Building Construction Costs with RSMeans data

Stephen C. Plotner, Senior Editor

D1103960

2018
76th annual edition

Chief Data Officer
Noam Reininger

Engineering Director
Bob Mewis, CCP

Contributing Editors
Christopher Babbitt
Sam Babbitt
Michelle Curran
Matthew Doheny (8)
Cheryl Elsmore
Linval Gentles
John Gomes (13, 41)
Derrick Hale, PE (2, 31, 32, 33, 34, 35, 44, 46)
Wafaa Hamitou (11, 12)

Joseph Kelble (14, 21, 22, 23, 25, 26, 27, 28, 48)
Charles Kibbee (1, 4)
Gerard Lafond, PE
Thomas Lane (6, 7)
Genevieve Medeiros
Elisa Mello
Ken Monty
Marilyn Phelan, AIA (9, 10)
Stephen C. Plotner (3, 5)
Callum Riley
Stephen Rosenberg
Jeff Sessions
Gabe Sirota

Matthew Sorrentino
Kevin Souza
Keegan Spraker
Tim Tonello
Jen Walsh
David Yazbek

Product Manager
Andrea Sillah

Production Manager
Debbie Panarelli

Production
Jonathan Forgit
Mary Lou Geary

Sharon Larsen
Sheryl Rose

Technical Support
Judy Abbruzzese
Gary L. Hoitt

Cover Design
Blaire Collins

Data Analytics
Tim Duggan
Todd Glowac
Matthew Kelliher-Gibson

Numbers in italics are the divisional responsibilities for each editor. Please contact the designated editor directly with any questions.

Gordian RSMeans data

Construction Publishers & Consultants
1099 Hingham Street, Suite 201
Rockland, MA 02370
United States of America
1-800-448-8182
www.RSMeans.com

Copyright 2017 by The Gordian Group Inc.
All rights reserved.
Cover photo © iStock.com/Ksene

Printed in the United States of America
ISSN 0068-3531
ISBN 978-1-946872-01-2

Gordian's authors, editors, and engineers apply diligence and judgment in locating and using reliable sources for the information published. However, Gordian makes no express or implied warranty or guarantee in connection with the content of the information contained herein, including the accuracy, correctness, value, sufficiency, or completeness of the data, methods, and other information contained herein. Gordian makes no express or implied warranty of merchantability or fitness for a particular purpose. Gordian shall have no liability to any customer or third party for any loss, expense, or damage, including consequential, incidental, special, or punitive damage, including lost profits or lost revenue, caused directly or indirectly by any error or omission, or arising out of, or in connection with, the information contained herein. For the purposes of this paragraph, "Gordian" shall include The Gordian Group, Inc., and its divisions, subsidiaries, successors, parent companies, and their employees, partners, principals, agents and representatives, and any third-party providers or sources of information or data. Gordian grants the purchaser of this publication limited license to use the cost data contained herein for purchaser's internal business purposes in connection with construction estimating and related work. The publication, and all cost data contained herein, may not be reproduced, integrated into any software or computer program, developed into a database or other electronic compilation, stored in an information storage or retrieval system, or transmitted or distributed to anyone in any form or by any means, electronic or mechanical, including photocopying or scanning, without prior written permission of Gordian. This publication is subject to protection under copyright law, trade secret law and other intellectual property laws of the United States, Canada and other jurisdictions. Gordian and its affiliates exclusively own and retain all rights, title and interest in and to this publication and the cost data contained herein including, without limitation, all copyright, patent, trademark and trade secret rights. Except for the limited license contained herein, your purchase of this publication does not grant you any intellectual property rights in the publication or the cost data.

| 0018 | $272.99 per copy (in United States) Price is subject to change without prior notice. |

Related Data and Services

2018 Building Construction Costs with RSMeans data has been tirelessly researched and carefully compiled to provide construction cost data for commercial and industrial projects or large multi-family housing projects costing $3,500,000 and up. For civil engineering structures such as bridges, dams, highways, or the like, please refer to Heavy Construction Costs with RSMeans data.

Our engineers recommend the following products and services to complement Building Construction Costs with RSMeans data:

Annual Cost Data Books
2018 Assemblies Costs with RSMeans data
2018 Square Foot Costs with RSMeans data

Reference Books
Estimating Building Costs
RSMeans Estimating Handbook
Green Building: Project Planning & Estimating
How to Estimate with RSMeans data
Plan Reading & Material Takeoff
Project Scheduling & Management for Construction

Seminars and In-House Training
Unit Price Estimating
Training for our online estimating solution
Practical Project Management for Construction Professionals
Scheduling with MSProject for Construction Professionals
Mechanical & Electrical Estimating

RSMeans data Online
For access to the latest cost data, an intuitive search, and an easy-to-use estimate builder, take advantage of the time savings available from our online application. To learn more visit: www.RSMeans.com/2018online.

Enterprise Solutions
Building owners, facility managers, building product manufacturers, and attorneys across the public and private sectors engage with RSMeans data Enterprise to solve unique challenges where trusted construction cost data is critical. To learn more visit: www.RSMeans.com/Enterprise.

Custom Built Data Sets
Building and Space Models: Quickly plan construction costs across multiple locations based on geography, project size, building system component, product options, and other variables for precise budgeting and cost control.

Predictive Analytics: Accurately plan future builds with custom graphical interactive dashboards, negotiate future costs of tenant build-outs, and identify and compare national account pricing.

Consulting
Building Product Manufacturing Analytics: Validate your claims and assist with new product launches.

Third-Party Legal Resources: Used in cases of construction cost or estimate disputes, construction product failure vs. installation failure, eminent domain, class action construction product liability, and more.

API
For resellers or internal application integration, RSMeans data is offered via API. Deliver Unit, Assembly, and Square Foot Model data within your interface. To learn more about how you can provide your customers with the latest in localized construction cost data visit: www.RSMeans.com/API.

Table of Contents

Foreword   iv

MasterFormat® Comparison Table   v

How the Cost Data Is Built: An Overview   vii

Estimating with RSMeans data: Unit Prices   ix

How to Use the Cost Data: The Details   xi

Unit Price Section   1

RSMeans data: Unit Prices—How They Work   4

Reference Section
  Construction Equipment Rental Costs   727
  Crew Listings   739
  Historical Cost Indexes   776
  City Cost Indexes   777
  Location Factors   820
  Reference Tables   826
  Change Orders   888
  Project Costs   891
  Abbreviations   896

Index   900

Other Data and Services   941

Labor Trade Rates including
Overhead & Profit   Inside Back Cover

Foreword

The Value of RSMeans data from Gordian

Since 1942, RSMeans data has been the industry-standard materials, labor, and equipment cost information database for contractors, facility owners and managers, architects, engineers, and anyone else that requires the latest localized construction cost information. Over 75 years later, the objective remains the same: to provide facility and construction professionals with the most current and comprehensive construction cost database possible.

With the constant influx of new construction methods and materials, in addition to ever-changing labor and material costs, last year's cost data is not reliable for today's designs, estimates, or budgets. The RSMeans data engineers invest over 22,000 hours in cost research annually and apply real-world construction experience to identify and quantify new building products and methodologies, adjust productivity rates, and adjust costs to local market conditions across the nation. This unparalleled construction cost expertise is why so many facility and construction professionals rely on RSMeans data year over year.

About Gordian

Gordian originated in the spirit of innovation and a strong commitment to helping clients reach and exceed their construction goals. In 1982, Gordian's Chairman and Founder, Harry H. Mellon, created Job Order Contracting while serving as Chief Engineer at the Supreme Headquarters Allied Powers Europe. Job Order Contracting is a unique indefinite delivery/indefinite quantity (IDIQ) process, which enables facility owners to complete a substantial number of repair, maintenance, and construction projects with a single, competitively awarded contract. Realizing facility and infrastructure owners across various industries could greatly benefit from the time and cost saving advantages of this innovative construction procurement solution, he established Gordian in 1990.

Continuing the commitment to providing the most relevant and accurate facility and construction data, software, and expertise in the industry, Gordian enhanced the fortitude of its data with the acquisition of RSMeans in 2014. And in an effort to expand its facility management capabilities, Gordian acquired Sightlines, the leading provider of facilities benchmarking data and analysis, in 2015.

Our Offerings

Gordian is the leader in facility and construction cost data, software, and expertise for all phases of the building life cycle. From planning to design, procurement, construction, and operations, Gordian's solutions help clients maximize efficiency, optimize cost savings, and increase building quality with its highly specialized data engineers, software, and unique proprietary data sets.

Our Commitment

At Gordian, we do more than talk about the quality of our data and the usefulness of its application. We stand behind all of our RSMeans data—from historical cost indexes to construction materials and techniques—to craft current costs and predict future trends. If you have any questions about our products or services, please call us toll-free at 800-448-8182 or visit our website at www.gordian.com.

MasterFormat® 2014/ MasterFormat® 2016 Comparison Table

This table compares the 2014 edition of the Construction Specifications Institute's MasterFormat® to the expanded 2016 edition. For your convenience, all revised 2014 numbers and titles are listed along with the corresponding 2016 numbers and titles. In some cases, a designation of RSMeans is used to identify sections of data numbered exclusively in RSMeans products.

CSI 2014 MF ID	CSI 2014 MF Description	2014 Designation	CSI 2016 MF ID	CSI 2016 MF Description	2016 Designation
35 20 23.23	Hydraulic Dredging	CSI	35 24 13.13	Cutter Suction Dredging	RSMeans
35 20 23.13	Mechanical Dredging	CSI	35 24 23.13	Mechanical Dredging	CSI
35 20 23	Dredging	CSI	35 24 23	Clamshell Dredging	RSMeans
35 20 16.73	Slide Gates	RSMeans	35 22 73.16	Slide Gates	CSI
35 20 16.69	Knife Gates	RSMeans	35 22 69.16	Knife Gates	RSMeans
35 20 16.66	Flap Gates	RSMeans	35 22 66.16	Flap Gates	RSMeans
35 20 16.63	Canal Gates	RSMeans	35 22 63.16	Canal Gates	RSMeans
35 20 16.26	Hydraulic Sluice Gates	CSI	35 22 26.16	Hydraulic Sluice Gates	RSMeans
35 20 16	Hydraulic Gates	CSI	35 22 26	Sluice Gates	RSMeans
35 20	Waterway and Marine Construction and Equipment	CSI	35 22	Hydraulic Gates	RSMeans
33 72 33.46	Substation Converter Stations	CSI	33 78 33.46	Substation Converter Stations	RSMeans
33 72 33.36	Cable Trays For Utility Substations	CSI	26 05 36.36	Cable Trays For Utility Substations	CSI
33 72 33.33	Raceway/Boxes For Utility Substations	CSI	26 05 33.33	Raceway/Boxes For Utility Substations	CSI
33 52 43.13	Aviation Fuel Piping	CSI	33 52 13.43	Aviation Fuel Piping	RSMeans
33 52 16.13	Gasoline Piping	CSI	33 52 13.16	Gasoline Piping	RSMeans
33 52 13.14	Petroleum Products	RSMeans	33 52 13.19	Petroleum Products	RSMeans
33 51 33.10	Piping, Valves & Meters, Gas Distribution	RSMeans	33 59 33.10	Piping, Valves & Meters, Gas Distribution	CSI
33 51 33	Natural-Gas Metering	CSI	33 59 33	Natural-Gas Metering	CSI
33 51 13.30	Piping, Gas Service & Distribution, Steel	RSMeans	33 52 16.16	Piping, Gas Service & Distribution, Steel	RSMeans
33 51 13.20	Piping, Gas Service & Distribution, Steel	RSMeans	33 52 16.13	Steel Natural Gas Piping	RSMeans
33 51 13.10	Piping, Gas Service and Distribution, Polyethylene	RSMeans	33 52 16.20	Piping, Gas Service And Distribution, Polyethylene	CSI
33 49 23	Storm Drainage Water Retention Structures	CSI	33 46 23	Modular Buried Stormwater Storage Units	CSI
33 49 13	Storm Drainage Manholes, Frames, and Covers	CSI	33 05 61	Concrete Manholes	RSMeans
33 47 19	Water Ponds and Reservoirs	CSI	33 46 11	Stormwater Ponds	CSI
33 47 13.54	Garden Ponds	RSMeans	13 12 13.54	Garden Ponds	CSI
33 47 13.53	Reservoir Liners HDPE	CSI	31 05 19.53	Reservoir Liners HDPE	RSMeans
33 47 13	Pond and Reservoir Liners	CSI	31 05 19	Geosynthetics for Earthwork	RSMeans
33 46 26.10	Geotextiles For Subsurface Drainage	RSMeans	33 41 23.19	Geosynthetic Drainage Layers	RSMeans
33 46 26	Geotextile Subsurface Drainage Filtration	CSI	33 41 23	Drainage Layers	CSI
33 46 16	Subdrainage Piping	CSI	33 41 16	Subdrainage Piping	CSI
33 46	Subdrainage	CSI	33 41	Subdrainage	CSI
33 44 16	Utility Trench Drains	CSI	33 42 36	Stormwater Trench Drains	CSI
33 44 13	Utility Area Drains	CSI	33 42 33	Stormwater Curbside Drains and Inlets	RSMeans
33 42 16.15	Oval Arch Culverts	RSMeans	33 42 13.15	Oval Arch Culverts	RSMeans
33 42 16.13	Culverts & Box Trench Sections	CSI	33 42 13.14	Culverts & Box Trench Sections	RSMeans
33 41 13	Public Storm Utility Drainage Piping	CSI	33 42 11	Stormwater Gravity Piping	CSI
33 41	Storm Utility Drainage Piping	CSI	33 42	Stormwater Conveyance	CSI
33 36 50	Drainage Field Systems	RSMeans	33 34 51	Drainage Field System	CSI
33 36 33.13	Utility Septic Tank Tile Drainage Field	CSI	33 34 51.13	Utility Septic Tank Tile Drainage Field	RSMeans
33 36 19	Utility Septic Tank Effluent Filter	CSI	33 34 16	Septic Tank Effluent Filters	CSI
33 36 13.19	Polyethylene Utility Septic Tank	CSI	33 34 13.33	Polyethylene Septic Tanks	CSI
33 36 13.13	Concrete Utility Septic Tank	CSI	33 34 13.13	Concrete Septic Tanks	RSMeans
33 36 13	Utility Septic Tank and Effluent Wet Wells	CSI	33 34 13	Septic Tanks	CSI
33 36	Utility Septic Tanks	CSI	33 34	Onsite Wastewater Disposal	CSI
33 31 13	Public Sanitary Utility Sewerage Piping	CSI	33 31 11	Public Sanitary Sewerage Gravity Piping	CSI
33 21 13	Public Water Supply Wells	CSI	33 11 13	Potable Water Supply Wells	RSMeans
33 21	Water Supply Wells	CSI	33 11	Groundwater Sources	CSI
33 16 19.50	Elevated Water Storage Tanks	RSMeans	33 16 11.50	Elevated Water Storage Tanks	CSI
33 16 19	Elevated Water Utility Storage Tanks	CSI	33 16 11	Elevated Composite Water Storage Tanks	CSI
33 16 13.29	Wood Water Storage Tanks	RSMeans	33 16 59.29	Wood Water Storage Tanks	CSI
33 16 13.23	Plastic-Coated Fabric Pillow Water Tanks	RSMeans	33 16 56.23	Plastic-Coated Fabric Pillow Water Tanks	CSI
33 16 13.19	Horizontal Plastic Water Tanks	CSI	33 16 56.19	Horizontal Plastic Water Tanks	CSI
33 16 13.16	Prestressed Conc. Water Storage Tanks	CSI	33 16 36.16	Prestressed Conc. Water Storage Tanks	RSMeans
33 16 13.13	Steel Water Storage Tanks	CSI	33 16 23.13	Steel Water Storage Tanks	RSMeans
33 16 13	Aboveground Water Utility Storage Tanks	CSI	33 16 23	Ground-Level Steel Water Storage Tanks	CSI
33 12 19.10	Fire Hydrants	RSMeans	33 14 19.30	Fire Hydrants	CSI
33 12 16.20	Valves	RSMeans	33 14 19.20	Valves	CSI

Number	Description	Source	Number	Description	Source
33 12 16.10	Valves	RSMeans	33 14 19.10	Valves	CSI
33 12 16	Water Utility Distribution Valves	CSI	33 14 19	Valves and Hydrants for Water Utility Service	CSI
33 12 13.15	Tapping, Crosses and Sleeves	RSMeans	33 14 17.15	Tapping, Crosses and Sleeves	CSI
33 12 13	Water Service Connections	CSI	33 14 17	Site Water Utility Service Laterals	RSMeans
33 11 13	Public Water Utility Distribution Piping	CSI	33 14 13	Public Water Utility Distribution Piping	RSMeans
33 11	Water Utility Distribution Piping	CSI	33 14	Water Utility Transmission and Distribution	CSI
33 05 26	Utility Identification	CSI	33 05 97	Identification and Signage for Utilities	RSMeans
33 05 23.22	Directional Drilling	RSMeans	33 05 07.13	Utility Directional Drilling	RSMeans
33 05 23.20	Horizontal Boring	RSMeans	33 05 07.23	Utility Boring and Jacking	CSI
33 05 23.19	Microtunneling	CSI	33 05 07.36	Microtunneling	CSI
33 05 23	Trenchless Utility Installation	CSI	33 05 07	Trenchless Installation of Utility Piping	RSMeans
33 05 16	Utility Structures	CSI	33 05 63	Concrete Vaults and Chambers	RSMeans
33 01 30.71	Rehabilitation of Sewer Utilities	CSI	33 01 30.23	Pipe Bursting	RSMeans
33 01 30.16	TV Inspection of Sewer Pipelines	CSI	33 01 30.11	Television Inspection of Sewers	CSI
32 31 13.30	Fence, Chain Link, Gates & Posts	RSMeans	32 31 11.10	Gate Operators	CSI
31 71 21.10	Cut and Cover Tunnels	RSMeans	31 71 23.10	Cut and Cover Tunnels	CSI
31 71 21	Tunnel Excavation by Cut and Cover	RSMeans	31 71 23	Tunneling by Cut and Cover	CSI
28 46 13	Hard-Wired Detention Monitoring & Control Systems	CSI	28 52 11	Detention Monitoring and Control Systems	CSI
28 46	Electronic Detention Monitoring & Control Systems	CSI	28 52	Detention Security Systems	CSI
28 41 13	Building Systems	RSMeans	28 33 11	Electronic Structural Monitoring Systems	RSMeans
28 39 10	Notification Systems	RSMeans	28 47 12	Notification Systems	CSI
28 39	Mass Notification Systems	CSI	28 47	Mass Notification	RSMeans
28 33 33	Gas Detection Sensors	CSI	28 42 15	Gas Detection Sensors	CSI
28 33	Gas Detection and Alarm	CSI	28 42	Gas Detection and Alarm	RSMeans
28 32 33	Radiation Detection Sensors	CSI	28 41 15	Radiation Detection Sensors	CSI
28 31 49.50	Carbon-Monoxide Detectors	RSMeans	28 46 11.21	Carbon-Monoxide Detection Sensors	CSI
28 31 46.50	Smoke Detectors	RSMeans	28 46 11.27	Other Sensors	CSI
28 31 43	Fire Detection Sensors	CSI	28 46 11	Fire Sensors and Detectors	CSI
28 31 23	Fire Det. & Alarm Annunciation Panels & Fire Station	CSI	28 46 21	Fire Alarm	CSI
28 23 19.10	Digital Video Recorder (DVR)	RSMeans	28 05 19.11	Digital Video Recorders	RSMeans
28 23 19	Digital Video Recorders & Analog Recording Devices	CSI	28 05 19	Storage Appliances for Electronic Safety & Security	RSMeans
28 16 16	Intrusion Detection Systems Infrastructure	CSI	28 31 16	Intrusion Detection Systems Infrastructure	RSMeans
28 13 53.39	Security Access Full Body Imaging Machine	RSMeans	28 18 15.39	Security Access Full Body Imaging Machine	CSI
28 13 53.36	Security Access Debugging Kit	RSMeans	28 18 53.36	Security Access Debugging Kit	CSI
28 13 53.33	Security Access Counterfeit Money Detector	RSMeans	28 18 53.33	Security Access Counterfeit Money Detector	CSI
28 13 53.23	Security Access Explosive Detection Equipment	CSI	28 18 15.23	Security Access Explosive Detection Equipment	CSI
28 13 53.16	Security Access X-Ray Equipment	CSI	28 18 13.16	Security Access X-Ray Equipment	CSI
28 13 53.13	Security Access Metal Detectors	CSI	28 18 11.13	Security Access Metal Detectors	RSMeans
28 13 53	Security Access Detection	CSI	28 18 11	Security Access Metal Detectors	CSI
28 13 23.50	Vehicle Barriers	RSMeans	28 19 15.50	Vehicle Barriers	CSI
28 13 23	Access Control Remote Devices	RSMeans	28 19 15	Perimeter Vehicle Access Management Systems	RSMeans
28 13	Access Control	CSI	28 19	Access Control Vehicle Identification Systems	CSI
28 05 13.23	Fire Alarm Communications Conductors and Cables	CSI	27 15 01.19	Fire Alarm Communications Conductors and Cables	RSMeans
28 05 13.10	Alarm & Communications Cable	RSMeans	27 15 01.11	Conductors & Cables For Electronic Safety & Security	CSI
28 01 30.51	Maint. and Admin. of Elec. Detection and Alarm	CSI	28 01 80.51	Maint. & Administration of Fire Detection & Alarm	RSMeans
28 01 30	Operation and Maint. of Elec. Detection and Alarm	CSI	28 01 80	Operation and Maint. of Fire Detection and Alarm	RSMeans
26 56 19.20	Roadway Luminaire	RSMeans	26 56 21.20	Roadway Luminaire	RSMeans
26 56 19.10	Roadway Lighting Fixtures	RSMeans	26 56 21.10	LED Exterior Lighting	RSMeans
26 56 16.55	Parking LED Lighting	RSMeans	26 56 19.60	Parking LED Lighting	RSMeans
26 53 13.10	Exit Lighting Fixtures	RSMeans	26 52 13.16	Exit Signs	CSI
26 26 13.10	Power Distribution Unit	RSMeans	26 27 33.20	Power Distribution Unit	RSMeans
13 34 23.35	Geodesic Domes	RSMeans	13 33 13.35	Geodesic Domes	CSI
13 34 23.15	Domes	RSMeans	13 34 56.15	Domes	CSI
11 14 13.13	Portable Posts and Railings	CSI	11 14 19.13	Portable Posts and Railings	CSI
10 21 13.20	Plastic Toilet Compartment Components	RSMeans	10 21 14.19	Plastic Toilet Compartment Components	CSI
10 21 13.17	Plastic-Laminate Clad Toilet Compartment Components	RSMeans	10 21 14.16	Plastic-Laminate Clad Toilet Compartment Components	CSI
10 21 13.14	Metal Toilet Compartment Components	RSMeans	10 21 14.13	Metal Toilet Compartment Components	CSI
08 74 23.50	Security Access Control Accessories	RSMeans	28 15 11.19	Security Access Control Accessories	CSI
08 74 19.50	Biometric Identity Access	RSMeans	28 15 11.15	Biometric Identity Devices	RSMeans
08 74 16.50	Keypad Access	RSMeans	28 15 11.13	Keypads	RSMeans
08 74 13	Card Key Access Control Hardware	CSI	28 15 11	Integrated Credential Readers & Field Entry Mgmt	RSMeans
08 74	Access Control Hardware	CSI	28 15	Access Control Hardware Devices	CSI
08 56 63	Detention Windows	CSI	11 98 21	Detention Windows	RSMeans
08 34 63	Detention Doors and Frames	CSI	11 98 12	Detention Doors and Frames	RSMeans
07 72 73.10	Pitch Pockets, Variable Sizes	RSMeans	07 71 16.20	Pitch Pockets, Variable Sizes	CSI
05 53 13.70	Expanded Steel Grating, at Ground	RSMeans	05 53 19.20	Expanded Grating, Steel	CSI
02 85 33	Removal and Disposal of Materials with Mold	CSI	02 87 13.33	Removal and Disposal of Materials with Mold	RSMeans
02 85 16	Mold Remediation Preparation and Containment	CSI	02 87 13	Mold Remediation	RSMeans
02 85	Mold Remediation	CSI	02 87	Biohazard Remediation	CSI

For additional tools that help with the utilization of the Construction Specifications Institute's 2016 edition of MasterFormat® please visit the following website: http://www.masterformat.com/revisions/

How the Cost Data Is Built: An Overview

Unit Prices*

All cost data have been divided into 50 divisions according to the MasterFormat® system of classification and numbering.

Assemblies*

The cost data in this section have been organized in an "Assemblies" format. These assemblies are the functional elements of a building and are arranged according to the 7 elements of the UNIFORMAT II classification system. For a complete explanation of a typical "Assembly", see "RSMeans data: Assemblies—How They Work."

Residential Models*

Model buildings for four classes of construction—economy, average, custom, and luxury—are developed and shown with complete costs per square foot.

Commercial/Industrial/ Institutional Models*

This section contains complete costs for 77 typical model buildings expressed as costs per square foot.

Green Commercial/Industrial/ Institutional Models*

This section contains complete costs for 25 green model buildings expressed as costs per square foot.

References*

This section includes information on Equipment Rental Costs, Crew Listings, Historical Cost Indexes, City Cost Indexes, Location Factors, Reference Tables, and Change Orders, as well as a listing of abbreviations.

- **Equipment Rental Costs:** Included are the average costs to rent and operate hundreds of pieces of construction equipment.
- **Crew Listings:** This section lists all the crews referenced in the cost data. A crew is composed of more than one trade classification and/or the addition of power equipment to any trade classification. Power equipment is included in the cost of the crew. Costs are shown both with bare labor rates and with the installing contractor's overhead and profit added. For each, the total crew cost per eight-hour day and the composite cost per labor-hour are listed.

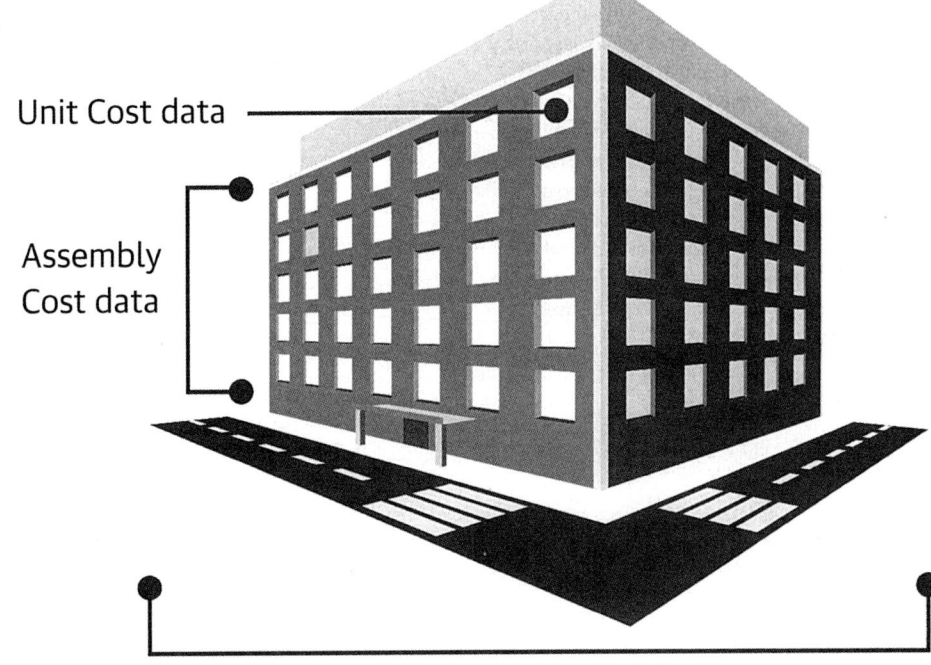

Unit Cost data

Assembly Cost data

Square Foot Models

- **Historical Cost Indexes:** These indexes provide you with data to adjust construction costs over time.
- **City Cost Indexes:** All costs in this data set are U.S. national averages. Costs vary by region. You can adjust for this by CSI Division to over 730 cities in 900+ 3-digit zip codes throughout the U.S. and Canada by using this data.
- **Location Factors:** You can adjust total project costs to over 730 cities in 900+ 3-digit zip codes throughout the U.S. and Canada by using the weighted number, which applies across all divisions.
- **Reference Tables:** At the beginning of selected major classifications in the Unit Prices are reference numbers indicators. These numbers refer you to related information in the Reference Section. In this section, you'll find reference tables, explanations, and estimating information that support how we develop the unit price data, technical data, and estimating procedures.
- **Change Orders:** This section includes information on the factors that influence the pricing of change orders.

- **Abbreviations:** A listing of abbreviations used throughout this information, along with the terms they represent, is included.

Index (printed versions only)

A comprehensive listing of all terms and subjects will help you quickly find what you need when you are not sure where it occurs in MasterFormat®.

Conclusion

This information is designed to be as comprehensive and easy to use as possible.

The Construction Specifications Institute (CSI) and Construction Specifications Canada (CSC) have produced the 2016 edition of MasterFormat®, a system of titles and numbers used extensively to organize construction information.

All unit prices in the RSMeans cost data are now arranged in the 50-division MasterFormat® 2016 system.

---

* Not all information is available in all data sets

Note: The material prices in RSMeans cost data are "contractor's prices." They are the prices that contractors can expect to pay at the lumberyards, suppliers'/distributors' warehouses, etc. Small orders of specialty items would be higher than the costs shown, while very large orders, such as truckload lots, would be less. The variation would depend on the size, timing, and negotiating power of the contractor. The labor costs are primarily for new construction or major renovation rather than repairs or minor alterations. With reasonable exercise of judgment, the figures can be used for any building work.

Estimating with RSMeans data: Unit Prices

Following these steps will allow you to complete an accurate estimate using RSMeans data Unit Prices.

1. Scope Out the Project

- Think through the project and identify the CSI divisions needed in your estimate.
- Identify the individual work tasks that will need to be covered in your estimate.
- The Unit Price Data have been divided into 50 divisions according to CSI MasterFormat® 2016.
- In printed versions, the Unit Price Section Table of Contents on page 1 may also be helpful when scoping out your project.
- Experienced estimators find it helpful to begin with Division 2 and continue through completion. Division 1 can be estimated after the full project scope is known.

2. Quantify

- Determine the number of units required for each work task that you identified.
- Experienced estimators include an allowance for waste in their quantities. (Waste is not included in our Unit Price line items unless otherwise stated.)

3. Price the Quantities

- Use the search tools available to locate individual Unit Price line items for your estimate.
- Reference Numbers indicated within a Unit Price section refer to additional information that you may find useful.
- The crew indicates who is performing the work for that task. Crew codes are expanded in the Crew Listings in the Reference Section to include all trades and equipment that comprise the crew.
- The Daily Output is the amount of work the crew is expected to complete in one day.
- The Labor-Hours value is the amount of time it will take for the crew to install one unit of work.
- The abbreviated Unit designation indicates the unit of measure upon which the crew, productivity, and prices are based.
- Bare Costs are shown for materials, labor, and equipment needed to complete the Unit Price line item. Bare costs do not include waste, project overhead, payroll insurance, payroll taxes, main office overhead, or profit.
- The Total Incl O&P cost is the billing rate or invoice amount of the installing contractor or subcontractor who performs the work for the Unit Price line item.

4. Multiply

- Multiply the total number of units needed for your project by the Total Incl O&P cost for each Unit Price line item.
- Be careful that your take off unit of measure matches the unit of measure in the Unit column.
- The price you calculate is an estimate for a completed item of work.
- Keep scoping individual tasks, determining the number of units required for those tasks, matching each task with individual Unit Price line items, and multiplying quantities by Total Incl O&P costs.
- An estimate completed in this manner is priced as if a subcontractor, or set of subcontractors, is performing the work. The estimate does not yet include Project Overhead or Estimate Summary components such as general contractor markups on subcontracted work, general contractor office overhead and profit, contingency, and location factors.

5. Project Overhead

- Include project overhead items from Division 1–General Requirements.
- These items are needed to make the job run. They are typically, but not always, provided by the general contractor. Items include, but are not limited to, field personnel, insurance, performance bond, permits, testing, temporary utilities, field office and storage facilities, temporary scaffolding and platforms, equipment mobilization and demobilization, temporary roads and sidewalks, winter protection, temporary barricades and fencing, temporary security, temporary signs, field engineering and layout, final cleaning, and commissioning.
- Each item should be quantified and matched to individual Unit Price line items in Division 1, then priced and added to your estimate.
- An alternate method of estimating project overhead costs is to apply a percentage of the total project cost, usually 5% to 15% with an average of 10% (see General Conditions).
- Include other project related expenses in your estimate such as:
  - Rented equipment not itemized in the Crew Listings
  - Rubbish handling throughout the project (see section 02 41 19.19)

6. Estimate Summary

- Include sales tax as required by laws of your state or county.
- Include the general contractor's markup on self-performed work, usually 5% to 15% with an average of 10%.
- Include the general contractor's markup on subcontracted work, usually 5% to 15% with an average of 10%.
- Include the general contractor's main office overhead and profit:
  - RSMeans data provides general guidelines on the general contractor's main office overhead (see section 01 31 13.60 and Reference Number R013113-50).
  - Markups will depend on the size of the general contractor's operations, projected annual revenue, the level of risk, and the level of competition in the local area and for this project in particular.
- Include a contingency, usually 3% to 5%, if appropriate.
- Adjust your estimate to the project's location by using the City Cost Indexes or the Location Factors in the Reference Section:
  - Look at the rules in "How to Use the City Cost Indexes" to see how to apply the Indexes for your location.
  - When the proper Index or Factor has been identified for the project's location, convert it to a multiplier by dividing it by 100, then multiply that multiplier by your estimated total cost. The original estimated total cost will now be adjusted up or down from the national average to a total that is appropriate for your location.

Editors' Note:
We urge you to spend time reading and understanding the supporting material. An accurate estimate requires experience, knowledge, and careful calculation. The more you know about how we at RSMeans developed the data, the more accurate your estimate will be. In addition, it is important to take into consideration the reference material such as Equipment Listings, Crew Listings, City Cost Indexes, Location Factors, and Reference Tables.

How to Use the Cost Data: The Details

What's Behind the Numbers? The Development of Cost data

RSMeans data engineers continually monitor developments in the construction industry in order to ensure reliable, thorough, and up-to-date cost information. While overall construction costs may vary relative to general economic conditions, price fluctuations within the industry are dependent upon many factors. Individual price variations may, in fact, be opposite to overall economic trends. Therefore, costs are constantly tracked, and complete updates are performed yearly. Also, new items are frequently added in response to changes in materials and methods.

Costs in U.S. Dollars

All costs represent U.S. national averages and are given in U.S. dollars. The City Cost Index (CCI) with RSMeans data can be used to adjust costs to a particular location. The CCI for Canada can be used to adjust U.S. national averages to local costs in Canadian dollars. No exchange rate conversion is necessary because it has already been factored in.

G The processes or products identified by the green symbol in our publications have been determined to be environmentally responsible and/or resource-efficient solely by RSMeans data engineering staff. The inclusion of the green symbol does not represent compliance with any specific industry association or standard.

Material Costs

RSMeans data engineers contact manufacturers, dealers, distributors, and contractors all across the U.S. and Canada to determine national average material costs. If you have access to current material costs for your specific location, you may wish to make adjustments to reflect differences from the national average. Included within material costs are fasteners for a normal installation. RSMeans data engineers use manufacturers' recommendations, written specifications, and/or standard construction practices for the sizing and spacing of fasteners. Adjustments to material costs may be required for your specific application or location. The manufacturer's warranty is assumed. Extended warranties are not included in the material costs. **Material costs do not include sales tax.**

Labor Costs

Labor costs are based upon a mathematical average of trade-specific wages in 30 major U.S. cities. The type of wage (union, open shop, or residential) is identified on the inside back cover of printed publications or is selected by the estimator when using the electronic products. Markups for the wages can also be found on the inside back cover of printed publications and/or under the labor references found in the electronic products.

- If wage rates in your area vary from those used, or if rate increases are expected within a given year, labor costs should be adjusted accordingly.

Labor costs reflect productivity based on actual working conditions. In addition to actual installation, these figures include time spent during a normal weekday on tasks, such as material receiving and handling, mobilization at site, site movement, breaks, and cleanup.

Productivity data is developed over an extended period so as not to be influenced by abnormal variations and reflects a typical average.

Equipment Costs

Equipment costs include not only rental but also operating costs for equipment under normal use. The operating costs include parts and labor for routine servicing, such as the repair and replacement of pumps, filters, and worn lines. Normal operating expendables, such as fuel, lubricants, tires, and electricity (where applicable), are also included. Extraordinary operating expendables with highly variable wear patterns, such as diamond bits and blades, are excluded. These costs are included under materials. Equipment rental rates are obtained from industry sources throughout North America—contractors, suppliers, dealers, manufacturers, and distributors.

Rental rates can also be treated as reimbursement costs for contractor-owned equipment. Owned equipment costs include depreciation, loan payments, interest, taxes, insurance, storage, and major repairs.

Equipment costs do not include operators' wages.

Equipment Cost/Day—The cost of equipment required for each crew is included in the Crew Listings in the Reference Section (small tools that are considered essential everyday tools are not listed out separately). The Crew Listings itemize specialized tools and heavy equipment along with labor trades. The daily cost of itemized equipment included in a crew is based on dividing the weekly bare rental rate by 5 (number of working days per week), then adding the hourly operating cost times 8 (the number of hours per day). This Equipment Cost/Day is shown in the last column of the Equipment Rental Costs in the Reference Section.

Mobilization, Demobilization—The cost to move construction equipment from an equipment yard or rental company to the job site and back again is not included in equipment costs. Mobilization (to the site) and demobilization (from the site) costs can be found in the Unit Price Section. If a piece of equipment is already at the job site, it is not appropriate to utilize mobilization or demobilization costs again in an estimate.

Overhead and Profit

Total Cost including O&P for the installing contractor is shown in the last column of the Unit Price and/or Assemblies. This figure is the sum of the bare material cost plus 10% for profit, the bare labor cost plus total overhead and profit, and the bare equipment cost plus 10% for profit. Details for the calculation of overhead and profit on labor are shown on the inside back cover of the printed product and in the Reference Section of the electronic product.

General Conditions

Cost data in this data set are presented in two ways: Bare Costs and Total Cost including O&P (Overhead and Profit). General Conditions, or General Requirements, of the contract should also be added to the Total Cost including O&P when applicable. Costs for General Conditions are listed in Division 1 of the Unit Price Section and in the Reference Section.

General Conditions for the installing contractor may range from 0% to 10% of the Total Cost including O&P. For the general or prime contractor, costs for General Conditions may range from 5% to 15% of the Total Cost including O&P, with a figure of 10% as the most typical allowance. If applicable, the Assemblies and Models sections use costs that include the installing contractor's overhead and profit (O&P).

Factors Affecting Costs

Costs can vary depending upon a number of variables. Here's a listing of some factors that affect costs and points to consider.

Quality—The prices for materials and the workmanship upon which productivity is based represent sound construction work. They are also in line with industry standard and manufacturer specifications and are frequently used by federal, state, and local governments.

Overtime—We have made no allowance for overtime. If you anticipate premium time or work beyond normal working hours, be sure to make an appropriate adjustment to your labor costs.

Productivity—The productivity, daily output, and labor-hour figures for each line item are based on an eight-hour work day in daylight hours in moderate temperatures, and up to a 14' working height unless otherwise indicated. For work that extends beyond normal work hours or is performed under adverse conditions, productivity may decrease.

Size of Project—The size, scope of work, and type of construction project will have a significant impact on cost. Economies of scale can reduce costs for large projects. Unit costs can often run higher for small projects.

Location—Material prices are for metropolitan areas. However, in dense urban areas, traffic and site storage limitations may increase costs. Beyond a 20-mile radius of metropolitan areas, extra trucking or transportation charges may also increase the material costs slightly. On the other hand, lower wage rates may be in effect. Be sure to consider both of these factors when preparing an estimate, particularly if the job site is located in a central city or remote rural location. In addition, highly specialized subcontract items may require travel and per-diem expenses for mechanics.

Other Factors—

- season of year
- contractor management
- weather conditions
- local union restrictions
- building code requirements
- availability of:
  - adequate energy
  - skilled labor
  - building materials
- owner's special requirements/restrictions
- safety requirements
- environmental considerations
- access

Unpredictable Factors—General business conditions influence "in-place" costs of all items. Substitute materials and construction methods may have to be employed. These may affect the installed cost and/or life cycle costs. Such factors may be difficult to evaluate and cannot necessarily be predicted on the basis of the job's location in a particular section of the country. Thus, where these factors apply, you may find significant but unavoidable cost variations for which you will have to apply a measure of judgment to your estimate.

Rounding of Costs

In printed publications only, all unit prices in excess of $5.00 have been rounded to make them easier to use and still maintain adequate precision of the results.

How Subcontracted Items Affect Costs

A considerable portion of all large construction jobs is usually subcontracted. In fact, the percentage done by subcontractors is constantly increasing and may run over 90%. Since the workers employed by these companies do nothing else but install their particular product, they soon become experts in that line. The result is, installation by these firms is accomplished so efficiently that the total in-place cost, even with the general contractor's overhead and profit, is no more, and often less, than if the principal contractor had handled the installation. Companies that deal with construction specialties are anxious to have their product perform well and, consequently, the installation will be the best possible.

Contingencies

The allowance for contingencies generally provides for unforeseen construction difficulties. On alterations or repair jobs, 20% is not too much. If drawings are final and only field contingencies are being considered, 2% or 3% is probably sufficient, and often nothing needs to be added. Contractually, changes in plans will be covered by extras. The contractor should consider inflationary price trends and possible material shortages during the course of the job. These escalation factors are dependent upon both economic conditions and the anticipated time between the estimate and actual construction. If drawings are not complete or approved, or a budget cost is wanted, it is wise to add 5% to 10%. Contingencies, then, are a matter of judgment.

Important Estimating Considerations

The productivity, or daily output, of each craftsman or crew assumes a well-managed job where tradesmen with the proper tools and equipment, along with the appropriate construction materials, are present. Included are daily set-up and cleanup time, break time, and plan layout time. Unless otherwise indicated, time for material movement on site (for items that can be transported by hand) of up to 200' into the building and to the first or second floor is also included. If material has to be transported by other means, over greater distances, or to higher floors, an additional allowance should be considered by the estimator.

While horizontal movement is typically a sole function of distances, vertical transport introduces other variables that can significantly impact productivity. In an occupied building, the use of elevators (assuming access, size, and required protective measures are acceptable) must be understood at the time of the estimate. For new construction, hoist wait and cycle times can easily be 15 minutes and may result in scheduled access extending beyond the normal work day. Finally, all vertical transport will impose strict weight limits likely to preclude the use of any motorized material handling.

The productivity, or daily output, also assumes installation that meets manufacturer/designer/ standard specifications. A time allowance for quality control checks, minor adjustments, and any task required to ensure the proper function or operation is also included. For items that require connections to services, time is included for positioning, leveling, securing the unit, and for making all the necessary connections (and start up where applicable), ensuring a complete installation. Estimating of the services themselves (electrical, plumbing, water, steam, hydraulics, dust collection, etc.) is separate.

In some cases, the estimator must consider the use of a crane and an appropriate crew for the installation of large or heavy items. For those situations where a crane is not included in the assigned crew and as part of the line item cost, then equipment rental costs, mobilization and demobilization costs, and operator and support personnel costs must be considered.

Labor-Hours

The labor-hours expressed in this publication are derived by dividing the total daily labor-hours for the crew by the daily output. Based on average installation time and the assumptions listed above, the labor-hours include: direct labor, indirect labor, and nonproductive time. A typical day for a craftsman might include, but is not limited to:

- Direct Work
  - ☐ Measuring and layout
  - ☐ Preparing materials
  - ☐ Actual installation
  - ☐ Quality assurance/quality control
- Indirect Work
  - ☐ Reading plans or specifications
  - ☐ Preparing space
  - ☐ Receiving materials
  - ☐ Material movement
  - ☐ Giving or receiving instruction
  - ☐ Miscellaneous
- Non-Work
  - ☐ Chatting
  - ☐ Personal issues
  - ☐ Breaks
  - ☐ Interruptions (i.e., sickness, weather, material or equipment shortages, etc.)

If any of the items for a typical day do not apply to the particular work or project situation, the estimator should make any necessary adjustments.

Final Checklist

Estimating can be a straightforward process provided you remember the basics. Here's a checklist of some of the steps you should remember to complete before finalizing your estimate.

Did you remember to:

- factor in the City Cost Index for your locale?
- take into consideration which items have been marked up and by how much?
- mark up the entire estimate sufficiently for your purposes?
- read the background information on techniques and technical matters that could impact your project time span and cost?
- include all components of your project in the final estimate?
- double check your figures for accuracy?
- call RSMeans data engineers if you have any questions about your estimate or the data you've used? Remember, Gordian stands behind all of our products, including our extensive RSMeans data solutions. If you have any questions about your estimate, about the costs you've used from our data, or even about the technical aspects of the job that may affect your estimate, feel free to call the Gordian RSMeans editors at 1-800-448-8182.

Access Quarterly Data Updates

rsmeans.com/2018books

Unit Price Section

Table of Contents

Sect. No.		Page
	General Requirements	
01 11	Summary of Work	10
01 21	Allowances	11
01 31	Project Management and Coordination	12
01 32	Construction Progress Documentation	14
01 41	Regulatory Requirements	14
01 45	Quality Control	15
01 51	Temporary Utilities	17
01 52	Construction Facilities	17
01 54	Construction Aids	18
01 55	Vehicular Access and Parking	22
01 56	Temporary Barriers and Enclosures	23
01 58	Project Identification	25
01 66	Product Storage and Handling Requirements	25
01 71	Examination and Preparation	25
01 74	Cleaning and Waste Management	25
01 76	Protecting Installed Construction	25
01 91	Commissioning	26
	Existing Conditions	
02 21	Surveys	28
02 32	Geotechnical Investigations	28
02 41	Demolition	29
02 42	Removal and Salvage of Construction Materials	35
02 43	Structure Moving	38
02 56	Site Containment	38
02 58	Snow Control	39
02 65	Underground Storage Tank Rmvl.	39
02 81	Transportation and Disposal of Hazardous Materials	40
02 82	Asbestos Remediation	40
02 83	Lead Remediation	44
02 87	Biohazard Remediation	46
02 91	Chemical Sampling, Testing and Analysis	47
	Concrete	
03 01	Maintenance of Concrete	50
03 05	Common Work Results for Concrete	50
03 11	Concrete Forming	52
03 15	Concrete Accessories	60
03 21	Reinforcement Bars	68
03 22	Fabric and Grid Reinforcing	72
03 23	Stressed Tendon Reinforcing	73
03 24	Fibrous Reinforcing	74
03 30	Cast-In-Place Concrete	75
03 31	Structural Concrete	77
03 35	Concrete Finishing	80
03 37	Specialty Placed Concrete	82
03 39	Concrete Curing	83
03 41	Precast Structural Concrete	83
03 45	Precast Architectural Concrete	86
03 47	Site-Cast Concrete	86
03 48	Precast Concrete Specialties	86
03 51	Cast Roof Decks	87
03 52	Lightweight Concrete Roof Insulation	87
03 53	Concrete Topping	87
03 54	Cast Underlayment	88
03 62	Non-Shrink Grouting	88
03 63	Epoxy Grouting	88
03 81	Concrete Cutting	89
03 82	Concrete Boring	89
	Masonry	
04 01	Maintenance of Masonry	94
04 05	Common Work Results for Masonry	95
04 21	Clay Unit Masonry	99
04 22	Concrete Unit Masonry	104
04 23	Glass Unit Masonry	110
04 24	Adobe Unit Masonry	111

Sect. No.		Page
04 25	Unit Masonry Panels	111
04 27	Multiple-Wythe Unit Masonry	111
04 41	Dry-Placed Stone	112
04 43	Stone Masonry	112
04 51	Flue Liner Masonry	116
04 54	Refractory Brick Masonry	116
04 57	Masonry Fireplaces	117
04 71	Manufactured Brick Masonry	117
04 72	Cast Stone Masonry	117
04 73	Manufactured Stone Masonry	118
	Metals	
05 01	Maintenance of Metals	120
05 05	Common Work Results for Metals	120
05 12	Structural Steel Framing	127
05 14	Structural Aluminum Framing	135
05 15	Wire Rope Assemblies	136
05 21	Steel Joist Framing	138
05 31	Steel Decking	141
05 35	Raceway Decking Assemblies	143
05 41	Structural Metal Stud Framing	143
05 42	Cold-Formed Metal Joist Framing	147
05 44	Cold-Formed Metal Trusses	153
05 51	Metal Stairs	154
05 52	Metal Railings	156
05 53	Metal Gratings	157
05 54	Metal Floor Plates	159
05 55	Metal Stair Treads and Nosings	160
05 56	Metal Castings	161
05 58	Formed Metal Fabrications	161
05 71	Decorative Metal Stairs	163
05 73	Decorative Metal Railings	164
05 75	Decorative Formed Metal	164
	Wood, Plastics & Composites	
06 05	Common Work Results for Wood, Plastics, and Composites	166
06 11	Wood Framing	180
06 12	Structural Panels	188
06 13	Heavy Timber Construction	189
06 15	Wood Decking	190
06 16	Sheathing	190
06 17	Shop-Fabricated Structural Wood	193
06 18	Glued-Laminated Construction	193
06 22	Millwork	196
06 25	Prefinished Paneling	208
06 26	Board Paneling	209
06 43	Wood Stairs and Railings	209
06 44	Ornamental Woodwork	212
06 48	Wood Frames	213
06 49	Wood Screens and Exterior Wood Shutters	214
06 51	Structural Plastic Shapes and Plates	214
06 52	Plastic Structural Assemblies	216
06 63	Plastic Railings	217
06 65	Plastic Trim	217
06 80	Composite Fabrications	218
06 81	Composite Railings	219
	Thermal & Moisture Protection	
07 01	Operation and Maint. of Thermal and Moisture Protection	222
07 05	Common Work Results for Thermal and Moisture Protection	222
07 11	Dampproofing	224
07 12	Built-up Bituminous Waterproofing	224
07 13	Sheet Waterproofing	225
07 16	Cementitious and Reactive Waterproofing	225
07 17	Bentonite Waterproofing	225
07 19	Water Repellents	226
07 21	Thermal Insulation	226
07 22	Roof and Deck Insulation	230
07 24	Exterior Insulation and Finish Systems	231
07 25	Weather Barriers	232
07 26	Vapor Retarders	232

Sect. No.		Page
07 27	Air Barriers	233
07 31	Shingles and Shakes	233
07 32	Roof Tiles	235
07 33	Natural Roof Coverings	236
07 41	Roof Panels	237
07 42	Wall Panels	238
07 44	Faced Panels	239
07 46	Siding	240
07 51	Built-Up Bituminous Roofing	243
07 52	Modified Bituminous Membrane Roofing	244
07 53	Elastomeric Membrane Roofing	245
07 54	Thermoplastic Membrane Roofing	246
07 55	Protected Membrane Roofing	247
07 56	Fluid-Applied Roofing	247
07 57	Coated Foamed Roofing	247
07 58	Roll Roofing	248
07 61	Sheet Metal Roofing	248
07 65	Flexible Flashing	249
07 71	Roof Specialties	251
07 72	Roof Accessories	256
07 76	Roof Pavers	257
07 81	Applied Fireproofing	257
07 84	Firestopping	258
07 91	Preformed Joint Seals	260
07 92	Joint Sealants	261
07 95	Expansion Control	262
	Openings	
08 05	Common Work Results for Openings	264
08 11	Metal Doors and Frames	265
08 12	Metal Frames	266
08 13	Metal Doors	267
08 14	Wood Doors	270
08 16	Composite Doors	276
08 17	Integrated Door Opening Assemblies	277
08 31	Access Doors and Panels	278
08 32	Sliding Glass Doors	279
08 33	Coiling Doors and Grilles	280
08 34	Special Function Doors	281
08 36	Panel Doors	283
08 38	Traffic Doors	285
08 41	Entrances and Storefronts	285
08 42	Entrances	287
08 43	Storefronts	287
08 44	Curtain Wall and Glazed Assemblies	288
08 45	Translucent Wall and Roof Assemblies	288
08 51	Metal Windows	289
08 52	Wood Windows	290
08 53	Plastic Windows	295
08 54	Composite Windows	298
08 62	Unit Skylights	298
08 63	Metal-Framed Skylights	298
08 71	Door Hardware	299
08 75	Window Hardware	310
08 79	Hardware Accessories	310
08 81	Glass Glazing	310
08 83	Mirrors	313
08 84	Plastic Glazing	313
08 87	Glazing Surface Films	314
08 88	Special Function Glazing	315
08 91	Louvers	315
08 95	Vents	316
	Finishes	
09 01	Maintenance of Finishes	318
09 05	Common Work Results for Finishes	318
09 21	Plaster and Gypsum Board Assemblies	320

Table of Contents (cont.)

Sect. No.		Page
09 22	Supports for Plaster and Gypsum Board	322
09 23	Gypsum Plastering	327
09 24	Cement Plastering	328
09 25	Other Plastering	328
09 26	Veneer Plastering	328
09 28	Backing Boards and Underlayments	329
09 29	Gypsum Board	329
09 30	Tiling	333
09 31	Thin-Set Tiling	334
09 32	Mortar-Bed Tiling	335
09 34	Waterproofing-Membrane Tiling	337
09 35	Chemical-Resistant Tiling	337
09 51	Acoustical Ceilings	338
09 53	Acoustical Ceiling Suspension Assemblies	339
09 54	Specialty Ceilings	340
09 57	Special Function Ceilings	341
09 61	Flooring Treatment	341
09 62	Specialty Flooring	341
09 63	Masonry Flooring	342
09 64	Wood Flooring	343
09 65	Resilient Flooring	344
09 66	Terrazzo Flooring	347
09 67	Fluid-Applied Flooring	349
09 68	Carpeting	350
09 69	Access Flooring	352
09 72	Wall Coverings	352
09 74	Flexible Wood Sheets	354
09 77	Special Wall Surfacing	354
09 81	Acoustic Insulation	355
09 84	Acoustic Room Components	355
09 91	Painting	356
09 93	Staining and Transparent Finishing	369
09 96	High-Performance Coatings	369
09 97	Special Coatings	371

Specialties

10 05	Common Work Results for Specialties	374
10 11	Visual Display Units	374
10 13	Directories	377
10 14	Signage	378
10 17	Telephone Specialties	380
10 21	Compartments and Cubicles	380
10 22	Partitions	384
10 26	Wall and Door Protection	387
10 28	Toilet, Bath, and Laundry Accessories	388
10 31	Manufactured Fireplaces	391
10 32	Fireplace Specialties	392
10 35	Stoves	393
10 43	Emergency Aid Specialties	393
10 44	Fire Protection Specialties	393
10 51	Lockers	394
10 55	Postal Specialties	396
10 56	Storage Assemblies	396
10 57	Wardrobe and Closet Specialties	397
10 71	Exterior Protection	398
10 73	Protective Covers	398
10 74	Manufactured Exterior Specialties	399
10 75	Flagpoles	400
10 81	Pest Control Devices	401
10 86	Security Mirrors and Domes	401
10 88	Scales	402

Equipment

11 05	Common Work Results for Equipment	404
11 11	Vehicle Service Equipment	406
11 12	Parking Control Equipment	406
11 13	Loading Dock Equipment	407
11 14	Pedestrian Control Equipment	408
11 21	Retail and Service Equipment	409
11 22	Banking Equipment	410
11 30	Residential Equipment	412

Sect. No.		Page
11 32	Unit Kitchens	414
11 41	Foodservice Storage Equipment	414
11 42	Food Preparation Equipment	416
11 43	Food Delivery Carts and Conveyors	417
11 44	Food Cooking Equipment	417
11 46	Food Dispensing Equipment	418
11 48	Foodservice Cleaning and Disposal Equipment	419
11 52	Audio-Visual Equipment	419
11 53	Laboratory Equipment	421
11 57	Vocational Shop Equipment	422
11 61	Broadcast, Theater, and Stage Equipment	422
11 62	Musical Equipment	423
11 66	Athletic Equipment	423
11 67	Recreational Equipment	425
11 68	Play Field Equipment and Structures	425
11 71	Medical Sterilizing Equipment	427
11 72	Examination and Treatment Equipment	427
11 73	Patient Care Equipment	428
11 74	Dental Equipment	428
11 76	Operating Room Equipment	429
11 77	Radiology Equipment	429
11 78	Mortuary Equipment	430
11 81	Facility Maintenance Equipment	430
11 82	Facility Solid Waste Handling Equipment	430
11 91	Religious Equipment	432
11 92	Agricultural Equipment	433
11 97	Security Equipment	433
11 98	Detention Equipment	433

Furnishings

12 05	Common Work Results for Furnishings	436
12 21	Window Blinds	436
12 22	Curtains and Drapes	437
12 23	Interior Shutters	438
12 24	Window Shades	438
12 32	Manufactured Wood Casework	439
12 35	Specialty Casework	441
12 36	Countertops	442
12 46	Furnishing Accessories	445
12 48	Rugs and Mats	445
12 51	Office Furniture	446
12 52	Seating	447
12 54	Hospitality Furniture	447
12 55	Detention Furniture	448
12 56	Institutional Furniture	448
12 61	Fixed Audience Seating	450
12 63	Stadium and Arena Seating	450
12 67	Pews and Benches	450
12 92	Interior Planters and Artificial Plants	450
12 93	Interior Public Space Furnishings	451

Special Construction

13 05	Common Work Results for Special Construction	454
13 11	Swimming Pools	457
13 12	Fountains	459
13 17	Tubs and Pools	459
13 18	Ice Rinks	459
13 21	Controlled Environment Rooms	460
13 24	Special Activity Rooms	462
13 28	Athletic and Recreational Special Construction	462
13 31	Fabric Structures	463
13 33	Geodesic Structures	464
13 34	Fabricated Engineered Structures	465
13 36	Towers	471
13 42	Building Modules	471
13 47	Facility Protection	471

Sect. No.		Page
13 48	Sound, Vibration, and Seismic Control	472
13 49	Radiation Protection	472
13 53	Meteorological Instrumentation	474

Conveying Equipment

14 11	Manual Dumbwaiters	476
14 12	Electric Dumbwaiters	476
14 21	Electric Traction Elevators	476
14 24	Hydraulic Elevators	477
14 27	Custom Elevator Cabs and Doors	478
14 28	Elevator Equipment and Controls	479
14 31	Escalators	480
14 32	Moving Walks	481
14 42	Wheelchair Lifts	481
14 45	Vehicle Lifts	481
14 91	Facility Chutes	482
14 92	Pneumatic Tube Systems	482

Fire Suppression

21 05	Common Work Results for Fire Suppression	486
21 11	Facility Fire-Suppression Water-Service Piping	486
21 12	Fire-Suppression Standpipes	487
21 13	Fire-Suppression Sprinkler Systems	488
21 21	Carbon-Dioxide Fire-Extinguishing Systems	489
21 22	Clean-Agent Fire-Extinguishing Systems	489
21 31	Centrifugal Fire Pumps	490

Plumbing

22 01	Operation and Maintenance of Plumbing	492
22 05	Common Work Results for Plumbing	493
22 07	Plumbing Insulation	499
22 11	Facility Water Distribution	500
22 13	Facility Sanitary Sewerage	508
22 14	Facility Storm Drainage	511
22 31	Domestic Water Softeners	512
22 33	Electric Domestic Water Heaters	513
22 34	Fuel-Fired Domestic Water Heaters	513
22 35	Domestic Water Heat Exchangers	514
22 41	Residential Plumbing Fixtures	515
22 42	Commercial Plumbing Fixtures	518
22 45	Emergency Plumbing Fixtures	522
22 47	Drinking Fountains and Water Coolers	522
22 51	Swimming Pool Plumbing Systems	524
22 52	Fountain Plumbing Systems	524
22 66	Chemical-Waste Systems for Lab. and Healthcare Facilities	524

Heating Ventilation Air Conditioning

23 05	Common Work Results for HVAC	528
23 07	HVAC Insulation	530
23 09	Instrumentation and Control for HVAC	531
23 13	Facility Fuel-Storage Tanks	532
23 21	Hydronic Piping and Pumps	534
23 22	Steam and Condensate Piping and Pumps	537
23 31	HVAC Ducts and Casings	538
23 33	Air Duct Accessories	539
23 34	HVAC Fans	541
23 37	Air Outlets and Inlets	544
23 38	Ventilation Hoods	547
23 41	Particulate Air Filtration	547
23 42	Gas-Phase Air Filtration	548
23 43	Electronic Air Cleaners	548
23 51	Breechings, Chimneys, and Stacks	548
23 52	Heating Boilers	549
23 54	Furnaces	551
23 55	Fuel-Fired Heaters	551
23 56	Solar Energy Heating Equipment	553

Table of Contents (cont.)

Sect. No.		Page
23 57	Heat Exchangers for HVAC	554
23 62	Packaged Compressor and Condenser Units	554
23 63	Refrigerant Condensers	554
23 64	Packaged Water Chillers	555
23 65	Cooling Towers	557
23 73	Indoor Central-Station Air-Handling Units	557
23 74	Packaged Outdoor HVAC Equipment	558
23 81	Decentralized Unitary HVAC Equipment	559
23 82	Convection Heating and Cooling Units	560
23 83	Radiant Heating Units	562
23 84	Humidity Control Equipment	564

Electrical

26 01	Operation and Maintenance of Electrical Systems	566
26 05	Common Work Results for Electrical	567
26 09	Instrumentation and Control for Electrical Systems	584
26 12	Medium-Voltage Transformers	585
26 22	Low-Voltage Transformers	585
26 24	Switchboards and Panelboards	586
26 25	Low-Voltage Enclosed Bus Assemblies	589
26 27	Low-Voltage Distribution Equipment	590
26 28	Low-Voltage Circuit Protective Devices	591
26 29	Low-Voltage Controllers	592
26 32	Packaged Generator Assemblies	592
26 33	Battery Equipment	593
26 35	Power Filters and Conditioners	593
26 51	Interior Lighting	594
26 52	Safety Lighting	596
26 54	Classified Location Lighting	597
26 55	Special Purpose Lighting	597
26 56	Exterior Lighting	597
26 61	Lighting Systems and Accessories	599
26 71	Electrical Machines	602

Communications

27 13	Communications Backbone Cabling	604
27 41	Audio-Video Systems	604
27 51	Distributed Audio-Video Communications Systems	605
27 52	Healthcare Communications and Monitoring Systems	606
27 53	Distributed Systems	606

Sect. No.		Page
Electronic Safety & Security		
28 15	Access Control Hardware Devices	608
28 18	Security Access Detection Equipment	608
28 23	Video Management System	609
28 31	Intrusion Detection	609
28 42	Gas Detection and Alarm	610
28 46	Fire Detection and Alarm	610
28 47	Mass Notification	611

Earthwork

31 05	Common Work Results for Earthwork	614
31 06	Schedules for Earthwork	614
31 11	Clearing and Grubbing	615
31 13	Selective Tree and Shrub Removal and Trimming	616
31 14	Earth Stripping and Stockpiling	617
31 22	Grading	617
31 23	Excavation and Fill	618
31 25	Erosion and Sedimentation Controls	639
31 31	Soil Treatment	639
31 32	Soil Stabilization	640
31 33	Rock Stabilization	640
31 36	Gabions	640
31 37	Riprap	640
31 41	Shoring	641
31 43	Concrete Raising	642
31 45	Vibroflotation and Densification	642
31 46	Needle Beams	642
31 48	Underpinning	643
31 52	Cofferdams	643
31 56	Slurry Walls	644
31 62	Driven Piles	644
31 63	Bored Piles	646

Exterior Improvements

32 01	Operation and Maintenance of Exterior Improvements	652
32 06	Schedules for Exterior Improvements	652
32 11	Base Courses	653
32 12	Flexible Paving	654
32 13	Rigid Paving	655
32 14	Unit Paving	656
32 16	Curbs, Gutters, Sidewalks, and Driveways	657
32 17	Paving Specialties	658
32 18	Athletic and Recreational Surfacing	660
32 31	Fences and Gates	661
32 32	Retaining Walls	665
32 33	Site Furnishings	667

Sect. No.		Page
32 34	Fabricated Bridges	668
32 35	Screening Devices	668
32 84	Planting Irrigation	669
32 91	Planting Preparation	670
32 92	Turf and Grasses	671
32 93	Plants	672
32 94	Planting Accessories	673
32 96	Transplanting	674

Utilities

33 01	Operation and Maintenance of Utilities	676
33 05	Common Work Results for Utilities	676
33 11	Groundwater Sources	678
33 14	Water Utility Transmission and Distribution	678
33 16	Water Utility Storage Tanks	681
33 31	Sanitary Sewerage Piping	682
33 34	Onsite Wastewater Disposal	682
33 41	Subdrainage	684
33 42	Stormwater Conveyance	684
33 52	Hydrocarbon Transmission and Distribution	687
33 71	Electrical Utility Transmission and Distribution	703
33 81	Communications Structures	704

Transportation

34 01	Operation and Maintenance of Transportation	706
34 11	Rail Tracks	706
34 41	Roadway Signaling and Control Equipment	707
34 71	Roadway Construction	707
34 72	Railway Construction	709

Waterway & Marine

35 24	Dredging	712
35 51	Floating Construction	712

Material Processing & Handling Equipment

41 21	Conveyors	714
41 22	Cranes and Hoists	714

Pollution Control Equipment

44 11	Particulate Control Equipment	718

Water and Wastewater Equipment

46 07	Packaged Water and Wastewater Treatment Equipment	722

Electrical Power Generation

48 15	Wind Energy Electrical Power Generation Equipment	724

RSMeans data: Unit Prices— How They Work

All RSMeans data: Unit Prices are organized in the same way.

03 30 Cast-In-Place Concrete

03 30 53 – Miscellaneous Cast-In-Place Concrete

03 30 53.40 Concrete In Place		Crew	Daily Output	Labor-Hours	Unit	Material	2018 Bare Costs Labor	Equipment	Total	Total Incl O&P
0010 **CONCRETE IN PLACE**	R033105-10									
0020 Including forms (4 uses), Grade 60 rebar, concrete (Portland cement	R033105-20									
0050 Type I), placement and finishing unless otherwise indicated	R033105-50									
0300 Beams (3500 psi), 5 kip/L.F., 10' span	R033105-65	C-14A	15.62	12.804	C.Y.	340	645	58	1,043	1,425
0350 25' span	R033105-70	"	18.55	10.782		355	545	49	949	1,275
0500 Chimney foundations (5000 psi), over 5 C.Y.	R033105-85	C-14C	32.22	3.476		171	167	.80	338.80	445
0510 (3500 psi), under 5 C.Y.	R033053-50	"	23.71	4.724		198	227	1.09	426.09	565
0700 Columns, square (4000 psi), 12" x 12", up to 1% reinforcing by area		C-14A	11.96	16.722		380	845	76	1,301	1,775
3540 Equipment pad (3000 psi), 3' x 3' x 6" thick		C-14H	45	1.067	Ea.	44.50	52.50	.57	97.57	130
3550 4' x 4' x 6" thick			30	1.600		69	79	.85	148.85	197
3560 5' x 5' x 8" thick			18	2.667		126	132	1.41	259.41	340
3570 6' x 6' x 8" thick			14	3.429		173	169	1.82	343.82	450
3580 8' x 8' x 10" thick			8	6		370	296	3.18	669.18	860
3590 10' x 10' x 12" thick			5	9.600		645	475	5.10	1,125.10	1,425

It is important to understand the structure of RSMeans data: Unit Prices, so that you can find information easily and use it correctly.

① Line Numbers

Line Numbers consist of 12 characters, which identify a unique location in the database for each task. The first 6 or 8 digits conform to the Construction Specifications Institute MasterFormat® 2016. The remainder of the digits are a further breakdown in order to arrange items in understandable groups of similar tasks. Line numbers are consistent across all of our publications, so a line number in any of our products will always refer to the same item of work.

② Descriptions

Descriptions are shown in a hierarchical structure to make them readable. In order to read a complete description, read up through the indents to the top of the section. Include everything that is above and to the left that is not contradicted by information below. For instance, the complete description for line 03 30 53.40 3550 is "Concrete in place, including forms (4 uses), Grade 60 rebar, concrete (Portland cement Type 1), placement and finishing unless otherwise indicated; Equipment pad (3000 psi), 4' × 4' × 6" thick."

③ RSMeans data

When using **RSMeans data**, it is important to read through an entire section to ensure that you use the data that most closely matches your work. Note that sometimes there is additional information shown in the section that may improve your price. There are frequently lines that further describe, add to, or adjust data for specific situations.

④ Reference Information

RSMeans data engineers have created **reference** information to assist you in your estimate. **If** there is information that applies to a section, it will be indicated at the start of the section.The Reference Section is located in the back of the data set.

⑤ Crews

Crews include labor and/or equipment necessary to accomplish each task. In this case, Crew C-14H is used. RSMeans data Engineers selects a crew to represent the workers and equipment that are typically used for that task. In this case, Crew C-14H consists of one carpenter foreman (outside), two carpenters, one rodman, one laborer, one cement finisher, and one gas engine vibrator. Details of all crews can be found in the Reference Section.

Crews - Standard

Crew No.	Bare Costs		Incl. Subs O&P		Cost Per Labor-Hour	
Crew C-14H	Hr.	Daily	Hr.	Daily	Bare Costs	Incl. O&P
1 Carpenter Foreman (outside)	$52.70	$421.60	$80.25	$642.00	$49.36	$74.88
2 Carpenters	50.70	811.20	77.20	1235.20		
1 Rodman (reinf.)	54.65	437.20	83.45	667.60		
1 Laborer	39.85	318.80	60.70	485.60		
1 Cement Finisher	47.55	380.40	70.45	563.60		
1 Gas Engine Vibrator		25.60		28.16	.53	.59
48 L.H., Daily Totals		$2394.80		$3622.16	$49.89	$75.46

6 Daily Output

The **Daily Output** is the amount of work that the crew can do in a normal 8-hour workday, including mobilization, layout, movement of materials, and cleanup. In this case, crew C-14H can install thirty 4' × 4' × 6" thick concrete pads in a day. Daily output is variable and based on many factors, including the size of the job, location, and environmental conditions. RSMeans data represents work done in daylight (or adequate lighting) and temperate conditions.

7 Labor-Hours

The figure in the **Labor-Hours** column is the amount of labor required to perform one unit of work–in this case the amount of labor required to construct one 4' × 4' equipment pad. This figure is calculated by dividing the number of hours of labor in the crew by the daily output (48 labor-hours divided by 30 pads = 1.6 hours of labor per pad). Multiply 1.600 times 60 to see the value in minutes: 60 × 1.6 = 96

minutes. Note: the labor-hour figure is not dependent on the crew size. A change in crew size will result in a corresponding change in daily output, but the labor-hours per unit of work will not change.

8 Unit of Measure

All RSMeans data: Unit Prices include the typical **Unit of Measure** used for estimating that item. For concrete-in-place the typical unit is cubic yards (C.Y.) or each (Ea.). For installing broadloom carpet it is square yard, and for gypsum board it is square foot. The estimator needs to take special care that the unit in the data matches the unit in the take-off. Unit conversions may be found in the Reference Section.

9 Bare Costs

Bare Costs are the costs of materials, labor, and equipment that the installing contractor pays. They represent the cost, in U.S. dollars, for one unit of work. They do not include any markups for profit or labor burden.

10 Bare Total

The **Total column** represents the total bare cost for the installing contractor in U.S. dollars. In this case, the sum of $69 for material + $79 for labor + $.85 for equipment is $148.85.

11 Total Incl O&P

The **Total Incl O&P column** is the total cost, including overhead and profit, that the installing contractor will charge the customer. This represents the cost of materials plus 10% profit, the cost of labor plus labor burden and 10% profit, and the cost of equipment plus 10% profit. It does not include the general contractor's overhead and profit. Note: See the inside back cover of the printed product or the Reference Section of the electronic product for details of how the labor burden is calculated.

National Average

*The RSMeans data in our print publications represents a "national average" cost. This data should be modified to the project location using the **City Cost Indexes** or **Location Factors** tables found in the Reference Section. Use the Location Factors to adjust estimate totals if the project covers multiple trades. Use the City Cost Indexes (CCI) for single trade*

projects or projects where a more detailed analysis is required. All figures in the two tables are derived from the same research. The last row of data in the CCI—the weighted average—is the same as the numbers reported for each location in the location factor table.

Project Name: Pre-Engineered Steel Building | **Architect: As Shown**

Location:	Anywhere, USA						01/01/18	STD
Line Number	Description	Qty	Unit	Material	Labor	Equipment	SubContract	Estimate Total
03 30 53.40 3940	Strip footing, 12" x 24", reinforced	34	C.Y.	$5,406.00	$3,808.00	$18.36	$0.00	
03 30 53.40 3950	Strip footing, 12" x 36", reinforced	15	C.Y.	$2,295.00	$1,350.00	$6.45	$0.00	
03 11 13.65 3000	Concrete slab edge forms	500	L.F.	$145.00	$1,280.00	$0.00	$0.00	
03 22 11.10 0200	Welded wire fabric reinforcing	150	C.S.F.	$2,940.00	$4,200.00	$0.00	$0.00	
03 31 13.35 0300	Ready mix concrete, 4000 psi for slab on grade	278	C.Y.	$35,584.00	$0.00	$0.00	$0.00	
03 31 13.70 4300	Place, strike off & consolidate concrete slab	278	C.Y.	$0.00	$5,031.80	$130.66	$0.00	
03 35 13.30 0250	Machine float & trowel concrete slab	15,000	S.F.	$0.00	$9,450.00	$300.00	$0.00	
03 15 16.20 0140	Cut control joints in concrete slab	950	L.F.	$47.50	$399.00	$57.00	$0.00	
03 39 23.13 0300	Sprayed concrete curing membrane	150	C.S.F.	$1,815.00	$1,005.00	$0.00	$0.00	
Division 03	**Subtotal**			**$48,232.50**	**$26,523.80**	**$512.47**	**$0.00**	**$75,268.77**
08 36 13.10 2650	Manual 10' x 10' steel sectional overhead door	8	Ea.	$10,400.00	$3,600.00	$0.00	$0.00	
08 36 13.10 2860	Insulation and steel back panel for OH door	800	S.F.	$4,000.00	$0.00	$0.00	$0.00	
Division 08	**Subtotal**			**$14,400.00**	**$3,600.00**	**$0.00**	**$0.00**	**$18,000.00**
13 34 19.50 1100	Pre-Engineered Steel Building, 100' x 150' x 24'	15,000	SF Flr.	$0.00	$0.00	$0.00	$367,500.00	
13 34 19.50 6050	Framing for PESB door opening, 3' x 7'	4	Opng.	$0.00	$0.00	$0.00	$2,240.00	
13 34 19.50 6100	Framing for PESB door opening, 10' x 10'	8	Opng.	$0.00	$0.00	$0.00	$9,200.00	
13 34 19.50 6200	Framing for PESB window opening, 4' x 3'	6	Opng.	$0.00	$0.00	$0.00	$3,330.00	
13 34 19.50 5750	PESB door, 3' x 7', single leaf	4	Opng.	$2,620.00	$700.00	$0.00	$0.00	
13 34 19.50 7750	PESB sliding window, 4' x 3' with screen	6	Opng.	$2,550.00	$600.00	$45.30	$0.00	
13 34 19.50 6550	PESB gutter, eave type, 26 ga., painted	300	L.F.	$2,220.00	$819.00	$0.00	$0.00	
13 34 19.50 8650	PESB roof vent, 12" wide x 10' long	15	Ea.	$555.00	$3,285.00	$0.00	$0.00	
13 34 19.50 6900	PESB insulation, vinyl faced, 4" thick	27,400	S.F.	$13,152.00	$9,590.00	$0.00	$0.00	
Division 13	**Subtotal**			**$21,097.00**	**$14,994.00**	**$45.30**	**$382,270.00**	**$418,406.30**
			Subtotal	$83,729.50	$45,117.80	$557.77	$382,270.00	$511,675.07
Division 01	General Requirements @ 7%			5,861.07	3,158.25	39.04	26,758.90	
			Estimate Subtotal	$89,590.57	$48,276.05	$596.81	$409,028.90	$511,675.07
			Sales Tax @ 5%	4,479.53		29.84	10,225.72	
			Subtotal A	94,070.09	48,276.05	626.65	419,254.62	
			GC O & P	9,407.01	25,634.58	62.67	41,925.46	
			Subtotal B	103,477.10	73,910.63	689.32	461,180.08	$639,257.13
			Contingency @ 5%					31,962.86
			Subtotal C					$671,219.99
			Bond @ $12/1000 +10% O&P					8,860.10
			Subtotal D					$680,080.09
			Location Adjustment Factor		102.30			15,641.84
			Grand Total					**$695,721.94**

This estimate is based on an interactive spreadsheet. You are free to download it and adjust it to your methodology.
A copy of this spreadsheet is available at **www.RSMeans.com/2018books.**

Sample Estimate

This sample demonstrates the elements of an estimate, including a tally of the RSMeans data lines and a summary of the markups on a contractor's work to arrive at a total cost to the owner. The Location Factor with RSMeans data is added at the bottom of the estimate to adjust the cost of the work to a specific location.

① Work Performed

The body of the estimate shows the RSMeans data selected, including the line number, a brief description of each item, its take-off unit and quantity, and the bare costs of materials, labor, and equipment. This estimate also includes a column titled "SubContract." This data is taken from the column "Total Incl O&P" and represents the total that a subcontractor would charge a general contractor for the work, including the sub's markup for overhead and profit.

② Division 1, General Requirements

This is the first division numerically but the last division estimated. Division 1 includes project-wide needs provided by the general contractor. These requirements vary by project but may include temporary facilities and utilities, security, testing, project cleanup, etc. For small projects a percentage can be used—typically between 5% and 15% of project cost. For large projects the costs may be itemized and priced individually.

③ Sales Tax

If the work is subject to state or local sales taxes, the amount must be added to the estimate. Sales tax may be added to material costs, equipment costs, and subcontracted work. In this case, sales tax was added in all three categories. It was assumed that approximately half the subcontracted work would be material cost, so the tax was applied to 50% of the subcontract total.

④ GC O&P

This entry represents the general contractor's markup on material, labor, equipment, and subcontractor costs. Our standard markup on materials, equipment, and subcontracted work is 10%. In this estimate, the markup on the labor performed by the GC's workers uses "Skilled Workers Average" shown in Column F on the table "Installing Contractor's Overhead & Profit," which can be found on the inside back cover of the printed product or in the Reference Section of the electronic product.

⑤ Contingency

A factor for contingency may be added to any estimate to represent the cost of unknowns that may occur between the time that the estimate is performed and the time the project is constructed. The amount of the allowance will depend on the stage of design at which the estimate is done and the contractor's assessment of the risk involved. Refer to section 01 21 16.50 for contingency allowances.

⑥ Bonds

Bond costs should be added to the estimate. The figures here represent a typical performance bond, ensuring the owner that if the general contractor does not complete the obligations in the construction contract the bonding company will pay the cost for completion of the work.

⑦ Location Adjustment

Published prices are based on national average costs. If necessary, adjust the total cost of the project using a location factor from the "Location Factor" table or the "City Cost Index" table. Use location factors if the work is general, covering multiple trades. If the work is by a single trade (e.g., masonry) use the more specific data found in the "City Cost Indexes."

Estimating Tips
01 20 00 Price and Payment Procedures

- Allowances that should be added to estimates to cover contingencies and job conditions that are not included in the national average material and labor costs are shown in Section 01 21.

- When estimating historic preservation projects (depending on the condition of the existing structure and the owner's requirements), a 15%–20% contingency or allowance is recommended, regardless of the stage of the drawings.

01 30 00 Administrative Requirements

- Before determining a final cost estimate, it is a good practice to review all the items listed in Subdivisions 01 31 and 01 32 to make final adjustments for items that may need customizing to specific job conditions.

- Requirements for initial and periodic submittals can represent a significant cost to the General Requirements of a job. Thoroughly check the submittal specifications when estimating a project to determine any costs that should be included.

01 40 00 Quality Requirements

- All projects will require some degree of quality control. This cost is not included in the unit cost of construction listed in each division. Depending upon the terms of the contract, the various costs of inspection and testing can be the responsibility of either the owner or the contractor. Be sure to include the required costs in your estimate.

01 50 00 Temporary Facilities and Controls

- Barricades, access roads, safety nets, scaffolding, security, and many more requirements for the execution of a safe project are elements of direct cost. These costs can easily be overlooked when preparing an estimate. When looking through the major classifications of this subdivision, determine which items apply to each division in your estimate.

- Construction equipment rental costs can be found in the Reference Section in Section 01 54 33. Operators' wages are not included in equipment rental costs.

- Equipment mobilization and demobilization costs are not included in equipment rental costs and must be considered separately.

- The cost of small tools provided by the installing contractor for his workers is covered in the "Overhead" column on the "Installing Contractor's Overhead and Profit" table that lists labor trades, base rates, and markups and, therefore, is included in the "Total Incl. O&P" cost of any unit price line item.

01 70 00 Execution and Closeout Requirements

- When preparing an estimate, thoroughly read the specifications to determine the requirements for Contract Closeout. Final cleaning, record documentation, operation and maintenance data, warranties and bonds, and spare parts and maintenance materials can all be elements of cost for the completion of a contract. Do not overlook these in your estimate.

Reference Numbers

Reference numbers are shown at the beginning of some major classifications. These numbers refer to related items in the Reference Section. The reference information may be an estimating procedure, an alternate pricing method, or technical information.

Note: Not all subdivisions listed here necessarily appear. ■

Did you know?

RSMeans data is available through our online application with 24/7 access:

- Search for unit prices by keyword
- Leverage the most up-to-date data
- Build and export estimates

Try it free for 30 days!
www.rsmeans.com/2018freetrial

No part of this cost data may be reproduced, stored in a retrieval system, or transmitted in any form or by any means without prior written permission of Gordian.

01 11 Summary of Work

01 11 31 – Professional Consultants

01 11 31.10 Architectural Fees

	01 11 31.10 Architectural Fees		Crew	Daily Output	Labor-Hours	Unit	Material	2018 Bare Costs Labor	Equipment	Total	Total Incl O&P
0010	**ARCHITECTURAL FEES**	R011110-10									
0020	For new construction										
0060	Minimum					Project				4.90%	4.90%
0090	Maximum									16%	16%
0100	For alteration work, to $500,000, add to new construction fee									50%	50%
0150	Over $500,000, add to new construction fee									25%	25%
2000	For "Greening" of building	G								3%	3%

01 11 31.20 Construction Management Fees

0010	**CONSTRUCTION MANAGEMENT FEES**										
0020	$1,000,000 job, minimum					Project				4.50%	4.50%
0050	Maximum									7.50%	7.50%
0060	For work to $100,000									10%	10%
0070	To $250,000									9%	9%
0090	To $1,000,000									6%	6%
0100	To $5,000,000									5%	5%
0110	To $10,000,000									4%	4%
0300	$50,000,000 job, minimum									2.50%	2.50%
0350	Maximum									4%	4%

01 11 31.30 Engineering Fees

0010	**ENGINEERING FEES**	R011110-30									
0020	Educational planning consultant, minimum					Project				.50%	.50%
0100	Maximum					"				2.50%	2.50%
0200	Electrical, minimum					Contrct				4.10%	4.10%
0300	Maximum									10.10%	10.10%
0400	Elevator & conveying systems, minimum									2.50%	2.50%
0500	Maximum									5%	5%
0600	Food service & kitchen equipment, minimum									8%	8%
0700	Maximum									12%	12%
0800	Landscaping & site development, minimum									2.50%	2.50%
0900	Maximum									6%	6%
1000	Mechanical (plumbing & HVAC), minimum									4.10%	4.10%
1100	Maximum									10.10%	10.10%
1200	Structural, minimum					Project				1%	1%
1300	Maximum					"				2.50%	2.50%

01 11 31.50 Models

0010	**MODELS**										
0500	2 story building, scaled 100' x 200', simple materials and details					Ea.	4,650			4,650	5,100
0510	Elaborate materials and details					"	30,300			30,300	33,300

01 11 31.75 Renderings

0010	**RENDERINGS** Color, matted, 20" x 30", eye level,										
0020	1 building, minimum					Ea.	2,225			2,225	2,450
0050	Average						3,125			3,125	3,450
0100	Maximum						5,050			5,050	5,550
1000	5 buildings, minimum						4,150			4,150	4,550
1100	Maximum						8,175			8,175	9,000
2000	Aerial perspective, color, 1 building, minimum						3,025			3,025	3,325
2100	Maximum						8,275			8,275	9,100
3000	5 buildings, minimum						6,050			6,050	6,675
3100	Maximum						12,200			12,200	13,400

For customer support on your Building Construction Costs with RSMeans data, call 800.448.8182.

01 21 Allowances

01 21 16 – Contingency Allowances

01 21 16.50 Contingencies	Crew	Daily Output	Labor-Hours	Unit	Material	2018 Bare Costs Labor	Equipment	Total	Total Incl O&P
0010 **CONTINGENCIES**, Add to estimate									
0020 Conceptual stage				Project				20%	20%
0050 Schematic stage								15%	15%
0100 Preliminary working drawing stage (Design Dev.)								10%	10%
0150 Final working drawing stage				▼				3%	3%

01 21 53 – Factors Allowance

01 21 53.60 Security Factors

	Crew	Daily Output	Labor-Hours	Unit	Material	2018 Bare Costs Labor	Equipment	Total	Total Incl O&P
0010 **SECURITY FACTORS** R012153-60									
0100 Additional costs due to security requirements									
0110 Daily search of personnel, supplies, equipment and vehicles									
0120 Physical search, inventory and doc of assets, at entry				Costs		30%			
0130 At entry and exit						50%			
0140 Physical search, at entry						6.25%			
0150 At entry and exit						12.50%			
0160 Electronic scan search, at entry						2%			
0170 At entry and exit						4%			
0180 Visual inspection only, at entry						.25%			
0190 At entry and exit						.50%			
0200 ID card or display sticker only, at entry						.12%			
0210 At entry and exit				▼		.25%			
0220 Day 1 as described below, then visual only for up to 5 day job duration									
0230 Physical search, inventory and doc of assets, at entry				Costs		5%			
0240 At entry and exit						10%			
0250 Physical search, at entry						1.25%			
0260 At entry and exit						2.50%			
0270 Electronic scan search, at entry						.42%			
0280 At entry and exit				▼		.83%			
0290 Day 1 as described below, then visual only for 6-10 day job duration									
0300 Physical search, inventory and doc of assets, at entry				Costs		2.50%			
0310 At entry and exit						5%			
0320 Physical search, at entry						.63%			
0330 At entry and exit						1.25%			
0340 Electronic scan search, at entry						.21%			
0350 At entry and exit				▼		.42%			
0360 Day 1 as described below, then visual only for 11-20 day job duration									
0370 Physical search, inventory and doc of assets, at entry				Costs		1.25%			
0380 At entry and exit						2.50%			
0390 Physical search, at entry						.31%			
0400 At entry and exit						.63%			
0410 Electronic scan search, at entry						.10%			
0420 At entry and exit				▼		.21%			
0430 Beyond 20 days, costs are negligible									
0440 Escort required to be with tradesperson during work effort				Costs		6.25%			

01 21 55 – Job Conditions Allowance

01 21 55.50 Job Conditions

	Crew	Daily Output	Labor-Hours	Unit	Material	2018 Bare Costs Labor	Equipment	Total	Total Incl O&P
0010 **JOB CONDITIONS** Modifications to applicable R012157-20									
0020 cost summaries									
0100 Economic conditions, favorable, deduct				Project				2%	2%
0200 Unfavorable, add								5%	5%
0300 Hoisting conditions, favorable, deduct								2%	2%
0400 Unfavorable, add								5%	5%
0500 General contractor management, experienced, deduct				▼				2%	2%

For customer support on your Building Construction Costs with RSMeans data, call 800.448.8182.

01 21 Allowances

01 21 55 – Job Conditions Allowance

01 21 55.50 Job Conditions	Crew	Daily Output	Labor-Hours	Unit	Material	2018 Bare Costs Labor	Equipment	Total	Total Incl O&P
0600 Inexperienced, add				Project				10%	10%
0700 Labor availability, surplus, deduct								1%	1%
0800 Shortage, add								10%	10%
0900 Material storage area, available, deduct								1%	1%
1000 Not available, add								2%	2%
1100 Subcontractor availability, surplus, deduct								5%	5%
1200 Shortage, add								12%	12%
1300 Work space, available, deduct								2%	2%
1400 Not available, add								5%	5%

01 21 57 – Overtime Allowance

01 21 57.50 Overtime

0010 **OVERTIME** for early completion of projects or where	R012909-90									
0020 labor shortages exist, add to usual labor, up to					Costs		100%			

01 21 63 – Taxes

01 21 63.10 Taxes

0010 **TAXES**	R012909-80									
0020 Sales tax, State, average					%	5.08%				
0050 Maximum	R012909-85					7.50%				
0200 Social Security, on first $118,500 of wages							7.65%			
0300 Unemployment, combined Federal and State, minimum	R012909-86						.60%			
0350 Average							9.60%			
0400 Maximum							12%			

01 31 Project Management and Coordination

01 31 13 – Project Coordination

01 31 13.20 Field Personnel

					Unit	Material	Labor	Equipment	Total	Total Incl O&P
0010 **FIELD PERSONNEL**										
0020 Clerk, average					Week		485		485	740
0100 Field engineer, junior engineer							1,150		1,150	1,750
0120 Engineer							1,500		1,500	2,300
0140 Senior engineer							1,700		1,700	2,600
0160 General purpose laborer, average							1,600		1,600	2,425
0180 Project manager, minimum							2,125		2,125	3,275
0200 Average							2,450		2,450	3,750
0220 Maximum							2,800		2,800	4,275
0240 Superintendent, minimum							2,075		2,075	3,175
0260 Average							2,275		2,275	3,500
0280 Maximum							2,600		2,600	3,975
0290 Timekeeper, average							1,325		1,325	2,025

01 31 13.30 Insurance

					Unit	Material	Labor	Equipment	Total	Total Incl O&P
0010 **INSURANCE**	R013113-40									
0020 Builders risk, standard, minimum					Job				.24%	.24%
0050 Maximum	R013113-50								.64%	.64%
0200 All-risk type, minimum									.25%	.25%
0250 Maximum	R013113-60								.62%	.62%
0400 Contractor's equipment floater, minimum					Value				.50%	.50%
0450 Maximum					"				1.50%	1.50%
0600 Public liability, average					Job				2.02%	2.02%
0800 Workers' compensation & employer's liability, average										
0850 by trade, carpentry, general					Payroll		13.05%			

For customer support on your Building Construction Costs with RSMeans data, call 800.448.8182.

01 31 Project Management and Coordination

01 31 13 – Project Coordination

01 31 13.30 Insurance

		Crew	Daily Output	Labor-Hours	Unit	Material	2018 Bare Costs Labor	Equipment	Total	Total Incl O&P
0900	Clerical				Payroll		.42%			
0950	Concrete						11.85%			
1000	Electrical						5.22%			
1050	Excavation						8.48%			
1100	Glazing						12.10%			
1150	Insulation						10.90%			
1200	Lathing						8.50%			
1250	Masonry						13.73%			
1300	Painting & decorating						11.19%			
1350	Pile driving						13.38%			
1400	Plastering						10.72%			
1450	Plumbing						6.94%			
1500	Roofing						28.99%			
1550	Sheet metal work (HVAC)						8.27%			
1600	Steel erection, structural						27.45%			
1650	Tile work, interior ceramic						8.85%			
1700	Waterproofing, brush or hand caulking						6.46%			
1800	Wrecking						18.01%			
2000	Range of 35 trades in 50 states, excl. wrecking & clerical, min.						1.30%			
2100	Average						11.53%			
2200	Maximum						120.29%			

01 31 13.40 Main Office Expense

		Crew	Daily Output	Labor-Hours	Unit	Material	Labor	Equipment	Total	Total Incl O&P
0010	**MAIN OFFICE EXPENSE** Average for General Contractors R013113-50									
0020	As a percentage of their annual volume									
0125	Annual volume under $1,000,000				% Vol.				17.50%	
0145	Up to $2,500,000								8%	
0150	Up to $4,000,000								6.80%	
0200	Up to $7,000,000								5.60%	
0250	Up to $10,000,000								5.10%	
0300	Over $10,000,000								3.90%	

01 31 13.50 General Contractor's Mark-Up

		Crew	Daily Output	Labor-Hours	Unit	Material	Labor	Equipment	Total	Total Incl O&P
0010	**GENERAL CONTRACTOR'S MARK-UP** on Change Orders									
0200	Extra work, by subcontractors, add				%				10%	10%
0250	By General Contractor, add								15%	15%
0400	Omitted work, by subcontractors, deduct all but								5%	5%
0450	By General Contractor, deduct all but								7.50%	7.50%
0600	Overtime work, by subcontractors, add								15%	15%
0650	By General Contractor, add								10%	10%

01 31 13.80 Overhead and Profit

		Crew	Daily Output	Labor-Hours	Unit	Material	Labor	Equipment	Total	Total Incl O&P
0010	**OVERHEAD & PROFIT** Allowance to add to items in this									
0020	book that do not include Subs O&P, average				%				25%	
0100	Allowance to add to items in this book that R013113-55									
0110	do include Subs O&P, minimum				%				5%	5%
0150	Average								10%	10%
0200	Maximum								15%	15%
0300	Typical, by size of project, under $100,000								30%	
0350	$500,000 project								25%	
0400	$2,000,000 project								20%	
0450	Over $10,000,000 project								15%	

For customer support on your Building Construction Costs with RSMeans data, call 800.448.8182.

01 31 Project Management and Coordination

01 31 13 – Project Coordination

01 31 13.90 Performance Bond		Crew	Daily Output	Labor-Hours	Unit	Material	2018 Bare Costs Labor	Equipment	Total	Total Incl O&P
0010	**PERFORMANCE BOND**	R013113-80									
0020	For buildings, minimum					Job				.60%	.60%
0100	Maximum					"				2.50%	2.50%

01 32 Construction Progress Documentation

01 32 13 – Scheduling of Work

01 32 13.50 Scheduling of Work

		Crew	Daily Output	Labor-Hours	Unit	Material	Labor	Equipment	Total	Total Incl O&P
0010	**SCHEDULING**									
0020	Critical path, as % of architectural fee, minimum				%				.50%	.50%
0100	Maximum				"				1%	1%
0300	Computer-update, micro, no plots, minimum				Ea.				455	500
0400	Including plots, maximum				"				1,450	1,600
0600	Rule of thumb, CPM scheduling, small job ($10 Million)				Job				.05%	.05%
0650	Large job ($50 Million +)								.03%	.03%
0700	Including cost control, small job								.08%	.08%
0750	Large job								.04%	.04%

01 32 33 – Photographic Documentation

01 32 33.50 Photographs

		Crew	Daily Output	Labor-Hours	Unit	Material	Labor	Equipment	Total	Total Incl O&P
0010	**PHOTOGRAPHS**									
0020	8" x 10", 4 shots, 2 prints ea., std. mounting				Set	540			540	595
0100	Hinged linen mounts					545			545	600
0200	8" x 10", 4 shots, 2 prints each, in color					500			500	550
0300	For I.D. slugs, add to all above					5.05			5.05	5.55
0500	Aerial photos, initial fly-over, 5 shots, digital images					415			415	455
0550	10 shots, digital images, 1 print					455			455	500
0600	For each additional print from fly-over					207			207	228
0700	For full color prints, add					40%				
0750	Add for traffic control area				Ea.	335			335	370
0900	For over 30 miles from airport, add per				Mile	6.75			6.75	7.40
1500	Time lapse equipment, camera and projector, buy				Ea.	2,700			2,700	2,975
1550	Rent per month				"	1,275			1,275	1,400
1700	Cameraman and processing, black & white				Day	1,275			1,275	1,400
1720	Color				"	1,475			1,475	1,625

01 41 Regulatory Requirements

01 41 26 – Permit Requirements

01 41 26.50 Permits

		Crew	Daily Output	Labor-Hours	Unit	Material	Labor	Equipment	Total	Total Incl O&P
0010	**PERMITS**									
0020	Rule of thumb, most cities, minimum				Job				.50%	.50%
0100	Maximum				"				2%	2%

For customer support on your Building Construction Costs with RSMeans data, call 800.448.8182.

01 45 Quality Control

01 45 23 – Testing and Inspecting Services

01 45 23.50 Testing	Crew	Daily Output	Labor-Hours	Unit	Material	2018 Bare Costs Labor	Equipment	Total	Total Incl O&P
0010 **TESTING** and Inspecting Services									
0015 For concrete building costing $1,000,000, minimum				Project				4,725	5,200
0020 Maximum								38,000	41,800
0050 Steel building, minimum								4,725	5,200
0070 Maximum								14,800	16,300
0100 For building costing $10,000,000, minimum								30,100	33,100
0150 Maximum				▼				48,200	53,000
0200 Asphalt testing, compressive strength Marshall stability, set of 3				Ea.				145	165
0220 Density, set of 3								86	95
0250 Extraction, individual tests on sample								136	150
0300 Penetration								41	45
0350 Mix design, 5 specimens								182	200
0360 Additional specimen								36	40
0400 Specific gravity								41	45
0420 Swell test								64	70
0450 Water effect and cohesion, set of 6								182	200
0470 Water effect and plastic flow								64	70
0600 Concrete testing, aggregates, abrasion, ASTM C 131								136	150
0650 Absorption, ASTM C 127								42	46
0800 Petrographic analysis, ASTM C 295								775	850
0900 Specific gravity, ASTM C 127								50	55
1000 Sieve analysis, washed, ASTM C 136								59	65
1050 Unwashed								59	65
1200 Sulfate soundness								114	125
1300 Weight per cubic foot								36	40
1500 Cement, physical tests, ASTM C 150								320	350
1600 Chemical tests, ASTM C 150								245	270
1800 Compressive test, cylinder, delivered to lab, ASTM C 39								12	13
1900 Picked up by lab, minimum								14	15
1950 Average								18	20
2000 Maximum								27	30
2200 Compressive strength, cores (not incl. drilling), ASTM C 42				▼				36	40
2250 Core drilling, 4" diameter (plus technician)				Inch				23	25
2260 Technician for core drilling				Hr.				45	50
2300 Patching core holes				Ea.				22	24
2400 Drying shrinkage at 28 days								236	260
2500 Flexural test beams, ASTM C 78								59	65
2600 Mix design, one batch mix								259	285
2650 Added trial batches								120	132
2800 Modulus of elasticity, ASTM C 469								164	180
2900 Tensile test, cylinders, ASTM C 496								45	50
3000 Water-Cement ratio curve, 3 batches								141	155
3100 4 batches								186	205
3300 Masonry testing, absorption, per 5 brick, ASTM C 67								45	50
3350 Chemical resistance, per 2 brick								50	55
3400 Compressive strength, per 5 brick, ASTM C 67								68	75
3420 Efflorescence, per 5 brick, ASTM C 67								68	75
3440 Imperviousness, per 5 brick								87	96
3470 Modulus of rupture, per 5 brick								86	95
3500 Moisture, block only								32	35
3550 Mortar, compressive strength, set of 3								23	25
4100 Reinforcing steel, bend test								55	61
4200 Tensile test, up to #8 bar				▼				36	40

For customer support on your Building Construction Costs with RSMeans data, call 800.448.8182.

15

01 45 23.50 Testing		Crew	Daily Output	Labor-Hours	Unit	Material	2018 Bare Costs Labor	Equipment	Total	Total Incl O&P
4220	#9 to #11 bar				Ea.				41	45
4240	#14 bar and larger								64	70
4400	Soil testing, Atterberg limits, liquid and plastic limits								59	65
4510	Hydrometer analysis								109	120
4530	Specific gravity, ASTM D 354								44	48
4600	Sieve analysis, washed, ASTM D 422								55	60
4700	Unwashed, ASTM D 422								59	65
4710	Consolidation test (ASTM D 2435), minimum								250	275
4715	Maximum								430	475
4720	Density and classification of undisturbed sample								73	80
4735	Soil density, nuclear method, ASTM D 2922								35	38.50
4740	Sand cone method, ASTM D 1556								27	30
4750	Moisture content, ASTM D 2216								9	10
4780	Permeability test, double ring infiltrometer								500	550
4800	Permeability, var. or constant head, undist., ASTM D 2434								227	250
4850	Recompacted								250	275
4900	Proctor compaction, 4" standard mold, ASTM D 698								123	135
4950	6" modified mold								68	75
5100	Shear tests, triaxial, minimum								410	450
5150	Maximum								545	600
5300	Direct shear, minimum, ASTM D 3080								320	350
5350	Maximum								410	450
5550	Technician for inspection, per day, earthwork								320	350
5650	Bolting								400	440
5750	Roofing								480	530
5790	Welding				▼				480	530
5820	Non-destructive metal testing, dye penetrant				Day				310	340
5840	Magnetic particle								310	340
5860	Radiography								450	495
5880	Ultrasonic				▼				310	340
6000	Welding certification, minimum				Ea.				91	100
6100	Maximum				"				250	275
7000	Underground storage tank									
7500	Volumetric tightness test, <=12,000 gal.				Ea.				435	480
7510	<=30,000 gal.				"				615	675
7600	Vadose zone (soil gas) sampling, 10-40 samples, min.				Day				1,375	1,500
7610	Maximum				"				2,275	2,500
7700	Ground water monitoring incl. drilling 3 wells, min.				Total				4,550	5,000
7710	Maximum				"				6,375	7,000
8000	X-ray concrete slabs				Ea.				182	200
9000	Thermographic testing, for bldg envelope heat loss, average 2,000 S.F. [G]				"				500	500

01 51 Temporary Utilities

01 51 13 – Temporary Electricity

01 51 13.80 Temporary Utilities

		Crew	Daily Output	Labor-Hours	Unit	Material	2018 Bare Costs Labor	2018 Bare Costs Equipment	Total	Total Incl O&P
0010	**TEMPORARY UTILITIES** R015113-65									
0350	Lighting, lamps, wiring, outlets, 40,000 S.F. building, 8 strings	1 Elec	34	.235	CSF Flr	5.60	13.70		19.30	26.50
0360	16 strings	"	17	.471		11.20	27.50		38.70	53.50
0400	Power for temp lighting only, 6.6 KWH, per month								.92	1.01
0430	11.8 KWH, per month								1.65	1.82
0450	23.6 KWH, per month								3.30	3.63
0600	Power for job duration incl. elevator, etc., minimum								47	51.50
0650	Maximum								110	121
1000	Toilet, portable, see Equip. Rental 01 54 33 in Reference Section									

01 52 Construction Facilities

01 52 13 – Field Offices and Sheds

01 52 13.20 Office and Storage Space

		Crew	Daily Output	Labor-Hours	Unit	Material	Labor	Equipment	Total	Total Incl O&P
0010	**OFFICE AND STORAGE SPACE**									
0020	Office trailer, furnished, no hookups, 20' x 8', buy	2 Skwk	1	16	Ea.	8,900	840		9,740	11,100
0250	Rent per month					198			198	218
0300	32' x 8', buy	2 Skwk	.70	22.857		14,200	1,200		15,400	17,400
0350	Rent per month					247			247	272
0400	50' x 10', buy	2 Skwk	.60	26.667		29,300	1,400		30,700	34,400
0450	Rent per month					355			355	395
0500	50' x 12', buy	2 Skwk	.50	32		25,900	1,675		27,575	31,100
0550	Rent per month					450			450	495
0700	For air conditioning, rent per month, add					50			50	55
0800	For delivery, add per mile				Mile	12			12	13.20
0890	Delivery each way				Ea.	2,725			2,725	3,000
0900	Bunk house trailer, 8' x 40' duplex dorm with kitchen, no hookups, buy	2 Carp	1	16		87,000	810		87,810	96,500
0910	9 man with kitchen and bath, no hookups, buy		1	16		89,000	810		89,810	99,000
0920	18 man sleeper with bath, no hookups, buy		1	16		96,000	810		96,810	106,500
1000	Portable buildings, prefab, on skids, economy, 8' x 8'		265	.060	S.F.	25	3.06		28.06	32
1100	Deluxe, 8' x 12'		150	.107	"	28	5.40		33.40	39.50
1200	Storage boxes, 20' x 8', buy	2 Skwk	1.80	8.889	Ea.	3,325	465		3,790	4,375
1250	Rent per month					84.50			84.50	93
1300	40' x 8', buy	2 Skwk	1.40	11.429		3,875	600		4,475	5,200
1350	Rent per month					110			110	121
5000	Air supported structures, see Section 13 31 13.13									

01 52 13.40 Field Office Expense

		Crew	Daily Output	Labor-Hours	Unit	Material	Labor	Equipment	Total	Total Incl O&P
0010	**FIELD OFFICE EXPENSE**									
0100	Office equipment rental average				Month	205			205	226
0120	Office supplies, average				"	82			82	90
0125	Office trailer rental, see Section 01 52 13.20									
0140	Telephone bill; avg. bill/month incl. long dist.				Month	86			86	94.50
0160	Lights & HVAC				"	161			161	177

For customer support on your Building Construction Costs with RSMeans data, call 800.448.8182.

17

01 54 Construction Aids

01 54 09 – Protection Equipment

01 54 09.50 Personnel Protective Equipment

	01 54 09.50 Personnel Protective Equipment	Crew	Daily Output	Labor-Hours	Unit	Material	2018 Bare Costs Labor	Equipment	Total	Total Incl O&P
0010	**PERSONNEL PROTECTIVE EQUIPMENT**									
0015	Hazardous waste protection									
0020	Respirator mask only, full face, silicone				Ea.	287			287	315
0030	Half face, silicone					49.50			49.50	54.50
0040	Respirator cartridges, 2 req'd/mask, dust or asbestos					4.76			4.76	5.25
0050	Chemical vapor					4.28			4.28	4.71
0060	Combination vapor and dust					11.70			11.70	12.90
0100	Emergency escape breathing apparatus, 5 minutes					705			705	780
0110	10 minutes					870			870	955
0150	Self contained breathing apparatus with full face piece, 30 minutes					2,275			2,275	2,525
0160	60 minutes					3,000			3,000	3,300
0200	Encapsulating suits, limited use, level A					1,775			1,775	1,975
0210	Level B				↓	405			405	445
0300	Over boots, latex				Pr.	8.05			8.05	8.85
0310	PVC					33			33	36.50
0320	Neoprene					31			31	34.50
0400	Gloves, nitrile/PVC					85.50			85.50	94
0410	Neoprene coated				↓	41.50			41.50	46

01 54 09.60 Safety Nets

	01 54 09.60 Safety Nets	Crew	Daily Output	Labor-Hours	Unit	Material	Labor	Equipment	Total	Total Incl O&P
0010	**SAFETY NETS**									
0020	No supports, stock sizes, nylon, 3-1/2" mesh				S.F.	3.11			3.11	3.42
0100	Polypropylene, 6" mesh					1.63			1.63	1.79
0200	Small mesh debris nets, 1/4" mesh, stock sizes					.54			.54	.59
0220	Combined 3-1/2" mesh and 1/4" mesh, stock sizes					4.85			4.85	5.35
0300	Rental, 4" mesh, stock sizes, 3 months					.73			.73	.80
0320	6 month rental					1.03			1.03	1.13
0340	12 months				↓	1.41			1.41	1.55

01 54 16 – Temporary Hoists

01 54 16.50 Weekly Forklift Crew

	01 54 16.50 Weekly Forklift Crew	Crew	Daily Output	Labor-Hours	Unit	Material	Labor	Equipment	Total	Total Incl O&P
0010	**WEEKLY FORKLIFT CREW**									
0100	All-terrain forklift, 45' lift, 35' reach, 9000 lb. capacity	A-3P	.20	40	Week		2,050	2,425	4,475	5,775

01 54 19 – Temporary Cranes

01 54 19.50 Daily Crane Crews

	01 54 19.50 Daily Crane Crews	Crew	Daily Output	Labor-Hours	Unit	Material	Labor	Equipment	Total	Total Incl O&P
0010	**DAILY CRANE CREWS** for small jobs, portal to portal R015433-15									
0100	12-ton truck-mounted hydraulic crane	A-3H	1	8	Day		450	630	1,080	1,375
0200	25-ton	A-3I	1	8			450	760	1,210	1,500
0300	40-ton	A-3J	1	8			450	1,325	1,775	2,125
0400	55-ton	A-3K	1	16			840	1,500	2,340	2,925
0500	80-ton	A-3L	1	16	↓		840	2,250	3,090	3,750
0900	If crane is needed on a Saturday, Sunday or Holiday									
0910	At time-and-a-half, add				Day		50%			
0920	At double time, add				"		100%			

01 54 19.60 Monthly Tower Crane Crew

	01 54 19.60 Monthly Tower Crane Crew	Crew	Daily Output	Labor-Hours	Unit	Material	Labor	Equipment	Total	Total Incl O&P
0010	**MONTHLY TOWER CRANE CREW**, excludes concrete footing									
0100	Static tower crane, 130' high, 106' jib, 6200 lb. capacity	A-3N	.05	176	Month		9,875	26,000	35,875	43,500

01 54 23 – Temporary Scaffolding and Platforms

01 54 23.60 Pump Staging

	01 54 23.60 Pump Staging	Crew	Daily Output	Labor-Hours	Unit	Material	Labor	Equipment	Total	Total Incl O&P
0010	**PUMP STAGING**, Aluminum R015423-20									
0200	24' long pole section, buy				Ea.	425			425	465
0300	18' long pole section, buy				↓	330			330	360

For customer support on your Building Construction Costs with RSMeans data, call 800.448.8182.

01 54 Construction Aids

01 54 23 – Temporary Scaffolding and Platforms

01 54 23.60 Pump Staging

		Crew	Daily Output	Labor-Hours	Unit	Material	2018 Bare Costs Labor	Equipment	Total	Total Incl O&P
0400	12' long pole section, buy				Ea.	222			222	244
0500	6' long pole section, buy					117			117	128
0600	6' long splice joint section, buy					86.50			86.50	95.50
0700	Pump jack, buy					174			174	192
0900	Foldable brace, buy					69.50			69.50	76
1000	Workbench/back safety rail support, buy					93.50			93.50	103
1100	Scaffolding planks/workbench, 14" wide x 24' long, buy					775			775	850
1200	Plank end safety rail, buy					355			355	390
1250	Safety net, 22' long, buy				↓	425			425	465
1300	System in place, 50' working height, per use based on 50 uses	2 Carp	84.80	.189	C.S.F.	7.05	9.55		16.60	22.50
1400	100 uses	↓	84.80	.189		3.53	9.55		13.08	18.45
1500	150 uses	↓	84.80	.189	↓	2.36	9.55		11.91	17.15

01 54 23.70 Scaffolding

		Crew	Daily Output	Labor-Hours	Unit	Material	2018 Bare Costs Labor	Equipment	Total	Total Incl O&P
0010	**SCAFFOLDING** R015423-10									
0015	Steel tube, regular, no plank, labor only to erect & dismantle									
0090	Building exterior, wall face, 1 to 5 stories, 6'-4" x 5' frames	3 Carp	8	3	C.S.F.		152		152	232
0200	6 to 12 stories	4 Carp	8	4			203		203	310
0301	13 to 20 stories	5 Clab	8	5			199		199	305
0460	Building interior, wall face area, up to 16' high	3 Carp	12	2			101		101	154
0560	16' to 40' high	↓	10	2.400	↓		122		122	185
0800	Building interior floor area, up to 30' high	↓	150	.160	C.C.F.		8.10		8.10	12.35
0900	Over 30' high	4 Carp	160	.200	"		10.15		10.15	15.45
0906	Complete system for face of walls, no plank, material only rent/mo				C.S.F.	33			33	36
0908	Interior spaces, no plank, material only rent/mo				C.C.F.	3.77			3.77	4.14
0910	Steel tubular, heavy duty shoring, buy									
0920	Frames 5' high 2' wide				Ea.	99.50			99.50	109
0925	5' high 4' wide					114			114	125
0930	6' high 2' wide					115			115	127
0935	6' high 4' wide				↓	126			126	139
0940	Accessories									
0945	Cross braces				Ea.	20			20	22
0950	U-head, 8" x 8"					22			22	24
0955	J-head, 4" x 8"					16			16	17.60
0960	Base plate, 8" x 8"					17.70			17.70	19.45
0965	Leveling jack				↓	39			39	42.50
1000	Steel tubular, regular, buy									
1100	Frames 3' high 5' wide				Ea.	91			91	100
1150	5' high 5' wide					107			107	117
1200	6'-4" high 5' wide					99.50			99.50	109
1350	7'-6" high 6' wide					170			170	187
1500	Accessories, cross braces					16			16	17.60
1550	Guardrail post					20			20	22
1600	Guardrail 7' section					8.05			8.05	8.85
1650	Screw jacks & plates					25.50			25.50	28
1700	Sidearm brackets					22.50			22.50	24.50
1750	8" casters					37			37	40.50
1800	Plank 2" x 10" x 16'-0"					66			66	72.50
1900	Stairway section					283			283	310
1910	Stairway starter bar					32			32	35
1920	Stairway inside handrail					53			53	58
1930	Stairway outside handrail					84			84	92.50
1940	Walk-thru frame guardrail				↓	41.50			41.50	46

For customer support on your Building Construction Costs with RSMeans data, call 800.448.8182.

19

01 54 23.70 Scaffolding		Crew	Daily Output	Labor-Hours	Unit	Material	2018 Bare Costs Labor	Equipment	Total	Total Incl O&P
2000	Steel tubular, regular, rent/mo.									
2100	Frames 3' high 5' wide				Ea.	4.38			4.38	4.82
2150	5' high 5' wide					4.38			4.38	4.82
2200	6'-4" high 5' wide					5.25			5.25	5.80
2250	7'-6" high 6' wide					9.75			9.75	10.75
2500	Accessories, cross braces					.88			.88	.97
2550	Guardrail post					.88			.88	.97
2600	Guardrail 7' section					.88			.88	.97
2650	Screw jacks & plates					1.75			1.75	1.93
2700	Sidearm brackets					1.75			1.75	1.93
2750	8" casters					7			7	7.70
2800	Outrigger for rolling tower					2.63			2.63	2.89
2850	Plank 2" x 10" x 16'-0"					9.75			9.75	10.75
2900	Stairway section					32.50			32.50	36
2940	Walk-thru frame guardrail				▼	2.19			2.19	2.41
3000	Steel tubular, heavy duty shoring, rent/mo.									
3250	5' high 2' & 4' wide				Ea.	8.30			8.30	9.15
3300	6' high 2' & 4' wide					8.30			8.30	9.15
3500	Accessories, cross braces					.88			.88	.97
3600	U-head, 8" x 8"					2.44			2.44	2.68
3650	J-head, 4" x 8"					2.44			2.44	2.68
3700	Base plate, 8" x 8"					.88			.88	.97
3750	Leveling jack					2.44			2.44	2.68
5700	Planks, 2" x 10" x 16'-0", labor only to erect & remove to 50' H	3 Carp	72	.333			16.90		16.90	25.50
5800	Over 50' high	4 Carp	80	.400	▼		20.50		20.50	31
6000	Heavy duty shoring for elevated slab forms to 8'-2" high, floor area									
6100	Labor only to erect & dismantle	4 Carp	16	2	C.S.F.		101		101	154
6110	Materials only, rent/mo.				"	43			43	47
6500	To 14'-8" high									
6600	Labor only to erect & dismantle	4 Carp	10	3.200	C.S.F.		162		162	247
6610	Materials only, rent/mo				"	62.50			62.50	69

01 54 23.75 Scaffolding Specialties

		Crew	Daily Output	Labor-Hours	Unit	Material	Labor	Equipment	Total	Total Incl O&P
0010	**SCAFFOLDING SPECIALTIES**									
1200	Sidewalk bridge, heavy duty steel posts & beams, including									
1210	parapet protection & waterproofing (material cost is rent/month)									
1220	8' to 10' wide, 2 posts	3 Carp	15	1.600	L.F.	45	81		126	173
1230	3 posts	"	10	2.400	"	69	122		191	261
1500	Sidewalk bridge using tubular steel scaffold frames including									
1510	planking (material cost is rent/month)	3 Carp	45	.533	L.F.	8.25	27		35.25	50
1600	For 2 uses per month, deduct from all above					50%				
1700	For 1 use every 2 months, add to all above					100%				
1900	Catwalks, 20" wide, no guardrails, 7' span, buy				Ea.	150			150	165
2000	10' span, buy					211			211	232
3720	Putlog, standard, 8' span, with hangers, buy					75.50			75.50	83
3730	Rent per month					16			16	17.60
3750	12' span, buy					99.50			99.50	109
3755	Rent per month					20			20	22
3760	Trussed type, 16' span, buy					255			255	281
3770	Rent per month					24			24	26.50
3790	22' span, buy					275			275	300
3795	Rent per month					32			32	35
3800	Rolling ladders with handrails, 30" wide, buy, 2 step					291			291	320

For customer support on your Building Construction Costs with RSMeans data, call 800.448.8182.

01 54 Construction Aids

01 54 23 – Temporary Scaffolding and Platforms

01 54 23.75 Scaffolding Specialties

		Crew	Daily Output	Labor-Hours	Unit	Material	2018 Bare Costs Labor	2018 Bare Costs Equipment	Total	Total Incl O&P
4000	7 step				Ea.	815			815	900
4050	10 step					1,175			1,175	1,275
4100	Rolling towers, buy, 5' wide, 7' long, 10' high					1,300			1,300	1,425
4200	For additional 5' high sections, to buy					245			245	270
4300	Complete incl. wheels, railings, outriggers,									
4350	21' high, to buy				Ea.	2,225			2,225	2,450
4400	Rent/month = 5% of purchase cost				"	196			196	216
5000	Motorized work platform, mast climber									
5050	Base unit, 50' W, less than 100' tall, rent/mo				Ea.	3,150			3,150	3,450
5100	Less than 200' tall, rent/mo					3,750			3,750	4,125
5150	Less than 300' tall, rent/mo					4,400			4,400	4,850
5200	Less than 400' tall, rent/mo					4,975			4,975	5,475
5250	Set up and demob, per unit, less than 100' tall	B-68F	16.60	1.446	C.S.F.		76.50	16.80	93.30	136
5300	Less than 200' tall		25	.960			51	11.15	62.15	90.50
5350	Less than 300' tall		30	.800			42.50	9.30	51.80	75.50
5400	Less than 400' tall		33	.727			38.50	8.45	46.95	68.50
5500	Mobilization (price includes freight in and out) per unit				Ea.				1,100	1,225

01 54 23.80 Staging Aids

		Crew	Daily Output	Labor-Hours	Unit	Material	2018 Bare Costs Labor	2018 Bare Costs Equipment	Total	Total Incl O&P
0010	**STAGING AIDS** and fall protection equipment									
0100	Sidewall staging bracket, tubular, buy				Ea.	57.50			57.50	63.50
0110	Cost each per day, based on 250 days use				Day	.23			.23	.25
0200	Guard post, buy				Ea.	54.50			54.50	60
0210	Cost each per day, based on 250 days use				Day	.22			.22	.24
0300	End guard chains, buy per pair				Pair	42			42	46
0310	Cost per set per day, based on 250 days use				Day	.23			.23	.25
1000	Roof shingling bracket, steel, buy				Ea.	11.25			11.25	12.35
1010	Cost each per day, based on 250 days use				Day	.04			.04	.05
1100	Wood bracket, buy				Ea.	24.50			24.50	27
1110	Cost each per day, based on 250 days use				Day	.10			.10	.11
2000	Ladder jack, aluminum, buy per pair				Pair	128			128	141
2010	Cost per pair per day, based on 250 days use				Day	.51			.51	.56
3000	Laminated wood plank, 2" x 10" x 16', buy				Ea.	57.50			57.50	63.50
3010	Cost each per day, based on 250 days use				Day	.23			.23	.25
3100	Aluminum scaffolding plank, 20" wide x 24' long, buy				Ea.	815			815	900
3110	Cost each per day, based on 250 days use				Day	3.27			3.27	3.59
4000	Nylon full body harness, lanyard and rope grab				Ea.	166			166	182
4010	Cost each per day, based on 250 days use				Day	.66			.66	.73
4100	Rope for safety line, 5/8" x 100' nylon, buy				Ea.	56			56	61.50
4110	Cost each per day, based on 250 days use				Day	.22			.22	.25
4200	Permanent U-Bolt roof anchor, buy				Ea.	30			30	33
4300	Temporary (one use) roof ridge anchor, buy				"	6.40			6.40	7
5000	Installation (setup and removal) of staging aids									
5010	Sidewall staging bracket	2 Carp	64	.250	Ea.		12.70		12.70	19.30
5020	Guard post with 2 wood rails	"	64	.250			12.70		12.70	19.30
5030	End guard chains, set	1 Carp	64	.125			6.35		6.35	9.65
5100	Roof shingling bracket		96	.083			4.22		4.22	6.45
5200	Ladder jack		64	.125			6.35		6.35	9.65
5300	Wood plank, 2" x 10" x 16'	2 Carp	80	.200			10.15		10.15	15.45
5310	Aluminum scaffold plank, 20" x 24'	"	40	.400			20.50		20.50	31
5410	Safety rope	1 Carp	40	.200			10.15		10.15	15.45
5420	Permanent U-Bolt roof anchor (install only)	2 Carp	40	.400			20.50		20.50	31
5430	Temporary roof ridge anchor (install only)	1 Carp	64	.125			6.35		6.35	9.65

For customer support on your Building Construction Costs with RSMeans data, call 800.448.8182.

21

01 54 Construction Aids

01 54 26 – Temporary Swing Staging

01 54 26.50 Swing Staging	Crew	Daily Output	Labor-Hours	Unit	Material	2018 Bare Costs Labor	2018 Bare Costs Equipment	Total	Total Incl O&P
0010 **SWING STAGING**, 500 lb. cap., 2' wide to 24' long, hand operated									
0020 steel cable type, with 60' cables, buy				Ea.	5,725			5,725	6,300
0030 Rent per month				"	570			570	630
0600 Lightweight (not for masons) 24' long for 150' height,									
0610 manual type, buy				Ea.	11,900			11,900	13,100
0620 Rent per month					1,200			1,200	1,325
0700 Powered, electric or air, to 150' high, buy					30,000			30,000	33,000
0710 Rent per month					2,100			2,100	2,300
0780 To 300' high, buy					30,500			30,500	33,500
0800 Rent per month					2,125			2,125	2,350
1000 Bosun's chair or work basket 3' x 3.5', to 300' high, electric, buy					12,200			12,200	13,400
1010 Rent per month					850			850	935
2200 Move swing staging (setup and remove)	E-4	2	16	Move		880	49.50	929.50	1,500

01 54 36 – Equipment Mobilization

01 54 36.50 Mobilization

01 54 36.50 Mobilization	Crew	Daily Output	Labor-Hours	Unit	Material	2018 Bare Costs Labor	2018 Bare Costs Equipment	Total	Total Incl O&P
0010 **MOBILIZATION** (Use line item again for demobilization) R015436-50									
0015 Up to 25 mi. haul dist. (50 mi. RT for mob/demob crew)									
1200 Small equipment, placed in rear of, or towed by pickup truck	A-3A	4	2	Ea.		103	31.50	134.50	190
1300 Equipment hauled on 3-ton capacity towed trailer	A-3Q	2.67	3			154	57	211	295
1400 20-ton capacity	B-34U	2	8			390	212	602	820
1500 40-ton capacity	B-34N	2	8			400	330	730	960
1600 50-ton capacity	B-34V	1	24			1,225	910	2,135	2,850
1700 Crane, truck-mounted, up to 75 ton (driver only)	1 Eqhv	4	2			112		112	169
1800 Over 75 ton (with chase vehicle)	A-3E	2.50	6.400			325	50.50	375.50	545
2400 Crane, large lattice boom, requiring assembly	B-34W	.50	144			7,025	6,725	13,750	18,000
2500 For each additional 5 miles haul distance, add						10%	10%		
3000 For large pieces of equipment, allow for assembly/knockdown									
3001 For mob/demob of vibrofloatation equip, see Section 31 45 13.10									
3100 For mob/demob of micro-tunneling equip, see Section 33 05 23.19									
3200 For mob/demob of pile driving equip, see Section 31 62 19.10									
3300 For mob/demob of caisson drilling equip, see Section 31 63 26.13									

01 55 Vehicular Access and Parking

01 55 23 – Temporary Roads

01 55 23.50 Roads and Sidewalks

	Crew	Daily Output	Labor-Hours	Unit	Material	2018 Bare Costs Labor	2018 Bare Costs Equipment	Total	Total Incl O&P
0010 **ROADS AND SIDEWALKS** Temporary									
0050 Roads, gravel fill, no surfacing, 4" gravel depth	B-14	715	.067	S.Y.	3.36	2.83	.44	6.63	8.50
0100 8" gravel depth	"	615	.078	"	6.70	3.29	.51	10.50	12.95
1000 Ramp, 3/4" plywood on 2" x 6" joists, 16" OC	2 Carp	300	.053	S.F.	1.56	2.70		4.26	5.85
1100 On 2" x 10" joists, 16" OC	"	275	.058	"	2.31	2.95		5.26	7.05

01 56 Temporary Barriers and Enclosures

01 56 13 – Temporary Air Barriers

01 56 13.60 Tarpaulins

		Crew	Daily Output	Labor-Hours	Unit	Material	2018 Bare Costs Labor	Equipment	Total	Total Incl O&P
0010	**TARPAULINS**									
0020	Cotton duck, 10-13.13 oz./S.Y., 6' x 8'				S.F.	.85			.85	.94
0050	30' x 30'					.59			.59	.65
0100	Polyvinyl coated nylon, 14-18 oz., minimum					1.44			1.44	1.58
0150	Maximum					1.44			1.44	1.58
0200	Reinforced polyethylene 3 mils thick, white					.04			.04	.04
0300	4 mils thick, white, clear or black					.13			.13	.14
0400	5.5 mils thick, clear					.19			.19	.21
0500	White, fire retardant					.61			.61	.67
0600	12 mils, oil resistant, fire retardant					.49			.49	.54
0700	8.5 mils, black					.18			.18	.20
0710	Woven polyethylene, 6 mils thick					.19			.19	.21
0730	Polyester reinforced w/integral fastening system, 11 mils thick					.19			.19	.21
0740	Polyethylene, reflective, 23 mils thick					1.35			1.35	1.49

01 56 13.90 Winter Protection

		Crew	Daily Output	Labor-Hours	Unit	Material	2018 Bare Costs Labor	Equipment	Total	Total Incl O&P
0010	**WINTER PROTECTION**									
0100	Framing to close openings	2 Clab	500	.032	S.F.	.45	1.28		1.73	2.43
0200	Tarpaulins hung over scaffolding, 8 uses, not incl. scaffolding		1500	.011		.25	.43		.68	.93
0250	Tarpaulin polyester reinf. w/integral fastening system, 11 mils thick		1600	.010		.21	.40		.61	.84
0300	Prefab fiberglass panels, steel frame, 8 uses		1200	.013		2.52	.53		3.05	3.58

01 56 16 – Temporary Dust Barriers

01 56 16.10 Dust Barriers, Temporary

		Crew	Daily Output	Labor-Hours	Unit	Material	2018 Bare Costs Labor	Equipment	Total	Total Incl O&P
0010	**DUST BARRIERS, TEMPORARY**									
0020	Spring loaded telescoping pole & head, to 12', erect and dismantle	1 Clab	240	.033	Ea.		1.33		1.33	2.02
0025	Cost per day (based upon 250 days)				Day	.29			.29	.31
0030	To 21', erect and dismantle	1 Clab	240	.033	Ea.		1.33		1.33	2.02
0035	Cost per day (based upon 250 days)				Day	.58			.58	.63
0040	Accessories, caution tape reel, erect and dismantle	1 Clab	480	.017	Ea.		.66		.66	1.01
0045	Cost per day (based upon 250 days)				Day	.34			.34	.37
0060	Foam rail and connector, erect and dismantle	1 Clab	240	.033	Ea.		1.33		1.33	2.02
0065	Cost per day (based upon 250 days)				Day	.12			.12	.13
0070	Caution tape	1 Clab	384	.021	C.L.F.	3.13	.83		3.96	4.71
0080	Zipper, standard duty		60	.133	Ea.	7.30	5.30		12.60	16.10
0090	Heavy duty		48	.167	"	9.75	6.65		16.40	21
0100	Polyethylene sheet, 4 mil		37	.216	Sq.	2.58	8.60		11.18	15.95
0110	6 mil		37	.216	"	3.73	8.60		12.33	17.20
1000	Dust partition, 6 mil polyethylene, 1" x 3" frame	2 Carp	2000	.008	S.F.	.32	.41		.73	.97
1080	2" x 4" frame	"	2000	.008	"	.35	.41		.76	1
1085	Negative air machine, 1800 CFM				Ea.	860			860	950
1090	Adhesive strip application, 2" width	1 Clab	192	.042	C.L.F.	6.55	1.66		8.21	9.75

01 56 23 – Temporary Barricades

01 56 23.10 Barricades

		Crew	Daily Output	Labor-Hours	Unit	Material	2018 Bare Costs Labor	Equipment	Total	Total Incl O&P
0010	**BARRICADES**									
0020	5' high, 3 rail @ 2" x 8", fixed	2 Carp	20	.800	L.F.	6.05	40.50		46.55	68.50
0150	Movable	"	30	.533	"	5	27		32	46.50
0300	Stock units, 58' high, 8' wide, reflective, buy				Ea.	211			211	232
0350	With reflective tape, buy				"	360			360	395
0400	Break-a-way 3" PVC pipe barricade									
0410	with 3 ea. 1' x 4' reflectorized panels, buy				Ea.	125			125	137
0500	Barricades, plastic, 8" x 24" wide, foldable					59			59	65
0800	Traffic cones, PVC, 18" high					11			11	12.10

01 56 Temporary Barriers and Enclosures

01 56 23 – Temporary Barricades

01 56 23.10 Barricades	Crew	Daily Output	Labor-Hours	Unit	Material	2018 Bare Costs Labor	Equipment	Total	Total Incl O&P
0850	28" high				Ea.	17.75			17.75	19.55
1000	Guardrail, wooden, 3' high, 1" x 6" on 2" x 4" posts	2 Carp	200	.080	L.F.	1.32	4.06		5.38	7.65
1100	2" x 6" on 4" x 4" posts	"	165	.097		2.54	4.92		7.46	10.30
1200	Portable metal with base pads, buy					14.25			14.25	15.65
1250	Typical installation, assume 10 reuses	2 Carp	600	.027		2.30	1.35		3.65	4.59
1300	Barricade tape, polyethylene, 7 mil, 3" wide x 500' long roll				Ea.	25			25	27.50
3000	Detour signs, set up and remove									
3010	Reflective aluminum, MUTCD, 24" x 24", post mounted	1 Clab	20	.400	Ea.	2.54	15.95		18.49	27.50
4000	Roof edge portable barrier stands and warning flags, 50 uses	1 Rohe	9100	.001	L.F.	.06	.03		.09	.12
4010	100 uses	"	9100	.001	"	.03	.03		.06	.08
5000	Barricades, see Section 01 54 33.40									

01 56 26 – Temporary Fencing

01 56 26.50 Temporary Fencing	Crew	Daily Output	Labor-Hours	Unit	Material	2018 Bare Costs Labor	Equipment	Total	Total Incl O&P
0010	**TEMPORARY FENCING**									
0020	Chain link, 11 ga., 4' high	2 Clab	400	.040	L.F.	1.63	1.59		3.22	4.22
0100	6' high		300	.053		4.55	2.13		6.68	8.25
0200	Rented chain link, 6' high, to 1000' (up to 12 mo.)		400	.040		2.99	1.59		4.58	5.70
0250	Over 1000' (up to 12 mo.)		300	.053		3.19	2.13		5.32	6.75
0350	Plywood, painted, 2" x 4" frame, 4' high	A-4	135	.178		6.35	8.55		14.90	19.90
0400	4" x 4" frame, 8' high	"	110	.218		12.10	10.45		22.55	29
0500	Wire mesh on 4" x 4" posts, 4' high	2 Carp	100	.160		10.35	8.10		18.45	24
0550	8' high	"	80	.200		15.60	10.15		25.75	32.50

01 56 29 – Temporary Protective Walkways

01 56 29.50 Protection	Crew	Daily Output	Labor-Hours	Unit	Material	2018 Bare Costs Labor	Equipment	Total	Total Incl O&P
0010	**PROTECTION**									
0020	Stair tread, 2" x 12" planks, 1 use	1 Carp	75	.107	Tread	5.35	5.40		10.75	14.10
0100	Exterior plywood, 1/2" thick, 1 use		65	.123		1.85	6.25		8.10	11.55
0200	3/4" thick, 1 use		60	.133		2.77	6.75		9.52	13.35
2200	Sidewalks, 2" x 12" planks, 2 uses		350	.023	S.F.	.89	1.16		2.05	2.74
2300	Exterior plywood, 2 uses, 1/2" thick		750	.011		.31	.54		.85	1.16
2400	5/8" thick		650	.012		.38	.62		1	1.37
2500	3/4" thick		600	.013		.46	.68		1.14	1.54

01 56 32 – Temporary Security

01 56 32.50 Watchman	Crew	Daily Output	Labor-Hours	Unit	Material	2018 Bare Costs Labor	Equipment	Total	Total Incl O&P
0010	**WATCHMAN**									
0020	Service, monthly basis, uniformed person, minimum				Hr.				25	27.50
0100	Maximum								45.50	50
0200	Person and command dog, minimum								31	34
0300	Maximum								54.50	60
0500	Sentry dog, leased, with job patrol (yard dog), 1 dog				Week				290	320
0600	2 dogs				"				390	430
0800	Purchase, trained sentry dog, minimum				Ea.				1,375	1,500
0900	Maximum				"				2,725	3,000

For customer support on your Building Construction Costs with RSMeans data, call 800.448.8182.

01 58 Project Identification

01 58 13 – Temporary Project Signage

01 58 13.50 Signs	Crew	Daily Output	Labor-Hours	Unit	Material	2018 Bare Costs Labor	2018 Bare Costs Equipment	Total	Total Incl O&P
0010 **SIGNS**									
0020 High intensity reflectorized, no posts, buy				Ea.	25			25	27.50

01 66 Product Storage and Handling Requirements

01 66 19 – Material Handling

01 66 19.10 Material Handling

	Crew	Daily Output	Labor-Hours	Unit	Material	Labor	Equipment	Total	Total Incl O&P
0010 **MATERIAL HANDLING**									
0020 Above 2nd story, via stairs, per C.Y. of material per floor	2 Clab	145	.110	C.Y.		4.40		4.40	6.70
0030 Via elevator, per C.Y. of material		240	.067			2.66		2.66	4.05
0050 Distances greater than 200', per C.Y. of material per each addl 200'	↓	300	.053	↓		2.13		2.13	3.24

01 71 Examination and Preparation

01 71 23 – Field Engineering

01 71 23.13 Construction Layout

	Crew	Daily Output	Labor-Hours	Unit	Material	Labor	Equipment	Total	Total Incl O&P
0010 **CONSTRUCTION LAYOUT**									
1100 Crew for layout of building, trenching or pipe laying, 2 person crew	A-6	1	16	Day		805	50.50	855.50	1,275
1200 3 person crew	A-7	1	24			1,325	50.50	1,375.50	2,050
1400 Crew for roadway layout, 4 person crew	A-8	1	32	↓		1,700	50.50	1,750.50	2,650

01 71 23.19 Surveyor Stakes

	Crew	Daily Output	Labor-Hours	Unit	Material	Labor	Equipment	Total	Total Incl O&P
0010 **SURVEYOR STAKES**									
0020 Hardwood, 1" x 1" x 48" long				C	70			70	77
0100 2" x 2" x 18" long					78			78	86
0150 2" x 2" x 24" long				↓	150			150	165

01 74 Cleaning and Waste Management

01 74 13 – Progress Cleaning

01 74 13.20 Cleaning Up

	Crew	Daily Output	Labor-Hours	Unit	Material	Labor	Equipment	Total	Total Incl O&P
0010 **CLEANING UP**									
0020 After job completion, allow, minimum				Job				.30%	.30%
0040 Maximum				"				1%	1%
0050 Cleanup of floor area, continuous, per day, during const.	A-5	24	.750	M.S.F.	2.41	30.50	1.97	34.88	51
0100 Final by GC at end of job	"	11.50	1.565	"	2.54	63	4.10	69.64	103
0200 Rubbish removal, see Section 02 41 19.19									

01 76 Protecting Installed Construction

01 76 13 – Temporary Protection of Installed Construction

01 76 13.20 Temporary Protection

	Crew	Daily Output	Labor-Hours	Unit	Material	Labor	Equipment	Total	Total Incl O&P
0010 **TEMPORARY PROTECTION**									
0020 Flooring, 1/8" tempered hardboard, taped seams	2 Carp	1500	.011	S.F.	.45	.54		.99	1.32
0030 Peel away carpet protection	1 Clab	3200	.003	"	.13	.10		.23	.29

For customer support on your Building Construction Costs with RSMeans data, call 800.448.8182.

25

01 91 Commissioning

01 91 13 – General Commissioning Requirements

01 91 13.50 Building Commissioning	Crew	Daily Output	Labor-Hours	Unit	Material	2018 Bare Costs Labor	Equipment	Total	Total Incl O&P
0010 **BUILDING COMMISSIONING**									
0100 Systems operation and verification during turnover				%				.25%	.25%
0150 Including all systems subcontractors								.50%	.50%
0200 Systems design assistance, operation, verification and training								.50%	.50%
0250 Including all systems subcontractors				↓				1%	1%

For customer support on your Building Construction Costs with RSMeans data, call 800.448.8182.

Estimating Tips
02 30 00 Subsurface Investigation

In preparing estimates on structures involving earthwork or foundations, all information concerning soil characteristics should be obtained. Look particularly for hazardous waste, evidence of prior dumping of debris, and previous stream beds.

02 40 00 Demolition and Structure Moving

The costs shown for selective demolition do not include rubbish handling or disposal. These items should be estimated separately using RSMeans data or other sources.

- Historic preservation often requires that the contractor remove materials from the existing structure, rehab them, and replace them. The estimator must be aware of any related measures and precautions that must be taken when doing selective demolition and cutting and patching. Requirements may include special handling and storage, as well as security.

- In addition to Subdivision 02 41 00, you can find selective demolition items in each division. Example: Roofing demolition is in Division 7.

- Absent of any other specific reference, an approximate demolish-in-place cost can be obtained by halving the new-install labor cost. To remove for reuse, allow the entire new-install labor figure.

02 40 00 Building Deconstruction

This section provides costs for the careful dismantling and recycling of most low-rise building materials.

02 50 00 Containment of Hazardous Waste

This section addresses on-site hazardous waste disposal costs.

02 80 00 Hazardous Material Disposal/Remediation

This subdivision includes information on hazardous waste handling, asbestos remediation, lead remediation, and mold remediation. See reference numbers R028213-20 and R028319-60 for further guidance in using these unit price lines.

02 90 00 Monitoring Chemical Sampling, Testing Analysis

This section provides costs for on-site sampling and testing hazardous waste.

Reference Numbers

Reference numbers are shown at the beginning of some major classifications. These numbers refer to related items in the Reference Section. The reference information may be an estimating procedure, an alternate pricing method, or technical information.

Note: Not all subdivisions listed here necessarily appear. ∎

Did you know?

RSMeans data is available through our online application with 24/7 access:

- Search for unit prices by keyword
- Leverage the most up-to-date data
- Build and export estimates

Try it free for 30 days!
www.rsmeans.com/2018freetrial

No part of this cost data may be reproduced, stored in a retrieval system, or transmitted in any form or by any means without prior written permission of Gordian.

02 21 Surveys

02 21 13 – Site Surveys

02 21 13.09 Topographical Surveys	Crew	Daily Output	Labor-Hours	Unit	Material	Labor	Equipment	Total	Total Incl O&P
0010 **TOPOGRAPHICAL SURVEYS**									
0020 Topographical surveying, conventional, minimum	A-7	3.30	7.273	Acre	21.50	400	15.25	436.75	645
0100 Maximum	A-8	.60	53.333	"	58.50	2,850	84.50	2,993	4,475

02 21 13.13 Boundary and Survey Markers

	Crew	Daily Output	Labor-Hours	Unit	Material	Labor	Equipment	Total	Total Incl O&P
0010 **BOUNDARY AND SURVEY MARKERS**									
0300 Lot location and lines, large quantities, minimum	A-7	2	12	Acre	34.50	660	25	719.50	1,075
0320 Average	"	1.25	19.200		58.50	1,050	40.50	1,149	1,700
0400 Small quantities, maximum	A-8	1	32	↓	73	1,700	50.50	1,823.50	2,725
0600 Monuments, 3' long	A-7	10	2.400	Ea.	31	132	5.05	168.05	240
0800 Property lines, perimeter, cleared land	"	1000	.024	L.F.	.07	1.32	.05	1.44	2.14
0900 Wooded land	A-8	875	.037	"	.09	1.95	.06	2.10	3.12

02 21 13.16 Aerial Surveys

	Crew	Daily Output	Labor-Hours	Unit	Material	Labor	Equipment	Total	Total Incl O&P
0010 **AERIAL SURVEYS**									
1500 Aerial surveying, including ground control, minimum fee, 10 acres				Total				4,700	4,700
1510 100 acres								9,400	9,400
1550 From existing photography, deduct				↓				1,625	1,625
1600 2' contours, 10 acres				Acre				470	470
1850 100 acres								94	94
2000 1000 acres								90	90
2050 10,000 acres				↓				85	85

02 32 Geotechnical Investigations

02 32 13 – Subsurface Drilling and Sampling

02 32 13.10 Boring and Exploratory Drilling

	Crew	Daily Output	Labor-Hours	Unit	Material	Labor	Equipment	Total	Total Incl O&P
0010 **BORING AND EXPLORATORY DRILLING**									
0020 Borings, initial field stake out & determination of elevations	A-6	1	16	Day		805	50.50	855.50	1,275
0100 Drawings showing boring details				Total		335		335	425
0200 Report and recommendations from P.E.						775		775	970
0300 Mobilization and demobilization	B-55	4	6	↓		248	249	497	650
0350 For over 100 miles, per added mile		450	.053	Mile		2.21	2.21	4.42	5.80
0600 Auger holes in earth, no samples, 2-1/2" diameter		78.60	.305	L.F.		12.65	12.70	25.35	33
0650 4" diameter		67.50	.356			14.70	14.75	29.45	39
0800 Cased borings in earth, with samples, 2-1/2" diameter		55.50	.432		16.65	17.90	17.95	52.50	65
0850 4" diameter		32.60	.736		28	30.50	30.50	89	111
1000 Drilling in rock, "BX" core, no sampling	B-56	34.90	.458			21	41.50	62.50	77.50
1050 With casing & sampling		31.70	.505		16.65	23	46	85.65	104
1200 "NX" core, no sampling		25.92	.617			28	56	84	104
1250 With casing and sampling	↓	25	.640	↓	19.75	29	58	106.75	130
1400 Borings, earth, drill rig and crew with truck mounted auger	B-55	1	24	Day		995	995	1,990	2,600
1450 Rock using crawler type drill	B-56	1	16	"		730	1,450	2,180	2,700
1500 For inner city borings add, minimum								10%	10%
1510 Maximum								20%	20%

02 32 19 – Exploratory Excavations

02 32 19.10 Test Pits

	Crew	Daily Output	Labor-Hours	Unit	Material	Labor	Equipment	Total	Total Incl O&P
0010 **TEST PITS**									
0020 Hand digging, light soil	1 Clab	4.50	1.778	C.Y.		71		71	108
0100 Heavy soil	"	2.50	3.200			128		128	194
0120 Loader-backhoe, light soil	B-11M	28	.571			26.50	13.75	40.25	55.50
0130 Heavy soil	"	20	.800	↓		37.50	19.25	56.75	77.50
1000 Subsurface exploration, mobilization				Mile				6.75	8.40

For customer support on your Building Construction Costs with RSMeans data, call 800.448.8182.

02 32 Geotechnical Investigations

02 32 19 – Exploratory Excavations

02 32 19.10 Test Pits	Crew	Daily Output	Labor-Hours	Unit	Material	2018 Bare Costs Labor	2018 Bare Costs Equipment	Total	Total Incl O&P
1010 Difficult access for rig, add				Hr.				260	320
1020 Auger borings, drill rig, incl. samples				L.F.				26.50	33
1030 Hand auger				↓				31.50	40
1050 Drill and sample every 5', split spoon								31.50	40
1060 Extra samples				Ea.				36	45.50

02 41 Demolition

02 41 13 – Selective Site Demolition

02 41 13.15 Hydrodemolition

		Crew	Daily Output	Labor-Hours	Unit	Material	Labor	Equipment	Total	Total Incl O&P
0010	**HYDRODEMOLITION** R024119-10									
0015	Hydrodemolition, concrete pavement									
0120	20,000 psi, Crew to include loader/vacuum truck as required									
0130	2" depth	B-5	1000	.056	S.F.		2.47	1.44	3.91	5.35
0410	4" depth		800	.070			3.09	1.80	4.89	6.65
0420	6" depth	↓	600	.093	↓		4.12	2.40	6.52	8.90

02 41 13.17 Demolish, Remove Pavement and Curb

		Crew	Daily Output	Labor-Hours	Unit	Material	Labor	Equipment	Total	Total Incl O&P
0010	**DEMOLISH, REMOVE PAVEMENT AND CURB** R024119-10									
5010	Pavement removal, bituminous roads, up to 3" thick	B-38	690	.058	S.Y.		2.63	1.63	4.26	5.75
5050	4"-6" thick		420	.095			4.32	2.68	7	9.50
5100	Bituminous driveways		640	.063			2.83	1.76	4.59	6.20
5200	Concrete to 6" thick, hydraulic hammer, mesh reinforced		255	.157			7.10	4.41	11.51	15.65
5300	Rod reinforced		200	.200	↓		9.05	5.65	14.70	19.95
5400	Concrete, 7"-24" thick, plain		33	1.212	C.Y.		55	34	89	121
5500	Reinforced	↓	24	1.667	"		75.50	47	122.50	167
5600	With hand held air equipment, bituminous, to 6" thick	B-39	1900	.025	S.F.		1.06	.12	1.18	1.76
5700	Concrete to 6" thick, no reinforcing		1600	.030			1.26	.15	1.41	2.08
5800	Mesh reinforced		1400	.034			1.44	.17	1.61	2.37
5900	Rod reinforced	↓	765	.063	↓		2.64	.31	2.95	4.35
6000	Curbs, concrete, plain	B-6	360	.067	L.F.		2.91	.87	3.78	5.35
6100	Reinforced		275	.087			3.81	1.14	4.95	7.05
6200	Granite		360	.067			2.91	.87	3.78	5.35
6300	Bituminous	↓	528	.045	↓		1.98	.59	2.57	3.66

02 41 13.23 Utility Line Removal

		Crew	Daily Output	Labor-Hours	Unit	Material	Labor	Equipment	Total	Total Incl O&P
0010	**UTILITY LINE REMOVAL**									
0015	No hauling, abandon catch basin or manhole	B-6	7	3.429	Ea.		150	44.50	194.50	276
0020	Remove existing catch basin or manhole, masonry		4	6			262	78	340	485
0030	Catch basin or manhole frames and covers, stored		13	1.846			80.50	24	104.50	149
0040	Remove and reset	↓	7	3.429			150	44.50	194.50	276
0900	Hydrants, fire, remove only	B-21A	5	8			400	72.50	472.50	685
0950	Remove and reset	"	2	20	↓		1,000	182	1,182	1,700
2900	Pipe removal, sewer/water, no excavation, 12" diameter	B-6	175	.137	L.F.		6	1.79	7.79	11.05
2930	15"-18" diameter	B-12Z	150	.160			7.25	9.40	16.65	21.50
2960	21"-24" diameter		120	.200			9.05	11.75	20.80	26.50
3000	27"-36" diameter	↓	90	.267			12.05	15.70	27.75	35.50
3200	Steel, welded connections, 4" diameter	B-6	160	.150			6.55	1.95	8.50	12.10
3300	10" diameter	"	80	.300	↓		13.10	3.91	17.01	24

02 41 13.30 Minor Site Demolition

		Crew	Daily Output	Labor-Hours	Unit	Material	Labor	Equipment	Total	Total Incl O&P
0010	**MINOR SITE DEMOLITION** R024119-10									
0100	Roadside delineators, remove only	B-80	175	.183	Ea.		8.10	3.50	11.60	16.15
0110	Remove and reset	"	100	.320	"		14.20	6.10	20.30	28.50

For customer support on your Building Construction Costs with RSMeans data, call 800.448.8182.

29

02 41 Demolition

02 41 13 – Selective Site Demolition

02 41 13.30 Minor Site Demolition

		Crew	Daily Output	Labor-Hours	Unit	Material	2018 Bare Costs Labor	2018 Bare Costs Equipment	Total	Total Incl O&P
0800	Guiderail, corrugated steel, remove only	B-80A	100	.240	L.F.		9.55	2.38	11.93	17.15
0850	Remove and reset	"	40	.600	"		24	5.95	29.95	43
0860	Guide posts, remove only	B-80B	120	.267	Ea.		11.40	1.97	13.37	19.45
0870	Remove and reset	B-55	50	.480	"		19.85	19.95	39.80	52
1000	Masonry walls, block, solid	B-5	1800	.031	C.F.		1.37	.80	2.17	2.96
1200	Brick, solid		900	.062			2.74	1.60	4.34	5.95
1400	Stone, with mortar		900	.062			2.74	1.60	4.34	5.95
1500	Dry set		1500	.037			1.65	.96	2.61	3.55
1600	Median barrier, precast concrete, remove and store	B-3	430	.112	L.F.		4.97	5.30	10.27	13.40
1610	Remove and reset	"	390	.123	"		5.50	5.85	11.35	14.75
4000	Sidewalk removal, bituminous, 2" thick	B-6	350	.069	S.Y.		2.99	.89	3.88	5.50
4010	2-1/2" thick		325	.074			3.23	.96	4.19	5.95
4050	Brick, set in mortar		185	.130			5.65	1.69	7.34	10.45
4100	Concrete, plain, 4"		160	.150			6.55	1.95	8.50	12.10
4110	Plain, 5"		140	.171			7.50	2.23	9.73	13.80
4120	Plain, 6"		120	.200			8.75	2.60	11.35	16.10
4200	Mesh reinforced, concrete, 4"		150	.160			7	2.08	9.08	12.90
4210	5" thick		131	.183			8	2.39	10.39	14.75
4220	6" thick		112	.214			9.35	2.79	12.14	17.25
4300	Slab on grade removal, plain	B-5	45	1.244	C.Y.		55	32	87	119
4310	Mesh reinforced		33	1.697			75	43.50	118.50	162
4320	Rod reinforced		25	2.240			99	57.50	156.50	214
4400	For congested sites or small quantities, add up to								200%	200%
4450	For disposal on site, add	B-11A	232	.069			3.23	5.50	8.73	10.95
4500	To 5 miles, add	B-34D	76	.105			4.84	8.20	13.04	16.35

02 41 13.33 Railtrack Removal

		Crew	Daily Output	Labor-Hours	Unit	Material	Labor	Equipment	Total	Total Incl O&P
0010	**RAILTRACK REMOVAL**									
3500	Railroad track removal, ties and track	B-13	330	.170	L.F.		7.40	1.79	9.19	13.20
3600	Ballast	B-14	500	.096	C.Y.		4.04	.63	4.67	6.85
3700	Remove and re-install, ties & track using new bolts & spikes		50	.960	L.F.		40.50	6.25	46.75	68.50
3800	Turnouts using new bolts and spikes		1	48	Ea.		2,025	310	2,335	3,425

02 41 13.60 Selective Demolition Fencing

		Crew	Daily Output	Labor-Hours	Unit	Material	Labor	Equipment	Total	Total Incl O&P
0010	**SELECTIVE DEMOLITION FENCING** R024119-10									
1600	Fencing, barbed wire, 3 strand	2 Clab	430	.037	L.F.		1.48		1.48	2.26
1650	5 strand	"	280	.057			2.28		2.28	3.47
1700	Chain link, posts & fabric, 8'-10' high, remove only	B-6	445	.054			2.36	.70	3.06	4.34

02 41 16 – Structure Demolition

02 41 16.13 Building Demolition

		Crew	Daily Output	Labor-Hours	Unit	Material	Labor	Equipment	Total	Total Incl O&P
0010	**BUILDING DEMOLITION** Large urban projects, incl. 20 mi. haul R024119-10									
0011	No foundation or dump fees, C.F. is vol. of building standing									
0020	Steel	B-8	21500	.003	C.F.		.14	.13	.27	.36
0050	Concrete		15300	.004			.19	.19	.38	.50
0080	Masonry		20100	.003			.15	.14	.29	.38
0100	Mixture of types		20100	.003			.15	.14	.29	.38
0500	Small bldgs, or single bldgs, no salvage included, steel	B-3	14800	.003			.14	.15	.29	.39
0600	Concrete		11300	.004			.19	.20	.39	.51
0650	Masonry		14800	.003			.14	.15	.29	.39
0700	Wood		14800	.003			.14	.15	.29	.39
0750	For buildings with no interior walls, deduct								30%	30%
1000	Demolition single family house, one story, wood 1600 S.F.	B-3	1	48	Ea.		2,150	2,300	4,450	5,775
1020	3200 S.F.		.50	96			4,275	4,575	8,850	11,500
1200	Demolition two family house, two story, wood 2400 S.F.		.67	71.964			3,200	3,425	6,625	8,625

For customer support on your Building Construction Costs with RSMeans data, call 800.448.8182.

02 41 Demolition

02 41 16 – Structure Demolition

02 41 16.13 Building Demolition

		Crew	Daily Output	Labor-Hours	Unit	Material	2018 Bare Costs Labor	2018 Bare Costs Equipment	Total	Total Incl O&P
1220	4200 S.F.	B-3	.38	128	Ea.		5,700	6,100	11,800	15,400
1300	Demolition three family house, three story, wood 3200 S.F.		.50	96			4,275	4,575	8,850	11,500
1320	5400 S.F.	▼	.30	160			7,125	7,625	14,750	19,200
5000	For buildings with no interior walls, deduct				▼				30%	30%

02 41 16.15 Explosive/Implosive Demolition

		Crew	Daily Output	Labor-Hours	Unit	Material	2018 Bare Costs Labor	2018 Bare Costs Equipment	Total	Total Incl O&P
0010	**EXPLOSIVE/IMPLOSIVE DEMOLITION** R024119-10									
0011	Large projects,									
0020	No disposal fee based on building volume, steel building	B-5B	16900	.003	C.F.		.14	.14	.28	.36
0100	Concrete building		16900	.003			.14	.14	.28	.36
0200	Masonry building	▼	16900	.003	▼		.14	.14	.28	.36
0400	Disposal of material, minimum	B-3	445	.108	C.Y.		4.81	5.15	9.96	12.95
0500	Maximum	"	365	.132	"		5.85	6.25	12.10	15.75

02 41 16.17 Building Demolition Footings and Foundations

		Crew	Daily Output	Labor-Hours	Unit	Material	2018 Bare Costs Labor	2018 Bare Costs Equipment	Total	Total Incl O&P
0010	**BUILDING DEMOLITION FOOTINGS AND FOUNDATIONS** R024119-10									
0200	Floors, concrete slab on grade,									
0240	4" thick, plain concrete	B-13L	5000	.003	S.F.		.18	.39	.57	.70
0280	Reinforced, wire mesh		4000	.004			.22	.49	.71	.88
0300	Rods		4500	.004			.20	.44	.64	.78
0400	6" thick, plain concrete		4000	.004			.22	.49	.71	.88
0420	Reinforced, wire mesh		3200	.005			.28	.61	.89	1.09
0440	Rods		3600	.004	▼		.25	.54	.79	.98
1000	Footings, concrete, 1' thick, 2' wide		300	.053	L.F.		2.99	6.50	9.49	11.70
1080	1'-6" thick, 2' wide		250	.064			3.59	7.85	11.44	14
1120	3' wide		200	.080			4.49	9.80	14.29	17.50
1140	2' thick, 3' wide	▼	175	.091			5.15	11.20	16.35	20
1200	Average reinforcing, add								10%	10%
1220	Heavy reinforcing, add				▼				20%	20%
2000	Walls, block, 4" thick	B-13L	8000	.002	S.F.		.11	.24	.35	.44
2040	6" thick		6000	.003			.15	.33	.48	.59
2080	8" thick		4000	.004			.22	.49	.71	.88
2100	12" thick	▼	3000	.005			.30	.65	.95	1.17
2200	For horizontal reinforcing, add								10%	10%
2220	For vertical reinforcing, add								20%	20%
2400	Concrete, plain concrete, 6" thick	B-13L	4000	.004			.22	.49	.71	.88
2420	8" thick		3500	.005			.26	.56	.82	1
2440	10" thick		3000	.005			.30	.65	.95	1.17
2500	12" thick	▼	2500	.006			.36	.78	1.14	1.40
2600	For average reinforcing, add								10%	10%
2620	For heavy reinforcing, add								20%	20%
4000	For congested sites or small quantities, add up to				▼				200%	200%
4200	Add for disposal, on site	B-11A	232	.069	C.Y.		3.23	5.50	8.73	10.95
4250	To five miles	B-30	220	.109	"		5.30	9	14.30	17.90

02 41 19 – Selective Demolition

02 41 19.13 Selective Building Demolition

		Crew	Daily Output	Labor-Hours	Unit	Material	2018 Bare Costs Labor	2018 Bare Costs Equipment	Total	Total Incl O&P
0010	**SELECTIVE BUILDING DEMOLITION**									
0020	Costs related to selective demolition of specific building components									
0025	are included under Common Work Results (XX 05)									
0030	in the component's appropriate division.									

For customer support on your Building Construction Costs with RSMeans data, call 800.448.8182.

31

02 41 Demolition

02 41 19 – Selective Demolition

02 41 19.16 Selective Demolition, Cutout

		Crew	Daily Output	Labor-Hours	Unit	Material	2018 Bare Costs Labor	Equipment	Total	Total Incl O&P
0010	**SELECTIVE DEMOLITION, CUTOUT** R024119-10									
0020	Concrete, elev. slab, light reinforcement, under 6 C.F.	B-9	65	.615	C.F.		25	3.60	28.60	41.50
0050	Light reinforcing, over 6 C.F.		75	.533	"		21.50	3.12	24.62	36
0200	Slab on grade to 6" thick, not reinforced, under 8 S.F.		85	.471	S.F.		18.95	2.75	21.70	32
0250	8-16 S.F.	↓	175	.229	"		9.20	1.34	10.54	15.45
0255	For over 16 S.F. see Line 02 41 16.17 0400									
0600	Walls, not reinforced, under 6 C.F.	B-9	60	.667	C.F.		27	3.90	30.90	45.50
0650	6-12 C.F.	"	80	.500	"		20	2.93	22.93	33.50
0655	For over 12 C.F. see Line 02 41 16.17 2500									
1000	Concrete, elevated slab, bar reinforced, under 6 C.F.	B-9	45	.889	C.F.		36	5.20	41.20	60
1050	Bar reinforced, over 6 C.F.		50	.800	"		32	4.68	36.68	54
1200	Slab on grade to 6" thick, bar reinforced, under 8 S.F.		75	.533	S.F.		21.50	3.12	24.62	36
1250	8-16 S.F.	↓	150	.267	"		10.75	1.56	12.31	18.05
1255	For over 16 S.F. see Line 02 41 16.17 0440									
1400	Walls, bar reinforced, under 6 C.F.	B-9	50	.800	C.F.		32	4.68	36.68	54
1450	6-12 C.F.	"	70	.571	"		23	3.34	26.34	38.50
1455	For over 12 C.F. see Lines 02 41 16.17 2500 and 2600									
2000	Brick, to 4 S.F. opening, not including toothing									
2040	4" thick	B-9	30	1.333	Ea.		53.50	7.80	61.30	90.50
2060	8" thick		18	2.222			89.50	13	102.50	150
2080	12" thick		10	4			161	23.50	184.50	271
2400	Concrete block, to 4 S.F. opening, 2" thick		35	1.143			46	6.70	52.70	77.50
2420	4" thick		30	1.333			53.50	7.80	61.30	90.50
2440	8" thick		27	1.481			59.50	8.65	68.15	101
2460	12" thick		24	1.667			67	9.75	76.75	113
2600	Gypsum block, to 4 S.F. opening, 2" thick		80	.500			20	2.93	22.93	33.50
2620	4" thick		70	.571			23	3.34	26.34	38.50
2640	8" thick		55	.727			29.50	4.25	33.75	49
2800	Terra cotta, to 4 S.F. opening, 4" thick		70	.571			23	3.34	26.34	38.50
2840	8" thick		65	.615			25	3.60	28.60	41.50
2880	12" thick	↓	50	.800	↓		32	4.68	36.68	54
3000	Toothing masonry cutouts, brick, soft old mortar	1 Brhe	40	.200	V.L.F.		7.85		7.85	12.05
3100	Hard mortar		30	.267			10.50		10.50	16.05
3200	Block, soft old mortar		70	.114			4.49		4.49	6.85
3400	Hard mortar	↓	50	.160	↓		6.30		6.30	9.60
6000	Walls, interior, not including re-framing,									
6010	openings to 5 S.F.									
6100	Drywall to 5/8" thick	1 Clab	24	.333	Ea.		13.30		13.30	20
6200	Paneling to 3/4" thick		20	.400			15.95		15.95	24.50
6300	Plaster, on gypsum lath		20	.400			15.95		15.95	24.50
6340	On wire lath	↓	14	.571	↓		23		23	34.50
7000	Wood frame, not including re-framing, openings to 5 S.F.									
7200	Floors, sheathing and flooring to 2" thick	1 Clab	5	1.600	Ea.		64		64	97
7310	Roofs, sheathing to 1" thick, not including roofing		6	1.333			53		53	81
7410	Walls, sheathing to 1" thick, not including siding	↓	7	1.143	↓		45.50		45.50	69.50

02 41 19.18 Selective Demolition, Disposal Only

		Crew	Daily Output	Labor-Hours	Unit	Material	2018 Bare Costs Labor	Equipment	Total	Total Incl O&P
0010	**SELECTIVE DEMOLITION, DISPOSAL ONLY** R024119-10									
0015	Urban bldg w/salvage value allowed									
0020	Including loading and 5 mile haul to dump									
0200	Steel frame	B-3	430	.112	C.Y.		4.97	5.30	10.27	13.40
0300	Concrete frame		365	.132			5.85	6.25	12.10	15.75
0400	Masonry construction	↓	445	.108	↓		4.81	5.15	9.96	12.95

For customer support on your Building Construction Costs with RSMeans data, call 800.448.8182.

02 41 Demolition

02 41 19 – Selective Demolition

02 41 19.18 Selective Demolition, Disposal Only	Crew	Daily Output	Labor-Hours	Unit	Material	2018 Bare Costs Labor	Equipment	Total	Total Incl O&P
0500 Wood frame	B-3	247	.194	C.Y.		8.65	9.25	17.90	23.50

02 41 19.19 Selective Demolition

		Crew	Daily Output	Labor-Hours	Unit	Material	Labor	Equipment	Total	Total Incl O&P
0010	**SELECTIVE DEMOLITION**, Rubbish Handling R024119-10									
0020	The following are to be added to the demolition prices									
0050	The following are components for a complete chute system									
0100	Top chute circular steel, 4' long, 18" diameter R024119-30	B-1C	15	1.600	Ea.	272	65	30.50	367.50	430
0102	23" diameter		15	1.600		295	65	30.50	390.50	460
0104	27" diameter		15	1.600		320	65	30.50	415.50	485
0106	30" diameter		15	1.600		340	65	30.50	435.50	510
0108	33" diameter		15	1.600		365	65	30.50	460.50	535
0110	36" diameter		15	1.600		385	65	30.50	480.50	560
0112	Regular chute, 18" diameter		15	1.600		204	65	30.50	299.50	360
0114	23" diameter		15	1.600		227	65	30.50	322.50	380
0116	27" diameter		15	1.600		249	65	30.50	344.50	405
0118	30" diameter		15	1.600		261	65	30.50	356.50	420
0120	33" diameter		15	1.600		295	65	30.50	390.50	460
0122	36" diameter		15	1.600		320	65	30.50	415.50	485
0124	Control door chute, 18" diameter		15	1.600		385	65	30.50	480.50	560
0126	23" diameter		15	1.600		410	65	30.50	505.50	585
0128	27" diameter		15	1.600		430	65	30.50	525.50	610
0130	30" diameter		15	1.600		455	65	30.50	550.50	635
0132	33" diameter		15	1.600		475	65	30.50	570.50	660
0134	36" diameter		15	1.600		500	65	30.50	595.50	685
0136	Chute liners, 14 ga., 18"-30" diameter		15	1.600		209	65	30.50	304.50	365
0138	33"-36" diameter		15	1.600		261	65	30.50	356.50	420
0140	17% thinner chute, 30" diameter		15	1.600		220	65	30.50	315.50	375
0142	33% thinner chute, 30" diameter	▼	15	1.600		166	65	30.50	261.50	315
0144	Top chute cover	1 Clab	24	.333		143	13.30		156.30	177
0146	Door chute cover	"	24	.333		143	13.30		156.30	177
0148	Top chute trough	2 Clab	12	1.333		475	53		528	605
0150	Bolt down frame & counter weights, 250 lb.	B-1	4	6		4,175	243		4,418	4,975
0152	500 lb.		4	6		6,175	243		6,418	7,175
0154	750 lb.		4	6		8,925	243		9,168	10,200
0156	1000 lb.		2.67	8.989		9,800	365		10,165	11,400
0158	1500 lb.	▼	2.67	8.989		12,600	365		12,965	14,400
0160	Chute warning light system, 5 stories	B-1C	4	6		8,900	243	114	9,257	10,300
0162	10 stories	"	2	12		14,200	485	227	14,912	16,600
0164	Dust control device for dumpsters	1 Clab	8	1		133	40		173	207
0166	Install or replace breakaway cord		8	1		25	40		65	88
0168	Install or replace warning sign	▼	16	.500	▼	10	19.95		29.95	41.50
0600	Dumpster, weekly rental, 1 dump/week, 6 C.Y. capacity (2 tons)				Week	415			415	455
0700	10 C.Y. capacity (3 tons)					480			480	530
0725	20 C.Y. capacity (5 tons) R024119-20					565			565	625
0800	30 C.Y. capacity (7 tons)					730			730	800
0840	40 C.Y. capacity (10 tons)				▼	775			775	850
2000	Load, haul, dump and return, 0'-50' haul, hand carried	2 Clab	24	.667	C.Y.		26.50		26.50	40.50
2005	Wheeled		37	.432			17.25		17.25	26.50
2040	0'-100' haul, hand carried		16.50	.970			38.50		38.50	59
2045	Wheeled		25	.640			25.50		25.50	39
2050	Forklift	A-3R	25	.320			16.40	5.95	22.35	31.50
2080	Haul and return, add per each extra 100' haul, hand carried	2 Clab	35.50	.451			17.95		17.95	27.50
2085	Wheeled	▼	54	.296	▼		11.80		11.80	18

02 41 19 – Selective Demolition

02 41 19.19 Selective Demolition

		Crew	Daily Output	Labor-Hours	Unit	Material	2018 Bare Costs Labor	Equipment	Total	Total Incl O&P
2120	For travel in elevators, up to 10 floors, add	2 Clab	140	.114	C.Y.		4.55		4.55	6.95
2130	0'-50' haul, incl. up to 5 riser stairs, hand carried		23	.696			27.50		27.50	42
2135	Wheeled		35	.457			18.20		18.20	28
2140	6-10 riser stairs, hand carried		22	.727			29		29	44
2145	Wheeled		34	.471			18.75		18.75	28.50
2150	11-20 riser stairs, hand carried		20	.800			32		32	48.50
2155	Wheeled		31	.516			20.50		20.50	31.50
2160	21-40 riser stairs, hand carried		16	1			40		40	60.50
2165	Wheeled		24	.667			26.50		26.50	40.50
2170	0-100' haul, incl. 5 riser stairs, hand carried		15	1.067			42.50		42.50	65
2175	Wheeled		23	.696			27.50		27.50	42
2180	6-10 riser stairs, hand carried		14	1.143			45.50		45.50	69.50
2185	Wheeled		21	.762			30.50		30.50	46.50
2190	11-20 riser stairs, hand carried		12	1.333			53		53	81
2195	Wheeled		18	.889			35.50		35.50	54
2200	21-40 riser stairs, hand carried		8	2			79.50		79.50	121
2205	Wheeled		12	1.333			53		53	81
2210	Haul and return, add per each extra 100' haul, hand carried		35.50	.451			17.95		17.95	27.50
2215	Wheeled		54	.296			11.80		11.80	18
2220	For each additional flight of stairs, up to 5 risers, add		550	.029	Flight		1.16		1.16	1.77
2225	6-10 risers, add		275	.058			2.32		2.32	3.53
2230	11-20 risers, add		138	.116			4.62		4.62	7.05
2235	21-40 risers, add		69	.232			9.25		9.25	14.10
3000	Loading & trucking, including 2 mile haul, chute loaded	B-16	45	.711	C.Y.		30	12.05	42.05	58.50
3040	Hand loading truck, 50' haul	"	48	.667			28	11.30	39.30	55
3080	Machine loading truck	B-17	120	.267			11.80	5.40	17.20	24
5000	Haul, per mile, up to 8 C.Y. truck	B-34B	1165	.007			.32	.47	.79	.99
5100	Over 8 C.Y. truck	"	1550	.005			.24	.35	.59	.75

02 41 19.20 Selective Demolition, Dump Charges

		Crew	Daily Output	Labor-Hours	Unit	Material	Labor	Equipment	Total	Total Incl O&P
0010	**SELECTIVE DEMOLITION, DUMP CHARGES** R024119-10									
0020	Dump charges, typical urban city, tipping fees only									
0100	Building construction materials				Ton	74			74	81
0200	Trees, brush, lumber					63			63	69.50
0300	Rubbish only					63			63	69.50
0500	Reclamation station, usual charge					74			74	81

02 41 19.21 Selective Demolition, Gutting

		Crew	Daily Output	Labor-Hours	Unit	Material	Labor	Equipment	Total	Total Incl O&P
0010	**SELECTIVE DEMOLITION, GUTTING** R024119-10									
0020	Building interior, including disposal, dumpster fees not included									
0500	Residential building									
0560	Minimum	B-16	400	.080	SF Flr.		3.35	1.36	4.71	6.60
0580	Maximum	"	360	.089	"		3.72	1.51	5.23	7.30
0900	Commercial building									
1000	Minimum	B-16	350	.091	SF Flr.		3.83	1.55	5.38	7.50
1020	Maximum	"	250	.128	"		5.35	2.17	7.52	10.55

02 41 19.25 Selective Demolition, Saw Cutting

		Crew	Daily Output	Labor-Hours	Unit	Material	Labor	Equipment	Total	Total Incl O&P
0010	**SELECTIVE DEMOLITION, SAW CUTTING** R024119-10									
0015	Asphalt, up to 3" deep	B-89	1050	.015	L.F.	.19	.73	.40	1.32	1.75
0020	Each additional inch of depth	"	1800	.009		.06	.43	.23	.72	.97
1200	Masonry walls, hydraulic saw, brick, per inch of depth	B-89B	300	.053		.06	2.55	2.24	4.85	6.40
1220	Block walls, solid, per inch of depth	"	250	.064		.06	3.07	2.69	5.82	7.65
2000	Brick or masonry w/hand held saw, per inch of depth	A-1	125	.064		.04	2.55	.57	3.16	4.56
5000	Wood sheathing to 1" thick, on walls	1 Carp	200	.040			2.03		2.03	3.09

02 41 Demolition

02 41 19 – Selective Demolition

02 41 19.25 Selective Demolition, Saw Cutting	Crew	Daily Output	Labor-Hours	Unit	Material	2018 Bare Costs Labor	Equipment	Total	Total Incl O&P
5020 On roof	1 Carp	250	.032	L.F.		1.62		1.62	2.47

02 41 19.27 Selective Demolition, Torch Cutting

		Crew	Daily Output	Labor-Hours	Unit	Material	2018 Bare Costs Labor	Equipment	Total	Total Incl O&P
0010	**SELECTIVE DEMOLITION, TORCH CUTTING** R024119-10									
0020	Steel, 1" thick plate	E-25	333	.024	L.F.	.87	1.36	.04	2.27	3.22
0040	1" diameter bar	"	600	.013	Ea.	.15	.76	.02	.93	1.41
1000	Oxygen lance cutting, reinforced concrete walls									
1040	12"-16" thick walls	1 Clab	10	.800	L.F.		32		32	48.50
1080	24" thick walls	"	6	1.333	"		53		53	81

02 42 Removal and Salvage of Construction Materials

02 42 10 – Building Deconstruction

02 42 10.10 Estimated Salvage Value or Savings

						Unit	Material	2018 Bare Costs Labor	Equipment	Total	Total Incl O&P
0010	**ESTIMATED SALVAGE VALUE OR SAVINGS**										
0015	Excludes material handling, packaging, container costs and										
0020	transportation for salvage or disposal										
0050	All items in Section 02 42 10.10 are credit deducts and not costs										
0100	Copper wire salvage value	G				Lb.				1.60	1.60
0110	Disposal savings	G								.04	.04
0200	Copper pipe salvage value	G								2.50	2.50
0210	Disposal savings	G								.05	.05
0300	Steel pipe salvage value	G								.06	.06
0310	Disposal savings	G								.03	.03
0400	Cast iron pipe salvage value	G								.03	.03
0410	Disposal savings	G								.01	.01
0500	Steel doors or windows salvage value	G								.06	.06
0510	Aluminum	G								.55	.55
0520	Disposal savings	G								.03	.03
0600	Aluminum siding salvage value	G								.49	.49
0630	Disposal savings	G				▼				.03	.03
0640	Wood siding (no lead or asbestos)	G				C.Y.				12	12
0800	Clean concrete disposal savings	G				Ton				62	62
0850	Asphalt shingles disposal savings	G				"				60	60
1000	Wood wall framing clean salvage value	G				M.B.F.				55	55
1010	Painted	G								44	44
1020	Floor framing	G								55	55
1030	Painted	G								44	44
1050	Roof framing	G								55	55
1060	Painted	G								44	44
1100	Wood beams salvage value	G				▼				55	55
1200	Wood framing and beams disposal savings	G				Ton				66	66
1220	Wood sheathing and sub-base flooring	G								72.50	72.50
1230	Wood wall paneling (1/4" thick)	G				▼				66	66
1300	Wood panel 3/4"-1" thick low salvage value	G				S.F.				.55	.55
1350	High salvage value	G				"				2.20	2.20
1400	Disposal savings	G				Ton				66	66
1500	Flooring tongue and groove 25/32" thick low salvage value	G				S.F.				.55	.55
1530	High salvage value	G				"				1.10	1.10
1560	Disposal savings	G				Ton				66	66
1600	Drywall or sheet rock salvage value	G								22	22
1650	Disposal savings	G				▼				66	66

02 42 10.20 Deconstruction of Building Components		Crew	Daily Output	Labor-Hours	Unit	Material	2018 Bare Costs Labor	Equipment	Total	Total Incl O&P
0010	**DECONSTRUCTION OF BUILDING COMPONENTS**									
0012	Buildings one or two stories only									
0015	Excludes material handling, packaging, container costs and									
0020	transportation for salvage or disposal									
0050	Deconstruction of plumbing fixtures									
0100	Wall hung or countertop lavatory	G	2 Clab	16	1	Ea.		40		40	60.50
0110	Single or double compartment kitchen sink	G		14	1.143			45.50		45.50	69.50
0120	Wall hung urinal	G		14	1.143			45.50		45.50	69.50
0130	Floor mounted	G		8	2			79.50		79.50	121
0140	Floor mounted water closet	G		16	1			40		40	60.50
0150	Wall hung	G		14	1.143			45.50		45.50	69.50
0160	Water fountain, free standing	G		16	1			40		40	60.50
0170	Wall hung or deck mounted	G		12	1.333			53		53	81
0180	Bathtub, steel or fiberglass	G		10	1.600			64		64	97
0190	Cast iron	G		8	2			79.50		79.50	121
0200	Shower, single	G		6	2.667			106		106	162
0210	Group	G		7	2.286			91		91	139
0300	Deconstruction of electrical fixtures									
0310	Surface mounted incandescent fixtures	G	2 Clab	48	.333	Ea.		13.30		13.30	20
0320	Fluorescent, 2 lamp	G		32	.500			19.95		19.95	30.50
0330	4 lamp	G		24	.667			26.50		26.50	40.50
0340	Strip fluorescent, 1 lamp	G		40	.400			15.95		15.95	24.50
0350	2 lamp	G		32	.500			19.95		19.95	30.50
0400	Recessed drop-in fluorescent fixture, 2 lamp	G		27	.593			23.50		23.50	36
0410	4 lamp	G		18	.889			35.50		35.50	54
0500	Deconstruction of appliances									
0510	Cooking stoves	G	2 Clab	26	.615	Ea.		24.50		24.50	37.50
0520	Dishwashers	G	"	26	.615	"		24.50		24.50	37.50
0600	Deconstruction of millwork and trim									
0610	Cabinets, wood	G	2 Carp	40	.400	L.F.		20.50		20.50	31
0620	Countertops	G		100	.160	"		8.10		8.10	12.35
0630	Wall paneling, 1" thick	G		500	.032	S.F.		1.62		1.62	2.47
0640	Ceiling trim	G		500	.032	L.F.		1.62		1.62	2.47
0650	Wainscoting	G		500	.032	S.F.		1.62		1.62	2.47
0660	Base, 3/4"-1" thick	G		600	.027	L.F.		1.35		1.35	2.06
0700	Deconstruction of doors and windows									
0710	Doors, wrap, interior, wood, single, no closers	G	2 Carp	21	.762	Ea.	4.75	38.50		43.25	64.50
0720	Double	G		13	1.231		9.50	62.50		72	105
0730	Solid core, single, exterior or interior	G		10	1.600		4.75	81		85.75	129
0740	Double	G		8	2		9.50	101		110.50	164
0810	Windows, wrap, wood, single									
0812	with no casement or cladding	G	2 Carp	21	.762	Ea.	4.75	38.50		43.25	64.50
0820	with casement and/or cladding	G	"	18	.889	"	4.75	45		49.75	74
0900	Deconstruction of interior finishes									
0910	Drywall for recycling	G	2 Clab	1775	.009	S.F.		.36		.36	.55
0920	Plaster wall, first floor	G		1775	.009			.36		.36	.55
0930	Second floor	G		1330	.012			.48		.48	.73
1000	Deconstruction of roofing and accessories									
1010	Built-up roofs	G	2 Clab	570	.028	S.F.		1.12		1.12	1.70
1020	Gutters, fascia and rakes	G	"	1140	.014	L.F.		.56		.56	.85
2000	Deconstruction of wood components									
2010	Roof sheeting	G	2 Clab	570	.028	S.F.		1.12		1.12	1.70

For customer support on your Building Construction Costs with RSMeans data, call 800.448.8182.

02 42 Removal and Salvage of Construction Materials

02 42 10 – Building Deconstruction

02 42 10.20 Deconstruction of Building Components

		Crew	Daily Output	Labor-Hours	Unit	Material	2018 Bare Costs Labor	Equipment	Total	Total Incl O&P
2020	Main roof framing	G 2 Clab	760	.021	L.F.		.84		.84	1.28
2030	Porch roof framing	G ↓	445	.036			1.43		1.43	2.18
2040	Beams 4" x 8"	G B-1	375	.064			2.59		2.59	3.95
2050	4" x 10"	G	300	.080			3.24		3.24	4.94
2055	4" x 12"	G	250	.096			3.89		3.89	5.95
2060	6" x 8"	G	250	.096			3.89		3.89	5.95
2065	6" x 10"	G	200	.120			4.86		4.86	7.40
2070	6" x 12"	G	170	.141			5.70		5.70	8.70
2075	8" x 12"	G	126	.190			7.70		7.70	11.75
2080	10" x 12"	G ↓	100	.240			9.70		9.70	14.80
2100	Ceiling joists	G 2 Clab	800	.020			.80		.80	1.21
2150	Wall framing, interior	G	1230	.013	↓		.52		.52	.79
2160	Sub-floor	G	2000	.008	S.F.		.32		.32	.49
2170	Floor joists	G	2000	.008	L.F.		.32		.32	.49
2200	Wood siding (no lead or asbestos)	G	1300	.012	S.F.		.49		.49	.75
2300	Wall framing, exterior	G	1600	.010	L.F.		.40		.40	.61
2400	Stair risers	G	53	.302	Ea.		12.05		12.05	18.30
2500	Posts	G ↓	800	.020	L.F.		.80		.80	1.21
3000	Deconstruction of exterior brick walls									
3010	Exterior brick walls, first floor	G 2 Clab	200	.080	S.F.		3.19		3.19	4.86
3020	Second floor	G	64	.250	"		9.95		9.95	15.20
3030	Brick chimney	G ↓	100	.160	C.F.		6.40		6.40	9.70
4000	Deconstruction of concrete									
4010	Slab on grade, 4" thick, plain concrete	G B-9	500	.080	S.F.		3.22	.47	3.69	5.40
4020	Wire mesh reinforced	G	470	.085			3.43	.50	3.93	5.75
4030	Rod reinforced	G	400	.100			4.03	.59	4.62	6.80
4110	Foundation wall, 6" thick, plain concrete	G	160	.250			10.05	1.46	11.51	16.95
4120	8" thick	G	140	.286			11.50	1.67	13.17	19.35
4130	10" thick	G ↓	120	.333	↓		13.40	1.95	15.35	22.50
9000	Deconstruction process, support equipment as needed									
9010	Daily use, portal to portal, 12-ton truck-mounted hydraulic crane crew	G A-3H	1	8	Day		450	630	1,080	1,375
9020	Daily use, skid steer and operator	G A-3C	1	8			410	365	775	1,025
9030	Daily use, backhoe 48 HP, operator and labor	G "	1	8	↓		410	365	775	1,025

02 42 10.30 Deconstruction Material Handling

		Crew	Daily Output	Labor-Hours	Unit	Material	2018 Bare Costs Labor	Equipment	Total	Total Incl O&P
0010	**DECONSTRUCTION MATERIAL HANDLING**									
0012	Buildings one or two stories only									
0100	Clean and stack brick on pallet	G 2 Clab	1200	.013	Ea.		.53		.53	.81
0200	Haul 50' and load rough lumber up to 2" x 8" size	G	2000	.008	"		.32		.32	.49
0210	Lumber larger than 2" x 8"	G	3200	.005	B.F.		.20		.20	.30
0300	Finish wood for recycling stack and wrap per pallet	G	8	2	Ea.	38	79.50		117.50	163
0350	Light fixtures		6	2.667		68.50	106		174.50	237
0375	Windows		6	2.667		64.50	106		170.50	233
0400	Miscellaneous materials	↓	8	2	↓	19	79.50		98.50	142
1000	See Section 02 41 19.19 for bulk material handling									

02 43 Structure Moving

02 43 13 – Structure Relocation

02 43 13.13 Building Relocation		Crew	Daily Output	Labor-Hours	Unit	Material	2018 Bare Costs Labor	Equipment	Total	Total Incl O&P
0010	**BUILDING RELOCATION**									
0011	One day move, up to 24' wide									
0020	Reset on existing foundation				Total				11,500	11,500
0040	Wood or steel frame bldg., based on ground floor area G	B-4	185	.259	S.F.		10.70	2.54	13.24	19.05
0060	Masonry bldg., based on ground floor area G	"	137	.350			14.45	3.43	17.88	26
0200	For 24'-42' wide, add								15%	15%

02 56 Site Containment

02 56 13 – Waste Containment

02 56 13.10 Containment of Hazardous Waste

		Crew	Daily Output	Labor-Hours	Unit	Material	2018 Bare Costs Labor	Equipment	Total	Total Incl O&P
0010	**CONTAINMENT OF HAZARDOUS WASTE**									
0020	OSHA Hazard level C									
0030	OSHA Hazard level D decrease labor and equipment, deduct						45%	45%		
0035	OSHA Hazard level B increase labor and equipment, add						22%	22%		
0040	OSHA Hazard level A increase labor and equipment, add						71%	71%		
0100	Excavation of contaminated soil & waste									
0105	Includes one respirator filter and two disposable suits per work day									
0110	3/4 C.Y. excavator to 10' deep	B-12F	51	.314	B.C.Y.	1.42	15.05	13.15	29.62	39
0120	Labor crew to 6' deep	B-2	19	2.105		9.55	84.50		94.05	140
0130	6'-12' deep	"	12	3.333		15.10	134		149.10	221
0200	Move contaminated soil/waste up to 150' on-site with 2.5 C.Y. loader	B-10T	300	.040	L.C.Y.	.24	1.96	1.74	3.94	5.15
0210	300'	"	186	.065	"	.39	3.17	2.81	6.37	8.30
0300	Secure burial cell construction									
0310	Various liner and cover materials									
0400	Very low density polyethylene (VLDPE)									
0410	50 mil top cover	B-47H	4000	.008	S.F.	.53	.42	.06	1.01	1.30
0420	80 mil liner	"	4000	.008	"	.68	.42	.06	1.16	1.47
0500	Chlorosulfonated polyethylene									
0510	36 mil hypalon top cover	B-47H	4000	.008	S.F.	2.32	.42	.06	2.80	3.27
0520	45 mil hypalon liner	"	4000	.008	"	2.76	.42	.06	3.24	3.76
0600	Polyvinyl chloride (PVC)									
0610	60 mil top cover	B-47H	4000	.008	S.F.	.96	.42	.06	1.44	1.77
0620	80 mil liner	"	4000	.008	"	1.26	.42	.06	1.74	2.11
0700	Rough textured H.D. polyethylene (HDPE)									
0710	40 mil top cover	B-47H	4000	.008	S.F.	.51	.42	.06	.99	1.28
0720	60 mil top cover		4000	.008		.61	.42	.06	1.09	1.39
0722	60 mil liner		4000	.008		.57	.42	.06	1.05	1.35
0730	80 mil liner		3800	.008		.71	.45	.06	1.22	1.53
1000	3/4" crushed stone, 6" deep ballast around liner	B-6	30	.800	L.C.Y.	28	35	10.40	73.40	95
1100	Hazardous waste, ballast cover with common borrow material	B-63	56	.714		12.75	30	3.11	45.86	63
1110	Mixture of common borrow & topsoil		56	.714		18.95	30	3.11	52.06	70
1120	Bank sand		56	.714		18.15	30	3.11	51.26	69
1130	Medium priced clay		44	.909		26.50	38.50	3.95	68.95	91.50
1140	Mixture of common borrow & medium priced clay		56	.714		19.60	30	3.11	52.71	70.50

38

02 58 Snow Control

02 58 13 – Snow Fencing

02 58 13.10 Snow Fencing System		Crew	Daily Output	Labor-Hours	Unit	Material	2018 Bare Costs Labor	Equipment	Total	Total Incl O&P
0010	**SNOW FENCING SYSTEM**									
7001	Snow fence on steel posts 10' OC, 4' high	B-1	500	.048	L.F.	.46	1.95		2.41	3.47

02 65 Underground Storage Tank Removal

02 65 10 – Underground Tank and Contaminated Soil Removal

02 65 10.30 Removal of Underground Storage Tanks

			Crew	Daily Output	Labor-Hours	Unit	Material	2018 Bare Costs Labor	Equipment	Total	Total Incl O&P
0010	**REMOVAL OF UNDERGROUND STORAGE TANKS**	R026510-20									
0011	Petroleum storage tanks, non-leaking										
0100	Excavate & load onto trailer										
0110	3,000 gal. to 5,000 gal. tank	G	B-14	4	12	Ea.		505	78	583	855
0120	6,000 gal. to 8,000 gal. tank	G	B-3A	3	13.333			570	298	868	1,200
0130	9,000 gal. to 12,000 gal. tank	G	"	2	20	↓		855	445	1,300	1,800
0190	Known leaking tank, add					%				100%	100%
0200	Remove sludge, water and remaining product from bottom										
0201	of tank with vacuum truck										
0300	3,000 gal. to 5,000 gal. tank	G	A-13	5	1.600	Ea.		82	144	226	282
0310	6,000 gal. to 8,000 gal. tank	G		4	2			103	180	283	355
0320	9,000 gal. to 12,000 gal. tank	G	↓	3	2.667	↓		137	240	377	470
0390	Dispose of sludge off-site, average					Gal.				6.25	6.80
0400	Insert inert solid CO_2 "dry ice" into tank										
0401	For cleaning/transporting tanks (1.5 lb./100 gal. cap)	G	1 Clab	500	.016	Lb.	1.19	.64		1.83	2.28
0403	Insert solid carbon dioxide, 1.5 lb./100 gal.	G	"	400	.020	"	1.19	.80		1.99	2.52
0503	Disconnect and remove piping	G	1 Plum	160	.050	L.F.		3.11		3.11	4.68
0603	Transfer liquids, 10% of volume	G	"	1600	.005	Gal.		.31		.31	.47
0703	Cut accessway into underground storage tank	G	1 Clab	5.33	1.501	Ea.		60		60	91
0813	Remove sludge, wash and wipe tank, 500 gal.	G	1 Plum	8	1			62		62	93.50
0823	3,000 gal.	G		6.67	1.199			74.50		74.50	112
0833	5,000 gal.	G		6.15	1.301			81		81	122
0843	8,000 gal.	G		5.33	1.501			93.50		93.50	140
0853	10,000 gal.	G		4.57	1.751			109		109	164
0863	12,000 gal.	G	↓	4.21	1.900	↓		118		118	178
1020	Haul tank to certified salvage dump, 100 miles round trip										
1023	3,000 gal. to 5,000 gal. tank					Ea.				760	830
1026	6,000 gal. to 8,000 gal. tank									880	960
1029	9,000 gal. to 12,000 gal. tank					↓				1,050	1,150
1100	Disposal of contaminated soil to landfill										
1110	Minimum					C.Y.				145	160
1111	Maximum					"				400	440
1120	Disposal of contaminated soil to										
1121	bituminous concrete batch plant										
1130	Minimum					C.Y.				80	88
1131	Maximum					"				115	125
1203	Excavate, pull, & load tank, backfill hole, 8,000 gal. +	G	B-12C	.50	32	Ea.		1,525	2,100	3,625	4,650
1213	Haul tank to certified dump, 100 miles rt, 8,000 gal. +	G	B-34K	1	8			370	835	1,205	1,475
1223	Excavate, pull, & load tank, backfill hole, 500 gal.	G	B-11C	1	16			750	310	1,060	1,475
1233	Excavate, pull, & load tank, backfill hole, 3,000-5,000 gal.	G	B-11M	.50	32			1,500	770	2,270	3,125
1243	Haul tank to certified dump, 100 miles rt, 500 gal.	G	B-34L	1	8			410	188	598	825
1253	Haul tank to certified dump, 100 miles rt, 3,000-5,000 gal.	G	B-34M	1	8	↓		410	238	648	880
2010	Decontamination of soil on site incl poly tarp on top/bottom										
2011	Soil containment berm and chemical treatment										
2020	Minimum	G	B-11C	100	.160	C.Y.	7.45	7.50	3.12	18.07	23

02 65 Underground Storage Tank Removal

02 65 10 – Underground Tank and Contaminated Soil Removal

02 65 10.30 Removal of Underground Storage Tanks		Crew	Daily Output	Labor-Hours	Unit	Material	2018 Bare Costs Labor	Equipment	Total	Total Incl O&P
2021	Maximum	G	B-11C	100	.160	C.Y.	9.65	7.50	3.12	20.27	25.50
2050	Disposal of decontaminated soil, minimum					↓				135	150
2055	Maximum					▼				400	440

02 81 Transportation and Disposal of Hazardous Materials

02 81 20 – Hazardous Waste Handling

02 81 20.10 Hazardous Waste Cleanup/Pickup/Disposal

		Unit	Total	Total Incl O&P
0010	**HAZARDOUS WASTE CLEANUP/PICKUP/DISPOSAL**			
0100	For contractor rental equipment, i.e., dozer,			
0110	Front end loader, dump truck, etc., see 01 54 33 Reference Section			
1000	Solid pickup			
1100	55 gal. drums	Ea.	240	265
1120	Bulk material, minimum	Ton	190	210
1130	Maximum	"	595	655
1200	Transportation to disposal site			
1220	Truckload = 80 drums or 25 C.Y. or 18 tons			
1260	Minimum	Mile	3.95	4.45
1270	Maximum	"	7.25	7.35
3000	Liquid pickup, vacuum truck, stainless steel tank			
3100	Minimum charge, 4 hours			
3110	1 compartment, 2200 gallon	Hr.	140	155
3120	2 compartment, 5000 gallon	"	200	225
3400	Transportation in 6900 gallon bulk truck	Mile	7.95	8.75
3410	In teflon lined truck	"	10.20	11.25
5000	Heavy sludge or dry vacuumable material	Hr.	140	160
6000	Dumpsite disposal charge, minimum	Ton	140	155
6020	Maximum	"	415	455

02 82 Asbestos Remediation

02 82 13 – Asbestos Abatement

02 82 13.39 Asbestos Remediation Plans and Methods

		Unit	Total	Total Incl O&P
0010	**ASBESTOS REMEDIATION PLANS AND METHODS**			
0100	Building Survey-Commercial Building	Ea.	2,200	2,400
0200	Asbestos Abatement Remediation Plan	"	1,350	1,475

02 82 13.41 Asbestos Abatement Equipment

			Unit	Material	Total	Total Incl O&P
0010	**ASBESTOS ABATEMENT EQUIPMENT**	R028213-20				
0011	Equipment and supplies, buy					
0200	Air filtration device, 2000 CFM		Ea.	820	820	900
0250	Large volume air sampling pump, minimum			315	315	345
0260	Maximum			360	360	395
0300	Airless sprayer unit, 2 gun			2,175	2,175	2,375
0350	Light stand, 500 watt		▼	38	38	42
0400	Personal respirators					
0410	Negative pressure, 1/2 face, dual operation, minimum		Ea.	27	27	29.50
0420	Maximum			30	30	33
0450	P.A.P.R., full face, minimum			675	675	740
0460	Maximum			1,400	1,400	1,550
0470	Supplied air, full face, including air line, minimum			380	380	420
0480	Maximum		▼	525	525	580

02 82 Asbestos Remediation

02 82 13 – Asbestos Abatement

02 82 13.41 Asbestos Abatement Equipment

		Crew	Daily Output	Labor-Hours	Unit	Material	2018 Bare Costs Labor	Equipment	Total	Total Incl O&P
0500	Personnel sampling pump				Ea.	245			245	270
1500	Power panel, 20 unit, including GFI					445			445	490
1600	Shower unit, including pump and filters					1,075			1,075	1,175
1700	Supplied air system (type C)					3,500			3,500	3,850
1750	Vacuum cleaner, HEPA, 16 gal., stainless steel, wet/dry					1,100			1,100	1,200
1760	55 gallon					1,300			1,300	1,425
1800	Vacuum loader, 9-18 ton/hr.					97,000			97,000	107,000
1900	Water atomizer unit, including 55 gal. drum					289			289	320
2000	Worker protection, whole body, foot, head cover & gloves, plastic					7.95			7.95	8.75
2500	Respirator, single use					26			26	28.50
2550	Cartridge for respirator					4.40			4.40	4.84
2570	Glove bag, 7 mil, 50" x 64"					7.60			7.60	8.35
2580	10 mil, 44" x 60"					8.90			8.90	9.75
2590	6 mil, 44" x 60"					5.45			5.45	6
3000	HEPA vacuum for work area, minimum					330			330	360
3050	Maximum					725			725	795
6000	Disposable polyethylene bags, 6 mil, 3 C.F.					.86			.86	.95
6300	Disposable fiber drums, 3 C.F.					18.70			18.70	20.50
6400	Pressure sensitive caution labels, 3" x 5"					2.42			2.42	2.66
6450	11" x 17"					7.60			7.60	8.35
6500	Negative air machine, 1800 CFM					860			860	950

02 82 13.42 Preparation of Asbestos Containment Area

		Crew	Daily Output	Labor-Hours	Unit	Material	2018 Bare Costs Labor	Equipment	Total	Total Incl O&P
0010	**PREPARATION OF ASBESTOS CONTAINMENT AREA**									
0100	Pre-cleaning, HEPA vacuum and wet wipe, flat surfaces	A-9	12000	.005	S.F.	.01	.30		.31	.48
0200	Protect carpeted area, 2 layers 6 mil poly on 3/4" plywood	"	1000	.064		2.02	3.61		5.63	7.80
0300	Separation barrier, 2" x 4" @ 16", 1/2" plywood ea. side, 8' high	2 Carp	400	.040		2.81	2.03		4.84	6.20
0310	12' high		320	.050		2.75	2.54		5.29	6.90
0320	16' high		200	.080		2.73	4.06		6.79	9.20
0400	Personnel decontam. chamber, 2" x 4" @ 16", 3/4" ply ea. side		280	.057		3.36	2.90		6.26	8.10
0450	Waste decontam. chamber, 2" x 4" studs @ 16", 3/4" ply ea. side		360	.044		3.36	2.25		5.61	7.15
0500	Cover surfaces with polyethylene sheeting									
0501	Including glue and tape									
0550	Floors, each layer, 6 mil	A-9	8000	.008	S.F.	.04	.45		.49	.74
0551	4 mil		9000	.007		.03	.40		.43	.65
0560	Walls, each layer, 6 mil		6000	.011		.04	.60		.64	.97
0561	4 mil		7000	.009		.03	.52		.55	.83
0570	For heights above 14', add						20%			
0575	For heights above 20', add						30%			
0580	For fire retardant poly, add					100%				
0590	For large open areas, deduct					10%	20%			
0600	Seal floor penetrations with foam firestop to 36 sq. in.	2 Carp	200	.080	Ea.	11.35	4.06		15.41	18.70
0610	36 sq. in. to 72 sq. in.		125	.128		22.50	6.50		29	35
0615	72 sq. in. to 144 sq. in.		80	.200		45.50	10.15		55.65	65.50
0620	Wall penetrations, to 36 sq. in.		180	.089		11.35	4.51		15.86	19.35
0630	36 sq. in. to 72 sq. in.		100	.160		22.50	8.10		30.60	37.50
0640	72 sq. in. to 144 sq. in.		60	.267		45.50	13.50		59	70.50
0800	Caulk seams with latex	1 Carp	230	.035	L.F.	.17	1.76		1.93	2.88
0900	Set up neg. air machine, 1-2k CFM/25 M.C.F. volume	1 Asbe	4.30	1.860	Ea.		105		105	163
0950	Set up and remove portable shower unit	2 Asbe	4	4	"		225		225	350

For customer support on your Building Construction Costs with RSMeans data, call 800.448.8182.

41

02 82 13.43 Bulk Asbestos Removal

		Crew	Daily Output	Labor-Hours	Unit	Material	2018 Bare Costs Labor	Equipment	Total	Total Incl O&P
0010	**BULK ASBESTOS REMOVAL**									
0020	Includes disposable tools and 2 suits and 1 respirator filter/day/worker									
0100	Beams, W 10 x 19	A-9	235	.272	L.F.	.69	15.35		16.04	25
0110	W 12 x 22		210	.305		.77	17.20		17.97	27.50
0120	W 14 x 26		180	.356		.90	20		20.90	32
0130	W 16 x 31		160	.400		1.02	22.50		23.52	36
0140	W 18 x 40		140	.457		1.16	26		27.16	41.50
0150	W 24 x 55		110	.582		1.48	33		34.48	52.50
0160	W 30 x 108		85	.753		1.91	42.50		44.41	68
0170	W 36 x 150		72	.889		2.26	50		52.26	80.50
0200	Boiler insulation		480	.133	S.F.	.41	7.50		7.91	12.10
0210	With metal lath, add				%				50%	50%
0300	Boiler breeching or flue insulation	A-9	520	.123	S.F.	.31	6.95		7.26	11.10
0310	For active boiler, add				%				100%	100%
0400	Duct or AHU insulation	A-10B	440	.073	S.F.	.18	4.10		4.28	6.55
0500	Duct vibration isolation joints, up to 24 sq. in. duct	A-9	56	1.143	Ea.	2.90	64.50		67.40	103
0520	25 sq. in. to 48 sq. in. duct		48	1.333		3.39	75		78.39	121
0530	49 sq. in. to 76 sq. in. duct		40	1.600		4.06	90		94.06	144
0600	Pipe insulation, air cell type, up to 4" diameter pipe		900	.071	L.F.	.18	4.01		4.19	6.40
0610	4" to 8" diameter pipe		800	.080		.20	4.51		4.71	7.20
0620	10" to 12" diameter pipe		700	.091		.23	5.15		5.38	8.25
0630	14" to 16" diameter pipe		550	.116		.30	6.55		6.85	10.55
0650	Over 16" diameter pipe		650	.098	S.F.	.25	5.55		5.80	8.90
0700	With glove bag up to 3" diameter pipe		200	.320	L.F.	9.50	18.05		27.55	38.50
1000	Pipe fitting insulation up to 4" diameter pipe		320	.200	Ea.	.51	11.25		11.76	18.05
1100	6" to 8" diameter pipe		304	.211		.53	11.85		12.38	19
1110	10" to 12" diameter pipe		192	.333		.85	18.80		19.65	30
1120	14" to 16" diameter pipe		128	.500		1.27	28		29.27	45.50
1130	Over 16" diameter pipe		176	.364	S.F.	.92	20.50		21.42	33
1200	With glove bag, up to 8" diameter pipe		75	.853	L.F.	6.45	48		54.45	81.50
2000	Scrape foam fireproofing from flat surface		2400	.027	S.F.	.07	1.50		1.57	2.40
2100	Irregular surfaces		1200	.053		.14	3.01		3.15	4.82
3000	Remove cementitious material from flat surface		1800	.036		.09	2		2.09	3.21
3100	Irregular surface		1000	.064		.12	3.61		3.73	5.75
4000	Scrape acoustical coating/fireproofing, from ceiling		3200	.020		.05	1.13		1.18	1.81
5000	Remove VAT and mastic from floor by hand		2400	.027		.07	1.50		1.57	2.40
5100	By machine	A-11	4800	.013		.03	.75	.01	.79	1.22
5150	For 2 layers, add				%				50%	50%
6000	Remove contaminated soil from crawl space by hand	A-9	400	.160	C.F.	.41	9		9.41	14.45
6100	With large production vacuum loader	A-12	700	.091	"	.23	5.15	1.03	6.41	9.40
7000	Radiator backing, not including radiator removal	A-9	1200	.053	S.F.	.14	3.01		3.15	4.82
8000	Cement-asbestos transite board and cement wall board	2 Asbe	1000	.016		.15	.90		1.05	1.57
8100	Transite shingle siding	A-10B	750	.043		.21	2.41		2.62	3.97
8200	Shingle roofing	"	2000	.016		.08	.90		.98	1.49
8250	Built-up, no gravel, non-friable	B-2	1400	.029		.08	1.15		1.23	1.84
8260	Bituminous flashing	1 Rofc	300	.027		.08	1.17		1.25	2.06
8300	Asbestos millboard, flat board and VAT contaminated plywood	2 Asbe	1000	.016		.08	.90		.98	1.49
9000	For type B (supplied air) respirator equipment, add				%				10%	10%

02 82 13.44 Demolition In Asbestos Contaminated Area

		Crew	Daily Output	Labor-Hours	Unit	Material	2018 Bare Costs Labor	Equipment	Total	Total Incl O&P
0010	**DEMOLITION IN ASBESTOS CONTAMINATED AREA**									
0200	Ceiling, including suspension system, plaster and lath	A-9	2100	.030	S.F.	.08	1.72		1.80	2.76
0210	Finished plaster, leaving wire lath		585	.109		.28	6.15		6.43	9.85

For customer support on your Building Construction Costs with RSMeans data, call 800.448.8182.

02 82 Asbestos Remediation

02 82 13 – Asbestos Abatement

02 82 13.44 Demolition In Asbestos Contaminated Area

		Crew	Daily Output	Labor-Hours	Unit	Material	2018 Bare Costs Labor	Equipment	Total	Total Incl O&P
0220	Suspended acoustical tile	A-9	3500	.018	S.F.	.05	1.03		1.08	1.65
0230	Concealed tile grid system		3000	.021		.05	1.20		1.25	1.93
0240	Metal pan grid system		1500	.043		.11	2.40		2.51	3.85
0250	Gypsum board		2500	.026		.07	1.44		1.51	2.31
0260	Lighting fixtures up to 2' x 4'		72	.889	Ea.	2.26	50		52.26	80.50
0400	Partitions, non load bearing									
0410	Plaster, lath, and studs	A-9	690	.093	S.F.	.88	5.25		6.13	9.05
0450	Gypsum board and studs	"	1390	.046	"	.12	2.59		2.71	4.16
9000	For type B (supplied air) respirator equipment, add				%				10%	10%

02 82 13.45 OSHA Testing

		Crew	Daily Output	Labor-Hours	Unit	Material	2018 Bare Costs Labor	Equipment	Total	Total Incl O&P
0010	**OSHA TESTING**									
0100	Certified technician, minimum				Day				200	220
0110	Maximum								300	330
0121	Industrial hygienist, minimum	1 Asbe	1.75	4.571			257		257	400
0130	Maximum								400	440
0200	Asbestos sampling and PCM analysis, NIOSH 7400, minimum	1 Asbe	8	1	Ea.	16.05	56.50		72.55	105
0210	Maximum		4	2		27.50	113		140.50	206
1000	Cleaned area samples		8	1		202	56.50		258.50	310
1100	PCM air sample analysis, NIOSH 7400, minimum		8	1		15.10	56.50		71.60	104
1110	Maximum		4	2		2.22	113		115.22	177
1200	TEM air sample analysis, NIOSH 7402, minimum								80	106
1210	Maximum								360	450

02 82 13.46 Decontamination of Asbestos Containment Area

		Crew	Daily Output	Labor-Hours	Unit	Material	2018 Bare Costs Labor	Equipment	Total	Total Incl O&P
0010	**DECONTAMINATION OF ASBESTOS CONTAINMENT AREA**									
0100	Spray exposed substrate with surfactant (bridging)									
0200	Flat surfaces	A-9	6000	.011	S.F.	.40	.60		1	1.37
0250	Irregular surfaces		4000	.016	"	.40	.90		1.30	1.84
0300	Pipes, beams, and columns		2000	.032	L.F.	.60	1.80		2.40	3.46
1000	Spray encapsulate polyethylene sheeting		8000	.008	S.F.	.40	.45		.85	1.14
1100	Roll down polyethylene sheeting		8000	.008	"		.45		.45	.70
1500	Bag polyethylene sheeting		400	.160	Ea.	1	9		10	15.10
2000	Fine clean exposed substrate, with nylon brush		2400	.027	S.F.		1.50		1.50	2.33
2500	Wet wipe substrate		4800	.013			.75		.75	1.17
2600	Vacuum surfaces, fine brush		6400	.010			.56		.56	.87
3000	Structural demolition									
3100	Wood stud walls	A-9	2800	.023	S.F.		1.29		1.29	2
3500	Window manifolds, not incl. window replacement		4200	.015			.86		.86	1.33
3600	Plywood carpet protection		2000	.032			1.80		1.80	2.80
4000	Remove custom decontamination facility	A-10A	8	3	Ea.	15.15	169		184.15	280
4100	Remove portable decontamination facility	3 Asbe	12	2	"	14.15	113		127.15	191
5000	HEPA vacuum, shampoo carpeting	A-9	4800	.013	S.F.	.10	.75		.85	1.28
9000	Final cleaning of protected surfaces	A-10A	8000	.003	"		.17		.17	.26

02 82 13.47 Asbestos Waste Pkg., Handling, and Disp.

		Crew	Daily Output	Labor-Hours	Unit	Material	2018 Bare Costs Labor	Equipment	Total	Total Incl O&P
0010	**ASBESTOS WASTE PACKAGING, HANDLING, AND DISPOSAL**									
0100	Collect and bag bulk material, 3 C.F. bags, by hand	A-9	400	.160	Ea.	.86	9		9.86	14.95
0200	Large production vacuum loader	A-12	880	.073		1	4.10	.82	5.92	8.35
1000	Double bag and decontaminate	A-9	960	.067		.86	3.76		4.62	6.80
2000	Containerize bagged material in drums, per 3 C.F. drum	"	800	.080		18.70	4.51		23.21	27.50
3000	Cart bags 50' to dumpster	2 Asbe	400	.040			2.25		2.25	3.50
5000	Disposal charges, not including haul, minimum				C.Y.				61	67
5020	Maximum				"				355	395
9000	For type B (supplied air) respirator equipment, add				%				10%	10%

For customer support on your Building Construction Costs with RSMeans data, call 800.448.8182.

43

02 82 Asbestos Remediation

02 82 13 – Asbestos Abatement

02 82 13.48 Asbestos Encapsulation With Sealants

	02 82 13.48 Asbestos Encapsulation With Sealants	Crew	Daily Output	Labor-Hours	Unit	Material	2018 Bare Costs Labor	Equipment	Total	Total Incl O&P
0010	**ASBESTOS ENCAPSULATION WITH SEALANTS**									
0100	Ceilings and walls, minimum	A-9	21000	.003	S.F.	.40	.17		.57	.71
0110	Maximum		10600	.006		.51	.34		.85	1.09
0200	Columns and beams, minimum		13300	.005		.40	.27		.67	.86
0210	Maximum		5325	.012		.61	.68		1.29	1.72
0300	Pipes to 12" diameter including minor repairs, minimum		800	.080	L.F.	.51	4.51		5.02	7.55
0310	Maximum		400	.160	"	1.21	9		10.21	15.35

02 83 Lead Remediation

02 83 19 – Lead-Based Paint Remediation

02 83 19.21 Lead Paint Remediation Plans and Methods

		Crew	Daily Output	Labor-Hours	Unit	Material	2018 Bare Costs Labor	Equipment	Total	Total Incl O&P
0010	**LEAD PAINT REMEDIATION PLANS AND METHODS**									
0100	Building Survey-Commercial Building				Ea.				2,050	2,250
0200	Lead Abatement Remediation Plan								1,225	1,350
0300	Lead Paint Testing, AAS Analysis								51	56
0400	Lead Paint Testing, X-Ray Fluorescence								51	56

02 83 19.23 Encapsulation of Lead-Based Paint

		Crew	Daily Output	Labor-Hours	Unit	Material	2018 Bare Costs Labor	Equipment	Total	Total Incl O&P
0010	**ENCAPSULATION OF LEAD-BASED PAINT**									
0020	Interior, brushwork, trim, under 6"	1 Pord	240	.033	L.F.	2.32	1.42		3.74	4.68
0030	6"-12" wide		180	.044		3.03	1.89		4.92	6.20
0040	Balustrades		300	.027		2.02	1.13		3.15	3.93
0050	Pipe to 4" diameter		500	.016		1.21	.68		1.89	2.35
0060	To 8" diameter		375	.021		1.52	.91		2.43	3.04
0070	To 12" diameter		250	.032		2.22	1.36		3.58	4.49
0080	To 16" diameter		170	.047		3.33	2		5.33	6.65
0090	Cabinets, ornate design		200	.040	S.F.	3.03	1.70		4.73	5.90
0100	Simple design		250	.032	"	2.32	1.36		3.68	4.60
0110	Doors, 3' x 7', both sides, incl. frame & trim									
0120	Flush	1 Pord	6	1.333	Ea.	28.50	56.50		85	117
0130	French, 10-15 lite		3	2.667		6.05	113		119.05	178
0140	Panel		4	2		35.50	85		120.50	167
0150	Louvered		2.75	2.909		32.50	124		156.50	222
0160	Windows, per interior side, per 15 S.F.									
0170	1-6 lite	1 Pord	14	.571	Ea.	20	24.50		44.50	58.50
0180	7-10 lite		7.50	1.067		22	45.50		67.50	93
0190	12 lite		5.75	1.391		30.50	59		89.50	123
0200	Radiators		8	1		70.50	42.50		113	142
0210	Grilles, vents		275	.029	S.F.	2.02	1.24		3.26	4.08
0220	Walls, roller, drywall or plaster		1000	.008		.61	.34		.95	1.18
0230	With spunbonded reinforcing fabric		720	.011		.71	.47		1.18	1.49
0240	Wood		800	.010		.71	.43		1.14	1.42
0250	Ceilings, roller, drywall or plaster		900	.009		.71	.38		1.09	1.35
0260	Wood		700	.011		.81	.49		1.30	1.62
0270	Exterior, brushwork, gutters and downspouts		300	.027	L.F.	1.92	1.13		3.05	3.82
0280	Columns		400	.020	S.F.	1.42	.85		2.27	2.84
0290	Spray, siding		600	.013	"	1.01	.57		1.58	1.96
0300	Miscellaneous									
0310	Electrical conduit, brushwork, to 2" diameter	1 Pord	500	.016	L.F.	1.21	.68		1.89	2.35
0320	Brick, block or concrete, spray		500	.016	S.F.	1.21	.68		1.89	2.35
0330	Steel, flat surfaces and tanks to 12"		500	.016		1.21	.68		1.89	2.35

For customer support on your Building Construction Costs with RSMeans data, call 800.448.8182.

02 83 Lead Remediation

02 83 19 – Lead-Based Paint Remediation

02 83 19.23 Encapsulation of Lead-Based Paint	Crew	Daily Output	Labor-Hours	Unit	Material	2018 Bare Costs Labor	Equipment	Total	Total Incl O&P
0340	Beams, brushwork	1 Pord	400	.020	S.F.	1.41	.85		2.26	2.83
0350	Trusses	↓	400	.020	↓	1.41	.85		2.26	2.83

02 83 19.26 Removal of Lead-Based Paint

		Crew	Daily Output	Labor-Hours	Unit	Material	Labor	Equipment	Total	Total Incl O&P
0010	**REMOVAL OF LEAD-BASED PAINT** R028319-60									
0011	By chemicals, per application									
0050	Baseboard, to 6" wide	1 Pord	64	.125	L.F.	.80	5.30		6.10	8.90
0070	To 12" wide		32	.250	"	1.47	10.65		12.12	17.60
0200	Balustrades, one side		28	.286	S.F.	1.50	12.15		13.65	19.95
1400	Cabinets, simple design		32	.250		1.37	10.65		12.02	17.50
1420	Ornate design		25	.320		1.60	13.60		15.20	22.50
1600	Cornice, simple design		60	.133		1.60	5.65		7.25	10.30
1620	Ornate design		20	.400		5.55	17		22.55	31.50
2800	Doors, one side, flush		84	.095		1.73	4.05		5.78	8
2820	Two panel		80	.100		1.37	4.26		5.63	7.90
2840	Four panel		45	.178	↓	1.43	7.55		8.98	12.95
2880	For trim, one side, add		64	.125	L.F.	.77	5.30		6.07	8.85
3000	Fence, picket, one side		30	.267	S.F.	1.37	11.35		12.72	18.60
3200	Grilles, one side, simple design		30	.267		1.37	11.35		12.72	18.60
3220	Ornate design		25	.320	↓	1.47	13.60		15.07	22
3240	Handrails		90	.089	L.F.	1.37	3.78		5.15	7.20
4400	Pipes, to 4" diameter		90	.089		1.73	3.78		5.51	7.60
4420	To 8" diameter		50	.160		3.47	6.80		10.27	14.05
4440	To 12" diameter		36	.222		5.20	9.45		14.65	19.95
4460	To 16" diameter		20	.400	↓	6.95	17		23.95	33
4500	For hangers, add		40	.200	Ea.	2.53	8.50		11.03	15.60
4800	Siding		90	.089	S.F.	1.32	3.78		5.10	7.15
5000	Trusses, open		55	.145	SF Face	2.02	6.20		8.22	11.50
6200	Windows, one side only, double-hung, 1/1 light, 24" x 48" high		4	2	Ea.	24	85		109	155
6220	30" x 60" high		3	2.667		32.50	113		145.50	207
6240	36" x 72" high		2.50	3.200		38.50	136		174.50	247
6280	40" x 80" high		2	4		47.50	170		217.50	310
6400	Colonial window, 6/6 light, 24" x 48" high		2	4		47.50	170		217.50	310
6420	30" x 60" high		1.50	5.333		62.50	227		289.50	410
6440	36" x 72" high		1	8		94	340		434	615
6480	40" x 80" high		1	8		94	340		434	615
6600	8/8 light, 24" x 48" high		2	4		47.50	170		217.50	310
6620	40" x 80" high		1	8		94	340		434	615
6800	12/12 light, 24" x 48" high		1	8		94	340		434	615
6820	40" x 80" high	↓	.75	10.667	↓	126	455		581	825
6840	Window frame & trim items, included in pricing above									
7000	Hand scraping and HEPA vacuum, less than 4 S.F.	1 Pord	8	1	Ea.	.86	42.50		43.36	65
8000	Collect and bag bulk material, 3 C.F. bags, by hand	"	30	.267	"	.86	11.35		12.21	18.05

For customer support on your Building Construction Costs with RSMeans data, call 800.448.8182.

45

02 87 Biohazard Remediation

02 87 13 – Mold Remediation

02 87 13.16 Mold Remediation Preparation and Containment

		Crew	Daily Output	Labor-Hours	Unit	Material	2018 Bare Costs Labor	2018 Bare Costs Equipment	Total	Total Incl O&P
0010	**MOLD REMEDIATION PREPARATION AND CONTAINMENT**									
0015	Mold remediation plans and methods									
0020	Initial inspection, areas to 2500 S.F.				Total				268	295
0030	Areas to 5000 S.F.								440	485
0032	Areas to 10000 S.F.								440	485
0040	Testing, air sample each								126	139
0050	Swab sample								115	127
0060	Tape sample								125	140
0070	Post remediation air test								126	139
0080	Mold abatement plan, area to 2500 S.F.								1,225	1,325
0090	Areas to 5000 S.F.								1,625	1,775
0095	Areas to 10000 S.F.								2,550	2,800
0100	Packup & removal of contents, average 3 bedroom home, excl storage								8,150	8,975
0110	Average 5 bedroom home, excl storage				↓				15,300	16,800
0600	For demolition in mold contaminated areas, see Section 02 85 33.50									
0610	For personal protection equipment, see Section 02 82 13.41									
6010	Preparation of mold containment area									
6100	Pre-cleaning, HEPA vacuum and wet wipe, flat surfaces	A-9	12000	.005	S.F.	.01	.30		.31	.48
6300	Separation barrier, 2" x 4" @ 16", 1/2" plywood ea. side, 8' high	2 Carp	400	.040		3.43	2.03		5.46	6.85
6310	12' high		320	.050		3.43	2.54		5.97	7.65
6320	16' high		200	.080		2.32	4.06		6.38	8.75
6400	Personnel decontam. chamber, 2" x 4" @ 16", 3/4" ply ea. side		280	.057		4.44	2.90		7.34	9.30
6450	Waste decontam. chamber, 2" x 4" studs @ 16", 3/4" ply each side	↓	360	.044	↓	4.44	2.25		6.69	8.30
6500	Cover surfaces with polyethylene sheeting									
6501	Including glue and tape									
6550	Floors, each layer, 6 mil	A-9	8000	.008	S.F.	.04	.45		.49	.74
6551	4 mil		9000	.007		.03	.40		.43	.65
6560	Walls, each layer, 6 mil		6000	.011		.04	.60		.64	.97
6561	4 mil	↓	7000	.009	↓	.03	.52		.55	.83
6570	For heights above 14', add						20%			
6575	For heights above 20', add						30%			
6580	For fire retardant poly, add					100%				
6590	For large open areas, deduct					10%	20%			
6600	Seal floor penetrations with foam firestop to 36 sq. in.	2 Carp	200	.080	Ea.	11.35	4.06		15.41	18.70
6610	36 sq. in. to 72 sq. in.		125	.128		22.50	6.50		29	35
6615	72 sq. in. to 144 sq. in.		80	.200		45.50	10.15		55.65	65.50
6620	Wall penetrations, to 36 sq. in.		180	.089		11.35	4.51		15.86	19.35
6630	36 sq. in. to 72 sq. in.		100	.160		22.50	8.10		30.60	37.50
6640	72 sq. in. to 144 sq. in.	↓	60	.267	↓	45.50	13.50		59	70.50
6800	Caulk seams with latex caulk	1 Carp	230	.035	L.F.	.17	1.76		1.93	2.88
6900	Set up neg. air machine, 1-2k CFM/25 M.C.F. volume	1 Asbe	4.30	1.860	Ea.		105		105	163

02 87 13.33 Removal and Disposal of Materials With Mold

		Crew	Daily Output	Labor-Hours	Unit	Material	2018 Bare Costs Labor	2018 Bare Costs Equipment	Total	Total Incl O&P
0010	**REMOVAL AND DISPOSAL OF MATERIALS WITH MOLD**									
0015	Demolition in mold contaminated area									
0200	Ceiling, including suspension system, plaster and lath	A-9	2100	.030	S.F.	.08	1.72		1.80	2.76
0210	Finished plaster, leaving wire lath		585	.109		.28	6.15		6.43	9.85
0220	Suspended acoustical tile		3500	.018		.05	1.03		1.08	1.65
0230	Concealed tile grid system		3000	.021		.05	1.20		1.25	1.93
0240	Metal pan grid system		1500	.043		.11	2.40		2.51	3.85
0250	Gypsum board		2500	.026		.07	1.44		1.51	2.31
0255	Plywood		2500	.026	↓	.07	1.44		1.51	2.31
0260	Lighting fixtures up to 2' x 4'	↓	72	.889	Ea.	2.26	50		52.26	80.50

For customer support on your Building Construction Costs with RSMeans data, call 800.448.8182.

02 87 Biohazard Remediation

02 87 13 – Mold Remediation

02 87 13.33 Removal and Disposal of Materials With Mold

	02 87 13.33 Removal and Disposal of Materials With Mold	Crew	Daily Output	Labor-Hours	Unit	Material	2018 Bare Costs Labor	Equipment	Total	Total Incl O&P
0400	Partitions, non load bearing									
0410	Plaster, lath, and studs	A-9	690	.093	S.F.	.88	5.25		6.13	9.05
0450	Gypsum board and studs		1390	.046		.12	2.59		2.71	4.16
0465	Carpet & pad		1390	.046	↓	.12	2.59		2.71	4.16
0600	Pipe insulation, air cell type, up to 4" diameter pipe		900	.071	L.F.	.18	4.01		4.19	6.40
0610	4" to 8" diameter pipe		800	.080		.20	4.51		4.71	7.20
0620	10" to 12" diameter pipe		700	.091		.23	5.15		5.38	8.25
0630	14" to 16" diameter pipe		550	.116	↓	.30	6.55		6.85	10.55
0650	Over 16" diameter pipe	↓	650	.098	S.F.	.25	5.55		5.80	8.90
9000	For type B (supplied air) respirator equipment, add				%				10%	10%

02 91 Chemical Sampling, Testing and Analysis

02 91 10 – Monitoring, Sampling, Testing and Analysis

02 91 10.10 Monitoring, Chem. Sampling, Testing & Analysis

	02 91 10.10 Monitoring, Chem. Sampling, Testing & Analysis	Crew	Daily Output	Labor-Hours	Unit	Material	2018 Bare Costs Labor	Equipment	Total	Total Incl O&P
0010	**MONITORING, CHEMICAL SAMPLING, TESTING AND ANALYSIS**									
0015	Field sampling of waste									
0100	Field samples, sample collection, sludge	1 Skwk	32	.250	Ea.		13.10		13.10	20
0110	Contaminated soils	"	32	.250	"		13.10		13.10	20
0200	Vials and bottles									
0210	32 oz. clear wide mouth jar (case of 12)				Ea.	36			36	40
0220	32 oz. Boston round bottle (case of 12)					37.50			37.50	41
0230	32 oz. HDPE bottle (case of 12)				↓	34			34	37.50
0300	Laboratory analytical services									
0310	Laboratory testing 13 metals				Ea.	214			214	236
0312	13 metals + mercury					214			214	236
0314	8 metals					163			163	179
0316	Mercury only					56			56	61.50
0318	Single metal (only Cs, Li, Sr, Ta)					46			46	50.50
0320	Single metal (excludes Hg, Cs, Li, Sr, Ta)					46			46	50.50
0400	Hydrocarbons standard					230			230	252
0410	Hydrocarbons fingerprint					255			255	281
0500	Radioactivity gross alpha					157			157	173
0510	Gross alpha & beta					157			157	173
0520	Radium 226					102			102	112
0530	Radium 228					133			133	146
0540	Radon					151			151	166
0550	Uranium					133			133	146
0600	Volatile organics without GC/MS					159			159	175
0610	Volatile organics including GC/MS					305			305	335
0630	Synthetic organic compounds					1,050			1,050	1,150
0640	Herbicides					220			220	242
0650	Pesticides					154			154	170
0660	PCB's				↓	143			143	157

Division Notes

	CREW	DAILY OUTPUT	LABOR-HOURS	UNIT	BARE COSTS				TOTAL INCL O&P
					MAT.	LABOR	EQUIP.	TOTAL	

Estimating Tips
General

- Carefully check all the plans and specifications. Concrete often appears on drawings other than structural drawings, including mechanical and electrical drawings for equipment pads. The cost of cutting and patching is often difficult to estimate. See Subdivision 03 81 for Concrete Cutting, Subdivision 02 41 19.16 for Cutout Demolition, Subdivision 03 05 05.10 for Concrete Demolition, and Subdivision 02 41 19.19 for Rubbish Handling (handling, loading, and hauling of debris).

- Always obtain concrete prices from suppliers near the job site. A volume discount can often be negotiated, depending upon competition in the area. Remember to add for waste, particularly for slabs and footings on grade.

03 10 00 Concrete Forming and Accessories

- A primary cost for concrete construction is forming. Most jobs today are constructed with prefabricated forms. The selection of the forms best suited for the job and the total square feet of forms required for efficient concrete forming and placing are key elements in estimating concrete construction. Enough forms must be available for erection to make efficient use of the concrete placing equipment and crew.

- Concrete accessories for forming and placing depend upon the systems used. Study the plans and specifications to ensure that all special accessory requirements have been included in the cost estimate, such as anchor bolts, inserts, and hangers.

- Included within costs for forms-in-place are all necessary bracing and shoring.

03 20 00 Concrete Reinforcing

- Ascertain that the reinforcing steel supplier has included all accessories, cutting, bending, and an allowance for lapping, splicing, and waste. A good rule of thumb is 10% for lapping, splicing, and waste. Also, 10% waste should be allowed for welded wire fabric.

- The unit price items in the subdivisions for Reinforcing In Place, Glass Fiber Reinforcing, and Welded Wire Fabric include the labor to install accessories such as beam and slab bolsters, high chairs, and bar ties and tie wire. The material cost for these accessories is not included; they may be obtained from the Accessories Subdivisions.

03 30 00 Cast-In-Place Concrete

- When estimating structural concrete, pay particular attention to requirements for concrete additives, curing methods, and surface treatments. Special consideration for climate, hot or cold, must be included in your estimate. Be sure to include requirements for concrete placing equipment and concrete finishing.

- For accurate concrete estimating, the estimator must consider each of the following major components individually: forms, reinforcing steel, ready-mix concrete, placement of the concrete, and finishing of the top surface. For faster estimating, Subdivision 03 30 53.40 for Concrete-In-Place can be used; here, various items of concrete work are presented that include the costs of all five major components (unless specifically stated otherwise).

03 40 00 Precast Concrete
03 50 00 Cast Decks and Underlayment

- The cost of hauling precast concrete structural members is often an important factor. For this reason, it is important to get a quote from the nearest supplier. It may become economically feasible to set up precasting beds on the site if the hauling costs are prohibitive.

Reference Numbers

Reference numbers are shown at the beginning of some major classifications. These numbers refer to related items in the Reference Section. The reference information may be an estimating procedure, an alternate pricing method, or technical information.

Note: Not all subdivisions listed here necessarily appear. ■

Did you know?

RSMeans data is available through our online application with 24/7 access:

- Search for unit prices by keyword
- Leverage the most up-to-date data
- Build and export estimates

Try it free for 30 days!
www.rsmeans.com/2018freetrial

No part of this cost data may be reproduced, stored in a retrieval system, or transmitted in any form or by any means without prior written permission of Gordian.

03 01 Maintenance of Concrete

03 01 30 – Maintenance of Cast-In-Place Concrete

03 01 30.62 Concrete Patching

03 01 30.62 Concrete Patching	Crew	Daily Output	Labor-Hours	Unit	Material	2018 Bare Costs Labor	Equipment	Total	Total Incl O&P
0010 **CONCRETE PATCHING**									
0100 Floors, 1/4" thick, small areas, regular grout	1 Cefi	170	.047	S.F.	1.54	2.24		3.78	5
0150 Epoxy grout	"	100	.080	"	8.65	3.80		12.45	15.20
2000 Walls, including chipping, cleaning and epoxy grout									
2100 1/4" deep	1 Cefi	65	.123	S.F.	7.45	5.85		13.30	16.80
2150 1/2" deep		50	.160		14.85	7.60		22.45	27.50
2200 3/4" deep	↓	40	.200	↓	22.50	9.50		32	38.50

03 05 Common Work Results for Concrete

03 05 05 – Selective Demolition for Concrete

03 05 05.10 Selective Demolition, Concrete

03 05 05.10 Selective Demolition, Concrete	Crew	Daily Output	Labor-Hours	Unit	Material	2018 Bare Costs Labor	Equipment	Total	Total Incl O&P
0010 **SELECTIVE DEMOLITION, CONCRETE** R024119-10									
0012 Excludes saw cutting, torch cutting, loading or hauling									
0050 Break into small pieces, reinf. less than 1% of cross-sectional area	B-9	24	1.667	C.Y.		67	9.75	76.75	113
0060 Reinforcing 1% to 2% of cross-sectional area		16	2.500			101	14.65	115.65	169
0070 Reinforcing more than 2% of cross-sectional area	↓	8	5	↓		201	29.50	230.50	335
0150 Remove whole pieces, up to 2 tons per piece	E-18	36	1.111	Ea.		61	30	91	131
0160 2-5 tons per piece		30	1.333			73	36	109	158
0170 5-10 tons per piece		24	1.667			91.50	45	136.50	197
0180 10-15 tons per piece	↓	18	2.222			122	60	182	262
0250 Precast unit embedded in masonry, up to 1 C.F.	D-1	16	1			45		45	68.50
0260 1-2 C.F.		12	1.333			59.50		59.50	91.50
0270 2-5 C.F.		10	1.600			71.50		71.50	110
0280 5-10 C.F.	↓	8	2	↓		89.50		89.50	137
0990 For hydrodemolition see Section 02 41 13.15									

03 05 13 – Basic Concrete Materials

03 05 13.20 Concrete Admixtures and Surface Treatments

03 05 13.20 Concrete Admixtures and Surface Treatments	Crew	Daily Output	Labor-Hours	Unit	Material	2018 Bare Costs Labor	Equipment	Total	Total Incl O&P
0010 **CONCRETE ADMIXTURES AND SURFACE TREATMENTS**									
0040 Abrasives, aluminum oxide, over 20 tons				Lb.	2.53			2.53	2.78
0050 1 to 20 tons					2.63			2.63	2.89
0070 Under 1 ton					2.73			2.73	3.01
0100 Silicon carbide, black, over 20 tons					4.32			4.32	4.75
0110 1 to 20 tons					4.49			4.49	4.94
0120 Under 1 ton				↓	4.67			4.67	5.15
0200 Air entraining agent, .7 to 1.5 oz. per bag, 55 gallon drum				Gal.	21			21	23
0220 5 gallon pail					29			29	32
0300 Bonding agent, acrylic latex, 250 S.F./gallon, 5 gallon pail					24.50			24.50	27
0320 Epoxy resin, 80 S.F./gallon, 4 gallon case				↓	65.50			65.50	72
0400 Calcium chloride, 50 lb. bags, T.L. lots				Ton	1,225			1,225	1,325
0420 Less than truckload lots				Bag	22.50			22.50	25
0500 Carbon black, liquid, 2 to 8 lb. per bag of cement				Lb.	13.70			13.70	15.05
0600 Concrete admixture, integral colors, dry pigment, 5 lb. bag				Ea.	28			28	31
0610 10 lb. bag					36.50			36.50	40
0620 25 lb. bag				↓	74			74	81.50
0920 Dustproofing compound, 250 S.F./gal., 5 gallon pail				Gal.	7.10			7.10	7.80
1010 Epoxy based, 125 S.F./gal., 5 gallon pail				"	59			59	65
1100 Hardeners, metallic, 55 lb. bags, natural (grey)				Lb.	.86			.86	.95
1200 Colors					1.48			1.48	1.63
1300 Non-metallic, 55 lb. bags, natural grey					.63			.63	.70
1320 Colors				↓	.80			.80	.88

For customer support on your Building Construction Costs with RSMeans data, call 800.448.8182.

03 05 13 – Basic Concrete Materials

03 05 13.20 Concrete Admixtures and Surface Treatments		Crew	Daily Output	Labor-Hours	Unit	Material	2018 Bare Costs Labor	Equipment	Total	Total Incl O&P
1550	Release agent, for tilt slabs, 5 gallon pail				Gal.	18.25			18.25	20
1570	For forms, 5 gallon pail					11.85			11.85	13
1590	Concrete release agent for forms, 100% biodegradable, zero VOC, 5 gal. pail	G				21.50			21.50	23.50
1595	55 gallon drum	G				18.15			18.15	19.95
1600	Sealer, hardener and dustproofer, epoxy-based, 125 S.F./gal., 5 gallon unit					59			59	65
1620	3 gallon unit					65			65	71.50
1630	Sealer, solvent-based, 250 S.F./gal., 55 gallon drum					21.50			21.50	23.50
1640	5 gallon pail					32			32	35
1650	Sealer, water based, 350 S.F., 55 gallon drum					21.50			21.50	23.50
1660	5 gallon pail					24			24	26.50
1900	Set retarder, 100 S.F./gal., 1 gallon pail				▼	6.20			6.20	6.80
2000	Waterproofing, integral 1 lb. per bag of cement				Lb.	3.19			3.19	3.51
2100	Powdered metallic, 40 lbs. per 100 S.F., standard colors					3.79			3.79	4.17
2120	Premium colors				▼	5.30			5.30	5.85
3000	For colored ready-mix concrete, add to prices in section 03 31 13.35									
3100	Subtle shades, 5 lb. dry pigment per C.Y., add				C.Y.	28			28	31
3400	Medium shades, 10 lb. dry pigment per C.Y., add					36.50			36.50	40
3700	Deep shades, 25 lb. dry pigment per C.Y., add				▼	74			74	81.50
6000	Concrete ready mix additives, recycled coal fly ash, mixed at plant	G			Ton	63			63	69
6010	Recycled blast furnace slag, mixed at plant	G			"	90.50			90.50	99.50

03 05 13.25 Aggregate

			Crew	Daily Output	Labor-Hours	Unit	Material	Labor	Equipment	Total	Total Incl O&P
0010	**AGGREGATE**	R033105-20									
0100	Lightweight vermiculite or perlite, 4 C.F. bag, C.L. lots	G				Bag	22.50			22.50	24.50
0150	L.C.L. lots	G				"	25			25	27.50
0250	Sand & stone, loaded at pit, crushed bank gravel					Ton	19.25			19.25	21
0350	Sand, washed, for concrete	R033105-50					21			21	23
0400	For plaster or brick						21			21	23
0450	Stone, 3/4" to 1-1/2"						17.55			17.55	19.30
0470	Round, river stone						40			40	44
0500	3/8" roofing stone & 1/2" pea stone						30			30	33
0550	For trucking 10-mile round trip, add to the above		B-34B	117	.068			3.15	4.64	7.79	9.85
0600	For trucking 30-mile round trip, add to the above		"	72	.111	▼		5.10	7.55	12.65	16
0850	Sand & stone, loaded at pit, crushed bank gravel					C.Y.	27			27	29.50
0950	Sand, washed, for concrete						29.50			29.50	32.50
1000	For plaster or brick						29.50			29.50	32.50
1050	Stone, 3/4" to 1-1/2"						33.50			33.50	37
1055	Round, river stone						42			42	46.50
1100	3/8" roofing stone & 1/2" pea stone						30			30	33.50
1150	For trucking 10-mile round trip, add to the above		B-34B	78	.103			4.72	6.95	11.67	14.75
1200	For trucking 30-mile round trip, add to the above		"	48	.167	▼		7.65	11.30	18.95	24
1310	Onyx chips, 50 lb. bags					Cwt.	32			32	35
1330	Quartz chips, 50 lb. bags						32			32	35
1410	White marble, 3/8" to 1/2", 50 lb. bags					▼	9.60			9.60	10.55
1430	3/4", bulk					Ton	166			166	182

03 05 13.30 Cement

			Crew	Daily Output	Labor-Hours	Unit	Material	Labor	Equipment	Total	Total Incl O&P
0010	**CEMENT**	R033105-20									
0240	Portland, Type I/II, T.L. lots, 94 lb. bags					Bag	12.60			12.60	13.85
0250	L.T.L./L.C.L. lots	R033105-30				"	13.80			13.80	15.15
0300	Trucked in bulk, per cwt.					Cwt.	7.80			7.80	8.60
0400	Type III, high early strength, T.L. lots, 94 lb. bags	R033105-40				Bag	13.55			13.55	14.90
0420	L.T.L. or L.C.L. lots					↓	19.20			19.20	21
0500	White, type III, high early strength, T.L. or C.L. lots, bags	R033105-50				▼	27			27	30

03 05 Common Work Results for Concrete

03 05 13 – Basic Concrete Materials

03 05 13.30 Cement

		Crew	Daily Output	Labor-Hours	Unit	Material	2018 Bare Costs Labor	Equipment	Total	Total Incl O&P
0520	L.T.L. or L.C.L. lots				Bag	38.50			38.50	42.50
0600	White, type I, T.L. or C.L. lots, bags					28.50			28.50	31
0620	L.T.L. or L.C.L. lots					30			30	32.50

03 05 13.80 Waterproofing and Dampproofing

		Crew	Daily Output	Labor-Hours	Unit	Material	2018 Bare Costs Labor	Equipment	Total	Total Incl O&P
0010	**WATERPROOFING AND DAMPPROOFING**									
0050	Integral waterproofing, add to cost of regular concrete				C.Y.	19.15			19.15	21

03 05 13.85 Winter Protection

		Crew	Daily Output	Labor-Hours	Unit	Material	2018 Bare Costs Labor	Equipment	Total	Total Incl O&P
0010	**WINTER PROTECTION**									
0012	For heated ready mix, add				C.Y.	5.35			5.35	5.90
0100	Temporary heat to protect concrete, 24 hours	2 Clab	50	.320	M.S.F.	200	12.75		212.75	239
0200	Temporary shelter for slab on grade, wood frame/polyethylene sheeting									
0201	Build or remove, light framing for short spans	2 Carp	10	1.600	M.S.F.	305	81		386	460
0210	Large framing for long spans	"	3	5.333	"	410	270		680	860
0500	Electrically heated pads, 110 volts, 15 watts/S.F., buy				S.F.	11.15			11.15	12.30
0600	20 watts/S.F., buy					14.85			14.85	16.35
0710	Electrically heated pads, 15 watts/S.F., 20 uses					.56			.56	.61

03 11 Concrete Forming

03 11 13 – Structural Cast-In-Place Concrete Forming

03 11 13.20 Forms In Place, Beams and Girders

			Crew	Daily Output	Labor-Hours	Unit	Material	2018 Bare Costs Labor	Equipment	Total	Total Incl O&P
0010	**FORMS IN PLACE, BEAMS AND GIRDERS**	R031113-40									
0500	Exterior spandrel, job-built plywood, 12" wide, 1 use		C-2	225	.213	SFCA	3.05	10.50		13.55	19.35
0550	2 use	R031113-60		275	.175		1.60	8.60		10.20	14.85
0600	3 use			295	.163		1.22	8		9.22	13.55
0650	4 use			310	.155		.99	7.60		8.59	12.70
1000	18" wide, 1 use			250	.192		2.70	9.45		12.15	17.35
1050	2 use			275	.175		1.48	8.60		10.08	14.75
1100	3 use			305	.157		1.08	7.75		8.83	13
1150	4 use			315	.152		.88	7.50		8.38	12.35
1500	24" wide, 1 use			265	.181		2.46	8.90		11.36	16.30
1550	2 use			290	.166		1.39	8.15		9.54	13.90
1600	3 use			315	.152		.98	7.50		8.48	12.50
1650	4 use			325	.148		.80	7.25		8.05	11.95
2000	Interior beam, job-built plywood, 12" wide, 1 use			300	.160		3.71	7.90		11.61	16.10
2050	2 use			340	.141		1.79	6.95		8.74	12.55
2100	3 use			364	.132		1.48	6.50		7.98	11.55
2150	4 use			377	.127		1.21	6.25		7.46	10.90
2500	24" wide, 1 use			320	.150		2.51	7.40		9.91	14
2550	2 use			365	.132		1.42	6.45		7.87	11.40
2600	3 use			385	.125		1	6.15		7.15	10.45
2650	4 use			395	.122		.81	6		6.81	10
3000	Encasing steel beam, hung, job-built plywood, 1 use			325	.148		3.15	7.25		10.40	14.50
3050	2 use			390	.123		1.73	6.05		7.78	11.15
3100	3 use			415	.116		1.26	5.70		6.96	10.05
3150	4 use			430	.112		1.02	5.50		6.52	9.50
3500	Bottoms only, to 30" wide, job-built plywood, 1 use			230	.209		4.14	10.25		14.39	20
3550	2 use			265	.181		2.32	8.90		11.22	16.15
3600	3 use			280	.171		1.66	8.45		10.11	14.65
3650	4 use			290	.166		1.35	8.15		9.50	13.90
4000	Sides only, vertical, 36" high, job-built plywood, 1 use			335	.143		5.15	7.05		12.20	16.45

For customer support on your Building Construction Costs with RSMeans data, call 800.448.8182.

03 11 Concrete Forming

03 11 13 – Structural Cast-In-Place Concrete Forming

03 11 13.20 Forms In Place, Beams and Girders

		Crew	Daily Output	Labor-Hours	Unit	Material	2018 Bare Costs Labor	Equipment	Total	Total Incl O&P
4050	2 use	C-2	405	.119	SFCA	2.84	5.85		8.69	12
4100	3 use		430	.112		2.06	5.50		7.56	10.60
4150	4 use		445	.108		1.68	5.30		6.98	9.95
4500	Sloped sides, 36" high, 1 use		305	.157		4.96	7.75		12.71	17.25
4550	2 use		370	.130		2.76	6.40		9.16	12.75
4600	3 use		405	.119		1.98	5.85		7.83	11.10
4650	4 use		425	.113		1.61	5.55		7.16	10.20
5000	Upstanding beams, 36" high, 1 use		225	.213		6.35	10.50		16.85	23
5050	2 use		255	.188		3.54	9.25		12.79	18
5100	3 use		275	.175		2.57	8.60		11.17	15.90
5150	4 use		280	.171		2.08	8.45		10.53	15.15

03 11 13.25 Forms In Place, Columns

			Crew	Daily Output	Labor-Hours	Unit	Material	2018 Bare Costs Labor	Equipment	Total	Total Incl O&P
0010	**FORMS IN PLACE, COLUMNS**	R031113-40									
0500	Round fiberglass, 4 use per mo., rent, 12" diameter		C-1	160	.200	L.F.	8.15	9.60		17.75	23.50
0550	16" diameter	R031113-60		150	.213		9.70	10.25		19.95	26.50
0600	18" diameter			140	.229		10.85	10.95		21.80	28.50
0650	24" diameter			135	.237		13.55	11.40		24.95	32
0700	28" diameter			130	.246		15.15	11.80		26.95	34.50
0800	30" diameter			125	.256		15.80	12.30		28.10	36
0850	36" diameter			120	.267		21	12.80		33.80	42.50
1500	Round fiber tube, recycled paper, 1 use, 8" diameter	G		155	.206		2.88	9.90		12.78	18.25
1550	10" diameter	G		155	.206		3.61	9.90		13.51	19.05
1600	12" diameter	G		150	.213		4.05	10.25		14.30	20
1650	14" diameter	G		145	.221		4.29	10.60		14.89	21
1700	16" diameter	G		140	.229		5.30	10.95		16.25	22.50
1720	18" diameter	G		140	.229		6.20	10.95		17.15	23.50
1750	20" diameter	G		135	.237		7.75	11.40		19.15	26
1800	24" diameter	G		130	.246		9.65	11.80		21.45	28.50
1850	30" diameter	G		125	.256		14.50	12.30		26.80	34.50
1900	36" diameter	G		115	.278		19.60	13.35		32.95	42
1950	42" diameter	G		100	.320		43.50	15.35		58.85	71.50
2000	48" diameter	G		85	.376		54.50	18.05		72.55	87.50
2200	For seamless type, add						15%				
3000	Round, steel, 4 use per mo., rent, regular duty, 14" diameter	G	C-1	145	.221	L.F.	18.10	10.60		28.70	36
3050	16" diameter	G		125	.256		18.45	12.30		30.75	39
3100	Heavy duty, 20" diameter	G		105	.305		20	14.65		34.65	44.50
3150	24" diameter	G		85	.376		22	18.05		40.05	51.50
3200	30" diameter	G		70	.457		25	22		47	61
3250	36" diameter	G		60	.533		27	25.50		52.50	68.50
3300	48" diameter	G		50	.640		40	30.50		70.50	91
3350	60" diameter	G		45	.711		49.50	34		83.50	107
4500	For second and succeeding months, deduct						50%				
5000	Job-built plywood, 8" x 8" columns, 1 use		C-1	165	.194	SFCA	2.66	9.30		11.96	17.10
5050	2 use			195	.164		1.52	7.90		9.42	13.70
5100	3 use			210	.152		1.06	7.30		8.36	12.30
5150	4 use			215	.149		.88	7.15		8.03	11.85
5500	12" x 12" columns, 1 use			180	.178		2.58	8.55		11.13	15.85
5550	2 use			210	.152		1.42	7.30		8.72	12.70
5600	3 use			220	.145		1.03	7		8.03	11.80
5650	4 use			225	.142		.84	6.85		7.69	11.30
6000	16" x 16" columns, 1 use			185	.173		2.59	8.30		10.89	15.50
6050	2 use			215	.149		1.38	7.15		8.53	12.40

03 11 13 – Structural Cast-In-Place Concrete Forming

03 11 13.25 Forms In Place, Columns

		Crew	Daily Output	Labor-Hours	Unit	Material	2018 Bare Costs Labor	Equipment	Total	Total Incl O&P
6100	3 use	C-1	230	.139	SFCA	1.04	6.70		7.74	11.30
6150	4 use		235	.136		.85	6.55		7.40	10.90
6500	24" x 24" columns, 1 use		190	.168		2.94	8.10		11.04	15.55
6550	2 use		216	.148		1.61	7.10		8.71	12.65
6600	3 use		230	.139		1.17	6.70		7.87	11.45
6650	4 use		238	.134		.95	6.45		7.40	10.90
7000	36" x 36" columns, 1 use		200	.160		2.16	7.70		9.86	14.10
7050	2 use		230	.139		1.22	6.70		7.92	11.50
7100	3 use		245	.131		.86	6.25		7.11	10.50
7150	4 use		250	.128		.70	6.15		6.85	10.10
7400	Steel framed plywood, based on 50 uses of purchased									
7420	forms, and 4 uses of bracing lumber									
7500	8" x 8" column	C-1	340	.094	SFCA	2.18	4.52		6.70	9.30
7550	10" x 10"		350	.091		1.89	4.39		6.28	8.80
7600	12" x 12"		370	.086		1.61	4.15		5.76	8.05
7650	16" x 16"		400	.080		1.25	3.84		5.09	7.25
7700	20" x 20"		420	.076		1.11	3.66		4.77	6.75
7750	24" x 24"		440	.073		.80	3.49		4.29	6.20
7755	30" x 30"		440	.073		1.02	3.49		4.51	6.40
7760	36" x 36"		460	.070		.89	3.34		4.23	6.10

03 11 13.30 Forms In Place, Culvert

		Crew	Daily Output	Labor-Hours	Unit	Material	2018 Bare Costs Labor	Equipment	Total	Total Incl O&P
0010	**FORMS IN PLACE, CULVERT** R031113-40									
0015	5' to 8' square or rectangular, 1 use	C-1	170	.188	SFCA	3.97	9.05		13.02	18.10
0050	2 use R031113-60		180	.178		2.39	8.55		10.94	15.65
0100	3 use		190	.168		1.87	8.10		9.97	14.35
0150	4 use		200	.160		1.61	7.70		9.31	13.45

03 11 13.35 Forms In Place, Elevated Slabs

		Crew	Daily Output	Labor-Hours	Unit	Material	2018 Bare Costs Labor	Equipment	Total	Total Incl O&P
0010	**FORMS IN PLACE, ELEVATED SLABS** R031113-40									
1000	Flat plate, job-built plywood, to 15' high, 1 use R031113-60	C-2	470	.102	S.F.	3.87	5.05		8.92	11.90
1050	2 use		520	.092		2.13	4.54		6.67	9.25
1100	3 use		545	.088		1.55	4.34		5.89	8.30
1150	4 use		560	.086		1.26	4.22		5.48	7.80
1500	15' to 20' high ceilings, 4 use		495	.097		1.26	4.77		6.03	8.65
1600	21' to 35' high ceilings, 4 use		450	.107		1.58	5.25		6.83	9.75
2000	Flat slab, drop panels, job-built plywood, to 15' high, 1 use		449	.107		4.45	5.25		9.70	12.90
2050	2 use		509	.094		2.45	4.64		7.09	9.75
2100	3 use		532	.090		1.78	4.44		6.22	8.70
2150	4 use		544	.088		1.45	4.34		5.79	8.20
2250	15' to 20' high ceilings, 4 use		480	.100		2.55	4.92		7.47	10.30
2350	20' to 35' high ceilings, 4 use		435	.110		2.86	5.45		8.31	11.40
3000	Floor slab hung from steel beams, 1 use		485	.099		2.68	4.87		7.55	10.35
3050	2 use		535	.090		2.12	4.42		6.54	9.10
3100	3 use		550	.087		1.94	4.30		6.24	8.70
3150	4 use		565	.085		1.84	4.18		6.02	8.40
3500	Floor slab, with 1-way joist pans, 1 use		415	.116		7.95	5.70		13.65	17.40
3550	2 use		445	.108		5.95	5.30		11.25	14.65
3600	3 use		475	.101		5.30	4.97		10.27	13.35
3650	4 use		500	.096		4.95	4.73		9.68	12.65
4500	With 2-way waffle domes, 1 use		405	.119		8.20	5.85		14.05	17.95
4520	2 use		450	.107		6.20	5.25		11.45	14.85
4530	3 use		460	.104		5.55	5.15		10.70	13.90
4550	4 use		470	.102		5.20	5.05		10.25	13.40

03 11 Concrete Forming

03 11 13 - Structural Cast-In-Place Concrete Forming

03 11 13.35 Forms In Place, Elevated Slabs

		Crew	Daily Output	Labor-Hours	Unit	Material	2018 Bare Costs Labor	Equipment	Total	Total Incl O&P
5000	Box out for slab openings, over 16" deep, 1 use	C-2	190	.253	SFCA	3.48	12.45		15.93	23
5050	2 use		240	.200	"	1.92	9.85		11.77	17.10
5500	Shallow slab box outs, to 10 S.F.		42	1.143	Ea.	11.60	56.50		68.10	98.50
5550	Over 10 S.F. (use perimeter)		600	.080	L.F.	1.55	3.94		5.49	7.70
6000	Bulkhead forms for slab, with keyway, 1 use, 2 piece		500	.096		1.72	4.73		6.45	9.10
6100	3 piece (see also edge forms)		460	.104		1.91	5.15		7.06	9.90
6200	Slab bulkhead form, 4-1/2" high, exp metal, w/keyway & stakes [G]	C-1	1200	.027		.91	1.28		2.19	2.95
6210	5-1/2" high [G]		1100	.029		1.14	1.40		2.54	3.38
6215	7-1/2" high [G]		960	.033		1.37	1.60		2.97	3.95
6220	9-1/2" high [G]		840	.038		1.49	1.83		3.32	4.42
6500	Curb forms, wood, 6" to 12" high, on elevated slabs, 1 use		180	.178	SFCA	1.71	8.55		10.26	14.90
6550	2 use		205	.156		.94	7.50		8.44	12.45
6600	3 use		220	.145		.69	7		7.69	11.40
6650	4 use		225	.142		.56	6.85		7.41	11
7000	Edge forms to 6" high, on elevated slab, 4 use		500	.064	L.F.	.21	3.07		3.28	4.91
7070	7" to 12" high, 1 use		162	.198	SFCA	1.30	9.50		10.80	15.90
7080	2 use		198	.162		.72	7.75		8.47	12.60
7090	3 use		222	.144		.52	6.90		7.42	11.10
7101	4 use		350	.091		.21	4.39		4.60	6.95
7500	Depressed area forms to 12" high, 4 use		300	.107	L.F.	.77	5.10		5.87	8.65
7550	12" to 24" high, 4 use		175	.183		1.05	8.80		9.85	14.50
8000	Perimeter deck and rail for elevated slabs, straight		90	.356		11.95	17.05		29	39
8050	Curved		65	.492		16.45	23.50		39.95	54
8500	Void forms, round plastic, 8" high x 3" diameter [G]		450	.071	Ea.	2.02	3.41		5.43	7.40
8550	4" diameter [G]		425	.075		2.29	3.61		5.90	8
8600	6" diameter [G]		400	.080		4.09	3.84		7.93	10.35
8650	8" diameter [G]		375	.085		6.75	4.10		10.85	13.70

03 11 13.40 Forms In Place, Equipment Foundations

		Crew	Daily Output	Labor-Hours	Unit	Material	2018 Bare Costs Labor	Equipment	Total	Total Incl O&P
0010	**FORMS IN PLACE, EQUIPMENT FOUNDATIONS** R031113-40									
0020	1 use	C-2	160	.300	SFCA	2.74	14.75		17.49	25.50
0050	2 use R031113-60		190	.253		1.51	12.45		13.96	20.50
0100	3 use		200	.240		1.10	11.80		12.90	19.20
0150	4 use		205	.234		.90	11.55		12.45	18.55

03 11 13.45 Forms In Place, Footings

		Crew	Daily Output	Labor-Hours	Unit	Material	2018 Bare Costs Labor	Equipment	Total	Total Incl O&P
0010	**FORMS IN PLACE, FOOTINGS** R031113-40									
0020	Continuous wall, plywood, 1 use	C-1	375	.085	SFCA	6.60	4.10		10.70	13.50
0050	2 use R031113-60		440	.073		3.62	3.49		7.11	9.30
0100	3 use		470	.068		2.63	3.27		5.90	7.90
0150	4 use		485	.066		2.15	3.17		5.32	7.20
0500	Dowel supports for footings or beams, 1 use		500	.064	L.F.	.93	3.07		4	5.70
1000	Integral starter wall, to 4" high, 1 use		400	.080		.95	3.84		4.79	6.90
1500	Keyway, 4 use, tapered wood, 2" x 4"	1 Carp	530	.015		.22	.77		.99	1.40
1550	2" x 6"		500	.016		.32	.81		1.13	1.59
2000	Tapered plastic		530	.015		1.34	.77		2.11	2.63
2250	For keyway hung from supports, add		150	.053		.93	2.70		3.63	5.15
3000	Pile cap, square or rectangular, job-built plywood, 1 use	C-1	290	.110	SFCA	2.92	5.30		8.22	11.25
3050	2 use		346	.092		1.61	4.44		6.05	8.50
3100	3 use		371	.086		1.17	4.14		5.31	7.60
3150	4 use		383	.084		.95	4.01		4.96	7.15
4000	Triangular or hexagonal, 1 use		225	.142		3.42	6.85		10.27	14.15
4050	2 use		280	.114		1.88	5.50		7.38	10.40
4100	3 use		305	.105		1.37	5.05		6.42	9.15

For customer support on your Building Construction Costs with RSMeans data, call 800.448.8182.

55

03 11 Concrete Forming

03 11 13 – Structural Cast-In-Place Concrete Forming

03 11 13.45 Forms In Place, Footings

		Crew	Daily Output	Labor-Hours	Unit	Material	Labor	2018 Bare Costs Equipment	Total	Total Incl O&P
4150	4 use	C-1	315	.102	SFCA	1.11	4.88		5.99	8.60
5000	Spread footings, job-built lumber, 1 use		305	.105		2.10	5.05		7.15	9.95
5050	2 use		371	.086		1.17	4.14		5.31	7.60
5100	3 use		401	.080		.84	3.83		4.67	6.75
5150	4 use		414	.077		.68	3.71		4.39	6.40
6000	Supports for dowels, plinths or templates, 2' x 2' footing		25	1.280	Ea.	6.10	61.50		67.60	100
6050	4' x 4' footing		22	1.455		12.25	70		82.25	119
6100	8' x 8' footing		20	1.600		24.50	77		101.50	144
6150	12' x 12' footing		17	1.882		30	90.50		120.50	171
7000	Plinths, job-built plywood, 1 use		250	.128	SFCA	3.11	6.15		9.26	12.75
7100	4 use		270	.119	"	1.02	5.70		6.72	9.75

03 11 13.47 Forms In Place, Gas Station Forms

			Crew	Daily Output	Labor-Hours	Unit	Material	Labor	2018 Bare Costs Equipment	Total	Total Incl O&P
0010	**FORMS IN PLACE, GAS STATION FORMS**										
0050	Curb fascia, with template, 12 ga. steel, left in place, 9" high	G	1 Carp	50	.160	L.F.	14.10	8.10		22.20	28
1000	Sign or light bases, 18" diameter, 9" high	G		9	.889	Ea.	89	45		134	166
1050	30" diameter, 13" high	G		8	1		141	50.50		191.50	232
2000	Island forms, 10' long, 9" high, 3'-6" wide	G	C-1	10	3.200		395	154		549	670
2050	4' wide	G		9	3.556		405	171		576	710
2500	20' long, 9" high, 4' wide	G		6	5.333		655	256		911	1,100
2550	5' wide	G		5	6.400		680	305		985	1,225

03 11 13.50 Forms In Place, Grade Beam

			Crew	Daily Output	Labor-Hours	Unit	Material	Labor	2018 Bare Costs Equipment	Total	Total Incl O&P
0010	**FORMS IN PLACE, GRADE BEAM**	R031113-40									
0020	Job-built plywood, 1 use		C-2	530	.091	SFCA	2.97	4.46		7.43	10.05
0050	2 use	R031113-60		580	.083		1.63	4.07		5.70	8
0100	3 use			600	.080		1.19	3.94		5.13	7.30
0150	4 use			605	.079		.96	3.91		4.87	7

03 11 13.55 Forms In Place, Mat Foundation

			Crew	Daily Output	Labor-Hours	Unit	Material	Labor	2018 Bare Costs Equipment	Total	Total Incl O&P
0010	**FORMS IN PLACE, MAT FOUNDATION**	R031113-40									
0020	Job-built plywood, 1 use		C-2	290	.166	SFCA	2.93	8.15		11.08	15.60
0050	2 use	R031113-60		310	.155		1.15	7.60		8.75	12.85
0100	3 use			330	.145		.74	7.15		7.89	11.70
0120	4 use			350	.137		.68	6.75		7.43	11.05

03 11 13.65 Forms In Place, Slab On Grade

			Crew	Daily Output	Labor-Hours	Unit	Material	Labor	2018 Bare Costs Equipment	Total	Total Incl O&P
0010	**FORMS IN PLACE, SLAB ON GRADE**	R031113-40									
1000	Bulkhead forms w/keyway, wood, 6" high, 1 use		C-1	510	.063	L.F.	1.05	3.01		4.06	5.75
1050	2 uses	R031113-60		400	.080		.58	3.84		4.42	6.50
1100	4 uses			350	.091		.34	4.39		4.73	7.10
1400	Bulkhead form for slab, 4-1/2" high, exp metal, incl keyway & stakes	G		1200	.027		.91	1.28		2.19	2.95
1410	5-1/2" high	G		1100	.029		1.14	1.40		2.54	3.38
1420	7-1/2" high	G		960	.033		1.37	1.60		2.97	3.95
1430	9-1/2" high	G		840	.038		1.49	1.83		3.32	4.42
2000	Curb forms, wood, 6" to 12" high, on grade, 1 use			215	.149	SFCA	1.90	7.15		9.05	13
2050	2 use			250	.128		1.05	6.15		7.20	10.50
2100	3 use			265	.121		.76	5.80		6.56	9.65
2150	4 use			275	.116		.62	5.60		6.22	9.20
3000	Edge forms, wood, 4 use, on grade, to 6" high			600	.053	L.F.	.29	2.56		2.85	4.22
3050	7" to 12" high			435	.074	SFCA	.69	3.53		4.22	6.15
3060	Over 12"			350	.091	"	.93	4.39		5.32	7.70
3500	For depressed slabs, 4 use, to 12" high			300	.107	L.F.	.76	5.10		5.86	8.65
3550	To 24" high			175	.183		1.01	8.80		9.81	14.45
4000	For slab blockouts, to 12" high, 1 use			200	.160		.82	7.70		8.52	12.60
4050	To 24" high, 1 use			120	.267		1.04	12.80		13.84	20.50

03 11 Concrete Forming

03 11 13 — Structural Cast-In-Place Concrete Forming

03 11 13.65 Forms In Place, Slab On Grade

		Crew	Daily Output	Labor-Hours	Unit	Material	2018 Bare Costs Labor	Equipment	Total	Total Incl O&P
4100	Plastic (extruded), to 6" high, multiple use, on grade	C-1	800	.040	L.F.	6.65	1.92		8.57	10.25
5000	Screed, 24 ga. metal key joint, see Section 03 15 16.30									
5020	Wood, incl. wood stakes, 1" x 3"	C-1	900	.036	L.F.	.85	1.71		2.56	3.54
5050	2" x 4"		900	.036	"	.84	1.71		2.55	3.52
6000	Trench forms in floor, wood, 1 use		160	.200	SFCA	1.89	9.60		11.49	16.70
6050	2 use		175	.183		1.04	8.80		9.84	14.50
6100	3 use		180	.178		.75	8.55		9.30	13.85
6150	4 use		185	.173		.61	8.30		8.91	13.30
8760	Void form, corrugated fiberboard, 4" x 12", 4' long [G]		3000	.011	S.F.	3.43	.51		3.94	4.55
8770	6" x 12", 4' long		3000	.011		4.10	.51		4.61	5.30
8780	1/4" thick hardboard protective cover for void form	2 Carp	1500	.011		.68	.54		1.22	1.57

03 11 13.85 Forms In Place, Walls

			Crew	Daily Output	Labor-Hours	Unit	Material	2018 Bare Costs Labor	Equipment	Total	Total Incl O&P
0010	**FORMS IN PLACE, WALLS**	R031113-10									
0100	Box out for wall openings, to 16" thick, to 10 S.F.		C-2	24	2	Ea.	26.50	98.50		125	179
0150	Over 10 S.F. (use perimeter)	R031113-40	"	280	.171	L.F.	2.28	8.45		10.73	15.35
0250	Brick shelf, 4" w, add to wall forms, use wall area above shelf										
0260	1 use	R031113-60	C-2	240	.200	SFCA	2.47	9.85		12.32	17.70
0300	2 use			275	.175		1.36	8.60		9.96	14.60
0350	4 use			300	.160		.99	7.90		8.89	13.10
0500	Bulkhead, wood with keyway, 1 use, 2 piece			265	.181	L.F.	2.19	8.90		11.09	16
0600	Bulkhead forms with keyway, 1 piece expanded metal, 8" wall [G]		C-1	1000	.032		1.37	1.54		2.91	3.85
0610	10" wall [G]			800	.040		1.49	1.92		3.41	4.56
0620	12" wall [G]			525	.061		1.79	2.93		4.72	6.40
0700	Buttress, to 8' high, 1 use		C-2	350	.137	SFCA	4.23	6.75		10.98	14.95
0750	2 use			430	.112		2.32	5.50		7.82	10.90
0800	3 use			460	.104		1.70	5.15		6.85	9.65
0850	4 use			480	.100		1.40	4.92		6.32	9.05
1000	Corbel or haunch, to 12" wide, add to wall forms, 1 use			150	.320	L.F.	2.38	15.75		18.13	26.50
1050	2 use			170	.282		1.31	13.90		15.21	22.50
1100	3 use			175	.274		.95	13.50		14.45	21.50
1150	4 use			180	.267		.77	13.15		13.92	21
2000	Wall, job-built plywood, to 8' high, 1 use			370	.130	SFCA	2.76	6.40		9.16	12.75
2050	2 use			435	.110		1.75	5.45		7.20	10.20
2100	3 use			495	.097		1.28	4.77		6.05	8.65
2150	4 use			505	.095		1.04	4.68		5.72	8.25
2400	Over 8' to 16' high, 1 use			280	.171		3.05	8.45		11.50	16.20
2450	2 use			345	.139		1.33	6.85		8.18	11.90
2500	3 use			375	.128		.95	6.30		7.25	10.65
2550	4 use			395	.122		.78	6		6.78	9.95
2700	Over 16' high, 1 use			235	.204		2.71	10.05		12.76	18.30
2750	2 use			290	.166		1.49	8.15		9.64	14.05
2800	3 use			315	.152		1.08	7.50		8.58	12.60
2850	4 use			330	.145		.88	7.15		8.03	11.85
4000	Radial, smooth curved, job-built plywood, 1 use			245	.196		2.55	9.65		12.20	17.50
4050	2 use			300	.160		1.40	7.90		9.30	13.55
4100	3 use			325	.148		1.02	7.25		8.27	12.15
4150	4 use			335	.143		.83	7.05		7.88	11.65
4200	Below grade, job-built plywood, 1 use			225	.213		2.71	10.50		13.21	19
4210	2 use			225	.213		1.50	10.50		12	17.65
4220	3 use			225	.213		1.24	10.50		11.74	17.35
4230	4 use			225	.213		.89	10.50		11.39	16.95
4300	Curved, 2' chords, job-built plywood, to 8' high, 1 use			290	.166		2.15	8.15		10.30	14.75

For customer support on your Building Construction Costs with RSMeans data, call 800.448.8182.

57

03 11 Concrete Forming

03 11 13 – Structural Cast-In-Place Concrete Forming

03 11 13.85 Forms In Place, Walls

		Crew	Daily Output	Labor-Hours	Unit	Material	Labor	Equipment	Total	Total Incl O&P
4350	2 use	C-2	355	.135	SFCA	1.18	6.65		7.83	11.45
4400	3 use		385	.125		.86	6.15		7.01	10.30
4450	4 use		400	.120		.70	5.90		6.60	9.75
4500	Over 8' to 16' high, 1 use		290	.166		.92	8.15		9.07	13.40
4525	2 use		355	.135		.51	6.65		7.16	10.70
4550	3 use		385	.125		.37	6.15		6.52	9.75
4575	4 use		400	.120		.30	5.90		6.20	9.35
4600	Retaining wall, battered, job-built plyw'd, to 8' high, 1 use		300	.160		2.03	7.90		9.93	14.25
4650	2 use		355	.135		1.11	6.65		7.76	11.40
4700	3 use		375	.128		.81	6.30		7.11	10.50
4750	4 use		390	.123		.66	6.05		6.71	9.95
4900	Over 8' to 16' high, 1 use		240	.200		2.21	9.85		12.06	17.45
4950	2 use		295	.163		1.22	8		9.22	13.55
5000	3 use		305	.157		.88	7.75		8.63	12.75
5050	4 use	↓	320	.150		.72	7.40		8.12	12.05
5500	For gang wall forming, 192 S.F. sections, deduct					10%	10%			
5550	384 S.F. sections, deduct					20%	20%			
7500	Lintel or sill forms, 1 use	1 Carp	30	.267		3.25	13.50		16.75	24
7520	2 use		34	.235		1.79	11.95		13.74	20
7540	3 use		36	.222		1.30	11.25		12.55	18.60
7560	4 use	↓	37	.216	↓	1.05	10.95		12	17.85
7800	Modular prefabricated plywood, based on 20 uses of purchased									
7820	forms, and 4 uses of bracing lumber									
7860	To 8' high	C-2	800	.060	SFCA	1.09	2.95		4.04	5.70
8060	Over 8' to 16' high		600	.080		1.15	3.94		5.09	7.25
8600	Pilasters, 1 use		270	.178		3.20	8.75		11.95	16.85
8620	2 use		330	.145		1.76	7.15		8.91	12.85
8640	3 use		370	.130		1.28	6.40		7.68	11.10
8660	4 use		385	.125	↓	1.04	6.15		7.19	10.50
9010	Steel framed plywood, based on 50 uses of purchased									
9020	forms, and 4 uses of bracing lumber									
9060	To 8' high	C-2	600	.080	SFCA	.73	3.94		4.67	6.80
9260	Over 8' to 16' high		450	.107		.73	5.25		5.98	8.80
9460	Over 16' to 20' high	↓	400	.120	↓	.73	5.90		6.63	9.80
9475	For elevated walls, add						10%			
9480	For battered walls, 1 side battered, add					10%	10%			
9485	For battered walls, 2 sides battered, add					15%	15%			

03 11 16 – Architectural Cast-in-Place Concrete Forming

03 11 16.13 Concrete Form Liners

		Crew	Daily Output	Labor-Hours	Unit	Material	Labor	Equipment	Total	Total Incl O&P
0010	**CONCRETE FORM LINERS**									
5750	Liners for forms (add to wall forms), ABS plastic									
5800	Aged wood, 4" wide, 1 use	1 Carp	256	.031	SFCA	3.47	1.58		5.05	6.25
5820	2 use		256	.031		1.91	1.58		3.49	4.51
5830	3 use		256	.031		1.39	1.58		2.97	3.94
5840	4 use		256	.031		1.13	1.58		2.71	3.65
5900	Fractured rope rib, 1 use		192	.042		5.05	2.11		7.16	8.75
5925	2 use		192	.042		2.77	2.11		4.88	6.25
5950	3 use		192	.042		2.01	2.11		4.12	5.45
6000	4 use		192	.042		1.63	2.11		3.74	5
6100	Ribbed, 3/4" deep x 1-1/2" OC, 1 use		224	.036		5.05	1.81		6.86	8.30
6125	2 use		224	.036		2.77	1.81		4.58	5.80
6150	3 use	↓	224	.036	↓	2.01	1.81		3.82	4.97

For customer support on your Building Construction Costs with RSMeans data, call 800.448.8182.

03 11 Concrete Forming

03 11 16 – Architectural Cast-in-Place Concrete Forming

03 11 16.13 Concrete Form Liners

		Crew	Daily Output	Labor-Hours	Unit	Material	2018 Bare Costs Labor	Equipment	Total	Total Incl O&P
6200	4 use	1 Carp	224	.036	SFCA	1.63	1.81		3.44	4.56
6300	Rustic brick pattern, 1 use		224	.036		3.47	1.81		5.28	6.60
6325	2 use		224	.036		1.91	1.81		3.72	4.86
6350	3 use		224	.036		1.39	1.81		3.20	4.29
6400	4 use		224	.036		1.13	1.81		2.94	4
6500	3/8" striated, random, 1 use		224	.036		3.47	1.81		5.28	6.60
6525	2 use		224	.036		1.91	1.81		3.72	4.86
6550	3 use		224	.036		1.39	1.81		3.20	4.29
6600	4 use		224	.036		1.13	1.81		2.94	4
6850	Random vertical rustication, 1 use		384	.021		6.60	1.06		7.66	8.85
6900	2 use		384	.021		3.62	1.06		4.68	5.60
6925	3 use		384	.021		2.64	1.06		3.70	4.51
6950	4 use		384	.021		2.14	1.06		3.20	3.97
7050	Wood, beveled edge, 3/4" deep, 1 use		384	.021	L.F.	.16	1.06		1.22	1.79
7100	1" deep, 1 use		384	.021	"	.29	1.06		1.35	1.93
7200	4" wide aged cedar, 1 use		256	.031	SFCA	3.47	1.58		5.05	6.25
7300	4" variable depth rough cedar		224	.036	"	5.05	1.81		6.86	8.30

03 11 19 – Insulating Concrete Forming

03 11 19.10 Insulating Forms, Left In Place

			Crew	Daily Output	Labor-Hours	Unit	Material	2018 Bare Costs Labor	Equipment	Total	Total Incl O&P
0010	**INSULATING FORMS, LEFT IN PLACE**										
0020	S.F. is for exterior face, but includes forms for both faces (total R22)										
2000	4" wall, straight block, 16" x 48" (5.33 S.F.)	G	2 Carp	90	.178	Ea.	21	9		30	37
2010	90 corner block, exterior 16" x 38" x 22" (6.67 S.F.)	G		75	.213		25.50	10.80		36.30	44.50
2020	45 corner block, exterior 16" x 34" x 18" (5.78 S.F.)	G		75	.213		25	10.80		35.80	44
2100	6" wall, straight block, 16" x 48" (5.33 S.F.)	G		90	.178		22	9		31	37.50
2110	90 corner block, exterior 16" x 32" x 24" (6.22 S.F.)	G		75	.213		25	10.80		35.80	44
2120	45 corner block, exterior 16" x 26" x 18" (4.89 S.F.)	G		75	.213		24.50	10.80		35.30	43.50
2130	Brick ledge block, 16" x 48" (5.33 S.F.)	G		80	.200		27	10.15		37.15	45.50
2140	Taper top block, 16" x 48" (5.33 S.F.)	G		80	.200		25.50	10.15		35.65	43.50
2200	8" wall, straight block, 16" x 48" (5.33 S.F.)	G		90	.178		23	9		32	39
2210	90 corner block, exterior 16" x 34" x 26" (6.67 S.F.)	G		75	.213		31	10.80		41.80	50.50
2220	45 corner block, exterior 16" x 28" x 20" (5.33 S.F.)	G		75	.213		25.50	10.80		36.30	44.50
2230	Brick ledge block, 16" x 48" (5.33 S.F.)	G		80	.200		28	10.15		38.15	46.50
2240	Taper top block, 16" x 48" (5.33 S.F.)	G		80	.200		26.50	10.15		36.65	44.50

03 11 19.60 Roof Deck Form Boards

			Crew	Daily Output	Labor-Hours	Unit	Material	2018 Bare Costs Labor	Equipment	Total	Total Incl O&P
0010	**ROOF DECK FORM BOARDS**	R051223-50									
0050	Includes bulb tee sub-purlins @ 32-5/8" OC										
0070	Non-asbestos fiber cement, 5/16" thick		C-13	2950	.008	S.F.	2.97	.43	.03	3.43	3.99
0100	Fiberglass, 1" thick			2700	.009		3.65	.47	.04	4.16	4.81
0500	Wood fiber, 1" thick	G		2700	.009		2.29	.47	.04	2.80	3.31

03 11 23 – Permanent Stair Forming

03 11 23.75 Forms In Place, Stairs

			Crew	Daily Output	Labor-Hours	Unit	Material	2018 Bare Costs Labor	Equipment	Total	Total Incl O&P
0010	**FORMS IN PLACE, STAIRS**	R031113-40									
0015	(Slant length x width), 1 use		C-2	165	.291	S.F.	5.75	14.30		20.05	28.50
0050	2 use	R031113-60		170	.282		3.26	13.90		17.16	24.50
0100	3 use			180	.267		2.44	13.15		15.59	22.50
0150	4 use			190	.253		2.03	12.45		14.48	21
1000	Alternate pricing method (1.0 L.F./S.F.), 1 use			100	.480	LF Rsr	5.75	23.50		29.25	42.50
1050	2 use			105	.457		3.26	22.50		25.76	38
1100	3 use			110	.436		2.44	21.50		23.94	35
1150	4 use			115	.417		2.03	20.50		22.53	33.50

03 11 Concrete Forming

03 11 23 – Permanent Stair Forming

03 11 23.75 Forms In Place, Stairs		Crew	Daily Output	Labor-Hours	Unit	Material	2018 Bare Costs Labor	Equipment	Total	Total Incl O&P
2000	Stairs, cast on sloping ground (length x width), 1 use	C-2	220	.218	S.F.	2.34	10.75		13.09	18.95
2025	2 use		232	.207		1.29	10.20		11.49	16.90
2050	3 use		244	.197		.94	9.70		10.64	15.80
2100	4 use	↓	256	.188	↓	.76	9.25		10.01	14.90

03 15 Concrete Accessories

03 15 05 – Concrete Forming Accessories

03 15 05.12 Chamfer Strips

		Crew	Daily Output	Labor-Hours	Unit	Material	2018 Bare Costs Labor	Equipment	Total	Total Incl O&P
0010	**CHAMFER STRIPS**									
2000	Polyvinyl chloride, 1/2" wide with leg	1 Carp	535	.015	L.F.	.68	.76		1.44	1.90
2200	3/4" wide with leg		525	.015		.75	.77		1.52	2.01
2400	1" radius with leg		515	.016		.78	.79		1.57	2.06
2800	2" radius with leg		500	.016		1.50	.81		2.31	2.89
5000	Wood, 1/2" wide		535	.015		.14	.76		.90	1.30
5200	3/4" wide		525	.015		.16	.77		.93	1.36
5400	1" wide	↓	515	.016	↓	.29	.79		1.08	1.52

03 15 05.15 Column Form Accessories

					Unit	Material	Labor	Equipment	Total	Total Incl O&P
0010	**COLUMN FORM ACCESSORIES**									
1000	Column clamps, adjustable to 24" x 24", buy	G			Set	172			172	189
1100	Rent per month	G				12.25			12.25	13.50
1300	For sizes to 30" x 30", buy	G				201			201	221
1400	Rent per month	G				14.05			14.05	15.50
1600	For sizes to 36" x 36", buy	G				248			248	273
1700	Rent per month	G				17.20			17.20	18.95
2000	Bar type with wedges, 36" x 36", buy	G				160			160	176
2100	Rent per month	G				11.20			11.20	12.30
2300	48" x 48", buy	G				220			220	242
2400	Rent per month	G				15.40			15.40	16.95
3000	Scissor type with wedges, 36" x 36", buy	G				138			138	151
3100	Rent per month	G				13.75			13.75	15.15
3300	60" x 60", buy	G				192			192	211
3400	Rent per month	G				19.25			19.25	21
4000	Friction collars 2'-6" diam., buy	G				2,700			2,700	2,975
4100	Rent per month	G				189			189	208
4300	4'-0" diam., buy	G				3,300			3,300	3,625
4400	Rent per month	G			↓	231			231	254

03 15 05.30 Hangers

					Unit	Material	Labor	Equipment	Total	Total Incl O&P
0010	**HANGERS**									
0020	Slab and beam form									
0500	Banding iron									
0550	3/4" x 22 ga., 14 L.F./lb. or 1/2" x 14 ga., 7 L.F./lb.	G			Lb.	1.37			1.37	1.51
1000	Fascia ties, coil type, to 24" long	G			C	465			465	515
1500	Frame ties to 8-1/8"	G				545			545	600
1550	8-1/8" to 10-1/8"	G				575			575	635
5000	Snap tie hanger, to 30" overall length, 3000#	G				465			465	515
5050	To 36" overall length	G				525			525	580
5100	To 48" overall length	G			↓	645			645	710
5500	Steel beam hanger									
5600	Flange to 8-1/8"	G			C	545			545	600
5650	8-1/8" to 10-1/8"	G			"	575			575	635

For customer support on your Building Construction Costs with RSMeans data, call 800.448.8182.

03 15 Concrete Accessories

03 15 05 – Concrete Forming Accessories

03 15 05.30 Hangers		Crew	Daily Output	Labor-Hours	Unit	Material	2018 Bare Costs Labor	Equipment	Total	Total Incl O&P
5900	Coil threaded rods, continuous, 1/2" diameter	G				L.F.	1.41			1.41	1.55
6000	Tie hangers to 30" overall length, 4000#	G				C	530			530	580
6100	To 36" overall length	G					590			590	645
6500	Tie back hanger, up to 12-1/8" flange	G				▼	1,475			1,475	1,625
8500	Wire, black annealed, 15 gauge	G				Cwt.	154			154	169
8600	16 gauge					"	170			170	187

03 15 05.70 Shores		Crew	Daily Output	Labor-Hours	Unit	Material	2018 Bare Costs Labor	Equipment	Total	Total Incl O&P
0010	**SHORES**										
0020	Erect and strip, by hand, horizontal members										
0500	Aluminum joists and stringers	G	2 Carp	60	.267	Ea.		13.50		13.50	20.50
0600	Steel, adjustable beams	G		45	.356			18.05		18.05	27.50
0700	Wood joists			50	.320			16.20		16.20	24.50
0800	Wood stringers			30	.533			27		27	41
1000	Vertical members to 10' high	G		55	.291			14.75		14.75	22.50
1050	To 13' high	G		50	.320			16.20		16.20	24.50
1100	To 16' high	G		45	.356	▼		18.05		18.05	27.50
1500	Reshoring	G	▼	1400	.011	S.F.	.61	.58		1.19	1.55
1600	Flying truss system	G	C-17D	9600	.009	SFCA		.46	.08	.54	.80
1760	Horizontal, aluminum joists, 6-1/4" high x 5' to 21' span, buy	G				L.F.	16.25			16.25	17.85
1770	Beams, 7-1/4" high x 4' to 30' span	G				"	19			19	21
1810	Horizontal, steel beam, W8x10, 7' span, buy	G				Ea.	61			61	67.50
1830	10' span	G					71			71	78
1920	15' span	G					122			122	135
1940	20' span	G				▼	172			172	189
1970	Steel stringer, W8x10, 4' to 16' span, buy	G				L.F.	7.10			7.10	7.85
3000	Rent for job duration, aluminum joist @ 2' OC, per mo.	G				SF Flr.	.41			.41	.45
3050	Steel W8x10	G					.18			.18	.20
3060	Steel adjustable	G				▼	.18			.18	.20
3500	#1 post shore, steel, 5'-7" to 9'-6" high, 10,000# cap., buy	G				Ea.	155			155	171
3550	#2 post shore, 7'-3" to 12'-10" high, 7800# capacity	G					180			180	198
3600	#3 post shore, 8'-10" to 16'-1" high, 3800# capacity	G				▼	196			196	216
5010	Frame shoring systems, steel, 12,000#/leg, buy										
5040	Frame, 2' wide x 6' high	G				Ea.	115			115	127
5250	X-brace	G					20			20	22
5550	Base plate	G					17.70			17.70	19.45
5600	Screw jack	G					39			39	42.50
5650	U-head, 8" x 8"	G				▼	22			22	24

03 15 05.75 Sleeves and Chases		Crew	Daily Output	Labor-Hours	Unit	Material	2018 Bare Costs Labor	Equipment	Total	Total Incl O&P
0010	**SLEEVES AND CHASES**										
0100	Plastic, 1 use, 12" long, 2" diameter	1 Carp	100	.080	Ea.	1.89	4.06		5.95	8.30
0150	4" diameter		90	.089		5.30	4.51		9.81	12.70
0200	6" diameter		75	.107		9.30	5.40		14.70	18.50
0250	12" diameter	▼	60	.133	▼	27.50	6.75		34.25	41

03 15 05.80 Snap Ties		Crew	Daily Output	Labor-Hours	Unit	Material	2018 Bare Costs Labor	Equipment	Total	Total Incl O&P
0010	**SNAP TIES**, 8-1/4" L&W (Lumber and wedge)										
0100	2250 lb., w/flat washer, 8" wall	G				C	93			93	102
0150	10" wall	G					135			135	149
0200	12" wall	G					140			140	154
0250	16" wall	G					155			155	171
0300	18" wall	G					161			161	177
0500	With plastic cone, 8" wall	G					82			82	90
0550	10" wall	G				▼	85			85	93.50

03 15 Concrete Accessories

03 15 05 – Concrete Forming Accessories

03 15 05.80 Snap Ties

			Crew	Daily Output	Labor-Hours	Unit	Material	2018 Bare Costs Labor	2018 Bare Costs Equipment	Total	Total Incl O&P
0600	12" wall	G				C	92			92	101
0650	16" wall	G					101			101	111
0700	18" wall	G					104			104	114
1000	3350 lb., w/flat washer, 8" wall	G					168			168	185
1100	10" wall	G					184			184	202
1150	12" wall	G					188			188	207
1200	16" wall	G					216			216	238
1250	18" wall	G					225			225	248
1500	With plastic cone, 8" wall	G					136			136	150
1550	10" wall	G					149			149	164
1600	12" wall	G					153			153	168
1650	16" wall	G					175			175	193
1700	18" wall	G					181			181	199

03 15 05.85 Stair Tread Inserts

		Crew	Daily Output	Labor-Hours	Unit	Material	2018 Bare Costs Labor	2018 Bare Costs Equipment	Total	Total Incl O&P
0010	**STAIR TREAD INSERTS**									
0105	Cast nosing insert, abrasive surface, pre-drilled, includes screws									
0110	Aluminum, 3" wide x 3' long	1 Cefi	32	.250	Ea.	52	11.90		63.90	75
0120	4' long		31	.258		69.50	12.25		81.75	94.50
0130	5' long		30	.267		86.50	12.70		99.20	114
0135	Extruded nosing insert, black abrasive strips, continuous anchor									
0140	Aluminum, 3" wide x 3' long	1 Cefi	64	.125	Ea.	33.50	5.95		39.45	46
0150	4' long		60	.133		45	6.35		51.35	59
0160	5' long		56	.143		56	6.80		62.80	71.50
0165	Extruded nosing insert, black abrasive strips, pre-drilled, incl. screws									
0170	Aluminum, 3" wide x 3' long	1 Cefi	32	.250	Ea.	42	11.90		53.90	63.50
0180	4' long		31	.258		56	12.25		68.25	79.50
0190	5' long		30	.267		70	12.70		82.70	96

03 15 05.95 Wall and Foundation Form Accessories

			Crew	Daily Output	Labor-Hours	Unit	Material	2018 Bare Costs Labor	2018 Bare Costs Equipment	Total	Total Incl O&P
0010	**WALL AND FOUNDATION FORM ACCESSORIES**										
2000	Footings, turnbuckle form aligner	G				Ea.	16.65			16.65	18.30
2050	Spreaders for footer, adjustable	G				"	25			25	27.50
3000	Form oil, up to 1200 S.F./gallon coverage					Gal.	14.15			14.15	15.55
3050	Up to 800 S.F./gallon					"	21.50			21.50	23.50
3500	Form patches, 1-3/4" diameter					C	28			28	31
3550	2-3/4" diameter					"	48			48	53
4000	Nail stakes, 3/4" diameter, 18" long	G				Ea.	1.98			1.98	2.18
4050	24" long	G					2.50			2.50	2.75
4200	30" long	G					3.16			3.16	3.48
4250	36" long	G					3.99			3.99	4.39

03 15 13 – Waterstops

03 15 13.50 Waterstops

		Crew	Daily Output	Labor-Hours	Unit	Material	2018 Bare Costs Labor	2018 Bare Costs Equipment	Total	Total Incl O&P
0010	**WATERSTOPS**, PVC and Rubber									
0020	PVC, ribbed 3/16" thick, 4" wide	1 Carp	155	.052	L.F.	1.43	2.62		4.05	5.55
0050	6" wide		145	.055		2.41	2.80		5.21	6.90
0500	With center bulb, 6" wide, 3/16" thick		135	.059		2.43	3		5.43	7.25
0550	3/8" thick		130	.062		4.43	3.12		7.55	9.60
0600	9" wide x 3/8" thick		125	.064		7.20	3.24		10.44	12.90
0800	Dumbbell type, 6" wide, 3/16" thick		150	.053		3.45	2.70		6.15	7.90
0850	3/8" thick		145	.055		3.71	2.80		6.51	8.35
1000	9" wide, 3/8" thick, plain		130	.062		6	3.12		9.12	11.35
1050	Center bulb		130	.062		8.50	3.12		11.62	14.10
1250	Ribbed type, split, 3/16" thick, 6" wide		145	.055		2	2.80		4.80	6.45

For customer support on your Building Construction Costs with RSMeans data, call 800.448.8182.

03 15 Concrete Accessories

03 15 13 – Waterstops

03 15 13.50 Waterstops

		Crew	Daily Output	Labor-Hours	Unit	Material	2018 Bare Costs Labor	Equipment	Total	Total Incl O&P
1300	3/8" thick	1 Carp	130	.062	L.F.	4.66	3.12		7.78	9.90
2000	Rubber, flat dumbbell, 3/8" thick, 6" wide		145	.055		4.55	2.80		7.35	9.25
2050	9" wide		135	.059		9.90	3		12.90	15.45
2500	Flat dumbbell split, 3/8" thick, 6" wide		145	.055		2	2.80		4.80	6.45
2550	9" wide		135	.059		4.66	3		7.66	9.70
3000	Center bulb, 1/4" thick, 6" wide		145	.055		6.90	2.80		9.70	11.85
3050	9" wide		135	.059		13.75	3		16.75	19.70
3500	Center bulb split, 3/8" thick, 6" wide		145	.055		5.70	2.80		8.50	10.50
3550	9" wide		135	.059		9.85	3		12.85	15.40
5000	Waterstop fittings, rubber, flat									
5010	Dumbbell or center bulb, 3/8" thick,									
5200	Field union, 6" wide	1 Carp	50	.160	Ea.	31.50	8.10		39.60	47.50
5250	9" wide		50	.160		35	8.10		43.10	51
5500	Flat cross, 6" wide		30	.267		49.50	13.50		63	75
5550	9" wide		30	.267		67.50	13.50		81	95
6000	Flat tee, 6" wide		30	.267		47.50	13.50		61	73
6050	9" wide		30	.267		62.50	13.50		76	89
6500	Flat ell, 6" wide		40	.200		46.50	10.15		56.65	67
6550	9" wide		40	.200		57.50	10.15		67.65	79
7000	Vertical tee, 6" wide		25	.320		20.50	16.20		36.70	47.50
7050	9" wide		25	.320		28.50	16.20		44.70	55.50
7500	Vertical ell, 6" wide		35	.229		20.50	11.60		32.10	40
7550	9" wide		35	.229		39.50	11.60		51.10	61

03 15 16 – Concrete Construction Joints

03 15 16.20 Control Joints, Saw Cut

		Crew	Daily Output	Labor-Hours	Unit	Material	2018 Bare Costs Labor	Equipment	Total	Total Incl O&P
0010	**CONTROL JOINTS, SAW CUT**									
0100	Sawcut control joints in green concrete									
0120	1" depth	C-27	2000	.008	L.F.	.03	.38	.05	.46	.66
0140	1-1/2" depth		1800	.009		.05	.42	.06	.53	.75
0160	2" depth		1600	.010		.07	.48	.06	.61	.85
0180	Sawcut joint reservoir in cured concrete									
0182	3/8" wide x 3/4" deep, with single saw blade	C-27	1000	.016	L.F.	.05	.76	.10	.91	1.30
0184	1/2" wide x 1" deep, with double saw blades		900	.018		.10	.85	.11	1.06	1.49
0186	3/4" wide x 1-1/2" deep, with double saw blades		800	.020		.21	.95	.13	1.29	1.78
0190	Water blast joint to wash away laitance, 2 passes	C-29	2500	.003			.13	.03	.16	.22
0200	Air blast joint to blow out debris and air dry, 2 passes	C-28	2000	.004			.19	.01	.20	.29
0300	For backer rod, see Section 07 91 23.10									
0340	For joint sealant, see Section 03 15 16.30 or 07 92 13.20									
0900	For replacement of joint sealant, see Section 07 01 90.81									

03 15 16.30 Expansion Joints

			Crew	Daily Output	Labor-Hours	Unit	Material	2018 Bare Costs Labor	Equipment	Total	Total Incl O&P
0010	**EXPANSION JOINTS**										
0020	Keyed, cold, 24 ga., incl. stakes, 3-1/2" high	G	1 Carp	200	.040	L.F.	.85	2.03		2.88	4.03
0050	4-1/2" high	G		200	.040		.91	2.03		2.94	4.09
0100	5-1/2" high	G		195	.041		1.14	2.08		3.22	4.42
0150	7-1/2" high	G		190	.042		1.37	2.14		3.51	4.76
0160	9-1/2" high	G		185	.043		1.49	2.19		3.68	4.98
0300	Poured asphalt, plain, 1/2" x 1"		1 Clab	450	.018		.24	.71		.95	1.34
0350	1" x 2"			400	.020		.95	.80		1.75	2.26
0500	Neoprene, liquid, cold applied, 1/2" x 1"			450	.018		2.28	.71		2.99	3.59
0550	1" x 2"			400	.020		9.10	.80		9.90	11.25
0700	Polyurethane, poured, 2 part, 1/2" x 1"			400	.020		1.39	.80		2.19	2.74
0750	1" x 2"			350	.023		5.55	.91		6.46	7.50

For customer support on your Building Construction Costs with RSMeans data, call 800.448.8182.

63

03 15 Concrete Accessories

03 15 16 – Concrete Construction Joints

03 15 16.30 Expansion Joints

		Crew	Daily Output	Labor-Hours	Unit	Material	2018 Bare Costs Labor	Equipment	Total	Total Incl O&P
0900	Rubberized asphalt, hot or cold applied, 1/2" x 1"	1 Clab	450	.018	L.F.	.30	.71		1.01	1.41
0950	1" x 2"		400	.020		1.19	.80		1.99	2.52
1100	Hot applied, fuel resistant, 1/2" x 1"		450	.018		.45	.71		1.16	1.58
1150	1" x 2"		400	.020		1.79	.80		2.59	3.17
2000	Premolded, bituminous fiber, 1/2" x 6"	1 Carp	375	.021		.42	1.08		1.50	2.11
2050	1" x 12"		300	.027		1.97	1.35		3.32	4.23
2140	Concrete expansion joint, recycled paper and fiber, 1/2" x 6" G		390	.021		.42	1.04		1.46	2.04
2150	1/2" x 12" G		360	.022		.84	1.13		1.97	2.65
2250	Cork with resin binder, 1/2" x 6"		375	.021		1.17	1.08		2.25	2.94
2300	1" x 12"		300	.027		3.20	1.35		4.55	5.60
2500	Neoprene sponge, closed cell, 1/2" x 6"		375	.021		2.31	1.08		3.39	4.19
2550	1" x 12"		300	.027		8.85	1.35		10.20	11.75
2750	Polyethylene foam, 1/2" x 6"		375	.021		.52	1.08		1.60	2.22
2800	1" x 12"		300	.027		2.78	1.35		4.13	5.10
3000	Polyethylene backer rod, 3/8" diameter		460	.017		.03	.88		.91	1.38
3050	3/4" diameter		460	.017		.07	.88		.95	1.41
3100	1" diameter		460	.017		.12	.88		1	1.47
3500	Polyurethane foam, with polybutylene, 1/2" x 1/2"		475	.017		1.19	.85		2.04	2.61
3550	1" x 1"		450	.018		3.03	.90		3.93	4.70
3750	Polyurethane foam, regular, closed cell, 1/2" x 6"		375	.021		.89	1.08		1.97	2.63
3800	1" x 12"		300	.027		3.15	1.35		4.50	5.55
4000	Polyvinyl chloride foam, closed cell, 1/2" x 6"		375	.021		2.40	1.08		3.48	4.29
4050	1" x 12"		300	.027		8.25	1.35		9.60	11.15
4250	Rubber, gray sponge, 1/2" x 6"		375	.021		2.02	1.08		3.10	3.87
4300	1" x 12"		300	.027		7.25	1.35		8.60	10.05
4400	Redwood heartwood, 1" x 4"		400	.020		1.16	1.01		2.17	2.82
4450	1" x 6"		375	.021		1.75	1.08		2.83	3.58
5000	For installation in walls, add						75%			
5250	For installation in boxouts, add						25%			

03 15 19 – Cast-In Concrete Anchors

03 15 19.05 Anchor Bolt Accessories

		Crew	Daily Output	Labor-Hours	Unit	Material	2018 Bare Costs Labor	Equipment	Total	Total Incl O&P
0010	**ANCHOR BOLT ACCESSORIES**									
0015	For anchor bolts set in fresh concrete, see Section 03 15 19.10									
8150	Anchor bolt sleeve, plastic, 1" diameter bolts	1 Carp	60	.133	Ea.	13.40	6.75		20.15	25
8500	1-1/2" diameter		28	.286		19.75	14.50		34.25	43.50
8600	2" diameter		24	.333		19.45	16.90		36.35	47
8650	3" diameter		20	.400		35.50	20.50		56	70

03 15 19.10 Anchor Bolts

		Crew	Daily Output	Labor-Hours	Unit	Material	2018 Bare Costs Labor	Equipment	Total	Total Incl O&P
0010	**ANCHOR BOLTS**									
0015	Made from recycled materials									
0025	Single bolts installed in fresh concrete, no templates									
0030	Hooked w/nut and washer, 1/2" diameter, 8" long G	1 Carp	132	.061	Ea.	1.42	3.07		4.49	6.25
0040	12" long G		131	.061		1.58	3.10		4.68	6.45
0050	5/8" diameter, 8" long G		129	.062		3.80	3.14		6.94	8.95
0060	12" long G		127	.063		4.67	3.19		7.86	10
0070	3/4" diameter, 8" long G		127	.063		4.67	3.19		7.86	10
0080	12" long G		125	.064		5.85	3.24		9.09	11.35
0090	2-bolt pattern, including job-built 2-hole template, per set									
0100	J-type, incl. hex nut & washer, 1/2" diameter x 6" long G	1 Carp	21	.381	Set	5.15	19.30		24.45	35
0110	12" long G		21	.381		5.75	19.30		25.05	36
0120	18" long G		21	.381		6.70	19.30		26	37
0130	3/4" diameter x 8" long G		20	.400		11.95	20.50		32.45	44

For customer support on your Building Construction Costs with RSMeans data, call 800.448.8182.

03 15 Concrete Accessories

03 15 19 – Cast-In Concrete Anchors

03 15 19.10 Anchor Bolts		Crew	Daily Output	Labor-Hours	Unit	Material	2018 Bare Costs Labor	Equipment	Total	Total Incl O&P
0140	12" long	G	1 Carp	20	.400	Set	14.30	20.50		34.80	46.50
0150	18" long	G		20	.400		17.80	20.50		38.30	50.50
0160	1" diameter x 12" long	G		19	.421		22.50	21.50		44	57.50
0170	18" long	G		19	.421		27	21.50		48.50	62
0180	24" long	G		19	.421		32	21.50		53.50	68
0190	36" long	G		18	.444		43	22.50		65.50	82
0200	1-1/2" diameter x 18" long	G		17	.471		41.50	24		65.50	82
0210	24" long	G		16	.500		48.50	25.50		74	92
0300	L-type, incl. hex nut & washer, 3/4" diameter x 12" long	G		20	.400		14.10	20.50		34.60	46.50
0310	18" long	G		20	.400		17.35	20.50		37.85	50
0320	24" long	G		20	.400		20.50	20.50		41	53.50
0330	30" long	G		20	.400		25.50	20.50		46	59
0340	36" long	G		20	.400		28.50	20.50		49	62.50
0350	1" diameter x 12" long	G		19	.421		22	21.50		43.50	56.50
0360	18" long	G		19	.421		26.50	21.50		48	61.50
0370	24" long	G		19	.421		32	21.50		53.50	68
0380	30" long	G		19	.421		37.50	21.50		59	73.50
0390	36" long	G		18	.444		42.50	22.50		65	81
0400	42" long	G		18	.444		51	22.50		73.50	90.50
0410	48" long	G		18	.444		57	22.50		79.50	97
0420	1-1/4" diameter x 18" long	G		18	.444		33.50	22.50		56	71.50
0430	24" long	G		18	.444		39	22.50		61.50	77.50
0440	30" long	G		17	.471		45	24		69	86
0450	36" long	G		17	.471		50.50	24		74.50	92.50
0460	42" long	G	2 Carp	32	.500		57	25.50		82.50	101
0470	48" long	G		32	.500		64.50	25.50		90	110
0480	54" long	G		31	.516		76	26		102	124
0490	60" long	G		31	.516		83	26		109	132
0500	1-1/2" diameter x 18" long	G		33	.485		49.50	24.50		74	92
0510	24" long	G		32	.500		57	25.50		82.50	102
0520	30" long	G		31	.516		64.50	26		90.50	111
0530	36" long	G		30	.533		73.50	27		100.50	122
0540	42" long	G		30	.533		83.50	27		110.50	133
0550	48" long	G		29	.552		93.50	28		121.50	146
0560	54" long	G		28	.571		113	29		142	168
0570	60" long	G		28	.571		124	29		153	180
0580	1-3/4" diameter x 18" long	G		31	.516		64	26		90	111
0590	24" long	G		30	.533		74.50	27		101.50	123
0600	30" long	G		29	.552		86.50	28		114.50	138
0610	36" long	G		28	.571		98	29		127	152
0620	42" long	G		27	.593		110	30		140	167
0630	48" long	G		26	.615		120	31		151	181
0640	54" long	G		26	.615		149	31		180	212
0650	60" long	G		25	.640		161	32.50		193.50	227
0660	2" diameter x 24" long	G		27	.593		121	30		151	179
0670	30" long	G		27	.593		136	30		166	196
0680	36" long	G		26	.615		149	31		180	212
0690	42" long	G		25	.640		166	32.50		198.50	233
0700	48" long	G		24	.667		190	34		224	261
0710	54" long	G		23	.696		226	35.50		261.50	305
0720	60" long	G		23	.696		243	35.50		278.50	320
0730	66" long	G		22	.727		260	37		297	340
0740	72" long	G		21	.762		284	38.50		322.50	375

03 15 19 – Cast-In Concrete Anchors

03 15 19.10 Anchor Bolts		Crew	Daily Output	Labor-Hours	Unit	Material	2018 Bare Costs Labor	Equipment	Total	Total Incl O&P
1000	4-bolt pattern, including job-built 4-hole template, per set									
1100	J-type, incl. hex nut & washer, 1/2" diameter x 6" long	G	1 Carp	19	.421	Set	7.65	21.50		29.15	41
1110	12" long	G		19	.421		8.95	21.50		30.45	42.50
1120	18" long	G		18	.444		10.85	22.50		33.35	46.50
1130	3/4" diameter x 8" long	G		17	.471		21.50	24		45.50	60
1140	12" long	G		17	.471		26	24		50	65
1150	18" long	G		17	.471		33	24		57	73
1160	1" diameter x 12" long	G		16	.500		42.50	25.50		68	85.50
1170	18" long	G		15	.533		51	27		78	97
1180	24" long	G		15	.533		61.50	27		88.50	109
1190	36" long	G		15	.533		83.50	27		110.50	133
1200	1-1/2" diameter x 18" long	G		13	.615		80	31		111	136
1210	24" long	G		12	.667		94.50	34		128.50	156
1300	L-type, incl. hex nut & washer, 3/4" diameter x 12" long	G		17	.471		25.50	24		49.50	64.50
1310	18" long	G		17	.471		32	24		56	72
1320	24" long	G		17	.471		38.50	24		62.50	79
1330	30" long	G		16	.500		48	25.50		73.50	91.50
1340	36" long	G		16	.500		54.50	25.50		80	98.50
1350	1" diameter x 12" long	G		16	.500		41	25.50		66.50	83.50
1360	18" long	G		15	.533		50.50	27		77.50	96.50
1370	24" long	G		15	.533		61.50	27		88.50	109
1380	30" long	G		15	.533		72	27		99	121
1390	36" long	G		15	.533		82	27		109	132
1400	42" long	G		14	.571		99.50	29		128.50	153
1410	48" long	G		14	.571		111	29		140	166
1420	1-1/4" diameter x 18" long	G		14	.571		64	29		93	115
1430	24" long	G		14	.571		76	29		105	128
1440	30" long	G		13	.615		87.50	31		118.50	144
1450	36" long	G		13	.615		99	31		130	157
1460	42" long	G	2 Carp	25	.640		111	32.50		143.50	173
1470	48" long	G		24	.667		127	34		161	191
1480	54" long	G		23	.696		149	35.50		184.50	218
1490	60" long	G		23	.696		163	35.50		198.50	234
1500	1-1/2" diameter x 18" long	G		25	.640		96	32.50		128.50	156
1510	24" long	G		24	.667		112	34		146	175
1520	30" long	G		23	.696		126	35.50		161.50	193
1530	36" long	G		22	.727		144	37		181	215
1540	42" long	G		22	.727		164	37		201	237
1550	48" long	G		21	.762		184	38.50		222.50	261
1560	54" long	G		20	.800		224	40.50		264.50	310
1570	60" long	G		20	.800		245	40.50		285.50	330
1580	1-3/4" diameter x 18" long	G		22	.727		125	37		162	194
1590	24" long	G		21	.762		146	38.50		184.50	220
1600	30" long	G		21	.762		170	38.50		208.50	246
1610	36" long	G		20	.800		194	40.50		234.50	275
1620	42" long	G		19	.842		217	42.50		259.50	305
1630	48" long	G		18	.889		238	45		283	330
1640	54" long	G		18	.889		295	45		340	395
1650	60" long	G		17	.941		320	47.50		367.50	425
1660	2" diameter x 24" long	G		19	.842		239	42.50		281.50	330
1670	30" long	G		18	.889		269	45		314	365
1680	36" long	G		18	.889		295	45		340	395
1690	42" long	G		17	.941		330	47.50		377.50	435

03 15 Concrete Accessories

03 15 19 – Cast-In Concrete Anchors

03 15 19.10 Anchor Bolts

			Crew	Daily Output	Labor-Hours	Unit	Material	2018 Bare Costs Labor	Equipment	Total	Total Incl O&P
1700	48" long	G	2 Carp	16	1	Set	380	50.50		430.50	490
1710	54" long	G		15	1.067		450	54		504	580
1720	60" long	G		15	1.067		485	54		539	615
1730	66" long	G		14	1.143		515	58		573	660
1740	72" long	G		14	1.143		565	58		623	710
1990	For galvanized, add					Ea.	75%				

03 15 19.20 Dovetail Anchor System

			Crew	Daily Output	Labor-Hours	Unit	Material	2018 Bare Costs Labor	Equipment	Total	Total Incl O&P
0010	**DOVETAIL ANCHOR SYSTEM**										
0500	Dovetail anchor slot, galvanized, foam-filled, 26 ga.	G	1 Carp	425	.019	L.F.	1.17	.95		2.12	2.74
0600	24 ga.	G		400	.020		1.61	1.01		2.62	3.31
0625	22 ga.	G		400	.020		1.75	1.01		2.76	3.47
0900	Stainless steel, foam-filled, 26 ga.	G		375	.021		1.87	1.08		2.95	3.71
1200	Dovetail brick anchor, corrugated, galvanized, 3-1/2" long, 16 ga.	G	1 Bric	10.50	.762	C	40.50	38.50		79	103
1300	12 ga.	G		10.50	.762		60	38.50		98.50	125
1500	Seismic, galvanized, 3-1/2" long, 16 ga.	G		10.50	.762		71	38.50		109.50	137
1600	12 ga.	G		10.50	.762		84	38.50		122.50	151
2000	Dovetail cavity wall, corrugated, galvanized, 5-1/2" long, 16 ga.	G		10.50	.762		40	38.50		78.50	103
2100	12 ga.	G		10.50	.762		57	38.50		95.50	121
3000	Dovetail furring anchors, corrugated, galvanized, 1-1/2" long, 16 ga.	G		10.50	.762		24	38.50		62.50	85
3100	12 ga.	G		10.50	.762		31	38.50		69.50	92.50
6000	Dovetail stone panel anchors, galvanized, 1/8" x 1" wide, 3-1/2" long	G		10.50	.762		97	38.50		135.50	166
6100	1/4" x 1" wide	G		10.50	.762		164	38.50		202.50	239

03 15 19.30 Inserts

			Crew	Daily Output	Labor-Hours	Unit	Material	2018 Bare Costs Labor	Equipment	Total	Total Incl O&P
0010	**INSERTS**										
1000	Inserts, slotted nut type for 3/4" bolts, 4" long	G	1 Carp	84	.095	Ea.	20	4.83		24.83	30
2100	6" long	G		84	.095		23	4.83		27.83	32.50
2150	8" long	G		84	.095		30.50	4.83		35.33	41
2200	Slotted, strap type, 4" long	G		84	.095		22	4.83		26.83	31.50
2250	6" long	G		84	.095		24.50	4.83		29.33	34.50
2300	8" long	G		84	.095		32.50	4.83		37.33	43.50
2350	Strap for slotted insert, 4" long	G		84	.095		11.90	4.83		16.73	20.50
4100	6" long	G		84	.095		12.95	4.83		17.78	21.50
4150	8" long	G		84	.095		16.15	4.83		20.98	25
4200	10" long	G		84	.095		19.60	4.83		24.43	29
7000	Loop ferrule type										
7100	1/4" diameter bolt	G	1 Carp	84	.095	Ea.	2.32	4.83		7.15	9.90
7350	7/8" diameter bolt	G	"	84	.095	"	10.80	4.83		15.63	19.20
9000	Wedge type										
9100	For 3/4" diameter bolt	G	1 Carp	60	.133	Ea.	9.70	6.75		16.45	21
9800	Cut washers, black										
9900	3/4" bolt	G				Ea.	1.21			1.21	1.33
9950	For galvanized inserts, add						30%				

03 15 19.45 Machinery Anchors

			Crew	Daily Output	Labor-Hours	Unit	Material	2018 Bare Costs Labor	Equipment	Total	Total Incl O&P
0010	**MACHINERY ANCHORS**, heavy duty, incl. sleeve, floating base nut,										
0020	lower stud & coupling nut, fiber plug, connecting stud, washer & nut.										
0030	For flush mounted embedment in poured concrete heavy equip. pads.										
0200	Stud & bolt, 1/2" diameter	G	E-16	40	.400	Ea.	52.50	22.50	2.46	77.46	97
0300	5/8" diameter	G		35	.457		61	25.50	2.81	89.31	112
0500	3/4" diameter	G		30	.533		72.50	29.50	3.28	105.28	132
0600	7/8" diameter	G		25	.640		82.50	35.50	3.94	121.94	153
0800	1" diameter	G		20	.800		88.50	44.50	4.92	137.92	176
0900	1-1/4" diameter	G		15	1.067		118	59.50	6.55	184.05	234

03 21 05.10 Rebar Accessories		Crew	Daily Output	Labor-Hours	Unit	Material	2018 Bare Costs Labor	Equipment	Total	Total Incl O&P
0010	**REBAR ACCESSORIES**									
0030	Steel & plastic made from recycled materials									
0100	Beam bolsters (BB), lower, 1-1/2" high, plain steel	G				C.L.F.	35			35	38.50
0102	Galvanized	G					42			42	46
0104	Stainless tipped legs	G					460			460	505
0106	Plastic tipped legs	G					51			51	56
0108	Epoxy dipped	G					88			88	97
0110	2" high, plain	G					44			44	48.50
0120	Galvanized	G					53			53	58
0140	Stainless tipped legs	G					470			470	515
0160	Plastic tipped legs	G					60			60	66
0162	Epoxy dipped	G					100			100	110
0200	Upper (BBU), 1-1/2" high, plain steel	G					85			85	93.50
0210	3" high	G					96			96	106
0500	Slab bolsters, continuous (SB), 1" high, plain steel	G					30			30	33
0502	Galvanized	G					36			36	39.50
0504	Stainless tipped legs	G					455			455	500
0506	Plastic tipped legs	G					44			44	48.50
0510	2" high, plain steel	G					38			38	42
0515	Galvanized	G					45.50			45.50	50
0520	Stainless tipped legs	G					465			465	510
0525	Plastic tipped legs	G					51			51	56
0530	For bolsters with wire runners (SBR), add	G					41			41	45
0540	For bolsters with plates (SBP), add	G				↓	97			97	107
0700	Bag ties, 16 ga., plain, 4" long	G				C	4			4	4.40
0710	5" long	G					5			5	5.50
0720	6" long	G					4			4	4.40
0730	7" long	G					5			5	5.50
1200	High chairs, individual (HC), 3" high, plain steel	G					62			62	68
1202	Galvanized	G					74.50			74.50	82
1204	Stainless tipped legs	G					485			485	535
1206	Plastic tipped legs	G					67			67	73.50
1210	5" high, plain	G					90			90	99
1212	Galvanized	G					108			108	119
1214	Stainless tipped legs	G					515			515	565
1216	Plastic tipped legs	G					99			99	109
1220	8" high, plain	G					132			132	145
1222	Galvanized	G					158			158	174
1224	Stainless tipped legs	G					555			555	615
1226	Plastic tipped legs	G					146			146	161
1230	12" high, plain	G					315			315	345
1232	Galvanized	G					380			380	415
1234	Stainless tipped legs	G					740			740	815
1236	Plastic tipped legs	G					345			345	380
1400	Individual high chairs, with plate (HCP), 5" high	G					188			188	207
1410	8" high	G					260			260	286
1500	Bar chair (BC), 1-1/2" high, plain steel	G					40			40	44
1520	Galvanized	G					45			45	49.50
1530	Stainless tipped legs	G					465			465	510
1540	Plastic tipped legs	G				↓	43			43	47.50
1700	Continuous high chairs (CHC), legs 8" OC, 4" high, plain steel	G				C.L.F.	54			54	59.50
1705	Galvanized	G					65			65	71.50
1710	Stainless tipped legs	G				↓	480			480	525

03 21 05 – Reinforcing Steel Accessories

03 21 05.10 Rebar Accessories		Crew	Daily Output	Labor-Hours	Unit	Material	2018 Bare Costs Labor	Equipment	Total	Total Incl O&P
1715	Plastic tipped legs	G				C.L.F.	72			72	79
1718	Epoxy dipped	G					93			93	102
1720	6" high, plain	G					74			74	81.50
1725	Galvanized	G					89			89	97.50
1730	Stainless tipped legs	G					500			500	550
1735	Plastic tipped legs	G					99			99	109
1738	Epoxy dipped	G					127			127	140
1740	8" high, plain	G					105			105	116
1745	Galvanized	G					126			126	139
1750	Stainless tipped legs	G					530			530	585
1755	Plastic tipped legs	G					125			125	137
1758	Epoxy dipped	G					161			161	177
1900	For continuous bottom wire runners, add	G					29			29	32
1940	For continuous bottom plate, add	G					199			199	219
2200	Screed chair base, 1/2" coil thread diam., 2-1/2" high, plain steel	G				C	335			335	370
2210	Galvanized	G					405			405	445
2220	5-1/2" high, plain	G					405			405	445
2250	Galvanized	G					485			485	535
2300	3/4" coil thread diam., 2-1/2" high, plain steel	G					420			420	465
2310	Galvanized	G					505			505	555
2320	5-1/2" high, plain steel	G					520			520	570
2350	Galvanized	G					625			625	685
2400	Screed holder, 1/2" coil thread diam. for pipe screed, plain steel, 6" long	G					405			405	445
2420	12" long	G					625			625	685
2500	3/4" coil thread diam. for pipe screed, plain steel, 6" long	G					585			585	640
2520	12" long	G					935			935	1,025
2700	Screw anchor for bolts, plain steel, 3/4" diameter x 4" long	G					565			565	620
2720	1" diameter x 6" long	G					930			930	1,025
2740	1-1/2" diameter x 8" long	G					1,175			1,175	1,300
2800	Screw anchor eye bolts, 3/4" x 3" long	G					2,950			2,950	3,250
2820	1" x 3-1/2" long	G					3,950			3,950	4,350
2840	1-1/2" x 6" long	G					12,100			12,100	13,300
2900	Screw anchor bolts, 3/4" x 9" long	G					1,700			1,700	1,850
2920	1" x 12" long	G					3,250			3,250	3,575
3001	Slab lifting inserts, single pickup, galv, 3/4" diam., 5" high	G					1,825			1,825	2,000
3010	6" high	G					1,850			1,850	2,025
3030	7" high	G					1,875			1,875	2,075
3100	1" diameter, 5-1/2" high	G					1,925			1,925	2,125
3120	7" high	G					1,975			1,975	2,175
3200	Double pickup lifting inserts, 1" diameter, 5-1/2" high	G					3,700			3,700	4,075
3220	7" high	G					4,100			4,100	4,525
3330	1-1/2" diameter, 8" high	G					5,175			5,175	5,675
3800	Subgrade chairs, #4 bar head, 3-1/2" high	G					40			40	44
3850	12" high	G					47			47	51.50
3900	#6 bar head, 3-1/2" high	G					40			40	44
3950	12" high	G					47			47	51.50
4200	Subgrade stakes, no nail holes, 3/4" diameter, 12" long	G					145			145	160
4250	24" long	G					240			240	264
4300	7/8" diameter, 12" long	G					390			390	430
4350	24" long	G					660			660	725
4500	Tie wire, 16 ga. annealed steel	G				Cwt.	170			170	187

03 21 05.75 Splicing Reinforcing Bars		Crew	Daily Output	Labor-Hours	Unit	Material	2018 Bare Costs Labor	Equipment	Total	Total Incl O&P
0010	**SPLICING REINFORCING BARS**	R032110-70								
0020	Including holding bars in place while splicing									
0100	Standard, self-aligning type, taper threaded, #4 bars	G	C-25	190	.168	Ea.	7.25	7.35		14.60	19.70
0105	#5 bars	G		170	.188		9.05	8.25		17.30	23
0110	#6 bars	G		150	.213		10.20	9.35		19.55	26
0120	#7 bars	G		130	.246		11.80	10.75		22.55	30
0300	#8 bars	G		115	.278		19.55	12.15		31.70	41
0305	#9 bars	G	C-5	105	.533		23	29	5.65	57.65	75
0310	#10 bars	G		95	.589		25.50	32	6.20	63.70	83.50
0320	#11 bars	G		85	.659		27	36	6.95	69.95	91.50
0330	#14 bars	G		65	.862		35	47	9.10	91.10	120
0340	#18 bars	G		45	1.244		53.50	67.50	13.15	134.15	176
0500	Transition self-aligning, taper threaded, #18-14	G		45	1.244		71	67.50	13.15	151.65	195
0510	#18-11	G		45	1.244		72	67.50	13.15	152.65	197
0520	#14-11	G		65	.862		47.50	47	9.10	103.60	134
0540	#11-10	G		85	.659		33	36	6.95	75.95	98.50
0550	#10-9	G		95	.589		31.50	32	6.20	69.70	90
0560	#9-8	G	C-25	105	.305		28.50	13.35		41.85	52.50
0580	#8-7	G		115	.278		26.50	12.15		38.65	48.50
0590	#7-6	G		130	.246		18.30	10.75		29.05	37
0600	Position coupler for curved bars, taper threaded, #4 bars	G		160	.200		34.50	8.75		43.25	52
0610	#5 bars	G		145	.221		36.50	9.65		46.15	55.50
0620	#6 bars	G		130	.246		44	10.75		54.75	65
0630	#7 bars	G		110	.291		46.50	12.75		59.25	71.50
0640	#8 bars	G		100	.320		48.50	14		62.50	75
0650	#9 bars	G	C-5	90	.622		53	34	6.55	93.55	117
0660	#10 bars	G		80	.700		57	38	7.40	102.40	129
0670	#11 bars	G		70	.800		59.50	43.50	8.45	111.45	140
0680	#14 bars	G		55	1.018		74	55.50	10.75	140.25	177
0690	#18 bars	G		40	1.400		91.50	76	14.75	182.25	233
0700	Transition position coupler for curved bars, taper threaded, #18-14	G		40	1.400		121	76	14.75	211.75	265
0710	#18-11	G		40	1.400		122	76	14.75	212.75	267
0720	#14-11	G		55	1.018		85	55.50	10.75	151.25	189
0730	#11-10	G		70	.800		69	43.50	8.45	120.95	151
0740	#10-9	G		80	.700		66	38	7.40	111.40	139
0750	#9-8	G	C-25	90	.356		61.50	15.55		77.05	92.50
0760	#8-7	G		100	.320		56.50	14		70.50	84
0770	#7-6	G		110	.291		55	12.75		67.75	80.50
0800	Sleeve type w/grout filler, for precast concrete, #6 bars	G		72	.444		29	19.45		48.45	62.50
0802	#7 bars	G		64	.500		34.50	22		56.50	72
0805	#8 bars	G		56	.571		41	25		66	84.50
0807	#9 bars	G		48	.667		47	29		76	98.50
0810	#10 bars	G	C-5	40	1.400		57.50	76	14.75	148.25	195
0900	#11 bars	G		32	1.750		63.50	95	18.45	176.95	236
0920	#14 bars	G		24	2.333		98.50	127	24.50	250	330
1000	Sleeve type w/ferrous filler, for critical structures, #6 bars	G	C-25	72	.444		91	19.45		110.45	131
1210	#7 bars	G		64	.500		92.50	22		114.50	137
1220	#8 bars	G		56	.571		97	25		122	147
1230	#9 bars	G	C-5	48	1.167		101	63.50	12.30	176.80	221
1240	#10 bars	G		40	1.400		106	76	14.75	196.75	249
1250	#11 bars	G		32	1.750		129	95	18.45	242.45	305
1260	#14 bars	G		24	2.333		161	127	24.50	312.50	395

For customer support on your Building Construction Costs with RSMeans data, call 800.448.8182.

03 21 Reinforcement Bars

03 21 05 – Reinforcing Steel Accessories

03 21 05.75 Splicing Reinforcing Bars

			Crew	Daily Output	Labor-Hours	Unit	Material	2018 Bare Costs Labor	Equipment	Total	Total Incl O&P
1270	#18 bars	G	C-5	16	3.500	Ea.	233	190	37	460	585
2000	Weldable half coupler, taper threaded, #4 bars	G	E-16	120	.133		11.35	7.40	.82	19.57	25.50
2100	#5 bars	G		112	.143		13.35	7.95	.88	22.18	28.50
2200	#6 bars	G		104	.154		21	8.55	.95	30.50	38.50
2300	#7 bars	G		96	.167		24.50	9.30	1.03	34.83	43.50
2400	#8 bars	G		88	.182		25.50	10.10	1.12	36.72	46.50
2500	#9 bars	G		80	.200		28.50	11.15	1.23	40.88	50.50
2600	#10 bars	G		72	.222		29	12.35	1.37	42.72	53.50
2700	#11 bars	G		64	.250		31	13.90	1.54	46.44	58.50
2800	#14 bars	G		56	.286		35.50	15.90	1.76	53.16	67.50
2900	#18 bars	G		48	.333		58	18.55	2.05	78.60	97

03 21 11 – Plain Steel Reinforcement Bars

03 21 11.60 Reinforcing In Place

			Crew	Daily Output	Labor-Hours	Unit	Material	2018 Bare Costs Labor	Equipment	Total	Total Incl O&P
0010	**REINFORCING IN PLACE**, 50-60 ton lots, A615 Grade 60 R032110-10										
0020	Includes labor, but not material cost, to install accessories										
0030	Made from recycled materials										
0100	Beams & Girders, #3 to #7	G	4 Rodm	1.60	20	Ton	960	1,100		2,060	2,725
0150	#8 to #18 R032110-20	G		2.70	11.852		960	650		1,610	2,050
0200	Columns, #3 to #7	G		1.50	21.333		960	1,175		2,135	2,825
0250	#8 to #18	G		2.30	13.913		960	760		1,720	2,200
0300	Spirals, hot rolled, 8" to 15" diameter	G		2.20	14.545		1,575	795		2,370	2,950
0320	15" to 24" diameter R032110-40	G		2.20	14.545		1,500	795		2,295	2,875
0330	24" to 36" diameter	G		2.30	13.913		1,425	760		2,185	2,725
0340	36" to 48" diameter R032110-50	G		2.40	13.333		1,350	730		2,080	2,625
0360	48" to 64" diameter	G		2.50	12.800		1,500	700		2,200	2,725
0380	64" to 84" diameter R032110-70	G		2.60	12.308		1,575	675		2,250	2,750
0390	84" to 96" diameter	G		2.70	11.852		1,650	650		2,300	2,800
0400	Elevated slabs, #4 to #7 R032110-80	G		2.90	11.034		960	605		1,565	1,975
0500	Footings, #4 to #7	G		2.10	15.238		960	835		1,795	2,325
0550	#8 to #18	G		3.60	8.889		960	485		1,445	1,800
0600	Slab on grade, #3 to #7	G		2.30	13.913		960	760		1,720	2,200
0700	Walls, #3 to #7	G		3	10.667		960	585		1,545	1,950
0750	#8 to #18	G		4	8		960	435		1,395	1,725
0900	For other than 50-60 ton lots										
1000	Under 10 ton job, #3 to #7, add						25%	10%			
1010	#8 to #18, add						20%	10%			
1050	10-50 ton job, #3 to #7, add						10%				
1060	#8 to #18, add						5%				
1100	60-100 ton job, #3 to #7, deduct						5%				
1110	#8 to #18, deduct						10%				
1150	Over 100 ton job, #3 to #7, deduct						10%				
1160	#8 to #18, deduct						15%				
1200	Reinforcing in place, A615 Grade 75, add	G				Ton	91.50			91.50	101
1220	Grade 90, add						122			122	134
2000	Unloading & sorting, add to above		C-5	100	.560			30.50	5.90	36.40	53
2200	Crane cost for handling, 90 picks/day, up to 1.5 tons/bundle, add to above			135	.415			22.50	4.38	26.88	39.50
2210	1.0 ton/bundle			92	.609			33	6.40	39.40	57.50
2220	0.5 ton/bundle			35	1.600			87	16.90	103.90	151
2400	Dowels, 2 feet long, deformed, #3	G	2 Rodm	520	.031	Ea.	.40	1.68		2.08	3.01
2410	#4	G		480	.033		.70	1.82		2.52	3.55
2420	#5	G		435	.037		1.10	2.01		3.11	4.28
2430	#6	G		360	.044		1.58	2.43		4.01	5.45

For customer support on your Building Construction Costs with RSMeans data, call 800.448.8182.

71

03 21 Reinforcement Bars

03 21 11 – Plain Steel Reinforcement Bars

03 21 11.60 Reinforcing In Place

		Crew	Daily Output	Labor-Hours	Unit	Material	2018 Bare Costs Labor	Equipment	Total	Total Incl O&P
2450	Longer and heavier dowels, add	G 2 Rodm	725	.022	Lb.	.53	1.21		1.74	2.42
2500	Smooth dowels, 12" long, 1/4" or 3/8" diameter	G	140	.114	Ea.	.73	6.25		6.98	10.35
2520	5/8" diameter	G	125	.128		1.28	7		8.28	12.10
2530	3/4" diameter	G	110	.145		1.58	7.95		9.53	13.90
2600	Dowel sleeves for CIP concrete, 2-part system									
2610	Sleeve base, plastic, for 5/8" smooth dowel sleeve, fasten to edge form	1 Rodm	200	.040	Ea.	.53	2.19		2.72	3.92
2615	Sleeve, plastic, 12" long, for 5/8" smooth dowel, snap onto base		400	.020		1.33	1.09		2.42	3.13
2620	Sleeve base, for 3/4" smooth dowel sleeve		175	.046		.53	2.50		3.03	4.39
2625	Sleeve, 12" long, for 3/4" smooth dowel		350	.023		1.21	1.25		2.46	3.24
2630	Sleeve base, for 1" smooth dowel sleeve		150	.053		.68	2.91		3.59	5.20
2635	Sleeve, 12" long, for 1" smooth dowel		300	.027		1.42	1.46		2.88	3.79
2700	Dowel caps, visual warning only, plastic, #3 to #8	2 Rodm	800	.020		.30	1.09		1.39	2
2720	#8 to #18		750	.021		.74	1.17		1.91	2.60
2750	Impalement protective, plastic, #4 to #9		800	.020		1.08	1.09		2.17	2.86

03 21 13 – Galvanized Reinforcement Steel Bars

03 21 13.10 Galvanized Reinforcing

		Crew	Daily Output	Labor-Hours	Unit	Material	2018 Bare Costs Labor	Equipment	Total	Total Incl O&P
0010	**GALVANIZED REINFORCING**									
0150	Add to plain steel rebar pricing for galvanized rebar				Ton	460			460	505

03 21 16 – Epoxy-Coated Reinforcement Steel Bars

03 21 16.10 Epoxy-Coated Reinforcing

		Crew	Daily Output	Labor-Hours	Unit	Material	2018 Bare Costs Labor	Equipment	Total	Total Incl O&P
0010	**EPOXY-COATED REINFORCING**									
0100	Add to plain steel rebar pricing for epoxy-coated rebar				Ton	785			785	865

03 21 21 – Composite Reinforcement Bars

03 21 21.11 Glass Fiber-Reinforced Polymer Reinf. Bars

		Crew	Daily Output	Labor-Hours	Unit	Material	2018 Bare Costs Labor	Equipment	Total	Total Incl O&P
0010	**GLASS FIBER-REINFORCED POLYMER REINFORCEMENT BARS**									
0020	Includes labor, but not material cost, to install accessories									
0050	#2 bar, .043 lb./L.F.	4 Rodm	9500	.003	L.F.	.40	.18		.58	.72
0100	#3 bar, .092 lb./L.F.		9300	.003		.63	.19		.82	.98
0150	#4 bar, .160 lb./L.F.		9100	.004		.91	.19		1.10	1.29
0200	#5 bar, .258 lb./L.F.		8700	.004		1.09	.20		1.29	1.51
0250	#6 bar, .372 lb./L.F.		8300	.004		1.81	.21		2.02	2.31
0300	#7 bar, .497 lb./L.F.		7900	.004		2.37	.22		2.59	2.95
0350	#8 bar, .620 lb./L.F.		7400	.004		2.38	.24		2.62	2.98
0400	#9 bar, .800 lb./L.F.		6800	.005		3.99	.26		4.25	4.78
0450	#10 bar, 1.08 lb./L.F.		5800	.006		3.75	.30		4.05	4.59
0500	For bends, add per bend				Ea.	1.54			1.54	1.69

03 22 Fabric and Grid Reinforcing

03 22 11 – Plain Welded Wire Fabric Reinforcing

03 22 11.10 Plain Welded Wire Fabric

		Crew	Daily Output	Labor-Hours	Unit	Material	2018 Bare Costs Labor	Equipment	Total	Total Incl O&P
0010	**PLAIN WELDED WIRE FABRIC** ASTM A185 R032205-30									
0020	Includes labor, but not material cost, to install accessories									
0030	Made from recycled materials									
0050	Sheets									
0100	6 x 6 - W1.4 x W1.4 (10 x 10) 21 lb./C.S.F.	G 2 Rodm	35	.457	C.S.F.	15.10	25		40.10	54.50
0200	6 x 6 - W2.1 x W2.1 (8 x 8) 30 lb./C.S.F.	G	31	.516		19.60	28		47.60	64.50
0300	6 x 6 - W2.9 x W2.9 (6 x 6) 42 lb./C.S.F.	G	29	.552		25.50	30		55.50	74
0400	6 x 6 - W4 x W4 (4 x 4) 58 lb./C.S.F.	G	27	.593		35	32.50		67.50	88
0500	4 x 4 - W1.4 x W1.4 (10 x 10) 31 lb./C.S.F.	G	31	.516		22.50	28		50.50	68

03 22 Fabric and Grid Reinforcing

03 22 11 – Plain Welded Wire Fabric Reinforcing

03 22 11.10 Plain Welded Wire Fabric		Crew	Daily Output	Labor-Hours	Unit	Material	2018 Bare Costs Labor	Equipment	Total	Total Incl O&P
0600	4 x 4 - W2.1 x W2.1 (8 x 8) 44 lb./C.S.F.	G 2 Rodm	29	.552	C.S.F.	28	30		58	77
0650	4 x 4 - W2.9 x W2.9 (6 x 6) 61 lb./C.S.F.	G	27	.593		45	32.50		77.50	99
0700	4 x 4 - W4 x W4 (4 x 4) 85 lb./C.S.F.	G	25	.640		56.50	35		91.50	116
0750	Rolls									
0800	2 x 2 - #14 galv., 21 lb./C.S.F., beam & column wrap	G 2 Rodm	6.50	2.462	C.S.F.	46.50	135		181.50	256
0900	2 x 2 - #12 galv. for gunite reinforcing	G "	6.50	2.462	"	70	135		205	282

03 22 13 – Galvanized Welded Wire Fabric Reinforcing

03 22 13.10 Galvanized Welded Wire Fabric

		Crew	Daily Output	Labor-Hours	Unit	Material	2018 Bare Costs Labor	Equipment	Total	Total Incl O&P
0010	**GALVANIZED WELDED WIRE FABRIC**									
0100	Add to plain welded wire pricing for galvanized welded wire				Lb.	.23			.23	.25

03 22 16 – Epoxy-Coated Welded Wire Fabric Reinforcing

03 22 16.10 Epoxy-Coated Welded Wire Fabric

		Crew	Daily Output	Labor-Hours	Unit	Material	2018 Bare Costs Labor	Equipment	Total	Total Incl O&P
0010	**EPOXY-COATED WELDED WIRE FABRIC**									
0100	Add to plain welded wire pricing for epoxy-coated welded wire				Lb.	.39			.39	.43

03 23 Stressed Tendon Reinforcing

03 23 05 – Prestressing Tendons

03 23 05.50 Prestressing Steel

03 23 05.50 Prestressing Steel			Crew	Daily Output	Labor-Hours	Unit	Material	2018 Bare Costs Labor	Equipment	Total	Total Incl O&P
0010	**PRESTRESSING STEEL**	R034136-90									
0100	Grouted strand, in beams, post-tensioned in field, 50' span, 100 kip	G	C-3	1200	.053	Lb.	2.85	2.71	.09	5.65	7.35
0150	300 kip	G		2700	.024		1.18	1.20	.04	2.42	3.18
0300	100' span, 100 kip	G		1700	.038		2.85	1.91	.07	4.83	6.10
0350	300 kip	G		3200	.020		2.44	1.02	.03	3.49	4.28
0500	200' span, 100 kip	G		2700	.024		2.85	1.20	.04	4.09	5
0550	300 kip	G		3500	.018		2.44	.93	.03	3.40	4.13
0800	Grouted bars, in beams, 50' span, 42 kip	G		2600	.025		1.10	1.25	.04	2.39	3.17
0850	143 kip	G		3200	.020		1.05	1.02	.03	2.10	2.75
1000	75' span, 42 kip	G		3200	.020		1.12	1.02	.03	2.17	2.82
1050	143 kip	G		4200	.015		.93	.77	.03	1.73	2.23
1200	Ungrouted strand, in beams, 50' span, 100 kip	G	C-4	1275	.025		.62	1.38	.02	2.02	2.82
1250	300 kip	G		1475	.022		.62	1.20	.02	1.84	2.53
1400	100' span, 100 kip	G		1500	.021		.62	1.18	.02	1.82	2.50
1450	300 kip	G		1650	.019		.62	1.07	.02	1.71	2.33
1600	200' span, 100 kip	G		1500	.021		.62	1.18	.02	1.82	2.50
1650	300 kip	G		1700	.019		.62	1.04	.02	1.68	2.28
1800	Ungrouted bars, in beams, 50' span, 42 kip	G		1400	.023		.48	1.26	.02	1.76	2.48
1850	143 kip	G		1700	.019		.48	1.04	.02	1.54	2.13
2000	75' span, 42 kip	G		1800	.018		.48	.98	.02	1.48	2.05
2050	143 kip	G		2200	.015		.48	.80	.01	1.29	1.78
2220	Ungrouted single strand, 100' elevated slab, 25 kip	G		1200	.027		.62	1.47	.03	2.12	2.96
2250	35 kip	G		1475	.022		.62	1.20	.02	1.84	2.53
3000	Slabs on grade, 0.5-inch diam. non-bonded strands, HDPE sheathed,										
3050	attached dead-end anchors, loose stressing-end anchors										
3100	25' x 30' slab, strands @ 36" OC, placing		2 Rodm	2940	.005	S.F.	.65	.30		.95	1.16
3105	Stressing		C-4A	3750	.004			.23	.01	.24	.37
3110	42" OC, placing		2 Rodm	3200	.005		.57	.27		.84	1.05
3115	Stressing		C-4A	4040	.004			.22	.01	.23	.34
3120	48" OC, placing		2 Rodm	3510	.005		.50	.25		.75	.93
3125	Stressing		C-4A	4390	.004			.20	.01	.21	.31
3150	25' x 40' slab, strands @ 36" OC, placing		2 Rodm	3370	.005		.63	.26		.89	1.09

For customer support on your Building Construction Costs with RSMeans data, call 800.448.8182.

73

03 23 Stressed Tendon Reinforcing

03 23 05 – Prestressing Tendons

03 23 05.50 Prestressing Steel

		Crew	Daily Output	Labor-Hours	Unit	Material	2018 Bare Costs Labor	Equipment	Total	Total Incl O&P
3155	Stressing	C-4A	4360	.004	S.F.		.20	.01	.21	.32
3160	42" OC, placing	2 Rodm	3760	.004		.54	.23		.77	.95
3165	Stressing	C-4A	4820	.003			.18	.01	.19	.29
3170	48" OC, placing	2 Rodm	4090	.004		.48	.21		.69	.86
3175	Stressing	C-4A	5190	.003			.17	.01	.18	.27
3200	30' x 30' slab, strands @ 36" OC, placing	2 Rodm	3260	.005		.62	.27		.89	1.09
3205	Stressing	C-4A	4190	.004			.21	.01	.22	.33
3210	42" OC, placing	2 Rodm	3530	.005		.56	.25		.81	1
3215	Stressing	C-4A	4500	.004			.19	.01	.20	.31
3220	48" OC, placing	2 Rodm	3840	.004		.50	.23		.73	.90
3225	Stressing	C-4A	4850	.003			.18	.01	.19	.29
3230	30' x 40' slab, strands @ 36" OC, placing	2 Rodm	3780	.004		.60	.23		.83	1.01
3235	Stressing	C-4A	4920	.003			.18	.01	.19	.28
3240	42" OC, placing	2 Rodm	4190	.004		.52	.21		.73	.90
3245	Stressing	C-4A	5410	.003			.16	.01	.17	.26
3250	48" OC, placing	2 Rodm	4520	.004		.47	.19		.66	.82
3255	Stressing	C-4A	5790	.003			.15	.01	.16	.24
3260	30' x 50' slab, strands @ 36" OC, placing	2 Rodm	4300	.004		.57	.20		.77	.93
3265	Stressing	C-4A	5650	.003			.15	.01	.16	.25
3270	42" OC, placing	2 Rodm	4720	.003		.50	.19		.69	.83
3275	Stressing	C-4A	6150	.003			.14	.01	.15	.23
3280	48" OC, placing	2 Rodm	5240	.003		.44	.17		.61	.74
3285	Stressing	C-4A	6760	.002			.13	.01	.14	.21

03 24 Fibrous Reinforcing

03 24 05 – Reinforcing Fibers

03 24 05.30 Synthetic Fibers

					Unit	Material	Labor	Equipment	Total	Total Incl O&P
0010	**SYNTHETIC FIBERS**									
0100	Synthetic fibers, add to concrete				Lb.	4.71			4.71	5.20
0110	1-1/2 lb./C.Y.				C.Y.	7.30			7.30	8

03 24 05.70 Steel Fibers

						Unit	Material	Labor	Equipment	Total	Total Incl O&P
0010	**STEEL FIBERS**										
0140	ASTM A850, Type V, continuously deformed, 1-1/2" long x 0.045" diam.										
0150	Add to price of ready mix concrete	G				Lb.	1.22			1.22	1.34
0205	Alternate pricing, dosing at 5 lb./C.Y., add to price of RMC	G				C.Y.	6.10			6.10	6.70
0210	10 lb./C.Y.	G					12.20			12.20	13.40
0215	15 lb./C.Y.	G					18.30			18.30	20
0220	20 lb./C.Y.	G					24.50			24.50	27
0225	25 lb./C.Y.	G					30.50			30.50	33.50
0230	30 lb./C.Y.	G					36.50			36.50	40.50
0235	35 lb./C.Y.	G					42.50			42.50	47
0240	40 lb./C.Y.	G					49			49	53.50
0250	50 lb./C.Y.	G					61			61	67
0275	75 lb./C.Y.	G					91.50			91.50	101
0300	100 lb./C.Y.	G					122			122	134

03 30 53 – Miscellaneous Cast-In-Place Concrete

03 30 53.40 Concrete In Place		Crew	Daily Output	Labor-Hours	Unit	Material	2018 Bare Costs Labor	Equipment	Total	Total Incl O&P
0010	**CONCRETE IN PLACE**	R033105-10								
0020	Including forms (4 uses), Grade 60 rebar, concrete (Portland cement	R033105-20								
0050	Type I), placement and finishing unless otherwise indicated	R033105-50								
0300	Beams (3500 psi), 5 kip/L.F., 10' span	R033105-65	C-14A	15.62	12.804	C.Y.	340	645	58	1,043	1,425
0350	25' span	R033105-70	"	18.55	10.782		355	545	49	949	1,275
0500	Chimney foundations (5000 psi), over 5 C.Y.	R033105-85	C-14C	32.22	3.476		171	167	.80	338.80	445
0510	(3500 psi), under 5 C.Y.	R033053-50	"	23.71	4.724		198	227	1.09	426.09	565
0700	Columns, square (4000 psi), 12" x 12", up to 1% reinforcing by area		C-14A	11.96	16.722		380	845	76	1,301	1,775
0720	Up to 2% reinforcing by area			10.13	19.743		580	1,000	89.50	1,669.50	2,250
0740	Up to 3% reinforcing by area			9.03	22.148		850	1,125	101	2,076	2,750
0800	16" x 16", up to 1% reinforcing by area			16.22	12.330		305	625	56	986	1,350
0820	Up to 2% reinforcing by area			12.57	15.911		495	805	72	1,372	1,850
0840	Up to 3% reinforcing by area			10.25	19.512		750	985	88.50	1,823.50	2,425
0900	24" x 24", up to 1% reinforcing by area			23.66	8.453		261	425	38.50	724.50	980
0920	Up to 2% reinforcing by area			17.71	11.293		440	570	51.50	1,061.50	1,400
0940	Up to 3% reinforcing by area			14.15	14.134		685	715	64	1,464	1,900
1000	36" x 36", up to 1% reinforcing by area			33.69	5.936		232	300	27	559	740
1020	Up to 2% reinforcing by area			23.32	8.576		390	435	39	864	1,125
1040	Up to 3% reinforcing by area			17.82	11.223		640	565	51	1,256	1,625
1100	Columns, round (4000 psi), tied, 12" diameter, up to 1% reinforcing by area			20.97	9.537		380	480	43.50	903.50	1,200
1120	Up to 2% reinforcing by area			15.27	13.098		575	660	59.50	1,294.50	1,700
1140	Up to 3% reinforcing by area			12.11	16.515		835	835	75	1,745	2,275
1200	16" diameter, up to 1% reinforcing by area			31.49	6.351		320	320	29	669	870
1220	Up to 2% reinforcing by area			19.12	10.460		520	530	47.50	1,097.50	1,425
1240	Up to 3% reinforcing by area			13.77	14.524		755	735	66	1,556	2,025
1300	20" diameter, up to 1% reinforcing by area			41.04	4.873		310	246	22	578	740
1320	Up to 2% reinforcing by area			24.05	8.316		495	420	38	953	1,225
1340	Up to 3% reinforcing by area			17.01	11.758		750	595	53.50	1,398.50	1,800
1400	24" diameter, up to 1% reinforcing by area			51.85	3.857		290	195	17.50	502.50	635
1420	Up to 2% reinforcing by area			27.06	7.391		490	375	33.50	898.50	1,150
1440	Up to 3% reinforcing by area			18.29	10.935		730	555	49.50	1,334.50	1,700
1500	36" diameter, up to 1% reinforcing by area			75.04	2.665		296	135	12.10	443.10	545
1520	Up to 2% reinforcing by area			37.49	5.335		475	270	24	769	955
1540	Up to 3% reinforcing by area			22.84	8.757		720	445	40	1,205	1,500
1900	Elevated slab (4000 psi), flat slab with drops, 125 psf Sup. Load, 20' span		C-14B	38.45	5.410		284	273	23.50	580.50	750
1950	30' span			50.99	4.079		298	206	17.80	521.80	660
2100	Flat plate, 125 psf Sup. Load, 15' span			30.24	6.878		261	345	30	636	845
2150	25' span			49.60	4.194		270	211	18.30	499.30	635
2300	Waffle const., 30" domes, 125 psf Sup. Load, 20' span			37.07	5.611		297	283	24.50	604.50	780
2350	30' span			44.07	4.720		275	238	20.50	533.50	690
2500	One way joists, 30" pans, 125 psf Sup. Load, 15' span			27.38	7.597		400	385	33	818	1,050
2550	25' span			31.15	6.677		365	335	29	729	945
2700	One way beam & slab, 125 psf Sup. Load, 15' span			20.59	10.102		279	510	44	833	1,125
2750	25' span			28.36	7.334		265	370	32	667	885
2900	Two way beam & slab, 125 psf Sup. Load, 15' span			24.04	8.652		270	435	37.50	742.50	1,000
2950	25' span			35.87	5.799		236	292	25.50	553.50	730
3100	Elevated slabs, flat plate, including finish, not										
3110	including forms or reinforcing										
3150	Regular concrete (4000 psi), 4" slab		C-8	2613	.021	S.F.	1.70	.95	.34	2.99	3.67
3200	6" slab			2585	.022		2.48	.96	.34	3.78	4.56
3250	2-1/2" thick floor fill			2685	.021		1.11	.92	.33	2.36	2.97
3300	Lightweight, 110 #/C.F., 2-1/2" thick floor fill			2585	.022		1.14	.96	.34	2.44	3.09
3400	Cellular concrete, 1-5/8" fill, under 5000 S.F.			2000	.028		.79	1.24	.44	2.47	3.22

For customer support on your Building Construction Costs with RSMeans data, call 800.448.8182.

75

03 30 Cast-In-Place Concrete

03 30 53 – Miscellaneous Cast-In-Place Concrete

03 30 53.40 Concrete In Place	Crew	Daily Output	Labor-Hours	Unit	Material	2018 Bare Costs Labor	2018 Bare Costs Equipment	Total	Total Incl O&P
3450	Over 10,000 S.F.	C-8	2200	.025	S.F.	.75	1.13	.40	2.28	2.97
3500	Add per floor for 3 to 6 stories high		31800	.002			.08	.03	.11	.15
3520	For 7 to 20 stories high		21200	.003			.12	.04	.16	.23
3540	Equipment pad (3000 psi), 3' x 3' x 6" thick	C-14H	45	1.067	Ea.	44.50	52.50	.57	97.57	130
3550	4' x 4' x 6" thick		30	1.600		69	79	.85	148.85	197
3560	5' x 5' x 8" thick		18	2.667		126	132	1.41	259.41	340
3570	6' x 6' x 8" thick		14	3.429		173	169	1.82	343.82	450
3580	8' x 8' x 10" thick		8	6		370	296	3.18	669.18	860
3590	10' x 10' x 12" thick		5	9.600		645	475	5.10	1,125.10	1,425
3800	Footings (3000 psi), spread under 1 C.Y.	C-14C	28	4	C.Y.	185	192	.92	377.92	495
3825	1 C.Y. to 5 C.Y.		43	2.605		219	125	.60	344.60	430
3850	Over 5 C.Y.		75	1.493		203	72	.34	275.34	330
3900	Footings, strip (3000 psi), 18" x 9", unreinforced	C-14L	40	2.400		146	113	.65	259.65	335
3920	18" x 9", reinforced	C-14C	35	3.200		169	154	.74	323.74	420
3925	20" x 10", unreinforced	C-14L	45	2.133		142	100	.58	242.58	310
3930	20" x 10", reinforced	C-14C	40	2.800		161	135	.64	296.64	385
3935	24" x 12", unreinforced	C-14L	55	1.745		140	82	.47	222.47	280
3940	24" x 12", reinforced	C-14C	48	2.333		159	112	.54	271.54	345
3945	36" x 12", unreinforced	C-14L	70	1.371		136	64.50	.37	200.87	248
3950	36" x 12", reinforced	C-14C	60	1.867		153	90	.43	243.43	305
4000	Foundation mat (3000 psi), under 10 C.Y.		38.67	2.896		221	139	.67	360.67	455
4050	Over 20 C.Y.		56.40	1.986		196	95.50	.46	291.96	360
4200	Wall, free-standing (3000 psi), 8" thick, 8' high	C-14D	45.83	4.364		182	219	19.80	420.80	555
4250	14' high		27.26	7.337		211	370	33.50	614.50	830
4260	12" thick, 8' high		64.32	3.109		165	156	14.10	335.10	435
4270	14' high		40.01	4.999		175	251	22.50	448.50	595
4300	15" thick, 8' high		80.02	2.499		160	126	11.35	297.35	380
4350	12' high		51.26	3.902		160	196	17.70	373.70	490
4500	18' high		48.85	4.094		178	206	18.60	402.60	530
4520	Handicap access ramp (4000 psi), railing both sides, 3' wide	C-14H	14.58	3.292	L.F.	360	163	1.74	524.74	645
4525	5' wide		12.22	3.928		370	194	2.08	566.08	705
4530	With 6" curb and rails both sides, 3' wide		8.55	5.614		370	277	2.98	649.98	835
4535	5' wide		7.31	6.566		375	325	3.48	703.48	910
4650	Slab on grade (3500 psi), not including finish, 4" thick	C-14E	60.75	1.449	C.Y.	145	71	.42	216.42	268
4700	6" thick	"	92	.957	"	140	47	.28	187.28	226
4701	Thickened slab edge (3500 psi), for slab on grade poured									
4702	monolithically with slab; depth is in addition to slab thickness;									
4703	formed vertical outside edge, earthen bottom and inside slope									
4705	8" deep x 8" wide bottom, unreinforced	C-14L	2190	.044	L.F.	3.97	2.06	.01	6.04	7.50
4710	8" x 8", reinforced	C-14C	1670	.067		6.20	3.22	.02	9.44	11.75
4715	12" deep x 12" wide bottom, unreinforced	C-14L	1800	.053		8.15	2.51	.01	10.67	12.80
4720	12" x 12", reinforced	C-14C	1310	.086		12.25	4.11	.02	16.38	19.75
4725	16" deep x 16" wide bottom, unreinforced	C-14L	1440	.067		13.80	3.13	.02	16.95	20
4730	16" x 16", reinforced	C-14C	1120	.100		18.70	4.81	.02	23.53	28
4735	20" deep x 20" wide bottom, unreinforced	C-14L	1150	.083		21	3.92	.02	24.94	29
4740	20" x 20", reinforced	C-14C	920	.122		27	5.85	.03	32.88	39
4745	24" deep x 24" wide bottom, unreinforced	C-14L	930	.103		29.50	4.85	.03	34.38	40
4750	24" x 24", reinforced	C-14C	740	.151		38	7.30	.03	45.33	52.50
4751	Slab on grade (3500 psi), incl. troweled finish, not incl. forms									
4760	or reinforcing, over 10,000 S.F., 4" thick	C-14F	3425	.021	S.F.	1.61	.95	.01	2.57	3.21
4820	6" thick		3350	.021		2.36	.97	.01	3.34	4.06
4840	8" thick		3184	.023		3.23	1.02	.01	4.26	5.10
4900	12" thick		2734	.026		4.84	1.19	.01	6.04	7.15

For customer support on your Building Construction Costs with RSMeans data, call 800.448.8182.

03 30 Cast-In-Place Concrete

03 30 53 – Miscellaneous Cast-In-Place Concrete

03 30 53.40 Concrete In Place

		Crew	Daily Output	Labor-Hours	Unit	Material	2018 Bare Costs Labor	Equipment	Total	Total Incl O&P
4950	15" thick	C-14F	2505	.029	S.F.	6.10	1.30	.01	7.41	8.65
5000	Slab on grade (3000 psi), incl. broom finish, not incl. forms									
5001	or reinforcing, 4" thick	C-14G	2873	.019	S.F.	1.57	.87	.01	2.45	3.04
5010	6" thick		2590	.022		2.46	.96	.01	3.43	4.15
5020	8" thick		2320	.024		3.20	1.08	.01	4.29	5.15
5200	Lift slab in place above the foundation, incl. forms, reinforcing,									
5210	concrete (4000 psi) and columns, over 20,000 S.F./floor	C-14B	2113	.098	S.F.	7.45	4.96	.43	12.84	16.20
5250	10,000 S.F. to 20,000 S.F./floor		1650	.126		8.10	6.35	.55	15	19.15
5300	Under 10,000 S.F./floor		1500	.139		8.75	7	.60	16.35	21
5500	Lightweight, ready mix, including screed finish only,									
5510	not including forms or reinforcing									
5550	1:4 (2500 psi) for structural roof decks	C-14B	260	.800	C.Y.	116	40.50	3.49	159.99	192
5600	1:6 (3000 psi) for ground slab with radiant heat	C-14F	92	.783		127	35.50	.28	162.78	192
5650	1:3:2 (2000 psi) with sand aggregate, roof deck	C-14B	260	.800		115	40.50	3.49	158.99	191
5700	Ground slab (2000 psi)	C-14F	107	.673		115	30.50	.24	145.74	172
5900	Pile caps (3000 psi), incl. forms and reinf., sq. or rect., under 10 C.Y.	C-14C	54.14	2.069		185	99.50	.48	284.98	355
5950	Over 10 C.Y.		75	1.493		175	72	.34	247.34	300
6000	Triangular or hexagonal, under 10 C.Y.		53	2.113		142	102	.49	244.49	315
6050	Over 10 C.Y.		85	1.318		158	63.50	.30	221.80	270
6200	Retaining walls (3000 psi), gravity, 4' high see Section 32 32	C-14D	66.20	3.021		160	152	13.70	325.70	420
6250	10' high		125	1.600		154	80.50	7.25	241.75	299
6300	Cantilever, level backfill loading, 8' high		70	2.857		171	143	12.95	326.95	420
6350	16' high		91	2.198		165	110	10	285	360
6800	Stairs (3500 psi), not including safety treads, free standing, 3'-6" wide	C-14H	83	.578	LF Nose	6.05	28.50	.31	34.86	50.50
6850	Cast on ground		125	.384	"	5.15	18.95	.20	24.30	35
7000	Stair landings, free standing		200	.240	S.F.	4.83	11.85	.13	16.81	23.50
7050	Cast on ground		475	.101	"	3.93	4.99	.05	8.97	11.95

03 31 Structural Concrete

03 31 13 – Heavyweight Structural Concrete

03 31 13.25 Concrete, Hand Mix

		Crew	Daily Output	Labor-Hours	Unit	Material	2018 Bare Costs Labor	Equipment	Total	Total Incl O&P
0010	**CONCRETE, HAND MIX** for small quantities or remote areas									
0050	Includes bulk local aggregate, bulk sand, bagged Portland									
0060	cement (Type I) and water, using gas powered cement mixer									
0125	2500 psi	C-30	135	.059	C.F.	3.80	2.36	1.19	7.35	9.10
0130	3000 psi		135	.059		4.11	2.36	1.19	7.66	9.45
0135	3500 psi		135	.059		4.29	2.36	1.19	7.84	9.65
0140	4000 psi		135	.059		4.50	2.36	1.19	8.05	9.85
0145	4500 psi		135	.059		4.75	2.36	1.19	8.30	10.10
0150	5000 psi		135	.059		5.05	2.36	1.19	8.60	10.50
0300	Using pre-bagged dry mix and wheelbarrow (80-lb. bag = 0.6 C.F.)									
0340	4000 psi	1 Clab	48	.167	C.F.	6.85	6.65		13.50	17.65

03 31 13.30 Concrete, Volumetric Site-Mixed

		Crew	Daily Output	Labor-Hours	Unit	Material	2018 Bare Costs Labor	Equipment	Total	Total Incl O&P
0010	**CONCRETE, VOLUMETRIC SITE-MIXED**									
0015	Mixed on-site in volumetric truck									
0020	Includes local aggregate, sand, Portland cement (Type I) and water									
0025	Excludes all additives and treatments									
0100	3000 psi, 1 C.Y. mixed and discharged				C.Y.	214			214	235
0110	2 C.Y.					158			158	174
0120	3 C.Y.					138			138	152
0130	4 C.Y.					125			125	137

For customer support on your Building Construction Costs with RSMeans data, call 800.448.8182.

77

03 31 Structural Concrete

03 31 13 – Heavyweight Structural Concrete

03 31 13.30 Concrete, Volumetric Site-Mixed	Crew	Daily Output	Labor-Hours	Unit	Material	2018 Bare Costs Labor	Equipment	Total	Total Incl O&P
0140 5 C.Y.				C.Y.	112			112	123
0200 For truck holding/waiting time past first 2 on-site hours, add				Hr.	93.50			93.50	103
0210 For trip charge beyond first 20 miles, each way, add				Mile	3.64			3.64	4
0220 For each additional increase of 500 psi, add				Ea.	4.17			4.17	4.59

03 31 13.35 Heavyweight Concrete, Ready Mix

	Crew	Daily Output	Labor-Hours	Unit	Material	2018 Bare Costs Labor	Equipment	Total	Total Incl O&P
0010 **HEAVYWEIGHT CONCRETE, READY MIX**, delivered R033105-10									
0012 Includes local aggregate, sand, Portland cement (Type I) and water									
0015 Excludes all additives and treatments R033105-20									
0020 2000 psi				C.Y.	115			115	127
0100 2500 psi R033105-30					119			119	130
0150 3000 psi					121			121	133
0200 3500 psi R033105-40					124			124	137
0300 4000 psi					128			128	140
0350 4500 psi R033105-50					131			131	144
0400 5000 psi					134			134	148
0411 6000 psi					138			138	152
0412 8000 psi					145			145	160
0413 10,000 psi					153			153	168
0414 12,000 psi					160			160	176
1000 For high early strength (Portland cement Type III), add					10%				
1010 For structural lightweight with regular sand, add					25%				
1300 For winter concrete (hot water), add					5.35			5.35	5.90
1410 For mid-range water reducer, add					3.59			3.59	3.95
1420 For high-range water reducer/superplasticizer, add					6.30			6.30	6.90
1430 For retarder, add					3.23			3.23	3.55
1440 For non-Chloride accelerator, add					6.40			6.40	7.05
1450 For Chloride accelerator, per 1%, add					3.86			3.86	4.25
1460 For fiber reinforcing, synthetic (1 lb./C.Y.), add					8			8	8.80
1500 For Saturday delivery, add					8.50			8.50	9.35
1510 For truck holding/waiting time past 1st hour per load, add				Hr.	105			105	116
1520 For short load (less than 4 C.Y.), add per load				Ea.	86			86	94.50
2000 For all lightweight aggregate, add				C.Y.	45%				

03 31 13.70 Placing Concrete

	Crew	Daily Output	Labor-Hours	Unit	Material	2018 Bare Costs Labor	Equipment	Total	Total Incl O&P
0010 **PLACING CONCRETE** R033105-70									
0020 Includes labor and equipment to place, level (strike off) and consolidate									
0050 Beams, elevated, small beams, pumped	C-20	60	1.067	C.Y.		45.50	15.55	61.05	86
0100 With crane and bucket	C-7	45	1.600			69.50	24	93.50	131
0200 Large beams, pumped	C-20	90	.711			30.50	10.35	40.85	57.50
0250 With crane and bucket	C-7	65	1.108			48	16.45	64.45	91
0400 Columns, square or round, 12" thick, pumped	C-20	60	1.067			45.50	15.55	61.05	86
0450 With crane and bucket	C-7	40	1.800			78	27	105	148
0600 18" thick, pumped	C-20	90	.711			30.50	10.35	40.85	57.50
0650 With crane and bucket	C-7	55	1.309			57	19.45	76.45	108
0800 24" thick, pumped	C-20	92	.696			30	10.15	40.15	56
0850 With crane and bucket	C-7	70	1.029			44.50	15.30	59.80	84.50
1000 36" thick, pumped	C-20	140	.457			19.55	6.65	26.20	37
1050 With crane and bucket	C-7	100	.720			31.50	10.70	42.20	59.50
1400 Elevated slabs, less than 6" thick, pumped	C-20	140	.457			19.55	6.65	26.20	37
1450 With crane and bucket	C-7	95	.758			33	11.25	44.25	62.50
1500 6" to 10" thick, pumped	C-20	160	.400			17.10	5.85	22.95	32.50
1550 With crane and bucket	C-7	110	.655			28.50	9.75	38.25	53.50
1600 Slabs over 10" thick, pumped	C-20	180	.356			15.20	5.20	20.40	28.50

03 31 13 – Heavyweight Structural Concrete

03 31 13.70 Placing Concrete	Crew	Daily Output	Labor-Hours	Unit	Material	2018 Bare Costs Labor	Equipment	Total	Total Incl O&P
1650	With crane and bucket	C-7	130	.554	C.Y.		24	8.25	32.25	45.50
1900	Footings, continuous, shallow, direct chute	C-6	120	.400			16.60	.43	17.03	25.50
1950	Pumped	C-20	150	.427			18.25	6.20	24.45	34.50
2000	With crane and bucket	C-7	90	.800			35	11.90	46.90	65.50
2100	Footings, continuous, deep, direct chute	C-6	140	.343			14.20	.37	14.57	22
2150	Pumped	C-20	160	.400			17.10	5.85	22.95	32.50
2200	With crane and bucket	C-7	110	.655			28.50	9.75	38.25	53.50
2400	Footings, spread, under 1 C.Y., direct chute	C-6	55	.873			36	.93	36.93	56
2450	Pumped	C-20	65	.985			42	14.35	56.35	80
2500	With crane and bucket	C-7	45	1.600			69.50	24	93.50	131
2600	Over 5 C.Y., direct chute	C-6	120	.400			16.60	.43	17.03	25.50
2650	Pumped	C-20	150	.427			18.25	6.20	24.45	34.50
2700	With crane and bucket	C-7	100	.720			31.50	10.70	42.20	59.50
2900	Foundation mats, over 20 C.Y., direct chute	C-6	350	.137			5.70	.15	5.85	8.75
2950	Pumped	C-20	400	.160			6.85	2.33	9.18	12.90
3000	With crane and bucket	C-7	300	.240			10.45	3.57	14.02	19.70
3200	Grade beams, direct chute	C-6	150	.320			13.25	.34	13.59	20.50
3250	Pumped	C-20	180	.356			15.20	5.20	20.40	28.50
3300	With crane and bucket	C-7	120	.600			26	8.90	34.90	49.50
3500	High rise, for more than 5 stories, pumped, add per story	C-20	2100	.030			1.30	.44	1.74	2.47
3510	With crane and bucket, add per story	C-7	2100	.034			1.49	.51	2	2.82
3700	Pile caps, under 5 C.Y., direct chute	C-6	90	.533			22	.57	22.57	34
3750	Pumped	C-20	110	.582			25	8.50	33.50	47
3800	With crane and bucket	C-7	80	.900			39	13.35	52.35	73.50
3850	Pile cap, 5 C.Y. to 10 C.Y., direct chute	C-6	175	.274			11.35	.29	11.64	17.55
3900	Pumped	C-20	200	.320			13.70	4.67	18.37	26
3950	With crane and bucket	C-7	150	.480			21	7.15	28.15	39.50
4000	Over 10 C.Y., direct chute	C-6	215	.223			9.25	.24	9.49	14.30
4050	Pumped	C-20	240	.267			11.40	3.89	15.29	21.50
4100	With crane and bucket	C-7	185	.389			16.90	5.80	22.70	32
4300	Slab on grade, up to 6" thick, direct chute	C-6	110	.436			18.10	.47	18.57	28
4350	Pumped	C-20	130	.492			21	7.20	28.20	40
4400	With crane and bucket	C-7	110	.655			28.50	9.75	38.25	53.50
4600	Over 6" thick, direct chute	C-6	165	.291			12.05	.31	12.36	18.65
4650	Pumped	C-20	185	.346			14.80	5.05	19.85	28
4700	With crane and bucket	C-7	145	.497			21.50	7.40	28.90	40.50
4900	Walls, 8" thick, direct chute	C-6	90	.533			22	.57	22.57	34
4950	Pumped	C-20	100	.640			27.50	9.35	36.85	52
5000	With crane and bucket	C-7	80	.900			39	13.35	52.35	73.50
5050	12" thick, direct chute	C-6	100	.480			19.90	.51	20.41	30.50
5100	Pumped	C-20	110	.582			25	8.50	33.50	47
5200	With crane and bucket	C-7	90	.800			35	11.90	46.90	65.50
5300	15" thick, direct chute	C-6	105	.457			18.95	.49	19.44	29
5350	Pumped	C-20	120	.533			23	7.80	30.80	43
5400	With crane and bucket	C-7	95	.758			33	11.25	44.25	62.50
5600	Wheeled concrete dumping, add to placing costs above									
5610	Walking cart, 50' haul, add	C-18	32	.281	C.Y.		11.25	1.81	13.06	19.15
5620	150' haul, add		24	.375			15.05	2.41	17.46	25.50
5700	250' haul, add		18	.500			20	3.21	23.21	34
5800	Riding cart, 50' haul, add	C-19	80	.113			4.51	1.22	5.73	8.20
5810	150' haul, add		60	.150			6	1.62	7.62	10.95
5900	250' haul, add		45	.200			8	2.16	10.16	14.60
6000	Concrete in-fill for pan-type metal stairs and landings. Manual placement									

03 31 Structural Concrete

03 31 13 – Heavyweight Structural Concrete

03 31 13.70 Placing Concrete

		Crew	Daily Output	Labor-Hours	Unit	Material	2018 Bare Costs Labor	2018 Bare Costs Equipment	Total	Total Incl O&P
6010	includes up to 50' horizontal haul from point of concrete discharge.									
6100	Stair pan treads, 2" deep									
6110	Flights in 1st floor level up/down from discharge point	C-8A	3200	.015	S.F.		.64		.64	.97
6120	2nd floor level		2500	.019			.82		.82	1.24
6130	3rd floor level		2000	.024			1.03		1.03	1.55
6140	4th floor level		1800	.027			1.14		1.14	1.72
6200	Intermediate stair landings, pan-type 4" deep									
6210	Flights in 1st floor level up/down from discharge point	C-8A	2000	.024	S.F.		1.03		1.03	1.55
6220	2nd floor level		1500	.032			1.37		1.37	2.06
6230	3rd floor level		1200	.040			1.71		1.71	2.58
6240	4th floor level		1000	.048			2.05		2.05	3.09

03 35 Concrete Finishing

03 35 13 – High-Tolerance Concrete Floor Finishing

03 35 13.30 Finishing Floors, High Tolerance

		Crew	Daily Output	Labor-Hours	Unit	Material	2018 Bare Costs Labor	2018 Bare Costs Equipment	Total	Total Incl O&P
0010	**FINISHING FLOORS, HIGH TOLERANCE**									
0012	Finishing of fresh concrete flatwork requires that concrete									
0013	first be placed, struck off & consolidated									
0015	Basic finishing for various unspecified flatwork									
0100	Bull float only	C-10	4000	.006	S.F.		.27		.27	.40
0125	Bull float & manual float		2000	.012			.54		.54	.81
0150	Bull float, manual float & broom finish, w/edging & joints		1850	.013			.58		.58	.87
0200	Bull float, manual float & manual steel trowel		1265	.019			.85		.85	1.27
0210	For specified Random Access Floors in ACI Classes 1, 2, 3 and 4 to achieve									
0215	Composite Overall Floor Flatness and Levelness values up to FF35/FL25									
0250	Bull float, machine float & machine trowel (walk-behind)	C-10C	1715	.014	S.F.		.63	.02	.65	.97
0300	Power screed, bull float, machine float & trowel (walk-behind)	C-10D	2400	.010			.45	.05	.50	.72
0350	Power screed, bull float, machine float & trowel (ride-on)	C-10E	4000	.006			.27	.06	.33	.47
0352	For specified Random Access Floors in ACI Classes 5, 6, 7 and 8 to achieve									
0354	Composite Overall Floor Flatness and Levelness values up to FF50/FL50									
0356	Add for two-dimensional restraightening after power float	C-10	6000	.004	S.F.		.18		.18	.27
0358	For specified Random or Defined Access Floors in ACI Class 9 to achieve									
0360	Composite Overall Floor Flatness and Levelness values up to FF100/FL100									
0362	Add for two-dimensional restraightening after bull float & power float	C-10	3000	.008	S.F.		.36		.36	.54
0364	For specified Superflat Defined Access Floors in ACI Class 9 to achieve									
0366	Minimum Floor Flatness and Levelness values of FF100/FL100									
0368	Add for 2-dim'l restraightening after bull float, power float, power trowel	C-10	2000	.012	S.F.		.54		.54	.81

03 35 16 – Heavy-Duty Concrete Floor Finishing

03 35 16.30 Finishing Floors, Heavy-Duty

		Crew	Daily Output	Labor-Hours	Unit	Material	2018 Bare Costs Labor	2018 Bare Costs Equipment	Total	Total Incl O&P
0010	**FINISHING FLOORS, HEAVY-DUTY**									
1800	Floor abrasives, dry shake on fresh concrete, .25 psf, aluminum oxide	1 Cefi	850	.009	S.F.	.55	.45		1	1.27
1850	Silicon carbide		850	.009		.83	.45		1.28	1.58
2000	Floor hardeners, dry shake, metallic, light service, .50 psf		850	.009		.43	.45		.88	1.13
2050	Medium service, .75 psf		750	.011		.65	.51		1.16	1.46
2100	Heavy service, 1.0 psf		650	.012		.86	.59		1.45	1.82
2150	Extra heavy, 1.5 psf		575	.014		1.29	.66		1.95	2.40
2300	Non-metallic, light service, .50 psf		850	.009		.16	.45		.61	.84
2350	Medium service, .75 psf		750	.011		.24	.51		.75	1.02
2400	Heavy service, 1.00 psf		650	.012		.33	.59		.92	1.23
2450	Extra heavy, 1.50 psf		575	.014		.49	.66		1.15	1.52

For customer support on your Building Construction Costs with RSMeans data, call 800.448.8182.

03 35 Concrete Finishing

03 35 16 – Heavy-Duty Concrete Floor Finishing

03 35 16.30 Finishing Floors, Heavy-Duty

		Crew	Daily Output	Labor-Hours	Unit	Material	2018 Bare Costs Labor	2018 Bare Costs Equipment	Total	Total Incl O&P
2800	Trap rock wearing surface, dry shake, for monolithic floors									
2810	2.0 psf	C-10B	1250	.032	S.F.	.02	1.37	.19	1.58	2.31
3800	Dustproofing, liquid, for cured concrete, solvent-based, 1 coat	1 Cefi	1900	.004		.16	.20		.36	.48
3850	2 coats		1300	.006		.59	.29		.88	1.08
4000	Epoxy-based, 1 coat		1500	.005		.15	.25		.40	.55
4050	2 coats		1500	.005		.30	.25		.55	.71

03 35 19 – Colored Concrete Finishing

03 35 19.30 Finishing Floors, Colored

		Crew	Daily Output	Labor-Hours	Unit	Material	2018 Bare Costs Labor	2018 Bare Costs Equipment	Total	Total Incl O&P
0010	**FINISHING FLOORS, COLORED**									
3000	Floor coloring, dry shake on fresh concrete (0.6 psf)	1 Cefi	1300	.006	S.F.	.41	.29		.70	.88
3050	(1.0 psf)	"	625	.013	"	.68	.61		1.29	1.65
3100	Colored dry shake powder only				Lb.	.68			.68	.75
3600	1/2" topping using 0.6 psf dry shake powdered color	C-10B	590	.068	S.F.	5.35	2.91	.41	8.67	10.75
3650	1.0 psf dry shake powdered color	"	590	.068	"	5.60	2.91	.41	8.92	11.05

03 35 23 – Exposed Aggregate Concrete Finishing

03 35 23.30 Finishing Floors, Exposed Aggregate

		Crew	Daily Output	Labor-Hours	Unit	Material	2018 Bare Costs Labor	2018 Bare Costs Equipment	Total	Total Incl O&P
0010	**FINISHING FLOORS, EXPOSED AGGREGATE**									
1600	Exposed local aggregate finish, seeded on fresh concrete, 3 lb./S.F.	1 Cefi	625	.013	S.F.	.20	.61		.81	1.12
1650	4 lb./S.F.	"	465	.017	"	.33	.82		1.15	1.58

03 35 29 – Tooled Concrete Finishing

03 35 29.30 Finishing Floors, Tooled

		Crew	Daily Output	Labor-Hours	Unit	Material	2018 Bare Costs Labor	2018 Bare Costs Equipment	Total	Total Incl O&P
0010	**FINISHING FLOORS, TOOLED**									
4400	Stair finish, fresh concrete, float finish	1 Cefi	275	.029	S.F.		1.38		1.38	2.05
4500	Steel trowel finish		200	.040			1.90		1.90	2.82
4600	Silicon carbide finish, dry shake on fresh concrete, .25 psf		150	.053		.55	2.54		3.09	4.37

03 35 29.60 Finishing Walls

		Crew	Daily Output	Labor-Hours	Unit	Material	2018 Bare Costs Labor	2018 Bare Costs Equipment	Total	Total Incl O&P
0010	**FINISHING WALLS**									
0020	Break ties and patch voids	1 Cefi	540	.015	S.F.	.04	.70		.74	1.09
0050	Burlap rub with grout		450	.018		.04	.85		.89	1.30
0100	Carborundum rub, dry		270	.030			1.41		1.41	2.09
0150	Wet rub		175	.046			2.17		2.17	3.22
0300	Bush hammer, green concrete	B-39	1000	.048			2.02	.23	2.25	3.33
0350	Cured concrete	"	650	.074			3.11	.36	3.47	5.10
0500	Acid etch	1 Cefi	575	.014		.13	.66		.79	1.13
0600	Float finish, 1/16" thick	"	300	.027		.39	1.27		1.66	2.30
0700	Sandblast, light penetration	E-11	1100	.029		.55	1.30	.18	2.03	2.86
0750	Heavy penetration	"	375	.085		1.09	3.80	.54	5.43	7.85
0850	Grind form fins flush	1 Clab	700	.011	L.F.		.46		.46	.69

03 35 33 – Stamped Concrete Finishing

03 35 33.50 Slab Texture Stamping

		Crew	Daily Output	Labor-Hours	Unit	Material	2018 Bare Costs Labor	2018 Bare Costs Equipment	Total	Total Incl O&P
0010	**SLAB TEXTURE STAMPING**									
0050	Stamping requires that concrete first be placed, struck off, consolidated,									
0060	bull floated and free of bleed water. Decorative stamping tasks include:									
0100	Step 1 - first application of dry shake colored hardener	1 Cefi	6400	.001	S.F.	.43	.06		.49	.56
0110	Step 2 - bull float		6400	.001			.06		.06	.09
0130	Step 3 - second application of dry shake colored hardener		6400	.001		.21	.06		.27	.32
0140	Step 4 - bull float, manual float & steel trowel	3 Cefi	1280	.019			.89		.89	1.32
0150	Step 5 - application of dry shake colored release agent	1 Cefi	6400	.001		.10	.06		.16	.20
0160	Step 6 - place, tamp & remove mats	3 Cefi	2400	.010		.81	.48		1.29	1.59
0170	Step 7 - touch up edges, mat joints & simulated grout lines	1 Cefi	1280	.006			.30		.30	.44

03 35 Concrete Finishing

03 35 33 – Stamped Concrete Finishing

03 35 33.50 Slab Texture Stamping	Crew	Daily Output	Labor-Hours	Unit	Material	2018 Bare Costs Labor	Equipment	Total	Total Incl O&P
0300	Alternate stamping estimating method includes all tasks above	4 Cefi	800	.040	S.F.	1.55	1.90		3.45	4.52
0400	Step 8 - pressure wash @ 3000 psi after 24 hours	1 Cefi	1600	.005			.24		.24	.35
0500	Step 9 - roll 2 coats cure/seal compound when dry	"	800	.010	↓	.63	.48		1.11	1.39

03 35 43 – Polished Concrete Finishing

03 35 43.10 Polished Concrete Floors

		Crew	Daily Output	Labor-Hours	Unit	Material	2018 Bare Costs Labor	Equipment	Total	Total Incl O&P
0010	**POLISHED CONCRETE FLOORS** R033543-10									
0015	Processing of cured concrete to include grinding, honing,									
0020	and polishing of interior floors with 22" segmented diamond									
0025	planetary floor grinder (2 passes in different directions per grit)									
0100	Removal of pre-existing coatings, dry, with carbide discs using									
0105	dry vacuum pick-up system, final hand sweeping									
0110	Glue, adhesive or tar	J-4	1.60	15	M.S.F.	21	675	152	848	1,200
0120	Paint, epoxy, 1 coat		3.60	6.667		21	300	67.50	388.50	550
0130	2 coats	↓	1.80	13.333	↓	21	600	135	756	1,075
0200	Grinding and edging, wet, including wet vac pick-up and auto									
0205	scrubbing between grit changes									
0210	40-grit diamond/metal matrix	J-4A	1.60	20	M.S.F.	41.50	875	310	1,226.50	1,675
0220	80-grit diamond/metal matrix		2	16		41.50	700	249	990.50	1,375
0230	120-grit diamond/metal matrix		2.40	13.333		41.50	585	207	833.50	1,150
0240	200-grit diamond/metal matrix	↓	2.80	11.429		41.50	500	178	719.50	990
0300	Spray on dye or stain (1 coat)	1 Cefi	16	.500		222	24		246	279
0400	Spray on densifier/hardener (2 coats)	"	8	1		365	47.50		412.50	470
0410	Auto scrubbing after 2nd coat, when dry	J-4B	16	.500	↓		19.95	15.90	35.85	48
0500	Honing and edging, wet, including wet vac pick-up and auto									
0505	scrubbing between grit changes									
0510	100-grit diamond/resin matrix	J-4A	2.80	11.429	M.S.F.	41.50	500	178	719.50	990
0520	200-grit diamond/resin matrix	"	2.80	11.429	"	41.50	500	178	719.50	990
0530	Dry, including dry vacuum pick-up system, final hand sweeping									
0540	400-grit diamond/resin matrix	J-4A	2.80	11.429	M.S.F.	41.50	500	178	719.50	990
0600	Polishing and edging, dry, including dry vac pick-up and hand									
0605	sweeping between grit changes									
0610	800-grit diamond/resin matrix	J-4A	2.80	11.429	M.S.F.	41.50	500	178	719.50	990
0620	1500-grit diamond/resin matrix		2.80	11.429		41.50	500	178	719.50	990
0630	3000-grit diamond/resin matrix	↓	2.80	11.429		41.50	500	178	719.50	990
0700	Auto scrubbing after final polishing step	J-4B	16	.500	↓		19.95	15.90	35.85	48

03 37 Specialty Placed Concrete

03 37 13 – Shotcrete

03 37 13.30 Gunite (Dry-Mix)

		Crew	Daily Output	Labor-Hours	Unit	Material	2018 Bare Costs Labor	Equipment	Total	Total Incl O&P
0010	**GUNITE (DRY-MIX)**									
0020	Typical in place, 1" layers, no mesh included	C-16	2000	.028	S.F.	.36	1.24	.19	1.79	2.48
0100	Mesh for gunite 2 x 2, #12	2 Rodm	800	.020		.70	1.09		1.79	2.44
0150	#4 reinforcing bars @ 6" each way	"	500	.032		1.60	1.75		3.35	4.43
0300	Typical in place, including mesh, 2" thick, flat surfaces	C-16	1000	.056		1.43	2.48	.39	4.30	5.75
0350	Curved surfaces		500	.112		1.43	4.96	.78	7.17	9.90
0500	4" thick, flat surfaces		750	.075		2.16	3.31	.52	5.99	7.95
0550	Curved surfaces	↓	350	.160		2.16	7.10	1.11	10.37	14.30
0900	Prepare old walls, no scaffolding, good condition	C-10	1000	.024			1.08		1.08	1.61
0950	Poor condition	"	275	.087			3.93		3.93	5.85
1100	For high finish requirement or close tolerance, add				↓		50%			

03 37 Specialty Placed Concrete

03 37 13 – Shotcrete

03 37 13.30 Gunite (Dry-Mix)

		Crew	Daily Output	Labor-Hours	Unit	Material	2018 Bare Costs Labor	Equipment	Total	Total Incl O&P
1150	Very high				S.F.		110%			

03 37 13.60 Shotcrete (Wet-Mix)

		Crew	Daily Output	Labor-Hours	Unit	Material	2018 Bare Costs Labor	Equipment	Total	Total Incl O&P
0010	**SHOTCRETE (WET-MIX)**									
0020	Wet mix, placed @ up to 12 C.Y./hour, 3000 psi	C-8C	80	.600	C.Y.	118	26.50	6.25	150.75	176
0100	Up to 35 C.Y./hour	C-8E	240	.200	"	106	8.70	2.40	117.10	133
1010	Fiber reinforced, 1" thick	C-8C	1740	.028	S.F.	.87	1.21	.29	2.37	3.10
1020	2" thick		900	.053		1.74	2.33	.56	4.63	6.05
1030	3" thick		825	.058		2.60	2.55	.61	5.76	7.40
1040	4" thick		750	.064		3.47	2.80	.67	6.94	8.80

03 39 Concrete Curing

03 39 13 – Water Concrete Curing

03 39 13.50 Water Curing

		Crew	Daily Output	Labor-Hours	Unit	Material	2018 Bare Costs Labor	Equipment	Total	Total Incl O&P
0010	**WATER CURING**									
0015	With burlap, 4 uses assumed, 7.5 oz.	2 Clab	55	.291	C.S.F.	14.75	11.60		26.35	34
0100	10 oz.	"	55	.291	"	26.50	11.60		38.10	46.50
0400	Curing blankets, 1" to 2" thick, buy				S.F.	.26			.26	.29

03 39 23 – Membrane Concrete Curing

03 39 23.13 Chemical Compound Membrane Concrete Curing

		Crew	Daily Output	Labor-Hours	Unit	Material	2018 Bare Costs Labor	Equipment	Total	Total Incl O&P
0010	**CHEMICAL COMPOUND MEMBRANE CONCRETE CURING**									
0300	Sprayed membrane curing compound	2 Clab	95	.168	C.S.F.	12.10	6.70		18.80	23.50
0700	Curing compound, solvent based, 400 S.F./gal., 55 gallon lots				Gal.	21.50			21.50	23.50
0720	5 gallon lots					32			32	35
0800	Curing compound, water based, 250 S.F./gal., 55 gallon lots					21.50			21.50	23.50
0820	5 gallon lots					24			24	26.50

03 39 23.23 Sheet Membrane Concrete Curing

		Crew	Daily Output	Labor-Hours	Unit	Material	2018 Bare Costs Labor	Equipment	Total	Total Incl O&P
0010	**SHEET MEMBRANE CONCRETE CURING**									
0200	Curing blanket, burlap/poly, 2-ply	2 Clab	70	.229	C.S.F.	25	9.10		34.10	41.50

03 41 Precast Structural Concrete

03 41 13 – Precast Concrete Hollow Core Planks

03 41 13.50 Precast Slab Planks

		Crew	Daily Output	Labor-Hours	Unit	Material	2018 Bare Costs Labor	Equipment	Total	Total Incl O&P
0010	**PRECAST SLAB PLANKS** R034105-30									
0020	Prestressed roof/floor members, grouted, solid, 4" thick	C-11	2400	.030	S.F.	8	1.63	.84	10.47	12.35
0050	6" thick		2800	.026		8.95	1.40	.72	11.07	12.90
0100	Hollow, 8" thick		3200	.023		10	1.22	.63	11.85	13.65
0150	10" thick		3600	.020		10.35	1.09	.56	12	13.75
0200	12" thick		4000	.018		10.75	.98	.50	12.23	13.90

03 41 16 – Precast Concrete Slabs

03 41 16.20 Precast Concrete Channel Slabs

		Crew	Daily Output	Labor-Hours	Unit	Material	2018 Bare Costs Labor	Equipment	Total	Total Incl O&P
0010	**PRECAST CONCRETE CHANNEL SLABS**									
0335	Lightweight concrete channel slab, long runs, 2-3/4" thick	C-12	1575	.030	S.F.	11.35	1.53	.31	13.19	15.15
0375	3-3/4" thick		1550	.031		11.70	1.55	.32	13.57	15.55
0475	4-3/4" thick		1525	.031		13	1.58	.33	14.91	17.05
1275	Short pieces, 2-3/4" thick		785	.061		17.05	3.07	.63	20.75	24
1375	3-3/4" thick		770	.062		17.50	3.13	.64	21.27	24.50
1475	4-3/4" thick		762	.063		19.55	3.16	.65	23.36	27

For customer support on your Building Construction Costs with RSMeans data, call 800.448.8182.

03 41 Precast Structural Concrete

03 41 16 – Precast Concrete Slabs

03 41 16.50 Precast Lightweight Concrete Plank	Crew	Daily Output	Labor-Hours	Unit	Material	2018 Bare Costs Labor	Equipment	Total	Total Incl O&P
0010 **PRECAST LIGHTWEIGHT CONCRETE PLANK**									
0015 Lightweight plank, nailable, T&G, 2" thick	C-12	1800	.027	S.F.	10.10	1.34	.28	11.72	13.45
0150 For premium ceiling finish, add				"	10%				
0200 For sloping roofs, slope over 4 in 12, add						25%			
0250 Slope over 6 in 12, add						150%			

03 41 23 – Precast Concrete Stairs

03 41 23.50 Precast Stairs

03 41 23.50 Precast Stairs	Crew	Daily Output	Labor-Hours	Unit	Material	2018 Bare Costs Labor	Equipment	Total	Total Incl O&P
0010 **PRECAST STAIRS**									
0020 Precast concrete treads on steel stringers, 3' wide	C-12	75	.640	Riser	153	32	6.60	191.60	224
0300 Front entrance, 5' wide with 48" platform, 2 risers		16	3	Flight	625	150	31	806	955
0350 5 risers		12	4		995	201	41.50	1,237.50	1,450
0500 6' wide, 2 risers		15	3.200		695	160	33	888	1,050
0550 5 risers		11	4.364		1,100	219	45	1,364	1,575
0700 7' wide, 2 risers		14	3.429		885	172	35.50	1,092.50	1,275
0750 5 risers		10	4.800		1,475	241	49.50	1,765.50	2,050
1200 Basement entrance stairwell, 6 steps, incl. steel bulkhead door	B-51	22	2.182		1,875	89.50	8.55	1,973.05	2,200
1250 14 steps	"	11	4.364		3,125	179	17.15	3,321.15	3,725

03 41 33 – Precast Structural Pretensioned Concrete

03 41 33.10 Precast Beams

03 41 33.10 Precast Beams	Crew	Daily Output	Labor-Hours	Unit	Material	2018 Bare Costs Labor	Equipment	Total	Total Incl O&P
0010 **PRECAST BEAMS** R034105-30									
0011 L-shaped, 20' span, 12" x 20"	C-11	32	2.250	Ea.	4,100	122	63	4,285	4,800
0060 18" x 36"		24	3		5,600	163	84	5,847	6,525
0100 24" x 44"		22	3.273		6,725	178	91.50	6,994.50	7,775
0150 30' span, 12" x 36"		24	3		7,575	163	84	7,822	8,675
0200 18" x 44"		20	3.600		9,225	196	101	9,522	10,600
0250 24" x 52"		16	4.500		11,100	245	126	11,471	12,700
0400 40' span, 12" x 52"		20	3.600		11,800	196	101	12,097	13,300
0450 18" x 52"		16	4.500		13,100	245	126	13,471	14,900
0500 24" x 52"		12	6		14,800	325	168	15,293	16,900
1200 Rectangular, 20' span, 12" x 20"		32	2.250		3,975	122	63	4,160	4,650
1250 18" x 36"		24	3		4,925	163	84	5,172	5,775
1300 24" x 44"		22	3.273		5,875	178	91.50	6,144.50	6,850
1400 30' span, 12" x 36"		24	3		6,625	163	84	6,872	7,650
1450 18" x 44"		20	3.600		8,000	196	101	8,297	9,225
1500 24" x 52"		16	4.500		9,625	245	126	9,996	11,100
1600 40' span, 12" x 52"		20	3.600		9,850	196	101	10,147	11,200
1650 18" x 52"		16	4.500		11,200	245	126	11,571	12,800
1700 24" x 52"		12	6		12,800	325	168	13,293	14,800
2000 "T" shaped, 20' span, 12" x 20"		32	2.250		4,675	122	63	4,860	5,400
2050 18" x 36"		24	3		6,275	163	84	6,522	7,250
2100 24" x 44"		22	3.273		7,525	178	91.50	7,794.50	8,650
2200 30' span, 12" x 36"		24	3		8,625	163	84	8,872	9,825
2250 18" x 44"		20	3.600		10,500	196	101	10,797	11,900
2300 24" x 52"		16	4.500		12,500	245	126	12,871	14,300
2500 40' span, 12" x 52"		20	3.600		14,200	196	101	14,497	16,000
2550 18" x 52"		16	4.500		15,100	245	126	15,471	17,100
2600 24" x 52"		12	6		16,700	325	168	17,193	19,000

03 41 33.15 Precast Columns

03 41 33.15 Precast Columns	Crew	Daily Output	Labor-Hours	Unit	Material	2018 Bare Costs Labor	Equipment	Total	Total Incl O&P
0010 **PRECAST COLUMNS** R034105-30									
0020 Rectangular to 12' high, 16" x 16"	C-11	120	.600	L.F.	234	32.50	16.80	283.30	330
0050 24" x 24"		96	.750		315	41	21	377	435

03 41 Precast Structural Concrete

03 41 33 – Precast Structural Pretensioned Concrete

03 41 33.15 Precast Columns

		Crew	Daily Output	Labor-Hours	Unit	Material	2018 Bare Costs Labor	Equipment	Total	Total Incl O&P
0300	24' high, 28" x 28"	C-11	192	.375	L.F.	365	20.50	10.50	396	445
0350	36" x 36"	↓	144	.500	↓	490	27	14	531	600

03 41 33.25 Precast Joists

		Crew	Daily Output	Labor-Hours	Unit	Material	2018 Bare Costs Labor	Equipment	Total	Total Incl O&P
0010	**PRECAST JOISTS** R034105-30									
0015	40 psf L.L., 6" deep for 12' spans	C-12	600	.080	L.F.	33	4.01	.83	37.84	43
0050	8" deep for 16' spans		575	.083		54.50	4.18	.86	59.54	67.50
0100	10" deep for 20' spans		550	.087		95.50	4.37	.90	100.77	113
0150	12" deep for 24' spans	↓	525	.091	↓	131	4.58	.94	136.52	152

03 41 33.60 Precast Tees

		Crew	Daily Output	Labor-Hours	Unit	Material	2018 Bare Costs Labor	Equipment	Total	Total Incl O&P
0010	**PRECAST TEES** R034105-30									
0020	Quad tee, short spans, roof	C-11	7200	.010	S.F.	10.40	.54	.28	11.22	12.65
0050	Floor		7200	.010		10.40	.54	.28	11.22	12.65
0200	Double tee, floor members, 60' span		8400	.009		11.70	.47	.24	12.41	13.90
0250	80' span		8000	.009		16.20	.49	.25	16.94	18.90
0300	Roof members, 30' span		4800	.015		11.25	.82	.42	12.49	14.10
0350	50' span		6400	.011		11.45	.61	.32	12.38	13.90
0400	Wall members, up to 55' high		3600	.020		15.55	1.09	.56	17.20	19.45
0500	Single tee roof members, 40' span		3200	.023		15.60	1.22	.63	17.45	19.85
0550	80' span		5120	.014		16.10	.76	.39	17.25	19.35
0600	100' span		6000	.012		24	.65	.34	24.99	28
0650	120' span	↓	6000	.012	↓	26	.65	.34	26.99	30
1000	Double tees, floor members									
1100	Lightweight, 20" x 8' wide, 45' span	C-11	20	3.600	Ea.	4,125	196	101	4,422	4,950
1150	24" x 8' wide, 50' span		18	4		4,575	217	112	4,904	5,500
1200	32" x 10' wide, 60' span		16	4.500		6,875	245	126	7,246	8,075
1250	Standard weight, 12" x 8' wide, 20' span		22	3.273		1,675	178	91.50	1,944.50	2,200
1300	16" x 8' wide, 25' span		20	3.600		2,075	196	101	2,372	2,725
1350	18" x 8' wide, 30' span		20	3.600		2,500	196	101	2,797	3,175
1400	20" x 8' wide, 45' span		18	4		3,750	217	112	4,079	4,600
1450	24" x 8' wide, 50' span		16	4.500		4,150	245	126	4,521	5,100
1500	32" x 10' wide, 60' span	↓	14	5.143	↓	6,250	280	144	6,674	7,475
2000	Roof members									
2050	Lightweight, 20" x 8' wide, 40' span	C-11	20	3.600	Ea.	3,650	196	101	3,947	4,450
2100	24" x 8' wide, 50' span		18	4		4,575	217	112	4,904	5,500
2150	32" x 10' wide, 60' span		16	4.500		6,875	245	126	7,246	8,075
2200	Standard weight, 12" x 8' wide, 30' span		22	3.273		2,500	178	91.50	2,769.50	3,125
2250	16" x 8' wide, 30' span		20	3.600		2,625	196	101	2,922	3,300
2300	18" x 8' wide, 30' span		20	3.600		2,750	196	101	3,047	3,450
2350	20" x 8' wide, 40' span		18	4		3,325	217	112	3,654	4,125
2400	24" x 8' wide, 50' span		16	4.500		4,150	245	126	4,521	5,100
2450	32" x 10' wide, 60' span	↓	14	5.143	↓	6,250	280	144	6,674	7,475

For customer support on your Building Construction Costs with RSMeans data, call 800.448.8182.

85

03 45 Precast Architectural Concrete

03 45 13 – Faced Architectural Precast Concrete

03 45 13.50 Precast Wall Panels

	03 45 13.50 Precast Wall Panels	Crew	Daily Output	Labor-Hours	Unit	Material	2018 Bare Costs Labor	Equipment	Total	Total Incl O&P
0010	**PRECAST WALL PANELS** R034513-10									
0050	Uninsulated, smooth gray									
0150	Low rise, 4' x 8' x 4" thick	C-11	320	.225	S.F.	27.50	12.25	6.30	46.05	56.50
0210	8' x 8', 4" thick		576	.125		27	6.80	3.50	37.30	45
0250	8' x 16' x 4" thick		1024	.070		27	3.82	1.97	32.79	38
0600	High rise, 4' x 8' x 4" thick		288	.250		27.50	13.60	7	48.10	59.50
0650	8' x 8' x 4" thick		512	.141		27	7.65	3.94	38.59	46.50
0700	8' x 16' x 4" thick		768	.094		27	5.10	2.63	34.73	40.50
0750	10' x 20', 6" thick		1400	.051		46	2.80	1.44	50.24	56.50
0800	Insulated panel, 2" polystyrene, add					1.14			1.14	1.25
0850	2" urethane, add					.90			.90	.99
1200	Finishes, white, add					3.31			3.31	3.64
1250	Exposed aggregate, add					.69			.69	.76
1300	Granite faced, domestic, add					30.50			30.50	33.50
1350	Brick faced, modular, red, add					4.66			4.66	5.15
2200	Fiberglass reinforced cement with urethane core									
2210	R20, 8' x 8', 5" plain finish	E-2	750	.075	S.F.	28	4.05	2.25	34.30	40
2220	Exposed aggregate or brick finish	"	600	.093	"	42.50	5.05	2.81	50.36	57.50

03 47 Site-Cast Concrete

03 47 13 – Tilt-Up Concrete

03 47 13.50 Tilt-Up Wall Panels

	03 47 13.50 Tilt-Up Wall Panels	Crew	Daily Output	Labor-Hours	Unit	Material	2018 Bare Costs Labor	Equipment	Total	Total Incl O&P
0010	**TILT-UP WALL PANELS** R034713-20									
0015	Wall panel construction, walls only, 5-1/2" thick	C-14	1600	.090	S.F.	5.95	4.42	.95	11.32	14.30
0100	7-1/2" thick		1550	.093		7.40	4.56	.98	12.94	16.10
0500	Walls and columns, 5-1/2" thick walls, 12" x 12" columns		1565	.092		8.75	4.52	.97	14.24	17.55
0550	7-1/2" thick wall, 12" x 12" columns		1370	.105		10.65	5.15	1.11	16.91	21
0800	Columns only, site precast, 12" x 12"		200	.720	L.F.	20.50	35.50	7.60	63.60	84.50
0850	16" x 16"		105	1.371	"	30	67.50	14.45	111.95	150

03 48 Precast Concrete Specialties

03 48 43 – Precast Concrete Trim

03 48 43.40 Precast Lintels

	03 48 43.40 Precast Lintels	Crew	Daily Output	Labor-Hours	Unit	Material	2018 Bare Costs Labor	Equipment	Total	Total Incl O&P
0010	**PRECAST LINTELS**, smooth gray, prestressed, stock units only									
0800	4" wide, 8" high, x 4' long	D-10	28	1.143	Ea.	32.50	56.50	13	102	136
0850	8' long		24	1.333		82	66	15.15	163.15	208
1000	6" wide, 8" high, x 4' long		26	1.231		48	61	14	123	161
1050	10' long		22	1.455		120	72	16.50	208.50	260
1200	8" wide, 8" high, x 4' long		24	1.333		56.50	66	15.15	137.65	180
1250	12' long		20	1.600		183	79	18.20	280.20	345
1275	For custom sizes, types, colors, or finishes of precast lintels, add					150%				

03 48 43.90 Precast Window Sills

	03 48 43.90 Precast Window Sills	Crew	Daily Output	Labor-Hours	Unit	Material	2018 Bare Costs Labor	Equipment	Total	Total Incl O&P
0010	**PRECAST WINDOW SILLS**									
0600	Precast concrete, 4" tapers to 3", 9" wide	D-1	70	.229	L.F.	19	10.25		29.25	36.50
0650	11" wide	"	60	.267	"	31	11.95		42.95	52.50

For customer support on your Building Construction Costs with RSMeans data, call 800.448.8182.

03 51 Cast Roof Decks

03 51 13 – Cementitious Wood Fiber Decks

03 51 13.50 Cementitious/Wood Fiber Planks		Crew	Daily Output	Labor- Hours	Unit	Material	2018 Bare Costs Labor	Equipment	Total	Total Incl O&P
0010	**CEMENTITIOUS/WOOD FIBER PLANKS** R051223-50									
0050	Plank, beveled edge, 1" thick	2 Carp	1000	.016	S.F.	3.41	.81		4.22	4.99
0100	1-1/2" thick		975	.016		4.42	.83		5.25	6.15
0150	T&G, 2" thick		950	.017		3.75	.85		4.60	5.45
0200	2-1/2" thick		925	.017		4.09	.88		4.97	5.85
0250	3" thick		900	.018		4.63	.90		5.53	6.45
1000	Bulb tee, sub-purlin and grout, 6' span, add	E-1	5000	.005		2.16	.26	.02	2.44	2.81
1100	8' span	"	4200	.006		2.16	.31	.02	2.49	2.90

03 51 16 – Gypsum Concrete Roof Decks

03 51 16.50 Gypsum Roof Deck		Crew	Daily Output	Labor- Hours	Unit	Material	2018 Bare Costs Labor	Equipment	Total	Total Incl O&P
0010	**GYPSUM ROOF DECK**									
1000	Poured gypsum, 2" thick	C-8	6000	.009	S.F.	1.74	.41	.15	2.30	2.69
1100	3" thick	"	4800	.012	"	2.61	.52	.18	3.31	3.85

03 52 Lightweight Concrete Roof Insulation

03 52 16 – Lightweight Insulating Concrete

03 52 16.13 Lightweight Cellular Insulating Concrete

		Crew	Daily Output	Labor- Hours	Unit	Material	2018 Bare Costs Labor	Equipment	Total	Total Incl O&P
0010	**LIGHTWEIGHT CELLULAR INSULATING CONCRETE** R035216-10									
0020	Portland cement and foaming agent [G]	C-8	50	1.120	C.Y.	128	49.50	17.65	195.15	234

03 52 16.16 Lightweight Aggregate Insulating Concrete

		Crew	Daily Output	Labor- Hours	Unit	Material	2018 Bare Costs Labor	Equipment	Total	Total Incl O&P
0010	**LIGHTWEIGHT AGGREGATE INSULATING CONCRETE** R035216-10									
0100	Poured vermiculite or perlite, field mix,									
0110	1:6 field mix [G]	C-8	50	1.120	C.Y.	255	49.50	17.65	322.15	375
0200	Ready mix, 1:6 mix, roof fill, 2" thick [G]		10000	.006	S.F.	1.42	.25	.09	1.76	2.03
0250	3" thick [G]		7700	.007		2.13	.32	.11	2.56	2.96
0400	Expanded volcanic glass rock, 1" thick [G]	2 Carp	1500	.011		.62	.54		1.16	1.50
0450	3" thick [G]	"	1200	.013		1.87	.68		2.55	3.08

03 53 Concrete Topping

03 53 16 – Iron-Aggregate Concrete Topping

03 53 16.50 Floor Topping		Crew	Daily Output	Labor- Hours	Unit	Material	2018 Bare Costs Labor	Equipment	Total	Total Incl O&P
0010	**FLOOR TOPPING**									
0400	Integral topping/finish, on fresh concrete, using 1:1:2 mix, 3/16" thick	C-10B	1000	.040	S.F.	.12	1.72	.24	2.08	2.99
0450	1/2" thick		950	.042		.33	1.81	.26	2.40	3.36
0500	3/4" thick		850	.047		.49	2.02	.29	2.80	3.89
0600	1" thick		750	.053		.66	2.29	.32	3.27	4.53
0800	Granolithic topping, on fresh or cured concrete, 1:1:1-1/2 mix, 1/2" thick		590	.068		.36	2.91	.41	3.68	5.25
0820	3/4" thick		580	.069		.55	2.96	.42	3.93	5.50
0850	1" thick		575	.070		.73	2.99	.42	4.14	5.75
0950	2" thick		500	.080		1.46	3.43	.49	5.38	7.30
1200	Heavy duty, 1:1:2, 3/4" thick, preshrunk, gray, 20 M.S.F.		320	.125		.89	5.35	.76	7	9.90
1300	100 M.S.F.		380	.105		.49	4.52	.64	5.65	8.05

03 54 Cast Underlayment

03 54 13 – Gypsum Cement Underlayment

03 54 13.50 Poured Gypsum Underlayment	Crew	Daily Output	Labor-Hours	Unit	Material	2018 Bare Costs Labor	Equipment	Total	Total Incl O&P
0010 **POURED GYPSUM UNDERLAYMENT**									
0400 Underlayment, gypsum based, self-leveling 2500 psi, pumped, 1/2" thick	C-8	24000	.002	S.F.	.43	.10	.04	.57	.68
0500 3/4" thick		20000	.003		.65	.12	.04	.81	.96
0600 1" thick	↓	16000	.004		.87	.16	.06	1.09	1.25
1400 Hand placed, 1/2" thick	C-18	450	.020		.43	.80	.13	1.36	1.84
1500 3/4" thick	"	300	.030	↓	.65	1.20	.19	2.04	2.76

03 54 16 – Hydraulic Cement Underlayment

03 54 16.50 Cement Underlayment	Crew	Daily Output	Labor-Hours	Unit	Material	2018 Bare Costs Labor	Equipment	Total	Total Incl O&P
0010 **CEMENT UNDERLAYMENT**									
2510 Underlayment, P.C. based, self-leveling, 4100 psi, pumped, 1/4" thick	C-8	20000	.003	S.F.	1.68	.12	.04	1.84	2.09
2520 1/2" thick		19000	.003		3.36	.13	.05	3.54	3.94
2530 3/4" thick		18000	.003		5.05	.14	.05	5.24	5.80
2540 1" thick		17000	.003		6.70	.15	.05	6.90	7.70
2550 1-1/2" thick	↓	15000	.004		10.10	.17	.06	10.33	11.40
2560 Hand placed, 1/2" thick	C-18	450	.020		3.36	.80	.13	4.29	5.05
2610 Topping, P.C. based, self-leveling, 6100 psi, pumped, 1/4" thick	C-8	20000	.003		2.47	.12	.04	2.63	2.95
2620 1/2" thick		19000	.003		4.93	.13	.05	5.11	5.70
2630 3/4" thick		18000	.003		7.40	.14	.05	7.59	8.40
2660 1" thick		17000	.003		9.85	.15	.05	10.05	11.15
2670 1-1/2" thick	↓	15000	.004		14.80	.17	.06	15.03	16.60
2680 Hand placed, 1/2" thick	C-18	450	.020	↓	4.93	.80	.13	5.86	6.80

03 62 Non-Shrink Grouting

03 62 13 – Non-Metallic Non-Shrink Grouting

03 62 13.50 Grout, Non-Metallic Non-Shrink	Crew	Daily Output	Labor-Hours	Unit	Material	2018 Bare Costs Labor	Equipment	Total	Total Incl O&P
0010 **GROUT, NON-METALLIC NON-SHRINK**									
0300 Non-shrink, non-metallic, 1" deep	1 Cefi	35	.229	S.F.	6.95	10.85		17.80	24
0350 2" deep	"	25	.320	"	13.85	15.20		29.05	38

03 62 16 – Metallic Non-Shrink Grouting

03 62 16.50 Grout, Metallic Non-Shrink	Crew	Daily Output	Labor-Hours	Unit	Material	2018 Bare Costs Labor	Equipment	Total	Total Incl O&P
0010 **GROUT, METALLIC NON-SHRINK**									
0020 Column & machine bases, non-shrink, metallic, 1" deep	1 Cefi	35	.229	S.F.	12.15	10.85		23	29.50
0050 2" deep	"	25	.320	"	24.50	15.20		39.70	49.50

03 63 Epoxy Grouting

03 63 05 – Grouting of Dowels and Fasteners

03 63 05.10 Epoxy Only	Crew	Daily Output	Labor-Hours	Unit	Material	2018 Bare Costs Labor	Equipment	Total	Total Incl O&P
0010 **EPOXY ONLY**									
1500 Chemical anchoring, epoxy cartridge, excludes layout, drilling, fastener									
1530 For fastener 3/4" diam. x 6" embedment	2 Skwk	72	.222	Ea.	5.20	11.65		16.85	23.50
1535 1" diam. x 8" embedment		66	.242		7.80	12.70		20.50	28
1540 1-1/4" diam. x 10" embedment		60	.267		15.65	13.95		29.60	38.50
1545 1-3/4" diam. x 12" embedment		54	.296		26	15.50		41.50	52.50
1550 14" embedment		48	.333		31.50	17.45		48.95	61
1555 2" diam. x 12" embedment		42	.381		41.50	19.95		61.45	76.50
1560 18" embedment	↓	32	.500	↓	52	26		78	97.50

For customer support on your Building Construction Costs with RSMeans data, call 800.448.8182.

03 81 Concrete Cutting

03 81 13 – Flat Concrete Sawing

03 81 13.50 Concrete Floor/Slab Cutting

		Crew	Daily Output	Labor-Hours	Unit	Material	2018 Bare Costs Labor	Equipment	Total	Total Incl O&P
0010	**CONCRETE FLOOR/SLAB CUTTING**									
0050	Includes blade cost, layout and set-up time									
0300	Saw cut concrete slabs, plain, up to 3" deep	B-89	1060	.015	L.F.	.12	.72	.40	1.24	1.67
0320	Each additional inch of depth		3180	.005		.04	.24	.13	.41	.56
0400	Mesh reinforced, up to 3" deep		980	.016		.14	.78	.43	1.35	1.80
0420	Each additional inch of depth		2940	.005		.05	.26	.14	.45	.60
0500	Rod reinforced, up to 3" deep		800	.020		.17	.96	.53	1.66	2.21
0520	Each additional inch of depth		2400	.007		.06	.32	.18	.56	.73

03 81 13.75 Concrete Saw Blades

		Crew	Daily Output	Labor-Hours	Unit	Material	2018 Bare Costs Labor	Equipment	Total	Total Incl O&P
0010	**CONCRETE SAW BLADES**									
3000	Blades for saw cutting, included in cutting line items									
3020	Diamond, 12" diameter				Ea.	220			220	242
3040	18" diameter					415			415	460
3080	24" diameter					695			695	765
3120	30" diameter					980			980	1,075
3160	36" diameter					1,300			1,300	1,425
3200	42" diameter					2,550			2,550	2,800

03 81 16 – Track Mounted Concrete Wall Sawing

03 81 16.50 Concrete Wall Cutting

		Crew	Daily Output	Labor-Hours	Unit	Material	2018 Bare Costs Labor	Equipment	Total	Total Incl O&P
0010	**CONCRETE WALL CUTTING**									
0750	Includes blade cost, layout and set-up time									
0800	Concrete walls, hydraulic saw, plain, per inch of depth	B-89B	250	.064	L.F.	.04	3.07	2.69	5.80	7.65
0820	Rod reinforcing, per inch of depth	"	150	.107	"	.06	5.10	4.48	9.64	12.70

03 82 Concrete Boring

03 82 13 – Concrete Core Drilling

03 82 13.10 Core Drilling

		Crew	Daily Output	Labor-Hours	Unit	Material	2018 Bare Costs Labor	Equipment	Total	Total Incl O&P
0010	**CORE DRILLING**									
0015	Includes bit cost, layout and set-up time									
0020	Reinforced concrete slab, up to 6" thick									
0100	1" diameter core	B-89A	17	.941	Ea.	.18	43.50	6.60	50.28	74
0150	For each additional inch of slab thickness in same hole, add		1440	.011		.03	.51	.08	.62	.90
0200	2" diameter core		16.50	.970		.29	44.50	6.80	51.59	76.50
0250	For each additional inch of slab thickness in same hole, add		1080	.015		.05	.68	.10	.83	1.20
0300	3" diameter core		16	1		.39	46	7	53.39	78.50
0350	For each additional inch of slab thickness in same hole, add		720	.022		.07	1.02	.16	1.25	1.81
0500	4" diameter core		15	1.067		.50	49	7.50	57	84
0550	For each additional inch of slab thickness in same hole, add		480	.033		.08	1.54	.23	1.85	2.70
0700	6" diameter core		14	1.143		.77	52.50	8	61.27	90
0750	For each additional inch of slab thickness in same hole, add		360	.044		.13	2.05	.31	2.49	3.61
0900	8" diameter core		13	1.231		1.18	56.50	8.65	66.33	97.50
0950	For each additional inch of slab thickness in same hole, add		288	.056		.20	2.56	.39	3.15	4.56
1100	10" diameter core		12	1.333		1.49	61.50	9.35	72.34	106
1150	For each additional inch of slab thickness in same hole, add		240	.067		.25	3.07	.47	3.79	5.50
1300	12" diameter core		11	1.455		1.83	67	10.20	79.03	115
1350	For each additional inch of slab thickness in same hole, add		206	.078		.30	3.58	.54	4.42	6.40
1500	14" diameter core		10	1.600		2.12	74	11.20	87.32	128
1550	For each additional inch of slab thickness in same hole, add		180	.089		.35	4.10	.62	5.07	7.35
1700	18" diameter core		9	1.778		2.92	82	12.45	97.37	142
1750	For each additional inch of slab thickness in same hole, add		144	.111		.49	5.10	.78	6.37	9.25

03 82 Concrete Boring

03 82 13 – Concrete Core Drilling

03 82 13.10 Core Drilling

		Crew	Daily Output	Labor-Hours	Unit	Material	2018 Bare Costs Labor	Equipment	Total	Total Incl O&P
1754	24" diameter core	B-89A	8	2	Ea.	4.14	92	14	110.14	161
1756	For each additional inch of slab thickness in same hole, add	▼	120	.133		.69	6.15	.93	7.77	11.20
1760	For horizontal holes, add to above				▼		20%	20%		
1770	Prestressed hollow core plank, 8" thick									
1780	1" diameter core	B-89A	17.50	.914	Ea.	.24	42	6.40	48.64	72
1790	For each additional inch of plank thickness in same hole, add		3840	.004		.03	.19	.03	.25	.35
1794	2" diameter core		17.25	.928		.39	43	6.50	49.89	73
1796	For each additional inch of plank thickness in same hole, add		2880	.006		.05	.26	.04	.35	.48
1800	3" diameter core		17	.941		.52	43.50	6.60	50.62	74.50
1810	For each additional inch of plank thickness in same hole, add		1920	.008		.07	.38	.06	.51	.72
1820	4" diameter core		16.50	.970		.67	44.50	6.80	51.97	76.50
1830	For each additional inch of plank thickness in same hole, add		1280	.013		.08	.58	.09	.75	1.07
1840	6" diameter core		15.50	1.032		1.03	47.50	7.25	55.78	81.50
1850	For each additional inch of plank thickness in same hole, add		960	.017		.13	.77	.12	1.02	1.44
1860	8" diameter core		15	1.067		1.57	49	7.50	58.07	85
1870	For each additional inch of plank thickness in same hole, add		768	.021		.20	.96	.15	1.31	1.85
1880	10" diameter core		14	1.143		1.99	52.50	8	62.49	91.50
1890	For each additional inch of plank thickness in same hole, add		640	.025		.25	1.15	.18	1.58	2.22
1900	12" diameter core		13.50	1.185		2.44	54.50	8.30	65.24	95.50
1910	For each additional inch of plank thickness in same hole, add	▼	548	.029	▼	.30	1.35	.20	1.85	2.63
3000	Bits for core drilling, included in drilling line items									
3010	Diamond, premium, 1" diameter				Ea.	73			73	80.50
3020	2" diameter					116			116	127
3030	3" diameter					156			156	172
3040	4" diameter					201			201	222
3060	6" diameter					310			310	340
3080	8" diameter					470			470	520
3110	10" diameter					600			600	660
3120	12" diameter					730			730	805
3140	14" diameter					850			850	935
3180	18" diameter					1,175			1,175	1,275
3240	24" diameter				▼	1,650			1,650	1,825

03 82 16 – Concrete Drilling

03 82 16.10 Concrete Impact Drilling

		Crew	Daily Output	Labor-Hours	Unit	Material	2018 Bare Costs Labor	Equipment	Total	Total Incl O&P
0010	**CONCRETE IMPACT DRILLING**									
0020	Includes bit cost, layout and set-up time, no anchors									
0050	Up to 4" deep in concrete/brick floors/walls									
0100	Holes, 1/4" diameter	1 Carp	75	.107	Ea.	.07	5.40		5.47	8.35
0150	For each additional inch of depth in same hole, add		430	.019		.02	.94		.96	1.46
0200	3/8" diameter		63	.127		.06	6.45		6.51	9.85
0250	For each additional inch of depth in same hole, add		340	.024		.01	1.19		1.20	1.84
0300	1/2" diameter		50	.160		.06	8.10		8.16	12.40
0350	For each additional inch of depth in same hole, add		250	.032		.01	1.62		1.63	2.49
0400	5/8" diameter		48	.167		.10	8.45		8.55	12.95
0450	For each additional inch of depth in same hole, add		240	.033		.03	1.69		1.72	2.60
0500	3/4" diameter		45	.178		.13	9		9.13	13.85
0550	For each additional inch of depth in same hole, add		220	.036		.03	1.84		1.87	2.85
0600	7/8" diameter		43	.186		.18	9.45		9.63	14.55
0650	For each additional inch of depth in same hole, add		210	.038		.05	1.93		1.98	2.99
0700	1" diameter		40	.200		.16	10.15		10.31	15.60
0750	For each additional inch of depth in same hole, add		190	.042		.04	2.14		2.18	3.29
0800	1-1/4" diameter	▼	38	.211		.29	10.65		10.94	16.55

03 82 Concrete Boring

03 82 16 – Concrete Drilling

03 82 16.10 Concrete Impact Drilling		Crew	Daily Output	Labor-Hours	Unit	Material	2018 Bare Costs Labor	Equipment	Total	Total Incl O&P
0850	For each additional inch of depth in same hole, add	1 Carp	180	.044	Ea.	.07	2.25		2.32	3.51
0900	1-1/2" diameter		35	.229		.50	11.60		12.10	18.20
0950	For each additional inch of depth in same hole, add	↓	165	.048	↓	.12	2.46		2.58	3.88
1000	For ceiling installations, add						40%			

For customer support on your Building Construction Costs with RSMeans data, call 800.448.8182.

Division Notes

	CREW	DAILY OUTPUT	LABOR-HOURS	UNIT	BARE COSTS				TOTAL INCL O&P
					MAT.	LABOR	EQUIP.	TOTAL	

Estimating Tips
04 05 00 Common Work Results for Masonry

- The terms mortar and grout are often used interchangeably—and incorrectly. Mortar is used to bed masonry units, seal the entry of air and moisture, provide architectural appearance, and allow for size variations in the units. Grout is used primarily in reinforced masonry construction and to bond the masonry to the reinforcing steel. Common mortar types are M (2500 psi), S (1800 psi), N (750 psi), and O (350 psi), and they conform to ASTM C270. Grout is either fine or coarse and conforms to ASTM C476, and in-place strengths generally exceed 2500 psi. Mortar and grout are different components of masonry construction and are placed by entirely different methods. An estimator should be aware of their unique uses and costs.

- Mortar is included in all assembled masonry line items. The mortar cost, part of the assembled masonry material cost, includes all ingredients, all labor, and all equipment required. Please see reference number R040513-10.

- Waste, specifically the loss/droppings of mortar and the breakage of brick and block, is included in all unit cost lines that include mortar and masonry units in this division. A factor of 25% is added for mortar and 3% for brick and concrete masonry units.

- Scaffolding or staging is not included in any of the Division 4 costs. Refer to Subdivision 01 54 23 for scaffolding and staging costs.

04 20 00 Unit Masonry

- The most common types of unit masonry are brick and concrete masonry. The major classifications of brick are building brick (ASTM C62), facing brick (ASTM C216), glazed brick, fire brick, and pavers. Many varieties of texture and appearance can exist within these classifications, and the estimator would be wise to check local custom and availability within the project area. For repair and remodeling jobs, matching the existing brick may be the most important criteria.

- Brick and concrete block are priced by the piece and then converted into a price per square foot of wall. Openings less than two square feet are generally ignored by the estimator because any savings in units used are offset by the cutting and trimming required.

- It is often difficult and expensive to find and purchase small lots of historic brick. Costs can vary widely. Many design issues affect costs, selection of mortar mix, and repairs or replacement of masonry materials. Cleaning techniques must be reflected in the estimate.

- All masonry walls, whether interior or exterior, require bracing. The cost of bracing walls during construction should be included by the estimator, and this bracing must remain in place until permanent bracing is complete. Permanent bracing of masonry walls is accomplished by masonry itself, in the form of pilasters or abutting wall corners, or by anchoring the walls to the structural frame. Accessories in the form of anchors, anchor slots, and ties are used, but their supply and installation can be by different trades. For instance, anchor slots on spandrel beams and columns are supplied and welded in place by the steel fabricator, but the ties from the slots into the masonry are installed by the bricklayer. Regardless of the installation method, the estimator must be certain that these accessories are accounted for in pricing.

Reference Numbers

Reference numbers are shown at the beginning of some major classifications. These numbers refer to related items in the Reference Section. The reference information may be an estimating procedure, an alternate pricing method, or technical information.

Note: Not all subdivisions listed here necessarily appear. ■

Did you know?

RSMeans data is available through our online application with 24/7 access:

- Search for unit prices by keyword
- Leverage the most up-to-date data
- Build and export estimates

Try it free for 30 days!
www.rsmeans.com/2018freetrial

No part of this cost data may be reproduced, stored in a retrieval system, or transmitted in any form or by any means without prior written permission of Gordian.

04 01 Maintenance of Masonry

04 01 20 – Maintenance of Unit Masonry

04 01 20.20 Pointing Masonry

04 01 20.20 Pointing Masonry	Crew	Daily Output	Labor-Hours	Unit	Material	2018 Bare Costs Labor	2018 Bare Costs Equipment	Total	Total Incl O&P
0010 POINTING MASONRY									
0300 Cut and repoint brick, hard mortar, running bond	1 Bric	80	.100	S.F.	.56	5.05		5.61	8.30
0320 Common bond		77	.104		.56	5.25		5.81	8.60
0360 Flemish bond		70	.114		.59	5.75		6.34	9.45
0400 English bond		65	.123		.59	6.20		6.79	10.15
0600 Soft old mortar, running bond		100	.080		.56	4.02		4.58	6.75
0620 Common bond		96	.083		.56	4.19		4.75	7
0640 Flemish bond		90	.089		.59	4.47		5.06	7.50
0680 English bond		82	.098		.59	4.91		5.50	8.15
0700 Stonework, hard mortar		140	.057	L.F.	.74	2.87		3.61	5.20
0720 Soft old mortar		160	.050	"	.74	2.52		3.26	4.67
1000 Repoint, mask and grout method, running bond		95	.084	S.F.	.74	4.24		4.98	7.30
1020 Common bond		90	.089		.74	4.47		5.21	7.65
1040 Flemish bond		86	.093		.78	4.68		5.46	8
1060 English bond		77	.104		.78	5.25		6.03	8.85
2000 Scrub coat, sand grout on walls, thin mix, brushed		120	.067		3.36	3.35		6.71	8.85
2020 Troweled		98	.082		4.68	4.11		8.79	11.45

04 01 20.30 Pointing CMU

04 01 20.30 Pointing CMU	Crew	Daily Output	Labor-Hours	Unit	Material	Labor	Equipment	Total	Total Incl O&P
0010 POINTING CMU									
0300 Cut and repoint block, hard mortar, running bond	1 Bric	190	.042	S.F.	.23	2.12		2.35	3.49
0310 Stacked bond		200	.040		.23	2.01		2.24	3.33
0600 Soft old mortar, running bond		230	.035		.23	1.75		1.98	2.93
0610 Stacked bond		245	.033		.23	1.64		1.87	2.76

04 01 20.40 Sawing Masonry

04 01 20.40 Sawing Masonry	Crew	Daily Output	Labor-Hours	Unit	Material	Labor	Equipment	Total	Total Incl O&P
0010 SAWING MASONRY									
0050 Brick or block by hand, per inch depth	A-1	125	.064	L.F.	.04	2.55	.57	3.16	4.56

04 01 20.41 Unit Masonry Stabilization

04 01 20.41 Unit Masonry Stabilization	Crew	Daily Output	Labor-Hours	Unit	Material	Labor	Equipment	Total	Total Incl O&P
0010 UNIT MASONRY STABILIZATION									
0100 Structural repointing method									
0110 Cut/grind mortar joint	1 Bric	240	.033	L.F.		1.68		1.68	2.57
0120 Clean and mask joint		2500	.003		.11	.16		.27	.37
0130 Epoxy paste and 1/4" FRP rod		240	.033		2.06	1.68		3.74	4.84
0132 3/8" FRP rod		160	.050		2.92	2.52		5.44	7.05
0140 Remove masking		14400	.001			.03		.03	.04
0300 Structural fabric method									
0310 Primer	1 Bric	600	.013	S.F.	.85	.67		1.52	1.97
0320 Apply filling/leveling paste		720	.011		.87	.56		1.43	1.82
0330 Epoxy, glass fiber fabric		720	.011		9.25	.56		9.81	11
0340 Carbon fiber fabric		720	.011		22.50	.56		23.06	26

04 01 30 – Unit Masonry Cleaning

04 01 30.20 Cleaning Masonry

04 01 30.20 Cleaning Masonry	Crew	Daily Output	Labor-Hours	Unit	Material	Labor	Equipment	Total	Total Incl O&P
0010 CLEANING MASONRY									
0200 By chemical, brush and rinse, new work, light construction dust	D-1	1000	.016	S.F.	.06	.72		.78	1.17
0220 Medium construction dust		800	.020		.09	.90		.99	1.47
0240 Heavy construction dust, drips or stains		600	.027		.12	1.19		1.31	1.96
0260 Low pressure wash and rinse, light restoration, light soil		800	.020		.14	.90		1.04	1.53
0270 Average soil, biological staining		400	.040		.22	1.79		2.01	2.98
0280 Heavy soil, biological and mineral staining, paint		330	.048		.29	2.17		2.46	3.64
0300 High pressure wash and rinse, heavy restoration, light soil		600	.027		.10	1.19		1.29	1.94
0310 Average soil, biological staining		400	.040		.16	1.79		1.95	2.91
0320 Heavy soil, biological and mineral staining, paint		250	.064		.21	2.87		3.08	4.62

For customer support on your Building Construction Costs with RSMeans data, call 800.448.8182.

04 01 Maintenance of Masonry

04 01 30 – Unit Masonry Cleaning

04 01 30.20 Cleaning Masonry

		Crew	Daily Output	Labor-Hours	Unit	Material	2018 Bare Costs Labor	Equipment	Total	Total Incl O&P
0400	High pressure wash, water only, light soil	C-29	500	.016	S.F.		.64	.13	.77	1.11
0420	Average soil, biological staining		375	.021			.85	.17	1.02	1.48
0440	Heavy soil, biological and mineral staining, paint		250	.032			1.28	.25	1.53	2.22
0800	High pressure water and chemical, light soil		450	.018		.17	.71	.14	1.02	1.43
0820	Average soil, biological staining		300	.027		.26	1.06	.21	1.53	2.13
0840	Heavy soil, biological and mineral staining, paint		200	.040		.34	1.59	.32	2.25	3.16
1200	Sandblast, wet system, light soil	J-6	1750	.018		.36	.81	.12	1.29	1.75
1220	Average soil, biological staining		1100	.029		.55	1.28	.18	2.01	2.74
1240	Heavy soil, biological and mineral staining, paint		700	.046		.73	2.01	.29	3.03	4.16
1400	Dry system, light soil		2500	.013		.36	.56	.08	1	1.34
1420	Average soil, biological staining		1750	.018		.55	.81	.12	1.48	1.95
1440	Heavy soil, biological and mineral staining, paint		1000	.032		.73	1.41	.20	2.34	3.15
1800	For walnut shells, add					.80			.80	.88
1820	For corn chips, add					.80			.80	.88
2000	Steam cleaning, light soil	A-1H	750	.011			.43	.10	.53	.76
2020	Average soil, biological staining		625	.013			.51	.12	.63	.91
2040	Heavy soil, biological and mineral staining		375	.021			.85	.20	1.05	1.51
4000	Add for masking doors and windows	1 Clab	800	.010		.07	.40		.47	.69
4200	Add for pedestrian protection				Job				10%	10%

04 01 30.60 Brick Washing

			Crew	Daily Output	Labor-Hours	Unit	Material	2018 Bare Costs Labor	Equipment	Total	Total Incl O&P
0010	**BRICK WASHING**	R040130-10									
0012	Acid cleanser, smooth brick surface		1 Bric	560	.014	S.F.	.05	.72		.77	1.16
0050	Rough brick			400	.020		.07	1.01		1.08	1.61
0060	Stone, acid wash			600	.013		.08	.67		.75	1.12
1000	Muriatic acid, price per gallon in 5 gallon lots					Gal.	10.10			10.10	11.10

04 05 Common Work Results for Masonry

04 05 05 – Selective Demolition for Masonry

04 05 05.10 Selective Demolition

			Crew	Daily Output	Labor-Hours	Unit	Material	2018 Bare Costs Labor	Equipment	Total	Total Incl O&P
0010	**SELECTIVE DEMOLITION**	R024119-10									
0200	Bond beams, 8" block with #4 bar		2 Clab	32	.500	L.F.		19.95		19.95	30.50
0300	Concrete block walls, unreinforced, 2" thick			1200	.013	S.F.		.53		.53	.81
0310	4" thick			1150	.014			.55		.55	.84
0320	6" thick			1100	.015			.58		.58	.88
0330	8" thick			1050	.015			.61		.61	.93
0340	10" thick			1000	.016			.64		.64	.97
0360	12" thick			950	.017			.67		.67	1.02
0380	Reinforced alternate courses, 2" thick			1130	.014			.56		.56	.86
0390	4" thick			1080	.015			.59		.59	.90
0400	6" thick			1035	.015			.62		.62	.94
0410	8" thick			990	.016			.64		.64	.98
0420	10" thick			940	.017			.68		.68	1.03
0430	12" thick			890	.018			.72		.72	1.09
0440	Reinforced alternate courses & vertically 48" OC, 4" thick			900	.018			.71		.71	1.08
0450	6" thick			850	.019			.75		.75	1.14
0460	8" thick			800	.020			.80		.80	1.21
0480	10" thick			750	.021			.85		.85	1.29
0490	12" thick			700	.023			.91		.91	1.39
1000	Chimney, 16" x 16", soft old mortar		1 Clab	55	.145	C.F.		5.80		5.80	8.85
1020	Hard mortar			40	.200			7.95		7.95	12.15
1030	16" x 20", soft old mortar			55	.145			5.80		5.80	8.85

04 05 05 – Selective Demolition for Masonry

04 05 05.10 Selective Demolition	Crew	Daily Output	Labor-Hours	Unit	Material	2018 Bare Costs Labor	Equipment	Total	Total Incl O&P
1040	Hard mortar	1 Clab	40	.200	C.F.		7.95		7.95	12.15
1050	16" x 24", soft old mortar		55	.145			5.80		5.80	8.85
1060	Hard mortar		40	.200			7.95		7.95	12.15
1080	20" x 20", soft old mortar		55	.145			5.80		5.80	8.85
1100	Hard mortar		40	.200			7.95		7.95	12.15
1110	20" x 24", soft old mortar		55	.145			5.80		5.80	8.85
1120	Hard mortar		40	.200			7.95		7.95	12.15
1140	20" x 32", soft old mortar		55	.145			5.80		5.80	8.85
1160	Hard mortar		40	.200			7.95		7.95	12.15
1200	48" x 48", soft old mortar		55	.145			5.80		5.80	8.85
1220	Hard mortar	▼	40	.200	▼		7.95		7.95	12.15
1250	Metal, high temp steel jacket, 24" diameter	E-2	130	.431	V.L.F.		23.50	13	36.50	52
1260	60" diameter	"	60	.933			50.50	28	78.50	112
1280	Flue lining, up to 12" x 12"	1 Clab	200	.040			1.59		1.59	2.43
1282	Up to 24" x 24"		150	.053			2.13		2.13	3.24
2000	Columns, 8" x 8", soft old mortar		48	.167			6.65		6.65	10.10
2020	Hard mortar		40	.200			7.95		7.95	12.15
2060	16" x 16", soft old mortar		16	.500			19.95		19.95	30.50
2100	Hard mortar		14	.571			23		23	34.50
2140	24" x 24", soft old mortar		8	1			40		40	60.50
2160	Hard mortar		6	1.333			53		53	81
2200	36" x 36", soft old mortar		4	2			79.50		79.50	121
2220	Hard mortar		3	2.667	▼		106		106	162
2230	Alternate pricing method, soft old mortar		30	.267	C.F.		10.65		10.65	16.20
2240	Hard mortar	▼	23	.348	"		13.85		13.85	21
3000	Copings, precast or masonry, to 8" wide									
3020	Soft old mortar	1 Clab	180	.044	L.F.		1.77		1.77	2.70
3040	Hard mortar	"	160	.050	"		1.99		1.99	3.04
3100	To 12" wide									
3120	Soft old mortar	1 Clab	160	.050	L.F.		1.99		1.99	3.04
3140	Hard mortar	"	140	.057	"		2.28		2.28	3.47
4000	Fireplace, brick, 30" x 24" opening									
4020	Soft old mortar	1 Clab	2	4	Ea.		159		159	243
4040	Hard mortar		1.25	6.400			255		255	390
4100	Stone, soft old mortar		1.50	5.333			213		213	325
4120	Hard mortar		1	8	▼		320		320	485
5000	Veneers, brick, soft old mortar		140	.057	S.F.		2.28		2.28	3.47
5020	Hard mortar		125	.064			2.55		2.55	3.88
5050	Glass block, up to 4" thick		500	.016			.64		.64	.97
5100	Granite and marble, 2" thick		180	.044			1.77		1.77	2.70
5120	4" thick		170	.047			1.88		1.88	2.86
5140	Stone, 4" thick		180	.044			1.77		1.77	2.70
5160	8" thick		175	.046	▼		1.82		1.82	2.77
5400	Alternate pricing method, stone, 4" thick		60	.133	C.F.		5.30		5.30	8.10
5420	8" thick	▼	85	.094	"		3.75		3.75	5.70

04 05 13 – Masonry Mortaring

04 05 13.10 Cement

0010	**CEMENT**									
0100	Masonry, 70 lb. bag, T.L. lots				Bag	13.80			13.80	15.20
0150	L.T.L. lots					14.65			14.65	16.10
0200	White, 70 lb. bag, T.L. lots					18.05			18.05	19.85
0250	L.T.L. lots				▼	19.95			19.95	22

96

For customer support on your Building Construction Costs with RSMeans data, call 800.448.8182.

04 05 Common Work Results for Masonry

04 05 13 – Masonry Mortaring

04 05 13.20 Lime

		Crew	Daily Output	Labor-Hours	Unit	Material	2018 Bare Costs Labor	Equipment	Total	Total Incl O&P
0010	**LIME**									
0020	Masons, hydrated, 50 lb. bag, T.L. lots				Bag	11.20			11.20	12.30
0050	L.T.L. lots					12.30			12.30	13.55
0200	Finish, double hydrated, 50 lb. bag, T.L. lots					10.40			10.40	11.45
0250	L.T.L. lots				▼	11.45			11.45	12.60

04 05 13.23 Surface Bonding Masonry Mortaring

		Crew	Daily Output	Labor-Hours	Unit	Material	Labor	Equipment	Total	Total Incl O&P
0010	**SURFACE BONDING MASONRY MORTARING**									
0020	Gray or white colors, not incl. block work	1 Bric	540	.015	S.F.	.15	.74		.89	1.31

04 05 13.30 Mortar

			Crew	Daily Output	Labor-Hours	Unit	Material	Labor	Equipment	Total	Total Incl O&P
0010	**MORTAR**	R040513-10									
0020	With masonry cement										
0100	Type M, 1:1:6 mix		1 Brhe	143	.056	C.F.	6.10	2.20		8.30	10.10
0200	Type N, 1:3 mix			143	.056		6.10	2.20		8.30	10.05
0300	Type O, 1:3 mix			143	.056		4.90	2.20		7.10	8.75
0400	Type PM, 1:1:6 mix, 2500 psi			143	.056		6	2.20		8.20	9.95
0500	Type S, 1/2:1:4 mix		▼	143	.056	▼	6.40	2.20		8.60	10.40
2000	With Portland cement and lime										
2100	Type M, 1:1/4:3 mix		1 Brhe	143	.056	C.F.	9.15	2.20		11.35	13.45
2200	Type N, 1:1:6 mix, 750 psi			143	.056		7.15	2.20		9.35	11.20
2300	Type O, 1:2:9 mix (Pointing Mortar)			143	.056		8.55	2.20		10.75	12.75
2400	Type PL, 1:1/2:4 mix, 2500 psi			143	.056		6	2.20		8.20	9.95
2600	Type S, 1:1/2:4 mix, 1800 psi		▼	143	.056		8.40	2.20		10.60	12.60
2650	Pre-mixed, type S or N						6.30			6.30	6.95
2700	Mortar for glass block		1 Brhe	143	.056	▼	12.60	2.20		14.80	17.20
2900	Mortar for fire brick, dry mix, 10 lb. pail					Ea.	22.50			22.50	25

04 05 13.91 Masonry Restoration Mortaring

		Crew	Daily Output	Labor-Hours	Unit	Material	Labor	Equipment	Total	Total Incl O&P
0010	**MASONRY RESTORATION MORTARING**									
0020	Masonry restoration mix				Lb.	.71			.71	.78
0050	White				"	1.18			1.18	1.30

04 05 13.93 Mortar Pigments

			Crew	Daily Output	Labor-Hours	Unit	Material	Labor	Equipment	Total	Total Incl O&P
0010	**MORTAR PIGMENTS**, 50 lb. bags (2 bags per M bricks)	R040513-10									
0020	Color admixture, range 2 to 10 lb. per bag of cement, light colors					Lb.	5.55			5.55	6.10
0050	Medium colors						7.40			7.40	8.15
0100	Dark colors						15.50			15.50	17.05

04 05 13.95 Sand

		Crew	Daily Output	Labor-Hours	Unit	Material	Labor	Equipment	Total	Total Incl O&P
0010	**SAND**, screened and washed at pit									
0020	For mortar, per ton				Ton	21			21	23
0050	With 10 mile haul					41.50			41.50	45.50
0100	With 30 mile haul					62.50			62.50	69
0200	Screened and washed, at the pit				C.Y.	29.50			29.50	32.50
0250	With 10 mile haul					57.50			57.50	63.50
0300	With 30 mile haul					87			87	95.50

04 05 13.98 Mortar Admixtures

		Crew	Daily Output	Labor-Hours	Unit	Material	Labor	Equipment	Total	Total Incl O&P
0010	**MORTAR ADMIXTURES**									
0020	Waterproofing admixture, per quart (1 qt. to 2 bags of masonry cement)				Qt.	4.98			4.98	5.50

04 05 16 – Masonry Grouting

04 05 16.30 Grouting

		Crew	Daily Output	Labor-Hours	Unit	Material	Labor	Equipment	Total	Total Incl O&P
0010	**GROUTING**									
0011	Bond beams & lintels, 8" deep, 6" thick, 0.15 C.F./L.F.	D-4	1480	.022	L.F.	.75	.97	.09	1.81	2.41
0020	8" thick, 0.2 C.F./L.F.		1400	.023		1.22	1.03	.09	2.34	3.01

04 05 Common Work Results for Masonry

04 05 16 – Masonry Grouting

04 05 16.30 Grouting

		Crew	Daily Output	Labor-Hours	Unit	Material	2018 Bare Costs Labor	Equipment	Total	Total Incl O&P
0050	10" thick, 0.25 C.F./L.F.	D-4	1200	.027	L.F.	1.26	1.20	.11	2.57	3.33
0060	12" thick, 0.3 C.F./L.F.	↓	1040	.031	↓	1.51	1.39	.12	3.02	3.91
0200	Concrete block cores, solid, 4" thk., by hand, 0.067 C.F./S.F. of wall	D-8	1100	.036	S.F.	.34	1.67		2.01	2.92
0210	6" thick, pumped, 0.175 C.F./S.F.	D-4	720	.044		.88	2	.18	3.06	4.22
0250	8" thick, pumped, 0.258 C.F./S.F.		680	.047		1.30	2.12	.19	3.61	4.87
0300	10" thick, pumped, 0.340 C.F./S.F.		660	.048		1.71	2.18	.19	4.08	5.40
0350	12" thick, pumped, 0.422 C.F./S.F.		640	.050		2.12	2.25	.20	4.57	6
0500	Cavity walls, 2" space, pumped, 0.167 C.F./S.F. of wall		1700	.019		.84	.85	.08	1.77	2.29
0550	3" space, 0.250 C.F./S.F.		1200	.027		1.26	1.20	.11	2.57	3.33
0600	4" space, 0.333 C.F./S.F.		1150	.028		1.67	1.25	.11	3.03	3.87
0700	6" space, 0.500 C.F./S.F.		800	.040	↓	2.51	1.80	.16	4.47	5.70
0800	Door frames, 3' x 7' opening, 2.5 C.F. per opening		60	.533	Opng.	12.55	24	2.14	38.69	52.50
0850	6' x 7' opening, 3.5 C.F. per opening		45	.711	"	17.60	32	2.86	52.46	71.50
2000	Grout, C476, for bond beams, lintels and CMU cores	↓	350	.091	C.F.	5.05	4.12	.37	9.54	12.25

04 05 19 – Masonry Anchorage and Reinforcing

04 05 19.05 Anchor Bolts

		Crew	Daily Output	Labor-Hours	Unit	Material	2018 Bare Costs Labor	Equipment	Total	Total Incl O&P
0010	**ANCHOR BOLTS**									
0015	Installed in fresh grout in CMU bond beams or filled cores, no templates									
0020	Hooked, with nut and washer, 1/2" diameter, 8" long	1 Bric	132	.061	Ea.	1.42	3.05		4.47	6.25
0030	12" long		131	.061		1.58	3.07		4.65	6.45
0040	5/8" diameter, 8" long		129	.062		3.80	3.12		6.92	8.95
0050	12" long		127	.063		4.67	3.17		7.84	10
0060	3/4" diameter, 8" long		127	.063		4.67	3.17		7.84	10
0070	12" long	↓	125	.064	↓	5.85	3.22		9.07	11.35

04 05 19.16 Masonry Anchors

		Crew	Daily Output	Labor-Hours	Unit	Material	2018 Bare Costs Labor	Equipment	Total	Total Incl O&P
0010	**MASONRY ANCHORS**									
0020	For brick veneer, galv., corrugated, 7/8" x 7", 22 ga.	1 Bric	10.50	.762	C	16.05	38.50		54.55	76
0100	24 ga.		10.50	.762		10.15	38.50		48.65	69.50
0150	16 ga.		10.50	.762		30	38.50		68.50	91.50
0200	Buck anchors, galv., corrugated, 16 ga., 2" bend, 8" x 2"		10.50	.762		65	38.50		103.50	130
0250	8" x 3"		10.50	.762		67.50	38.50		106	133
0660	Cavity wall, Z-type, galvanized, 6" long, 1/8" diam.		10.50	.762		24	38.50		62.50	84.50
0670	3/16" diameter		10.50	.762		33	38.50		71.50	94.50
0680	1/4" diameter		10.50	.762		40.50	38.50		79	104
0850	8" long, 3/16" diameter		10.50	.762		26	38.50		64.50	87
0855	1/4" diameter		10.50	.762		47.50	38.50		86	111
1000	Rectangular type, galvanized, 1/4" diameter, 2" x 6"		10.50	.762		75.50	38.50		114	142
1050	4" x 6"		10.50	.762		91	38.50		129.50	159
1100	3/16" diameter, 2" x 6"		10.50	.762		48	38.50		86.50	112
1150	4" x 6"		10.50	.762		55	38.50		93.50	119
1500	Rigid partition anchors, plain, 8" long, 1" x 1/8"		10.50	.762		237	38.50		275.50	320
1550	1" x 1/4"		10.50	.762		279	38.50		317.50	365
1580	1-1/2" x 1/8"		10.50	.762		261	38.50		299.50	345
1600	1-1/2" x 1/4"		10.50	.762		325	38.50		363.50	420
1650	2" x 1/8"		10.50	.762		310	38.50		348.50	400
1700	2" x 1/4"	↓	10.50	.762	↓	405	38.50		443.50	505

04 05 19.26 Masonry Reinforcing Bars

			Crew	Daily Output	Labor-Hours	Unit	Material	2018 Bare Costs Labor	Equipment	Total	Total Incl O&P
0010	**MASONRY REINFORCING BARS**	R040519-50									
0015	Steel bars A615, placed horiz., #3 & #4 bars		1 Bric	450	.018	Lb.	.48	.89		1.37	1.90
0020	#5 & #6 bars			800	.010		.48	.50		.98	1.30
0050	Placed vertical, #3 & #4 bars			350	.023		.48	1.15		1.63	2.29
0060	#5 & #6 bars		↓	650	.012		.48	.62		1.10	1.48

For customer support on your Building Construction Costs with RSMeans data, call 800.448.8182.

04 05 Common Work Results for Masonry

04 05 19 – Masonry Anchorage and Reinforcing

04 05 19.26 Masonry Reinforcing Bars

	04 05 19.26 Masonry Reinforcing Bars	Crew	Daily Output	Labor-Hours	Unit	Material	2018 Bare Costs Labor	Equipment	Total	Total Incl O&P
0200	Joint reinforcing, regular truss, to 6" wide, mill std galvanized	1 Bric	30	.267	C.L.F.	24	13.40		37.40	47
0250	12" wide		20	.400		28	20		48	62
0400	Cavity truss with drip section, to 6" wide		30	.267		22.50	13.40		35.90	45.50
0450	12" wide		20	.400		26	20		46	59.50

04 05 23 – Masonry Accessories

04 05 23.13 Masonry Control and Expansion Joints

	04 05 23.13 Masonry Control and Expansion Joints	Crew	Daily Output	Labor-Hours	Unit	Material	2018 Bare Costs Labor	Equipment	Total	Total Incl O&P
0010	**MASONRY CONTROL AND EXPANSION JOINTS**									
0020	Rubber, for double wythe 8" minimum wall (Brick/CMU)	1 Bric	400	.020	L.F.	2.24	1.01		3.25	4
0025	"T" shaped		320	.025		1.25	1.26		2.51	3.31
0030	Cross-shaped for CMU units		280	.029		1.66	1.44		3.10	4.03
0050	PVC, for double wythe 8" minimum wall (Brick/CMU)		400	.020		1.66	1.01		2.67	3.37
0120	"T" shaped		320	.025		.84	1.26		2.10	2.85
0160	Cross-shaped for CMU units		280	.029		1.16	1.44		2.60	3.48

04 05 23.19 Masonry Cavity Drainage, Weepholes, and Vents

	04 05 23.19 Masonry Cavity Drainage, Weepholes, and Vents	Crew	Daily Output	Labor-Hours	Unit	Material	2018 Bare Costs Labor	Equipment	Total	Total Incl O&P
0010	**MASONRY CAVITY DRAINAGE, WEEPHOLES, AND VENTS**									
0020	Extruded aluminum, 4" deep, 2-3/8" x 8-1/8"	1 Bric	30	.267	Ea.	38.50	13.40		51.90	63
0050	5" x 8-1/8"		25	.320		50	16.10		66.10	80
0100	2-1/4" x 25"		25	.320		87	16.10		103.10	120
0150	5" x 16-1/2"		22	.364		60	18.30		78.30	94
0175	5" x 24"		22	.364		90.50	18.30		108.80	128
0200	6" x 16-1/2"		22	.364		97	18.30		115.30	135
0250	7-3/4" x 16-1/2"		20	.400		88	20		108	128
0400	For baked enamel finish, add					35%				
0500	For cast aluminum, painted, add					60%				
1000	Stainless steel ventilators, 6" x 6"	1 Bric	25	.320		222	16.10		238.10	270
1050	8" x 8"		24	.333		248	16.75		264.75	299
1100	12" x 12"		23	.348		285	17.50		302.50	340
1150	12" x 6"		24	.333		269	16.75		285.75	320
1200	Foundation block vent, galv., 1-1/4" thk, 8" high, 16" long, no damper		30	.267		16.20	13.40		29.60	38.50
1250	For damper, add					3.28			3.28	3.61

04 05 23.95 Wall Plugs

	04 05 23.95 Wall Plugs	Crew	Daily Output	Labor-Hours	Unit	Material	2018 Bare Costs Labor	Equipment	Total	Total Incl O&P
0010	**WALL PLUGS** (for nailing to brickwork)									
0020	25 ga., galvanized, plain	1 Bric	10.50	.762	C	26.50	38.50		65	87.50
0050	Wood filled	"	10.50	.762	"	67.50	38.50		106	133

04 21 Clay Unit Masonry

04 21 13 – Brick Masonry

04 21 13.13 Brick Veneer Masonry

	04 21 13.13 Brick Veneer Masonry	Crew	Daily Output	Labor-Hours	Unit	Material	2018 Bare Costs Labor	Equipment	Total	Total Incl O&P
0010	**BRICK VENEER MASONRY**, T.L. lots, excl. scaff., grout & reinforcing R042110-20									
0015	Material costs incl. 3% brick and 25% mortar waste									
0020	Standard, select common, 4" x 2-2/3" x 8" (6.75/S.F.)	D-8	1.50	26.667	M	640	1,225		1,865	2,575
0050	Red, 4" x 2-2/3" x 8", running bond		1.50	26.667		600	1,225		1,825	2,525
0100	Full header every 6th course (7.88/S.F.) R042110-50		1.45	27.586		600	1,275		1,875	2,600
0150	English, full header every 2nd course (10.13/S.F.)		1.40	28.571		600	1,300		1,900	2,650
0200	Flemish, alternate header every course (9.00/S.F.)		1.40	28.571		600	1,300		1,900	2,650
0250	Flemish, alt. header every 6th course (7.13/S.F.)		1.45	27.586		600	1,275		1,875	2,600
0300	Full headers throughout (13.50/S.F.)		1.40	28.571		595	1,300		1,895	2,650
0350	Rowlock course (13.50/S.F.)		1.35	29.630		595	1,350		1,945	2,725
0400	Rowlock stretcher (4.50/S.F.)		1.40	28.571		610	1,300		1,910	2,675
0450	Soldier course (6.75/S.F.)		1.40	28.571		600	1,300		1,900	2,650

For customer support on your Building Construction Costs with RSMeans data, call 800.448.8182.

99

04 21 13.13 Brick Veneer Masonry	Crew	Daily Output	Labor-Hours	Unit	Material	2018 Bare Costs Labor	Equipment	Total	Total Incl O&P
0500 Sailor course (4.50/S.F.)	D-8	1.30	30.769	M	610	1,400		2,010	2,825
0601 Buff or gray face, running bond (6.75/S.F.)		1.50	26.667		600	1,225		1,825	2,525
0700 Glazed face, 4" x 2-2/3" x 8", running bond		1.40	28.571		2,025	1,300		3,325	4,225
0750 Full header every 6th course (7.88/S.F.)		1.35	29.630		1,925	1,350		3,275	4,200
1000 Jumbo, 6" x 4" x 12" (3.00/S.F.)		1.30	30.769		1,975	1,400		3,375	4,325
1051 Norman, 4" x 2-2/3" x 12" (4.50/S.F.)		1.45	27.586		1,150	1,275		2,425	3,225
1100 Norwegian, 4" x 3-1/5" x 12" (3.75/S.F.)		1.40	28.571		1,550	1,300		2,850	3,700
1150 Economy, 4" x 4" x 8" (4.50/S.F.)		1.40	28.571		1,000	1,300		2,300	3,100
1201 Engineer, 4" x 3-1/5" x 8" (5.63/S.F.)		1.45	27.586		700	1,275		1,975	2,725
1251 Roman, 4" x 2" x 12" (6.00/S.F.)		1.50	26.667		1,250	1,225		2,475	3,225
1300 S.C.R., 6" x 2-2/3" x 12" (4.50/S.F.)		1.40	28.571		1,400	1,300		2,700	3,550
1350 Utility, 4" x 4" x 12" (3.00/S.F.)	▼	1.08	37.037		1,725	1,700		3,425	4,500
1360 For less than truck load lots, add					15%				
1400 For battered walls, add						30%			
1450 For corbels, add						75%			
1500 For curved walls, add						30%			
1550 For pits and trenches, deduct				▼		20%			
1999 Alternate method of figuring by square foot									
2000 Standard, sel. common, 4" x 2-2/3" x 8" (6.75/S.F.)	D-8	230	.174	S.F.	4.32	8		12.32	16.95
2020 Red, 4" x 2-2/3" x 8", running bond		220	.182		4.06	8.35		12.41	17.20
2050 Full header every 6th course (7.88/S.F.)		185	.216		4.73	9.90		14.63	20.50
2100 English, full header every 2nd course (10.13/S.F.)		140	.286		6.05	13.10		19.15	26.50
2150 Flemish, alternate header every course (9.00/S.F.)		150	.267		5.40	12.25		17.65	24.50
2200 Flemish, alt. header every 6th course (7.13/S.F.)		205	.195		4.29	8.95		13.24	18.40
2250 Full headers throughout (13.50/S.F.)		105	.381		8.05	17.50		25.55	36
2300 Rowlock course (13.50/S.F.)		100	.400		8.05	18.35		26.40	37
2350 Rowlock stretcher (4.50/S.F.)		310	.129		2.74	5.90		8.64	12.05
2400 Soldier course (6.75/S.F.)		200	.200		4.06	9.20		13.26	18.50
2450 Sailor course (4.50/S.F.)		290	.138		2.74	6.35		9.09	12.70
2600 Buff or gray face, running bond (6.75/S.F.)		220	.182		4.30	8.35		12.65	17.50
2700 Glazed face brick, running bond		210	.190		13	8.75		21.75	27.50
2750 Full header every 6th course (7.88/S.F.)		170	.235		15.15	10.80		25.95	33
3000 Jumbo, 6" x 4" x 12" running bond (3.00/S.F.)		435	.092		5.45	4.22		9.67	12.45
3050 Norman, 4" x 2-2/3" x 12" running bond (4.5/S.F.)		320	.125		6.40	5.75		12.15	15.85
3100 Norwegian, 4" x 3-1/5" x 12" (3.75/S.F.)		375	.107		5.65	4.90		10.55	13.75
3150 Economy, 4" x 4" x 8" (4.50/S.F.)		310	.129		4.46	5.90		10.36	13.95
3200 Engineer, 4" x 3-1/5" x 8" (5.63/S.F.)		260	.154		3.92	7.05		10.97	15.10
3250 Roman, 4" x 2" x 12" (6.00/S.F.)		250	.160		7.30	7.35		14.65	19.25
3300 S.C.R., 6" x 2-2/3" x 12" (4.50/S.F.)		310	.129		6.30	5.90		12.20	16
3350 Utility, 4" x 4" x 12" (3.00/S.F.)	▼	360	.111		5.10	5.10		10.20	13.40
3360 For less than truck load lots, add					.10%				
3370 For battered walls, add						30%			
3380 For corbels, add						75%			
3400 For cavity wall construction, add						15%			
3450 For stacked bond, add						10%			
3500 For interior veneer construction, add						15%			
3510 For pits and trenches, deduct						20%			
3550 For curved walls, add				▼		30%			

04 21 13.14 Thin Brick Veneer

		Crew	Daily Output	Labor-Hours	Unit	Material	2018 Bare Costs Labor	Equipment	Total	Total Incl O&P
0010	**THIN BRICK VENEER**									
0015	Material costs incl. 3% brick and 25% mortar waste									
0020	On & incl. metal panel support sys, modular, 2-2/3" x 5/8" x 8", red	D-7	92	.174	S.F.	9.60	7.30		16.90	21.50
0100	Closure, 4" x 5/8" x 8"		110	.145		9.35	6.10		15.45	19.35
0110	Norman, 2-2/3" x 5/8" x 12"		110	.145		9.40	6.10		15.50	19.40
0120	Utility, 4" x 5/8" x 12"		125	.128		9.15	5.40		14.55	18
0130	Emperor, 4" x 3/4" x 16"		175	.091		10.30	3.84		14.14	17.05
0140	Super emperor, 8" x 3/4" x 16"	↓	195	.082	↓	10.60	3.45		14.05	16.80
0150	For L shaped corners with 4" return, add				L.F.	9.25			9.25	10.20
0200	On masonry/plaster back-up, modular, 2-2/3" x 5/8" x 8", red	D-7	137	.117	S.F.	4.49	4.91		9.40	12.20
0210	Closure, 4" x 5/8" x 8"		165	.097		4.24	4.08		8.32	10.70
0220	Norman, 2-2/3" x 5/8" x 12"		165	.097		4.28	4.08		8.36	10.75
0230	Utility, 4" x 5/8" x 12"		185	.086		4	3.64		7.64	9.80
0240	Emperor, 4" x 3/4" x 16"		260	.062		5.15	2.59		7.74	9.55
0250	Super emperor, 8" x 3/4" x 16"	↓	285	.056	↓	5.50	2.36		7.86	9.55
0260	For L shaped corners with 4" return, add				L.F.	9.25			9.25	10.20
0270	For embedment into pre-cast concrete panels, add				S.F.	14.40			14.40	15.85

04 21 13.15 Chimney

		Crew	Daily Output	Labor-Hours	Unit	Material	2018 Bare Costs Labor	Equipment	Total	Total Incl O&P
0010	**CHIMNEY**, excludes foundation, scaffolding, grout and reinforcing									
0100	Brick, 16" x 16", 8" flue	D-1	18.20	.879	V.L.F.	25.50	39.50		65	88.50
0150	16" x 20" with one 8" x 12" flue		16	1		40	45		85	113
0200	16" x 24" with two 8" x 8" flues		14	1.143		58.50	51		109.50	143
0250	20" x 20" with one 12" x 12" flue		13.70	1.168		48	52.50		100.50	133
0300	20" x 24" with two 8" x 12" flues		12	1.333		66.50	59.50		126	165
0350	20" x 32" with two 12" x 12" flues	↓	10	1.600	↓	84.50	71.50		156	203

04 21 13.18 Columns

			Crew	Daily Output	Labor-Hours	Unit	Material	2018 Bare Costs Labor	Equipment	Total	Total Incl O&P
0010	**COLUMNS**, solid, excludes scaffolding, grout and reinforcing	R042110-10									
0050	Brick, 8" x 8", 9 brick/V.L.F.		D-1	56	.286	V.L.F.	5.30	12.80		18.10	25.50
0100	12" x 8", 13.5 brick/V.L.F.			37	.432		7.95	19.35		27.30	38.50
0200	12" x 12", 20 brick/V.L.F.			25	.640		11.80	28.50		40.30	57
0300	16" x 12", 27 brick/V.L.F.			19	.842		15.95	37.50		53.45	75.50
0400	16" x 16", 36 brick/V.L.F.			14	1.143		21.50	51		72.50	102
0500	20" x 16", 45 brick/V.L.F.			11	1.455		26.50	65		91.50	130
0600	20" x 20", 56 brick/V.L.F.			9	1.778		33	79.50		112.50	159
0700	24" x 20", 68 brick/V.L.F.			7	2.286		40	102		142	201
0800	24" x 24", 81 brick/V.L.F.			6	2.667		48	119		167	236
1000	36" x 36", 182 brick/V.L.F.		↓	3	5.333		108	239		347	485

04 21 13.30 Oversized Brick

		Crew	Daily Output	Labor-Hours	Unit	Material	2018 Bare Costs Labor	Equipment	Total	Total Incl O&P
0010	**OVERSIZED BRICK**, excludes scaffolding, grout and reinforcing									
0100	Veneer, 4" x 2.25" x 16"	D-8	387	.103	S.F.	5.40	4.74		10.14	13.20
0102	8" x 2.25" x 16", multicell		265	.151		17.10	6.95		24.05	29.50
0105	4" x 2.75" x 16"		412	.097		5.60	4.46		10.06	12.95
0107	8" x 2.75" x 16", multicell		295	.136		17	6.20		23.20	28.50
0110	4" x 4" x 16"		460	.087		3.61	3.99		7.60	10.10
0120	4" x 8" x 16"		533	.075		4.26	3.44		7.70	9.95
0122	4" x 8" x 16" multicell		327	.122		16.05	5.60		21.65	26.50
0125	Loadbearing, 6" x 4" x 16", grouted and reinforced		387	.103		11.15	4.74		15.89	19.50
0130	8" x 4" x 16", grouted and reinforced		327	.122		12.25	5.60		17.85	22
0132	10" x 4" x 16", grouted and reinforced		327	.122		25	5.60		30.60	36
0135	6" x 8" x 16", grouted and reinforced		440	.091		14.40	4.17		18.57	22.50
0140	8" x 8" x 16", grouted and reinforced	↓	400	.100	↓	15.45	4.59		20.04	24

04 21 Clay Unit Masonry

04 21 13 – Brick Masonry

04 21 13.30 Oversized Brick

		Crew	Daily Output	Labor-Hours	Unit	Material	2018 Bare Costs Labor	Equipment	Total	Total Incl O&P
0145	Curtainwall/reinforced veneer, 6" x 4" x 16"	D-8	387	.103	S.F.	16.15	4.74		20.89	25
0150	8" x 4" x 16"		327	.122		19.65	5.60		25.25	30
0152	10" x 4" x 16"		327	.122		27.50	5.60		33.10	39
0155	6" x 8" x 16"		440	.091		19.90	4.17		24.07	28.50
0160	8" x 8" x 16"		400	.100		28	4.59		32.59	37.50
0200	For 1 to 3 slots in face, add					15%				
0210	For 4 to 7 slots in face, add					25%				
0220	For bond beams, add					20%				
0230	For bullnose shapes, add					20%				
0240	For open end knockout, add					10%				
0250	For white or gray color group, add					10%				
0260	For 135 degree corner, add					250%				

04 21 13.35 Common Building Brick

		Crew	Daily Output	Labor-Hours	Unit	Material	2018 Bare Costs Labor	Equipment	Total	Total Incl O&P
0010	**COMMON BUILDING BRICK**, C62, T.L. lots, material only R042110-20									
0020	Standard				M	570			570	625
0050	Select				"	535			535	585

04 21 13.40 Structural Brick

		Crew	Daily Output	Labor-Hours	Unit	Material	2018 Bare Costs Labor	Equipment	Total	Total Incl O&P
0010	**STRUCTURAL BRICK** C652, Grade SW, incl. mortar, scaffolding not incl.									
0100	Standard unit, 4-5/8" x 2-3/4" x 9-5/8"	D-8	245	.163	S.F.	4.27	7.50		11.77	16.15
0120	Bond beam		225	.178		4.27	8.15		12.42	17.20
0140	V cut bond beam		225	.178		4.27	8.15		12.42	17.20
0160	Stretcher quoin, 5-5/8" x 2-3/4" x 9-5/8"		245	.163		7.90	7.50		15.40	20
0180	Corner quoin		245	.163		7.90	7.50		15.40	20
0200	Corner, 45 degree, 4-5/8" x 2-3/4" x 10-7/16"		235	.170		7.85	7.80		15.65	20.50

04 21 13.45 Face Brick

		Crew	Daily Output	Labor-Hours	Unit	Material	2018 Bare Costs Labor	Equipment	Total	Total Incl O&P
0010	**FACE BRICK** Material Only, C216, T.L. lots R042110-20									
0300	Standard modular, 4" x 2-2/3" x 8"				M	495			495	545
0450	Economy, 4" x 4" x 8"					870			870	955
0510	Economy, 4" x 4" x 12"					1,375			1,375	1,525
0550	Jumbo, 6" x 4" x 12"					1,650			1,650	1,825
0610	Jumbo, 8" x 4" x 12"					1,650			1,650	1,825
0650	Norwegian, 4" x 3-1/5" x 12"					1,375			1,375	1,500
0710	Norwegian, 6" x 3-1/5" x 12"					1,700			1,700	1,875
0850	Standard glazed, plain colors, 4" x 2-2/3" x 8"					1,775			1,775	1,950
1000	Deep trim shades, 4" x 2-2/3" x 8"					2,275			2,275	2,500
1080	Jumbo utility, 4" x 4" x 12"					1,550			1,550	1,700
1120	4" x 8" x 8"					1,975			1,975	2,175
1140	4" x 8" x 16"					5,850			5,850	6,425
1260	Engineer, 4" x 3-1/5" x 8"					585			585	645
1350	King, 4" x 2-3/4" x 10"					540			540	595
1770	Standard modular, double glazed, 4" x 2-2/3" x 8"					2,700			2,700	2,950
1850	Jumbo, colored glazed ceramic, 6" x 4" x 12"					2,825			2,825	3,125
2050	Jumbo utility, glazed, 4" x 4" x 12"					5,325			5,325	5,850
2100	4" x 8" x 8"					6,250			6,250	6,875
2150	4" x 16" x 8"					7,325			7,325	8,050
2170	For less than truck load lots, add					15			15	16.50
2180	For buff or gray brick, add					16			16	17.60

For customer support on your Building Construction Costs with RSMeans data, call 800.448.8182.

04 21 Clay Unit Masonry

04 21 26 – Glazed Structural Clay Tile Masonry

04 21 26.10 Structural Facing Tile

		Crew	Daily Output	Labor-Hours	Unit	Material	2018 Bare Costs Labor	Equipment	Total	Total Incl O&P
0010	**STRUCTURAL FACING TILE**, std. colors, excl. scaffolding, grout, reinforcing									
0020	6T series, 5-1/3" x 12", 2.3 pieces per S.F., glazed 1 side, 2" thick	D-8	225	.178	S.F.	8.65	8.15		16.80	22
0100	4" thick		220	.182		13.10	8.35		21.45	27
0150	Glazed 2 sides		195	.205		17.40	9.40		26.80	33.50
0250	6" thick		210	.190		17.60	8.75		26.35	33
0300	Glazed 2 sides		185	.216		21	9.90		30.90	38
0400	8" thick		180	.222		23	10.20		33.20	41
0500	Special shapes, group 1		400	.100	Ea.	8.35	4.59		12.94	16.20
0550	Group 2		375	.107		13.50	4.90		18.40	22.50
0600	Group 3		350	.114		17.40	5.25		22.65	27
0650	Group 4		325	.123		36	5.65		41.65	48
0700	Group 5		300	.133		43	6.10		49.10	57
0750	Group 6		275	.145		58	6.70		64.70	74
1000	Fire rated, 4" thick, 1 hr. rating		210	.190	S.F.	18.65	8.75		27.40	34
1300	Acoustic, 4" thick		210	.190	"	39.50	8.75		48.25	56.50
2000	8W series, 8" x 16", 1.125 pieces per S.F.									
2050	2" thick, glazed 1 side	D-8	360	.111	S.F.	10.10	5.10		15.20	18.95
2100	4" thick, glazed 1 side		345	.116		13.85	5.30		19.15	23.50
2150	Glazed 2 sides		325	.123		16.50	5.65		22.15	27
2200	6" thick, glazed 1 side		330	.121		28	5.55		33.55	39.50
2250	8" thick, glazed 1 side		310	.129		29	5.90		34.90	41
2500	Special shapes, group 1		300	.133	Ea.	17	6.10		23.10	28
2550	Group 2		280	.143		22.50	6.55		29.05	35
2600	Group 3		260	.154		24	7.05		31.05	37.50
2650	Group 4		250	.160		50.50	7.35		57.85	67
2700	Group 5		240	.167		43.50	7.65		51.15	59.50
2750	Group 6		230	.174		94.50	8		102.50	116
3000	4" thick, glazed 1 side		345	.116	S.F.	15.90	5.30		21.20	25.50
3100	Acoustic, 4" thick		345	.116	"	19.10	5.30		24.40	29
3120	4W series, 8" x 8", 2.25 pieces per S.F.									
3125	2" thick, glazed 1 side	D-8	360	.111	S.F.	9.90	5.10		15	18.70
3130	4" thick, glazed 1 side		345	.116		11.65	5.30		16.95	21
3135	Glazed 2 sides		325	.123		16.05	5.65		21.70	26.50
3140	6" thick, glazed 1 side		330	.121		16.60	5.55		22.15	27
3150	8" thick, glazed 1 side		310	.129		24	5.90		29.90	35.50
3155	Special shapes, group 1		300	.133	Ea.	7.75	6.10		13.85	17.90
3160	Group 2		280	.143	"	8.95	6.55		15.50	19.90
3200	For designer colors, add					25%				
3300	For epoxy mortar joints, add				S.F.	1.79			1.79	1.97

04 21 29 – Terra Cotta Masonry

04 21 29.10 Terra Cotta Masonry Components

		Crew	Daily Output	Labor-Hours	Unit	Material	2018 Bare Costs Labor	Equipment	Total	Total Incl O&P
0010	**TERRA COTTA MASONRY COMPONENTS**									
0020	Coping, split type, not glazed, 9" wide	D-1	90	.178	L.F.	10.50	7.95		18.45	24
0100	13" wide		80	.200		15.35	8.95		24.30	30.50
0200	Coping, split type, glazed, 9" wide		90	.178		17.85	7.95		25.80	32
0250	13" wide		80	.200		23.50	8.95		32.45	39.50
0500	Partition or back-up blocks, scored, in C.L. lots									
0700	Non-load bearing 12" x 12", 3" thick, special order	D-8	550	.073	S.F.	19.50	3.34		22.84	26.50
0750	4" thick, standard		500	.080		6.10	3.67		9.77	12.30
0800	6" thick		450	.089		8.20	4.08		12.28	15.30
0850	8" thick		400	.100		10.30	4.59		14.89	18.40
1000	Load bearing, 12" x 12", 4" thick, in walls		500	.080		5.75	3.67		9.42	11.90

04 21 Clay Unit Masonry

04 21 29 – Terra Cotta Masonry

04 21 29.10 Terra Cotta Masonry Components	Crew	Daily Output	Labor-Hours	Unit	Material	2018 Bare Costs Labor	Equipment	Total	Total Incl O&P
1050	In floors	D-8	750	.053	S.F.	5.75	2.45		8.20	10.05
1200	6" thick, in walls		450	.089		9.30	4.08		13.38	16.50
1250	In floors		675	.059		9.30	2.72		12.02	14.40
1400	8" thick, in walls		400	.100		11.55	4.59		16.14	19.80
1450	In floors		575	.070		11.55	3.19		14.74	17.65
1600	10" thick, in walls, special order		350	.114		27	5.25		32.25	38
1650	In floors, special order		500	.080		27	3.67		30.67	35.50
1800	12" thick, in walls, special order		300	.133		26.50	6.10		32.60	38.50
1850	In floors, special order	▼	450	.089		26.50	4.08		30.58	35.50
2000	For reinforcing with steel rods, add to above					15%	5%			
2100	For smooth tile instead of scored, add					2.83			2.83	3.11
2200	For L.C.L. quantities, add				▼	10%	10%			

04 21 29.20 Terra Cotta Tile

		Crew	Daily Output	Labor-Hours	Unit	Material	Labor	Equipment	Total	Total Incl O&P
0010	**TERRA COTTA TILE**, on walls, dry set, 1/2" thick									
0100	Square, hexagonal or lattice shapes, unglazed	1 Tilf	135	.059	S.F.	4.74	2.78		7.52	9.30
0300	Glazed, plain colors		130	.062		7.25	2.88		10.13	12.20
0400	Intense colors	▼	125	.064	▼	8.50	3		11.50	13.80

04 22 Concrete Unit Masonry

04 22 10 – Concrete Masonry Units

04 22 10.11 Autoclave Aerated Concrete Block

			Crew	Daily Output	Labor-Hours	Unit	Material	Labor	Equipment	Total	Total Incl O&P
0010	**AUTOCLAVE AERATED CONCRETE BLOCK**, excl. scaffolding, grout & reinforcing										
0050	Solid, 4" x 8" x 24", incl. mortar	G	D-8	600	.067	S.F.	1.50	3.06		4.56	6.35
0060	6" x 8" x 24"	G		600	.067		2.25	3.06		5.31	7.15
0070	8" x 8" x 24"	G		575	.070		3	3.19		6.19	8.20
0080	10" x 8" x 24"	G		575	.070		3.66	3.19		6.85	8.90
0090	12" x 8" x 24"	G	▼	550	.073	▼	4.50	3.34		7.84	10.05

04 22 10.12 Chimney Block

		Crew	Daily Output	Labor-Hours	Unit	Material	Labor	Equipment	Total	Total Incl O&P
0010	**CHIMNEY BLOCK**, excludes scaffolding, grout and reinforcing									
0220	1 piece, with 8" x 8" flue, 16" x 16"	D-1	28	.571	V.L.F.	19.70	25.50		45.20	60.50
0230	2 piece, 16" x 16"		26	.615		21	27.50		48.50	65
0240	2 piece, with 8" x 12" flue, 16" x 20"	▼	24	.667	▼	33.50	30		63.50	82.50

04 22 10.14 Concrete Block, Back-Up

			Crew	Daily Output	Labor-Hours	Unit	Material	Labor	Equipment	Total	Total Incl O&P
0010	**CONCRETE BLOCK, BACK-UP**, C90, 2000 psi	R042210-20									
0020	Normal weight, 8" x 16" units, tooled joint 1 side										
0050	Not-reinforced, 2000 psi, 2" thick		D-8	475	.084	S.F.	1.60	3.87		5.47	7.65
0200	4" thick			460	.087		1.92	3.99		5.91	8.20
0300	6" thick			440	.091		2.50	4.17		6.67	9.15
0350	8" thick			400	.100		2.66	4.59		7.25	10
0400	10" thick		▼	330	.121		3.15	5.55		8.70	11.95
0450	12" thick		D-9	310	.155		4.36	6.95		11.31	15.40
1000	Reinforced, alternate courses, 4" thick		D-8	450	.089		2.10	4.08		6.18	8.55
1100	6" thick			430	.093		2.68	4.27		6.95	9.50
1150	8" thick			395	.101		2.88	4.65		7.53	10.25
1200	10" thick		▼	320	.125		3.32	5.75		9.07	12.45
1250	12" thick		D-9	300	.160	▼	4.54	7.15		11.69	15.95

For customer support on your Building Construction Costs with RSMeans data, call 800.448.8182.

04 22 Concrete Unit Masonry

04 22 10 – Concrete Masonry Units

04 22 10.16 Concrete Block, Bond Beam

	04 22 10.16 Concrete Block, Bond Beam	Crew	Daily Output	Labor-Hours	Unit	Material	2018 Bare Costs Labor	Equipment	Total	Total Incl O&P
0010	**CONCRETE BLOCK, BOND BEAM**, C90, 2000 psi									
0020	Not including grout or reinforcing									
0125	Regular block, 6" thick	D-8	584	.068	L.F.	2.75	3.14		5.89	7.85
0130	8" high, 8" thick	"	565	.071		2.86	3.25		6.11	8.10
0150	12" thick	D-9	510	.094		4.05	4.22		8.27	10.90
0525	Lightweight, 6" thick	D-8	592	.068		2.98	3.10		6.08	8
0530	8" high, 8" thick	"	575	.070		3.59	3.19		6.78	8.85
0550	12" thick	D-9	520	.092	↓	4.84	4.14		8.98	11.65
2000	Including grout and 2 #5 bars									
2100	Regular block, 8" high, 8" thick	D-8	300	.133	L.F.	5.05	6.10		11.15	14.90
2150	12" thick	D-9	250	.192		6.90	8.60		15.50	21
2500	Lightweight, 8" high, 8" thick	D-8	305	.131		5.80	6		11.80	15.60
2550	12" thick	D-9	255	.188	↓	7.70	8.45		16.15	21.50

04 22 10.18 Concrete Block, Column

	04 22 10.18 Concrete Block, Column	Crew	Daily Output	Labor-Hours	Unit	Material	2018 Bare Costs Labor	Equipment	Total	Total Incl O&P
0010	**CONCRETE BLOCK, COLUMN** or pilaster									
0050	Including vertical reinforcing (4-#4 bars) and grout									
0160	1 piece unit, 16" x 16"	D-1	26	.615	V.L.F.	15.65	27.50		43.15	59
0170	2 piece units, 16" x 20"		24	.667		21.50	30		51.50	69.50
0180	20" x 20"		22	.727		28	32.50		60.50	80.50
0190	22" x 24"		18	.889		42	40		82	108
0200	20" x 32"	↓	14	1.143	↓	49	51		100	132

04 22 10.19 Concrete Block, Insulation Inserts

	04 22 10.19 Concrete Block, Insulation Inserts	Crew	Daily Output	Labor-Hours	Unit	Material	2018 Bare Costs Labor	Equipment	Total	Total Incl O&P
0010	**CONCRETE BLOCK, INSULATION INSERTS**									
0100	Styrofoam, plant installed, add to block prices									
0200	8" x 16" units, 6" thick				S.F.	1.22			1.22	1.34
0250	8" thick					1.37			1.37	1.51
0300	10" thick					1.42			1.42	1.56
0350	12" thick					1.57			1.57	1.73
0500	8" x 8" units, 8" thick					1.22			1.22	1.34
0550	12" thick				↓	1.42			1.42	1.56

04 22 10.23 Concrete Block, Decorative

	04 22 10.23 Concrete Block, Decorative	Crew	Daily Output	Labor-Hours	Unit	Material	2018 Bare Costs Labor	Equipment	Total	Total Incl O&P
0010	**CONCRETE BLOCK, DECORATIVE**, C90, 2000 psi									
0020	Embossed, simulated brick face									
0100	8" x 16" units, 4" thick	D-8	400	.100	S.F.	2.88	4.59		7.47	10.20
0200	8" thick		340	.118		3.13	5.40		8.53	11.70
0250	12" thick	↓	300	.133	↓	5.30	6.10		11.40	15.15
0400	Embossed both sides									
0500	8" thick	D-8	300	.133	S.F.	4.23	6.10		10.33	14
0550	12" thick	"	275	.145	"	5.55	6.70		12.25	16.30
1000	Fluted high strength									
1100	8" x 16" x 4" thick, flutes 1 side,	D-8	345	.116	S.F.	4.09	5.30		9.39	12.65
1150	Flutes 2 sides		335	.119		4.69	5.50		10.19	13.55
1200	8" thick	↓	300	.133		5.50	6.10		11.60	15.40
1250	For special colors, add				↓	.67			.67	.73
1400	Deep grooved, smooth face									
1450	8" x 16" x 4" thick	D-8	345	.116	S.F.	2.63	5.30		7.93	11.05
1500	8" thick	"	300	.133	"	4.14	6.10		10.24	13.90
2000	Formblock, incl. inserts & reinforcing									
2100	8" x 16" x 8" thick	D-8	345	.116	S.F.	3.65	5.30		8.95	12.15
2150	12" thick	"	310	.129	"	4.83	5.90		10.73	14.35
2500	Ground face									

04 22 Concrete Unit Masonry

04 22 10 – Concrete Masonry Units

04 22 10.23 Concrete Block, Decorative		Crew	Daily Output	Labor-Hours	Unit	Material	2018 Bare Costs Labor	Equipment	Total	Total Incl O&P
2600	8" x 16" x 4" thick	D-8	345	.116	S.F.	3.97	5.30		9.27	12.50
2650	6" thick		325	.123		4.74	5.65		10.39	13.85
2700	8" thick	↓	300	.133		5.20	6.10		11.30	15.10
2750	12" thick	D-9	265	.181	↓	5.70	8.10		13.80	18.70
2900	For special colors, add, minimum					15%				
2950	For special colors, add, maximum					45%				
4000	Slump block									
4100	4" face height x 16" x 4" thick	D-1	165	.097	S.F.	4.37	4.34		8.71	11.45
4150	6" thick		160	.100		6.25	4.48		10.73	13.75
4200	8" thick		155	.103		6.20	4.62		10.82	13.95
4250	10" thick		140	.114		12.95	5.10		18.05	22
4300	12" thick		130	.123		13.10	5.50		18.60	23
4400	6" face height x 16" x 6" thick		155	.103		5.80	4.62		10.42	13.45
4450	8" thick		150	.107		8.75	4.78		13.53	16.95
4500	10" thick		130	.123		13.45	5.50		18.95	23.50
4550	12" thick	↓	120	.133	↓	13.95	5.95		19.90	24.50
5000	Split rib profile units, 1" deep ribs, 8 ribs									
5100	8" x 16" x 4" thick	D-8	345	.116	S.F.	4.06	5.30		9.36	12.60
5150	6" thick		325	.123		4.61	5.65		10.26	13.70
5200	8" thick	↓	300	.133		5.20	6.10		11.30	15.05
5250	12" thick	D-9	275	.175		6.05	7.80		13.85	18.60
5400	For special deeper colors, 4" thick, add					1.37			1.37	1.50
5450	12" thick, add					1.42			1.42	1.57
5600	For white, 4" thick, add					1.37			1.37	1.50
5650	6" thick, add					1.39			1.39	1.53
5700	8" thick, add					1.44			1.44	1.58
5750	12" thick, add				↓	1.48			1.48	1.63
6000	Split face									
6100	8" x 16" x 4" thick	D-8	350	.114	S.F.	3.77	5.25		9.02	12.20
6150	6" thick		325	.123		4.28	5.65		9.93	13.35
6200	8" thick	↓	300	.133		4.82	6.10		10.92	14.65
6250	12" thick	D-9	270	.178		5.85	7.95		13.80	18.60
6300	For scored, add					.40			.40	.44
6400	For special deeper colors, 4" thick, add					.67			.67	.73
6450	6" thick, add					.79			.79	.87
6500	8" thick, add					.81			.81	.89
6550	12" thick, add					.84			.84	.92
6650	For white, 4" thick, add					1.36			1.36	1.49
6700	6" thick, add					1.37			1.37	1.50
6750	8" thick, add					1.38			1.38	1.52
6800	12" thick, add				↓	1.42			1.42	1.57
7000	Scored ground face, 2 to 5 scores									
7100	8" x 16" x 4" thick	D-8	340	.118	S.F.	8.80	5.40		14.20	17.90
7150	6" thick		310	.129		9.70	5.90		15.60	19.75
7200	8" thick	↓	290	.138		10.90	6.35		17.25	21.50
7250	12" thick	D-9	265	.181	↓	14.70	8.10		22.80	28.50
8000	Hexagonal face profile units, 8" x 16" units									
8100	4" thick, hollow	D-8	340	.118	S.F.	3.83	5.40		9.23	12.45
8200	Solid		340	.118		4.90	5.40		10.30	13.65
8300	6" thick, hollow		310	.129		4.07	5.90		9.97	13.50
8350	8" thick, hollow	↓	290	.138	↓	4.64	6.35		10.99	14.80
8500	For stacked bond, add						26%			
8550	For high rise construction, add per story	D-8	67.80	.590	M.S.F.		27		27	41.50

106

04 22 Concrete Unit Masonry

04 22 10 – Concrete Masonry Units

04 22 10.23 Concrete Block, Decorative

		Crew	Daily Output	Labor-Hours	Unit	Material	2018 Bare Costs Labor	Equipment	Total	Total Incl O&P
8600	For scored block, add					10%				
8650	For honed or ground face, per face, add				Ea.	1.21			1.21	1.33
8700	For honed or ground end, per end, add				"	1.21			1.21	1.33
8750	For bullnose block, add					10%				
8800	For special color, add					13%				

04 22 10.24 Concrete Block, Exterior

		Crew	Daily Output	Labor-Hours	Unit	Material	2018 Bare Costs Labor	Equipment	Total	Total Incl O&P
0010	**CONCRETE BLOCK, EXTERIOR**, C90, 2000 psi									
0020	Reinforced alt courses, tooled joints 2 sides									
0100	Normal weight, 8" x 16" x 6" thick	D-8	395	.101	S.F.	2.46	4.65		7.11	9.80
0200	8" thick		360	.111		3.80	5.10		8.90	12
0250	10" thick		290	.138		4.41	6.35		10.76	14.55
0300	12" thick	D-9	250	.192		5.15	8.60		13.75	18.80
0500	Lightweight, 8" x 16" x 6" thick	D-8	450	.089		3.41	4.08		7.49	10
0600	8" thick		430	.093		3.84	4.27		8.11	10.75
0650	10" thick		395	.101		4.24	4.65		8.89	11.75
0700	12" thick	D-9	350	.137		4.37	6.15		10.52	14.20

04 22 10.26 Concrete Block Foundation Wall

		Crew	Daily Output	Labor-Hours	Unit	Material	2018 Bare Costs Labor	Equipment	Total	Total Incl O&P
0010	**CONCRETE BLOCK FOUNDATION WALL**, C90/C145									
0050	Normal-weight, cut joints, horiz joint reinf, no vert reinf.									
0200	Hollow, 8" x 16" x 6" thick	D-8	455	.088	S.F.	2.93	4.04		6.97	9.40
0250	8" thick		425	.094		3.11	4.32		7.43	10
0300	10" thick		350	.114		3.57	5.25		8.82	11.95
0350	12" thick	D-9	300	.160		4.78	7.15		11.93	16.20
0500	Solid, 8" x 16" block, 6" thick	D-8	440	.091		2.99	4.17		7.16	9.70
0550	8" thick	"	415	.096		4.23	4.42		8.65	11.40
0600	12" thick	D-9	350	.137		6.10	6.15		12.25	16.10

04 22 10.28 Concrete Block, High Strength

		Crew	Daily Output	Labor-Hours	Unit	Material	2018 Bare Costs Labor	Equipment	Total	Total Incl O&P
0010	**CONCRETE BLOCK, HIGH STRENGTH**									
0050	Hollow, reinforced alternate courses, 8" x 16" units									
0200	3500 psi, 4" thick	D-8	440	.091	S.F.	2.46	4.17		6.63	9.10
0250	6" thick		395	.101		2.37	4.65		7.02	9.70
0300	8" thick		360	.111		3.72	5.10		8.82	11.90
0350	12" thick	D-9	250	.192		5	8.60		13.60	18.65
0500	5000 psi, 4" thick	D-8	440	.091		2.16	4.17		6.33	8.75
0550	6" thick		395	.101		3.06	4.65		7.71	10.45
0600	8" thick		360	.111		4.35	5.10		9.45	12.60
0650	12" thick	D-9	300	.160		5	7.15		12.15	16.45
1000	For 75% solid block, add					30%				
1050	For 100% solid block, add					50%				

04 22 10.30 Concrete Block, Interlocking

		Crew	Daily Output	Labor-Hours	Unit	Material	2018 Bare Costs Labor	Equipment	Total	Total Incl O&P
0010	**CONCRETE BLOCK, INTERLOCKING**									
0100	Not including grout or reinforcing									
0200	8" x 16" units, 2,000 psi, 8" thick	D-1	245	.065	S.F.	3.15	2.93		6.08	7.95
0300	12" thick		220	.073		4.69	3.26		7.95	10.15
0350	16" thick		185	.086		7.05	3.87		10.92	13.70
0400	Including grout & reinforcing, 8" thick	D-4	245	.131		8.60	5.90	.53	15.03	19
0450	12" thick		220	.145		10.30	6.55	.58	17.43	22
0500	16" thick		185	.173		12.85	7.80	.70	21.35	27

For customer support on your Building Construction Costs with RSMeans data, call 800.448.8182.

107

04 22 Concrete Unit Masonry

04 22 10 – Concrete Masonry Units

04 22 10.32 Concrete Block, Lintels

		Crew	Daily Output	Labor-Hours	Unit	Material	2018 Bare Costs Labor	2018 Bare Costs Equipment	Total	Total Incl O&P
0010	**CONCRETE BLOCK, LINTELS**, C90, normal weight									
0100	Including grout and horizontal reinforcing									
0200	8" x 8" x 8", 1 #4 bar	D-4	300	.107	L.F.	4.03	4.81	.43	9.27	12.20
0250	2 #4 bars		295	.108		4.24	4.89	.44	9.57	12.60
0400	8" x 16" x 8", 1 #4 bar		275	.116		3.69	5.25	.47	9.41	12.55
0450	2 #4 bars		270	.119		3.91	5.35	.48	9.74	12.95
1000	12" x 8" x 8", 1 #4 bar		275	.116		5.50	5.25	.47	11.22	14.55
1100	2 #4 bars		270	.119		5.75	5.35	.48	11.58	14.95
1150	2 #5 bars		270	.119		6	5.35	.48	11.83	15.20
1200	2 #6 bars		265	.121		6.25	5.45	.49	12.19	15.75
1500	12" x 16" x 8", 1 #4 bar		250	.128		6.60	5.75	.51	12.86	16.60
1600	2 #3 bars		245	.131		6.60	5.90	.53	13.03	16.80
1650	2 #4 bars		245	.131		6.80	5.90	.53	13.23	17
1700	2 #5 bars	↓	240	.133	↓	7.05	6	.54	13.59	17.50

04 22 10.33 Lintel Block

		Crew	Daily Output	Labor-Hours	Unit	Material	2018 Bare Costs Labor	2018 Bare Costs Equipment	Total	Total Incl O&P
0010	**LINTEL BLOCK**									
3481	Lintel block 6" x 8" x 8"	D-1	300	.053	Ea.	1.36	2.39		3.75	5.15
3501	6" x 16" x 8"		275	.058		2.10	2.61		4.71	6.30
3521	8" x 8" x 8"		275	.058		1.21	2.61		3.82	5.30
3561	8" x 16" x 8"	↓	250	.064	↓	1.90	2.87		4.77	6.50

04 22 10.34 Concrete Block, Partitions

		Crew	Daily Output	Labor-Hours	Unit	Material	2018 Bare Costs Labor	2018 Bare Costs Equipment	Total	Total Incl O&P
0010	**CONCRETE BLOCK, PARTITIONS**, excludes scaffolding R042210-20									
0100	Acoustical slotted block									
0200	4" thick, type A-1	D-8	315	.127	S.F.	6.45	5.85		12.30	16
0210	8" thick		275	.145		7.35	6.70		14.05	18.30
0250	8" thick, type Q		275	.145		14.40	6.70		21.10	26
0260	4" thick, type RSC		315	.127		10.05	5.85		15.90	19.95
0270	6" thick		295	.136		10.05	6.20		16.25	20.50
0280	8" thick		275	.145		10.05	6.70		16.75	21.50
0290	12" thick		250	.160		10.05	7.35		17.40	22.50
0300	8" thick, type RSR		275	.145		10.05	6.70		16.75	21.50
0400	8" thick, type RSC/RF		275	.145		7.75	6.70		14.45	18.70
0410	10" thick		260	.154		10	7.05		17.05	22
0420	12" thick		250	.160		8.90	7.35		16.25	21
0430	12" thick, type RSC/RF-4		250	.160		15.85	7.35		23.20	28.50
0500	NRC .60 type R, 8" thick		265	.151		12.35	6.95		19.30	24
0600	NRC .65 type RR, 8" thick		265	.151		8.65	6.95		15.60	20
0700	NRC .65 type 4R-RF, 8" thick		265	.151		12.65	6.95		19.60	24.50
0710	NRC .70 type R, 12" thick	↓	245	.163	↓	13.25	7.50		20.75	26
1000	Lightweight block, tooled joints, 2 sides, hollow									
1100	Not reinforced, 8" x 16" x 4" thick	D-8	440	.091	S.F.	1.95	4.17		6.12	8.55
1150	6" thick		410	.098		2.75	4.48		7.23	9.90
1200	8" thick		385	.104		3.36	4.77		8.13	11
1250	10" thick	↓	370	.108		4.07	4.96		9.03	12.10
1300	12" thick	D-9	350	.137		4.27	6.15		10.42	14.10
2000	Not reinforced, 8" x 24" x 4" thick, hollow		460	.104		1.42	4.67		6.09	8.70
2100	6" thick		440	.109		1.97	4.89		6.86	9.65
2150	8" thick		415	.116		2.45	5.20		7.65	10.65
2200	10" thick		385	.125		2.96	5.60		8.56	11.80
2250	12" thick	↓	365	.132		3.14	5.90		9.04	12.45
2800	Solid, not reinforced, 8" x 16" x 2" thick	D-8	440	.091		1.76	4.17		5.93	8.35
2900	4" thick	↓	420	.095	↓	1.72	4.37		6.09	8.60

04 22 10.34 Concrete Block, Partitions

		Crew	Daily Output	Labor-Hours	Unit	Material	2018 Bare Costs Labor	Equipment	Total	Total Incl O&P
2950	6" thick	D-8	390	.103	S.F.	3.26	4.71		7.97	10.80
3000	8" thick		365	.110		3.67	5.05		8.72	11.75
3050	10" thick	▼	350	.114		4.06	5.25		9.31	12.50
3100	12" thick	D-9	330	.145	▼	4.18	6.50		10.68	14.60
4000	Regular block, tooled joints, 2 sides, hollow									
4100	Not reinforced, 8" x 16" x 4" thick	D-8	430	.093	S.F.	1.82	4.27		6.09	8.55
4150	6" thick		400	.100		2.40	4.59		6.99	9.70
4200	8" thick		375	.107		2.57	4.90		7.47	10.30
4250	10" thick	▼	360	.111		3.05	5.10		8.15	11.15
4300	12" thick	D-9	340	.141		4.26	6.30		10.56	14.40
4500	Reinforced alternate courses, 8" x 16" x 4" thick	D-8	425	.094		2	4.32		6.32	8.80
4550	6" thick		395	.101		2.58	4.65		7.23	9.95
4600	8" thick		370	.108		2.79	4.96		7.75	10.65
4650	10" thick	▼	355	.113		4.30	5.15		9.45	12.65
4700	12" thick	D-9	335	.143		4.45	6.40		10.85	14.75
4900	Solid, not reinforced, 2" thick	D-8	435	.092		1.54	4.22		5.76	8.15
5000	3" thick		430	.093		1.43	4.27		5.70	8.10
5050	4" thick		415	.096		2.08	4.42		6.50	9.05
5100	6" thick		385	.104		2.47	4.77		7.24	10
5150	8" thick	▼	360	.111		3.69	5.10		8.79	11.85
5200	12" thick	D-9	325	.148		5.60	6.60		12.20	16.30
5500	Solid, reinforced alternate courses, 4" thick	D-8	420	.095		2.26	4.37		6.63	9.20
5550	6" thick		380	.105		2.62	4.83		7.45	10.30
5600	8" thick	▼	355	.113		3.85	5.15		9	12.15
5650	12" thick	D-9	320	.150	▼	4.55	6.70		11.25	15.30

04 22 10.38 Concrete Brick

		Crew	Daily Output	Labor-Hours	Unit	Material	2018 Bare Costs Labor	Equipment	Total	Total Incl O&P
0010	**CONCRETE BRICK**, C55, grade N, type 1									
0100	Regular, 4" x 2-1/4" x 8"	D-8	660	.061	Ea.	.49	2.78		3.27	4.80
0125	Rusticated, 4" x 2-1/4" x 8"		660	.061		.55	2.78		3.33	4.86
0150	Frog, 4" x 2-1/4" x 8"		660	.061		.53	2.78		3.31	4.84
0200	Double, 4" x 4-7/8" x 8"	▼	535	.075	▼	.86	3.43		4.29	6.20

04 22 10.42 Concrete Block, Screen Block

		Crew	Daily Output	Labor-Hours	Unit	Material	2018 Bare Costs Labor	Equipment	Total	Total Incl O&P
0010	**CONCRETE BLOCK, SCREEN BLOCK**									
0200	8" x 16", 4" thick	D-8	330	.121	S.F.	9.10	5.55		14.65	18.50
0300	8" thick		270	.148		9.90	6.80		16.70	21.50
0350	12" x 12", 4" thick		290	.138		6.90	6.35		13.25	17.30
0500	8" thick	▼	250	.160	▼	7.10	7.35		14.45	19.05

04 22 10.44 Glazed Concrete Block

		Crew	Daily Output	Labor-Hours	Unit	Material	2018 Bare Costs Labor	Equipment	Total	Total Incl O&P
0010	**GLAZED CONCRETE BLOCK** C744									
0100	Single face, 8" x 16" units, 2" thick	D-8	360	.111	S.F.	10.80	5.10		15.90	19.70
0200	4" thick		345	.116		11.10	5.30		16.40	20.50
0250	6" thick		330	.121		12	5.55		17.55	21.50
0300	8" thick		310	.129		12.60	5.90		18.50	23
0350	10" thick	▼	295	.136		14.65	6.20		20.85	25.50
0400	12" thick	D-9	280	.171		15.60	7.70		23.30	29
0700	Double face, 8" x 16" units, 4" thick	D-8	340	.118		15.95	5.40		21.35	26
0750	6" thick		320	.125		19.05	5.75		24.80	30
0800	8" thick	▼	300	.133	▼	20	6.10		26.10	31.50
1000	Jambs, bullnose or square, single face, 8" x 16", 2" thick		315	.127	Ea.	20.50	5.85		26.35	31.50
1050	4" thick		285	.140	"	21	6.45		27.45	33
1200	Caps, bullnose or square, 8" x 16", 2" thick		420	.095	L.F.	19.15	4.37		23.52	27.50
1250	4" thick	▼	380	.105	"	21.50	4.83		26.33	31

04 22 Concrete Unit Masonry

04 22 10 – Concrete Masonry Units

04 22 10.44 Glazed Concrete Block		Crew	Daily Output	Labor-Hours	Unit	Material	2018 Bare Costs Labor	Equipment	Total	Total Incl O&P
1256	Corner, bullnose or square, 2" thick	D-8	280	.143	Ea.	22.50	6.55		29.05	34.50
1258	4" thick		270	.148		25	6.80		31.80	38
1260	6" thick		260	.154		30	7.05		37.05	44
1270	8" thick		250	.160		36.50	7.35		43.85	52
1280	10" thick		240	.167		37	7.65		44.65	52.50
1290	12" thick		230	.174	↓	39	8		47	55
1500	Cove base, 8" x 16", 2" thick		315	.127	L.F.	9.90	5.85		15.75	19.80
1550	4" thick		285	.140		10	6.45		16.45	21
1600	6" thick		265	.151		10.70	6.95		17.65	22.50
1650	8" thick	↓	245	.163	↓	11.20	7.50		18.70	24

04 23 Glass Unit Masonry

04 23 13 – Vertical Glass Unit Masonry

04 23 13.10 Glass Block

		Crew	Daily Output	Labor-Hours	Unit	Material	2018 Bare Costs Labor	Equipment	Total	Total Incl O&P
0010	**GLASS BLOCK**									
0100	Plain, 4" thick, under 1,000 S.F., 6" x 6"	D-8	115	.348	S.F.	26	15.95		41.95	53.50
0150	8" x 8"		160	.250		16.40	11.50		27.90	35.50
0160	end block		160	.250		60.50	11.50		72	84
0170	90 degree corner		160	.250		63.50	11.50		75	87
0180	45 degree corner		160	.250		54	11.50		65.50	76.50
0200	12" x 12"		175	.229		24	10.50		34.50	42.50
0210	4" x 8"		160	.250		30.50	11.50		42	51
0220	6" x 8"		160	.250		19.90	11.50		31.40	39.50
0300	1,000 to 5,000 S.F., 6" x 6"		135	.296		25.50	13.60		39.10	49
0350	8" x 8"		190	.211		16.05	9.65		25.70	32.50
0400	12" x 12"		215	.186		24	8.55		32.55	39
0410	4" x 8"		215	.186		29.50	8.55		38.05	45.50
0420	6" x 8"		215	.186		19.50	8.55		28.05	34.50
0500	Over 5,000 S.F., 6" x 6"		145	.276		25	12.65		37.65	47
0550	8" x 8"		215	.186		15.60	8.55		24.15	30
0600	12" x 12"		240	.167		23	7.65		30.65	37
0610	4" x 8"		240	.167		29	7.65		36.65	43
0620	6" x 8"	↓	240	.167	↓	18.90	7.65		26.55	32.50
0700	For solar reflective blocks, add					100%				
1000	Thinline, plain, 3-1/8" thick, under 1,000 S.F., 6" x 6"	D-8	115	.348	S.F.	23.50	15.95		39.45	50.50
1050	8" x 8"		160	.250		13.15	11.50		24.65	32
1200	Over 5,000 S.F., 6" x 6"		145	.276		22.50	12.65		35.15	44.50
1250	8" x 8"		215	.186		12.60	8.55		21.15	27
1400	For cleaning block after installation (both sides), add	↓	1000	.040	↓	.16	1.84		2	2.99
4000	Accessories									
4100	Anchors, 20 ga. galv., 1-3/4" wide x 24" long				Ea.	5.35			5.35	5.90
4200	Emulsion asphalt				Gal.	11.85			11.85	13.05
4300	Expansion joint, fiberglass				L.F.	.70			.70	.77
4400	Steel mesh, double galvanized				"	1.01			1.01	1.11

For customer support on your Building Construction Costs with RSMeans data, call 800.448.8182.

04 24 Adobe Unit Masonry

04 24 16 – Manufactured Adobe Unit Masonry

04 24 16.06 Adobe Brick

		Crew	Daily Output	Labor-Hours	Unit	Material	2018 Bare Costs Labor	2018 Bare Costs Equipment	Total	Total Incl O&P
0010	**ADOBE BRICK**, Semi-stabilized, with cement mortar									
0060	Brick, 10" x 4" x 14", 2.6/S.F. [G]	D-8	560	.071	S.F.	4.78	3.28		8.06	10.25
0080	12" x 4" x 16", 2.3/S.F. [G]		580	.069		7	3.17		10.17	12.55
0100	10" x 4" x 16", 2.3/S.F. [G]		590	.068		6.55	3.11		9.66	11.95
0120	8" x 4" x 16", 2.3/S.F. [G]		560	.071		4.99	3.28		8.27	10.50
0140	4" x 4" x 16", 2.3/S.F. [G]		540	.074		4.88	3.40		8.28	10.55
0160	6" x 4" x 16", 2.3/S.F. [G]		540	.074		4.53	3.40		7.93	10.20
0180	4" x 4" x 12", 3.0/S.F. [G]		520	.077		5.25	3.53		8.78	11.20
0200	8" x 4" x 12", 3.0/S.F. [G]		520	.077		4.34	3.53		7.87	10.20

04 25 Unit Masonry Panels

04 25 20 – Pre-Fabricated Masonry Panels

04 25 20.10 Brick and Epoxy Mortar Panels

		Crew	Daily Output	Labor-Hours	Unit	Material	2018 Bare Costs Labor	2018 Bare Costs Equipment	Total	Total Incl O&P
0010	**BRICK AND EPOXY MORTAR PANELS**									
0020	Prefabricated brick & epoxy mortar, 4" thick, minimum	C-11	775	.093	S.F.	8	5.05	2.60	15.65	19.75
0100	Maximum	"	500	.144		10	7.85	4.03	21.88	28
0200	For 2" concrete back-up, add					50%				
0300	For 1" urethane & 3" concrete back-up, add					70%				

04 27 Multiple-Wythe Unit Masonry

04 27 10 – Multiple-Wythe Masonry

04 27 10.20 Cavity Walls

		Crew	Daily Output	Labor-Hours	Unit	Material	2018 Bare Costs Labor	2018 Bare Costs Equipment	Total	Total Incl O&P
0010	**CAVITY WALLS**, brick and CMU, includes joint reinforcing and ties									
0200	4" face brick, 4" block	D-8	165	.242	S.F.	6.10	11.15		17.25	24
0400	6" block		145	.276		6.50	12.65		19.15	26.50
0600	8" block		125	.320		6.50	14.70		21.20	29.50

04 27 10.30 Brick Walls

		Crew	Daily Output	Labor-Hours	Unit	Material	2018 Bare Costs Labor	2018 Bare Costs Equipment	Total	Total Incl O&P
0010	**BRICK WALLS**, including mortar, excludes scaffolding R042110-20									
0020	Estimating by number of brick									
0140	Face brick, 4" thick wall, 6.75 brick/S.F.	D-8	1.45	27.586	M	590	1,275		1,865	2,600
0150	Common brick, 4" thick wall, 6.75 brick/S.F.		1.60	25		665	1,150		1,815	2,475
0204	8" thick, 13.50 brick/S.F.		1.80	22.222		685	1,025		1,710	2,300
0250	12" thick, 20.25 brick/S.F.		1.90	21.053		690	965		1,655	2,225
0304	16" thick, 27.00 brick/S.F.		2	20		695	920		1,615	2,175
0500	Reinforced, face brick, 4" thick wall, 6.75 brick/S.F.		1.40	28.571		615	1,300		1,915	2,675
0520	Common brick, 4" thick wall, 6.75 brick/S.F.		1.55	25.806		690	1,175		1,865	2,575
0550	8" thick, 13.50 brick/S.F.		1.75	22.857		710	1,050		1,760	2,375
0600	12" thick, 20.25 brick/S.F.		1.85	21.622		710	990		1,700	2,300
0650	16" thick, 27.00 brick/S.F.		1.95	20.513		720	940		1,660	2,250
0790	Alternate method of figuring by square foot									
0800	Face brick, 4" thick wall, 6.75 brick/S.F.	D-8	215	.186	S.F.	3.99	8.55		12.54	17.45
0850	Common brick, 4" thick wall, 6.75 brick/S.F.		240	.167		4.50	7.65		12.15	16.65
0900	8" thick, 13.50 brick/S.F.		135	.296		9.25	13.60		22.85	31
1000	12" thick, 20.25 brick/S.F.		95	.421		13.90	19.35		33.25	45
1050	16" thick, 27.00 brick/S.F.		75	.533		18.75	24.50		43.25	58
1200	Reinforced, face brick, 4" thick wall, 6.75 brick/S.F.		210	.190		4.15	8.75		12.90	17.95
1220	Common brick, 4" thick wall, 6.75 brick/S.F.		235	.170		4.66	7.80		12.46	17.10
1250	8" thick, 13.50 brick/S.F.		130	.308		9.55	14.10		23.65	32
1300	12" thick, 20.25 brick/S.F.		90	.444		14.40	20.50		34.90	47

04 27 Multiple-Wythe Unit Masonry

04 27 10 – Multiple-Wythe Masonry

04 27 10.30 Brick Walls	Crew	Daily Output	Labor-Hours	Unit	Material	2018 Bare Costs Labor	2018 Bare Costs Equipment	Total	Total Incl O&P
1350	16" thick, 27.00 brick/S.F.	D-8	70	.571	S.F.	19.40	26		45.40	61.50

04 27 10.40 Steps
0010	**STEPS**									
0012	Entry steps, select common brick	D-1	.30	53.333	M	535	2,400		2,935	4,225

04 41 Dry-Placed Stone

04 41 10 – Dry Placed Stone

04 41 10.10 Rough Stone Wall

			Crew	Daily Output	Labor-Hours	Unit	Material	Labor	Equipment	Total	Total Incl O&P
0011	**ROUGH STONE WALL,** Dry										
0012	Dry laid (no mortar), under 18" thick	G	D-1	60	.267	C.F.	13.50	11.95		25.45	33
0100	Random fieldstone, under 18" thick	G	D-12	60	.533		13.50	24		37.50	52
0150	Over 18" thick	G	"	63	.508		16.20	23		39.20	53
0500	Field stone veneer	G	D-8	120	.333	S.F.	12.65	15.30		27.95	37.50
0510	Valley stone veneer	G		120	.333		12.65	15.30		27.95	37.50
0520	River stone veneer	G		120	.333		12.65	15.30		27.95	37.50
0600	Rubble stone walls, in mortar bed, up to 18" thick	G	D-11	75	.320	C.F.	16.30	15.15		31.45	41

04 43 Stone Masonry

04 43 10 – Masonry with Natural and Processed Stone

04 43 10.05 Ashlar Veneer

		Crew	Daily Output	Labor-Hours	Unit	Material	Labor	Equipment	Total	Total Incl O&P
0011	**ASHLAR VENEER** +/- 4" thk, random or random rectangular									
0150	Sawn face, split joints, low priced stone	D-8	140	.286	S.F.	12.20	13.10		25.30	33.50
0200	Medium priced stone		130	.308		13.85	14.10		27.95	36.50
0300	High priced stone		120	.333		18.40	15.30		33.70	43.50
0600	Seam face, split joints, medium price stone		125	.320		19.35	14.70		34.05	44
0700	High price stone		120	.333		19.10	15.30		34.40	44.50
1000	Split or rock face, split joints, medium price stone		125	.320		11.55	14.70		26.25	35
1100	High price stone		120	.333		17.55	15.30		32.85	43

04 43 10.10 Bluestone

		Crew	Daily Output	Labor-Hours	Unit	Material	Labor	Equipment	Total	Total Incl O&P
0010	**BLUESTONE,** cut to size									
0500	Sills, natural cleft, 10" wide to 6' long, 1-1/2" thick	D-11	70	.343	L.F.	13.75	16.20		29.95	40
0550	2" thick	"	63	.381		15.05	18		33.05	44
1000	Stair treads, natural cleft, 12" wide, 6' long, 1-1/2" thick	D-10	115	.278		12.85	13.75	3.16	29.76	38.50
1050	2" thick		105	.305		13.90	15.10	3.46	32.46	42
1100	Smooth finish, 1-1/2" thick		115	.278		12.85	13.75	3.16	29.76	38.50
1150	2" thick		105	.305		13.90	15.10	3.46	32.46	42
1300	Thermal finish, 1-1/2" thick		115	.278		12.85	13.75	3.16	29.76	38.50
1350	2" thick		105	.305		13.90	15.10	3.46	32.46	42

04 43 10.45 Granite

		Crew	Daily Output	Labor-Hours	Unit	Material	Labor	Equipment	Total	Total Incl O&P
0010	**GRANITE,** cut to size									
0050	Veneer, polished face, 3/4" to 1-1/2" thick									
0150	Low price, gray, light gray, etc.	D-10	130	.246	S.F.	27.50	12.20	2.80	42.50	52
0180	Medium price, pink, brown, etc.		130	.246		30.50	12.20	2.80	45.50	55
0220	High price, red, black, etc.		130	.246		43	12.20	2.80	58	68.50
0300	1-1/2" to 2-1/2" thick, veneer									
0350	Low price, gray, light gray, etc.	D-10	130	.246	S.F.	29	12.20	2.80	44	53.50
0500	Medium price, pink, brown, etc.		130	.246		34	12.20	2.80	49	59
0550	High price, red, black, etc.		130	.246		53	12.20	2.80	68	80

For customer support on your Building Construction Costs with RSMeans data, call 800.448.8182.

04 43 Stone Masonry

04 43 10 – Masonry with Natural and Processed Stone

04 43 10.45 Granite

		Crew	Daily Output	Labor-Hours	Unit	Material	2018 Bare Costs Labor	2018 Bare Costs Equipment	Total	Total Incl O&P
0700	2-1/2" to 4" thick, veneer									
0750	Low price, gray, light gray, etc.	D-10	110	.291	S.F.	39	14.40	3.30	56.70	68.50
0850	Medium price, pink, brown, etc.		110	.291		45	14.40	3.30	62.70	75
0950	High price, red, black, etc.	▼	110	.291		64	14.40	3.30	81.70	96
1000	For bush hammered finish, deduct					5%				
1050	Coarse rubbed finish, deduct					10%				
1100	Honed finish, deduct					5%				
1150	Thermal finish, deduct				▼	18%				
2450	For radius under 5', add				L.F.	100%				
2500	Steps, copings, etc., finished on more than one surface									
2550	Low price, gray, light gray, etc.	D-10	50	.640	C.F.	94	31.50	7.25	132.75	160
2575	Medium price, pink, brown, etc.		50	.640		122	31.50	7.25	160.75	191
2600	High price, red, black, etc.	▼	50	.640	▼	150	31.50	7.25	188.75	222
2800	Pavers, 4" x 4" x 4" blocks, split face and joints									
2850	Low price, gray, light gray, etc.	D-11	80	.300	S.F.	13.30	14.20		27.50	36
2875	Medium price, pink, brown, etc.		80	.300		21.50	14.20		35.70	45
2900	High price, red, black, etc.	▼	80	.300		29.50	14.20		43.70	54
4000	Soffits, 2" thick, low price, gray, light gray	D-13	35	1.371		39	66	10.40	115.40	154
4050	Medium price, pink, brown, etc.		35	1.371		65.50	66	10.40	141.90	183
4100	High price, red, black, etc.		35	1.371		92.50	66	10.40	168.90	213
4200	Low price, gray, light gray, etc.		35	1.371		64	66	10.40	140.40	182
4250	Medium price, pink, brown, etc.		35	1.371		92	66	10.40	168.40	212
4300	High price, red, black, etc.	▼	35	1.371	▼	120	66	10.40	196.40	243
5000	Reclaimed or antique									
5010	Treads, up to 12" wide	D-10	100	.320	L.F.	42.50	15.85	3.64	61.99	75
5020	Up to 18" wide		100	.320		38.50	15.85	3.64	57.99	70.50
5030	Capstone, size varies		50	.640	▼	30.50	31.50	7.25	69.25	90
5040	Posts	▼	30	1.067	V.L.F.	30.50	53	12.10	95.60	127

04 43 10.50 Lightweight Natural Stone

		Crew	Daily Output	Labor-Hours	Unit	Material	2018 Bare Costs Labor	2018 Bare Costs Equipment	Total	Total Incl O&P
0011	**LIGHTWEIGHT NATURAL STONE** Lava type									
0100	Veneer, rubble face, sawed back, irregular shapes G	D-10	130	.246	S.F.	8.95	12.20	2.80	23.95	31.50
0200	Sawed face and back, irregular shapes G		130	.246	"	8.95	12.20	2.80	23.95	31.50
1000	Reclaimed or antique, barn or foundation stone	▼	1	32	Ton	395	1,575	365	2,335	3,250

04 43 10.55 Limestone

		Crew	Daily Output	Labor-Hours	Unit	Material	2018 Bare Costs Labor	2018 Bare Costs Equipment	Total	Total Incl O&P
0010	**LIMESTONE**, cut to size									
0020	Veneer facing panels									
0500	Texture finish, light stick, 4-1/2" thick, 5' x 12'	D-4	300	.107	S.F.	20.50	4.81	.43	25.74	30.50
0750	5" thick, 5' x 14' panels	D-10	275	.116		21.50	5.75	1.32	28.57	34.50
1000	Sugarcube finish, 2" thick, 3' x 5' panels		275	.116		29	5.75	1.32	36.07	42.50
1050	3" thick, 4' x 9' panels		275	.116		25.50	5.75	1.32	32.57	39
1200	4" thick, 5' x 11' panels		275	.116		31.50	5.75	1.32	38.57	45
1400	Sugarcube, textured finish, 4-1/2" thick, 5' x 12'		275	.116		32.50	5.75	1.32	39.57	46.50
1450	5" thick, 5' x 14' panels		275	.116	▼	34	5.75	1.32	41.07	47.50
2000	Coping, sugarcube finish, top & 2 sides		30	1.067	C.F.	68.50	53	12.10	133.60	169
2100	Sills, lintels, jambs, trim, stops, sugarcube finish, simple		20	1.600		68.50	79	18.20	165.70	217
2150	Detailed		20	1.600	▼	68.50	79	18.20	165.70	217
2300	Steps, extra hard, 14" wide, 6" rise	▼	50	.640	L.F.	25.50	31.50	7.25	64.25	84.50
3000	Quoins, plain finish, 6" x 12" x 12"	D-12	25	1.280	Ea.	41	58		99	135
3050	6" x 16" x 24"	"	25	1.280	"	55	58		113	150

For customer support on your Building Construction Costs with RSMeans data, call 800.448.8182.

113

04 43 10.60 Marble

		Crew	Daily Output	Labor-Hours	Unit	Material	2018 Bare Costs Labor	Equipment	Total	Total Incl O&P
0011	**MARBLE**, ashlar, split face, +/- 4" thick, random									
0040	Lengths 1' to 4' & heights 2" to 7-1/2", average	D-8	175	.229	S.F.	18.30	10.50		28.80	36
0100	Base, polished, 3/4" or 7/8" thick, polished, 6" high	D-10	65	.492	L.F.	12	24.50	5.60	42.10	56.50
0300	Carvings or bas-relief, from templates, simple design		80	.400	S.F.	150	19.80	4.54	174.34	200
0350	Intricate design	↓	80	.400	"	350	19.80	4.54	374.34	415
0600	Columns, cornices, mouldings, etc.									
0650	Hand or special machine cut, simple design	D-10	35	.914	C.F.	55	45.50	10.40	110.90	141
0700	Intricate design	"	35	.914	"	288	45.50	10.40	343.90	395
1000	Facing, polished finish, cut to size, 3/4" to 7/8" thick									
1050	Carrara or equal	D-10	130	.246	S.F.	22.50	12.20	2.80	37.50	46.50
1100	Arabescato or equal		130	.246		39.50	12.20	2.80	54.50	65
1300	1-1/4" thick, Botticino Classico or equal		125	.256		24	12.65	2.91	39.56	49
1350	Statuarietto or equal		125	.256		41.50	12.65	2.91	57.06	68
1500	2" thick, Crema Marfil or equal		120	.267		48.50	13.20	3.03	64.73	77
1550	Cafe Pinta or equal	↓	120	.267	↓	70	13.20	3.03	86.23	100
1700	Rubbed finish, cut to size, 4" thick									
1740	Average	D-10	100	.320	S.F.	41.50	15.85	3.64	60.99	73.50
1780	Maximum	"	100	.320	"	71.50	15.85	3.64	90.99	107
2200	Window sills, 6" x 3/4" thick	D-1	85	.188	L.F.	11.10	8.45		19.55	25
2500	Flooring, polished tiles, 12" x 12" x 3/8" thick									
2510	Thin set, Giallo Solare or equal	D-11	90	.267	S.F.	17.50	12.60		30.10	38.50
2600	Sky Blue or equal		90	.267		16	12.60		28.60	37
2700	Mortar bed, Giallo Solare or equal		65	.369		17.50	17.45		34.95	46
2740	Sky Blue or equal	↓	65	.369		16	17.45		33.45	44
2780	Travertine, 3/8" thick, Sierra or equal	D-10	130	.246		9.40	12.20	2.80	24.40	32
2790	Silver or equal	"	130	.246		26	12.20	2.80	41	50
2800	Patio tile, non-slip, 1/2" thick, flame finish	D-11	75	.320	↓	11.15	15.15		26.30	35.50
2900	Shower or toilet partitions, 7/8" thick partitions									
3050	3/4" or 1-1/4" thick stiles, polished 2 sides, average	D-11	75	.320	S.F.	46.50	15.15		61.65	74
3201	Soffits, add to above prices				"	20%	100%			
3210	Stairs, risers, 7/8" thick x 6" high	D-10	115	.278	L.F.	15.60	13.75	3.16	32.51	41.50
3360	Treads, 12" wide x 1-1/4" thick	"	115	.278	"	44	13.75	3.16	60.91	73
3500	Thresholds, 3' long, 7/8" thick, 4" to 5" wide, plain	D-12	24	1.333	Ea.	35.50	60.50		96	132
3550	Beveled		24	1.333	"	71.50	60.50		132	172
3700	Window stools, polished, 7/8" thick, 5" wide	↓	85	.376	L.F.	21.50	17.05		38.55	50

04 43 10.75 Sandstone or Brownstone

		Crew	Daily Output	Labor-Hours	Unit	Material	2018 Bare Costs Labor	Equipment	Total	Total Incl O&P
0011	**SANDSTONE OR BROWNSTONE**									
0100	Sawed face veneer, 2-1/2" thick, to 2' x 4' panels	D-10	130	.246	S.F.	21.50	12.20	2.80	36.50	45
0150	4" thick, to 3'-6" x 8' panels		100	.320		21.50	15.85	3.64	40.99	51.50
0300	Split face, random sizes	↓	100	.320	↓	13.40	15.85	3.64	32.89	43
0350	Cut stone trim (limestone)									
0360	Ribbon stone, 4" thick, 5' pieces	D-8	120	.333	Ea.	160	15.30		175.30	200
0370	Cove stone, 4" thick, 5' pieces		105	.381		161	17.50		178.50	204
0380	Cornice stone, 10" to 12" wide		90	.444		199	20.50		219.50	250
0390	Band stone, 4" thick, 5' pieces		145	.276		109	12.65		121.65	139
0410	Window and door trim, 3" to 4" wide		160	.250		93.50	11.50		105	121
0420	Key stone, 18" long	↓	60	.667	↓	93.50	30.50		124	150

04 43 10 – Masonry with Natural and Processed Stone

04 43 10.80 Slate	Crew	Daily Output	Labor-Hours	Unit	Material	2018 Bare Costs Labor	Equipment	Total	Total Incl O&P
0010 **SLATE**									
0040 Pennsylvania - blue gray to black									
0050 Vermont - unfading green, mottled green & purple, gray & purple									
0100 Virginia - blue black									
0200 Exterior paving, natural cleft, 1" thick									
0250 6" x 6" Pennsylvania	D-12	100	.320	S.F.	7.15	14.50		21.65	30
0300 Vermont		100	.320		10.70	14.50		25.20	34
0350 Virginia		100	.320		14.95	14.50		29.45	38.50
0500 24" x 24", Pennsylvania		120	.267		13.80	12.10		25.90	33.50
0550 Vermont		120	.267		26.50	12.10		38.60	47.50
0600 Virginia		120	.267		21.50	12.10		33.60	42.50
0700 18" x 30" Pennsylvania		120	.267		15.65	12.10		27.75	35.50
0750 Vermont		120	.267		26.50	12.10		38.60	47.50
0800 Virginia		120	.267		19.30	12.10		31.40	39.50
1000 Interior flooring, natural cleft, 1/2" thick									
1100 6" x 6" Pennsylvania	D-12	100	.320	S.F.	4.24	14.50		18.74	26.50
1150 Vermont		100	.320		9.40	14.50		23.90	32.50
1200 Virginia		100	.320		11.80	14.50		26.30	35
1300 24" x 24" Pennsylvania		120	.267		8.20	12.10		20.30	27.50
1350 Vermont		120	.267		21.50	12.10		33.60	42
1400 Virginia		120	.267		15.60	12.10		27.70	35.50
1500 18" x 24" Pennsylvania		120	.267		8.20	12.10		20.30	27.50
1550 Vermont		120	.267		17.10	12.10		29.20	37.50
1600 Virginia		120	.267		15.85	12.10		27.95	36
2000 Facing panels, 1-1/4" thick, to 4' x 4' panels									
2100 Natural cleft finish, Pennsylvania	D-10	180	.178	S.F.	36	8.80	2.02	46.82	55
2110 Vermont		180	.178		29	8.80	2.02	39.82	47
2120 Virginia		180	.178		35	8.80	2.02	45.82	54.50
2150 Sand rubbed finish, surface, add					10.80			10.80	11.85
2200 Honed finish, add					7.80			7.80	8.55
2500 Ribbon, natural cleft finish, 1" thick, to 9 S.F.	D-10	80	.400		13.90	19.80	4.54	38.24	50.50
2550 Sand rubbed finish		80	.400		18.80	19.80	4.54	43.14	55.50
2600 Honed finish		80	.400		17.50	19.80	4.54	41.84	54.50
2700 1-1/2" thick		78	.410		18.05	20.50	4.66	43.21	56
2750 Sand rubbed finish		78	.410		24	20.50	4.66	49.16	62
2800 Honed finish		78	.410		22.50	20.50	4.66	47.66	61
2850 2" thick		76	.421		21.50	21	4.78	47.28	61.50
2900 Sand rubbed finish		76	.421		30	21	4.78	55.78	70.50
2950 Honed finish		76	.421		27.50	21	4.78	53.28	68
3100 Stair landings, 1" thick, black, clear	D-1	65	.246		21	11.05		32.05	40.50
3200 Ribbon	"	65	.246		23.50	11.05		34.55	42.50
3500 Stair treads, sand finish, 1" thick x 12" wide									
3550 Under 3 L.F.	D-10	85	.376	L.F.	23.50	18.65	4.28	46.43	58.50
3600 3 L.F. to 6 L.F.	"	120	.267	"	25	13.20	3.03	41.23	51
3700 Ribbon, sand finish, 1" thick x 12" wide									
3750 To 6 L.F.	D-10	120	.267	L.F.	21	13.20	3.03	37.23	47
4000 Stools or sills, sand finish, 1" thick, 6" wide	D-12	160	.200		12.20	9.05		21.25	27.50
4100 Honed finish		160	.200		11.65	9.05		20.70	26.50
4200 10" wide		90	.356		18.80	16.10		34.90	45
4250 Honed finish		90	.356		17.50	16.10		33.60	44
4400 2" thick, 6" wide		140	.229		19.60	10.35		29.95	37.50
4450 Honed finish		140	.229		18.65	10.35		29	36.50

04 43 Stone Masonry

04 43 10 – Masonry with Natural and Processed Stone

04 43 10.80 Slate

		Crew	Daily Output	Labor-Hours	Unit	Material	2018 Bare Costs Labor	2018 Bare Costs Equipment	Total	Total Incl O&P
4600	10" wide	D-12	90	.356	L.F.	30.50	16.10		46.60	58.50
4650	Honed finish	↓	90	.356		29	16.10		45.10	56.50
4800	For lengths over 3', add				↓	25%				

04 43 10.85 Window Sill

		Crew	Daily Output	Labor-Hours	Unit	Material	2018 Bare Costs Labor	2018 Bare Costs Equipment	Total	Total Incl O&P
0010	**WINDOW SILL**									
0020	Bluestone, thermal top, 10" wide, 1-1/2" thick	D-1	85	.188	S.F.	9.60	8.45		18.05	23.50
0050	2" thick		75	.213	"	9.60	9.55		19.15	25
0100	Cut stone, 5" x 8" plain		48	.333	L.F.	12.55	14.95		27.50	37
0200	Face brick on edge, brick, 8" wide		80	.200		5.50	8.95		14.45	19.75
0400	Marble, 9" wide, 1" thick		85	.188		8.95	8.45		17.40	23
0900	Slate, colored, unfading, honed, 12" wide, 1" thick		85	.188		8.95	8.45		17.40	22.50
0950	2" thick	↓	70	.229	↓	8.95	10.25		19.20	25.50

04 51 Flue Liner Masonry

04 51 10 – Clay Flue Lining

04 51 10.10 Flue Lining

		Crew	Daily Output	Labor-Hours	Unit	Material	2018 Bare Costs Labor	2018 Bare Costs Equipment	Total	Total Incl O&P
0010	**FLUE LINING**, including mortar									
0020	Clay, 8" x 8"	D-1	125	.128	V.L.F.	6	5.75		11.75	15.40
0100	8" x 12"		103	.155		8.75	6.95		15.70	20.50
0200	12" x 12"		93	.172		11.30	7.70		19	24.50
0300	12" x 18"		84	.190		22.50	8.55		31.05	38
0400	18" x 18"		75	.213		29	9.55		38.55	46.50
0500	20" x 20"		66	.242		43.50	10.85		54.35	64
0600	24" x 24"		56	.286		55.50	12.80		68.30	80.50
1000	Round, 18" diameter		66	.242		39.50	10.85		50.35	60
1100	24" diameter	↓	47	.340	↓	76	15.25		91.25	107

04 54 Refractory Brick Masonry

04 54 10 – Refractory Brick Work

04 54 10.10 Fire Brick

		Crew	Daily Output	Labor-Hours	Unit	Material	2018 Bare Costs Labor	2018 Bare Costs Equipment	Total	Total Incl O&P
0010	**FIRE BRICK**									
0012	Low duty, 2000°F, 9" x 2-1/2" x 4-1/2"	D-1	.60	26.667	M	1,650	1,200		2,850	3,625
0050	High duty, 3000°F	"	.60	26.667	"	2,800	1,200		4,000	4,900

04 54 10.20 Fire Clay

		Crew	Daily Output	Labor-Hours	Unit	Material	2018 Bare Costs Labor	2018 Bare Costs Equipment	Total	Total Incl O&P
0010	**FIRE CLAY**									
0020	Gray, high duty, 100 lb. bag				Bag	38			38	42
0050	100 lb. drum, premixed (400 brick per drum)				Drum	43			43	47.50

04 57 Masonry Fireplaces

04 57 10 – Brick or Stone Fireplaces

04 57 10.10 Fireplace

		Crew	Daily Output	Labor-Hours	Unit	Material	2018 Bare Costs Labor	Equipment	Total	Total Incl O&P
0010	**FIREPLACE**									
0100	Brick fireplace, not incl. foundations or chimneys									
0110	30" x 29" opening, incl. chamber, plain brickwork	D-1	.40	40	Ea.	590	1,800		2,390	3,400
0200	Fireplace box only (110 brick)	"	2	8	"	162	360		522	730
0300	For elaborate brickwork and details, add					35%	35%			
0400	For hearth, brick & stone, add	D-1	2	8	Ea.	213	360		573	785
0410	For steel, damper, cleanouts, add		4	4		17.55	179		196.55	293
0600	Plain brickwork, incl. metal circulator		.50	32		900	1,425		2,325	3,200
0800	Face brick only, standard size, 8" x 2-2/3" x 4"		.30	53.333	M	620	2,400		3,020	4,325

04 71 Manufactured Brick Masonry

04 71 10 – Simulated or Manufactured Brick

04 71 10.10 Simulated Brick

		Crew	Daily Output	Labor-Hours	Unit	Material	2018 Bare Costs Labor	Equipment	Total	Total Incl O&P
0010	**SIMULATED BRICK**									
0020	Aluminum, baked on colors	1 Carp	200	.040	S.F.	4.50	2.03		6.53	8.05
0050	Fiberglass panels		200	.040		10	2.03		12.03	14.10
0100	Urethane pieces cemented in mastic		150	.053		8.60	2.70		11.30	13.55
0150	Vinyl siding panels		200	.040		10.90	2.03		12.93	15.10
0160	Cement base, brick, incl. mastic	D-1	100	.160		9.75	7.15		16.90	21.50
0170	Corner		50	.320	V.L.F.	21.50	14.35		35.85	45.50
0180	Stone face, incl. mastic		100	.160	S.F.	10.40	7.15		17.55	22.50
0190	Corner		50	.320	V.L.F.	24.50	14.35		38.85	49

04 72 Cast Stone Masonry

04 72 10 – Cast Stone Masonry Features

04 72 10.10 Coping

		Crew	Daily Output	Labor-Hours	Unit	Material	2018 Bare Costs Labor	Equipment	Total	Total Incl O&P
0010	**COPING**, stock units									
0050	Precast concrete, 10" wide, 4" tapers to 3-1/2", 8" wall	D-1	75	.213	L.F.	22.50	9.55		32.05	39.50
0100	12" wide, 3-1/2" tapers to 3", 10" wall		70	.229		24.50	10.25		34.75	42.50
0110	14" wide, 4" tapers to 3-1/2", 12" wall		65	.246		28	11.05		39.05	47.50
0150	16" wide, 4" tapers to 3-1/2", 14" wall		60	.267		30	11.95		41.95	51.50
0250	Precast concrete corners		40	.400	Ea.	39	17.90		56.90	70.50
0300	Limestone for 12" wall, 4" thick		90	.178	L.F.	14.80	7.95		22.75	28.50
0350	6" thick		80	.200		22	8.95		30.95	38
0500	Marble, to 4" thick, no wash, 9" wide		90	.178		12.90	7.95		20.85	26.50
0550	12" wide		80	.200		18	8.95		26.95	33.50
0700	Terra cotta, 9" wide		90	.178		6.45	7.95		14.40	19.30
0750	12" wide		80	.200		8.80	8.95		17.75	23.50
0800	Aluminum, for 12" wall		80	.200		9.20	8.95		18.15	24

04 72 20 – Cultured Stone Veneer

04 72 20.10 Cultured Stone Veneer Components

		Crew	Daily Output	Labor-Hours	Unit	Material	2018 Bare Costs Labor	Equipment	Total	Total Incl O&P
0010	**CULTURED STONE VENEER COMPONENTS**									
0110	On wood frame and sheathing substrate, random sized cobbles, corner stones	D-8	70	.571	V.L.F.	10.40	26		36.40	51.50
0120	Field stones		140	.286	S.F.	7.55	13.10		20.65	28.50
0130	Random sized flats, corner stones		70	.571	V.L.F.	10.70	26		36.70	52
0140	Field stones		140	.286	S.F.	8.90	13.10		22	30
0150	Horizontal lined ledgestones, corner stones		75	.533	V.L.F.	10.40	24.50		34.90	49
0160	Field stones		150	.267	S.F.	7.55	12.25		19.80	27
0170	Random shaped flats, corner stones		65	.615	V.L.F.	10.40	28.50		38.90	54.50

04 72 Cast Stone Masonry

04 72 20 – Cultured Stone Veneer

04 72 20.10 Cultured Stone Veneer Components	Crew	Daily Output	Labor-Hours	Unit	Material	2018 Bare Costs Labor	Equipment	Total	Total Incl O&P
0180	Field stones	D-8	150	.267	S.F.	7.55	12.25		19.80	27
0190	Random shaped/textured face, corner stones		65	.615	V.L.F.	10.40	28.50		38.90	54.50
0200	Field stones		130	.308	S.F.	7.55	14.10		21.65	30
0210	Random shaped river rock, corner stones		65	.615	V.L.F.	10.40	28.50		38.90	54.50
0220	Field stones		130	.308	S.F.	7.55	14.10		21.65	30
0240	On concrete or CMU substrate, random sized cobbles, corner stones		70	.571	V.L.F.	9.70	26		35.70	50.50
0250	Field stones		140	.286	S.F.	7.20	13.10		20.30	28
0260	Random sized flats, corner stones		70	.571	V.L.F.	10	26		36	51
0270	Field stones		140	.286	S.F.	8.55	13.10		21.65	29.50
0280	Horizontal lined ledgestones, corner stones		75	.533	V.L.F.	9.70	24.50		34.20	48
0290	Field stones		150	.267	S.F.	7.20	12.25		19.45	26.50
0300	Random shaped flats, corner stones		70	.571	V.L.F.	9.70	26		35.70	50.50
0310	Field stones		140	.286	S.F.	7.20	13.10		20.30	28
0320	Random shaped/textured face, corner stones		65	.615	V.L.F.	9.70	28.50		38.20	53.50
0330	Field stones		130	.308	S.F.	7.20	14.10		21.30	29.50
0340	Random shaped river rock, corner stones		65	.615	V.L.F.	9.70	28.50		38.20	53.50
0350	Field stones		130	.308	S.F.	7.20	14.10		21.30	29.50
0360	Cultured stone veneer, #15 felt weather resistant barrier	1 Clab	3700	.002	Sq.	5.35	.09		5.44	6.05
0370	Expanded metal lath, diamond, 2.5 lb./S.Y., galvanized	1 Lath	85	.094	S.Y.	3.10	4.63		7.73	10.25
0390	Water table or window sill, 18" long	1 Bric	80	.100	Ea.	10	5.05		15.05	18.70

04 73 Manufactured Stone Masonry

04 73 20 – Simulated or Manufactured Stone

04 73 20.10 Simulated Stone

		Crew	Daily Output	Labor-Hours	Unit	Material	2018 Bare Costs Labor	Equipment	Total	Total Incl O&P
0010	**SIMULATED STONE**									
0100	Insulated fiberglass panels, 5/8" ply backer	L-4	200	.120	S.F.	10	5.70		15.70	19.75

For customer support on your Building Construction Costs with RSMeans data, call 800.448.8182.

Estimating Tips

05 05 00 Common Work Results for Metals

- Nuts, bolts, washers, connection angles, and plates can add a significant amount to both the tonnage of a structural steel job and the estimated cost. As a rule of thumb, add 10% to the total weight to account for these accessories.

- Type 2 steel construction, commonly referred to as "simple construction," consists generally of field-bolted connections with lateral bracing supplied by other elements of the building, such as masonry walls or x-bracing. The estimator should be aware, however, that shop connections may be accomplished by welding or bolting. The method may be particular to the fabrication shop and may have an impact on the estimated cost.

05 10 00 Structural Steel

- Steel items can be obtained from two sources: a fabrication shop or a metals service center. Fabrication shops can fabricate items under more controlled conditions than crews in the field can. They are also more efficient and can produce items more economically. Metal service centers serve as a source of long mill shapes to both fabrication shops and contractors.

- Most line items in this structural steel subdivision, and most items in 05 50 00 Metal Fabrications, are indicated as being shop fabricated. The bare material cost for these shop fabricated items is the "Invoice Cost" from the shop and includes the mill base price of steel plus mill extras, transportation to the shop, shop drawings and detailing where warranted, shop fabrication and handling, sandblasting and a shop coat of primer paint, all necessary structural bolts, and delivery to the job site. The bare labor cost and bare equipment cost for these shop fabricated items are for field installation or erection.

- Line items in Subdivision 05 12 23.40 Lightweight Framing, and other items scattered in Division 5, are indicated as being field fabricated. The bare material cost for these field fabricated items is the "Invoice Cost" from the metals service center and includes the mill base price of steel plus mill extras, transportation to the metals service center, material handling, and delivery of long lengths of mill shapes to the job site. Material costs for structural bolts and welding rods should be added to the estimate. The bare labor cost and bare equipment cost for these items are for both field fabrication and field installation or erection, and include time for cutting, welding, and drilling in the fabricated metal items. Drilling into concrete and fasteners to fasten field fabricated items to other work are not included and should be added to the estimate.

05 20 00 Steel Joist Framing

- In any given project the total weight of open web steel joists is determined by the loads to be supported and the design. However, economies can be realized in minimizing the amount of labor used to place the joists. This is done by maximizing the joist spacing, and therefore minimizing the number of joists required to be installed on the job. Certain spacings and locations may be required by the design, but in other cases maximizing the spacing and keeping it as uniform as possible will keep the costs down.

05 30 00 Steel Decking

- The takeoff and estimating of a metal deck involves more than the area of the floor or roof and the type of deck specified or shown on the drawings. Many different sizes and types of openings may exist. Small openings for individual pipes or conduits may be drilled after the floor/roof is installed, but larger openings may require special deck lengths as well as reinforcing or structural support. The estimator should determine who will be supplying this reinforcing. Additionally, some deck terminations are part of the deck package, such as screed angles and pour stops, and others will be part of the steel contract, such as angles attached to structural members and cast-in-place angles and plates. The estimator must ensure that all pieces are accounted for in the complete estimate.

05 50 00 Metal Fabrications

- The most economical steel stairs are those that use common materials, standard details, and most importantly, a uniform and relatively simple method of field assembly. Commonly available A36/A992 channels and plates are very good choices for the main stringers of the stairs, as are angles and tees for the carrier members. Risers and treads are usually made by specialty shops, and it is most economical to use a typical detail in as many places as possible. The stairs should be pre-assembled and shipped directly to the site. The field connections should be simple and straightforward enough to be accomplished efficiently, and with minimum equipment and labor.

Reference Numbers

Reference numbers are shown at the beginning of some major classifications. These numbers refer to related items in the Reference Section. The reference information may be an estimating procedure, an alternate pricing method, or technical information.

Note: Not all subdivisions listed here necessarily appear. ■

No part of this cost data may be reproduced, stored in a retrieval system, or transmitted in any form or by any means without prior written permission of Gordian.

Did you know?

RSMeans data is available through our online application with 24/7 access:

- Search for unit prices by keyword
- Leverage the most up-to-date data
- Build and export estimates

Try it free for 30 days!
www.rsmeans.com/2018freetrial

05 01 Maintenance of Metals

05 01 10 – Maintenance of Structural Metal Framing

05 01 10.51 Cleaning of Structural Metal Framing

	05 01 10.51 Cleaning of Structural Metal Framing	Crew	Daily Output	Labor-Hours	Unit	Material	2018 Bare Costs Labor	Equipment	Total	Total Incl O&P
0010	**CLEANING OF STRUCTURAL METAL FRAMING**									
6125	Steel surface treatments, PDCA guidelines									
6170	Wire brush, hand (SSPC-SP2)	1 Psst	400	.020	S.F.	.02	.87		.89	1.47
6180	Power tool (SSPC-SP3)	"	700	.011		.08	.50		.58	.92
6215	Pressure washing, up to 5,000 psi, 5,000-15,000 S.F./day	1 Pord	10000	.001			.03		.03	.05
6220	Steam cleaning, 600 psi @ 300 degree F, 1,250-2,500 S.F./day		2000	.004			.17		.17	.26
6225	Water blasting, up to 25,000 psi, 1,750-3,500 S.F./day	↓	2500	.003			.14		.14	.21
6230	Brush-off blast (SSPC-SP7)	E-11	1750	.018		.18	.81	.12	1.11	1.62
6235	Com'l blast (SSPC-SP6), loose scale, fine pwder rust, 2.0#/S.F. sand		1200	.027		.36	1.19	.17	1.72	2.48
6240	Tight mill scale, little/no rust, 3.0#/S.F. sand		1000	.032		.55	1.43	.20	2.18	3.08
6245	Exist coat blistered/pitted, 4.0#/S.F. sand		875	.037		.73	1.63	.23	2.59	3.65
6250	Exist coat badly pitted/nodules, 6.7#/S.F. sand		825	.039		1.22	1.73	.25	3.20	4.35
6255	Near white blast (SSPC-SP10), loose scale, fine rust, 5.6#/S.F. sand		450	.071		1.02	3.17	.45	4.64	6.65
6260	Tight mill scale, little/no rust, 6.9#/S.F. sand		325	.098		1.25	4.39	.63	6.27	9
6265	Exist coat blistered/pitted, 9.0#/S.F. sand		225	.142		1.64	6.35	.90	8.89	12.85
6270	Exist coat badly pitted/nodules, 11.3#/S.F. sand	↓	150	.213	↓	2.05	9.50	1.35	12.90	18.85

05 05 Common Work Results for Metals

05 05 05 – Selective Demolition for Metals

05 05 05.10 Selective Demolition, Metals

	05 05 05.10 Selective Demolition, Metals	Crew	Daily Output	Labor-Hours	Unit	Material	2018 Bare Costs Labor	Equipment	Total	Total Incl O&P
0010	**SELECTIVE DEMOLITION, METALS** R024119-10									
0015	Excludes shores, bracing, cutting, loading, hauling, dumping									
0020	Remove nuts only up to 3/4" diameter	1 Sswk	480	.017	Ea.		.91		.91	1.49
0030	7/8" to 1-1/4" diameter		240	.033			1.82		1.82	2.98
0040	1-3/8" to 2" diameter		160	.050			2.73		2.73	4.47
0060	Unbolt and remove structural bolts up to 3/4" diameter		240	.033			1.82		1.82	2.98
0070	7/8" to 2" diameter		160	.050			2.73		2.73	4.47
0140	Light weight framing members, remove whole or cut up, up to 20 lb.	↓	240	.033			1.82		1.82	2.98
0150	21-40 lb.	2 Sswk	210	.076			4.16		4.16	6.80
0160	41-80 lb.	3 Sswk	180	.133			7.30		7.30	11.90
0170	81-120 lb.	4 Sswk	150	.213			11.65		11.65	19.05
0230	Structural members, remove whole or cut up, up to 500 lb.	E-19	48	.500			27	22.50	49.50	68
0240	1/4-2 tons	E-18	36	1.111			61	30	91	131
0250	2-5 tons	E-24	30	1.067			58	19.70	77.70	115
0260	5-10 tons	E-20	24	2.667			145	55	200	293
0270	10-15 tons	E-2	18	3.111			169	93.50	262.50	375
0340	Fabricated item, remove whole or cut up, up to 20 lb.	1 Sswk	96	.083			4.55		4.55	7.45
0350	21-40 lb.	2 Sswk	84	.190			10.40		10.40	17
0360	41-80 lb.	3 Sswk	72	.333			18.20		18.20	30
0370	81-120 lb.	4 Sswk	60	.533			29		29	47.50
0380	121-500 lb.	E-19	48	.500			27	22.50	49.50	68
0390	501-1000 lb.	"	36	.667	↓		36	30	66	90.50
0500	Steel roof decking, uncovered, bare	B-2	5000	.008	S.F.		.32		.32	.49

05 05 13 – Shop-Applied Coatings for Metal

05 05 13.50 Paints and Protective Coatings

	05 05 13.50 Paints and Protective Coatings	Crew	Daily Output	Labor-Hours	Unit	Material	2018 Bare Costs Labor	Equipment	Total	Total Incl O&P
0010	**PAINTS AND PROTECTIVE COATINGS**									
5900	Galvanizing structural steel in shop, under 1 ton R050516-30				Ton	550			550	605
5950	1 ton to 20 tons					510			510	560
6000	Over 20 tons				↓	460			460	505

For customer support on your Building Construction Costs with RSMeans data, call 800.448.8182.

05 05 19 – Post-Installed Concrete Anchors

05 05 19.10 Chemical Anchors		Crew	Daily Output	Labor-Hours	Unit	Material	2018 Bare Costs Labor	Equipment	Total	Total Incl O&P
0010	**CHEMICAL ANCHORS**									
0020	Includes layout & drilling									
1430	Chemical anchor, w/rod & epoxy cartridge, 3/4" diameter x 9-1/2" long	B-89A	27	.593	Ea.	10.85	27.50	4.15	42.50	58
1435	1" diameter x 11-3/4" long		24	.667		20	30.50	4.67	55.17	74.50
1440	1-1/4" diameter x 14" long		21	.762		38	35	5.35	78.35	101
1445	1-3/4" diameter x 15" long		20	.800		69.50	37	5.60	112.10	139
1450	18" long		17	.941		83	43.50	6.60	133.10	165
1455	2" diameter x 18" long		16	1		113	46	7	166	202
1460	24" long		15	1.067		147	49	7.50	203.50	244

05 05 19.20 Expansion Anchors		Crew	Daily Output	Labor-Hours	Unit	Material	2018 Bare Costs Labor	Equipment	Total	Total Incl O&P
0010	**EXPANSION ANCHORS**									
0100	Anchors for concrete, brick or stone, no layout and drilling									
0200	Expansion shields, zinc, 1/4" diameter, 1-5/16" long, single G	1 Carp	90	.089	Ea.	.45	4.51		4.96	7.35
0300	1-3/8" long, double G		85	.094		.58	4.77		5.35	7.90
0400	3/8" diameter, 1-1/2" long, single G		85	.094		.69	4.77		5.46	8
0500	2" long, double G		80	.100		1.15	5.05		6.20	8.95
0600	1/2" diameter, 2-1/16" long, single G		80	.100		1.26	5.05		6.31	9.10
0700	2-1/2" long, double G		75	.107		2.13	5.40		7.53	10.60
0800	5/8" diameter, 2-5/8" long, single G		75	.107		2.18	5.40		7.58	10.65
0900	2-3/4" long, double G		70	.114		2.92	5.80		8.72	12
1000	3/4" diameter, 2-3/4" long, single G		70	.114		3.29	5.80		9.09	12.40
1100	3-15/16" long, double G		65	.123		5.45	6.25		11.70	15.50
2100	Hollow wall anchors for gypsum wall board, plaster or tile									
2300	1/8" diameter, short G	1 Carp	160	.050	Ea.	.33	2.54		2.87	4.22
2400	Long G		150	.053		.32	2.70		3.02	4.47
2500	3/16" diameter, short G		150	.053		.51	2.70		3.21	4.68
2600	Long G		140	.057		.67	2.90		3.57	5.15
2700	1/4" diameter, short G		140	.057		.73	2.90		3.63	5.20
2800	Long G		130	.062		.80	3.12		3.92	5.65
3000	Toggle bolts, bright steel, 1/8" diameter, 2" long G		85	.094		.25	4.77		5.02	7.55
3100	4" long G		80	.100		.29	5.05		5.34	8
3200	3/16" diameter, 3" long G		80	.100		.33	5.05		5.38	8.05
3300	6" long G		75	.107		.45	5.40		5.85	8.75
3400	1/4" diameter, 3" long G		75	.107		.41	5.40		5.81	8.70
3500	6" long G		70	.114		.59	5.80		6.39	9.45
3600	3/8" diameter, 3" long G		70	.114		.86	5.80		6.66	9.75
3700	6" long G		60	.133		1.50	6.75		8.25	11.95
3800	1/2" diameter, 4" long G		60	.133		1.65	6.75		8.40	12.10
3900	6" long G		50	.160		2.08	8.10		10.18	14.65
4000	Nailing anchors									
4100	Nylon nailing anchor, 1/4" diameter, 1" long	1 Carp	3.20	2.500	C	23	127		150	218
4200	1-1/2" long		2.80	2.857		25.50	145		170.50	250
4300	2" long		2.40	3.333		27.50	169		196.50	287
4400	Metal nailing anchor, 1/4" diameter, 1" long G		3.20	2.500		22.50	127		149.50	218
4500	1-1/2" long G		2.80	2.857		29	145		174	253
4600	2" long G		2.40	3.333		35.50	169		204.50	296
5000	Screw anchors for concrete, masonry,									
5100	stone & tile, no layout or drilling included									
5700	Lag screw shields, 1/4" diameter, short G	1 Carp	90	.089	Ea.	.47	4.51		4.98	7.35
5800	Long G		85	.094		.54	4.77		5.31	7.85
5900	3/8" diameter, short G		85	.094		.72	4.77		5.49	8.05
6000	Long G		80	.100		.92	5.05		5.97	8.70

05 05 Common Work Results for Metals

05 05 19 – Post-Installed Concrete Anchors

05 05 19.20 Expansion Anchors		Crew	Daily Output	Labor-Hours	Unit	Material	2018 Bare Costs Labor	Equipment	Total	Total Incl O&P
6100	1/2" diameter, short	G 1 Carp	80	.100	Ea.	1.03	5.05		6.08	8.85
6200	Long	G	75	.107		1.37	5.40		6.77	9.75
6300	5/8" diameter, short	G	70	.114		1.42	5.80		7.22	10.35
6400	Long	G	65	.123		1.96	6.25		8.21	11.65
6600	Lead, #6 & #8, 3/4" long	G	260	.031		.19	1.56		1.75	2.59
6700	#10 - #14, 1-1/2" long	G	200	.040		.38	2.03		2.41	3.51
6800	#16 & #18, 1-1/2" long	G	160	.050		.41	2.54		2.95	4.31
6900	Plastic, #6 & #8, 3/4" long		260	.031		.05	1.56		1.61	2.44
7000	#8 & #10, 7/8" long		240	.033		.06	1.69		1.75	2.64
7100	#10 & #12, 1" long		220	.036		.07	1.84		1.91	2.89
7200	#14 & #16, 1-1/2" long		160	.050		.07	2.54		2.61	3.94
8000	Wedge anchors, not including layout or drilling									
8050	Carbon steel, 1/4" diameter, 1-3/4" long	G 1 Carp	150	.053	Ea.	.51	2.70		3.21	4.68
8100	3-1/4" long	G	140	.057		.67	2.90		3.57	5.15
8150	3/8" diameter, 2-1/4" long	G	145	.055		.54	2.80		3.34	4.86
8200	5" long	G	140	.057		.95	2.90		3.85	5.45
8250	1/2" diameter, 2-3/4" long	G	140	.057		1.07	2.90		3.97	5.60
8300	7" long	G	125	.064		1.82	3.24		5.06	6.95
8350	5/8" diameter, 3-1/2" long	G	130	.062		1.81	3.12		4.93	6.75
8400	8-1/2" long	G	115	.070		3.86	3.53		7.39	9.60
8450	3/4" diameter, 4-1/4" long	G	115	.070		2.75	3.53		6.28	8.35
8500	10" long	G	95	.084		6.25	4.27		10.52	13.35
8550	1" diameter, 6" long	G	100	.080		8.70	4.06		12.76	15.80
8575	9" long	G	85	.094		11.35	4.77		16.12	19.70
8600	12" long	G	75	.107		12.25	5.40		17.65	21.50
8650	1-1/4" diameter, 9" long	G	70	.114		22	5.80		27.80	33
8700	12" long	G	60	.133		28	6.75		34.75	41
8750	For type 303 stainless steel, add					350%				
8800	For type 316 stainless steel, add					450%				
8950	Self-drilling concrete screw, hex washer head, 3/16" diam. x 1-3/4" long	G 1 Carp	300	.027	Ea.	.19	1.35		1.54	2.27
8960	2-1/4" long	G	250	.032		.21	1.62		1.83	2.70
8970	Phillips flat head, 3/16" diam. x 1-3/4" long	G	300	.027		.19	1.35		1.54	2.27
8980	2-1/4" long	G	250	.032		.21	1.62		1.83	2.70

05 05 21 – Fastening Methods for Metal

05 05 21.10 Cutting Steel

		Crew	Daily Output	Labor-Hours	Unit	Material	Labor	Equipment	Total	Total Incl O&P
0010	**CUTTING STEEL**									
0020	Hand burning, incl. preparation, torch cutting & grinding, no staging									
0050	Steel to 1/4" thick	E-25	400	.020	L.F.	.20	1.13	.03	1.36	2.10
0100	1/2" thick		320	.025		.37	1.42	.04	1.83	2.77
0150	3/4" thick		260	.031		.61	1.74	.05	2.40	3.57
0200	1" thick		200	.040		.87	2.27	.06	3.20	4.73

05 05 21.15 Drilling Steel

		Crew	Daily Output	Labor-Hours	Unit	Material	Labor	Equipment	Total	Total Incl O&P
0010	**DRILLING STEEL**									
1910	Drilling & layout for steel, up to 1/4" deep, no anchor									
1920	Holes, 1/4" diameter	1 Sswk	112	.071	Ea.	.10	3.90		4	6.50
1925	For each additional 1/4" depth, add		336	.024		.10	1.30		1.40	2.24
1930	3/8" diameter		104	.077		.09	4.20		4.29	6.95
1935	For each additional 1/4" depth, add		312	.026		.09	1.40		1.49	2.39
1940	1/2" diameter		96	.083		.11	4.55		4.66	7.55
1945	For each additional 1/4" depth, add		288	.028		.11	1.52		1.63	2.60
1950	5/8" diameter		88	.091		.16	4.97		5.13	8.30
1955	For each additional 1/4" depth, add		264	.030		.16	1.66		1.82	2.89

For customer support on your Building Construction Costs with RSMeans data, call 800.448.8182.

05 05 Common Work Results for Metals

05 05 21 – Fastening Methods for Metal

05 05 21.15 Drilling Steel

		Crew	Daily Output	Labor-Hours	Unit	Material	2018 Bare Costs Labor	Equipment	Total	Total Incl O&P
1960	3/4" diameter	1 Sswk	80	.100	Ea.	.20	5.45		5.65	9.15
1965	For each additional 1/4" depth, add		240	.033		.20	1.82		2.02	3.20
1970	7/8" diameter		72	.111		.27	6.05		6.32	10.25
1975	For each additional 1/4" depth, add		216	.037		.27	2.02		2.29	3.60
1980	1" diameter		64	.125		.23	6.85		7.08	11.40
1985	For each additional 1/4" depth, add		192	.042		.23	2.28		2.51	3.97
1990	For drilling up, add						40%			

05 05 21.90 Welding Steel

			Crew	Daily Output	Labor-Hours	Unit	Material	2018 Bare Costs Labor	Equipment	Total	Total Incl O&P
0010	**WELDING STEEL**, Structural	R050521-20									
0020	Field welding, 1/8" E6011, cost per welder, no operating engineer		E-14	8	1	Hr.	4.71	56.50	12.30	73.51	111
0200	With 1/2 operating engineer		E-13	8	1.500		4.71	82.50	12.30	99.51	150
0300	With 1 operating engineer		E-12	8	2		4.71	108	12.30	125.01	189
0500	With no operating engineer, 2# weld rod per ton		E-14	8	1	Ton	4.71	56.50	12.30	73.51	111
0600	8# E6011 per ton		"	2	4		18.85	227	49	294.85	445
0800	With one operating engineer per welder, 2# E6011 per ton		E-12	8	2		4.71	108	12.30	125.01	189
0900	8# E6011 per ton		"	2	8		18.85	430	49	497.85	755
1200	Continuous fillet, down welding										
1300	Single pass, 1/8" thick, 0.1#/L.F.		E-14	150	.053	L.F.	.24	3.02	.66	3.92	5.90
1400	3/16" thick, 0.2#/L.F.			75	.107		.47	6.05	1.31	7.83	11.85
1500	1/4" thick, 0.3#/L.F.			50	.160		.71	9.05	1.97	11.73	17.75
1610	5/16" thick, 0.4#/L.F.			38	.211		.94	11.95	2.59	15.48	23.50
1800	3 passes, 3/8" thick, 0.5#/L.F.			30	.267		1.18	15.10	3.28	19.56	29.50
2010	4 passes, 1/2" thick, 0.7#/L.F.			22	.364		1.65	20.50	4.47	26.62	40
2200	5 to 6 passes, 3/4" thick, 1.3#/L.F.			12	.667		3.06	38	8.20	49.26	74
2400	8 to 11 passes, 1" thick, 2.4#/L.F.			6	1.333		5.65	75.50	16.40	97.55	147
2600	For vertical joint welding, add							20%			
2700	Overhead joint welding, add							300%			
2900	For semi-automatic welding, obstructed joints, deduct							5%			
3000	Exposed joints, deduct							15%			
4000	Cleaning and welding plates, bars, or rods										
4010	to existing beams, columns, or trusses		E-14	12	.667	L.F.	1.18	38	8.20	47.38	72

05 05 23 – Metal Fastenings

05 05 23.10 Bolts and Hex Nuts

			Crew	Daily Output	Labor-Hours	Unit	Material	2018 Bare Costs Labor	Equipment	Total	Total Incl O&P
0010	**BOLTS & HEX NUTS**, Steel, A307										
0100	1/4" diameter, 1/2" long	G	1 Sswk	140	.057	Ea.	.06	3.12		3.18	5.15
0200	1" long	G		140	.057		.07	3.12		3.19	5.20
0300	2" long	G		130	.062		.10	3.36		3.46	5.60
0400	3" long	G		130	.062		.15	3.36		3.51	5.65
0500	4" long	G		120	.067		.17	3.64		3.81	6.15
0600	3/8" diameter, 1" long	G		130	.062		.14	3.36		3.50	5.65
0700	2" long	G		130	.062		.18	3.36		3.54	5.70
0800	3" long	G		120	.067		.24	3.64		3.88	6.20
0900	4" long	G		120	.067		.30	3.64		3.94	6.30
1000	5" long	G		115	.070		.38	3.80		4.18	6.60
1100	1/2" diameter, 1-1/2" long	G		120	.067		.40	3.64		4.04	6.40
1200	2" long	G		120	.067		.46	3.64		4.10	6.45
1300	4" long	G		115	.070		.75	3.80		4.55	7.05
1400	6" long	G		110	.073		1.05	3.97		5.02	7.65
1500	8" long	G		105	.076		1.38	4.16		5.54	8.30
1600	5/8" diameter, 1-1/2" long	G		120	.067		.85	3.64		4.49	6.90
1700	2" long	G		120	.067		.94	3.64		4.58	7
1800	4" long	G		115	.070		1.34	3.80		5.14	7.70

For customer support on your Building Construction Costs with RSMeans data, call 800.448.8182.

123

05 05 23.10 Bolts and Hex Nuts

			Crew	Daily Output	Labor-Hours	Unit	Material	2018 Bare Costs Labor	Equipment	Total	Total Incl O&P
1900	6" long	G	1 Sswk	110	.073	Ea.	1.72	3.97		5.69	8.40
2000	8" long	G		105	.076		2.55	4.16		6.71	9.60
2100	10" long	G		100	.080		3.20	4.37		7.57	10.65
2200	3/4" diameter, 2" long	G		120	.067		1.15	3.64		4.79	7.20
2300	4" long	G		110	.073		1.65	3.97		5.62	8.30
2400	6" long	G		105	.076		2.12	4.16		6.28	9.15
2500	8" long	G		95	.084		3.20	4.60		7.80	11
2600	10" long	G		85	.094		4.20	5.15		9.35	13
2700	12" long	G		80	.100		4.92	5.45		10.37	14.35
2800	1" diameter, 3" long	G		105	.076		2.89	4.16		7.05	9.95
2900	6" long	G		90	.089		4.20	4.86		9.06	12.55
3000	12" long	G	↓	75	.107		7.50	5.85		13.35	17.80
3100	For galvanized, add						75%				
3200	For stainless, add				↓		350%				

05 05 23.25 High Strength Bolts

			Crew	Daily Output	Labor-Hours	Unit	Material	2018 Bare Costs Labor	Equipment	Total	Total Incl O&P
0010	**HIGH STRENGTH BOLTS** R050523-10										
0020	A325 Type 1, structural steel, bolt-nut-washer set										
0100	1/2" diameter x 1-1/2" long	G	1 Sswk	130	.062	Ea.	.94	3.36		4.30	6.55
0120	2" long	G		125	.064		1.02	3.50		4.52	6.80
0150	3" long	G		120	.067		1.42	3.64		5.06	7.50
0170	5/8" diameter x 1-1/2" long	G		125	.064		1.78	3.50		5.28	7.65
0180	2" long	G		120	.067		1.92	3.64		5.56	8.05
0190	3" long	G		115	.070		2.38	3.80		6.18	8.80
0200	3/4" diameter x 2" long	G		120	.067		3	3.64		6.64	9.25
0220	3" long	G		115	.070		3.61	3.80		7.41	10.15
0250	4" long	G		110	.073		4.44	3.97		8.41	11.40
0300	6" long	G		105	.076		5.80	4.16		9.96	13.15
0350	8" long	G		95	.084		11.55	4.60		16.15	20
0360	7/8" diameter x 2" long	G		115	.070		4.01	3.80		7.81	10.60
0365	3" long	G		110	.073		4.75	3.97		8.72	11.75
0370	4" long	G		105	.076		5.75	4.16		9.91	13.10
0380	6" long	G		100	.080		7.30	4.37		11.67	15.20
0390	8" long	G		90	.089		11.65	4.86		16.51	21
0400	1" diameter x 2" long	G		105	.076		3.92	4.16		8.08	11.10
0420	3" long	G		100	.080		4.39	4.37		8.76	12
0450	4" long	G		95	.084		4.95	4.60		9.55	12.95
0500	6" long	G		90	.089		6.50	4.86		11.36	15.10
0550	8" long	G		85	.094		11.10	5.15		16.25	20.50
0600	1-1/4" diameter x 3" long	G		85	.094		10.90	5.15		16.05	20.50
0650	4" long	G		80	.100		11.90	5.45		17.35	22
0700	6" long	G		75	.107		15.50	5.85		21.35	26.50
0750	8" long	G	↓	70	.114	↓	19.80	6.25		26.05	32
1020	A490, bolt-nut-washer set										
1170	5/8" diameter x 1-1/2" long	G	1 Sswk	125	.064	Ea.	4.49	3.50		7.99	10.65
1180	2" long	G		120	.067		5.35	3.64		8.99	11.85
1190	3" long	G		115	.070		6.55	3.80		10.35	13.40
1200	3/4" diameter x 2" long	G		120	.067		4.11	3.64		7.75	10.45
1220	3" long	G		115	.070		4.87	3.80		8.67	11.55
1250	4" long	G		110	.073		5.65	3.97		9.62	12.75
1300	6" long	G		105	.076		8.35	4.16		12.51	16
1350	8" long	G		95	.084		14.20	4.60		18.80	23
1360	7/8" diameter x 2" long	G	↓	115	.070	↓	6.20	3.80		10	13

05 05 23.25 High Strength Bolts		Crew	Daily Output	Labor-Hours	Unit	Material	2018 Bare Costs Labor	Equipment	Total	Total Incl O&P
1365	3" long	G	1 Sswk	110	.073	Ea.	7.30	3.97		11.27	14.55
1370	4" long	G		105	.076		9.05	4.16		13.21	16.80
1380	6" long	G		100	.080		12.80	4.37		17.17	21
1390	8" long	G		90	.089		18.60	4.86		23.46	28.50
1400	1" diameter x 2" long	G		105	.076		8.15	4.16		12.31	15.80
1420	3" long	G		100	.080		9.90	4.37		14.27	18.05
1450	4" long	G		95	.084		11.45	4.60		16.05	20
1500	6" long	G		90	.089		15.35	4.86		20.21	25
1550	8" long	G		85	.094		24.50	5.15		29.65	35.50
1600	1-1/4" diameter x 3" long	G		85	.094		45	5.15		50.15	58
1650	4" long	G		80	.100		52.50	5.45		57.95	66.50
1700	6" long	G		75	.107		73	5.85		78.85	90
1750	8" long	G		70	.114		96	6.25		102.25	116

05 05 23.30 Lag Screws		Crew	Daily Output	Labor-Hours	Unit	Material	Labor	Equipment	Total	Total Incl O&P
0010	**LAG SCREWS**									
0020	Steel, 1/4" diameter, 2" long	G	1 Carp	200	.040	Ea.	.10	2.03		2.13	3.20
0100	3/8" diameter, 3" long	G		150	.053		.28	2.70		2.98	4.43
0200	1/2" diameter, 3" long	G		130	.062		.66	3.12		3.78	5.50
0300	5/8" diameter, 3" long	G		120	.067		1.29	3.38		4.67	6.55

05 05 23.35 Machine Screws		Crew	Daily Output	Labor-Hours	Unit	Material	Labor	Equipment	Total	Total Incl O&P
0010	**MACHINE SCREWS**									
0020	Steel, round head, #8 x 1" long	G	1 Carp	4.80	1.667	C	4.06	84.50		88.56	133
0110	#8 x 2" long	G		2.40	3.333		8.10	169		177.10	266
0200	#10 x 1" long	G		4	2		5.05	101		106.05	160
0300	#10 x 2" long	G		2	4		8.75	203		211.75	320

05 05 23.50 Powder Actuated Tools and Fasteners		Crew	Daily Output	Labor-Hours	Unit	Material	Labor	Equipment	Total	Total Incl O&P
0010	**POWDER ACTUATED TOOLS & FASTENERS**									
0020	Stud driver, .22 caliber, single shot					Ea.	152			152	167
0100	.27 caliber, semi automatic, strip					"	460			460	505
0300	Powder load, single shot, .22 cal, power level 2, brown					C	5.65			5.65	6.25
0400	Strip, .27 cal, power level 4, red						8.15			8.15	9
0600	Drive pin, .300 x 3/4" long	G	1 Carp	4.80	1.667		4.39	84.50		88.89	134
0700	.300 x 3" long with washer	G	"	4	2		13.40	101		114.40	169

05 05 23.55 Rivets		Crew	Daily Output	Labor-Hours	Unit	Material	Labor	Equipment	Total	Total Incl O&P
0010	**RIVETS**									
0100	Aluminum rivet & mandrel, 1/2" grip length x 1/8" diameter	G	1 Carp	4.80	1.667	C	7.70	84.50		92.20	137
0200	3/16" diameter	G		4	2		10.80	101		111.80	166
0300	Aluminum rivet, steel mandrel, 1/8" diameter	G		4.80	1.667		10.40	84.50		94.90	140
0400	3/16" diameter	G		4	2		16.65	101		117.65	172
0500	Copper rivet, steel mandrel, 1/8" diameter	G		4.80	1.667		9.95	84.50		94.45	140
0800	Stainless rivet & mandrel, 1/8" diameter	G		4.80	1.667		23	84.50		107.50	154
0900	3/16" diameter	G		4	2		37.50	101		138.50	196
1000	Stainless rivet, steel mandrel, 1/8" diameter	G		4.80	1.667		15.95	84.50		100.45	147
1100	3/16" diameter	G		4	2		26	101		127	183
1200	Steel rivet and mandrel, 1/8" diameter	G		4.80	1.667		7.55	84.50		92.05	137
1300	3/16" diameter	G		4	2		10.95	101		111.95	166
1400	Hand riveting tool, standard					Ea.	70			70	77
1500	Deluxe						395			395	435
1600	Power riveting tool, standard						540			540	595
1700	Deluxe						1,650			1,650	1,825

05 05 Common Work Results for Metals

05 05 23 – Metal Fastenings

05 05 23.70 Structural Blind Bolts

		Crew	Daily Output	Labor-Hours	Unit	Material	2018 Bare Costs Labor	2018 Bare Costs Equipment	Total	Total Incl O&P
0010	**STRUCTURAL BLIND BOLTS**									
0100	1/4" diameter x 1/4" grip	G 1 Sswk	240	.033	Ea.	1.73	1.82		3.55	4.88
0150	1/2" grip	G	216	.037		1.86	2.02		3.88	5.35
0200	3/8" diameter x 1/2" grip	G	232	.034		3.12	1.88		5	6.50
0250	3/4" grip	G	208	.038		3.27	2.10		5.37	7.05
0300	1/2" diameter x 1/2" grip	G	224	.036		6.05	1.95		8	9.85
0350	3/4" grip	G	200	.040		8.20	2.19		10.39	12.55
0400	5/8" diameter x 3/4" grip	G	216	.037		10.90	2.02		12.92	15.30
0450	1" grip	G	192	.042		14.80	2.28		17.08	19.95

05 05 23.80 Vibration and Bearing Pads

		Crew	Daily Output	Labor-Hours	Unit	Material	2018 Bare Costs Labor	2018 Bare Costs Equipment	Total	Total Incl O&P
0010	**VIBRATION & BEARING PADS**									
0300	Laminated synthetic rubber impregnated cotton duck, 1/2" thick	2 Sswk	24	.667	S.F.	74	36.50		110.50	141
0400	1" thick		20	.800		145	43.50		188.50	231
0600	Neoprene bearing pads, 1/2" thick		24	.667		28	36.50		64.50	90.50
0700	1" thick		20	.800		56	43.50		99.50	133
0900	Fabric reinforced neoprene, 5000 psi, 1/2" thick		24	.667		12.20	36.50		48.70	73
1000	1" thick		20	.800		24.50	43.50		68	98.50
1200	Felt surfaced vinyl pads, cork and sisal, 5/8" thick		24	.667		5.60	36.50		42.10	65.50
1300	1" thick		20	.800		10.10	43.50		53.60	82.50
1500	Teflon bonded to 10 ga. carbon steel, 1/32" layer		24	.667		55	36.50		91.50	120
1600	3/32" layer		24	.667		82.50	36.50		119	150
1800	Bonded to 10 ga. stainless steel, 1/32" layer		24	.667		97	36.50		133.50	167
1900	3/32" layer		24	.667		128	36.50		164.50	201
2100	Circular machine leveling pad & stud				Kip	7.35			7.35	8.05

05 05 23.85 Weld Shear Connectors

		Crew	Daily Output	Labor-Hours	Unit	Material	2018 Bare Costs Labor	2018 Bare Costs Equipment	Total	Total Incl O&P
0010	**WELD SHEAR CONNECTORS**									
0020	3/4" diameter, 3-3/16" long	G E-10	960	.017	Ea.	.55	.93	.35	1.83	2.51
0030	3-3/8" long	G	950	.017		.57	.94	.35	1.86	2.55
0200	3-7/8" long	G	945	.017		.61	.94	.36	1.91	2.61
0300	4-3/16" long	G	935	.017		.64	.95	.36	1.95	2.67
0500	4-7/8" long	G	930	.017		.72	.96	.36	2.04	2.75
0600	5-3/16" long	G	920	.017		.75	.97	.37	2.09	2.80
0800	5-3/8" long	G	910	.018		.76	.98	.37	2.11	2.84
0900	6-3/16" long	G	905	.018		.83	.98	.37	2.18	2.93
1000	7-3/16" long	G	895	.018		1.03	1	.38	2.41	3.17
1100	8-3/16" long	G	890	.018		1.13	1	.38	2.51	3.30
1500	7/8" diameter, 3-11/16" long	G	920	.017		.89	.97	.37	2.23	2.96
1600	4-3/16" long	G	910	.018		.95	.98	.37	2.30	3.06
1700	5-3/16" long	G	905	.018		1.08	.98	.37	2.43	3.21
1800	6-3/16" long	G	895	.018		1.21	1	.38	2.59	3.37
1900	7-3/16" long	G	890	.018		1.34	1	.38	2.72	3.54
2000	8-3/16" long	G	880	.018		1.47	1.01	.38	2.86	3.68

05 05 23.87 Weld Studs

		Crew	Daily Output	Labor-Hours	Unit	Material	2018 Bare Costs Labor	2018 Bare Costs Equipment	Total	Total Incl O&P
0010	**WELD STUDS**									
0020	1/4" diameter, 2-11/16" long	G E-10	1120	.014	Ea.	.37	.80	.30	1.47	2.04
0100	4-1/8" long	G	1080	.015		.35	.82	.31	1.48	2.08
0200	3/8" diameter, 4-1/8" long	G	1080	.015		.40	.82	.31	1.53	2.13
0300	6-1/8" long	G	1040	.015		.52	.86	.32	1.70	2.33
0400	1/2" diameter, 2-1/8" long	G	1040	.015		.37	.86	.32	1.55	2.16
0500	3-1/8" long	G	1025	.016		.44	.87	.33	1.64	2.27
0600	4-1/8" long	G	1010	.016		.52	.88	.33	1.73	2.38

05 05 Common Work Results for Metals

05 05 23 – Metal Fastenings

05 05 23.87 Weld Studs

			Crew	Daily Output	Labor-Hours	Unit	Material	2018 Bare Costs Labor	Equipment	Total	Total Incl O&P
0700	5-5/16" long	G	E-10	990	.016	Ea.	.64	.90	.34	1.88	2.54
0800	6-1/8" long	G		975	.016		.69	.91	.35	1.95	2.63
0900	8-1/8" long	G		960	.017		.98	.93	.35	2.26	2.98
1000	5/8" diameter, 2-11/16" long	G		1000	.016		.63	.89	.34	1.86	2.53
1010	4-3/16" long	G		990	.016		.78	.90	.34	2.02	2.70
1100	6-9/16" long	G		975	.016		1.02	.91	.35	2.28	2.99
1200	8-3/16" long	G		960	.017		1.37	.93	.35	2.65	3.41

05 05 23.90 Welding Rod

			Crew	Daily Output	Labor-Hours	Unit	Material	2018 Bare Costs Labor	Equipment	Total	Total Incl O&P
0010	**WELDING ROD**										
0020	Steel, type 6011, 1/8" diam., less than 500#					Lb.	2.35			2.35	2.59
0100	500# to 2,000#						2.12			2.12	2.33
0200	2,000# to 5,000#						1.99			1.99	2.19
0300	5/32" diam., less than 500#						2.42			2.42	2.66
0310	500# to 2,000#						2.18			2.18	2.40
0320	2,000# to 5,000#						2.05			2.05	2.25
0400	3/16" diam., less than 500#						2.39			2.39	2.63
0500	500# to 2,000#						2.15			2.15	2.37
0600	2,000# to 5,000#						2.02			2.02	2.22
0620	Steel, type 6010, 1/8" diam., less than 500#						2.41			2.41	2.65
0630	500# to 2,000#						2.17			2.17	2.39
0640	2,000# to 5,000#						2.04			2.04	2.24
0650	Steel, type 7018 Low Hydrogen, 1/8" diam., less than 500#						2.44			2.44	2.69
0660	500# to 2,000#						2.20			2.20	2.42
0670	2,000# to 5,000#						2.07			2.07	2.27
0700	Steel, type 7024 Jet Weld, 1/8" diam., less than 500#						2.50			2.50	2.75
0710	500# to 2,000#						2.25			2.25	2.48
0720	2,000# to 5,000#						2.12			2.12	2.33
1550	Aluminum, type 4043 TIG, 1/8" diam., less than 10#						5.10			5.10	5.65
1560	10# to 60#						4.61			4.61	5.05
1570	Over 60#						4.33			4.33	4.77
1600	Aluminum, type 5356 TIG, 1/8" diam., less than 10#						5.45			5.45	6
1610	10# to 60#						4.90			4.90	5.40
1620	Over 60#						4.61			4.61	5.05
1900	Cast iron, type 8 Nickel, 1/8" diam., less than 500#						22			22	24
1910	500# to 1,000#						19.75			19.75	21.50
1920	Over 1,000#						18.55			18.55	20.50
2000	Stainless steel, type 316/316L, 1/8" diam., less than 500#						7.05			7.05	7.75
2100	500# to 1,000#						6.35			6.35	6.95
2220	Over 1,000#						5.95			5.95	6.55

05 12 Structural Steel Framing

05 12 23 – Structural Steel for Buildings

05 12 23.05 Canopy Framing

			Crew	Daily Output	Labor-Hours	Unit	Material	2018 Bare Costs Labor	Equipment	Total	Total Incl O&P
0010	**CANOPY FRAMING**										
0020	6" and 8" members, shop fabricated	G	E-4	3000	.011	Lb.	1.62	.59	.03	2.24	2.78

05 12 23.10 Ceiling Supports

			Crew	Daily Output	Labor-Hours	Unit	Material	2018 Bare Costs Labor	Equipment	Total	Total Incl O&P
0010	**CEILING SUPPORTS**										
1000	Entrance door/folding partition supports, shop fabricated	G	E-4	60	.533	L.F.	27	29.50	1.64	58.14	79.50
1100	Linear accelerator door supports	G		14	2.286		123	126	7.05	256.05	350
1200	Lintels or shelf angles, hung, exterior hot dipped galv.	G		267	.120		18.40	6.60	.37	25.37	31

For customer support on your Building Construction Costs with RSMeans data, call 800.448.8182.

127

05 12 23.10 Ceiling Supports

		Crew	Daily Output	Labor-Hours	Unit	Material	2018 Bare Costs Labor	Equipment	Total	Total Incl O&P
1250	Two coats primer paint instead of galv.	E-4	267	.120	L.F.	15.95	6.60	.37	22.92	28.50
1400	Monitor support, ceiling hung, expansion bolted	G	4	8	Ea.	425	440	24.50	889.50	1,225
1450	Hung from pre-set inserts	G	6	5.333		460	294	16.45	770.45	1,000
1600	Motor supports for overhead doors	G	4	8		217	440	24.50	681.50	985
1700	Partition support for heavy folding partitions, without pocket	G	24	1.333	L.F.	61.50	73.50	4.11	139.11	192
1750	Supports at pocket only	G	12	2.667		123	147	8.20	278.20	385
2000	Rolling grilles & fire door supports	G	34	.941		52.50	52	2.90	107.40	146
2100	Spider-leg light supports, expansion bolted to ceiling slab	G	8	4	Ea.	175	221	12.30	408.30	565
2150	Hung from pre-set inserts	G	12	2.667	"	188	147	8.20	343.20	455
2400	Toilet partition support	G	36	.889	L.F.	61.50	49	2.74	113.24	151
2500	X-ray travel gantry support	G	12	2.667	"	210	147	8.20	365.20	480

05 12 23.15 Columns, Lightweight

		Crew	Daily Output	Labor-Hours	Unit	Material	2018 Bare Costs Labor	Equipment	Total	Total Incl O&P
0010	**COLUMNS, LIGHTWEIGHT**									
1000	Lightweight units (lally), 3-1/2" diameter	E-2	780	.072	L.F.	5.70	3.90	2.16	11.76	14.95
1050	4" diameter	"	900	.062	"	7.60	3.38	1.87	12.85	15.85
5800	Adjustable jack post, 8' maximum height, 2-3/4" diameter	G			Ea.	52.50			52.50	57.50
5850	4" diameter	G			"	84			84	92

05 12 23.17 Columns, Structural

		Crew	Daily Output	Labor-Hours	Unit	Material	2018 Bare Costs Labor	Equipment	Total	Total Incl O&P
0010	**COLUMNS, STRUCTURAL**	R051223-10								
0015	Made from recycled materials									
0020	Shop fab'd for 100-ton, 1-2 story project, bolted connections									
0800	Steel, concrete filled, extra strong pipe, 3-1/2" diameter	E-2	660	.085	L.F.	44.50	4.61	2.56	51.67	59
0830	4" diameter		780	.072		49.50	3.90	2.16	55.56	63
0890	5" diameter		1020	.055		59	2.98	1.65	63.63	71.50
0930	6" diameter		1200	.047		78	2.53	1.41	81.94	91.50
0940	8" diameter		1100	.051		78	2.76	1.53	82.29	92
1100	For galvanizing, add				Lb.	.25			.25	.28
1300	For web ties, angles, etc., add per added lb.	1 Sswk	945	.008		1.35	.46		1.81	2.24
1500	Steel pipe, extra strong, no concrete, 3" to 5" diameter	G	E-2	16000	.004		1.35	.19	.11	1.65	1.90
1600	6" to 12" diameter	G	14000	.004		1.35	.22	.12	1.69	1.96
1700	Steel pipe, extra strong, no concrete, 3" diameter x 12'-0"	G	60	.933	Ea.	166	50.50	28	244.50	294
1750	4" diameter x 12'-0"	G	58	.966		242	52.50	29	323.50	380
1800	6" diameter x 12'-0"	G	54	1.037		460	56.50	31.50	548	635
1850	8" diameter x 14'-0"	G	50	1.120		820	61	34	915	1,025
1900	10" diameter x 16'-0"	G	48	1.167		1,175	63.50	35	1,273.50	1,450
1950	12" diameter x 18'-0"	G	45	1.244		1,575	67.50	37.50	1,680	1,900
3300	Structural tubing, square, A500GrB, 4" to 6" square, light section	G	11270	.005	Lb.	1.35	.27	.15	1.77	2.07
3600	Heavy section	G	32000	.002	"	1.35	.10	.05	1.50	1.69
4000	Concrete filled, add				L.F.	4.78			4.78	5.25
4500	Structural tubing, square, 4" x 4" x 1/4" x 12'-0"	G	E-2	58	.966	Ea.	222	52.50	29	303.50	360
4550	6" x 6" x 1/4" x 12'-0"	G	54	1.037		365	56.50	31.50	453	525
4600	8" x 8" x 3/8" x 14'-0"	G	50	1.120		790	61	34	885	1,000
4650	10" x 10" x 1/2" x 16'-0"	G	48	1.167		1,450	63.50	35	1,548.50	1,750
5100	Structural tubing, rect., 5" to 6" wide, light section	G	8000	.007	Lb.	1.35	.38	.21	1.94	2.32
5200	Heavy section	G	12000	.005		1.35	.25	.14	1.74	2.04
5300	7" to 10" wide, light section	G	15000	.004		1.35	.20	.11	1.66	1.92
5400	Heavy section	G	18000	.003		1.35	.17	.09	1.61	1.85
5500	Structural tubing, rect., 5" x 3" x 1/4" x 12'-0"	G	58	.966	Ea.	215	52.50	29	296.50	355
5550	6" x 4" x 5/16" x 12'-0"	G	54	1.037		335	56.50	31.50	423	495
5600	8" x 4" x 3/8" x 12'-0"	G	54	1.037		490	56.50	31.50	578	665
5650	10" x 6" x 3/8" x 14'-0"	G	50	1.120		790	61	34	885	1,000
5700	12" x 8" x 1/2" x 16'-0"	G	48	1.167		1,450	63.50	35	1,548.50	1,750

05 12 Structural Steel Framing

05 12 23 – Structural Steel for Buildings

05 12 23.17 Columns, Structural

		Crew	Daily Output	Labor-Hours	Unit	Material	2018 Bare Costs Labor	Equipment	Total	Total Incl O&P
6800	W Shape, A992 steel, 2 tier, W8 x 24 **G**	E-2	1080	.052	L.F.	35.50	2.81	1.56	39.87	45
6850	W8 x 31 **G**		1080	.052		46	2.81	1.56	50.37	56.50
6900	W8 x 48 **G**		1032	.054		71	2.95	1.63	75.58	84.50
6950	W8 x 67 **G**		984	.057		99	3.09	1.71	103.80	116
7000	W10 x 45 **G**		1032	.054		66.50	2.95	1.63	71.08	80
7050	W10 x 68 **G**		984	.057		101	3.09	1.71	105.80	118
7100	W10 x 112 **G**		960	.058		166	3.17	1.76	170.93	189
7150	W12 x 50 **G**		1032	.054		74	2.95	1.63	78.58	88
7200	W12 x 87 **G**		984	.057		129	3.09	1.71	133.80	149
7250	W12 x 120 **G**		960	.058		178	3.17	1.76	182.93	202
7300	W12 x 190 **G**		912	.061		281	3.33	1.85	286.18	315
7350	W14 x 74 **G**		984	.057		110	3.09	1.71	114.80	128
7400	W14 x 120 **G**		960	.058		178	3.17	1.76	182.93	202
7450	W14 x 176 **G**		912	.061	%	261	3.33	1.85	266.18	294
8090	For projects 75 to 99 tons, add				%	10%				
8092	50 to 74 tons, add					20%				
8094	25 to 49 tons, add					30%	10%			
8096	10 to 24 tons, add					50%	25%			
8098	2 to 9 tons, add					75%	50%			
8099	Less than 2 tons, add					100%	100%			

05 12 23.18 Corner Guards

		Crew	Daily Output	Labor-Hours	Unit	Material	2018 Bare Costs Labor	Equipment	Total	Total Incl O&P
0010	**CORNER GUARDS**									
0020	Steel angle w/anchors, 1" x 1" x 1/4", 1.5#/L.F.	2 Carp	160	.100	L.F.	8.25	5.05		13.30	16.80
0100	2" x 2" x 1/4" angles, 3.2#/L.F.		150	.107		10.70	5.40		16.10	20
0200	3" x 3" x 5/16" angles, 6.1#/L.F.		140	.114		16.55	5.80		22.35	27
0300	4" x 4" x 5/16" angles, 8.2#/L.F.		120	.133		16.55	6.75		23.30	28.50
0350	For angles drilled and anchored to masonry, add					15%	120%			
0370	Drilled and anchored to concrete, add					20%	170%			
0400	For galvanized angles, add					35%				
0450	For stainless steel angles, add					100%				

05 12 23.20 Curb Edging

		Crew	Daily Output	Labor-Hours	Unit	Material	2018 Bare Costs Labor	Equipment	Total	Total Incl O&P
0010	**CURB EDGING**									
0020	Steel angle w/anchors, shop fabricated, on forms, 1" x 1", 0.8#/L.F. **G**	E-4	350	.091	L.F.	1.67	5.05	.28	7	10.40
0100	2" x 2" angles, 3.92#/L.F. **G**		330	.097		6.70	5.35	.30	12.35	16.50
0200	3" x 3" angles, 6.1#/L.F. **G**		300	.107		10.85	5.90	.33	17.08	22
0300	4" x 4" angles, 8.2#/L.F. **G**		275	.116		14.25	6.40	.36	21.01	26.50
1000	6" x 4" angles, 12.3#/L.F. **G**		250	.128		21	7.05	.39	28.44	35
1050	Steel channels with anchors, on forms, 3" channel, 5#/L.F. **G**		290	.110		8.45	6.10	.34	14.89	19.60
1100	4" channel, 5.4#/L.F. **G**		270	.119		9.10	6.55	.37	16.02	21
1200	6" channel, 8.2#/L.F. **G**		255	.125		14.25	6.90	.39	21.54	27.50
1300	8" channel, 11.5#/L.F. **G**		225	.142		19.60	7.85	.44	27.89	35
1400	10" channel, 15.3#/L.F. **G**		180	.178		25.50	9.80	.55	35.85	45
1500	12" channel, 20.7#/L.F. **G**		140	.229		34.50	12.60	.70	47.80	59.50
2000	For curved edging, add					35%	10%			

05 12 23.40 Lightweight Framing

		Crew	Daily Output	Labor-Hours	Unit	Material	2018 Bare Costs Labor	Equipment	Total	Total Incl O&P
0010	**LIGHTWEIGHT FRAMING** R051223-35									
0015	Made from recycled materials									
0200	For load-bearing steel studs see Section 05 41 13.30									
0400	Angle framing, field fabricated, 4" and larger R051223-45 **G**	E-3	440	.055	Lb.	.78	3.02	.22	4.02	6.05
0450	Less than 4" angles **G**		265	.091	"	.81	5	.37	6.18	9.50
0460	1/2" x 1/2" x 1/8" **G**		200	.120	L.F.	.16	6.65	.49	7.30	11.55
0462	3/4" x 3/4" x 1/8" **G**		160	.150		.45	8.30	.62	9.37	14.75

05 12 23 – Structural Steel for Buildings

05 12 23.40 Lightweight Framing		Crew	Daily Output	Labor-Hours	Unit	Material	2018 Bare Costs Labor	Equipment	Total	Total Incl O&P
0464	1" x 1" x 1/8"	G	E-3	135	.178	L.F.	.65	9.85	.73	11.23	17.60
0466	1-1/4" x 1-1/4" x 3/16"	G		115	.209		1.20	11.55	.86	13.61	21
0468	1-1/2" x 1-1/2" x 3/16"	G		100	.240		1.45	13.30	.98	15.73	24
0470	2" x 2" x 1/4"	G		90	.267		2.58	14.75	1.09	18.42	28
0472	2-1/2" x 2-1/2" x 1/4"	G		72	.333		3.31	18.45	1.37	23.13	35
0474	3" x 2" x 3/8"	G		65	.369		4.77	20.50	1.51	26.78	40.50
0476	3" x 3" x 3/8"	G		57	.421	↓	5.80	23.50	1.73	31.03	46.50
0600	Channel framing, field fabricated, 8" and larger	G		500	.048	Lb.	.81	2.66	.20	3.67	5.45
0650	Less than 8" channels	G		335	.072	"	.81	3.96	.29	5.06	7.70
0660	C2 x 1.78	G		115	.209	L.F.	1.44	11.55	.86	13.85	21.50
0662	C3 x 4.1	G		80	.300		3.31	16.60	1.23	21.14	32
0664	C4 x 5.4	G		66	.364		4.36	20	1.49	25.85	39.50
0666	C5 x 6.7	G		57	.421		5.40	23.50	1.73	30.63	46
0668	C6 x 8.2	G		55	.436		6.40	24	1.79	32.19	48.50
0670	C7 x 9.8	G		40	.600		7.90	33	2.46	43.36	66
0672	C8 x 11.5	G		36	.667		9.30	37	2.73	49.03	73.50
0710	Structural bar tee, field fabricated, 3/4" x 3/4" x 1/8"	G		160	.150		.45	8.30	.62	9.37	14.75
0712	1" x 1" x 1/8"	G		135	.178		.65	9.85	.73	11.23	17.60
0714	1-1/2" x 1-1/2" x 1/4"	G		114	.211		1.89	11.65	.86	14.40	22
0716	2" x 2" x 1/4"	G		89	.270		2.58	14.90	1.11	18.59	28.50
0718	2-1/2" x 2-1/2" x 3/8"	G		72	.333		4.77	18.45	1.37	24.59	37
0720	3" x 3" x 3/8"	G		57	.421		5.80	23.50	1.73	31.03	46.50
0730	Structural zee, field fabricated, 1-1/4" x 1-3/4" x 1-3/4"	G		114	.211		.61	11.65	.86	13.12	20.50
0732	2-11/16" x 3" x 2-11/16"	G		114	.211		1.44	11.65	.86	13.95	21.50
0734	3-1/16" x 4" x 3-1/16"	G		133	.180		2.17	10	.74	12.91	19.50
0736	3-1/4" x 5" x 3-1/4"	G		133	.180		2.96	10	.74	13.70	20.50
0738	3-1/2" x 6" x 3-1/2"	G		160	.150		4.47	8.30	.62	13.39	19.15
0740	Junior beam, field fabricated, 3"	G		80	.300		4.60	16.60	1.23	22.43	33.50
0742	4"	G		72	.333		6.20	18.45	1.37	26.02	38.50
0744	5"	G		67	.358		8.10	19.80	1.47	29.37	43
0746	6"	G		62	.387		10.10	21.50	1.59	33.19	48
0748	7"	G		57	.421		12.35	23.50	1.73	37.58	53.50
0750	8"	G		53	.453	↓	14.85	25	1.86	41.71	59.50
1000	Continuous slotted channel framing system, shop fab, simple framing	G	2 Sswk	2400	.007	Lb.	4.17	.36		4.53	5.20
1200	Complex framing	G	"	1600	.010		4.71	.55		5.26	6.10
1300	Cross bracing, rods, shop fabricated, 3/4" diameter	G	E-3	700	.034		1.62	1.90	.14	3.66	5.05
1310	7/8" diameter	G		850	.028		1.62	1.56	.12	3.30	4.46
1320	1" diameter	G		1000	.024		1.62	1.33	.10	3.05	4.06
1330	Angle, 5" x 5" x 3/8"	G		2800	.009		1.62	.47	.04	2.13	2.60
1350	Hanging lintels, shop fabricated	G	↓	850	.028		1.62	1.56	.12	3.30	4.46
1380	Roof frames, shop fabricated, 3'-0" square, 5' span	G	E-2	4200	.013		1.62	.72	.40	2.74	3.38
1400	Tie rod, not upset, 1-1/2" to 4" diameter, with turnbuckle	G	2 Sswk	800	.020		1.75	1.09		2.84	3.72
1420	No turnbuckle	G		700	.023		1.68	1.25		2.93	3.89
1500	Upset, 1-3/4" to 4" diameter, with turnbuckle	G		800	.020		1.75	1.09		2.84	3.72
1520	No turnbuckle	G	↓	700	.023	↓	1.68	1.25		2.93	3.89

05 12 23.45 Lintels

0010	**LINTELS**										
0015	Made from recycled materials										
0020	Plain steel angles, shop fabricated, under 500 lb.	G	1 Bric	550	.015	Lb.	1.04	.73		1.77	2.26
0100	500 to 1,000 lb.	G		640	.013		1.01	.63		1.64	2.07
0200	1,000 to 2,000 lb.	G		640	.013		.98	.63		1.61	2.04
0300	2,000 to 4,000 lb.	G	↓	640	.013	↓	.96	.63		1.59	2.01

05 12 Structural Steel Framing

05 12 23 – Structural Steel for Buildings

05 12 23.45 Lintels

		Crew	Daily Output	Labor-Hours	Unit	Material	2018 Bare Costs Labor	Equipment	Total	Total Incl O&P
0500	For built-up angles and plates, add to above	G			Lb.	1.35			1.35	1.48
0700	For engineering, add to above					.13			.13	.15
0900	For galvanizing, add to above, under 500 lb.					.30			.30	.33
0950	500 to 2,000 lb.					.27			.27	.30
1000	Over 2,000 lb.					.25			.25	.28
2000	Steel angles, 3-1/2" x 3", 1/4" thick, 2'-6" long	G	1 Bric	47	.170	Ea.	14.55	8.55		23.10	29
2100	4'-6" long	G		26	.308		26	15.50		41.50	52.50
2600	4" x 3-1/2", 1/4" thick, 5'-0" long	G		21	.381		33.50	19.15		52.65	66
2700	9'-0" long	G		12	.667		60	33.50		93.50	118

05 12 23.60 Pipe Support Framing

		Crew	Daily Output	Labor-Hours	Unit	Material	2018 Bare Costs Labor	Equipment	Total	Total Incl O&P
0010	**PIPE SUPPORT FRAMING**									
0020	Under 10#/L.F., shop fabricated	G	E-4	3900	.008	Lb.	1.80	.45	.03	2.28	2.75
0200	10.1 to 15#/L.F.	G		4300	.007		1.78	.41	.02	2.21	2.65
0400	15.1 to 20#/L.F.	G		4800	.007		1.75	.37	.02	2.14	2.55
0600	Over 20#/L.F.	G		5400	.006		1.72	.33	.02	2.07	2.45

05 12 23.65 Plates

			Crew	Daily Output	Labor-Hours	Unit	Material	2018 Bare Costs Labor	Equipment	Total	Total Incl O&P
0010	**PLATES**	R051223-80									
0015	Made from recycled materials										
0020	For connections & stiffener plates, shop fabricated										
0050	1/8" thick (5.1 lb./S.F.)		G				S.F.	6.85			6.85	7.55
0100	1/4" thick (10.2 lb./S.F.)		G					13.75			13.75	15.10
0300	3/8" thick (15.3 lb./S.F.)		G					20.50			20.50	22.50
0400	1/2" thick (20.4 lb./S.F.)		G					27.50			27.50	30
0450	3/4" thick (30.6 lb./S.F.)		G					41			41	45.50
0500	1" thick (40.8 lb./S.F.)		G					55			55	60.50
2000	Steel plate, warehouse prices, no shop fabrication										
2100	1/4" thick (10.2 lb./S.F.)		G				S.F.	7.30			7.30	8

05 12 23.70 Stressed Skin Steel Roof and Ceiling System

		Crew	Daily Output	Labor-Hours	Unit	Material	2018 Bare Costs Labor	Equipment	Total	Total Incl O&P
0010	**STRESSED SKIN STEEL ROOF & CEILING SYSTEM**										
0020	Double panel flat roof, spans to 100'	G	E-2	1150	.049	S.F.	10.75	2.64	1.47	14.86	17.70
0100	Double panel convex roof, spans to 200'	G		960	.058		17.50	3.17	1.76	22.43	26
0200	Double panel arched roof, spans to 300'	G		760	.074		27	4	2.22	33.22	38.50

05 12 23.75 Structural Steel Members

			Crew	Daily Output	Labor-Hours	Unit	Material	2018 Bare Costs Labor	Equipment	Total	Total Incl O&P
0010	**STRUCTURAL STEEL MEMBERS**	R051223-10									
0015	Made from recycled materials										
0020	Shop fab'd for 100-ton, 1-2 story project, bolted connections										
0100	Beam or girder, W 6 x 9		G	E-2	600	.093	L.F.	13.35	5.05	2.81	21.21	26
0120	x 15		G		600	.093		22	5.05	2.81	29.86	35.50
0140	x 20		G		600	.093		29.50	5.05	2.81	37.36	43.50
0300	W 8 x 10		G		600	.093		14.80	5.05	2.81	22.66	27.50
0320	x 15		G		600	.093		22	5.05	2.81	29.86	35.50
0350	x 21		G		600	.093		31	5.05	2.81	38.86	45
0360	x 24		G		550	.102		35.50	5.55	3.07	44.12	51
0370	x 28		G		550	.102		41.50	5.55	3.07	50.12	57.50
0500	x 31		G		550	.102		46	5.55	3.07	54.62	62.50
0520	x 35		G		550	.102		52	5.55	3.07	60.62	69
0540	x 48		G		550	.102		71	5.55	3.07	79.62	90
0600	W 10 x 12		G		600	.093		17.75	5.05	2.81	25.61	30.50
0620	x 15		G		600	.093		22	5.05	2.81	29.86	35.50
0700	x 22		G		600	.093		32.50	5.05	2.81	40.36	47
0720	x 26		G		600	.093		38.50	5.05	2.81	46.36	53.50
0740	x 33		G		550	.102		49	5.55	3.07	57.62	66

For customer support on your Building Construction Costs with RSMeans data, call 800.448.8182.

131

05 12 Structural Steel Framing

05 12 23 – Structural Steel for Buildings

05 12 23.75 Structural Steel Members		Crew	Daily Output	Labor-Hours	Unit	Material	2018 Bare Costs Labor	Equipment	Total	Total Incl O&P
0900	x 49	G	E-2	550	.102	L.F.	72.50	5.55	3.07	81.12	92
1100	W 12 x 16	G		880	.064		23.50	3.45	1.92	28.87	33.50
1300	x 22	G		880	.064		32.50	3.45	1.92	37.87	43.50
1500	x 26	G		880	.064		38.50	3.45	1.92	43.87	50
1520	x 35	G		810	.069		52	3.75	2.08	57.83	65.50
1560	x 50	G		750	.075		74	4.05	2.25	80.30	90.50
1580	x 58	G		750	.075		86	4.05	2.25	92.30	103
1700	x 72	G		640	.088		107	4.75	2.64	114.39	128
1740	x 87	G		640	.088		129	4.75	2.64	136.39	153
1900	W 14 x 26	G		990	.057		38.50	3.07	1.70	43.27	49.50
2100	x 30	G		900	.062		44.50	3.38	1.87	49.75	56.50
2300	x 34	G		810	.069		50.50	3.75	2.08	56.33	64
2320	x 43	G		810	.069		63.50	3.75	2.08	69.33	78.50
2340	x 53	G		800	.070		78.50	3.80	2.11	84.41	95
2360	x 74	G		760	.074		110	4	2.22	116.22	130
2380	x 90	G		740	.076		133	4.11	2.28	139.39	156
2500	x 120	G		720	.078		178	4.22	2.34	184.56	204
2700	W 16 x 26	G		1000	.056		38.50	3.04	1.69	43.23	49
2900	x 31	G		900	.062		46	3.38	1.87	51.25	58
3100	x 40	G		800	.070		59	3.80	2.11	64.91	73.50
3120	x 50	G		800	.070		74	3.80	2.11	79.91	90
3140	x 67	G	▼	760	.074		99	4	2.22	105.22	118
3300	W 18 x 35	G	E-5	960	.083		52	4.55	1.86	58.41	66.50
3500	x 40	G		960	.083		59	4.55	1.86	65.41	74.50
3520	x 46	G		960	.083		68	4.55	1.86	74.41	84.50
3700	x 50	G		912	.088		74	4.79	1.96	80.75	91.50
3900	x 55	G		912	.088		81.50	4.79	1.96	88.25	99.50
3920	x 65	G		900	.089		96.50	4.85	1.98	103.33	116
3940	x 76	G		900	.089		113	4.85	1.98	119.83	134
3960	x 86	G		900	.089		127	4.85	1.98	133.83	150
3980	x 106	G		900	.089		157	4.85	1.98	163.83	183
4100	W 21 x 44	G		1064	.075		65	4.10	1.68	70.78	80
4300	x 50	G		1064	.075		74	4.10	1.68	79.78	90
4500	x 62	G		1036	.077		92	4.22	1.72	97.94	110
4700	x 68	G		1036	.077		101	4.22	1.72	106.94	120
4720	x 83	G		1000	.080		123	4.37	1.79	129.16	144
4740	x 93	G		1000	.080		138	4.37	1.79	144.16	160
4760	x 101	G		1000	.080		150	4.37	1.79	156.16	174
4780	x 122	G		1000	.080		181	4.37	1.79	187.16	208
4900	W 24 x 55	G		1110	.072		81.50	3.93	1.61	87.04	97.50
5100	x 62	G		1110	.072		92	3.93	1.61	97.54	109
5300	x 68	G		1110	.072		101	3.93	1.61	106.54	119
5500	x 76	G		1110	.072		113	3.93	1.61	118.54	132
5700	x 84	G		1080	.074		124	4.04	1.65	129.69	145
5720	x 94	G		1080	.074		139	4.04	1.65	144.69	161
5740	x 104	G	▼	1050	.076	▼	154	4.16	1.70	159.86	178
5760	x 117	G	E-5	1050	.076	L.F.	173	4.16	1.70	178.86	200
5780	x 146	G		1050	.076		216	4.16	1.70	221.86	247
5800	W 27 x 84	G		1190	.067		124	3.67	1.50	129.17	145
5900	x 94	G		1190	.067		139	3.67	1.50	144.17	161
5920	x 114	G		1150	.070		169	3.80	1.55	174.35	194
5940	x 146	G		1150	.070		216	3.80	1.55	221.35	246
5960	x 161	G	▼	1150	.070	▼	238	3.80	1.55	243.35	270

For customer support on your Building Construction Costs with RSMeans data, call 800.448.8182.

05 12 23 – Structural Steel for Buildings

05 12 23.75 Structural Steel Members

			Crew	Daily Output	Labor-Hours	Unit	Material	2018 Bare Costs Labor	Equipment	Total	Total Incl O&P
6100	W 30 x 99	G	E-5	1200	.067	L.F.	147	3.64	1.49	152.13	168
6300	x 108	G		1200	.067		160	3.64	1.49	165.13	183
6500	x 116	G		1160	.069		172	3.77	1.54	177.31	197
6520	x 132	G		1160	.069		195	3.77	1.54	200.31	223
6540	x 148	G		1160	.069		219	3.77	1.54	224.31	249
6560	x 173	G		1120	.071		256	3.90	1.59	261.49	290
6580	x 191	G		1120	.071		283	3.90	1.59	288.49	320
6700	W 33 x 118	G		1176	.068		175	3.71	1.52	180.23	200
6900	x 130	G		1134	.071		193	3.85	1.57	198.42	220
7100	x 141	G		1134	.071		209	3.85	1.57	214.42	238
7120	x 169	G		1100	.073		250	3.97	1.62	255.59	283
7140	x 201	G		1100	.073		298	3.97	1.62	303.59	335
7300	W 36 x 135	G		1170	.068		200	3.73	1.53	205.26	228
7500	x 150	G		1170	.068		222	3.73	1.53	227.26	252
7600	x 170	G		1150	.070		252	3.80	1.55	257.35	285
7700	x 194	G		1125	.071		287	3.88	1.59	292.47	325
7900	x 231	G		1125	.071		340	3.88	1.59	345.47	385
7920	x 262	G		1035	.077		390	4.22	1.73	395.95	435
8100	x 302	G		1035	.077		445	4.22	1.73	450.95	500
8490	For projects 75 to 99 tons, add						10%				
8492	50 to 74 tons, add						20%				
8494	25 to 49 tons, add						30%	10%			
8496	10 to 24 tons, add						50%	25%			
8498	2 to 9 tons, add						75%	50%			
8499	Less than 2 tons, add						100%	100%			

05 12 23.77 Structural Steel Projects

				Crew	Daily Output	Labor-Hours	Unit	Material	Labor	Equipment	Total	Total Incl O&P
0010	**STRUCTURAL STEEL PROJECTS**		R050516-30									
0015	Made from recycled materials											
0020	Shop fab'd for 100-ton, 1-2 story project, bolted connections											
0200	Apartments, nursing homes, etc., 1 to 2 stories	R050523-10	G	E-5	10.30	7.767	Ton	2,700	425	173	3,298	3,825
0300	3 to 6 stories		G	"	10.10	7.921		2,750	430	177	3,357	3,925
0400	7 to 15 stories	R051223-10	G	E-6	14.20	9.014		2,800	490	137	3,427	4,025
0500	Over 15 stories		G	"	13.90	9.209		2,900	500	140	3,540	4,175
0700	Offices, hospitals, etc., steel bearing, 1 to 2 stories	R051223-20	G	E-5	10.30	7.767		2,700	425	173	3,298	3,825
0800	3 to 6 stories		G	E-6	14.40	8.889		2,750	485	135	3,370	3,950
0900	7 to 15 stories	R051223-25	G		14.20	9.014		2,800	490	137	3,427	4,025
1000	Over 15 stories		G		13.90	9.209		2,900	500	140	3,540	4,175
1100	For multi-story masonry wall bearing construction, add	R051223-30							30%			
1300	Industrial bldgs., 1 story, beams & girders, steel bearing		G	E-5	12.90	6.202		2,700	340	138	3,178	3,650
1400	Masonry bearing		G	"	10	8		2,700	435	179	3,314	3,850
1500	Industrial bldgs., 1 story, under 10 tons,											
1510	steel from warehouse, trucked		G	E-2	7.50	7.467	Ton	3,225	405	225	3,855	4,450
1600	1 story with roof trusses, steel bearing		G	E-5	10.60	7.547		3,175	410	168	3,753	4,350
1700	Masonry bearing		G	"	8.30	9.639		3,175	525	215	3,915	4,575
1900	Monumental structures, banks, stores, etc., simple connections		G	E-6	13	9.846		2,700	535	149	3,384	3,975
2000	Moment/composite connections		G	"	9	14.222		4,475	775	215	5,465	6,400
2200	Churches, simple connections		G	E-5	11.60	6.897		2,500	375	154	3,029	3,525
2300	Moment/composite connections		G	"	5.20	15.385		3,350	840	345	4,535	5,400
2800	Power stations, fossil fuels, simple connections		G	E-6	11	11.636		2,700	635	176	3,511	4,175
2900	Moment/composite connections		G		5.70	22.456		4,050	1,225	340	5,615	6,800
2950	Nuclear fuels, non-safety steel, simple connections		G		7	18.286		2,700	995	277	3,972	4,850
3000	Moment/composite connections		G		5.50	23.273		4,050	1,275	355	5,680	6,900

For customer support on your Building Construction Costs with RSMeans data, call 800.448.8182.

133

05 12 Structural Steel Framing

05 12 23 – Structural Steel for Buildings

05 12 23.77 Structural Steel Projects

		Crew	Daily Output	Labor-Hours	Unit	Material	2018 Bare Costs Labor	Equipment	Total	Total Incl O&P
3040	Safety steel, simple connections G	E-6	2.50	51.200	Ton	3,925	2,800	775	7,500	9,675
3070	Moment/composite connections G	↓	1.50	85.333		5,175	4,650	1,300	11,125	14,600
3100	Roof trusses, simple connections G	E-5	13	6.154		3,775	335	137	4,247	4,850
3200	Moment/composite connections G		8.30	9.639		4,575	525	215	5,315	6,100
3210	Schools, simple connections G		14.50	5.517		2,700	300	123	3,123	3,575
3220	Moment/composite connections G	↓	8.30	9.639		3,925	525	215	4,665	5,400
3400	Welded construction, simple commercial bldgs., 1 to 2 stories G	E-7	7.60	10.526		2,750	575	248	3,573	4,225
3500	7 to 15 stories G	E-9	8.30	15.422		3,175	840	262	4,277	5,150
3700	Welded rigid frame, 1 story, simple connections G	E-7	15.80	5.063		2,800	276	119	3,195	3,650
3800	Moment/composite connections G	"	5.50	14.545	↓	3,625	795	345	4,765	5,650
3810	Fabrication shop costs (incl in project bare material cost, above)									
3820	Mini mill base price, Grade A992 G				Ton	730			730	800
3830	Mill extras plus delivery to warehouse					280			280	310
3835	Delivery from warehouse to fabrication shop					108			108	118
3840	Shop extra for shop drawings and detailing					325			325	355
3850	Shop fabricating and handling					965			965	1,050
3860	Shop sandblasting and primer coat of paint					148			148	163
3870	Shop delivery to the job site					139			139	153
3880	Total material cost, shop fabricated, primed, delivered				↓	2,700			2,700	2,950
3900	High strength steel mill spec extras:									
3950	A529, A572 (50 ksi) and A36: same as A992 steel (no extra)									
4000	Add to A992 price for A572 (60, 65 ksi) G				Ton	82			82	90.50
4100	A242 and A588 Weathering G				"	82			82	90.50
4200	Mill size extras for W-Shapes: 0 to 30 plf: no extra charge									
4210	Member sizes 31 to 65 plf, add G				Ton	5.10			5.10	5.65
4220	Member sizes 66 to 100 plf, add G					61.50			61.50	68
4230	Member sizes 101 to 387 plf, add G				↓	149			149	164
4300	Column base plates, light, up to 150 lb. G	2 Sswk	2000	.008	Lb.	1.48	.44		1.92	2.34
4400	Heavy, over 150 lb. G	E-2	7500	.007	"	1.55	.41	.23	2.19	2.60
4600	Castellated beams, light sections, to 50#/L.F., simple connections G		10.70	5.234	Ton	2,825	284	158	3,267	3,725
4700	Moment/composite connections G		7	8		3,100	435	241	3,776	4,350
4900	Heavy sections, over 50 plf, simple connections G		11.70	4.786		2,950	260	144	3,354	3,825
5000	Moment/composite connections G	↓	7.80	7.179		3,225	390	216	3,831	4,425
5390	For projects 75 to 99 tons, add					10%				
5392	50 to 74 tons, add					20%				
5394	25 to 49 tons, add					30%	10%			
5396	10 to 24 tons, add					50%	25%			
5398	2 to 9 tons, add					75%	50%			
5399	Less than 2 tons, add				↓	100%	100%			

05 12 23.78 Structural Steel Secondary Members

		Crew	Daily Output	Labor-Hours	Unit	Material	2018 Bare Costs Labor	Equipment	Total	Total Incl O&P
0010	**STRUCTURAL STEEL SECONDARY MEMBERS**									
0015	Made from recycled materials									
0020	Shop fabricated for 20-ton girt/purlin framing package, materials only									
0100	Girts/purlins, C/Z-shapes, includes clips and bolts									
0110	6" x 2-1/2" x 2-1/2", 16 ga., 3.0 lb./L.F.				L.F.	3.63			3.63	4
0115	14 ga., 3.5 lb./L.F.					4.24			4.24	4.66
0120	8" x 2-3/4" x 2-3/4", 16 ga., 3.4 lb./L.F.					4.12			4.12	4.53
0125	14 ga., 4.1 lb./L.F.					4.97			4.97	5.45
0130	12 ga., 5.6 lb./L.F.					6.80			6.80	7.45
0135	10" x 3-1/2" x 3-1/2", 14 ga., 4.7 lb./L.F.					5.70			5.70	6.25
0140	12 ga., 6.7 lb./L.F.					8.10			8.10	8.95
0145	12" x 3-1/2" x 3-1/2", 14 ga., 5.3 lb./L.F.				↓	6.40			6.40	7.05

For customer support on your Building Construction Costs with RSMeans data, call 800.448.8182.

05 12 Structural Steel Framing

05 12 23 – Structural Steel for Buildings

05 12 23.78 Structural Steel Secondary Members

		Crew	Daily Output	Labor-Hours	Unit	Material	2018 Bare Costs Labor	2018 Bare Costs Equipment	Total	Total Incl O&P
0150	12 ga., 7.4 lb./L.F.				L.F.	8.95			8.95	9.85
0200	Eave struts, C-shape, includes clips and bolts									
0210	6" x 4" x 3", 16 ga., 3.1 lb./L.F.				L.F.	3.76			3.76	4.13
0215	14 ga., 3.9 lb./L.F.					4.73			4.73	5.20
0220	8" x 4" x 3", 16 ga., 3.5 lb./L.F.					4.24			4.24	4.66
0225	14 ga., 4.4 lb./L.F.					5.35			5.35	5.85
0230	12 ga., 6.2 lb./L.F.					7.50			7.50	8.25
0235	10" x 5" x 3", 14 ga., 5.2 lb./L.F.					6.30			6.30	6.95
0240	12 ga., 7.3 lb./L.F.					8.85			8.85	9.75
0245	12" x 5" x 4", 14 ga., 6.0 lb./L.F.					7.25			7.25	8
0250	12 ga., 8.4 lb./L.F.					10.20			10.20	11.20
0300	Rake/base angle, excludes concrete drilling and expansion anchors									
0310	2" x 2", 14 ga., 1.0 lb./L.F.	2 Sswk	640	.025	L.F.	1.21	1.37		2.58	3.56
0315	3" x 2", 14 ga., 1.3 lb./L.F.		535	.030		1.58	1.63		3.21	4.40
0320	3" x 3", 14 ga., 1.6 lb./L.F.		500	.032		1.94	1.75		3.69	4.99
0325	4" x 3", 14 ga., 1.8 lb./L.F.		480	.033		2.18	1.82		4	5.40
0600	Installation of secondary members, erection only									
0610	Girts, purlins, eave struts, 16 ga., 6" deep	E-18	100	.400	Ea.		22	10.85	32.85	47.50
0615	8" deep		80	.500			27.50	13.55	41.05	59
0620	14 ga., 6" deep		80	.500			27.50	13.55	41.05	59
0625	8" deep		65	.615			34	16.65	50.65	73
0630	10" deep		55	.727			40	19.70	59.70	86
0635	12" deep		50	.800			44	21.50	65.50	94.50
0640	12 ga., 8" deep		50	.800			44	21.50	65.50	94.50
0645	10" deep		45	.889			49	24	73	105
0650	12" deep		40	1			55	27	82	119
0900	For less than 20-ton job lots									
0905	For 15 to 19 tons, add				%	10%				
0910	For 10 to 14 tons, add					25%				
0915	For 5 to 9 tons, add					50%	50%	50%		
0920	For 1 to 4 tons, add					75%	75%	75%		
0925	For less than 1 ton, add					100%	100%	100%		

05 12 23.80 Subpurlins

			Crew	Daily Output	Labor-Hours	Unit	Material	2018 Bare Costs Labor	2018 Bare Costs Equipment	Total	Total Incl O&P
0010	**SUBPURLINS**	R051223-50									
0015	Made from recycled materials										
0020	Bulb tees, shop fabricated, painted, 32-5/8" OC, 40 psf L.L.										
0200	Type 218, max 10'-2" span, 3.19 plf, 2-1/8" high x 2-1/8" wide [G]		E-1	3100	.008	S.F.	1.73	.42	.03	2.18	2.60
1420	For 24-5/8" spacing, add						33%	33%			
1430	For 48-5/8" spacing, deduct						33%	33%			

05 14 Structural Aluminum Framing

05 14 23 – Non-Exposed Structural Aluminum Framing

05 14 23.05 Aluminum Shapes

			Crew	Daily Output	Labor-Hours	Unit	Material	2018 Bare Costs Labor	2018 Bare Costs Equipment	Total	Total Incl O&P
0010	**ALUMINUM SHAPES**										
0015	Made from recycled materials										
0020	Structural shapes, 1" to 10" members, under 1 ton	[G]	E-2	4000	.014	Lb.	3.71	.76	.42	4.89	5.75
0050	1 to 5 tons	[G]		4300	.013		3.40	.71	.39	4.50	5.30
0100	Over 5 tons	[G]		4600	.012		3.24	.66	.37	4.27	5
0300	Extrusions, over 5 tons, stock shapes	[G]		1330	.042		3.41	2.29	1.27	6.97	8.80
0400	Custom shapes	[G]		1330	.042		4.52	2.29	1.27	8.08	10.05

05 15 Wire Rope Assemblies

05 15 16 – Steel Wire Rope Assemblies

05 15 16.05 Accessories for Steel Wire Rope		Crew	Daily Output	Labor-Hours	Unit	Material	2018 Bare Costs Labor	Equipment	Total	Total Incl O&P
0010	**ACCESSORIES FOR STEEL WIRE ROPE**									
0015	Made from recycled materials									
1500	Thimbles, heavy duty, 1/4"	G	E-17	160	.100	Ea.	.37	5.55		5.92	9.50
1510	1/2"	G		160	.100		1.63	5.55		7.18	10.90
1520	3/4"	G		105	.152		3.72	8.50		12.22	17.95
1530	1"	G		52	.308		7.45	17.10		24.55	36
1540	1-1/4"	G		38	.421		11.45	23.50		34.95	51
1550	1-1/2"	G		13	1.231		32	68.50		100.50	148
1560	1-3/4"	G		8	2		66.50	111		177.50	255
1570	2"	G		6	2.667		96.50	148		244.50	350
1580	2-1/4"	G		4	4		131	223		354	510
1600	Clips, 1/4" diameter	G		160	.100		1.98	5.55		7.53	11.30
1610	3/8" diameter	G		160	.100		2.17	5.55		7.72	11.50
1620	1/2" diameter	G		160	.100		3.49	5.55		9.04	12.95
1630	3/4" diameter	G		102	.157		5.65	8.75		14.40	20.50
1640	1" diameter	G		64	.250		9.45	13.90		23.35	33.50
1650	1-1/4" diameter	G		35	.457		15.45	25.50		40.95	58.50
1670	1-1/2" diameter	G		26	.615		21	34.50		55.50	79
1680	1-3/4" diameter	G		16	1		48.50	55.50		104	145
1690	2" diameter	G		12	1.333		54	74		128	181
1700	2-1/4" diameter	G		10	1.600		79.50	89		168.50	234
1800	Sockets, open swage, 1/4" diameter	G		160	.100		26	5.55		31.55	37.50
1810	1/2" diameter	G		77	.208		37.50	11.55		49.05	60
1820	3/4" diameter	G		19	.842		58	47		105	141
1830	1" diameter	G		9	1.778		104	99		203	276
1840	1-1/4" diameter	G		5	3.200		144	178		322	450
1850	1-1/2" diameter	G		3	5.333		315	297		612	835
1860	1-3/4" diameter	G		3	5.333		560	297		857	1,100
1870	2" diameter	G		1.50	10.667		850	595		1,445	1,900
1900	Closed swage, 1/4" diameter	G		160	.100		15.60	5.55		21.15	26.50
1910	1/2" diameter	G		104	.154		27	8.55		35.55	43.50
1920	3/4" diameter	G		32	.500		40.50	28		68.50	90
1930	1" diameter	G		15	1.067		71	59.50		130.50	175
1940	1-1/4" diameter	G		7	2.286		106	127		233	325
1950	1-1/2" diameter	G		4	4		193	223		416	575
1960	1-3/4" diameter	G		3	5.333		284	297		581	800
1970	2" diameter	G		2	8		555	445		1,000	1,350
2000	Open spelter, galv., 1/4" diameter	G		160	.100		35	5.55		40.55	47.50
2010	1/2" diameter	G		70	.229		36.50	12.70		49.20	61
2020	3/4" diameter	G		26	.615		55	34.50		89.50	117
2030	1" diameter	G		10	1.600		152	89		241	315
2040	1-1/4" diameter	G		5	3.200		218	178		396	530
2050	1-1/2" diameter	G		4	4		460	223		683	870
2060	1-3/4" diameter	G		2	8		805	445		1,250	1,625
2070	2" diameter	G		1.20	13.333		925	740		1,665	2,250
2080	2-1/2" diameter	G		1	16		1,700	890		2,590	3,325
2100	Closed spelter, galv., 1/4" diameter	G		160	.100		29	5.55		34.55	41
2110	1/2" diameter	G		88	.182		31	10.10		41.10	51
2120	3/4" diameter	G		30	.533		47	29.50		76.50	101
2130	1" diameter	G		13	1.231		101	68.50		169.50	223
2140	1-1/4" diameter	G		7	2.286		161	127		288	385
2150	1-1/2" diameter	G		6	2.667		345	148		493	625
2160	1-3/4" diameter	G		2.80	5.714		460	320		780	1,025

For customer support on your Building Construction Costs with RSMeans data, call 800.448.8182.

05 15 Wire Rope Assemblies

05 15 16 – Steel Wire Rope Assemblies

05 15 16.05 Accessories for Steel Wire Rope

		Crew	Daily Output	Labor-Hours	Unit	Material	2018 Bare Costs Labor	Equipment	Total	Total Incl O&P
2170	2" diameter	G E-17	2	8	Ea.	570	445		1,015	1,350
2200	Jaw & jaw turnbuckles, 1/4" x 4"	G	160	.100		12	5.55		17.55	22.50
2250	1/2" x 6"	G	96	.167		15.15	9.30		24.45	32
2260	1/2" x 9"	G	77	.208		20	11.55		31.55	41
2270	1/2" x 12"	G	66	.242		22.50	13.50		36	47
2300	3/4" x 6"	G	38	.421		29.50	23.50		53	71
2310	3/4" x 9"	G	30	.533		33	29.50		62.50	84.50
2320	3/4" x 12"	G	28	.571		42.50	32		74.50	98.50
2330	3/4" x 18"	G	23	.696		50.50	38.50		89	119
2350	1" x 6"	G	17	.941		57.50	52.50		110	149
2360	1" x 12"	G	13	1.231		63	68.50		131.50	182
2370	1" x 18"	G	10	1.600		94.50	89		183.50	250
2380	1" x 24"	G	9	1.778		104	99		203	276
2400	1-1/4" x 12"	G	7	2.286		106	127		233	325
2410	1-1/4" x 18"	G	6.50	2.462		131	137		268	370
2420	1-1/4" x 24"	G	5.60	2.857		177	159		336	455
2450	1-1/2" x 12"	G	5.20	3.077		232	171		403	535
2460	1-1/2" x 18"	G	4	4		248	223		471	640
2470	1-1/2" x 24"	G	3.20	5		335	278		613	820
2500	1-3/4" x 18"	G	3.20	5		505	278		783	1,000
2510	1-3/4" x 24"	G	2.80	5.714		575	320		895	1,150
2550	2" x 24"	G	1.60	10		775	555		1,330	1,750

05 15 16.50 Steel Wire Rope

		Crew	Daily Output	Labor-Hours	Unit	Material	2018 Bare Costs Labor	Equipment	Total	Total Incl O&P
0010	**STEEL WIRE ROPE**									
0015	Made from recycled materials									
0020	6 x 19, bright, fiber core, 5000' rolls, 1/2" diameter	G			L.F.	.86			.86	.94
0050	Steel core	G				1.13			1.13	1.24
0100	Fiber core, 1" diameter	G				2.89			2.89	3.18
0150	Steel core	G				3.30			3.30	3.63
0300	6 x 19, galvanized, fiber core, 1/2" diameter	G				1.27			1.27	1.39
0350	Steel core	G				1.45			1.45	1.59
0400	Fiber core, 1" diameter	G				3.71			3.71	4.08
0450	Steel core	G				3.89			3.89	4.28
0500	6 x 7, bright, IPS, fiber core, <500 L.F. w/acc., 1/4" diameter	G E-17	6400	.003		1.11	.14		1.25	1.45
0510	1/2" diameter	G	2100	.008		2.70	.42		3.12	3.66
0520	3/4" diameter	G	960	.017		4.89	.93		5.82	6.90
0550	6 x 19, bright, IPS, IWRC, <500 L.F. w/acc., 1/4" diameter	G	5760	.003		.95	.15		1.10	1.29
0560	1/2" diameter	G	1730	.009		1.54	.51		2.05	2.53
0570	3/4" diameter	G	770	.021		2.67	1.16		3.83	4.82
0580	1" diameter	G	420	.038		4.52	2.12		6.64	8.45
0590	1-1/4" diameter	G	290	.055		7.50	3.07		10.57	13.25
0600	1-1/2" diameter	G	192	.083		9.20	4.64		13.84	17.75
0610	1-3/4" diameter	G E-18	240	.167		14.70	9.15	4.51	28.36	36
0620	2" diameter	G	160	.250		18.85	13.70	6.75	39.30	50
0630	2-1/4" diameter	G	160	.250		25	13.70	6.75	45.45	57
0650	6 x 37, bright, IPS, IWRC, <500 L.F. w/acc., 1/4" diameter	G E-17	6400	.003		1.11	.14		1.25	1.45
0660	1/2" diameter	G	1730	.009		1.88	.51		2.39	2.90
0670	3/4" diameter	G	770	.021		3.03	1.16		4.19	5.20
0680	1" diameter	G	430	.037		4.81	2.07		6.88	8.70
0690	1-1/4" diameter	G	290	.055		7.25	3.07		10.32	13
0700	1-1/2" diameter	G	190	.084		10.40	4.69		15.09	19.10
0710	1-3/4" diameter	G E-18	260	.154		16.50	8.45	4.17	29.12	36.50

05 15 16 – Steel Wire Rope Assemblies

05 15 16.50 Steel Wire Rope		Crew	Daily Output	Labor-Hours	Unit	Material	2018 Bare Costs Labor	Equipment	Total	Total Incl O&P
0720	2" diameter	G	E-18	200	.200	L.F.	21.50	10.95	5.40	37.85	47
0730	2-1/4" diameter	G	↓	160	.250		28.50	13.70	6.75	48.95	60.50
0800	6 x 19 & 6 x 37, swaged, 1/2" diameter	G	E-17	1220	.013		2.53	.73		3.26	3.97
0810	9/16" diameter	G		1120	.014		2.94	.80		3.74	4.54
0820	5/8" diameter	G		930	.017		3.49	.96		4.45	5.40
0830	3/4" diameter	G		640	.025		4.45	1.39		5.84	7.15
0840	7/8" diameter	G		480	.033		5.60	1.85		7.45	9.20
0850	1" diameter	G		350	.046		6.85	2.54		9.39	11.65
0860	1-1/8" diameter	G		288	.056		8.40	3.09		11.49	14.30
0870	1-1/4" diameter	G		230	.070		10.20	3.87		14.07	17.55
0880	1-3/8" diameter	G	↓	192	.083		11.75	4.64		16.39	20.50
0890	1-1/2" diameter	G	E-18	300	.133	↓	14.30	7.30	3.61	25.21	31.50

05 15 16.60 Galvanized Steel Wire Rope and Accessories

			Crew	Daily Output	Labor-Hours	Unit	Material	Labor	Equipment	Total	Total Incl O&P
0010	**GALVANIZED STEEL WIRE ROPE & ACCESSORIES**										
0015	Made from recycled materials										
3000	Aircraft cable, galvanized, 7 x 7 x 1/8"	G	E-17	5000	.003	L.F.	.20	.18		.38	.51
3100	Clamps, 1/8"	G	"	125	.128	Ea.	1.41	7.10		8.51	13.20

05 15 16.70 Temporary Cable Safety Railing

			Crew	Daily Output	Labor-Hours	Unit	Material	Labor	Equipment	Total	Total Incl O&P
0010	**TEMPORARY CABLE SAFETY RAILING**, Each 100' strand incl.										
0020	2 eyebolts, 1 turnbuckle, 100' cable, 2 thimbles, 6 clips										
0025	Made from recycled materials										
0100	One strand using 1/4" cable & accessories	G	2 Sswk	4	4	C.L.F.	153	219		372	525
0200	1/2" cable & accessories	G	"	2	8	"	325	435		760	1,075

05 21 13 – Deep Longspan Steel Joist Framing

05 21 13.50 Deep Longspan Joists

			Crew	Daily Output	Labor-Hours	Unit	Material	Labor	Equipment	Total	Total Incl O&P
0010	**DEEP LONGSPAN JOISTS**										
3010	DLH series, 40-ton job lots, bolted cross bridging, shop primer										
3015	Made from recycled materials										
3040	Spans to 144' (shipped in 2 pieces)	G	E-7	13	6.154	Ton	2,125	335	145	2,605	3,025
3200	52DLH11, 26 lb./L.F.	G		2000	.040	L.F.	26.50	2.18	.94	29.62	33.50
3220	52DLH16, 45 lb./L.F.	G		2000	.040		48	2.18	.94	51.12	57
3240	56DLH11, 26 lb./L.F.	G		2000	.040		27.50	2.18	.94	30.62	35
3260	56DLH16, 46 lb./L.F.	G		2000	.040		49	2.18	.94	52.12	58.50
3280	60DLH12, 29 lb./L.F.	G		2000	.040		31	2.18	.94	34.12	38.50
3300	60DLH17, 52 lb./L.F.	G		2000	.040		55	2.18	.94	58.12	65.50
3320	64DLH12, 31 lb./L.F.	G		2200	.036		33	1.98	.86	35.84	40
3340	64DLH17, 52 lb./L.F.	G		2200	.036		55	1.98	.86	57.84	65
3360	68DLH13, 37 lb./L.F.	G		2200	.036		39.50	1.98	.86	42.34	47
3380	68DLH18, 61 lb./L.F.	G		2200	.036		65	1.98	.86	67.84	75.50
3400	72DLH14, 41 lb./L.F.	G		2200	.036		43.50	1.98	.86	46.34	52
3420	72DLH19, 70 lb./L.F.	G	↓	2200	.036	↓	74.50	1.98	.86	77.34	86
3500	For less than 40-ton job lots										
3502	For 30 to 39 tons, add					%	10%				
3504	20 to 29 tons, add						20%				
3506	10 to 19 tons, add						30%				
3507	5 to 9 tons, add						50%	25%			
3508	1 to 4 tons, add						75%	50%			
3509	Less than 1 ton, add						100%	100%			

For customer support on your Building Construction Costs with RSMeans data, call 800.448.8182.

05 21 Steel Joist Framing

05 21 13 – Deep Longspan Steel Joist Framing

05 21 13.50 Deep Longspan Joists

		Crew	Daily Output	Labor-Hours	Unit	Material	2018 Bare Costs Labor	Equipment	Total	Total Incl O&P
4010	SLH series, 40-ton job lots, bolted cross bridging, shop primer									
4040	Spans to 200' (shipped in 3 pieces) [G]	E-7	13	6.154	Ton	2,200	335	145	2,680	3,125
4200	80SLH15, 40 lb./L.F. [G]		1500	.053	L.F.	44	2.91	1.26	48.17	54.50
4220	80SLH20, 75 lb./L.F. [G]		1500	.053		82.50	2.91	1.26	86.67	96.50
4240	88SLH16, 46 lb./L.F. [G]		1500	.053		50.50	2.91	1.26	54.67	61.50
4260	88SLH21, 89 lb./L.F. [G]		1500	.053		97.50	2.91	1.26	101.67	113
4280	96SLH17, 52 lb./L.F. [G]		1500	.053		57	2.91	1.26	61.17	69
4300	96SLH22, 102 lb./L.F. [G]		1500	.053		112	2.91	1.26	116.17	129
4320	104SLH18, 59 lb./L.F. [G]		1800	.044		65	2.43	1.05	68.48	76
4340	104SLH23, 109 lb./L.F. [G]		1800	.044		120	2.43	1.05	123.48	137
4360	112SLH19, 67 lb./L.F. [G]		1800	.044		73.50	2.43	1.05	76.98	86
4380	112SLH24, 131 lb./L.F. [G]		1800	.044		144	2.43	1.05	147.48	163
4400	120SLH20, 77 lb./L.F. [G]		1800	.044		84.50	2.43	1.05	87.98	98
4420	120SLH25, 152 lb./L.F. [G]		1800	.044		167	2.43	1.05	170.48	189
6100	For less than 40-ton job lots									
6102	For 30 to 39 tons, add				%	10%				
6104	20 to 29 tons, add					20%				
6106	10 to 19 tons, add					30%				
6107	5 to 9 tons, add					50%	25%			
6108	1 to 4 tons, add					75%	50%			
6109	Less than 1 ton, add					100%	100%			

05 21 16 – Longspan Steel Joist Framing

05 21 16.50 Longspan Joists

		Crew	Daily Output	Labor-Hours	Unit	Material	2018 Bare Costs Labor	Equipment	Total	Total Incl O&P
0010	**LONGSPAN JOISTS**									
2000	LH series, 40-ton job lots, bolted cross bridging, shop primer									
2015	Made from recycled materials									
2040	Longspan joists, LH series, up to 96' [G]	E-7	13	6.154	Ton	2,025	335	145	2,505	2,950
2200	18LH04, 12 lb./L.F. [G]		1400	.057	L.F.	12.20	3.12	1.35	16.67	19.95
2220	18LH08, 19 lb./L.F. [G]		1400	.057		19.35	3.12	1.35	23.82	28
2240	20LH04, 12 lb./L.F. [G]		1400	.057		12.20	3.12	1.35	16.67	19.95
2260	20LH08, 19 lb./L.F. [G]		1400	.057		19.35	3.12	1.35	23.82	28
2280	24LH05, 13 lb./L.F. [G]		1400	.057		13.25	3.12	1.35	17.72	21
2300	24LH10, 23 lb./L.F. [G]		1400	.057		23.50	3.12	1.35	27.97	32.50
2320	28LH06, 16 lb./L.F. [G]		1800	.044		16.30	2.43	1.05	19.78	23
2340	28LH11, 25 lb./L.F. [G]		1800	.044		25.50	2.43	1.05	28.98	33
2360	32LH08, 17 lb./L.F. [G]		1800	.044		17.30	2.43	1.05	20.78	24
2380	32LH13, 30 lb./L.F. [G]		1800	.044		30.50	2.43	1.05	33.98	38.50
2400	36LH09, 21 lb./L.F. [G]		1800	.044		21.50	2.43	1.05	24.98	28.50
2420	36LH14, 36 lb./L.F. [G]		1800	.044		36.50	2.43	1.05	39.98	45.50
2440	40LH10, 21 lb./L.F. [G]		2200	.036		21.50	1.98	.86	24.34	27.50
2460	40LH15, 36 lb./L.F. [G]		2200	.036		36.50	1.98	.86	39.34	44.50
2480	44LH11, 22 lb./L.F. [G]		2200	.036		22.50	1.98	.86	25.34	28.50
2500	44LH16, 42 lb./L.F. [G]		2200	.036		43	1.98	.86	45.84	51
2520	48LH11, 22 lb./L.F. [G]		2200	.036		22.50	1.98	.86	25.34	28.50
2540	48LH16, 42 lb./L.F. [G]		2200	.036		43	1.98	.86	45.84	51
2600	For less than 40-ton job lots									
2602	For 30 to 39 tons, add				%	10%				
2604	20 to 29 tons, add					20%				
2606	10 to 19 tons, add					30%				
2607	5 to 9 tons, add					50%	25%			
2608	1 to 4 tons, add					75%	50%			
2609	Less than 1 ton, add					100%	100%			

05 21 Steel Joist Framing

05 21 16 – Longspan Steel Joist Framing

05 21 16.50 Longspan Joists		Crew	Daily Output	Labor-Hours	Unit	Material	2018 Bare Costs Labor	Equipment	Total	Total Incl O&P
6000	For welded cross bridging, add						30%			

05 21 19 – Open Web Steel Joist Framing

05 21 19.10 Open Web Joists

			Crew	Daily Output	Labor-Hours	Unit	Material	2018 Bare Costs Labor	Equipment	Total	Total Incl O&P
0010	**OPEN WEB JOISTS**										
0015	Made from recycled materials										
0050	K series, 40-ton lots, horiz. bridging, spans to 30', shop primer	G	E-7	12	6.667	Ton	1,800	365	157	2,322	2,725
0130	8K1, 5.1 lb./L.F.	G		1200	.067	L.F.	4.60	3.64	1.57	9.81	12.65
0140	10K1, 5.0 lb./L.F.	G		1200	.067		4.51	3.64	1.57	9.72	12.55
0160	12K3, 5.7 lb./L.F.	G		1500	.053		5.15	2.91	1.26	9.32	11.70
0180	14K3, 6.0 lb./L.F.	G		1500	.053		5.40	2.91	1.26	9.57	12
0200	16K3, 6.3 lb./L.F.	G		1800	.044		5.70	2.43	1.05	9.18	11.30
0220	16K6, 8.1 lb./L.F.	G		1800	.044		7.30	2.43	1.05	10.78	13.10
0240	18K5, 7.7 lb./L.F.	G		2000	.040		6.95	2.18	.94	10.07	12.20
0260	18K9, 10.2 lb./L.F.	G		2000	.040		9.20	2.18	.94	12.32	14.70
0440	K series, 30' to 50' spans	G		17	4.706	Ton	1,775	257	111	2,143	2,475
0500	20K5, 8.2 lb./L.F.	G		2000	.040	L.F.	7.25	2.18	.94	10.37	12.55
0520	20K9, 10.8 lb./L.F.	G		2000	.040		9.55	2.18	.94	12.67	15.05
0540	22K5, 8.8 lb./L.F.	G		2000	.040		7.80	2.18	.94	10.92	13.10
0560	22K9, 11.3 lb./L.F.	G		2000	.040		10	2.18	.94	13.12	15.55
0580	24K6, 9.7 lb./L.F.	G		2200	.036		8.60	1.98	.86	11.44	13.60
0600	24K10, 13.1 lb./L.F.	G		2200	.036		11.60	1.98	.86	14.44	16.90
0620	26K6, 10.6 lb./L.F.	G		2200	.036		9.40	1.98	.86	12.24	14.45
0640	26K10, 13.8 lb./L.F.	G		2200	.036		12.20	1.98	.86	15.04	17.60
0660	28K8, 12.7 lb./L.F.	G		2400	.033		11.25	1.82	.78	13.85	16.15
0680	28K12, 17.1 lb./L.F.	G		2400	.033		15.15	1.82	.78	17.75	20.50
0700	30K8, 13.2 lb./L.F.	G		2400	.033		11.70	1.82	.78	14.30	16.65
0720	30K12, 17.6 lb./L.F.	G		2400	.033		15.60	1.82	.78	18.20	21
0800	For less than 40-ton job lots										
0802	For 30 to 39 tons, add					%	10%				
0804	20 to 29 tons, add						20%				
0806	10 to 19 tons, add						30%				
0807	5 to 9 tons, add						50%	25%			
0808	1 to 4 tons, add						75%	50%			
0809	Less than 1 ton, add						100%	100%			
1010	CS series, 40-ton job lots, horizontal bridging, shop primer										
1040	Spans to 30'	G	E-7	12	6.667	Ton	1,850	365	157	2,372	2,800
1100	10CS2, 7.5 lb./L.F.	G		1200	.067	L.F.	6.95	3.64	1.57	12.16	15.25
1120	12CS2, 8.0 lb./L.F.	G		1500	.053		7.45	2.91	1.26	11.62	14.25
1140	14CS2, 8.0 lb./L.F.	G		1500	.053		7.45	2.91	1.26	11.62	14.25
1160	16CS2, 8.5 lb./L.F.	G		1800	.044		7.90	2.43	1.05	11.38	13.75
1180	16CS4, 14.5 lb./L.F.	G		1800	.044		13.50	2.43	1.05	16.98	19.85
1200	18CS2, 9.0 lb./L.F.	G		2000	.040		8.35	2.18	.94	11.47	13.75
1220	18CS4, 15.0 lb./L.F.	G		2000	.040		13.95	2.18	.94	17.07	19.90
1240	20CS2, 9.5 lb./L.F.	G		2000	.040		8.85	2.18	.94	11.97	14.25
1260	20CS4, 16.5 lb./L.F.	G		2000	.040		15.35	2.18	.94	18.47	21.50
1280	22CS2, 10.0 lb./L.F.	G		2000	.040		9.30	2.18	.94	12.42	14.75
1300	22CS4, 16.5 lb./L.F.	G		2000	.040		15.35	2.18	.94	18.47	21.50
1320	24CS2, 10.0 lb./L.F.	G		2200	.036		9.30	1.98	.86	12.14	14.35
1340	24CS4, 16.5 lb./L.F.	G		2200	.036		15.35	1.98	.86	18.19	21
1360	26CS2, 10.0 lb./L.F.	G		2200	.036		9.30	1.98	.86	12.14	14.35
1380	26CS4, 16.5 lb./L.F.	G		2200	.036		15.35	1.98	.86	18.19	21
1400	28CS2, 10.5 lb./L.F.	G		2400	.033		9.75	1.82	.78	12.35	14.55

For customer support on your Building Construction Costs with RSMeans data, call 800.448.8182.

05 21 Steel Joist Framing

05 21 19 – Open Web Steel Joist Framing

05 21 19.10 Open Web Joists

		Crew	Daily Output	Labor-Hours	Unit	Material	2018 Bare Costs Labor	Equipment	Total	Total Incl O&P
1420	28CS4, 16.5 lb./L.F.	E-7	2400	.033	L.F.	15.35	1.82	.78	17.95	20.50
1440	30CS2, 11.0 lb./L.F.		2400	.033		10.20	1.82	.78	12.80	15.05
1460	30CS4, 16.5 lb./L.F.	▼	2400	.033	▼	15.35	1.82	.78	17.95	20.50
1500	For less than 40-ton job lots									
1502	For 30 to 39 tons, add				%	10%				
1504	20 to 29 tons, add					20%				
1506	10 to 19 tons, add					30%				
1507	5 to 9 tons, add					50%	25%			
1508	1 to 4 tons, add					75%	50%			
1509	Less than 1 ton, add				▼	100%	100%			
6200	For shop prime paint other than mfrs. standard, add					20%				
6300	For bottom chord extensions, add per chord				Ea.	40			40	44
6400	Individual steel bearing plate, 6" x 6" x 1/4" with J-hook	1 Bric	160	.050	"	8.10	2.52		10.62	12.75

05 21 23 – Steel Joist Girder Framing

05 21 23.50 Joist Girders

		Crew	Daily Output	Labor-Hours	Unit	Material	2018 Bare Costs Labor	Equipment	Total	Total Incl O&P
0010	**JOIST GIRDERS**									
0015	Made from recycled materials									
7020	Joist girders, 40-ton job lots, shop primer	E-5	13	6.154	Ton	1,800	335	137	2,272	2,675
7100	For less than 40-ton job lots									
7102	For 30 to 39 tons, add				Ton	10%				
7104	20 to 29 tons, add					20%				
7106	10 to 19 tons, add					30%				
7107	5 to 9 tons, add					50%	25%			
7108	1 to 4 tons, add					75%	50%			
7109	Less than 1 ton, add					100%	100%			
8000	Trusses, 40-ton job lots, shop fabricated WT chords, shop primer	E-5	11	7.273	▼	5,925	395	162	6,482	7,350
8100	For less than 40-ton job lots									
8102	For 30 to 39 tons, add				Ton	10%				
8104	20 to 29 tons, add					20%				
8106	10 to 19 tons, add					30%				
8107	5 to 9 tons, add					50%	25%			
8108	1 to 4 tons, add					75%	50%			
8109	Less than 1 ton, add				▼	100%	100%			

05 31 Steel Decking

05 31 13 – Steel Floor Decking

05 31 13.50 Floor Decking

		Crew	Daily Output	Labor-Hours	Unit	Material	2018 Bare Costs Labor	Equipment	Total	Total Incl O&P
0010	**FLOOR DECKING** R053100-10									
0015	Made from recycled materials									
5100	Non-cellular composite decking, galvanized, 1-1/2" deep, 16 ga.	E-4	3500	.009	S.F.	4.01	.50	.03	4.54	5.25
5120	18 ga.		3650	.009		3.24	.48	.03	3.75	4.38
5140	20 ga.		3800	.008		2.58	.46	.03	3.07	3.63
5200	2" deep, 22 ga.		3860	.008		2.22	.46	.03	2.71	3.22
5300	20 ga.		3600	.009		2.48	.49	.03	3	3.56
5400	18 ga.		3380	.009		3.17	.52	.03	3.72	4.36
5500	16 ga.		3200	.010		3.96	.55	.03	4.54	5.30
5700	3" deep, 22 ga.		3200	.010		2.42	.55	.03	3	3.59
5800	20 ga.		3000	.011		2.73	.59	.03	3.35	4
5900	18 ga.		2850	.011		3.37	.62	.03	4.02	4.76
6000	16 ga.	▼	2700	.012	▼	4.50	.65	.04	5.19	6.05

For customer support on your Building Construction Costs with RSMeans data, call 800.448.8182.

141

05 31 Steel Decking

05 31 23 – Steel Roof Decking

05 31 23.50 Roof Decking		Crew	Daily Output	Labor-Hours	Unit	Material	2018 Bare Costs Labor	Equipment	Total	Total Incl O&P
0010	**ROOF DECKING**									
0015	Made from recycled materials									
2100	Open type, 1-1/2" deep, Type B, wide rib, galv., 22 ga., under 50 sq.	[G] E-4	4500	.007	S.F.	2.38	.39	.02	2.79	3.28
2200	50-500 squares	[G]	4900	.007		1.85	.36	.02	2.23	2.65
2400	Over 500 squares	[G]	5100	.006		1.71	.35	.02	2.08	2.48
2600	20 ga., under 50 squares	[G]	3865	.008		2.78	.46	.03	3.27	3.84
2650	50-500 squares	[G]	4170	.008		2.22	.42	.02	2.66	3.17
2700	Over 500 squares	[G]	4300	.007		2	.41	.02	2.43	2.90
2900	18 ga., under 50 squares	[G]	3800	.008		3.58	.46	.03	4.07	4.73
2950	50-500 squares	[G]	4100	.008		2.86	.43	.02	3.31	3.88
3000	Over 500 squares	[G]	4300	.007		2.58	.41	.02	3.01	3.53
3050	16 ga., under 50 squares	[G]	3700	.009		4.84	.48	.03	5.35	6.15
3060	50-500 squares	[G]	4000	.008		3.87	.44	.02	4.33	5
3100	Over 500 squares	[G]	4200	.008		3.49	.42	.02	3.93	4.55
3200	3" deep, Type N, 22 ga., under 50 squares	[G]	3600	.009		3.50	.49	.03	4.02	4.68
3250	50-500 squares	[G]	3800	.008		2.80	.46	.03	3.29	3.87
3260	over 500 squares	[G]	4000	.008		2.52	.44	.02	2.98	3.52
3300	20 ga., under 50 squares	[G]	3400	.009		3.76	.52	.03	4.31	5
3350	50-500 squares	[G]	3600	.009		3.01	.49	.03	3.53	4.14
3360	over 500 squares	[G]	3800	.008		2.71	.46	.03	3.20	3.77
3400	18 ga., under 50 squares	[G]	3200	.010		4.88	.55	.03	5.46	6.30
3450	50-500 squares	[G]	3400	.009		3.90	.52	.03	4.45	5.15
3460	over 500 squares	[G]	3600	.009		3.51	.49	.03	4.03	4.69
3500	16 ga., under 50 squares	[G]	3000	.011		6.45	.59	.03	7.07	8.10
3550	50-500 squares	[G]	3200	.010		5.15	.55	.03	5.73	6.65
3560	over 500 squares	[G]	3400	.009		4.64	.52	.03	5.19	6
3700	4-1/2" deep, Type J, 20 ga., over 50 squares	[G]	2700	.012		4.24	.65	.04	4.93	5.75
3800	18 ga.	[G]	2460	.013		5.55	.72	.04	6.31	7.35
3900	16 ga.	[G]	2350	.014		7.25	.75	.04	8.04	9.30
4100	6" deep, Type H, 18 ga., over 50 squares	[G]	2000	.016		6.70	.88	.05	7.63	8.85
4200	16 ga.	[G]	1930	.017		8.35	.91	.05	9.31	10.75
4300	14 ga.	[G]	1860	.017		10.75	.95	.05	11.75	13.40
4500	7-1/2" deep, Type H, 18 ga., over 50 squares	[G]	1690	.019		7.95	1.04	.06	9.05	10.50
4600	16 ga.	[G]	1590	.020		9.90	1.11	.06	11.07	12.80
4700	14 ga.	[G]	1490	.021		12.30	1.18	.07	13.55	15.55
4800	For painted instead of galvanized, deduct					5%				
5000	For acoustical perforated with fiberglass insulation, add				S.F.	25%				
5100	For type F intermediate rib instead of type B wide rib, add	[G]				25%				
5150	For type A narrow rib instead of type B wide rib, add	[G]				25%				

05 31 33 – Steel Form Decking

05 31 33.50 Form Decking

		Crew	Daily Output	Labor-Hours	Unit	Material	Labor	Equipment	Total	Total Incl O&P
0010	**FORM DECKING**									
0015	Made from recycled materials									
6100	Slab form, steel, 28 ga., 9/16" deep, Type UFS, uncoated	[G] E-4	4000	.008	S.F.	1.80	.44	.02	2.26	2.73
6200	Galvanized	[G]	4000	.008		1.59	.44	.02	2.05	2.50
6220	24 ga., 1" deep, Type UF1X, uncoated	[G]	3900	.008		1.72	.45	.03	2.20	2.66
6240	Galvanized	[G]	3900	.008		2.02	.45	.03	2.50	2.99
6300	24 ga., 1-5/16" deep, Type UFX, uncoated	[G]	3800	.008		1.83	.46	.03	2.32	2.80
6400	Galvanized	[G]	3800	.008		2.15	.46	.03	2.64	3.16
6500	22 ga., 1-5/16" deep, uncoated	[G]	3700	.009		2.32	.48	.03	2.83	3.36
6600	Galvanized	[G]	3700	.009		2.37	.48	.03	2.88	3.42
6700	22 ga., 2" deep, uncoated	[G]	3600	.009		3.02	.49	.03	3.54	4.15

142

05 31 Steel Decking

05 31 33 – Steel Form Decking

05 31 33.50 Form Decking		Crew	Daily Output	Labor-Hours	Unit	Material	2018 Bare Costs Labor	Equipment	Total	Total Incl O&P
6800	Galvanized	G	E-4	3600	.009	S.F.	2.96	.49	.03	3.48	4.09
7000	Sheet metal edge closure form, 12" wide with 2 bends, galvanized										
7100	18 ga.	G	E-14	360	.022	L.F.	4.88	1.26	.27	6.41	7.70
7200	16 ga.	G	"	360	.022	"	6.60	1.26	.27	8.13	9.60

05 35 Raceway Decking Assemblies

05 35 13 – Steel Cellular Decking

05 35 13.50 Cellular Decking		Crew	Daily Output	Labor-Hours	Unit	Material	2018 Bare Costs Labor	Equipment	Total	Total Incl O&P
0010	**CELLULAR DECKING**										
0015	Made from recycled materials										
0200	Cellular units, galv, 1-1/2" deep, Type BC, 20-20 ga., over 15 squares	G	E-4	1460	.022	S.F.	10.10	1.21	.07	11.38	13.15
0250	18-20 ga.	G		1420	.023		11.45	1.24	.07	12.76	14.70
0300	18-18 ga.	G		1390	.023		11.75	1.27	.07	13.09	15.05
0320	16-18 ga.	G		1360	.024		14	1.30	.07	15.37	17.60
0340	16-16 ga.	G		1330	.024		15.60	1.33	.07	17	19.40
0400	3" deep, Type NC, galvanized, 20-20 ga.	G		1375	.023		11.10	1.28	.07	12.45	14.40
0500	18-20 ga.	G		1350	.024		13.40	1.31	.07	14.78	16.95
0600	18-18 ga.	G		1290	.025		13.35	1.37	.08	14.80	17
0700	16-18 ga.	G		1230	.026		15.05	1.44	.08	16.57	19
0800	16-16 ga.	G		1150	.028		16.40	1.53	.09	18.02	20.50
1000	4-1/2" deep, Type JC, galvanized, 18-20 ga.	G		1100	.029		15.45	1.60	.09	17.14	19.70
1100	18-18 ga.	G		1040	.031		15.35	1.70	.09	17.14	19.75
1200	16-18 ga.	G		980	.033		17.30	1.80	.10	19.20	22
1300	16-16 ga.	G		935	.034		18.85	1.89	.11	20.85	24
1500	For acoustical deck, add						15%				
1700	For cells used for ventilation, add						15%				
1900	For multi-story or congested site, add							50%			
8000	Metal deck and trench, 2" thick, 20 ga., combination										
8010	60% cellular, 40% non-cellular, inserts and trench	G	R-4	1100	.036	S.F.	20	2.03	.09	22.12	25.50

05 41 Structural Metal Stud Framing

05 41 13 – Load-Bearing Metal Stud Framing

05 41 13.05 Bracing		Crew	Daily Output	Labor-Hours	Unit	Material	2018 Bare Costs Labor	Equipment	Total	Total Incl O&P
0010	**BRACING**, shear wall X-bracing, per 10' x 10' bay, one face										
0015	Made of recycled materials										
0120	Metal strap, 20 ga. x 4" wide	G	2 Carp	18	.889	Ea.	18.90	45		63.90	89.50
0130	6" wide	G		18	.889		31.50	45		76.50	103
0160	18 ga. x 4" wide	G		16	1		33	50.50		83.50	113
0170	6" wide	G		16	1		48.50	50.50		99	131
0410	Continuous strap bracing, per horizontal row on both faces										
0420	Metal strap, 20 ga. x 2" wide, studs 12" OC	G	1 Carp	7	1.143	C.L.F.	57	58		115	151
0430	16" OC	G		8	1		57	50.50		107.50	140
0440	24" OC	G		10	.800		57	40.50		97.50	125
0450	18 ga. x 2" wide, studs 12" OC	G		6	1.333		79	67.50		146.50	190
0460	16" OC	G		7	1.143		79	58		137	175
0470	24" OC	G		8	1		79	50.50		129.50	164

143

For customer support on your Building Construction Costs with RSMeans data, call 800.448.8182.

05 41 Structural Metal Stud Framing

05 41 13 – Load-Bearing Metal Stud Framing

05 41 13.10 Bridging

		Crew	Daily Output	Labor-Hours	Unit	Material	2018 Bare Costs Labor	2018 Bare Costs Equipment	Total	Total Incl O&P
0010	**BRIDGING,** solid between studs w/1-1/4" leg track, per stud bay									
0015	Made from recycled materials									
0200	Studs 12" OC, 18 ga. x 2-1/2" wide	G	1 Carp	125	.064	Ea.	.95	3.24		4.19	6
0210	3-5/8" wide	G		120	.067		1.16	3.38		4.54	6.40
0220	4" wide	G		120	.067		1.22	3.38		4.60	6.50
0230	6" wide	G		115	.070		1.60	3.53		5.13	7.10
0240	8" wide	G		110	.073		1.97	3.69		5.66	7.75
0300	16 ga. x 2-1/2" wide	G		115	.070		1.21	3.53		4.74	6.70
0310	3-5/8" wide	G		110	.073		1.47	3.69		5.16	7.20
0320	4" wide	G		110	.073		1.57	3.69		5.26	7.35
0330	6" wide	G		105	.076		2.01	3.86		5.87	8.10
0340	8" wide	G		100	.080		2.51	4.06		6.57	8.95
1200	Studs 16" OC, 18 ga. x 2-1/2" wide	G		125	.064		1.22	3.24		4.46	6.30
1210	3-5/8" wide	G		120	.067		1.49	3.38		4.87	6.80
1220	4" wide	G		120	.067		1.57	3.38		4.95	6.85
1230	6" wide	G		115	.070		2.05	3.53		5.58	7.60
1240	8" wide	G		110	.073		2.52	3.69		6.21	8.40
1300	16 ga. x 2-1/2" wide	G		115	.070		1.55	3.53		5.08	7.05
1310	3-5/8" wide	G		110	.073		1.88	3.69		5.57	7.65
1320	4" wide	G		110	.073		2.01	3.69		5.70	7.80
1330	6" wide	G		105	.076		2.57	3.86		6.43	8.75
1340	8" wide	G		100	.080		3.22	4.06		7.28	9.75
2200	Studs 24" OC, 18 ga. x 2-1/2" wide	G		125	.064		1.77	3.24		5.01	6.90
2210	3-5/8" wide	G		120	.067		2.15	3.38		5.53	7.50
2220	4" wide	G		120	.067		2.27	3.38		5.65	7.65
2230	6" wide	G		115	.070		2.96	3.53		6.49	8.60
2240	8" wide	G		110	.073		3.65	3.69		7.34	9.60
2300	16 ga. x 2-1/2" wide	G		115	.070		2.24	3.53		5.77	7.80
2310	3-5/8" wide	G		110	.073		2.72	3.69		6.41	8.60
2320	4" wide	G		110	.073		2.91	3.69		6.60	8.80
2330	6" wide	G		105	.076		3.72	3.86		7.58	10
2340	8" wide	G		100	.080		4.65	4.06		8.71	11.30
3000	Continuous bridging, per row									
3100	16 ga. x 1-1/2" channel thru studs 12" OC	G	1 Carp	6	1.333	C.L.F.	51.50	67.50		119	160
3110	16" OC	G		7	1.143		51.50	58		109.50	145
3120	24" OC	G		8.80	.909		51.50	46		97.50	127
4100	2" x 2" angle x 18 ga., studs 12" OC	G		7	1.143		80.50	58		138.50	177
4110	16" OC	G		9	.889		80.50	45		125.50	157
4120	24" OC	G		12	.667		80.50	34		114.50	140
4200	16 ga., studs 12" OC	G		5	1.600		101	81		182	235
4210	16" OC	G		7	1.143		101	58		159	199
4220	24" OC	G		10	.800		101	40.50		141.50	173

05 41 13.25 Framing, Boxed Headers/Beams

		Crew	Daily Output	Labor-Hours	Unit	Material	2018 Bare Costs Labor	2018 Bare Costs Equipment	Total	Total Incl O&P
0010	**FRAMING, BOXED HEADERS/BEAMS**									
0015	Made from recycled materials									
0200	Double, 18 ga. x 6" deep	G	2 Carp	220	.073	L.F.	5.50	3.69		9.19	11.65
0210	8" deep	G		210	.076		6.05	3.86		9.91	12.55
0220	10" deep	G		200	.080		7.40	4.06		11.46	14.35
0230	12" deep	G		190	.084		8.05	4.27		12.32	15.35
0300	16 ga. x 8" deep	G		180	.089		7	4.51		11.51	14.55
0310	10" deep	G		170	.094		8.45	4.77		13.22	16.50
0320	12" deep	G		160	.100		9.15	5.05		14.20	17.80

For customer support on your Building Construction Costs with RSMeans data, call 800.448.8182.

05 41 Structural Metal Stud Framing

05 41 13 – Load-Bearing Metal Stud Framing

05 41 13.25 Framing, Boxed Headers/Beams

		Crew	Daily Output	Labor-Hours	Unit	Material	2018 Bare Costs Labor	Equipment	Total	Total Incl O&P
0400	14 ga. x 10" deep	2 Carp	140	.114	L.F.	9.75	5.80		15.55	19.50
0410	12" deep		130	.123		10.65	6.25		16.90	21.50
1210	Triple, 18 ga. x 8" deep		170	.094		8.75	4.77		13.52	16.90
1220	10" deep		165	.097		10.60	4.92		15.52	19.15
1230	12" deep		160	.100		11.60	5.05		16.65	20.50
1300	16 ga. x 8" deep		145	.110		10.15	5.60		15.75	19.65
1310	10" deep		140	.114		12.15	5.80		17.95	22
1320	12" deep		135	.119		13.25	6		19.25	24
1400	14 ga. x 10" deep		115	.139		13.25	7.05		20.30	25.50
1410	12" deep		110	.145		14.65	7.35		22	27.50

05 41 13.30 Framing, Stud Walls

		Crew	Daily Output	Labor-Hours	Unit	Material	2018 Bare Costs Labor	Equipment	Total	Total Incl O&P
0010	**FRAMING, STUD WALLS** w/top & bottom track, no openings,									
0020	Headers, beams, bridging or bracing									
0025	Made from recycled materials									
4100	8' high walls, 18 ga. x 2-1/2" wide, studs 12" OC	2 Carp	54	.296	L.F.	9.15	15		24.15	33
4110	16" OC		77	.208		7.35	10.55		17.90	24
4120	24" OC		107	.150		5.50	7.60		13.10	17.60
4130	3-5/8" wide, studs 12" OC		53	.302		10.85	15.30		26.15	35.50
4140	16" OC		76	.211		8.70	10.65		19.35	26
4150	24" OC		105	.152		6.55	7.75		14.30	18.95
4160	4" wide, studs 12" OC		52	.308		11.30	15.60		26.90	36.50
4170	16" OC		74	.216		9.05	10.95		20	26.50
4180	24" OC		103	.155		6.80	7.90		14.70	19.50
4190	6" wide, studs 12" OC		51	.314		14.40	15.90		30.30	40
4200	16" OC		73	.219		11.55	11.10		22.65	29.50
4210	24" OC		101	.158		8.70	8.05		16.75	22
4220	8" wide, studs 12" OC		50	.320		17.45	16.20		33.65	43.50
4230	16" OC		72	.222		14	11.25		25.25	32.50
4240	24" OC		100	.160		10.60	8.10		18.70	24
4300	16 ga. x 2-1/2" wide, studs 12" OC		47	.340		10.85	17.25		28.10	38.50
4310	16" OC		68	.235		8.60	11.95		20.55	27.50
4320	24" OC		94	.170		6.35	8.65		15	20
4330	3-5/8" wide, studs 12" OC		46	.348		12.95	17.65		30.60	41
4340	16" OC		66	.242		10.25	12.30		22.55	30
4350	24" OC		92	.174		7.55	8.80		16.35	22
4360	4" wide, studs 12" OC		45	.356		13.55	18.05		31.60	42.50
4370	16" OC		65	.246		10.75	12.50		23.25	31
4380	24" OC		90	.178		7.95	9		16.95	22.50
4390	6" wide, studs 12" OC		44	.364		17.05	18.45		35.50	47
4400	16" OC		64	.250		13.55	12.70		26.25	34
4410	24" OC		88	.182		10.05	9.20		19.25	25
4420	8" wide, studs 12" OC		43	.372		21	18.85		39.85	51.50
4430	16" OC		63	.254		16.60	12.90		29.50	38
4440	24" OC		86	.186		12.30	9.45		21.75	28
5100	10' high walls, 18 ga. x 2-1/2" wide, studs 12" OC		54	.296		11	15		26	35
5110	16" OC		77	.208		8.70	10.55		19.25	25.50
5120	24" OC		107	.150		6.40	7.60		14	18.60
5130	3-5/8" wide, studs 12" OC		53	.302		13	15.30		28.30	38
5140	16" OC		76	.211		10.30	10.65		20.95	27.50
5150	24" OC		105	.152		7.60	7.75		15.35	20
5160	4" wide, studs 12" OC		52	.308		13.55	15.60		29.15	39
5170	16" OC		74	.216		10.75	10.95		21.70	28.50

For customer support on your Building Construction Costs with RSMeans data, call 800.448.8182.

145

05 41 13 – Load-Bearing Metal Stud Framing

05 41 13.30 Framing, Stud Walls

		Crew	Daily Output	Labor-Hours	Unit	Material	2018 Bare Costs Labor	Equipment	Total	Total Incl O&P
5180	24" OC	G	2 Carp	103	.155	L.F.	7.95	7.90		15.85	20.50
5190	6" wide, studs 12" OC	G		51	.314		17.25	15.90		33.15	43
5200	16" OC	G		73	.219		13.70	11.10		24.80	32
5210	24" OC	G		101	.158		10.15	8.05		18.20	23.50
5220	8" wide, studs 12" OC	G		50	.320		21	16.20		37.20	47.50
5230	16" OC	G		72	.222		16.55	11.25		27.80	35.50
5240	24" OC	G		100	.160		12.30	8.10		20.40	26
5300	16 ga. x 2-1/2" wide, studs 12" OC	G		47	.340		13.10	17.25		30.35	41
5310	16" OC	G		68	.235		10.30	11.95		22.25	29.50
5320	24" OC	G		94	.170		7.45	8.65		16.10	21.50
5330	3-5/8" wide, studs 12" OC	G		46	.348		15.60	17.65		33.25	44
5340	16" OC	G		66	.242		12.25	12.30		24.55	32
5350	24" OC	G		92	.174		8.90	8.80		17.70	23.50
5360	4" wide, studs 12" OC	G		45	.356		16.35	18.05		34.40	45.50
5370	16" OC	G		65	.246		12.85	12.50		25.35	33
5380	24" OC	G		90	.178		9.35	9		18.35	24
5390	6" wide, studs 12" OC	G		44	.364		20.50	18.45		38.95	50.50
5400	16" OC	G		64	.250		16.15	12.70		28.85	37
5410	24" OC	G		88	.182		11.80	9.20		21	27
5420	8" wide, studs 12" OC	G		43	.372		25	18.85		43.85	56
5430	16" OC	G		63	.254		19.80	12.90		32.70	41.50
5440	24" OC	G		86	.186		14.45	9.45		23.90	30.50
6190	12' high walls, 18 ga. x 6" wide, studs 12" OC	G		41	.390		20	19.80		39.80	52
6200	16" OC	G		58	.276		15.80	14		29.80	39
6210	24" OC	G		81	.198		11.55	10		21.55	28
6220	8" wide, studs 12" OC	G		40	.400		24.50	20.50		45	57.50
6230	16" OC	G		57	.281		19.15	14.25		33.40	42.50
6240	24" OC	G		80	.200		14	10.15		24.15	31
6390	16 ga. x 6" wide, studs 12" OC	G		35	.457		24	23		47	62
6400	16" OC	G		51	.314		18.80	15.90		34.70	44.50
6410	24" OC	G		70	.229		13.55	11.60		25.15	32.50
6420	8" wide, studs 12" OC	G		34	.471		29.50	24		53.50	69
6430	16" OC	G		50	.320		23	16.20		39.20	50
6440	24" OC	G		69	.232		16.60	11.75		28.35	36
6530	14 ga. x 3-5/8" wide, studs 12" OC	G		34	.471		22.50	24		46.50	61.50
6540	16" OC	G		48	.333		17.75	16.90		34.65	45
6550	24" OC	G		65	.246		12.75	12.50		25.25	33
6560	4" wide, studs 12" OC	G		33	.485		24	24.50		48.50	64
6570	16" OC	G		47	.340		18.75	17.25		36	47
6580	24" OC	G		64	.250		13.50	12.70		26.20	34
6730	12 ga. x 3-5/8" wide, studs 12" OC	G		31	.516		31.50	26		57.50	74.50
6740	16" OC	G		43	.372		24.50	18.85		43.35	55
6750	24" OC	G		59	.271		17.15	13.75		30.90	40
6760	4" wide, studs 12" OC	G		30	.533		33.50	27		60.50	78
6770	16" OC	G		42	.381		26	19.30		45.30	58
6780	24" OC	G		58	.276		18.30	14		32.30	41.50
7390	16' high walls, 16 ga. x 6" wide, studs 12" OC	G		33	.485		31	24.50		55.50	71.50
7400	16" OC	G		48	.333		24	16.90		40.90	52
7410	24" OC	G		67	.239		17.05	12.10		29.15	37
7420	8" wide, studs 12" OC	G		32	.500		38	25.50		63.50	80.50
7430	16" OC	G		47	.340		29.50	17.25		46.75	59
7440	24" OC	G		66	.242		21	12.30		33.30	41.50
7560	14 ga. x 4" wide, studs 12" OC	G		31	.516		31	26		57	74

146

05 41 Structural Metal Stud Framing

05 41 13 – Load-Bearing Metal Stud Framing

05 41 13.30 Framing, Stud Walls

			Crew	Daily Output	Labor-Hours	Unit	Material	2018 Bare Costs Labor	Equipment	Total	Total Incl O&P
7570	16" OC	G	2 Carp	45	.356	L.F.	24	18.05		42.05	54
7580	24" OC	G		61	.262		17	13.30		30.30	39
7590	6" wide, studs 12" OC	G		30	.533		39	27		66	84
7600	16" OC	G		44	.364		30	18.45		48.45	61
7610	24" OC	G		60	.267		21.50	13.50		35	44
7760	12 ga. x 4" wide, studs 12" OC	G		29	.552		44	28		72	90.50
7770	16" OC	G		40	.400		33.50	20.50		54	68
7780	24" OC	G		55	.291		23.50	14.75		38.25	48
7790	6" wide, studs 12" OC	G		28	.571		55.50	29		84.50	105
7800	16" OC	G		39	.410		42.50	21		63.50	78
7810	24" OC	G		54	.296		29.50	15		44.50	55.50
8590	20' high walls, 14 ga. x 6" wide, studs 12" OC	G		29	.552		48	28		76	95
8600	16" OC	G		42	.381		37	19.30		56.30	70
8610	24" OC	G		57	.281		26	14.25		40.25	50
8620	8" wide, studs 12" OC	G		28	.571		52	29		81	101
8630	16" OC	G		41	.390		40	19.80		59.80	74
8640	24" OC	G		56	.286		28.50	14.50		43	53
8790	12 ga. x 6" wide, studs 12" OC	G		27	.593		68	30		98	121
8800	16" OC	G		37	.432		52	22		74	91
8810	24" OC	G		51	.314		36	15.90		51.90	63.50
8820	8" wide, studs 12" OC	G		26	.615		82.50	31		113.50	139
8830	16" OC	G		36	.444		63	22.50		85.50	104
8840	24" OC	G		50	.320		43.50	16.20		59.70	72.50

05 42 Cold-Formed Metal Joist Framing

05 42 13 – Cold-Formed Metal Floor Joist Framing

05 42 13.05 Bracing

			Crew	Daily Output	Labor-Hours	Unit	Material	2018 Bare Costs Labor	Equipment	Total	Total Incl O&P
0010	**BRACING**, continuous, per row, top & bottom										
0015	Made from recycled materials										
0120	Flat strap, 20 ga. x 2" wide, joists at 12" OC	G	1 Carp	4.67	1.713	C.L.F.	60	87		147	198
0130	16" OC	G		5.33	1.501		57.50	76		133.50	180
0140	24" OC	G		6.66	1.201		55.50	61		116.50	154
0150	18 ga. x 2" wide, joists at 12" OC	G		4	2		78.50	101		179.50	241
0160	16" OC	G		4.67	1.713		77	87		164	217
0170	24" OC	G		5.33	1.501		75.50	76		151.50	199

05 42 13.10 Bridging

			Crew	Daily Output	Labor-Hours	Unit	Material	2018 Bare Costs Labor	Equipment	Total	Total Incl O&P
0010	**BRIDGING**, solid between joists w/1-1/4" leg track, per joist bay										
0015	Made from recycled materials										
0230	Joists 12" OC, 18 ga. track x 6" wide	G	1 Carp	80	.100	Ea.	1.60	5.05		6.65	9.45
0240	8" wide	G		75	.107		1.97	5.40		7.37	10.40
0250	10" wide	G		70	.114		2.46	5.80		8.26	11.50
0260	12" wide	G		65	.123		2.78	6.25		9.03	12.55
0330	16 ga. track x 6" wide	G		70	.114		2.01	5.80		7.81	11
0340	8" wide	G		65	.123		2.51	6.25		8.76	12.25
0350	10" wide	G		60	.133		3.10	6.75		9.85	13.70
0360	12" wide	G		55	.145		3.58	7.35		10.93	15.20
0440	14 ga. track x 8" wide	G		60	.133		3.13	6.75		9.88	13.75
0450	10" wide	G		55	.145		3.89	7.35		11.24	15.55
0460	12" wide	G		50	.160		4.49	8.10		12.59	17.30
0550	12 ga. track x 10" wide	G		45	.178		5.70	9		14.70	19.95
0560	12" wide	G		40	.200		5.80	10.15		15.95	22

05 42 13 – Cold-Formed Metal Floor Joist Framing

05 42 13.10 Bridging

			Crew	Daily Output	Labor-Hours	Unit	Material	2018 Bare Costs Labor	Equipment	Total	Total Incl O&P
1230	16" OC, 18 ga. track x 6" wide	G	1 Carp	80	.100	Ea.	2.05	5.05		7.10	9.95
1240	8" wide	G		75	.107		2.52	5.40		7.92	11.05
1250	10" wide	G		70	.114		3.15	5.80		8.95	12.25
1260	12" wide	G		65	.123		3.56	6.25		9.81	13.40
1330	16 ga. track x 6" wide	G		70	.114		2.57	5.80		8.37	11.65
1340	8" wide	G		65	.123		3.22	6.25		9.47	13.05
1350	10" wide	G		60	.133		3.98	6.75		10.73	14.65
1360	12" wide	G		55	.145		4.59	7.35		11.94	16.30
1440	14 ga. track x 8" wide	G		60	.133		4.01	6.75		10.76	14.70
1450	10" wide	G		55	.145		4.98	7.35		12.33	16.75
1460	12" wide	G		50	.160		5.75	8.10		13.85	18.70
1550	12 ga. track x 10" wide	G		45	.178		7.30	9		16.30	21.50
1560	12" wide	G		40	.200		7.45	10.15		17.60	23.50
2230	24" OC, 18 ga. track x 6" wide	G		80	.100		2.96	5.05		8.01	10.95
2240	8" wide	G		75	.107		3.65	5.40		9.05	12.25
2250	10" wide	G		70	.114		4.56	5.80		10.36	13.80
2260	12" wide	G		65	.123		5.15	6.25		11.40	15.15
2330	16 ga. track x 6" wide	G		70	.114		3.72	5.80		9.52	12.90
2340	8" wide	G		65	.123		4.65	6.25		10.90	14.60
2350	10" wide	G		60	.133		5.75	6.75		12.50	16.65
2360	12" wide	G		55	.145		6.65	7.35		14	18.55
2440	14 ga. track x 8" wide	G		60	.133		5.80	6.75		12.55	16.70
2450	10" wide	G		55	.145		7.20	7.35		14.55	19.20
2460	12" wide	G		50	.160		8.35	8.10		16.45	21.50
2550	12 ga. track x 10" wide	G		45	.178		10.55	9		19.55	25.50
2560	12" wide	G	▼	40	.200	▼	10.80	10.15		20.95	27.50

05 42 13.25 Framing, Band Joist

			Crew	Daily Output	Labor-Hours	Unit	Material	2018 Bare Costs Labor	Equipment	Total	Total Incl O&P
0010	**FRAMING, BAND JOIST** (track) fastened to bearing wall										
0015	Made from recycled materials										
0220	18 ga. track x 6" deep	G	2 Carp	1000	.016	L.F.	1.30	.81		2.11	2.67
0230	8" deep	G		920	.017		1.61	.88		2.49	3.11
0240	10" deep	G		860	.019		2.01	.94		2.95	3.65
0320	16 ga. track x 6" deep	G		900	.018		1.64	.90		2.54	3.17
0330	8" deep	G		840	.019		2.05	.97		3.02	3.72
0340	10" deep	G		780	.021		2.53	1.04		3.57	4.36
0350	12" deep	G		740	.022		2.92	1.10		4.02	4.88
0430	14 ga. track x 8" deep	G		750	.021		2.55	1.08		3.63	4.46
0440	10" deep	G		720	.022		3.17	1.13		4.30	5.20
0450	12" deep	G		700	.023		3.66	1.16		4.82	5.80
0540	12 ga. track x 10" deep	G		670	.024		4.64	1.21		5.85	6.95
0550	12" deep	G	▼	650	.025	▼	4.75	1.25		6	7.10

05 42 13.30 Framing, Boxed Headers/Beams

			Crew	Daily Output	Labor-Hours	Unit	Material	2018 Bare Costs Labor	Equipment	Total	Total Incl O&P
0010	**FRAMING, BOXED HEADERS/BEAMS**										
0015	Made from recycled materials										
0200	Double, 18 ga. x 6" deep	G	2 Carp	220	.073	L.F.	5.50	3.69		9.19	11.65
0210	8" deep	G		210	.076		6.05	3.86		9.91	12.55
0220	10" deep	G		200	.080		7.40	4.06		11.46	14.35
0230	12" deep	G		190	.084		8.05	4.27		12.32	15.35
0300	16 ga. x 8" deep	G		180	.089		7	4.51		11.51	14.55
0310	10" deep	G		170	.094		8.45	4.77		13.22	16.50
0320	12" deep	G		160	.100		9.15	5.05		14.20	17.80
0400	14 ga. x 10" deep	G	▼	140	.114	▼	9.75	5.80		15.55	19.50

For customer support on your Building Construction Costs with RSMeans data, call 800.448.8182.

05 42 Cold-Formed Metal Joist Framing

05 42 13 – Cold-Formed Metal Floor Joist Framing

05 42 13.30 Framing, Boxed Headers/Beams

			Crew	Daily Output	Labor-Hours	Unit	Material	2018 Bare Costs Labor	Equipment	Total	Total Incl O&P
0410	12" deep	G	2 Carp	130	.123	L.F.	10.65	6.25		16.90	21.50
0500	12 ga. x 10" deep	G		110	.145		12.85	7.35		20.20	25.50
0510	12" deep	G		100	.160		14.15	8.10		22.25	28
1210	Triple, 18 ga. x 8" deep	G		170	.094		8.75	4.77		13.52	16.90
1220	10" deep	G		165	.097		10.60	4.92		15.52	19.15
1230	12" deep	G		160	.100		11.60	5.05		16.65	20.50
1300	16 ga. x 8" deep	G		145	.110		10.15	5.60		15.75	19.65
1310	10" deep	G		140	.114		12.15	5.80		17.95	22
1320	12" deep	G		135	.119		13.25	6		19.25	24
1400	14 ga. x 10" deep	G		115	.139		14.10	7.05		21.15	26.50
1410	12" deep	G		110	.145		15.50	7.35		22.85	28.50
1500	12 ga. x 10" deep	G		90	.178		18.75	9		27.75	34
1510	12" deep	G		85	.188		21	9.55		30.55	37.50

05 42 13.40 Framing, Joists

			Crew	Daily Output	Labor-Hours	Unit	Material	2018 Bare Costs Labor	Equipment	Total	Total Incl O&P
0010	**FRAMING, JOISTS**, no band joists (track), web stiffeners, headers,										
0020	Beams, bridging or bracing										
0025	Made from recycled materials										
0030	Joists (2" flange) and fasteners, materials only										
0220	18 ga. x 6" deep	G				L.F.	1.69			1.69	1.86
0230	8" deep	G					1.98			1.98	2.18
0240	10" deep	G					2.34			2.34	2.58
0320	16 ga. x 6" deep	G					2.07			2.07	2.28
0330	8" deep	G					2.47			2.47	2.71
0340	10" deep	G					2.89			2.89	3.18
0350	12" deep	G					3.28			3.28	3.60
0430	14 ga. x 8" deep	G					3.10			3.10	3.41
0440	10" deep	G					3.57			3.57	3.93
0450	12" deep	G					4.06			4.06	4.47
0540	12 ga. x 10" deep	G					5.20			5.20	5.70
0550	12" deep	G					5.90			5.90	6.50
1010	Installation of joists to band joists, beams & headers, labor only										
1220	18 ga. x 6" deep		2 Carp	110	.145	Ea.		7.35		7.35	11.25
1230	8" deep			90	.178			9		9	13.70
1240	10" deep			80	.200			10.15		10.15	15.45
1320	16 ga. x 6" deep			95	.168			8.55		8.55	13
1330	8" deep			70	.229			11.60		11.60	17.65
1340	10" deep			60	.267			13.50		13.50	20.50
1350	12" deep			55	.291			14.75		14.75	22.50
1430	14 ga. x 8" deep			65	.246			12.50		12.50	19
1440	10" deep			45	.356			18.05		18.05	27.50
1450	12" deep			35	.457			23		23	35.50
1540	12 ga. x 10" deep			40	.400			20.50		20.50	31
1550	12" deep			30	.533			27		27	41

05 42 13.45 Framing, Web Stiffeners

			Crew	Daily Output	Labor-Hours	Unit	Material	2018 Bare Costs Labor	Equipment	Total	Total Incl O&P
0010	**FRAMING, WEB STIFFENERS** at joist bearing, fabricated from										
0020	Stud piece (1-5/8" flange) to stiffen joist (2" flange)										
0025	Made from recycled materials										
2120	For 6" deep joist, with 18 ga. x 2-1/2" stud	G	1 Carp	120	.067	Ea.	.92	3.38		4.30	6.15
2130	3-5/8" stud	G		110	.073		1.08	3.69		4.77	6.80
2140	4" stud	G		105	.076		1.12	3.86		4.98	7.15
2150	6" stud	G		100	.080		1.42	4.06		5.48	7.75
2160	8" stud	G		95	.084		1.71	4.27		5.98	8.40

05 42 Cold-Formed Metal Joist Framing

05 42 13 – Cold-Formed Metal Floor Joist Framing

05 42 13.45 Framing, Web Stiffeners		Crew	Daily Output	Labor-Hours	Unit	Material	2018 Bare Costs Labor	Equipment	Total	Total Incl O&P
2220	8" deep joist, with 2-1/2" stud	G	1 Carp	120	.067	Ea.	1.23	3.38		4.61	6.50
2230	3-5/8" stud	G		110	.073		1.45	3.69		5.14	7.20
2240	4" stud	G		105	.076		1.50	3.86		5.36	7.55
2250	6" stud	G		100	.080		1.90	4.06		5.96	8.30
2260	8" stud	G		95	.084		2.29	4.27		6.56	9
2320	10" deep joist, with 2-1/2" stud	G		110	.073		1.53	3.69		5.22	7.30
2330	3-5/8" stud	G		100	.080		1.79	4.06		5.85	8.15
2340	4" stud	G		95	.084		1.86	4.27		6.13	8.55
2350	6" stud	G		90	.089		2.36	4.51		6.87	9.45
2360	8" stud	G		85	.094		2.84	4.77		7.61	10.35
2420	12" deep joist, with 2-1/2" stud	G		110	.073		1.84	3.69		5.53	7.60
2430	3-5/8" stud	G		100	.080		2.16	4.06		6.22	8.60
2440	4" stud	G		95	.084		2.24	4.27		6.51	8.95
2450	6" stud	G		90	.089		2.84	4.51		7.35	9.95
2460	8" stud	G		85	.094		3.42	4.77		8.19	11
3130	For 6" deep joist, with 16 ga. x 3-5/8" stud	G		100	.080		1.34	4.06		5.40	7.65
3140	4" stud	G		95	.084		1.40	4.27		5.67	8.05
3150	6" stud	G		90	.089		1.75	4.51		6.26	8.80
3160	8" stud	G		85	.094		2.14	4.77		6.91	9.60
3230	8" deep joist, with 3-5/8" stud	G		100	.080		1.80	4.06		5.86	8.20
3240	4" stud	G		95	.084		1.88	4.27		6.15	8.55
3250	6" stud	G		90	.089		2.35	4.51		6.86	9.45
3260	8" stud	G		85	.094		2.87	4.77		7.64	10.40
3330	10" deep joist, with 3-5/8" stud	G		85	.094		2.22	4.77		6.99	9.70
3340	4" stud	G		80	.100		2.32	5.05		7.37	10.25
3350	6" stud	G		75	.107		2.91	5.40		8.31	11.45
3360	8" stud	G		70	.114		3.55	5.80		9.35	12.70
3430	12" deep joist, with 3-5/8" stud	G		85	.094		2.68	4.77		7.45	10.20
3440	4" stud	G		80	.100		2.80	5.05		7.85	10.80
3450	6" stud	G		75	.107		3.50	5.40		8.90	12.10
3460	8" stud	G		70	.114		4.28	5.80		10.08	13.50
4230	For 8" deep joist, with 14 ga. x 3-5/8" stud	G		90	.089		2.22	4.51		6.73	9.30
4240	4" stud	G		85	.094		2.35	4.77		7.12	9.85
4250	6" stud	G		80	.100		2.95	5.05		8	10.95
4260	8" stud	G		75	.107		3.15	5.40		8.55	11.70
4330	10" deep joist, with 3-5/8" stud	G		75	.107		2.76	5.40		8.16	11.30
4340	4" stud	G		70	.114		2.91	5.80		8.71	12
4350	6" stud	G		65	.123		3.65	6.25		9.90	13.50
4360	8" stud	G		60	.133		3.90	6.75		10.65	14.60
4430	12" deep joist, with 3-5/8" stud	G		75	.107		3.32	5.40		8.72	11.90
4440	4" stud	G		70	.114		3.50	5.80		9.30	12.65
4450	6" stud	G		65	.123		4.40	6.25		10.65	14.35
4460	8" stud	G		60	.133		4.70	6.75		11.45	15.45
5330	For 10" deep joist, with 12 ga. x 3-5/8" stud	G		65	.123		3.97	6.25		10.22	13.85
5340	4" stud	G		60	.133		4.23	6.75		10.98	14.95
5350	6" stud	G		55	.145		5.35	7.35		12.70	17.15
5360	8" stud	G		50	.160		6.45	8.10		14.55	19.45
5430	12" deep joist, with 3-5/8" stud	G		65	.123		4.78	6.25		11.03	14.75
5440	4" stud	G		60	.133		5.10	6.75		11.85	15.90
5450	6" stud	G		55	.145		6.45	7.35		13.80	18.35
5460	8" stud	G		50	.160		7.80	8.10		15.90	21

For customer support on your Building Construction Costs with RSMeans data, call 800.448.8182.

05 42 23 – Cold-Formed Metal Roof Joist Framing

05 42 23.05 Framing, Bracing

05 42 23.05 Framing, Bracing		Crew	Daily Output	Labor-Hours	Unit	Material	2018 Bare Costs Labor	Equipment	Total	Total Incl O&P
0010	**FRAMING, BRACING**									
0015	Made from recycled materials									
0020	Continuous bracing, per row									
0100	16 ga. x 1-1/2" channel thru rafters/trusses @ 16" OC	G	1 Carp	4.50	1.778	C.L.F.	51.50	90		141.50	194
0120	24" OC	G		6	1.333		51.50	67.50		119	160
0300	2" x 2" angle x 18 ga., rafters/trusses @ 16" OC	G		6	1.333		80.50	67.50		148	192
0320	24" OC	G		8	1		80.50	50.50		131	166
0400	16 ga., rafters/trusses @ 16" OC	G		4.50	1.778		101	90		191	248
0420	24" OC	G		6.50	1.231		101	62.50		163.50	206

05 42 23.10 Framing, Bridging

05 42 23.10 Framing, Bridging		Crew	Daily Output	Labor-Hours	Unit	Material	2018 Bare Costs Labor	Equipment	Total	Total Incl O&P
0010	**FRAMING, BRIDGING**									
0015	Made from recycled materials									
0020	Solid, between rafters w/1-1/4" leg track, per rafter bay									
1200	Rafters 16" OC, 18 ga. x 4" deep	G	1 Carp	60	.133	Ea.	1.57	6.75		8.32	12
1210	6" deep	G		57	.140		2.05	7.10		9.15	13.10
1220	8" deep	G		55	.145		2.52	7.35		9.87	14.05
1230	10" deep	G		52	.154		3.15	7.80		10.95	15.35
1240	12" deep	G		50	.160		3.56	8.10		11.66	16.25
2200	24" OC, 18 ga. x 4" deep	G		60	.133		2.27	6.75		9.02	12.80
2210	6" deep	G		57	.140		2.96	7.10		10.06	14.10
2220	8" deep	G		55	.145		3.65	7.35		11	15.25
2230	10" deep	G		52	.154		4.56	7.80		12.36	16.90
2240	12" deep	G		50	.160		5.15	8.10		13.25	18

05 42 23.50 Framing, Parapets

05 42 23.50 Framing, Parapets		Crew	Daily Output	Labor-Hours	Unit	Material	2018 Bare Costs Labor	Equipment	Total	Total Incl O&P
0010	**FRAMING, PARAPETS**									
0015	Made from recycled materials									
0100	3' high installed on 1st story, 18 ga. x 4" wide studs, 12" OC	G	2 Carp	100	.160	L.F.	5.70	8.10		13.80	18.60
0110	16" OC	G		150	.107		4.85	5.40		10.25	13.60
0120	24" OC	G		200	.080		4.01	4.06		8.07	10.60
0200	6" wide studs, 12" OC	G		100	.160		7.30	8.10		15.40	20.50
0210	16" OC	G		150	.107		6.25	5.40		11.65	15.10
0220	24" OC	G		200	.080		5.15	4.06		9.21	11.90
1100	Installed on 2nd story, 18 ga. x 4" wide studs, 12" OC	G		95	.168		5.70	8.55		14.25	19.25
1110	16" OC	G		145	.110		4.85	5.60		10.45	13.85
1120	24" OC	G		190	.084		4.01	4.27		8.28	10.90
1200	6" wide studs, 12" OC	G		95	.168		7.30	8.55		15.85	21
1210	16" OC	G		145	.110		6.25	5.60		11.85	15.35
1220	24" OC	G		190	.084		5.15	4.27		9.42	12.20
2100	Installed on gable, 18 ga. x 4" wide studs, 12" OC	G		85	.188		5.70	9.55		15.25	21
2110	16" OC	G		130	.123		4.85	6.25		11.10	14.85
2120	24" OC	G		170	.094		4.01	4.77		8.78	11.65
2200	6" wide studs, 12" OC	G		85	.188		7.30	9.55		16.85	22.50
2210	16" OC	G		130	.123		6.25	6.25		12.50	16.35
2220	24" OC	G		170	.094		5.15	4.77		9.92	12.95

05 42 23.60 Framing, Roof Rafters

05 42 23.60 Framing, Roof Rafters		Crew	Daily Output	Labor-Hours	Unit	Material	2018 Bare Costs Labor	Equipment	Total	Total Incl O&P
0010	**FRAMING, ROOF RAFTERS**									
0015	Made from recycled materials									
0100	Boxed ridge beam, double, 18 ga. x 6" deep	G	2 Carp	160	.100	L.F.	5.50	5.05		10.55	13.75
0110	8" deep	G		150	.107		6.05	5.40		11.45	14.90
0120	10" deep	G		140	.114		7.40	5.80		13.20	16.95
0130	12" deep	G		130	.123		8.05	6.25		14.30	18.35

For customer support on your Building Construction Costs with RSMeans data, call 800.448.8182.

151

05 42 Cold-Formed Metal Joist Framing

05 42 23 – Cold-Formed Metal Roof Joist Framing

05 42 23.60 Framing, Roof Rafters

			Crew	Daily Output	Labor-Hours	Unit	Material	2018 Bare Costs Labor	Equipment	Total	Total Incl O&P
0200	16 ga. x 6" deep	G	2 Carp	150	.107	L.F.	6.20	5.40		11.60	15.10
0210	8" deep	G		140	.114		7	5.80		12.80	16.50
0220	10" deep	G		130	.123		8.45	6.25		14.70	18.75
0230	12" deep	G		120	.133		9.15	6.75		15.90	20.50
1100	Rafters, 2" flange, material only, 18 ga. x 6" deep	G					1.69			1.69	1.86
1110	8" deep	G					1.98			1.98	2.18
1120	10" deep	G					2.34			2.34	2.58
1130	12" deep	G					2.70			2.70	2.97
1200	16 ga. x 6" deep	G					2.07			2.07	2.28
1210	8" deep	G					2.47			2.47	2.71
1220	10" deep	G					2.89			2.89	3.18
1230	12" deep	G					3.28			3.28	3.60
2100	Installation only, ordinary rafter to 4:12 pitch, 18 ga. x 6" deep		2 Carp	35	.457	Ea.		23		23	35.50
2110	8" deep			30	.533			27		27	41
2120	10" deep			25	.640			32.50		32.50	49.50
2130	12" deep			20	.800			40.50		40.50	62
2200	16 ga. x 6" deep			30	.533			27		27	41
2210	8" deep			25	.640			32.50		32.50	49.50
2220	10" deep			20	.800			40.50		40.50	62
2230	12" deep			15	1.067			54		54	82.50
8100	Add to labor, ordinary rafters on steep roofs							25%			
8110	Dormers & complex roofs							50%			
8200	Hip & valley rafters to 4:12 pitch							25%			
8210	Steep roofs							50%			
8220	Dormers & complex roofs							75%			
8300	Hip & valley jack rafters to 4:12 pitch							50%			
8310	Steep roofs							75%			
8320	Dormers & complex roofs							100%			

05 42 23.70 Framing, Soffits and Canopies

			Crew	Daily Output	Labor-Hours	Unit	Material	2018 Bare Costs Labor	Equipment	Total	Total Incl O&P
0010	**FRAMING, SOFFITS & CANOPIES**										
0015	Made from recycled materials										
0130	Continuous ledger track @ wall, studs @ 16" OC, 18 ga. x 4" wide	G	2 Carp	535	.030	L.F.	1.05	1.52		2.57	3.46
0140	6" wide	G		500	.032		1.36	1.62		2.98	3.97
0150	8" wide	G		465	.034		1.68	1.74		3.42	4.51
0160	10" wide	G		430	.037		2.10	1.89		3.99	5.20
0230	Studs @ 24" OC, 18 ga. x 4" wide	G		800	.020		1	1.01		2.01	2.64
0240	6" wide	G		750	.021		1.30	1.08		2.38	3.08
0250	8" wide	G		700	.023		1.61	1.16		2.77	3.53
0260	10" wide	G		650	.025		2.01	1.25		3.26	4.11
1000	Horizontal soffit and canopy members, material only										
1030	1-5/8" flange studs, 18 ga. x 4" deep	G				L.F.	1.34			1.34	1.48
1040	6" deep	G					1.70			1.70	1.87
1050	8" deep	G					2.05			2.05	2.26
1140	2" flange joists, 18 ga. x 6" deep	G					1.93			1.93	2.13
1150	8" deep	G					2.27			2.27	2.49
1160	10" deep	G					2.68			2.68	2.94
4030	Installation only, 18 ga., 1-5/8" flange x 4" deep		2 Carp	130	.123	Ea.		6.25		6.25	9.50
4040	6" deep			110	.145			7.35		7.35	11.25
4050	8" deep			90	.178			9		9	13.70
4140	2" flange, 18 ga. x 6" deep			110	.145			7.35		7.35	11.25
4150	8" deep			90	.178			9		9	13.70
4160	10" deep			80	.200			10.15		10.15	15.45

For customer support on your Building Construction Costs with RSMeans data, call 800.448.8182.

05 42 Cold-Formed Metal Joist Framing

05 42 23 – Cold-Formed Metal Roof Joist Framing

05 42 23.70 Framing, Soffits and Canopies

		Crew	Daily Output	Labor-Hours	Unit	Material	2018 Bare Costs Labor	Equipment	Total	Total Incl O&P
6010	Clips to attach fascia to rafter tails, 2" x 2" x 18 ga. angle	G	1 Carp	120	.067	Ea.	.95	3.38		4.33	6.20
6020	16 ga. angle	G	"	100	.080	↓	1.20	4.06		5.26	7.50

05 44 Cold-Formed Metal Trusses

05 44 13 – Cold-Formed Metal Roof Trusses

05 44 13.60 Framing, Roof Trusses

			Crew	Daily Output	Labor-Hours	Unit	Material	2018 Bare Costs Labor	Equipment	Total	Total Incl O&P
0010	**FRAMING, ROOF TRUSSES**										
0015	Made from recycled materials										
0020	Fabrication of trusses on ground, Fink (W) or King Post, to 4:12 pitch										
0120	18 ga. x 4" chords, 16' span	G	2 Carp	12	1.333	Ea.	62.50	67.50		130	172
0130	20' span	G		11	1.455		78.50	74		152.50	198
0140	24' span	G		11	1.455		94	74		168	215
0150	28' span	G		10	1.600		110	81		191	245
0160	32' span	G		10	1.600		125	81		206	262
0250	6" chords, 28' span	G		9	1.778		139	90		229	290
0260	32' span	G		9	1.778		159	90		249	310
0270	36' span	G		8	2		179	101		280	350
0280	40' span	G		8	2		199	101		300	375
1120	5:12 to 8:12 pitch, 18 ga. x 4" chords, 16' span	G		10	1.600		71.50	81		152.50	203
1130	20' span	G		9	1.778		89.50	90		179.50	236
1140	24' span	G		9	1.778		108	90		198	255
1150	28' span	G		8	2		125	101		226	292
1160	32' span	G		8	2		143	101		244	310
1250	6" chords, 28' span	G		7	2.286		159	116		275	350
1260	32' span	G		7	2.286		182	116		298	375
1270	36' span	G		6	2.667		204	135		339	430
1280	40' span	G		6	2.667		227	135		362	455
2120	9:12 to 12:12 pitch, 18 ga. x 4" chords, 16' span	G		8	2		89.50	101		190.50	253
2130	20' span	G		7	2.286		112	116		228	299
2140	24' span	G		7	2.286		134	116		250	325
2150	28' span	G		6	2.667		157	135		292	380
2160	32' span	G		6	2.667		179	135		314	405
2250	6" chords, 28' span	G		5	3.200		199	162		361	465
2260	32' span	G		5	3.200		227	162		389	495
2270	36' span	G		4	4		256	203		459	590
2280	40' span	G	↓	4	4		284	203		487	620
5120	Erection only of roof trusses, to 4:12 pitch, 16' span		F-6	48	.833			39.50	10.35	49.85	71.50
5130	20' span			46	.870			41.50	10.80	52.30	74.50
5140	24' span			44	.909			43	11.25	54.25	78
5150	28' span			42	.952			45	11.80	56.80	81.50
5160	32' span			40	1			47.50	12.40	59.90	85.50
5170	36' span			38	1.053			50	13.05	63.05	90.50
5180	40' span			36	1.111			52.50	13.80	66.30	95
5220	5:12 to 8:12 pitch, 16' span			42	.952			45	11.80	56.80	81.50
5230	20' span			40	1			47.50	12.40	59.90	85.50
5240	24' span			38	1.053			50	13.05	63.05	90.50
5250	28' span			36	1.111			52.50	13.80	66.30	95
5260	32' span			34	1.176			56	14.60	70.60	101
5270	36' span			32	1.250			59.50	15.50	75	107
5280	40' span			30	1.333			63.50	16.55	80.05	114
5320	9:12 to 12:12 pitch, 16' span		↓	36	1.111	↓		52.50	13.80	66.30	95

05 44 Cold-Formed Metal Trusses

05 44 13 – Cold-Formed Metal Roof Trusses

05 44 13.60 Framing, Roof Trusses

		Crew	Daily Output	Labor-Hours	Unit	Material	2018 Bare Costs Labor	Equipment	Total	Total Incl O&P
5330	20' span	F-6	34	1.176	Ea.		56	14.60	70.60	101
5340	24' span		32	1.250			59.50	15.50	75	107
5350	28' span		30	1.333			63.50	16.55	80.05	114
5360	32' span		28	1.429			68	17.70	85.70	123
5370	36' span		26	1.538			73	19.10	92.10	132
5380	40' span	↓	24	1.667	↓		79	20.50	99.50	143

05 51 Metal Stairs

05 51 13 – Metal Pan Stairs

05 51 13.50 Pan Stairs

			Crew	Daily Output	Labor-Hours	Unit	Material	2018 Bare Costs Labor	Equipment	Total	Total Incl O&P
0010	**PAN STAIRS**, shop fabricated, steel stringers										
0015	Made from recycled materials										
0200	Metal pan tread for concrete in-fill, picket rail, 3'-6" wide	G	E-4	35	.914	Riser	560	50.50	2.82	613.32	700
0300	4'-0" wide	G		30	1.067		625	59	3.29	687.29	785
0350	Wall rail, both sides, 3'-6" wide	G		53	.604	↓	425	33.50	1.86	460.36	525
1500	Landing, steel pan, conventional	G		160	.200	S.F.	74.50	11.05	.62	86.17	101
1600	Pre-erected	G	↓	255	.125	"	132	6.90	.39	139.29	157
1700	Pre-erected, steel pan tread, 3'-6" wide, 2 line pipe rail	G	E-2	87	.644	Riser	555	35	19.40	609.40	690

05 51 16 – Metal Floor Plate Stairs

05 51 16.50 Floor Plate Stairs

			Crew	Daily Output	Labor-Hours	Unit	Material	2018 Bare Costs Labor	Equipment	Total	Total Incl O&P
0010	**FLOOR PLATE STAIRS**, shop fabricated, steel stringers										
0015	Made from recycled materials										
0400	Cast iron tread and pipe rail, 3'-6" wide	G	E-4	35	.914	Riser	600	50.50	2.82	653.32	745
0500	Checkered plate tread, industrial, 3'-6" wide	G		28	1.143		365	63	3.52	431.52	505
0550	Circular, for tanks, 3'-0" wide	G	↓	33	.970		415	53.50	2.99	471.49	550
0600	For isolated stairs, add							100%			
0800	Custom steel stairs, 3'-6" wide, economy	G	E-4	35	.914		560	50.50	2.82	613.32	700
0810	Medium priced	G		30	1.067		740	59	3.29	802.29	915
0900	Deluxe	G	↓	20	1.600		920	88	4.93	1,012.93	1,175
1100	For 4' wide stairs, add							5%	5%		
1300	For 5' wide stairs, add					↓		10%	10%		

05 51 19 – Metal Grating Stairs

05 51 19.50 Grating Stairs

			Crew	Daily Output	Labor-Hours	Unit	Material	2018 Bare Costs Labor	Equipment	Total	Total Incl O&P
0010	**GRATING STAIRS**, shop fabricated, steel stringers, safety nosing on treads										
0015	Made from recycled materials										
0020	Grating tread and pipe railing, 3'-6" wide	G	E-4	35	.914	Riser	365	50.50	2.82	418.32	485
0100	4'-0" wide	G	"	30	1.067	"	420	59	3.29	482.29	565

05 51 23 – Metal Fire Escapes

05 51 23.25 Fire Escapes

			Crew	Daily Output	Labor-Hours	Unit	Material	2018 Bare Costs Labor	Equipment	Total	Total Incl O&P
0010	**FIRE ESCAPES**, shop fabricated										
0200	2' wide balcony, 1" x 1/4" bars 1-1/2" OC, with railing	G	2 Sswk	10	1.600	L.F.	60	87.50		147.50	209
0400	1st story cantilevered stair, standard, with railing	G		.50	32	Ea.	2,500	1,750		4,250	5,600
0500	Cable counterweighted, with railing	G		.40	40	"	2,325	2,175		4,500	6,125
0700	36" x 40" platform & fixed stair, with railing	G	↓	.40	40	Flight	1,100	2,175		3,275	4,800
0900	For 3'-6" wide escapes, add to above							100%	150%		

For customer support on your Building Construction Costs with RSMeans data, call 800.448.8182.

05 51 Metal Stairs

05 51 23 – Metal Fire Escapes

05 51 23.50 Fire Escape Stairs		Crew	Daily Output	Labor-Hours	Unit	Material	2018 Bare Costs Labor	Equipment	Total	Total Incl O&P
0010	**FIRE ESCAPE STAIRS**, portable									
0100	Portable ladder					Ea.	128			128	141

05 51 33 – Metal Ladders

05 51 33.13 Vertical Metal Ladders

05 51 33.13 Vertical Metal Ladders			Crew	Daily Output	Labor-Hours	Unit	Material	2018 Bare Costs Labor	Equipment	Total	Total Incl O&P
0010	**VERTICAL METAL LADDERS**, shop fabricated										
0015	Made from recycled materials										
0020	Steel, 20" wide, bolted to concrete, with cage	G	E-4	50	.640	V.L.F.	67	35.50	1.97	104.47	133
0100	Without cage	G		85	.376		36.50	21	1.16	58.66	75.50
0300	Aluminum, bolted to concrete, with cage	G		50	.640		133	35.50	1.97	170.47	206
0400	Without cage	G		85	.376		52	21	1.16	74.16	92.50

05 51 33.16 Inclined Metal Ladders

05 51 33.16 Inclined Metal Ladders			Crew	Daily Output	Labor-Hours	Unit	Material	2018 Bare Costs Labor	Equipment	Total	Total Incl O&P
0010	**INCLINED METAL LADDERS**, shop fabricated										
0015	Made from recycled materials										
3900	Industrial ships ladder, steel, 24" W, grating treads, 2 line pipe rail	G	E-4	30	1.067	Riser	214	59	3.29	276.29	335
4000	Aluminum	G	"	30	1.067	"	299	59	3.29	361.29	430

05 51 33.23 Alternating Tread Ladders

05 51 33.23 Alternating Tread Ladders		Crew	Daily Output	Labor-Hours	Unit	Material	2018 Bare Costs Labor	Equipment	Total	Total Incl O&P
0010	**ALTERNATING TREAD LADDERS**, shop fabricated									
0015	Made from recycled materials									
0800	Alternating tread ladders, 68-degree angle of incline									
0810	8' vertical rise, steel, 149 lb., standard paint color	B-68G	3	5.333	Ea.	2,400	291	93	2,784	3,225
0820	Non-standard paint color		3	5.333		2,850	291	93	3,234	3,700
0830	Galvanized		3	5.333		2,825	291	93	3,209	3,675
0840	Stainless		3	5.333		4,075	291	93	4,459	5,075
0850	Aluminum, 87 lb.		3	5.333		2,975	291	93	3,359	3,850
1010	10' vertical rise, steel, 181 lb., standard paint color		2.75	5.818		2,925	320	101	3,346	3,825
1020	Non-standard paint color		2.75	5.818		3,400	320	101	3,821	4,350
1030	Galvanized		2.75	5.818		3,400	320	101	3,821	4,375
1040	Stainless		2.75	5.818		4,925	320	101	5,346	6,025
1050	Aluminum, 103 lb.		2.75	5.818		3,600	320	101	4,021	4,575
1210	12' vertical rise, steel, 245 lb., standard paint color		2.50	6.400		3,425	350	112	3,887	4,475
1220	Non-standard paint color		2.50	6.400		3,925	350	112	4,387	5,025
1230	Galvanized		2.50	6.400		4,000	350	112	4,462	5,100
1240	Stainless		2.50	6.400		5,750	350	112	6,212	7,025
1250	Aluminum, 103 lb.		2.50	6.400		4,200	350	112	4,662	5,325
1410	14' vertical rise, steel, 281 lb., standard paint color		2.25	7.111		3,925	390	124	4,439	5,100
1420	Non-standard paint color		2.25	7.111		4,475	390	124	4,989	5,700
1430	Galvanized		2.25	7.111		4,600	390	124	5,114	5,825
1440	Stainless		2.25	7.111		6,575	390	124	7,089	8,025
1450	Aluminum, 136 lb.		2.25	7.111		4,825	390	124	5,339	6,075
1610	16' vertical rise, steel, 317 lb., standard paint color		2	8		4,450	435	140	5,025	5,750
1620	Non-standard paint color		2	8		5,025	435	140	5,600	6,400
1630	Galvanized		2	8		5,175	435	140	5,750	6,575
1640	Stainless		2	8		7,425	435	140	8,000	9,025
1650	Aluminum, 153 lb.		2	8		5,425	435	140	6,000	6,850

05 52 13.50 Railings, Pipe

		Crew	Daily Output	Labor-Hours	Unit	Material	2018 Bare Costs Labor	2018 Bare Costs Equipment	Total	Total Incl O&P
0010	**RAILINGS, PIPE**, shop fab'd, 3'-6" high, posts @ 5' OC									
0015	Made from recycled materials									
0020	Aluminum, 2 rail, satin finish, 1-1/4" diameter	G E-4	160	.200	L.F.	52.50	11.05	.62	64.17	76.50
0030	Clear anodized	G	160	.200		64.50	11.05	.62	76.17	89.50
0040	Dark anodized	G	160	.200		71.50	11.05	.62	83.17	97
0080	1-1/2" diameter, satin finish	G	160	.200		61.50	11.05	.62	73.17	86.50
0090	Clear anodized	G	160	.200		69.50	11.05	.62	81.17	94.50
0100	Dark anodized	G	160	.200		76.50	11.05	.62	88.17	103
0140	Aluminum, 3 rail, 1-1/4" diam., satin finish	G	137	.234		68	12.90	.72	81.62	96.50
0150	Clear anodized	G	137	.234		84.50	12.90	.72	98.12	115
0160	Dark anodized	G	137	.234		93	12.90	.72	106.62	124
0200	1-1/2" diameter, satin finish	G	137	.234		80.50	12.90	.72	94.12	110
0210	Clear anodized	G	137	.234		91.50	12.90	.72	105.12	123
0220	Dark anodized	G	137	.234		101	12.90	.72	114.62	133
0500	Steel, 2 rail, on stairs, primed, 1-1/4" diameter	G	160	.200		29	11.05	.62	40.67	50.50
0520	1-1/2" diameter	G	160	.200		31.50	11.05	.62	43.17	53
0540	Galvanized, 1-1/4" diameter	G	160	.200		39	11.05	.62	50.67	61.50
0560	1-1/2" diameter	G	160	.200		44	11.05	.62	55.67	67
0580	Steel, 3 rail, primed, 1-1/4" diameter	G	137	.234		43	12.90	.72	56.62	69.50
0600	1-1/2" diameter	G	137	.234		45.50	12.90	.72	59.12	72
0620	Galvanized, 1-1/4" diameter	G	137	.234		60.50	12.90	.72	74.12	88.50
0640	1-1/2" diameter	G	137	.234		70	12.90	.72	83.62	99
0700	Stainless steel, 2 rail, 1-1/4" diam., #4 finish	G	137	.234		128	12.90	.72	141.62	162
0720	High polish	G	137	.234		206	12.90	.72	219.62	249
0740	Mirror polish	G	137	.234		258	12.90	.72	271.62	305
0760	Stainless steel, 3 rail, 1-1/2" diam., #4 finish	G	120	.267		192	14.70	.82	207.52	236
0770	High polish	G	120	.267		320	14.70	.82	335.52	375
0780	Mirror finish	G	120	.267		390	14.70	.82	405.52	450
0900	Wall rail, alum. pipe, 1-1/4" diam., satin finish	G	213	.150		25	8.30	.46	33.76	41.50
0905	Clear anodized	G	213	.150		31.50	8.30	.46	40.26	48.50
0910	Dark anodized	G	213	.150		37	8.30	.46	45.76	55
0915	1-1/2" diameter, satin finish	G	213	.150		28	8.30	.46	36.76	45
0920	Clear anodized	G	213	.150		35	8.30	.46	43.76	52.50
0925	Dark anodized	G	213	.150		44	8.30	.46	52.76	62.50
0930	Steel pipe, 1-1/4" diameter, primed	G	213	.150		17.35	8.30	.46	26.11	33
0935	Galvanized	G	213	.150		25	8.30	.46	33.76	41.50
0940	1-1/2" diameter	G	176	.182		17.90	10.05	.56	28.51	36.50
0945	Galvanized	G	213	.150		25	8.30	.46	33.76	41.50
0955	Stainless steel pipe, 1-1/2" diam., #4 finish	G	107	.299		102	16.50	.92	119.42	140
0960	High polish	G	107	.299		207	16.50	.92	224.42	256
0965	Mirror polish	G	107	.299		245	16.50	.92	262.42	298
2000	2-line pipe rail (1-1/2" T&B) with 1/2" pickets @ 4-1/2" OC,									
2005	attached handrail on brackets									
2010	42" high aluminum, satin finish, straight & level	G E-4	120	.267	L.F.	268	14.70	.82	283.52	320
2050	42" high steel, primed, straight & level	G "	120	.267		147	14.70	.82	162.52	186
4000	For curved and level rails, add						10%	10%		
4100	For sloped rails for stairs, add						30%	30%		

05 52 16 - Industrial Railings

05 52 16.50 Railings, Industrial

		Crew	Daily Output	Labor-Hours	Unit	Material	2018 Bare Costs Labor	2018 Bare Costs Equipment	Total	Total Incl O&P
0010	**RAILINGS, INDUSTRIAL**, shop fab'd, 3'-6" high, posts @ 5' OC									
0020	2 rail, 3'-6" high, 1-1/2" pipe	G E-4	255	.125	L.F.	41.50	6.90	.39	48.79	57
0100	2" angle rail	G "	255	.125		37.50	6.90	.39	44.79	52.50

For customer support on your Building Construction Costs with RSMeans data, call 800.448.8182.

05 52 Metal Railings

05 52 16 – Industrial Railings

05 52 16.50 Railings, Industrial		Crew	Daily Output	Labor-Hours	Unit	Material	2018 Bare Costs Labor	Equipment	Total	Total Incl O&P
0200	For 4" high kick plate, 10 ga., add	G				L.F.	5.90			5.90	6.45
0300	1/4" thick, add	G					7.70			7.70	8.45
0500	For curved level rails, add						10%	10%			
0550	For sloped rails for stairs, add						30%	30%			

05 53 Metal Gratings

05 53 13 – Bar Gratings

05 53 13.10 Floor Grating, Aluminum

		Crew	Daily Output	Labor-Hours	Unit	Material	2018 Bare Costs Labor	Equipment	Total	Total Incl O&P
0010	**FLOOR GRATING, ALUMINUM**, field fabricated from panels										
0015	Made from recycled materials										
0110	Bearing bars @ 1-3/16" OC, cross bars @ 4" OC,										
0111	Up to 300 S.F., 1" x 1/8" bar	G	E-4	900	.036	S.F.	17.40	1.96	.11	19.47	22.50
0112	Over 300 S.F.	G		850	.038		15.80	2.08	.12	18	21
0113	1-1/4" x 1/8" bar, up to 300 S.F.	G		800	.040		17.65	2.21	.12	19.98	23
0114	Over 300 S.F.	G		1000	.032		16.05	1.76	.10	17.91	20.50
0122	1-1/4" x 3/16" bar, up to 300 S.F.	G		750	.043		28	2.35	.13	30.48	34.50
0124	Over 300 S.F.	G		1000	.032		25.50	1.76	.10	27.36	31
0132	1-1/2" x 3/16" bar, up to 300 S.F.	G		700	.046		31.50	2.52	.14	34.16	39.50
0134	Over 300 S.F.	G		1000	.032		29	1.76	.10	30.86	34.50
0136	1-3/4" x 3/16" bar, up to 300 S.F.	G		500	.064		34.50	3.53	.20	38.23	44
0138	Over 300 S.F.	G		1000	.032		31.50	1.76	.10	33.36	37.50
0146	2-1/4" x 3/16" bar, up to 300 S.F.	G		600	.053		44	2.94	.16	47.10	53.50
0148	Over 300 S.F.	G		1000	.032		40	1.76	.10	41.86	47
0162	Cross bars @ 2" OC, 1" x 1/8", up to 300 S.F.	G		600	.053		31	2.94	.16	34.10	39.50
0164	Over 300 S.F.	G		1000	.032		28.50	1.76	.10	30.36	34
0172	1-1/4" x 3/16" bar, up to 300 S.F.	G		600	.053		50	2.94	.16	53.10	60
0174	Over 300 S.F.	G		1000	.032		45.50	1.76	.10	47.36	53
0182	1-1/2" x 3/16" bar, up to 300 S.F.	G		600	.053		60	2.94	.16	63.10	71
0184	Over 300 S.F.	G		1000	.032		54.50	1.76	.10	56.36	63
0186	1-3/4" x 3/16" bar, up to 300 S.F.	G		600	.053		63	2.94	.16	66.10	74.50
0188	Over 300 S.F.	G		1000	.032		57.50	1.76	.10	59.36	66
0200	For straight cuts, add					L.F.	4.40			4.40	4.84
0300	For curved cuts, add						5.40			5.40	5.95
0400	For straight banding, add	G					5.60			5.60	6.20
0500	For curved banding, add	G					6.75			6.75	7.45
0600	For aluminum checkered plate nosings, add	G					7.30			7.30	8.05
0700	For straight toe plate, add	G					11.05			11.05	12.15
0800	For curved toe plate, add	G					12.90			12.90	14.15
1000	For cast aluminum abrasive nosings, add	G					10.75			10.75	11.80
1400	Extruded I bars are 10% less than 3/16" bars										
1600	Heavy duty, all extruded plank, 3/4" deep, 1.8 #/S.F.	G	E-4	1100	.029	S.F.	26.50	1.60	.09	28.19	31.50
1700	1-1/4" deep, 2.9 #/S.F.	G		1000	.032		31	1.76	.10	32.86	37
1800	1-3/4" deep, 4.2 #/S.F.	G		925	.035		43.50	1.91	.11	45.52	51
1900	2-1/4" deep, 5.0 #/S.F.	G		875	.037		64.50	2.02	.11	66.63	74.50
2100	For safety serrated surface, add						15%				

05 53 13.70 Floor Grating, Steel

		Crew	Daily Output	Labor-Hours	Unit	Material	2018 Bare Costs Labor	Equipment	Total	Total Incl O&P
0010	**FLOOR GRATING, STEEL**, field fabricated from panels										
0015	Made from recycled materials										
0300	Platforms, to 12' high, rectangular	G	E-4	3150	.010	Lb.	3.21	.56	.03	3.80	4.48
0400	Circular	G	"	2300	.014	"	4.01	.77	.04	4.82	5.70
0410	Painted bearing bars @ 1-3/16"										

05 53 Metal Gratings

05 53 13 – Bar Gratings

05 53 13.70 Floor Grating, Steel

		Crew	Daily Output	Labor-Hours	Unit	Material	2018 Bare Costs Labor	Equipment	Total	Total Incl O&P
0412	Cross bars @ 4" OC, 3/4" x 1/8" bar, up to 300 S.F.	G	E-2	500	.112	S.F.	7.80	6.10	3.37	17.27	22
0414	Over 300 S.F.	G		750	.075		7.10	4.05	2.25	13.40	16.75
0422	1-1/4" x 3/16", up to 300 S.F.	G		400	.140		12.10	7.60	4.22	23.92	30
0424	Over 300 S.F.	G		600	.093		11	5.05	2.81	18.86	23.50
0432	1-1/2" x 3/16", up to 300 S.F.	G		400	.140		14.05	7.60	4.22	25.87	32.50
0434	Over 300 S.F.	G		600	.093		12.80	5.05	2.81	20.66	25
0436	1-3/4" x 3/16", up to 300 S.F.	G		400	.140		17.70	7.60	4.22	29.52	36
0438	Over 300 S.F.	G		600	.093		16.10	5.05	2.81	23.96	29
0452	2-1/4" x 3/16", up to 300 S.F.	G		300	.187		21	10.15	5.60	36.75	45.50
0454	Over 300 S.F.	G		450	.124		19.10	6.75	3.75	29.60	36
0462	Cross bars @ 2" OC, 3/4" x 1/8", up to 300 S.F.	G		500	.112		15.15	6.10	3.37	24.62	30
0464	Over 300 S.F.	G		750	.075		12.60	4.05	2.25	18.90	23
0472	1-1/4" x 3/16", up to 300 S.F.	G		400	.140		19.80	7.60	4.22	31.62	39
0474	Over 300 S.F.	G		600	.093		16.50	5.05	2.81	24.36	29.50
0482	1-1/2" x 3/16", up to 300 S.F.	G		400	.140		22.50	7.60	4.22	34.32	41.50
0484	Over 300 S.F.	G		600	.093		18.55	5.05	2.81	26.41	31.50
0486	1-3/4" x 3/16", up to 300 S.F.	G		400	.140		33.50	7.60	4.22	45.32	54
0488	Over 300 S.F.	G		600	.093		28	5.05	2.81	35.86	42
0502	2-1/4" x 3/16", up to 300 S.F.	G		300	.187		33	10.15	5.60	48.75	58.50
0504	Over 300 S.F.	G		450	.124		27.50	6.75	3.75	38	45
0690	For galvanized grating, add						25%				
0800	For straight cuts, add					L.F.	6.30			6.30	6.90
0900	For curved cuts, add						8			8	8.80
1000	For straight banding, add	G					6.75			6.75	7.45
1100	For curved banding, add	G					8.90			8.90	9.80
1200	For checkered plate nosings, add	G					7.85			7.85	8.60
1300	For straight toe or kick plate, add	G					14			14	15.40
1400	For curved toe or kick plate, add	G					15.85			15.85	17.45
1500	For abrasive nosings, add	G					10.40			10.40	11.40
1510	For stair treads, see Section 05 55 13.50										
1600	For safety serrated surface, bearing bars @ 1-3/16" OC, add						15%				
1700	Bearing bars @ 15/16" OC, add						25%				
2000	Stainless steel gratings, close spaced, 1" x 1/8" bars, up to 300 S.F.	G	E-4	450	.071	S.F.	76.50	3.92	.22	80.64	90.50
2100	Standard spacing, 3/4" x 1/8" bars	G		500	.064		84	3.53	.20	87.73	98
2200	1-1/4" x 3/16" bars	G		400	.080		82	4.41	.25	86.66	98

05 53 16 – Plank Gratings

05 53 16.50 Grating Planks

		Crew	Daily Output	Labor-Hours	Unit	Material	2018 Bare Costs Labor	Equipment	Total	Total Incl O&P
0010	**GRATING PLANKS**, field fabricated from planks										
0020	Aluminum, 9-1/2" wide, 14 ga., 2" rib	G	E-4	950	.034	L.F.	32	1.86	.10	33.96	38
0200	Galvanized steel, 9-1/2" wide, 14 ga., 2-1/2" rib	G		950	.034		16.55	1.86	.10	18.51	21.50
0300	4" rib	G		950	.034		18.10	1.86	.10	20.06	23
0500	12 ga., 2-1/2" rib	G		950	.034		16.75	1.86	.10	18.71	21.50
0600	3" rib	G		950	.034		21.50	1.86	.10	23.46	27
0800	Stainless steel, type 304, 16 ga., 2" rib	G		950	.034		41	1.86	.10	42.96	48
0900	Type 316	G		950	.034		56.50	1.86	.10	58.46	65

For customer support on your Building Construction Costs with RSMeans data, call 800.448.8182.

05 53 19.10 Expanded Grating, Aluminum		Crew	Daily Output	Labor-Hours	Unit	Material	2018 Bare Costs Labor	Equipment	Total	Total Incl O&P
0010	**EXPANDED GRATING, ALUMINUM**									
1200	Expanded aluminum, .65 #/S.F.	G	E-4	1050	.030	S.F.	21.50	1.68	.09	23.27	26.50

05 53 19.20 Expanded Grating, Steel

		Crew	Daily Output	Labor-Hours	Unit	Material	Labor	Equipment	Total	Total Incl O&P
0010	**EXPANDED GRATING, STEEL**									
2400	Expanded steel grating, at ground, 3.0 #/S.F.	G	E-4	900	.036	S.F.	8.05	1.96	.11	10.12	12.25
2500	3.14 #/S.F.	G		900	.036		7.05	1.96	.11	9.12	11.10
2600	4.0 #/S.F.	G		850	.038		8.35	2.08	.12	10.55	12.65
2650	4.27 #/S.F.	G		850	.038		9.50	2.08	.12	11.70	13.95
2700	5.0 #/S.F.	G		800	.040		13.80	2.21	.12	16.13	18.90
2800	6.25 #/S.F.	G		750	.043		17.80	2.35	.13	20.28	23.50
2900	7.0 #/S.F.	G		700	.046		19.65	2.52	.14	22.31	26
3100	For flattened expanded steel grating, add						8%				
3300	For elevated installation above 15', add							15%			

05 53 19.30 Grating Frame

		Crew	Daily Output	Labor-Hours	Unit	Material	Labor	Equipment	Total	Total Incl O&P
0010	**GRATING FRAME**, field fabricated									
0020	Aluminum, for gratings 1" to 1-1/2" deep	G	1 Sswk	70	.114	L.F.	3.83	6.25		10.08	14.40
0100	For each corner, add	G				Ea.	5.75			5.75	6.30

05 54 13.20 Checkered Plates

		Crew	Daily Output	Labor-Hours	Unit	Material	Labor	Equipment	Total	Total Incl O&P
0010	**CHECKERED PLATES**, steel, field fabricated									
0015	Made from recycled materials									
0020	1/4" & 3/8", 2000 to 5000 S.F., bolted	G	E-4	2900	.011	Lb.	.74	.61	.03	1.38	1.84
0100	Welded	G		4400	.007	"	.70	.40	.02	1.12	1.45
0300	Pit or trench cover and frame, 1/4" plate, 2' to 3' wide	G		100	.320	S.F.	9.35	17.65	.99	27.99	40.50
0400	For galvanizing, add	G				Lb.	.28			.28	.31
0500	Platforms, 1/4" plate, no handrails included, rectangular	G	E-4	4200	.008		3.52	.42	.02	3.96	4.59
0600	Circular	G	"	2500	.013		4.40	.71	.04	5.15	6.05

05 54 13.70 Trench Covers

		Crew	Daily Output	Labor-Hours	Unit	Material	Labor	Equipment	Total	Total Incl O&P
0010	**TRENCH COVERS**, field fabricated									
0020	Cast iron grating with bar stops and angle frame, to 18" wide	G	1 Sswk	20	.400	L.F.	213	22		235	270
0100	Frame only (both sides of trench), 1" grating	G		45	.178		1.30	9.70		11	17.35
0150	2" grating	G		35	.229		2.86	12.50		15.36	23.50
0200	Aluminum, stock units, including frames and									
0210	3/8" plain cover plate, 4" opening	G	E-4	205	.156	L.F.	17.80	8.60	.48	26.88	34
0300	6" opening	G		185	.173		22.50	9.55	.53	32.58	41
0400	10" opening	G		170	.188		32	10.40	.58	42.98	53
0500	16" opening	G		155	.206		46.50	11.40	.64	58.54	70.50
0700	Add per inch for additional widths to 24"	G					1.98			1.98	2.18
0900	For custom fabrication, add						50%				
1100	For 1/4" plain cover plate, deduct						12%				
1500	For cover recessed for tile, 1/4" thick, deduct						12%				
1600	3/8" thick, add						5%				
1800	For checkered plate cover, 1/4" thick, deduct						12%				
1900	3/8" thick, add						2%				
2100	For slotted or round holes in cover, 1/4" thick, add						3%				
2200	3/8" thick, add						4%				
2300	For abrasive cover, add						12%				

05 55 13.50 Stair Treads		Crew	Daily Output	Labor-Hours	Unit	Material	2018 Bare Costs Labor	2018 Bare Costs Equipment	Total	Total Incl O&P
0010	**STAIR TREADS**, stringers and bolts not included									
3000	Diamond plate treads, steel, 1/8" thick									
3005	Open riser, black enamel									
3010	9" deep x 36" long	G 2 Sswk	48	.333	Ea.	91.50	18.20		109.70	130
3020	42" long	G	48	.333		96.50	18.20		114.70	136
3030	48" long	G	48	.333		101	18.20		119.20	141
3040	11" deep x 36" long	G	44	.364		96.50	19.85		116.35	139
3050	42" long	G	44	.364		102	19.85		121.85	145
3060	48" long	G	44	.364		108	19.85		127.85	152
3110	Galvanized, 9" deep x 36" long	G	48	.333		144	18.20		162.20	189
3120	42" long	G	48	.333		153	18.20		171.20	199
3130	48" long	G	48	.333		163	18.20		181.20	209
3140	11" deep x 36" long	G	44	.364		149	19.85		168.85	197
3150	42" long	G	44	.364		163	19.85		182.85	212
3160	48" long	G	44	.364		169	19.85		188.85	219
3200	Closed riser, black enamel									
3210	12" deep x 36" long	G 2 Sswk	40	.400	Ea.	110	22		132	157
3220	42" long	G	40	.400		119	22		141	167
3230	48" long	G	40	.400		126	22		148	174
3240	Galvanized, 12" deep x 36" long	G	40	.400		173	22		195	227
3250	42" long	G	40	.400		189	22		211	244
3260	48" long	G	40	.400		198	22		220	254
4000	Bar grating treads									
4005	Steel, 1-1/4" x 3/16" bars, anti-skid nosing, black enamel									
4010	8-5/8" deep x 30" long	G 2 Sswk	48	.333	Ea.	54	18.20		72.20	89.50
4020	36" long	G	48	.333		63.50	18.20		81.70	99.50
4030	48" long	G	48	.333		97.50	18.20		115.70	137
4040	10-15/16" deep x 36" long	G	44	.364		70	19.85		89.85	110
4050	48" long	G	44	.364		100	19.85		119.85	143
4060	Galvanized, 8-5/8" deep x 30" long	G	48	.333		62	18.20		80.20	98.50
4070	36" long	G	48	.333		73.50	18.20		91.70	111
4080	48" long	G	48	.333		108	18.20		126.20	149
4090	10-15/16" deep x 36" long	G	44	.364		86.50	19.85		106.35	128
4100	48" long	G	44	.364		112	19.85		131.85	156
4200	Aluminum, 1-1/4" x 3/16" bars, serrated, with nosing									
4210	7-5/8" deep x 18" long	G 2 Sswk	52	.308	Ea.	50	16.80		66.80	82.50
4220	24" long	G	52	.308		59	16.80		75.80	92.50
4230	30" long	G	52	.308		68.50	16.80		85.30	103
4240	36" long	G	52	.308		177	16.80		193.80	223
4250	8-13/16" deep x 18" long	G	48	.333		67	18.20		85.20	104
4260	24" long	G	48	.333		99.50	18.20		117.70	139
4270	30" long	G	48	.333		115	18.20		133.20	156
4280	36" long	G	48	.333		194	18.20		212.20	244
4290	10" deep x 18" long	G	44	.364		130	19.85		149.85	176
4300	30" long	G	44	.364		173	19.85		192.85	223
4310	36" long	G	44	.364		210	19.85		229.85	264
5000	Channel grating treads									
5005	Steel, 14 ga., 2-1/2" thick, galvanized									
5010	9" deep x 36" long	G 2 Sswk	48	.333	Ea.	103	18.20		121.20	143
5020	48" long	G "	48	.333	"	139	18.20		157.20	182

For customer support on your Building Construction Costs with RSMeans data, call 800.448.8182.

05 55 Metal Stair Treads and Nosings

05 55 19 – Metal Stair Tread Covers

05 55 19.50 Stair Tread Covers for Renovation

		Crew	Daily Output	Labor-Hours	Unit	Material	2018 Bare Costs Labor	2018 Bare Costs Equipment	Total	Total Incl O&P
0010	**STAIR TREAD COVERS FOR RENOVATION**									
0205	Extruded tread cover with nosing, pre-drilled, includes screws									
0210	Aluminum with black abrasive strips, 9" wide x 3' long	1 Carp	24	.333	Ea.	104	16.90		120.90	141
0220	4' long		22	.364		139	18.45		157.45	181
0230	5' long		20	.400		179	20.50		199.50	228
0240	11" wide x 3' long		24	.333		135	16.90		151.90	174
0250	4' long		22	.364		180	18.45		198.45	226
0260	5' long		20	.400		221	20.50		241.50	274
0305	Black abrasive strips with yellow front strips									
0310	Aluminum, 9" wide x 3' long	1 Carp	24	.333	Ea.	117	16.90		133.90	155
0320	4' long		22	.364		156	18.45		174.45	200
0330	5' long		20	.400		195	20.50		215.50	246
0340	11" wide x 3' long		24	.333		141	16.90		157.90	181
0350	4' long		22	.364		186	18.45		204.45	232
0360	5' long		20	.400		236	20.50		256.50	290
0405	Black abrasive strips with photoluminescent front strips									
0410	Aluminum, 9" wide x 3' long	1 Carp	24	.333	Ea.	150	16.90		166.90	191
0420	4' long		22	.364		183	18.45		201.45	230
0430	5' long		20	.400		229	20.50		249.50	283
0440	11" wide x 3' long		24	.333		163	16.90		179.90	205
0450	4' long		22	.364		218	18.45		236.45	267
0460	5' long		20	.400		272	20.50		292.50	330

05 56 Metal Castings

05 56 13 – Metal Construction Castings

05 56 13.50 Construction Castings

			Crew	Daily Output	Labor-Hours	Unit	Material	2018 Bare Costs Labor	2018 Bare Costs Equipment	Total	Total Incl O&P
0010	**CONSTRUCTION CASTINGS**										
0020	Manhole covers and frames, see Section 33 44 13.13										
0100	Column bases, cast iron, 16" x 16", approx. 65 lb.	G	E-4	46	.696	Ea.	142	38.50	2.14	182.64	221
0200	32" x 32", approx. 256 lb.	G	"	23	1.391		520	76.50	4.29	600.79	705
0400	Cast aluminum for wood columns, 8" x 8"	G	1 Carp	32	.250		46	12.70		58.70	70
0500	12" x 12"	G	"	32	.250		70	12.70		82.70	97
0600	Miscellaneous C.I. castings, light sections, less than 150 lb.	G	E-4	3200	.010	Lb.	8.95	.55	.03	9.53	10.80
1100	Heavy sections, more than 150 lb.	G		4200	.008		5.15	.42	.02	5.59	6.40
1300	Special low volume items	G		3200	.010		11.25	.55	.03	11.83	13.30
1500	For ductile iron, add						100%				

05 58 Formed Metal Fabrications

05 58 13 – Column Covers

05 58 13.05 Column Covers

			Crew	Daily Output	Labor-Hours	Unit	Material	2018 Bare Costs Labor	2018 Bare Costs Equipment	Total	Total Incl O&P
0010	**COLUMN COVERS**										
0015	Made from recycled materials										
0020	Excludes structural steel, light ga. metal framing, misc. metals, sealants										
0100	Round covers, 2 halves with 2 vertical joints for backer rod and sealant										
0110	Up to 12' high, no horizontal joints										
0120	12" diameter, 0.125" aluminum, anodized/painted finish	G	2 Sswk	32	.500	V.L.F.	31	27.50		58.50	78.50
0130	Type 304 stainless steel, 16 gauge, #4 brushed finish	G		32	.500		44	27.50		71.50	93
0140	Type 316 stainless steel, 16 gauge, #4 brushed finish	G		32	.500		55	27.50		82.50	105
0150	18" diameter, aluminum	G		32	.500		46.50	27.50		74	96

05 58 Formed Metal Fabrications

05 58 13 – Column Covers

05 58 13.05 Column Covers

			Crew	Daily Output	Labor-Hours	Unit	Material	2018 Bare Costs Labor	Equipment	Total	Total Incl O&P
0160		Type 304 stainless steel	G	2 Sswk	32	.500	V.L.F.	66	27.50		93.50	117
0170		Type 316 stainless steel	G		32	.500		82.50	27.50		110	136
0180		24" diameter, aluminum	G		32	.500		62	27.50		89.50	113
0190		Type 304 stainless steel	G		32	.500		88	27.50		115.50	142
0200		Type 316 stainless steel	G		32	.500		110	27.50		137.50	166
0210		30" diameter, aluminum	G		30	.533		78	29		107	133
0220		Type 304 stainless steel	G		30	.533		110	29		139	169
0230		Type 316 stainless steel	G		30	.533		138	29		167	199
0240		36" diameter, aluminum	G		30	.533		93.50	29		122.50	151
0250		Type 304 stainless steel	G		30	.533		132	29		161	193
0260		Type 316 stainless steel	G		30	.533		165	29		194	230
0400		Up to 24' high, 2 stacked sections with 1 horizontal joint										
0410		18" diameter, aluminum	G	2 Sswk	28	.571	V.L.F.	49	31		80	105
0450		Type 304 stainless steel	G		28	.571		69.50	31		100.50	128
0460		Type 316 stainless steel	G		28	.571		86.50	31		117.50	147
0470		24" diameter, aluminum	G		28	.571		65.50	31		96.50	123
0480		Type 304 stainless steel	G		28	.571		92.50	31		123.50	153
0490		Type 316 stainless steel	G		28	.571		116	31		147	178
0500		30" diameter, aluminum	G		24	.667		81.50	36.50		118	150
0510		Type 304 stainless steel	G		24	.667		116	36.50		152.50	187
0520		Type 316 stainless steel	G		24	.667		145	36.50		181.50	219
0530		36" diameter, aluminum	G		24	.667		98	36.50		134.50	168
0540		Type 304 stainless steel	G		24	.667		139	36.50		175.50	213
0550		Type 316 stainless steel	G		24	.667		173	36.50		209.50	251

05 58 21 – Formed Chain

05 58 21.05 Alloy Steel Chain

			Crew	Daily Output	Labor-Hours	Unit	Material	2018 Bare Costs Labor	Equipment	Total	Total Incl O&P
0010	**ALLOY STEEL CHAIN**, Grade 80, for lifting											
0015	Self-colored, cut lengths, 1/4"		G	E-17	4	4	C.L.F.	900	223		1,123	1,350
0020	3/8"		G		2	8		1,125	445		1,570	1,950
0030	1/2"		G		1.20	13.333		1,750	740		2,490	3,150
0040	5/8"		G		.72	22.222		2,850	1,225		4,075	5,150
0050	3/4"		G	E-18	.48	83.333		3,675	4,575	2,250	10,500	13,900
0060	7/8"		G		.40	100		6,675	5,475	2,700	14,850	19,200
0070	1"		G		.35	114		8,550	6,275	3,100	17,925	22,900
0080	1-1/4"		G		.24	167		13,100	9,150	4,525	26,775	34,200
0110	Hook, Grade 80, Clevis slip, 1/4"		G				Ea.	28.50			28.50	31.50
0120	3/8"		G					42.50			42.50	47
0130	1/2"		G					67			67	73.50
0140	5/8"		G					102			102	112
0150	3/4"		G					129			129	142
0160	Hook, Grade 80, eye/sling w/hammerlock coupling, 15 ton		G					395			395	435
0170	22 ton		G					965			965	1,075
0180	37 ton		G					3,125			3,125	3,450

05 58 23 – Formed Metal Guards

05 58 23.90 Window Guards

			Crew	Daily Output	Labor-Hours	Unit	Material	2018 Bare Costs Labor	Equipment	Total	Total Incl O&P
0010	**WINDOW GUARDS**, shop fabricated											
0015	Expanded metal, steel angle frame, permanent		G	E-4	350	.091	S.F.	23.50	5.05	.28	28.83	34
0025	Steel bars, 1/2" x 1/2", spaced 5" OC		G	"	290	.110	"	16.25	6.10	.34	22.69	28
0030	Hinge mounted, add		G				Opng.	47			47	52
0040	Removable type, add		G				"	29.50			29.50	32.50
0050	For galvanized guards, add						S.F.	35%				
0070	For pivoted or projected type, add							105%	40%			

For customer support on your Building Construction Costs with RSMeans data, call 800.448.8182.

05 58 Formed Metal Fabrications

05 58 23 – Formed Metal Guards

05 58 23.90 Window Guards		Crew	Daily Output	Labor-Hours	Unit	Material	2018 Bare Costs Labor	Equipment	Total	Total Incl O&P
0100	Mild steel, stock units, economy	E-4	405	.079	S.F.	6.40	4.36	.24	11	14.35
0200	Deluxe		405	.079		12.95	4.36	.24	17.55	21.50
0400	Woven wire, stock units, 3/8" channel frame, 3' x 5' opening		40	.800	Opng.	172	44	2.46	218.46	265
0500	4' x 6' opening		38	.842		276	46.50	2.59	325.09	385
0800	Basket guards for above, add					238			238	262
1000	Swinging guards for above, add					81			81	89

05 58 25 – Formed Lamp Posts

05 58 25.40 Lamp Posts

		Crew	Daily Output	Labor-Hours	Unit	Material	2018 Bare Costs Labor	Equipment	Total	Total Incl O&P
0010	**LAMP POSTS**									
0020	Aluminum, 7' high, stock units, post only	1 Carp	16	.500	Ea.	85.50	25.50		111	133
0100	Mild steel, plain	"	16	.500	"	63.50	25.50		89	108

05 71 Decorative Metal Stairs

05 71 13 – Fabricated Metal Spiral Stairs

05 71 13.50 Spiral Stairs

		Crew	Daily Output	Labor-Hours	Unit	Material	2018 Bare Costs Labor	Equipment	Total	Total Incl O&P
0010	**SPIRAL STAIRS**									
1805	Shop fabricated, custom ordered									
1810	Aluminum, 5'-0" diameter, plain units	E-4	45	.711	Riser	580	39	2.19	621.19	700
1820	Fancy units		45	.711		1,000	39	2.19	1,041.19	1,175
1900	Cast iron, 4'-0" diameter, plain units		45	.711		745	39	2.19	786.19	885
1920	Fancy units		25	1.280		1,300	70.50	3.94	1,374.44	1,550
2000	Steel, industrial checkered plate, 4' diameter		45	.711		490	39	2.19	531.19	605
2200	6' diameter		40	.800		695	44	2.46	741.46	835
3100	Spiral stair kits, 12 stacking risers to fit exact floor height									
3110	Steel, flat metal treads, primed, 3'-6" diameter	2 Carp	1.60	10	Flight	1,350	505		1,855	2,250
3120	4'-0" diameter		1.45	11.034		1,525	560		2,085	2,525
3130	4'-6" diameter		1.35	11.852		1,700	600		2,300	2,775
3140	5'-0" diameter		1.25	12.800		1,850	650		2,500	3,050
3210	Galvanized, 3'-6" diameter		1.60	10		1,850	505		2,355	2,825
3220	4'-0" diameter		1.45	11.034		2,200	560		2,760	3,275
3230	4'-6" diameter		1.35	11.852		2,425	600		3,025	3,600
3240	5'-0" diameter		1.25	12.800		2,650	650		3,300	3,925
3310	Checkered plate tread, primed, 3'-6" diameter		1.45	11.034		1,600	560		2,160	2,600
3320	4'-0" diameter		1.35	11.852		1,800	600		2,400	2,900
3330	4'-6" diameter		1.25	12.800		1,975	650		2,625	3,150
3340	5'-0" diameter		1.15	13.913		2,150	705		2,855	3,450
3410	Galvanized, 3'-6" diameter		1.45	11.034		2,200	560		2,760	3,250
3420	4'-0" diameter		1.35	11.852		2,525	600		3,125	3,700
3430	4'-6" diameter		1.25	12.800		2,775	650		3,425	4,050
3440	5'-0" diameter		1.15	13.913		3,000	705		3,705	4,375
3510	Red oak covers on flat metal treads, 3'-6" diameter		1.35	11.852		2,450	600		3,050	3,600
3520	4'-0" diameter		1.25	12.800		2,925	650		3,575	4,200
3530	4'-6" diameter		1.15	13.913		3,150	705		3,855	4,550
3540	5'-0" diameter		1.05	15.238		3,400	775		4,175	4,900

05 73 Decorative Metal Railings

05 73 16 – Wire Rope Decorative Metal Railings

05 73 16.10 Cable Railings

		Crew	Daily Output	Labor- Hours	Unit	Material	2018 Bare Costs Labor	2018 Bare Costs Equipment	Total	Total Incl O&P
0010	**CABLE RAILINGS**, with 316 stainless steel 1 x 19 cable, 3/16" diameter									
0015	Made from recycled materials									
0100	1-3/4" diameter stainless steel posts x 42" high, cables 4" OC	G 2 Sswk	25	.640	L.F.	45.50	35		80.50	107

05 73 23 – Ornamental Railings

05 73 23.50 Railings, Ornamental

		Crew	Daily Output	Labor- Hours	Unit	Material	2018 Bare Costs Labor	2018 Bare Costs Equipment	Total	Total Incl O&P
0010	**RAILINGS, ORNAMENTAL**, 3'-6" high, posts @ 6' OC									
0020	Bronze or stainless, hand forged, plain	G 2 Sswk	24	.667	L.F.	260	36.50		296.50	345
0100	Fancy	G	18	.889		520	48.50		568.50	650
0200	Aluminum, panelized, plain	G	24	.667		12.20	36.50		48.70	73
0300	Fancy	G	18	.889		25.50	48.50		74	108
0400	Wrought iron, hand forged, plain	G	24	.667		94	36.50		130.50	164
0500	Fancy	G	18	.889		231	48.50		279.50	335
0550	Steel, panelized, plain	G	24	.667		20.50	36.50		57	82
0560	Fancy	G	18	.889		30	48.50		78.50	113
0600	Composite metal/wood/glass, plain	G	18	.889		142	48.50		190.50	236
0700	Fancy	↓	12	1.333	↓	284	73		357	435

05 75 Decorative Formed Metal

05 75 13 – Columns

05 75 13.10 Aluminum Columns

		Crew	Daily Output	Labor- Hours	Unit	Material	2018 Bare Costs Labor	2018 Bare Costs Equipment	Total	Total Incl O&P
0010	**ALUMINUM COLUMNS**									
0015	Made from recycled materials									
0020	Aluminum, extruded, stock units, no cap or base, 6" diameter	G E-4	240	.133	L.F.	12.85	7.35	.41	20.61	26.50
0100	8" diameter	G	170	.188		16.75	10.40	.58	27.73	36
0200	10" diameter	G	150	.213		22	11.75	.66	34.41	44.50
0300	12" diameter	G	140	.229		35.50	12.60	.70	48.80	61
0400	15" diameter	G	120	.267	↓	49	14.70	.82	64.52	79
0410	Caps and bases, plain, 6" diameter	G			Set	25			25	27.50
0420	8" diameter	G				31			31	34.50
0430	10" diameter	G				47			47	51.50
0440	12" diameter	G				77			77	84.50
0450	15" diameter	G				138			138	152
0460	Caps, ornamental, plain	G				310			310	340
0470	Fancy	G			↓	1,550			1,550	1,725
0500	For square columns, add to column prices above				L.F.	50%				
0700	Residential, flat, 8' high, plain	G E-4	20	1.600	Ea.	101	88	4.93	193.93	260
0720	Fancy	G	20	1.600		196	88	4.93	288.93	365
0740	Corner type, plain	G	20	1.600		174	88	4.93	266.93	340
0760	Fancy	↓	20	1.600	↓	345	88	4.93	437.93	530

05 75 13.20 Columns, Ornamental

		Crew	Daily Output	Labor- Hours	Unit	Material	2018 Bare Costs Labor	2018 Bare Costs Equipment	Total	Total Incl O&P
0010	**COLUMNS, ORNAMENTAL**, shop fabricated [R051223-10]									
6400	Mild steel, flat, 9" wide, stock units, painted, plain	G E-4	160	.200	V.L.F.	9.45	11.05	.62	21.12	29
6450	Fancy	G	160	.200		18.40	11.05	.62	30.07	39
6500	Corner columns, painted, plain	G	160	.200		16.35	11.05	.62	28.02	36.50
6550	Fancy	G ↓	160	.200	↓	32.50	11.05	.62	44.17	54

For customer support on your Building Construction Costs with RSMeans data, call 800.448.8182.

Estimating Tips
06 05 00 Common Work Results for Wood, Plastics, and Composites

- Common to any wood-framed structure are the accessory connector items such as screws, nails, adhesives, hangers, connector plates, straps, angles, and hold-downs. For typical wood-framed buildings, such as residential projects, the aggregate total for these items can be significant, especially in areas where seismic loading is a concern. For floor and wall framing, the material cost is based on 10 to 25 lbs. of accessory connectors per MBF. Hold-downs, hangers, and other connectors should be taken off by the piece.

  Included with material costs are fasteners for a normal installation. Gordian's RSMeans engineers use manufacturers' recommendations, written specifications, and/or standard construction practice for the sizing and spacing of fasteners. Prices for various fasteners are shown for informational purposes only. Adjustments should be made if unusual fastening conditions exist.

06 10 00 Carpentry

- Lumber is a traded commodity and therefore sensitive to supply and demand in the marketplace. Even with "budgetary" estimating of wood-framed projects, it is advisable to call local suppliers for the latest market pricing.

- The common quantity unit for wood-framed projects is "thousand board feet" (MBF). A board foot is a volume of wood—1" x 1' x 1' or 144 cubic inches. Board-foot quantities are generally calculated using nominal material dimensions— dressed sizes are ignored. Board foot per lineal foot of any stick of lumber can be calculated by dividing the nominal cross-sectional area by 12. As an example, 2,000 lineal feet of 2 x 12 equates to 4 MBF by dividing the nominal area, 2 x 12, by 12, which equals 2, and multiplying by 2,000 to give 4,000 board feet. This simple rule applies to all nominal dimensioned lumber.

- Waste is an issue of concern at the quantity takeoff for any area of construction. Framing lumber is sold in even foot lengths, i.e., 8', 10', 12', 14', 16' and depending on spans, wall heights, and the grade of lumber, waste is inevitable. A rule of thumb for lumber waste is 5%–10% depending on material quality and the complexity of the framing.

- Wood in various forms and shapes is used in many projects, even where the main structural framing is steel, concrete, or masonry. Plywood as a back-up partition material and 2x boards used as blocking and cant strips around roof edges are two common examples. The estimator should ensure that the costs of all wood materials are included in the final estimate.

06 20 00 Finish Carpentry

- It is necessary to consider the grade of workmanship when estimating labor costs for erecting millwork and an interior finish. In practice, there are three grades: premium, custom, and economy. The RSMeans daily output for base and case moldings is in the range of 200 to 250 L.F. per carpenter per day. This is appropriate for most average custom-grade projects. For premium projects, an adjustment to productivity of 25%–50% should be made, depending on the complexity of the job.

Reference Numbers

Reference numbers are shown at the beginning of some major classifications. These numbers refer to related items in the Reference Section. The reference information may be an estimating procedure, an alternate pricing method, or technical information.

Note: Not all subdivisions listed here necessarily appear. ■

Did you know?

RSMeans data is available through our online application with 24/7 access:

- Search for unit prices by keyword
- Leverage the most up-to-date data
- Build and export estimates

Try it free for 30 days!
www.rsmeans.com/2018freetrial

No part of this cost data may be reproduced, stored in a retrieval system, or transmitted in any form or by any means without prior written permission of Gordian.

06 05 05.10 Selective Demolition Wood Framing	Crew	Daily Output	Labor-Hours	Unit	Material	2018 Bare Costs Labor	2018 Bare Costs Equipment	Total	Total Incl O&P
0010 **SELECTIVE DEMOLITION WOOD FRAMING** R024119-10									
0100 Timber connector, nailed, small	1 Clab	96	.083	Ea.		3.32		3.32	5.05
0110 Medium		60	.133			5.30		5.30	8.10
0120 Large		48	.167			6.65		6.65	10.10
0130 Bolted, small		48	.167			6.65		6.65	10.10
0140 Medium		32	.250			9.95		9.95	15.20
0150 Large		24	.333			13.30		13.30	20
2958 Beams, 2" x 6"	2 Clab	1100	.015	L.F.		.58		.58	.88
2960 2" x 8"		825	.019			.77		.77	1.18
2965 2" x 10"		665	.024			.96		.96	1.46
2970 2" x 12"		550	.029			1.16		1.16	1.77
2972 2" x 14"		470	.034			1.36		1.36	2.07
2975 4" x 8"	B-1	413	.058			2.35		2.35	3.59
2980 4" x 10"		330	.073			2.95		2.95	4.49
2985 4" x 12"		275	.087			3.54		3.54	5.40
3000 6" x 8"		275	.087			3.54		3.54	5.40
3040 6" x 10"		220	.109			4.42		4.42	6.75
3080 6" x 12"		185	.130			5.25		5.25	8
3120 8" x 12"		140	.171			6.95		6.95	10.60
3160 10" x 12"		110	.218			8.85		8.85	13.45
3162 Alternate pricing method		1.10	21.818	M.B.F.		885		885	1,350
3170 Blocking, in 16" OC wall framing, 2" x 4"	1 Clab	600	.013	L.F.		.53		.53	.81
3172 2" x 6"		400	.020			.80		.80	1.21
3174 In 24" OC wall framing, 2" x 4"		600	.013			.53		.53	.81
3176 2" x 6"		400	.020			.80		.80	1.21
3178 Alt method, wood blocking removal from wood framing		.40	20	M.B.F.		795		795	1,225
3179 Wood blocking removal from steel framing		.36	22.222	"		885		885	1,350
3180 Bracing, let in, 1" x 3", studs 16" OC		1050	.008	L.F.		.30		.30	.46
3181 Studs 24" OC		1080	.007			.30		.30	.45
3182 1" x 4", studs 16" OC		1050	.008			.30		.30	.46
3183 Studs 24" OC		1080	.007			.30		.30	.45
3184 1" x 6", studs 16" OC		1050	.008			.30		.30	.46
3185 Studs 24" OC		1080	.007			.30		.30	.45
3186 2" x 3", studs 16" OC		800	.010			.40		.40	.61
3187 Studs 24" OC		830	.010			.38		.38	.59
3188 2" x 4", studs 16" OC		800	.010			.40		.40	.61
3189 Studs 24" OC		830	.010			.38		.38	.59
3190 2" x 6", studs 16" OC		800	.010			.40		.40	.61
3191 Studs 24" OC		830	.010			.38		.38	.59
3192 2" x 8", studs 16" OC		800	.010			.40		.40	.61
3193 Studs 24" OC		830	.010			.38		.38	.59
3194 "T" shaped metal bracing, studs at 16" OC		1060	.008			.30		.30	.46
3195 Studs at 24" OC		1200	.007			.27		.27	.40
3196 Metal straps, studs at 16" OC		1200	.007			.27		.27	.40
3197 Studs at 24" OC		1240	.006			.26		.26	.39
3200 Columns, round, 8' to 14' tall		40	.200	Ea.		7.95		7.95	12.15
3202 Dimensional lumber sizes	2 Clab	1.10	14.545	M.B.F.		580		580	885
3250 Blocking, between joists	1 Clab	320	.025	Ea.		1		1	1.52
3252 Bridging, metal strap, between joists		320	.025	Pr.		1		1	1.52
3254 Wood, between joists		320	.025	"		1		1	1.52
3260 Door buck, studs, header & access., 8' high 2" x 4" wall, 3' wide		32	.250	Ea.		9.95		9.95	15.20
3261 4' wide		32	.250			9.95		9.95	15.20
3262 5' wide		32	.250			9.95		9.95	15.20

06 05 05 – Selective Demolition for Wood, Plastics, and Composites

06 05 05.10 Selective Demolition Wood Framing		Crew	Daily Output	Labor-Hours	Unit	Material	2018 Bare Costs Labor	Equipment	Total	Total Incl O&P
3263	6' wide	1 Clab	32	.250	Ea.		9.95		9.95	15.20
3264	8' wide		30	.267			10.65		10.65	16.20
3265	10' wide		30	.267			10.65		10.65	16.20
3266	12' wide		30	.267			10.65		10.65	16.20
3267	2" x 6" wall, 3' wide		32	.250			9.95		9.95	15.20
3268	4' wide		32	.250			9.95		9.95	15.20
3269	5' wide		32	.250			9.95		9.95	15.20
3270	6' wide		32	.250			9.95		9.95	15.20
3271	8' wide		30	.267			10.65		10.65	16.20
3272	10' wide		30	.267			10.65		10.65	16.20
3273	12' wide		30	.267			10.65		10.65	16.20
3274	Window buck, studs, header & access., 8' high 2" x 4" wall, 2' wide		24	.333			13.30		13.30	20
3275	3' wide		24	.333			13.30		13.30	20
3276	4' wide		24	.333			13.30		13.30	20
3277	5' wide		24	.333			13.30		13.30	20
3278	6' wide		24	.333			13.30		13.30	20
3279	7' wide		24	.333			13.30		13.30	20
3280	8' wide		22	.364			14.50		14.50	22
3281	10' wide		22	.364			14.50		14.50	22
3282	12' wide		22	.364			14.50		14.50	22
3283	2" x 6" wall, 2' wide		24	.333			13.30		13.30	20
3284	3' wide		24	.333			13.30		13.30	20
3285	4' wide		24	.333			13.30		13.30	20
3286	5' wide		24	.333			13.30		13.30	20
3287	6' wide		24	.333			13.30		13.30	20
3288	7' wide		24	.333			13.30		13.30	20
3289	8' wide		22	.364			14.50		14.50	22
3290	10' wide		22	.364			14.50		14.50	22
3291	12' wide		22	.364			14.50		14.50	22
3360	Deck or porch decking		825	.010	L.F.		.39		.39	.59
3400	Fascia boards, 1" x 6"		500	.016			.64		.64	.97
3440	1" x 8"		450	.018			.71		.71	1.08
3480	1" x 10"		400	.020			.80		.80	1.21
3490	2" x 6"		450	.018			.71		.71	1.08
3500	2" x 8"		400	.020			.80		.80	1.21
3510	2" x 10"		350	.023			.91		.91	1.39
3610	Furring, on wood walls or ceiling		4000	.002	S.F.		.08		.08	.12
3620	On masonry or concrete walls or ceiling		1200	.007	"		.27		.27	.40
3800	Headers over openings, 2 @ 2" x 6"		110	.073	L.F.		2.90		2.90	4.41
3840	2 @ 2" x 8"		100	.080			3.19		3.19	4.86
3880	2 @ 2" x 10"		90	.089			3.54		3.54	5.40
3885	Alternate pricing method		.26	30.651	M.B.F.		1,225		1,225	1,850
3920	Joists, 1" x 4"		1250	.006	L.F.		.26		.26	.39
3930	1" x 6"		1135	.007			.28		.28	.43
3940	1" x 8"		1000	.008			.32		.32	.49
3950	1" x 10"		895	.009			.36		.36	.54
3960	1" x 12"		765	.010			.42		.42	.63
4200	2" x 4"	2 Clab	1000	.016			.64		.64	.97
4230	2" x 6"		970	.016			.66		.66	1
4240	2" x 8"		940	.017			.68		.68	1.03
4250	2" x 10"		910	.018			.70		.70	1.07
4280	2" x 12"		880	.018			.72		.72	1.10
4281	2" x 14"		850	.019			.75		.75	1.14

For customer support on your Building Construction Costs with RSMeans data, call 800.448.8182.

167

06 05 05.10 Selective Demolition Wood Framing		Crew	Daily Output	Labor-Hours	Unit	Material	2018 Bare Costs Labor	Equipment	Total	Total Incl O&P
4282	Composite joists, 9-1/2"	2 Clab	960	.017	L.F.		.66		.66	1.01
4283	11-7/8"		930	.017			.69		.69	1.04
4284	14"		897	.018			.71		.71	1.08
4285	16"		865	.019			.74		.74	1.12
4290	Wood joists, alternate pricing method		1.50	10.667	M.B.F.		425		425	645
4500	Open web joist, 12" deep		500	.032	L.F.		1.28		1.28	1.94
4505	14" deep		475	.034			1.34		1.34	2.04
4510	16" deep		450	.036			1.42		1.42	2.16
4520	18" deep		425	.038			1.50		1.50	2.29
4530	24" deep		400	.040			1.59		1.59	2.43
4550	Ledger strips, 1" x 2"	1 Clab	1200	.007			.27		.27	.40
4560	1" x 3"		1200	.007			.27		.27	.40
4570	1" x 4"		1200	.007			.27		.27	.40
4580	2" x 2"		1100	.007			.29		.29	.44
4590	2" x 4"		1000	.008			.32		.32	.49
4600	2" x 6"		1000	.008			.32		.32	.49
4601	2" x 8" or 2" x 10"		800	.010			.40		.40	.61
4602	4" x 6"		600	.013			.53		.53	.81
4604	4" x 8"		450	.018			.71		.71	1.08
5400	Posts, 4" x 4"	2 Clab	800	.020			.80		.80	1.21
5405	4" x 6"		550	.029			1.16		1.16	1.77
5410	4" x 8"		440	.036			1.45		1.45	2.21
5425	4" x 10"		390	.041			1.64		1.64	2.49
5430	4" x 12"		350	.046			1.82		1.82	2.77
5440	6" x 6"		400	.040			1.59		1.59	2.43
5445	6" x 8"		350	.046			1.82		1.82	2.77
5450	6" x 10"		320	.050			1.99		1.99	3.04
5455	6" x 12"		290	.055			2.20		2.20	3.35
5480	8" x 8"		300	.053			2.13		2.13	3.24
5500	10" x 10"		240	.067			2.66		2.66	4.05
5660	T&G floor planks		2	8	M.B.F.		320		320	485
5682	Rafters, ordinary, 16" OC, 2" x 4"		880	.018	S.F.		.72		.72	1.10
5683	2" x 6"		840	.019			.76		.76	1.16
5684	2" x 8"		820	.020			.78		.78	1.18
5685	2" x 10"		820	.020			.78		.78	1.18
5686	2" x 12"		810	.020			.79		.79	1.20
5687	24" OC, 2" x 4"		1170	.014			.55		.55	.83
5688	2" x 6"		1117	.014			.57		.57	.87
5689	2" x 8"		1091	.015			.58		.58	.89
5690	2" x 10"		1091	.015			.58		.58	.89
5691	2" x 12"		1077	.015			.59		.59	.90
5795	Rafters, ordinary, 2" x 4" (alternate method)		862	.019	L.F.		.74		.74	1.13
5800	2" x 6" (alternate method)		850	.019			.75		.75	1.14
5840	2" x 8" (alternate method)		837	.019			.76		.76	1.16
5855	2" x 10" (alternate method)		825	.019			.77		.77	1.18
5865	2" x 12" (alternate method)		812	.020			.79		.79	1.20
5870	Sill plate, 2" x 4"	1 Clab	1170	.007			.27		.27	.42
5871	2" x 6"		780	.010			.41		.41	.62
5872	2" x 8"		586	.014			.54		.54	.83
5873	Alternate pricing method		.78	10.256	M.B.F.		410		410	625
5885	Ridge board, 1" x 4"	2 Clab	900	.018	L.F.		.71		.71	1.08
5886	1" x 6"		875	.018			.73		.73	1.11
5887	1" x 8"		850	.019			.75		.75	1.14

06 05 05.10 Selective Demolition Wood Framing	Crew	Daily Output	Labor-Hours	Unit	Material	2018 Bare Costs Labor	2018 Bare Costs Equipment	Total	Total Incl O&P
5888	1" x 10"	2 Clab	825	.019	L.F.		.77		.77	1.18
5889	1" x 12"		800	.020			.80		.80	1.21
5890	2" x 4"		900	.018			.71		.71	1.08
5892	2" x 6"		875	.018			.73		.73	1.11
5894	2" x 8"		850	.019			.75		.75	1.14
5896	2" x 10"		825	.019			.77		.77	1.18
5898	2" x 12"		800	.020			.80		.80	1.21
6050	Rafter tie, 1" x 4"		1250	.013			.51		.51	.78
6052	1" x 6"		1135	.014			.56		.56	.86
6054	2" x 4"		1000	.016			.64		.64	.97
6056	2" x 6"	▼	970	.016			.66		.66	1
6070	Sleepers, on concrete, 1" x 2"	1 Clab	4700	.002			.07		.07	.10
6075	1" x 3"		4000	.002			.08		.08	.12
6080	2" x 4"		3000	.003			.11		.11	.16
6085	2" x 6"	▼	2600	.003	▼		.12		.12	.19
6086	Sheathing from roof, 5/16"	2 Clab	1600	.010	S.F.		.40		.40	.61
6088	3/8"		1525	.010			.42		.42	.64
6090	1/2"		1400	.011			.46		.46	.69
6092	5/8"		1300	.012			.49		.49	.75
6094	3/4"		1200	.013			.53		.53	.81
6096	Board sheathing from roof		1400	.011			.46		.46	.69
6100	Sheathing, from walls, 1/4"		1200	.013			.53		.53	.81
6110	5/16"		1175	.014			.54		.54	.83
6120	3/8"		1150	.014			.55		.55	.84
6130	1/2"		1125	.014			.57		.57	.86
6140	5/8"		1100	.015			.58		.58	.88
6150	3/4"		1075	.015			.59		.59	.90
6152	Board sheathing from walls		1500	.011			.43		.43	.65
6158	Subfloor/roof deck, with boards		2200	.007			.29		.29	.44
6159	Subfloor/roof deck, with tongue & groove boards		2000	.008			.32		.32	.49
6160	Plywood, 1/2" thick		768	.021			.83		.83	1.26
6162	5/8" thick		760	.021			.84		.84	1.28
6164	3/4" thick		750	.021			.85		.85	1.29
6165	1-1/8" thick	▼	720	.022			.89		.89	1.35
6166	Underlayment, particle board, 3/8" thick	1 Clab	780	.010			.41		.41	.62
6168	1/2" thick		768	.010			.42		.42	.63
6170	5/8" thick		760	.011			.42		.42	.64
6172	3/4" thick	▼	750	.011	▼		.43		.43	.65
6200	Stairs and stringers, straight run	2 Clab	40	.400	Riser		15.95		15.95	24.50
6240	With platforms, winders or curves	"	26	.615	"		24.50		24.50	37.50
6300	Components, tread	1 Clab	110	.073	Ea.		2.90		2.90	4.41
6320	Riser		80	.100	"		3.99		3.99	6.05
6390	Stringer, 2" x 10"		260	.031	L.F.		1.23		1.23	1.87
6400	2" x 12"		260	.031			1.23		1.23	1.87
6410	3" x 10"		250	.032			1.28		1.28	1.94
6420	3" x 12"	▼	250	.032			1.28		1.28	1.94
6590	Wood studs, 2" x 3"	2 Clab	3076	.005			.21		.21	.32
6600	2" x 4"		2000	.008			.32		.32	.49
6640	2" x 6"	▼	1600	.010	▼		.40		.40	.61
6720	Wall framing, including studs, plates and blocking, 2" x 4"	1 Clab	600	.013	S.F.		.53		.53	.81
6740	2" x 6"		480	.017	"		.66		.66	1.01
6750	Headers, 2" x 4"		1125	.007	L.F.		.28		.28	.43
6755	2" x 6"	▼	1125	.007	▼		.28		.28	.43

06 05 05 – Selective Demolition for Wood, Plastics, and Composites

06 05 05.10 Selective Demolition Wood Framing		Crew	Daily Output	Labor-Hours	Unit	Material	2018 Bare Costs Labor	Equipment	Total	Total Incl O&P
6760	2" x 8"	1 Clab	1050	.008	L.F.		.30		.30	.46
6765	2" x 10"		1050	.008			.30		.30	.46
6770	2" x 12"		1000	.008			.32		.32	.49
6780	4" x 10"		525	.015			.61		.61	.93
6785	4" x 12"		500	.016			.64		.64	.97
6790	6" x 8"		560	.014			.57		.57	.87
6795	6" x 10"		525	.015			.61		.61	.93
6797	6" x 12"		500	.016			.64		.64	.97
7000	Trusses									
7050	12' span	2 Clab	74	.216	Ea.		8.60		8.60	13.10
7150	24' span	F-3	66	.606			31.50	7.50	39	56
7200	26' span		64	.625			32.50	7.75	40.25	57.50
7250	28' span		62	.645			33.50	8	41.50	60
7300	30' span		58	.690			35.50	8.55	44.05	64
7350	32' span		56	.714			37	8.85	45.85	66
7400	34' span		54	.741			38.50	9.20	47.70	68.50
7450	36' span		52	.769			40	9.55	49.55	71
8000	Soffit, T&G wood	1 Clab	520	.015	S.F.		.61		.61	.93
8010	Hardboard, vinyl or aluminum	"	640	.013			.50		.50	.76
8030	Plywood	2 Carp	315	.051			2.58		2.58	3.92
9500	See Section 02 41 19.19 for rubbish handling									

06 05 05.20 Selective Demolition Millwork and Trim

		Crew	Daily Output	Labor-Hours	Unit	Material	Labor	Equipment	Total	Total Incl O&P
0010	**SELECTIVE DEMOLITION MILLWORK AND TRIM** R024119-10									
1000	Cabinets, wood, base cabinets, per L.F.	2 Clab	80	.200	L.F.		7.95		7.95	12.15
1020	Wall cabinets, per L.F.	"	80	.200	"		7.95		7.95	12.15
1060	Remove and reset, base cabinets	2 Carp	18	.889	Ea.		45		45	68.50
1070	Wall cabinets		20	.800			40.50		40.50	62
1072	Oven cabinet, 7' high		11	1.455			74		74	112
1074	Cabinet door, up to 2' high	1 Clab	66	.121			4.83		4.83	7.35
1076	2' - 4' high	"	46	.174			6.95		6.95	10.55
1100	Steel, painted, base cabinets	2 Clab	60	.267	L.F.		10.65		10.65	16.20
1120	Wall cabinets		60	.267	"		10.65		10.65	16.20
1200	Casework, large area		320	.050	S.F.		1.99		1.99	3.04
1220	Selective		200	.080	"		3.19		3.19	4.86
1500	Counter top, straight runs		200	.080	L.F.		3.19		3.19	4.86
1510	L, U or C shapes		120	.133			5.30		5.30	8.10
1550	Remove and reset, straight runs	2 Carp	50	.320			16.20		16.20	24.50
1560	L, U or C shape	"	40	.400			20.50		20.50	31
2000	Paneling, 4' x 8' sheets	2 Clab	2000	.008	S.F.		.32		.32	.49
2100	Boards, 1" x 4"		700	.023			.91		.91	1.39
2120	1" x 6"		750	.021			.85		.85	1.29
2140	1" x 8"		800	.020			.80		.80	1.21
3000	Trim, baseboard, to 6" wide		1200	.013	L.F.		.53		.53	.81
3040	Greater than 6" and up to 12" wide		1000	.016			.64		.64	.97
3080	Remove and reset, minimum	2 Carp	400	.040			2.03		2.03	3.09
3090	Maximum	"	300	.053			2.70		2.70	4.12
3100	Ceiling trim	2 Clab	1000	.016			.64		.64	.97
3120	Chair rail		1200	.013			.53		.53	.81
3140	Railings with balusters		240	.067			2.66		2.66	4.05
3160	Wainscoting		700	.023	S.F.		.91		.91	1.39

For customer support on your Building Construction Costs with RSMeans data, call 800.448.8182.

06 05 23 – Wood, Plastic, and Composite Fastenings

06 05 23.10 Nails

		Crew	Daily Output	Labor-Hours	Unit	Material	2018 Bare Costs Labor	Equipment	Total	Total Incl O&P
0010	**NAILS**, material only, based upon 50# box purchase									
0020	Copper nails, plain				Lb.	11.15			11.15	12.30
0400	Stainless steel, plain					8.55			8.55	9.40
0500	Box, 3d to 20d, bright					1.43			1.43	1.57
0520	Galvanized					2.34			2.34	2.57
0600	Common, 3d to 60d, plain					1.15			1.15	1.27
0700	Galvanized					2.19			2.19	2.41
0800	Aluminum					10.80			10.80	11.90
1000	Annular or spiral thread, 4d to 60d, plain					3.19			3.19	3.51
1200	Galvanized					3.02			3.02	3.32
1400	Drywall nails, plain					1.78			1.78	1.96
1600	Galvanized					1.84			1.84	2.02
1800	Finish nails, 4d to 10d, plain					1.24			1.24	1.36
2000	Galvanized					1.81			1.81	1.99
2100	Aluminum					7.95			7.95	8.75
2300	Flooring nails, hardened steel, 2d to 10d, plain					3.54			3.54	3.89
2400	Galvanized					3.75			3.75	4.13
2500	Gypsum lath nails, 1-1/8", 13 ga. flathead, blued					3.32			3.32	3.65
2600	Masonry nails, hardened steel, 3/4" to 3" long, plain					2.34			2.34	2.57
2700	Galvanized					3.90			3.90	4.29
2900	Roofing nails, threaded, galvanized					2.76			2.76	3.04
3100	Aluminum					7.25			7.25	8
3300	Compressed lead head, threaded, galvanized					2.97			2.97	3.27
3600	Siding nails, plain shank, galvanized					2.53			2.53	2.78
3800	Aluminum					5.75			5.75	6.35
5000	Add to prices above for cement coating					.15			.15	.17
5200	Zinc or tin plating					.25			.25	.28
5500	Vinyl coated sinkers, 8d to 16d					2.56			2.56	2.82

06 05 23.40 Sheet Metal Screws

		Crew	Daily Output	Labor-Hours	Unit	Material	2018 Bare Costs Labor	Equipment	Total	Total Incl O&P
0010	**SHEET METAL SCREWS**									
0020	Steel, standard, #8 x 3/4", plain				C	2.24			2.24	2.46
0100	Galvanized					2.24			2.24	2.46
0300	#10 x 1", plain					3.12			3.12	3.43
0400	Galvanized					3.12			3.12	3.43
0600	With washers, #14 x 1", plain					7.25			7.25	8
0700	Galvanized					7.25			7.25	8
0900	#14 x 2", plain					10.75			10.75	11.85
1000	Galvanized					10.75			10.75	11.85
1500	Self-drilling, with washers (pinch point), #8 x 3/4", plain					2.92			2.92	3.21
1600	Galvanized					2.92			2.92	3.21
1800	#10 x 3/4", plain					3.44			3.44	3.78
1900	Galvanized					3.44			3.44	3.78
3000	Stainless steel w/aluminum or neoprene washers, #14 x 1", plain					12.95			12.95	14.25
3100	#14 x 2", plain					21.50			21.50	24

06 05 23.50 Wood Screws

		Crew	Daily Output	Labor-Hours	Unit	Material	2018 Bare Costs Labor	Equipment	Total	Total Incl O&P
0010	**WOOD SCREWS**									
0020	#8, 1" long, steel				C	4.60			4.60	5.05
0100	Brass					12.20			12.20	13.40
0200	#8, 2" long, steel					8			8	8.80
0300	Brass					24			24	26.50
0400	#10, 1" long, steel					3.30			3.30	3.63
0500	Brass					15.60			15.60	17.15

For customer support on your Building Construction Costs with RSMeans data, call 800.448.8182.

171

06 05 23 – Wood, Plastic, and Composite Fastenings

06 05 23.50 Wood Screws	Crew	Daily Output	Labor-Hours	Unit	Material	2018 Bare Costs Labor	Equipment	Total	Total Incl O&P
0600	#10, 2" long, steel				C	5.30			5.30	5.85
0700	Brass					27			27	29.50
0800	#10, 3" long, steel					8.85			8.85	9.75
1000	#12, 2" long, steel					7.30			7.30	8.05
1100	Brass					34.50			34.50	38
1500	#12, 3" long, steel					11.15			11.15	12.30
2000	#12, 4" long, steel					24.50			24.50	27

06 05 23.60 Timber Connectors	Crew	Daily Output	Labor-Hours	Unit	Material	2018 Bare Costs Labor	Equipment	Total	Total Incl O&P
0010	**TIMBER CONNECTORS**									
0020	Add up cost of each part for total cost of connection									
0100	Connector plates, steel, with bolts, straight	2 Carp	75	.213	Ea.	27.50	10.80		38.30	47
0110	Tee, 7 ga.		50	.320		34	16.20		50.20	62
0120	T- Strap, 14 ga., 12" x 8" x 2"		50	.320		34	16.20		50.20	62
0150	Anchor plates, 7 ga., 9" x 7"		75	.213		27.50	10.80		38.30	47
0200	Bolts, machine, sq. hd. with nut & washer, 1/2" diameter, 4" long	1 Carp	140	.057		.75	2.90		3.65	5.25
0300	7-1/2" long		130	.062		1.37	3.12		4.49	6.25
0500	3/4" diameter, 7-1/2" long		130	.062		3.20	3.12		6.32	8.25
0610	Machine bolts, w/nut, washer, 3/4" diameter, 15" L, HD's & beam hangers		95	.084		5.95	4.27		10.22	13.05
0800	Drilling bolt holes in timber, 1/2" diameter		450	.018	Inch		.90		.90	1.37
0900	1" diameter		350	.023	"		1.16		1.16	1.76
1100	Framing anchor, angle, 3" x 3" x 1-1/2", 12 ga.		175	.046	Ea.	2.43	2.32		4.75	6.20
1150	Framing anchors, 18 ga., 4-1/2" x 2-3/4"		175	.046		2.43	2.32		4.75	6.20
1160	Framing anchors, 18 ga., 4-1/2" x 3"		175	.046		2.43	2.32		4.75	6.20
1170	Clip anchors plates, 18 ga., 12" x 1-1/8"		175	.046		2.43	2.32		4.75	6.20
1250	Holdowns, 3 ga. base, 10 ga. body		8	1		38	50.50		88.50	119
1260	Holdowns, 7 ga. 11-1/16" x 3-1/4"		8	1		38	50.50		88.50	119
1270	Holdowns, 7 ga. 14-3/8" x 3-1/8"		8	1		38	50.50		88.50	119
1275	Holdowns, 12 ga. 8" x 2-1/2"		8	1		38	50.50		88.50	119
1300	Joist and beam hangers, 18 ga. galv., for 2" x 4" joist		175	.046		.75	2.32		3.07	4.36
1400	2" x 6" to 2" x 10" joist		165	.048		1.27	2.46		3.73	5.15
1600	16 ga. galv., 3" x 6" to 3" x 10" joist		160	.050		2.92	2.54		5.46	7.05
1700	3" x 10" to 3" x 14" joist		160	.050		4.32	2.54		6.86	8.60
1800	4" x 6" to 4" x 10" joist		155	.052		2.89	2.62		5.51	7.15
1900	4" x 10" to 4" x 14" joist		155	.052		4.89	2.62		7.51	9.40
2000	Two-2" x 6" to two-2" x 10" joists		150	.053		3.83	2.70		6.53	8.35
2100	Two-2" x 10" to two-2" x 14" joists		150	.053		4.29	2.70		6.99	8.85
2300	3/16" thick, 6" x 8" joist		145	.055		59.50	2.80		62.30	70
2400	6" x 10" joist		140	.057		62	2.90		64.90	72.50
2500	6" x 12" joist		135	.059		64.50	3		67.50	75.50
2700	1/4" thick, 6" x 14" joist		130	.062		67	3.12		70.12	78.50
2900	Plywood clips, extruded aluminum H clip, for 3/4" panels					.22			.22	.24
3000	Galvanized 18 ga. back-up clip					.18			.18	.20
3200	Post framing, 16 ga. galv. for 4" x 4" base, 2 piece	1 Carp	130	.062		15.15	3.12		18.27	21.50
3300	Cap		130	.062		21	3.12		24.12	28
3500	Rafter anchors, 18 ga. galv., 1-1/2" wide, 5-1/4" long		145	.055		.45	2.80		3.25	4.76
3600	10-3/4" long		145	.055		1.37	2.80		4.17	5.75
3800	Shear plates, 2-5/8" diameter		120	.067		2.47	3.38		5.85	7.85
3900	4" diameter		115	.070		2.40	3.53		5.93	8
4000	Sill anchors, embedded in concrete or block, 25-1/2" long		115	.070		11.60	3.53		15.13	18.10
4100	Spike grids, 3" x 6"		120	.067		1.09	3.38		4.47	6.35
4400	Split rings, 2-1/2" diameter		120	.067		2.44	3.38		5.82	7.85
4500	4" diameter		110	.073		2.78	3.69		6.47	8.65

06 05 23.60 Timber Connectors	Crew	Daily Output	Labor-Hours	Unit	Material	2018 Bare Costs Labor	Equipment	Total	Total Incl O&P
4550	Tie plate, 20 ga., 7" x 3-1/8"	1 Carp	110	.073	Ea.	2.78	3.69		6.47	8.65
4560	5" x 4-1/8"		110	.073		2.78	3.69		6.47	8.65
4575	Twist straps, 18 ga., 12" x 1-1/4"		110	.073		2.78	3.69		6.47	8.65
4580	16" x 1-1/4"		110	.073		2.78	3.69		6.47	8.65
4600	Strap ties, 20 ga., 2-1/16" wide, 12-13/16" long		180	.044		.87	2.25		3.12	4.39
4700	16 ga., 1-3/8" wide, 12" long		180	.044		.87	2.25		3.12	4.39
4800	1-1/4" wide, 21-5/8" long		160	.050		2.70	2.54		5.24	6.85
5000	Toothed rings, 2-5/8" or 4" diameter		90	.089		2.05	4.51		6.56	9.10
5200	Truss plates, nailed, 20 ga., up to 32' span	↓	17	.471	Truss	12.85	24		36.85	50.50
5400	Washers, 2" x 2" x 1/8"				Ea.	.46			.46	.51
5500	3" x 3" x 3/16"				"	1.32			1.32	1.45
6000	Angles and gussets, painted									
6012	7 ga., 3-1/4" x 3-1/4" x 2-1/2" long	1 Carp	1.90	4.211	C	1,075	213		1,288	1,500
6014	3-1/4" x 3-1/4" x 5" long		1.90	4.211		2,100	213		2,313	2,625
6016	3-1/4" x 3-1/4" x 7-1/2" long		1.85	4.324		3,975	219		4,194	4,675
6018	5-3/4" x 5-3/4" x 2-1/2" long		1.85	4.324		2,550	219		2,769	3,150
6020	5-3/4" x 5-3/4" x 5" long		1.85	4.324		4,075	219		4,294	4,800
6022	5-3/4" x 5-3/4" x 7-1/2" long		1.80	4.444		6,025	225		6,250	6,975
6024	3 ga., 4-1/4" x 4-1/4" x 3" long		1.85	4.324		2,725	219		2,944	3,300
6026	4-1/4" x 4-1/4" x 6" long		1.85	4.324		5,850	219		6,069	6,750
6028	4-1/4" x 4-1/4" x 9" long		1.80	4.444		6,575	225		6,800	7,575
6030	7-1/4" x 7-1/4" x 3" long		1.80	4.444		4,700	225		4,925	5,525
6032	7-1/4" x 7-1/4" x 6" long		1.80	4.444		6,350	225		6,575	7,325
6034	7-1/4" x 7-1/4" x 9" long	↓	1.75	4.571	↓	14,200	232		14,432	16,000
6036	Gussets									
6038	7 ga., 8-1/8" x 8-1/8" x 2-3/4" long	1 Carp	1.80	4.444	C	4,850	225		5,075	5,675
6040	3 ga., 9-3/4" x 9-3/4" x 3-1/4" long	"	1.80	4.444	"	7,275	225		7,500	8,350
6101	Beam hangers, polymer painted									
6102	Bolted, 3 ga. (W x H x L)									
6104	3-1/4" x 9" x 12" top flange	1 Carp	1	8	C	19,100	405		19,505	21,600
6106	5-1/4" x 9" x 12" top flange		1	8		20,000	405		20,405	22,600
6108	5-1/4" x 11" x 11-3/4" top flange		1	8		20,800	405		21,205	23,400
6110	6-7/8" x 9" x 12" top flange		1	8		20,600	405		21,005	23,300
6112	6-7/8" x 11" x 13-1/2" top flange		1	8		24,300	405		24,705	27,400
6114	8-7/8" x 11" x 15-1/2" top flange	↓	1	8	↓	23,600	405		24,005	26,500
6116	Nailed, 3 ga. (W x H x L)									
6118	3-1/4" x 10-1/2" x 10" top flange	1 Carp	1.80	4.444	C	18,700	225		18,925	20,800
6120	3-1/4" x 10-1/2" x 12" top flange		1.80	4.444		21,400	225		21,625	23,900
6122	5-1/4" x 9-1/2" x 10" top flange		1.80	4.444		17,500	225		17,725	19,500
6124	5-1/4" x 9-1/2" x 12" top flange		1.80	4.444		29,800	225		30,025	33,100
6128	6-7/8" x 8-1/2" x 12" top flange	↓	1.80	4.444	↓	20,600	225		20,825	23,000
6134	Saddle hangers, glu-lam (W x H x L)									
6136	3-1/4" x 10-1/2" x 5-1/4" x 6" saddle	1 Carp	.50	16	C	14,900	810		15,710	17,600
6138	3-1/4" x 10-1/2" x 6-7/8" x 6" saddle		.50	16		15,700	810		16,510	18,400
6140	3-1/4" x 10-1/2" x 8-7/8" x 6" saddle		.50	16		17,400	810		18,210	20,400
6142	3-1/4" x 19-1/2" x 5-1/4" x 10-1/8" saddle		.40	20		14,900	1,025		15,925	18,000
6144	3-1/4" x 19-1/2" x 6-7/8" x 10-1/8" saddle		.40	20		15,700	1,025		16,725	18,800
6146	3-1/4" x 19-1/2" x 8-7/8" x 10-1/8" saddle		.40	20		16,500	1,025		17,525	19,700
6148	5-1/4" x 9-1/2" x 5-1/4" x 12" saddle		.50	16		17,800	810		18,610	20,700
6150	5-1/4" x 9-1/2" x 6-7/8" x 9" saddle		.50	16		19,500	810		20,310	22,600
6152	5-1/4" x 10-1/2" x spec x 12" saddle		.50	16		21,200	810		22,010	24,500
6154	5-1/4" x 18" x 5-1/4" x 12-1/8" saddle		.40	20		17,800	1,025		18,825	21,100
6156	5-1/4" x 18" x 6-7/8" x 12-1/8" saddle	↓	.40	20		19,500	1,025		20,525	23,000

06 05 23.60 Timber Connectors	Crew	Daily Output	Labor-Hours	Unit	Material	2018 Bare Costs Labor	Equipment	Total	Total Incl O&P
6158	5-1/4" x 18" x spec x 12-1/8" saddle	1 Carp	.40	20	C	21,200	1,025		22,225	24,900
6160	6-7/8" x 8-1/2" x 6-7/8" x 12" saddle		.50	16		20,000	810		20,810	23,200
6162	6-7/8" x 8-1/2" x 8-7/8" x 12" saddle		.50	16		20,700	810		21,510	23,900
6164	6-7/8" x 10-1/2" x spec x 12" saddle		.50	16		20,000	810		20,810	23,200
6166	6-7/8" x 18" x 6-7/8" x 13-3/4" saddle		.40	20		20,000	1,025		21,025	23,600
6168	6-7/8" x 18" x 8-7/8" x 13-3/4" saddle		.40	20		22,000	1,025		23,025	25,800
6170	6-7/8" x 18" x spec x 13-3/4" saddle		.40	20		22,300	1,025		23,325	26,100
6172	8-7/8" x 18" x spec x 15-3/4" saddle	▼	.40	20	▼	34,800	1,025		35,825	39,900
6201	Beam and purlin hangers, galvanized, 12 ga.									
6202	Purlin or joist size, 3" x 8"	1 Carp	1.70	4.706	C	1,850	239		2,089	2,425
6204	3" x 10"		1.70	4.706		2,125	239		2,364	2,725
6206	3" x 12"		1.65	4.848		2,300	246		2,546	2,900
6208	3" x 14"		1.65	4.848		2,450	246		2,696	3,050
6210	3" x 16"		1.65	4.848		2,575	246		2,821	3,225
6212	4" x 8"		1.65	4.848		1,875	246		2,121	2,425
6214	4" x 10"		1.65	4.848		2,125	246		2,371	2,725
6216	4" x 12"		1.60	5		2,525	254		2,779	3,150
6218	4" x 14"		1.60	5		2,525	254		2,779	3,150
6220	4" x 16"		1.60	5		2,850	254		3,104	3,500
6222	6" x 8"		1.60	5		2,275	254		2,529	2,875
6224	6" x 10"		1.55	5.161		2,425	262		2,687	3,075
6226	6" x 12"		1.55	5.161		4,450	262		4,712	5,300
6228	6" x 14"		1.50	5.333		4,725	270		4,995	5,600
6230	6" x 16"	▼	1.50	5.333	▼	5,000	270		5,270	5,900
6250	Beam seats									
6252	Beam size, 5-1/4" wide									
6254	5" x 7" x 1/4"	1 Carp	1.80	4.444	C	7,500	225		7,725	8,600
6256	6" x 7" x 3/8"		1.80	4.444		8,425	225		8,650	9,625
6258	7" x 7" x 3/8"		1.80	4.444		9,000	225		9,225	10,200
6260	8" x 7" x 3/8"	▼	1.80	4.444	▼	10,600	225		10,825	12,000
6262	Beam size, 6-7/8" wide									
6264	5" x 9" x 1/4"	1 Carp	1.80	4.444	C	8,975	225		9,200	10,200
6266	6" x 9" x 3/8"		1.80	4.444		11,700	225		11,925	13,200
6268	7" x 9" x 3/8"		1.80	4.444		11,800	225		12,025	13,300
6270	8" x 9" x 3/8"	▼	1.80	4.444	▼	14,000	225		14,225	15,700
6272	Special beams, over 6-7/8" wide									
6274	5" x 10" x 3/8"	1 Carp	1.80	4.444	C	12,100	225		12,325	13,700
6276	6" x 10" x 3/8"		1.80	4.444		14,200	225		14,425	15,900
6278	7" x 10" x 3/8"		1.80	4.444		14,800	225		15,025	16,600
6280	8" x 10" x 3/8"		1.75	4.571		15,900	232		16,132	17,900
6282	5-1/4" x 12" x 5/16"		1.75	4.571		12,300	232		12,532	13,900
6284	6-1/2" x 12" x 3/8"		1.75	4.571		20,400	232		20,632	22,800
6286	5-1/4" x 16" x 5/16"		1.70	4.706		18,100	239		18,339	20,300
6288	6-1/2" x 16" x 3/8"		1.70	4.706		23,600	239		23,839	26,300
6290	5-1/4" x 20" x 5/16"		1.70	4.706		21,200	239		21,439	23,800
6292	6-1/2" x 20" x 3/8"	▼	1.65	4.848	▼	27,700	246		27,946	30,900
6300	Column bases									
6302	4" x 4", 16 ga.	1 Carp	1.80	4.444	C	715	225		940	1,125
6306	7 ga.		1.80	4.444		2,900	225		3,125	3,525
6308	4" x 6", 16 ga.		1.80	4.444		1,650	225		1,875	2,175
6312	7 ga.		1.80	4.444		2,850	225		3,075	3,475
6314	6" x 6", 16 ga.		1.75	4.571		1,875	232		2,107	2,425
6318	7 ga.	▼	1.75	4.571		3,650	232		3,882	4,350

06 05 23.60 Timber Connectors		Crew	Daily Output	Labor-Hours	Unit	Material	2018 Bare Costs Labor	Equipment	Total	Total Incl O&P
6320	6" x 8", 7 ga.	1 Carp	1.70	4.706	C	2,900	239		3,139	3,550
6322	6" x 10", 7 ga.		1.70	4.706		3,300	239		3,539	4,000
6324	6" x 12', 7 ga.		1.70	4.706		3,575	239		3,814	4,325
6326	8" x 8', 7 ga.		1.65	4.848		5,700	246		5,946	6,650
6330	8" x 10", 7 ga		1.65	4.848		7,225	246		7,471	8,325
6332	8" x 12", 7 ga.		1.60	5		7,850	254		8,104	9,000
6334	10" x 10", 3 ga.		1.60	5		8,000	254		8,254	9,200
6336	10" x 12", 3 ga.		1.60	5		9,200	254		9,454	10,500
6338	12" x 12", 3 ga.	↓	1.55	5.161	↓	10,000	262		10,262	11,400
6350	Column caps, painted, 3 ga.									
6352	3-1/4" x 3-5/8"	1 Carp	1.80	4.444	C	9,550	225		9,775	10,800
6354	3-1/4" x 5-1/2"		1.80	4.444		9,550	225		9,775	10,800
6356	3-5/8" x 3-5/8"		1.80	4.444		7,750	225		7,975	8,850
6358	3-5/8" x 5-1/2"		1.80	4.444		7,750	225		7,975	8,850
6360	5-1/4" x 5-1/2"		1.75	4.571		10,200	232		10,432	11,600
6362	5-1/4" x 7-1/2"		1.75	4.571		10,900	232		11,132	12,400
6364	5-1/2" x 3-5/8"		1.75	4.571		10,200	232		10,432	11,600
6366	5-1/2" x 5-1/2"		1.75	4.571		10,200	232		10,432	11,600
6368	5-1/2" x 7-1/2"		1.70	4.706		11,700	239		11,939	13,300
6370	6-7/8" x 5-1/2"		1.70	4.706		12,200	239		12,439	13,900
6372	6-7/8" x 6-7/8"		1.70	4.706		12,200	239		12,439	13,900
6374	6-7/8" x 7-1/2"		1.70	4.706		12,200	239		12,439	13,900
6376	7-1/2" x 5-1/2"		1.65	4.848		12,800	246		13,046	14,500
6378	7-1/2" x 7-1/2"		1.65	4.848		12,800	246		13,046	14,500
6380	8-7/8" x 5-1/2"		1.60	5		13,500	254		13,754	15,300
6382	8-7/8" x 7-1/2"		1.60	5		13,500	254		13,754	15,300
6384	9-1/2" x 5-1/2"	↓	1.60	5	↓	18,300	254		18,554	20,500
6400	Floor tie anchors, polymer paint									
6402	10 ga., 3" x 37-1/2"	1 Carp	1.80	4.444	C	4,875	225		5,100	5,700
6404	3-1/2" x 45-1/2"		1.75	4.571		5,100	232		5,332	5,950
6406	3 ga., 3-1/2" x 56"	↓	1.70	4.706	↓	8,975	239		9,214	10,200
6410	Girder hangers									
6412	6" wall thickness, 4" x 6"	1 Carp	1.80	4.444	C	2,725	225		2,950	3,350
6414	4" x 8"		1.80	4.444		3,050	225		3,275	3,700
6416	8" wall thickness, 4" x 6"		1.80	4.444		3,075	225		3,300	3,725
6418	4" x 8"	↓	1.80	4.444	↓	3,475	225		3,700	4,175
6420	Hinge connections, polymer painted									
6422	3/4" thick top plate									
6424	5-1/4" x 12" w/5" x 5" top	1 Carp	1	8	C	32,900	405		33,305	36,700
6426	5-1/4" x 15" w/6" x 6" top		.80	10		34,900	505		35,405	39,200
6428	5-1/4" x 18" w/7" x 7" top		.70	11.429		36,700	580		37,280	41,300
6430	5-1/4" x 26" w/9" x 9" top	↓	.60	13.333		39,100	675		39,775	44,000
6432	1" thick top plate									
6434	6-7/8" x 14" w/5" x 5" top	1 Carp	.80	10	C	40,000	505		40,505	44,800
6436	6-7/8" x 17" w/6" x 6" top		.80	10		44,500	505		45,005	49,700
6438	6-7/8" x 21" w/7" x 7" top		.70	11.429		48,500	580		49,080	54,500
6440	6-7/8" x 31" w/9" x 9" top	↓	.60	13.333	↓	53,000	675		53,675	59,500
6442	1-1/4" thick top plate									
6444	8-7/8" x 16" w/5" x 5" top	1 Carp	.60	13.333	C	49,900	675		50,575	56,000
6446	8-7/8" x 21" w/6" x 6" top		.50	16		55,000	810		55,810	61,500
6448	8-7/8" x 26" w/7" x 7" top		.40	20		62,000	1,025		63,025	70,000
6450	8-7/8" x 39" w/9" x 9" top	↓	.30	26.667		77,500	1,350		78,850	87,000
6460	Holddowns									

06 05 23.60 Timber Connectors	Crew	Daily Output	Labor-Hours	Unit	Material	2018 Bare Costs Labor	Equipment	Total	Total Incl O&P
6462	Embedded along edge									
6464	26" long, 12 ga.	1 Carp	.90	8.889	C	1,175	450		1,625	1,975
6466	35" long, 12 ga.		.85	9.412		1,625	475		2,100	2,525
6468	35" long, 10 ga.	↓	.85	9.412	↓	1,800	475		2,275	2,700
6470	Embedded away from edge									
6472	Medium duty, 12 ga.									
6474	18-1/2" long	1 Carp	.95	8.421	C	705	425		1,130	1,425
6476	23-3/4" long		.90	8.889		880	450		1,330	1,650
6478	28" long		.85	9.412		895	475		1,370	1,700
6480	35" long	↓	.85	9.412	↓	1,225	475		1,700	2,075
6482	Heavy duty, 10 ga.									
6484	28" long	1 Carp	.85	9.412	C	1,725	475		2,200	2,600
6486	35" long	"	.85	9.412	"	1,875	475		2,350	2,775
6490	Surface mounted (W x H)									
6492	2-1/2" x 5-3/4", 7 ga.	1 Carp	1	8	C	2,325	405		2,730	3,175
6494	2-1/2" x 8", 12 ga.		1	8		1,250	405		1,655	2,000
6496	2-7/8" x 6-3/8", 7 ga.		1	8		4,775	405		5,180	5,875
6498	2-7/8" x 12-1/2", 3 ga.		1	8		4,050	405		4,455	5,075
6500	3-3/16" x 9-3/8", 10 ga.		1	8		4,325	405		4,730	5,375
6502	3-1/2" x 11-5/8", 3 ga.		1	8		4,875	405		5,280	6,000
6504	3-1/2" x 14-3/4", 3 ga.		1	8		12,100	405		12,505	13,900
6506	3-1/2" x 16-1/2", 3 ga.		1	8		14,100	405		14,505	16,100
6508	3-1/2" x 20-1/2", 3 ga.		.90	8.889		16,100	450		16,550	18,400
6510	3-1/2" x 24-1/2", 3 ga.		.90	8.889		22,400	450		22,850	25,400
6512	4-1/4" x 20-3/4", 3 ga.	↓	.90	8.889	↓	22,000	450		22,450	25,000
6520	Joist hangers									
6522	Sloped, field adjustable, 18 ga.									
6524	2" x 6"	1 Carp	1.65	4.848	C	455	246		701	875
6526	2" x 8"		1.65	4.848		970	246		1,216	1,450
6528	2" x 10" and up		1.65	4.848		1,125	246		1,371	1,600
6530	3" x 10" and up		1.60	5		1,175	254		1,429	1,675
6532	4" x 10" and up	↓	1.55	5.161	↓	1,350	262		1,612	1,875
6536	Skewed 45°, 16 ga.									
6538	2" x 4"	1 Carp	1.75	4.571	C	900	232		1,132	1,350
6540	2" x 6" or 2" x 8"		1.65	4.848		910	246		1,156	1,375
6542	2" x 10" or 2" x 12"		1.65	4.848		955	246		1,201	1,425
6544	2" x 14" or 2" x 16"		1.60	5		1,650	254		1,904	2,200
6546	(2) 2" x 6" or (2) 2" x 8"		1.60	5		1,500	254		1,754	2,025
6548	(2) 2" x 10" or (2) 2" x 12"		1.55	5.161		1,625	262		1,887	2,175
6550	(2) 2" x 14" or (2) 2" x 16"		1.50	5.333		2,550	270		2,820	3,200
6552	4" x 6" or 4" x 8"		1.60	5		1,400	254		1,654	1,925
6554	4" x 10" or 4" x 12"		1.55	5.161		1,500	262		1,762	2,050
6556	4" x 14" or 4" x 16"	↓	1.55	5.161	↓	2,300	262		2,562	2,925
6560	Skewed 45°, 14 ga.									
6562	(2) 2" x 6" or (2) 2" x 8"	1 Carp	1.60	5	C	1,725	254		1,979	2,275
6564	(2) 2" x 10" or (2) 2" x 12"		1.55	5.161		2,400	262		2,662	3,025
6566	(2) 2" x 14" or (2) 2" x 16"		1.50	5.333		3,525	270		3,795	4,275
6568	4" x 6" or 4" x 8"		1.60	5		2,050	254		2,304	2,625
6570	4" x 10" or 4" x 12"		1.55	5.161		2,175	262		2,437	2,775
6572	4" x 14" or 4" x 16"	↓	1.55	5.161	↓	2,900	262		3,162	3,600
6590	Joist hangers, heavy duty 12 ga., galvanized									
6592	2" x 4"	1 Carp	1.75	4.571	C	1,175	232		1,407	1,625
6594	2" x 6"	↓	1.65	4.848	↓	1,275	246		1,521	1,775

06 05 23.60 Timber Connectors		Crew	Daily Output	Labor-Hours	Unit	Material	2018 Bare Costs Labor	Equipment	Total	Total Incl O&P
6595	2" x 6", 16 ga.	1 Carp	1.65	4.848	C	1,225	246		1,471	1,725
6596	2" x 8"		1.65	4.848		1,950	246		2,196	2,500
6597	2" x 8", 16 ga.		1.65	4.848		1,850	246		2,096	2,400
6598	2" x 10"		1.65	4.848		2,000	246		2,246	2,575
6600	2" x 12"		1.65	4.848		2,450	246		2,696	3,075
6602	2" x 14"		1.65	4.848		2,700	246		2,946	3,350
6604	2" x 16"		1.65	4.848		2,850	246		3,096	3,525
6606	3" x 4"		1.65	4.848		1,725	246		1,971	2,275
6608	3" x 6"		1.65	4.848		2,300	246		2,546	2,900
6610	3" x 8"		1.65	4.848		2,350	246		2,596	2,950
6612	3" x 10"		1.60	5		2,700	254		2,954	3,350
6614	3" x 12"		1.60	5		3,250	254		3,504	3,950
6616	3" x 14"		1.60	5		3,800	254		4,054	4,550
6618	3" x 16"		1.60	5		4,200	254		4,454	4,975
6620	(2) 2" x 4"		1.75	4.571		1,900	232		2,132	2,450
6622	(2) 2" x 6"		1.60	5		2,300	254		2,554	2,900
6624	(2) 2" x 8"		1.60	5		2,350	254		2,604	2,950
6626	(2) 2" x 10"		1.55	5.161		2,550	262		2,812	3,200
6628	(2) 2" x 12"		1.55	5.161		3,250	262		3,512	3,975
6630	(2) 2" x 14"		1.50	5.333		3,475	270		3,745	4,200
6632	(2) 2" x 16"		1.50	5.333		3,500	270		3,770	4,250
6634	4" x 4"		1.65	4.848		1,450	246		1,696	1,975
6636	4" x 6"		1.60	5		1,600	254		1,854	2,150
6638	4" x 8"		1.60	5		1,850	254		2,104	2,425
6640	4" x 10"		1.55	5.161		2,275	262		2,537	2,900
6642	4" x 12"		1.55	5.161		2,425	262		2,687	3,075
6644	4" x 14"		1.55	5.161		2,900	262		3,162	3,600
6646	4" x 16"		1.55	5.161		3,175	262		3,437	3,900
6648	(3) 2" x 10"		1.50	5.333		3,325	270		3,595	4,050
6650	(3) 2" x 12"		1.50	5.333		3,700	270		3,970	4,475
6652	(3) 2" x 14"		1.45	5.517		4,200	280		4,480	5,050
6654	(3) 2" x 16"		1.45	5.517		4,275	280		4,555	5,125
6656	6" x 6"		1.60	5		1,725	254		1,979	2,275
6658	6" x 8"		1.60	5		1,950	254		2,204	2,525
6660	6" x 10"		1.55	5.161		2,325	262		2,587	2,975
6662	6" x 12"		1.55	5.161		2,650	262		2,912	3,325
6664	6" x 14"		1.50	5.333		3,325	270		3,595	4,050
6666	6" x 16"		1.50	5.333		3,925	270		4,195	4,700
6690	Knee braces, galvanized, 12 ga.									
6692	Beam depth, 10" x 15" x 5' long	1 Carp	1.80	4.444	C	5,100	225		5,325	5,950
6694	15" x 22-1/2" x 7' long		1.70	4.706		5,850	239		6,089	6,800
6696	22-1/2" x 28-1/2" x 8' long		1.60	5		6,275	254		6,529	7,275
6698	28-1/2" x 36" x 10' long		1.55	5.161		6,550	262		6,812	7,600
6700	36" x 42" x 12' long		1.50	5.333		7,225	270		7,495	8,350
6710	Mudsill anchors									
6714	2" x 4" or 3" x 4"	1 Carp	115	.070	C	177	3.53		180.53	200
6716	2" x 6" or 3" x 6"		115	.070		177	3.53		180.53	200
6718	Block wall, 13-1/4" long		115	.070		78.50	3.53		82.03	91.50
6720	21-1/4" long		115	.070		127	3.53		130.53	145
6730	Post bases, 12 ga. galvanized									
6732	Adjustable, 3-9/16" x 3-9/16"	1 Carp	1.30	6.154	C	975	310		1,285	1,550
6734	3-9/16" x 5-1/2"		1.30	6.154		1,950	310		2,260	2,625
6736	4" x 4"		1.30	6.154		915	310		1,225	1,475

06 05 23.60 Timber Connectors		Crew	Daily Output	Labor-Hours	Unit	Material	2018 Bare Costs Labor	Equipment	Total	Total Incl O&P
6738	4" x 6"	1 Carp	1.30	6.154	C	1,325	310		1,635	1,925
6740	5-1/2" x 5-1/2"		1.30	6.154		1,725	310		2,035	2,375
6742	6" x 6"		1.30	6.154		3,425	310		3,735	4,225
6744	Elevated, 3-9/16" x 3-1/4"		1.30	6.154		1,100	310		1,410	1,675
6746	5-1/2" x 3-5/16"		1.30	6.154		1,475	310		1,785	2,100
6748	5-1/2" x 5"		1.30	6.154		2,250	310		2,560	2,950
6750	Regular, 3-9/16" x 3-3/8"		1.30	6.154		810	310		1,120	1,375
6752	4" x 3-3/8"		1.30	6.154		1,200	310		1,510	1,800
6754	18 ga., 5-1/4" x 3-1/8"		1.30	6.154		1,200	310		1,510	1,800
6755	5-1/2" x 3-3/8"		1.30	6.154		1,200	310		1,510	1,800
6756	5-1/2" x 5-3/8"		1.30	6.154		1,700	310		2,010	2,350
6758	6" x 3-3/8"		1.30	6.154		1,925	310		2,235	2,600
6760	6" x 5-3/8"		1.30	6.154		2,450	310		2,760	3,175
6762	Post combination cap/bases									
6764	3-9/16" x 3-9/16"	1 Carp	1.20	6.667	C	420	340		760	980
6766	3-9/16" x 5-1/2"		1.20	6.667		1,100	340		1,440	1,750
6768	4" x 4"		1.20	6.667		2,225	340		2,565	2,950
6770	5-1/2" x 5-1/2"		1.20	6.667		1,100	340		1,440	1,750
6772	6" x 6"		1.20	6.667		4,050	340		4,390	5,000
6774	7-1/2" x 7-1/2"		1.20	6.667		4,225	340		4,565	5,175
6776	8" x 8"		1.20	6.667		4,450	340		4,790	5,425
6790	Post-beam connection caps									
6792	Beam size 3-9/16"									
6794	12 ga. post, 4" x 4"	1 Carp	1	8	C	2,500	405		2,905	3,375
6796	4" x 6"		1	8		3,750	405		4,155	4,750
6798	4" x 8"		1	8		5,550	405		5,955	6,725
6800	16 ga. post, 4" x 4"		1	8		1,100	405		1,505	1,850
6802	4" x 6"		1	8		1,925	405		2,330	2,750
6804	4" x 8"		1	8		3,575	405		3,980	4,550
6805	18 ga. post, 2-7/8" x 3"		1	8		3,575	405		3,980	4,550
6806	Beam size 5-1/2"									
6808	12 ga. post, 6" x 4"	1 Carp	1	8	C	3,675	405		4,080	4,675
6810	6" x 6"		1	8		4,650	405		5,055	5,725
6812	6" x 8"		1	8		4,925	405		5,330	6,025
6816	16 ga. post, 6" x 4"		1	8		2,000	405		2,405	2,825
6818	6" x 6"		1	8		2,025	405		2,430	2,850
6820	Beam size 7-1/2"									
6822	12 ga. post, 8" x 4"	1 Carp	1	8	C	5,275	405		5,680	6,425
6824	8" x 6"		1	8		5,325	405		5,730	6,500
6826	8" x 8"		1	8		8,025	405		8,430	9,450
6840	Purlin anchors, embedded									
6842	Heavy duty, 10 ga.									
6844	Straight, 28" long	1 Carp	1.60	5	C	1,625	254		1,879	2,150
6846	35" long		1.50	5.333		1,700	270		1,970	2,275
6848	Twisted, 28" long		1.60	5		1,625	254		1,879	2,150
6850	35" long		1.50	5.333		1,700	270		1,970	2,275
6852	Regular duty, 12 ga.									
6854	Straight, 18-1/2" long	1 Carp	1.80	4.444	C	820	225		1,045	1,250
6856	23-3/4" long		1.70	4.706		1,025	239		1,264	1,500
6858	29" long		1.60	5		1,050	254		1,304	1,525
6860	35" long		1.50	5.333		1,450	270		1,720	2,000
6862	Twisted, 18" long		1.80	4.444		820	225		1,045	1,250
6866	28" long		1.60	5		980	254		1,234	1,450

For customer support on your Building Construction Costs with RSMeans data, call 800.448.8182.

06 05 23.60 Timber Connectors	Crew	Daily Output	Labor-Hours	Unit	Material	2018 Bare Costs Labor	2018 Bare Costs Equipment	Total	Total Incl O&P
6868	35" long	1 Carp	1.50	5.333	C	1,450	270		1,720	2,000
6870	Straight, plastic coated									
6872	23-1/2" long	1 Carp	1.60	5	C	2,025	254		2,279	2,600
6874	26-7/8" long		1.60	5		2,375	254		2,629	2,975
6876	32-1/2" long		1.50	5.333		2,525	270		2,795	3,175
6878	35-7/8" long		1.50	5.333		2,625	270		2,895	3,300
6890	Purlin hangers, painted									
6892	12 ga., 2" x 6"	1 Carp	1.80	4.444	C	1,700	225		1,925	2,225
6894	2" x 8"		1.80	4.444		1,850	225		2,075	2,375
6896	2" x 10"		1.80	4.444		2,000	225		2,225	2,550
6898	2" x 12"		1.75	4.571		2,150	232		2,382	2,700
6900	2" x 14"		1.75	4.571		2,275	232		2,507	2,875
6902	2" x 16"		1.75	4.571		2,425	232		2,657	3,025
6904	3" x 6"		1.70	4.706		1,700	239		1,939	2,250
6906	3" x 8"		1.70	4.706		1,850	239		2,089	2,425
6908	3" x 10"		1.70	4.706		2,000	239		2,239	2,575
6910	3" x 12"		1.65	4.848		2,300	246		2,546	2,900
6912	3" x 14"		1.65	4.848		2,450	246		2,696	3,050
6914	3" x 16"		1.65	4.848		2,575	246		2,821	3,225
6916	4" x 6"		1.65	4.848		1,725	246		1,971	2,275
6918	4" x 8"		1.65	4.848		1,875	246		2,121	2,425
6920	4" x 10"		1.65	4.848		2,000	246		2,246	2,575
6922	4" x 12"		1.60	5		2,375	254		2,629	3,000
6924	4" x 14"		1.60	5		2,525	254		2,779	3,150
6926	4" x 16"		1.60	5		2,675	254		2,929	3,325
6928	6" x 6"		1.60	5		2,250	254		2,504	2,850
6930	6" x 8"		1.60	5		2,425	254		2,679	3,025
6932	6" x 10"		1.55	5.161		2,575	262		2,837	3,225
6934	double 2" x 6"		1.70	4.706		1,850	239		2,089	2,425
6936	double 2" x 8"		1.70	4.706		2,000	239		2,239	2,575
6938	double 2" x 10"		1.70	4.706		2,150	239		2,389	2,725
6940	double 2" x 12"		1.65	4.848		2,300	246		2,546	2,900
6942	double 2" x 14"		1.65	4.848		2,450	246		2,696	3,050
6944	double 2" x 16"		1.65	4.848		2,575	246		2,821	3,225
6960	11 ga., 4" x 6"		1.65	4.848		3,400	246		3,646	4,100
6962	4" x 8"		1.65	4.848		3,650	246		3,896	4,400
6964	4" x 10"		1.65	4.848		3,900	246		4,146	4,675
6966	6" x 6"		1.60	5		3,450	254		3,704	4,150
6968	6" x 8"		1.60	5		3,700	254		3,954	4,425
6970	6" x 10"		1.55	5.161		3,950	262		4,212	4,725
6972	6" x 12"		1.55	5.161		4,200	262		4,462	5,000
6974	6" x 14"		1.55	5.161		4,450	262		4,712	5,300
6976	6" x 16"		1.50	5.333		4,700	270		4,970	5,575
6978	7 ga., 8" x 6"		1.60	5		3,725	254		3,979	4,475
6980	8" x 8"		1.60	5		3,975	254		4,229	4,750
6982	8" x 10"		1.55	5.161		4,225	262		4,487	5,050
6984	8" x 12"		1.55	5.161		4,475	262		4,737	5,325
6986	8" x 14"		1.50	5.333		4,725	270		4,995	5,600
6988	8" x 16"		1.50	5.333		4,975	270		5,245	5,875
7000	Strap connectors, galvanized									
7002	12 ga., 2-1/16" x 36"	1 Carp	1.55	5.161	C	1,075	262		1,337	1,600
7004	2-1/16" x 47"		1.50	5.333		1,500	270		1,770	2,050
7005	10 ga., 2-1/16" x 72"		1.50	5.333		1,575	270		1,845	2,125

06 05 Common Work Results for Wood, Plastics, and Composites

06 05 23 – Wood, Plastic, and Composite Fastenings

06 05 23.60 Timber Connectors

		Crew	Daily Output	Labor-Hours	Unit	Material	2018 Bare Costs Labor	Equipment	Total	Total Incl O&P
7006	7 ga., 2-1/16" x 34"	1 Carp	1.55	5.161	C	2,675	262		2,937	3,325
7008	2-1/16" x 45"		1.50	5.333		3,500	270		3,770	4,250
7010	3 ga., 3" x 32"		1.55	5.161		4,525	262		4,787	5,375
7012	3" x 41"		1.55	5.161		4,700	262		4,962	5,550
7014	3" x 50"		1.50	5.333		7,150	270		7,420	8,275
7016	3" x 59"		1.50	5.333		8,725	270		8,995	9,975
7018	3-1/2" x 68"	↓	1.45	5.517	↓	8,850	280		9,130	10,200
7030	Tension ties									
7032	19-1/8" long, 16 ga., 3/4" anchor bolt	1 Carp	1.80	4.444	C	1,250	225		1,475	1,725
7034	20" long, 12 ga., 1/2" anchor bolt		1.80	4.444		1,600	225		1,825	2,125
7036	20" long, 12 ga., 3/4" anchor bolt		1.80	4.444		1,600	225		1,825	2,125
7038	27-3/4" long, 12 ga., 3/4" anchor bolt	↓	1.75	4.571	↓	2,850	232		3,082	3,475
7050	Truss connectors, galvanized									
7052	Adjustable hanger									
7054	18 ga., 2" x 6"	1 Carp	1.65	4.848	C	485	246		731	910
7056	4" x 6"		1.65	4.848		640	246		886	1,075
7058	16 ga., 4" x 10"		1.60	5		930	254		1,184	1,400
7060	(2) 2" x 10"	↓	1.60	5	↓	930	254		1,184	1,400
7062	Connectors to plate									
7064	16 ga., 2" x 4" plate	1 Carp	1.80	4.444	C	480	225		705	870
7066	2" x 6" plate	"	1.80	4.444	"	620	225		845	1,025
7068	Hip jack connector									
7070	14 ga.	1 Carp	1.50	5.333	C	2,950	270		3,220	3,650

06 05 23.80 Metal Bracing

		Crew	Daily Output	Labor-Hours	Unit	Material	2018 Bare Costs Labor	Equipment	Total	Total Incl O&P
0010	**METAL BRACING**									
0302	Let-in, "T" shaped, 22 ga. galv. steel, studs at 16" OC	1 Carp	580	.014	L.F.	.81	.70		1.51	1.95
0402	Studs at 24" OC		600	.013		.81	.68		1.49	1.92
0502	Steel straps, 16 ga. galv. steel, studs at 16" OC		600	.013		1.05	.68		1.73	2.19
0602	Studs at 24" OC	↓	620	.013	↓	1.05	.65		1.70	2.16

06 11 Wood Framing

06 11 10 – Framing with Dimensional, Engineered or Composite Lumber

06 11 10.01 Forest Stewardship Council Certification

		Crew	Daily Output	Labor-Hours	Unit	Material	2018 Bare Costs Labor	Equipment	Total	Total Incl O&P
0010	**FOREST STEWARDSHIP COUNCIL CERTIFICATION**									
0020	For Forest Stewardship Council (FSC) cert dimension lumber, add [G]						65%			

06 11 10.02 Blocking

		Crew	Daily Output	Labor-Hours	Unit	Material	2018 Bare Costs Labor	Equipment	Total	Total Incl O&P
0010	**BLOCKING**									
2600	Miscellaneous, to wood construction									
2620	2" x 4"	1 Carp	.17	47.059	M.B.F.	630	2,375		3,005	4,325
2625	Pneumatic nailed		.21	38.095		635	1,925		2,560	3,650
2660	2" x 8"		.27	29.630		670	1,500		2,170	3,000
2665	Pneumatic nailed	↓	.33	24.242	↓	680	1,225		1,905	2,625
2720	To steel construction									
2740	2" x 4"	1 Carp	.14	57.143	M.B.F.	630	2,900		3,530	5,100
2780	2" x 8"	"	.21	38.095	"	670	1,925		2,595	3,700

06 11 10.04 Wood Bracing

		Crew	Daily Output	Labor-Hours	Unit	Material	2018 Bare Costs Labor	Equipment	Total	Total Incl O&P
0010	**WOOD BRACING**									
0012	Let-in, with 1" x 6" boards, studs @ 16" OC	1 Carp	150	.053	L.F.	.78	2.70		3.48	4.98
0202	Studs @ 24" OC	"	230	.035	"	.78	1.76		2.54	3.55

For customer support on your Building Construction Costs with RSMeans data, call 800.448.8182.

06 11 Wood Framing

06 11 10 – Framing with Dimensional, Engineered or Composite Lumber

06 11 10.06 Bridging

		Crew	Daily Output	Labor-Hours	Unit	Material	2018 Bare Costs Labor	Equipment	Total	Total Incl O&P
0010	**BRIDGING**									
0012	Wood, for joists 16" OC, 1" x 3"	1 Carp	130	.062	Pr.	.69	3.12		3.81	5.50
0017	Pneumatic nailed		170	.047		.76	2.39		3.15	4.47
0102	2" x 3" bridging		130	.062		.72	3.12		3.84	5.55
0107	Pneumatic nailed		170	.047		.75	2.39		3.14	4.46
0302	Steel, galvanized, 18 ga., for 2" x 10" joists at 12" OC		130	.062		1.71	3.12		4.83	6.65
0352	16" OC		135	.059		1.71	3		4.71	6.45
0402	24" OC		140	.057		2.57	2.90		5.47	7.25
0602	For 2" x 14" joists at 16" OC		130	.062		1.86	3.12		4.98	6.80
0902	Compression type, 16" OC, 2" x 8" joists		200	.040		1.26	2.03		3.29	4.47
1002	2" x 12" joists		200	.040		1.25	2.03		3.28	4.47

06 11 10.10 Beam and Girder Framing

		Crew	Daily Output	Labor-Hours	Unit	Material	2018 Bare Costs Labor	Equipment	Total	Total Incl O&P
0010	**BEAM AND GIRDER FRAMING** R061110-30									
3500	Single, 2" x 6"	2 Carp	.70	22.857	M.B.F.	645	1,150		1,795	2,475
3505	Pneumatic nailed		.81	19.704		650	1,000		1,650	2,250
3520	2" x 8"		.86	18.605		670	945		1,615	2,150
3525	Pneumatic nailed		1	16.048		680	815		1,495	2,000
3540	2" x 10"		1	16		845	810		1,655	2,150
3545	Pneumatic nailed		1.16	13.793		850	700		1,550	2,000
3560	2" x 12"		1.10	14.545		895	735		1,630	2,100
3565	Pneumatic nailed		1.28	12.539		905	635		1,540	1,975
3580	2" x 14"		1.17	13.675		970	695		1,665	2,125
3585	Pneumatic nailed		1.36	11.791		980	600		1,580	1,975
3600	3" x 8"		1.10	14.545		1,425	735		2,160	2,700
3620	3" x 10"		1.25	12.800		1,450	650		2,100	2,600
3640	3" x 12"		1.35	11.852		1,450	600		2,050	2,525
3660	3" x 14"		1.40	11.429		1,450	580		2,030	2,475
3680	4" x 8"	F-3	2.66	15.038		1,550	780	186	2,516	3,100
3700	4" x 10"		3.16	12.658		1,500	655	157	2,312	2,825
3720	4" x 12"		3.60	11.111		1,375	575	138	2,088	2,525
3740	4" x 14"		3.96	10.101		1,375	525	125	2,025	2,425
4000	Double, 2" x 6"	2 Carp	1.25	12.800		645	650		1,295	1,700
4005	Pneumatic nailed		1.45	11.034		650	560		1,210	1,575
4020	2" x 8"		1.60	10		670	505		1,175	1,500
4025	Pneumatic nailed		1.86	8.621		680	435		1,115	1,400
4040	2" x 10"		1.92	8.333		845	425		1,270	1,575
4045	Pneumatic nailed		2.23	7.185		850	365		1,215	1,500
4060	2" x 12"		2.20	7.273		895	370		1,265	1,550
4065	Pneumatic nailed		2.55	6.275		905	320		1,225	1,475
4080	2" x 14"		2.45	6.531		970	330		1,300	1,575
4085	Pneumatic nailed		2.84	5.634		980	286		1,266	1,500
5000	Triple, 2" x 6"		1.65	9.697		645	490		1,135	1,450
5005	Pneumatic nailed		1.91	8.377		650	425		1,075	1,350
5020	2" x 8"		2.10	7.619		670	385		1,055	1,325
5025	Pneumatic nailed		2.44	6.568		680	335		1,015	1,250
5040	2" x 10"		2.50	6.400		845	325		1,170	1,425
5045	Pneumatic nailed		2.90	5.517		850	280		1,130	1,350
5060	2" x 12"		2.85	5.614		895	285		1,180	1,425
5065	Pneumatic nailed		3.31	4.840		905	245		1,150	1,375
5080	2" x 14"		3.15	5.079		970	258		1,228	1,475
5085	Pneumatic nailed		3.35	4.770		980	242		1,222	1,450

For customer support on your Building Construction Costs with RSMeans data, call 800.448.8182.

181

06 11 10 - Framing with Dimensional, Engineered or Composite Lumber

06 11 10.12 Ceiling Framing

		Crew	Daily Output	Labor-Hours	Unit	Material	2018 Bare Costs Labor	Equipment	Total	Total Incl O&P
0010	**CEILING FRAMING**									
6400	Suspended, 2" x 3"	2 Carp	.50	32	M.B.F.	845	1,625		2,470	3,400
6450	2" x 4"		.59	27.119		630	1,375		2,005	2,800
6500	2" x 6"		.80	20		645	1,025		1,670	2,250
6550	2" x 8"	↓	.86	18.605	↓	670	945		1,615	2,150

06 11 10.14 Posts and Columns

		Crew	Daily Output	Labor-Hours	Unit	Material	2018 Bare Costs Labor	Equipment	Total	Total Incl O&P
0010	**POSTS AND COLUMNS**									
0400	4" x 4"	2 Carp	.52	30.769	M.B.F.	1,375	1,550		2,925	3,900
0420	4" x 6"		.55	29.091		1,550	1,475		3,025	3,950
0440	4" x 8"		.59	27.119		1,575	1,375		2,950	3,825
0460	6" x 6"		.65	24.615		1,800	1,250		3,050	3,875
0480	6" x 8"		.70	22.857		1,975	1,150		3,125	3,950
0500	6" x 10"	↓	.75	21.333	↓	1,400	1,075		2,475	3,175

06 11 10.18 Joist Framing

		Crew	Daily Output	Labor-Hours	Unit	Material	2018 Bare Costs Labor	Equipment	Total	Total Incl O&P
0010	**JOIST FRAMING** R061110-30									
2650	Joists, 2" x 4"	2 Carp	.83	19.277	M.B.F.	630	975		1,605	2,200
2655	Pneumatic nailed		.96	16.667		635	845		1,480	1,975
2680	2" x 6"		1.25	12.800		645	650		1,295	1,700
2685	Pneumatic nailed		1.44	11.111		650	565		1,215	1,575
2700	2" x 8"		1.46	10.959		670	555		1,225	1,575
2705	Pneumatic nailed		1.68	9.524		680	485		1,165	1,475
2720	2" x 10"		1.49	10.738		845	545		1,390	1,750
2725	Pneumatic nailed		1.71	9.357		850	475		1,325	1,650
2740	2" x 12"		1.75	9.143		895	465		1,360	1,700
2745	Pneumatic nailed		2.01	7.960		905	405		1,310	1,600
2760	2" x 14"		1.79	8.939		970	455		1,425	1,775
2765	Pneumatic nailed		2.06	7.767		980	395		1,375	1,675
2780	3" x 6"		1.39	11.511		1,325	585		1,910	2,375
2790	3" x 8"		1.90	8.421		1,425	425		1,850	2,225
2800	3" x 10"		1.95	8.205		1,450	415		1,865	2,225
2820	3" x 12"		1.80	8.889		1,450	450		1,900	2,275
2840	4" x 6"		1.60	10		1,525	505		2,030	2,450
2860	4" x 10"		2	8		1,500	405		1,905	2,275
2880	4" x 12"		1.80	8.889	↓	1,375	450		1,825	2,175
3000	Composite wood joist 9-1/2" deep		.90	17.778	M.L.F.	1,775	900		2,675	3,325
3010	11-1/2" deep		.88	18.182		2,050	920		2,970	3,650
3020	14" deep		.82	19.512		2,550	990		3,540	4,325
3030	16" deep		.78	20.513		4,075	1,050		5,125	6,050
4000	Open web joist 12" deep		.88	18.182	↓	3,600	920		4,520	5,375
4002	Per linear foot		880	.018	L.F.	3.60	.92		4.52	5.35
4004	Treated, per linear foot		880	.018	"	4.53	.92		5.45	6.40
4010	14" deep		.82	19.512	M.L.F.	3,900	990		4,890	5,800
4012	Per linear foot		820	.020	L.F.	3.90	.99		4.89	5.80
4014	Treated, per linear foot		820	.020	"	4.98	.99		5.97	7
4020	16" deep		.78	20.513	M.L.F.	3,850	1,050		4,900	5,825
4022	Per linear foot		780	.021	L.F.	3.86	1.04		4.90	5.85
4024	Treated, per linear foot		780	.021	"	5.10	1.04		6.14	7.20
4030	18" deep		.74	21.622	M.L.F.	4,225	1,100		5,325	6,325
4032	Per linear foot		740	.022	L.F.	4.22	1.10		5.32	6.30
4034	Treated, per linear foot		740	.022	"	5.60	1.10		6.70	7.85
6000	Composite rim joist, 1-1/4" x 9-1/2"		.90	17.778	M.L.F.	2,075	900		2,975	3,650
6010	1-1/4" x 11-1/2"	↓	.88	18.182	↓	2,275	920		3,195	3,900

06 11 Wood Framing

06 11 10 – Framing with Dimensional, Engineered or Composite Lumber

06 11 10.18 Joist Framing

		Crew	Daily Output	Labor-Hours	Unit	Material	2018 Bare Costs Labor	Equipment	Total	Total Incl O&P
6020	1-1/4" x 14-1/2"	2 Carp	.82	19.512	M.L.F.	3,000	990		3,990	4,800
6030	1-1/4" x 16-1/2"	↓	.78	20.513	↓	2,750	1,050		3,800	4,600

06 11 10.24 Miscellaneous Framing

		Crew	Daily Output	Labor-Hours	Unit	Material	2018 Bare Costs Labor	Equipment	Total	Total Incl O&P
0010	**MISCELLANEOUS FRAMING**									
8500	Firestops, 2" x 4"	2 Carp	.51	31.373	M.B.F.	630	1,600		2,230	3,125
8505	Pneumatic nailed		.62	25.806		635	1,300		1,935	2,700
8520	2" x 6"		.60	26.667		645	1,350		1,995	2,750
8525	Pneumatic nailed		.73	21.858		650	1,100		1,750	2,400
8540	2" x 8"		.60	26.667		670	1,350		2,020	2,775
8560	2" x 12"		.70	22.857		895	1,150		2,045	2,750
8600	Nailers, treated, wood construction, 2" x 4"		.53	30.189		880	1,525		2,405	3,300
8605	Pneumatic nailed		.64	25.157		890	1,275		2,165	2,925
8620	2" x 6"		.75	21.333		735	1,075		1,810	2,450
8625	Pneumatic nailed		.90	17.778		740	900		1,640	2,200
8640	2" x 8"		.93	17.204		840	870		1,710	2,250
8645	Pneumatic nailed		1.12	14.337		845	725		1,570	2,025
8660	Steel construction, 2" x 4"		.50	32		880	1,625		2,505	3,450
8680	2" x 6"		.70	22.857		735	1,150		1,885	2,575
8700	2" x 8"		.87	18.391		840	930		1,770	2,350
8760	Rough bucks, treated, for doors or windows, 2" x 6"		.40	40		735	2,025		2,760	3,900
8765	Pneumatic nailed		.48	33.333		740	1,700		2,440	3,400
8780	2" x 8"		.51	31.373		840	1,600		2,440	3,350
8785	Pneumatic nailed		.61	26.144		845	1,325		2,170	2,950
8800	Stair stringers, 2" x 10"		.22	72.727		845	3,675		4,520	6,550
8820	2" x 12"		.26	61.538		895	3,125		4,020	5,725
8840	3" x 10"		.31	51.613		1,450	2,625		4,075	5,575
8860	3" x 12"		.38	42.105	↓	1,450	2,125		3,575	4,850
8870	Laminated structural lumber, 1-1/4" x 11-1/2"		130	.123	L.F.	2.28	6.25		8.53	12
8880	1-1/4" x 14-1/2"	↓	130	.123	"	2.97	6.25		9.22	12.75

06 11 10.26 Partitions

		Crew	Daily Output	Labor-Hours	Unit	Material	2018 Bare Costs Labor	Equipment	Total	Total Incl O&P
0010	**PARTITIONS**									
0020	Single bottom and double top plate, no waste, std. & better lumber									
0180	2" x 4" studs, 8' high, studs 12" OC	2 Carp	80	.200	L.F.	4.60	10.15		14.75	20.50
0185	12" OC, pneumatic nailed		96	.167		4.66	8.45		13.11	17.95
0200	16" OC		100	.160		3.77	8.10		11.87	16.50
0205	16" OC, pneumatic nailed		120	.133		3.81	6.75		10.56	14.50
0300	24" OC		125	.128		2.93	6.50		9.43	13.10
0305	24" OC, pneumatic nailed		150	.107		2.96	5.40		8.36	11.50
0380	10' high, studs 12" OC		80	.200		5.45	10.15		15.60	21.50
0385	12" OC, pneumatic nailed		96	.167		5.50	8.45		13.95	18.90
0400	16" OC		100	.160		4.39	8.10		12.49	17.20
0405	16" OC, pneumatic nailed		120	.133		4.44	6.75		11.19	15.20
0500	24" OC		125	.128		3.35	6.50		9.85	13.60
0505	24" OC, pneumatic nailed		150	.107		3.39	5.40		8.79	11.95
0580	12' high, studs 12" OC		65	.246		6.30	12.50		18.80	26
0585	12" OC, pneumatic nailed		78	.205		6.35	10.40		16.75	23
0600	16" OC		80	.200		5	10.15		15.15	21
0605	16" OC, pneumatic nailed		96	.167		5.10	8.45		13.55	18.45
0700	24" OC		100	.160		3.77	8.10		11.87	16.50
0705	24" OC, pneumatic nailed		120	.133		3.81	6.75		10.56	14.50
0780	2" x 6" studs, 8' high, studs 12" OC		70	.229		7.10	11.60		18.70	25.50
0785	12" OC, pneumatic nailed	↓	84	.190		7.15	9.65		16.80	22.50

For customer support on your Building Construction Costs with RSMeans data, call 800.448.8182.

183

06 11 10 – Framing with Dimensional, Engineered or Composite Lumber

06 11 10.26 Partitions

		Crew	Daily Output	Labor-Hours	Unit	Material	2018 Bare Costs Labor	Equipment	Total	Total Incl O&P
0800	16" OC	2 Carp	90	.178	L.F.	5.80	9		14.80	20
0805	16" OC, pneumatic nailed		108	.148		5.85	7.50		13.35	17.90
0900	24" OC		115	.139		4.50	7.05		11.55	15.70
0905	24" OC, pneumatic nailed		138	.116		4.55	5.90		10.45	13.95
0980	10' high, studs 12" OC		70	.229		8.35	11.60		19.95	27
0985	12" OC, pneumatic nailed		84	.190		8.45	9.65		18.10	24
1000	16" OC		90	.178		6.75	9		15.75	21
1005	16" OC, pneumatic nailed		108	.148		6.85	7.50		14.35	18.95
1100	24" OC		115	.139		5.15	7.05		12.20	16.40
1105	24" OC, pneumatic nailed		138	.116		5.20	5.90		11.10	14.65
1180	12' high, studs 12" OC		55	.291		9.65	14.75		24.40	33
1185	12" OC, pneumatic nailed		66	.242		9.75	12.30		22.05	29.50
1200	16" OC		70	.229		7.70	11.60		19.30	26
1205	16" OC, pneumatic nailed		84	.190		7.80	9.65		17.45	23.50
1300	24" OC		90	.178		5.80	9		14.80	20
1305	24" OC, pneumatic nailed		108	.148		5.85	7.50		13.35	17.90
1400	For horizontal blocking, 2" x 4", add		600	.027		.42	1.35		1.77	2.52
1500	2" x 6", add		600	.027		.64	1.35		1.99	2.77
1600	For openings, add		250	.064			3.24		3.24	4.94
1700	Headers for above openings, material only, add				M.B.F.	675			675	740

06 11 10.28 Porch or Deck Framing

		Crew	Daily Output	Labor-Hours	Unit	Material	2018 Bare Costs Labor	Equipment	Total	Total Incl O&P
0010	**PORCH OR DECK FRAMING**									
0100	Treated lumber, posts or columns, 4" x 4"	2 Carp	390	.041	L.F.	1.20	2.08		3.28	4.49
0110	4" x 6"		275	.058		1.94	2.95		4.89	6.65
0120	4" x 8"		220	.073		3.96	3.69		7.65	9.95
0130	Girder, single, 4" x 4"		675	.024		1.20	1.20		2.40	3.15
0140	4" x 6"		600	.027		1.94	1.35		3.29	4.20
0150	4" x 8"		525	.030		3.96	1.55		5.51	6.70
0160	Double, 2" x 4"		625	.026		1.21	1.30		2.51	3.31
0170	2" x 6"		600	.027		1.52	1.35		2.87	3.73
0180	2" x 8"		575	.028		2.30	1.41		3.71	4.68
0190	2" x 10"		550	.029		2.88	1.47		4.35	5.40
0200	2" x 12"		525	.030		4.15	1.55		5.70	6.90
0210	Triple, 2" x 4"		575	.028		1.81	1.41		3.22	4.14
0220	2" x 6"		550	.029		2.28	1.47		3.75	4.75
0230	2" x 8"		525	.030		3.45	1.55		5	6.15
0240	2" x 10"		500	.032		4.32	1.62		5.94	7.20
0250	2" x 12"		475	.034		6.20	1.71		7.91	9.45
0260	Ledger, bolted 4' OC, 2" x 4"		400	.040		.75	2.03		2.78	3.91
0270	2" x 6"		395	.041		.89	2.05		2.94	4.11
0280	2" x 8"		390	.041		1.27	2.08		3.35	4.57
0290	2" x 10"		385	.042		1.55	2.11		3.66	4.91
0300	2" x 12"		380	.042		2.17	2.14		4.31	5.65
0310	Joists, 2" x 4"		1250	.013		.60	.65		1.25	1.65
0320	2" x 6"		1250	.013		.76	.65		1.41	1.83
0330	2" x 8"		1100	.015		1.15	.74		1.89	2.39
0340	2" x 10"		900	.018		1.44	.90		2.34	2.96
0350	2" x 12"		875	.018		1.73	.93		2.66	3.31
0360	Railings and trim, 1" x 4"	1 Carp	300	.027		.51	1.35		1.86	2.62
0370	2" x 2"		300	.027		.43	1.35		1.78	2.53
0380	2" x 4"		300	.027		.59	1.35		1.94	2.71
0390	2" x 6"		300	.027		.74	1.35		2.09	2.88

06 11 10.28 Porch or Deck Framing

		Crew	Daily Output	Labor-Hours	Unit	Material	2018 Bare Costs Labor	Equipment	Total	Total Incl O&P
0400	Decking, 1" x 4"	1 Carp	275	.029	S.F.	2.76	1.47		4.23	5.30
0410	2" x 4"		300	.027		2	1.35		3.35	4.26
0420	2" x 6"		320	.025		1.60	1.27		2.87	3.69
0430	5/4" x 6"		320	.025		2.15	1.27		3.42	4.29
0440	Balusters, square, 2" x 2"	2 Carp	660	.024	L.F.	.43	1.23		1.66	2.34
0450	Turned, 2" x 2"		420	.038		.57	1.93		2.50	3.57
0460	Stair stringer, 2" x 10"		130	.123		1.44	6.25		7.69	11.10
0470	2" x 12"		130	.123		1.73	6.25		7.98	11.40
0480	Stair treads, 1" x 4"		140	.114		2.76	5.80		8.56	11.85
0490	2" x 4"		140	.114		.60	5.80		6.40	9.45
0500	2" x 6"		160	.100		.91	5.05		5.96	8.70
0510	5/4" x 6"		160	.100		1	5.05		6.05	8.80
0520	Turned handrail post, 4" x 4"		64	.250	Ea.	34	12.70		46.70	57
0530	Lattice panel, 4' x 8', 1/2"		1600	.010	S.F.	.72	.51		1.23	1.56
0535	3/4"		1600	.010	"	1.07	.51		1.58	1.94
0540	Cedar, posts or columns, 4" x 4"		390	.041	L.F.	3.69	2.08		5.77	7.25
0550	4" x 6"		275	.058		6.90	2.95		9.85	12.05
0560	4" x 8"		220	.073		11	3.69		14.69	17.70
0800	Decking, 1" x 4"		550	.029		2.75	1.47		4.22	5.25
0810	2" x 4"		600	.027		5.60	1.35		6.95	8.20
0820	2" x 6"		640	.025		10.15	1.27		11.42	13.10
0830	5/4" x 6"		640	.025		6.30	1.27		7.57	8.85
0840	Railings and trim, 1" x 4"		600	.027		2.75	1.35		4.10	5.10
0860	2" x 4"		600	.027		5.60	1.35		6.95	8.20
0870	2" x 6"		600	.027		10.15	1.35		11.50	13.20
0920	Stair treads, 1" x 4"		140	.114		2.75	5.80		8.55	11.80
0930	2" x 4"		140	.114		5.60	5.80		11.40	14.95
0940	2" x 6"		160	.100		10.15	5.05		15.20	18.85
0950	5/4" x 6"		160	.100		6.30	5.05		11.35	14.60
0980	Redwood, posts or columns, 4" x 4"		390	.041		6.45	2.08		8.53	10.25
0990	4" x 6"		275	.058		12.60	2.95		15.55	18.35
1000	4" x 8"		220	.073		23.50	3.69		27.19	31.50
1240	Decking, 1" x 4"	1 Carp	275	.029	S.F.	3.97	1.47		5.44	6.60
1260	2" x 6"		340	.024		7.50	1.19		8.69	10.05
1270	5/4" x 6"		320	.025		4.77	1.27		6.04	7.20
1280	Railings and trim, 1" x 4"	2 Carp	600	.027	L.F.	1.17	1.35		2.52	3.35
1310	2" x 6"		600	.027		7.50	1.35		8.85	10.30
1420	Alternative decking, wood/plastic composite, 5/4" x 6" [G]		640	.025		3.27	1.27		4.54	5.55
1440	1" x 4" square edge fir		550	.029		2.77	1.47		4.24	5.30
1450	1" x 4" tongue and groove fir		450	.036		1.51	1.80		3.31	4.41
1460	1" x 4" mahogany		550	.029		2.07	1.47		3.54	4.52
1462	5/4" x 6" PVC		550	.029		3.36	1.47		4.83	5.95
1465	Framing, porch or deck, alt deck fastening, screws, add	1 Carp	240	.033	S.F.		1.69		1.69	2.57
1470	Accessories, joist hangers, 2" x 4"		160	.050	Ea.	.75	2.54		3.29	4.69
1480	2" x 6" through 2" x 12"		150	.053		1.27	2.70		3.97	5.50
1530	Post footing, incl excav, backfill, tube form & concrete, 4' deep, 8" diam.	F-7	12	2.667		17.55	121		138.55	203
1540	10" diameter		11	2.909		24	132		156	228
1550	12" diameter		10	3.200		30.50	145		175.50	255

06 11 10.30 Roof Framing

		Crew	Daily Output	Labor-Hours	Unit	Material	2018 Bare Costs Labor	Equipment	Total	Total Incl O&P
0010	**ROOF FRAMING**									
5250	Composite rafter, 9-1/2" deep	2 Carp	575	.028	L.F.	1.76	1.41		3.17	4.09
5260	11-1/2" deep		575	.028	"	2.05	1.41		3.46	4.40

06 11 Wood Framing

06 11 10 – Framing with Dimensional, Engineered or Composite Lumber

06 11 10.30 Roof Framing

		Crew	Daily Output	Labor-Hours	Unit	Material	2018 Bare Costs Labor	Equipment	Total	Total Incl O&P
6070	Fascia boards, 2" x 8"	2 Carp	.30	53.333	M.B.F.	670	2,700		3,370	4,850
6080	2" x 10"		.30	53.333		845	2,700		3,545	5,050
7000	Rafters, to 4 in 12 pitch, 2" x 6"		1	16		645	810		1,455	1,925
7060	2" x 8"		1.26	12.698		670	645		1,315	1,725
7300	Hip and valley rafters, 2" x 6"		.76	21.053		645	1,075		1,720	2,325
7360	2" x 8"		.96	16.667		670	845		1,515	2,000
7540	Hip and valley jacks, 2" x 6"		.60	26.667		645	1,350		1,995	2,750
7600	2" x 8"		.65	24.615		670	1,250		1,920	2,625
7780	For slopes steeper than 4 in 12, add						30%			
7790	For dormers or complex roofs, add						50%			
7800	Rafter tie, 1" x 4", #3	2 Carp	.27	59.259	M.B.F.	1,525	3,000		4,525	6,250
7820	Ridge board, #2 or better, 1" x 6"		.30	53.333		1,575	2,700		4,275	5,850
7840	1" x 8"		.37	43.243		1,950	2,200		4,150	5,500
7860	1" x 10"		.42	38.095		2,025	1,925		3,950	5,175
7880	2" x 6"		.50	32		645	1,625		2,270	3,175
7900	2" x 8"		.60	26.667		670	1,350		2,020	2,775
7920	2" x 10"		.66	24.242		845	1,225		2,070	2,800
7940	Roof cants, split, 4" x 4"		.86	18.605		1,375	945		2,320	2,925
7960	6" x 6"		1.80	8.889		1,775	450		2,225	2,650
7980	Roof curbs, untreated, 2" x 6"		.52	30.769		645	1,550		2,195	3,075
8000	2" x 12"		.80	20		895	1,025		1,920	2,525

06 11 10.32 Sill and Ledger Framing

		Crew	Daily Output	Labor-Hours	Unit	Material	2018 Bare Costs Labor	Equipment	Total	Total Incl O&P
0010	**SILL AND LEDGER FRAMING**									
0020	Extruded polystyrene sill sealer, 5-1/2" wide	1 Carp	1600	.005	L.F.	.16	.25		.41	.57
4482	Ledgers, nailed, 2" x 4"	2 Carp	.50	32	M.B.F.	630	1,625		2,255	3,175
4484	2" x 6"		.60	26.667		645	1,350		1,995	2,750
4486	Bolted, not including bolts, 3" x 8"		.65	24.615		1,425	1,250		2,675	3,475
4488	3" x 12"		.70	22.857		1,450	1,150		2,600	3,350
4490	Mud sills, redwood, construction grade, 2" x 4"		.59	27.119		3,400	1,375		4,775	5,825
4492	2" x 6"		.78	20.513		3,400	1,050		4,450	5,325
4500	Sills, 2" x 4"		.40	40		620	2,025		2,645	3,775
4520	2" x 6"		.55	29.091		635	1,475		2,110	2,950
4540	2" x 8"		.67	23.881		660	1,200		1,860	2,575
4600	Treated, 2" x 4"		.36	44.444		870	2,250		3,120	4,375
4620	2" x 6"		.50	32		725	1,625		2,350	3,275
4640	2" x 8"		.60	26.667		830	1,350		2,180	2,975
4700	4" x 4"		.60	26.667		870	1,350		2,220	3,000
4720	4" x 6"		.70	22.857		940	1,150		2,090	2,800
4740	4" x 8"		.80	20		1,450	1,025		2,475	3,150
4760	4" x 10"		.87	18.391		1,675	930		2,605	3,275

06 11 10.34 Sleepers

		Crew	Daily Output	Labor-Hours	Unit	Material	2018 Bare Costs Labor	Equipment	Total	Total Incl O&P
0010	**SLEEPERS**									
0300	On concrete, treated, 1" x 2"	2 Carp	.39	41.026	M.B.F.	1,775	2,075		3,850	5,150
0320	1" x 3"		.50	32		1,950	1,625		3,575	4,625
0340	2" x 4"		.99	16.162		1,025	820		1,845	2,375
0360	2" x 6"		1.30	12.308		870	625		1,495	1,900

06 11 10.36 Soffit and Canopy Framing

		Crew	Daily Output	Labor-Hours	Unit	Material	2018 Bare Costs Labor	Equipment	Total	Total Incl O&P
0010	**SOFFIT AND CANOPY FRAMING**									
1300	Canopy or soffit framing, 1" x 4"	2 Carp	.30	53.333	M.B.F.	1,525	2,700		4,225	5,800
1340	1" x 8"		.50	32		1,950	1,625		3,575	4,625
1360	2" x 4"		.41	39.024		630	1,975		2,605	3,725
1400	2" x 8"		.67	23.881		670	1,200		1,870	2,575

For customer support on your Building Construction Costs with RSMeans data, call 800.448.8182.

06 11 Wood Framing

06 11 10 – Framing with Dimensional, Engineered or Composite Lumber

06 11 10.36 Soffit and Canopy Framing		Crew	Daily Output	Labor-Hours	Unit	Material	2018 Bare Costs Labor	Equipment	Total	Total Incl O&P
1420	3" x 4"	2 Carp	.50	32	M.B.F.	1,200	1,625		2,825	3,800
1460	3" x 8"	↓	.60	26.667	↓	1,425	1,350		2,775	3,625

06 11 10.38 Treated Lumber Framing Material

	06 11 10.38 Treated Lumber Framing Material				Unit	Material			Total	Total Incl O&P
0010	**TREATED LUMBER FRAMING MATERIAL**									
0100	2" x 4"				M.B.F.	870			870	960
0110	2" x 6"					725			725	800
0120	2" x 8"					830			830	915
0130	2" x 10"					830			830	915
0140	2" x 12"					1,000			1,000	1,100
0200	4" x 4"					870			870	955
0210	4" x 6"					940			940	1,025
0220	4" x 8"				↓	1,450			1,450	1,600

06 11 10.40 Wall Framing

	06 11 10.40 Wall Framing		Crew	Daily Output	Labor-Hours	Unit	Material	Labor	Equipment	Total	Total Incl O&P
0010	**WALL FRAMING**	R061110-30									
5860	Headers over openings, 2" x 6"		2 Carp	.36	44.444	M.B.F.	645	2,250		2,895	4,125
5865	2" x 6", pneumatic nailed			.43	37.209		650	1,875		2,525	3,600
5880	2" x 8"			.45	35.556		670	1,800		2,470	3,475
5885	2" x 8", pneumatic nailed			.54	29.630		680	1,500		2,180	3,025
5900	2" x 10"			.53	30.189		845	1,525		2,370	3,250
5905	2" x 10", pneumatic nailed			.67	23.881		850	1,200		2,050	2,775
5920	2" x 12"			.60	26.667		895	1,350		2,245	3,025
5925	2" x 12", pneumatic nailed			.72	22.222		905	1,125		2,030	2,725
5940	4" x 12"			.76	21.053		1,375	1,075		2,450	3,125
5945	4" x 12", pneumatic nailed			.92	17.391		1,375	880		2,255	2,875
5960	6" x 12"			.84	19.048		1,450	965		2,415	3,075
5965	6" x 12", pneumatic nailed			1.01	15.873		1,450	805		2,255	2,825
6000	Plates, untreated, 2" x 3"			.43	37.209		845	1,875		2,720	3,800
6005	2" x 3", pneumatic nailed			.52	30.769		855	1,550		2,405	3,325
6020	2" x 4"			.53	30.189		630	1,525		2,155	3,025
6025	2" x 4", pneumatic nailed			.67	23.881		635	1,200		1,835	2,550
6040	2" x 6"			.75	21.333		645	1,075		1,720	2,350
6045	2" x 6", pneumatic nailed			.90	17.778		650	900		1,550	2,100
6120	Studs, 8' high wall, 2" x 3"			.60	26.667		845	1,350		2,195	2,975
6125	2" x 3", pneumatic nailed			.72	22.222		855	1,125		1,980	2,675
6140	2" x 4"			.92	17.391		630	880		1,510	2,050
6145	2" x 4", pneumatic nailed			1.10	14.493		635	735		1,370	1,825
6160	2" x 6"			1	16		645	810		1,455	1,925
6165	2" x 6", pneumatic nailed			1.20	13.333		650	675		1,325	1,750
6180	3" x 4"			.80	20		1,200	1,025		2,225	2,875
6185	3" x 4", pneumatic nailed		↓	.96	16.667	↓	1,200	845		2,045	2,600
8200	For 12' high walls, deduct							5%			
8220	For stub wall, 6' high, add							20%			
8240	3' high, add							40%			
8250	For second story & above, add							5%			
8300	For dormer & gable, add							15%			

06 11 10.42 Furring

	06 11 10.42 Furring	Crew	Daily Output	Labor-Hours	Unit	Material	Labor	Equipment	Total	Total Incl O&P
0010	**FURRING**									
0012	Wood strips, 1" x 2", on walls, on wood	1 Carp	550	.015	L.F.	.27	.74		1.01	1.41
0015	On wood, pneumatic nailed		710	.011		.27	.57		.84	1.16
0300	On masonry		495	.016		.29	.82		1.11	1.57
0400	On concrete		260	.031		.29	1.56		1.85	2.70
0600	1" x 3", on walls, on wood	↓	550	.015	↓	.43	.74		1.17	1.59

06 11 Wood Framing

06 11 10 – Framing with Dimensional, Engineered or Composite Lumber

06 11 10.42 Furring		Crew	Daily Output	Labor-Hours	Unit	Material	2018 Bare Costs Labor	Equipment	Total	Total Incl O&P
0605	On wood, pneumatic nailed	1 Carp	710	.011	L.F.	.43	.57		1	1.34
0700	On masonry		495	.016		.46	.82		1.28	1.76
0800	On concrete		260	.031		.46	1.56		2.02	2.89
0850	On ceilings, on wood		350	.023		.43	1.16		1.59	2.23
0855	On wood, pneumatic nailed		450	.018		.43	.90		1.33	1.84
0900	On masonry		320	.025		.46	1.27		1.73	2.44
0950	On concrete		210	.038		.46	1.93		2.39	3.45

06 11 10.44 Grounds

		Crew	Daily Output	Labor-Hours	Unit	Material	Labor	Equipment	Total	Total Incl O&P
0010	**GROUNDS**									
0020	For casework, 1" x 2" wood strips, on wood	1 Carp	330	.024	L.F.	.27	1.23		1.50	2.16
0100	On masonry		285	.028		.29	1.42		1.71	2.49
0200	On concrete		250	.032		.29	1.62		1.91	2.79
0400	For plaster, 3/4" deep, on wood		450	.018		.27	.90		1.17	1.66
0500	On masonry		225	.036		.29	1.80		2.09	3.07
0600	On concrete		175	.046		.29	2.32		2.61	3.85
0700	On metal lath		200	.040		.29	2.03		2.32	3.41

06 12 Structural Panels

06 12 10 – Structural Insulated Panels

06 12 10.10 OSB Faced Panels

			Crew	Daily Output	Labor-Hours	Unit	Material	Labor	Equipment	Total	Total Incl O&P
0010	**OSB FACED PANELS**										
0100	Structural insul. panels, 7/16" OSB both faces, EPS insul., 3-5/8" T	G	F-3	2075	.019	S.F.	3.65	1	.24	4.89	5.80
0110	5-5/8" thick	G		1725	.023		4.10	1.20	.29	5.59	6.65
0120	7-3/8" thick	G		1425	.028		4.45	1.45	.35	6.25	7.50
0130	9-3/8" thick	G		1125	.036		4.75	1.84	.44	7.03	8.55
0140	7/16" OSB one face, EPS insul., 3-5/8" thick	G		2175	.018		3.75	.95	.23	4.93	5.85
0150	5-5/8" thick	G		1825	.022		4.35	1.14	.27	5.76	6.80
0160	7-3/8" thick	G		1525	.026		4.85	1.36	.33	6.54	7.75
0170	9-3/8" thick	G		1225	.033		5.35	1.69	.40	7.44	8.90
0190	7/16" OSB - 1/2" GWB faces, EPS insul., 3-5/8" T	G		2075	.019		3.45	1	.24	4.69	5.60
0200	5-5/8" thick	G		1725	.023		4.10	1.20	.29	5.59	6.65
0210	7-3/8" thick	G		1425	.028		4.65	1.45	.35	6.45	7.70
0220	9-3/8" thick	G		1125	.036		5.25	1.84	.44	7.53	9.10
0240	7/16" OSB - 1/2" MRGWB faces, EPS insul., 3-5/8" T	G		2075	.019		3.55	1	.24	4.79	5.70
0250	5-5/8" thick	G		1725	.023		4.25	1.20	.29	5.74	6.80
0260	7-3/8" thick	G		1425	.028		4.65	1.45	.35	6.45	7.70
0270	9-3/8" thick	G		1125	.036		5.35	1.84	.44	7.63	9.20
0300	For 1/2" GWB added to OSB skin, add	G					1.40			1.40	1.54
0310	For 1/2" MRGWB added to OSB skin, add	G					1.40			1.40	1.54
0320	For one T1-11 skin, add to OSB-OSB	G					1.95			1.95	2.15
0330	For one 19/32" CDX skin, add to OSB-OSB	G					1.50			1.50	1.65
0500	Structural insulated panel, 7/16" OSB both sides, straw core										
0510	4-3/8" T, walls (w/sill, splines, plates)	G	F-6	2400	.017	S.F.	7.55	.79	.21	8.55	9.75
0520	Floors (w/splines)	G		2400	.017		7.55	.79	.21	8.55	9.75
0530	Roof (w/splines)	G		2400	.017		7.55	.79	.21	8.55	9.75
0550	7-7/8" T, walls (w/sill, splines, plates)	G		2400	.017		11.40	.79	.21	12.40	14
0560	Floors (w/splines)	G		2400	.017		11.40	.79	.21	12.40	14
0570	Roof (w/splines)	G		2400	.017		11.40	.79	.21	12.40	14

For customer support on your Building Construction Costs with RSMeans data, call 800.448.8182.

06 12 Structural Panels

06 12 19 – Composite Shearwall Panels

06 12 19.10 Steel and Wood Composite Shearwall Panels	Crew	Daily Output	Labor-Hours	Unit	Material	2018 Bare Costs Labor	Equipment	Total	Total Incl O&P
0010	**STEEL & WOOD COMPOSITE SHEARWALL PANELS**									
0020	Anchor bolts, 36" long (must be placed in wet concrete)	1 Carp	150	.053	Ea.	37	2.70		39.70	44.50
0030	On concrete, 2" x 4" & 2" x 6" walls, 7'-10' high, 360 lb. shear, 12" wide	2 Carp	8	2		460	101		561	660
0040	715 lb. shear, 15" wide		8	2		510	101		611	715
0050	1860 lb. shear, 18" wide		8	2		525	101		626	735
0060	2780 lb. shear, 21" wide		8	2		560	101		661	770
0070	3790 lb. shear, 24" wide		8	2		640	101		741	855
0080	2" x 6" walls, 11'-13' high, 1180 lb. shear, 18" wide		6	2.667		645	135		780	915
0090	1555 lb. shear, 21" wide		6	2.667		720	135		855	995
0100	2280 lb. shear, 24" wide		6	2.667		805	135		940	1,100
0110	For installing above on wood floor frame, add									
0120	Coupler nuts, threaded rods, bolts, shear transfer plate kit	1 Carp	16	.500	Ea.	65	25.50		90.50	110
0130	Framing anchors, angle (2 required)	"	96	.083	"	2.43	4.22		6.65	9.10
0140	For blocking see Section 06 11 10.02									
0150	For installing above, first floor to second floor, wood floor frame, add									
0160	Add stack option to first floor wall panel				Ea.	71.50			71.50	78.50
0170	Threaded rods, bolts, shear transfer plate kit	1 Carp	16	.500		75	25.50		100.50	121
0180	Framing anchors, angle (2 required)	"	96	.083		2.43	4.22		6.65	9.10
0190	For blocking see section 06 11 10.02									
0200	For installing stacked panels, balloon framing									
0210	Add stack option to first floor wall panel				Ea.	71.50			71.50	78.50
0220	Threaded rods, bolts kit	1 Carp	16	.500	"	45	25.50		70.50	88

06 13 Heavy Timber Construction

06 13 23 – Heavy Timber Framing

06 13 23.10 Heavy Framing

		Crew	Daily Output	Labor-Hours	Unit	Material	2018 Bare Costs Labor	Equipment	Total	Total Incl O&P
0010	**HEAVY FRAMING**									
0020	Beams, single 6" x 10"	2 Carp	1.10	14.545	M.B.F.	1,550	735		2,285	2,850
0100	Single 8" x 16"		1.20	13.333		1,925	675		2,600	3,150
0200	Built from 2" lumber, multiple 2" x 14"		.90	17.778		960	900		1,860	2,425
0210	Built from 3" lumber, multiple 3" x 6"		.70	22.857		1,325	1,150		2,475	3,225
0220	Multiple 3" x 8"		.80	20		1,425	1,025		2,450	3,125
0230	Multiple 3" x 10"		.90	17.778		1,450	900		2,350	2,950
0240	Multiple 3" x 12"		1	16		1,450	810		2,260	2,800
0250	Built from 4" lumber, multiple 4" x 6"		.80	20		1,525	1,025		2,550	3,225
0260	Multiple 4" x 8"		.90	17.778		1,550	900		2,450	3,075
0270	Multiple 4" x 10"		1	16		1,475	810		2,285	2,850
0280	Multiple 4" x 12"		1.10	14.545		1,375	735		2,110	2,625
0290	Columns, structural grade, 1500f, 4" x 4"		.60	26.667		1,325	1,350		2,675	3,525
0300	6" x 6"		.65	24.615		1,400	1,250		2,650	3,425
0400	8" x 8"		.70	22.857		1,500	1,150		2,650	3,425
0500	10" x 10"		.75	21.333		1,625	1,075		2,700	3,425
0600	12" x 12"		.80	20		1,550	1,025		2,575	3,250
0800	Floor planks, 2" thick, T&G, 2" x 6"		1.05	15.238		1,600	775		2,375	2,950
0900	2" x 10"		1.10	14.545		1,625	735		2,360	2,925
1100	3" thick, 3" x 6"		1.05	15.238		1,625	775		2,400	2,950
1200	3" x 10"		1.10	14.545		1,650	735		2,385	2,950
1400	Girders, structural grade, 12" x 12"		.80	20		1,550	1,025		2,575	3,250
1500	10" x 16"		1	16		2,550	810		3,360	4,025
2300	Roof purlins, 4" thick, structural grade		1.05	15.238		1,550	775		2,325	2,875

06 15 Wood Decking

06 15 16 – Wood Roof Decking

06 15 16.10 Solid Wood Roof Decking

		Crew	Daily Output	Labor-Hours	Unit	Material	2018 Bare Costs Labor	Equipment	Total	Total Incl O&P
0010	**SOLID WOOD ROOF DECKING**									
0350	Cedar planks, 2" thick	2 Carp	350	.046	S.F.	6.65	2.32		8.97	10.90
0400	3" thick		320	.050		10	2.54		12.54	14.85
0500	4" thick		250	.064		13.35	3.24		16.59	19.60
0550	6" thick		200	.080		20	4.06		24.06	28
0650	Douglas fir, 2" thick		350	.046		2.78	2.32		5.10	6.60
0700	3" thick		320	.050		4.17	2.54		6.71	8.45
0800	4" thick		250	.064		5.55	3.24		8.79	11.05
0850	6" thick		200	.080		8.35	4.06		12.41	15.35
0950	Hemlock, 2" thick		350	.046		2.83	2.32		5.15	6.65
1000	3" thick		320	.050		4.25	2.54		6.79	8.55
1100	4" thick		250	.064		5.65	3.24		8.89	11.20
1150	6" thick		200	.080		8.50	4.06		12.56	15.55
1250	Western white spruce, 2" thick		350	.046		1.81	2.32		4.13	5.50
1300	3" thick		320	.050		2.71	2.54		5.25	6.85
1400	4" thick		250	.064		3.61	3.24		6.85	8.90
1450	6" thick		200	.080		5.40	4.06		9.46	12.15

06 15 23 – Laminated Wood Decking

06 15 23.10 Laminated Roof Deck

		Crew	Daily Output	Labor-Hours	Unit	Material	2018 Bare Costs Labor	Equipment	Total	Total Incl O&P
0010	**LAMINATED ROOF DECK**									
0020	Pine or hemlock, 3" thick	2 Carp	425	.038	S.F.	5.75	1.91		7.66	9.25
0100	4" thick		325	.049		7.50	2.50		10	12.05
0300	Cedar, 3" thick		425	.038		7.05	1.91		8.96	10.65
0400	4" thick		325	.049		9.45	2.50		11.95	14.15
0600	Fir, 3" thick		425	.038		6	1.91		7.91	9.50
0700	4" thick		325	.049		7.55	2.50		10.05	12.10

06 16 Sheathing

06 16 13 – Insulating Sheathing

06 16 13.10 Insulating Sheathing

			Crew	Daily Output	Labor-Hours	Unit	Material	2018 Bare Costs Labor	Equipment	Total	Total Incl O&P
0010	**INSULATING SHEATHING**										
0020	Expanded polystyrene, 1#/C.F. density, 3/4" thick, R2.89	G	2 Carp	1400	.011	S.F.	.36	.58		.94	1.28
0030	1" thick, R3.85	G		1300	.012		.43	.62		1.05	1.42
0040	2" thick, R7.69	G		1200	.013		.70	.68		1.38	1.80
0050	Extruded polystyrene, 15 psi compressive strength, 1" thick, R5	G		1300	.012		.71	.62		1.33	1.73
0060	2" thick, R10	G		1200	.013		.87	.68		1.55	1.99
0070	Polyisocyanurate, 2#/C.F. density, 3/4" thick	G		1400	.011		.58	.58		1.16	1.52
0080	1" thick	G		1300	.012		.59	.62		1.21	1.60
0090	1-1/2" thick	G		1250	.013		.74	.65		1.39	1.80
0100	2" thick	G		1200	.013		.90	.68		1.58	2.02

06 16 23 – Subflooring

06 16 23.10 Subfloor

			Crew	Daily Output	Labor-Hours	Unit	Material	2018 Bare Costs Labor	Equipment	Total	Total Incl O&P
0010	**SUBFLOOR**	R061636-20									
0011	Plywood, CDX, 1/2" thick		2 Carp	1500	.011	SF Flr.	.62	.54		1.16	1.50
0015	Pneumatic nailed			1860	.009		.62	.44		1.06	1.34
0100	5/8" thick			1350	.012		.77	.60		1.37	1.75
0105	Pneumatic nailed			1674	.010		.77	.48		1.25	1.58
0200	3/4" thick			1250	.013		.92	.65		1.57	2.01
0205	Pneumatic nailed			1550	.010		.92	.52		1.44	1.82
0300	1-1/8" thick, 2-4-1 including underlayment			1050	.015		2.07	.77		2.84	3.46

For customer support on your Building Construction Costs with RSMeans data, call 800.448.8182.

06 16 Sheathing

06 16 23 – Subflooring

06 16 23.10 Subfloor

		Crew	Daily Output	Labor-Hours	Unit	Material	2018 Bare Costs Labor	Equipment	Total	Total Incl O&P
0440	With boards, 1" x 6", S4S, laid regular	2 Carp	900	.018	SF Flr.	1.70	.90		2.60	3.23
0450	1" x 8", laid regular		1000	.016		2.07	.81		2.88	3.52
0460	Laid diagonal		850	.019		2.07	.95		3.02	3.73
0500	1" x 10", laid regular		1100	.015		2.12	.74		2.86	3.45
0600	Laid diagonal	↓	900	.018	↓	2.12	.90		3.02	3.70
8990	Subfloor adhesive, 3/8" bead	1 Carp	2300	.003	L.F.	.11	.18		.29	.39

06 16 26 – Underlayment

06 16 26.10 Wood Product Underlayment

		Crew	Daily Output	Labor-Hours	Unit	Material	2018 Bare Costs Labor	Equipment	Total	Total Incl O&P
0010	**WOOD PRODUCT UNDERLAYMENT** R061636-20									
0015	Plywood, underlayment grade, 1/4" thick	2 Carp	1500	.011	S.F.	.94	.54		1.48	1.85
0018	Pneumatic nailed		1860	.009		.94	.44		1.38	1.69
0030	3/8" thick		1500	.011		1.04	.54		1.58	1.96
0070	Pneumatic nailed		1860	.009		1.04	.44		1.48	1.80
0100	1/2" thick		1450	.011		1.23	.56		1.79	2.20
0105	Pneumatic nailed		1798	.009		1.23	.45		1.68	2.04
0200	5/8" thick		1400	.011		1.36	.58		1.94	2.38
0205	Pneumatic nailed		1736	.009		1.36	.47		1.83	2.21
0300	3/4" thick		1300	.012		1.47	.62		2.09	2.57
0305	Pneumatic nailed		1612	.010		1.47	.50		1.97	2.39
0500	Particle board, 3/8" thick G		1500	.011		.41	.54		.95	1.27
0505	Pneumatic nailed G		1860	.009		.41	.44		.85	1.11
0600	1/2" thick G		1450	.011		.43	.56		.99	1.32
0605	Pneumatic nailed G		1798	.009		.43	.45		.88	1.16
0800	5/8" thick G		1400	.011		.55	.58		1.13	1.49
0805	Pneumatic nailed G		1736	.009		.55	.47		1.02	1.32
0900	3/4" thick G		1300	.012		.67	.62		1.29	1.69
0905	Pneumatic nailed G		1612	.010		.67	.50		1.17	1.51
1100	Hardboard, underlayment grade, 4' x 4', .215" thick G	↓	1500	.011	↓	.68	.54		1.22	1.57

06 16 33 – Wood Board Sheathing

06 16 33.10 Board Sheathing

		Crew	Daily Output	Labor-Hours	Unit	Material	2018 Bare Costs Labor	Equipment	Total	Total Incl O&P
0009	**BOARD SHEATHING**									
0010	Roof, 1" x 6" boards, laid horizontal	2 Carp	725	.022	S.F.	1.70	1.12		2.82	3.56
0020	On steep roof		520	.031		1.70	1.56		3.26	4.24
0040	On dormers, hips, & valleys		480	.033		1.70	1.69		3.39	4.43
0050	Laid diagonal		650	.025		1.70	1.25		2.95	3.76
0070	1" x 8" boards, laid horizontal		875	.018		2.07	.93		3	3.69
0080	On steep roof		635	.025		2.07	1.28		3.35	4.23
0090	On dormers, hips, & valleys		580	.028		2.12	1.40		3.52	4.46
0100	Laid diagonal	↓	725	.022		2.07	1.12		3.19	3.98
0110	Skip sheathing, 1" x 4", 7" OC	1 Carp	1200	.007		.63	.34		.97	1.21
0120	1" x 6", 9" OC		1450	.006		.78	.28		1.06	1.29
0180	T&G sheathing/decking, 1" x 6"		1000	.008		1.80	.41		2.21	2.60
0190	2" x 6"	↓	1000	.008		3.86	.41		4.27	4.87
0200	Walls, 1" x 6" boards, laid regular	2 Carp	650	.025		1.70	1.25		2.95	3.76
0210	Laid diagonal		585	.027		1.70	1.39		3.09	3.97
0220	1" x 8" boards, laid regular		765	.021		2.07	1.06		3.13	3.90
0230	Laid diagonal	↓	650	.025	↓	2.07	1.25		3.32	4.18

For customer support on your Building Construction Costs with RSMeans data, call 800.448.8182.

191

06 16 36.10 Sheathing		Crew	Daily Output	Labor-Hours	Unit	Material	2018 Bare Costs Labor	Equipment	Total	Total Incl O&P
0010	**SHEATHING** R061636-20									
0012	Plywood on roofs, CDX									
0030	5/16" thick	2 Carp	1600	.010	S.F.	.59	.51		1.10	1.41
0035	Pneumatic nailed R061110-30		1952	.008		.59	.42		1.01	1.27
0050	3/8" thick		1525	.010		.60	.53		1.13	1.47
0055	Pneumatic nailed		1860	.009		.60	.44		1.04	1.32
0100	1/2" thick		1400	.011		.62	.58		1.20	1.56
0105	Pneumatic nailed		1708	.009		.62	.48		1.10	1.40
0200	5/8" thick		1300	.012		.77	.62		1.39	1.79
0205	Pneumatic nailed		1586	.010		.77	.51		1.28	1.62
0300	3/4" thick		1200	.013		.92	.68		1.60	2.05
0305	Pneumatic nailed		1464	.011		.92	.55		1.47	1.86
0500	Plywood on walls, with exterior CDX, 3/8" thick		1200	.013		.60	.68		1.28	1.69
0505	Pneumatic nailed		1488	.011		.60	.55		1.15	1.49
0600	1/2" thick		1125	.014		.62	.72		1.34	1.78
0605	Pneumatic nailed		1395	.011		.62	.58		1.20	1.57
0700	5/8" thick		1050	.015		.77	.77		1.54	2.02
0705	Pneumatic nailed		1302	.012		.77	.62		1.39	1.79
0800	3/4" thick		975	.016		.92	.83		1.75	2.29
0805	Pneumatic nailed		1209	.013		.92	.67		1.59	2.04
1000	For shear wall construction, add						20%			
1200	For structural 1 exterior plywood, add				S.F.	10%				
3000	Wood fiber, regular, no vapor barrier, 1/2" thick	2 Carp	1200	.013		.64	.68		1.32	1.73
3100	5/8" thick		1200	.013		.70	.68		1.38	1.80
3300	No vapor barrier, in colors, 1/2" thick		1200	.013		.81	.68		1.49	1.92
3400	5/8" thick		1200	.013		.85	.68		1.53	1.97
3600	With vapor barrier one side, white, 1/2" thick		1200	.013		.63	.68		1.31	1.72
3700	Vapor barrier 2 sides, 1/2" thick		1200	.013		.84	.68		1.52	1.95
3800	Asphalt impregnated, 25/32" thick		1200	.013		.32	.68		1	1.38
3850	Intermediate, 1/2" thick		1200	.013		.32	.68		1	1.38
4500	Oriented strand board, on roof, 7/16" thick G		1460	.011		.48	.56		1.04	1.38
4505	Pneumatic nailed G		1780	.009		.48	.46		.94	1.22
4550	1/2" thick G		1400	.011		.48	.58		1.06	1.41
4555	Pneumatic nailed G		1736	.009		.48	.47		.95	1.24
4600	5/8" thick G		1300	.012		.54	.62		1.16	1.54
4605	Pneumatic nailed G		1586	.010		.54	.51		1.05	1.37
4610	On walls, 7/16" thick G		1200	.013		.48	.68		1.16	1.56
4615	Pneumatic nailed G		1488	.011		.48	.55		1.03	1.36
4620	1/2" thick G		1195	.013		.48	.68		1.16	1.56
4625	Pneumatic nailed G		1325	.012		.48	.61		1.09	1.46
4630	5/8" thick G		1050	.015		.54	.77		1.31	1.77
4635	Pneumatic nailed G		1302	.012		.54	.62		1.16	1.54
4700	Oriented strand board, factory laminated W.R. barrier, on roof, 1/2" thick G		1400	.011		.75	.58		1.33	1.71
4705	Pneumatic nailed G		1736	.009		.75	.47		1.22	1.54
4720	5/8" thick G		1300	.012		.91	.62		1.53	1.95
4725	Pneumatic nailed G		1586	.010		.91	.51		1.42	1.78
4730	5/8" thick, T&G G		1150	.014		1.09	.71		1.80	2.27
4735	Pneumatic nailed, T&G G		1400	.011		1.09	.58		1.67	2.08
4740	On walls, 7/16" thick G		1200	.013		.69	.68		1.37	1.79
4745	Pneumatic nailed G		1488	.011		.69	.55		1.24	1.59
4750	1/2" thick G		1195	.013		.75	.68		1.43	1.86
4755	Pneumatic nailed G		1325	.012		.75	.61		1.36	1.76

192

06 16 Sheathing

06 16 36 – Wood Panel Product Sheathing

06 16 36.10 Sheathing	Crew	Daily Output	Labor-Hours	Unit	Material	2018 Bare Costs Labor	2018 Bare Costs Equipment	Total	Total Incl O&P
4800	Joint sealant tape, 3-1/2"	2 Carp	7600	.002	L.F.	.28	.11		.39	.47
4810	Joint sealant tape, 6"	↓	7600	.002	"	.41	.11		.52	.61

06 16 43 – Gypsum Sheathing

06 16 43.10 Gypsum Sheathing	Crew	Daily Output	Labor-Hours	Unit	Material	2018 Bare Costs Labor	2018 Bare Costs Equipment	Total	Total Incl O&P
0010	**GYPSUM SHEATHING**									
0020	Gypsum, weatherproof, 1/2" thick	2 Carp	1125	.014	S.F.	.46	.72		1.18	1.61
0040	With embedded glass mats	"	1100	.015	"	.71	.74		1.45	1.90

06 17 Shop-Fabricated Structural Wood

06 17 33 – Wood I-Joists

06 17 33.10 Wood and Composite I-Joists	Crew	Daily Output	Labor-Hours	Unit	Material	2018 Bare Costs Labor	2018 Bare Costs Equipment	Total	Total Incl O&P
0010	**WOOD AND COMPOSITE I-JOISTS**									
0100	Plywood webs, incl. bridging & blocking, panels 24" OC									
1200	15' to 24' span, 50 psf live load	F-5	2400	.013	SF Flr.	1.99	.68		2.67	3.23
1300	55 psf live load		2250	.014		2.31	.73		3.04	3.65
1400	24' to 30' span, 45 psf live load		2600	.012		2.90	.63		3.53	4.15
1500	55 psf live load	↓	2400	.013	↓	4.62	.68		5.30	6.15

06 17 53 – Shop-Fabricated Wood Trusses

06 17 53.10 Roof Trusses	Crew	Daily Output	Labor-Hours	Unit	Material	2018 Bare Costs Labor	2018 Bare Costs Equipment	Total	Total Incl O&P
0010	**ROOF TRUSSES**									
0100	Fink (W) or King post type, 2'-0" OC									
0200	Metal plate connected, 4 in 12 slope									
0210	24' to 29' span	F-3	3000	.013	SF Flr.	1.68	.69	.17	2.54	3.08
0300	30' to 43' span		3000	.013		2.39	.69	.17	3.25	3.85
0400	44' to 60' span	↓	3000	.013	↓	2.39	.69	.17	3.25	3.86
0700	Glued and nailed, add					50%				

06 18 Glued-Laminated Construction

06 18 13 – Glued-Laminated Beams

06 18 13.20 Laminated Framing	Crew	Daily Output	Labor-Hours	Unit	Material	2018 Bare Costs Labor	2018 Bare Costs Equipment	Total	Total Incl O&P
0010	**LAMINATED FRAMING**									
0020	30 lb., short term live load, 15 lb. dead load									
0200	Straight roof beams, 20' clear span, beams 8' OC	F-3	2560	.016	SF Flr.	2.26	.81	.19	3.26	3.93
0300	Beams 16' OC		3200	.013		1.65	.65	.16	2.46	2.97
0500	40' clear span, beams 8' OC		3200	.013		4.29	.65	.16	5.10	5.85
0600	Beams 16' OC	↓	3840	.010		3.54	.54	.13	4.21	4.85
0800	60' clear span, beams 8' OC	F-4	2880	.017		7.35	.85	.35	8.55	9.80
0900	Beams 16' OC	"	3840	.013		5.50	.64	.26	6.40	7.30
1100	Tudor arches, 30' to 40' clear span, frames 8' OC	F-3	1680	.024		9.60	1.23	.30	11.13	12.75
1200	Frames 16' OC	"	2240	.018		7.50	.92	.22	8.64	9.90
1400	50' to 60' clear span, frames 8' OC	F-4	2200	.022		10.30	1.12	.45	11.87	13.55
1500	Frames 16' OC		2640	.018		8.80	.93	.38	10.11	11.50
1700	Radial arches, 60' clear span, frames 8' OC		1920	.025		9.65	1.28	.52	11.45	13.15
1800	Frames 16' OC		2880	.017		7.65	.85	.35	8.85	10.10
2000	100' clear span, frames 8' OC		1600	.030		9.95	1.54	.62	12.11	13.95
2100	Frames 16' OC		2400	.020		8.80	1.03	.41	10.24	11.65
2300	120' clear span, frames 8' OC		1440	.033		13.25	1.71	.69	15.65	17.90
2400	Frames 16' OC	↓	1920	.025		12.10	1.28	.52	13.90	15.80

06 18 Glued-Laminated Construction

06 18 13 – Glued-Laminated Beams

06 18 13.20 Laminated Framing	Crew	Daily Output	Labor-Hours	Unit	Material	2018 Bare Costs Labor	Equipment	Total	Total Incl O&P
2600 Bowstring trusses, 20' OC, 40' clear span	F-3	2400	.017	SF Flr.	6	.86	.21	7.07	8.15
2700 60' clear span	F-4	3600	.013		5.40	.68	.28	6.36	7.30
2800 100' clear span		4000	.012		7.65	.62	.25	8.52	9.60
2900 120' clear span		3600	.013		8.10	.68	.28	9.06	10.25
3000 For less than 1000 B.F., add					20%				
3050 For over 5000 B.F., deduct					10%				
3100 For premium appearance, add to S.F. prices					5%				
3300 For industrial type, deduct					15%				
3500 For stain and varnish, add					5%				
3900 For 3/4" laminations, add to straight					25%				
4100 Add to curved					15%				
4300 Alternate pricing method: (use nominal footage of									
4310 components). Straight beams, camber less than 6"	F-3	3.50	11.429	M.B.F.	3,175	590	142	3,907	4,550
4400 Columns, including hardware		2	20		3,425	1,025	248	4,698	5,600
4600 Curved members, radius over 32'		2.50	16		3,500	830	198	4,528	5,325
4700 Radius 10' to 32'		3	13.333		3,475	690	165	4,330	5,025
4900 For complicated shapes, add maximum					100%				
5100 For pressure treating, add to straight					35%				
5200 Add to curved					45%				
6000 Laminated veneer members, southern pine or western species									
6050 1-3/4" wide x 5-1/2" deep	2 Carp	480	.033	L.F.	3.47	1.69		5.16	6.40
6100 9-1/2" deep		480	.033		4.25	1.69		5.94	7.25
6150 14" deep		450	.036		7.35	1.80		9.15	10.85
6200 18" deep		450	.036		10.30	1.80		12.10	14.05
6300 Parallel strand members, southern pine or western species									
6350 1-3/4" wide x 9-1/4" deep	2 Carp	480	.033	L.F.	4.77	1.69		6.46	7.80
6400 11-1/4" deep		450	.036		4.94	1.80		6.74	8.20
6450 14" deep		400	.040		7.60	2.03		9.63	11.45
6500 3-1/2" wide x 9-1/4" deep		480	.033		16.10	1.69		17.79	20.50
6550 11-1/4" deep		450	.036		19.90	1.80		21.70	25
6600 14" deep		400	.040		22.50	2.03		24.53	27.50
6650 7" wide x 9-1/4" deep		450	.036		33.50	1.80		35.30	40
6700 11-1/4" deep		420	.038		42	1.93		43.93	49.50
6750 14" deep		400	.040		50	2.03		52.03	58
8000 Straight beams									
8102 20' span									
8104 3-1/8" x 9"	F-3	30	1.333	Ea.	149	69	16.55	234.55	287
8106 x 10-1/2"		30	1.333		174	69	16.55	259.55	315
8108 x 12"		30	1.333		199	69	16.55	284.55	340
8110 x 13-1/2"		30	1.333		224	69	16.55	309.55	370
8112 x 15"		29	1.379		249	71.50	17.10	337.60	400
8114 5-1/8" x 10-1/2"		30	1.333		285	69	16.55	370.55	440
8116 x 12"		30	1.333		325	69	16.55	410.55	485
8118 x 13-1/2"		30	1.333		365	69	16.55	450.55	530
8120 x 15"		29	1.379		410	71.50	17.10	498.60	580
8122 x 16-1/2"		29	1.379		450	71.50	17.10	538.60	625
8124 x 18"		29	1.379		490	71.50	17.10	578.60	670
8126 x 19-1/2"		29	1.379		530	71.50	17.10	618.60	715
8128 x 21"		28	1.429		570	74	17.70	661.70	760
8130 x 22-1/2"		28	1.429		610	74	17.70	701.70	800
8132 x 24"		28	1.429		650	74	17.70	741.70	845
8134 6-3/4" x 12"		29	1.379		430	71.50	17.10	518.60	600
8136 x 13-1/2"		29	1.379		485	71.50	17.10	573.60	660

06 18 13 – Glued-Laminated Beams

06 18 13.20 Laminated Framing		Crew	Daily Output	Labor-Hours	Unit	Material	2018 Bare Costs Labor	Equipment	Total	Total Incl O&P
8138	x 15"	F-3	29	1.379	Ea.	535	71.50	17.10	623.60	720
8140	x 16-1/2"		28	1.429		590	74	17.70	681.70	780
8142	x 18"		28	1.429		645	74	17.70	736.70	840
8144	x 19-1/2"		28	1.429		700	74	17.70	791.70	900
8146	x 21"		27	1.481		750	76.50	18.35	844.85	960
8148	x 22-1/2"		27	1.481		805	76.50	18.35	899.85	1,025
8150	x 24"		27	1.481		860	76.50	18.35	954.85	1,075
8152	x 25-1/2"		27	1.481		915	76.50	18.35	1,009.85	1,125
8154	x 27"		26	1.538		965	79.50	19.10	1,063.60	1,225
8156	x 28-1/2"		26	1.538		1,025	79.50	19.10	1,123.60	1,275
8158	x 30"	↓	26	1.538	↓	1,075	79.50	19.10	1,173.60	1,325
8200	30' span									
8250	3-1/8" x 9"	F-3	30	1.333	Ea.	224	69	16.55	309.55	370
8252	x 10-1/2"		30	1.333		261	69	16.55	346.55	410
8254	x 12"		30	1.333		298	69	16.55	383.55	455
8256	x 13-1/2"		30	1.333		335	69	16.55	420.55	495
8258	x 15"		29	1.379		375	71.50	17.10	463.60	540
8260	5-1/8" x 10-1/2"		30	1.333		430	69	16.55	515.55	595
8262	x 12"		30	1.333		490	69	16.55	575.55	665
8264	x 13-1/2"		30	1.333		550	69	16.55	635.55	730
8266	x 15"		29	1.379		610	71.50	17.10	698.60	800
8268	x 16-1/2"		29	1.379		670	71.50	17.10	758.60	870
8270	x 18"		29	1.379		735	71.50	17.10	823.60	935
8272	x 19-1/2"		29	1.379		795	71.50	17.10	883.60	1,000
8274	x 21"		28	1.429		855	74	17.70	946.70	1,075
8276	x 22-1/2"		28	1.429		915	74	17.70	1,006.70	1,125
8278	x 24"		28	1.429		980	74	17.70	1,071.70	1,200
8280	6-3/4" x 12"		29	1.379		645	71.50	17.10	733.60	840
8282	x 13-1/2"		29	1.379		725	71.50	17.10	813.60	925
8284	x 15"		29	1.379		805	71.50	17.10	893.60	1,025
8286	x 16-1/2"		28	1.429		885	74	17.70	976.70	1,100
8288	x 18"		28	1.429		965	74	17.70	1,056.70	1,200
8290	x 19-1/2"		28	1.429		1,050	74	17.70	1,141.70	1,275
8292	x 21"		27	1.481		1,125	76.50	18.35	1,219.85	1,375
8294	x 22-1/2"		27	1.481		1,200	76.50	18.35	1,294.85	1,450
8296	x 24"		27	1.481		1,300	76.50	18.35	1,394.85	1,550
8298	x 25-1/2"		27	1.481		1,375	76.50	18.35	1,469.85	1,625
8300	x 27"		26	1.538		1,450	79.50	19.10	1,548.60	1,750
8302	x 28-1/2"		26	1.538		1,525	79.50	19.10	1,623.60	1,825
8304	x 30"	↓	26	1.538	↓	1,600	79.50	19.10	1,698.60	1,925
8400	40' span									
8402	3-1/8" x 9"	F-3	30	1.333	Ea.	298	69	16.55	383.55	455
8404	x 10-1/2"		30	1.333		350	69	16.55	435.55	510
8406	x 12"		30	1.333		400	69	16.55	485.55	560
8408	x 13-1/2"		30	1.333		445	69	16.55	530.55	615
8410	x 15"		29	1.379		495	71.50	17.10	583.60	675
8412	5-1/8" x 10-1/2"		30	1.333		570	69	16.55	655.55	755
8414	x 12"		30	1.333		650	69	16.55	735.55	840
8416	x 13-1/2"		30	1.333		735	69	16.55	820.55	930
8418	x 15"		29	1.379		815	71.50	17.10	903.60	1,025
8420	x 16-1/2"		29	1.379		895	71.50	17.10	983.60	1,125
8422	x 18"		29	1.379		980	71.50	17.10	1,068.60	1,200
8424	x 19-1/2"		29	1.379		1,050	71.50	17.10	1,138.60	1,300

06 18 Glued-Laminated Construction

06 18 13 – Glued-Laminated Beams

06 18 13.20 Laminated Framing		Crew	Daily Output	Labor-Hours	Unit	Material	2018 Bare Costs Labor	Equipment	Total	Total Incl O&P
8426	x 21"	F-3	28	1.429	Ea.	1,150	74	17.70	1,241.70	1,375
8428	x 22-1/2"		28	1.429		1,225	74	17.70	1,316.70	1,475
8430	x 24"		28	1.429		1,300	74	17.70	1,391.70	1,550
8432	6-3/4" x 12"		29	1.379		860	71.50	17.10	948.60	1,075
8434	x 13-1/2"		29	1.379		965	71.50	17.10	1,053.60	1,200
8436	x 15"		29	1.379		1,075	71.50	17.10	1,163.60	1,300
8438	x 16-1/2"		28	1.429		1,175	74	17.70	1,266.70	1,425
8440	x 18"		28	1.429		1,300	74	17.70	1,391.70	1,550
8442	x 19-1/2"		28	1.429		1,400	74	17.70	1,491.70	1,650
8444	x 21"		27	1.481		1,500	76.50	18.35	1,594.85	1,775
8446	x 22-1/2"		27	1.481		1,600	76.50	18.35	1,694.85	1,900
8448	x 24"		27	1.481		1,725	76.50	18.35	1,819.85	2,025
8450	x 25-1/2"		27	1.481		1,825	76.50	18.35	1,919.85	2,125
8452	x 27"		26	1.538		1,925	79.50	19.10	2,023.60	2,275
8454	x 28-1/2"		26	1.538		2,050	79.50	19.10	2,148.60	2,400
8456	x 30"		26	1.538		2,150	79.50	19.10	2,248.60	2,500

06 22 Millwork

06 22 13 – Standard Pattern Wood Trim

06 22 13.15 Moldings, Base

		Crew	Daily Output	Labor-Hours	Unit	Material	2018 Bare Costs Labor	Equipment	Total	Total Incl O&P
0010	**MOLDINGS, BASE**									
5100	Classic profile, 5/8" x 5-1/2", finger jointed and primed	1 Carp	250	.032	L.F.	1.59	1.62		3.21	4.22
5105	Poplar		240	.033		1.62	1.69		3.31	4.35
5110	Red oak		220	.036		2.49	1.84		4.33	5.55
5115	Maple		220	.036		3.75	1.84		5.59	6.95
5120	Cherry		220	.036		4.45	1.84		6.29	7.70
5125	3/4" x 7-1/2", finger jointed and primed		250	.032		2	1.62		3.62	4.67
5130	Poplar		240	.033		2.59	1.69		4.28	5.40
5135	Red oak		220	.036		3.61	1.84		5.45	6.80
5140	Maple		220	.036		4.99	1.84		6.83	8.30
5145	Cherry		220	.036		5.95	1.84		7.79	9.35
5150	Modern profile, 5/8" x 3-1/2", finger jointed and primed		250	.032		.96	1.62		2.58	3.52
5155	Poplar		240	.033		1.03	1.69		2.72	3.70
5160	Red oak		220	.036		1.70	1.84		3.54	4.68
5165	Maple		220	.036		2.63	1.84		4.47	5.70
5170	Cherry		220	.036		2.84	1.84		4.68	5.95
5175	Ogee profile, 7/16" x 3", finger jointed and primed		250	.032		.66	1.62		2.28	3.19
5180	Poplar		240	.033		.67	1.69		2.36	3.31
5185	Red oak		220	.036		.95	1.84		2.79	3.85
5200	9/16" x 3-1/2", finger jointed and primed		250	.032		.66	1.62		2.28	3.19
5205	Pine		240	.033		1.17	1.69		2.86	3.86
5210	Red oak		220	.036		2.80	1.84		4.64	5.90
5215	9/16" x 4-1/2", red oak		220	.036		4.19	1.84		6.03	7.40
5220	5/8" x 3-1/2", finger jointed and primed		250	.032		.98	1.62		2.60	3.55
5225	Poplar		240	.033		1.03	1.69		2.72	3.70
5230	Red oak		220	.036		1.70	1.84		3.54	4.68
5235	Maple		220	.036		2.63	1.84		4.47	5.70
5240	Cherry		220	.036		2.84	1.84		4.68	5.95
5245	5/8" x 4", finger jointed and primed		250	.032		1.25	1.62		2.87	3.84
5250	Poplar		240	.033		1.33	1.69		3.02	4.03
5255	Red oak		220	.036		1.91	1.84		3.75	4.91

For customer support on your Building Construction Costs with RSMeans data, call 800.448.8182.

06 22 Millwork

06 22 13 – Standard Pattern Wood Trim

06 22 13.15 Moldings, Base

		Crew	Daily Output	Labor-Hours	Unit	Material	2018 Bare Costs Labor	2018 Bare Costs Equipment	Total	Total Incl O&P
5260	Maple	1 Carp	220	.036	L.F.	2.94	1.84		4.78	6.05
5265	Cherry		220	.036		3.08	1.84		4.92	6.20
5270	Rectangular profile, oak, 3/8" x 1-1/4"		260	.031		1.24	1.56		2.80	3.75
5275	1/2" x 2-1/2"		255	.031		2.35	1.59		3.94	5
5280	1/2" x 3-1/2"		250	.032		2.78	1.62		4.40	5.55
5285	1" x 6"		240	.033		4.14	1.69		5.83	7.10
5290	1" x 8"		240	.033		5.15	1.69		6.84	8.20
5295	Pine, 3/8" x 1-3/4"		260	.031		.50	1.56		2.06	2.93
5300	7/16" x 2-1/2"		255	.031		.78	1.59		2.37	3.28
5305	1" x 6"		240	.033		.72	1.69		2.41	3.36
5310	1" x 8"		240	.033		.93	1.69		2.62	3.59
5315	Shoe, 1/2" x 3/4", primed		260	.031		.51	1.56		2.07	2.94
5320	Pine		240	.033		.34	1.69		2.03	2.94
5325	Poplar		240	.033		.41	1.69		2.10	3.02
5330	Red oak		220	.036		.55	1.84		2.39	3.41
5335	Maple		220	.036		.73	1.84		2.57	3.61
5340	Cherry		220	.036		.83	1.84		2.67	3.72
5345	11/16" x 1-1/2", pine		240	.033		.73	1.69		2.42	3.37
5350	Caps, 11/16" x 1-3/8", pine		240	.033		.58	1.69		2.27	3.21
5355	3/4" x 1-3/4", finger jointed and primed		260	.031		.77	1.56		2.33	3.23
5360	Poplar		240	.033		1.04	1.69		2.73	3.71
5365	Red oak		220	.036		1.25	1.84		3.09	4.18
5370	Maple		220	.036		1.63	1.84		3.47	4.60
5375	Cherry		220	.036		3.30	1.84		5.14	6.45
5380	Combination base & shoe, 9/16" x 3-1/2" & 1/2" x 3/4", pine		125	.064		1.51	3.24		4.75	6.60
5385	Three piece oak, 6" high		80	.100		5.95	5.05		11	14.25
5390	Including 3/4" x 1" base shoe		70	.114		6.45	5.80		12.25	15.90
5395	Flooring cant strip, 3/4" x 3/4", pre-finished pine		260	.031		.44	1.56		2	2.86
5400	For pre-finished, stain and clear coat, add						.55		.55	.61
5405	Clear coat only, add						.45		.45	.50

06 22 13.30 Moldings, Casings

		Crew	Daily Output	Labor-Hours	Unit	Material	2018 Bare Costs Labor	2018 Bare Costs Equipment	Total	Total Incl O&P
0010	**MOLDINGS, CASINGS**									
0085	Apron, 9/16" x 2-1/2", pine	1 Carp	250	.032	L.F.	1.43	1.62		3.05	4.04
0090	5/8" x 2-1/2", pine		250	.032		1.71	1.62		3.33	4.35
0110	5/8" x 3-1/2", pine		220	.036		1.94	1.84		3.78	4.94
0300	Band, 11/16" x 1-1/8", pine		270	.030		.75	1.50		2.25	3.11
0310	11/16" x 1-1/2", finger jointed and primed		270	.030		.76	1.50		2.26	3.12
0320	Pine		270	.030		1	1.50		2.50	3.39
0330	11/16" x 1-3/4", finger jointed and primed		270	.030		.95	1.50		2.45	3.33
0350	Pine		270	.030		.95	1.50		2.45	3.33
0355	Beaded, 3/4" x 3-1/2", finger jointed and primed		220	.036		.98	1.84		2.82	3.89
0360	Poplar		220	.036		1.15	1.84		2.99	4.07
0365	Red oak		220	.036		1.70	1.84		3.54	4.68
0370	Maple		220	.036		2.63	1.84		4.47	5.70
0375	Cherry		220	.036		3.18	1.84		5.02	6.30
0380	3/4" x 4", finger jointed and primed		220	.036		1.22	1.84		3.06	4.15
0385	Poplar		220	.036		1.44	1.84		3.28	4.39
0390	Red oak		220	.036		2.02	1.84		3.86	5.05
0395	Maple		220	.036		2.65	1.84		4.49	5.70
0400	Cherry		220	.036		3.36	1.84		5.20	6.50
0405	3/4" x 5-1/2", finger jointed and primed		200	.040		1.35	2.03		3.38	4.57
0410	Poplar		200	.040		1.99	2.03		4.02	5.30

06 22 13.30 Moldings, Casings		Crew	Daily Output	Labor-Hours	Unit	Material	2018 Bare Costs Labor	Equipment	Total	Total Incl O&P
0415	Red oak	1 Carp	200	.040	L.F.	2.93	2.03		4.96	6.30
0420	Maple		200	.040		3.85	2.03		5.88	7.30
0425	Cherry		200	.040		4.43	2.03		6.46	7.95
0430	Classic profile, 3/4" x 2-3/4", finger jointed and primed		250	.032		.86	1.62		2.48	3.41
0435	Poplar		250	.032		1.03	1.62		2.65	3.60
0440	Red oak		250	.032		1.51	1.62		3.13	4.13
0445	Maple		250	.032		2.16	1.62		3.78	4.84
0450	Cherry		250	.032		2.45	1.62		4.07	5.15
0455	Fluted, 3/4" x 3-1/2", poplar		220	.036		1.15	1.84		2.99	4.07
0460	Red oak		220	.036		1.70	1.84		3.54	4.68
0465	Maple		220	.036		2.63	1.84		4.47	5.70
0470	Cherry		220	.036		3.18	1.84		5.02	6.30
0475	3/4" x 4", poplar		220	.036		1.44	1.84		3.28	4.39
0480	Red oak		220	.036		2.02	1.84		3.86	5.05
0485	Maple		220	.036		2.65	1.84		4.49	5.70
0490	Cherry		220	.036		3.36	1.84		5.20	6.50
0495	3/4" x 5-1/2", poplar		200	.040		1.99	2.03		4.02	5.30
0500	Red oak		200	.040		2.93	2.03		4.96	6.30
0505	Maple		200	.040		3.85	2.03		5.88	7.30
0510	Cherry		200	.040		4.43	2.03		6.46	7.95
0515	3/4" x 7-1/2", poplar		190	.042		1.26	2.14		3.40	4.63
0520	Red oak		190	.042		3.93	2.14		6.07	7.55
0525	Maple		190	.042		5.65	2.14		7.79	9.45
0530	Cherry		190	.042		6.75	2.14		8.89	10.70
0535	3/4" x 9-1/2", poplar		180	.044		4.12	2.25		6.37	7.95
0540	Red oak		180	.044		6.55	2.25		8.80	10.65
0545	Maple		180	.044		9	2.25		11.25	13.35
0550	Cherry		180	.044		9.80	2.25		12.05	14.25
0555	Modern profile, 9/16" x 2-1/4", poplar		250	.032		.78	1.62		2.40	3.33
0560	Red oak		250	.032		.81	1.62		2.43	3.36
0565	11/16" x 2-1/2", finger jointed & primed		250	.032		.84	1.62		2.46	3.39
0570	Pine		250	.032		1.34	1.62		2.96	3.94
0575	3/4" x 2-1/2", poplar		250	.032		.91	1.62		2.53	3.47
0580	Red oak		250	.032		1.23	1.62		2.85	3.82
0585	Maple		250	.032		1.86	1.62		3.48	4.51
0590	Cherry		250	.032		2.56	1.62		4.18	5.30
0595	Mullion, 5/16" x 2", pine		270	.030		.90	1.50		2.40	3.28
0600	9/16" x 2-1/2", finger jointed and primed		250	.032		.98	1.62		2.60	3.55
0605	Pine		250	.032		1.34	1.62		2.96	3.94
0610	Red oak		250	.032		3.28	1.62		4.90	6.10
0615	1-1/16" x 3-3/4", red oak		220	.036		6.95	1.84		8.79	10.40
0620	Ogee, 7/16" x 2-1/2", poplar		250	.032		.74	1.62		2.36	3.28
0625	Red oak		250	.032		.94	1.62		2.56	3.50
0630	9/16" x 2-1/4", finger jointed and primed		250	.032		.55	1.62		2.17	3.07
0635	Poplar		250	.032		.59	1.62		2.21	3.12
0640	Red oak		250	.032		.81	1.62		2.43	3.36
0645	11/16" x 2-1/2", finger jointed and primed		250	.032		.72	1.62		2.34	3.26
0700	Pine		250	.032		1.42	1.62		3.04	4.03
0701	Red oak		250	.032		3.25	1.62		4.87	6.05
0730	11/16" x 3-1/2", finger jointed and primed		220	.036		1.37	1.84		3.21	4.32
0750	Pine		220	.036		1.79	1.84		3.63	4.78
0755	3/4" x 2-1/2", finger jointed and primed		250	.032		.77	1.62		2.39	3.32
0760	Poplar		250	.032		.91	1.62		2.53	3.47

06 22 Millwork

06 22 13 – Standard Pattern Wood Trim

06 22 13.30 Moldings, Casings

		Crew	Daily Output	Labor-Hours	Unit	Material	2018 Bare Costs Labor	Equipment	Total	Total Incl O&P
0765	Red oak	1 Carp	250	.032	L.F.	1.23	1.62		2.85	3.82
0770	Maple		250	.032		1.86	1.62		3.48	4.51
0775	Cherry		250	.032		2.34	1.62		3.96	5.05
0780	3/4" x 3-1/2", finger jointed and primed		220	.036		.98	1.84		2.82	3.89
0785	Poplar		220	.036		1.15	1.84		2.99	4.07
0790	Red oak		220	.036		1.70	1.84		3.54	4.68
0795	Maple		220	.036		2.63	1.84		4.47	5.70
0800	Cherry		220	.036		3.18	1.84		5.02	6.30
4700	Square profile, 1" x 1", teak		215	.037		2.24	1.89		4.13	5.35
4800	Rectangular profile, 1" x 3", teak		200	.040		6.70	2.03		8.73	10.45

06 22 13.35 Moldings, Ceilings

		Crew	Daily Output	Labor-Hours	Unit	Material	2018 Bare Costs Labor	Equipment	Total	Total Incl O&P
0010	**MOLDINGS, CEILINGS**									
0600	Bed, 9/16" x 1-3/4", pine	1 Carp	270	.030	L.F.	1.20	1.50		2.70	3.61
0650	9/16" x 2", pine		270	.030		1.19	1.50		2.69	3.60
0710	9/16" x 1-3/4", oak		270	.030		2.35	1.50		3.85	4.87
1200	Cornice, 9/16" x 1-3/4", pine		270	.030		1	1.50		2.50	3.39
1300	9/16" x 2-1/4", pine		265	.030		1.31	1.53		2.84	3.77
1350	Cove, 1/2" x 2-1/4", poplar		265	.030		1.07	1.53		2.60	3.50
1360	Red oak		265	.030		1.37	1.53		2.90	3.83
1370	Hard maple		265	.030		1.64	1.53		3.17	4.13
1380	Cherry		265	.030		2.27	1.53		3.80	4.82
2400	9/16" x 1-3/4", pine		270	.030		1	1.50		2.50	3.39
2500	11/16" x 2-3/4", pine		265	.030		1.87	1.53		3.40	4.38
2510	Crown, 5/8" x 5/8", poplar		300	.027		.46	1.35		1.81	2.56
2520	Red oak		300	.027		.54	1.35		1.89	2.65
2530	Hard maple		300	.027		.73	1.35		2.08	2.86
2540	Cherry		300	.027		.77	1.35		2.12	2.91
2600	9/16" x 3-5/8", pine		250	.032		2.10	1.62		3.72	4.78
2700	11/16" x 4-1/4", pine		250	.032		2.97	1.62		4.59	5.75
2705	Oak		250	.032		6.30	1.62		7.92	9.40
2710	3/4" x 1-3/4", poplar		270	.030		.74	1.50		2.24	3.10
2720	Red oak		270	.030		1.09	1.50		2.59	3.49
2730	Hard maple		270	.030		1.44	1.50		2.94	3.87
2740	Cherry		270	.030		1.77	1.50		3.27	4.24
2750	3/4" x 2", poplar		270	.030		.97	1.50		2.47	3.36
2760	Red oak		270	.030		1.30	1.50		2.80	3.72
2770	Hard maple		270	.030		1.74	1.50		3.24	4.20
2780	Cherry		270	.030		1.94	1.50		3.44	4.42
2790	3/4" x 2-3/4", poplar		265	.030		1.05	1.53		2.58	3.48
2800	Red oak		265	.030		1.62	1.53		3.15	4.11
2810	Hard maple		265	.030		2.16	1.53		3.69	4.70
2820	Cherry		265	.030		2.48	1.53		4.01	5.05
2830	3/4" x 3-1/2", poplar		250	.032		1.34	1.62		2.96	3.94
2840	Red oak		250	.032		2.02	1.62		3.64	4.69
2850	Hard maple		250	.032		2.64	1.62		4.26	5.35
2860	Cherry		250	.032		3.09	1.62		4.71	5.85
2870	FJP poplar		250	.032		.99	1.62		2.61	3.56
2880	3/4" x 5", poplar		245	.033		1.99	1.66		3.65	4.71
2890	Red oak		245	.033		2.94	1.66		4.60	5.75
2900	Hard maple		245	.033		3.86	1.66		5.52	6.75
2910	Cherry		245	.033		4.45	1.66		6.11	7.40
2920	FJP poplar		245	.033		1.41	1.66		3.07	4.07

06 22 13.35 Moldings, Ceilings

		Crew	Daily Output	Labor-Hours	Unit	Material	2018 Bare Costs Labor	Equipment	Total	Total Incl O&P
2930	3/4" x 6-1/4", poplar	1 Carp	240	.033	L.F.	2.36	1.69		4.05	5.15
2940	Red oak		240	.033		3.57	1.69		5.26	6.50
2950	Hard maple		240	.033		4.66	1.69		6.35	7.70
2960	Cherry		240	.033		5.55	1.69		7.24	8.65
2970	7/8" x 8-3/4", poplar		220	.036		3.89	1.84		5.73	7.10
2980	Red oak		220	.036		6.05	1.84		7.89	9.45
2990	Hard maple		220	.036		8.40	1.84		10.24	12
3000	Cherry		220	.036		9.80	1.84		11.64	13.55
3010	1" x 7-1/4", poplar		220	.036		3.89	1.84		5.73	7.10
3020	Red oak		220	.036		6.05	1.84		7.89	9.45
3030	Hard maple		220	.036		8.40	1.84		10.24	12
3040	Cherry		220	.036		9.80	1.84		11.64	13.55
3050	1-1/16" x 4-1/4", poplar		250	.032		2.36	1.62		3.98	5.05
3060	Red oak		250	.032		2.78	1.62		4.40	5.50
3070	Hard maple		250	.032		3.67	1.62		5.29	6.50
3080	Cherry		250	.032		5.25	1.62		6.87	8.20
3090	Dentil crown, 3/4" x 5", poplar		250	.032		1.99	1.62		3.61	4.66
3100	Red oak		250	.032		2.94	1.62		4.56	5.70
3110	Hard maple		250	.032		3.86	1.62		5.48	6.70
3120	Cherry		250	.032		4.45	1.62		6.07	7.35
3130	Dentil piece for above, 1/2" x 1/2", poplar		300	.027		2.82	1.35		4.17	5.15
3140	Red oak		300	.027		3.24	1.35		4.59	5.60
3150	Hard maple		300	.027		3.68	1.35		5.03	6.10
3160	Cherry		300	.027		4.01	1.35		5.36	6.45

06 22 13.40 Moldings, Exterior

		Crew	Daily Output	Labor-Hours	Unit	Material	2018 Bare Costs Labor	Equipment	Total	Total Incl O&P
0010	**MOLDINGS, EXTERIOR**									
0100	Band board, cedar, rough sawn, 1" x 2"	1 Carp	300	.027	L.F.	.69	1.35		2.04	2.82
0110	1" x 3"		300	.027		1.02	1.35		2.37	3.19
0120	1" x 4"		250	.032		1.36	1.62		2.98	3.97
0130	1" x 6"		250	.032		2.05	1.62		3.67	4.72
0140	1" x 8"		225	.036		2.73	1.80		4.53	5.75
0150	1" x 10"		225	.036		3.39	1.80		5.19	6.50
0160	1" x 12"		200	.040		4.08	2.03		6.11	7.55
0240	STK, 1" x 2"		300	.027		.45	1.35		1.80	2.55
0250	1" x 3"		300	.027		.49	1.35		1.84	2.60
0260	1" x 4"		250	.032		.81	1.62		2.43	3.36
0270	1" x 6"		250	.032		1.32	1.62		2.94	3.92
0280	1" x 8"		225	.036		2.01	1.80		3.81	4.96
0290	1" x 10"		225	.036		2.77	1.80		4.57	5.80
0300	1" x 12"		200	.040		4.29	2.03		6.32	7.80
0310	Pine, #2, 1" x 2"		300	.027		.28	1.35		1.63	2.37
0320	1" x 3"		300	.027		.45	1.35		1.80	2.55
0330	1" x 4"		250	.032		.54	1.62		2.16	3.07
0340	1" x 6"		250	.032		.82	1.62		2.44	3.37
0350	1" x 8"		225	.036		1.33	1.80		3.13	4.21
0360	1" x 10"		225	.036		1.74	1.80		3.54	4.66
0370	1" x 12"		200	.040		2.13	2.03		4.16	5.45
0380	D & better, 1" x 2"		300	.027		.50	1.35		1.85	2.61
0390	1" x 3"		300	.027		.64	1.35		1.99	2.76
0400	1" x 4"		250	.032		.82	1.62		2.44	3.37
0410	1" x 6"		250	.032		1.07	1.62		2.69	3.64
0420	1" x 8"		225	.036		1.62	1.80		3.42	4.53

06 22 13 – Standard Pattern Wood Trim

06 22 13.40 Moldings, Exterior	Crew	Daily Output	Labor-Hours	Unit	Material	2018 Bare Costs Labor	Equipment	Total	Total Incl O&P
0430	1" x 10"	1 Carp	225	.036	L.F.	2.24	1.80		4.04	5.20
0440	1" x 12"		200	.040		2.69	2.03		4.72	6.05
0450	Redwood, clear all heart, 1" x 2"		300	.027		.64	1.35		1.99	2.76
0460	1" x 3"		300	.027		.95	1.35		2.30	3.10
0470	1" x 4"		250	.032		1.20	1.62		2.82	3.79
0480	1" x 6"		252	.032		1.79	1.61		3.40	4.42
0490	1" x 8"		225	.036		2.37	1.80		4.17	5.35
0500	1" x 10"		225	.036		3.98	1.80		5.78	7.10
0510	1" x 12"		200	.040		4.77	2.03		6.80	8.35
0530	Corner board, cedar, rough sawn, 1" x 2"		225	.036		.69	1.80		2.49	3.51
0540	1" x 3"		225	.036		1.02	1.80		2.82	3.88
0550	1" x 4"		200	.040		1.36	2.03		3.39	4.59
0560	1" x 6"		200	.040		2.05	2.03		4.08	5.35
0570	1" x 8"		200	.040		2.73	2.03		4.76	6.10
0580	1" x 10"		175	.046		3.39	2.32		5.71	7.25
0590	1" x 12"		175	.046		4.08	2.32		6.40	8
0670	STK, 1" x 2"		225	.036		.45	1.80		2.25	3.24
0680	1" x 3"		225	.036		.49	1.80		2.29	3.29
0690	1" x 4"		200	.040		.79	2.03		2.82	3.96
0700	1" x 6"		200	.040		1.32	2.03		3.35	4.54
0710	1" x 8"		200	.040		2.01	2.03		4.04	5.30
0720	1" x 10"		175	.046		2.77	2.32		5.09	6.55
0730	1" x 12"		175	.046		4.29	2.32		6.61	8.25
0740	Pine, #2, 1" x 2"		225	.036		.28	1.80		2.08	3.06
0750	1" x 3"		225	.036		.45	1.80		2.25	3.24
0760	1" x 4"		200	.040		.54	2.03		2.57	3.69
0770	1" x 6"		200	.040		.82	2.03		2.85	3.99
0780	1" x 8"		200	.040		1.33	2.03		3.36	4.55
0790	1" x 10"		175	.046		1.74	2.32		4.06	5.45
0800	1" x 12"		175	.046		2.13	2.32		4.45	5.85
0810	D & better, 1" x 2"		225	.036		.50	1.80		2.30	3.30
0820	1" x 3"		225	.036		.64	1.80		2.44	3.45
0830	1" x 4"		200	.040		.82	2.03		2.85	3.99
0840	1" x 6"		200	.040		1.07	2.03		3.10	4.26
0850	1" x 8"		200	.040		1.62	2.03		3.65	4.87
0860	1" x 10"		175	.046		2.24	2.32		4.56	6
0870	1" x 12"		175	.046		2.69	2.32		5.01	6.50
0880	Redwood, clear all heart, 1" x 2"		225	.036		.64	1.80		2.44	3.45
0890	1" x 3"		225	.036		.95	1.80		2.75	3.79
0900	1" x 4"		200	.040		1.20	2.03		3.23	4.41
0910	1" x 6"		200	.040		1.79	2.03		3.82	5.05
0920	1" x 8"		200	.040		2.37	2.03		4.40	5.70
0930	1" x 10"		175	.046		3.98	2.32		6.30	7.90
0940	1" x 12"		175	.046		4.77	2.32		7.09	8.80
0950	Cornice board, cedar, rough sawn, 1" x 2"		330	.024		.69	1.23		1.92	2.63
0960	1" x 3"		290	.028		1.02	1.40		2.42	3.26
0970	1" x 4"		250	.032		1.36	1.62		2.98	3.97
0980	1" x 6"		250	.032		2.05	1.62		3.67	4.72
0990	1" x 8"		200	.040		2.73	2.03		4.76	6.10
1000	1" x 10"		180	.044		3.39	2.25		5.64	7.15
1010	1" x 12"		180	.044		4.08	2.25		6.33	7.90
1020	STK, 1" x 2"		330	.024		.45	1.23		1.68	2.36
1030	1" x 3"		290	.028		.49	1.40		1.89	2.67

06 22 13.40 Moldings, Exterior

		Crew	Daily Output	Labor-Hours	Unit	Material	Labor	Equipment	Total	Total Incl O&P
1040	1" x 4"	1 Carp	250	.032	L.F.	.81	1.62		2.43	3.36
1050	1" x 6"		250	.032		1.32	1.62		2.94	3.92
1060	1" x 8"		200	.040		2.01	2.03		4.04	5.30
1070	1" x 10"		180	.044		2.77	2.25		5.02	6.45
1080	1" x 12"		180	.044		4.29	2.25		6.54	8.15
1500	Pine, #2, 1" x 2"		330	.024		.28	1.23		1.51	2.18
1510	1" x 3"		290	.028		.30	1.40		1.70	2.46
1600	1" x 4"		250	.032		.54	1.62		2.16	3.07
1700	1" x 6"		250	.032		.82	1.62		2.44	3.37
1800	1" x 8"		200	.040		1.33	2.03		3.36	4.55
1900	1" x 10"		180	.044		1.74	2.25		3.99	5.35
2000	1" x 12"		180	.044		2.13	2.25		4.38	5.75
2020	D & better, 1" x 2"		330	.024		.50	1.23		1.73	2.42
2030	1" x 3"		290	.028		.64	1.40		2.04	2.83
2040	1" x 4"		250	.032		.82	1.62		2.44	3.37
2050	1" x 6"		250	.032		1.07	1.62		2.69	3.64
2060	1" x 8"		200	.040		1.62	2.03		3.65	4.87
2070	1" x 10"		180	.044		2.24	2.25		4.49	5.90
2080	1" x 12"		180	.044		2.69	2.25		4.94	6.40
2090	Redwood, clear all heart, 1" x 2"		330	.024		.64	1.23		1.87	2.57
2100	1" x 3"		290	.028		.95	1.40		2.35	3.17
2110	1" x 4"		250	.032		1.20	1.62		2.82	3.79
2120	1" x 6"		250	.032		1.79	1.62		3.41	4.44
2130	1" x 8"		200	.040		2.37	2.03		4.40	5.70
2140	1" x 10"		180	.044		3.98	2.25		6.23	7.80
2150	1" x 12"		180	.044		4.77	2.25		7.02	8.70
2160	3 piece, 1" x 2", 1" x 4", 1" x 6", rough sawn cedar		80	.100		4.11	5.05		9.16	12.20
2180	STK cedar		80	.100		2.57	5.05		7.62	10.55
2200	#2 pine		80	.100		1.64	5.05		6.69	9.50
2210	D & better pine		80	.100		2.38	5.05		7.43	10.30
2220	Clear all heart redwood		80	.100		3.62	5.05		8.67	11.70
2230	1" x 8", 1" x 10", 1" x 12", rough sawn cedar		65	.123		10.15	6.25		16.40	20.50
2240	STK cedar		65	.123		9	6.25		15.25	19.40
2300	#2 pine		65	.123		5.15	6.25		11.40	15.20
2320	D & better pine		65	.123		6.50	6.25		12.75	16.65
2330	Clear all heart redwood		65	.123		11.05	6.25		17.30	21.50
2340	Door/window casing, cedar, rough sawn, 1" x 2"		275	.029		.69	1.47		2.16	3.01
2350	1" x 3"		275	.029		1.02	1.47		2.49	3.38
2360	1" x 4"		250	.032		1.36	1.62		2.98	3.97
2370	1" x 6"		250	.032		2.05	1.62		3.67	4.72
2380	1" x 8"		230	.035		2.73	1.76		4.49	5.70
2390	1" x 10"		230	.035		3.39	1.76		5.15	6.40
2395	1" x 12"		210	.038		4.08	1.93		6.01	7.40
2410	STK, 1" x 2"		275	.029		.45	1.47		1.92	2.74
2420	1" x 3"		275	.029		.49	1.47		1.96	2.79
2430	1" x 4"		250	.032		.81	1.62		2.43	3.36
2440	1" x 6"		250	.032		1.32	1.62		2.94	3.92
2450	1" x 8"		230	.035		2.01	1.76		3.77	4.90
2460	1" x 10"		230	.035		2.77	1.76		4.53	5.75
2470	1" x 12"		210	.038		4.29	1.93		6.22	7.65
2550	Pine, #2, 1" x 2"		275	.029		.28	1.47		1.75	2.56
2560	1" x 3"		275	.029		.45	1.47		1.92	2.74
2570	1" x 4"		250	.032		.54	1.62		2.16	3.07

06 22 13.40 Moldings, Exterior		Crew	Daily Output	Labor-Hours	Unit	Material	2018 Bare Costs Labor	Equipment	Total	Total Incl O&P
2580	1" x 6"	1 Carp	250	.032	L.F.	.82	1.62		2.44	3.37
2590	1" x 8"		230	.035		1.33	1.76		3.09	4.15
2600	1" x 10"		230	.035		1.74	1.76		3.50	4.60
2610	1" x 12"		210	.038		2.13	1.93		4.06	5.30
2620	Pine, D & better, 1" x 2"		275	.029		.50	1.47		1.97	2.80
2630	1" x 3"		275	.029		.64	1.47		2.11	2.95
2640	1" x 4"		250	.032		.82	1.62		2.44	3.37
2650	1" x 6"		250	.032		1.07	1.62		2.69	3.64
2660	1" x 8"		230	.035		1.62	1.76		3.38	4.47
2670	1" x 10"		230	.035		2.24	1.76		4	5.15
2680	1" x 12"		210	.038		2.69	1.93		4.62	5.90
2690	Redwood, clear all heart, 1" x 2"		275	.029		.64	1.47		2.11	2.95
2695	1" x 3"		275	.029		.95	1.47		2.42	3.29
2710	1" x 4"		250	.032		1.20	1.62		2.82	3.79
2715	1" x 6"		250	.032		1.79	1.62		3.41	4.44
2730	1" x 8"		230	.035		2.37	1.76		4.13	5.30
2740	1" x 10"		230	.035		3.98	1.76		5.74	7.05
2750	1" x 12"		210	.038		4.77	1.93		6.70	8.20
3500	Bellyband, pine, 11/16" x 4-1/4"		250	.032		2.82	1.62		4.44	5.55
3610	Brickmold, pine, 1-1/4" x 2"		200	.040		2.10	2.03		4.13	5.40
3620	FJP, 1-1/4" x 2"		200	.040		.97	2.03		3	4.16
5100	Fascia, cedar, rough sawn, 1" x 2"		275	.029		.69	1.47		2.16	3.01
5110	1" x 3"		275	.029		1.02	1.47		2.49	3.38
5120	1" x 4"		250	.032		1.36	1.62		2.98	3.97
5200	1" x 6"		250	.032		2.05	1.62		3.67	4.72
5300	1" x 8"		230	.035		2.73	1.76		4.49	5.70
5310	1" x 10"		230	.035		3.39	1.76		5.15	6.40
5320	1" x 12"		210	.038		4.08	1.93		6.01	7.40
5400	2" x 4"		220	.036		1.07	1.84		2.91	3.98
5500	2" x 6"		220	.036		1.59	1.84		3.43	4.56
5600	2" x 8"		200	.040		2.12	2.03		4.15	5.45
5700	2" x 10"		180	.044		2.64	2.25		4.89	6.35
5800	2" x 12"		170	.047		6.30	2.39		8.69	10.55
6120	STK, 1" x 2"		275	.029		.45	1.47		1.92	2.74
6130	1" x 3"		275	.029		.49	1.47		1.96	2.79
6140	1" x 4"		250	.032		.81	1.62		2.43	3.36
6150	1" x 6"		250	.032		1.32	1.62		2.94	3.92
6160	1" x 8"		230	.035		2.01	1.76		3.77	4.90
6170	1" x 10"		230	.035		2.77	1.76		4.53	5.75
6180	1" x 12"		210	.038		4.29	1.93		6.22	7.65
6185	2" x 2"		260	.031		.74	1.56		2.30	3.19
6190	Pine, #2, 1" x 2"		275	.029		.28	1.47		1.75	2.56
6200	1" x 3"		275	.029		.45	1.47		1.92	2.74
6210	1" x 4"		250	.032		.54	1.62		2.16	3.07
6220	1" x 6"		250	.032		.82	1.62		2.44	3.37
6230	1" x 8"		230	.035		1.33	1.76		3.09	4.15
6240	1" x 10"		230	.035		1.74	1.76		3.50	4.60
6250	1" x 12"		210	.038		2.13	1.93		4.06	5.30
6260	D & better, 1" x 2"		275	.029		.50	1.47		1.97	2.80
6270	1" x 3"		275	.029		.64	1.47		2.11	2.95
6280	1" x 4"		250	.032		.82	1.62		2.44	3.37
6290	1" x 6"		250	.032		1.07	1.62		2.69	3.64
6300	1" x 8"		230	.035		1.62	1.76		3.38	4.47

06 22 13 – Standard Pattern Wood Trim

06 22 13.40 Moldings, Exterior	Crew	Daily Output	Labor-Hours	Unit	Material	2018 Bare Costs Labor	Equipment	Total	Total Incl O&P
6310	1" x 10"	1 Carp	230	.035	L.F.	2.24	1.76		4	5.15
6312	1" x 12"		210	.038		2.69	1.93		4.62	5.90
6330	Southern yellow, 1-1/4" x 5"		240	.033		2.88	1.69		4.57	5.75
6340	1-1/4" x 6"		240	.033		2.57	1.69		4.26	5.40
6350	1-1/4" x 8"		215	.037		3.65	1.89		5.54	6.90
6360	1-1/4" x 12"		190	.042		5.35	2.14		7.49	9.15
6370	Redwood, clear all heart, 1" x 2"		275	.029		.64	1.47		2.11	2.95
6380	1" x 3"		275	.029		1.20	1.47		2.67	3.57
6390	1" x 4"		250	.032		1.20	1.62		2.82	3.79
6400	1" x 6"		250	.032		1.79	1.62		3.41	4.44
6410	1" x 8"		230	.035		2.37	1.76		4.13	5.30
6420	1" x 10"		230	.035		3.98	1.76		5.74	7.05
6430	1" x 12"		210	.038		4.77	1.93		6.70	8.20
6440	1-1/4" x 5"		240	.033		1.87	1.69		3.56	4.63
6450	1-1/4" x 6"		240	.033		2.23	1.69		3.92	5.05
6460	1-1/4" x 8"		215	.037		3.54	1.89		5.43	6.75
6470	1-1/4" x 12"		190	.042		7.15	2.14		9.29	11.10
6580	Frieze, cedar, rough sawn, 1" x 2"		275	.029		.69	1.47		2.16	3.01
6590	1" x 3"		275	.029		1.02	1.47		2.49	3.38
6600	1" x 4"		250	.032		1.36	1.62		2.98	3.97
6610	1" x 6"		250	.032		2.05	1.62		3.67	4.72
6620	1" x 8"		250	.032		2.73	1.62		4.35	5.45
6630	1" x 10"		225	.036		3.39	1.80		5.19	6.50
6640	1" x 12"		200	.040		4.04	2.03		6.07	7.55
6650	STK, 1" x 2"		275	.029		.45	1.47		1.92	2.74
6660	1" x 3"		275	.029		.49	1.47		1.96	2.79
6670	1" x 4"		250	.032		.81	1.62		2.43	3.36
6680	1" x 6"		250	.032		1.32	1.62		2.94	3.92
6690	1" x 8"		250	.032		2.01	1.62		3.63	4.68
6700	1" x 10"		225	.036		2.77	1.80		4.57	5.80
6710	1" x 12"		200	.040		4.29	2.03		6.32	7.80
6790	Pine, #2, 1" x 2"		275	.029		.28	1.47		1.75	2.56
6800	1" x 3"		275	.029		.45	1.47		1.92	2.74
6810	1" x 4"		250	.032		.54	1.62		2.16	3.07
6820	1" x 6"		250	.032		.82	1.62		2.44	3.37
6830	1" x 8"		250	.032		1.33	1.62		2.95	3.93
6840	1" x 10"		225	.036		1.74	1.80		3.54	4.66
6850	1" x 12"		200	.040		2.13	2.03		4.16	5.45
6860	D & better, 1" x 2"		275	.029		.50	1.47		1.97	2.80
6870	1" x 3"		275	.029		.64	1.47		2.11	2.95
6880	1" x 4"		250	.032		.82	1.62		2.44	3.37
6890	1" x 6"		250	.032		1.07	1.62		2.69	3.64
6900	1" x 8"		250	.032		1.62	1.62		3.24	4.25
6910	1" x 10"		225	.036		2.24	1.80		4.04	5.20
6920	1" x 12"		200	.040		2.69	2.03		4.72	6.05
6930	Redwood, clear all heart, 1" x 2"		275	.029		.64	1.47		2.11	2.95
6940	1" x 3"		275	.029		.95	1.47		2.42	3.29
6950	1" x 4"		250	.032		1.20	1.62		2.82	3.79
6960	1" x 6"		250	.032		1.79	1.62		3.41	4.44
6970	1" x 8"		250	.032		2.37	1.62		3.99	5.05
6980	1" x 10"		225	.036		3.98	1.80		5.78	7.10
6990	1" x 12"		200	.040		4.77	2.03		6.80	8.35
7000	Grounds, 1" x 1", cedar, rough sawn		300	.027		.35	1.35		1.70	2.45

For customer support on your Building Construction Costs with RSMeans data, call 800.448.8182.

06 22 Millwork

06 22 13 – Standard Pattern Wood Trim

06 22 13.40 Moldings, Exterior		Crew	Daily Output	Labor-Hours	Unit	Material	2018 Bare Costs Labor	Equipment	Total	Total Incl O&P
7010	STK	1 Carp	300	.027	L.F.	.28	1.35		1.63	2.36
7020	Pine, #2		300	.027		.18	1.35		1.53	2.26
7030	D & better		300	.027		.31	1.35		1.66	2.40
7050	Redwood		300	.027		.39	1.35		1.74	2.49
7060	Rake/verge board, cedar, rough sawn, 1" x 2"		225	.036		.69	1.80		2.49	3.51
7070	1" x 3"		225	.036		1.02	1.80		2.82	3.88
7080	1" x 4"		200	.040		1.36	2.03		3.39	4.59
7090	1" x 6"		200	.040		2.05	2.03		4.08	5.35
7100	1" x 8"		190	.042		2.73	2.14		4.87	6.25
7110	1" x 10"		190	.042		3.39	2.14		5.53	7
7120	1" x 12"		180	.044		4.08	2.25		6.33	7.90
7130	STK, 1" x 2"		225	.036		.45	1.80		2.25	3.24
7140	1" x 3"		225	.036		.49	1.80		2.29	3.29
7150	1" x 4"		200	.040		.81	2.03		2.84	3.98
7160	1" x 6"		200	.040		1.32	2.03		3.35	4.54
7170	1" x 8"		190	.042		2.01	2.14		4.15	5.45
7180	1" x 10"		190	.042		2.77	2.14		4.91	6.30
7190	1" x 12"		180	.044		4.29	2.25		6.54	8.15
7200	Pine, #2, 1" x 2"		225	.036		.28	1.80		2.08	3.06
7210	1" x 3"		225	.036		.45	1.80		2.25	3.24
7220	1" x 4"		200	.040		.54	2.03		2.57	3.69
7230	1" x 6"		200	.040		.82	2.03		2.85	3.99
7240	1" x 8"		190	.042		1.33	2.14		3.47	4.71
7250	1" x 10"		190	.042		1.74	2.14		3.88	5.15
7260	1" x 12"		180	.044		2.13	2.25		4.38	5.75
7340	D & better, 1" x 2"		225	.036		.50	1.80		2.30	3.30
7350	1" x 3"		225	.036		.64	1.80		2.44	3.45
7360	1" x 4"		200	.040		.82	2.03		2.85	3.99
7370	1" x 6"		200	.040		1.07	2.03		3.10	4.26
7380	1" x 8"		190	.042		1.62	2.14		3.76	5.05
7390	1" x 10"		190	.042		2.24	2.14		4.38	5.70
7400	1" x 12"		180	.044		2.69	2.25		4.94	6.40
7410	Redwood, clear all heart, 1" x 2"		225	.036		.64	1.80		2.44	3.45
7420	1" x 3"		225	.036		.95	1.80		2.75	3.79
7430	1" x 4"		200	.040		1.20	2.03		3.23	4.41
7440	1" x 6"		200	.040		1.79	2.03		3.82	5.05
7450	1" x 8"		190	.042		2.37	2.14		4.51	5.85
7460	1" x 10"		190	.042		3.98	2.14		6.12	7.60
7470	1" x 12"		180	.044		4.77	2.25		7.02	8.70
7480	2" x 4"		200	.040		2.30	2.03		4.33	5.60
7490	2" x 6"		182	.044		3.44	2.23		5.67	7.15
7500	2" x 8"		165	.048		4.57	2.46		7.03	8.75
7630	Soffit, cedar, rough sawn, 1" x 2"	2 Carp	440	.036		.69	1.84		2.53	3.57
7640	1" x 3"		440	.036		1.02	1.84		2.86	3.94
7650	1" x 4"		420	.038		1.36	1.93		3.29	4.44
7660	1" x 6"		420	.038		2.05	1.93		3.98	5.20
7670	1" x 8"		420	.038		2.73	1.93		4.66	5.95
7680	1" x 10"		400	.040		3.39	2.03		5.42	6.80
7690	1" x 12"		400	.040		4.08	2.03		6.11	7.55
7700	STK, 1" x 2"		440	.036		.45	1.84		2.29	3.30
7710	1" x 3"		440	.036		.49	1.84		2.33	3.35
7720	1" x 4"		420	.038		.81	1.93		2.74	3.83
7730	1" x 6"		420	.038		1.32	1.93		3.25	4.39

06 22 13 – Standard Pattern Wood Trim

06 22 13.40 Moldings, Exterior

		Crew	Daily Output	Labor-Hours	Unit	Material	2018 Bare Costs Labor	Equipment	Total	Total Incl O&P
7740	1" x 8"	2 Carp	420	.038	L.F.	2.01	1.93		3.94	5.15
7750	1" x 10"		400	.040		2.77	2.03		4.80	6.15
7760	1" x 12"		400	.040		4.29	2.03		6.32	7.80
7770	Pine, #2, 1" x 2"		440	.036		.28	1.84		2.12	3.12
7780	1" x 3"		440	.036		.45	1.84		2.29	3.30
7790	1" x 4"		420	.038		.54	1.93		2.47	3.54
7800	1" x 6"		420	.038		.82	1.93		2.75	3.84
7810	1" x 8"		420	.038		1.33	1.93		3.26	4.40
7820	1" x 10"		400	.040		1.74	2.03		3.77	5
7830	1" x 12"		400	.040		2.13	2.03		4.16	5.45
7840	D & better, 1" x 2"		440	.036		.50	1.84		2.34	3.36
7850	1" x 3"		440	.036		.64	1.84		2.48	3.51
7860	1" x 4"		420	.038		.82	1.93		2.75	3.84
7870	1" x 6"		420	.038		1.07	1.93		3	4.11
7880	1" x 8"		420	.038		1.62	1.93		3.55	4.72
7890	1" x 10"		400	.040		2.24	2.03		4.27	5.55
7900	1" x 12"		400	.040		2.69	2.03		4.72	6.05
7910	Redwood, clear all heart, 1" x 2"		440	.036		.64	1.84		2.48	3.51
7920	1" x 3"		440	.036		.95	1.84		2.79	3.85
7930	1" x 4"		420	.038		1.20	1.93		3.13	4.26
7940	1" x 6"		420	.038		1.79	1.93		3.72	4.91
7950	1" x 8"		420	.038		2.37	1.93		4.30	5.55
7960	1" x 10"		400	.040		3.98	2.03		6.01	7.45
7970	1" x 12"		400	.040		4.77	2.03		6.80	8.35
8050	Trim, crown molding, pine, 11/16" x 4-1/4"	1 Carp	250	.032		3.93	1.62		5.55	6.80
8060	Back band, 11/16" x 1-1/16"		250	.032		.90	1.62		2.52	3.46
8070	Insect screen frame stock, 1-1/16" x 1-3/4"		395	.020		2.23	1.03		3.26	4.01
8080	Dentils, 2-1/2" x 2-1/2" x 4", 6" OC		30	.267		1.22	13.50		14.72	22
8100	Fluted, 5-1/2"		165	.048		5.05	2.46		7.51	9.30
8110	Stucco bead, 1-3/8" x 1-5/8"		250	.032		2.33	1.62		3.95	5.05

06 22 13.45 Moldings, Trim

		Crew	Daily Output	Labor-Hours	Unit	Material	2018 Bare Costs Labor	Equipment	Total	Total Incl O&P
0010	**MOLDINGS, TRIM**									
0200	Astragal, stock pine, 11/16" x 1-3/4"	1 Carp	255	.031	L.F.	1.24	1.59		2.83	3.78
0250	1-5/16" x 2-3/16"		240	.033		2.62	1.69		4.31	5.45
0800	Chair rail, stock pine, 5/8" x 2-1/2"		270	.030		1.54	1.50		3.04	3.98
0900	5/8" x 3-1/2"		240	.033		2.33	1.69		4.02	5.15
1000	Closet pole, stock pine, 1-1/8" diameter		200	.040		1.16	2.03		3.19	4.37
1100	Fir, 1-5/8" diameter		200	.040		2.05	2.03		4.08	5.35
3300	Half round, stock pine, 1/4" x 1/2"		270	.030		.25	1.50		1.75	2.57
3350	1/2" x 1"		255	.031		.64	1.59		2.23	3.12
3400	Handrail, fir, single piece, stock, hardware not included									
3450	1-1/2" x 1-3/4"	1 Carp	80	.100	L.F.	2.10	5.05		7.15	10
3470	Pine, 1-1/2" x 1-3/4"		80	.100		1.97	5.05		7.02	9.85
3500	1-1/2" x 2-1/2"		76	.105		2.57	5.35		7.92	11
3600	Lattice, stock pine, 1/4" x 1-1/8"		270	.030		.36	1.50		1.86	2.69
3700	1/4" x 1-3/4"		250	.032		.61	1.62		2.23	3.14
3800	Miscellaneous, custom, pine, 1" x 1"		270	.030		.44	1.50		1.94	2.77
3850	1" x 2"		265	.030		.88	1.53		2.41	3.30
3900	1" x 3"		240	.033		1.32	1.69		3.01	4.02
4100	Birch or oak, nominal 1" x 1"		240	.033		.39	1.69		2.08	2.99
4200	Nominal 1" x 3"		215	.037		1.16	1.89		3.05	4.14
4400	Walnut, nominal 1" x 1"		215	.037		.60	1.89		2.49	3.53

06 22 Millwork

06 22 13 – Standard Pattern Wood Trim

06 22 13.45 Moldings, Trim

		Crew	Daily Output	Labor-Hours	Unit	Material	2018 Bare Costs Labor	2018 Bare Costs Equipment	Total	Total Incl O&P
4500	Nominal 1" x 3"	1 Carp	200	.040	L.F.	1.79	2.03		3.82	5.05
4700	Teak, nominal 1" x 1"		215	.037		2.79	1.89		4.68	5.95
4800	Nominal 1" x 3"		200	.040		8.35	2.03		10.38	12.30
4900	Quarter round, stock pine, 1/4" x 1/4"		275	.029		.25	1.47		1.72	2.53
4950	3/4" x 3/4"		255	.031	▼	.52	1.59		2.11	2.99
5600	Wainscot moldings, 1-1/8" x 9/16", 2' high, minimum		76	.105	S.F.	12.35	5.35		17.70	21.50
5700	Maximum	▼	65	.123	"	16.25	6.25		22.50	27.50

06 22 13.50 Moldings, Window and Door

		Crew	Daily Output	Labor-Hours	Unit	Material	2018 Bare Costs Labor	2018 Bare Costs Equipment	Total	Total Incl O&P
0010	**MOLDINGS, WINDOW AND DOOR**									
2800	Door moldings, stock, decorative, 1-1/8" wide, plain	1 Carp	17	.471	Set	48	24		72	89
2900	Detailed		17	.471	"	110	24		134	158
2960	Clear pine door jamb, no stops, 11/16" x 4-9/16"		240	.033	L.F.	4.93	1.69		6.62	7.95
3150	Door trim set, 1 head and 2 sides, pine, 2-1/2" wide		12	.667	Opng.	24	34		58	78
3170	3-1/2" wide		11	.727	"	30.50	37		67.50	89.50
3250	Glass beads, stock pine, 3/8" x 1/2"		275	.029	L.F.	.31	1.47		1.78	2.59
3270	3/8" x 7/8"		270	.030		.40	1.50		1.90	2.73
4850	Parting bead, stock pine, 3/8" x 3/4"		275	.029		.47	1.47		1.94	2.77
4870	1/2" x 3/4"		255	.031		.41	1.59		2	2.87
5000	Stool caps, stock pine, 11/16" x 3-1/2"		200	.040		2.17	2.03		4.20	5.50
5100	1-1/16" x 3-1/4"		150	.053	▼	3.40	2.70		6.10	7.85
5300	Threshold, oak, 3' long, inside, 5/8" x 3-5/8"		32	.250	Ea.	11.20	12.70		23.90	31.50
5400	Outside, 1-1/2" x 7-5/8"	▼	16	.500	"	47	25.50		72.50	90
5900	Window trim sets, including casings, header, stops,									
5910	stool and apron, 2-1/2" wide, FJP	1 Carp	13	.615	Opng.	32	31		63	83
5950	Pine		10	.800		38	40.50		78.50	104
6000	Oak	▼	6	1.333	▼	77	67.50		144.50	188

06 22 13.60 Moldings, Soffits

		Crew	Daily Output	Labor-Hours	Unit	Material	2018 Bare Costs Labor	2018 Bare Costs Equipment	Total	Total Incl O&P
0010	**MOLDINGS, SOFFITS**									
0200	Soffits, pine, 1" x 4"	2 Carp	420	.038	L.F.	.51	1.93		2.44	3.50
0210	1" x 6"		420	.038		.78	1.93		2.71	3.80
0220	1" x 8"		420	.038		1.29	1.93		3.22	4.36
0230	1" x 10"		400	.040		1.68	2.03		3.71	4.94
0240	1" x 12"		400	.040		2.07	2.03		4.10	5.35
0250	STK cedar, 1" x 4"		420	.038		.77	1.93		2.70	3.79
0260	1" x 6"		420	.038		1.28	1.93		3.21	4.35
0270	1" x 8"		420	.038		1.97	1.93		3.90	5.10
0280	1" x 10"		400	.040		2.71	2.03		4.74	6.05
0290	1" x 12"		400	.040	▼	4.23	2.03		6.26	7.75
1000	Exterior AC plywood, 1/4" thick		400	.040	S.F.	1	2.03		3.03	4.19
1050	3/8" thick		400	.040		1.04	2.03		3.07	4.23
1100	1/2" thick	▼	400	.040		1.23	2.03		3.26	4.44
1150	Polyvinyl chloride, white, solid	1 Carp	230	.035		2.17	1.76		3.93	5.10
1160	Perforated	"	230	.035	▼	2.17	1.76		3.93	5.10
1170	Accessories, "J" channel 5/8"	2 Carp	700	.023	L.F.	.50	1.16		1.66	2.31

06 25 Prefinished Paneling

06 25 13 – Prefinished Hardboard Paneling

06 25 13.10 Paneling, Hardboard

		Crew	Daily Output	Labor-Hours	Unit	Material	2018 Bare Costs Labor	2018 Bare Costs Equipment	Total	Total Incl O&P
0010	**PANELING, HARDBOARD**									
0050	Not incl. furring or trim, hardboard, tempered, 1/8" thick	G 2 Carp	500	.032	S.F.	.45	1.62		2.07	2.97
0100	1/4" thick	G	500	.032		.64	1.62		2.26	3.17
0300	Tempered pegboard, 1/8" thick	G	500	.032		.44	1.62		2.06	2.95
0400	1/4" thick	G	500	.032		.70	1.62		2.32	3.24
0600	Untempered hardboard, natural finish, 1/8" thick	G	500	.032		.44	1.62		2.06	2.95
0700	1/4" thick	G	500	.032		.53	1.62		2.15	3.05
0900	Untempered pegboard, 1/8" thick	G	500	.032		.48	1.62		2.10	3
1000	1/4" thick	G	500	.032		.48	1.62		2.10	3
1200	Plastic faced hardboard, 1/8" thick	G	500	.032		.62	1.62		2.24	3.15
1300	1/4" thick	G	500	.032		.84	1.62		2.46	3.39
1500	Plastic faced pegboard, 1/8" thick	G	500	.032		.62	1.62		2.24	3.15
1600	1/4" thick	G	500	.032		.80	1.62		2.42	3.35
1800	Wood grained, plain or grooved, 1/8" thick	G	500	.032		.65	1.62		2.27	3.19
1900	1/4" thick	G	425	.038		1.42	1.91		3.33	4.47
2100	Moldings, wood grained MDF		500	.032	L.F.	.41	1.62		2.03	2.92
2200	Pine		425	.038	"	1.41	1.91		3.32	4.46

06 25 16 – Prefinished Plywood Paneling

06 25 16.10 Paneling, Plywood

		Crew	Daily Output	Labor-Hours	Unit	Material	2018 Bare Costs Labor	2018 Bare Costs Equipment	Total	Total Incl O&P
0010	**PANELING, PLYWOOD** R061636-20									
2400	Plywood, prefinished, 1/4" thick, 4' x 8' sheets									
2410	with vertical grooves. Birch faced, economy	2 Carp	500	.032	S.F.	1.47	1.62		3.09	4.09
2420	Average		420	.038		1.20	1.93		3.13	4.26
2430	Custom		350	.046		1.10	2.32		3.42	4.74
2600	Mahogany, African		400	.040		2.45	2.03		4.48	5.80
2700	Philippine (Lauan)		500	.032		.65	1.62		2.27	3.19
2900	Oak		500	.032		1.34	1.62		2.96	3.94
3000	Cherry		400	.040		2	2.03		4.03	5.30
3200	Rosewood		320	.050		3.10	2.54		5.64	7.25
3400	Teak		400	.040		3.12	2.03		5.15	6.50
3600	Chestnut		375	.043		5.30	2.16		7.46	9.10
3800	Pecan		400	.040		2.50	2.03		4.53	5.85
3900	Walnut, average		500	.032		2.47	1.62		4.09	5.20
3950	Custom		400	.040		2.46	2.03		4.49	5.80
4000	Plywood, prefinished, 3/4" thick, stock grades, economy		320	.050		1.44	2.54		3.98	5.45
4100	Average		224	.071		4.93	3.62		8.55	10.90
4300	Architectural grade, custom		224	.071		5.40	3.62		9.02	11.40
4400	Luxury		160	.100		5.30	5.05		10.35	13.50
4600	Plywood, "A" face, birch, VC, 1/2" thick, natural		450	.036		1.38	1.80		3.18	4.27
4700	Select		450	.036		1.95	1.80		3.75	4.90
4900	Veneer core, 3/4" thick, natural		320	.050		2.25	2.54		4.79	6.35
5000	Select		320	.050		2.53	2.54		5.07	6.65
5200	Lumber core, 3/4" thick, natural		320	.050		3.14	2.54		5.68	7.30
5500	Plywood, knotty pine, 1/4" thick, A2 grade		450	.036		1.53	1.80		3.33	4.43
5600	A3 grade		450	.036		2.09	1.80		3.89	5.05
5800	3/4" thick, veneer core, A2 grade		320	.050		2.31	2.54		4.85	6.40
5900	A3 grade		320	.050		2.39	2.54		4.93	6.50
6100	Aromatic cedar, 1/4" thick, plywood		400	.040		2.25	2.03		4.28	5.55
6200	1/4" thick, particle board		400	.040		1.15	2.03		3.18	4.36

For customer support on your Building Construction Costs with RSMeans data, call 800.448.8182.

06 25 Prefinished Paneling

06 25 26 – Panel System

06 25 26.10 Panel Systems

		Crew	Daily Output	Labor-Hours	Unit	Material	2018 Bare Costs Labor	Equipment	Total	Total Incl O&P
0010	**PANEL SYSTEMS**									
0100	Raised panel, eng. wood core w/wood veneer, std., paint grade	2 Carp	300	.053	S.F.	12.15	2.70		14.85	17.45
0110	Oak veneer		300	.053		24	2.70		26.70	30.50
0120	Maple veneer		300	.053		30	2.70		32.70	37
0130	Cherry veneer		300	.053		37.50	2.70		40.20	45.50
0300	Class I fire rated, paint grade		300	.053		13.50	2.70		16.20	18.95
0310	Oak veneer		300	.053		30	2.70		32.70	37
0320	Maple veneer		300	.053		41	2.70		43.70	49
0330	Cherry veneer		300	.053		49	2.70		51.70	58
0510	Beadboard, 5/8" MDF, standard, primed		300	.053		9.35	2.70		12.05	14.40
0520	Oak veneer, unfinished		300	.053		14.70	2.70		17.40	20.50
0530	Maple veneer, unfinished		300	.053		15.80	2.70		18.50	21.50
0610	Rustic paneling, 5/8" MDF, standard, maple veneer, unfinished	▼	300	.053	▼	18.50	2.70		21.20	24.50

06 26 Board Paneling

06 26 13 – Profile Board Paneling

06 26 13.10 Paneling, Boards

		Crew	Daily Output	Labor-Hours	Unit	Material	2018 Bare Costs Labor	Equipment	Total	Total Incl O&P
0010	**PANELING, BOARDS**									
6400	Wood board paneling, 3/4" thick, knotty pine	2 Carp	300	.053	S.F.	2.05	2.70		4.75	6.40
6500	Rough sawn cedar		300	.053		3.30	2.70		6	7.75
6700	Redwood, clear, 1" x 4" boards		300	.053		4.96	2.70		7.66	9.55
6900	Aromatic cedar, closet lining, boards	▼	275	.058	▼	2.42	2.95		5.37	7.15

06 43 Wood Stairs and Railings

06 43 13 – Wood Stairs

06 43 13.20 Prefabricated Wood Stairs

		Crew	Daily Output	Labor-Hours	Unit	Material	2018 Bare Costs Labor	Equipment	Total	Total Incl O&P
0010	**PREFABRICATED WOOD STAIRS**									
0100	Box stairs, prefabricated, 3'-0" wide									
0110	Oak treads, up to 14 risers	2 Carp	39	.410	Riser	99.50	21		120.50	141
0600	With pine treads for carpet, up to 14 risers	"	39	.410	"	64	21		85	102
1100	For 4' wide stairs, add				Flight	25%				
1550	Stairs, prefabricated stair handrail with balusters	1 Carp	30	.267	L.F.	82	13.50		95.50	111
1700	Basement stairs, prefabricated, pine treads									
1710	Pine risers, 3' wide, up to 14 risers	2 Carp	52	.308	Riser	64	15.60		79.60	94.50
4000	Residential, wood, oak treads, prefabricated		1.50	10.667	Flight	1,300	540		1,840	2,250
4200	Built in place	▼	.44	36.364	"	2,300	1,850		4,150	5,325
4400	Spiral, oak, 4'-6" diameter, unfinished, prefabricated,									
4500	incl. railing, 9' high	2 Carp	1.50	10.667	Flight	2,750	540		3,290	3,850

06 43 13.40 Wood Stair Parts

		Crew	Daily Output	Labor-Hours	Unit	Material	2018 Bare Costs Labor	Equipment	Total	Total Incl O&P
0010	**WOOD STAIR PARTS**									
0020	Pin top balusters, 1-1/4", oak, 34"	1 Carp	96	.083	Ea.	5.20	4.22		9.42	12.15
0030	38"		96	.083		5.85	4.22		10.07	12.90
0040	42"		96	.083		6.35	4.22		10.57	13.45
0050	Poplar, 34"		96	.083		3.37	4.22		7.59	10.15
0060	38"		96	.083		3.99	4.22		8.21	10.85
0070	42"		96	.083		8.80	4.22		13.02	16.10
0080	Maple, 34"		96	.083		4.90	4.22		9.12	11.85
0090	38"		96	.083	▼	5.60	4.22		9.82	12.65

06 43 13.40 Wood Stair Parts	Crew	Daily Output	Labor-Hours	Unit	Material	2018 Bare Costs Labor	2018 Bare Costs Equipment	Total	Total Incl O&P
0100	42"	1 Carp	96	.083	Ea.	6.50	4.22		10.72	13.60
0130	Primed, 34"		96	.083		3.70	4.22		7.92	10.50
0140	38"		96	.083		4.25	4.22		8.47	11.15
0150	42"		96	.083		4.89	4.22		9.11	11.85
0180	Box top balusters, 1-1/4", oak, 34"		60	.133		8.75	6.75		15.50	19.95
0190	38"		60	.133		11.85	6.75		18.60	23.50
0200	42"		60	.133		13	6.75		19.75	24.50
0210	Poplar, 34"		60	.133		7.50	6.75		14.25	18.55
0220	38"		60	.133		8.10	6.75		14.85	19.20
0230	42"		60	.133		8.95	6.75		15.70	20
0240	Maple, 34"		60	.133		9.80	6.75		16.55	21
0250	38"		60	.133		10.75	6.75		17.50	22
0260	42"		60	.133		11.85	6.75		18.60	23.50
0290	Primed, 34"		60	.133		8.25	6.75		15	19.40
0300	38"		60	.133		9.50	6.75		16.25	21
0310	42"		60	.133		9.85	6.75		16.60	21
0340	Square balusters, cut from lineal stock, pine, 1-1/16" x 1-1/16"		180	.044	L.F.	1.31	2.25		3.56	4.87
0350	1-5/16" x 1-5/16"		180	.044		1.98	2.25		4.23	5.60
0360	1-5/8" x 1-5/8"		180	.044		2.68	2.25		4.93	6.40
0370	Turned newel, oak, 3-1/2" square, 48" high		8	1	Ea.	105	50.50		155.50	193
0380	62" high		8	1		101	50.50		151.50	188
0390	Poplar, 3-1/2" square, 48" high		8	1		46.50	50.50		97	129
0400	62" high		8	1		58	50.50		108.50	141
0410	Maple, 3-1/2" square, 48" high		8	1		65	50.50		115.50	149
0420	62" high		8	1		80	50.50		130.50	165
0430	Square newel, oak, 3-1/2" square, 48" high		8	1		55	50.50		105.50	138
0440	58" high		8	1		68.50	50.50		119	152
0450	Poplar, 3-1/2" square, 48" high		8	1		35.50	50.50		86	116
0460	58" high		8	1		43.50	50.50		94	125
0470	Maple, 3" square, 48" high		8	1		54	50.50		104.50	137
0480	58" high		8	1		68	50.50		118.50	152
0490	Railings, oak, economy		96	.083	L.F.	9.45	4.22		13.67	16.85
0500	Average		96	.083		13.50	4.22		17.72	21.50
0510	Custom		96	.083		17.95	4.22		22.17	26
0520	Maple, economy		96	.083		11.90	4.22		16.12	19.50
0530	Average		96	.083		13.50	4.22		17.72	21.50
0540	Custom		96	.083		20.50	4.22		24.72	29
0550	Oak, for bending rail, economy		48	.167		26	8.45		34.45	41.50
0560	Average		48	.167		29.50	8.45		37.95	45.50
0570	Custom		48	.167		33	8.45		41.45	49.50
0580	Maple, for bending rail, economy		48	.167		29	8.45		37.45	44.50
0590	Average		48	.167		31.50	8.45		39.95	47.50
0600	Custom		48	.167		35	8.45		43.45	51
0610	Risers, oak, 3/4" x 8", 36" long		80	.100	Ea.	13	5.05		18.05	22
0620	42" long		70	.114		15.15	5.80		20.95	25.50
0630	48" long		63	.127		17.30	6.45		23.75	29
0640	54" long		56	.143		19.50	7.25		26.75	32.50
0650	60" long		50	.160		21.50	8.10		29.60	36.50
0660	72" long		42	.190		26	9.65		35.65	43
0670	Poplar, 3/4" x 8", 36" long		80	.100		12.75	5.05		17.80	22
0680	42" long		71	.113		14.90	5.70		20.60	25
0690	48" long		63	.127		17	6.45		23.45	28.50
0700	54" long		56	.143		19.15	7.25		26.40	32

For customer support on your Building Construction Costs with RSMeans data, call 800.448.8182.

06 43 Wood Stairs and Railings

06 43 13 – Wood Stairs

06 43 13.40 Wood Stair Parts		Crew	Daily Output	Labor-Hours	Unit	Material	2018 Bare Costs Labor	Equipment	Total	Total Incl O&P
0710	60" long	1 Carp	50	.160	Ea.	21.50	8.10		29.60	36
0720	72" long		42	.190		25.50	9.65		35.15	42.50
0730	Pine, 1" x 8", 36" long		80	.100		3.88	5.05		8.93	11.95
0740	42" long		70	.114		4.53	5.80		10.33	13.80
0750	48" long		63	.127		5.20	6.45		11.65	15.50
0760	54" long		56	.143		5.80	7.25		13.05	17.45
0770	60" long		50	.160		6.45	8.10		14.55	19.45
0780	72" long		42	.190		7.75	9.65		17.40	23.50
0790	Treads, oak, no returns, 1-1/32" x 11-1/2" x 36" long		32	.250		29.50	12.70		42.20	52
0800	42" long		32	.250		34.50	12.70		47.20	57.50
0810	48" long		32	.250		39.50	12.70		52.20	62.50
0820	54" long		32	.250		44	12.70		56.70	68
0830	60" long		32	.250		49	12.70		61.70	73.50
0840	72" long		32	.250		59	12.70		71.70	84.50
0850	Mitred return one end, 1-1/32" x 11-1/2" x 36" long		24	.333		39	16.90		55.90	68.50
0860	42" long		24	.333		45.50	16.90		62.40	75.50
0870	48" long		24	.333		52	16.90		68.90	82.50
0880	54" long		24	.333		58.50	16.90		75.40	90
0890	60" long		24	.333		65	16.90		81.90	97
0900	72" long		24	.333		78	16.90		94.90	111
0910	Mitred return two ends, 1-1/32" x 11-1/2" x 36" long		12	.667		49	34		83	106
0920	42" long		12	.667		57	34		91	115
0930	48" long		12	.667		65.50	34		99.50	124
0940	54" long		12	.667		73.50	34		107.50	133
0950	60" long		12	.667		81.50	34		115.50	142
0960	72" long		12	.667		98	34		132	160
0970	Starting step, oak, 48", bullnose		8	1		172	50.50		222.50	266
0980	Double end bullnose		8	1		254	50.50		304.50	355
1030	Skirt board, pine, 1" x 10"		55	.145	L.F.	1.68	7.35		9.03	13.10
1040	1" x 12"		52	.154	"	2.07	7.80		9.87	14.20
1050	Oak landing tread, 1-1/16" thick		54	.148	S.F.	7.95	7.50		15.45	20
1060	Oak cove molding		96	.083	L.F.	1	4.22		5.22	7.55
1070	Oak stringer molding		96	.083	"	4	4.22		8.22	10.85
1090	Rail bolt, 5/16" x 3-1/2"		48	.167	Ea.	2.75	8.45		11.20	15.90
1100	5/16" x 4-1/2"		48	.167		2.75	8.45		11.20	15.90
1120	Newel post anchor		16	.500		13	25.50		38.50	53
1130	Tapered plug, 1/2"		240	.033		1	1.69		2.69	3.67
1140	1"		240	.033		.99	1.69		2.68	3.66

06 43 16 – Wood Railings

06 43 16.10 Wood Handrails and Railings

06 43 16.10 Wood Handrails and Railings		Crew	Daily Output	Labor-Hours	Unit	Material	2018 Bare Costs Labor	Equipment	Total	Total Incl O&P
0010	**WOOD HANDRAILS AND RAILINGS**									
0020	Custom design, architectural grade, hardwood, plain	1 Carp	38	.211	L.F.	12.65	10.65		23.30	30
0100	Shaped		30	.267		53	13.50		66.50	79
0300	Stock interior railing with spindles 4" OC, 4' long		40	.200		46.50	10.15		56.65	67
0400	8' long		48	.167		46.50	8.45		54.95	64.50

06 44 Ornamental Woodwork

06 44 19 – Wood Grilles

06 44 19.10 Grilles	Crew	Daily Output	Labor-Hours	Unit	Material	2018 Bare Costs Labor	Equipment	Total	Total Incl O&P
0010 **GRILLES** and panels, hardwood, sanded									
0020 2' x 4' to 4' x 8', custom designs, unfinished, economy	1 Carp	38	.211	S.F.	58	10.65		68.65	80.50
0050 Average		30	.267		68.50	13.50		82	96
0100 Custom		19	.421		72	21.50		93.50	112

06 44 33 – Wood Mantels

06 44 33.10 Fireplace Mantels

06 44 33.10 Fireplace Mantels	Crew	Daily Output	Labor-Hours	Unit	Material	Labor	Equipment	Total	Total Incl O&P
0010 **FIREPLACE MANTELS**									
0015 6" molding, 6' x 3'-6" opening, plain, paint grade	1 Carp	5	1.600	Opng.	450	81		531	620
0100 Ornate, oak		5	1.600		600	81		681	780
0300 Prefabricated pine, colonial type, stock, deluxe		2	4		1,825	203		2,028	2,300
0400 Economy		3	2.667		790	135		925	1,075

06 44 33.20 Fireplace Mantel Beam

06 44 33.20 Fireplace Mantel Beam	Crew	Daily Output	Labor-Hours	Unit	Material	Labor	Equipment	Total	Total Incl O&P
0010 **FIREPLACE MANTEL BEAM**									
0020 Rough texture wood, 4" x 8"	1 Carp	36	.222	L.F.	8.05	11.25		19.30	26
0100 4" x 10"		35	.229	"	10.80	11.60		22.40	29.50
0300 Laminated hardwood, 2-1/4" x 10-1/2" wide, 6' long		5	1.600	Ea.	105	81		186	239
0400 8' long		5	1.600	"	146	81		227	284
0600 Brackets for above, rough sawn		12	.667	Pr.	10.10	34		44.10	62.50
0700 Laminated		12	.667	"	19.50	34		53.50	73

06 44 39 – Wood Posts and Columns

06 44 39.10 Decorative Beams

06 44 39.10 Decorative Beams	Crew	Daily Output	Labor-Hours	Unit	Material	Labor	Equipment	Total	Total Incl O&P
0010 **DECORATIVE BEAMS**									
0020 Rough sawn cedar, non-load bearing, 4" x 4"	2 Carp	180	.089	L.F.	1.62	4.51		6.13	8.65
0100 4" x 6"		170	.094		1.77	4.77		6.54	9.20
0200 4" x 8"		160	.100		2.32	5.05		7.37	10.25
0300 4" x 10"		150	.107		4.04	5.40		9.44	12.70
0400 4" x 12"		140	.114		4.66	5.80		10.46	13.95
0500 8" x 8"		130	.123		4.77	6.25		11.02	14.75

06 44 39.20 Columns

06 44 39.20 Columns	Crew	Daily Output	Labor-Hours	Unit	Material	Labor	Equipment	Total	Total Incl O&P
0010 **COLUMNS**									
0050 Aluminum, round colonial, 6" diameter	2 Carp	80	.200	V.L.F.	17.75	10.15		27.90	35
0100 8" diameter		62.25	.257		18.55	13.05		31.60	40.50
0200 10" diameter		55	.291		21	14.75		35.75	46
0250 Fir, stock units, hollow round, 6" diameter		80	.200		26	10.15		36.15	44
0300 8" diameter		80	.200		33.50	10.15		43.65	52
0350 10" diameter		70	.229		41	11.60		52.60	62.50
0360 12" diameter		65	.246		59	12.50		71.50	84
0400 Solid turned, to 8' high, 3-1/2" diameter		80	.200		10.15	10.15		20.30	26.50
0500 4-1/2" diameter		75	.213		12.10	10.80		22.90	30
0600 5-1/2" diameter		70	.229		16.35	11.60		27.95	35.50
0800 Square columns, built-up, 5" x 5"		65	.246		33.50	12.50		46	56
0900 Solid, 3-1/2" x 3-1/2"		130	.123		10.15	6.25		16.40	20.50
1600 Hemlock, tapered, T&G, 12" diam., 10' high		100	.160		47.50	8.10		55.60	64.50
1700 16' high		65	.246		79	12.50		91.50	106
1900 14" diameter, 10' high		100	.160		96	8.10		104.10	117
2000 18' high		65	.246		106	12.50		118.50	135
2200 18" diameter, 12' high		65	.246		173	12.50		185.50	210
2300 20' high		50	.320		124	16.20		140.20	161
2500 20" diameter, 14' high		40	.400		187	20.50		207.50	237
2600 20' high		35	.457		177	23		200	231
2800 For flat pilasters, deduct					33%				

For customer support on your Building Construction Costs with RSMeans data, call 800.448.8182.

06 44 Ornamental Woodwork

06 44 39 - Wood Posts and Columns

06 44 39.20 Columns		Crew	Daily Output	Labor-Hours	Unit	Material	2018 Bare Costs Labor	Equipment	Total	Total Incl O&P
3000	For splitting into halves, add				Ea.	110			110	121
4000	Rough sawn cedar posts, 4" x 4"	2 Carp	250	.064	V.L.F.	4.24	3.24		7.48	9.60
4100	4" x 6"		235	.068		8.30	3.45		11.75	14.35
4200	6" x 6"		220	.073		14	3.69		17.69	21
4300	8" x 8"		200	.080		18.70	4.06		22.76	26.50

06 48 Wood Frames

06 48 13 - Exterior Wood Door Frames

06 48 13.10 Exterior Wood Door Frames and Accessories

06 48 13.10 Exterior Wood Door Frames and Accessories		Crew	Daily Output	Labor-Hours	Unit	Material	2018 Bare Costs Labor	Equipment	Total	Total Incl O&P
0010	**EXTERIOR WOOD DOOR FRAMES AND ACCESSORIES**									
0400	Exterior frame, incl. ext. trim, pine, 5/4 x 4-9/16" deep	2 Carp	375	.043	L.F.	6.75	2.16		8.91	10.75
0420	5-3/16" deep		375	.043		8.30	2.16		10.46	12.40
0440	6-9/16" deep		375	.043		9.75	2.16		11.91	14.05
0600	Oak, 5/4 x 4-9/16" deep		350	.046		12.55	2.32		14.87	17.35
0620	5-3/16" deep		350	.046		13.80	2.32		16.12	18.70
0640	6-9/16" deep		350	.046		18.55	2.32		20.87	24
1000	Sills, 8/4 x 8" deep, oak, no horns		100	.160		7.15	8.10		15.25	20
1020	2" horns		100	.160		21.50	8.10		29.60	36
1040	3" horns		100	.160		21.50	8.10		29.60	36
1100	8/4 x 10" deep, oak, no horns		90	.178		6.55	9		15.55	21
1120	2" horns		90	.178		27	9		36	43
1140	3" horns		90	.178		27	9		36	43
2000	Wood frame & trim, ext., colonial, 3' opng., fluted pilasters, flat head		22	.727	Ea.	505	37		542	610
2010	Dentil head		21	.762		600	38.50		638.50	720
2020	Ram's head		20	.800		650	40.50		690.50	775
2100	5'-4" opening, in-swing, fluted pilasters, flat head		17	.941		445	47.50		492.50	565
2120	Ram's head		15	1.067		1,325	54		1,379	1,525
2140	Out-swing, fluted pilasters, flat head		17	.941		480	47.50		527.50	600
2160	Ram's head		15	1.067		1,500	54		1,554	1,725
2400	6'-0" opening, in-swing, fluted pilasters, flat head		16	1		480	50.50		530.50	600
2420	Ram's head		10	1.600		1,500	81		1,581	1,775
2460	Out-swing, fluted pilasters, flat head		16	1		480	50.50		530.50	605
2480	Ram's head		10	1.600		1,500	81		1,581	1,775
2600	For two sidelights, flat head, add		30	.533	Opng.	277	27		304	345
2620	Ram's head, add		20	.800	"	920	40.50		960.50	1,050
2700	Custom birch frame, 3'-0" opening		16	1	Ea.	232	50.50		282.50	330
2750	6'-0" opng.		16	1		365	50.50		415.50	475
2900	Exterior, modern, plain trim, 3' opng., in-swing, FJP		26	.615		48	31		79	101
2920	Fir		24	.667		56.50	34		90.50	114
2940	Oak		22	.727		66	37		103	129

06 48 16 - Interior Wood Door Frames

06 48 16.10 Interior Wood Door Jamb and Frames

06 48 16.10 Interior Wood Door Jamb and Frames		Crew	Daily Output	Labor-Hours	Unit	Material	2018 Bare Costs Labor	Equipment	Total	Total Incl O&P
0010	**INTERIOR WOOD DOOR JAMB AND FRAMES**									
3000	Interior frame, pine, 11/16" x 3-5/8" deep	2 Carp	375	.043	L.F.	4.15	2.16		6.31	7.85
3020	4-9/16" deep		375	.043		4.80	2.16		6.96	8.60
3200	Oak, 11/16" x 3-5/8" deep		350	.046		2.76	2.32		5.08	6.55
3220	4-9/16" deep		350	.046		11	2.32		13.32	15.65
3240	5-3/16" deep		350	.046		17.85	2.32		20.17	23
3400	Walnut, 11/16" x 3-5/8" deep		350	.046		9.10	2.32		11.42	13.55
3420	4-9/16" deep		350	.046		10.80	2.32		13.12	15.40

For customer support on your Building Construction Costs with RSMeans data, call 800.448.8182.

213

06 48 Wood Frames

06 48 16 – Interior Wood Door Frames

06 48 16.10 Interior Wood Door Jamb and Frames		Crew	Daily Output	Labor-Hours	Unit	Material	2018 Bare Costs Labor	Equipment	Total	Total Incl O&P
3440	5-3/16" deep	2 Carp	350	.046	L.F.	9.65	2.32		11.97	14.20
3600	Pocket door frame		16	1	Ea.	86	50.50		136.50	172
3800	Threshold, oak, 5/8" x 3-5/8" deep		200	.080	L.F.	3.57	4.06		7.63	10.15
3820	4-5/8" deep		190	.084		3.95	4.27		8.22	10.85
3840	5-5/8" deep	↓	180	.089	↓	6.65	4.51		11.16	14.15

06 49 Wood Screens and Exterior Wood Shutters

06 49 19 – Exterior Wood Shutters

06 49 19.10 Shutters, Exterior

		Crew	Daily Output	Labor-Hours	Unit	Material	2018 Bare Costs Labor	Equipment	Total	Total Incl O&P
0010	**SHUTTERS, EXTERIOR**									
0012	Aluminum, louvered, 1'-4" wide, 3'-0" long	1 Carp	10	.800	Pr.	192	40.50		232.50	273
0400	6'-8" long		9	.889		340	45		385	445
1000	Pine, louvered, primed, each 1'-2" wide, 3'-3" long		10	.800		235	40.50		275.50	320
1100	4'-7" long		10	.800		270	40.50		310.50	360
1500	Each 1'-6" wide, 3'-3" long		10	.800		255	40.50		295.50	340
1600	4'-7" long		10	.800		315	40.50		355.50	410
1620	Cedar, louvered, 1'-2" wide, 5'-7" long		10	.800		320	40.50		360.50	410
1630	Each 1'-4" wide, 2'-2" long		10	.800		171	40.50		211.50	250
1670	4'-3" long		10	.800		294	40.50		334.50	385
1690	5'-11" long		10	.800		350	40.50		390.50	445
1700	Door blinds, 6'-9" long, each 1'-3" wide		9	.889		385	45		430	495
1710	1'-6" wide		9	.889		415	45		460	525
1720	Cedar, solid raised panel, each 1'-4" wide, 3'-3" long		10	.800		325	40.50		365.50	415
1740	4'-3" long		10	.800		335	40.50		375.50	430
1770	5'-11" long		10	.800		515	40.50		555.50	630
1800	Door blinds, 6'-9" long, each 1'-3" wide		9	.889		490	45		535	610
1900	1'-6" wide		9	.889		580	45		625	710
2500	Polystyrene, solid raised panel, each 1'-4" wide, 3'-3" long		10	.800		82	40.50		122.50	152
2700	4'-7" long		10	.800		102	40.50		142.50	174
4500	Polystyrene, louvered, each 1'-2" wide, 3'-3" long		10	.800		38.50	40.50		79	104
4750	5'-3" long		10	.800		53.50	40.50		94	121
6000	Vinyl, louvered, each 1'-2" x 4'-7" long		10	.800		58	40.50		98.50	126
6200	Each 1'-4" x 6'-8" long	↓	9	.889	↓	81.50	45		126.50	158

06 51 Structural Plastic Shapes and Plates

06 51 13 – Plastic Lumber

06 51 13.10 Recycled Plastic Lumber

			Crew	Daily Output	Labor-Hours	Unit	Material	2018 Bare Costs Labor	Equipment	Total	Total Incl O&P
0010	**RECYCLED PLASTIC LUMBER**										
4000	Sheeting, recycled plastic, black or white, 4' x 8' x 1/8"	G	2 Carp	1100	.015	S.F.	1.09	.74		1.83	2.32
4010	4' x 8' x 3/16"	G		1100	.015		1.51	.74		2.25	2.78
4020	4' x 8' x 1/4"	G		950	.017		1.81	.85		2.66	3.29
4030	4' x 8' x 3/8"	G		950	.017		3.02	.85		3.87	4.62
4040	4' x 8' x 1/2"	G		900	.018		4.03	.90		4.93	5.80
4050	4' x 8' x 5/8"	G		900	.018		6	.90		6.90	7.95
4060	4' x 8' x 3/4"	G	↓	850	.019	↓	7.50	.95		8.45	9.70
4070	Add for colors	G				Ea.	5%				
8500	100% recycled plastic, var colors, NLB, 2" x 2"	G				L.F.	1.90			1.90	2.09
8510	2" x 4"	G					3.90			3.90	4.29
8520	2" x 6"	G					6.10			6.10	6.75

For customer support on your Building Construction Costs with RSMeans data, call 800.448.8182.

06 51 Structural Plastic Shapes and Plates

06 51 13 – Plastic Lumber

06 51 13.10 Recycled Plastic Lumber

		Crew	Daily Output	Labor-Hours	Unit	Material	2018 Bare Costs Labor	Equipment	Total	Total Incl O&P
8530	2" x 8" G				L.F.	8.60			8.60	9.50
8540	2" x 10" G					12.25			12.25	13.50
8550	5/4" x 4" G					4.58			4.58	5.05
8560	5/4" x 6" G					5.70			5.70	6.25
8570	1" x 6" G					2.83			2.83	3.11
8580	1/2" x 8" G					3.32			3.32	3.65
8590	2" x 10" T&G G					12.65			12.65	13.90
8600	3" x 10" T&G G					18.95			18.95	21
8610	Add for premium colors G					20%				

06 51 13.12 Structural Plastic Lumber

		Crew	Daily Output	Labor-Hours	Unit	Material	2018 Bare Costs Labor	Equipment	Total	Total Incl O&P
0010	**STRUCTURAL PLASTIC LUMBER**									
1320	Plastic lumber, posts or columns, 4" x 4"	2 Carp	390	.041	L.F.	11.10	2.08		13.18	15.40
1325	4" x 6"		275	.058		13.30	2.95		16.25	19.15
1330	4" x 8"		220	.073		19.55	3.69		23.24	27
1340	Girder, single, 4" x 4"		675	.024		11.10	1.20		12.30	14.10
1345	4" x 6"		600	.027		13.30	1.35		14.65	16.70
1350	4" x 8"		525	.030		19.55	1.55		21.10	24
1352	Double, 2" x 4"		625	.026		7.95	1.30		9.25	10.75
1354	2" x 6"		600	.027		9.65	1.35		11	12.65
1356	2" x 8"		575	.028		16.85	1.41		18.26	20.50
1358	2" x 10"		550	.029		23.50	1.47		24.97	28.50
1360	2" x 12"		525	.030		25	1.55		26.55	30
1362	Triple, 2" x 4"		575	.028		11.90	1.41		13.31	15.25
1364	2" x 6"		550	.029		14.45	1.47		15.92	18.15
1366	2" x 8"		525	.030		25	1.55		26.55	30.50
1368	2" x 10"		500	.032		35.50	1.62		37.12	41.50
1370	2" x 12"		475	.034		37.50	1.71		39.21	43.50
1372	Ledger, bolted 4' OC, 2" x 4"		400	.040		4.12	2.03		6.15	7.60
1374	2" x 6"		550	.029		4.89	1.47		6.36	7.65
1376	2" x 8"		390	.041		8.55	2.08		10.63	12.55
1378	2" x 10"		385	.042		11.90	2.11		14.01	16.30
1380	2" x 12"		380	.042		12.60	2.14		14.74	17.10
1382	Joists, 2" x 4"		1250	.013		3.97	.65		4.62	5.35
1384	2" x 6"		1250	.013		4.83	.65		5.48	6.30
1386	2" x 8"		1100	.015		8.40	.74		9.14	10.35
1388	2" x 10"		500	.032		11.90	1.62		13.52	15.55
1390	2" x 12"		875	.018		12.50	.93		13.43	15.15
1392	Railings and trim, 5/4" x 4"	1 Carp	300	.027		4.35	1.35		5.70	6.85
1394	2" x 2"		300	.027		1.91	1.35		3.26	4.16
1396	2" x 4"		300	.027		3.95	1.35		5.30	6.40
1398	2" x 6"		300	.027		4.79	1.35		6.14	7.30

For customer support on your Building Construction Costs with RSMeans data, call 800.448.8182.

215

06 52 Plastic Structural Assemblies

06 52 10 – Fiberglass Structural Assemblies

06 52 10.10 Castings, Fiberglass

		Crew	Daily Output	Labor-Hours	Unit	Material	2018 Bare Costs Labor	Equipment	Total	Total Incl O&P
0010	**CASTINGS, FIBERGLASS**									
0100	Angle, 1" x 1" x 1/8" thick	2 Sswk	240	.067	L.F.	1.68	3.64		5.32	7.80
0120	3" x 3" x 1/4" thick		200	.080		7.45	4.37		11.82	15.35
0140	4" x 4" x 1/4" thick		200	.080		9.30	4.37		13.67	17.40
0160	4" x 4" x 3/8" thick		200	.080		10.70	4.37		15.07	18.90
0180	6" x 6" x 1/2" thick		160	.100		26	5.45		31.45	37.50
1000	Flat sheet, 1/8" thick		140	.114	S.F.	6	6.25		12.25	16.80
1020	1/4" thick		120	.133		9.70	7.30		17	22.50
1040	3/8" thick		100	.160		15.50	8.75		24.25	31.50
1060	1/2" thick		80	.200		19.55	10.95		30.50	39.50
2000	Handrail, 42" high, 2" diam. rails pickets 5' OC		32	.500	L.F.	51.50	27.50		79	101
3000	Round bar, 1/4" diam.		240	.067		.41	3.64		4.05	6.40
3020	1/2" diam.		200	.080		1.17	4.37		5.54	8.45
3040	3/4" diam.		200	.080		2.91	4.37		7.28	10.35
3060	1" diam.		160	.100		5.15	5.45		10.60	14.60
3080	1-1/4" diam.		160	.100		8.90	5.45		14.35	18.75
3100	1-1/2" diam.		140	.114		11.50	6.25		17.75	23
3500	Round tube, 1" diam. x 1/8" thick		240	.067		3	3.64		6.64	9.25
3520	2" diam. x 1/4" thick		200	.080		7.55	4.37		11.92	15.45
3540	3" diam. x 1/4" thick		160	.100		11.50	5.45		16.95	21.50
4000	Square bar, 1/2" square		240	.067		4.45	3.64		8.09	10.85
4020	1" square		200	.080		5.90	4.37		10.27	13.60
4040	1-1/2" square		160	.100		11.10	5.45		16.55	21
4500	Square tube, 1" x 1" x 1/8" thick		240	.067		3.31	3.64		6.95	9.60
4520	2" x 2" x 1/8" thick		200	.080		5.10	4.37		9.47	12.75
4540	3" x 3" x 1/4" thick		160	.100		13.85	5.45		19.30	24
5000	Threaded rod, 3/8" diam.		320	.050		4.57	2.73		7.30	9.50
5020	1/2" diam.		320	.050		5.75	2.73		8.48	10.75
5040	5/8" diam.		280	.057		5.90	3.12		9.02	11.60
5060	3/4" diam.		280	.057		6.85	3.12		9.97	12.65
6000	Wide flange beam, 4" x 4" x 1/4" thick		120	.133		10.75	7.30		18.05	24
6020	6" x 6" x 1/4" thick		100	.160		17.95	8.75		26.70	34
6040	8" x 8" x 3/8" thick		80	.200		37	10.95		47.95	58.50

06 52 10.20 Fiberglass Stair Treads

		Crew	Daily Output	Labor-Hours	Unit	Material	2018 Bare Costs Labor	Equipment	Total	Total Incl O&P
0010	**FIBERGLASS STAIR TREADS**									
0100	24" wide	2 Sswk	52	.308	Ea.	36.50	16.80		53.30	67.50
0140	30" wide		52	.308		45	16.80		61.80	77
0180	36" wide		52	.308		54.50	16.80		71.30	87.50
0220	42" wide		52	.308		63.50	16.80		80.30	97

06 52 10.30 Fiberglass Grating

		Crew	Daily Output	Labor-Hours	Unit	Material	2018 Bare Costs Labor	Equipment	Total	Total Incl O&P
0010	**FIBERGLASS GRATING**									
0100	Molded, green (for mod. corrosive environment)									
0140	1" x 4" mesh, 1" thick	2 Sswk	400	.040	S.F.	7.40	2.19		9.59	11.70
0180	1-1/2" square mesh, 1" thick		400	.040		15.75	2.19		17.94	21
0220	1-1/4" thick		400	.040		9.05	2.19		11.24	13.50
0260	1-1/2" thick		400	.040		26.50	2.19		28.69	32.50
0300	2" square mesh, 2" thick		320	.050		27.50	2.73		30.23	34.50
1000	Orange (for highly corrosive environment)									
1040	1" x 4" mesh, 1" thick	2 Sswk	400	.040	S.F.	16.95	2.19		19.14	22
1080	1-1/2" square mesh, 1" thick		400	.040		18.95	2.19		21.14	24.50
1120	1-1/4" thick		400	.040		19.95	2.19		22.14	25.50
1160	1-1/2" thick		400	.040		21	2.19		23.19	26.50

For customer support on your Building Construction Costs with RSMeans data, call 800.448.8182.

06 52 Plastic Structural Assemblies

06 52 10 – Fiberglass Structural Assemblies

06 52 10.30 Fiberglass Grating

		Crew	Daily Output	Labor-Hours	Unit	Material	2018 Bare Costs Labor	Equipment	Total	Total Incl O&P
1200	2" square mesh, 2" thick	2 Sswk	320	.050	S.F.	21	2.73		23.73	27.50
3000	Pultruded, green (for mod. corrosive environment)									
3040	1" OC bar spacing, 1" thick	2 Sswk	400	.040	S.F.	13.95	2.19		16.14	18.90
3080	1-1/2" thick		320	.050		14.65	2.73		17.38	20.50
3120	1-1/2" OC bar spacing, 1" thick		400	.040		14	2.19		16.19	18.95
3160	1-1/2" thick	↓	400	.040	↓	20.50	2.19		22.69	26.50
4000	Grating support legs, fixed height, no base				Ea.	46			46	50.50
4040	With base					40.50			40.50	45
4080	Adjustable to 60"				↓	61.50			61.50	67.50

06 52 10.40 Fiberglass Floor Grating

		Crew	Daily Output	Labor-Hours	Unit	Material	2018 Bare Costs Labor	Equipment	Total	Total Incl O&P
0010	**FIBERGLASS FLOOR GRATING**									
0100	Reinforced polyester, fire retardant, 1" x 4" grid, 1" thick	E-4	510	.063	S.F.	14.50	3.46	.19	18.15	22
0200	1-1/2" x 6" mesh, 1-1/2" thick		500	.064		16.85	3.53	.20	20.58	24.50
0300	With grit surface, 1-1/2" x 6" grid, 1-1/2" thick	↓	500	.064	↓	17.20	3.53	.20	20.93	25

06 63 Plastic Railings

06 63 10 – Plastic (PVC) Railings

06 63 10.10 Plastic Railings

		Crew	Daily Output	Labor-Hours	Unit	Material	2018 Bare Costs Labor	Equipment	Total	Total Incl O&P
0010	**PLASTIC RAILINGS**									
0100	Horizontal PVC handrail with balusters, 3-1/2" wide, 36" high	1 Carp	96	.083	L.F.	29	4.22		33.22	38
0150	42" high		96	.083		28.50	4.22		32.72	38
0200	Angled PVC handrail with balusters, 3-1/2" wide, 36" high		72	.111		18.85	5.65		24.50	29.50
0250	42" high		72	.111		28.50	5.65		34.15	40
0300	Post sleeve for 4 x 4 post		96	.083	↓	14.70	4.22		18.92	22.50
0400	Post cap for 4 x 4 post, flat profile		48	.167	Ea.	13	8.45		21.45	27
0450	Newel post style profile		48	.167		25.50	8.45		33.95	41
0500	Raised corbeled profile		48	.167		36.50	8.45		44.95	53.50
0550	Post base trim for 4 x 4 post	↓	96	.083	↓	20.50	4.22		24.72	29

06 65 Plastic Trim

06 65 10 – PVC Trim

06 65 10.10 PVC Trim, Exterior

		Crew	Daily Output	Labor-Hours	Unit	Material	2018 Bare Costs Labor	Equipment	Total	Total Incl O&P
0010	**PVC TRIM, EXTERIOR**									
0100	Cornerboards, 5/4" x 6" x 6"	1 Carp	240	.033	L.F.	7.20	1.69		8.89	10.45
0110	Door/window casing, 1" x 4"		200	.040		1.39	2.03		3.42	4.62
0120	1" x 6"		200	.040		2.06	2.03		4.09	5.35
0130	1" x 8"		195	.041		2.72	2.08		4.80	6.15
0140	1" x 10"		195	.041		3.44	2.08		5.52	6.95
0150	1" x 12"		190	.042		4.22	2.14		6.36	7.90
0160	5/4" x 4"		195	.041		1.69	2.08		3.77	5.05
0170	5/4" x 6"		195	.041		2.67	2.08		4.75	6.10
0180	5/4" x 8"		190	.042		3.50	2.14		5.64	7.10
0190	5/4" x 10"		190	.042		4.50	2.14		6.64	8.20
0200	5/4" x 12"		185	.043		5.35	2.19		7.54	9.25
0210	Fascia, 1" x 4"		250	.032		1.39	1.62		3.01	4
0220	1" x 6"		250	.032		2.06	1.62		3.68	4.74
0230	1" x 8"		225	.036		2.72	1.80		4.52	5.75
0240	1" x 10"		225	.036		3.44	1.80		5.24	6.55
0250	1" x 12"		200	.040		4.22	2.03		6.25	7.75

For customer support on your Building Construction Costs with RSMeans data, call 800.448.8182.

217

06 65 10 − PVC Trim

06 65 10.10 PVC Trim, Exterior		Crew	Daily Output	Labor-Hours	Unit	Material	2018 Bare Costs Labor	Equipment	Total	Total Incl O&P
0260	5/4" x 4"	1 Carp	240	.033	L.F.	1.69	1.69		3.38	4.43
0270	5/4" x 6"		240	.033		2.67	1.69		4.36	5.50
0280	5/4" x 8"		215	.037		3.50	1.89		5.39	6.70
0290	5/4" x 10"		215	.037		4.50	1.89		6.39	7.80
0300	5/4" x 12"		190	.042		5.35	2.14		7.49	9.15
0310	Frieze, 1" x 4"		250	.032		1.39	1.62		3.01	4
0320	1" x 6"		250	.032		2.06	1.62		3.68	4.74
0330	1" x 8"		225	.036		2.72	1.80		4.52	5.75
0340	1" x 10"		225	.036		3.44	1.80		5.24	6.55
0350	1" x 12"		200	.040		4.22	2.03		6.25	7.75
0360	5/4" x 4"		240	.033		1.69	1.69		3.38	4.43
0370	5/4" x 6"		240	.033		2.67	1.69		4.36	5.50
0380	5/4" x 8"		215	.037		3.50	1.89		5.39	6.70
0390	5/4" x 10"		215	.037		4.50	1.89		6.39	7.80
0400	5/4" x 12"		190	.042		5.35	2.14		7.49	9.15
0410	Rake, 1" x 4"		200	.040		1.39	2.03		3.42	4.62
0420	1" x 6"		200	.040		2.06	2.03		4.09	5.35
0430	1" x 8"		190	.042		2.72	2.14		4.86	6.25
0440	1" x 10"		190	.042		3.44	2.14		5.58	7.05
0450	1" x 12"		180	.044		4.22	2.25		6.47	8.05
0460	5/4" x 4"		195	.041		1.69	2.08		3.77	5.05
0470	5/4" x 6"		195	.041		2.67	2.08		4.75	6.10
0480	5/4" x 8"		185	.043		3.50	2.19		5.69	7.20
0490	5/4" x 10"		185	.043		4.50	2.19		6.69	8.30
0500	5/4" x 12"		175	.046		5.35	2.32		7.67	9.45
0510	Rake trim, 1" x 4"		225	.036		1.39	1.80		3.19	4.28
0520	1" x 6"		225	.036		2.06	1.80		3.86	5
0560	5/4" x 4"		220	.036		1.69	1.84		3.53	4.67
0570	5/4" x 6"		220	.036		2.67	1.84		4.51	5.75
0610	Soffit, 1" x 4"	2 Carp	420	.038		1.39	1.93		3.32	4.47
0620	1" x 6"		420	.038		2.06	1.93		3.99	5.20
0630	1" x 8"		420	.038		2.72	1.93		4.65	5.95
0640	1" x 10"		400	.040		3.44	2.03		5.47	6.85
0650	1" x 12"		400	.040		4.22	2.03		6.25	7.75
0660	5/4" x 4"		410	.039		1.69	1.98		3.67	4.87
0670	5/4" x 6"		410	.039		2.67	1.98		4.65	5.95
0680	5/4" x 8"		410	.039		3.50	1.98		5.48	6.85
0690	5/4" x 10"		390	.041		4.50	2.08		6.58	8.10
0700	5/4" x 12"		390	.041		5.35	2.08		7.43	9.05

06 80 Composite Fabrications

06 80 10 − Composite Decking

06 80 10.10 Woodgrained Composite Decking

0010	**WOODGRAINED COMPOSITE DECKING**									
0100	Woodgrained composite decking, 1" x 6"	2 Carp	640	.025	L.F.	3.72	1.27		4.99	6
0110	Grooved edge		660	.024		3.87	1.23		5.10	6.15
0120	2" x 6"		640	.025		3.81	1.27		5.08	6.10
0130	Encased, 1" x 6"		640	.025		3.81	1.27		5.08	6.10
0140	Grooved edge		660	.024		3.96	1.23		5.19	6.25
0150	2" x 6"		640	.025		5.35	1.27		6.62	7.80

218

For customer support on your Building Construction Costs with RSMeans data, call 800.448.8182.

06 81 Composite Railings

06 81 10 – Encased Railings

06 81 10.10 Encased Composite Railings	Crew	Daily Output	Labor-Hours	Unit	Material	2018 Bare Costs Labor	Equipment	Total	Total Incl O&P
0010 **ENCASED COMPOSITE RAILINGS**									
0100 Encased composite railing, 6' long, 36" high, incl. balusters	1 Carp	16	.500	Ea.	143	25.50		168.50	197
0110 42" high, incl. balusters		16	.500		207	25.50		232.50	267
0120 8' long, 36" high, incl. balusters		12	.667		165	34		199	234
0130 42" high, incl. balusters		12	.667		156	34		190	223
0140 Accessories, post sleeve, 4" x 4", 39" long		32	.250		25	12.70		37.70	47
0150 96" long		24	.333		77	16.90		93.90	110
0160 6" x 6", 39" long		32	.250		54	12.70		66.70	79
0170 96" long		24	.333		155	16.90		171.90	196
0180 Accessories, post skirt, 4" x 4"		96	.083		3.99	4.22		8.21	10.85
0190 6" x 6"		96	.083		4.72	4.22		8.94	11.65
0200 Post cap, 4" x 4", flat		48	.167		6.65	8.45		15.10	20
0210 Pyramid		48	.167		5.70	8.45		14.15	19.10
0220 Post cap, 6" x 6", flat		48	.167		11.30	8.45		19.75	25.50
0230 Pyramid		48	.167		10.80	8.45		19.25	25

For customer support on your Building Construction Costs with RSMeans data, call 800.448.8182.

Division Notes

	CREW	DAILY OUTPUT	LABOR-HOURS	UNIT	BARE COSTS				TOTAL INCL O&P
					MAT.	LABOR	EQUIP.	TOTAL	

Estimating Tips
07 10 00 Dampproofing and Waterproofing

- Be sure of the job specifications before pricing this subdivision. The difference in cost between waterproofing and dampproofing can be great. Waterproofing will hold back standing water. Dampproofing prevents the transmission of water vapor. Also included in this section are vapor retarding membranes.

07 20 00 Thermal Protection

- Insulation and fireproofing products are measured by area, thickness, volume or R-value. Specifications may give only what the specific R-value should be in a certain situation. The estimator may need to choose the type of insulation to meet that R-value.

07 30 00 Steep Slope Roofing
07 40 00 Roofing and Siding Panels

- Many roofing and siding products are bought and sold by the square. One square is equal to an area that measures 100 square feet.

  This simple change in unit of measure could create a large error if the estimator is not observant. Accessories necessary for a complete installation must be figured into any calculations for both material and labor.

07 50 00 Membrane Roofing
07 60 00 Flashing and Sheet Metal
07 70 00 Roofing and Wall Specialties and Accessories

- The items in these subdivisions compose a roofing system. No one component completes the installation, and all must be estimated. Built-up or single-ply membrane roofing systems are made up of many products and installation trades. Wood blocking at roof perimeters or penetrations, parapet coverings, reglets, roof drains, gutters, downspouts, sheet metal flashing, skylights, smoke vents, and roof hatches all need to be considered along with the roofing material. Several different installation trades will need to work together on the roofing system. Inherent difficulties in the scheduling and coordination of various trades must be accounted for when estimating labor costs.

07 90 00 Joint Protection

- To complete the weather-tight shell, the sealants and caulkings must be estimated. Where different materials meet—at expansion joints, at flashing penetrations, and at hundreds of other locations throughout a construction project—caulking and sealants provide another line of defense against water penetration. Often, an entire system is based on the proper location and placement of caulking or sealants. The detailed drawings that are included as part of a set of architectural plans show typical locations for these materials. When caulking or sealants are shown at typical locations, this means the estimator must include them for all the locations where this detail is applicable. Be careful to keep different types of sealants separate, and remember to consider backer rods and primers if necessary.

Reference Numbers

Reference numbers are shown at the beginning of some major classifications. These numbers refer to related items in the Reference Section. The reference information may be an estimating procedure, an alternate pricing method, or technical information.

Note: Not all subdivisions listed here necessarily appear. ■

Did you know?

RSMeans data is available through our online application with 24/7 access:

- Search for unit prices by keyword
- Leverage the most up-to-date data
- Build and export estimates

Try it free for 30 days!
www.rsmeans.com/2018freetrial

No part of this cost data may be reproduced, stored in a retrieval system, or transmitted in any form or by any means without prior written permission of Gordian.

07 01 Operation and Maint. of Thermal and Moisture Protection

07 01 50 – Maintenance of Membrane Roofing

07 01 50.10 Roof Coatings

		Crew	Daily Output	Labor-Hours	Unit	Material	2018 Bare Costs Labor	Equipment	Total	Total Incl O&P
0010	**ROOF COATINGS**									
0012	Asphalt, brush grade, material only				Gal.	8.85			8.85	9.75
0200	Asphalt base, fibered aluminum coating ⬚G					8.40			8.40	9.25
0300	Asphalt primer, 5 gal.					7.50			7.50	8.25
0600	Coal tar pitch, 200 lb. barrels				Ton	1,250			1,250	1,375
0700	Tar roof cement, 5 gal. lots				Gal.	14.25			14.25	15.65
0800	Glass fibered roof & patching cement, 5 gal.				"	8.40			8.40	9.25
0900	Reinforcing glass membrane, 450 S.F./roll				Ea.	59.50			59.50	65.50
1000	Neoprene roof coating, 5 gal., 2 gal./sq.				Gal.	30.50			30.50	33.50
1100	Roof patch & flashing cement, 5 gal.					8.90			8.90	9.80
1200	Roof resaturant, glass fibered, 3 gal./sq.					9.65			9.65	10.60

07 01 90 – Maintenance of Joint Protection

07 01 90.81 Joint Sealant Replacement

		Crew	Daily Output	Labor-Hours	Unit	Material	2018 Bare Costs Labor	Equipment	Total	Total Incl O&P
0010	**JOINT SEALANT REPLACEMENT**									
0050	Control joints in concrete floors/slabs									
0100	Option 1 for joints with hard dry sealant									
0110	Step 1: Sawcut to remove 95% of old sealant									
0112	1/4" wide x 1/2" deep, with single saw blade	C-27	4800	.003	L.F.	.02	.16	.02	.20	.27
0114	3/8" wide x 3/4" deep, with single saw blade		4000	.004		.03	.19	.03	.25	.34
0116	1/2" wide x 1" deep, with double saw blades		3600	.004		.06	.21	.03	.30	.40
0118	3/4" wide x 1-1/2" deep, with double saw blades		3200	.005		.12	.24	.03	.39	.52
0120	Step 2: Water blast joint faces and edges	C-29	2500	.003			.13	.03	.16	.22
0130	Step 3: Air blast joint faces and edges	C-28	2000	.004			.19	.01	.20	.29
0140	Step 4: Sand blast joint faces and edges	E-11	2000	.016			.71	.10	.81	1.24
0150	Step 5: Air blast joint faces and edges	C-28	2000	.004			.19	.01	.20	.29
0200	Option 2 for joints with soft pliable sealant									
0210	Step 1: Plow joint with rectangular blade	B-62	2600	.009	L.F.		.40	.07	.47	.68
0220	Step 2: Sawcut to re-face joint faces									
0222	1/4" wide x 1/2" deep, with single saw blade	C-27	2400	.007	L.F.	.02	.32	.04	.38	.54
0224	3/8" wide x 3/4" deep, with single saw blade		2000	.008		.04	.38	.05	.47	.66
0226	1/2" wide x 1" deep, with double saw blades		1800	.009		.08	.42	.06	.56	.77
0228	3/4" wide x 1-1/2" deep, with double saw blades		1600	.010		.16	.48	.06	.70	.94
0230	Step 3: Water blast joint faces and edges	C-29	2500	.003			.13	.03	.16	.22
0240	Step 4: Air blast joint faces and edges	C-28	2000	.004			.19	.01	.20	.29
0250	Step 5: Sand blast joint faces and edges	E-11	2000	.016			.71	.10	.81	1.24
0260	Step 6: Air blast joint faces and edges	C-28	2000	.004			.19	.01	.20	.29
0290	For saw cutting new control joints, see Section 03 15 16.20									
8910	For backer rod, see Section 07 91 23.10									
8920	For joint sealant, see Section 03 15 16.30 or 07 92 13.20									

07 05 Common Work Results for Thermal and Moisture Protection

07 05 05 – Selective Demolition for Thermal and Moisture Protection

07 05 05.10 Selective Demo., Thermal and Moist. Protection

			Crew	Daily Output	Labor-Hours	Unit	Material	2018 Bare Costs Labor	Equipment	Total	Total Incl O&P
0010	**SELECTIVE DEMO., THERMAL AND MOISTURE PROTECTION**										
0020	Caulking/sealant, to 1" x 1" joint	R024119-10	1 Clab	600	.013	L.F.		.53		.53	.81
0120	Downspouts, including hangers			350	.023	"		.91		.91	1.39
0220	Flashing, sheet metal			290	.028	S.F.		1.10		1.10	1.67
0420	Gutters, aluminum or wood, edge hung			240	.033	L.F.		1.33		1.33	2.02
0520	Built-in			100	.080	"		3.19		3.19	4.86
0620	Insulation, air/vapor barrier			3500	.002	S.F.		.09		.09	.14

For customer support on your Building Construction Costs with RSMeans data, call 800.448.8182.

07 05 Common Work Results for Thermal and Moisture Protection

07 05 05 – Selective Demolition for Thermal and Moisture Protection

07 05 05.10 Selective Demo., Thermal and Moist. Protection		Crew	Daily Output	Labor-Hours	Unit	Material	2018 Bare Costs Labor	Equipment	Total	Total Incl O&P
0670	Batts or blankets	1 Clab	1400	.006	C.F.		.23		.23	.35
0720	Foamed or sprayed in place	2 Clab	1000	.016	B.F.		.64		.64	.97
0770	Loose fitting	1 Clab	3000	.003	C.F.		.11		.11	.16
0870	Rigid board		3450	.002	B.F.		.09		.09	.14
1120	Roll roofing, cold adhesive		12	.667	Sq.		26.50		26.50	40.50
1170	Roof accessories, adjustable metal chimney flashing		9	.889	Ea.		35.50		35.50	54
1325	Plumbing vent flashing		32	.250	"		9.95		9.95	15.20
1375	Ridge vent strip, aluminum		310	.026	L.F.		1.03		1.03	1.57
1620	Skylight to 10 S.F.		8	1	Ea.		40		40	60.50
2120	Roof edge, aluminum soffit and fascia	▼	570	.014	L.F.		.56		.56	.85
2170	Concrete coping, up to 12" wide	2 Clab	160	.100			3.99		3.99	6.05
2220	Drip edge	1 Clab	1000	.008			.32		.32	.49
2270	Gravel stop		950	.008			.34		.34	.51
2370	Sheet metal coping, up to 12" wide	▼	240	.033	▼		1.33		1.33	2.02
2470	Roof insulation board, over 2" thick	B-2	7800	.005	B.F.		.21		.21	.31
2520	Up to 2" thick	"	3900	.010	S.F.		.41		.41	.63
2620	Roof ventilation, louvered gable vent	1 Clab	16	.500	Ea.		19.95		19.95	30.50
2670	Remove, roof hatch	G-3	15	2.133			106		106	162
2675	Rafter vents	1 Clab	960	.008	▼		.33		.33	.51
2720	Soffit vent and/or fascia vent		575	.014	L.F.		.55		.55	.84
2775	Soffit vent strip, aluminum, 3" to 4" wide		160	.050			1.99		1.99	3.04
2820	Roofing accessories, shingle moulding, to 1" x 4"	▼	1600	.005			.20		.20	.30
2870	Cant strip	B-2	2000	.020			.81		.81	1.23
2920	Concrete block walkway	1 Clab	230	.035	▼		1.39		1.39	2.11
3070	Roofing, felt paper, #15		70	.114	Sq.		4.55		4.55	6.95
3125	#30 felt	▼	30	.267	"		10.65		10.65	16.20
3170	Asphalt shingles, 1 layer	B-2	3500	.011	S.F.		.46		.46	.70
3180	2 layers		1750	.023	"		.92		.92	1.40
3370	Modified bitumen		26	1.538	Sq.		62		62	94.50
3420	Built-up, no gravel, 3 ply		25	1.600			64.50		64.50	98
3470	4 ply		21	1.905	▼		76.50		76.50	117
3620	5 ply		1600	.025	S.F.		1.01		1.01	1.53
3720	5 ply, with gravel		890	.045			1.81		1.81	2.76
3725	Loose gravel removal		5000	.008			.32		.32	.49
3730	Embedded gravel removal		2000	.020			.81		.81	1.23
3870	Fiberglass sheet		1200	.033			1.34		1.34	2.04
4120	Slate shingles		1900	.021	▼		.85		.85	1.29
4170	Ridge shingles, clay or slate		2000	.020	L.F.		.81		.81	1.23
4320	Single ply membrane, attached at seams		52	.769	Sq.		31		31	47
4370	Ballasted		75	.533			21.50		21.50	32.50
4420	Fully adhered	▼	39	1.026	▼		41.50		41.50	63
4550	Roof hatch, 2'-6" x 3'-0"	1 Clab	10	.800	Ea.		32		32	48.50
4670	Wood shingles	B-2	2200	.018	S.F.		.73		.73	1.11
4820	Sheet metal roofing	"	2150	.019			.75		.75	1.14
4970	Siding, horizontal wood clapboards	1 Clab	380	.021			.84		.84	1.28
5025	Exterior insulation finish system	"	120	.067			2.66		2.66	4.05
5070	Tempered hardboard, remove and reset	1 Carp	380	.021			1.07		1.07	1.63
5120	Tempered hardboard sheet siding	"	375	.021	▼		1.08		1.08	1.65
5170	Metal, corner strips	1 Clab	850	.009	L.F.		.38		.38	.57
5225	Horizontal strips		444	.018	S.F.		.72		.72	1.09
5320	Vertical strips		400	.020			.80		.80	1.21
5520	Wood shingles		350	.023			.91		.91	1.39
5620	Stucco siding	▼	360	.022			.89		.89	1.35

For customer support on your Building Construction Costs with RSMeans data, call 800.448.8182.

223

07 05 Common Work Results for Thermal and Moisture Protection

07 05 05 – Selective Demolition for Thermal and Moisture Protection

07 05 05.10 Selective Demo., Thermal and Moist. Protection	Crew	Daily Output	Labor-Hours	Unit	Material	2018 Bare Costs Labor	Equipment	Total	Total Incl O&P
5670	Textured plywood	1 Clab	725	.011	S.F.		.44		.44	.67
5720	Vinyl siding		510	.016	↓		.63		.63	.95
5770	Corner strips		900	.009	L.F.		.35		.35	.54
5870	Wood, boards, vertical		400	.020	S.F.		.80		.80	1.21
5880	Steel siding, corrugated/ribbed	↓	402.50	.020	"		.79		.79	1.21
5920	Waterproofing, protection/drain board	2 Clab	3900	.004	B.F.		.16		.16	.25
5970	Over 1/2" thick		1750	.009	S.F.		.36		.36	.55
6020	To 1/2" thick	↓	2000	.008	"		.32		.32	.49

07 11 Dampproofing

07 11 13 – Bituminous Dampproofing

07 11 13.10 Bituminous Asphalt Coating

		Crew	Daily Output	Labor-Hours	Unit	Material	2018 Bare Costs Labor	Equipment	Total	Total Incl O&P
0010	**BITUMINOUS ASPHALT COATING**									
0030	Brushed on, below grade, 1 coat	1 Rofc	665	.012	S.F.	.22	.53		.75	1.13
0100	2 coat		500	.016		.44	.70		1.14	1.67
0300	Sprayed on, below grade, 1 coat		830	.010		.22	.42		.64	.95
0400	2 coat	↓	500	.016	↓	.43	.70		1.13	1.66
0500	Asphalt coating, with fibers				Gal.	8.40			8.40	9.25
0600	Troweled on, asphalt with fibers, 1/16" thick	1 Rofc	500	.016	S.F.	.37	.70		1.07	1.58
0700	1/8" thick		400	.020		.65	.88		1.53	2.19
1000	1/2" thick		350	.023	↓	2.11	1		3.11	4.01

07 11 16 – Cementitious Dampproofing

07 11 16.20 Cementitious Parging

		Crew	Daily Output	Labor-Hours	Unit	Material	2018 Bare Costs Labor	Equipment	Total	Total Incl O&P
0010	**CEMENTITIOUS PARGING**									
0020	Portland cement, 2 coats, 1/2" thick	D-1	250	.064	S.F.	.37	2.87		3.24	4.80
0100	Waterproofed Portland cement, 1/2" thick, 2 coats	"	250	.064	"	5.80	2.87		8.67	10.75

07 12 Built-up Bituminous Waterproofing

07 12 13 – Built-Up Asphalt Waterproofing

07 12 13.20 Membrane Waterproofing

		Crew	Daily Output	Labor-Hours	Unit	Material	2018 Bare Costs Labor	Equipment	Total	Total Incl O&P
0010	**MEMBRANE WATERPROOFING**									
0012	On slabs, 1 ply, felt, mopped	G-1	3000	.019	S.F.	.43	.77	.17	1.37	1.96
0100	On slabs, 1 ply, glass fiber fabric, mopped		2100	.027		.47	1.10	.24	1.81	2.63
0300	On slabs, 2 ply, felt, mopped		2500	.022		.86	.92	.21	1.99	2.73
0400	On slabs, 2 ply, glass fiber fabric, mopped		1650	.034		1.04	1.39	.31	2.74	3.83
0600	On slabs, 3 ply, felt, mopped		2100	.027		1.30	1.10	.24	2.64	3.54
0700	On slabs, 3 ply, glass fiber fabric, mopped	↓	1550	.036		1.42	1.48	.33	3.23	4.42
0710	Asphaltic hardboard protection board, 1/8" thick	2 Rofc	500	.032		.67	1.41		2.08	3.11
1000	EPS membrane protection board, 1/4" thick		3500	.005		.34	.20		.54	.71
1050	3/8" thick		3500	.005		.37	.20		.57	.75
1060	1/2" thick		3500	.005	↓	.41	.20		.61	.79
1070	Fiberglass fabric, black, 20/10 mesh	↓	116	.138	Sq.	14.70	6.05		20.75	26.50

For customer support on your Building Construction Costs with RSMeans data, call 800.448.8182.

07 13 Sheet Waterproofing

07 13 53 – Elastomeric Sheet Waterproofing

07 13 53.10 Elastomeric Sheet Waterproofing and Access.	Crew	Daily Output	Labor-Hours	Unit	Material	2018 Bare Costs Labor	Equipment	Total	Total Incl O&P
0010	**ELASTOMERIC SHEET WATERPROOFING AND ACCESS.**								
0090	EPDM, plain, 45 mils thick	2 Rofc	580	.028	S.F.	1.47	1.21		2.68	3.66
0100	60 mils thick		570	.028		1.58	1.23		2.81	3.81
0300	Nylon reinforced sheets, 45 mils thick		580	.028		1.66	1.21		2.87	3.86
0400	60 mils thick	↓	570	.028		1.69	1.23		2.92	3.93
0600	Vulcanizing splicing tape for above, 2" wide				C.L.F.	63			63	69
0700	4" wide				"	121			121	134
0900	Adhesive, bonding, 60 S.F./gal.				Gal.	26.50			26.50	29
1000	Splicing, 75 S.F./gal.				"	38			38	42
1200	Neoprene sheets, plain, 45 mils thick	2 Rofc	580	.028	S.F.	2.11	1.21		3.32	4.36
1300	60 mils thick		570	.028		2.23	1.23		3.46	4.53
1500	Nylon reinforced, 45 mils thick		580	.028		2.15	1.21		3.36	4.41
1600	60 mils thick		570	.028		2.78	1.23		4.01	5.15
1800	120 mils thick	↓	500	.032	↓	5.65	1.41		7.06	8.55
1900	Adhesive, splicing, 150 S.F./gal. per coat				Gal.	38			38	42
2100	Fiberglass reinforced, fluid applied, 1/8" thick	2 Rofc	500	.032	S.F.	1.63	1.41		3.04	4.16
2200	Polyethylene and rubberized asphalt sheets, 60 mils thick		550	.029		.94	1.28		2.22	3.18
2400	Polyvinyl chloride sheets, plain, 10 mils thick		580	.028		.15	1.21		1.36	2.21
2500	20 mils thick		570	.028		.20	1.23		1.43	2.30
2700	30 mils thick	↓	560	.029	↓	.25	1.26		1.51	2.39
3000	Adhesives, trowel grade, 40-100 S.F./gal.				Gal.	24			24	26.50
3100	Brush grade, 100-250 S.F./gal.				"	21.50			21.50	24
3300	Bitumen modified polyurethane, fluid applied, 55 mils thick	2 Rofc	665	.024	S.F.	.99	1.06		2.05	2.87

07 16 Cementitious and Reactive Waterproofing

07 16 16 – Crystalline Waterproofing

07 16 16.20 Cementitious Waterproofing

		Crew	Daily Output	Labor-Hours	Unit	Material	Labor	Equipment	Total	Total Incl O&P
0010	**CEMENTITIOUS WATERPROOFING**									
0020	1/8" application, sprayed on	G-2A	1000	.024	S.F.	.68	.93	.61	2.22	2.94
0050	4 coat cementitious metallic slurry	1 Cefi	1.20	6.667	C.S.F.	39	315		354	515

07 17 Bentonite Waterproofing

07 17 13 – Bentonite Panel Waterproofing

07 17 13.10 Bentonite

		Crew	Daily Output	Labor-Hours	Unit	Material	Labor	Equipment	Total	Total Incl O&P
0010	**BENTONITE**									
0020	Panels, 4' x 4', 3/16" thick	1 Rofc	625	.013	S.F.	1.72	.56		2.28	2.84
0100	Rolls, 3/8" thick, with geotextile fabric both sides	"	550	.015	"	1.82	.64		2.46	3.08
0300	Granular bentonite, 50 lb. bags (.625 C.F.)				Bag	17.95			17.95	19.75
0400	3/8" thick, troweled on	1 Rofc	475	.017	S.F.	.90	.74		1.64	2.24
0500	Drain board, expanded polystyrene, 1-1/2" thick	1 Rohe	1600	.005		.41	.16		.57	.73
0510	2" thick		1600	.005		.54	.16		.70	.87
0520	3" thick		1600	.005		.81	.16		.97	1.17
0530	4" thick		1600	.005		1.08	.16		1.24	1.47
0600	With filter fabric, 1-1/2" thick		1600	.005		.47	.16		.63	.79
0625	2" thick		1600	.005		.60	.16		.76	.94
0650	3" thick		1600	.005		.87	.16		1.03	1.24
0675	4" thick	↓	1600	.005	↓	1.14	.16		1.30	1.54

07 19 Water Repellents

07 19 19 – Silicone Water Repellents

07 19 19.10 Silicone Based Water Repellents		Crew	Daily Output	Labor-Hours	Unit	Material	2018 Bare Costs Labor	Equipment	Total	Total Incl O&P
0010	**SILICONE BASED WATER REPELLENTS**									
0020	Water base liquid, roller applied	2 Rofc	7000	.002	S.F.	.42	.10		.52	.63
0200	Silicone or stearate, sprayed on CMU, 1 coat	1 Rofc	4000	.002		.39	.09		.48	.58
0300	2 coats	"	3000	.003		.79	.12		.91	1.07

07 21 Thermal Insulation

07 21 13 – Board Insulation

07 21 13.10 Rigid Insulation

07 21 13.10 Rigid Insulation			Crew	Daily Output	Labor-Hours	Unit	Material	2018 Bare Costs Labor	Equipment	Total	Total Incl O&P
0010	**RIGID INSULATION**, for walls										
0040	Fiberglass, 1.5#/C.F., unfaced, 1" thick, R4.1	G	1 Carp	1000	.008	S.F.	.34	.41		.75	.99
0060	1-1/2" thick, R6.2	G		1000	.008		.40	.41		.81	1.06
0080	2" thick, R8.3	G		1000	.008		.50	.41		.91	1.17
0120	3" thick, R12.4	G		800	.010		.60	.51		1.11	1.43
0370	3#/C.F., unfaced, 1" thick, R4.3	G		1000	.008		.54	.41		.95	1.21
0390	1-1/2" thick, R6.5	G		1000	.008		.79	.41		1.20	1.49
0400	2" thick, R8.7	G		890	.009		1.07	.46		1.53	1.87
0420	2-1/2" thick, R10.9	G		800	.010		1.11	.51		1.62	1.99
0440	3" thick, R13	G		800	.010		1.62	.51		2.13	2.55
0520	Foil faced, 1" thick, R4.3	G		1000	.008		.84	.41		1.25	1.54
0540	1-1/2" thick, R6.5	G		1000	.008		1.25	.41		1.66	2
0560	2" thick, R8.7	G		890	.009		1.58	.46		2.04	2.43
0580	2-1/2" thick, R10.9	G		800	.010		1.86	.51		2.37	2.82
0600	3" thick, R13	G		800	.010		2.09	.51		2.60	3.07
1600	Isocyanurate, 4' x 8' sheet, foil faced, both sides										
1610	1/2" thick	G	1 Carp	800	.010	S.F.	.31	.51		.82	1.11
1620	5/8" thick	G		800	.010		.48	.51		.99	1.30
1630	3/4" thick	G		800	.010		.45	.51		.96	1.27
1640	1" thick	G		800	.010		.61	.51		1.12	1.44
1650	1-1/2" thick	G		730	.011		.67	.56		1.23	1.59
1660	2" thick	G		730	.011		.90	.56		1.46	1.84
1670	3" thick	G		730	.011		2.76	.56		3.32	3.89
1680	4" thick	G		730	.011		2.52	.56		3.08	3.62
1700	Perlite, 1" thick, R2.77	G		800	.010		.47	.51		.98	1.29
1750	2" thick, R5.55	G		730	.011		.78	.56		1.34	1.71
1900	Extruded polystyrene, 25 psi compressive strength, 1" thick, R5	G		800	.010		.57	.51		1.08	1.40
1940	2" thick, R10	G		730	.011		1.14	.56		1.70	2.10
1960	3" thick, R15	G		730	.011		1.59	.56		2.15	2.60
2100	Expanded polystyrene, 1" thick, R3.85	G		800	.010		.27	.51		.78	1.07
2120	2" thick, R7.69	G		730	.011		.54	.56		1.10	1.44
2140	3" thick, R11.49	G		730	.011		.81	.56		1.37	1.74

07 21 13.13 Foam Board Insulation

07 21 13.13 Foam Board Insulation			Crew	Daily Output	Labor-Hours	Unit	Material	2018 Bare Costs Labor	Equipment	Total	Total Incl O&P
0010	**FOAM BOARD INSULATION**										
0600	Polystyrene, expanded, 1" thick, R4	G	1 Carp	680	.012	S.F.	.27	.60		.87	1.21
0700	2" thick, R8	G	"	675	.012	"	.54	.60		1.14	1.50

07 21 16 – Blanket Insulation

07 21 16.10 Blanket Insulation for Floors/Ceilings

07 21 16.10 Blanket Insulation for Floors/Ceilings			Crew	Daily Output	Labor-Hours	Unit	Material	2018 Bare Costs Labor	Equipment	Total	Total Incl O&P
0010	**BLANKET INSULATION FOR FLOORS/CEILINGS**										
0020	Including spring type wire fasteners										
2000	Fiberglass, blankets or batts, paper or foil backing										
2100	3-1/2" thick, R13	G	1 Carp	700	.011	S.F.	.41	.58		.99	1.33

07 21 Thermal Insulation

07 21 16 – Blanket Insulation

07 21 16.10 Blanket Insulation for Floors/Ceilings

		Crew	Daily Output	Labor-Hours	Unit	Material	2018 Bare Costs Labor	Equipment	Total	Total Incl O&P
2150	6-1/4" thick, R19	G 1 Carp	600	.013	S.F.	.51	.68		1.19	1.59
2210	9-1/2" thick, R30	G	500	.016		.73	.81		1.54	2.04
2220	12" thick, R38	G	475	.017		1.05	.85		1.90	2.46
3000	Unfaced, 3-1/2" thick, R13	G	600	.013		.33	.68		1.01	1.39
3010	6-1/4" thick, R19	G	500	.016		.39	.81		1.20	1.67
3020	9-1/2" thick, R30	G	450	.018		.60	.90		1.50	2.03
3030	12" thick, R38	G	425	.019		.75	.95		1.70	2.28

07 21 16.20 Blanket Insulation for Walls

		Crew	Daily Output	Labor-Hours	Unit	Material	2018 Bare Costs Labor	Equipment	Total	Total Incl O&P
0010	**BLANKET INSULATION FOR WALLS**									
0020	Kraft faced fiberglass, 3-1/2" thick, R11, 15" wide	G 1 Carp	1350	.006	S.F.	.32	.30		.62	.81
0030	23" wide	G	1600	.005		.32	.25		.57	.74
0060	R13, 11" wide	G	1150	.007		.34	.35		.69	.91
0080	15" wide	G	1350	.006		.34	.30		.64	.83
0100	23" wide	G	1600	.005		.34	.25		.59	.76
0110	R15, 11" wide	G	1150	.007		.49	.35		.84	1.08
0120	15" wide	G	1350	.006		.49	.30		.79	1
0130	23" wide	G	1600	.005		.49	.25		.74	.93
0140	6" thick, R19, 11" wide	G	1150	.007		.44	.35		.79	1.02
0160	15" wide	G	1350	.006		.44	.30		.74	.94
0180	23" wide	G	1600	.005		.44	.25		.69	.87
0182	R21, 11" wide	G	1150	.007		.66	.35		1.01	1.27
0184	15" wide	G	1350	.006		.66	.30		.96	1.19
0186	23" wide	G	1600	.005		.66	.25		.91	1.12
0188	9" thick, R30, 11" wide	G	985	.008		.73	.41		1.14	1.43
0200	15" wide	G	1150	.007		.73	.35		1.08	1.34
0220	23" wide	G	1350	.006		.73	.30		1.03	1.26
0230	12" thick, R38, 11" wide	G	985	.008		1.05	.41		1.46	1.79
0240	15" wide	G	1150	.007		1.05	.35		1.40	1.70
0260	23" wide	G	1350	.006		1.05	.30		1.35	1.62
0410	Foil faced fiberglass, 3-1/2" thick, R13, 11" wide	G	1150	.007		.47	.35		.82	1.06
0420	15" wide	G	1350	.006		.47	.30		.77	.98
0440	23" wide	G	1600	.005		.47	.25		.72	.91
0442	R15, 11" wide	G	1150	.007		.48	.35		.83	1.07
0444	15" wide	G	1350	.006		.48	.30		.78	.99
0446	23" wide	G	1600	.005		.48	.25		.73	.92
0448	6" thick, R19, 11" wide	G	1150	.007		.62	.35		.97	1.22
0460	15" wide	G	1350	.006		.62	.30		.92	1.14
0480	23" wide	G	1600	.005		.62	.25		.87	1.07
0482	R21, 11" wide	G	1150	.007		.64	.35		.99	1.24
0484	15" wide	G	1350	.006		.64	.30		.94	1.16
0486	23" wide	G	1600	.005		.64	.25		.89	1.09
0488	9" thick, R30, 11" wide	G	985	.008		.94	.41		1.35	1.66
0500	15" wide	G	1150	.007		.94	.35		1.29	1.57
0550	23" wide	G	1350	.006		.94	.30		1.24	1.49
0560	12" thick, R38, 11" wide	G	985	.008		1.15	.41		1.56	1.90
0570	15" wide	G	1150	.007		1.15	.35		1.50	1.81
0580	23" wide	G	1350	.006		1.15	.30		1.45	1.73
0620	Unfaced fiberglass, 3-1/2" thick, R13, 11" wide	G	1150	.007		.33	.35		.68	.90
0820	15" wide	G	1350	.006		.33	.30		.63	.82
0830	23" wide	G	1600	.005		.33	.25		.58	.75
0832	R15, 11" wide	G	1150	.007		.46	.35		.81	1.05
0834	15" wide	G	1350	.006		.46	.30		.76	.97

07 21 Thermal Insulation

07 21 16 – Blanket Insulation

07 21 16.20 Blanket Insulation for Walls

		Crew	Daily Output	Labor-Hours	Unit	Material	2018 Bare Costs Labor	Equipment	Total	Total Incl O&P
0836	23" wide	1 Carp	1600	.005	S.F.	.46	.25		.71	.90
0838	6" thick, R19, 11" wide		1150	.007		.39	.35		.74	.97
0860	15" wide		1150	.007		.39	.35		.74	.97
0880	23" wide		1350	.006		.39	.30		.69	.89
0882	R21, 11" wide		1150	.007		.69	.35		1.04	1.30
0886	15" wide		1350	.006		.69	.30		.99	1.22
0888	23" wide		1600	.005		.69	.25		.94	1.15
0890	9" thick, R30, 11" wide		985	.008		.60	.41		1.01	1.29
0900	15" wide		1150	.007		.60	.35		.95	1.20
0920	23" wide		1350	.006		.60	.30		.90	1.12
0930	12" thick, R38, 11" wide		985	.008		.75	.41		1.16	1.46
0940	15" wide		1000	.008		.75	.41		1.16	1.45
0960	23" wide		1150	.007		.75	.35		1.10	1.37
1300	Wall or ceiling insulation, mineral wool batts									
1320	3-1/2" thick, R15	1 Carp	1600	.005	S.F.	.77	.25		1.02	1.24
1340	5-1/2" thick, R23		1600	.005		1.21	.25		1.46	1.72
1380	7-1/4" thick, R30		1350	.006		1.60	.30		1.90	2.21
1700	Non-rigid insul., recycled blue cotton fiber, unfaced batts, R13, 16" wide		1600	.005		.99	.25		1.24	1.48
1710	R19, 16" wide		1600	.005		1.35	.25		1.60	1.88
1850	Friction fit wire insulation supports, 16" OC		960	.008	Ea.	.07	.42		.49	.72

07 21 19 – Foamed In Place Insulation

07 21 19.10 Masonry Foamed In Place Insulation

		Crew	Daily Output	Labor-Hours	Unit	Material	2018 Bare Costs Labor	Equipment	Total	Total Incl O&P
0010	**MASONRY FOAMED IN PLACE INSULATION**									
0100	Amino-plast foam, injected into block core, 6" block	G-2A	6000	.004	Ea.	.17	.16	.10	.43	.55
0110	8" block		5000	.005		.20	.19	.12	.51	.66
0120	10" block		4000	.006		.25	.23	.15	.63	.83
0130	12" block		3000	.008		.34	.31	.20	.85	1.12
0140	Injected into cavity wall		13000	.002	B.F.	.06	.07	.05	.18	.24
0150	Preparation, drill holes into mortar joint every 4 V.L.F., 5/8" diameter	1 Clab	960	.008	Ea.		.33		.33	.51
0160	7/8" diameter		680	.012			.47		.47	.71
0170	Patch drilled holes, 5/8" diameter		1800	.004		.04	.18		.22	.31
0180	7/8" diameter		1200	.007		.05	.27		.32	.46

07 21 23 – Loose-Fill Insulation

07 21 23.10 Poured Loose-Fill Insulation

		Crew	Daily Output	Labor-Hours	Unit	Material	2018 Bare Costs Labor	Equipment	Total	Total Incl O&P
0010	**POURED LOOSE-FILL INSULATION**									
0020	Cellulose fiber, R3.8 per inch	1 Carp	200	.040	C.F.	.69	2.03		2.72	3.85
0021	4" thick		1000	.008	S.F.	.17	.41		.58	.80
0022	6" thick		800	.010	"	.28	.51		.79	1.08
0080	Fiberglass wool, R4 per inch		200	.040	C.F.	.62	2.03		2.65	3.77
0081	4" thick		600	.013	S.F.	.21	.68		.89	1.26
0082	6" thick		400	.020	"	.30	1.01		1.31	1.87
0100	Mineral wool, R3 per inch		200	.040	C.F.	.49	2.03		2.52	3.63
0101	4" thick		600	.013	S.F.	.16	.68		.84	1.21
0102	6" thick		400	.020	"	.25	1.01		1.26	1.81
0300	Polystyrene, R4 per inch		200	.040	C.F.	1.42	2.03		3.45	4.65
0301	4" thick		600	.013	S.F.	.47	.68		1.15	1.55
0302	6" thick		400	.020	"	.71	1.01		1.72	2.32
0400	Perlite, R2.78 per inch		200	.040	C.F.	5.30	2.03		7.33	8.90
0401	4" thick		1000	.008	S.F.	1.76	.41		2.17	2.56
0402	6" thick		800	.010	"	2.65	.51		3.16	3.68

For customer support on your Building Construction Costs with RSMeans data, call 800.448.8182.

07 21 Thermal Insulation

07 21 23 – Loose-Fill Insulation

07 21 23.20 Masonry Loose-Fill Insulation		Crew	Daily Output	Labor-Hours	Unit	Material	2018 Bare Costs Labor	Equipment	Total	Total Incl O&P
0010	**MASONRY LOOSE-FILL INSULATION**, vermiculite or perlite									
0100	In cores of concrete block, 4" thick wall, .115 C.F./S.F.	G	D-1	4800	.003	S.F.	.61	.15		.76	.90
0200	6" thick wall, .175 C.F./S.F.	G		3000	.005		.93	.24		1.17	1.39
0300	8" thick wall, .258 C.F./S.F.	G		2400	.007		1.36	.30		1.66	1.96
0400	10" thick wall, .340 C.F./S.F.	G		1850	.009		1.80	.39		2.19	2.57
0500	12" thick wall, .422 C.F./S.F.	G		1200	.013		2.23	.60		2.83	3.37
0600	Poured cavity wall, vermiculite or perlite, water repellent	G		250	.064	C.F.	5.30	2.87		8.17	10.20
0700	Foamed in place, urethane in 2-5/8" cavity	G	G-2A	1035	.023	S.F.	1.35	.90	.59	2.84	3.60
0800	For each 1" added thickness, add	G	"	2372	.010	"	.51	.39	.26	1.16	1.49

07 21 26 – Blown Insulation

07 21 26.10 Blown Insulation

		Crew	Daily Output	Labor-Hours	Unit	Material	2018 Bare Costs Labor	Equipment	Total	Total Incl O&P
0010	**BLOWN INSULATION** Ceilings, with open access										
0020	Cellulose, 3-1/2" thick, R13	G	G-4	5000	.005	S.F.	.24	.19	.06	.49	.63
0030	5-3/16" thick, R19	G		3800	.006		.35	.26	.08	.69	.87
0050	6-1/2" thick, R22	G		3000	.008		.45	.32	.10	.87	1.10
0100	8-11/16" thick, R30	G		2600	.009		.61	.37	.12	1.10	1.37
0120	10-7/8" thick, R38	G		1800	.013		.78	.54	.17	1.49	1.86
1000	Fiberglass, 5.5" thick, R11	G		3800	.006		.21	.26	.08	.55	.71
1050	6" thick, R12	G		3000	.008		.30	.32	.10	.72	.93
1100	8.8" thick, R19	G		2200	.011		.37	.44	.14	.95	1.23
1200	10" thick, R22	G		1800	.013		.43	.54	.17	1.14	1.49
1300	11.5" thick, R26	G		1500	.016		.52	.65	.20	1.37	1.78
1350	13" thick, R30	G		1400	.017		.60	.69	.22	1.51	1.96
1450	16" thick, R38	G		1145	.021		.77	.85	.27	1.89	2.42
1500	20" thick, R49	G		920	.026		1.01	1.06	.33	2.40	3.09

07 21 27 – Reflective Insulation

07 21 27.10 Reflective Insulation Options

		Crew	Daily Output	Labor-Hours	Unit	Material	2018 Bare Costs Labor	Equipment	Total	Total Incl O&P
0010	**REFLECTIVE INSULATION OPTIONS**										
0020	Aluminum foil on reinforced scrim	G	1 Carp	19	.421	C.S.F.	15	21.50		36.50	49
0100	Reinforced with woven polyolefin	G		19	.421		22	21.50		43.50	57
0500	With single bubble air space, R8.8	G		15	.533		26	27		53	70
0600	With double bubble air space, R9.8	G		15	.533		31.50	27		58.50	75.50

07 21 29 – Sprayed Insulation

07 21 29.10 Sprayed-On Insulation

		Crew	Daily Output	Labor-Hours	Unit	Material	2018 Bare Costs Labor	Equipment	Total	Total Incl O&P
0010	**SPRAYED-ON INSULATION**										
0020	Fibrous/cementitious, finished wall, 1" thick, R3.7	G	G-2	2050	.012	S.F.	.36	.49	.06	.91	1.21
0100	Attic, 5.2" thick, R19	G		1550	.015	"	.42	.65	.08	1.15	1.53
0200	Fiberglass, R4 per inch, vertical	G		1600	.015	B.F.	.19	.63	.08	.90	1.25
0210	Horizontal	G		1200	.020	"	.19	.84	.11	1.14	1.60
0300	Closed cell, spray polyurethane foam, 2 lb./C.F. density										
0310	1" thick	G	G-2A	6000	.004	S.F.	.51	.16	.10	.77	.92
0320	2" thick	G		3000	.008		1.03	.31	.20	1.54	1.87
0330	3" thick	G		2000	.012		1.54	.47	.31	2.32	2.79
0335	3-1/2" thick	G		1715	.014		1.80	.54	.36	2.70	3.26
0340	4" thick	G		1500	.016		2.05	.62	.41	3.08	3.72
0350	5" thick	G		1200	.020		2.57	.78	.51	3.86	4.65
0355	5-1/2" thick	G		1090	.022		2.82	.86	.56	4.24	5.10
0360	6" thick	G		1000	.024		3.08	.93	.61	4.62	5.60

For customer support on your Building Construction Costs with RSMeans data, call 800.448.8182.

229

07 22 Roof and Deck Insulation

07 22 16 – Roof Board Insulation

07 22 16.10 Roof Deck Insulation		Crew	Daily Output	Labor-Hours	Unit	Material	2018 Bare Costs Labor	Equipment	Total	Total Incl O&P
0010	**ROOF DECK INSULATION**, fastening excluded									
0016	Asphaltic cover board, fiberglass lined, 1/8" thick	1 Rofc	1400	.006	S.F.	.48	.25		.73	.95
0018	1/4" thick		1400	.006		.96	.25		1.21	1.48
0020	Fiberboard low density, 1/2" thick, R1.39	G	1300	.006		.35	.27		.62	.84
0030	1" thick, R2.78	G	1040	.008		.58	.34		.92	1.21
0080	1-1/2" thick, R4.17	G	1040	.008		.89	.34		1.23	1.55
0100	2" thick, R5.56	G	1040	.008		1.15	.34		1.49	1.84
0110	Fiberboard high density, 1/2" thick, R1.3	G	1300	.006		.27	.27		.54	.75
0120	1" thick, R2.5	G	1040	.008		.56	.34		.90	1.19
0130	1-1/2" thick, R3.8	G	1040	.008		.83	.34		1.17	1.48
0200	Fiberglass, 3/4" thick, R2.78	G	1300	.006		.62	.27		.89	1.13
0400	15/16" thick, R3.70	G	1300	.006		.82	.27		1.09	1.35
0460	1-1/16" thick, R4.17	G	1300	.006		1.07	.27		1.34	1.63
0600	1-5/16" thick, R5.26	G	1300	.006		1.42	.27		1.69	2.01
0650	2-1/16" thick, R8.33	G	1040	.008		1.52	.34		1.86	2.24
0700	2-7/16" thick, R10	G	1040	.008		1.72	.34		2.06	2.46
0800	Gypsum cover board, fiberglass mat facer, 1/4" thick		1400	.006		.48	.25		.73	.95
0810	1/2" thick		1300	.006		.59	.27		.86	1.10
0820	5/8" thick		1200	.007		.62	.29		.91	1.17
0830	Primed fiberglass mat facer, 1/4" thick		1400	.006		.50	.25		.75	.97
0840	1/2" thick		1300	.006		.58	.27		.85	1.09
0850	5/8" thick		1200	.007		.61	.29		.90	1.16
1650	Perlite, 1/2" thick, R1.32	G	1365	.006		.27	.26		.53	.73
1655	3/4" thick, R2.08	G	1040	.008		.38	.34		.72	.99
1660	1" thick, R2.78	G	1040	.008		.54	.34		.88	1.16
1670	1-1/2" thick, R4.17	G	1040	.008		.80	.34		1.14	1.45
1680	2" thick, R5.56	G	910	.009		1.08	.39		1.47	1.84
1685	2-1/2" thick, R6.67	G	910	.009		1.45	.39		1.84	2.25
1690	Tapered for drainage	G	1040	.008	B.F.	1.05	.34		1.39	1.73
1700	Polyisocyanurate, 2#/C.F. density, 3/4" thick	G	1950	.004	S.F.	.42	.18		.60	.76
1705	1" thick	G	1820	.004		.43	.19		.62	.80
1715	1-1/2" thick	G	1625	.005		.58	.22		.80	1
1725	2" thick	G	1430	.006		.74	.25		.99	1.22
1735	2-1/2" thick	G	1365	.006		.99	.26		1.25	1.52
1745	3" thick	G	1300	.006		1.10	.27		1.37	1.66
1755	3-1/2" thick	G	1300	.006		1.71	.27		1.98	2.33
1765	Tapered for drainage	G	1820	.004	B.F.	.52	.19		.71	.90
1900	Extruded polystyrene									
1910	15 psi compressive strength, 1" thick, R5	G	1 Rofc	1950	.004	S.F.	.55	.18		.73	.91
1920	2" thick, R10	G	1625	.005		.72	.22		.94	1.15
1930	3" thick, R15	G	1300	.006		1.43	.27		1.70	2.02
1932	4" thick, R20	G	1300	.006		1.93	.27		2.20	2.57
1934	Tapered for drainage	G	1950	.004	B.F.	.51	.18		.69	.86
1940	25 psi compressive strength, 1" thick, R5	G	1950	.004	S.F.	1.11	.18		1.29	1.52
1942	2" thick, R10	G	1625	.005		2.11	.22		2.33	2.68
1944	3" thick, R15	G	1300	.006		3.22	.27		3.49	3.99
1946	4" thick, R20	G	1300	.006		4.44	.27		4.71	5.35
1948	Tapered for drainage	G	1950	.004	B.F.	.56	.18		.74	.92
1950	40 psi compressive strength, 1" thick, R5	G	1950	.004	S.F.	.87	.18		1.05	1.26
1952	2" thick, R10	G	1625	.005		1.65	.22		1.87	2.18
1954	3" thick, R15	G	1300	.006		2.39	.27		2.66	3.08
1956	4" thick, R20	G	1300	.006		3.13	.27		3.40	3.90
1958	Tapered for drainage	G	1820	.004	B.F.	.87	.19		1.06	1.29

For customer support on your Building Construction Costs with RSMeans data, call 800.448.8182.

07 22 Roof and Deck Insulation

07 22 16 - Roof Board Insulation

07 22 16.10 Roof Deck Insulation

		Crew	Daily Output	Labor-Hours	Unit	Material	2018 Bare Costs Labor	Equipment	Total	Total Incl O&P
1960	60 psi compressive strength, 1" thick, R5	**G** 1 Rofc	1885	.004	S.F.	1.05	.19		1.24	1.47
1962	2" thick, R10	**G**	1560	.005		2	.23		2.23	2.57
1964	3" thick, R15	**G**	1270	.006		3.26	.28		3.54	4.05
1966	4" thick, R20	**G**	1235	.006		4.04	.28		4.32	4.93
1968	Tapered for drainage	**G**	1820	.004	B.F.	1	.19		1.19	1.43
2010	Expanded polystyrene, 1#/C.F. density, 3/4" thick, R2.89	**G**	1950	.004	S.F.	.20	.18		.38	.52
2020	1" thick, R3.85	**G**	1950	.004		.27	.18		.45	.60
2100	2" thick, R7.69	**G**	1625	.005		.54	.22		.76	.95
2110	3" thick, R11.49	**G**	1625	.005		.81	.22		1.03	1.25
2120	4" thick, R15.38	**G**	1625	.005		1.08	.22		1.30	1.55
2130	5" thick, R19.23	**G**	1495	.005		1.35	.24		1.59	1.89
2140	6" thick, R23.26	**G**	1495	.005		1.62	.24		1.86	2.18
2150	Tapered for drainage	**G**	1950	.004	B.F.	.53	.18		.71	.88
2400	Composites with 2" EPS									
2410	1" fiberboard	**G** 1 Rofc	1325	.006	S.F.	1.45	.27		1.72	2.05
2420	7/16" oriented strand board	**G**	1040	.008		1.15	.34		1.49	1.84
2430	1/2" plywood	**G**	1040	.008		1.44	.34		1.78	2.15
2440	1" perlite	**G**	1040	.008		1.17	.34		1.51	1.86
2450	Composites with 1-1/2" polyisocyanurate									
2460	1" fiberboard	**G** 1 Rofc	1040	.008	S.F.	1.07	.34		1.41	1.75
2470	1" perlite	**G**	1105	.007		.96	.32		1.28	1.60
2480	7/16" oriented strand board	**G**	1040	.008		.84	.34		1.18	1.49
3000	Fastening alternatives, coated screws, 2" long		3744	.002	Ea.	.06	.09		.15	.23
3010	4" long		3120	.003		.11	.11		.22	.31
3020	6" long		2675	.003		.19	.13		.32	.43
3030	8" long		2340	.003		.28	.15		.43	.56
3040	10" long		1872	.004		.47	.19		.66	.84
3050	Pre-drill and drive wedge spike, 2-1/2"		1248	.006		.40	.28		.68	.91
3060	3-1/2"		1101	.007		.58	.32		.90	1.18
3070	4-1/2"		936	.009		.65	.38		1.03	1.35
3075	3" galvanized deck plates		7488	.001		.08	.05		.13	.17
3080	Spot mop asphalt	G-1	295	.190	Sq.	5.80	7.80	1.74	15.34	21.50
3090	Full mop asphalt	"	192	.292		11.60	12	2.68	26.28	35.50
3110	Low-rise polyurethane adhesive, from 5 gallon kit, 12" OC beads	1 Rofc	45	.178		33	7.80		40.80	49.50
3120	6" OC beads		32	.250		66	11		77	91
3130	4" OC beads		30	.267		99	11.70		110.70	129

07 24 Exterior Insulation and Finish Systems

07 24 13 - Polymer-Based Exterior Insulation and Finish System

07 24 13.10 Exterior Insulation and Finish Systems

		Crew	Daily Output	Labor-Hours	Unit	Material	2018 Bare Costs Labor	Equipment	Total	Total Incl O&P
0010	**EXTERIOR INSULATION AND FINISH SYSTEMS**									
0095	Field applied, 1" EPS insulation	**G** J-1	390	.103	S.F.	1.80	4.49	.34	6.63	9.10
0100	With 1/2" cement board sheathing	**G**	268	.149		2.59	6.55	.50	9.64	13.20
0105	2" EPS insulation	**G**	390	.103		2.07	4.49	.34	6.90	9.40
0110	With 1/2" cement board sheathing	**G**	268	.149		2.86	6.55	.50	9.91	13.50
0115	3" EPS insulation	**G**	390	.103		2.34	4.49	.34	7.17	9.70
0120	With 1/2" cement board sheathing	**G**	268	.149		3.13	6.55	.50	10.18	13.80
0125	4" EPS insulation	**G**	390	.103		2.61	4.49	.34	7.44	10
0130	With 1/2" cement board sheathing	**G**	268	.149		4.18	6.55	.50	11.23	14.95
0140	Premium finish add		1265	.032		.36	1.38	.11	1.85	2.60
0150	Heavy duty reinforcement add		914	.044		.60	1.92	.15	2.67	3.69

07 24 Exterior Insulation and Finish Systems

07 24 13 – Polymer-Based Exterior Insulation and Finish System

07 24 13.10 Exterior Insulation and Finish Systems	Crew	Daily Output	Labor-Hours	Unit	Material	2018 Bare Costs Labor	Equipment	Total	Total Incl O&P
0160	2.5#/S.Y. metal lath substrate add	1 Lath	75	.107	S.Y.	3.10	5.25		8.35	11.15
0170	3.4#/S.Y. metal lath substrate add	"	75	.107	"	4.44	5.25		9.69	12.65
0180	Color or texture change	J-1	1265	.032	S.F.	.82	1.38	.11	2.31	3.10
0190	With substrate leveling base coat	1 Plas	530	.015		.83	.70		1.53	1.96
0210	With substrate sealing base coat	1 Pord	1224	.007	↓	.12	.28		.40	.55
0370	V groove shape in panel face				L.F.	.68			.68	.75
0380	U groove shape in panel face				"	.85			.85	.94
0440	For higher than one story, add						25%			

07 25 Weather Barriers

07 25 10 – Weather Barriers or Wraps

07 25 10.10 Weather Barriers

		Crew	Daily Output	Labor-Hours	Unit	Material	2018 Bare Costs Labor	Equipment	Total	Total Incl O&P
0010	**WEATHER BARRIERS**									
0400	Asphalt felt paper, #15	1 Carp	37	.216	Sq.	5.35	10.95		16.30	22.50
0401	Per square foot	"	3700	.002	S.F.	.05	.11		.16	.23
0450	Housewrap, exterior, spun bonded polypropylene									
0470	Small roll	1 Carp	3800	.002	S.F.	.15	.11		.26	.33
0480	Large roll	"	4000	.002	"	.14	.10		.24	.30
2100	Asphalt felt roof deck vapor barrier, class 1 metal decks	1 Rofc	37	.216	Sq.	22.50	9.50		32	40.50
2200	For all other decks	"	37	.216		16.95	9.50		26.45	34.50
2800	Asphalt felt, 50% recycled content, 15 lb., 4 sq./roll	1 Carp	36	.222		5.65	11.25		16.90	23.50
2810	30 lb., 2 sq./roll	"	36	.222	↓	9	11.25		20.25	27
3000	Building wrap, spun bonded polyethylene	2 Carp	8000	.002	S.F.	.18	.10		.28	.35

07 26 Vapor Retarders

07 26 10 – Above-Grade Vapor Retarders

07 26 10.10 Vapor Retarders

			Crew	Daily Output	Labor-Hours	Unit	Material	2018 Bare Costs Labor	Equipment	Total	Total Incl O&P
0010	**VAPOR RETARDERS**										
0020	Aluminum and kraft laminated, foil 1 side	G	1 Carp	37	.216	Sq.	13.65	10.95		24.60	31.50
0100	Foil 2 sides	G		37	.216		14.15	10.95		25.10	32.50
0600	Polyethylene vapor barrier, standard, 2 mil	G		37	.216		1.50	10.95		12.45	18.35
0700	4 mil	G		37	.216		2.58	10.95		13.53	19.55
0900	6 mil	G		37	.216		3.73	10.95		14.68	21
1200	10 mil	G		37	.216		8.75	10.95		19.70	26.50
1300	Clear reinforced, fire retardant, 8 mil	G		37	.216		26.50	10.95		37.45	45.50
1350	Cross laminated type, 3 mil	G		37	.216		12.35	10.95		23.30	30.50
1400	4 mil	G		37	.216		13.10	10.95		24.05	31
1800	Reinf. waterproof, 2 mil polyethylene backing, 1 side			37	.216		9.95	10.95		20.90	27.50
1900	2 sides			37	.216		12.90	10.95		23.85	31
2400	Waterproofed kraft with sisal or fiberglass fibers		↓	37	.216	↓	21	10.95		31.95	39.50

For customer support on your Building Construction Costs with RSMeans data, call 800.448.8182.

07 27 Air Barriers

07 27 13 – Modified Bituminous Sheet Air Barriers

07 27 13.10 Modified Bituminous Sheet Air Barrier	Crew	Daily Output	Labor-Hours	Unit	Material	2018 Bare Costs Labor	Equipment	Total	Total Incl O&P
0010 **MODIFIED BITUMINOUS SHEET AIR BARRIER**									
0100 SBS modified sheet laminated to polyethylene sheet, 40 mils, 4" wide	1 Carp	1200	.007	L.F.	.33	.34		.67	.87
0120 6" wide		1100	.007		.45	.37		.82	1.05
0140 9" wide		1000	.008		.62	.41		1.03	1.31
0160 12" wide	↓	900	.009	↓	.80	.45		1.25	1.57
0180 18" wide	2 Carp	1700	.009	S.F.	.75	.48		1.23	1.55
0200 36" wide	"	1800	.009		.73	.45		1.18	1.49
0220 Adhesive for above	1 Carp	1400	.006	↓	.31	.29		.60	.79

07 27 26 – Fluid-Applied Membrane Air Barriers

07 27 26.10 Fluid Applied Membrane Air Barrier

	Crew	Daily Output	Labor-Hours	Unit	Material	Labor	Equipment	Total	Total Incl O&P
0010 **FLUID APPLIED MEMBRANE AIR BARRIER**									
0100 Spray applied vapor barrier, 25 S.F./gallon	1 Pord	1375	.006	S.F.	.01	.25		.26	.39

07 31 Shingles and Shakes

07 31 13 – Asphalt Shingles

07 31 13.10 Asphalt Roof Shingles

	Crew	Daily Output	Labor-Hours	Unit	Material	Labor	Equipment	Total	Total Incl O&P
0010 **ASPHALT ROOF SHINGLES**									
0100 Standard strip shingles									
0150 Inorganic, class A, 25 year	1 Rofc	5.50	1.455	Sq.	75	64		139	191
0155 Pneumatic nailed		7	1.143		75	50		125	167
0200 30 year		5	1.600		95.50	70.50		166	223
0205 Pneumatic nailed	↓	6.25	1.280	↓	95.50	56.50		152	200
0250 Standard laminated multi-layered shingles									
0300 Class A, 240-260 lb./square	1 Rofc	4.50	1.778	Sq.	110	78		188	252
0305 Pneumatic nailed		5.63	1.422		110	62.50		172.50	226
0350 Class A, 250-270 lb./square		4	2		110	88		198	269
0355 Pneumatic nailed	↓	5	1.600	↓	110	70.50		180.50	239
0400 Premium, laminated multi-layered shingles									
0450 Class A, 260-300 lb./square	1 Rofc	3.50	2.286	Sq.	144	100		244	325
0455 Pneumatic nailed		4.37	1.831		144	80.50		224.50	293
0500 Class A, 300-385 lb./square		3	2.667		260	117		377	485
0505 Pneumatic nailed		3.75	2.133		260	94		354	445
0800 #15 felt underlayment		64	.125		5.35	5.50		10.85	15.15
0825 #30 felt underlayment		58	.138		10.30	6.05		16.35	21.50
0850 Self adhering polyethylene and rubberized asphalt underlayment		22	.364	↓	78	16		94	113
0900 Ridge shingles		330	.024	L.F.	2.27	1.07		3.34	4.29
0905 Pneumatic nailed	↓	412.50	.019	"	2.27	.85		3.12	3.93
1000 For steep roofs (7 to 12 pitch or greater), add						50%			

07 31 16 – Metal Shingles

07 31 16.10 Aluminum Shingles

	Crew	Daily Output	Labor-Hours	Unit	Material	Labor	Equipment	Total	Total Incl O&P
0010 **ALUMINUM SHINGLES**									
0020 Mill finish, .019" thick	1 Carp	5	1.600	Sq.	228	81		309	375
0100 .020" thick	"	5	1.600		257	81		338	405
0300 For colors, add				↓	21.50			21.50	24
0600 Ridge cap, .024" thick	1 Carp	170	.047	L.F.	4	2.39		6.39	8.05
0700 End wall flashing, .024" thick		170	.047		2.27	2.39		4.66	6.15
0900 Valley section, .024" thick		170	.047		3.97	2.39		6.36	8
1000 Starter strip, .024" thick		400	.020		1.82	1.01		2.83	3.54
1200 Side wall flashing, .024" thick		170	.047		2.22	2.39		4.61	6.05
1500 Gable flashing, .024" thick		400	.020	↓	1.77	1.01		2.78	3.49

07 31 Shingles and Shakes

07 31 16 – Metal Shingles

07 31 16.20 Steel Shingles

07 31 16.20 Steel Shingles	Crew	Daily Output	Labor-Hours	Unit	Material	2018 Bare Costs Labor	Equipment	Total	Total Incl O&P
0010 **STEEL SHINGLES**									
0012 Galvanized, 26 ga.	1 Rots	2.20	3.636	Sq.	360	161		521	665
0200 24 ga.	"	2.20	3.636		355	161		516	660
0300 For colored galvanized shingles, add					58			58	64

07 31 26 – Slate Shingles

07 31 26.10 Slate Roof Shingles

07 31 26.10 Slate Roof Shingles		Crew	Daily Output	Labor-Hours	Unit	Material	2018 Bare Costs Labor	Equipment	Total	Total Incl O&P
0010 **SLATE ROOF SHINGLES** R073126-20										
0100 Buckingham Virginia black, 3/16" - 1/4" thick	G	1 Rots	1.75	4.571	Sq.	555	202		757	950
0200 1/4" thick	G		1.75	4.571		555	202		757	950
0900 Pennsylvania black, Bangor, #1 clear	G		1.75	4.571		495	202		697	885
1200 Vermont, unfading, green, mottled green	G		1.75	4.571		505	202		707	895
1300 Semi-weathering green & gray	G		1.75	4.571		360	202		562	735
1400 Purple	G		1.75	4.571		435	202		637	820
1500 Black or gray	G		1.75	4.571		485	202		687	875
1600 Red	G		1.75	4.571		1,175	202		1,377	1,625
1700 Variegated purple			1.75	4.571		425	202		627	810
2700 Ridge shingles, slate			200	.040	L.F.	10.10	1.77		11.87	14.05

07 31 29 – Wood Shingles and Shakes

07 31 29.13 Wood Shingles

07 31 29.13 Wood Shingles	Crew	Daily Output	Labor-Hours	Unit	Material	2018 Bare Costs Labor	Equipment	Total	Total Incl O&P
0010 **WOOD SHINGLES**									
0012 16" No. 1 red cedar shingles, 5" exposure, on roof	1 Carp	2.50	3.200	Sq.	295	162		457	570
0015 Pneumatic nailed		3.25	2.462		295	125		420	515
0200 7-1/2" exposure, on walls		2.05	3.902		196	198		394	515
0205 Pneumatic nailed		2.67	2.996		196	152		348	445
0300 18" No. 1 red cedar perfections, 5-1/2" exposure, on roof		2.75	2.909		253	147		400	505
0305 Pneumatic nailed		3.57	2.241		253	114		367	450
0500 7-1/2" exposure, on walls		2.25	3.556		186	180		366	480
0505 Pneumatic nailed		2.92	2.740		186	139		325	415
0600 Resquared and rebutted, 5-1/2" exposure, on roof		3	2.667		290	135		425	525
0605 Pneumatic nailed		3.90	2.051		290	104		394	480
0900 7-1/2" exposure, on walls		2.45	3.265		214	166		380	485
0905 Pneumatic nailed		3.18	2.516		214	128		342	430
1000 Add to above for fire retardant shingles					55			55	60.50
1060 Preformed ridge shingles	1 Carp	400	.020	L.F.	3.89	1.01		4.90	5.80
2000 White cedar shingles, 16" long, extras, 5" exposure, on roof		2.40	3.333	Sq.	197	169		366	475
2005 Pneumatic nailed		3.12	2.564		197	130		327	415
2050 5" exposure on walls		2	4		197	203		400	525
2055 Pneumatic nailed		2.60	3.077		197	156		353	455
2100 7-1/2" exposure, on walls		2	4		141	203		344	465
2105 Pneumatic nailed		2.60	3.077		141	156		297	395
2150 "B" grade, 5" exposure on walls		2	4		172	203		375	500
2155 Pneumatic nailed		2.60	3.077		172	156		328	425
2300 For #15 organic felt underlayment on roof, 1 layer, add		64	.125		5.35	6.35		11.70	15.55
2400 2 layers, add		32	.250		10.75	12.70		23.45	31
2600 For steep roofs (7/12 pitch or greater), add to above						50%			
2700 Panelized systems, No.1 cedar shingles on 5/16" CDX plywood									
2800 On walls, 8' strips, 7" or 14" exposure	2 Carp	700	.023	S.F.	6.40	1.16		7.56	8.80
3500 On roofs, 8' strips, 7" or 14" exposure	1 Carp	3	2.667	Sq.	665	135		800	935
3505 Pneumatic nailed	"	4	2	"	665	101		766	885

For customer support on your Building Construction Costs with RSMeans data, call 800.448.8182.

07 31 Shingles and Shakes

07 31 29 – Wood Shingles and Shakes

07 31 29.16 Wood Shakes

		Crew	Daily Output	Labor-Hours	Unit	Material	2018 Bare Costs Labor	Equipment	Total	Total Incl O&P
0010	**WOOD SHAKES**									
1100	Hand-split red cedar shakes, 1/2" thick x 24" long, 10" exp. on roof	1 Carp	2.50	3.200	Sq.	300	162		462	575
1105	Pneumatic nailed		3.25	2.462		300	125		425	520
1110	3/4" thick x 24" long, 10" exp. on roof		2.25	3.556		300	180		480	605
1115	Pneumatic nailed		2.92	2.740		300	139		439	540
1200	1/2" thick, 18" long, 8-1/2" exp. on roof		2	4		275	203		478	610
1205	Pneumatic nailed		2.60	3.077		275	156		431	540
1210	3/4" thick x 18" long, 8-1/2" exp. on roof		1.80	4.444		275	225		500	645
1215	Pneumatic nailed		2.34	3.419		275	173		448	565
1255	10" exposure on walls		2	4		265	203		468	600
1260	10" exposure on walls, pneumatic nailed	▼	2.60	3.077		265	156		421	530
1700	Add to above for fire retardant shakes, 24" long					55			55	60.50
1800	18" long				▼	55			55	60.50
1810	Ridge shakes	1 Carp	350	.023	L.F.	5.75	1.16		6.91	8.10

07 32 Roof Tiles

07 32 13 – Clay Roof Tiles

07 32 13.10 Clay Tiles

		Crew	Daily Output	Labor-Hours	Unit	Material	Labor	Equipment	Total	Total Incl O&P
0010	**CLAY TILES**, including accessories									
0300	Flat shingle, interlocking, 15", 166 pcs./sq., fireflashed blend	3 Rots	6	4	Sq.	470	177		647	810
0500	Terra cotta red		6	4		520	177		697	865
0600	Roman pan and top, 18", 102 pcs./sq., fireflashed blend	▼	5.50	4.364		505	193		698	880
0640	Terra cotta red	1 Rots	2.40	3.333		570	147		717	875
1100	Barrel mission tile, 18", 166 pcs./sq., fireflashed blend	3 Rots	5.50	4.364		415	193		608	780
1140	Terra cotta red		5.50	4.364		420	193		613	785
1700	Scalloped edge flat shingle, 14", 145 pcs./sq., fireflashed blend		6	4		1,150	177		1,327	1,575
1800	Terra cotta red	▼	6	4		1,050	177		1,227	1,450
3010	#15 felt underlayment	1 Rofc	64	.125		5.35	5.50		10.85	15.15
3020	#30 felt underlayment		58	.138		10.30	6.05		16.35	21.50
3040	Polyethylene and rubberized asph. underlayment	▼	22	.364	▼	78	16		94	113

07 32 16 – Concrete Roof Tiles

07 32 16.10 Concrete Tiles

		Crew	Daily Output	Labor-Hours	Unit	Material	Labor	Equipment	Total	Total Incl O&P
0010	**CONCRETE TILES**									
0020	Corrugated, 13" x 16-1/2", 90 per sq., 950 lb./sq.									
0050	Earthtone colors, nailed to wood deck	1 Rots	1.35	5.926	Sq.	105	262		367	555
0150	Blues		1.35	5.926		105	262		367	555
0200	Greens		1.35	5.926		106	262		368	555
0250	Premium colors	▼	1.35	5.926	▼	106	262		368	555
0500	Shakes, 13" x 16-1/2", 90 per sq., 950 lb./sq.									
0600	All colors, nailed to wood deck	1 Rots	1.50	5.333	Sq.	126	235		361	535
1500	Accessory pieces, ridge & hip, 10" x 16-1/2", 8 lb. each	"	120	.067	Ea.	3.80	2.94		6.74	9.15
1700	Rake, 6-1/2" x 16-3/4", 9 lb. each					3.80			3.80	4.18
1800	Mansard hip, 10" x 16-1/2", 9.2 lb. each					3.80			3.80	4.18
1900	Hip starter, 10" x 16-1/2", 10.5 lb. each					10.50			10.50	11.55
2000	3 or 4 way apex, 10" each side, 11.5 lb. each				▼	12			12	13.20

07 32 Roof Tiles

07 32 19 – Metal Roof Tiles

07 32 19.10 Metal Roof Tiles		Crew	Daily Output	Labor-Hours	Unit	Material	2018 Bare Costs Labor	Equipment	Total	Total Incl O&P
0010	**METAL ROOF TILES**									
0020	Accessories included, .032" thick aluminum, mission tile	1 Carp	2.50	3.200	Sq.	830	162		992	1,150
0200	Spanish tiles	"	3	2.667	"	550	135		685	810

07 33 Natural Roof Coverings

07 33 63 – Vegetated Roofing

07 33 63.10 Green Roof Systems

			Crew	Daily Output	Labor-Hours	Unit	Material	2018 Bare Costs Labor	Equipment	Total	Total Incl O&P
0010	**GREEN ROOF SYSTEMS**										
0020	Soil mixture for green roof 30% sand, 55% gravel, 15% soil										
0100	Hoist and spread soil mixture 4" depth up to 5 stories tall roof	G	B-13B	4000	.014	S.F.	.23	.61	.25	1.09	1.45
0150	6" depth	G		2667	.021		.35	.92	.37	1.64	2.18
0200	8" depth	G		2000	.028		.46	1.22	.50	2.18	2.92
0250	10" depth	G		1600	.035		.58	1.53	.62	2.73	3.63
0300	12" depth	G		1335	.042		.69	1.83	.74	3.26	4.36
0310	Alt. man-made soil mix, hoist & spread, 4" deep up to 5 stories tall roof	G		4000	.014		1.92	.61	.25	2.78	3.31
0350	Mobilization 55 ton crane to site	G	1 Eqhv	3.60	2.222	Ea.		125		125	188
0355	Hoisting cost to 5 stories per day (Avg. 28 picks per day)	G	B-13B	1	56	Day		2,450	995	3,445	4,825
0360	Mobilization or demobilization, 100 ton crane to site driver & escort	G	A-3E	2.50	6.400	Ea.		325	50.50	375.50	545
0365	Hoisting cost 6-10 stories per day (Avg. 21 picks per day)	G	B-13C	1	56	Day		2,450	1,875	4,325	5,775
0370	Hoist and spread soil mixture 4" depth 6-10 stories tall roof	G		4000	.014	S.F.	.23	.61	.47	1.31	1.69
0375	6" depth	G		2667	.021		.35	.92	.70	1.97	2.54
0380	8" depth	G		2000	.028		.46	1.22	.94	2.62	3.40
0385	10" depth	G		1600	.035		.58	1.53	1.17	3.28	4.24
0390	12" depth	G		1335	.042		.69	1.83	1.40	3.92	5.10
0400	Green roof edging treated lumber 4" x 4", no hoisting included	G	2 Carp	400	.040	L.F.	1.20	2.03		3.23	4.41
0410	4" x 6"	G		400	.040		1.94	2.03		3.97	5.25
0420	4" x 8"	G		360	.044		3.96	2.25		6.21	7.80
0430	4" x 6" double stacked	G		300	.053		3.89	2.70		6.59	8.40
0500	Green roof edging redwood lumber 4" x 4", no hoisting included	G		400	.040		6.45	2.03		8.48	10.20
0510	4" x 6"	G		400	.040		12.60	2.03		14.63	16.95
0520	4" x 8"	G		360	.044		23.50	2.25		25.75	29.50
0530	4" x 6" double stacked	G		300	.053		25	2.70		27.70	31.50
0550	Components, not including membrane or insulation:										
0560	Fluid applied rubber membrane, reinforced, 215 mil thick	G	G-5	350	.114	S.F.	.30	4.56	.52	5.38	8.60
0570	Root barrier	G	2 Rofc	775	.021		.70	.91		1.61	2.30
0580	Moisture retention barrier and reservoir	G	"	900	.018		2.66	.78		3.44	4.24
0600	Planting sedum, light soil, potted, 2-1/4" diameter, 2 per S.F.	G	1 Clab	420	.019		6.10	.76		6.86	7.85
0610	1 per S.F.	G	"	840	.010		3.05	.38		3.43	3.94
0630	Planting sedum mat per S.F. including shipping (4000 S.F. min)	G	4 Clab	4000	.008		7.30	.32		7.62	8.55
0640	Installation sedum mat system (no soil required) per S.F. (4000 S.F. min)	G	"	4000	.008		10.20	.32		10.52	11.70
0645	Note: pricing of sedum mats shipped in full truck loads (4000-5000 S.F.)										

For customer support on your Building Construction Costs with RSMeans data, call 800.448.8182.

07 41 Roof Panels

07 41 13 – Metal Roof Panels

07 41 13.10 Aluminum Roof Panels

		Crew	Daily Output	Labor-Hours	Unit	Material	2018 Bare Costs Labor	Equipment	Total	Total Incl O&P
0010	**ALUMINUM ROOF PANELS**									
0020	Corrugated or ribbed, .0155" thick, natural	G-3	1200	.027	S.F.	1	1.33		2.33	3.13
0300	Painted		1200	.027		1.45	1.33		2.78	3.63
0400	Corrugated, .018" thick, on steel frame, natural finish		1200	.027		1.24	1.33		2.57	3.39
0600	Painted		1200	.027		1.55	1.33		2.88	3.74
0700	Corrugated, on steel frame, natural, .024" thick		1200	.027		1.80	1.33		3.13	4.01
0800	Painted		1200	.027		2.20	1.33		3.53	4.45
0900	.032" thick, natural		1200	.027		3.04	1.33		4.37	5.35
1200	Painted		1200	.027		3.28	1.33		4.61	5.65
1300	V-Beam, on steel frame construction, .032" thick, natural		1200	.027		2.62	1.33		3.95	4.91
1500	Painted		1200	.027		3.91	1.33		5.24	6.35
1600	.040" thick, natural		1200	.027		3.82	1.33		5.15	6.25
1800	Painted		1200	.027		4.61	1.33		5.94	7.10
1900	.050" thick, natural		1200	.027		3.95	1.33		5.28	6.40
2100	Painted		1200	.027		4.75	1.33		6.08	7.30
2200	For roofing on wood frame, deduct		4600	.007		.08	.35		.43	.62
2400	Ridge cap, .032" thick, natural		800	.040	L.F.	3.18	1.99		5.17	6.55

07 41 13.20 Steel Roofing Panels

		Crew	Daily Output	Labor-Hours	Unit	Material	2018 Bare Costs Labor	Equipment	Total	Total Incl O&P
0010	**STEEL ROOFING PANELS**									
0012	Corrugated or ribbed, on steel framing, 30 ga. galv	G-3	1100	.029	S.F.	1.67	1.45		3.12	4.05
0100	28 ga.		1050	.030		1.44	1.52		2.96	3.90
0300	26 ga.		1000	.032		2.02	1.59		3.61	4.65
0400	24 ga.		950	.034		3	1.68		4.68	5.85
0600	Colored, 28 ga.		1050	.030		1.81	1.52		3.33	4.31
0700	26 ga.		1000	.032		2.12	1.59		3.71	4.76
0710	Flat profile, 1-3/4" standing seams, 10" wide, standard finish, 26 ga.		1000	.032		3.93	1.59		5.52	6.75
0715	24 ga.		950	.034		4.60	1.68		6.28	7.60
0720	22 ga.		900	.036		5.70	1.77		7.47	8.95
0725	Zinc aluminum alloy finish, 26 ga.		1000	.032		3.20	1.59		4.79	5.95
0730	24 ga.		950	.034		3.72	1.68		5.40	6.65
0735	22 ga.		900	.036		4.24	1.77		6.01	7.35
0740	12" wide, standard finish, 26 ga.		1000	.032		3.98	1.59		5.57	6.80
0745	24 ga.		950	.034		5.15	1.68		6.83	8.20
0750	Zinc aluminum alloy finish, 26 ga.		1000	.032		4.44	1.59		6.03	7.30
0755	24 ga.		950	.034		3.72	1.68		5.40	6.65
0840	Flat profile, 1" x 3/8" batten, 12" wide, standard finish, 26 ga.		1000	.032		3.46	1.59		5.05	6.25
0845	24 ga.		950	.034		4.08	1.68		5.76	7.05
0850	22 ga.		900	.036		4.91	1.77		6.68	8.10
0855	Zinc aluminum alloy finish, 26 ga.		1000	.032		3.36	1.59		4.95	6.15
0860	24 ga.		950	.034		3.82	1.68		5.50	6.75
0865	22 ga.		900	.036		4.34	1.77		6.11	7.45
0870	16-1/2" wide, standard finish, 24 ga.		950	.034		4.03	1.68		5.71	7
0875	22 ga.		900	.036		4.55	1.77		6.32	7.70
0880	Zinc aluminum alloy finish, 24 ga.		950	.034		3.51	1.68		5.19	6.40
0885	22 ga.		900	.036		3.98	1.77		5.75	7.10
0890	Flat profile, 2" x 2" batten, 12" wide, standard finish, 26 ga.		1000	.032		4.03	1.59		5.62	6.85
0895	24 ga.		950	.034		4.86	1.68		6.54	7.90
0900	22 ga.		900	.036		5.90	1.77		7.67	9.20
0905	Zinc aluminum alloy finish, 26 ga.		1000	.032		3.82	1.59		5.41	6.65
0910	24 ga.		950	.034		4.29	1.68		5.97	7.30
0915	22 ga.		900	.036		4.96	1.77		6.73	8.15
0920	16-1/2" wide, standard finish, 24 ga.		950	.034		4.44	1.68		6.12	7.45

07 41 Roof Panels

07 41 13 – Metal Roof Panels

07 41 13.20 Steel Roofing Panels		Crew	Daily Output	Labor-Hours	Unit	Material	2018 Bare Costs Labor	Equipment	Total	Total Incl O&P
0925	22 ga.	G-3	900	.036	S.F.	5.10	1.77		6.87	8.35
0930	Zinc aluminum alloy finish, 24 ga.		950	.034		3.98	1.68		5.66	6.95
0935	22 ga.		900	.036		4.55	1.77		6.32	7.70
1200	Ridge, galvanized, 10" wide	G	800	.040	L.F.	3.23	1.99		5.22	6.60
1203	14" wide	G	2 Shee 316	.051		3.88	3.03		6.91	8.90
1205	18" wide	G	" 308	.052		4.52	3.11		7.63	9.70
1210	20" wide	G	G-3 750	.043		4.25	2.13		6.38	7.90

07 41 33 – Plastic Roof Panels

07 41 33.10 Fiberglass Panels

		Crew	Daily Output	Labor-Hours	Unit	Material	Labor	Equipment	Total	Total Incl O&P
0010	**FIBERGLASS PANELS**									
0012	Corrugated panels, roofing, 8 oz./S.F.	G-3	1000	.032	S.F.	2.47	1.59		4.06	5.15
0100	12 oz./S.F.		1000	.032		4.52	1.59		6.11	7.40
0300	Corrugated siding, 6 oz./S.F.		880	.036		1.99	1.81		3.80	4.95
0400	8 oz./S.F.		880	.036		2.47	1.81		4.28	5.50
0500	Fire retardant		880	.036		3.98	1.81		5.79	7.15
0600	12 oz. siding, textured		880	.036		3.87	1.81		5.68	7
0700	Fire retardant		880	.036		4.48	1.81		6.29	7.70
0900	Flat panels, 6 oz./S.F., clear or colors		880	.036		2.52	1.81		4.33	5.55
1100	Fire retardant, class A		880	.036		3.56	1.81		5.37	6.70
1300	8 oz./S.F., clear or colors		880	.036		2.47	1.81		4.28	5.50
1700	Sandwich panels, fiberglass, 1-9/16" thick, panels to 20 S.F.		180	.178		35.50	8.85		44.35	52.50
1900	As above, but 2-3/4" thick, panels to 100 S.F.		265	.121		25	6		31	36.50

07 42 Wall Panels

07 42 13 – Metal Wall Panels

07 42 13.10 Mansard Panels

		Crew	Daily Output	Labor-Hours	Unit	Material	Labor	Equipment	Total	Total Incl O&P
0010	**MANSARD PANELS**									
0600	Aluminum, stock units, straight surfaces	1 Shee	115	.070	S.F.	4.17	4.16		8.33	10.95
0700	Concave or convex surfaces, add		75	.107	"	2.20	6.40		8.60	12.15
0800	For framing, to 5' high, add		115	.070	L.F.	4	4.16		8.16	10.75
0900	Soffits, to 1' wide		125	.064	S.F.	2.75	3.83		6.58	8.90

07 42 13.20 Aluminum Siding Panels

		Crew	Daily Output	Labor-Hours	Unit	Material	Labor	Equipment	Total	Total Incl O&P
0010	**ALUMINUM SIDING PANELS**									
0012	Corrugated, on steel framing, .019" thick, natural finish	G-3	775	.041	S.F.	1.66	2.06		3.72	4.97
0100	Painted		775	.041		1.82	2.06		3.88	5.15
0400	Farm type, .021" thick on steel frame, natural		775	.041		1.72	2.06		3.78	5.05
0600	Painted		775	.041		1.82	2.06		3.88	5.15
0700	Industrial type, corrugated, on steel, .024" thick, mill		775	.041		2.37	2.06		4.43	5.75
0900	Painted		775	.041		2.52	2.06		4.58	5.90
1000	.032" thick, mill		775	.041		2.64	2.06		4.70	6.05
1200	Painted		775	.041		3.17	2.06		5.23	6.65
1300	V-Beam, on steel frame, .032" thick, mill		775	.041		2.99	2.06		5.05	6.45
1500	Painted		775	.041		3.33	2.06		5.39	6.80
1600	.040" thick, mill		775	.041		3.59	2.06		5.65	7.10
1800	Painted		775	.041		4.17	2.06		6.23	7.75
1900	.050" thick, mill		775	.041		4.22	2.06		6.28	7.80
2100	Painted		775	.041		4.99	2.06		7.05	8.65
2200	Ribbed, 3" profile, on steel frame, .032" thick, natural		775	.041		2.62	2.06		4.68	6
2400	Painted		775	.041		3.22	2.06		5.28	6.70
2500	.040" thick, natural		775	.041		3.03	2.06		5.09	6.45

For customer support on your Building Construction Costs with RSMeans data, call 800.448.8182.

07 42 Wall Panels

07 42 13 – Metal Wall Panels

07 42 13.20 Aluminum Siding Panels

		Crew	Daily Output	Labor-Hours	Unit	Material	2018 Bare Costs Labor	Equipment	Total	Total Incl O&P
2700	Painted	G-3	775	.041	S.F.	3.50	2.06		5.56	7
2750	.050" thick, natural		775	.041		3.45	2.06		5.51	6.95
2760	Painted		775	.041		3.97	2.06		6.03	7.50
3300	For siding on wood frame, deduct from above		2800	.011		.10	.57		.67	.98
3400	Screw fasteners, aluminum, self tapping, neoprene washer, 1"				M	222			222	244
3600	Stitch screws, self tapping, with neoprene washer, 5/8"				"	167			167	183
3630	Flashing, sidewall, .032" thick	G-3	800	.040	L.F.	3.03	1.99		5.02	6.35
3650	End wall, .040" thick		800	.040		3.54	1.99		5.53	6.95
3670	Closure strips, corrugated, .032" thick		800	.040		.96	1.99		2.95	4.10
3680	Ribbed, 4" or 8", .032" thick		800	.040		.95	1.99		2.94	4.09
3690	V-beam, .040" thick		800	.040		1.26	1.99		3.25	4.43
3800	Horizontal, colored clapboard, 8" wide, plain	2 Carp	515	.031	S.F.	2.50	1.58		4.08	5.15
3810	Insulated		515	.031		2.88	1.58		4.46	5.55
4000	Vertical board & batten, colored, non-insulated		515	.031		2.10	1.58		3.68	4.71
4200	For simulated wood design, add					.15			.15	.17
4300	Corners for above, outside	2 Carp	515	.031	V.L.F.	3.64	1.58		5.22	6.40
4500	Inside corners	"	515	.031	"	1.77	1.58		3.35	4.35

07 42 13.30 Steel Siding

		Crew	Daily Output	Labor-Hours	Unit	Material	2018 Bare Costs Labor	Equipment	Total	Total Incl O&P
0010	**STEEL SIDING**									
0020	Beveled, vinyl coated, 8" wide	1 Carp	265	.030	S.F.	1.80	1.53		3.33	4.31
0050	10" wide	"	275	.029		1.95	1.47		3.42	4.40
0080	Galv, corrugated or ribbed, on steel frame, 30 ga.	G-3	800	.040		1.25	1.99		3.24	4.42
0100	28 ga.		795	.040		1.35	2.01		3.36	4.55
0300	26 ga.		790	.041		1.78	2.02		3.80	5.05
0400	24 ga.		785	.041		2.25	2.03		4.28	5.60
0600	22 ga.		770	.042		2.35	2.07		4.42	5.75
0700	Colored, corrugated/ribbed, on steel frame, 10 yr. finish, 28 ga.		800	.040		2.05	1.99		4.04	5.30
0900	26 ga.		795	.040		2.11	2.01		4.12	5.40
1000	24 ga.		790	.041		2.36	2.02		4.38	5.70
1020	20 ga.		785	.041		3	2.03		5.03	6.40
1200	Factory sandwich panel, 26 ga., 1" insulation, galvanized		380	.084		5.40	4.20		9.60	12.30
1300	Colored 1 side		380	.084		6.95	4.20		11.15	14.05
1500	Galvanized 2 sides		380	.084		8.25	4.20		12.45	15.50
1600	Colored 2 sides		380	.084		8.50	4.20		12.70	15.75
1800	Acrylic paint face, regular paint liner		380	.084		6.25	4.20		10.45	13.30
1900	For 2" thick polystyrene, add					1			1	1.10
2000	22 ga., galv, 2" insulation, baked enamel exterior	G-3	360	.089		12.25	4.43		16.68	20.50
2100	Polyvinylidene exterior finish	"	360	.089		13	4.43		17.43	21

07 44 Faced Panels

07 44 73 – Metal Faced Panels

07 44 73.10 Metal Faced Panels and Accessories

		Crew	Daily Output	Labor-Hours	Unit	Material	2018 Bare Costs Labor	Equipment	Total	Total Incl O&P
0010	**METAL FACED PANELS AND ACCESSORIES**									
0400	Textured aluminum, 4' x 8' x 5/16" plywood backing, single face	2 Shee	375	.043	S.F.	4.13	2.55		6.68	8.45
0600	Double face		375	.043		5.40	2.55		7.95	9.85
0700	4' x 10' x 5/16" plywood backing, single face		375	.043		4.40	2.55		6.95	8.75
0900	Double face		375	.043		5.90	2.55		8.45	10.40
1000	4' x 12' x 5/16" plywood backing, single face		375	.043		4.40	2.55		6.95	8.75
1300	Smooth aluminum, 1/4" plywood panel, fluoropolymer finish, double face		375	.043		6.15	2.55		8.70	10.65
1350	Clear anodized finish, double face		375	.043		10.20	2.55		12.75	15.10
1400	Double face textured aluminum, structural panel, 1" EPS insulation		375	.043		5.80	2.55		8.35	10.30

07 44 Faced Panels

07 44 73 – Metal Faced Panels

07 44 73.10 Metal Faced Panels and Accessories	Crew	Daily Output	Labor-Hours	Unit	Material	2018 Bare Costs Labor	Equipment	Total	Total Incl O&P
1500	Accessories, outside corner	1 Shee	175	.046	L.F.	1.92	2.73		4.65	6.30
1600	Inside corner		175	.046		1.38	2.73		4.11	5.70
1800	Batten mounting clip		200	.040		.50	2.39		2.89	4.20
1900	Low profile batten		480	.017		.62	1		1.62	2.20
2100	High profile batten		480	.017		1.43	1		2.43	3.09
2200	Water table		200	.040		2.16	2.39		4.55	6.05
2400	Horizontal joint connector		200	.040		1.70	2.39		4.09	5.50
2500	Corner cap		200	.040		1.88	2.39		4.27	5.70
2700	H - moulding	▼	480	.017	▼	1.27	1		2.27	2.92

07 46 Siding

07 46 23 – Wood Siding

07 46 23.10 Wood Board Siding

		Crew	Daily Output	Labor-Hours	Unit	Material	2018 Bare Costs Labor	Equipment	Total	Total Incl O&P
0010	**WOOD BOARD SIDING**									
3200	Wood, cedar bevel, A grade, 1/2" x 6"	1 Carp	295	.027	S.F.	4.42	1.38		5.80	6.95
3300	1/2" x 8"		330	.024		7.50	1.23		8.73	10.10
3500	3/4" x 10", clear grade		375	.021		7.30	1.08		8.38	9.65
3600	"B" grade		375	.021		3.99	1.08		5.07	6.05
3800	Cedar, rough sawn, 1" x 4", A grade, natural		220	.036		7.30	1.84		9.14	10.85
3900	Stained		220	.036		7.45	1.84		9.29	11
4100	1" x 12", board & batten, #3 & Btr., natural		420	.019		4.76	.97		5.73	6.70
4200	Stained		420	.019		5.10	.97		6.07	7.05
4400	1" x 8" channel siding, #3 & Btr., natural		330	.024		4.74	1.23		5.97	7.05
4500	Stained		330	.024		5	1.23		6.23	7.35
4700	Redwood, clear, beveled, vertical grain, 1/2" x 4"		220	.036		4.81	1.84		6.65	8.10
4750	1/2" x 6"		295	.027		4.80	1.38		6.18	7.40
4800	1/2" x 8"		330	.024		5.20	1.23		6.43	7.55
5000	3/4" x 10"		375	.021		4.95	1.08		6.03	7.10
5200	Channel siding, 1" x 10", B grade		375	.021		4.60	1.08		5.68	6.70
5250	Redwood, T&G boards, B grade, 1" x 4"		220	.036		7.50	1.84		9.34	11.05
5270	1" x 8"		330	.024		7.90	1.23		9.13	10.55
5400	White pine, rough sawn, 1" x 8", natural		330	.024		2.48	1.23		3.71	4.60
5500	Stained	▼	330	.024	▼	2.38	1.23		3.61	4.49

07 46 29 – Plywood Siding

07 46 29.10 Plywood Siding Options

		Crew	Daily Output	Labor-Hours	Unit	Material	2018 Bare Costs Labor	Equipment	Total	Total Incl O&P
0010	**PLYWOOD SIDING OPTIONS**									
0900	Plywood, medium density overlaid, 3/8" thick	2 Carp	750	.021	S.F.	1.35	1.08		2.43	3.14
1000	1/2" thick		700	.023		1.55	1.16		2.71	3.47
1100	3/4" thick		650	.025		1.90	1.25		3.15	3.99
1600	Texture 1-11, cedar, 5/8" thick, natural		675	.024		2.60	1.20		3.80	4.69
1700	Factory stained		675	.024		2.87	1.20		4.07	4.99
1900	Texture 1-11, fir, 5/8" thick, natural		675	.024		1.37	1.20		2.57	3.34
2000	Factory stained		675	.024		1.91	1.20		3.11	3.93
2050	Texture 1-11, S.Y.P., 5/8" thick, natural		675	.024		1.44	1.20		2.64	3.41
2100	Factory stained		675	.024		1.51	1.20		2.71	3.49
2200	Rough sawn cedar, 3/8" thick, natural		675	.024		1.27	1.20		2.47	3.23
2300	Factory stained		675	.024		1.57	1.20		2.77	3.56
2500	Rough sawn fir, 3/8" thick, natural		675	.024		.92	1.20		2.12	2.84
2600	Factory stained		675	.024		1.09	1.20		2.29	3.03
2800	Redwood, textured siding, 5/8" thick	▼	675	.024	▼	2.01	1.20		3.21	4.04

For customer support on your Building Construction Costs with RSMeans data, call 800.448.8182.

07 46 Siding

07 46 29 – Plywood Siding

07 46 29.10 Plywood Siding Options	Crew	Daily Output	Labor-Hours	Unit	Material	2018 Bare Costs Labor	Equipment	Total	Total Incl O&P
3000 Polyvinyl chloride coated, 3/8" thick	2 Carp	750	.021	S.F.	1.16	1.08		2.24	2.93

07 46 33 – Plastic Siding

07 46 33.10 Vinyl Siding

	Crew	Daily Output	Labor-Hours	Unit	Material	Labor	Equipment	Total	Total Incl O&P
0010 **VINYL SIDING**									
3995 Clapboard profile, woodgrain texture, .048 thick, double 4	2 Carp	495	.032	S.F.	1.07	1.64		2.71	3.68
4000 Double 5		550	.029		1.07	1.47		2.54	3.43
4005 Single 8		495	.032		1.37	1.64		3.01	4.01
4010 Single 10		550	.029		1.65	1.47		3.12	4.06
4015 .044 thick, double 4		495	.032		1.05	1.64		2.69	3.66
4020 Double 5		550	.029		1.07	1.47		2.54	3.43
4025 .042 thick, double 4		495	.032		1.07	1.64		2.71	3.68
4030 Double 5		550	.029		1.07	1.47		2.54	3.43
4035 Cross sawn texture, .040 thick, double 4		495	.032		.73	1.64		2.37	3.31
4040 Double 5		550	.029		.67	1.47		2.14	2.99
4045 Smooth texture, .042 thick, double 4		495	.032		.80	1.64		2.44	3.38
4050 Double 5		550	.029		.80	1.47		2.27	3.13
4055 Single 8		495	.032		.80	1.64		2.44	3.38
4060 Cedar texture, .044 thick, double 4		495	.032		1.14	1.64		2.78	3.76
4065 Double 6		600	.027		1.15	1.35		2.50	3.33
4070 Dutch lap profile, woodgrain texture, .048 thick, double 5		550	.029		1.09	1.47		2.56	3.45
4075 .044 thick, double 4.5		525	.030		1.09	1.55		2.64	3.55
4080 .042 thick, double 4.5		525	.030		.92	1.55		2.47	3.37
4085 .040 thick, double 4.5		525	.030		.73	1.55		2.28	3.16
4100 Shake profile, 10" wide		400	.040		3.71	2.03		5.74	7.20
4105 Vertical pattern, .046 thick, double 5		550	.029		1.56	1.47		3.03	3.97
4110 .044 thick, triple 3		550	.029		1.78	1.47		3.25	4.21
4115 .040 thick, triple 4		550	.029		1.68	1.47		3.15	4.10
4120 .040 thick, triple 2.66		550	.029		1.78	1.47		3.25	4.21
4125 Insulation, fan folded extruded polystyrene, 1/4"		2000	.008		.29	.41		.70	.94
4130 3/8"		2000	.008		.32	.41		.73	.97
4135 Accessories, J channel, 5/8" pocket		700	.023	L.F.	.51	1.16		1.67	2.32
4140 3/4" pocket		695	.023		.56	1.17		1.73	2.39
4145 1-1/4" pocket		680	.024		.86	1.19		2.05	2.76
4150 Flexible, 3/4" pocket		600	.027		2.39	1.35		3.74	4.69
4155 Under sill finish trim		500	.032		.56	1.62		2.18	3.08
4160 Vinyl starter strip		700	.023		.66	1.16		1.82	2.48
4165 Aluminum starter strip		700	.023		.29	1.16		1.45	2.08
4170 Window casing, 2-1/2" wide, 3/4" pocket		510	.031		1.71	1.59		3.30	4.30
4175 Outside corner, woodgrain finish, 4" face, 3/4" pocket		700	.023		2.15	1.16		3.31	4.13
4180 5/8" pocket		700	.023		2.13	1.16		3.29	4.11
4185 Smooth finish, 4" face, 3/4" pocket		700	.023		2.13	1.16		3.29	4.11
4190 7/8" pocket		690	.023		2.03	1.18		3.21	4.03
4195 1-1/4" pocket		700	.023		1.42	1.16		2.58	3.33
4200 Soffit and fascia, 1' overhang, solid		120	.133		4.72	6.75		11.47	15.50
4205 Vented		120	.133		4.72	6.75		11.47	15.50
4207 18" overhang, solid		110	.145		5.50	7.35		12.85	17.30
4208 Vented		110	.145		5.50	7.35		12.85	17.30
4210 2' overhang, solid		100	.160		6.30	8.10		14.40	19.25
4215 Vented		100	.160		6.30	8.10		14.40	19.25
4217 3' overhang, solid		100	.160		7.85	8.10		15.95	21
4218 Vented		100	.160		7.85	8.10		15.95	21
4220 Colors for siding and soffits, add				S.F.	.15			.15	.17

07 46 Siding

07 46 33 – Plastic Siding

07 46 33.10 Vinyl Siding

		Crew	Daily Output	Labor-Hours	Unit	Material	2018 Bare Costs Labor	Equipment	Total	Total Incl O&P
4225	Colors for accessories and trim, add				L.F.	.31			.31	.34

07 46 33.20 Polypropylene Siding

		Crew	Daily Output	Labor-Hours	Unit	Material	2018 Bare Costs Labor	Equipment	Total	Total Incl O&P
0010	**POLYPROPYLENE SIDING**									
4090	Shingle profile, random grooves, double 7	2 Carp	400	.040	S.F.	3.24	2.03		5.27	6.65
4092	Cornerpost for above	1 Carp	365	.022	L.F.	13.10	1.11		14.21	16.10
4095	Triple 5	2 Carp	400	.040	S.F.	3.24	2.03		5.27	6.65
4097	Cornerpost for above	1 Carp	365	.022	L.F.	12.30	1.11		13.41	15.25
5000	Staggered butt, double 7"	2 Carp	400	.040	S.F.	3.70	2.03		5.73	7.15
5002	Cornerpost for above	1 Carp	365	.022	L.F.	13.10	1.11		14.21	16.10
5010	Half round, double 6-1/4"	2 Carp	360	.044	S.F.	3.70	2.25		5.95	7.50
5020	Shake profile, staggered butt, double 9"	"	510	.031	"	3.70	1.59		5.29	6.50
5022	Cornerpost for above	1 Carp	365	.022	L.F.	9.80	1.11		10.91	12.45
5030	Straight butt, double 7"	2 Carp	400	.040	S.F.	3.70	2.03		5.73	7.15
5032	Cornerpost for above	1 Carp	365	.022	L.F.	13	1.11		14.11	16
6000	Accessories, J channel, 5/8" pocket	2 Carp	700	.023		.51	1.16		1.67	2.32
6010	3/4" pocket		695	.023		.56	1.17		1.73	2.39
6020	1-1/4" pocket		680	.024		.86	1.19		2.05	2.76
6030	Aluminum starter strip	↓	700	.023	↓	.29	1.16		1.45	2.08

07 46 46 – Fiber Cement Siding

07 46 46.10 Fiber Cement Siding

		Crew	Daily Output	Labor-Hours	Unit	Material	2018 Bare Costs Labor	Equipment	Total	Total Incl O&P
0010	**FIBER CEMENT SIDING**									
0020	Lap siding, 5/16" thick, 6" wide, 4-3/4" exposure, smooth texture	2 Carp	415	.039	S.F.	1.31	1.95		3.26	4.42
0025	Woodgrain texture		415	.039		1.31	1.95		3.26	4.42
0030	7-1/2" wide, 6-1/4" exposure, smooth texture		425	.038		1.63	1.91		3.54	4.71
0035	Woodgrain texture		425	.038		1.63	1.91		3.54	4.71
0040	8" wide, 6-3/4" exposure, smooth texture		425	.038		1.23	1.91		3.14	4.26
0045	Rough sawn texture		425	.038		1.23	1.91		3.14	4.26
0050	9-1/2" wide, 8-1/4" exposure, smooth texture		440	.036		1.29	1.84		3.13	4.23
0055	Woodgrain texture		440	.036		1.29	1.84		3.13	4.23
0060	12" wide, 10-3/8" exposure, smooth texture		455	.035		2.09	1.78		3.87	5
0065	Woodgrain texture		455	.035		2.09	1.78		3.87	5
0070	Panel siding, 5/16" thick, smooth texture		750	.021		1.33	1.08		2.41	3.11
0075	Stucco texture		750	.021		1.33	1.08		2.41	3.11
0080	Grooved woodgrain texture		750	.021		1.33	1.08		2.41	3.11
0085	V - grooved woodgrain texture		750	.021		1.33	1.08		2.41	3.11
0088	Shingle siding, 48" x 15-1/4" panels, 7" exposure		700	.023	↓	4.19	1.16		5.35	6.35
0090	Wood starter strip	↓	400	.040	L.F.	.45	2.03		2.48	3.59

07 46 73 – Soffit

07 46 73.10 Soffit Options

		Crew	Daily Output	Labor-Hours	Unit	Material	2018 Bare Costs Labor	Equipment	Total	Total Incl O&P
0010	**SOFFIT OPTIONS**									
0012	Aluminum, residential, .020" thick	1 Carp	210	.038	S.F.	2.07	1.93		4	5.20
0100	Baked enamel on steel, 16 or 18 ga.		105	.076		6.05	3.86		9.91	12.55
0300	Polyvinyl chloride, white, solid		230	.035		2.17	1.76		3.93	5.10
0400	Perforated	↓	230	.035		2.17	1.76		3.93	5.10
0500	For colors, add				↓	.15			.15	.17

07 51 Built-Up Bituminous Roofing

07 51 13 – Built-Up Asphalt Roofing

07 51 13.10 Built-Up Roofing Components

		Crew	Daily Output	Labor-Hours	Unit	Material	2018 Bare Costs Labor	Equipment	Total	Total Incl O&P
0010	**BUILT-UP ROOFING COMPONENTS**									
0012	Asphalt saturated felt, #30, 2 sq./roll	1 Rofc	58	.138	Sq.	10.30	6.05		16.35	21.50
0200	#15, 4 sq./roll, plain or perforated, not mopped		58	.138		5.35	6.05		11.40	16.10
0300	Roll roofing, smooth, #65		15	.533		10.40	23.50		33.90	51
0500	#90		12	.667		37	29.50		66.50	90
0520	Mineralized		12	.667		36	29.50		65.50	89
0540	D.C. (double coverage), 19" selvage edge	↓	10	.800	↓	48	35		83	112
0580	Adhesive (lap cement)				Gal.	8.45			8.45	9.30
0800	Steep, flat or dead level asphalt, 10 ton lots, packaged				Ton	965			965	1,075

07 51 13.13 Cold-Applied Built-Up Asphalt Roofing

		Crew	Daily Output	Labor-Hours	Unit	Material	2018 Bare Costs Labor	Equipment	Total	Total Incl O&P
0010	**COLD-APPLIED BUILT-UP ASPHALT ROOFING**									
0020	3 ply system, installation only (components listed below)	G-5	50	.800	Sq.		32	3.62	35.62	57.50
0100	Spunbond poly. fabric, 1.35 oz./S.Y., 36"W, 10.8 sq./roll				Ea.	133			133	147
0500	Base & finish coat, 3 gal./sq., 5 gal./can				Gal.	8			8	8.80
0600	Coating, ceramic granules, 1/2 sq./bag				Ea.	23.50			23.50	26
0700	Aluminum, 2 gal./sq.				Gal.	11.80			11.80	13
0800	Emulsion, fibered or non-fibered, 4 gal./sq.				"	6.30			6.30	6.95

07 51 13.20 Built-Up Roofing Systems

		Crew	Daily Output	Labor-Hours	Unit	Material	2018 Bare Costs Labor	Equipment	Total	Total Incl O&P
0010	**BUILT-UP ROOFING SYSTEMS** R075113-20									
0120	Asphalt flood coat with gravel/slag surfacing, not including									
0140	Insulation, flashing or wood nailers									
0200	Asphalt base sheet, 3 plies #15 asphalt felt, mopped	G-1	22	2.545	Sq.	104	105	23.50	232.50	315
0350	On nailable decks		21	2.667		107	109	24.50	240.50	330
0500	4 plies #15 asphalt felt, mopped		20	2.800		142	115	25.50	282.50	380
0550	On nailable decks		19	2.947		126	121	27	274	370
0700	Coated glass base sheet, 2 plies glass (type IV), mopped		22	2.545		111	105	23.50	239.50	325
0850	3 plies glass, mopped		20	2.800		134	115	25.50	274.50	370
0950	On nailable decks		19	2.947		126	121	27	274	370
1100	4 plies glass fiber felt (type IV), mopped		20	2.800		165	115	25.50	305.50	405
1150	On nailable decks		19	2.947		149	121	27	297	400
1200	Coated & saturated base sheet, 3 plies #15 asph. felt, mopped		20	2.800		116	115	25.50	256.50	350
1250	On nailable decks		19	2.947		108	121	27	256	355
1300	4 plies #15 asphalt felt, mopped	↓	22	2.545	↓	135	105	23.50	263.50	350
2000	Asphalt flood coat, smooth surface									
2200	Asphalt base sheet & 3 plies #15 asphalt felt, mopped	G-1	24	2.333	Sq.	110	96	21.50	227.50	305
2400	On nailable decks		23	2.435		102	100	22.50	224.50	305
2600	4 plies #15 asphalt felt, mopped		24	2.333		129	96	21.50	246.50	325
2700	On nailable decks	↓	23	2.435	↓	121	100	22.50	243.50	325
2900	Coated glass fiber base sheet, mopped, and 2 plies of									
2910	glass fiber felt (type IV)	G-1	25	2.240	Sq.	106	92	20.50	218.50	295
3100	On nailable decks		24	2.333		100	96	21.50	217.50	295
3200	3 plies, mopped		23	2.435		129	100	22.50	251.50	335
3300	On nailable decks		22	2.545		121	105	23.50	249.50	335
3800	4 plies glass fiber felt (type IV), mopped		23	2.435		152	100	22.50	274.50	360
3900	On nailable decks		22	2.545		144	105	23.50	272.50	360
4000	Coated & saturated base sheet, 3 plies #15 asph. felt, mopped		24	2.333		111	96	21.50	228.50	305
4200	On nailable decks		23	2.435		103	100	22.50	225.50	305
4300	4 plies #15 organic felt, mopped	↓	22	2.545	↓	130	105	23.50	258.50	345
4500	Coal tar pitch with gravel/slag surfacing									
4600	4 plies #15 tarred felt, mopped	G-1	21	2.667	Sq.	199	109	24.50	332.50	430
4800	3 plies glass fiber felt (type IV), mopped	"	19	2.947	"	164	121	27	312	415
5000	Coated glass fiber base sheet, and 2 plies of									

07 51 Built-Up Bituminous Roofing

07 51 13 – Built-Up Asphalt Roofing

07 51 13.20 Built-Up Roofing Systems

		Crew	Daily Output	Labor-Hours	Unit	Material	2018 Bare Costs Labor	Equipment	Total	Total Incl O&P
5010	glass fiber felt (type IV), mopped	G-1	19	2.947	Sq.	170	121	27	318	420
5300	On nailable decks		18	3.111		148	128	28.50	304.50	410
5600	4 plies glass fiber felt (type IV), mopped		21	2.667		228	109	24.50	361.50	460
5800	On nailable decks		20	2.800		207	115	25.50	347.50	450

07 51 13.30 Cants

		Crew	Daily Output	Labor-Hours	Unit	Material	2018 Bare Costs Labor	Equipment	Total	Total Incl O&P
0010	**CANTS**									
0012	Lumber, treated, 4" x 4" cut diagonally	1 Rofc	325	.025	L.F.	1.91	1.08		2.99	3.92
0300	Mineral or fiber, trapezoidal, 1" x 4" x 48"		325	.025		.30	1.08		1.38	2.15
0400	1-1/2" x 5-5/8" x 48"		325	.025		.48	1.08		1.56	2.35

07 51 13.40 Felts

		Crew	Daily Output	Labor-Hours	Unit	Material	2018 Bare Costs Labor	Equipment	Total	Total Incl O&P
0010	**FELTS**									
0012	Glass fibered roofing felt, #15, not mopped	1 Rofc	58	.138	Sq.	9.40	6.05		15.45	20.50
0300	Base sheet, #80, channel vented		58	.138		43.50	6.05		49.55	58
0400	#70, coated		58	.138		18.20	6.05		24.25	30
0500	Cap, #87, mineral surfaced		58	.138		83	6.05		89.05	101
0600	Flashing membrane, #65		16	.500		10.40	22		32.40	48.50
0800	Coal tar fibered, #15, no mopping		58	.138		17.75	6.05		23.80	30
0900	Asphalt felt, #15, 4 sq./roll, no mopping		58	.138		5.35	6.05		11.40	16.10
1100	#30, 2 sq./roll		58	.138		10.30	6.05		16.35	21.50
1200	Double coated, #33		58	.138		11.20	6.05		17.25	22.50
1400	#40, base sheet		58	.138		11.35	6.05		17.40	22.50
1450	Coated and saturated		58	.138		12.35	6.05		18.40	24
1500	Tarred felt, organic, #15, 4 sq. rolls		58	.138		12.90	6.05		18.95	24.50
1550	#30, 2 sq. roll		58	.138		25.50	6.05		31.55	38
1700	Add for mopping above felts, per ply, asphalt, 24 lb./sq.	G-1	192	.292		11.60	12	2.68	26.28	35.50
1800	Coal tar mopping, 30 lb./sq.		186	.301		18.90	12.35	2.76	34.01	45
1900	Flood coat, with asphalt, 60 lb./sq.		60	.933		29	38.50	8.55	76.05	106
2000	With coal tar, 75 lb./sq.		56	1		47.50	41	9.20	97.70	131

07 51 13.50 Walkways for Built-Up Roofs

		Crew	Daily Output	Labor-Hours	Unit	Material	2018 Bare Costs Labor	Equipment	Total	Total Incl O&P
0010	**WALKWAYS FOR BUILT-UP ROOFS**									
0020	Asphalt impregnated, 3' x 6' x 1/2" thick	1 Rofc	400	.020	S.F.	1.87	.88		2.75	3.53
0100	3' x 3' x 3/4" thick	"	400	.020		5.35	.88		6.23	7.40
0300	Concrete patio blocks, 2" thick, natural	1 Clab	115	.070		3.51	2.77		6.28	8.10
0400	Colors	"	115	.070		3.77	2.77		6.54	8.35

07 52 Modified Bituminous Membrane Roofing

07 52 13 – Atactic-Polypropylene-Modified Bituminous Membrane Roofing

07 52 13.10 APP Modified Bituminous Membrane

		Crew	Daily Output	Labor-Hours	Unit	Material	2018 Bare Costs Labor	Equipment	Total	Total Incl O&P
0010	**APP MODIFIED BITUMINOUS MEMBRANE** R075213-30									
0020	Base sheet, #15 glass fiber felt, nailed to deck	1 Rofc	58	.138	Sq.	10.95	6.05		17	22.50
0030	Spot mopped to deck	G-1	295	.190		15.20	7.80	1.74	24.74	32
0040	Fully mopped to deck	"	192	.292		21	12	2.68	35.68	46
0050	#15 organic felt, nailed to deck	1 Rofc	58	.138		6.90	6.05		12.95	17.80
0060	Spot mopped to deck	G-1	295	.190		11.15	7.80	1.74	20.69	27.50
0070	Fully mopped to deck	"	192	.292		16.95	12	2.68	31.63	41.50
2100	APP mod., smooth surf. cap sheet, poly. reinf., torched, 160 mils	G-5	2100	.019	S.F.	.76	.76	.09	1.61	2.21
2150	170 mils		2100	.019		.77	.76	.09	1.62	2.22
2200	Granule surface cap sheet, poly. reinf., torched, 180 mils		2000	.020		.96	.80	.09	1.85	2.50
2250	Smooth surface flashing, torched, 160 mils		1260	.032		.76	1.27	.14	2.17	3.13
2300	170 mils		1260	.032		.77	1.27	.14	2.18	3.14

07 52 Modified Bituminous Membrane Roofing

07 52 13 – Atactic-Polypropylene-Modified Bituminous Membrane Roofing

07 52 13.10 APP Modified Bituminous Membrane	Crew	Daily Output	Labor-Hours	Unit	Material	2018 Bare Costs Labor	Equipment	Total	Total Incl O&P
2350	Granule surface flashing, torched, 180 mils	G-5	1260	.032	S.F.	.96	1.27	.14	2.37	3.35
2400	Fibrated aluminum coating	1 Rofc	3800	.002	↓	.08	.09		.17	.25
2450	Seam heat welding	"	205	.039	L.F.	.08	1.71		1.79	2.98

07 52 16 – Styrene-Butadiene-Styrene Modified Bituminous Membrane Roofing

07 52 16.10 SBS Modified Bituminous Membrane

		Crew	Daily Output	Labor-Hours	Unit	Material	Labor	Equipment	Total	Total Incl O&P
0010	**SBS MODIFIED BITUMINOUS MEMBRANE**									
0080	Mod. bit. rfng., SBS mod, gran surf. cap sheet, poly. reinf.									
0650	120 to 149 mils thick	G-1	2000	.028	S.F.	1.34	1.15	.26	2.75	3.69
0750	150 to 160 mils	"	2000	.028		1.81	1.15	.26	3.22	4.21
1150	For reflective granules, add					.71	1.02	.28	2.01	2.82
1600	Smooth surface cap sheet, mopped, 145 mils	G-1	2100	.027		.82	1.10	.24	2.16	3.01
1620	Lightweight base sheet, fiberglass reinforced, 35 to 47 mil		2100	.027		.29	1.10	.24	1.63	2.43
1625	Heavyweight base/ply sheet, reinforced, 87 to 120 mil thick	↓	2100	.027		.93	1.10	.24	2.27	3.13
1650	Granulated walkpad, 180 to 220 mils	1 Rofc	400	.020		1.88	.88		2.76	3.55
1700	Smooth surface flashing, 145 mils	G-1	1260	.044		.82	1.82	.41	3.05	4.42
1800	150 mils		1260	.044		.51	1.82	.41	2.74	4.08
1900	Granular surface flashing, 150 mils		1260	.044		.72	1.82	.41	2.95	4.31
2000	160 mils	↓	1260	.044		.75	1.82	.41	2.98	4.35
2010	Elastomeric asphalt primer	1 Rofc	2600	.003		.17	.14		.31	.42
2015	Roofing asphalt, 30 lb./square	G-1	19000	.003		.15	.12	.03	.30	.39
2020	Cold process adhesive, 20 to 30 mils thick	1 Rofc	750	.011		.25	.47		.72	1.07
2025	Self adhering vapor retarder, 30 to 45 mils thick	G-5	2150	.019	↓	1.07	.74	.08	1.89	2.52
2050	Seam heat welding	1 Rofc	205	.039	L.F.	.08	1.71		1.79	2.98

07 53 Elastomeric Membrane Roofing

07 53 16 – Chlorosulfonate-Polyethylene Roofing

07 53 16.10 Chlorosulfonated Polyethylene Roofing

		Crew	Daily Output	Labor-Hours	Unit	Material	Labor	Equipment	Total	Total Incl O&P
0010	**CHLOROSULFONATED POLYETHYLENE ROOFING**									
0800	Chlorosulfonated polyethylene (CSPE)									
0900	45 mils, heat welded seams, plate attachment	G-5	35	1.143	Sq.	241	45.50	5.20	291.70	350
1100	Heat welded seams, plate attachment and ballasted		26	1.538		252	61.50	6.95	320.45	390
1200	60 mils, heat welded seams, plate attachment		35	1.143		310	45.50	5.20	360.70	430
1300	Heat welded seams, plate attachment and ballasted	↓	26	1.538	↓	320	61.50	6.95	388.45	465

07 53 23 – Ethylene-Propylene-Diene-Monomer Roofing

07 53 23.20 Ethylene-Propylene-Diene-Monomer Roofing

		Crew	Daily Output	Labor-Hours	Unit	Material	Labor	Equipment	Total	Total Incl O&P
0010	**ETHYLENE-PROPYLENE-DIENE-MONOMER ROOFING (EPDM)**									
3500	Ethylene-propylene-diene-monomer (EPDM), 45 mils, 0.28 psf									
3600	Loose-laid & ballasted with stone (10 psf)	G-5	51	.784	Sq.	85	31.50	3.55	120.05	150
3700	Mechanically attached		35	1.143		79	45.50	5.20	129.70	170
3800	Fully adhered with adhesive	↓	26	1.538	↓	112	61.50	6.95	180.45	235
4500	60 mils, 0.40 psf									
4600	Loose-laid & ballasted with stone (10 psf)	G-5	51	.784	Sq.	101	31.50	3.55	136.05	167
4700	Mechanically attached		35	1.143		94	45.50	5.20	144.70	186
4800	Fully adhered with adhesive	↓	26	1.538	↓	127	61.50	6.95	195.45	251
4810	45 mil, 0.28 psf, membrane only					49			49	54
4820	60 mil, 0.40 psf, membrane only				↓	62.50			62.50	68.50
4850	Seam tape for membrane, 3" x 100' roll				Ea.	44.50			44.50	49
4900	Batten strips, 10' sections					3.94			3.94	4.33
4910	Cover tape for batten strips, 6" x 100' roll				↓	187			187	206
4930	Plate anchors				M	81			81	89

For customer support on your Building Construction Costs with RSMeans data, call 800.448.8182.

245

07 53 Elastomeric Membrane Roofing

07 53 23 – Ethylene-Propylene-Diene-Monomer Roofing

07 53 23.20 Ethylene-Propylene-Diene-Monomer Roofing	Crew	Daily Output	Labor-Hours	Unit	Material	2018 Bare Costs Labor	Equipment	Total	Total Incl O&P
4970	Adhesive for fully adhered systems, 60 S.F./gal.				Gal.	20.50			20.50	22.50

07 53 29 – Polyisobutylene Roofing

07 53 29.10 Polyisobutylene Roofing

		Crew	Daily Output	Labor-Hours	Unit	Material	Labor	Equipment	Total	Total Incl O&P
0010	**POLYISOBUTYLENE ROOFING**									
7500	Polyisobutylene (PIB), 100 mils, 0.57 psf									
7600	Loose-laid & ballasted with stone/gravel (10 psf)	G-5	51	.784	Sq.	211	31.50	3.55	246.05	288
7700	Partially adhered with adhesive		35	1.143		251	45.50	5.20	301.70	360
7800	Hot asphalt attachment		35	1.143		241	45.50	5.20	291.70	350
7900	Fully adhered with contact cement		26	1.538		262	61.50	6.95	330.45	400

07 54 Thermoplastic Membrane Roofing

07 54 19 – Polyvinyl-Chloride Roofing

07 54 19.10 Polyvinyl-Chloride Roofing (PVC)

		Crew	Daily Output	Labor-Hours	Unit	Material	Labor	Equipment	Total	Total Incl O&P
0010	**POLYVINYL-CHLORIDE ROOFING (PVC)**									
8200	Heat welded seams									
8700	Reinforced, 48 mils, 0.33 psf									
8750	Loose-laid & ballasted with stone/gravel (12 psf)	G-5	51	.784	Sq.	113	31.50	3.55	148.05	180
8800	Mechanically attached		35	1.143		106	45.50	5.20	156.70	200
8850	Fully adhered with adhesive		26	1.538		153	61.50	6.95	221.45	279
8860	Reinforced, 60 mils, 0.40 psf									
8870	Loose-laid & ballasted with stone/gravel (12 psf)	G-5	51	.784	Sq.	114	31.50	3.55	149.05	181
8880	Mechanically attached		35	1.143		107	45.50	5.20	157.70	201
8890	Fully adhered with adhesive		26	1.538		154	61.50	6.95	222.45	280

07 54 23 – Thermoplastic-Polyolefin Roofing

07 54 23.10 Thermoplastic Polyolefin Roofing (T.P.O.)

		Crew	Daily Output	Labor-Hours	Unit	Material	Labor	Equipment	Total	Total Incl O&P
0010	**THERMOPLASTIC POLYOLEFIN ROOFING (T.P.O.)**									
0100	45 mil, loose laid & ballasted with stone (1/2 ton/sq.)	G-5	51	.784	Sq.	88	31.50	3.55	123.05	153
0120	Fully adhered		25	1.600		78.50	64	7.25	149.75	201
0140	Mechanically attached		34	1.176		78	47	5.35	130.35	170
0160	Self adhered		35	1.143		78.50	45.50	5.20	129.20	169
0180	60 mil membrane, heat welded seams, ballasted		50	.800		100	32	3.62	135.62	167
0200	Fully adhered		25	1.600		90	64	7.25	161.25	214
0220	Mechanically attached		34	1.176		94	47	5.35	146.35	189
0240	Self adhered		35	1.143		106	45.50	5.20	156.70	200

07 54 30 – Ketone Ethylene Ester Roofing

07 54 30.10 Ketone Ethylene Ester Roofing

		Crew	Daily Output	Labor-Hours	Unit	Material	Labor	Equipment	Total	Total Incl O&P
0010	**KETONE ETHYLENE ESTER ROOFING**									
0100	Ketone ethylene ester roofing, 50 mil, fully adhered	G-5	26	1.538	Sq.	202	61.50	6.95	270.45	335
0120	Mechanically attached		35	1.143		131	45.50	5.20	181.70	227
0140	Ballasted with stone		51	.784		138	31.50	3.55	173.05	208
0160	50 mil, fleece backed, adhered w/hot asphalt	G-1	26	2.154		162	88.50	19.75	270.25	350
0180	Accessories, pipe boot	1 Rofc	32	.250	Ea.	25.50	11		36.50	47
0200	Pre-formed corners		32	.250	"	8.90	11		19.90	28.50
0220	Ketone clad metal, including up to 4 bends		330	.024	S.F.	4.12	1.07		5.19	6.30
0240	Walkway pad	2 Rofc	800	.020	"	4.33	.88		5.21	6.25
0260	Stripping material	1 Rofc	310	.026	L.F.	1	1.13		2.13	3.01

For customer support on your Building Construction Costs with RSMeans data, call 800.448.8182.

07 55 Protected Membrane Roofing

07 55 10 – Protected Membrane Roofing Components

07 55 10.10 Protected Membrane Roofing Components	Crew	Daily Output	Labor-Hours	Unit	Material	2018 Bare Costs Labor	Equipment	Total	Total Incl O&P
0010	**PROTECTED MEMBRANE ROOFING COMPONENTS**								
0100	Choose roofing membrane from 07 50								
0120	Then choose roof deck insulation from 07 22								
0130	Filter fabric	2 Rofc	10000	.002	S.F.	.09	.07		.16	.22
0140	Ballast, 3/8" - 1/2" in place	G-1	36	1.556	Ton	20.50	64	14.30	98.80	146
0150	3/4" - 1-1/2" in place	"	36	1.556	"	20.50	64	14.30	98.80	146
0200	2" concrete blocks, natural	1 Clab	115	.070	S.F.	3.51	2.77		6.28	8.10
0210	Colors	"	115	.070	"	3.77	2.77		6.54	8.35

07 56 Fluid-Applied Roofing

07 56 10 – Fluid-Applied Roofing Elastomers

07 56 10.10 Elastomeric Roofing

		Crew	Daily Output	Labor-Hours	Unit	Material	2018 Bare Costs Labor	Equipment	Total	Total Incl O&P
0010	**ELASTOMERIC ROOFING**									
0020	Acrylic, 44% solids, 2 coats, on corrugated metal	2 Rofc	2400	.007	S.F.	.56	.29		.85	1.11
0025	On smooth metal		3000	.005		.45	.23		.68	.89
0030	On foam or modified bitumen		1500	.011		.90	.47		1.37	1.78
0035	On concrete		1500	.011		.90	.47		1.37	1.78
0040	On tar and gravel		1500	.011		.90	.47		1.37	1.78
0045	36% solids, 2 coats, on corrugated metal		2400	.007		.52	.29		.81	1.07
0050	On smooth metal		3000	.005		.42	.23		.65	.85
0055	On foam or modified bitumen		1500	.011		.84	.47		1.31	1.71
0060	On concrete		1500	.011		.84	.47		1.31	1.71
0065	On tar and gravel		1500	.011		.84	.47		1.31	1.71
0070	Primer if required, 2 coats on corrugated metal		2400	.007		.51	.29		.80	1.05
0075	On smooth metal		3000	.005		.51	.23		.74	.95
0080	On foam or modified bitumen		1500	.011		.74	.47		1.21	1.61
0085	On concrete		1500	.011		.56	.47		1.03	1.40
0090	On tar & gravel/rolled roof		1500	.011		.74	.47		1.21	1.61
0110	Acrylic rubber, fluid applied, 20 mils thick	G-5	2000	.020		2.22	.80	.09	3.11	3.88
0120	50 mils, reinforced		1200	.033		3.28	1.33	.15	4.76	6
0130	For walking surface, add		900	.044		1.22	1.77	.20	3.19	4.55
0300	Neoprene, fluid applied, 20 mil thick, not reinforced	G-1	1135	.049		1.44	2.03	.45	3.92	5.50
0600	Non-woven polyester, reinforced		960	.058		1.57	2.40	.54	4.51	6.35
0700	5 coat neoprene deck, 60 mil thick, under 10,000 S.F.		325	.172		4.81	7.10	1.58	13.49	18.95
0900	Over 10,000 S.F.		625	.090		4.81	3.68	.82	9.31	12.40

07 57 Coated Foamed Roofing

07 57 13 – Sprayed Polyurethane Foam Roofing

07 57 13.10 Sprayed Polyurethane Foam Roofing (S.P.F.)

		Crew	Daily Output	Labor-Hours	Unit	Material	2018 Bare Costs Labor	Equipment	Total	Total Incl O&P
0010	**SPRAYED POLYURETHANE FOAM ROOFING (S.P.F.)**									
0100	Primer for metal substrate (when required)	G-2A	3000	.008	S.F.	.49	.31	.20	1	1.28
0200	Primer for non-metal substrate (when required)		3000	.008		.19	.31	.20	.70	.95
0300	Closed cell spray, polyurethane foam, 3 lb./C.F. density, 1", R6.7		15000	.002		.64	.06	.04	.74	.86
0400	2", R13.4		13125	.002		1.29	.07	.05	1.41	1.58
0500	3", R18.6		11485	.002		1.93	.08	.05	2.06	2.31
0550	4", R24.8		10080	.002		2.57	.09	.06	2.72	3.05
0700	Spray-on silicone coating		2500	.010		1.24	.37	.25	1.86	2.24
0800	Warranty 5-20 year manufacturer's								.15	.15
0900	Warranty 20 year, no dollar limit								.20	.20

For customer support on your Building Construction Costs with RSMeans data, call 800.448.8182.

247

07 58 Roll Roofing

07 58 10 – Asphalt Roll Roofing

07 58 10.10 Roll Roofing	Crew	Daily Output	Labor-Hours	Unit	Material	2018 Bare Costs Labor	Equipment	Total	Total Incl O&P
0010 **ROLL ROOFING**									
0100 Asphalt, mineral surface									
0200 1 ply #15 organic felt, 1 ply mineral surfaced									
0300 Selvage roofing, lap 19", nailed & mopped	G-1	27	2.074	Sq.	71	85	19.05	175.05	242
0400 3 plies glass fiber felt (type IV), 1 ply mineral surfaced									
0500 Selvage roofing, lapped 19", mopped	G-1	25	2.240	Sq.	123	92	20.50	235.50	315
0600 Coated glass fiber base sheet, 2 plies of glass fiber									
0700 Felt (type IV), 1 ply mineral surfaced selvage									
0800 Roofing, lapped 19", mopped	G-1	25	2.240	Sq.	131	92	20.50	243.50	320
0900 On nailable decks	"	24	2.333	"	120	96	21.50	237.50	315
1000 3 plies glass fiber felt (type III), 1 ply mineral surfaced									
1100 Selvage roofing, lapped 19", mopped	G-1	25	2.240	Sq.	123	92	20.50	235.50	315

07 61 Sheet Metal Roofing

07 61 13 – Standing Seam Sheet Metal Roofing

07 61 13.10 Standing Seam Sheet Metal Roofing, Field Fab.

	Crew	Daily Output	Labor-Hours	Unit	Material	2018 Bare Costs Labor	Equipment	Total	Total Incl O&P
0010 **STANDING SEAM SHEET METAL ROOFING, FIELD FABRICATED**									
0400 Copper standing seam roofing, over 10 squares, 16 oz., 125 lb./sq.	1 Shee	1.30	6.154	Sq.	1,025	370		1,395	1,675
0600 18 oz., 140 lb./sq.		1.20	6.667		1,125	400		1,525	1,850
0700 20 oz., 150 lb./sq.	↓	1.10	7.273		1,350	435		1,785	2,150
1200 For abnormal conditions or small areas, add					25%	100%			
1300 For lead-coated copper, add				↓	25%				

07 61 16 – Batten Seam Sheet Metal Roofing

07 61 16.10 Batten Seam Sheet Metal Roofing, Field Fab.

	Crew	Daily Output	Labor-Hours	Unit	Material	2018 Bare Costs Labor	Equipment	Total	Total Incl O&P
0010 **BATTEN SEAM SHEET METAL ROOFING, FIELD FABRICATED**									
0012 Copper batten seam roofing, over 10 sq., 16 oz., 130 lb./sq.	1 Shee	1.10	7.273	Sq.	1,275	435		1,710	2,075
0020 Lead batten seam roofing, 5 lb./S.F.		1.20	6.667		1,650	400		2,050	2,400
0100 Zinc/copper alloy batten seam roofing, .020" thick		1.20	6.667		1,325	400		1,725	2,050
0200 Copper roofing, batten seam, over 10 sq., 18 oz., 145 lb./sq.		1	8		1,425	480		1,905	2,300
0300 20 oz., 160 lb./sq.		1	8		1,700	480		2,180	2,600
0500 Stainless steel batten seam roofing, type 304, 28 ga.		1.20	6.667		655	400		1,055	1,325
0600 26 ga.		1.15	6.957		755	415		1,170	1,475
0800 Zinc, copper alloy roofing, batten seam, .027" thick		1.15	6.957		1,675	415		2,090	2,450
0900 .032" thick		1.10	7.273		1,900	435		2,335	2,775
1000 .040" thick	↓	1.05	7.619	↓	2,450	455		2,905	3,400

07 61 19 – Flat Seam Sheet Metal Roofing

07 61 19.10 Flat Seam Sheet Metal Roofing, Field Fabricated

	Crew	Daily Output	Labor-Hours	Unit	Material	2018 Bare Costs Labor	Equipment	Total	Total Incl O&P
0010 **FLAT SEAM SHEET METAL ROOFING, FIELD FABRICATED**									
0900 Copper flat seam roofing, over 10 squares, 16 oz., 115 lb./sq.	1 Shee	1.20	6.667	Sq.	940	400		1,340	1,625
0950 18 oz., 130 lb./sq.		1.15	6.957		1,050	415		1,465	1,775
1000 20 oz., 145 lb./sq.		1.10	7.273		1,250	435		1,685	2,050
1008 Zinc flat seam roofing, .020" thick		1.20	6.667		1,125	400		1,525	1,850
1010 .027" thick		1.15	6.957		1,425	415		1,840	2,200
1020 .032" thick		1.12	7.143		1,625	425		2,050	2,425
1030 .040" thick		1.05	7.619		2,100	455		2,555	3,000
1100 Lead flat seam roofing, 5 lb./S.F.	↓	1.30	6.154	↓	1,400	370		1,770	2,100

For customer support on your Building Construction Costs with RSMeans data, call 800.448.8182.

07 65 Flexible Flashing

07 65 10 – Sheet Metal Flashing

07 65 10.10 Sheet Metal Flashing and Counter Flashing		Crew	Daily Output	Labor-Hours	Unit	Material	2018 Bare Costs Labor	Equipment	Total	Total Incl O&P
0010	**SHEET METAL FLASHING AND COUNTER FLASHING**									
0011	Including up to 4 bends									
0020	Aluminum, mill finish, .013" thick	1 Rofc	145	.055	S.F.	.83	2.42		3.25	4.99
0030	.016" thick		145	.055		.98	2.42		3.40	5.15
0060	.019" thick		145	.055		1.40	2.42		3.82	5.60
0100	.032" thick		145	.055		1.35	2.42		3.77	5.55
0200	.040" thick		145	.055		2.29	2.42		4.71	6.60
0300	.050" thick		145	.055		2.75	2.42		5.17	7.10
0325	Mill finish 5" x 7" step flashing, .016" thick		1920	.004	Ea.	.15	.18		.33	.48
0350	Mill finish 12" x 12" step flashing, .016" thick		1600	.005	"	.55	.22		.77	.98
0400	Painted finish, add				S.F.	.33			.33	.36
1000	Mastic-coated 2 sides, .005" thick	1 Rofc	330	.024		1.81	1.07		2.88	3.78
1100	.016" thick		330	.024		2	1.07		3.07	3.99
1600	Copper, 16 oz. sheets, under 1000 lb.		115	.070		8.10	3.06		11.16	14.05
1700	Over 4000 lb.		155	.052		8.10	2.27		10.37	12.70
1900	20 oz. sheets, under 1000 lb.		110	.073		10.70	3.20		13.90	17.20
2000	Over 4000 lb.		145	.055		10.15	2.42		12.57	15.30
2200	24 oz. sheets, under 1000 lb.		105	.076		14.75	3.35		18.10	22
2300	Over 4000 lb.		135	.059		14	2.60		16.60	19.80
2500	32 oz. sheets, under 1000 lb.		100	.080		19	3.52		22.52	27
2600	Over 4000 lb.		130	.062		18.05	2.70		20.75	24.50
2700	W shape for valleys, 16 oz., 24" wide		100	.080	L.F.	16.40	3.52		19.92	24
5800	Lead, 2.5 lb./S.F., up to 12" wide		135	.059	S.F.	6.15	2.60		8.75	11.15
5900	Over 12" wide		135	.059		4.04	2.60		6.64	8.80
8900	Stainless steel sheets, 32 ga.		155	.052		3.35	2.27		5.62	7.50
9000	28 ga.		155	.052		4.66	2.27		6.93	8.95
9100	26 ga.		155	.052		4.50	2.27		6.77	8.75
9200	24 ga.		155	.052		5	2.27		7.27	9.30
9290	For mechanically keyed flashing, add					40%				
9320	Steel sheets, galvanized, 20 ga.	1 Rofc	130	.062	S.F.	1.27	2.70		3.97	5.95
9322	22 ga.		135	.059		1.25	2.60		3.85	5.75
9324	24 ga.		140	.057		.95	2.51		3.46	5.30
9326	26 ga.		148	.054		.83	2.38		3.21	4.91
9328	28 ga.		155	.052		.72	2.27		2.99	4.61
9340	30 ga.		160	.050		.60	2.20		2.80	4.36
9400	Terne coated stainless steel, .015" thick, 28 ga.		155	.052		8.10	2.27		10.37	12.70
9500	.018" thick, 26 ga.		155	.052		9.05	2.27		11.32	13.75
9600	Zinc and copper alloy (brass), .020" thick		155	.052		10.25	2.27		12.52	15.10
9700	.027" thick		155	.052		12.25	2.27		14.52	17.30
9800	.032" thick		155	.052		15.50	2.27		17.77	21
9900	.040" thick		155	.052		20.50	2.27		22.77	26.50

07 65 12 – Fabric and Mastic Flashings

07 65 12.10 Fabric and Mastic Flashing and Counter Flashing

07 65 12.10 Fabric and Mastic Flashing and Counter Flashing		Crew	Daily Output	Labor-Hours	Unit	Material	2018 Bare Costs Labor	Equipment	Total	Total Incl O&P
0010	**FABRIC AND MASTIC FLASHING AND COUNTER FLASHING**									
1300	Asphalt flashing cement, 5 gallon				Gal.	9.65			9.65	10.60
4900	Fabric, asphalt-saturated cotton, specification grade	1 Rofc	35	.229	S.Y.	3.20	10.05		13.25	20.50
5000	Utility grade		35	.229		1.50	10.05		11.55	18.55
5300	Close-mesh fabric, saturated, 17 oz./S.Y.		35	.229		2.17	10.05		12.22	19.30
5500	Fiberglass, resin-coated		35	.229		1.07	10.05		11.12	18.10
8500	Shower pan, bituminous membrane, 7 oz.		155	.052	S.F.	1.75	2.27		4.02	5.75

07 65 13 – Laminated Sheet Flashing

07 65 13.10 Laminated Sheet Flashing	Crew	Daily Output	Labor-Hours	Unit	Material	2018 Bare Costs Labor	Equipment	Total	Total Incl O&P
0010	**LAMINATED SHEET FLASHING**, Including up to 4 bends								
0500	Aluminum, fabric-backed 2 sides, mill finish, .004" thick	1 Rofc	330	.024	S.F.	1.60	1.07		2.67	3.55
0700	.005" thick		330	.024		1.80	1.07		2.87	3.77
0750	Mastic-backed, self adhesive		460	.017		3.35	.76		4.11	4.98
0800	Mastic-coated 2 sides, .004" thick		330	.024		1.60	1.07		2.67	3.55
2800	Copper, paperbacked 1 side, 2 oz.		330	.024		2.18	1.07		3.25	4.19
2900	3 oz.		330	.024		3.12	1.07		4.19	5.20
3100	Paperbacked 2 sides, 2 oz.		330	.024		2.35	1.07		3.42	4.38
3150	3 oz.		330	.024		2.20	1.07		3.27	4.21
3200	5 oz.		330	.024		3.38	1.07		4.45	5.50
3250	7 oz.		330	.024		6.70	1.07		7.77	9.15
3400	Mastic-backed 2 sides, copper, 2 oz.		330	.024		2	1.07		3.07	3.99
3500	3 oz.		330	.024		2.50	1.07		3.57	4.54
3700	5 oz.		330	.024		3.80	1.07		4.87	5.95
3800	Fabric-backed 2 sides, copper, 2 oz.		330	.024		2.01	1.07		3.08	4
4000	3 oz.		330	.024		2.80	1.07		3.87	4.87
4100	5 oz.		330	.024		3.90	1.07		4.97	6.10
4300	Copper-clad stainless steel, .015" thick, under 500 lb.		115	.070		6.65	3.06		9.71	12.45
4400	Over 2000 lb.		155	.052		6.75	2.27		9.02	11.25
4600	.018" thick, under 500 lb.		100	.080		7.95	3.52		11.47	14.65
4700	Over 2000 lb.		145	.055		7.75	2.42		10.17	12.65
8550	Shower pan, 3 ply copper and fabric, 3 oz.		155	.052		4	2.27		6.27	8.20
8600	7 oz.		155	.052		4.73	2.27		7	9
9300	Stainless steel, paperbacked 2 sides, .005" thick		330	.024		3.92	1.07		4.99	6.10

07 65 19 – Plastic Sheet Flashing

07 65 19.10 Plastic Sheet Flashing and Counter Flashing	Crew	Daily Output	Labor-Hours	Unit	Material	2018 Bare Costs Labor	Equipment	Total	Total Incl O&P
0010	**PLASTIC SHEET FLASHING AND COUNTER FLASHING**								
7300	Polyvinyl chloride, black, 10 mil	1 Rofc	285	.028	S.F.	.26	1.23		1.49	2.37
7400	20 mil		285	.028		.26	1.23		1.49	2.37
7600	30 mil		285	.028		.33	1.23		1.56	2.44
7700	60 mil		285	.028		.85	1.23		2.08	3.02
7900	Black or white for exposed roofs, 60 mil		285	.028		1.22	1.23		2.45	3.42
8060	PVC tape, 5" x 45 mils, for joint covers, 100 L.F./roll				Ea.	177			177	194
8850	Polyvinyl chloride, 30 mil	1 Rofc	160	.050	S.F.	1.50	2.20		3.70	5.35

07 65 23 – Rubber Sheet Flashing

07 65 23.10 Rubber Sheet Flashing and Counter Flashing	Crew	Daily Output	Labor-Hours	Unit	Material	2018 Bare Costs Labor	Equipment	Total	Total Incl O&P
0010	**RUBBER SHEET FLASHING AND COUNTER FLASHING**								
4810	EPDM 90 mils, 1" diameter pipe flashing	1 Rofc	32	.250	Ea.	19.95	11		30.95	40.50
4820	2" diameter		30	.267		19.90	11.70		31.60	41.50
4830	3" diameter		28	.286		21	12.55		33.55	44.50
4840	4" diameter		24	.333		28.50	14.65		43.15	56
4850	6" diameter		22	.364		28.50	16		44.50	58.50
8100	Rubber, butyl, 1/32" thick		285	.028	S.F.	2.22	1.23		3.45	4.52
8200	1/16" thick		285	.028		3.27	1.23		4.50	5.70
8300	Neoprene, cured, 1/16" thick		285	.028		2.60	1.23		3.83	4.94
8400	1/8" thick		285	.028		6.10	1.23		7.33	8.80

07 65 Flexible Flashing

07 65 26 – Self-Adhering Sheet Flashing

07 65 26.10 Self-Adhering Sheet or Roll Flashing	Crew	Daily Output	Labor-Hours	Unit	Material	2018 Bare Costs Labor	Equipment	Total	Total Incl O&P
0010 **SELF-ADHERING SHEET OR ROLL FLASHING**									
0020 Self-adhered flashing, 25 mil cross laminated HDPE, 4" wide	1 Rofc	960	.008	L.F.	.20	.37		.57	.84
0040 6" wide		896	.009		.30	.39		.69	.99
0060 9" wide		832	.010		.45	.42		.87	1.21
0080 12" wide	↓	768	.010	↓	.60	.46		1.06	1.43

07 71 Roof Specialties

07 71 16 – Manufactured Counterflashing Systems

07 71 16.20 Pitch Pockets, Variable Sizes

	Crew	Daily Output	Labor-Hours	Unit	Material	Labor	Equipment	Total	Total Incl O&P
0010 **PITCH POCKETS, VARIABLE SIZES**									
0100 Adjustable, 4" to 7", welded corners, 4" deep	1 Rofc	48	.167	Ea.	16.65	7.35		24	30.50
0200 Side extenders, 6"	"	240	.033	"	2.78	1.46		4.24	5.50

07 71 19 – Manufactured Gravel Stops and Fasciae

07 71 19.10 Gravel Stop

	Crew	Daily Output	Labor-Hours	Unit	Material	Labor	Equipment	Total	Total Incl O&P
0010 **GRAVEL STOP**									
0020 Aluminum, .050" thick, 4" face height, mill finish	1 Shee	145	.055	L.F.	6.55	3.30		9.85	12.25
0080 Duranodic finish		145	.055		7.40	3.30		10.70	13.20
0100 Painted		145	.055		7.50	3.30		10.80	13.30
0300 6" face height		135	.059		6.85	3.54		10.39	12.90
0350 Duranodic finish		135	.059		7.95	3.54		11.49	14.15
0400 Painted		135	.059		8.75	3.54		12.29	15.05
0600 8" face height		125	.064		7.75	3.83		11.58	14.40
0650 Duranodic finish		125	.064		9.15	3.83		12.98	15.90
0700 Painted		125	.064		9.75	3.83		13.58	16.60
0900 12" face height, .080" thick, 2 piece		100	.080		11.05	4.78		15.83	19.45
0950 Duranodic finish		100	.080		11	4.78		15.78	19.40
1000 Painted		100	.080		13.50	4.78		18.28	22
1200 Copper, 16 oz., 3" face height		145	.055		24.50	3.30		27.80	32
1300 6" face height		135	.059		33.50	3.54		37.04	42.50
1350 Galv steel, 24 ga., 4" leg, plain, with continuous cleat, 4" face		145	.055		6.50	3.30		9.80	12.20
1360 6" face height		145	.055		6.55	3.30		9.85	12.25
1500 Polyvinyl chloride, 6" face height		135	.059		5.75	3.54		9.29	11.75
1600 9" face height		125	.064		6.90	3.83		10.73	13.45
1800 Stainless steel, 24 ga., 6" face height		135	.059		15.80	3.54		19.34	23
1900 12" face height		100	.080		23	4.78		27.78	33
2100 20 ga., 6" face height		135	.059		19.50	3.54		23.04	27
2200 12" face height	↓	100	.080	↓	28	4.78		32.78	38

07 71 19.30 Fascia

	Crew	Daily Output	Labor-Hours	Unit	Material	Labor	Equipment	Total	Total Incl O&P
0010 **FASCIA**									
0100 Aluminum, reverse board and batten, .032" thick, colored, no furring incl.	1 Shee	145	.055	S.F.	7	3.30		10.30	12.75
0300 Steel, galv and enameled, stock, no furring, long panels		145	.055		5.20	3.30		8.50	10.75
0600 Short panels	↓	115	.070	↓	5.20	4.16		9.36	12.05

07 71 23 – Manufactured Gutters and Downspouts

07 71 23.10 Downspouts

	Crew	Daily Output	Labor-Hours	Unit	Material	Labor	Equipment	Total	Total Incl O&P
0010 **DOWNSPOUTS**									
0020 Aluminum, embossed, .020" thick, 2" x 3"	1 Shee	190	.042	L.F.	.91	2.52		3.43	4.84
0100 Enameled		190	.042		1.36	2.52		3.88	5.35
0300 .024" thick, 2" x 3"		180	.044		2.12	2.66		4.78	6.40
0400 3" x 4"	↓	140	.057	↓	1.95	3.42		5.37	7.35

07 71 23 – Manufactured Gutters and Downspouts

07 71 23.10 Downspouts		Crew	Daily Output	Labor-Hours	Unit	Material	2018 Bare Costs Labor	Equipment	Total	Total Incl O&P
0600	Round, corrugated aluminum, 3" diameter, .020" thick	1 Shee	190	.042	L.F.	2.08	2.52		4.60	6.15
0700	4" diameter, .025" thick	↓	140	.057	↓	3.19	3.42		6.61	8.70
0900	Wire strainer, round, 2" diameter		155	.052	Ea.	1.74	3.09		4.83	6.60
1000	4" diameter		155	.052		2.38	3.09		5.47	7.35
1200	Rectangular, perforated, 2" x 3"		145	.055		2.32	3.30		5.62	7.60
1300	3" x 4"		145	.055	↓	3.33	3.30		6.63	8.70
1500	Copper, round, 16 oz., stock, 2" diameter		190	.042	L.F.	8.35	2.52		10.87	13
1600	3" diameter		190	.042		8.60	2.52		11.12	13.30
1800	4" diameter		145	.055		9.80	3.30		13.10	15.85
1900	5" diameter		130	.062		14.65	3.68		18.33	22
2100	Rectangular, corrugated copper, stock, 2" x 3"		190	.042		8.30	2.52		10.82	13
2200	3" x 4"		145	.055		9.35	3.30		12.65	15.35
2400	Rectangular, plain copper, stock, 2" x 3"		190	.042		11.25	2.52		13.77	16.25
2500	3" x 4"		145	.055	↓	14	3.30		17.30	20.50
2700	Wire strainers, rectangular, 2" x 3"		145	.055	Ea.	17.20	3.30		20.50	24
2800	3" x 4"		145	.055		17.95	3.30		21.25	25
3000	Round, 2" diameter		145	.055		6.50	3.30		9.80	12.20
3100	3" diameter		145	.055		7.25	3.30		10.55	13.05
3300	4" diameter		145	.055		12.25	3.30		15.55	18.55
3400	5" diameter		115	.070	↓	22.50	4.16		26.66	31.50
3600	Lead-coated copper, round, stock, 2" diameter		190	.042	L.F.	22.50	2.52		25.02	29
3700	3" diameter		190	.042		23	2.52		25.52	29.50
3900	4" diameter		145	.055		24	3.30		27.30	31.50
4000	5" diameter, corrugated		130	.062		24	3.68		27.68	31.50
4200	6" diameter, corrugated		105	.076		31.50	4.56		36.06	41.50
4300	Rectangular, corrugated, stock, 2" x 3"		190	.042		15.15	2.52		17.67	20.50
4500	Plain, stock, 2" x 3"		190	.042		24.50	2.52		27.02	31
4600	3" x 4"		145	.055		33	3.30		36.30	41.50
4800	Steel, galvanized, round, corrugated, 2" or 3" diameter, 28 ga.		190	.042		2.07	2.52		4.59	6.10
4900	4" diameter, 28 ga.		145	.055		2.08	3.30		5.38	7.35
5100	5" diameter, 26 ga.		130	.062		3.53	3.68		7.21	9.50
5400	6" diameter, 28 ga.		105	.076		3.48	4.56		8.04	10.80
5500	26 ga.		105	.076		3.78	4.56		8.34	11.10
5700	Rectangular, corrugated, 28 ga., 2" x 3"		190	.042		2.04	2.52		4.56	6.10
5800	3" x 4"		145	.055		2.20	3.30		5.50	7.45
6000	Rectangular, plain, 28 ga., galvanized, 2" x 3"		190	.042		3.83	2.52		6.35	8.05
6100	3" x 4"		145	.055		4.32	3.30		7.62	9.80
6300	Epoxy painted, 24 ga., corrugated, 2" x 3"		190	.042		2.40	2.52		4.92	6.50
6400	3" x 4"		145	.055	↓	2.91	3.30		6.21	8.25
6600	Wire strainers, rectangular, 2" x 3"		145	.055	Ea.	17.30	3.30		20.60	24
6700	3" x 4"		145	.055		19	3.30		22.30	26
6900	Round strainers, 2" or 3" diameter		145	.055		4.23	3.30		7.53	9.70
7000	4" diameter		145	.055	↓	6.25	3.30		9.55	11.95
7800	Stainless steel tubing, schedule 5, 2" x 3" or 3" diameter		190	.042	L.F.	84	2.52		86.52	96.50
7900	3" x 4" or 4" diameter		145	.055		114	3.30		117.30	130
8100	4" x 5" or 5" diameter		135	.059		126	3.54		129.54	144
8200	Vinyl, rectangular, 2" x 3"		210	.038		1.90	2.28		4.18	5.55
8300	Round, 2-1/2"	↓	220	.036	↓	1.38	2.17		3.55	4.84

07 71 23.20 Downspout Elbows

		Crew	Daily Output	Labor-Hours	Unit	Material	2018 Bare Costs Labor	Equipment	Total	Total Incl O&P
0010	**DOWNSPOUT ELBOWS**									
0020	Aluminum, embossed, 2" x 3", .020" thick	1 Shee	100	.080	Ea.	.93	4.78		5.71	8.30
0100	Enameled	↓	100	.080	↓	1.78	4.78		6.56	9.25

07 71 Roof Specialties

07 71 23 – Manufactured Gutters and Downspouts

07 71 23.20 Downspout Elbows		Crew	Daily Output	Labor-Hours	Unit	Material	2018 Bare Costs Labor	Equipment	Total	Total Incl O&P
0200	Embossed, 3" x 4", .025" thick	1 Shee	100	.080	Ea.	3.10	4.78		7.88	10.70
0300	Enameled		100	.080		3.93	4.78		8.71	11.60
0400	Embossed, corrugated, 3" diameter, .020" thick		100	.080		3.10	4.78		7.88	10.70
0500	4" diameter, .025" thick		100	.080		6.55	4.78		11.33	14.50
0600	Copper, 16 oz., 2" diameter		100	.080		9.45	4.78		14.23	17.70
0700	3" diameter		100	.080		9.10	4.78		13.88	17.30
0800	4" diameter		100	.080		13.95	4.78		18.73	22.50
1000	Rectangular, 2" x 3" corrugated		100	.080		9.25	4.78		14.03	17.45
1100	3" x 4" corrugated		100	.080		15.10	4.78		19.88	24
1300	Vinyl, 2-1/2" diameter, 45 or 75 degree bend		100	.080		3.98	4.78		8.76	11.70
1400	Tee Y junction		75	.107		12.95	6.40		19.35	24

07 71 23.30 Gutters

07 71 23.30 Gutters		Crew	Daily Output	Labor-Hours	Unit	Material	2018 Bare Costs Labor	Equipment	Total	Total Incl O&P
0010	**GUTTERS**									
0012	Aluminum, stock units, 5" K type, .027" thick, plain	1 Shee	125	.064	L.F.	2.80	3.83		6.63	8.95
0100	Enameled		125	.064		2.89	3.83		6.72	9.05
0300	5" K type, .032" thick, plain		125	.064		3.53	3.83		7.36	9.75
0400	Enameled		125	.064		3.52	3.83		7.35	9.70
0700	Copper, half round, 16 oz., stock units, 4" wide		125	.064		8.50	3.83		12.33	15.20
0900	5" wide		125	.064		7.10	3.83		10.93	13.65
1000	6" wide		118	.068		11.25	4.05		15.30	18.60
1200	K type, 16 oz., stock, 5" wide		125	.064		8.25	3.83		12.08	14.95
1300	6" wide		125	.064		9.05	3.83		12.88	15.80
1500	Lead coated copper, 16 oz., half round, stock, 4" wide		125	.064		15.55	3.83		19.38	23
1600	6" wide		118	.068		18.50	4.05		22.55	26.50
1800	K type, stock, 5" wide		125	.064		18.50	3.83		22.33	26.50
1900	6" wide		125	.064		18.50	3.83		22.33	26.50
2100	Copper clad stainless steel, K type, 5" wide		125	.064		7.75	3.83		11.58	14.40
2200	6" wide		125	.064		9.80	3.83		13.63	16.65
2400	Steel, galv, half round or box, 28 ga., 5" wide, plain		125	.064		2.18	3.83		6.01	8.25
2500	Enameled		125	.064		2.28	3.83		6.11	8.35
2700	26 ga., stock, 5" wide		125	.064		2.47	3.83		6.30	8.55
2800	6" wide		125	.064		2.50	3.83		6.33	8.60
3000	Vinyl, O.G., 4" wide	1 Carp	115	.070		1.34	3.53		4.87	6.80
3100	5" wide		115	.070		1.64	3.53		5.17	7.15
3200	4" half round, stock units		115	.070		1.39	3.53		4.92	6.90
3250	Joint connectors				Ea.	3.03			3.03	3.33
3300	Wood, clear treated cedar, fir or hemlock, 3" x 4"	1 Carp	100	.080	L.F.	11	4.06		15.06	18.30
3400	4" x 5"	"	100	.080	"	22	4.06		26.06	30
5000	Accessories, end cap, K type, aluminum 5"	1 Shee	625	.013	Ea.	.72	.77		1.49	1.96
5010	6"		625	.013		1.56	.77		2.33	2.89
5020	Copper, 5"		625	.013		3.49	.77		4.26	5
5030	6"		625	.013		3.54	.77		4.31	5.05
5040	Lead coated copper, 5"		625	.013		13	.77		13.77	15.45
5050	6"		625	.013		13.90	.77		14.67	16.40
5060	Copper clad stainless steel, 5"		625	.013		3.52	.77		4.29	5.05
5070	6"		625	.013		3.52	.77		4.29	5.05
5080	Galvanized steel, 5"		625	.013		1.36	.77		2.13	2.67
5090	6"		625	.013		2.34	.77		3.11	3.74
5100	Vinyl, 4"	1 Carp	625	.013		6.20	.65		6.85	7.80
5110	5"	"	625	.013		6.55	.65		7.20	8.20
5120	Half round, copper, 4"	1 Shee	625	.013		4.40	.77		5.17	6
5130	5"		625	.013		4.72	.77		5.49	6.35

07 71 23 – Manufactured Gutters and Downspouts

07 71 23.30 Gutters

07 71 23.30 Gutters		Crew	Daily Output	Labor-Hours	Unit	Material	2018 Bare Costs Labor	Equipment	Total	Total Incl O&P
5140	6"	1 Shee	625	.013	Ea.	7.65	.77		8.42	9.60
5150	Lead coated copper, 5"		625	.013		14.65	.77		15.42	17.25
5160	6"		625	.013		22	.77		22.77	25
5170	Copper clad stainless steel, 5"		625	.013		4.53	.77		5.30	6.15
5180	6"		625	.013		4.53	.77		5.30	6.15
5190	Galvanized steel, 5"		625	.013		2.40	.77		3.17	3.81
5200	6"		625	.013		3	.77		3.77	4.47
5210	Outlet, aluminum, 2" x 3"		420	.019		.62	1.14		1.76	2.42
5220	3" x 4"		420	.019		1.05	1.14		2.19	2.90
5230	2-3/8" round		420	.019		.56	1.14		1.70	2.36
5240	Copper, 2" x 3"		420	.019		7.20	1.14		8.34	9.65
5250	3" x 4"		420	.019		8.15	1.14		9.29	10.75
5260	2-3/8" round		420	.019		4.55	1.14		5.69	6.75
5270	Lead coated copper, 2" x 3"		420	.019		26	1.14		27.14	30
5280	3" x 4"		420	.019		29.50	1.14		30.64	33.50
5290	2-3/8" round		420	.019		26	1.14		27.14	30
5300	Copper clad stainless steel, 2" x 3"		420	.019		7.20	1.14		8.34	9.65
5310	3" x 4"		420	.019		8.15	1.14		9.29	10.75
5320	2-3/8" round		420	.019		4.55	1.14		5.69	6.75
5330	Galvanized steel, 2" x 3"		420	.019		3.40	1.14		4.54	5.50
5340	3" x 4"		420	.019		5.25	1.14		6.39	7.50
5350	2-3/8" round		420	.019		4.25	1.14		5.39	6.40
5360	K type mitres, aluminum		65	.123		3.38	7.35		10.73	14.95
5370	Copper		65	.123		13.15	7.35		20.50	26
5380	Lead coated copper		65	.123		55	7.35		62.35	72
5390	Copper clad stainless steel		65	.123		27.50	7.35		34.85	41.50
5400	Galvanized steel		65	.123		24	7.35		31.35	38
5420	Half round mitres, copper		65	.123		63.50	7.35		70.85	81.50
5430	Lead coated copper		65	.123		90	7.35		97.35	110
5440	Copper clad stainless steel		65	.123		56.50	7.35		63.85	73.50
5450	Galvanized steel		65	.123		29	7.35		36.35	43.50
5460	Vinyl mitres and outlets		65	.123		10.45	7.35		17.80	23
5470	Sealant		940	.009	L.F.	.01	.51		.52	.79
5480	Soldering		96	.083	"	.26	4.98		5.24	7.90

07 71 23.35 Gutter Guard

		Crew	Daily Output	Labor-Hours	Unit	Material	2018 Bare Costs Labor	Equipment	Total	Total Incl O&P
0010	**GUTTER GUARD**									
0020	6" wide strip, aluminum mesh	1 Carp	500	.016	L.F.	2.58	.81		3.39	4.08
0100	Vinyl mesh	"	500	.016	"	2.78	.81		3.59	4.30

07 71 26 – Reglets

07 71 26.10 Reglets and Accessories

		Crew	Daily Output	Labor-Hours	Unit	Material	2018 Bare Costs Labor	Equipment	Total	Total Incl O&P
0010	**REGLETS AND ACCESSORIES**									
0020	Reglet, aluminum, .025" thick, in parapet	1 Carp	225	.036	L.F.	1.57	1.80		3.37	4.48
0300	16 oz. copper		225	.036		6.50	1.80		8.30	9.90
0400	Galvanized steel, 24 ga.		225	.036		1.23	1.80		3.03	4.10
0600	Stainless steel, .020" thick		225	.036		3.85	1.80		5.65	7
0900	Counter flashing for above, 12" wide, .032" aluminum	1 Shee	150	.053		2.16	3.19		5.35	7.25
1200	16 oz. copper		150	.053		6.25	3.19		9.44	11.75
1300	Galvanized steel, 26 ga.		150	.053		1.25	3.19		4.44	6.25
1500	Stainless steel, .020" thick		150	.053		6.30	3.19		9.49	11.80

For customer support on your Building Construction Costs with RSMeans data, call 800.448.8182.

07 71 Roof Specialties

07 71 29 – Manufactured Roof Expansion Joints

07 71 29.10 Expansion Joints

	Crew	Daily Output	Labor-Hours	Unit	Material	2018 Bare Costs Labor	Equipment	Total	Total Incl O&P
0010 **EXPANSION JOINTS**									
0300 Butyl or neoprene center with foam insulation, metal flanges									
0400 Aluminum, .032" thick for openings to 2-1/2"	1 Rofc	165	.048	L.F.	12.10	2.13		14.23	16.90
0600 For joint openings to 3-1/2"		165	.048		12.10	2.13		14.23	16.90
0610 For joint openings to 5"		165	.048		14.15	2.13		16.28	19.20
0620 For joint openings to 8"		165	.048		17	2.13		19.13	22.50
0700 Copper, 16 oz. for openings to 2-1/2"		165	.048		19.05	2.13		21.18	24.50
0900 For joint openings to 3-1/2"		165	.048		19.70	2.13		21.83	25
0910 For joint openings to 5"		165	.048		22	2.13		24.13	28
0920 For joint openings to 8"		165	.048		25	2.13		27.13	31.50
1000 Galvanized steel, 26 ga. for openings to 2-1/2"		165	.048		10.30	2.13		12.43	14.95
1200 For joint openings to 3-1/2"		165	.048		10.30	2.13		12.43	14.95
1210 For joint openings to 5"		165	.048		11.85	2.13		13.98	16.65
1220 For joint openings to 8"		165	.048		15.45	2.13		17.58	20.50
1300 Lead-coated copper, 16 oz. for openings to 2-1/2"		165	.048		34.50	2.13		36.63	41.50
1500 For joint openings to 3-1/2"		165	.048		34.50	2.13		36.63	41.50
1600 Stainless steel, .018", for openings to 2-1/2"		165	.048		13.90	2.13		16.03	18.90
1800 For joint openings to 3-1/2"		165	.048		13.90	2.13		16.03	18.90
1810 For joint openings to 5"		165	.048		15.20	2.13		17.33	20.50
1820 For joint openings to 8"		165	.048		21.50	2.13		23.63	27.50
1900 Neoprene, double-seal type with thick center, 4-1/2" wide		125	.064		15.15	2.81		17.96	21.50
1950 Polyethylene bellows, with galv steel flat flanges		100	.080		6.55	3.52		10.07	13.15
1960 With galvanized angle flanges	▼	100	.080		7.05	3.52		10.57	13.70
2000 Roof joint with extruded aluminum cover, 2"	1 Shee	115	.070		29	4.16		33.16	38
2100 Roof joint, plastic curbs, foam center, standard	1 Rofc	100	.080		13.65	3.52		17.17	21
2200 Large	"	100	.080		17.70	3.52		21.22	25.50
2500 Roof to wall joint with extruded aluminum cover	1 Shee	115	.070		29	4.16		33.16	38
2700 Wall joint, closed cell foam on PVC cover, 9" wide	1 Rofc	125	.064		5.55	2.81		8.36	10.85
2800 12" wide	"	115	.070	▼	6.60	3.06		9.66	12.45

07 71 43 – Drip Edge

07 71 43.10 Drip Edge, Rake Edge, Ice Belts

	Crew	Daily Output	Labor-Hours	Unit	Material	2018 Bare Costs Labor	Equipment	Total	Total Incl O&P
0010 **DRIP EDGE, RAKE EDGE, ICE BELTS**									
0020 Aluminum, .016" thick, 5" wide, mill finish	1 Carp	400	.020	L.F.	.58	1.01		1.59	2.18
0100 White finish		400	.020		.64	1.01		1.65	2.24
0200 8" wide, mill finish		400	.020		1.51	1.01		2.52	3.20
0300 Ice belt, 28" wide, mill finish		100	.080		7.85	4.06		11.91	14.80
0310 Vented, mill finish		400	.020		2.26	1.01		3.27	4.03
0320 Painted finish		400	.020		2.54	1.01		3.55	4.33
0400 Galvanized, 5" wide		400	.020		.61	1.01		1.62	2.21
0500 8" wide, mill finish		400	.020		.83	1.01		1.84	2.45
0510 Rake edge, aluminum, 1-1/2" x 1-1/2"		400	.020		.33	1.01		1.34	1.90
0520 3-1/2" x 1-1/2"	▼	400	.020	▼	.45	1.01		1.46	2.04

For customer support on your Building Construction Costs with RSMeans data, call 800.448.8182.

255

07 72 Roof Accessories

07 72 23 – Relief Vents

07 72 23.10 Roof Vents

07 72 23.10 Roof Vents	Crew	Daily Output	Labor-Hours	Unit	Material	2018 Bare Costs Labor	2018 Bare Costs Equipment	Total	Total Incl O&P
0010 **ROOF VENTS**									
0020 Mushroom shape, for built-up roofs, aluminum	1 Rofc	30	.267	Ea.	65	11.70		76.70	91
0100 PVC, 6" high	"	30	.267	"	21.50	11.70		33.20	43.50

07 72 26 – Ridge Vents

07 72 26.10 Ridge Vents and Accessories

	Crew	Daily Output	Labor-Hours	Unit	Material	Labor	Equipment	Total	Total Incl O&P
0010 **RIDGE VENTS AND ACCESSORIES**									
0100 Aluminum strips, mill finish	1 Rofc	160	.050	L.F.	2.40	2.20		4.60	6.35
0150 Painted finish		160	.050	"	4.17	2.20		6.37	8.30
0200 Connectors		48	.167	Ea.	4.95	7.35		12.30	17.80
0300 End caps		48	.167	"	2.25	7.35		9.60	14.85
0400 Galvanized strips		160	.050	L.F.	3.71	2.20		5.91	7.80
0430 Molded polyethylene, shingles not included		160	.050	"	2.78	2.20		4.98	6.75
0440 End plugs		48	.167	Ea.	2.25	7.35		9.60	14.85
0450 Flexible roll, shingles not included	▼	160	.050	L.F.	2.37	2.20		4.57	6.30
2300 Ridge vent strip, mill finish	1 Shee	155	.052	"	3.93	3.09		7.02	9.05

07 72 33 – Roof Hatches

07 72 33.10 Roof Hatch Options

	Crew	Daily Output	Labor-Hours	Unit	Material	Labor	Equipment	Total	Total Incl O&P
0010 **ROOF HATCH OPTIONS**									
0500 2'-6" x 3', aluminum curb and cover	G-3	10	3.200	Ea.	885	159		1,044	1,225
0520 Galvanized steel curb and aluminum cover		10	3.200		845	159		1,004	1,175
0540 Galvanized steel curb and cover		10	3.200		615	159		774	920
0600 2'-6" x 4'-6", aluminum curb and cover		9	3.556		1,075	177		1,252	1,475
0800 Galvanized steel curb and aluminum cover		9	3.556		950	177		1,127	1,325
0900 Galvanized steel curb and cover		9	3.556		995	177		1,172	1,375
1100 4' x 4' aluminum curb and cover		8	4		1,775	199		1,974	2,250
1120 Galvanized steel curb and aluminum cover		8	4		1,700	199		1,899	2,175
1140 Galvanized steel curb and cover		8	4		1,100	199		1,299	1,500
1200 2'-6" x 8'-0", aluminum curb and cover		6.60	4.848		2,050	242		2,292	2,650
1400 Galvanized steel curb and aluminum cover		6.60	4.848		1,900	242		2,142	2,450
1500 Galvanized steel curb and cover	▼	6.60	4.848		1,200	242		1,442	1,700
1800 For plexiglass panels, 2'-6" x 3'-0", add to above				▼	475			475	525

07 72 36 – Smoke Vents

07 72 36.10 Smoke Hatches

	Crew	Daily Output	Labor-Hours	Unit	Material	Labor	Equipment	Total	Total Incl O&P
0010 **SMOKE HATCHES**									
0200 For 3'-0" long, add to roof hatches from Section 07 72 33.10				Ea.	25%	5%			
0250 For 4'-0" long, add to roof hatches from Section 07 72 33.10					20%	5%			
0300 For 8'-0" long, add to roof hatches from Section 07 72 33.10				▼	10%	5%			

07 72 36.20 Smoke Vent Options

	Crew	Daily Output	Labor-Hours	Unit	Material	Labor	Equipment	Total	Total Incl O&P
0010 **SMOKE VENT OPTIONS**									
0100 4' x 4' aluminum cover and frame	G-3	13	2.462	Ea.	2,200	123		2,323	2,575
0200 Galvanized steel cover and frame		13	2.462		1,650	123		1,773	2,000
0300 4' x 8' aluminum cover and frame		8	4		2,825	199		3,024	3,400
0400 Galvanized steel cover and frame	▼	8	4	▼	2,350	199		2,549	2,875

07 72 53 – Snow Guards

07 72 53.10 Snow Guard Options

	Crew	Daily Output	Labor-Hours	Unit	Material	Labor	Equipment	Total	Total Incl O&P
0010 **SNOW GUARD OPTIONS**									
0100 Slate & asphalt shingle roofs, fastened with nails	1 Rofc	160	.050	Ea.	12.35	2.20		14.55	17.30
0200 Standing seam metal roofs, fastened with set screws		48	.167		17.55	7.35		24.90	31.50
0300 Surface mount for metal roofs, fastened with solder		48	.167	▼	7.45	7.35		14.80	20.50
0400 Double rail pipe type, including pipe	▼	130	.062	L.F.	34	2.70		36.70	42

For customer support on your Building Construction Costs with RSMeans data, call 800.448.8182.

07 72 Roof Accessories

07 72 80 – Vents

07 72 80.30 Vent Options

		Crew	Daily Output	Labor-Hours	Unit	Material	2018 Bare Costs Labor	Equipment	Total	Total Incl O&P
0010	**VENT OPTIONS**									
0020	Plastic, for insulated decks, 1 per M.S.F.	1 Rofc	40	.200	Ea.	19.85	8.80		28.65	37
0100	Heavy duty		20	.400		50	17.60		67.60	84.50
0300	Aluminum		30	.267		21	11.70		32.70	43
0800	Polystyrene baffles, 12" wide for 16" OC rafter spacing	1 Carp	90	.089		.44	4.51		4.95	7.35
0900	For 24" OC rafter spacing	"	110	.073		.79	3.69		4.48	6.45

07 76 Roof Pavers

07 76 16 – Roof Decking Pavers

07 76 16.10 Roof Pavers and Supports

		Crew	Daily Output	Labor-Hours	Unit	Material	2018 Bare Costs Labor	Equipment	Total	Total Incl O&P
0010	**ROOF PAVERS AND SUPPORTS**									
1000	Roof decking pavers, concrete blocks, 2" thick, natural	1 Clab	115	.070	S.F.	3.51	2.77		6.28	8.10
1100	Colors		115	.070	"	3.77	2.77		6.54	8.35
1200	Support pedestal, bottom cap		960	.008	Ea.	3	.33		3.33	3.81
1300	Top cap		960	.008		4.80	.33		5.13	5.80
1400	Leveling shims, 1/16"		1920	.004		1.20	.17		1.37	1.57
1500	1/8"		1920	.004		1.20	.17		1.37	1.57
1600	Buffer pad		960	.008		2.50	.33		2.83	3.26
1700	PVC legs (4" SDR 35)		2880	.003	Inch	.14	.11		.25	.32
2000	Alternate pricing method, system in place		101	.079	S.F.	7.15	3.16		10.31	12.70

07 81 Applied Fireproofing

07 81 16 – Cementitious Fireproofing

07 81 16.10 Sprayed Cementitious Fireproofing

		Crew	Daily Output	Labor-Hours	Unit	Material	2018 Bare Costs Labor	Equipment	Total	Total Incl O&P
0010	**SPRAYED CEMENTITIOUS FIREPROOFING**									
0050	Not including canvas protection, normal density									
0100	Per 1" thick, on flat plate steel	G-2	3000	.008	S.F.	.57	.34	.04	.95	1.18
0200	Flat decking		2400	.010		.57	.42	.05	1.04	1.31
0400	Beams		1500	.016		.57	.67	.09	1.33	1.72
0500	Corrugated or fluted decks		1250	.019		.85	.81	.10	1.76	2.27
0700	Columns, 1-1/8" thick		1100	.022		.64	.92	.12	1.68	2.21
0800	2-3/16" thick		700	.034		1.37	1.44	.18	2.99	3.88
0900	For canvas protection, add		5000	.005		.10	.20	.03	.33	.44
1000	Not including canvas protection, high density									
1100	Per 1" thick, on flat plate steel	G-2	3000	.008	S.F.	2.23	.34	.04	2.61	3.01
1110	On flat decking		2400	.010		2.23	.42	.05	2.70	3.14
1120	On beams		1500	.016		2.23	.67	.09	2.99	3.55
1130	Corrugated or fluted decks		1250	.019		2.23	.81	.10	3.14	3.78
1140	Columns, 1-1/8" thick		1100	.022		2.51	.92	.12	3.55	4.27
1150	2-3/16" thick		1100	.022		5	.92	.12	6.04	7
1170	For canvas protection, add		5000	.005		.10	.20	.03	.33	.44
1200	Not including canvas protection, retrofitting									
1210	Per 1" thick, on flat plate steel	G-2	1500	.016	S.F.	.50	.67	.09	1.26	1.65
1220	On flat decking		1200	.020		.50	.84	.11	1.45	1.94
1230	On beams		750	.032		.50	1.34	.17	2.01	2.77
1240	Corrugated or fluted decks		625	.038		.75	1.61	.21	2.57	3.49
1250	Columns, 1-1/8" thick		550	.044		.56	1.83	.23	2.62	3.64
1260	2-3/16" thick		500	.048		1.13	2.02	.26	3.41	4.56
1400	Accessories, preliminary spattered texture coat		4500	.005		.05	.22	.03	.30	.42

For customer support on your Building Construction Costs with RSMeans data, call 800.448.8182.

07 81 Applied Fireproofing

07 81 16 – Cementitious Fireproofing

07 81 16.10 Sprayed Cementitious Fireproofing	Crew	Daily Output	Labor-Hours	Unit	Material	2018 Bare Costs Labor	Equipment	Total	Total Incl O&P
1410	Bonding agent	1 Plas	1000	.008	S.F.	.10	.37		.47	.67
1500	Intumescent epoxy fireproofing on wire mesh, 3/16" thick									
1550	1 hour rating, exterior use	G-2	136	.176	S.F.	7.50	7.40	.95	15.85	20.50
1551	2 hour rating, exterior use		136	.176		7.50	7.40	.95	15.85	20.50
1600	Magnesium oxychloride, 35# to 40# density, 1/4" thick		3000	.008		1.60	.34	.04	1.98	2.32
1650	1/2" thick		2000	.012		3.20	.50	.06	3.76	4.35
1700	60# to 70# density, 1/4" thick		3000	.008		2.10	.34	.04	2.48	2.87
1750	1/2" thick		2000	.012		4.25	.50	.06	4.81	5.50
2000	Vermiculite cement, troweled or sprayed, 1/4" thick		3000	.008		1.45	.34	.04	1.83	2.16
2050	1/2" thick	↓	2000	.012	↓	2.85	.50	.06	3.41	3.97

07 84 Firestopping

07 84 13 – Penetration Firestopping

07 84 13.10 Firestopping

		Crew	Daily Output	Labor-Hours	Unit	Material	2018 Bare Costs Labor	Equipment	Total	Total Incl O&P
0010	**FIRESTOPPING** R078413-30									
0100	Metallic piping, non insulated									
0110	Through walls, 2" diameter	1 Carp	16	.500	Ea.	3.63	25.50		29.13	42.50
0120	4" diameter		14	.571		6.35	29		35.35	51
0130	6" diameter		12	.667		9.30	34		43.30	62
0140	12" diameter		10	.800		18.60	40.50		59.10	82.50
0150	Through floors, 2" diameter		32	.250		1.85	12.70		14.55	21.50
0160	4" diameter		28	.286		3.25	14.50		17.75	25.50
0170	6" diameter		24	.333		4.74	16.90		21.64	30.50
0180	12" diameter	↓	20	.400	↓	9.55	20.50		30.05	41.50
0190	Metallic piping, insulated									
0200	Through walls, 2" diameter	1 Carp	16	.500	Ea.	6.35	25.50		31.85	45.50
0210	4" diameter		14	.571		9.30	29		38.30	54.50
0220	6" diameter		12	.667		12.50	34		46.50	65.50
0230	12" diameter		10	.800		21.50	40.50		62	85.50
0240	Through floors, 2" diameter		32	.250		3.25	12.70		15.95	23
0250	4" diameter		28	.286		4.74	14.50		19.24	27
0260	6" diameter		24	.333		6.40	16.90		23.30	32.50
0270	12" diameter	↓	20	.400	↓	9.75	20.50		30.25	42
0280	Non metallic piping, non insulated									
0290	Through walls, 2" diameter	1 Carp	12	.667	Ea.	46	34		80	102
0300	4" diameter		10	.800		89.50	40.50		130	160
0310	6" diameter		8	1		172	50.50		222.50	266
0330	Through floors, 2" diameter		16	.500		30	25.50		55.50	71.50
0340	4" diameter		6	1.333		57.50	67.50		125	166
0350	6" diameter	↓	6	1.333	↓	111	67.50		178.50	225
0370	Ductwork, insulated & non insulated, round									
0380	Through walls, 6" diameter	1 Carp	12	.667	Ea.	12.50	34		46.50	65.50
0390	12" diameter		10	.800		21.50	40.50		62	85.50
0400	18" diameter		8	1		26	50.50		76.50	106
0410	Through floors, 6" diameter		16	.500		6.40	25.50		31.90	45.50
0420	12" diameter		14	.571		10.90	29		39.90	56
0430	18" diameter		12	.667		13.25	34		47.25	66
0440	Ductwork, insulated & non insulated, rectangular									
0450	With stiffener/closure angle, through walls, 6" x 12"	1 Carp	8	1	Ea.	16.70	50.50		67.20	95.50
0460	12" x 24"		6	1.333		33.50	67.50		101	140
0470	24" x 48"	↓	4	2		67	101		168	228

For customer support on your Building Construction Costs with RSMeans data, call 800.448.8182.

07 84 Firestopping

07 84 13 – Penetration Firestopping

07 84 13.10 Firestopping	Crew	Daily Output	Labor-Hours	Unit	Material	2018 Bare Costs Labor	Equipment	Total	Total Incl O&P
0480	With stiffener/closure angle, through floors, 6" x 12"	1 Carp	10	.800	Ea.	8.35	40.50		48.85	71
0490	12" x 24"		8	1		16.70	50.50		67.20	95.50
0500	24" x 48"		6	1.333		33.50	67.50		101	140
0510	Multi trade openings									
0520	Through walls, 6" x 12"	1 Carp	2	4	Ea.	46.50	203		249.50	360
0530	12" x 24"	"	1	8		152	405		557	785
0540	24" x 48"	2 Carp	1	16		540	810		1,350	1,825
0550	48" x 96"	"	.75	21.333		2,025	1,075		3,100	3,875
0560	Through floors, 6" x 12"	1 Carp	2	4		42	203		245	355
0570	12" x 24"	"	1	8		58	405		463	685
0580	24" x 48"	2 Carp	.75	21.333		139	1,075		1,214	1,800
0590	48" x 96"	"	.50	32		330	1,625		1,955	2,850
0600	Structural penetrations, through walls									
0610	Steel beams, W8 x 10	1 Carp	8	1	Ea.	70.50	50.50		121	155
0620	W12 x 14		6	1.333		102	67.50		169.50	215
0630	W21 x 44		5	1.600		146	81		227	284
0640	W36 x 135		3	2.667		226	135		361	455
0650	Bar joists, 18" deep		6	1.333		41.50	67.50		109	149
0660	24" deep		6	1.333		55.50	67.50		123	164
0670	36" deep		5	1.600		83	81		164	216
0680	48" deep		4	2		111	101		212	276
0690	Construction joints, floor slab at exterior wall									
0700	Precast, brick, block or drywall exterior									
0710	2" wide joint	1 Carp	125	.064	L.F.	9.80	3.24		13.04	15.70
0720	4" wide joint	"	75	.107	"	15.30	5.40		20.70	25
0730	Metal panel, glass or curtain wall exterior									
0740	2" wide joint	1 Carp	40	.200	L.F.	9.80	10.15		19.95	26
0750	4" wide joint	"	25	.320	"	15.30	16.20		31.50	41.50
0760	Floor slab to drywall partition									
0770	Flat joint	1 Carp	100	.080	L.F.	13.60	4.06		17.66	21
0780	Fluted joint		50	.160		15.15	8.10		23.25	29
0790	Etched fluted joint		75	.107		19.05	5.40		24.45	29.50
0800	Floor slab to concrete/masonry partition									
0810	Flat joint	1 Carp	75	.107	L.F.	13.60	5.40		19	23
0820	Fluted joint	"	50	.160	"	15.50	8.10		23.60	29.50
0830	Concrete/CMU wall joints									
0840	1" wide	1 Carp	100	.080	L.F.	19.40	4.06		23.46	27.50
0850	2" wide		75	.107		24	5.40		29.40	35
0860	4" wide		50	.160		33.50	8.10		41.60	49.50
0870	Concrete/CMU floor joints									
0880	1" wide	1 Carp	200	.040	L.F.	19.40	2.03		21.43	24.50
0890	2" wide		150	.053		24	2.70		26.70	30.50
0900	4" wide		100	.080		34.50	4.06		38.56	43.50

07 91 Preformed Joint Seals

07 91 13 – Compression Seals

07 91 13.10 Compression Seals

		Crew	Daily Output	Labor-Hours	Unit	Material	2018 Bare Costs Labor	Equipment	Total	Total Incl O&P
0010	**COMPRESSION SEALS**									
4900	O-ring type cord, 1/4"	1 Bric	472	.017	L.F.	.44	.85		1.29	1.79
4910	1/2"		440	.018		1.20	.91		2.11	2.72
4920	3/4"		424	.019		2.35	.95		3.30	4.04
4930	1"		408	.020		4.07	.99		5.06	6
4940	1-1/4"		384	.021		7.75	1.05		8.80	10.10
4950	1-1/2"		368	.022		9.50	1.09		10.59	12.10
4960	1-3/4"		352	.023		16.20	1.14		17.34	19.60
4970	2"		344	.023		22	1.17		23.17	26

07 91 16 – Joint Gaskets

07 91 16.10 Joint Gaskets

		Crew	Daily Output	Labor-Hours	Unit	Material	2018 Bare Costs Labor	Equipment	Total	Total Incl O&P
0010	**JOINT GASKETS**									
4400	Joint gaskets, neoprene, closed cell w/adh, 1/8" x 3/8"	1 Bric	240	.033	L.F.	.33	1.68		2.01	2.93
4500	1/4" x 3/4"		215	.037		.63	1.87		2.50	3.56
4700	1/2" x 1"		200	.040		1.50	2.01		3.51	4.73
4800	3/4" x 1-1/2"		165	.048		1.63	2.44		4.07	5.50

07 91 23 – Backer Rods

07 91 23.10 Backer Rods

		Crew	Daily Output	Labor-Hours	Unit	Material	2018 Bare Costs Labor	Equipment	Total	Total Incl O&P
0010	**BACKER RODS**									
0030	Backer rod, polyethylene, 1/4" diameter	1 Bric	4.60	1.739	C.L.F.	2.37	87.50		89.87	137
0050	1/2" diameter		4.60	1.739		4.28	87.50		91.78	139
0070	3/4" diameter		4.60	1.739		6.75	87.50		94.25	141
0090	1" diameter		4.60	1.739		12.10	87.50		99.60	147

07 91 26 – Joint Fillers

07 91 26.10 Joint Fillers

		Crew	Daily Output	Labor-Hours	Unit	Material	2018 Bare Costs Labor	Equipment	Total	Total Incl O&P
0010	**JOINT FILLERS**									
4360	Butyl rubber filler, 1/4" x 1/4"	1 Bric	290	.028	L.F.	.22	1.39		1.61	2.37
4365	1/2" x 1/2"		250	.032		.89	1.61		2.50	3.44
4370	1/2" x 3/4"		210	.038		1.34	1.92		3.26	4.40
4375	3/4" x 3/4"		230	.035		2.01	1.75		3.76	4.89
4380	1" x 1"		180	.044		2.68	2.24		4.92	6.35
4390	For coloring, add					12%				
4980	Polyethylene joint backing, 1/4" x 2"	1 Bric	2.08	3.846	C.L.F.	13.50	193		206.50	310
4990	1/4" x 6"		1.28	6.250	"	29	315		344	510
5600	Silicone, room temp vulcanizing foam seal, 1/4" x 1/2"		1312	.006	L.F.	.45	.31		.76	.97
5610	1/2" x 1/2"		656	.012		.90	.61		1.51	1.93
5620	1/2" x 3/4"		442	.018		1.35	.91		2.26	2.88
5630	3/4" x 3/4"		328	.024		2.03	1.23		3.26	4.12
5640	1/8" x 1"		1312	.006		.45	.31		.76	.97
5650	1/8" x 3"		442	.018		1.35	.91		2.26	2.88
5670	1/4" x 3"		295	.027		2.71	1.36		4.07	5.05
5680	1/4" x 6"		148	.054		5.40	2.72		8.12	10.10
5690	1/2" x 6"		82	.098		10.85	4.91		15.76	19.40
5700	1/2" x 9"		52.50	.152		16.25	7.65		23.90	29.50
5710	1/2" x 12"		33	.242		21.50	12.20		33.70	42.50

For customer support on your Building Construction Costs with RSMeans data, call 800.448.8182.

07 92 Joint Sealants

07 92 13 – Elastomeric Joint Sealants

07 92 13.20 Caulking and Sealant Options

07 92 13.20 Caulking and Sealant Options	Crew	Daily Output	Labor-Hours	Unit	Material	2018 Bare Costs Labor	Equipment	Total	Total Incl O&P
0010 **CAULKING AND SEALANT OPTIONS**									
0050 Latex acrylic based, bulk				Gal.	29.50			29.50	32.50
0055 Bulk in place 1/4" x 1/4" bead	1 Bric	300	.027	L.F.	.09	1.34		1.43	2.15
0060 1/4" x 3/8"		294	.027		.16	1.37		1.53	2.27
0065 1/4" x 1/2"		288	.028		.21	1.40		1.61	2.37
0075 3/8" x 3/8"		284	.028		.23	1.42		1.65	2.43
0080 3/8" x 1/2"		280	.029		.31	1.44		1.75	2.54
0085 3/8" x 5/8"		276	.029		.39	1.46		1.85	2.66
0095 3/8" x 3/4"		272	.029		.47	1.48		1.95	2.77
0100 1/2" x 1/2"		275	.029		.42	1.46		1.88	2.70
0105 1/2" x 5/8"		269	.030		.52	1.50		2.02	2.86
0110 1/2" x 3/4"		263	.030		.62	1.53		2.15	3.03
0115 1/2" x 7/8"		256	.031		.73	1.57		2.30	3.21
0120 1/2" x 1"		250	.032		.83	1.61		2.44	3.37
0125 3/4" x 3/4"		244	.033		.94	1.65		2.59	3.55
0130 3/4" x 1"		225	.036		1.25	1.79		3.04	4.11
0135 1" x 1"		200	.040		1.66	2.01		3.67	4.91
0190 Cartridges				Gal.	32.50			32.50	36
0200 11 fl. oz. cartridge				Ea.	2.81			2.81	3.09
0500 1/4" x 1/2"	1 Bric	288	.028	L.F.	.23	1.40		1.63	2.39
0600 1/2" x 1/2"		275	.029		.46	1.46		1.92	2.74
0800 3/4" x 3/4"		244	.033		1.03	1.65		2.68	3.66
0900 3/4" x 1"		225	.036		1.38	1.79		3.17	4.26
1000 1" x 1"		200	.040		1.72	2.01		3.73	4.98
1400 Butyl based, bulk				Gal.	37			37	40.50
1500 Cartridges				"	40.50			40.50	44.50
1700 1/4" x 1/2", 154 L.F./gal.	1 Bric	288	.028	L.F.	.24	1.40		1.64	2.40
1800 1/2" x 1/2", 77 L.F./gal.	"	275	.029	"	.48	1.46		1.94	2.77
2300 Polysulfide compounds, 1 component, bulk				Gal.	83			83	91
2600 1 or 2 component, in place, 1/4" x 1/4", 308 L.F./gal.	1 Bric	300	.027	L.F.	.27	1.34		1.61	2.35
2700 1/2" x 1/4", 154 L.F./gal.		288	.028		.54	1.40		1.94	2.73
2900 3/4" x 3/8", 68 L.F./gal.		272	.029		1.22	1.48		2.70	3.60
3000 1" x 1/2", 38 L.F./gal.		250	.032		2.18	1.61		3.79	4.86
3200 Polyurethane, 1 or 2 component				Gal.	53.50			53.50	58.50
3500 Bulk, in place, 1/4" x 1/4"	1 Bric	300	.027	L.F.	.17	1.34		1.51	2.24
3655 1/2" x 1/4"		288	.028		.35	1.40		1.75	2.52
3800 3/4" x 3/8"		272	.029		.79	1.48		2.27	3.12
3900 1" x 1/2"		250	.032		1.39	1.61		3	3.99
4100 Silicone rubber, bulk				Gal.	57			57	62.50
4200 Cartridges				"	52.50			52.50	57.50

07 92 16 – Rigid Joint Sealants

07 92 16.10 Rigid Joint Sealants

	Crew	Daily Output	Labor-Hours	Unit	Material	Labor	Equipment	Total	Total Incl O&P
0010 **RIGID JOINT SEALANTS**									
5800 Tapes, sealant, PVC foam adhesive, 1/16" x 1/4"				C.L.F.	9			9	9.90
5900 1/16" x 1/2"					9			9	9.90
5950 1/16" x 1"					16			16	17.60
6000 1/8" x 1/2"					9			9	9.90

07 92 19 – Acoustical Joint Sealants

07 92 19.10 Acoustical Sealant

	Crew	Daily Output	Labor-Hours	Unit	Material	Labor	Equipment	Total	Total Incl O&P
0010 **ACOUSTICAL SEALANT**									
0020 Acoustical sealant, elastomeric, cartridges				Ea.	8.45			8.45	9.25
0025 In place, 1/4" x 1/4"	1 Bric	300	.027	L.F.	.34	1.34		1.68	2.43

07 92 Joint Sealants

07 92 19 – Acoustical Joint Sealants

07 92 19.10 Acoustical Sealant		Crew	Daily Output	Labor-Hours	Unit	Material	2018 Bare Costs Labor	Equipment	Total	Total Incl O&P
0030	1/4" x 1/2"	1 Bric	288	.028	L.F.	.69	1.40		2.09	2.90
0035	1/2" x 1/2"		275	.029		1.38	1.46		2.84	3.75
0040	1/2" x 3/4"		263	.030		2.07	1.53		3.60	4.61
0045	3/4" x 3/4"		244	.033		3.10	1.65		4.75	5.95
0050	1" x 1"		200	.040		5.50	2.01		7.51	9.15

07 95 Expansion Control

07 95 13 – Expansion Joint Cover Assemblies

07 95 13.50 Expansion Joint Assemblies		Crew	Daily Output	Labor-Hours	Unit	Material	2018 Bare Costs Labor	Equipment	Total	Total Incl O&P
0010	**EXPANSION JOINT ASSEMBLIES**									
0200	Floor cover assemblies, 1" space, aluminum	1 Sswk	38	.211	L.F.	19	11.50		30.50	40
0300	Bronze		38	.211		60.50	11.50		72	85.50
0500	2" space, aluminum		38	.211		18.50	11.50		30	39.50
0600	Bronze		38	.211		60.50	11.50		72	85.50
0800	Wall and ceiling assemblies, 1" space, aluminum		38	.211		16.75	11.50		28.25	37.50
0900	Bronze		38	.211		53.50	11.50		65	78
1100	2" space, aluminum		38	.211		17.10	11.50		28.60	37.50
1200	Bronze		38	.211		52.50	11.50		64	76.50
1400	Floor to wall assemblies, 1" space, aluminum		38	.211		19.50	11.50		31	40.50
1500	Bronze or stainless		38	.211		64.50	11.50		76	90
1700	Gym floor angle covers, aluminum, 3" x 3" angle		46	.174		19.50	9.50		29	37
1800	3" x 4" angle		46	.174		23	9.50		32.50	41
2000	Roof closures, aluminum, flat roof, low profile, 1" space		57	.140		24	7.65		31.65	39
2100	High profile		57	.140		26	7.65		33.65	41
2300	Roof to wall, low profile, 1" space		57	.140		23.50	7.65		31.15	38
2400	High profile		57	.140		26	7.65		33.65	41

Estimating Tips
08 10 00 Doors and Frames

All exterior doors should be addressed for their energy conservation (insulation and seals).

- Most metal doors and frames look alike, but there may be significant differences among them. When estimating these items, be sure to choose the line item that most closely compares to the specification or door schedule requirements regarding:
  - ☐ type of metal
  - ☐ metal gauge
  - ☐ door core material
  - ☐ fire rating
  - ☐ finish
- Wood and plastic doors vary considerably in price. The primary determinant is the veneer material. Lauan, birch, and oak are the most common veneers. Other variables include the following:
  - ☐ hollow or solid core
  - ☐ fire rating
  - ☐ flush or raised panel
  - ☐ finish
- Door pricing includes bore for cylindrical locksets and mortise for hinges.

08 30 00 Specialty Doors and Frames

- There are many varieties of special doors, and they are usually priced per each. Add frames, hardware, or operators required for a complete installation.

08 40 00 Entrances, Storefronts, and Curtain Walls

- Glazed curtain walls consist of the metal tube framing and the glazing material. The cost data in this subdivision is presented for the metal tube framing alone or the composite wall. If your estimate requires a detailed takeoff of the framing, be sure to add the glazing cost and any tints.

08 50 00 Windows

- Steel windows are unglazed and aluminum can be glazed or unglazed. Some metal windows are priced without glass. Refer to 08 80 00 Glazing for glass pricing. The grade C indicates commercial grade windows, usually ASTM C-35.
- All wood windows and vinyl are priced preglazed. The glazing is insulating glass. Add the cost of screens and grills if required and not already included.

08 70 00 Hardware

- Hardware costs add considerably to the cost of a door. The most efficient method to determine the hardware requirements for a project is to review the door and hardware schedule together. One type of door may have different hardware, depending on the door usage.
- Door hinges are priced by the pair, with most doors requiring 1-1/2 pairs per door. The hinge prices do not include installation labor, because it is included in door installation.

Hinges are classified according to the frequency of use, base material, and finish.

08 80 00 Glazing

- Different openings require different types of glass. The most common types are:
  - ☐ float
  - ☐ tempered
  - ☐ insulating
  - ☐ impact-resistant
  - ☐ ballistic-resistant
- Most exterior windows are glazed with insulating glass. Entrance doors and window walls, where the glass is less than 18" from the floor, are generally glazed with tempered glass. Interior windows and some residential windows are glazed with float glass.
- Coastal communities require the use of impact-resistant glass, dependent on wind speed.
- The insulation or 'u' value is a strong consideration, along with solar heat gain, to determine total energy efficiency.

Reference Numbers

Reference numbers are shown at the beginning of some major classifications. These numbers refer to related items in the Reference Section. The reference information may be an estimating procedure, an alternate pricing method, or technical information.

Note: Not all subdivisions listed here necessarily appear. ■

No part of this cost data may be reproduced, stored in a retrieval system, or transmitted in any form or by any means without prior written permission of Gordian.

Did you know?

RSMeans data is available through our online application with 24/7 access:

- Search for unit prices by keyword
- Leverage the most up-to-date data
- Build and export estimates

Try it free for 30 days!
www.rsmeans.com/2018freetrial

08 05 Common Work Results for Openings

08 05 05 – Selective Demolition for Openings

08 05 05.10 Selective Demolition Doors

		Crew	Daily Output	Labor-Hours	Unit	Material	2018 Bare Costs Labor	Equipment	Total	Total Incl O&P
0010	**SELECTIVE DEMOLITION DOORS** R024119-10									
0200	Doors, exterior, 1-3/4" thick, single, 3' x 7' high	1 Clab	16	.500	Ea.		19.95		19.95	30.50
0220	Double, 6' x 7' high		12	.667			26.50		26.50	40.50
0500	Interior, 1-3/8" thick, single, 3' x 7' high		20	.400			15.95		15.95	24.50
0520	Double, 6' x 7' high		16	.500			19.95		19.95	30.50
0700	Bi-folding, 3' x 6'-8" high		20	.400			15.95		15.95	24.50
0720	6' x 6'-8" high		18	.444			17.70		17.70	27
0900	Bi-passing, 3' x 6'-8" high		16	.500			19.95		19.95	30.50
0940	6' x 6'-8" high		14	.571			23		23	34.50
0960	Interior metal door 1-3/4" thick, 3'-0" x 6'-8"		18	.444			17.70		17.70	27
0980	Interior metal door 1-3/4" thick, 3'-0" x 7'-0"		18	.444			17.70		17.70	27
1000	Interior wood door 1-3/4" thick, 3'-0" x 6'-8"		20	.400			15.95		15.95	24.50
1020	Interior wood door 1-3/4" thick, 3'-0" x 7'-0"		20	.400			15.95		15.95	24.50
1100	Door demo, floor door	2 Sswk	5	3.200			175		175	286
1500	Remove and reset, hollow core	1 Carp	8	1			50.50		50.50	77
1520	Solid		6	1.333			67.50		67.50	103
2000	Frames, including trim, metal		8	1			50.50		50.50	77
2200	Wood	2 Carp	32	.500			25.50		25.50	38.50
3000	Special doors, counter doors		6	2.667			135		135	206
3100	Double acting		10	1.600			81		81	124
3200	Floor door (trap type), or access type		8	2			101		101	154
3300	Glass, sliding, including frames		12	1.333			67.50		67.50	103
3400	Overhead, commercial, 12' x 12' high		4	4			203		203	310
3440	up to 20' x 16' high		3	5.333			270		270	410
3445	up to 35' x 30' high		1	16			810		810	1,225
3500	Residential, 9' x 7' high		8	2			101		101	154
3540	16' x 7' high		7	2.286			116		116	176
3620	Large		2.50	6.400			325		325	495
3700	Roll-up grille		5	3.200			162		162	247
3800	Revolving door, no wire or connections		2	8			405		405	620
3900	Storefront swing door		3	5.333			270		270	410
6600	Demo flexible transparent strip entrance	3 Shee	115	.209	SF Surf		12.50		12.50	19.05
7100	Remove double swing pneumatic doors, openers and sensors	2 Skwk	.50	32	Opng.		1,675		1,675	2,575
7110	Remove automatic operators, industrial, sliding doors, to 12' wide	"	.40	40	"		2,100		2,100	3,200
7550	Hangar door demo	2 Sswk	330	.048	S.F.		2.65		2.65	4.33
7570	Remove shock absorbing door	"	1.90	8.421	Opng.		460		460	750

08 05 05.20 Selective Demolition of Windows

		Crew	Daily Output	Labor-Hours	Unit	Material	2018 Bare Costs Labor	Equipment	Total	Total Incl O&P
0010	**SELECTIVE DEMOLITION OF WINDOWS** R024119-10									
0200	Aluminum, including trim, to 12 S.F.	1 Clab	16	.500	Ea.		19.95		19.95	30.50
0240	To 25 S.F.		11	.727			29		29	44
0280	To 50 S.F.		5	1.600			64		64	97
0320	Storm windows/screens, to 12 S.F.		27	.296			11.80		11.80	18
0360	To 25 S.F.		21	.381			15.20		15.20	23
0400	To 50 S.F.		16	.500			19.95		19.95	30.50
0600	Glass, up to 10 S.F./window		200	.040	S.F.		1.59		1.59	2.43
0620	Over 10 S.F./window		150	.053	"		2.13		2.13	3.24
1000	Steel, including trim, to 12 S.F.		13	.615	Ea.		24.50		24.50	37.50
1020	To 25 S.F.		9	.889			35.50		35.50	54
1040	To 50 S.F.		4	2			79.50		79.50	121
2000	Wood, including trim, to 12 S.F.		22	.364			14.50		14.50	22
2020	To 25 S.F.		18	.444			17.70		17.70	27
2060	To 50 S.F.		13	.615			24.50		24.50	37.50

For customer support on your Building Construction Costs with RSMeans data, call 800.448.8182.

08 05 Common Work Results for Openings

08 05 05 – Selective Demolition for Openings

08 05 05.20 Selective Demolition of Windows

		Crew	Daily Output	Labor-Hours	Unit	Material	2018 Bare Costs Labor	Equipment	Total	Total Incl O&P
2065	To 180 S.F.	1 Clab	8	1	Ea.		40		40	60.50
4300	Remove bay/bow window	2 Carp	6	2.667	↓		135		135	206
4410	Remove skylight, plstc domes, flush/curb mtd	"	395	.041	S.F.		2.05		2.05	3.13
4420	Remove skylight, plstc/glass up to 2' x 3'	1 Carp	15	.533	Ea.		27		27	41
4440	Remove skylight, plstc/glass up to 4' x 6'	2 Carp	10	1.600			81		81	124
4480	Remove roof window up to 3' x 4'	1 Carp	8	1			50.50		50.50	77
4500	Remove roof window up to 4' x 6'	2 Carp	6	2.667			135		135	206
5020	Remove and reset window, up to a 2 'x 2' window	1 Carp	6	1.333			67.50		67.50	103
5040	Up to a 3' x 3' window	↓	4	2			101		101	154
5080	Up to a 4' x 5' window	↓	2	4	↓		203		203	310

08 11 Metal Doors and Frames

08 11 16 – Aluminum Doors and Frames

08 11 16.10 Entrance Doors

		Crew	Daily Output	Labor-Hours	Unit	Material	2018 Bare Costs Labor	Equipment	Total	Total Incl O&P
0010	**ENTRANCE DOORS** and frame, aluminum, narrow stile									
0011	Including standard hardware, clear finish, no glass									
0012	Top and bottom offset pivots, 1/4" beveled glass stops, threshold									
0013	Dead bolt lock with inside thumb screw, standard push pull									
0020	3'-0" x 7'-0" opening	2 Sswk	2	8	Ea.	945	435		1,380	1,775
0025	Anodizing aluminum entr. door & frame, add					113			113	124
0030	3'-6" x 7'-0" opening	2 Sswk	2	8		870	435		1,305	1,675
0100	3'-0" x 10'-0" opening, 3' high transom		1.80	8.889		1,350	485		1,835	2,275
0200	3'-6" x 10'-0" opening, 3' high transom		1.80	8.889		1,375	485		1,860	2,325
0280	5'-0" x 7'-0" opening		2	8		1,600	435		2,035	2,475
0300	6'-0" x 7'-0" opening		1.30	12.308	↓	1,150	675		1,825	2,375
0301	6'-0" x 7'-0" opening		1.30	12.308	Pr.	1,150	675		1,825	2,375
0400	6'-0" x 10'-0" opening, 3' high transom		1.10	14.545	"	1,475	795		2,270	2,925
0520	3'-0" x 7'-0" opening, wide stile		2	8	Ea.	925	435		1,360	1,750
0540	3'-6" x 7'-0" opening		2	8		1,000	435		1,435	1,825
0560	5'-0" x 7'-0" opening		2	8	↓	1,350	435		1,785	2,225
0580	6'-0" x 7'-0" opening		1.30	12.308	Pr.	1,550	675		2,225	2,800
0600	7'-0" x 7'-0" opening	↓	1	16	"	2,275	875		3,150	3,925
1200	For non-standard size, add				Leaf	80%				
1250	For installation of non-standard size, add						20%			
1300	Light bronze finish, add				Leaf	36%				
1400	Dark bronze finish, add					25%				
1500	For black finish, add					40%				
1600	Concealed panic device, add				↓	970			970	1,075
1700	Electric striker release, add				Opng.	320			320	350
1800	Floor check, add				Leaf	670			670	740
1900	Concealed closer, add				"	550			550	600

08 11 63 – Metal Screen and Storm Doors and Frames

08 11 63.23 Aluminum Screen and Storm Doors and Frames

		Crew	Daily Output	Labor-Hours	Unit	Material	2018 Bare Costs Labor	Equipment	Total	Total Incl O&P
0010	**ALUMINUM SCREEN AND STORM DOORS AND FRAMES**									
0020	Combination storm and screen									
0420	Clear anodic coating, 2'-8" wide	2 Carp	14	1.143	Ea.	220	58		278	330
0440	3'-0" wide	"	14	1.143	"	191	58		249	298
0500	For 7'-0" door height, add					8%				
1020	Mill finish, 2'-8" wide	2 Carp	14	1.143	Ea.	250	58		308	365
1040	3'-0" wide	"	14	1.143	↓	270	58		328	385

08 11 Metal Doors and Frames

08 11 63 – Metal Screen and Storm Doors and Frames

08 11 63.23 Aluminum Screen and Storm Doors and Frames	Crew	Daily Output	Labor-Hours	Unit	Material	2018 Bare Costs Labor	Equipment	Total	Total Incl O&P
1100	For 7'-0" door, add				Ea.	8%				
1520	White painted, 2'-8" wide	2 Carp	14	1.143		239	58		297	350
1540	3'-0" wide	"	14	1.143		315	58		373	435
1600	For 7'-0" door, add					8%				
2000	Wood door & screen, see Section 08 14 33.20									

08 12 Metal Frames

08 12 13 – Hollow Metal Frames

08 12 13.13 Standard Hollow Metal Frames

			Crew	Daily Output	Labor-Hours	Unit	Material	Labor	Equipment	Total	Total Incl O&P
0010	**STANDARD HOLLOW METAL FRAMES**										
0020	16 ga., up to 5-3/4" jamb depth										
0025	3'-0" x 6'-8" single	G	2 Carp	16	1	Ea.	121	50.50		171.50	210
0028	3'-6" wide, single	G		16	1		232	50.50		282.50	330
0030	4'-0" wide, single	G		16	1		268	50.50		318.50	370
0040	6'-0" wide, double	G		14	1.143		227	58		285	340
0045	8'-0" wide, double	G		14	1.143		227	58		285	340
0100	3'-0" x 7'-0" single	G		16	1		192	50.50		242.50	288
0110	3'-6" wide, single	G		16	1		194	50.50		244.50	290
0112	4'-0" wide, single	G		16	1		227	50.50		277.50	325
0140	6'-0" wide, double	G		14	1.143		227	58		285	340
0145	8'-0" wide, double	G		14	1.143		237	58		295	350
1000	16 ga., up to 4-7/8" deep, 3'-0" x 7'-0" single	G		16	1		181	50.50		231.50	276
1140	6'-0" wide, double	G		14	1.143		204	58		262	315
1200	16 ga., 8-3/4" deep, 3'-0" x 7'-0" single	G		16	1		192	50.50		242.50	288
1240	6'-0" wide, double	G		14	1.143		256	58		314	370
2800	14 ga., up to 3-7/8" deep, 3'-0" x 7'-0" single	G		16	1		247	50.50		297.50	350
2840	6'-0" wide, double	G		14	1.143		278	58		336	395
3000	14 ga., up to 5-3/4" deep, 3'-0" x 6'-8" single	G		16	1		154	50.50		204.50	246
3002	3'-6" wide, single	G		16	1		239	50.50		289.50	340
3005	4'-0" wide, single	G		16	1		155	50.50		205.50	248
3600	up to 5-3/4" jamb depth, 4'-0" x 7'-0" single	G		15	1.067		237	54		291	345
3620	6'-0" wide, double	G		12	1.333		177	67.50		244.50	297
3640	8'-0" wide, double	G		12	1.333		282	67.50		349.50	415
3700	8'-0" high, 4'-0" wide, single	G		15	1.067		270	54		324	380
3740	8'-0" wide, double	G		12	1.333		320	67.50		387.50	455
4000	6-3/4" deep, 4'-0" x 7'-0" single	G		15	1.067		266	54		320	375
4020	6'-0" wide, double	G		12	1.333		300	67.50		367.50	435
4040	8'-0" wide, double	G		12	1.333		224	67.50		291.50	350
4100	8'-0" high, 4'-0" wide, single	G		15	1.067		176	54		230	276
4140	8'-0" wide, double	G		12	1.333		415	67.50		482.50	560
4400	8-3/4" deep, 4'-0" x 7'-0", single	G		15	1.067		385	54		439	505
4440	8'-0" wide, double	G		12	1.333		405	67.50		472.50	555
4500	4'-0" x 8'-0", single	G		15	1.067		440	54		494	570
4540	8'-0" wide, double	G		12	1.333		450	67.50		517.50	600
4900	For welded frames, add						51			51	56
5400	14 ga., "B" label, up to 5-3/4" deep, 4'-0" x 7'-0" single	G	2 Carp	15	1.067		167	54		221	267
5440	8'-0" wide, double	G		12	1.333		219	67.50		286.50	345
5800	6-3/4" deep, 7'-0" high, 4'-0" wide, single	G		15	1.067		173	54		227	274
5840	8'-0" wide, double	G		12	1.333		305	67.50		372.50	445
6200	8-3/4" deep, 4'-0" x 7'-0" single	G		15	1.067		247	54		301	355
6240	8'-0" wide, double	G		12	1.333		380	67.50		447.50	525

For customer support on your Building Construction Costs with RSMeans data, call 800.448.8182.

08 12 Metal Frames

08 12 13 – Hollow Metal Frames

08 12 13.13 Standard Hollow Metal Frames

		Crew	Daily Output	Labor-Hours	Unit	Material	2018 Bare Costs Labor	Equipment	Total	Total Incl O&P
6300	For "A" label use same price as "B" label									
6400	For baked enamel finish, add					30%	15%			
6500	For galvanizing, add					20%				
6600	For hospital stop, add				Ea.	235			235	259
6620	For hospital stop, stainless steel, add				"	109			109	120
7900	Transom lite frames, fixed, add	2 Carp	155	.103	S.F.	43	5.25		48.25	55.50
8000	Movable, add	"	130	.123	"	55	6.25		61.25	70

08 12 13.25 Channel Metal Frames

			Crew	Daily Output	Labor-Hours	Unit	Material	2018 Bare Costs Labor	Equipment	Total	Total Incl O&P
0010	**CHANNEL METAL FRAMES**										
0020	Steel channels with anchors and bar stops										
0100	6" channel @ 8.2#/L.F., 3' x 7' door, weighs 150#	G	E-4	13	2.462	Ea.	242	136	7.60	385.60	495
0200	8" channel @ 11.5#/L.F., 6' x 8' door, weighs 275#	G		9	3.556		445	196	10.95	651.95	820
0300	8' x 12' door, weighs 400#	G		6.50	4.923		645	272	15.15	932.15	1,175
0400	10" channel @ 15.3#/L.F., 10' x 10' door, weighs 500#	G		6	5.333		810	294	16.45	1,120.45	1,400
0500	12' x 12' door, weighs 600#	G		5.50	5.818		970	320	17.90	1,307.90	1,625
0600	12" channel @ 20.7#/L.F., 12' x 12' door, weighs 825#	G		4.50	7.111		1,325	390	22	1,737	2,150
0700	12' x 16' door, weighs 1000#	G		4	8		1,625	440	24.50	2,089.50	2,525
0800	For frames without bar stops, light sections, deduct						15%				
0900	Heavy sections, deduct						10%				

08 13 Metal Doors

08 13 13 – Hollow Metal Doors

08 13 13.13 Standard Hollow Metal Doors

			Crew	Daily Output	Labor-Hours	Unit	Material	2018 Bare Costs Labor	Equipment	Total	Total Incl O&P
0010	**STANDARD HOLLOW METAL DOORS**	R081313-20									
0015	Flush, full panel, hollow core										
0017	When noted doors are prepared but do not include glass or louvers										
0020	1-3/8" thick, 20 ga., 2'-0" x 6'-8"	G	2 Carp	20	.800	Ea.	340	40.50		380.50	435
0040	2'-8" x 6'-8"	G		18	.889		350	45		395	455
0060	3'-0" x 6'-8"	G		17	.941		355	47.50		402.50	465
0100	3'-0" x 7'-0"	G		17	.941		365	47.50		412.50	480
0120	For vision lite, add						108			108	119
0140	For narrow lite, add						116			116	128
0320	Half glass, 20 ga., 2'-0" x 6'-8"	G	2 Carp	20	.800		600	40.50		640.50	720
0340	2'-8" x 6'-8"	G		18	.889		600	45		645	730
0360	3'-0" x 6'-8"	G		17	.941		600	47.50		647.50	735
0400	3'-0" x 7'-0"	G		17	.941		400	47.50		447.50	515
0410	1-3/8" thick, 18 ga., 2'-0" x 6'-8"	G		20	.800		405	40.50		445.50	505
0420	3'-0" x 6'-8"	G		17	.941		410	47.50		457.50	525
0425	3'-0" x 7'-0"	G		17	.941		430	47.50		477.50	550
0450	For vision lite, add						108			108	119
0452	For narrow lite, add						116			116	128
0460	Half glass, 18 ga., 2'-0" x 6'-8"	G	2 Carp	20	.800		560	40.50		600.50	680
0465	2'-8" x 6'-8"	G		18	.889		580	45		625	710
0470	3'-0" x 6'-8"	G		17	.941		570	47.50		617.50	700
0475	3'-0" x 7'-0"	G		17	.941		585	47.50		632.50	720
0500	Hollow core, 1-3/4" thick, full panel, 20 ga., 2'-8" x 6'-8"	G		18	.889		435	45		480	550
0520	3'-0" x 6'-8"	G		17	.941		435	47.50		482.50	555
0640	3'-0" x 7'-0"	G		17	.941		465	47.50		512.50	585
0680	4'-0" x 7'-0"	G		15	1.067		635	54		689	780
0700	4'-0" x 8'-0"	G		13	1.231		750	62.50		812.50	915
1000	18 ga., 2'-8" x 6'-8"	G		17	.941		485	47.50		532.50	605

For customer support on your Building Construction Costs with RSMeans data, call 800.448.8182.

267

08 13 13.13 Standard Hollow Metal Doors

			Crew	Daily Output	Labor-Hours	Unit	Material	2018 Bare Costs Labor	Equipment	Total	Total Incl O&P
1020	3'-0" x 6'-8"	G	2 Carp	16	1	Ea.	485	50.50		535.50	605
1120	3'-0" x 7'-0"	G		17	.941		495	47.50		542.50	620
1180	4'-0" x 7'-0"	G		14	1.143		695	58		753	855
1200	4'-0" x 8'-0"	G		17	.941		850	47.50		897.50	1,000
1212	For vision lite, add						108			108	119
1214	For narrow lite, add						116			116	128
1230	Half glass, 20 ga., 2'-8" x 6'-8"	G	2 Carp	20	.800		580	40.50		620.50	700
1240	3'-0" x 6'-8"	G		18	.889		580	45		625	705
1260	3'-0" x 7'-0"	G		18	.889		595	45		640	725
1280	Embossed panel, 1-3/4" thick, poly core, 20 ga., 3'-0" x 7'-0"			18	.889		420	45		465	530
1290	Half glass, 1-3/4" thick, poly core, 20 ga., 3'-0" x 7'-0"	G		18	.889		615	45		660	745
1320	18 ga., 2'-8" x 6'-8"	G		18	.889		640	45		685	775
1340	3'-0" x 6'-8"	G		17	.941		635	47.50		682.50	770
1360	3'-0" x 7'-0"	G		17	.941		645	47.50		692.50	785
1380	4'-0" x 7'-0"	G		15	1.067		795	54		849	960
1400	4'-0" x 8'-0"	G		14	1.143		895	58		953	1,075
1500	Flush full panel, 16 ga., steel hollow core										
1520	2'-0" x 6'-8"	G	2 Carp	20	.800	Ea.	570	40.50		610.50	685
1530	2'-8" x 6'-8"	G		20	.800		570	40.50		610.50	685
1540	3'-0" x 6'-8"	G		20	.800		555	40.50		595.50	670
1560	2'-8" x 7'-0"	G		18	.889		580	45		625	710
1570	3'-0" x 7'-0"	G		18	.889		560	45		605	690
1580	3'-6" x 7'-0"	G		18	.889		665	45		710	805
1590	4'-0" x 7'-0"	G		18	.889		730	45		775	875
1600	2'-8" x 8'-0"	G		18	.889		715	45		760	855
1620	3'-0" x 8'-0"	G		18	.889		730	45		775	870
1630	3'-6" x 8'-0"	G		18	.889		810	45		855	960
1640	4'-0" x 8'-0"	G		18	.889		850	45		895	1,000
1720	Insulated, 1-3/4" thick, full panel, 18 ga., 3'-0" x 6'-8"	G		15	1.067		485	54		539	620
1740	2'-8" x 7'-0"	G		16	1		505	50.50		555.50	630
1760	3'-0" x 7'-0"	G		15	1.067		505	54		559	640
1800	4'-0" x 8'-0"	G		13	1.231		780	62.50		842.50	955
1820	Half glass, 18 ga., 3'-0" x 6'-8"	G		16	1		635	50.50		685.50	770
1840	2'-8" x 7'-0"	G		17	.941		665	47.50		712.50	805
1860	3'-0" x 7'-0"	G		16	1		690	50.50		740.50	835
1900	4'-0" x 8'-0"	G		14	1.143		555	58		613	700
2000	For vision lite, add						108			108	119
2010	For narrow lite, add						116			116	128
8100	For bottom louver, add						224			224	246
8110	For baked enamel finish, add						30%	15%			
8120	For galvanizing, add						20%				

08 13 13.15 Metal Fire Doors

			Crew	Daily Output	Labor-Hours	Unit	Material	2018 Bare Costs Labor	Equipment	Total	Total Incl O&P
0010	**METAL FIRE DOORS** R081313-20										
0015	Steel, flush, "B" label, 90 minutes										
0020	Full panel, 20 ga., 2'-0" x 6'-8"		2 Carp	20	.800	Ea.	425	40.50		465.50	525
0040	2'-8" x 6'-8"			18	.889		440	45		485	555
0060	3'-0" x 6'-8"			17	.941		440	47.50		487.50	560
0080	3'-0" x 7'-0"			17	.941		455	47.50		502.50	575
0140	18 ga., 3'-0" x 6'-8"			16	1		505	50.50		555.50	630
0160	2'-8" x 7'-0"			17	.941		535	47.50		582.50	660
0180	3'-0" x 7'-0"			16	1		515	50.50		565.50	645
0200	4'-0" x 7'-0"			15	1.067		650	54		704	800

08 13 Metal Doors

08 13 13 – Hollow Metal Doors

08 13 13.15 Metal Fire Doors	Crew	Daily Output	Labor-Hours	Unit	Material	2018 Bare Costs Labor	Equipment	Total	Total Incl O&P
0220	For "A" label, 3 hour, 18 ga., use same price as "B" label									
0240	For vision lite, add				Ea.	163			163	179
0300	Full panel, 16 ga., 2'-0" x 6'-8"	2 Carp	20	.800		315	40.50		355.50	405
0310	2'-8" x 6'-8"		18	.889		550	45		595	675
0320	3'-0" x 6'-8"		17	.941		440	47.50		487.50	555
0350	2'-8" x 7'-0"		17	.941		570	47.50		617.50	700
0360	3'-0" x 7'-0"		16	1		555	50.50		605.50	685
0370	4'-0" x 7'-0"		15	1.067		715	54		769	875
0520	Flush, "B" label, 90 minutes, egress core, 20 ga., 2'-0" x 6'-8"		18	.889		665	45		710	805
0540	2'-8" x 6'-8"		17	.941		670	47.50		717.50	810
0560	3'-0" x 6'-8"		16	1		675	50.50		725.50	815
0580	3'-0" x 7'-0"		16	1		700	50.50		750.50	845
0640	Flush, "A" label, 3 hour, egress core, 18 ga., 3'-0" x 6'-8"		15	1.067		730	54		784	885
0660	2'-8" x 7'-0"		16	1		765	50.50		815.50	915
0680	3'-0" x 7'-0"		15	1.067		750	54		804	905
0700	4'-0" x 7'-0"	▼	14	1.143	▼	910	58		968	1,100

08 13 13.20 Residential Steel Doors

			Crew	Daily Output	Labor-Hours	Unit	Material	Labor	Equipment	Total	Total Incl O&P
0010	**RESIDENTIAL STEEL DOORS**											
0020	Prehung, insulated, exterior											
0030	Embossed, full panel, 2'-8" x 6'-8"	G	2 Carp	17	.941	Ea.	405	47.50		452.50	525
0040	3'-0" x 6'-8"	G		15	1.067		290	54		344	405
0060	3'-0" x 7'-0"	G		15	1.067		350	54		404	470
0070	5'-4" x 6'-8", double	G		8	2		575	101		676	790
0220	Half glass, 2'-8" x 6'-8"	G		17	.941		330	47.50		377.50	440
0240	3'-0" x 6'-8"	G		16	1		330	50.50		380.50	440
0260	3'-0" x 7'-0"	G		16	1		400	50.50		450.50	515
0270	5'-4" x 6'-8", double	G		8	2		990	101		1,091	1,225
1320	Flush face, full panel, 2'-8" x 6'-8"	G		16	1		291	50.50		341.50	395
1340	3'-0" x 6'-8"	G		15	1.067		291	54		345	405
1360	3'-0" x 7'-0"	G		15	1.067		310	54		364	425
1380	5'-4" x 6'-8", double	G		8	2		720	101		821	945
1420	Half glass, 2'-8" x 6'-8"	G		17	.941		350	47.50		397.50	460
1440	3'-0" x 6'-8"	G		16	1		350	50.50		400.50	460
1460	3'-0" x 7'-0"	G		16	1		435	50.50		485.50	555
1480	5'-4" x 6'-8", double	G	▼	8	2		665	101		766	885
1500	Sidelight, full lite, 1'-0" x 6'-8" with grille	G					345			345	375
1510	1'-0" x 6'-8", low E	G					365			365	405
1520	1'-0" x 6'-8", half lite	G					305			305	335
1530	1'-0" x 6'-8", half lite, low E	G					296			296	325
2300	Interior, residential, closet, bi-fold, 2'-0" x 6'-8"	G	2 Carp	16	1		222	50.50		272.50	320
2330	3'-0" wide	G		16	1		260	50.50		310.50	365
2360	4'-0" wide	G		15	1.067		280	54		334	395
2400	5'-0" wide	G		14	1.143		340	58		398	465
2420	6'-0" wide	G	▼	13	1.231		300	62.50		362.50	425

08 13 13.25 Doors Hollow Metal

			Crew	Daily Output	Labor-Hours	Unit	Material	Labor	Equipment	Total	Total Incl O&P
0010	**DOORS HOLLOW METAL**											
0500	Exterior, commercial, flush, 20 ga., 1-3/4" x 7'-0" x 2'-6" wide	G	2 Carp	15	1.067	Ea.	291	54		345	405
0530	2'-8" wide	G		15	1.067		291	54		345	405
0560	3'-0" wide	G		14	1.143		292	58		350	410
1000	18 ga., 1-3/4" x 7'-0" x 2'-6" wide	G		15	1.067		320	54		374	435
1030	2'-8" wide	G		15	1.067		320	54		374	440
1060	3'-0" wide	G	▼	14	1.143		315	58		373	435

For customer support on your Building Construction Costs with RSMeans data, call 800.448.8182.

269

08 13 Metal Doors

08 13 13 – Hollow Metal Doors

08 13 13.25 Doors Hollow Metal		Crew	Daily Output	Labor-Hours	Unit	Material	2018 Bare Costs Labor	Equipment	Total	Total Incl O&P
1500	16 ga., 1-3/4" x 7'-0" x 2'-6" wide	G 2 Carp	15	1.067	Ea.	370	54		424	495
1530	2'-8" wide	G	15	1.067		370	54		424	495
1560	3'-0" wide	G	14	1.143		370	58		428	500
1590	3'-6" wide	G	14	1.143		690	58		748	850
2900	Fire door, "A" label, 18 gauge, 1-3/4" x 2'-6" x 7'-0"	G	15	1.067		795	54		849	960
2930	2'-8" wide	G	15	1.067		690	54		744	845
2960	3'-0" wide	G	14	1.143		690	58		748	850
2990	3'-6" wide	G	14	1.143		470	58		528	610
3100	"B" label, 2'-6" wide	G	15	1.067		585	54		639	730
3130	2'-8" wide	G	15	1.067		585	54		639	730
3160	3'-0" wide	G	14	1.143		585	58		643	735

08 13 16 – Aluminum Doors

08 13 16.10 Commercial Aluminum Doors		Crew	Daily Output	Labor-Hours	Unit	Material	2018 Bare Costs Labor	Equipment	Total	Total Incl O&P
0010	**COMMERCIAL ALUMINUM DOORS**, flush, no glazing									
5000	Flush panel doors, pair of 2'-6" x 7'-0"	2 Sswk	2	8	Pr.	1,300	435		1,735	2,150
5050	3'-0" x 7'-0", single		2.50	6.400	Ea.	790	350		1,140	1,425
5100	Pair of 3'-0" x 7'-0"		2	8	Pr.	1,400	435		1,835	2,250
5150	3'-6" x 7'-0", single		2.50	6.400	Ea.	915	350		1,265	1,575

08 14 Wood Doors

08 14 13 – Carved Wood Doors

08 14 13.10 Types of Wood Doors, Carved		Crew	Daily Output	Labor-Hours	Unit	Material	2018 Bare Costs Labor	Equipment	Total	Total Incl O&P
0010	**TYPES OF WOOD DOORS, CARVED**									
3000	Solid wood, 1-3/4" thick stile and rail									
3020	Mahogany, 3'-0" x 7'-0", six panel	2 Carp	14	1.143	Ea.	1,200	58		1,258	1,400
3030	With two lites		10	1.600		3,050	81		3,131	3,475
3040	3'-6" x 8'-0", six panel		10	1.600		1,625	81		1,706	1,925
3050	With two lites		8	2		2,850	101		2,951	3,300
3100	Pine, 3'-0" x 7'-0", six panel		14	1.143		510	58		568	650
3110	With two lites		10	1.600		805	81		886	1,000
3120	3'-6" x 8'-0", six panel		10	1.600		1,025	81		1,106	1,250
3130	With two lites		8	2		1,950	101		2,051	2,300
3200	Red oak, 3'-0" x 7'-0", six panel		14	1.143		1,325	58		1,383	1,550
3210	With two lites		10	1.600		2,525	81		2,606	2,900
3220	3'-6" x 8'-0", six panel		10	1.600		2,700	81		2,781	3,100
3230	With two lites		8	2		3,500	101		3,601	4,000
4000	Hand carved door, mahogany									
4020	3'-0" x 7'-0", simple design	2 Carp	14	1.143	Ea.	1,750	58		1,808	2,025
4030	Intricate design		11	1.455		3,675	74		3,749	4,125
4040	3'-6" x 8'-0", simple design		10	1.600		2,775	81		2,856	3,175
4050	Intricate design		8	2		3,675	101		3,776	4,175
4400	For custom finish, add					575			575	635
4600	Side light, mahogany, 7'-0" x 1'-6" wide, 4 lites	2 Carp	18	.889		1,100	45		1,145	1,275
4610	6 lites		14	1.143		2,525	58		2,583	2,875
4620	8'-0" x 1'-6" wide, 4 lites		14	1.143		2,075	58		2,133	2,375
4630	6 lites		10	1.600		2,075	81		2,156	2,400
4640	Side light, oak, 7'-0" x 1'-6" wide, 4 lites		18	.889		1,200	45		1,245	1,400
4650	6 lites		14	1.143		2,100	58		2,158	2,400
4660	8'-0" x 1'-6" wide, 4 lites		14	1.143		1,250	58		1,308	1,475
4670	6 lites		10	1.600		2,100	81		2,181	2,425

For customer support on your Building Construction Costs with RSMeans data, call 800.448.8182.

08 14 Wood Doors

08 14 16 – Flush Wood Doors

08 14 16.09 Smooth Wood Doors		Crew	Daily Output	Labor- Hours	Unit	Material	2018 Bare Costs Labor	Equipment	Total	Total Incl O&P
0010	**SMOOTH WOOD DOORS**									
0015	Flush, interior, hollow core									
0025	Lauan face, 1-3/8", 3'-0" x 6'-8"	2 Carp	17	.941	Ea.	55	47.50		102.50	133
0030	4'-0" x 6'-8"		16	1		126	50.50		176.50	216
0080	1-3/4", 2'-0" x 6'-8"		17	.941		55	47.50		102.50	133
0108	3'-0" x 7'-0"		16	1		206	50.50		256.50	305
0112	Pair of 3'-0" x 7'-0"		9	1.778	Pr.	132	90		222	282
0140	Birch face, 1-3/8", 2'-6" x 6'-8"		17	.941	Ea.	113	47.50		160.50	198
0180	3'-0" x 6'-8"		17	.941		97	47.50		144.50	180
0200	4'-0" x 6'-8"		16	1		158	50.50		208.50	251
0202	1-3/4", 2'-0" x 6'-8"		17	.941		57.50	47.50		105	136
0210	3'-0" x 7'-0"		16	1		158	50.50		208.50	250
0214	Pair of 3'-0" x 7'-0"		9	1.778	Pr.	259	90		349	420
0220	Oak face, 1-3/8", 2'-0" x 6'-8"		17	.941	Ea.	110	47.50		157.50	194
0280	3'-0" x 6'-8"		17	.941		115	47.50		162.50	199
0300	4'-0" x 6'-8"		16	1		141	50.50		191.50	232
0305	1-3/4", 2'-6" x 6'-8"		17	.941		125	47.50		172.50	210
0310	3'-0" x 7'-0"		16	1		250	50.50		300.50	350
0320	Walnut face, 1-3/8", 2'-0" x 6'-8"		17	.941		185	47.50		232.50	277
0340	2'-6" x 6'-8"		17	.941		192	47.50		239.50	284
0380	3'-0" x 6'-8"		17	.941		200	47.50		247.50	293
0400	4'-0" x 6'-8"		16	1		219	50.50		269.50	320
0430	For 7'-0" high, add					27			27	30
0440	For 8'-0" high, add					39.50			39.50	43.50
0480	For prefinishing, clear, add					48.50			48.50	53
0500	For prefinishing, stain, add					60.50			60.50	66.50
1320	M.D. overlay on hardboard, 1-3/8", 2'-0" x 6'-8"	2 Carp	17	.941		123	47.50		170.50	208
1340	2'-6" x 6'-8"		17	.941		124	47.50		171.50	209
1380	3'-0" x 6'-8"		17	.941		130	47.50		177.50	216
1400	4'-0" x 6'-8"		16	1		185	50.50		235.50	281
1420	For 7'-0" high, add					17.90			17.90	19.65
1440	For 8'-0" high, add					35.50			35.50	39
1720	H.P. plastic laminate, 1-3/8", 2'-0" x 6'-8"	2 Carp	16	1		275	50.50		325.50	380
1740	2'-6" x 6'-8"		16	1		270	50.50		320.50	375
1780	3'-0" x 6'-8"		15	1.067		299	54		353	415
1800	4'-0" x 6'-8"		14	1.143		395	58		453	525
1820	For 7'-0" high, add					17.90			17.90	19.70
1840	For 8'-0" high, add					35			35	39
2020	Particle core, lauan face, 1-3/8", 2'-6" x 6'-8"	2 Carp	15	1.067		97	54		151	190
2040	3'-0" x 6'-8"		14	1.143		101	58		159	199
2080	3'-0" x 7'-0"		13	1.231		120	62.50		182.50	227
2085	4'-0" x 7'-0"		12	1.333		129	67.50		196.50	245
2110	1-3/4", 3'-0" x 7'-0"		13	1.231		206	62.50		268.50	320
2120	Birch face, 1-3/8", 2'-6" x 6'-8"		15	1.067		110	54		164	204
2140	3'-0" x 6'-8"		14	1.143		121	58		179	221
2180	3'-0" x 7'-0"		13	1.231		132	62.50		194.50	240
2200	4'-0" x 7'-0"		12	1.333		148	67.50		215.50	266
2205	1-3/4", 3'-0" x 7'-0"		13	1.231		195	62.50		257.50	310
2220	Oak face, 1-3/8", 2'-6" x 6'-8"		15	1.067		122	54		176	217
2240	3'-0" x 6'-8"		14	1.143		134	58		192	235
2280	3'-0" x 7'-0"		13	1.231		140	62.50		202.50	249
2300	4'-0" x 7'-0"		12	1.333		166	67.50		233.50	286

08 14 Wood Doors

08 14 16 – Flush Wood Doors

08 14 16.09 Smooth Wood Doors

		Crew	Daily Output	Labor-Hours	Unit	Material	2018 Bare Costs Labor	Equipment	Total	Total Incl O&P
2305	1-3/4", 3'-0" x 7'-0"	2 Carp	13	1.231	Ea.	236	62.50		298.50	355
2320	Walnut face, 1-3/8", 2'-0" x 6'-8"		15	1.067		131	54		185	227
2340	2'-6" x 6'-8"		14	1.143		148	58		206	251
2380	3'-0" x 6'-8"		13	1.231		166	62.50		228.50	278
2400	4'-0" x 6'-8"	▼	12	1.333		219	67.50		286.50	345
2440	For 8'-0" high, add					44.50			44.50	49
2460	For 8'-0" high walnut, add					40.50			40.50	44.50
2720	For prefinishing, clear, add					39.50			39.50	43.50
2740	For prefinishing, stain, add					57			57	62.50
3320	M.D. overlay on hardboard, 1-3/8", 2'-6" x 6'-8"	2 Carp	14	1.143		178	58		236	284
3340	3'-0" x 6'-8"		13	1.231		193	62.50		255.50	305
3380	3'-0" x 7'-0"		12	1.333		196	67.50		263.50	320
3400	4'-0" x 7'-0"	▼	10	1.600		265	81		346	415
3440	For 8'-0" height, add					39.50			39.50	43.50
3460	For solid wood core, add					70.50			70.50	77.50
3720	H.P. plastic laminate, 1-3/8", 2'-6" x 6'-8"	2 Carp	13	1.231		160	62.50		222.50	271
3740	3'-0" x 6'-8"		12	1.333		188	67.50		255.50	310
3780	3'-0" x 7'-0"		11	1.455		193	74		267	325
3800	4'-0" x 7'-0"	▼	8	2		380	101		481	570
3840	For 8'-0" height, add					39.50			39.50	43.50
3860	For solid wood core, add					43			43	47.50
4000	Exterior, flush, solid core, birch, 1-3/4" x 2'-6" x 7'-0"	2 Carp	15	1.067		166	54		220	266
4020	2'-8" wide		15	1.067		247	54		301	355
4040	3'-0" wide		14	1.143		217	58		275	325
4100	Oak faced 1-3/4" x 2'-6" x 7'-0"		15	1.067		222	54		276	325
4120	2'-8" wide		15	1.067		227	54		281	335
4140	3'-0" wide		14	1.143		215	58		273	325
4200	Walnut faced, 1-3/4" x 2'-6" x 7'-0"		15	1.067		325	54		379	445
4220	2'-8" wide		15	1.067		300	54		354	415
4240	3'-0" wide	▼	14	1.143		291	58		349	410
4300	For 6'-8" high door, deduct from 7'-0" door					18.05			18.05	19.90
5000	Wood doors, for vision lite, add					108			108	119
5010	Wood doors, for narrow lite, add					116			116	128
5015	Wood doors, for bottom (or top) louver, add			▼		224			224	246

08 14 16.20 Wood Fire Doors

		Crew	Daily Output	Labor-Hours	Unit	Material	2018 Bare Costs Labor	Equipment	Total	Total Incl O&P
0010	**WOOD FIRE DOORS**									
0020	Particle core, 7 face plys, "B" label,									
0040	1 hour, birch face, 1-3/4" x 2'-6" x 6'-8"	2 Carp	14	1.143	Ea.	435	58		493	565
0080	3'-0" x 6'-8"		13	1.231		490	62.50		552.50	630
0090	3'-0" x 7'-0"		12	1.333		455	67.50		522.50	605
0100	4'-0" x 7'-0"		12	1.333		655	67.50		722.50	825
0140	Oak face, 2'-6" x 6'-8"		14	1.143		490	58		548	630
0180	3'-0" x 6'-8"		13	1.231		500	62.50		562.50	645
0190	3'-0" x 7'-0"		12	1.333		500	67.50		567.50	655
0200	4'-0" x 7'-0"		12	1.333		635	67.50		702.50	805
0240	Walnut face, 2'-6" x 6'-8"		14	1.143		500	58		558	640
0280	3'-0" x 6'-8"		13	1.231		515	62.50		577.50	665
0290	3'-0" x 7'-0"		12	1.333		550	67.50		617.50	715
0300	4'-0" x 7'-0"		12	1.333		820	67.50		887.50	1,000
0440	M.D. overlay on hardboard, 2'-6" x 6'-8"		15	1.067		340	54		394	455
0480	3'-0" x 6'-8"		14	1.143		400	58		458	530
0490	3'-0" x 7'-0"	▼	13	1.231		425	62.50		487.50	565

For customer support on your Building Construction Costs with RSMeans data, call 800.448.8182.

08 14 Wood Doors

08 14 16 - Flush Wood Doors

08 14 16.20 Wood Fire Doors

		Crew	Daily Output	Labor-Hours	Unit	Material	2018 Bare Costs Labor	2018 Bare Costs Equipment	Total	Total Incl O&P
0500	4'-0" x 7'-0"	2 Carp	12	1.333	Ea.	450	67.50		517.50	600
0740	90 minutes, birch face, 1-3/4" x 2'-6" x 6'-8"		14	1.143		335	58		393	455
0780	3'-0" x 6'-8"		13	1.231		380	62.50		442.50	515
0790	3'-0" x 7'-0"		12	1.333		385	67.50		452.50	530
0800	4'-0" x 7'-0"		12	1.333		520	67.50		587.50	680
0840	Oak face, 2'-6" x 6'-8"		14	1.143		495	58		553	635
0880	3'-0" x 6'-8"		13	1.231		505	62.50		567.50	650
0890	3'-0" x 7'-0"		12	1.333		560	67.50		627.50	720
0900	4'-0" x 7'-0"		12	1.333		660	67.50		727.50	830
0940	Walnut face, 2'-6" x 6'-8"		14	1.143		500	58		558	640
0980	3'-0" x 6'-8"		13	1.231		440	62.50		502.50	580
0990	3'-0" x 7'-0"		12	1.333		505	67.50		572.50	665
1000	4'-0" x 7'-0"		12	1.333		655	67.50		722.50	830
1140	M.D. overlay on hardboard, 2'-6" x 6'-8"		15	1.067		405	54		459	530
1180	3'-0" x 6'-8"		14	1.143		415	58		473	545
1190	3'-0" x 7'-0"		13	1.231		495	62.50		557.50	640
1200	4'-0" x 7'-0"		12	1.333		560	67.50		627.50	720
1240	For 8'-0" height, add					82			82	90
1260	For 8'-0" height walnut, add					96.50			96.50	106
2200	Custom architectural "B" label, flush, 1-3/4" thick, birch,									
2210	Solid core									
2220	2'-6" x 7'-0"	2 Carp	15	1.067	Ea.	395	54		449	515
2260	3'-0" x 7'-0"		14	1.143		340	58		398	465
2300	4'-0" x 7'-0"		13	1.231		520	62.50		582.50	665
2420	4'-0" x 8'-0"		11	1.455		525	74		599	690
2480	For oak veneer, add					50%				
2500	For walnut veneer, add					75%				

08 14 33 - Stile and Rail Wood Doors

08 14 33.10 Wood Doors Paneled

		Crew	Daily Output	Labor-Hours	Unit	Material	2018 Bare Costs Labor	2018 Bare Costs Equipment	Total	Total Incl O&P
0010	**WOOD DOORS PANELED**									
0020	Interior, six panel, hollow core, 1-3/8" thick									
0040	Molded hardboard, 2'-0" x 6'-8"	2 Carp	17	.941	Ea.	62	47.50		109.50	141
0060	2'-6" x 6'-8"		17	.941		64	47.50		111.50	143
0070	2'-8" x 6'-8"		17	.941		67	47.50		114.50	146
0080	3'-0" x 6'-8"		17	.941		73.50	47.50		121	154
0140	Embossed print, molded hardboard, 2'-0" x 6'-8"		17	.941		64	47.50		111.50	143
0160	2'-6" x 6'-8"		17	.941		64	47.50		111.50	143
0180	3'-0" x 6'-8"		17	.941		73.50	47.50		121	154
0540	Six panel, solid, 1-3/8" thick, pine, 2'-0" x 6'-8"		15	1.067		155	54		209	254
0560	2'-6" x 6'-8"		14	1.143		154	58		212	258
0580	3'-0" x 6'-8"		13	1.231		148	62.50		210.50	258
1020	Two panel, bored rail, solid, 1-3/8" thick, pine, 1'-6" x 6'-8"		16	1		245	50.50		295.50	345
1040	2'-0" x 6'-8"		15	1.067		320	54		374	440
1060	2'-6" x 6'-8"		14	1.143		365	58		423	490
1340	Two panel, solid, 1-3/8" thick, fir, 2'-0" x 6'-8"		15	1.067		160	54		214	259
1360	2'-6" x 6'-8"		14	1.143		215	58		273	325
1380	3'-0" x 6'-8"		13	1.231		375	62.50		437.50	510
1740	Five panel, solid, 1-3/8" thick, fir, 2'-0" x 6'-8"		15	1.067		286	54		340	400
1760	2'-6" x 6'-8"		14	1.143		380	58		438	510
1780	3'-0" x 6'-8"		13	1.231		380	62.50		442.50	515

08 14 33.20 Wood Doors Residential	Crew	Daily Output	Labor-Hours	Unit	Material	Labor	2018 Bare Costs Equipment	Total	Total Incl O&P
0010	**WOOD DOORS RESIDENTIAL**								
0200	Exterior, combination storm & screen, pine								
0260	2'-8" wide	2 Carp	10	1.600	Ea.	310	81		391	470
0280	3'-0" wide		9	1.778		325	90		415	495
0300	7'-1" x 3'-0" wide		9	1.778		355	90		445	525
0400	Full lite, 6'-9" x 2'-6" wide		11	1.455		325	74		399	470
0420	2'-8" wide		10	1.600		330	81		411	490
0440	3'-0" wide		9	1.778		335	90		425	500
0500	7'-1" x 3'-0" wide		9	1.778		365	90		455	535
0700	Dutch door, pine, 1-3/4" x 2'-8" x 6'-8", 6 panel		12	1.333		695	67.50		762.50	870
0720	Half glass		10	1.600		985	81		1,066	1,200
0800	3'-0" wide, 6 panel		12	1.333		590	67.50		657.50	755
0820	Half glass		10	1.600		1,050	81		1,131	1,275
1000	Entrance door, colonial, 1-3/4" x 6'-8" x 2'-8" wide		16	1		570	50.50		620.50	700
1020	6 panel pine, 3'-0" wide		15	1.067		560	54		614	700
1100	8 panel pine, 2'-8" wide		16	1		685	50.50		735.50	825
1120	3'-0" wide	▼	15	1.067		635	54		689	785
1200	For tempered safety glass lites (min. of 2), add					85.50			85.50	94
1300	Flush, birch, solid core, 1-3/4" x 6'-8" x 2'-8" wide	2 Carp	16	1		142	50.50		192.50	234
1320	3'-0" wide		15	1.067		152	54		206	250
1350	7'-0" x 2'-8" wide		16	1		133	50.50		183.50	224
1360	3'-0" wide	▼	15	1.067		155	54		209	254
1380	For tempered safety glass lites, add				▼	114			114	126
1930	For dutch door with shelf, add					140%				
2700	Interior, closet, bi-fold, w/hardware, no frame or trim incl.									
2720	Flush, birch, 2'-6" x 6'-8"	2 Carp	13	1.231	Ea.	73	62.50		135.50	175
2740	3'-0" wide		13	1.231		74.50	62.50		137	177
2760	4'-0" wide		12	1.333		115	67.50		182.50	230
2780	5'-0" wide		11	1.455		114	74		188	238
2800	6'-0" wide		10	1.600		135	81		216	272
3000	Raised panel pine, 6'-6" or 6'-8" x 2'-6" wide		13	1.231		179	62.50		241.50	292
3020	3'-0" wide		13	1.231		289	62.50		351.50	415
3040	4'-0" wide		12	1.333		315	67.50		382.50	450
3060	5'-0" wide		11	1.455		405	74		479	555
3080	6'-0" wide		10	1.600		445	81		526	615
3200	Louvered, pine, 6'-6" or 6'-8" x 2'-6" wide		13	1.231		154	62.50		216.50	264
3220	3'-0" wide		13	1.231		239	62.50		301.50	360
3240	4'-0" wide		12	1.333		251	67.50		318.50	380
3260	5'-0" wide		11	1.455		278	74		352	415
3280	6'-0" wide	▼	10	1.600	▼	310	81		391	465
4400	Bi-passing closet, incl. hardware and frame, no trim incl.									
4420	Flush, lauan, 6'-8" x 4'-0" wide	2 Carp	12	1.333	Opng.	167	67.50		234.50	287
4440	5'-0" wide		11	1.455		183	74		257	315
4460	6'-0" wide		10	1.600		169	81		250	310
4600	Flush, birch, 6'-8" x 4'-0" wide		12	1.333		254	67.50		321.50	385
4620	5'-0" wide		11	1.455		223	74		297	355
4640	6'-0" wide		10	1.600		335	81		416	495
4800	Louvered, pine, 6'-8" x 4'-0" wide		12	1.333		485	67.50		552.50	635
4820	5'-0" wide		11	1.455		610	74		684	780
4840	6'-0" wide		10	1.600		720	81		801	915
5000	Paneled, pine, 6'-8" x 4'-0" wide		12	1.333		490	67.50		557.50	645
5020	5'-0" wide	▼	11	1.455		630	74		704	800

08 14 Wood Doors

08 14 33 – Stile and Rail Wood Doors

08 14 33.20 Wood Doors Residential	Crew	Daily Output	Labor-Hours	Unit	Material	2018 Bare Costs Labor	Equipment	Total	Total Incl O&P
5040	6'-0" wide	2 Carp	10	1.600	Opng.	840	81		921	1,050
5042	8'-0" wide	↓	12	1.333	↓	980	67.50		1,047.50	1,175
6100	Folding accordion, closet, including track and frame									
6120	Vinyl, 2 layer, stock	2 Carp	10	1.600	Ea.	68	81		149	199
6140	Woven mahogany and vinyl, stock		10	1.600		58.50	81		139.50	188
6160	Wood slats with vinyl overlay, stock		10	1.600		159	81		240	299
6180	Economy vinyl, stock		10	1.600		39	81		120	167
6200	Rigid PVC	↓	10	1.600	↓	108	81		189	243
7310	Passage doors, flush, no frame included									
7320	Hardboard, hollow core, 1-3/8" x 6'-8" x 1'-6" wide	2 Carp	18	.889	Ea.	41.50	45		86.50	115
7330	2'-0" wide		18	.889		45	45		90	118
7340	2'-6" wide		18	.889		52.50	45		97.50	126
7350	2'-8" wide		18	.889		53	45		98	127
7360	3'-0" wide		17	.941		55	47.50		102.50	133
7420	Lauan, hollow core, 1-3/8" x 6'-8" x 1'-6" wide		18	.889		53	45		98	127
7440	2'-0" wide		18	.889		60	45		105	135
7450	2'-4" wide		18	.889		68	45		113	144
7460	2'-6" wide		18	.889		68	45		113	144
7480	2'-8" wide		18	.889		70	45		115	146
7500	3'-0" wide		17	.941		75	47.50		122.50	155
7700	Birch, hollow core, 1-3/8" x 6'-8" x 1'-6" wide		18	.889		62	45		107	137
7720	2'-0" wide		18	.889		69	45		114	145
7740	2'-6" wide		18	.889		79	45		124	156
7760	2'-8" wide		18	.889		82	45		127	159
7780	3'-0" wide		17	.941		88	47.50		135.50	170
8000	Pine louvered, 1-3/8" x 6'-8" x 1'-6" wide		19	.842		148	42.50		190.50	228
8020	2'-0" wide		18	.889		164	45		209	249
8040	2'-6" wide		18	.889		199	45		244	288
8060	2'-8" wide		18	.889		209	45		254	299
8080	3'-0" wide		17	.941		219	47.50		266.50	315
8300	Pine paneled, 1-3/8" x 6'-8" x 1'-6" wide		19	.842		191	42.50		233.50	275
8320	2'-0" wide		18	.889		223	45		268	315
8330	2'-4" wide		18	.889		241	45		286	335
8340	2'-6" wide		18	.889		250	45		295	345
8360	2'-8" wide		18	.889		258	45		303	355
8380	3'-0" wide	↓	17	.941	↓	271	47.50		318.50	370

08 14 35 – Torrified Doors

08 14 35.10 Torrified Exterior Doors

		Crew	Daily Output	Labor-Hours	Unit	Material	Labor	Equipment	Total	Total Incl O&P
0010	**TORRIFIED EXTERIOR DOORS**									
0020	Wood doors made from torrified wood, exterior									
0030	All doors require a finish be applied, all glass is insulated									
0040	All doors require pilot holes for all fasteners									
0100	6 panel, paint grade poplar, 1-3/4" x 3'-0" x 6'-8"	2 Carp	12	1.333	Ea.	1,325	67.50		1,392.50	1,550
0120	Half glass 3'-0" x 6'-8"	"	12	1.333		1,425	67.50		1,492.50	1,650
0200	Side lite, full glass, 1-3/4" x 1'-2" x 6'-8"					930			930	1,025
0220	Side lite, half glass, 1-3/4" x 1'-2" x 6'-8"					910			910	1,000
0300	Raised face, 2 panel, paint grade poplar, 1-3/4" x 3'-0" x 7'-0"	2 Carp	12	1.333		1,325	67.50		1,392.50	1,550
0320	Side lite, raised face, half glass, 1-3/4" x 1'-2" x 7'-0"					1,050			1,050	1,175
0500	6 panel, Fir, 1-3/4" x 3'-0" x 6'-8"	2 Carp	12	1.333		1,875	67.50		1,942.50	2,150
0520	Half glass 3'-0" x 6'-8"	"	12	1.333		1,850	67.50		1,917.50	2,150
0600	Side lite, full glass, 1-3/4" x 1'-2" x 6'-8"					940			940	1,025
0620	Side lite, half glass, 1-3/4" x 1'-2" x 6'-8"			↓		970			970	1,075

08 14 Wood Doors

08 14 35 – Torrified Doors

08 14 35.10 Torrified Exterior Doors		Crew	Daily Output	Labor-Hours	Unit	Material	2018 Bare Costs Labor	Equipment	Total	Total Incl O&P
0700	6 panel, Mahogany, 1-3/4" x 3'-0" x 6'-8"	2 Carp	12	1.333	Ea.	1,975	67.50		2,042.50	2,275
0800	Side lite, full glass, 1-3/4" x 1'-2" x 6'-8"					1,050			1,050	1,150
0820	Side lite, half glass, 1-3/4" x 1'-2" x 6'-8"		↓		↓	1,050			1,050	1,150

08 14 40 – Interior Cafe Doors

08 14 40.10 Cafe Style Doors

		Crew	Daily Output	Labor-Hours	Unit	Material	Labor	Equipment	Total	Total Incl O&P
0010	**CAFE STYLE DOORS**									
6520	Interior cafe doors, 2'-6" opening, stock, panel pine	2 Carp	16	1	Ea.	400	50.50		450.50	515
6540	3'-0" opening	"	16	1	"	445	50.50		495.50	560
6550	Louvered pine									
6560	2'-6" opening	2 Carp	16	1	Ea.	320	50.50		370.50	425
8000	3'-0" opening		16	1		335	50.50		385.50	445
8010	2'-6" opening, hardwood		16	1		360	50.50		410.50	470
8020	3'-0" opening	↓	16	1	↓	395	50.50		445.50	510

08 16 Composite Doors

08 16 13 – Fiberglass Doors

08 16 13.10 Entrance Doors, Fibrous Glass

			Crew	Daily Output	Labor-Hours	Unit	Material	Labor	Equipment	Total	Total Incl O&P
0010	**ENTRANCE DOORS, FIBROUS GLASS**										
0020	Exterior, fiberglass, door, 2'-8" wide x 6'-8" high	G	2 Carp	15	1.067	Ea.	281	54		335	395
0040	3'-0" wide x 6'-8" high	G		15	1.067		274	54		328	385
0060	3'-0" wide x 7'-0" high	G		15	1.067		490	54		544	620
0080	3'-0" wide x 6'-8" high, with two lites	G		15	1.067		340	54		394	460
0100	3'-0" wide x 8'-0" high, with two lites	G		15	1.067		525	54		579	665
0110	Half glass, 3'-0" wide x 6'-8" high	G		15	1.067		450	54		504	580
0120	3'-0" wide x 6'-8" high, low E	G		15	1.067		490	54		544	620
0130	3'-0" wide x 8'-0" high	G		15	1.067		585	54		639	730
0140	3'-0" wide x 8'-0" high, low E	G	↓	15	1.067		675	54		729	830
0150	Side lights, 1'-0" wide x 6'-8" high	G					281			281	310
0160	1'-0" wide x 6'-8" high, low E	G					289			289	320
0180	1'-0" wide x 6'-8" high, full glass	G					340			340	370
0190	1'-0" wide x 6'-8" high, low E	G				↓	360			360	395

08 16 14 – French Doors

08 16 14.10 Exterior Doors With Glass Lites

		Crew	Daily Output	Labor-Hours	Unit	Material	Labor	Equipment	Total	Total Incl O&P
0010	**EXTERIOR DOORS WITH GLASS LITES**									
0020	French, Fir, 1-3/4", 3'-0" wide x 6'-8" high	2 Carp	12	1.333	Ea.	620	67.50		687.50	785
0025	Double		12	1.333		1,250	67.50		1,317.50	1,450
0030	Maple, 1-3/4", 3'-0" wide x 6'-8" high		12	1.333		695	67.50		762.50	870
0035	Double		12	1.333		1,400	67.50		1,467.50	1,625
0040	Cherry, 1-3/4", 3'-0" wide x 6'-8" high		12	1.333		810	67.50		877.50	995
0045	Double		12	1.333		1,625	67.50		1,692.50	1,875
0100	Mahogany, 1-3/4", 3'-0" wide x 8'-0" high		10	1.600		825	81		906	1,025
0105	Double		10	1.600		1,650	81		1,731	1,950
0110	Fir, 1-3/4", 3'-0" wide x 8'-0" high		10	1.600		1,225	81		1,306	1,475
0115	Double		10	1.600		2,475	81		2,556	2,850
0120	Oak, 1-3/4", 3'-0" wide x 8'-0" high		10	1.600		1,875	81		1,956	2,200
0125	Double	↓	10	1.600	↓	3,750	81		3,831	4,250

For customer support on your Building Construction Costs with RSMeans data, call 800.448.8182.

08 17 Integrated Door Opening Assemblies

08 17 13 – Integrated Metal Door Opening Assemblies

08 17 13.20 Stainless Steel Doors and Frames		Crew	Daily Output	Labor-Hours	Unit	Material	2018 Bare Costs Labor	Equipment	Total	Total Incl O&P
0010	**STAINLESS STEEL DOORS AND FRAMES**									
0500	Stainless steel, prehung door, foam core, 14 ga., 3'-0" x 7'-0"	G	2 Carp	5	3.200	Ea.	3,150	162		3,312	3,725
0600	Stainless steel, prehung double door, foam core, 14 ga., 3'-0" x 7'-0"	G	"	4	4	"	6,200	203		6,403	7,125

08 17 23 – Integrated Wood Door Opening Assemblies

08 17 23.10 Pre-Hung Doors

		Crew	Daily Output	Labor-Hours	Unit	Material	2018 Bare Costs Labor	Equipment	Total	Total Incl O&P
0010	**PRE-HUNG DOORS**									
0300	Exterior, wood, comb. storm & screen, 6'-9" x 2'-6" wide	2 Carp	15	1.067	Ea.	248	54		302	355
0320	2'-8" wide		15	1.067		330	54		384	450
0340	3'-0" wide		15	1.067		310	54		364	425
0360	For 7'-0" high door, add					40			40	44
1600	Entrance door, flush, birch, solid core									
1620	4-5/8" solid jamb, 1-3/4" x 6'-8" x 2'-8" wide	2 Carp	16	1	Ea.	300	50.50		350.50	405
1640	3'-0" wide		16	1		405	50.50		455.50	520
1642	5-5/8" jamb		16	1		345	50.50		395.50	455
1680	For 7'-0" high door, add					25.50			25.50	28
2000	Entrance door, colonial, 6 panel pine									
2020	4-5/8" solid jamb, 1-3/4" x 6'-8" x 2'-8" wide	2 Carp	16	1	Ea.	660	50.50		710.50	800
2040	3'-0" wide	"	16	1		690	50.50		740.50	835
2060	For 7'-0" high door, add					56.50			56.50	62
2200	For 5-5/8" solid jamb, add					44.50			44.50	49
4000	Interior, passage door, 4-5/8" solid jamb									
4400	Lauan, flush, solid core, 1-3/8" x 6'-8" x 2'-6" wide	2 Carp	17	.941	Ea.	194	47.50		241.50	286
4420	2'-8" wide		17	.941		194	47.50		241.50	286
4440	3'-0" wide		16	1		211	50.50		261.50	310
4600	Hollow core, 1-3/8" x 6'-8" x 2'-6" wide		17	.941		136	47.50		183.50	223
4620	2'-8" wide		17	.941		139	47.50		186.50	225
4640	3'-0" wide		16	1		142	50.50		192.50	234
4700	For 7'-0" high door, add					41			41	45.50
5000	Birch, flush, solid core, 1-3/8" x 6'-8" x 2'-6" wide	2 Carp	17	.941		294	47.50		341.50	400
5020	2'-8" wide		17	.941		209	47.50		256.50	305
5040	3'-0" wide		16	1		320	50.50		370.50	430
5200	Hollow core, 1-3/8" x 6'-8" x 2'-6" wide		17	.941		242	47.50		289.50	340
5220	2'-8" wide		17	.941		273	47.50		320.50	375
5240	3'-0" wide		16	1		264	50.50		314.50	370
5280	For 7'-0" high door, add					34.50			34.50	38
5500	Hardboard paneled, 1-3/8" x 6'-8" x 2'-6" wide	2 Carp	17	.941		152	47.50		199.50	240
5520	2'-8" wide		17	.941		163	47.50		210.50	252
5540	3'-0" wide		16	1		161	50.50		211.50	255
6000	Pine paneled, 1-3/8" x 6'-8" x 2'-6" wide		17	.941		276	47.50		323.50	380
6020	2'-8" wide		17	.941		293	47.50		340.50	400
6040	3'-0" wide		16	1		300	50.50		350.50	405

08 31 13 – Access Doors and Frames

08 31 13.10 Types of Framed Access Doors

08 31 13.10 Types of Framed Access Doors	Crew	Daily Output	Labor-Hours	Unit	Material	2018 Bare Costs Labor	2018 Bare Costs Equipment	Total	Total Incl O&P
0010 **TYPES OF FRAMED ACCESS DOORS**									
1000 Fire rated door with lock									
1100 Metal, 12" x 12"	1 Carp	10	.800	Ea.	162	40.50		202.50	240
1150 18" x 18"		9	.889		203	45		248	292
1200 24" x 24"		9	.889		350	45		395	450
1250 24" x 36"		8	1		350	50.50		400.50	460
1300 24" x 48"		8	1		400	50.50		450.50	510
1350 36" x 36"		7.50	1.067		520	54		574	655
1400 48" x 48"		7.50	1.067		670	54		724	825
1600 Stainless steel, 12" x 12"		10	.800		345	40.50		385.50	435
1650 18" x 18"		9	.889		375	45		420	485
1700 24" x 24"		9	.889		550	45		595	680
1750 24" x 36"	↓	8	1	↓	705	50.50		755.50	850
2000 Flush door for finishing									
2100 Metal 8" x 8"	1 Carp	10	.800	Ea.	36	40.50		76.50	102
2150 12" x 12"	"	10	.800	"	37.50	40.50		78	103
3000 Recessed door for acoustic tile									
3100 Metal, 12" x 12"	1 Carp	4.50	1.778	Ea.	52	90		142	194
3150 12" x 24"		4.50	1.778		107	90		197	255
3200 24" x 24"		4	2		99	101		200	263
3250 24" x 36"	↓	4	2	↓	164	101		265	335
4000 Recessed door for drywall									
4100 Metal 12" x 12"	1 Carp	6	1.333	Ea.	74.50	67.50		142	185
4150 12" x 24"		5.50	1.455		119	74		193	243
4200 24" x 36"	↓	5	1.600	↓	168	81		249	310
6000 Standard door									
6100 Metal, 8" x 8"	1 Carp	10	.800	Ea.	28.50	40.50		69	93.50
6150 12" x 12"		10	.800		31	40.50		71.50	96
6200 18" x 18"		9	.889		37	45		82	109
6250 24" x 24"		9	.889		51	45		96	125
6300 24" x 36"		8	1		88	50.50		138.50	174
6350 36" x 36"		8	1		100	50.50		150.50	187
6500 Stainless steel, 8" x 8"		10	.800		80	40.50		120.50	150
6550 12" x 12"		10	.800		99	40.50		139.50	171
6600 18" x 18"		9	.889		190	45		235	277
6650 24" x 24"	↓	9	.889	↓	281	45		326	380

08 31 13.20 Bulkhead/Cellar Doors

	Crew	Daily Output	Labor-Hours	Unit	Material	Labor	Equipment	Total	Total Incl O&P
0010 **BULKHEAD/CELLAR DOORS**									
0020 Steel, not incl. sides, 44" x 62"	1 Carp	5.50	1.455	Ea.	655	74		729	830
0100 52" x 73"		5.10	1.569		825	79.50		904.50	1,025
0500 With sides and foundation plates, 57" x 45" x 24"		4.70	1.702		870	86.50		956.50	1,075
0600 42" x 49" x 51"	↓	4.30	1.860	↓	595	94.50		689.50	800

08 31 13.30 Commercial Floor Doors

	Crew	Daily Output	Labor-Hours	Unit	Material	Labor	Equipment	Total	Total Incl O&P
0010 **COMMERCIAL FLOOR DOORS**									
0020 Aluminum tile, steel frame, one leaf, 2' x 2' opng.	2 Sswk	3.50	4.571	Opng.	630	250		880	1,100
0050 3'-6" x 3'-6" opening		3.50	4.571		1,150	250		1,400	1,650
0500 Double leaf, 4' x 4' opening		3	5.333		1,625	291		1,916	2,275
0550 5' x 5' opening	↓	3	5.333	↓	2,875	291		3,166	3,625

08 31 13.35 Industrial Floor Doors

	Crew	Daily Output	Labor-Hours	Unit	Material	Labor	Equipment	Total	Total Incl O&P
0010 **INDUSTRIAL FLOOR DOORS**									
0020 Steel 300 psf L.L., single leaf, 2' x 2', 175#	2 Sswk	6	2.667	Opng.	750	146		896	1,075
0050 3' x 3' opening, 300#	↓	5.50	2.909	↓	1,100	159		1,259	1,450

08 31 Access Doors and Panels

08 31 13 – Access Doors and Frames

08 31 13.35 Industrial Floor Doors

		Crew	Daily Output	Labor-Hours	Unit	Material	2018 Bare Costs Labor	Equipment	Total	Total Incl O&P
0300	Double leaf, 4' x 4' opening, 455#	2 Sswk	5	3.200	Opng.	2,500	175		2,675	3,025
0350	5' x 5' opening, 645#		4.50	3.556		2,875	194		3,069	3,475
1000	Aluminum, 300 psf L.L., single leaf, 2' x 2', 60#		6	2.667		750	146		896	1,075
1050	3' x 3' opening, 100#		5.50	2.909		935	159		1,094	1,275
1500	Double leaf, 4' x 4' opening, 160#		5	3.200		2,150	175		2,325	2,625
1550	5' x 5' opening, 235#		4.50	3.556		2,875	194		3,069	3,475
2000	Aluminum, 150 psf L.L., single leaf, 2' x 2', 60#		6	2.667		710	146		856	1,025
2050	3' x 3' opening, 95#		5.50	2.909		935	159		1,094	1,275
2500	Double leaf, 4' x 4' opening, 150#		5	3.200		1,475	175		1,650	1,900
2550	5' x 5' opening, 230#	↓	4.50	3.556	↓	1,975	194		2,169	2,500

08 31 13.40 Kennel Doors

		Crew	Daily Output	Labor-Hours	Unit	Material	2018 Bare Costs Labor	Equipment	Total	Total Incl O&P
0010	**KENNEL DOORS**									
0020	2 way, swinging type, 13" x 19" opening	2 Carp	11	1.455	Opng.	90	74		164	211
0100	17" x 29" opening		11	1.455		130	74		204	255
0200	9" x 9" opening, electronic with accessories	↓	11	1.455	↓	149	74		223	276

08 32 Sliding Glass Doors

08 32 13 – Sliding Aluminum-Framed Glass Doors

08 32 13.10 Sliding Aluminum Doors

		Crew	Daily Output	Labor-Hours	Unit	Material	2018 Bare Costs Labor	Equipment	Total	Total Incl O&P
0010	**SLIDING ALUMINUM DOORS**									
0350	Aluminum, 5/8" tempered insulated glass, 6' wide									
0400	Premium	2 Carp	4	4	Ea.	1,550	203		1,753	2,000
0450	Economy		4	4		820	203		1,023	1,225
0500	8' wide, premium		3	5.333		1,700	270		1,970	2,275
0550	Economy		3	5.333		1,475	270		1,745	2,025
0600	12' wide, premium		2.50	6.400		3,000	325		3,325	3,800
0650	Economy		2.50	6.400		1,575	325		1,900	2,225
4000	Aluminum, baked on enamel, temp glass, 6'-8" x 10'-0" wide		4	4		1,100	203		1,303	1,500
4020	Insulating glass, 6'-8" x 6'-0" wide		4	4		955	203		1,158	1,350
4040	8'-0" wide		3	5.333		1,125	270		1,395	1,625
4060	10'-0" wide		2	8		1,400	405		1,805	2,175
4080	Anodized, temp glass, 6'-8" x 6'-0" wide		4	4		455	203		658	810
4100	8'-0" wide		3	5.333		575	270		845	1,050
4120	10'-0" wide	↓	2	8	↓	660	405		1,065	1,350
5000	Aluminum sliding glass door system									
5010	Sliding door 4' wide opening single side	2 Carp	2	8	Ea.	6,000	405		6,405	7,225
5015	8' wide opening single side		2	8		9,000	405		9,405	10,500
5020	Telescoping glass door system, 4' wide opening biparting		2	8		5,100	405		5,505	6,225
5025	8' wide opening biparting		2	8		6,000	405		6,405	7,225
5030	Folding glass door, 4' wide opening biparting		2	8		8,000	405		8,405	9,425
5035	8' wide opening biparting		2	8		10,000	405		10,405	11,600
5040	ICU-CCU sliding telescoping glass door, 4' x 7', single side opening		2	8		3,050	405		3,455	3,975
5045	8' x 7', single side opening	↓	2	8	↓	4,625	405		5,030	5,725

08 32 19 – Sliding Wood-Framed Glass Doors

08 32 19.15 Sliding Glass Vinyl-Clad Wood Doors

			Crew	Daily Output	Labor-Hours	Unit	Material	2018 Bare Costs Labor	Equipment	Total	Total Incl O&P
0010	**SLIDING GLASS VINYL-CLAD WOOD DOORS**										
0020	Glass, sliding vinyl-clad, insul. glass, 6'-0" x 6'-8"	G	2 Carp	4	4	Opng.	1,525	203		1,728	1,975
0025	6'-0" x 6'-10" high	G		4	4		1,700	203		1,903	2,150
0030	6'-0" x 8'-0" high	G		4	4		2,050	203		2,253	2,550
0050	5'-0" x 6'-8" high	G	↓	4	4	↓	1,575	203		1,778	2,050

08 32 Sliding Glass Doors

08 32 19 – Sliding Wood-Framed Glass Doors

08 32 19.15 Sliding Glass Vinyl-Clad Wood Doors		Crew	Daily Output	Labor-Hours	Unit	Material	2018 Bare Costs Labor	Equipment	Total	Total Incl O&P
0100	8'-0" x 6'-10" high	G 2 Carp	4	4	Opng.	2,025	203		2,228	2,525
0500	4 leaf, 9'-0" x 6'-10" high	G	3	5.333		3,350	270		3,620	4,100
0600	12'-0" x 6'-10" high	G	3	5.333		4,025	270		4,295	4,825

08 33 Coiling Doors and Grilles

08 33 13 – Coiling Counter Doors

08 33 13.10 Counter Doors, Coiling Type

		Crew	Daily Output	Labor-Hours	Unit	Material	2018 Bare Costs Labor	Equipment	Total	Total Incl O&P
0010	**COUNTER DOORS, COILING TYPE**									
0020	Manual, incl. frame and hardware, galv. stl., 4' roll-up, 6' long	2 Carp	2	8	Opng.	1,300	405		1,705	2,050
0300	Galvanized steel, UL label		1.80	8.889		1,325	450		1,775	2,150
0600	Stainless steel, 4' high roll-up, 6' long		2	8		2,225	405		2,630	3,050
0700	10' long		1.80	8.889		2,650	450		3,100	3,600
2000	Aluminum, 4' high, 4' long		2.20	7.273		1,675	370		2,045	2,375
2020	6' long		2	8		2,000	405		2,405	2,825
2040	8' long		1.90	8.421		2,275	425		2,700	3,150
2060	10' long		1.80	8.889		2,275	450		2,725	3,200
2080	14' long		1.40	11.429		2,925	580		3,505	4,100
2100	6' high, 4' long		2	8		2,025	405		2,430	2,850
2120	6' long		1.60	10		1,800	505		2,305	2,750
2140	10' long		1.40	11.429		2,375	580		2,955	3,475

08 33 16 – Coiling Counter Grilles

08 33 16.10 Coiling Grilles

		Crew	Daily Output	Labor-Hours	Unit	Material	2018 Bare Costs Labor	Equipment	Total	Total Incl O&P
0010	**COILING GRILLES**									
0015	Aluminum, manual, incl. frame, mill finish									
0020	Top coiling, 4' high, 4' long	2 Sswk	3.20	5	Opng.	1,525	273		1,798	2,125
0030	6' long		3.20	5		1,775	273		2,048	2,400
0040	8' long		2.40	6.667		1,975	365		2,340	2,750
0050	12' long		2.40	6.667		2,375	365		2,740	3,225
0060	16' long		1.60	10		2,550	545		3,095	3,725
0070	6' high, 4' long		3.20	5		1,775	273		2,048	2,400
0080	6' long		3.20	5		1,725	273		1,998	2,350
0090	8' long		2.40	6.667		1,750	365		2,115	2,525
0100	12' long		1.60	10		2,275	545		2,820	3,425
0110	16' long		1.20	13.333		2,725	730		3,455	4,200
0200	Side coiling, 8' high, 12' long		.60	26.667		2,225	1,450		3,675	4,800
0220	18' long		.50	32		4,750	1,750		6,500	8,075
0240	24' long		.40	40		3,575	2,175		5,750	7,500
0260	12' high, 12' long		.50	32		3,225	1,750		4,975	6,375
0280	18' long		.40	40		5,025	2,175		7,200	9,100
0300	24' long		.28	57.143		6,350	3,125		9,475	12,100

08 33 23 – Overhead Coiling Doors

08 33 23.10 Coiling Service Doors

		Crew	Daily Output	Labor-Hours	Unit	Material	2018 Bare Costs Labor	Equipment	Total	Total Incl O&P
0010	**COILING SERVICE DOORS** Steel, manual, 20 ga., incl. hardware									
0050	8' x 8' high	2 Sswk	1.60	10	Ea.	725	545		1,270	1,700
0100	10' x 10' high		1.40	11.429		1,075	625		1,700	2,200
0200	20' x 10' high		1	16		3,000	875		3,875	4,700
0300	12' x 12' high		1.20	13.333		1,200	730		1,930	2,525
0400	20' x 12' high		.90	17.778		2,100	970		3,070	3,900
0500	14' x 14' high		.80	20		3,000	1,100		4,100	5,075
0600	20' x 16' high		.60	26.667		3,650	1,450		5,100	6,400

For customer support on your Building Construction Costs with RSMeans data, call 800.448.8182.

08 33 Coiling Doors and Grilles

08 33 23 – Overhead Coiling Doors

08 33 23.10 Coiling Service Doors

08 33 23.10 Coiling Service Doors		Crew	Daily Output	Labor-Hours	Unit	Material	2018 Bare Costs Labor	Equipment	Total	Total Incl O&P
0700	10' x 20' high	2 Sswk	.50	32	Ea.	2,575	1,750		4,325	5,700
1000	12' x 12', crank operated, crank on door side		.80	20		1,750	1,100		2,850	3,700
1100	Crank thru wall	↓	.70	22.857		2,025	1,250		3,275	4,275
1300	For vision panel, add					355			355	390
1600	3' x 7' pass door within rolling steel door, new construction					1,650			1,650	1,825
1700	Existing construction	2 Sswk	2	8		2,000	435		2,435	2,925
2000	Class A fire doors, manual, 20 ga., 8' x 8' high		1.40	11.429		1,575	625		2,200	2,750
2100	10' x 10' high		1.10	14.545		2,150	795		2,945	3,650
2200	20' x 10' high		.80	20		4,400	1,100		5,500	6,600
2300	12' x 12' high		1	16		3,375	875		4,250	5,125
2400	20' x 12' high		.80	20		4,750	1,100		5,850	7,000
2500	14' x 14' high		.60	26.667		3,700	1,450		5,150	6,450
2600	20' x 16' high		.50	32		5,775	1,750		7,525	9,200
2700	10' x 20' high	↓	.40	40	↓	4,575	2,175		6,750	8,600
3000	For 18 ga. doors, add				S.F.	1.10			1.10	1.21
3300	For enamel finish, add				"	2.60			2.60	2.86
3600	For safety edge bottom bar, pneumatic, add				L.F.	22			22	24.50
3700	Electric, add					42			42	46
4000	For weatherstripping, extruded rubber, jambs, add					18			18	19.80
4100	Hood, add					9			9	9.90
4200	Sill, add				↓	9			9	9.90
4500	Motor operators, to 14' x 14' opening	2 Sswk	5	3.200	Ea.	1,175	175		1,350	1,575
4600	Over 14' x 14', jack shaft type	"	5	3.200		1,275	175		1,450	1,675
4700	For fire door, additional fusible link, add				↓	75			75	82.50

08 34 Special Function Doors

08 34 13 – Cold Storage Doors

08 34 13.10 Doors for Cold Area Storage

08 34 13.10 Doors for Cold Area Storage		Crew	Daily Output	Labor-Hours	Unit	Material	2018 Bare Costs Labor	Equipment	Total	Total Incl O&P
0010	**DOORS FOR COLD AREA STORAGE**									
0020	Single, 20 ga. galvanized steel									
0300	Horizontal sliding, 5' x 7', manual operation, 3.5" thick	2 Carp	2	8	Ea.	3,050	405		3,455	3,975
0400	4" thick		2	8		3,350	405		3,755	4,325
0500	6" thick		2	8		3,425	405		3,830	4,400
0800	5' x 7', power operation, 2" thick		1.90	8.421		5,600	425		6,025	6,800
0900	4" thick		1.90	8.421		5,700	425		6,125	6,925
1000	6" thick		1.90	8.421		6,475	425		6,900	7,775
1300	9' x 10', manual operation, 2" insulation		1.70	9.412		4,450	475		4,925	5,625
1400	4" insulation		1.70	9.412		4,675	475		5,150	5,875
1500	6" insulation		1.70	9.412		5,600	475		6,075	6,875
1800	Power operation, 2" insulation		1.60	10		7,700	505		8,205	9,250
1900	4" insulation		1.60	10		7,850	505		8,355	9,425
2000	6" insulation	↓	1.70	9.412	↓	8,250	475		8,725	9,800
2300	For stainless steel face, add					25%				
3000	Hinged, lightweight, 3' x 7'-0", 2" thick	2 Carp	2	8	Ea.	1,325	405		1,730	2,075
3050	4" thick		1.90	8.421		1,650	425		2,075	2,475
3300	Polymer doors, 3' x 7'-0"		1.90	8.421		1,350	425		1,775	2,125
3350	6" thick		1.40	11.429		2,350	580		2,930	3,450
3600	Stainless steel, 3' x 7'-0", 4" thick		1.90	8.421		1,725	425		2,150	2,550
3650	6" thick		1.40	11.429		2,750	580		3,330	3,900
3900	Painted, 3' x 7'-0", 4" thick		1.90	8.421		1,325	425		1,750	2,100
3950	6" thick	↓	1.40	11.429	↓	2,350	580		2,930	3,450

08 34 Special Function Doors

08 34 13 – Cold Storage Doors

08 34 13.10 Doors for Cold Area Storage

		Crew	Daily Output	Labor-Hours	Unit	Material	2018 Bare Costs Labor	Equipment	Total	Total Incl O&P
5000	Bi-parting, electric operated									
5010	6' x 8' opening, galv. faces, 4" thick for cooler	2 Carp	.80	20	Opng.	7,225	1,025		8,250	9,500
5050	For freezer, 4" thick		.80	20		7,950	1,025		8,975	10,300
5300	For door buck framing and door protection, add		2.50	6.400		670	325		995	1,225
6000	Galvanized batten door, galvanized hinges, 4' x 7'		2	8		1,825	405		2,230	2,650
6050	6' x 8'		1.80	8.889		2,475	450		2,925	3,400
6500	Fire door, 3 hr., 6' x 8', single slide		.80	20		8,250	1,025		9,275	10,600
6550	Double, bi-parting		.70	22.857		11,400	1,150		12,550	14,300

08 34 36 – Darkroom Doors

08 34 36.10 Various Types of Darkroom Doors

		Crew	Daily Output	Labor-Hours	Unit	Material	2018 Bare Costs Labor	Equipment	Total	Total Incl O&P
0010	**VARIOUS TYPES OF DARKROOM DOORS**									
0015	Revolving, standard, 2 way, 36" diameter	2 Carp	3.10	5.161	Opng.	3,025	262		3,287	3,725
0020	41" diameter		3.10	5.161		3,150	262		3,412	3,875
0050	3 way, 51" diameter		1.40	11.429		4,000	580		4,580	5,275
1000	4 way, 49" diameter		1.40	11.429		4,125	580		4,705	5,425
2000	Hinged safety, 2 way, 41" diameter		2.30	6.957		3,925	355		4,280	4,825
2500	3 way, 51" diameter		1.40	11.429		4,225	580		4,805	5,525
3000	Pop out safety, 2 way, 41" diameter		3.10	5.161		4,225	262		4,487	5,050
4000	3 way, 51" diameter		1.40	11.429		5,200	580		5,780	6,575
5000	Wheelchair-type, pop out, 51" diameter		1.40	11.429		5,800	580		6,380	7,275
5020	72" diameter		.90	17.778		9,400	900		10,300	11,700
9300	For complete darkrooms, see Section 13 21 53.50									

08 34 53 – Security Doors and Frames

08 34 53.20 Steel Door

		Crew	Daily Output	Labor-Hours	Unit	Material	2018 Bare Costs Labor	Equipment	Total	Total Incl O&P
0010	**STEEL DOOR** flush with ballistic core and welded frame both 14 ga.									
0120	UL 752 Level 3, 1-3/4", 3'-0" x 7'-0"	2 Carp	1.50	10.667	Opng.	2,600	540		3,140	3,675
0125	1-3/4", 3'-6" x 7'-0"		1.50	10.667		2,850	540		3,390	3,950
0130	1-3/4", 4'-0" x 7'-0"		1.20	13.333		3,050	675		3,725	4,375
0150	UL 752 Level 8, 1-3/4", 3'-0" x 7'-0"		1.50	10.667		13,500	540		14,040	15,600
0155	1-3/4", 3'-6" x 7'-0"		1.50	10.667		14,700	540		15,240	17,000
0160	1-3/4", 4'-0" x 7'-0"		1.20	13.333		14,700	675		15,375	17,200
1000	Safe room sliding door and hardware, 1-3/4", 3'-0" x 7'-0" UL 752 Level 3		.50	32		23,700	1,625		25,325	28,600
1050	Safe room swinging door and hardware, 1-3/4", 3'-0" x 7'-0" UL 752 Level 3		.50	32		28,400	1,625		30,025	33,800

08 34 53.30 Wood Ballistic Doors

		Crew	Daily Output	Labor-Hours	Unit	Material	2018 Bare Costs Labor	Equipment	Total	Total Incl O&P
0010	**WOOD BALLISTIC DOORS** with frames and hardware									
0050	Wood, 1-3/4", 3'-0" x 7'-0" UL 752 Level 3	2 Carp	1.50	10.667	Opng.	2,575	540		3,115	3,675

08 34 56 – Security Gates

08 34 56.10 Gates

		Crew	Daily Output	Labor-Hours	Unit	Material	2018 Bare Costs Labor	Equipment	Total	Total Incl O&P
0010	**GATES**									
0015	Driveway gates include mounting hardware									
0500	Wood, security gate, driveway, dual, 10' wide	H-4	.80	25	Opng.	3,775	1,175		4,950	5,950
0505	12' wide		.80	25		3,650	1,175		4,825	5,800
0510	15' wide		.80	25		4,325	1,175		5,500	6,575
0600	Steel, security gate, driveway, single, 10' wide		.80	25		1,700	1,175		2,875	3,675
0605	12' wide		.80	25		1,900	1,175		3,075	3,900
0620	Steel, security gate, driveway, dual, 12' wide		.80	25		2,050	1,175		3,225	4,075
0625	14' wide		.80	25		2,250	1,175		3,425	4,275
0630	16' wide		.80	25		2,450	1,175		3,625	4,500
0700	Aluminum, security gate, driveway, dual, 10' wide		.80	25		3,200	1,175		4,375	5,325
0705	12' wide		.80	25		3,850	1,175		5,025	6,025
0710	16' wide		.80	25		4,375	1,175		5,550	6,625

08 34 Special Function Doors

08 34 56 – Security Gates

08 34 56.10 Gates

		Crew	Daily Output	Labor-Hours	Unit	Material	2018 Bare Costs Labor	Equipment	Total	Total Incl O&P
1000	Security gate, driveway, opener 12 VDC				Ea.	830			830	910
1010	Wireless					1,900			1,900	2,075
1020	Security gate, driveway, opener 24 VDC					1,425			1,425	1,550
1030	Wireless					1,575			1,575	1,750
1040	Security gate, driveway, opener 12 VDC, solar panel 10 watt	1 Elec	2	4		445	233		678	840
1050	20 watt	"	2	4		585	233		818	995

08 34 59 – Vault Doors and Day Gates

08 34 59.10 Secure Storage Doors

		Crew	Daily Output	Labor-Hours	Unit	Material	2018 Bare Costs Labor	Equipment	Total	Total Incl O&P
0010	**SECURE STORAGE DOORS**									
0020	Door and frame, 32" x 78", clear opening									
0100	1 hour test, 32" door, weighs 750 lb.	2 Sswk	1.50	10.667	Opng.	6,400	585		6,985	8,000
0200	2 hour test, 32" door, weighs 950 lb.		1.30	12.308		7,925	675		8,600	9,800
0250	40" door, weighs 1130 lb.		1	16		8,950	875		9,825	11,300
0300	4 hour test, 32" door, weighs 1025 lb.		1.20	13.333		9,900	730		10,630	12,100
0350	40" door, weighs 1140 lb.		.90	17.778		11,100	970		12,070	13,800
0600	For time lock, two movement, add	1 Elec	2	4	Ea.	1,850	233		2,083	2,400
0800	Day gate, painted, steel, 32" wide	2 Sswk	1.50	10.667		2,025	585		2,610	3,175
0850	40" wide		1.40	11.429		1,925	625		2,550	3,150
0900	Aluminum, 32" wide		1.50	10.667		2,950	585		3,535	4,200
0950	40" wide		1.40	11.429		3,125	625		3,750	4,475
2050	Security vault door, class I, 3' wide, 3-1/2" thick	E-24	.19	167	Opng.	14,500	9,075	3,075	26,650	33,800
2100	Class II, 3' wide, 7" thick		.19	167		17,100	9,075	3,075	29,250	36,700
2150	Class III, 9R, 3' wide, 10" thick		.13	250		21,600	13,600	4,625	39,825	50,500
2160	Class V, type 1, 40" door		2.48	12.903	Ea.	6,550	700	238	7,488	8,575
2170	Class V, type 2, 40" door		2.48	12.903		6,300	700	238	7,238	8,300
2180	Day gate for class V vault	2 Sswk	2	8		1,725	435		2,160	2,625

08 34 73 – Sound Control Door Assemblies

08 34 73.10 Acoustical Doors

		Crew	Daily Output	Labor-Hours	Unit	Material	2018 Bare Costs Labor	Equipment	Total	Total Incl O&P
0010	**ACOUSTICAL DOORS**									
0020	Including framed seals, 3' x 7', wood, 40 STC rating	2 Carp	1.50	10.667	Ea.	3,350	540		3,890	4,500
0100	Steel, 41 STC rating		1.50	10.667		3,600	540		4,140	4,775
0200	45 STC rating		1.50	10.667		3,975	540		4,515	5,200
0300	48 STC rating		1.50	10.667		4,650	540		5,190	5,950
0400	52 STC rating		1.50	10.667		5,200	540		5,740	6,525

08 36 Panel Doors

08 36 13 – Sectional Doors

08 36 13.10 Overhead Commercial Doors

		Crew	Daily Output	Labor-Hours	Unit	Material	2018 Bare Costs Labor	Equipment	Total	Total Incl O&P
0010	**OVERHEAD COMMERCIAL DOORS**									
1000	Stock, sectional, heavy duty, wood, 1-3/4" thick, 8' x 8' high	2 Carp	2	8	Ea.	1,225	405		1,630	1,975
1100	10' x 10' high		1.80	8.889		1,700	450		2,150	2,550
1200	12' x 12' high		1.50	10.667		2,350	540		2,890	3,400
1300	Chain hoist, 14' x 14' high		1.30	12.308		3,650	625		4,275	4,950
1400	12' x 16' high		1	16		3,600	810		4,410	5,200
1500	20' x 8' high		1.30	12.270		2,925	620		3,545	4,175
1600	20' x 16' high		.65	24.615		5,975	1,250		7,225	8,475
1800	Center mullion openings, 8' high		4	4		1,300	203		1,503	1,725
1900	20' high		2	8		2,150	405		2,555	2,975
2100	For medium duty custom door, deduct					5%	5%			
2150	For medium duty stock doors, deduct					10%	5%			

For customer support on your Building Construction Costs with RSMeans data, call 800.448.8182.

283

08 36 13 – Sectional Doors

08 36 13.10 Overhead Commercial Doors		Crew	Daily Output	Labor-Hours	Unit	Material	2018 Bare Costs Labor	Equipment	Total	Total Incl O&P
2300	Fiberglass and aluminum, heavy duty, sectional, 12' x 12' high	2 Carp	1.50	10.667	Ea.	3,350	540		3,890	4,500
2450	Chain hoist, 20' x 20' high		.50	32		7,450	1,625		9,075	10,700
2600	Steel, 24 ga. sectional, manual, 8' x 8' high		2	8		1,000	405		1,405	1,725
2650	10' x 10' high		1.80	8.889		1,300	450		1,750	2,100
2700	12' x 12' high		1.50	10.667		1,525	540		2,065	2,500
2800	Chain hoist, 20' x 14' high		.70	22.857		3,600	1,150		4,750	5,725
2850	For 1-1/4" rigid insulation and 26 ga. galv.									
2860	back panel, add				S.F.	5			5	5.50
2900	For electric trolley operator, 1/3 HP, to 12' x 12', add	1 Carp	2	4	Ea.	1,100	203		1,303	1,525
2950	Over 12' x 12', 1/2 HP, add		1	8		1,200	405		1,605	1,950
2980	Overhead, for row of clear lites, add		1	8		164	405		569	800

08 36 13.20 Residential Garage Doors

		Crew	Daily Output	Labor-Hours	Unit	Material	Labor	Equipment	Total	Total Incl O&P
0010	**RESIDENTIAL GARAGE DOORS**									
0050	Hinged, wood, custom, double door, 9' x 7'	2 Carp	4	4	Ea.	875	203		1,078	1,275
0070	16' x 7'		3	5.333		1,300	270		1,570	1,825
0200	Overhead, sectional, incl. hardware, fiberglass, 9' x 7', standard		5	3.200		1,025	162		1,187	1,375
0220	Deluxe		5	3.200		1,200	162		1,362	1,575
0300	16' x 7', standard		6	2.667		1,625	135		1,760	2,000
0320	Deluxe		6	2.667		2,250	135		2,385	2,675
0500	Hardboard, 9' x 7', standard		8	2		715	101		816	940
0520	Deluxe		8	2		860	101		961	1,100
0600	16' x 7', standard		6	2.667		1,300	135		1,435	1,625
0620	Deluxe		6	2.667		1,500	135		1,635	1,850
0700	Metal, 9' x 7', standard		8	2		885	101		986	1,125
0720	Deluxe		6	2.667		995	135		1,130	1,300
0800	16' x 7', standard		6	2.667		1,075	135		1,210	1,375
0820	Deluxe		5	3.200		1,450	162		1,612	1,850
0900	Wood, 9' x 7', standard		8	2		1,025	101		1,126	1,275
0920	Deluxe		8	2		2,250	101		2,351	2,625
1000	16' x 7', standard		6	2.667		1,675	135		1,810	2,050
1020	Deluxe		6	2.667		3,125	135		3,260	3,650
1800	Door hardware, sectional	1 Carp	4	2		360	101		461	555
1810	Door tracks only		4	2		170	101		271	340
1820	One side only		7	1.143		128	58		186	228
4000	For electric operator, economy, add		8	1		440	50.50		490.50	560
4100	Deluxe, including remote control		8	1		635	50.50		685.50	770
4500	For transmitter/receiver control, add to operator				Total	113			113	124
4600	Transmitters, additional				"	63.50			63.50	70

08 36 19 – Multi-Leaf Vertical Lift Doors

08 36 19.10 Sectional Vertical Lift Doors

		Crew	Daily Output	Labor-Hours	Unit	Material	Labor	Equipment	Total	Total Incl O&P
0010	**SECTIONAL VERTICAL LIFT DOORS**									
0020	Motorized, 14 ga. steel, incl. frame and control panel									
0050	16' x 16' high	L-10	.50	48	Ea.	21,800	2,675	990	25,465	29,400
0100	10' x 20' high		1.30	18.462		36,600	1,025	380	38,005	42,400
0120	15' x 20' high		1.30	18.462		44,800	1,025	380	46,205	51,500
0140	20' x 20' high		1	24		52,500	1,350	495	54,345	60,000
0160	25' x 20' high		1	24		59,000	1,350	495	60,845	67,000
0170	32' x 24' high		.75	32		50,000	1,775	660	52,435	58,500
0180	20' x 25' high		1	24		60,500	1,350	495	62,345	69,000
0200	25' x 25' high		.70	34.286		73,500	1,925	710	76,135	85,000
0220	25' x 30' high		.70	34.286		79,500	1,925	710	82,135	91,000
0240	30' x 30' high		.70	34.286		90,500	1,925	710	93,135	103,500

For customer support on your Building Construction Costs with RSMeans data, call 800.448.8182.

08 36 Panel Doors

08 36 19 – Multi-Leaf Vertical Lift Doors

08 36 19.10 Sectional Vertical Lift Doors	Crew	Daily Output	Labor-Hours	Unit	Material	2018 Bare Costs Labor	Equipment	Total	Total Incl O&P
0260	35' x 30' high	L-10	.70	34.286	Ea.	95,500	1,925	710	98,135	109,000

08 38 Traffic Doors

08 38 13 – Flexible Strip Doors

08 38 13.10 Flexible Transparent Strip Doors

		Crew	Daily Output	Labor-Hours	Unit	Material	Labor	Equipment	Total	Total Incl O&P
0010	**FLEXIBLE TRANSPARENT STRIP DOORS**									
0100	12" strip width, 2/3 overlap	3 Shee	135	.178	SF Surf	7.65	10.65		18.30	24.50
0200	Full overlap		115	.209		9.60	12.50		22.10	29.50
0220	8" strip width, 1/2 overlap		140	.171		6.25	10.25		16.50	22.50
0240	Full overlap	▼	120	.200	▼	7.85	11.95		19.80	27
0300	Add for suspension system, header mount				L.F.	9.20			9.20	10.10
0400	Wall mount				"	9.50			9.50	10.45

08 38 19 – Rigid Traffic Doors

08 38 19.20 Double Acting Swing Doors

		Crew	Daily Output	Labor-Hours	Unit	Material	Labor	Equipment	Total	Total Incl O&P
0010	**DOUBLE ACTING SWING DOORS**									
0020	Including frame, closer, hardware and vision panel									
1000	Polymer, 7'-0" high, 4'-0" wide	2 Carp	4.20	3.810	Pr.	2,100	193		2,293	2,600
1025	6'-0" wide		4	4		2,200	203		2,403	2,725
1050	6'-8" wide	▼	4	4	▼	2,400	203		2,603	2,950
2000	3/4" thick, stainless steel									
2010	Stainless steel, 7' high opening, 4' wide	2 Carp	4	4	Pr.	2,600	203		2,803	3,150
2050	7' wide	"	3.80	4.211	"	2,800	213		3,013	3,400

08 38 19.30 Shock Absorbing Doors

		Crew	Daily Output	Labor-Hours	Unit	Material	Labor	Equipment	Total	Total Incl O&P
0010	**SHOCK ABSORBING DOORS**									
0020	Rigid, no frame, 1-1/2" thick, 5' x 7'	2 Sswk	1.90	8.421	Opng.	1,525	460		1,985	2,425
0100	8' x 8'		1.80	8.889		2,000	485		2,485	3,000
0500	Flexible, no frame, insulated, .16" thick, economy, 5' x 7'		2	8		1,750	435		2,185	2,650
0600	Deluxe		1.90	8.421		2,625	460		3,085	3,625
1000	8' x 8' opening, economy		2	8		2,750	435		3,185	3,750
1100	Deluxe	▼	1.90	8.421	▼	3,500	460		3,960	4,600

08 41 Entrances and Storefronts

08 41 13 – Aluminum-Framed Entrances and Storefronts

08 41 13.20 Tube Framing

		Crew	Daily Output	Labor-Hours	Unit	Material	Labor	Equipment	Total	Total Incl O&P
0010	**TUBE FRAMING**, For window walls and storefronts, aluminum stock									
0050	Plain tube frame, mill finish, 1-3/4" x 1-3/4"	2 Glaz	103	.155	L.F.	10.50	7.55		18.05	23
0150	1-3/4" x 4"		98	.163		14.05	7.90		21.95	27.50
0200	1-3/4" x 4-1/2"		95	.168		16.90	8.15		25.05	31
0250	2" x 6"		89	.180		25	8.70		33.70	40.50
0350	4" x 4"		87	.184		28.50	8.90		37.40	44.50
0400	4-1/2" x 4-1/2"		85	.188		30	9.15		39.15	47
0450	Glass bead		240	.067		3.27	3.23		6.50	8.50
1000	Flush tube frame, mill finish, 1/4" glass, 1-3/4" x 4", open header		80	.200		14.05	9.70		23.75	30
1050	Open sill		82	.195		11.45	9.45		20.90	27
1100	Closed back header		83	.193		20	9.35		29.35	36
1150	Closed back sill	▼	85	.188	▼	19.20	9.15		28.35	35
1160	Tube fmg., spandrel cover both sides, alum 1" wide	1 Sswk	85	.094	S.F.	96	5.15		101.15	114
1170	Tube fmg., spandrel cover both sides, alum 2" wide	"	85	.094	"	38	5.15		43.15	50.50

08 41 Entrances and Storefronts

08 41 13 – Aluminum-Framed Entrances and Storefronts

08 41 13.20 Tube Framing

		Crew	Daily Output	Labor-Hours	Unit	Material	2018 Bare Costs Labor	Equipment	Total	Total Incl O&P
1200	Vertical mullion, one piece	2 Glaz	75	.213	L.F.	20.50	10.35		30.85	38
1250	Two piece		73	.219		21.50	10.65		32.15	39.50
1300	90° or 180° vertical corner post		75	.213		33	10.35		43.35	52
1400	1-3/4" x 4-1/2", open header		80	.200		16.75	9.70		26.45	33
1450	Open sill		82	.195		13.90	9.45		23.35	29.50
1500	Closed back header		83	.193		20.50	9.35		29.85	36.50
1550	Closed back sill		85	.188		19.50	9.15		28.65	35.50
1600	Vertical mullion, one piece		75	.213		22	10.35		32.35	39.50
1650	Two piece		73	.219		23	10.65		33.65	41.50
1700	90° or 180° vertical corner post		75	.213		23.50	10.35		33.85	41.50
2000	Flush tube frame, mil fin.,ins. glass w/thml brk, 2" x 4-1/2", open header		75	.213		16	10.35		26.35	33.50
2050	Open sill		77	.208		13.45	10.10		23.55	30
2100	Closed back header		78	.205		15.05	9.95		25	31.50
2150	Closed back sill		80	.200		15.70	9.70		25.40	32
2200	Vertical mullion, one piece		70	.229		16.95	11.10		28.05	35.50
2250	Two piece		68	.235		18.30	11.40		29.70	37.50
2300	90° or 180° vertical corner post		70	.229		18.65	11.10		29.75	37.50
5000	Flush tube frame, mill fin., thermal brk., 2-1/4" x 4-1/2", open header		74	.216		16.80	10.50		27.30	34.50
5050	Open sill		75	.213		14.85	10.35		25.20	32
5100	Vertical mullion, one piece		69	.232		18.75	11.25		30	37.50
5150	Two piece		67	.239		21	11.60		32.60	40.50
5200	90° or 180° vertical corner post		69	.232		18.80	11.25		30.05	37.50
6980	Door stop (snap in)	▼	380	.042	▼	3.66	2.04		5.70	7.10
7000	For joints, 90°, clip type, add				Ea.	25.50			25.50	28
7050	Screw spline joint, add					23.50			23.50	26
7100	For joint other than 90°, add				▼	49.50			49.50	54.50
8000	For bronze anodized aluminum, add					15%				
8020	For black finish, add					30%				
8050	For stainless steel materials, add					350%				
8100	For monumental grade, add					53%				
8150	For steel stiffener, add	2 Glaz	200	.080	L.F.	11.30	3.88		15.18	18.30
8200	For 2 to 5 stories, add per story				Story		8%			

08 41 19 – Stainless-Steel-Framed Entrances and Storefronts

08 41 19.10 Stainless-Steel and Glass Entrance Unit

		Crew	Daily Output	Labor-Hours	Unit	Material	2018 Bare Costs Labor	Equipment	Total	Total Incl O&P
0010	**STAINLESS-STEEL AND GLASS ENTRANCE UNIT**, narrow stiles									
0020	3' x 7' opening, including hardware, minimum	2 Sswk	1.60	10	Opng.	7,450	545		7,995	9,100
0050	Average		1.40	11.429		7,950	625		8,575	9,775
0100	Maximum	▼	1.20	13.333		8,475	730		9,205	10,500
1000	For solid bronze entrance units, statuary finish, add					64%				
1100	Without statuary finish, add				▼	45%				
2000	Balanced doors, 3' x 7', economy	2 Sswk	.90	17.778	Ea.	10,100	970		11,070	12,700
2100	Premium	"	.70	22.857	"	16,500	1,250		17,750	20,300

08 41 26 – All-Glass Entrances and Storefronts

08 41 26.10 Window Walls Aluminum, Stock

		Crew	Daily Output	Labor-Hours	Unit	Material	2018 Bare Costs Labor	Equipment	Total	Total Incl O&P
0010	**WINDOW WALLS ALUMINUM, STOCK**, including glazing									
0020	Minimum	H-2	160	.150	S.F.	50	6.85		56.85	65.50
0050	Average		140	.171		69	7.80		76.80	87.50
0100	Maximum	▼	110	.218	▼	183	9.95		192.95	216
0500	For translucent sandwich wall systems, see Section 07 41 33.10									
0850	Cost of the above walls depends on material,									
0860	finish, repetition, and size of units.									
0870	The larger the opening, the lower the S.F. cost									

For customer support on your Building Construction Costs with RSMeans data, call 800.448.8182.

08 42 Entrances

08 42 26 – All-Glass Entrances

08 42 26.10 Swinging Glass Doors

		Crew	Daily Output	Labor-Hours	Unit	Material	2018 Bare Costs Labor	Equipment	Total	Total Incl O&P
0010	**SWINGING GLASS DOORS**									
0020	Including hardware, 1/2" thick, tempered, 3' x 7' opening	2 Glaz	2	8	Opng.	2,325	390		2,715	3,175
0100	6' x 7' opening	"	1.40	11.429	"	4,575	555		5,130	5,900

08 42 33 – Revolving Door Entrances

08 42 33.10 Circular Rotating Entrance Doors

		Crew	Daily Output	Labor-Hours	Unit	Material	2018 Bare Costs Labor	Equipment	Total	Total Incl O&P
0010	**CIRCULAR ROTATING ENTRANCE DOORS**, Aluminum									
0020	6'-10" to 7' high, stock units, minimum	4 Sswk	.75	42.667	Opng.	31,500	2,325		33,825	38,500
0050	Average		.60	53.333		36,500	2,925		39,425	45,000
0100	Maximum		.45	71.111		43,700	3,875		47,575	54,500
1000	Stainless steel		.30	106		49,300	5,800		55,100	63,500
1100	Solid bronze		.15	213		49,800	11,700		61,500	74,000
1500	For automatic controls, add	2 Elec	2	8		15,400	465		15,865	17,600

08 42 36 – Balanced Door Entrances

08 42 36.10 Balanced Entrance Doors

		Crew	Daily Output	Labor-Hours	Unit	Material	2018 Bare Costs Labor	Equipment	Total	Total Incl O&P
0010	**BALANCED ENTRANCE DOORS**									
0020	Hardware & frame, alum. & glass, 3' x 7', econ.	2 Sswk	.90	17.778	Ea.	7,025	970		7,995	9,325
0150	Premium	"	.70	22.857	"	8,425	1,250		9,675	11,300

08 43 Storefronts

08 43 13 – Aluminum-Framed Storefronts

08 43 13.10 Aluminum-Framed Entrance Doors and Frames

		Crew	Daily Output	Labor-Hours	Unit	Material	2018 Bare Costs Labor	Equipment	Total	Total Incl O&P
0010	**ALUMINUM-FRAMED ENTRANCE DOORS AND FRAMES**									
0015	Standard hardware and glass stops but no glass									
0020	Entrance door, 3' x 7' opening, clear anodized finish	2 Sswk	7	2.286	Opng.	625	125		750	890
0040	Bronze finish		7	2.286		675	125		800	950
0060	Black finish		7	2.286		690	125		815	965
0200	3'-6" x 7'-0", mill finish		7	2.286		730	125		855	1,000
0220	Bronze finish		7	2.286		865	125		990	1,150
0240	Black finish		7	2.286		915	125		1,040	1,200
0500	6' x 7' opening, clear finish		6	2.667		980	146		1,126	1,325
0520	Bronze finish		6	2.667		1,075	146		1,221	1,425
0540	Black finish		6	2.667		1,150	146		1,296	1,525
0600	Door frame for above doors 3'-0" x 7'-0", mill finish		6	2.667		445	146		591	725
0620	Bronze finish		6	2.667		500	146		646	790
0640	Black finish		6	2.667		550	146		696	845
0700	3'-6" x 7'-0", mill finish		6	2.667		365	146		511	640
0720	Bronze finish		6	2.667		365	146		511	640
0740	Black finish		6	2.667		365	146		511	640
0800	6'-0" x 7'-0", mill finish		6	2.667		365	146		511	640
0820	Bronze finish		6	2.667		370	146		516	645
0840	Black finish		6	2.667		400	146		546	680
1000	With 3' high transom above, 3' x 7' opening, clear finish		5.50	2.909		540	159		699	855
1050	Bronze finish		5.50	2.909		550	159		709	865
1100	Black finish		5.50	2.909		630	159		789	950
1300	3'-6" x 7'-0" opening, clear finish		5.50	2.909		375	159		534	675
1320	Bronze finish		5.50	2.909		395	159		554	695
1340	Black finish		5.50	2.909		405	159		564	705
1500	6' x 7' opening, clear finish		5.50	2.909		645	159		804	970
1550	Bronze finish		5.50	2.909		670	159		829	1,000
1600	Black finish		5.50	2.909		740	159		899	1,075

For customer support on your Building Construction Costs with RSMeans data, call 800.448.8182.

287

08 43 Storefronts

08 43 13 – Aluminum-Framed Storefronts

08 43 13.20 Storefront Systems	Crew	Daily Output	Labor-Hours	Unit	Material	2018 Bare Costs Labor	Equipment	Total	Total Incl O&P
0010 **STOREFRONT SYSTEMS**, aluminum frame clear 3/8" plate glass									
0020 incl. 3' x 7' door with hardware (400 sq. ft. max. wall)									
0500 Wall height to 12' high, commercial grade	2 Glaz	150	.107	S.F.	24	5.15		29.15	34.50
0600 Institutional grade		130	.123		29.50	5.95		35.45	41.50
0700 Monumental grade		115	.139		82	6.75		88.75	100
1000 6' x 7' door with hardware, commercial grade		135	.119		75	5.75		80.75	91
1100 Institutional grade		115	.139		58.50	6.75		65.25	74.50
1200 Monumental grade		100	.160		113	7.75		120.75	136
1500 For bronze anodized finish, add					15%				
1600 For black anodized finish, add					36%				
1700 For stainless steel framing, add to monumental					78%				

08 43 29 – Sliding Storefronts

08 43 29.10 Sliding Panels

	Crew	Daily Output	Labor-Hours	Unit	Material	Labor	Equipment	Total	Total Incl O&P
0010 **SLIDING PANELS**									
0020 Mall fronts, aluminum & glass, 15' x 9' high	2 Glaz	1.30	12.308	Opng.	3,825	595		4,420	5,125
0100 24' x 9' high		.70	22.857		5,500	1,100		6,600	7,725
0200 48' x 9' high, with fixed panels		.90	17.778		9,950	860		10,810	12,200
0500 For bronze finish, add					17%				

08 44 Curtain Wall and Glazed Assemblies

08 44 13 – Glazed Aluminum Curtain Walls

08 44 13.10 Glazed Curtain Walls

	Crew	Daily Output	Labor-Hours	Unit	Material	Labor	Equipment	Total	Total Incl O&P
0010 **GLAZED CURTAIN WALLS**, aluminum, stock, including glazing									
0020 Minimum	H-1	205	.156	S.F.	40	8.05		48.05	56.50
0050 Average, single glazed		195	.164		57	8.45		65.45	76.50
0150 Average, double glazed		180	.178		73.50	9.15		82.65	95.50
0200 Maximum		160	.200		192	10.30		202.30	228

08 45 Translucent Wall and Roof Assemblies

08 45 10 – Translucent Roof Assemblies

08 45 10.10 Skyroofs

	Crew	Daily Output	Labor-Hours	Unit	Material	Labor	Equipment	Total	Total Incl O&P
0010 **SKYROOFS**									
1200 Skylights, circular, clear, double glazed acrylic									
1230 30" diameter	2 Carp	3	5.333	Ea.	3,000	270		3,270	3,700
1250 60" diameter		3	5.333		4,000	270		4,270	4,800
1290 96" diameter		2	8		5,000	405		5,405	6,125
1300 Skylight Barrel Vault, clear, double glazed, acrylic									
1330 3'-0" x 12'-0"	G-3	3	10.667	Ea.	5,000	530		5,530	6,300
1350 4'-0" x 12'-0"		3	10.667		5,500	530		6,030	6,850
1390 5'-0" x 12'-0"		2	16		6,000	795		6,795	7,825
1400 Skylight Pyramid, aluminum frame, clear low E laminated glass									
1410 The glass is installed in the frame except where noted									
1430 Square, 3' x 3'	G-3	3	10.667	Ea.	6,000	530		6,530	7,400
1440 4' x 4'		3	10.667		7,000	530		7,530	8,500
1450 5' x 5', glass must be field installed		3	10.667		8,000	530		8,530	9,600
1460 6' x 6', glass must be field installed		2	16		10,000	795		10,795	12,200
1550 Install pre-cut laminated glass in aluminum frame on a flat roof	2 Glaz	55	.291	SF Surf		14.10		14.10	21.50
1560 Install pre-cut laminated glass in aluminum frame on a sloped roof	"	40	.400	"		19.40		19.40	29.50

For customer support on your Building Construction Costs with RSMeans data, call 800.448.8182.

08 51 Metal Windows

08 51 13 – Aluminum Windows

08 51 13.10 Aluminum Sash

		Crew	Daily Output	Labor-Hours	Unit	Material	2018 Bare Costs Labor	2018 Bare Costs Equipment	Total	Total Incl O&P
0010	**ALUMINUM SASH**									
0020	Stock, grade C, glaze & trim not incl., casement	2 Sswk	200	.080	S.F.	41	4.37		45.37	52.50
0050	Double-hung		200	.080		41.50	4.37		45.87	52.50
0100	Fixed casement		200	.080		18.20	4.37		22.57	27
0150	Picture window		200	.080		19.55	4.37		23.92	28.50
0200	Projected window		200	.080		37.50	4.37		41.87	48
0250	Single-hung		200	.080		17.20	4.37		21.57	26
0300	Sliding		200	.080		22.50	4.37		26.87	31.50
1000	Mullions for above, tubular		240	.067	L.F.	6.50	3.64		10.14	13.10
2000	Custom aluminum sash, grade HC, glazing not included		140	.114	S.F.	41	6.25		47.25	55.50

08 51 13.20 Aluminum Windows

		Crew	Daily Output	Labor-Hours	Unit	Material	2018 Bare Costs Labor	2018 Bare Costs Equipment	Total	Total Incl O&P
0010	**ALUMINUM WINDOWS**, incl. frame and glazing, commercial grade									
1000	Stock units, casement, 3'-1" x 3'-2" opening	2 Sswk	10	1.600	Ea.	380	87.50		467.50	565
1050	Add for storms					122			122	134
1600	Projected, with screen, 3'-1" x 3'-2" opening	2 Sswk	10	1.600		360	87.50		447.50	540
1700	Add for storms					119			119	131
2000	4'-5" x 5'-3" opening	2 Sswk	8	2		410	109		519	630
2100	Add for storms					128			128	141
2500	Enamel finish windows, 3'-1" x 3'-2"	2 Sswk	10	1.600		365	87.50		452.50	550
2600	4'-5" x 5'-3"		8	2		415	109		524	635
3000	Single-hung, 2' x 3' opening, enameled, standard glazed		10	1.600		212	87.50		299.50	375
3100	Insulating glass		10	1.600		257	87.50		344.50	425
3300	2'-8" x 6'-8" opening, standard glazed		8	2		370	109		479	585
3400	Insulating glass		8	2		480	109		589	710
3700	3'-4" x 5'-0" opening, standard glazed		9	1.778		305	97		402	495
3800	Insulating glass		9	1.778		340	97		437	535
3890	Awning type, 3' x 3' opening, standard glass		14	1.143		430	62.50		492.50	575
3900	Insulating glass		14	1.143		460	62.50		522.50	605
3910	3' x 4' opening, standard glass		10	1.600		500	87.50		587.50	695
3920	Insulating glass		10	1.600		575	87.50		662.50	780
3930	3' x 5'-4" opening, standard glass		10	1.600		600	87.50		687.50	805
3940	Insulating glass		10	1.600		710	87.50		797.50	925
3950	4' x 5'-4" opening, standard glass		9	1.778		660	97		757	885
3960	Insulating glass		9	1.778		785	97		882	1,025
4000	Sliding aluminum, 3' x 2' opening, standard glazed		10	1.600		220	87.50		307.50	385
4100	Insulating glass		10	1.600		236	87.50		323.50	405
4300	5' x 3' opening, standard glazed		9	1.778		335	97		432	530
4400	Insulating glass		9	1.778		390	97		487	590
4600	8' x 4' opening, standard glazed		6	2.667		355	146		501	635
4700	Insulating glass		6	2.667		575	146		721	875
5000	9' x 5' opening, standard glazed		4	4		540	219		759	950
5100	Insulating glass		4	4		850	219		1,069	1,300
5500	Sliding, with thermal barrier and screen, 6' x 4', 2 track		8	2		725	109		834	975
5700	4 track		8	2		910	109		1,019	1,175
6000	For above units with bronze finish, add					15%				
6200	For installation in concrete openings, add					8%				

08 51 13.30 Impact Resistant Aluminum Windows

		Crew	Daily Output	Labor-Hours	Unit	Material	2018 Bare Costs Labor	2018 Bare Costs Equipment	Total	Total Incl O&P
0010	**IMPACT RESISTANT ALUMINUM WINDOWS**, incl. frame and glazing									
0100	Single-hung, impact resistant, 2'-8" x 5'-0"	2 Carp	9	1.778	Ea.	1,250	90		1,340	1,500
0120	3'-0" x 5'-0"		9	1.778		1,350	90		1,440	1,625
0130	4'-0" x 5'-0"		9	1.778		1,450	90		1,540	1,725
0250	Horizontal slider, impact resistant, 5'-5" x 5'-2"		9	1.778		1,625	90		1,715	1,900

08 51 Metal Windows

08 51 23 – Steel Windows

08 51 23.10 Steel Sash

08 51 23.10 Steel Sash	Crew	Daily Output	Labor-Hours	Unit	Material	2018 Bare Costs Labor	Equipment	Total	Total Incl O&P
0010 **STEEL SASH** Custom units, glazing and trim not included	R085123-10								
0100 Casement, 100% vented	2 Sswk	200	.080	S.F.	67.50	4.37		71.87	81
0200 50% vented		200	.080		55.50	4.37		59.87	68
0300 Fixed		200	.080		29.50	4.37		33.87	39.50
1000 Projected, commercial, 40% vented		200	.080		52.50	4.37		56.87	64.50
1100 Intermediate, 50% vented		200	.080		60	4.37		64.37	73
1500 Industrial, horizontally pivoted		200	.080		55.50	4.37		59.87	68
1600 Fixed		200	.080		32	4.37		36.37	42.50
2000 Industrial security sash, 50% vented		200	.080		60	4.37		64.37	73
2100 Fixed		200	.080		48.50	4.37		52.87	60.50
2500 Picture window		200	.080		31	4.37		35.37	41
3000 Double-hung		200	.080		60.50	4.37		64.87	73.50
5000 Mullions for above, open interior face		240	.067	L.F.	10.65	3.64		14.29	17.65
5100 With interior cover		240	.067	"	17.60	3.64		21.24	25.50

08 51 23.20 Steel Windows

08 51 23.20 Steel Windows	Crew	Daily Output	Labor-Hours	Unit	Material	2018 Bare Costs Labor	Equipment	Total	Total Incl O&P
0010 **STEEL WINDOWS** Stock, including frame, trim and insul. glass									
0020 See Section 13 34 19.50									
1000 Custom units, double-hung, 2'-8" x 4'-6" opening	R085123-10 2 Sswk	12	1.333	Ea.	725	73		798	915
1100 2'-4" x 3'-9" opening		12	1.333		595	73		668	775
1500 Commercial projected, 3'-9" x 5'-5" opening		10	1.600		1,250	87.50		1,337.50	1,525
1600 6'-9" x 4'-1" opening		7	2.286		1,650	125		1,775	2,025
2000 Intermediate projected, 2'-9" x 4'-1" opening		12	1.333		700	73		773	890
2100 4'-1" x 5'-5" opening		10	1.600		1,425	87.50		1,512.50	1,725

08 51 23.40 Basement Utility Windows

08 51 23.40 Basement Utility Windows	Crew	Daily Output	Labor-Hours	Unit	Material	2018 Bare Costs Labor	Equipment	Total	Total Incl O&P
0010 **BASEMENT UTILITY WINDOWS**									
0015 1'-3" x 2'-8"	1 Carp	16	.500	Ea.	143	25.50		168.50	197
1100 1'-7" x 2'-8"	"	16	.500	"	147	25.50		172.50	201

08 51 66 – Metal Window Screens

08 51 66.10 Screens

08 51 66.10 Screens	Crew	Daily Output	Labor-Hours	Unit	Material	2018 Bare Costs Labor	Equipment	Total	Total Incl O&P
0010 **SCREENS**									
0020 For metal sash, aluminum or bronze mesh, flat screen	2 Sswk	1200	.013	S.F.	4.50	.73		5.23	6.15
0500 Wicket screen, inside window		1000	.016		6.85	.87		7.72	9
0800 Security screen, aluminum frame with stainless steel cloth		1200	.013		24.50	.73		25.23	28
0900 Steel grate, painted, on steel frame		1600	.010		13.50	.55		14.05	15.75
1000 Screens for solar louvers		160	.100		25.50	5.45		30.95	37
4000 See Section 05 58 23.90 for window guards									

08 52 Wood Windows

08 52 10 – Plain Wood Windows

08 52 10.20 Awning Window

08 52 10.20 Awning Window	Crew	Daily Output	Labor-Hours	Unit	Material	2018 Bare Costs Labor	Equipment	Total	Total Incl O&P
0010 **AWNING WINDOW**, Including frame, screens and grilles									
0100 34" x 22", insulated glass	1 Carp	10	.800	Ea.	276	40.50		316.50	365
0200 Low E glass		10	.800		298	40.50		338.50	390
0300 40" x 28", insulated glass		9	.889		315	45		360	415
0400 Low E glass		9	.889		345	45		390	450
0500 48" x 36", insulated glass		8	1		470	50.50		520.50	590
0600 Low E glass		8	1		495	50.50		545.50	620

For customer support on your Building Construction Costs with RSMeans data, call 800.448.8182.

08 52 Wood Windows

08 52 10 – Plain Wood Windows

08 52 10.40 Casement Window

		Crew	Daily Output	Labor-Hours	Unit	Material	2018 Bare Costs Labor	Equipment	Total	Total Incl O&P
0010	**CASEMENT WINDOW**, including frame, screen and grilles									
0100	2'-0" x 3'-0" H, double insulated glass	G	1 Carp	10	.800	Ea.	282	40.50		322.50	370
0150	Low E glass	G		10	.800		275	40.50		315.50	360
0200	2'-0" x 4'-6" high, double insulated glass	G		9	.889		385	45		430	495
0250	Low E glass	G		9	.889		385	45		430	495
0260	Casement 4'-2" x 4'-2" double insulated glass	G		11	.727		920	37		957	1,050
0270	4'-0" x 4'-0" Low E glass	G		11	.727		550	37		587	655
0290	6'-4" x 5'-7" Low E glass	G		9	.889		1,175	45		1,220	1,375
0300	2'-4" x 6'-0" high, double insulated glass	G		8	1		445	50.50		495.50	565
0350	Low E glass	G		8	1		440	50.50		490.50	560
0522	Vinyl-clad, premium, double insulated glass, 2'-0" x 3'-0"	G		10	.800		285	40.50		325.50	375
0524	2'-0" x 4'-0"	G		9	.889		330	45		375	435
0525	2'-0" x 5'-0"	G		8	1		380	50.50		430.50	495
0528	2'-0" x 6'-0"	G		8	1		400	50.50		450.50	520
0600	3'-0" x 5'-0"	G		8	1		700	50.50		750.50	840
0700	4'-0" x 3'-0"	G		8	1		765	50.50		815.50	915
0710	4'-0" x 4'-0"	G		8	1		655	50.50		705.50	795
0720	4'-8" x 4'-0"	G		8	1		720	50.50		770.50	870
0730	4'-8" x 5'-0"	G		6	1.333		825	67.50		892.50	1,000
0740	4'-8" x 6'-0"	G		6	1.333		930	67.50		997.50	1,125
0750	6'-0" x 4'-0"	G		6	1.333		845	67.50		912.50	1,025
0800	6'-0" x 5'-0"	G		6	1.333		930	67.50		997.50	1,125
0900	5'-6" x 5'-6"	G	2 Carp	15	1.067		1,475	54		1,529	1,700
2000	Bay, casement units, 8' x 5', w/screens, double insulated glass			2.50	6.400	Opng.	1,625	325		1,950	2,300
2100	Low E glass			2.50	6.400	"	1,725	325		2,050	2,375
8190	For installation, add per leaf					Ea.		15%			
8200	For multiple leaf units, deduct for stationary sash										
8220	2' high					Ea.	24			24	26.50
8240	4'-6" high						27			27	30
8260	6' high						36.50			36.50	40

08 52 10.50 Double-Hung

		Crew	Daily Output	Labor-Hours	Unit	Material	Labor	Equipment	Total	Total Incl O&P
0010	**DOUBLE-HUNG**, Including frame, screens and grilles	R085216-10									
0100	2'-0" x 3'-0" high, low E insul. glass	G	1 Carp	10	.800	Ea.	193	40.50		233.50	274
0200	3'-0" x 4'-0" high, double insulated glass	G		9	.889		288	45		333	385
0300	4'-0" x 4'-6" high, low E insulated glass	G		8	1		335	50.50		385.50	445

08 52 10.55 Picture Window

		Crew	Daily Output	Labor-Hours	Unit	Material	Labor	Equipment	Total	Total Incl O&P
0010	**PICTURE WINDOW**, Including frame and grilles										
0100	3'-6" x 4'-0" high, double insulated glass		2 Carp	12	1.333	Ea.	435	67.50		502.50	585
0150	Low E glass			12	1.333		445	67.50		512.50	595
0200	4'-0" x 4'-6" high, double insulated glass			11	1.455		550	74		624	715
0250	Low E glass			11	1.455		530	74		604	695
0300	5'-0" x 4'-0" high, double insulated glass			11	1.455		580	74		654	750
0350	Low E glass			11	1.455		605	74		679	775
0400	6'-0" x 4'-6" high, double insulated glass			10	1.600		640	81		721	825
0450	Low E glass			10	1.600		640	81		721	830

08 52 10.65 Wood Sash

		Crew	Daily Output	Labor-Hours	Unit	Material	Labor	Equipment	Total	Total Incl O&P
0010	**WOOD SASH**, Including glazing but not trim										
0050	Custom, 5'-0" x 4'-0", 1" double glazed, 3/16" thick lites		2 Carp	3.20	5	Ea.	233	254		487	640
0100	1/4" thick lites			5	3.200		263	162		425	535
0200	1" thick, triple glazed			5	3.200		430	162		592	720
0300	7'-0" x 4'-6" high, 1" double glazed, 3/16" thick lites			4.30	3.721		430	189		619	755

For customer support on your Building Construction Costs with RSMeans data, call 800.448.8182.

291

08 52 Wood Windows

08 52 10 – Plain Wood Windows

08 52 10.65 Wood Sash

		Crew	Daily Output	Labor-Hours	Unit	Material	2018 Bare Costs Labor	2018 Bare Costs Equipment	Total	Total Incl O&P
0400	1/4" thick lites	2 Carp	4.30	3.721	Ea.	490	189		679	820
0500	1" thick, triple glazed		4.30	3.721		560	189		749	900
0600	8'-6" x 5'-0" high, 1" double glazed, 3/16" thick lites		3.50	4.571		580	232		812	995
0700	1/4" thick lites		3.50	4.571		640	232		872	1,050
0800	1" thick, triple glazed		3.50	4.571		675	232		907	1,100
0900	Window frames only, based on perimeter length				L.F.	4.13			4.13	4.54
1200	Window sill, stock, per lineal foot					8.65			8.65	9.50
1250	Casing, stock					3.38			3.38	3.72

08 52 10.70 Sliding Windows

			Crew	Daily Output	Labor-Hours	Unit	Material	2018 Bare Costs Labor	2018 Bare Costs Equipment	Total	Total Incl O&P
0010	**SLIDING WINDOWS**										
0100	3'-0" x 3'-0" high, double insulated	G	1 Carp	10	.800	Ea.	293	40.50		333.50	380
0120	Low E glass	G		10	.800		320	40.50		360.50	410
0200	4'-0" x 3'-6" high, double insulated	G		9	.889		380	45		425	490
0220	Low E glass	G		9	.889		385	45		430	490
0300	6'-0" x 5'-0" high, double insulated	G		8	1		505	50.50		555.50	630
0320	Low E glass	G		8	1		550	50.50		600.50	680

08 52 13 – Metal-Clad Wood Windows

08 52 13.10 Awning Windows, Metal-Clad

		Crew	Daily Output	Labor-Hours	Unit	Material	2018 Bare Costs Labor	2018 Bare Costs Equipment	Total	Total Incl O&P
0010	**AWNING WINDOWS, METAL-CLAD**									
2000	Metal-clad, awning deluxe, double insulated glass, 34" x 22"	1 Carp	9	.889	Ea.	255	45		300	350
2050	36" x 25"		9	.889		276	45		321	375
2100	40" x 22"		9	.889		295	45		340	395
2150	40" x 30"		9	.889		340	45		385	445
2200	48" x 28"		8	1		350	50.50		400.50	460
2250	60" x 36"		8	1		375	50.50		425.50	485

08 52 13.20 Casement Windows, Metal-Clad

			Crew	Daily Output	Labor-Hours	Unit	Material	2018 Bare Costs Labor	2018 Bare Costs Equipment	Total	Total Incl O&P
0010	**CASEMENT WINDOWS, METAL-CLAD**										
0100	Metal-clad, deluxe, dbl. insul. glass, 2'-0" x 3'-0" high	G	1 Carp	10	.800	Ea.	294	40.50		334.50	385
0120	2'-0" x 4'-0" high	G		9	.889		320	45		365	420
0130	2'-0" x 5'-0" high	G		8	1		340	50.50		390.50	450
0140	2'-0" x 6'-0" high	G		8	1		380	50.50		430.50	495
0300	Metal-clad, casement, bldrs mdl, 6'-0" x 4'-0", dbl. insul. glass, 3 panels		2 Carp	10	1.600		1,200	81		1,281	1,450
0310	9'-0" x 4'-0", 4 panels			8	2		1,550	101		1,651	1,850
0320	10'-0" x 5'-0", 5 panels			7	2.286		2,100	116		2,216	2,500
0330	12'-0" x 6'-0", 6 panels			6	2.667		2,700	135		2,835	3,150

08 52 13.30 Double-Hung Windows, Metal-Clad

			Crew	Daily Output	Labor-Hours	Unit	Material	2018 Bare Costs Labor	2018 Bare Costs Equipment	Total	Total Incl O&P
0010	**DOUBLE-HUNG WINDOWS, METAL-CLAD**										
0100	Metal-clad, deluxe, dbl. insul. glass, 2'-6" x 3'-0" high	G	1 Carp	10	.800	Ea.	285	40.50		325.50	375
0120	3'-0" x 3'-6" high	G		10	.800		325	40.50		365.50	420
0140	3'-0" x 4'-0" high	G		9	.889		340	45		385	445
0160	3'-0" x 4'-6" high	G		9	.889		355	45		400	460
0180	3'-0" x 5'-0" high	G		8	1		385	50.50		435.50	500
0200	3'-6" x 6'-0" high	G		8	1		465	50.50		515.50	585

08 52 13.35 Picture and Sliding Windows Metal-Clad

			Crew	Daily Output	Labor-Hours	Unit	Material	2018 Bare Costs Labor	2018 Bare Costs Equipment	Total	Total Incl O&P
0010	**PICTURE AND SLIDING WINDOWS METAL-CLAD**										
2000	Metal-clad, dlx picture, dbl. insul. glass, 4'-0" x 4'-0" high		2 Carp	12	1.333	Ea.	380	67.50		447.50	525
2100	4'-0" x 6'-0" high			11	1.455		560	74		634	725
2200	5'-0" x 6'-0" high			10	1.600		620	81		701	805
2300	6'-0" x 6'-0" high			10	1.600		710	81		791	905
2400	Metal-clad, dlx sliding, dbl. insul. glass, 3'-0" x 3'-0" high	G	1 Carp	10	.800		330	40.50		370.50	420
2420	4'-0" x 3'-6" high	G		9	.889		400	45		445	510

For customer support on your Building Construction Costs with RSMeans data, call 800.448.8182.

08 52 Wood Windows

08 52 13 – Metal-Clad Wood Windows

08 52 13.35 Picture and Sliding Windows Metal-Clad

		Crew	Daily Output	Labor-Hours	Unit	Material	2018 Bare Costs Labor	Equipment	Total	Total Incl O&P
2440	5'-0" x 4'-0" high [G]	1 Carp	9	.889	Ea.	480	45		525	595
2460	6'-0" x 5'-0" high [G]	↓	8	1	↓	745	50.50		795.50	895

08 52 13.40 Bow and Bay Windows, Metal-Clad

		Crew	Daily Output	Labor-Hours	Unit	Material	2018 Bare Costs Labor	Equipment	Total	Total Incl O&P
0010	**BOW AND BAY WINDOWS, METAL-CLAD**									
0100	Metal-clad, deluxe, dbl. insul. glass, 8'-0" x 5'-0" high, 4 panels	2 Carp	10	1.600	Ea.	1,700	81		1,781	2,000
0120	10'-0" x 5'-0" high, 5 panels		8	2		1,825	101		1,926	2,150
0140	10'-0" x 6'-0" high, 5 panels		7	2.286		2,150	116		2,266	2,550
0160	12'-0" x 6'-0" high, 6 panels		6	2.667		2,925	135		3,060	3,425
0400	Double-hung, bldrs. model, bay, 8' x 4' high, dbl. insul. glass		10	1.600		1,350	81		1,431	1,600
0440	Low E glass		10	1.600		1,450	81		1,531	1,725
0460	9'-0" x 5'-0" high, dbl. insul. glass		6	2.667		1,450	135		1,585	1,800
0480	Low E glass		6	2.667		1,525	135		1,660	1,875
0500	Metal-clad, deluxe, dbl. insul. glass, 7'-0" x 4'-0" high		10	1.600		1,300	81		1,381	1,550
0520	8'-0" x 4'-0" high		8	2		1,350	101		1,451	1,625
0540	8'-0" x 5'-0" high		7	2.286		1,375	116		1,491	1,700
0560	9'-0" x 5'-0" high	↓	6	2.667	↓	1,475	135		1,610	1,825

08 52 16 – Plastic-Clad Wood Windows

08 52 16.10 Bow Window

		Crew	Daily Output	Labor-Hours	Unit	Material	2018 Bare Costs Labor	Equipment	Total	Total Incl O&P
0010	**BOW WINDOW** including frames, screens, and grilles									
0020	End panels operable									
1000	Bow type, casement, wood, bldrs. mdl., 8' x 5' dbl. insul. glass, 4 panel	2 Carp	10	1.600	Ea.	1,575	81		1,656	1,850
1050	Low E glass		10	1.600		1,325	81		1,406	1,575
1100	10'-0" x 5'-0", dbl. insul. glass, 6 panels		6	2.667		1,350	135		1,485	1,700
1200	Low E glass, 6 panels		6	2.667		1,450	135		1,585	1,800
1300	Vinyl-clad, bldrs. model, dbl. insul. glass, 6'-0" x 4'-0", 3 panel		10	1.600		1,050	81		1,131	1,275
1340	9'-0" x 4'-0", 4 panel		8	2		1,425	101		1,526	1,725
1380	10'-0" x 6'-0", 5 panels		7	2.286		2,350	116		2,466	2,750
1420	12'-0" x 6'-0", 6 panels		6	2.667		3,075	135		3,210	3,575
2000	Bay window, 8' x 5', dbl. insul. glass		10	1.600		1,925	81		2,006	2,225
2050	Low E glass		10	1.600		2,325	81		2,406	2,675
2100	12'-0" x 6'-0", dbl. insul. glass, 6 panels		6	2.667		2,400	135		2,535	2,825
2200	Low E glass		6	2.667		3,250	135		3,385	3,775
2280	6'-0" x 4'-0"		11	1.455		1,250	74		1,324	1,475
2300	Vinyl-clad, premium, dbl. insul. glass, 8'-0" x 5'-0"		10	1.600		1,800	81		1,881	2,100
2340	10'-0" x 5'-0"		8	2		2,400	101		2,501	2,800
2380	10'-0" x 6'-0"		7	2.286		2,800	116		2,916	3,250
2420	12'-0" x 6'-0"		6	2.667		3,350	135		3,485	3,900
3300	Vinyl-clad, premium, dbl. insul. glass, 7'-0" x 4'-6"		10	1.600		1,400	81		1,481	1,650
3340	8'-0" x 4'-6"		8	2		1,425	101		1,526	1,700
3380	8'-0" x 5'-0"		7	2.286		1,500	116		1,616	1,825
3420	9'-0" x 5'-0"	↓	6	2.667	↓	1,525	135		1,660	1,875

08 52 16.15 Awning Window Vinyl-Clad

		Crew	Daily Output	Labor-Hours	Unit	Material	2018 Bare Costs Labor	Equipment	Total	Total Incl O&P
0010	**AWNING WINDOW VINYL-CLAD** including frames, screens, and grilles									
0240	Vinyl-clad, 34" x 22"	1 Carp	10	.800	Ea.	267	40.50		307.50	355
0280	36" x 28"		9	.889		305	45		350	410
0300	36" x 36"		9	.889		350	45		395	455
0340	40" x 22"		10	.800		295	40.50		335.50	385
0360	48" x 28"		8	1		370	50.50		420.50	485
0380	60" x 36"	↓	8	1	↓	520	50.50		570.50	645

08 52 Wood Windows

08 52 16 – Plastic-Clad Wood Windows

08 52 16.30 Palladian Windows		Crew	Daily Output	Labor-Hours	Unit	Material	2018 Bare Costs Labor	Equipment	Total	Total Incl O&P
0010	**PALLADIAN WINDOWS**									
0020	Vinyl-clad, double insulated glass, including frame and grilles									
0040	3'-2" x 2'-6" high	2 Carp	11	1.455	Ea.	1,275	74		1,349	1,500
0060	3'-2" x 4'-10"		11	1.455		1,750	74		1,824	2,025
0080	3'-2" x 6'-4"		10	1.600		1,750	81		1,831	2,050
0100	4'-0" x 4'-0"		10	1.600		1,575	81		1,656	1,850
0120	4'-0" x 5'-4"	3 Carp	10	2.400		1,875	122		1,997	2,250
0140	4'-0" x 6'-0"		9	2.667		1,875	135		2,010	2,275
0160	4'-0" x 7'-4"		9	2.667		2,125	135		2,260	2,550
0180	5'-5" x 4'-10"		9	2.667		2,300	135		2,435	2,725
0200	5'-5" x 6'-10"		9	2.667		2,650	135		2,785	3,100
0220	5'-5" x 7'-9"		9	2.667		2,800	135		2,935	3,300
0240	6'-0" x 7'-11"		8	3		3,575	152		3,727	4,175
0260	8'-0" x 6'-0"		8	3		3,200	152		3,352	3,725

08 52 16.35 Double-Hung Window

			Crew	Daily Output	Labor-Hours	Unit	Material	Labor	Equipment	Total	Total Incl O&P
0010	**DOUBLE-HUNG WINDOW** including frames, screens, and grilles										
0300	Vinyl-clad, premium, double insulated glass, 2'-6" x 3'-0"	G	1 Carp	10	.800	Ea.	340	40.50		380.50	430
0305	2'-6" x 4'-0"	G		10	.800		375	40.50		415.50	475
0400	3'-0" x 3'-6"	G		10	.800		345	40.50		385.50	435
0500	3'-0" x 4'-0"	G		9	.889		400	45		445	510
0600	3'-0" x 4'-6"	G		9	.889		425	45		470	540
0700	3'-0" x 5'-0"	G		8	1		460	50.50		510.50	580
0790	3'-4" x 5'-0"	G		8	1		450	50.50		500.50	570
0800	3'-6" x 6'-0"	G		8	1		515	50.50		565.50	640
0820	4'-0" x 5'-0"	G		7	1.143		570	58		628	720
0830	4'-0" x 6'-0"	G		7	1.143		725	58		783	885

08 52 16.40 Transom Windows

			Crew	Daily Output	Labor-Hours	Unit	Material	Labor	Equipment	Total	Total Incl O&P
0010	**TRANSOM WINDOWS**										
0050	Vinyl-clad, premium, dbl. insul. glass, 32" x 8"	1 Carp	16	.500	Ea.	191	25.50		216.50	249
0100	36" x 8"		16	.500		206	25.50		231.50	266
0110	36" x 12"		16	.500		221	25.50		246.50	282
0200	44" x 48"		12	.667		590	34		624	700
1000	Vinyl-clad, premium, dbl. insul. glass, 4'-0" x 4'-0"	2 Carp	12	1.333		530	67.50		597.50	690
1100	4'-0" x 6'-0"		11	1.455		975	74		1,049	1,175
1200	5'-0" x 6'-0"		10	1.600		1,075	81		1,156	1,325
1300	6'-0" x 6'-0"		10	1.600		1,150	81		1,231	1,375

08 52 16.70 Vinyl-Clad, Premium, DBL. Insul. Glass

			Crew	Daily Output	Labor-Hours	Unit	Material	Labor	Equipment	Total	Total Incl O&P
0010	**VINYL-CLAD, PREMIUM, DBL. INSUL. GLASS**										
1000	Sliding, 3'-0" x 3'-0"	G	1 Carp	10	.800	Ea.	625	40.50		665.50	745
1050	4'-0" x 3'-6"	G		9	.889		690	45		735	830
1100	5'-0" x 4'-0"	G		9	.889		920	45		965	1,075
1150	6'-0" x 5'-0"	G		8	1		1,175	50.50		1,225.50	1,350

08 52 50 – Window Accessories

08 52 50.10 Window Grille or Muntin

			Crew	Daily Output	Labor-Hours	Unit	Material	Labor	Equipment	Total	Total Incl O&P
0010	**WINDOW GRILLE OR MUNTIN**, snap in type										
0020	Standard pattern interior grilles										
2000	Wood, awning window, glass size, 28" x 16" high	1 Carp	30	.267	Ea.	30.50	13.50		44	54
2060	44" x 24" high		32	.250		44	12.70		56.70	68
2100	Casement, glass size, 20" x 36" high		30	.267		33.50	13.50		47	57.50
2180	20" x 56" high		32	.250		46.50	12.70		59.20	70.50
2200	Double-hung, glass size, 16" x 24" high		24	.333	Set	56	16.90		72.90	87

For customer support on your Building Construction Costs with RSMeans data, call 800.448.8182.

08 52 Wood Windows

08 52 50 – Window Accessories

08 52 50.10 Window Grille or Muntin

		Crew	Daily Output	Labor-Hours	Unit	Material	2018 Bare Costs Labor	Equipment	Total	Total Incl O&P
2280	32" x 32" high	1 Carp	34	.235	Set	140	11.95		151.95	172
2500	Picture, glass size, 48" x 48" high		30	.267	Ea.	128	13.50		141.50	162
2580	60" x 68" high		28	.286	"	196	14.50		210.50	238
2600	Sliding, glass size, 14" x 36" high		24	.333	Set	37.50	16.90		54.40	67
2680	36" x 36" high		22	.364	"	45	18.45		63.45	77.50

08 52 66 – Wood Window Screens

08 52 66.10 Wood Screens

0010	**WOOD SCREENS**									
0020	Over 3 S.F., 3/4" frames	2 Carp	375	.043	S.F.	4.75	2.16		6.91	8.55
0100	1-1/8" frames	"	375	.043	"	8.45	2.16		10.61	12.60

08 52 69 – Wood Storm Windows

08 52 69.10 Storm Windows

0010	**STORM WINDOWS**, aluminum residential										
0300	Basement, mill finish, incl. fiberglass screen										
0320	1'-10" x 1'-0" high	G	2 Carp	30	.533	Ea.	36	27		63	80.50
0340	2'-9" x 1'-6" high	G		30	.533		39	27		66	84
0360	3'-4" x 2'-0" high	G		30	.533		46	27		73	91.50
1600	Double-hung, combination, storm & screen										
2000	Clear anodic coating, 2'-0" x 3'-5" high	G	2 Carp	30	.533	Ea.	97	27		124	148
2020	2'-6" x 5'-0" high	G		28	.571		124	29		153	180
2040	4'-0" x 6'-0" high	G		25	.640		135	32.50		167.50	199
2400	White painted, 2'-0" x 3'-5" high	G		30	.533		95	27		122	146
2420	2'-6" x 5'-0" high	G		28	.571		100	29		129	154
2440	4'-0" x 6'-0" high	G		25	.640		115	32.50		147.50	177
2600	Mill finish, 2'-0" x 3'-5" high	G		30	.533		90	27		117	140
2620	2'-6" x 5'-0" high	G		28	.571		95	29		124	149
2640	4'-0" x 6'-8" high	G		25	.640		115	32.50		147.50	177

08 53 Plastic Windows

08 53 13 – Vinyl Windows

08 53 13.20 Vinyl Single-Hung Windows

0010	**VINYL SINGLE-HUNG WINDOWS**, insulated glass										
0020	Grids, low E, J fin, extension jambs										
0130	25" x 41"	G	2 Carp	20	.800	Ea.	210	40.50		250.50	293
0140	25" x 49"	G		18	.889		220	45		265	310
0150	25" x 57"	G		17	.941		230	47.50		277.50	325
0160	25" x 65"	G		16	1		250	50.50		300.50	350
0170	29" x 41"	G		18	.889		210	45		255	300
0180	29" x 53"	G		18	.889		230	45		275	320
0190	29" x 57"	G		17	.941		240	47.50		287.50	335
0200	29" x 65"	G		16	1		250	50.50		300.50	350
0210	33" x 41"	G		20	.800		225	40.50		265.50	310
0220	33" x 53"	G		18	.889		245	45		290	340
0230	33" x 57"	G		17	.941		250	47.50		297.50	350
0240	33" x 65"	G		16	1		260	50.50		310.50	365
0250	37" x 41"	G		20	.800		250	40.50		290.50	335
0260	37" x 53"	G		18	.889		275	45		320	375
0270	37" x 57"	G		17	.941		285	47.50		332.50	390
0280	37" x 65"	G		16	1		300	50.50		350.50	405

08 53 Plastic Windows

08 53 13 – Vinyl Windows

08 53 13.30 Vinyl Double-Hung Windows		Crew	Daily Output	Labor-Hours	Unit	Material	2018 Bare Costs Labor	Equipment	Total	Total Incl O&P
0010	**VINYL DOUBLE-HUNG WINDOWS**, insulated glass									
0100	Grids, low E, J fin, ext. jambs, 21" x 53"	G	2 Carp	18	.889	Ea.	220	45		265	310
0102	21" x 37"	G		18	.889		230	45		275	320
0104	21" x 41"	G		18	.889		240	45		285	335
0106	21" x 49"	G		18	.889		250	45		295	345
0110	21" x 57"	G		17	.941		270	47.50		317.50	370
0120	21" x 65"	G		16	1		285	50.50		335.50	390
0128	25" x 37"	G		20	.800		240	40.50		280.50	325
0130	25" x 41"	G		20	.800		250	40.50		290.50	335
0140	25" x 49"	G		18	.889		260	45		305	355
0145	25" x 53"	G		18	.889		275	45		320	375
0150	25" x 57"	G		17	.941		280	47.50		327.50	385
0160	25" x 65"	G		16	1		295	50.50		345.50	400
0162	25" x 69"	G		16	1		300	50.50		350.50	405
0164	25" x 77"	G		16	1		325	50.50		375.50	435
0168	29" x 37"	G		18	.889		250	45		295	345
0170	29" x 41"	G		18	.889		260	45		305	355
0172	29" x 49"	G		18	.889		270	45		315	365
0180	29" x 53"	G		18	.889		280	45		325	380
0190	29" x 57"	G		17	.941		290	47.50		337.50	395
0200	29" x 65"	G		16	1		310	50.50		360.50	415
0202	29" x 69"	G		16	1		315	50.50		365.50	420
0205	29" x 77"	G		16	1		340	50.50		390.50	450
0208	33" x 37"	G		20	.800		265	40.50		305.50	355
0210	33" x 41"	G		20	.800		270	40.50		310.50	360
0215	33" x 49"	G		20	.800		285	40.50		325.50	375
0220	33" x 53"	G		18	.889		290	45		335	390
0230	33" x 57"	G		17	.941		300	47.50		347.50	405
0240	33" x 65"	G		16	1		320	50.50		370.50	425
0242	33" x 69"	G		16	1		330	50.50		380.50	440
0246	33" x 77"	G		16	1		350	50.50		400.50	460
0250	37" x 41"	G		20	.800		295	40.50		335.50	385
0255	37" x 49"	G		20	.800		297	40.50		337.50	385
0260	37" x 53"	G		18	.889		320	45		365	420
0270	37" x 57"	G		17	.941		340	47.50		387.50	450
0280	37" x 65"	G		16	1		360	50.50		410.50	470
0282	37" x 69"	G		16	1		365	50.50		415.50	475
0286	37" x 77"	G		16	1		380	50.50		430.50	495
0300	Solid vinyl, average quality, double insulated glass, 2'-0" x 3'-0"	G	1 Carp	10	.800		291	40.50		331.50	380
0310	3'-0" x 4'-0"	G		9	.889		214	45		259	305
0330	Premium, double insulated glass, 2'-6" x 3'-0"	G		10	.800		276	40.50		316.50	365
0340	3'-0" x 3'-6"	G		9	.889		310	45		355	410
0350	3'-0" x 4'-0"	G		9	.889		330	45		375	435
0360	3'-0" x 4'-6"	G		9	.889		335	45		380	440
0370	3'-0" x 5'-0"	G		8	1		360	50.50		410.50	470
0380	3'-6" x 6'-0"	G		8	1		395	50.50		445.50	510

08 53 13.40 Vinyl Casement Windows

		Crew	Daily Output	Labor-Hours	Unit	Material	2018 Bare Costs Labor	Equipment	Total	Total Incl O&P
0010	**VINYL CASEMENT WINDOWS**, insulated glass									
0015	Grids, low E, J fin, extension jambs, screens									
0100	One lite, 21" x 41"	G	2 Carp	20	.800	Ea.	315	40.50		355.50	405
0110	21" x 47"	G		20	.800		330	40.50		370.50	425
0120	21" x 53"	G		20	.800		365	40.50		405.50	460

296

08 53 Plastic Windows

08 53 13 - Vinyl Windows

08 53 13.40 Vinyl Casement Windows

			Crew	Daily Output	Labor-Hours	Unit	Material	2018 Bare Costs Labor	Equipment	Total	Total Incl O&P
0128	24" x 35"	G	2 Carp	19	.842	Ea.	297	42.50		339.50	390
0130	24" x 41"	G		19	.842		320	42.50		362.50	415
0140	24" x 47"	G		19	.842		355	42.50		397.50	455
0150	24" x 53"	G		19	.842		380	42.50		422.50	485
0158	28" x 35"	G		19	.842		310	42.50		352.50	405
0160	28" x 41"	G		19	.842		340	42.50		382.50	440
0170	28" x 47"	G		19	.842		375	42.50		417.50	480
0180	28" x 53"	G		19	.842		405	42.50		447.50	510
0184	28" x 59"	G		19	.842		415	42.50		457.50	525
0188	Two lites, 33" x 35"	G		18	.889		495	45		540	615
0190	33" x 41"	G		18	.889		530	45		575	655
0200	33" x 47"	G		18	.889		565	45		610	690
0210	33" x 53"	G		18	.889		580	45		625	710
0212	33" x 59"	G		18	.889		635	45		680	770
0215	33" x 72"	G		18	.889		665	45		710	805
0220	41" x 41"	G		18	.889		550	45		595	675
0230	41" x 47"	G		18	.889		610	45		655	740
0240	41" x 53"	G		17	.941		650	47.50		697.50	790
0242	41" x 59"	G		17	.941		690	47.50		737.50	830
0246	41" x 72"	G		17	.941		720	47.50		767.50	865
0250	47" x 41"	G		17	.941		585	47.50		632.50	715
0260	47" x 47"	G		17	.941		625	47.50		672.50	765
0270	47" x 53"	G		17	.941		665	47.50		712.50	810
0272	47" x 59"	G		17	.941		725	47.50		772.50	870
0280	56" x 41"	G		15	1.067		625	54		679	775
0290	56" x 47"	G		15	1.067		665	54		719	820
0300	56" x 53"	G		15	1.067		725	54		779	880
0302	56" x 59"	G		15	1.067		750	54		804	910
0310	56" x 72"	G		15	1.067		810	54		864	975
0340	Solid vinyl, premium, double insulated glass, 2'-0" x 3'-0" high	G	1 Carp	10	.800		270	40.50		310.50	360
0360	2'-0" x 4'-0" high	G		9	.889		299	45		344	400
0380	2'-0" x 5'-0" high	G		8	1		335	50.50		385.50	445

08 53 13.50 Vinyl Picture Windows

			Crew	Daily Output	Labor-Hours	Unit	Material	2018 Bare Costs Labor	Equipment	Total	Total Incl O&P
0010	**VINYL PICTURE WINDOWS**, insulated glass										
0120	Grids, low E, J fin, ext. jambs, 47" x 35"		2 Carp	12	1.333	Ea.	305	67.50		372.50	440
0130	47" x 41"			12	1.333		400	67.50		467.50	545
0140	47" x 47"			12	1.333		350	67.50		417.50	490
0150	47" x 53"			11	1.455		375	74		449	525
0160	71" x 35"			11	1.455		400	74		474	550
0170	71" x 41"			11	1.455		420	74		494	570
0180	71" x 47"			11	1.455		450	74		524	605

For customer support on your Building Construction Costs with RSMeans data, call 800.448.8182.

297

08 54 Composite Windows

08 54 13 – Fiberglass Windows

08 54 13.10 Fiberglass Single-Hung Windows		Crew	Daily Output	Labor-Hours	Unit	Material	2018 Bare Costs Labor	Equipment	Total	Total Incl O&P
0010	**FIBERGLASS SINGLE-HUNG WINDOWS**									
0100	Grids, low E, 18" x 24"	G	2 Carp	18	.889	Ea.	335	45		380	440
0110	18" x 40"	G		17	.941		340	47.50		387.50	450
0130	24" x 40"	G		20	.800		360	40.50		400.50	455
0230	36" x 36"	G		17	.941		370	47.50		417.50	480
0250	36" x 48"	G		20	.800		405	40.50		445.50	505
0260	36" x 60"	G		18	.889		445	45		490	560
0280	36" x 72"	G		16	1		470	50.50		520.50	590
0290	48" x 40"	G		16	1		470	50.50		520.50	590

08 62 Unit Skylights

08 62 13 – Domed Unit Skylights

08 62 13.20 Skylights

		Crew	Daily Output	Labor-Hours	Unit	Material	2018 Bare Costs Labor	Equipment	Total	Total Incl O&P
0010	**SKYLIGHTS**, flush or curb mounted									
2120	Ventilating insulated plexiglass dome with									
2130	curb mounting, 36" x 36"	G	G-3	12	2.667	Ea.	480	133		613	735
2150	52" x 52"	G		12	2.667		670	133		803	940
2160	28" x 52"	G		10	3.200		495	159		654	790
2170	36" x 52"	G		10	3.200		545	159		704	840
2180	For electric opening system, add	G					315			315	350
2300	Insulated safety glass with aluminum frame	G	G-3	160	.200	S.F.	95	9.95		104.95	120

08 63 Metal-Framed Skylights

08 63 13 – Domed Metal-Framed Skylights

08 63 13.20 Skylight Rigid Metal-Framed

		Crew	Daily Output	Labor-Hours	Unit	Material	2018 Bare Costs Labor	Equipment	Total	Total Incl O&P
0010	**SKYLIGHT RIGID METAL-FRAMED** skylight framing is aluminum									
0050	Fixed acrylic double domes, curb mount, 25-1/2" x 25-1/2"		G-3	10	3.200	Ea.	210	159		369	475
0060	25-1/2" x 33-1/2"			10	3.200		242	159		401	510
0070	25-1/2" x 49-1/2"			6	5.333		279	266		545	710
0080	33-1/2" x 33-1/2"			8	4		305	199		504	640
0090	37-1/2" x 25-1/2"			8	4		250	199		449	580
0100	37-1/2" x 37-1/2"			8	4		245	199		444	575
0110	37-1/2" x 49-1/2"			6	5.333		420	266		686	865
0120	49-1/2" x 33-1/2"			6	5.333		375	266		641	815
0130	49-1/2" x 49-1/2"			6	5.333		470	266		736	920
1000	Fixed tempered glass, curb mount, 17-1/2" x 33-1/2"			10	3.200		170	159		329	430
1020	17-1/2" x 49-1/2"			6	5.333		190	266		456	615
1030	25-1/2" x 25-1/2"			10	3.200		170	159		329	430
1040	25-1/2" x 33-1/2"			10	3.200		200	159		359	465
1050	25-1/2" x 37-1/2"			8	4		210	199		409	535
1060	25-1/2" x 49-1/2"			6	5.333		222	266		488	650
1070	25-1/2" x 73-1/2"			6	5.333		375	266		641	815
1080	33-1/2" x 33-1/2"			8	4		235	199		434	565
2000	Manual vent tempered glass & screen, curb, 25-1/2" x 25-1/2"			10	3.200		410	159		569	695
2020	25-1/2" x 37-1/2"			10	3.200		465	159		624	755
2030	25-1/2" x 49-1/2"			8	4		505	199		704	860
2040	33-1/2" x 33-1/2"			8	4		535	199		734	895
2050	33-1/2" x 49-1/2"			6	5.333		675	266		941	1,150
2060	37-1/2" x 37-1/2"			6	5.333		615	266		881	1,075

For customer support on your Building Construction Costs with RSMeans data, call 800.448.8182.

08 63 Metal-Framed Skylights

08 63 13 – Domed Metal-Framed Skylights

08 63 13.20 Skylight Rigid Metal-Framed	Crew	Daily Output	Labor-Hours	Unit	Material	2018 Bare Costs Labor	Equipment	Total	Total Incl O&P
2070	49-1/2" x 49-1/2"	G-3	6	5.333	Ea.	790	266		1,056	1,275
3000	Electric vent tempered glass, curb mount, 25-1/2" x 25-1/2"		10	3.200		960	159		1,119	1,300
3020	25-1/2" x 37-1/2"		10	3.200		1,050	159		1,209	1,400
3030	25-1/2" x 49-1/2"		8	4		1,125	199		1,324	1,525
3040	33-1/2" x 33-1/2"		8	4		1,125	199		1,324	1,550
3050	33-1/2" x 49-1/2"		6	5.333		1,225	266		1,491	1,725
3060	37-1/2" x 37-1/2"		6	5.333		1,200	266		1,466	1,700
3070	49-1/2" x 49-1/2"		6	5.333		1,325	266		1,591	1,850

08 71 Door Hardware

08 71 13 – Automatic Door Operators

08 71 13.10 Automatic Openers Commercial

		Crew	Daily Output	Labor-Hours	Unit	Material	2018 Bare Costs Labor	Equipment	Total	Total Incl O&P
0010	**AUTOMATIC OPENERS COMMERCIAL**									
0020	Pneumatic door opener incl. motion sens, control box, tubing, compressor									
0050	For single swing door, per opening	2 Skwk	.80	20	Ea.	4,625	1,050		5,675	6,675
0100	Pair, per opening		.50	32	Opng.	7,500	1,675		9,175	10,800
1000	For single sliding door, per opening		.60	26.667		5,075	1,400		6,475	7,700
1300	Bi-parting pair		.50	32		7,600	1,675		9,275	11,000
1420	Electronic door opener incl. motion sens, 12 V control box, motor									
1450	For single swing door, per opening	2 Skwk	.80	20	Opng.	3,750	1,050		4,800	5,725
1500	Pair, per opening		.50	32		6,925	1,675		8,600	10,200
1600	For single sliding door, per opening		.60	26.667		4,575	1,400		5,975	7,150
1700	Bi-parting pair		.50	32		5,475	1,675		7,150	8,600
1750	Handicap actuator buttons, 2, incl. 12 V DC wiring, add	1 Carp	1.50	5.333	Pr.	480	270		750	935

08 71 13.20 Automatic Openers Industrial

		Crew	Daily Output	Labor-Hours	Unit	Material	2018 Bare Costs Labor	Equipment	Total	Total Incl O&P
0010	**AUTOMATIC OPENERS INDUSTRIAL**									
0015	Sliding doors up to 6' wide	2 Skwk	.60	26.667	Opng.	6,000	1,400		7,400	8,725
0200	To 12' wide	"	.40	40	"	7,175	2,100		9,275	11,100
0400	Over 12' wide, add per L.F. of excess				L.F.	815			815	895
1000	Swing doors, to 5' wide	2 Skwk	.80	20	Ea.	3,550	1,050		4,600	5,500
1860	Add for controls, wall pushbutton, 3 button		4	4		245	209		454	590
1870	Control pull cord		4.30	3.721		198	195		393	515
1880	For addl elec eye for sliding door operation, one side, add					9%				

08 71 20 – Hardware

08 71 20.10 Bolts, Flush

		Crew	Daily Output	Labor-Hours	Unit	Material	2018 Bare Costs Labor	Equipment	Total	Total Incl O&P
0010	**BOLTS, FLUSH**									
0020	Standard, concealed	1 Carp	7	1.143	Ea.	22.50	58		80.50	113
0800	Automatic fire exit	"	5	1.600		170	81		251	310
1600	Electrified dead bolt	1 Elec	3	2.667		165	155		320	415
3000	Barrel, brass, 2" long	1 Carp	40	.200		4.69	10.15		14.84	20.50
3020	4" long		40	.200		13.10	10.15		23.25	30
3060	6" long		40	.200		23.50	10.15		33.65	41.50

08 71 20.15 Hardware

		Crew	Daily Output	Labor-Hours	Unit	Material	2018 Bare Costs Labor	Equipment	Total	Total Incl O&P
0010	**HARDWARE**									
0020	Average hardware cost									
1000	Door hardware, apartment, interior	1 Carp	4	2	Door	505	101		606	710
1300	Average, door hardware, motel/hotel interior, with access card		4	2		660	101		761	880
1500	Hospital bedroom, average quality		4	2		685	101		786	905
2000	High quality		3	2.667		670	135		805	945
2100	Pocket door		6	1.333	Ea.	111	67.50		178.50	225

08 71 Door Hardware

08 71 20 – Hardware

08 71 20.15 Hardware

		Crew	Daily Output	Labor-Hours	Unit	Material	2018 Bare Costs Labor	Equipment	Total	Total Incl O&P
2250	School, single exterior, incl. lever, incl. panic device	1 Carp	3	2.667	Door	1,450	135		1,585	1,800
2500	Single interior, regular use, lever included		3	2.667		550	135		685	810
2550	Avg., door hdwe., school, classroom, ANSI F84, lever handle		3	2.667		785	135		920	1,075
2600	Avg., door hdwe. set, school, classroom, ANSI F88, incl. lever		3	2.667		850	135		985	1,150
2850	Stairway, single interior		3	2.667		620	135		755	890
3100	Double exterior, with panic device		2	4	Pr.	2,725	203		2,928	3,300
6020	Add for fire alarm door holder, electro-magnetic	1 Elec	4	2	Ea.	104	116		220	288

08 71 20.20 Door Protectors

		Crew	Daily Output	Labor-Hours	Unit	Material	2018 Bare Costs Labor	Equipment	Total	Total Incl O&P
0010	**DOOR PROTECTORS**									
0020	1-3/4" x 3/4" U channel	2 Carp	80	.200	L.F.	22.50	10.15		32.65	40.50
0021	1-3/4" x 1-1/4" U channel		80	.200	"	20	10.15		30.15	37.50
1000	Tear drop, spring stl., 8" high x 19" long		15	1.067	Ea.	104	54		158	197
1010	8" high x 32" long		15	1.067		129	54		183	225
1100	Tear drop, stainless stl., 8" high x 19" long		15	1.067		273	54		327	385
1200	8" high x 32" long		15	1.067		340	54		394	460

08 71 20.30 Door Closers

		Crew	Daily Output	Labor-Hours	Unit	Material	2018 Bare Costs Labor	Equipment	Total	Total Incl O&P
0010	**DOOR CLOSERS** adjustable backcheck, multiple mounting									
0015	and rack and pinion									
0020	Standard Regular Arm	1 Carp	6	1.333	Ea.	201	67.50		268.50	325
0040	Hold open arm		6	1.333		96.50	67.50		164	209
0100	Fusible link		6.50	1.231		172	62.50		234.50	284
0210	Light duty, regular arm		6	1.333		117	67.50		184.50	231
0220	Parallel arm		6	1.333		143	67.50		210.50	260
0230	Hold open arm		6	1.333		124	67.50		191.50	240
0240	Fusible link arm		6	1.333		156	67.50		223.50	275
0250	Medium duty, regular arm		6	1.333		124	67.50		191.50	239
0500	Surface mount regular arm		6.50	1.231		166	62.50		228.50	278
0550	Fusible link		6.50	1.231		150	62.50		212.50	260
1520	Overhead concealed, all sizes, regular arm		5.50	1.455		210	74		284	345
1525	Concealed arm		5	1.600		335	81		416	490
1530	Concealed in door, all sizes, regular arm		5.50	1.455		350	74		424	495
1535	Concealed arm		5	1.600		274	81		355	425
1560	Floor concealed, all sizes, single acting		2.20	3.636		555	184		739	890
1565	Double acting		2.20	3.636		535	184		719	865
1610	Hold open arm		6	1.333		450	67.50		517.50	600
1620	Double acting, standard arm		6	1.333		830	67.50		897.50	1,025
1630	Hold open arm		6	1.333		840	67.50		907.50	1,025
1640	Floor, center hung, single acting, bottom arm		6	1.333		440	67.50		507.50	590
1650	Double acting		6	1.333		500	67.50		567.50	660
1660	Offset hung, single acting, bottom arm		6	1.333		600	67.50		667.50	765
2000	Backcheck and adjustable power, hinge face mount									
5000	For cast aluminum cylinder, deduct				Ea.	37			37	40.50
5010	For delayed action surface mounted, add	1 Carp	6	1.333		38	67.50		105.50	145
5040	For delayed action, add					52			52	57.50
5080	For fusible link arm, add					43.50			43.50	48
5120	For shock absorbing arm, add					51.50			51.50	57
5160	For spring power adjustment, add					41.50			41.50	45.50
6000	Closer-holder, hinge face mount, all sizes, exposed arm	1 Carp	6.50	1.231		219	62.50		281.50	335
6500	Electro magnetic closer/holder									
6510	Single point, no detector	1 Carp	4	2	Ea.	550	101		651	760
6515	Including detector		4	2		705	101		806	930
6520	Multi-point, no detector		4	2		965	101		1,066	1,200

For customer support on your Building Construction Costs with RSMeans data, call 800.448.8182.

08 71 Door Hardware

08 71 20 – Hardware

08 71 20.30 Door Closers

		Crew	Daily Output	Labor-Hours	Unit	Material	2018 Bare Costs Labor	Equipment	Total	Total Incl O&P
6524	Including detector	1 Carp	4	2	Ea.	1,450	101		1,551	1,725
6550	Electric automatic operators									
6555	Operator	1 Carp	4	2	Ea.	2,050	101		2,151	2,400
6570	Wall plate actuator		4	2		228	101		329	405
7000	Electronic closer-holder, hinge facemount, concealed arm		5	1.600		445	81		526	615
7400	With built-in detector		5	1.600		615	81		696	800
8000	Surface mounted, standard duty, parallel arm, primed, traditional		6	1.333		197	67.50		264.50	320
8030	Light duty		6	1.333		160	67.50		227.50	279
8042	Extra duty parallel arm		6	1.333		124	67.50		191.50	240
8044	Hold open arm		6	1.333		167	67.50		234.50	286
8046	Positive stop arm		6	1.333		229	67.50		296.50	355
8050	Heavy duty		6	1.333		200	67.50		267.50	325
8052	Heavy duty, regular arm		6	1.333		231	67.50		298.50	360
8054	Top jamb mount		6	1.333		231	67.50		298.50	360
8056	Extra duty parallel arm		6	1.333		232	67.50		299.50	360
8058	Hold open arm		6	1.333		284	67.50		351.50	420
8060	Positive stop arm		6	1.333		260	67.50		327.50	390
8062	Fusible link arm		6	1.333		276	67.50		343.50	410
8080	Universal heavy duty, regular arm		6	1.333		243	67.50		310.50	370
8084	Parallel arm		6	1.333		243	67.50		310.50	370
8088	Extra duty, parallel arm		6	1.333		272	67.50		339.50	400
8090	Hold open arm		6	1.333		279	67.50		346.50	410
8094	Positive stop arm		6	1.333		299	67.50		366.50	435
8100	Standard duty, parallel arm, modern		6	1.333		271	67.50		338.50	400
8150	Heavy duty		6	1.333		305	67.50		372.50	440

08 71 20.36 Panic Devices

		Crew	Daily Output	Labor-Hours	Unit	Material	2018 Bare Costs Labor	Equipment	Total	Total Incl O&P
0010	**PANIC DEVICES** R087110-10									
0015	Touch bars various styles									
0040	Single door exit only, rim device, wide stile									
0050	Economy US28	1 Carp	5	1.600	Ea.	279	81		360	430
0060	Standard duty US28		5	1.600		715	81		796	915
0065	US26D		5	1.600		805	81		886	1,025
0070	US10		5	1.600		775	81		856	980
0075	US3		5	1.600		885	81		966	1,100
0080	Night latch, economy US28		5	1.600		380	81		461	545
0085	Standard duty, night latch US28		5	1.600		815	81		896	1,025
0090	US26D		5	1.600		1,125	81		1,206	1,375
0095	US10		5	1.600		1,075	81		1,156	1,325
0100	US3		5	1.600		1,225	81		1,306	1,475
0500	Single door exit only, rim device, narrow stile									
0540	Economy US28	1 Carp	5	1.600	Ea.	465	81		546	640
0550	Standard duty US28		5	1.600		875	81		956	1,075
0565	US26D		5	1.600		780	81		861	980
0570	US10		5	1.600		795	81		876	1,000
0575	US3		5	1.600		1,000	81		1,081	1,225
0580	Economy with night latch US28		5	1.600		335	81		416	495
0585	Standard duty US28		5	1.600		940	81		1,021	1,150
0590	US26D		5	1.600		1,125	81		1,206	1,375
0595	US10		5	1.600		945	81		1,026	1,175
0600	US32D		5	1.600		1,250	81		1,331	1,500
1040	Single door exit only, surface vertical rod									
1050	Surface vertical rod, economy US28	1 Carp	4	2	Ea.	490	101		591	695

For customer support on your Building Construction Costs with RSMeans data, call 800.448.8182.

08 71 20.36 Panic Devices

		Crew	Daily Output	Labor-Hours	Unit	Material	2018 Bare Costs Labor	2018 Bare Costs Equipment	Total	Total Incl O&P
1060	Surface vertical rod, standard duty US28	1 Carp	4	2	Ea.	1,125	101		1,226	1,400
1065	US26D		4	2		1,250	101		1,351	1,500
1070	US10		4	2		1,150	101		1,251	1,425
1075	US3		4	2		1,200	101		1,301	1,475
1080	US32D		4	2		1,250	101		1,351	1,550
1500	Single door exit, rim, narrow stile, economy, surface vertical rod									
1540	Surface vertical rod, narrow, economy US28	1 Carp	4	2	Ea.	500	101		601	705
2040	Single door exit only, concealed vertical rod									
2042	These devices will require separate and extra door preparation									
2060	Concealed rod, standard duty US28	1 Carp	4	2	Ea.	1,050	101		1,151	1,300
2065	US26D		4	2		1,175	101		1,276	1,425
2070	US10		4	2		1,275	101		1,376	1,550
2560	Concealed rod, narrow, standard duty US28		4	2		1,100	101		1,201	1,375
2565	US26D		4	2		1,325	101		1,426	1,600
2567	US26		4	2		1,400	101		1,501	1,700
2570	US10		4	2		1,150	101		1,251	1,425
2575	US3		4	2		1,300	101		1,401	1,600
3040	Single door exit only, mortise device, wide stile									
3042	These devices must be combined with a mortise lock									
3060	Mortise, standard duty US28	1 Carp	4	2	Ea.	1,125	101		1,226	1,375
3065	US26D		4	2		1,075	101		1,176	1,350
3067	US26		4	2		1,250	101		1,351	1,525
3070	US10		4	2		1,150	101		1,251	1,425
3075	US3		4	2		1,425	101		1,526	1,725
3080	US32D		4	2		1,125	101		1,226	1,375
3500	Single door exit only, mortise device, narrow stile									
3510	These devices must be combined with a mortise lock									
3540	Mortise, economy US28	1 Carp	4	2	Ea.	510	101		611	715

08 71 20.40 Lockset

		Crew	Daily Output	Labor-Hours	Unit	Material	2018 Bare Costs Labor	2018 Bare Costs Equipment	Total	Total Incl O&P
0010	**LOCKSET**, Standard duty									
0020	Non-keyed, passage, w/sect. trim	1 Carp	12	.667	Ea.	78	34		112	138
0100	Privacy		12	.667		80	34		114	140
0400	Keyed, single cylinder function		10	.800		151	40.50		191.50	229
0420	Hotel (see also Section 08 71 20.15)		8	1		210	50.50		260.50	310
0500	Lever handled, keyed, single cylinder function		10	.800		127	40.50		167.50	202
1000	Heavy duty with sectional trim, non-keyed, passages		12	.667		127	34		161	191
1100	Privacy		12	.667		157	34		191	225
1400	Keyed, single cylinder function		10	.800		188	40.50		228.50	268
1420	Hotel		8	1		525	50.50		575.50	655
1600	Communicating		10	.800		310	40.50		350.50	400
1690	For re-core cylinder, add					52			52	57
3800	Cipher lockset w/key pad (security item)	1 Carp	13	.615		1,000	31		1,031	1,150
3900	Cipher lockset with dial for swinging doors (security item)		13	.615		1,850	31		1,881	2,075
3920	with dial for swinging doors & drill resistant plate (security item)		12	.667		2,250	34		2,284	2,525
3950	Cipher lockset with dial for safe/vault door (security item)		12	.667		1,650	34		1,684	1,850
3980	Keyless, pushbutton type									
4000	Residential/light commercial, deadbolt, standard	1 Carp	9	.889	Ea.	144	45		189	228
4010	Heavy duty		9	.889		241	45		286	335
4020	Industrial, heavy duty, with deadbolt		9	.889		425	45		470	535
4030	Key override		9	.889		425	45		470	540
4040	Lever activated handle		9	.889		435	45		480	545
4050	Key override		9	.889		445	45		490	560

For customer support on your Building Construction Costs with RSMeans data, call 800.448.8182.

08 71 Door Hardware

08 71 20 – Hardware

08 71 20.40 Lockset

		Crew	Daily Output	Labor-Hours	Unit	Material	2018 Bare Costs Labor	Equipment	Total	Total Incl O&P
4060	Double sided pushbutton type	1 Carp	8	1	Ea.	800	50.50		850.50	955
4070	Key override	▼	8	1	▼	820	50.50		870.50	980

08 71 20.41 Dead Locks

		Crew	Daily Output	Labor-Hours	Unit	Material	2018 Bare Costs Labor	Equipment	Total	Total Incl O&P
0010	**DEAD LOCKS**									
0011	Mortise heavy duty outside key (security item)	1 Carp	9	.889	Ea.	182	45		227	269
0020	Double cylinder		9	.889		182	45		227	269
0100	Medium duty, outside key		10	.800		112	40.50		152.50	185
0110	Double cylinder		10	.800		121	40.50		161.50	195
1000	Tubular, standard duty, outside key		10	.800		50.50	40.50		91	118
1010	Double cylinder		10	.800		65.50	40.50		106	135
1200	Night latch, outside key	▼	10	.800	▼	50	40.50		90.50	117

08 71 20.42 Mortise Locksets

		Crew	Daily Output	Labor-Hours	Unit	Material	2018 Bare Costs Labor	Equipment	Total	Total Incl O&P
0010	**MORTISE LOCKSETS**, Comm., wrought knobs & full escutcheon trim									
0015	Assumes mortise is cut									
0020	Non-keyed, passage, Grade 3	1 Carp	9	.889	Ea.	157	45		202	242
0030	Grade 1		8	1		425	50.50		475.50	540
0040	Privacy set, Grade 3		9	.889		172	45		217	259
0050	Grade 1		8	1		465	50.50		515.50	585
0100	Keyed, office/entrance/apartment, Grade 2		8	1		201	50.50		251.50	298
0110	Grade 1		7	1.143		530	58		588	670
0120	Single cylinder, typical, Grade 3		8	1		193	50.50		243.50	289
0130	Grade 1		7	1.143		525	58		583	670
0200	Hotel, room, Grade 3		7	1.143		192	58		250	299
0210	Grade 1 (see also Section 08 71 20.15)		6	1.333		530	67.50		597.50	685
0300	Double cylinder, Grade 3		8	1		229	50.50		279.50	330
0310	Grade 1		7	1.143		545	58		603	690
1000	Wrought knobs and sectional trim, non-keyed, passage, Grade 3		10	.800		133	40.50		173.50	208
1010	Grade 1		9	.889		425	45		470	540
1040	Privacy, Grade 3		10	.800		153	40.50		193.50	231
1050	Grade 1		9	.889		480	45		525	600
1100	Keyed, entrance, office/apartment, Grade 3		9	.889		225	45		270	315
1103	Install lockset		6.92	1.156		225	58.50		283.50	335
1110	Grade 1		8	1		550	50.50		600.50	675
1120	Single cylinder, Grade 3		9	.889		230	45		275	320
1130	Grade 1	▼	8	1	▼	510	50.50		560.50	635
2000	Cast knobs and full escutcheon trim									
2010	Non-keyed, passage, Grade 3	1 Carp	9	.889	Ea.	280	45		325	380
2020	Grade 1		8	1		385	50.50		435.50	500
2040	Privacy, Grade 3		9	.889		325	45		370	430
2050	Grade 1		8	1		450	50.50		500.50	570
2120	Keyed, single cylinder, Grade 3		8	1		335	50.50		385.50	445
2123	Mortise lock		6.15	1.301		335	66		401	470
2130	Grade 1		7	1.143		535	58		593	680
3000	Cast knob and sectional trim, non-keyed, passage, Grade 3		10	.800		215	40.50		255.50	298
3010	Grade 1		10	.800		395	40.50		435.50	490
3040	Privacy, Grade 3		10	.800		229	40.50		269.50	315
3050	Grade 1		10	.800		450	40.50		490.50	555
3100	Keyed, office/entrance/apartment, Grade 3		9	.889		260	45		305	355
3110	Grade 1		9	.889		590	45		635	715
3120	Single cylinder, Grade 3		9	.889		265	45		310	360
3130	Grade 1	▼	9	.889		535	45		580	660
3190	For re-core cylinder, add				▼	86			86	94.50

08 71 20 – Hardware

08 71 20.44 Anti-Ligature Locksets Grade 1

		Crew	Daily Output	Labor-Hours	Unit	Material	2018 Bare Costs Labor	Equipment	Total	Total Incl O&P
0010	**ANTI-LIGATURE LOCKSETS GRADE 1**									
0100	Anti-ligature cylindrical locksets									
0500	Arch handle passage set, US32D	1 Carp	8	1	Ea.	500	50.50		550.50	625
0510	privacy set		8	1		600	50.50		650.50	735
0520	classroom set		8	1		615	50.50		665.50	755
0530	storeroom set		8	1		615	50.50		665.50	755
0550	Lever handle passage set US32D		8	1		475	50.50		525.50	600
0560	Hospital/privacy set		8	1		495	50.50		545.50	620
0570	Office set		8	1		445	50.50		495.50	565
0580	Classroom set		8	1		500	50.50		550.50	625
0590	Storeroom set		8	1		445	50.50		495.50	565
0600	Exit set		8	1		485	50.50		535.50	610
0610	Entry set		8	1		490	50.50		540.50	615
0620	Asylum set		8	1		460	50.50		510.50	580
0630	Back to back dummy set		8	1		295	50.50		345.50	400
0690	Anti-ligature mortise locksets									
0700	Knob handle privacy set, US32D	1 Carp	8	1	Ea.	620	50.50		670.50	760
0710	Office set		8	1		725	50.50		775.50	875
0720	Classroom set		8	1		725	50.50		775.50	875
0730	Hotel set		8	1		740	50.50		790.50	890
0740	Apartment set		8	1		740	50.50		790.50	890
0750	Institutional privacy set		8	1		785	50.50		835.50	940
0760	Lever handle office set US32D		8	1		600	50.50		650.50	735
0770	Classroom set sectional trim		8	1		600	50.50		650.50	735
0780	Hotel set		8	1		615	50.50		665.50	755
0790	Apartment set		8	1		615	50.50		665.50	755
0800	Institutional privacy set		8	1		665	50.50		715.50	810
0810	Office set		8	1		765	50.50		815.50	920
0820	Classroom set escutcheon trim		8	1		765	50.50		815.50	920
0830	Hotel set escutcheon trim		8	1		785	50.50		835.50	935
0840	Apartment set escutcheon trim		8	1		785	50.50		835.50	935
0850	Institutional privacy set escutcheon trim		8	1		785	50.50		835.50	935

08 71 20.45 Peepholes

		Crew	Daily Output	Labor-Hours	Unit	Material	Labor	Equipment	Total	Total Incl O&P
0010	**PEEPHOLES**									
2010	Peephole	1 Carp	32	.250	Ea.	8.95	12.70		21.65	29
2020	Peephole, wide view	"	32	.250	"	12.55	12.70		25.25	33

08 71 20.50 Door Stops

		Crew	Daily Output	Labor-Hours	Unit	Material	Labor	Equipment	Total	Total Incl O&P
0010	**DOOR STOPS**									
0020	Holder & bumper, floor or wall	1 Carp	32	.250	Ea.	36	12.70		48.70	59
1300	Wall bumper, 4" diameter, with rubber pad, aluminum		32	.250		12.70	12.70		25.40	33.50
1600	Door bumper, floor type, aluminum		32	.250		3.32	12.70		16.02	23
1900	Plunger type, door mounted		32	.250		30.50	12.70		43.20	53.50

08 71 20.55 Push-Pull Plates

		Crew	Daily Output	Labor-Hours	Unit	Material	Labor	Equipment	Total	Total Incl O&P
0010	**PUSH-PULL PLATES**									
0090	Push plate, 0.050 thick, 3" x 12", aluminum	1 Carp	12	.667	Ea.	5.95	34		39.95	58
0100	4" x 16"		12	.667		14.50	34		48.50	67.50
0110	6" x 16"		12	.667		8	34		42	60.50
0120	8" x 16"		12	.667		9.40	34		43.40	62
0200	Push plate, 0.050 thick, 3" x 12", brass		12	.667		13.70	34		47.70	66.50
0210	4" x 16"		12	.667		17.10	34		51.10	70.50
0220	6" x 16"		12	.667		26.50	34		60.50	80.50

For customer support on your Building Construction Costs with RSMeans data, call 800.448.8182.

08 71 Door Hardware

08 71 20 – Hardware

08 71 20.55 Push-Pull Plates

		Crew	Daily Output	Labor-Hours	Unit	Material	2018 Bare Costs Labor	2018 Bare Costs Equipment	Total	Total Incl O&P
0230	8" x 16"	1 Carp	12	.667	Ea.	34.50	34		68.50	89.50
0250	Push plate, 0.050 thick, 3" x 12", satin brass		12	.667		13.90	34		47.90	67
0260	4" x 16"		12	.667		17.30	34		51.30	70.50
0270	6" x 16"		12	.667		26.50	34		60.50	80.50
0280	8" x 16"		12	.667		34.50	34		68.50	89.50
0490	Push plate, 0.050 thick, 3" x 12", bronze		12	.667		16.50	34		50.50	69.50
0500	4" x 16"		12	.667		26	34		60	80.50
0510	6" x 16"		12	.667		31.50	34		65.50	86
0520	8" x 16"		12	.667		38	34		72	93
0740	Push plate, 0.050 thick, 3" x 12", stainless steel		12	.667		11.65	34		45.65	64.50
0760	6" x 16"		12	.667		18.30	34		52.30	71.50
0780	8" x 16"		12	.667		23.50	34		57.50	77
0790	Push plate, 0.050 thick, 3" x 12", satin stainless steel		12	.667		6.60	34		40.60	59
0820	6" x 16"		12	.667		11.30	34		45.30	64
0830	8" x 16"		12	.667		15.60	34		49.60	68.50
0980	Pull plate, 0.050 thick, 3" x 12", aluminum		12	.667		25.50	34		59.50	79.50
1050	Pull plate, 0.050 thick, 3" x 12", brass		12	.667		39.50	34		73.50	95
1080	Pull plate, 0.050 thick, 3" x 12", bronze		12	.667		50	34		84	107
1180	Pull plate, 0.050 thick, 3" x 12", stainless steel		12	.667		45.50	34		79.50	102
1250	Pull plate, 0.050 thick, 3" x 12", chrome		12	.667		45.50	34		79.50	102
1500	Pull handle and push bar, aluminum		11	.727		123	37		160	191
2000	Bronze		10	.800		166	40.50		206.50	245

08 71 20.60 Entrance Locks

		Crew	Daily Output	Labor-Hours	Unit	Material	Labor	Equipment	Total	Total Incl O&P
0010	**ENTRANCE LOCKS**									
0015	Cylinder, grip handle deadlocking latch	1 Carp	9	.889	Ea.	181	45		226	268
0020	Deadbolt		8	1		195	50.50		245.50	291
0100	Push and pull plate, dead bolt		8	1		234	50.50		284.50	335
0900	For handicapped lever, add					152			152	167

08 71 20.65 Thresholds

		Crew	Daily Output	Labor-Hours	Unit	Material	Labor	Equipment	Total	Total Incl O&P
0010	**THRESHOLDS**									
0011	Threshold 3' long saddles aluminum	1 Carp	48	.167	L.F.	10.40	8.45		18.85	24.50
0100	Aluminum, 8" wide, 1/2" thick		12	.667	Ea.	50	34		84	107
0500	Bronze		60	.133	L.F.	45.50	6.75		52.25	60.50
0600	Bronze, panic threshold, 5" wide, 1/2" thick		12	.667	Ea.	162	34		196	230
0700	Rubber, 1/2" thick, 5-1/2" wide		20	.400		41	20.50		61.50	76
0800	2-3/4" wide		20	.400		49.50	20.50		70	85
1950	ADA compliant thresholds									
2300	Threshold, aluminum 4" wide x 36" long	1 Carp	12	.667	Ea.	25	34		59	79
2310	4" wide x 48" long		12	.667		30	34		64	84.50
2360	6" wide x 36" long		12	.667		39	34		73	94.50
2370	6" wide x 48" long		12	.667		48	34		82	105

08 71 20.70 Floor Checks

		Crew	Daily Output	Labor-Hours	Unit	Material	Labor	Equipment	Total	Total Incl O&P
0010	**FLOOR CHECKS**									
0020	For over 3' wide doors single acting	1 Carp	2.50	3.200	Ea.	775	162		937	1,100
0500	Double acting	"	2.50	3.200	"	885	162		1,047	1,225

08 71 20.75 Door Hardware Accessories

		Crew	Daily Output	Labor-Hours	Unit	Material	Labor	Equipment	Total	Total Incl O&P
0010	**DOOR HARDWARE ACCESSORIES**									
0050	Door closing coordinator, 36" (for paired openings up to 56")	1 Carp	8	1	Ea.	113	50.50		163.50	202
0060	48" (for paired openings up to 84")		8	1		120	50.50		170.50	209
0070	56" (for paired openings up to 96")		8	1		129	50.50		179.50	219

For customer support on your Building Construction Costs with RSMeans data, call 800.448.8182.

305

08 71 Door Hardware

08 71 20 – Hardware

08 71 20.80 Hasps	Crew	Daily Output	Labor-Hours	Unit	Material	2018 Bare Costs Labor	Equipment	Total	Total Incl O&P
0010 **HASPS**, steel assembly									
0015 3"	1 Carp	26	.308	Ea.	5.45	15.60		21.05	30
0020 4-1/2"		13	.615		6.70	31		37.70	55
0040 6"		12.50	.640		9.95	32.50		42.45	60.50

08 71 20.90 Hinges	Crew	Daily Output	Labor-Hours	Unit	Material	2018 Bare Costs Labor	Equipment	Total	Total Incl O&P
0010 **HINGES**									
0012 Full mortise, avg. freq., steel base, USP, 4-1/2" x 4-1/2"				Pr.	39.50			39.50	43.50
0100 5" x 5", USP					59			59	65
0200 6" x 6", USP					124			124	137
0400 Brass base, 4-1/2" x 4-1/2", US10					64			64	70
0500 5" x 5", US10					71.50			71.50	78.50
0600 6" x 6", US10					164			164	180
0800 Stainless steel base, 4-1/2" x 4-1/2", US32					77.50			77.50	85
0900 For non removable pin, add (security item)				Ea.	5.55			5.55	6.10
0910 For floating pin, driven tips, add					3.40			3.40	3.74
0930 For hospital type tip on pin, add					14.20			14.20	15.60
0940 For steeple type tip on pin, add					19.30			19.30	21
0950 Full mortise, high frequency, steel base, 3-1/2" x 3-1/2", US26D				Pr.	31			31	34
1000 4-1/2" x 4-1/2", USP					66.50			66.50	73.50
1100 5" x 5", USP					52.50			52.50	58
1200 6" x 6", USP					139			139	153
1400 Brass base, 3-1/2" x 3-1/2", US4					52.50			52.50	57.50
1430 4-1/2" x 4-1/2", US10					76			76	83.50
1500 5" x 5", US10					127			127	139
1600 6" x 6", US10					171			171	188
1800 Stainless steel base, 4-1/2" x 4-1/2", US32					104			104	115
1810 5" x 4-1/2", US32					140			140	154
1930 For hospital type tip on pin, add				Ea.	14.60			14.60	16.05
1950 Full mortise, low frequency, steel base, 3-1/2" x 3-1/2", US26D				Pr.	28			28	31
2000 4-1/2" x 4-1/2", USP					23.50			23.50	25.50
2100 5" x 5", USP					49			49	54
2200 6" x 6", USP					94			94	104
2300 4-1/2" x 4-1/2", US3					17.70			17.70	19.45
2310 5" x 5", US3					42.50			42.50	46.50
2400 Brass bass, 4-1/2" x 4-1/2", US10					55			55	60.50
2500 5" x 5", US10					79			79	86.50
2800 Stainless steel base, 4-1/2" x 4-1/2", US32					76.50			76.50	84.50

08 71 20.91 Special Hinges	Crew	Daily Output	Labor-Hours	Unit	Material	2018 Bare Costs Labor	Equipment	Total	Total Incl O&P
0010 **SPECIAL HINGES**									
0015 Paumelle, high frequency									
0020 Steel base, 6" x 4-1/2", US10				Pr.	138			138	152
0100 Brass base, 5" x 4-1/2", US10				Ea.	248			248	273
0200 Paumelle, average frequency, steel base, 4-1/2" x 3-1/2", US10				Pr.	93.50			93.50	103
0400 Olive knuckle, low frequency, brass base, 6" x 4-1/2", US10				Ea.	143			143	158
1000 Electric hinge with concealed conductor, average frequency									
1010 Steel base, 4-1/2" x 4-1/2", US26D				Pr.	315			315	345
1100 Bronze base, 4-1/2" x 4-1/2", US26D				"	335			335	365
1200 Electric hinge with concealed conductor, high frequency									
1210 Steel base, 4-1/2" x 4-1/2", US26D				Pr.	252			252	277
1600 Double weight, 800 lb., steel base, removable pin, 5" x 6", USP					485			485	535
1700 Steel base-welded pin, 5" x 6", USP					170			170	187
1800 Triple weight, 2000 lb., steel base, welded pin, 5" x 6", USP					630			630	690

For customer support on your Building Construction Costs with RSMeans data, call 800.448.8182.

08 71 Door Hardware

08 71 20 – Hardware

08 71 20.91 Special Hinges

		Crew	Daily Output	Labor-Hours	Unit	Material	2018 Bare Costs Labor	Equipment	Total	Total Incl O&P
2000	Pivot reinf., high frequency, steel base, 7-3/4" door plate, USP				Pr.	151			151	167
2200	Bronze base, 7-3/4" door plate, US10				▼	238			238	262
3000	Swing clear, full mortise, full or half surface, high frequency,									
3010	Steel base, 5" high, USP				Pr.	144			144	159
3200	Swing clear, full mortise, average frequency									
3210	Steel base, 4-1/2" high, USP				Pr.	130			130	143
4000	Wide throw, average frequency, steel base, 4-1/2" x 6", USP					90.50			90.50	99.50
4200	High frequency, steel base, 4-1/2" x 6", USP				▼	113			113	125
4600	Spring hinge, single acting, 6" flange, steel				Ea.	53			53	58
4700	Brass					92.50			92.50	102
4900	Double acting, 6" flange, steel					84			84	92.50
4950	Brass				▼	130			130	143
8000	Continuous hinges									
8010	Steel, piano, 2" x 72"	1 Carp	20	.400	Ea.	23	20.50		43.50	56.50
8020	Brass, piano, 1-1/16" x 30"		30	.267		9	13.50		22.50	30.50
8030	Acrylic, piano, 1-3/4" x 12"		40	.200		16	10.15		26.15	33
9000	Continuous hinge, steel, full mortise, heavy duty, 96"	▼	2	4	▼	380	203		583	730

08 71 20.95 Kick Plates

		Crew	Daily Output	Labor-Hours	Unit	Material	2018 Bare Costs Labor	Equipment	Total	Total Incl O&P
0010	**KICK PLATES**									
0020	Stainless steel, .050", 16 ga., 8" x 28", US32	1 Carp	15	.533	Ea.	41	27		68	86
0030	8" x 30"		15	.533		44.50	27		71.50	89.50
0040	8" x 34"		15	.533		50	27		77	96
0050	10" x 28"		15	.533		82	27		109	131
0060	10" x 30"		15	.533		88.50	27		115.50	139
0070	10" x 34"		15	.533		99	27		126	150
0080	Mop/Kick, 4" x 28"		15	.533		37	27		64	81.50
0090	4" x 30"		15	.533		39	27		66	84
0100	4" x 34"		15	.533		44	27		71	89.50
0110	6" x 28"		15	.533		42	27		69	87
0120	6" x 30"		15	.533		51	27		78	97
0130	6" x 34"		15	.533		57	27		84	104
0500	Bronze, .050", 8" x 28"		15	.533		72	27		99	121
0510	8" x 30"		15	.533		69	27		96	117
0520	8" x 34"		15	.533		77.50	27		104.50	127
0530	10" x 28"		15	.533		80.50	27		107.50	130
0540	10" x 30"		15	.533		81.50	27		108.50	131
0550	10" x 34"		15	.533		98	27		125	149
0560	Mop/Kick, 4" x 28"		15	.533		35	27		62	80
0570	4" x 30"		15	.533		38.50	27		65.50	83.50
0580	4" x 34"		15	.533		39.50	27		66.50	84.50
0590	6" x 28"		15	.533		49	27		76	95
0600	6" x 30"		15	.533		50.50	27		77.50	96.50
0610	6" x 34"		15	.533		60	27		87	107
1000	Acrylic, .125", 8" x 26"		15	.533		27	27		54	71
1010	8" x 36"		15	.533		37.50	27		64.50	82
1020	8" x 42"		15	.533		44	27		71	89.50
1030	10" x 26"		15	.533		34	27		61	78.50
1040	10" x 36"		15	.533		47.50	27		74.50	93
1050	10" x 42"		15	.533		55	27		82	102
1060	Mop/Kick, 4" x 26"		15	.533		16.40	27		43.40	59
1070	4" x 36"		15	.533		22.50	27		49.50	65.50
1080	4" x 42"		15	.533		26	27		53	69.50

08 71 20 – Hardware

08 71 20.95 Kick Plates		Crew	Daily Output	Labor-Hours	Unit	Material	2018 Bare Costs Labor	Equipment	Total	Total Incl O&P
1090	6" x 26"	1 Carp	15	.533	Ea.	24	27		51	67.50
1100	6" x 36"		15	.533		33	27		60	77.50
1110	6" x 42"		15	.533		38.50	27		65.50	83.50
1220	Brass, .050", 8" x 26"		15	.533		69.50	27		96.50	117
1230	8" x 36"		15	.533		73.50	27		100.50	122
1240	8" x 42"		15	.533		84.50	27		111.50	134
1250	10" x 26"		15	.533		70	27		97	118
1260	10" x 36"		15	.533		89.50	27		116.50	140
1270	10" x 42"		15	.533		103	27		130	155
1320	Mop/Kick, 4" x 26"		15	.533		23	27		50	66.50
1330	4" x 36"		15	.533		32	27		59	76
1340	4" x 42"		15	.533		37	27		64	82
1350	6" x 26"		15	.533		34.50	27		61.50	79
1360	6" x 36"		15	.533		48	27		75	93.50
1370	6" x 42"		15	.533		55	27		82	102
1800	Aluminum, .050", 8" x 26"		15	.533		38	27		65	82.50
1810	8" x 36"		15	.533		36	27		63	80.50
1820	8" x 42"		15	.533		42.50	27		69.50	87.50
1830	10" x 26"		15	.533		32.50	27		59.50	77
1840	10" x 36"		15	.533		45	27		72	90.50
1850	10" x 42"		15	.533		62	27		89	110
1860	Mop/Kick, 4" x 26"		15	.533		15.85	27		42.85	58.50
1870	4" x 36"		15	.533		21	27		48	64
1880	4" x 42"		15	.533		25.50	27		52.50	69
1890	6" x 26"		15	.533		23	27		50	66.50
1900	6" x 36"		15	.533		31.50	27		58.50	75.50
1910	6" x 42"		15	.533		36.50	27		63.50	81.50

08 71 21 – Astragals

08 71 21.10 Exterior Mouldings, Astragals

	08 71 21.10 Exterior Mouldings, Astragals	Crew	Daily Output	Labor-Hours	Unit	Material	2018 Bare Costs Labor	Equipment	Total	Total Incl O&P
0010	**EXTERIOR MOULDINGS, ASTRAGALS**									
0400	One piece, overlapping cadmium plated steel, flat, 3/16" x 2"	1 Carp	90	.089	L.F.	3.88	4.51		8.39	11.10
0600	Prime coated steel, flat, 1/8" x 3"		90	.089		5.90	4.51		10.41	13.35
0800	Stainless steel, flat, 3/32" x 1-5/8"		90	.089		15.55	4.51		20.06	24
1000	Aluminum, flat, 1/8" x 2"		90	.089		3.98	4.51		8.49	11.25
1200	Nail on, "T" extrusion		120	.067		2.05	3.38		5.43	7.40
1300	Vinyl bulb insert		105	.076		2.66	3.86		6.52	8.85
1600	Screw on, "T" extrusion		90	.089		3.79	4.51		8.30	11
1700	Vinyl insert		75	.107		4.37	5.40		9.77	13.05
2000	"L" extrusion, neoprene bulbs		75	.107		4.38	5.40		9.78	13.05
2100	Neoprene sponge insert		75	.107		6.70	5.40		12.10	15.60
2200	Magnetic		75	.107		10.30	5.40		15.70	19.55
2400	Spring hinged security seal, with cam		75	.107		6.60	5.40		12	15.50
2600	Spring loaded locking bolt, vinyl insert		45	.178		8.95	9		17.95	23.50
2800	Neoprene sponge strip, "Z" shaped, aluminum		60	.133		8.05	6.75		14.80	19.15
2900	Solid neoprene strip, nail on aluminum strip		90	.089		3.93	4.51		8.44	11.15
3000	One piece stile protection									
3020	Neoprene fabric loop, nail on aluminum strips	1 Carp	60	.133	L.F.	1.15	6.75		7.90	11.55
3110	Flush mounted aluminum extrusion, 1/2" x 1-1/4"		60	.133		6.70	6.75		13.45	17.65
3140	3/4" x 1-3/8"		60	.133		3.98	6.75		10.73	14.70
3160	1-1/8" x 1-3/4"		60	.133		4.56	6.75		11.31	15.30
3300	Mortise, 9/16" x 3/4"		60	.133		3.98	6.75		10.73	14.70
3320	13/16" x 1-3/8"		60	.133		4.17	6.75		10.92	14.90

For customer support on your Building Construction Costs with RSMeans data, call 800.448.8182.

08 71 Door Hardware

08 71 21 – Astragals

08 71 21.10 Exterior Mouldings, Astragals

		Crew	Daily Output	Labor-Hours	Unit	Material	2018 Bare Costs Labor	Equipment	Total	Total Incl O&P
3600	Spring bronze strip, nail on type	1 Carp	105	.076	L.F.	1.90	3.86		5.76	8
3620	Screw on, with retainer		75	.107		2.62	5.40		8.02	11.15
3800	Flexible stainless steel housing, pile insert, 1/2" door		105	.076		7.05	3.86		10.91	13.65
3820	3/4" door		105	.076		7.85	3.86		11.71	14.55
4000	Extruded aluminum retainer, flush mount, pile insert		105	.076		2.18	3.86		6.04	8.30
4080	Mortise, felt insert		90	.089		4.42	4.51		8.93	11.70
4160	Mortise with spring, pile insert		90	.089		3.30	4.51		7.81	10.50
4400	Rigid vinyl retainer, mortise, pile insert		105	.076		2.62	3.86		6.48	8.80
4600	Wool pile filler strip, aluminum backing		105	.076		2.77	3.86		6.63	8.95
5000	Two piece overlapping astragal, extruded aluminum retainer									
5010	Pile insert	1 Carp	60	.133	L.F.	3.25	6.75		10	13.90
5020	Vinyl bulb insert		60	.133		1.88	6.75		8.63	12.35
5040	Vinyl flap insert		60	.133		3.54	6.75		10.29	14.20
5060	Solid neoprene flap insert		60	.133		6.35	6.75		13.10	17.30
5080	Hypalon rubber flap insert		60	.133		6.45	6.75		13.20	17.40
5090	Snap on cover, pile insert		60	.133		9.15	6.75		15.90	20.50
5400	Magnetic aluminum, surface mounted		60	.133		23.50	6.75		30.25	36
5500	Interlocking aluminum, 5/8" x 1" neoprene bulb insert		45	.178		5.55	9		14.55	19.80
5600	Adjustable aluminum, 9/16" x 21/32", pile insert		45	.178		17.05	9		26.05	32.50
5800	Magnetic, adjustable, 9/16" x 21/32"		45	.178		22	9		31	38
6000	Two piece stile protection									
6010	Cloth backed rubber loop, 1" gap, nail on aluminum strips	1 Carp	45	.178	L.F.	4.22	9		13.22	18.35
6040	Screw on aluminum strips		45	.178		6.35	9		15.35	20.50
6100	1-1/2" gap, screw on aluminum extrusion		45	.178		5.70	9		14.70	19.95
6240	Vinyl fabric loop, slotted aluminum extrusion, 1" gap		45	.178		2.14	9		11.14	16.05
6300	1-1/4" gap		45	.178		6	9		15	20.50

08 71 25 – Weatherstripping

08 71 25.10 Mechanical Seals, Weatherstripping

		Crew	Daily Output	Labor-Hours	Unit	Material	2018 Bare Costs Labor	Equipment	Total	Total Incl O&P
0010	**MECHANICAL SEALS, WEATHERSTRIPPING**									
1000	Doors, wood frame, interlocking, for 3' x 7' door, zinc	1 Carp	3	2.667	Opng.	47.50	135		182.50	259
1100	Bronze		3	2.667		60.50	135		195.50	273
1300	6' x 7' opening, zinc		2	4		58.50	203		261.50	375
1400	Bronze		2	4		69.50	203		272.50	385
1700	Wood frame, spring type, bronze									
1800	3' x 7' door	1 Carp	7.60	1.053	Opng.	25	53.50		78.50	109
1900	6' x 7' door	"	7	1.143	"	31	58		89	122
2200	Metal frame, spring type, bronze									
2300	3' x 7' door	1 Carp	3	2.667	Opng.	49.50	135		184.50	260
2400	6' x 7' door	"	2.50	3.200	"	53	162		215	305
2500	For stainless steel, spring type, add					133%				
2700	Metal frame, extruded sections, 3' x 7' door, aluminum	1 Carp	3	2.667	Opng.	28	135		163	237
2800	Bronze		3	2.667		82.50	135		217.50	297
3100	6' x 7' door, aluminum		1.50	5.333		35	270		305	450
3200	Bronze		1.50	5.333		140	270		410	565
3500	Threshold weatherstripping									
3650	Door sweep, flush mounted, aluminum	1 Carp	25	.320	Ea.	20.50	16.20		36.70	47
3700	Vinyl		25	.320		18.35	16.20		34.55	44.50
5000	Garage door bottom weatherstrip, 12' aluminum, clear		14	.571		25.50	29		54.50	72
5010	Bronze		14	.571		91.50	29		120.50	145
5050	Bottom protection, rubber		14	.571		41.50	29		70.50	89.50
5100	Threshold		14	.571		74.50	29		103.50	126

For customer support on your Building Construction Costs with RSMeans data, call 800.448.8182.

08 75 Window Hardware

08 75 30 – Weatherstripping

08 75 30.10 Mechanical Weather Seals	Crew	Daily Output	Labor-Hours	Unit	Material	2018 Bare Costs Labor	Equipment	Total	Total Incl O&P
0010 **MECHANICAL WEATHER SEALS**, Window, double-hung, 3' x 5'									
0020 Zinc	1 Carp	7.20	1.111	Opng.	22	56.50		78.50	110
0100 Bronze		7.20	1.111		42	56.50		98.50	132
0500 As above but heavy duty, zinc		4.60	1.739		20.50	88		108.50	157
0600 Bronze	↓	4.60	1.739	↓	75.50	88		163.50	217

08 79 Hardware Accessories

08 79 13 – Key Storage Equipment

08 79 13.10 Key Cabinets

	Crew	Daily Output	Labor-Hours	Unit	Material	2018 Bare Costs Labor	Equipment	Total	Total Incl O&P
0010 **KEY CABINETS**									
0020 Wall mounted, 60 key capacity	1 Carp	20	.400	Ea.	96	20.50		116.50	137
0400 Wall mounted, 30 key capacity	1 Clab	50	.160		74	6.40		80.40	91
0500 Wall mounted, 80 key capacity	"	40	.200	↓	118	7.95		125.95	142

08 79 20 – Door Accessories

08 79 20.10 Door Hardware Accessories

	Crew	Daily Output	Labor-Hours	Unit	Material	2018 Bare Costs Labor	Equipment	Total	Total Incl O&P
0010 **DOOR HARDWARE ACCESSORIES**									
0140 Door bolt, surface, 4"	1 Carp	32	.250	Ea.	15.55	12.70		28.25	36.50
0160 Door latch	"	12	.667	"	8.75	34		42.75	61
0200 Sliding closet door									
0220 Track and hanger, single	1 Carp	10	.800	Ea.	67	40.50		107.50	136
0240 Double		8	1		81.50	50.50		132	167
0260 Door guide, single		48	.167		31	8.45		39.45	47
0280 Double		48	.167		41	8.45		49.45	58
0600 Deadbolt and lock cover plate, brass or stainless steel		30	.267		30	13.50		43.50	53.50
0620 Hole cover plate, brass or chrome		35	.229		8.20	11.60		19.80	26.50
2240 Mortise lockset, passage, lever handle		9	.889		161	45		206	246
4000 Security chain, standard	↓	18	.444	↓	12.95	22.50		35.45	49

08 81 Glass Glazing

08 81 10 – Float Glass

08 81 10.10 Various Types and Thickness of Float Glass

	Crew	Daily Output	Labor-Hours	Unit	Material	2018 Bare Costs Labor	Equipment	Total	Total Incl O&P
0010 **VARIOUS TYPES AND THICKNESS OF FLOAT GLASS** R088110-10									
0020 3/16" plain	2 Glaz	130	.123	S.F.	6.80	5.95		12.75	16.55
0200 Tempered, clear		130	.123		7.25	5.95		13.20	17.05
0300 Tinted		130	.123		6.80	5.95		12.75	16.55
0600 1/4" thick, clear, plain		120	.133		9.90	6.45		16.35	20.50
0700 Tinted		120	.133		9.45	6.45		15.90	20
0800 Tempered, clear		120	.133		6.80	6.45		13.25	17.30
0900 Tinted		120	.133		11.70	6.45		18.15	22.50
1600 3/8" thick, clear, plain		75	.213		11.20	10.35		21.55	28
1700 Tinted		75	.213		16.55	10.35		26.90	34
1800 Tempered, clear		75	.213		17.70	10.35		28.05	35
1900 Tinted		75	.213		20	10.35		30.35	37.50
2200 1/2" thick, clear, plain		55	.291		19.05	14.10		33.15	42.50
2300 Tinted		55	.291		29	14.10		43.10	53.50
2400 Tempered, clear		55	.291		26.50	14.10		40.60	50.50
2500 Tinted		55	.291		28	14.10		42.10	52
2800 5/8" thick, clear, plain		45	.356		29	17.25		46.25	58
2900 Tempered, clear	↓	45	.356	↓	33	17.25		50.25	62.50

310

For customer support on your Building Construction Costs with RSMeans data, call 800.448.8182.

08 81 Glass Glazing

08 81 10 – Float Glass

08 81 10.10 Various Types and Thickness of Float Glass

		Crew	Daily Output	Labor-Hours	Unit	Material	2018 Bare Costs Labor	Equipment	Total	Total Incl O&P
3200	3/4" thick, clear, plain	2 Glaz	35	.457	S.F.	37.50	22		59.50	74.50
3300	Tempered, clear		35	.457		43.50	22		65.50	81.50
3600	1" thick, clear, plain		30	.533		62	26		88	108
8900	For low emissivity coating for 3/16" & 1/4" only, add to above					18%				

08 81 13 – Decorative Glass Glazing

08 81 13.10 Beveled Glass

		Crew	Daily Output	Labor-Hours	Unit	Material	2018 Bare Costs Labor	Equipment	Total	Total Incl O&P
0010	**BEVELED GLASS**, with design patterns									
0020	Simple pattern	2 Glaz	150	.107	S.F.	64	5.15		69.15	78.50
0050	Intricate pattern	"	125	.128	"	139	6.20		145.20	162

08 81 13.30 Sandblasted Glass

		Crew	Daily Output	Labor-Hours	Unit	Material	2018 Bare Costs Labor	Equipment	Total	Total Incl O&P
0010	**SANDBLASTED GLASS**, float glass									
0020	1/8" thick	2 Glaz	160	.100	S.F.	11.70	4.85		16.55	20.50
0100	3/16" thick		130	.123		12.90	5.95		18.85	23.50
0500	1/4" thick		120	.133		13.45	6.45		19.90	24.50
0600	3/8" thick		75	.213		14.40	10.35		24.75	31.50

08 81 17 – Fire Glass

08 81 17.10 Fire Resistant Glass

		Crew	Daily Output	Labor-Hours	Unit	Material	2018 Bare Costs Labor	Equipment	Total	Total Incl O&P
0010	**FIRE RESISTANT GLASS**									
0020	Fire glass minimum	2 Glaz	40	.400	S.F.	40.50	19.40		59.90	74
0030	Mid range		40	.400		82.50	19.40		101.90	120
0050	High end		40	.400		365	19.40		384.40	430

08 81 20 – Vision Panels

08 81 20.10 Full Vision

		Crew	Daily Output	Labor-Hours	Unit	Material	2018 Bare Costs Labor	Equipment	Total	Total Incl O&P
0010	**FULL VISION**, window system with 3/4" glass mullions									
0020	Up to 10' high	H-2	130	.185	S.F.	67.50	8.40		75.90	87.50
0100	10' to 20' high, minimum		110	.218		72	9.95		81.95	94
0150	Average		100	.240		77.50	10.95		88.45	102
0200	Maximum		80	.300		86.50	13.70		100.20	117

08 81 25 – Glazing Variables

08 81 25.10 Applications of Glazing

			Crew	Daily Output	Labor-Hours	Unit	Material	2018 Bare Costs Labor	Equipment	Total	Total Incl O&P
0010	**APPLICATIONS OF GLAZING**	R088110-10									
0600	For glass replacement, add					S.F.		100%			
0700	For gasket settings, add					L.F.	6.10			6.10	6.75
0900	For sloped glazing, add					S.F.		26%			
2000	Fabrication, polished edges, 1/4" thick					Inch	.59			.59	.65
2100	1/2" thick						1.38			1.38	1.52
2500	Mitered edges, 1/4" thick						1.38			1.38	1.52
2600	1/2" thick						2.29			2.29	2.52

08 81 30 – Insulating Glass

08 81 30.10 Reduce Heat Transfer Glass

				Crew	Daily Output	Labor-Hours	Unit	Material	2018 Bare Costs Labor	Equipment	Total	Total Incl O&P
0010	**REDUCE HEAT TRANSFER GLASS**		R088110-10									
0015	2 lites 1/8" float, 1/2" thk under 15 S.F.											
0020	Clear	G		2 Glaz	95	.168	S.F.	10.35	8.15		18.50	24
0100	Tinted	G			95	.168		14.50	8.15		22.65	28.50
0200	2 lites 3/16" float, for 5/8" thk unit, 15 to 30 S.F., clear	G			90	.178		13.80	8.60		22.40	28
0300	Tinted	G			90	.178		13.85	8.60		22.45	28.50
0400	1" thk, dbl. glazed, 1/4" float, 30 to 70 S.F., clear	G			75	.213		17.35	10.35		27.70	35
0500	Tinted	G			75	.213		23.50	10.35		33.85	41.50
0600	1" thk, dbl. glazed, 1/4" float, 1/4" wire				75	.213		24.50	10.35		34.85	42.50
0700	1/4" float, 1/4" tempered				75	.213		33.50	10.35		43.85	52

For customer support on your Building Construction Costs with RSMeans data, call 800.448.8182.

311

08 81 Glass Glazing

08 81 30 - Insulating Glass

08 81 30.10 Reduce Heat Transfer Glass

		Crew	Daily Output	Labor-Hours	Unit	Material	2018 Bare Costs Labor	Equipment	Total	Total Incl O&P
0800	1/4" wire, 1/4" tempered	2 Glaz	75	.213	S.F.	32	10.35		42.35	51
2000	Both lites, light & heat reflective [G]		85	.188		31.50	9.15		40.65	49
2500	Heat reflective, film inside, 1" thick unit, clear [G]		85	.188		27.50	9.15		36.65	44.50
2600	Tinted [G]		85	.188		30	9.15		39.15	47.50
3000	Film on weatherside, clear, 1/2" thick unit [G]		95	.168		19.75	8.15		27.90	34
3100	5/8" thick unit [G]		90	.178		20.50	8.60		29.10	35.50
3200	1" thick unit [G]		85	.188		27.50	9.15		36.65	44

08 81 35 - Translucent Glass

08 81 35.10 Obscure Glass

		Crew	Daily Output	Labor-Hours	Unit	Material	2018 Bare Costs Labor	Equipment	Total	Total Incl O&P
0010	**OBSCURE GLASS**									
0020	1/8" thick, textured	2 Glaz	140	.114	S.F.	12.30	5.55		17.85	22
0100	Color		125	.128		14.60	6.20		20.80	25.50
0300	7/32" thick, textured		120	.133		13.50	6.45		19.95	24.50
0400	Color		105	.152		16.95	7.40		24.35	30

08 81 35.20 Patterned Glass

		Crew	Daily Output	Labor-Hours	Unit	Material	2018 Bare Costs Labor	Equipment	Total	Total Incl O&P
0010	**PATTERNED GLASS**, colored									
0020	1/8" thick	2 Glaz	140	.114	S.F.	9.95	5.55		15.50	19.30
0300	7/32" thick	"	120	.133	"	12.45	6.45		18.90	23.50

08 81 45 - Sheet Glass

08 81 45.10 Window Glass, Sheet

		Crew	Daily Output	Labor-Hours	Unit	Material	2018 Bare Costs Labor	Equipment	Total	Total Incl O&P
0010	**WINDOW GLASS, SHEET** gray									
0020	1/8" thick	2 Glaz	160	.100	S.F.	6.45	4.85		11.30	14.45
0200	1/4" thick	"	130	.123	"	7.65	5.95		13.60	17.50

08 81 50 - Spandrel Glass

08 81 50.10 Glass for Non Vision Areas

		Crew	Daily Output	Labor-Hours	Unit	Material	2018 Bare Costs Labor	Equipment	Total	Total Incl O&P
0010	**GLASS FOR NON VISION AREAS**, 1/4" thick standard colors									
0020	Up to 1,000 S.F.	2 Glaz	110	.145	S.F.	18.55	7.05		25.60	31
0200	1,000 to 2,000 S.F.	"	120	.133	"	16.10	6.45		22.55	27.50
0300	For custom colors, add				Total	10%				
0500	For 3/8" thick, add				S.F.	12.65			12.65	13.90
1000	For double coated, 1/4" thick, add					4.34			4.34	4.77
1200	For insulation on panels, add					7.10			7.10	7.80
2000	Panels, insulated, with aluminum backed fiberglass, 1" thick	2 Glaz	120	.133		17.20	6.45		23.65	28.50
2100	2" thick	"	120	.133		21.50	6.45		27.95	33.50

08 81 55 - Window Glass

08 81 55.10 Sheet Glass

		Crew	Daily Output	Labor-Hours	Unit	Material	2018 Bare Costs Labor	Equipment	Total	Total Incl O&P
0010	**SHEET GLASS** (window), clear float, stops, putty bed									
0015	1/8" thick, clear float	2 Glaz	480	.033	S.F.	3.92	1.62		5.54	6.75
0500	3/16" thick, clear		480	.033		6.30	1.62		7.92	9.40
0600	Tinted		480	.033		7.85	1.62		9.47	11.10
0700	Tempered		480	.033		9.75	1.62		11.37	13.15

08 81 65 - Wire Glass

08 81 65.10 Glass Reinforced With Wire

		Crew	Daily Output	Labor-Hours	Unit	Material	2018 Bare Costs Labor	Equipment	Total	Total Incl O&P
0010	**GLASS REINFORCED WITH WIRE**									
0012	1/4" thick rough obscure	2 Glaz	135	.119	S.F.	25.50	5.75		31.25	36.50
1000	Polished wire, 1/4" thick, diamond, clear		135	.119		29.50	5.75		35.25	41
1500	Pinstripe, obscure		135	.119		54	5.75		59.75	68

For customer support on your Building Construction Costs with RSMeans data, call 800.448.8182.

08 83 Mirrors

08 83 13 – Mirrored Glass Glazing

08 83 13.10 Mirrors

		Crew	Daily Output	Labor-Hours	Unit	Material	2018 Bare Costs Labor	Equipment	Total	Total Incl O&P
0010	**MIRRORS**, No frames, wall type, 1/4" plate glass, polished edge									
0100	Up to 5 S.F.	2 Glaz	125	.128	S.F.	9.80	6.20		16	20
0200	Over 5 S.F.		160	.100		9.55	4.85		14.40	17.85
0500	Door type, 1/4" plate glass, up to 12 S.F.		160	.100		9.15	4.85		14	17.40
1000	Float glass, up to 10 S.F., 1/8" thick		160	.100		6.10	4.85		10.95	14.05
1100	3/16" thick		150	.107		7.50	5.15		12.65	16.10
1500	12" x 12" wall tiles, square edge, clear		195	.082		2.61	3.98		6.59	8.90
1600	Veined		195	.082		6.55	3.98		10.53	13.25
2000	1/4" thick, stock sizes, one way transparent		125	.128		20.50	6.20		26.70	32
2010	Bathroom, unframed, laminated		160	.100		14.75	4.85		19.60	23.50

08 83 13.15 Reflective Glass

			Crew	Daily Output	Labor-Hours	Unit	Material	2018 Bare Costs Labor	Equipment	Total	Total Incl O&P
0010	**REFLECTIVE GLASS**										
0100	1/4" float with fused metallic oxide fixed	G	2 Glaz	115	.139	S.F.	17.90	6.75		24.65	30
0500	1/4" float glass with reflective applied coating	G	"	115	.139	"	14.50	6.75		21.25	26

08 84 Plastic Glazing

08 84 10 – Plexiglass Glazing

08 84 10.10 Plexiglass Acrylic

		Crew	Daily Output	Labor-Hours	Unit	Material	2018 Bare Costs Labor	Equipment	Total	Total Incl O&P
0010	**PLEXIGLASS ACRYLIC**, clear, masked,									
0020	1/8" thick, cut sheets	2 Glaz	170	.094	S.F.	11.35	4.56		15.91	19.40
0200	Full sheets		195	.082		5.75	3.98		9.73	12.35
0500	1/4" thick, cut sheets		165	.097		13.35	4.70		18.05	22
0600	Full sheets		185	.086		9.30	4.19		13.49	16.60
0900	3/8" thick, cut sheets		155	.103		21	5		26	30.50
1000	Full sheets		180	.089		16.75	4.31		21.06	25
1300	1/2" thick, cut sheets		135	.119		31.50	5.75		37.25	43
1400	Full sheets		150	.107		18.85	5.15		24	28.50
1700	3/4" thick, cut sheets		115	.139		31	6.75		37.75	44.50
1800	Full sheets		130	.123		18	5.95		23.95	29
2100	1" thick, cut sheets		105	.152		35	7.40		42.40	49.50
2200	Full sheets		125	.128		22	6.20		28.20	33.50
3000	Colored, 1/8" thick, cut sheets		170	.094		20	4.56		24.56	29
3200	Full sheets		195	.082		11.40	3.98		15.38	18.60
3500	1/4" thick, cut sheets		165	.097		22	4.70		26.70	31.50
3600	Full sheets		185	.086		14.80	4.19		18.99	22.50
4000	Mirrors, untinted, cut sheets, 1/8" thick		185	.086		13.30	4.19		17.49	21
4200	1/4" thick		180	.089		17.15	4.31		21.46	25.50

08 84 20 – Polycarbonate

08 84 20.10 Thermoplastic

		Crew	Daily Output	Labor-Hours	Unit	Material	2018 Bare Costs Labor	Equipment	Total	Total Incl O&P
0010	**THERMOPLASTIC**, clear, masked, cut sheets									
0020	1/8" thick	2 Glaz	170	.094	S.F.	15.20	4.56		19.76	23.50
0500	3/16" thick		165	.097		18.60	4.70		23.30	27.50
1000	1/4" thick		155	.103		18.25	5		23.25	27.50
1500	3/8" thick		150	.107		28.50	5.15		33.65	39.50

08 87 Glazing Surface Films

08 87 13 - Solar Control Films

08 87 13.10 Solar Films On Glass

		Crew	Daily Output	Labor-Hours	Unit	Material	2018 Bare Costs Labor	Equipment	Total	Total Incl O&P
0010	**SOLAR FILMS ON GLASS** (glass not included)									
1000	Bronze, 20% VLT	H-2	180	.133	S.F.	1.60	6.10		7.70	11
1020	50% VLT		180	.133		1.80	6.10		7.90	11.25
1050	Neutral, 20% VLT		180	.133		1.80	6.10		7.90	11.25
1100	Silver, 15% VLT		180	.133		1.04	6.10		7.14	10.40
1120	35% VLT		180	.133		3.02	6.10		9.12	12.55
1150	68% VLT		180	.133		.40	6.10		6.50	9.70
3000	One way mirror with night vision 5% in and 15% out VLT		180	.133		3.03	6.10		9.13	12.60

08 87 16 - Glass Safety Films

08 87 16.10 Safety Films On Glass

		Crew	Daily Output	Labor-Hours	Unit	Material	2018 Bare Costs Labor	Equipment	Total	Total Incl O&P
0010	**SAFETY FILMS ON GLASS** (glass not included)									
0015	Safety film helps hold glass together when broken									
0050	Safety film, clear, 2 mil	H-2	180	.133	S.F.	1.21	6.10		7.31	10.60
0100	Safety film, clear, 4 mil		180	.133		1.81	6.10		7.91	11.25
0110	Safety film, tinted, 4 mil		180	.133		3.46	6.10		9.56	13.05
0150	Safety film, black tint, 8 mil		180	.133		2.23	6.10		8.33	11.70

08 87 26 - Bird Control Film

08 87 26.10 Bird Control Film

		Crew	Daily Output	Labor-Hours	Unit	Material	2018 Bare Costs Labor	Equipment	Total	Total Incl O&P
0010	**BIRD CONTROL FILM**									
0050	Patterned, adhered to glass	2 Glaz	180	.089	S.F.	6.75	4.31		11.06	14
0200	Decals small, adhered to glass	1 Glaz	50	.160	Ea.	2.48	7.75		10.23	14.50
0250	Decals large, adhered to glass		50	.160	"	6.95	7.75		14.70	19.40
0300	Decals set of 4, adhered to glass		25	.320	Set	6.95	15.50		22.45	31
0400	Bird control film tape		10	.800	Roll	19.95	39		58.95	81

08 87 33 - Electrically Tinted Window Film

08 87 33.20 Window Film

		Crew	Daily Output	Labor-Hours	Unit	Material	2018 Bare Costs Labor	Equipment	Total	Total Incl O&P
0010	**WINDOW FILM** adhered on glass (glass not included)									
0015	Film is pre-wired and can be trimmed									
0100	Window film 4 S.F. adhered on glass	1 Glaz	200	.040	S.F.	224	1.94		225.94	250
0120	6 S.F.		180	.044		335	2.16		337.16	375
0160	9 S.F.		170	.047		505	2.28		507.28	560
0180	10 S.F.		150	.053		560	2.59		562.59	620
0200	12 S.F.		144	.056		670	2.69		672.69	740
0220	14 S.F.		140	.057		780	2.77		782.77	860
0240	16 S.F.		128	.063		890	3.03		893.03	985
0260	18 S.F.		126	.063		1,000	3.08		1,003.08	1,100
0280	20 S.F.		120	.067		1,125	3.23		1,128.23	1,225

08 87 53 - Security Films On Glass

08 87 53.10 Security Film Adhered On Glass

		Crew	Daily Output	Labor-Hours	Unit	Material	2018 Bare Costs Labor	Equipment	Total	Total Incl O&P
0010	**SECURITY FILM ADHERED ON GLASS** (glass not included)									
0020	Security film, clear, 7 mil	1 Glaz	200	.040	S.F.	1.09	1.94		3.03	4.14
0030	8 mil		200	.040		1.97	1.94		3.91	5.10
0040	9 mil		200	.040		1.08	1.94		3.02	4.13
0050	10 mil		200	.040		1.87	1.94		3.81	5
0060	12 mil		200	.040		1.84	1.94		3.78	4.96
0075	15 mil		200	.040		2.32	1.94		4.26	5.50
0100	Security film, sealed with structural adhesive, 7 mil		180	.044		5.55	2.16		7.71	9.35
0110	8 mil		180	.044		6.45	2.16		8.61	10.30
0140	14 mil		180	.044		7.05	2.16		9.21	11

For customer support on your Building Construction Costs with RSMeans data, call 800.448.8182.

08 88 Special Function Glazing

08 88 40 – Acoustical Glass Units

08 88 40.10 Sound Reduction Units		Crew	Daily Output	Labor-Hours	Unit	Material	2018 Bare Costs Labor	Equipment	Total	Total Incl O&P
0010	**SOUND REDUCTION UNITS**, 1 lite at 3/8", 1 lite at 3/16"									
0020	For 1" thick	2 Glaz	100	.160	S.F.	36.50	7.75		44.25	52
0100	For 4" thick	"	80	.200	"	62	9.70		71.70	82.50

08 88 56 – Ballistics-Resistant Glazing

08 88 56.10 Laminated Glass		Crew	Daily Output	Labor-Hours	Unit	Material	2018 Bare Costs Labor	Equipment	Total	Total Incl O&P
0010	**LAMINATED GLASS**									
0020	Clear float .03" vinyl 1/4" thick	2 Glaz	90	.178	S.F.	12.75	8.60		21.35	27
0100	3/8" thick		78	.205		27.50	9.95		37.45	45.50
0200	.06" vinyl, 1/2" thick		65	.246		28	11.95		39.95	49
1000	5/8" thick		90	.178		30.50	8.60		39.10	46.50
2000	Bullet-resisting, 1-3/16" thick, to 15 S.F.		16	1		117	48.50		165.50	202
2100	Over 15 S.F.		16	1		150	48.50		198.50	239
2500	2-1/4" thick, to 15 S.F.		12	1.333		207	64.50		271.50	325
2600	Over 15 S.F.		12	1.333		196	64.50		260.50	315
2700	Level 2 (.357 magnum), NIJ and UL		12	1.333		58	64.50		122.50	162
2750	Level 3A (.44 magnum) NIJ, UL 3		12	1.333		63	64.50		127.50	168
2800	Level 4 (AK-47) NIJ, UL 7 & 8		12	1.333		83	64.50		147.50	190
2850	Level 5 (M-16) UL		12	1.333		93	64.50		157.50	200
2900	Level 3 (7.62 Armor Piercing) NIJ, UL 4 & 5		12	1.333		108	64.50		172.50	217

08 91 Louvers

08 91 19 – Fixed Louvers

08 91 19.10 Aluminum Louvers		Crew	Daily Output	Labor-Hours	Unit	Material	2018 Bare Costs Labor	Equipment	Total	Total Incl O&P
0010	**ALUMINUM LOUVERS**									
0020	Aluminum with screen, residential, 8" x 8"	1 Carp	38	.211	Ea.	20	10.65		30.65	38.50
0100	12" x 12"		38	.211		16	10.65		26.65	34
0200	12" x 18"		35	.229		21	11.60		32.60	41
0250	14" x 24"		30	.267		29	13.50		42.50	52.50
0300	18" x 24"		27	.296		32	15		47	58
0500	24" x 30"		24	.333		61.50	16.90		78.40	93
0700	Triangle, adjustable, small		20	.400		58.50	20.50		79	95.50
0800	Large		15	.533		80.50	27		107.50	130
1200	Extruded aluminum, see Section 23 37 15.40									
2100	Midget, aluminum, 3/4" deep, 1" diameter	1 Carp	85	.094	Ea.	.86	4.77		5.63	8.20
2150	3" diameter		60	.133		2.94	6.75		9.69	13.55
2200	4" diameter		50	.160		5.60	8.10		13.70	18.50
2250	6" diameter		30	.267		3.78	13.50		17.28	24.50

08 91 26 – Door Louvers

08 91 26.10 Steel Louvers, 18 Gauge, Fixed Blade		Crew	Daily Output	Labor-Hours	Unit	Material	2018 Bare Costs Labor	Equipment	Total	Total Incl O&P
0010	**STEEL LOUVERS, 18 GAUGE, FIXED BLADE**									
0050	12" x 12", with powder coat	1 Carp	20	.400	Ea.	85	20.50		105.50	125
0055	18" x 12"		20	.400		87.50	20.50		108	127
0060	18" x 18"		20	.400		91.50	20.50		112	132
0065	24" x 12"		20	.400		113	20.50		133.50	156
0070	24" x 18"		20	.400		124	20.50		144.50	167
0075	24" x 24"		20	.400		152	20.50		172.50	198
0100	12" x 12", galvanized		20	.400		77	20.50		97.50	116
0105	18" x 12"		20	.400		79.50	20.50		100	119
0115	24" x 12"		20	.400		103	20.50		123.50	145
0125	24" x 24"		20	.400		143	20.50		163.50	189

For customer support on your Building Construction Costs with RSMeans data, call 800.448.8182.

315

08 91 Louvers

08 91 26 – Door Louvers

08 91 26.10 Steel Louvers, 18 Gauge, Fixed Blade		Crew	Daily Output	Labor-Hours	Unit	Material	2018 Bare Costs Labor	Equipment	Total	Total Incl O&P
0300	6" x 12", painted	1 Carp	20	.400	Ea.	51	20.50		71.50	87
0320	10" x 16"		20	.400		79	20.50		99.50	118
0340	12" x 22"		20	.400		93.50	20.50		114	134
0360	14" x 14"		20	.400		91	20.50		111.50	131
0400	18" x 18"		20	.400		80	20.50		100.50	119
0420	20" x 26"		20	.400		139	20.50		159.50	184
0440	22" x 22"		20	.400		133	20.50		153.50	177
0460	26" x 26"		20	.400		208	20.50		228.50	260
3000	Fire rated									
3010	12" x 12"	1 Carp	10	.800	Ea.	159	40.50		199.50	237
3050	18" x 18"		10	.800		227	40.50		267.50	310
3100	24" x 12"		10	.800		242	40.50		282.50	330
3120	24" x 18"		10	.800		255	40.50		295.50	340
3130	24" x 24"		10	.800		283	40.50		323.50	370

08 91 26.20 Aluminum Louvers

		Crew	Daily Output	Labor-Hours	Unit	Material	Labor	Equipment	Total	Total Incl O&P
0010	**ALUMINUM LOUVERS**, 18 ga., fixed blade, clear anodized									
0050	6" x 12"	1 Carp	20	.400	Ea.	62	20.50		82.50	99
0060	10" x 10"		20	.400		73	20.50		93.50	112
0065	10" x 14"		20	.400		87.50	20.50		108	127
0080	12" x 22"		20	.400		116	20.50		136.50	158
0090	14" x 14"		20	.400		104	20.50		124.50	146
0100	18" x 10"		20	.400		86.50	20.50		107	126
0110	18" x 18"		20	.400		134	20.50		154.50	179
0120	22" x 22"		20	.400		190	20.50		210.50	240
0130	26" x 26"		20	.400		243	20.50		263.50	298

08 95 Vents

08 95 13 – Soffit Vents

08 95 13.10 Wall Louvers

		Crew	Daily Output	Labor-Hours	Unit	Material	Labor	Equipment	Total	Total Incl O&P
0010	**WALL LOUVERS**									
2340	Baked enamel finish	1 Carp	200	.040	L.F.	5.70	2.03		7.73	9.40
2400	Under eaves vent, aluminum, mill finish, 16" x 4"		48	.167	Ea.	1.89	8.45		10.34	14.95
2500	16" x 8"		48	.167	"	2.61	8.45		11.06	15.70

08 95 16 – Wall Vents

08 95 16.10 Louvers

		Crew	Daily Output	Labor-Hours	Unit	Material	Labor	Equipment	Total	Total Incl O&P
0010	**LOUVERS**									
0020	Redwood, 2'-0" diameter, full circle	1 Carp	16	.500	Ea.	275	25.50		300.50	345
0100	Half circle		16	.500		214	25.50		239.50	274
0200	Octagonal		16	.500		194	25.50		219.50	252
0300	Triangular, 5/12 pitch, 5'-0" at base		16	.500		555	25.50		580.50	650
7000	Vinyl gable vent, 8" x 8"		38	.211		17	10.65		27.65	35
7020	12" x 12"		38	.211		30	10.65		40.65	49.50
7080	12" x 18"		35	.229		38.50	11.60		50.10	60
7200	18" x 24"		30	.267		55.50	13.50		69	81.50

For customer support on your Building Construction Costs with RSMeans data, call 800.448.8182.

Estimating Tips
General
- Room Finish Schedule: A complete set of plans should contain a room finish schedule. If one is not available, it would be well worth the time and effort to obtain one.

09 20 00 Plaster and Gypsum Board
- Lath is estimated by the square yard plus a 5% allowance for waste. Furring, channels, and accessories are measured by the linear foot. An extra foot should be allowed for each accessory miter or stop.
- Plaster is also estimated by the square yard. Deductions for openings vary by preference, from zero deduction to 50% of all openings over 2 feet in width. The estimator should allow one extra square foot for each linear foot of horizontal interior or exterior angle located below the ceiling level. Also, double the areas of small radius work.
- Drywall accessories, studs, track, and acoustical caulking are all measured by the linear foot. Drywall taping is figured by the square foot. Gypsum wallboard is estimated by the square foot. No material deductions should be made for door or window openings under 32 S.F.

09 60 00 Flooring
- Tile and terrazzo areas are taken off on a square foot basis. Trim and base materials are measured by the linear foot. Accent tiles are listed per each. Two basic methods of installation are used. Mud set is approximately 30% more expensive than thin set. The cost of grout is included with tile unit price lines unless otherwise noted. In terrazzo work, be sure to include the linear footage of embedded decorative strips, grounds, machine rubbing, and power cleanup.
- Wood flooring is available in strip, parquet, or block configuration. The latter two types are set in adhesives with quantities estimated by the square foot. The laying pattern will influence labor costs and material waste. In addition to the material and labor for laying wood floors, the estimator must make allowances for sanding and finishing these areas, unless the flooring is prefinished.
- Sheet flooring is measured by the square yard. Roll widths vary, so consideration should be given to use the most economical width, as waste must be figured into the total quantity. Consider also the installation methods available—direct glue down or stretched. Direct glue-down installation is assumed with sheet carpet unit price lines unless otherwise noted.

09 70 00 Wall Finishes
- Wall coverings are estimated by the square foot. The area to be covered is measured—length by height of the wall above the baseboards—to calculate the square footage of each wall. This figure is divided by the number of square feet in the single roll which is being used. Deduct, in full, the areas of openings such as doors and windows. Where a pattern match is required allow 25%–30% waste.

09 80 00 Acoustic Treatment
- Acoustical systems fall into several categories. The takeoff of these materials should be by the square foot of area with a 5% allowance for waste. Do not forget about scaffolding, if applicable, when estimating these systems.

09 90 00 Painting and Coating
- A major portion of the work in painting involves surface preparation. Be sure to include cleaning, sanding, filling, and masking costs in the estimate.
- Protection of adjacent surfaces is not included in painting costs. When considering the method of paint application, an important factor is the amount of protection and masking required. These must be estimated separately and may be the determining factor in choosing the method of application.

Reference Numbers
Reference numbers are shown at the beginning of some major classifications. These numbers refer to related items in the Reference Section. The reference information may be an estimating procedure, an alternate pricing method, or technical information.

Note: Not all subdivisions listed here necessarily appear. ■

Did you know?

RSMeans data is available through our online application with 24/7 access:
- Search for unit prices by keyword
- Leverage the most up-to-date data
- Build and export estimates

Try it free for 30 days!
www.rsmeans.com/2018freetrial

No part of this cost data may be reproduced, stored in a retrieval system, or transmitted in any form or by any means without prior written permission of Gordian.

09 01 Maintenance of Finishes

09 01 60 – Maintenance of Flooring

09 01 60.10 Carpet Maintenance

		Crew	Daily Output	Labor-Hours	Unit	Material	2018 Bare Costs Labor	Equipment	Total	Total Incl O&P
0010	**CARPET MAINTENANCE**									
0020	Steam clean, per cleaning, routine maintenance	1 Clab	3000	.003	S.F.	.09	.11		.20	.26
0500	Stain removal	"	2000	.004	"	.13	.16		.29	.38

09 01 70 – Maintenance of Wall Finishes

09 01 70.10 Gypsum Wallboard Repairs

		Crew	Daily Output	Labor-Hours	Unit	Material	2018 Bare Costs Labor	Equipment	Total	Total Incl O&P
0010	**GYPSUM WALLBOARD REPAIRS**									
0100	Fill and sand, pin/nail holes	1 Carp	960	.008	Ea.		.42		.42	.64
0110	Screw head pops		480	.017			.85		.85	1.29
0120	Dents, up to 2" square		48	.167		.01	8.45		8.46	12.85
0130	2" to 4" square		24	.333		.03	16.90		16.93	25.50
0140	Cut square, patch, sand and finish, holes, up to 2" square		12	.667		.03	34		34.03	51.50
0150	2" to 4" square		11	.727		.09	37		37.09	56
0160	4" to 8" square		10	.800		.24	40.50		40.74	62.50
0170	8" to 12" square		8	1		.48	50.50		50.98	77.50
0180	12" to 32" square		6	1.333		1.60	67.50		69.10	105
0210	16" by 48"		5	1.600		2.76	81		83.76	127
0220	32" by 48"		4	2		4.52	101		105.52	159
0230	48" square		3.50	2.286		6.40	116		122.40	183
0240	60" square		3.20	2.500		9.95	127		136.95	204
0500	Skim coat surface with joint compound		1600	.005	S.F.	.03	.25		.28	.42
0510	Prepare, retape and refinish joints		60	.133	L.F.	.63	6.75		7.38	11

09 05 Common Work Results for Finishes

09 05 05 – Selective Demolition for Finishes

09 05 05.10 Selective Demolition, Ceilings

		Crew	Daily Output	Labor-Hours	Unit	Material	2018 Bare Costs Labor	Equipment	Total	Total Incl O&P
0010	**SELECTIVE DEMOLITION, CEILINGS** R024119-10									
0200	Ceiling, gypsum wall board, furred and nailed or screwed	2 Clab	800	.020	S.F.		.80		.80	1.21
0220	On metal frame		760	.021			.84		.84	1.28
0240	On suspension system, including system		720	.022			.89		.89	1.35
1000	Plaster, lime and horse hair, on wood lath, incl. lath		700	.023			.91		.91	1.39
1020	On metal lath		570	.028			1.12		1.12	1.70
1100	Gypsum, on gypsum lath		720	.022			.89		.89	1.35
1120	On metal lath		500	.032			1.28		1.28	1.94
1200	Suspended ceiling, mineral fiber, 2' x 2' or 2' x 4'		1500	.011			.43		.43	.65
1250	On suspension system, incl. system		1200	.013			.53		.53	.81
1500	Tile, wood fiber, 12" x 12", glued		900	.018			.71		.71	1.08
1540	Stapled		1500	.011			.43		.43	.65
1580	On suspension system, incl. system		760	.021			.84		.84	1.28
2000	Wood, tongue and groove, 1" x 4"		1000	.016			.64		.64	.97
2040	1" x 8"		1100	.015			.58		.58	.88
2400	Plywood or wood fiberboard, 4' x 8' sheets		1200	.013			.53		.53	.81
2500	Remove & refinish textured ceiling	1 Plas	222	.036		.03	1.67		1.70	2.54

09 05 05.20 Selective Demolition, Flooring

		Crew	Daily Output	Labor-Hours	Unit	Material	2018 Bare Costs Labor	Equipment	Total	Total Incl O&P
0010	**SELECTIVE DEMOLITION, FLOORING** R024119-10									
0200	Brick with mortar	2 Clab	475	.034	S.F.		1.34		1.34	2.04
0400	Carpet, bonded, including surface scraping		2000	.008			.32		.32	.49
0440	Scrim applied		8000	.002			.08		.08	.12
0480	Tackless		9000	.002			.07		.07	.11
0550	Carpet tile, releasable adhesive		5000	.003			.13		.13	.19
0560	Permanent adhesive		1850	.009			.34		.34	.53

For customer support on your Building Construction Costs with RSMeans data, call 800.448.8182.

09 05 Common Work Results for Finishes

09 05 05 – Selective Demolition for Finishes

09 05 05.20 Selective Demolition, Flooring

		Crew	Daily Output	Labor-Hours	Unit	Material	2018 Bare Costs Labor	Equipment	Total	Total Incl O&P
0600	Composition, acrylic or epoxy	2 Clab	400	.040	S.F.		1.59		1.59	2.43
0700	Concrete, scarify skin	A-1A	225	.036			1.86	.95	2.81	3.90
0800	Resilient, sheet goods	2 Clab	1400	.011			.46		.46	.69
0820	For gym floors	"	900	.018			.71		.71	1.08
0850	Vinyl or rubber cove base	1 Clab	1000	.008	L.F.		.32		.32	.49
0860	Vinyl or rubber cove base, molded corner	"	1000	.008	Ea.		.32		.32	.49
0870	For glued and caulked installation, add to labor						50%			
0900	Vinyl composition tile, 12" x 12"	2 Clab	1000	.016	S.F.		.64		.64	.97
2000	Tile, ceramic, thin set		675	.024			.94		.94	1.44
2020	Mud set		625	.026			1.02		1.02	1.55
2200	Marble, slate, thin set		675	.024			.94		.94	1.44
2220	Mud set		625	.026			1.02		1.02	1.55
2600	Terrazzo, thin set		450	.036			1.42		1.42	2.16
2620	Mud set		425	.038			1.50		1.50	2.29
2640	Terrazzo, cast in place		300	.053			2.13		2.13	3.24
3000	Wood, block, on end	1 Carp	400	.020			1.01		1.01	1.54
3200	Parquet		450	.018			.90		.90	1.37
3400	Strip flooring, interior, 2-1/4" x 25/32" thick		325	.025			1.25		1.25	1.90
3500	Exterior, porch flooring, 1" x 4"		220	.036			1.84		1.84	2.81
3800	Subfloor, tongue and groove, 1" x 6"		325	.025			1.25		1.25	1.90
3820	1" x 8"		430	.019			.94		.94	1.44
3840	1" x 10"		520	.015			.78		.78	1.19
4000	Plywood, nailed		600	.013			.68		.68	1.03
4100	Glued and nailed		400	.020			1.01		1.01	1.54
4200	Hardboard, 1/4" thick		760	.011			.53		.53	.81
8000	Remove flooring, bead blast, simple floor plan	A-1A	1000	.008		.05	.42	.21	.68	.94
8100	Complex floor plan		400	.020		.05	1.05	.54	1.64	2.25
8150	Mastic only		1500	.005		.05	.28	.14	.47	.65
9050	For grinding concrete floors, see Section 03 35 43.10									

09 05 05.30 Selective Demolition, Walls and Partitions

		Crew	Daily Output	Labor-Hours	Unit	Material	2018 Bare Costs Labor	Equipment	Total	Total Incl O&P
0010	**SELECTIVE DEMOLITION, WALLS AND PARTITIONS** R024119-10									
0020	Walls, concrete, reinforced	B-39	120	.400	C.F.		16.85	1.95	18.80	27.50
0025	Plain	"	160	.300			12.65	1.46	14.11	21
0100	Brick, 4" to 12" thick	B-9	220	.182			7.30	1.06	8.36	12.30
0200	Concrete block, 4" thick		1150	.035	S.F.		1.40	.20	1.60	2.35
0280	8" thick		1050	.038			1.53	.22	1.75	2.59
0300	Exterior stucco 1" thick over mesh		3200	.013			.50	.07	.57	.85
1000	Gypsum wallboard, nailed or screwed	1 Clab	1000	.008			.32		.32	.49
1010	2 layers		400	.020			.80		.80	1.21
1020	Glued and nailed		900	.009			.35		.35	.54
1500	Fiberboard, nailed		900	.009			.35		.35	.54
1520	Glued and nailed		800	.010			.40		.40	.61
1568	Plenum barrier, sheet lead		300	.027			1.06		1.06	1.62
2000	Movable walls, metal, 5' high		300	.027			1.06		1.06	1.62
2020	8' high		400	.020			.80		.80	1.21
2200	Metal or wood studs, finish 2 sides, fiberboard	B-1	520	.046			1.87		1.87	2.85
2250	Lath and plaster		260	.092			3.74		3.74	5.70
2300	Gypsum wallboard		520	.046			1.87		1.87	2.85
2350	Plywood		450	.053			2.16		2.16	3.29
2800	Paneling, 4' x 8' sheets	1 Clab	475	.017			.67		.67	1.02
3000	Plaster, lime and horsehair, on wood lath		400	.020			.80		.80	1.21
3020	On metal lath		335	.024			.95		.95	1.45

For customer support on your Building Construction Costs with RSMeans data, call 800.448.8182.

319

09 05 Common Work Results for Finishes

09 05 05 – Selective Demolition for Finishes

09 05 05.30 Selective Demolition, Walls and Partitions

		Crew	Daily Output	Labor-Hours	Unit	Material	2018 Bare Costs Labor	Equipment	Total	Total Incl O&P
3400	Gypsum or perlite, on gypsum lath	1 Clab	410	.020	S.F.		.78		.78	1.18
3420	On metal lath		300	.027	↓		1.06		1.06	1.62
3450	Plaster, interior gypsum, acoustic, or cement		60	.133	S.Y.		5.30		5.30	8.10
3500	Stucco, on masonry		145	.055			2.20		2.20	3.35
3510	Commercial 3-coat		80	.100			3.99		3.99	6.05
3520	Interior stucco	↓	25	.320	↓		12.75		12.75	19.40
3600	Plywood, one side	B-1	1500	.016	S.F.		.65		.65	.99
3750	Terra cotta block and plaster, to 6" thick	"	175	.137			5.55		5.55	8.45
3760	Tile, ceramic, on walls, thin set	1 Clab	300	.027			1.06		1.06	1.62
3765	Mud set		250	.032			1.28		1.28	1.94
3800	Toilet partitions, slate or marble		5	1.600	Ea.		64		64	97
3820	Metal or plastic	↓	8	1	"		40		40	60.50
5000	Wallcovering, vinyl	1 Pape	700	.011	S.F.		.49		.49	.73
5010	With release agent		1500	.005			.23		.23	.34
5025	Wallpaper, 2 layers or less, by hand		250	.032			1.36		1.36	2.05
5035	3 layers or more	↓	165	.048	↓		2.07		2.07	3.11

09 05 71 – Acoustic Underlayment

09 05 71.10 Acoustical Underlayment

		Crew	Daily Output	Labor-Hours	Unit	Material	2018 Bare Costs Labor	Equipment	Total	Total Incl O&P
0010	**ACOUSTICAL UNDERLAYMENT**									
0100	Rubber underlayment, 5/64"	1 Tilf	275	.029	S.F.	.32	1.36		1.68	2.37
0110	1/8" thk.		275	.029		.51	1.36		1.87	2.58
0120	13/64" thk.		270	.030		.69	1.39		2.08	2.81
0130	15/64" thk.		270	.030		.92	1.39		2.31	3.06
0140	23/64" thk.		265	.030		1.26	1.41		2.67	3.48
0150	15/32" thk.		265	.030		1.60	1.41		3.01	3.85
0400	Cork underlayment, 3/32"		275	.029		.39	1.36		1.75	2.45
0410	1/8" thk.		275	.029		.40	1.36		1.76	2.46
0420	5/32" thk.		275	.029		.47	1.36		1.83	2.54
0430	15/64" thk.		270	.030		.61	1.39		2	2.72
0440	15/32" thk.		265	.030		1.27	1.41		2.68	3.49
0600	Rubber cork underlayment, 3/64"		275	.029		.31	1.36		1.67	2.36
0610	13/64" thk.		270	.030		1.27	1.39		2.66	3.45
0620	3/8" thk.		270	.030		2.42	1.39		3.81	4.71
0630	25/64" thk.		270	.030		2.55	1.39		3.94	4.86
0800	Foam underlayment, 5/64"		275	.029		.54	1.36		1.90	2.61
0810	1/8" thk.		275	.029		.59	1.36		1.95	2.67
0820	15/32" thk.	↓	265	.030	↓	.49	1.41		1.90	2.63
4000	Nylon matting 0.4" thick, with carbon black spinerette									
4010	plus polyester fabric, on floor	D-7	1600	.010	S.F.	1.41	.42		1.83	2.17

09 21 Plaster and Gypsum Board Assemblies

09 21 13 – Plaster Assemblies

09 21 13.10 Plaster Partition Wall

		Crew	Daily Output	Labor-Hours	Unit	Material	2018 Bare Costs Labor	Equipment	Total	Total Incl O&P
0010	**PLASTER PARTITION WALL**									
0400	Stud walls, 3.4 lb. metal lath, 3 coat gypsum plaster, 2 sides									
0600	2" x 4" wood studs, 16" OC	J-2	315	.152	S.F.	2.72	6.80	.42	9.94	13.65
0700	2-1/2" metal studs, 25 ga., 12" OC		325	.148		2.52	6.60	.41	9.53	13.05
0800	3-5/8" metal studs, 25 ga., 16" OC	↓	320	.150		2.47	6.70	.42	9.59	13.25
0900	Gypsum lath, 2 coat vermiculite plaster, 2 sides									
1000	2" x 4" wood studs, 16" OC	J-2	355	.135	S.F.	2.49	6.05	.38	8.92	12.20

For customer support on your Building Construction Costs with RSMeans data, call 800.448.8182.

09 21 Plaster and Gypsum Board Assemblies

09 21 13 – Plaster Assemblies

09 21 13.10 Plaster Partition Wall

		Crew	Daily Output	Labor-Hours	Unit	Material	2018 Bare Costs Labor	Equipment	Total	Total Incl O&P
1200	2-1/2" metal studs, 25 ga., 12" OC	J-2	365	.132	S.F.	2.45	5.85	.37	8.67	11.90
1300	3-5/8" metal studs, 25 ga., 16" OC	▼	360	.133	▼	2.40	5.95	.37	8.72	11.95

09 21 16 – Gypsum Board Assemblies

09 21 16.23 Gypsum Board Shaft Wall Assemblies

		Crew	Daily Output	Labor-Hours	Unit	Material	2018 Bare Costs Labor	Equipment	Total	Total Incl O&P
0010	**GYPSUM BOARD SHAFT WALL ASSEMBLIES**									
0020	Cavity type on 25 ga. J-track & C-H studs, 24" OC									
0030	1" thick coreboard wall liner on shaft side									
0040	2-hour assembly with double layer									
0060	5/8" f.r. gyp bd on rm side, 2-1/2" J-track & C-H studs	2 Carp	220	.073	S.F.	2.95	3.69		6.64	8.85
0065	4" J-track & C-H studs	↓	220	.073	↓	3.11	3.69		6.80	9
0070	6" J-track & C-H studs	▼	220	.073	▼	3.33	3.69		7.02	9.25
0100	3-hour assembly with triple layer									
0300	5/8" f.r. gyp bd on rm side, 2-1/2" J-track & C-H studs	2 Carp	180	.089	S.F.	2.60	4.51		7.11	9.70
0305	4" J-track & C-H studs		180	.089		2.76	4.51		7.27	9.90
0310	6" J-track & C-H studs	▼	180	.089	▼	2.98	4.51		7.49	10.10
0400	4-hour assembly, 1" coreboard, 5/8" fire rated gypsum board									
0600	and 3/4" galv. metal furring channels, 24" OC, with									
0700	Dbl. layer 5/8" f.r. gyp bd on rm side, 2-1/2" trk. & C-H studs	2 Carp	110	.145	S.F.	3.33	7.35		10.68	14.90
0705	4" J-track & C-H studs		110	.145		3.11	7.35		10.46	14.65
0710	6" J-track & C-H studs	↓	110	.145		3.33	7.35		10.68	14.90
0900	For taping & finishing, add per side	1 Carp	1050	.008	▼	.05	.39		.44	.64
1000	For insulation, see Section 07 21									
5200	For work over 8' high, add	2 Carp	3060	.005	S.F.		.27		.27	.40
5300	For distribution cost 3 stories and above, add per story	"	6100	.003	"		.13		.13	.20

09 21 16.33 Partition Wall

		Crew	Daily Output	Labor-Hours	Unit	Material	2018 Bare Costs Labor	Equipment	Total	Total Incl O&P
0010	**PARTITION WALL** Stud wall, 8' to 12' high									
0050	1/2", interior, gypsum board, std, tape & finish 2 sides									
0500	Installed on and incl., 2" x 4" wood studs, 16" OC	2 Carp	310	.052	S.F.	1.24	2.62		3.86	5.35
1000	Metal studs, NLB, 25 ga., 16" OC, 3-5/8" wide		350	.046		1.15	2.32		3.47	4.80
1200	6" wide		330	.048		1.28	2.46		3.74	5.15
1400	Water resistant, on 2" x 4" wood studs, 16" OC		310	.052		1.40	2.62		4.02	5.50
1600	Metal studs, NLB, 25 ga., 16" OC, 3-5/8" wide		350	.046		1.31	2.32		3.63	4.97
1800	6" wide		330	.048		1.44	2.46		3.90	5.30
2000	Fire res., 2 layers, 1-1/2 hr., on 2" x 4" wood studs, 16" OC		210	.076		2.04	3.86		5.90	8.15
2200	Metal studs, NLB, 25 ga., 16" OC, 3-5/8" wide		250	.064		1.95	3.24		5.19	7.10
2400	6" wide		230	.070		2.08	3.53		5.61	7.65
2600	Fire & water res., 2 layers, 1-1/2 hr., 2" x 4" studs, 16" OC		210	.076		2.04	3.86		5.90	8.15
2800	Metal studs, NLB, 25 ga., 16" OC, 3-5/8" wide		250	.064		1.95	3.24		5.19	7.10
3000	6" wide	▼	230	.070	▼	2.08	3.53		5.61	7.65
3200	5/8", interior, gypsum board, standard, tape & finish 2 sides									
3400	Installed on and including 2" x 4" wood studs, 16" OC	2 Carp	300	.053	S.F.	1.26	2.70		3.96	5.50
3600	24" OC		330	.048		1.15	2.46		3.61	5
3800	Metal studs, NLB, 25 ga., 16" OC, 3-5/8" wide		340	.047		1.17	2.39		3.56	4.92
4000	6" wide		320	.050		1.30	2.54		3.84	5.30
4200	24" OC, 3-5/8" wide		360	.044		1.07	2.25		3.32	4.61
4400	6" wide		340	.047		1.16	2.39		3.55	4.91
4800	Water resistant, on 2" x 4" wood studs, 16" OC		300	.053		1.44	2.70		4.14	5.70
5000	24" OC		330	.048		1.33	2.46		3.79	5.20
5200	Metal studs, NLB, 25 ga. 16" OC, 3-5/8" wide		340	.047		1.35	2.39		3.74	5.10
5400	6" wide		320	.050		1.48	2.54		4.02	5.50
5600	24" OC, 3-5/8" wide		360	.044		1.25	2.25		3.50	4.81
5800	6" wide	▼	340	.047		1.34	2.39		3.73	5.10

For customer support on your Building Construction Costs with RSMeans data, call 800.448.8182.

321

09 21 Plaster and Gypsum Board Assemblies

09 21 16 – Gypsum Board Assemblies

09 21 16.33 Partition Wall

		Crew	Daily Output	Labor-Hours	Unit	Material	2018 Bare Costs Labor	Equipment	Total	Total Incl O&P
6000	Fire resistant, 2 layers, 2 hr., on 2" x 4" wood studs, 16" OC	2 Carp	205	.078	S.F.	2	3.96		5.96	8.25
6200	24" OC		235	.068		1.89	3.45		5.34	7.35
6400	Metal studs, NLB, 25 ga., 16" OC, 3-5/8" wide		245	.065		1.95	3.31		5.26	7.20
6600	6" wide		225	.071		2.04	3.61		5.65	7.75
6800	24" OC, 3-5/8" wide		265	.060		1.81	3.06		4.87	6.65
7000	6" wide		245	.065		1.90	3.31		5.21	7.15
7200	Fire & water resistant, 2 layers, 2 hr., 2" x 4" studs, 16" OC		205	.078		2	3.96		5.96	8.25
7400	24" OC		235	.068		1.89	3.45		5.34	7.35
7600	Metal studs, NLB, 25 ga., 16" OC, 3-5/8" wide		245	.065		1.91	3.31		5.22	7.15
7800	6" wide		225	.071		2.04	3.61		5.65	7.75
8000	24" OC, 3-5/8" wide		265	.060		1.81	3.06		4.87	6.65
8200	6" wide		245	.065		1.90	3.31		5.21	7.15
8600	1/2" blueboard, mesh tape both sides									
8620	Installed on and including 2" x 4" wood studs, 16" OC	2 Carp	300	.053	S.F.	1.40	2.70		4.10	5.65
8640	Metal studs, NLB, 25 ga., 16" OC, 3-5/8" wide		340	.047		1.31	2.39		3.70	5.05
8660	6" wide		320	.050		1.44	2.54		3.98	5.45
8800	Hospital security partition, 5/8" fiber reinf. high abuse gyp. bd.									
8810	Mtl. studs, NLB, 20 ga., 16" OC, 3-5/8" wide, w/sec. mesh, gyp. bd.	2 Carp	208	.077	S.F.	4.25	3.90		8.15	10.65
9000	Exterior, 1/2" gypsum sheathing, 1/2" gypsum finished, interior,									
9100	including foil faced insulation, metal studs, 20 ga.									
9200	16" OC, 3-5/8" wide	2 Carp	290	.055	S.F.	1.79	2.80		4.59	6.25
9400	6" wide		270	.059		1.99	3		4.99	6.75
9600	Partitions, for work over 8' high, add		1530	.010			.53		.53	.81

09 22 Supports for Plaster and Gypsum Board

09 22 03 – Fastening Methods for Finishes

09 22 03.20 Drilling Plaster/Drywall

		Crew	Daily Output	Labor-Hours	Unit	Material	2018 Bare Costs Labor	Equipment	Total	Total Incl O&P
0010	**DRILLING PLASTER/DRYWALL**									
1100	Drilling & layout for drywall/plaster walls, up to 1" deep, no anchor									
1200	Holes, 1/4" diameter	1 Carp	150	.053	Ea.	.01	2.70		2.71	4.13
1300	3/8" diameter		140	.057		.01	2.90		2.91	4.42
1400	1/2" diameter		130	.062		.01	3.12		3.13	4.76
1500	3/4" diameter		120	.067		.02	3.38		3.40	5.15
1600	1" diameter		110	.073		.02	3.69		3.71	5.60
1700	1-1/4" diameter		100	.080		.04	4.06		4.10	6.25
1800	1-1/2" diameter		90	.089		.06	4.51		4.57	6.90
1900	For ceiling installations, add						40%			

09 22 13 – Metal Furring

09 22 13.13 Metal Channel Furring

		Crew	Daily Output	Labor-Hours	Unit	Material	2018 Bare Costs Labor	Equipment	Total	Total Incl O&P
0010	**METAL CHANNEL FURRING**									
0030	Beams and columns, 7/8" hat channels, galvanized, 12" OC	1 Lath	155	.052	S.F.	.44	2.54		2.98	4.23
0050	16" OC		170	.047		.36	2.31		2.67	3.81
0070	24" OC		185	.043		.24	2.13		2.37	3.40
0100	Ceilings, on steel, 7/8" hat channels, galvanized, 12" OC		210	.038		.40	1.87		2.27	3.21
0300	16" OC		290	.028		.36	1.36		1.72	2.39
0400	24" OC		420	.019		.24	.94		1.18	1.64
0600	1-5/8" hat channels, galvanized, 12" OC		190	.042		.53	2.07		2.60	3.64
0700	16" OC		260	.031		.47	1.51		1.98	2.76
0900	24" OC		390	.021		.32	1.01		1.33	1.84
0930	7/8" hat channels with sound isolation clips, 12" OC		120	.067		1.72	3.28		5	6.75

For customer support on your Building Construction Costs with RSMeans data, call 800.448.8182.

09 22 Supports for Plaster and Gypsum Board

09 22 13 – Metal Furring

09 22 13.13 Metal Channel Furring

		Crew	Daily Output	Labor-Hours	Unit	Material	2018 Bare Costs Labor	Equipment	Total	Total Incl O&P
0940	16" OC	1 Lath	100	.080	S.F.	1.29	3.93		5.22	7.20
0950	24" OC		165	.048		.86	2.38		3.24	4.46
0960	1-5/8" hat channels, galvanized, 12" OC		110	.073		1.85	3.57		5.42	7.35
0970	16" OC		100	.080		1.38	3.93		5.31	7.30
0980	24" OC		155	.052		.92	2.54		3.46	4.76
1000	Walls, 7/8" hat channels, galvanized, 12" OC		235	.034		.40	1.67		2.07	2.91
1200	16" OC		265	.030		.36	1.48		1.84	2.58
1300	24" OC		350	.023		.24	1.12		1.36	1.92
1500	1-5/8" hat channels, galvanized, 12" OC		210	.038		.53	1.87		2.40	3.35
1600	16" OC		240	.033		.47	1.64		2.11	2.94
1800	24" OC		305	.026		.32	1.29		1.61	2.26
1920	7/8" hat channels with sound isolation clips, 12" OC		125	.064		1.72	3.15		4.87	6.55
1940	16" OC		100	.080		1.29	3.93		5.22	7.20
1950	24" OC		150	.053		.86	2.62		3.48	4.81
1960	1-5/8" hat channels, galvanized, 12" OC		115	.070		1.85	3.42		5.27	7.10
1970	16" OC		95	.084		1.38	4.14		5.52	7.60
1980	24" OC		140	.057		.92	2.81		3.73	5.15
3000	Z Furring, walls, 1" deep, 25 ga., 24" OC		350	.023		1.38	1.12		2.50	3.17
3010	48" OC		700	.011		.69	.56		1.25	1.59
3020	1-1/2" deep, 24" OC		345	.023		1.61	1.14		2.75	3.45
3030	48" OC		695	.012		.80	.57		1.37	1.72
3040	2" deep, 24" OC		340	.024		1.93	1.16		3.09	3.83
3050	48" OC		690	.012		.97	.57		1.54	1.90
3060	1" deep, 20 ga., 24" OC		350	.023		2.28	1.12		3.40	4.16
3070	48" OC		700	.011		1.14	.56		1.70	2.08
3080	1-1/2" deep, 24" OC		345	.023		2.62	1.14		3.76	4.56
3090	48" OC		695	.012		1.31	.57		1.88	2.28
4000	2" deep, 24" OC		340	.024		3.22	1.16		4.38	5.25
4010	48" OC		690	.012		1.61	.57		2.18	2.61

09 22 16 – Non-Structural Metal Framing

09 22 16.13 Non-Structural Metal Stud Framing

		Crew	Daily Output	Labor-Hours	Unit	Material	2018 Bare Costs Labor	Equipment	Total	Total Incl O&P
0010	**NON-STRUCTURAL METAL STUD FRAMING**									
1600	Non-load bearing, galv., 8' high, 25 ga. 1-5/8" wide, 16" OC	1 Carp	619	.013	S.F.	.27	.66		.93	1.30
1610	24" OC		950	.008		.20	.43		.63	.87
1620	2-1/2" wide, 16" OC		613	.013		.36	.66		1.02	1.40
1630	24" OC		938	.009		.27	.43		.70	.95
1640	3-5/8" wide, 16" OC		600	.013		.40	.68		1.08	1.47
1650	24" OC		925	.009		.30	.44		.74	1
1660	4" wide, 16" OC		594	.013		.45	.68		1.13	1.53
1670	24" OC		925	.009		.33	.44		.77	1.04
1680	6" wide, 16" OC		588	.014		.53	.69		1.22	1.63
1690	24" OC		906	.009		.40	.45		.85	1.12
1700	20 ga. studs, 1-5/8" wide, 16" OC		494	.016		.34	.82		1.16	1.63
1710	24" OC		763	.010		.26	.53		.79	1.09
1720	2-1/2" wide, 16" OC		488	.016		.44	.83		1.27	1.75
1730	24" OC		750	.011		.33	.54		.87	1.18
1740	3-5/8" wide, 16" OC		481	.017		.50	.84		1.34	1.83
1750	24" OC		738	.011		.37	.55		.92	1.25
1760	4" wide, 16" OC		475	.017		.60	.85		1.45	1.96
1770	24" OC		738	.011		.45	.55		1	1.34
1780	6" wide, 16" OC		469	.017		.71	.86		1.57	2.11
1790	24" OC		725	.011		.54	.56		1.10	1.44

09 22 16.13 Non-Structural Metal Stud Framing

		Crew	Daily Output	Labor-Hours	Unit	Material	2018 Bare Costs Labor	2018 Bare Costs Equipment	Total	Total Incl O&P
2000	Non-load bearing, galv., 10' high, 25 ga. 1-5/8" wide, 16" OC	1 Carp	495	.016	S.F.	.25	.82		1.07	1.53
2100	24" OC		760	.011		.19	.53		.72	1.02
2200	2-1/2" wide, 16" OC		490	.016		.34	.83		1.17	1.63
2250	24" OC		750	.011		.25	.54		.79	1.09
2300	3-5/8" wide, 16" OC		480	.017		.38	.85		1.23	1.71
2350	24" OC		740	.011		.28	.55		.83	1.14
2400	4" wide, 16" OC		475	.017		.42	.85		1.27	1.77
2450	24" OC		740	.011		.31	.55		.86	1.17
2500	6" wide, 16" OC		470	.017		.50	.86		1.36	1.86
2550	24" OC		725	.011		.37	.56		.93	1.26
2600	20 ga. studs, 1-5/8" wide, 16" OC		395	.020		.32	1.03		1.35	1.92
2650	24" OC		610	.013		.24	.66		.90	1.27
2700	2-1/2" wide, 16" OC		390	.021		.41	1.04		1.45	2.03
2750	24" OC		600	.013		.30	.68		.98	1.36
2800	3-5/8" wide, 16" OC		385	.021		.47	1.05		1.52	2.12
2850	24" OC		590	.014		.35	.69		1.04	1.43
2900	4" wide, 16" OC		380	.021		.57	1.07		1.64	2.26
2950	24" OC		590	.014		.42	.69		1.11	1.51
3000	6" wide, 16" OC		375	.021		.68	1.08		1.76	2.39
3050	24" OC		580	.014		.50	.70		1.20	1.61
3060	Non-load bearing, galv., 12' high, 25 ga. 1-5/8" wide, 16" OC		413	.019		.24	.98		1.22	1.77
3070	24" OC		633	.013		.18	.64		.82	1.18
3080	2-1/2" wide, 16" OC		408	.020		.32	.99		1.31	1.87
3090	24" OC		625	.013		.24	.65		.89	1.25
3100	3-5/8" wide, 16" OC		400	.020		.36	1.01		1.37	1.94
3110	24" OC		617	.013		.26	.66		.92	1.29
3120	4" wide, 16" OC		396	.020		.40	1.02		1.42	2
3130	24" OC		617	.013		.30	.66		.96	1.32
3140	6" wide, 16" OC		392	.020		.48	1.03		1.51	2.11
3150	24" OC		604	.013		.35	.67		1.02	1.41
3160	20 ga. studs, 1-5/8" wide, 16" OC		329	.024		.31	1.23		1.54	2.22
3170	24" OC		508	.016		.23	.80		1.03	1.47
3180	2-1/2" wide, 16" OC		325	.025		.39	1.25		1.64	2.33
3190	24" OC		500	.016		.29	.81		1.10	1.56
3200	3-5/8" wide, 16" OC		321	.025		.45	1.26		1.71	2.42
3210	24" OC		492	.016		.33	.82		1.15	1.62
3220	4" wide, 16" OC		317	.025		.55	1.28		1.83	2.55
3230	24" OC		492	.016		.40	.82		1.22	1.70
3240	6" wide, 16" OC		313	.026		.65	1.30		1.95	2.68
3250	24" OC		483	.017		.47	.84		1.31	1.80
3260	Non-load bearing, galv., 16' high, 25 ga. 4" wide, 12" OC		195	.041		.52	2.08		2.60	3.74
3270	16" OC		275	.029		.41	1.47		1.88	2.70
3280	24" OC		400	.020		.30	1.01		1.31	1.87
3290	6" wide, 12" OC		190	.042		.62	2.14		2.76	3.93
3300	16" OC		280	.029		.49	1.45		1.94	2.74
3310	24" OC		400	.020		.35	1.01		1.36	1.93
3320	20 ga. studs, 2-1/2" wide, 12" OC		180	.044		.50	2.25		2.75	3.98
3330	16" OC		254	.032		.40	1.60		2	2.87
3340	24" OC		390	.021		.29	1.04		1.33	1.90
3350	3-5/8" wide, 12" OC		170	.047		.58	2.39		2.97	4.26
3360	16" OC		251	.032		.45	1.62		2.07	2.96
3370	24" OC		384	.021		.33	1.06		1.39	1.97
3380	4" wide, 12" OC		170	.047		.70	2.39		3.09	4.40

324

09 22 16 – Non-Structural Metal Framing

09 22 16.13 Non-Structural Metal Stud Framing

		Crew	Daily Output	Labor-Hours	Unit	Material	2018 Bare Costs Labor	Equipment	Total	Total Incl O&P
3390	16" OC	1 Carp	247	.032	S.F.	.55	1.64		2.19	3.11
3400	24" OC		384	.021		.40	1.06		1.46	2.05
3410	6" wide, 12" OC		175	.046		.83	2.32		3.15	4.44
3420	16" OC		245	.033		.65	1.66		2.31	3.24
3430	24" OC		400	.020		.48	1.01		1.49	2.07
3440	Non-load bearing, galv., 20' high, 25 ga. 6" wide, 12" OC		125	.064		.60	3.24		3.84	5.60
3450	16" OC		220	.036		.47	1.84		2.31	3.33
3460	24" OC		360	.022		.34	1.13		1.47	2.10
3470	20 ga. studs, 4" wide, 12" OC		120	.067		.69	3.38		4.07	5.90
3480	16" OC		215	.037		.54	1.89		2.43	3.46
3490	6" wide, 12" OC		115	.070		.81	3.53		4.34	6.25
3500	16" OC		215	.037		.64	1.89		2.53	3.57
3510	24" OC	▼	331	.024	▼	.46	1.23		1.69	2.38
5000	For load bearing studs, see Section 05 41 13.30									

09 22 26 – Suspension Systems

09 22 26.13 Ceiling Suspension Systems

		Crew	Daily Output	Labor-Hours	Unit	Material	2018 Bare Costs Labor	Equipment	Total	Total Incl O&P
0010	**CEILING SUSPENSION SYSTEMS** for gypsum board or plaster									
8000	Suspended ceilings, including carriers									
8200	1-1/2" carriers, 24" OC with:									
8300	7/8" channels, 16" OC	1 Lath	275	.029	S.F.	.58	1.43		2.01	2.74
8320	24" OC		310	.026		.46	1.27		1.73	2.38
8400	1-5/8" channels, 16" OC		205	.039		.69	1.92		2.61	3.59
8420	24" OC	▼	250	.032	▼	.53	1.57		2.10	2.91
8600	2" carriers, 24" OC with:									
8700	7/8" channels, 16" OC	1 Lath	250	.032	S.F.	.66	1.57		2.23	3.04
8720	24" OC		285	.028		.54	1.38		1.92	2.63
8800	1-5/8" channels, 16" OC		190	.042		.77	2.07		2.84	3.91
8820	24" OC	▼	225	.036		.62	1.75		2.37	3.26

09 22 36 – Lath

09 22 36.13 Gypsum Lath

		Crew	Daily Output	Labor-Hours	Unit	Material	2018 Bare Costs Labor	Equipment	Total	Total Incl O&P
0010	**GYPSUM LATH** R092000-50									
0020	Plain or perforated, nailed, 3/8" thick	1 Lath	85	.094	S.Y.	3.15	4.63		7.78	10.30
0100	1/2" thick		80	.100		2.52	4.92		7.44	10
0300	Clipped to steel studs, 3/8" thick		75	.107		3.15	5.25		8.40	11.20
0400	1/2" thick		70	.114		2.52	5.60		8.12	11.05
1500	For ceiling installations, add		216	.037			1.82		1.82	2.69
1600	For columns and beams, add	▼	170	.047	▼		2.31		2.31	3.42

09 22 36.23 Metal Lath

		Crew	Daily Output	Labor-Hours	Unit	Material	2018 Bare Costs Labor	Equipment	Total	Total Incl O&P
0010	**METAL LATH** R092000-50									
0020	Diamond, expanded, 2.5 lb./S.Y., painted				S.Y.	4.15			4.15	4.57
0100	Galvanized					3.10			3.10	3.41
0300	3.4 lb./S.Y., painted					4.33			4.33	4.76
0400	Galvanized					4.44			4.44	4.88
0600	For #15 asphalt sheathing paper, add					.48			.48	.53
0900	Flat rib, 1/8" high, 2.75 lb., painted					3.60			3.60	3.96
1000	Foil backed					3.81			3.81	4.19
1200	3.4 lb./S.Y., painted					4.54			4.54	4.99
1300	Galvanized					4.62			4.62	5.10
1500	For #15 asphalt sheathing paper, add					.48			.48	.53
1800	High rib, 3/8" high, 3.4 lb./S.Y., painted					4.72			4.72	5.20
1900	Galvanized				▼	4.28			4.28	4.71

09 22 36.23 Metal Lath

	Crew	Daily Output	Labor-Hours	Unit	Material	2018 Bare Costs Labor	Equipment	Total	Total Incl O&P
2400	3/4" high, painted, .60 lb./S.F.				S.F.	.73			.73	.80
2500	.75 lb./S.F.				"	1.58			1.58	1.74
2800	Stucco mesh, painted, 3.6 lb.				S.Y.	4.29			4.29	4.72
3000	K-lath, perforated, absorbent paper, regular					4.42			4.42	4.86
3100	Heavy duty					5.20			5.20	5.75
3300	Waterproof, heavy duty, grade B backing					5.10			5.10	5.60
3400	Fire resistant backing					5.65			5.65	6.20
3600	2.5 lb. diamond painted, on wood framing, on walls	1 Lath	85	.094		4.15	4.63		8.78	11.40
3700	On ceilings		75	.107		4.15	5.25		9.40	12.30
3900	3.4 lb. diamond painted, on wood framing, on walls		80	.100		4.54	4.92		9.46	12.25
4000	On ceilings		70	.114		4.54	5.60		10.14	13.30
4200	3.4 lb. diamond painted, wired to steel framing		75	.107		4.54	5.25		9.79	12.75
4300	On ceilings		60	.133		4.54	6.55		11.09	14.70
4500	Columns and beams, wired to steel		40	.200		4.54	9.85		14.39	19.55
4600	Cornices, wired to steel		35	.229		4.54	11.25		15.79	21.50
4800	Screwed to steel studs, 2.5 lb.		80	.100		4.15	4.92		9.07	11.80
4900	3.4 lb.		75	.107		4.33	5.25		9.58	12.50
5100	Rib lath, painted, wired to steel, on walls, 2.5 lb.		75	.107		3.60	5.25		8.85	11.70
5200	3.4 lb.		70	.114		4.72	5.60		10.32	13.50
5400	4.0 lb.		65	.123		5.75	6.05		11.80	15.25
5500	For self-furring lath, add					.12			.12	.13
5700	Suspended ceiling system, incl. 3.4 lb. diamond lath, painted	1 Lath	15	.533		4.36	26		30.36	44
5800	Galvanized	"	15	.533		4.54	26		30.54	44
6000	Hollow metal stud partitions, 3.4 lb. painted lath both sides									
6010	Non-load bearing, 25 ga., w/rib lath, 2-1/2" studs, 12" OC	1 Lath	20.30	.394	S.Y.	13.30	19.35		32.65	43
6300	16" OC		21.10	.379		12.45	18.65		31.10	41
6350	24" OC		22.70	.352		11.65	17.30		28.95	38.50
6400	3-5/8" studs, 16" OC		19.50	.410		12.85	20		32.85	44
6600	24" OC		20.40	.392		11.95	19.25		31.20	41.50
6700	4" studs, 16" OC		20.40	.392		13.25	19.25		32.50	43
6900	24" OC		21.60	.370		12.25	18.20		30.45	40.50
7000	6" studs, 16" OC		19.50	.410		13.95	20		33.95	45.50
7100	24" OC		21.10	.379		12.75	18.65		31.40	41.50
7200	L.B. partitions, 16 ga., w/rib lath, 2-1/2" studs, 16" OC		20	.400		13.70	19.65		33.35	44
7300	3-5/8" studs, 16 ga.		19.70	.406		15.45	19.95		35.40	46.50
7500	4" studs, 16 ga.		19.50	.410		16	20		36	47.50
7600	6" studs, 16 ga.		18.70	.428		19	21		40	52

09 22 36.43 Security Mesh

	Crew	Daily Output	Labor-Hours	Unit	Material	2018 Bare Costs Labor	Equipment	Total	Total Incl O&P
0010	SECURITY MESH, expanded metal, flat, screwed to framing									
0100	On walls, 3/4", 1.76 lb./S.F.	2 Carp	1500	.011	S.F.	2.07	.54		2.61	3.10
0110	1-1/2", 1.14 lb./S.F.		1600	.010		1.59	.51		2.10	2.52
0200	On ceilings, 3/4", 1.76 lb./S.F.		1350	.012		2.07	.60		2.67	3.19
0210	1-1/2", 1.14 lb./S.F.		1450	.011		1.59	.56		2.15	2.60

09 22 36.83 Accessories, Plaster

	Crew	Daily Output	Labor-Hours	Unit	Material	2018 Bare Costs Labor	Equipment	Total	Total Incl O&P
0010	ACCESSORIES, PLASTER									
0020	Casing bead, expanded flange, galvanized	1 Lath	2.70	2.963	C.L.F.	52.50	146		198.50	273
0200	Foundation weep screed, galvanized	"	2.70	2.963		53	146		199	274
0900	Channels, cold rolled, 16 ga., 3/4" deep, galvanized					39.50			39.50	43.50
1200	1-1/2" deep, 16 ga., galvanized					52.50			52.50	58
1620	Corner bead, expanded bullnose, 3/4" radius, #10, galvanized	1 Lath	2.60	3.077		26.50	151		177.50	253
1650	#1, galvanized		2.55	3.137		49	154		203	282
1670	Expanded wing, 2-3/4" wide, #1, galvanized		2.65	3.019		40.50	148		188.50	264

09 22 Supports for Plaster and Gypsum Board

09 22 36 – Lath

09 22 36.83 Accessories, Plaster	Crew	Daily Output	Labor-Hours	Unit	Material	2018 Bare Costs Labor	Equipment	Total	Total Incl O&P
1700	Inside corner (corner rite), 3" x 3", painted	1 Lath	2.60	3.077	C.L.F.	21	151		172	247
1750	Strip-ex, 4" wide, painted		2.55	3.137		23.50	154		177.50	254
1800	Expansion joint, 3/4" grounds, limited expansion, galv., 1 piece		2.70	2.963		73	146		219	296
2100	Extreme expansion, galvanized, 2 piece		2.60	3.077		141	151		292	380

09 23 Gypsum Plastering

09 23 13 – Acoustical Gypsum Plastering

09 23 13.10 Perlite or Vermiculite Plaster

		Crew	Daily Output	Labor-Hours	Unit	Material	2018 Bare Costs Labor	Equipment	Total	Total Incl O&P
0010	**PERLITE OR VERMICULITE PLASTER** R092000-50									
0020	In 100 lb. bags, under 200 bags				Bag	18.20			18.20	20
0100	Over 200 bags				"	17.45			17.45	19.15
0300	2 coats, no lath included, on walls	J-1	92	.435	S.Y.	5.95	19.05	1.45	26.45	36.50
0400	On ceilings	"	79	.506		5.95	22	1.69	29.64	42
0600	On and incl. 3/8" gypsum lath, on metal studs	J-2	84	.571		10.20	25.50	1.59	37.29	51
0700	On ceilings	"	70	.686		10.20	30.50	1.91	42.61	59.50
0900	3 coats, no lath included, on walls	J-1	74	.541		6.40	23.50	1.80	31.70	44.50
1000	On ceilings	"	63	.635		6.40	28	2.11	36.51	51
1200	On and incl. painted metal lath, on metal studs	J-2	72	.667		11.10	30	1.85	42.95	58.50
1300	On ceilings		61	.787		11.10	35	2.19	48.29	67
1500	On and incl. suspended metal lath ceiling		37	1.297		10.75	58	3.61	72.36	102
1700	For irregular or curved surfaces, add to above						30%			
1800	For columns and beams, add to above						50%			
1900	For soffits, add to ceiling prices						40%			

09 23 20 – Gypsum Plaster

09 23 20.10 Gypsum Plaster On Walls and Ceilings

		Crew	Daily Output	Labor-Hours	Unit	Material	2018 Bare Costs Labor	Equipment	Total	Total Incl O&P
0010	**GYPSUM PLASTER ON WALLS AND CEILINGS** R092000-50									
0020	80# bag, less than 1 ton				Bag	15.80			15.80	17.40
0100	Over 1 ton				"	13.80			13.80	15.20
0300	2 coats, no lath included, on walls	J-1	105	.381	S.Y.	3.76	16.65	1.27	21.68	30.50
0400	On ceilings	"	92	.435		3.76	19.05	1.45	24.26	34
0600	On and incl. 3/8" gypsum lath on steel, on walls	J-2	97	.495		6.90	22	1.38	30.28	42
0700	On ceilings	"	83	.578		6.90	26	1.61	34.51	48
0900	3 coats, no lath included, on walls	J-1	87	.460		5.40	20	1.53	26.93	37.50
1000	On ceilings	"	78	.513		5.40	22.50	1.71	29.61	41.50
1200	On and including painted metal lath, on wood studs	J-2	86	.558		10.40	25	1.55	36.95	50.50
1300	On ceilings	"	76.50	.627		10.40	28	1.74	40.14	55.50
1600	For irregular or curved surfaces, add						30%			
1800	For columns & beams, add						50%			

09 23 20.20 Gauging Plaster

		Crew	Daily Output	Labor-Hours	Unit	Material	2018 Bare Costs Labor	Equipment	Total	Total Incl O&P
0010	**GAUGING PLASTER** R092000-50									
0020	100 lb. bags, less than 1 ton				Bag	19.75			19.75	21.50
0100	Over 1 ton				"	18.70			18.70	20.50

09 23 20.30 Keenes Cement

		Crew	Daily Output	Labor-Hours	Unit	Material	2018 Bare Costs Labor	Equipment	Total	Total Incl O&P
0010	**KEENES CEMENT** R092000-50									
0020	In 100 lb. bags, less than 1 ton				Bag	22.50			22.50	24.50
0100	Over 1 ton				"	20.50			20.50	22.50
0300	Finish only, add to plaster prices, standard	J-1	215	.186	S.Y.	1.97	8.15	.62	10.74	15.05
0400	High quality	"	144	.278	"	1.99	12.15	.93	15.07	21.50

For customer support on your Building Construction Costs with RSMeans data, call 800.448.8182.

327

09 24 Cement Plastering

09 24 23 – Cement Stucco

09 24 23.40 Stucco		Crew	Daily Output	Labor-Hours	Unit	Material	2018 Bare Costs Labor	Equipment	Total	Total Incl O&P
0010	**STUCCO** R092000-50									
0015	3 coats 7/8" thick, float finish, with mesh, on wood frame	J-2	63	.762	S.Y.	8.35	34	2.12	44.47	62.50
0100	On masonry construction, no mesh incl.	J-1	67	.597		3.43	26	1.99	31.42	45
0300	For trowel finish, add	1 Plas	170	.047			2.19		2.19	3.28
0600	For coloring, add	J-1	685	.058		.42	2.56	.19	3.17	4.51
0700	For special texture, add	"	200	.200		1.46	8.75	.67	10.88	15.50
0900	For soffits, add	J-2	155	.310		2.25	13.85	.86	16.96	24
1000	Stucco, with bonding agent, 3 coats, on walls, no mesh incl.	J-1	200	.200		4.36	8.75	.67	13.78	18.70
1200	Ceilings		180	.222		3.77	9.75	.74	14.26	19.55
1300	Beams		80	.500		3.77	22	1.67	27.44	39
1500	Columns		100	.400		3.77	17.50	1.33	22.60	32
1600	Mesh, galvanized, nailed to wood, 1.8 lb.	1 Lath	60	.133		7.35	6.55		13.90	17.80
1800	3.6 lb.		55	.145		4.29	7.15		11.44	15.25
1900	Wired to steel, galvanized, 1.8 lb.		53	.151		7.35	7.40		14.75	19.05
2100	3.6 lb.		50	.160		4.29	7.85		12.14	16.30

09 25 Other Plastering

09 25 23 – Lime Based Plastering

09 25 23.10 Venetian Plaster		Crew	Daily Output	Labor-Hours	Unit	Material	2018 Bare Costs Labor	Equipment	Total	Total Incl O&P
0010	**VENETIAN PLASTER**									
0100	Walls, 1 coat primer, roller applied	1 Plas	950	.008	S.F.	.17	.39		.56	.78
0200	Plaster, 3 coats, incl. sanding	2 Plas	700	.023		.42	1.06		1.48	2.05
0210	For pigment, light colors add per S.F. plaster					.02			.02	.02
0220	For pigment, dark colors add per S.F. plaster					.04			.04	.04
0300	For sealer/wax coat incl. burnishing, add	1 Plas	300	.027		.38	1.24		1.62	2.27

09 26 Veneer Plastering

09 26 13 – Gypsum Veneer Plastering

09 26 13.20 Blueboard		Crew	Daily Output	Labor-Hours	Unit	Material	2018 Bare Costs Labor	Equipment	Total	Total Incl O&P
0010	**BLUEBOARD** For use with thin coat									
0100	plaster application see Section 09 26 13.80									
1000	3/8" thick, on walls or ceilings, standard, no finish included	2 Carp	1900	.008	S.F.	.34	.43		.77	1.02
1100	With thin coat plaster finish		875	.018		.45	.93		1.38	1.91
1400	On beams, columns, or soffits, standard, no finish included		675	.024		.39	1.20		1.59	2.26
1450	With thin coat plaster finish		475	.034		.50	1.71		2.21	3.15
3000	1/2" thick, on walls or ceilings, standard, no finish included		1900	.008		.35	.43		.78	1.04
3100	With thin coat plaster finish		875	.018		.46	.93		1.39	1.92
3300	Fire resistant, no finish included		1900	.008		.35	.43		.78	1.04
3400	With thin coat plaster finish		875	.018		.46	.93		1.39	1.92
3450	On beams, columns, or soffits, standard, no finish included		675	.024		.40	1.20		1.60	2.27
3500	With thin coat plaster finish		475	.034		.52	1.71		2.23	3.17
3700	Fire resistant, no finish included		675	.024		.40	1.20		1.60	2.27
3800	With thin coat plaster finish		475	.034		.52	1.71		2.23	3.17
5000	5/8" thick, on walls or ceilings, fire resistant, no finish included		1900	.008		.36	.43		.79	1.05
5100	With thin coat plaster finish		875	.018		.47	.93		1.40	1.93
5500	On beams, columns, or soffits, no finish included		675	.024		.41	1.20		1.61	2.29
5600	With thin coat plaster finish		475	.034		.53	1.71		2.24	3.18
6000	For high ceilings, over 8' high, add		3060	.005			.27		.27	.40
6500	For distribution costs 3 stories and above, add per story		6100	.003			.13		.13	.20

For customer support on your Building Construction Costs with RSMeans data, call 800.448.8182.

09 26 Veneer Plastering

09 26 13 – Gypsum Veneer Plastering

09 26 13.80 Thin Coat Plaster

		Crew	Daily Output	Labor-Hours	Unit	Material	2018 Bare Costs Labor	Equipment	Total	Total Incl O&P
0010	**THIN COAT PLASTER**									
0012	1 coat veneer, not incl. lath	J-1	3600	.011	S.F.	.11	.49	.04	.64	.89
1000	In 50 lb. bags				Bag	15.20			15.20	16.75

09 28 Backing Boards and Underlayments

09 28 13 – Cementitious Backing Boards

09 28 13.10 Cementitious Backerboard

		Crew	Daily Output	Labor-Hours	Unit	Material	2018 Bare Costs Labor	Equipment	Total	Total Incl O&P
0010	**CEMENTITIOUS BACKERBOARD**									
0070	Cementitious backerboard, on floor, 3' x 4' x 1/2" sheets	2 Carp	525	.030	S.F.	.86	1.55		2.41	3.30
0080	3' x 5' x 1/2" sheets		525	.030		.79	1.55		2.34	3.22
0090	3' x 6' x 1/2" sheets		525	.030		.78	1.55		2.33	3.21
0100	3' x 4' x 5/8" sheets		525	.030		.95	1.55		2.50	3.40
0110	3' x 5' x 5/8" sheets		525	.030		.95	1.55		2.50	3.39
0120	3' x 6' x 5/8" sheets		525	.030		.95	1.55		2.50	3.40
0150	On wall, 3' x 4' x 1/2" sheets		350	.046		.86	2.32		3.18	4.48
0160	3' x 5' x 1/2" sheets		350	.046		.79	2.32		3.11	4.40
0170	3' x 6' x 1/2" sheets		350	.046		.78	2.32		3.10	4.39
0180	3' x 4' x 5/8" sheets		350	.046		.95	2.32		3.27	4.58
0190	3' x 5' x 5/8" sheets		350	.046		.95	2.32		3.27	4.57
0200	3' x 6' x 5/8" sheets		350	.046		.95	2.32		3.27	4.58
0250	On counter, 3' x 4' x 1/2" sheets		180	.089		.86	4.51		5.37	7.80
0260	3' x 5' x 1/2" sheets		180	.089		.79	4.51		5.30	7.70
0270	3' x 6' x 1/2" sheets		180	.089		.78	4.51		5.29	7.70
0300	3' x 4' x 5/8" sheets		180	.089		.95	4.51		5.46	7.90
0310	3' x 5' x 5/8" sheets		180	.089		.95	4.51		5.46	7.90
0320	3' x 6' x 5/8" sheets		180	.089		.95	4.51		5.46	7.90

09 29 Gypsum Board

09 29 10 – Gypsum Board Panels

09 29 10.30 Gypsum Board

		Crew	Daily Output	Labor-Hours	Unit	Material	2018 Bare Costs Labor	Equipment	Total	Total Incl O&P
0010	**GYPSUM BOARD** on walls & ceilings R092910-10									
0100	Nailed or screwed to studs unless otherwise noted									
0110	1/4" thick, on walls or ceilings, standard, no finish included	2 Carp	1330	.012	S.F.	.37	.61		.98	1.34
0115	1/4" thick, on walls or ceilings, flexible, no finish included		1050	.015		.53	.77		1.30	1.76
0117	1/4" thick, on columns or soffits, flexible, no finish included		1050	.015		.53	.77		1.30	1.76
0130	1/4" thick, standard, no finish included, less than 800 S.F.		510	.031		.37	1.59		1.96	2.83
0150	3/8" thick, on walls, standard, no finish included		2000	.008		.36	.41		.77	1.02
0200	On ceilings, standard, no finish included		1800	.009		.36	.45		.81	1.09
0250	On beams, columns, or soffits, no finish included		675	.024		.36	1.20		1.56	2.23
0300	1/2" thick, on walls, standard, no finish included		2000	.008		.34	.41		.75	.99
0350	Taped and finished (level 4 finish)		965	.017		.39	.84		1.23	1.71
0390	With compound skim coat (level 5 finish)		775	.021		.44	1.05		1.49	2.07
0400	Fire resistant, no finish included		2000	.008		.37	.41		.78	1.03
0450	Taped and finished (level 4 finish)		965	.017		.42	.84		1.26	1.74
0490	With compound skim coat (level 5 finish)		775	.021		.47	1.05		1.52	2.11
0500	Water resistant, no finish included		2000	.008		.42	.41		.83	1.08
0550	Taped and finished (level 4 finish)		965	.017		.47	.84		1.31	1.79
0590	With compound skim coat (level 5 finish)		775	.021		.52	1.05		1.57	2.16
0600	Prefinished, vinyl, clipped to studs		900	.018		.51	.90		1.41	1.93

For customer support on your Building Construction Costs with RSMeans data, call 800.448.8182.

329

09 29 10.30 Gypsum Board	Crew	Daily Output	Labor-Hours	Unit	Material	2018 Bare Costs Labor	Equipment	Total	Total Incl O&P
0700 Mold resistant, no finish included	2 Carp	2000	.008	S.F.	.44	.41		.85	1.10
0710 Taped and finished (level 4 finish)		965	.017		.49	.84		1.33	1.82
0720 With compound skim coat (level 5 finish)		775	.021		.54	1.05		1.59	2.18
1000 On ceilings, standard, no finish included		1800	.009		.34	.45		.79	1.06
1050 Taped and finished (level 4 finish)		765	.021		.39	1.06		1.45	2.05
1090 With compound skim coat (level 5 finish)		610	.026		.44	1.33		1.77	2.51
1100 Fire resistant, no finish included		1800	.009		.37	.45		.82	1.10
1150 Taped and finished (level 4 finish)		765	.021		.42	1.06		1.48	2.08
1195 With compound skim coat (level 5 finish)		610	.026		.47	1.33		1.80	2.55
1200 Water resistant, no finish included		1800	.009		.42	.45		.87	1.15
1250 Taped and finished (level 4 finish)		765	.021		.47	1.06		1.53	2.13
1290 With compound skim coat (level 5 finish)		610	.026		.52	1.33		1.85	2.60
1310 Mold resistant, no finish included		1800	.009		.44	.45		.89	1.17
1320 Taped and finished (level 4 finish)		765	.021		.49	1.06		1.55	2.16
1330 With compound skim coat (level 5 finish)		610	.026		.54	1.33		1.87	2.62
1350 Sag resistant, no finish included		1600	.010		.36	.51		.87	1.17
1360 Taped and finished (level 4 finish)		765	.021		.41	1.06		1.47	2.07
1370 With compound skim coat (level 5 finish)		610	.026		.46	1.33		1.79	2.53
1500 On beams, columns, or soffits, standard, no finish included		675	.024		.39	1.20		1.59	2.26
1550 Taped and finished (level 4 finish)		540	.030		.39	1.50		1.89	2.72
1590 With compound skim coat (level 5 finish)		475	.034		.44	1.71		2.15	3.08
1600 Fire resistant, no finish included		675	.024		.37	1.20		1.57	2.24
1650 Taped and finished (level 4 finish)		540	.030		.42	1.50		1.92	2.75
1690 With compound skim coat (level 5 finish)		475	.034		.47	1.71		2.18	3.12
1700 Water resistant, no finish included		675	.024		.48	1.20		1.68	2.36
1750 Taped and finished (level 4 finish)		540	.030		.47	1.50		1.97	2.80
1790 With compound skim coat (level 5 finish)		475	.034		.52	1.71		2.23	3.17
1800 Mold resistant, no finish included		675	.024		.51	1.20		1.71	2.39
1810 Taped and finished (level 4 finish)		540	.030		.49	1.50		1.99	2.83
1820 With compound skim coat (level 5 finish)		475	.034		.54	1.71		2.25	3.19
1850 Sag resistant, no finish included		675	.024		.41	1.20		1.61	2.29
1860 Taped and finished (level 4 finish)		540	.030		.41	1.50		1.91	2.74
1870 With compound skim coat (level 5 finish)		475	.034		.46	1.71		2.17	3.10
2000 5/8" thick, on walls, standard, no finish included		2000	.008		.35	.41		.76	1.01
2050 Taped and finished (level 4 finish)		965	.017		.40	.84		1.24	1.72
2090 With compound skim coat (level 5 finish)		775	.021		.45	1.05		1.50	2.08
2100 Fire resistant, no finish included		2000	.008		.36	.41		.77	1.02
2150 Taped and finished (level 4 finish)		965	.017		.41	.84		1.25	1.73
2195 With compound skim coat (level 5 finish)		775	.021		.46	1.05		1.51	2.09
2200 Water resistant, no finish included		2000	.008		.44	.41		.85	1.10
2250 Taped and finished (level 4 finish)		965	.017		.49	.84		1.33	1.82
2290 With compound skim coat (level 5 finish)		775	.021		.54	1.05		1.59	2.18
2300 Prefinished, vinyl, clipped to studs		900	.018		.79	.90		1.69	2.24
2510 Mold resistant, no finish included		2000	.008		.49	.41		.90	1.16
2520 Taped and finished (level 4 finish)		965	.017		.54	.84		1.38	1.87
2530 With compound skim coat (level 5 finish)		775	.021		.59	1.05		1.64	2.24
3000 On ceilings, standard, no finish included		1800	.009		.35	.45		.80	1.08
3050 Taped and finished (level 4 finish)		765	.021		.40	1.06		1.46	2.06
3090 With compound skim coat (level 5 finish)		615	.026		.45	1.32		1.77	2.50
3100 Fire resistant, no finish included		1800	.009		.36	.45		.81	1.09
3150 Taped and finished (level 4 finish)		765	.021		.41	1.06		1.47	2.07
3190 With compound skim coat (level 5 finish)		615	.026		.46	1.32		1.78	2.51
3200 Water resistant, no finish included		1800	.009		.44	.45		.89	1.17

330

09 29 10 – Gypsum Board Panels

09 29 10.30 Gypsum Board		Crew	Daily Output	Labor-Hours	Unit	Material	2018 Bare Costs Labor	Equipment	Total	Total Incl O&P
3250	Taped and finished (level 4 finish)	2 Carp	765	.021	S.F.	.49	1.06		1.55	2.16
3290	With compound skim coat (level 5 finish)		615	.026		.54	1.32		1.86	2.60
3300	Mold resistant, no finish included		1800	.009		.49	.45		.94	1.23
3310	Taped and finished (level 4 finish)		765	.021		.54	1.06		1.60	2.21
3320	With compound skim coat (level 5 finish)		615	.026		.59	1.32		1.91	2.66
3500	On beams, columns, or soffits, no finish included		675	.024		.40	1.20		1.60	2.27
3550	Taped and finished (level 4 finish)		475	.034		.46	1.71		2.17	3.10
3590	With compound skim coat (level 5 finish)		380	.042		.52	2.14		2.66	3.82
3600	Fire resistant, no finish included		675	.024		.41	1.20		1.61	2.29
3650	Taped and finished (level 4 finish)		475	.034		.47	1.71		2.18	3.11
3690	With compound skim coat (level 5 finish)		380	.042		.46	2.14		2.60	3.75
3700	Water resistant, no finish included		675	.024		.51	1.20		1.71	2.39
3750	Taped and finished (level 4 finish)		475	.034		.54	1.71		2.25	3.19
3790	With compound skim coat (level 5 finish)		380	.042		.56	2.14		2.70	3.87
3800	Mold resistant, no finish included		675	.024		.56	1.20		1.76	2.45
3810	Taped and finished (level 4 finish)		475	.034		.59	1.71		2.30	3.25
3820	With compound skim coat (level 5 finish)		380	.042		.62	2.14		2.76	3.93
4000	Fireproofing, beams or columns, 2 layers, 1/2" thick, incl finish		330	.048		.83	2.46		3.29	4.66
4010	Mold resistant		330	.048		.97	2.46		3.43	4.81
4050	5/8" thick		300	.053		.81	2.70		3.51	5
4060	Mold resistant		300	.053		1.07	2.70		3.77	5.30
4100	3 layers, 1/2" thick		225	.071		1.25	3.61		4.86	6.85
4110	Mold resistant		225	.071		1.46	3.61		5.07	7.10
4150	5/8" thick		210	.076		1.22	3.86		5.08	7.25
4160	Mold resistant		210	.076		1.61	3.86		5.47	7.65
5050	For 1" thick coreboard on columns		480	.033		.77	1.69		2.46	3.42
5100	For foil-backed board, add					.18			.18	.20
5200	For work over 8' high, add	2 Carp	3060	.005			.27		.27	.40
5270	For textured spray, add	2 Lath	1600	.010		.04	.49		.53	.77
5300	For distribution cost 3 stories and above, add per story	2 Carp	6100	.003			.13		.13	.20
5350	For finishing inner corners, add		950	.017	L.F.	.10	.85		.95	1.41
5355	For finishing outer corners, add		1250	.013		.23	.65		.88	1.24
5500	For acoustical sealant, add per bead	1 Carp	500	.016		.04	.81		.85	1.29
5550	Sealant, 1 quart tube				Ea.	7.05			7.05	7.80
6000	Gypsum sound dampening panels									
6010	1/2" thick on walls, multi-layer, lightweight, no finish included	2 Carp	1500	.011	S.F.	2.16	.54		2.70	3.20
6015	Taped and finished (level 4 finish)		725	.022		2.21	1.12		3.33	4.13
6020	With compound skim coat (level 5 finish)		580	.028		2.26	1.40		3.66	4.61
6025	5/8" thick on walls, for wood studs, no finish included		1500	.011		2.21	.54		2.75	3.25
6030	Taped and finished (level 4 finish)		725	.022		2.26	1.12		3.38	4.18
6035	With compound skim coat (level 5 finish)		580	.028		2.31	1.40		3.71	4.67
6040	For metal stud, no finish included		1500	.011		2.28	.54		2.82	3.33
6045	Taped and finished (level 4 finish)		725	.022		2.33	1.12		3.45	4.26
6050	With compound skim coat (level 5 finish)		580	.028		2.38	1.40		3.78	4.75
6055	Abuse resist, no finish included		1500	.011		4.10	.54		4.64	5.35
6060	Taped and finished (level 4 finish)		725	.022		4.15	1.12		5.27	6.25
6065	With compound skim coat (level 5 finish)		580	.028		4.20	1.40		5.60	6.75
6070	Shear rated, no finish included		1500	.011		5.15	.54		5.69	6.45
6075	Taped and finished (level 4 finish)		725	.022		5.20	1.12		6.32	7.40
6080	With compound skim coat (level 5 finish)		580	.028		5.25	1.40		6.65	7.90
6085	For SCIF applications, no finish included		1500	.011		5.15	.54		5.69	6.45
6090	Taped and finished (level 4 finish)		725	.022		5.20	1.12		6.32	7.40
6095	With compound skim coat (level 5 finish)		580	.028		5.25	1.40		6.65	7.90

09 29 10.30 Gypsum Board

		Crew	Daily Output	Labor-Hours	Unit	Material	2018 Bare Costs Labor	Equipment	Total	Total Incl O&P
6100	1-3/8" thick on walls, THX certified, no finish included	2 Carp	1500	.011	S.F.	9.15	.54		9.69	10.85
6105	Taped and finished (level 4 finish)		725	.022		9.20	1.12		10.32	11.80
6110	With compound skim coat (level 5 finish)		580	.028		9.25	1.40		10.65	12.30
6115	5/8" thick on walls, score & snap installation, no finish included		2000	.008		1.96	.41		2.37	2.78
6120	Taped and finished (level 4 finish)		965	.017		2.01	.84		2.85	3.49
6125	With compound skim coat (level 5 finish)		775	.021		2.06	1.05		3.11	3.85
7020	5/8" thick on ceilings, for wood joists, no finish included		1200	.013		2.21	.68		2.89	3.46
7025	Taped and finished (level 4 finish)		510	.031		2.26	1.59		3.85	4.90
7030	With compound skim coat (level 5 finish)		410	.039		2.31	1.98		4.29	5.55
7035	For metal joists, no finish included		1200	.013		2.28	.68		2.96	3.54
7040	Taped and finished (level 4 finish)		510	.031		2.33	1.59		3.92	4.98
7045	With compound skim coat (level 5 finish)		410	.039		2.38	1.98		4.36	5.65
7050	Abuse resist, no finish included		1200	.013		4.10	.68		4.78	5.55
7055	Taped and finished (level 4 finish)		510	.031		4.15	1.59		5.74	7
7060	With compound skim coat (level 5 finish)		410	.039		4.20	1.98		6.18	7.65
7065	Shear rated, no finish included		1200	.013		5.15	.68		5.83	6.70
7070	Taped and finished (level 4 finish)		510	.031		5.20	1.59		6.79	8.10
7075	With compound skim coat (level 5 finish)		410	.039		5.25	1.98		7.23	8.75
7080	For SCIF applications, no finish included		1200	.013		5.15	.68		5.83	6.70
7085	Taped and finished (level 4 finish)		510	.031		5.20	1.59		6.79	8.10
7090	With compound skim coat (level 5 finish)		410	.039		5.25	1.98		7.23	8.75
8010	5/8" thick on ceilings, score & snap installation, no finish included		1600	.010		1.96	.51		2.47	2.93
8015	Taped and finished (level 4 finish)		680	.024		2.01	1.19		3.20	4.03
8020	With compound skim coat (level 5 finish)	▼	545	.029	▼	2.06	1.49		3.55	4.53

09 29 10.50 High Abuse Gypsum Board

		Crew	Daily Output	Labor-Hours	Unit	Material	2018 Bare Costs Labor	Equipment	Total	Total Incl O&P
0010	**HIGH ABUSE GYPSUM BOARD**, fiber reinforced, nailed or									
0100	screwed to studs unless otherwise noted									
0110	1/2" thick, on walls, no finish included	2 Carp	1800	.009	S.F.	.78	.45		1.23	1.55
0120	Taped and finished (level 4 finish)		870	.018		.83	.93		1.76	2.33
0130	With compound skim coat (level 5 finish)		700	.023		.88	1.16		2.04	2.73
0150	On ceilings, no finish included		1620	.010		.78	.50		1.28	1.62
0160	Taped and finished (level 4 finish)		690	.023		.83	1.18		2.01	2.70
0170	With compound skim coat (level 5 finish)		550	.029		.88	1.47		2.35	3.22
0210	5/8" thick, on walls, no finish included		1800	.009		.81	.45		1.26	1.58
0220	Taped and finished (level 4 finish)		870	.018		.86	.93		1.79	2.36
0230	With compound skim coat (level 5 finish)		700	.023		.91	1.16		2.07	2.76
0250	On ceilings, no finish included		1620	.010		.81	.50		1.31	1.65
0260	Taped and finished (level 4 finish)		690	.023		.86	1.18		2.04	2.73
0270	With compound skim coat (level 5 finish)		550	.029		.91	1.47		2.38	3.25
0272	5/8" thick, on roof sheathing for 1-hour rating		1620	.010		.81	.50		1.31	1.65
0310	5/8" thick, on walls, very high impact, no finish included		1800	.009		.90	.45		1.35	1.68
0320	Taped and finished (level 4 finish)		870	.018		.95	.93		1.88	2.46
0330	With compound skim coat (level 5 finish)		700	.023		1	1.16		2.16	2.86
0350	On ceilings, no finish included		1620	.010		.90	.50		1.40	1.75
0360	Taped and finished (level 4 finish)		690	.023		.95	1.18		2.13	2.83
0370	With compound skim coat (level 5 finish)	▼	550	.029	▼	1	1.47		2.47	3.35
0400	High abuse, gypsum core, paper face									
0410	1/2" thick, on walls, no finish included	2 Carp	1800	.009	S.F.	.64	.45		1.09	1.39
0420	Taped and finished (level 4 finish)		870	.018		.69	.93		1.62	2.18
0430	With compound skim coat (level 5 finish)		700	.023		.74	1.16		1.90	2.57
0450	On ceilings, no finish included		1620	.010		.64	.50		1.14	1.46
0460	Taped and finished (level 4 finish)		690	.023		.69	1.18		1.87	2.55

09 29 Gypsum Board

09 29 10 – Gypsum Board Panels

09 29 10.50 High Abuse Gypsum Board	Crew	Daily Output	Labor-Hours	Unit	Material	2018 Bare Costs Labor	Equipment	Total	Total Incl O&P
0470	With compound skim coat (level 5 finish)	2 Carp	550	.029	S.F.	.74	1.47		2.21	3.06
0510	5/8" thick, on walls, no finish included		1800	.009		.69	.45		1.14	1.45
0520	Taped and finished (level 4 finish)		870	.018		.74	.93		1.67	2.23
0530	With compound skim coat (level 5 finish)		700	.023		.79	1.16		1.95	2.63
0550	On ceilings, no finish included		1620	.010		.69	.50		1.19	1.52
0560	Taped and finished (level 4 finish)		690	.023		.74	1.18		1.92	2.60
0570	With compound skim coat (level 5 finish)		550	.029		.79	1.47		2.26	3.12
1000	For high ceilings, over 8' high, add		2750	.006			.30		.30	.45
1010	For distribution cost 3 stories and above, add per story		5500	.003			.15		.15	.22

09 29 15 – Gypsum Board Accessories

09 29 15.10 Accessories, Gypsum Board

		Crew	Daily Output	Labor-Hours	Unit	Material	2018 Bare Costs Labor	Equipment	Total	Total Incl O&P
0010	**ACCESSORIES, GYPSUM BOARD**									
0020	Casing bead, galvanized steel	1 Carp	2.90	2.759	C.L.F.	24	140		164	240
0100	Vinyl		3	2.667		23	135		158	232
0300	Corner bead, galvanized steel, 1" x 1"		4	2		15.50	101		116.50	171
0400	1-1/4" x 1-1/4"		3.50	2.286		16.95	116		132.95	195
0600	Vinyl		4	2		20.50	101		121.50	177
0900	Furring channel, galv. steel, 7/8" deep, standard		2.60	3.077		34.50	156		190.50	276
1000	Resilient		2.55	3.137		26	159		185	271
1100	J trim, galvanized steel, 1/2" wide		3	2.667		22.50	135		157.50	231
1120	5/8" wide		2.95	2.712		32.50	137		169.50	245
1140	L trim, galvanized		3	2.667		18.75	135		153.75	227
1150	U trim, galvanized		2.95	2.712		23.50	137		160.50	235
1160	Screws #6 x 1" A				M	10.85			10.85	11.95
1170	#6 x 1-5/8" A				"	14.40			14.40	15.80
1200	For stud partitions, see Sections 05 41 13.30 and 09 22 16.13									
1500	Z stud, galvanized steel, 1-1/2" wide	1 Carp	2.60	3.077	C.L.F.	41.50	156		197.50	284
1600	2" wide	"	2.55	3.137	"	65	159		224	315

09 30 Tiling

09 30 13 – Ceramic Tiling

09 30 13.20 Ceramic Tile Repairs

		Crew	Daily Output	Labor-Hours	Unit	Material	2018 Bare Costs Labor	Equipment	Total	Total Incl O&P
0010	**CERAMIC TILE REPAIRS**									
1000	Grout removal, carbide tipped, rotary grinder	1 Clab	240	.033	L.F.		1.33		1.33	2.02
1100	Regrout tile 4-1/2" x 4-1/2" or larger, wall	1 Tilf	100	.080	S.F.	.15	3.75		3.90	5.70
1150	Floor		125	.064		.16	3		3.16	4.62
1200	Seal tile and grout		360	.022			1.04		1.04	1.54

09 30 13.45 Ceramic Tile Accessories

		Crew	Daily Output	Labor-Hours	Unit	Material	2018 Bare Costs Labor	Equipment	Total	Total Incl O&P
0010	**CERAMIC TILE ACCESSORIES**									
0100	Spacers, 1/8"				C	1.98			1.98	2.18
1310	Sealer for natural stone tile, installed	1 Tilf	650	.012	S.F.	.05	.58		.63	.91

09 30 29 – Metal Tiling

09 30 29.10 Metal Tile

		Crew	Daily Output	Labor-Hours	Unit	Material	2018 Bare Costs Labor	Equipment	Total	Total Incl O&P
0010	**METAL TILE** 4' x 4' sheet, 24 ga., tile pattern, nailed									
0200	Stainless steel	2 Carp	512	.031	S.F.	28	1.58		29.58	33.50
0400	Aluminized steel	"	512	.031	"	19	1.58		20.58	23.50

09 30 Tiling

09 30 95 – Tile & Stone Setting Materials and Specialties

09 30 95.10 Moisture Resistant, Anti-Fracture Membrane	Crew	Daily Output	Labor-Hours	Unit	Material	2018 Bare Costs Labor	2018 Bare Costs Equipment	Total	Total Incl O&P
0010 **MOISTURE RESISTANT, ANTI-FRACTURE MEMBRANE**									
0200 Elastomeric membrane, 1/16" thick	D-7	275	.058	S.F.	1.12	2.45		3.57	4.85

09 31 Thin-Set Tiling

09 31 13 – Thin-Set Ceramic Tiling

09 31 13.10 Thin-Set Ceramic Tile

		Crew	Daily Output	Labor-Hours	Unit	Material	Labor	Equipment	Total	Total Incl O&P
0010	**THIN-SET CERAMIC TILE**									
0020	Backsplash, average grade tiles	1 Tilf	50	.160	S.F.	2.79	7.50		10.29	14.15
0022	Custom grade tiles		50	.160		5.60	7.50		13.10	17.25
0024	Luxury grade tiles		50	.160		11.15	7.50		18.65	23.50
0026	Economy grade tiles		50	.160		2.56	7.50		10.06	13.90
0100	Base, using 1' x 4" high piece with 1" x 1" tiles	D-7	128	.125	L.F.	4.99	5.25		10.24	13.30
0300	For 6" high base, 1" x 1" tile face, add					1.22			1.22	1.35
0400	For 2" x 2" tile face, add to above					.72			.72	.79
0700	Cove base, 4-1/4" x 4-1/4"	D-7	128	.125		4.19	5.25		9.44	12.40
1000	6" x 4-1/4" high		137	.117		4.32	4.91		9.23	12
1300	Sanitary cove base, 6" x 4-1/4" high		124	.129		4.64	5.40		10.04	13.15
1600	6" x 6" high		117	.137		5.30	5.75		11.05	14.30
1800	Bathroom accessories, average (soap dish, toothbrush holder)		82	.195	Ea.	9.80	8.20		18	23
1900	Bathtub, 5', rec. 4-1/4" x 4-1/4" tile wainscot, adhesive set 6' high		2.90	5.517		175	232		407	535
2100	7' high wainscot		2.50	6.400		204	269		473	625
2200	8' high wainscot		2.20	7.273		233	305		538	705
2500	Bullnose trim, 4-1/4" x 4-1/4"		128	.125	L.F.	4.17	5.25		9.42	12.40
2800	2" x 6"		124	.129	"	4.22	5.40		9.62	12.70
3300	Ceramic tile, porcelain type, 1 color, color group 2, 1" x 1"		183	.087	S.F.	6.40	3.67		10.07	12.50
3310	2" x 2" or 2" x 1"		190	.084		6.25	3.54		9.79	12.15
3350	For random blend, 2 colors, add					1			1	1.10
3360	4 colors, add					1.50			1.50	1.65
3370	For color group 3, add					.65			.65	.72
3380	For abrasive non-slip tile, add					.45			.45	.50
4300	Specialty tile, 4-1/4" x 4-1/4" x 1/2", decorator finish	D-7	183	.087		12.70	3.67		16.37	19.45
4500	Add for epoxy grout, 1/16" joint, 1" x 1" tile		800	.020		.68	.84		1.52	1.99
4600	2" x 2" tile		820	.020		.62	.82		1.44	1.89
4610	Add for epoxy grout, 1/8" joint, 8" x 8" x 3/8" tile, add		900	.018		1.42	.75		2.17	2.67
4800	Pregrouted sheets, walls, 4-1/4" x 4-1/4", 6" x 4-1/4"									
4810	and 8-1/2" x 4-1/4", 4 S.F. sheets, silicone grout	D-7	240	.067	S.F.	5.60	2.80		8.40	10.35
5100	Floors, unglazed, 2 S.F. sheets,									
5110	urethane adhesive	D-7	180	.089	S.F.	2.03	3.74		5.77	7.80
5400	Walls, interior, 4-1/4" x 4-1/4" tile		190	.084		2.61	3.54		6.15	8.10
5500	6" x 4-1/4" tile		190	.084		3.15	3.54		6.69	8.70
5700	8-1/2" x 4-1/4" tile		190	.084		5.65	3.54		9.19	11.50
5800	6" x 6" tile		175	.091		3.57	3.84		7.41	9.65
5810	8" x 8" tile		170	.094		5.30	3.96		9.26	11.70
5820	12" x 12" tile		160	.100		4.66	4.20		8.86	11.35
5830	16" x 16" tile		150	.107		5.15	4.48		9.63	12.30
6000	Decorated wall tile, 4-1/4" x 4-1/4", color group 1		270	.059		3.68	2.49		6.17	7.75
6100	Color group 4		180	.089		52.50	3.74		56.24	63
9300	Ceramic tiles, recycled glass, standard colors, 2" x 2" thru 6" x 6" [G]		190	.084		22	3.54		25.54	30
9310	6" x 6" [G]		175	.091		22	3.84		25.84	30
9320	8" x 8" [G]		170	.094		23.50	3.96		27.46	32

For customer support on your Building Construction Costs with RSMeans data, call 800.448.8182.

09 31 Thin-Set Tiling

09 31 13 – Thin-Set Ceramic Tiling

09 31 13.10 Thin-Set Ceramic Tile

		Crew	Daily Output	Labor-Hours	Unit	Material	2018 Bare Costs Labor	Equipment	Total	Total Incl O&P
9330	12" x 12" [G]	D-7	160	.100	S.F.	23.50	4.20		27.70	32
9340	Earthtones, 2" x 2" to 4" x 8" [G]		190	.084		26	3.54		29.54	34
9350	6" x 6" [G]		175	.091		26	3.84		29.84	34
9360	8" x 8" [G]		170	.094		27	3.96		30.96	35.50
9370	12" x 12" [G]		160	.100		27	4.20		31.20	35.50
9380	Deep colors, 2" x 2" to 4" x 8" [G]		190	.084		30.50	3.54		34.04	39.50
9390	6" x 6" [G]		175	.091		30.50	3.84		34.34	39.50
9400	8" x 8" [G]		170	.094		32	3.96		35.96	41.50
9410	12" x 12" [G]	▼	160	.100	▼	32	4.20		36.20	41.50

09 31 33 – Thin-Set Stone Tiling

09 31 33.10 Tiling, Thin-Set Stone

		Crew	Daily Output	Labor-Hours	Unit	Material	2018 Bare Costs Labor	Equipment	Total	Total Incl O&P
0010	**TILING, THIN-SET STONE**									
3000	Floors, natural clay, random or uniform, color group 1	D-7	183	.087	S.F.	4.31	3.67		7.98	10.20
3100	Color group 2		183	.087		5.95	3.67		9.62	12
3255	Floors, glazed, 6" x 6", color group 1		300	.053		5.25	2.24		7.49	9.10
3260	8" x 8" tile		300	.053		5.25	2.24		7.49	9.10
3270	12" x 12" tile		290	.055		5.85	2.32		8.17	9.85
3280	16" x 16" tile		280	.057		7.50	2.40		9.90	11.80
3281	18" x 18" tile		270	.059		7.10	2.49		9.59	11.55
3282	20" x 20" tile		260	.062		7.05	2.59		9.64	11.60
3283	24" x 24" tile		250	.064		9	2.69		11.69	13.90
3285	Border, 6" x 12" tile		200	.080		11.05	3.36		14.41	17.20
3290	3" x 12" tile	▼	200	.080	▼	10.60	3.36		13.96	16.65

09 32 Mortar-Bed Tiling

09 32 13 – Mortar-Bed Ceramic Tiling

09 32 13.10 Ceramic Tile

		Crew	Daily Output	Labor-Hours	Unit	Material	2018 Bare Costs Labor	Equipment	Total	Total Incl O&P
0010	**CERAMIC TILE**									
0050	Base, using 1' x 4" high pc. with 1" x 1" tiles	D-7	82	.195	L.F.	5.25	8.20		13.45	17.95
0600	Cove base, 4-1/4" x 4-1/4" high		91	.176		4.33	7.40		11.73	15.70
0900	6" x 4-1/4" high		100	.160		4.46	6.70		11.16	14.85
1200	Sanitary cove base, 6" x 4-1/4" high		93	.172		4.78	7.25		12.03	15.95
1500	6" x 6" high		84	.190		5.45	8		13.45	17.80
2400	Bullnose trim, 4-1/4" x 4-1/4"		82	.195		4.27	8.20		12.47	16.85
2700	2" x 6" bullnose trim	▼	84	.190	▼	4.29	8		12.29	16.55
6210	Wall tile, 4-1/4" x 4-1/4", better grade	1 Tilf	50	.160	S.F.	9.50	7.50		17	21.50
6240	2" x 2"		50	.160		6.50	7.50		14	18.25
6250	6" x 6"		55	.145		10.15	6.80		16.95	21.50
6260	8" x 8"	▼	60	.133		9.40	6.25		15.65	19.60
6300	Exterior walls, frostproof, 4-1/4" x 4-1/4"	D-7	102	.157		7.25	6.60		13.85	17.70
6400	1-3/8" x 1-3/8"		93	.172		6.60	7.25		13.85	18
6600	Crystalline glazed, 4-1/4" x 4-1/4", plain		100	.160		4.52	6.70		11.22	14.90
6700	4-1/4" x 4-1/4", scored tile		100	.160		6.50	6.70		13.20	17.10
6900	6" x 6" plain		93	.172		5.75	7.25		13	17.05
7000	For epoxy grout, 1/16" joints, 4-1/4" tile, add		800	.020		.41	.84		1.25	1.69
7200	For tile set in dry mortar, add		1735	.009			.39		.39	.57
7300	For tile set in Portland cement mortar, add	▼	290	.055	▼	.18	2.32		2.50	3.63

For customer support on your Building Construction Costs with RSMeans data, call 800.448.8182.

335

09 32 Mortar-Bed Tiling

09 32 16 – Mortar-Bed Quarry Tiling

09 32 16.10 Quarry Tile	Crew	Daily Output	Labor-Hours	Unit	Material	2018 Bare Costs Labor	Equipment	Total	Total Incl O&P
0010 **QUARRY TILE**									
0100 Base, cove or sanitary, to 5" high, 1/2" thick	D-7	110	.145	L.F.	6.40	6.10		12.50	16.10
0300 Bullnose trim, red, 6" x 6" x 1/2" thick		120	.133		5.40	5.60		11	14.20
0400 4" x 4" x 1/2" thick		110	.145		5	6.10		11.10	14.55
0600 4" x 8" x 1/2" thick, using 8" as edge		130	.123		5.35	5.15		10.50	13.55
0700 Floors, 1,000 S.F. lots, red, 4" x 4" x 1/2" thick		120	.133	S.F.	8.70	5.60		14.30	17.90
0900 6" x 6" x 1/2" thick		140	.114		8.25	4.80		13.05	16.20
1000 4" x 8" x 1/2" thick		130	.123		6.60	5.15		11.75	14.95
1300 For waxed coating, add					.76			.76	.84
1500 For non-standard colors, add					.46			.46	.51
1600 For abrasive surface, add					.52			.52	.57
1800 Brown tile, imported, 6" x 6" x 3/4"	D-7	120	.133		7.20	5.60		12.80	16.20
1900 8" x 8" x 1"		110	.145		9.55	6.10		15.65	19.55
2100 For thin set mortar application, deduct		700	.023			.96		.96	1.42
2200 For epoxy grout & mortar, 6" x 6" x 1/2", add		350	.046		2.32	1.92		4.24	5.40
2700 Stair tread, 6" x 6" x 3/4", plain		50	.320		7.20	13.45		20.65	28
2800 Abrasive		47	.340		8.55	14.30		22.85	30.50
3000 Wainscot, 6" x 6" x 1/2", thin set, red		105	.152		6.30	6.40		12.70	16.40
3100 Non-standard colors		105	.152		6.40	6.40		12.80	16.50
3300 Window sill, 6" wide, 3/4" thick		90	.178	L.F.	8.65	7.45		16.10	20.50
3400 Corners		80	.200	Ea.	6	8.40		14.40	19.05

09 32 23 – Mortar-Bed Glass Mosaic Tiling

09 32 23.10 Glass Mosaics

	Crew	Daily Output	Labor-Hours	Unit	Material	2018 Bare Costs Labor	Equipment	Total	Total Incl O&P
0010 **GLASS MOSAICS** 3/4" tile on 12" sheets, standard grout									
0300 Color group 1 & 2	D-7	73	.219	S.F.	20	9.20		29.20	35.50
0350 Color group 3		73	.219		21	9.20		30.20	37
0400 Color group 4		73	.219		27	9.20		36.20	43.50
0450 Color group 5		73	.219		30.50	9.20		39.70	47
0500 Color group 6		73	.219		40.50	9.20		49.70	58
0600 Color group 7		73	.219		41	9.20		50.20	58.50
0700 Color group 8, golds, silvers & specialties		64	.250		41.50	10.50		52	61.50
1020 1" tile on 12" sheets, opalescent finish		73	.219		18.40	9.20		27.60	34
1040 1" x 2" tile on 12" sheet, blend		73	.219		21	9.20		30.20	36.50
1060 2" tile on 12" sheet, blend		73	.219		17.60	9.20		26.80	33
1080 5/8" x random tile, linear, on 12" sheet, blend		73	.219		25	9.20		34.20	40.50
1600 Dots on 12" sheet		73	.219		25	9.20		34.20	41
1700 For glass mosaic tiles set in dry mortar, add		290	.055		.45	2.32		2.77	3.93
1720 For glass mosaic tiles set in Portland cement mortar, add		290	.055		.01	2.32		2.33	3.44
1730 For polyblend sanded tile grout		96.15	.166	Lb.	2.19	7		9.19	12.75

For customer support on your Building Construction Costs with RSMeans data, call 800.448.8182.

09 34 Waterproofing-Membrane Tiling

09 34 13 – Waterproofing-Membrane Ceramic Tiling

09 34 13.10 Ceramic Tile Waterproofing Membrane

09 34 13.10 Ceramic Tile Waterproofing Membrane	Crew	Daily Output	Labor-Hours	Unit	Material	2018 Bare Costs Labor	Equipment	Total	Total Incl O&P
0010 **CERAMIC TILE WATERPROOFING MEMBRANE**									
0020 On floors, including thinset									
0030 Fleece laminated polyethylene grid, 1/8" thick	D-7	250	.064	S.F.	2.28	2.69		4.97	6.50
0040 5/16" thick	"	250	.064	"	2.60	2.69		5.29	6.85
0050 On walls, including thinset									
0060 Fleece laminated polyethylene sheet, 8 mil thick	D-7	480	.033	S.F.	2.28	1.40		3.68	4.57
0070 Accessories, including thinset									
0080 Joint and corner sheet, 4 mils thick, 5" wide	1 Tilf	240	.033	L.F.	1.35	1.56		2.91	3.79
0090 7-1/4" wide		180	.044		1.71	2.08		3.79	4.96
0100 10" wide		120	.067	↓	2.08	3.12		5.20	6.90
0110 Pre-formed corners, inside		32	.250	Ea.	7.85	11.70		19.55	26
0120 Outside		32	.250		7.65	11.70		19.35	26
0130 2" flanged floor drain with 6" stainless steel grate		16	.500	↓	370	23.50		393.50	445
0140 EPS, sloped shower floor		480	.017	S.F.	5.55	.78		6.33	7.25
0150 Curb	↓	32	.250	L.F.	14.05	11.70		25.75	33

09 35 Chemical-Resistant Tiling

09 35 13 – Chemical-Resistant Ceramic Tiling

09 35 13.10 Chemical-Resistant Ceramic Tiling

09 35 13.10 Chemical-Resistant Ceramic Tiling	Crew	Daily Output	Labor-Hours	Unit	Material	2018 Bare Costs Labor	Equipment	Total	Total Incl O&P
0010 **CHEMICAL-RESISTANT CERAMIC TILING**									
0100 4-1/4" x 4-1/4" x 1/4", 1/8" joint	D-7	130	.123	S.F.	12.05	5.15		17.20	21
0200 6"x 6" x 1/2" thick		120	.133		9.75	5.60		15.35	19
0300 8"x 8" x 1/2" thick		110	.145		10.90	6.10		17	21
0400 4-1/4" x 4-1/4" x 1/4", 1/4" joint		130	.123		12.80	5.15		17.95	22
0500 6"x 6" x 1/2" thick		120	.133		10.85	5.60		16.45	20.50
0600 8"x 8" x 1/2" thick		110	.145		11.55	6.10		17.65	22
0700 4-1/4" x 4-1/4" x 1/4", 3/8" joint		130	.123		13.50	5.15		18.65	22.50
0800 6"x 6" x 1/2" thick		120	.133		11.80	5.60		17.40	21.50
0900 8"x 8" x 1/2" thick	↓	110	.145	↓	12.70	6.10		18.80	23

09 35 16 – Chemical-Resistant Quarry Tiling

09 35 16.10 Chemical-Resistant Quarry Tiling

09 35 16.10 Chemical-Resistant Quarry Tiling	Crew	Daily Output	Labor-Hours	Unit	Material	2018 Bare Costs Labor	Equipment	Total	Total Incl O&P
0010 **CHEMICAL-RESISTANT QUARRY TILING**									
0100 4"x 8" x 1/2" thick, 1/8" joint	D-7	130	.123	S.F.	11.20	5.15		16.35	19.95
0200 6"x 6" x 1/2" thick		120	.133		11.25	5.60		16.85	20.50
0300 8"x 8" x 1/2" thick		110	.145		10.35	6.10		16.45	20.50
0400 4"x 8" x 1/2" thick, 1/4" joint		130	.123		12.40	5.15		17.55	21.50
0500 6"x 6" x 1/2" thick		120	.133		12.35	5.60		17.95	22
0600 8"x 8" x 1/2" thick		110	.145		11	6.10		17.10	21
0700 4"x 8" x 1/2" thick, 3/8" joint		130	.123		13.50	5.15		18.65	22.50
0800 6"x 6" x 1/2" thick		120	.133		13.30	5.60		18.90	23
0900 8"x 8" x 1/2" thick	↓	110	.145	↓	12.15	6.10		18.25	22.50

For customer support on your Building Construction Costs with RSMeans data, call 800.448.8182.

337

09 51 13 – Acoustical Panel Ceilings

09 51 13.10 Ceiling, Acoustical Panel	Crew	Daily Output	Labor-Hours	Unit	Material	2018 Bare Costs Labor	Equipment	Total	Total Incl O&P
0010 **CEILING, ACOUSTICAL PANEL**									
0100 Fiberglass boards, film faced, 2' x 2' or 2' x 4', 5/8" thick	1 Carp	625	.013	S.F.	1.26	.65		1.91	2.38
0120 3/4" thick		600	.013		3	.68		3.68	4.33
0130 3" thick, thermal, R11		450	.018		3.50	.90		4.40	5.20

09 51 14 – Acoustical Fabric-Faced Panel Ceilings

09 51 14.10 Ceiling, Acoustical Fabric-Faced Panel

	Crew	Daily Output	Labor-Hours	Unit	Material	2018 Bare Costs Labor	Equipment	Total	Total Incl O&P
0010 **CEILING, ACOUSTICAL FABRIC-FACED PANEL**									
0100 Glass cloth faced fiberglass, 3/4" thick	1 Carp	500	.016	S.F.	2.99	.81		3.80	4.53
0120 1" thick		485	.016		3.61	.84		4.45	5.25
0130 1-1/2" thick, nubby face		475	.017		2.73	.85		3.58	4.30

09 51 23 – Acoustical Tile Ceilings

09 51 23.10 Suspended Acoustic Ceiling Tiles

	Crew	Daily Output	Labor-Hours	Unit	Material	2018 Bare Costs Labor	Equipment	Total	Total Incl O&P
0010 **SUSPENDED ACOUSTIC CEILING TILES**, not including									
0100 suspension system									
1110 Mineral fiber tile, lay-in, 2' x 2' or 2' x 4', 5/8" thick, fine texture	1 Carp	625	.013	S.F.	.79	.65		1.44	1.86
1115 Rough textured		625	.013		.75	.65		1.40	1.82
1125 3/4" thick, fine textured		600	.013		2.11	.68		2.79	3.35
1130 Rough textured		600	.013		1.73	.68		2.41	2.93
1135 Fissured		600	.013		2.17	.68		2.85	3.42
1150 Tegular, 5/8" thick, fine textured		470	.017		1.12	.86		1.98	2.54
1155 Rough textured		470	.017		1.34	.86		2.20	2.78
1165 3/4" thick, fine textured		450	.018		2.33	.90		3.23	3.93
1170 Rough textured		450	.018		1.53	.90		2.43	3.05
1175 Fissured		450	.018		2.21	.90		3.11	3.80
1185 For plastic film face, add					.81			.81	.89
1190 For fire rating, add					.50			.50	.55
3720 Mineral fiber, 24" x 24" or 48", reveal edge, painted, 5/8" thick	1 Carp	600	.013		1.36	.68		2.04	2.53
3740 3/4" thick		575	.014		1.67	.71		2.38	2.91
5020 66-78% recycled content, 3/4" thick [G]		600	.013		2.08	.68		2.76	3.32
5040 Mylar, 42% recycled content, 3/4" thick [G]		600	.013		5	.68		5.68	6.55
6000 Remove & install new ceiling tiles, min fiber, 2' x 2' or 2' x 4', 5/8"thk.		335	.024		.79	1.21		2	2.71

09 51 23.30 Suspended Ceilings, Complete

	Crew	Daily Output	Labor-Hours	Unit	Material	2018 Bare Costs Labor	Equipment	Total	Total Incl O&P
0010 **SUSPENDED CEILINGS, COMPLETE**, incl. standard									
0100 suspension system but not incl. 1-1/2" carrier channels									
0600 Fiberglass ceiling board, 2' x 4' x 5/8", plain faced	1 Carp	500	.016	S.F.	2.05	.81		2.86	3.50
0700 Offices, 2' x 4' x 3/4"		380	.021		3.79	1.07		4.86	5.80
0800 Mineral fiber, on 15/16" T bar susp. 2' x 2' x 3/4" lay-in board		345	.023		3.13	1.18		4.31	5.25
0810 2' x 4' x 5/8" tile		380	.021		2.21	1.07		3.28	4.06
0820 Tegular, 2' x 2' x 5/8" tile on 9/16" grid		250	.032		2.58	1.62		4.20	5.30
0830 2' x 4' x 3/4" tile		275	.029		2.82	1.47		4.29	5.35
0900 Luminous panels, prismatic, acrylic		255	.031		3.72	1.59		5.31	6.50
1200 Metal pan with acoustic pad, steel		75	.107		4.93	5.40		10.33	13.65
1300 Painted aluminum		75	.107		3.49	5.40		8.89	12.10
1500 Aluminum, degreased finish		75	.107		4.94	5.40		10.34	13.70
1600 Stainless steel		75	.107		9.30	5.40		14.70	18.50
1800 Tile, Z bar suspension, 5/8" mineral fiber tile		150	.053		2.14	2.70		4.84	6.45
1900 3/4" mineral fiber tile		150	.053		2.42	2.70		5.12	6.80
2400 For strip lighting, see Section 26 51 13.50									
2500 For rooms under 500 S.F., add				S.F.		25%			

338

For customer support on your Building Construction Costs with RSMeans data, call 800.448.8182.

09 51 Acoustical Ceilings

09 51 33 – Acoustical Metal Pan Ceilings

09 51 33.10 Ceiling, Acoustical Metal Pan	Crew	Daily Output	Labor-Hours	Unit	Material	2018 Bare Costs Labor	Equipment	Total	Total Incl O&P
0010 **CEILING, ACOUSTICAL METAL PAN**									
0100 Metal panel, lay-in, 2' x 2', sq. edge	1 Carp	500	.016	S.F.	10.90	.81		11.71	13.25
0110 Tegular edge		500	.016		14	.81		14.81	16.65
0120 2' x 4', sq. edge		500	.016		13.75	.81		14.56	16.40
0130 Tegular edge		500	.016		14	.81		14.81	16.65
0140 Perforated alum. clip-in, 2' x 2'		500	.016		14	.81		14.81	16.65
0150 2' x 4'	▼	500	.016	▼	11.40	.81		12.21	13.80

09 51 53 – Direct-Applied Acoustical Ceilings

09 51 53.10 Ceiling Tile

	Crew	Daily Output	Labor-Hours	Unit	Material	2018 Bare Costs Labor	Equipment	Total	Total Incl O&P
0010 **CEILING TILE**, stapled or cemented									
0100 12" x 12" or 12" x 24", not including furring									
0600 Mineral fiber, vinyl coated, 5/8" thick	1 Carp	300	.027	S.F.	2.34	1.35		3.69	4.63
0700 3/4" thick		300	.027		3.05	1.35		4.40	5.40
0900 Fire rated, 3/4" thick, plain faced		300	.027		1.42	1.35		2.77	3.62
1000 Plastic coated face		300	.027		2.10	1.35		3.45	4.37
1200 Aluminum faced, 5/8" thick, plain		300	.027		1.83	1.35		3.18	4.07
3700 Wall application of above, add	▼	1000	.008			.41		.41	.62
3900 For ceiling primer, add					.12			.12	.13
4000 For ceiling cement, add				▼	.40			.40	.44

09 53 Acoustical Ceiling Suspension Assemblies

09 53 23 – Metal Acoustical Ceiling Suspension Assemblies

09 53 23.30 Ceiling Suspension Systems

	Crew	Daily Output	Labor-Hours	Unit	Material	2018 Bare Costs Labor	Equipment	Total	Total Incl O&P
0010 **CEILING SUSPENSION SYSTEMS** for boards and tile									
0050 Class A suspension system, 15/16" T bar, 2' x 4' grid	1 Carp	800	.010	S.F.	.79	.51		1.30	1.64
0300 2' x 2' grid		650	.012		1.02	.62		1.64	2.07
0310 25% recycled steel, 2' x 4' grid [G]		800	.010		.83	.51		1.34	1.68
0320 2' x 2' grid [G]	▼	650	.012		1.04	.62		1.66	2.09
0350 For 9/16" grid, add					.16			.16	.18
0360 For fire rated grid, add					.09			.09	.10
0370 For colored grid, add					.21			.21	.23
0400 Concealed Z bar suspension system, 12" module	1 Carp	520	.015		.93	.78		1.71	2.21
0600 1-1/2" carrier channels, 4' OC, add	"	470	.017	▼	.12	.86		.98	1.44
0700 Carrier channels for ceilings with									
0900 recessed lighting fixtures, add	1 Carp	460	.017	S.F.	.22	.88		1.10	1.58
1040 Hanging wire, 12 ga., 4' long		65	.123	C.S.F.	.43	6.25		6.68	10
1080 8' long	▼	65	.123	"	.87	6.25		7.12	10.45
3000 Seismic ceiling bracing, IBC Site Class D, Occupancy Category II									
3050 For ceilings less than 2500 S.F.									
3060 Seismic clips at attached walls	1 Carp	180	.044	Ea.	1.13	2.25		3.38	4.67
3100 For ceilings greater than 2500 S.F., add									
3120 Seismic clips, joints at cross tees	1 Carp	120	.067	Ea.	1.85	3.38		5.23	7.20
3140 At cross tees and mains, mains field cut	"	60	.133	"	1.85	6.75		8.60	12.35
3200 Compression posts, telescopic, attached to structure above									
3210 To 30" high	1 Carp	26	.308	Ea.	38.50	15.60		54.10	66.50
3220 30" to 48" high		25.50	.314		42	15.90		57.90	70.50
3230 48" to 84" high		25	.320		51	16.20		67.20	80.50
3240 84" to 102" high		24.50	.327		58	16.55		74.55	89
3250 102" to 120" high		24	.333		83	16.90		99.90	117
3260 120" to 144" high	▼	24	.333	▼	92	16.90		108.90	127

For customer support on your Building Construction Costs with RSMeans data, call 800.448.8182.

09 53 Acoustical Ceiling Suspension Assemblies

09 53 23 – Metal Acoustical Ceiling Suspension Assemblies

09 53 23.30 Ceiling Suspension Systems	Crew	Daily Output	Labor-Hours	Unit	Material	2018 Bare Costs Labor	Equipment	Total	Total Incl O&P
3300	Stabilizer bars									
3310	12" long	1 Carp	240	.033	Ea.	1.05	1.69		2.74	3.72
3320	24" long		235	.034		.99	1.73		2.72	3.72
3330	36" long		230	.035		.96	1.76		2.72	3.74
3340	48" long		220	.036		.79	1.84		2.63	3.68
3400	Wire support for light fixtures, per L.F. height to structure above									
3410	Less than 10 lb.	1 Carp	400	.020	L.F.	.29	1.01		1.30	1.86
3420	10 lb. to 56 lb.	"	240	.033	"	.58	1.69		2.27	3.21

09 54 Specialty Ceilings

09 54 16 – Luminous Ceilings

09 54 16.10 Ceiling, Luminous

		Crew	Daily Output	Labor-Hours	Unit	Material	Labor	Equipment	Total	Total Incl O&P
0010	**CEILING, LUMINOUS**									
0020	Translucent lay-in panels, 2' x 2'	1 Carp	500	.016	S.F.	23	.81		23.81	26
0030	2' x 6'	"	500	.016	"	17.80	.81		18.61	21

09 54 23 – Linear Metal Ceilings

09 54 23.10 Metal Ceilings

		Crew	Daily Output	Labor-Hours	Unit	Material	Labor	Equipment	Total	Total Incl O&P
0010	**METAL CEILINGS**									
0015	Solid alum. planks, 3-1/4" x 12', open reveal	1 Carp	500	.016	S.F.	2.35	.81		3.16	3.83
0020	Closed reveal		500	.016		3	.81		3.81	4.54
0030	7-1/4" x 12', open reveal		500	.016		4	.81		4.81	5.65
0040	Closed reveal		500	.016		5.10	.81		5.91	6.85
0050	Metal, open cell, 2' x 2', 6" cell		500	.016		8.70	.81		9.51	10.80
0060	8" cell		500	.016		9.60	.81		10.41	11.80
0070	2' x 4', 6" cell		500	.016		5.70	.81		6.51	7.50
0080	8" cell		500	.016		5.70	.81		6.51	7.50

09 54 26 – Suspended Wood Ceilings

09 54 26.10 Wood Ceilings

		Crew	Daily Output	Labor-Hours	Unit	Material	Labor	Equipment	Total	Total Incl O&P
0010	**WOOD CEILINGS**									
1000	4"-6" wood slats on heavy duty 15/16" T-bar grid	2 Carp	250	.064	S.F.	26	3.24		29.24	34

09 54 33 – Decorative Panel Ceilings

09 54 33.20 Metal Panel Ceilings

		Crew	Daily Output	Labor-Hours	Unit	Material	Labor	Equipment	Total	Total Incl O&P
0010	**METAL PANEL CEILINGS**									
0020	Lay-in or screwed to furring, not including grid									
0100	Tin ceilings, 2' x 2' or 2' x 4', bare steel finish	2 Carp	300	.053	S.F.	2.14	2.70		4.84	6.50
0120	Painted white finish		300	.053	"	3.56	2.70		6.26	8.05
0140	Copper, chrome or brass finish		300	.053	L.F.	6.30	2.70		9	11.05
0200	Cornice molding, 2-1/2" to 3-1/2" wide, 4' long, bare steel finish		200	.080	S.F.	2.21	4.06		6.27	8.65
0220	Painted white finish		200	.080		2.81	4.06		6.87	9.30
0240	Copper, chrome or brass finish		200	.080		4.01	4.06		8.07	10.60
0320	5" to 6-1/2" wide, 4' long, bare steel finish		150	.107		3.40	5.40		8.80	12
0340	Painted white finish		150	.107		3.87	5.40		9.27	12.50
0360	Copper, chrome or brass finish		150	.107		5.60	5.40		11	14.40
0420	Flat molding, 3-1/2" to 5" wide, 4' long, bare steel finish		250	.064		3.79	3.24		7.03	9.10
0440	Painted white finish		250	.064		3.96	3.24		7.20	9.30
0460	Copper, chrome or brass finish		250	.064		7.20	3.24		10.44	12.85

For customer support on your Building Construction Costs with RSMeans data, call 800.448.8182.

09 57 Special Function Ceilings

09 57 53 – Security Ceiling Assemblies

09 57 53.10 Ceiling Assem., Security, Radio Freq. Shielding

09 57 53.10 Ceiling Assem., Security, Radio Freq. Shielding	Crew	Daily Output	Labor-Hours	Unit	Material	2018 Bare Costs Labor	Equipment	Total	Total Incl O&P
0010 **CEILING ASSEMBLY, SECURITY, RADIO FREQUENCY SHIELDING**									
0020 Prefabricated, galvanized steel	2 Carp	375	.043	SF Surf	5.30	2.16		7.46	9.10
0050 5 oz., copper ceiling panel		155	.103		4.48	5.25		9.73	12.90
0110 12 oz., copper ceiling panel		140	.114		9.05	5.80		14.85	18.75
0250 Ceiling hangers	E-1	45	.533	Ea.	33.50	29	2.19	64.69	85
0300 Shielding transition, 11 ga. preformed angles	"	1365	.018	L.F.	33.50	.95	.07	34.52	38

09 61 Flooring Treatment

09 61 19 – Concrete Floor Staining

09 61 19.40 Floors, Interior

09 61 19.40 Floors, Interior	Crew	Daily Output	Labor-Hours	Unit	Material	2018 Bare Costs Labor	Equipment	Total	Total Incl O&P
0010 **FLOORS, INTERIOR**									
0300 Acid stain and sealer									
0310 Stain, one coat	1 Pord	650	.012	S.F.	.14	.52		.66	.94
0320 Two coats		570	.014		.28	.60		.88	1.21
0330 Acrylic sealer, one coat		2600	.003		.22	.13		.35	.45
0340 Two coats		1400	.006		.45	.24		.69	.86

09 62 Specialty Flooring

09 62 19 – Laminate Flooring

09 62 19.10 Floating Floor

09 62 19.10 Floating Floor	Crew	Daily Output	Labor-Hours	Unit	Material	2018 Bare Costs Labor	Equipment	Total	Total Incl O&P
0010 **FLOATING FLOOR**									
8300 Floating floor, laminate, wood pattern strip, complete	1 Clab	133	.060	S.F.	4.32	2.40		6.72	8.40
8310 Components, T&G wood composite strips					3.83			3.83	4.22
8320 Film					.17			.17	.18
8330 Foam					.26			.26	.29
8340 Adhesive					.43			.43	.47
8350 Installation kit					.19			.19	.21
8360 Trim, 2" wide x 3' long				L.F.	4.30			4.30	4.73
8370 Reducer moulding				"	5.70			5.70	6.25

09 62 23 – Bamboo Flooring

09 62 23.10 Flooring, Bamboo

09 62 23.10 Flooring, Bamboo		Crew	Daily Output	Labor-Hours	Unit	Material	2018 Bare Costs Labor	Equipment	Total	Total Incl O&P
0010 **FLOORING, BAMBOO**										
8600 Flooring, wood, bamboo strips, unfinished, 5/8" x 4" x 3'	G	1 Carp	255	.031	S.F.	5.70	1.59		7.29	8.65
8610 5/8" x 4" x 4'	G		275	.029		5.90	1.47		7.37	8.75
8620 5/8" x 4" x 6'	G		295	.027		6.45	1.38		7.83	9.20
8630 Finished, 5/8" x 4" x 3'	G		255	.031		6.25	1.59		7.84	9.30
8640 5/8" x 4" x 4'	G		275	.029		6.55	1.47		8.02	9.45
8650 5/8" x 4" x 6'	G		295	.027		4.99	1.38		6.37	7.60
8660 Stair treads, unfinished, 1-1/16" x 11-1/2" x 4'	G		18	.444	Ea.	54.50	22.50		77	94.50
8670 Finished, 1-1/16" x 11-1/2" x 4'	G		18	.444		83	22.50		105.50	126
8680 Stair risers, unfinished, 5/8" x 7-1/2" x 4'	G		18	.444		20	22.50		42.50	56.50
8690 Finished, 5/8" x 7-1/2" x 4'	G		18	.444		38	22.50		60.50	76.50
8700 Stair nosing, unfinished, 6' long	G		16	.500		44.50	25.50		70	87.50
8710 Finished, 6' long	G		16	.500		42	25.50		67.50	85

For customer support on your Building Construction Costs with RSMeans data, call 800.448.8182.

341

09 62 Specialty Flooring

09 62 29 – Cork Flooring

09 62 29.10 Cork Tile Flooring

09 62 29.10 Cork Tile Flooring		Crew	Daily Output	Labor-Hours	Unit	Material	2018 Bare Costs Labor	Equipment	Total	Total Incl O&P
0010	**CORK TILE FLOORING**									
2200	Cork tile, standard finish, 1/8" thick	G	1 Tilf	315	.025	S.F.	5.25	1.19		6.44	7.50
2250	3/16" thick	G		315	.025		5.80	1.19		6.99	8.10
2300	5/16" thick	G		315	.025		6.45	1.19		7.64	8.85
2350	1/2" thick	G		315	.025		7	1.19		8.19	9.45
2500	Urethane finish, 1/8" thick	G		315	.025		5.25	1.19		6.44	7.55
2550	3/16" thick	G		315	.025		7.30	1.19		8.49	9.80
2600	5/16" thick	G		315	.025		7.25	1.19		8.44	9.70
2650	1/2" thick	G		315	.025		7.20	1.19		8.39	9.65

09 63 Masonry Flooring

09 63 13 – Brick Flooring

09 63 13.10 Miscellaneous Brick Flooring

09 63 13.10 Miscellaneous Brick Flooring		Crew	Daily Output	Labor-Hours	Unit	Material	2018 Bare Costs Labor	Equipment	Total	Total Incl O&P
0010	**MISCELLANEOUS BRICK FLOORING**									
0020	Acid-proof shales, red, 8" x 3-3/4" x 1-1/4" thick	D-7	.43	37.209	M	715	1,575		2,290	3,100
0050	2-1/4" thick	D-1	.40	40		1,050	1,800		2,850	3,900
0200	Acid-proof clay brick, 8" x 3-3/4" x 2-1/4" thick	G		.40	40		1,000	1,800		2,800	3,850
0250	9" x 4-1/2" x 3"	G		95	.168	S.F.	4.41	7.55		11.96	16.40
0260	Cast ceramic, pressed, 4" x 8" x 1/2", unglazed	D-7	100	.160		6.85	6.70		13.55	17.50
0270	Glazed		100	.160		9.15	6.70		15.85	20
0280	Hand molded flooring, 4" x 8" x 3/4", unglazed		95	.168		9.05	7.10		16.15	20.50
0290	Glazed		95	.168		11.40	7.10		18.50	23
0300	8" hexagonal, 3/4" thick, unglazed		85	.188		9.95	7.90		17.85	22.50
0310	Glazed		85	.188		17.95	7.90		25.85	31.50
0400	Heavy duty industrial, cement mortar bed, 2" thick, not incl. brick	D-1	80	.200		1.13	8.95		10.08	14.95
0450	Acid-proof joints, 1/4" wide	"	65	.246		1.58	11.05		12.63	18.65
0500	Pavers, 8" x 4", 1" to 1-1/4" thick, red	D-7	95	.168		3.99	7.10		11.09	14.90
0510	Ironspot	"	95	.168		5.65	7.10		12.75	16.70
0540	1-3/8" to 1-3/4" thick, red	D-1	95	.168		3.85	7.55		11.40	15.80
0560	Ironspot		95	.168		5.60	7.55		13.15	17.70
0580	2-1/4" thick, red		90	.178		3.92	7.95		11.87	16.50
0590	Ironspot		90	.178		6.10	7.95		14.05	18.90
0700	Paver, adobe brick, 6" x 12", 1/2" joint	G		42	.381		1.44	17.05		18.49	27.50
0710	Mexican red, 12" x 12"	G	1 Tilf	48	.167		1.85	7.80		9.65	13.60
0720	Saltillo, 12" x 12"	G	"	48	.167		1.49	7.80		9.29	13.20
0800	For sidewalks and patios with pavers, see Section 32 14 16.10									
0870	For epoxy joints, add	D-1	600	.027	S.F.	3.05	1.19		4.24	5.20
0880	For Furan underlayment, add	"	600	.027		2.53	1.19		3.72	4.61
0890	For waxed surface, steam cleaned, add	A-1H	1000	.008		.21	.32	.07	.60	.80

09 63 40 – Stone Flooring

09 63 40.10 Marble

09 63 40.10 Marble		Crew	Daily Output	Labor-Hours	Unit	Material	2018 Bare Costs Labor	Equipment	Total	Total Incl O&P
0010	**MARBLE**									
0020	Thin gauge tile, 12" x 6", 3/8", white Carara	D-7	60	.267	S.F.	17	11.20		28.20	35.50
0100	Travertine		60	.267		8.80	11.20		20	26.50
0200	12" x 12" x 3/8", thin set, floors		60	.267		11.10	11.20		22.30	29
0300	On walls		52	.308		9.85	12.95		22.80	30
1000	Marble threshold, 4" wide x 36" long x 5/8" thick, white		60	.267	Ea.	11.10	11.20		22.30	29

For customer support on your Building Construction Costs with RSMeans data, call 800.448.8182.

09 63 Masonry Flooring

09 63 40 – Stone Flooring

09 63 40.20 Slate Tile

09 63 40.20 Slate Tile	Crew	Daily Output	Labor-Hours	Unit	Material	2018 Bare Costs Labor	Equipment	Total	Total Incl O&P
0010 **SLATE TILE**									
0020 Vermont, 6" x 6" x 1/4" thick, thin set	D-7	180	.089	S.F.	7.70	3.74		11.44	14
0200 See also Section 32 14 40.10									

09 64 Wood Flooring

09 64 16 – Wood Block Flooring

09 64 16.10 End Grain Block Flooring

	Crew	Daily Output	Labor-Hours	Unit	Material	2018 Bare Costs Labor	Equipment	Total	Total Incl O&P
0010 **END GRAIN BLOCK FLOORING**									
0020 End grain flooring, coated, 2" thick	1 Carp	295	.027	S.F.	3.65	1.38		5.03	6.10
0400 Natural finish, 1" thick, fir		125	.064		3.77	3.24		7.01	9.10
0600 1-1/2" thick, pine		125	.064		3.70	3.24		6.94	9
0700 2" thick, pine	↓	125	.064	↓	4.91	3.24		8.15	10.35

09 64 19 – Wood Composition Flooring

09 64 19.10 Wood Composition

	Crew	Daily Output	Labor-Hours	Unit	Material	2018 Bare Costs Labor	Equipment	Total	Total Incl O&P
0010 **WOOD COMPOSITION** Gym floors									
0100 2-1/4" x 6-7/8" x 3/8", on adh, corkbd & bond coat	D-7	150	.107	S.F.	8.05	4.48		12.53	15.50
0200 Thin set, on concrete	"	250	.064		5.75	2.69		8.44	10.35
0300 Sanding and finishing, add	1 Carp	200	.040	↓	.87	2.03		2.90	4.05

09 64 23 – Wood Parquet Flooring

09 64 23.10 Wood Parquet

	Crew	Daily Output	Labor-Hours	Unit	Material	2018 Bare Costs Labor	Equipment	Total	Total Incl O&P
0010 **WOOD PARQUET** flooring									
5200 Parquetry, 5/16" thk, no finish, oak, plain pattern	1 Carp	160	.050	S.F.	5.45	2.54		7.99	9.85
5300 Intricate pattern		100	.080		9.90	4.06		13.96	17.05
5500 Teak, plain pattern		160	.050		6.35	2.54		8.89	10.80
5600 Intricate pattern		100	.080		10.75	4.06		14.81	18.05
5650 13/16" thick, select grade oak, plain pattern		160	.050		11.85	2.54		14.39	16.85
5700 Intricate pattern		100	.080		17.35	4.06		21.41	25.50
5800 Custom parquetry, including finish, plain pattern		100	.080		17.10	4.06		21.16	25
5900 Intricate pattern		50	.160		25	8.10		33.10	40
6700 Parquetry, prefinished white oak, 5/16" thick, plain pattern		160	.050		8	2.54		10.54	12.65
6800 Intricate pattern		100	.080		8.65	4.06		12.71	15.70
7000 Walnut or teak, parquetry, plain pattern		160	.050		8.50	2.54		11.04	13.20
7100 Intricate pattern	↓	100	.080	↓	13.65	4.06		17.71	21
7200 Acrylic wood parquet blocks, 12" x 12" x 5/16",									
7210 Irradiated, set in epoxy	1 Carp	160	.050	S.F.	10.45	2.54		12.99	15.35

09 64 29 – Wood Strip and Plank Flooring

09 64 29.10 Wood

	Crew	Daily Output	Labor-Hours	Unit	Material	2018 Bare Costs Labor	Equipment	Total	Total Incl O&P
0010 **WOOD**									
0020 Fir, vertical grain, 1" x 4", not incl. finish, grade B & better	1 Carp	255	.031	S.F.	3.45	1.59		5.04	6.20
0100 Grade C & better		255	.031		3.25	1.59		4.84	6
4000 Maple, strip, 25/32" x 2-1/4", not incl. finish, select		170	.047		4.93	2.39		7.32	9.05
4100 #2 & better		170	.047		4.84	2.39		7.23	8.95
4300 33/32" x 3-1/4", not incl. finish, #1 grade		170	.047		5.65	2.39		8.04	9.85
4400 #2 & better	↓	170	.047	↓	5	2.39		7.39	9.15
4600 Oak, white or red, 25/32" x 2-1/4", not incl. finish									
4700 #1 common	1 Carp	170	.047	S.F.	3.42	2.39		5.81	7.40
4900 Select quartered, 2-1/4" wide		170	.047		4.29	2.39		6.68	8.35
5000 Clear		170	.047		4.23	2.39		6.62	8.30
6100 Prefinished, white oak, prime grade, 2-1/4" wide	↓	170	.047		5.25	2.39		7.64	9.40

09 64 Wood Flooring

09 64 29 – Wood Strip and Plank Flooring

09 64 29.10 Wood		Crew	Daily Output	Labor-Hours	Unit	Material	2018 Bare Costs Labor	Equipment	Total	Total Incl O&P
6200	3-1/4" wide	1 Carp	185	.043	S.F.	5.70	2.19		7.89	9.65
6400	Ranch plank		145	.055		7.30	2.80		10.10	12.25
6500	Hardwood blocks, 9" x 9", 25/32" thick		160	.050		7.45	2.54		9.99	12
7400	Yellow pine, 3/4" x 3-1/8", T&G, C & better, not incl. finish		200	.040		1.67	2.03		3.70	4.93
7500	Refinish wood floor, sand, 2 coats poly, wax, soft wood	1 Clab	400	.020		.21	.80		1.01	1.44
7600	Hardwood		130	.062		.21	2.45		2.66	3.97
7800	Sanding and finishing, 2 coats polyurethane		295	.027		.21	1.08		1.29	1.88
7900	Subfloor and underlayment, see Section 06 16									
8015	Transition molding, 2-1/4" wide, 5' long	1 Carp	19.20	.417	Ea.	18.65	21		39.65	52.50

09 64 66 – Wood Athletic Flooring

09 64 66.10 Gymnasium Flooring

		Crew	Daily Output	Labor-Hours	Unit	Material	2018 Bare Costs Labor	Equipment	Total	Total Incl O&P
0010	**GYMNASIUM FLOORING**									
0600	Gym floor, in mastic, over 2 ply felt, #2 & better									
0700	25/32" thick maple	1 Carp	100	.080	S.F.	4.13	4.06		8.19	10.75
0900	33/32" thick maple		98	.082		5.55	4.14		9.69	12.45
1000	For 1/2" corkboard underlayment, add		750	.011		1.20	.54		1.74	2.14
1300	For #1 grade maple, add					.63			.63	.69
1600	Maple flooring, over sleepers, #2 & better									
1700	25/32" thick	1 Carp	85	.094	S.F.	5.80	4.77		10.57	13.65
1900	33/32" thick	"	83	.096		6.75	4.89		11.64	14.90
2000	For #1 grade, add					.68			.68	.75
2200	For 3/4" subfloor, add	1 Carp	350	.023		1.50	1.16		2.66	3.41
2300	With two 1/2" subfloors, 25/32" thick	"	69	.116		7.25	5.90		13.15	16.95
2500	Maple, incl. finish, #2 & btr., 25/32" thick, on rubber									
2600	Sleepers, with two 1/2" subfloors	1 Carp	76	.105	S.F.	7.80	5.35		13.15	16.70
2800	With steel spline, double connection to channels	"	73	.110		8.30	5.55		13.85	17.60
2900	For 33/32" maple, add					.88			.88	.97
3100	For #1 grade maple, add					.68			.68	.75
3500	For termite proofing all of the above, add					.36			.36	.40
3700	Portable hardwood, prefinished panels	1 Carp	83	.096		10.40	4.89		15.29	18.90
3720	Insulated with polystyrene, 1" thick, add		165	.048		.88	2.46		3.34	4.71
3750	Running tracks, Sitka spruce surface, 25/32" x 2-1/4"		62	.129		19.05	6.55		25.60	31
3770	3/4" plywood surface, finished		100	.080		4.51	4.06		8.57	11.15

09 65 Resilient Flooring

09 65 10 – Resilient Tile Underlayment

09 65 10.10 Latex Underlayment

		Crew	Daily Output	Labor-Hours	Unit	Material	2018 Bare Costs Labor	Equipment	Total	Total Incl O&P
0010	**LATEX UNDERLAYMENT**									
3600	Latex underlayment, 1/8" thk., cementitious for resilient flooring	1 Tilf	160	.050	S.F.	1.25	2.34		3.59	4.85
4000	Liquid, fortified				Gal.	32			32	35

09 65 13 – Resilient Base and Accessories

09 65 13.13 Resilient Base

		Crew	Daily Output	Labor-Hours	Unit	Material	2018 Bare Costs Labor	Equipment	Total	Total Incl O&P
0010	**RESILIENT BASE**									
0690	1/8" vinyl base, 2-1/2" H, straight or cove, standard colors	1 Tilf	315	.025	L.F.	.70	1.19		1.89	2.53
0700	4" high		315	.025		1.19	1.19		2.38	3.07
0710	6" high		315	.025		1.47	1.19		2.66	3.38
0720	Corners, 2-1/2" high		315	.025	Ea.	2.21	1.19		3.40	4.19
0730	4" high		315	.025		2.52	1.19		3.71	4.53
0740	6" high		315	.025		2.84	1.19		4.03	4.88
0800	1/8" rubber base, 2-1/2" H, straight or cove, standard colors		315	.025	L.F.	1.10	1.19		2.29	2.97

09 65 Resilient Flooring

09 65 13 – Resilient Base and Accessories

09 65 13.13 Resilient Base

		Crew	Daily Output	Labor-Hours	Unit	Material	2018 Bare Costs Labor	Equipment	Total	Total Incl O&P
1100	4" high	1 Tilf	315	.025	L.F.	1.28	1.19		2.47	3.17
1110	6" high		315	.025		1.90	1.19		3.09	3.85
1150	Corners, 2-1/2" high		315	.025	Ea.	2.46	1.19		3.65	4.47
1153	4" high		315	.025		2.54	1.19		3.73	4.55
1155	6" high	↓	315	.025	↓	3.11	1.19		4.30	5.20
1450	For premium color/finish add					50%				
1500	Millwork profile	1 Tilf	315	.025	L.F.	6.25	1.19		7.44	8.65

09 65 13.23 Resilient Stair Treads and Risers

		Crew	Daily Output	Labor-Hours	Unit	Material	2018 Bare Costs Labor	Equipment	Total	Total Incl O&P
0010	**RESILIENT STAIR TREADS AND RISERS**									
0300	Rubber, molded tread, 12" wide, 5/16" thick, black	1 Tilf	115	.070	L.F.	14.30	3.26		17.56	20.50
0400	Colors		115	.070		15.70	3.26		18.96	22
0600	1/4" thick, black		115	.070		13.70	3.26		16.96	19.85
0700	Colors		115	.070		15.35	3.26		18.61	21.50
0900	Grip strip safety tread, colors, 5/16" thick		115	.070		20	3.26		23.26	27
1000	3/16" thick		120	.067	↓	14.75	3.12		17.87	21
1200	Landings, smooth sheet rubber, 1/8" thick		120	.067	S.F.	8.20	3.12		11.32	13.60
1300	3/16" thick		120	.067	"	9.05	3.12		12.17	14.55
1500	Nosings, 3" wide, 3/16" thick, black		140	.057	L.F.	4.61	2.68		7.29	9
1600	Colors		140	.057		6.05	2.68		8.73	10.60
1800	Risers, 7" high, 1/8" thick, flat		250	.032		8.55	1.50		10.05	11.60
1900	Coved		250	.032		9.35	1.50		10.85	12.50
2100	Vinyl, molded tread, 12" wide, colors, 1/8" thick		115	.070		5.95	3.26		9.21	11.35
2200	1/4" thick		115	.070	↓	8.05	3.26		11.31	13.65
2300	Landing material, 1/8" thick		200	.040	S.F.	6.15	1.87		8.02	9.55
2400	Riser, 7" high, 1/8" thick, coved		175	.046	L.F.	2.94	2.14		5.08	6.40
2500	Tread and riser combined, 1/8" thick	↓	80	.100	"	10.05	4.69		14.74	18

09 65 13.37 Vinyl Transition Strips

		Crew	Daily Output	Labor-Hours	Unit	Material	2018 Bare Costs Labor	Equipment	Total	Total Incl O&P
0010	**VINYL TRANSITION STRIPS**									
0100	Various mats. to various mats., adhesive applied, 1/4" to 1/8"	1 Tilf	315	.025	L.F.	1.47	1.19		2.66	3.38
0105	0.08" to 1/8"		315	.025		1.33	1.19		2.52	3.22
0110	0.08" to 1/4"		315	.025		1.45	1.19		2.64	3.36
0115	1/4" to 3/8"		315	.025		1.35	1.19		2.54	3.25
0120	1/4" to 1/2"		315	.025		1.35	1.19		2.54	3.25
0125	1/4" to 0.08"		315	.025		1.45	1.19		2.64	3.36
0200	Vinyl wheeled trans. strips, carpet to var. mats., 1/4" to 1/8" x 2-1/2"		315	.025		4.71	1.19		5.90	6.95
0205	1/4" to 1/8" x 4"		315	.025		6.10	1.19		7.29	8.45
0210	Various mats. to various mats. 1/4" to 0.08" x 2-1/2"		315	.025		5.05	1.19		6.24	7.30
0215	Carpet to various materials, 1/4" to flush x 2-1/2"		315	.025		3.93	1.19		5.12	6.10
0220	1/4" to flush x 4"		315	.025		6.05	1.19		7.24	8.40
0225	Various materials to resilient, 3/8" to 1/8" x 2-1/2"		315	.025		3.93	1.19		5.12	6.10
0230	Carpet to various materials, 3/8" to 1/4" x 2-1/2"		315	.025		5.40	1.19		6.59	7.70
0235	1/4" to 1/4" x 2-1/2"		315	.025		6.05	1.19		7.24	8.40
0240	Various materials to resilient, 1/8" to 1/8" x 2-1/2"		315	.025		4.57	1.19		5.76	6.80
0245	Various materials to var. mats., 1/8" to flush x 2-1/2"		315	.025		3.06	1.19		4.25	5.15
0250	3/8" to flush x 4"		315	.025		6.15	1.19		7.34	8.55
0255	1/2" to flush x 4"		315	.025		8.25	1.19		9.44	10.85
0260	Various materials to resilient, 1/8" to 0.08" x 2-1/2"		315	.025		3.55	1.19		4.74	5.65
0265	0.08" to 0.08" x 2-1/2"		315	.025		3.37	1.19		4.56	5.45
0270	3/8" to 0.08" x 2-1/2"	↓	315	.025	↓	3.37	1.19		4.56	5.45

For customer support on your Building Construction Costs with RSMeans data, call 800.448.8182.

09 65 16 – Resilient Sheet Flooring

09 65 16.10 Rubber and Vinyl Sheet Flooring		Crew	Daily Output	Labor-Hours	Unit	Material	2018 Bare Costs Labor	Equipment	Total	Total Incl O&P
0010	**RUBBER AND VINYL SHEET FLOORING**									
5500	Linoleum, sheet goods [G]	1 Tilf	360	.022	S.F.	3.59	1.04		4.63	5.50
5900	Rubber, sheet goods, 36" wide, 1/8" thick		120	.067		8.25	3.12		11.37	13.70
5950	3/16" thick		100	.080		10.20	3.75		13.95	16.80
6000	1/4" thick		90	.089		12.05	4.16		16.21	19.40
8000	Vinyl sheet goods, backed, .065" thick, plain pattern/colors		250	.032		4.30	1.50		5.80	6.95
8050	Intricate pattern/colors		200	.040		3.94	1.87		5.81	7.10
8100	.080" thick, plain pattern/colors		230	.035		4.22	1.63		5.85	7.05
8150	Intricate pattern/colors		200	.040		6.45	1.87		8.32	9.85
8200	.125" thick, plain pattern/colors		230	.035		3.87	1.63		5.50	6.65
8250	Intricate pattern/colors		200	.040		7.50	1.87		9.37	11
8400	For welding seams, add		100	.080	L.F.	.24	3.75		3.99	5.80
8450	For integral cove base, add		175	.046	"	.81	2.14		2.95	4.06
8700	Adhesive cement, 1 gallon per 200 to 300 S.F.				Gal.	31.50			31.50	34.50
8800	Asphalt primer, 1 gallon per 300 S.F.					14.95			14.95	16.45
8900	Emulsion, 1 gallon per 140 S.F.					18.95			18.95	21

09 65 19 – Resilient Tile Flooring

09 65 19.19 Vinyl Composition Tile Flooring

		Crew	Daily Output	Labor-Hours	Unit	Material	Labor	Equipment	Total	Total Incl O&P
0010	**VINYL COMPOSITION TILE FLOORING**									
7000	Vinyl composition tile, 12" x 12", 1/16" thick	1 Tilf	500	.016	S.F.	1.22	.75		1.97	2.45
7050	Embossed		500	.016		2.66	.75		3.41	4.04
7100	Marbleized		500	.016		2.66	.75		3.41	4.04
7150	Solid		500	.016		3.44	.75		4.19	4.89
7200	3/32" thick, embossed		500	.016		1.55	.75		2.30	2.82
7250	Marbleized		500	.016		3.06	.75		3.81	4.48
7300	Solid		500	.016		2.85	.75		3.60	4.25
7350	1/8" thick, marbleized		500	.016		2.44	.75		3.19	3.79
7400	Solid		500	.016		1.77	.75		2.52	3.06
7450	Conductive		500	.016		5.90	.75		6.65	7.60

09 65 19.23 Vinyl Tile Flooring

		Crew	Daily Output	Labor-Hours	Unit	Material	Labor	Equipment	Total	Total Incl O&P
0010	**VINYL TILE FLOORING**									
7500	Vinyl tile, 12" x 12", 3/32" thick, standard colors/patterns	1 Tilf	500	.016	S.F.	3.71	.75		4.46	5.20
7550	1/8" thick, standard colors/patterns		500	.016		5.20	.75		5.95	6.80
7600	1/8" thick, premium colors/patterns		500	.016		7	.75		7.75	8.80
7650	Solid colors		500	.016		3.23	.75		3.98	4.66
7700	Marbleized or Travertine pattern		500	.016		6.20	.75		6.95	7.90
7750	Florentine pattern		500	.016		6.60	.75		7.35	8.35
7800	Premium colors/patterns		500	.016		6.25	.75		7	7.95

09 65 19.33 Rubber Tile Flooring

		Crew	Daily Output	Labor-Hours	Unit	Material	Labor	Equipment	Total	Total Incl O&P
0010	**RUBBER TILE FLOORING**									
6050	Rubber tile, marbleized colors, 12" x 12", 1/8" thick	1 Tilf	400	.020	S.F.	5.80	.94		6.74	7.75
6100	3/16" thick		400	.020		8	.94		8.94	10.20
6300	Special tile, plain colors, 1/8" thick		400	.020		8.15	.94		9.09	10.35
6350	3/16" thick		400	.020		9.75	.94		10.69	12.10
6410	Raised, radial or square, .5 mm black		400	.020		7.75	.94		8.69	9.95
6430	.5 mm colored		400	.020		7.75	.94		8.69	9.95
6450	For golf course, skating rink, etc., 1/4" thick		275	.029		10.45	1.36		11.81	13.45

09 65 Resilient Flooring

09 65 33 – Conductive Resilient Flooring

09 65 33.10 Conductive Rubber and Vinyl Flooring	Crew	Daily Output	Labor-Hours	Unit	Material	2018 Bare Costs Labor	Equipment	Total	Total Incl O&P
0010 **CONDUCTIVE RUBBER AND VINYL FLOORING**									
1700 Conductive flooring, rubber tile, 1/8" thick	1 Tilf	315	.025	S.F.	7.05	1.19		8.24	9.50
1800 Homogeneous vinyl tile, 1/8" thick	"	315	.025	"	6.80	1.19		7.99	9.25

09 65 66 – Resilient Athletic Flooring

09 65 66.10 Resilient Athletic Flooring

	Crew	Daily Output	Labor-Hours	Unit	Material	2018 Bare Costs Labor	Equipment	Total	Total Incl O&P
0010 **RESILIENT ATHLETIC FLOORING**									
1000 Recycled rubber rolled goods, for weight rooms, 3/8" thk.	1 Tilf	315	.025	S.F.	2.66	1.19		3.85	4.69
1050 Interlocking 2'x2' squares, rubber, 1/4" thk.		310	.026		1.89	1.21		3.10	3.87
1055 5/16" thk.		310	.026		2.29	1.21		3.50	4.31
1060 3/8" thk.		300	.027		2.98	1.25		4.23	5.15
1065 1/2" thk.		310	.026		3.08	1.21		4.29	5.20
2000 Vinyl sheet flooring, 1/4" thk.		315	.025		4.28	1.19		5.47	6.45

09 66 Terrazzo Flooring

09 66 13 – Portland Cement Terrazzo Flooring

09 66 13.10 Portland Cement Terrazzo

	Crew	Daily Output	Labor-Hours	Unit	Material	2018 Bare Costs Labor	Equipment	Total	Total Incl O&P
0010 **PORTLAND CEMENT TERRAZZO**, cast-in-place R096613-10									
0020 Cove base, 6" high, 16 ga. zinc	1 Mstz	20	.400	L.F.	3.30	18.80		22.10	31.50
0100 Curb, 6" high and 6" wide		6	1.333		6.25	62.50		68.75	99.50
0300 Divider strip for floors, 14 ga., 1-1/4" deep, zinc		375	.021		1.47	1		2.47	3.10
0400 Brass		375	.021		2.49	1		3.49	4.22
0600 Heavy top strip 1/4" thick, 1-1/4" deep, zinc		300	.027		2.16	1.25		3.41	4.23
0900 Galv. bottoms, brass		300	.027		2.68	1.25		3.93	4.80
1200 For thin set floors, 16 ga., 1/2" x 1/2", zinc		350	.023		1.29	1.07		2.36	3.01
1300 Brass		350	.023		2.75	1.07		3.82	4.62
1500 Floor, bonded to concrete, 1-3/4" thick, gray cement	J-3	75	.213	S.F.	3.57	9.20	4.15	16.92	22
1600 White cement, mud set		75	.213		4.16	9.20	4.15	17.51	23
1800 Not bonded, 3" total thickness, gray cement		70	.229		4.40	9.85	4.45	18.70	24.50
1900 White cement, mud set		70	.229		5.40	9.85	4.45	19.70	25.50
2100 For Venetian terrazzo, 1" topping, add					50%	50%			
2200 For heavy duty abrasive terrazzo, add					50%	50%			
2700 Monolithic terrazzo, 1/2" thick									
2710 10' panels	J-3	125	.128	S.F.	3.36	5.50	2.49	11.35	14.60
3000 Stairs, cast-in-place, pan filled treads		30	.533	L.F.	3.61	23	10.40	37.01	49.50
3100 Treads and risers		14	1.143	"	6.25	49	22	77.25	104
3300 For stair landings, add to floor prices						50%			
3400 Stair stringers and fascia	J-3	30	.533	S.F.	5.30	23	10.40	38.70	51.50
3600 For abrasive metal nosings on stairs, add		150	.107	L.F.	9.50	4.59	2.08	16.17	19.55
3700 For abrasive surface finish, add		600	.027	S.F.	1.75	1.15	.52	3.42	4.20
3900 For raised abrasive strips, add		150	.107	L.F.	1.37	4.59	2.08	8.04	10.60
4000 Wainscot, bonded, 1-1/2" thick		30	.533	S.F.	4	23	10.40	37.40	50
4200 1/4" thick		40	.400	"	6.20	17.20	7.80	31.20	41
4300 Stone chips, onyx gemstone, per 50 lb. bag				Bag	18.60			18.60	20.50

09 66 16 – Terrazzo Floor Tile

09 66 16.10 Tile or Terrazzo Base

	Crew	Daily Output	Labor-Hours	Unit	Material	2018 Bare Costs Labor	Equipment	Total	Total Incl O&P
0010 **TILE OR TERRAZZO BASE**									
0020 Scratch coat only	1 Mstz	150	.053	S.F.	.48	2.51		2.99	4.24
0500 Scratch and brown coat only	"	75	.107	"	.96	5		5.96	8.45

For customer support on your Building Construction Costs with RSMeans data, call 800.448.8182.

09 66 Terrazzo Flooring

09 66 16 – Terrazzo Floor Tile

09 66 16.13 Portland Cement Terrazzo Floor Tile

		Crew	Daily Output	Labor-Hours	Unit	Material	2018 Bare Costs Labor	Equipment	Total	Total Incl O&P
0010	**PORTLAND CEMENT TERRAZZO FLOOR TILE**									
1200	Floor tiles, non-slip, 1" thick, 12" x 12"	D-1	60	.267	S.F.	24.50	11.95		36.45	45.50
1300	1-1/4" thick, 12" x 12"		60	.267		25.50	11.95		37.45	46.50
1500	16" x 16"		50	.320		27.50	14.35		41.85	52.50
1600	1-1/2" thick, 16" x 16"	↓	45	.356		25	15.95		40.95	52.50
1800	For Venetian terrazzo, add					7.55			7.55	8.35
1900	For white cement, add				↓	.71			.71	.78

09 66 16.16 Plastic Matrix Terrazzo Floor Tile

		Crew	Daily Output	Labor-Hours	Unit	Material	Labor	Equipment	Total	Total Incl O&P
0010	**PLASTIC MATRIX TERRAZZO FLOOR TILE**									
0100	12" x 12", 3/16" thick, floor tiles w/marble chips	1 Tilf	500	.016	S.F.	7.30	.75		8.05	9.15
0200	12" x 12", 3/16" thick, floor tiles w/glass chips		500	.016		7.90	.75		8.65	9.75
0300	12" x 12", 3/16" thick, floor tiles w/recycled content	↓	500	.016	↓	6.20	.75		6.95	7.95

09 66 16.30 Terrazzo, Precast

		Crew	Daily Output	Labor-Hours	Unit	Material	Labor	Equipment	Total	Total Incl O&P
0010	**TERRAZZO, PRECAST**									
0020	Base, 6" high, straight	1 Mstz	70	.114	L.F.	12.35	5.35		17.70	21.50
0100	Cove		60	.133		16.50	6.25		22.75	27.50
0300	8" high, straight		60	.133		14.75	6.25		21	25.50
0400	Cove	↓	50	.160		21.50	7.50		29	35
0600	For white cement, add					.55			.55	.61
0700	For 16 ga. zinc toe strip, add					2.18			2.18	2.40
0900	Curbs, 4" x 4" high	1 Mstz	40	.200		40	9.40		49.40	58
1000	8" x 8" high	"	30	.267		46.50	12.55		59.05	69.50
2400	Stair treads, 1-1/2" thick, non-slip, three line pattern	2 Mstz	70	.229		50	10.75		60.75	71.50
2500	Nosing and two lines		70	.229		50	10.75		60.75	71.50
2700	2" thick treads, straight		60	.267		58	12.55		70.55	82
2800	Curved		50	.320		75.50	15.05		90.55	106
3000	Stair risers, 1" thick, to 6" high, straight sections		60	.267		13.75	12.55		26.30	33.50
3100	Cove		50	.320		19.15	15.05		34.20	43.50
3300	Curved, 1" thick, to 6" high, vertical		48	.333		26	15.65		41.65	52
3400	Cove		38	.421		42	19.80		61.80	76
3600	Stair tread and riser, single piece, straight, smooth surface		60	.267		64.50	12.55		77.05	89.50
3700	Non skid surface		40	.400		83.50	18.80		102.30	120
3900	Curved tread and riser, smooth surface		40	.400		90.50	18.80		109.30	128
4000	Non skid surface		32	.500		114	23.50		137.50	161
4200	Stair stringers, notched, 1" thick		25	.640		37	30		67	85
4300	2" thick		22	.727	↓	43.50	34		77.50	98.50
4500	Stair landings, structural, non-slip, 1-1/2" thick		85	.188	S.F.	40.50	8.85		49.35	57.50
4600	3" thick	↓	75	.213		57	10.05		67.05	78
4800	Wainscot, 12" x 12" x 1" tiles	1 Mstz	12	.667		8.75	31.50		40.25	56
4900	16" x 16" x 1-1/2" tiles	"	8	1	↓	17.45	47		64.45	88.50

09 66 23 – Resinous Matrix Terrazzo Flooring

09 66 23.13 Polyacrylate Mod. Cementitious Terrazzo Flr.

		Crew	Daily Output	Labor-Hours	Unit	Material	Labor	Equipment	Total	Total Incl O&P
0010	**POLYACRYLATE MODIFIED CEMENTITIOUS TERRAZZO FLOORING**									
3150	Polyacrylate, 1/4" thick, granite chips	C-6	735	.065	S.F.	3.73	2.71	.07	6.51	8.30
3170	Recycled porcelain		480	.100		4.84	4.15	.11	9.10	11.70
3200	3/8" thick, granite chips		620	.077		4.90	3.21	.08	8.19	10.35
3220	Recycled porcelain	↓	480	.100	↓	6.85	4.15	.11	11.11	13.95

09 66 23.16 Epoxy-Resin Terrazzo Flooring

		Crew	Daily Output	Labor-Hours	Unit	Material	Labor	Equipment	Total	Total Incl O&P
0010	**EPOXY-RESIN TERRAZZO FLOORING**									
1800	Epoxy terrazzo, 1/4" thick, chemical resistant, granite chips	J-3	200	.080	S.F.	6.35	3.44	1.56	11.35	13.80
1900	Recycled porcelain	↓	150	.107		9.70	4.59	2.08	16.37	19.75

For customer support on your Building Construction Costs with RSMeans data, call 800.448.8182.

09 66 Terrazzo Flooring

09 66 23 – Resinous Matrix Terrazzo Flooring

	09 66 23.16 Epoxy-Resin Terrazzo Flooring	Crew	Daily Output	Labor-Hours	Unit	Material	2018 Bare Costs Labor	Equipment	Total	Total Incl O&P
2500	Epoxy terrazzo, 1/4" thick, granite chips	J-3	200	.080	S.F.	5.80	3.44	1.56	10.80	13.20
2550	Average		175	.091		5.60	3.94	1.78	11.32	13.95
2600	Recycled aggregate		150	.107		5.90	4.59	2.08	12.57	15.60
2650	Epoxy terrazzo, 3/8" thick, marble chips		200	.080		5.90	3.44	1.56	10.90	13.30
2675	Glass or mother of pearl		200	.080		6.90	3.44	1.56	11.90	14.40

09 66 33 – Conductive Terrazzo Flooring

09 66 33.10 Conductive Terrazzo

		Crew	Daily Output	Labor-Hours	Unit	Material	Labor	Equipment	Total	Total Incl O&P
0010	**CONDUCTIVE TERRAZZO**									
2400	Bonded conductive floor for hospitals	J-3	90	.178	S.F.	5.40	7.65	3.46	16.51	21

09 66 33.13 Conductive Epoxy-Resin Terrazzo

		Crew	Daily Output	Labor-Hours	Unit	Material	Labor	Equipment	Total	Total Incl O&P
0010	**CONDUCTIVE EPOXY-RESIN TERRAZZO**									
2100	Epoxy terrazzo, 1/4" thick, conductive, granite chips	J-3	100	.160	S.F.	8.95	6.90	3.11	18.96	23.50
2200	Recycled porcelain	"	90	.178	"	11.15	7.65	3.46	22.26	27.50

09 66 33.19 Conductive Plastic-Matrix Terrazzo Flooring

		Crew	Daily Output	Labor-Hours	Unit	Material	Labor	Equipment	Total	Total Incl O&P
0010	**CONDUCTIVE PLASTIC-MATRIX TERRAZZO FLOORING**									
3300	Conductive, 1/4" thick, granite chips	C-6	450	.107	S.F.	8.10	4.42	.11	12.63	15.70
3330	Recycled porcelain		305	.157		10.70	6.55	.17	17.42	22
3350	3/8" thick, granite chips		365	.132		11.05	5.45	.14	16.64	20.50
3370	Recycled porcelain		255	.188		13.65	7.80	.20	21.65	27
3450	Granite, conductive, 1/4" thick, 20% chip		695	.069		9.85	2.86	.07	12.78	15.25
3470	50% chip		420	.114		12.75	4.74	.12	17.61	21.50
3500	3/8" thick, 20% chip		695	.069		14.40	2.86	.07	17.33	20.50
3520	50% chip		380	.126		17.45	5.25	.14	22.84	27.50

09 67 Fluid-Applied Flooring

09 67 13 – Elastomeric Liquid Flooring

09 67 13.13 Elastomeric Liquid Flooring

		Crew	Daily Output	Labor-Hours	Unit	Material	Labor	Equipment	Total	Total Incl O&P
0010	**ELASTOMERIC LIQUID FLOORING**									
0020	Cementitious acrylic, 1/4" thick	C-6	520	.092	S.F.	1.75	3.83	.10	5.68	7.85
0100	3/8" thick	"	450	.107		2.30	4.42	.11	6.83	9.35
0200	Methyl methachrylate, 1/4" thick	C-8A	3000	.016		7.15	.68		7.83	8.90
0210	1/8" thick	"	3000	.016		6.05	.68		6.73	7.70
0300	Cupric oxychloride, on bond coat, simple configs and patterns	C-6	480	.100		3.93	4.15	.11	8.19	10.75
0400	Complex configurations and patterns		420	.114		6.55	4.74	.12	11.41	14.55
2400	Mastic, hot laid, 2 coat, 1-1/2" thick, std., simple configs and patterns		690	.070		4.55	2.89	.07	7.51	9.45
2500	Maximum		520	.092		5.85	3.83	.10	9.78	12.30
2700	Acid-proof, minimum		605	.079		5.85	3.29	.08	9.22	11.45
2800	Maximum		350	.137		8.10	5.70	.15	13.95	17.65
3000	Neoprene, troweled on, 1/4" thick, minimum		545	.088		4.48	3.65	.09	8.22	10.60
3100	Maximum		430	.112		6.10	4.63	.12	10.85	13.85
4300	Polyurethane, with suspended vinyl chips, clear		1065	.045		7.60	1.87	.05	9.52	11.25
4500	Pigmented		860	.056		11.05	2.31	.06	13.42	15.75

09 67 23 – Resinous Flooring

09 67 23.23 Resinous Flooring

		Crew	Daily Output	Labor-Hours	Unit	Material	Labor	Equipment	Total	Total Incl O&P
0010	**RESINOUS FLOORING**									
1200	Heavy duty epoxy topping, 1/4" thick,									
1300	500 to 1,000 S.F.	C-6	420	.114	S.F.	5.95	4.74	.12	10.81	13.90
1500	1,000 to 2,000 S.F.		450	.107		5.15	4.42	.11	9.68	12.45
1600	Over 10,000 S.F.		480	.100		4.75	4.15	.11	9.01	11.65

09 67 Fluid-Applied Flooring

09 67 26 – Quartz Flooring

09 67 26.26 Quartz Flooring

		Crew	Daily Output	Labor-Hours	Unit	Material	2018 Bare Costs Labor	2018 Bare Costs Equipment	Total	Total Incl O&P
0010	**QUARTZ FLOORING**									
0600	Epoxy, with colored quartz chips, broadcast, 3/8" thick	C-6	675	.071	S.F.	3.08	2.95	.08	6.11	7.95
0700	1/2" thick		490	.098		4.44	4.06	.10	8.60	11.15
0900	Troweled, minimum		560	.086		3.57	3.55	.09	7.21	9.45
1000	Maximum		480	.100		5.90	4.15	.11	10.16	12.90
3600	Polyester, with colored quartz chips, 1/16" thick, minimum		1065	.045		3.38	1.87	.05	5.30	6.60
3700	Maximum		560	.086		4.65	3.55	.09	8.29	10.60
3900	1/8" thick, minimum		810	.059		3.93	2.46	.06	6.45	8.10
4000	Maximum		675	.071		5.25	2.95	.08	8.28	10.35
4200	Polyester, heavy duty, compared to epoxy, add		2590	.019		1.59	.77	.02	2.38	2.93

09 67 66 – Fluid-Applied Athletic Flooring

09 67 66.10 Polyurethane

		Crew	Daily Output	Labor-Hours	Unit	Material	2018 Bare Costs Labor	2018 Bare Costs Equipment	Total	Total Incl O&P
0010	**POLYURETHANE**									
4400	Thermoset, prefabricated in place, indoor									
4500	3/8" thick for basketball, gyms, etc.	1 Tilf	100	.080	S.F.	5.65	3.75		9.40	11.75
4600	1/2" thick for professional sports		95	.084		7.55	3.95		11.50	14.15
4700	Outdoor, 1/4" thick, smooth, for tennis		100	.080		5.85	3.75		9.60	12
5000	Poured in place, indoor, with finish, 1/4" thick		80	.100		4.17	4.69		8.86	11.55
5050	3/8" thick		65	.123		5.05	5.75		10.80	14.10
5100	1/2" thick		50	.160		6.60	7.50		14.10	18.35

09 68 Carpeting

09 68 05 – Carpet Accessories

09 68 05.11 Flooring Transition Strip

		Crew	Daily Output	Labor-Hours	Unit	Material	2018 Bare Costs Labor	2018 Bare Costs Equipment	Total	Total Incl O&P
0010	**FLOORING TRANSITION STRIP**									
0107	Clamp down brass divider, 12' strip, vinyl to carpet	1 Tilf	31.25	.256	Ea.	14.65	12		26.65	34
0117	Vinyl to hard surface	"	31.25	.256	"	14.65	12		26.65	34

09 68 10 – Carpet Pad

09 68 10.10 Commercial Grade Carpet Pad

		Crew	Daily Output	Labor-Hours	Unit	Material	2018 Bare Costs Labor	2018 Bare Costs Equipment	Total	Total Incl O&P
0010	**COMMERCIAL GRADE CARPET PAD**									
9000	Sponge rubber pad, 20 oz./sq. yd.	1 Tilf	150	.053	S.Y.	4.69	2.50		7.19	8.85
9100	40 to 62 oz./sq. yd.		150	.053		8.75	2.50		11.25	13.35
9200	Felt pad, 20 oz./sq. yd.		150	.053		5.80	2.50		8.30	10.05
9300	32 to 56 oz./sq. yd.		150	.053		11.10	2.50		13.60	15.95
9400	Bonded urethane pad, 2.7 density		150	.053		5.95	2.50		8.45	10.25
9500	13.0 density		150	.053		8	2.50		10.50	12.50
9600	Prime urethane pad, 2.7 density		150	.053		3.51	2.50		6.01	7.55
9700	13.0 density		150	.053		6.55	2.50		9.05	10.95

09 68 13 – Tile Carpeting

09 68 13.10 Carpet Tile

		Crew	Daily Output	Labor-Hours	Unit	Material	2018 Bare Costs Labor	2018 Bare Costs Equipment	Total	Total Incl O&P
0010	**CARPET TILE**									
0100	Tufted nylon, 18" x 18", hard back, 20 oz.	1 Tilf	80	.100	S.Y.	27	4.69		31.69	36.50
0110	26 oz.		80	.100		25.50	4.69		30.19	35
0200	Cushion back, 20 oz.		80	.100		25	4.69		29.69	34.50
0210	26 oz.		80	.100		30	4.69		34.69	40.50
1100	Tufted, 24" x 24", hard back, 24 oz. nylon		80	.100		31	4.69		35.69	41
1180	35 oz.		80	.100		36	4.69		40.69	47
5060	42 oz.		80	.100		47	4.69		51.69	59
6000	Electrostatic dissapative carpet tile, 24" x 24", 24 oz.		80	.100		37.50	4.69		42.19	48

For customer support on your Building Construction Costs with RSMeans data, call 800.448.8182.

09 68 Carpeting

09 68 13 – Tile Carpeting

09 68 13.10 Carpet Tile

	Crew	Daily Output	Labor-Hours	Unit	Material	2018 Bare Costs Labor	2018 Bare Costs Equipment	Total	Total Incl O&P
6100 Electrostatic dissapative carpet tile for access floors, 24" x 24", 24 oz.	1 Tilf	80	.100	S.Y.	47	4.69		51.69	58.50

09 68 16 – Sheet Carpeting

09 68 16.10 Sheet Carpet

	Crew	Daily Output	Labor-Hours	Unit	Material	2018 Bare Costs Labor	2018 Bare Costs Equipment	Total	Total Incl O&P
0010 **SHEET CARPET**									
0700 Nylon, level loop, 26 oz., light to medium traffic	1 Tilf	75	.107	S.Y.	22	5		27	31.50
0720 28 oz., light to medium traffic		75	.107		30	5		35	40.50
0900 32 oz., medium traffic		75	.107		39.50	5		44.50	51
1100 40 oz., medium to heavy traffic		75	.107		49	5		54	61
2920 Nylon plush, 30 oz., medium traffic		75	.107		28.50	5		33.50	39
3000 36 oz., medium traffic		75	.107		36	5		41	47.50
3100 42 oz., medium to heavy traffic		70	.114		45.50	5.35		50.85	58
3200 46 oz., medium to heavy traffic		70	.114		52.50	5.35		57.85	66
3300 54 oz., heavy traffic		70	.114		59.50	5.35		64.85	73
3340 60 oz., heavy traffic		70	.114		66.50	5.35		71.85	81
3665 Olefin, 24 oz., light to medium traffic		75	.107		20.50	5		25.50	30
3670 26 oz., medium traffic		75	.107		12.60	5		17.60	21.50
3680 28 oz., medium to heavy traffic		75	.107		25	5		30	35
3700 32 oz., medium to heavy traffic		75	.107		29.50	5		34.50	39.50
3730 42 oz., heavy traffic		70	.114		28	5.35		33.35	39
4110 Wool, level loop, 40 oz., medium traffic		70	.114		114	5.35		119.35	133
4500 50 oz., medium to heavy traffic		70	.114		109	5.35		114.35	128
4700 Patterned, 32 oz., medium to heavy traffic		70	.114		98.50	5.35		103.85	116
4900 48 oz., heavy traffic		70	.114		109	5.35		114.35	128
5000 For less than full roll (approx. 1500 S.F.), add					25%				
5100 For small rooms, less than 12' wide, add						25%			
5200 For large open areas (no cuts), deduct						25%			
5600 For bound carpet baseboard, add	1 Tilf	300	.027	L.F.	1.81	1.25		3.06	3.84
5610 For stairs, not incl. price of carpet, add	"	30	.267	Riser		12.50		12.50	18.50
5620 For borders and patterns, add to labor						18%			
8950 For tackless, stretched installation, add padding from 09 68 10.10 to above									
9850 For brand-named specific fiber, add				S.Y.	25%				

09 68 20 – Athletic Carpet

09 68 20.10 Indoor Athletic Carpet

	Crew	Daily Output	Labor-Hours	Unit	Material	2018 Bare Costs Labor	2018 Bare Costs Equipment	Total	Total Incl O&P
0010 **INDOOR ATHLETIC CARPET**									
3700 Polyethylene, in rolls, no base incl., landscape surfaces	1 Tilf	275	.029	S.F.	4.08	1.36		5.44	6.50
3800 Nylon action surface, 1/8" thick		275	.029		3.94	1.36		5.30	6.35
3900 1/4" thick		275	.029		5.70	1.36		7.06	8.25
4000 3/8" thick		275	.029		7.15	1.36		8.51	9.85
4100 Golf tee surface with foam back		235	.034		7.05	1.59		8.64	10.15
4200 Practice putting, knitted nylon surface		235	.034		6	1.59		7.59	8.95
4300 Synthetic turf, 1/2" ht.		90	.089		4.44	4.16		8.60	11.05
4350 3/4" ht.		210	.038		4.46	1.79		6.25	7.55
4400 1" ht.		190	.042		4.69	1.97		6.66	8.05
5500 Polyvinyl chloride, sheet goods for gyms, 1/4" thick		80	.100		6.85	4.69		11.54	14.50
5600 3/8" thick		60	.133		10.05	6.25		16.30	20.50

09 69 Access Flooring

09 69 13 – Rigid-Grid Access Flooring

09 69 13.10 Access Floors

		Crew	Daily Output	Labor-Hours	Unit	Material	2018 Bare Costs Labor	Equipment	Total	Total Incl O&P
0010	**ACCESS FLOORS**									
0015	Access floor pkg. including panel, pedestal, & stringers 1500 lbs. load									
0100	Package pricing, conc. fill panels, no fin., 6" ht.	4 Carp	750	.043	S.F.	16.70	2.16		18.86	21.50
0105	12" ht.		750	.043		17.10	2.16		19.26	22
0110	18" ht.		750	.043		17.15	2.16		19.31	22
0115	24" ht.		750	.043		17.55	2.16		19.71	22.50
0120	Package pricing, steel panels, no fin., 6" ht.		750	.043		13.35	2.16		15.51	17.95
0125	12" ht.		750	.043		16.75	2.16		18.91	21.50
0130	18" ht.		750	.043		16.85	2.16		19.01	22
0135	24" ht.		750	.043		16.95	2.16		19.11	22
0140	Package pricing, wood core panels, no fin., 6" ht.		750	.043		13.50	2.16		15.66	18.15
0145	12" ht.		750	.043		13.75	2.16		15.91	18.40
0150	18" ht.		750	.043		13.75	2.16		15.91	18.40
0155	24" ht.		750	.043		14	2.16		16.16	18.70
0160	Pkg. pricing, conc. fill pnls, no stringers, no fin., 6" ht, 1250 lbs. load		700	.046		10.80	2.32		13.12	15.40
0165	12" ht.		700	.046		10	2.32		12.32	14.55
0170	18" ht.		700	.046		10.95	2.32		13.27	15.55
0175	24" ht.		700	.046		11.90	2.32		14.22	16.60
0250	Panels, 2' x 2' conc. fill, no fin.	2 Carp	500	.032		23	1.62		24.62	27.50
0255	With 1/8" high pressure laminate		500	.032		56	1.62		57.62	64.50
0260	Metal panels with 1/8" high pressure laminate		500	.032		66.50	1.62		68.12	75.50
0265	Wood core panels with 1/8" high pressure laminate		500	.032		43.50	1.62		45.12	50.50
0400	Aluminum panels, no fin.		500	.032		33.50	1.62		35.12	39
0600	For carpet covering, add					9.05			9.05	10
0700	For vinyl floor covering, add					9.35			9.35	10.30
0900	For high pressure laminate covering, add					7.70			7.70	8.45
0910	For snap on stringer system, add	2 Carp	1000	.016		1.59	.81		2.40	2.99
1000	Machine cutouts after initial installation	1 Carp	50	.160	Ea.	20	8.10		28.10	34.50
1050	Pedestals, 6" to 12"	2 Carp	85	.188		8.65	9.55		18.20	24
1100	Air conditioning grilles, 4" x 12"	1 Carp	17	.471		72.50	24		96.50	117
1150	4" x 18"	"	14	.571		99.50	29		128.50	154
1200	Approach ramps, steel	2 Carp	60	.267	S.F.	26.50	13.50		40	49.50
1300	Aluminum	"	40	.400	"	34	20.50		54.50	68.50
1500	Handrail, 2 rail, aluminum	1 Carp	15	.533	L.F.	117	27		144	170

09 72 Wall Coverings

09 72 13 – Cork Wall Coverings

09 72 13.10 Covering, Cork Wall

		Crew	Daily Output	Labor-Hours	Unit	Material	2018 Bare Costs Labor	Equipment	Total	Total Incl O&P
0010	**COVERING, CORK WALL**									
0600	Cork tiles, light or dark, 12" x 12" x 3/16"	1 Pape	240	.033	S.F.	4.26	1.42		5.68	6.85
0700	5/16" thick		235	.034		3.33	1.45		4.78	5.85
0900	1/4" basket weave		240	.033		3.39	1.42		4.81	5.85
1000	1/2" natural, non-directional pattern		240	.033		6.80	1.42		8.22	9.60
1100	3/4" natural, non-directional pattern		240	.033		11.60	1.42		13.02	14.90
1200	Granular surface, 12" x 36", 1/2" thick		385	.021		1.30	.89		2.19	2.76
1300	1" thick		370	.022		1.66	.92		2.58	3.22
1500	Polyurethane coated, 12" x 12" x 3/16" thick		240	.033		4.03	1.42		5.45	6.55
1600	5/16" thick		235	.034		6	1.45		7.45	8.80
1800	Cork wallpaper, paperbacked, natural		480	.017		1.60	.71		2.31	2.83
1900	Colors		480	.017		2.84	.71		3.55	4.19

For customer support on your Building Construction Costs with RSMeans data, call 800.448.8182.

09 72 Wall Coverings

09 72 16 – Vinyl-Coated Fabric Wall Coverings

09 72 16.13 Flexible Vinyl Wall Coverings

09 72 16.13 Flexible Vinyl Wall Coverings	Crew	Daily Output	Labor-Hours	Unit	Material	2018 Bare Costs Labor	Equipment	Total	Total Incl O&P
0010 **FLEXIBLE VINYL WALL COVERINGS**									
3000 Vinyl wall covering, fabric-backed, lightweight, type 1 (12-15 oz./S.Y.)	1 Pape	640	.013	S.F.	1.07	.53		1.60	1.98
3300 Medium weight, type 2 (20-24 oz./S.Y.)		480	.017		.91	.71		1.62	2.07
3400 Heavy weight, type 3 (28 oz./S.Y.)		435	.018		1.40	.78		2.18	2.72
3600 Adhesive, 5 gal. lots (18 S.Y./gal.)				Gal.	12.05			12.05	13.25

09 72 16.16 Rigid-Sheet Vinyl Wall Coverings

	Crew	Daily Output	Labor-Hours	Unit	Material	Labor	Equipment	Total	Total Incl O&P
0010 **RIGID-SHEET VINYL WALL COVERINGS**									
0100 Acrylic, modified, semi-rigid PVC, .028" thick	2 Carp	330	.048	S.F.	1.30	2.46		3.76	5.15
0110 .040" thick	"	320	.050	"	1.85	2.54		4.39	5.90

09 72 19 – Textile Wall Coverings

09 72 19.10 Textile Wall Covering

	Crew	Daily Output	Labor-Hours	Unit	Material	Labor	Equipment	Total	Total Incl O&P
0010 **TEXTILE WALL COVERING**, including sizing; add 10-30% waste @ takeoff									
0020 Silk	1 Pape	640	.013	S.F.	4.72	.53		5.25	6
0030 Cotton		640	.013		7.05	.53		7.58	8.60
0040 Linen		640	.013		1.92	.53		2.45	2.91
0050 Blend		640	.013		3.26	.53		3.79	4.39
0060 Linen wall covering, paper backed									
0070 Flame treatment				S.F.	1.07			1.07	1.18
0080 Stain resistance treatment					1.92			1.92	2.11
0090 Grass cloth, natural fabric [G]	1 Pape	400	.020		1.79	.85		2.64	3.25
0100 Grass cloths with lining paper [G]		400	.020		1.38	.85		2.23	2.80
0110 Premium texture/color [G]		350	.023		3.15	.97		4.12	4.94

09 72 20 – Natural Fiber Wall Covering

09 72 20.10 Natural Fiber Wall Covering

	Crew	Daily Output	Labor-Hours	Unit	Material	Labor	Equipment	Total	Total Incl O&P
0010 **NATURAL FIBER WALL COVERING**, including sizing; add 10-30% waste @ takeoff									
0015 Bamboo	1 Pape	640	.013	S.F.	2.22	.53		2.75	3.24
0030 Burlap		640	.013		1.88	.53		2.41	2.87
0045 Jute		640	.013		1.33	.53		1.86	2.26
0060 Sisal		640	.013		1.53	.53		2.06	2.48

09 72 23 – Wallpapering

09 72 23.10 Wallpaper

	Crew	Daily Output	Labor-Hours	Unit	Material	Labor	Equipment	Total	Total Incl O&P
0010 **WALLPAPER** including sizing; add 10-30% waste @ takeoff R097223-10									
0050 Aluminum foil	1 Pape	275	.029	S.F.	1.03	1.24		2.27	2.99
0100 Copper sheets, .025" thick, vinyl backing		240	.033		5.50	1.42		6.92	8.20
0300 Phenolic backing		240	.033		7.10	1.42		8.52	10
2400 Gypsum-based, fabric-backed, fire resistant									
2500 for masonry walls, 21 oz./S.Y.	1 Pape	800	.010	S.F.	.77	.43		1.20	1.49
2700 Small quantities		640	.013		.68	.53		1.21	1.55
3700 Wallpaper, average workmanship, solid pattern, low cost paper		640	.013		.61	.53		1.14	1.47
3900 Basic patterns (matching required), avg. cost paper		535	.015		1.19	.64		1.83	2.27
4000 Paper at $85 per double roll, quality workmanship		435	.018		2.11	.78		2.89	3.50

For customer support on your Building Construction Costs with RSMeans data, call 800.448.8182.

353

09 74 Flexible Wood Sheets

09 74 16 – Flexible Wood Veneers

09 74 16.10 Veneer, Flexible Wood	Crew	Daily Output	Labor-Hours	Unit	2018 Bare Costs Material	Labor	Equipment	Total	Total Incl O&P
0010 **VENEER, FLEXIBLE WOOD**									
0100 Flexible wood veneer, 1/32" thick, plain woods	1 Pape	100	.080	S.F.	2.41	3.41		5.82	7.80
0110 Exotic woods	"	95	.084	"	3.64	3.59		7.23	9.40

09 77 Special Wall Surfacing

09 77 30 – Fiberglass Reinforced Panels

09 77 30.10 Fiberglass Reinforced Plastic Panels

		Crew	Daily Output	Labor-Hours	Unit	Material	Labor	Equipment	Total	Total Incl O&P
0010	**FIBERGLASS REINFORCED PLASTIC PANELS**, .090" thick									
0020	On walls, adhesive mounted, embossed surface	2 Carp	640	.025	S.F.	1.20	1.27		2.47	3.25
0030	Smooth surface		640	.025		1.48	1.27		2.75	3.56
0040	Fire rated, embossed surface		640	.025		2.15	1.27		3.42	4.30
0050	Nylon rivet mounted, on drywall, embossed surface		480	.033		1.20	1.69		2.89	3.89
0060	Smooth surface		480	.033		1.48	1.69		3.17	4.20
0070	Fire rated, embossed surface		480	.033		2.15	1.69		3.84	4.94
0080	On masonry, embossed surface		320	.050		1.20	2.54		3.74	5.20
0090	Smooth surface		320	.050		1.48	2.54		4.02	5.50
0100	Fire rated, embossed surface		320	.050		2.15	2.54		4.69	6.25
0110	Nylon rivet and adhesive mounted, on drywall, embossed surface		240	.067		1.36	3.38		4.74	6.65
0120	Smooth surface		240	.067		1.36	3.38		4.74	6.65
0130	Fire rated, embossed surface		240	.067		2.37	3.38		5.75	7.75
0140	On masonry, embossed surface		190	.084		1.36	4.27		5.63	8
0150	Smooth surface		190	.084		1.36	4.27		5.63	8
0160	Fire rated, embossed surface		190	.084		2.37	4.27		6.64	9.10
0170	For moldings, add	1 Carp	250	.032	L.F.	.28	1.62		1.90	2.78
0180	On ceilings, for lay in grid system, embossed surface		400	.020	S.F.	1.20	1.01		2.21	2.86
0190	Smooth surface		400	.020		1.48	1.01		2.49	3.17
0200	Fire rated, embossed surface		400	.020		2.15	1.01		3.16	3.91

09 77 43 – Panel Systems

09 77 43.20 Slatwall Panels and Accessories

		Crew	Daily Output	Labor-Hours	Unit	Material	Labor	Equipment	Total	Total Incl O&P
0010	**SLATWALL PANELS AND ACCESSORIES**									
0100	Slatwall panel, 4' x 8' x 3/4" T, MDF, paint grade	1 Carp	500	.016	S.F.	1.56	.81		2.37	2.96
0110	Melamine finish		500	.016		2.04	.81		2.85	3.48
0120	High pressure plastic laminate finish		500	.016		3.41	.81		4.22	4.99
0125	Wood veneer		500	.016		3.80	.81		4.61	5.40
0130	Aluminum channel inserts, add					2.64			2.64	2.90
0200	Accessories, corner forms, 8' L				L.F.	5.30			5.30	5.85
0210	T-connector, 8' L					7.50			7.50	8.25
0220	J-mold, 8' L					1.28			1.28	1.41
0230	Edge cap, 8' L					1.72			1.72	1.89
0240	Finish end cap, 8' L					3.24			3.24	3.56
0300	Display hook, metal, 4" L				Ea.	.42			.42	.46
0310	6" L					.48			.48	.53
0320	8" L					.51			.51	.56
0330	10" L					.57			.57	.63
0340	12" L					.64			.64	.70
0350	Acrylic, 4" L					1.23			1.23	1.35
0360	6" L					.99			.99	1.09
0400	Waterfall hanger, metal, 12"-16"					3.75			3.75	4.13
0410	Acrylic					11.50			11.50	12.65
0500	Shelf bracket, metal, 8"					1.90			1.90	2.09

For customer support on your Building Construction Costs with RSMeans data, call 800.448.8182.

09 77 Special Wall Surfacing

09 77 43 – Panel Systems

09 77 43.20 Slatwall Panels and Accessories	Crew	Daily Output	Labor-Hours	Unit	Material	2018 Bare Costs Labor	Equipment	Total	Total Incl O&P
0510	10"				Ea.	2.06			2.06	2.27
0520	12"					2.28			2.28	2.51
0530	14"					2.50			2.50	2.75
0540	16"					2.85			2.85	3.14
0550	Acrylic, 8"					3.82			3.82	4.20
0560	10"					4.11			4.11	4.52
0570	12"					4.61			4.61	5.05
0580	14"					6.05			6.05	6.65
0600	Shelf, acrylic, 12" x 16" x 1/4"					22			22	24.50
0610	12" x 24" x 1/4"					25			25	27.50

09 81 Acoustic Insulation

09 81 13 – Acoustic Board Insulation

09 81 13.10 Acoustic Board Insulation

		Crew	Daily Output	Labor-Hours	Unit	Material	Labor	Equipment	Total	Total Incl O&P
0010	**ACOUSTIC BOARD INSULATION**									
0020	Cellulose fiber board, 1/2" thk.	1 Carp	800	.010	S.F.	.88	.51		1.39	1.74

09 81 16 – Acoustic Blanket Insulation

09 81 16.10 Sound Attenuation Blanket

		Crew	Daily Output	Labor-Hours	Unit	Material	Labor	Equipment	Total	Total Incl O&P
0010	**SOUND ATTENUATION BLANKET**									
0020	Blanket, 1" thick	1 Carp	925	.009	S.F.	.26	.44		.70	.96
0500	1-1/2" thick		920	.009		.30	.44		.74	1
1000	2" thick		915	.009		.42	.44		.86	1.13
1500	3" thick		910	.009		.61	.45		1.06	1.35
2000	Wall hung, STC 18-21, 1" thick, 4' x 20'	2 Carp	22	.727	Ea.	525	37		562	635
2010	10' x 20'	"	19	.842		1,325	42.50		1,367.50	1,525
2020	Wall hung, STC 27-28, 3" thick, 4' x 20'	3 Carp	12	2		575	101		676	785
2030	10' x 20'	"	9	2.667		1,425	135		1,560	1,775
3000	Thermal or acoustical batt above ceiling, 2" thick	1 Carp	900	.009	S.F.	.53	.45		.98	1.27
3100	3" thick		900	.009		.81	.45		1.26	1.58
3200	4" thick		900	.009		1.01	.45		1.46	1.80
3400	Urethane plastic foam, open cell, on wall, 2" thick	2 Carp	2050	.008		3.32	.40		3.72	4.25
3500	3" thick		1550	.010		4.41	.52		4.93	5.65
3600	4" thick		1050	.015		6.20	.77		6.97	8
3700	On ceiling, 2" thick		1700	.009		3.31	.48		3.79	4.37
3800	3" thick		1300	.012		4.41	.62		5.03	5.80
3900	4" thick		900	.018		6.20	.90		7.10	8.15

09 84 Acoustic Room Components

09 84 13 – Fixed Sound-Absorptive Panels

09 84 13.10 Fixed Panels

		Crew	Daily Output	Labor-Hours	Unit	Material	Labor	Equipment	Total	Total Incl O&P
0010	**FIXED PANELS** Perforated steel facing, painted with									
0100	Fiberglass or mineral filler, no backs, 2-1/4" thick, modular									
0200	space units, ceiling or wall hung, white or colored	1 Carp	100	.080	S.F.	9.45	4.06		13.51	16.60
0300	Fiberboard sound deadening panels, 1/2" thick	"	600	.013	"	.34	.68		1.02	1.40
0500	Fiberglass panels, 4' x 8' x 1" thick, with									
0600	glass cloth face for walls, cemented	1 Carp	155	.052	S.F.	9.30	2.62		11.92	14.25
0700	1-1/2" thick, dacron covered, inner aluminum frame,									
0710	wall mounted	1 Carp	300	.027	S.F.	9.10	1.35		10.45	12.05
0900	Mineral fiberboard panels, fabric covered, 30" x 108",									

For customer support on your Building Construction Costs with RSMeans data, call 800.448.8182.

355

09 84 Acoustic Room Components

09 84 13 – Fixed Sound-Absorptive Panels

09 84 13.10 Fixed Panels	Crew	Daily Output	Labor-Hours	Unit	Material	2018 Bare Costs Labor	Equipment	Total	Total Incl O&P
1000 3/4" thick, concealed spline, wall mounted	1 Carp	150	.053	S.F.	6.55	2.70		9.25	11.30

09 84 36 – Sound-Absorbing Ceiling Units

09 84 36.10 Barriers

		Crew	Daily Output	Labor-Hours	Unit	Material	2018 Bare Costs Labor	Equipment	Total	Total Incl O&P
0010	**BARRIERS** Plenum									
0600	Aluminum foil, fiberglass reinf., parallel with joists	1 Carp	275	.029	S.F.	1.13	1.47		2.60	3.49
0700	Perpendicular to joists		180	.044		1.13	2.25		3.38	4.67
0900	Aluminum mesh, kraft paperbacked		275	.029		.81	1.47		2.28	3.14
0970	Fiberglass batts, kraft faced, 3-1/2" thick		1400	.006		.38	.29		.67	.86
0980	6" thick		1300	.006		.67	.31		.98	1.21
1000	Sheet lead, 1 lb., 1/64" thick, perpendicular to joists		150	.053		6.90	2.70		9.60	11.70
1100	Vinyl foam reinforced, 1/8" thick, 1.0 lb./S.F.		150	.053		5.95	2.70		8.65	10.65

09 91 Painting

09 91 03 – Paint Restoration

09 91 03.20 Sanding

		Crew	Daily Output	Labor-Hours	Unit	Material	2018 Bare Costs Labor	Equipment	Total	Total Incl O&P
0010	**SANDING** and puttying interior trim, compared to R099100-10									
0100	Painting 1 coat, on quality work				L.F.		100%			
0300	Medium work						50%			
0400	Industrial grade						25%			
0500	Surface protection, placement and removal									
0510	Surface protection, placement and removal, basic drop cloths	1 Pord	6400	.001	S.F.		.05		.05	.08
0520	Masking with paper		800	.010		.07	.43		.50	.72
0530	Volume cover up (using plastic sheathing or building paper)		16000	.001			.02		.02	.03

09 91 03.30 Exterior Surface Preparation

		Crew	Daily Output	Labor-Hours	Unit	Material	2018 Bare Costs Labor	Equipment	Total	Total Incl O&P
0010	**EXTERIOR SURFACE PREPARATION** R099100-10									
0015	Doors, per side, not incl. frames or trim									
0020	Scrape & sand									
0030	Wood, flush	1 Pord	616	.013	S.F.		.55		.55	.83
0040	Wood, detail		496	.016			.69		.69	1.03
0050	Wood, louvered		280	.029			1.22		1.22	1.83
0060	Wood, overhead		616	.013			.55		.55	.83
0070	Wire brush									
0080	Metal, flush	1 Pord	640	.013	S.F.		.53		.53	.80
0090	Metal, detail		520	.015			.65		.65	.99
0100	Metal, louvered		360	.022			.95		.95	1.42
0110	Metal or fibr., overhead		640	.013			.53		.53	.80
0120	Metal, roll up		560	.014			.61		.61	.92
0130	Metal, bulkhead		640	.013			.53		.53	.80
0140	Power wash, based on 2500 lb. operating pressure									
0150	Metal, flush	A-1H	2240	.004	S.F.		.14	.03	.17	.26
0160	Metal, detail		2120	.004			.15	.04	.19	.27
0170	Metal, louvered		2000	.004			.16	.04	.20	.28
0180	Metal or fibr., overhead		2400	.003			.13	.03	.16	.23
0190	Metal, roll up		2400	.003			.13	.03	.16	.23
0200	Metal, bulkhead		2200	.004			.15	.03	.18	.26
0400	Windows, per side, not incl. trim									
0410	Scrape & sand									
0420	Wood, 1-2 lite	1 Pord	320	.025	S.F.		1.06		1.06	1.60
0430	Wood, 3-6 lite		280	.029			1.22		1.22	1.83
0440	Wood, 7-10 lite		240	.033			1.42		1.42	2.13

For customer support on your Building Construction Costs with RSMeans data, call 800.448.8182.

09 91 Painting

09 91 03 – Paint Restoration

09 91 03.30 Exterior Surface Preparation

		Crew	Daily Output	Labor-Hours	Unit	Material	2018 Bare Costs Labor	Equipment	Total	Total Incl O&P
0450	Wood, 12 lite	1 Pord	200	.040	S.F.		1.70		1.70	2.56
0460	Wood, Bay/Bow	↓	320	.025	↓		1.06		1.06	1.60
0470	Wire brush									
0480	Metal, 1-2 lite	1 Pord	480	.017	S.F.		.71		.71	1.07
0490	Metal, 3-6 lite		400	.020			.85		.85	1.28
0500	Metal, Bay/Bow	↓	480	.017	↓		.71		.71	1.07
0510	Power wash, based on 2500 lb. operating pressure									
0520	1-2 lite	A-1H	4400	.002	S.F.		.07	.02	.09	.13
0530	3-6 lite		4320	.002			.07	.02	.09	.13
0540	7-10 lite		4240	.002			.08	.02	.10	.13
0550	12 lite		4160	.002			.08	.02	.10	.14
0560	Bay/Bow	↓	4400	.002	↓		.07	.02	.09	.13
0600	Siding, scrape and sand, light=10-30%, med.=30-70%									
0610	Heavy=70-100% of surface to sand									
0650	Texture 1-11, light	1 Pord	480	.017	S.F.		.71		.71	1.07
0660	Med.		440	.018			.77		.77	1.16
0670	Heavy		360	.022			.95		.95	1.42
0680	Wood shingles, shakes, light		440	.018			.77		.77	1.16
0690	Med.		360	.022			.95		.95	1.42
0700	Heavy		280	.029			1.22		1.22	1.83
0710	Clapboard, light		520	.015			.65		.65	.99
0720	Med.		480	.017			.71		.71	1.07
0730	Heavy	↓	400	.020	↓		.85		.85	1.28
0740	Wire brush									
0750	Aluminum, light	1 Pord	600	.013	S.F.		.57		.57	.85
0760	Med.		520	.015			.65		.65	.99
0770	Heavy	↓	440	.018	↓		.77		.77	1.16
0780	Pressure wash, based on 2500 lb. operating pressure									
0790	Stucco	A-1H	3080	.003	S.F.		.10	.02	.12	.19
0800	Aluminum or vinyl		3200	.003			.10	.02	.12	.18
0810	Siding, masonry, brick & block	↓	2400	.003	↓		.13	.03	.16	.23
1300	Miscellaneous, wire brush									
1310	Metal, pedestrian gate	1 Pord	100	.080	S.F.		3.40		3.40	5.10
1320	Aluminum chain link, both sides		250	.032			1.36		1.36	2.05
1400	Existing galvanized surface, clean and prime, prep for painting	↓	380	.021	↓	.13	.90		1.03	1.49
8000	For chemical washing, see Section 04 01 30									
8010	For steam cleaning, see Section 04 01 30.20									
8020	For sand blasting, see Sections 03 35 29.60 and 05 01 10.51									

09 91 03.40 Interior Surface Preparation

		Crew	Daily Output	Labor-Hours	Unit	Material	2018 Bare Costs Labor	Equipment	Total	Total Incl O&P
0010	**INTERIOR SURFACE PREPARATION** R099100-10									
0020	Doors, per side, not incl. frames or trim									
0030	Scrape & sand									
0040	Wood, flush	1 Pord	616	.013	S.F.		.55		.55	.83
0050	Wood, detail		496	.016			.69		.69	1.03
0060	Wood, louvered	↓	280	.029	↓		1.22		1.22	1.83
0070	Wire brush									
0080	Metal, flush	1 Pord	640	.013	S.F.		.53		.53	.80
0090	Metal, detail		520	.015			.65		.65	.99
0100	Metal, louvered	↓	360	.022	↓		.95		.95	1.42
0110	Hand wash									
0120	Wood, flush	1 Pord	2160	.004	S.F.		.16		.16	.24
0130	Wood, detail	↓	2000	.004	↓		.17		.17	.26

For customer support on your Building Construction Costs with RSMeans data, call 800.448.8182.

357

09 91 Painting

09 91 03 – Paint Restoration

09 91 03.40 Interior Surface Preparation

		Crew	Daily Output	Labor-Hours	Unit	Material	2018 Bare Costs Labor	Equipment	Total	Total Incl O&P
0140	Wood, louvered	1 Pord	1360	.006	S.F.		.25		.25	.38
0150	Metal, flush		2160	.004			.16		.16	.24
0160	Metal, detail		2000	.004			.17		.17	.26
0170	Metal, louvered		1360	.006			.25		.25	.38
0400	Windows, per side, not incl. trim									
0410	Scrape & sand									
0420	Wood, 1-2 lite	1 Pord	360	.022	S.F.		.95		.95	1.42
0430	Wood, 3-6 lite		320	.025			1.06		1.06	1.60
0440	Wood, 7-10 lite		280	.029			1.22		1.22	1.83
0450	Wood, 12 lite		240	.033			1.42		1.42	2.13
0460	Wood, Bay/Bow		360	.022			.95		.95	1.42
0470	Wire brush									
0480	Metal, 1-2 lite	1 Pord	520	.015	S.F.		.65		.65	.99
0490	Metal, 3-6 lite		440	.018			.77		.77	1.16
0500	Metal, Bay/Bow		520	.015			.65		.65	.99
0600	Walls, sanding, light=10-30%, medium=30-70%,									
0610	heavy=70-100% of surface to sand									
0650	Walls, sand									
0660	Gypsum board or plaster, light	1 Pord	3077	.003	S.F.		.11		.11	.17
0670	Gypsum board or plaster, medium		2160	.004			.16		.16	.24
0680	Gypsum board or plaster, heavy		923	.009			.37		.37	.56
0690	Wood, T&G, light		2400	.003			.14		.14	.21
0700	Wood, T&G, medium		1600	.005			.21		.21	.32
0710	Wood, T&G, heavy		800	.010			.43		.43	.64
0720	Walls, wash									
0730	Gypsum board or plaster	1 Pord	3200	.003	S.F.		.11		.11	.16
0740	Wood, T&G		3200	.003			.11		.11	.16
0750	Masonry, brick & block, smooth		2800	.003			.12		.12	.18
0760	Masonry, brick & block, coarse		2000	.004			.17		.17	.26
8000	For chemical washing, see Section 04 01 30									
8010	For steam cleaning, see Section 04 01 30.20									
8020	For sand blasting, see Sections 03 35 29.60 and 05 01 10.51									

09 91 13 – Exterior Painting

09 91 13.30 Fences

		Crew	Daily Output	Labor-Hours	Unit	Material	2018 Bare Costs Labor	Equipment	Total	Total Incl O&P
0010	**FENCES** R099100-20									
0100	Chain link or wire metal, one side, water base									
0110	Roll & brush, first coat	1 Pord	960	.008	S.F.	.07	.35		.42	.61
0120	Second coat		1280	.006		.07	.27		.34	.47
0130	Spray, first coat		2275	.004		.07	.15		.22	.31
0140	Second coat		2600	.003		.07	.13		.20	.28
0150	Picket, water base									
0160	Roll & brush, first coat	1 Pord	865	.009	S.F.	.08	.39		.47	.67
0170	Second coat		1050	.008		.08	.32		.40	.57
0180	Spray, first coat		2275	.004		.08	.15		.23	.31
0190	Second coat		2600	.003		.08	.13		.21	.28
0200	Stockade, water base									
0210	Roll & brush, first coat	1 Pord	1040	.008	S.F.	.08	.33		.41	.57
0220	Second coat		1200	.007		.08	.28		.36	.51
0230	Spray, first coat		2275	.004		.08	.15		.23	.31
0240	Second coat		2600	.003		.08	.13		.21	.28

09 91 13.42 Miscellaneous, Exterior

		Crew	Daily Output	Labor-Hours	Unit	Material	2018 Bare Costs Labor	Equipment	Total	Total Incl O&P
0010	**MISCELLANEOUS, EXTERIOR** R099100-20									
0015	For painting metals, see Section 09 97 13.23									
0100	Railing, ext., decorative wood, incl. cap & baluster									
0110	Newels & spindles @ 12" OC									
0120	Brushwork, stain, sand, seal & varnish									
0130	First coat	1 Pord	90	.089	L.F.	.88	3.78		4.66	6.65
0140	Second coat	"	120	.067	"	.88	2.84		3.72	5.25
0150	Rough sawn wood, 42" high, 2" x 2" verticals, 6" OC									
0160	Brushwork, stain, each coat	1 Pord	90	.089	L.F.	.31	3.78		4.09	6.05
0170	Wrought iron, 1" rail, 1/2" sq. verticals									
0180	Brushwork, zinc chromate, 60" high, bars 6" OC									
0190	Primer	1 Pord	130	.062	L.F.	.88	2.62		3.50	4.91
0200	Finish coat		130	.062		1.13	2.62		3.75	5.20
0210	Additional coat	↓	190	.042	↓	1.32	1.79		3.11	4.15
0220	Shutters or blinds, single panel, 2' x 4', paint all sides									
0230	Brushwork, primer	1 Pord	20	.400	Ea.	.69	17		17.69	26.50
0240	Finish coat, exterior latex		20	.400		.57	17		17.57	26
0250	Primer & 1 coat, exterior latex		13	.615		1.11	26		27.11	40.50
0260	Spray, primer		35	.229		1.01	9.75		10.76	15.75
0270	Finish coat, exterior latex		35	.229		1.21	9.75		10.96	16
0280	Primer & 1 coat, exterior latex	↓	20	.400	↓	1.09	17		18.09	26.50
0290	For louvered shutters, add				S.F.	10%				
0300	Stair stringers, exterior, metal									
0310	Roll & brush, zinc chromate, to 14", each coat	1 Pord	320	.025	L.F.	.38	1.06		1.44	2.01
0320	Rough sawn wood, 4" x 12"									
0330	Roll & brush, exterior latex, each coat	1 Pord	215	.037	L.F.	.08	1.58		1.66	2.47
0340	Trellis/lattice, 2" x 2" @ 3" OC with 2" x 8" supports									
0350	Spray, latex, per side, each coat	1 Pord	475	.017	S.F.	.08	.72		.80	1.17
0450	Decking, ext., sealer, alkyd, brushwork, sealer coat		1140	.007		.10	.30		.40	.56
0460	1st coat		1140	.007		.12	.30		.42	.58
0470	2nd coat		1300	.006		.09	.26		.35	.48
0500	Paint, alkyd, brushwork, primer coat		1140	.007		.11	.30		.41	.57
0510	1st coat		1140	.007		.14	.30		.44	.60
0520	2nd coat		1300	.006		.10	.26		.36	.50
0600	Sand paint, alkyd, brushwork, 1 coat	↓	150	.053	↓	.14	2.27		2.41	3.57

09 91 13.60 Siding Exterior

		Crew	Daily Output	Labor-Hours	Unit	Material	2018 Bare Costs Labor	Equipment	Total	Total Incl O&P
0010	**SIDING EXTERIOR**, Alkyd (oil base) R099100-10									
0450	Steel siding, oil base, paint 1 coat, brushwork	2 Pord	2015	.008	S.F.	.11	.34		.45	.63
0500	Spray R099100-20		4550	.004		.17	.15		.32	.42
0800	Paint 2 coats, brushwork		1300	.012		.23	.52		.75	1.04
1000	Spray		2750	.006		.15	.25		.40	.54
1200	Stucco, rough, oil base, paint 2 coats, brushwork		1300	.012		.23	.52		.75	1.04
1400	Roller		1625	.010		.24	.42		.66	.89
1600	Spray		2925	.005		.25	.23		.48	.63
1800	Texture 1-11 or clapboard, oil base, primer coat, brushwork		1300	.012		.14	.52		.66	.95
2000	Spray		4550	.004		.14	.15		.29	.39
2400	Paint 2 coats, brushwork		810	.020		.33	.84		1.17	1.63
2600	Spray		2600	.006		.36	.26		.62	.79
3400	Stain 2 coats, brushwork		950	.017		.20	.72		.92	1.31
4000	Spray		3050	.005		.23	.22		.45	.59
4200	Wood shingles, oil base primer coat, brushwork		1300	.012		.13	.52		.65	.94
4400	Spray	↓	3900	.004		.12	.17		.29	.40

For customer support on your Building Construction Costs with RSMeans data, call 800.448.8182.

359

09 91 13.60 Siding Exterior

		Crew	Daily Output	Labor-Hours	Unit	Material	2018 Bare Costs Labor	2018 Bare Costs Equipment	Total	Total Incl O&P
5000	Paint 2 coats, brushwork	2 Pord	810	.020	S.F.	.27	.84		1.11	1.57
5200	Spray		2275	.007		.26	.30		.56	.74
6500	Stain 2 coats, brushwork		950	.017		.20	.72		.92	1.31
7000	Spray		2660	.006		.28	.26		.54	.70
8000	For latex paint, deduct					10%				
8100	For work over 12' H, from pipe scaffolding, add						15%			
8200	For work over 12' H, from extension ladder, add						25%			
8300	For work over 12' H, from swing staging, add						35%			

09 91 13.62 Siding, Misc.

		Crew	Daily Output	Labor-Hours	Unit	Material	2018 Bare Costs Labor	2018 Bare Costs Equipment	Total	Total Incl O&P
0010	**SIDING, MISC.,** latex paint R099100-10									
0100	Aluminum siding									
0110	Brushwork, primer R099100-20	2 Pord	2275	.007	S.F.	.06	.30		.36	.52
0120	Finish coat, exterior latex		2275	.007		.06	.30		.36	.51
0130	Primer & 1 coat exterior latex		1300	.012		.13	.52		.65	.94
0140	Primer & 2 coats exterior latex		975	.016		.19	.70		.89	1.26
0150	Mineral fiber shingles									
0160	Brushwork, primer	2 Pord	1495	.011	S.F.	.14	.46		.60	.85
0170	Finish coat, industrial enamel		1495	.011		.18	.46		.64	.89
0180	Primer & 1 coat enamel		810	.020		.32	.84		1.16	1.62
0190	Primer & 2 coats enamel		540	.030		.50	1.26		1.76	2.45
0200	Roll, primer		1625	.010		.16	.42		.58	.80
0210	Finish coat, industrial enamel		1625	.010		.19	.42		.61	.84
0220	Primer & 1 coat enamel		975	.016		.35	.70		1.05	1.44
0230	Primer & 2 coats enamel		650	.025		.55	1.05		1.60	2.18
0240	Spray, primer		3900	.004		.12	.17		.29	.40
0250	Finish coat, industrial enamel		3900	.004		.16	.17		.33	.44
0260	Primer & 1 coat enamel		2275	.007		.28	.30		.58	.76
0270	Primer & 2 coats enamel		1625	.010		.45	.42		.87	1.12
0280	Waterproof sealer, first coat		4485	.004		.12	.15		.27	.36
0290	Second coat		5235	.003		.11	.13		.24	.32
0300	Rough wood incl. shingles, shakes or rough sawn siding									
0310	Brushwork, primer	2 Pord	1280	.013	S.F.	.14	.53		.67	.95
0320	Finish coat, exterior latex		1280	.013		.10	.53		.63	.91
0330	Primer & 1 coat exterior latex		960	.017		.24	.71		.95	1.34
0340	Primer & 2 coats exterior latex		700	.023		.34	.97		1.31	1.84
0350	Roll, primer		2925	.005		.19	.23		.42	.56
0360	Finish coat, exterior latex		2925	.005		.12	.23		.35	.48
0370	Primer & 1 coat exterior latex		1790	.009		.31	.38		.69	.91
0380	Primer & 2 coats exterior latex		1300	.012		.43	.52		.95	1.27
0390	Spray, primer		3900	.004		.16	.17		.33	.43
0400	Finish coat, exterior latex		3900	.004		.09	.17		.26	.36
0410	Primer & 1 coat exterior latex		2600	.006		.25	.26		.51	.67
0420	Primer & 2 coats exterior latex		2080	.008		.35	.33		.68	.87
0430	Waterproof sealer, first coat		4485	.004		.21	.15		.36	.46
0440	Second coat		4485	.004		.12	.15		.27	.36
0450	Smooth wood incl. butt, T&G, beveled, drop or B&B siding									
0460	Brushwork, primer	2 Pord	2325	.007	S.F.	.10	.29		.39	.55
0470	Finish coat, exterior latex		1280	.013		.10	.53		.63	.91
0480	Primer & 1 coat exterior latex		800	.020		.20	.85		1.05	1.50
0490	Primer & 2 coats exterior latex		630	.025		.31	1.08		1.39	1.97
0500	Roll, primer		2275	.007		.11	.30		.41	.57
0510	Finish coat, exterior latex		2275	.007		.11	.30		.41	.57

09 91 Painting

09 91 13 - Exterior Painting

09 91 13.62 Siding, Misc.

		Crew	Daily Output	Labor-Hours	Unit	Material	2018 Bare Costs Labor	Equipment	Total	Total Incl O&P
0520	Primer & 1 coat exterior latex	2 Pord	1300	.012	S.F.	.22	.52		.74	1.03
0530	Primer & 2 coats exterior latex		975	.016		.33	.70		1.03	1.42
0540	Spray, primer		4550	.004		.09	.15		.24	.33
0550	Finish coat, exterior latex		4550	.004		.09	.15		.24	.33
0560	Primer & 1 coat exterior latex		2600	.006		.18	.26		.44	.59
0570	Primer & 2 coats exterior latex		1950	.008		.28	.35		.63	.83
0580	Waterproof sealer, first coat		5230	.003		.12	.13		.25	.33
0590	Second coat	▼	5980	.003		.12	.11		.23	.30
0600	For oil base paint, add				▼	10%				

09 91 13.70 Doors and Windows, Exterior

		Crew	Daily Output	Labor-Hours	Unit	Material	2018 Bare Costs Labor	Equipment	Total	Total Incl O&P
0010	**DOORS AND WINDOWS, EXTERIOR** R099100-10									
0100	Door frames & trim, only									
0110	Brushwork, primer R099100-20	1 Pord	512	.016	L.F.	.06	.67		.73	1.07
0120	Finish coat, exterior latex		512	.016		.07	.67		.74	1.08
0130	Primer & 1 coat, exterior latex		300	.027		.13	1.13		1.26	1.86
0140	Primer & 2 coats, exterior latex	▼	265	.030	▼	.21	1.28		1.49	2.16
0150	Doors, flush, both sides, incl. frame & trim									
0160	Roll & brush, primer	1 Pord	10	.800	Ea.	4.80	34		38.80	56.50
0170	Finish coat, exterior latex		10	.800		5.40	34		39.40	57
0180	Primer & 1 coat, exterior latex		7	1.143		10.20	48.50		58.70	84.50
0190	Primer & 2 coats, exterior latex		5	1.600		15.65	68		83.65	119
0200	Brushwork, stain, sealer & 2 coats polyurethane	▼	4	2	▼	30	85		115	161
0210	Doors, French, both sides, 10-15 lite, incl. frame & trim									
0220	Brushwork, primer	1 Pord	6	1.333	Ea.	2.40	56.50		58.90	88
0230	Finish coat, exterior latex		6	1.333		2.71	56.50		59.21	88.50
0240	Primer & 1 coat, exterior latex		3	2.667		5.10	113		118.10	177
0250	Primer & 2 coats, exterior latex		2	4		7.65	170		177.65	264
0260	Brushwork, stain, sealer & 2 coats polyurethane	▼	2.50	3.200		11.10	136		147.10	217
0270	Doors, louvered, both sides, incl. frame & trim									
0280	Brushwork, primer	1 Pord	7	1.143	Ea.	4.80	48.50		53.30	78.50
0290	Finish coat, exterior latex		7	1.143		5.40	48.50		53.90	79
0300	Primer & 1 coat, exterior latex		4	2		10.20	85		95.20	139
0310	Primer & 2 coats, exterior latex		3	2.667		15.35	113		128.35	188
0320	Brushwork, stain, sealer & 2 coats polyurethane	▼	4.50	1.778		30	75.50		105.50	147
0330	Doors, panel, both sides, incl. frame & trim									
0340	Roll & brush, primer	1 Pord	6	1.333	Ea.	4.80	56.50		61.30	91
0350	Finish coat, exterior latex		6	1.333		5.40	56.50		61.90	91.50
0360	Primer & 1 coat, exterior latex		3	2.667		10.20	113		123.20	182
0370	Primer & 2 coats, exterior latex		2.50	3.200		15.35	136		151.35	222
0380	Brushwork, stain, sealer & 2 coats polyurethane	▼	3	2.667	▼	30	113		143	204
0400	Windows, per ext. side, based on 15 S.F.									
0410	1 to 6 lite									
0420	Brushwork, primer	1 Pord	13	.615	Ea.	.95	26		26.95	40.50
0430	Finish coat, exterior latex		13	.615		1.07	26		27.07	40.50
0440	Primer & 1 coat, exterior latex		8	1		2.02	42.50		44.52	66
0450	Primer & 2 coats, exterior latex		6	1.333		3.03	56.50		59.53	89
0460	Stain, sealer & 1 coat varnish	▼	7	1.143	▼	4.38	48.50		52.88	78
0470	7 to 10 lite									
0480	Brushwork, primer	1 Pord	11	.727	Ea.	.95	31		31.95	47.50
0490	Finish coat, exterior latex		11	.727		1.07	31		32.07	47.50
0500	Primer & 1 coat, exterior latex		7	1.143		2.02	48.50		50.52	75
0510	Primer & 2 coats, exterior latex		5	1.600		3.03	68		71.03	105

For customer support on your Building Construction Costs with RSMeans data, call 800.448.8182.

361

09 91 Painting

09 91 13 – Exterior Painting

09 91 13.70 Doors and Windows, Exterior

		Crew	Daily Output	Labor-Hours	Unit	Material	2018 Bare Costs Labor	2018 Bare Costs Equipment	Total	Total Incl O&P
0520	Stain, sealer & 1 coat varnish	1 Pord	6	1.333	Ea.	4.38	56.50		60.88	90.50
0530	12 lite									
0540	Brushwork, primer	1 Pord	10	.800	Ea.	.95	34		34.95	52
0550	Finish coat, exterior latex		10	.800		1.07	34		35.07	52
0560	Primer & 1 coat, exterior latex		6	1.333		2.02	56.50		58.52	87.50
0570	Primer & 2 coats, exterior latex		5	1.600		3.03	68		71.03	105
0580	Stain, sealer & 1 coat varnish		6	1.333		4.28	56.50		60.78	90
0590	For oil base paint, add					10%				

09 91 13.80 Trim, Exterior

		Crew	Daily Output	Labor-Hours	Unit	Material	2018 Bare Costs Labor	2018 Bare Costs Equipment	Total	Total Incl O&P
0010	**TRIM, EXTERIOR** R099100-10									
0100	Door frames & trim (see Doors, interior or exterior)									
0110	Fascia, latex paint, one coat coverage R099100-20									
0120	1" x 4", brushwork	1 Pord	640	.013	L.F.	.02	.53		.55	.82
0130	Roll		1280	.006		.02	.27		.29	.43
0140	Spray		2080	.004		.02	.16		.18	.27
0150	1" x 6" to 1" x 10", brushwork		640	.013		.08	.53		.61	.88
0160	Roll		1230	.007		.08	.28		.36	.51
0170	Spray		2100	.004		.06	.16		.22	.31
0180	1" x 12", brushwork		640	.013		.08	.53		.61	.88
0190	Roll		1050	.008		.08	.32		.40	.58
0200	Spray		2200	.004		.06	.15		.21	.30
0210	Gutters & downspouts, metal, zinc chromate paint									
0220	Brushwork, gutters, 5", first coat	1 Pord	640	.013	L.F.	.40	.53		.93	1.24
0230	Second coat		960	.008		.38	.35		.73	.94
0240	Third coat		1280	.006		.30	.27		.57	.74
0250	Downspouts, 4", first coat		640	.013		.40	.53		.93	1.24
0260	Second coat		960	.008		.38	.35		.73	.94
0270	Third coat		1280	.006		.30	.27		.57	.74
0280	Gutters & downspouts, wood									
0290	Brushwork, gutters, 5", primer	1 Pord	640	.013	L.F.	.06	.53		.59	.87
0300	Finish coat, exterior latex		640	.013		.06	.53		.59	.87
0310	Primer & 1 coat exterior latex		400	.020		.13	.85		.98	1.43
0320	Primer & 2 coats exterior latex		325	.025		.21	1.05		1.26	1.81
0330	Downspouts, 4", primer		640	.013		.06	.53		.59	.87
0340	Finish coat, exterior latex		640	.013		.06	.53		.59	.87
0350	Primer & 1 coat exterior latex		400	.020		.13	.85		.98	1.43
0360	Primer & 2 coats exterior latex		325	.025		.10	1.05		1.15	1.69
0370	Molding, exterior, up to 14" wide									
0380	Brushwork, primer	1 Pord	640	.013	L.F.	.08	.53		.61	.88
0390	Finish coat, exterior latex		640	.013		.08	.53		.61	.88
0400	Primer & 1 coat exterior latex		400	.020		.16	.85		1.01	1.46
0410	Primer & 2 coats exterior latex		315	.025		.16	1.08		1.24	1.81
0420	Stain & fill		1050	.008		.12	.32		.44	.63
0430	Shellac		1850	.004		.15	.18		.33	.45
0440	Varnish		1275	.006		.11	.27		.38	.52

09 91 13.90 Walls, Masonry (CMU), Exterior

		Crew	Daily Output	Labor-Hours	Unit	Material	2018 Bare Costs Labor	2018 Bare Costs Equipment	Total	Total Incl O&P
0010	**WALLS, MASONRY (CMU), EXTERIOR** R099100-10									
0360	Concrete masonry units (CMU), smooth surface									
0370	Brushwork, latex, first coat	1 Pord	640	.013	S.F.	.06	.53		.59	.87
0380	Second coat		960	.008		.05	.35		.40	.58
0390	Waterproof sealer, first coat		736	.011		.27	.46		.73	1
0400	Second coat		1104	.007		.27	.31		.58	.76

362

09 91 Painting

09 91 13 – Exterior Painting

09 91 13.90 Walls, Masonry (CMU), Exterior

		Crew	Daily Output	Labor-Hours	Unit	Material	2018 Bare Costs Labor	Equipment	Total	Total Incl O&P
0410	Roll, latex, paint, first coat	1 Pord	1465	.005	S.F.	.07	.23		.30	.43
0420	Second coat		1790	.004		.06	.19		.25	.35
0430	Waterproof sealer, first coat		1680	.005		.27	.20		.47	.60
0440	Second coat		2060	.004		.27	.17		.44	.55
0450	Spray, latex, paint, first coat		1950	.004		.06	.17		.23	.32
0460	Second coat		2600	.003		.05	.13		.18	.25
0470	Waterproof sealer, first coat		2245	.004		.27	.15		.42	.53
0480	Second coat		2990	.003		.27	.11		.38	.47
0490	Concrete masonry unit (CMU), porous									
0500	Brushwork, latex, first coat	1 Pord	640	.013	S.F.	.12	.53		.65	.93
0510	Second coat		960	.008		.06	.35		.41	.60
0520	Waterproof sealer, first coat		736	.011		.27	.46		.73	1
0530	Second coat		1104	.007		.27	.31		.58	.76
0540	Roll latex, first coat		1465	.005		.09	.23		.32	.45
0550	Second coat		1790	.004		.06	.19		.25	.35
0560	Waterproof sealer, first coat		1680	.005		.27	.20		.47	.60
0570	Second coat		2060	.004		.27	.17		.44	.55
0580	Spray latex, first coat		1950	.004		.07	.17		.24	.33
0590	Second coat		2600	.003		.05	.13		.18	.25
0600	Waterproof sealer, first coat		2245	.004		.27	.15		.42	.53
0610	Second coat		2990	.003		.27	.11		.38	.47

09 91 23 – Interior Painting

09 91 23.20 Cabinets and Casework

			Crew	Daily Output	Labor-Hours	Unit	Material	2018 Bare Costs Labor	Equipment	Total	Total Incl O&P
0010	**CABINETS AND CASEWORK**	R099100-10									
1000	Primer coat, oil base, brushwork		1 Pord	650	.012	S.F.	.07	.52		.59	.86
2000	Paint, oil base, brushwork, 1 coat	R099100-20		650	.012		.12	.52		.64	.92
3000	Stain, brushwork, wipe off			650	.012		.10	.52		.62	.90
4000	Shellac, 1 coat, brushwork			650	.012		.13	.52		.65	.93
4500	Varnish, 3 coats, brushwork, sand after 1st coat			325	.025		.27	1.05		1.32	1.88
5000	For latex paint, deduct						10%				
6300	Strip, prep and refinish wood furniture										
6310	Remove paint using chemicals, wood furniture		1 Pord	28	.286	S.F.	1.60	12.15		13.75	20
6320	Prep for painting, sanding			75	.107		.24	4.54		4.78	7.10
6350	Stain and wipe, brushwork			600	.013		.10	.57		.67	.96
6355	Spray applied			900	.009		.09	.38		.47	.67
6360	Sealer or varnish, brushwork			1080	.007		.09	.32		.41	.57
6365	Spray applied			2100	.004		.09	.16		.25	.34
6370	Paint, primer, brushwork			720	.011		.07	.47		.54	.78
6375	Spray applied			2100	.004		.06	.16		.22	.31
6380	Finish coat, brushwork			810	.010		.12	.42		.54	.76
6385	Spray applied			2100	.004		.11	.16		.27	.36

09 91 23.33 Doors and Windows, Interior Alkyd (Oil Base)

			Crew	Daily Output	Labor-Hours	Unit	Material	2018 Bare Costs Labor	Equipment	Total	Total Incl O&P
0010	**DOORS AND WINDOWS, INTERIOR ALKYD (OIL BASE)**	R099100-10									
0500	Flush door & frame, 3' x 7', oil, primer, brushwork		1 Pord	10	.800	Ea.	3.98	34		37.98	55.50
1000	Paint, 1 coat	R099100-20		10	.800		4.16	34		38.16	55.50
1400	Stain, brushwork, wipe off			18	.444		2.15	18.90		21.05	31
1600	Shellac, 1 coat, brushwork			25	.320		2.64	13.60		16.24	23.50
1800	Varnish, 3 coats, brushwork, sand after 1st coat			9	.889		5.75	38		43.75	63.50
2000	Panel door & frame, 3' x 7', oil, primer, brushwork			6	1.333		2.50	56.50		59	88.50
2200	Paint, 1 coat			6	1.333		4.16	56.50		60.66	90
2600	Stain, brushwork, panel door, 3' x 7', not incl. frame			16	.500		2.15	21.50		23.65	34.50
2800	Shellac, 1 coat, brushwork			22	.364		2.64	15.45		18.09	26.50

09 91 23.33 Doors and Windows, Interior Alkyd (Oil Base)

		Crew	Daily Output	Labor-Hours	Unit	Material	2018 Bare Costs Labor	Equipment	Total	Total Incl O&P
3000	Varnish, 3 coats, brushwork, sand after 1st coat	1 Pord	7.50	1.067	Ea.	5.75	45.50		51.25	75
4400	Windows, including frame and trim, per side									
4600	Colonial type, 6/6 lites, 2' x 3', oil, primer, brushwork	1 Pord	14	.571	Ea.	.39	24.50		24.89	37
5800	Paint, 1 coat		14	.571		.66	24.50		25.16	37
6200	3' x 5' opening, 6/6 lites, primer coat, brushwork		12	.667		.99	28.50		29.49	43.50
6400	Paint, 1 coat		12	.667		1.64	28.50		30.14	44.50
6800	4' x 8' opening, 6/6 lites, primer coat, brushwork		8	1		2.11	42.50		44.61	66.50
7000	Paint, 1 coat		8	1		3.50	42.50		46	68
8000	Single lite type, 2' x 3', oil base, primer coat, brushwork		33	.242		.39	10.30		10.69	16
8200	Paint, 1 coat		33	.242		.66	10.30		10.96	16.25
8600	3' x 5' opening, primer coat, brushwork		20	.400		.99	17		17.99	26.50
8800	Paint, 1 coat		20	.400		1.64	17		18.64	27.50
9200	4' x 8' opening, primer coat, brushwork		14	.571		2.11	24.50		26.61	39
9400	Paint, 1 coat		14	.571		3.50	24.50		28	40.50

09 91 23.35 Doors and Windows, Interior Latex

		Crew	Daily Output	Labor-Hours	Unit	Material	2018 Bare Costs Labor	Equipment	Total	Total Incl O&P
0010	**DOORS & WINDOWS, INTERIOR LATEX** R099100-10									
0100	Doors, flush, both sides, incl. frame & trim									
0110	Roll & brush, primer R099100-20	1 Pord	10	.800	Ea.	4.08	34		38.08	55.50
0120	Finish coat, latex		10	.800		5.55	34		39.55	57
0130	Primer & 1 coat latex		7	1.143		9.65	48.50		58.15	83.50
0140	Primer & 2 coats latex		5	1.600		14.90	68		82.90	118
0160	Spray, both sides, primer		20	.400		4.30	17		21.30	30
0170	Finish coat, latex		20	.400		5.85	17		22.85	32
0180	Primer & 1 coat latex		11	.727		10.20	31		41.20	57.50
0190	Primer & 2 coats latex		8	1		15.75	42.50		58.25	81.50
0200	Doors, French, both sides, 10-15 lite, incl. frame & trim									
0210	Roll & brush, primer	1 Pord	6	1.333	Ea.	2.04	56.50		58.54	88
0220	Finish coat, latex		6	1.333		2.78	56.50		59.28	88.50
0230	Primer & 1 coat latex		3	2.667		4.82	113		117.82	176
0240	Primer & 2 coats latex		2	4		7.45	170		177.45	264
0260	Doors, louvered, both sides, incl. frame & trim									
0270	Roll & brush, primer	1 Pord	7	1.143	Ea.	4.08	48.50		52.58	77.50
0280	Finish coat, latex		7	1.143		5.55	48.50		54.05	79
0290	Primer & 1 coat, latex		4	2		9.40	85		94.40	138
0300	Primer & 2 coats, latex		3	2.667		15.20	113		128.20	188
0320	Spray, both sides, primer		20	.400		4.30	17		21.30	30
0330	Finish coat, latex		20	.400		5.85	17		22.85	32
0340	Primer & 1 coat, latex		11	.727		10.20	31		41.20	57.50
0350	Primer & 2 coats, latex		8	1		16.10	42.50		58.60	81.50
0360	Doors, panel, both sides, incl. frame & trim									
0370	Roll & brush, primer	1 Pord	6	1.333	Ea.	4.30	56.50		60.80	90
0380	Finish coat, latex		6	1.333		5.55	56.50		62.05	91.50
0390	Primer & 1 coat, latex		3	2.667		9.65	113		122.65	182
0400	Primer & 2 coats, latex		2.50	3.200		15.20	136		151.20	222
0420	Spray, both sides, primer		10	.800		4.30	34		38.30	55.50
0430	Finish coat, latex		10	.800		5.85	34		39.85	57.50
0440	Primer & 1 coat, latex		5	1.600		10.20	68		78.20	113
0450	Primer & 2 coats, latex		4	2		16.10	85		101.10	146
0460	Windows, per interior side, based on 15 S.F.									
0470	1 to 6 lite									
0480	Brushwork, primer	1 Pord	13	.615	Ea.	.81	26		26.81	40.50
0490	Finish coat, enamel		13	.615		1.10	26		27.10	40.50

09 91 Painting

09 91 23 – Interior Painting

09 91 23.35 Doors and Windows, Interior Latex

		Crew	Daily Output	Labor-Hours	Unit	Material	2018 Bare Costs Labor	Equipment	Total	Total Incl O&P
0500	Primer & 1 coat enamel	1 Pord	8	1	Ea.	1.90	42.50		44.40	66
0510	Primer & 2 coats enamel	↓	6	1.333	↓	3	56.50		59.50	89
0530	7 to 10 lite									
0540	Brushwork, primer	1 Pord	11	.727	Ea.	.81	31		31.81	47.50
0550	Finish coat, enamel		11	.727		1.10	31		32.10	47.50
0560	Primer & 1 coat enamel		7	1.143		1.90	48.50		50.40	75
0570	Primer & 2 coats enamel	↓	5	1.600	↓	3	68		71	105
0590	12 lite									
0600	Brushwork, primer	1 Pord	10	.800	Ea.	.81	34		34.81	52
0610	Finish coat, enamel		10	.800		1.10	34		35.10	52
0620	Primer & 1 coat enamel		6	1.333		1.90	56.50		58.40	87.50
0630	Primer & 2 coats enamel	↓	5	1.600		3	68		71	105
0650	For oil base paint, add				↓	10%				

09 91 23.39 Doors and Windows, Interior Latex, Zero Voc

			Crew	Daily Output	Labor-Hours	Unit	Material	2018 Bare Costs Labor	Equipment	Total	Total Incl O&P
0010	**DOORS & WINDOWS, INTERIOR LATEX, ZERO VOC**										
0100	Doors flush, both sides, incl. frame & trim										
0110	Roll & brush, primer	G	1 Pord	10	.800	Ea.	6.30	34		40.30	58
0120	Finish coat, latex	G		10	.800		6.95	34		40.95	58.50
0130	Primer & 1 coat latex	G		7	1.143		13.25	48.50		61.75	87.50
0140	Primer & 2 coats latex	G		5	1.600		19.80	68		87.80	124
0160	Spray, both sides, primer	G		20	.400		6.65	17		23.65	33
0170	Finish coat, latex	G		20	.400		7.30	17		24.30	33.50
0180	Primer & 1 coat latex	G		11	.727		14.05	31		45.05	62
0190	Primer & 2 coats latex	G	↓	8	1	↓	21	42.50		63.50	87
0200	Doors, French, both sides, 10-15 lite, incl. frame & trim										
0210	Roll & brush, primer	G	1 Pord	6	1.333	Ea.	3.15	56.50		59.65	89
0220	Finish coat, latex	G		6	1.333		3.48	56.50		59.98	89.50
0230	Primer & 1 coat latex	G		3	2.667		6.65	113		119.65	178
0240	Primer & 2 coats latex	G	↓	2	4	↓	9.90	170		179.90	267
0360	Doors, panel, both sides, incl. frame & trim										
0370	Roll & brush, primer	G	1 Pord	6	1.333	Ea.	6.65	56.50		63.15	93
0380	Finish coat, latex	G		6	1.333		6.95	56.50		63.45	93
0390	Primer & 1 coat, latex	G		3	2.667		13.25	113		126.25	186
0400	Primer & 2 coats, latex	G		2.50	3.200		20	136		156	227
0420	Spray, both sides, primer	G		10	.800		6.65	34		40.65	58.50
0430	Finish coat, latex	G		10	.800		7.30	34		41.30	59
0440	Primer & 1 coat, latex	G		5	1.600		14.05	68		82.05	117
0450	Primer & 2 coats, latex	G	↓	4	2	↓	21.50	85		106.50	152
0460	Windows, per interior side, based on 15 S.F.										
0470	1 to 6 lite										
0480	Brushwork, primer	G	1 Pord	13	.615	Ea.	1.24	26		27.24	41
0490	Finish coat, enamel	G		13	.615		1.37	26		27.37	41
0500	Primer & 1 coat enamel	G		8	1		2.62	42.50		45.12	67
0510	Primer & 2 coats enamel	G	↓	6	1.333	↓	3.99	56.50		60.49	90

09 91 23.40 Floors, Interior

			Crew	Daily Output	Labor-Hours	Unit	Material	2018 Bare Costs Labor	Equipment	Total	Total Incl O&P
0010	**FLOORS, INTERIOR**	R099100-10									
0100	Concrete paint, latex										
0110	Brushwork										
0120	1st coat		1 Pord	975	.008	S.F.	.16	.35		.51	.70
0130	2nd coat			1150	.007		.10	.30		.40	.56
0140	3rd coat		↓	1300	.006	↓	.08	.26		.34	.48
0150	Roll										

For customer support on your Building Construction Costs with RSMeans data, call 800.448.8182.

365

09 91 Painting

09 91 23 – Interior Painting

09 91 23.40 Floors, Interior

		Crew	Daily Output	Labor-Hours	Unit	Material	2018 Bare Costs Labor	2018 Bare Costs Equipment	Total	Total Incl O&P
0160	1st coat	1 Pord	2600	.003	S.F.	.21	.13		.34	.43
0170	2nd coat		3250	.002		.12	.10		.22	.30
0180	3rd coat		3900	.002		.09	.09		.18	.23
0190	Spray									
0200	1st coat	1 Pord	2600	.003	S.F.	.18	.13		.31	.40
0210	2nd coat		3250	.002		.10	.10		.20	.27
0220	3rd coat		3900	.002		.08	.09		.17	.22

09 91 23.44 Anti-Slip Floor Treatments

		Crew	Daily Output	Labor-Hours	Unit	Material	2018 Bare Costs Labor	2018 Bare Costs Equipment	Total	Total Incl O&P
0010	**ANTI-SLIP FLOOR TREATMENTS**									
1000	Walking surface treatment, ADA compliant, mop on and rinse									
1100	For tile, terrazzo, stone or smooth concrete	1 Pord	4000	.002	S.F.	.33	.09		.42	.50
1110	For marble		4000	.002		.24	.09		.33	.39
1120	For wood		4000	.002		.22	.09		.31	.38
1130	For baths and showers		500	.016		.38	.68		1.06	1.44
2000	Granular additive for paint or sealer, add to paint cost					.02			.02	.02

09 91 23.52 Miscellaneous, Interior

		Crew	Daily Output	Labor-Hours	Unit	Material	2018 Bare Costs Labor	2018 Bare Costs Equipment	Total	Total Incl O&P
0010	**MISCELLANEOUS, INTERIOR** R099100-10									
2400	Floors, conc./wood, oil base, primer/sealer coat, brushwork	2 Pord	1950	.008	S.F.	.09	.35		.44	.63
2450	Roller		5200	.003		.09	.13		.22	.30
2600	Spray		6000	.003		.09	.11		.20	.27
2650	Paint 1 coat, brushwork		1950	.008		.11	.35		.46	.65
2800	Roller		5200	.003		.11	.13		.24	.32
2850	Spray		6000	.003		.12	.11		.23	.30
3000	Stain, wood floor, brushwork, 1 coat		4550	.004		.10	.15		.25	.34
3200	Roller		5200	.003		.11	.13		.24	.32
3250	Spray		6000	.003		.11	.11		.22	.29
3400	Varnish, wood floor, brushwork		4550	.004		.09	.15		.24	.33
3450	Roller		5200	.003		.10	.13		.23	.31
3600	Spray		6000	.003		.10	.11		.21	.28
3650	For anti-skid, see Section 09 91 23.44									
3800	Grilles, per side, oil base, primer coat, brushwork	1 Pord	520	.015	S.F.	.13	.65		.78	1.13
3850	Spray		1140	.007		.14	.30		.44	.60
3920	Paint 2 coats, brushwork		325	.025		.43	1.05		1.48	2.05
3940	Spray		650	.012		.49	.52		1.01	1.33
4600	Miscellaneous surfaces, metallic paint, spray applied									
4610	Water based, non-tintable, warm silver	1 Pord	1140	.007	S.F.	.78	.30		1.08	1.31
4620	Rusted iron		1140	.007		1.01	.30		1.31	1.56
4630	Low VOC, tintable		1140	.007		.35	.30		.65	.83
5000	Pipe, 1"-4" diameter, primer or sealer coat, oil base, brushwork	2 Pord	1250	.013	L.F.	.09	.54		.63	.92
5100	Spray		2165	.007		.09	.31		.40	.57
5350	Paint 2 coats, brushwork		775	.021		.20	.88		1.08	1.54
5400	Spray		1240	.013		.23	.55		.78	1.08
6300	13"-16" diameter, primer or sealer coat, brushwork		310	.052		.37	2.20		2.57	3.72
6350	Spray		540	.030		.42	1.26		1.68	2.36
6500	Paint 2 coats, brushwork		195	.082		.81	3.49		4.30	6.15
6550	Spray		310	.052		.90	2.20		3.10	4.31
7000	Trim, wood, incl. puttying, under 6" wide									
7200	Primer coat, oil base, brushwork	1 Pord	650	.012	L.F.	.03	.52		.55	.83
7250	Paint, 1 coat, brushwork		650	.012		.05	.52		.57	.85
7450	3 coats		325	.025		.16	1.05		1.21	1.75
7500	Over 6" wide, primer coat, brushwork		650	.012		.07	.52		.59	.86
7550	Paint, 1 coat, brushwork		650	.012		.11	.52		.63	.91

For customer support on your Building Construction Costs with RSMeans data, call 800.448.8182.

09 91 23.52 Miscellaneous, Interior

		Crew	Daily Output	Labor-Hours	Unit	Material	2018 Bare Costs Labor	Equipment	Total	Total Incl O&P
7650	3 coats	1 Pord	325	.025	L.F.	.32	1.05		1.37	1.93
8000	Cornice, simple design, primer coat, oil base, brushwork		650	.012	S.F.	.07	.52		.59	.86
8250	Paint, 1 coat		650	.012		.11	.52		.63	.91
8350	Ornate design, primer coat		350	.023		.07	.97		1.04	1.53
8400	Paint, 1 coat		350	.023		.11	.97		1.08	1.58
8600	Balustrades, primer coat, oil base, brushwork		520	.015		.07	.65		.72	1.06
8650	Paint, 1 coat		520	.015		.11	.65		.76	1.11
8900	Trusses and wood frames, primer coat, oil base, brushwork		800	.010		.07	.43		.50	.71
8950	Spray		1200	.007		.07	.28		.35	.51
9220	Paint 2 coats, brushwork		500	.016		.21	.68		.89	1.25
9240	Spray		600	.013		.24	.57		.81	1.11
9260	Stain, brushwork, wipe off		600	.013		.10	.57		.67	.96
9280	Varnish, 3 coats, brushwork	▼	275	.029		.27	1.24		1.51	2.16
9350	For latex paint, deduct				▼	10%				

09 91 23.62 Electrostatic Painting

		Crew	Daily Output	Labor-Hours	Unit	Material	2018 Bare Costs Labor	Equipment	Total	Total Incl O&P
0010	**ELECTROSTATIC PAINTING**									
0100	In shop									
0200	Flat surfaces (lockers, casework, elevator doors, etc.)									
0300	One coat	1 Pord	200	.040	S.F.	.63	1.70		2.33	3.25
0400	Two coats	"	120	.067	"	.91	2.84		3.75	5.25
0500	Irregular surfaces (furniture, door frames, etc.)									
0600	One coat	1 Pord	150	.053	S.F.	.63	2.27		2.90	4.11
0700	Two coats	"	100	.080	"	.91	3.40		4.31	6.10
0800	On site									
0900	Flat surfaces (lockers, casework, elevator doors, etc.)									
1000	One coat	1 Pord	150	.053	S.F.	.63	2.27		2.90	4.11
1100	Two coats	"	100	.080	"	.91	3.40		4.31	6.10
1200	Irregular surfaces (furniture, door frames, etc.)									
1300	One coat	1 Pord	115	.070	S.F.	.63	2.96		3.59	5.15
1400	Two coats		70	.114		.91	4.86		5.77	8.30
2000	Anti-microbial coating, hospital application	▼	150	.053	▼	.05	2.27		2.32	3.48

09 91 23.72 Walls and Ceilings, Interior

		Crew	Daily Output	Labor-Hours	Unit	Material	2018 Bare Costs Labor	Equipment	Total	Total Incl O&P
0010	**WALLS AND CEILINGS, INTERIOR** R099100-10									
0100	Concrete, drywall or plaster, latex, primer or sealer coat R099100-20									
0200	Smooth finish, brushwork	1 Pord	1150	.007	S.F.	.06	.30		.36	.52
0240	Roller		1350	.006		.06	.25		.31	.45
0280	Spray		2750	.003		.05	.12		.17	.25
0300	Sand finish, brushwork		975	.008		.06	.35		.41	.60
0340	Roller		1150	.007		.06	.30		.36	.52
0380	Spray		2275	.004		.05	.15		.20	.29
0800	Paint 2 coats, smooth finish, brushwork		680	.012		.14	.50		.64	.91
0840	Roller		800	.010		.14	.43		.57	.80
0880	Spray		1625	.005		.13	.21		.34	.47
0900	Sand finish, brushwork		605	.013		.14	.56		.70	1.01
0940	Roller		1020	.008		.14	.33		.47	.66
0980	Spray		1700	.005		.13	.20		.33	.45
1200	Paint 3 coats, smooth finish, brushwork		510	.016		.22	.67		.89	1.24
1240	Roller		650	.012		.22	.52		.74	1.03
1280	Spray		850	.009		.20	.40		.60	.82
1600	Glaze coating, 2 coats, spray, clear		1200	.007		.56	.28		.84	1.05
1640	Multicolor	▼	1200	.007	▼	.85	.28		1.13	1.37
1660	Painting walls, complete, including surface prep, primer &									

For customer support on your Building Construction Costs with RSMeans data, call 800.448.8182.

367

09 91 23.72 Walls and Ceilings, Interior

		Crew	Daily Output	Labor-Hours	Unit	Material	2018 Bare Costs Labor	2018 Bare Costs Equipment	Total	Total Incl O&P
1670	2 coats finish, on drywall or plaster, with roller	1 Pord	325	.025	S.F.	.21	1.05		1.26	1.81
1700	For oil base paint, add					10%				
1800	For ceiling installations, add						25%			
2000	Masonry or concrete block, primer/sealer, latex paint									
2100	Primer, smooth finish, brushwork	1 Pord	1000	.008	S.F.	.15	.34		.49	.68
2110	Roller		1150	.007		.11	.30		.41	.57
2180	Spray		2400	.003		.10	.14		.24	.32
2200	Sand finish, brushwork		850	.009		.11	.40		.51	.72
2210	Roller		975	.008		.11	.35		.46	.65
2280	Spray		2050	.004		.10	.17		.27	.36
2400	Finish coat, smooth finish, brush		1100	.007		.09	.31		.40	.56
2410	Roller		1300	.006		.09	.26		.35	.48
2480	Spray		2400	.003		.08	.14		.22	.29
2500	Sand finish, brushwork		950	.008		.09	.36		.45	.63
2510	Roller		1090	.007		.09	.31		.40	.56
2580	Spray		2040	.004		.08	.17		.25	.33
2800	Primer plus one finish coat, smooth brush		525	.015		.30	.65		.95	1.31
2810	Roller		615	.013		.19	.55		.74	1.04
2880	Spray		1200	.007		.17	.28		.45	.62
2900	Sand finish, brushwork		450	.018		.19	.76		.95	1.35
2910	Roller		515	.016		.19	.66		.85	1.20
2980	Spray		1025	.008		.17	.33		.50	.69
3200	Primer plus 2 finish coats, smooth, brush		355	.023		.28	.96		1.24	1.75
3210	Roller		415	.019		.28	.82		1.10	1.54
3280	Spray		800	.010		.25	.43		.68	.91
3300	Sand finish, brushwork		305	.026		.28	1.12		1.40	1.99
3310	Roller		350	.023		.28	.97		1.25	1.77
3380	Spray		675	.012		.25	.50		.75	1.03
3600	Glaze coating, 3 coats, spray, clear		900	.009		.80	.38		1.18	1.45
3620	Multicolor		900	.009		1.05	.38		1.43	1.72
4000	Block filler, 1 coat, brushwork		425	.019		.13	.80		.93	1.35
4100	Silicone, water repellent, 2 coats, spray		2000	.004		.45	.17		.62	.75
4120	For oil base paint, add					10%				
8200	For work 8'-15' H, add						10%			
8300	For work over 15' H, add						20%			
8400	For light textured surfaces, add						10%			
8410	Heavy textured, add						25%			

09 91 23.74 Walls and Ceilings, Interior, Zero VOC Latex

			Crew	Daily Output	Labor-Hours	Unit	Material	2018 Bare Costs Labor	2018 Bare Costs Equipment	Total	Total Incl O&P
0010	**WALLS AND CEILINGS, INTERIOR, ZERO VOC LATEX**										
0100	Concrete, dry wall or plaster, latex, primer or sealer coat										
0200	Smooth finish, brushwork	G	1 Pord	1150	.007	S.F.	.08	.30		.38	.54
0240	Roller	G		1350	.006		.08	.25		.33	.47
0280	Spray	G		2750	.003		.06	.12		.18	.26
0300	Sand finish, brushwork	G		975	.008		.08	.35		.43	.62
0340	Roller	G		1150	.007		.09	.30		.39	.55
0380	Spray	G		2275	.004		.07	.15		.22	.30
0800	Paint 2 coats, smooth finish, brushwork	G		680	.012		.18	.50		.68	.95
0840	Roller	G		800	.010		.19	.43		.62	.84
0880	Spray	G		1625	.005		.16	.21		.37	.50
0900	Sand finish, brushwork	G		605	.013		.18	.56		.74	1.04
0940	Roller	G		1020	.008		.19	.33		.52	.70
0980	Spray	G		1700	.005		.16	.20		.36	.48

For customer support on your Building Construction Costs with RSMeans data, call 800.448.8182.

09 91 Painting

09 91 23 – Interior Painting

09 91 23.74 Walls and Ceilings, Interior, Zero VOC Latex

			Crew	Daily Output	Labor-Hours	Unit	Material	2018 Bare Costs Labor	2018 Bare Costs Equipment	Total	Total Incl O&P
1200	Paint 3 coats, smooth finish, brushwork	G	1 Pord	510	.016	S.F.	.26	.67		.93	1.29
1240	Roller	G		650	.012		.28	.52		.80	1.10
1280	Spray	G		850	.009		.24	.40		.64	.86
1800	For ceiling installations, add	G						25%			
8200	For work 8'- 15' H, add							10%			
8300	For work over 15' H, add							20%			

09 91 23.75 Dry Fall Painting

			Crew	Daily Output	Labor-Hours	Unit	Material	Labor	Equipment	Total	Total Incl O&P
0010	**DRY FALL PAINTING**	R099100-10									
0100	Sprayed on walls, gypsum board or plaster										
0220	One coat	R099100-20	1 Pord	2600	.003	S.F.	.09	.13		.22	.29
0250	Two coats			1560	.005		.17	.22		.39	.52
0280	Concrete or textured plaster, one coat			1560	.005		.09	.22		.31	.42
0310	Two coats			1300	.006		.17	.26		.43	.58
0340	Concrete block, one coat			1560	.005		.09	.22		.31	.42
0370	Two coats			1300	.006		.17	.26		.43	.58
0400	Wood, one coat			877	.009		.09	.39		.48	.67
0430	Two coats			650	.012		.17	.52		.69	.98
0440	On ceilings, gypsum board or plaster										
0470	One coat		1 Pord	1560	.005	S.F.	.09	.22		.31	.42
0500	Two coats			1300	.006		.17	.26		.43	.58
0530	Concrete or textured plaster, one coat			1560	.005		.09	.22		.31	.42
0560	Two coats			1300	.006		.17	.26		.43	.58
0570	Structural steel, bar joists or metal deck, one coat			1560	.005		.09	.22		.31	.42
0580	Two coats			1040	.008		.17	.33		.50	.68

09 93 Staining and Transparent Finishing

09 93 23 – Interior Staining and Finishing

09 93 23.10 Varnish

			Crew	Daily Output	Labor-Hours	Unit	Material	Labor	Equipment	Total	Total Incl O&P
0010	**VARNISH**										
0012	1 coat + sealer, on wood trim, brush, no sanding included		1 Pord	400	.020	S.F.	.07	.85		.92	1.36
0020	1 coat + sealer, on wood trim, brush, no sanding included, no VOC			400	.020		.22	.85		1.07	1.52
0100	Hardwood floors, 2 coats, no sanding included, roller			1890	.004		.15	.18		.33	.44

09 96 High-Performance Coatings

09 96 23 – Graffiti-Resistant Coatings

09 96 23.10 Graffiti-Resistant Treatments

			Crew	Daily Output	Labor-Hours	Unit	Material	Labor	Equipment	Total	Total Incl O&P
0010	**GRAFFITI-RESISTANT TREATMENTS**, sprayed on walls										
0100	Non-sacrificial, permanent non-stick coating, clear, on metals		1 Pord	2000	.004	S.F.	2.06	.17		2.23	2.53
0200	Concrete			2000	.004		2.35	.17		2.52	2.84
0300	Concrete block			2000	.004		3.03	.17		3.20	3.60
0400	Brick			2000	.004		3.44	.17		3.61	4.04
0500	Stone			2000	.004		3.44	.17		3.61	4.04
0600	Unpainted wood			2000	.004		3.97	.17		4.14	4.63
2000	Semi-permanent cross linking polymer primer, on metals			2000	.004		.65	.17		.82	.97
2100	Concrete			2000	.004		.78	.17		.95	1.12
2200	Concrete block			2000	.004		.97	.17		1.14	1.33
2300	Brick			2000	.004		.78	.17		.95	1.12
2400	Stone			2000	.004		.78	.17		.95	1.12
2500	Unpainted wood			2000	.004		1.08	.17		1.25	1.45

09 96 23 – Graffiti-Resistant Coatings

09 96 23.10 Graffiti-Resistant Treatments

		Crew	Daily Output	Labor-Hours	Unit	Material	2018 Bare Costs Labor	Equipment	Total	Total Incl O&P
3000	Top coat, on metals	1 Pord	2000	.004	S.F.	.55	.17		.72	.86
3100	Concrete		2000	.004		.62	.17		.79	.95
3200	Concrete block		2000	.004		.87	.17		1.04	1.22
3300	Brick		2000	.004		.73	.17		.90	1.06
3400	Stone		2000	.004		.73	.17		.90	1.06
3500	Unpainted wood		2000	.004		.87	.17		1.04	1.22
5000	Sacrificial, water based, on metal		2000	.004		.32	.17		.49	.62
5100	Concrete		2000	.004		.32	.17		.49	.62
5200	Concrete block		2000	.004		.32	.17		.49	.62
5300	Brick		2000	.004		.32	.17		.49	.62
5400	Stone		2000	.004		.32	.17		.49	.62
5500	Unpainted wood		2000	.004		.32	.17		.49	.62
8000	Cleaner for use after treatment									
8100	Towels or wipes, per package of 30				Ea.	.63			.63	.70
8200	Aerosol spray, 24 oz. can				"	18			18	19.80
8500	Graffiti removal with chemicals									
8510	Brush on, spray rinse off, on brick, masonry or stone	A-1H	200	.040	S.F.	.33	1.59	.37	2.29	3.20
8520	On smooth concrete	"	400	.020		.26	.80	.19	1.25	1.71
8530	Wipe on, wipe off, on plastic or painted metal	1 Clab	1500	.005		.63	.21		.84	1.02
8540	Brush on, wipe off, on painted wood	"	500	.016		.35	.64		.99	1.35

09 96 46 – Intumescent Painting

09 96 46.10 Coatings, Intumescent

		Crew	Daily Output	Labor-Hours	Unit	Material	2018 Bare Costs Labor	Equipment	Total	Total Incl O&P
0010	**COATINGS, INTUMESCENT**, spray applied									
0100	On exterior structural steel, 0.25" d.f.t.	1 Pord	475	.017	S.F.	.47	.72		1.19	1.59
0150	0.51" d.f.t.		350	.023		.47	.97		1.44	1.97
0200	0.98" d.f.t.		280	.029		.47	1.22		1.69	2.34
0300	On interior structural steel, 0.108" d.f.t.		300	.027		.46	1.13		1.59	2.21
0350	0.310" d.f.t.		150	.053		.46	2.27		2.73	3.92
0400	0.670" d.f.t.		100	.080		.46	3.40		3.86	5.60

09 96 53 – Elastomeric Coatings

09 96 53.10 Coatings, Elastomeric

		Crew	Daily Output	Labor-Hours	Unit	Material	2018 Bare Costs Labor	Equipment	Total	Total Incl O&P
0010	**COATINGS, ELASTOMERIC**									
0020	High build, water proof, one coat system									
0100	Concrete, brush	1 Pord	650	.012	S.F.	.29	.52		.81	1.11
0110	Roll		1650	.005		.29	.21		.50	.63
0120	Spray		2600	.003		.29	.13		.42	.52
0200	Concrete block, brush		600	.013		.34	.57		.91	1.23
0210	Roll		1400	.006		.34	.24		.58	.75
0220	Spray		1900	.004		.34	.18		.52	.65
0300	Stucco, brush		400	.020		.47	.85		1.32	1.80
0310	Roll		1000	.008		.47	.34		.81	1.03
0320	Spray		1500	.005		.47	.23		.70	.86

09 96 56 – Epoxy Coatings

09 96 56.20 Wall Coatings

		Crew	Daily Output	Labor-Hours	Unit	Material	2018 Bare Costs Labor	Equipment	Total	Total Incl O&P
0010	**WALL COATINGS**									
0100	Acrylic glazed coatings, matte	1 Pord	525	.015	S.F.	.35	.65		1	1.37
0200	Gloss		305	.026		.73	1.12		1.85	2.48
0300	Epoxy coatings, solvent based		525	.015		.45	.65		1.10	1.48
0400	Water based		170	.047		.32	2		2.32	3.36
0600	Exposed aggregate, troweled on, 1/16" to 1/4", solvent based		235	.034		.70	1.45		2.15	2.95
0700	Water based (epoxy or polyacrylate)		130	.062		1.50	2.62		4.12	5.60

For customer support on your Building Construction Costs with RSMeans data, call 800.448.8182.

09 96 High-Performance Coatings

09 96 56 – Epoxy Coatings

09 96 56.20 Wall Coatings	Crew	Daily Output	Labor-Hours	Unit	Material	2018 Bare Costs Labor	Equipment	Total	Total Incl O&P
0900	1/2" to 5/8" aggregate, solvent based	1 Pord	130	.062	S.F.	1.35	2.62		3.97	5.45
1000	Water based		80	.100		2.35	4.26		6.61	9
1500	Exposed aggregate, sprayed on, 1/8" aggregate, solvent based		295	.027		.57	1.15		1.72	2.37
1600	Water based		145	.055		1.18	2.35		3.53	4.83
1800	High build epoxy, 50 mil, solvent based		390	.021		.76	.87		1.63	2.15
1900	Water based		95	.084		1.29	3.58		4.87	6.80
2100	Laminated epoxy with fiberglass, solvent based		295	.027		.82	1.15		1.97	2.64
2200	Water based		145	.055		1.49	2.35		3.84	5.15
2400	Sprayed perlite or vermiculite, 1/16" thick, solvent based		2935	.003		.28	.12		.40	.48
2500	Water based		640	.013		.83	.53		1.36	1.71
2700	Vinyl plastic wall coating, solvent based		735	.011		.38	.46		.84	1.12
2800	Water based		240	.033		.93	1.42		2.35	3.15
3000	Urethane on smooth surface, 2 coats, solvent based		1135	.007		.34	.30		.64	.82
3100	Water based		665	.012		.60	.51		1.11	1.43
3600	Ceramic-like glazed coating, cementitious, solvent based		440	.018		.48	.77		1.25	1.69
3700	Water based		345	.023		.92	.99		1.91	2.50
3900	Resin base, solvent based		640	.013		.33	.53		.86	1.16
4000	Water based		330	.024		.58	1.03		1.61	2.19

09 97 Special Coatings

09 97 13 – Steel Coatings

09 97 13.23 Exterior Steel Coatings

0010	EXTERIOR STEEL COATINGS	R050516-30								
6100	Cold galvanizing, brush in field		1 Psst	1100	.007	S.F.	.21	.32		.53	.76
6510	Paints & protective coatings, sprayed in field										
6520	Alkyds, primer		2 Psst	3600	.004	S.F.	.09	.19		.28	.42
6540	Gloss topcoats			3200	.005		.08	.22		.30	.45
6560	Silicone alkyd			3200	.005		.17	.22		.39	.55
6610	Epoxy, primer			3000	.005		.26	.23		.49	.67
6630	Intermediate or topcoat			2800	.006		.28	.25		.53	.72
6650	Enamel coat			2800	.006		.35	.25		.60	.79
6700	Epoxy ester, primer			2800	.006		.49	.25		.74	.95
6720	Topcoats			2800	.006		.23	.25		.48	.67
6810	Latex primer			3600	.004		.06	.19		.25	.39
6830	Topcoats			3200	.005		.08	.22		.30	.44
6910	Universal primers, one part, phenolic, modified alkyd			2000	.008		.38	.35		.73	1
6940	Two part, epoxy spray			2000	.008		.40	.35		.75	1.02
7000	Zinc rich primers, self cure, spray, inorganic			1800	.009		.88	.39		1.27	1.61
7010	Epoxy, spray, organic			1800	.009		.28	.39		.67	.95
7020	Above one story, spray painting simple structures, add							25%			
7030	Intricate structures, add							50%			

09 97 35 – Dry Erase Coatings

09 97 35.10 Dry Erase Coatings

0010	DRY ERASE COATINGS									
0020	Dry erase coatings, clear, roller applied	1 Pord	1325	.006	S.F.	2.08	.26		2.34	2.68

For customer support on your Building Construction Costs with RSMeans data, call 800.448.8182.

371

Division Notes

	CREW	DAILY OUTPUT	LABOR-HOURS	UNIT	BARE COSTS				TOTAL INCL O&P
					MAT.	LABOR	EQUIP.	TOTAL	

Estimating Tips
General

- The items in this division are usually priced per square foot or each.

- Many items in Division 10 require some type of support system or special anchors that are not usually furnished with the item. The required anchors must be added to the estimate in the appropriate division.

- Some items in Division 10, such as lockers, may require assembly before installation. Verify the amount of assembly required. Assembly can often exceed installation time.

10 20 00 Interior Specialties

- Support angles and blocking are not included in the installation of toilet compartments, shower/dressing compartments, or cubicles. Appropriate line items from Division 5 or 6 may need to be added to support the installations.

- Toilet partitions are priced by the stall. A stall consists of a side wall, pilaster, and door with hardware. Toilet tissue holders and grab bars are extra.

- The required acoustical rating of a folding partition can have a significant impact on costs. Verify the sound transmission coefficient rating of the panel priced against the specification requirements.

- Grab bar installation does not include supplemental blocking or backing to support the required load. When grab bars are installed at an existing facility, provisions must be made to attach the grab bars to a solid structure.

Reference Numbers

Reference numbers are shown at the beginning of some major classifications. These numbers refer to related items in the Reference Section. The reference information may be an estimating procedure, an alternate pricing method, or technical information.

Note: Not all subdivisions listed here necessarily appear. ■

Did you know?

RSMeans data is available through our online application with 24/7 access:

- Search for unit prices by keyword
- Leverage the most up-to-date data
- Build and export estimates

Try it free for 30 days!
www.rsmeans.com/2018freetrial

No part of this cost data may be reproduced, stored in a retrieval system, or transmitted in any form or by any means without prior written permission of Gordian.

10 05 05.10 Selective Demolition, Specialties

	10 05 05.10 Selective Demolition, Specialties	Crew	Daily Output	Labor-Hours	Unit	Material	2018 Bare Costs Labor	Equipment	Total	Total Incl O&P
0010	**SELECTIVE DEMOLITION, SPECIALTIES**									
1100	Boards and panels, wall mounted	2 Clab	15	1.067	Ea.		42.50		42.50	65
1105	Demolition, mirror, wall mounted	1 Clab	30	.267			10.65		10.65	16.20
1200	Cases, for directory and/or bulletin boards, including doors	2 Clab	24	.667			26.50		26.50	40.50
1850	Shower partitions, cabinet or stall, including base and door	"	8	2			79.50		79.50	121
1855	Shower receptor, terrazzo or concrete	1 Clab	14	.571			23		23	34.50
1900	Curtain track or rod, hospital type, ceiling mounted or suspended	"	220	.036	L.F.		1.45		1.45	2.21
1910	Toilet cubicles, remove	2 Clab	8	2	Ea.		79.50		79.50	121
1930	Urinal screen, remove	1 Clab	12	.667	"		26.50		26.50	40.50
2650	Wall guard, misc. wall or corner protection	"	320	.025	L.F.		1		1	1.52
2750	Access floor, metal panel system, including pedestals, covering	2 Clab	850	.019	S.F.		.75		.75	1.14
3050	Fireplace, prefab, freestanding or wall hung, including hood and screen	1 Clab	2	4	Ea.		159		159	243
3054	Chimney top, simulated brick, 4' high	"	15	.533			21.50		21.50	32.50
3200	Stove, woodburning, cast iron	2 Clab	2	8			320		320	485
3440	Weathervane, residential	1 Clab	12	.667			26.50		26.50	40.50
3500	Flagpole, groundset, to 70' high, excluding base/foundation	K-1	1	16			760	238	998	1,400
3555	To 30' high	"	2.50	6.400			305	95	400	565
4300	Letter, signs or plaques, exterior on wall	1 Clab	20	.400			15.95		15.95	24.50
4310	Signs, street, reflective aluminum, including post and bracket		60	.133			5.30		5.30	8.10
4320	Door signs interior on door 6" x 6", selective demolition		20	.400			15.95		15.95	24.50
4550	Turnstiles, manual or electric	2 Clab	2	8			320		320	485
5050	Lockers	1 Clab	15	.533	Opng.		21.50		21.50	32.50
5250	Cabinets, recessed	Q-12	12	1.333	Ea.		72		72	109
5260	Mail boxes, horiz., key lock, front loading, remove	1 Carp	34	.235	"		11.95		11.95	18.15
5350	Awning, fabric, including frame	2 Clab	100	.160	S.F.		6.40		6.40	9.70
6050	Partition, woven wire		1400	.011			.46		.46	.69
6100	Folding gate, security, door or window		500	.032			1.28		1.28	1.94
6580	Acoustic air wall		650	.025			.98		.98	1.49
7550	Telephone enclosure, exterior, post mounted		3	5.333	Ea.		213		213	325
8850	Scale, platform, excludes foundation or pit		.25	64	"		2,550		2,550	3,875

10 11 Visual Display Units
10 11 13 – Chalkboards

10 11 13.13 Fixed Chalkboards

	10 11 13.13 Fixed Chalkboards	Crew	Daily Output	Labor-Hours	Unit	Material	2018 Bare Costs Labor	Equipment	Total	Total Incl O&P
0010	**FIXED CHALKBOARDS** Porcelain enamel steel									
3900	Wall hung									
4000	Aluminum frame and chalktrough									
4200	3' x 4'	2 Carp	16	1	Ea.	241	50.50		291.50	340
4300	3' x 5'		15	1.067		310	54		364	425
4500	4' x 8'		14	1.143		405	58		463	535
4600	4' x 12'		13	1.231		560	62.50		622.50	710
4700	Wood frame and chalktrough									
4800	3' x 4'	2 Carp	16	1	Ea.	212	50.50		262.50	310
5000	3' x 5'		15	1.067		265	54		319	375
5100	4' x 5'		14	1.143		277	58		335	395
5300	4' x 8'		13	1.231		365	62.50		427.50	495
5400	Liquid chalk, white porcelain enamel, wall hung									
5420	Deluxe units, aluminum trim and chalktrough									
5450	4' x 4'	2 Carp	16	1	Ea.	300	50.50		350.50	405
5500	4' x 8'		14	1.143		450	58		508	585
5550	4' x 12'		12	1.333		620	67.50		687.50	785

For customer support on your Building Construction Costs with RSMeans data, call 800.448.8182.

10 11 Visual Display Units

10 11 13 – Chalkboards

10 11 13.13 Fixed Chalkboards

		Crew	Daily Output	Labor-Hours	Unit	Material	2018 Bare Costs Labor	2018 Bare Costs Equipment	Total	Total Incl O&P
5700	Wood trim and chalktrough									
5900	4' x 4'	2 Carp	16	1	Ea.	750	50.50		800.50	900
6000	4' x 6'		15	1.067		850	54		904	1,025
6200	4' x 8'	↓	14	1.143		1,025	58		1,083	1,225
6300	Liquid chalk, felt tip markers					2.18			2.18	2.40
6500	Erasers					2.42			2.42	2.66
6600	Board cleaner, 8 oz. bottle				↓	6			6	6.60

10 11 13.23 Modular-Support-Mounted Chalkboards

		Crew	Daily Output	Labor-Hours	Unit	Material	2018 Bare Costs Labor	2018 Bare Costs Equipment	Total	Total Incl O&P
0010	**MODULAR-SUPPORT-MOUNTED CHALKBOARDS**									
0400	Sliding chalkboards									
0450	Vertical, one sliding board with back panel, wall mounted									
0500	8' x 4'	2 Carp	8	2	Ea.	3,000	101		3,101	3,450
0520	8' x 8'	↓	7.50	2.133		3,725	108		3,833	4,275
0540	8' x 12'	↓	7	2.286	↓	4,575	116		4,691	5,200
0600	Two sliding boards, with back panel									
0620	8' x 4'	2 Carp	8	2	Ea.	4,425	101		4,526	5,025
0640	8' x 8'		7.50	2.133		5,525	108		5,633	6,250
0660	8' x 12'	↓	7	2.286	↓	8,925	116		9,041	9,975
0700	Horizontal, two track									
0800	4' x 8', 2 sliding panels	2 Carp	8	2	Ea.	2,050	101		2,151	2,400
0820	4' x 12', 2 sliding panels	↓	7.50	2.133		2,675	108		2,783	3,100
0840	4' x 16', 4 sliding panels	↓	7	2.286	↓	3,600	116		3,716	4,125
0900	Four track, four sliding panels									
0920	4' x 8'	2 Carp	8	2	Ea.	3,300	101		3,401	3,775
0940	4' x 12'	↓	7.50	2.133		4,300	108		4,408	4,925
0960	4' x 16'	↓	7	2.286	↓	5,600	116		5,716	6,350
1200	Vertical, motor operated									
1400	One sliding panel with back panel									
1450	10' x 4'	2 Carp	4	4	Ea.	5,350	203		5,553	6,175
1500	10' x 10'	↓	3.75	4.267	↓	6,450	216		6,666	7,400
1550	10' x 16'	↓	3.50	4.571	↓	7,575	232		7,807	8,700
1700	Two sliding panels with back panel									
1750	10' x 4'	2 Carp	4	4	Ea.	9,600	203		9,803	10,900
1800	10' x 10'	↓	3.75	4.267		10,800	216		11,016	12,200
1850	10' x 16'	↓	3.50	4.571	↓	12,600	232		12,832	14,300
2000	Three sliding panels with back panel									
2100	10' x 4'	2 Carp	4	4	Ea.	15,700	203		15,903	17,600
2150	10' x 10'	↓	3.75	4.267		16,900	216		17,116	18,900
2200	10' x 16'	↓	3.50	4.571	↓	19,200	232		19,432	21,500
2400	For projection screen, glass beaded, add				S.F.	4.77			4.77	5.25
2500	For remote control, 1 panel control, add				Ea.	360			360	395
2600	2 panel control, add				"	620			620	680
2800	For units without back panels, deduct				S.F.	5.05			5.05	5.55
2850	For liquid chalk porcelain panels, add				"	5.15			5.15	5.70
3000	Swing leaf, any comb. of chalkboard & cork, aluminum frame									
3100	Floor style, 6 panels									
3150	30" x 40" panels				Ea.	1,550			1,550	1,700
3200	48" x 40" panels				"	2,600			2,600	2,850
3300	Wall mounted, 6 panels									
3400	30" x 40" panels	2 Carp	16	1	Ea.	1,450	50.50		1,500.50	1,675
3450	48" x 40" panels	"	16	1	"	1,800	50.50		1,850.50	2,075
3600	Extra panels for swing leaf units									

10 11 Visual Display Units

10 11 13 – Chalkboards

10 11 13.23 Modular-Support-Mounted Chalkboards

		Crew	Daily Output	Labor-Hours	Unit	Material	2018 Bare Costs Labor	Equipment	Total	Total Incl O&P
3700	30" x 40" panels				Ea.	294			294	325
3750	48" x 40" panels				"	365			365	400

10 11 13.43 Portable Chalkboards

		Crew	Daily Output	Labor-Hours	Unit	Material	2018 Bare Costs Labor	Equipment	Total	Total Incl O&P
0010	**PORTABLE CHALKBOARDS**									
0100	Freestanding, reversible									
0120	Economy, wood frame, 4' x 6'									
0140	Chalkboard both sides				Ea.	710			710	780
0160	Chalkboard one side, cork other side				"	615			615	680
0200	Standard, lightweight satin finished aluminum, 4' x 6'									
0220	Chalkboard both sides				Ea.	700			700	770
0240	Chalkboard one side, cork other side				"	620			620	685
0300	Deluxe, heavy duty extruded aluminum, 4' x 6'									
0320	Chalkboard both sides				Ea.	1,150			1,150	1,250
0340	Chalkboard one side, cork other side				"	1,150			1,150	1,250

10 11 16 – Markerboards

10 11 16.53 Electronic Markerboards

		Crew	Daily Output	Labor-Hours	Unit	Material	2018 Bare Costs Labor	Equipment	Total	Total Incl O&P
0010	**ELECTRONIC MARKERBOARDS**									
0100	Wall hung or free standing, 3' x 4' to 4' x 6'	2 Carp	8	2	S.F.	85.50	101		186.50	248
0150	5' x 6' to 4' x 8'		8	2	"	61	101		162	221
0500	Interactive projection module for existing whiteboards		8	2	Ea.	1,300	101		1,401	1,575

10 11 23 – Tackboards

10 11 23.10 Fixed Tackboards

		Crew	Daily Output	Labor-Hours	Unit	Material	2018 Bare Costs Labor	Equipment	Total	Total Incl O&P
0010	**FIXED TACKBOARDS**									
0020	Cork sheets, unbacked, no frame, 1/4" thick	2 Carp	290	.055	S.F.	1.58	2.80		4.38	6
0100	1/2" thick		290	.055		4.34	2.80		7.14	9.05
0300	Fabric-face, no frame, on 7/32" cork underlay		290	.055		7.25	2.80		10.05	12.20
0400	On 1/4" cork on 1/4" hardboard		290	.055		9.05	2.80		11.85	14.20
0600	With edges wrapped		290	.055		11	2.80		13.80	16.35
0700	On 7/16" fire retardant core		290	.055		6.40	2.80		9.20	11.30
0900	With edges wrapped		290	.055		8.50	2.80		11.30	13.60
1000	Designer fabric only, cut to size					2.88			2.88	3.17
1200	1/4" vinyl cork, on 1/4" hardboard, no frame	2 Carp	290	.055		9.10	2.80		11.90	14.25
1300	On 1/4" coreboard		290	.055		5.05	2.80		7.85	9.80
2000	For map and display rail, economy, add		385	.042	L.F.	3.21	2.11		5.32	6.75
2100	Deluxe, add		350	.046	"	4.94	2.32		7.26	9
2120	Prefabricated, 1/4" cork, 3' x 5' with aluminum frame		16	1	Ea.	140	50.50		190.50	231
2140	Wood frame		16	1		164	50.50		214.50	258
2160	4' x 4' with aluminum frame		16	1		161	50.50		211.50	254
2180	Wood frame		16	1		213	50.50		263.50	310
2200	4' x 8' with aluminum frame		14	1.143		300	58		358	420
2210	Wood frame		14	1.143		290	58		348	410
2220	4' x 12' with aluminum frame		12	1.333		365	67.50		432.50	505
2230	Bulletin board case, single glass door, with lock									
2240	36" x 24", economy	2 Carp	12	1.333	Ea.	355	67.50		422.50	495
2250	Deluxe		12	1.333		390	67.50		457.50	530
2260	42" x 30", economy		12	1.333		415	67.50		482.50	560
2270	Deluxe		12	1.333		635	67.50		702.50	800
2300	Glass enclosed cabinets, alum., cork panel, hinged doors									
2400	3' x 3', 1 door	2 Carp	12	1.333	Ea.	635	67.50		702.50	805
2500	4' x 4', 2 door		11	1.455		1,050	74		1,124	1,275
2600	4' x 7', 3 door		10	1.600		1,875	81		1,956	2,175

For customer support on your Building Construction Costs with RSMeans data, call 800.448.8182.

10 11 Visual Display Units

10 11 23 – Tackboards

10 11 23.10 Fixed Tackboards

		Crew	Daily Output	Labor-Hours	Unit	Material	2018 Bare Costs Labor	2018 Bare Costs Equipment	Total	Total Incl O&P
2800	4' x 10', 4 door	2 Carp	8	2	Ea.	2,475	101		2,576	2,875
2900	For lights, add per door opening	1 Elec	13	.615		173	36		209	245
3100	Horizontal sliding units, 4 doors, 4' x 8', 8' x 4'	2 Carp	9	1.778		2,050	90		2,140	2,375
3200	4' x 12'		7	2.286		2,675	116		2,791	3,100
3400	8 doors, 4' x 16'		5	3.200		3,625	162		3,787	4,225
3500	4' x 24'	▼	4	4	▼	4,875	203		5,078	5,675

10 11 23.20 Control Boards

		Crew	Daily Output	Labor-Hours	Unit	Material	2018 Bare Costs Labor	2018 Bare Costs Equipment	Total	Total Incl O&P
0010	**CONTROL BOARDS**									
0020	Magnetic, porcelain finish, 18" x 24", framed	2 Carp	8	2	Ea.	199	101		300	375
0100	24" x 36"		7.50	2.133		275	108		383	465
0200	36" x 48"		7	2.286		335	116		451	545
0300	48" x 72"		6	2.667		655	135		790	930
0400	48" x 96"	▼	5	3.200	▼	1,075	162		1,237	1,425
1000	Hospital patient display board, 4-color custom design									
1010	Porcelain steel dry erase board, 36" x 24"	2 Carp	7.50	2.133	Ea.	235	108		343	425

10 13 Directories

10 13 10 – Building Directories

10 13 10.10 Directory Boards

		Crew	Daily Output	Labor-Hours	Unit	Material	2018 Bare Costs Labor	2018 Bare Costs Equipment	Total	Total Incl O&P
0010	**DIRECTORY BOARDS**									
0050	Plastic, glass covered, 30" x 20"	2 Carp	3	5.333	Ea.	207	270		477	640
0100	36" x 48"		2	8		915	405		1,320	1,625
0300	Grooved cork, 30" x 20"		3	5.333		425	270		695	880
0400	36" x 48"		2	8		585	405		990	1,250
0600	Black felt, 30" x 20"		3	5.333		239	270		509	675
0700	36" x 48"		2	8		465	405		870	1,125
0900	Outdoor, weatherproof, black plastic, 36" x 24"		2	8		700	405		1,105	1,400
1000	36" x 36"		1.50	10.667		810	540		1,350	1,725
1800	Indoor, economy, open face, 18" x 24"		7	2.286		173	116		289	365
1900	24" x 36"		7	2.286		161	116		277	355
2000	36" x 24"		6	2.667		161	135		296	385
2100	36" x 48"		6	2.667		263	135		398	495
2400	Building directory, alum., black felt panels, 1 door, 24" x 18"		4	4		289	203		492	630
2500	36" x 24"		3.50	4.571		395	232		627	790
2600	48" x 32"		3	5.333		560	270		830	1,025
2700	2 door, 36" x 48"		2.50	6.400		675	325		1,000	1,225
2800	36" x 60"		2	8		790	405		1,195	1,475
2900	48" x 60"	▼	1	16		890	810		1,700	2,200
3100	For bronze enamel finish, add					15%				
3200	For bronze anodized finish, add					25%				
3400	For illuminated directory, single door unit, add				▼	145			145	159
3500	For 6" header panel, 6 letters per foot, add				L.F.	19.70			19.70	21.50
6050	Building directory, electronic display, alum. frame, wall mounted	2 Carp	32	.500	Ea.	4,175	25.50		4,200.50	4,625
6100	Free standing	"	60	.267	"	3,675	13.50		3,688.50	4,075

For customer support on your Building Construction Costs with RSMeans data, call 800.448.8182.

377

10 14 Signage

10 14 19 – Dimensional Letter Signage

10 14 19.10 Exterior Signs

		Crew	Daily Output	Labor-Hours	Unit	Material	2018 Bare Costs Labor	2018 Bare Costs Equipment	Total	Total Incl O&P
0010	**EXTERIOR SIGNS**									
0020	Letters, 2" high, 3/8" deep, cast bronze	1 Carp	24	.333	Ea.	27	16.90		43.90	55.50
0140	1/2" deep, cast aluminum		18	.444		27	22.50		49.50	64.50
0160	Cast bronze		32	.250		35	12.70		47.70	58
0300	6" high, 5/8" deep, cast aluminum		24	.333		30	16.90		46.90	58.50
0400	Cast bronze		24	.333		60	16.90		76.90	91.50
0600	8" high, 3/4" deep, cast aluminum		14	.571		32.50	29		61.50	79.50
0700	Cast bronze		20	.400		91	20.50		111.50	131
0900	10" high, 1" deep, cast aluminum		18	.444		60	22.50		82.50	101
1000	Cast bronze		18	.444		114	22.50		136.50	161
1200	12" high, 1-1/4" deep, cast aluminum		12	.667		59.50	34		93.50	117
1500	Cast bronze		18	.444		167	22.50		189.50	219
1600	14" high, 2-5/16" deep, cast aluminum		12	.667		101	34		135	163
1800	Fabricated stainless steel, 6" high, 2" deep		20	.400		47.50	20.50		68	83
1900	12" high, 3" deep		18	.444		72	22.50		94.50	114
2100	18" high, 3" deep		12	.667		97.50	34		131.50	159
2200	24" high, 4" deep		10	.800		193	40.50		233.50	274
2700	Acrylic, on high density foam, 12" high, 2" deep		20	.400		18.90	20.50		39.40	52
2800	18" high, 2" deep		18	.444		38	22.50		60.50	76
3900	Plaques, custom, 20" x 30", for up to 450 letters, cast aluminum	2 Carp	4	4		1,875	203		2,078	2,350
4000	Cast bronze		4	4		1,900	203		2,103	2,375
4200	30" x 36", up to 900 letters, cast aluminum		3	5.333		2,775	270		3,045	3,450
4300	Cast bronze		3	5.333		4,325	270		4,595	5,175
4500	36" x 48", for up to 1300 letters, cast bronze		2	8		4,875	405		5,280	6,000
4800	Signs, reflective alum. directional signs, dbl. face, 2-way, w/bracket		30	.533		108	27		135	160
4900	4-way		30	.533		189	27		216	248
5100	Exit signs, 24 ga. alum., 14" x 12" surface mounted	1 Carp	30	.267		51.50	13.50		65	77.50
5200	10" x 7"		20	.400		30.50	20.50		51	64.50
5400	Bracket mounted, double face, 12" x 10"		30	.267		54	13.50		67.50	80
5500	Sticky back, stock decals, 14" x 10"	1 Clab	50	.160		26	6.40		32.40	38
6400	Replacement sign faces, 6" or 8"	"	50	.160		66	6.40		72.40	82
8000	Internally illuminated, custom									
8100	On pedestal, 84" x 30" x 12"	L-7	.50	56	Ea.	11,500	2,725		14,225	16,900

10 14 23 – Panel Signage

10 14 23.13 Engraved Panel Signage

		Crew	Daily Output	Labor-Hours	Unit	Material	2018 Bare Costs Labor	2018 Bare Costs Equipment	Total	Total Incl O&P
0010	**ENGRAVED PANEL SIGNAGE**, interior									
1010	Flexible door sign, adhesive back, w/Braille, 5/8" letters, 4" x 4"	1 Clab	32	.250	Ea.	34	9.95		43.95	52
1050	6" x 6"		32	.250		49.50	9.95		59.45	69.50
1100	8" x 2"		32	.250		35	9.95		44.95	53.50
1150	8" x 4"		32	.250		44	9.95		53.95	63.50
1200	8" x 8"		32	.250		69.50	9.95		79.45	91.50
1250	12" x 2"		32	.250		35	9.95		44.95	53.50
1300	12" x 6"		32	.250		41	9.95		50.95	60
1350	12" x 12"		32	.250		155	9.95		164.95	185
1500	Graphic symbols, 2" x 2"		32	.250		12	9.95		21.95	28.50
1550	6" x 6"		32	.250		31	9.95		40.95	49
1600	8" x 8"		32	.250		39	9.95		48.95	57.50
2010	Corridor, stock acrylic, 2-sided, with mounting bracket, 2" x 8"	1 Carp	24	.333		29.50	16.90		46.40	58
2020	2" x 10"		24	.333		35	16.90		51.90	64
2050	3" x 8"		24	.333		27	16.90		43.90	55
2060	3" x 10"		24	.333		36.50	16.90		53.40	65.50
2070	3" x 12"		24	.333		34.50	16.90		51.40	63.50

For customer support on your Building Construction Costs with RSMeans data, call 800.448.8182.

10 14 Signage

10 14 23 – Panel Signage

10 14 23.13 Engraved Panel Signage

		Crew	Daily Output	Labor-Hours	Unit	Material	2018 Bare Costs Labor	2018 Bare Costs Equipment	Total	Total Incl O&P
2100	4" x 8"	1 Carp	24	.333	Ea.	22	16.90		38.90	49.50
2110	4" x 10"		24	.333		40	16.90		56.90	69.50
2120	4" x 12"		24	.333		50	16.90		66.90	80.50
7000	Wayfinding signage, custom									
7010	Plastic, flexible, 1/8" thick, incl. mounting									
7020	6" x 6"	1 Carp	32	.250	Ea.	16.90	12.70		29.60	38
7030	6" x 12"		32	.250		36.50	12.70		49.20	59.50
7040	6" x 24"		28	.286		60	14.50		74.50	88
7050	8" x 8"		32	.250		27	12.70		39.70	49
7060	8" x 16"		28	.286		59.50	14.50		74	87
7070	8" x 24"		26	.308		90.50	15.60		106.10	124
7080	12" x 12"		28	.286		71	14.50		85.50	101
7090	12" x 24"		24	.333		131	16.90		147.90	170
7100	12" x 36"		20	.400		199	20.50		219.50	250
7210	Weather resistant, engraved and color filled									
7220	8" x 8"	1 Carp	32	.250	Ea.	39	12.70		51.70	62
7230	8" x 16"		28	.286		61	14.50		75.50	89
7240	8" x 24"		26	.308		90.50	15.60		106.10	124
7250	12" x 12"		28	.286		69	14.50		83.50	98
7260	12" x 24"		24	.333		122	16.90		138.90	161
7270	12" x 36"		20	.400		202	20.50		222.50	254
7280	16" x 32"	2 Carp	60	.267		223	13.50		236.50	266
7290	16" x 48"	"	60	.267		350	13.50		363.50	405
7320	For engraved letters, 1/2" high, add					.70			.70	.77
7330	1" high, add					.83			.83	.91
7340	2" high, add					1.65			1.65	1.82
7350	3" high, add					2.25			2.25	2.48
7360	For engraved graphic symbols, 3" high, add					14.30			14.30	15.75
7370	7" high, add					17.70			17.70	19.45
7380	10" high, add					24			24	26.50
9990	Replace interior labels	1 Carp	12	.667		26	34		60	80.50
9991	Replace interior plaques	"	12	.667		84.50	34		118.50	145

10 14 26 – Post and Panel/Pylon Signage

10 14 26.10 Post and Panel Signage

		Crew	Daily Output	Labor-Hours	Unit	Material	2018 Bare Costs Labor	2018 Bare Costs Equipment	Total	Total Incl O&P
0010	**POST AND PANEL SIGNAGE**, Alum., incl. 2 posts,									
0011	2-sided panel (blank), concrete footings per post									
0025	24" W x 18" H	B-1	25	.960	Ea.	905	39		944	1,050
0050	36" W x 24" H		25	.960		970	39		1,009	1,125
0100	57" W x 36" H		25	.960		1,825	39		1,864	2,050
0150	81" W x 46" H		25	.960		2,225	39		2,264	2,500

10 14 43 – Photoluminescent Signage

10 14 43.10 Photoluminescent Signage

		Crew	Daily Output	Labor-Hours	Unit	Material	2018 Bare Costs Labor	2018 Bare Costs Equipment	Total	Total Incl O&P
0010	**PHOTOLUMINESCENT SIGNAGE**									
0011	Photoluminescent exit sign, plastic, 7"x10"	1 Carp	32	.250	Ea.	27.50	12.70		40.20	50
0022	Polyester		32	.250		14.80	12.70		27.50	35.50
0033	Aluminum		32	.250		52.50	12.70		65.20	77
0044	Plastic, 10"x14"		32	.250		58.50	12.70		71.20	84
0055	Polyester		32	.250		35	12.70		47.70	58
0066	Aluminum		32	.250		69.50	12.70		82.20	96
0111	Directional exit sign, plastic, 10"x14"		32	.250		57.50	12.70		70.20	82.50
0122	Polyester		32	.250		45	12.70		57.70	69
0133	Aluminum		32	.250		67	12.70		79.70	93.50

For customer support on your Building Construction Costs with RSMeans data, call 800.448.8182.

379

10 14 Signage

10 14 43 – Photoluminescent Signage

10 14 43.10 Photoluminescent Signage	Crew	Daily Output	Labor-Hours	Unit	Material	2018 Bare Costs Labor	Equipment	Total	Total Incl O&P
0153 Kick plate exit sign, 10"x34"	1 Carp	28	.286	Ea.	258	14.50		272.50	305
0173 Photoluminescent tape, 1"x60'		300	.027	L.F.	.55	1.35		1.90	2.66
0183 2"x60'		290	.028	"	1.16	1.40		2.56	3.41
0211 Photoluminescent directional arrows		42	.190	Ea.	1.28	9.65		10.93	16.10

10 14 53 – Traffic Signage

10 14 53.20 Traffic Signs

	Crew	Daily Output	Labor-Hours	Unit	Material	2018 Bare Costs Labor	Equipment	Total	Total Incl O&P
0010 **TRAFFIC SIGNS**									
0012 Stock, 24" x 24", no posts, .080" alum. reflectorized	B-80	70	.457	Ea.	88	20.50	8.75	117.25	137
0100 High intensity		70	.457		101	20.50	8.75	130.25	151
0300 30" x 30", reflectorized		70	.457		122	20.50	8.75	151.25	174
0400 High intensity		70	.457		132	20.50	8.75	161.25	185
0600 Guide and directional signs, 12" x 18", reflectorized		70	.457		41	20.50	8.75	70.25	85
0700 High intensity		70	.457		54.50	20.50	8.75	83.75	100
0900 18" x 24", stock signs, reflectorized		70	.457		48	20.50	8.75	77.25	93
1000 High intensity		70	.457		53.50	20.50	8.75	82.75	98.50
1200 24" x 24", stock signs, reflectorized		70	.457		58.50	20.50	8.75	87.75	104
1300 High intensity		70	.457		63	20.50	8.75	92.25	109
1500 Add to above for steel posts, galvanized, 10'-0" upright, bolted		200	.160		32.50	7.10	3.06	42.66	50
1600 12'-0" upright, bolted		140	.229		39	10.15	4.37	53.52	63
1800 Highway road signs, aluminum, over 20 S.F., reflectorized		350	.091	S.F.	16.50	4.06	1.75	22.31	26
2000 High intensity		350	.091		16.50	4.06	1.75	22.31	26
2200 Highway, suspended over road, 80 S.F. min., reflectorized		165	.194		16.50	8.60	3.71	28.81	35.50
2300 High intensity		165	.194		18.50	8.60	3.71	30.81	37.50
9000 Replace directional sign		6	5.333	Ea.	132	237	102	471	615

10 17 Telephone Specialties

10 17 16 – Telephone Enclosures

10 17 16.10 Commercial Telephone Enclosures

	Crew	Daily Output	Labor-Hours	Unit	Material	2018 Bare Costs Labor	Equipment	Total	Total Incl O&P
0010 **COMMERCIAL TELEPHONE ENCLOSURES**									
0300 Shelf type, wall hung, recessed	2 Carp	5	3.200	Ea.	810	162		972	1,125
0400 Surface mount	"	5	3.200	"	1,675	162		1,837	2,100

10 21 Compartments and Cubicles

10 21 13 – Toilet Compartments

10 21 13.13 Metal Toilet Compartments

	Crew	Daily Output	Labor-Hours	Unit	Material	2018 Bare Costs Labor	Equipment	Total	Total Incl O&P
0010 **METAL TOILET COMPARTMENTS**									
0110 Cubicles, ceiling hung									
0200 Powder coated steel	2 Carp	4	4	Ea.	555	203		758	920
0500 Stainless steel	"	4	4		1,150	203		1,353	1,550
0600 For handicap units, add					465			465	515
0900 Floor and ceiling anchored									
1000 Powder coated steel	2 Carp	5	3.200	Ea.	600	162		762	905
1300 Stainless steel	"	5	3.200		1,250	162		1,412	1,625
1400 For handicap units, add					330			330	360
1610 Floor anchored									
1700 Powder coated steel	2 Carp	7	2.286	Ea.	625	116		741	860
2000 Stainless steel	"	7	2.286		1,225	116		1,341	1,525
2100 For handicap units, add					330			330	360
2200 For juvenile units, deduct					43.50			43.50	48

For customer support on your Building Construction Costs with RSMeans data, call 800.448.8182.

10 21 Compartments and Cubicles

10 21 13 – Toilet Compartments

10 21 13.13 Metal Toilet Compartments

		Crew	Daily Output	Labor-Hours	Unit	Material	2018 Bare Costs Labor	Equipment	Total	Total Incl O&P
2450	Floor anchored, headrail braced									
2500	Powder coated steel	2 Carp	6	2.667	Ea.	390	135		525	635
2804	Stainless steel	"	6	2.667		1,075	135		1,210	1,375
2900	For handicap units, add ♿					330			330	360
3000	Wall hung partitions, powder coated steel	2 Carp	7	2.286		670	116		786	910
3300	Stainless steel	"	7	2.286		1,725	116		1,841	2,075
3400	For handicap units, add ♿				▼	330			330	360
4000	Screens, entrance, floor mounted, 58" high, 48" wide									
4200	Powder coated steel	2 Carp	15	1.067	Ea.	242	54		296	350
4500	Stainless steel	"	15	1.067	"	960	54		1,014	1,125
4650	Urinal screen, 18" wide									
4704	Powder coated steel	2 Carp	6.15	2.602	Ea.	230	132		362	455
5004	Stainless steel	"	6.15	2.602	"	535	132		667	790
5100	Floor mounted, headrail braced									
5300	Powder coated steel	2 Carp	8	2	Ea.	230	101		331	405
5600	Stainless steel	"	8	2	"	590	101		691	800
5750	Pilaster, flush									
5800	Powder coated steel	2 Carp	10	1.600	Ea.	276	81		357	430
6100	Stainless steel		10	1.600		610	81		691	795
6300	Post braced, powder coated steel		10	1.600		169	81		250	310
6800	Powder coated steel		10	1.600		170	81		251	310
7800	Wedge type, powder coated steel		10	1.600		139	81		220	277
8100	Stainless steel	▼	10	1.600	▼	610	81		691	795

10 21 13.16 Plastic-Laminate-Clad Toilet Compartments

		Crew	Daily Output	Labor-Hours	Unit	Material	2018 Bare Costs Labor	Equipment	Total	Total Incl O&P
0010	**PLASTIC-LAMINATE-CLAD TOILET COMPARTMENTS**									
0110	Cubicles, ceiling hung									
0300	Plastic laminate on particle board ♿	2 Carp	4	4	Ea.	550	203		753	915
0600	For handicap units, add				"	465			465	515
0900	Floor and ceiling anchored									
1100	Plastic laminate on particle board	2 Carp	5	3.200	Ea.	765	162		927	1,075
1400	For handicap units, add ♿				"	330			330	360
1610	Floor mounted									
1800	Plastic laminate on particle board	2 Carp	7	2.286	Ea.	550	116		666	780
2450	Floor mounted, headrail braced									
2600	Plastic laminate on particle board	2 Carp	6	2.667	Ea.	835	135		970	1,125
3400	For handicap units, add ♿					330			330	360
4300	Entrance screen, floor mtd., plas. lam., 58" high, 48" wide	2 Carp	15	1.067		515	54		569	650
4800	Urinal screen, 18" wide, ceiling braced, plastic laminate		8	2		211	101		312	385
5400	Floor mounted, headrail braced		8	2		211	101		312	385
5900	Pilaster, flush, plastic laminate		10	1.600		425	81		506	595
6400	Post braced, plastic laminate	▼	10	1.600	▼	256	81		337	405
6700	Wall hung, bracket supported									
6900	Plastic laminate on particle board	2 Carp	10	1.600	Ea.	97.50	81		178.50	231
7450	Flange supported									
7500	Plastic laminate on particle board	2 Carp	10	1.600	Ea.	245	81		326	395

10 21 13.19 Plastic Toilet Compartments

		Crew	Daily Output	Labor-Hours	Unit	Material	2018 Bare Costs Labor	Equipment	Total	Total Incl O&P
0010	**PLASTIC TOILET COMPARTMENTS**									
0110	Cubicles, ceiling hung									
0250	Phenolic ♿	2 Carp	4	4	Ea.	910	203		1,113	1,300
0260	Polymer plastic	"	4	4	▼	995	203		1,198	1,400
0600	For handicap units, add					465			465	515
0900	Floor and ceiling anchored									

For customer support on your Building Construction Costs with RSMeans data, call 800.448.8182.

381

10 21 Compartments and Cubicles

10 21 13 – Toilet Compartments

10 21 13.19 Plastic Toilet Compartments

		Crew	Daily Output	Labor-Hours	Unit	Material	2018 Bare Costs Labor	Equipment	Total	Total Incl O&P
1050	Phenolic	2 Carp	5	3.200	Ea.	860	162		1,022	1,200
1060	Polymer plastic	"	5	3.200		995	162		1,157	1,350
1400	For handicap units, add					330			330	360
1610	Floor mounted									
1750	Phenolic	2 Carp	7	2.286	Ea.	750	116		866	1,000
1760	Polymer plastic	"	7	2.286		885	116		1,001	1,150
2100	For handicap units, add					330			330	360
2200	For juvenile units, deduct					43.50			43.50	48
2450	Floor mounted, headrail braced									
2550	Phenolic	2 Carp	6	2.667	Ea.	750	135		885	1,025
3600	Polymer plastic		6	2.667		860	135		995	1,150
3810	Entrance screen, polymer plastic, flr. mtd., 48"x58"		6	2.667		505	135		640	760
3820	Entrance screen, polymer plastic, flr. to clg pilaster, 48"x58"		6	2.667		600	135		735	865
6110	Urinal screen, polymer plastic, pilaster flush, 18" W		6	2.667		435	135		570	685
7110	Wall hung		6	2.667		745	135		880	1,025
7710	Flange mounted		6	2.667		930	135		1,065	1,225

10 21 13.40 Stone Toilet Compartments

		Crew	Daily Output	Labor-Hours	Unit	Material	2018 Bare Costs Labor	Equipment	Total	Total Incl O&P
0010	**STONE TOILET COMPARTMENTS**									
0100	Cubicles, ceiling hung, marble	2 Marb	2	8	Ea.	1,650	390		2,040	2,425
0600	For handicap units, add					465			465	515
0800	Floor & ceiling anchored, marble	2 Marb	2.50	6.400		1,750	315		2,065	2,400
1400	For handicap units, add					330			330	360
1600	Floor mounted, marble	2 Marb	3	5.333		1,075	262		1,337	1,600
2400	Floor mounted, headrail braced, marble	"	3	5.333		1,200	262		1,462	1,725
2900	For handicap units, add					330			330	360
4100	Entrance screen, floor mounted marble, 58" high, 48" wide	2 Marb	9	1.778		700	87		787	905
4600	Urinal screen, 18" wide, ceiling braced, marble	D-1	6	2.667		795	119		914	1,050
5100	Floor mounted, headrail braced									
5200	Marble	D-1	6	2.667	Ea.	675	119		794	930
5700	Pilaster, flush, marble		9	1.778		885	79.50		964.50	1,100
6200	Post braced, marble		9	1.778		865	79.50		944.50	1,075

10 21 14 – Toilet Compartment Components

10 21 14.13 Metal Toilet Compartment Components

		Crew	Daily Output	Labor-Hours	Unit	Material	2018 Bare Costs Labor	Equipment	Total	Total Incl O&P
0010	**METAL TOILET COMPARTMENT COMPONENTS**									
0100	Pilasters									
0110	Overhead braced, powder coated steel, 7" wide x 82" high	2 Carp	22.20	.721	Ea.	66.50	36.50		103	129
0120	Stainless steel		22.20	.721		132	36.50		168.50	201
0130	Floor braced, powder coated steel, 7" wide x 70" high		23.30	.687		104	35		139	167
0140	Stainless steel		23.30	.687		218	35		253	293
0150	Ceiling hung, powder coated steel, 7" wide x 83" high		13.30	1.203		139	61		200	246
0160	Stainless steel		13.30	1.203		285	61		346	410
0170	Wall hung, powder coated steel, 3" wide x 58" high		18.90	.847		133	43		176	212
0180	Stainless steel		18.90	.847		169	43		212	252
0200	Panels									
0210	Powder coated steel, 31" wide x 58" high	2 Carp	18.90	.847	Ea.	123	43		166	202
0220	Stainless steel		18.90	.847		330	43		373	425
0230	Powder coated steel, 53" wide x 58" high		18.90	.847		151	43		194	232
0240	Stainless steel		18.90	.847		400	43		443	505
0250	Powder coated steel, 63" wide x 58" high		18.90	.847		180	43		223	264
0260	Stainless steel		18.90	.847		495	43		538	610
0300	Doors									
0310	Powder coated steel, 24" wide x 58" high	2 Carp	14.10	1.135	Ea.	127	57.50		184.50	228

For customer support on your Building Construction Costs with RSMeans data, call 800.448.8182.

10 21 Compartments and Cubicles

10 21 14 – Toilet Compartment Components

10 21 14.13 Metal Toilet Compartment Components

		Crew	Daily Output	Labor-Hours	Unit	Material	2018 Bare Costs Labor	Equipment	Total	Total Incl O&P
0320	Stainless steel	2 Carp	14.10	1.135	Ea.	310	57.50		367.50	430
0330	Powder coated steel, 26" wide x 58" high		14.10	1.135		151	57.50		208.50	254
0340	Stainless steel		14.10	1.135		320	57.50		377.50	440
0350	Powder coated steel, 28" wide x 58" high		14.10	1.135		171	57.50		228.50	276
0360	Stainless steel		14.10	1.135		335	57.50		392.50	460
0370	Powder coated steel, 36" wide x 58" high		14.10	1.135		185	57.50		242.50	292
0380	Stainless steel	▼	14.10	1.135	▼	395	57.50		452.50	525
0400	Headrails									
0410	For powder coated steel, 62" long	2 Carp	65	.246	Ea.	21	12.50		33.50	42
0420	Stainless steel		65	.246		21	12.50		33.50	42
0430	For powder coated steel, 84" long		50	.320		30	16.20		46.20	57.50
0440	Stainless steel		50	.320		30	16.20		46.20	57.50
0450	For powder coated steel, 120" long		30	.533		40	27		67	84.50
0460	Stainless steel	▼	30	.533	▼	40	27		67	84.50

10 21 14.16 Plastic-Laminate Clad Toilet Compartment Components

		Crew	Daily Output	Labor-Hours	Unit	Material	2018 Bare Costs Labor	Equipment	Total	Total Incl O&P
0010	**PLASTIC-LAMINATE CLAD TOILET COMPARTMENT COMPONENTS**									
0100	Pilasters									
0110	Overhead braced, 7" wide x 82" high	2 Carp	22.20	.721	Ea.	103	36.50		139.50	169
0130	Floor anchored, 7" wide x 70" high		23.30	.687		103	35		138	166
0150	Ceiling hung, 7" wide x 83" high		13.30	1.203		107	61		168	211
0180	Wall hung, 3" wide x 58" high	▼	18.90	.847	▼	96	43		139	171
0200	Panels									
0210	31" wide x 58" high	2 Carp	18.90	.847	Ea.	153	43		196	234
0230	51" wide x 58" high		18.90	.847		209	43		252	295
0250	63" wide x 58" high	▼	18.90	.847	▼	242	43		285	330
0300	Doors									
0310	24" wide x 58" high	2 Carp	14.10	1.135	Ea.	151	57.50		208.50	254
0330	26" wide x 58" high		14.10	1.135		156	57.50		213.50	259
0350	28" wide x 58" high		14.10	1.135		161	57.50		218.50	265
0370	36" wide x 58" high	▼	14.10	1.135	▼	185	57.50		242.50	291
0400	Headrails									
0410	62" long	2 Carp	65	.246	Ea.	22.50	12.50		35	44
0430	84" long		60	.267		29.50	13.50		43	53
0450	120" long	▼	30	.533	▼	39.50	27		66.50	84.50

10 21 14.19 Plastic Toilet Compartment Components

		Crew	Daily Output	Labor-Hours	Unit	Material	2018 Bare Costs Labor	Equipment	Total	Total Incl O&P
0010	**PLASTIC TOILET COMPARTMENT COMPONENTS**									
0100	Pilasters									
0110	Overhead braced, polymer plastic, 7" wide x 82" high	2 Carp	22.20	.721	Ea.	114	36.50		150.50	182
0120	Phenolic		22.20	.721		141	36.50		177.50	211
0130	Floor braced, polymer plastic, 7" wide x 70" high		23.30	.687		183	35		218	254
0140	Phenolic		23.30	.687		135	35		170	201
0150	Ceiling hung, polymer plastic, 7" wide x 83" high		13.30	1.203		171	61		232	281
0160	Phenolic		13.30	1.203		164	61		225	274
0180	Wall hung, phenolic, 3" wide x 58" high	▼	18.90	.847		105	43		148	182
0200	Panels									
0203	Polymer plastic, 18" wide x 55" high	2 Carp	18.90	.847	Ea.	290	43		333	385
0206	Phenolic, 18" wide x 58" high		18.90	.847		251	43		294	340
0210	Polymer plastic, 31" wide x 55" high		18.90	.847		297	43		340	390
0220	Phenolic, 31" wide x 58" high		18.90	.847		278	43		321	370
0223	Polymer plastic, 48" wide x 55" high		18.90	.847		535	43		578	655
0226	Phenolic, 48" wide x 58" high		18.90	.847		470	43		513	580
0230	Polymer plastic, 51" wide x 55" high	▼	18.90	.847	▼	450	43		493	560

For customer support on your Building Construction Costs with RSMeans data, call 800.448.8182.

383

10 21 Compartments and Cubicles

10 21 14 – Toilet Compartment Components

10 21 14.19 Plastic Toilet Compartment Components

		Crew	Daily Output	Labor-Hours	Unit	Material	2018 Bare Costs Labor	Equipment	Total	Total Incl O&P
0240	Phenolic, 51" wide x 58" high	2 Carp	18.90	.847	Ea.	400	43		443	505
0250	Polymer plastic, 63" wide x 55" high		18.90	.847		545	43		588	665
0260	Phenolic, 63" wide x 58" high		18.90	.847		420	43		463	525
0300	Doors									
0310	Polymer plastic, 24" wide x 55" high	2 Carp	14.10	1.135	Ea.	218	57.50		275.50	330
0320	Phenolic, 24" wide x 58" high		14.10	1.135		310	57.50		367.50	430
0330	Polymer plastic, 26" wide x 55" high		14.10	1.135		242	57.50		299.50	355
0340	Phenolic, 26" wide x 58" high		14.10	1.135		330	57.50		387.50	450
0350	Polymer plastic, 28" wide x 55" high		14.10	1.135		260	57.50		317.50	375
0360	Phenolic, 28" wide x 58" high		14.10	1.135		345	57.50		402.50	470
0370	Polymer plastic, 36" wide x 55" high		14.10	1.135		310	57.50		367.50	435
0380	Phenolic, 36" wide x 58" high		14.10	1.135		430	57.50		487.50	565
0400	Headrails									
0410	For polymer plastic, 62" long	2 Carp	65	.246	Ea.	23	12.50		35.50	44.50
0420	Phenolic		65	.246		22.50	12.50		35	44
0430	For polymer plastic, 84" long		50	.320		31.50	16.20		47.70	59
0440	Phenolic		50	.320		34.50	16.20		50.70	62.50
0450	For polymer plastic, 120" long		30	.533		41	27		68	86.50
0460	Phenolic		30	.533		41	27		68	86

10 21 23 – Cubicle Curtains and Track

10 21 23.16 Cubicle Track and Hardware

		Crew	Daily Output	Labor-Hours	Unit	Material	2018 Bare Costs Labor	Equipment	Total	Total Incl O&P
0010	**CUBICLE TRACK AND HARDWARE**									
0020	Curtain track, box channel, ceiling mounted	1 Carp	135	.059	L.F.	6.35	3		9.35	11.55
0100	Suspended	"	100	.080	"	8.55	4.06		12.61	15.60
0300	Curtains, nylon mesh tops, fire resistant, 11 oz. per lineal yard									
0310	Polyester oxford cloth, 9' ceiling height	1 Carp	425	.019	L.F.	18	.95		18.95	21.50
0500	8' ceiling height		425	.019		15.05	.95		16	18
0550	Polyester, antimicrobial, 9' ceiling height		425	.019		19.65	.95		20.60	23
0560	8' ceiling height		425	.019		17.50	.95		18.45	20.50
0700	Designer oxford cloth		425	.019		6.65	.95		7.60	8.75
0800	I.V. track systems									
0820	I.V. track, oval	1 Carp	135	.059	L.F.	7.95	3		10.95	13.30
0830	I.V. trolley		32	.250	Ea.	41	12.70		53.70	65
0840	I.V. pendant (tree, 5 hook)		32	.250	"	182	12.70		194.70	219

10 22 Partitions

10 22 13 – Wire Mesh Partitions

10 22 13.10 Partitions, Woven Wire

		Crew	Daily Output	Labor-Hours	Unit	Material	2018 Bare Costs Labor	Equipment	Total	Total Incl O&P
0010	**PARTITIONS, WOVEN WIRE** for tool or stockroom enclosures									
0100	Channel frame, 1-1/2" diamond mesh, 10 ga. wire, painted									
0300	Wall panels, 4'-0" wide, 7' high	2 Carp	25	.640	Ea.	144	32.50		176.50	208
0400	8' high		23	.696		155	35.50		190.50	225
0600	10' high		18	.889		195	45		240	284
0700	For 5' wide panels, add					5%				
0900	Ceiling panels, 10' long, 2' wide	2 Carp	25	.640		140	32.50		172.50	204
1000	4' wide		15	1.067		170	54		224	270
1200	Panel with service window & shelf, 5' wide, 7' high		20	.800		340	40.50		380.50	430
1300	8' high		15	1.067		460	54		514	590
1500	Sliding doors, full height, 3' wide, 7' high		6	2.667		425	135		560	675
1600	10' high		5	3.200		565	162		727	870

For customer support on your Building Construction Costs with RSMeans data, call 800.448.8182.

10 22 Partitions

10 22 13 – Wire Mesh Partitions

10 22 13.10 Partitions, Woven Wire

		Crew	Daily Output	Labor-Hours	Unit	Material	2018 Bare Costs Labor	Equipment	Total	Total Incl O&P
1800	6' wide sliding door, 7' full height	2 Carp	5	3.200	Ea.	590	162		752	895
1900	10' high		4	4		775	203		978	1,150
2100	Swinging doors, 3' wide, 7' high, no transom		6	2.667		286	135		421	520
2200	7' high, 3' transom		5	3.200		360	162		522	640

10 22 16 – Folding Gates

10 22 16.10 Security Gates

		Crew	Daily Output	Labor-Hours	Unit	Material	2018 Bare Costs Labor	Equipment	Total	Total Incl O&P
0010	**SECURITY GATES**									
0015	For roll up type, see Section 08 33 13.10									
0300	Scissors type folding gate, ptd. steel, single, 6-1/2' high, 5-1/2' wide	2 Sswk	4	4	Opng.	225	219		444	605
0350	6-1/2' wide		4	4		238	219		457	615
0400	7-1/2' wide		4	4		237	219		456	615
0600	Double gate, 8' high, 8' wide		2.50	6.400		390	350		740	1,000
0650	10' wide		2.50	6.400		435	350		785	1,050
0700	12' wide		2	8		590	435		1,025	1,375
0750	14' wide		2	8		620	435		1,055	1,400
0900	Door gate, folding steel, 4' wide, 61" high		4	4		134	219		353	500
1000	71" high		4	4		176	219		395	550
1200	81" high		4	4		182	219		401	555
1300	Window gates, 2' to 4' wide, 31" high		4	4		85	219		304	450
1500	55" high		3.75	4.267		122	233		355	515
1600	79" high		3.50	4.571		144	250		394	570

10 22 19 – Demountable Partitions

10 22 19.43 Demountable Composite Partitions

		Crew	Daily Output	Labor-Hours	Unit	Material	2018 Bare Costs Labor	Equipment	Total	Total Incl O&P
0010	**DEMOUNTABLE COMPOSITE PARTITIONS**, add for doors									
0100	Do not deduct door openings from total L.F.									
0900	Demountable gypsum system on 2" to 2-1/2"									
1000	Steel studs, 9' high, 3" to 3-3/4" thick									
1200	Vinyl-clad gypsum	2 Carp	48	.333	L.F.	60	16.90		76.90	91.50
1300	Fabric clad gypsum		44	.364		150	18.45		168.45	193
1500	Steel clad gypsum		40	.400		167	20.50		187.50	215
1600	1.75 system, aluminum framing, vinyl-clad hardboard,									
1800	Paper honeycomb core panel, 1-3/4" to 2-1/2" thick									
1900	9' high	2 Carp	48	.333	L.F.	101	16.90		117.90	137
2100	7' high		60	.267		90.50	13.50		104	120
2200	5' high		80	.200		76.50	10.15		86.65	99.50
2250	Unitized gypsum system									
2300	Unitized panel, 9' high, 2" to 2-1/2" thick									
2350	Vinyl-clad gypsum	2 Carp	48	.333	L.F.	130	16.90		146.90	169
2400	Fabric clad gypsum	"	44	.364	"	214	18.45		232.45	263
2500	Unitized mineral fiber system									
2510	Unitized panel, 9' high, 2-1/4" thick, aluminum frame									
2550	Vinyl-clad mineral fiber	2 Carp	48	.333	L.F.	129	16.90		145.90	168
2600	Fabric clad mineral fiber	"	44	.364	"	193	18.45		211.45	240
2800	Movable steel walls, modular system									
2900	Unitized panels, 9' high, 48" wide									
3100	Baked enamel, pre-finished	2 Carp	60	.267	L.F.	146	13.50		159.50	182
3200	Fabric clad steel		56	.286	"	212	14.50		226.50	255
5310	Trackless wall, cork finish, semi-acoustic, 1-5/8" thick, unsealed		325	.049	S.F.	38.50	2.50		41	46.50
5320	Sealed		190	.084		42.50	4.27		46.77	53.50
5330	Acoustic, 2" thick, unsealed		305	.052		36.50	2.66		39.16	44
5340	Sealed		225	.071		56	3.61		59.61	67
5500	For acoustical partitions, add, unsealed					2.36			2.36	2.60

For customer support on your Building Construction Costs with RSMeans data, call 800.448.8182.

385

10 22 Partitions

10 22 19 – Demountable Partitions

10 22 19.43 Demountable Composite Partitions	Crew	Daily Output	Labor-Hours	Unit	Material	2018 Bare Costs Labor	Equipment	Total	Total Incl O&P
5550	Sealed				S.F.	11			11	12.10
5700	For doors, see Section 08 16									
5800	For door hardware, see Section 08 71									
6100	In-plant modular office system, w/prehung hollow core door									
6200	3" thick polystyrene core panels									
6250	12' x 12', 2 wall	2 Clab	3.80	4.211	Ea.	4,825	168		4,993	5,550
6300	4 wall		1.90	8.421		7,550	335		7,885	8,800
6350	16' x 16', 2 wall		3.60	4.444		7,850	177		8,027	8,900
6400	4 wall		1.80	8.889		10,000	355		10,355	11,500

10 22 23 – Portable Partitions, Screens, and Panels

10 22 23.13 Wall Screens

		Crew	Daily Output	Labor-Hours	Unit	Material	2018 Bare Costs Labor	Equipment	Total	Total Incl O&P
0010	**WALL SCREENS**, divider panels, free standing, fiber core									
0020	Fabric face straight									
0100	3'-0" long, 4'-0" high	2 Carp	100	.160	L.F.	134	8.10		142.10	160
0200	5'-0" high		90	.178		117	9		126	143
0500	6'-0" high		75	.213		122	10.80		132.80	150
0900	5'-0" long, 4'-0" high		175	.091		65.50	4.64		70.14	79
1000	5'-0" high		150	.107		88	5.40		93.40	105
1500	6'-0" high		125	.128		110	6.50		116.50	131
1600	6'-0" long, 5'-0" high		162	.099		110	5		115	129
3200	Economical panels, fabric face, 4'-0" long, 5'-0" high		132	.121		59	6.15		65.15	74.50
3250	6'-0" high		112	.143		66.50	7.25		73.75	84
3300	5'-0" long, 5'-0" high		150	.107		60	5.40		65.40	74.50
3350	6'-0" high		125	.128		54	6.50		60.50	69
3450	Acoustical panels, 60 to 90 NRC, 3'-0" long, 5'-0" high		90	.178		78	9		87	99.50
3550	6'-0" high		75	.213		79.50	10.80		90.30	104
3600	5'-0" long, 5'-0" high		150	.107		64.50	5.40		69.90	79.50
3650	6'-0" high		125	.128		69	6.50		75.50	86
3700	6'-0" long, 5'-0" high		162	.099		54.50	5		59.50	67.50
3750	6'-0" high		138	.116		82.50	5.90		88.40	100
3800	Economy acoustical panels, 40 NRC, 4'-0" long, 5'-0" high		132	.121		59	6.15		65.15	74.50
3850	6'-0" high		112	.143		66.50	7.25		73.75	84
3900	5'-0" long, 6'-0" high		125	.128		54	6.50		60.50	69
3950	6'-0" long, 5'-0" high		162	.099		49.50	5		54.50	62
4000	Metal chalkboard, 6'-6" high, chalkboard, 1 side		125	.128		124	6.50		130.50	146
4100	Metal chalkboard, 2 sides		120	.133		141	6.75		147.75	166
4300	Tackboard, both sides		123	.130		112	6.60		118.60	133

10 22 33 – Accordion Folding Partitions

10 22 33.10 Partitions, Accordion Folding

		Crew	Daily Output	Labor-Hours	Unit	Material	2018 Bare Costs Labor	Equipment	Total	Total Incl O&P
0010	**PARTITIONS, ACCORDION FOLDING**									
0100	Vinyl covered, over 150 S.F., frame not included									
0300	Residential, 1.25 lb./S.F., 8' maximum height	2 Carp	300	.053	S.F.	22.50	2.70		25.20	29
0400	Commercial, 1.75 lb./S.F., 8' maximum height		225	.071		26	3.61		29.61	34
0600	2 lb./S.F., 17' maximum height		150	.107		27	5.40		32.40	38
0700	Industrial, 4 lb./S.F., 20' maximum height		75	.213		40	10.80		50.80	60.50
0900	Acoustical, 3 lb./S.F., 17' maximum height		100	.160		27.50	8.10		35.60	42.50
1200	5 lb./S.F., 20' maximum height		95	.168		38	8.55		46.55	55
1300	5.5 lb./S.F., 17' maximum height		90	.178		44.50	9		53.50	62.50
1400	Fire rated, 4.5 psf, 20' maximum height		160	.100		44.50	5.05		49.55	56.50
1500	Vinyl-clad wood or steel, electric operation, 5.0 psf		160	.100		56.50	5.05		61.55	70
1900	Wood, non-acoustic, birch or mahogany, to 10' high		300	.053		30.50	2.70		33.20	37.50

For customer support on your Building Construction Costs with RSMeans data, call 800.448.8182.

10 22 Partitions

10 22 39 – Folding Panel Partitions

10 22 39.10 Partitions, Folding Panel	Crew	Daily Output	Labor-Hours	Unit	Material	2018 Bare Costs Labor	Equipment	Total	Total Incl O&P
0010 **PARTITIONS, FOLDING PANEL**, acoustic, wood									
0100 Vinyl faced, to 18' high, 6 psf, economy trim	2 Carp	60	.267	S.F.	58.50	13.50		72	85
0150 Standard trim		45	.356		70	18.05		88.05	105
0200 Premium trim		30	.533		90	27		117	141
0400 Plastic laminate or hardwood finish, standard trim		60	.267		60.50	13.50		74	87
0500 Premium trim		30	.533		64.50	27		91.50	112
0600 Wood, low acoustical type, 4.5 psf, to 14' high		50	.320		44	16.20		60.20	72.50
1100 Steel, acoustical, 9 to 12 lb./S.F., vinyl faced, standard trim		60	.267		62.50	13.50		76	89
1200 Premium trim		30	.533		76	27		103	125
1700 Aluminum framed, acoustical, to 12' high, 5.5 psf, standard trim		60	.267		42	13.50		55.50	66.50
1800 Premium trim		30	.533		50.50	27		77.50	97
2000 6.5 lb./S.F., standard trim		60	.267		44	13.50		57.50	69
2100 Premium trim	↓	30	.533	↓	54.50	27		81.50	101

10 22 43 – Sliding Partitions

10 22 43.10 Partitions, Sliding	Crew	Daily Output	Labor-Hours	Unit	Material	2018 Bare Costs Labor	Equipment	Total	Total Incl O&P
0010 **PARTITIONS, SLIDING**									
0020 Acoustic air wall, 1-5/8" thick, standard trim	2 Carp	375	.043	S.F.	34	2.16		36.16	41
0100 Premium trim		365	.044		58.50	2.22		60.72	67.50
0300 2-1/4" thick, standard trim		360	.044		38	2.25		40.25	45.50
0400 Premium trim	↓	330	.048	↓	67	2.46		69.46	77
0600 For track type, add to above				L.F.	124			124	137
0700 Overhead track type, acoustical, 3" thick, 11 psf, standard trim	2 Carp	350	.046	S.F.	86	2.32		88.32	98
0800 Premium trim	"	300	.053	"	103	2.70		105.70	118

10 26 Wall and Door Protection

10 26 13 – Corner Guards

10 26 13.20 Corner Protection	Crew	Daily Output	Labor-Hours	Unit	Material	2018 Bare Costs Labor	Equipment	Total	Total Incl O&P
0010 **CORNER PROTECTION**									
0100 Stainless steel, 16 ga., adhesive mount, 3-1/2" leg	1 Carp	80	.100	L.F.	24.50	5.05		29.55	34.50
0200 12 ga. stainless, adhesive mount	"	80	.100		20.50	5.05		25.55	30
0300 For screw mount, add						10%			
0500 Vinyl acrylic, adhesive mount, 3" leg	1 Carp	128	.063		9.90	3.17		13.07	15.75
0550 1-1/2" leg		160	.050		5.05	2.54		7.59	9.45
0600 Screw mounted, 3" leg		80	.100		9.80	5.05		14.85	18.50
0650 1-1/2" leg		100	.080		4.52	4.06		8.58	11.15
0700 Clear plastic, screw mounted, 2-1/2"		60	.133		4.68	6.75		11.43	15.45
1000 Vinyl cover, alum. retainer, surface mount, 3" x 3"		48	.167		10.90	8.45		19.35	25
1050 2" x 2"		48	.167		9.75	8.45		18.20	23.50
1100 Flush mounted, 3" x 3"		32	.250		21	12.70		33.70	42.50
1150 2" x 2"	↓	32	.250	↓	16.70	12.70		29.40	37.50

10 26 16 – Bumper Guards

10 26 16.10 Guard, Bumper	Crew	Daily Output	Labor-Hours	Unit	Material	2018 Bare Costs Labor	Equipment	Total	Total Incl O&P
0010 **GUARD, BUMPER**									
1200 Bed bumper, vinyl acrylic, alum. retainer, 21" long	1 Carp	10	.800	Ea.	42	40.50		82.50	108
1300 53" long with aligner		9	.889	"	105	45		150	185
1400 Bumper, vinyl cover, alum. retain., cush. mnt., 1-1/2" x 2-3/4"		80	.100	L.F.	14.20	5.05		19.25	23.50
1500 2" x 4-1/4"		80	.100		20.50	5.05		25.55	30.50
1600 Surface mounted, 1-3/4" x 3-5/8"	↓	80	.100	↓	11.95	5.05		17	21

10 26 Wall and Door Protection

10 26 16 – Bumper Guards

10 26 16.16 Protective Corridor Handrails

10 26 16.16 Protective Corridor Handrails	Crew	Daily Output	Labor-Hours	Unit	Material	2018 Bare Costs Labor	Equipment	Total	Total Incl O&P
0010 **PROTECTIVE CORRIDOR HANDRAILS**									
3000 Handrail/bumper, vinyl cover, alum. retainer									
3010 Bracket mounted, flat rail, 5-1/2"	1 Carp	80	.100	L.F.	18.30	5.05		23.35	27.50
3100 6-1/2"		80	.100		23	5.05		28.05	32.50
3200 Bronze bracket, 1-3/4" diam. rail	↓	80	.100	↓	16.50	5.05		21.55	26

10 26 23 – Protective Wall Covering

10 26 23.10 Wall Covering, Protective

	Crew	Daily Output	Labor-Hours	Unit	Material	2018 Bare Costs Labor	Equipment	Total	Total Incl O&P
0010 **WALL COVERING, PROTECTIVE**									
0400 Rub rail, vinyl, adhesive mounted	1 Carp	185	.043	L.F.	9.65	2.19		11.84	13.95
0500 Neoprene, aluminum backing, 1-1/2" x 2"		110	.073		9.10	3.69		12.79	15.60
1000 Trolley rail, PVC, clipped to wall, 5" high		185	.043		8.85	2.19		11.04	13.10
1050 8" high	↓	180	.044		15.10	2.25		17.35	20
1700 Bumper rail, stainless steel, flat bar on brackets, 4" x 1/4"	2 Skwk	120	.133	↓	39	7		46	53
1775 Wall end guard, stainless steel, 16 ga., 36" tall, screwed to studs	1 Skwk	30	.267	Ea.	25.50	13.95		39.45	50
2000 Crash rail, vinyl cover, alum. retainer, 1" x 4"	1 Carp	110	.073	L.F.	11.05	3.69		14.74	17.75
2100 1" x 8"		90	.089		18.70	4.51		23.21	27.50
2150 Vinyl inserts, aluminum plate, 1" x 2-1/2"		110	.073		14.90	3.69		18.59	22
2200 1" x 5"	↓	90	.089	↓	24	4.51		28.51	33.50

10 26 23.13 Impact Resistant Wall Protection

	Crew	Daily Output	Labor-Hours	Unit	Material	2018 Bare Costs Labor	Equipment	Total	Total Incl O&P
0010 **IMPACT RESISTANT WALL PROTECTION**									
0100 Vinyl wall protection, complete instl. incl. panels, trim, adhesive									
0110 .040" thk., std. colors	2 Carp	320	.050	S.F.	16.05	2.54		18.59	21.50
0120 .040" thk., element patterns		300	.053		17.25	2.70		19.95	23
0130 .060" thk., std. colors		300	.053		17.25	2.70		19.95	23
0140 .060" thk., element patterns	↓	280	.057	↓	19.05	2.90		21.95	25.50
1750 Wallguard stainless steel baseboard, 12" tall, adhesive applied	2 Skwk	260	.062	L.F.	33	3.22		36.22	41.50
1775 Wall protection, stainless steel, 16 ga., 48" x 36" tall, screwed to studs	"	500	.032	S.F.	7.95	1.68		9.63	11.30

10 26 33 – Door and Frame Protection

10 26 33.10 Protection, Door and Frame

	Crew	Daily Output	Labor-Hours	Unit	Material	2018 Bare Costs Labor	Equipment	Total	Total Incl O&P
0010 **PROTECTION, DOOR AND FRAME**									
0100 Door frame guard, vinyl, 3" x 3" x 4'	1 Carp	30	.267	Ea.	83	13.50		96.50	112
0110 Door frame guard, vinyl, 3" x 3" x 8'		26	.308		129	15.60		144.60	166
0120 Door frame guard, stainless steel, 3" x 3" x 4'	↓	30	.267		420	13.50		433.50	485
0500 Steel door track/wheel guards, 4'-0" high	E-4	22	1.455	↓	109	80	4.48	193.48	256

10 28 Toilet, Bath, and Laundry Accessories

10 28 13 – Toilet Accessories

10 28 13.13 Commercial Toilet Accessories

	Crew	Daily Output	Labor-Hours	Unit	Material	2018 Bare Costs Labor	Equipment	Total	Total Incl O&P
0010 **COMMERCIAL TOILET ACCESSORIES**									
0200 Curtain rod, stainless steel, 5' long, 1" diameter	1 Carp	13	.615	Ea.	27.50	31		58.50	77.50
0300 1-1/4" diameter		13	.615		30	31		61	80.50
0350 Chrome, 1" diameter		13	.615	↓	28	31		59	78.50
0360 For vinyl curtain, add		1950	.004	S.F.	.95	.21		1.16	1.37
0400 Diaper changing station, horizontal, wall mounted, plastic		10	.800	Ea.	248	40.50		288.50	335
0420 Vertical		10	.800		222	40.50		262.50	305
0430 Oval shaped		10	.800		222	40.50		262.50	305
0440 Recessed, with stainless steel flange	↓	6	1.333	↓	510	67.50		577.50	665
0500 Dispenser units, combined soap & towel dispensers,									
0510 Mirror and shelf, flush mounted	1 Carp	10	.800	Ea.	365	40.50		405.50	460

For customer support on your Building Construction Costs with RSMeans data, call 800.448.8182.

10 28 13.13 Commercial Toilet Accessories	Crew	Daily Output	Labor-Hours	Unit	Material	2018 Bare Costs Labor	Equipment	Total	Total Incl O&P
0600	Towel dispenser and waste receptacle,									
0610	18 gallon capacity	1 Carp	10	.800	Ea.	330	40.50		370.50	425
0800	Grab bar, straight, 1-1/4" diameter, stainless steel, 18" long		24	.333		30	16.90		46.90	58.50
0900	24" long		23	.348		29.50	17.65		47.15	59.50
1000	30" long		22	.364		32.50	18.45		50.95	64
1100	36" long		20	.400		35	20.50		55.50	69.50
1105	42" long		20	.400		36.50	20.50		57	71
1120	Corner, 36" long		20	.400		92.50	20.50		113	133
1200	1-1/2" diameter, 24" long		23	.348		31.50	17.65		49.15	61.50
1300	36" long		20	.400		34	20.50		54.50	68.50
1310	42" long		18	.444		37.50	22.50		60	76
1500	Tub bar, 1-1/4" diameter, 24" x 36"		14	.571		88.50	29		117.50	142
1600	Plus vertical arm		12	.667		93.50	34		127.50	155
1900	End tub bar, 1" diameter, 90° angle, 16" x 32"		12	.667		104	34		138	166
2010	Tub/shower/toilet, 2-wall, 36" x 24"		12	.667		94	34		128	156
2300	Hand dryer, surface mounted, electric, 115 volt, 20 amp		4	2		490	101		591	695
2400	230 volt, 10 amp		4	2		655	101		756	875
2450	Hand dryer, touch free, 1400 watt, 81,000 rpm		4	2		1,375	101		1,476	1,650
2600	Hat and coat strip, stainless steel, 4 hook, 36" long		24	.333		68	16.90		84.90	101
2700	6 hook, 60" long		20	.400		124	20.50		144.50	168
3000	Mirror, with stainless steel 3/4" square frame, 18" x 24"		20	.400		47	20.50		67.50	83
3100	36" x 24"		15	.533		123	27		150	176
3200	48" x 24"		10	.800		179	40.50		219.50	259
3300	72" x 24"		6	1.333		282	67.50		349.50	415
3500	With 5" stainless steel shelf, 18" x 24"		20	.400		189	20.50		209.50	239
3600	36" x 24"		15	.533		238	27		265	305
3700	48" x 24"		10	.800		257	40.50		297.50	345
3800	72" x 24"		6	1.333		259	67.50		326.50	390
4100	Mop holder strip, stainless steel, 5 holders, 48" long		20	.400		80	20.50		100.50	119
4200	Napkin/tampon dispenser, recessed		15	.533		620	27		647	720
4220	Semi-recessed		6.50	1.231		370	62.50		432.50	500
4250	Napkin receptacle, recessed		6.50	1.231		170	62.50		232.50	282
4300	Robe hook, single, regular		96	.083		19.45	4.22		23.67	28
4400	Heavy duty, concealed mounting		56	.143		22.50	7.25		29.75	36
4600	Soap dispenser, chrome, surface mounted, liquid		20	.400		51	20.50		71.50	87
5000	Recessed stainless steel, liquid		10	.800		161	40.50		201.50	239
5600	Shelf, stainless steel, 5" wide, 18 ga., 24" long		24	.333		82	16.90		98.90	116
5700	48" long		16	.500		157	25.50		182.50	211
5800	8" wide shelf, 18 ga., 24" long		22	.364		69.50	18.45		87.95	105
5900	48" long		14	.571		145	29		174	203
6000	Toilet seat cover dispenser, stainless steel, recessed		20	.400		173	20.50		193.50	222
6050	Surface mounted		15	.533		30	27		57	74
6100	Toilet tissue dispenser, surface mounted, SS, single roll		30	.267		18.05	13.50		31.55	40.50
6200	Double roll		24	.333		22	16.90		38.90	50
6240	Plastic, twin/jumbo dbl. roll		24	.333		28.50	16.90		45.40	57
6400	Towel bar, stainless steel, 18" long		23	.348		42.50	17.65		60.15	74
6500	30" long		21	.381		52	19.30		71.30	86.50
6700	Towel dispenser, stainless steel, surface mounted		16	.500		43	25.50		68.50	85.50
6800	Flush mounted, recessed		10	.800		112	40.50		152.50	186
6900	Plastic, touchless, battery operated		16	.500		91	25.50		116.50	139
7000	Towel holder, hotel type, 2 guest size		20	.400		56	20.50		76.50	92.50
7200	Towel shelf, stainless steel, 24" long, 8" wide		20	.400		64.50	20.50		85	102
7400	Tumbler holder, for tumbler only		30	.267		25	13.50		38.50	48

10 28 Toilet, Bath, and Laundry Accessories

10 28 13 – Toilet Accessories

10 28 13.13 Commercial Toilet Accessories

		Crew	Daily Output	Labor-Hours	Unit	Material	2018 Bare Costs Labor	Equipment	Total	Total Incl O&P
7410	Tumbler holder, recessed	1 Carp	20	.400	Ea.	8	20.50		28.50	40
7500	Soap, tumbler & toothbrush		30	.267		19.60	13.50		33.10	42
7510	Tumbler & toothbrush holder		20	.400		13.45	20.50		33.95	46
8000	Waste receptacles, stainless steel, with top, 13 gallon		10	.800		310	40.50		350.50	400
8100	36 gallon		8	1		410	50.50		460.50	525
9996	Bathroom access., grab bar, straight, 1-1/2" diam., SS, 42" L install only	▼	18	.444	▼		22.50		22.50	34.50

10 28 16 – Bath Accessories

10 28 16.20 Medicine Cabinets

		Crew	Daily Output	Labor-Hours	Unit	Material	2018 Bare Costs Labor	Equipment	Total	Total Incl O&P
0010	**MEDICINE CABINETS**									
0020	With mirror, sst frame, 16" x 22", unlighted	1 Carp	14	.571	Ea.	99	29		128	153
0100	Wood frame		14	.571		136	29		165	194
0300	Sliding mirror doors, 20" x 16" x 4-3/4", unlighted		7	1.143		124	58		182	225
0400	24" x 19" x 8-1/2", lighted		5	1.600		206	81		287	350
0600	Triple door, 30" x 32", unlighted, plywood body		7	1.143		350	58		408	475
0700	Steel body		7	1.143		375	58		433	500
0900	Oak door, wood body, beveled mirror, single door		7	1.143		175	58		233	280
1000	Double door		6	1.333		370	67.50		437.50	510
1200	Hotel cabinets, stainless, with lower shelf, unlighted		10	.800		218	40.50		258.50	300
1300	Lighted	▼	5	1.600	▼	340	81		421	500

10 28 19 – Tub and Shower Enclosures

10 28 19.10 Partitions, Shower

		Crew	Daily Output	Labor-Hours	Unit	Material	2018 Bare Costs Labor	Equipment	Total	Total Incl O&P
0010	**PARTITIONS, SHOWER** floor mounted, no plumbing									
0400	Cabinet, one piece, fiberglass, 32" x 32"	2 Carp	5	3.200	Ea.	625	162		787	935
0420	36" x 36"		5	3.200		695	162		857	1,000
0440	36" x 48"		5	3.200		1,400	162		1,562	1,775
0460	Acrylic, 32" x 32"		5	3.200		360	162		522	640
0480	36" x 36"		5	3.200		1,075	162		1,237	1,425
0500	36" x 48"	▼	5	3.200		1,300	162		1,462	1,675
0520	Shower door for above, clear plastic, 24" wide	1 Carp	8	1		182	50.50		232.50	277
0540	28" wide		8	1		249	50.50		299.50	350
0560	Tempered glass, 24" wide		8	1		239	50.50		289.50	340
0580	28" wide	▼	8	1		281	50.50		331.50	385
2400	Glass stalls, with doors, no receptors, chrome on brass	2 Shee	3	5.333		1,650	320		1,970	2,275
2700	Anodized aluminum	"	4	4		1,325	239		1,564	1,825
2900	Marble shower stall, stock design, with shower door	2 Marb	1.20	13.333		2,300	655		2,955	3,525
3000	With curtain		1.30	12.308		2,025	605		2,630	3,150
3200	Receptors, precast terrazzo, 32" x 32"		14	1.143		355	56		411	475
3300	48" x 34"		9.50	1.684		470	82.50		552.50	645
3500	Plastic, simulated terrazzo receptor, 32" x 32"		14	1.143		167	56		223	270
3600	32" x 48"		12	1.333		278	65.50		343.50	405
3800	Precast concrete, colors, 32" x 32"		14	1.143		251	56		307	360
3900	48" x 48"	▼	8	2		273	98		371	450
4100	Shower doors, economy plastic, 24" wide	1 Shee	9	.889		137	53		190	232
4200	Tempered glass door, economy		8	1		293	60		353	410
4400	Folding, tempered glass, aluminum frame		6	1.333		450	79.50		529.50	615
4500	Sliding, tempered glass, 48" opening		6	1.333		580	79.50		659.50	760
4700	Deluxe, tempered glass, chrome on brass frame, 42" to 44"		8	1		440	60		500	570
4800	39" to 48" wide		1	8		670	480		1,150	1,475
4850	On anodized aluminum frame, obscure glass		2	4		580	239		819	1,000
4900	Clear glass	▼	1	8	▼	670	480		1,150	1,475
5100	Shower enclosure, tempered glass, anodized alum. frame									
5120	2 panel & door, corner unit, 32" x 32"	1 Shee	2	4	Ea.	965	239		1,204	1,450

For customer support on your Building Construction Costs with RSMeans data, call 800.448.8182.

10 28 Toilet, Bath, and Laundry Accessories

10 28 19 – Tub and Shower Enclosures

10 28 19.10 Partitions, Shower

	Crew	Daily Output	Labor-Hours	Unit	Material	2018 Bare Costs Labor	Equipment	Total	Total Incl O&P
5140	Neo-angle corner unit, 16" x 24" x 16"	1 Shee	2	4	Ea.	1,125	239		1,364	1,625
5200	Shower surround, 3 wall, polypropylene, 32" x 32"	1 Carp	4	2		620	101		721	835
5220	PVC, 32" x 32"		4	2		395	101		496	590
5240	Fiberglass		4	2		400	101		501	595
5250	2 wall, polypropylene, 32" x 32"		4	2		310	101		411	500
5270	PVC		4	2		380	101		481	570
5290	Fiberglass		4	2		380	101		481	575
5300	Tub doors, tempered glass & frame, obscure glass	1 Shee	8	1		224	60		284	340
5400	Clear glass		6	1.333		520	79.50		599.50	695
5600	Chrome plated, brass frame, obscure glass		8	1		295	60		355	415
5700	Clear glass		6	1.333		725	79.50		804.50	915
5900	Tub/shower enclosure, temp. glass, alum. frame, obscure glass		2	4		400	239		639	805
6200	Clear glass		1.50	5.333		835	320		1,155	1,400
6500	On chrome-plated brass frame, obscure glass		2	4		550	239		789	970
6600	Clear glass		1.50	5.333		1,175	320		1,495	1,775
6800	Tub surround, 3 wall, polypropylene	1 Carp	4	2		252	101		353	430
6900	PVC		4	2		375	101		476	565
7000	Fiberglass, obscure glass		4	2		390	101		491	585
7100	Clear glass		3	2.667		665	135		800	935

10 28 23 – Laundry Accessories

10 28 23.13 Built-In Ironing Boards

		Crew	Daily Output	Labor-Hours	Unit	Material	Labor	Equipment	Total	Total Incl O&P
0010	**BUILT-IN IRONING BOARDS**									
0020	Including cabinet, board & light, 42"	1 Carp	2	4	Ea.	445	203		648	800
0100	46"	"	1.50	5.333	"	520	270		790	985

10 31 Manufactured Fireplaces

10 31 13 – Manufactured Fireplace Chimneys

10 31 13.10 Fireplace Chimneys

		Crew	Daily Output	Labor-Hours	Unit	Material	Labor	Equipment	Total	Total Incl O&P
0010	**FIREPLACE CHIMNEYS**									
0500	Chimney dbl. wall, all stainless, over 8'-6", 7" diam., add to fireplace	1 Carp	33	.242	V.L.F.	87.50	12.30		99.80	115
0600	10" diameter, add to fireplace		32	.250		112	12.70		124.70	143
0700	12" diameter, add to fireplace		31	.258		172	13.10		185.10	210
0800	14" diameter, add to fireplace		30	.267		227	13.50		240.50	271
1000	Simulated brick chimney top, 4' high, 16" x 16"		10	.800	Ea.	455	40.50		495.50	565
1100	24" x 24"		7	1.143	"	560	58		618	705

10 31 13.20 Chimney Accessories

		Crew	Daily Output	Labor-Hours	Unit	Material	Labor	Equipment	Total	Total Incl O&P
0010	**CHIMNEY ACCESSORIES**									
0020	Chimney screens, galv., 13" x 13" flue	1 Bric	8	1	Ea.	58.50	50.50		109	141
0050	24" x 24" flue		5	1.600		124	80.50		204.50	260
0200	Stainless steel, 13" x 13" flue		8	1		97.50	50.50		148	184
0250	20" x 20" flue		5	1.600		152	80.50		232.50	290
2400	Squirrel and bird screens, galvanized, 8" x 8" flue		16	.500		56	25		81	100
2450	13" x 13" flue		12	.667		66.50	33.50		100	125

10 31 16 – Manufactured Fireplace Forms

10 31 16.10 Fireplace Forms

		Crew	Daily Output	Labor-Hours	Unit	Material	Labor	Equipment	Total	Total Incl O&P
0010	**FIREPLACE FORMS**									
1800	Fireplace forms, no accessories, 32" opening	1 Bric	3	2.667	Ea.	745	134		879	1,025
1900	36" opening		2.50	3.200		950	161		1,111	1,300
2000	40" opening		2	4		1,275	201		1,476	1,700
2100	78" opening		1.50	5.333		1,850	268		2,118	2,425

For customer support on your Building Construction Costs with RSMeans data, call 800.448.8182.

391

10 31 Manufactured Fireplaces

10 31 23 – Prefabricated Fireplaces

10 31 23.10 Fireplace, Prefabricated	Crew	Daily Output	Labor-Hours	Unit	Material	2018 Bare Costs Labor	Equipment	Total	Total Incl O&P
0010	**FIREPLACE, PREFABRICATED**, free standing or wall hung								
0100	With hood & screen, painted	1 Carp	1.30	6.154	Ea.	1,625	310		1,935	2,250
0150	Average		1	8		1,825	405		2,230	2,625
0200	Stainless steel		.90	8.889		3,250	450		3,700	4,250
1500	Simulated logs, gas fired, 40,000 BTU, 2' long, manual safety pilot		7	1.143	Set	545	58		603	690
1600	Adjustable flame remote pilot		6	1.333		1,275	67.50		1,342.50	1,500
1700	Electric, 1,500 BTU, 1'-6" long, incandescent flame		7	1.143		260	58		318	375
1800	1,500 BTU, LED flame		6	1.333		345	67.50		412.50	485
2000	Fireplace, built-in, 36" hearth, radiant		1.30	6.154	Ea.	720	310		1,030	1,275
2100	Recirculating, small fan		1	8		925	405		1,330	1,650
2150	Large fan		.90	8.889		2,075	450		2,525	2,975
2200	42" hearth, radiant		1.20	6.667		970	340		1,310	1,600
2300	Recirculating, small fan		.90	8.889		1,250	450		1,700	2,075
2350	Large fan		.80	10		1,450	505		1,955	2,375
2400	48" hearth, radiant		1.10	7.273		2,350	370		2,720	3,150
2500	Recirculating, small fan		.80	10		2,475	505		2,980	3,500
2550	Large fan		.70	11.429		2,575	580		3,155	3,700
3000	See through, including doors		.80	10		2,375	505		2,880	3,375
3200	Corner (2 wall)		1	8		3,150	405		3,555	4,100

10 32 Fireplace Specialties

10 32 13 – Fireplace Dampers

10 32 13.10 Dampers

		Crew	Daily Output	Labor-Hours	Unit	Material	2018 Bare Costs Labor	Equipment	Total	Total Incl O&P
0010	**DAMPERS**									
0800	Damper, rotary control, steel, 30" opening	1 Bric	6	1.333	Ea.	119	67		186	234
0850	Cast iron, 30" opening		6	1.333		125	67		192	240
0880	36" opening		6	1.333		127	67		194	243
0900	48" opening		6	1.333		167	67		234	287
0920	60" opening		6	1.333		355	67		422	495
0950	72" opening		5	1.600		425	80.50		505.50	590
1000	84" opening, special order		5	1.600		910	80.50		990.50	1,125
1050	96" opening, special order		4	2		925	101		1,026	1,175
1200	Steel plate, poker control, 60" opening		8	1		320	50.50		370.50	430
1250	84" opening, special order		5	1.600		585	80.50		665.50	770
1400	"Universal" type, chain operated, 32" x 20" opening		8	1		250	50.50		300.50	355
1450	48" x 24" opening		5	1.600		375	80.50		455.50	535

10 32 23 – Fireplace Doors

10 32 23.10 Doors

		Crew	Daily Output	Labor-Hours	Unit	Material	2018 Bare Costs Labor	Equipment	Total	Total Incl O&P
0010	**DOORS**									
0400	Cleanout doors and frames, cast iron, 8" x 8"	1 Bric	12	.667	Ea.	54.50	33.50		88	112
0450	12" x 12"		10	.800		89	40		129	160
0500	18" x 24"		8	1		150	50.50		200.50	242
0550	Cast iron frame, steel door, 24" x 30"		5	1.600		315	80.50		395.50	470
1600	Dutch oven door and frame, cast iron, 12" x 15" opening		13	.615		131	31		162	193
1650	Copper plated, 12" x 15" opening		13	.615		257	31		288	330

For customer support on your Building Construction Costs with RSMeans data, call 800.448.8182.

10 35 Stoves

10 35 13 – Heating Stoves

10 35 13.10 Wood Burning Stoves	Crew	Daily Output	Labor-Hours	Unit	Material	2018 Bare Costs Labor	Equipment	Total	Total Incl O&P
0010 **WOOD BURNING STOVES**									
0015 Cast iron, less than 1,500 S.F.	2 Carp	1.30	12.308	Ea.	1,350	625		1,975	2,425
0020 1,500 to 2,000 S.F.		1	16		2,075	810		2,885	3,500
0030 greater than 2,000 S.F.	↓	.80	20		2,800	1,025		3,825	4,625
0050 For gas log lighter, add				↓	47			47	52

10 43 Emergency Aid Specialties

10 43 13 – Defibrillator Cabinets

10 43 13.05 Defibrillator Cabinets

	Crew	Daily Output	Labor-Hours	Unit	Material	2018 Bare Costs Labor	Equipment	Total	Total Incl O&P
0010 **DEFIBRILLATOR CABINETS**, not equipped, stainless steel									
0050 Defibrillator cabinet, stainless steel with strobe & alarm 12" x 27"	1 Carp	10	.800	Ea.	455	40.50		495.50	560
0100 Automatic External Defibrillator	"	30	.267	"	1,375	13.50		1,388.50	1,550

10 44 Fire Protection Specialties

10 44 13 – Fire Protection Cabinets

10 44 13.53 Fire Equipment Cabinets

	Crew	Daily Output	Labor-Hours	Unit	Material	2018 Bare Costs Labor	Equipment	Total	Total Incl O&P
0010 **FIRE EQUIPMENT CABINETS**, not equipped, 20 ga. steel box									
0040 Recessed, D.S. glass in door, box size given									
1000 Portable extinguisher, single, 8" x 12" x 27", alum. door & frame	Q-12	8	2	Ea.	134	108		242	310
1100 Steel door and frame		8	2		128	108		236	305
2700 Fire blanket & extinguisher cab, inc blanket, rec stl., 14" x 40" x 8"		7	2.286		197	123		320	405
2800 Fire blanket cab, inc blanket, surf mtd, stl, 15"x10"x5", w/pwdr coat fin	↓	8	2	↓	99	108		207	272
3000 Hose rack assy., 1-1/2" valve & 100' hose, 24" x 40" x 5-1/2"									
3100 Aluminum door and frame	Q-12	6	2.667	Ea.	485	144		629	750
3200 Steel door and frame		6	2.667		271	144		415	515
3300 Stainless steel door and frame	↓	6	2.667	↓	505	144		649	770
4000 Hose rack assy., 2-1/2" x 1-1/2" valve, 100' hose, 24" x 40" x 8"									
4100 Aluminum door and frame	Q-12	6	2.667	Ea.	495	144		639	755
4200 Steel door and frame		6	2.667		335	144		479	580
4300 Stainless steel door and frame	↓	6	2.667	↓	655	144		799	935
5000 Hose rack assy., 2-1/2" x 1-1/2" valve, 100' hose									
5010 and extinguisher, 30" x 40" x 8"									
5100 Aluminum door and frame	Q-12	5	3.200	Ea.	625	172		797	950
5200 Steel door and frame		5	3.200		291	172		463	580
5300 Stainless steel door and frame	↓	5	3.200	↓	635	172		807	960
8000 Valve cabinet for 2-1/2" FD angle valve, 18" x 18" x 8"									
8100 Aluminum door and frame	Q-12	12	1.333	Ea.	216	72		288	345
8200 Steel door and frame		12	1.333		178	72		250	305
8300 Stainless steel door and frame	↓	12	1.333	↓	291	72		363	430

10 44 16 – Fire Extinguishers

10 44 16.13 Portable Fire Extinguishers

	Crew	Daily Output	Labor-Hours	Unit	Material	2018 Bare Costs Labor	Equipment	Total	Total Incl O&P
0010 **PORTABLE FIRE EXTINGUISHERS**									
0140 CO_2, with hose and "H" horn, 10 lb.				Ea.	300			300	330
0160 15 lb.					385			385	425
0180 20 lb.				↓	430			430	470
1000 Dry chemical, pressurized									
1040 Standard type, portable, painted, 2-1/2 lb.				Ea.	41			41	45
1060 5 lb.					56			56	61.50
1080 10 lb.				↓	85.50			85.50	94

10 44 Fire Protection Specialties

10 44 16 – Fire Extinguishers

10 44 16.13 Portable Fire Extinguishers

		Crew	Daily Output	Labor-Hours	Unit	Material	2018 Bare Costs Labor	Equipment	Total	Total Incl O&P
1100	20 lb.				Ea.	138			138	152
1120	30 lb.					425			425	465
1300	Standard type, wheeled, 150 lb.					1,250			1,250	1,375
2000	ABC all purpose type, portable, 2-1/2 lb.					23.50			23.50	26
2060	5 lb.					28			28	31
2080	9-1/2 lb.					51.50			51.50	56.50
2100	20 lb.					85			85	93.50
3500	Halotron 1, 2-1/2 lb.					132			132	145
3600	5 lb.					223			223	246
3700	11 lb.					445			445	490
5000	Pressurized water, 2-1/2 gallon, stainless steel					106			106	116
5060	With anti-freeze					118			118	130
9400	Installation of extinguishers, 12 or more, on nailable surface	1 Carp	30	.267			13.50		13.50	20.50
9420	On masonry or concrete	"	15	.533			27		27	41

10 44 16.16 Wheeled Fire Extinguisher Units

		Crew	Daily Output	Labor-Hours	Unit	Material	2018 Bare Costs Labor	Equipment	Total	Total Incl O&P
0010	**WHEELED FIRE EXTINGUISHER UNITS**									
0350	CO_2, portable, with swivel horn									
0360	Wheeled type, cart mounted, 50 lb.				Ea.	1,200			1,200	1,325
0400	100 lb.				"	4,450			4,450	4,900
2200	ABC all purpose type									
2300	Wheeled, 45 lb.				Ea.	775			775	850
2360	150 lb.				"	1,975			1,975	2,150

10 51 Lockers

10 51 13 – Metal Lockers

10 51 13.10 Lockers

		Crew	Daily Output	Labor-Hours	Unit	Material	2018 Bare Costs Labor	Equipment	Total	Total Incl O&P
0011	**LOCKERS** steel, baked enamel, knock down construction									
0012	Body- 24ga, door- 16 or 18ga									
0110	1-tier locker, 12" x 15" x 72"	1 Shee	20	.400	Ea.	271	24		295	335
0120	18" x 15" x 72"		20	.400		233	24		257	293
0130	12" x 18" x 72"		20	.400		197	24		221	254
0140	18" x 18" x 72"		20	.400		270	24		294	335
0410	2- tier, 12" x 15" x 36"		30	.267		252	15.95		267.95	300
0420	18" x 15" x 36"		30	.267		235	15.95		250.95	284
0430	12" x 18" x 36"		30	.267		300	15.95		315.95	355
0440	18" x 18" x 36"		30	.267		256	15.95		271.95	305
0450	3- tier 12" x 15" x 24"		30	.267		285	15.95		300.95	340
0455	12" x 18" x 24"		30	.267		294	15.95		309.95	350
0480	4- tier 12" x 15" x 18"		30	.267		365	15.95		380.95	425
0485	12" x 18" x 18"		30	.267		380	15.95		395.95	445
0500	Two person, 18" x 15" x 72"		20	.400		335	24		359	400
0510	18" x 18" x 72"		20	.400		300	24		324	365
0520	Duplex, 15" x 15" x 72"		20	.400		335	24		359	405
0530	15" x 21" x 72"		20	.400		375	24		399	450
0600	5 tier box lockers, unassembled		30	.267	Opng.	44	15.95		59.95	73
0700	Set up		24	.333		51.50	19.95		71.45	87
0900	6 tier box lockers, unassembled		36	.222		39	13.30		52.30	63
1000	Set up		30	.267		45	15.95		60.95	73.50
1098	All welded, ventilated athletic lockers, body- 16 ga., door- 14 ga.									
1100	1-tier, 12" x 18" x 72"	1 Shee	7.50	1.067	Ea.	335	64		399	470
1110	18" x 15" x 72"		7.50	1.067		545	64		609	695

For customer support on your Building Construction Costs with RSMeans data, call 800.448.8182.

10 51 Lockers

10 51 13 – Metal Lockers

10 51 13.10 Lockers

		Crew	Daily Output	Labor-Hours	Unit	Material	2018 Bare Costs Labor	Equipment	Total	Total Incl O&P
1115	18" x 18" x 72"	1 Shee	7.50	1.067	Ea.	555	64		619	710
1130	2-tier, 12" x 15" x 36"		7.50	1.067		565	64		629	725
1135	12" x 18" x 36"		7.50	1.067		605	64		669	765
1140	18" x 15" x 36"		7.50	1.067		660	64		724	825
1145	18" x 18" x 36"		7.50	1.067		670	64		734	835
1160	3-tier, 12" x 18" x 24"		7.50	1.067		775	64		839	950
1180	4-tier, 12" x 18" x 18"		7.50	1.067		590	64		654	745
1190	5H box, 12" x 18" x 14.4"		7.50	1.067		605	64		669	765
1195	18" x 18" x 14.4"		7.50	1.067		775	64		839	950
1196	6H box, 12" x 15" x 12"		7.50	1.067		485	64		549	635
1197	12" x 18" x 12"		7.50	1.067		595	64		659	755
1198	18" x 18" x 12"	▼	7.50	1.067	▼	735	64		799	905
2399	Standard duty lockers, body- 24 ga., doors- 16-18 ga.									
2400	16-person locker unit with clothing rack									
2500	72" wide x 15" deep x 72" high	1 Shee	15	.533	Ea.	500	32		532	600
2550	18" deep	"	15	.533	"	660	32		692	780
3000	Wall mounted lockers, 4 person, with coat bar									
3100	48" wide x 18" deep x 12" high	1 Shee	20	.400	Ea.	330	24		354	400
3250	Rack w/24 wire mesh baskets		1.50	5.333	Set	400	320		720	925
3260	30 baskets		1.25	6.400		390	385		775	1,025
3270	36 baskets		.95	8.421		445	505		950	1,250
3280	42 baskets	▼	.80	10	▼	455	600		1,055	1,425
3300	For built-in lock with 2 keys, add				Ea.	11.90			11.90	13.05
3600	For hanger rods, add					2.48			2.48	2.73
3650	For number plate kit, 100 plates #1 - #100, add	1 Shee	4	2		82	120		202	273
3700	For locker base, closed front panel		90	.089		5.10	5.30		10.40	13.70
3710	End panel, bolted		36	.222		8.40	13.30		21.70	30
3800	For sloping top, 12" wide		24	.333		31.50	19.95		51.45	65
3810	15" wide		24	.333		31	19.95		50.95	64.50
3820	18" wide		24	.333		34	19.95		53.95	68
3850	Sloping top end panel, 12" deep		72	.111		16.05	6.65		22.70	28
3860	15" deep		72	.111		16.60	6.65		23.25	28.50
3870	18" deep		72	.111		17.10	6.65		23.75	29
3900	For finish end panels, steel, 60" high, 15" deep		12	.667		30	40		70	94
3910	72" high, 12" deep		12	.667		27	40		67	90.50
3920	18" deep	▼	12	.667	▼	41.50	40		81.50	107
5000	For "ready to assemble" lockers,									
5010	Add to labor						75%			
5020	Deduct from material					20%				
6000	Heavy duty for detention facility, tamper proof, 14 ga. welded steel, solid									
6100	24" W x 24" D x 74" H, single tier	1 Shee	18	.444	Ea.	540	26.50		566.50	635
6110	Double tier		18	.444		535	26.50		561.50	630
6120	Triple tier	▼	18	.444	▼	555	26.50		581.50	650

10 51 26 – Plastic Lockers

10 51 26.13 Recycled Plastic Lockers

			Crew	Daily Output	Labor-Hours	Unit	Material	Labor	Equipment	Total	Total Incl O&P
0011	**RECYCLED PLASTIC LOCKERS**, 30% recycled										
0110	Single tier box locker, 12" x 12" x 72"	G	1 Shee	8	1	Ea.	420	60		480	555
0120	12" x 15" x 72"	G		8	1		450	60		510	585
0130	12" x 18" x 72"	G		8	1		425	60		485	560
0410	Double tier, 12" x 12" x 72"	G		21	.381		455	23		478	535
0420	12" x 15" x 72"	G		21	.381		485	23		508	570
0430	12" x 18" x 72"	G	▼	21	.381	▼	450	23		473	530

10 51 Lockers

10 51 53 – Locker Room Benches

10 51 53.10 Benches	Crew	Daily Output	Labor-Hours	Unit	Material	2018 Bare Costs Labor	Equipment	Total	Total Incl O&P
0010 **BENCHES**									
2100 Locker bench, laminated maple, top only	1 Shee	100	.080	L.F.	26	4.78		30.78	36.50
2200 Pedestals, steel pipe		25	.320	Ea.	45.50	19.15		64.65	79
2250 Plastic, 9.5" top with PVC pedestals		80	.100	L.F.	62	6		68	77

10 55 Postal Specialties

10 55 23 – Mail Boxes

10 55 23.10 Commercial Mail Boxes

	Crew	Daily Output	Labor-Hours	Unit	Material	2018 Bare Costs Labor	Equipment	Total	Total Incl O&P
0010 **COMMERCIAL MAIL BOXES**									
0020 Horiz., key lock, 5"H x 6"W x 15"D, alum., rear load	1 Carp	34	.235	Ea.	42.50	11.95		54.45	65
0100 Front loading		34	.235		42.50	11.95		54.45	65
0200 Double, 5"H x 12"W x 15"D, rear loading		26	.308		66.50	15.60		82.10	97
0300 Front loading		26	.308		74.50	15.60		90.10	106
0500 Quadruple, 10"H x 12"W x 15"D, rear loading		20	.400		115	20.50		135.50	158
0600 Front loading		20	.400		95.50	20.50		116	136
0800 Vertical, front load, 15"H x 5"W x 6"D, alum., per compartment		34	.235		45	11.95		56.95	67.50
0900 Bronze, duranodic finish		34	.235		48.50	11.95		60.45	71.50
1000 Steel, enameled		34	.235		45	11.95		56.95	67.50
1700 Alphabetical directories, 120 names		10	.800		104	40.50		144.50	176
1800 Letter collection box		6	1.333		725	67.50		792.50	905
1830 Lobby collection boxes, aluminum	2 Shee	5	3.200		1,750	191		1,941	2,225
1840 Bronze or stainless	"	4.50	3.556		1,950	213		2,163	2,475
1900 Letter slot, residential	1 Carp	20	.400		80.50	20.50		101	120
2000 Post office type		8	1		120	50.50		170.50	209
2250 Key keeper, single key, aluminum		26	.308		45	15.60		60.60	73.50
2300 Steel, enameled		26	.308		77	15.60		92.60	109

10 56 Storage Assemblies

10 56 13 – Metal Storage Shelving

10 56 13.10 Shelving

	Crew	Daily Output	Labor-Hours	Unit	Material	2018 Bare Costs Labor	Equipment	Total	Total Incl O&P
0010 **SHELVING**									
0020 Metal, industrial, cross-braced, 3' W, 12" D	1 Sswk	175	.046	SF Shlf	8.10	2.50		10.60	13
0100 24" D		330	.024		6.10	1.32		7.42	8.85
0300 4' W, 12" D		185	.043		7.55	2.36		9.91	12.15
0400 24" D		380	.021		5.45	1.15		6.60	7.90
1200 Enclosed sides, cross-braced back, 3' W, 12" D		175	.046		13.35	2.50		15.85	18.80
1300 24" D		290	.028		7.80	1.51		9.31	11
1500 Fully enclosed, sides and back, 3' W, 12" D		150	.053		16.55	2.91		19.46	23
1600 24" D		255	.031		11.15	1.71		12.86	15.05
1800 4' W, 12" D		150	.053		10.30	2.91		13.21	16.05
1900 24" D		290	.028		9	1.51		10.51	12.35
2200 Wide span, 1600 lb. capacity per shelf, 6' W, 24" D		380	.021		7.15	1.15		8.30	9.80
2400 36" D		440	.018		6.10	.99		7.09	8.30
2600 8' W, 24" D		440	.018		6.85	.99		7.84	9.15
2800 36" D		520	.015		4.61	.84		5.45	6.40
4000 Pallet racks, steel frame 5,000 lb. capacity, 8' long, 36" D	2 Sswk	450	.036		9.35	1.94		11.29	13.45
4200 42" D		500	.032		8.20	1.75		9.95	11.90
4400 48" D		520	.031		7.35	1.68		9.03	10.80

10 56 Storage Assemblies

10 56 13 – Metal Storage Shelving

10 56 13.20 Parts Bins

		Crew	Daily Output	Labor-Hours	Unit	Material	2018 Bare Costs Labor	Equipment	Total	Total Incl O&P
0010	**PARTS BINS** metal, gray baked enamel finish									
0100	6'-3" high, 3' wide									
0300	12 bins, 18" wide x 12" high, 12" deep	2 Clab	10	1.600	Ea.	350	64		414	480
0400	24" deep		10	1.600		455	64		519	595
0600	72 bins, 6" wide x 6" high, 12" deep		8	2		535	79.50		614.50	710
0700	18" deep	↓	8	2	↓	845	79.50		924.50	1,050
1000	7'-3" high, 3' wide									
1200	14 bins, 18" wide x 12" high, 12" deep	2 Clab	10	1.600	Ea.	350	64		414	480
1300	24" deep		10	1.600		440	64		504	580
1500	84 bins, 6" wide x 6" high, 12" deep		8	2		890	79.50		969.50	1,100
1600	24" deep	↓	8	2	↓	1,075	79.50		1,154.50	1,325

10 57 Wardrobe and Closet Specialties

10 57 13 – Hat and Coat Racks

10 57 13.10 Coat Racks and Wardrobes

		Crew	Daily Output	Labor-Hours	Unit	Material	2018 Bare Costs Labor	Equipment	Total	Total Incl O&P
0010	**COAT RACKS AND WARDROBES**									
0020	Hat & coat rack, floor model, 6 hangers									
0050	Standing, beech wood, 21" x 21" x 72", chrome				Ea.	260			260	286
0100	18 ga. tubular steel, 21" x 21" x 69", wood walnut				"	350			350	385
0500	16 ga. steel frame, 22 ga. steel shelves									
0650	Single pedestal, 30" x 18" x 63"				Ea.	345			345	380
0800	Single face rack, 29" x 18-1/2" x 62"					315			315	345
0900	51" x 18-1/2" x 70"					435			435	480
0910	Double face rack, 39" x 26" x 70"					405			405	445
0920	63" x 26" x 70"				↓	540			540	595
0940	For 2" ball casters, add				Set	83			83	91.50
1400	Utility hook strips, 3/8" x 2-1/2" x 18", 6 hooks	1 Carp	48	.167	Ea.	56	8.45		64.45	74.50
1500	34" long, 12 hooks	"	48	.167	"	68	8.45		76.45	88
1650	Wall mounted racks, 16 ga. steel frame, 22 ga. steel shelves									
1850	12" x 15" x 26", 6 hangers	1 Carp	32	.250	Ea.	140	12.70		152.70	173
2000	12" x 15" x 50", 12 hangers	"	32	.250	"	162	12.70		174.70	197
2150	Wardrobe cabinet, steel, baked enamel finish									
2300	36" x 21" x 78", incl. top shelf & hanger rod				Ea.	330			330	365
2400	Wardrobe, 24" x 24" x 76", KD, w/door, hospital, baked enamel steel	1 Carp	2	4		685	203		888	1,075
2500	Hardwood	"	2	4	↓	1,225	203		1,428	1,650

10 57 23 – Closet and Utility Shelving

10 57 23.19 Wood Closet and Utility Shelving

		Crew	Daily Output	Labor-Hours	Unit	Material	2018 Bare Costs Labor	Equipment	Total	Total Incl O&P
0010	**WOOD CLOSET AND UTILITY SHELVING**									
0020	Pine, clear grade, no edge band, 1" x 8"	1 Carp	115	.070	L.F.	3.52	3.53		7.05	9.20
0100	1" x 10"		110	.073		4.38	3.69		8.07	10.40
0200	1" x 12"		105	.076		5.30	3.86		9.16	11.70
0600	Plywood, 3/4" thick with lumber edge, 12" wide		75	.107		1.91	5.40		7.31	10.35
0700	24" wide		70	.114	↓	3.38	5.80		9.18	12.50
0900	Bookcase, clear grade pine, shelves 12" OC, 8" deep, per S.F. shelf		70	.114	S.F.	11.45	5.80		17.25	21.50
1000	12" deep shelves		65	.123	"	17.15	6.25		23.40	28.50
1200	Adjustable closet rod and shelf, 12" wide, 3' long		20	.400	Ea.	12.60	20.50		33.10	45
1300	8' long		15	.533	"	24.50	27		51.50	68
1500	Prefinished shelves with supports, stock, 8" wide		75	.107	L.F.	5.60	5.40		11	14.40
1600	10" wide	↓	70	.114	"	5.40	5.80		11.20	14.75

For customer support on your Building Construction Costs with RSMeans data, call 800.448.8182.

10 71 Exterior Protection

10 71 13 – Exterior Sun Control Devices

10 71 13.19 Rolling Exterior Shutters	Crew	Daily Output	Labor-Hours	Unit	Material	2018 Bare Costs Labor	Equipment	Total	Total Incl O&P
0010 **ROLLING EXTERIOR SHUTTERS**									
0020 Roll-up, manual operation, aluminum, 3' x 4', incl. frame	2 Carp	8	2	Ea.	625	101		726	840
0030 6' x 7'	"	8	2	"	1,300	101		1,401	1,575

10 73 Protective Covers

10 73 13 – Awnings

10 73 13.10 Awnings, Fabric

		Crew	Daily Output	Labor-Hours	Unit	Material	2018 Bare Costs Labor	Equipment	Total	Total Incl O&P
0010	**AWNINGS, FABRIC**									
0020	Including acrylic canvas and frame, standard design									
0100	Door and window, slope, 3' high, 4' wide	1 Carp	4.50	1.778	Ea.	755	90		845	965
0110	6' wide		3.50	2.286		970	116		1,086	1,250
0120	8' wide		3	2.667		1,200	135		1,335	1,500
0200	Quarter round convex, 4' wide		3	2.667		1,175	135		1,310	1,500
0210	6' wide		2.25	3.556		1,525	180		1,705	1,950
0220	8' wide		1.80	4.444		1,875	225		2,100	2,400
0300	Dome, 4' wide		7.50	1.067		455	54		509	585
0310	6' wide		3.50	2.286		1,025	116		1,141	1,300
0320	8' wide		2	4		1,800	203		2,003	2,300
0350	Elongated dome, 4' wide		1.33	6.015		1,700	305		2,005	2,350
0360	6' wide		1.11	7.207		2,025	365		2,390	2,775
0370	8' wide		1	8		2,375	405		2,780	3,250
1000	Entry or walkway, peak, 12' long, 4' wide	2 Carp	.90	17.778		5,475	900		6,375	7,400
1010	6' wide		.60	26.667		8,450	1,350		9,800	11,300
1020	8' wide		.40	40		11,700	2,025		13,725	15,900
1100	Radius with dome end, 4' wide		1.10	14.545		4,150	735		4,885	5,700
1110	6' wide		.70	22.857		6,675	1,150		7,825	9,125
1120	8' wide		.50	32		9,500	1,625		11,125	12,900
2000	Retractable lateral arm awning, manual									
2010	To 12' wide, 8'-6" projection	2 Carp	1.70	9.412	Ea.	1,250	475		1,725	2,075
2020	To 14' wide, 8'-6" projection		1.10	14.545		1,450	735		2,185	2,725
2030	To 19' wide, 8'-6" projection		.85	18.824		1,950	955		2,905	3,600
2040	To 24' wide, 8'-6" projection		.67	23.881		2,475	1,200		3,675	4,575
2050	Motor for above, add	1 Carp	2.67	3		1,050	152		1,202	1,400
3000	Patio/deck canopy with frame									
3010	12' wide, 12' projection	2 Carp	2	8	Ea.	1,750	405		2,155	2,550
3020	16' wide, 14' projection	"	1.20	13.333		2,725	675		3,400	4,025
9000	For fire retardant canvas, add					7%				
9010	For lettering or graphics, add					35%				
9020	For painted or coated acrylic canvas, deduct					8%				
9030	For translucent or opaque vinyl canvas, add					10%				
9040	For 6 or more units, deduct					20%	15%			

10 73 16 – Canopies

10 73 16.20 Metal Canopies

		Crew	Daily Output	Labor-Hours	Unit	Material	2018 Bare Costs Labor	Equipment	Total	Total Incl O&P
0010	**METAL CANOPIES**									
0020	Wall hung, .032", aluminum, prefinished, 8' x 10'	K-2	1.30	18.462	Ea.	2,500	960	183	3,643	4,475
0300	8' x 20'		1.10	21.818		4,150	1,125	216	5,491	6,600
0500	10' x 10'		1.30	18.462		2,950	960	183	4,093	4,950
0700	10' x 20'		1.10	21.818		4,700	1,125	216	6,041	7,200

For customer support on your Building Construction Costs with RSMeans data, call 800.448.8182.

10 73 Protective Covers

10 73 16 – Canopies

10 73 16.20 Metal Canopies

		Crew	Daily Output	Labor-Hours	Unit	Material	2018 Bare Costs Labor	Equipment	Total	Total Incl O&P
1000	12' x 20'	K-2	1	24	Ea.	5,625	1,250	238	7,113	8,450
1360	12' x 30'		.80	30		8,450	1,550	298	10,298	12,100
1700	12' x 40'		.60	40		11,300	2,075	395	13,770	16,200
1900	For free standing units, add					20%	10%			
2300	Aluminum entrance canopies, flat soffit, .032"									
2500	3'-6" x 4'-0", clear anodized	2 Carp	4	4	Ea.	1,050	203		1,253	1,450
2700	Bronze anodized		4	4		1,825	203		2,028	2,325
3000	Polyurethane painted		4	4		1,475	203		1,678	1,925
3300	4'-6" x 10'-0", clear anodized		2	8		2,875	405		3,280	3,775
3500	Bronze anodized		2	8		3,675	405		4,080	4,650
3700	Polyurethane painted		2	8		3,075	405		3,480	4,000
4000	Wall downspout, 10 L.F., clear anodized	1 Carp	7	1.143		178	58		236	284
4300	Bronze anodized		7	1.143		310	58		368	430
4500	Polyurethane painted		7	1.143		267	58		325	380
7000	Carport, baked vinyl finish, .032", 20' x 10', no foundations, flat panel	K-2	4	6	Car	4,325	310	59.50	4,694.50	5,325
7250	Insulated flat panel		2	12	"	4,825	625	119	5,569	6,450
7500	Walkway cover, to 12' wide, stl., vinyl finish, .032", no fndtns., flat		250	.096	S.F.	22.50	4.99	.95	28.44	34
7750	Arched		200	.120	"	55	6.25	1.19	62.44	72

10 74 Manufactured Exterior Specialties

10 74 23 – Cupolas

10 74 23.10 Wood Cupolas

		Crew	Daily Output	Labor-Hours	Unit	Material	2018 Bare Costs Labor	Equipment	Total	Total Incl O&P
0010	**WOOD CUPOLAS**									
0020	Stock units, pine, painted, 18" sq., 28" high, alum. roof	1 Carp	4.10	1.951	Ea.	285	99		384	465
0100	Copper roof		3.80	2.105		291	107		398	485
0300	23" square, 33" high, aluminum roof		3.70	2.162		440	110		550	650
0400	Copper roof		3.30	2.424		590	123		713	835
0600	30" square, 37" high, aluminum roof		3.70	2.162		600	110		710	825
0700	Copper roof		3.30	2.424		715	123		838	975
0900	Hexagonal, 31" wide, 46" high, copper roof		4	2		945	101		1,046	1,200
1000	36" wide, 50" high, copper roof		3.50	2.286		1,675	116		1,791	2,000
1200	For deluxe stock units, add to above					25%				
1400	For custom built units, add to above					50%	50%			

10 74 29 – Steeples

10 74 29.10 Prefabricated Steeples

		Crew	Daily Output	Labor-Hours	Unit	Material	2018 Bare Costs Labor	Equipment	Total	Total Incl O&P
0010	**PREFABRICATED STEEPLES**									
4000	Steeples, translucent fiberglass, 30" square, 15' high	F-3	2	20	Ea.	9,300	1,025	248	10,573	12,000
4150	25' high		1.80	22.222		10,500	1,150	276	11,926	13,700
4350	Opaque fiberglass, 24" square, 14' high		2	20		7,575	1,025	248	8,848	10,200
4500	28' high		1.80	22.222		6,325	1,150	276	7,751	9,000
4600	Aluminum, baked finish, 16" square, 14' high					6,150			6,150	6,750
4620	20' high, 3'-6" base					10,700			10,700	11,800
4640	35' high, 8' base					38,400			38,400	42,200
4660	60' high, 14' base					76,500			76,500	84,000
4680	152' high, custom					628,500			628,500	691,500
4700	Porcelain enamel steeples, custom, 40' high	F-3	.50	80		14,100	4,150	990	19,240	23,000
4800	60' high	"	.30	133		20,000	6,900	1,650	28,550	34,300

For customer support on your Building Construction Costs with RSMeans data, call 800.448.8182.

399

10 74 Manufactured Exterior Specialties

10 74 46 – Window Wells

10 74 46.10 Area Window Wells	Crew	Daily Output	Labor-Hours	Unit	Material	Labor	Equipment	Total	Total Incl O&P
0010 **AREA WINDOW WELLS**, Galvanized steel									
0020 20 ga., 3'-2" wide, 1' deep	1 Sswk	29	.276	Ea.	16.95	15.10		32.05	43
0100 2' deep		23	.348		31	19		50	65.50
0300 16 ga., 3'-2" wide, 1' deep		29	.276		23	15.10		38.10	50
0400 3' deep		23	.348		47	19		66	82.50
0600 Welded grating for above, 15 lb., painted		45	.178		91.50	9.70		101.20	117
0700 Galvanized		45	.178		124	9.70		133.70	153
0900 Translucent plastic cap for above		60	.133		20.50	7.30		27.80	34.50

10 75 Flagpoles

10 75 16 – Ground-Set Flagpoles

10 75 16.10 Flagpoles

	Crew	Daily Output	Labor-Hours	Unit	Material	Labor	Equipment	Total	Total Incl O&P
0010 **FLAGPOLES**, ground set									
0050 Not including base or foundation									
0100 Aluminum, tapered, ground set 20' high	K-1	2	8	Ea.	1,100	380	119	1,599	1,900
0200 25' high		1.70	9.412		1,175	450	140	1,765	2,100
0300 30' high		1.50	10.667		1,400	510	159	2,069	2,475
0400 35' high		1.40	11.429		1,875	545	170	2,590	3,075
0500 40' high		1.20	13.333		3,025	635	198	3,858	4,500
0600 50' high		1	16		3,525	760	238	4,523	5,275
0700 60' high		.90	17.778		4,975	845	265	6,085	7,050
0800 70' high		.80	20		8,950	950	298	10,198	11,600
1100 Counterbalanced, internal halyard, 20' high		1.80	8.889		2,800	425	132	3,357	3,850
1200 30' high		1.50	10.667		2,975	510	159	3,644	4,225
1300 40' high		1.30	12.308		7,125	585	183	7,893	8,900
1400 50' high		1	16		9,875	760	238	10,873	12,300
2820 Aluminum, electronically operated, 30' high		1.40	11.429		4,500	545	170	5,215	5,950
2840 35' high		1.30	12.308		5,350	585	183	6,118	6,950
2860 39' high		1.10	14.545		6,500	690	216	7,406	8,450
2880 45' high		1	16		6,850	760	238	7,848	8,950
2900 50' high		.90	17.778		8,800	845	265	9,910	11,200
3000 Fiberglass, tapered, ground set, 23' high		2	8		675	380	119	1,174	1,450
3100 29'-7" high		1.50	10.667		1,325	510	159	1,994	2,425
3200 36'-1" high		1.40	11.429		1,625	545	170	2,340	2,800
3300 39'-5" high		1.20	13.333		1,950	635	198	2,783	3,325
3400 49'-2" high		1	16		4,575	760	238	5,573	6,425
3500 59' high		.90	17.778		5,200	845	265	6,310	7,300
4300 Steel, direct imbedded installation									
4400 Internal halyard, 20' high	K-1	2.50	6.400	Ea.	1,375	305	95	1,775	2,100
4500 25' high		2.50	6.400		2,075	305	95	2,475	2,850
4600 30' high		2.30	6.957		2,475	330	104	2,909	3,325
4700 40' high		2.10	7.619		3,750	365	113	4,228	4,800
4800 50' high		1.90	8.421		4,375	400	125	4,900	5,550
5000 60' high		1.80	8.889		7,275	425	132	7,832	8,775
5100 70' high		1.60	10		7,925	475	149	8,549	9,600
5200 80' high		1.40	11.429		10,200	545	170	10,915	12,200
5300 90' high		1.20	13.333		16,400	635	198	17,233	19,300
5500 100' high		1	16		18,100	760	238	19,098	21,300
6400 Wood poles, tapered, clear vertical grain fir with tilting									
6410 base, not incl. foundation, 4" butt, 25' high	K-1	1.90	8.421	Ea.	1,625	400	125	2,150	2,550

For customer support on your Building Construction Costs with RSMeans data, call 800.448.8182.

10 75 Flagpoles

10 75 16 – Ground-Set Flagpoles

10 75 16.10 Flagpoles	Crew	Daily Output	Labor-Hours	Unit	Material	2018 Bare Costs Labor	Equipment	Total	Total Incl O&P
6800 6" butt, 30' high	K-1	1.30	12.308	Ea.	2,950	585	183	3,718	4,325
7300 Foundations for flagpoles, including									
7400 excavation and concrete, to 35' high poles	C-1	10	3.200	Ea.	710	154		864	1,025
7600 40' to 50' high		3.50	9.143		1,325	440		1,765	2,125
7700 Over 60' high	↓	2	16	↓	1,650	770		2,420	2,975

10 75 23 – Wall-Mounted Flagpoles

10 75 23.10 Flagpoles	Crew	Daily Output	Labor-Hours	Unit	Material	2018 Bare Costs Labor	Equipment	Total	Total Incl O&P
0010 **FLAGPOLES**, structure mounted									
0100 Fiberglass, vertical wall set, 19'-8" long	K-1	1.50	10.667	Ea.	1,050	510	159	1,719	2,100
0200 23' long		1.40	11.429		1,575	545	170	2,290	2,725
0300 26'-3" long		1.30	12.308		1,775	585	183	2,543	3,025
0800 19'-8" long outrigger		1.30	12.308		1,375	585	183	2,143	2,600
1300 Aluminum, vertical wall set, tapered, with base, 20' high		1.20	13.333		1,200	635	198	2,033	2,500
1400 29'-6" high		1	16		2,925	760	238	3,923	4,600
2400 Outrigger poles with base, 12' long		1.30	12.308		1,300	585	183	2,068	2,500
2500 14' long	↓	1	16	↓	1,550	760	238	2,548	3,100

10 81 Pest Control Devices

10 81 13 – Bird Control Devices

10 81 13.10 Bird Control Netting	Crew	Daily Output	Labor-Hours	Unit	Material	2018 Bare Costs Labor	Equipment	Total	Total Incl O&P
0010 **BIRD CONTROL NETTING**									
0020 1/8" square mesh	4 Clab	4000	.008	S.F.	.56	.32		.88	1.11
0100 1/4" square mesh		4000	.008		.64	.32		.96	1.19
0120 1/2" square mesh		4000	.008		.14	.32		.46	.64
0140 5/8" x 3/4" mesh		4000	.008		.08	.32		.40	.58
0160 1-1/4" x 1-1/2" mesh		4000	.008		.16	.32		.48	.67
0200 4" square mesh	↓	4000	.008	↓	.12	.32		.44	.62
1000 Poly clips				Ea.	.24			.24	.26

10 86 Security Mirrors and Domes

10 86 10 – Security Mirrors

10 86 10.10 Exterior Traffic Control Mirrors	Crew	Daily Output	Labor-Hours	Unit	Material	2018 Bare Costs Labor	Equipment	Total	Total Incl O&P
0010 **EXTERIOR TRAFFIC CONTROL MIRRORS**									
0100 Convex, stainless steel, 20 ga., 26" diameter	1 Carp	12	.667	Ea.	172	34		206	241

10 86 20 – Security Domes

10 86 20.10 Domes	Crew	Daily Output	Labor-Hours	Unit	Material	2018 Bare Costs Labor	Equipment	Total	Total Incl O&P
0010 **DOMES** for security cameras (CCTV)									
0100 Ceiling mounted, 10" diameter	1 Carp	30	.267	Ea.	12.50	13.50		26	34.50
0110 12" diameter	"	30	.267	"	9	13.50		22.50	30.50

For customer support on your Building Construction Costs with RSMeans data, call 800.448.8182.

401

10 88 05.10 Scales	Crew	Daily Output	Labor-Hours	Unit	Material	2018 Bare Costs Labor	Equipment	Total	Total Incl O&P
0010 **SCALES**									
0700 Truck scales, incl. steel weigh bridge,									
0800 not including foundation, pits									
1550 Digital, electronic, 100 ton capacity, steel deck 12' x 10' platform	3 Carp	.20	120	Ea.	14,900	6,075		20,975	25,700
1600 40' x 10' platform		.14	171		29,700	8,700		38,400	45,900
1640 60' x 10' platform		.13	185		39,400	9,350		48,750	57,500
1680 70' x 10' platform		.12	200		41,800	10,100		51,900	61,500
2000 For standard automatic printing device, add					1,400			1,400	1,525
2100 For remote reading electronic system, add					2,800			2,800	3,075
2300 Concrete foundation pits for above, 8' x 6', 5 C.Y. required	C-1	.50	64		1,125	3,075		4,200	5,925
2400 14' x 6' platform, 10 C.Y. required		.35	91.429		1,675	4,400		6,075	8,500
2600 50' x 10' platform, 30 C.Y. required		.25	128		2,250	6,150		8,400	11,800
2700 70' x 10' platform, 40 C.Y. required		.15	213		4,900	10,200		15,100	21,000
2750 Crane scales, dial, 1 ton capacity					1,150			1,150	1,250
2780 5 ton capacity					1,475			1,475	1,600
2800 Digital, 1 ton capacity					1,900			1,900	2,075
2850 10 ton capacity					4,800			4,800	5,300
2900 Low profile electronic warehouse scale,									
3000 not incl. printer, 4' x 4' platform, 10,000 lb. capacity	2 Carp	.30	53.333	Ea.	1,425	2,700		4,125	5,675
3300 5' x 7' platform, 10,000 lb. capacity		.25	64		5,100	3,250		8,350	10,600
3400 20,000 lb. capacity		.20	80		6,275	4,050		10,325	13,100
3500 For printers, incl. time, date & numbering, add					655			655	720
3800 Portable, beam type, capacity 1000 lb., platform 18" x 24"					860			860	945
3900 Dial type, capacity 2000 lb., platform 24" x 24"					1,500			1,500	1,650
4000 Digital type, capacity 1000 lb., platform 24" x 30"					2,400			2,400	2,650
4100 Portable contractor truck scales, 50 ton cap., 40' x 10' platform					27,800			27,800	30,600
4200 60' x 10' platform					33,500			33,500	36,900
4400 Heavy-Duty Steel Deck Truck Scales 20' x 10'	3 Carp	.20	120		18,000	6,075		24,075	29,100
4500 Heavy-Duty Steel Deck Truck Scales 80' x 10'		.10	240		51,500	12,200		63,700	75,000
4600 Heavy-Duty Steel Deck Truck Scales 160' x 10'		.04	600		104,000	30,400		134,400	161,000
4700 Heavy-Duty Steel Deck Truck Scales 20' x 12'		.18	137		21,000	6,950		27,950	33,700
4800 Heavy-Duty Steel Deck Truck Scales 80' x 12'		.09	267		60,000	13,500		73,500	86,500
4900 Heavy-Duty Steel Deck Truck Scales 160' x 12'		.04	600		121,000	30,400		151,400	179,500
5000 Heavy-Duty Steel Deck Truck Scales 20' x 14'		.16	150		27,900	7,600		35,500	42,300
5100 Heavy-Duty Steel Deck Truck Scales 80' x 14'		.06	400		66,000	20,300		86,300	103,500
5200 Heavy-Duty Steel Deck Truck Scales 160' x 14'		.03	800		129,000	40,600		169,600	203,500

Estimating Tips
General

- The items in this division are usually priced per square foot or each. Many of these items are purchased by the owner for installation by the contractor. Check the specifications for responsibilities and include time for receiving, storage, installation, and mechanical and electrical hookups in the appropriate divisions.

- Many items in Division 11 require some type of support system that is not usually furnished with the item. Examples of these systems include blocking for the attachment of casework and support angles for ceiling-hung projection screens. The required blocking or supports must be added to the estimate in the appropriate division.

- Some items in Division 11 may require assembly or electrical hookups. Verify the amount of assembly required or the need for a hard electrical connection and add the appropriate costs.

Reference Numbers

Reference numbers are shown at the beginning of some major classifications. These numbers refer to related items in the Reference Section. The reference information may be an estimating procedure, an alternate pricing method, or technical information.

Note: Not all subdivisions listed here necessarily appear. ■

Did you know?

RSMeans data is available through our online application with 24/7 access:

- Search for unit prices by keyword
- Leverage the most up-to-date data
- Build and export estimates

Try it free for 30 days!
www.rsmeans.com/2018freetrial

No part of this cost data may be reproduced, stored in a retrieval system, or transmitted in any form or by any means without prior written permission of Gordian.

11 05 Common Work Results for Equipment

11 05 05 – Selective Demolition for Equipment

11 05 05.10 Selective Demolition	Crew	Daily Output	Labor-Hours	Unit	Material	2018 Bare Costs Labor	2018 Bare Costs Equipment	Total	Total Incl O&P
0010 **SELECTIVE DEMOLITION**									
0130 Central vacuum, motor unit, residential or commercial	1 Clab	2	4	Ea.		159		159	243
0210 Vault door and frame	2 Skwk	2	8			420		420	640
0215 Day gate, for vault	"	3	5.333			279		279	425
0380 Bank equipment, teller window, bullet resistant	1 Clab	1.20	6.667			266		266	405
0381 Counter	2 Clab	1.50	10.667	Station		425		425	645
0382 Drive-up window, including drawer and glass		1.50	10.667	"		425		425	645
0383 Thru-wall boxes and chests, selective demolition		2.50	6.400	Ea.		255		255	390
0384 Bullet resistant partitions		20	.800	L.F.		32		32	48.50
0385 Pneumatic tube system, 2 lane drive-up	L-3	.45	35.556	Ea.		1,950		1,950	2,950
0386 Safety deposit box	1 Clab	50	.160	Opng.		6.40		6.40	9.70
0387 Surveillance system, video, complete	2 Elec	2	8	Ea.		465		465	695
0410 Church equipment, misc movable fixtures	2 Clab	1	16			640		640	970
0412 Steeple, to 28' high	F-3	3	13.333			690	165	855	1,225
0414 40' to 60' high	"	.80	50			2,600	620	3,220	4,600
0510 Library equipment, bookshelves, wood, to 90" high	1 Clab	20	.400	L.F.		15.95		15.95	24.50
0515 Carrels, hardwood, 36" x 24"	"	9	.889	Ea.		35.50		35.50	54
0630 Stage equipment, light control panel	1 Elec	1	8	"		465		465	695
0632 Border lights		40	.200	L.F.		11.65		11.65	17.40
0634 Spotlights		8	1	Ea.		58		58	87
0636 Telescoping platforms and risers	2 Clab	175	.091	SF Stg.		3.64		3.64	5.55
1020 Barber equipment, hydraulic chair	1 Clab	40	.200	Ea.		7.95		7.95	12.15
1030 Checkout counter, supermarket or warehouse conveyor	2 Clab	18	.889			35.50		35.50	54
1040 Food cases, refrigerated or frozen	Q-5	6	2.667			151		151	228
1190 Laundry equipment, commercial	L-6	3	4			243		243	365
1360 Movie equipment, lamphouse, to 4000 watt, incl. rectifier	1 Elec	4	2			116		116	174
1365 Sound system, incl. amplifier	"	1.25	6.400			370		370	555
1410 Air compressor, to 5 HP	2 Clab	2.50	6.400			255		255	390
1412 Lubrication equipment, automotive, 3 reel type, incl. pump, excl. piping	L-4	1	24	Set		1,150		1,150	1,750
1414 Booth, spray paint, complete, to 26' long	"	.80	30	"		1,425		1,425	2,175
1560 Parking equipment, cashier booth	B-22	2	15	Ea.		705	97	802	1,175
1600 Loading dock equipment, dock bumpers, rubber	1 Clab	50	.160	"		6.40		6.40	9.70
1610 Door seal for door perimeter	"	50	.160	L.F.		6.40		6.40	9.70
1611 Loading dock equipment, dock seal for perimeter, selective demolition	2 Clab	13	1.231	"		49		49	74.50
1620 Platform lifter, fixed, 6' x 8', 5000 lb. capacity	E-16	1.50	10.667	Ea.		595	65.50	660.50	1,050
1630 Dock leveller	"	2	8			445	49	494	785
1640 Lights, single or double arm	1 Elec	8	1			58		58	87
1650 Shelter, fabric, truck or train	1 Clab	1.50	5.333			213		213	325
1790 Waste handling equipment, commercial compactor	L-4	2	12			570		570	875
1792 Commercial or municipal incinerator, gas	"	2	12			570		570	875
1795 Crematory, excluding building	Q-3	.25	128			7,575		7,575	11,400
1910 Detection equipment, cell bar front	E-4	4	8			440	24.50	464.50	745
1912 Cell door and frame		8	4			221	12.30	233.30	375
1914 Prefab cell, 4' to 5' wide, 7' to 8' high, 7' deep		8	4			221	12.30	233.30	375
1916 Cot, bolted, single		40	.800			44	2.46	46.46	74.50
1918 Visitor cubicle		4	8			440	24.50	464.50	745
2850 Hydraulic gates, canal, flap, knife, slide or sluice, to 18" diameter	L-5A	8	4			222	83.50	305.50	445
2852 19" to 36" diameter		6	5.333			296	111	407	600
2854 37" to 48" diameter		2	16			890	335	1,225	1,800
2856 49" to 60" diameter		1	32			1,775	670	2,445	3,575
2858 Over 60" diameter		.30	107			5,925	2,225	8,150	12,000
3100 Sewage pumping system, prefabricated, to 1000 GPM	C-17D	.20	420			22,200	3,800	26,000	38,200
3110 Sewage treatment, holding tank for recirc chemical water closet	1 Plum	8	1			62		62	93.50

404

11 05 Common Work Results for Equipment

11 05 05 – Selective Demolition for Equipment

11 05 05.10 Selective Demolition	Crew	Daily Output	Labor-Hours	Unit	Material	2018 Bare Costs Labor	Equipment	Total	Total Incl O&P
3900	Wastewater treatment system, to 1500 gal.	B-21	2	14	Ea.		650	65	715	1,050
4050	Food storage equipment, walk-in refrigerator/freezer	2 Clab	64	.250	S.F.		9.95		9.95	15.20
4052	Shelving, stainless steel, 4 tier or dunnage rack	1 Clab	12	.667	Ea.		26.50		26.50	40.50
4100	Food preparation equipment, small countertop		18	.444			17.70		17.70	27
4150	Food delivery carts, heated cabinets		18	.444			17.70		17.70	27
4200	Cooking equipment, commercial range	Q-1	12	1.333			74.50		74.50	112
4250	Hood and ventilation equipment, kitchen exhaust hood, excl. fire prot	1 Clab	3	2.667			106		106	162
4255	Fire protection system	Q-1	3	5.333			298		298	450
4300	Food dispensing equipment, countertop items	1 Clab	15	.533			21.50		21.50	32.50
4310	Serving counter	"	65	.123	L.F.		4.90		4.90	7.45
4350	Ice machine, ice cube maker, flakers and storage bins, to 2000 lb./day	Q-1	1.60	10	Ea.		560		560	840
4400	Cleaning and disposal, commercial dishwasher, to 50 racks/hour	L-6	1	12			730		730	1,100
4405	To 275 racks/hour	L-4	1	24			1,150		1,150	1,750
4410	Dishwasher hood	2 Clab	5	3.200			128		128	194
4420	Garbage disposal, commercial, to 5 HP	L-1	8	2			120		120	181
4540	Water heater, residential, to 80 gal./day	"	5	3.200			193		193	289
4542	Water softener, automatic	2 Plum	10	1.600			99.50		99.50	150
4544	Disappearing stairway, to 15' floor height	2 Clab	6	2.667			106		106	162
4710	Darkroom equipment, light	L-7	10	2.800			136		136	207
4712	Heavy	"	1.50	18.667			910		910	1,375
4720	Doors	2 Clab	3.50	4.571	Opng.		182		182	277
4830	Bowling alley, complete, incl. pinsetter, scorer, counters, misc supplies	4 Clab	.40	80	Lane		3,200		3,200	4,850
4840	Health club equipment, circuit training apparatus	2 Clab	2	8	Set		320		320	485
4842	Squat racks	"	10	1.600	Ea.		64		64	97
4860	School equipment, basketball backstop	L-2	2	8			355		355	540
4862	Table and benches, folding, in wall, 14' long	L-4	4	6			285		285	435
4864	Bleachers, telescoping, to 30 tier	F-5	120	.267	Seat		13.65		13.65	21
4865	to 15 tier	"	160	.200	"		10.25		10.25	15.60
4866	Boxing ring, elevated	L-4	.20	120	Ea.		5,700		5,700	8,750
4867	Boxing ring, floor level	"	2	12			570		570	875
4868	Exercise equipment	1 Clab	6	1.333			53		53	81
4870	Gym divider	L-4	1000	.024	S.F.		1.14		1.14	1.75
4875	Scoreboard	R-3	2	10	Ea.		580	65	645	940
4880	Shooting range, incl. bullet traps, targets, excl. structure	L-9	1	36	Point		1,625		1,625	2,525
5200	Vocational shop equipment	2 Clab	8	2	Ea.		79.50		79.50	121
6200	Fume hood, incl. countertop, excl. HVAC	"	6	2.667	L.F.		106		106	162
7100	Medical sterilizing, distiller, water, steam heated, 50 gal. capacity	1 Plum	2.80	2.857	Ea.		178		178	267
7200	Medical equipment, surgery table, minor	1 Clab	1	8			320		320	485
7210	Surgical lights, doctor's office, single or double arm	2 Elec	3	5.333			310		310	465
7300	Physical therapy, table	2 Clab	4	4			159		159	243
7310	Whirlpool bath, fixed, incl. mixing valves	1 Plum	4	2			124		124	187
7400	Dental equipment, chair, electric or hydraulic	1 Clab	.75	10.667			425		425	645
7410	Central suction system	1 Plum	2	4			249		249	375
7420	Drill console with accessories	1 Clab	3.20	2.500			99.50		99.50	152
7430	X-ray unit	"	4	2			79.50		79.50	121
7440	X-ray developer	1 Plum	10	.800			49.50		49.50	75

11 05 10 – Equipment Installation

11 05 10.10 Industrial Equipment Installation

		Crew	Daily Output	Labor-Hours	Unit	Material	Labor	Equipment	Total	Total Incl O&P
0010	**INDUSTRIAL EQUIPMENT INSTALLATION**									
0020	Industrial equipment, minimum	E-2	12	4.667	Ton		253	141	394	560
0200	Maximum	"	2	28	"		1,525	845	2,370	3,350

For customer support on your Building Construction Costs with RSMeans data, call 800.448.8182.

405

11 11 Vehicle Service Equipment

11 11 13 – Compressed-Air Vehicle Service Equipment

11 11 13.10 Compressed Air Equipment	Crew	Daily Output	Labor-Hours	Unit	Material	2018 Bare Costs Labor	Equipment	Total	Total Incl O&P
0010 **COMPRESSED AIR EQUIPMENT**									
0030 Compressors, electric, 1-1/2 HP, standard controls	L-4	1.50	16	Ea.	485	760		1,245	1,700
0550 Dual controls		1.50	16		1,025	760		1,785	2,300
0600 5 HP, 115/230 volt, standard controls		1	24		2,025	1,150		3,175	4,000
0650 Dual controls	↓	1	24	↓	3,225	1,150		4,375	5,300

11 11 19 – Vehicle Lubrication Equipment

11 11 19.10 Lubrication Equipment	Crew	Daily Output	Labor-Hours	Unit	Material	2018 Bare Costs Labor	Equipment	Total	Total Incl O&P
0010 **LUBRICATION EQUIPMENT**									
3000 Lube equipment, 3 reel type, with pumps, not including piping	L-4	.50	48	Set	10,200	2,275		12,475	14,700
3100 Hose reel, including hose, oil/lube, 1000 psi	2 Sswk	2	8	Ea.	465	435		900	1,225
3200 Grease, 5000 psi		2	8		510	435		945	1,275
3300 Air, 50', 160 psi		2	8		610	435		1,045	1,375
3350 25', 160 psi	↓	2	8	↓	365	435		800	1,125

11 11 33 – Vehicle Spray Painting Equipment

11 11 33.10 Spray Painting Equipment	Crew	Daily Output	Labor-Hours	Unit	Material	2018 Bare Costs Labor	Equipment	Total	Total Incl O&P
0010 **SPRAY PAINTING EQUIPMENT**									
4000 Spray painting booth, 26' long, complete	L-4	.40	60	Ea.	10,500	2,850		13,350	15,900

11 12 Parking Control Equipment

11 12 13 – Parking Key and Card Control Units

11 12 13.10 Parking Control Units	Crew	Daily Output	Labor-Hours	Unit	Material	2018 Bare Costs Labor	Equipment	Total	Total Incl O&P
0010 **PARKING CONTROL UNITS**									
5100 Card reader	1 Elec	2	4	Ea.	1,400	233		1,633	1,900
5120 Proximity with customer display	2 Elec	1	16		6,000	930		6,930	8,000
6000 Parking control software, basic functionality	1 Elec	.50	16		24,000	930		24,930	27,800
6020 Multi-function	"	.20	40	↓	98,000	2,325		100,325	111,500

11 12 16 – Parking Ticket Dispensers

11 12 16.10 Ticket Dispensers	Crew	Daily Output	Labor-Hours	Unit	Material	2018 Bare Costs Labor	Equipment	Total	Total Incl O&P
0010 **TICKET DISPENSERS**									
5900 Ticket spitter with time/date stamp, standard	2 Elec	2	8	Ea.	6,675	465		7,140	8,025
5920 Mag stripe encoding	"	2	8	"	19,000	465		19,465	21,500

11 12 26 – Parking Fee Collection Equipment

11 12 26.13 Parking Fee Coin Collection Equipment	Crew	Daily Output	Labor-Hours	Unit	Material	2018 Bare Costs Labor	Equipment	Total	Total Incl O&P
0010 **PARKING FEE COIN COLLECTION EQUIPMENT**									
5200 Cashier booth, average	B-22	1	30	Ea.	10,600	1,400	194	12,194	14,100
5300 Collector station, pay on foot	2 Elec	.20	80		112,000	4,650		116,650	130,000
5320 Credit card only	"	.50	32	↓	20,700	1,850		22,550	25,500

11 12 26.23 Fee Equipment

	Crew	Daily Output	Labor-Hours	Unit	Material	2018 Bare Costs Labor	Equipment	Total	Total Incl O&P
0010 **FEE EQUIPMENT**									
5600 Fee computer	1 Elec	1.50	5.333	Ea.	12,200	310		12,510	13,900

11 12 33 – Parking Gates

11 12 33.13 Lift Arm Parking Gates

	Crew	Daily Output	Labor-Hours	Unit	Material	2018 Bare Costs Labor	Equipment	Total	Total Incl O&P
0010 **LIFT ARM PARKING GATES**									
5000 Barrier gate with programmable controller	2 Elec	3	5.333	Ea.	3,050	310		3,360	3,825
5020 Industrial		3	5.333		5,250	310		5,560	6,250
5050 Non-programmable, with reader and 12' arm		3	5.333		1,925	310		2,235	2,600
5500 Exit verifier	↓	1	16		19,500	930		20,430	22,800
5700 Full sign, 4" letters	1 Elec	2	4	↓	1,250	233		1,483	1,725

For customer support on your Building Construction Costs with RSMeans data, call 800.448.8182.

11 12 Parking Control Equipment

11 12 33 – Parking Gates

11 12 33.13 Lift Arm Parking Gates		Crew	Daily Output	Labor-Hours	Unit	Material	2018 Bare Costs Labor	Equipment	Total	Total Incl O&P
5800	Inductive loop	2 Elec	4	4	Ea.	210	233		443	580
5950	Vehicle detector, microprocessor based	1 Elec	3	2.667		440	155		595	715
7100	Traffic spike unit, flush mount, spring loaded, 72" L	B-89	4	4		1,275	192	105	1,572	1,800
7200	Surface mount, 72" L	2 Skwk	10	1.600		2,375	84		2,459	2,725

11 13 Loading Dock Equipment

11 13 13 – Loading Dock Bumpers

11 13 13.10 Dock Bumpers

		Crew	Daily Output	Labor-Hours	Unit	Material	2018 Bare Costs Labor	Equipment	Total	Total Incl O&P
0010	**DOCK BUMPERS** Bolts not included									
0020	2" x 6" to 4" x 8", average	1 Carp	.30	26.667	M.B.F.	1,400	1,350		2,750	3,575
0050	Bumpers, lam. rubber blocks 4-1/2" thick, 10" high, 14" long		26	.308	Ea.	54.50	15.60		70.10	84
0200	24" long		22	.364		98.50	18.45		116.95	136
0300	36" long		17	.471		271	24		295	335
0500	12" high, 14" long		25	.320		74.50	16.20		90.70	106
0550	24" long		20	.400		74.50	20.50		95	113
0600	36" long		15	.533		117	27		144	170
0800	Laminated rubber blocks 6" thick, 10" high, 14" long		22	.364		70	18.45		88.45	105
0850	24" long		18	.444		125	22.50		147.50	172
0900	36" long		13	.615		201	31		232	269
0910	20" high, 11" long		13	.615		125	31		156	185
0920	Extruded rubber bumpers, T section, 22" x 22" x 3" thick		41	.195		60.50	9.90		70.40	81.50
0940	Molded rubber bumpers, 24" x 12" x 3" thick		20	.400		59	20.50		79.50	96
1000	Welded installation of above bumpers	E-14	8	1		3.90	56.50	12.30	72.70	110
1100	For drilled anchors, add per anchor	1 Carp	36	.222		7	11.25		18.25	25
1300	Steel bumpers, see Section 10 26 13.10									

11 13 16 – Loading Dock Seals and Shelters

11 13 16.10 Dock Seals and Shelters

		Crew	Daily Output	Labor-Hours	Unit	Material	2018 Bare Costs Labor	Equipment	Total	Total Incl O&P
0010	**DOCK SEALS AND SHELTERS**									
3600	Door seal for door perimeter, 12" x 12", vinyl covered	1 Carp	26	.308	L.F.	48.50	15.60		64.10	77.50
3700	Loading dock, seal for perimeter, 9' x 8', with 12" vinyl	2 Carp	6	2.667	Ea.	1,450	135		1,585	1,800
3900	Folding gates, see Section 10 22 16.10									
6200	Shelters, fabric, for truck or train, scissor arms, minimum	1 Carp	1	8	Ea.	2,125	405		2,530	2,950
6300	Maximum	"	.50	16	"	2,600	810		3,410	4,100

11 13 19 – Stationary Loading Dock Equipment

11 13 19.10 Dock Equipment

		Crew	Daily Output	Labor-Hours	Unit	Material	2018 Bare Costs Labor	Equipment	Total	Total Incl O&P
0010	**DOCK EQUIPMENT**									
2200	Dock boards, heavy duty, 60" x 60", aluminum, 5,000 lb. capacity				Ea.	1,400			1,400	1,550
2700	9,000 lb. capacity					1,350			1,350	1,475
3200	15,000 lb. capacity					1,275			1,275	1,400
4200	Platform lifter, 6' x 6', portable, 3,000 lb. capacity					9,550			9,550	10,500
4250	4,000 lb. capacity					11,800			11,800	12,900
4400	Fixed, 6' x 8', 5,000 lb. capacity	E-16	.70	22.857		9,600	1,275	141	11,016	12,800
4500	Levelers, hinged for trucks, 10 ton capacity, 6' x 8'		1.08	14.815		4,375	825	91	5,291	6,250
4650	7' x 8'		1.08	14.815		6,525	825	91	7,441	8,625
4670	Air bag power operated, 10 ton cap., 6' x 8'		1.08	14.815		4,875	825	91	5,791	6,825
4680	7' x 8'		1.08	14.815		5,350	825	91	6,266	7,350
4700	Hydraulic, 10 ton capacity, 6' x 8'		1.08	14.815		5,350	825	91	6,266	7,325
4800	7' x 8'		1.08	14.815		5,750	825	91	6,666	7,775
5800	Loading dock safety restraints, manual style		1.08	14.815		3,350	825	91	4,266	5,125
5900	Automatic style		1.08	14.815		5,625	825	91	6,541	7,650

11 13 Loading Dock Equipment

11 13 19 – Stationary Loading Dock Equipment

11 13 19.10 Dock Equipment	Crew	Daily Output	Labor-Hours	Unit	Material	2018 Bare Costs Labor	Equipment	Total	Total Incl O&P
6000	Dock leveler, 15 ton capacity									
6100	I-beam construction, mechanical, 6' x 8'	E-16	.50	32	Ea.	5,100	1,775	197	7,072	8,750
6150	Hydraulic		.50	32		6,000	1,775	197	7,972	9,725
6200	Formed beam deck construction, mechanical, 6' x 8'		.50	32		3,825	1,775	197	5,797	7,325
6250	Hydraulic		.50	32		5,225	1,775	197	7,197	8,875
6300	Edge of dock leveler, mechanical, 15 ton capacity		2	8		1,150	445	49	1,644	2,025
6301	Edge of dock leveler, mechanical, 10 ton capacity		2	8		1,150	445	49	1,644	2,025
7000	22.5 ton capacity									
7100	Vertical storing dock leveler, hydraulic, 6' x 6'	E-16	.40	40	Ea.	8,125	2,225	246	10,596	12,800

11 13 26 – Loading Dock Lights

11 13 26.10 Dock Lights

		Crew	Daily Output	Labor-Hours	Unit	Material	Labor	Equipment	Total	Total Incl O&P
0010	**DOCK LIGHTS**									
5000	Lights for loading docks, single arm, 24" long	1 Elec	3.80	2.105	Ea.	205	123		328	410
5700	Double arm, 60" long	"	3.80	2.105	"	199	123		322	400

11 14 Pedestrian Control Equipment

11 14 13 – Pedestrian Gates

11 14 13.19 Turnstiles

		Crew	Daily Output	Labor-Hours	Unit	Material	Labor	Equipment	Total	Total Incl O&P
0010	**TURNSTILES**									
0020	One way, 4 arm, 46" diameter, economy, manual	2 Carp	5	3.200	Ea.	1,850	162		2,012	2,300
0100	Electric		1.20	13.333		2,200	675		2,875	3,425
0300	High security, galv., 5'-5" diameter, 7' high, manual		1	16		5,950	810		6,760	7,775
0350	Electric		.60	26.667		7,925	1,350		9,275	10,800
0420	Three arm, 24" opening, light duty, manual		2	8		3,225	405		3,630	4,175
0450	Heavy duty		1.50	10.667		4,825	540		5,365	6,125
0460	Manual, with registering & controls, light duty		2	8		3,900	405		4,305	4,900
0470	Heavy duty		1.50	10.667		4,550	540		5,090	5,825
0480	Electric, heavy duty		1.10	14.545		4,625	735		5,360	6,200
0500	For coin or token operating, add					705			705	775
1200	One way gate with horizontal bars, 5'-5" diameter									
1300	7' high, recreation or transit type	2 Carp	.80	20	Ea.	5,375	1,025		6,400	7,450
1500	For electronic counter, add				"	257			257	283

11 14 19 – Portable Posts and Railings

11 14 19.13 Portable Posts and Railings

		Crew	Daily Output	Labor-Hours	Unit	Material	Labor	Equipment	Total	Total Incl O&P
0010	**PORTABLE POSTS AND RAILINGS**									
0020	Portable for pedestrian traffic control, standard				Ea.	143			143	158
0300	Deluxe posts				"	235			235	258
0600	Ropes for above posts, plastic covered, 1-1/2" diameter				L.F.	18.80			18.80	20.50
0700	Chain core				"	14			14	15.40
1500	Portable security or safety barrier, black with 7' yellow strap				Ea.	235			235	258
1510	12' yellow strap					260			260	286
1550	Sign holder, standard design					85			85	93.50

For customer support on your Building Construction Costs with RSMeans data, call 800.448.8182.

11 21 Retail and Service Equipment

11 21 13 – Cash Registers and Checking Equipment

11 21 13.10 Checkout Counter

11 21 13.10 Checkout Counter	Crew	Daily Output	Labor-Hours	Unit	Material	2018 Bare Costs Labor	Equipment	Total	Total Incl O&P
0010 **CHECKOUT COUNTER**									
0020 Supermarket conveyor, single belt	2 Clab	10	1.600	Ea.	3,400	64		3,464	3,850
0100 Double belt, power take-away		9	1.778		4,900	71		4,971	5,475
0400 Double belt, power take-away, incl. side scanning		7	2.286		5,750	91		5,841	6,475
0800 Warehouse or bulk type	↓	6	2.667	↓	6,800	106		6,906	7,625
1000 Scanning system, 2 lanes, w/registers, scan gun & memory				System	17,800			17,800	19,600
1100 10 lanes, single processor, full scan, with scales				"	169,000			169,000	186,000
2000 Register, restaurant, minimum				Ea.	790			790	870
2100 Maximum					3,350			3,350	3,675
2150 Store, minimum					790			790	870
2200 Maximum				↓	3,350			3,350	3,675

11 21 33 – Checkroom Equipment

11 21 33.10 Clothes Check Equipment

11 21 33.10 Clothes Check Equipment	Crew	Daily Output	Labor-Hours	Unit	Material	2018 Bare Costs Labor	Equipment	Total	Total Incl O&P
0010 **CLOTHES CHECK EQUIPMENT**									
0030 Clothes check rack, free standing, st. stl., 2-tier, 90 bag capacity				Ea.	1,550			1,550	1,700
0050 Wall mounted, 45 bag capacity	L-2	8	2		895	88.50		983.50	1,125
0100 Garment checking bag, green mesh fabric, 21" H x 17" W with 4.5" hook				↓	27			27	30

11 21 53 – Barber and Beauty Shop Equipment

11 21 53.10 Barber Equipment

11 21 53.10 Barber Equipment	Crew	Daily Output	Labor-Hours	Unit	Material	2018 Bare Costs Labor	Equipment	Total	Total Incl O&P
0010 **BARBER EQUIPMENT**									
0020 Chair, hydraulic, movable, minimum	1 Carp	24	.333	Ea.	595	16.90		611.90	680
0050 Maximum	"	16	.500		3,775	25.50		3,800.50	4,200
0200 Wall hung styling station with mirrors, minimum	L-2	8	2		575	88.50		663.50	770
0300 Maximum	"	4	4		2,525	177		2,702	3,050
0500 Sink, hair washing basin, rough plumbing not incl.	1 Plum	8	1		495	62		557	640
1000 Sterilizer, liquid solution for tools					161			161	177
1100 Total equipment, rule of thumb, per chair, minimum	L-8	1	20		2,100	1,050		3,150	3,900
1150 Maximum	"	1	20		5,600	1,050		6,650	7,750

11 21 73 – Commercial Laundry and Dry Cleaning Equipment

11 21 73.13 Dry Cleaning Equipment

11 21 73.13 Dry Cleaning Equipment	Crew	Daily Output	Labor-Hours	Unit	Material	2018 Bare Costs Labor	Equipment	Total	Total Incl O&P
0010 **DRY CLEANING EQUIPMENT**									
2000 Dry cleaners, electric, 20 lb. capacity, not incl. rough-in	L-1	.20	80	Ea.	33,700	4,825		38,525	44,300
2050 25 lb. capacity		.17	94.118		51,500	5,675		57,175	65,000
2100 30 lb. capacity		.15	107		54,000	6,425		60,425	69,000
2150 60 lb. capacity	↓	.09	178	↓	78,500	10,700		89,200	102,000

11 21 73.16 Drying and Conditioning Equipment

11 21 73.16 Drying and Conditioning Equipment	Crew	Daily Output	Labor-Hours	Unit	Material	2018 Bare Costs Labor	Equipment	Total	Total Incl O&P
0010 **DRYING AND CONDITIONING EQUIPMENT**									
0100 Dryers, not including rough-in									
1500 Industrial, 30 lb. capacity	1 Plum	2	4	Ea.	3,375	249		3,624	4,075
1600 50 lb. capacity	"	1.70	4.706		3,825	292		4,117	4,650
4700 Lint collector, ductwork not included, 8,000 to 10,000 CFM	Q-10	.30	80	↓	9,750	4,475		14,225	17,500

11 21 73.19 Finishing Equipment

11 21 73.19 Finishing Equipment	Crew	Daily Output	Labor-Hours	Unit	Material	2018 Bare Costs Labor	Equipment	Total	Total Incl O&P
0010 **FINISHING EQUIPMENT**									
3500 Folders, blankets & sheets, minimum	1 Elec	.17	47.059	Ea.	35,100	2,750		37,850	42,700
3700 King size with automatic stacker		.10	80		63,500	4,650		68,150	77,000
3800 For conveyor delivery, add	↓	.45	17.778		16,100	1,025		17,125	19,400
4900 Spreader feeders, 240V, 2 station	L-6	.70	17.143		59,000	1,050		60,050	66,500
4920 4 station	"	.35	34.286	↓	71,500	2,075		73,575	81,500

For customer support on your Building Construction Costs with RSMeans data, call 800.448.8182.

409

11 21 Retail and Service Equipment

11 21 73 – Commercial Laundry and Dry Cleaning Equipment

11 21 73.23 Commercial Ironing Equipment

11 21 73.23 Commercial Ironing Equipment	Crew	Daily Output	Labor-Hours	Unit	Material	2018 Bare Costs Labor	Equipment	Total	Total Incl O&P
0010	**COMMERCIAL IRONING EQUIPMENT**								
4500	Ironers, institutional, 110", single roll	1 Elec	.20	40	Ea.	34,000	2,325		36,325	40,900
4800	Pressers, low capacity air operated	L-6	1.75	6.857		10,100	415		10,515	11,800
4820	Hand operated		1.75	6.857		9,450	415		9,865	11,000
4840	Ironer 48", 240V		3.50	3.429		116,000	209		116,209	128,000
6600	Hand operated presser		.70	17.143		6,225	1,050		7,275	8,425
6620	Mushroom press 115V		.70	17.143		7,850	1,050		8,900	10,200

11 21 73.26 Commercial Washers and Extractors

		Crew	Daily Output	Labor-Hours	Unit	Material	Labor	Equipment	Total	Total Incl O&P
0010	**COMMERCIAL WASHERS AND EXTRACTORS**, not including rough-in									
6000	Combination washer/extractor, 20 lb. capacity	L-6	1.50	8	Ea.	6,575	485		7,060	7,950
6100	30 lb. capacity		.80	15		10,000	910		10,910	12,400
6200	50 lb. capacity		.68	17.647		12,700	1,075		13,775	15,600
6300	75 lb. capacity		.30	40		18,700	2,425		21,125	24,300
6350	125 lb. capacity		.16	75		29,100	4,550		33,650	39,000
6380	Washer extractor/dryer, 110 lb., 240V		1	12		19,300	730		20,030	22,300
6400	Washer extractor, 135 lb., 240V		1	12		32,500	730		33,230	36,800
6450	Pass through		1	12		55,000	730		55,730	61,500
6500	200 lb. washer extractor		1	12		73,000	730		73,730	81,000
6550	Pass through		1	12		77,000	730		77,730	85,500
6600	Extractor, low capacity		1.75	6.857		7,475	415		7,890	8,850

11 21 73.33 Coin-Operated Laundry Equipment

		Crew	Daily Output	Labor-Hours	Unit	Material	Labor	Equipment	Total	Total Incl O&P
0010	**COIN-OPERATED LAUNDRY EQUIPMENT**									
0990	Dryer, gas fired									
1000	Commercial, 30 lb. capacity, coin operated, single	1 Plum	3	2.667	Ea.	3,500	166		3,666	4,100
1100	Double stacked	"	2	4		8,200	249		8,449	9,400
4860	Coin dry cleaner 20 lb.	L-6	1.75	6.857		29,100	415		29,515	32,600
5290	Clothes washer									
5300	Commercial, coin operated, average	1 Plum	3	2.667	Ea.	1,325	166		1,491	1,700

11 21 83 – Photo Processing Equipment

11 21 83.13 Darkroom Equipment

		Crew	Daily Output	Labor-Hours	Unit	Material	Labor	Equipment	Total	Total Incl O&P
0010	**DARKROOM EQUIPMENT**									
0020	Developing sink, 5" deep, 24" x 48"	Q-1	2	8	Ea.	580	445		1,025	1,325
0050	48" x 52"		1.70	9.412		1,400	525		1,925	2,325
0200	10" deep, 24" x 48"		1.70	9.412		1,700	525		2,225	2,650
0250	24" x 108"		1.50	10.667		3,550	595		4,145	4,800
0500	Dryers, dehumidified filtered air, 36" x 25" x 68" high	L-7	6	4.667		2,325	227		2,552	2,900
3000	Viewing lights, 20" x 24"		6	4.667		286	227		513	660
3100	20" x 24" with color correction		6	4.667		415	227		642	800

11 22 Banking Equipment

11 22 13 – Vault Equipment

11 22 13.16 Safes

		Crew	Daily Output	Labor-Hours	Unit	Material	Labor	Equipment	Total	Total Incl O&P
0010	**SAFES**									
0200	Office, 1 hr. rating, 30" x 18" x 18"				Ea.	2,250			2,250	2,475
0250	40" x 18" x 18"					4,900			4,900	5,375
0300	60" x 36" x 18", double door					9,225			9,225	10,200
0600	Data, 1 hr. rating, 27" x 19" x 16"					5,375			5,375	5,900
0700	63" x 34" x 16"					16,600			16,600	18,200
0750	Diskette, 1 hr., 14" x 12" x 11", inside					4,375			4,375	4,825

For customer support on your Building Construction Costs with RSMeans data, call 800.448.8182.

11 22 Banking Equipment

11 22 13 – Vault Equipment

11 22 13.16 Safes

		Crew	Daily Output	Labor-Hours	Unit	Material	2018 Bare Costs Labor	2018 Bare Costs Equipment	Total	Total Incl O&P
0800	Money, "B" label, 9" x 14" x 14"				Ea.	565			565	620
0900	Tool resistive, 24" x 24" x 20"					4,125			4,125	4,550
1050	Tool and torch resistive, 24" x 24" x 20"					8,350			8,350	9,200
1150	Jewelers, 23" x 20" x 18"					9,425			9,425	10,400
1200	63" x 25" x 18"					14,200			14,200	15,600
1300	For handling into building, add, minimum	A-2	8.50	2.824			117	22	139	202
1400	Maximum	"	.78	30.769			1,275	242	1,517	2,200

11 22 16 – Teller and Service Equipment

11 22 16.13 Teller Equipment Systems

		Crew	Daily Output	Labor-Hours	Unit	Material	2018 Bare Costs Labor	2018 Bare Costs Equipment	Total	Total Incl O&P
0010	**TELLER EQUIPMENT SYSTEMS**									
0020	Alarm system, police	2 Elec	1.60	10	Ea.	4,975	580		5,555	6,350
0100	With vault alarm	"	.40	40		19,800	2,325		22,125	25,300
0400	Bullet resistant teller window, 44" x 60"	1 Glaz	.60	13.333		4,425	645		5,070	5,850
0500	48" x 60"	"	.60	13.333		6,100	645		6,745	7,700
3000	Counters for banks, frontal only	2 Carp	1	16	Station	1,950	810		2,760	3,350
3100	Complete with steel undercounter	"	.50	32	"	3,625	1,625		5,250	6,475
4600	Door and frame, bullet-resistant, with vision panel, minimum	2 Sswk	1.10	14.545	Ea.	5,525	795		6,320	7,375
4700	Maximum		1.10	14.545		7,475	795		8,270	9,525
4800	Drive-up window, drawer & mike, not incl. glass, minimum		1	16		7,350	875		8,225	9,525
4900	Maximum		.50	32		9,475	1,750		11,225	13,300
5000	Night depository, with chest, minimum		1	16		7,800	875		8,675	10,000
5100	Maximum		.50	32		11,100	1,750		12,850	15,200
5200	Package receiver, painted		3.20	5		1,400	273		1,673	1,975
5300	Stainless steel		3.20	5		2,350	273		2,623	3,050
5400	Partitions, bullet-resistant, 1-3/16" glass, 8' high	2 Carp	10	1.600	L.F.	201	81		282	345
5450	Acrylic	"	10	1.600	"	380	81		461	545
5500	Pneumatic tube systems, 2 lane drive-up, complete	L-3	.25	64	Total	26,200	3,500		29,700	34,200
5550	With T.V. viewer	"	.20	80	"	51,000	4,400		55,400	62,500
5570	Safety deposit boxes, minimum	1 Sswk	44	.182	Opng.	58.50	9.95		68.45	80.50
5580	Maximum, 10" x 15" opening		19	.421		124	23		147	174
5590	Teller locker, average		15	.533		1,600	29		1,629	1,800
5600	Pass thru, bullet-res. window, painted steel, 24" x 36"	2 Sswk	1.60	10	Ea.	2,925	545		3,470	4,100
5700	48" x 48"		1.20	13.333		2,700	730		3,430	4,175
5800	72" x 40"		.80	20		4,450	1,100		5,550	6,650
5900	For stainless steel frames, add					20%				
6100	Surveillance system, video camera, complete	2 Elec	1	16	Ea.	9,700	930		10,630	12,100
6110	For each additional camera, add				"	1,025			1,025	1,125
6120	CCTV system, see Section 28 23 13.10									
6200	24 hour teller, single unit,									
6300	automated deposit, cash and memo	L-3	.25	64	Ea.	44,700	3,500		48,200	54,500
7000	Vault front, see Section 08 34 59.10									

For customer support on your Building Construction Costs with RSMeans data, call 800.448.8182.

411

11 30 13.15 Cooking Equipment

		Crew	Daily Output	Labor-Hours	Unit	Material	2018 Bare Costs Labor	Equipment	Total	Total Incl O&P
0010	**COOKING EQUIPMENT**									
0020	Cooking range, 30" free standing, 1 oven, minimum	2 Clab	10	1.600	Ea.	470	64		534	610
0050	Maximum		4	4		2,300	159		2,459	2,775
0150	2 oven, minimum		10	1.600		1,000	64		1,064	1,200
0200	Maximum		10	1.600		3,300	64		3,364	3,725
0350	Built-in, 30" wide, 1 oven, minimum	1 Elec	6	1.333		800	77.50		877.50	995
0400	Maximum	2 Carp	2	8		1,750	405		2,155	2,550
0500	2 oven, conventional, minimum		4	4		1,150	203		1,353	1,575
0550	1 conventional, 1 microwave, maximum		2	8		2,275	405		2,680	3,125
0700	Free standing, 1 oven, 21" wide range, minimum	2 Clab	10	1.600		475	64		539	615
0750	21" wide, maximum	"	4	4		660	159		819	970
0900	Countertop cooktops, 4 burner, standard, minimum	1 Elec	6	1.333		325	77.50		402.50	475
0950	Maximum		3	2.667		1,750	155		1,905	2,150
1050	As above, but with grill and griddle attachment, minimum		6	1.333		1,450	77.50		1,527.50	1,700
1100	Maximum		3	2.667		3,625	155		3,780	4,200
1200	Induction cooktop, 30" wide		3	2.667		1,475	155		1,630	1,850
1250	Microwave oven, minimum		4	2		97	116		213	281
1300	Maximum		2	4		460	233		693	855

11 30 13.16 Refrigeration Equipment

		Crew	Daily Output	Labor-Hours	Unit	Material	2018 Bare Costs Labor	Equipment	Total	Total Incl O&P
0010	**REFRIGERATION EQUIPMENT**									
2000	Deep freeze, 15 to 23 C.F., minimum	2 Clab	10	1.600	Ea.	595	64		659	750
2050	Maximum		5	3.200		805	128		933	1,075
2200	30 C.F., minimum		8	2		800	79.50		879.50	1,000
2250	Maximum		3	5.333		860	213		1,073	1,275
5200	Icemaker, automatic, 20 lbs./day	1 Plum	7	1.143		1,250	71		1,321	1,475
5350	51 lbs./day	"	2	4		1,375	249		1,624	1,875
5500	Refrigerator, no frost, 10 C.F. to 12 C.F., minimum	2 Clab	10	1.600		440	64		504	580
5600	Maximum		6	2.667		535	106		641	750
5750	14 C.F. to 16 C.F., minimum		9	1.778		555	71		626	720
5800	Maximum		5	3.200		765	128		893	1,025
5950	18 C.F. to 20 C.F., minimum		8	2		705	79.50		784.50	895
6000	Maximum		4	4		1,650	159		1,809	2,075
6150	21 C.F. to 29 C.F., minimum		7	2.286		1,075	91		1,166	1,325
6200	Maximum		3	5.333		2,350	213		2,563	2,900
6790	Energy-star qualified, 18 C.F., minimum [G]	2 Carp	4	4		535	203		738	900
6795	Maximum [G]		2	8		1,600	405		2,005	2,375
6797	21.7 C.F., minimum [G]		4	4		1,400	203		1,603	1,850
6799	Maximum [G]		4	4		1,875	203		2,078	2,375

11 30 13.17 Kitchen Cleaning Equipment

		Crew	Daily Output	Labor-Hours	Unit	Material	2018 Bare Costs Labor	Equipment	Total	Total Incl O&P
0010	**KITCHEN CLEANING EQUIPMENT**									
2750	Dishwasher, built-in, 2 cycles, minimum	L-1	4	4	Ea.	296	241		537	685
2800	Maximum		2	8		455	480		935	1,225
2950	4 or more cycles, minimum		4	4		400	241		641	800
2960	Average		4	4		535	241		776	945
3000	Maximum		2	8		1,175	480		1,655	2,025
3100	Energy-star qualified, minimum [G]		4	4		380	241		621	780
3110	Maximum [G]		2	8		1,700	480		2,180	2,600

11 30 13.18 Waste Disposal Equipment

		Crew	Daily Output	Labor-Hours	Unit	Material	2018 Bare Costs Labor	Equipment	Total	Total Incl O&P
0010	**WASTE DISPOSAL EQUIPMENT**									
1750	Compactor, residential size, 4 to 1 compaction, minimum	1 Carp	5	1.600	Ea.	690	81		771	885
1800	Maximum	"	3	2.667		1,150	135		1,285	1,475
3300	Garbage disposal, sink type, minimum	L-1	10	1.600		106	96.50		202.50	260

For customer support on your Building Construction Costs with RSMeans data, call 800.448.8182.

11 30 Residential Equipment

11 30 13 – Residential Appliances

11 30 13.18 Waste Disposal Equipment

		Crew	Daily Output	Labor-Hours	Unit	Material	2018 Bare Costs Labor	Equipment	Total	Total Incl O&P
3350	Maximum	L-1	10	1.600	Ea.	204	96.50		300.50	370

11 30 13.19 Kitchen Ventilation Equipment

		Crew	Daily Output	Labor-Hours	Unit	Material	Labor	Equipment	Total	Total Incl O&P
0010	**KITCHEN VENTILATION EQUIPMENT**									
4150	Hood for range, 2 speed, vented, 30" wide, minimum	L-3	5	3.200	Ea.	93	176		269	370
4200	Maximum		3	5.333		910	293		1,203	1,450
4300	42" wide, minimum		5	3.200		170	176		346	450
4330	Custom		5	3.200		1,650	176		1,826	2,075
4350	Maximum		3	5.333		2,000	293		2,293	2,650
4500	For ventless hood, 2 speed, add					17.45			17.45	19.20
4650	For vented 1 speed, deduct from maximum					63			63	69

11 30 13.24 Washers

		Crew	Daily Output	Labor-Hours	Unit	Material	Labor	Equipment	Total	Total Incl O&P
0010	**WASHERS**									
5000	Residential, 4 cycle, average	1 Plum	3	2.667	Ea.	975	166		1,141	1,325
6650	Washing machine, automatic, minimum		3	2.667		645	166		811	960
6700	Maximum		1	8		1,350	495		1,845	2,225
6750	Energy star qualified, front loading, minimum G		3	2.667		700	166		866	1,025
6760	Maximum G		1	8		1,800	495		2,295	2,725
6764	Top loading, minimum G		3	2.667		640	166		806	955
6766	Maximum G		3	2.667		1,000	166		1,166	1,350

11 30 13.25 Dryers

		Crew	Daily Output	Labor-Hours	Unit	Material	Labor	Equipment	Total	Total Incl O&P
0010	**DRYERS**									
0500	Gas fired residential, 16 lb. capacity, average	1 Plum	3	2.667	Ea.	715	166		881	1,025
6770	Electric, front loading, energy-star qualified, minimum G	L-2	3	5.333		405	236		641	805
6780	Maximum G	"	2	8		1,475	355		1,830	2,150
7450	Vent kits for dryers	1 Carp	10	.800		43.50	40.50		84	110

11 30 15 – Miscellaneous Residential Appliances

11 30 15.13 Sump Pumps

		Crew	Daily Output	Labor-Hours	Unit	Material	Labor	Equipment	Total	Total Incl O&P
0010	**SUMP PUMPS**									
6400	Cellar drainer, pedestal, 1/3 HP, molded PVC base	1 Plum	3	2.667	Ea.	140	166		306	405
6450	Solid brass	"	2	4	"	240	249		489	640
6460	Sump pump, see also Section 22 14 29.16									

11 30 15.23 Water Heaters

		Crew	Daily Output	Labor-Hours	Unit	Material	Labor	Equipment	Total	Total Incl O&P
0010	**WATER HEATERS**									
6900	Electric, glass lined, 30 gallon, minimum	L-1	5	3.200	Ea.	900	193		1,093	1,275
6950	Maximum		3	5.333		1,250	320		1,570	1,850
7100	80 gallon, minimum		2	8		1,675	480		2,155	2,550
7150	Maximum		1	16		2,325	965		3,290	4,000
7180	Gas, glass lined, 30 gallon, minimum	2 Plum	5	3.200		1,600	199		1,799	2,075
7220	Maximum		3	5.333		2,225	330		2,555	2,950
7260	50 gallon, minimum		2.50	6.400		1,725	400		2,125	2,500
7300	Maximum		1.50	10.667		2,400	665		3,065	3,650
7310	Water heater, see also Section 22 33 30.13									

11 30 15.43 Air Quality

		Crew	Daily Output	Labor-Hours	Unit	Material	Labor	Equipment	Total	Total Incl O&P
0010	**AIR QUALITY**									
2450	Dehumidifier, portable, automatic, 15 pint	1 Elec	4	2	Ea.	205	116		321	400
2550	40 pint		3.75	2.133		250	124		374	460
3550	Heater, electric, built-in, 1250 watt, ceiling type, minimum		4	2		117	116		233	300
3600	Maximum		3	2.667		213	155		368	465
3700	Wall type, minimum		4	2		197	116		313	390
3750	Maximum		3	2.667		194	155		349	445
3900	1500 watt wall type, with blower		4	2		188	116		304	380

For customer support on your Building Construction Costs with RSMeans data, call 800.448.8182.

413

11 30 Residential Equipment

11 30 15 – Miscellaneous Residential Appliances

11 30 15.43 Air Quality

		Crew	Daily Output	Labor-Hours	Unit	Material	2018 Bare Costs Labor	Equipment	Total	Total Incl O&P
3950	3000 watt	1 Elec	3	2.667	Ea.	480	155		635	760
4850	Humidifier, portable, 8 gallons/day					148			148	162
5000	15 gallons/day					191			191	210

11 30 33 – Retractable Stairs

11 30 33.10 Disappearing Stairway

		Crew	Daily Output	Labor-Hours	Unit	Material	2018 Bare Costs Labor	Equipment	Total	Total Incl O&P
0010	**DISAPPEARING STAIRWAY** No trim included									
0100	Custom grade, pine, 8'-6" ceiling, minimum	1 Carp	4	2	Ea.	219	101		320	395
0150	Average		3.50	2.286		243	116		359	445
0200	Maximum		3	2.667		291	135		426	525
0500	Heavy duty, pivoted, from 7'-7" to 12'-10" floor to floor		3	2.667		1,375	135		1,510	1,725
0600	16'-0" ceiling		2	4		1,575	203		1,778	2,025
0800	Economy folding, pine, 8'-6" ceiling		4	2		177	101		278	350
0900	9'-6" ceiling		4	2		205	101		306	380
1100	Automatic electric, aluminum, floor to floor height, 8' to 9'	2 Carp	1	16		9,350	810		10,160	11,500
1400	11' to 12'		.90	17.778		9,925	900		10,825	12,300
1700	14' to 15'		.70	22.857		10,600	1,150		11,750	13,500

11 32 Unit Kitchens

11 32 13 – Metal Unit Kitchens

11 32 13.10 Commercial Unit Kitchens

		Crew	Daily Output	Labor-Hours	Unit	Material	2018 Bare Costs Labor	Equipment	Total	Total Incl O&P
0010	**COMMERCIAL UNIT KITCHENS**									
1500	Combination range, refrigerator and sink, 30" wide, minimum	L-1	2	8	Ea.	1,225	480		1,705	2,075
1550	Maximum		1	16		1,075	965		2,040	2,625
1570	60" wide, average		1.40	11.429		1,275	690		1,965	2,425
1590	72" wide, average		1.20	13.333		1,700	800		2,500	3,075
1600	Office model, 48" wide		2	8		1,550	480		2,030	2,450
1620	Refrigerator and sink only		2.40	6.667		2,625	400		3,025	3,475
1640	Combination range, refrigerator, sink, microwave									
1660	Oven and ice maker	L-1	.80	20	Ea.	4,825	1,200		6,025	7,100

11 41 Foodservice Storage Equipment

11 41 13 – Refrigerated Food Storage Cases

11 41 13.10 Refrigerated Food Cases

		Crew	Daily Output	Labor-Hours	Unit	Material	2018 Bare Costs Labor	Equipment	Total	Total Incl O&P
0010	**REFRIGERATED FOOD CASES**									
0030	Dairy, multi-deck, 12' long	Q-5	3	5.333	Ea.	10,700	300		11,000	12,300
0100	For rear sliding doors, add					1,925			1,925	2,125
0200	Delicatessen case, service deli, 12' long, single deck	Q-5	3.90	4.103		7,975	233		8,208	9,125
0300	Multi-deck, 18 S.F. shelf display		3	5.333		7,950	300		8,250	9,200
0400	Freezer, self-contained, chest-type, 30 C.F.		3.90	4.103		4,375	233		4,608	5,150
0500	Glass door, upright, 78 C.F.		3.30	4.848		9,250	275		9,525	10,600
0600	Frozen food, chest type, 12' long		3.30	4.848		7,825	275		8,100	9,025
0700	Glass door, reach-in, 5 door		3	5.333		9,800	300		10,100	11,300
0800	Island case, 12' long, single deck		3.30	4.848		7,650	275		7,925	8,825
0900	Multi-deck		3	5.333		8,925	300		9,225	10,300
1000	Meat case, 12' long, single deck		3.30	4.848		7,650	275		7,925	8,825
1050	Multi-deck		3.10	5.161		9,875	293		10,168	11,300
1100	Produce, 12' long, single deck		3.30	4.848		6,550	275		6,825	7,625
1200	Multi-deck		3.10	5.161		8,300	293		8,593	9,575

For customer support on your Building Construction Costs with RSMeans data, call 800.448.8182.

11 41 Foodservice Storage Equipment

11 41 13 – Refrigerated Food Storage Cases

11 41 13.20 Refrigerated Food Storage Equipment		Crew	Daily Output	Labor-Hours	Unit	Material	2018 Bare Costs Labor	Equipment	Total	Total Incl O&P
0010	**REFRIGERATED FOOD STORAGE EQUIPMENT**									
2350	Cooler, reach-in, beverage, 6' long	Q-1	6	2.667	Ea.	3,200	149		3,349	3,750
4300	Freezers, reach-in, 44 C.F.		4	4		4,725	224		4,949	5,525
4500	68 C.F.		3	5.333		5,800	298		6,098	6,825
4600	Freezer, pre-fab, 8' x 8' w/refrigeration	2 Carp	.45	35.556		10,500	1,800		12,300	14,400
4620	8' x 12'		.35	45.714		11,400	2,325		13,725	16,000
4640	8' x 16'		.25	64		13,800	3,250		17,050	20,200
4660	8' x 20'		.17	94.118		19,700	4,775		24,475	29,000
4680	Reach-in, 1 compartment	Q-1	4	4		3,225	224		3,449	3,875
4685	Energy star rated [G]	R-18	7.80	3.333		2,900	170		3,070	3,450
4700	2 compartment	Q-1	3	5.333		4,450	298		4,748	5,325
4705	Energy star rated [G]	R-18	6.20	4.194		3,575	214		3,789	4,275
4710	3 compartment	Q-1	3	5.333		6,150	298		6,448	7,200
4715	Energy star rated [G]	R-18	5.60	4.643		4,925	237		5,162	5,750
8320	Refrigerator, reach-in, 1 compartment		7.80	3.333		2,925	170		3,095	3,450
8325	Energy star rated [G]		7.80	3.333		2,900	170		3,070	3,450
8330	2 compartment		6.20	4.194		4,175	214		4,389	4,925
8335	Energy star rated [G]		6.20	4.194		3,575	214		3,789	4,275
8340	3 compartment		5.60	4.643		5,650	237		5,887	6,550
8345	Energy star rated [G]		5.60	4.643		4,925	237		5,162	5,750
8350	Pre-fab, with refrigeration, 8' x 8'	2 Carp	.45	35.556		6,900	1,800		8,700	10,300
8360	8' x 12'		.35	45.714		8,350	2,325		10,675	12,700
8370	8' x 16'		.25	64		10,400	3,250		13,650	16,500
8380	8' x 20'		.17	94.118		13,200	4,775		17,975	21,800
8390	Pass-thru/roll-in, 1 compartment	R-18	7.80	3.333		3,775	170		3,945	4,400
8400	2 compartment		6.24	4.167		5,325	213		5,538	6,200
8410	3 compartment		5.60	4.643		7,950	237		8,187	9,100
8420	Walk-in, alum, door & floor only, no refrig, 6' x 6' x 7'-6"	2 Carp	1.40	11.429		6,600	580		7,180	8,125
8430	10' x 6' x 7'-6"		.55	29.091		9,725	1,475		11,200	13,000
8440	12' x 14' x 7'-6"		.25	64		13,200	3,250		16,450	19,600
8450	12' x 20' x 7'-6"		.17	94.118		12,200	4,775		16,975	20,700
8460	Refrigerated cabinets, mobile					4,350			4,350	4,775
8470	Refrigerator/freezer, reach-in, 1 compartment	R-18	5.60	4.643		6,450	237		6,687	7,425
8480	2 compartment	"	4.80	5.417		7,175	277		7,452	8,325

11 41 13.30 Wine Cellar

		Crew	Daily Output	Labor-Hours	Unit	Material	2018 Bare Costs Labor	Equipment	Total	Total Incl O&P
0010	**WINE CELLAR**, refrigerated, Redwood interior, carpeted, walk-in type									
0020	6'-8" high, including racks									
0200	80" W x 48" D for 900 bottles	2 Carp	1.50	10.667	Ea.	4,200	540		4,740	5,450
0250	80" W x 72" D for 1300 bottles		1.33	12.030		5,125	610		5,735	6,550
0300	80" W x 94" D for 1900 bottles		1.17	13.675		6,200	695		6,895	7,875
0400	80" W x 124" D for 2500 bottles		1	16		7,300	810		8,110	9,250
0600	Portable cabinets, red oak, reach-in temp. & humidity controlled									
0650	26-5/8" W x 26-1/2" D x 68" H for 235 bottles				Ea.	4,350			4,350	4,800
0660	32" W x 21-1/2" D x 73-1/2" H for 144 bottles					3,525			3,525	3,875
0670	32" W x 29-1/2" D x 73-1/2" H for 288 bottles					4,525			4,525	4,975
0680	39-1/2" W x 29-1/2" D x 86-1/2" H for 440 bottles					4,550			4,550	5,000
0690	52-1/2" W x 29-1/2" D x 73-1/2" H for 468 bottles					5,000			5,000	5,500
0700	52-1/2" W x 29-1/2" D x 86-1/2" H for 572 bottles					5,125			5,125	5,625
0730	Portable, red oak, can be built-in with glass door									
0750	23-7/8" W x 24" D x 34-1/2" H for 50 bottles				Ea.	1,050			1,050	1,150

For customer support on your Building Construction Costs with RSMeans data, call 800.448.8182.

415

11 41 Foodservice Storage Equipment

11 41 33 – Foodservice Shelving

11 41 33.20 Metal Food Storage Shelving

		Crew	Daily Output	Labor-Hours	Unit	Material	2018 Bare Costs Labor	2018 Bare Costs Equipment	Total	Total Incl O&P
0010	**METAL FOOD STORAGE SHELVING**									
8600	Stainless steel shelving, louvered 4-tier, 20" x 3'	1 Clab	6	1.333	Ea.	1,450	53		1,503	1,675
8605	20" x 4'		6	1.333		1,650	53		1,703	1,875
8610	20" x 6'		6	1.333		1,725	53		1,778	1,975
8615	24" x 3'		6	1.333		2,050	53		2,103	2,350
8620	24" x 4'		6	1.333		2,450	53		2,503	2,775
8625	24" x 6'		6	1.333		3,400	53		3,453	3,825
8630	Flat 4-tier, 20" x 3'		6	1.333		1,175	53		1,228	1,375
8635	20" x 4'		6	1.333		1,400	53		1,453	1,625
8640	20" x 5'		6	1.333		1,600	53		1,653	1,825
8645	24" x 3'		6	1.333		905	53		958	1,075
8650	24" x 4'		6	1.333		2,300	53		2,353	2,600
8655	24" x 6'		6	1.333		2,750	53		2,803	3,100
8700	Galvanized shelving, louvered 4-tier, 20" x 3'		6	1.333		790	53		843	950
8705	20" x 4'		6	1.333		895	53		948	1,075
8710	20" x 6'		6	1.333		950	53		1,003	1,125
8715	24" x 3'		6	1.333		725	53		778	875
8720	24" x 4'		6	1.333		955	53		1,008	1,125
8725	24" x 6'		6	1.333		1,375	53		1,428	1,575
8730	Flat 4-tier, 20" x 3'		6	1.333		730	53		783	885
8735	20" x 4'		6	1.333		640	53		693	785
8740	20" x 6'		6	1.333		950	53		1,003	1,125
8745	24" x 3'		6	1.333		715	53		768	865
8750	24" x 4'		6	1.333		550	53		603	685
8755	24" x 6'		6	1.333		850	53		903	1,025
8760	Stainless steel dunnage rack, 24" x 3'		8	1		350	40		390	445
8765	24" x 4'		8	1		445	40		485	550
8770	Galvanized dunnage rack, 24" x 3'		8	1		131	40		171	205
8775	24" x 4'		8	1		192	40		232	272

11 42 Food Preparation Equipment

11 42 10 – Commercial Food Preparation Equipment

11 42 10.10 Choppers, Mixers and Misc. Equipment

		Crew	Daily Output	Labor-Hours	Unit	Material	2018 Bare Costs Labor	2018 Bare Costs Equipment	Total	Total Incl O&P
0010	**CHOPPERS, MIXERS AND MISC. EQUIPMENT**									
1700	Choppers, 5 pounds	R-18	7	3.714	Ea.	2,725	190		2,915	3,275
1720	16 pounds		5	5.200		2,525	266		2,791	3,175
1740	35 to 40 pounds		4	6.500		3,575	330		3,905	4,425
1840	Coffee brewer, 5 burners	1 Plum	3	2.667		915	166		1,081	1,250
1850	Coffee urn, twin 6 gallon urns		2	4		2,300	249		2,549	2,900
1860	Single, 3 gallon		3	2.667		1,600	166		1,766	2,025
3000	Fast food equipment, total package, minimum	6 Skwk	.08	600		211,500	31,400		242,900	281,000
3100	Maximum	"	.07	686		288,500	35,900		324,400	372,000
3800	Food mixers, bench type, 20 quarts	L-7	7	4		3,175	195		3,370	3,800
3850	40 quarts		5.40	5.185		9,450	252		9,702	10,800
3900	60 quarts		5	5.600		12,000	273		12,273	13,600
4040	80 quarts		3.90	7.179		18,900	350		19,250	21,200
4100	Floor type, 20 quarts		15	1.867		3,625	91		3,716	4,150
4120	60 quarts		14	2		10,700	97.50		10,797.50	11,900
4140	80 quarts		12	2.333		17,400	114		17,514	19,400
4160	140 quarts		8.60	3.256		26,700	158		26,858	29,600

For customer support on your Building Construction Costs with RSMeans data, call 800.448.8182.

11 42 Food Preparation Equipment

11 42 10 - Commercial Food Preparation Equipment

11 42 10.10 Choppers, Mixers and Misc. Equipment

		Crew	Daily Output	Labor-Hours	Unit	Material	2018 Bare Costs Labor	2018 Bare Costs Equipment	Total	Total Incl O&P
6700	Peelers, small	R-18	8	3.250	Ea.	1,825	166		1,991	2,250
6720	Large	"	6	4.333		3,150	221		3,371	3,800
6800	Pulper/extractor, close coupled, 5 HP	1 Plum	1.90	4.211		2,475	262		2,737	3,125
8580	Slicer with table	R-18	9	2.889		3,925	148		4,073	4,550

11 43 Food Delivery Carts and Conveyors

11 43 13 - Food Delivery Carts

11 43 13.10 Mobile Carts, Racks and Trays

		Crew	Daily Output	Labor-Hours	Unit	Material	2018 Bare Costs Labor	2018 Bare Costs Equipment	Total	Total Incl O&P
0010	**MOBILE CARTS, RACKS AND TRAYS**									
1650	Cabinet, heated, 1 compartment, reach-in	R-18	5.60	4.643	Ea.	2,850	237		3,087	3,500
1655	Pass-thru roll-in		5.60	4.643		3,575	237		3,812	4,275
1660	2 compartment, reach-in		4.80	5.417		6,250	277		6,527	7,300
1670	Mobile					3,725			3,725	4,100
2000	Hospital food cart, hot and cold service, 20 tray capacity					15,900			15,900	17,500
6850	Mobile rack w/pan slide					1,225			1,225	1,350
9180	Tray and silver dispenser, mobile	1 Clab	16	.500		650	19.95		669.95	745

11 44 Food Cooking Equipment

11 44 13 - Commercial Ranges

11 44 13.10 Cooking Equipment

		Crew	Daily Output	Labor-Hours	Unit	Material	2018 Bare Costs Labor	2018 Bare Costs Equipment	Total	Total Incl O&P
0010	**COOKING EQUIPMENT**									
0020	Bake oven, gas, one section	Q-1	8	2	Ea.	5,725	112		5,837	6,475
0300	Two sections		7	2.286		9,425	128		9,553	10,600
0600	Three sections		6	2.667		12,000	149		12,149	13,400
0900	Electric convection, single deck	L-7	4	7		5,525	340		5,865	6,600
1300	Broiler, without oven, standard	Q-1	8	2		3,875	112		3,987	4,425
1550	Infrared	L-7	4	7		7,425	340		7,765	8,700
4750	Fryer, with twin baskets, modular model	Q-1	7	2.286		1,375	128		1,503	1,725
5000	Floor model, on 6" legs	"	5	3.200		2,575	179		2,754	3,100
5100	Extra single basket, large					50			50	55
5170	Energy star rated, 50 lb. capacity [G]	R-18	4	6.500		4,750	330		5,080	5,725
5175	85 lb. capacity [G]	"	4	6.500		9,900	330		10,230	11,400
5300	Griddle, SS, 24" plate, w/4" legs, elec, 208 V, 3 phase, 3' long	Q-1	7	2.286		2,200	128		2,328	2,600
5550	4' long	"	6	2.667		2,025	149		2,174	2,450
6200	Iced tea brewer	1 Plum	3.44	2.326		630	145		775	910
6350	Kettle, w/steam jacket, tilting, w/positive lock, SS, 20 gallons	L-7	7	4		8,700	195		8,895	9,875
6600	60 gallons	"	6	4.667		20,000	227		20,227	22,300
6900	Range, restaurant type, 6 burners and 1 standard oven, 36" wide	Q-1	7	2.286		3,075	128		3,203	3,575
6950	Convection		7	2.286		4,550	128		4,678	5,225
7150	2 standard ovens, 24" griddle, 60" wide		6	2.667		5,325	149		5,474	6,100
7200	1 standard, 1 convection oven		6	2.667		9,500	149		9,649	10,600
7450	Heavy duty, single 34" standard oven, open top		5	3.200		5,550	179		5,729	6,375
7500	Convection oven		5	3.200		6,000	179		6,179	6,875
7700	Griddle top		6	2.667		2,175	149		2,324	2,625
7750	Convection oven		6	2.667		3,175	149		3,324	3,725
7760	Induction cooker, electric	L-7	7	4		1,825	195		2,020	2,300
8850	Steamer, electric 27 KW		7	4		11,800	195		11,995	13,200
9100	Electric, 10 KW or gas 100,000 BTU		5	5.600		7,200	273		7,473	8,325
9150	Toaster, conveyor type, 16-22 slices/minute					1,250			1,250	1,375

For customer support on your Building Construction Costs with RSMeans data, call 800.448.8182.

417

11 44 Food Cooking Equipment

11 44 13 – Commercial Ranges

11 44 13.10 Cooking Equipment	Crew	Daily Output	Labor-Hours	Unit	Material	2018 Bare Costs Labor	Equipment	Total	Total Incl O&P
9160	Pop-up, 2 slot				Ea.	645			645	705
9200	For deluxe models of above equipment, add					75%				
9400	Rule of thumb: Equipment cost based									
9410	on kitchen work area									
9420	Office buildings, minimum	L-7	77	.364	S.F.	94	17.70		111.70	131
9450	Maximum		58	.483		159	23.50		182.50	211
9550	Public eating facilities, minimum		77	.364		123	17.70		140.70	163
9600	Maximum		46	.609		201	29.50		230.50	266
9750	Hospitals, minimum		58	.483		127	23.50		150.50	176
9800	Maximum		39	.718		234	35		269	310

11 46 Food Dispensing Equipment

11 46 16 – Service Line Equipment

11 46 16.10 Commercial Food Dispensing Equipment

		Crew	Daily Output	Labor-Hours	Unit	Material	Labor	Equipment	Total	Total Incl O&P
0010	**COMMERCIAL FOOD DISPENSING EQUIPMENT**									
1050	Butter pat dispenser	1 Clab	13	.615	Ea.	835	24.50		859.50	955
1100	Bread dispenser, counter top		13	.615		750	24.50		774.50	865
1900	Cup and glass dispenser, drop in		4	2		565	79.50		644.50	745
1920	Disposable cup, drop in		16	.500		655	19.95		674.95	755
2650	Dish dispenser, drop in, 12"		11	.727		2,575	29		2,604	2,875
2660	Mobile		10	.800		2,250	32		2,282	2,525
3300	Food warmer, counter, 1.2 KW					720			720	790
3550	1.6 KW					2,200			2,200	2,425
3600	Well, hot food, built-in, rectangular, 12" x 20"	R-30	10	2.600		740	122		862	1,000
3610	Circular, 7 qt.		10	2.600		405	122		527	630
3620	Refrigerated, 2 compartments		10	2.600		3,200	122		3,322	3,700
3630	3 compartments		9	2.889		3,950	136		4,086	4,550
3640	4 compartments		8	3.250		4,600	153		4,753	5,275
4720	Frost cold plate		9	2.889		21,400	136		21,536	23,700
5700	Hot chocolate dispenser	1 Plum	4	2		1,150	124		1,274	1,450
5750	Ice dispenser 567 pound	Q-1	6	2.667		5,450	149		5,599	6,225
6250	Jet spray dispenser	R-18	4.50	5.778		1,850	295		2,145	2,475
6300	Juice dispenser, concentrate	"	4.50	5.778		1,925	295		2,220	2,575
6690	Milk dispenser, bulk, 2 flavor	R-30	8	3.250		1,850	153		2,003	2,275
6695	3 flavor	"	8	3.250		2,375	153		2,528	2,850
8800	Serving counter, straight	1 Carp	40	.200	L.F.	1,425	10.15		1,435.15	1,575
8820	Curved section	"	30	.267	"	2,100	13.50		2,113.50	2,350
8825	Solid surface, see Section 12 36 61.16									
8860	Sneeze guard with lights, 60" L	1 Clab	16	.500	Ea.	345	19.95		364.95	410
8900	Sneeze guard, stainless steel and glass, single sided									
8910	Portable, 48" W				Ea.	435			435	480
8920	Portable, 72" W					415			415	460
8930	Adjustable, 36" W	1 Carp	24	.333		250	16.90		266.90	300
8940	Adjustable, 48" W	"	20	.400		315	20.50		335.50	380
9100	Soft serve ice cream machine, medium	R-18	11	2.364		7,700	121		7,821	8,650
9110	Large	"	9	2.889		21,900	148		22,048	24,200

For customer support on your Building Construction Costs with RSMeans data, call 800.448.8182.

11 46 Food Dispensing Equipment

11 46 83 – Ice Machines

11 46 83.10 Commercial Ice Equipment	Crew	Daily Output	Labor-Hours	Unit	Material	2018 Bare Costs Labor	Equipment	Total	Total Incl O&P
0010 **COMMERCIAL ICE EQUIPMENT**									
5800 Ice cube maker, 50 lbs./day	Q-1	6	2.667	Ea.	1,700	149		1,849	2,100
5810 65 lbs./day, energy star rated		6	2.667		1,625	149		1,774	2,000
5900 250 lbs./day		1.20	13.333		2,525	745		3,270	3,900
5950 300 lbs./day, remote condensing		1.20	13.333		2,400	745		3,145	3,750
6050 500 lbs./day		4	4		2,850	224		3,074	3,450
6060 With bin		1.20	13.333		3,975	745		4,720	5,500
6070 Modular, with bin and condenser		1.20	13.333		4,025	745		4,770	5,550
6090 1000 lbs./day, with bin		1	16		5,150	895		6,045	7,025
6100 Ice flakers, 300 lbs./day		1.60	10		3,300	560		3,860	4,475
6120 600 lbs./day		.95	16.842		3,975	940		4,915	5,800
6130 1000 lbs./day		.75	21.333		4,875	1,200		6,075	7,175
6140 2000 lbs./day	↓	.65	24.615		22,400	1,375		23,775	26,800
6160 Ice storage bin, 500 pound capacity	Q-5	1	16		1,125	905		2,030	2,625
6180 1000 pound	"	.56	28.571	↓	2,900	1,625		4,525	5,650

11 48 Foodservice Cleaning and Disposal Equipment

11 48 13 – Commercial Dishwashers

11 48 13.10 Dishwashers

	Crew	Daily Output	Labor-Hours	Unit	Material	2018 Bare Costs Labor	Equipment	Total	Total Incl O&P
0010 **DISHWASHERS**									
2700 Dishwasher, commercial, rack type									
2720 10 to 12 racks/hour	Q-1	3.20	5	Ea.	3,400	280		3,680	4,150
2730 Energy star rated, 35 to 40 racks/hour [G]		1.30	12.308		4,250	690		4,940	5,700
2740 50 to 60 racks/hour [G]	↓	1.30	12.308		11,100	690		11,790	13,200
2800 Automatic, 190 to 230 racks/hour	L-6	.35	34.286		12,600	2,075		14,675	16,900
2820 235 to 275 racks/hour		.25	48		26,200	2,925		29,125	33,200
2840 8,750 to 12,500 dishes/hour	↓	.10	120	↓	47,200	7,300		54,500	63,000
2950 Dishwasher hood, canopy type	L-3A	10	1.200	L.F.	1,025	66		1,091	1,250
2960 Pant leg type	"	2.50	4.800	Ea.	7,925	264		8,189	9,125
5200 Garbage disposal 1.5 HP, 100 GPH	L-1	4.80	3.333		1,500	201		1,701	1,950
5210 3 HP, 120 GPH		4.60	3.478		2,125	209		2,334	2,675
5220 5 HP, 250 GPH	↓	4.50	3.556	↓	2,825	214		3,039	3,425
6750 Pot sink, 3 compartment	1 Plum	7.25	1.103	L.F.	970	68.50		1,038.50	1,175
6760 Pot washer, low temp wash/rinse		1.60	5	Ea.	5,475	310		5,785	6,500
6770 High pressure wash, high temperature rinse	↓	1.20	6.667		38,000	415		38,415	42,400
9170 Trash compactor, small, up to 125 lb. compacted weight	L-4	4	6		24,300	285		24,585	27,200
9175 Large, up to 175 lb. compacted weight	"	3	8	↓	28,900	380		29,280	32,400

11 52 Audio-Visual Equipment

11 52 13 – Projection Screens

11 52 13.10 Projection Screens, Wall or Ceiling Hung

	Crew	Daily Output	Labor-Hours	Unit	Material	2018 Bare Costs Labor	Equipment	Total	Total Incl O&P
0010 **PROJECTION SCREENS, WALL OR CEILING HUNG**, matte white									
0100 Manually operated, economy	2 Carp	500	.032	S.F.	6.45	1.62		8.07	9.50
0300 Intermediate		450	.036		7.85	1.80		9.65	11.40
0400 Deluxe		400	.040		9.75	2.03		11.78	13.85
0600 Electric operated, matte white, 25 S.F., economy		5	3.200	Ea.	1,125	162		1,287	1,500
0700 Deluxe		4	4		1,325	203		1,528	1,775
0900 50 S.F., economy		3	5.333		1,100	270		1,370	1,600
1000 Deluxe	↓	2	8	↓	2,650	405		3,055	3,550

For customer support on your Building Construction Costs with RSMeans data, call 800.448.8182.

419

11 52 Audio-Visual Equipment

11 52 13 – Projection Screens

11 52 13.10 Projection Screens, Wall or Ceiling Hung	Crew	Daily Output	Labor-Hours	Unit	Material	2018 Bare Costs Labor	Equipment	Total	Total Incl O&P
1200	Heavy duty, electric operated, 200 S.F.	2 Carp	1.50	10.667	Ea.	4,150	540		4,690	5,400
1300	400 S.F.	↓	1	16	↓	4,950	810		5,760	6,675
1500	Rigid acrylic in wall, for rear projection, 1/4" thick	2 Glaz	30	.533	S.F.	70	26		96	116
1600	1/2" thick (maximum size 10' x 20')	"	25	.640	"	82	31		113	137

11 52 16 – Projectors

11 52 16.10 Movie Equipment

		Crew	Daily Output	Labor-Hours	Unit	Material	2018 Bare Costs Labor	Equipment	Total	Total Incl O&P
0010	**MOVIE EQUIPMENT**									
0020	Changeover, minimum				Ea.	515			515	565
0100	Maximum					975			975	1,075
0400	Film transport, incl. platters and autowind, minimum					5,425			5,425	5,975
0500	Maximum					15,400			15,400	17,000
0800	Lamphouses, incl. rectifiers, xenon, 1,000 watt	1 Elec	2	4		7,200	233		7,433	8,250
0900	1,600 watt		2	4		7,675	233		7,908	8,800
1000	2,000 watt		1.50	5.333		8,350	310		8,660	9,650
1100	4,000 watt	↓	1.50	5.333		11,500	310		11,810	13,100
1400	Lenses, anamorphic, minimum					1,375			1,375	1,525
1500	Maximum					3,100			3,100	3,425
1800	Flat 35 mm, minimum					1,200			1,200	1,325
1900	Maximum					1,875			1,875	2,050
2200	Pedestals, for projectors					1,675			1,675	1,825
2300	Console type					11,900			11,900	13,100
2600	Projector mechanisms, incl. soundhead, 35 mm, minimum					12,300			12,300	13,600
2700	Maximum				↓	16,900			16,900	18,600
3000	Projection screens, rigid, in wall, acrylic, 1/4" thick	2 Glaz	195	.082	S.F.	46.50	3.98		50.48	57
3100	1/2" thick	"	130	.123	"	54	5.95		59.95	68.50
3300	Electric operated, heavy duty, 400 S.F.	2 Carp	1	16	Ea.	3,300	810		4,110	4,850
3320	Theater projection screens, matte white, including frames	"	200	.080	S.F.	7.40	4.06		11.46	14.35
3400	Also see Section 11 52 13.10									
3700	Sound systems, incl. amplifier, mono, minimum	1 Elec	.90	8.889	Ea.	3,725	515		4,240	4,875
3800	Dolby/Super Sound, maximum		.40	20		20,100	1,175		21,275	23,900
4100	Dual system, 2 channel, front surround, minimum		.70	11.429		5,150	665		5,815	6,675
4200	Dolby/Super Sound, 4 channel, maximum	↓	.40	20		18,800	1,175		19,975	22,400
4500	Sound heads, 35 mm					5,975			5,975	6,575
4900	Splicer, tape system, minimum					830			830	915
5000	Tape type, maximum					1,475			1,475	1,625
5300	Speakers, recessed behind screen, minimum	1 Elec	2	4		1,175	233		1,408	1,650
5400	Maximum	"	1	8		3,450	465		3,915	4,475
5700	Seating, painted steel, upholstered, minimum	2 Carp	35	.457		152	23		175	203
5800	Maximum	"	28	.571		500	29		529	595
6100	Rewind tables, minimum					2,975			2,975	3,275
6200	Maximum				↓	5,275			5,275	5,825
7000	For automation, varying sophistication, minimum	1 Elec	1	8	System	2,650	465		3,115	3,625
7100	Maximum	2 Elec	.30	53.333	"	6,200	3,100		9,300	11,500

11 52 16.20 Movie Equipment- Digital

		Crew	Daily Output	Labor-Hours	Unit	Material	2018 Bare Costs Labor	Equipment	Total	Total Incl O&P
0010	**MOVIE EQUIPMENT- DIGITAL**									
1000	Digital 2K projection system, 98" DMD	1 Elec	2	4	Ea.	49,000	233		49,233	54,500
1100	OEM lens		2	4		5,975	233		6,208	6,925
2000	Pedestal with power distribution		2	4		2,300	233		2,533	2,875
3000	Software	↓	2	4	↓	1,950	233		2,183	2,500

11 53 Laboratory Equipment

11 53 03 - Laboratory Test Equipment

11 53 03.13 Test Equipment	Crew	Daily Output	Labor-Hours	Unit	Material	2018 Bare Costs Labor	Equipment	Total	Total Incl O&P
0010	**TEST EQUIPMENT**								
1700	Thermometer, electric, portable				Ea.	330			330	365
1800	Titration unit, four 2000 ml reservoirs				"	6,050			6,050	6,650

11 53 13 - Laboratory Fume Hoods

11 53 13.13 Recirculating Laboratory Fume Hoods

		Crew	Daily Output	Labor-Hours	Unit	Material	Labor	Equipment	Total	Total Incl O&P
0010	**RECIRCULATING LABORATORY FUME HOODS**									
0600	Fume hood, with countertop & base, not including HVAC									
0610	Simple, minimum	2 Carp	5.40	2.963	L.F.	560	150		710	845
0620	Complex, including fixtures		2.40	6.667		1,050	340		1,390	1,675
0630	Special, maximum		1.70	9.412		1,175	475		1,650	2,000
0670	Service fixtures, average				Ea.	335			335	365
0680	For sink assembly with hot and cold water, add	1 Plum	1.40	5.714		760	355		1,115	1,375
0750	Glove box, fiberglass, bacteriological					17,300			17,300	19,000
0760	Controlled atmosphere					20,500			20,500	22,500
0770	Radioisotope					17,300			17,300	19,000
0780	Carcinogenic					17,300			17,300	19,000

11 53 16 - Laboratory Incubators

11 53 16.13 Incubators

		Crew	Daily Output	Labor-Hours	Unit	Material	Labor	Equipment	Total	Total Incl O&P
0010	**INCUBATORS**									
1000	Incubators, minimum				Ea.	3,350			3,350	3,675
1010	Maximum				"	13,800			13,800	15,200

11 53 19 - Laboratory Sterilizers

11 53 19.13 Sterilizers

		Crew	Daily Output	Labor-Hours	Unit	Material	Labor	Equipment	Total	Total Incl O&P
0010	**STERILIZERS**									
0700	Glassware washer, undercounter, minimum	L-1	1.80	8.889	Ea.	6,675	535		7,210	8,150
0710	Maximum	"	1	16		14,600	965		15,565	17,500
1850	Utensil washer-sanitizer	1 Plum	2	4		8,975	249		9,224	10,300

11 53 23 - Laboratory Refrigerators

11 53 23.13 Refrigerators

		Crew	Daily Output	Labor-Hours	Unit	Material	Labor	Equipment	Total	Total Incl O&P
0010	**REFRIGERATORS**									
1200	Blood bank, 28.6 C.F. emergency signal				Ea.	13,000			13,000	14,400
1210	Reach-in, 16.9 C.F.				"	9,250			9,250	10,200

11 53 33 - Emergency Safety Appliances

11 53 33.13 Emergency Equipment

		Crew	Daily Output	Labor-Hours	Unit	Material	Labor	Equipment	Total	Total Incl O&P
0010	**EMERGENCY EQUIPMENT**									
1400	Safety equipment, eye wash, hand held				Ea.	415			415	455
1450	Deluge shower				"	820			820	900

11 53 43 - Service Fittings and Accessories

11 53 43.13 Fittings

		Crew	Daily Output	Labor-Hours	Unit	Material	Labor	Equipment	Total	Total Incl O&P
0010	**FITTINGS**									
1600	Sink, one piece plastic, flask wash, hose, free standing	1 Plum	1.60	5	Ea.	2,025	310		2,335	2,700
1610	Epoxy resin sink, 25" x 16" x 10"	"	2	4	"	229	249		478	625
1950	Utility table, acid resistant top with drawers	2 Carp	30	.533	L.F.	171	27		198	229
8000	Alternate pricing method: as percent of lab furniture									
8050	Installation, not incl. plumbing & duct work				% Furn.				22%	22%
8100	Plumbing, final connections, simple system								10%	10%
8110	Moderately complex system								15%	15%
8120	Complex system								20%	20%
8150	Electrical, simple system								10%	10%

For customer support on your Building Construction Costs with RSMeans data, call 800.448.8182.

421

11 53 Laboratory Equipment

11 53 43 – Service Fittings and Accessories

11 53 43.13 Fittings

	11 53 43.13 Fittings	Crew	Daily Output	Labor-Hours	Unit	Material	2018 Bare Costs Labor	Equipment	Total	Total Incl O&P
8160	Moderately complex system				% Furn.				20%	20%
8170	Complex system								35%	35%

11 53 53 – Biological Safety Cabinets

11 53 53.10 Pharmacy Cabinets

		Crew	Daily Output	Labor-Hours	Unit	Material	2018 Bare Costs Labor	Equipment	Total	Total Incl O&P
0010	**PHARMACY CABINETS**, vertical flow									
0100	Class II, type B2, 6' L	2 Carp	1.50	10.667	Ea.	14,700	540		15,240	17,000

11 57 Vocational Shop Equipment

11 57 10 – Shop Equipment

11 57 10.10 Vocational School Shop Equipment

		Crew	Daily Output	Labor-Hours	Unit	Material	2018 Bare Costs Labor	Equipment	Total	Total Incl O&P
0010	**VOCATIONAL SCHOOL SHOP EQUIPMENT**									
0020	Benches, work, wood, average	2 Carp	5	3.200	Ea.	715	162		877	1,025
0100	Metal, average		5	3.200		450	162		612	740
0400	Combination belt & disc sander, 6"		4	4		1,700	203		1,903	2,175
0700	Drill press, floor mounted, 12", 1/2 HP		4	4		400	203		603	750
0800	Dust collector, not incl. ductwork, 6" diameter	1 Shee	1.10	7.273		5,250	435		5,685	6,450
0810	Dust collector bag, 20" diameter	"	5	1.600		585	95.50		680.50	785
1000	Grinders, double wheel, 1/2 HP	2 Carp	5	3.200		215	162		377	485
1300	Jointer, 4", 3/4 HP		4	4		1,325	203		1,528	1,775
1600	Kilns, 16 C.F., to 2000°		4	4		1,575	203		1,778	2,025
1900	Lathe, woodworking, 10", 1/2 HP		4	4		575	203		778	940
2200	Planer, 13" x 6"		4	4		1,175	203		1,378	1,600
2500	Potter's wheel, motorized		4	4		1,200	203		1,403	1,600
2800	Saws, band, 14", 3/4 HP		4	4		1,100	203		1,303	1,525
3100	Metal cutting band saw, 14"		4	4		2,475	203		2,678	3,025
3400	Radial arm saw, 10", 2 HP		4	4		1,275	203		1,478	1,725
3700	Scroll saw, 24"		4	4		625	203		828	995
4000	Table saw, 10", 3 HP		4	4		2,900	203		3,103	3,500
4300	Welder AC arc, 30 amp capacity		4	4		3,175	203		3,378	3,800

11 61 Broadcast, Theater, and Stage Equipment

11 61 23 – Folding and Portable Stages

11 61 23.10 Portable Stages

		Crew	Daily Output	Labor-Hours	Unit	Material	2018 Bare Costs Labor	Equipment	Total	Total Incl O&P
0010	**PORTABLE STAGES**									
1500	Flooring, portable oak parquet, 3' x 3' sections				S.F.	14.10			14.10	15.55
1600	Cart to carry 225 S.F. of flooring				Ea.	435			435	480
5000	Stages, portable with steps, folding legs, stock, 8" high				SF Stg.	41			41	45.50
5100	16" high					59.50			59.50	65.50
5200	32" high					61			61	67
5300	40" high					62.50			62.50	69
6000	Telescoping platforms, extruded alum., straight, minimum	4 Carp	157	.204		37.50	10.35		47.85	57
6100	Maximum		77	.416		53	21		74	90
6500	Pie-shaped, minimum		150	.213		81	10.80		91.80	105
6600	Maximum		70	.457		90.50	23		113.50	135
6800	For 3/4" plywood covered deck, deduct					5.40			5.40	5.95
7000	Band risers, steel frame, plywood deck, minimum	4 Carp	275	.116		32	5.90		37.90	44.50
7100	Maximum	"	138	.232		77	11.75		88.75	103
7500	Chairs for above, self-storing, minimum	2 Carp	43	.372	Ea.	116	18.85		134.85	157
7600	Maximum	"	40	.400	"	206	20.50		226.50	258

11 61 Broadcast, Theater, and Stage Equipment

11 61 33 – Rigging Systems and Controls

11 61 33.10 Controls		Crew	Daily Output	Labor-Hours	Unit	Material	2018 Bare Costs Labor	Equipment	Total	Total Incl O&P
0010	**CONTROLS**									
0050	Control boards with dimmers and breakers, minimum	1 Elec	1	8	Ea.	14,600	465		15,065	16,800
0100	Average		.50	16		49,300	930		50,230	55,500
0150	Maximum		.20	40		140,000	2,325		142,325	157,500
8000	Rule of thumb: total stage equipment, minimum	4 Carp	100	.320	SF Stg.	103	16.20		119.20	138
8100	Maximum	"	25	1.280	"	575	65		640	735

11 61 43 – Stage Curtains

11 61 43.10 Curtains		Crew	Daily Output	Labor-Hours	Unit	Material	2018 Bare Costs Labor	Equipment	Total	Total Incl O&P
0010	**CURTAINS**									
0500	Curtain track, straight, light duty	2 Carp	20	.800	L.F.	30.50	40.50		71	95.50
0600	Heavy duty		18	.889		66	45		111	141
0700	Curved sections		12	1.333		196	67.50		263.50	320
1000	Curtains, velour, medium weight		600	.027	S.F.	8.60	1.35		9.95	11.50
1150	Silica based yarn, inherently fire retardant		50	.320	"	16.05	16.20		32.25	42

11 62 Musical Equipment

11 62 16 – Carillons

11 62 16.10 Bell Tower Equipment		Crew	Daily Output	Labor-Hours	Unit	Material	2018 Bare Costs Labor	Equipment	Total	Total Incl O&P
0010	**BELL TOWER EQUIPMENT**									
0300	Carillon, 4 octave (48 bells), with keyboard				System	1,062,500			1,062,500	1,168,500
0320	2 octave (24 bells)					500,000			500,000	550,000
0340	3 to 4 bell peal, minimum					125,000			125,000	137,500
0360	Maximum					750,000			750,000	825,000
0380	Cast bronze bell, average				Ea.	112,500			112,500	123,500
0400	Electronic, digital, minimum					18,700			18,700	20,600
0410	With keyboard, maximum					93,500			93,500	103,000

11 66 Athletic Equipment

11 66 13 – Exercise Equipment

11 66 13.10 Physical Training Equipment		Crew	Daily Output	Labor-Hours	Unit	Material	2018 Bare Costs Labor	Equipment	Total	Total Incl O&P
0010	**PHYSICAL TRAINING EQUIPMENT**									
0020	Abdominal rack, 2 board capacity				Ea.	520			520	575
0050	Abdominal board, upholstered					735			735	805
0200	Bicycle trainer, minimum					520			520	570
0300	Deluxe, electric					4,875			4,875	5,350
0400	Barbell set, chrome plated steel, 25 lb.					291			291	320
0420	100 lb.					585			585	645
0450	200 lb.					775			775	850
0500	Weight plates, cast iron, per lb.				Lb.	3.62			3.62	3.98
0520	Storage rack, 10 station				Ea.	1,025			1,025	1,125
0600	Circuit training apparatus, 12 machines minimum	2 Clab	1.25	12.800	Set	32,500	510		33,010	36,500
0700	Average		1	16		40,500	640		41,140	45,600
0800	Maximum		.75	21.333		48,600	850		49,450	55,000
0820	Dumbbell set, cast iron, with rack and 5 pair					445			445	490
0900	Squat racks	2 Clab	5	3.200	Ea.	900	128		1,028	1,175
1200	Multi-station gym machine, 5 station					5,000			5,000	5,500
1250	9 station					11,400			11,400	12,500
1280	Rowing machine, hydraulic					1,750			1,750	1,925

11 66 Athletic Equipment

11 66 13 – Exercise Equipment

11 66 13.10 Physical Training Equipment

	11 66 13.10 Physical Training Equipment	Crew	Daily Output	Labor-Hours	Unit	Material	2018 Bare Costs Labor	Equipment	Total	Total Incl O&P
1300	Treadmill, manual				Ea.	1,050			1,050	1,175
1320	Motorized					3,475			3,475	3,825
1340	Electronic					4,050			4,050	4,450
1360	Cardio-testing					5,200			5,200	5,700
1400	Treatment/massage tables, minimum					575			575	635
1420	Deluxe, with accessories					795			795	875
4150	Exercise equipment, bicycle trainer					965			965	1,075
4180	Chinning bar, adjustable, wall mounted	1 Carp	5	1.600		218	81		299	365
4200	Exercise ladder, 16' x 1'-7", suspended	L-2	3	5.333		1,425	236		1,661	1,925
4210	High bar, floor plate attached	1 Carp	4	2		2,525	101		2,626	2,925
4240	Parallel bars, adjustable		4	2		1,650	101		1,751	1,975
4270	Uneven parallel bars, adjustable		4	2		3,725	101		3,826	4,250
4280	Wall mounted, adjustable	L-2	1.50	10.667	Set	975	470		1,445	1,800
4300	Rope, ceiling mounted, 18' long	1 Carp	3.66	2.186	Ea.	201	111		312	390
4330	Side horse, vaulting		5	1.600		1,475	81		1,556	1,725
4360	Treadmill, motorized, deluxe, training type		5	1.600		3,925	81		4,006	4,425
4390	Weight lifting multi-station, minimum	2 Clab	1	16		280	640		920	1,275
4450	Maximum	"	.50	32		15,000	1,275		16,275	18,500

11 66 23 – Gymnasium Equipment

11 66 23.13 Basketball Equipment

		Crew	Daily Output	Labor-Hours	Unit	Material	Labor	Equipment	Total	Total Incl O&P
0010	**BASKETBALL EQUIPMENT**									
1000	Backstops, wall mtd., 6' extended, fixed, minimum	L-2	1	16	Ea.	1,600	710		2,310	2,825
1100	Maximum		1	16		1,875	710		2,585	3,150
1200	Swing up, minimum		1	16		1,750	710		2,460	3,000
1250	Maximum		1	16		3,050	710		3,760	4,425
1300	Portable, manual, heavy duty, spring operated		1.90	8.421		13,100	375		13,475	15,100
1400	Ceiling suspended, stationary, minimum		.78	20.513		4,275	910		5,185	6,100
1450	Fold up, with accessories, maximum		.40	40		6,350	1,775		8,125	9,700
1600	For electrically operated, add	1 Elec	1	8		2,550	465		3,015	3,525
5800	Wall pads, 1-1/2" thick, standard (not fire rated)	2 Carp	640	.025	S.F.	6.25	1.27		7.52	8.85

11 66 23.19 Boxing Ring

		Crew	Daily Output	Labor-Hours	Unit	Material	Labor	Equipment	Total	Total Incl O&P
0010	**BOXING RING**									
4100	Elevated, 22' x 22'	L-4	.10	240	Ea.	5,725	11,400		17,125	23,800
4110	For cellular plastic foam padding, add		.10	240		1,225	11,400		12,625	18,900
4120	Floor level, including posts and ropes only, 20' x 20'		.80	30		4,175	1,425		5,600	6,775
4130	Canvas, 30' x 30'		5	4.800		1,525	228		1,753	2,025

11 66 23.47 Gym Mats

		Crew	Daily Output	Labor-Hours	Unit	Material	Labor	Equipment	Total	Total Incl O&P
0010	**GYM MATS**									
5500	2" thick, naugahyde covered				S.F.	5.20			5.20	5.70
5600	Vinyl/nylon covered					8.30			8.30	9.10
6000	Wrestling mats, 1" thick, heavy duty					5			5	5.50

11 66 43 – Interior Scoreboards

11 66 43.10 Scoreboards

		Crew	Daily Output	Labor-Hours	Unit	Material	Labor	Equipment	Total	Total Incl O&P
0010	**SCOREBOARDS**									
7000	Baseball, minimum	R-3	1.30	15.385	Ea.	3,700	890	99.50	4,689.50	5,500
7200	Maximum		.05	400		20,700	23,200	2,600	46,500	60,500
7300	Football, minimum		.86	23.256		5,125	1,350	151	6,626	7,850
7400	Maximum		.20	100		19,000	5,800	650	25,450	30,300
7500	Basketball (one side), minimum		2.07	9.662		2,475	560	62.50	3,097.50	3,625
7600	Maximum		.30	66.667		7,550	3,875	430	11,855	14,600
7700	Hockey-basketball (four sides), minimum		.25	80		9,300	4,650	520	14,470	17,700

11 66 Athletic Equipment

11 66 43 – Interior Scoreboards

11 66 43.10 Scoreboards		Crew	Daily Output	Labor-Hours	Unit	Material	2018 Bare Costs Labor	Equipment	Total	Total Incl O&P
7800	Maximum	R-3	.15	133	Ea.	16,200	7,725	865	24,790	30,400

11 66 53 – Gymnasium Dividers

11 66 53.10 Divider Curtains

		Crew	Daily Output	Labor-Hours	Unit	Material	Labor	Equipment	Total	Total Incl O&P
0010	**DIVIDER CURTAINS**									
4500	Gym divider curtain, mesh top, vinyl bottom, manual	L-4	500	.048	S.F.	10.25	2.28		12.53	14.80
4700	Electric roll up	L-7	400	.070	"	13.60	3.41		17.01	20

11 67 Recreational Equipment

11 67 13 – Bowling Alley Equipment

11 67 13.10 Bowling Alleys

		Crew	Daily Output	Labor-Hours	Unit	Material	Labor	Equipment	Total	Total Incl O&P
0010	**BOWLING ALLEYS** Including alley, pinsetter, scorer,									
0020	Counters and misc. supplies, minimum	4 Carp	.20	160	Lane	40,600	8,100		48,700	57,000
0150	Average		.19	168		49,400	8,550		57,950	67,500
0300	Maximum		.18	178		57,000	9,025		66,025	76,000
0400	Combo table ball rack, add					1,350			1,350	1,475
0600	For automatic scorer, add, minimum					6,125			6,125	6,750
0700	Maximum					10,200			10,200	11,200

11 67 23 – Shooting Range Equipment

11 67 23.10 Shooting Range

		Crew	Daily Output	Labor-Hours	Unit	Material	Labor	Equipment	Total	Total Incl O&P
0010	**SHOOTING RANGE** Incl. bullet traps, target provisions, controls,									
0100	Separators, ceiling system, etc. Not incl. structural shell									
0200	Commercial	L-9	.64	56.250	Point	33,500	2,550		36,050	40,800
0300	Law enforcement		.28	129		44,000	5,825		49,825	57,500
0400	National Guard armories		.71	50.704		24,100	2,300		26,400	30,100
0500	Reserve training centers		.71	50.704		18,800	2,300		21,100	24,200
0600	Schools and colleges		.32	113		43,400	5,100		48,500	55,500
0700	Major academies		.19	189		61,000	8,575		69,575	80,500
0800	For acoustical treatment, add					10%	10%			
0900	For lighting, add					28%	25%			
1000	For plumbing, add					5%	5%			
1100	For ventilating system, add, minimum					40%	40%			
1200	Add, average					25%	25%			
1300	Add, maximum					35%	35%			

11 68 Play Field Equipment and Structures

11 68 13 – Playground Equipment

11 68 13.10 Free-Standing Playground Equipment

		Crew	Daily Output	Labor-Hours	Unit	Material	Labor	Equipment	Total	Total Incl O&P
0010	**FREE-STANDING PLAYGROUND EQUIPMENT** See also individual items									
0200	Bike rack, 10' long, permanent G	B-1	12	2	Ea.	570	81		651	750
0392	Upper body warm-up station		2.60	9.231		3,125	375		3,500	4,000
0394	Bench stepper station		2.60	9.231		1,350	375		1,725	2,075
0396	Standing push up station		2.60	9.231		1,175	375		1,550	1,875
0398	Upper body stretch station		2.60	9.231		2,275	375		2,650	3,075
0400	Horizontal monkey ladder, 14' long, 6' high		4	6		1,400	243		1,643	1,925
0590	Parallel bars, 10' long		4	6		565	243		808	995
0600	Posts, tether ball set, 2-3/8" OD		12	2		420	81		501	585
0800	Poles, multiple purpose, 10'-6" long		12	2	Pr.	211	81		292	355
1000	Ground socket for movable posts, 2-3/8" post		10	2.400		109	97.50		206.50	268

11 68 Play Field Equipment and Structures

11 68 13 – Playground Equipment

11 68 13.10 Free-Standing Playground Equipment

		Crew	Daily Output	Labor-Hours	Unit	Material	2018 Bare Costs Labor	Equipment	Total	Total Incl O&P
1100	3-1/2" post	B-1	10	2.400	Pr.	205	97.50		302.50	375
1300	See-saw, spring, steel, 2 units		6	4	Ea.	900	162		1,062	1,225
1400	4 units		4	6		1,400	243		1,643	1,925
1500	6 units		3	8		1,900	325		2,225	2,600
1700	Shelter, fiberglass golf tee, 3 person		4.60	5.217		4,500	211		4,711	5,275
1900	Slides, stainless steel bed, 12' long, 6' high		3	8		4,250	325		4,575	5,175
2000	20' long, 10' high		2	12		7,525	485		8,010	9,025
2200	Swings, plain seats, 8' high, 4 seats		2	12		1,600	485		2,085	2,525
2300	8 seats		1.30	18.462		2,775	750		3,525	4,200
2500	12' high, 4 seats		2	12		2,125	485		2,610	3,100
2600	8 seats		1.30	18.462		3,675	750		4,425	5,200
2800	Whirlers, 8' diameter		3	8		3,225	325		3,550	4,025
2900	10' diameter		3	8		5,900	325		6,225	7,000

11 68 13.20 Modular Playground

		Crew	Daily Output	Labor-Hours	Unit	Material	2018 Bare Costs Labor	Equipment	Total	Total Incl O&P
0010	**MODULAR PLAYGROUND** Basic components									
0100	Deck, square, steel, 48" x 48"	B-1	1	24	Ea.	685	970		1,655	2,225
0110	Recycled polyurethane		1	24		630	970		1,600	2,175
0120	Triangular, steel, 48" side		1	24		760	970		1,730	2,300
0130	Post, steel, 5" square		18	1.333	L.F.	47.50	54		101.50	135
0140	Aluminum, 2-3/8" square		20	1.200		53.50	48.50		102	133
0150	5" square		18	1.333		46.50	54		100.50	134
0160	Roof, square poly, 54" side		18	1.333	Ea.	1,650	54		1,704	1,875
0170	Wheelchair transfer module, for 3' high deck		3	8	"	3,200	325		3,525	4,025
0180	Guardrail, pipe, 36" high		60	.400	L.F.	291	16.20		307.20	345
0190	Steps, deck-to-deck, three 8" steps		8	3	Ea.	1,225	122		1,347	1,525
0200	Activity panel, crawl through panel		2	12		555	485		1,040	1,350
0210	Alphabet/spelling panel		2	12		655	485		1,140	1,450
0360	With guardrails		3	8		1,900	325		2,225	2,600
0370	Crawl tunnel, straight, 56" long		4	6		1,250	243		1,493	1,750
0380	90°, 4' long		4	6		1,650	243		1,893	2,175
1200	Slide, tunnel, for 56" high deck		8	3		2,200	122		2,322	2,600
1210	Straight, poly		8	3		440	122		562	665
1220	Stainless steel, 54" high deck		6	4		845	162		1,007	1,175
1230	Curved, poly, 40" high deck		6	4		845	162		1,007	1,175
1240	Spiral slide, 56"-72" high		5	4.800		5,000	195		5,195	5,800
1300	Ladder, vertical, for 24"-72" high deck		5	4.800		535	195		730	885
1310	Horizontal, 8' long		5	4.800		905	195		1,100	1,300
1320	Corkscrew climber, 6' high		3	8		1,150	325		1,475	1,775
1330	Fire pole for 72" high deck		6	4		292	162		454	565
1340	Bridge, ring climber, 8' long		4	6		2,400	243		2,643	3,000
1350	Suspension		4	6	L.F.	325	243		568	725

11 68 16 – Play Structures

11 68 16.10 Handball/Squash Court

		Crew	Daily Output	Labor-Hours	Unit	Material	2018 Bare Costs Labor	Equipment	Total	Total Incl O&P
0010	**HANDBALL/SQUASH COURT**, outdoor									
0900	Handball or squash court, outdoor, wood	2 Carp	.50	32	Ea.	4,475	1,625		6,100	7,400
1000	Masonry handball/squash court	D-1	.30	53.333	"	25,500	2,400		27,900	31,700

11 68 16.30 Platform/Paddle Tennis Court

		Crew	Daily Output	Labor-Hours	Unit	Material	2018 Bare Costs Labor	Equipment	Total	Total Incl O&P
0010	**PLATFORM/PADDLE TENNIS COURT** Complete with lighting, etc.									
0100	Aluminum slat deck with aluminum frame	B-1	.08	300	Court	65,000	12,200		77,200	90,000
0500	Aluminum slat deck with wood frame	C-1	.12	267		78,000	12,800		90,800	105,000
0800	Aluminum deck heater, add	B-1	1.18	20.339		2,750	825		3,575	4,275
0900	Douglas fir planking with wood frame 2" x 6" x 30'	C-1	.12	267		73,000	12,800		85,800	100,000

426

11 68 Play Field Equipment and Structures

11 68 16 – Play Structures

11 68 16.30 Platform/Paddle Tennis Court	Crew	Daily Output	Labor-Hours	Unit	Material	2018 Bare Costs Labor	Equipment	Total	Total Incl O&P
1000	Plywood deck with steel frame	C-1	.12	267	Court	73,000	12,800		85,800	100,000
1100	Steel slat deck with wood frame	↓	.12	267	↓	44,600	12,800		57,400	68,500

11 68 33 – Athletic Field Equipment

11 68 33.13 Football Field Equipment

		Crew	Daily Output	Labor-Hours	Unit	Material	2018 Bare Costs Labor	Equipment	Total	Total Incl O&P
0010	**FOOTBALL FIELD EQUIPMENT**									
0020	Goal posts, steel, football, double post	B-1	1.50	16	Pr.	4,225	650		4,875	5,650
0100	Deluxe, single post		1.50	16		3,175	650		3,825	4,500
0300	Football, convertible to soccer		1.50	16		3,000	650		3,650	4,300
0500	Soccer, regulation	↓	2	12	↓	2,050	485		2,535	3,000

11 71 Medical Sterilizing Equipment

11 71 10 – Medical Sterilizers & Distillers

11 71 10.10 Sterilizers and Distillers

		Crew	Daily Output	Labor-Hours	Unit	Material	2018 Bare Costs Labor	Equipment	Total	Total Incl O&P
0010	**STERILIZERS AND DISTILLERS**									
0700	Distiller, water, steam heated, 50 gal. capacity	1 Plum	1.40	5.714	Ea.	24,300	355		24,655	27,300
3010	Portable, top loading, 105-135 degree C, 3 to 30 psi, 50 L chamber					9,625			9,625	10,600
3020	Stainless steel basket, 10.7" diam. x 11.8" H					212			212	233
3025	Stainless steel pail, 10.7" diam. x 10.7" H					315			315	345
3050	85 L chamber					17,300			17,300	19,000
3060	Stainless steel basket, 15.3" diam. x 11.5" H					445			445	485
3065	Stainless steel pail, 15.3" diam. x 11" H					665			665	730
5600	Sterilizers, floor loading, 26" x 62" x 42", single door, steam					135,000			135,000	148,500
5650	Double door, steam					224,000			224,000	246,000
5800	General purpose, 20" x 20" x 38", single door					12,400			12,400	13,700
6000	Portable, counter top, steam, minimum					2,725			2,725	3,000
6020	Maximum					4,950			4,950	5,450
6050	Portable, counter top, gas, 17" x 15" x 32-1/2"					42,700			42,700	47,000
6150	Manual washer/sterilizer, 16" x 16" x 26"	1 Plum	2	4	↓	58,500	249		58,749	65,000
6200	Steam generators, electric 10 kW to 180 kW, freestanding									
6250	Minimum	1 Elec	3	2.667	Ea.	11,100	155		11,255	12,500
6300	Maximum	"	.70	11.429		23,400	665		24,065	26,700
8200	Bed pan washer-sanitizer	1 Plum	2	4	↓	9,500	249		9,749	10,800

11 72 Examination and Treatment Equipment

11 72 13 – Examination Equipment

11 72 13.13 Examination Equipment

		Crew	Daily Output	Labor-Hours	Unit	Material	2018 Bare Costs Labor	Equipment	Total	Total Incl O&P
0010	**EXAMINATION EQUIPMENT**									
0300	Blood pressure unit, mercurial, wall				Ea.	150			150	165
0400	Diagnostic set, wall					630			630	690
4400	Scale, physician's, with height rod				↓	430			430	470

11 72 53 – Treatment Equipment

11 72 53.13 Medical Treatment Equipment

		Crew	Daily Output	Labor-Hours	Unit	Material	2018 Bare Costs Labor	Equipment	Total	Total Incl O&P
0010	**MEDICAL TREATMENT EQUIPMENT**									
6300	Exam light, portable, 14" flexible arm				Ea.	193			193	212
6500	Surgery table, minor minimum	1 Sswk	.70	11.429		14,900	625		15,525	17,400
6520	Maximum	"	.50	16		18,300	875		19,175	21,500
6700	Surgical lights, doctor's office, single arm	2 Elec	2	8	↓	3,025	465		3,490	4,025
6750	Dual arm	"	1	16	↓	5,225	930		6,155	7,150

11 73 Patient Care Equipment

11 73 10 – Patient Treatment Equipment

11 73 10.10 Treatment Equipment

		Crew	Daily Output	Labor-Hours	Unit	Material	2018 Bare Costs Labor	Equipment	Total	Total Incl O&P
0010	**TREATMENT EQUIPMENT**									
0750	Exam room furnishings, average per room				Ea.	5,775			5,775	6,350
1800	Heat therapy unit, humidified, 26" x 78" x 28"				"	3,350			3,350	3,675
2100	Hubbard tank with accessories, stainless steel,									
2110	125 GPM at 45 psi water pressure				Ea.	15,000			15,000	16,500
2150	For electric overhead hoist, add					3,250			3,250	3,575
2900	K-Module for heat therapy, 20 oz. capacity, 75°F to 110°F					550			550	605
3600	Paraffin bath, 126°F, auto controlled					255			255	281
3900	Parallel bars for walking training, 12'-0"					1,875			1,875	2,075
4600	Station, dietary, medium, with ice					21,100			21,100	23,300
4700	Medicine					9,550			9,550	10,500
7000	Tables, physical therapy, walk off, electric	2 Carp	3	5.333		2,500	270		2,770	3,150
7150	Standard, vinyl top with base cabinets, minimum		3	5.333		1,075	270		1,345	1,575
7200	Maximum	↓	2	8		5,375	405		5,780	6,525
7250	Table, hospital, adjustable height					1,525			1,525	1,675
8400	Whirlpool bath, mobile, sst, 18" x 24" x 60"					5,425			5,425	5,975
8450	Fixed, incl. mixing valves	1 Plum	2	4	↓	4,850	249		5,099	5,700

11 73 10.20 Bariatric Equipment

		Crew	Daily Output	Labor-Hours	Unit	Material	2018 Bare Costs Labor	Equipment	Total	Total Incl O&P
0010	**BARIATRIC EQUIPMENT**									
5000	Patient lift, electric operated, arm style									
5110	400 lb. capacity				Ea.	1,250			1,250	1,375
5120	450 lb. capacity					2,275			2,275	2,525
5130	600 lb. capacity					3,450			3,450	3,775
5140	700 lb. capacity					4,725			4,725	5,175
5150	1,000 lb. capacity					7,025			7,025	7,725
5200	Overhead, 4-post, 1,000 lb. capacity					10,200			10,200	11,200
5300	Overhead, track type, 450 lb. capacity, not including track					3,350			3,350	3,675
5500	For fabric sling, add					305			305	335
5550	For digital scale, add				↓	810			810	890

11 74 Dental Equipment

11 74 10 – Dental Office Equipment

11 74 10.10 Diagnostic and Treatment Equipment

		Crew	Daily Output	Labor-Hours	Unit	Material	2018 Bare Costs Labor	Equipment	Total	Total Incl O&P
0010	**DIAGNOSTIC AND TREATMENT EQUIPMENT**									
0020	Central suction system, minimum	1 Plum	1.20	6.667	Ea.	1,075	415		1,490	1,800
0100	Maximum	"	.90	8.889		4,900	550		5,450	6,225
0300	Air compressor, minimum	1 Skwk	.80	10		2,800	525		3,325	3,875
0400	Maximum	↓	.50	16		8,750	840		9,590	10,900
0600	Chair, electric or hydraulic, minimum		.50	16		2,450	840		3,290	3,975
0700	Maximum	↓	.25	32		4,250	1,675		5,925	7,250
0800	Doctor's/assistant's stool, minimum					252			252	277
0850	Maximum					690			690	760
1000	Drill console with accessories, minimum	1 Skwk	1.60	5		2,475	262		2,737	3,100
1100	Maximum		1.60	5		5,175	262		5,437	6,100
2000	Light, ceiling mounted, minimum		8	1		900	52.50		952.50	1,075
2100	Maximum	↓	8	1		2,100	52.50		2,152.50	2,375
2200	Unit light, minimum	2 Skwk	5.33	3.002		760	157		917	1,075
2210	Maximum		5.33	3.002		1,550	157		1,707	1,950
2220	Track light, minimum		3.20	5		1,700	262		1,962	2,250
2230	Maximum	↓	3.20	5		2,050	262		2,312	2,650
2300	Sterilizers, steam portable, minimum					1,900			1,900	2,075

For customer support on your Building Construction Costs with RSMeans data, call 800.448.8182.

11 74 Dental Equipment

11 74 10 - Dental Office Equipment

| | | Daily | Labor- | | | 2018 Bare Costs | | | Total |
11 74 10.10 Diagnostic and Treatment Equipment	Crew	Output	Hours	Unit	Material	Labor	Equipment	Total	Incl O&P
2350	Maximum				Ea.	6,175			6,175	6,800
2600	Steam, institutional					2,750			2,750	3,025
2650	Dry heat, electric, portable, 3 trays					1,375			1,375	1,500
2700	Ultra-sonic cleaner, portable, minimum					465			465	510
2750	Maximum (institutional)					1,225			1,225	1,350
3000	X-ray unit, wall, minimum	1 Skwk	4	2		2,450	105		2,555	2,850
3010	Maximum		4	2		5,200	105		5,305	5,875
3100	Panoramic unit	↓	.60	13.333		17,000	700		17,700	19,800
3105	Deluxe, minimum	2 Skwk	1.60	10		11,700	525		12,225	13,700
3110	Maximum	"	1.60	10		32,500	525		33,025	36,500
3500	Developers, X-ray, average	1 Plum	5.33	1.501		5,500	93.50		5,593.50	6,200
3600	Maximum	"	5.33	1.501	↓	10,500	93.50		10,593.50	11,600

11 76 Operating Room Equipment

11 76 10 - Operating Room Equipment

11 76 10.10 Surgical Equipment

		Crew	Daily Output	Labor-Hours	Unit	Material	Labor	Equipment	Total	Total Incl O&P
0010	**SURGICAL EQUIPMENT**									
5000	Scrub, surgical, stainless steel, single station, minimum	1 Plum	3	2.667	Ea.	5,350	166		5,516	6,150
5100	Maximum					7,925			7,925	8,725
6550	Major surgery table, minimum	1 Sswk	.50	16		25,100	875		25,975	29,000
6570	Maximum		.50	16		35,200	875		36,075	40,100
6600	Hydraulic, hand-held control, general surgery		.60	13.333		37,900	730		38,630	42,900
6650	Stationary, universal	↓	.50	16		49,300	875		50,175	56,000
6800	Surgical lights, major operating room, dual head, minimum	2 Elec	1	16		5,050	930		5,980	6,975
6850	Maximum		1	16		37,500	930		38,430	42,600
6900	Ceiling mount articulation, single arm	↓	1	16	↓	4,650	930		5,580	6,525

11 77 Radiology Equipment

11 77 10 - Radiology Equipment

11 77 10.10 X-Ray Equipment

		Crew	Daily Output	Labor-Hours	Unit	Material	Labor	Equipment	Total	Total Incl O&P
0010	**X-RAY EQUIPMENT**									
8700	X-ray, mobile, minimum				Ea.	18,900			18,900	20,800
8750	Maximum					95,500			95,500	105,000
8900	Stationary, minimum					37,100			37,100	40,900
8950	Maximum					276,000			276,000	303,500
9150	Developing processors, minimum					4,525			4,525	4,975
9200	Maximum				↓	12,200			12,200	13,400

For customer support on your Building Construction Costs with RSMeans data, call 800.448.8182.

429

11 78 Mortuary Equipment

11 78 13 – Mortuary Refrigerators

11 78 13.10 Mortuary and Autopsy Equipment	Crew	Daily Output	Labor-Hours	Unit	Material	2018 Bare Costs Labor	2018 Bare Costs Equipment	Total	Total Incl O&P
0010 **MORTUARY AND AUTOPSY EQUIPMENT**									
0015 Autopsy table, standard	1 Plum	1	8	Ea.	9,825	495		10,320	11,600
0020 Deluxe	"	.60	13.333		16,500	830		17,330	19,400
3200 Mortuary refrigerator, end operated, 2 capacity					8,900			8,900	9,800
3300 6 capacity					16,700			16,700	18,400

11 78 16 – Crematorium Equipment

11 78 16.10 Crematory

	Crew	Daily Output	Labor-Hours	Unit	Material	Labor	Equipment	Total	Total Incl O&P
0010 **CREMATORY**									
1500 Crematory, not including building, 1 place	Q-3	.20	160	Ea.	77,000	9,475		86,475	99,000
1750 2 place	"	.10	320	"	110,000	18,900		128,900	149,000

11 81 Facility Maintenance Equipment

11 81 19 – Vacuum Cleaning Systems

11 81 19.10 Vacuum Cleaning

	Crew	Daily Output	Labor-Hours	Unit	Material	Labor	Equipment	Total	Total Incl O&P
0010 **VACUUM CLEANING**									
0020 Central, 3 inlet, residential	1 Skwk	.90	8.889	Total	1,150	465		1,615	1,975
0200 Commercial		.70	11.429		1,350	600		1,950	2,425
0400 5 inlet system, residential		.50	16		1,675	840		2,515	3,100
0600 7 inlet system, commercial		.40	20		1,875	1,050		2,925	3,675
0800 9 inlet system, residential		.30	26.667		4,025	1,400		5,425	6,550
4010 Rule of thumb: First 1200 S.F., installed								1,425	1,575
4020 For each additional S.F., add				S.F.				.26	.26

11 82 Facility Solid Waste Handling Equipment

11 82 19 – Packaged Incinerators

11 82 19.10 Packaged Gas Fired Incinerators

	Crew	Daily Output	Labor-Hours	Unit	Material	Labor	Equipment	Total	Total Incl O&P
0010 **PACKAGED GAS FIRED INCINERATORS**									
4400 Incinerator, gas, not incl. chimney, elec. or pipe, 50 lbs./hr., minimum	Q-3	.80	40	Ea.	40,800	2,375		43,175	48,400
4420 Maximum		.70	45.714		42,900	2,700		45,600	51,500
4440 200 lbs./hr., minimum (batch type)		.60	53.333		72,000	3,150		75,150	84,000
4460 Maximum (with feeder)		.50	64		79,500	3,775		83,275	92,500
4480 400 lbs./hr., minimum (batch type)		.30	107		78,500	6,300		84,800	96,000
4500 Maximum (with feeder)		.25	128		94,000	7,575		101,575	114,500
4520 800 lbs./hr., with feeder, minimum		.20	160		120,000	9,475		129,475	146,000
4540 Maximum		.17	188		201,000	11,100		212,100	238,000
4560 1,200 lbs./hr., with feeder, minimum		.15	213		146,500	12,600		159,100	180,500
4580 Maximum		.11	291		183,500	17,200		200,700	228,000
4600 2,000 lbs./hr., with feeder, minimum		.10	320		396,500	18,900		415,400	464,500
4620 Maximum		.05	640		640,500	37,900		678,400	761,500
4700 For heat recovery system, add, minimum		.25	128		84,500	7,575		92,075	104,500
4710 Add, maximum		.11	291		264,500	17,200		281,700	316,500
4720 For automatic ash conveyer, add		.50	64		35,200	3,775		38,975	44,400
4750 Large municipal incinerators, incl. stack, minimum		.25	128	Ton/day	21,400	7,575		28,975	34,900
4850 Maximum		.10	320	"	57,000	18,900		75,900	91,000

For customer support on your Building Construction Costs with RSMeans data, call 800.448.8182.

11 82 Facility Solid Waste Handling Equipment

11 82 26 – Facility Waste Compactors

11 82 26.10 Compactors

		Crew	Daily Output	Labor-Hours	Unit	Material	2018 Bare Costs Labor	Equipment	Total	Total Incl O&P
0010	**COMPACTORS**									
0020	Compactors, 115 volt, 250 lbs./hr., chute fed	L-4	1	24	Ea.	12,500	1,150		13,650	15,500
0100	Hand fed		2.40	10		16,000	475		16,475	18,300
0300	Multi-bag, 230 volt, 600 lbs./hr., chute fed		1	24		16,000	1,150		17,150	19,400
0400	Hand fed		1	24		16,100	1,150		17,250	19,500
0500	Containerized, hand fed, 2 to 6 C.Y. containers, 250 lbs./hr.		1	24		16,000	1,150		17,150	19,400
0550	For chute fed, add per floor		1	24		1,475	1,150		2,625	3,375
1000	Heavy duty industrial compactor, 0.5 C.Y. capacity		1	24		10,700	1,150		11,850	13,600
1050	1.0 C.Y. capacity		1	24		16,400	1,150		17,550	19,800
1100	3.0 C.Y. capacity		.50	48		27,400	2,275		29,675	33,600
1150	5.0 C.Y. capacity		.50	48		34,500	2,275		36,775	41,500
1200	Combination shredder/compactor (5,000 lbs./hr.)		.50	48		67,000	2,275		69,275	77,500
1400	For handling hazardous waste materials, 55 gallon drum packer, std.					20,900			20,900	23,000
1410	55 gallon drum packer w/HEPA filter					26,100			26,100	28,700
1420	55 gallon drum packer w/charcoal & HEPA filter					34,800			34,800	38,200
1430	All of the above made explosion proof, add					1,525			1,525	1,675
5500	Shredder, municipal use, 35 tons/hour					318,000			318,000	350,000
5600	60 tons/hour					677,500			677,500	745,000
5750	Shredder & baler, 50 tons/day					635,000			635,000	698,500
5800	Shredder, industrial, minimum					25,200			25,200	27,700
5850	Maximum					134,000			134,000	147,500
5900	Baler, industrial, minimum					10,000			10,000	11,000
5950	Maximum					585,500			585,500	644,500
6000	Transfer station compactor, with power unit									
6050	and pedestal, not including pit, 50 tons/hour				Ea.	201,000			201,000	221,000

11 82 39 – Medical Waste Disposal Systems

11 82 39.10 Off-Site Disposal

		Crew	Daily Output	Labor-Hours	Unit	Material	2018 Bare Costs Labor	Equipment	Total	Total Incl O&P
0010	**OFF-SITE DISPOSAL**									
0100	Medical waste disposal, Red Bag system, pick up & treat, 200 lbs./week				Week	190			190	209
0110	Per month				Month	730			730	800
0150	Red bags, 7-10 gal., 1.2 mil, pkg of 500				Ea.	70			70	77
0200	15 gal., package of 250					66			66	72.50
0250	33 gal., package of 250					69			69	76
0300	45 gal., package of 100					58			58	64

11 82 39.20 Disposal Carts

		Crew	Daily Output	Labor-Hours	Unit	Material	2018 Bare Costs Labor	Equipment	Total	Total Incl O&P
0010	**DISPOSAL CARTS**									
2010	Medical waste disposal cart, HDPE, w/lid, 28 gal. capacity				Ea.	278			278	305
2020	96 gal. capacity					320			320	355
2030	150 gal. capacity, low profile					500			500	550
2040	200 gal. capacity					965			965	1,050

11 82 39.30 Medical Waste Sanitizers

		Crew	Daily Output	Labor-Hours	Unit	Material	2018 Bare Costs Labor	Equipment	Total	Total Incl O&P
0010	**MEDICAL WASTE SANITIZERS**									
2010	Small, hand loaded, 1.5 C.Y., 225 lb. capacity				Ea.	83,000			83,000	91,000
2020	Medium, cart loaded, 6.25 C.Y., 938 lb. capacity					110,500			110,500	121,500
2030	Large, cart loaded, 15 C.Y., 2250 lb. capacity					131,500			131,500	144,500
3010	Cart, aluminum, 75 lb. capacity					2,175			2,175	2,400
3020	95 lb. capacity					2,375			2,375	2,600
4010	Stainless steel, 173 lb. capacity					3,100			3,100	3,425
4020	232 lb. capacity					3,475			3,475	3,825
4030	Cart lift, hydraulic scissor type					6,150			6,150	6,775
4040	Portable aluminum ramp					1,725			1,725	1,900

11 82 Facility Solid Waste Handling Equipment

11 82 39 – Medical Waste Disposal Systems

11 82 39.30 Medical Waste Sanitizers		Crew	Daily Output	Labor-Hours	Unit	2018 Bare Costs Material	Labor	Equipment	Total	Total Incl O&P
4050	Fold-down steel tracks				Ea.	1,400			1,400	1,550
4060	Pull-out drawer, small					6,625			6,625	7,300
4070	Medium					9,575			9,575	10,500
4080	Large					13,500			13,500	14,900
5000	Medical waste treatment, sanitize, on-site									
5010	Less than 15,000 lbs./month				Lb.	.20			.20	.22
5020	Over 15,000 lbs./month				"	.16			.16	.18

11 91 Religious Equipment

11 91 13 – Baptisteries

11 91 13.10 Baptistry		Crew	Daily Output	Labor-Hours	Unit	Material	Labor	Equipment	Total	Total Incl O&P
0010	**BAPTISTRY**									
0150	Fiberglass, 3'-6" deep, x 13'-7" long,									
0160	steps at both ends, incl. plumbing, minimum	L-8	1	20	Ea.	5,550	1,050		6,600	7,700
0200	Maximum	"	.70	28.571		9,325	1,525		10,850	12,600
0250	Add for filter, heater and lights					1,850			1,850	2,025

11 91 23 – Sanctuary Equipment

11 91 23.10 Sanctuary Furnishings		Crew	Daily Output	Labor-Hours	Unit	Material	Labor	Equipment	Total	Total Incl O&P
0010	**SANCTUARY FURNISHINGS**									
0020	Altar, wood, custom design, plain	1 Carp	1.40	5.714	Ea.	2,575	290		2,865	3,300
0050	Deluxe	"	.20	40		12,400	2,025		14,425	16,800
0070	Granite or marble, average	2 Marb	.50	32		13,400	1,575		14,975	17,100
0090	Deluxe	"	.20	80		36,500	3,925		40,425	46,100
0100	Arks, prefabricated, plain	2 Carp	.80	20		9,275	1,025		10,300	11,800
0130	Deluxe, maximum	"	.20	80		131,500	4,050		135,550	151,000
0500	Reconciliation room, wood, prefabricated, single, plain	1 Carp	.60	13.333		3,200	675		3,875	4,550
0550	Deluxe		.40	20		8,600	1,025		9,625	11,000
0650	Double, plain		.40	20		6,300	1,025		7,325	8,475
0700	Deluxe		.20	40		18,800	2,025		20,825	23,800
1000	Lecterns, wood, plain		5	1.600		740	81		821	940
1100	Deluxe		2	4		5,875	203		6,078	6,775
2000	Pulpits, hardwood, prefabricated, plain		2	4		1,550	203		1,753	2,000
2100	Deluxe		1.60	5		10,200	254		10,454	11,600
2500	Railing, hardwood, average		25	.320	L.F.	211	16.20		227.20	257
3000	Seating, individual, oak, contour, laminated		21	.381	Person	180	19.30		199.30	228
3100	Cushion seat		21	.381		164	19.30		183.30	210
3200	Fully upholstered		21	.381		178	19.30		197.30	226
3300	Combination, self-rising		21	.381		315	19.30		334.30	375
3500	For cherry, add					30%				
5000	Wall cross, aluminum, extruded, 2" x 2" section	1 Carp	34	.235	L.F.	242	11.95		253.95	284
5150	4" x 4" section		29	.276		350	14		364	405
5300	Bronze, extruded, 1" x 2" section		31	.258		475	13.10		488.10	545
5350	2-1/2" x 2-1/2" section		34	.235		720	11.95		731.95	815
5450	Solid bar stock, 1/2" x 3" section		29	.276		950	14		964	1,075
5600	Fiberglass, stock		34	.235		166	11.95		177.95	201
5700	Stainless steel, 4" deep, channel section		29	.276		765	14		779	860
5800	4" deep box section		29	.276		1,050	14		1,064	1,200

For customer support on your Building Construction Costs with RSMeans data, call 800.448.8182.

11 92 Agricultural Equipment

11 92 16 – Stock Feeders

11 92 16.16 Barns

		Crew	Daily Output	Labor-Hours	Unit	Material	2018 Bare Costs Labor	2018 Bare Costs Equipment	Total	Total Incl O&P
0010	**BARNS**									
0015	Swine barn, farrowing pens and equipment	B-1	1000	.024	S.F.	13.85	.97		14.82	16.75
0020	Gestation pens and equipment		800	.030		12.20	1.22		13.42	15.25
0030	Nursery pens and equipment		1150	.021		10.65	.85		11.50	13
0040	Finishing pens and equipment		1400	.017		8.10	.69		8.79	10
0120	Poultry barn, cages and equipment		700	.034		17.80	1.39		19.19	21.50
0220	Animal barn stall, feed and water equipment		1000	.024		15.85	.97		16.82	18.90
0230	Manure floor scraper system		1500	.016		2.97	.65		3.62	4.26
0240	Below slab manure gutter and shuttle stroker		400	.060		10.15	2.43		12.58	14.85
0250	Exhaust system		1400	.017		1.23	.69		1.92	2.41
0260	Milking barn, milking equipment		220	.109		89	4.42		93.42	105
0270	Milk storage equipment		450	.053		26.50	2.16		28.66	33
0280	Sheep barn, equipment, sheep shear street gates		1000	.024		1,025	.97		1,025.97	1,125
0290	Gate in frame		1000	.024		105	.97		105.97	116
0295	Maternity fence with drinking trough	▼	1000	.024	▼	61.50	.97		62.47	69
0300	Tobacco barn, fruit & tobacco auto. dryer machine/heat pump dryer	Q-20	2.90	6.897	Ea.	6,350	375		6,725	7,550
0310	Tobacco curing generator		4.90	4.082		4,450	223		4,673	5,250
0320	Automatic system controller		6.90	2.899		1,625	159		1,784	2,025
0330	Humidity & temp. transmitters for humidity measurement		6.90	2.899		900	159		1,059	1,225
0340	Handling system for leaf tobacco	▼	5.90	3.390	▼	560	185		745	895

11 97 Security Equipment

11 97 30 – Security Drawers

11 97 30.10 Pass Through Drawer

		Crew	Daily Output	Labor-Hours	Unit	Material	2018 Bare Costs Labor	2018 Bare Costs Equipment	Total	Total Incl O&P
0010	**PASS THROUGH DRAWER**									
0100	Pass-thru drawer for personal items, 18" x 15" x 24"	1 Skwk	2	4	Ea.	2,900	209		3,109	3,500
0110	Including speakers	"	1.50	5.333	"	3,250	279		3,529	4,000

11 98 Detention Equipment

11 98 21 – Detention Windows

11 98 21.13 Visitor Cubicle Windows

		Crew	Daily Output	Labor-Hours	Unit	Material	2018 Bare Costs Labor	2018 Bare Costs Equipment	Total	Total Incl O&P
0010	**VISITOR CUBICLE WINDOWS**									
4000	Visitor cubicle, vision panel, no intercom	E-4	2	16	Ea.	3,350	880	49.50	4,279.50	5,200

11 98 30 – Detention Cell Equipment

11 98 30.10 Cell Equipment

		Crew	Daily Output	Labor-Hours	Unit	Material	2018 Bare Costs Labor	2018 Bare Costs Equipment	Total	Total Incl O&P
0010	**CELL EQUIPMENT**									
3000	Toilet apparatus including wash basin, average	L-8	1.50	13.333	Ea.	3,400	705		4,105	4,825

For customer support on your Building Construction Costs with RSMeans data, call 800.448.8182.

433

Division Notes

	CREW	DAILY OUTPUT	LABOR-HOURS	UNIT	BARE COSTS				TOTAL INCL O&P
					MAT.	LABOR	EQUIP.	TOTAL	

Estimating Tips
General

- The items in this division are usually priced per square foot or each. Most of these items are purchased by the owner and installed by the contractor. Do not assume the items in Division 12 will be purchased and installed by the contractor. Check the specifications for responsibilities and include receiving, storage, installation, and mechanical and electrical hookups in the appropriate divisions.

- Some items in this division require some type of support system that is not usually furnished with the item. Examples of these systems include blocking for the attachment of casework and heavy drapery rods. The required blocking must be added to the estimate in the appropriate division.

Reference Numbers

Reference numbers are shown at the beginning of some major classifications. These numbers refer to related items in the Reference Section. The reference information may be an estimating procedure, an alternate pricing method, or technical information.

Note: Not all subdivisions listed here necessarily appear. ■

Did you know?

RSMeans data is available through our online application with 24/7 access:

- Search for unit prices by keyword
- Leverage the most up-to-date data
- Build and export estimates

Try it free for 30 days!
www.rsmeans.com/2018freetrial

No part of this cost data may be reproduced, stored in a retrieval system, or transmitted in any form or by any means without prior written permission of Gordian.

12 05 Common Work Results for Furnishings

12 05 05 – Selective Demolition for Furnishings

12 05 05.10 Selective Demolition, Interiors

		Crew	Daily Output	Labor-Hours	Unit	Material	2018 Bare Costs Labor	Equipment	Total	Total Incl O&P
0010	**SELECTIVE DEMOLITION, INTERIORS**									
1100	Casework, wood base cabinets	2 Clab	24	.667	L.F.		26.50		26.50	40.50
1200	Countertop		96	.167			6.65		6.65	10.10
3100	Casework, metal base cabinets		20	.800			32		32	48.50
3110	Cabinet base trim		400	.040			1.59		1.59	2.43
3120	Countertop, stainless steel or acid proof		80	.200			7.95		7.95	12.15
3122	custom	1 Carp	48	.167	S.F.		8.45		8.45	12.85
3125	Ceramic tile	1 Clab	120	.067	"		2.66		2.66	4.05
3127	Trim	1 Tilf	72	.111	L.F.		5.20		5.20	7.70
3130	Wall cabinets, wood, 84" high	2 Clab	30	.533			21.50		21.50	32.50
3500	Laboratory casework, tall storage cabinets, 84" high		40	.400			15.95		15.95	24.50
3510	Wall cabinets, metal		40	.400			15.95		15.95	24.50
4830	Floor mats, recessed or link mats	1 Clab	300	.027	S.F.		1.06		1.06	1.62
4832	Skate lock tile		200	.040			1.59		1.59	2.43
4834	Duckboard		300	.027			1.06		1.06	1.62
4920	Blinds, interior, horizontal or vertical		150	.053	L.F.		2.13		2.13	3.24
4922	Wood folding panels		35	.229	Pr.		9.10		9.10	13.85
4924	Shades, interior		700	.011	S.F.		.46		.46	.69
4930	Drapery hardware, traverse rods		35	.229	Ea.		9.10		9.10	13.85
4950	Blast curtains, including hardware		25	.320			12.75		12.75	19.40
5200	Fixed seating, per seat	2 Carp	44	.364			18.45		18.45	28
6400	Booth, restaurant	2 Clab	80	.200	L.F.		7.95		7.95	12.15
7400	Office systems, furniture, cubicle	"	3000	.005	S.F.		.21		.21	.32

12 21 Window Blinds

12 21 13 – Horizontal Louver Blinds

12 21 13.13 Metal Horizontal Louver Blinds

		Crew	Daily Output	Labor-Hours	Unit	Material	2018 Bare Costs Labor	Equipment	Total	Total Incl O&P
0010	**METAL HORIZONTAL LOUVER BLINDS**									
0020	Horizontal, 1" aluminum slats, solid color, stock	1 Carp	590	.014	S.F.	5.90	.69		6.59	7.55
0250	2" aluminum slats, solid color, stock	"	590	.014	"	4.85	.69		5.54	6.40

12 21 13.33 Vinyl Horizontal Louver Blinds

		Crew	Daily Output	Labor-Hours	Unit	Material	2018 Bare Costs Labor	Equipment	Total	Total Incl O&P
0010	**VINYL HORIZONTAL LOUVER BLINDS**									
0100	2" composite, 48" wide, 48" high	1 Carp	30	.267	Ea.	97	13.50		110.50	128
0120	72" high		29	.276		131	14		145	166
0140	96" high		28	.286		179	14.50		193.50	219
0200	60" wide, 60" high		27	.296		135	15		150	172
0220	72" high		25	.320		156	16.20		172.20	196
0240	96" high		24	.333		233	16.90		249.90	282
0300	72" wide, 72" high		25	.320		209	16.20		225.20	254
0320	96" high		23	.348		288	17.65		305.65	340
0400	96" wide, 96" high		20	.400		425	20.50		445.50	495
1000	2" faux wood, 48" wide, 48" high		30	.267		76	13.50		89.50	105
1020	72" high		29	.276		99	14		113	131
1040	96" high		28	.286		118	14.50		132.50	152
1300	72" wide, 72" high		25	.320		128	16.20		144.20	166
1320	96" high		23	.348		192	17.65		209.65	238
1400	96" wide, 96" high		20	.400		256	20.50		276.50	315

For customer support on your Building Construction Costs with RSMeans data, call 800.448.8182.

12 21 Window Blinds

12 21 16 – Vertical Louver Blinds

	12 21 16.13 Metal Vertical Louver Blinds	Crew	Daily Output	Labor-Hours	Unit	Material	2018 Bare Costs Labor	Equipment	Total	Total Incl O&P
0010	**METAL VERTICAL LOUVER BLINDS**									
1500	Vertical, 3" PVC strips, minimum	1 Carp	460	.017	S.F.	8.65	.88		9.53	10.85
1600	Maximum		400	.020		16	1.01		17.01	19.15
1800	4" aluminum slats, minimum		460	.017		8.10	.88		8.98	10.30
1900	Maximum		400	.020		18	1.01		19.01	21.50

12 22 Curtains and Drapes

12 22 16 – Drapery Track and Accessories

12 22 16.10 Drapery Hardware

		Crew	Daily Output	Labor-Hours	Unit	Material	2018 Bare Costs Labor	Equipment	Total	Total Incl O&P
0010	**DRAPERY HARDWARE**									
0030	Standard traverse, per foot, minimum	1 Carp	59	.136	L.F.	6.15	6.85		13	17.25
0100	Maximum		51	.157	"	15	7.95		22.95	28.50
4000	Traverse rods, adjustable, 28" to 48"		22	.364	Ea.	32	18.45		50.45	63
4020	48" to 84"		20	.400		40.50	20.50		61	75.50
4040	66" to 120"		18	.444		49.50	22.50		72	88.50
4060	84" to 156"		16	.500		56	25.50		81.50	100
4080	100" to 180"		14	.571		65.50	29		94.50	116
4090	156" to 228"		13	.615		78.50	31		109.50	134
4100	228" to 312"		13	.615		90.50	31		121.50	147
4200	Double rods, adjustable, 30" to 48"		9	.889		54	45		99	128
4220	48" to 86"		9	.889		75	45		120	151
4240	86" to 150"		8	1		83	50.50		133.50	168
4260	100" to 180"		7	1.143		87.50	58		145.50	184
4300	Curtain rod & brackets, adjustable, 30" to 48"		9	.889		31.50	45		76.50	103
4320	48" to 86"		9	.889		44.50	45		89.50	118
4340	86" to 150"		8	1		55	50.50		105.50	138
4360	100" to 180"		7	1.143		68	58		126	163
4600	Valance, pinch pleated fabric, 12" deep, up to 54" long, minimum					41			41	45
4610	Maximum					102			102	112
4620	Up to 77" long, minimum					72.50			72.50	79.50
4630	Maximum					165			165	181
5000	Stationary rods, first 2'					8.35			8.35	9.20
5020	Each additional foot, add				L.F.	4.02			4.02	4.42

12 22 16.20 Blast Curtains

		Crew	Daily Output	Labor-Hours	Unit	Material	2018 Bare Costs Labor	Equipment	Total	Total Incl O&P
0010	**BLAST CURTAINS** per L.F. horizontal opening width, off-white or gray fabric									
0100	Blast curtains, drapery system, complete, including hardware, minimum	1 Carp	10.25	.780	L.F.	244	39.50		283.50	330
0120	Average		10.25	.780		258	39.50		297.50	345
0140	Maximum		10.25	.780		273	39.50		312.50	360

12 23 Interior Shutters

12 23 10 – Wood Interior Shutters

12 23 10.10 Wood Interior Shutters

		Crew	Daily Output	Labor-Hours	Unit	Material	2018 Bare Costs Labor	2018 Bare Costs Equipment	Total	Total Incl O&P
0010	**WOOD INTERIOR SHUTTERS**, louvered									
0200	Two panel, 27" wide, 36" high	1 Carp	5	1.600	Set	160	81		241	300
0300	33" wide, 36" high		5	1.600		206	81		287	350
0500	47" wide, 36" high		5	1.600		276	81		357	430
1000	Four panel, 27" wide, 36" high		5	1.600		158	81		239	298
1100	33" wide, 36" high		5	1.600		203	81		284	345
1300	47" wide, 36" high		5	1.600		271	81		352	420
1400	Plantation shutters, 16" x 48"		5	1.600	Ea.	181	81		262	325
1450	16" x 96"		4	2		297	101		398	480
1460	36" x 96"		3	2.667		655	135		790	925

12 23 10.13 Wood Panels

		Crew	Daily Output	Labor-Hours	Unit	Material	2018 Bare Costs Labor	2018 Bare Costs Equipment	Total	Total Incl O&P
0010	**WOOD PANELS**									
3000	Wood folding panels with movable louvers, 7" x 20" each	1 Carp	17	.471	Pr.	92	24		116	138
3300	8" x 28" each		17	.471		92	24		116	138
3450	9" x 36" each		17	.471		106	24		130	154
3600	10" x 40" each		17	.471		115	24		139	164
4000	Fixed louver type, stock units, 8" x 20" each		17	.471		96.50	24		120.50	143
4150	10" x 28" each		17	.471		81.50	24		105.50	127
4300	12" x 36" each		17	.471		96.50	24		120.50	143
4450	18" x 40" each		17	.471		138	24		162	189
5000	Insert panel type, stock, 7" x 20" each		17	.471		27	24		51	66.50
5150	8" x 28" each		17	.471		50	24		74	91
5300	9" x 36" each		17	.471		63	24		87	106
5450	10" x 40" each		17	.471		67.50	24		91.50	111
5600	Raised panel type, stock, 10" x 24" each		17	.471		115	24		139	164
5650	12" x 26" each		17	.471		115	24		139	164
5700	14" x 30" each		17	.471		127	24		151	177
5750	16" x 36" each		17	.471		141	24		165	192
6000	For custom built pine, add					22%				
6500	For custom built hardwood blinds, add					42%				

12 24 Window Shades

12 24 13 – Roller Window Shades

12 24 13.10 Shades

		Crew	Daily Output	Labor-Hours	Unit	Material	2018 Bare Costs Labor	2018 Bare Costs Equipment	Total	Total Incl O&P
0010	**SHADES**									
0020	Basswood, roll-up, stain finish, 3/8" slats	1 Carp	300	.027	S.F.	19.75	1.35		21.10	23.50
0200	7/8" slats		300	.027		19.60	1.35		20.95	23.50
0300	Vertical side slide, stain finish, 3/8" slats		300	.027		22.50	1.35		23.85	26.50
0400	7/8" slats		300	.027		24	1.35		25.35	28.50
0500	For fire retardant finishes, add					16%				
0600	For "B" rated finishes, add					20%				
0900	Mylar, single layer, non-heat reflective	1 Carp	685	.012		3	.59		3.59	4.20
0910	Mylar, single layer, heat reflective		685	.012		2.54	.59		3.13	3.70
1000	Double layered, heat reflective		685	.012		6.05	.59		6.64	7.55
1100	Triple layered, heat reflective		685	.012		7.60	.59		8.19	9.25
1200	For metal roller instead of wood, add per				Shade	6.35			6.35	6.95
1300	Vinyl coated cotton, standard	1 Carp	685	.012	S.F.	4.28	.59		4.87	5.60
1400	Lightproof decorator shades		685	.012		5.60	.59		6.19	7.05
1500	Vinyl, lightweight, 4 ga.		685	.012		.81	.59		1.40	1.79
1600	Heavyweight, 6 ga.		685	.012		2.48	.59		3.07	3.63
1700	Vinyl laminated fiberglass, 6 ga., translucent		685	.012		4.72	.59		5.31	6.10

For customer support on your Building Construction Costs with RSMeans data, call 800.448.8182.

12 24 Window Shades

12 24 13 – Roller Window Shades

12 24 13.10 Shades		Crew	Daily Output	Labor- Hours	Unit	Material	2018 Bare Costs Labor	Equipment	Total	Total Incl O&P
1800	Lightproof	1 Carp	685	.012	S.F.	5.55	.59		6.14	7
2000	Polyester, room darkening, with continuous cord, GEI									
2010	36" x 72"	1 Carp	38	.211	Ea.	142	10.65		152.65	172
2020	48" x 72"		28	.286		164	14.50		178.50	202
2030	60" x 72"		23	.348		192	17.65		209.65	238
2040	72" x 72"		19	.421		237	21.50		258.50	293
6011	Solar screening, fiberglass [G]		85	.094	S.F.	7.65	4.77		12.42	15.65

12 32 Manufactured Wood Casework

12 32 23 – Hardwood Casework

12 32 23.10 Manufactured Wood Casework, Stock Units

		Crew	Daily Output	Labor- Hours	Unit	Material	2018 Bare Costs Labor	Equipment	Total	Total Incl O&P
0010	**MANUFACTURED WOOD CASEWORK, STOCK UNITS**									
0300	Built-in drawer units, pine, 18" deep, 32" high, unfinished									
0400	Minimum	2 Carp	53	.302	L.F.	137	15.30		152.30	175
0500	Maximum	"	40	.400	"	161	20.50		181.50	209
0700	Kitchen base cabinets, hardwood, not incl. counter tops,									
0710	24" deep, 35" high, prefinished									
0800	One top drawer, one door below, 12" wide	2 Carp	24.80	.645	Ea.	300	32.50		332.50	380
0840	18" wide		23.30	.687		340	35		375	430
0880	24" wide		22.30	.717		415	36.50		451.50	510
1000	Four drawers, 12" wide		24.80	.645		315	32.50		347.50	400
1040	18" wide		23.30	.687		355	35		390	445
1060	24" wide		22.30	.717		390	36.50		426.50	485
1200	Two top drawers, two doors below, 27" wide		22	.727		445	37		482	545
1260	36" wide		20.30	.788		525	40		565	635
1300	48" wide		18.90	.847		590	43		633	715
1500	Range or sink base, two doors below, 30" wide		21.40	.748		405	38		443	505
1540	36" wide		20.30	.788		455	40		495	560
1580	48" wide		18.90	.847		500	43		543	615
1800	For sink front units, deduct					179			179	197
2000	Corner base cabinets, 36" wide, standard	2 Carp	18	.889		745	45		790	890
2100	Lazy Susan with revolving door	"	16.50	.970		955	49		1,004	1,125
4000	Kitchen wall cabinets, hardwood, 12" deep with two doors									
4050	12" high, 30" wide	2 Carp	24.80	.645	Ea.	272	32.50		304.50	350
4100	36" wide		24	.667		75.50	34		109.50	135
4400	15" high, 30" wide		24	.667		277	34		311	355
4440	36" wide		22.70	.705		330	35.50		365.50	420
4700	24" high, 30" wide		23.30	.687		370	35		405	465
4720	36" wide		22.70	.705		405	35.50		440.50	505
5000	30" high, one door, 12" wide		22	.727		261	37		298	345
5040	18" wide		20.90	.766		300	39		339	390
5060	24" wide		20.30	.788		350	40		390	445
5300	Two doors, 27" wide		19.80	.808		390	41		431	490
5340	36" wide		18.80	.851		465	43		508	580
5380	48" wide		18.40	.870		570	44		614	690
6000	Corner wall, 30" high, 24" wide		18	.889		400	45		445	510
6050	30" wide		17.20	.930		425	47		472	535
6100	36" wide		16.50	.970		485	49		534	610
6500	Revolving Lazy Susan		15.20	1.053		127	53.50		180.50	222
7000	Broom cabinet, 84" high, 24" deep, 18" wide		10	1.600		745	81		826	945
7500	Oven cabinets, 84" high, 24" deep, 27" wide		8	2		1,125	101		1,226	1,400

For customer support on your Building Construction Costs with RSMeans data, call 800.448.8182.

439

12 32 Manufactured Wood Casework

12 32 23 – Hardwood Casework

12 32 23.10 Manufactured Wood Casework, Stock Units		Crew	Daily Output	Labor-Hours	Unit	Material	2018 Bare Costs Labor	Equipment	Total	Total Incl O&P
7750	Valance board trim	2 Carp	396	.040	L.F.	18.05	2.05		20.10	23
7780	Toe kick trim	1 Carp	256	.031	"	3.45	1.58		5.03	6.20
7790	Base cabinet corner filler		16	.500	Ea.	48.50	25.50		74	91.50
7800	Cabinet filler, 3" x 24"		20	.400		19.05	20.50		39.55	52
7810	3" x 30"		20	.400		24	20.50		44.50	57
7820	3" x 42"		18	.444		33.50	22.50		56	71
7830	3" x 80"		16	.500		63.50	25.50		89	109
7850	Cabinet panel		50	.160	S.F.	10.35	8.10		18.45	24
9000	For deluxe models of all cabinets, add					40%				
9500	For custom built in place, add					25%	10%			
9558	Rule of thumb, kitchen cabinets not including									
9560	appliances & counter top, minimum	2 Carp	30	.533	L.F.	202	27		229	263
9600	Maximum	"	25	.640	"	440	32.50		472.50	535
9610	For metal cabinets, see Section 12 35 70.13									

12 32 23.30 Manufactured Wood Casework Vanities

		Crew	Daily Output	Labor-Hours	Unit	Material	2018 Bare Costs Labor	Equipment	Total	Total Incl O&P
0010	**MANUFACTURED WOOD CASEWORK VANITIES**									
8000	Vanity bases, 2 doors, 30" high, 21" deep, 24" wide	2 Carp	20	.800	Ea.	350	40.50		390.50	445
8050	30" wide		16	1		420	50.50		470.50	535
8100	36" wide		13.33	1.200		405	61		466	540
8150	48" wide		11.43	1.400		530	71		601	695
9000	For deluxe models of all vanities, add to above					40%				
9500	For custom built in place, add to above					25%	10%			

12 32 23.35 Manufactured Wood Casework Hardware

		Crew	Daily Output	Labor-Hours	Unit	Material	2018 Bare Costs Labor	Equipment	Total	Total Incl O&P
0010	**MANUFACTURED WOOD CASEWORK HARDWARE**									
1000	Catches, minimum	1 Carp	235	.034	Ea.	1.40	1.73		3.13	4.17
1040	Maximum	"	80	.100	"	8.05	5.05		13.10	16.55
2000	Door/drawer pulls, handles									
2200	Handles and pulls, projecting, metal, minimum	1 Carp	48	.167	Ea.	5.20	8.45		13.65	18.55
2240	Maximum		36	.222		10.65	11.25		21.90	29
2300	Wood, minimum		48	.167		5.30	8.45		13.75	18.70
2340	Maximum		36	.222		9.80	11.25		21.05	28
2400	Drawer pulls, antimicrobial copper alloy finish		50	.160		19.55	8.10		27.65	34
2600	Flush, metal, minimum		48	.167		5.30	8.45		13.75	18.70
2640	Maximum		36	.222		9.90	11.25		21.15	28
2900	Drawer knobs, antimicrobial copper alloy finish		50	.160		12.95	8.10		21.05	26.50
3000	Drawer tracks/glides, minimum		48	.167	Pr.	8.95	8.45		17.40	22.50
3040	Maximum		24	.333		26	16.90		42.90	54
4000	Cabinet hinges, minimum		160	.050		3.10	2.54		5.64	7.25
4040	Maximum		68	.118		14.10	5.95		20.05	24.50
7000	Appliance pulls, antimicrobial copper alloy finish		50	.160	L.F.	70	8.10		78.10	89.50

For customer support on your Building Construction Costs with RSMeans data, call 800.448.8182.

12 35 Specialty Casework

12 35 50 – Educational/Library Casework

12 35 50.13 Educational Casework

		Crew	Daily Output	Labor-Hours	Unit	Material	2018 Bare Costs Labor	Equipment	Total	Total Incl O&P
0010	**EDUCATIONAL CASEWORK**									
5000	School, 24" deep, metal, 84" high units	2 Carp	15	1.067	L.F.	575	54		629	715
5150	Counter height units		20	.800		410	40.50		450.50	515
5450	Wood, custom fabricated, 32" high counter		20	.800		252	40.50		292.50	340
5600	Add for counter top		56	.286		27	14.50		41.50	51.50
5800	84" high wall units	↓	15	1.067	↓	625	54		679	775
6000	Laminated plastic finish is same price as wood									

12 35 53 – Laboratory Casework

12 35 53.13 Metal Laboratory Casework

		Crew	Daily Output	Labor-Hours	Unit	Material	2018 Bare Costs Labor	Equipment	Total	Total Incl O&P
0010	**METAL LABORATORY CASEWORK**									
0020	Cabinets, base, door units, metal	2 Carp	18	.889	L.F.	253	45		298	350
0300	Drawer units		18	.889		560	45		605	690
0700	Tall storage cabinets, open, 7' high		20	.800		525	40.50		565.50	640
0900	With glazed doors		20	.800		870	40.50		910.50	1,025
1300	Wall cabinets, metal, 12-1/2" deep, open		20	.800		211	40.50		251.50	294
1500	With doors	↓	20	.800	↓	415	40.50		455.50	520
6300	Rule of thumb: lab furniture including installation & connection									
6320	High school				S.F.				35	39
6340	College								52	57
6360	Clinical, health care								45	49.50
6380	Industrial				↓				72.50	79.50

12 35 59 – Display Casework

12 35 59.10 Display Cases

		Crew	Daily Output	Labor-Hours	Unit	Material	2018 Bare Costs Labor	Equipment	Total	Total Incl O&P
0010	**DISPLAY CASES** Free standing, all glass									
0020	Aluminum frame, 42" high x 36" wide x 12" deep	2 Carp	8	2	Ea.	1,425	101		1,526	1,725
0100	70" high x 48" wide x 18" deep	"	6	2.667		4,175	135		4,310	4,800
0500	For wood bases, add					9%				
0600	For hardwood frames, deduct					8%				
0700	For bronze, baked enamel finish, add				↓	10%				
2000	Wall mounted, glass front, aluminum frame									
2010	Non-illuminated, one section 3' x 4' x 1'-4"	2 Carp	5	3.200	Ea.	2,325	162		2,487	2,800
2100	5' x 4' x 1'-4"		5	3.200		2,475	162		2,637	2,950
2200	6' x 4' x 1'-4"		4	4		3,225	203		3,428	3,825
2500	Two sections, 8' x 4' x 1'-4"		2	8		2,550	405		2,955	3,425
2600	10' x 4' x 1'-4"		2	8		4,200	405		4,605	5,250
3000	Three sections, 16' x 4' x 1'-4"	↓	1.50	10.667	↓	4,725	540		5,265	6,025
3500	For fluorescent lights, add				Section	415			415	455
4000	Table exhibit cases, 2' wide, 3' high, 4' long, flat top	2 Carp	5	3.200	Ea.	1,225	162		1,387	1,600
4100	3' wide, 3' high, 4' long, sloping top	"	3	5.333	"	730	270		1,000	1,200

12 35 70 – Healthcare Casework

12 35 70.13 Hospital Casework

		Crew	Daily Output	Labor-Hours	Unit	Material	2018 Bare Costs Labor	Equipment	Total	Total Incl O&P
0010	**HOSPITAL CASEWORK**									
0500	Base cabinets, laminated plastic	2 Carp	10	1.600	L.F.	370	81		451	530
1000	Stainless steel	"	10	1.600		680	81		761	870
1200	For all drawers, add					38.50			38.50	42.50
1300	Cabinet base trim, 4" high, enameled steel	2 Carp	200	.080		62	4.06		66.06	74
1400	Stainless steel		200	.080		124	4.06		128.06	142
1450	Countertop, laminated plastic, no backsplash		40	.400		51	20.50		71.50	87.50
1650	With backsplash		40	.400		63.50	20.50		84	101
1800	For sink cutout, add		12.20	1.311	Ea.		66.50		66.50	101
1900	Stainless steel counter top	↓	40	.400	L.F.	166	20.50		186.50	213

For customer support on your Building Construction Costs with RSMeans data, call 800.448.8182.

441

12 35 Specialty Casework

12 35 70 – Healthcare Casework

12 35 70.13 Hospital Casework

		Crew	Daily Output	Labor-Hours	Unit	Material	2018 Bare Costs Labor	2018 Bare Costs Equipment	Total	Total Incl O&P
2000	For drop-in stainless 43" x 21" sink, add				Ea.	1,350			1,350	1,500
2050	Laminate with antimicrobial finish #4	2 Carp	40	.400	L.F.	33.50	20.50		54	68
2500	Wall cabinets, laminated plastic		15	1.067		276	54		330	390
2600	Enameled steel		15	1.067		340	54		394	460
2700	Stainless steel		15	1.067		675	54		729	825
3000	Hospital cabinets, stainless steel with glass door(s), lockable									
3010	One door, 24" W x 18" D x 60" H	2 Clab	18	.889	Ea.	2,900	35.50		2,935.50	3,250
3020	Two doors, 36" W x 18" D x 60" H		15	1.067		4,175	42.50		4,217.50	4,675
3030	36" W x 24" D x 67" H		15	1.067		5,925	42.50		5,967.50	6,600
3040	48" W x 24" D x 66" H		12	1.333		4,850	53		4,903	5,400
3050	48" W x 24" D x 72" H		12	1.333		6,025	53		6,078	6,700
3060	60" W x 24" D x 72" H		9	1.778		6,050	71		6,121	6,750

12 35 70.16 Nurse Station Casework

		Crew	Daily Output	Labor-Hours	Unit	Material	2018 Bare Costs Labor	2018 Bare Costs Equipment	Total	Total Incl O&P
0010	NURSE STATION CASEWORK									
2100	Door type, laminated plastic	2 Carp	10	1.600	L.F.	425	81		506	595
2200	Enameled steel		10	1.600		410	81		491	575
2300	Stainless steel		10	1.600		815	81		896	1,025
2400	For drawer type, add					345			345	380

12 35 80 – Commercial Kitchen Casework

12 35 80.13 Metal Kitchen Casework

		Crew	Daily Output	Labor-Hours	Unit	Material	2018 Bare Costs Labor	2018 Bare Costs Equipment	Total	Total Incl O&P
0010	METAL KITCHEN CASEWORK									
3500	Base cabinets, metal, minimum	2 Carp	30	.533	L.F.	99.50	27		126.50	150
3600	Maximum		25	.640		252	32.50		284.50	330
3700	Wall cabinets, metal, minimum		30	.533		99.50	27		126.50	150
3800	Maximum		25	.640		228	32.50		260.50	300

12 36 Countertops

12 36 16 – Metal Countertops

12 36 16.10 Stainless Steel Countertops

		Crew	Daily Output	Labor-Hours	Unit	Material	2018 Bare Costs Labor	2018 Bare Costs Equipment	Total	Total Incl O&P
0010	STAINLESS STEEL COUNTERTOPS									
3200	Stainless steel, custom	1 Carp	24	.333	S.F.	179	16.90		195.90	223
3210	Stainless steel	"	12	.667	L.F.	360	34		394	445

12 36 19 – Wood Countertops

12 36 19.10 Maple Countertops

		Crew	Daily Output	Labor-Hours	Unit	Material	2018 Bare Costs Labor	2018 Bare Costs Equipment	Total	Total Incl O&P
0010	MAPLE COUNTERTOPS									
2900	Solid, laminated, 1-1/2" thick, no splash	1 Carp	28	.286	L.F.	88.50	14.50		103	119
3000	With square splash		28	.286	"	105	14.50		119.50	138
3400	Recessed cutting block with trim, 16" x 20" x 1"		8	1	Ea.	108	50.50		158.50	195

12 36 23 – Plastic Countertops

12 36 23.13 Plastic-Laminate-Clad Countertops

		Crew	Daily Output	Labor-Hours	Unit	Material	2018 Bare Costs Labor	2018 Bare Costs Equipment	Total	Total Incl O&P
0010	PLASTIC-LAMINATE-CLAD COUNTERTOPS									
0020	Stock, 24" wide w/backsplash, minimum	1 Carp	30	.267	L.F.	17.25	13.50		30.75	39.50
0100	Maximum		25	.320		39	16.20		55.20	67.50
0300	Custom plastic, 7/8" thick, aluminum molding, no splash		30	.267		34.50	13.50		48	58.50
0400	Cove splash		30	.267		41	13.50		54.50	65.50
0600	1-1/4" thick, no splash		28	.286		39	14.50		53.50	64.50
0700	Square splash		28	.286		43.50	14.50		58	69.50
0900	Square edge, plastic face, 7/8" thick, no splash		30	.267		34	13.50		47.50	58
1000	With splash		30	.267		41	13.50		54.50	66

12 36 Countertops

12 36 23 – Plastic Countertops

12 36 23.13 Plastic-Laminate-Clad Countertops	Crew	Daily Output	Labor-Hours	Unit	Material	2018 Bare Costs Labor	Equipment	Total	Total Incl O&P
1200	For stainless channel edge, 7/8" thick, add				L.F.	3.66			3.66	4.03
1300	1-1/4" thick, add					4.36			4.36	4.80
1500	For solid color suede finish, add					5.80			5.80	6.40
1700	For end splash, add				Ea.	21			21	23
1900	For cut outs, standard, add, minimum	1 Carp	32	.250		17.40	12.70		30.10	38.50
2000	Maximum		8	1		7	50.50		57.50	84.50
2100	Postformed, including backsplash and front edge		30	.267	L.F.	12.60	13.50		26.10	34.50
2110	Mitred, add		12	.667	Ea.		34		34	51.50
2200	Built-in place, 25" wide, plastic laminate		25	.320	L.F.	60	16.20		76.20	90.50

12 36 33 – Tile Countertops

12 36 33.10 Ceramic Tile Countertops

		Crew	Daily Output	Labor-Hours	Unit	Material	Labor	Equipment	Total	Total Incl O&P
0010	**CERAMIC TILE COUNTERTOPS**									
2300	Ceramic tile mosaic	1 Carp	25	.320	L.F.	39.50	16.20		55.70	68

12 36 40 – Stone Countertops

12 36 40.10 Natural Stone Countertops

		Crew	Daily Output	Labor-Hours	Unit	Material	Labor	Equipment	Total	Total Incl O&P
0010	**NATURAL STONE COUNTERTOPS**									
2500	Marble, stock, with splash, 1/2" thick, minimum	1 Bric	17	.471	L.F.	49	23.50		72.50	90
2700	3/4" thick, maximum		13	.615		123	31		154	183
2800	Granite, average, 1-1/4" thick, 24" wide, no splash		13.01	.615		165	31		196	230

12 36 53 – Laboratory Countertops

12 36 53.10 Laboratory Countertops and Sinks

		Crew	Daily Output	Labor-Hours	Unit	Material	Labor	Equipment	Total	Total Incl O&P
0010	**LABORATORY COUNTERTOPS AND SINKS**									
0020	Countertops, epoxy resin, not incl. base cabinets, acid-proof, minimum	2 Carp	82	.195	S.F.	47.50	9.90		57.40	67
0030	Maximum		70	.229		47.50	11.60		59.10	70
0040	Stainless steel		82	.195		165	9.90		174.90	196

12 36 61 – Simulated Stone Countertops

12 36 61.16 Solid Surface Countertops

		Crew	Daily Output	Labor-Hours	Unit	Material	Labor	Equipment	Total	Total Incl O&P
0010	**SOLID SURFACE COUNTERTOPS**, Acrylic polymer									
0020	Pricing for orders of 100 L.F. or greater									
0100	25" wide, solid colors	2 Carp	28	.571	L.F.	52.50	29		81.50	102
0200	Patterned colors		28	.571		66.50	29		95.50	117
0300	Premium patterned colors		28	.571		88.50	29		117.50	142
0400	With silicone attached 4" backsplash, solid colors		27	.593		61	30		91	114
0500	Patterned colors		27	.593		77.50	30		107.50	132
0600	Premium patterned colors		27	.593		96.50	30		126.50	152
0700	With hard seam attached 4" backsplash, solid colors		23	.696		61	35.50		96.50	121
0800	Patterned colors		23	.696		77.50	35.50		113	139
0900	Premium patterned colors		23	.696		96.50	35.50		132	160
1000	Pricing for order of 51-99 L.F.									
1100	25" wide, solid colors	2 Carp	24	.667	L.F.	60	34		94	118
1200	Patterned colors		24	.667		76.50	34		110.50	136
1300	Premium patterned colors		24	.667		102	34		136	164
1400	With silicone attached 4" backsplash, solid colors		23	.696		70.50	35.50		106	131
1500	Patterned colors		23	.696		89	35.50		124.50	152
1600	Premium patterned colors		23	.696		111	35.50		146.50	176
1700	With hard seam attached 4" backsplash, solid colors		20	.800		70.50	40.50		111	140
1800	Patterned colors		20	.800		89	40.50		129.50	160
1900	Premium patterned colors		20	.800		111	40.50		151.50	184
2000	Pricing for order of 1-50 L.F.									
2100	25" wide, solid colors	2 Carp	20	.800	L.F.	70.50	40.50		111	140

For customer support on your Building Construction Costs with RSMeans data, call 800.448.8182.

443

12 36 Countertops

12 36 61 – Simulated Stone Countertops

12 36 61.16 Solid Surface Countertops

		Crew	Daily Output	Labor-Hours	Unit	Material	2018 Bare Costs Labor	Equipment	Total	Total Incl O&P
2200	Patterned colors	2 Carp	20	.800	L.F.	90	40.50		130.50	161
2300	Premium patterned colors		20	.800		119	40.50		159.50	193
2400	With silicone attached 4" backsplash, solid colors		19	.842		82.50	42.50		125	156
2500	Patterned colors		19	.842		105	42.50		147.50	180
2600	Premium patterned colors		19	.842		130	42.50		172.50	208
2700	With hard seam attached 4" backsplash, solid colors		15	1.067		82.50	54		136.50	174
2800	Patterned colors		15	1.067		105	54		159	198
2900	Premium patterned colors	↓	15	1.067	↓	130	54		184	226
3000	Sinks, pricing for order of 100 or greater units									
3100	Single bowl, hard seamed, solid colors, 13" x 17"	1 Carp	3	2.667	Ea.	375	135		510	620
3200	10" x 15"		7	1.143		174	58		232	280
3300	Cutouts for sinks	↓	8	1	↓		50.50		50.50	77
3400	Sinks, pricing for order of 51-99 units									
3500	Single bowl, hard seamed, solid colors, 13" x 17"	1 Carp	2.55	3.137	Ea.	435	159		594	715
3600	10" x 15"		6	1.333		200	67.50		267.50	325
3700	Cutouts for sinks	↓	7	1.143	↓		58		58	88
3800	Sinks, pricing for order of 1-50 units									
3900	Single bowl, hard seamed, solid colors, 13" x 17"	1 Carp	2	4	Ea.	510	203		713	870
4000	10" x 15"		4.55	1.758		235	89		324	395
4100	Cutouts for sinks		5.25	1.524			77.50		77.50	118
4200	Cooktop cutouts, pricing for 100 or greater units		4	2		25.50	101		126.50	182
4300	51-99 units		3.40	2.353		29.50	119		148.50	215
4400	1-50 units	↓	3	2.667	↓	34.50	135		169.50	244

12 36 61.17 Solid Surface Vanity Tops

		Crew	Daily Output	Labor-Hours	Unit	Material	2018 Bare Costs Labor	Equipment	Total	Total Incl O&P
0010	**SOLID SURFACE VANITY TOPS**									
0015	Solid surface, center bowl, 17" x 19"	1 Carp	12	.667	Ea.	181	34		215	251
0020	19" x 25"		12	.667		185	34		219	255
0030	19" x 31"		12	.667		216	34		250	290
0040	19" x 37"		12	.667		252	34		286	330
0050	22" x 25"		10	.800		330	40.50		370.50	425
0060	22" x 31"		10	.800		385	40.50		425.50	485
0070	22" x 37"		10	.800		450	40.50		490.50	555
0080	22" x 43"		10	.800		520	40.50		560.50	630
0090	22" x 49"		10	.800		560	40.50		600.50	680
0110	22" x 55"		8	1		405	50.50		455.50	520
0120	22" x 61"		8	1		460	50.50		510.50	580
0220	Double bowl, 22" x 61"		8	1		515	50.50		565.50	645
0230	Double bowl, 22" x 73"	↓	8	1	↓	980	50.50		1,030.50	1,150
0240	For aggregate colors, add					35%				
0250	For faucets and fittings, see Section 22 41 39.10									

12 36 61.19 Quartz Agglomerate Countertops

		Crew	Daily Output	Labor-Hours	Unit	Material	2018 Bare Costs Labor	Equipment	Total	Total Incl O&P
0010	**QUARTZ AGGLOMERATE COUNTERTOPS**									
0100	25" wide, 4" backsplash, color group A, minimum	2 Carp	15	1.067	L.F.	67.50	54		121.50	157
0110	Maximum		15	1.067		94	54		148	187
0120	Color group B, minimum		15	1.067		72	54		126	162
0130	Maximum		15	1.067		98.50	54		152.50	191
0140	Color group C, minimum		15	1.067		81	54		135	172
0150	Maximum		15	1.067		115	54		169	210
0160	Color group D, minimum		15	1.067		88	54		142	180
0170	Maximum	↓	15	1.067	↓	119	54		173	214

For customer support on your Building Construction Costs with RSMeans data, call 800.448.8182.

12 46 Furnishing Accessories

12 46 13 – Ash Receptacles

12 46 13.10 Ash/Trash Receivers

		Crew	Daily Output	Labor-Hours	Unit	Material	2018 Bare Costs Labor	Equipment	Total	Total Incl O&P
0010	**ASH/TRASH RECEIVERS**									
1000	Ash urn, cylindrical metal									
1020	8" diameter, 20" high	1 Clab	60	.133	Ea.	197	5.30		202.30	225
1060	10" diameter, 26" high	"	60	.133	"	158	5.30		163.30	181
2000	Combination ash/trash urn, metal									
2020	8" diameter, 20" high	1 Clab	60	.133	Ea.	254	5.30		259.30	288
2050	10" diameter, 26" high	"	60	.133	"	204	5.30		209.30	232

12 46 19 – Clocks

12 46 19.50 Wall Clocks

		Crew	Daily Output	Labor-Hours	Unit	Material	2018 Bare Costs Labor	Equipment	Total	Total Incl O&P
0010	**WALL CLOCKS**									
0080	12" diameter, single face	1 Elec	8	1	Ea.	102	58		160	200
0100	Double face	"	6.20	1.290	"	159	75		234	287

12 46 33 – Waste Receptacles

12 46 33.13 Trash Receptacles

		Crew	Daily Output	Labor-Hours	Unit	Material	2018 Bare Costs Labor	Equipment	Total	Total Incl O&P
0010	**TRASH RECEPTACLES**									
4000	Trash receptacle, metal									
4020	8" diameter, 15" high	1 Clab	60	.133	Ea.	71	5.30		76.30	86
4040	10" diameter, 18" high		60	.133		133	5.30		138.30	154
5040	16" x 8" x 14" high		60	.133		58.50	5.30		63.80	72
5500	Plastic, with lid									
5520	35 gal.	1 Clab	60	.133	Ea.	190	5.30		195.30	217
5540	45 gal.		60	.133		305	5.30		310.30	345
5550	Plastic recycling barrel, w/lid & wheels, 32 gal. [G]		60	.133		94	5.30		99.30	111
5560	65 gal. [G]		60	.133		605	5.30		610.30	675
5570	95 gal. [G]		60	.133		1,125	5.30		1,130.30	1,250

12 48 Rugs and Mats

12 48 13 – Entrance Floor Mats and Frames

12 48 13.13 Entrance Floor Mats

		Crew	Daily Output	Labor-Hours	Unit	Material	2018 Bare Costs Labor	Equipment	Total	Total Incl O&P
0010	**ENTRANCE FLOOR MATS**									
0020	Recessed, black rubber, 3/8" thick, solid	1 Clab	155	.052	S.F.	26	2.06		28.06	31.50
0050	Perforated		155	.052		21.50	2.06		23.56	26.50
0100	1/2" thick, solid		155	.052		25.50	2.06		27.56	31
0150	Perforated		155	.052		26.50	2.06		28.56	32
0200	In colors, 3/8" thick, solid		155	.052		27.50	2.06		29.56	33.50
0250	Perforated		155	.052		29.50	2.06		31.56	35.50
0300	1/2" thick, solid		155	.052		35.50	2.06		37.56	42.50
0350	Perforated		155	.052		36.50	2.06		38.56	43
1225	Recessed, alum. rail, hinged mat, 7/16" thick									
1250	Carpet insert	1 Clab	360	.022	S.F.	66	.89		66.89	74
1275	Vinyl insert		360	.022		66	.89		66.89	74
1300	Abrasive insert		360	.022		66	.89		66.89	74
1325	Recessed, vinyl rail, hinged mat, 7/16" thick									
1350	Carpet insert	1 Clab	360	.022	S.F.	73	.89		73.89	82
1375	Vinyl insert		360	.022		73	.89		73.89	82
1400	Abrasive insert		360	.022		73	.89		73.89	82
2000	Recycled rubber tire tile, 12" x 12" x 3/8" thick [G]		125	.064		9.95	2.55		12.50	14.85
2510	Natural cocoa fiber, 1/2" thick [G]		125	.064		7.10	2.55		9.65	11.70
2520	3/4" thick [G]		125	.064		6.45	2.55		9	10.95
2530	1" thick [G]		125	.064		12.90	2.55		15.45	18.10

For customer support on your Building Construction Costs with RSMeans data, call 800.448.8182.

445

12 48 Rugs and Mats

12 48 13 – Entrance Floor Mats and Frames

12 48 13.13 Entrance Floor Mats	Crew	Daily Output	Labor-Hours	Unit	Material	2018 Bare Costs Labor	Equipment	Total	Total Incl O&P
3000	Hospital tacky mats, package of 30 with frame				Ea.	52			52	57.50
3010	4 packages of 30				"	85			85	93.50

12 51 Office Furniture

12 51 16 – Case Goods

12 51 16.13 Metal Case Goods

		Crew	Daily Output	Labor-Hours	Unit	Material	Labor	Equipment	Total	Total Incl O&P
0010	**METAL CASE GOODS**									
0020	Desks, 29" high, double pedestal, 30" x 60", metal, minimum				Ea.	575			575	630
0030	Maximum					895			895	985
0600	Desks, single pedestal, 30" x 60", metal, minimum					505			505	555
0620	Maximum					965			965	1,075
0720	Desks, secretarial, 30" x 60", metal, minimum					455			455	500
0730	Maximum					875			875	960
0740	Return, 20" x 42", minimum					420			420	465
0750	Maximum					675			675	745
0940	59" x 12" x 23" high, steel, minimum					350			350	385
0960	Maximum					450			450	495

12 51 16.16 Wood Case Goods

		Crew	Daily Output	Labor-Hours	Unit	Material	Labor	Equipment	Total	Total Incl O&P
0010	**WOOD CASE GOODS**									
0150	Desk, 29" high, double pedestal, 30" x 60"									
0160	Wood, minimum				Ea.	740			740	815
0180	Maximum				"	1,375			1,375	1,500
0630	Single pedestal, 30" x 60"									
0640	Wood, minimum				Ea.	625			625	690
0650	Maximum					950			950	1,050
0670	Executive return, 24" x 42", with box, file, wood, minimum					460			460	505
0680	Maximum					890			890	980
0790	Desk, 29" high, secretarial, 30" x 60"									
0800	Wood, minimum				Ea.	465			465	510
0810	Maximum					1,400			1,400	1,550
0820	Return, 20" x 42", minimum					400			400	440
0830	Maximum					700			700	775
0900	Desktop organizer, 72" x 14" x 36" high, wood, minimum					183			183	201
0920	Maximum					440			440	485
1110	Furniture, credenza, 29" high, 18" to 22" x 60" to 72"									
1120	Wood, minimum				Ea.	810			810	890
1140	Maximum				"	2,900			2,900	3,175

12 51 23 – Office Tables

12 51 23.33 Conference Tables

		Crew	Daily Output	Labor-Hours	Unit	Material	Labor	Equipment	Total	Total Incl O&P
0010	**CONFERENCE TABLES**									
6050	Boat, 96" x 42", minimum				Ea.	850			850	935
6150	Maximum					3,975			3,975	4,375
6720	Rectangle, 96" x 42", minimum					1,375			1,375	1,525
6740	Maximum					3,175			3,175	3,500

For customer support on your Building Construction Costs with RSMeans data, call 800.448.8182.

12 52 Seating

12 52 23 – Office Seating

12 52 23.13 Office Chairs	Crew	Daily Output	Labor-Hours	Unit	Material	2018 Bare Costs Labor	Equipment	Total	Total Incl O&P
0010 **OFFICE CHAIRS**									
2000 Standard office chair, executive, minimum				Ea.	291			291	320
2150 Maximum					2,075			2,075	2,275
2200 Management, minimum					259			259	285
2250 Maximum					1,850			1,850	2,025
2280 Task, minimum					198			198	218
2290 Maximum					625			625	685
2300 Arm kit, minimum					77			77	85
2320 Maximum				▼	123			123	135

12 54 Hospitality Furniture

12 54 13 – Hotel and Motel Furniture

12 54 13.10 Hotel Furniture

	Crew	Daily Output	Labor-Hours	Unit	Material	Labor	Equipment	Total	Total Incl O&P
0010 **HOTEL FURNITURE**									
0020 Standard quality set, minimum				Room	2,550			2,550	2,800
0200 Maximum				"	8,525			8,525	9,375

12 54 16 – Restaurant Furniture

12 54 16.10 Tables, Folding

	Crew	Daily Output	Labor-Hours	Unit	Material	Labor	Equipment	Total	Total Incl O&P
0010 **TABLES, FOLDING** Laminated plastic tops									
1000 Tubular steel legs with glides									
1020 18" x 60", minimum				Ea.	289			289	320
1040 Maximum					1,275			1,275	1,400
1840 36" x 96", minimum					310			310	340
1860 Maximum					2,275			2,275	2,500
2000 Round, wood stained, plywood top, 60" diameter, minimum					223			223	245
2020 Maximum				▼	305			305	335

12 54 16.20 Furniture, Restaurant

	Crew	Daily Output	Labor-Hours	Unit	Material	Labor	Equipment	Total	Total Incl O&P
0010 **FURNITURE, RESTAURANT**									
0020 Bars, built-in, front bar	1 Carp	5	1.600	L.F.	297	81		378	450
0200 Back bar	"	5	1.600	"	216	81		297	360
0300 Booth seating, see Section 12 54 16.70									
2000 Chair, bentwood side chair, metal, minimum				Ea.	106			106	116
2020 Maximum					124			124	136
2600 Upholstered seat & back, arms, minimum					172			172	189
2620 Maximum				▼	490			490	535

12 54 16.70 Booths

	Crew	Daily Output	Labor-Hours	Unit	Material	Labor	Equipment	Total	Total Incl O&P
0010 **BOOTHS**									
1000 Banquet, upholstered seat and back, custom									
1500 Straight, minimum	2 Carp	40	.400	L.F.	206	20.50		226.50	258
1520 Maximum		36	.444		400	22.50		422.50	475
1600 "L" or "U" shape, minimum		35	.457		210	23		233	267
1620 Maximum	▼	30	.533	▼	370	27		397	450
1800 Upholstered outside finished backs for									
1810 single booths and custom banquets									
1820 Minimum	2 Carp	44	.364	L.F.	24.50	18.45		42.95	55
1840 Maximum	"	40	.400	"	71.50	20.50		92	110
3000 Fixed seating, one piece plastic chair and									
3010 plastic laminate table top									
3100 Two seat, 24" x 24" table, minimum	F-7	30	1.067	Ea.	825	48.50		873.50	985
3120 Maximum	↓	26	1.231	↓	1,175	55.50		1,230.50	1,375

For customer support on your Building Construction Costs with RSMeans data, call 800.448.8182.

447

12 54 Hospitality Furniture

12 54 16 – Restaurant Furniture

12 54 16.70 Booths

12 54 16.70 Booths		Crew	Daily Output	Labor-Hours	Unit	Material	2018 Bare Costs Labor	Equipment	Total	Total Incl O&P
3200	Four seat, 24" x 48" table, minimum	F-7	28	1.143	Ea.	820	52		872	980
3220	Maximum	↓	24	1.333	↓	1,400	60.50		1,460.50	1,650
5000	Mount in floor, wood fiber core with									
5010	plastic laminate face, single booth									
5050	24" wide	F-7	30	1.067	Ea.	315	48.50		363.50	425
5100	48" wide	"	28	1.143	"	405	52		457	525

12 55 Detention Furniture

12 55 13 – Detention Bunks

12 55 13.13 Cots

	COTS	Crew	Daily Output	Labor-Hours	Unit	Material	2018 Bare Costs Labor	Equipment	Total	Total Incl O&P
0010	**COTS**									
2500	Bolted, single, painted steel	E-4	20	1.600	Ea.	340	88	4.93	432.93	520
2700	Stainless steel	"	20	1.600	"	975	88	4.93	1,067.93	1,225

12 56 Institutional Furniture

12 56 33 – Classroom Furniture

12 56 33.10 Furniture, School

	FURNITURE, SCHOOL	Crew	Daily Output	Labor-Hours	Unit	Material	2018 Bare Costs Labor	Equipment	Total	Total Incl O&P
0010	**FURNITURE, SCHOOL**									
0500	Classroom, movable chair & desk type, minimum				Set				73.50	81
0600	Maximum				"				155	171
1000	Chair, molded plastic									
1100	Integral tablet arm, minimum				Ea.	101			101	112
1150	Maximum					199			199	219
2000	Desk, single pedestal, top book compartment, minimum					102			102	112
2020	Maximum					199			199	219
2200	Flip top, minimum					256			256	282
2220	Maximum				↓	305			305	335

12 56 43 – Dormitory Furniture

12 56 43.10 Dormitory Furnishings

	DORMITORY FURNISHINGS	Crew	Daily Output	Labor-Hours	Unit	Material	2018 Bare Costs Labor	Equipment	Total	Total Incl O&P
0010	**DORMITORY FURNISHINGS**									
0300	Bunkable bed, twin, minimum				Ea.	395			395	435
0320	Maximum					650			650	715
1000	Chest, four drawer, minimum					395			395	430
1020	Maximum				↓	700			700	770
1050	Built-in, minimum	2 Carp	13	1.231	L.F.	137	62.50		199.50	246
1150	Maximum		10	1.600		254	81		335	405
1200	Desk top, built-in, laminated plastic, 24" deep, minimum		50	.320		48.50	16.20		64.70	78
1300	Maximum		40	.400		145	20.50		165.50	191
1450	30" deep, minimum		50	.320		62	16.20		78.20	92.50
1550	Maximum		40	.400		271	20.50		291.50	330
1750	Dressing unit, built-in, minimum		12	1.333		206	67.50		273.50	330
1850	Maximum	↓	8	2	↓	620	101		721	835
8000	Rule of thumb: total cost for furniture, minimum				Student				2,525	2,800
8050	Maximum				"				4,850	5,350

For customer support on your Building Construction Costs with RSMeans data, call 800.448.8182.

12 56 Institutional Furniture

12 56 51 – Library Furniture

12 56 51.10 Library Furnishings

		Crew	Daily Output	Labor-Hours	Unit	Material	2018 Bare Costs Labor	2018 Bare Costs Equipment	Total	Total Incl O&P
0010	**LIBRARY FURNISHINGS**									
0100	Attendant desk, 36" x 62" x 29" high	1 Carp	16	.500	Ea.	2,200	25.50		2,225.50	2,475
0200	Book display, "A" frame display, both sides, 42" x 42" x 60" high		16	.500		1,150	25.50		1,175.50	1,325
0220	Table with bulletin board, 42" x 24" x 49" high		16	.500		710	25.50		735.50	820
0800	Card catalog, 30 tray unit		16	.500		3,500	25.50		3,525.50	3,900
0840	60 tray unit		16	.500		7,025	25.50		7,050.50	7,775
0880	72 tray unit	2 Carp	16	1		8,200	50.50		8,250.50	9,100
1000	Carrels, single face, initial unit	1 Carp	16	.500		880	25.50		905.50	1,000
1500	Double face, initial unit	2 Carp	16	1		1,475	50.50		1,525.50	1,675
1710	Carrels, hardwood, 36" x 24", minimum	1 Carp	5	1.600		560	81		641	740
1720	Maximum	"	4	2		1,625	101		1,726	1,925
2700	Card catalog file, 60 trays, complete					8,625			8,625	9,500
2720	Alternate method: each tray					144			144	158
3800	Charging desk, built-in, with counter, plastic laminated top	1 Carp	7	1.143	L.F.	465	58		523	600
4000	Dictionary stand, stationary		16	.500	Ea.	710	25.50		735.50	820
4020	Revolving		16	.500		217	25.50		242.50	278
4200	Exhibit case, table style, 60" x 28" x 36"		11	.727		2,700	37		2,737	3,000
6010	Bookshelf, metal, 90" high, 10" shelf, double face		11.50	.696	L.F.	187	35.50		222.50	260
6020	Single face		12	.667	"	117	34		151	181
6050	For 8" shelving, subtract from above					10%				
6060	For 12" shelving, add to above					10%				
6070	For 42" high with countertop, subtract from above					20%				
6100	Mobile compacted shelving, hand crank, 9'-0" high									
6110	Double face, including track, 3' section				Ea.	2,175			2,175	2,375
6150	For electrical operation, add					25%				
6200	Magazine shelving, 82" high, 12" deep, single face	1 Carp	11.50	.696	L.F.	149	35.50		184.50	218
6210	Double face	"	11.50	.696	"	283	35.50		318.50	365
7200	Reading table, laminated top, 60" x 36"				Ea.	460			460	505

12 56 70 – Healthcare Furniture

12 56 70.10 Furniture, Hospital

		Crew	Daily Output	Labor-Hours	Unit	Material	2018 Bare Costs Labor	2018 Bare Costs Equipment	Total	Total Incl O&P
0010	**FURNITURE, HOSPITAL**									
0020	Beds, manual, minimum				Ea.	855			855	940
0100	Maximum					2,825			2,825	3,100
0600	All electric hospital beds, minimum					1,875			1,875	2,075
0700	Maximum					3,725			3,725	4,100
0900	Manual, nursing home beds, minimum					685			685	755
1000	Maximum					1,825			1,825	2,000
1020	Overbed table, laminated top, minimum					355			355	395
1040	Maximum					715			715	790
1100	Patient wall systems, not incl. plumbing, minimum				Room	1,400			1,400	1,550
1200	Maximum				"	2,000			2,000	2,200
2000	Geriatric chairs, minimum				Ea.	500			500	555
2020	Maximum				"	760			760	835

12 61 Fixed Audience Seating

12 61 13 – Upholstered Audience Seating

12 61 13.13 Auditorium Chairs

		Crew	Daily Output	Labor-Hours	Unit	Material	2018 Bare Costs Labor	Equipment	Total	Total Incl O&P
0010	**AUDITORIUM CHAIRS**									
2000	All veneer construction	2 Carp	22	.727	Ea.	274	37		311	355
2200	Veneer back, padded seat		22	.727		297	37		334	380
2350	Fully upholstered, spring seat		22	.727		254	37		291	335
2450	For tablet arms, add					81.50			81.50	90
2500	For fire retardancy, CATB-133, add					29			29	32

12 61 13.23 Lecture Hall Seating

		Crew	Daily Output	Labor-Hours	Unit	Material	2018 Bare Costs Labor	Equipment	Total	Total Incl O&P
0010	**LECTURE HALL SEATING**									
1000	Pedestal type, minimum	2 Carp	22	.727	Ea.	265	37		302	350
1200	Maximum	"	14.50	1.103	"	515	56		571	650

12 63 Stadium and Arena Seating

12 63 13 – Stadium and Arena Bench Seating

12 63 13.13 Bleachers

		Crew	Daily Output	Labor-Hours	Unit	Material	2018 Bare Costs Labor	Equipment	Total	Total Incl O&P
0010	**BLEACHERS**									
3000	Telescoping, manual to 15 tier, minimum	F-5	65	.492	Seat	114	25		139	164
3100	Maximum		60	.533		163	27.50		190.50	222
3300	16 to 20 tier, minimum		60	.533		274	27.50		301.50	340
3400	Maximum		55	.582		325	30		355	405
3600	21 to 30 tier, minimum		50	.640		262	33		295	340
3700	Maximum		40	.800		360	41		401	460
3900	For integral power operation, add, minimum	2 Elec	300	.053		53	3.10		56.10	63
4000	Maximum	"	250	.064		86.50	3.72		90.22	101
5000	Benches, folding, in wall, 14' table, 2 benches	L-4	2	12	Set	965	570		1,535	1,925

12 67 Pews and Benches

12 67 13 – Pews

12 67 13.13 Sanctuary Pews

		Crew	Daily Output	Labor-Hours	Unit	Material	2018 Bare Costs Labor	Equipment	Total	Total Incl O&P
0010	**SANCTUARY PEWS**									
1500	Bench type, hardwood, minimum	1 Carp	20	.400	L.F.	100	20.50		120.50	141
1550	Maximum	"	15	.533		175	27		202	233
1570	For kneeler, add					22.50			22.50	24.50

12 92 Interior Planters and Artificial Plants

12 92 33 – Interior Planters

12 92 33.10 Planters

		Crew	Daily Output	Labor-Hours	Unit	Material	2018 Bare Costs Labor	Equipment	Total	Total Incl O&P
0010	**PLANTERS**									
1000	Fiberglass, hanging, 12" diameter, 7" high				Ea.	112			112	123
1500	Rectangular, 48" long, 16" high, 15" wide					605			605	665
1650	60" long, 30" high, 28" wide					1,025			1,025	1,125
2000	Round, 12" diameter, 13" high					161			161	177
2050	25" high					212			212	233
5000	Square, 10" side, 20" high					199			199	219
5100	14" side, 15" high					214			214	235
6000	Metal bowl, 32" diameter, 8" high, minimum					570			570	630
6050	Maximum					795			795	875
8750	Wood, fiberglass liner, square									

For customer support on your Building Construction Costs with RSMeans data, call 800.448.8182.

12 92 Interior Planters and Artificial Plants

12 92 33 – Interior Planters

12 92 33.10 Planters	Crew	Daily Output	Labor-Hours	Unit	Material	2018 Bare Costs Labor	Equipment	Total	Total Incl O&P
8780	14" square, 15" high, minimum				Ea.	430			430	475
8800	Maximum					530			530	580
9400	Plastic cylinder, molded, 10" diameter, 10" high					47			47	51.50
9500	11" diameter, 11" high					89.50			89.50	98

12 93 Interior Public Space Furnishings

12 93 13 – Bicycle Racks

12 93 13.10 Bicycle Racks

		Crew	Daily Output	Labor-Hours	Unit	Material	2018 Bare Costs Labor	Equipment	Total	Total Incl O&P
0010	**BICYCLE RACKS**									
0020	Single side, grid, 1-5/8" OD stl. pipe, w/1/2" bars, galv, 5 bike cap	2 Clab	10	1.600	Ea.	315	64		379	440
0025	Powder coat finish		10	1.600		320	64		384	445
0030	Single side, grid, 1-5/8" OD stl. pipe, w/1/2" bars, galv, 9 bike cap		8	2		400	79.50		479.50	560
0035	Powder coat finish		8	2		415	79.50		494.50	575
0040	Single side, grid, 1-5/8" OD stl. pipe, w/1/2" bars, galv, 18 bike cap		4	4		465	159		624	755
0045	Powder coat finish		4	4		525	159		684	820
0050	S curve, 1-7/8" OD stl. pipe, 11 ga., galv, 5 bike cap		10	1.600		157	64		221	270
0055	Powder coat finish		10	1.600		157	64		221	270
0060	S curve, 1-7/8" OD stl. pipe, 11 ga., galv, 7 bike cap		9	1.778		167	71		238	292
0065	Powder coat finish		9	1.778		212	71		283	340
0070	S curve, 1-7/8" OD stl. pipe, 11 ga., galv, 9 bike cap		8	2		320	79.50		399.50	470
0075	Powder coat finish		8	2		310	79.50		389.50	460
0080	S curve, 1-7/8" OD stl. pipe, 11 ga., galv, 11 bike cap		6	2.667		430	106		536	635
0085	Powder coat finish		6	2.667		415	106		521	615

12 93 23 – Trash and Litter Receptacles

12 93 23.10 Trash Receptacles

		Crew	Daily Output	Labor-Hours	Unit	Material	2018 Bare Costs Labor	Equipment	Total	Total Incl O&P
0010	**TRASH RECEPTACLES**									
0020	Fiberglass, 2' square, 18" high	2 Clab	30	.533	Ea.	600	21.50		621.50	695
0100	2' square, 2'-6" high		30	.533		930	21.50		951.50	1,050
0300	Circular, 2' diameter, 18" high		30	.533		530	21.50		551.50	615
0400	2' diameter, 2'-6" high		30	.533		625	21.50		646.50	725
0500	Recycled plastic, var colors, round, 32 gal., 28" x 38" high [G]		5	3.200		660	128		788	920
0510	32 gal., 31" x 32" high [G]		5	3.200		770	128		898	1,050
9110	Plastic, with dome lid, 32 gal. capacity		35	.457		62.50	18.20		80.70	96.50
9120	Recycled plastic slats, plastic dome lid, 32 gal. capacity		35	.457		297	18.20		315.20	355

12 93 23.20 Trash Closure

		Crew	Daily Output	Labor-Hours	Unit	Material	2018 Bare Costs Labor	Equipment	Total	Total Incl O&P
0010	**TRASH CLOSURE**									
0020	Steel with pullover cover, 2'-3" wide, 4'-7" high, 6'-2" long	2 Clab	5	3.200	Ea.	2,100	128		2,228	2,525
0100	10'-1" long		4	4		2,700	159		2,859	3,225
0300	Wood, 10' wide, 6' high, 10' long		1.20	13.333		2,100	530		2,630	3,125

Division Notes

	CREW	DAILY OUTPUT	LABOR-HOURS	UNIT	BARE COSTS				TOTAL INCL O&P
					MAT.	LABOR	EQUIP.	TOTAL	

Estimating Tips
General

- The items and systems in this division are usually estimated, purchased, supplied, and installed as a unit by one or more subcontractors. The estimator must ensure that all parties are operating from the same set of specifications and assumptions, and that all necessary items are estimated and will be provided. Many times the complex items and systems are covered, but the more common ones, such as excavation or a crane, are overlooked for the very reason that everyone assumes nobody could miss them. The estimator should be the central focus and be able to ensure that all systems are complete.

- Another area where problems can develop in this division is at the interface between systems. The estimator must ensure, for instance, that anchor bolts, nuts, and washers are estimated and included for the air-supported structures and pre-engineered buildings to be bolted to their foundations. Utility supply is a common area where essential items or pieces of equipment can be missed or overlooked, because each subcontractor may feel it is another's responsibility. The estimator should also be aware of certain items which may be supplied as part of a package but installed by others, and ensure that the installing contractor's estimate includes the cost of installation. Conversely, the estimator must also ensure that items are not costed by two different subcontractors, resulting in an inflated overall estimate.

13 30 00 Special Structures

- The foundations and floor slab, as well as rough mechanical and electrical, should be estimated, as this work is required for the assembly and erection of the structure. Generally, as noted in the data set, the pre-engineered building comes as a shell. Pricing is based on the size and structural design parameters stated in the reference section. Additional features, such as windows and doors with their related structural framing, must also be included by the estimator. Here again, the estimator must have a clear understanding of the scope of each portion of the work and all the necessary interfaces.

Reference Numbers

Reference numbers are shown at the beginning of some major classifications. These numbers refer to related items in the Reference Section. The reference information may be an estimating procedure, an alternate pricing method, or technical information.

Note: Not all subdivisions listed here necessarily appear. ∎

Did you know?

RSMeans data is available through our online application with 24/7 access:

- Search for unit prices by keyword
- Leverage the most up-to-date data
- Build and export estimates

Try it free for 30 days!
www.rsmeans.com/2018freetrial

No part of this cost data may be reproduced, stored in a retrieval system, or transmitted in any form or by any means without prior written permission of Gordian.

13 05 Common Work Results for Special Construction

13 05 05 – Selective Demolition for Special Construction

13 05 05.10 Selective Demolition, Air Supported Structures

		Crew	Daily Output	Labor-Hours	Unit	Material	2018 Bare Costs Labor	Equipment	Total	Total Incl O&P
0010	**SELECTIVE DEMOLITION, AIR SUPPORTED STRUCTURES**									
0020	Tank covers, scrim, dbl. layer, vinyl poly w/hdwe., blower & controls									
0050	Round and rectangular R024119-10	B-2	9000	.004	S.F.		.18		.18	.27
0100	Warehouse structures									
0120	Poly/vinyl fabric, 28 oz., incl. tension cables & inflation system	4 Clab	9000	.004	SF Flr.		.14		.14	.22
0150	Reinforced vinyl, 12 oz., 3,000 S.F.	"	5000	.006			.26		.26	.39
0200	12,000 to 24,000 S.F.	8 Clab	20000	.003			.13		.13	.19
0250	Tedlar vinyl fabric, 28 oz. w/liner, to 3,000 S.F.	4 Clab	5000	.006			.26		.26	.39
0300	12,000 to 24,000 S.F.	8 Clab	20000	.003	↓		.13		.13	.19
0350	Greenhouse/shelter, woven polyethylene with liner									
0400	3,000 S.F.	4 Clab	5000	.006	SF Flr.		.26		.26	.39
0450	12,000 to 24,000 S.F.	8 Clab	20000	.003			.13		.13	.19
0500	Tennis/gymnasium, poly/vinyl fabric, 28 oz., incl. thermal liner	4 Clab	9000	.004			.14		.14	.22
0600	Stadium/convention center, teflon coated fiberglass, incl. thermal liner	9 Clab	40000	.002	↓		.07		.07	.11
0700	Doors, air lock, 15' long, 10' x 10'	2 Carp	1.50	10.667	Ea.		540		540	825
0720	15' x 15'		.80	20			1,025		1,025	1,550
0750	Revolving personnel door, 6' diam. x 6'-6" high	↓	1.50	10.667			540		540	825

13 05 05.20 Selective Demolition, Garden Houses

		Crew	Daily Output	Labor-Hours	Unit	Material	2018 Bare Costs Labor	Equipment	Total	Total Incl O&P
0010	**SELECTIVE DEMOLITION, GARDEN HOUSES** R024119-10									
0020	Prefab, wood, excl. foundation, average	2 Clab	400	.040	SF Flr.		1.59		1.59	2.43

13 05 05.25 Selective Demolition, Geodesic Domes

		Crew	Daily Output	Labor-Hours	Unit	Material	2018 Bare Costs Labor	Equipment	Total	Total Incl O&P
0010	**SELECTIVE DEMOLITION, GEODESIC DOMES**									
0050	Shell only, interlocking plywood panels, 30' diameter	F-5	3.20	10	Ea.		510		510	780
0060	34' diameter		2.30	13.913			710		710	1,075
0070	39' diameter	↓	2	16			820		820	1,250
0080	45' diameter	F-3	2.20	18.182			940	225	1,165	1,675
0090	55' diameter		2	20			1,025	248	1,273	1,850
0100	60' diameter		2	20			1,025	248	1,273	1,850
0110	65' diameter	↓	1.60	25	↓		1,300	310	1,610	2,325

13 05 05.30 Selective Demolition, Greenhouses

		Crew	Daily Output	Labor-Hours	Unit	Material	2018 Bare Costs Labor	Equipment	Total	Total Incl O&P
0010	**SELECTIVE DEMOLITION, GREENHOUSES** R024119-10									
0020	Resi-type, free standing, excl. foundations, 9' long x 8' wide	2 Clab	160	.100	SF Flr.		3.99		3.99	6.05
0030	9' long x 11' wide		170	.094			3.75		3.75	5.70
0040	9' long x 14' wide		220	.073			2.90		2.90	4.41
0050	9' long x 17' wide		320	.050			1.99		1.99	3.04
0060	Lean-to type, 4' wide		64	.250			9.95		9.95	15.20
0070	7' wide		120	.133	↓		5.30		5.30	8.10
0080	Geodesic hemisphere, 1/8" plexiglass glazing, 8' diam.		4	4	Ea.		159		159	243
0090	24' diam.		.80	20			795		795	1,225
0100	48' diam.	↓	.40	40	↓		1,600		1,600	2,425

13 05 05.35 Selective Demolition, Hangars

		Crew	Daily Output	Labor-Hours	Unit	Material	2018 Bare Costs Labor	Equipment	Total	Total Incl O&P
0010	**SELECTIVE DEMOLITION, HANGARS**									
0020	T type hangars, prefab, steel, galv roof & walls, incl doors, excl fndtn	E-2	2550	.022	SF Flr.		1.19	.66	1.85	2.64
0030	Circular type, prefab, steel frame, plastic skin, incl foundation, 80' diam	"	.50	112	Total		6,075	3,375	9,450	13,400

13 05 05.45 Selective Demolition, Lightning Protection

		Crew	Daily Output	Labor-Hours	Unit	Material	2018 Bare Costs Labor	Equipment	Total	Total Incl O&P
0010	**SELECTIVE DEMOLITION, LIGHTNING PROTECTION**									
0020	Air terminal & base, copper, 3/8" diam. x 10", to 75' H	1 Clab	16	.500	Ea.		19.95		19.95	30.50
0030	1/2" diam. x 12", over 75' H		16	.500			19.95		19.95	30.50
0050	Aluminum, 1/2" diam. x 12", to 75' H		16	.500			19.95		19.95	30.50
0060	5/8" diam. x 12", over 75' H		16	.500	↓		19.95		19.95	30.50
0070	Cable, copper, 220 lb. per thousand feet, to 75' H	↓	640	.013	L.F.		.50		.50	.76

454

13 05 Common Work Results for Special Construction

13 05 05 – Selective Demolition for Special Construction

13 05 05.45 Selective Demolition, Lightning Protection	Crew	Daily Output	Labor-Hours	Unit	Material	2018 Bare Costs Labor	Equipment	Total	Total Incl O&P
0080	375 lb. per thousand feet, over 75' H	1 Clab	460	.017	L.F.		.69		.69	1.06
0090	Aluminum, 101 lb. per thousand feet, to 75' H		560	.014			.57		.57	.87
0100	199 lb. per thousand feet, over 75' H		480	.017	▼		.66		.66	1.01
0110	Arrester, 175 V AC, to ground		16	.500	Ea.		19.95		19.95	30.50
0120	650 V AC, to ground	▼	13	.615	"		24.50		24.50	37.50

13 05 05.50 Selective Demolition, Pre-Engineered Steel Buildings

		Crew	Daily Output	Labor-Hours	Unit	Material	Labor	Equipment	Total	Total Incl O&P
0010	**SELECTIVE DEMOLITION, PRE-ENGINEERED STEEL BUILDINGS**									
0500	Pre-engd. steel bldgs., rigid frame, clear span & multi post, excl. salvage									
0550	3,500 to 7,500 S.F.	L-10	1000	.024	SF Flr.		1.34	.50	1.84	2.68
0600	7,501 to 12,500 S.F.		1500	.016			.89	.33	1.22	1.78
0650	12,501 S.F. or greater	▼	1650	.015	▼		.81	.30	1.11	1.62
0700	Pre-engd. steel building components									
0710	Entrance canopy, including frame 4' x 4'	E-24	8	4	Ea.		218	74	292	430
0720	4' x 8'	"	7	4.571			249	84.50	333.50	495
0730	HM doors, self framing, single leaf	2 Skwk	8	2			105		105	160
0740	Double leaf		5	3.200	▼		168		168	256
0760	Gutter, eave type		600	.027	L.F.		1.40		1.40	2.14
0770	Sash, single slide, double slide or fixed		24	.667	Ea.		35		35	53.50
0780	Skylight, fiberglass, to 30 S.F.		16	1			52.50		52.50	80
0785	Roof vents, circular, 12" to 24" diameter		12	1.333			70		70	107
0790	Continuous, 10' long	▼	8	2	▼		105		105	160
0900	Shelters, aluminum frame									
0910	Acrylic glazing, 3' x 9' x 8' high	2 Skwk	2	8	Ea.		420		420	640
0920	9' x 12' x 8' high	"	1.50	10.667	"		560		560	855

13 05 05.60 Selective Demolition, Silos

		Crew	Daily Output	Labor-Hours	Unit	Material	Labor	Equipment	Total	Total Incl O&P
0010	**SELECTIVE DEMOLITION, SILOS**									
0020	Conc stave, indstrl, conical/sloping bott, excl fndtn, 12' diam., 35' H	E-24	.18	178	Ea.		9,675	3,275	12,950	19,100
0030	16' diam., 45' H		.12	267			14,500	4,925	19,425	28,700
0040	25' diam., 75' H	▼	.08	400			21,800	7,375	29,175	43,000
0050	Steel, factory fabricated, 30,000 gal. cap, painted or epoxy lined	L-5	2	28	▼		1,550	295	1,845	2,825

13 05 05.65 Selective Demolition, Sound Control

		Crew	Daily Output	Labor-Hours	Unit	Material	Labor	Equipment	Total	Total Incl O&P
0010	**SELECTIVE DEMOLITION, SOUND CONTROL** R024119-10									
0120	Acoustical enclosure, 4" thick walls & ceiling panels, 8 lb./S.F.	3 Carp	144	.167	SF Surf		8.45		8.45	12.85
0130	10.5 lb./S.F.		128	.188			9.50		9.50	14.50
0140	Reverb chamber, parallel walls, 4" thick		120	.200			10.15		10.15	15.45
0150	Skewed walls, parallel roof, 4" thick		110	.218			11.05		11.05	16.85
0160	Skewed walls/roof, 4" layer/air space		96	.250			12.70		12.70	19.30
0170	Sound-absorbing panels, painted metal, 2'-6" x 8', under 1,000 S.F.		430	.056			2.83		2.83	4.31
0180	Over 1,000 S.F.	▼	480	.050			2.54		2.54	3.86
0190	Flexible transparent curtain, clear	3 Shee	430	.056			3.34		3.34	5.10
0192	50% clear, 50% foam		430	.056			3.34		3.34	5.10
0194	25% clear, 75% foam		430	.056			3.34		3.34	5.10
0196	100% foam		430	.056	▼		3.34		3.34	5.10
0200	Audio-masking sys., incl. speakers, amplfr., signal gnrtr.									
0205	Ceiling mounted, 5,000 S.F.	2 Elec	4800	.003	S.F.		.19		.19	.29
0210	10,000 S.F.		5600	.003			.17		.17	.25
0220	Plenum mounted, 5,000 S.F.		7600	.002			.12		.12	.18
0230	10,000 S.F.		8800	.002			.11		.11	.16

13 05 05.70 Selective Demolition, Special Purpose Rooms

		Crew	Daily Output	Labor-Hours	Unit	Material	Labor	Equipment	Total	Total Incl O&P
0010	**SELECTIVE DEMOLITION, SPECIAL PURPOSE ROOMS** R024119-10									
0100	Audiometric rooms, under 500 S.F. surface	4 Carp	200	.160	SF Surf		8.10		8.10	12.35
0110	Over 500 S.F. surface	"	240	.133	"		6.75		6.75	10.30

For customer support on your Building Construction Costs with RSMeans data, call 800.448.8182.

455

13 05 05.70 Selective Demolition, Special Purpose Rooms		Crew	Daily Output	Labor-Hours	Unit	Material	2018 Bare Costs Labor	Equipment	Total	Total Incl O&P
0200	Clean rooms, 12' x 12' soft wall, class 100	1 Carp	.30	26.667	Ea.		1,350		1,350	2,050
0210	Class 1,000		.30	26.667			1,350		1,350	2,050
0220	Class 10,000		.35	22.857			1,150		1,150	1,775
0230	Class 100,000		.35	22.857			1,150		1,150	1,775
0300	Darkrooms, shell complete, 8' high	2 Carp	220	.073	SF Flr.		3.69		3.69	5.60
0310	12' high		110	.145	"		7.35		7.35	11.25
0350	Darkrooms doors, mini-cylindrical, revolving		4	4	Ea.		203		203	310
0400	Music room, practice modular		140	.114	SF Surf		5.80		5.80	8.80
0500	Refrigeration structures and finishes									
0510	Wall finish, 2 coat Portland cement plaster, 1/2" thick	1 Clab	200	.040	S.F.		1.59		1.59	2.43
0520	Fiberglass panels, 1/8" thick		400	.020			.80		.80	1.21
0530	Ceiling finish, polystyrene plastic, 1" to 2" thick		500	.016			.64		.64	.97
0540	4" thick		450	.018			.71		.71	1.08
0550	Refrigerator, prefab aluminum walk-in, 7'-6" high, 6' x 6' OD	2 Carp	100	.160	SF Flr.		8.10		8.10	12.35
0560	10' x 10' OD		160	.100			5.05		5.05	7.70
0570	Over 150 S.F.		200	.080			4.06		4.06	6.20
0600	Sauna, prefabricated, including heater & controls, 7' high, to 30 S.F.		120	.133			6.75		6.75	10.30
0610	To 40 S.F.		140	.114			5.80		5.80	8.80
0620	To 60 S.F.		175	.091			4.64		4.64	7.05
0630	To 100 S.F.		220	.073			3.69		3.69	5.60
0640	To 130 S.F.		250	.064			3.24		3.24	4.94
0650	Steam bath, heater, timer, head, single, to 140 C.F.	1 Plum	2.20	3.636	Ea.		226		226	340
0660	To 300 C.F.		2.20	3.636			226		226	340
0670	Steam bath, comm. size, w/blow-down assembly, to 800 C.F.		1.80	4.444			276		276	415
0680	To 2500 C.F.		1.60	5			310		310	470
0690	Steam bath, comm. size, multiple, for motels, apts, 500 C.F., 2 baths		2	4			249		249	375
0700	1,000 C.F., 4 baths		1.40	5.714			355		355	535

13 05 05.75 Selective Demolition, Storage Tanks

		Crew	Daily Output	Labor-Hours	Unit	Material	Labor	Equipment	Total	Total Incl O&P
0010	**SELECTIVE DEMOLITION, STORAGE TANKS**									
0500	Steel tank, single wall, above ground, not incl. fdn., pumps or piping									
0510	Single wall, 275 gallon R024119-10	Q-1	3	5.333	Ea.		298		298	450
0520	550 thru 2,000 gallon	B-34P	2	12			645	275	920	1,275
0530	5,000 thru 10,000 gallon	B-34Q	2	12			655	530	1,185	1,575
0540	15,000 thru 30,000 gallon	B-34S	2	16			910	1,625	2,535	3,150
0600	Steel tank, double wall, above ground not incl. fdn., pumps & piping									
0620	500 thru 2,000 gallon	B-34P	2	12	Ea.		645	275	920	1,275

13 05 05.85 Selective Demolition, Swimming Pool Equip

		Crew	Daily Output	Labor-Hours	Unit	Material	Labor	Equipment	Total	Total Incl O&P
0010	**SELECTIVE DEMOLITION, SWIMMING POOL EQUIP**									
0020	Diving stand, stainless steel, 3 meter	2 Clab	3	5.333	Ea.		213		213	325
0030	1 meter		5	3.200			128		128	194
0040	Diving board, 16' long, aluminum		5.40	2.963			118		118	180
0050	Fiberglass		5.40	2.963			118		118	180
0070	Ladders, heavy duty, stainless steel, 2 tread		14	1.143			45.50		45.50	69.50
0080	4 tread		12	1.333			53		53	81
0090	Lifeguard chair, stainless steel, fixed		5	3.200			128		128	194
0100	Slide, tubular, fiberglass, aluminum handrails & ladder, 5', straight		4	4			159		159	243
0110	8', curved		6	2.667			106		106	162
0120	10', curved		3	5.333			213		213	325
0130	12' straight, with platform		2.50	6.400			255		255	390
0140	Removable access ramp, stainless steel		4	4			159		159	243
0150	Removable stairs, stainless steel, collapsible		4	4			159		159	243

456

For customer support on your Building Construction Costs with RSMeans data, call 800.448.8182.

13 05 Common Work Results for Special Construction

13 05 05 – Selective Demolition for Special Construction

13 05 05.90 Selective Demolition, Tension Structures

		Crew	Daily Output	Labor-Hours	Unit	Material	2018 Bare Costs Labor	Equipment	Total	Total Incl O&P
0010	**SELECTIVE DEMOLITION, TENSION STRUCTURES**									
0020	Steel/alum. frame, fabric shell, 60' clear span, 6,000 S.F.	B-41	2000	.022	SF Flr.		.91	.17	1.08	1.57
0030	12,000 S.F.		2200	.020			.83	.16	.99	1.43
0040	80' clear span, 20,800 S.F.	↓	2440	.018			.75	.14	.89	1.28
0050	100' clear span, 10,000 S.F.	L-5	4350	.013			.71	.14	.85	1.30
0060	26,000 S.F.		4600	.012			.67	.13	.80	1.22
0070	36,000 S.F.	↓	5000	.011	↓		.62	.12	.74	1.13

13 05 05.95 Selective Demo, X-Ray/Radio Freq Protection

		Crew	Daily Output	Labor-Hours	Unit	Material	2018 Bare Costs Labor	Equipment	Total	Total Incl O&P
0010	**SELECTIVE DEMO, X-RAY/RADIO FREQ PROTECTION**									
0020	Shielding lead, lined door frame, excl. hdwe., 1/16" thick	1 Clab	4.80	1.667	Ea.		66.50		66.50	101
0030	Lead sheets, 1/16" thick R024119-10	2 Clab	270	.059	S.F.		2.36		2.36	3.60
0040	1/8" thick		240	.067			2.66		2.66	4.05
0050	Lead shielding, 1/4" thick		270	.059			2.36		2.36	3.60
0060	1/2" thick	↓	240	.067	↓		2.66		2.66	4.05
0070	Lead glass, 1/4" thick, 2.0 mm LE, 12" x 16"	2 Glaz	16	1	Ea.		48.50		48.50	73.50
0080	24" x 36"		8	2			97		97	147
0090	36" x 60"		4	4			194		194	294
0100	Lead glass window frame, with 1/16" lead & voice passage, 36" x 60"		4	4			194		194	294
0110	Lead glass window frame, 24" x 36"	↓	8	2			97		97	147
0120	Lead gypsum board, 5/8" thick with 1/16" lead	2 Clab	320	.050	S.F.		1.99		1.99	3.04
0130	1/8" lead		280	.057			2.28		2.28	3.47
0140	1/32" lead		400	.040	↓		1.59		1.59	2.43
0150	Butt joints, 1/8" lead or thicker, 2" x 7' long batten strip		480	.033	Ea.		1.33		1.33	2.02
0160	X-ray protection, average radiography room, up to 300 S.F., 1/16" lead, min		.50	32	Total		1,275		1,275	1,950
0170	Maximum		.30	53.333			2,125		2,125	3,225
0180	Deep therapy X-ray room, 250 kV cap, up to 300 S.F., 1/4" lead, min		.20	80			3,200		3,200	4,850
0190	Maximum		.12	133	↓		5,325		5,325	8,100
0880	Radio frequency shielding, prefab or screen-type copper or steel, minimum		360	.044	SF Surf		1.77		1.77	2.70
0890	Average		310	.052			2.06		2.06	3.13
0895	Maximum	↓	290	.055	↓		2.20		2.20	3.35

13 11 Swimming Pools

13 11 13 – Below-Grade Swimming Pools

13 11 13.50 Swimming Pools

		Crew	Daily Output	Labor-Hours	Unit	Material	2018 Bare Costs Labor	Equipment	Total	Total Incl O&P
0010	**SWIMMING POOLS** Residential in-ground, vinyl lined									
0020	Concrete sides, w/equip, sand bottom	B-52	300	.187	SF Surf	26	8.65	2.01	36.66	44.50
0100	Metal or polystyrene sides R131113-20	B-14	410	.117		22	4.93	.76	27.69	32.50
0200	Add for vermiculite bottom				↓	1.67			1.67	1.84
0500	Gunite bottom and sides, white plaster finish									
0600	12' x 30' pool	B-52	145	.386	SF Surf	49	17.90	4.15	71.05	85
0720	16' x 32' pool		155	.361		44	16.75	3.88	64.63	78.50
0750	20' x 40' pool	↓	250	.224	↓	39.50	10.40	2.41	52.31	61.50
0810	Concrete bottom and sides, tile finish									
0820	12' x 30' pool	B-52	80	.700	SF Surf	49.50	32.50	7.55	89.55	111
0830	16' x 32' pool		95	.589		40.50	27.50	6.35	74.35	93.50
0840	20' x 40' pool	↓	130	.431	↓	32.50	20	4.63	57.13	71
1100	Motel, gunite with plaster finish, incl. medium									
1150	capacity filtration & chlorination	B-52	115	.487	SF Surf	60	22.50	5.25	87.75	106
1200	Municipal, gunite with plaster finish, incl. high									
1250	capacity filtration & chlorination	B-52	100	.560	SF Surf	77.50	26	6	109.50	132

For customer support on your Building Construction Costs with RSMeans data, call 800.448.8182.

457

13 11 Swimming Pools

13 11 13 – Below-Grade Swimming Pools

13 11 13.50 Swimming Pools

		Crew	Daily Output	Labor-Hours	Unit	Material	2018 Bare Costs Labor	Equipment	Total	Total Incl O&P
1350	Add for formed gutters				L.F.	114			114	126
1360	Add for stainless steel gutters				"	340			340	370
1600	For water heating system, see Section 23 52 28.10									
1700	Filtration and deck equipment only, as % of total				Total				20%	20%
1800	Deck equipment, rule of thumb, 20' x 40' pool				SF Pool				1.18	1.30
1900	5,000 S.F. pool				"				1.73	1.90
3000	Painting pools, preparation + 3 coats, 20' x 40' pool, epoxy	2 Pord	.33	48.485	Total	1,725	2,075		3,800	5,000
3100	Rubber base paint, 18 gallons	"	.33	48.485		1,325	2,075		3,400	4,550
3500	42' x 82' pool, 75 gallons, epoxy paint	3 Pord	.14	171		7,300	7,300		14,600	19,100
3600	Rubber base paint	"	.14	171		5,500	7,300		12,800	17,000

13 11 23 – On-Grade Swimming Pools

13 11 23.50 Swimming Pools

		Crew	Daily Output	Labor-Hours	Unit	Material	2018 Bare Costs Labor	Equipment	Total	Total Incl O&P
0010	**SWIMMING POOLS** Residential above ground, steel construction									
0100	Round, 15' diam.	B-80A	3	8	Ea.	785	320	79.50	1,184.50	1,450
0120	18' diam.		2.50	9.600		930	385	95	1,410	1,725
0140	21' diam.		2	12		1,075	480	119	1,674	2,025
0160	24' diam.		1.80	13.333		1,175	530	132	1,837	2,225
0180	27' diam.		1.50	16		1,500	640	159	2,299	2,825
0200	30' diam.		1	24		1,650	955	238	2,843	3,500
0220	Oval, 12' x 24'		2.30	10.435		1,600	415	104	2,119	2,525
0240	15' x 30'		1.80	13.333		2,525	530	132	3,187	3,725
0260	18' x 33'		1	24		2,775	955	238	3,968	4,750

13 11 46 – Swimming Pool Accessories

13 11 46.50 Swimming Pool Equipment

		Crew	Daily Output	Labor-Hours	Unit	Material	2018 Bare Costs Labor	Equipment	Total	Total Incl O&P
0010	**SWIMMING POOL EQUIPMENT**									
0020	Diving stand, stainless steel, 3 meter	2 Carp	.40	40	Ea.	16,700	2,025		18,725	21,500
0300	1 meter		2.70	5.926		10,100	300		10,400	11,600
0600	Diving boards, 16' long, aluminum		2.70	5.926		4,325	300		4,625	5,200
0700	Fiberglass		2.70	5.926		3,375	300		3,675	4,150
0800	14' long, aluminum		2.70	5.926		3,975	300		4,275	4,825
0850	Fiberglass		2.70	5.926		3,350	300		3,650	4,125
1100	Bulkhead, movable, PVC, 8'-2" wide	2 Clab	8	2		2,425	79.50		2,504.50	2,800
1120	7'-9" wide		8	2		2,075	79.50		2,154.50	2,425
1140	7'-3" wide		8	2		2,125	79.50		2,204.50	2,450
1160	6'-9" wide		8	2		2,125	79.50		2,204.50	2,450
1200	Ladders, heavy duty, stainless steel, 2 tread	2 Carp	7	2.286		880	116		996	1,150
1500	4 tread		6	2.667		775	135		910	1,050
1800	Lifeguard chair, stainless steel, fixed		2.70	5.926		3,400	300		3,700	4,200
1900	Portable					2,875			2,875	3,175
2100	Lights, underwater, 12 volt, with transformer, 300 watt	1 Elec	1	8		360	465		825	1,100
2200	110 volt, 500 watt, standard		1	8		295	465		760	1,025
2400	Low water cutoff type		1	8		282	465		747	1,000
2800	Heaters, see Section 23 52 28.10									
3000	Pool covers, reinforced vinyl	3 Clab	1800	.013	S.F.	1.20	.53		1.73	2.13
3050	Automatic, electric								8.75	9.65
3100	Vinyl, for winter, 400 S.F. max pool surface	3 Clab	3200	.008		.22	.30		.52	.70
3200	With water tubes, 400 S.F. max pool surface	"	3000	.008		.37	.32		.69	.90
3250	Sealed air bubble polyethylene solar blanket, 16 mils					.37			.37	.41
3300	Slides, tubular, fiberglass, aluminum handrails & ladder, 5'-0", straight	2 Carp	1.60	10	Ea.	3,700	505		4,205	4,825
3320	8'-0", curved		3	5.333		7,225	270		7,495	8,350
3400	10'-0", curved		1	16		27,100	810		27,910	31,000
3420	12'-0", straight with platform		1.20	13.333		15,300	675		15,975	17,900

For customer support on your Building Construction Costs with RSMeans data, call 800.448.8182.

13 11 Swimming Pools

13 11 46 – Swimming Pool Accessories

	13 11 46.50 Swimming Pool Equipment	Crew	Daily Output	Labor-Hours	Unit	Material	2018 Bare Costs Labor	Equipment	Total	Total Incl O&P
4500	Hydraulic lift, movable pool bottom, single ram									
4520	Under 1,000 S.F. area	L-9	72	.500	S.F.	180	22.50		202.50	233
4600	Four ram lift, over 1,000 S.F.	"	109	.330	"	150	14.95		164.95	188
5000	Removable access ramp, stainless steel	2 Clab	2	8	Ea.	5,400	320		5,720	6,425

13 12 Fountains

13 12 13 – Exterior Fountains

13 12 13.10 Outdoor Fountains

		Crew	Daily Output	Labor-Hours	Unit	Material	Labor	Equipment	Total	Total Incl O&P
0010	**OUTDOOR FOUNTAINS**									
0100	Outdoor fountain, 48" high with bowl and figures	2 Clab	2	8	Ea.	570	320		890	1,100
0200	Commercial, concrete or cast stone, 40-60" H, simple		2	8		775	320		1,095	1,350
0220	Average		2	8		1,350	320		1,670	1,950
0240	Ornate		2	8		2,625	320		2,945	3,350
0260	Metal, 72" high		2	8		1,200	320		1,520	1,800
0280	90" high		2	8		1,875	320		2,195	2,550
0300	120" high		2	8		5,000	320		5,320	5,975
0320	Resin or fiberglass, 40-60" H, wall type		2	8		480	320		800	1,025
0340	Waterfall type		2	8		1,000	320		1,320	1,575

13 12 23 – Interior Fountains

13 12 23.10 Indoor Fountains

		Crew	Daily Output	Labor-Hours	Unit	Material	Labor	Equipment	Total	Total Incl O&P
0010	**INDOOR FOUNTAINS**									
0100	Commercial, floor type, resin or fiberglass, lighted, cascade type	2 Clab	2	8	Ea.	286	320		606	800
0120	Tiered type		2	8		285	320		605	800
0140	Waterfall type		2	8		275	320		595	785

13 17 Tubs and Pools

13 17 33 – Whirlpool Tubs

13 17 33.10 Whirlpool Bath

		Crew	Daily Output	Labor-Hours	Unit	Material	Labor	Equipment	Total	Total Incl O&P
0010	**WHIRLPOOL BATH**									
6000	Whirlpool, bath with vented overflow, molded fiberglass									
6100	66" x 36" x 24"	Q-1	1	16	Ea.	1,025	895		1,920	2,475

13 18 Ice Rinks

13 18 13 – Ice Rink Floor Systems

13 18 13.50 Ice Skating

		Crew	Daily Output	Labor-Hours	Unit	Material	Labor	Equipment	Total	Total Incl O&P
0010	**ICE SKATING** Equipment incl. refrigeration, plumbing & cooling									
0020	coils & concrete slab, 85' x 200' rink									
0300	55° system, 5 mos., 100 ton				Total	604,000			604,000	664,000
0700	90° system, 12 mos., 135 ton				"	682,500			682,500	751,000
1200	Subsoil heating system (recycled from compressor), 85' x 200'	Q-7	.27	119	Ea.	42,000	7,100		49,100	57,000
1300	Subsoil insulation, 2 lb. polystyrene with vapor barrier, 85' x 200'	2 Carp	.14	114	"	31,500	5,800		37,300	43,500

13 18 16 – Ice Rink Dasher Boards

13 18 16.50 Ice Rink Dasher Boards

		Crew	Daily Output	Labor-Hours	Unit	Material	Labor	Equipment	Total	Total Incl O&P
0010	**ICE RINK DASHER BOARDS**									
1000	Dasher boards, 1/2" H.D. polyethylene faced steel frame, 3' acrylic									
1020	screen at sides, 5' acrylic ends, 85' x 200'	F-5	.06	533	Ea.	142,000	27,300		169,300	197,500

13 18 Ice Rinks

13 18 16 – Ice Rink Dasher Boards

13 18 16.50 Ice Rink Dasher Boards	Crew	Daily Output	Labor-Hours	Unit	Material	2018 Bare Costs Labor	Equipment	Total	Total Incl O&P
1100	Fiberglass & aluminum construction, same sides and ends	F-5	.06	533	Ea.	163,000	27,300		190,300	220,500

13 21 Controlled Environment Rooms

13 21 13 – Clean Rooms

13 21 13.50 Clean Room Components

		Crew	Daily Output	Labor-Hours	Unit	Material	2018 Bare Costs Labor	Equipment	Total	Total Incl O&P
0010	**CLEAN ROOM COMPONENTS**									
1100	Clean room, soft wall, 12' x 12', Class 100	1 Carp	.18	44.444	Ea.	16,600	2,250		18,850	21,700
1110	Class 1,000		.18	44.444		15,700	2,250		17,950	20,600
1120	Class 10,000		.21	38.095		13,600	1,925		15,525	17,900
1130	Class 100,000		.21	38.095		13,000	1,925		14,925	17,300
2800	Ceiling grid support, slotted channel struts 4'-0" OC, ea. way				S.F.				5.90	6.50
3000	Ceiling panel, vinyl coated foil on mineral substrate									
3020	Sealed, non-perforated				S.F.				1.27	1.40
4000	Ceiling panel seal, silicone sealant, 150 L.F./gal.	1 Carp	150	.053	L.F.	.29	2.70		2.99	4.44
4100	Two sided adhesive tape	"	240	.033	"	.13	1.69		1.82	2.71
4200	Clips, one per panel				Ea.	1.02			1.02	1.12
6000	HEPA filter, 2' x 4', 99.97% eff., 3" dp beveled frame (silicone seal)					390			390	425
6040	6" deep skirted frame (channel seal)					475			475	525
6100	99.99% efficient, 3" deep beveled frame (silicone seal)					495			495	545
6140	6" deep skirted frame (channel seal)					480			480	525
6200	99.999% efficient, 3" deep beveled frame (silicone seal)					535			535	590
6240	6" deep skirted frame (channel seal)					515			515	565
7000	Wall panel systems, including channel strut framing									
7020	Polyester coated aluminum, particle board				S.F.				18.20	18.20
7100	Porcelain coated aluminum, particle board								32	32
7400	Wall panel support, slotted channel struts, to 12' high								16.35	16.35

13 21 26 – Cold Storage Rooms

13 21 26.50 Refrigeration

		Crew	Daily Output	Labor-Hours	Unit	Material	2018 Bare Costs Labor	Equipment	Total	Total Incl O&P
0010	**REFRIGERATION**									
0020	Curbs, 12" high, 4" thick, concrete	2 Carp	58	.276	L.F.	3.48	14		17.48	25.50
1000	Doors, see Section 08 34 13.10									
2400	Finishes, 2 coat Portland cement plaster, 1/2" thick	1 Plas	48	.167	S.F.	.94	7.75		8.69	12.65
2500	For galvanized reinforcing mesh, add	1 Lath	335	.024		.64	1.17		1.81	2.43
2700	3/16" thick latex cement	1 Plas	88	.091		1.62	4.22		5.84	8.15
2900	For glass cloth reinforced ceilings, add	"	450	.018		.62	.83		1.45	1.92
3100	Fiberglass panels, 1/8" thick	1 Carp	149.45	.054		2.16	2.71		4.87	6.50
3200	Polystyrene, plastic finish ceiling, 1" thick		274	.029		2.77	1.48		4.25	5.30
3400	2" thick		274	.029		3.14	1.48		4.62	5.70
3500	4" thick		219	.037		4.08	1.85		5.93	7.30
3800	Floors, concrete, 4" thick	1 Cefi	93	.086		1.57	4.09		5.66	7.80
3900	6" thick	"	85	.094		2.46	4.48		6.94	9.35
4000	Insulation, 1" to 6" thick, cork				B.F.	1.49			1.49	1.64
4100	Urethane					.61			.61	.67
4300	Polystyrene, regular					.57			.57	.63
4400	Bead board					.27			.27	.30
4600	Installation of above, add per layer	2 Carp	657.60	.024	S.F.	.46	1.23		1.69	2.39
4700	Wall and ceiling juncture		298.90	.054	L.F.	1.49	2.71		4.20	5.75
4900	Partitions, galvanized sandwich panels, 4" thick, stock		219.20	.073	S.F.	10.15	3.70		13.85	16.85
5000	Aluminum or fiberglass		219.20	.073	"	6.80	3.70		10.50	13.15
5200	Prefab walk-in, 7'-6" high, aluminum, incl. refrigeration, door & floor									

For customer support on your Building Construction Costs with RSMeans data, call 800.448.8182.

13 21 Controlled Environment Rooms

13 21 26 – Cold Storage Rooms

13 21 26.50 Refrigeration

		Crew	Daily Output	Labor-Hours	Unit	Material	2018 Bare Costs Labor	Equipment	Total	Total Incl O&P
5210	not incl. partitions, 6' x 6'	2 Carp	54.80	.292	SF Flr.	111	14.80		125.80	145
5500	10' x 10'		82.20	.195		89.50	9.85		99.35	114
5700	12' x 14'		109.60	.146		80	7.40		87.40	99.50
5800	12' x 20'		109.60	.146		114	7.40		121.40	137
6100	For 8'-6" high, add					5%				
6300	Rule of thumb for complete units, w/o doors & refrigeration, cooler	2 Carp	146	.110		165	5.55		170.55	189
6400	Freezer		109.60	.146		119	7.40		126.40	142
6600	Shelving, plated or galvanized, steel wire type		360	.044	SF Hor.	8.80	2.25		11.05	13.15
6700	Slat shelf type		375	.043		17.75	2.16		19.91	23
6900	For stainless steel shelving, add					300%				
7000	Vapor barrier, on wood walls	2 Carp	1644	.010	S.F.	.19	.49		.68	.96
7200	On masonry walls	"	1315	.012	"	.34	.62		.96	1.31
7500	For air curtain doors, see Section 23 34 33.10									

13 21 48 – Sound-Conditioned Rooms

13 21 48.10 Anechoic Chambers

		Crew	Daily Output	Labor-Hours	Unit	Material	2018 Bare Costs Labor	Equipment	Total	Total Incl O&P
0010	**ANECHOIC CHAMBERS** Standard units, 7' ceiling heights									
0100	Area for pricing is net inside dimensions									
0300	200 cycles per second cutoff, 25 S.F. floor area				SF Flr.	1,850			1,850	2,050
0400	50 S.F.								1,050	1,150
0600	75 S.F.								1,000	1,100
0700	100 S.F.					1,400			1,400	1,550
0900	For 150 cycles per second cutoff, add to 100 S.F. room								30%	30%
1000	For 100 cycles per second cutoff, add to 100 S.F. room								45%	45%

13 21 48.15 Audiometric Rooms

		Crew	Daily Output	Labor-Hours	Unit	Material	2018 Bare Costs Labor	Equipment	Total	Total Incl O&P
0010	**AUDIOMETRIC ROOMS**									
0020	Under 500 S.F. surface	4 Carp	98	.327	SF Surf	60	16.55		76.55	91
0100	Over 500 S.F. surface	"	120	.267	"	58.50	13.50		72	85

13 21 53 – Darkrooms

13 21 53.50 Darkrooms

		Crew	Daily Output	Labor-Hours	Unit	Material	2018 Bare Costs Labor	Equipment	Total	Total Incl O&P
0010	**DARKROOMS**									
0020	Shell, complete except for door, 64 S.F., 8' high	2 Carp	128	.125	SF Flr.	55.50	6.35		61.85	70.50
0100	12' high		64	.250		72.50	12.70		85.20	99
0500	120 S.F. floor, 8' high		120	.133		40.50	6.75		47.25	55
0600	12' high		60	.267		54.50	13.50		68	80.50
0800	240 S.F. floor, 8' high		120	.133		28	6.75		34.75	41.50
0900	12' high		60	.267		40	13.50		53.50	64.50
1200	Mini-cylindrical, revolving, unlined, 4' diameter		3.50	4.571	Ea.	3,075	232		3,307	3,725
1400	5'-6" diameter		2.50	6.400		4,875	325		5,200	5,850
1600	Add for lead lining, inner cylinder, 1/32" thick					1,650			1,650	1,825
1700	1/16" thick					4,450			4,450	4,875
1800	Add for lead lining, inner and outer cylinder, 1/32" thick					3,050			3,050	3,375
1900	1/16" thick					6,825			6,825	7,500
2000	For darkroom door, see Section 08 34 36.10									

13 21 56 – Music Rooms

13 21 56.50 Music Rooms

		Crew	Daily Output	Labor-Hours	Unit	Material	2018 Bare Costs Labor	Equipment	Total	Total Incl O&P
0010	**MUSIC ROOMS**									
0020	Practice room, modular, perforated steel, under 500 S.F.	2 Carp	70	.229	SF Surf	38.50	11.60		50.10	60
0100	Over 500 S.F.	"	80	.200	"	32.50	10.15		42.65	51.50

13 24 Special Activity Rooms

13 24 16 – Saunas

13 24 16.50 Saunas and Heaters

		Crew	Daily Output	Labor-Hours	Unit	Material	2018 Bare Costs Labor	Equipment	Total	Total Incl O&P
0010	**SAUNAS AND HEATERS**									
0020	Prefabricated, incl. heater & controls, 7' high, 6' x 4', C/C	L-7	2.20	12.727	Ea.	5,125	620		5,745	6,600
0050	6' x 4', C/P		2	14		4,325	680		5,005	5,775
0400	6' x 5', C/C		2	14		5,950	680		6,630	7,575
0450	6' x 5', C/P		2	14		5,175	680		5,855	6,725
0600	6' x 6', C/C		1.80	15.556		7,000	755		7,755	8,850
0650	6' x 6', C/P		1.80	15.556		5,425	755		6,180	7,100
0800	6' x 9', C/C		1.60	17.500		7,900	850		8,750	10,000
0850	6' x 9', C/P		1.60	17.500		6,900	850		7,750	8,900
1000	8' x 12', C/C		1.10	25.455		10,500	1,250		11,750	13,500
1050	8' x 12', C/P		1.10	25.455		9,300	1,250		10,550	12,100
1200	8' x 8', C/C		1.40	20		8,275	975		9,250	10,600
1250	8' x 8', C/P		1.40	20		7,375	975		8,350	9,575
1400	8' x 10', C/C		1.20	23.333		9,050	1,125		10,175	11,700
1450	8' x 10', C/P		1.20	23.333		7,975	1,125		9,100	10,500
1600	10' x 12', C/C		1	28		12,200	1,375		13,575	15,500
1650	10' x 12', C/P	▼	1	28		10,800	1,375		12,175	14,000
1700	Door only, cedar, 2'x6', with 1' x 4' tempered insulated glass window	2 Carp	3.40	4.706		620	239		859	1,050
1800	Prehung, incl. jambs, pulls & hardware	"	12	1.333		680	67.50		747.50	855
2500	Heaters only (incl. above), wall mounted, to 200 C.F.					775			775	855
2750	To 300 C.F.					1,050			1,050	1,150
3000	Floor standing, to 720 C.F., 10,000 watts, w/controls	1 Elec	3	2.667		2,500	155		2,655	2,975
3250	To 1,000 C.F., 16,000 watts	"	3	2.667	▼	3,625	155		3,780	4,200

13 24 26 – Steam Baths

13 24 26.50 Steam Baths and Components

		Crew	Daily Output	Labor-Hours	Unit	Material	2018 Bare Costs Labor	Equipment	Total	Total Incl O&P
0010	**STEAM BATHS AND COMPONENTS**									
0020	Heater, timer & head, single, to 140 C.F.	1 Plum	1.20	6.667	Ea.	2,325	415		2,740	3,175
0500	To 300 C.F.		1.10	7.273		2,600	450		3,050	3,525
1000	Commercial size, with blow-down assembly, to 800 C.F.		.90	8.889		5,725	550		6,275	7,125
1500	To 2,500 C.F.	▼	.80	10		8,000	620		8,620	9,725
2000	Multiple, motels, apts., 2 baths, w/blow-down assm., 500 C.F.	Q-1	1.30	12.308		6,675	690		7,365	8,375
2500	4 baths	"	.70	22.857		10,700	1,275		11,975	13,600
2700	Conversion unit for residential tub, including door				▼	3,750			3,750	4,100

13 28 Athletic and Recreational Special Construction

13 28 33 – Athletic and Recreational Court Walls

13 28 33.50 Sport Court

		Crew	Daily Output	Labor-Hours	Unit	Material	2018 Bare Costs Labor	Equipment	Total	Total Incl O&P
0010	**SPORT COURT**									
0020	Floors, No. 2 & better maple, 25/32" thick				SF Flr.				6.05	6.65
0100	Walls, laminated plastic bonded to galv. steel studs				SF Wall				7.70	8.45
0300	Squash, regulation court in existing building, minimum				Court	40,300			40,300	44,400
0400	Maximum				"	44,900			44,900	49,400
0450	Rule of thumb for components:									
0470	Walls	3 Carp	.15	160	Court	12,100	8,100		20,200	25,700
0500	Floor	"	.25	96		9,575	4,875		14,450	17,900
0550	Lighting	2 Elec	.60	26.667		2,300	1,550		3,850	4,850
0600	Handball, racquetball court in existing building, minimum	C-1	.20	160		43,700	7,675		51,375	59,500
0800	Maximum	"	.10	320		47,200	15,400		62,600	75,500
0900	Rule of thumb for components: walls	3 Carp	.12	200		13,800	10,100		23,900	30,600
1000	Floor	▼	.25	96		9,575	4,875		14,450	17,900

For customer support on your Building Construction Costs with RSMeans data, call 800.448.8182.

13 28 Athletic and Recreational Special Construction

13 28 33 – Athletic and Recreational Court Walls

13 28 33.50 Sport Court	Crew	Daily Output	Labor-Hours	Unit	Material	2018 Bare Costs Labor	Equipment	Total	Total Incl O&P
1100 Ceiling	3 Carp	.33	72.727	Court	4,600	3,675		8,275	10,700
1200 Lighting	2 Elec	.60	26.667	↓	2,425	1,550		3,975	4,975

13 31 Fabric Structures

13 31 13 – Air-Supported Fabric Structures

13 31 13.09 Air Supported Tank Covers

		Crew	Daily Output	Labor-Hours	Unit	Material	2018 Bare Costs Labor	Equipment	Total	Total Incl O&P
0010	**AIR SUPPORTED TANK COVERS**, vinyl polyester									
0100	Scrim, double layer, with hardware, blower, standby & controls									
0200	Round, 75' diameter	B-2	4500	.009	S.F.	12.15	.36		12.51	13.95
0300	100' diameter		5000	.008		11.05	.32		11.37	12.70
0400	150' diameter		5000	.008		8.75	.32		9.07	10.15
0500	Rectangular, 20' x 20'		4500	.009		24	.36		24.36	27
0600	30' x 40'		4500	.009		24	.36		24.36	27
0700	50' x 60'	↓	4500	.009		24	.36		24.36	27
0800	For single wall construction, deduct, minimum					.82			.82	.90
0900	Maximum					2.43			2.43	2.67
1000	For maximum resistance to atmosphere or cold, add				↓	1.18			1.18	1.30
1100	For average shipping charges, add				Total	2,025			2,025	2,225

13 31 13.13 Single-Walled Air-Supported Structures

		Crew	Daily Output	Labor-Hours	Unit	Material	2018 Bare Costs Labor	Equipment	Total	Total Incl O&P
0010	**SINGLE-WALLED AIR-SUPPORTED STRUCTURES** R133113-10									
0020	Site preparation, incl. anchor placement and utilities	B-11B	1000	.016	SF Flr.	1.18	.73	.29	2.20	2.71
0030	For concrete, see Section 03 30 53.40									
0050	Warehouse, polyester/vinyl fabric, 28 oz., over 10 yr. life, welded									
0060	Seams, tension cables, primary & auxiliary inflation system,									
0070	airlock, personnel doors and liner									
0100	5,000 S.F.	4 Clab	5000	.006	SF Flr.	27	.26		27.26	30.50
0250	12,000 S.F.	"	6000	.005		19.55	.21		19.76	22
0400	24,000 S.F.	8 Clab	12000	.005		13.75	.21		13.96	15.45
0500	50,000 S.F.	"	12500	.005	↓	12.75	.20		12.95	14.35
0700	12 oz. reinforced vinyl fabric, 5 yr. life, sewn seams,									
0710	accordion door, including liner									
0750	3,000 S.F.	4 Clab	3000	.011	SF Flr.	13.45	.43		13.88	15.45
0800	12,000 S.F.	"	6000	.005		11.45	.21		11.66	12.90
0850	24,000 S.F.	8 Clab	12000	.005		9.70	.21		9.91	11
0950	Deduct for single layer					1.06			1.06	1.17
1000	Add for welded seams					1.53			1.53	1.68
1050	Add for double layer, welded seams included				↓	3.06			3.06	3.37
1250	Tedlar/vinyl fabric, 28 oz., with liner, over 10 yr. life,									
1260	incl. overhead and personnel doors									
1300	3,000 S.F.	4 Clab	3000	.011	SF Flr.	25.50	.43		25.93	28.50
1450	12,000 S.F.	"	6000	.005		17.80	.21		18.01	19.90
1550	24,000 S.F.	8 Clab	12000	.005		13.75	.21		13.96	15.45
1700	Deduct for single layer				↓	2.04			2.04	2.24
2250	Greenhouse/shelter, woven polyethylene with liner, 2 yr. life,									
2260	sewn seams, including doors									
2300	3,000 S.F.	4 Clab	3000	.011	SF Flr.	16.30	.43		16.73	18.60
2350	12,000 S.F.	"	6000	.005		14.30	.21		14.51	16
2450	24,000 S.F.	8 Clab	12000	.005		12.25	.21		12.46	13.75
2550	Deduct for single layer				↓	1			1	1.10
2600	Tennis/gymnasium, polyester/vinyl fabric, 28 oz., over 10 yr. life,									
2610	including thermal liner, heat and lights									

For customer support on your Building Construction Costs with RSMeans data, call 800.448.8182.

463

13 31 Fabric Structures

13 31 13 – Air-Supported Fabric Structures

13 31 13.13 Single-Walled Air-Supported Structures

		Crew	Daily Output	Labor-Hours	Unit	Material	2018 Bare Costs Labor	Equipment	Total	Total Incl O&P
2650	7,200 S.F.	4 Clab	6000	.005	SF Flr.	24.50	.21		24.71	27
2750	13,000 S.F.	"	6500	.005		18.60	.20		18.80	21
2850	Over 24,000 S.F.	8 Clab	12000	.005		17	.21		17.21	19
2860	For low temperature conditions, add					1.17			1.17	1.29
2870	For average shipping charges, add				Total	5,700			5,700	6,275
2900	Thermal liner, translucent reinforced vinyl				SF Flr.	1.17			1.17	1.29
2950	Metalized mylar fabric and mesh, double liner				"	2.42			2.42	2.66
3050	Stadium/convention center, teflon coated fiberglass, heavy weight,									
3060	over 20 yr. life, incl. thermal liner and heating system									
3100	Minimum	9 Clab	26000	.003	SF Flr.	59.50	.11		59.61	65
3110	Maximum	"	19000	.004	"	70.50	.15		70.65	77.50
3400	Doors, air lock, 15' long, 10' x 10'	2 Carp	.80	20	Ea.	21,100	1,025		22,125	24,800
3600	15' x 15'	"	.50	32		31,400	1,625		33,025	37,000
3700	For each added 5' length, add					5,600			5,600	6,175
3900	Revolving personnel door, 6' diameter, 6'-6" high	2 Carp	.80	20		15,700	1,025		16,725	18,900

13 31 23 – Tensioned Fabric Structures

13 31 23.50 Tension Structures

		Crew	Daily Output	Labor-Hours	Unit	Material	2018 Bare Costs Labor	Equipment	Total	Total Incl O&P
0010	**TENSION STRUCTURES** Rigid steel/alum. frame, vinyl coated poly									
0100	Fabric shell, 60' clear span, not incl. foundations or floors									
0200	6,000 S.F.	B-41	1000	.044	SF Flr.	17.60	1.82	.34	19.76	22.50
0300	12,000 S.F.		1100	.040		16.75	1.65	.31	18.71	21.50
0400	80' to 99' clear span, 20,800 S.F.		1220	.036		16.45	1.49	.28	18.22	20.50
0410	100' to 119' clear span, 10,000 S.F.	L-5	2175	.026		17.40	1.42	.27	19.09	22
0430	26,000 S.F.		2300	.024		16.10	1.34	.26	17.70	20
0450	36,000 S.F.		2500	.022		15.90	1.24	.24	17.38	19.75
0460	120' to 149' clear span, 24,000 S.F.		3000	.019		17.65	1.03	.20	18.88	21.50
0470	150' to 199' clear span, 30,000 S.F.		6000	.009		18.40	.51	.10	19.01	21
0480	200' clear span, 40,000 S.F.	E-6	8000	.016		23	.87	.24	24.11	27
0500	For roll-up door, 12' x 14', add	L-2	1	16	Ea.	6,975	710		7,685	8,750

13 33 Geodesic Structures

13 33 13 – Geodesic Domes

13 33 13.35 Geodesic Domes

		Crew	Daily Output	Labor-Hours	Unit	Material	2018 Bare Costs Labor	Equipment	Total	Total Incl O&P
0010	**GEODESIC DOMES** Shell only, interlocking plywood panels R133423-30									
0400	30' diameter	F-5	1.60	20	Ea.	24,400	1,025		25,425	28,400
0500	33' diameter		1.14	28.070		25,800	1,425		27,225	30,500
0600	40' diameter		1	32		29,900	1,650		31,550	35,400
0700	45' diameter	F-3	1.13	35.556		31,400	1,850	440	33,690	37,800
0750	56' diameter		1	40		55,000	2,075	495	57,570	64,000
0800	60' diameter		1	40		62,500	2,075	495	65,070	72,000
0850	67' diameter		.80	50		87,500	2,600	620	90,720	101,000
1100	Aluminum panel, with 6" insulation									
1200	100' diameter				SF Flr.	26			26	29
1300	500' diameter				"	25.50			25.50	28
1600	Aluminum framed, plexiglass closure panels									
1700	40' diameter				SF Flr.	64.50			64.50	71
1800	200' diameter				"	60			60	65.50
2100	Aluminum framed, aluminum closure panels									
2200	40' diameter				SF Flr.	21			21	23.50
2300	100' diameter				"	20.50			20.50	22.50

For customer support on your Building Construction Costs with RSMeans data, call 800.448.8182.

13 33 Geodesic Structures

13 33 13 – Geodesic Domes

13 33 13.35 Geodesic Domes

		Crew	Daily Output	Labor-Hours	Unit	Material	2018 Bare Costs Labor	Equipment	Total	Total Incl O&P
2400	200' diameter				SF Flr.	20.50			20.50	22.50
2500	For VRP faced bonded fiberglass insulation, add.				▼				10	10
2700	Aluminum framed, fiberglass sandwich panel closure									
2800	6' diameter	2 Carp	150	.107	SF Flr.	28.50	5.40		33.90	39.50
2900	28' diameter	"	350	.046	"	25.50	2.32		27.82	32

13 34 Fabricated Engineered Structures

13 34 13 – Glazed Structures

13 34 13.13 Greenhouses

		Crew	Daily Output	Labor-Hours	Unit	Material	2018 Bare Costs Labor	Equipment	Total	Total Incl O&P
0010	**GREENHOUSES**, Shell only, stock units, not incl. 2' stub walls,									
0020	foundation, floors, heat or compartments									
0300	Residential type, free standing, 8'-6" long x 7'-6" wide	2 Carp	59	.271	SF Flr.	24	13.75		37.75	47.50
0400	10'-6" wide		85	.188		44.50	9.55		54.05	63.50
0600	13'-6" wide		108	.148		46	7.50		53.50	62
0700	17'-0" wide		160	.100		46	5.05		51.05	58
0900	Lean-to type, 3'-10" wide		34	.471		44	24		68	84.50
1000	6'-10" wide	▼	58	.276	▼	29	14		43	53.50
1500	Commercial, custom, truss frame, incl. equip., plumbing, elec.,									
1550	benches and controls, under 2,000 S.F.				SF Flr.	13.40			13.40	14.75
1700	Over 5,000 S.F.				"	12.35			12.35	13.55
2000	Institutional, custom, rigid frame, including compartments and									
2050	multi-controls, under 500 S.F.				SF Flr.	30			30	33
2150	Over 2,000 S.F.				"	11.20			11.20	12.35
3700	For 1/4" tempered glass, add				SF Surf	1.57			1.57	1.73
3900	Cooling, 1,200 CFM exhaust fan, add				Ea.	330			330	360
4000	7,850 CFM					1,050			1,050	1,150
4200	For heaters, 10 MBH, add					230			230	253
4300	60 MBH, add					800			800	880
4500	For benches, 2' x 8', add					175			175	192
4600	4' x 10', add				▼	213			213	234
4800	For ventilation & humidity control w/4 integrated outlets, add				Total	262			262	288
4900	For environmental controls and automation, 8 outputs, 9 stages, add				"	1,050			1,050	1,150
5100	For humidification equipment, add				Ea.	325			325	360
5200	For vinyl shading, add				S.F.	.42			.42	.46
6000	Geodesic hemisphere, 1/8" plexiglass glazing									
6050	8' diameter	2 Carp	2	8	Ea.	6,450	405		6,855	7,725
6150	24' diameter	▼	.35	45.714		15,000	2,325		17,325	19,900
6250	48' diameter	▼	.20	80	▼	35,000	4,050		39,050	44,600

13 34 13.19 Swimming Pool Enclosures

		Crew	Daily Output	Labor-Hours	Unit	Material	2018 Bare Costs Labor	Equipment	Total	Total Incl O&P
0010	**SWIMMING POOL ENCLOSURES** Translucent, free standing									
0020	not including foundations, heat or light									
0200	Economy	2 Carp	200	.080	SF Hor.	51	4.06		55.06	62
0600	Deluxe	"	70	.229		93	11.60		104.60	120
0700	For motorized roof, 40% opening, solid roof, add					21			21	23
0800	Skylight type roof, add				▼	13.50			13.50	14.85

13 34 16 – Grandstands and Bleachers

13 34 16.13 Grandstands

		Crew	Daily Output	Labor-Hours	Unit	Material	2018 Bare Costs Labor	Equipment	Total	Total Incl O&P
0010	**GRANDSTANDS** Permanent, municipal, including foundation									
0300	Steel, economy				Seat	27			27	29.50
0400	Steel, deluxe					30.50			30.50	33.50

For customer support on your Building Construction Costs with RSMeans data, call 800.448.8182.

465

13 34 Fabricated Engineered Structures

13 34 16 – Grandstands and Bleachers

13 34 16.13 Grandstands

		Crew	Daily Output	Labor-Hours	Unit	Material	2018 Bare Costs Labor	Equipment	Total	Total Incl O&P
0900	Composite, steel, wood and plastic, stock design, economy				Seat	37			37	41
1000	Deluxe			↓		90.50			90.50	100

13 34 16.53 Bleachers

		Crew	Daily Output	Labor-Hours	Unit	Material	2018 Bare Costs Labor	Equipment	Total	Total Incl O&P
0010	**BLEACHERS**									
0020	Bleachers, outdoor, portable, 5 tiers, 42 seats	2 Sswk	120	.133	Seat	71.50	7.30		78.80	91
0100	5 tiers, 54 seats		80	.200		74	10.95		84.95	99.50
0200	10 tiers, 104 seats		120	.133		84.50	7.30		91.80	105
0300	10 tiers, 144 seats	↓	80	.200	↓	69	10.95		79.95	93.50
0500	Permanent bleachers, aluminum seat, steel frame, 24" row									
0600	8 tiers, 80 seats	2 Sswk	60	.267	Seat	73	14.55		87.55	104
0700	8 tiers, 160 seats		48	.333		68.50	18.20		86.70	106
0925	15 tiers, 154 to 165 seats		60	.267		77	14.55		91.55	109
0975	15 tiers, 214 to 225 seats		60	.267		69	14.55		83.55	100
1050	15 tiers, 274 to 285 seats		60	.267		69.50	14.55		84.05	101
1200	Seat backs only, 30" row, fiberglass		160	.100		25	5.45		30.45	36.50
1300	Steel and wood	↓	160	.100	↓	25	5.45		30.45	36.50
1400	NOTE: average seating is 1.5' in width									

13 34 19 – Metal Building Systems

13 34 19.50 Pre-Engineered Steel Buildings

		Crew	Daily Output	Labor-Hours	Unit	Material	2018 Bare Costs Labor	Equipment	Total	Total Incl O&P
0010	**PRE-ENGINEERED STEEL BUILDINGS** R133419-10									
0100	Clear span rigid frame, 26 ga. colored roofing and siding									
0150	20' to 29' wide, 10' eave height	E-2	425	.132	SF Flr.	9.35	7.15	3.97	20.47	26
0160	14' eave height		350	.160		10.10	8.70	4.82	23.62	30.50
0170	16' eave height		320	.175		10.80	9.50	5.25	25.55	33
0180	20' eave height		275	.204		11.80	11.05	6.15	29	37.50
0190	24' eave height		240	.233		13.05	12.65	7.05	32.75	42.50
0200	30' to 49' wide, 10' eave height		535	.105		7.15	5.70	3.15	16	20.50
0300	14' eave height		450	.124		7.70	6.75	3.75	18.20	23.50
0400	16' eave height		415	.135		8.20	7.30	4.07	19.57	25
0500	20' eave height		360	.156		8.90	8.45	4.69	22.04	28.50
0600	24' eave height		320	.175		9.80	9.50	5.25	24.55	32
0700	50' to 100' wide, 10' eave height		770	.073		6.05	3.95	2.19	12.19	15.35
0900	16' eave height		600	.093		6.95	5.05	2.81	14.81	18.85
1000	20' eave height		490	.114		7.55	6.20	3.44	17.19	22
1100	24' eave height	↓	435	.129	↓	8.30	7	3.88	19.18	24.50
1200	Clear span tapered beam frame, 26 ga. colored roofing/siding									
1300	30' to 39' wide, 10' eave height	E-2	535	.105	SF Flr.	8.10	5.70	3.15	16.95	21.50
1400	14' eave height		450	.124		8.95	6.75	3.75	19.45	24.50
1500	16' eave height		415	.135		9.35	7.30	4.07	20.72	26.50
1600	20' eave height		360	.156		10.35	8.45	4.69	23.49	30
1700	40' wide, 10' eave height		600	.093		7.20	5.05	2.81	15.06	19.10
1800	14' eave height		510	.110		7.95	5.95	3.31	17.21	22
1900	16' eave height		475	.118		8.30	6.40	3.55	18.25	23.50
2000	20' eave height		415	.135		9.15	7.30	4.07	20.52	26
2100	50' to 79' wide, 10' eave height		770	.073		6.70	3.95	2.19	12.84	16.10
2200	14' eave height		675	.083		7.30	4.50	2.50	14.30	17.95
2300	16' eave height		635	.088		7.55	4.79	2.66	15	18.90
2400	20' eave height		490	.114		8.95	6.20	3.44	18.59	23.50
2410	80' to 100' wide, 10' eave height		935	.060		5.95	3.25	1.80	11	13.75
2420	14' eave height		750	.075		6.55	4.05	2.25	12.85	16.15
2430	16' eave height		685	.082		6.85	4.44	2.46	13.75	17.35
2440	20' eave height	↓	560	.100		7.35	5.45	3.01	15.81	20

For customer support on your Building Construction Costs with RSMeans data, call 800.448.8182.

13 34 19 – Metal Building Systems

13 34 19.50 Pre-Engineered Steel Buildings	Crew	Daily Output	Labor-Hours	Unit	Material	2018 Bare Costs Labor	Equipment	Total	Total Incl O&P
2460	101' to 120' wide, 10' eave height	E-2	950	.059	SF Flr.	5.45	3.20	1.78	10.43	13.05
2470	14' eave height		770	.073		6.05	3.95	2.19	12.19	15.35
2480	16' eave height		675	.083		6.45	4.50	2.50	13.45	17.05
2490	20' eave height	▼	560	.100	▼	6.90	5.45	3.01	15.36	19.60
2500	Single post 2-span frame, 26 ga. colored roofing and siding									
2600	80' wide, 14' eave height	E-2	740	.076	SF Flr.	6.05	4.11	2.28	12.44	15.70
2700	16' eave height		695	.081		6.45	4.37	2.43	13.25	16.75
2800	20' eave height		625	.090		6.95	4.86	2.70	14.51	18.40
2900	24' eave height		570	.098		7.65	5.35	2.96	15.96	20
3000	100' wide, 14' eave height		835	.067		5.90	3.64	2.02	11.56	14.45
3100	16' eave height		795	.070		5.45	3.82	2.12	11.39	14.45
3200	20' eave height		730	.077		6.65	4.16	2.31	13.12	16.50
3300	24' eave height		670	.084		7.35	4.54	2.52	14.41	18.10
3400	120' wide, 14' eave height		870	.064		6.85	3.49	1.94	12.28	15.30
3500	16' eave height		830	.067		6.10	3.66	2.03	11.79	14.80
3600	20' eave height		765	.073		6.65	3.97	2.21	12.83	16.10
3700	24' eave height	▼	705	.079	▼	7.30	4.31	2.39	14	17.55
3800	Double post 3-span frame, 26 ga. colored roofing and siding									
3900	150' wide, 14' eave height	E-2	925	.061	SF Flr.	4.81	3.29	1.82	9.92	12.55
4000	16' eave height		890	.063		5	3.42	1.90	10.32	13.05
4100	20' eave height		820	.068		5.45	3.71	2.06	11.22	14.20
4200	24' eave height	▼	765	.073	▼	6.05	3.97	2.21	12.23	15.45
4300	Triple post 4-span frame, 26 ga. colored roofing and siding									
4400	160' wide, 14' eave height	E-2	970	.058	SF Flr.	4.68	3.13	1.74	9.55	12.05
4500	16' eave height		930	.060		4.92	3.27	1.81	10	12.65
4600	20' eave height		870	.064		4.84	3.49	1.94	10.27	13.05
4700	24' eave height		815	.069		5.50	3.73	2.07	11.30	14.30
4800	200' wide, 14' eave height		1030	.054		4.33	2.95	1.64	8.92	11.30
4900	16' eave height		995	.056		4.48	3.05	1.70	9.23	11.70
5000	20' eave height		935	.060		4.93	3.25	1.80	9.98	12.60
5100	24' eave height	▼	885	.063	▼	5.55	3.43	1.91	10.89	13.70
5200	Accessory items: add to the basic building cost above									
5250	Eave overhang, 2' wide, 26 ga., with soffit	E-2	360	.156	L.F.	33	8.45	4.69	46.14	55
5300	4' wide, without soffit		300	.187		29	10.15	5.60	44.75	54
5350	With soffit		250	.224		42	12.15	6.75	60.90	73
5400	6' wide, without soffit		250	.224		37.50	12.15	6.75	56.40	68.50
5450	With soffit		200	.280	▼	50	15.20	8.45	73.65	89
5500	Entrance canopy, incl. frame, 4' x 4'		25	2.240	Ea.	515	122	67.50	704.50	840
5550	4' x 8'		19	2.947	"	600	160	89	849	1,025
5600	End wall roof overhang, 4' wide, without soffit		850	.066	L.F.	18.35	3.58	1.99	23.92	28
5650	With soffit	▼	500	.112	"	30	6.10	3.37	39.47	46
5700	Doors, HM self-framing, incl. butts, lockset and trim									
5750	Single leaf, 3070 (3' x 7'), economy	2 Sswk	5	3.200	Opng.	655	175		830	1,000
5800	Deluxe		4	4		695	219		914	1,125
5825	Glazed		4	4		785	219		1,004	1,225
5850	3670 (3'-6" x 7')		4	4		870	219		1,089	1,325
5900	4070 (4' x 7')		3	5.333		1,075	291		1,366	1,650
5950	Double leaf, 6070 (6' x 7')		2	8		1,200	435		1,635	2,050
6000	Glazed		2	8		1,475	435		1,910	2,350
6050	Framing only, for openings, 3' x 7'		4	4		188	219		407	560
6100	10' x 10'		3	5.333		620	291		911	1,150
6150	For windows below, 2020 (2' x 2')		6	2.667		199	146		345	455
6200	4030 (4' x 3')	▼	5	3.200	▼	243	175		418	555

13 34 19.50 Pre-Engineered Steel Buildings	Crew	Daily Output	Labor-Hours	Unit	Material	2018 Bare Costs Labor	Equipment	Total	Total Incl O&P
6250	Flashings, 26 ga., corner or eave, painted	2 Sswk	240	.067	L.F.	4.73	3.64		8.37	11.15
6300	Galvanized		240	.067		4.47	3.64		8.11	10.85
6350	Rake flashing, painted		240	.067		5.10	3.64		8.74	11.55
6400	Galvanized		240	.067		4.79	3.64		8.43	11.20
6450	Ridge flashing, 18" wide, painted		240	.067		6.80	3.64		10.44	13.45
6500	Galvanized		240	.067		7.75	3.64		11.39	14.45
6550	Gutter, eave type, 26 ga., painted		320	.050		7.40	2.73		10.13	12.60
6650	Valley type, between buildings, painted		120	.133		14	7.30		21.30	27.50
6710	Insulation, rated .6 lb. density, unfaced 4" thick, R13	2 Carp	2300	.007	S.F.	.47	.35		.82	1.06
6720	6" thick, R19		2300	.007		.63	.35		.98	1.23
6730	10" thick, R30		2300	.007		1.16	.35		1.51	1.82
6750	Insulation, rated .6 lb. density, poly/scrim/foil (PSF) faced									
6760	4" thick, R13	2 Carp	2300	.007	S.F.	.61	.35		.96	1.21
6770	6" thick, R19		2300	.007		.82	.35		1.17	1.44
6780	9-1/2" thick, R30		2300	.007		1.05	.35		1.40	1.70
6800	Insulation, rated .6 lb. density, vinyl faced 1-1/2" thick, R5		2300	.007		.38	.35		.73	.96
6850	3" thick, R10		2300	.007		.36	.35		.71	.94
6900	4" thick, R13		2300	.007		.48	.35		.83	1.07
6920	6" thick, R19		2300	.007		.62	.35		.97	1.22
6930	10" thick, R30		2300	.007		1.70	.35		2.05	2.41
6950	Foil/scrim/kraft (FSK) faced, 1-1/2" thick, R5		2300	.007		.41	.35		.76	.99
7000	2" thick, R6		2300	.007		.53	.35		.88	1.12
7050	3" thick, R10		2300	.007		.49	.35		.84	1.08
7100	4" thick, R13		2300	.007		.50	.35		.85	1.09
7110	6" thick, R19		2300	.007		.62	.35		.97	1.22
7120	10" thick, R30		2300	.007		.96	.35		1.31	1.60
7150	Metalized polyester/scrim/kraft (PSK) facing,1-1/2" thk, R5		2300	.007		.66	.35		1.01	1.27
7200	2" thick, R6		2300	.007		.78	.35		1.13	1.40
7250	3" thick, R11		2300	.007		.88	.35		1.23	1.51
7300	4" thick, R13		2300	.007		1	.35		1.35	1.64
7310	6" thick, R19		2300	.007		1.25	.35		1.60	1.92
7320	10" thick, R30		2300	.007		1.39	.35		1.74	2.07
7350	Vinyl/scrim/foil (VSF), 1-1/2" thick, R5		2300	.007		.58	.35		.93	1.18
7400	2" thick, R6		2300	.007		.74	.35		1.09	1.35
7450	3" thick, R10		2300	.007		.80	.35		1.15	1.42
7500	4" thick, R13		2300	.007		.99	.35		1.34	1.63
7510	Vinyl/scrim/vinyl (VSV), 4" thick, R13		2300	.007		.60	.35		.95	1.20
7520	6" thick, R19		2300	.007		.76	.35		1.11	1.38
7530	9-1/2" thick, R30		2300	.007		1.02	.35		1.37	1.66
7540	Polyprop/scrim/polyester (PSP), 4" thick, R13		2300	.007		.66	.35		1.01	1.27
7550	6" thick, R19		2300	.007		.82	.35		1.17	1.44
7555	10" thick, R30		2300	.007		1.51	.35		1.86	2.20
7560	Polyprop/scrim/kraft/polyester (PSKP), 4" thick, R13		2300	.007		.63	.35		.98	1.23
7570	6" thick, R19		2300	.007		.88	.35		1.23	1.51
7580	10" thick, R19		2300	.007		1.68	.35		2.03	2.39
7585	Vinyl/scrim/polyester (VSP), 4" thick, R13		2300	.007		.63	.35		.98	1.23
7590	6" thick, R19		2300	.007		.78	.35		1.13	1.40
7600	10" thick, R30		2300	.007		1.22	.35		1.57	1.88
7635	Insulation installation, over the purlin, second layer, up to 4" thick, add						90%			
7640	Insulation installation, between the purlins, up to 4" thick, add						100%			
7650	Sash, single slide, glazed, with screens, 2020 (2' x 2')	E-1	22	1.091	Opng.	141	59	4.47	204.47	255
7700	3030 (3' x 3')		14	1.714		320	93	7.05	420.05	505
7750	4030 (4' x 3')		13	1.846		425	100	7.55	532.55	635

For customer support on your Building Construction Costs with RSMeans data, call 800.448.8182.

13 34 Fabricated Engineered Structures

13 34 19 – Metal Building Systems

13 34 19.50 Pre-Engineered Steel Buildings	Crew	Daily Output	Labor-Hours	Unit	Material	2018 Bare Costs Labor	Equipment	Total	Total Incl O&P
7800	6040 (6' x 4')	E-1	12	2	Opng.	850	108	8.20	966.20	1,125
7850	Double slide sash, 3030 (3' x 3')		14	1.714		232	93	7.05	332.05	410
7900	6040 (6' x 4')		12	2		620	108	8.20	736.20	860
7950	Fixed glass, no screens, 3030 (3' x 3')		14	1.714		223	93	7.05	323.05	400
8000	6040 (6' x 4')		12	2		595	108	8.20	711.20	835
8050	Prefinished storm sash, 3030 (3' x 3')	▼	70	.343	▼	85	18.60	1.41	105.01	125
8100	Siding and roofing, see Sections 07 41 13 & 07 42 13									
8200	Skylight, fiberglass panels, to 30 S.F.	E-1	10	2.400	Ea.	136	130	9.85	275.85	365
8250	Larger sizes, add for excess over 30 S.F.	"	300	.080	S.F.	4.52	4.34	.33	9.19	12.25
8300	Roof vents, turbine ventilator, wind driven									
8350	No damper, includes base, galvanized									
8400	12" diameter	Q-9	10	1.600	Ea.	67	86		153	205
8450	20" diameter		8	2		189	108		297	370
8500	24" diameter	▼	8	2		299	108		407	495
8600	Continuous, 26 ga., 10' long, 9" wide	2 Sswk	4	4		37	219		256	395
8650	12" wide	"	4	4	▼	37	219		256	395

13 34 23 – Fabricated Structures

13 34 23.10 Comfort Stations

		Crew	Daily Output	Labor-Hours	Unit	Material	Labor	Equipment	Total	Total Incl O&P
0010	**COMFORT STATIONS** Prefab., stock, w/doors, windows & fixt.									
0100	Not incl. interior finish or electrical									
0300	Mobile, on steel frame, 2 unit				S.F.	191			191	211
0350	7 unit					320			320	350
0400	Permanent, including concrete slab, 2 unit	B-12J	50	.320		254	15.35	17.40	286.75	320
0500	6 unit	"	43	.372	▼	188	17.85	20	225.85	256
0600	Alternate pricing method, mobile, 2 fixture				Fixture	6,625			6,625	7,300
0650	7 fixture					11,400			11,400	12,600
0700	Permanent, 2 unit	B-12J	.70	22.857		20,900	1,100	1,250	23,250	26,000
0750	6 unit	"	.50	32	▼	17,800	1,525	1,750	21,075	23,900

13 34 23.15 Domes

		Crew	Daily Output	Labor-Hours	Unit	Material	Labor	Equipment	Total	Total Incl O&P
0010	**DOMES**									
1500	Domes, bulk storage, shell only, dual radius hemisphere, arch, steel									
1600	framing, corrugated steel covering, 150' diameter	E-2	550	.102	SF Flr.	34	5.55	3.07	42.62	49.50
1700	400' diameter	"	720	.078		27.50	4.22	2.34	34.06	40
1800	Wood framing, wood decking, to 400' diameter	F-4	400	.120	▼	37	6.15	2.48	45.63	52.50
1900	Radial framed wood (2" x 6"), 1/2" thick									
2000	plywood, asphalt shingles, 50' diameter	F-3	2000	.020	SF Flr.	68.50	1.04	.25	69.79	77.50
2100	60' diameter		1900	.021		59	1.09	.26	60.35	66.50
2200	72' diameter		1800	.022		49	1.15	.28	50.43	56
2300	116' diameter		1730	.023		33.50	1.20	.29	34.99	38.50
2400	150' diameter	▼	1500	.027	▼	36	1.38	.33	37.71	42

13 34 23.16 Fabricated Control Booths

		Crew	Daily Output	Labor-Hours	Unit	Material	Labor	Equipment	Total	Total Incl O&P
0010	**FABRICATED CONTROL BOOTHS**									
0100	Guard House, prefab conc. w/bullet resistant doors & windows, roof & wiring									
0110	8' x 8', Level III	L-10	1	24	Ea.	54,000	1,350	495	55,845	62,000
0120	8' x 8', Level IV	"	1	24	"	61,500	1,350	495	63,345	70,000

13 34 23.25 Garage Costs

		Crew	Daily Output	Labor-Hours	Unit	Material	Labor	Equipment	Total	Total Incl O&P
0010	**GARAGE COSTS**									
0020	Public parking, average				Car				18,400	20,200
0300	Residential, wood, 12' x 20', one car prefab shell, stock, economy	2 Carp	1	16	Total	6,225	810		7,035	8,075
0350	Custom		.67	23.881		7,200	1,200		8,400	9,775
0400	Two car, 24' x 20', economy	▼	.67	23.881	▼	12,500	1,200		13,700	15,700

For customer support on your Building Construction Costs with RSMeans data, call 800.448.8182.

469

13 34 Fabricated Engineered Structures

13 34 23 – Fabricated Structures

13 34 23.25 Garage Costs

		Crew	Daily Output	Labor-Hours	Unit	Material	2018 Bare Costs Labor	2018 Bare Costs Equipment	Total	Total Incl O&P
0450	Custom	2 Carp	.50	32	Total	15,600	1,625		17,225	19,700

13 34 23.30 Garden House

		Crew	Daily Output	Labor-Hours	Unit	Material	Labor	Equipment	Total	Total Incl O&P
0010	**GARDEN HOUSE** Prefab wood, no floors or foundations									
0100	6' x 6'	2 Carp	200	.080	SF Flr.	51	4.06		55.06	62
0300	8' x 12'	"	48	.333	"	31	16.90		47.90	59.50

13 34 23.45 Kiosks

		Crew	Daily Output	Labor-Hours	Unit	Material	Labor	Equipment	Total	Total Incl O&P
0010	**KIOSKS**									
0020	Round, advertising type, 5' diameter, 7' high, aluminum wall, illuminated				Ea.	23,600			23,600	26,000
0100	Aluminum wall, non-illuminated					22,600			22,600	24,900
0500	Rectangular, 5' x 9', 7'-6" high, aluminum wall, illuminated					25,600			25,600	28,200
0600	Aluminum wall, non-illuminated					24,100			24,100	26,500

13 34 23.60 Portable Booths

		Crew	Daily Output	Labor-Hours	Unit	Material	Labor	Equipment	Total	Total Incl O&P
0010	**PORTABLE BOOTHS** Prefab. aluminum with doors, windows, ext. roof									
0100	lights wiring & insulation, 15 S.F. building, OD, painted				S.F.	284			284	310
0300	30 S.F. building					228			228	251
0400	50 S.F. building					180			180	198
0600	80 S.F. building					156			156	172
0700	100 S.F. building					133			133	146
0900	Acoustical booth, 27 Db @ 1,000 Hz, 15 S.F. floor				Ea.	3,850			3,850	4,225
1000	7' x 7'-6", including light & ventilation					7,925			7,925	8,700
1200	Ticket booth, galv. steel, not incl. foundations., 4' x 4'					5,100			5,100	5,600
1300	4' x 6'					7,475			7,475	8,225

13 34 23.70 Shelters

		Crew	Daily Output	Labor-Hours	Unit	Material	Labor	Equipment	Total	Total Incl O&P
0010	**SHELTERS**									
0020	Aluminum frame, acrylic glazing, 3' x 9' x 8' high	2 Sswk	1.14	14.035	Ea.	3,175	765		3,940	4,725
0100	9' x 12' x 8' high	"	.73	21.918	"	7,450	1,200		8,650	10,200

13 34 43 – Aircraft Hangars

13 34 43.50 Hangars

		Crew	Daily Output	Labor-Hours	Unit	Material	Labor	Equipment	Total	Total Incl O&P
0010	**HANGARS** Prefabricated steel T hangars, galv. steel roof &									
0100	walls, incl. electric bi-folding doors									
0110	not including floors or foundations, 4 unit	E-2	1275	.044	SF Flr.	15.05	2.38	1.32	18.75	22
0130	8 unit		1063	.053		12.75	2.86	1.59	17.20	20.50
0900	With bottom rolling doors, 4 unit		1386	.040		13.85	2.19	1.22	17.26	20
1000	8 unit		966	.058		11.10	3.15	1.75	16	19.20
1200	Alternate pricing method:									
1300	Galv. roof and walls, electric bi-folding doors, 4 plane	E-2	1.06	52.830	Plane	20,000	2,875	1,600	24,475	28,400
1500	8 plane		.91	61.538		12,700	3,350	1,850	17,900	21,400
1600	With bottom rolling doors, 4 plane		1.25	44.800		17,100	2,425	1,350	20,875	24,200
1800	8 plane		.97	57.732		15,000	3,125	1,750	19,875	23,500
2000	Circular type, prefab., steel frame, plastic skin, electric									
2010	door, including foundations, 80' diameter									

13 34 53 – Agricultural Structures

13 34 53.50 Silos

		Crew	Daily Output	Labor-Hours	Unit	Material	Labor	Equipment	Total	Total Incl O&P
0010	**SILOS**									
0500	Steel, factory fab., 30,000 gallon cap., painted, economy	L-5	1	56	Ea.	22,500	3,100	590	26,190	30,400
0700	Deluxe		.50	112		35,800	6,175	1,175	43,150	50,500
0800	Epoxy lined, economy		1	56		36,800	3,100	590	40,490	46,200
1000	Deluxe		.50	112		46,700	6,175	1,175	54,050	63,000

13 34 Fabricated Engineered Structures

13 34 56 – Observatories

13 34 56.15 Domes		Crew	Daily Output	Labor-Hours	Unit	Material	2018 Bare Costs Labor	Equipment	Total	Total Incl O&P
0010	**DOMES**									
0020	Domes, rev. alum., elec. drive, for astronomy obsv. shell only, stock units									
0600	10'-6" diameter	2 Carp	.25	64	Ea.	48,400	3,250		51,650	58,000
0900	18'-6" diameter		.17	94.118		82,000	4,775		86,775	97,500
1200	24'-6" diameter		.08	200		118,500	10,100		128,600	145,500

13 34 63 – Natural Fiber Construction

13 34 63.50 Straw Bale Construction

0010	**STRAW BALE CONSTRUCTION**										
2020	Straw bales in walls w/modified post and beam frame	G	2 Carp	320	.050	S.F.	6.35	2.54		8.89	10.80

13 36 Towers

13 36 13 – Metal Towers

13 36 13.50 Control Towers

0010	**CONTROL TOWERS**									
0020	Modular 12' x 10', incl. instruments				Ea.	801,500			801,500	882,000
0500	With standard 40' tower				"	1,260,000			1,260,000	1,386,000
1000	Temporary portable control towers, 8' x 12',									
1010	complete with one position communications				Ea.				266,000	293,000

13 42 Building Modules

13 42 63 – Detention Cell Modules

13 42 63.16 Steel Detention Cell Modules

0010	**STEEL DETENTION CELL MODULES**									
2000	Cells, prefab., 5' to 6' wide, 7' to 8' high, 7' to 8' deep,									
2010	bar front, cot, not incl. plumbing	E-4	1.50	21.333	Ea.	9,700	1,175	65.50	10,940.50	12,700

13 47 Facility Protection

13 47 13 – Cathodic Protection

13 47 13.16 Cathodic Prot. for Underground Storage Tanks

0010	**CATHODIC PROTECTION FOR UNDERGROUND STORAGE TANKS**									
1000	Anodes, magnesium type, 9 #	R-15	18.50	2.595	Ea.	35	148	15.30	198.30	277
1010	17 #		13	3.692		74.50	211	22	307.50	420
1020	32 #		10	4.800		117	274	28.50	419.50	570
1030	48 #		7.20	6.667		161	380	39.50	580.50	790
1100	Graphite type w/epoxy cap, 3" x 60" (32 #)	R-22	8.40	4.438		120	236		356	485
1110	4" x 80" (68 #)		6	6.213		232	330		562	750
1120	6" x 72" (80 #)		5.20	7.169		1,500	380		1,880	2,225
1130	6" x 36" (45 #)		9.60	3.883		760	207		967	1,150
2000	Rectifiers, silicon type, air cooled, 28 V/10 A	R-19	3.50	5.714		2,050	335		2,385	2,750
2010	20 V/20 A		3.50	5.714		2,100	335		2,435	2,825
2100	Oil immersed, 28 V/10 A		3	6.667		2,825	390		3,215	3,675
2110	20 V/20 A		3	6.667		3,000	390		3,390	3,875
3000	Anode backfill, coke breeze	R-22	3850	.010	Lb.	.28	.52		.80	1.08
4000	Cable, HMWPE, No. 8		2.40	15.533	M.L.F.	570	825		1,395	1,850
4010	No. 6		2.40	15.533		845	825		1,670	2,150
4020	No. 4		2.40	15.533		1,275	825		2,100	2,625

For customer support on your Building Construction Costs with RSMeans data, call 800.448.8182.

471

13 47 Facility Protection

13 47 13 – Cathodic Protection

13 47 13.16 Cathodic Prot. for Underground Storage Tanks

		Crew	Daily Output	Labor-Hours	Unit	Material	2018 Bare Costs Labor	Equipment	Total	Total Incl O&P
4030	No. 2	R-22	2.40	15.533	M.L.F.	1,975	825		2,800	3,400
4040	No. 1		2.20	16.945		2,700	905		3,605	4,325
4050	No. 1/0		2.20	16.945		3,500	905		4,405	5,200
4060	No. 2/0		2.20	16.945		4,975	905		5,880	6,825
4070	No. 4/0		2	18.640		6,275	995		7,270	8,400
5000	Test station, 7 terminal box, flush curb type w/lockable cover	R-19	12	1.667	Ea.	76.50	97		173.50	230
5010	Reference cell, 2" diam. PVC conduit, cplg., plug, set flush	"	4.80	4.167	"	151	243		394	530

13 48 Sound, Vibration, and Seismic Control

13 48 13 – Manufactured Sound and Vibration Control Components

13 48 13.50 Audio Masking

		Crew	Daily Output	Labor-Hours	Unit	Material	2018 Bare Costs Labor	Equipment	Total	Total Incl O&P
0010	**AUDIO MASKING**, acoustical enclosure, 4" thick wall and ceiling									
0020	8 lb./S.F., up to 12' span	3 Carp	72	.333	SF Surf	31.50	16.90		48.40	60
0300	Better quality panels, 10.5 #/S.F		64	.375		36	19		55	68.50
0400	Reverb-chamber, 4" thick, parallel walls		60	.400		44.50	20.50		65	80
0600	Skewed wall, parallel roof, 4" thick panels		55	.436		51	22		73	89.50
0700	Skewed walls, skewed roof, 4" layers, 4" air space		48	.500		57.50	25.50		83	102
0900	Sound-absorbing panels, pntd. mtl., 2'-6" x 8', under 1,000 S.F.		215	.112		11.85	5.65		17.50	21.50
1100	Over 1,000 S.F.		240	.100		11.40	5.05		16.45	20.50
1200	Fabric faced		240	.100		9.25	5.05		14.30	17.85
1500	Flexible transparent curtain, clear	3 Shee	215	.112		7.15	6.70		13.85	18.05
1600	50% foam		215	.112		10	6.70		16.70	21
1700	75% foam		215	.112		10	6.70		16.70	21
1800	100% foam		215	.112		10	6.70		16.70	21
3100	Audio masking system, including speakers, amplification									
3110	and signal generator									
3200	Ceiling mounted, 5,000 S.F.	2 Elec	2400	.007	S.F.	1.21	.39		1.60	1.91
3300	10,000 S.F.		2800	.006		.97	.33		1.30	1.57
3400	Plenum mounted, 5,000 S.F.		3800	.004		1.04	.25		1.29	1.51
3500	10,000 S.F.		4400	.004		.70	.21		.91	1.09

13 49 Radiation Protection

13 49 13 – Integrated X-Ray Shielding Assemblies

13 49 13.50 Lead Sheets

		Crew	Daily Output	Labor-Hours	Unit	Material	2018 Bare Costs Labor	Equipment	Total	Total Incl O&P
0010	**LEAD SHEETS**									
0300	Lead sheets, 1/16" thick	2 Lath	135	.119	S.F.	10.90	5.85		16.75	20.50
0400	1/8" thick		120	.133		22	6.55		28.55	34
0500	Lead shielding, 1/4" thick		135	.119		42.50	5.85		48.35	55
0550	1/2" thick		120	.133		79.50	6.55		86.05	97
0950	Lead headed nails (average 1 lb. per sheet)				Lb.	8.40			8.40	9.25
1000	Butt joints in 1/8" lead or thicker, 2" batten strip x 7' long	2 Lath	240	.067	Ea.	31	3.28		34.28	39
1200	X-ray protection, average radiography or fluoroscopy									
1210	room, up to 300 S.F. floor, 1/16" lead, economy	2 Lath	.25	64	Total	11,100	3,150		14,250	17,000
1500	7'-0" walls, deluxe	"	.15	107	"	13,400	5,250		18,650	22,500
1600	Deep therapy X-ray room, 250 kV capacity,									
1800	up to 300 S.F. floor, 1/4" lead, economy	2 Lath	.08	200	Total	31,100	9,825		40,925	48,700
1900	7'-0" walls, deluxe	"	.06	267	"	38,400	13,100		51,500	61,500

472

For customer support on your Building Construction Costs with RSMeans data, call 800.448.8182.

13 49 Radiation Protection

13 49 19 – Lead-Lined Materials

13 49 19.50 Shielding Lead

		Crew	Daily Output	Labor-Hours	Unit	Material	2018 Bare Costs Labor	Equipment	Total	Total Incl O&P
0010	**SHIELDING LEAD**									
0100	Laminated lead in wood doors, 1/16" thick, no hardware				S.F.	53.50			53.50	59
0200	Lead lined door frame, not incl. hardware,									
0210	1/16" thick lead, butt prepared for hardware	1 Lath	2.40	3.333	Ea.	860	164		1,024	1,175
0850	Window frame with 1/16" lead and voice passage, 36" x 60"	2 Glaz	2	8		4,525	390		4,915	5,575
0870	24" x 36" frame		4	4		2,350	194		2,544	2,875
0900	Lead gypsum board, 5/8" thick with 1/16" lead		160	.100	S.F.	12.10	4.85		16.95	20.50
0910	1/8" lead		140	.114		24.50	5.55		30.05	35.50
0930	1/32" lead	2 Lath	200	.080		8.50	3.93		12.43	15.15

13 49 21 – Lead Glazing

13 49 21.50 Lead Glazing

		Crew	Daily Output	Labor-Hours	Unit	Material	2018 Bare Costs Labor	Equipment	Total	Total Incl O&P
0010	**LEAD GLAZING**									
0600	Lead glass, 1/4" thick, 2.0 mm LE, 12" x 16"	2 Glaz	13	1.231	Ea.	390	59.50		449.50	520
0700	24" x 36"		8	2		1,375	97		1,472	1,650
0800	36" x 60"		2	8		3,775	390		4,165	4,775
2000	X-ray viewing panels, clear lead plastic									
2010	7 mm thick, 0.3 mm LE, 2.3 lb./S.F.	H-3	139	.115	S.F.	271	4.97		275.97	305
2020	12 mm thick, 0.5 mm LE, 3.9 lb./S.F.		82	.195		385	8.40		393.40	435
2030	18 mm thick, 0.8 mm LE, 5.9 lb./S.F.		54	.296		440	12.80		452.80	505
2040	22 mm thick, 1.0 mm LE, 7.2 lb./S.F.		44	.364		575	15.70		590.70	655
2050	35 mm thick, 1.5 mm LE, 11.5 lb./S.F.		28	.571		880	24.50		904.50	1,000
2060	46 mm thick, 2.0 mm LE, 15.0 lb./S.F.		21	.762		1,150	33		1,183	1,300
2090	For panels 12 S.F. to 48 S.F., add crating charge				Ea.				50	50

13 49 23 – Integrated RFI/EMI Shielding Assemblies

13 49 23.50 Modular Shielding Partitions

		Crew	Daily Output	Labor-Hours	Unit	Material	2018 Bare Costs Labor	Equipment	Total	Total Incl O&P
0010	**MODULAR SHIELDING PARTITIONS**									
4000	X-ray barriers, modular, panels mounted within framework for									
4002	attaching to floor, wall or ceiling, upper portion is clear lead									
4005	plastic window panels 48"H, lower portion is opaque leaded									
4008	steel panels 36"H, structural supports not incl.									
4010	1-section barrier, 36"W x 84"H overall									
4020	0.5 mm LE panels	H-3	6.40	2.500	Ea.	8,800	108		8,908	9,850
4030	0.8 mm LE panels		6.40	2.500		9,500	108		9,608	10,600
4040	1.0 mm LE panels		5.33	3.002		11,100	130		11,230	12,400
4050	1.5 mm LE panels		5.33	3.002		14,800	130		14,930	16,500
4060	2-section barrier, 72"W x 84"H overall									
4070	0.5 mm LE panels	H-3	4	4	Ea.	12,800	173		12,973	14,300
4080	0.8 mm LE panels		4	4		14,100	173		14,273	15,800
4090	1.0 mm LE panels		3.56	4.494		17,300	194		17,494	19,400
5000	1.5 mm LE panels		3.20	5		24,700	216		24,916	27,400
5010	3-section barrier, 108"W x 84"H overall									
5020	0.5 mm LE panels	H-3	3.20	5	Ea.	19,100	216		19,316	21,300
5030	0.8 mm LE panels		3.20	5		21,200	216		21,416	23,600
5040	1.0 mm LE panels		2.67	5.993		26,000	259		26,259	29,000
5050	1.5 mm LE panels		2.46	6.504		37,000	281		37,281	41,100
7000	X-ray barriers, mobile, mounted within framework w/casters on									
7005	bottom, clear lead plastic window panels on upper portion,									
7010	opaque on lower, 30"W x 75"H overall, incl. framework									
7020	24"H upper w/0.5 mm LE, 48"H lower w/0.8 mm LE	1 Carp	16	.500	Ea.	4,225	25.50		4,250.50	4,700
7030	48"W x 75"H overall, incl. framework									
7040	36"H upper w/0.5 mm LE, 36"H lower w/0.8 mm LE	1 Carp	16	.500	Ea.	6,675	25.50		6,700.50	7,400

For customer support on your Building Construction Costs with RSMeans data, call 800.448.8182.

473

13 49 Radiation Protection

13 49 23 – Integrated RFI/EMI Shielding Assemblies

13 49 23.50 Modular Shielding Partitions	Crew	Daily Output	Labor-Hours	Unit	Material	2018 Bare Costs Labor	Equipment	Total	Total Incl O&P
7050 36"H upper w/1.0 mm LE, 36"H lower w/1.5 mm LE	1 Carp	16	.500	Ea.	7,900	25.50		7,925.50	8,750
7060 72"W x 75"H overall, incl. framework									
7070 36"H upper w/0.5 mm LE, 36"H lower w/0.8 mm LE	1 Carp	16	.500	Ea.	7,900	25.50		7,925.50	8,750
7080 36"H upper w/1.0 mm LE, 36"H lower w/1.5 mm LE	"	16	.500	"	9,900	25.50		9,925.50	10,900

13 49 33 – Radio Frequency Shielding

13 49 33.50 Shielding, Radio Frequency	Crew	Daily Output	Labor-Hours	Unit	Material	2018 Bare Costs Labor	Equipment	Total	Total Incl O&P
0010 **SHIELDING, RADIO FREQUENCY**									
0020 Prefabricated, galvanized steel	2 Carp	375	.043	SF Surf	5.30	2.16		7.46	9.10
0040 5 oz., copper floor panel		480	.033		4.48	1.69		6.17	7.50
0050 5 oz., copper wall/ceiling panel		155	.103		4.48	5.25		9.73	12.90
0100 12 oz., copper floor panel		470	.034		9.05	1.73		10.78	12.60
0110 12 oz., copper wall/ceiling panel		140	.114		9.05	5.80		14.85	18.75
0150 Door, copper/wood laminate, 4' x 7'		1.50	10.667	Ea.	8,400	540		8,940	10,100

13 53 Meteorological Instrumentation

13 53 09 – Weather Instrumentation

13 53 09.50 Weather Station	Crew	Daily Output	Labor-Hours	Unit	Material	2018 Bare Costs Labor	Equipment	Total	Total Incl O&P
0010 **WEATHER STATION**									
0020 Remote recording, solar powered, with rain gauge & display, 400' range				Ea.	775			775	850
0100 1 mile range				"	1,900			1,900	2,075

For customer support on your Building Construction Costs with RSMeans data, call 800.448.8182.

Estimating Tips
General

- Many products in Division 14 will require some type of support or blocking for installation not included with the item itself. Examples are supports for conveyors or tube systems, attachment points for lifts, and footings for hoists or cranes. Add these supports in the appropriate division.

14 10 00 Dumbwaiters
14 20 00 Elevators

- Dumbwaiters and elevators are estimated and purchased in a method similar to buying a car. The manufacturer has a base unit with standard features. Added to this base unit price will be whatever options the owner or specifications require. Increased load capacity, additional vertical travel, additional stops, higher speed, and cab finish options are items to be considered. When developing an estimate for dumbwaiters and elevators, remember that some items needed by the installers may have to be included as part of the general contract.

Examples are:
- □ shaftway
- □ rail support brackets
- □ machine room
- □ electrical supply
- □ sill angles
- □ electrical connections
- □ pits
- □ roof penthouses
- □ pit ladders

Check the job specifications and drawings before pricing.

- Installation of elevators and handicapped lifts in historic structures can require significant additional costs. The associated structural requirements may involve cutting into and repairing finishes, moldings, flooring, etc. The estimator must account for these special conditions.

14 30 00 Escalators and Moving Walks

- Escalators and moving walks are specialty items installed by specialty contractors. There are numerous options associated with these items. For specific options, contact a manufacturer or contractor. In a method similar to estimating dumbwaiters and elevators, you should verify the extent of general contract work and add items as necessary.

14 40 00 Lifts
14 90 00 Other Conveying Equipment

- Products such as correspondence lifts, chutes, and pneumatic tube systems, as well as other items specified in this subdivision, may require trained installers. The general contractor might not have any choice as to who will perform the installation or when it will be performed. Long lead times are often required for these products, making early decisions in scheduling necessary.

Reference Numbers

Reference numbers are shown at the beginning of some major classifications. These numbers refer to related items in the Reference Section. The reference information may be an estimating procedure, an alternate pricing method, or technical information.

Note: Not all subdivisions listed here necessarily appear. ■

Did you know?

RSMeans data is available through our online application with 24/7 access:

- Search for unit prices by keyword
- Leverage the most up-to-date data
- Build and export estimates

Try it free for 30 days!
www.rsmeans.com/2018freetrial

No part of this cost data may be reproduced, stored in a retrieval system, or transmitted in any form or by any means without prior written permission of Gordian.

14 11 Manual Dumbwaiters

14 11 10 – Hand Operated Dumbwaiters

14 11 10.20 Manual Dumbwaiters	Crew	Daily Output	Labor-Hours	Unit	Material	2018 Bare Costs Labor	Equipment	Total	Total Incl O&P
0010 **MANUAL DUMBWAITERS**									
0020 2 stop, hand powered, up to 75 lb. capacity	2 Elev	.75	21.333	Ea.	3,100	1,750		4,850	6,025
0100 76 lb. capacity and up		.50	32	"	6,900	2,625		9,525	11,500
0300 For each additional stop, add		.75	21.333	Stop	1,125	1,750		2,875	3,875

14 12 Electric Dumbwaiters

14 12 10 – Dumbwaiters

14 12 10.10 Electric Dumbwaiters

	Crew	Daily Output	Labor-Hours	Unit	Material	2018 Bare Costs Labor	Equipment	Total	Total Incl O&P
0010 **ELECTRIC DUMBWAITERS**									
0020 2 stop, up to 75 lb. capacity	2 Elev	.13	123	Ea.	7,625	10,100		17,725	23,500
0100 76 lb. capacity and up		.11	145	"	23,000	12,000		35,000	43,100
0600 For each additional stop, add		.54	29.630	Stop	3,400	2,425		5,825	7,375

14 21 Electric Traction Elevators

14 21 13 – Electric Traction Freight Elevators

14 21 13.10 Electric Traction Freight Elevators and Options

		Crew	Daily Output	Labor-Hours	Unit	Material	2018 Bare Costs Labor	Equipment	Total	Total Incl O&P
0010 **ELECTRIC TRACTION FREIGHT ELEVATORS AND OPTIONS**	R142000-10									
0425 Electric freight, base unit, 4000 lb., 200 fpm, 4 stop, std. fin.	R142000-20	2 Elev	.05	320	Ea.	122,000	26,300		148,300	173,500
0450 For 5000 lb. capacity, add	R142000-30					6,375			6,375	7,000
0500 For 6000 lb. capacity, add	R142000-40					15,000			15,000	16,500
0525 For 7000 lb. capacity, add						20,100			20,100	22,100
0550 For 8000 lb. capacity, add						24,300			24,300	26,800
0575 For 10000 lb. capacity, add						33,700			33,700	37,000
0600 For 12000 lb. capacity, add						40,900			40,900	45,000
0625 For 16000 lb. capacity, add						49,600			49,600	54,500
0650 For 20000 lb. capacity, add						55,000			55,000	61,000
0675 For increased speed, 250 fpm, add						14,600			14,600	16,000
0700 300 fpm, geared electric, add						18,200			18,200	20,000
0725 350 fpm, geared electric, add						22,200			22,200	24,400
0750 400 fpm, geared electric, add						26,200			26,200	28,800
0775 500 fpm, gearless electric, add						36,100			36,100	39,700
0800 600 fpm, gearless electric, add						40,400			40,400	44,400
0825 700 fpm, gearless electric, add						47,700			47,700	52,500
0850 800 fpm, gearless electric, add						53,000			53,000	58,000
0875 For class "B" loading, add						2,900			2,900	3,175
0900 For class "C-1" loading, add						7,250			7,250	7,975
0925 For class "C-2" loading, add						8,650			8,650	9,500
0950 For class "C-3" loading, add						12,000			12,000	13,200
0975 For travel over 40 V.L.F., add		2 Elev	7.25	2.207	V.L.F.	640	181		821	975
1000 For number of stops over 4, add		"	.27	59.259	Stop	4,675	4,875		9,550	12,400

14 21 23 – Electric Traction Passenger Elevators

14 21 23.10 Electric Traction Passenger Elevators and Options

		Crew	Daily Output	Labor-Hours	Unit	Material	2018 Bare Costs Labor	Equipment	Total	Total Incl O&P
0010 **ELECTRIC TRACTION PASSENGER ELEVATORS AND OPTIONS**	R142000-10									
1625 Electric pass., base unit, 2,000 lb., 200 fpm, 4 stop, std. fin.	R142000-20	2 Elev	.05	320	Ea.	105,000	26,300		131,300	154,500
1650 For 2,500 lb. capacity, add	R142000-30					4,200			4,200	4,600
1675 For 3,000 lb. capacity, add	R142000-40					4,675			4,675	5,150
1700 For 3,500 lb. capacity, add						6,250			6,250	6,900
1725 For 4,000 lb. capacity, add						7,275			7,275	8,000
1750 For 4,500 lb. capacity, add						10,100			10,100	11,100

For customer support on your Building Construction Costs with RSMeans data, call 800.448.8182.

14 21 23 – Electric Traction Passenger Elevators

14 21 23.10 Electric Traction Passenger Elevators and Options

		Crew	Daily Output	Labor-Hours	Unit	Material	2018 Bare Costs Labor	2018 Bare Costs Equipment	Total	Total Incl O&P
1775	For 5000 lb. capacity, add				Ea.	12,900			12,900	14,200
1800	For increased speed, 250 fpm, geared electric, add					3,475			3,475	3,825
1825	300 fpm, geared electric, add					6,900			6,900	7,600
1850	350 fpm, geared electric, add					8,600			8,600	9,475
1875	400 fpm, geared electric, add					12,700			12,700	14,000
1900	500 fpm, gearless electric, add					31,100			31,100	34,200
1925	600 fpm, gearless electric, add					49,400			49,400	54,500
1950	700 fpm, gearless electric, add					55,500			55,500	61,500
1975	800 fpm, gearless electric, add					61,500			61,500	67,500
2000	For travel over 40 V.L.F., add	2 Elev	7.25	2.207	V.L.F.	770	181		951	1,125
2025	For number of stops over 4, add		.27	59.259	Stop	3,275	4,875		8,150	10,900
2400	Electric hospital, base unit, 4,000 lb., 200 fpm, 4 stop, std fin.		.05	320	Ea.	90,000	26,300		116,300	138,000
2425	For 4,500 lb. capacity, add					6,225			6,225	6,825
2450	For 5,000 lb. capacity, add					8,150			8,150	8,950
2475	For increased speed, 250 fpm, geared electric, add					3,900			3,900	4,275
2500	300 fpm, geared electric, add					7,400			7,400	8,150
2525	350 fpm, geared electric, add					8,950			8,950	9,825
2550	400 fpm, geared electric, add					13,100			13,100	14,400
2575	500 fpm, gearless electric, add					36,600			36,600	40,300
2600	600 fpm, gearless electric, add					53,500			53,500	59,000
2625	700 fpm, gearless electric, add					59,500			59,500	65,500
2650	800 fpm, gearless electric, add					66,500			66,500	73,500
2675	For travel over 40 V.L.F., add	2 Elev	7.25	2.207	V.L.F.	185	181		366	475
2700	For number of stops over 4, add	"	.27	59.259	Stop	4,850	4,875		9,725	12,600

14 21 33 – Electric Traction Residential Elevators

14 21 33.20 Residential Elevators

		Crew	Daily Output	Labor-Hours	Unit	Material	2018 Bare Costs Labor	2018 Bare Costs Equipment	Total	Total Incl O&P
0010	**RESIDENTIAL ELEVATORS**									
7000	Residential, cab type, 1 floor, 2 stop, economy model	2 Elev	.20	80	Ea.	10,800	6,575		17,375	21,600
7100	Custom model		.10	160		18,200	13,200		31,400	39,600
7200	2 floor, 3 stop, economy model		.12	133		16,000	11,000		27,000	33,900
7300	Custom model		.06	267		26,100	21,900		48,000	61,500

14 24 13 – Hydraulic Freight Elevators

14 24 13.10 Hydraulic Freight Elevators and Options

			Crew	Daily Output	Labor-Hours	Unit	Material	2018 Bare Costs Labor	2018 Bare Costs Equipment	Total	Total Incl O&P
0010	**HYDRAULIC FREIGHT ELEVATORS AND OPTIONS**	R142000-10									
1025	Hydraulic freight, base unit, 2,000 lb., 50 fpm, 2 stop, std. fin.	R142000-20	2 Elev	.10	160	Ea.	84,500	13,200		97,700	112,500
1050	For 2,500 lb. capacity, add	R142000-30					4,225			4,225	4,625
1075	For 3,000 lb. capacity, add	R142000-40					5,575			5,575	6,125
1100	For 3,500 lb. capacity, add						9,125			9,125	10,000
1125	For 4,000 lb. capacity, add						10,400			10,400	11,400
1150	For 4,500 lb. capacity, add						12,100			12,100	13,300
1175	For 5,000 lb. capacity, add						16,400			16,400	18,100
1200	For 6,000 lb. capacity, add						16,800			16,800	18,400
1225	For 7,000 lb. capacity, add						25,700			25,700	28,300
1250	For 8,000 lb. capacity, add						30,500			30,500	33,500
1275	For 10,000 lb. capacity, add						32,900			32,900	36,200
1300	For 12,000 lb. capacity, add						40,000			40,000	44,000
1325	For 16,000 lb. capacity, add						53,000			53,000	58,000
1350	For 20,000 lb. capacity, add						59,000			59,000	65,000

14 24 Hydraulic Elevators

14 24 13 – Hydraulic Freight Elevators

14 24 13.10 Hydraulic Freight Elevators and Options

		Crew	Daily Output	Labor-Hours	Unit	Material	2018 Bare Costs Labor	Equipment	Total	Total Incl O&P
1375	For increased speed, 100 fpm, add				Ea.	1,200			1,200	1,325
1400	125 fpm, add					3,100			3,100	3,425
1425	150 fpm, add					4,950			4,950	5,450
1450	175 fpm, add					6,950			6,950	7,650
1475	For class "B" loading, add					2,775			2,775	3,050
1500	For class "C-1" loading, add					7,100			7,100	7,800
1525	For class "C-2" loading, add					8,500			8,500	9,350
1550	For class "C-3" loading, add					11,700			11,700	12,900
1575	For travel over 20 V.L.F., add	2 Elev	7.25	2.207	V.L.F.	900	181		1,081	1,250
1600	For number of stops over 2, add	"	.27	59.259	Stop	2,250	4,875		7,125	9,750

14 24 23 – Hydraulic Passenger Elevators

14 24 23.10 Hydraulic Passenger Elevators and Options

		Crew	Daily Output	Labor-Hours	Unit	Material	2018 Bare Costs Labor	Equipment	Total	Total Incl O&P
0010	**HYDRAULIC PASSENGER ELEVATORS AND OPTIONS** R142000-10									
2050	Hyd. pass., base unit, 1,500 lb., 100 fpm, 2 stop, std. fin. R142000-20	2 Elev	.10	160	Ea.	42,900	13,200		56,100	67,000
2075	For 2,000 lb. capacity, add R142000-30					865			865	955
2100	For 2,500 lb. capacity, add R142000-40					3,200			3,200	3,500
2125	For 3,000 lb. capacity, add					4,525			4,525	4,975
2150	For 3,500 lb. capacity, add					7,625			7,625	8,400
2175	For 4,000 lb. capacity, add					9,200			9,200	10,100
2200	For 4,500 lb. capacity, add					12,100			12,100	13,300
2225	For 5,000 lb. capacity, add					17,100			17,100	18,800
2250	For increased speed, 125 fpm, add					1,200			1,200	1,325
2275	150 fpm, add					2,750			2,750	3,025
2300	175 fpm, add					5,450			5,450	6,000
2325	200 fpm, add					10,300			10,300	11,300
2350	For travel over 12 V.L.F., add	2 Elev	7.25	2.207	V.L.F.	730	181		911	1,075
2375	For number of stops over 2, add		.27	59.259	Stop	1,025	4,875		5,900	8,375
2725	Hydraulic hospital, base unit, 4,000 lb., 100 fpm, 2 stop, std. fin.		.10	160	Ea.	66,500	13,200		79,700	93,000
2775	For 4,500 lb. capacity, add					7,500			7,500	8,250
2800	For 5,000 lb. capacity, add					11,000			11,000	12,100
2825	For increased speed, 125 fpm, add					2,250			2,250	2,475
2850	150 fpm, add					3,775			3,775	4,175
2875	175 fpm, add					6,350			6,350	6,975
2900	200 fpm, add					9,300			9,300	10,200
2925	For travel over 12 V.L.F., add	2 Elev	7.25	2.207	V.L.F.	405	181		586	715
2950	For number of stops over 2, add	"	.27	59.259	Stop	4,650	4,875		9,525	12,400

14 27 Custom Elevator Cabs and Doors

14 27 13 – Custom Elevator Cab Finishes

14 27 13.10 Cab Finishes

		Crew	Daily Output	Labor-Hours	Unit	Material	2018 Bare Costs Labor	Equipment	Total	Total Incl O&P
0010	**CAB FINISHES**									
3325	Passenger elevator cab finishes (based on 3,500 lb. cab size)									
3350	Acrylic panel ceiling				Ea.	800			800	880
3375	Aluminum eggcrate ceiling					720			720	795
3400	Stainless steel doors					4,150			4,150	4,550
3425	Carpet flooring					635			635	695
3450	Epoxy flooring					480			480	530
3475	Quarry tile flooring					920			920	1,000
3500	Slate flooring					1,675			1,675	1,850
3525	Textured rubber flooring					670			670	735

For customer support on your Building Construction Costs with RSMeans data, call 800.448.8182.

14 27 Custom Elevator Cabs and Doors

14 27 13 – Custom Elevator Cab Finishes

14 27 13.10 Cab Finishes

14 27 13.10 Cab Finishes	Crew	Daily Output	Labor-Hours	Unit	Material	2018 Bare Costs Labor	Equipment	Total	Total Incl O&P
3550	Stainless steel walls				Ea.	4,275			4,275	4,700
3575	Stainless steel returns at door					1,225			1,225	1,350
4450	Hospital elevator cab finishes (based on 3,500 lb. cab size)									
4475	Aluminum eggcrate ceiling				Ea.	715			715	785
4500	Stainless steel doors					4,150			4,150	4,550
4525	Epoxy flooring					480			480	530
4550	Quarry tile flooring					920			920	1,000
4575	Textured rubber flooring					670			670	735
4600	Stainless steel walls					4,775			4,775	5,250
4625	Stainless steel returns at door					930			930	1,025

14 28 Elevator Equipment and Controls

14 28 10 – Elevator Equipment and Control Options

14 28 10.10 Elevator Controls and Doors

	14 28 10.10 Elevator Controls and Doors	Crew	Daily Output	Labor-Hours	Unit	Material	2018 Bare Costs Labor	Equipment	Total	Total Incl O&P
0010	**ELEVATOR CONTROLS AND DOORS**									
2975	Passenger elevator options									
3000	2 car group automatic controls	2 Elev	.66	24.242	Ea.	3,575	2,000		5,575	6,900
3025	3 car group automatic controls		.44	36.364		5,425	3,000		8,425	10,400
3050	4 car group automatic controls		.33	48.485		9,250	3,975		13,225	16,200
3075	5 car group automatic controls		.26	61.538		13,700	5,050		18,750	22,700
3100	6 car group automatic controls		.22	72.727		20,900	5,975		26,875	31,900
3125	Intercom service		3	5.333		570	440		1,010	1,275
3150	Duplex car selective collective		.66	24.242		4,000	2,000		6,000	7,375
3175	Center opening 1 speed doors		2	8		2,075	660		2,735	3,275
3200	Center opening 2 speed doors		2	8		2,925	660		3,585	4,175
3225	Rear opening doors (opposite front)		2	8		4,500	660		5,160	5,925
3250	Side opening 2 speed doors		2	8		7,625	660		8,285	9,375
3275	Automatic emergency power switching		.66	24.242		1,275	2,000		3,275	4,375
3300	Manual emergency power switching		8	2		540	164		704	840
3625	Hall finishes, stainless steel doors					1,500			1,500	1,625
3650	Stainless steel frames					1,525			1,525	1,675
3675	12 month maintenance contract								3,750	4,125
3700	Signal devices, hall lanterns	2 Elev	8	2		545	164		709	845
3725	Position indicators, up to 3		9.40	1.702		345	140		485	590
3750	Position indicators, per each over 3		32	.500		96.50	41		137.50	168
3775	High speed heavy duty door opener					2,725			2,725	3,000
3800	Variable voltage, O.H. gearless machine, min.	2 Elev	.16	100		35,200	8,225		43,425	51,000
3815	Maximum		.07	229		77,500	18,800		96,300	113,000
3825	Basement installed geared machine		.33	48.485		13,800	3,975		17,775	21,200
3850	Freight elevator options									
3875	Doors, bi-parting	2 Elev	.66	24.242	Ea.	6,175	2,000		8,175	9,750
3900	Power operated door and gate	"	.66	24.242		25,300	2,000		27,300	30,800
3925	Finishes, steel plate floor					1,125			1,125	1,225
3950	14 ga. 1/4" x 4' steel plate walls					2,075			2,075	2,275
3975	12 month maintenance contract								3,750	4,125
4000	Signal devices, hall lanterns	2 Elev	8	2		530	164		694	830
4025	Position indicators, up to 3		9.40	1.702		370	140		510	620
4050	Position indicators, per each over 3		32	.500		105	41		146	177
4075	Variable voltage basement installed geared machine		.66	24.242		20,900	2,000		22,900	26,000
4100	Hospital elevator options									
4125	2 car group automatic controls	2 Elev	.66	24.242	Ea.	3,575	2,000		5,575	6,900

For customer support on your Building Construction Costs with RSMeans data, call 800.448.8182.

479

14 28 10.10 Elevator Controls and Doors	Crew	Daily Output	Labor-Hours	Unit	Material	2018 Bare Costs Labor	Equipment	Total	Total Incl O&P
4150	3 car group automatic controls	2 Elev	.44	36.364	Ea.	5,375	3,000		8,375	10,400
4175	4 car group automatic controls		.33	48.485		13,400	3,975		17,375	20,700
4200	5 car group automatic controls		.26	61.538		13,200	5,050		18,250	22,200
4225	6 car group automatic controls		.22	72.727		20,300	5,975		26,275	31,300
4250	Intercom service		3	5.333		540	440		980	1,250
4275	Duplex car selective collective		.66	24.242		3,900	2,000		5,900	7,250
4300	Center opening 1 speed doors		2	8		2,075	660		2,735	3,250
4325	Center opening 2 speed doors		2	8		2,725	660		3,385	3,975
4350	Rear opening doors (opposite front)		2	8		4,475	660		5,135	5,900
4375	Side opening 2 speed doors		2	8		6,725	660		7,385	8,375
4400	Automatic emergency power switching		.66	24.242		1,225	2,000		3,225	4,325
4425	Manual emergency power switching	▼	8	2		525	164		689	820
4675	Hall finishes, stainless steel doors					1,550			1,550	1,700
4700	Stainless steel frames					1,550			1,550	1,700
4725	12 month maintenance contract								3,750	4,125
4750	Signal devices, hall lanterns	2 Elev	8	2		520	164		684	815
4775	Position indicators, up to 3		9.40	1.702		350	140		490	595
4800	Position indicators, per each over 3	▼	32	.500		96.50	41		137.50	168
4825	High speed heavy duty door opener					2,775			2,775	3,050
4850	Variable voltage, O.H. gearless machine, min.	2 Elev	.16	100		35,800	8,225		44,025	51,500
4865	Maximum		.07	229		78,000	18,800		96,800	114,000
4875	Basement installed geared machine	▼	.33	48.485	▼	18,000	3,975		21,975	25,800
5000	Drilling for piston, casing included, 18" diameter	B-48	80	.700	V.L.F.	64.50	31.50	34.50	130.50	157

14 31 10.10 Escalators

		Crew	Daily Output	Labor-Hours	Unit	Material	2018 Bare Costs Labor	Equipment	Total	Total Incl O&P
0010	**ESCALATORS**									
1000	Glass, 32" wide x 10' floor to floor height	M-1	.07	457	Ea.	90,500	35,700	715	126,915	153,500
1010	48" wide x 10' floor to floor height		.07	457		97,500	35,700	715	133,915	161,000
1020	32" wide x 15' floor to floor height		.06	533		95,500	41,600	830	137,930	168,000
1030	48" wide x 15' floor to floor height		.06	533		101,000	41,600	830	143,430	174,000
1040	32" wide x 20' floor to floor height		.05	653		101,500	51,000	1,025	153,525	188,500
1050	48" wide x 20' floor to floor height		.05	653		110,000	51,000	1,025	162,025	198,000
1060	32" wide x 25' floor to floor height		.04	800		111,500	62,500	1,250	175,250	217,500
1070	48" wide x 25' floor to floor height		.04	800		129,000	62,500	1,250	192,750	236,000
1080	Enameled steel, 32" wide x 10' floor to floor height		.07	457		98,000	35,700	715	134,415	162,000
1090	48" wide x 10' floor to floor height		.07	457		106,500	35,700	715	142,915	171,000
1110	32" wide x 15' floor to floor height		.06	533		104,000	41,600	830	146,430	177,500
1120	48" wide x 15' floor to floor height		.06	533		110,500	41,600	830	152,930	184,500
1130	32" wide x 20' floor to floor height		.05	653		111,000	51,000	1,025	163,025	199,000
1140	48" wide x 20' floor to floor height		.05	653		120,000	51,000	1,025	172,025	209,000
1150	32" wide x 25' floor to floor height		.04	800		121,000	62,500	1,250	184,750	227,500
1160	48" wide x 25' floor to floor height		.04	800		139,500	62,500	1,250	203,250	248,000
1170	Stainless steel, 32" wide x 10' floor to floor height		.07	457		104,000	35,700	715	140,415	168,000
1180	48" wide x 10' floor to floor height		.07	457		111,500	35,700	715	147,915	177,000
1500	32" wide x 15' floor to floor height		.06	533		110,000	41,600	830	152,430	184,000
1700	48" wide x 15' floor to floor height		.06	533		116,000	41,600	830	158,430	190,500
1750	32" wide x 18' floor to floor height		.05	615		134,500	48,100	960	183,560	220,500
1775	48" wide x 18' floor to floor height		.05	615		146,500	48,100	960	195,560	233,500
2300	32" wide x 25' floor to floor height	▼	.04	800	▼	136,500	62,500	1,250	200,250	245,000

14 31 Escalators

14 31 10 - Glass and Steel Escalators

14 31 10.10 Escalators

		Crew	Daily Output	Labor-Hours	Unit	Material	2018 Bare Costs Labor	2018 Bare Costs Equipment	Total	Total Incl O&P
2500	48" wide x 25' floor to floor height	M-1	.04	800	Ea.	146,500	62,500	1,250	210,250	255,500

14 32 Moving Walks

14 32 10 - Moving Walkways

14 32 10.10 Moving Walks

			Crew	Daily Output	Labor-Hours	Unit	Material	2018 Bare Costs Labor	2018 Bare Costs Equipment	Total	Total Incl O&P
0010	**MOVING WALKS**	R143210-20									
0020	Walk, 27" tread width, minimum		M-1	6.50	4.923	L.F.	925	385	7.70	1,317.70	1,600
0100	300' to 500', maximum			4.43	7.223		1,275	565	11.25	1,851.25	2,250
0300	48" tread width walk, minimum			4.43	7.223		2,100	565	11.25	2,676.25	3,150
0400	100' to 350', maximum			3.82	8.377		2,425	655	13.05	3,093.05	3,675
0600	Ramp, 12° incline, 36" tread width, minimum			5.27	6.072		1,700	475	9.45	2,184.45	2,600
0700	70' to 90' maximum			3.82	8.377		2,525	655	13.05	3,193.05	3,775
0900	48" tread width, minimum			3.57	8.964		2,500	700	14	3,214	3,825
1000	40' to 70', maximum			2.91	10.997		3,125	860	17.15	4,002.15	4,750

14 42 Wheelchair Lifts

14 42 13 - Inclined Wheelchair Lifts

14 42 13.10 Inclined Wheelchair Lifts and Stairclimbers

		Crew	Daily Output	Labor-Hours	Unit	Material	2018 Bare Costs Labor	2018 Bare Costs Equipment	Total	Total Incl O&P
0010	**INCLINED WHEELCHAIR LIFTS AND STAIRCLIMBERS**									
7700	Stair climber (chair lift), single seat, minimum	2 Elev	1	16	Ea.	5,025	1,325		6,350	7,475
7800	Maximum		.20	80		6,875	6,575		13,450	17,400
8700	Stair lift, minimum		1	16		13,600	1,325		14,925	17,000
8900	Maximum		.20	80		21,600	6,575		28,175	33,500

14 42 16 - Vertical Wheelchair Lifts

14 42 16.10 Wheelchair Lifts

		Crew	Daily Output	Labor-Hours	Unit	Material	2018 Bare Costs Labor	2018 Bare Costs Equipment	Total	Total Incl O&P
0010	**WHEELCHAIR LIFTS**									
8000	Wheelchair lift, minimum	2 Elev	1	16	Ea.	6,875	1,325		8,200	9,500
8500	Maximum	"	.50	32	"	16,300	2,625		18,925	21,800

14 45 Vehicle Lifts

14 45 10 - Hydraulic Vehicle Lifts

14 45 10.10 Hydraulic Lifts

		Crew	Daily Output	Labor-Hours	Unit	Material	2018 Bare Costs Labor	2018 Bare Costs Equipment	Total	Total Incl O&P
0010	**HYDRAULIC LIFTS**									
2200	Single post, 8,000 lb. capacity	L-4	.40	60	Ea.	6,625	2,850		9,475	11,700
2810	Double post, 6,000 lb. capacity		2.67	8.989		9,350	425		9,775	11,000
2815	9,000 lb. capacity		2.29	10.480		22,200	500		22,700	25,200
2820	15,000 lb. capacity		2	12		25,100	570		25,670	28,500
2822	Four post, 26,000 lb. capacity		1.80	13.333		13,900	635		14,535	16,300
2825	30,000 lb. capacity		1.60	15		55,500	715		56,215	62,000
2830	Ramp style, 4 post, 25,000 lb. capacity		2	12		21,700	570		22,270	24,700
2835	35,000 lb. capacity		1	24		101,000	1,150		102,150	113,000
2840	50,000 lb. capacity		1	24		113,000	1,150		114,150	126,500
2845	75,000 lb. capacity		1	24		131,000	1,150		132,150	146,500
2850	For drive thru tracks, add, minimum					1,125			1,125	1,225
2855	Maximum					1,925			1,925	2,125
2860	Ramp extensions, 3' (set of 2)					1,150			1,150	1,250
2865	Rolling jack platform					3,975			3,975	4,375

For customer support on your Building Construction Costs with RSMeans data, call 800.448.8182.

481

14 45 Vehicle Lifts

14 45 10 - Hydraulic Vehicle Lifts

14 45 10.10 Hydraulic Lifts	Crew	Daily Output	Labor-Hours	Unit	Material	2018 Bare Costs Labor	Equipment	Total	Total Incl O&P
2870	Electric/hydraulic jacking beam				Ea.	10,600			10,600	11,700
2880	Scissor lift, portable, 6,000 lb. capacity				↓	10,400			10,400	11,500

14 91 Facility Chutes

14 91 33 - Laundry and Linen Chutes

14 91 33.10 Chutes

		Crew	Daily Output	Labor-Hours	Unit	Material	2018 Bare Costs Labor	Equipment	Total	Total Incl O&P
0011	**CHUTES**, linen, trash or refuse									
0050	Aluminized steel, 16 ga., 18" diameter	2 Shee	3.50	4.571	Floor	1,700	273		1,973	2,300
0100	24" diameter		3.20	5		1,875	299		2,174	2,525
0200	30" diameter		3	5.333		2,125	320		2,445	2,800
0300	36" diameter		2.80	5.714		2,625	340		2,965	3,425
0400	Galvanized steel, 16 ga., 18" diameter		3.50	4.571		1,025	273		1,298	1,550
0500	24" diameter		3.20	5		1,150	299		1,449	1,700
0600	30" diameter		3	5.333		1,275	320		1,595	1,875
0700	36" diameter		2.80	5.714		1,525	340		1,865	2,200
0800	Stainless steel, 18" diameter		3.50	4.571		3,000	273		3,273	3,725
0900	24" diameter		3.20	5		3,150	299		3,449	3,900
1000	30" diameter		3	5.333		3,750	320		4,070	4,600
1005	36" diameter		2.80	5.714	↓	3,950	340		4,290	4,850
1200	Linen chute bottom collector, aluminized steel		4	4	Ea.	1,425	239		1,664	1,950
1300	Stainless steel		4	4		1,825	239		2,064	2,375
1500	Refuse, bottom hopper, aluminized steel, 18" diameter		3	5.333		1,050	320		1,370	1,625
1600	24" diameter		3	5.333		1,250	320		1,570	1,850
1800	36" diameter		3	5.333	↓	2,500	320		2,820	3,225

14 91 82 - Trash Chutes

14 91 82.10 Trash Chutes and Accessories

		Crew	Daily Output	Labor-Hours	Unit	Material	2018 Bare Costs Labor	Equipment	Total	Total Incl O&P
0010	**TRASH CHUTES AND ACCESSORIES**									
2900	Package chutes, spiral type, minimum	2 Shee	4.50	3.556	Floor	2,425	213		2,638	3,000
3000	Maximum	"	1.50	10.667	"	6,325	640		6,965	7,950

14 92 Pneumatic Tube Systems

14 92 10 - Conventional, Automatic and Computer Controlled Pneumatic Tube Systems

14 92 10.10 Pneumatic Tube Systems

		Crew	Daily Output	Labor-Hours	Unit	Material	2018 Bare Costs Labor	Equipment	Total	Total Incl O&P
0010	**PNEUMATIC TUBE SYSTEMS**									
0020	100' long, single tube, 2 stations, stock									
0100	3" diameter	2 Stpi	.12	133	Total	3,375	8,400		11,775	16,400
0300	4" diameter	"	.09	178	"	4,275	11,200		15,475	21,600
0400	Twin tube, two stations or more, conventional system									
0600	2-1/2" round	2 Stpi	62.50	.256	L.F.	42	16.15		58.15	70.50
0700	3" round		46	.348		38	22		60	75
0900	4" round		49.60	.323		48	20.50		68.50	83.50
1000	4" x 7" oval		37.60	.426	↓	92	27		119	142
1050	Add for blower		2	8	System	5,200	505		5,705	6,475
1110	Plus for each round station, add		7.50	2.133	Ea.	1,350	134		1,484	1,700
1150	Plus for each oval station, add		7.50	2.133	"	1,350	134		1,484	1,700
1200	Alternate pricing method: base cost, economy model		.75	21.333	Total	5,750	1,350		7,100	8,350
1300	Custom model		.25	64	"	11,500	4,025		15,525	18,800
1500	Plus total system length, add, for economy model		93.40	.171	L.F.	8.25	10.80		19.05	25.50
1600	For custom model		37.60	.426	"	25	27		52	68

For customer support on your Building Construction Costs with RSMeans data, call 800.448.8182.

14 92 Pneumatic Tube Systems

14 92 10 – Conventional, Automatic and Computer Controlled Pneumatic Tube Systems

14 92 10.10 Pneumatic Tube Systems	Crew	Daily Output	Labor-Hours	Unit	Material	2018 Bare Costs Labor	Equipment	Total	Total Incl O&P
1800	Completely automatic system, 4" round, 15 to 50 stations	2 Stpi	.29	55.172	Station	22,600	3,475		26,075	30,000
2200	51 to 144 stations		.32	50		15,600	3,150		18,750	22,000
2400	6" round or 4" x 7" oval, 15 to 50 stations		.24	66.667		27,600	4,200		31,800	36,700
2800	51 to 144 stations		.23	69.565		21,100	4,375		25,475	29,800

For customer support on your Building Construction Costs with RSMeans data, call 800.448.8182.

483

Division Notes

	CREW	DAILY OUTPUT	LABOR-HOURS	UNIT	BARE COSTS				TOTAL INCL O&P
					MAT.	LABOR	EQUIP.	TOTAL	

Estimating Tips

Pipe for fire protection and all uses is located in Subdivisions 21 11 13 and 22 11 13.

The labor adjustment factors listed in Subdivision 22 01 02.20 also apply to Division 21.

Many, but not all, areas in the U.S. require backflow protection in the fire system. Insurance underwriters may have specific requirements for the type of materials to be installed or design requirements based on the hazard to be protected. Local jurisdictions may have requirements not covered by code. It is advisable to be aware of any special conditions.

For your reference, the following is a list of the most applicable Fire Codes and Standards which may be purchased from the NFPA, 1 Batterymarch Park, Quincy, MA 02169-7471.

- NFPA 1: Uniform Fire Code
- NFPA 10: Portable Fire Extinguishers
- NFPA 11: Low-, Medium-, and High-Expansion Foam
- NFPA 12: Carbon Dioxide Extinguishing Systems (Also companion 12A)
- NFPA 13: Installation of Sprinkler Systems (Also companion 13D, 13E, and 13R)
- NFPA 14: Installation of Standpipe and Hose Systems
- NFPA 15: Water Spray Fixed Systems for Fire Protection
- NFPA 16: Installation of Foam-Water Sprinkler and Foam-Water Spray Systems
- NFPA 17: Dry Chemical Extinguishing Systems (Also companion 17A)
- NFPA 18: Wetting Agents
- NFPA 20: Installation of Stationary Pumps for Fire Protection

- NFPA 22: Water Tanks for Private Fire Protection
- NFPA 24: Installation of Private Fire Service Mains and their Appurtenances
- NFPA 25: Inspection, Testing and Maintenance of Water-Based Fire Protection

Reference Numbers

Reference numbers are shown at the beginning of some major classifications. These numbers refer to related items in the Reference Section. The reference information may be an estimating procedure, an alternate pricing method, or technical information.

Note: Not all subdivisions listed here necessarily appear. ∎

Did you know?

RSMeans data is available through our online application with 24/7 access:

- Search for unit prices by keyword
- Leverage the most up-to-date data
- Build and export estimates

Try it free for 30 days!
www.rsmeans.com/2018freetrial

No part of this cost data may be reproduced, stored in a retrieval system, or transmitted in any form or by any means without prior written permission of Gordian.

21 05 Common Work Results for Fire Suppression

21 05 23 – General-Duty Valves for Water-Based Fire-Suppression Piping

21 05 23.50 General-Duty Valves	Crew	Daily Output	Labor-Hours	Unit	Material	2018 Bare Costs Labor	Equipment	Total	Total Incl O&P
0010 **GENERAL-DUTY VALVES**, for water-based fire suppression									
6200 Valves and components									
6500 Check, swing, C.I. body, brass fittings, auto. ball drip									
6520 4" size	Q-12	3	5.333	Ea.	380	287		667	850
6800 Check, wafer, butterfly type, C.I. body, bronze fittings									
6820 4" size	Q-12	4	4	Ea.	1,275	216		1,491	1,725

21 11 Facility Fire-Suppression Water-Service Piping

21 11 16 – Facility Fire Hydrants

21 11 16.50 Fire Hydrants for Buildings

21 11 16.50 Fire Hydrants for Buildings	Crew	Daily Output	Labor-Hours	Unit	Material	2018 Bare Costs Labor	Equipment	Total	Total Incl O&P
0010 **FIRE HYDRANTS FOR BUILDINGS**									
3750 Hydrants, wall, w/caps, single, flush, polished brass									
3800 2-1/2" x 2-1/2"	Q-12	5	3.200	Ea.	248	172		420	535
3840 2-1/2" x 3"	"	5	3.200	↓	500	172		672	810
3900 For polished chrome, add				↓	20%				
3950 Double, flush, polished brass									
4000 2-1/2" x 2-1/2" x 4"	Q-12	5	3.200	Ea.	765	172		937	1,100
4040 2-1/2" x 2-1/2" x 6"	"	4.60	3.478	↓	875	187		1,062	1,250
4200 For polished chrome, add				↓	10%				
4350 Double, projecting, polished brass									
4400 2-1/2" x 2-1/2" x 4"	Q-12	5	3.200	Ea.	296	172		468	585
4450 2-1/2" x 2-1/2" x 6"	"	4.60	3.478	"	605	187		792	950
4460 Valve control, dbl. flush/projecting hydrant, cap &									
4470 chain, extension rod & cplg., escutcheon, polished brass	Q-12	8	2	Ea.	272	108		380	460

21 11 19 – Fire-Department Connections

21 11 19.50 Connections for the Fire-Department

21 11 19.50 Connections for the Fire-Department	Crew	Daily Output	Labor-Hours	Unit	Material	2018 Bare Costs Labor	Equipment	Total	Total Incl O&P
0010 **CONNECTIONS FOR THE FIRE-DEPARTMENT**									
0020 For fire pro. cabinets, see Section 10 44 13.53									
7140 Standpipe connections, wall, w/plugs & chains									
7160 Single, flush, brass, 2-1/2" x 2-1/2"	Q-12	5	3.200	Ea.	188	172		360	465
7180 2-1/2" x 3"	"	5	3.200	"	196	172		368	475
7240 For polished chrome, add					15%				
7280 Double, flush, polished brass									
7300 2-1/2" x 2-1/2" x 4"	Q-12	5	3.200	Ea.	760	172		932	1,100
7330 2-1/2" x 2-1/2" x 6"	"	4.60	3.478	"	890	187		1,077	1,275
7400 For polished chrome, add					15%				
7440 For sill cock combination, add				Ea.	96			96	105
7900 Three way, flush, polished brass									
7920 2-1/2" (3) x 4"	Q-12	4.80	3.333	Ea.	1,950	180		2,130	2,425
7930 2-1/2" (3) x 6"	"	4.60	3.478	↓	2,000	187		2,187	2,475
8000 For polished chrome, add				↓	9%				
8020 Three way, projecting, polished brass									
8040 2-1/2" (3) x 4"	Q-12	4.80	3.333	Ea.	850	180		1,030	1,200

For customer support on your Building Construction Costs with RSMeans data, call 800.448.8182.

21 12 13 – Fire-Suppression Hoses and Nozzles

21 12 13.50 Fire Hoses and Nozzles	Crew	Daily Output	Labor-Hours	Unit	Material	2018 Bare Costs Labor	Equipment	Total	Total Incl O&P
0010 **FIRE HOSES AND NOZZLES**									
0200 Adapters, rough brass, straight hose threads									
0220 One piece, female to male, rocker lugs									
0240 1" x 1"				Ea.	58			58	63.50
0320 2" x 2"					42			42	46
0380 2-1/2" x 2-1/2"					19			19	21
2200 Hose, less couplings									
2260 Synthetic jacket, lined, 300 lb. test, 1-1/2" diameter	Q-12	2600	.006	L.F.	3	.33		3.33	3.80
2270 2" diameter		2200	.007		2.45	.39		2.84	3.29
2280 2-1/2" diameter		2200	.007		5.55	.39		5.94	6.70
2290 3" diameter		2200	.007		3.10	.39		3.49	4
2360 High strength, 500 lb. test, 1-1/2" diameter		2600	.006		2.40	.33		2.73	3.14
2380 2-1/2" diameter		2200	.007		5.35	.39		5.74	6.50
5600 Nozzles, brass									
5620 Adjustable fog, 3/4" booster line				Ea.	140			140	154
5630 1" booster line					137			137	151
5640 1-1/2" leader line					103			103	113
5660 2-1/2" direct connection					184			184	202
5680 2-1/2" playpipe nozzle					230			230	253
5780 For chrome plated, add					8%				
5850 Electrical fire, adjustable fog, no shock									
5900 1-1/2"				Ea.	495			495	545
5920 2-1/2"					840			840	925
5980 For polished chrome, add					6%				
6200 Heavy duty, comb. adj. fog and str. stream, with handle									
6210 1" booster line				Ea.	191			191	210

21 12 19 – Fire-Suppression Hose Racks

21 12 19.50 Fire Hose Racks

	Crew	Daily Output	Labor-Hours	Unit	Material	2018 Bare Costs Labor	Equipment	Total	Total Incl O&P
0010 **FIRE HOSE RACKS**									
2600 Hose rack, swinging, for 1-1/2" diameter hose,									
2620 Enameled steel, 50' and 75' lengths of hose	Q-12	20	.800	Ea.	69	43		112	141
2640 100' and 125' lengths of hose	"	20	.800	"	90	43		133	164

21 12 23 – Fire-Suppression Hose Valves

21 12 23.70 Fire Hose Valves

	Crew	Daily Output	Labor-Hours	Unit	Material	2018 Bare Costs Labor	Equipment	Total	Total Incl O&P
0010 **FIRE HOSE VALVES**									
0080 Wheel handle, 300 lb., 1-1/2"	1 Spri	12	.667	Ea.	114	40		154	186
0090 2-1/2"	"	7	1.143	"	206	68.50		274.50	330
0100 For polished brass, add					35%				
0110 For polished chrome, add					50%				

21 13 13 – Wet-Pipe Sprinkler Systems

21 13 13.50 Wet-Pipe Sprinkler System Components

21 13 13.50 Wet-Pipe Sprinkler System Components	Crew	Daily Output	Labor-Hours	Unit	Material	2018 Bare Costs Labor	Equipment	Total	Total Incl O&P
0010 **WET-PIPE SPRINKLER SYSTEM COMPONENTS**									
2600 Sprinkler heads, not including supply piping									
3700 Standard spray, pendent or upright, brass, 135°F to 286°F									
3730 1/2" NPT, 7/16" orifice	1 Spri	16	.500	Ea.	16	30		46	62.50
3740 1/2" NPT, 1/2" orifice	"	16	.500		10.40	30		40.40	56.50
3860 For wax and lead coating, add					41.50			41.50	46
3880 For wax coating, add					26			26	28.50
3900 For lead coating, add					31.50			31.50	34.50
3920 For 360°F, same cost									
3930 For 400°F	1 Spri	16	.500	Ea.	99	30		129	154
3940 For 500°F	"	16	.500	"	99	30		129	154
4500 Sidewall, horizontal, brass, 135°F to 286°F									
4520 1/2" NPT, 1/2" orifice	1 Spri	16	.500	Ea.	26.50	30		56.50	74.50
4540 For 360°F, same cost									
4800 Recessed pendent, brass, 135°F to 286°F									
4820 1/2" NPT, 3/8" orifice	1 Spri	10	.800	Ea.	45	48		93	122
4830 1/2" NPT, 7/16" orifice		10	.800		19.55	48		67.55	94
4840 1/2" NPT, 1/2" orifice		10	.800		15.15	48		63.15	89

21 13 16 – Dry-Pipe Sprinkler Systems

21 13 16.50 Dry-Pipe Sprinkler System Components

21 13 16.50 Dry-Pipe Sprinkler System Components	Crew	Daily Output	Labor-Hours	Unit	Material	Labor	Equipment	Total	Total Incl O&P
0010 **DRY-PIPE SPRINKLER SYSTEM COMPONENTS**									
0600 Accelerator	1 Spri	8	1	Ea.	820	60		880	990
2600 Sprinkler heads, not including supply piping									
2640 Dry, pendent, 1/2" orifice, 3/4" or 1" NPT									
2700 15-1/4" to 18" length	1 Spri	14	.571	Ea.	154	34		188	221
2710 18-1/4" to 21" length		13	.615		160	37		197	232
2720 21-1/4" to 24" length		13	.615		166	37		203	238
2730 24-1/4" to 27" length		13	.615		172	37		209	245

21 13 19 – Preaction Sprinkler Systems

21 13 19.50 Preaction Sprinkler System Components

21 13 19.50 Preaction Sprinkler System Components	Crew	Daily Output	Labor-Hours	Unit	Material	Labor	Equipment	Total	Total Incl O&P
0010 **PREACTION SPRINKLER SYSTEM COMPONENTS**									
3000 Preaction valve cabinet									
3100 Single interlock, pneum. release, panel, 1/2 HP comp. regul. air trim									
3110 1-1/2"	Q-12	3	5.333	Ea.	44,400	287		44,687	49,200
3120 2"		3	5.333		44,400	287		44,687	49,200
3130 2-1/2"		3	5.333		44,400	287		44,687	49,200
3140 3"		3	5.333		44,500	287		44,787	49,300
3150 4"		2	8		46,100	430		46,530	51,000
3160 6"	Q-13	4	8		48,400	455		48,855	54,000
3200 Double interlock, pneum. release, panel, 1/2 HP comp. regul. air trim									
3210 1-1/2"	Q-12	3	5.333	Ea.	43,200	287		43,487	47,900
3220 2"		3	5.333		43,200	287		43,487	47,900
3230 2-1/2"		3	5.333		43,200	287		43,487	47,900
3240 3"		3	5.333		43,200	287		43,487	48,000
3250 4"		2	8		45,400	430		45,830	50,500
3260 6"	Q-13	4	8		47,200	455		47,655	52,500

21 13 26 – Deluge Fire-Suppression Sprinkler Systems

21 13 26.50 Deluge Fire-Suppression Sprinkler Sys. Comp.

21 13 26.50 Deluge Fire-Suppression Sprinkler Sys. Comp.	Crew	Daily Output	Labor-Hours	Unit	Material	Labor	Equipment	Total	Total Incl O&P
0010 **DELUGE FIRE-SUPPRESSION SPRINKLER SYSTEM COMPONENTS**									
1400 Deluge system, monitoring panel w/deluge valve & trim	1 Spri	18	.444	Ea.	6,950	26.50		6,976.50	7,675
6200 Valves and components									

488

For customer support on your Building Construction Costs with RSMeans data, call 800.448.8182.

21 13 Fire-Suppression Sprinkler Systems

21 13 26 – Deluge Fire-Suppression Sprinkler Systems

21 13 26.50 Deluge Fire-Suppression Sprinkler Sys. Comp.	Crew	Daily Output	Labor-Hours	Unit	Material	2018 Bare Costs Labor	Equipment	Total	Total Incl O&P
7000	Deluge, assembly, incl. trim, pressure									
7020	operated relief, emergency release, gauges									
7040	2" size	Q-12	2	8	Ea.	3,975	430		4,405	5,000
7060	3" size	"	1.50	10.667	"	4,450	575		5,025	5,775

21 13 39 – Foam-Water Systems

21 13 39.50 Foam-Water System Components

		Crew	Daily Output	Labor-Hours	Unit	Material	Labor	Equipment	Total	Total Incl O&P
0010	**FOAM-WATER SYSTEM COMPONENTS**									
2600	Sprinkler heads, not including supply piping									
3600	Foam-water, pendent or upright, 1/2" NPT	1 Spri	12	.667	Ea.	216	40		256	299

21 21 Carbon-Dioxide Fire-Extinguishing Systems

21 21 16 – Carbon-Dioxide Fire-Extinguishing Equipment

21 21 16.50 CO2 Fire Extinguishing System

		Crew	Daily Output	Labor-Hours	Unit	Material	Labor	Equipment	Total	Total Incl O&P
0010	**CO$_2$ FIRE EXTINGUISHING SYSTEM**									
0042	For detectors and control stations, see Section 28 31 23.50									
0100	Control panel, single zone with batteries (2 zones det., 1 suppr.)	1 Elec	1	8	Ea.	920	465		1,385	1,725
0150	Multizone (4) with batteries (8 zones det., 4 suppr.)	"	.50	16		2,775	930		3,705	4,450
1000	Dispersion nozzle, CO$_2$, 3" x 5"	1 Plum	18	.444		144	27.50		171.50	200
2000	Extinguisher, CO$_2$ system, high pressure, 75 lb. cylinder	Q-1	6	2.667		1,300	149		1,449	1,675
2100	100 lb. cylinder	"	5	3.200		1,675	179		1,854	2,125
3000	Electro/mechanical release	L-1	4	4		935	241		1,176	1,375
3400	Manual pull station	1 Plum	6	1.333		79	83		162	212
4000	Pneumatic damper release	"	8	1		155	62		217	264

21 22 Clean-Agent Fire-Extinguishing Systems

21 22 16 – Clean-Agent Fire-Extinguishing Equipment

21 22 16.50 Clean-Agent Extinguishing Systems

		Crew	Daily Output	Labor-Hours	Unit	Material	Labor	Equipment	Total	Total Incl O&P
0010	**CLEAN-AGENT EXTINGUISHING SYSTEMS**									
0020	FM200 fire extinguishing system									
1100	Dispersion nozzle FM200, 1-1/2"	1 Plum	14	.571	Ea.	182	35.50		217.50	254
2400	Extinguisher, FM200 system, filled, with mounting bracket									
2460	26 lb. container	Q-1	8	2	Ea.	1,975	112		2,087	2,325
2480	44 lb. container		7	2.286		2,700	128		2,828	3,175
2500	63 lb. container		6	2.667		3,100	149		3,249	3,650
2520	101 lb. container		5	3.200		4,900	179		5,079	5,650
2540	196 lb. container		4	4		6,625	224		6,849	7,600
6000	FM200 system, simple nozzle layout, with broad dispersion				C.F.	1.76			1.76	1.94
6010	Extinguisher, FM200 system, filled, with mounting bracket									
6020	Complex nozzle layout and/or including underfloor dispersion				C.F.	3.50			3.50	3.85
6100	20,000 C.F. 2 exits, 8' clng					2			2	2.20
6200	100,000 C.F. 4 exits, 8' clng					1.80			1.80	1.98
6300	250,000 C.F. 6 exits, 8' clng					1.52			1.52	1.67
7010	HFC-227ea fire extinguishing system									
7100	Cylinders with clean-agent									
7110	Does not include pallete jack/fork lift rental fees									
7120	70 lb. cyl, w/35 lb. agent, no solenoid	Q-12	14	1.143	Ea.	2,400	61.50		2,461.50	2,750
7130	70 lb. cyl w/70 lb. agent, no solenoid		10	1.600		3,075	86		3,161	3,525
7140	70 lb. cyl w/35 lb. agent, w/solenoid		14	1.143		3,275	61.50		3,336.50	3,700
7150	70 lb. cyl w/70 lb. agent, w/solenoid		10	1.600		4,150	86		4,236	4,675

For customer support on your Building Construction Costs with RSMeans data, call 800.448.8182.

489

21 22 Clean-Agent Fire-Extinguishing Systems

21 22 16 – Clean-Agent Fire-Extinguishing Equipment

21 22 16.50 Clean-Agent Extinguishing Systems	Crew	Daily Output	Labor-Hours	Unit	Material	2018 Bare Costs Labor	Equipment	Total	Total Incl O&P
7220	250 lb. cyl, w/125 lb. agent, no solenoid	Q-12	8	2	Ea.	5,425	108		5,533	6,125
7230	250 lb. cyl, w/250 lb. agent, no solenoid		5	3.200		5,425	172		5,597	6,200
7240	250 lb. cyl, w/125 lb. agent, w/solenoid		8	2		7,100	108		7,208	7,975
7250	250 lb. cyl, w/250 lb. agent, w/solenoid		5	3.200		10,200	172		10,372	11,500
7320	560 lb. cyl, w/300 lb. agent, no solenoid		4	4		10,200	216		10,416	11,500
7330	560 lb. cyl, w/560 lb. agent, no solenoid		2.50	6.400		15,200	345		15,545	17,300
7340	560 lb. cyl, w/300 lb. agent, w/solenoid		4	4		13,100	216		13,316	14,700
7350	560 lb. cyl, w/560 lb. agent, w/solenoid		2.50	6.400		19,500	345		19,845	22,000
7420	1,200 lb. cyl, w/600 lb. agent, no solenoid	Q-13	4	8		19,200	455		19,655	21,800
7430	1,200 lb. cyl, w/1,200 lb. agent, no solenoid		3	10.667		30,900	610		31,510	34,900
7440	1,200 lb. cyl, w/600 lb. agent, w/solenoid		4	8		45,700	455		46,155	50,500
7450	1,200 lb. cyl, w/1,200 lb. agent, w/solenoid		3	10.667		71,500	610		72,110	79,500
7500	Accessories									
7510	Dispersion nozzle	1 Spri	16	.500	Ea.	95	30		125	150
7520	Agent release panel	1 Elec	4	2		60	116		176	240
7530	Maintenance switch	"	6	1.333		37.50	77.50		115	157
7540	Solenoid valve, 12v dc	1 Spri	8	1		232	60		292	345
7550	12v ac		8	1		470	60		530	610
7560	12v dc, explosion proof		8	1		390	60		450	520

21 31 Centrifugal Fire Pumps

21 31 13 – Electric-Drive, Centrifugal Fire Pumps

21 31 13.50 Electric-Drive Fire Pumps

		Crew	Daily Output	Labor-Hours	Unit	Material	2018 Bare Costs Labor	Equipment	Total	Total Incl O&P
0010	**ELECTRIC-DRIVE FIRE PUMPS** Including controller, fittings and relief valve									
3100	250 GPM, 55 psi, 15 HP, 3550 RPM, 2" pump	Q-13	.70	45.714	Ea.	15,100	2,600		17,700	20,500
3200	500 GPM, 50 psi, 27 HP, 1770 RPM, 4" pump		.68	47.059		15,500	2,675		18,175	21,100
3350	750 GPM, 50 psi, 44 HP, 1770 RPM, 5" pump		.64	50		16,200	2,850		19,050	22,100
3400	750 GPM, 100 psi, 66 HP, 3550 RPM, 4" pump		.58	55.172		18,700	3,150		21,850	25,300
5000	For jockey pump 1", 3 HP, with control, add	Q-12	2	8		2,875	430		3,305	3,800

21 31 16 – Diesel-Drive, Centrifugal Fire Pumps

21 31 16.50 Diesel-Drive Fire Pumps

		Crew	Daily Output	Labor-Hours	Unit	Material	2018 Bare Costs Labor	Equipment	Total	Total Incl O&P
0010	**DIESEL-DRIVE FIRE PUMPS** Including controller, fittings and relief valve									
0050	500 GPM, 50 psi, 27 HP, 4" pump	Q-13	.64	50	Ea.	39,200	2,850		42,050	47,400
0200	750 GPM, 50 psi, 44 HP, 5" pump		.60	53.333		40,400	3,050		43,450	49,000
0400	1000 GPM, 100 psi, 89 HP, 4" pump		.56	57.143		45,800	3,250		49,050	55,500
0700	2000 GPM, 100 psi, 167 HP, 6" pump		.34	94.118		58,000	5,375		63,375	72,000
0950	3500 GPM, 100 psi, 300 HP, 10" pump		.24	133		80,500	7,600		88,100	100,000

For customer support on your Building Construction Costs with RSMeans data, call 800.448.8182.

Estimating Tips
22 10 00 Plumbing Piping and Pumps

This subdivision is primarily basic pipe and related materials. The pipe may be used by any of the mechanical disciplines, i.e., plumbing, fire protection, heating, and air conditioning.

Note: CPVC plastic piping approved for fire protection is located in 21 11 13.

- The labor adjustment factors listed in Subdivision 22 01 02.20 apply throughout Divisions 21, 22, and 23. CAUTION: the correct percentage may vary for the same items. For example, the percentage add for the basic pipe installation should be based on the maximum height that the installer must install for that particular section. If the pipe is to be located 14' above the floor but it is suspended on threaded rod from beams, the bottom flange of which is 18' high (4' rods), then the height is actually 18' and the add is 20%. The pipe cover, however, does not have to go above the 14', and so the add should be 10%.

- Most pipe is priced first as straight pipe with a joint (coupling, weld, etc.) every 10' and a hanger usually every 10'. There are exceptions with hanger spacing such as for cast iron pipe (5')

and plastic pipe (3 per 10'). Following each type of pipe there are several lines listing sizes and the amount to be subtracted to delete couplings and hangers. This is for pipe that is to be buried or supported together on trapeze hangers. The reason that the couplings are deleted is that these runs are usually long, and frequently longer lengths of pipe are used. By deleting the couplings, the estimator is expected to look up and add back the correct reduced number of couplings.

- When preparing an estimate, it may be necessary to approximate the fittings. Fittings usually run between 25% and 50% of the cost of the pipe. The lower percentage is for simpler runs, and the higher number is for complex areas, such as mechanical rooms.

- For historic restoration projects, the systems must be as invisible as possible, and pathways must be sought for pipes, conduit, and ductwork. While installations in accessible spaces (such as basements and attics) are relatively straightforward to estimate, labor costs may be more difficult to determine when delivery systems must be concealed.

22 40 00 Plumbing Fixtures
- Plumbing fixture costs usually require two lines: the fixture itself and its "rough-in, supply, and waste."
- In the Assemblies Section (Plumbing D2010) for the desired fixture, the System Components Group at the center of the page shows the fixture on the first line. The rest of the list (fittings, pipe, tubing, etc.) will total up to what we refer to in the Unit Price section as "Rough-in, supply, waste, and vent." Note that for most fixtures we allow a nominal 5' of tubing to reach from the fixture to a main or riser.
- Remember that gas- and oil-fired units need venting.

Reference Numbers
Reference numbers are shown at the beginning of some major classifications. These numbers refer to related items in the Reference Section. The reference information may be an estimating procedure, an alternate pricing method, or technical information.

Note: Not all subdivisions listed here necessarily appear. ■

No part of this cost data may be reproduced, stored in a retrieval system, or transmitted in any form or by any means without prior written permission of Gordian.

Did you know?
RSMeans data is available through our online application with 24/7 access:
- Search for unit prices by keyword
- Leverage the most up-to-date data
- Build and export estimates

Try it free for 30 days!
www.rsmeans.com/2018freetrial

22 01 Operation and Maintenance of Plumbing

22 01 02 – Labor Adjustments

22 01 02.10 Boilers, General

		Crew	Daily Output	Labor-Hours	Unit	Material	2018 Bare Costs Labor	2018 Bare Costs Equipment	Total	Total Incl O&P
0010	**BOILERS, GENERAL**, Prices do not include flue piping, elec. wiring,									
0020	gas or oil piping, boiler base, pad, or tankless unless noted									
0100	Boiler H.P.: 10 KW = 34 lb./steam/hr. = 33,475 BTU/hr.									
0150	To convert SFR to BTU rating: Hot water, 150 x SFR;									
0160	Forced hot water, 180 x SFR; steam, 240 x SFR									

22 01 02.20 Labor Adjustment Factors

		Crew	Daily Output	Labor-Hours	Unit	Material	2018 Bare Costs Labor	2018 Bare Costs Equipment	Total	Total Incl O&P
0010	**LABOR ADJUSTMENT FACTORS** (For Div. 21, 22 and 23) R220102-20									
0100	Labor factors, The below are reasonable suggestions, however									
0110	each project must be evaluated for its own peculiarities, and									
0120	the adjustments be increased or decreased depending on the									
0130	severity of the special conditions.									
1000	Add to labor for elevated installation (Above floor level)									
1080	10' to 14.5' high						10%			
1100	15' to 19.5' high						20%			
1120	20' to 24.5' high						25%			
1140	25' to 29.5' high						35%			
1160	30' to 34.5' high						40%			
1180	35' to 39.5' high						50%			
1200	40' and higher						55%			
2000	Add to labor for crawl space									
2100	3' high						40%			
2140	4' high						30%			
3000	Add to labor for multi-story building									
3010	For new construction (No elevator available)									
3100	Add for floors 3 thru 10						5%			
3110	Add for floors 11 thru 15						10%			
3120	Add for floors 16 thru 20						15%			
3130	Add for floors 21 thru 30						20%			
3140	Add for floors 31 and up						30%			
3170	For existing structure (Elevator available)									
3180	Add for work on floor 3 and above						2%			
4000	Add to labor for working in existing occupied buildings									
4100	Hospital						35%			
4140	Office building						25%			
4180	School						20%			
4220	Factory or warehouse						15%			
4260	Multi dwelling						15%			
5000	Add to labor, miscellaneous									
5100	Cramped shaft						35%			
5140	Congested area						15%			
5180	Excessive heat or cold						30%			
9000	Labor factors, The above are reasonable suggestions, however									
9010	each project should be evaluated for its own peculiarities.									
9100	Other factors to be considered are:									
9140	Movement of material and equipment through finished areas									
9180	Equipment room									
9220	Attic space									
9260	No service road									
9300	Poor unloading/storage area									
9340	Congested site area/heavy traffic									

For customer support on your Building Construction Costs with RSMeans data, call 800.448.8182.

22 05 05 – Selective Demolition for Plumbing

22 05 05.10 Plumbing Demolition

	Crew	Daily Output	Labor-Hours	Unit	Material	2018 Bare Costs Labor	Equipment	Total	Total Incl O&P
0010 **PLUMBING DEMOLITION** R220105-10									
1020 Fixtures, including 10' piping									
1100 Bathtubs, cast iron	1 Plum	4	2	Ea.		124		124	187
1120 Fiberglass		6	1.333			83		83	125
1140 Steel		5	1.600			99.50		99.50	150
1200 Lavatory, wall hung		10	.800			49.50		49.50	75
1220 Counter top		8	1			62		62	93.50
1300 Sink, single compartment		8	1			62		62	93.50
1320 Double compartment		7	1.143			71		71	107
1400 Water closet, floor mounted		8	1			62		62	93.50
1420 Wall mounted		7	1.143			71		71	107
1500 Urinal, floor mounted		4	2			124		124	187
1520 Wall mounted		7	1.143			71		71	107
1600 Water fountains, free standing		8	1			62		62	93.50
1620 Wall or deck mounted		6	1.333	▼		83		83	125
2000 Piping, metal, up thru 1-1/2" diameter		200	.040	L.F.		2.49		2.49	3.74
2050 2" thru 3-1/2" diameter	▼	150	.053			3.31		3.31	4.99
2100 4" thru 6" diameter	2 Plum	100	.160			9.95		9.95	15
2150 8" thru 14" diameter	"	60	.267			16.55		16.55	25
2153 16" thru 20" diameter	Q-18	70	.343			20	.85	20.85	31.50
2155 24" thru 26" diameter		55	.436			25.50	1.08	26.58	39.50
2156 30" thru 36" diameter	▼	40	.600			35.50	1.48	36.98	54.50
2160 Plastic pipe with fittings, up thru 1-1/2" diameter	1 Plum	250	.032			1.99		1.99	3
2162 2" thru 3" diameter	"	200	.040			2.49		2.49	3.74
2164 4" thru 6" diameter	Q-1	200	.080			4.47		4.47	6.75
2166 8" thru 14" diameter		150	.107			5.95		5.95	9
2168 16" diameter	▼	100	.160	▼		8.95		8.95	13.50
2212 Deduct for salvage, aluminum scrap				Ton				700	770
2214 Brass scrap								2,675	2,675
2216 Copper scrap								3,525	3,525
2218 Lead scrap								570	570
2220 Steel scrap				▼				200	200
2250 Water heater, 40 gal.	1 Plum	6	1.333	Ea.		83		83	125
9470 Water softener	Q-1	2	8	"		445		445	675

22 05 23 – General-Duty Valves for Plumbing Piping

22 05 23.10 Valves, Brass

	Crew	Daily Output	Labor-Hours	Unit	Material	2018 Bare Costs Labor	Equipment	Total	Total Incl O&P
0010 **VALVES, BRASS**									
0500 Gas cocks, threaded									
0530 1/2"	1 Plum	24	.333	Ea.	13.15	20.50		33.65	45.50
0540 3/4"		22	.364		15.80	22.50		38.30	51.50
0550 1"		19	.421		31	26		57	74
0560 1-1/4"	▼	15	.533	▼	44.50	33		77.50	99

22 05 23.20 Valves, Bronze

	Crew	Daily Output	Labor-Hours	Unit	Material	2018 Bare Costs Labor	Equipment	Total	Total Incl O&P
0010 **VALVES, BRONZE**									
1020 Angle, 150 lb., rising stem, threaded									
1030 1/8"	1 Plum	24	.333	Ea.	137	20.50		157.50	181
1040 1/4"		24	.333		137	20.50		157.50	181
1050 3/8"		24	.333		137	20.50		157.50	181
1060 1/2"		22	.364		137	22.50		159.50	184
1070 3/4"		20	.400		211	25		236	270
1080 1"		19	.421		300	26		326	375
1100 1-1/2"	▼	13	.615	▼	450	38.50		488.50	555

22 05 23.20 Valves, Bronze		Crew	Daily Output	Labor-Hours	Unit	Material	2018 Bare Costs Labor	Equipment	Total	Total Incl O&P
1102	Soldered same price as threaded									
1110	2"	1 Plum	11	.727	Ea.	725	45		770	870
1300	Ball									
1398	Threaded, 150 psi									
1400	1/4"	1 Plum	24	.333	Ea.	14.50	20.50		35	47
1430	3/8"		24	.333		14.50	20.50		35	47
1450	1/2"		22	.364		14.50	22.50		37	50
1460	3/4"		20	.400		24	25		49	64
1470	1"		19	.421		30	26		56	72.50
1480	1-1/4"		15	.533		52.50	33		85.50	108
1490	1-1/2"		13	.615		68.50	38.50		107	133
1500	2"		11	.727		82.50	45		127.50	159
1522	Solder the same price as threaded									
1750	Check, swing, class 150, regrinding disc, threaded									
1800	1/8"	1 Plum	24	.333	Ea.	70	20.50		90.50	108
1830	1/4"		24	.333		65	20.50		85.50	103
1840	3/8"		24	.333		69	20.50		89.50	107
1850	1/2"		24	.333		66.50	20.50		87	104
1860	3/4"		20	.400		86.50	25		111.50	133
1870	1"		19	.421		141	26		167	195
1880	1-1/4"		15	.533		203	33		236	274
1890	1-1/2"		13	.615		263	38.50		301.50	345
1900	2"		11	.727		305	45		350	405
1910	2-1/2"	Q-1	15	1.067		780	59.50		839.50	950
2000	For 200 lb., add					5%	10%			
2040	For 300 lb., add					15%	15%			
2850	Gate, N.R.S., soldered, 125 psi									
2900	3/8"	1 Plum	24	.333	Ea.	62.50	20.50		83	99.50
2920	1/2"		24	.333		56	20.50		76.50	92.50
2940	3/4"		20	.400		64	25		89	108
2950	1"		19	.421		74.50	26		100.50	122
2960	1-1/4"		15	.533		137	33		170	201
2970	1-1/2"		13	.615		154	38.50		192.50	228
2980	2"		11	.727		181	45		226	267
2990	2-1/2"	Q-1	15	1.067		495	59.50		554.50	635
3000	3"	"	13	1.231		550	69		619	710
3850	Rising stem, soldered, 300 psi									
3950	1"	1 Plum	19	.421	Ea.	191	26		217	250
3980	2"	"	11	.727		520	45		565	645
4000	3"	Q-1	13	1.231		1,725	69		1,794	2,000
4250	Threaded, class 150									
4310	1/4"	1 Plum	24	.333	Ea.	74	20.50		94.50	113
4320	3/8"		24	.333		74	20.50		94.50	113
4330	1/2"		24	.333		67	20.50		87.50	105
4340	3/4"		20	.400		78.50	25		103.50	124
4350	1"		19	.421		105	26		131	156
4360	1-1/4"		15	.533		143	33		176	207
4370	1-1/2"		13	.615		181	38.50		219.50	257
4380	2"		11	.727		243	45		288	335
4390	2-1/2"	Q-1	15	1.067		565	59.50		624.50	715
4400	3"	"	13	1.231		790	69		859	975
4500	For 300 psi, threaded, add					100%	15%			
4540	For chain operated type, add					15%				

494

For customer support on your Building Construction Costs with RSMeans data, call 800.448.8182.

22 05 23.20 Valves, Bronze		Crew	Daily Output	Labor-Hours	Unit	Material	2018 Bare Costs Labor	Equipment	Total	Total Incl O&P
4850	Globe, class 150, rising stem, threaded									
4920	1/4"	1 Plum	24	.333	Ea.	104	20.50		124.50	145
4940	3/8"		24	.333		102	20.50		122.50	144
4950	1/2"		24	.333		102	20.50		122.50	144
4960	3/4"		20	.400		137	25		162	188
4970	1"		19	.421		206	26		232	266
4980	1-1/4"		15	.533		299	33		332	380
4990	1-1/2"		13	.615		420	38.50		458.50	525
5000	2"		11	.727		550	45		595	675
5010	2-1/2"	Q-1	15	1.067		1,250	59.50		1,309.50	1,475
5020	3"	"	13	1.231		1,775	69		1,844	2,050
5120	For 300 lb. threaded, add					50%	15%			
5600	Relief, pressure & temperature, self-closing, ASME, threaded									
5640	3/4"	1 Plum	28	.286	Ea.	209	17.75		226.75	257
5650	1"		24	.333		335	20.50		355.50	400
5660	1-1/4"		20	.400		685	25		710	795
5670	1-1/2"		18	.444		1,325	27.50		1,352.50	1,500
5680	2"		16	.500		1,425	31		1,456	1,625
5950	Pressure, poppet type, threaded									
6000	1/2"	1 Plum	30	.267	Ea.	75.50	16.55		92.05	108
6040	3/4"	"	28	.286	"	85.50	17.75		103.25	121
6400	Pressure, water, ASME, threaded									
6440	3/4"	1 Plum	28	.286	Ea.	144	17.75		161.75	185
6450	1"		24	.333		296	20.50		316.50	355
6460	1-1/4"		20	.400		450	25		475	535
6470	1-1/2"		18	.444		650	27.50		677.50	750
6480	2"		16	.500		935	31		966	1,075
6490	2-1/2"		15	.533		3,500	33		3,533	3,900
6900	Reducing, water pressure									
6920	300 psi to 25-75 psi, threaded or sweat									
6940	1/2"	1 Plum	24	.333	Ea.	440	20.50		460.50	515
6950	3/4"		20	.400		450	25		475	535
6960	1"		19	.421		700	26		726	810
6970	1-1/4"		15	.533		1,225	33		1,258	1,400
6980	1-1/2"		13	.615		1,850	38.50		1,888.50	2,075
8350	Tempering, water, sweat connections									
8400	1/2"	1 Plum	24	.333	Ea.	106	20.50		126.50	147
8440	3/4"	"	20	.400	"	139	25		164	191
8650	Threaded connections									
8700	1/2"	1 Plum	24	.333	Ea.	148	20.50		168.50	194
8740	3/4"		20	.400		875	25		900	1,000
8750	1"		19	.421		985	26		1,011	1,125
8760	1-1/4"		15	.533		1,525	33		1,558	1,725
8770	1-1/2"		13	.615		1,650	38.50		1,688.50	1,875
8780	2"		11	.727		2,500	45		2,545	2,825
8800	Water heater water & gas safety shut off									
8810	Protection against a leaking water heater									
8814	Shut off valve	1 Plum	16	.500	Ea.	185	31		216	251
8818	Water heater dam		32	.250		31.50	15.55		47.05	58
8822	Gas control wiring harness		32	.250		19.75	15.55		35.30	45
8830	Whole house flood safety shut off									
8834	Connections									
8838	3/4" NPT	1 Plum	12	.667	Ea.	935	41.50		976.50	1,100

22 05 23.20 Valves, Bronze

		Crew	Daily Output	Labor-Hours	Unit	Material	2018 Bare Costs Labor	2018 Bare Costs Equipment	Total	Total Incl O&P
8842	1" NPT	1 Plum	11	.727	Ea.	960	45		1,005	1,125
8846	1-1/4" NPT	↓	10	.800	↓	995	49.50		1,044.50	1,175

22 05 23.60 Valves, Plastic

		Crew	Daily Output	Labor-Hours	Unit	Material	2018 Bare Costs Labor	2018 Bare Costs Equipment	Total	Total Incl O&P
0010	**VALVES, PLASTIC**									
1100	Angle, PVC, threaded									
1110	1/4"	1 Plum	26	.308	Ea.	51	19.10		70.10	85
1120	1/2"		26	.308		73	19.10		92.10	110
1130	3/4"		25	.320		87	19.90		106.90	126
1140	1"	↓	23	.348	↓	105	21.50		126.50	149
1150	Ball, PVC, socket or threaded, true union									
1230	1/2"	1 Plum	26	.308	Ea.	42.50	19.10		61.60	76
1240	3/4"		25	.320		42.50	19.90		62.40	76.50
1250	1"		23	.348		50.50	21.50		72	88
1260	1-1/4"		21	.381		80	23.50		103.50	124
1270	1-1/2"		20	.400		80	25		105	126
1280	2"	↓	17	.471		127	29.50		156.50	183
1360	For PVC, flanged, add				↓	100%	15%			
1650	CPVC, socket or threaded, single union									
1700	1/2"	1 Plum	26	.308	Ea.	55.50	19.10		74.60	90
1720	3/4"		25	.320		69	19.90		88.90	106
1730	1"		23	.348		84	21.50		105.50	125
1750	1-1/4"		21	.381		146	23.50		169.50	196
1760	1-1/2"	↓	20	.400		134	25		159	185
1840	For CPVC, flanged, add					65%	15%			
1880	For true union, socket or threaded, add				↓	50%	5%			
2050	Polypropylene, threaded									
2100	1/4"	1 Plum	26	.308	Ea.	37	19.10		56.10	70
2120	3/8"		26	.308		37	19.10		56.10	70
2130	1/2"		26	.308		37	19.10		56.10	69.50
2140	3/4"		25	.320		44	19.90		63.90	78.50
2150	1"		23	.348		51.50	21.50		73	89
2160	1-1/4"		21	.381		68.50	23.50		92	111
2170	1-1/2"		20	.400		84.50	25		109.50	131
2180	2"	↓	17	.471	↓	114	29.50		143.50	169
4850	Foot valve, PVC, socket or threaded									
4900	1/2"	1 Plum	34	.235	Ea.	59	14.60		73.60	87
4930	3/4"		32	.250		66.50	15.55		82.05	97
4940	1"		28	.286		86.50	17.75		104.25	122
4950	1-1/4"		27	.296		164	18.40		182.40	208
4960	1-1/2"	↓	26	.308	↓	166	19.10		185.10	212
6350	Y sediment strainer, PVC, socket or threaded									
6400	1/2"	1 Plum	26	.308	Ea.	51	19.10		70.10	85
6440	3/4"		24	.333		54.50	20.50		75	91
6450	1"		23	.348		64.50	21.50		86	104
6460	1-1/4"		21	.381		109	23.50		132.50	155
6470	1-1/2"	↓	20	.400	↓	109	25		134	157

22 05 48.10 Seismic Bracing Supports

		Crew	Daily Output	Labor-Hours	Unit	Material	2018 Bare Costs Labor	2018 Bare Costs Equipment	Total	Total Incl O&P
0010	**SEISMIC BRACING SUPPORTS**	R133113-90								
0020	Clamps									
0030	C-clamp, for mounting on steel beam									
0040	3/8" threaded rod	1 Skwk	160	.050	Ea.	3.93	2.62		6.55	8.35

496

For customer support on your Building Construction Costs with RSMeans data, call 800.448.8182.

22 05 48.10 Seismic Bracing Supports	Crew	Daily Output	Labor-Hours	Unit	Material	2018 Bare Costs Labor	Equipment	Total	Total Incl O&P
0050	1/2" threaded rod	1 Skwk	160	.050	Ea.	4.88	2.62		7.50	9.35
0060	5/8" threaded rod		160	.050		5.30	2.62		7.92	9.85
0070	3/4" threaded rod		160	.050		6.80	2.62		9.42	11.50
0100	Brackets									
0110	Beam side or wall malleable iron									
0120	3/8" threaded rod	1 Skwk	48	.167	Ea.	4.07	8.75		12.82	17.85
0130	1/2" threaded rod		48	.167		3.78	8.75		12.53	17.50
0140	5/8" threaded rod		48	.167		11	8.75		19.75	25.50
0150	3/4" threaded rod		48	.167		18.90	8.75		27.65	34.50
0160	7/8" threaded rod		48	.167		12.50	8.75		21.25	27
0170	For concrete installation, add						30%			
0180	Wall, welded steel									
0190	0 size 12" wide 18" deep	1 Skwk	34	.235	Ea.	178	12.30		190.30	214
0200	1 size 18" wide 24" deep		34	.235		211	12.30		223.30	251
0210	2 size 24" wide 30" deep		34	.235		279	12.30		291.30	325
0300	Rod, carbon steel									
0310	Continuous thread									
0320	1/4" thread	1 Skwk	144	.056	L.F.	2.18	2.91		5.09	6.85
0330	3/8" thread		144	.056		2.33	2.91		5.24	7
0340	1/2" thread		144	.056		3.67	2.91		6.58	8.50
0350	5/8" thread		144	.056		5.20	2.91		8.11	10.15
0360	3/4" thread		144	.056		9.15	2.91		12.06	14.55
0370	7/8" thread		144	.056		11.50	2.91		14.41	17.10
0380	For galvanized, add						30%			
0400	Channel, steel									
0410	3/4" x 1-1/2"	1 Skwk	80	.100	L.F.	2.61	5.25		7.86	10.85
0420	1-1/2" x 1-1/2"		70	.114		3.41	6		9.41	12.90
0430	1-7/8" x 1-1/2"		60	.133		16.30	7		23.30	28.50
0440	3" x 1-1/2"		50	.160		18.25	8.40		26.65	33
0450	Spring nuts									
0460	3/8"	1 Skwk	100	.080	Ea.	1.59	4.19		5.78	8.15
0470	1/2"	"	80	.100	"	1.67	5.25		6.92	9.85
0500	Welding, field									
0510	Cleaning and welding plates, bars, or rods									
0520	To existing beams, columns, or trusses									
0530	1" weld	1 Skwk	144	.056	Ea.	.23	2.91		3.14	4.70
0540	2" weld		72	.111		.46	5.80		6.26	9.40
0550	3" weld		54	.148		.69	7.75		8.44	12.60
0560	4" weld		36	.222		.92	11.65		12.57	18.80
0570	5" weld		30	.267		1.15	13.95		15.10	23
0580	6" weld		24	.333		1.38	17.45		18.83	28
0600	Vibration absorbers									
0610	Hangers, neoprene flex									
0620	10-120 lb. capacity	1 Skwk	8	1	Ea.	25.50	52.50		78	108
0630	75-550 lb. capacity		8	1		40.50	52.50		93	125
0640	250-1,100 lb. capacity		6	1.333		82.50	70		152.50	198
0650	1,000-4,000 lb. capacity		6	1.333		147	70		217	269
0660	Spring flex									
0670	60-450 lb. capacity	1 Skwk	8	1	Ea.	82.50	52.50		135	171
0680	85-450 lb. capacity		8	1		100	52.50		152.50	190
0690	600-900 lb. capacity		8	1		148	52.50		200.50	242
0700	1,100-1,300 lb. capacity		6	1.333		174	70		244	298
0710	Mounts, neoprene									

22 05 Common Work Results for Plumbing

22 05 48 – Vibration and Seismic Controls for Plumbing Piping and Equipment

22 05 48.10 Seismic Bracing Supports

		Crew	Daily Output	Labor-Hours	Unit	Material	2018 Bare Costs Labor	Equipment	Total	Total Incl O&P
0720	135-380 lb. capacity	1 Skwk	7	1.143	Ea.	18.20	60		78.20	112
0730	250-1,100 lb. capacity		7	1.143		60	60		120	158
0740	1,000-4,000 lb. capacity		5	1.600		128	84		212	269

22 05 76 – Facility Drainage Piping Cleanouts

22 05 76.10 Cleanouts

		Crew	Daily Output	Labor-Hours	Unit	Material	2018 Bare Costs Labor	Equipment	Total	Total Incl O&P
0010	**CLEANOUTS**									
0060	Floor type									
0080	Round or square, scoriated nickel bronze top									
0100	2" pipe size	1 Plum	10	.800	Ea.	184	49.50		233.50	278
0120	3" pipe size		8	1		249	62		311	365
0140	4" pipe size		6	1.333		277	83		360	430
0980	Round top, recessed for terrazzo									
1000	2" pipe size	1 Plum	9	.889	Ea.	515	55		570	650
1080	3" pipe size		6	1.333		530	83		613	710
1100	4" pipe size		4	2		650	124		774	900
1120	5" pipe size	Q-1	6	2.667		940	149		1,089	1,250

22 05 76.20 Cleanout Tees

		Crew	Daily Output	Labor-Hours	Unit	Material	2018 Bare Costs Labor	Equipment	Total	Total Incl O&P
0010	**CLEANOUT TEES**									
0100	Cast iron, B&S, with countersunk plug									
0200	2" pipe size	1 Plum	4	2	Ea.	120	124		244	320
0220	3" pipe size		3.60	2.222		150	138		288	375
0240	4" pipe size		3.30	2.424		232	151		383	480
0280	6" pipe size	Q-1	5	3.200		660	179		839	995
0500	For round smooth access cover, same price									
4000	Plastic, tees and adapters. Add plugs									
4010	ABS, DWV									
4020	Cleanout tee, 1-1/2" pipe size	1 Plum	15	.533	Ea.	12.20	33		45.20	63.50
4030	2" pipe size	Q-1	27	.593		13.30	33		46.30	64.50
4040	3" pipe size		21	.762		25	42.50		67.50	91.50
4050	4" pipe size		16	1		54	56		110	144
4100	Cleanout plug, 1-1/2" pipe size	1 Plum	32	.250		2.11	15.55		17.66	26
4110	2" pipe size	Q-1	56	.286		2.75	16		18.75	27
4120	3" pipe size		36	.444		4.50	25		29.50	42.50
4130	4" pipe size		30	.533		7.80	30		37.80	53.50
4180	Cleanout adapter fitting, 1-1/2" pipe size	1 Plum	32	.250		3.36	15.55		18.91	27
4190	2" pipe size	Q-1	56	.286		5.10	16		21.10	29.50
4200	3" pipe size		36	.444		13.05	25		38.05	52
4210	4" pipe size		30	.533		24.50	30		54.50	71.50
5000	PVC, DWV									
5010	Cleanout tee, 1-1/2" pipe size	1 Plum	15	.533	Ea.	6.90	33		39.90	57.50
5020	2" pipe size	Q-1	27	.593		8.95	33		41.95	60
5030	3" pipe size		21	.762		15.70	42.50		58.20	81.50
5040	4" pipe size		16	1		31	56		87	118
5090	Cleanout plug, 1-1/2" pipe size	1 Plum	32	.250		1.81	15.55		17.36	25.50
5100	2" pipe size	Q-1	56	.286		1.84	16		17.84	26
5110	3" pipe size		36	.444		3.63	25		28.63	41.50
5120	4" pipe size		30	.533		5.35	30		35.35	51
5130	6" pipe size		24	.667		17.10	37.50		54.60	75
5170	Cleanout adapter fitting, 1-1/2" pipe size	1 Plum	32	.250		2.22	15.55		17.77	26
5180	2" pipe size	Q-1	56	.286		2.93	16		18.93	27
5190	3" pipe size		36	.444		8	25		33	46.50
5200	4" pipe size		30	.533		13.10	30		43.10	59.50

For customer support on your Building Construction Costs with RSMeans data, call 800.448.8182.

22 05 Common Work Results for Plumbing

22 05 76 – Facility Drainage Piping Cleanouts

22 05 76.20 Cleanout Tees

		Crew	Daily Output	Labor-Hours	Unit	Material	2018 Bare Costs Labor	Equipment	Total	Total Incl O&P
5210	6" pipe size	Q-1	24	.667	Ea.	37.50	37.50		75	97

22 07 Plumbing Insulation

22 07 19 – Plumbing Piping Insulation

22 07 19.10 Piping Insulation

			Crew	Daily Output	Labor-Hours	Unit	Material	2018 Bare Costs Labor	Equipment	Total	Total Incl O&P
0010	**PIPING INSULATION**										
0110	Insulation req'd. is based on the surface size/area to be covered										
0600	Pipe covering (price copper tube one size less than IPS)										
6600	Fiberglass, with all service jacket										
6840	1" wall, 1/2" iron pipe size	G	Q-14	240	.067	L.F.	.85	3.38		4.23	6.20
6870	1" iron pipe size	G		220	.073		.99	3.69		4.68	6.80
6900	2" iron pipe size	G		200	.080		1.53	4.05		5.58	8
6920	3" iron pipe size	G		180	.089		1.86	4.50		6.36	9.05
6940	4" iron pipe size	G		150	.107		2.48	5.40		7.88	11.15
7320	2" wall, 1/2" iron pipe size	G		220	.073		3.07	3.69		6.76	9.10
7440	6" iron pipe size	G		100	.160		6.40	8.10		14.50	19.60
7460	8" iron pipe size	G		80	.200		7.50	10.15		17.65	24
7480	10" iron pipe size	G		70	.229		8.95	11.60		20.55	28
7490	12" iron pipe size	G	↓	65	.246	↓	10.05	12.45		22.50	30.50
7800	For fiberglass with standard canvas jacket, deduct						5%				
7802	For fittings, add 3 L.F. for each fitting										
7804	plus 4 L.F. for each flange of the fitting										
7810	Finishes										
7812	For .016" aluminum jacket, add	G	Q-14	200	.080	S.F.	1.01	4.05		5.06	7.40
7813	For .010" stainless steel, add	G	"	160	.100	"	3.15	5.05		8.20	11.30
7879	Rubber tubing, flexible closed cell foam										
7880	3/8" wall, 1/4" iron pipe size	G	1 Asbe	120	.067	L.F.	.31	3.75		4.06	6.20
7910	1/2" iron pipe size	G		115	.070		.47	3.92		4.39	6.60
7920	3/4" iron pipe size	G		115	.070		.54	3.92		4.46	6.70
7930	1" iron pipe size	G		110	.073		.57	4.09		4.66	7
7950	1-1/2" iron pipe size	G		110	.073		.75	4.09		4.84	7.20
8100	1/2" wall, 1/4" iron pipe size	G		90	.089		.72	5		5.72	8.55
8130	1/2" iron pipe size	G		89	.090		.84	5.05		5.89	8.75
8140	3/4" iron pipe size	G		89	.090		.95	5.05		6	8.90
8150	1" iron pipe size	G		88	.091		.82	5.10		5.92	8.85
8170	1-1/2" iron pipe size	G		87	.092		1.49	5.20		6.69	9.70
8180	2" iron pipe size	G		86	.093		1.91	5.25		7.16	10.25
8200	3" iron pipe size	G		85	.094		2.37	5.30		7.67	10.85
8300	3/4" wall, 1/4" iron pipe size	G		90	.089		.89	5		5.89	8.75
8330	1/2" iron pipe size	G		89	.090		1.09	5.05		6.14	9.05
8340	3/4" iron pipe size	G		89	.090		1.54	5.05		6.59	9.55
8350	1" iron pipe size	G		88	.091		1.77	5.10		6.87	9.90
8370	1-1/2" iron pipe size	G		87	.092		2.62	5.20		7.82	10.95
8380	2" iron pipe size	G		86	.093		3.04	5.25		8.29	11.50
8400	3" iron pipe size	G		85	.094		4.77	5.30		10.07	13.50
8444	1" wall, 1/2" iron pipe size	G		86	.093		2.32	5.25		7.57	10.70
8445	3/4" iron pipe size	G		84	.095		2.81	5.35		8.16	11.40
8446	1" iron pipe size	G		84	.095		3.39	5.35		8.74	12.05
8447	1-1/4" iron pipe size	G		82	.098		3.75	5.50		9.25	12.70
8448	1-1/2" iron pipe size	G		82	.098		4.88	5.50		10.38	13.90
8449	2" iron pipe size	G	↓	80	.100	↓	6.10	5.65		11.75	15.45

For customer support on your Building Construction Costs with RSMeans data, call 800.448.8182.

499

22 07 Plumbing Insulation

22 07 19 – Plumbing Piping Insulation

22 07 19.10 Piping Insulation	Crew	Daily Output	Labor-Hours	Unit	Material	2018 Bare Costs Labor	Equipment	Total	Total Incl O&P
8450	2-1/2" iron pipe size	**G** 1 Asbe	80	.100	L.F.	7.45	5.65		13.10	16.95
8456	Rubber insulation tape, 1/8" x 2" x 30'	**G**			Ea.	21			21	23

22 11 Facility Water Distribution

22 11 13 – Facility Water Distribution Piping

22 11 13.14 Pipe, Brass

		Crew	Daily Output	Labor-Hours	Unit	Material	2018 Bare Costs Labor	Equipment	Total	Total Incl O&P
0010	**PIPE, BRASS**, Plain end									
0900	Field threaded, coupling & clevis hanger assembly 10' OC									
0920	Regular weight									
1120	1/2" diameter	1 Plum	48	.167	L.F.	7.65	10.35		18	24
1140	3/4" diameter		46	.174		10.05	10.80		20.85	27.50
1160	1" diameter	↓	43	.186		6.90	11.55		18.45	25
1180	1-1/4" diameter	Q-1	72	.222		22.50	12.45		34.95	43.50
1200	1-1/2" diameter		65	.246		27	13.75		40.75	50
1220	2" diameter	↓	53	.302	↓	18.45	16.90		35.35	46

22 11 13.23 Pipe/Tube, Copper

		Crew	Daily Output	Labor-Hours	Unit	Material	2018 Bare Costs Labor	Equipment	Total	Total Incl O&P
0010	**PIPE/TUBE, COPPER**, Solder joints									
1000	Type K tubing, couplings & clevis hanger assemblies 10' OC									
1100	1/4" diameter	1 Plum	84	.095	L.F.	4.11	5.90		10.01	13.40
1200	1" diameter		66	.121		12.30	7.55		19.85	25
1260	2" diameter	↓	40	.200	↓	26	12.45		38.45	47
2000	Type L tubing, couplings & clevis hanger assemblies 10' OC									
2100	1/4" diameter	1 Plum	88	.091	L.F.	2.64	5.65		8.29	11.40
2120	3/8" diameter		84	.095		3.25	5.90		9.15	12.45
2140	1/2" diameter		81	.099		3.34	6.15		9.49	12.95
2160	5/8" diameter		79	.101		5.75	6.30		12.05	15.85
2180	3/4" diameter		76	.105		4.49	6.55		11.04	14.80
2200	1" diameter		68	.118		6.90	7.30		14.20	18.60
2220	1-1/4" diameter		58	.138		11.45	8.55		20	25.50
2240	1-1/2" diameter		52	.154		10.30	9.55		19.85	26
2260	2" diameter	↓	42	.190		17.90	11.85		29.75	37.50
2280	2-1/2" diameter	Q-1	62	.258		25.50	14.45		39.95	49.50
2300	3" diameter		56	.286		43	16		59	71
2320	3-1/2" diameter		43	.372		50.50	21		71.50	87
2340	4" diameter		39	.410		62	23		85	103
2360	5" diameter	↓	34	.471		143	26.50		169.50	198
2380	6" diameter	Q-2	40	.600		145	35		180	212
2400	8" diameter	"	36	.667		241	38.50		279.50	325
2410	For other than full hard temper, add				↓	21%				
2590	For silver solder, add						15%			
4000	Type DWV tubing, couplings & clevis hanger assemblies 10' OC									
4100	1-1/4" diameter	1 Plum	60	.133	L.F.	11.60	8.30		19.90	25.50
4120	1-1/2" diameter		54	.148		11.25	9.20		20.45	26
4140	2" diameter	↓	44	.182		18.05	11.30		29.35	37
4160	3" diameter	Q-1	58	.276		24.50	15.45		39.95	50
4180	4" diameter		40	.400		55	22.50		77.50	94
4200	5" diameter	↓	36	.444		97.50	25		122.50	145
4220	6" diameter	Q-2	42	.571	↓	142	33		175	207

For customer support on your Building Construction Costs with RSMeans data, call 800.448.8182.

22 11 13 – Facility Water Distribution Piping

22 11 13.44 Pipe, Steel	Crew	Daily Output	Labor-Hours	Unit	Material	2018 Bare Costs Labor	Equipment	Total	Total Incl O&P
0010 **PIPE, STEEL**									
0012 The steel pipe in this section does not include fittings such as ells, tees									
0014 For fittings either add a % (usually 25 to 35%) or see									
0015 the Mechanical or Plumbing Costs									
0020 All pipe sizes are to Spec. A-53 unless noted otherwise　　R221113-50									
0050 Schedule 40, threaded, with couplings, and clevis hanger									
0060 assemblies sized for covering, 10' OC									
0540 Black, 1/4" diameter	1 Plum	66	.121	L.F.	5.55	7.55		13.10	17.45
0550 3/8" diameter		65	.123		6.15	7.65		13.80	18.25
0560 1/2" diameter		63	.127		3.82	7.90		11.72	16.10
0570 3/4" diameter		61	.131		4.18	8.15		12.33	16.90
0580 1" diameter		53	.151		4.38	9.40		13.78	18.95
0590 1-1/4" diameter	Q-1	89	.180		5.15	10.05		15.20	21
0600 1-1/2" diameter		80	.200		5.70	11.20		16.90	23
0610 2" diameter		64	.250		11.95	14		25.95	34
0620 2-1/2" diameter		50	.320		16.05	17.90		33.95	44.50
0630 3" diameter		43	.372		18.80	21		39.80	52
0640 3-1/2" diameter		40	.400		24.50	22.50		47	60.50
0650 4" diameter		36	.444		21.50	25		46.50	61
1280 All pipe sizes are to Spec. A-53 unless noted otherwise									
1281 Schedule 40, threaded, with couplings and clevis hanger									
1282 assemblies sized for covering, 10' OC									
1290 Galvanized, 1/4" diameter	1 Plum	66	.121	L.F.	7.25	7.55		14.80	19.35
1300 3/8" diameter		65	.123		7.70	7.65		15.35	19.95
1310 1/2" diameter		63	.127		3.86	7.90		11.76	16.15
1320 3/4" diameter		61	.131		4.43	8.15		12.58	17.15
1330 1" diameter		53	.151		4.87	9.40		14.27	19.50
1340 1-1/4" diameter	Q-1	89	.180		5.65	10.05		15.70	21.50
1350 1-1/2" diameter		80	.200		6.40	11.20		17.60	24
1360 2" diameter		64	.250		12.95	14		26.95	35.50
1370 2-1/2" diameter		50	.320		17.80	17.90		35.70	46.50
1380 3" diameter		43	.372		21	21		42	54.50
1390 3-1/2" diameter		40	.400		26	22.50		48.50	62
1400 4" diameter		36	.444		24	25		49	64
2000 Welded, sch. 40, on yoke & roll hanger assy's, sized for covering, 10' OC									
2040 Black, 1" diameter	Q-15	93	.172	L.F.	4.18	9.60	.64	14.42	19.80
2070 2" diameter		61	.262		11.50	14.65	.97	27.12	36
2090 3" diameter		43	.372		17	21	1.38	39.38	51.50
2110 4" diameter		37	.432		16.85	24	1.60	42.45	57
2120 5" diameter		32	.500		34.50	28	1.85	64.35	82
2130 6" diameter	Q-16	36	.667		42.50	38.50	1.65	82.65	106
2140 8" diameter		29	.828		68.50	48	2.04	118.54	150
2150 10" diameter		24	1		87.50	58	2.47	147.97	187
2160 12" diameter		19	1.263		104	73.50	3.12	180.62	228

22 11 13.48 Pipe, Fittings and Valves, Steel, Grooved-Joint

	Crew	Daily Output	Labor-Hours	Unit	Material	Labor	Equipment	Total	Total Incl O&P
0010 **PIPE, FITTINGS AND VALVES, STEEL, GROOVED-JOINT**									
0012 Fittings are ductile iron. Steel fittings noted.									
0020 Pipe includes coupling & clevis type hanger assemblies, 10' OC									
1000 Schedule 40, black									
1040 3/4" diameter	1 Plum	71	.113	L.F.	6.05	7		13.05	17.20
1050 1" diameter		63	.127		5.85	7.90		13.75	18.35
1060 1-1/4" diameter		58	.138		6.95	8.55		15.50	20.50

22 11 13.48 Pipe, Fittings and Valves, Steel, Grooved-Joint	Crew	Daily Output	Labor-Hours	Unit	Material	2018 Bare Costs Labor	Equipment	Total	Total Incl O&P
1070	1-1/2" diameter	1 Plum	51	.157	L.F.	7.55	9.75		17.30	23
1080	2" diameter	▼	40	.200		8.75	12.45		21.20	28.50
1090	2-1/2" diameter	Q-1	57	.281		11.45	15.70		27.15	36
1100	3" diameter		50	.320		13.85	17.90		31.75	42.50
1110	4" diameter		45	.356		24	19.90		43.90	56.50
1120	5" diameter	▼	37	.432		43.50	24		67.50	84
1130	6" diameter	Q-2	42	.571	▼	49	33		82	104
1800	Galvanized									
1840	3/4" diameter	1 Plum	71	.113	L.F.	6.25	7		13.25	17.45
1850	1" diameter		63	.127		6.50	7.90		14.40	19.05
1860	1-1/4" diameter		58	.138		7.80	8.55		16.35	21.50
1870	1-1/2" diameter		51	.157		8.65	9.75		18.40	24
1880	2" diameter	▼	40	.200		10	12.45		22.45	29.50
1890	2-1/2" diameter	Q-1	57	.281		12.95	15.70		28.65	38
1900	3" diameter		50	.320		15.75	17.90		33.65	44.50
1910	4" diameter		45	.356		26.50	19.90		46.40	59.50
1920	5" diameter	▼	37	.432		31	24		55	71
1930	6" diameter	Q-2	42	.571	▼	33	33		66	86.50
3990	Fittings: coupling material required at joints not incl. in fitting price.									
3994	Add 1 selected coupling, material only, per joint for installed price.									
4000	Elbow, 90° or 45°, painted									
4030	3/4" diameter	1 Plum	50	.160	Ea.	72	9.95		81.95	94
4040	1" diameter		50	.160		38.50	9.95		48.45	57
4050	1-1/4" diameter		40	.200		38.50	12.45		50.95	60.50
4060	1-1/2" diameter		33	.242		38.50	15.05		53.55	64.50
4070	2" diameter	▼	25	.320		38.50	19.90		58.40	72
4080	2-1/2" diameter	Q-1	40	.400		38.50	22.50		61	75.50
4090	3" diameter		33	.485		67.50	27		94.50	116
4100	4" diameter		25	.640		73.50	36		109.50	135
4110	5" diameter	▼	20	.800		175	44.50		219.50	260
4120	6" diameter	Q-2	25	.960		205	55.50		260.50	310
4250	For galvanized elbows, add				▼	26%				
4690	Tee, painted									
4700	3/4" diameter	1 Plum	38	.211	Ea.	77	13.10		90.10	104
4740	1" diameter		33	.242		59.50	15.05		74.55	88
4750	1-1/4" diameter		27	.296		59.50	18.40		77.90	93
4760	1-1/2" diameter		22	.364		59.50	22.50		82	99.50
4770	2" diameter	▼	17	.471		59.50	29.50		89	110
4780	2-1/2" diameter	Q-1	27	.593		59.50	33		92.50	116
4790	3" diameter		22	.727		81	40.50		121.50	151
4800	4" diameter		17	.941		123	52.50		175.50	216
4810	5" diameter	▼	13	1.231		287	69		356	420
4820	6" diameter	Q-2	17	1.412		330	82		412	490
4900	For galvanized tees, add				▼	24%				
4906	Couplings, rigid style, painted									
4908	1" diameter	1 Plum	100	.080	Ea.	30	4.97		34.97	40.50
4909	1-1/4" diameter		100	.080		28	4.97		32.97	38
4910	1-1/2" diameter		67	.119		30	7.40		37.40	44
4912	2" diameter	▼	50	.160		38	9.95		47.95	56.50
4914	2-1/2" diameter	Q-1	80	.200		43	11.20		54.20	64.50
4916	3" diameter		67	.239		50	13.35		63.35	75
4918	4" diameter		50	.320		69.50	17.90		87.40	104
4920	5" diameter	▼	40	.400		89.50	22.50		112	132

502

For customer support on your Building Construction Costs with RSMeans data, call 800.448.8182.

22 11 13.48 Pipe, Fittings and Valves, Steel, Grooved-Joint	Crew	Daily Output	Labor-Hours	Unit	Material	2018 Bare Costs Labor	Equipment	Total	Total Incl O&P
4922	6" diameter	Q-2	50	.480	Ea.	118	28		146	172
4940	Flexible, standard, painted									
4950	3/4" diameter	1 Plum	100	.080	Ea.	21.50	4.97		26.47	31.50
4960	1" diameter		100	.080		21.50	4.97		26.47	31.50
4970	1-1/4" diameter		80	.100		28	6.20		34.20	40.50
4980	1-1/2" diameter		67	.119		30.50	7.40		37.90	44.50
4990	2" diameter		50	.160		33	9.95		42.95	51
5000	2-1/2" diameter	Q-1	80	.200		38	11.20		49.20	59
5010	3" diameter		67	.239		42	13.35		55.35	66.50
5020	3-1/2" diameter		57	.281		60.50	15.70		76.20	90
5030	4" diameter		50	.320		61	17.90		78.90	94
5040	5" diameter		40	.400		91.50	22.50		114	135
5050	6" diameter	Q-2	50	.480		108	28		136	161
5200	For galvanized couplings, add					33%				
7790	Valves: coupling material required at joints not incl. in valve price.									
7794	Add 1 selected coupling, material only, per joint for installed price.									

22 11 13.64 Pipe, Stainless Steel

		Crew	Daily Output	Labor-Hours	Unit	Material	2018 Bare Costs Labor	Equipment	Total	Total Incl O&P
0010	**PIPE, STAINLESS STEEL**									
3500	Threaded, couplings and clevis hanger assemblies, 10' OC									
3520	Schedule 40, type 304									
3540	1/4" diameter	1 Plum	54	.148	L.F.	9.15	9.20		18.35	24
3550	3/8" diameter		53	.151		10.65	9.40		20.05	26
3560	1/2" diameter		52	.154		11.90	9.55		21.45	27.50
3580	1" diameter		45	.178		23	11.05		34.05	42
3610	2" diameter	Q-1	57	.281		47.50	15.70		63.20	76
3640	4" diameter	Q-2	51	.471		128	27.50		155.50	182
3740	For small quantities, add					10%				
4250	Schedule 40, type 316									
4290	1/4" diameter	1 Plum	54	.148	L.F.	21.50	9.20		30.70	37.50
4300	3/8" diameter		53	.151		23.50	9.40		32.90	40
4310	1/2" diameter		52	.154		29.50	9.55		39.05	47
4320	3/4" diameter		51	.157		34	9.75		43.75	52
4330	1" diameter		45	.178		47	11.05		58.05	68.50
4360	2" diameter	Q-1	57	.281		95	15.70		110.70	129
4390	4" diameter	Q-2	51	.471		219	27.50		246.50	282
4490	For small quantities, add					10%				

22 11 13.74 Pipe, Plastic

		Crew	Daily Output	Labor-Hours	Unit	Material	2018 Bare Costs Labor	Equipment	Total	Total Incl O&P
0010	**PIPE, PLASTIC**									
1800	PVC, couplings 10' OC, clevis hanger assemblies, 3 per 10'									
1820	Schedule 40									
1860	1/2" diameter	1 Plum	54	.148	L.F.	5.10	9.20		14.30	19.45
1870	3/4" diameter		51	.157		5.60	9.75		15.35	21
1880	1" diameter		46	.174		8.75	10.80		19.55	26
1890	1-1/4" diameter		42	.190		9.50	11.85		21.35	28.50
1900	1-1/2" diameter		36	.222		9.60	13.80		23.40	31.50
1910	2" diameter	Q-1	59	.271		11.10	15.15		26.25	35
1920	2-1/2" diameter		56	.286		10.60	16		26.60	35.50
1930	3" diameter		53	.302		12.90	16.90		29.80	39.50
1940	4" diameter		48	.333		30.50	18.65		49.15	61.50
1950	5" diameter		43	.372		39	21		60	74
1960	6" diameter		39	.410		27	23		50	64.50
4100	DWV type, schedule 40, couplings 10' OC, clevis hanger assy's, 3 per 10'									

22 11 13.74 Pipe, Plastic

		Crew	Daily Output	Labor-Hours	Unit	Material	2018 Bare Costs Labor	Equipment	Total	Total Incl O&P
4210	ABS, schedule 40, foam core type									
4212	Plain end black									
4214	1-1/2" diameter	1 Plum	39	.205	L.F.	8	12.75		20.75	28
4216	2" diameter	Q-1	62	.258		8.55	14.45		23	31
4218	3" diameter		56	.286		8.35	16		24.35	33
4220	4" diameter		51	.314		25	17.55		42.55	54
4222	6" diameter		42	.381		18.95	21.50		40.45	53
4240	To delete coupling & hangers, subtract									
4244	1-1/2" diam. to 6" diam.					43%	48%			
4400	PVC									
4410	1-1/4" diameter	1 Plum	42	.190	L.F.	8.50	11.85		20.35	27
4420	1-1/2" diameter	"	36	.222		7.75	13.80		21.55	29.50
4460	2" diameter	Q-1	59	.271		8.65	15.15		23.80	32.50
4470	3" diameter		53	.302		8.55	16.90		25.45	35
4480	4" diameter		48	.333		10.20	18.65		28.85	39
4490	6" diameter		39	.410		17.35	23		40.35	53.50
5300	CPVC, socket joint, couplings 10' OC, clevis hanger assemblies, 3 per 10'									
5302	Schedule 40									
5304	1/2" diameter	1 Plum	54	.148	L.F.	5.85	9.20		15.05	20.50
5305	3/4" diameter		51	.157		6.90	9.75		16.65	22.50
5306	1" diameter		46	.174		10.50	10.80		21.30	28
5307	1-1/4" diameter		42	.190		11.85	11.85		23.70	31
5308	1-1/2" diameter		36	.222		11.75	13.80		25.55	34
5309	2" diameter	Q-1	59	.271		14.80	15.15		29.95	39.50
5310	2-1/2" diameter		56	.286		18.75	16		34.75	44.50
5311	3" diameter		53	.302		21.50	16.90		38.40	49
5360	CPVC, threaded, couplings 10' OC, clevis hanger assemblies, 3 per 10'									
5380	Schedule 40									
5460	1/2" diameter	1 Plum	54	.148	L.F.	6.70	9.20		15.90	21
5470	3/4" diameter		51	.157		8.35	9.75		18.10	24
5480	1" diameter		46	.174		12	10.80		22.80	29.50
5490	1-1/4" diameter		42	.190		13	11.85		24.85	32
5500	1-1/2" diameter		36	.222		12.75	13.80		26.55	35
5510	2" diameter	Q-1	59	.271		16.05	15.15		31.20	40.50
5520	2-1/2" diameter		56	.286		20	16		36	46
5530	3" diameter		53	.302		23	16.90		39.90	51
7280	PEX, flexible, no couplings or hangers									
7282	Note: For labor costs add 25% to the couplings and fittings labor total.									
7285	For fittings see section 23 83 16.10 7000									
7300	Non-barrier type, hot/cold tubing rolls									
7310	1/4" diameter x 100'				L.F.	.49			.49	.54
7350	3/8" diameter x 100'					.55			.55	.61
7360	1/2" diameter x 100'					.61			.61	.67
7370	1/2" diameter x 500'					.61			.61	.67
7380	1/2" diameter x 1000'					.61			.61	.67
7400	3/4" diameter x 100'					.92			.92	1.01
7410	3/4" diameter x 500'					1.11			1.11	1.22
7420	3/4" diameter x 1000'					1.11			1.11	1.22
7460	1" diameter x 100'					1.90			1.90	2.09
7470	1" diameter x 300'					1.90			1.90	2.09
7480	1" diameter x 500'					1.90			1.90	2.09
7500	1-1/4" diameter x 100'					3.23			3.23	3.55
7510	1-1/4" diameter x 300'					3.23			3.23	3.55

For customer support on your Building Construction Costs with RSMeans data, call 800.448.8182.

22 11 Facility Water Distribution

22 11 13 – Facility Water Distribution Piping

22 11 13.74 Pipe, Plastic

		Crew	Daily Output	Labor-Hours	Unit	Material	2018 Bare Costs Labor	Equipment	Total	Total Incl O&P
7540	1-1/2" diameter x 100'				L.F.	4.40			4.40	4.84
7550	1-1/2" diameter x 300'				↓	4.40			4.40	4.84
7596	Most sizes available in red or blue									
7700	Non-barrier type, hot/cold tubing straight lengths									
7710	1/2" diameter x 20'				L.F.	.61			.61	.67
7750	3/4" diameter x 20'					1.11			1.11	1.22
7760	1" diameter x 20'					1.90			1.90	2.09
7770	1-1/4" diameter x 20'					3.23			3.23	3.55
7780	1-1/2" diameter x 20'					4.40			4.40	4.84
7790	2" diameter				↓	8.60			8.60	9.45
7796	Most sizes available in red or blue									

22 11 19 – Domestic Water Piping Specialties

22 11 19.10 Flexible Connectors

		Crew	Daily Output	Labor-Hours	Unit	Material	2018 Bare Costs Labor	Equipment	Total	Total Incl O&P
0010	**FLEXIBLE CONNECTORS**, Corrugated, 7/8" OD, 1/2" ID									
0050	Gas, seamless brass, steel fittings									
0200	12" long	1 Plum	36	.222	Ea.	6.40	13.80		20.20	28
0220	18" long		36	.222		7.95	13.80		21.75	30
0240	24" long		34	.235		9.40	14.60		24	32.50
0280	36" long		32	.250		11.25	15.55		26.80	36
0340	60" long	↓	30	.267	↓	16.95	16.55		33.50	43.50
2000	Water, copper tubing, dielectric separators									
2100	12" long	1 Plum	36	.222	Ea.	6.40	13.80		20.20	28
2260	24" long	"	34	.235	"	9.50	14.60		24.10	32.50

22 11 19.14 Flexible Metal Hose

		Crew	Daily Output	Labor-Hours	Unit	Material	2018 Bare Costs Labor	Equipment	Total	Total Incl O&P
0010	**FLEXIBLE METAL HOSE**, Connectors, standard lengths									
0100	Bronze braided, bronze ends									
0120	3/8" diameter x 12"	1 Stpi	26	.308	Ea.	25.50	19.40		44.90	57
0160	3/4" diameter x 12"		20	.400		35.50	25		60.50	77
0180	1" diameter x 18"		19	.421		46	26.50		72.50	90.50
0200	1-1/2" diameter x 18"		13	.615		60	39		99	125
0220	2" diameter x 18"	↓	11	.727	↓	78	46		124	155

22 11 19.26 Pressure Regulators

		Crew	Daily Output	Labor-Hours	Unit	Material	2018 Bare Costs Labor	Equipment	Total	Total Incl O&P
0010	**PRESSURE REGULATORS**									
3000	Steam, high capacity, bronze body, stainless steel trim									
3020	Threaded, 1/2" diameter	1 Stpi	24	.333	Ea.	2,550	21		2,571	2,825
3030	3/4" diameter		24	.333		2,550	21		2,571	2,825
3040	1" diameter		19	.421		2,825	26.50		2,851.50	3,175
3060	1-1/4" diameter		15	.533		3,125	33.50		3,158.50	3,500
3080	1-1/2" diameter		13	.615		3,575	39		3,614	4,000
3100	2" diameter	↓	11	.727		4,375	46		4,421	4,900
3120	2-1/2" diameter	Q-5	12	1.333		5,500	75.50		5,575.50	6,175
3140	3" diameter	"	11	1.455	↓	6,275	82.50		6,357.50	7,025
3500	Flanged connection, iron body, 125 lb. W.S.P.									
3520	3" diameter	Q-5	11	1.455	Ea.	6,875	82.50		6,957.50	7,675
3540	4" diameter	"	5	3.200	"	8,650	181		8,831	9,800

22 11 19.38 Water Supply Meters

		Crew	Daily Output	Labor-Hours	Unit	Material	2018 Bare Costs Labor	Equipment	Total	Total Incl O&P
0010	**WATER SUPPLY METERS**									
2000	Domestic/commercial, bronze									
2020	Threaded									
2060	5/8" diameter, to 20 GPM	1 Plum	16	.500	Ea.	50.50	31		81.50	103
2080	3/4" diameter, to 30 GPM	↓	14	.571	↓	92	35.50		127.50	155

For customer support on your Building Construction Costs with RSMeans data, call 800.448.8182.

505

22 11 19.38 Water Supply Meters

		Crew	Daily Output	Labor-Hours	Unit	Material	2018 Bare Costs Labor	2018 Bare Costs Equipment	Total	Total Incl O&P
2100	1" diameter, to 50 GPM	1 Plum	12	.667	Ea.	140	41.50		181.50	217
2300	Threaded/flanged									
2340	1-1/2" diameter, to 100 GPM	1 Plum	8	1	Ea.	340	62		402	470
2360	2" diameter, to 160 GPM	"	6	1.333	"	465	83		548	635
2600	Flanged, compound									
2640	3" diameter, 320 GPM	Q-1	3	5.333	Ea.	2,175	298		2,473	2,850
2660	4" diameter, to 500 GPM		1.50	10.667		3,475	595		4,070	4,725
2680	6" diameter, to 1,000 GPM		1	16		5,700	895		6,595	7,625
2700	8" diameter, to 1,800 GPM		.80	20		8,850	1,125		9,975	11,400

22 11 19.42 Backflow Preventers

		Crew	Daily Output	Labor-Hours	Unit	Material	2018 Bare Costs Labor	2018 Bare Costs Equipment	Total	Total Incl O&P
0010	**BACKFLOW PREVENTERS**, Includes valves									
0020	and four test cocks, corrosion resistant, automatic operation									
4000	Reduced pressure principle									
4100	Threaded, bronze, valves are ball									
4120	3/4" pipe size	1 Plum	16	.500	Ea.	455	31		486	545
4140	1" pipe size		14	.571		480	35.50		515.50	585
4150	1-1/4" pipe size		12	.667		895	41.50		936.50	1,050
4160	1-1/2" pipe size		10	.800		985	49.50		1,034.50	1,150
4180	2" pipe size		7	1.143		1,100	71		1,171	1,325
5000	Flanged, bronze, valves are OS&Y									
5060	2-1/2" pipe size	Q-1	5	3.200	Ea.	5,325	179		5,504	6,125
5080	3" pipe size		4.50	3.556		6,050	199		6,249	6,950
5100	4" pipe size		3	5.333		7,550	298		7,848	8,775
5120	6" pipe size	Q-2	3	8		11,000	465		11,465	12,800
5600	Flanged, iron, valves are OS&Y									
5660	2-1/2" pipe size	Q-1	5	3.200	Ea.	3,025	179		3,204	3,600
5680	3" pipe size		4.50	3.556		3,175	199		3,374	3,800
5700	4" pipe size		3	5.333		4,000	298		4,298	4,850
5720	6" pipe size	Q-2	3	8		5,800	465		6,265	7,075
5740	8" pipe size		2	12		10,200	695		10,895	12,300
5760	10" pipe size		1	24		19,500	1,400		20,900	23,500

22 11 19.50 Vacuum Breakers

		Crew	Daily Output	Labor-Hours	Unit	Material	2018 Bare Costs Labor	2018 Bare Costs Equipment	Total	Total Incl O&P
0010	**VACUUM BREAKERS**									
0013	See also backflow preventers Section 22 11 19.42									
1000	Anti-siphon continuous pressure type									
1010	Max. 150 psi - 210°F									
1020	Bronze body									
1030	1/2" size	1 Stpi	24	.333	Ea.	171	21		192	220
1040	3/4" size		20	.400		171	25		196	226
1050	1" size		19	.421		176	26.50		202.50	234
1060	1-1/4" size		15	.533		345	33.50		378.50	430
1070	1-1/2" size		13	.615		425	39		464	530
1080	2" size		11	.727		440	46		486	555
1200	Max. 125 psi with atmospheric vent									
1210	Brass, in-line construction									
1220	1/4" size	1 Stpi	24	.333	Ea.	133	21		154	178
1230	3/8" size	"	24	.333		133	21		154	178
1260	For polished chrome finish, add					13%				
2000	Anti-siphon, non-continuous pressure type									
2010	Hot or cold water 125 psi - 210°F									
2020	Bronze body									
2030	1/4" size	1 Stpi	24	.333	Ea.	73	21		94	112

For customer support on your Building Construction Costs with RSMeans data, call 800.448.8182.

22 11 19.50 Vacuum Breakers

		Crew	Daily Output	Labor-Hours	Unit	Material	2018 Bare Costs Labor	Equipment	Total	Total Incl O&P
2040	3/8" size	1 Stpi	24	.333	Ea.	73	21		94	112
2050	1/2" size		24	.333		82.50	21		103.50	122
2060	3/4" size		20	.400		98.50	25		123.50	146
2070	1" size		19	.421		152	26.50		178.50	208
2080	1-1/4" size		15	.533		267	33.50		300.50	345
2090	1-1/2" size		13	.615		315	39		354	405
2100	2" size		11	.727		485	46		531	605
2110	2-1/2" size		8	1		1,400	63		1,463	1,650
2120	3" size	▼	6	1.333	▼	1,850	84		1,934	2,175
2150	For polished chrome finish, add					50%				

22 11 19.54 Water Hammer Arresters/Shock Absorbers

		Crew	Daily Output	Labor-Hours	Unit	Material	2018 Bare Costs Labor	Equipment	Total	Total Incl O&P
0010	**WATER HAMMER ARRESTERS/SHOCK ABSORBERS**									
0490	Copper									
0500	3/4" male IPS For 1 to 11 fixtures	1 Plum	12	.667	Ea.	28	41.50		69.50	93.50
0600	1" male IPS For 12 to 32 fixtures		8	1		47.50	62		109.50	146
0700	1-1/4" male IPS For 33 to 60 fixtures		8	1		48	62		110	147
0800	1-1/2" male IPS For 61 to 113 fixtures		8	1		69	62		131	170
0900	2" male IPS For 114 to 154 fixtures		8	1		101	62		163	205
1000	2-1/2" male IPS For 155 to 330 fixtures	▼	4	2	▼	305	124		429	520

22 11 19.64 Hydrants

		Crew	Daily Output	Labor-Hours	Unit	Material	2018 Bare Costs Labor	Equipment	Total	Total Incl O&P
0010	**HYDRANTS**									
0050	Wall type, moderate climate, bronze, encased									
0200	3/4" IPS connection	1 Plum	16	.500	Ea.	850	31		881	980
0300	1" IPS connection		14	.571		920	35.50		955.50	1,050
0500	Anti-siphon type, 3/4" connection	▼	16	.500	▼	750	31		781	865
1000	Non-freeze, bronze, exposed									
1100	3/4" IPS connection, 4" to 9" thick wall	1 Plum	14	.571	Ea.	380	35.50		415.50	475
1120	10" to 14" thick wall		12	.667		410	41.50		451.50	515
1140	15" to 19" thick wall		12	.667		450	41.50		491.50	560
1160	20" to 24" thick wall	▼	10	.800	▼	755	49.50		804.50	905
1200	For 1" IPS connection, add					15%	10%			
1240	For 3/4" adapter type vacuum breaker, add				Ea.	72			72	79.50
1280	For anti-siphon type, add				"	153			153	168
2000	Non-freeze bronze, encased, anti-siphon type									
2100	3/4" IPS connection, 5" to 9" thick wall	1 Plum	14	.571	Ea.	1,225	35.50		1,260.50	1,400
2120	10" to 14" thick wall		12	.667		1,500	41.50		1,541.50	1,725
2140	15" to 19" thick wall	▼	12	.667	▼	1,575	41.50		1,616.50	1,800
3000	Ground box type, bronze frame, 3/4" IPS connection									
3080	Non-freeze, all bronze, polished face, set flush									
3100	2' depth of bury	1 Plum	8	1	Ea.	1,100	62		1,162	1,325
3140	4' depth of bury		8	1		1,275	62		1,337	1,500
3180	6' depth of bury		7	1.143		1,450	71		1,521	1,700
3220	8' depth of bury	▼	5	1.600		1,625	99.50		1,724.50	1,925
3400	For 1" IPS connection, add					15%	10%			
3550	For 2" IPS connection, add					445%	24%			
3600	For tapped drain port in box, add				▼	93			93	102
5000	Moderate climate, all bronze, polished face									
5020	and scoriated cover, set flush									
5100	3/4" IPS connection	1 Plum	16	.500	Ea.	755	31		786	875
5120	1" IPS connection	"	14	.571	▼	1,825	35.50		1,860.50	2,075
5200	For tapped drain port in box, add					98.50			98.50	109

22 11 Facility Water Distribution

22 11 23 – Domestic Water Pumps

22 11 23.10 General Utility Pumps

22 11 23.10 General Utility Pumps		Crew	Daily Output	Labor-Hours	Unit	Material	2018 Bare Costs Labor	Equipment	Total	Total Incl O&P
0010	**GENERAL UTILITY PUMPS**									
2000	Single stage									
3000	Double suction,									
3190	75 HP, to 2,500 GPM	Q-3	.28	114	Ea.	14,600	6,750		21,350	26,300
3220	100 HP, to 3,000 GPM		.26	123		18,200	7,275		25,475	31,000
3240	150 HP, to 4,000 GPM		.24	133		21,400	7,900		29,300	35,400

22 13 Facility Sanitary Sewerage

22 13 16 – Sanitary Waste and Vent Piping

22 13 16.20 Pipe, Cast Iron

22 13 16.20 Pipe, Cast Iron		Crew	Daily Output	Labor-Hours	Unit	Material	2018 Bare Costs Labor	Equipment	Total	Total Incl O&P
0010	**PIPE, CAST IRON**, Soil, on clevis hanger assemblies, 5' OC									
0020	Single hub, service wt., lead & oakum joints 10' OC									
2120	2" diameter	Q-1	63	.254	L.F.	16.90	14.20		31.10	40
2140	3" diameter		60	.267		20	14.90		34.90	44.50
2160	4" diameter		55	.291		35	16.25		51.25	63
2180	5" diameter	Q-2	76	.316		43.50	18.30		61.80	75
2200	6" diameter	"	73	.329		41.50	19.05		60.55	74.50
2220	8" diameter	Q-3	59	.542		63.50	32		95.50	119
2240	10" diameter		54	.593		101	35		136	164
2260	12" diameter		48	.667		144	39.50		183.50	219
2320	For service weight, double hub, add					10%				
2340	For extra heavy, single hub, add					48%	4%			
2360	For extra heavy, double hub, add					71%	4%			
2400	Lead for caulking (1#/diam. in.)	Q-1	160	.100	Lb.	1.04	5.60		6.64	9.55
2420	Oakum for caulking (1/8#/diam. in.)	"	40	.400	"	4.22	22.50		26.72	38
4000	No hub, couplings 10' OC									
4100	1-1/2" diameter	Q-1	71	.225	L.F.	16	12.60		28.60	36.50
4120	2" diameter		67	.239		17.10	13.35		30.45	39
4140	3" diameter		64	.250		19.60	14		33.60	42.50
4160	4" diameter		58	.276		34.50	15.45		49.95	60.50

22 13 16.50 Shower Drains

22 13 16.50 Shower Drains		Crew	Daily Output	Labor-Hours	Unit	Material	2018 Bare Costs Labor	Equipment	Total	Total Incl O&P
0010	**SHOWER DRAINS**									
2780	Shower, with strainer, uniform diam. trap, bronze top									
2800	2" and 3" pipe size	Q-1	8	2	Ea.	350	112		462	555
2820	4" pipe size	"	7	2.286		395	128		523	630
2840	For galvanized body, add					190			190	208

22 13 16.60 Traps

22 13 16.60 Traps		Crew	Daily Output	Labor-Hours	Unit	Material	2018 Bare Costs Labor	Equipment	Total	Total Incl O&P
0010	**TRAPS**									
0030	Cast iron, service weight									
0050	Running P trap, without vent									
1100	2"	Q-1	16	1	Ea.	148	56		204	247
1140	3"		14	1.143		148	64		212	260
1150	4"		13	1.231		148	69		217	267
1160	6"	Q-2	17	1.412		720	82		802	915
1180	Running trap, single hub, with vent									
2080	3" pipe size, 3" vent	Q-1	14	1.143	Ea.	117	64		181	226
2120	4" pipe size, 4" vent	"	13	1.231		175	69		244	296
2300	For double hub, vent, add					10%	20%			
3000	P trap, B&S, 2" pipe size	Q-1	16	1		38.50	56		94.50	127
3040	3" pipe size	"	14	1.143		57.50	64		121.50	160

For customer support on your Building Construction Costs with RSMeans data, call 800.448.8182.

22 13 16 – Sanitary Waste and Vent Piping

22 13 16.60 Traps

		Crew	Daily Output	Labor-Hours	Unit	Material	2018 Bare Costs Labor	Equipment	Total	Total Incl O&P
3350	Deep seal trap, B&S									
3400	1-1/4" pipe size	Q-1	14	1.143	Ea.	60	64		124	163
3410	1-1/2" pipe size		14	1.143		60	64		124	163
3420	2" pipe size		14	1.143		55.50	64		119.50	158
3440	3" pipe size	↓	12	1.333	↓	70.50	74.50		145	190
4700	Copper, drainage, drum trap									
4800	3" x 5" solid, 1-1/2" pipe size	1 Plum	16	.500	Ea.	136	31		167	197
4840	3" x 6" swivel, 1-1/2" pipe size	"	16	.500	"	275	31		306	345
5100	P trap, standard pattern									
5200	1-1/4" pipe size	1 Plum	18	.444	Ea.	88.50	27.50		116	139
5240	1-1/2" pipe size		17	.471		98.50	29.50		128	152
5260	2" pipe size		15	.533		152	33		185	217
5280	3" pipe size	↓	11	.727	↓	475	45		520	595
5340	With cleanout, swivel joint and slip joint									
5360	1-1/4" pipe size	1 Plum	18	.444	Ea.	117	27.50		144.50	171
5400	1-1/2" pipe size	"	17	.471	"	178	29.50		207.50	240

22 13 16.80 Vent Flashing and Caps

		Crew	Daily Output	Labor-Hours	Unit	Material	2018 Bare Costs Labor	Equipment	Total	Total Incl O&P
0010	**VENT FLASHING AND CAPS**									
0120	Vent caps									
0140	Cast iron									
0180	2-1/2" to 3-5/8" pipe	1 Plum	21	.381	Ea.	49.50	23.50		73	90
0190	4" to 4-1/8" pipe	"	19	.421	"	69	26		95	116
0900	Vent flashing									
1000	Aluminum with lead ring									
1020	1-1/4" pipe	1 Plum	20	.400	Ea.	5.25	25		30.25	43.50
1030	1-1/2" pipe		20	.400		4.95	25		29.95	43
1040	2" pipe		18	.444		4.97	27.50		32.47	47
1050	3" pipe		17	.471		5.50	29.50		35	50
1060	4" pipe	↓	16	.500	↓	6.65	31		37.65	54.50
1350	Copper with neoprene ring									
1400	1-1/4" pipe	1 Plum	20	.400	Ea.	70.50	25		95.50	115
1430	1-1/2" pipe		20	.400		70.50	25		95.50	115
1440	2" pipe		18	.444		70.50	27.50		98	119
1450	3" pipe		17	.471		85.50	29.50		115	138
1460	4" pipe	↓	16	.500	↓	85.50	31		116.50	141

22 13 19 – Sanitary Waste Piping Specialties

22 13 19.13 Sanitary Drains

		Crew	Daily Output	Labor-Hours	Unit	Material	2018 Bare Costs Labor	Equipment	Total	Total Incl O&P
0010	**SANITARY DRAINS**									
0400	Deck, auto park, CI, 13" top									
0440	3", 4", 5", and 6" pipe size	Q-1	8	2	Ea.	1,650	112		1,762	1,975
0480	For galvanized body, add				"	890			890	980
2000	Floor, medium duty, CI, deep flange, 7" diam. top									
2040	2" and 3" pipe size	Q-1	12	1.333	Ea.	230	74.50		304.50	365
2080	For galvanized body, add					111			111	122
2120	With polished bronze top				↓	360			360	395
2400	Heavy duty, with sediment bucket, CI, 12" diam. loose grate									
2420	2", 3", 4", 5", and 6" pipe size	Q-1	9	1.778	Ea.	790	99.50		889.50	1,025
2460	With polished bronze top				"	1,125			1,125	1,225
2500	Heavy duty, cleanout & trap w/bucket, CI, 15" top									
2540	2", 3", and 4" pipe size	Q-1	6	2.667	Ea.	7,525	149		7,674	8,500
2560	For galvanized body, add					1,925			1,925	2,125
2580	With polished bronze top				↓	8,350			8,350	9,200

22 13 23 – Sanitary Waste Interceptors

22 13 23.10 Interceptors	Crew	Daily Output	Labor-Hours	Unit	Material	2018 Bare Costs Labor	Equipment	Total	Total Incl O&P
0010	**INTERCEPTORS**								
0150	Grease, fabricated steel, 4 GPM, 8 lb. fat capacity	1 Plum	4	2	Ea.	1,325	124		1,449	1,625
0200	7 GPM, 14 lb. fat capacity		4	2		1,825	124		1,949	2,200
1000	10 GPM, 20 lb. fat capacity		4	2		2,150	124		2,274	2,550
1040	15 GPM, 30 lb. fat capacity		4	2		3,200	124		3,324	3,700
1060	20 GPM, 40 lb. fat capacity		3	2.667		3,925	166		4,091	4,550
1120	50 GPM, 100 lb. fat capacity	Q-1	2	8		7,200	445		7,645	8,600
1160	100 GPM, 200 lb. fat capacity	"	2	8		16,300	445		16,745	18,700
1580	For seepage pan, add					7%				
3000	Hair, cast iron, 1-1/4" and 1-1/2" pipe connection	1 Plum	8	1	Ea.	500	62		562	640
3100	For chrome-plated cast iron, add					305			305	335
4000	Oil, fabricated steel, 10 GPM, 2" pipe size	1 Plum	4	2		2,950	124		3,074	3,425
4100	15 GPM, 2" or 3" pipe size		4	2		4,050	124		4,174	4,625
4120	20 GPM, 2" or 3" pipe size		3	2.667		4,900	166		5,066	5,625
4220	100 GPM, 3" pipe size	Q-1	2	8		16,300	445		16,745	18,700
6000	Solids, precious metals recovery, CI, 1-1/4" to 2" pipe	1 Plum	4	2		750	124		874	1,000
6100	Dental lab., large, CI, 1-1/2" to 2" pipe	"	3	2.667		2,625	166		2,791	3,125

22 13 29 – Sanitary Sewerage Pumps

22 13 29.13 Wet-Pit-Mounted, Vertical Sewerage Pumps

		Crew	Daily Output	Labor-Hours	Unit	Material	2018 Bare Costs Labor	Equipment	Total	Total Incl O&P
0010	**WET-PIT-MOUNTED, VERTICAL SEWERAGE PUMPS**									
0020	Controls incl. alarm/disconnect panel w/wire. Excavation not included									
0260	Simplex, 9 GPM at 60 PSIG, 91 gal. tank				Ea.	3,375			3,375	3,725
0300	Unit with manway, 26" ID, 18" high					3,925			3,925	4,325
0340	26" ID, 36" high					3,975			3,975	4,375
0380	43" ID, 4' high					4,100			4,100	4,500
3000	Indoor residential type installation									
3020	Simplex, 9 GPM at 60 PSIG, 91 gal. HDPE tank				Ea.	3,450			3,450	3,800

22 13 29.14 Sewage Ejector Pumps

		Crew	Daily Output	Labor-Hours	Unit	Material	2018 Bare Costs Labor	Equipment	Total	Total Incl O&P
0010	**SEWAGE EJECTOR PUMPS**, With operating and level controls									
0100	Simplex system incl. tank, cover, pump 15' head									
0500	37 gal. PE tank, 12 GPM, 1/2 HP, 2" discharge	Q-1	3.20	5	Ea.	490	280		770	960
0510	3" discharge		3.10	5.161		535	289		824	1,025
0530	87 GPM, .7 HP, 2" discharge		3.20	5		755	280		1,035	1,250
0540	3" discharge		3.10	5.161		820	289		1,109	1,325
0600	45 gal. coated stl. tank, 12 GPM, 1/2 HP, 2" discharge		3	5.333		880	298		1,178	1,425
0610	3" discharge		2.90	5.517		915	310		1,225	1,475
0630	87 GPM, .7 HP, 2" discharge		3	5.333		1,125	298		1,423	1,700
0640	3" discharge		2.90	5.517		1,200	310		1,510	1,775
0660	134 GPM, 1 HP, 2" discharge		2.80	5.714		1,225	320		1,545	1,825
0680	3" discharge		2.70	5.926		1,275	330		1,605	1,925
0700	70 gal. PE tank, 12 GPM, 1/2 HP, 2" discharge		2.60	6.154		950	345		1,295	1,575
0710	3" discharge		2.40	6.667		1,000	375		1,375	1,650
0730	87 GPM, .7 HP, 2" discharge		2.50	6.400		1,225	360		1,585	1,900
0740	3" discharge		2.30	6.957		1,300	390		1,690	2,000
0760	134 GPM, 1 HP, 2" discharge		2.20	7.273		1,325	405		1,730	2,100
0770	3" discharge		2	8		1,425	445		1,870	2,250

510

For customer support on your Building Construction Costs with RSMeans data, call 800.448.8182.

22 14 Facility Storm Drainage

22 14 23 – Storm Drainage Piping Specialties

22 14 23.33 Backwater Valves

		Crew	Daily Output	Labor-Hours	Unit	Material	2018 Bare Costs Labor	Equipment	Total	Total Incl O&P
0010	**BACKWATER VALVES**, CI Body									
6980	Bronze gate and automatic flapper valves									
7000	3" and 4" pipe size	Q-1	13	1.231	Ea.	2,325	69		2,394	2,675
7100	5" and 6" pipe size	"	13	1.231	"	3,575	69		3,644	4,050
7240	Bronze flapper valve, bolted cover									
7260	2" pipe size	Q-1	16	1	Ea.	685	56		741	835
7300	4" pipe size	"	13	1.231		1,325	69		1,394	1,550
7340	6" pipe size	Q-2	17	1.412		1,900	82		1,982	2,200

22 14 26 – Facility Storm Drains

22 14 26.13 Roof Drains

		Crew	Daily Output	Labor-Hours	Unit	Material	2018 Bare Costs Labor	Equipment	Total	Total Incl O&P
0010	**ROOF DRAINS**									
0140	Cornice, CI, 45° or 90° outlet									
0200	3" and 4" pipe size	Q-1	12	1.333	Ea.	380	74.50		454.50	525
0260	For galvanized body, add					87.50			87.50	96.50
0280	For polished bronze dome, add					108			108	119
3860	Roof, flat metal deck, CI body, 12" CI dome									
3890	3" pipe size	Q-1	14	1.143	Ea.	330	64		394	455
3920	6" pipe size	"	10	1.600	"	785	89.50		874.50	995
4620	Main, all aluminum, 12" low profile dome									
4640	2", 3" and 4" pipe size	Q-1	14	1.143	Ea.	500	64		564	645

22 14 26.16 Facility Area Drains

		Crew	Daily Output	Labor-Hours	Unit	Material	2018 Bare Costs Labor	Equipment	Total	Total Incl O&P
0010	**FACILITY AREA DRAINS**									
4980	Scupper floor, oblique strainer, CI									
5000	6" x 7" top, 2", 3" and 4" pipe size	Q-1	16	1	Ea.	325	56		381	440
5100	8" x 12" top, 5" and 6" pipe size	"	14	1.143		630	64		694	790
5160	For galvanized body, add					40%				
5200	For polished bronze strainer, add					85%				

22 14 26.19 Facility Trench Drains

		Crew	Daily Output	Labor-Hours	Unit	Material	2018 Bare Costs Labor	Equipment	Total	Total Incl O&P
0010	**FACILITY TRENCH DRAINS**									
5980	Trench, floor, heavy duty, modular, CI, 12" x 12" top									
6000	2", 3", 4", 5", & 6" pipe size	Q-1	8	2	Ea.	1,025	112		1,137	1,300
6100	For unit with polished bronze top	"	8	2	"	1,525	112		1,637	1,850
6600	Trench, floor, for cement concrete encasement									
6610	Not including trenching or concrete									
6640	Polyester polymer concrete									
6650	4" internal width, with grate									
6660	Light duty steel grate	Q-1	120	.133	L.F.	63	7.45		70.45	81
6670	Medium duty steel grate		115	.139		124	7.80		131.80	148
6680	Heavy duty iron grate		110	.145		113	8.15		121.15	137
6700	12" internal width, with grate									
6770	Heavy duty galvanized grate	Q-1	80	.200	L.F.	176	11.20		187.20	210
6800	Fiberglass									
6810	8" internal width, with grate									
6820	Medium duty galvanized grate	Q-1	115	.139	L.F.	109	7.80		116.80	132
6830	Heavy duty iron grate	"	110	.145	"	171	8.15		179.15	200

22 14 29 – Sump Pumps

22 14 29.13 Wet-Pit-Mounted, Vertical Sump Pumps

		Crew	Daily Output	Labor-Hours	Unit	Material	2018 Bare Costs Labor	Equipment	Total	Total Incl O&P
0010	**WET-PIT-MOUNTED, VERTICAL SUMP PUMPS**									
0400	Molded PVC base, 21 GPM at 15' head, 1/3 HP	1 Plum	5	1.600	Ea.	140	99.50		239.50	305
0800	Iron base, 21 GPM at 15' head, 1/3 HP		5	1.600		155	99.50		254.50	320
1200	Solid brass, 21 GPM at 15' head, 1/3 HP		5	1.600		240	99.50		339.50	415

22 14 Facility Storm Drainage

22 14 29 – Sump Pumps

22 14 29.13 Wet-Pit-Mounted, Vertical Sump Pumps

		Crew	Daily Output	Labor-Hours	Unit	Material	2018 Bare Costs Labor	Equipment	Total	Total Incl O&P
2000	Sump pump, single stage									
2010	25 GPM, 1 HP, 1-1/2" discharge	Q-1	1.80	8.889	Ea.	3,875	495		4,370	5,000
2020	75 GPM, 1-1/2 HP, 2" discharge		1.50	10.667		4,100	595		4,695	5,400
2030	100 GPM, 2 HP, 2-1/2" discharge		1.30	12.308		4,175	690		4,865	5,600
2040	150 GPM, 3 HP, 3" discharge		1.10	14.545		4,175	815		4,990	5,800
2050	200 GPM, 3 HP, 3" discharge		1	16		4,425	895		5,320	6,225
2060	300 GPM, 10 HP, 4" discharge	Q-2	1.20	20		4,775	1,150		5,925	7,000
2070	500 GPM, 15 HP, 5" discharge		1.10	21.818		5,425	1,275		6,700	7,875
2080	800 GPM, 20 HP, 6" discharge		1	24		6,425	1,400		7,825	9,150
2090	1,000 GPM, 30 HP, 6" discharge		.85	28.235		7,050	1,650		8,700	10,200
2100	1,600 GPM, 50 HP, 8" discharge		.72	33.333		11,000	1,925		12,925	15,000
2110	2,000 GPM, 60 HP, 8" discharge	Q-3	.85	37.647		11,200	2,225		13,425	15,800
2202	For general purpose float switch, copper coated float, add	Q-1	5	3.200		103	179		282	385

22 14 29.16 Submersible Sump Pumps

		Crew	Daily Output	Labor-Hours	Unit	Material	2018 Bare Costs Labor	Equipment	Total	Total Incl O&P
0010	**SUBMERSIBLE SUMP PUMPS**									
7000	Sump pump, automatic									
7100	Plastic, 1-1/4" discharge, 1/4 HP	1 Plum	6.40	1.250	Ea.	150	77.50		227.50	282
7140	1/3 HP		6	1.333		217	83		300	365
7160	1/2 HP		5.40	1.481		266	92		358	430
7180	1-1/2" discharge, 1/2 HP		5.20	1.538		305	95.50		400.50	480
7500	Cast iron, 1-1/4" discharge, 1/4 HP		6	1.333		211	83		294	355
7540	1/3 HP		6	1.333		248	83		331	400
7560	1/2 HP		5	1.600		300	99.50		399.50	480

22 31 Domestic Water Softeners

22 31 13 – Residential Domestic Water Softeners

22 31 13.10 Residential Water Softeners

		Crew	Daily Output	Labor-Hours	Unit	Material	2018 Bare Costs Labor	Equipment	Total	Total Incl O&P
0010	**RESIDENTIAL WATER SOFTENERS**									
7350	Water softener, automatic, to 30 grains per gallon	2 Plum	5	3.200	Ea.	370	199		569	705
7400	To 100 grains per gallon	"	4	4	"	840	249		1,089	1,300

22 31 16 – Commercial Domestic Water Softeners

22 31 16.10 Water Softeners

		Crew	Daily Output	Labor-Hours	Unit	Material	2018 Bare Costs Labor	Equipment	Total	Total Incl O&P
0010	**WATER SOFTENERS**									
5800	Softener systems, automatic, intermediate sizes									
5820	available, may be used in multiples.									
6000	Hardness capacity between regenerations and flow									
6100	150,000 grains, 37 GPM cont., 51 GPM peak	Q-1	1.20	13.333	Ea.	5,175	745		5,920	6,825
6200	300,000 grains, 81 GPM cont., 113 GPM peak		1	16		8,450	895		9,345	10,700
6300	750,000 grains, 160 GPM cont., 230 GPM peak		.80	20		12,500	1,125		13,625	15,400
6400	900,000 grains, 185 GPM cont., 270 GPM peak		.70	22.857		20,000	1,275		21,275	23,900

For customer support on your Building Construction Costs with RSMeans data, call 800.448.8182.

22 33 Electric Domestic Water Heaters

22 33 13 – Instantaneous Electric Domestic Water Heaters

22 33 13.10 Hot Water Dispensers

		Crew	Daily Output	Labor-Hours	Unit	Material	2018 Bare Costs Labor	2018 Bare Costs Equipment	Total	Total Incl O&P
0010	**HOT WATER DISPENSERS**									
0160	Commercial, 100 cup, 11.3 amp	1 Plum	14	.571	Ea.	455	35.50		490.50	555
3180	Household, 60 cup	"	14	.571	"	239	35.50		274.50	315

22 33 30 – Residential, Electric Domestic Water Heaters

22 33 30.13 Residential, Small-Capacity Elec. Water Heaters

		Crew	Daily Output	Labor-Hours	Unit	Material	2018 Bare Costs Labor	2018 Bare Costs Equipment	Total	Total Incl O&P
0010	**RESIDENTIAL, SMALL-CAPACITY ELECTRIC DOMESTIC WATER HEATERS**									
1000	Residential, electric, glass lined tank, 5 yr., 10 gal., single element	1 Plum	2.30	3.478	Ea.	430	216		646	795
1040	20 gallon, single element		2.20	3.636		570	226		796	965
1060	30 gallon, double element		2.20	3.636		1,000	226		1,226	1,450
1080	40 gallon, double element		2	4		1,075	249		1,324	1,550
1100	52 gallon, double element		2	4		1,200	249		1,449	1,700
1180	120 gallon, double element	▼	1.40	5.714	▼	2,575	355		2,930	3,350

22 33 33 – Light-Commercial Electric Domestic Water Heaters

22 33 33.10 Commercial Electric Water Heaters

		Crew	Daily Output	Labor-Hours	Unit	Material	2018 Bare Costs Labor	2018 Bare Costs Equipment	Total	Total Incl O&P
0010	**COMMERCIAL ELECTRIC WATER HEATERS**									
4000	Commercial, 100° rise. NOTE: for each size tank, a range of									
4010	heaters between the ones shown are available									
4020	Electric									
4100	5 gal., 3 kW, 12 GPH, 208 volt	1 Plum	2	4	Ea.	4,775	249		5,024	5,625
4120	10 gal., 6 kW, 25 GPH, 208 volt		2	4		5,275	249		5,524	6,175
4130	30 gal., 24 kW, 98 GPH, 208 volt		1.92	4.167		8,700	259		8,959	9,975
4136	40 gal., 36 kW, 148 GPH, 208 volt		1.88	4.255		10,400	264		10,664	11,900
4140	50 gal., 9 kW, 37 GPH, 208 volt		1.80	4.444		7,275	276		7,551	8,425
4160	50 gal., 36 kW, 148 GPH, 208 volt	▼	1.80	4.444		11,200	276		11,476	12,700
4300	200 gal., 15 kW, 61 GPH, 480 volt	Q-1	1.70	9.412		34,200	525		34,725	38,400
4320	200 gal., 120 kW, 490 GPH, 480 volt		1.70	9.412		46,800	525		47,325	52,500
4460	400 gal., 30 kW, 123 GPH, 480 volt	▼	1	16		46,600	895		47,495	52,500
5400	Modulating step control for under 90 kW, 2-5 steps	1 Elec	5.30	1.509		790	88		878	1,000
5440	For above 90 kW, 1 through 5 steps beyond standard, add		3.20	2.500		230	146		376	470
5460	For above 90 kW, 6 through 10 steps beyond standard, add		2.70	2.963		450	172		622	750
5480	For above 90 kW, 11 through 18 steps beyond standard, add	▼	1.60	5	▼	670	291		961	1,175

22 34 Fuel-Fired Domestic Water Heaters

22 34 13 – Instantaneous, Tankless, Gas Domestic Water Heaters

22 34 13.10 Instantaneous, Tankless, Gas Water Heaters

			Crew	Daily Output	Labor-Hours	Unit	Material	2018 Bare Costs Labor	2018 Bare Costs Equipment	Total	Total Incl O&P
0010	**INSTANTANEOUS, TANKLESS, GAS WATER HEATERS**										
9410	Natural gas/propane, 3.2 GPM	G	1 Plum	2	4	Ea.	500	249		749	925
9420	6.4 GPM	G		1.90	4.211		715	262		977	1,175
9430	8.4 GPM	G		1.80	4.444		855	276		1,131	1,350
9440	9.5 GPM	G	▼	1.60	5	▼	1,000	310		1,310	1,575

22 34 30 – Residential Gas Domestic Water Heaters

22 34 30.13 Residential, Atmos, Gas Domestic Wtr Heaters

		Crew	Daily Output	Labor-Hours	Unit	Material	2018 Bare Costs Labor	2018 Bare Costs Equipment	Total	Total Incl O&P
0010	**RESIDENTIAL, ATMOSPHERIC, GAS DOMESTIC WATER HEATERS**									
2000	Gas fired, foam lined tank, 10 yr., vent not incl.									
2040	30 gallon	1 Plum	2	4	Ea.	1,775	249		2,024	2,325
2100	75 gallon		1.50	5.333		2,700	330		3,030	3,475
2120	100 gallon		1.30	6.154		2,850	380		3,230	3,700
2900	Water heater, safety-drain pan, 26" round	▼	20	.400	▼	19.90	25		44.90	59.50
3000	Tank leak safety, water & gas shut off see 22 05 23.20 8800									

For customer support on your Building Construction Costs with RSMeans data, call 800.448.8182.

513

22 34 Fuel-Fired Domestic Water Heaters

22 34 36 – Commercial Gas Domestic Water Heaters

22 34 36.13 Commercial, Atmos., Gas Domestic Water Htrs.	Crew	Daily Output	Labor-Hours	Unit	Material	2018 Bare Costs Labor	Equipment	Total	Total Incl O&P
0010 **COMMERCIAL, ATMOSPHERIC, GAS DOMESTIC WATER HEATERS**									
6000 Gas fired, flush jacket, std. controls, vent not incl.									
6040 75 MBH input, 73 GPH	1 Plum	1.40	5.714	Ea.	3,775	355		4,130	4,675
6060 98 MBH input, 95 GPH		1.40	5.714		7,475	355		7,830	8,750
6080 120 MBH input, 110 GPH		1.20	6.667		7,700	415		8,115	9,100
6180 200 MBH input, 192 GPH		.60	13.333		10,200	830		11,030	12,500
6200 250 MBH input, 245 GPH		.50	16		10,900	995		11,895	13,500
6900 For low water cutoff, add		8	1		365	62		427	495
6960 For bronze body hot water circulator, add		4	2		2,150	124		2,274	2,550

22 34 46 – Oil-Fired Domestic Water Heaters

22 34 46.10 Residential Oil-Fired Water Heaters

	Crew	Daily Output	Labor-Hours	Unit	Material	Labor	Equipment	Total	Total Incl O&P
0010 **RESIDENTIAL OIL-FIRED WATER HEATERS**									
3000 Oil fired, glass lined tank, 5 yr., vent not included, 30 gallon	1 Plum	2	4	Ea.	1,275	249		1,524	1,775
3040 50 gallon		1.80	4.444		1,575	276		1,851	2,150
3060 70 gallon		1.50	5.333		2,200	330		2,530	2,925

22 34 46.20 Commercial Oil-Fired Water Heaters

	Crew	Daily Output	Labor-Hours	Unit	Material	Labor	Equipment	Total	Total Incl O&P
0010 **COMMERCIAL OIL-FIRED WATER HEATERS**									
8000 Oil fired, glass lined, UL listed, std. controls, vent not incl.									
8060 140 gal., 140 MBH input, 134 GPH	Q-1	2.13	7.512	Ea.	26,000	420		26,420	29,200
8080 140 gal., 199 MBH input, 191 GPH		2	8		26,900	445		27,345	30,300
8100 140 gal., 255 MBH input, 247 GPH		1.60	10		27,700	560		28,260	31,200
8160 140 gal., 540 MBH input, 519 GPH		.96	16.667		36,700	930		37,630	41,800
8180 140 gal., 720 MBH input, 691 GPH		.92	17.391		37,400	975		38,375	42,600
8280 201 gal., 1,250 MBH input, 1,200 GPH	Q-2	1.22	19.672		58,000	1,150		59,150	65,000
8300 201 gal., 1,500 MBH input, 1,441 GPH	"	1.16	20.690		62,500	1,200		63,700	70,500
8900 For low water cutoff, add	1 Plum	8	1		365	62		427	495
8960 For bronze body hot water circulator, add	"	4	2		875	124		999	1,150

22 35 Domestic Water Heat Exchangers

22 35 30 – Water Heating by Steam

22 35 30.10 Water Heating Transfer Package

	Crew	Daily Output	Labor-Hours	Unit	Material	Labor	Equipment	Total	Total Incl O&P
0010 **WATER HEATING TRANSFER PACKAGE**, Complete controls,									
0020 expansion tank, converter, air separator									
1000 Hot water, 180°F enter, 200°F leaving, 15# steam									
1010 One pump system, 28 GPM	Q-6	.75	32	Ea.	21,100	1,875		22,975	26,000
1020 35 GPM		.70	34.286		22,900	2,025		24,925	28,200
1040 55 GPM		.65	36.923		27,100	2,175		29,275	33,100
1060 130 GPM		.55	43.636		34,200	2,575		36,775	41,500
1080 255 GPM		.40	60		45,000	3,525		48,525	55,000
1100 550 GPM		.30	80		61,000	4,700		65,700	74,000

For customer support on your Building Construction Costs with RSMeans data, call 800.448.8182.

22 41 Residential Plumbing Fixtures

22 41 06 – Plumbing Fixtures General

22 41 06.10 Plumbing Fixture Notes

	Crew	Daily Output	Labor-Hours	Unit	Material	2018 Bare Costs Labor	2018 Bare Costs Equipment	Total	Total Incl O&P
0010 **PLUMBING FIXTURE NOTES**, Incl. trim fittings unless otherwise noted									
0080 For rough-in, supply, waste, and vent, see add for each type									
0122 For electric water coolers, see Section 22 47 16.10									
0160 For color, unless otherwise noted, add				Ea.	20%				

22 41 13 – Residential Water Closets, Urinals, and Bidets

22 41 13.13 Water Closets

	Crew	Daily Output	Labor-Hours	Unit	Material	2018 Bare Costs Labor	2018 Bare Costs Equipment	Total	Total Incl O&P
0010 **WATER CLOSETS**									
0032 For automatic flush, see Line 22 42 39.10 0972									
0150 Tank type, vitreous china, incl. seat, supply pipe w/stop, 1.6 gpf or noted									
0200 Wall hung									
0400 Two piece, close coupled	Q-1	5.30	3.019	Ea.	400	169		569	695
0960 For rough-in, supply, waste, vent and carrier	"	2.73	5.861	"	1,150	330		1,480	1,775
0999 Floor mounted									
1100 Two piece, close coupled	Q-1	5.30	3.019	Ea.	230	169		399	505
1102 Economy		5.30	3.019		126	169		295	390
1110 Two piece, close coupled, dual flush		5.30	3.019		320	169		489	610
1140 Two piece, close coupled, 1.28 gpf, ADA [G]		5.30	3.019		310	169		479	595
1960 For color, add					30%				
1980 For rough-in, supply, waste and vent	Q-1	3.05	5.246	Ea.	360	293		653	835

22 41 16 – Residential Lavatories and Sinks

22 41 16.13 Lavatories

	Crew	Daily Output	Labor-Hours	Unit	Material	2018 Bare Costs Labor	2018 Bare Costs Equipment	Total	Total Incl O&P
0010 **LAVATORIES**, With trim, white unless noted otherwise									
0500 Vanity top, porcelain enamel on cast iron									
0600 20" x 18"	Q-1	6.40	2.500	Ea.	300	140		440	540
0640 33" x 19" oval		6.40	2.500		520	140		660	785
0720 19" round		6.40	2.500		390	140		530	640
0860 For color, add					25%				
1000 Cultured marble, 19" x 17", single bowl	Q-1	6.40	2.500	Ea.	127	140		267	350
1040 25" x 19", single bowl	"	6.40	2.500	"	150	140		290	375
1580 For color, same price									
1900 Stainless steel, self-rimming, 25" x 22", single bowl, ledge	Q-1	6.40	2.500	Ea.	320	140		460	560
1960 17" x 22", single bowl		6.40	2.500		305	140		445	550
2600 Steel, enameled, 20" x 17", single bowl		5.80	2.759		129	154		283	375
2660 19" round		5.80	2.759		157	154		311	405
2900 Vitreous china, 20" x 16", single bowl		5.40	2.963		213	166		379	485
2960 20" x 17", single bowl		5.40	2.963		116	166		282	375
3020 19" round, single bowl		5.40	2.963		114	166		280	375
3200 22" x 13", single bowl		5.40	2.963		220	166		386	490
3560 For color, add					50%				
3580 Rough-in, supply, waste and vent for all above lavatories	Q-1	2.30	6.957	Ea.	238	390		628	845
4000 Wall hung									
4040 Porcelain enamel on cast iron, 16" x 14", single bowl	Q-1	8	2	Ea.	445	112		557	660
4180 20" x 18", single bowl		8	2		246	112		358	440
4240 22" x 19", single bowl		8	2		690	112		802	925
4580 For color, add					30%				
6000 Vitreous china, 18" x 15", single bowl with backsplash	Q-1	7	2.286	Ea.	161	128		289	370
6500 For color, add					30%				
6960 Rough-in, supply, waste and vent for above lavatories	Q-1	1.66	9.639	Ea.	465	540		1,005	1,325
7000 Pedestal type									
7600 Vitreous china, 27" x 21", white	Q-1	6.60	2.424	Ea.	670	136		806	945
7610 27" x 21", colored		6.60	2.424		855	136		991	1,150
7620 27" x 21", premium color		6.60	2.424		975	136		1,111	1,275

22 41 16 – Residential Lavatories and Sinks

22 41 16.13 Lavatories

		Crew	Daily Output	Labor-Hours	Unit	Material	2018 Bare Costs Labor	Equipment	Total	Total Incl O&P
7660	26" x 20", white	Q-1	6.60	2.424	Ea.	660	136		796	930
7670	26" x 20", colored		6.60	2.424		840	136		976	1,125
7680	26" x 20", premium color		6.60	2.424		955	136		1,091	1,250
7700	24" x 20", white		6.60	2.424		470	136		606	725
7710	24" x 20", colored		6.60	2.424		595	136		731	860
7720	24" x 20", premium color		6.60	2.424		675	136		811	950
7760	21" x 18", white		6.60	2.424		258	136		394	485
7770	21" x 18", colored		6.60	2.424		278	136		414	510
7990	Rough-in, supply, waste and vent for pedestal lavatories	▼	1.66	9.639	▼	465	540		1,005	1,325

22 41 16.16 Sinks

		Crew	Daily Output	Labor-Hours	Unit	Material	2018 Bare Costs Labor	Equipment	Total	Total Incl O&P
0010	**SINKS**, With faucets and drain									
2000	Kitchen, counter top style, PE on CI, 24" x 21" single bowl	Q-1	5.60	2.857	Ea.	305	160		465	575
2100	31" x 22" single bowl		5.60	2.857		660	160		820	965
2200	32" x 21" double bowl		4.80	3.333		375	186		561	695
3000	Stainless steel, self rimming, 19" x 18" single bowl		5.60	2.857		615	160		775	915
3100	25" x 22" single bowl		5.60	2.857		680	160		840	990
4000	Steel, enameled, with ledge, 24" x 21" single bowl		5.60	2.857		525	160		685	820
4100	32" x 21" double bowl	▼	4.80	3.333		540	186		726	875
4960	For color sinks except stainless steel, add					10%				
4980	For rough-in, supply, waste and vent, counter top sinks	Q-1	2.14	7.477	▼	275	420		695	935
5000	Kitchen, raised deck, PE on CI									
5100	32" x 21", dual level, double bowl	Q-1	2.60	6.154	Ea.	450	345		795	1,000
5700	For color, add					20%				
5790	For rough-in, supply, waste & vent, sinks	Q-1	1.85	8.649	▼	275	485		760	1,025

22 41 19 – Residential Bathtubs

22 41 19.10 Baths

		Crew	Daily Output	Labor-Hours	Unit	Material	2018 Bare Costs Labor	Equipment	Total	Total Incl O&P
0010	**BATHS**									
0100	Tubs, recessed porcelain enamel on cast iron, with trim									
0180	48" x 42"	Q-1	4	4	Ea.	2,800	224		3,024	3,400
0220	72" x 36"		3	5.333		2,850	298		3,148	3,600
2000	Enameled formed steel, 4'-6" long		5.80	2.759		490	154		644	770
4000	Soaking, acrylic, w/pop-up drain 66" x 36" x 20" deep		5.50	2.909		1,675	163		1,838	2,100
4100	60" x 42" x 20" deep		5	3.200		1,250	179		1,429	1,650
9600	Rough-in, supply, waste and vent, for all above tubs, add	▼	2.07	7.729	▼	430	430		860	1,125

22 41 23 – Residential Showers

22 41 23.20 Showers

		Crew	Daily Output	Labor-Hours	Unit	Material	2018 Bare Costs Labor	Equipment	Total	Total Incl O&P
0010	**SHOWERS**									
1500	Stall, with drain only. Add for valve and door/curtain									
3000	Fiberglass, one piece, with 3 walls, 32" x 32" square	Q-1	5.50	2.909	Ea.	345	163		508	625
3100	36" x 36" square		5.50	2.909		400	163		563	685
3250	64" x 65-3/4" x 81-1/2" fold. seat, ADA		3.80	4.211		1,425	235		1,660	1,900
4000	Polypropylene, stall only, w/molded-stone floor, 30" x 30"		2	8		635	445		1,080	1,375
4200	Rough-in, supply, waste and vent for above showers	▼	2.05	7.805	▼	370	435		805	1,075

22 41 23.40 Shower System Components

		Crew	Daily Output	Labor-Hours	Unit	Material	2018 Bare Costs Labor	Equipment	Total	Total Incl O&P
0010	**SHOWER SYSTEM COMPONENTS**									
4500	Receptor only									
4510	For tile, 36" x 36"	1 Plum	4	2	Ea.	375	124		499	595
4520	Fiberglass receptor only, 32" x 32"		8	1		102	62		164	206
4530	34" x 34"		7.80	1.026		123	63.50		186.50	231
4540	36" x 36"	▼	7.60	1.053	▼	119	65.50		184.50	230
4600	Rectangular									

For customer support on your Building Construction Costs with RSMeans data, call 800.448.8182.

22 41 Residential Plumbing Fixtures

22 41 23 – Residential Showers

22 41 23.40 Shower System Components

		Crew	Daily Output	Labor-Hours	Unit	Material	2018 Bare Costs Labor	Equipment	Total	Total Incl O&P
4620	32" x 48"	1 Plum	7.40	1.081	Ea.	141	67		208	257
4630	34" x 54"		7.20	1.111		166	69		235	287
4640	34" x 60"		7	1.143		180	71		251	305
5000	Built-in, head, arm, 2.5 GPM valve		4	2		91	124		215	287
5200	Head, arm, by-pass, integral stops, handles	↓	3.60	2.222	↓	286	138		424	525

22 41 36 – Residential Laundry Trays

22 41 36.10 Laundry Sinks

		Crew	Daily Output	Labor-Hours	Unit	Material	2018 Bare Costs Labor	Equipment	Total	Total Incl O&P
0010	**LAUNDRY SINKS**, With trim									
0020	Porcelain enamel on cast iron, black iron frame									
0050	24" x 21", single compartment	Q-1	6	2.667	Ea.	595	149		744	875
0100	26" x 21", single compartment	"	6	2.667	"	620	149		769	910
2000	Molded stone, on wall hanger or legs									
2020	22" x 23", single compartment	Q-1	6	2.667	Ea.	185	149		334	430
2100	45" x 21", double compartment	"	5	3.200	"	310	179		489	615
3000	Plastic, on wall hanger or legs									
3020	18" x 23", single compartment	Q-1	6.50	2.462	Ea.	145	138		283	365
3300	40" x 24", double compartment		5.50	2.909		287	163		450	560
5000	Stainless steel, counter top, 22" x 17" single compartment		6	2.667		76.50	149		225.50	310
5200	33" x 22", double compartment		5	3.200		91.50	179		270.50	370
9600	Rough-in, supply, waste and vent, for all laundry sinks	↓	2.14	7.477		275	420		695	935

22 41 39 – Residential Faucets, Supplies and Trim

22 41 39.10 Faucets and Fittings

		Crew	Daily Output	Labor-Hours	Unit	Material	2018 Bare Costs Labor	Equipment	Total	Total Incl O&P
0010	**FAUCETS AND FITTINGS**									
0150	Bath, faucets, diverter spout combination, sweat	1 Plum	8	1	Ea.	86	62		148	188
0200	For integral stops, IPS unions, add					111			111	122
0420	Bath, press-bal mix valve w/diverter, spout, shower head, arm/flange	1 Plum	8	1	↓	175	62		237	286
0810	Bidet									
0812	Fitting, over the rim, swivel spray/pop-up drain	1 Plum	8	1	Ea.	260	62		322	380
1000	Kitchen sink faucets, top mount, cast spout		10	.800	↓	83.50	49.50		133	167
1100	For spray, add	↓	24	.333	↓	17.10	20.50		37.60	50
1300	Single control lever handle									
1310	With pull out spray									
1320	Polished chrome	1 Plum	10	.800	Ea.	196	49.50		245.50	291
2000	Laundry faucets, shelf type, IPS or copper unions		12	.667		61.50	41.50		103	130
2100	Lavatory faucet, centerset, without drain	↓	10	.800	↓	67	49.50		116.50	149
2210	Porcelain cross handles and pop-up drain									
2220	Polished chrome	1 Plum	6.66	1.201	Ea.	207	74.50		281.50	340
2230	Polished brass	"	6.66	1.201	"	310	74.50		384.50	455
2260	Single lever handle and pop-up drain									
2280	Satin nickel	1 Plum	6.66	1.201	Ea.	273	74.50		347.50	410
2290	Polished chrome		6.66	1.201		197	74.50		271.50	330
2810	Automatic sensor and operator, with faucet head [G]		6.15	1.301		480	81		561	650
4000	Shower by-pass valve with union		18	.444		57	27.50		84.50	105
4200	Shower thermostatic mixing valve, concealed, with shower head trim kit	↓	8	1	↓	350	62		412	480
4220	Shower pressure balancing mixing valve									
4230	With shower head, arm, flange and diverter tub spout									
4240	Chrome	1 Plum	6.14	1.303	Ea.	410	81		491	575
4250	Satin nickel		6.14	1.303		555	81		636	730
4260	Polished graphite		6.14	1.303		555	81		636	730
5000	Sillcock, compact, brass, IPS or copper to hose	↓	24	.333	↓	10.55	20.50		31.05	42.50

For customer support on your Building Construction Costs with RSMeans data, call 800.448.8182.

517

22 42 Commercial Plumbing Fixtures

22 42 13 – Commercial Water Closets, Urinals, and Bidets

22 42 13.13 Water Closets

	22 42 13.13 Water Closets	Crew	Daily Output	Labor-Hours	Unit	Material	2018 Bare Costs Labor	Equipment	Total	Total Incl O&P
0010	**WATER CLOSETS**									
3000	Bowl only, with flush valve, seat, 1.6 gpf unless noted									
3100	Wall hung	Q-1	5.80	2.759	Ea.	970	154		1,124	1,300
3200	For rough-in, supply, waste and vent, single WC		2.56	6.250		1,200	350		1,550	1,825
3300	Floor mounted		5.80	2.759		325	154		479	585
3350	With wall outlet		5.80	2.759		580	154		734	870
3360	With floor outlet, 1.28 gpf G		5.80	2.759		550	154		704	835
3362	With floor outlet, 1.28 gpf, ADA G		5.80	2.759		575	154		729	860
3370	For rough-in, supply, waste and vent, single WC		2.84	5.634		395	315		710	910
3390	Floor mounted children's size, 10-3/4" high									
3392	With automatic flush sensor, 1.6 gpf	Q-1	6.20	2.581	Ea.	620	144		764	900
3396	With automatic flush sensor, 1.28 gpf		6.20	2.581		620	144		764	900
3400	For rough-in, supply, waste and vent, single WC		2.84	5.634		395	315		710	910

22 42 13.16 Urinals

	22 42 13.16 Urinals	Crew	Daily Output	Labor-Hours	Unit	Material	Labor	Equipment	Total	Total Incl O&P
0010	**URINALS**									
3000	Wall hung, vitreous china, with self-closing valve									
3100	Siphon jet type	Q-1	3	5.333	Ea.	281	298		579	760
3120	Blowout type		3	5.333		470	298		768	965
3140	Water saving .5 gpf G		3	5.333		555	298		853	1,050
3300	Rough-in, supply, waste & vent		2.83	5.654		670	315		985	1,225
5000	Stall type, vitreous china, includes valve		2.50	6.400		770	360		1,130	1,375
6980	Rough-in, supply, waste and vent		1.99	8.040		430	450		880	1,150
8000	Waterless (no flush) urinal									
8010	Wall hung									
8014	Fiberglass reinforced polyester									
8020	Standard unit G	Q-1	21.30	.751	Ea.	385	42		427	485
8030	ADA compliant unit G	"	21.30	.751		400	42		442	505
8070	For solid color, add G					60			60	66
8080	For 2" brass flange (new const.), add G	Q-1	96	.167		19.20	9.30		28.50	35
8200	Vitreous china									
8220	ADA compliant unit, 14" G	Q-1	21.30	.751	Ea.	198	42		240	282
8250	ADA compliant unit, 15.5"	"	21.30	.751		230	42		272	315
8270	For solid color, add G					60			60	66
8290	Rough-in, supply, waste & vent G	Q-1	2.92	5.479		635	305		940	1,150
8400	Trap liquid									
8410	1 quart G				Ea.	17.35			17.35	19.10
8420	1 gallon G				"	61			61	67

22 42 16 – Commercial Lavatories and Sinks

22 42 16.13 Lavatories

0010	**LAVATORIES**, With trim, white unless noted otherwise									
0020	Commercial lavatories same as residential. See Section 22 41 16									

22 42 16.40 Service Sinks

	22 42 16.40 Service Sinks	Crew	Daily Output	Labor-Hours	Unit	Material	Labor	Equipment	Total	Total Incl O&P
0010	**SERVICE SINKS**									
6650	Service, floor, corner, PE on CI, 28" x 28"	Q-1	4.40	3.636	Ea.	1,025	203		1,228	1,425
6790	For rough-in, supply, waste & vent, floor service sinks		1.64	9.756		960	545		1,505	1,875
7000	Service, wall, PE on CI, roll rim, 22" x 18"		4	4		800	224		1,024	1,225
7100	24" x 20"		4	4		885	224		1,109	1,300
8600	Vitreous china, 22" x 20"		4	4		800	224		1,024	1,225
8960	For stainless steel rim guard, front or one side, add					53.50			53.50	59
8980	For rough-in, supply, waste & vent, wall service sinks	Q-1	1.30	12.308		1,375	690		2,065	2,550

For customer support on your Building Construction Costs with RSMeans data, call 800.448.8182.

22 42 Commercial Plumbing Fixtures

22 42 23 – Commercial Showers

22 42 23.30 Group Showers

		Crew	Daily Output	Labor-Hours	Unit	Material	2018 Bare Costs Labor	Equipment	Total	Total Incl O&P
0010	**GROUP SHOWERS**									
6000	Group, w/pressure balancing valve, rough-in and rigging not included									
6800	Column, 6 heads, no receptors, less partitions	Q-1	3	5.333	Ea.	9,400	298		9,698	10,800
6900	With stainless steel partitions		1	16		12,100	895		12,995	14,800
7600	5 heads, no receptors, less partitions		3	5.333		6,475	298		6,773	7,575
7620	4 heads (1 ADA compliant) no receptors, less partitions ♿		3	5.333		5,750	298		6,048	6,775
7700	With stainless steel partitions		1	16		6,725	895		7,620	8,750
8000	Wall, 2 heads, no receptors, less partitions		4	4		2,775	224		2,999	3,375
8100	With stainless steel partitions		2	8		6,075	445		6,520	7,375

22 42 33 – Wash Fountains

22 42 33.20 Commercial Wash Fountains

		Crew	Daily Output	Labor-Hours	Unit	Material	2018 Bare Costs Labor	Equipment	Total	Total Incl O&P
0010	**COMMERCIAL WASH FOUNTAINS**									
1900	Group, foot control									
2000	Precast terrazzo, circular, 36" diam., 5 or 6 persons	Q-2	3	8	Ea.	7,425	465		7,890	8,850
2100	54" diam. for 8 or 10 persons		2.50	9.600		9,250	555		9,805	11,000
2400	Semi-circular, 36" diam. for 3 persons		3	8		6,500	465		6,965	7,850
2500	54" diam. for 4 or 5 persons		2.50	9.600		8,750	555		9,305	10,500
2700	Quarter circle (corner), 54" diam. for 3 persons		3.50	6.857		8,025	400		8,425	9,425
3000	Stainless steel, circular, 36" diameter		3.50	6.857		6,675	400		7,075	7,950
3100	54" diameter		2.80	8.571		8,200	495		8,695	9,775
3400	Semi-circular, 36" diameter		3.50	6.857		5,125	400		5,525	6,250
3500	54" diameter ♿		2.80	8.571		7,150	495		7,645	8,625
5610	Group, infrared control, barrier free									
5614	Precast terrazzo									
5620	Semi-circular 36" diam. for 3 persons	Q-2	3	8	Ea.	7,400	465		7,865	8,825
5630	46" diam. for 4 persons		2.80	8.571		7,975	495		8,470	9,525
5640	Circular, 54" diam. for 8 persons, button control		2.50	9.600		9,675	555		10,230	11,400
5700	Rough-in, supply, waste and vent for above wash fountains	Q-1	1.82	8.791		455	490		945	1,250
6200	Duo for small washrooms, stainless steel		2	8		2,750	445		3,195	3,700
6500	Rough-in, supply, waste & vent for duo fountains		2.02	7.921		238	445		683	925

22 42 39 – Commercial Faucets, Supplies, and Trim

22 42 39.10 Faucets and Fittings

		Crew	Daily Output	Labor-Hours	Unit	Material	2018 Bare Costs Labor	Equipment	Total	Total Incl O&P
0010	**FAUCETS AND FITTINGS**									
0840	Flush valves, with vacuum breaker									
0850	Water closet									
0860	Exposed, rear spud	1 Plum	8	1	Ea.	156	62		218	266
0870	Top spud		8	1		182	62		244	294
0880	Concealed, rear spud		8	1		210	62		272	325
0890	Top spud		8	1		171	62		233	283
0900	Wall hung		8	1		191	62		253	305
0920	Urinal									
0930	Exposed, stall	1 Plum	8	1	Ea.	182	62		244	294
0940	Wall (washout)		8	1		144	62		206	252
0950	Pedestal, top spud		8	1		146	62		208	255
0960	Concealed, stall		8	1		168	62		230	279
0970	Wall (washout) ♿		8	1		181	62		243	294
0971	Automatic flush sensor and operator for									
0972	urinals or water closets, standard Ⓖ	1 Plum	8	1	Ea.	475	62		537	615
0980	High efficiency water saving									
0984	Water closets, 1.28 gpf Ⓖ	1 Plum	8	1	Ea.	415	62		477	555
0988	Urinals, .5 gpf Ⓖ	"	8	1	"	415	62		477	555

For customer support on your Building Construction Costs with RSMeans data, call 800.448.8182.

519

22 42 39.10 Faucets and Fittings

	22 42 39.10 Faucets and Fittings	Crew	Daily Output	Labor-Hours	Unit	Material	2018 Bare Costs Labor	Equipment	Total	Total Incl O&P
2790	Faucets for lavatories									
2800	Self-closing, center set	1 Plum	10	.800	Ea.	149	49.50		198.50	239
2810	Automatic sensor and operator, with faucet head		6.15	1.301		480	81		561	650
3000	Service sink faucet, cast spout, pail hook, hose end	↓	14	.571	↓	76	35.50		111.50	137

22 42 39.30 Carriers and Supports

	22 42 39.30 Carriers and Supports	Crew	Daily Output	Labor-Hours	Unit	Material	2018 Bare Costs Labor	Equipment	Total	Total Incl O&P
0010	**CARRIERS AND SUPPORTS**, For plumbing fixtures									
0500	Drinking fountain, wall mounted									
0600	Plate type with studs, top back plate	1 Plum	7	1.143	Ea.	57	71		128	170
0700	Top front and back plate		7	1.143		152	71		223	274
0800	Top & bottom, front & back plates, w/bearing jacks	↓	7	1.143	↓	152	71		223	274
3000	Lavatory, concealed arm									
3050	Floor mounted, single									
3100	High back fixture	1 Plum	6	1.333	Ea.	605	83		688	790
3200	Flat slab fixture		6	1.333		520	83		603	695
3220	ADA compliant	↓	6	1.333	↓	555	83		638	735
3250	Floor mounted, back to back									
3300	High back fixtures	1 Plum	5	1.600	Ea.	860	99.50		959.50	1,100
3400	Flat slab fixtures		5	1.600		1,050	99.50		1,149.50	1,325
3430	ADA compliant	↓	5	1.600	↓	775	99.50		874.50	1,000
3500	Wall mounted, in stud or masonry									
3600	High back fixture	1 Plum	6	1.333	Ea.	355	83		438	520
3700	Flat slab fixture	"	6	1.333	"	295	83		378	450
4600	Sink, floor mounted									
4650	Exposed arm system									
4700	Single heavy fixture	1 Plum	5	1.600	Ea.	530	99.50		629.50	730
4750	Single heavy sink with slab		5	1.600		1,325	99.50		1,424.50	1,600
4800	Back to back, standard fixtures		5	1.600		740	99.50		839.50	965
4850	Back to back, heavy fixtures		5	1.600		1,150	99.50		1,249.50	1,400
4900	Back to back, heavy sink with slab	↓	5	1.600	↓	1,625	99.50		1,724.50	1,950
4950	Exposed offset arm system									
5000	Single heavy deep fixture	1 Plum	5	1.600	Ea.	960	99.50		1,059.50	1,200
5100	Plate type system									
5200	With bearing jacks, single fixture	1 Plum	5	1.600	Ea.	1,425	99.50		1,524.50	1,700
5300	With exposed arms, single heavy fixture		5	1.600		1,425	99.50		1,524.50	1,700
5400	Wall mounted, exposed arms, single heavy fixture		5	1.600		515	99.50		614.50	715
6000	Urinal, floor mounted, 2" or 3" coupling, blowout type		6	1.333		640	83		723	830
6100	With fixture or hanger bolts, blowout or washout		6	1.333		460	83		543	630
6200	With bearing plate		6	1.333		510	83		593	685
6300	Wall mounted, plate type system	↓	6	1.333	↓	395	83		478	560
6980	Water closet, siphon jet									
7000	Horizontal, adjustable, caulk									
7040	Single, 4" pipe size	1 Plum	5.33	1.501	Ea.	965	93.50		1,058.50	1,225
7050	4" pipe size, ADA compliant		5.33	1.501		775	93.50		868.50	990
7060	5" pipe size		5.33	1.501		1,100	93.50		1,193.50	1,375
7100	Double, 4" pipe size		5	1.600		1,600	99.50		1,699.50	1,925
7110	4" pipe size, ADA compliant		5	1.600		1,450	99.50		1,549.50	1,725
7120	5" pipe size	↓	5	1.600	↓	1,950	99.50		2,049.50	2,275
7160	Horizontal, adjustable, extended, caulk									
7180	Single, 4" pipe size	1 Plum	5.33	1.501	Ea.	1,150	93.50		1,243.50	1,425
7200	5" pipe size		5.33	1.501		1,475	93.50		1,568.50	1,775
7240	Double, 4" pipe size		5	1.600		2,000	99.50		2,099.50	2,375
7260	5" pipe size	↓	5	1.600	↓	2,450	99.50		2,549.50	2,825

22 42 39.30 Carriers and Supports	Crew	Daily Output	Labor-Hours	Unit	Material	2018 Bare Costs Labor	Equipment	Total	Total Incl O&P
7400	Vertical, adjustable, caulk or thread									
7440	Single, 4" pipe size	1 Plum	5.33	1.501	Ea.	1,025	93.50		1,118.50	1,275
7460	5" pipe size		5.33	1.501		1,275	93.50		1,368.50	1,550
7480	6" pipe size		5	1.600		1,500	99.50		1,599.50	1,800
7520	Double, 4" pipe size		5	1.600		1,750	99.50		1,849.50	2,075
7540	5" pipe size		5	1.600		2,025	99.50		2,124.50	2,375
7560	6" pipe size	↓	4	2	↓	2,250	124		2,374	2,650
7600	Vertical, adjustable, extended, caulk									
7620	Single, 4" pipe size	1 Plum	5.33	1.501	Ea.	1,175	93.50		1,268.50	1,450
7640	5" pipe size		5.33	1.501		855	93.50		948.50	1,075
7680	6" pipe size		5	1.600		1,675	99.50		1,774.50	1,975
7720	Double, 4" pipe size		5	1.600		1,925	99.50		2,024.50	2,275
7740	5" pipe size		5	1.600		1,000	99.50		1,099.50	1,275
7760	6" pipe size	↓	4	2	↓	1,100	124		1,224	1,400
7780	Water closet, blow out									
7800	Vertical offset, caulk or thread									
7820	Single, 4" pipe size	1 Plum	5.33	1.501	Ea.	1,075	93.50		1,168.50	1,325
7840	Double, 4" pipe size	"	5	1.600	"	1,850	99.50		1,949.50	2,175
7880	Vertical offset, extended, caulk									
7900	Single, 4" pipe size	1 Plum	5.33	1.501	Ea.	1,350	93.50		1,443.50	1,625
7920	Double, 4" pipe size	"	5	1.600	"	2,125	99.50		2,224.50	2,475
7960	Vertical, for floor mounted back-outlet									
7980	Single, 4" thread, 2" vent	1 Plum	5.33	1.501	Ea.	765	93.50		858.50	980
8000	Double, 4" thread, 2" vent	"	6	1.333	"	2,250	83		2,333	2,600
8040	Vertical, for floor mounted back-outlet, extended									
8060	Single, 4" caulk, 2" vent	1 Plum	6	1.333	Ea.	765	83		848	965
8080	Double, 4" caulk, 2" vent	"	6	1.333	"	2,250	83		2,333	2,600
8200	Water closet, residential									
8220	Vertical centerline, floor mount									
8240	Single, 3" caulk, 2" or 3" vent	1 Plum	6	1.333	Ea.	695	83		778	890
8260	4" caulk, 2" or 4" vent		6	1.333		895	83		978	1,100
8280	3" copper sweat, 3" vent		6	1.333		625	83		708	815
8300	4" copper sweat, 4" vent	↓	6	1.333	↓	755	83		838	955
8400	Vertical offset, floor mount									
8420	Single, 3" or 4" caulk, vent	1 Plum	4	2	Ea.	870	124		994	1,150
8440	3" or 4" copper sweat, vent		5	1.600		870	99.50		969.50	1,100
8460	Double, 3" or 4" caulk, vent		4	2		1,500	124		1,624	1,825
8480	3" or 4" copper sweat, vent	↓	5	1.600	↓	1,500	99.50		1,599.50	1,800
9000	Water cooler (electric), floor mounted									
9100	Plate type with bearing plate, single	1 Plum	6	1.333	Ea.	440	83		523	610

For customer support on your Building Construction Costs with RSMeans data, call 800.448.8182.

521

22 45 Emergency Plumbing Fixtures

22 45 13 – Emergency Showers

22 45 13.10 Emergency Showers	Crew	Daily Output	Labor-Hours	Unit	Material	2018 Bare Costs Labor	Equipment	Total	Total Incl O&P
0010 **EMERGENCY SHOWERS**, Rough-in not included									
5000 Shower, single head, drench, ball valve, pull, freestanding	Q-1	4	4	Ea.	385	224		609	755
5200 Horizontal or vertical supply		4	4		560	224		784	950
6000 Multi-nozzle, eye/face wash combination		4	4		690	224		914	1,100
6400 Multi-nozzle, 12 spray, shower only		4	4		2,050	224		2,274	2,575
6600 For freeze-proof, add		6	2.667		485	149		634	760

22 45 16 – Eyewash Equipment

22 45 16.10 Eyewash Safety Equipment

	Crew	Daily Output	Labor-Hours	Unit	Material	2018 Bare Costs Labor	Equipment	Total	Total Incl O&P
0010 **EYEWASH SAFETY EQUIPMENT**, Rough-in not included									
1000 Eye wash fountain									
1400 Plastic bowl, pedestal mounted	Q-1	4	4	Ea.	294	224		518	660
1600 Unmounted		4	4		214	224		438	570
1800 Wall mounted		4	4		460	224		684	840
2000 Stainless steel, pedestal mounted		4	4		370	224		594	740
2200 Unmounted		4	4		282	224		506	645
2400 Wall mounted		4	4		300	224		524	665

22 45 19 – Self-Contained Eyewash Equipment

22 45 19.10 Self-Contained Eyewash Safety Equipment

	Crew	Daily Output	Labor-Hours	Unit	Material	2018 Bare Costs Labor	Equipment	Total	Total Incl O&P
0010 **SELF-CONTAINED EYEWASH SAFETY EQUIPMENT**									
3000 Eye wash, portable, self-contained				Ea.	1,675			1,675	1,850

22 45 26 – Eye/Face Wash Equipment

22 45 26.10 Eye/Face Wash Safety Equipment

	Crew	Daily Output	Labor-Hours	Unit	Material	2018 Bare Costs Labor	Equipment	Total	Total Incl O&P
0010 **EYE/FACE WASH SAFETY EQUIPMENT**, Rough-in not included									
4000 Eye and face wash, combination fountain									
4200 Stainless steel, pedestal mounted	Q-1	4	4	Ea.	1,075	224		1,299	1,525
4400 Unmounted		4	4		282	224		506	645
4600 Wall mounted		4	4		256	224		480	615

22 47 Drinking Fountains and Water Coolers

22 47 13 – Drinking Fountains

22 47 13.10 Drinking Water Fountains

	Crew	Daily Output	Labor-Hours	Unit	Material	2018 Bare Costs Labor	Equipment	Total	Total Incl O&P
0010 **DRINKING WATER FOUNTAINS**, For connection to cold water supply									
1000 Wall mounted, non-recessed									
1400 Bronze, with no back	1 Plum	4	2	Ea.	1,000	124		1,124	1,275
1800 Cast aluminum, enameled, for correctional institutions		4	2		1,400	124		1,524	1,725
2000 Fiberglass, 12" back, single bubbler unit		4	2		2,025	124		2,149	2,400
2040 Dual bubbler		3.20	2.500		2,375	155		2,530	2,825
2400 Precast stone, no back		4	2		980	124		1,104	1,250
2700 Stainless steel, single bubbler, no back		4	2		1,000	124		1,124	1,275
2740 With back		4	2		1,025	124		1,149	1,300
2780 Dual handle, ADA compliant		4	2		720	124		844	975
2820 Dual level, ADA compliant		3.20	2.500		1,575	155		1,730	1,950
3300 Vitreous china									
3340 7" back	1 Plum	4	2	Ea.	575	124		699	820
3940 For vandal-resistant bottom plate, add					77			77	84.50
3960 For freeze-proof valve system, add	1 Plum	2	4		705	249		954	1,150
3980 For rough-in, supply and waste, add	"	2.21	3.620		205	225		430	565
4000 Wall mounted, semi-recessed									
4200 Poly-marble, single bubbler	1 Plum	4	2	Ea.	955	124		1,079	1,225

22 47 Drinking Fountains and Water Coolers

22 47 13 – Drinking Fountains

22 47 13.10 Drinking Water Fountains

22 47 13.10 Drinking Water Fountains	Crew	Daily Output	Labor-Hours	Unit	Material	2018 Bare Costs Labor	Equipment	Total	Total Incl O&P
4600	Stainless steel, satin finish, single bubbler	1 Plum	4	2	Ea.	1,300	124		1,424	1,625
4900	Vitreous china, single bubbler		4	2		895	124		1,019	1,175
5980	For rough-in, supply and waste, add		1.83	4.372		205	272		477	635
6000	Wall mounted, fully recessed									
6400	Poly-marble, single bubbler	1 Plum	4	2	Ea.	1,675	124		1,799	2,000
6800	Stainless steel, single bubbler		4	2		1,525	124		1,649	1,850
7560	For freeze-proof valve system, add		2	4		895	249		1,144	1,350
7580	For rough-in, supply and waste, add		1.83	4.372		205	272		477	635
7600	Floor mounted, pedestal type									
7700	Aluminum, architectural style, CI base	1 Plum	2	4	Ea.	2,400	249		2,649	3,000
7780	ADA compliant unit		2	4		1,675	249		1,924	2,200
8400	Stainless steel, architectural style		2	4		1,950	249		2,199	2,525
8600	Enameled iron, heavy duty service, 2 bubblers		2	4		2,850	249		3,099	3,500
8660	4 bubblers		2	4		4,175	249		4,424	4,975
8880	For freeze-proof valve system, add		2	4		705	249		954	1,150
8900	For rough-in, supply and waste, add		1.83	4.372		205	272		477	635
9100	Deck mounted									
9500	Stainless steel, circular receptor	1 Plum	4	2	Ea.	435	124		559	665
9760	White enameled steel, 14" x 9" receptor		4	2		380	124		504	600
9860	White enameled cast iron, 24" x 16" receptor		3	2.667		485	166		651	785
9980	For rough-in, supply and waste, add		1.83	4.372		205	272		477	635

22 47 16 – Pressure Water Coolers

22 47 16.10 Electric Water Coolers

	22 47 16.10 Electric Water Coolers	Crew	Daily Output	Labor-Hours	Unit	Material	2018 Bare Costs Labor	Equipment	Total	Total Incl O&P
0010	**ELECTRIC WATER COOLERS**									
0100	Wall mounted, non-recessed									
0140	4 GPH	Q-1	4	4	Ea.	680	224		904	1,075
0160	8 GPH, barrier free, sensor operated		4	4		1,075	224		1,299	1,500
0180	8.2 GPH		4	4		910	224		1,134	1,325
0600	8 GPH hot and cold water		4	4		1,050	224		1,274	1,500
0640	For stainless steel cabinet, add					90.50			90.50	99.50
1000	Dual height, 8.2 GPH	Q-1	3.80	4.211		1,975	235		2,210	2,525
1040	14.3 GPH	"	3.80	4.211		1,475	235		1,710	1,975
1240	For stainless steel cabinet, add					188			188	207
2600	ADA compliant, 8 GPH	Q-1	4	4		990	224		1,214	1,425
3300	Semi-recessed, 8.1 GPH		4	4		825	224		1,049	1,250
3320	12 GPH		4	4		950	224		1,174	1,375
4600	Floor mounted, flush-to-wall									
4640	4 GPH	1 Plum	3	2.667	Ea.	800	166		966	1,125
4680	8.2 GPH		3	2.667		840	166		1,006	1,175
4720	14.3 GPH		3	2.667		965	166		1,131	1,300
4960	14 GPH hot and cold water		3	2.667		1,100	166		1,266	1,475
4980	For stainless steel cabinet, add					138			138	152
5000	Dual height, 8.2 GPH	1 Plum	2	4		1,200	249		1,449	1,700
5040	14.3 GPH	"	2	4		1,250	249		1,499	1,750
5120	For stainless steel cabinet, add					201			201	221
9800	For supply, waste & vent, all coolers	1 Plum	2.21	3.620		205	225		430	565

For customer support on your Building Construction Costs with RSMeans data, call 800.448.8182.

523

22 51 Swimming Pool Plumbing Systems

22 51 19 – Swimming Pool Water Treatment Equipment

22 51 19.50 Swimming Pool Filtration Equipment	Crew	Daily Output	Labor-Hours	Unit	Material	2018 Bare Costs Labor	Equipment	Total	Total Incl O&P
0010 **SWIMMING POOL FILTRATION EQUIPMENT**									
0900 Filter system, sand or diatomite type, incl. pump, 6,000 gal./hr.	2 Plum	1.80	8.889	Total	2,125	550		2,675	3,175
1020 Add for chlorination system, 800 S.F. pool		3	5.333	Ea.	177	330		507	695
1040 5,000 S.F. pool		3	5.333	"	1,800	330		2,130	2,475

22 52 Fountain Plumbing Systems

22 52 16 – Fountain Pumps

22 52 16.10 Fountain Water Pumps

	Crew	Daily Output	Labor-Hours	Unit	Material	2018 Bare Costs Labor	Equipment	Total	Total Incl O&P
0010 **FOUNTAIN WATER PUMPS**									
0100 Pump w/controls									
0200 Single phase, 100' cord, 1/2 HP pump	2 Skwk	4.40	3.636	Ea.	1,200	190		1,390	1,625
0300 3/4 HP pump		4.30	3.721		1,275	195		1,470	1,700
0400 1 HP pump		4.20	3.810		1,875	199		2,074	2,375
0500 1-1/2 HP pump		4.10	3.902		2,550	204		2,754	3,125
0600 2 HP pump		4	4		3,925	209		4,134	4,650
0700 Three phase, 200' cord, 5 HP pump		3.90	4.103		5,525	215		5,740	6,400
0800 7-1/2 HP pump		3.80	4.211		11,000	220		11,220	12,400
0900 10 HP pump		3.70	4.324		14,000	226		14,226	15,700
1000 15 HP pump		3.60	4.444		21,000	233		21,233	23,500
2000 DESIGN NOTE: Use two horsepower per surface acre.									

22 52 33 – Fountain Ancillary

22 52 33.10 Fountain Miscellaneous

	Crew	Daily Output	Labor-Hours	Unit	Material	2018 Bare Costs Labor	Equipment	Total	Total Incl O&P
0010 **FOUNTAIN MISCELLANEOUS**									
1300 Lights w/mounting kits, 200 watt	2 Skwk	18	.889	Ea.	1,175	46.50		1,221.50	1,375
1400 300 watt		18	.889		1,325	46.50		1,371.50	1,550
1500 500 watt		18	.889		1,500	46.50		1,546.50	1,725
1600 Color blender		12	1.333		585	70		655	745

22 66 Chemical-Waste Systems for Lab. and Healthcare Facilities

22 66 53 – Laboratory Chemical-Waste and Vent Piping

22 66 53.30 Glass Pipe

	Crew	Daily Output	Labor-Hours	Unit	Material	2018 Bare Costs Labor	Equipment	Total	Total Incl O&P
0010 **GLASS PIPE**, Borosilicate, couplings & clevis hanger assemblies, 10' OC									
0020 Drainage									
1100 1-1/2" diameter	Q-1	52	.308	L.F.	13.70	17.20		30.90	41
1120 2" diameter		44	.364		17.35	20.50		37.85	49.50
1140 3" diameter		39	.410		20.50	23		43.50	57
1160 4" diameter		30	.533		43.50	30		73.50	93
1180 6" diameter		26	.615		71	34.50		105.50	130

22 66 53.60 Corrosion Resistant Pipe

	Crew	Daily Output	Labor-Hours	Unit	Material	2018 Bare Costs Labor	Equipment	Total	Total Incl O&P
0010 **CORROSION RESISTANT PIPE**, No couplings or hangers									
0020 Iron alloy, drain, mechanical joint									
1000 1-1/2" diameter	Q-1	70	.229	L.F.	78	12.80		90.80	105
1100 2" diameter		66	.242		80.50	13.55		94.05	109
1120 3" diameter		60	.267		87	14.90		101.90	119
1140 4" diameter		52	.308		112	17.20		129.20	149
2980 Plastic, epoxy, fiberglass filament wound, B&S joint									
3000 2" diameter	Q-1	62	.258	L.F.	12.10	14.45		26.55	35
3100 3" diameter		51	.314		14.15	17.55		31.70	42

524

For customer support on your Building Construction Costs with RSMeans data, call 800.448.8182.

22 66 53.60 Corrosion Resistant Pipe		Crew	Daily Output	Labor-Hours	Unit	Material	2018 Bare Costs Labor	Equipment	Total	Total Incl O&P
3120	4" diameter	Q-1	45	.356	L.F.	20	19.90		39.90	52
3140	6" diameter	↓	32	.500	↓	28.50	28		56.50	73
3980	Polyester, fiberglass filament wound, B&S joint									
4000	2" diameter	Q-1	62	.258	L.F.	13.20	14.45		27.65	36
4100	3" diameter		51	.314		17.20	17.55		34.75	45.50
4120	4" diameter		45	.356		25	19.90		44.90	58
4140	6" diameter	↓	32	.500	↓	36.50	28		64.50	82
4980	Polypropylene, acid resistant, fire retardant, Schedule 40									
5000	1-1/2" diameter	Q-1	68	.235	L.F.	8.25	13.15		21.40	29
5100	2" diameter		62	.258		11.40	14.45		25.85	34
5120	3" diameter		51	.314		22.50	17.55		40.05	51.50
5140	4" diameter	↓	45	.356	↓	29	19.90		48.90	62
5980	Proxylene, fire retardant, Schedule 40									
6000	1-1/2" diameter	Q-1	68	.235	L.F.	12.80	13.15		25.95	34
6100	2" diameter		62	.258		17.50	14.45		31.95	41
6120	3" diameter		51	.314		31.50	17.55		49.05	61.50
6140	4" diameter	↓	45	.356	↓	44.50	19.90		64.40	79

Division Notes

	CREW	DAILY OUTPUT	LABOR-HOURS	UNIT	BARE COSTS				TOTAL INCL O&P
					MAT.	LABOR	EQUIP.	TOTAL	

Estimating Tips

The labor adjustment factors listed in Subdivision 22 01 02.20 also apply to Division 23.

23 10 00 Facility Fuel Systems

- The prices in this subdivision for above- and below-ground storage tanks do not include foundations or hold-down slabs, unless noted. The estimator should refer to Divisions 3 and 31 for foundation system pricing. In addition to the foundations, required tank accessories, such as tank gauges, leak detection devices, and additional manholes and piping, must be added to the tank prices.

23 50 00 Central Heating Equipment

- When estimating the cost of an HVAC system, check to see who is responsible for providing and installing the temperature control system. It is possible to overlook controls, assuming that they would be included in the electrical estimate.

- When looking up a boiler, be careful on specified capacity. Some manufacturers rate their products on output while others use input.

- Include HVAC insulation for pipe, boiler, and duct (wrap and liner).

- Be careful when looking up mechanical items to get the correct pressure rating and connection type (thread, weld, flange).

23 70 00 Central HVAC Equipment

- Combination heating and cooling units are sized by the air conditioning requirements. (See Reference No. R236000-20 for the preliminary sizing guide.)

- A ton of air conditioning is nominally 400 CFM.

- Rectangular duct is taken off by the linear foot for each size, but its cost is usually estimated by the pound. Remember that SMACNA standards now base duct on internal pressure.

- Prefabricated duct is estimated and purchased like pipe: straight sections and fittings.

- Note that cranes or other lifting equipment are not included on any lines in Division 23. For example, if a crane is required to lift a heavy piece of pipe into place high above a gym floor, or to put a rooftop unit on the roof of a four-story building, etc., it must be added. Due to the potential for extreme variation—from nothing additional required to a major crane or helicopter—we feel that including a nominal amount for "lifting contingency" would be useless and detract from the accuracy of the estimate. When using equipment rental cost data from RSMeans, do not forget to include the cost of the operator(s).

Reference Numbers

Reference numbers are shown at the beginning of some major classifications. These numbers refer to related items in the Reference Section. The reference information may be an estimating procedure, an alternate pricing method, or technical information.

Note: Not all subdivisions listed here necessarily appear. ∎

No part of this cost data may be reproduced, stored in a retrieval system, or transmitted in any form or by any means without prior written permission of Gordian.

Note: Trade Service, in part, has been used as a reference source for some of the material prices used in Division 23.

Did you know?

RSMeans data is available through our online application with 24/7 access:

- Search for unit prices by keyword
- Leverage the most up-to-date data
- Build and export estimates

Try it free for 30 days!
www.rsmeans.com/2018freetrial

23 05 Common Work Results for HVAC

23 05 02 – HVAC General

23 05 02.10 Air Conditioning, General	Crew	Daily Output	Labor-Hours	Unit	Material	2018 Bare Costs Labor	Equipment	Total	Total Incl O&P
0010 **AIR CONDITIONING, GENERAL** Prices are for standard efficiencies (SEER 13)									
0020 for upgrade to SEER 14 add					10%				

23 05 05 – Selective Demolition for HVAC

23 05 05.10 HVAC Demolition

		Crew	Daily Output	Labor-Hours	Unit	Material	Labor	Equipment	Total	Total Incl O&P
0010	**HVAC DEMOLITION**									
0100	Air conditioner, split unit, 3 ton	Q-5	2	8	Ea.		455		455	685
0150	Package unit, 3 ton	Q-6	3	8	"		470		470	710
0298	Boilers									
0300	Electric, up thru 148 kW	Q-19	2	12	Ea.		685		685	1,025
0310	150 thru 518 kW	"	1	24			1,375		1,375	2,050
0320	550 thru 2,000 kW	Q-21	.40	80			4,700		4,700	7,050
0330	2,070 kW and up	"	.30	107			6,250		6,250	9,400
0340	Gas and/or oil, up thru 150 MBH	Q-7	2.20	14.545			870		870	1,325
0350	160 thru 2,000 MBH		.80	40			2,400		2,400	3,625
0360	2,100 thru 4,500 MBH		.50	64			3,850		3,850	5,775
0370	4,600 thru 7,000 MBH		.30	107			6,400		6,400	9,625
0380	7,100 thru 12,000 MBH		.16	200			12,000		12,000	18,100
0390	12,200 thru 25,000 MBH	▼	.12	267	▼		16,000		16,000	24,100
1000	Ductwork, 4" high, 8" wide	1 Clab	200	.040	L.F.		1.59		1.59	2.43
1100	6" high, 8" wide		165	.048			1.93		1.93	2.94
1200	10" high, 12" wide		125	.064			2.55		2.55	3.88
1300	12"-14" high, 16"-18" wide		85	.094			3.75		3.75	5.70
1400	18" high, 24" wide		67	.119			4.76		4.76	7.25
1500	30" high, 36" wide		56	.143			5.70		5.70	8.65
1540	72" wide	▼	50	.160	▼		6.40		6.40	9.70
3000	Mechanical equipment, light items. Unit is weight, not cooling.	Q-5	.90	17.778	Ton		1,000		1,000	1,525
3600	Heavy items	"	1.10	14.545	"		825		825	1,250
5090	Remove refrigerant from system	1 Stpi	40	.200	Lb.		12.60		12.60	19

23 05 23 – General-Duty Valves for HVAC Piping

23 05 23.30 Valves, Iron Body

		Crew	Daily Output	Labor-Hours	Unit	Material	Labor	Equipment	Total	Total Incl O&P
0010	**VALVES, IRON BODY**									
1020	Butterfly, wafer type, gear actuator, 200 lb.									
1030	2"	1 Plum	14	.571	Ea.	105	35.50		140.50	169
1040	2-1/2"	Q-1	9	1.778		120	99.50		219.50	282
1050	3"		8	2		110	112		222	289
1060	4"	▼	5	3.200		123	179		302	405
1070	5"	Q-2	5	4.800	▼	138	278		416	570
1080	6"	"	5	4.800		156	278		434	590
1650	Gate, 125 lb., N.R.S.									
2150	Flanged									
2200	2"	1 Plum	5	1.600	Ea.	590	99.50		689.50	800
2240	2-1/2"	Q-1	5	3.200		605	179		784	935
2260	3"		4.50	3.556		680	199		879	1,050
2280	4"	▼	3	5.333		975	298		1,273	1,525
2300	6"	Q-2	3	8	▼	1,650	465		2,115	2,525
3550	OS&Y, 125 lb., flanged									
3600	2"	1 Plum	5	1.600	Ea.	400	99.50		499.50	590
3660	3"	Q-1	4.50	3.556		455	199		654	800
3680	4"	"	3	5.333		665	298		963	1,175
3700	6"	Q-2	3	8	▼	1,050	465		1,515	1,850
3900	For 175 lb., flanged, add					200%	10%			
5450	Swing check, 125 lb., threaded									

For customer support on your Building Construction Costs with RSMeans data, call 800.448.8182.

23 05 23 – General-Duty Valves for HVAC Piping

23 05 23.30 Valves, Iron Body

		Crew	Daily Output	Labor-Hours	Unit	Material	2018 Bare Costs Labor	Equipment	Total	Total Incl O&P
5500	2"	1 Plum	11	.727	Ea.	485	45		530	600
5540	2-1/2"	Q-1	15	1.067		620	59.50		679.50	770
5550	3"		13	1.231		670	69		739	845
5560	4"	↓	10	1.600	↓	1,075	89.50		1,164.50	1,300
5950	Flanged									
6000	2"	1 Plum	5	1.600	Ea.	425	99.50		524.50	615
6040	2-1/2"	Q-1	5	3.200		330	179		509	630
6050	3"		4.50	3.556		415	199		614	755
6060	4"	↓	3	5.333		675	298		973	1,200
6070	6"	Q-2	3	8	↓	1,100	465		1,565	1,925

23 05 23.80 Valves, Steel

		Crew	Daily Output	Labor-Hours	Unit	Material	2018 Bare Costs Labor	Equipment	Total	Total Incl O&P
0010	**VALVES, STEEL**									
0800	Cast									
1350	Check valve, swing type, 150 lb., flanged									
1370	1"	1 Plum	10	.800	Ea.	360	49.50		409.50	470
1400	2"	"	8	1		610	62		672	765
1440	2-1/2"	Q-1	5	3.200		705	179		884	1,050
1450	3"		4.50	3.556		720	199		919	1,100
1460	4"	↓	3	5.333		1,025	298		1,323	1,575
1540	For 300 lb., flanged, add						50%	15%		
1548	For 600 lb., flanged, add				↓		110%	20%		
1950	Gate valve, 150 lb., flanged									
2000	2"	1 Plum	8	1	Ea.	630	62		692	785
2040	2-1/2"	Q-1	5	3.200		890	179		1,069	1,250
2050	3"		4.50	3.556		890	199		1,089	1,275
2060	4"	↓	3	5.333		1,125	298		1,423	1,700
2070	6"	Q-2	3	8	↓	1,800	465		2,265	2,675
3650	Globe valve, 150 lb., flanged									
3700	2"	1 Plum	8	1	Ea.	790	62		852	965
3740	2-1/2"	Q-1	5	3.200		1,000	179		1,179	1,375
3750	3"		4.50	3.556		1,000	199		1,199	1,400
3760	4"	↓	3	5.333	↓	1,475	298		1,773	2,050
3770	6"	Q-2	3	8	↓	2,300	465		2,765	3,225
5150	Forged									
5650	Check valve, class 800, horizontal									
5698	Threaded									
5700	1/4"	1 Plum	24	.333	Ea.	90.50	20.50		111	131
5720	3/8"		24	.333		90.50	20.50		111	131
5730	1/2"		24	.333		90.50	20.50		111	131
5740	3/4"		20	.400		96.50	25		121.50	144
5750	1"		19	.421		114	26		140	165
5760	1-1/4"	↓	15	.533	↓	223	33		256	295

23 05 93 – Testing, Adjusting, and Balancing for HVAC

23 05 93.10 Balancing, Air

		Crew	Daily Output	Labor-Hours	Unit	Material	2018 Bare Costs Labor	Equipment	Total	Total Incl O&P
0010	**BALANCING, AIR** (Subcontractor's quote incl. material and labor)									
0900	Heating and ventilating equipment									
1000	Centrifugal fans, utility sets				Ea.				420	420
1100	Heating and ventilating unit								630	630
1200	In-line fan								630	630
1300	Propeller and wall fan								119	119
1400	Roof exhaust fan								280	280
2000	Air conditioning equipment, central station				↓				910	910

23 05 Common Work Results for HVAC

23 05 93 – Testing, Adjusting, and Balancing for HVAC

23 05 93.10 Balancing, Air

		Crew	Daily Output	Labor-Hours	Unit	Material	2018 Bare Costs Labor	Equipment	Total	Total Incl O&P
2100	Built-up low pressure unit				Ea.				840	840
2200	Built-up high pressure unit								980	980
2500	Multi-zone A.C. and heating unit								630	630
2600	For each zone over one, add								140	140
2700	Package A.C. unit								350	350
2800	Rooftop heating and cooling unit								490	490
3000	Supply, return, exhaust, registers & diffusers, avg. height ceiling								84	84
3100	High ceiling								126	126
3200	Floor height				▼				70	70

23 05 93.20 Balancing, Water

		Crew	Daily Output	Labor-Hours	Unit	Material	2018 Bare Costs Labor	Equipment	Total	Total Incl O&P
0010	**BALANCING, WATER** (Subcontractor's quote incl. material and labor)									
0050	Air cooled condenser				Ea.				256	256
0080	Boiler								515	515
0100	Cabinet unit heater								88	88
0200	Chiller								620	620
0300	Convector								73	73
0500	Cooling tower								475	475
0600	Fan coil unit, unit ventilator								132	132
0700	Fin tube and radiant panels								146	146
0800	Main and duct re-heat coils								135	135
0810	Heat exchanger								135	135
1000	Pumps								320	320
1100	Unit heater				▼				102	102

23 07 HVAC Insulation

23 07 13 – Duct Insulation

23 07 13.10 Duct Thermal Insulation

			Crew	Daily Output	Labor-Hours	Unit	Material	2018 Bare Costs Labor	Equipment	Total	Total Incl O&P
0010	**DUCT THERMAL INSULATION**										
0110	Insulation req'd. is based on the surface size/area to be covered										
3000	Ductwork										
3020	Blanket type, fiberglass, flexible										
3030	Fire rated for grease and hazardous exhaust ducts										
3060	1-1/2" thick		Q-14	84	.190	S.F.	4.54	9.65		14.19	20
3090	Fire rated for plenums										
3100	1/2" x 24" x 25'		Q-14	1.94	8.247	Roll	167	420		587	835
3110	1/2" x 24" x 25'			98	.163	S.F.	3.35	8.25		11.60	16.55
3120	1/2" x 48" x 25'			1.04	15.385	Roll	335	780		1,115	1,575
3126	1/2" x 48" x 25'		▼	104	.154	S.F.	3.35	7.80		11.15	15.80
3140	FSK vapor barrier wrap, .75 lb. density										
3160	1" thick	G	Q-14	350	.046	S.F.	.22	2.32		2.54	3.84
3170	1-1/2" thick	G		320	.050		.27	2.53		2.80	4.23
3180	2" thick	G		300	.053		.32	2.70		3.02	4.54
3190	3" thick	G		260	.062		.45	3.12		3.57	5.35
3200	4" thick	G	▼	242	.066	▼	.64	3.35		3.99	5.90
3210	Vinyl jacket, same as FSK										
3280	Unfaced, 1 lb. density										
3310	1" thick	G	Q-14	360	.044	S.F.	.24	2.25		2.49	3.76
3320	1-1/2" thick	G		330	.048		.36	2.46		2.82	4.21
3330	2" thick	G	▼	310	.052	▼	.44	2.62		3.06	4.54
3400	FSK facing, 1 lb. density										
3420	1-1/2" thick	G	Q-14	310	.052	S.F.	.36	2.62		2.98	4.46

530

23 07 HVAC Insulation

23 07 13 – Duct Insulation

23 07 13.10 Duct Thermal Insulation

23 07 13.10 Duct Thermal Insulation		Crew	Daily Output	Labor-Hours	Unit	Material	2018 Bare Costs Labor	Equipment	Total	Total Incl O&P
3430	2" thick [G]	Q-14	300	.053	S.F.	.44	2.70		3.14	4.67
3450	FSK facing, 1.5 lb. density									
3470	1-1/2" thick [G]	Q-14	300	.053	S.F.	.45	2.70		3.15	4.69
3480	2" thick [G]	"	290	.055	"	.54	2.80		3.34	4.93
3795	Finishes									
3800	Stainless steel woven mesh	Q-14	100	.160	S.F.	.93	8.10		9.03	13.60
3810	For .010" stainless steel, add		160	.100		3.15	5.05		8.20	11.30
3820	18 oz. fiberglass cloth, pasted on		170	.094		.84	4.77		5.61	8.30
3900	8 oz. canvas, pasted on		180	.089		.26	4.50		4.76	7.30
3940	For .016" aluminum jacket, add	↓	200	.080	↓	1.01	4.05		5.06	7.40
7878	Contact cement, quart can				Ea.	12.50			12.50	13.75

23 07 16 – HVAC Equipment Insulation

23 07 16.10 HVAC Equipment Thermal Insulation

23 07 16.10 HVAC Equipment Thermal Insulation		Crew	Daily Output	Labor-Hours	Unit	Material	2018 Bare Costs Labor	Equipment	Total	Total Incl O&P
0010	**HVAC EQUIPMENT THERMAL INSULATION**									
0110	Insulation req'd. is based on the surface size/area to be covered									
1000	Boiler, 1-1/2" calcium silicate only [G]	Q-14	110	.145	S.F.	3.65	7.35		11	15.45
1020	Plus 2" fiberglass [G]	"	80	.200	"	5.15	10.15		15.30	21.50
2000	Breeching, 2" calcium silicate									
2020	Rectangular [G]	Q-14	42	.381	S.F.	7.15	19.30		26.45	38
2040	Round [G]	"	38.70	.413	"	7.45	21		28.45	40.50

23 09 Instrumentation and Control for HVAC

23 09 33 – Electric and Electronic Control System for HVAC

23 09 33.10 Electronic Control Systems

23 09 33.10 Electronic Control Systems		Crew	Daily Output	Labor-Hours	Unit	Material	2018 Bare Costs Labor	Equipment	Total	Total Incl O&P
0010	**ELECTRONIC CONTROL SYSTEMS**									
0020	For electronic costs, add to Section 23 09 43.10				Ea.				15%	15%

23 09 43 – Pneumatic Control System for HVAC

23 09 43.10 Pneumatic Control Systems

23 09 43.10 Pneumatic Control Systems		Crew	Daily Output	Labor-Hours	Unit	Material	2018 Bare Costs Labor	Equipment	Total	Total Incl O&P
0010	**PNEUMATIC CONTROL SYSTEMS**									
0011	Including a nominal 50' of tubing. Add control panelboard if req'd.									
0100	Heating and ventilating, split system									
0200	Mixed air control, economizer cycle, panel readout, tubing									
0220	Up to 10 tons [G]	Q-19	.68	35.294	Ea.	4,450	2,025		6,475	7,925
0240	For 10 to 20 tons [G]		.63	37.915		4,750	2,175		6,925	8,450
0260	For over 20 tons [G]	↓	.58	41.096	↓	5,125	2,350		7,475	9,175
0300	Heating coil, hot water, 3 way valve,									
0320	Freezestat, limit control on discharge, readout	Q-5	.69	23.088	Ea.	3,300	1,300		4,600	5,625
0500	Cooling coil, chilled water, room									
0520	Thermostat, 3 way valve	Q-5	2	8	Ea.	1,475	455		1,930	2,300
0600	Cooling tower, fan cycle, damper control,									
0620	Control system including water readout in/out at panel	Q-19	.67	35.821	Ea.	5,850	2,050		7,900	9,525
1000	Unit ventilator, day/night operation,									
1100	freezestat, ASHRAE, cycle 2	Q-19	.91	26.374	Ea.	3,250	1,500		4,750	5,850
2000	Compensated hot water from boiler, valve control,									
2100	readout and reset at panel, up to 60 GPM	Q-19	.55	43.956	Ea.	6,025	2,525		8,550	10,400
2120	For 120 GPM		.51	47.059		6,500	2,700		9,200	11,200
2140	For 240 GPM		.49	49.180		6,750	2,825		9,575	11,700
3000	Boiler room combustion air, damper to 5 S.F., controls		1.37	17.582		2,925	1,000		3,925	4,725
3500	Fan coil, heating and cooling valves, 4 pipe control system		3	8		1,300	460		1,760	2,125
3600	Heat exchanger system controls	↓	.86	27.907	↓	2,850	1,600		4,450	5,525

23 09 Instrumentation and Control for HVAC

23 09 43 – Pneumatic Control System for HVAC

23 09 43.10 Pneumatic Control Systems	Crew	Daily Output	Labor-Hours	Unit	Material	2018 Bare Costs Labor	2018 Bare Costs Equipment	Total	Total Incl O&P
4000	Pneumatic thermostat, including controlling room radiator valve	Q-5	2.43	6.593	Ea.	880	375		1,255	1,525
4060	Pump control system	Q-19	3	8	↓	1,350	460		1,810	2,175
4500	Air supply for pneumatic control system									
4600	Tank mounted duplex compressor, starter, alternator,									
4620	piping, dryer, PRV station and filter									
4630	1/2 HP	Q-19	.68	35.139	Ea.	10,800	2,000		12,800	14,900
4660	1-1/2 HP	↓	.58	41.739	↓	13,300	2,375		15,675	18,200
4690	5 HP	↓	.42	57.143	↓	31,400	3,275		34,675	39,400

23 13 Facility Fuel-Storage Tanks

23 13 13 – Facility Underground Fuel-Oil, Storage Tanks

23 13 13.09 Single-Wall Steel Fuel-Oil Tanks

		Crew	Daily Output	Labor-Hours	Unit	Material	2018 Bare Costs Labor	2018 Bare Costs Equipment	Total	Total Incl O&P
0010	**SINGLE-WALL STEEL FUEL-OIL TANKS**									
5000	Tanks, steel ugnd., sti-p3, not incl. hold-down bars									
5500	Excavation, pad, pumps and piping not included									
5510	Single wall, 500 gallon capacity, 7 ga. shell	Q-5	2.70	5.926	Ea.	1,800	335		2,135	2,500
5520	1,000 gallon capacity, 7 ga. shell	"	2.50	6.400		2,525	365		2,890	3,325
5530	2,000 gallon capacity, 1/4" thick shell	Q-7	4.60	6.957		3,275	415		3,690	4,225
5535	2,500 gallon capacity, 7 ga. shell	Q-5	3	5.333		4,900	300		5,200	5,825
5540	5,000 gallon capacity, 1/4" thick shell	Q-7	3.20	10		6,425	600		7,025	7,975
5580	15,000 gallon capacity, 5/16" thick shell		1.70	18.824		11,500	1,125		12,625	14,400
5600	20,000 gallon capacity, 5/16" thick shell		1.50	21.333		20,200	1,275		21,475	24,200
5610	25,000 gallon capacity, 3/8" thick shell		1.30	24.615		26,500	1,475		27,975	31,400
5620	30,000 gallon capacity, 3/8" thick shell		1.10	29.091		30,900	1,750		32,650	36,600
5630	40,000 gallon capacity, 3/8" thick shell		.90	35.556		42,400	2,125		44,525	49,900
5640	50,000 gallon capacity, 3/8" thick shell	↓	.80	40	↓	47,100	2,400		49,500	55,500

23 13 13.23 Glass-Fiber-Reinfcd-Plastic, Fuel-Oil, Storage

		Crew	Daily Output	Labor-Hours	Unit	Material	2018 Bare Costs Labor	2018 Bare Costs Equipment	Total	Total Incl O&P
0010	**GLASS-FIBER-REINFCD-PLASTIC, UNDERGRND. FUEL-OIL, STORAGE**									
0210	Fiberglass, underground, single wall, UL listed, not including									
0220	manway or hold-down strap									
0240	2,000 gallon capacity	Q-7	4.57	7.002	Ea.	6,900	420		7,320	8,200
0245	3,000 gallon capacity		3.90	8.205		8,250	490		8,740	9,825
0250	4,000 gallon capacity		3.55	9.014		9,575	540		10,115	11,300
0255	5,000 gallon capacity		3.20	10		10,600	600		11,200	12,600
0260	6,000 gallon capacity		2.67	11.985		10,900	720		11,620	13,100
0280	10,000 gallon capacity		2	16		14,800	960		15,760	17,800
0284	15,000 gallon capacity		1.68	19.048		23,700	1,150		24,850	27,800
0290	20,000 gallon capacity	↓	1.45	22.069	↓	30,500	1,325		31,825	35,500
0500	For manway, fittings and hold-downs, add					20%	15%			
1020	Fiberglass, underground, double wall, UL listed									
1030	includes manways, not incl. hold-down straps									
1040	600 gallon capacity	Q-5	2.42	6.612	Ea.	8,325	375		8,700	9,725
1050	1,000 gallon capacity	"	2.25	7.111		11,400	405		11,805	13,100
1060	2,500 gallon capacity	Q-7	4.16	7.692		16,500	460		16,960	18,800
1070	3,000 gallon capacity		3.90	8.205		18,500	490		18,990	21,100
1080	4,000 gallon capacity		3.64	8.791		18,800	525		19,325	21,500
1090	6,000 gallon capacity		2.42	13.223		24,500	795		25,295	28,100
1100	8,000 gallon capacity		2.08	15.385		27,300	925		28,225	31,400
1110	10,000 gallon capacity		1.82	17.582		31,600	1,050		32,650	36,300
1120	12,000 gallon capacity	↓	1.70	18.824	↓	39,400	1,125		40,525	45,100
2210	Fiberglass, underground, single wall, UL listed, including									

For customer support on your Building Construction Costs with RSMeans data, call 800.448.8182.

23 13 Facility Fuel-Storage Tanks

23 13 13 – Facility Underground Fuel-Oil, Storage Tanks

23 13 13.23 Glass-Fiber-Reinfcd-Plastic, Fuel-Oil, Storage

		Crew	Daily Output	Labor-Hours	Unit	Material	2018 Bare Costs Labor	2018 Bare Costs Equipment	Total	Total Incl O&P
2220	hold-down straps, no manways									
2240	2,000 gallon capacity	Q-7	3.55	9.014	Ea.	7,375	540		7,915	8,950
2250	4,000 gallon capacity		2.90	11.034		10,100	660		10,760	12,100
2260	6,000 gallon capacity		2	16		11,900	960		12,860	14,600
2280	10,000 gallon capacity		1.60	20		15,800	1,200		17,000	19,200
2284	15,000 gallon capacity		1.39	23.022		24,700	1,375		26,075	29,200
2290	20,000 gallon capacity		1.14	28.070		32,000	1,675		33,675	37,700
3020	Fiberglass, underground, double wall, UL listed									
3030	includes manways and hold-down straps									
3040	600 gallon capacity	Q-5	1.86	8.602	Ea.	8,825	490		9,315	10,400
3050	1,000 gallon capacity	"	1.70	9.412		11,900	535		12,435	13,900
3060	2,500 gallon capacity	Q-7	3.29	9.726		16,900	585		17,485	19,500
3070	3,000 gallon capacity		3.13	10.224		19,000	615		19,615	21,800
3080	4,000 gallon capacity		2.93	10.922		19,300	655		19,955	22,200
3090	6,000 gallon capacity		1.86	17.204		25,500	1,025		26,525	29,600
3100	8,000 gallon capacity		1.65	19.394		28,200	1,175		29,375	32,900
3110	10,000 gallon capacity		1.48	21.622		32,600	1,300		33,900	37,800
3120	12,000 gallon capacity		1.40	22.857		40,400	1,375		41,775	46,600

23 13 23 – Facility Aboveground Fuel-Oil, Storage Tanks

23 13 23.13 Vertical, Steel, Abvground Fuel-Oil, Stor. Tanks

		Crew	Daily Output	Labor-Hours	Unit	Material	2018 Bare Costs Labor	2018 Bare Costs Equipment	Total	Total Incl O&P
0010	**VERTICAL, STEEL, ABOVEGROUND FUEL-OIL, STORAGE TANKS**									
4000	Fixed roof oil storage tanks, steel (1 BBL=42 gal. w/foundation 3'D x 1'W)									
4200	5,000 barrels				Ea.				194,000	213,500
4300	24,000 barrels								333,500	367,000
4500	56,000 barrels								729,000	802,000
4600	110,000 barrels								1,060,000	1,166,000
4800	143,000 barrels								1,250,000	1,375,000
4900	225,000 barrels								1,360,000	1,496,000
5100	Floating roof gasoline tanks, steel, 5,000 barrels (w/foundation 3'D x 1'W)								204,000	225,000
5200	25,000 barrels								381,000	419,000
5400	55,000 barrels								839,000	923,000
5500	100,000 barrels								1,253,000	1,379,000
5700	150,000 barrels								1,532,000	1,685,000
5800	225,000 barrels								2,300,000	2,783,000

23 13 23.16 Horizontal, Stl, Abvgrd Fuel-Oil, Storage Tanks

		Crew	Daily Output	Labor-Hours	Unit	Material	2018 Bare Costs Labor	2018 Bare Costs Equipment	Total	Total Incl O&P
0010	**HORIZONTAL, STEEL, ABOVEGROUND FUEL-OIL, STORAGE TANKS**									
3000	Steel, storage, aboveground, including cradles, coating,									
3020	fittings, not including foundation, pumps or piping									
3040	Single wall, 275 gallon	Q-5	5	3.200	Ea.	515	181		696	840
3060	550 gallon	"	2.70	5.926		4,325	335		4,660	5,275
3080	1,000 gallon	Q-7	5	6.400		7,125	385		7,510	8,400
3100	1,500 gallon		4.75	6.737		10,200	405		10,605	11,800
3120	2,000 gallon		4.60	6.957		12,300	415		12,715	14,100
3140	5,000 gallon		3.20	10		21,000	600		21,600	24,000
3150	10,000 gallon		2	16		40,600	960		41,560	46,200
3160	15,000 gallon		1.70	18.824		52,000	1,125		53,125	59,000
3170	20,000 gallon		1.45	22.069		67,500	1,325		68,825	76,500
3180	25,000 gallon		1.30	24.615		79,000	1,475		80,475	89,000
3190	30,000 gallon		1.10	29.091		94,500	1,750		96,250	106,500
3320	Double wall, 500 gallon capacity	Q-5	2.40	6.667		1,750	380		2,130	2,500
3330	2,000 gallon capacity	Q-7	4.15	7.711		5,700	465		6,165	6,975
3340	4,000 gallon capacity		3.60	8.889		12,500	535		13,035	14,600

23 13 23 – Facility Aboveground Fuel-Oil, Storage Tanks

23 13 23.16 Horizontal, Stl, Abvgrd Fuel-Oil, Storage Tanks		Crew	Daily Output	Labor-Hours	Unit	Material	2018 Bare Costs Labor	Equipment	Total	Total Incl O&P
3350	6,000 gallon capacity	Q-7	2.40	13.333	Ea.	14,200	800		15,000	16,800
3360	8,000 gallon capacity		2	16		16,800	960		17,760	20,000
3370	10,000 gallon capacity		1.80	17.778		30,000	1,075		31,075	34,600
3380	15,000 gallon capacity		1.50	21.333		40,000	1,275		41,275	45,900
3390	20,000 gallon capacity		1.30	24.615		43,000	1,475		44,475	49,500
3400	25,000 gallon capacity		1.15	27.826		53,500	1,675		55,175	61,500
3410	30,000 gallon capacity	▼	1	32	▼	64,500	1,925		66,425	74,000

23 13 23.26 Horizontal, Conc., Abvgrd Fuel-Oil, Stor. Tanks

		Crew	Daily Output	Labor-Hours	Unit	Material	Labor	Equipment	Total	Total Incl O&P
0010	**HORIZONTAL, CONCRETE, ABOVEGROUND FUEL-OIL, STORAGE TANKS**									
0050	Concrete, storage, aboveground, including pad & pump									
0100	500 gallon	F-3	2	20	Ea.	9,950	1,025	248	11,223	12,700
0200	1,000 gallon	"	2	20		13,900	1,025	248	15,173	17,100
0300	2,000 gallon	F-4	2	24		17,900	1,225	495	19,620	22,100
0400	4,000 gallon		2	24		22,900	1,225	495	24,620	27,600
0500	8,000 gallon		2	24		35,800	1,225	495	37,520	41,800
0600	12,000 gallon	▼	2	24	▼	47,700	1,225	495	49,420	55,000

23 21 20 – Hydronic HVAC Piping Specialties

23 21 20.10 Air Control

		Crew	Daily Output	Labor-Hours	Unit	Material	Labor	Equipment	Total	Total Incl O&P
0010	**AIR CONTROL**									
0030	Air separator, with strainer									
0040	2" diameter	Q-5	6	2.667	Ea.	1,375	151		1,526	1,750
0080	2-1/2" diameter		5	3.200		1,550	181		1,731	1,975
0100	3" diameter		4	4		2,375	227		2,602	2,975
0120	4" diameter	▼	3	5.333		3,425	300		3,725	4,225
0130	5" diameter	Q-6	3.60	6.667		4,375	390		4,765	5,400
0140	6" diameter	"	3.40	7.059	▼	5,225	415		5,640	6,375

23 21 20.18 Automatic Air Vent

		Crew	Daily Output	Labor-Hours	Unit	Material	Labor	Equipment	Total	Total Incl O&P
0010	**AUTOMATIC AIR VENT**									
0020	Cast iron body, stainless steel internals, float type									
0060	1/2" NPT inlet, 300 psi	1 Stpi	12	.667	Ea.	163	42		205	243
0220	3/4" NPT inlet, 250 psi	"	10	.800		370	50.50		420.50	480
0340	1-1/2" NPT inlet, 250 psi	Q-5	12	1.333	▼	1,150	75.50		1,225.50	1,375

23 21 20.42 Expansion Joints

		Crew	Daily Output	Labor-Hours	Unit	Material	Labor	Equipment	Total	Total Incl O&P
0010	**EXPANSION JOINTS**									
0100	Bellows type, neoprene cover, flanged spool									
0140	6" face to face, 1-1/4" diameter	1 Stpi	11	.727	Ea.	255	46		301	350
0160	1-1/2" diameter	"	10.60	.755		255	47.50		302.50	355
0180	2" diameter	Q-5	13.30	1.203		258	68		326	385
0190	2-1/2" diameter		12.40	1.290		267	73		340	405
0200	3" diameter		11.40	1.404		299	79.50		378.50	450
0480	10" face to face, 2" diameter		13	1.231		370	70		440	515
0500	2-1/2" diameter		12	1.333		390	75.50		465.50	545
0520	3" diameter		11	1.455		400	82.50		482.50	565
0540	4" diameter		8	2		455	113		568	670
0560	5" diameter		7	2.286		540	130		670	790
0580	6" diameter	▼	6	2.667	▼	560	151		711	845

534

For customer support on your Building Construction Costs with RSMeans data, call 800.448.8182.

23 21 Hydronic Piping and Pumps

23 21 20 – Hydronic HVAC Piping Specialties

23 21 20.46 Expansion Tanks

		Crew	Daily Output	Labor-Hours	Unit	Material	2018 Bare Costs Labor	2018 Bare Costs Equipment	Total	Total Incl O&P
0010	**EXPANSION TANKS**									
1507	Underground fuel-oil storage tanks, see Section 23 13 13									
1512	Tank leak detection systems, see Section 28 33 33.50									
2000	Steel, liquid expansion, ASME, painted, 15 gallon capacity	Q-5	17	.941	Ea.	720	53.50		773.50	870
2020	24 gallon capacity		14	1.143		830	65		895	1,025
2040	30 gallon capacity		12	1.333		830	75.50		905.50	1,025
2060	40 gallon capacity		10	1.600		970	90.50		1,060.50	1,200
2080	60 gallon capacity		8	2		1,175	113		1,288	1,450
2100	80 gallon capacity		7	2.286		1,250	130		1,380	1,575
2120	100 gallon capacity		6	2.667		1,700	151		1,851	2,075
3000	Steel ASME expansion, rubber diaphragm, 19 gal. cap. accept.		12	1.333		2,725	75.50		2,800.50	3,125
3020	31 gallon capacity		8	2		3,025	113		3,138	3,525
3040	61 gallon capacity		6	2.667		4,425	151		4,576	5,100
3080	119 gallon capacity		4	4		4,600	227		4,827	5,400
3100	158 gallon capacity		3.80	4.211		6,400	239		6,639	7,375
3140	317 gallon capacity		2.80	5.714		9,650	325		9,975	11,100
3180	528 gallon capacity	▼	2.40	6.667	▼	15,700	380		16,080	17,900

23 21 20.58 Hydronic Heating Control Valves

		Crew	Daily Output	Labor-Hours	Unit	Material	2018 Bare Costs Labor	2018 Bare Costs Equipment	Total	Total Incl O&P
0010	**HYDRONIC HEATING CONTROL VALVES**									
0050	Hot water, nonelectric, thermostatic									
0100	Radiator supply, 1/2" diameter	1 Stpi	24	.333	Ea.	68.50	21		89.50	107
0120	3/4" diameter		20	.400		71.50	25		96.50	117
0140	1" diameter		19	.421		88	26.50		114.50	137
0160	1-1/4" diameter	▼	15	.533	▼	125	33.50		158.50	188
0500	For low pressure steam, add					25%				

23 21 20.70 Steam Traps

		Crew	Daily Output	Labor-Hours	Unit	Material	2018 Bare Costs Labor	2018 Bare Costs Equipment	Total	Total Incl O&P
0010	**STEAM TRAPS**									
0030	Cast iron body, threaded									
0040	Inverted bucket									
0050	1/2" pipe size	1 Stpi	12	.667	Ea.	157	42		199	237
0070	3/4" pipe size		10	.800		226	50.50		276.50	325
0100	1" pipe size		9	.889		350	56		406	470
0120	1-1/4" pipe size	▼	8	1	▼	525	63		588	675
1000	Float & thermostatic, 15 psi									
1010	3/4" pipe size	1 Stpi	16	.500	Ea.	149	31.50		180.50	212
1020	1" pipe size		15	.533		180	33.50		213.50	249
1040	1-1/2" pipe size		9	.889		315	56		371	435
1060	2" pipe size	▼	6	1.333	▼	905	84		989	1,125

23 21 20.76 Strainers, Y Type, Bronze Body

		Crew	Daily Output	Labor-Hours	Unit	Material	2018 Bare Costs Labor	2018 Bare Costs Equipment	Total	Total Incl O&P
0010	**STRAINERS, Y TYPE, BRONZE BODY**									
0050	Screwed, 125 lb., 1/4" pipe size	1 Stpi	24	.333	Ea.	33.50	21		54.50	68
0070	3/8" pipe size		24	.333		36.50	21		57.50	72
0100	1/2" pipe size		20	.400		36.50	25		61.50	78.50
0140	1" pipe size		17	.471		60.50	29.50		90	112
0160	1-1/2" pipe size		14	.571		131	36		167	198
0180	2" pipe size		13	.615		173	39		212	249
0182	3" pipe size	▼	12	.667		1,100	42		1,142	1,275
0200	300 lb., 2-1/2" pipe size	Q-5	17	.941		720	53.50		773.50	875
0220	3" pipe size		16	1		1,275	56.50		1,331.50	1,475
0240	4" pipe size	▼	15	1.067	▼	2,325	60.50		2,385.50	2,650
0500	For 300 lb. rating 1/4" thru 2", add					15%				

23 21 20 – Hydronic HVAC Piping Specialties

23 21 20.76 Strainers, Y Type, Bronze Body

		Crew	Daily Output	Labor-Hours	Unit	Material	2018 Bare Costs Labor	Equipment	Total	Total Incl O&P
1000	Flanged, 150 lb., 1-1/2" pipe size	1 Stpi	11	.727	Ea.	475	46		521	595
1020	2" pipe size	"	8	1		645	63		708	805
1030	2-1/2" pipe size	Q-5	5	3.200		945	181		1,126	1,325
1040	3" pipe size		4.50	3.556		1,175	202		1,377	1,600
1060	4" pipe size		3	5.333		1,800	300		2,100	2,425
1100	6" pipe size	Q-6	3	8		3,475	470		3,945	4,525
1106	8" pipe size	"	2.60	9.231		3,775	545		4,320	4,975
1500	For 300 lb. rating, add					40%				

23 21 20.78 Strainers, Y Type, Iron Body

		Crew	Daily Output	Labor-Hours	Unit	Material	2018 Bare Costs Labor	Equipment	Total	Total Incl O&P
0010	**STRAINERS, Y TYPE, IRON BODY**									
0050	Screwed, 250 lb., 1/4" pipe size	1 Stpi	20	.400	Ea.	13.70	25		38.70	53
0070	3/8" pipe size		20	.400		13.70	25		38.70	53
0100	1/2" pipe size		20	.400		13.70	25		38.70	53
0140	1" pipe size		16	.500		21	31.50		52.50	70.50
0160	1-1/2" pipe size		12	.667		40.50	42		82.50	108
0180	2" pipe size		8	1		60.50	63		123.50	162
0220	3" pipe size	Q-5	11	1.455		291	82.50		373.50	445
0240	4" pipe size	"	5	3.200		495	181		676	820
0500	For galvanized body, add					50%				
1000	Flanged, 125 lb., 1-1/2" pipe size	1 Stpi	11	.727	Ea.	163	46		209	249
1020	2" pipe size	"	8	1		136	63		199	244
1040	3" pipe size	Q-5	4.50	3.556		179	202		381	500
1060	4" pipe size	"	3	5.333		296	300		596	780
1080	5" pipe size	Q-6	3.40	7.059		375	415		790	1,025
1100	6" pipe size	"	3	8		595	470		1,065	1,375
1500	For 250 lb. rating, add					20%				
2000	For galvanized body, add					50%				
2500	For steel body, add					40%				

23 21 20.88 Venturi Flow

		Crew	Daily Output	Labor-Hours	Unit	Material	2018 Bare Costs Labor	Equipment	Total	Total Incl O&P
0010	**VENTURI FLOW**, Measuring device									
0050	1/2" diameter	1 Stpi	24	.333	Ea.	292	21		313	350
0120	1" diameter		19	.421		286	26.50		312.50	355
0140	1-1/4" diameter		15	.533		355	33.50		388.50	440
0160	1-1/2" diameter		13	.615		370	39		409	465
0180	2" diameter		11	.727		380	46		426	490
0220	3" diameter	Q-5	14	1.143		535	65		600	690
0240	4" diameter	"	11	1.455		800	82.50		882.50	1,000
0280	6" diameter	Q-6	3.50	6.857		1,175	405		1,580	1,900
0500	For meter, add					2,250			2,250	2,500

23 21 23 – Hydronic Pumps

23 21 23.13 In-Line Centrifugal Hydronic Pumps

		Crew	Daily Output	Labor-Hours	Unit	Material	2018 Bare Costs Labor	Equipment	Total	Total Incl O&P
0010	**IN-LINE CENTRIFUGAL HYDRONIC PUMPS**									
0600	Bronze, sweat connections, 1/40 HP, in line									
0640	3/4" size	Q-1	16	1	Ea.	246	56		302	355
1000	Flange connection, 3/4" to 1-1/2" size									
1040	1/12 HP	Q-1	6	2.667	Ea.	645	149		794	935
1060	1/8 HP		6	2.667		1,100	149		1,249	1,450
1100	1/3 HP		6	2.667		1,250	149		1,399	1,600
1140	2" size, 1/6 HP		5	3.200		1,600	179		1,779	2,025
1180	2-1/2" size, 1/4 HP		5	3.200		2,025	179		2,204	2,500
2000	Cast iron, flange connection									
2040	3/4" to 1-1/2" size, in line, 1/12 HP	Q-1	6	2.667	Ea.	430	149		579	695

For customer support on your Building Construction Costs with RSMeans data, call 800.448.8182.

23 21 Hydronic Piping and Pumps

23 21 23 – Hydronic Pumps

23 21 23.13 In-Line Centrifugal Hydronic Pumps	Crew	Daily Output	Labor-Hours	Unit	Material	2018 Bare Costs Labor	Equipment	Total	Total Incl O&P
2100	1/3 HP	Q-1	6	2.667	Ea.	775	149		924	1,075
2101	Pumps, circulating, 3/4" to 1-1/2" size, 1/3 HP		6	2.667		930	149		1,079	1,250
2140	2" size, 1/6 HP		5	3.200		850	179		1,029	1,200
2180	2-1/2" size, 1/4 HP		5	3.200		1,075	179		1,254	1,450
2220	3" size, 1/4 HP	↓	4	4	↓	1,100	224		1,324	1,525
2600	For nonferrous impeller, add					3%				

23 21 29 – Automatic Condensate Pump Units

23 21 29.10 Condensate Removal Pump System

			Crew	Daily Output	Labor-Hours	Unit	Material	Labor	Equipment	Total	Total Incl O&P
0010	**CONDENSATE REMOVAL PUMP SYSTEM**										
0020	Pump with 1 gal. ABS tank										
0100	115 V										
0120	1/50 HP, 200 GPH	G	1 Stpi	12	.667	Ea.	187	42		229	269
0140	1/18 HP, 270 GPH	G		10	.800		192	50.50		242.50	287
0160	1/5 HP, 450 GPH	G	↓	8	1	↓	435	63		498	575
0200	230 V										
0240	1/18 HP, 270 GPH		1 Stpi	10	.800	Ea.	206	50.50		256.50	305
0260	1/5 HP, 450 GPH	G	"	8	1	"	490	63		553	635

23 22 Steam and Condensate Piping and Pumps

23 22 13 – Steam and Condensate Heating Piping

23 22 13.23 Aboveground Steam and Condensate Piping

		Crew	Daily Output	Labor-Hours	Unit	Material	Labor	Equipment	Total	Total Incl O&P
0010	**ABOVEGROUND STEAM AND CONDENSATE HEATING PIPING**									
0020	Condensate meter									
0100	500 lb. per hour	1 Stpi	14	.571	Ea.	3,625	36		3,661	4,050
0140	1500 lb. per hour	"	7	1.143	"	4,250	72		4,322	4,775

23 22 23 – Steam Condensate Pumps

23 22 23.10 Condensate Return System

		Crew	Daily Output	Labor-Hours	Unit	Material	Labor	Equipment	Total	Total Incl O&P
0010	**CONDENSATE RETURN SYSTEM**									
2000	Simplex									
2010	With pump, motor, CI receiver, float switch									
2020	3/4 HP, 15 GPM	Q-1	1.80	8.889	Ea.	6,850	495		7,345	8,275
2100	Duplex									
2110	With 2 pumps and motors, CI receiver, float switch, alternator									
2120	3/4 HP, 15 GPM, 15 gal. CI receiver	Q-1	1.40	11.429	Ea.	7,450	640		8,090	9,175
2130	1 HP, 25 GPM		1.20	13.333		8,975	745		9,720	11,000
2140	1-1/2 HP, 45 GPM		1	16		10,400	895		11,295	12,800
2150	1-1/2 HP, 60 GPM	↓	1	16	↓	11,700	895		12,595	14,200

For customer support on your Building Construction Costs with RSMeans data, call 800.448.8182.

537

23 31 HVAC Ducts and Casings

23 31 13 – Metal Ducts

23 31 13.13 Rectangular Metal Ducts

23 31 13.13 Rectangular Metal Ducts	Crew	Daily Output	Labor-Hours	Unit	Material	2018 Bare Costs Labor	Equipment	Total	Total Incl O&P
0010 **RECTANGULAR METAL DUCTS**									
0020 Fabricated rectangular, includes fittings, joints, supports,									
0021 allowance for flexible connections and field sketches.									
0030 Does not include "as-built dwgs." or insulation.									
0031 NOTE: Fabrication and installation are combined									
0040 as LABOR cost. Approx. 25% fittings assumed.									
0042 Fabrication/Inst. is to commercial quality standards									
0043 (SMACNA or equiv.) for structure, sealing, leak testing, etc.									
0050 Add to labor for elevated installation									
0051 of fabricated ductwork									
0052 10' to 15' high						6%			
0053 15' to 20' high						12%			
0054 20' to 25' high						15%			
0055 25' to 30' high						21%			
0056 30' to 35' high						24%			
0057 35' to 40' high						30%			
0058 Over 40' high						33%			
0072 For duct insulation see Line 23 07 13.10 3000									
0100 Aluminum, alloy 3003-H14, under 100 lb.	Q-10	75	.320	Lb.	3.19	17.85		21.04	31
0110 100 to 500 lb.		80	.300		1.88	16.75		18.63	27.50
0120 500 to 1,000 lb.		95	.253		1.81	14.10		15.91	23.50
0140 1,000 to 2,000 lb.		120	.200		1.77	11.15		12.92	19
0150 2,000 to 5,000 lb.		130	.185		1.77	10.30		12.07	17.65
0160 Over 5,000 lb.		145	.166		1.77	9.25		11.02	16.05
0500 Galvanized steel, under 200 lb.		235	.102		.57	5.70		6.27	9.35
0520 200 to 500 lb.		245	.098		.56	5.45		6.01	8.95
0540 500 to 1,000 lb.		255	.094		.55	5.25		5.80	8.60
0560 1,000 to 2,000 lb.		265	.091		.54	5.05		5.59	8.30
0570 2,000 to 5,000 lb.		275	.087		.53	4.87		5.40	8.05
0580 Over 5,000 lb.		285	.084		.53	4.70		5.23	7.75
1000 Stainless steel, type 304, under 100 lb.		165	.145		4.71	8.10		12.81	17.60
1020 100 to 500 lb.		175	.137		3.87	7.65		11.52	15.95
1030 500 to 1,000 lb.		190	.126		2.58	7.05		9.63	13.60
1040 1,000 to 2,000 lb.		200	.120		2.29	6.70		8.99	12.70
1050 2,000 to 5,000 lb.		225	.107		2.21	5.95		8.16	11.55
1060 Over 5,000 lb.		235	.102		2.38	5.70		8.08	11.30
1100 For medium pressure ductwork, add						15%			
1200 For high pressure ductwork, add						40%			
1210 For welded ductwork, add						85%			
1220 For 30% fittings, add						11%			
1224 For 40% fittings, add						34%			
1228 For 50% fittings, add						56%			
1232 For 60% fittings, add						79%			
1236 For 70% fittings, add						101%			
1240 For 80% fittings, add						124%			
1244 For 90% fittings, add						147%			
1248 For 100% fittings, add						169%			
1252 Note: Fittings add includes time for detailing and installation.									

23 31 13.19 Metal Duct Fittings

	Crew	Daily Output	Labor-Hours	Unit	Material	Labor	Equipment	Total	Total Incl O&P
0010 **METAL DUCT FITTINGS**									
2000 Fabrics for flexible connections, with metal edge	1 Shee	100	.080	L.F.	3.66	4.78		8.44	11.35
2100 Without metal edge	"	160	.050	"	3.26	2.99		6.25	8.15

538

For customer support on your Building Construction Costs with RSMeans data, call 800.448.8182.

23 31 HVAC Ducts and Casings

23 31 16 – Nonmetal Ducts

23 31 16.13 Fibrous-Glass Ducts

		Crew	Daily Output	Labor-Hours	Unit	Material	2018 Bare Costs Labor	Equipment	Total	Total Incl O&P
0010	**FIBROUS-GLASS DUCTS**									
3490	Rigid fiberglass duct board, foil reinf. kraft facing									
3500	Rectangular, 1" thick, alum. faced, (FRK), std. weight	Q-10	350	.069	SF Surf	.90	3.83		4.73	6.85

23 33 Air Duct Accessories

23 33 13 – Dampers

23 33 13.13 Volume-Control Dampers

		Crew	Daily Output	Labor-Hours	Unit	Material	2018 Bare Costs Labor	Equipment	Total	Total Incl O&P
0010	**VOLUME-CONTROL DAMPERS**									
5990	Multi-blade dampers, opposed blade, 8" x 6"	1 Shee	24	.333	Ea.	29	19.95		48.95	62.50
5994	8" x 8"		22	.364		30	22		52	66
5996	10" x 10"		21	.381		35.50	23		58.50	74
6000	12" x 12"		21	.381		39.50	23		62.50	78.50
6020	12" x 18"		18	.444		53	26.50		79.50	99
6030	14" x 10"		20	.400		38.50	24		62.50	79
6031	14" x 14"		17	.471		47	28		75	95
6033	16" x 12"		17	.471		47	28		75	95
6035	16" x 16"		16	.500		58.50	30		88.50	110
6037	18" x 16"		15	.533		64.50	32		96.50	120
6038	18" x 18"		15	.533		69.50	32		101.50	125
6070	20" x 16"		14	.571		69.50	34		103.50	128
6072	20" x 20"		13	.615		84	37		121	148
6074	22" x 18"		14	.571		84	34		118	144
6076	24" x 16"		11	.727		82	43.50		125.50	157
6078	24" x 20"		8	1		97	60		157	199
6080	24" x 24"		8	1		114	60		174	217
6110	26" x 26"		6	1.333		127	79.50		206.50	262
6133	30" x 30"	Q-9	6.60	2.424		183	131		314	400
6135	32" x 32"		6.40	2.500		206	135		341	430
6180	48" x 36"		5.60	2.857		340	154		494	610
8000	Multi-blade dampers, parallel blade									
8100	8" x 8"	1 Shee	24	.333	Ea.	110	19.95		129.95	152
8140	16" x 10"		20	.400		138	24		162	189
8200	24" x 16"		11	.727		179	43.50		222.50	264
8260	30" x 18"		7	1.143		249	68.50		317.50	375

23 33 13.16 Fire Dampers

		Crew	Daily Output	Labor-Hours	Unit	Material	2018 Bare Costs Labor	Equipment	Total	Total Incl O&P
0010	**FIRE DAMPERS**									
3000	Fire damper, curtain type, 1-1/2 hr. rated, vertical, 6" x 6"	1 Shee	24	.333	Ea.	32.50	19.95		52.45	66
3020	8" x 6"		22	.364		32.50	22		54.50	68.50
3240	16" x 14"		18	.444		59	26.50		85.50	105
3400	24" x 20"		8	1		58.50	60		118.50	156

23 33 13.28 Splitter Damper Assembly

		Crew	Daily Output	Labor-Hours	Unit	Material	2018 Bare Costs Labor	Equipment	Total	Total Incl O&P
0010	**SPLITTER DAMPER ASSEMBLY**									
7000	Self locking, 1' rod	1 Shee	24	.333	Ea.	28	19.95		47.95	61.50
7020	3' rod		22	.364		40	22		62	77
7040	4' rod		20	.400		44.50	24		68.50	85.50
7060	6' rod		18	.444		54	26.50		80.50	100

23 33 Air Duct Accessories

23 33 19 – Duct Silencers

23 33 19.10 Duct Silencers

23 33 19.10 Duct Silencers	Crew	Daily Output	Labor-Hours	Unit	Material	2018 Bare Costs Labor	Equipment	Total	Total Incl O&P
0010 **DUCT SILENCERS**									
9000 Silencers, noise control for air flow, duct				MCFM	76.50			76.50	84

23 33 33 – Duct-Mounting Access Doors

23 33 33.13 Duct Access Doors

23 33 33.13 Duct Access Doors	Crew	Daily Output	Labor-Hours	Unit	Material	2018 Bare Costs Labor	Equipment	Total	Total Incl O&P
0010 **DUCT ACCESS DOORS**									
1000 Duct access door, insulated, 6" x 6"	1 Shee	14	.571	Ea.	23.50	34		57.50	78
1020 10" x 10"		11	.727		20.50	43.50		64	89
1040 12" x 12"		10	.800		29.50	48		77.50	106
1050 12" x 18"		9	.889		51	53		104	138
1070 18" x 18"		8	1		35.50	60		95.50	131
1074 24" x 18"		8	1		61	60		121	159

23 33 46 – Flexible Ducts

23 33 46.10 Flexible Air Ducts

23 33 46.10 Flexible Air Ducts	Crew	Daily Output	Labor-Hours	Unit	Material	2018 Bare Costs Labor	Equipment	Total	Total Incl O&P
0010 **FLEXIBLE AIR DUCTS**									
1280 Add to labor for elevated installation									
1282 of prefabricated (purchased) ductwork									
1283 10' to 15' high						10%			
1284 15' to 20' high						20%			
1285 20' to 25' high						25%			
1286 25' to 30' high						35%			
1287 30' to 35' high						40%			
1288 35' to 40' high						50%			
1289 Over 40' high						55%			
1300 Flexible, coated fiberglass fabric on corr. resist. metal helix									
1400 pressure to 12" (WG) UL-181									
1500 Noninsulated, 3" diameter	Q-9	400	.040	L.F.	1.20	2.15		3.35	4.61
1540 5" diameter		320	.050		1.38	2.69		4.07	5.65
1560 6" diameter		280	.057		1.60	3.08		4.68	6.45
1580 7" diameter		240	.067		1.78	3.59		5.37	7.45
1600 8" diameter		200	.080		2.02	4.31		6.33	8.75
1640 10" diameter		160	.100		2.61	5.40		8.01	11.05
1660 12" diameter		120	.133		3.10	7.20		10.30	14.35
1900 Insulated, 1" thick, PE jacket, 3" diameter G		380	.042		2.62	2.27		4.89	6.35
1910 4" diameter G		340	.047		2.80	2.53		5.33	6.95
1920 5" diameter G		300	.053		2.93	2.87		5.80	7.60
1940 6" diameter G		260	.062		3.11	3.31		6.42	8.45
1960 7" diameter G		220	.073		3.60	3.92		7.52	9.90
1980 8" diameter G		180	.089		3.80	4.78		8.58	11.50
2020 10" diameter G		140	.114		4.55	6.15		10.70	14.40
2040 12" diameter G		100	.160		5.25	8.60		13.85	18.95

23 33 53 – Duct Liners

23 33 53.10 Duct Liner Board

23 33 53.10 Duct Liner Board	Crew	Daily Output	Labor-Hours	Unit	Material	2018 Bare Costs Labor	Equipment	Total	Total Incl O&P
0010 **DUCT LINER BOARD**									
3340 Board type fiberglass liner, FSK, 1-1/2 lb. density									
3344 1" thick G	Q-14	150	.107	S.F.	.97	5.40		6.37	9.45
3345 1-1/2" thick G		130	.123		1.06	6.25		7.31	10.85
3346 2" thick G		120	.133		1.24	6.75		7.99	11.85
3348 3" thick G		110	.145		1.60	7.35		8.95	13.20
3350 4" thick G		100	.160		1.96	8.10		10.06	14.75
3356 3 lb. density, 1" thick G		150	.107		1.24	5.40		6.64	9.75

For customer support on your Building Construction Costs with RSMeans data, call 800.448.8182.

23 33 Air Duct Accessories

23 33 53 – Duct Liners

23 33 53.10 Duct Liner Board

23 33 53.10 Duct Liner Board		Crew	Daily Output	Labor-Hours	Unit	Material	2018 Bare Costs Labor	Equipment	Total	Total Incl O&P
3358	1-1/2" thick	G Q-14	130	.123	S.F.	1.56	6.25		7.81	11.40
3360	2" thick	G	120	.133		1.90	6.75		8.65	12.60
3362	2-1/2" thick	G	110	.145		2.23	7.35		9.58	13.90
3364	3" thick	G	100	.160		2.57	8.10		10.67	15.45
3366	4" thick	G	90	.178		3.22	9		12.22	17.55
3370	6 lb. density, 1" thick	G	140	.114		1.75	5.80		7.55	10.95
3374	1-1/2" thick	G	120	.133		2.35	6.75		9.10	13.10
3378	2" thick	G	100	.160		2.94	8.10		11.04	15.85
3490	Board type, fiberglass liner, 3 lb. density									
3680	No finish									
3700	1" thick	G Q-14	170	.094	S.F.	.69	4.77		5.46	8.15
3710	1-1/2" thick	G	140	.114		1.03	5.80		6.83	10.15
3720	2" thick	G	130	.123		1.38	6.25		7.63	11.20
3940	Board type, non-fibrous foam									
3950	Temperature, bacteria and fungi resistant									
3960	1" thick	G Q-14	150	.107	S.F.	2.57	5.40		7.97	11.25
3970	1-1/2" thick	G	130	.123		4.03	6.25		10.28	14.15
3980	2" thick	G	120	.133		4.02	6.75		10.77	14.90

23 34 HVAC Fans

23 34 13 – Axial HVAC Fans

23 34 13.10 Axial Flow HVAC Fans

		Crew	Daily Output	Labor-Hours	Unit	Material	2018 Bare Costs Labor	Equipment	Total	Total Incl O&P
0010	**AXIAL FLOW HVAC FANS**									
0020	Air conditioning and process air handling									
1500	Vaneaxial, low pressure, 2,000 CFM, 1/2 HP	Q-20	3.60	5.556	Ea.	2,375	305		2,680	3,050
1520	4,000 CFM, 1 HP		3.20	6.250		2,425	340		2,765	3,200
1540	8,000 CFM, 2 HP		2.80	7.143		3,500	390		3,890	4,450

23 34 14 – Blower HVAC Fans

23 34 14.10 Blower Type HVAC Fans

		Crew	Daily Output	Labor-Hours	Unit	Material	2018 Bare Costs Labor	Equipment	Total	Total Incl O&P
0010	**BLOWER TYPE HVAC FANS**									
2500	Ceiling fan, right angle, extra quiet, 0.10" S.P.									
2520	95 CFM	Q-20	20	1	Ea.	291	54.50		345.50	405
2540	210 CFM		19	1.053		345	57.50		402.50	470
2560	385 CFM		18	1.111		435	61		496	575
2580	885 CFM		16	1.250		860	68.50		928.50	1,050
2600	1,650 CFM		13	1.538		1,200	84		1,284	1,425
2620	2,960 CFM		11	1.818		1,575	99.50		1,674.50	1,900
2640	For wall or roof cap, add	1 Shee	16	.500		291	30		321	365
2660	For straight thru fan, add					10%				
2680	For speed control switch, add	1 Elec	16	.500		159	29		188	218
7500	Utility set, steel construction, pedestal, 1/4" S.P.									
7520	Direct drive, 150 CFM, 1/8 HP	Q-20	6.40	3.125	Ea.	925	171		1,096	1,275
7540	485 CFM, 1/6 HP		5.80	3.448		1,050	189		1,239	1,425
7560	1,950 CFM, 1/2 HP		4.80	4.167		1,350	228		1,578	1,850
7580	2,410 CFM, 3/4 HP		4.40	4.545		2,525	249		2,774	3,150
7600	3,328 CFM, 1-1/2 HP		3	6.667		2,500	365		2,865	3,300
7680	V-belt drive, drive cover, 3 phase									
7700	800 CFM, 1/4 HP	Q-20	6	3.333	Ea.	940	182		1,122	1,300
7720	1,300 CFM, 1/3 HP		5	4		985	219		1,204	1,400
7740	2,000 CFM, 1 HP		4.60	4.348		1,175	238		1,413	1,625

For customer support on your Building Construction Costs with RSMeans data, call 800.448.8182.

541

23 34 HVAC Fans

23 34 14 – Blower HVAC Fans

23 34 14.10 Blower Type HVAC Fans

		Crew	Daily Output	Labor-Hours	Unit	Material	2018 Bare Costs Labor	Equipment	Total	Total Incl O&P
7760	2,900 CFM, 3/4 HP	Q-20	4.20	4.762	Ea.	1,575	260		1,835	2,125

23 34 16 – Centrifugal HVAC Fans

23 34 16.10 Centrifugal Type HVAC Fans

		Crew	Daily Output	Labor-Hours	Unit	Material	2018 Bare Costs Labor	Equipment	Total	Total Incl O&P
0010	**CENTRIFUGAL TYPE HVAC FANS**									
0200	In-line centrifugal, supply/exhaust booster									
0220	aluminum wheel/hub, disconnect switch, 1/4" S.P.									
0240	500 CFM, 10" diameter connection	Q-20	3	6.667	Ea.	1,575	365		1,940	2,275
0260	1,380 CFM, 12" diameter connection		2	10		1,600	545		2,145	2,575
0280	1,520 CFM, 16" diameter connection		2	10		1,650	545		2,195	2,625
0300	2,560 CFM, 18" diameter connection		1	20		1,800	1,100		2,900	3,625
0320	3,480 CFM, 20" diameter connection		.80	25		2,075	1,375		3,450	4,375
0326	5,080 CFM, 20" diameter connection		.75	26.667		2,175	1,450		3,625	4,625
3500	Centrifugal, airfoil, motor and drive, complete									
3520	1,000 CFM, 1/2 HP	Q-20	2.50	8	Ea.	2,025	440		2,465	2,900
3540	2,000 CFM, 1 HP		2	10		2,300	545		2,845	3,350
3560	4,000 CFM, 3 HP		1.80	11.111		2,900	610		3,510	4,125
3580	8,000 CFM, 7-1/2 HP		1.40	14.286		4,275	780		5,055	5,875
3600	12,000 CFM, 10 HP		1	20		5,800	1,100		6,900	8,050
5000	Utility set, centrifugal, V belt drive, motor									
5020	1/4" S.P., 1,200 CFM, 1/4 HP	Q-20	6	3.333	Ea.	1,875	182		2,057	2,325
5040	1,520 CFM, 1/3 HP		5	4		2,450	219		2,669	3,025
5060	1,850 CFM, 1/2 HP		4	5		2,425	274		2,699	3,075
5080	2,180 CFM, 3/4 HP		3	6.667		2,875	365		3,240	3,725
5100	1/2" S.P., 3,600 CFM, 1 HP		2	10		2,975	545		3,520	4,100
5120	4,250 CFM, 1-1/2 HP		1.60	12.500		3,600	685		4,285	5,025
5140	4,800 CFM, 2 HP		1.40	14.286		4,375	780		5,155	5,975
7000	Roof exhauster, centrifugal, aluminum housing, 12" galvanized									
7020	curb, bird screen, back draft damper, 1/4" S.P.									
7100	Direct drive, 320 CFM, 11" sq. damper	Q-20	7	2.857	Ea.	770	156		926	1,075
7120	600 CFM, 11" sq. damper		6	3.333		985	182		1,167	1,350
7140	815 CFM, 13" sq. damper		5	4		985	219		1,204	1,400
7160	1,450 CFM, 13" sq. damper		4.20	4.762		1,575	260		1,835	2,125
7180	2,050 CFM, 16" sq. damper		4	5		1,925	274		2,199	2,550
7200	V-belt drive, 1,650 CFM, 12" sq. damper		6	3.333		1,425	182		1,607	1,825
7220	2,750 CFM, 21" sq. damper		5	4		1,675	219		1,894	2,175
7230	3,500 CFM, 21" sq. damper		4.50	4.444		1,900	243		2,143	2,475
7240	4,910 CFM, 23" sq. damper		4	5		2,300	274		2,574	2,950
7260	8,525 CFM, 28" sq. damper		3	6.667		3,075	365		3,440	3,925
7280	13,760 CFM, 35" sq. damper		2	10		4,250	545		4,795	5,500
7300	20,558 CFM, 43" sq. damper		1	20		8,425	1,100		9,525	10,900
7320	For 2 speed winding, add					15%				
7340	For explosion proof motor, add					755			755	835
7360	For belt driven, top discharge, add					15%				
8500	Wall exhausters, centrifugal, auto damper, 1/8" S.P.									
8520	Direct drive, 610 CFM, 1/20 HP	Q-20	14	1.429	Ea.	410	78		488	570
8540	796 CFM, 1/12 HP		13	1.538		970	84		1,054	1,200
8560	822 CFM, 1/6 HP		12	1.667		1,025	91		1,116	1,300
8580	1,320 CFM, 1/4 HP		12	1.667		1,375	91		1,466	1,675
9500	V-belt drive, 3 phase									
9520	2,800 CFM, 1/4 HP	Q-20	9	2.222	Ea.	1,850	122		1,972	2,225
9540	3,740 CFM, 1/2 HP	"	8	2.500	"	1,925	137		2,062	2,325

For customer support on your Building Construction Costs with RSMeans data, call 800.448.8182.

23 34 HVAC Fans

23 34 23 – HVAC Power Ventilators

23 34 23.10 HVAC Power Circulators and Ventilators		Crew	Daily Output	Labor-Hours	Unit	Material	2018 Bare Costs Labor	Equipment	Total	Total Incl O&P
0010	**HVAC POWER CIRCULATORS AND VENTILATORS**									
3000	Paddle blade air circulator, 3 speed switch									
3020	42", 5,000 CFM high, 3,000 CFM low Ⓖ	1 Elec	2.40	3.333	Ea.	123	194		317	425
3040	52", 6,500 CFM high, 4,000 CFM low Ⓖ	"	2.20	3.636	"	156	212		368	485
3100	For antique white motor, same cost									
3200	For brass plated motor, same cost									
3300	For light adaptor kit, add Ⓖ				Ea.	39.50			39.50	43.50
6000	Propeller exhaust, wall shutter									
6020	Direct drive, one speed, 0.075" S.P.									
6100	653 CFM, 1/30 HP	Q-20	10	2	Ea.	202	109		311	390
6120	1,033 CFM, 1/20 HP		9	2.222		300	122		422	515
6140	1,323 CFM, 1/15 HP	▼	8	2.500	▼	335	137		472	580
6300	V-belt drive, 3 phase									
6320	6,175 CFM, 3/4 HP	Q-20	5	4	Ea.	1,275	219		1,494	1,725
6340	7,500 CFM, 3/4 HP		5	4		1,350	219		1,569	1,800
6360	10,100 CFM, 1 HP		4.50	4.444		1,550	243		1,793	2,075
6380	14,300 CFM, 1-1/2 HP	▼	4	5	▼	1,825	274		2,099	2,425
6650	Residential, bath exhaust, grille, back draft damper									
6660	50 CFM	Q-20	24	.833	Ea.	52.50	45.50		98	127
6670	110 CFM		22	.909		104	49.50		153.50	190
6680	Light combination, squirrel cage, 100 watt, 70 CFM	▼	24	.833	▼	114	45.50		159.50	195
6700	Light/heater combination, ceiling mounted									
6710	70 CFM, 1,450 watt	Q-20	24	.833	Ea.	140	45.50		185.50	224
6800	Heater combination, recessed, 70 CFM		24	.833		69.50	45.50		115	146
6820	With 2 infrared bulbs		23	.870		97	47.50		144.50	180
6900	Kitchen exhaust, grille, complete, 160 CFM		22	.909		105	49.50		154.50	192
6910	180 CFM		20	1		106	54.50		160.50	200
6920	270 CFM		18	1.111		230	61		291	345
6930	350 CFM	▼	16	1.250	▼	145	68.50		213.50	264
6940	Residential roof jacks and wall caps									
6944	Wall cap with back draft damper									
6946	3" & 4" diam. round duct	1 Shee	11	.727	Ea.	25	43.50		68.50	94
6948	6" diam. round duct	"	11	.727	"	72.50	43.50		116	146
6958	Roof jack with bird screen and back draft damper									
6960	3" & 4" diam. round duct	1 Shee	11	.727	Ea.	24.50	43.50		68	93.50
6962	3-1/4" x 10" rectangular duct	"	10	.800	"	44	48		92	122
6980	Transition									
6982	3-1/4" x 10" to 6" diam. round	1 Shee	20	.400	Ea.	32	24		56	71.50

23 34 33 – Air Curtains

23 34 33.10 Air Barrier Curtains

		Crew	Daily Output	Labor-Hours	Unit	Material	Labor	Equipment	Total	Total Incl O&P
0010	**AIR BARRIER CURTAINS**, Incl. motor starters, transformers,									
0050	and door switches									
2450	Conveyor openings or service windows									
3000	Service window, 5' high x 25" wide	2 Shee	5	3.200	Ea.	350	191		541	675
3100	Environmental separation									
3110	Door heights up to 8', low profile, super quiet									
3120	Unheated, variable speed									
3130	36" wide	2 Shee	4	4	Ea.	660	239		899	1,100
3134	42" wide		3.80	4.211		690	252		942	1,150
3138	48" wide		3.60	4.444		715	266		981	1,200
3142	60" wide	▼	3.40	4.706		740	281		1,021	1,250
3146	72" wide	Q-3	4.60	6.957		915	410		1,325	1,625

For customer support on your Building Construction Costs with RSMeans data, call 800.448.8182.

543

23 34 HVAC Fans

23 34 33 – Air Curtains

23 34 33.10 Air Barrier Curtains

	23 34 33.10 Air Barrier Curtains	Crew	Daily Output	Labor-Hours	Unit	Material	2018 Bare Costs Labor	Equipment	Total	Total Incl O&P
3150	96" wide	Q-3	4.40	7.273	Ea.	1,375	430		1,805	2,175
3154	120" wide	↓	4.20	7.619		1,475	450		1,925	2,300
3158	144" wide	▼	4	8	▼	1,800	475		2,275	2,725
3200	Door heights up to 10'									
3210	Unheated									
3230	36" wide	2 Shee	3.80	4.211	Ea.	700	252		952	1,150
3234	42" wide		3.60	4.444		720	266		986	1,200
3238	48" wide		3.40	4.706		740	281		1,021	1,250
3242	60" wide	▼	3.20	5		1,075	299		1,374	1,625
3246	72" wide	Q-3	4.40	7.273		1,150	430		1,580	1,925
3250	96" wide		4.20	7.619		1,375	450		1,825	2,175
3254	120" wide		4	8		1,875	475		2,350	2,800
3258	144" wide	▼	3.80	8.421	▼	2,025	500		2,525	3,000
3300	Door heights up to 12'									
3310	Unheated									
3334	42" wide	2 Shee	3.40	4.706	Ea.	1,025	281		1,306	1,550
3338	48" wide		3.20	5		1,050	299		1,349	1,600
3342	60" wide	▼	3	5.333		1,050	320		1,370	1,650
3346	72" wide	Q-3	4.20	7.619		1,850	450		2,300	2,725
3350	96" wide		4	8		2,025	475		2,500	2,950
3354	120" wide		3.80	8.421		2,300	500		2,800	3,275
3358	144" wide	▼	3.60	8.889	▼	2,650	525		3,175	3,700
3400	Door heights up to 16'									
3410	Unheated									
3438	48" wide	2 Shee	3	5.333	Ea.	1,375	320		1,695	2,000
3442	60" wide	"	2.80	5.714		1,450	340		1,790	2,125
3446	72" wide	Q-3	3.80	8.421		2,475	500		2,975	3,475
3450	96" wide		3.60	8.889		2,400	525		2,925	3,450
3454	120" wide		3.40	9.412		3,250	555		3,805	4,425
3458	144" wide	▼	3.20	10	▼	3,375	590		3,965	4,600
3470	Heated, electric									
3474	48" wide	2 Shee	2.90	5.517	Ea.	2,300	330		2,630	3,025
3478	60" wide	"	2.70	5.926		2,350	355		2,705	3,125
3482	72" wide	Q-3	3.70	8.649		4,050	510		4,560	5,225
3486	96" wide		3.50	9.143		4,250	540		4,790	5,500
3490	120" wide		3.30	9.697		5,700	575		6,275	7,150
3494	144" wide	▼	3.10	10.323	▼	5,800	610		6,410	7,325

23 37 Air Outlets and Inlets

23 37 13 – Diffusers, Registers, and Grilles

23 37 13.10 Diffusers

	23 37 13.10 Diffusers	Crew	Daily Output	Labor-Hours	Unit	Material	2018 Bare Costs Labor	Equipment	Total	Total Incl O&P
0010	**DIFFUSERS**, Aluminum, opposed blade damper unless noted									
0100	Ceiling, linear, also for sidewall									
0500	Perforated, 24" x 24" lay-in panel size, 6" x 6"	1 Shee	16	.500	Ea.	158	30		188	220
0520	8" x 8"		15	.533		166	32		198	232
0530	9" x 9"		14	.571		169	34		203	238
0540	10" x 10"		14	.571		169	34		203	238
0560	12" x 12"		12	.667		176	40		216	255
0590	16" x 16"		11	.727		197	43.50		240.50	284
0600	18" x 18"		10	.800		211	48		259	305
0610	20" x 20"	▼	10	.800	▼	228	48		276	325

23 37 Air Outlets and Inlets

23 37 13 – Diffusers, Registers, and Grilles

23 37 13.10 Diffusers

		Crew	Daily Output	Labor-Hours	Unit	Material	2018 Bare Costs Labor	Equipment	Total	Total Incl O&P
0620	24" x 24"	1 Shee	9	.889	Ea.	250	53		303	355
1000	Rectangular, 1 to 4 way blow, 6" x 6"		16	.500		42	30		72	91.50
1010	8" x 8"		15	.533		58.50	32		90.50	113
1014	9" x 9"		15	.533		52.50	32		84.50	106
1016	10" x 10"		15	.533		80.50	32		112.50	137
1020	12" x 6"		15	.533		72	32		104	128
1040	12" x 9"		14	.571		76.50	34		110.50	136
1060	12" x 12"		12	.667		70.50	40		110.50	139
1070	14" x 6"		13	.615		78.50	37		115.50	142
1074	14" x 14"		12	.667		129	40		169	203
1150	18" x 18"		9	.889		114	53		167	206
1160	21" x 21"		8	1		217	60		277	330
1170	24" x 12"		10	.800		165	48		213	254
1500	Round, butterfly damper, steel, diffuser size, 6" diameter		18	.444		11	26.50		37.50	52.50
1520	8" diameter		16	.500		11.70	30		41.70	58.50
1540	10" diameter		14	.571		14.55	34		48.55	68
1560	12" diameter		12	.667		19.20	40		59.20	82
1580	14" diameter		10	.800		24	48		72	99.50
2000	T-bar mounting, 24" x 24" lay-in frame, 6" x 6"		16	.500		69.50	30		99.50	122
2020	8" x 8"		14	.571		69.50	34		103.50	129
2040	12" x 12"		12	.667		85.50	40		125.50	155
2060	16" x 16"		11	.727		106	43.50		149.50	183
2080	18" x 18"	▼	10	.800		117	48		165	202
6000	For steel diffusers instead of aluminum, deduct				▼	10%				

23 37 13.30 Grilles

		Crew	Daily Output	Labor-Hours	Unit	Material	2018 Bare Costs Labor	Equipment	Total	Total Incl O&P
0010	**GRILLES**									
0020	Aluminum, unless noted otherwise									
1000	Air return, steel, 6" x 6"	1 Shee	26	.308	Ea.	19.80	18.40		38.20	50
1020	10" x 6"		24	.333		19.80	19.95		39.75	52.50
1080	16" x 8"		22	.364		28	22		50	64
1100	12" x 12"		22	.364		28	22		50	64
1120	24" x 12"		18	.444		37	26.50		63.50	81
1220	24" x 18"		16	.500		45.50	30		75.50	95.50
1280	36" x 24"		14	.571		78.50	34		112.50	139
3000	Filter grille with filter, 12" x 12"		24	.333		55	19.95		74.95	91
3020	18" x 12"		20	.400		71.50	24		95.50	116
3040	24" x 18"		18	.444		87	26.50		113.50	136
3060	24" x 24"	▼	16	.500		101	30		131	157
6000	For steel grilles instead of aluminum in above, deduct				▼	10%				

23 37 13.60 Registers

		Crew	Daily Output	Labor-Hours	Unit	Material	2018 Bare Costs Labor	Equipment	Total	Total Incl O&P
0010	**REGISTERS**									
0980	Air supply									
1000	Ceiling/wall, O.B. damper, anodized aluminum									
1010	One or two way deflection, adj. curved face bars									
1020	8" x 4"	1 Shee	26	.308	Ea.	17.30	18.40		35.70	47
1120	12" x 12"		18	.444		29	26.50		55.50	72.50
1240	20" x 6"		18	.444		26.50	26.50		53	69.50
1340	24" x 8"		13	.615		35.50	37		72.50	95
1350	24" x 18"	▼	12	.667		63.50	40		103.50	131
2700	Above registers in steel instead of aluminum, deduct				▼	10%				
4000	Floor, toe operated damper, enameled steel									
4020	4" x 8"	1 Shee	32	.250	Ea.	12.70	14.95		27.65	37

For customer support on your Building Construction Costs with RSMeans data, call 800.448.8182.

545

23 37 13.60 Registers

		Crew	Daily Output	Labor-Hours	Unit	Material	2018 Bare Costs Labor	Equipment	Total	Total Incl O&P
4100	8" x 10"	1 Shee	22	.364	Ea.	15.50	22		37.50	50
4140	10" x 10"		20	.400		18.60	24		42.60	57
4220	14" x 14"		16	.500		44.50	30		74.50	94.50
4240	14" x 20"	↓	15	.533	↓	52	32		84	106
4980	Air return									
5000	Ceiling or wall, fixed 45° face blades									
5010	Adjustable O.B. damper, anodized aluminum									
5020	4" x 8"	1 Shee	26	.308	Ea.	16.25	18.40		34.65	46
5060	6" x 10"		19	.421		19.45	25		44.45	60
5280	24" x 24"		11	.727		89	43.50		132.50	165
5300	24" x 36"	↓	8	1	↓	142	60		202	248
6000	For steel construction instead of aluminum, deduct					10%				

		Crew	Daily Output	Labor-Hours	Unit	Material	2018 Bare Costs Labor	Equipment	Total	Total Incl O&P
0010	**HVAC LOUVERS**									
0100	Aluminum, extruded, with screen, mill finish									
1002	Brick vent, see also Section 04 05 23.19									
1100	Standard, 4" deep, 8" wide, 5" high	1 Shee	24	.333	Ea.	35.50	19.95		55.45	69.50
1200	Modular, 4" deep, 7-3/4" wide, 5" high		24	.333		37.50	19.95		57.45	72
1300	Speed brick, 4" deep, 11-5/8" wide, 3-7/8" high		24	.333		37.50	19.95		57.45	72
1400	Fuel oil brick, 4" deep, 8" wide, 5" high		24	.333	↓	65	19.95		84.95	102
2000	Cooling tower and mechanical equip., screens, light weight		40	.200	S.F.	16.25	11.95		28.20	36
2020	Standard weight		35	.229		43	13.65		56.65	68.50
2500	Dual combination, automatic, intake or exhaust		20	.400		59	24		83	102
2520	Manual operation		20	.400		44	24		68	85
2540	Electric or pneumatic operation		20	.400	↓	44	24		68	85
2560	Motor, for electric or pneumatic	↓	14	.571	Ea.	505	34		539	610
3000	Fixed blade, continuous line									
3100	Mullion type, stormproof	1 Shee	28	.286	S.F.	44	17.10		61.10	74.50
3200	Stormproof		28	.286		44	17.10		61.10	74.50
3300	Vertical line	↓	28	.286	↓	52	17.10		69.10	83
3500	For damper to use with above, add					50%	30%			
3520	Motor, for damper, electric or pneumatic	1 Shee	14	.571	Ea.	505	34		539	610
4000	Operating, 45°, manual, electric or pneumatic		24	.333	S.F.	53	19.95		72.95	89
4100	Motor, for electric or pneumatic		14	.571	Ea.	505	34		539	610
4200	Penthouse, roof		56	.143	S.F.	26	8.55		34.55	41.50
4300	Walls		40	.200		61	11.95		72.95	85.50
5000	Thinline, under 4" thick, fixed blade	↓	40	.200	↓	25.50	11.95		37.45	46.50
5010	Finishes, applied by mfr. at additional cost, available in colors									
5020	Prime coat only, add				S.F.	3.48			3.48	3.83
5040	Baked enamel finish coating, add					6.40			6.40	7.05
5060	Anodized finish, add					6.95			6.95	7.65
5080	Duranodic finish, add					12.65			12.65	13.90
5100	Fluoropolymer finish coating, add					19.90			19.90	22
9980	For small orders (under 10 pieces), add				↓	25%				

		Crew	Daily Output	Labor-Hours	Unit	Material	2018 Bare Costs Labor	Equipment	Total	Total Incl O&P
0010	**HVAC GRAVITY AIR VENTILATORS**, Includes base									
1280	Rotary ventilators, wind driven, galvanized									
1300	4" neck diameter	Q-9	20	.800	Ea.	47.50	43		90.50	118
1340	6" neck diameter		16	1		48	54		102	135
1400	12" neck diameter	↓	10	1.600		67	86		153	205

For customer support on your Building Construction Costs with RSMeans data, call 800.448.8182.

23 37 Air Outlets and Inlets

23 37 23 – HVAC Gravity Ventilators

	23 37 23.10 HVAC Gravity Air Ventilators	Crew	Daily Output	Labor-Hours	Unit	Material	2018 Bare Costs Labor	Equipment	Total	Total Incl O&P
1500	24" neck diameter	Q-9	8	2	Ea.	299	108		407	495
1540	36" neck diameter	↓	6	2.667	↓	575	144		719	850
2000	Stationary, gravity, syphon, galvanized									
2160	6" neck diameter, 66 CFM	Q-9	16	1	Ea.	34	54		88	119
2240	12" neck diameter, 160 CFM		10	1.600		78.50	86		164.50	217
2340	24" neck diameter, 900 CFM		8	2		266	108		374	455
2380	36" neck diameter, 2,000 CFM		6	2.667		415	144		559	675
4200	Stationary mushroom, aluminum, 16" orifice diameter		10	1.600		650	86		736	845
4220	26" orifice diameter		6.15	2.602		960	140		1,100	1,275
4230	30" orifice diameter		5.71	2.802		1,400	151		1,551	1,775
4240	38" orifice diameter		5	3.200		2,025	172		2,197	2,500
4250	42" orifice diameter		4.70	3.404		2,675	183		2,858	3,225
4260	50" orifice diameter	↓	4.44	3.604	↓	3,175	194		3,369	3,800
5000	Relief vent									
5500	Rectangular, aluminum, galvanized curb									
5510	intake/exhaust, 0.033" SP									
5580	500 CFM, 12" x 12"	Q-9	8.60	1.860	Ea.	715	100		815	940
5600	600 CFM, 12" x 16"		8	2		800	108		908	1,050
5640	1,000 CFM, 12" x 24"		6.60	2.424		895	131		1,026	1,175
5680	3,000 CFM, 20" x 42"	↓	4	4	↓	1,575	215		1,790	2,050
5880	Size is throat area, volume is at 500 fpm									
7000	Note: sizes based on exhaust. Intake, with 0.125" SP									
7100	loss, approximately twice listed capacity.									

23 38 Ventilation Hoods

23 38 13 – Commercial-Kitchen Hoods

23 38 13.10 Hood and Ventilation Equipment

		Crew	Daily Output	Labor-Hours	Unit	Material	Labor	Equipment	Total	Total Incl O&P
0010	**HOOD AND VENTILATION EQUIPMENT**									
2970	Exhaust hood, sst, gutter on all sides, 4' x 4' x 2'	1 Carp	1.80	4.444	Ea.	4,300	225		4,525	5,075
2980	4' x 4' x 7'	"	1.60	5	"	6,850	254		7,104	7,900

23 41 Particulate Air Filtration

23 41 13 – Panel Air Filters

23 41 13.10 Panel Type Air Filters

			Crew	Daily Output	Labor-Hours	Unit	Material	Labor	Equipment	Total	Total Incl O&P
0010	**PANEL TYPE AIR FILTERS**										
2950	Mechanical media filtration units										
3000	High efficiency type, with frame, non-supported	G				MCFM	35			35	38.50
3100	Supported type	G				"	55			55	60.50
5500	Throwaway glass or paper media type, 12" x 36" x 1"					Ea.	2.39			2.39	2.63

23 41 16 – Renewable-Media Air Filters

23 41 16.10 Disposable Media Air Filters

		Unit	Material	Labor	Equipment	Total	Total Incl O&P
0010	**DISPOSABLE MEDIA AIR FILTERS**						
5000	Renewable disposable roll	C.S.F.	5.25			5.25	5.75

23 41 19 – Washable Air Filters

23 41 19.10 Permanent Air Filters

			Unit	Material	Labor	Equipment	Total	Total Incl O&P
0010	**PERMANENT AIR FILTERS**							
4500	Permanent washable	G	MCFM	25			25	27.50

For customer support on your Building Construction Costs with RSMeans data, call 800.448.8182.

547

23 41 Particulate Air Filtration

23 41 23 – Extended Surface Filters

23 41 23.10 Expanded Surface Filters	Crew	Daily Output	Labor-Hours	Unit	Material	2018 Bare Costs Labor	Equipment	Total	Total Incl O&P
0010 **EXPANDED SURFACE FILTERS**									
4000　Medium efficiency, extended surface	G			MCFM	6			6	6.60

23 42 Gas-Phase Air Filtration

23 42 13 – Activated-Carbon Air Filtration

23 42 13.10 Charcoal Type Air Filtration

	Crew	Daily Output	Labor-Hours	Unit	Material	Labor	Equipment	Total	Total Incl O&P
0010 **CHARCOAL TYPE AIR FILTRATION**									
0050　Activated charcoal type, full flow				MCFM	650			650	715
0060　Full flow, impregnated media 12" deep					225			225	248
0070　HEPA filter & frame for field erection					450			450	495
0080　HEPA filter-diffuser, ceiling install.					350			350	385

23 43 Electronic Air Cleaners

23 43 13 – Washable Electronic Air Cleaners

23 43 13.10 Electronic Air Cleaners

	Crew	Daily Output	Labor-Hours	Unit	Material	Labor	Equipment	Total	Total Incl O&P
0010 **ELECTRONIC AIR CLEANERS**									
2000　Electronic air cleaner, duct mounted									
2150　　1,000 CFM	1 Shee	4	2	Ea.	430	120		550	660
2200　　1,200 CFM		3.80	2.105		505	126		631	745
2250　　1,400 CFM		3.60	2.222		530	133		663	790

23 51 Breechings, Chimneys, and Stacks

23 51 13 – Draft Control Devices

23 51 13.13 Draft-Induction Fans

	Crew	Daily Output	Labor-Hours	Unit	Material	Labor	Equipment	Total	Total Incl O&P
0010 **DRAFT-INDUCTION FANS**									
1000　Breeching installation									
1800　　Hot gas, 600°F, variable pitch pulley and motor									
1860　　　8" diam. inlet, 1/4 HP, 1 phase, 1,120 CFM	Q-9	4	4	Ea.	2,150	215		2,365	2,675
1900　　　12" diam. inlet, 3/4 HP, 3 phase, 2,960 CFM		3	5.333		2,925	287		3,212	3,650
1980　　　24" diam. inlet, 7-1/2 HP, 3 phase, 17,760 CFM		.80	20		8,375	1,075		9,450	10,900
2300　　For multi-blade damper at fan inlet, add					20%				

23 51 23 – Gas Vents

23 51 23.10 Gas Chimney Vents

	Crew	Daily Output	Labor-Hours	Unit	Material	Labor	Equipment	Total	Total Incl O&P
0010 **GAS CHIMNEY VENTS**, Prefab metal, UL listed									
0020　Gas, double wall, galvanized steel									
0080　　3" diameter	Q-9	72	.222	V.L.F.	6.60	11.95		18.55	25.50
0100　　4" diameter		68	.235		8.15	12.65		20.80	28.50
0120　　5" diameter		64	.250		9.25	13.45		22.70	30.50
0140　　6" diameter		60	.267		11.05	14.35		25.40	34
0160　　7" diameter		56	.286		21	15.40		36.40	46.50
0180　　8" diameter		52	.308		22.50	16.55		39.05	50.50
0200　　10" diameter		48	.333		43	17.95		60.95	75
0220　　12" diameter		44	.364		51.50	19.55		71.05	86.50
0260　　16" diameter		40	.400		123	21.50		144.50	168
0300　　20" diameter	Q-10	36	.667		179	37		216	253

For customer support on your Building Construction Costs with RSMeans data, call 800.448.8182.

23 51 Breechings, Chimneys, and Stacks

23 51 26 – All-Fuel Vent Chimneys

23 51 26.30 All-Fuel Vent Chimneys, Double Wall, St. Stl.	Crew	Daily Output	Labor-Hours	Unit	Material	2018 Bare Costs Labor	Equipment	Total	Total Incl O&P
0010 **ALL-FUEL VENT CHIMNEYS, DOUBLE WALL, STAINLESS STEEL**									
7780 All fuel, pressure tight, double wall, 4" insulation, UL listed, 1,400°F.									
7790 304 stainless steel liner, aluminized steel outer jacket									
7800 6" diameter	Q-9	60	.267	V.L.F.	65	14.35		79.35	93.50
7804 8" diameter		52	.308		75	16.55		91.55	108
7806 10" diameter		48	.333		83.50	17.95		101.45	119
7808 12" diameter		44	.364		95.50	19.55		115.05	135
7810 14" diameter		42	.381		107	20.50		127.50	150
7880 For 316 stainless steel liner, add				L.F.	30%				

23 52 Heating Boilers

23 52 13 – Electric Boilers

23 52 13.10 Electric Boilers, ASME

	Crew	Daily Output	Labor-Hours	Unit	Material	2018 Bare Costs Labor	Equipment	Total	Total Incl O&P
0010 **ELECTRIC BOILERS, ASME**, Standard controls and trim									
1000 Steam, 6 KW, 20.5 MBH	Q-19	1.20	20	Ea.	4,250	1,150		5,400	6,400
1160 60 KW, 205 MBH		1	24		6,925	1,375		8,300	9,650
1220 112 KW, 382 MBH		.75	32		9,475	1,825		11,300	13,200
1280 222 KW, 758 MBH		.55	43.636		23,800	2,500		26,300	30,000
1380 518 KW, 1,768 MBH	Q-21	.36	88.889		32,600	5,225		37,825	43,800
1480 814 KW, 2,778 MBH		.25	128		40,600	7,500		48,100	56,000
1600 2,340 KW, 7,984 MBH		.16	200		86,500	11,700		98,200	112,500
2000 Hot water, 7.5 KW, 25.6 MBH	Q-19	1.30	18.462		5,125	1,050		6,175	7,225
2100 90 KW, 307 MBH		1.10	21.818		6,100	1,250		7,350	8,600
2220 296 KW, 1,010 MBH		.55	43.636		15,700	2,500		18,200	21,000
2500 1,036 KW, 3,536 MBH	Q-21	.34	94.118		35,200	5,525		40,725	47,100
2680 2,400 KW, 8,191 MBH		.25	128		68,000	7,500		75,500	86,000
2820 3,600 KW, 12,283 MBH		.16	200		94,000	11,700		105,700	121,000

23 52 23 – Cast-Iron Boilers

23 52 23.20 Gas-Fired Boilers

	Crew	Daily Output	Labor-Hours	Unit	Material	2018 Bare Costs Labor	Equipment	Total	Total Incl O&P
0010 **GAS-FIRED BOILERS**, Natural or propane, standard controls, packaged									
1000 Cast iron, with insulated jacket									
2000 Steam, gross output, 81 MBH	Q-7	1.40	22.857	Ea.	2,300	1,375		3,675	4,625
2080 203 MBH		.90	35.556		3,700	2,125		5,825	7,275
2180 400 MBH		.56	56.838		5,675	3,400		9,075	11,400
2240 765 MBH		.43	74.419		11,800	4,475		16,275	19,700
2320 1,875 MBH		.30	107		24,200	6,400		30,600	36,300
2440 4,720 MBH		.15	208		65,500	12,500		78,000	91,000
2480 6,100 MBH		.13	246		85,000	14,800		99,800	115,500
2540 6,970 MBH		.10	320		95,000	19,200		114,200	133,500
3000 Hot water, gross output, 80 MBH		1.46	21.918		1,925	1,325		3,250	4,100
3140 320 MBH		.80	40		4,525	2,400		6,925	8,600
3260 1,088 MBH		.40	80		13,000	4,800		17,800	21,500
3360 2,856 MBH		.20	160		29,700	9,600		39,300	47,100
3380 3,264 MBH		.18	180		31,500	10,800		42,300	51,000
3480 6,100 MBH		.13	250		112,500	15,000		127,500	146,000
3540 6,970 MBH		.09	360		126,500	21,600		148,100	172,000
7000 For tankless water heater, add					10%				

For customer support on your Building Construction Costs with RSMeans data, call 800.448.8182.

549

23 52 Heating Boilers

23 52 23 – Cast-Iron Boilers

23 52 23.30 Gas/Oil Fired Boilers

23 52 23.30 Gas/Oil Fired Boilers	Crew	Daily Output	Labor-Hours	Unit	Material	2018 Bare Costs Labor	Equipment	Total	Total Incl O&P
0010 **GAS/OIL FIRED BOILERS**, Combination with burners and controls, packaged									
1000 Cast iron with insulated jacket									
2000 Steam, gross output, 720 MBH	Q-7	.43	74.074	Ea.	14,200	4,450		18,650	22,300
2080 1,600 MBH		.30	107		20,100	6,425		26,525	31,800
2140 2,700 MBH		.19	166		27,000	9,950		36,950	44,700
2280 5,520 MBH		.14	235		87,000	14,100		101,100	117,000
2340 6,390 MBH		.11	296		93,000	17,800		110,800	129,500
2380 6,970 MBH	↓	.09	372	↓	98,500	22,300		120,800	142,000
2900 Hot water, gross output									
2910 200 MBH	Q-6	.62	39.024	Ea.	10,200	2,300		12,500	14,800
2920 300 MBH		.49	49.080		10,200	2,875		13,075	15,700
2930 400 MBH		.41	57.971		12,000	3,400		15,400	18,300
2940 500 MBH	↓	.36	67.039		12,900	3,950		16,850	20,200
3000 584 MBH	Q-7	.44	72.072		14,300	4,325		18,625	22,200
3060 1,460 MBH		.28	113		35,400	6,775		42,175	49,100
3160 4,088 MBH		.16	195		60,000	11,700		71,700	83,500
3300 13,500 MBH, 403.3 BHP	↓	.04	727	↓	185,000	43,600		228,600	269,000

23 52 23.40 Oil-Fired Boilers

23 52 23.40 Oil-Fired Boilers	Crew	Daily Output	Labor-Hours	Unit	Material	Labor	Equipment	Total	Total Incl O&P
0010 **OIL-FIRED BOILERS**, Standard controls, flame retention burner, packaged									
1000 Cast iron, with insulated flush jacket									
2000 Steam, gross output, 109 MBH	Q-7	1.20	26.667	Ea.	2,225	1,600		3,825	4,850
2060 207 MBH		.90	35.556		3,025	2,125		5,150	6,525
2180 1,084 MBH		.38	85.106		10,500	5,100		15,600	19,300
2280 3,000 MBH		.19	170		22,800	10,200		33,000	40,500
2380 5,520 MBH		.14	235		76,000	14,100		90,100	105,000
2460 6,970 MBH	↓	.09	364	↓	97,500	21,800		119,300	140,000
3000 Hot water, same price as steam									
4000 For tankless coil in smaller sizes, add				Ea.	15%				

23 52 26 – Steel Boilers

23 52 26.40 Oil-Fired Boilers

23 52 26.40 Oil-Fired Boilers	Crew	Daily Output	Labor-Hours	Unit	Material	Labor	Equipment	Total	Total Incl O&P
0010 **OIL-FIRED BOILERS**, Standard controls, flame retention burner									
5000 Steel, with insulated flush jacket									
7000 Hot water, gross output, 103 MBH	Q-6	1.60	15	Ea.	1,875	880		2,755	3,400
7120 420 MBH		.70	34.483		5,775	2,025		7,800	9,400
7320 3,150 MBH	↓	.13	185		39,100	10,900		50,000	59,500
7340 For tankless coil in steam or hot water, add				↓	7%				

23 52 28 – Swimming Pool Boilers

23 52 28.10 Swimming Pool Heaters

23 52 28.10 Swimming Pool Heaters	Crew	Daily Output	Labor-Hours	Unit	Material	Labor	Equipment	Total	Total Incl O&P
0010 **SWIMMING POOL HEATERS**, Not including wiring, external									
0020 piping, base or pad									
0160 Gas fired, input, 155 MBH	Q-6	1.50	16	Ea.	1,975	940		2,915	3,600
0200 199 MBH		1	24		2,100	1,400		3,500	4,425
0280 500 MBH		.40	60		8,725	3,525		12,250	14,900
0400 1,800 MBH	↓	.14	171		21,100	10,100		31,200	38,400
2000 Electric, 12 KW, 4,800 gallon pool	Q-19	3	8		2,100	460		2,560	3,000
2020 15 KW, 7,200 gallon pool		2.80	8.571		2,150	490		2,640	3,075
2040 24 KW, 9,600 gallon pool		2.40	10		2,475	570		3,045	3,575
2100 57 KW, 24,000 gallon pool	↓	1.20	20	↓	3,650	1,150		4,800	5,725

For customer support on your Building Construction Costs with RSMeans data, call 800.448.8182.

23 54 Furnaces

23 54 13 – Electric-Resistance Furnaces

23 54 13.10 Electric Furnaces	Crew	Daily Output	Labor-Hours	Unit	Material	2018 Bare Costs Labor	Equipment	Total	Total Incl O&P
0010 **ELECTRIC FURNACES**, Hot air, blowers, std. controls									
0011 not including gas, oil or flue piping									
1000 Electric, UL listed									
1070 10.2 MBH	Q-20	4.40	4.545	Ea.	335	249		584	750
1080 17.1 MBH		4.60	4.348		415	238		653	815
1100 34.1 MBH		4.40	4.545		520	249		769	950

23 54 16 – Fuel-Fired Furnaces

23 54 16.13 Gas-Fired Furnaces

	Crew	Daily Output	Labor-Hours	Unit	Material	Labor	Equipment	Total	Total Incl O&P
0010 **GAS-FIRED FURNACES**									
3000 Gas, AGA certified, upflow, direct drive models									
3020 45 MBH input	Q-9	4	4	Ea.	645	215		860	1,050
3040 60 MBH input		3.80	4.211		640	227		867	1,050
3060 75 MBH input		3.60	4.444		710	239		949	1,150
3100 100 MBH input		3.20	5		740	269		1,009	1,225

23 54 16.16 Oil-Fired Furnaces

	Crew	Daily Output	Labor-Hours	Unit	Material	Labor	Equipment	Total	Total Incl O&P
0010 **OIL-FIRED FURNACES**									
6000 Oil, UL listed, atomizing gun type burner									
6020 56 MBH output	Q-9	3.60	4.444	Ea.	2,825	239		3,064	3,475
6030 84 MBH output		3.50	4.571		2,850	246		3,096	3,525
6040 95 MBH output		3.40	4.706		2,875	253		3,128	3,525
6060 134 MBH output		3.20	5		2,825	269		3,094	3,500
6080 151 MBH output		3	5.333		3,100	287		3,387	3,875

23 55 Fuel-Fired Heaters

23 55 13 – Fuel-Fired Duct Heaters

23 55 13.16 Gas-Fired Duct Heaters

	Crew	Daily Output	Labor-Hours	Unit	Material	Labor	Equipment	Total	Total Incl O&P
0010 **GAS-FIRED DUCT HEATERS**, Includes burner, controls, stainless steel									
0020 heat exchanger. Gas fired, electric ignition									
0030 Indoor installation									
0100 120 MBH output	Q-5	4	4	Ea.	3,325	227		3,552	4,025
0130 200 MBH output		2.70	5.926		4,225	335		4,560	5,150
0140 240 MBH output		2.30	6.957		4,425	395		4,820	5,475
0180 320 MBH output		1.60	10		5,175	565		5,740	6,550
0300 For powered venter and adapter, add					540			540	595
0502 For required flue pipe, see Section 23 51 23.10									
1000 Outdoor installation, with power venter									
1020 75 MBH output	Q-5	4	4	Ea.	3,700	227		3,927	4,425
1060 120 MBH output		4	4		4,075	227		4,302	4,850
1100 187 MBH output		3	5.333		5,025	300		5,325	6,000
1140 300 MBH output		1.80	8.889		6,475	505		6,980	7,850
1180 450 MBH output		1.40	11.429		9,450	650		10,100	11,400

23 55 23 – Gas-Fired Radiant Heaters

23 55 23.10 Infrared Type Heating Units

	Crew	Daily Output	Labor-Hours	Unit	Material	Labor	Equipment	Total	Total Incl O&P
0010 **INFRARED TYPE HEATING UNITS**									
0020 Gas fired, unvented, electric ignition, 100% shutoff.									
0030 Piping and wiring not included									
0120 45 MBH	Q-5	5	3.200	Ea.	955	181		1,136	1,325
0160 60 MBH		4	4		955	227		1,182	1,400
0240 120 MBH		2	8		1,575	455		2,030	2,400
1000 Gas fired, vented, electric ignition, tubular									

For customer support on your Building Construction Costs with RSMeans data, call 800.448.8182.

551

23 55 Fuel-Fired Heaters

23 55 23 – Gas-Fired Radiant Heaters

23 55 23.10 Infrared Type Heating Units	Crew	Daily Output	Labor-Hours	Unit	Material	2018 Bare Costs Labor	Equipment	Total	Total Incl O&P
1020 Piping and wiring not included, 20' to 80' lengths									
1030 Single stage, input, 60 MBH	Q-6	4.50	5.333	Ea.	1,450	315		1,765	2,075
1040 80 MBH		3.90	6.154		1,450	360		1,810	2,150
1050 100 MBH		3.40	7.059		1,450	415		1,865	2,225
1060 125 MBH		2.90	8.276		1,450	485		1,935	2,325
1070 150 MBH		2.70	8.889		1,450	525		1,975	2,375
1080 170 MBH		2.50	9.600		1,450	565		2,015	2,450
1090 200 MBH		2.20	10.909		1,650	640		2,290	2,800
1100 Note: Final pricing may vary due to									
1110 tube length and configuration package selected									
1130 Two stage, input, 60 MBH high, 45 MBH low	Q-6	4.50	5.333	Ea.	1,775	315		2,090	2,425
1140 80 MBH high, 60 MBH low		3.90	6.154		1,775	360		2,135	2,500
1150 100 MBH high, 65 MBH low		3.40	7.059		1,775	415		2,190	2,575
1160 125 MBH high, 95 MBH low		2.90	8.276		1,775	485		2,260	2,675
1170 150 MBH high, 100 MBH low		2.70	8.889		1,775	525		2,300	2,725
1180 170 MBH high, 125 MBH low		2.50	9.600		2,000	565		2,565	3,050
1190 200 MBH high, 150 MBH low		2.20	10.909		2,000	640		2,640	3,175
1220 Note: Final pricing may vary due to									
1230 tube length and configuration package selected									

23 55 33 – Fuel-Fired Unit Heaters

23 55 33.13 Oil-Fired Unit Heaters

	Crew	Daily Output	Labor-Hours	Unit	Material	Labor	Equipment	Total	Total Incl O&P
0010 **OIL-FIRED UNIT HEATERS**, Cabinet, grilles, fan, ctrl., burner, no piping									
6000 Oil fired, suspension mounted, 94 MBH output	Q-5	4	4	Ea.	5,100	227		5,327	5,975
6040 140 MBH output		3	5.333		5,375	300		5,675	6,350
6060 184 MBH output		3	5.333		5,700	300		6,000	6,725

23 55 33.16 Gas-Fired Unit Heaters

	Crew	Daily Output	Labor-Hours	Unit	Material	Labor	Equipment	Total	Total Incl O&P
0010 **GAS-FIRED UNIT HEATERS**, Cabinet, grilles, fan, ctrls., burner, no piping									
0022 thermostat, no piping. For flue see Section 23 51 23.10									
1000 Gas fired, floor mounted									
1100 60 MBH output	Q-5	10	1.600	Ea.	725	90.50		815.50	935
1140 100 MBH output		8	2		800	113		913	1,050
1180 180 MBH output		6	2.667		1,150	151		1,301	1,475
2000 Suspension mounted, propeller fan, 20 MBH output		8.50	1.882		1,575	107		1,682	1,875
2040 60 MBH output		7	2.286		1,800	130		1,930	2,200
2060 80 MBH output		6	2.667		1,975	151		2,126	2,400
2100 130 MBH output		5	3.200		2,350	181		2,531	2,850
2240 320 MBH output		2	8		4,350	455		4,805	5,475
2500 For powered venter and adapter, add					405			405	445
5000 Wall furnace, 17.5 MBH output	Q-5	6	2.667		635	151		786	930
5020 24 MBH output		5	3.200		635	181		816	975
5040 35 MBH output		4	4		980	227		1,207	1,425

For customer support on your Building Construction Costs with RSMeans data, call 800.448.8182.

23 56 Solar Energy Heating Equipment

23 56 16 – Packaged Solar Heating Equipment

23 56 16.40 Solar Heating Systems

		Crew	Daily Output	Labor-Hours	Unit	Material	2018 Bare Costs Labor	2018 Bare Costs Equipment	Total	Total Incl O&P
0010	**SOLAR HEATING SYSTEMS** R235616-60									
0020	System/package prices, not including connecting									
0030	pipe, insulation, or special heating/plumbing fixtures									
0500	Hot water, standard package, low temperature									
0540	1 collector, circulator, fittings, 65 gal. tank	G	Q-1	.50	32	Ea.	3,950	1,800		5,750	7,050
0580	2 collectors, circulator, fittings, 120 gal. tank	G		.40	40		5,400	2,225		7,625	9,325
0620	3 collectors, circulator, fittings, 120 gal. tank	G		.34	47.059		7,375	2,625		10,000	12,100
0700	Medium temperature package									
0720	1 collector, circulator, fittings, 80 gal. tank	G	Q-1	.50	32	Ea.	5,350	1,800		7,150	8,600
0740	2 collectors, circulator, fittings, 120 gal. tank	G		.40	40		6,900	2,225		9,125	11,000
0780	3 collectors, circulator, fittings, 120 gal. tank	G		.30	53.333		7,825	2,975		10,800	13,100
0980	For each additional 120 gal. tank, add	G					1,850			1,850	2,050

23 56 19 – Solar Heating Components

23 56 19.50 Solar Heating Ancillary

		Crew	Daily Output	Labor-Hours	Unit	Material	2018 Bare Costs Labor	2018 Bare Costs Equipment	Total	Total Incl O&P
0010	**SOLAR HEATING ANCILLARY**									
2300	Circulators, air									
2310	Blowers									
2400	Reversible fan, 20" diameter, 2 speed	G	Q-9	18	.889	Ea.	114	48		162	199
2870	1/12 HP, 30 GPM	G	Q-1	10	1.600	"	350	89.50		439.50	520
3000	Collector panels, air with aluminum absorber plate									
3010	Wall or roof mount									
3040	Flat black, plastic glazing									
3080	4' x 8'	G	Q-9	6	2.667	Ea.	670	144		814	960
3200	Flush roof mount, 10' to 16' x 22" wide	G	"	96	.167	L.F.	142	8.95		150.95	170
3300	Collector panels, liquid with copper absorber plate									
3330	Alum. frame, 4' x 8', 5/32" single glazing	G	Q-1	9.50	1.684	Ea.	1,025	94		1,119	1,275
3390	Alum. frame, 4' x 10', 5/32" single glazing	G		6	2.667		1,175	149		1,324	1,500
3450	Flat black, alum. frame, 3.5' x 7.5'	G		9	1.778		880	99.50		979.50	1,125
3500	4' x 8'	G		5.50	2.909		1,050	163		1,213	1,400
3520	4' x 10'	G		10	1.600		1,250	89.50		1,339.50	1,500
3540	4' x 12.5'	G		5	3.200		1,275	179		1,454	1,675
3600	Liquid, full wetted, plastic, alum. frame, 4' x 10'	G		5	3.200		330	179		509	630
3650	Collector panel mounting, flat roof or ground rack	G		7	2.286		248	128		376	465
3670	Roof clamps	G		70	.229	Set	2.87	12.80		15.67	22.50
3700	Roof strap, teflon	G	1 Plum	205	.039	L.F.	25	2.43		27.43	31
3900	Differential controller with two sensors									
3930	Thermostat, hard wired	G	1 Plum	8	1	Ea.	103	62		165	207
4100	Five station with digital read-out	G	"	3	2.667	"	267	166		433	545
4300	Heat exchanger									
4580	Fluid to fluid package includes two circulating pumps									
4590	expansion tank, check valve, relief valve									
4600	controller, high temperature cutoff and sensors	G	Q-1	2.50	6.400	Ea.	810	360		1,170	1,425
4650	Heat transfer fluid									
4700	Propylene glycol, inhibited anti-freeze	G	1 Plum	28	.286	Gal.	23.50	17.75		41.25	52.50
8250	Water storage tank with heat exchanger and electric element									
8300	80 gal. with 2" x 2 lb. density insulation	G	1 Plum	1.60	5	Ea.	1,675	310		1,985	2,325
8380	120 gal. with 2" x 2 lb. density insulation	G		1.40	5.714		1,925	355		2,280	2,650
8400	120 gal. with 2" x 2 lb. density insul., 40 S.F. heat coil	G		1.40	5.714		2,425	355		2,780	3,200

For customer support on your Building Construction Costs with RSMeans data, call 800.448.8182.

553

23 57 Heat Exchangers for HVAC

23 57 16 – Steam-to-Water Heat Exchangers

23 57 16.10 Shell/Tube Type Steam-to-Water Heat Exch.

		Crew	Daily Output	Labor-Hours	Unit	Material	2018 Bare Costs Labor	Equipment	Total	Total Incl O&P
0010	**SHELL AND TUBE TYPE STEAM-TO-WATER HEAT EXCHANGERS**									
0016	Shell & tube type, 2 or 4 pass, 3/4" OD copper tubes,									
0020	C.I. heads, C.I. tube sheet, steel shell									
0100	Hot water 40°F to 180°F, by steam at 10 psi									
0120	8 GPM	Q-5	6	2.667	Ea.	2,375	151		2,526	2,850
0140	10 GPM		5	3.200		3,575	181		3,756	4,225
0160	40 GPM		4	4		5,550	227		5,777	6,450
0180	64 GPM		2	8		8,475	455		8,930	10,000
0200	96 GPM	↓	1	16		11,400	905		12,305	13,900
0220	120 GPM	Q-6	1.50	16	↓	14,900	940		15,840	17,800

23 57 19 – Liquid-to-Liquid Heat Exchangers

23 57 19.13 Plate-Type, Liquid-to-Liquid Heat Exchangers

		Crew	Daily Output	Labor-Hours	Unit	Material	2018 Bare Costs Labor	Equipment	Total	Total Incl O&P
0010	**PLATE-TYPE, LIQUID-TO-LIQUID HEAT EXCHANGERS**									
3000	Plate type,									
3100	400 GPM	Q-6	.80	30	Ea.	42,500	1,775		44,275	49,500
3120	800 GPM	"	.50	48		73,000	2,825		75,825	85,000
3140	1,200 GPM	Q-7	.34	94.118		109,000	5,650		114,650	128,500
3160	1,800 GPM	"	.24	133	↓	144,500	8,000		152,500	171,000

23 57 19.16 Shell-Type, Liquid-to-Liquid Heat Exchangers

		Crew	Daily Output	Labor-Hours	Unit	Material	2018 Bare Costs Labor	Equipment	Total	Total Incl O&P
0010	**SHELL-TYPE, LIQUID-TO-LIQUID HEAT EXCHANGERS**									
1000	Hot water 40°F to 140°F, by water at 200°F									
1020	7 GPM	Q-5	6	2.667	Ea.	2,925	151		3,076	3,425
1040	16 GPM		5	3.200		4,125	181		4,306	4,825
1060	34 GPM		4	4		6,250	227		6,477	7,225
1100	74 GPM	↓	1.50	10.667	↓	11,300	605		11,905	13,400

23 62 Packaged Compressor and Condenser Units

23 62 13 – Packaged Air-Cooled Refrigerant Compressor and Condenser Units

23 62 13.10 Packaged Air-Cooled Refrig. Condensing Units

		Crew	Daily Output	Labor-Hours	Unit	Material	2018 Bare Costs Labor	Equipment	Total	Total Incl O&P
0010	**PACKAGED AIR-COOLED REFRIGERANT CONDENSING UNITS**									
0020	Condensing unit									
0030	Air cooled, compressor, standard controls									
0050	1.5 ton	Q-5	2.50	6.400	Ea.	1,025	365		1,390	1,675
0500	5 ton		.60	26.667		2,150	1,500		3,650	4,650
0600	10 ton	↓	.50	32		3,700	1,825		5,525	6,800
0700	20 ton	Q-6	.40	60	↓	8,275	3,525		11,800	14,400

23 63 Refrigerant Condensers

23 63 13 – Air-Cooled Refrigerant Condensers

23 63 13.10 Air-Cooled Refrig. Condensers

		Crew	Daily Output	Labor-Hours	Unit	Material	2018 Bare Costs Labor	Equipment	Total	Total Incl O&P
0010	**AIR-COOLED REFRIG. CONDENSERS**									
0080	Air cooled, belt drive, propeller fan									
0240	50 ton	Q-6	.69	34.985	Ea.	10,800	2,050		12,850	14,900
0280	59 ton		.58	41.308		12,900	2,425		15,325	17,900
0320	73 ton		.47	51.173		16,900	3,000		19,900	23,100
0360	86 ton		.40	60.302		19,600	3,550		23,150	26,900
0380	88 ton	↓	.39	61.697	↓	21,000	3,625		24,625	28,600
1550	Air cooled, direct drive, propeller fan									

For customer support on your Building Construction Costs with RSMeans data, call 800.448.8182.

23 63 Refrigerant Condensers

23 63 13 – Air-Cooled Refrigerant Condensers

23 63 13.10 Air-Cooled Refrig. Condensers

	23 63 13.10 Air-Cooled Refrig. Condensers	Crew	Daily Output	Labor-Hours	Unit	Material	2018 Bare Costs Labor	Equipment	Total	Total Incl O&P
1590	1 ton	Q-5	3.80	4.211	Ea.	1,650	239		1,889	2,175
1600	1-1/2 ton		3.60	4.444		1,975	252		2,227	2,550
1620	2 ton		3.20	5		2,175	284		2,459	2,825
1640	5 ton		2	8		5,025	455		5,480	6,200
1660	10 ton		1.40	11.429		6,850	650		7,500	8,525
1690	16 ton		1.10	14.545		10,200	825		11,025	12,500
1720	26 ton		.84	19.002		12,500	1,075		13,575	15,400
1760	41 ton	Q-6	.77	31.008		17,800	1,825		19,625	22,400
1800	63 ton	"	.55	44.037		28,000	2,600		30,600	34,700

23 64 Packaged Water Chillers

23 64 13 – Absorption Water Chillers

23 64 13.16 Indirect-Fired Absorption Water Chillers

		Crew	Daily Output	Labor-Hours	Unit	Material	2018 Bare Costs Labor	Equipment	Total	Total Incl O&P
0010	**INDIRECT-FIRED ABSORPTION WATER CHILLERS**									
0020	Steam or hot water, water cooled									
0050	100 ton	Q-7	.13	240	Ea.	131,500	14,400		145,900	166,000
0400	420 ton	"	.10	323	"	394,000	19,400		413,400	462,500

23 64 16 – Centrifugal Water Chillers

23 64 16.10 Centrifugal Type Water Chillers

		Crew	Daily Output	Labor-Hours	Unit	Material	2018 Bare Costs Labor	Equipment	Total	Total Incl O&P
0010	**CENTRIFUGAL TYPE WATER CHILLERS**, With standard controls									
0020	Centrifugal liquid chiller, water cooled									
0030	not including water tower									
0100	2,000 ton (twin 1,000 ton units)	Q-7	.07	478	Ea.	735,000	28,600		763,600	851,500

23 64 19 – Reciprocating Water Chillers

23 64 19.10 Reciprocating Type Water Chillers

		Crew	Daily Output	Labor-Hours	Unit	Material	2018 Bare Costs Labor	Equipment	Total	Total Incl O&P
0010	**RECIPROCATING TYPE WATER CHILLERS**, With standard controls									
0494	Water chillers, integral air cooled condenser									
0980	Water cooled, multiple compressor, semi-hermetic, tower not incl.									
1090	45 ton cooling	Q-7	.29	111	Ea.	22,700	6,675		29,375	35,000
1160	100 ton cooling		.18	180		50,500	10,800		61,300	71,500
1180	125 ton cooling		.16	196		53,000	11,800		64,800	76,000
1200	145 ton cooling		.16	203		59,500	12,100		71,600	84,000
1451	Water cooled, dual compressors, semi-hermetic, tower not incl.									

23 64 23 – Scroll Water Chillers

23 64 23.10 Scroll Water Chillers

		Crew	Daily Output	Labor-Hours	Unit	Material	2018 Bare Costs Labor	Equipment	Total	Total Incl O&P
0010	**SCROLL WATER CHILLERS**, With standard controls									
0480	Packaged w/integral air cooled condenser									
0482	10 ton cooling	Q-7	.34	94.118	Ea.	15,000	5,650		20,650	24,900
0490	15 ton cooling		.37	86.486		18,000	5,175		23,175	27,600
0500	20 ton cooling		.34	94.118		21,800	5,650		27,450	32,500
0520	40 ton cooling		.30	108		32,800	6,475		39,275	45,900
0540	70 ton cooling		.27	119		51,000	7,100		58,100	66,500
0590	100 ton cooling		.25	128		68,000	7,675		75,675	86,000
0600	110 ton cooling		.24	133		73,000	8,000		81,000	92,000
0630	130 ton cooling		.24	133		82,000	8,000		90,000	102,500
0670	170 ton cooling		.23	139		103,500	8,350		111,850	126,000
0671	210 ton cooling		.22	145		121,000	8,725		129,725	146,000
0675	250 ton cooling		.21	152		140,000	9,150		149,150	168,000
0676	275 ton cooling		.21	152		151,500	9,150		160,650	181,000

For customer support on your Building Construction Costs with RSMeans data, call 800.448.8182.

555

23 64 Packaged Water Chillers

23 64 23 – Scroll Water Chillers

23 64 23.10 Scroll Water Chillers

		Crew	Daily Output	Labor-Hours	Unit	Material	2018 Bare Costs Labor	Equipment	Total	Total Incl O&P
0677	300 ton cooling	Q-7	.20	160	Ea.	163,500	9,600		173,100	194,500
0678	330 ton cooling		.20	160		184,500	9,600		194,100	217,500
0679	390 ton cooling	↓	.19	168	↓	227,000	10,100		237,100	264,500
0680	Scroll water cooled, single compressor, hermetic, tower not incl.									
0700	2 ton cooling	Q-5	.57	28.070	Ea.	3,525	1,600		5,125	6,275
0710	5 ton cooling		.57	28.070		4,225	1,600		5,825	7,025
0740	8 ton cooling	↓	.31	52.117		5,800	2,950		8,750	10,900
0760	10 ton cooling	Q-6	.36	67.039		6,700	3,950		10,650	13,300
0800	20 ton cooling	Q-7	.38	83.990		11,900	5,050		16,950	20,700
0820	30 ton cooling	"	.33	96.096	↓	13,400	5,775		19,175	23,400
0900	Scroll water cooled, multiple compressor, hermetic, tower not incl.									
0930	30 ton cooling	Q-7	.36	88.889	Ea.	22,800	5,325		28,125	33,100
0935	35 ton cooling		.31	103		28,200	6,200		34,400	40,400
0940	40 ton cooling		.30	107		27,100	6,400		33,500	39,400
0950	50 ton cooling		.28	114		30,200	6,850		37,050	43,500
0960	60 ton cooling		.25	128		32,600	7,675		40,275	47,500
0962	75 ton cooling		.23	139		53,500	8,350		61,850	71,000
0964	85 ton cooling		.23	139		55,000	8,350		63,350	73,000
0970	100 ton cooling		.18	178		60,500	10,700		71,200	83,000
0980	115 ton cooling		.17	188		62,500	11,300		73,800	86,000
0990	125 ton cooling		.16	200		63,000	12,000		75,000	87,500
0995	145 ton cooling	↓	.16	200	↓	70,000	12,000		82,000	95,000

23 64 26 – Rotary-Screw Water Chillers

23 64 26.10 Rotary-Screw Type Water Chillers

		Crew	Daily Output	Labor-Hours	Unit	Material	2018 Bare Costs Labor	Equipment	Total	Total Incl O&P
0010	**ROTARY-SCREW TYPE WATER CHILLERS**, With standard controls									
0110	Screw, liquid chiller, air cooled, insulated evaporator									
0120	130 ton	Q-7	.14	229	Ea.	94,500	13,700		108,200	124,500
0124	160 ton		.13	246		120,500	14,800		135,300	154,500
0128	180 ton		.13	250		135,500	15,000		150,500	172,000
0132	210 ton		.12	258		144,000	15,500		159,500	182,000
0136	270 ton		.12	267		171,000	16,000		187,000	212,000
0140	320 ton	↓	.12	276	↓	214,000	16,500		230,500	260,500
0200	Packaged unit, water cooled, not incl. tower									
0210	80 ton	Q-7	.14	224	Ea.	47,100	13,400		60,500	72,000
0240	200 ton		.13	252		83,000	15,100		98,100	114,000
0270	350 ton	↓	.12	276		140,000	16,500		156,500	179,500
1450	Water cooled, tower not included									
1580	150 ton cooling, screw compressors	Q-7	.13	241	Ea.	69,500	14,400		83,900	98,000
1620	200 ton cooling, screw compressors		.13	250		96,000	15,000		111,000	128,000
1660	291 ton cooling, screw compressors	↓	.12	260	↓	100,000	15,600		115,600	133,500

For customer support on your Building Construction Costs with RSMeans data, call 800.448.8182.

23 65 Cooling Towers

23 65 13 - Forced-Draft Cooling Towers

23 65 13.10 Forced-Draft Type Cooling Towers	Crew	Daily Output	Labor-Hours	Unit	Material	2018 Bare Costs Labor	Equipment	Total	Total Incl O&P
0010 **FORCED-DRAFT TYPE COOLING TOWERS**, Packaged units									
0070 Galvanized steel									
0080 Induced draft, crossflow									
0100 Vertical, belt drive, 61 tons	Q-6	90	.267	TonAC	192	15.70		207.70	235
0150 100 ton		100	.240		185	14.10		199.10	226
0200 115 ton		109	.220		161	12.95		173.95	197
0250 131 ton		120	.200		193	11.75		204.75	231
0260 162 ton		132	.182		156	10.70		166.70	188
1000 For higher capacities, use multiples									
1500 Induced air, double flow									
1900 Vertical, gear drive, 167 ton	Q-6	126	.190	TonAC	152	11.20		163.20	184
2000 297 ton		129	.186		93.50	10.95		104.45	120
2100 582 ton		132	.182		52	10.70		62.70	73
2150 849 ton		142	.169		70	9.95		79.95	92
2200 1,016 ton		150	.160		70.50	9.40		79.90	91.50
3000 For higher capacities, use multiples									
3500 For pumps and piping, add	Q-6	38	.632	TonAC	96	37		133	162
4000 For absorption systems, add				"	75%	75%			
4100 Cooling water chemical feeder	Q-5	3	5.333	Ea.	410	300		710	905
5000 Fiberglass tower on galvanized steel support structure									
5010 Draw thru									
5100 100 ton	Q-6	1.40	17.143	Ea.	13,900	1,000		14,900	16,800
5120 120 ton		1.20	20		16,200	1,175		17,375	19,600
5140 140 ton		1	24		17,500	1,400		18,900	21,300
5160 160 ton		.80	30		19,400	1,775		21,175	24,100
5180 180 ton		.65	36.923		22,100	2,175		24,275	27,600
5251 600 ton		.16	150		75,000	8,825		83,825	96,000
5252 1,000 ton		.10	250		125,000	14,700		139,700	159,500
5300 For stainless steel support structure, add					30%				
5360 For higher capacities, use multiples of each size									
6000 Stainless steel									
6010 Induced draft, crossflow, horizontal, belt drive									
6100 57 ton	Q-6	1.50	16	Ea.	29,600	940		30,540	33,900
6120 91 ton		.99	24.242		35,800	1,425		37,225	41,500
6140 111 ton		.43	55.814		46,800	3,275		50,075	56,500
6160 126 ton		.22	109		48,600	6,425		55,025	63,000

23 73 Indoor Central-Station Air-Handling Units

23 73 13 - Modular Indoor Central-Station Air-Handling Units

23 73 13.10 Air-Handling Units

	Crew	Daily Output	Labor-Hours	Unit	Material	2018 Bare Costs Labor	Equipment	Total	Total Incl O&P
0010 **AIR-HANDLING UNITS**, Built-Up									
0100 With cooling/heating coil section, filters, mixing box									
0880 Single zone, horizontal/vertical									
0890 Constant volume									
0900 1,600 CFM	Q-5	1.20	13.333	Ea.	5,450	755		6,205	7,150
0920 5,000 CFM	Q-6	1.40	17.143		15,300	1,000		16,300	18,300
0940 11,500 CFM		1	24		27,300	1,400		28,700	32,100
0970 22,000 CFM		.60	40		53,500	2,350		55,850	62,500
1000 40,000 CFM		.30	80		98,500	4,700		103,200	115,000

23 73 Indoor Central-Station Air-Handling Units

23 73 39 – Indoor, Direct Gas-Fired Heating and Ventilating Units

23 73 39.10 Make-Up Air Unit	Crew	Daily Output	Labor-Hours	Unit	Material	2018 Bare Costs Labor	2018 Bare Costs Equipment	Total	Total Incl O&P
0010 **MAKE-UP AIR UNIT**									
0020 Indoor suspension, natural/LP gas, direct fired,									
0032 standard control. For flue see Section 23 51 23.10									
0040 70°F temperature rise, MBH is input									
0100 75 MBH input	Q-6	3.60	6.667	Ea.	5,150	390		5,540	6,275
0160 150 MBH input		3	8		6,250	470		6,720	7,575
0220 225 MBH input		2.40	10		7,000	590		7,590	8,575
0300 400 MBH input		1.60	15		16,500	880		17,380	19,500
0600 For discharge louver assembly, add					5%				
0700 For filters, add					10%				
0800 For air shut-off damper section, add					30%				

23 74 Packaged Outdoor HVAC Equipment

23 74 33 – Dedicated Outdoor-Air Units

23 74 33.10 Rooftop Air Conditioners

23 74 33.10 Rooftop Air Conditioners		Crew	Daily Output	Labor-Hours	Unit	Material	2018 Bare Costs Labor	2018 Bare Costs Equipment	Total	Total Incl O&P
0010 **ROOFTOP AIR CONDITIONERS**, Standard controls, curb, economizer										
1000 Single zone, electric cool, gas heat										
1100 3 ton cooling, 60 MBH heating	R236000-20	Q-5	.70	22.857	Ea.	2,800	1,300		4,100	5,025
1120 4 ton cooling, 95 MBH heating			.61	26.403		3,325	1,500		4,825	5,900
1140 5 ton cooling, 112 MBH heating			.56	28.521		4,375	1,625		6,000	7,225
1145 6 ton cooling, 140 MBH heating			.52	30.769		5,000	1,750		6,750	8,100
1150 7.5 ton cooling, 170 MBH heating			.50	32.258		5,825	1,825		7,650	9,150
1156 8.5 ton cooling, 170 MBH heating			.46	34.783		7,075	1,975		9,050	10,800
1160 10 ton cooling, 200 MBH heating		Q-6	.67	35.982		9,150	2,125		11,275	13,300
1170 12.5 ton cooling, 230 MBH heating			.63	37.975		10,400	2,225		12,625	14,800
1190 17.5 ton cooling, 330 MBH heating			.52	45.889		13,900	2,700		16,600	19,400
1200 20 ton cooling, 360 MBH heating		Q-7	.67	47.976		29,500	2,875		32,375	36,800
1210 25 ton cooling, 450 MBH heating			.56	57.554		32,000	3,450		35,450	40,400
1220 30 ton cooling, 540 MBH heating			.47	68.376		35,700	4,100		39,800	45,500
1240 40 ton cooling, 675 MBH heating			.35	91.168		43,900	5,475		49,375	56,500
2000 Multizone, electric cool, gas heat, economizer										
2100 15 ton cooling, 360 MBH heating		Q-7	.61	52.545	Ea.	65,500	3,150		68,650	77,000
2120 20 ton cooling, 360 MBH heating			.53	60.038		70,500	3,600		74,100	83,000
2200 40 ton cooling, 540 MBH heating			.28	114		124,500	6,825		131,325	147,500
2210 50 ton cooling, 540 MBH heating			.23	142		156,000	8,525		164,525	184,500
2220 70 ton cooling, 1,500 MBH heating			.16	199		167,500	11,900		179,400	202,000
2240 80 ton cooling, 1,500 MBH heating			.14	229		191,000	13,700		204,700	231,000
2260 90 ton cooling, 1,500 MBH heating			.13	256		196,500	15,400		211,900	239,000
2280 105 ton cooling, 1,500 MBH heating			.11	291		218,000	17,400		235,400	266,500
2400 For hot water heat coil, deduct						5%				
2500 For steam heat coil, deduct						2%				
2600 For electric heat, deduct						3%	5%			

For customer support on your Building Construction Costs with RSMeans data, call 800.448.8182.

23 81 Decentralized Unitary HVAC Equipment

23 81 13 – Packaged Terminal Air-Conditioners

23 81 13.10 Packaged Cabinet Type Air-Conditioners

		Daily Output	Labor-Hours	Unit	Material	2018 Bare Costs Labor	2018 Bare Costs Equipment	Total	Total Incl O&P
		Crew							
0010	**PACKAGED CABINET TYPE AIR-CONDITIONERS**, Cabinet, wall sleeve,								
0100	louver, electric heat, thermostat, manual changeover, 208 V								
0200	6,000 BTUH cooling, 8,800 BTU heat	Q-5	6	2.667	Ea.	720	151		871	1,025
0220	9,000 BTUH cooling, 13,900 BTU heat		5	3.200		940	181		1,121	1,300
0240	12,000 BTUH cooling, 13,900 BTU heat		4	4		1,450	227		1,677	1,950
0260	15,000 BTUH cooling, 13,900 BTU heat		3	5.333		1,400	300		1,700	2,000
0500	For hot water coil, increase heat by 10%, add					5%	10%			
1000	For steam, increase heat output by 30%, add					8%	10%			

23 81 19 – Self-Contained Air-Conditioners

23 81 19.20 Self-Contained Single Package

		Crew	Daily Output	Labor-Hours	Unit	Material	2018 Bare Costs Labor	2018 Bare Costs Equipment	Total	Total Incl O&P
0010	**SELF-CONTAINED SINGLE PACKAGE**									
0100	Air cooled, for free blow or duct, not incl. remote condenser									
0110	Constant volume									
0200	3 ton cooling	Q-5	1	16	Ea.	3,750	905		4,655	5,500
0220	5 ton cooling	Q-6	1.20	20		4,325	1,175		5,500	6,550
0240	10 ton cooling	Q-7	1	32		7,675	1,925		9,600	11,400
0260	20 ton cooling		.90	35.556		12,600	2,125		14,725	17,000
0280	30 ton cooling		.80	40		25,500	2,400		27,900	31,600
0340	60 ton cooling	Q-8	.40	80		56,500	4,800	148	61,448	70,000
0490	For duct mounting, no price change									
0500	For steam heating coils, add				Ea.	10%	10%			
1000	Water cooled for free blow or duct, not including tower									
1010	Constant volume									
1100	3 ton cooling	Q-6	1	24	Ea.	3,700	1,400		5,100	6,200
1120	5 ton cooling	"	1	24		4,525	1,400		5,925	7,100
1140	10 ton cooling	Q-7	.90	35.556		8,550	2,125		10,675	12,600
1160	20 ton cooling		.80	40		27,700	2,400		30,100	34,100
1180	30 ton cooling		.70	45.714		36,700	2,750		39,450	44,500

23 81 23 – Computer-Room Air-Conditioners

23 81 23.10 Computer Room Units

		Crew	Daily Output	Labor-Hours	Unit	Material	2018 Bare Costs Labor	2018 Bare Costs Equipment	Total	Total Incl O&P
0010	**COMPUTER ROOM UNITS**									
1000	Air cooled, includes remote condenser but not									
1020	interconnecting tubing or refrigerant									
1080	3 ton	Q-5	.50	32	Ea.	24,700	1,825		26,525	29,800
1120	5 ton		.45	35.556		26,300	2,025		28,325	32,000
1160	6 ton		.30	53.333		48,600	3,025		51,625	58,000
1200	8 ton		.27	59.259		49,200	3,350		52,550	59,000
1240	10 ton		.25	64		51,500	3,625		55,125	62,000
1260	12 ton		.24	66.667		53,000	3,775		56,775	64,000
1280	15 ton		.22	72.727		56,500	4,125		60,625	68,000
1290	18 ton		.20	80		65,000	4,525		69,525	78,000
1300	20 ton	Q-6	.26	92.308		67,500	5,425		72,925	82,500
1320	22 ton		.24	100		68,500	5,875		74,375	84,500
1360	30 ton		.21	114		84,500	6,725		91,225	103,000
2200	Chilled water, for connection to									
2220	existing chiller system of adequate capacity									
2260	5 ton	Q-5	.74	21.622	Ea.	18,800	1,225		20,025	22,600

23 81 Decentralized Unitary HVAC Equipment

23 81 43 – Air-Source Unitary Heat Pumps

23 81 43.10 Air-Source Heat Pumps

	Crew	Daily Output	Labor-Hours	Unit	Material	2018 Bare Costs Labor	Equipment	Total	Total Incl O&P
0010 **AIR-SOURCE HEAT PUMPS**, Not including interconnecting tubing									
1000 Air to air, split system, not including curbs, pads, fan coil and ductwork									
1012 Outside condensing unit only, for fan coil see Section 23 82 19.10									
1020 2 ton cooling, 8.5 MBH heat @ 0°F	Q-5	2	8	Ea.	1,650	455		2,105	2,475
1060 5 ton cooling, 27 MBH heat @ 0°F		.50	32		2,675	1,825		4,500	5,650
1080 7.5 ton cooling, 33 MBH heat @ 0°F	↓	.45	35.556		3,875	2,025		5,900	7,275
1100 10 ton cooling, 50 MBH heat @ 0°F	Q-6	.64	37.500		6,275	2,200		8,475	10,300
1120 15 ton cooling, 64 MBH heat @ 0°F		.50	48		8,725	2,825		11,550	13,800
1130 20 ton cooling, 85 MBH heat @ 0°F		.35	68.571		17,900	4,025		21,925	25,800
1140 25 ton cooling, 119 MBH heat @ 0°F	↓	.25	96	↓	21,100	5,650		26,750	31,700
1500 Single package, not including curbs, pads, or plenums									
1520 2 ton cooling, 6.5 MBH heat @ 0°F	Q-5	1.50	10.667	Ea.	3,200	605		3,805	4,425
1580 4 ton cooling, 13 MBH heat @ 0°F		.96	16.667		4,300	945		5,245	6,150
1640 7.5 ton cooling, 35 MBH heat @ 0°F	↓	.40	40	↓	7,225	2,275		9,500	11,400

23 81 46 – Water-Source Unitary Heat Pumps

23 81 46.10 Water Source Heat Pumps

	Crew	Daily Output	Labor-Hours	Unit	Material	2018 Bare Costs Labor	Equipment	Total	Total Incl O&P
0010 **WATER SOURCE HEAT PUMPS**, Not incl. connecting tubing or water source									
2000 Water source to air, single package									
2100 1 ton cooling, 13 MBH heat @ 75°F	Q-5	2	8	Ea.	1,900	455		2,355	2,750
2140 2 ton cooling, 19 MBH heat @ 75°F		1.70	9.412		2,225	535		2,760	3,250
2220 5 ton cooling, 29 MBH heat @ 75°F	↓	.90	17.778		3,300	1,000		4,300	5,175
3960 For supplementary heat coil, add				↓	10%				
4000 For increase in capacity thru use									
4020 of solar collector, size boiler at 60%									

23 82 Convection Heating and Cooling Units

23 82 16 – Air Coils

23 82 16.10 Flanged Coils

	Crew	Daily Output	Labor-Hours	Unit	Material	2018 Bare Costs Labor	Equipment	Total	Total Incl O&P
0010 **FLANGED COILS**									
0500 Chilled water cooling, 6 rows, 24" x 48"	Q-5	3.20	5	Ea.	4,400	284		4,684	5,275
1000 Direct expansion cooling, 6 rows, 24" x 48"		2.80	5.714		4,775	325		5,100	5,750
1500 Hot water heating, 1 row, 24" x 48"		4	4		1,750	227		1,977	2,275
2000 Steam heating, 1 row, 24" x 48"	↓	3.06	5.229	↓	2,475	296		2,771	3,175

23 82 16.20 Duct Heaters

	Crew	Daily Output	Labor-Hours	Unit	Material	2018 Bare Costs Labor	Equipment	Total	Total Incl O&P
0010 **DUCT HEATERS**, Electric, 480 V, 3 Ph.									
0020 Finned tubular insert, 500°F									
0100 8" wide x 6" high, 4.0 kW	Q-20	16	1.250	Ea.	795	68.50		863.50	980
0120 12" high, 8.0 kW		15	1.333		1,325	73		1,398	1,550
0140 18" high, 12.0 kW		14	1.429		1,850	78		1,928	2,150
0160 24" high, 16.0 kW		13	1.538		2,375	84		2,459	2,750
0180 30" high, 20.0 kW		12	1.667		2,900	91		2,991	3,350
0300 12" wide x 6" high, 6.7 kW		15	1.333		845	73		918	1,050
0360 24" high, 26.7 kW		12	1.667		2,450	91		2,541	2,850
0700 24" wide x 6" high, 17.8 kW		13	1.538		1,000	84		1,084	1,225
0760 24" high, 71.1 kW	↓	10	2	↓	3,075	109		3,184	3,550
8000 To obtain BTU multiply kW by 3413									

For customer support on your Building Construction Costs with RSMeans data, call 800.448.8182.

23 82 Convection Heating and Cooling Units

23 82 19 – Fan Coil Units

23 82 19.10 Fan Coil Air Conditioning

23 82 19.10 Fan Coil Air Conditioning	Crew	Daily Output	Labor-Hours	Unit	Material	2018 Bare Costs Labor	2018 Bare Costs Equipment	Total	Total Incl O&P
0010 **FAN COIL AIR CONDITIONING**									
0030 Fan coil AC, cabinet mounted, filters and controls									
0100 Chilled water, 1/2 ton cooling	Q-5	8	2	Ea.	555	113		668	780
0120 1 ton cooling		6	2.667		830	151		981	1,150
0140 1-1/2 ton cooling		5.50	2.909		835	165		1,000	1,175
0150 2 ton cooling		5.25	3.048		1,200	173		1,373	1,575
0180 3 ton cooling	▼	4	4	▼	1,950	227		2,177	2,500
0262 For hot water coil, add					40%	10%			
0320 1 ton cooling	Q-5	6	2.667	Ea.	1,625	151		1,776	2,000
0940 Direct expansion, for use w/air cooled condensing unit, 1-1/2 ton cooling		5	3.200		610	181		791	950
1000 5 ton cooling	▼	3	5.333		1,075	300		1,375	1,625
1040 10 ton cooling	Q-6	2.60	9.231		2,000	545		2,545	3,025
1060 20 ton cooling	"	.70	34.286		3,850	2,025		5,875	7,250
1500 For hot water coil, add				▼	40%	10%			

23 82 19.20 Heating and Ventilating Units

23 82 19.20 Heating and Ventilating Units	Crew	Daily Output	Labor-Hours	Unit	Material	2018 Bare Costs Labor	2018 Bare Costs Equipment	Total	Total Incl O&P
0010 **HEATING AND VENTILATING UNITS**, Classroom units									
0020 Includes filter, heating/cooling coils, standard controls									
0080 750 CFM, 2 tons cooling	Q-6	2	12	Ea.	4,350	705		5,055	5,875
0120 1,250 CFM, 3 tons cooling		1.40	17.143		5,300	1,000		6,300	7,375
0140 1,500 CFM, 4 tons cooling	▼	.80	30	▼	5,675	1,775		7,450	8,900
0500 For electric heat, add					35%				
1000 For no cooling, deduct					25%	10%			

23 82 29 – Radiators

23 82 29.10 Hydronic Heating

23 82 29.10 Hydronic Heating	Crew	Daily Output	Labor-Hours	Unit	Material	2018 Bare Costs Labor	2018 Bare Costs Equipment	Total	Total Incl O&P
0010 **HYDRONIC HEATING**, Terminal units, not incl. main supply pipe									
1000 Radiation									
1100 Panel, baseboard, C.I., including supports, no covers	Q-5	46	.348	L.F.	44.50	19.70		64.20	78.50
3000 Radiators, cast iron									
3100 Free standing or wall hung, 6 tube, 25" high	Q-5	96	.167	Section	62	9.45		71.45	82.50
3200 4 tube, 19" high	"	96	.167	"	37.50	9.45		46.95	56
3250 Adj. brackets, 2 per wall radiator up to 30 sections	1 Stpi	32	.250	Ea.	65	15.75		80.75	94.50
9500 To convert SFR to BTU rating: Hot water, 150 x SFR									
9510 Forced hot water, 180 x SFR; steam, 240 x SFR									

23 82 33 – Convectors

23 82 33.10 Convector Units

23 82 33.10 Convector Units	Crew	Daily Output	Labor-Hours	Unit	Material	2018 Bare Costs Labor	2018 Bare Costs Equipment	Total	Total Incl O&P
0010 **CONVECTOR UNITS**, Terminal units, not incl. main supply pipe									
2204 Convector, multifin, 2 pipe w/cabinet									
2210 17" H x 24" L	Q-5	10	1.600	Ea.	105	90.50		195.50	252
2214 17" H x 36" L		8.60	1.860		157	105		262	330
2218 17" H x 48" L		7.40	2.162		209	123		332	415
2222 21" H x 24" L		9	1.778		113	101		214	276
2226 21" H x 36" L		8.20	1.951		170	111		281	355
2228 21" H x 48" L	▼	6.80	2.353	▼	226	133		359	450
2240 For knob operated damper, add					140%				
2241 For metal trim strips, add	Q-5	64	.250	Ea.	13.55	14.20		27.75	36.50
2243 For snap-on inlet grille, add					10%	10%			
2245 For hinged access door, add	Q-5	64	.250	Ea.	37.50	14.20		51.70	63
2246 For air chamber, auto-venting, add	"	58	.276	"	8.35	15.65		24	32.50

23 82 Convection Heating and Cooling Units

23 82 36 – Finned-Tube Radiation Heaters

23 82 36.10 Finned Tube Radiation

	23 82 36.10 Finned Tube Radiation	Crew	Daily Output	Labor-Hours	Unit	Material	2018 Bare Costs Labor	Equipment	Total	Total Incl O&P
0010	**FINNED TUBE RADIATION**, Terminal units, not incl. main supply pipe									
1150	Fin tube, wall hung, 14" slope top cover, with damper									
1200	1-1/4" copper tube, 4-1/4" alum. fin	Q-5	38	.421	L.F.	48	24		72	89
1250	1-1/4" steel tube, 4-1/4" steel fin	"	36	.444	"	41	25		66	83
1500	Note: fin tube may also require corners, caps, etc.									

23 82 39 – Unit Heaters

23 82 39.16 Propeller Unit Heaters

	23 82 39.16 Propeller Unit Heaters	Crew	Daily Output	Labor-Hours	Unit	Material	Labor	Equipment	Total	Total Incl O&P
0010	**PROPELLER UNIT HEATERS**									
3950	Unit heaters, propeller, 115 V 2 psi steam, 60°F entering air									
4000	Horizontal, 12 MBH	Q-5	12	1.333	Ea.	380	75.50		455.50	535
4060	43.9 MBH		8	2		580	113		693	805
4140	96.8 MBH		6	2.667		845	151		996	1,150
4180	157.6 MBH		4	4		1,125	227		1,352	1,575
4240	286.9 MBH		2	8		1,750	455		2,205	2,600
4260	364 MBH		1.80	8.889		2,175	505		2,680	3,150
4270	404 MBH		1.60	10		2,275	565		2,840	3,350
4300	Vertical diffuser same price									
4310	Vertical flow, 40 MBH	Q-5	11	1.455	Ea.	570	82.50		652.50	755
4314	58.5 MBH		8	2		590	113		703	820
4326	131 MBH		4	4		890	227		1,117	1,325
4346	297 MBH		1.80	8.889		1,700	505		2,205	2,625
4354	420 MBH (460 V)	Q-6	1.80	13.333		2,275	785		3,060	3,675
4358	500 MBH (460 V)		1.71	14.035		3,025	825		3,850	4,575
4362	570 MBH (460 V)		1.40	17.143		4,150	1,000		5,150	6,075
4366	620 MBH (460 V)		1.30	18.462		4,150	1,075		5,225	6,175
4370	960 MBH (460 V)		1.10	21.818		7,875	1,275		9,150	10,600

23 83 Radiant Heating Units

23 83 16 – Radiant-Heating Hydronic Piping

23 83 16.10 Radiant Floor Heating

	23 83 16.10 Radiant Floor Heating	Crew	Daily Output	Labor-Hours	Unit	Material	Labor	Equipment	Total	Total Incl O&P
0010	**RADIANT FLOOR HEATING**									
0100	Tubing, PEX (cross-linked polyethylene)									
0110	Oxygen barrier type for systems with ferrous materials									
0120	1/2"	Q-5	800	.020	L.F.	1.05	1.13		2.18	2.87
0130	3/4"		535	.030		1.48	1.70		3.18	4.18
0140	1"		400	.040		2.31	2.27		4.58	5.95
0200	Non barrier type for ferrous free systems									
0210	1/2"	Q-5	800	.020	L.F.	.50	1.13		1.63	2.26
0220	3/4"		535	.030		.92	1.70		2.62	3.56
0230	1"		400	.040		1.58	2.27		3.85	5.15
1000	Manifolds									
1110	Brass									
1120	With supply and return valves, flow meter, thermometer,									
1122	auto air vent and drain/fill valve.									
1130	1", 2 circuit	Q-5	14	1.143	Ea.	298	65		363	430
1140	1", 3 circuit		13.50	1.185		340	67		407	475
1150	1", 4 circuit		13	1.231		370	70		440	510
1154	1", 5 circuit		12.50	1.280		440	72.50		512.50	595
1158	1", 6 circuit		12	1.333		480	75.50		555.50	645
1162	1", 7 circuit		11.50	1.391		525	79		604	700

For customer support on your Building Construction Costs with RSMeans data, call 800.448.8182.

23 83 Radiant Heating Units

23 83 16 – Radiant-Heating Hydronic Piping

23 83 16.10 Radiant Floor Heating		Crew	Daily Output	Labor-Hours	Unit	Material	2018 Bare Costs Labor	Equipment	Total	Total Incl O&P
1166	1", 8 circuit	Q-5	11	1.455	Ea.	565	82.50		647.50	750
1172	1", 9 circuit		10.50	1.524		625	86.50		711.50	820
1174	1", 10 circuit		10	1.600		670	90.50		760.50	875
1178	1", 11 circuit		9.50	1.684		700	95.50		795.50	915
1182	1", 12 circuit	▼	9	1.778	▼	775	101		876	1,000
1610	Copper manifold header (cut to size)									
1620	1" header, 12 circuit 1/2" sweat outlets	Q-5	3.33	4.805	Ea.	102	272		374	520
1630	1-1/4" header, 12 circuit 1/2" sweat outlets		3.20	5		118	284		402	555
1640	1-1/4" header, 12 circuit 3/4" sweat outlets		3	5.333		128	300		428	595
1650	1-1/2" header, 12 circuit 3/4" sweat outlets		3.10	5.161		153	293		446	610
1660	2" header, 12 circuit 3/4" sweat outlets	▼	2.90	5.517	▼	226	315		541	720
3000	Valves									
3110	Thermostatic zone valve actuator with end switch	Q-5	40	.400	Ea.	44.50	22.50		67	83
3114	Thermostatic zone valve actuator	"	36	.444	"	91	25		116	138
3120	Motorized straight zone valve with operator complete									
3130	3/4"	Q-5	35	.457	Ea.	146	26		172	200
3140	1"		32	.500		158	28.50		186.50	217
3150	1-1/4"	▼	29.60	.541	▼	200	30.50		230.50	266
3500	4 way mixing valve, manual, brass									
3530	1"	Q-5	13.30	1.203	Ea.	203	68		271	325
3540	1-1/4"		11.40	1.404		220	79.50		299.50	360
3550	1-1/2"		11	1.455		281	82.50		363.50	435
3560	2"		10.60	1.509		395	85.50		480.50	565
3800	Mixing valve motor, 4 way for valves, 1" and 1-1/4"		34	.471		405	26.50		431.50	485
3810	Mixing valve motor, 4 way for valves, 1-1/2" and 2"	▼	30	.533	▼	400	30		430	490
5000	Radiant floor heating, zone control panel									
5120	4 zone actuator valve control, expandable	Q-5	20	.800	Ea.	159	45.50		204.50	244
5130	6 zone actuator valve control, expandable		18	.889		253	50.50		303.50	355
6070	Thermal track, straight panel for long continuous runs, 5.333 S.F.		40	.400		31	22.50		53.50	68
6080	Thermal track, utility panel, for direction reverse at run end, 5.333 S.F.		40	.400		31	22.50		53.50	68
6090	Combination panel, for direction reverse plus straight run, 5.333 S.F.	▼	40	.400	▼	31	22.50		53.50	68
7000	PEX tubing fittings									
7100	Compression type									
7116	Coupling									
7120	1/2" x 1/2"	1 Stpi	27	.296	Ea.	6.75	18.65		25.40	35.50
7124	3/4" x 3/4"	"	23	.348	"	13.05	22		35.05	47.50
7130	Adapter									
7132	1/2" x female sweat 1/2"	1 Stpi	27	.296	Ea.	4.37	18.65		23.02	33
7134	1/2" x female sweat 3/4"		26	.308		4.89	19.40		24.29	34.50
7136	5/8" x female sweat 3/4"	▼	24	.333	▼	7	21		28	39
7140	Elbow									
7142	1/2" x female sweat 1/2"	1 Stpi	27	.296	Ea.	6.70	18.65		25.35	35.50
7144	1/2" x female sweat 3/4"		26	.308		7.80	19.40		27.20	37.50
7146	5/8" x female sweat 3/4"	▼	24	.333	▼	8.75	21		29.75	41
7200	Insert type									
7206	PEX x male NPT									
7210	1/2" x 1/2"	1 Stpi	29	.276	Ea.	2.73	17.40		20.13	29
7220	3/4" x 3/4"		27	.296		4.02	18.65		22.67	32.50
7230	1" x 1"	▼	26	.308		6.80	19.40		26.20	36.50
7300	PEX coupling									
7310	1/2" x 1/2"	1 Stpi	30	.267	Ea.	.54	16.80		17.34	26
7320	3/4" x 3/4"		29	.276		.76	17.40		18.16	27
7330	1" x 1"	▼	28	.286		1.35	18		19.35	28.50

23 83 Radiant Heating Units

23 83 16 – Radiant-Heating Hydronic Piping

23 83 16.10 Radiant Floor Heating

		Crew	Daily Output	Labor-Hours	Unit	Material	2018 Bare Costs Labor	Equipment	Total	Total Incl O&P
7400	PEX stainless crimp ring									
7410	1/2" x 1/2"	1 Stpi	86	.093	Ea.	.46	5.85		6.31	9.35
7420	3/4" x 3/4"		84	.095		.63	6		6.63	9.75
7430	1" x 1"		82	.098		.90	6.15		7.05	10.25

23 83 33 – Electric Radiant Heaters

23 83 33.10 Electric Heating

		Crew	Daily Output	Labor-Hours	Unit	Material	2018 Bare Costs Labor	Equipment	Total	Total Incl O&P
0010	**ELECTRIC HEATING**, not incl. conduit or feed wiring									
1100	Rule of thumb: Baseboard units, including control	1 Elec	4.40	1.818	kW	109	106		215	277
1300	Baseboard heaters, 2' long, 350 watt		8	1	Ea.	27.50	58		85.50	118
1400	3' long, 750 watt		8	1		31.50	58		89.50	122
1600	4' long, 1,000 watt		6.70	1.194		37	69.50		106.50	145
1800	5' long, 935 watt		5.70	1.404		46.50	81.50		128	173
2000	6' long, 1,500 watt		5	1.600		51.50	93		144.50	196
2400	8' long, 2,000 watt		4	2		62.50	116		178.50	243
2950	Wall heaters with fan, 120 to 277 volt									
3170	1,000 watt	1 Elec	6	1.333	Ea.	93	77.50		170.50	218
3180	1,250 watt		5	1.600		100	93		193	249
3190	1,500 watt		4	2		100	116		216	284
3600	Thermostats, integral		16	.500		27.50	29		56.50	73.50
3800	Line voltage, 1 pole		8	1		15.70	58		73.70	104

23 84 Humidity Control Equipment

23 84 13 – Humidifiers

23 84 13.10 Humidifier Units

		Crew	Daily Output	Labor-Hours	Unit	Material	2018 Bare Costs Labor	Equipment	Total	Total Incl O&P
0010	**HUMIDIFIER UNITS**									
0520	Steam, room or duct, filter, regulators, auto. controls, 220 V									
0540	11 lb./hr.	Q-5	6	2.667	Ea.	2,725	151		2,876	3,225
0560	22 lb./hr.		5	3.200		3,000	181		3,181	3,575
0580	33 lb./hr.		4	4		3,100	227		3,327	3,750
0600	50 lb./hr.		4	4		3,575	227		3,802	4,300
0620	100 lb./hr.		3	5.333		4,525	300		4,825	5,425

For customer support on your Building Construction Costs with RSMeans data, call 800.448.8182.

Estimating Tips
26 05 00 Common Work Results for Electrical

- Conduit should be taken off in three main categories—power distribution, branch power, and branch lighting—so the estimator can concentrate on systems and components, therefore making it easier to ensure all items have been accounted for.

- For cost modifications for elevated conduit installation, add the percentages to labor according to the height of installation and only to the quantities exceeding the different height levels, not to the total conduit quantities. Refer to 26 01 02.20 for labor adjustment factors.

- Remember that aluminum wiring of equal ampacity is larger in diameter than copper and may require larger conduit.

- If more than three wires at a time are being pulled, deduct percentages from the labor hours of that grouping of wires.

- When taking off grounding systems, identify separately the type and size of wire, and list each unique type of ground connection.

- The estimator should take the weights of materials into consideration when completing a takeoff. Topics to consider include: How will the materials be supported? What methods of support are available? How high will the support structure have to reach? Will the final support structure be able to withstand the total burden? Is the support material included or separate from the fixture, equipment, and material specified?

- Do not overlook the costs for equipment used in the installation. If scaffolding or highlifts are available in the field, contractors may use them in lieu of the proposed ladders and rolling staging.

26 20 00 Low-Voltage Electrical Transmission

- Supports and concrete pads may be shown on drawings for the larger equipment, or the support system may be only a piece of plywood for the back of a panelboard. In either case, they must be included in the costs.

26 40 00 Electrical and Cathodic Protection

- When taking off cathodic protection systems, identify the type and size of cable, and list each unique type of anode connection.

26 50 00 Lighting

- Fixtures should be taken off room by room using the fixture schedule, specifications, and the ceiling plan. For large concentrations of lighting fixtures in the same area, deduct percentages from labor hours.

Reference Numbers

Reference numbers are shown at the beginning of some major classifications. These numbers refer to related items in the Reference Section. The reference information may be an estimating procedure, an alternate pricing method, or technical information.

Note: Not all subdivisions listed here necessarily appear. ■

No part of this cost data may be reproduced, stored in a retrieval system, or transmitted in any form or by any means without prior written permission of Gordian.

Note: Trade Service, in part, has been used as a reference source for some of the material prices used in Division 26.

Did you know?

RSMeans data is available through our online application with 24/7 access:

- Search for unit prices by keyword
- Leverage the most up-to-date data
- Build and export estimates

Try it free for 30 days!
www.rsmeans.com/2018freetrial

26 01 02.20 Labor Adjustment Factors	Crew	Daily Output	Labor-Hours	Unit	Material	2018 Bare Costs Labor	Equipment	Total	Total Incl O&P
0010 **LABOR ADJUSTMENT FACTORS** (For Div. 26, 27 and 28) R260519-90									
0100 Subtract from labor for Economy of Scale for Wire									
0110 4-5 wires						25%			
0120 6-10 wires						30%			
0130 11-15 wires						35%			
0140 over 15 wires						40%			
0150 Labor adjustment factors (For Div. 26, 27, 28 and 48)									
0200 Labor factors: The below are reasonable suggestions, but									
0210 each project must be evaluated for its own peculiarities, and									
0220 the adjustments be increased or decreased depending on the									
0230 severity of the special conditions.									
1000 Add to labor for elevated installation (above floor level)									
1010 10' to 14.5' high						10%			
1020 15' to 19.5' high						20%			
1030 20' to 24.5' high						25%			
1040 25' to 29.5' high						35%			
1050 30' to 34.5' high						40%			
1060 35' to 39.5' high						50%			
1070 40' and higher						55%			
2000 Add to labor for crawl space									
2010 3' high						40%			
2020 4' high						30%			
3000 Add to labor for multi-story building									
3100 For new construction (No elevator available)									
3110 Add for floors 3 thru 10						5%			
3120 Add for floors 11 thru 15						10%			
3130 Add for floors 16 thru 20						15%			
3140 Add for floors 21 thru 30						20%			
3150 Add for floors 31 and up						30%			
3200 For existing structure (Elevator available)									
3210 Add for work on floor 3 and above						2%			
4000 Add to labor for working in existing occupied buildings									
4010 Hospital						35%			
4020 Office building						25%			
4030 School						20%			
4040 Factory or warehouse						15%			
4050 Multi-dwelling						15%			
5000 Add to labor, miscellaneous									
5010 Cramped shaft						35%			
5020 Congested area						15%			
5030 Excessive heat or cold						30%			
6000 Labor factors: the above are reasonable suggestions, but									
6100 each project should be evaluated for its own peculiarities									
6200 Other factors to be considered are:									
6210 Movement of material and equipment through finished areas						10%			
6220 Equipment room min security direct access w/authorization						15%			
6230 Attic space						25%			
6240 No service road						25%			
6250 Poor unloading/storage area, no hydraulic lifts or jacks						20%			
6260 Congested site area/heavy traffic						20%			
7000 Correctional facilities (no compounding division 1 adjustment factors)									
7010 Minimum security w/facilities escort						30%			
7020 Medium security w/facilities and correctional officer escort						40%			

26 01 Operation and Maintenance of Electrical Systems

26 01 02 – Labor Adjustment

26 01 02.20 Labor Adjustment Factors	Crew	Daily Output	Labor-Hours	Unit	Material	2018 Bare Costs Labor	Equipment	Total	Total Incl O&P
7030	Max security w/facilities & correctional officer escort (no inmate contact)						50%			

26 05 Common Work Results for Electrical

26 05 05 – Selective Demolition for Electrical

26 05 05.10 Electrical Demolition

		Crew	Daily Output	Labor-Hours	Unit	Material	2018 Bare Costs Labor	Equipment	Total	Total Incl O&P
0010	**ELECTRICAL DEMOLITION**									
0020	Electrical demolition, conduit to 10' high, incl. fittings & hangers									
0100	Rigid galvanized steel, 1/2" to 1" diameter	1 Elec	242	.033	L.F.		1.92		1.92	2.88
0120	1-1/4" to 2"	"	200	.040			2.33		2.33	3.48
0140	2-1/2" to 3-1/2"	2 Elec	302	.053			3.08		3.08	4.61
0160	4" to 6"	"	160	.100			5.80		5.80	8.70
0200	Electric metallic tubing (EMT), 1/2" to 1"	1 Elec	394	.020			1.18		1.18	1.77
0220	1-1/4" to 1-1/2"		326	.025			1.43		1.43	2.14
0240	2" to 3"		236	.034			1.97		1.97	2.95
0260	3-1/2" to 4"	2 Elec	310	.052			3		3	4.49
0270	Armored cable (BX) avg. 50' runs									
0280	#14, 2 wire	1 Elec	690	.012	L.F.		.67		.67	1.01
0290	#14, 3 wire		571	.014			.82		.82	1.22
0300	#12, 2 wire		605	.013			.77		.77	1.15
0310	#12, 3 wire		514	.016			.91		.91	1.35
0320	#10, 2 wire		514	.016			.91		.91	1.35
0330	#10, 3 wire		425	.019			1.10		1.10	1.64
0340	#8, 3 wire		342	.023			1.36		1.36	2.03
0350	Non metallic sheathed cable (Romex)									
0360	#14, 2 wire	1 Elec	720	.011	L.F.		.65		.65	.97
0370	#14, 3 wire		657	.012			.71		.71	1.06
0380	#12, 2 wire		629	.013			.74		.74	1.11
0390	#10, 3 wire		450	.018			1.03		1.03	1.55
0400	Wiremold raceway, including fittings & hangers									
0420	No. 3000	1 Elec	250	.032	L.F.		1.86		1.86	2.78
0440	No. 4000		217	.037			2.15		2.15	3.21
0460	No. 6000		166	.048			2.80		2.80	4.19
0462	Plugmold with receptacle		114	.070			4.08		4.08	6.10
0465	Telephone/power pole		12	.667	Ea.		39		39	58
0470	Non-metallic, straight section		480	.017	L.F.		.97		.97	1.45
0500	Channels, steel, including fittings & hangers									
0520	3/4" x 1-1/2"	1 Elec	308	.026	L.F.		1.51		1.51	2.26
0540	1-1/2" x 1-1/2"		269	.030			1.73		1.73	2.59
0560	1-1/2" x 1-7/8"		229	.035			2.03		2.03	3.04
0600	Copper bus duct, indoor, 3 phase									
0610	Including hangers & supports									
0620	225 amp	2 Elec	135	.119	L.F.		6.90		6.90	10.30
0640	400 amp		106	.151			8.80		8.80	13.15
0660	600 amp		86	.186			10.85		10.85	16.20
0680	1,000 amp		60	.267			15.50		15.50	23
0700	1,600 amp		40	.400			23.50		23.50	35
0720	3,000 amp		10	1.600			93		93	139
1300	Transformer, dry type, 1 phase, incl. removal of									
1320	supports, wire & conduit terminations									
1340	1 kVA	1 Elec	7.70	1.039	Ea.		60.50		60.50	90.50
1420	75 kVA	2 Elec	2.50	6.400	"		370		370	555

26 05 05.10 Electrical Demolition	Crew	Daily Output	Labor-Hours	Unit	Material	2018 Bare Costs Labor	2018 Bare Costs Equipment	Total	Total Incl O&P
1440	3 phase to 600 V, primary									
1460	3 kVA	1 Elec	3.87	2.067	Ea.		120		120	180
1520	75 kVA	2 Elec	2.69	5.948			345		345	515
1550	300 kVA	R-3	1.80	11.111			645	72	717	1,050
1570	750 kVA	"	1.10	18.182			1,050	118	1,168	1,700
1800	Wire, THW-THWN-THHN, removed from									
1810	in place conduit, to 10' high									
1830	#14	1 Elec	65	.123	C.L.F.		7.15		7.15	10.70
1840	#12		55	.145			8.45		8.45	12.65
1850	#10		45.50	.176			10.25		10.25	15.30
1860	#8		40.40	.198			11.50		11.50	17.25
1870	#6		32.60	.245			14.30		14.30	21.50
1880	#4	2 Elec	53	.302			17.55		17.55	26.50
1890	#3		50	.320			18.60		18.60	28
1900	#2		44.60	.359			21		21	31
1910	1/0		33.20	.482			28		28	42
1920	2/0		29.20	.548			32		32	47.50
1930	3/0		25	.640			37.50		37.50	55.50
1940	4/0		22	.727			42.50		42.50	63.50
1950	250 kcmil		20	.800			46.50		46.50	69.50
1960	300 kcmil		19	.842			49		49	73.50
1970	350 kcmil		18	.889			51.50		51.50	77.50
1980	400 kcmil		17	.941			55		55	82
1990	500 kcmil		16.20	.988			57.50		57.50	86
2000	Interior fluorescent fixtures, incl. supports									
2010	& whips, to 10' high									
2100	Recessed drop-in 2' x 2', 2 lamp	2 Elec	35	.457	Ea.		26.50		26.50	40
2110	2' x 2', 4 lamp		30	.533			31		31	46.50
2120	2' x 4', 2 lamp		33	.485			28		28	42
2140	2' x 4', 4 lamp		30	.533			31		31	46.50
2160	4' x 4', 4 lamp		20	.800			46.50		46.50	69.50
2180	Surface mount, acrylic lens & hinged frame									
2200	1' x 4', 2 lamp	2 Elec	44	.364	Ea.		21		21	31.50
2210	6" x 4', 2 lamp		44	.364			21		21	31.50
2220	2' x 2', 2 lamp		44	.364			21		21	31.50
2260	2' x 4', 4 lamp		33	.485			28		28	42
2280	4' x 4', 4 lamp		23	.696			40.50		40.50	60.50
2281	4' x 4', 6 lamp		23	.696			40.50		40.50	60.50
2300	Strip fixtures, surface mount									
2320	4' long, 1 lamp	2 Elec	53	.302	Ea.		17.55		17.55	26.50
2340	4' long, 2 lamp		50	.320			18.60		18.60	28
2360	8' long, 1 lamp		42	.381			22		22	33
2380	8' long, 2 lamp		40	.400			23.50		23.50	35
2400	Pendant mount, industrial, incl. removal									
2410	of chain or rod hangers, to 10' high									
2420	4' long, 2 lamp	2 Elec	35	.457	Ea.		26.50		26.50	40
2421	4' long, 4 lamp		35	.457			26.50		26.50	40
2440	8' long, 2 lamp		27	.593			34.50		34.50	51.50

26 05 Common Work Results for Electrical

26 05 13 – Medium-Voltage Cables

26 05 13.16 Medium-Voltage, Single Cable

		Crew	Daily Output	Labor-Hours	Unit	Material	2018 Bare Costs Labor	2018 Bare Costs Equipment	Total	Total Incl O&P
0010	**MEDIUM-VOLTAGE, SINGLE CABLE** Splicing & terminations not included									
0040	Copper, XLP shielding, 5 kV, #6	2 Elec	4.40	3.636	C.L.F.	165	212		377	495
0050	#4		4.40	3.636		214	212		426	550
0100	#2		4	4		262	233		495	640
0200	#1		4	4		315	233		548	695
0400	1/0		3.80	4.211		345	245		590	745
0600	2/0		3.60	4.444		420	259		679	845
0800	4/0		3.20	5		565	291		856	1,050
1000	250 kcmil	3 Elec	4.50	5.333		675	310		985	1,200
1200	350 kcmil		3.90	6.154		875	360		1,235	1,500
1400	500 kcmil		3.60	6.667		1,125	390		1,515	1,825
1600	15 kV, ungrounded neutral, #1	2 Elec	4	4		355	233		588	740
1800	1/0		3.80	4.211		425	245		670	830
2000	2/0		3.60	4.444		480	259		739	915
2200	4/0		3.20	5		640	291		931	1,150
2400	250 kcmil	3 Elec	4.50	5.333		710	310		1,020	1,250
2600	350 kcmil		3.90	6.154		895	360		1,255	1,525
2800	500 kcmil		3.60	6.667		1,100	390		1,490	1,775

26 05 19 – Low-Voltage Electrical Power Conductors and Cables

26 05 19.20 Armored Cable

		Crew	Daily Output	Labor-Hours	Unit	Material	2018 Bare Costs Labor	2018 Bare Costs Equipment	Total	Total Incl O&P
0010	**ARMORED CABLE**									
0050	600 volt, copper (BX), #14, 2 conductor, solid	1 Elec	2.40	3.333	C.L.F.	41.50	194		235.50	335
0100	3 conductor, solid		2.20	3.636		64.50	212		276.50	385
0150	#12, 2 conductor, solid		2.30	3.478		41.50	202		243.50	350
0200	3 conductor, solid		2	4		71	233		304	430
0250	#10, 2 conductor, solid		2	4		80.50	233		313.50	440
0300	3 conductor, solid		1.60	5		110	291		401	555
0340	#8, 2 conductor, stranded		1.50	5.333		244	310		554	735
0350	3 conductor, stranded		1.30	6.154		244	360		604	805
0400	3 conductor with PVC jacket, in cable tray, #6		3.10	2.581		545	150		695	825
0450	#4	2 Elec	5.40	2.963		665	172		837	990
0500	#2		4.60	3.478		800	202		1,002	1,175
0550	#1		4	4		1,025	233		1,258	1,475
0600	1/0		3.60	4.444		1,100	259		1,359	1,575
0650	2/0		3.40	4.706		1,325	274		1,599	1,850
0700	3/0		3.20	5		1,775	291		2,066	2,375
0750	4/0		3	5.333		2,125	310		2,435	2,800
0800	250 kcmil	3 Elec	3.60	6.667		2,625	390		3,015	3,475
0850	350 kcmil		3.30	7.273		3,300	425		3,725	4,275
0900	500 kcmil		3	8		4,575	465		5,040	5,725
1050	5 kV, copper, 3 conductor with PVC jacket,									
1060	non-shielded, in cable tray, #4	2 Elec	380	.042	L.F.	7.60	2.45		10.05	12
1100	#2		360	.044		9.85	2.59		12.44	14.70
1200	#1		300	.053		12.55	3.10		15.65	18.45
1400	1/0		290	.055		14.50	3.21		17.71	21
1600	2/0		260	.062		16.75	3.58		20.33	24
2000	4/0		240	.067		22.50	3.88		26.38	30.50
2100	250 kcmil	3 Elec	330	.073		30.50	4.23		34.73	40
2150	350 kcmil		315	.076		37.50	4.43		41.93	48
2200	500 kcmil		270	.089		51.50	5.15		56.65	64.50
2400	15 kV, copper, 3 conductor with PVC jacket galv., steel armored									
2500	grounded neutral, in cable tray, #2	2 Elec	300	.053	L.F.	16.25	3.10		19.35	22.50

26 05 19.20 Armored Cable

		Crew	Daily Output	Labor-Hours	Unit	Material	2018 Bare Costs Labor	Equipment	Total	Total Incl O&P
2600	#1	2 Elec	280	.057	L.F.	17	3.33		20.33	23.50
2800	1/0		260	.062		19.75	3.58		23.33	27.50
2900	2/0		220	.073		26	4.23		30.23	35
3000	4/0		190	.084		29.50	4.90		34.40	39.50
3100	250 kcmil	3 Elec	270	.089		32.50	5.15		37.65	44
3150	350 kcmil		240	.100		38.50	5.80		44.30	50.50
3200	500 kcmil		210	.114		51.50	6.65		58.15	67
3400	15 kV, copper, 3 conductor with PVC jacket,									
3450	ungrounded neutral, in cable tray, #2	2 Elec	260	.062	L.F.	17.20	3.58		20.78	24.50
3500	#1		230	.070		19.05	4.05		23.10	27
3600	1/0		200	.080		22	4.66		26.66	31
3700	2/0		190	.084		27	4.90		31.90	37
3800	4/0		160	.100		32.50	5.80		38.30	44
4000	250 kcmil	3 Elec	210	.114		37.50	6.65		44.15	51.50
4050	350 kcmil		195	.123		49.50	7.15		56.65	65
4100	500 kcmil		180	.133		60.50	7.75		68.25	78
9010	600 volt, copper (MC) steel clad, #14, 2 wire	R-1A	8.16	1.961	C.L.F.	42.50	103		145.50	201
9020	3 wire		7.21	2.219		75.50	116		191.50	257
9030	4 wire		6.69	2.392		184	125		309	390
9040	#12, 2 wire		7.21	2.219		54.50	116		170.50	234
9050	3 wire		6.69	2.392		70.50	125		195.50	265
9070	#10, 2 wire		6.04	2.649		103	139		242	320
9080	3 wire		5.65	2.832		133	148		281	370
9100	#8, 2 wire, stranded		3.94	4.061		158	213		371	495
9110	3 wire, stranded		3.40	4.706		193	246		439	580
9200	600 volt, copper (MC) aluminum clad, #14, 2 wire		8.16	1.961		41.50	103		144.50	200
9210	3 wire		7.21	2.219		73	116		189	255
9220	4 wire		6.69	2.392		88.50	125		213.50	285
9230	#12, 2 wire		7.21	2.219		46.50	116		162.50	225
9240	3 wire		6.69	2.392		7.35	125		132.35	195
9250	4 wire		6.04	2.649		131	139		270	350
9260	#10, 2 wire		6.04	2.649		85	139		224	300
9270	3 wire		5.65	2.832		116	148		264	350
9280	4 wire		5.23	3.059		208	160		368	470

26 05 19.35 Cable Terminations

		Crew	Daily Output	Labor-Hours	Unit	Material	2018 Bare Costs Labor	Equipment	Total	Total Incl O&P
0010	**CABLE TERMINATIONS**									
0015	Wire connectors, screw type, #22 to #14	1 Elec	260	.031	Ea.	.06	1.79		1.85	2.75
0020	#18 to #12		240	.033		.07	1.94		2.01	2.98
0035	#16 to #10		230	.035		.27	2.02		2.29	3.33
0040	#14 to #8		210	.038		.30	2.22		2.52	3.64
0045	#12 to #6		180	.044		.45	2.59		3.04	4.37
0050	Terminal lugs, solderless, #16 to #10		50	.160		.34	9.30		9.64	14.25
0100	#8 to #4		30	.267		.68	15.50		16.18	24
0150	#2 to #1		22	.364		.89	21		21.89	32.50
0200	1/0 to 2/0		16	.500		1.56	29		30.56	45
0250	3/0		12	.667		4.16	39		43.16	62.50
0300	4/0		11	.727		4.37	42.50		46.87	68.50
0350	250 kcmil		9	.889		3.29	51.50		54.79	81
0400	350 kcmil		7	1.143		4.24	66.50		70.74	104
0450	500 kcmil		6	1.333		7.35	77.50		84.85	124
1600	Crimp 1 hole lugs, copper or aluminum, 600 volt									
1620	#14	1 Elec	60	.133	Ea.	.55	7.75		8.30	12.20

570

For customer support on your Building Construction Costs with RSMeans data, call 800.448.8182.

26 05 Common Work Results for Electrical

26 05 19 – Low-Voltage Electrical Power Conductors and Cables

26 05 19.35 Cable Terminations

		Crew	Daily Output	Labor-Hours	Unit	Material	2018 Bare Costs Labor	Equipment	Total	Total Incl O&P
1630	#12	1 Elec	50	.160	Ea.	1.09	9.30		10.39	15.10
1640	#10		45	.178		1.09	10.35		11.44	16.65
1780	#8		36	.222		1.43	12.95		14.38	21
1800	#6		30	.267		1.82	15.50		17.32	25
2000	#4		27	.296		3.14	17.25		20.39	29.50
2200	#2		24	.333		2.90	19.40		22.30	32
2400	#1		20	.400		5.25	23.50		28.75	41
2500	1/0		17.50	.457		5.95	26.50		32.45	46.50
2600	2/0		15	.533		7.65	31		38.65	55
2800	3/0		12	.667		6.85	39		45.85	65.50
3000	4/0		11	.727		6	42.50		48.50	70
3200	250 kcmil		9	.889		7.40	51.50		58.90	85.50
3400	300 kcmil		8	1		11.35	58		69.35	99.50
3500	350 kcmil		7	1.143		11.85	66.50		78.35	113
3600	400 kcmil		6.50	1.231		16.10	71.50		87.60	125
3800	500 kcmil		6	1.333		17.75	77.50		95.25	136

26 05 19.50 Mineral Insulated Cable

		Crew	Daily Output	Labor-Hours	Unit	Material	2018 Bare Costs Labor	Equipment	Total	Total Incl O&P
0010	**MINERAL INSULATED CABLE** 600 volt									
0100	1 conductor, #12	1 Elec	1.60	5	C.L.F.	350	291		641	820
0200	#10		1.60	5		575	291		866	1,075
0400	#8		1.50	5.333		640	310		950	1,175
0500	#6		1.40	5.714		760	335		1,095	1,325
0600	#4	2 Elec	2.40	6.667		1,025	390		1,415	1,700
0800	#2		2.20	7.273		1,450	425		1,875	2,225
0900	#1		2.10	7.619		1,675	445		2,120	2,500
1000	1/0		2	8		1,950	465		2,415	2,850
1100	2/0		1.90	8.421		2,325	490		2,815	3,300
1200	3/0		1.80	8.889		2,800	515		3,315	3,850
1400	4/0		1.60	10		3,225	580		3,805	4,425
1410	250 kcmil	3 Elec	2.40	10		3,625	580		4,205	4,875
1420	350 kcmil		1.95	12.308		4,150	715		4,865	5,650
1430	500 kcmil		1.95	12.308		5,350	715		6,065	6,950

26 05 19.55 Non-Metallic Sheathed Cable

		Crew	Daily Output	Labor-Hours	Unit	Material	2018 Bare Costs Labor	Equipment	Total	Total Incl O&P
0010	**NON-METALLIC SHEATHED CABLE** 600 volt									
0100	Copper with ground wire (Romex)									
0150	#14, 2 conductor	1 Elec	2.70	2.963	C.L.F.	18.35	172		190.35	278
0200	3 conductor		2.40	3.333		25.50	194		219.50	320
0220	4 conductor		2.20	3.636		39.50	212		251.50	360
0250	#12, 2 conductor		2.50	3.200		24	186		210	305
0300	3 conductor		2.20	3.636		40	212		252	360
0320	4 conductor		2	4		60	233		293	415
0350	#10, 2 conductor		2.20	3.636		41	212		253	360
0400	3 conductor		1.80	4.444		62.50	259		321.50	455
0430	#8, 2 conductor		1.60	5		75.50	291		366.50	520
0450	3 conductor		1.50	5.333		117	310		427	595
0500	#6, 3 conductor		1.40	5.714		197	335		532	710
0550	SE type SER aluminum cable, 3 RHW and									
0600	1 bare neutral, 3 #8 & 1 #8	1 Elec	1.60	5	C.L.F.	56.50	291		347.50	495
0650	3 #6 & 1 #6	"	1.40	5.714		67	335		402	570
0700	3 #4 & 1 #6	2 Elec	2.40	6.667		71.50	390		461.50	660
0750	3 #2 & 1 #4		2.20	7.273		124	425		549	770
0800	3 #1/0 & 1 #2		2	8		182	465		647	895

For customer support on your Building Construction Costs with RSMeans data, call 800.448.8182.

571

26 05 19 – Low-Voltage Electrical Power Conductors and Cables

26 05 19.55 Non-Metallic Sheathed Cable		Crew	Daily Output	Labor-Hours	Unit	Material	2018 Bare Costs Labor	Equipment	Total	Total Incl O&P
0850	3 #2/0 & 1 #1	2 Elec	1.80	8.889	C.L.F.	212	515		727	1,000
0900	3 #4/0 & 1 #2/0	↓	1.60	10	↓	295	580		875	1,200

26 05 19.90 Wire

		Crew	Daily Output	Labor-Hours	Unit	Material	2018 Bare Costs Labor	Equipment	Total	Total Incl O&P
0010	**WIRE**, normal installation conditions in wireway, conduit, cable tray									
0020	600 volt, copper type THW, solid, #14 R260519-92	1 Elec	13	.615	C.L.F.	5.90	36		41.90	60
0030	#12		11	.727		9.65	42.50		52.15	74
0040	#10		10	.800		15.75	46.50		62.25	87
0050	Stranded, #14 R260533-22		13	.615		8.60	36		44.60	63
0100	#12		11	.727		13.20	42.50		55.70	78
0120	#10		10	.800		21	46.50		67.50	92.50
0140	#8		8	1		34.50	58		92.50	125
0160	#6	↓	6.50	1.231		43.50	71.50		115	155
0180	#4	2 Elec	10.60	1.509		69.50	88		157.50	207
0200	#3		10	1.600		105	93		198	254
0220	#2		9	1.778		102	103		205	267
0240	#1		8	2		136	116		252	325
0260	1/0		6.60	2.424		216	141		357	450
0280	2/0		5.80	2.759		205	161		366	465
0300	3/0		5	3.200		235	186		421	535
0350	4/0	↓	4.40	3.636		435	212		647	790
0400	250 kcmil	3 Elec	6	4		350	233		583	735
0420	300 kcmil		5.70	4.211		380	245		625	780
0450	350 kcmil		5.40	4.444		465	259		724	895
0480	400 kcmil		5.10	4.706		540	274		814	1,000
0490	500 kcmil	↓	4.80	5		690	291		981	1,200
0540	600 volt, aluminum type THHN, stranded, #6	1 Elec	8	1		24	58		82	114
0560	#4	2 Elec	13	1.231		30	71.50		101.50	140
0580	#2		10.60	1.509		41	88		129	176
0600	#1		9	1.778		60	103		163	221
0620	1/0		8	2		72	116		188	254
0640	2/0		7.20	2.222		86	129		215	288
0680	3/0		6.60	2.424		107	141		248	330
0700	4/0	↓	6.20	2.581		119	150		269	355
0720	250 kcmil	3 Elec	8.70	2.759		144	161		305	400
0740	300 kcmil		8.10	2.963		199	172		371	475
0760	350 kcmil		7.50	3.200		201	186		387	500
0780	400 kcmil		6.90	3.478		229	202		431	555
0800	500 kcmil		6	4		258	233		491	635
0850	600 kcmil		5.70	4.211		325	245		570	725
0880	700 kcmil		5.10	4.706		400	274		674	850
0900	750 kcmil		4.80	5		400	291		691	875
0910	1,000 kcmil	↓	3.78	6.349		595	370		965	1,200
0920	600 volt, copper type THWN-THHN, solid, #14	1 Elec	13	.615		6.55	36		42.55	60.50
0940	#12		11	.727		9.60	42.50		52.10	74
0960	#10		10	.800		14.25	46.50		60.75	85
1000	Stranded, #14		13	.615		7.35	36		43.35	61.50
1200	#12		11	.727		10.50	42.50		53	75
1250	#10		10	.800		16.10	46.50		62.60	87
1300	#8		8	1		28.50	58		86.50	118
1350	#6	↓	6.50	1.231		48.50	71.50		120	160
1400	#4	2 Elec	10.60	1.509	↓	69	88		157	207

For customer support on your Building Construction Costs with RSMeans data, call 800.448.8182.

26 05 Common Work Results for Electrical

26 05 23 – Control-Voltage Electrical Power Cables

26 05 23.10 Control Cable

26 05 23.10 Control Cable	Crew	Daily Output	Labor-Hours	Unit	Material	2018 Bare Costs Labor	2018 Bare Costs Equipment	Total	Total Incl O&P
0010 **CONTROL CABLE**									
0020 600 volt, copper, #14 THWN wire with PVC jacket, 2 wires	1 Elec	9	.889	C.L.F.	37.50	51.50		89	119
0030 3 wires		8	1		51	58		109	143
0100 4 wires		7	1.143		54.50	66.50		121	160
0150 5 wires		6.50	1.231		75	71.50		146.50	189
0200 6 wires		6	1.333		103	77.50		180.50	229
0300 8 wires		5.30	1.509		133	88		221	277
0400 10 wires		4.80	1.667		154	97		251	315
0500 12 wires		4.30	1.860		199	108		307	380
0600 14 wires		3.80	2.105		214	123		337	420
0700 16 wires		3.50	2.286		242	133		375	465
0800 18 wires		3.30	2.424		272	141		413	510
0810 19 wires		3.10	2.581		287	150		437	540
0900 20 wires		3	2.667		300	155		455	565
1000 22 wires		2.80	2.857		335	166		501	615

26 05 26 – Grounding and Bonding for Electrical Systems

26 05 26.80 Grounding

26 05 26.80 Grounding	Crew	Daily Output	Labor-Hours	Unit	Material	2018 Bare Costs Labor	2018 Bare Costs Equipment	Total	Total Incl O&P
0010 **GROUNDING**									
0030 Rod, copper clad, 8' long, 1/2" diameter	1 Elec	5.50	1.455	Ea.	19.90	84.50		104.40	149
0050 3/4" diameter		5.30	1.509		35	88		123	170
0080 10' long, 1/2" diameter		4.80	1.667		22.50	97		119.50	170
0100 3/4" diameter		4.40	1.818		42	106		148	205
0130 15' long, 3/4" diameter		4	2		49	116		165	228
0390 Bare copper wire, stranded, #8		11	.727	C.L.F.	43	42.50		85.50	111
0400 #6		10	.800		41	46.50		87.50	115
0600 #2	2 Elec	10	1.600		89	93		182	237
0800 3/0		6.60	2.424		315	141		456	555
1000 4/0		5.70	2.807		440	163		603	730
1200 250 kcmil	3 Elec	7.20	3.333		445	194		639	780
1800 Water pipe ground clamps, heavy duty									
2000 Bronze, 1/2" to 1" diameter	1 Elec	8	1	Ea.	29.50	58		87.50	120
2100 1-1/4" to 2" diameter		8	1		28.50	58		86.50	119
2200 2-1/2" to 3" diameter		6	1.333		68	77.50		145.50	191
2800 Brazed connections, #6 wire		12	.667		16.15	39		55.15	76
3000 #2 wire		10	.800		21.50	46.50		68	93.50
3100 3/0 wire		8	1		32	58		90	123
3200 4/0 wire		7	1.143		36.50	66.50		103	140
3400 250 kcmil wire		5	1.600		43	93		136	186
3600 500 kcmil wire		4	2		53	116		169	232

26 05 33 – Raceway and Boxes for Electrical Systems

26 05 33.13 Conduit

26 05 33.13 Conduit	Crew	Daily Output	Labor-Hours	Unit	Material	2018 Bare Costs Labor	2018 Bare Costs Equipment	Total	Total Incl O&P
0010 **CONDUIT** To 10' high, includes 2 terminations, 2 elbows, R260533-22									
0020 11 beam clamps, and 11 couplings per 100 L.F.									
0300 Aluminum, 1/2" diameter	1 Elec	100	.080	L.F.	1.72	4.66		6.38	8.85
0500 3/4" diameter		90	.089		2.28	5.15		7.43	10.25
0700 1" diameter		80	.100		2.82	5.80		8.62	11.80
1000 1-1/4" diameter		70	.114		3.91	6.65		10.56	14.25
1030 1-1/2" diameter		65	.123		4.70	7.15		11.85	15.85
1050 2" diameter		60	.133		6.40	7.75		14.15	18.60
1070 2-1/2" diameter		50	.160		9.25	9.30		18.55	24
1100 3" diameter	2 Elec	90	.178		12.15	10.35		22.50	29

573

For customer support on your Building Construction Costs with RSMeans data, call 800.448.8182.

26 05 33.13 Conduit

		Crew	Daily Output	Labor-Hours	Unit	Material	2018 Bare Costs Labor	2018 Bare Costs Equipment	Total	Total Incl O&P
1130	3-1/2" diameter	2 Elec	80	.200	L.F.	16	11.65		27.65	35
1140	4" diameter	▼	70	.229		18.05	13.30		31.35	40
1750	Rigid galvanized steel, 1/2" diameter	1 Elec	90	.089		2.31	5.15		7.46	10.30
1770	3/4" diameter		80	.100		4.57	5.80		10.37	13.75
1800	1" diameter		65	.123		6.75	7.15		13.90	18.15
1830	1-1/4" diameter		60	.133		4.74	7.75		12.49	16.80
1850	1-1/2" diameter		55	.145		7.90	8.45		16.35	21.50
1870	2" diameter		45	.178		9.95	10.35		20.30	26.50
1900	2-1/2" diameter	▼	35	.229		11.80	13.30		25.10	33
1930	3" diameter	2 Elec	50	.320		13.70	18.60		32.30	43
1950	3-1/2" diameter		44	.364		17.40	21		38.40	50.50
1970	4" diameter	▼	40	.400		20	23.50		43.50	57
2500	Steel, intermediate conduit (IMC), 1/2" diameter	1 Elec	100	.080		1.69	4.66		6.35	8.80
2530	3/4" diameter		90	.089		2.05	5.15		7.20	10
2550	1" diameter		70	.114		2.58	6.65		9.23	12.80
2570	1-1/4" diameter		65	.123		3.58	7.15		10.73	14.65
2600	1-1/2" diameter		60	.133		4.76	7.75		12.51	16.85
2630	2" diameter		50	.160		6.15	9.30		15.45	20.50
2650	2-1/2" diameter	▼	40	.200		9.15	11.65		20.80	27.50
2670	3" diameter	2 Elec	60	.267		12.10	15.50		27.60	36.50
2700	3-1/2" diameter		54	.296		16.90	17.25		34.15	44.50
2730	4" diameter		50	.320		11.70	18.60		30.30	41
5000	Electric metallic tubing (EMT), 1/2" diameter	1 Elec	170	.047		.80	2.74		3.54	4.96
5020	3/4" diameter		130	.062		1.13	3.58		4.71	6.60
5040	1" diameter		115	.070		1.84	4.05		5.89	8.10
5060	1-1/4" diameter		100	.080		2.99	4.66		7.65	10.25
5080	1-1/2" diameter		90	.089		2.82	5.15		7.97	10.85
5100	2" diameter		80	.100		3.85	5.80		9.65	12.95
5120	2-1/2" diameter	▼	60	.133		4.43	7.75		12.18	16.45
5140	3" diameter	2 Elec	100	.160		5.55	9.30		14.85	20
5160	3-1/2" diameter		90	.178		6.95	10.35		17.30	23
5180	4" diameter	▼	80	.200	▼	12.10	11.65		23.75	30.50
9900	Add to labor for higher elevated installation									
9905	10' to 14.5' high, add						10%			
9910	15' to 20' high, add						20%			
9920	20' to 25' high, add						25%			
9930	25' to 30' high, add						35%			
9940	30' to 35' high, add						40%			
9950	35' to 40' high, add						50%			
9960	Over 40' high, add						55%			

26 05 33.16 Boxes for Electrical Systems

		Crew	Daily Output	Labor-Hours	Unit	Material	2018 Bare Costs Labor	2018 Bare Costs Equipment	Total	Total Incl O&P
0010	**BOXES FOR ELECTRICAL SYSTEMS**									
0020	Pressed steel, octagon, 4"	1 Elec	20	.400	Ea.	2.86	23.50		26.36	38
0060	Covers, blank		64	.125		.88	7.30		8.18	11.85
0100	Extension rings		40	.200		3.68	11.65		15.33	21.50
0150	Square, 4"		20	.400		5.25	23.50		28.75	41
0200	Extension rings		40	.200		3.79	11.65		15.44	21.50
0250	Covers, blank		64	.125		.67	7.30		7.97	11.65
0300	Plaster rings		64	.125		1.54	7.30		8.84	12.60
0650	Switchbox		27	.296		4.61	17.25		21.86	31
1100	Concrete, floor, 1 gang		5.30	1.509		94.50	88		182.50	235
2000	Poke-thru fitting, fire rated, for 3-3/4" floor	▼	6.80	1.176	▼	142	68.50		210.50	258

26 05 33.16 Boxes for Electrical Systems	Crew	Daily Output	Labor-Hours	Unit	Material	2018 Bare Costs Labor	Equipment	Total	Total Incl O&P
2040 For 7" floor	1 Elec	6.80	1.176	Ea.	194	68.50		262.50	315
2100 Pedestal, 15 amp, duplex receptacle & blank plate		5.25	1.524		140	88.50		228.50	287
2120 Duplex receptacle and telephone plate		5.25	1.524		140	88.50		228.50	287
2140 Pedestal, 20 amp, duplex recept. & phone plate		5	1.600		141	93		234	295
2200 Abandonment plate		32	.250		42	14.55		56.55	68.50

26 05 33.18 Pull Boxes

		Crew	Daily Output	Labor-Hours	Unit	Material	2018 Bare Costs Labor	Equipment	Total	Total Incl O&P
0010	**PULL BOXES**									
0100	Steel, pull box, NEMA 1, type SC, 6" W x 6" H x 4" D	1 Elec	8	1	Ea.	9.75	58		67.75	97.50
0200	8" W x 8" H x 4" D		8	1		12.55	58		70.55	101
0300	10" W x 12" H x 6" D		5.30	1.509		28.50	88		116.50	162
0400	16" W x 20" H x 8" D		4	2		82	116		198	264
0500	20" W x 24" H x 8" D		3.20	2.500		96	146		242	325
0600	24" W x 36" H x 8" D		2.70	2.963		151	172		323	425
0650	Pull box, hinged, NEMA 1, 6" W x 6" H x 4" D		8	1		13	58		71	101
0800	12" W x 16" H x 6" D		4.70	1.702		51.50	99		150.50	205
1000	20" W x 20" H x 6" D		3.60	2.222		121	129		250	325
1200	20" W x 20" H x 8" D		3.20	2.500		151	146		297	385
1400	24" W x 36" H x 8" D		2.70	2.963		237	172		409	520
1600	24" W x 42" H x 8" D		2	4		355	233		588	740
2100	Pull box, NEMA 3R, type SC, raintight & weatherproof									
2150	6" L x 6" W x 6" D	1 Elec	10	.800	Ea.	16.65	46.50		63.15	88
2200	8" L x 6" W x 6" D		8	1		24	58		82	114
2250	10" L x 6" W x 6" D		7	1.143		32	66.50		98.50	135
2300	12" L x 12" W x 6" D		5	1.600		48	93		141	192
2350	16" L x 16" W x 6" D		4.50	1.778		89	103		192	253
2400	20" L x 20" W x 6" D		4	2		89	116		205	272
2450	24" L x 18" W x 8" D		3	2.667		146	155		301	395
2500	24" L x 24" W x 10" D		2.50	3.200		310	186		496	620
2550	30" L x 24" W x 12" D		2	4		385	233		618	775
2600	36" L x 36" W x 12" D		1.50	5.333		450	310		760	960
2800	Cast iron, pull boxes for surface mounting									
3000	NEMA 4, watertight & dust tight									
3050	6" L x 6" W x 6" D	1 Elec	4	2	Ea.	273	116		389	475
3100	8" L x 6" W x 6" D		3.20	2.500		380	146		526	635
3150	10" L x 6" W x 6" D		2.50	3.200		425	186		611	750
3200	12" L x 12" W x 6" D		2.30	3.478		775	202		977	1,150
3250	16" L x 16" W x 6" D		1.30	6.154		710	360		1,070	1,325
3300	20" L x 20" W x 6" D		.80	10		865	580		1,445	1,825
3350	24" L x 18" W x 8" D		.70	11.429		2,600	665		3,265	3,850
3400	24" L x 24" W x 10" D		.50	16		4,900	930		5,830	6,800
3450	30" L x 24" W x 12" D		.40	20		6,275	1,175		7,450	8,650
3500	36" L x 36" W x 12" D		.20	40		3,575	2,325		5,900	7,400
6000	J.I.C. wiring boxes, NEMA 12, dust tight & drip tight									
6050	6" L x 8" W x 4" D	1 Elec	10	.800	Ea.	40.50	46.50		87	115
6100	8" L x 10" W x 4" D		8	1		50.50	58		108.50	143
6150	12" L x 14" W x 6" D		5.30	1.509		86	88		174	226
6200	14" L x 16" W x 6" D		4.70	1.702		100	99		199	258
6250	16" L x 20" W x 6" D		4.40	1.818		219	106		325	400
6300	24" L x 30" W x 6" D		3.20	2.500		320	146		466	575
6350	24" L x 30" W x 8" D		2.90	2.759		470	161		631	755
6400	24" L x 36" W x 8" D		2.70	2.963		365	172		537	665
6450	24" L x 42" W x 8" D		2.30	3.478		415	202		617	760

26 05 Common Work Results for Electrical

26 05 33 – Raceway and Boxes for Electrical Systems

26 05 33.18 Pull Boxes

		Crew	Daily Output	Labor-Hours	Unit	Material	2018 Bare Costs Labor	2018 Bare Costs Equipment	Total	Total Incl O&P
6500	24" L x 48" W x 8" D	1 Elec	2	4	Ea.	450	233		683	845

26 05 33.23 Wireway

		Crew	Daily Output	Labor-Hours	Unit	Material	2018 Bare Costs Labor	2018 Bare Costs Equipment	Total	Total Incl O&P
0010	**WIREWAY** to 10' high									
0100	NEMA 1, screw cover w/fittings and supports, 2-1/2" x 2-1/2"	1 Elec	45	.178	L.F.	12.20	10.35		22.55	29
0200	4" x 4"	"	40	.200		13.15	11.65		24.80	32
0400	6" x 6"	2 Elec	60	.267		21	15.50		36.50	46
0600	8" x 8"	"	40	.400		28	23.50		51.50	65.50
4475	NEMA 3R, screw cover w/fittings and supports, 4" x 4"	1 Elec	36	.222		16.30	12.95		29.25	37.50
4480	6" x 6"	2 Elec	55	.291		18.10	16.95		35.05	45.50
4485	8" x 8"		36	.444		32	26		58	73.50
4490	12" x 12"		18	.889		57.50	51.50		109	141

26 05 36 – Cable Trays for Electrical Systems

26 05 36.10 Cable Tray Ladder Type

		Crew	Daily Output	Labor-Hours	Unit	Material	2018 Bare Costs Labor	2018 Bare Costs Equipment	Total	Total Incl O&P
0010	**CABLE TRAY LADDER TYPE** w/ftngs. & supports, 4" dp., to 15' elev.									
0160	Galvanized steel tray									
0170	4" rung spacing, 6" wide	2 Elec	98	.163	L.F.	13.60	9.50		23.10	29
0200	12" wide		86	.186		16.40	10.85		27.25	34
0400	18" wide		82	.195		19	11.35		30.35	38
0600	24" wide		78	.205		22	11.95		33.95	42
3200	Aluminum tray, 4" deep, 6" rung spacing, 6" wide		134	.119		13.40	6.95		20.35	25
3220	12" wide		124	.129		15	7.50		22.50	28
3230	18" wide		114	.140		16.70	8.15		24.85	30.50
3240	24" wide		106	.151		19.40	8.80		28.20	34.50

26 05 39 – Underfloor Raceways for Electrical Systems

26 05 39.30 Conduit In Concrete Slab

		Crew	Daily Output	Labor-Hours	Unit	Material	2018 Bare Costs Labor	2018 Bare Costs Equipment	Total	Total Incl O&P
0010	**CONDUIT IN CONCRETE SLAB** Including terminations,									
0020	fittings and supports									
3230	PVC, schedule 40, 1/2" diameter	1 Elec	270	.030	L.F.	.49	1.72		2.21	3.11
3250	3/4" diameter		230	.035		.56	2.02		2.58	3.65
3270	1" diameter		200	.040		.77	2.33		3.10	4.32
3300	1-1/4" diameter		170	.047		1.01	2.74		3.75	5.20
3330	1-1/2" diameter		140	.057		1.21	3.33		4.54	6.30
3350	2" diameter		120	.067		1.52	3.88		5.40	7.45
4350	Rigid galvanized steel, 1/2" diameter		200	.040		1.99	2.33		4.32	5.65
4400	3/4" diameter		170	.047		4.27	2.74		7.01	8.80
4450	1" diameter		130	.062		6.60	3.58		10.18	12.60
4500	1-1/4" diameter		110	.073		4.21	4.23		8.44	11
4600	1-1/2" diameter		100	.080		7.30	4.66		11.96	15
4800	2" diameter		90	.089		9.05	5.15		14.20	17.70

26 05 39.40 Conduit In Trench

		Crew	Daily Output	Labor-Hours	Unit	Material	2018 Bare Costs Labor	2018 Bare Costs Equipment	Total	Total Incl O&P
0010	**CONDUIT IN TRENCH** Includes terminations and fittings									
0020	Does not include excavation or backfill, see Section 31 23 16									
0200	Rigid galvanized steel, 2" diameter	1 Elec	150	.053	L.F.	8.70	3.10		11.80	14.25
0400	2-1/2" diameter	"	100	.080		10.70	4.66		15.36	18.70
0600	3" diameter	2 Elec	160	.100		12.55	5.80		18.35	22.50
0800	3-1/2" diameter		140	.114		16.55	6.65		23.20	28
1000	4" diameter		100	.160		18.90	9.30		28.20	35
1200	5" diameter		80	.200		36.50	11.65		48.15	57.50
1400	6" diameter		60	.267		51.50	15.50		67	79.50

For customer support on your Building Construction Costs with RSMeans data, call 800.448.8182.

26 05 43.10 Trench Duct

26 05 43.10 Trench Duct	Crew	Daily Output	Labor-Hours	Unit	Material	2018 Bare Costs Labor	Equipment	Total	Total Incl O&P
0010 **TRENCH DUCT** Steel with cover									
0020 Standard adjustable, depths to 4"									
0100 Straight, single compartment, 9" wide	2 Elec	40	.400	L.F.	254	23.50		277.50	315
0200 12" wide		32	.500		330	29		359	405
0400 18" wide		26	.615		365	36		401	455
0600 24" wide		22	.727		430	42.50		472.50	540
0800 30" wide		20	.800		230	46.50		276.50	325
1000 36" wide		16	1		264	58		322	375
1200 Horizontal elbow, 9" wide		5.40	2.963	Ea.	415	172		587	720
1400 12" wide		4.60	3.478		445	202		647	795
1600 18" wide		4	4		590	233		823	995
1800 24" wide		3.20	5		840	291		1,131	1,350
2000 30" wide		2.60	6.154		1,125	360		1,485	1,750
2200 36" wide		2.40	6.667		1,475	390		1,865	2,200
2400 Vertical elbow, 9" wide		5.40	2.963		168	172		340	440
2600 12" wide		4.60	3.478		158	202		360	480
2800 18" wide		4	4		181	233		414	550
3000 24" wide		3.20	5		225	291		516	685
3200 30" wide		2.60	6.154		248	360		608	810
3400 36" wide		2.40	6.667		273	390		663	880
3600 Cross, 9" wide		4	4		680	233		913	1,100
3800 12" wide		3.20	5		720	291		1,011	1,225
4000 18" wide		2.60	6.154		865	360		1,225	1,475
4200 24" wide		2.20	7.273		1,075	425		1,500	1,800
4400 30" wide		2	8		1,375	465		1,840	2,200
4600 36" wide		1.80	8.889		1,700	515		2,215	2,650
4800 End closure, 9" wide		14.40	1.111		41	64.50		105.50	142
5000 12" wide		12	1.333		47	77.50		124.50	168
5200 18" wide		10	1.600		72	93		165	219
5400 24" wide		8	2		95	116		211	278
5600 30" wide		6.60	2.424		145	141		286	370
5800 36" wide		5.80	2.759		141	161		302	395
6000 Tees, 9" wide		4	4		400	233		633	790
6200 12" wide		3.60	4.444		465	259		724	895
6400 18" wide		3.20	5		595	291		886	1,100
6600 24" wide		3	5.333		855	310		1,165	1,400
6800 30" wide		2.60	6.154		1,100	360		1,460	1,750
7000 36" wide		2	8		1,450	465		1,915	2,300
7200 Riser, and cabinet connector, 9" wide		5.40	2.963		175	172		347	450
7400 12" wide		4.60	3.478		204	202		406	530
7600 18" wide		4	4		205	233		438	575
7800 24" wide		3.20	5		305	291		596	770
8000 30" wide		2.60	6.154		350	360		710	920
8200 36" wide		2	8		405	465		870	1,150
8400 Insert assembly, cell to conduit adapter, 1-1/4"	1 Elec	16	.500		69.50	29		98.50	120

26 05 43.20 Underfloor Duct

26 05 43.20 Underfloor Duct	Crew	Daily Output	Labor-Hours	Unit	Material	2018 Bare Costs Labor	Equipment	Total	Total Incl O&P
0010 **UNDERFLOOR DUCT**									
0100 Duct, 1-3/8" x 3-1/8" blank, standard	2 Elec	160	.100	L.F.	12.85	5.80		18.65	23
0200 1-3/8" x 7-1/4" blank, super duct		120	.133		30	7.75		37.75	44.50
0400 7/8" or 1-3/8" insert type, 24" OC, 1-3/8" x 3-1/8", std.		140	.114		20	6.65		26.65	32
0600 1-3/8" x 7-1/4", super duct		100	.160		35	9.30		44.30	52.50
0800 Junction box, single duct, 1 level, 3-1/8"	1 Elec	4	2	Ea.	450	116		566	670

26 05 43 – Underground Ducts and Raceways for Electrical Systems

26 05 43.20 Underfloor Duct		Crew	Daily Output	Labor-Hours	Unit	Material	2018 Bare Costs Labor	Equipment	Total	Total Incl O&P
1000	Junction box, single duct, 1 level, 7-1/4"	1 Elec	2.70	2.963	Ea.	525	172		697	835
1200	1 level, 2 duct, 3-1/8"		3.20	2.500		590	146		736	870
1400	Junction box, 1 level, 2 duct, 7-1/4"		2.30	3.478		1,450	202		1,652	1,900
1580	Junction box, 1 level, one 3-1/8" + one 7-1/4" x same		2.30	3.478		1,000	202		1,202	1,400
1600	Triple duct, 3-1/8"		2.30	3.478		1,000	202		1,202	1,400
1800	Insert to conduit adapter, 3/4" & 1"		32	.250		37	14.55		51.55	63
2000	Support, single cell		27	.296		55	17.25		72.25	86.50
2200	Super duct		16	.500		56	29		85	105
2400	Double cell		16	.500		55.50	29		84.50	105
2600	Triple cell		11	.727		53.50	42.50		96	123
2800	Vertical elbow, standard duct		10	.800		99.50	46.50		146	179
3000	Super duct		8	1		99.50	58		157.50	196
3200	Cabinet connector, standard duct		32	.250		79	14.55		93.55	109
3400	Super duct		27	.296		77	17.25		94.25	111
3600	Conduit adapter, 1" to 1-1/4"		32	.250		74.50	14.55		89.05	104
3800	2" to 1-1/4"		27	.296		89.50	17.25		106.75	125
4000	Outlet, low tension (tele, computer, etc.)		8	1		105	58		163	202
4200	High tension, receptacle (120 volt)		8	1		108	58		166	206

26 05 80 – Wiring Connections

26 05 80.10 Motor Connections

		Crew	Daily Output	Labor-Hours	Unit	Material	2018 Bare Costs Labor	Equipment	Total	Total Incl O&P
0010	**MOTOR CONNECTIONS**									
0020	Flexible conduit and fittings, 115 volt, 1 phase, up to 1 HP motor	1 Elec	8	1	Ea.	4.87	58		62.87	92.50
0110	230 volt, 3 phase, 3 HP motor		6.78	1.180		5.90	68.50		74.40	110
0112	5 HP motor		5.47	1.463		5	85		90	133
0114	7-1/2 HP motor		4.61	1.735		8.55	101		109.55	160
0120	10 HP motor		4.20	1.905		15.90	111		126.90	184
0200	25 HP motor		2.70	2.963		24.50	172		196.50	285
0400	50 HP motor		2.20	3.636		49	212		261	370
0600	100 HP motor		1.50	5.333		103	310		413	580

26 05 90 – Residential Applications

26 05 90.10 Residential Wiring

		Crew	Daily Output	Labor-Hours	Unit	Material	2018 Bare Costs Labor	Equipment	Total	Total Incl O&P
0010	**RESIDENTIAL WIRING**									
0020	20' avg. runs and #14/2 wiring incl. unless otherwise noted									
1000	Service & panel, includes 24' SE-AL cable, service eye, meter,									
1010	Socket, panel board, main bkr., ground rod, 15 or 20 amp									
1020	1-pole circuit breakers, and misc. hardware									
1100	100 amp, with 10 branch breakers	1 Elec	1.19	6.723	Ea.	325	390		715	945
1110	With PVC conduit and wire		.92	8.696		360	505		865	1,150
1120	With RGS conduit and wire		.73	10.959		560	640		1,200	1,575
1150	150 amp, with 14 branch breakers		1.03	7.767		850	450		1,300	1,600
1170	With PVC conduit and wire		.82	9.756		920	570		1,490	1,850
1180	With RGS conduit and wire		.67	11.940		1,175	695		1,870	2,350
1200	200 amp, with 18 branch breakers	2 Elec	1.80	8.889		1,000	515		1,515	1,875
1220	With PVC conduit and wire		1.46	10.959		1,075	640		1,715	2,125
1230	With RGS conduit and wire		1.24	12.903		1,400	750		2,150	2,675
1800	Lightning surge suppressor	1 Elec	32	.250		76	14.55		90.55	106
2000	Switch devices									
2100	Single pole, 15 amp, ivory, with a 1-gang box, cover plate,									
2110	Type NM (Romex) cable	1 Elec	17.10	.468	Ea.	15.50	27		42.50	57.50
2120	Type MC cable		14.30	.559		25	32.50		57.50	76
2130	EMT & wire		5.71	1.401		36	81.50		117.50	162
2150	3-way, #14/3, type NM cable		14.55	.550		9.80	32		41.80	59

For customer support on your Building Construction Costs with RSMeans data, call 800.448.8182.

26 05 90.10 Residential Wiring	Crew	Daily Output	Labor-Hours	Unit	Material	2018 Bare Costs Labor	Equipment	Total	Total Incl O&P
2170	Type MC cable	1 Elec	12.31	.650	Ea.	22.50	38		60.50	81.50
2180	EMT & wire		5	1.600		30	93		123	172
2200	4-way, #14/3, type NM cable		14.55	.550		18.45	32		50.45	68.50
2220	Type MC cable		12.31	.650		31.50	38		69.50	91
2230	EMT & wire		5	1.600		38.50	93		131.50	182
2250	S.P., 20 amp, #12/2, type NM cable		13.33	.600		11.10	35		46.10	64
2270	Type MC cable		11.43	.700		19.65	40.50		60.15	82.50
2280	EMT & wire		4.85	1.649		33.50	96		129.50	181
2290	S.P. rotary dimmer, 600 W, no wiring		17	.471		30	27.50		57.50	74
2300	S.P. rotary dimmer, 600 W, type NM cable		14.55	.550		33.50	32		65.50	85
2320	Type MC cable		12.31	.650		43.50	38		81.50	104
2330	EMT & wire		5	1.600		55	93		148	200
2350	3-way rotary dimmer, type NM cable		13.33	.600		21.50	35		56.50	75.50
2370	Type MC cable		11.43	.700		31	40.50		71.50	95
2380	EMT & wire		4.85	1.649		43	96		139	191
2400	Interval timer wall switch, 20 amp, 1-30 min., #12/2									
2410	Type NM cable	1 Elec	14.55	.550	Ea.	59.50	32		91.50	114
2420	Type MC cable		12.31	.650		65.50	38		103.50	129
2430	EMT & wire		5	1.600		82	93		175	229
2500	Decorator style									
2510	S.P., 15 amp, type NM cable	1 Elec	17.10	.468	Ea.	19.35	27		46.35	62
2520	Type MC cable		14.30	.559		29	32.50		61.50	80.50
2530	EMT & wire		5.71	1.401		39.50	81.50		121	166
2550	3-way, #14/3, type NM cable		14.55	.550		13.70	32		45.70	63
2570	Type MC cable		12.31	.650		26.50	38		64.50	85.50
2580	EMT & wire		5	1.600		33.50	93		126.50	176
2600	4-way, #14/3, type NM cable		14.55	.550		22.50	32		54.50	72.50
2620	Type MC cable		12.31	.650		35	38		73	95
2630	EMT & wire		5	1.600		42.50	93		135.50	186
2650	S.P., 20 amp, #12/2, type NM cable		13.33	.600		14.95	35		49.95	68.50
2670	Type MC cable		11.43	.700		23.50	40.50		64	87
2680	EMT & wire		4.85	1.649		37.50	96		133.50	186
2700	S.P., slide dimmer, type NM cable		17.10	.468		30.50	27		57.50	74
2720	Type MC cable		14.30	.559		40	32.50		72.50	92.P
2730	EMT & wire		5.71	1.401		52	81.50		133.50	179
2750	S.P., touch dimmer, type NM cable		17.10	.468		46.50	27		73.50	92
2770	Type MC cable		14.30	.559		56	32.50		88.50	111
2780	EMT & wire		5.71	1.401		68	81.50		149.50	197
2800	3-way touch dimmer, type NM cable		13.33	.600		49.50	35		84.50	106
2820	Type MC cable		11.43	.700		59	40.50		99.50	126
2830	EMT & wire		4.85	1.649		71	96		167	222
3000	Combination devices									
3100	S.P. switch/15 amp recpt., ivory, 1-gang box, plate									
3110	Type NM cable	1 Elec	11.43	.700	Ea.	20.50	40.50		61	83.50
3120	Type MC cable		10	.800		30	46.50		76.50	103
3130	EMT & wire		4.40	1.818		42	106		148	204
3150	S.P. switch/pilot light, type NM cable		11.43	.700		21	40.50		61.50	84
3170	Type MC cable		10	.800		30.50	46.50		77	103
3180	EMT & wire		4.43	1.806		42.50	105		147.50	204
3190	2-S.P. switches, 2-#14/2, no wiring		14	.571		12.85	33.50		46.35	63.50
3200	2-S.P. switches, 2-#14/2, type NM cables		10	.800		22	46.50		68.50	94
3220	Type MC cable		8.89	.900		36.50	52.50		89	119
3230	EMT & wire		4.10	1.951		43.50	114		157.50	218

26 05 90.10 Residential Wiring	Crew	Daily Output	Labor-Hours	Unit	Material	2018 Bare Costs Labor	Equipment	Total	Total Incl O&P
3250	3-way switch/15 amp recpt., #14/3, type NM cable	1 Elec	10	.800	Ea.	26.50	46.50		73	99
3270	Type MC cable		8.89	.900		39.50	52.50		92	122
3280	EMT & wire		4.10	1.951		46.50	114		160.50	222
3300	2-3 way switches, 2-#14/3, type NM cables		8.89	.900		34.50	52.50		87	117
3320	Type MC cable		8	1		55.50	58		113.50	148
3330	EMT & wire		4	2		53	116		169	233
3350	S.P. switch/20 amp recpt., #12/2, type NM cable		10	.800		28	46.50		74.50	101
3370	Type MC cable		8.89	.900		34	52.50		86.50	116
3380	EMT & wire	▼	4.10	1.951	▼	50.50	114		164.50	226
3400	Decorator style									
3410	S.P. switch/15 amp recpt., type NM cable	1 Elec	11.43	.700	Ea.	24	40.50		64.50	87.50
3420	Type MC cable		10	.800		34	46.50		80.50	107
3430	EMT & wire		4.40	1.818		45.50	106		151.50	209
3450	S.P. switch/pilot light, type NM cable		11.43	.700		25	40.50		65.50	88.50
3470	Type MC cable		10	.800		34.50	46.50		81	108
3480	EMT & wire		4.40	1.818		46.50	106		152.50	209
3500	2-S.P. switches, 2-#14/2, type NM cables		10	.800		26	46.50		72.50	98
3520	Type MC cable		8.89	.900		40	52.50		92.50	123
3530	EMT & wire		4.10	1.951		47.50	114		161.50	222
3550	3-way/15 amp recpt., #14/3, type NM cable		10	.800		30.50	46.50		77	103
3570	Type MC cable		8.89	.900		43.50	52.50		96	126
3580	EMT & wire		4.10	1.951		50.50	114		164.50	226
3650	2-3 way switches, 2-#14/3, type NM cables		8.89	.900		38.50	52.50		91	121
3670	Type MC cable		8	1		59	58		117	152
3680	EMT & wire		4	2		57	116		173	237
3700	S.P. switch/20 amp recpt., #12/2, type NM cable		10	.800		32	46.50		78.50	105
3720	Type MC cable		8.89	.900		38	52.50		90.50	120
3730	EMT & wire	▼	4.10	1.951	▼	54.50	114		168.50	230
4000	Receptacle devices									
4010	Duplex outlet, 15 amp recpt., ivory, 1-gang box, plate									
4015	Type NM cable	1 Elec	14.55	.550	Ea.	8.40	32		40.40	57
4020	Type MC cable		12.31	.650		17.95	38		55.95	76.50
4030	EMT & wire		5.33	1.501		28.50	87.50		116	163
4050	With #12/2, type NM cable		12.31	.650		9.50	38		47.50	67
4070	Type MC cable		10.67	.750		18.05	43.50		61.55	85
4080	EMT & wire		4.71	1.699		32	99		131	184
4100	20 amp recpt., #12/2, type NM cable		12.31	.650		17.25	38		55.25	75.50
4120	Type MC cable		10.67	.750		26	43.50		69.50	93.50
4130	EMT & wire	▼	4.71	1.699	▼	40	99		139	192
4140	For GFI see Section 26 05 90.10 line 4300 below									
4150	Decorator style, 15 amp recpt., type NM cable	1 Elec	14.55	.550	Ea.	12.25	32		44.25	61.50
4170	Type MC cable		12.31	.650		22	38		60	80.50
4180	EMT & wire		5.33	1.501		32.50	87.50		120	167
4200	With #12/2, type NM cable		12.31	.650		13.35	38		51.35	71
4220	Type MC cable		10.67	.750		22	43.50		65.50	89
4230	EMT & wire		4.71	1.699		36	99		135	188
4250	20 amp recpt., #12/2, type NM cable		12.31	.650		21	38		59	80
4270	Type MC cable		10.67	.750		29.50	43.50		73	97.50
4280	EMT & wire		4.71	1.699		44	99		143	196
4300	GFI, 15 amp recpt., type NM cable		12.31	.650		21	38		59	79.50
4320	Type MC cable		10.67	.750		30.50	43.50		74	98.50
4330	EMT & wire		4.71	1.699		41	99		140	193
4350	GFI with #12/2, type NM cable	▼	10.67	.750		22	43.50		65.50	89

26 05 90 – Residential Applications

26 05 90.10 Residential Wiring	Crew	Daily Output	Labor-Hours	Unit	Material	2018 Bare Costs Labor	Equipment	Total	Total Incl O&P
4370 Type MC cable	1 Elec	9.20	.870	Ea.	30.50	50.50		81	109
4380 EMT & wire		4.21	1.900		44.50	111		155.50	214
4400 20 amp recpt., #12/2, type NM cable		10.67	.750		48.50	43.50		92	118
4420 Type MC cable		9.20	.870		57	50.50		107.50	138
4430 EMT & wire		4.21	1.900		71	111		182	243
4500 Weather-proof cover for above receptacles, add	↓	32	.250	↓	2.10	14.55		16.65	24.50
4550 Air conditioner outlet, 20 amp-240 volt recpt.									
4560 30' of #12/2, 2 pole circuit breaker									
4570 Type NM cable	1 Elec	10	.800	Ea.	58	46.50		104.50	134
4580 Type MC cable		9	.889		68.50	51.50		120	153
4590 EMT & wire		4	2		81	116		197	264
4600 Decorator style, type NM cable		10	.800		63	46.50		109.50	139
4620 Type MC cable		9	.889		73	51.50		124.50	158
4630 EMT & wire	↓	4	2	↓	86	116		202	269
4650 Dryer outlet, 30 amp-240 volt recpt., 20' of #10/3									
4660 2 pole circuit breaker									
4670 Type NM cable	1 Elec	6.41	1.248	Ea.	53.50	72.50		126	168
4680 Type MC cable		5.71	1.401		62.50	81.50		144	191
4690 EMT & wire	↓	3.48	2.299	↓	75	134		209	282
4700 Range outlet, 50 amp-240 volt recpt., 30' of #8/3									
4710 Type NM cable	1 Elec	4.21	1.900	Ea.	82	111		193	255
4720 Type MC cable		4	2		128	116		244	315
4730 EMT & wire		2.96	2.703		106	157		263	350
4750 Central vacuum outlet, type NM cable		6.40	1.250		50	73		123	164
4770 Type MC cable		5.71	1.401		64.50	81.50		146	193
4780 EMT & wire	↓	3.48	2.299	↓	83	134		217	291
4800 30 amp-110 volt locking recpt., #10/2 circ. bkr.									
4810 Type NM cable	1 Elec	6.20	1.290	Ea.	58	75		133	176
4820 Type MC cable		5.40	1.481		76	86		162	213
4830 EMT & wire	↓	3.20	2.500	↓	93.50	146		239.50	320
4900 Low voltage outlets									
4910 Telephone recpt., 20' of 4/C phone wire	1 Elec	26	.308	Ea.	8	17.90		25.90	36
4920 TV recpt., 20' of RG59U coax wire, F type connector	"	16	.500	"	15.55	29		44.55	60.50
4950 Door bell chime, transformer, 2 buttons, 60' of bellwire									
4970 Economy model	1 Elec	11.50	.696	Ea.	52	40.50		92.50	118
4980 Custom model		11.50	.696		94.50	40.50		135	165
4990 Luxury model, 3 buttons	↓	9.50	.842	↓	185	49		234	278
6000 Lighting outlets									
6050 Wire only (for fixture), type NM cable	1 Elec	32	.250	Ea.	5.75	14.55		20.30	28.50
6070 Type MC cable		24	.333		11.70	19.40		31.10	42
6080 EMT & wire		10	.800		21.50	46.50		68	93.50
6100 Box (4"), and wire (for fixture), type NM cable		25	.320		14.55	18.60		33.15	44
6120 Type MC cable		20	.400		20.50	23.50		44	57.50
6130 EMT & wire	↓	11	.727		30.50	42.50		73	97
6200 Fixtures (use with line 6050 or 6100 above)									
6210 Canopy style, economy grade	1 Elec	40	.200	Ea.	23	11.65		34.65	43
6220 Custom grade		40	.200		53.50	11.65		65.15	76.50
6250 Dining room chandelier, economy grade		19	.421		72	24.50		96.50	116
6260 Custom grade		19	.421		305	24.50		329.50	370
6270 Luxury grade		15	.533		955	31		986	1,100
6310 Kitchen fixture (fluorescent), economy grade		30	.267		71.50	15.50		87	102
6320 Custom grade		25	.320		137	18.60		155.60	179
6350 Outdoor, wall mounted, economy grade		30	.267		28.50	15.50		44	54.50

26 05 90.10 Residential Wiring	Crew	Daily Output	Labor-Hours	Unit	Material	2018 Bare Costs Labor	Equipment	Total	Total Incl O&P
6360	Custom grade	1 Elec	30	.267	Ea.	121	15.50		136.50	156
6370	Luxury grade		25	.320		248	18.60		266.60	300
6410	Outdoor PAR floodlights, 1 lamp, 150 watt		20	.400		26	23.50		49.50	63.50
6420	2 lamp, 150 watt each		20	.400		45.50	23.50		69	85
6425	Motion sensing, 2 lamp, 150 watt each		20	.400		88.50	23.50		112	133
6430	For infrared security sensor, add		32	.250		99.50	14.55		114.05	132
6450	Outdoor, quartz-halogen, 300 watt flood		20	.400		33	23.50		56.50	71.50
6600	Recessed downlight, round, pre-wired, 50 or 75 watt trim		30	.267		62.50	15.50		78	91.50
6610	With shower light trim		30	.267		86	15.50		101.50	118
6620	With wall washer trim		28	.286		79.50	16.65		96.15	113
6630	With eye-ball trim		28	.286		66.50	16.65		83.15	98
6700	Porcelain lamp holder		40	.200		2.92	11.65		14.57	20.50
6710	With pull switch		40	.200		11.30	11.65		22.95	30
6750	Fluorescent strip, 2-20 watt tube, wrap around diffuser, 24"		24	.333		46	19.40		65.40	80
6760	1-34 watt tube, 48"		24	.333		105	19.40		124.40	145
6770	2-34 watt tubes, 48"		20	.400		122	23.50		145.50	169
6800	Bathroom heat lamp, 1-250 watt		28	.286		31	16.65		47.65	59.50
6810	2-250 watt lamps	↓	28	.286	↓	73	16.65		89.65	106
6820	For timer switch, see Section 26 05 90.10 line 2400									
6900	Outdoor post lamp, incl. post, fixture, 35' of #14/2									
6910	Type NM cable	1 Elec	3.50	2.286	Ea.	300	133		433	530
6920	Photo-eye, add		27	.296		27.50	17.25		44.75	56
6950	Clock dial time switch, 24 hr., w/enclosure, type NM cable		11.43	.700		68.50	40.50		109	136
6970	Type MC cable		11	.727		78	42.50		120.50	150
6980	EMT & wire	↓	4.85	1.649	↓	88.50	96		184.50	242
7000	Alarm systems									
7050	Smoke detectors, box, #14/3, type NM cable	1 Elec	14.55	.550	Ea.	33.50	32		65.50	85
7070	Type MC cable		12.31	.650		44	38		82	105
7080	EMT & wire	↓	5	1.600		51	93		144	195
7090	For relay output to security system, add				↓	10.75			10.75	11.85
8000	Residential equipment									
8050	Disposal hook-up, incl. switch, outlet box, 3' of flex									
8060	20 amp-1 pole circ. bkr., and 25' of #12/2									
8070	Type NM cable	1 Elec	10	.800	Ea.	28	46.50		74.50	101
8080	Type MC cable		8	1		37.50	58		95.50	128
8090	EMT & wire	↓	5	1.600	↓	54.50	93		147.50	199
8100	Trash compactor or dishwasher hook-up, incl. outlet box,									
8110	3' of flex, 15 amp-1 pole circ. bkr., and 25' of #14/2									
8120	Type NM cable	1 Elec	10	.800	Ea.	15.50	46.50		62	86.50
8130	Type MC cable		8	1		26.50	58		84.50	116
8140	EMT & wire	↓	5	1.600		41	93		134	184
8150	Hot water sink dispenser hook-up, use line 8100									
8200	Vent/exhaust fan hook-up, type NM cable	1 Elec	32	.250	Ea.	5.75	14.55		20.30	28.50
8220	Type MC cable		24	.333		11.70	19.40		31.10	42
8230	EMT & wire	↓	10	.800	↓	21.50	46.50		68	93.50
8250	Bathroom vent fan, 50 CFM (use with above hook-up)									
8260	Economy model	1 Elec	15	.533	Ea.	19.40	31		50.40	68
8270	Low noise model		15	.533		44	31		75	95
8280	Custom model	↓	12	.667	↓	121	39		160	191
8300	Bathroom or kitchen vent fan, 110 CFM									
8310	Economy model	1 Elec	15	.533	Ea.	68	31		99	122
8320	Low noise model	"	15	.533	"	92	31		123	148
8350	Paddle fan, variable speed (w/o lights)									

26 05 90.10 Residential Wiring	Crew	Daily Output	Labor-Hours	Unit	Material	2018 Bare Costs Labor	Equipment	Total	Total Incl O&P
8360	Economy model (AC motor)	1 Elec	10	.800	Ea.	110	46.50		156.50	191
8362	With light kit		10	.800		149	46.50		195.50	234
8370	Custom model (AC motor)		10	.800		264	46.50		310.50	360
8372	With light kit		10	.800		305	46.50		351.50	405
8380	Luxury model (DC motor)		8	1		315	58		373	435
8382	With light kit		8	1		355	58		413	475
8390	Remote speed switch for above, add		12	.667		33.50	39		72.50	95
8500	Whole house exhaust fan, ceiling mount, 36", variable speed									
8510	Remote switch, incl. shutters, 20 amp-1 pole circ. bkr.									
8520	30' of #12/2, type NM cable	1 Elec	4	2	Ea.	1,225	116		1,341	1,525
8530	Type MC cable		3.50	2.286		1,250	133		1,383	1,575
8540	EMT & wire		3	2.667		1,250	155		1,405	1,600
8600	Whirlpool tub hook-up, incl. timer switch, outlet box									
8610	3' of flex, 20 amp-1 pole GFI circ. bkr.									
8620	30' of #12/2, type NM cable	1 Elec	5	1.600	Ea.	125	93		218	277
8630	Type MC cable		4.20	1.905		131	111		242	310
8640	EMT & wire		3.40	2.353		145	137		282	365
8650	Hot water heater hook-up, incl. 1-2 pole circ. bkr., box;									
8660	3' of flex, 20' of #10/2, type NM cable	1 Elec	5	1.600	Ea.	28.50	93		121.50	171
8670	Type MC cable		4.20	1.905		41	111		152	211
8680	EMT & wire		3.40	2.353		47.50	137		184.50	258
9000	Heating/air conditioning									
9050	Furnace/boiler hook-up, incl. firestat, local on-off switch									
9060	Emergency switch, and 40' of type NM cable	1 Elec	4	2	Ea.	52	116		168	231
9070	Type MC cable		3.50	2.286		66	133		199	272
9080	EMT & wire		1.50	5.333		89	310		399	565
9100	Air conditioner hook-up, incl. local 60 amp disc. switch									
9110	3' sealtite, 40 amp, 2 pole circuit breaker									
9130	40' of #8/2, type NM cable	1 Elec	3.50	2.286	Ea.	134	133		267	345
9140	Type MC cable		3	2.667		201	155		356	455
9150	EMT & wire		1.30	6.154		182	360		542	735
9200	Heat pump hook-up, 1-40 & 1-100 amp 2 pole circ. bkr.									
9210	Local disconnect switch, 3' sealtite									
9220	40' of #8/2 & 30' of #3/2									
9230	Type NM cable	1 Elec	1.30	6.154	Ea.	475	360		835	1,050
9240	Type MC cable		1.08	7.407		525	430		955	1,225
9250	EMT & wire		.94	8.511		490	495		985	1,275
9500	Thermostat hook-up, using low voltage wire									
9520	Heating only, 25' of #18-3	1 Elec	24	.333	Ea.	5.70	19.40		25.10	35.50
9530	Heating/cooling, 25' of #18-4	"	20	.400	"	7.15	23.50		30.65	43

For customer support on your Building Construction Costs with RSMeans data, call 800.448.8182.

583

26 09 13 - Electrical Power Monitoring

26 09 13.10 Switchboard Instruments

26 09 13.10 Switchboard Instruments		Crew	Daily Output	Labor-Hours	Unit	Material	2018 Bare Costs Labor	Equipment	Total	Total Incl O&P
0010	**SWITCHBOARD INSTRUMENTS** 3 phase, 4 wire									
0100	AC indicating, ammeter & switch	1 Elec	8	1	Ea.	2,875	58		2,933	3,250
0200	Voltmeter & switch		8	1		2,900	58		2,958	3,250
0300	Wattmeter		8	1		4,100	58		4,158	4,600
0400	AC recording, ammeter		4	2		7,300	116		7,416	8,225
0500	Voltmeter		4	2		7,300	116		7,416	8,225
0600	Ground fault protection, zero sequence		2.70	2.963		6,450	172		6,622	7,350
0700	Ground return path		2.70	2.963		6,450	172		6,622	7,350
0800	3 current transformers, 5 to 800 amp		2	4		3,000	233		3,233	3,650
0900	1,000 to 1,500 amp		1.30	6.154		4,325	360		4,685	5,275
1200	2,000 to 4,000 amp		1	8		5,100	465		5,565	6,300
1300	Fused potential transformer, maximum 600 volt		8	1		1,125	58		1,183	1,325

26 09 13.30 Smart Metering

26 09 13.30 Smart Metering			Crew	Daily Output	Labor-Hours	Unit	Material	2018 Bare Costs Labor	Equipment	Total	Total Incl O&P
0010	**SMART METERING,** In panel										
0100	Single phase, 120/208 volt, 100 amp	G	1 Elec	8.78	.911	Ea.	325	53		378	435
0120	200 amp	G		8.78	.911		335	53		388	450
0200	277 volt, 100 amp	G		8.78	.911		360	53		413	475
0220	200 amp	G		8.78	.911		395	53		448	510
1100	Three phase, 120/208 volt, 100 amp	G		4.69	1.706		675	99.50		774.50	890
1120	200 amp	G		4.69	1.706		800	99.50		899.50	1,025
1130	400 amp	G		4.69	1.706		800	99.50		899.50	1,025
1140	800 amp	G		4.69	1.706		805	99.50		904.50	1,025
1150	1,600 amp	G		4.69	1.706		830	99.50		929.50	1,050
1200	277/480 volt, 100 amp	G		4.69	1.706		870	99.50		969.50	1,100
1220	200 amp	G		4.69	1.706		880	99.50		979.50	1,125
1230	400 amp	G		4.69	1.706		880	99.50		979.50	1,125
1240	800 amp	G		4.69	1.706		850	99.50		949.50	1,075
1250	1,600 amp	G		4.69	1.706		865	99.50		964.50	1,100
2000	Data recorder, 8 meters	G		10.97	.729		1,450	42.50		1,492.50	1,650
2100	16 meters	G		8.53	.938		3,600	54.50		3,654.50	4,025
3000	Software package, per meter, basic	G					254			254	279
3100	Premium	G					665			665	735

26 09 23 - Lighting Control Devices

26 09 23.10 Energy Saving Lighting Devices

26 09 23.10 Energy Saving Lighting Devices			Crew	Daily Output	Labor-Hours	Unit	Material	2018 Bare Costs Labor	Equipment	Total	Total Incl O&P
0010	**ENERGY SAVING LIGHTING DEVICES**										
0100	Occupancy sensors, passive infrared ceiling mounted	G	1 Elec	7	1.143	Ea.	75	66.50		141.50	182
0110	Ultrasonic ceiling mounted	G		7	1.143		85.50	66.50		152	194
0120	Dual technology ceiling mounted	G		6.50	1.231		125	71.50		196.50	244
0150	Automatic wall switches	G		24	.333		75	19.40		94.40	112
0160	Daylighting sensor, manual control, ceiling mounted	G		7	1.143		108	66.50		174.50	218
0170	Remote and dimming control with remote controller	G		6.50	1.231		168	71.50		239.50	292
0200	Passive infrared ceiling mounted			6.50	1.231		33	71.50		104.50	143
0400	Remote power pack	G		10	.800		33	46.50		79.50	106
0450	Photoelectric control, S.P.S.T. 120 V	G		8	1		21	58		79	110
0500	S.P.S.T. 208 V/277 V	G		8	1		27	58		85	117
0550	D.P.S.T. 120 V	G		6	1.333		194	77.50		271.50	330
0600	D.P.S.T. 208 V/277 V	G		6	1.333		204	77.50		281.50	340
0650	S.P.D.T. 208 V/277 V	G		6	1.333		212	77.50		289.50	350
0660	Daylight level sensor, wall mounted, on/off or dimming	G		8	1		197	58		255	305

For customer support on your Building Construction Costs with RSMeans data, call 800.448.8182.

26 12 Medium-Voltage Transformers

26 12 19 – Pad-Mounted, Liquid-Filled, Medium-Voltage Transformers

26 12 19.10 Transformer, Oil-Filled		Crew	Daily Output	Labor-Hours	Unit	Material	2018 Bare Costs Labor	Equipment	Total	Total Incl O&P
0010	**TRANSFORMER, OIL-FILLED** primary delta or Y,									
0050	Pad mounted 5 kV or 15 kV, with taps, 277/480 V secondary, 3 phase									
0100	150 kVA	R-3	.65	30.769	Ea.	9,750	1,775	199	11,724	13,600
0200	300 kVA		.45	44.444		13,900	2,575	288	16,763	19,500
0300	500 kVA		.40	50		19,700	2,900	325	22,925	26,400
0400	750 kVA		.38	52.632		25,000	3,050	340	28,390	32,500
0500	1,000 kVA		.26	76.923		29,600	4,450	500	34,550	39,800
0600	1,500 kVA		.23	86.957		35,200	5,050	565	40,815	46,900
0700	2,000 kVA		.20	100		44,400	5,800	650	50,850	58,500
0800	3,750 kVA		.16	125		83,500	7,250	810	91,560	103,500

26 22 Low-Voltage Transformers

26 22 13 – Low-Voltage Distribution Transformers

26 22 13.10 Transformer, Dry-Type		Crew	Daily Output	Labor-Hours	Unit	Material	2018 Bare Costs Labor	Equipment	Total	Total Incl O&P
0010	**TRANSFORMER, DRY-TYPE**									
0050	Single phase, 240/480 volt primary, 120/240 volt secondary									
0100	1 kVA	1 Elec	2	4	Ea.	380	233		613	770
0300	2 kVA		1.60	5		590	291		881	1,075
0500	3 kVA		1.40	5.714		735	335		1,070	1,300
0700	5 kVA		1.20	6.667		1,000	390		1,390	1,675
0900	7.5 kVA	2 Elec	2.20	7.273		1,375	425		1,800	2,150
1100	10 kVA		1.60	10		1,575	580		2,155	2,600
1300	15 kVA		1.20	13.333		2,125	775		2,900	3,475
1500	25 kVA		1	16		2,375	930		3,305	4,025
1700	37.5 kVA		.80	20		2,875	1,175		4,050	4,925
1900	50 kVA		.70	22.857		3,500	1,325		4,825	5,850
2100	75 kVA		.65	24.615		4,500	1,425		5,925	7,100
2190	480 V primary, 120/240 V secondary, nonvent., 15 kVA		1.20	13.333		1,675	775		2,450	2,975
2200	25 kVA		.90	17.778		2,475	1,025		3,500	4,275
2210	37 kVA		.75	21.333		2,925	1,250		4,175	5,075
2220	50 kVA		.65	24.615		3,450	1,425		4,875	5,950
2300	3 phase, 480 volt primary, 120/208 volt secondary									
2310	Ventilated, 3 kVA	1 Elec	1	8	Ea.	1,125	465		1,590	1,950
2700	6 kVA		.80	10		1,275	580		1,855	2,275
2900	9 kVA		.70	11.429		1,275	665		1,940	2,400
3100	15 kVA	2 Elec	1.10	14.545		1,625	845		2,470	3,050
3300	30 kVA		.90	17.778		1,600	1,025		2,625	3,325
3500	45 kVA		.80	20		1,900	1,175		3,075	3,850
3700	75 kVA		.70	22.857		2,550	1,325		3,875	4,800
3900	112.5 kVA	R-3	.90	22.222		4,225	1,300	144	5,669	6,725
4100	150 kVA		.85	23.529		5,125	1,375	152	6,652	7,875
4300	225 kVA		.65	30.769		7,475	1,775	199	9,449	11,100
4500	300 kVA		.55	36.364		9,375	2,100	236	11,711	13,700
4700	500 kVA		.45	44.444		15,600	2,575	288	18,463	21,400
4800	750 kVA		.35	57.143		22,700	3,325	370	26,395	30,400

For customer support on your Building Construction Costs with RSMeans data, call 800.448.8182.

585

26 24 13.10 Incoming Switchboards

		Crew	Daily Output	Labor-Hours	Unit	Material	2018 Bare Costs Labor	Equipment	Total	Total Incl O&P
0010	**INCOMING SWITCHBOARDS** main service section									
0100	Aluminum bus bars, not including CT's or PT's									
0200	No main disconnect, includes CT compartment									
0300	120/208 volt, 4 wire, 600 amp	2 Elec	1	16	Ea.	4,000	930		4,930	5,800
0400	800 amp		.88	18.182		4,000	1,050		5,050	5,975
0500	1,000 amp		.80	20		4,800	1,175		5,975	7,025
0600	1,200 amp		.72	22.222		4,800	1,300		6,100	7,200
0700	1,600 amp		.66	24.242		4,800	1,400		6,200	7,375
0800	2,000 amp		.62	25.806		5,175	1,500		6,675	7,925
1000	3,000 amp		.56	28.571		6,825	1,675		8,500	9,975
2000	Fused switch & CT compartment									
2100	120/208 volt, 4 wire, 400 amp	2 Elec	1.12	14.286	Ea.	2,675	830		3,505	4,200
2200	600 amp		.94	17.021		3,175	990		4,165	4,975
2300	800 amp		.84	19.048		10,800	1,100		11,900	13,600
2400	1,200 amp		.68	23.529		14,100	1,375		15,475	17,600
2900	Pressure switch & CT compartment									
3000	120/208 volt, 4 wire, 800 amp	2 Elec	.80	20	Ea.	9,700	1,175		10,875	12,500
3100	1,200 amp		.66	24.242		18,800	1,400		20,200	22,800
3200	1,600 amp		.62	25.806		20,000	1,500		21,500	24,300
3300	2,000 amp		.56	28.571		21,200	1,675		22,875	25,800
4400	Circuit breaker, molded case & CT compartment									
4600	3 pole, 4 wire, 600 amp	2 Elec	.94	17.021	Ea.	8,350	990		9,340	10,700
4800	800 amp		.84	19.048		10,000	1,100		11,100	12,700
5000	1,200 amp		.68	23.529		13,600	1,375		14,975	17,100
5100	Copper bus bars, not incl. CT's or PT's, add, minimum					15%				

26 24 13.30 Distribution Switchboards Section

		Crew	Daily Output	Labor-Hours	Unit	Material	2018 Bare Costs Labor	Equipment	Total	Total Incl O&P
0010	**DISTRIBUTION SWITCHBOARDS SECTION**									
0100	Aluminum bus bars, not including breakers									
0195	120/208 or 277/480 volt, 4 wire, 400 amp	2 Elec	1.10	14.545	Ea.	1,450	845		2,295	2,875
0200	600 amp		1	16		1,525	930		2,455	3,075
0300	800 amp		.88	18.182		1,925	1,050		2,975	3,675
0400	1,000 amp		.80	20		2,550	1,175		3,725	4,550
0500	1,200 amp		.72	22.222		3,000	1,300		4,300	5,225
0600	1,600 amp		.66	24.242		4,000	1,400		5,400	6,475
0700	2,000 amp		.62	25.806		4,975	1,500		6,475	7,725
0800	2,500 amp		.60	26.667		5,975	1,550		7,525	8,900
0900	3,000 amp		.56	28.571		6,975	1,675		8,650	10,200
0950	4,000 amp		.52	30.769		7,975	1,800		9,775	11,500

26 24 13.40 Switchboards Feeder Section

		Crew	Daily Output	Labor-Hours	Unit	Material	2018 Bare Costs Labor	Equipment	Total	Total Incl O&P
0010	**SWITCHBOARDS FEEDER SECTION** group mounted devices									
0030	Circuit breakers									
0160	FA frame, 15 to 60 amp, 240 volt, 1 pole	1 Elec	8	1	Ea.	114	58		172	212
0280	FA frame, 70 to 100 amp, 240 volt, 1 pole		7	1.143		217	66.50		283.50	340
0420	KA frame, 70 to 225 amp		3.20	2.500		1,525	146		1,671	1,925
0430	LA frame, 125 to 400 amp		2.30	3.478		3,500	202		3,702	4,175
0460	MA frame, 450 to 600 amp		1.60	5		6,175	291		6,466	7,200
0470	700 to 800 amp		1.30	6.154		8,000	360		8,360	9,325
0480	MAL frame, 1,000 amp		1	8		8,300	465		8,765	9,850
0490	PA frame, 1,200 amp		.80	10		16,900	580		17,480	19,500
0500	Branch circuit, fusible switch, 600 volt, double 30/30 amp		4	2		845	116		961	1,100
0550	60/60 amp		3.20	2.500		870	146		1,016	1,175
0600	100/100 amp		2.70	2.963		1,100	172		1,272	1,450

For customer support on your Building Construction Costs with RSMeans data, call 800.448.8182.

26 24 Switchboards and Panelboards

26 24 13 – Switchboards

26 24 13.40 Switchboards Feeder Section

	26 24 13.40 Switchboards Feeder Section	Crew	Daily Output	Labor-Hours	Unit	Material	2018 Bare Costs Labor	Equipment	Total	Total Incl O&P
0650	Single, 30 amp	1 Elec	5.30	1.509	Ea.	695	88		783	895
0700	60 amp		4.70	1.702		825	99		924	1,050
0750	100 amp		4	2		1,050	116		1,166	1,325
0800	200 amp		2.70	2.963		1,350	172		1,522	1,750
0850	400 amp		2.30	3.478		2,475	202		2,677	3,025
0900	600 amp		1.80	4.444		3,025	259		3,284	3,700
0950	800 amp		1.30	6.154		5,100	360		5,460	6,125
1000	1,200 amp	▼	.80	10	▼	5,825	580		6,405	7,275

26 24 16 – Panelboards

26 24 16.20 Panelboard and Load Center Circuit Breakers

	26 24 16.20 Panelboard and Load Center Circuit Breakers	Crew	Daily Output	Labor-Hours	Unit	Material	2018 Bare Costs Labor	Equipment	Total	Total Incl O&P
0010	**PANELBOARD AND LOAD CENTER CIRCUIT BREAKERS**									
0050	Bolt-on, 10,000 amp I.C., 120 volt, 1 pole									
0100	15-50 amp	1 Elec	10	.800	Ea.	19.20	46.50		65.70	90.50
0200	60 amp		8	1		19.20	58		77.20	108
0300	70 amp	▼	8	1	▼	28	58		86	118
0350	240 volt, 2 pole									
0400	15-50 amp	1 Elec	8	1	Ea.	51	58		109	144
0500	60 amp		7.50	1.067		34.50	62		96.50	131
0600	80-100 amp		5	1.600		116	93		209	266
0700	3 pole, 15-60 amp		6.20	1.290		140	75		215	266
0800	70 amp		5	1.600		180	93		273	335
0900	80-100 amp		3.60	2.222		195	129		324	405
1000	22,000 amp I.C., 240 volt, 2 pole, 70-225 amp		2.70	2.963		555	172		727	875
1100	3 pole, 70-225 amp		2.30	3.478		275	202		477	605
1200	14,000 amp I.C., 277 volts, 1 pole, 15-30 amp		8	1		36	58		94	127
1300	22,000 amp I.C., 480 volts, 2 pole, 70-225 amp		2.70	2.963		505	172		677	815
1400	3 pole, 70-225 amp	▼	2.30	3.478	▼	770	202		972	1,150

26 24 16.30 Panelboards Commercial Applications

	26 24 16.30 Panelboards Commercial Applications	Crew	Daily Output	Labor-Hours	Unit	Material	2018 Bare Costs Labor	Equipment	Total	Total Incl O&P
0010	**PANELBOARDS COMMERCIAL APPLICATIONS**									
0050	NQOD, w/20 amp 1 pole bolt-on circuit breakers									
0100	3 wire, 120/240 volts, 100 amp main lugs									
0150	10 circuits	1 Elec	1	8	Ea.	925	465		1,390	1,725
0200	14 circuits		.88	9.091		1,025	530		1,555	1,925
0250	18 circuits		.75	10.667		1,125	620		1,745	2,150
0300	20 circuits	▼	.65	12.308		1,250	715		1,965	2,450
0350	225 amp main lugs, 24 circuits	2 Elec	1.20	13.333		1,425	775		2,200	2,700
0400	30 circuits		.90	17.778		1,625	1,025		2,650	3,350
0450	36 circuits		.80	20		1,875	1,175		3,050	3,800
0500	38 circuits		.72	22.222		2,025	1,300		3,325	4,150
0550	42 circuits	▼	.66	24.242		2,075	1,400		3,475	4,400
0600	4 wire, 120/208 volts, 100 amp main lugs, 12 circuits	1 Elec	1	8		995	465		1,460	1,800
0650	16 circuits		.75	10.667		1,150	620		1,770	2,175
0700	20 circuits		.65	12.308		1,325	715		2,040	2,525
0750	24 circuits		.60	13.333		1,275	775		2,050	2,550
0800	30 circuits	▼	.53	15.094		1,625	880		2,505	3,125
0850	225 amp main lugs, 32 circuits	2 Elec	.90	17.778		1,825	1,025		2,850	3,575
0900	34 circuits		.84	19.048		1,875	1,100		2,975	3,700
0950	36 circuits		.80	20		1,925	1,175		3,100	3,850
1000	42 circuits		.68	23.529		2,150	1,375		3,525	4,400
1010	400 amp main lugs, 42 circs	▼	.68	23.529	▼	2,150	1,375		3,525	4,400
1200	NEHB, w/20 amp, 1 pole bolt-on circuit-breakers									
1250	4 wire, 277/480 volts, 100 amp main lugs, 12 circuits	1 Elec	.88	9.091	Ea.	1,350	530		1,880	2,300

26 24 16.30 Panelboards Commercial Applications		Crew	Daily Output	Labor-Hours	Unit	Material	2018 Bare Costs Labor	Equipment	Total	Total Incl O&P
1300	20 circuits	1 Elec	.60	13.333	Ea.	2,025	775		2,800	3,375
1350	225 amp main lugs, 24 circuits	2 Elec	.90	17.778		2,300	1,025		3,325	4,075
1400	30 circuits		.80	20		2,750	1,175		3,925	4,775
1448	32 circuits		4.90	3.265		3,200	190		3,390	3,775
1450	36 circuits	▼	.72	22.222	▼	3,200	1,300		4,500	5,425
1600	NQOD panel, w/20 amp, 1 pole, circuit breakers									
1650	3 wire, 120/240 volt with main circuit breaker									
1700	100 amp main, 12 circuits	1 Elec	.80	10	Ea.	1,250	580		1,830	2,250
1750	20 circuits	"	.60	13.333		1,575	775		2,350	2,900
1800	225 amp main, 30 circuits	2 Elec	.68	23.529		2,975	1,375		4,350	5,325
1801	225 amp main, 32 circuits		5	3.200		2,975	186		3,161	3,550
1850	42 circuits		.52	30.769		3,450	1,800		5,250	6,450
1900	400 amp main, 30 circuits		.54	29.630		4,125	1,725		5,850	7,100
1950	42 circuits	▼	.50	32	▼	4,600	1,850		6,450	7,825
2000	4 wire, 120/208 volts with main circuit breaker									
2050	100 amp main, 24 circuits	1 Elec	.47	17.021	Ea.	1,925	990		2,915	3,575
2100	30 circuits	"	.40	20		2,075	1,175		3,250	4,050
2200	225 amp main, 32 circuits	2 Elec	.72	22.222		3,475	1,300		4,775	5,725
2250	42 circuits		.56	28.571		3,575	1,675		5,250	6,400
2300	400 amp main, 42 circuits		.48	33.333		5,075	1,950		7,025	8,500
2350	600 amp main, 42 circuits	▼	.40	40	▼	7,500	2,325		9,825	11,700
2400	NEHB, with 20 amp, 1 pole circuit breaker									
2450	4 wire, 277/480 volts with main circuit breaker									
2500	100 amp main, 24 circuits	1 Elec	.42	19.048	Ea.	2,650	1,100		3,750	4,550
2550	30 circuits	"	.38	21.053		3,100	1,225		4,325	5,225
2600	225 amp main, 30 circuits	2 Elec	.72	22.222		3,875	1,300		5,175	6,200
2650	42 circuits	"	.56	28.571	▼	4,775	1,675		6,450	7,725

26 24 19.40 Motor Starters and Controls		Crew	Daily Output	Labor-Hours	Unit	Material	2018 Bare Costs Labor	Equipment	Total	Total Incl O&P
0010	**MOTOR STARTERS AND CONTROLS**									
0050	Magnetic, FVNR, with enclosure and heaters, 480 volt									
0080	2 HP, size 00	1 Elec	3.50	2.286	Ea.	197	133		330	415
0100	5 HP, size 0		2.30	3.478		295	202		497	630
0200	10 HP, size 1	▼	1.60	5		267	291		558	730
0300	25 HP, size 2	2 Elec	2.20	7.273		500	425		925	1,175
0400	50 HP, size 3		1.80	8.889		820	515		1,335	1,675
0500	100 HP, size 4		1.20	13.333		1,825	775		2,600	3,150
0600	200 HP, size 5	▼	.90	17.778		4,250	1,025		5,275	6,225
0700	Combination, with motor circuit protectors, 5 HP, size 0	1 Elec	1.80	4.444		1,025	259		1,284	1,500
0800	10 HP, size 1	"	1.30	6.154		1,050	360		1,410	1,675
0900	25 HP, size 2	2 Elec	2	8		1,475	465		1,940	2,325
1000	50 HP, size 3		1.32	12.121		2,150	705		2,855	3,400
1200	100 HP, size 4	▼	.80	20		4,625	1,175		5,800	6,825
1400	Combination, with fused switch, 5 HP, size 0	1 Elec	1.80	4.444		590	259		849	1,025
1600	10 HP, size 1	"	1.30	6.154		630	360		990	1,225
1800	25 HP, size 2	2 Elec	2	8		1,025	465		1,490	1,825
2000	50 HP, size 3		1.32	12.121		1,725	705		2,430	2,950
2200	100 HP, size 4	▼	.80	20	▼	3,025	1,175		4,200	5,075

26 25 13.40 Copper Bus Duct	Crew	Daily Output	Labor-Hours	Unit	Material	2018 Bare Costs Labor	Equipment	Total	Total Incl O&P
0010 **COPPER BUS DUCT** 10' long									
0050 Indoor 3 pole 4 wire, plug-in, straight section, 225 amp	2 Elec	40	.400	L.F.	222	23.50		245.50	279
1000 400 amp		32	.500		375	29		404	455
1500 600 amp		26	.615		425	36		461	520
2400 800 amp		20	.800		620	46.50		666.50	750
2450 1,000 amp		18	.889		300	51.50		351.50	410
2500 1,350 amp		16	1		415	58		473	540
2510 1,600 amp		12	1.333		470	77.50		547.50	630
2520 2,000 amp		10	1.600		595	93		688	795
2550 Feeder, 600 amp		28	.571		204	33.50		237.50	274
2600 800 amp		22	.727		750	42.50		792.50	890
2700 1,000 amp		20	.800		885	46.50		931.50	1,050
2800 1,350 amp		18	.889		385	51.50		436.50	505
2900 1,600 amp		14	1.143		1,425	66.50		1,491.50	1,675
3000 2,000 amp		12	1.333	▼	1,775	77.50		1,852.50	2,075
3100 Elbows, 225 amp		4	4	Ea.	1,575	233		1,808	2,075
3200 400 amp		3.60	4.444		1,575	259		1,834	2,100
3300 600 amp		3.20	5		1,575	291		1,866	2,150
3400 800 amp		2.80	5.714		1,700	335		2,035	2,375
3500 1,000 amp		2.60	6.154		1,900	360		2,260	2,625
3600 1,350 amp		2.40	6.667		1,875	390		2,265	2,650
3700 1,600 amp		2.20	7.273		2,050	425		2,475	2,875
3800 2,000 amp		1.80	8.889		2,525	515		3,040	3,550
4000 End box, 225 amp		34	.471		170	27.50		197.50	228
4100 400 amp		32	.500		188	29		217	251
4200 600 amp		28	.571		188	33.50		221.50	257
4300 800 amp		26	.615		188	36		224	261
4400 1,000 amp		24	.667		179	39		218	255
4500 1,350 amp		22	.727		179	42.50		221.50	260
4600 1,600 amp		20	.800		179	46.50		225.50	266
4700 2,000 amp		18	.889		219	51.50		270.50	320
4800 Cable tap box end, 225 amp		3.20	5		1,175	291		1,466	1,725
5000 400 amp		2.60	6.154		1,200	360		1,560	1,850
5100 600 amp		2.20	7.273		1,625	425		2,050	2,400
5200 800 amp		2	8		1,700	465		2,165	2,575
5300 1,000 amp		1.60	10		1,725	580		2,305	2,775
5400 1,350 amp		1.40	11.429		2,250	665		2,915	3,475
5500 1,600 amp		1.20	13.333		2,500	775		3,275	3,900
5600 2,000 amp		1	16		2,800	930		3,730	4,475
5700 Switchboard stub, 225 amp		5.40	2.963		1,250	172		1,422	1,650
5800 400 amp		4.60	3.478		1,325	202		1,527	1,750
5900 600 amp		4	4		1,375	233		1,608	1,875
6000 800 amp		3.20	5		1,675	291		1,966	2,250
6100 1,000 amp		3	5.333		1,925	310		2,235	2,600
6200 1,350 amp		2.60	6.154		2,400	360		2,760	3,150
6300 1,600 amp		2.40	6.667		2,700	390		3,090	3,550
6400 2,000 amp		2	8		3,275	465		3,740	4,300
6490 Tee fittings, 225 amp		2.40	6.667		2,175	390		2,565	2,975
6500 400 amp		2	8		2,175	465		2,640	3,100
6600 600 amp		1.80	8.889		2,175	515		2,690	3,175
6700 800 amp		1.60	10		2,250	580		2,830	3,375
6800 1,350 amp		1.20	13.333		3,025	775		3,800	4,500
7000 1,600 amp	▼	1	16	▼	3,450	930		4,380	5,200

26 25 Low-Voltage Enclosed Bus Assemblies

26 25 13 – Low-Voltage Busways

26 25 13.40 Copper Bus Duct

		Crew	Daily Output	Labor-Hours	Unit	Material	2018 Bare Costs Labor	2018 Bare Costs Equipment	Total	Total Incl O&P
7100	2,000 amp	2 Elec	.80	20	Ea.	4,100	1,175		5,275	6,250
7200	Plug-in fusible switches w/3 fuses, 600 volt, 3 pole, 30 amp	1 Elec	4	2		860	116		976	1,125
7300	60 amp		3.60	2.222		970	129		1,099	1,275
7400	100 amp		2.70	2.963		1,325	172		1,497	1,700
7500	200 amp	2 Elec	3.20	5		2,325	291		2,616	3,000
7600	400 amp		1.40	11.429		6,500	665		7,165	8,150
7700	600 amp		.90	17.778		7,850	1,025		8,875	10,200
7800	800 amp		.66	24.242		12,600	1,400		14,000	15,900
7900	1,200 amp		.50	32		23,600	1,850		25,450	28,700
8000	Plug-in circuit breakers, molded case, 15 to 50 amp	1 Elec	4.40	1.818		815	106		921	1,050
8100	70 to 100 amp	"	3.10	2.581		905	150		1,055	1,225
8200	150 to 225 amp	2 Elec	3.40	4.706		2,450	274		2,724	3,100
8300	250 to 400 amp		1.40	11.429		4,300	665		4,965	5,750
8400	500 to 600 amp		1	16		5,825	930		6,755	7,800
8500	700 to 800 amp		.64	25		7,175	1,450		8,625	10,100
8600	900 to 1,000 amp		.56	28.571		10,300	1,675		11,975	13,800
8700	1,200 amp		.44	36.364		12,400	2,125		14,525	16,800

26 27 Low-Voltage Distribution Equipment

26 27 16 – Electrical Cabinets and Enclosures

26 27 16.10 Cabinets

		Crew	Daily Output	Labor-Hours	Unit	Material	2018 Bare Costs Labor	2018 Bare Costs Equipment	Total	Total Incl O&P
0010	**CABINETS**									
7000	Cabinets, current transformer									
7050	Single door, 24" H x 24" W x 10" D	1 Elec	1.60	5	Ea.	165	291		456	615
7100	30" H x 24" W x 10" D		1.30	6.154		179	360		539	730
7150	36" H x 24" W x 10" D		1.10	7.273		380	425		805	1,050
7200	30" H x 30" W x 10" D		1	8		246	465		711	965
7250	36" H x 30" W x 10" D		.90	8.889		310	515		825	1,125
7300	36" H x 36" W x 10" D		.80	10		315	580		895	1,225
7500	Double door, 48" H x 36" W x 10" D		.60	13.333		575	775		1,350	1,775
7550	24" H x 24" W x 12" D		1	8		179	465		644	890

26 27 23 – Indoor Service Poles

26 27 23.40 Surface Raceway

		Crew	Daily Output	Labor-Hours	Unit	Material	2018 Bare Costs Labor	2018 Bare Costs Equipment	Total	Total Incl O&P
0010	**SURFACE RACEWAY**									
0090	Metal, straight section									
0100	No. 500	1 Elec	100	.080	L.F.	1.07	4.66		5.73	8.15
0110	No. 700		100	.080		1.19	4.66		5.85	8.25
0400	No. 1500, small pancake		90	.089		2.20	5.15		7.35	10.15
0600	No. 2000, base & cover, blank		90	.089		2.24	5.15		7.39	10.20
0800	No. 3000, base & cover, blank		75	.107		4.28	6.20		10.48	14
1000	No. 4000, base & cover, blank		65	.123		6.95	7.15		14.10	18.35
1200	No. 6000, base & cover, blank		50	.160		11.75	9.30		21.05	27
2400	Fittings, elbows, No. 500		40	.200	Ea.	1.94	11.65		13.59	19.55
2800	Elbow cover, No. 2000		40	.200		3.64	11.65		15.29	21.50
2880	Tee, No. 500		42	.190		3.74	11.10		14.84	20.50
2900	No. 2000		27	.296		12.10	17.25		29.35	39.50
3000	Switch box, No. 500		16	.500		12.50	29		41.50	57.50
3400	Telephone outlet, No. 1500		16	.500		14.50	29		43.50	59.50
3600	Junction box, No. 1500		16	.500		10.30	29		39.30	55
3800	Plugmold wired sections, No. 2000									

For customer support on your Building Construction Costs with RSMeans data, call 800.448.8182.

26 27 Low-Voltage Distribution Equipment

26 27 23 – Indoor Service Poles

26 27 23.40 Surface Raceway

	Crew	Daily Output	Labor-Hours	Unit	Material	2018 Bare Costs Labor	Equipment	Total	Total Incl O&P
4000	1 circuit, 6 outlets, 3' long	1 Elec	8	1	Ea.	38	58		96	129
4100	2 circuits, 8 outlets, 6' long	"	5.30	1.509	"	62.50	88		150.50	200

26 27 26 – Wiring Devices

26 27 26.10 Low Voltage Switching

	Crew	Daily Output	Labor-Hours	Unit	Material	2018 Bare Costs Labor	Equipment	Total	Total Incl O&P
0010	**LOW VOLTAGE SWITCHING**									
3600	Relays, 120 V or 277 V standard	1 Elec	12	.667	Ea.	41	39		80	103
3800	Flush switch, standard		40	.200		11.50	11.65		23.15	30
4000	Interchangeable		40	.200		15.05	11.65		26.70	34
4100	Surface switch, standard		40	.200		8.20	11.65		19.85	26.50
4200	Transformer 115 V to 25 V		12	.667		128	39		167	199
4400	Master control, 12 circuit, manual		4	2		128	116		244	315
4500	25 circuit, motorized		4	2		141	116		257	330
4600	Rectifier, silicon		12	.667		46	39		85	109
4800	Switchplates, 1 gang, 1, 2 or 3 switch, plastic		80	.100		5	5.80		10.80	14.20
5000	Stainless steel		80	.100		11.35	5.80		17.15	21
5400	2 gang, 3 switch, stainless steel		53	.151		23	8.80		31.80	38
5500	4 switch, plastic		53	.151		10.45	8.80		19.25	24.50
5800	3 gang, 9 switch, stainless steel		32	.250		65.50	14.55		80.05	94

26 27 26.20 Wiring Devices Elements

	Crew	Daily Output	Labor-Hours	Unit	Material	2018 Bare Costs Labor	Equipment	Total	Total Incl O&P
0010	**WIRING DEVICES ELEMENTS**									
0200	Toggle switch, quiet type, single pole, 15 amp	1 Elec	40	.200	Ea.	.49	11.65		12.14	17.95
0600	3 way, 15 amp		23	.348		1.52	20		21.52	32
0900	4 way, 15 amp		15	.533		10.70	31		41.70	58.50
1650	Dimmer switch, 120 volt, incandescent, 600 watt, 1 pole [G]		16	.500		22.50	29		51.50	68.50
2460	Receptacle, duplex, 120 volt, grounded, 15 amp		40	.200		1.30	11.65		12.95	18.85
2470	20 amp		27	.296		9.05	17.25		26.30	36
2490	Dryer, 30 amp		15	.533		4.44	31		35.44	51.50
2500	Range, 50 amp		11	.727		10.60	42.50		53.10	75
2600	Wall plates, stainless steel, 1 gang		80	.100		2.56	5.80		8.36	11.50
2800	2 gang		53	.151		4.33	8.80		13.13	17.90
3200	Lampholder, keyless		26	.308		16.85	17.90		34.75	45.50
3400	Pullchain with receptacle		22	.364		20.50	21		41.50	54

26 27 73 – Door Chimes

26 27 73.10 Doorbell System

	Crew	Daily Output	Labor-Hours	Unit	Material	2018 Bare Costs Labor	Equipment	Total	Total Incl O&P
0010	**DOORBELL SYSTEM**, incl. transformer, button & signal									
0100	6" bell	1 Elec	4	2	Ea.	139	116		255	325
0200	Buzzer	"	4	2	"	109	116		225	294

26 28 Low-Voltage Circuit Protective Devices

26 28 16 – Enclosed Switches and Circuit Breakers

26 28 16.10 Circuit Breakers

	Crew	Daily Output	Labor-Hours	Unit	Material	2018 Bare Costs Labor	Equipment	Total	Total Incl O&P
0010	**CIRCUIT BREAKERS** (in enclosure)									
0100	Enclosed (NEMA 1), 600 volt, 3 pole, 30 amp	1 Elec	3.20	2.500	Ea.	520	146		666	790
0200	60 amp		2.80	2.857		635	166		801	950
0400	100 amp		2.30	3.478		730	202		932	1,100
0500	200 amp		1.50	5.333		1,525	310		1,835	2,150
0600	225 amp		1.50	5.333		1,675	310		1,985	2,325
0700	400 amp	2 Elec	1.60	10		2,875	580		3,455	4,050
0800	600 amp		1.20	13.333		4,175	775		4,950	5,725
1000	800 amp		.94	17.021		5,425	990		6,415	7,450

For customer support on your Building Construction Costs with RSMeans data, call 800.448.8182.

591

26 28 Low-Voltage Circuit Protective Devices

26 28 16 – Enclosed Switches and Circuit Breakers

26 28 16.20 Safety Switches

		Crew	Daily Output	Labor-Hours	Unit	Material	2018 Bare Costs Labor	2018 Bare Costs Equipment	Total	Total Incl O&P
0010	**SAFETY SWITCHES**									
0100	General duty 240 volt, 3 pole NEMA 1, fusible, 30 amp	1 Elec	3.20	2.500	Ea.	55.50	146		201.50	279
0200	60 amp		2.30	3.478		94	202		296	410
0300	100 amp		1.90	4.211		159	245		404	540
0400	200 amp		1.30	6.154		345	360		705	915
0500	400 amp	2 Elec	1.80	8.889		905	515		1,420	1,775
0600	600 amp	"	1.20	13.333		1,500	775		2,275	2,800
2900	Heavy duty, 240 volt, 3 pole NEMA 1 fusible									
2910	30 amp	1 Elec	3.20	2.500	Ea.	87.50	146		233.50	315
3000	60 amp		2.30	3.478		145	202		347	465
3300	100 amp		1.90	4.211		228	245		473	615
3500	200 amp		1.30	6.154		390	360		750	960
3700	400 amp	2 Elec	1.80	8.889		990	515		1,505	1,875
3900	600 amp	"	1.20	13.333		1,750	775		2,525	3,075

26 29 Low-Voltage Controllers

26 29 13 – Enclosed Controllers

26 29 13.20 Control Stations

		Crew	Daily Output	Labor-Hours	Unit	Material	2018 Bare Costs Labor	2018 Bare Costs Equipment	Total	Total Incl O&P
0010	**CONTROL STATIONS**									
0050	NEMA 1, heavy duty, stop/start	1 Elec	8	1	Ea.	146	58		204	248
0100	Stop/start, pilot light		6.20	1.290		199	75		274	330
0200	Hand/off/automatic		6.20	1.290		108	75		183	231
0400	Stop/start/reverse		5.30	1.509		197	88		285	350

26 32 Packaged Generator Assemblies

26 32 13 – Engine Generators

26 32 13.13 Diesel-Engine-Driven Generator Sets

		Crew	Daily Output	Labor-Hours	Unit	Material	2018 Bare Costs Labor	2018 Bare Costs Equipment	Total	Total Incl O&P
0010	**DIESEL-ENGINE-DRIVEN GENERATOR SETS**									
2000	Diesel engine, including battery, charger,									
2010	muffler, & day tank, 30 kW	R-3	.55	36.364	Ea.	11,000	2,100	236	13,336	15,500
2100	50 kW		.42	47.619		20,000	2,750	310	23,060	26,500
2200	75 kW		.35	57.143		23,100	3,325	370	26,795	30,800
2300	100 kW		.31	64.516		27,900	3,750	420	32,070	36,800
2400	125 kW		.29	68.966		29,300	4,000	445	33,745	38,800
2500	150 kW		.26	76.923		37,600	4,450	500	42,550	48,600
2501	Generator set, dsl eng in alum encl, incl btry, chgr, muf & day tank,150 kW		.26	76.923		41,400	4,450	500	46,350	52,500
2600	175 kW		.25	80		41,900	4,650	520	47,070	53,500
2700	200 kW		.24	83.333		45,200	4,825	540	50,565	57,500
2800	250 kW		.23	86.957		48,400	5,050	565	54,015	61,500
2900	300 kW		.22	90.909		51,500	5,275	590	57,365	65,000
3000	350 kW		.20	100		59,000	5,800	650	65,450	74,000
3100	400 kW		.19	105		72,000	6,100	680	78,780	89,000
3200	500 kW		.18	111		91,500	6,450	720	98,670	111,000

26 32 13.16 Gas-Engine-Driven Generator Sets

		Crew	Daily Output	Labor-Hours	Unit	Material	2018 Bare Costs Labor	2018 Bare Costs Equipment	Total	Total Incl O&P
0010	**GAS-ENGINE-DRIVEN GENERATOR SETS**									
0020	Gas or gasoline operated, includes battery,									
0050	charger, & muffler									
0200	7.5 kW	R-3	.83	24.096	Ea.	8,550	1,400	156	10,106	11,700

26 32 Packaged Generator Assemblies

26 32 13 – Engine Generators

26 32 13.16 Gas-Engine-Driven Generator Sets

		Crew	Daily Output	Labor-Hours	Unit	Material	2018 Bare Costs Labor	Equipment	Total	Total Incl O&P
0300	11.5 kW	R-3	.71	28.169	Ea.	12,100	1,625	183	13,908	16,000
0400	20 kW		.63	31.746		14,300	1,850	206	16,356	18,700
0500	35 kW	↓	.55	36.364		17,000	2,100	236	19,336	22,100
0600	80 kW	R-13	.40	105		27,900	5,925	590	34,415	40,200
0700	100 kW		.33	127		30,600	7,175	715	38,490	45,100
0800	125 kW		.28	150		62,500	8,450	840	71,790	82,500
0900	185 kW	↓	.25	168	↓	83,000	9,475	940	93,415	106,000

26 33 Battery Equipment

26 33 43 – Battery Chargers

26 33 43.55 Electric Vehicle Charging

			Crew	Daily Output	Labor-Hours	Unit	Material	2018 Bare Costs Labor	Equipment	Total	Total Incl O&P
0010	**ELECTRIC VEHICLE CHARGING**										
0020	Level 2, wall mounted										
2200	Heavy duty	G	R-1A	15.36	1.042	Ea.	1,575	54.50		1,629.50	1,825
2210	with RFID	G		12.29	1.302		3,175	68		3,243	3,600
2300	Free standing, single connector	G		10.24	1.563		1,800	82		1,882	2,125
2310	with RFID	G		8.78	1.822		3,625	95.50		3,720.50	4,125
2320	Double connector	G		7.68	2.083		3,350	109		3,459	3,850
2330	with RFID	G	↓	6.83	2.343	↓	6,725	123		6,848	7,575

26 35 Power Filters and Conditioners

26 35 13 – Capacitors

26 35 13.10 Capacitors Indoor

		Crew	Daily Output	Labor-Hours	Unit	Material	2018 Bare Costs Labor	Equipment	Total	Total Incl O&P
0010	**CAPACITORS INDOOR**									
0020	240 volts, single & 3 phase, 0.5 kVAR	1 Elec	2.70	2.963	Ea.	550	172		722	865
0100	1.0 kVAR		2.70	2.963		665	172		837	995
0150	2.5 kVAR		2	4		750	233		983	1,175
0200	5.0 kVAR		1.80	4.444		790	259		1,049	1,250
0250	7.5 kVAR		1.60	5		1,350	291		1,641	1,925
0300	10 kVAR		1.50	5.333		1,575	310		1,885	2,200
0350	15 kVAR		1.30	6.154		1,975	360		2,335	2,700
0400	20 kVAR		1.10	7.273		2,375	425		2,800	3,250
0450	25 kVAR		1	8		2,825	465		3,290	3,800
1000	480 volts, single & 3 phase, 1 kVAR		2.70	2.963		505	172		677	815
1050	2 kVAR		2.70	2.963		580	172		752	895
1100	5 kVAR		2	4		730	233		963	1,150
1150	7.5 kVAR		2	4		785	233		1,018	1,225
1200	10 kVAR		2	4		1,050	233		1,283	1,500
1250	15 kVAR		2	4		1,250	233		1,483	1,725
1300	20 kVAR		1.60	5		1,400	291		1,691	1,950
1350	30 kVAR		1.50	5.333		1,600	310		1,910	2,250
1400	40 kVAR		1.20	6.667		2,025	390		2,415	2,800
1450	50 kVAR	↓	1.10	7.273	↓	2,400	425		2,825	3,275

For customer support on your Building Construction Costs with RSMeans data, call 800.448.8182.

593

26 51 Interior Lighting

26 51 13 – Interior Lighting Fixtures, Lamps, and Ballasts

26 51 13.50 Interior Lighting Fixtures		Crew	Daily Output	Labor-Hours	Unit	Material	2018 Bare Costs Labor	Equipment	Total	Total Incl O&P
0010	**INTERIOR LIGHTING FIXTURES** Including lamps, mounting									
0030	hardware and connections									
0100	Fluorescent, C.W. lamps, troffer, recess mounted in grid, RS									
0130	Grid ceiling mount									
0200	Acrylic lens, 1' W x 4' L, two 40 watt	1 Elec	5.70	1.404	Ea.	51.50	81.50		133	179
0210	1' W x 4' L, three 40 watt		5.40	1.481		58.50	86		144.50	194
0300	2' W x 2' L, two U40 watt		5.70	1.404		55.50	81.50		137	183
0400	2' W x 4' L, two 40 watt		5.30	1.509		54	88		142	191
0500	2' W x 4' L, three 40 watt		5	1.600		59.50	93		152.50	205
0600	2' W x 4' L, four 40 watt		4.70	1.702		62	99		161	217
0700	4' W x 4' L, four 40 watt	2 Elec	6.40	2.500		276	146		422	525
0800	4' W x 4' L, six 40 watt		6.20	2.581		297	150		447	550
0900	4' W x 4' L, eight 40 watt		5.80	2.759		340	161		501	615
0910	Acrylic lens, 1' W x 4' L, two 32 watt T8 [G]	1 Elec	5.70	1.404		61.50	81.50		143	190
0930	2' W x 2' L, two U32 watt T8 [G]		5.70	1.404		96	81.50		177.50	227
0940	2' W x 4' L, two 32 watt T8 [G]		5.30	1.509		79	88		167	218
0950	2' W x 4' L, three 32 watt T8 [G]		5	1.600		70	93		163	216
0960	2' W x 4' L, four 32 watt T8 [G]		4.70	1.702		74.50	99		173.50	230
1000	Surface mounted, RS									
1030	Acrylic lens with hinged & latched door frame									
1100	1' W x 4' L, two 40 watt	1 Elec	7	1.143	Ea.	65.50	66.50		132	172
1110	1' W x 4' L, three 40 watt		6.70	1.194		68	69.50		137.50	179
1200	2' W x 2' L, two U40 watt		7	1.143		70.50	66.50		137	177
1300	2' W x 4' L, two 40 watt		6.20	1.290		80	75		155	200
1400	2' W x 4' L, three 40 watt		5.70	1.404		83	81.50		164.50	213
1500	2' W x 4' L, four 40 watt		5.30	1.509		85	88		173	225
1501	2' W x 4' L, six 40 watt T8		5.20	1.538		85	89.50		174.50	228
1600	4' W x 4' L, four 40 watt	2 Elec	7.20	2.222		420	129		549	655
1700	4' W x 4' L, six 40 watt		6.60	2.424		455	141		596	710
1800	4' W x 4' L, eight 40 watt		6.20	2.581		475	150		625	745
1900	2' W x 8' L, four 40 watt		6.40	2.500		167	146		313	400
2000	2' W x 8' L, eight 40 watt		6.20	2.581		179	150		329	420
2100	Strip fixture									
2130	Surface mounted									
2200	4' long, one 40 watt, RS	1 Elec	8.50	.941	Ea.	31	55		86	116
2300	4' long, two 40 watt, RS		8	1		43	58		101	135
2400	4' long, one 40 watt, SL		8	1		53.50	58		111.50	146
2500	4' long, two 40 watt, SL		7	1.143		72.50	66.50		139	179
2600	8' long, one 75 watt, SL	2 Elec	13.40	1.194		55	69.50		124.50	165
2700	8' long, two 75 watt, SL	"	12.40	1.290		66.50	75		141.50	185
2800	4' long, two 60 watt, HO	1 Elec	6.70	1.194		107	69.50		176.50	222
2900	8' long, two 110 watt, HO	2 Elec	10.60	1.509		108	88		196	250
2950	High bay pendent mounted, 16" W x 4' L, four 54 watt, T5HO [G]		8.90	1.798		244	105		349	425
2952	2' W x 4' L, six 54 watt, T5HO [G]		8.50	1.882		310	110		420	505
2954	2' W x 4' L, six 32 watt, T8 [G]		8.50	1.882		179	110		289	360
3000	Strip, pendent mounted, industrial, white porcelain enamel									
3100	4' long, two 40 watt, RS	1 Elec	5.70	1.404	Ea.	53	81.50		134.50	180
3200	4' long, two 60 watt, HO	"	5	1.600		83	93		176	231
3300	8' long, two 75 watt, SL	2 Elec	8.80	1.818		98.50	106		204.50	266
3400	8' long, two 110 watt, HO	"	8	2		126	116		242	315
3470	Troffer, air handling, 2' W x 4' L with four 32 watt T8 [G]	1 Elec	4	2		116	116		232	300
3480	2' W x 2' L with two U32 watt T8 [G]		5.50	1.455		111	84.50		195.50	250
3490	Air connector insulated, 5" diameter		20	.400		68	23.50		91.50	110

594

26 51 Interior Lighting

26 51 13 - Interior Lighting Fixtures, Lamps, and Ballasts

26 51 13.50 Interior Lighting Fixtures

		Crew	Daily Output	Labor-Hours	Unit	Material	2018 Bare Costs Labor	Equipment	Total	Total Incl O&P
3500	6" diameter	1 Elec	20	.400	Ea.	69	23.50		92.50	111
3510	Troffer parabolic lay-in, 1' W x 4' L with one 32 W T8 G		5.70	1.404		117	81.50		198.50	251
3520	1' W x 4' L with two 32 W T8 G		5.30	1.509		140	88		228	285
3525	2' W x 2' L with two U32 W T8 G		5.70	1.404		118	81.50		199.50	252
3530	2' W x 4' L with three 32 W T8 G		5	1.600		131	93		224	283
3531	Intr fxtr, fluor, troffer prismatic lay-in, 2' W x 4'l w/three 32 W T8		5	1.600		152	93		245	305
4450	Incandescent, high hat can, round alzak reflector, prewired									
4470	100 watt	1 Elec	8	1	Ea.	64.50	58		122.50	158
4480	150 watt		8	1		105	58		163	202
4500	300 watt		6.70	1.194		242	69.50		311.50	370
4600	Square glass lens with metal trim, prewired									
4630	100 watt	1 Elec	6.70	1.194	Ea.	55.50	69.50		125	165
4680	150 watt		6.70	1.194		98	69.50		167.50	212
4700	200 watt		6.70	1.194		98	69.50		167.50	212
4800	300 watt		5.70	1.404		146	81.50		227.50	282
4900	Ceiling/wall, surface mounted, metal cylinder, 75 watt		10	.800		50.50	46.50		97	126
4920	150 watt		10	.800		81	46.50		127.50	159
5200	Ceiling, surface mounted, opal glass drum									
5300	8", one 60 watt lamp	1 Elec	10	.800	Ea.	65.50	46.50		112	142
5400	10", two 60 watt lamps		8	1		72.50	58		130.50	167
5500	12", four 60 watt lamps		6.70	1.194		103	69.50		172.50	217
6010	Vapor tight, incandescent, ceiling mounted, 200 watt		6.20	1.290		78.50	75		153.50	198
6100	Fluorescent, surface mounted, 2 lamps, 4' L, RS, 40 watt		3.20	2.500		116	146		262	345
6850	Vandalproof, surface mounted, fluorescent, two 32 watt T8 G		3.20	2.500		252	146		398	495
6860	Incandescent, one 150 watt		8	1		90	58		148	186
7500	Ballast replacement, by weight of ballast, to 15' high									
7520	Indoor fluorescent, less than 2 lb.	1 Elec	10	.800	Ea.	23.50	46.50		70	95.50
7540	Two 40W, watt reducer, 2 to 5 lb.		9.40	.851		71	49.50		120.50	153
7560	Two F96 slimline, over 5 lb.		8	1		110	58		168	208
7580	Vaportite ballast, less than 2 lb.		9.40	.851		23.50	49.50		73	100
7600	2 lb. to 5 lb.		8.90	.899		71	52.50		123.50	157
7620	Over 5 lb.		7.60	1.053		110	61.50		171.50	213
7630	Electronic ballast for two tubes		8	1		43.50	58		101.50	135
7640	Dimmable ballast one-lamp G		8	1		108	58		166	206
7650	Dimmable ballast two-lamp G		7.60	1.053		108	61.50		169.50	210

26 51 13.55 Interior LED Fixtures

		Crew	Daily Output	Labor-Hours	Unit	Material	2018 Bare Costs Labor	Equipment	Total	Total Incl O&P
0010	**INTERIOR LED FIXTURES** Incl. lamps and mounting hardware									
0100	Downlight, recess mounted, 7.5" diameter, 25 watt G	1 Elec	8	1	Ea.	335	58		393	450
0120	10" diameter, 36 watt G		8	1		360	58		418	480
0160	cylinder, 10 watts G		8	1		103	58		161	200
0180	20 watts G		8	1		129	58		187	229
1000	Troffer, recess mounted, 2' x 4', 3,200 lumens G		5.30	1.509		138	88		226	282
1010	4,800 lumens G		5	1.600		179	93		272	335
1020	6,400 lumens G		4.70	1.702		198	99		297	365
1100	Troffer retrofit lamp, 38 watt G		21	.381		69.50	22		91.50	110
1110	60 watt G		20	.400		141	23.50		164.50	191
1120	100 watt G		18	.444		206	26		232	265
1200	Troffer, volumetric recess mounted, 2' x 2' G		5.70	1.404		305	81.50		386.50	455
2000	Strip, surface mounted, one light bar 4' long, 3,500 K G		8.50	.941		260	55		315	370
2010	5,000 K G		8	1		260	58		318	375
2020	Two light bar 4' long, 5,000 K G		7	1.143		410	66.50		476.50	550
3000	Linear, suspended mounted, one light bar 4' long, 37 watt G		6.70	1.194		153	69.50		222.50	272

For customer support on your Building Construction Costs with RSMeans data, call 800.448.8182.

595

26 51 Interior Lighting

26 51 13 – Interior Lighting Fixtures, Lamps, and Ballasts

26 51 13.55 Interior LED Fixtures		Crew	Daily Output	Labor-Hours	Unit	Material	2018 Bare Costs Labor	Equipment	Total	Total Incl O&P
3010	One light bar 8' long, 74 watt	G 2 Elec	12.20	1.311	Ea.	283	76.50		359.50	425
3020	Two light bar 4' long, 74 watt	G 1 Elec	5.70	1.404		305	81.50		386.50	455
3030	Two light bar 8' long, 148 watt	G 2 Elec	8.80	1.818		350	106		456	545
4000	High bay, surface mounted, round, 150 watts	G	5.41	2.959		425	172		597	725
4010	2 bars, 164 watts	G	5.41	2.959		385	172		557	680
4020	3 bars, 246 watts	G	5.01	3.197		515	186		701	845
4030	4 bars, 328 watts	G	4.60	3.478		715	202		917	1,100
4040	5 bars, 410 watts	G 3 Elec	4.20	5.716		810	335		1,145	1,400
4050	6 bars, 492 watts	G	3.80	6.324		910	370		1,280	1,550
4060	7 bars, 574 watts	G	3.39	7.075		990	410		1,400	1,725
4070	8 bars, 656 watts	G	2.99	8.029		1,050	465		1,515	1,850
5000	Track, lighthead, 6 watt	G 1 Elec	32	.250		54.50	14.55		69.05	82
5010	9 watt	G "	32	.250		61.50	14.55		76.05	89.50
6000	Garage, surface mounted, 103 watts	G 2 Elec	6.50	2.462		970	143		1,113	1,300
6100	pendent mounted, 80 watts	G	6.50	2.462		660	143		803	940
6200	95 watts	G	6.50	2.462		780	143		923	1,075
6300	125 watts	G	6.50	2.462		820	143		963	1,125

26 52 Safety Lighting

26 52 13 – Emergency and Exit Lighting

26 52 13.10 Emergency Lighting and Battery Units

		Crew	Daily Output	Labor-Hours	Unit	Material	Labor	Equipment	Total	Total Incl O&P
0010	**EMERGENCY LIGHTING AND BATTERY UNITS**									
0300	Emergency light units, battery operated									
0350	Twin sealed beam light, 25 W, 6 V each									
0500	Lead battery operated	1 Elec	4	2	Ea.	144	116		260	330
0700	Nickel cadmium battery operated		4	2		345	116		461	550
0900	Self-contained fluorescent lamp pack		10	.800		191	46.50		237.50	280

26 52 13.16 Exit Signs

		Crew	Daily Output	Labor-Hours	Unit	Material	Labor	Equipment	Total	Total Incl O&P
0010	**EXIT SIGNS**									
0080	Exit light ceiling or wall mount, incandescent, single face	1 Elec	8	1	Ea.	70	58		128	164
0100	Double face		6.70	1.194		50	69.50		119.50	159
0200	LED standard, single face	G	8	1		47.50	58		105.50	139
0220	Double face	G	6.70	1.194		51	69.50		120.50	160
0230	LED vandal-resistant, single face	G	7.27	1.100		212	64		276	330
0240	LED w/battery unit, single face	G	4.40	1.818		184	106		290	360
0260	Double face	G	4	2		188	116		304	380
0262	LED w/battery unit, vandal-resistant, single face	G	4.40	1.818		245	106		351	430
0270	Combination emergency light units and exit sign		4	2		179	116		295	370

For customer support on your Building Construction Costs with RSMeans data, call 800.448.8182.

26 54 Classified Location Lighting

26 54 13 - Incandescent Classified Location Lighting

26 54 13.20 Explosionproof	Crew	Daily Output	Labor-Hours	Unit	Material	2018 Bare Costs Labor	Equipment	Total	Total Incl O&P
0010 **EXPLOSION PROOF**, incl. lamps, mounting hardware and connections									
6510 Incandescent, ceiling mounted, 200 watt	1 Elec	4	2	Ea.	1,550	116		1,666	1,875
6600 Fluorescent, RS, 4' long, ceiling mounted, two 40 watt	"	2.70	2.963	"	4,075	172		4,247	4,725

26 55 Special Purpose Lighting

26 55 33 - Hazard Warning Lighting

26 55 33.10 Warning Beacons

	Crew	Daily Output	Labor-Hours	Unit	Material	2018 Bare Costs Labor	Equipment	Total	Total Incl O&P
0010 **WARNING BEACONS**									
0015 Surface mount with colored or clear lens									
0100 Rotating beacon									
0110 120V, 40 watt halogen	1 Elec	3.50	2.286	Ea.	106	133		239	315
0120 24V, 20 watt halogen	"	3.50	2.286	"	228	133		361	450
0200 Steady beacon									
0210 120V, 40 watt halogen	1 Elec	3.50	2.286	Ea.	108	133		241	320
0220 24V, 20 watt		3.50	2.286		112	133		245	320
0230 12V DC, incandescent		3.50	2.286		103	133		236	315
0300 Flashing beacon									
0310 120V, 40 watt halogen	1 Elec	3.50	2.286	Ea.	106	133		239	315
0320 24V, 20 watt halogen		3.50	2.286		106	133		239	315
0410 12V DC with two 6V lantern batteries		7	1.143		107	66.50		173.50	218

26 55 61 - Theatrical Lighting

26 55 61.10 Lights

	Crew	Daily Output	Labor-Hours	Unit	Material	2018 Bare Costs Labor	Equipment	Total	Total Incl O&P
0010 **LIGHTS**									
2000 Lights, border, quartz, reflector, vented,									
2100 colored or white	1 Elec	20	.400	L.F.	182	23.50		205.50	235
2500 Spotlight, follow spot, with transformer, 2,100 watt	"	4	2	Ea.	3,450	116		3,566	3,950
2600 For no transformer, deduct					960			960	1,050
3000 Stationary spot, fresnel quartz, 6" lens	1 Elec	4	2		196	116		312	390
3100 8" lens		4	2		244	116		360	440
3500 Ellipsoidal quartz, 1,000 watt, 6" lens		4	2		360	116		476	575
3600 12" lens		4	2		655	116		771	895
4000 Strobe light, 1 to 15 flashes per second, quartz		3	2.667		845	155		1,000	1,150
4500 Color wheel, portable, five hole, motorized		4	2		217	116		333	415

26 56 Exterior Lighting

26 56 13 - Lighting Poles and Standards

26 56 13.10 Lighting Poles

	Crew	Daily Output	Labor-Hours	Unit	Material	2018 Bare Costs Labor	Equipment	Total	Total Incl O&P
0010 **LIGHTING POLES**									
2800 Light poles, anchor base									
2820 not including concrete bases									
2840 Aluminum pole, 8' high	1 Elec	4	2	Ea.	785	116		901	1,025
3000 20' high	R-3	2.90	6.897		1,050	400	44.50	1,494.50	1,800
3200 30' high		2.60	7.692		1,975	445	50	2,470	2,900
3400 35' high		2.30	8.696		2,150	505	56.50	2,711.50	3,175
3600 40' high		2	10		2,450	580	65	3,095	3,650
3800 Bracket arms, 1 arm	1 Elec	8	1		134	58		192	235
4000 2 arms		8	1		270	58		328	385
4200 3 arms		5.30	1.509		405	88		493	575
4400 4 arms		4.80	1.667		540	97		637	740

26 56 Exterior Lighting

26 56 13 – Lighting Poles and Standards

26 56 13.10 Lighting Poles

		Crew	Daily Output	Labor-Hours	Unit	Material	2018 Bare Costs Labor	Equipment	Total	Total Incl O&P
4500	Steel pole, galvanized, 8' high	1 Elec	3.80	2.105	Ea.	680	123		803	930
4600	20' high	R-3	2.60	7.692		1,050	445	50	1,545	1,900
4800	30' high		2.30	8.696		1,450	505	56.50	2,011.50	2,425
5000	35' high		2.20	9.091		1,600	525	59	2,184	2,600
5200	40' high		1.70	11.765		1,975	680	76	2,731	3,275
5400	Bracket arms, 1 arm	1 Elec	8	1		201	58		259	310
5600	2 arms		8	1		310	58		368	425
5800	3 arms		5.30	1.509		225	88		313	380
6000	4 arms		5.30	1.509		320	88		408	480
6462	20' high	R-3	2.90	6.897		1,050	400	44.50	1,494.50	1,825
6463	30' high		2.30	8.696		1,450	505	56.50	2,011.50	2,425
6464	35' high		2.40	8.333		1,600	485	54	2,139	2,525
6465	25' high		2.70	7.407		1,050	430	48	1,528	1,875

26 56 19 – LED Exterior Lighting

26 56 19.55 Roadway LED Luminaire

			Crew	Daily Output	Labor-Hours	Unit	Material	2018 Bare Costs Labor	Equipment	Total	Total Incl O&P
0010	**ROADWAY LED LUMINAIRE**										
0100	LED fixture, 72 LEDs, 120 V AC or 12 V DC, equal to 60 watt	G	1 Elec	2.70	2.963	Ea.	585	172		757	905
0110	108 LEDs, 120 V AC or 12 V DC, equal to 90 watt	G		2.70	2.963		690	172		862	1,025
0120	144 LEDs, 120 V AC or 12 V DC, equal to 120 watt	G		2.70	2.963		845	172		1,017	1,200
0130	252 LEDs, 120 V AC or 12 V DC, equal to 210 watt	G	2 Elec	4.40	3.636		1,175	212		1,387	1,625
0140	Replaces high pressure sodium fixture, 75 watt	G	1 Elec	2.70	2.963		395	172		567	695
0150	125 watt	G		2.70	2.963		450	172		622	755
0160	150 watt	G		2.70	2.963		530	172		702	840
0170	175 watt	G		2.70	2.963		780	172		952	1,125
0180	200 watt	G		2.70	2.963		745	172		917	1,075
0190	250 watt	G	2 Elec	4.40	3.636		850	212		1,062	1,250
0200	320 watt	G	"	4.40	3.636		930	212		1,142	1,350

26 56 19.60 Parking LED Lighting

			Crew	Daily Output	Labor-Hours	Unit	Material	2018 Bare Costs Labor	Equipment	Total	Total Incl O&P
0010	**PARKING LED LIGHTING**										
0100	Round pole mounting, 88 lamp watts	G	1 Elec	2	4	Ea.	1,150	233		1,383	1,600

26 56 21 – HID Exterior Lighting

26 56 21.20 Roadway Luminaire

		Crew	Daily Output	Labor-Hours	Unit	Material	2018 Bare Costs Labor	Equipment	Total	Total Incl O&P
0010	**ROADWAY LUMINAIRE**									
2650	Roadway area luminaire, low pressure sodium, 135 watt	1 Elec	2	4	Ea.	840	233		1,073	1,275
2700	180 watt	"	2	4		900	233		1,133	1,350
2750	Metal halide, 400 watt	2 Elec	4.40	3.636		650	212		862	1,025
2760	1,000 watt		4	4		730	233		963	1,150
2780	High pressure sodium, 400 watt		4.40	3.636		775	212		987	1,175
2790	1,000 watt		4	4		880	233		1,113	1,325

26 56 23 – Area Lighting

26 56 23.10 Exterior Fixtures

		Crew	Daily Output	Labor-Hours	Unit	Material	2018 Bare Costs Labor	Equipment	Total	Total Incl O&P
0010	**EXTERIOR FIXTURES** With lamps									
0200	Wall mounted, incandescent, 100 watt	1 Elec	8	1	Ea.	40.50	58		98.50	132
0400	Quartz, 500 watt		5.30	1.509		61.50	88		149.50	199
1100	Wall pack, low pressure sodium, 35 watt		4	2		191	116		307	385
1150	55 watt		4	2		227	116		343	425

26 56 23.55 Exterior LED Fixtures

			Crew	Daily Output	Labor-Hours	Unit	Material	2018 Bare Costs Labor	Equipment	Total	Total Incl O&P
0010	**EXTERIOR LED FIXTURES**										
0100	Wall mounted, indoor/outdoor, 12 watt	G	1 Elec	10	.800	Ea.	216	46.50		262.50	310
0110	32 watt	G		10	.800		465	46.50		511.50	580

For customer support on your Building Construction Costs with RSMeans data, call 800.448.8182.

26 56 Exterior Lighting

26 56 23 – Area Lighting

26 56 23.55 Exterior LED Fixtures		Crew	Daily Output	Labor-Hours	Unit	Material	2018 Bare Costs Labor	Equipment	Total	Total Incl O&P
0120	66 watt	G	1 Elec	10	.800	Ea.	470	46.50		516.50	585
0200	outdoor, 110 watt	G		10	.800		1,200	46.50		1,246.50	1,400
0210	220 watt	G		10	.800		1,750	46.50		1,796.50	2,000
0300	modular, type IV, 120 V, 50 lamp watts	G		9	.889		1,150	51.50		1,201.50	1,350
0310	101 lamp watts	G		9	.889		1,300	51.50		1,351.50	1,500
0320	126 lamp watts	G		9	.889		1,625	51.50		1,676.50	1,850
0330	202 lamp watts	G		9	.889		1,850	51.50		1,901.50	2,100
0340	240 V, 50 lamp watts	G		8	1		1,200	58		1,258	1,400
0350	101 lamp watts	G		8	1		1,350	58		1,408	1,550
0360	126 lamp watts	G		8	1		1,450	58		1,508	1,675
0370	202 lamp watts	G		8	1		1,900	58		1,958	2,150
0400	wall pack, glass, 13 lamp watts	G		4	2		445	116		561	665
0410	poly w/photocell, 26 lamp watts	G		4	2		340	116		456	550
0420	50 lamp watts	G		4	2		775	116		891	1,025
0430	replacement, 40 watts	G		4	2		390	116		506	605
0440	60 watts	G		4	2		400	116		516	615

26 56 36 – Flood Lighting

26 56 36.20 Floodlights

		Crew	Daily Output	Labor-Hours	Unit	Material	Labor	Equipment	Total	Total Incl O&P
0010	**FLOODLIGHTS** with ballast and lamp,									
1400	Pole mounted, pole not included									
1950	Metal halide, 175 watt	1 Elec	2.70	2.963	Ea.	210	172		382	490
2000	400 watt	2 Elec	4.40	3.636		224	212		436	560
2200	1,000 watt	"	4	4		790	233		1,023	1,225
2340	High pressure sodium, 70 watt	1 Elec	2.70	2.963		250	172		422	535
2400	400 watt	2 Elec	4.40	3.636		315	212		527	665
2600	1,000 watt	"	4	4		580	233		813	985

26 56 36.55 LED Floodlights

			Crew	Daily Output	Labor-Hours	Unit	Material	Labor	Equipment	Total	Total Incl O&P
0010	**LED FLOODLIGHTS** with ballast and lamp,										
0020	Pole mounted, pole not included										
0100	11 watt	G	1 Elec	4	2	Ea.	435	116		551	655
0110	46 watt	G		4	2		1,450	116		1,566	1,775
0120	90 watt	G		4	2		2,000	116		2,116	2,375
0130	288 watt	G		4	2		1,800	116		1,916	2,150

26 61 Lighting Systems and Accessories

26 61 23 – Lamps Applications

26 61 23.10 Lamps

			Crew	Daily Output	Labor-Hours	Unit	Material	Labor	Equipment	Total	Total Incl O&P
0010	**LAMPS**										
0080	Fluorescent, rapid start, cool white, 2' long, 20 watt		1 Elec	1	8	C	305	465		770	1,025
0100	4' long, 40 watt			.90	8.889		256	515		771	1,050
0200	Slimline, 4' long, 40 watt			.90	8.889		840	515		1,355	1,700
0210	4' long, 30 watt energy saver	G		.90	8.889		840	515		1,355	1,700
0400	High output, 4' long, 60 watt			.90	8.889		545	515		1,060	1,375
0410	8' long, 95 watt energy saver	G		.80	10		505	580		1,085	1,425
0500	8' long, 110 watt			.80	10		505	580		1,085	1,425
0512	2' long, T5, 14 watt energy saver	G		1	8		159	465		624	870
0514	3' long, T5, 21 watt energy saver	G		.90	8.889		180	515		695	975
0516	4' long, T5, 28 watt energy saver	G		.90	8.889		180	515		695	975
0517	4' long, T5, 54 watt energy saver	G		.90	8.889		555	515		1,070	1,375
0560	Twin tube compact lamp	G		.90	8.889		450	515		965	1,275

For customer support on your Building Construction Costs with RSMeans data, call 800.448.8182.

599

26 61 23 – Lamps Applications

26 61 23.10 Lamps

		Crew	Daily Output	Labor-Hours	Unit	Material	2018 Bare Costs Labor	Equipment	Total	Total Incl O&P
0570	Double twin tube compact lamp	G	1 Elec	.80	10	C	985	580		1,565	1,950
0600	Mercury vapor, mogul base, deluxe white, 100 watt			.30	26.667		5,325	1,550		6,875	8,175
0700	250 watt			.30	26.667		5,200	1,550		6,750	8,025
0800	400 watt			.30	26.667		5,200	1,550		6,750	8,050
0900	1,000 watt			.20	40		13,800	2,325		16,125	18,700
1000	Metal halide, mogul base, 175 watt			.30	26.667		1,075	1,550		2,625	3,525
1200	400 watt			.30	26.667		1,900	1,550		3,450	4,400
1300	1,000 watt			.20	40		3,350	2,325		5,675	7,150
1350	High pressure sodium, 70 watt			.30	26.667		1,650	1,550		3,200	4,125
1380	250 watt			.30	26.667		2,225	1,550		3,775	4,775
1400	400 watt			.30	26.667		1,675	1,550		3,225	4,150
1450	1,000 watt			.20	40		4,450	2,325		6,775	8,350
3000	Guards, fluorescent lamp, 4' long			1	8		1,325	465		1,790	2,150
3200	8' long			.90	8.889	Ea.	13.20	515		528.20	790

26 61 23.55 LED Lamps

		Crew	Daily Output	Labor-Hours	Unit	Material	2018 Bare Costs Labor	Equipment	Total	Total Incl O&P
0010	**LED LAMPS**										
0100	LED lamp, interior, shape A60, equal to 60 W	G	1 Elec	160	.050	Ea.	19	2.91		21.91	25.50
0110	7 W LED decorative c, ca, f, g shape			160	.050		34.50	2.91		37.41	42
0120	12V mini lamp LED			160	.050		5	2.91		7.91	9.85
0200	Globe frosted A60, equal to 60 W	G		160	.050		11.40	2.91		14.31	16.90
0205	LED lamp, interior, globe			160	.050		16.15	2.91		19.06	22
0210	2.2 W LED LMP			160	.050		66	2.91		68.91	77.50
0220	2.2 W LED replacement decorative lamp			160	.050		8.70	2.91		11.61	13.90
0230	3.5 W LED replacement decorative lamp			160	.050		10.20	2.91		13.11	15.55
0240	4.5 W 120V LED replacement decorative lamp			160	.050		12.40	2.91		15.31	18
0250	4.5 W 120V LED, 2700k replacement decorative lamp			160	.050		10.10	2.91		13.01	15.45
0260	4.9 W 120V LED, 2700k replacement decorative lamp candelabra base			160	.050		10.05	2.91		12.96	15.40
0270	4.9 W 120V LED, 3000k replacement decorative lamp candelabra base			160	.050		9.40	2.91		12.31	14.70
0280	5 W LED PAR 20 parabolic reflector lamp FL			140	.057		60	3.33		63.33	71
0300	Globe earth, equal to 100 W	G		140	.057		27.50	3.33		30.83	35.50
0305	7 W LED reflector lamp 3000k DIM			140	.057		58	3.33		61.33	69
0310	10 W omni LED warm white light bulb E26 medium base 120 volt card			140	.057		16.75	3.33		20.08	23.50
0315	7 W LED reflector lamp WFL 2700k DIM			140	.057		58	3.33		61.33	69
0320	9 W omni LED warm white light bulb E26 medium base 120 volt card			140	.057		9.05	3.33		12.38	14.90
0500	8 W LED, A19 lamp, equal to 40 W			140	.057		5.90	3.33		9.23	11.45
0505	9 W LED, A19 lamp, 5000K dimmable, equal to 60 W			140	.057		7.20	3.33		10.53	12.85
0510	10.5 W LED, A19 lamp, dimmable, equal to 60 W			140	.057		6.35	3.33		9.68	11.90
0515	10 W LED, A19 lamp, frosted, dimmable, equal to 60 W			140	.057		10.85	3.33		14.18	16.85
0520	10 W LED, A19 lamp, omni-directional, dimmable, equal to 60 W			140	.057		8.40	3.33		11.73	14.20
0525	12 W LED, A19 lamp, dimmable, equal to 75 W			140	.057		10.80	3.33		14.13	16.80
1100	MR16, 3 W, replacement of halogen lamp 25 W	G		130	.062		19	3.58		22.58	26.50
1200	6 W replacement of halogen lamp 45 W	G		130	.062		20	3.58		23.58	27.50
2100	10 W, PAR20, equal to 60 W	G		130	.062		26.50	3.58		30.08	35
2200	15 W, PAR30, equal to 100 W	G		130	.062		47.50	3.58		51.08	57.50
2210	LED lamp 50 W w/6ft pri			130	.062		610	3.58		613.58	675
2220	11 Watt reflector dimmable warm white LED light bulb with medium base			130	.062		41.50	3.58		45.08	51
2221	12 W A-Line LED lamp DIM			130	.062		78	3.58		81.58	91.50
2225	13 Watt reflector LED warm white e26 with medium base 120 volt box			130	.062		39	3.58		42.58	48.50
2226	13 W br30 LED lamp			130	.062		68	3.58		71.58	80
2227	15 W 120V br30 inc LED lamp, 2700k			130	.062		63.50	3.58		67.08	75
2228	15 W 120V br30 inc LED lamp, 4000k			130	.062		60	3.58		63.58	71.50
2230	3 Watt dimmable warm white decorative LED lamp with medium base			160	.050		12.60	2.91		15.51	18.20

For customer support on your Building Construction Costs with RSMeans data, call 800.448.8182.

26 61 23 – Lamps Applications

26 61 23.55 LED Lamps		Crew	Daily Output	Labor-Hours	Unit	Material	2018 Bare Costs Labor	Equipment	Total	Total Incl O&P
2240	.43 W night light LED daylight bulb E12 candelabra base 120 volt 2 pack	1 Elec	160	.050	Ea.	4.72	2.91		7.63	9.55
2250	15 W omni-directional LED warm white e26 medium base 120 volt box		160	.050		40	2.91		42.91	48.50
2251	11 W omni-directional LED warm white e26 medium base 120 volt box		160	.050		28	2.91		30.91	35
2252	10 W omni-directional LED warm white e26 medium base 120 volt box		160	.050		28	2.91		30.91	35
2253	7 W omni A19 LED warm white e26 medium base 120 volt box		160	.050		20.50	2.91		23.41	27
2255	8 PAR 20 parabolic reflector 2700 LED lamp		160	.050		18.80	2.91		21.71	25
2256	8 PAR 20 parabolic reflector 3000 LED lamp		160	.050		18.95	2.91		21.86	25.50
2260	12 W PAR 38 120V LED 15 degree directional lamp		160	.050		203	2.91		205.91	227
2270	12 W PAR 38 120V LED 25 degree directional lamp		160	.050		31.50	2.91		34.41	39
2280	12 W PAR 38 120V LED 40 degree directional lamp		160	.050		203	2.91		205.91	227
2285	16 W PAR 38 120V 2700k LED parabolic reflector lamp		160	.050		23	2.91		25.91	30
2290	16 W PAR 38 120V 3000k LED parabolic reflector lamp		160	.050		23	2.91		25.91	30
3000	15 W PAR 30 LED daylight E26 medium base 120V box		160	.050		60	2.91		62.91	70.50
3100	17 W LED 3000k PAR 38 100 W replacement		160	.050		47	2.91		49.91	56
3110	17 W LED PAR 38 100 W replacement		160	.050		35	2.91		37.91	43
3120	17 W LED PAR 38 100 W replacement parabolic reflector		160	.050		35	2.91		37.91	43
3130	24 W LED T8 PW straight fluorescent lamp		8.89	.900		289	52.50		341.50	400
3135	Linear fluorescent LED lamp 120V		8.89	.900		111	52.50		163.50	201
3200	30 W LED 2700K recessed 8055E PAR 38 high power		160	.050		105	2.91		107.91	120
3210	30 W LED 4200K 120 degree 8055E PAR 38 high power		160	.050		105	2.91		107.91	120
3220	30 W LED 5700K 8055E PAR 38 high power		160	.050		105	2.91		107.91	120
3230	50 W LED 2700K 8045M PAR 38 277V high power		160	.050		340	2.91		342.91	380
3240	50 W LED 4200K 8045M PAR 38 277V high power retro fit		160	.050		129	2.91		131.91	145
3250	50 W LED 5700K 8045M PAR 38 277V high power retro fit		160	.050		340	2.91		342.91	380
8000	10 W LED PAR 30/fl 10 pk		160	.050		96	2.91		98.91	110
8010	Gen 3 PAR 30 15 W short neck power LED 120 VAC E26 80 +cri 300k dimm		160	.050		12.95	2.91		15.86	18.60
8020	12 PAR 30 2700K parabolic reflector LED lamp		160	.050		43.50	2.91		46.41	52.50
8030	12 PAR 30 3000K parabolic reflector LED lamp		160	.050		28	2.91		30.91	35.50
8040	3500K LED advantage T8 9 W 800LM 2ft linear 2 BD frosted		69	.116		9	6.75		15.75	20
8050	3500K LED litespan T8 9 W 900LM 2ft linear frosted		69	.116		9.90	6.75		16.65	21
8060	4000K LED advantage T8 9 W 800LM 2ft linear 2BD frosted		69	.116		9	6.75		15.75	20
8070	4000K LED litespan T8 9 W 900LM 2ft linear frosted		69	.116		9.90	6.75		16.65	21
8080	5000K LED advantage T8 9 W 800LM 2ft linear 2BD frosted		69	.116		9	6.75		15.75	20
8090	5000K LED litespan T8 9 W 900LM 2ft linear frosted		69	.116		9.90	6.75		16.65	21
8100	3500K LED advantage T8 18 W 1600LM 4ft linear 2 BD frosted		69	.116		9.90	6.75		16.65	21
8105	18 W LED 4ft T8 4000K frost 1600L linear lamp		69	.116		18.50	6.75		25.25	30.50
8108	18 W LED 48 inch T8 4100K 1890LM linear lamp		69	.116		57	6.75		63.75	73
8110	3500K LED advantage T8 18 W 1800LM 4ft linear 2 BD frosted		69	.116		21	6.75		27.75	33
8120	4000K LED advantage T8 18 W 1600LM 4ft linear 2 BD frosted		69	.116		16.30	6.75		23.05	28
8130	4000K LED litespan T8 18 W 1800LM 4ft linear frosted		69	.116		21	6.75		27.75	33
8140	5000K LED advantage T8 18 W 1600LM 4ft linear 2BD frosted		69	.116		9.90	6.75		16.65	21
8150	5000K LED litespan T8 18 W 1800LM 4ft linear 2BD frosted		69	.116		21	6.75		27.75	33
8200	16.5 T8 3000 IF-6U U-shape fluorescent LED lamp		65	.123		21.50	7.15		28.65	34
8210	16.5 T8 3500 IF-6U U-shape fluorescent LED lamp		65	.123		21.50	7.15		28.65	34.50
8220	16.5 T8 4000 IF-6U U-shape fluorescent LED lamp		65	.123		21	7.15		28.15	33.50
8230	16.5 T8 5000 IF-6U U-shape fluorescent LED lamp		65	.123		21	7.15		28.15	34
8240	18 W 6 inch T8 4100K U-shape frosted fluorescent LED lamp		60	.133		11.70	7.75		19.45	24.50
8250	18 W 6 inch T8 5000K U-shape frosted fluorescent LED lamp		60	.133		11.70	7.75		19.45	24.50
8260	Circular 12 W linear fluorescent LED 2700K MOD lamp		62.40	.128		300	7.45		307.45	340
8270	Circular 12 W linear fluorescent LED 3000K MOD lamp		62.40	.128		300	7.45		307.45	340
8280	Circular 12 W linear fluorescent LED 3500K MOD lamp		62.40	.128		273	7.45		280.45	310
8290	Circular 18 W linear fluorescent LED 3000K MOD lamp		62.40	.128		375	7.45		382.45	420
8300	Circular 18 W linear fluorescent LED 3500K MOD lamp		62.40	.128		410	7.45		417.45	465

26 61 Lighting Systems and Accessories

26 61 23 – Lamps Applications

26 61 23.55 LED Lamps		Crew	Daily Output	Labor-Hours	Unit	Material	2018 Bare Costs Labor	Equipment	Total	Total Incl O&P
8310	Circular 18 W linear fluorescent LED 5000K MOD lamp	1 Elec	62.40	.128	Ea.	375	7.45		382.45	420
8320	Circular 18 W linear fluorescent LED 4100K MOD lamp	↓	62.40	.128	↓	395	7.45		402.45	445

26 71 Electrical Machines

26 71 13 – Motors Applications

26 71 13.20 Motors

		Crew	Daily Output	Labor-Hours	Unit	Material	2018 Bare Costs Labor	Equipment	Total	Total Incl O&P
0010	**MOTORS** 230/460 V, 60 HZ									
0050	Dripproof, premium efficiency, 1.15 service factor									
0060	1,800 RPM, 1/4 HP	1 Elec	5.33	1.501	Ea.	261	87.50		348.50	420
0070	1/3 HP		5.33	1.501		241	87.50		328.50	395
0080	1/2 HP		5.33	1.501		201	87.50		288.50	350
0090	3/4 HP		5.33	1.501		281	87.50		368.50	440
0100	1 HP		4.50	1.778		300	103		403	485
0250	5 HP		4.50	1.778		640	103		743	860
0350	10 HP	↓	4	2		1,200	116		1,316	1,475
0450	20 HP	2 Elec	5.20	3.077	↓	2,025	179		2,204	2,500

26 71 13.40 Motors Explosion Proof

		Crew	Daily Output	Labor-Hours	Unit	Material	2018 Bare Costs Labor	Equipment	Total	Total Incl O&P
0010	**MOTORS EXPLOSION PROOF**, 208-230/460 V, 60 HZ									
0020	1,800 RPM, 1/4 HP	1 Elec	5	1.600	Ea.	565	93		658	765
0030	1/3 HP		5	1.600		272	93		365	440
0040	1/2 HP		5	1.600		440	93		533	625
0050	3/4 HP		5.33	1.501		350	87.50		437.50	515
0060	1 HP		4.20	1.905		555	111		666	775
0090	5 HP		4.20	1.905		890	111		1,001	1,150
0110	10 HP	↓	3.70	2.162		1,300	126		1,426	1,625
0130	20 HP	2 Elec	5	3.200		2,125	186		2,311	2,600
2000	3,600 RPM, 1/4 HP	1 Elec	5	1.600		405	93		498	585
2010	1/3 HP		5	1.600		425	93		518	605
2020	1/2 HP		5	1.600		320	93		413	490
2030	3/4 HP		5	1.600		430	93		523	615
2040	1 HP		4.20	1.905		610	111		721	835
2070	5 HP		4.20	1.905		975	111		1,086	1,250
2090	10 HP	↓	3.70	2.162		1,275	126		1,401	1,600
2110	20 HP	2 Elec	5	3.200	↓	2,150	186		2,336	2,625

For customer support on your Building Construction Costs with RSMeans data, call 800.448.8182.

Estimating Tips

27 20 00 Data Communications
27 30 00 Voice Communications
27 40 00 Audio-Video Communications

- When estimating material costs for special systems, it is always prudent to obtain manufacturers' quotations for equipment prices and special installation requirements that may affect the total cost.

- For cost modifications for elevated tray installation, add the percentages to labor according to the height of the installation and only to the quantities exceeding the different height levels, not to the total tray quantities. Refer to 26 01 02.20 for labor adjustment factors.

- Do not overlook the costs for equipment used in the installation. If scissor lifts and boom lifts are available in the field, contractors may use them in lieu of the proposed ladders and rolling staging.

Reference Numbers

Reference numbers are shown at the beginning of some major classifications. These numbers refer to related items in the Reference Section. The reference information may be an estimating procedure, an alternate pricing method, or technical information.

Note: Not all subdivisions listed here necessarily appear. ■

No part of this cost data may be reproduced, stored in a retrieval system, or transmitted in any form or by any means without prior written permission of Gordian.

Note: Trade Service, in part, has been used as a reference source for some of the material prices used in Division 27.

Did you know?

RSMeans data is available through our online application with 24/7 access:

- Search for unit prices by keyword
- Leverage the most up-to-date data
- Build and export estimates

Try it free for 30 days!
www.rsmeans.com/2018freetrial

27 13 Communications Backbone Cabling

27 13 23 – Communications Optical Fiber Backbone Cabling

27 13 23.13 Communications Optical Fiber	Crew	Daily Output	Labor-Hours	Unit	Material	2018 Bare Costs Labor	Equipment	Total	Total Incl O&P
0010 **COMMUNICATIONS OPTICAL FIBER**									
0040 Specialized tools & techniques cause installation costs to vary.									
0070 Fiber optic, cable, bulk simplex, single mode	1 Elec	8	1	C.L.F.	22.50	58		80.50	112
0080 Multi mode		8	1		29.50	58		87.50	120
0090 4 strand, single mode		7.34	1.090		38	63.50		101.50	137
0095 Multi mode		7.34	1.090		50.50	63.50		114	151
0100 12 strand, single mode		6.67	1.199		72.50	70		142.50	184
0105 Multi mode		6.67	1.199		96.50	70		166.50	210
0150 Jumper				Ea.	33			33	36.50
0200 Pigtail					29			29	32
0300 Connector	1 Elec	24	.333		26	19.40		45.40	57.50
0350 Finger splice		32	.250		38	14.55		52.55	64
0400 Transceiver (low cost bi-directional)		8	1		430	58		488	560
0450 Rack housing, 4 rack spaces, 12 panels (144 fibers)		2	4		560	233		793	965
0500 Patch panel, 12 ports		6	1.333		300	77.50		377.50	445
1000 Cable, 62.5 microns, direct burial, 4 fiber	R-15	1200	.040	L.F.	.91	2.29	.24	3.44	4.68
1020 Indoor, 2 fiber	R-19	1000	.020		.42	1.17		1.59	2.20
1040 Outdoor, aerial/duct	"	1670	.012		.65	.70		1.35	1.76
1060 50 microns, direct burial, 8 fiber	R-22	4000	.009		1.27	.50		1.77	2.14
1080 12 fiber		4000	.009		2.08	.50		2.58	3.03
1100 Indoor, 12 fiber		759	.049		1.98	2.62		4.60	6.10
1120 Connectors, 62.5 micron cable, transmission	R-19	40	.500	Ea.	14.35	29		43.35	59.50
1140 Cable splice		40	.500		17.50	29		46.50	63
1160 125 micron cable, transmission		16	1.250		15.10	73		88.10	126
1180 Receiver, 1.2 mile range		20	1		240	58.50		298.50	350
1200 1.9 mile range		20	1		226	58.50		284.50	335
1220 6.2 mile range		5	4		282	233		515	660
1240 Transmitter, 1.2 mile range		20	1		264	58.50		322.50	375
1260 1.9 mile range		20	1		299	58.50		357.50	415
1280 6.2 mile range		5	4		380	233		613	770
1300 Modem, 1.2 mile range		5	4		173	233		406	540
1320 6.2 mile range		5	4		315	233		548	700
1340 1.9 mile range, 12 channel		5	4		2,000	233		2,233	2,550
1360 Repeater, 1.2 mile range		10	2		370	117		487	585
1380 1.9 mile range		10	2		475	117		592	700
1400 6.2 mile range		5	4		920	233		1,153	1,350
1420 1.2 mile range, digital		5	4		440	233		673	830

27 41 Audio-Video Systems

27 41 33 – Master Antenna Television Systems

27 41 33.10 TV Systems

	Crew	Daily Output	Labor-Hours	Unit	Material	2018 Bare Costs Labor	Equipment	Total	Total Incl O&P
0010 **TV SYSTEMS,** not including rough-in wires, cables & conduits									
0100 Master TV antenna system									
0200 VHF reception & distribution, 12 outlets	1 Elec	6	1.333	Outlet	114	77.50		191.50	242
0400 30 outlets		10	.800		240	46.50		286.50	335
0600 100 outlets		13	.615		288	36		324	370
0800 VHF & UHF reception & distribution, 12 outlets		6	1.333		249	77.50		326.50	390
1000 30 outlets		10	.800		165	46.50		211.50	251
1200 100 outlets		13	.615		168	36		204	239
1400 School and deluxe systems, 12 outlets		2.40	3.333		330	194		524	650
1600 30 outlets		4	2		288	116		404	490

For customer support on your Building Construction Costs with RSMeans data, call 800.448.8182.

27 41 Audio-Video Systems

27 41 33 – Master Antenna Television Systems

27 41 33.10 TV Systems

		Crew	Daily Output	Labor-Hours	Unit	Material	2018 Bare Costs Labor	2018 Bare Costs Equipment	Total	Total Incl O&P
1800	80 outlets	1 Elec	5.30	1.509	Outlet	276	88		364	435

27 51 Distributed Audio-Video Communications Systems

27 51 16 – Public Address Systems

27 51 16.10 Public Address System

		Crew	Daily Output	Labor-Hours	Unit	Material	Labor	Equipment	Total	Total Incl O&P
0010	**PUBLIC ADDRESS SYSTEM**									
0100	Conventional, office	1 Elec	5.33	1.501	Speaker	159	87.50		246.50	305
0200	Industrial	"	2.70	2.963	"	305	172		477	600

27 51 19 – Sound Masking Systems

27 51 19.10 Sound System

		Crew	Daily Output	Labor-Hours	Unit	Material	Labor	Equipment	Total	Total Incl O&P
0010	**SOUND SYSTEM**, not including rough-in wires, cables & conduits									
0100	Components, projector outlet	1 Elec	8	1	Ea.	46	58		104	138
0200	Microphone		4	2		95.50	116		211.50	279
0400	Speakers, ceiling or wall		8	1		139	58		197	240
0600	Trumpets		4	2		259	116		375	460
0800	Privacy switch		8	1		103	58		161	201
1000	Monitor panel		4	2		460	116		576	680
1200	Antenna, AM/FM		4	2		130	116		246	315
1400	Volume control		8	1		52	58		110	144
1600	Amplifier, 250 W		1	8		1,200	465		1,665	2,025
1800	Cabinets		1	8		1,000	465		1,465	1,800
2000	Intercom, 30 station capacity, master station	2 Elec	2	8		2,300	465		2,765	3,225
2200	Remote station	1 Elec	8	1		194	58		252	300
2400	Intercom outlets		8	1		114	58		172	212
2600	Handset		4	2		375	116		491	590
2800	Emergency call system, 12 zones, annunciator		1.30	6.154		1,125	360		1,485	1,775
3000	Bell		5.30	1.509		117	88		205	260
3200	Light or relay		8	1		58.50	58		116.50	152
3400	Transformer		4	2		257	116		373	455
3600	House telephone, talking station		1.60	5		555	291		846	1,050
3800	Press to talk, release to listen		5.30	1.509		129	88		217	273
4000	System-on button					77			77	85
4200	Door release	1 Elec	4	2		138	116		254	325
4400	Combination speaker and microphone		8	1		235	58		293	345
4600	Termination box		3.20	2.500		74	146		220	299
4800	Amplifier or power supply		5.30	1.509		850	88		938	1,075
5000	Vestibule door unit		16	.500	Name	156	29		185	215
5200	Strip cabinet		27	.296	Ea.	294	17.25		311.25	350
5400	Directory		16	.500	"	139	29		168	196

For customer support on your Building Construction Costs with RSMeans data, call 800.448.8182.

605

27 52 Healthcare Communications and Monitoring Systems

27 52 23 – Nurse Call/Code Blue Systems

27 52 23.10 Nurse Call Systems

27 52 23.10 Nurse Call Systems	Crew	Daily Output	Labor-Hours	Unit	Material	2018 Bare Costs Labor	Equipment	Total	Total Incl O&P
0010 **NURSE CALL SYSTEMS**									
0100 Single bedside call station	1 Elec	8	1	Ea.	163	58		221	266
0200 Ceiling speaker station		8	1		83	58		141	178
0400 Emergency call station		8	1		89	58		147	185
0600 Pillow speaker		8	1		218	58		276	325
0800 Double bedside call station		4	2		176	116		292	370
1000 Duty station		4	2		131	116		247	320
1200 Standard call button		8	1		107	58		165	205
1400 Lights, corridor, dome or zone indicator		8	1		60.50	58		118.50	154
1600 Master control station for 20 stations	2 Elec	.65	24.615	Total	3,525	1,425		4,950	6,025

27 53 Distributed Systems

27 53 13 – Clock Systems

27 53 13.50 Clock Equipments

	Crew	Daily Output	Labor-Hours	Unit	Material	2018 Bare Costs Labor	Equipment	Total	Total Incl O&P
0010 **CLOCK EQUIPMENTS**, not including wires & conduits									
0100 Time system components, master controller	1 Elec	.33	24.242	Ea.	1,500	1,400		2,900	3,750
0200 Program bell		8	1		111	58		169	209
0400 Combination clock & speaker		3.20	2.500		198	146		344	435
0600 Frequency generator		2	4		2,425	233		2,658	3,025
0800 Job time automatic stamp recorder		4	2		565	116		681	795
1600 Master time clock system, clocks & bells, 20 room	4 Elec	.20	160		5,750	9,300		15,050	20,200
1800 50 room	"	.08	400		12,200	23,300		35,500	48,200
1900 Time clock	1 Elec	3.20	2.500		390	146		536	650
2000 100 cards in & out, 1 color					9			9	9.90
2200 2 colors					9			9	9.90
2800 Metal rack for 25 cards	1 Elec	7	1.143		46.50	66.50		113	151

For customer support on your Building Construction Costs with RSMeans data, call 800.448.8182.

Estimating Tips

- When estimating material costs for electronic safety and security systems, it is always prudent to obtain manufacturers' quotations for equipment prices and special installation requirements that may affect the total cost.

- Fire alarm systems consist of control panels, annunciator panels, batteries with rack, charger, and fire alarm actuating and indicating devices. Some fire alarm systems include speakers, telephone lines, door closer controls, and other components. Be careful not to overlook the costs related to installation for these items. Also be aware of costs for integrated automation instrumentation and terminal devices, control equipment, control wiring, and programming. Insurance underwriters may have specific requirements for the type of materials to be installed or design requirements based on the hazard to be protected. Local jurisdictions may have requirements not covered by code. It is advisable to be aware of any special conditions.

- Security equipment includes items such as CCTV, access control, and other detection and identification systems to perform alert and alarm functions. Be sure to consider the costs related to installation for this security equipment, such as for integrated automation instrumentation and terminal devices, control equipment, control wiring, and programming.

Reference Numbers

Reference numbers are shown at the beginning of some major classifications. These numbers refer to related items in the Reference Section. The reference information may be an estimating procedure, an alternate pricing method, or technical information.

Note: Not all subdivisions listed here necessarily appear. ■

No part of this cost data may be reproduced, stored in a retrieval system, or transmitted in any form or by any means without prior written permission of Gordian.

Note: Trade Service, in part, has been used as a reference source for some of the material prices used in Division 28.

Did you know?

RSMeans data is available through our online application with 24/7 access:

- Search for unit prices by keyword
- Leverage the most up-to-date data
- Build and export estimates

Try it free for 30 days!
www.rsmeans.com/2018freetrial

28 15 Access Control Hardware Devices

28 15 11 – Integrated Credential Readers and Field Entry Management

28 15 11.11 Standard Card Readers

		Crew	Daily Output	Labor-Hours	Unit	Material	2018 Bare Costs Labor	2018 Bare Costs Equipment	Total	Total Incl O&P
0010	**STANDARD CARD READERS**									
0015	Card key access									
0020	Computerized system, processor, proximity reader and cards									
0030	Does not include door hardware, lockset or wiring									
0040	Card key system for 1 door				Ea.	1,425			1,425	1,550
0060	Card key system for 2 doors					2,025			2,025	2,225
0080	Card key system for 4 doors					2,500			2,500	2,750
0100	Processor for card key access system					940			940	1,025
0160	Magnetic lock for electric access, 600 pound holding force					190			190	209
0170	Magnetic lock for electric access, 1200 pound holding force					190			190	209
0200	Proximity card reader					158			158	173

28 15 11.15 Biometric Identity Devices

		Crew	Daily Output	Labor-Hours	Unit	Material	Labor	Equipment	Total	Total Incl O&P
0010	**BIOMETRIC IDENTITY DEVICES**									
0220	Hand geometry scanner, mem of 512 users, excl. striker/power	1 Elec	3	2.667	Ea.	2,300	155		2,455	2,750
0230	Memory upgrade for, adds 9,700 user profiles		8	1		330	58		388	450
0240	Adds 32,500 user profiles		8	1		665	58		723	815
0250	Prison type, memory of 256 users, excl. striker/power		3	2.667		2,700	155		2,855	3,200
0260	Memory upgrade for, adds 3,300 user profiles		8	1		315	58		373	435
0270	Adds 9,700 user profiles		8	1		500	58		558	635
0280	Adds 27,900 user profiles		8	1		680	58		738	830
0290	All weather, mem of 512 users, excl. striker/power		3	2.667		4,075	155		4,230	4,700
0300	Facial & fingerprint scanner, combination unit, excl. striker/power		3	2.667		895	155		1,050	1,225
0310	Access for, for initial setup, excl. striker/power		3	2.667		1,200	155		1,355	1,550

28 18 Security Access Detection Equipment

28 18 11 – Security Access Metal Detectors

28 18 11.13 Security Access Metal Detectors

		Crew	Daily Output	Labor-Hours	Unit	Material	Labor	Equipment	Total	Total Incl O&P
0010	**SECURITY ACCESS METAL DETECTORS**									
0240	Metal detector, hand-held, wand type, unit only				Ea.	129			129	138
0250	Metal detector, walk through portal type, single zone	1 Elec	2	4		3,175	233		3,408	3,825
0260	Multi-zone	"	2	4		4,625	233		4,858	5,425

28 18 13 – Security Access X-Ray Equipment

28 18 13.16 Security Access X-Ray Equipment

		Crew	Daily Output	Labor-Hours	Unit	Material	Labor	Equipment	Total	Total Incl O&P
0010	**SECURITY ACCESS X-RAY EQUIPMENT**									
0290	X-ray machine, desk top, for mail/small packages/letters	1 Elec	4	2	Ea.	3,475	116		3,591	4,000
0300	Conveyor type, incl. monitor		2	4		16,000	233		16,233	18,000
0310	Includes additional features		2	4		28,000	233		28,233	31,200
0320	X-ray machine, large unit, for airports, incl. monitor	2 Elec	1	16		40,000	930		40,930	45,400
0330	Full console	"	.50	32		68,500	1,850		70,350	78,500

28 18 15 – Security Access Explosive Detection Equipment

28 18 15.23 Security Access Explosive Detection Equipment

		Crew	Daily Output	Labor-Hours	Unit	Material	Labor	Equipment	Total	Total Incl O&P
0010	**SECURITY ACCESS EXPLOSIVE DETECTION EQUIPMENT**									
0270	Explosives detector, walk through portal type	1 Elec	2	4	Ea.	44,200	233		44,433	49,000
0280	Hand-held, battery operated				"				25,500	28,100

28 23 Video Management System

28 23 13 – Video Management System Interfaces

28 23 13.10 Closed Circuit Television System	Crew	Daily Output	Labor-Hours	Unit	Material	2018 Bare Costs Labor	Equipment	Total	Total Incl O&P
0010 **CLOSED CIRCUIT TELEVISION SYSTEM**									
2000 Surveillance, one station (camera & monitor)	2 Elec	2.60	6.154	Total	850	360		1,210	1,475
2200 For additional camera stations, add	1 Elec	2.70	2.963	Ea.	370	172		542	665
2400 Industrial quality, one station (camera & monitor)	2 Elec	2.60	6.154	Total	2,000	360		2,360	2,725
2600 For additional camera stations, add	1 Elec	2.70	2.963	Ea.	800	172		972	1,150
2610 For low light, add		2.70	2.963		600	172		772	920
2620 For very low light, add		2.70	2.963		3,900	172		4,072	4,550
2800 For weatherproof camera station, add		1.30	6.154		640	360		1,000	1,250
3000 For pan and tilt, add		1.30	6.154		1,875	360		2,235	2,600
3200 For zoom lens - remote control, add		2	4		2,200	233		2,433	2,775
3400 Extended zoom lens		2	4		6,000	233		6,233	6,950
3410 For automatic iris for low light, add		2	4		1,200	233		1,433	1,675
3600 Educational TV studio, basic 3 camera system, black & white,									
3800 electrical & electronic equip. only	4 Elec	.80	40	Total	10,000	2,325		12,325	14,500
4000 Full console		.28	114		29,700	6,650		36,350	42,700
4100 As above, but color system		.28	114		61,500	6,650		68,150	77,500
4120 Full console		.12	267		228,500	15,500		244,000	274,000
4200 For film chain, black & white, add	1 Elec	1	8	Ea.	15,400	465		15,865	17,600
4250 Color, add		.25	32		10,500	1,850		12,350	14,400
4400 For video recorders, add		1	8		3,000	465		3,465	4,000
4600 Premium	4 Elec	.40	80		19,400	4,650		24,050	28,300

28 23 23 – Video Surveillance Systems Infrastructure

28 23 23.50 Video Surveillance Equipments	Crew	Daily Output	Labor-Hours	Unit	Material	2018 Bare Costs Labor	Equipment	Total	Total Incl O&P
0010 **VIDEO SURVEILLANCE EQUIPMENTS**									
0200 Video cameras, wireless, hidden in exit signs, clocks, etc., incl. receiver	1 Elec	3	2.667	Ea.	187	155		342	435
0210 Accessories for video recorder, single camera		3	2.667		183	155		338	435
0220 For multiple cameras		3	2.667		1,750	155		1,905	2,150
0230 Video cameras, wireless, for under vehicle searching, complete		2	4		16,000	233		16,233	18,000
0234 Master monitor station, 3 doors x 5 color monitor with tilt feature		2	4		800	233		1,033	1,225

28 31 Intrusion Detection

28 31 16 – Intrusion Detection Systems Infrastructure

28 31 16.50 Intrusion Detection	Crew	Daily Output	Labor-Hours	Unit	Material	2018 Bare Costs Labor	Equipment	Total	Total Incl O&P
0010 **INTRUSION DETECTION**, not including wires & conduits									
0100 Burglar alarm, battery operated, mechanical trigger	1 Elec	4	2	Ea.	280	116		396	485
0200 Electrical trigger		4	2		335	116		451	545
0400 For outside key control, add		8	1		87	58		145	183
0600 For remote signaling circuitry, add		8	1		138	58		196	239
0800 Card reader, flush type, standard		2.70	2.963		810	172		982	1,150
1000 Multi-code		2.70	2.963		1,200	172		1,372	1,575
1200 Door switches, hinge switch		5.30	1.509		62.50	88		150.50	200
1400 Magnetic switch		5.30	1.509		100	88		188	241
1600 Exit control locks, horn alarm		4	2		219	116		335	415
1800 Flashing light alarm		4	2		247	116		363	445
2000 Indicating panels, 1 channel		2.70	2.963		271	172		443	555
2200 10 channel	2 Elec	3.20	5		1,075	291		1,366	1,600
2400 20 channel		2	8		2,500	465		2,965	3,425
2600 40 channel		1.14	14.035		4,500	815		5,315	6,175
2800 Ultrasonic motion detector, 12 V	1 Elec	2.30	3.478		194	202		396	520
3000 Infrared photoelectric detector	"	4	2		146	116		262	335

28 42 Gas Detection and Alarm

28 42 15 – Gas Detection Sensors

28 42 15.50 Tank Leak Detection Systems	Crew	Daily Output	Labor-Hours	Unit	Material	2018 Bare Costs Labor	Equipment	Total	Total Incl O&P
0010	**TANK LEAK DETECTION SYSTEMS** Liquid and vapor									
0100	For hydrocarbons and hazardous liquids/vapors									
0120	Controller, data acquisition, incl. printer, modem, RS232 port									
0140	24 channel, for use with all probes				Ea.	2,925			2,925	3,200
0160	9 channel, for external monitoring				"	1,175			1,175	1,300
0200	Probes									
0210	Well monitoring									
0220	Liquid phase detection				Ea.	660			660	725
0230	Hydrocarbon vapor, fixed position					630			630	690
0240	Hydrocarbon vapor, float mounted					550			550	605
0250	Both liquid and vapor hydrocarbon					540			540	595
0300	Secondary containment, liquid phase									
0310	Pipe trench/manway sump				Ea.	595			595	655
0320	Double wall pipe and manual sump					500			500	550
0330	Double wall fiberglass annular space					375			375	410
0340	Double wall steel tank annular space					299			299	330
0500	Accessories									
0510	Modem, non-dedicated phone line				Ea.	274			274	300
0600	Monitoring, internal									
0610	Automatic tank gauge, incl. overfill				Ea.	1,225			1,225	1,350
0620	Product line				"	1,275			1,275	1,400
0700	Monitoring, special									
0710	Cathodic protection				Ea.	690			690	760
0720	Annular space chemical monitor				"	940			940	1,025

28 46 Fire Detection and Alarm

28 46 11 – Fire Sensors and Detectors

28 46 11.21 Carbon-Monoxide Detection Sensors

		Crew	Daily Output	Labor-Hours	Unit	Material	2018 Bare Costs Labor	Equipment	Total	Total Incl O&P
0010	**CARBON-MONOXIDE DETECTION SENSORS**									
8400	Smoke and carbon monoxide alarm battery operated photoelectric low profile	1 Elec	24	.333	Ea.	54.50	19.40		73.90	88.50
8410	low profile photoelectric battery powered		24	.333		28.50	19.40		47.90	60.50
8420	photoelectric low profile sealed lithium		24	.333		52	19.40		71.40	86
8430	Photoelectric low profile sealed lithium smoke and CO with voice combo		24	.333		38	19.40		57.40	71
8500	Carbon monoxide sensor, wall mount 1Mod 1 relay output smoke & heat		24	.333		490	19.40		509.40	570
8700	Carbon monoxide detector, battery operated, wall mounted		16	.500		52	29		81	101
8710	Hardwired, wall and ceiling mounted		8	1		99.50	58		157.50	196
8720	Duct mounted		8	1	"	345	58		403	460

28 46 11.27 Other Sensors

		Crew	Daily Output	Labor-Hours	Unit	Material	2018 Bare Costs Labor	Equipment	Total	Total Incl O&P
0010	**OTHER SENSORS**									
5200	Smoke detector, ceiling type	1 Elec	6.20	1.290	Ea.	120	75		195	244
5240	Smoke detector addressable type		6	1.333		224	77.50		301.50	360
5400	Duct type		3.20	2.500		330	146		476	585
5420	Duct addressable type		3.20	2.500		515	146		661	785
8300	Smoke alarm with integrated strobe light 120 V 16DB 60 fpm flash rate		16	.500		103	29		132	157
8310	Photoelectric smoke detector with strobe 120 V 90 DB ceiling mount		12	.667		158	39		197	232
8320	120 V, 90 DB wall mount		12	.667		166	39		205	240
8440	Fire alarm beam detector, motorised reflective, infrared optical beam	2 Elec	1.60	10		800	580		1,380	1,750
8450	Fire alarm beam detector, motorised reflective, IR/UV optical beam	"	1.60	10		1,025	580		1,605	2,000

For customer support on your Building Construction Costs with RSMeans data, call 800.448.8182.

28 46 Fire Detection and Alarm

28 46 11 – Fire Sensors and Detectors

28 46 11.50 Fire and Heat Detectors

28 46 11.50 Fire and Heat Detectors	Crew	Daily Output	Labor-Hours	Unit	Material	2018 Bare Costs Labor	Equipment	Total	Total Incl O&P
0010 **FIRE & HEAT DETECTORS**									
5000 Detector, rate of rise	1 Elec	8	1	Ea.	51.50	58		109.50	144

28 46 21 – Fire Alarm

28 46 21.50 Alarm Panels and Devices

28 46 21.50 Alarm Panels and Devices	Crew	Daily Output	Labor-Hours	Unit	Material	2018 Bare Costs Labor	Equipment	Total	Total Incl O&P
0010 **ALARM PANELS AND DEVICES**, not including wires & conduits									
2200 Intercom remote station	1 Elec	8	1	Ea.	71	58		129	165
2400 Intercom outlet		8	1		53	58		111	146
2600 Sound system, intercom handset		8	1		475	58		533	610
3600 4 zone	2 Elec	2	8		380	465		845	1,100
3800 8 zone		1	16		960	930		1,890	2,450
3810 5 zone		1.50	10.667		770	620		1,390	1,775
3900 10 zone		1.25	12.800		945	745		1,690	2,175
4000 12 zone		.67	23.988		2,100	1,400		3,500	4,375
4020 Alarm device, tamper, flow	1 Elec	8	1		240	58		298	350
4025 Fire alarm, loop expander card		16	.500		740	29		769	855
4050 Actuating device		8	1		340	58		398	460
4200 Battery and rack		4	2		415	116		531	630
4400 Automatic charger		8	1		590	58		648	735
4600 Signal bell		8	1		124	58		182	224
4610 Fire alarm signal bell 10" red 20-24 V P		8	1		136	58		194	236
4800 Trouble buzzer or manual station		8	1		84	58		142	180
5425 Duct smoke and heat detector 2 wire		8	1		107	58		165	204
5430 Fire alarm duct detector controller		3	2.667		235	155		390	490
5435 Fire alarm duct detector sensor kit		8	1		75	58		133	170
5440 Remote test station for smoke detector duct type		5.30	1.509		65	88		153	203
5460 Remote fire alarm indicator light		5.30	1.509		18.55	88		106.55	152
5600 Strobe and horn		5.30	1.509		154	88		242	300
5610 Strobe and horn (ADA type)		5.30	1.509		154	88		242	300
5620 Visual alarm (ADA type)		6.70	1.194		104	69.50		173.50	218
5800 Fire alarm horn		6.70	1.194		61.50	69.50		131	172
6000 Door holder, electro-magnetic		4	2		104	116		220	288
6200 Combination holder and closer		3.20	2.500		124	146		270	355
6600 Drill switch		8	1		375	58		433	495
6800 Master box		2.70	2.963		6,475	172		6,647	7,375
7000 Break glass station		8	1		57.50	58		115.50	151
7800 Remote annunciator, 8 zone lamp		1.80	4.444		213	259		472	620
8000 12 zone lamp	2 Elec	2.60	6.154		415	360		775	995
8200 16 zone lamp	"	2.20	7.273		375	425		800	1,050

28 47 Mass Notification

28 47 12 – Notification Systems

28 47 12.10 Mass Notification System

28 47 12.10 Mass Notification System	Crew	Daily Output	Labor-Hours	Unit	Material	2018 Bare Costs Labor	Equipment	Total	Total Incl O&P
0010 **MASS NOTIFICATION SYSTEM**									
0100 Wireless command center, 10,000 devices	2 Elec	1.33	12.030	Ea.	2,275	700		2,975	3,550
0200 Option, email notification					1,775			1,775	1,950
0210 Remote device supervision & monitor					2,525			2,525	2,800
0300 Antenna VHF or UHF, for medium range	1 Elec	4	2		87	116		203	270
0310 For high-power transmitter		2	4		1,100	233		1,333	1,550
0400 Transmitter, 25 watt		4	2		1,725	116		1,841	2,075
0410 40 watt		2.66	3.008		2,000	175		2,175	2,450

For customer support on your Building Construction Costs with RSMeans data, call 800.448.8182.

611

28 47 Mass Notification

28 47 12 – Notification Systems

28 47 12.10 Mass Notification System	Crew	Daily Output	Labor-Hours	Unit	Material	2018 Bare Costs Labor	Equipment	Total	Total Incl O&P
0420 100 watt	1 Elec	1.33	6.015	Ea.	6,475	350		6,825	7,650
0500 Wireless receiver/control module for speaker		8	1		279	58		337	390
0600 Desktop paging controller, stand alone		4	2		835	116		951	1,100

For customer support on your Building Construction Costs with RSMeans data, call 800.448.8182.

Estimating Tips
31 05 00 Common Work Results for Earthwork

- Estimating the actual cost of performing earthwork requires careful consideration of the variables involved. This includes items such as type of soil, whether water will be encountered, dewatering, whether banks need bracing, disposal of excavated earth, and length of haul to fill or spoil sites, etc. If the project has large quantities of cut or fill, consider raising or lowering the site to reduce costs, while paying close attention to the effect on site drainage and utilities.

- If the project has large quantities of fill, creating a borrow pit on the site can significantly lower the costs.

- It is very important to consider what time of year the project is scheduled for completion. Bad weather can create large cost overruns from dewatering, site repair, and lost productivity from cold weather.

Reference Numbers
Reference numbers are shown at the beginning of some major classifications. These numbers refer to related items in the Reference Section. The reference information may be an estimating procedure, an alternate pricing method, or technical information.

Note: Not all subdivisions listed here necessarily appear. ■

Did you know?

RSMeans data is available through our online application with 24/7 access:

- Search for unit prices by keyword
- Leverage the most up-to-date data
- Build and export estimates

Try it free for 30 days!
www.rsmeans.com/2018freetrial

No part of this cost data may be reproduced, stored in a retrieval system, or transmitted in any form or by any means without prior written permission of Gordian.

31 05 Common Work Results for Earthwork

31 05 13 – Soils for Earthwork

31 05 13.10 Borrow

		Crew	Daily Output	Labor-Hours	Unit	Material	2018 Bare Costs Labor	2018 Bare Costs Equipment	Total	Total Incl O&P
0010	**BORROW** R312316-40									
0020	Spread, 200 HP dozer, no compaction, 2 mile RT haul									
0200	Common borrow	B-15	600	.047	C.Y.	12.75	2.21	3.93	18.89	21.50
0700	Screened loam		600	.047		28	2.21	3.93	34.14	38
0800	Topsoil, weed free		600	.047		25	2.21	3.93	31.14	35
0900	For 5 mile haul, add	B-34B	200	.040			1.84	2.71	4.55	5.75

31 05 16 – Aggregates for Earthwork

31 05 16.10 Borrow

		Crew	Daily Output	Labor-Hours	Unit	Material	2018 Bare Costs Labor	2018 Bare Costs Equipment	Total	Total Incl O&P
0010	**BORROW** R312316-40									
0020	Spread, with 200 HP dozer, no compaction, 2 mile RT haul									
0100	Bank run gravel	B-15	600	.047	L.C.Y.	18.50	2.21	3.93	24.64	28
0300	Crushed stone (1.40 tons per C.Y.), 1-1/2"		600	.047		28	2.21	3.93	34.14	38
0320	3/4"		600	.047		28	2.21	3.93	34.14	38
0340	1/2"		600	.047		29.50	2.21	3.93	35.64	40
0360	3/8"		600	.047		32.50	2.21	3.93	38.64	43.50
0400	Sand, washed, concrete		600	.047		36	2.21	3.93	42.14	47
0500	Dead or bank sand		600	.047		18.15	2.21	3.93	24.29	27.50
0600	Select structural fill		600	.047		20	2.21	3.93	26.14	29.50
0900	For 5 mile haul, add	B-34B	200	.040			1.84	2.71	4.55	5.75

31 05 23 – Cement and Concrete for Earthwork

31 05 23.30 Plant Mixed Bituminous Concrete

		Crew	Daily Output	Labor-Hours	Unit	Material	2018 Bare Costs Labor	2018 Bare Costs Equipment	Total	Total Incl O&P
0010	**PLANT MIXED BITUMINOUS CONCRETE**									
0020	Asphaltic concrete plant mix (145 lb./C.F.)				Ton	65			65	71.50
0040	Asphaltic concrete less than 300 tons add trucking costs									
0050	See Section 31 23 23.20 for hauling costs									
0200	All weather patching mix, hot				Ton	65.50			65.50	72
0250	Cold patch					73.50			73.50	80.50
0300	Berm mix					64			64	70

31 06 Schedules for Earthwork

31 06 60 – Schedules for Special Foundations and Load Bearing Elements

31 06 60.14 Piling Special Costs

		Crew	Daily Output	Labor-Hours	Unit	Material	2018 Bare Costs Labor	2018 Bare Costs Equipment	Total	Total Incl O&P
0010	**PILING SPECIAL COSTS**									
0011	Piling special costs, pile caps, see Section 03 30 53.40									
0500	Cutoffs, concrete piles, plain	1 Pile	5.50	1.455	Ea.		74.50		74.50	118
0600	With steel thin shell, add		38	.211			10.80		10.80	17.05
0700	Steel pile or "H" piles		19	.421			21.50		21.50	34
0800	Wood piles		38	.211			10.80		10.80	17.05
0900	Pre-augering up to 30' deep, average soil, 24" diameter	B-43	180	.267	L.F.		11.85	13.60	25.45	33
0920	36" diameter		115	.417			18.50	21.50	40	51.50
0960	48" diameter		70	.686			30.50	35	65.50	84.50
0980	60" diameter		50	.960			42.50	49	91.50	119
1000	Testing, any type piles, test load is twice the design load									
1050	50 ton design load, 100 ton test				Ea.				14,000	15,500
1100	100 ton design load, 200 ton test								20,000	22,000
1150	150 ton design load, 300 ton test								26,000	28,500
1200	200 ton design load, 400 ton test								28,000	31,000
1250	400 ton design load, 800 ton test								32,000	35,000
1500	Wet conditions, soft damp ground									
1600	Requiring mats for crane, add								40%	40%

For customer support on your Building Construction Costs with RSMeans data, call 800.448.8182.

31 06 Schedules for Earthwork

31 06 60 – Schedules for Special Foundations and Load Bearing Elements

31 06 60.14 Piling Special Costs

		Crew	Daily Output	Labor-Hours	Unit	Material	2018 Bare Costs Labor	Equipment	Total	Total Incl O&P
1700	Barge mounted driving rig, add								30%	30%

31 06 60.15 Mobilization

		Crew	Daily Output	Labor-Hours	Unit	Material	2018 Bare Costs Labor	Equipment	Total	Total Incl O&P
0010	**MOBILIZATION**									
0020	Set up & remove, air compressor, 600 CFM	A-5	3.30	5.455	Ea.		220	14.30	234.30	350
0100	1,200 CFM	"	2.20	8.182			330	21.50	351.50	525
0200	Crane, with pile leads and pile hammer, 75 ton	B-19	.60	107			5,600	3,225	8,825	12,200
0300	150 ton	"	.36	178			9,325	5,350	14,675	20,300
0500	Drill rig, for caissons, to 36", minimum	B-43	2	24			1,075	1,225	2,300	2,975
0520	Maximum		.50	96			4,250	4,900	9,150	11,800
0600	Up to 84"		1	48			2,125	2,450	4,575	5,925
0800	Auxiliary boiler, for steam small	A-5	1.66	10.843			440	28.50	468.50	695
0900	Large	"	.83	21.687			875	57	932	1,400
1100	Rule of thumb: complete pile driving set up, small	B-19	.45	142			7,450	4,275	11,725	16,300
1200	Large	"	.27	237			12,400	7,150	19,550	27,200
1500	Mobilization, barge, by tug boat	B-83	25	.640	Mile		30	27	57	75.50

31 11 Clearing and Grubbing

31 11 10 – Clearing and Grubbing Land

31 11 10.10 Clear and Grub Site

		Crew	Daily Output	Labor-Hours	Unit	Material	2018 Bare Costs Labor	Equipment	Total	Total Incl O&P
0010	**CLEAR AND GRUB SITE**									
0020	Cut & chip light trees to 6" diam.	B-7	1	48	Acre		2,050	1,700	3,750	4,950
0150	Grub stumps and remove	B-30	2	12			585	990	1,575	1,975
0200	Cut & chip medium trees to 12" diam.	B-7	.70	68.571			2,925	2,425	5,350	7,075
0250	Grub stumps and remove	B-30	1	24			1,175	1,975	3,150	3,925
0300	Cut & chip heavy trees to 24" diam.	B-7	.30	160			6,800	5,625	12,425	16,500
0350	Grub stumps and remove	B-30	.50	48			2,325	3,950	6,275	7,875
0400	If burning is allowed, deduct cut & chip								40%	40%
3000	Chipping stumps, to 18" deep, 12" diam.	B-86	20	.400	Ea.		21.50	9.20	30.70	42.50
3040	18" diameter		16	.500			27	11.50	38.50	53
3080	24" diameter		14	.571			30.50	13.15	43.65	61
3100	30" diameter		12	.667			36	15.35	51.35	71
3120	36" diameter		10	.800			43	18.40	61.40	85.50
3160	48" diameter		8	1			54	23	77	107
5000	Tree thinning, feller buncher, conifer									
5080	Up to 8" diameter	B-93	240	.033	Ea.		1.79	3.38	5.17	6.40
5120	12" diameter		160	.050			2.69	5.05	7.74	9.60
5240	Hardwood, up to 4" diameter		240	.033			1.79	3.38	5.17	6.40
5280	8" diameter		180	.044			2.39	4.50	6.89	8.55
5320	12" diameter		120	.067			3.58	6.75	10.33	12.85
7000	Tree removal, congested area, aerial lift truck									
7040	8" diameter	B-85	7	5.714	Ea.		251	135	386	530
7080	12" diameter		6	6.667			292	158	450	620
7120	18" diameter		5	8			350	190	540	740
7160	24" diameter		4	10			440	237	677	925
7240	36" diameter		3	13.333			585	315	900	1,225
7280	48" diameter		2	20			875	475	1,350	1,850

For customer support on your Building Construction Costs with RSMeans data, call 800.448.8182.

615

31 13 Selective Tree and Shrub Removal and Trimming

31 13 13 – Selective Tree and Shrub Removal

31 13 13.10 Selective Clearing

		Crew	Daily Output	Labor-Hours	Unit	Material	2018 Bare Costs Labor	Equipment	Total	Total Incl O&P
0010	**SELECTIVE CLEARING**									
0020	Clearing brush with brush saw	A-1C	.25	32	Acre		1,275	111	1,386	2,075
0100	By hand	1 Clab	.12	66.667			2,650		2,650	4,050
0300	With dozer, ball and chain, light clearing	B-11A	2	8			375	635	1,010	1,275
0400	Medium clearing		1.50	10.667			500	850	1,350	1,700
0500	With dozer and brush rake, light		10	1.600			75	127	202	253
0550	Medium brush to 4" diameter		8	2			93.50	159	252.50	315
0600	Heavy brush to 4" diameter		6.40	2.500			117	199	316	395
1000	Brush mowing, tractor w/rotary mower, no removal									
1020	Light density	B-84	2	4	Acre		215	179	394	520
1040	Medium density		1.50	5.333			287	238	525	690
1080	Heavy density		1	8			430	355	785	1,050

31 13 13.20 Selective Tree Removal

		Crew	Daily Output	Labor-Hours	Unit	Material	2018 Bare Costs Labor	Equipment	Total	Total Incl O&P
0010	**SELECTIVE TREE REMOVAL**									
0011	With tractor, large tract, firm									
0020	level terrain, no boulders, less than 12" diam. trees									
0300	300 HP dozer, up to 400 trees/acre, 0 to 25% hardwoods	B-10M	.75	16	Acre		785	2,450	3,235	3,875
0340	25% to 50% hardwoods		.60	20			980	3,050	4,030	4,825
0370	75% to 100% hardwoods		.45	26.667			1,300	4,075	5,375	6,450
0400	500 trees/acre, 0% to 25% hardwoods		.60	20			980	3,050	4,030	4,825
0440	25% to 50% hardwoods		.48	25			1,225	3,800	5,025	6,050
0470	75% to 100% hardwoods		.36	33.333			1,625	5,075	6,700	8,075
0500	More than 600 trees/acre, 0 to 25% hardwoods		.52	23.077			1,125	3,525	4,650	5,600
0540	25% to 50% hardwoods		.42	28.571			1,400	4,350	5,750	6,925
0570	75% to 100% hardwoods		.31	38.710			1,900	5,900	7,800	9,375
0900	Large tract clearing per tree									
1500	300 HP dozer, to 12" diameter, softwood	B-10M	320	.038	Ea.		1.84	5.70	7.54	9.10
1550	Hardwood		100	.120			5.90	18.30	24.20	29
1600	12" to 24" diameter, softwood		200	.060			2.95	9.15	12.10	14.50
1650	Hardwood		80	.150			7.35	23	30.35	36
1700	24" to 36" diameter, softwood		100	.120			5.90	18.30	24.20	29
1750	Hardwood		50	.240			11.80	36.50	48.30	58
1800	36" to 48" diameter, softwood		70	.171			8.40	26	34.40	41.50
1850	Hardwood		35	.343			16.85	52.50	69.35	83
2000	Stump removal on site by hydraulic backhoe, 1-1/2 C.Y.									
2040	4" to 6" diameter	B-17	60	.533	Ea.		23.50	10.85	34.35	47.50
2050	8" to 12" diameter	B-30	33	.727			35.50	60	95.50	120
2100	14" to 24" diameter		25	.960			46.50	79	125.50	158
2150	26" to 36" diameter		16	1.500			73	124	197	246
3000	Remove selective trees, on site using chain saws and chipper,									
3050	not incl. stumps, up to 6" diameter	B-7	18	2.667	Ea.		113	94	207	275
3100	8" to 12" diameter		12	4			170	141	311	415
3150	14" to 24" diameter		10	4.800			204	169	373	495
3200	26" to 36" diameter		8	6			255	211	466	620
3300	Machine load, 2 mile haul to dump, 12" diam. tree	A-3B	8	2			100	133	233	297

For customer support on your Building Construction Costs with RSMeans data, call 800.448.8182.

31 14 Earth Stripping and Stockpiling

31 14 13 – Soil Stripping and Stockpiling

31 14 13.23 Topsoil Stripping and Stockpiling	Crew	Daily Output	Labor-Hours	Unit	Material	2018 Bare Costs Labor	Equipment	Total	Total Incl O&P
0010 **TOPSOIL STRIPPING AND STOCKPILING**									
0020 200 HP dozer, ideal conditions	B-10B	2300	.005	C.Y.		.26	.55	.81	1
0100 Adverse conditions	"	1150	.010			.51	1.11	1.62	1.99
0200 300 HP dozer, ideal conditions	B-10M	3000	.004			.20	.61	.81	.97
0300 Adverse conditions	"	1650	.007			.36	1.11	1.47	1.76
0400 400 HP dozer, ideal conditions	B-10X	3900	.003			.15	.59	.74	.88
0500 Adverse conditions	"	2000	.006			.29	1.15	1.44	1.71
0600 Clay, dry and soft, 200 HP dozer, ideal conditions	B-10B	1600	.008			.37	.80	1.17	1.44
0700 Adverse conditions	"	800	.015			.74	1.59	2.33	2.86
1000 Medium hard, 300 HP dozer, ideal conditions	B-10M	2000	.006			.29	.91	1.20	1.46
1100 Adverse conditions	"	1100	.011			.54	1.66	2.20	2.64
1200 Very hard, 400 HP dozer, ideal conditions	B-10X	2600	.005			.23	.88	1.11	1.31
1300 Adverse conditions	"	1340	.009	▼		.44	1.71	2.15	2.55
1400 Loam or topsoil, remove and stockpile on site									
1420 6" deep, 200' haul	B-10B	865	.014	C.Y.		.68	1.47	2.15	2.65
1430 300' haul		520	.023			1.13	2.45	3.58	4.40
1440 500' haul		225	.053	▼		2.62	5.65	8.27	10.15
1450 Alternate method: 6" deep, 200' haul		5090	.002	S.Y.		.12	.25	.37	.46
1460 500' haul	▼	1325	.009	"		.45	.96	1.41	1.73
1500 Loam or topsoil, remove/stockpile on site									
1510 By hand, 6" deep, 50' haul, less than 100 S.Y.	B-1	100	.240	S.Y.		9.70		9.70	14.80
1520 By skid steer, 6" deep, 100' haul, 101-500 S.Y.	B-62	500	.048			2.10	.35	2.45	3.56
1530 100' haul, 501-900 S.Y.	"	900	.027			1.16	.19	1.35	1.98
1540 200' haul, 901-1,100 S.Y.	B-63	1000	.040			1.69	.17	1.86	2.75
1550 By dozer, 200' haul, 1,101-4,000 S.Y.	B-10B	4000	.003	▼		.15	.32	.47	.57

31 22 Grading

31 22 13 – Rough Grading

31 22 13.20 Rough Grading Sites	Crew	Daily Output	Labor-Hours	Unit	Material	2018 Bare Costs Labor	Equipment	Total	Total Incl O&P
0010 **ROUGH GRADING SITES**									
0100 Rough grade sites 400 S.F. or less, hand labor	B-1	2	12	Ea.		485		485	740
0120 410-1,000 S.F.	"	1	24			970		970	1,475
0130 1,100-3,000 S.F., skid steer & labor	B-62	1.50	16			700	116	816	1,175
0140 3,100-5,000 S.F.	"	1	24			1,050	174	1,224	1,800
0150 5,100-8,000 S.F.	B-63	1	40			1,675	174	1,849	2,750
0160 8,100-10,000 S.F.	"	.75	53.333			2,250	232	2,482	3,675
0170 8,100-10,000 S.F., dozer	B-10L	1	12			590	465	1,055	1,400
0200 Rough grade open sites 10,000-20,000 S.F., grader	B-11L	1.80	8.889			415	360	775	1,025
0210 20,100-25,000 S.F.		1.40	11.429			535	460	995	1,325
0220 25,100-30,000 S.F.		1.20	13.333			625	535	1,160	1,525
0230 30,100-35,000 S.F.		1	16			750	645	1,395	1,825
0240 35,100-40,000 S.F.		.90	17.778			830	715	1,545	2,025
0250 40,100-45,000 S.F.		.80	20			935	805	1,740	2,300
0260 45,100-50,000 S.F.		.72	22.222			1,050	895	1,945	2,550
0270 50,100-75,000 S.F.		.50	32			1,500	1,300	2,800	3,700
0280 75,100-100,000 S.F.	▼	.36	44.444	▼		2,075	1,800	3,875	5,125

For customer support on your Building Construction Costs with RSMeans data, call 800.448.8182.

617

31 22 16.10 Finish Grading		Crew	Daily Output	Labor-Hours	Unit	Material	2018 Bare Costs Labor	Equipment	Total	Total Incl O&P
0010	**FINISH GRADING**									
0012	Finish grading area to be paved with grader, small area	B-11L	400	.040	S.Y.		1.87	1.61	3.48	4.61
0100	Large area		2000	.008			.37	.32	.69	.92
1100	Fine grade for slab on grade, machine		1040	.015			.72	.62	1.34	1.77
1150	Hand grading	B-18	700	.034			1.39	.06	1.45	2.18
3500	Finish grading lagoon bottoms	B-11L	4	4	M.S.F.		187	161	348	460

31 23 16.13 Excavating, Trench

		Crew	Daily Output	Labor-Hours	Unit	Material	2018 Bare Costs Labor	Equipment	Total	Total Incl O&P
0010	**EXCAVATING, TRENCH**									
0011	Or continuous footing									
0020	Common earth with no sheeting or dewatering included									
0050	1' to 4' deep, 3/8 C.Y. excavator	B-11C	150	.107	B.C.Y.		4.99	2.08	7.07	9.85
0060	1/2 C.Y. excavator	B-11M	200	.080			3.74	1.92	5.66	7.75
0090	4' to 6' deep, 1/2 C.Y. excavator	"	200	.080			3.74	1.92	5.66	7.75
0100	5/8 C.Y. excavator	B-12Q	250	.064			3.07	2.31	5.38	7.20
0110	3/4 C.Y. excavator	B-12F	300	.053			2.56	2.23	4.79	6.35
0300	1/2 C.Y. excavator, truck mounted	B-12J	200	.080			3.84	4.35	8.19	10.60
0500	6' to 10' deep, 3/4 C.Y. excavator	B-12F	225	.071			3.41	2.98	6.39	8.40
0510	1 C.Y. excavator	B-12A	400	.040			1.92	1.86	3.78	4.95
0600	1 C.Y. excavator, truck mounted	B-12K	400	.040			1.92	3.11	5.03	6.35
0610	1-1/2 C.Y. excavator	B-12B	600	.027			1.28	1.49	2.77	3.58
0900	10' to 14' deep, 3/4 C.Y. excavator	B-12F	200	.080			3.84	3.35	7.19	9.50
0910	1 C.Y. excavator	B-12A	360	.044			2.13	2.06	4.19	5.50
1000	1-1/2 C.Y. excavator	B-12B	540	.030			1.42	1.65	3.07	3.97
1300	14' to 20' deep, 1 C.Y. excavator	B-12A	320	.050			2.40	2.32	4.72	6.20
1310	1-1/2 C.Y. excavator	B-12B	480	.033			1.60	1.86	3.46	4.47
1320	2-1/2 C.Y. excavator	B-12S	765	.021			1	1.84	2.84	3.55
1340	20' to 24' deep, 1 C.Y. excavator	B-12A	288	.056			2.67	2.58	5.25	6.90
1342	1-1/2 C.Y. excavator	B-12B	432	.037			1.78	2.07	3.85	4.97
1344	2-1/2 C.Y. excavator	B-12S	685	.023			1.12	2.06	3.18	3.97
1352	4' to 6' deep, 1/2 C.Y. excavator w/trench box	B-13H	188	.085			4.08	5.05	9.13	11.75
1354	5/8 C.Y. excavator	"	235	.068			3.27	4.05	7.32	9.40
1356	3/4 C.Y. excavator	B-13G	282	.057			2.72	2.66	5.38	7.05
1362	6' to 10' deep, 3/4 C.Y. excavator w/trench box	"	212	.075			3.62	3.54	7.16	9.40
1370	1 C.Y. excavator	B-13D	376	.043			2.04	2.19	4.23	5.50
1371	1-1/2 C.Y. excavator	B-13E	564	.028			1.36	1.73	3.09	3.96
1374	10' to 14' deep, 3/4 C.Y. excavator w/trench box	B-13G	188	.085			4.08	3.99	8.07	10.60
1375	1 C.Y. excavator	B-13D	338	.047			2.27	2.44	4.71	6.10
1376	1-1/2 C.Y. excavator	B-13E	508	.032			1.51	1.92	3.43	4.40
1381	14' to 20' deep, 1 C.Y. excavator w/trench box	B-13D	301	.053			2.55	2.74	5.29	6.85
1382	1-1/2 C.Y. excavator	B-13E	451	.035			1.70	2.16	3.86	4.96
1383	2-1/2 C.Y. excavator	B-13J	720	.022			1.07	2.07	3.14	3.89
1386	20' to 24' deep, 1 C.Y. excavator w/trench box	B-13D	271	.059			2.83	3.04	5.87	7.65
1387	1-1/2 C.Y. excavator	B-13E	406	.039			1.89	2.40	4.29	5.50
1388	2-1/2 C.Y. excavator	B-13J	645	.025			1.19	2.31	3.50	4.35
1391	Shoring by S.F./day trench wall protected loose mat., 4' W	B-6	3200	.008	SF Wall	.49	.33	.10	.92	1.14
1392	Rent shoring per week per S.F. wall protected, loose mat., 4' W					1.34			1.34	1.48
1395	Hydraulic shoring, S.F. trench wall protected stable mat., 4' W	2 Clab	2700	.006		.19	.24		.43	.57

31 23 16.13 Excavating, Trench

		Crew	Daily Output	Labor-Hours	Unit	Material	2018 Bare Costs Labor	2018 Bare Costs Equipment	Total	Total Incl O&P
1397	semi-stable material, 4' W	2 Clab	2400	.007	SF Wall	.27	.27		.54	.70
1398	Rent hydraulic shoring per day/S.F. wall, stable mat., 4' W					.32			.32	.35
1399	semi-stable material					.39			.39	.43
1400	By hand with pick and shovel 2' to 6' deep, light soil	1 Clab	8	1	B.C.Y.		40		40	60.50
1500	Heavy soil	"	4	2	"		79.50		79.50	121
1700	For tamping backfilled trenches, air tamp, add	A-1G	100	.080	E.C.Y.		3.19	.52	3.71	5.45
1900	Vibrating plate, add	B-18	180	.133	"		5.40	.23	5.63	8.50
2100	Trim sides and bottom for concrete pours, common earth		1500	.016	S.F.		.65	.03	.68	1.02
2300	Hardpan		600	.040	"		1.62	.07	1.69	2.54
2400	Pier and spread footing excavation, add to above				B.C.Y.				30%	30%
3000	Backfill trench, F.E. loader, wheel mtd., 1 C.Y. bucket									
3020	Minimal haul	B-10R	400	.030	L.C.Y.		1.47	.73	2.20	3.03
3040	100' haul	"	200	.060			2.95	1.45	4.40	6.05
3080	2-1/4 C.Y. bucket, minimum haul	B-10T	600	.020			.98	.87	1.85	2.45
3090	100' haul	"	300	.040			1.96	1.74	3.70	4.89
5020	Loam & sandy clay with no sheeting or dewatering included									
5050	1' to 4' deep, 3/8 C.Y. tractor loader/backhoe	B-11C	162	.099	B.C.Y.		4.62	1.93	6.55	9.10
5060	1/2 C.Y. excavator	B-11M	216	.074			3.47	1.78	5.25	7.20
5080	4' to 6' deep, 1/2 C.Y. excavator	"	216	.074			3.47	1.78	5.25	7.20
5090	5/8 C.Y. excavator	B-12Q	276	.058			2.78	2.09	4.87	6.50
5100	3/4 C.Y. excavator	B-12F	324	.049			2.37	2.07	4.44	5.85
5130	1/2 C.Y. excavator, truck mounted	B-12J	216	.074			3.55	4.03	7.58	9.85
5140	6' to 10' deep, 3/4 C.Y. excavator	B-12F	243	.066			3.16	2.76	5.92	7.80
5150	1 C.Y. excavator	B-12A	432	.037			1.78	1.72	3.50	4.58
5160	1 C.Y. excavator, truck mounted	B-12K	432	.037			1.78	2.88	4.66	5.85
5170	1-1/2 C.Y. excavator	B-12B	648	.025			1.18	1.38	2.56	3.31
5190	10' to 14' deep, 3/4 C.Y. excavator	B-12F	216	.074			3.55	3.10	6.65	8.80
5200	1 C.Y. excavator	B-12A	389	.041			1.97	1.91	3.88	5.10
5210	1-1/2 C.Y. excavator	B-12B	583	.027			1.32	1.53	2.85	3.68
5250	14' to 20' deep, 1 C.Y. excavator	B-12A	346	.046			2.22	2.15	4.37	5.70
5260	1-1/2 C.Y. excavator	B-12B	518	.031			1.48	1.73	3.21	4.14
5270	2-1/2 C.Y. excavator	B-12S	826	.019			.93	1.71	2.64	3.29
5300	20' to 24' deep, 1 C.Y. excavator	B-12A	311	.051			2.47	2.39	4.86	6.35
5310	1-1/2 C.Y. excavator	B-12B	467	.034			1.64	1.91	3.55	4.59
5320	2-1/2 C.Y. excavator	B-12S	740	.022			1.04	1.91	2.95	3.67
5352	4' to 6' deep, 1/2 C.Y. excavator w/trench box	B-13H	205	.078			3.74	4.64	8.38	10.75
5354	5/8 C.Y. excavator	"	257	.062			2.99	3.70	6.69	8.60
5356	3/4 C.Y. excavator	B-13G	308	.052			2.49	2.44	4.93	6.45
5362	6' to 10' deep, 3/4 C.Y. excavator w/trench box	"	231	.069			3.32	3.25	6.57	8.60
5364	1 C.Y. excavator	B-13D	410	.039			1.87	2.01	3.88	5.05
5366	1-1/2 C.Y. excavator	B-13E	616	.026			1.25	1.58	2.83	3.63
5370	10' to 14' deep, 3/4 C.Y. excavator w/trench box	B-13G	205	.078			3.74	3.66	7.40	9.70
5372	1 C.Y. excavator	B-13D	370	.043			2.07	2.22	4.29	5.60
5374	1-1/2 C.Y. excavator	B-13E	554	.029			1.39	1.76	3.15	4.04
5382	14' to 20' deep, 1 C.Y. excavator w/trench box	B-13D	329	.049			2.33	2.50	4.83	6.30
5384	1-1/2 C.Y. excavator	B-13E	492	.033			1.56	1.98	3.54	4.54
5386	2-1/2 C.Y. excavator	B-13J	780	.021			.98	1.91	2.89	3.59
5392	20' to 24' deep, 1 C.Y. excavator w/trench box	B-13D	295	.054			2.60	2.79	5.39	7
5394	1-1/2 C.Y. excavator	B-13E	444	.036			1.73	2.20	3.93	5.05
5396	2-1/2 C.Y. excavator	B-13J	695	.023			1.10	2.15	3.25	4.03
6020	Sand & gravel with no sheeting or dewatering included									
6050	1' to 4' deep, 3/8 C.Y. excavator	B-11C	165	.097	B.C.Y.		4.54	1.89	6.43	8.95
6060	1/2 C.Y. excavator	B-11M	220	.073			3.40	1.75	5.15	7.10

31 23 16.13 Excavating, Trench

		Crew	Daily Output	Labor-Hours	Unit	Material	2018 Bare Costs Labor	Equipment	Total	Total Incl O&P
6080	4' to 6' deep, 1/2 C.Y. excavator	B-11M	220	.073	B.C.Y.		3.40	1.75	5.15	7.10
6090	5/8 C.Y. excavator	B-12Q	275	.058			2.79	2.10	4.89	6.55
6100	3/4 C.Y. excavator	B-12F	330	.048			2.33	2.03	4.36	5.75
6130	1/2 C.Y. excavator, truck mounted	B-12J	220	.073			3.49	3.95	7.44	9.65
6140	6' to 10' deep, 3/4 C.Y. excavator	B-12F	248	.065			3.10	2.70	5.80	7.65
6150	1 C.Y. excavator	B-12A	440	.036			1.74	1.69	3.43	4.50
6160	1 C.Y. excavator, truck mounted	B-12K	440	.036			1.74	2.82	4.56	5.75
6170	1-1/2 C.Y. excavator	B-12B	660	.024			1.16	1.35	2.51	3.25
6190	10' to 14' deep, 3/4 C.Y. excavator	B-12F	220	.073			3.49	3.04	6.53	8.65
6200	1 C.Y. excavator	B-12A	396	.040			1.94	1.87	3.81	5
6210	1-1/2 C.Y. excavator	B-12B	594	.027			1.29	1.50	2.79	3.62
6250	14' to 20' deep, 1 C.Y. excavator	B-12A	352	.045			2.18	2.11	4.29	5.60
6260	1-1/2 C.Y. excavator	B-12B	528	.030			1.45	1.69	3.14	4.06
6270	2-1/2 C.Y. excavator	B-12S	840	.019			.91	1.68	2.59	3.23
6300	20' to 24' deep, 1 C.Y. excavator	B-12A	317	.050			2.42	2.34	4.76	6.25
6310	1-1/2 C.Y. excavator	B-12B	475	.034			1.62	1.88	3.50	4.52
6320	2-1/2 C.Y. excavator	B-12S	755	.021			1.02	1.87	2.89	3.60
6352	4' to 6' deep, 1/2 C.Y. excavator w/trench box	B-13H	209	.077			3.67	4.55	8.22	10.55
6354	5/8 C.Y. excavator	"	261	.061			2.94	3.64	6.58	8.45
6356	3/4 C.Y. excavator	B-13G	314	.051			2.45	2.39	4.84	6.35
6362	6' to 10' deep, 3/4 C.Y. excavator w/trench box	"	236	.068			3.25	3.18	6.43	8.45
6364	1 C.Y. excavator	B-13D	418	.038			1.84	1.97	3.81	4.95
6366	1-1/2 C.Y. excavator	B-13E	627	.026			1.22	1.55	2.77	3.56
6370	10' to 14' deep, 3/4 C.Y. excavator w/trench box	B-13G	209	.077			3.67	3.59	7.26	9.50
6372	1 C.Y. excavator	B-13D	376	.043			2.04	2.19	4.23	5.50
6374	1-1/2 C.Y. excavator	B-13E	564	.028			1.36	1.73	3.09	3.96
6382	14' to 20' deep, 1 C.Y. excavator w/trench box	B-13D	334	.048			2.30	2.46	4.76	6.20
6384	1-1/2 C.Y. excavator	B-13E	502	.032			1.53	1.94	3.47	4.46
6386	2-1/2 C.Y. excavator	B-13J	790	.020			.97	1.89	2.86	3.55
6392	20' to 24' deep, 1 C.Y. excavator w/trench box	B-13D	301	.053			2.55	2.74	5.29	6.85
6394	1-1/2 C.Y. excavator	B-13E	452	.035			1.70	2.16	3.86	4.94
6396	2-1/2 C.Y. excavator	B-13J	710	.023			1.08	2.10	3.18	3.95
7020	Dense hard clay with no sheeting or dewatering included									
7050	1' to 4' deep, 3/8 C.Y. excavator	B-11C	132	.121	B.C.Y.		5.65	2.37	8.02	11.20
7060	1/2 C.Y. excavator	B-11M	176	.091			4.25	2.19	6.44	8.85
7080	4' to 6' deep, 1/2 C.Y. excavator	"	176	.091			4.25	2.19	6.44	8.85
7090	5/8 C.Y. excavator	B-12Q	220	.073			3.49	2.62	6.11	8.20
7100	3/4 C.Y. excavator	B-12F	264	.061			2.91	2.54	5.45	7.20
7130	1/2 C.Y. excavator, truck mounted	B-12J	176	.091			4.36	4.94	9.30	12.05
7140	6' to 10' deep, 3/4 C.Y. excavator	B-12F	198	.081			3.88	3.38	7.26	9.55
7150	1 C.Y. excavator	B-12A	352	.045			2.18	2.11	4.29	5.60
7160	1 C.Y. excavator, truck mounted	B-12K	352	.045			2.18	3.53	5.71	7.20
7170	1-1/2 C.Y. excavator	B-12B	528	.030			1.45	1.69	3.14	4.06
7190	10' to 14' deep, 3/4 C.Y. excavator	B-12F	176	.091			4.36	3.80	8.16	10.80
7200	1 C.Y. excavator	B-12A	317	.050			2.42	2.34	4.76	6.25
7210	1-1/2 C.Y. excavator	B-12B	475	.034			1.62	1.88	3.50	4.52
7250	14' to 20' deep, 1 C.Y. excavator	B-12A	282	.057			2.72	2.63	5.35	7
7260	1-1/2 C.Y. excavator	B-12B	422	.038			1.82	2.12	3.94	5.10
7270	2-1/2 C.Y. excavator	B-12S	675	.024			1.14	2.09	3.23	4.02
7300	20' to 24' deep, 1 C.Y. excavator	B-12A	254	.063			3.02	2.92	5.94	7.80
7310	1-1/2 C.Y. excavator	B-12B	380	.042			2.02	2.35	4.37	5.65
7320	2-1/2 C.Y. excavator	B-12S	605	.026			1.27	2.33	3.60	4.49

31 23 16.14 Excavating, Utility Trench

31 23 16.14 Excavating, Utility Trench	Crew	Daily Output	Labor-Hours	Unit	Material	2018 Bare Costs Labor	Equipment	Total	Total Incl O&P
0010 EXCAVATING, UTILITY TRENCH									
0011 Common earth									
0050 Trenching with chain trencher, 12 HP, operator walking									
0100 4" wide trench, 12" deep	B-53	800	.010	L.F.		.51	.08	.59	.86
0150 18" deep		750	.011			.55	.08	.63	.92
0200 24" deep		700	.011			.59	.09	.68	.98
0300 6" wide trench, 12" deep		650	.012			.63	.10	.73	1.05
0350 18" deep		600	.013			.68	.10	.78	1.14
0400 24" deep		550	.015			.75	.11	.86	1.25
0450 36" deep		450	.018			.91	.14	1.05	1.53
0600 8" wide trench, 12" deep		475	.017			.86	.13	.99	1.44
0650 18" deep		400	.020			1.03	.15	1.18	1.72
0700 24" deep		350	.023			1.17	.18	1.35	1.96
0750 36" deep		300	.027			1.37	.21	1.58	2.29
1000 Backfill by hand including compaction, add									
1050 4" wide trench, 12" deep	A-1G	800	.010	L.F.		.40	.07	.47	.68
1100 18" deep		530	.015			.60	.10	.70	1.03
1150 24" deep		400	.020			.80	.13	.93	1.35
1300 6" wide trench, 12" deep		540	.015			.59	.10	.69	1.01
1350 18" deep		405	.020			.79	.13	.92	1.34
1400 24" deep		270	.030			1.18	.19	1.37	2.01
1450 36" deep		180	.044			1.77	.29	2.06	3.02
1600 8" wide trench, 12" deep		400	.020			.80	.13	.93	1.35
1650 18" deep		265	.030			1.20	.20	1.40	2.05
1700 24" deep		200	.040			1.59	.26	1.85	2.72
1750 36" deep		135	.059			2.36	.39	2.75	4.02
2000 Chain trencher, 40 HP operator riding									
2050 6" wide trench and backfill, 12" deep	B-54	1200	.007	L.F.		.34	.28	.62	.83
2100 18" deep		1000	.008			.41	.34	.75	.99
2150 24" deep		975	.008			.42	.35	.77	1.02
2200 36" deep		900	.009			.46	.37	.83	1.10
2250 48" deep		750	.011			.55	.45	1	1.32
2300 60" deep		650	.012			.63	.52	1.15	1.52
2400 8" wide trench and backfill, 12" deep		1000	.008			.41	.34	.75	.99
2450 18" deep		950	.008			.43	.35	.78	1.04
2500 24" deep		900	.009			.46	.37	.83	1.10
2550 36" deep		800	.010			.51	.42	.93	1.23
2600 48" deep		650	.012			.63	.52	1.15	1.52
2700 12" wide trench and backfill, 12" deep		975	.008			.42	.35	.77	1.02
2750 18" deep		860	.009			.48	.39	.87	1.15
2800 24" deep		800	.010			.51	.42	.93	1.23
2850 36" deep		725	.011			.57	.46	1.03	1.36
3000 16" wide trench and backfill, 12" deep		835	.010			.49	.40	.89	1.18
3050 18" deep		750	.011			.55	.45	1	1.32
3100 24" deep		700	.011			.59	.48	1.07	1.41
3200 Compaction with vibratory plate, add								35%	35%
5100 Hand excavate and trim for pipe bells after trench excavation									
5200 8" pipe	1 Clab	155	.052	L.F.		2.06		2.06	3.13
5300 18" pipe	"	130	.062	"		2.45		2.45	3.74

For customer support on your Building Construction Costs with RSMeans data, call 800.448.8182.

621

31 23 16.16 Structural Excavation for Minor Structures

		Crew	Daily Output	Labor-Hours	Unit	Material	2018 Bare Costs Labor	2018 Bare Costs Equipment	Total	Total Incl O&P
0010	**STRUCTURAL EXCAVATION FOR MINOR STRUCTURES** R312316-40									
0015	Hand, pits to 6' deep, sandy soil	1 Clab	8	1	B.C.Y.		40		40	60.50
0100	Heavy soil or clay		4	2			79.50		79.50	121
0300	Pits 6' to 12' deep, sandy soil		5	1.600			64		64	97
0500	Heavy soil or clay		3	2.667			106		106	162
0700	Pits 12' to 18' deep, sandy soil		4	2			79.50		79.50	121
0900	Heavy soil or clay		2	4			159		159	243
1100	Hand loading trucks from stock pile, sandy soil		12	.667			26.50		26.50	40.50
1300	Heavy soil or clay		8	1			40		40	60.50
1500	For wet or muck hand excavation, add to above						50%			50%
6000	Machine excavation, for spread and mat footings, elevator pits,									
6001	and small building foundations									
6030	Common earth, hydraulic backhoe, 1/2 C.Y. bucket	B-12E	55	.291	B.C.Y.		13.95	7.90	21.85	29.50
6035	3/4 C.Y. bucket	B-12F	90	.178			8.55	7.45	16	21
6040	1 C.Y. bucket	B-12A	108	.148			7.10	6.85	13.95	18.30
6050	1-1/2 C.Y. bucket	B-12B	144	.111			5.35	6.20	11.55	14.90
6060	2 C.Y. bucket	B-12C	200	.080			3.84	5.25	9.09	11.60
6070	Sand and gravel, 3/4 C.Y. bucket	B-12F	100	.160			7.70	6.70	14.40	18.95
6080	1 C.Y. bucket	B-12A	120	.133			6.40	6.20	12.60	16.50
6090	1-1/2 C.Y. bucket	B-12B	160	.100			4.80	5.60	10.40	13.40
6100	2 C.Y. bucket	B-12C	220	.073			3.49	4.78	8.27	10.55
6110	Clay, till, or blasted rock, 3/4 C.Y. bucket	B-12F	80	.200			9.60	8.35	17.95	24
6120	1 C.Y. bucket	B-12A	95	.168			8.10	7.80	15.90	21
6130	1-1/2 C.Y. bucket	B-12B	130	.123			5.90	6.85	12.75	16.50
6140	2 C.Y. bucket	B-12C	175	.091			4.39	6	10.39	13.25
6230	Sandy clay & loam, hydraulic backhoe, 1/2 C.Y. bucket	B-12E	60	.267			12.80	7.25	20.05	27.50
6235	3/4 C.Y. bucket	B-12F	98	.163			7.85	6.85	14.70	19.35
6240	1 C.Y. bucket	B-12A	116	.138			6.60	6.40	13	17.05
6250	1-1/2 C.Y. bucket	B-12B	156	.103			4.92	5.75	10.67	13.75
9010	For mobilization or demobilization, see Section 01 54 36.50									
9020	For dewatering, see Section 31 23 19.20									
9022	For larger structures, see Bulk Excavation, Section 31 23 16.42									
9024	For loading onto trucks, add								15%	15%
9026	For hauling, see Section 31 23 23.20									
9030	For sheeting or soldier bms/lagging, see Section 31 52 16.10									
9040	For trench excavation of strip ftgs, see Section 31 23 16.13									

31 23 16.26 Rock Removal

		Crew	Daily Output	Labor-Hours	Unit	Material	2018 Bare Costs Labor	2018 Bare Costs Equipment	Total	Total Incl O&P
0010	**ROCK REMOVAL**									
0015	Drilling only rock, 2" hole for rock bolts	B-47	316	.076	L.F.		3.37	4.74	8.11	10.30
0800	2-1/2" hole for pre-splitting		250	.096			4.26	6	10.26	13.05
4600	Quarry operations, 2-1/2" to 3-1/2" diameter		240	.100			4.43	6.25	10.68	13.60

31 23 16.30 Drilling and Blasting Rock

		Crew	Daily Output	Labor-Hours	Unit	Material	2018 Bare Costs Labor	2018 Bare Costs Equipment	Total	Total Incl O&P
0010	**DRILLING AND BLASTING ROCK**									
0020	Rock, open face, under 1,500 C.Y.	B-47	225	.107	B.C.Y.	4.45	4.73	6.65	15.83	19.40
0100	Over 1,500 C.Y.		300	.080		4.45	3.55	4.99	12.99	15.80
0200	Areas where blasting mats are required, under 1,500 C.Y.		175	.137		4.45	6.10	8.55	19.10	23.50
0250	Over 1,500 C.Y.		250	.096		4.45	4.26	6	14.71	17.95
0300	Bulk drilling and blasting, can vary greatly, average								9.65	12.20
0500	Pits, average								25.50	31.50
1300	Deep hole method, up to 1,500 C.Y.	B-47	50	.480		4.45	21.50	30	55.95	70.50
1400	Over 1,500 C.Y.		66	.364		4.45	16.10	22.50	43.05	54.50
1900	Restricted areas, up to 1,500 C.Y.		13	1.846		4.45	82	115	201.45	256

31 23 Excavation and Fill

31 23 16 – Excavation

31 23 16.30 Drilling and Blasting Rock

		Crew	Daily Output	Labor-Hours	Unit	Material	2018 Bare Costs Labor	Equipment	Total	Total Incl O&P
2000	Over 1,500 C.Y.	B-47	20	1.200	B.C.Y.	4.45	53	75	132.45	168
2200	Trenches, up to 1,500 C.Y.		22	1.091		12.90	48.50	68	129.40	163
2300	Over 1,500 C.Y.		26	.923		12.90	41	57.50	111.40	140
2500	Pier holes, up to 1,500 C.Y.		22	1.091		4.45	48.50	68	120.95	153
2600	Over 1,500 C.Y.		31	.774		4.45	34.50	48.50	87.45	110
2800	Boulders under 1/2 C.Y., loaded on truck, no hauling	B-100	80	.150			7.35	12.25	19.60	24.50
2900	Boulders, drilled, blasted	B-47	100	.240		4.45	10.65	14.95	30.05	37.50
3100	Jackhammer operators with foreman compressor, air tools	B-9	1	40	Day		1,600	234	1,834	2,700
3300	Track drill, compressor, operator and foreman	B-47	1	24	"		1,075	1,500	2,575	3,275
3500	Blasting caps				Ea.	6.65			6.65	7.30
3700	Explosives					.53			.53	.58
3800	Blasting mats, for purchase, no mobilization, 10' x 15' x 12"					1,250			1,250	1,375
3900	Blasting mats, rent, for first day					214			214	235
4000	Per added day					64			64	70.50
4200	Preblast survey for 6 room house, individual lot, minimum	A-6	2.40	6.667			335	21	356	535
4300	Maximum	"	1.35	11.852			600	37.50	637.50	950
4500	City block within zone of influence, minimum	A-8	25200	.001	S.F.		.07		.07	.10
4600	Maximum	"	15100	.002	"		.11		.11	.17

31 23 16.42 Excavating, Bulk Bank Measure

		Crew	Daily Output	Labor-Hours	Unit	Material	2018 Bare Costs Labor	Equipment	Total	Total Incl O&P
0010	**EXCAVATING, BULK BANK MEASURE** R312316-40									
0011	Common earth piled									
0020	For loading onto trucks, add								15%	15%
0050	For mobilization and demobilization, see Section 01 54 36.50 R312316-45									
0100	For hauling, see Section 31 23 23.20									
0200	Excavator, hydraulic, crawler mtd., 1 C.Y. cap. = 100 C.Y./hr.	B-12A	800	.020	B.C.Y.		.96	.93	1.89	2.47
0250	1-1/2 C.Y. cap. = 125 C.Y./hr.	B-12B	1000	.016			.77	.89	1.66	2.14
0260	2 C.Y. cap. = 165 C.Y./hr.	B-12C	1320	.012			.58	.80	1.38	1.76
0300	3 C.Y. cap. = 260 C.Y./hr.	B-12D	2080	.008			.37	1.06	1.43	1.72
0305	3-1/2 C.Y. cap. = 300 C.Y./hr.	"	2400	.007			.32	.92	1.24	1.49
0310	Wheel mounted, 1/2 C.Y. cap. = 40 C.Y./hr.	B-12E	320	.050			2.40	1.36	3.76	5.15
0360	3/4 C.Y. cap. = 60 C.Y./hr.	B-12F	480	.033			1.60	1.39	2.99	3.95
0500	Clamshell, 1/2 C.Y. cap. = 20 C.Y./hr.	B-12G	160	.100			4.80	5.45	10.25	13.25
0550	1 C.Y. cap. = 35 C.Y./hr.	B-12H	280	.057			2.74	4.99	7.73	9.65
0950	Dragline, 1/2 C.Y. cap. = 30 C.Y./hr.	B-12I	240	.067			3.20	4.51	7.71	9.80
1000	3/4 C.Y. cap. = 35 C.Y./hr.	"	280	.057			2.74	3.86	6.60	8.40
1050	1-1/2 C.Y. cap. = 65 C.Y./hr.	B-12P	520	.031			1.48	2.70	4.18	5.20
1200	Front end loader, track mtd., 1-1/2 C.Y. cap. = 70 C.Y./hr.	B-10N	560	.021			1.05	1.04	2.09	2.73
1250	2-1/2 C.Y. cap. = 95 C.Y./hr.	B-100	760	.016			.78	1.29	2.07	2.59
1300	3 C.Y. cap. = 130 C.Y./hr.	B-10P	1040	.012			.57	1.16	1.73	2.13
1350	5 C.Y. cap. = 160 C.Y./hr.	B-10Q	1280	.009			.46	1.14	1.60	1.95
1500	Wheel mounted, 3/4 C.Y. cap. = 45 C.Y./hr.	B-10R	360	.033			1.64	.81	2.45	3.37
1550	1-1/2 C.Y. cap. = 80 C.Y./hr.	B-10S	640	.019			.92	.54	1.46	1.98
1600	2-1/4 C.Y. cap. = 100 C.Y./hr.	B-10T	800	.015			.74	.65	1.39	1.83
1650	5 C.Y. cap. = 185 C.Y./hr.	B-10U	1480	.008			.40	.66	1.06	1.33
1800	Hydraulic excavator, truck mtd. 1/2 C.Y. = 30 C.Y./hr.	B-12J	240	.067			3.20	3.62	6.82	8.85
1850	48" bucket, 1 C.Y. = 45 C.Y./hr.	B-12K	360	.044			2.13	3.45	5.58	7.05
3700	Shovel, 1/2 C.Y. cap. = 55 C.Y./hr.	B-12L	440	.036			1.74	2.03	3.77	4.87
3750	3/4 C.Y. cap. = 85 C.Y./hr.	B-12M	680	.024			1.13	1.66	2.79	3.54
3800	1 C.Y. cap. = 120 C.Y./hr.	B-12N	960	.017			.80	1.48	2.28	2.84
3850	1-1/2 C.Y. cap. = 160 C.Y./hr.	B-120	1280	.013			.60	1.14	1.74	2.16
3900	3 C.Y. cap. = 250 C.Y./hr.	B-12T	2000	.008			.38	.91	1.29	1.58
4000	For soft soil or sand, deduct								15%	15%

For customer support on your Building Construction Costs with RSMeans data, call 800.448.8182.

623

31 23 16.42 Excavating, Bulk Bank Measure	Crew	Daily Output	Labor-Hours	Unit	Material	2018 Bare Costs Labor	2018 Bare Costs Equipment	Total	Total Incl O&P
4100	For heavy soil or stiff clay, add				B.C.Y.			60%	60%
4200	For wet excavation with clamshell or dragline, add							100%	100%
4250	All other equipment, add							50%	50%
4400	Clamshell in sheeting or cofferdam, minimum	B-12H	160	.100			4.80	8.75	13.55	16.85
4450	Maximum	"	60	.267			12.80	23.50	36.30	45
5000	Excavating, bulk bank measure, sandy clay & loam piled								
5020	For loading onto trucks, add							15%	15%
5100	Excavator, hydraulic, crawler mtd., 1 C.Y. cap. = 120 C.Y./hr.	B-12A	960	.017	B.C.Y.		.80	.77	1.57	2.06
5150	1-1/2 C.Y. cap. = 150 C.Y./hr.	B-12B	1200	.013			.64	.74	1.38	1.79
5300	2 C.Y. cap. = 195 C.Y./hr.	B-12C	1560	.010			.49	.67	1.16	1.49
5400	3 C.Y. cap. = 300 C.Y./hr.	B-12D	2400	.007			.32	.92	1.24	1.49
5500	3-1/2 C.Y. cap. = 350 C.Y./hr.	"	2800	.006			.27	.78	1.05	1.27
5610	Wheel mounted, 1/2 C.Y. cap. = 44 C.Y./hr.	B-12E	352	.045			2.18	1.24	3.42	4.66
5660	3/4 C.Y. cap. = 66 C.Y./hr.	B-12F	528	.030			1.45	1.27	2.72	3.60
8000	For hauling excavated material, see Section 31 23 23.20								

31 23 16.46 Excavating, Bulk, Dozer

	EXCAVATING, BULK, DOZER	Crew	Daily Output	Labor-Hours	Unit	Material	Labor	Equipment	Total	Total Incl O&P
0010	**EXCAVATING, BULK, DOZER**									
0011	Open site									
2000	80 HP, 50' haul, sand & gravel	B-10L	460	.026	B.C.Y.		1.28	1.01	2.29	3.05
2010	Sandy clay & loam		440	.027			1.34	1.06	2.40	3.19
2020	Common earth		400	.030			1.47	1.16	2.63	3.51
2040	Clay		250	.048			2.36	1.86	4.22	5.60
2200	150' haul, sand & gravel		230	.052			2.56	2.02	4.58	6.10
2210	Sandy clay & loam		220	.055			2.68	2.11	4.79	6.35
2220	Common earth		200	.060			2.95	2.32	5.27	7
2240	Clay		125	.096			4.72	3.72	8.44	11.25
2400	300' haul, sand & gravel		120	.100			4.91	3.87	8.78	11.70
2410	Sandy clay & loam		115	.104			5.15	4.04	9.19	12.20
2420	Common earth		100	.120			5.90	4.64	10.54	14
2440	Clay		65	.185			9.05	7.15	16.20	21.50
3000	105 HP, 50' haul, sand & gravel	B-10W	700	.017			.84	.87	1.71	2.23
3010	Sandy clay & loam		680	.018			.87	.89	1.76	2.29
3020	Common earth		610	.020			.97	1	1.97	2.56
3040	Clay		385	.031			1.53	1.58	3.11	4.06
3200	150' haul, sand & gravel		310	.039			1.90	1.96	3.86	5.05
3210	Sandy clay & loam		300	.040			1.96	2.03	3.99	5.20
3220	Common earth		270	.044			2.18	2.25	4.43	5.80
3240	Clay		170	.071			3.47	3.58	7.05	9.20
3300	300' haul, sand & gravel		140	.086			4.21	4.34	8.55	11.15
3310	Sandy clay & loam		135	.089			4.37	4.50	8.87	11.55
3320	Common earth		120	.100			4.91	5.05	9.96	13
3340	Clay		100	.120			5.90	6.10	12	15.60
4000	200 HP, 50' haul, sand & gravel	B-10B	1400	.009			.42	.91	1.33	1.64
4010	Sandy clay & loam		1360	.009			.43	.94	1.37	1.69
4020	Common earth		1230	.010			.48	1.04	1.52	1.86
4040	Clay		770	.016			.77	1.65	2.42	2.98
4200	150' haul, sand & gravel		595	.020			.99	2.14	3.13	3.85
4210	Sandy clay & loam		580	.021			1.02	2.19	3.21	3.95
4220	Common earth		516	.023			1.14	2.47	3.61	4.44
4240	Clay		325	.037			1.81	3.92	5.73	7.05
4400	300' haul, sand & gravel		310	.039			1.90	4.11	6.01	7.40
4410	Sandy clay & loam		300	.040			1.96	4.24	6.20	7.65

31 23 16.46 Excavating, Bulk, Dozer

		Crew	Daily Output	Labor-Hours	Unit	Material	2018 Bare Costs Labor	Equipment	Total	Total Incl O&P
4420	Common earth	B-10B	270	.044	B.C.Y.		2.18	4.71	6.89	8.50
4440	Clay	↓	170	.071			3.47	7.50	10.97	13.50
5000	300 HP, 50' haul, sand & gravel	B-10M	1900	.006			.31	.96	1.27	1.53
5010	Sandy clay & loam		1850	.006			.32	.99	1.31	1.57
5020	Common earth		1650	.007			.36	1.11	1.47	1.76
5040	Clay		1025	.012			.58	1.78	2.36	2.83
5200	150' haul, sand & gravel		920	.013			.64	1.99	2.63	3.16
5210	Sandy clay & loam		895	.013			.66	2.04	2.70	3.25
5220	Common earth		800	.015			.74	2.29	3.03	3.62
5240	Clay		500	.024			1.18	3.66	4.84	5.80
5400	300' haul, sand & gravel		470	.026			1.25	3.89	5.14	6.20
5410	Sandy clay & loam		455	.026			1.30	4.02	5.32	6.40
5420	Common earth		410	.029			1.44	4.46	5.90	7.10
5440	Clay	↓	250	.048			2.36	7.30	9.66	11.60
5500	460 HP, 50' haul, sand & gravel	B-10X	1930	.006			.31	1.19	1.50	1.77
5506	Sandy clay & loam		1880	.006			.31	1.22	1.53	1.81
5510	Common earth		1680	.007			.35	1.36	1.71	2.03
5520	Clay		1050	.011			.56	2.18	2.74	3.25
5530	150' haul, sand & gravel		1290	.009			.46	1.78	2.24	2.64
5535	Sandy clay & loam		1250	.010			.47	1.83	2.30	2.73
5540	Common earth		1120	.011			.53	2.04	2.57	3.05
5550	Clay		700	.017			.84	3.27	4.11	4.87
5560	300' haul, sand & gravel		660	.018			.89	3.47	4.36	5.15
5565	Sandy clay & loam		640	.019			.92	3.58	4.50	5.35
5570	Common earth		575	.021			1.03	3.98	5.01	5.95
5580	Clay	↓	350	.034			1.68	6.55	8.23	9.75
6000	700 HP, 50' haul, sand & gravel	B-10V	3500	.003			.17	1.43	1.60	1.82
6006	Sandy clay & loam		3400	.004			.17	1.47	1.64	1.88
6010	Common earth		3035	.004			.19	1.65	1.84	2.10
6020	Clay		1925	.006			.31	2.60	2.91	3.31
6030	150' haul, sand & gravel		2025	.006			.29	2.47	2.76	3.16
6035	Sandy clay & loam		1960	.006			.30	2.55	2.85	3.25
6040	Common earth		1750	.007			.34	2.86	3.20	3.65
6050	Clay		1100	.011			.54	4.54	5.08	5.80
6060	300' haul, sand & gravel		1030	.012			.57	4.85	5.42	6.20
6065	Sandy clay & loam		1005	.012			.59	4.97	5.56	6.35
6070	Common earth		900	.013			.65	5.55	6.20	7.10
6080	Clay	↓	550	.022	↓		1.07	9.10	10.17	11.60

31 23 16.50 Excavation, Bulk, Scrapers

		Crew	Daily Output	Labor-Hours	Unit	Material	2018 Bare Costs Labor	Equipment	Total	Total Incl O&P
0010	**EXCAVATION, BULK, SCRAPERS** R312316-40									
0100	Elev. scraper 11 C.Y., sand & gravel 1,500' haul, 1/4 dozer	B-33F	690	.020	B.C.Y.		1.01	2.34	3.35	4.10
0150	3,000' haul		610	.023			1.14	2.64	3.78	4.64
0200	5,000' haul		505	.028			1.38	3.19	4.57	5.60
0300	Common earth, 1,500' haul		600	.023			1.16	2.69	3.85	4.72
0350	3,000' haul		530	.026			1.32	3.04	4.36	5.35
0400	5,000' haul		440	.032			1.58	3.66	5.24	6.40
0410	Sandy clay & loam, 1,500' haul		648	.022			1.08	2.49	3.57	4.37
0420	3,000' haul		572	.024			1.22	2.82	4.04	4.94
0430	5,000' haul		475	.029			1.47	3.39	4.86	5.95
0500	Clay, 1,500' haul		375	.037			1.86	4.30	6.16	7.55
0550	3,000' haul		330	.042			2.11	4.89	7	8.55
0600	5,000' haul	↓	275	.051	↓		2.53	5.85	8.38	10.30

31 23 16.50 Excavation, Bulk, Scrapers	Crew	Daily Output	Labor-Hours	Unit	Material	2018 Bare Costs Labor	Equipment	Total	Total Incl O&P
1000	Self propelled scraper, 14 C.Y., 1/4 push dozer									
1050	Sand and gravel, 1,500' haul	B-33D	920	.015	B.C.Y.		.76	3.20	3.96	4.67
1100	3,000' haul		805	.017			.87	3.66	4.53	5.35
1200	5,000' haul		645	.022			1.08	4.57	5.65	6.70
1300	Common earth, 1,500' haul		800	.018			.87	3.68	4.55	5.35
1350	3,000' haul		700	.020			1	4.21	5.21	6.15
1400	5,000' haul		560	.025			1.24	5.25	6.49	7.70
1420	Sandy clay & loam, 1,500' haul		864	.016			.81	3.41	4.22	4.97
1430	3,000' haul		786	.018			.89	3.75	4.64	5.45
1440	5,000' haul		605	.023			1.15	4.87	6.02	7.10
1500	Clay, 1,500' haul		500	.028			1.39	5.90	7.29	8.60
1550	3,000' haul		440	.032			1.58	6.70	8.28	9.75
1600	5,000' haul		350	.040			1.99	8.40	10.39	12.25
2000	21 C.Y., 1/4 push dozer, sand & gravel, 1,500' haul	B-33E	1180	.012			.59	2.48	3.07	3.62
2100	3,000' haul		910	.015			.77	3.21	3.98	4.69
2200	5,000' haul		750	.019			.93	3.90	4.83	5.70
2300	Common earth, 1,500' haul		1030	.014			.68	2.84	3.52	4.14
2350	3,000' haul		790	.018			.88	3.70	4.58	5.40
2400	5,000' haul		650	.022			1.07	4.50	5.57	6.55
2420	Sandy clay & loam, 1,500' haul		1112	.013			.63	2.63	3.26	3.84
2430	3,000' haul		854	.016			.82	3.42	4.24	5
2440	5,000' haul		702	.020			.99	4.17	5.16	6.10
2500	Clay, 1,500' haul		645	.022			1.08	4.54	5.62	6.60
2550	3,000' haul		495	.028			1.41	5.90	7.31	8.65
2600	5,000' haul		405	.035			1.72	7.20	8.92	10.55
2700	Towed, 10 C.Y., 1/4 push dozer, sand & gravel, 1,500' haul	B-33B	560	.025			1.24	4.36	5.60	6.70
2720	3,000' haul		450	.031			1.55	5.45	7	8.30
2730	5,000' haul		365	.038			1.91	6.70	8.61	10.25
2750	Common earth, 1,500' haul		420	.033			1.66	5.80	7.46	8.90
2770	3,000' haul		400	.035			1.74	6.10	7.84	9.35
2780	5,000' haul		310	.045			2.25	7.90	10.15	12.05
2785	Sandy clay & loam, 1,500' haul		454	.031			1.54	5.40	6.94	8.20
2790	3,000' haul		432	.032			1.61	5.65	7.26	8.65
2795	5,000' haul		340	.041			2.05	7.20	9.25	11
2800	Clay, 1,500' haul		315	.044			2.21	7.75	9.96	11.90
2820	3,000' haul		300	.047			2.32	8.15	10.47	12.45
2840	5,000' haul		225	.062			3.10	10.85	13.95	16.65
2900	15 C.Y., 1/4 push dozer, sand & gravel, 1,500' haul	B-33C	800	.018			.87	3.08	3.95	4.71
2920	3,000' haul		640	.022			1.09	3.85	4.94	5.90
2940	5,000' haul		520	.027			1.34	4.73	6.07	7.25
2960	Common earth, 1,500' haul		600	.023			1.16	4.10	5.26	6.25
2980	3,000' haul		560	.025			1.24	4.40	5.64	6.70
3000	5,000' haul		440	.032			1.58	5.60	7.18	8.55
3005	Sandy clay & loam, 1,500' haul		648	.022			1.08	3.80	4.88	5.80
3010	3,000' haul		605	.023			1.15	4.07	5.22	6.20
3015	5,000' haul		475	.029			1.47	5.20	6.67	7.90
3020	Clay, 1,500' haul		450	.031			1.55	5.45	7	8.35
3040	3,000' haul		420	.033			1.66	5.85	7.51	8.95
3060	5,000' haul		320	.044			2.18	7.70	9.88	11.75

31 23 19.20 Dewatering Systems

		Crew	Daily Output	Labor-Hours	Unit	Material	2018 Bare Costs Labor	Equipment	Total	Total Incl O&P
0010	**DEWATERING SYSTEMS**									
0020	Excavate drainage trench, 2' wide, 2' deep	B-11C	90	.178	C.Y.		8.30	3.47	11.77	16.40
0100	2' wide, 3' deep, with backhoe loader	"	135	.119			5.55	2.31	7.86	10.95
0200	Excavate sump pits by hand, light soil	1 Clab	7.10	1.127			45		45	68.50
0300	Heavy soil	"	3.50	2.286	↓		91		91	139
0500	Pumping 8 hrs., attended 2 hrs./day, incl. 20 L.F.									
0550	of suction hose & 100 L.F. discharge hose									
0600	2" diaphragm pump used for 8 hrs.	B-10H	4	3	Day		147	19.20	166.20	244
0650	4" diaphragm pump used for 8 hrs.	B-10I	4	3			147	30.50	177.50	257
0800	8 hrs. attended, 2" diaphragm pump	B-10H	1	12			590	77	667	975
0900	3" centrifugal pump	B-10J	1	12			590	85	675	985
1000	4" diaphragm pump	B-10I	1	12			590	123	713	1,025
1100	6" centrifugal pump	B-10K	1	12	↓		590	320	910	1,250
1300	CMP, incl. excavation 3' deep, 12" diameter	B-6	115	.209	L.F.	10.55	9.10	2.72	22.37	28.50
1400	18" diameter		100	.240	"	18.70	10.50	3.12	32.32	40
1600	Sump hole construction, incl. excavation and gravel, pit		1250	.019	C.F.	1.09	.84	.25	2.18	2.74
1700	With 12" gravel collar, 12" pipe, corrugated, 16 ga.		70	.343	L.F.	21	14.95	4.46	40.41	51
1800	15" pipe, corrugated, 16 ga.		55	.436		27.50	19.05	5.70	52.25	66
1900	18" pipe, corrugated, 16 ga.		50	.480		32	21	6.25	59.25	74.50
2000	24" pipe, corrugated, 14 ga.		40	.600	↓	38.50	26	7.80	72.30	91
2200	Wood lining, up to 4' x 4', add	↓	300	.080	SFCA	16.15	3.49	1.04	20.68	24
9950	See Section 31 23 19.40 for wellpoints									
9960	See Section 31 23 19.30 for deep well systems									

31 23 19.30 Wells

		Crew	Daily Output	Labor-Hours	Unit	Material	2018 Bare Costs Labor	Equipment	Total	Total Incl O&P
0010	**WELLS**									
0011	For dewatering 10' to 20' deep, 2' diameter									
0020	with steel casing, minimum	B-6	165	.145	V.L.F.	38	6.35	1.89	46.24	53.50
0050	Average		98	.245		43	10.70	3.19	56.89	67
0100	Maximum	↓	49	.490	↓	48	21.50	6.40	75.90	92.50
0300	For dewatering pumps see 01 54 33 in Reference Section									
0500	For domestic water wells, see Section 33 21 13.10									

31 23 19.40 Wellpoints

		Crew	Daily Output	Labor-Hours	Unit	Material	2018 Bare Costs Labor	Equipment	Total	Total Incl O&P
0010	**WELLPOINTS** R312319-90									
0011	For equipment rental, see 01 54 33 in Reference Section									
0100	Installation and removal of single stage system									
0110	Labor only, 0.75 labor-hours per L.F.	1 Clab	10.70	.748	LF Hdr		30		30	45.50
0200	2.0 labor-hours per L.F.	"	4	2	"		79.50		79.50	121
0400	Pump operation, 4 @ 6 hr. shifts									
0410	Per 24 hr. day	4 Eqlt	1.27	25.197	Day		1,300		1,300	1,950
0500	Per 168 hr. week, 160 hr. straight, 8 hr. double time		.18	178	Week		9,125		9,125	13,800
0550	Per 4.3 week month	↓	.04	800	Month		41,000		41,000	62,000
0600	Complete installation, operation, equipment rental, fuel &									
0610	removal of system with 2" wellpoints 5' OC									
0700	100' long header, 6" diameter, first month	4 Eqlt	3.23	9.907	LF Hdr	159	510		669	940
0800	Thereafter, per month		4.13	7.748		127	395		522	740
1000	200' long header, 8" diameter, first month		6	5.333		145	274		419	575
1100	Thereafter, per month		8.39	3.814		71.50	196		267.50	375
1300	500' long header, 8" diameter, first month		10.63	3.010		55.50	154		209.50	295
1400	Thereafter, per month		20.91	1.530		40	78.50		118.50	162
1600	1,000' long header, 10" diameter, first month		11.62	2.754		47.50	141		188.50	266
1700	Thereafter, per month	↓	41.81	.765	↓	24	39.50		63.50	85.50
1900	Note: above figures include pumping 168 hrs. per week,									

31 23 Excavation and Fill

31 23 19 – Dewatering

31 23 19.40 Wellpoints	Crew	Daily Output	Labor-Hours	Unit	Material	2018 Bare Costs Labor	2018 Bare Costs Equipment	Total	Total Incl O&P
1910	the pump operator, and one stand-by pump.									

31 23 23 – Fill

31 23 23.13 Backfill

	31 23 23.13 Backfill	Crew	Daily Output	Labor-Hours	Unit	Material	Labor	Equipment	Total	Total Incl O&P
0010	**BACKFILL** R312323-30									
0015	By hand, no compaction, light soil	1 Clab	14	.571	L.C.Y.		23		23	34.50
0100	Heavy soil		11	.727	"		29		29	44
0300	Compaction in 6" layers, hand tamp, add to above	↓	20.60	.388	E.C.Y.		15.50		15.50	23.50
0400	Roller compaction operator walking, add	B-10A	100	.120			5.90	1.80	7.70	10.90
0500	Air tamp, add	B-9D	190	.211			8.45	1.42	9.87	14.45
0600	Vibrating plate, add	A-1D	60	.133			5.30	.53	5.83	8.70
0800	Compaction in 12" layers, hand tamp, add to above	1 Clab	34	.235			9.40		9.40	14.30
0900	Roller compaction operator walking, add	B-10A	150	.080			3.93	1.20	5.13	7.25
1000	Air tamp, add	B-9	285	.140			5.65	.82	6.47	9.50
1100	Vibrating plate, add	A-1E	90	.089			3.54	.45	3.99	5.90
1300	Dozer backfilling, bulk, up to 300' haul, no compaction	B-10B	1200	.010	L.C.Y.		.49	1.06	1.55	1.91
1400	Air tamped, add	B-11B	80	.200	E.C.Y.		9.10	3.56	12.66	17.70
1600	Compacting backfill, 6" to 12" lifts, vibrating roller	B-10C	800	.015			.74	2.12	2.86	3.44
1700	Sheepsfoot roller	B-10D	750	.016	↓		.79	2.27	3.06	3.68
1900	Dozer backfilling, trench, up to 300' haul, no compaction	B-10B	900	.013	L.C.Y.		.65	1.41	2.06	2.55
2000	Air tamped, add	B-11B	80	.200	E.C.Y.		9.10	3.56	12.66	17.70
2200	Compacting backfill, 6" to 12" lifts, vibrating roller	B-10C	700	.017			.84	2.42	3.26	3.93
2300	Sheepsfoot roller	B-10D	650	.018	↓		.91	2.61	3.52	4.24
2350	Spreading in 8" layers, small dozer	B-10B	1060	.011	L.C.Y.		.56	1.20	1.76	2.16

31 23 23.14 Backfill, Structural

	31 23 23.14 Backfill, Structural	Crew	Daily Output	Labor-Hours	Unit	Material	Labor	Equipment	Total	Total Incl O&P
0010	**BACKFILL, STRUCTURAL**									
0011	Dozer or F.E. loader									
0020	From existing stockpile, no compaction									
1000	55 HP wheeled loader, 50' haul, common earth	B-11C	200	.080	L.C.Y.		3.74	1.56	5.30	7.35
2000	80 HP, 50' haul, sand & gravel	B-10L	1100	.011			.54	.42	.96	1.27
2010	Sandy clay & loam		1070	.011			.55	.43	.98	1.31
2020	Common earth		975	.012			.60	.48	1.08	1.43
2040	Clay		850	.014			.69	.55	1.24	1.65
2400	300' haul, sand & gravel		370	.032			1.59	1.26	2.85	3.79
2410	Sandy clay & loam		360	.033			1.64	1.29	2.93	3.90
2420	Common earth		330	.036			1.79	1.41	3.20	4.25
2440	Clay	↓	290	.041			2.03	1.60	3.63	4.83
3000	105 HP, 50' haul, sand & gravel	B-10W	1350	.009			.44	.45	.89	1.16
3010	Sandy clay & loam		1325	.009			.45	.46	.91	1.17
3020	Common earth		1225	.010			.48	.50	.98	1.28
3040	Clay		1100	.011			.54	.55	1.09	1.42
3300	300' haul, sand & gravel		465	.026			1.27	1.31	2.58	3.36
3310	Sandy clay & loam		455	.026			1.30	1.34	2.64	3.43
3320	Common earth		415	.029			1.42	1.47	2.89	3.76
3340	Clay		370	.032			1.59	1.64	3.23	4.22
4000	200 HP, 50' haul, sand & gravel	B-10B	2500	.005			.24	.51	.75	.92
4010	Sandy clay & loam		2435	.005			.24	.52	.76	.95
4020	Common earth		2200	.005			.27	.58	.85	1.04
4040	Clay		1950	.006			.30	.65	.95	1.18
4400	300' haul, sand & gravel		805	.015			.73	1.58	2.31	2.85
4410	Sandy clay & loam		790	.015			.75	1.61	2.36	2.90
4420	Common earth		735	.016			.80	1.73	2.53	3.12
4440	Clay	↓	660	.018			.89	1.93	2.82	3.47

For customer support on your Building Construction Costs with RSMeans data, call 800.448.8182.

31 23 Excavation and Fill

31 23 23 – Fill

31 23 23.14 Backfill, Structural

		Crew	Daily Output	Labor-Hours	Unit	Material	2018 Bare Costs Labor	Equipment	Total	Total Incl O&P
5000	300 HP, 50' haul, sand & gravel	B-10M	3170	.004	L.C.Y.		.19	.58	.77	.92
5010	Sandy clay & loam		3110	.004			.19	.59	.78	.94
5020	Common earth		2900	.004			.20	.63	.83	1
5040	Clay		2700	.004			.22	.68	.90	1.07
5400	300' haul, sand & gravel		1500	.008			.39	1.22	1.61	1.93
5410	Sandy clay & loam		1470	.008			.40	1.24	1.64	1.98
5420	Common earth		1350	.009			.44	1.36	1.80	2.15
5440	Clay		1225	.010			.48	1.49	1.97	2.37
6010	For trench backfill, see Sections 31 23 16.13 and 31 23 16.14									
6100	For compaction, see Section 31 23 23.24									

31 23 23.16 Fill By Borrow and Utility Bedding

		Crew	Daily Output	Labor-Hours	Unit	Material	2018 Bare Costs Labor	Equipment	Total	Total Incl O&P
0010	**FILL BY BORROW AND UTILITY BEDDING**									
0015	Fill by borrow, load, 1 mile haul, spread with dozer									
0020	for embankments	B-15	1200	.023	L.C.Y.	12.75	1.10	1.97	15.82	17.90
0035	Select fill for shoulders & embankments	"	1200	.023	"	20	1.10	1.97	23.07	26
0040	Fill, for hauling over 1 mile, add to above per C.Y., see Section 31 23 23.20				Mile				1.41	1.73
0049	Utility bedding, for pipe & conduit, not incl. compaction									
0050	Crushed or screened bank run gravel	B-6	150	.160	L.C.Y.	21	7	2.08	30.08	36
0100	Crushed stone 3/4" to 1/2"		150	.160		28	7	2.08	37.08	43.50
0200	Sand, dead or bank		150	.160		18.15	7	2.08	27.23	33
0500	Compacting bedding in trench	A-1D	90	.089	E.C.Y.		3.54	.35	3.89	5.80
0600	If material source exceeds 2 miles, add for extra mileage.									
0610	See Section 31 23 23.20 for hauling mileage add.									

31 23 23.17 General Fill

		Crew	Daily Output	Labor-Hours	Unit	Material	2018 Bare Costs Labor	Equipment	Total	Total Incl O&P
0010	**GENERAL FILL**									
0011	Spread dumped material, no compaction									
0020	By dozer	B-10B	1000	.012	L.C.Y.		.59	1.27	1.86	2.29
0100	By hand	1 Clab	12	.667	"		26.50		26.50	40.50
0500	Gravel fill, compacted, under floor slabs, 4" deep	B-37	10000	.005	S.F.	.40	.20	.02	.62	.77
0600	6" deep		8600	.006		.60	.23	.02	.85	1.04
0700	9" deep		7200	.007		1	.28	.02	1.30	1.55
0800	12" deep		6000	.008		1.40	.34	.03	1.77	2.08
1000	Alternate pricing method, 4" deep		120	.400	E.C.Y.	30	16.85	1.26	48.11	60
1100	6" deep		160	.300		30	12.65	.95	43.60	53
1200	9" deep		200	.240		30	10.10	.76	40.86	49
1300	12" deep		220	.218		30	9.20	.69	39.89	47.50
1500	For fill under exterior paving, see Section 32 11 23.23									

31 23 23.20 Hauling

		Crew	Daily Output	Labor-Hours	Unit	Material	2018 Bare Costs Labor	Equipment	Total	Total Incl O&P
0010	**HAULING**									
0011	Excavated or borrow, loose cubic yards									
0012	no loading equipment, including hauling, waiting, loading/dumping									
0013	time per cycle (wait, load, travel, unload or dump & return)									
0014	8 C.Y. truck, 15 MPH avg., cycle 0.5 miles, 10 min. wait/ld./uld.	B-34A	320	.025	L.C.Y.		1.15	1.06	2.21	2.89
0016	cycle 1 mile		272	.029			1.35	1.24	2.59	3.41
0018	cycle 2 miles		208	.038			1.77	1.63	3.40	4.46
0020	cycle 4 miles		144	.056			2.56	2.35	4.91	6.45
0022	cycle 6 miles		112	.071			3.29	3.02	6.31	8.30
0024	cycle 8 miles		88	.091			4.18	3.85	8.03	10.55
0026	20 MPH avg., cycle 0.5 mile		336	.024			1.10	1.01	2.11	2.76
0028	cycle 1 mile		296	.027			1.24	1.14	2.38	3.13
0030	cycle 2 miles		240	.033			1.53	1.41	2.94	3.86
0032	cycle 4 miles		176	.045			2.09	1.92	4.01	5.25

31 23 23.20 Hauling		Crew	Daily Output	Labor-Hours	Unit	Material	2018 Bare Costs Labor	2018 Bare Costs Equipment	Total	Total Incl O&P
0034	cycle 6 miles	B-34A	136	.059	L.C.Y.		2.71	2.49	5.20	6.80
0036	cycle 8 miles		112	.071			3.29	3.02	6.31	8.30
0044	25 MPH avg., cycle 4 miles		192	.042			1.92	1.76	3.68	4.83
0046	cycle 6 miles		160	.050			2.30	2.12	4.42	5.80
0048	cycle 8 miles		128	.063			2.88	2.65	5.53	7.25
0050	30 MPH avg., cycle 4 miles		216	.037			1.70	1.57	3.27	4.29
0052	cycle 6 miles		176	.045			2.09	1.92	4.01	5.25
0054	cycle 8 miles		144	.056	*		2.56	2.35	4.91	6.45
0114	15 MPH avg., cycle 0.5 mile, 15 min. wait/ld./uld.		224	.036			1.64	1.51	3.15	4.13
0116	cycle 1 mile		200	.040			1.84	1.69	3.53	4.63
0118	cycle 2 miles		168	.048			2.19	2.02	4.21	5.50
0120	cycle 4 miles		120	.067			3.07	2.82	5.89	7.70
0122	cycle 6 miles		96	.083			3.83	3.53	7.36	9.65
0124	cycle 8 miles		80	.100			4.60	4.23	8.83	11.60
0126	20 MPH avg., cycle 0.5 mile		232	.034			1.59	1.46	3.05	4
0128	cycle 1 mile		208	.038			1.77	1.63	3.40	4.46
0130	cycle 2 miles		184	.043			2	1.84	3.84	5.05
0132	cycle 4 miles		144	.056			2.56	2.35	4.91	6.45
0134	cycle 6 miles		112	.071			3.29	3.02	6.31	8.30
0136	cycle 8 miles		96	.083			3.83	3.53	7.36	9.65
0144	25 MPH avg., cycle 4 miles		152	.053			2.42	2.23	4.65	6.10
0146	cycle 6 miles		128	.063			2.88	2.65	5.53	7.25
0148	cycle 8 miles		112	.071			3.29	3.02	6.31	8.30
0150	30 MPH avg., cycle 4 miles		168	.048			2.19	2.02	4.21	5.50
0152	cycle 6 miles		144	.056			2.56	2.35	4.91	6.45
0154	cycle 8 miles		120	.067			3.07	2.82	5.89	7.70
0214	15 MPH avg., cycle 0.5 mile, 20 min. wait/ld./uld.		176	.045			2.09	1.92	4.01	5.25
0216	cycle 1 mile		160	.050			2.30	2.12	4.42	5.80
0218	cycle 2 miles		136	.059			2.71	2.49	5.20	6.80
0220	cycle 4 miles		104	.077			3.54	3.26	6.80	8.95
0222	cycle 6 miles		88	.091			4.18	3.85	8.03	10.55
0224	cycle 8 miles		72	.111			5.10	4.70	9.80	12.85
0226	20 MPH avg., cycle 0.5 mile		176	.045			2.09	1.92	4.01	5.25
0228	cycle 1 mile		168	.048			2.19	2.02	4.21	5.50
0230	cycle 2 miles		144	.056			2.56	2.35	4.91	6.45
0232	cycle 4 miles		120	.067			3.07	2.82	5.89	7.70
0234	cycle 6 miles		96	.083			3.83	3.53	7.36	9.65
0236	cycle 8 miles		88	.091			4.18	3.85	8.03	10.55
0244	25 MPH avg., cycle 4 miles		128	.063			2.88	2.65	5.53	7.25
0246	cycle 6 miles		112	.071			3.29	3.02	6.31	8.30
0248	cycle 8 miles		96	.083			3.83	3.53	7.36	9.65
0250	30 MPH avg., cycle 4 miles		136	.059			2.71	2.49	5.20	6.80
0252	cycle 6 miles		120	.067			3.07	2.82	5.89	7.70
0254	cycle 8 miles		104	.077			3.54	3.26	6.80	8.95
0314	15 MPH avg., cycle 0.5 mile, 25 min. wait/ld./uld.		144	.056			2.56	2.35	4.91	6.45
0316	cycle 1 mile		128	.063			2.88	2.65	5.53	7.25
0318	cycle 2 miles		112	.071			3.29	3.02	6.31	8.30
0320	cycle 4 miles		96	.083			3.83	3.53	7.36	9.65
0322	cycle 6 miles		80	.100			4.60	4.23	8.83	11.60
0324	cycle 8 miles		64	.125			5.75	5.30	11.05	14.45
0326	20 MPH avg., cycle 0.5 mile		144	.056			2.56	2.35	4.91	6.45
0328	cycle 1 mile		136	.059			2.71	2.49	5.20	6.80
0330	cycle 2 miles		120	.067			3.07	2.82	5.89	7.70

For customer support on your Building Construction Costs with RSMeans data, call 800.448.8182.

31 23 23.20 Hauling	Crew	Daily Output	Labor-Hours	Unit	Material	2018 Bare Costs Labor	2018 Bare Costs Equipment	2018 Bare Costs Total	Total Incl O&P
0332	cycle 4 miles	B-34A	104	.077	L.C.Y.		3.54	3.26	6.80	8.95
0334	cycle 6 miles		88	.091			4.18	3.85	8.03	10.55
0336	cycle 8 miles		80	.100			4.60	4.23	8.83	11.60
0344	25 MPH avg., cycle 4 miles		112	.071			3.29	3.02	6.31	8.30
0346	cycle 6 miles		96	.083			3.83	3.53	7.36	9.65
0348	cycle 8 miles		88	.091			4.18	3.85	8.03	10.55
0350	30 MPH avg., cycle 4 miles		112	.071			3.29	3.02	6.31	8.30
0352	cycle 6 miles		104	.077			3.54	3.26	6.80	8.95
0354	cycle 8 miles		96	.083			3.83	3.53	7.36	9.65
0414	15 MPH avg., cycle 0.5 mile, 30 min. wait/ld./uld.		120	.067			3.07	2.82	5.89	7.70
0416	cycle 1 mile		112	.071			3.29	3.02	6.31	8.30
0418	cycle 2 miles		96	.083			3.83	3.53	7.36	9.65
0420	cycle 4 miles		80	.100			4.60	4.23	8.83	11.60
0422	cycle 6 miles		72	.111			5.10	4.70	9.80	12.85
0424	cycle 8 miles		64	.125			5.75	5.30	11.05	14.45
0426	20 MPH avg., cycle 0.5 mile		120	.067			3.07	2.82	5.89	7.70
0428	cycle 1 mile		112	.071			3.29	3.02	6.31	8.30
0430	cycle 2 miles		104	.077			3.54	3.26	6.80	8.95
0432	cycle 4 miles		88	.091			4.18	3.85	8.03	10.55
0434	cycle 6 miles		80	.100			4.60	4.23	8.83	11.60
0436	cycle 8 miles		72	.111			5.10	4.70	9.80	12.85
0444	25 MPH avg., cycle 4 miles		96	.083			3.83	3.53	7.36	9.65
0446	cycle 6 miles		88	.091			4.18	3.85	8.03	10.55
0448	cycle 8 miles		80	.100			4.60	4.23	8.83	11.60
0450	30 MPH avg., cycle 4 miles		96	.083			3.83	3.53	7.36	9.65
0452	cycle 6 miles		88	.091			4.18	3.85	8.03	10.55
0454	cycle 8 miles		80	.100			4.60	4.23	8.83	11.60
0514	15 MPH avg., cycle 0.5 mile, 35 min. wait/ld./uld.		104	.077			3.54	3.26	6.80	8.95
0516	cycle 1 mile		96	.083			3.83	3.53	7.36	9.65
0518	cycle 2 miles		88	.091			4.18	3.85	8.03	10.55
0520	cycle 4 miles		72	.111			5.10	4.70	9.80	12.85
0522	cycle 6 miles		64	.125			5.75	5.30	11.05	14.45
0524	cycle 8 miles		56	.143			6.55	6.05	12.60	16.55
0526	20 MPH avg., cycle 0.5 mile		104	.077			3.54	3.26	6.80	8.95
0528	cycle 1 mile		96	.083			3.83	3.53	7.36	9.65
0530	cycle 2 miles		96	.083			3.83	3.53	7.36	9.65
0532	cycle 4 miles		80	.100			4.60	4.23	8.83	11.60
0534	cycle 6 miles		72	.111			5.10	4.70	9.80	12.85
0536	cycle 8 miles		64	.125			5.75	5.30	11.05	14.45
0544	25 MPH avg., cycle 4 miles		88	.091			4.18	3.85	8.03	10.55
0546	cycle 6 miles		80	.100			4.60	4.23	8.83	11.60
0548	cycle 8 miles		72	.111			5.10	4.70	9.80	12.85
0550	30 MPH avg., cycle 4 miles		88	.091			4.18	3.85	8.03	10.55
0552	cycle 6 miles		80	.100			4.60	4.23	8.83	11.60
0554	cycle 8 miles		72	.111			5.10	4.70	9.80	12.85
1014	12 C.Y. truck, cycle 0.5 mile, 15 MPH avg., 15 min. wait/ld./uld.	B-34B	336	.024			1.10	1.62	2.72	3.43
1016	cycle 1 mile		300	.027			1.23	1.81	3.04	3.84
1018	cycle 2 miles		252	.032			1.46	2.15	3.61	4.57
1020	cycle 4 miles		180	.044			2.04	3.02	5.06	6.40
1022	cycle 6 miles		144	.056			2.56	3.77	6.33	8
1024	cycle 8 miles		120	.067			3.07	4.52	7.59	9.60
1025	cycle 10 miles		96	.083			3.83	5.65	9.48	11.95
1026	20 MPH avg., cycle 0.5 mile		348	.023			1.06	1.56	2.62	3.31

31 23 23.20 Hauling		Crew	Daily Output	Labor-Hours	Unit	Material	2018 Bare Costs Labor	2018 Bare Costs Equipment	Total	Total Incl O&P
1028	cycle 1 mile	B-34B	312	.026	L.C.Y.		1.18	1.74	2.92	3.69
1030	cycle 2 miles		276	.029			1.33	1.97	3.30	4.17
1032	cycle 4 miles		216	.037			1.70	2.51	4.21	5.35
1034	cycle 6 miles		168	.048			2.19	3.23	5.42	6.85
1036	cycle 8 miles		144	.056			2.56	3.77	6.33	8
1038	cycle 10 miles		120	.067			3.07	4.52	7.59	9.60
1040	25 MPH avg., cycle 4 miles		228	.035			1.61	2.38	3.99	5.05
1042	cycle 6 miles		192	.042			1.92	2.83	4.75	6
1044	cycle 8 miles		168	.048			2.19	3.23	5.42	6.85
1046	cycle 10 miles		144	.056			2.56	3.77	6.33	8
1050	30 MPH avg., cycle 4 miles		252	.032			1.46	2.15	3.61	4.57
1052	cycle 6 miles		216	.037			1.70	2.51	4.21	5.35
1054	cycle 8 miles		180	.044			2.04	3.02	5.06	6.40
1056	cycle 10 miles		156	.051			2.36	3.48	5.84	7.40
1060	35 MPH avg., cycle 4 miles		264	.030			1.39	2.06	3.45	4.36
1062	cycle 6 miles		228	.035			1.61	2.38	3.99	5.05
1064	cycle 8 miles		204	.039			1.80	2.66	4.46	5.65
1066	cycle 10 miles		180	.044			2.04	3.02	5.06	6.40
1068	cycle 20 miles		120	.067			3.07	4.52	7.59	9.60
1069	cycle 30 miles		84	.095			4.38	6.45	10.83	13.70
1070	cycle 40 miles		72	.111			5.10	7.55	12.65	16
1072	40 MPH avg., cycle 6 miles		240	.033			1.53	2.26	3.79	4.80
1074	cycle 8 miles		216	.037			1.70	2.51	4.21	5.35
1076	cycle 10 miles		192	.042			1.92	2.83	4.75	6
1078	cycle 20 miles		120	.067			3.07	4.52	7.59	9.60
1080	cycle 30 miles		96	.083			3.83	5.65	9.48	11.95
1082	cycle 40 miles		72	.111			5.10	7.55	12.65	16
1084	cycle 50 miles		60	.133			6.15	9.05	15.20	19.20
1094	45 MPH avg., cycle 8 miles		216	.037			1.70	2.51	4.21	5.35
1096	cycle 10 miles		204	.039			1.80	2.66	4.46	5.65
1098	cycle 20 miles		132	.061			2.79	4.11	6.90	8.70
1100	cycle 30 miles		108	.074			3.41	5.05	8.46	10.70
1102	cycle 40 miles		84	.095			4.38	6.45	10.83	13.70
1104	cycle 50 miles		72	.111			5.10	7.55	12.65	16
1106	50 MPH avg., cycle 10 miles		216	.037			1.70	2.51	4.21	5.35
1108	cycle 20 miles		144	.056			2.56	3.77	6.33	8
1110	cycle 30 miles		108	.074			3.41	5.05	8.46	10.70
1112	cycle 40 miles		84	.095			4.38	6.45	10.83	13.70
1114	cycle 50 miles		72	.111			5.10	7.55	12.65	16
1214	15 MPH avg., cycle 0.5 mile, 20 min. wait/ld./uld.		264	.030			1.39	2.06	3.45	4.36
1216	cycle 1 mile		240	.033			1.53	2.26	3.79	4.80
1218	cycle 2 miles		204	.039			1.80	2.66	4.46	5.65
1220	cycle 4 miles		156	.051			2.36	3.48	5.84	7.40
1222	cycle 6 miles		132	.061			2.79	4.11	6.90	8.70
1224	cycle 8 miles		108	.074			3.41	5.05	8.46	10.70
1225	cycle 10 miles		96	.083			3.83	5.65	9.48	11.95
1226	20 MPH avg., cycle 0.5 mile		264	.030			1.39	2.06	3.45	4.36
1228	cycle 1 mile		252	.032			1.46	2.15	3.61	4.57
1230	cycle 2 miles		216	.037			1.70	2.51	4.21	5.35
1232	cycle 4 miles		180	.044			2.04	3.02	5.06	6.40
1234	cycle 6 miles		144	.056			2.56	3.77	6.33	8
1236	cycle 8 miles		132	.061			2.79	4.11	6.90	8.70
1238	cycle 10 miles		108	.074			3.41	5.05	8.46	10.70

For customer support on your Building Construction Costs with RSMeans data, call 800.448.8182.

31 23 23 – Fill

31 23 23.20 Hauling		Crew	Daily Output	Labor-Hours	Unit	Material	2018 Bare Costs Labor	2018 Bare Costs Equipment	Total	Total Incl O&P
1240	25 MPH avg., cycle 4 miles	B-34B	192	.042	L.C.Y.		1.92	2.83	4.75	6
1242	cycle 6 miles		168	.048			2.19	3.23	5.42	6.85
1244	cycle 8 miles		144	.056			2.56	3.77	6.33	8
1246	cycle 10 miles		132	.061			2.79	4.11	6.90	8.70
1250	30 MPH avg., cycle 4 miles		204	.039			1.80	2.66	4.46	5.65
1252	cycle 6 miles		180	.044			2.04	3.02	5.06	6.40
1254	cycle 8 miles		156	.051			2.36	3.48	5.84	7.40
1256	cycle 10 miles		144	.056			2.56	3.77	6.33	8
1260	35 MPH avg., cycle 4 miles		216	.037			1.70	2.51	4.21	5.35
1262	cycle 6 miles		192	.042			1.92	2.83	4.75	6
1264	cycle 8 miles		168	.048			2.19	3.23	5.42	6.85
1266	cycle 10 miles		156	.051			2.36	3.48	5.84	7.40
1268	cycle 20 miles		108	.074			3.41	5.05	8.46	10.70
1269	cycle 30 miles		72	.111			5.10	7.55	12.65	16
1270	cycle 40 miles		60	.133			6.15	9.05	15.20	19.20
1272	40 MPH avg., cycle 6 miles		192	.042			1.92	2.83	4.75	6
1274	cycle 8 miles		180	.044			2.04	3.02	5.06	6.40
1276	cycle 10 miles		156	.051			2.36	3.48	5.84	7.40
1278	cycle 20 miles		108	.074			3.41	5.05	8.46	10.70
1280	cycle 30 miles		84	.095			4.38	6.45	10.83	13.70
1282	cycle 40 miles		72	.111			5.10	7.55	12.65	16
1284	cycle 50 miles		60	.133			6.15	9.05	15.20	19.20
1294	45 MPH avg., cycle 8 miles		180	.044			2.04	3.02	5.06	6.40
1296	cycle 10 miles		168	.048			2.19	3.23	5.42	6.85
1298	cycle 20 miles		120	.067			3.07	4.52	7.59	9.60
1300	cycle 30 miles		96	.083			3.83	5.65	9.48	11.95
1302	cycle 40 miles		72	.111			5.10	7.55	12.65	16
1304	cycle 50 miles		60	.133			6.15	9.05	15.20	19.20
1306	50 MPH avg., cycle 10 miles		180	.044			2.04	3.02	5.06	6.40
1308	cycle 20 miles		132	.061			2.79	4.11	6.90	8.70
1310	cycle 30 miles		96	.083			3.83	5.65	9.48	11.95
1312	cycle 40 miles		84	.095			4.38	6.45	10.83	13.70
1314	cycle 50 miles		72	.111			5.10	7.55	12.65	16
1414	15 MPH avg., cycle 0.5 mile, 25 min. wait/ld./uld.		204	.039			1.80	2.66	4.46	5.65
1416	cycle 1 mile		192	.042			1.92	2.83	4.75	6
1418	cycle 2 miles		168	.048			2.19	3.23	5.42	6.85
1420	cycle 4 miles		132	.061			2.79	4.11	6.90	8.70
1422	cycle 6 miles		120	.067			3.07	4.52	7.59	9.60
1424	cycle 8 miles		96	.083			3.83	5.65	9.48	11.95
1425	cycle 10 miles		84	.095			4.38	6.45	10.83	13.70
1426	20 MPH avg., cycle 0.5 mile		216	.037			1.70	2.51	4.21	5.35
1428	cycle 1 mile		204	.039			1.80	2.66	4.46	5.65
1430	cycle 2 miles		180	.044			2.04	3.02	5.06	6.40
1432	cycle 4 miles		156	.051			2.36	3.48	5.84	7.40
1434	cycle 6 miles		132	.061			2.79	4.11	6.90	8.70
1436	cycle 8 miles		120	.067			3.07	4.52	7.59	9.60
1438	cycle 10 miles		96	.083			3.83	5.65	9.48	11.95
1440	25 MPH avg., cycle 4 miles		168	.048			2.19	3.23	5.42	6.85
1442	cycle 6 miles		144	.056			2.56	3.77	6.33	8
1444	cycle 8 miles		132	.061			2.79	4.11	6.90	8.70
1446	cycle 10 miles		108	.074			3.41	5.05	8.46	10.70
1450	30 MPH avg., cycle 4 miles		168	.048			2.19	3.23	5.42	6.85
1452	cycle 6 miles		156	.051			2.36	3.48	5.84	7.40

For customer support on your Building Construction Costs with RSMeans data, call 800.448.8182.

633

31 23 23.20 Hauling		Crew	Daily Output	Labor-Hours	Unit	Material	2018 Bare Costs			Total Incl O&P
							Labor	Equipment	Total	
1454	cycle 8 miles	B-34B	132	.061	L.C.Y.		2.79	4.11	6.90	8.70
1456	cycle 10 miles		120	.067			3.07	4.52	7.59	9.60
1460	35 MPH avg., cycle 4 miles		180	.044			2.04	3.02	5.06	6.40
1462	cycle 6 miles		156	.051			2.36	3.48	5.84	7.40
1464	cycle 8 miles		144	.056			2.56	3.77	6.33	8
1466	cycle 10 miles		132	.061			2.79	4.11	6.90	8.70
1468	cycle 20 miles		96	.083			3.83	5.65	9.48	11.95
1469	cycle 30 miles		72	.111			5.10	7.55	12.65	16
1470	cycle 40 miles		60	.133			6.15	9.05	15.20	19.20
1472	40 MPH avg., cycle 6 miles		168	.048			2.19	3.23	5.42	6.85
1474	cycle 8 miles		156	.051			2.36	3.48	5.84	7.40
1476	cycle 10 miles		144	.056			2.56	3.77	6.33	8
1478	cycle 20 miles		96	.083			3.83	5.65	9.48	11.95
1480	cycle 30 miles		84	.095			4.38	6.45	10.83	13.70
1482	cycle 40 miles		60	.133			6.15	9.05	15.20	19.20
1484	cycle 50 miles		60	.133			6.15	9.05	15.20	19.20
1494	45 MPH avg., cycle 8 miles		156	.051			2.36	3.48	5.84	7.40
1496	cycle 10 miles		144	.056			2.56	3.77	6.33	8
1498	cycle 20 miles		108	.074			3.41	5.05	8.46	10.70
1500	cycle 30 miles		84	.095			4.38	6.45	10.83	13.70
1502	cycle 40 miles		72	.111			5.10	7.55	12.65	16
1504	cycle 50 miles		60	.133			6.15	9.05	15.20	19.20
1506	50 MPH avg., cycle 10 miles		156	.051			2.36	3.48	5.84	7.40
1508	cycle 20 miles		120	.067			3.07	4.52	7.59	9.60
1510	cycle 30 miles		96	.083			3.83	5.65	9.48	11.95
1512	cycle 40 miles		72	.111			5.10	7.55	12.65	16
1514	cycle 50 miles		60	.133			6.15	9.05	15.20	19.20
1614	15 MPH avg., cycle 0.5 mile, 30 min. wait/ld./uld.		180	.044			2.04	3.02	5.06	6.40
1616	cycle 1 mile		168	.048			2.19	3.23	5.42	6.85
1618	cycle 2 miles		144	.056			2.56	3.77	6.33	8
1620	cycle 4 miles		120	.067			3.07	4.52	7.59	9.60
1622	cycle 6 miles		108	.074			3.41	5.05	8.46	10.70
1624	cycle 8 miles		84	.095			4.38	6.45	10.83	13.70
1625	cycle 10 miles		84	.095			4.38	6.45	10.83	13.70
1626	20 MPH avg., cycle 0.5 mile		180	.044			2.04	3.02	5.06	6.40
1628	cycle 1 mile		168	.048			2.19	3.23	5.42	6.85
1630	cycle 2 miles		156	.051			2.36	3.48	5.84	7.40
1632	cycle 4 miles		132	.061			2.79	4.11	6.90	8.70
1634	cycle 6 miles		120	.067			3.07	4.52	7.59	9.60
1636	cycle 8 miles		108	.074			3.41	5.05	8.46	10.70
1638	cycle 10 miles		96	.083			3.83	5.65	9.48	11.95
1640	25 MPH avg., cycle 4 miles		144	.056			2.56	3.77	6.33	8
1642	cycle 6 miles		132	.061			2.79	4.11	6.90	8.70
1644	cycle 8 miles		108	.074			3.41	5.05	8.46	10.70
1646	cycle 10 miles		108	.074			3.41	5.05	8.46	10.70
1650	30 MPH avg., cycle 4 miles		144	.056			2.56	3.77	6.33	8
1652	cycle 6 miles		132	.061			2.79	4.11	6.90	8.70
1654	cycle 8 miles		120	.067			3.07	4.52	7.59	9.60
1656	cycle 10 miles		108	.074			3.41	5.05	8.46	10.70
1660	35 MPH avg., cycle 4 miles		156	.051			2.36	3.48	5.84	7.40
1662	cycle 6 miles		144	.056			2.56	3.77	6.33	8
1664	cycle 8 miles		132	.061			2.79	4.11	6.90	8.70
1666	cycle 10 miles		120	.067			3.07	4.52	7.59	9.60

For customer support on your Building Construction Costs with RSMeans data, call 800.448.8182.

31 23 23.20 Hauling		Crew	Daily Output	Labor-Hours	Unit	Material	2018 Bare Costs Labor	Equipment	Total	Total Incl O&P
1668	cycle 20 miles	B-34B	84	.095	L.C.Y.		4.38	6.45	10.83	13.70
1669	cycle 30 miles		72	.111			5.10	7.55	12.65	16
1670	cycle 40 miles		60	.133			6.15	9.05	15.20	19.20
1672	40 MPH avg., cycle 6 miles		144	.056			2.56	3.77	6.33	8
1674	cycle 8 miles		132	.061			2.79	4.11	6.90	8.70
1676	cycle 10 miles		120	.067			3.07	4.52	7.59	9.60
1678	cycle 20 miles		96	.083			3.83	5.65	9.48	11.95
1680	cycle 30 miles		72	.111			5.10	7.55	12.65	16
1682	cycle 40 miles		60	.133			6.15	9.05	15.20	19.20
1684	cycle 50 miles		48	.167			7.65	11.30	18.95	24
1694	45 MPH avg., cycle 8 miles		144	.056			2.56	3.77	6.33	8
1696	cycle 10 miles		132	.061			2.79	4.11	6.90	8.70
1698	cycle 20 miles		96	.083			3.83	5.65	9.48	11.95
1700	cycle 30 miles		84	.095			4.38	6.45	10.83	13.70
1702	cycle 40 miles		60	.133			6.15	9.05	15.20	19.20
1704	cycle 50 miles		60	.133			6.15	9.05	15.20	19.20
1706	50 MPH avg., cycle 10 miles		132	.061			2.79	4.11	6.90	8.70
1708	cycle 20 miles		108	.074			3.41	5.05	8.46	10.70
1710	cycle 30 miles		84	.095			4.38	6.45	10.83	13.70
1712	cycle 40 miles		72	.111			5.10	7.55	12.65	16
1714	cycle 50 miles		60	.133			6.15	9.05	15.20	19.20
2000	Hauling, 8 C.Y. truck, small project cost per hour	B-34A	8	1	Hr.		46	42.50	88.50	116
2100	12 C.Y. truck	B-34B	8	1			46	68	114	144
2150	16.5 C.Y. truck	B-34C	8	1			46	76	122	154
2175	18 C.Y. 8 wheel truck	B-34I	8	1			46	88.50	134.50	167
2200	20 C.Y. truck	B-34D	8	1			46	78	124	156
2300	Grading at dump, or embankment if required, by dozer	B-10B	1000	.012	L.C.Y.		.59	1.27	1.86	2.29
2310	Spotter at fill or cut, if required	1 Clab	8	1	Hr.		40		40	60.50
9014	18 C.Y. truck, 8 wheels,15 min. wait/ld./uld.,15 MPH, cycle 0.5 mi.	B-34I	504	.016	L.C.Y.		.73	1.40	2.13	2.64
9016	cycle 1 mile		450	.018			.82	1.57	2.39	2.96
9018	cycle 2 miles		378	.021			.97	1.87	2.84	3.52
9020	cycle 4 miles		270	.030			1.36	2.61	3.97	4.93
9022	cycle 6 miles		216	.037			1.70	3.27	4.97	6.15
9024	cycle 8 miles		180	.044			2.04	3.92	5.96	7.40
9025	cycle 10 miles		144	.056			2.56	4.90	7.46	9.25
9026	20 MPH avg., cycle 0.5 mile		522	.015			.71	1.35	2.06	2.55
9028	cycle 1 mile		468	.017			.79	1.51	2.30	2.84
9030	cycle 2 miles		414	.019			.89	1.71	2.60	3.22
9032	cycle 4 miles		324	.025			1.14	2.18	3.32	4.11
9034	cycle 6 miles		252	.032			1.46	2.80	4.26	5.30
9036	cycle 8 miles		216	.037			1.70	3.27	4.97	6.15
9038	cycle 10 miles		180	.044			2.04	3.92	5.96	7.40
9040	25 MPH avg., cycle 4 miles		342	.023			1.08	2.06	3.14	3.89
9042	cycle 6 miles		288	.028			1.28	2.45	3.73	4.63
9044	cycle 8 miles		252	.032			1.46	2.80	4.26	5.30
9046	cycle 10 miles		216	.037			1.70	3.27	4.97	6.15
9050	30 MPH avg., cycle 4 miles		378	.021			.97	1.87	2.84	3.52
9052	cycle 6 miles		324	.025			1.14	2.18	3.32	4.11
9054	cycle 8 miles		270	.030			1.36	2.61	3.97	4.93
9056	cycle 10 miles		234	.034			1.57	3.02	4.59	5.70
9060	35 MPH avg., cycle 4 miles		396	.020			.93	1.78	2.71	3.36
9062	cycle 6 miles		342	.023			1.08	2.06	3.14	3.89
9064	cycle 8 miles		288	.028			1.28	2.45	3.73	4.63

31 23 23.20 Hauling		Crew	Daily Output	Labor-Hours	Unit	Material	2018 Bare Costs Labor	2018 Bare Costs Equipment	Total	Total Incl O&P
9066	cycle 10 miles	B-34I	270	.030	L.C.Y.		1.36	2.61	3.97	4.93
9068	cycle 20 miles		162	.049			2.27	4.36	6.63	8.20
9070	cycle 30 miles		126	.063			2.92	5.60	8.52	10.55
9072	cycle 40 miles		90	.089			4.09	7.85	11.94	14.80
9074	40 MPH avg., cycle 6 miles		360	.022			1.02	1.96	2.98	3.70
9076	cycle 8 miles		324	.025			1.14	2.18	3.32	4.11
9078	cycle 10 miles		288	.028			1.28	2.45	3.73	4.63
9080	cycle 20 miles		180	.044			2.04	3.92	5.96	7.40
9082	cycle 30 miles		144	.056			2.56	4.90	7.46	9.25
9084	cycle 40 miles		108	.074			3.41	6.55	9.96	12.35
9086	cycle 50 miles		90	.089			4.09	7.85	11.94	14.80
9094	45 MPH avg., cycle 8 miles		324	.025			1.14	2.18	3.32	4.11
9096	cycle 10 miles		306	.026			1.20	2.31	3.51	4.35
9098	cycle 20 miles		198	.040			1.86	3.57	5.43	6.70
9100	cycle 30 miles		144	.056			2.56	4.90	7.46	9.25
9102	cycle 40 miles		126	.063			2.92	5.60	8.52	10.55
9104	cycle 50 miles		108	.074			3.41	6.55	9.96	12.35
9106	50 MPH avg., cycle 10 miles		324	.025			1.14	2.18	3.32	4.11
9108	cycle 20 miles		216	.037			1.70	3.27	4.97	6.15
9110	cycle 30 miles		162	.049			2.27	4.36	6.63	8.20
9112	cycle 40 miles		126	.063			2.92	5.60	8.52	10.55
9114	cycle 50 miles		108	.074			3.41	6.55	9.96	12.35
9214	20 min. wait/ld./uld.,15 MPH, cycle 0.5 mi.		396	.020			.93	1.78	2.71	3.36
9216	cycle 1 mile		360	.022			1.02	1.96	2.98	3.70
9218	cycle 2 miles		306	.026			1.20	2.31	3.51	4.35
9220	cycle 4 miles		234	.034			1.57	3.02	4.59	5.70
9222	cycle 6 miles		198	.040			1.86	3.57	5.43	6.70
9224	cycle 8 miles		162	.049			2.27	4.36	6.63	8.20
9225	cycle 10 miles		144	.056			2.56	4.90	7.46	9.25
9226	20 MPH avg., cycle 0.5 mile		396	.020			.93	1.78	2.71	3.36
9228	cycle 1 mile		378	.021			.97	1.87	2.84	3.52
9230	cycle 2 miles		324	.025			1.14	2.18	3.32	4.11
9232	cycle 4 miles		270	.030			1.36	2.61	3.97	4.93
9234	cycle 6 miles		216	.037			1.70	3.27	4.97	6.15
9236	cycle 8 miles		198	.040			1.86	3.57	5.43	6.70
9238	cycle 10 miles		162	.049			2.27	4.36	6.63	8.20
9240	25 MPH avg., cycle 4 miles		288	.028			1.28	2.45	3.73	4.63
9242	cycle 6 miles		252	.032			1.46	2.80	4.26	5.30
9244	cycle 8 miles		216	.037			1.70	3.27	4.97	6.15
9246	cycle 10 miles		198	.040			1.86	3.57	5.43	6.70
9250	30 MPH avg., cycle 4 miles		306	.026			1.20	2.31	3.51	4.35
9252	cycle 6 miles		270	.030			1.36	2.61	3.97	4.93
9254	cycle 8 miles		234	.034			1.57	3.02	4.59	5.70
9256	cycle 10 miles		216	.037			1.70	3.27	4.97	6.15
9260	35 MPH avg., cycle 4 miles		324	.025			1.14	2.18	3.32	4.11
9262	cycle 6 miles		288	.028			1.28	2.45	3.73	4.63
9264	cycle 8 miles		252	.032			1.46	2.80	4.26	5.30
9266	cycle 10 miles		234	.034			1.57	3.02	4.59	5.70
9268	cycle 20 miles		162	.049			2.27	4.36	6.63	8.20
9270	cycle 30 miles		108	.074			3.41	6.55	9.96	12.35
9272	cycle 40 miles		90	.089			4.09	7.85	11.94	14.80
9274	40 MPH avg., cycle 6 miles		288	.028			1.28	2.45	3.73	4.63
9276	cycle 8 miles		270	.030			1.36	2.61	3.97	4.93

31 23 Excavation and Fill

31 23 23 – Fill

31 23 23.20 Hauling		Crew	Daily Output	Labor-Hours	Unit	Material	2018 Bare Costs Labor	Equipment	Total	Total Incl O&P
9278	cycle 10 miles	B-34I	234	.034	L.C.Y.		1.57	3.02	4.59	5.70
9280	cycle 20 miles		162	.049			2.27	4.36	6.63	8.20
9282	cycle 30 miles		126	.063			2.92	5.60	8.52	10.55
9284	cycle 40 miles		108	.074			3.41	6.55	9.96	12.35
9286	cycle 50 miles		90	.089			4.09	7.85	11.94	14.80
9294	45 MPH avg., cycle 8 miles		270	.030			1.36	2.61	3.97	4.93
9296	cycle 10 miles		252	.032			1.46	2.80	4.26	5.30
9298	cycle 20 miles		180	.044			2.04	3.92	5.96	7.40
9300	cycle 30 miles		144	.056			2.56	4.90	7.46	9.25
9302	cycle 40 miles		108	.074			3.41	6.55	9.96	12.35
9304	cycle 50 miles		90	.089			4.09	7.85	11.94	14.80
9306	50 MPH avg., cycle 10 miles		270	.030			1.36	2.61	3.97	4.93
9308	cycle 20 miles		198	.040			1.86	3.57	5.43	6.70
9310	cycle 30 miles		144	.056			2.56	4.90	7.46	9.25
9312	cycle 40 miles		126	.063			2.92	5.60	8.52	10.55
9314	cycle 50 miles		108	.074			3.41	6.55	9.96	12.35
9414	25 min. wait/ld./uld.,15 MPH, cycle 0.5 mi.		306	.026			1.20	2.31	3.51	4.35
9416	cycle 1 mile		288	.028			1.28	2.45	3.73	4.63
9418	cycle 2 miles		252	.032			1.46	2.80	4.26	5.30
9420	cycle 4 miles		198	.040			1.86	3.57	5.43	6.70
9422	cycle 6 miles		180	.044			2.04	3.92	5.96	7.40
9424	cycle 8 miles		144	.056			2.56	4.90	7.46	9.25
9425	cycle 10 miles		126	.063			2.92	5.60	8.52	10.55
9426	20 MPH avg., cycle 0.5 mile		324	.025			1.14	2.18	3.32	4.11
9428	cycle 1 mile		306	.026			1.20	2.31	3.51	4.35
9430	cycle 2 miles		270	.030			1.36	2.61	3.97	4.93
9432	cycle 4 miles		234	.034			1.57	3.02	4.59	5.70
9434	cycle 6 miles		198	.040			1.86	3.57	5.43	6.70
9436	cycle 8 miles		180	.044			2.04	3.92	5.96	7.40
9438	cycle 10 miles		144	.056			2.56	4.90	7.46	9.25
9440	25 MPH avg., cycle 4 miles		252	.032			1.46	2.80	4.26	5.30
9442	cycle 6 miles		216	.037			1.70	3.27	4.97	6.15
9444	cycle 8 miles		198	.040			1.86	3.57	5.43	6.70
9446	cycle 10 miles		180	.044			2.04	3.92	5.96	7.40
9450	30 MPH avg., cycle 4 miles		252	.032			1.46	2.80	4.26	5.30
9452	cycle 6 miles		234	.034			1.57	3.02	4.59	5.70
9454	cycle 8 miles		198	.040			1.86	3.57	5.43	6.70
9456	cycle 10 miles		180	.044			2.04	3.92	5.96	7.40
9460	35 MPH avg., cycle 4 miles		270	.030			1.36	2.61	3.97	4.93
9462	cycle 6 miles		234	.034			1.57	3.02	4.59	5.70
9464	cycle 8 miles		216	.037			1.70	3.27	4.97	6.15
9466	cycle 10 miles		198	.040			1.86	3.57	5.43	6.70
9468	cycle 20 miles		144	.056			2.56	4.90	7.46	9.25
9470	cycle 30 miles		108	.074			3.41	6.55	9.96	12.35
9472	cycle 40 miles		90	.089			4.09	7.85	11.94	14.80
9474	40 MPH avg., cycle 6 miles		252	.032			1.46	2.80	4.26	5.30
9476	cycle 8 miles		234	.034			1.57	3.02	4.59	5.70
9478	cycle 10 miles		216	.037			1.70	3.27	4.97	6.15
9480	cycle 20 miles		144	.056			2.56	4.90	7.46	9.25
9482	cycle 30 miles		126	.063			2.92	5.60	8.52	10.55
9484	cycle 40 miles		90	.089			4.09	7.85	11.94	14.80
9486	cycle 50 miles		90	.089			4.09	7.85	11.94	14.80
9494	45 MPH avg., cycle 8 miles		234	.034			1.57	3.02	4.59	5.70

31 23 23.20 Hauling		Crew	Daily Output	Labor-Hours	Unit	Material	2018 Bare Costs Labor	Equipment	Total	Total Incl O&P
9496	cycle 10 miles	B-34I	216	.037	L.C.Y.		1.70	3.27	4.97	6.15
9498	cycle 20 miles		162	.049			2.27	4.36	6.63	8.20
9500	cycle 30 miles		126	.063			2.92	5.60	8.52	10.55
9502	cycle 40 miles		108	.074			3.41	6.55	9.96	12.35
9504	cycle 50 miles		90	.089			4.09	7.85	11.94	14.80
9506	50 MPH avg., cycle 10 miles		234	.034			1.57	3.02	4.59	5.70
9508	cycle 20 miles		180	.044			2.04	3.92	5.96	7.40
9510	cycle 30 miles		144	.056			2.56	4.90	7.46	9.25
9512	cycle 40 miles		108	.074			3.41	6.55	9.96	12.35
9514	cycle 50 miles		90	.089			4.09	7.85	11.94	14.80
9614	30 min. wait/ld./uld.,15 MPH, cycle 0.5 mi.		270	.030			1.36	2.61	3.97	4.93
9616	cycle 1 mile		252	.032			1.46	2.80	4.26	5.30
9618	cycle 2 miles		216	.037			1.70	3.27	4.97	6.15
9620	cycle 4 miles		180	.044			2.04	3.92	5.96	7.40
9622	cycle 6 miles		162	.049			2.27	4.36	6.63	8.20
9624	cycle 8 miles		126	.063			2.92	5.60	8.52	10.55
9625	cycle 10 miles		126	.063			2.92	5.60	8.52	10.55
9626	20 MPH avg., cycle 0.5 mile		270	.030			1.36	2.61	3.97	4.93
9628	cycle 1 mile		252	.032			1.46	2.80	4.26	5.30
9630	cycle 2 miles		234	.034			1.57	3.02	4.59	5.70
9632	cycle 4 miles		198	.040			1.86	3.57	5.43	6.70
9634	cycle 6 miles		180	.044			2.04	3.92	5.96	7.40
9636	cycle 8 miles		162	.049			2.27	4.36	6.63	8.20
9638	cycle 10 miles		144	.056			2.56	4.90	7.46	9.25
9640	25 MPH avg., cycle 4 miles		216	.037			1.70	3.27	4.97	6.15
9642	cycle 6 miles		198	.040			1.86	3.57	5.43	6.70
9644	cycle 8 miles		180	.044			2.04	3.92	5.96	7.40
9646	cycle 10 miles		162	.049			2.27	4.36	6.63	8.20
9650	30 MPH avg., cycle 4 miles		216	.037			1.70	3.27	4.97	6.15
9652	cycle 6 miles		198	.040			1.86	3.57	5.43	6.70
9654	cycle 8 miles		180	.044			2.04	3.92	5.96	7.40
9656	cycle 10 miles		162	.049			2.27	4.36	6.63	8.20
9660	35 MPH avg., cycle 4 miles		234	.034			1.57	3.02	4.59	5.70
9662	cycle 6 miles		216	.037			1.70	3.27	4.97	6.15
9664	cycle 8 miles		198	.040			1.86	3.57	5.43	6.70
9666	cycle 10 miles		180	.044			2.04	3.92	5.96	7.40
9668	cycle 20 miles		126	.063			2.92	5.60	8.52	10.55
9670	cycle 30 miles		108	.074			3.41	6.55	9.96	12.35
9672	cycle 40 miles		90	.089			4.09	7.85	11.94	14.80
9674	40 MPH avg., cycle 6 miles		216	.037			1.70	3.27	4.97	6.15
9676	cycle 8 miles		198	.040			1.86	3.57	5.43	6.70
9678	cycle 10 miles		180	.044			2.04	3.92	5.96	7.40
9680	cycle 20 miles		144	.056			2.56	4.90	7.46	9.25
9682	cycle 30 miles		108	.074			3.41	6.55	9.96	12.35
9684	cycle 40 miles		90	.089			4.09	7.85	11.94	14.80
9686	cycle 50 miles		72	.111			5.10	9.80	14.90	18.50
9694	45 MPH avg., cycle 8 miles		216	.037			1.70	3.27	4.97	6.15
9696	cycle 10 miles		198	.040			1.86	3.57	5.43	6.70
9698	cycle 20 miles		144	.056			2.56	4.90	7.46	9.25
9700	cycle 30 miles		126	.063			2.92	5.60	8.52	10.55
9702	cycle 40 miles		108	.074			3.41	6.55	9.96	12.35
9704	cycle 50 miles		90	.089			4.09	7.85	11.94	14.80
9706	50 MPH avg., cycle 10 miles		198	.040			1.86	3.57	5.43	6.70

31 23 Excavation and Fill

31 23 23 – Fill

31 23 23.20 Hauling

		Crew	Daily Output	Labor-Hours	Unit	Material	2018 Bare Costs Labor	Equipment	Total	Total Incl O&P
9708	cycle 20 miles	B-34I	162	.049	L.C.Y.		2.27	4.36	6.63	8.20
9710	cycle 30 miles		126	.063			2.92	5.60	8.52	10.55
9712	cycle 40 miles		108	.074			3.41	6.55	9.96	12.35
9714	cycle 50 miles		90	.089			4.09	7.85	11.94	14.80

31 23 23.24 Compaction, Structural

		Crew	Daily Output	Labor-Hours	Unit	Material	2018 Bare Costs Labor	Equipment	Total	Total Incl O&P
0010	**COMPACTION, STRUCTURAL** R312323-30									
0020	Steel wheel tandem roller, 5 tons	B-10E	8	1.500	Hr.		73.50	18.90	92.40	132
0100	10 tons	B-10F	8	1.500	"		73.50	29	102.50	143
0300	Sheepsfoot or wobbly wheel roller, 8" lifts, common fill	B-10G	1300	.009	E.C.Y.		.45	1.01	1.46	1.80
0400	Select fill	"	1500	.008			.39	.88	1.27	1.56
0600	Vibratory plate, 8" lifts, common fill	A-1D	200	.040			1.59	.16	1.75	2.60
0700	Select fill	"	216	.037			1.48	.15	1.63	2.41

31 25 Erosion and Sedimentation Controls

31 25 14 – Stabilization Measures for Erosion and Sedimentation Control

31 25 14.16 Rolled Erosion Control Mats and Blankets

			Crew	Daily Output	Labor-Hours	Unit	Material	2018 Bare Costs Labor	Equipment	Total	Total Incl O&P
0010	**ROLLED EROSION CONTROL MATS AND BLANKETS**										
0020	Jute mesh, 100 S.Y. per roll, 4' wide, stapled	G	B-80A	2400	.010	S.Y.	.91	.40	.10	1.41	1.72
0070	Paper biodegradable mesh	G	B-1	2500	.010		.18	.39		.57	.79
0080	Paper mulch	G	B-64	20000	.001		.14	.03	.02	.19	.22
0100	Plastic netting, stapled, 2" x 1" mesh, 20 mil	G	B-1	2500	.010		.29	.39		.68	.91
0200	Polypropylene mesh, stapled, 6.5 oz./S.Y.	G		2500	.010		1.63	.39		2.02	2.38
0300	Tobacco netting, or jute mesh #2, stapled	G		2500	.010		.25	.39		.64	.87
1000	Silt fence, install and maintain, remove	G	B-62	1300	.018	L.F.	.46	.81	.13	1.40	1.88
1100	Allow 10% per month for maintenance; 6-month max life										
1200	Place and remove hay bales, staked	G	A-2	3	8	Ton	288	330	63	681	885
1250	Place and remove hay bales, staked (alt. pricing)	G	"	2500	.010	L.F.	4.78	.40	.08	5.26	5.95

31 31 Soil Treatment

31 31 16 – Termite Control

31 31 16.13 Chemical Termite Control

		Crew	Daily Output	Labor-Hours	Unit	Material	2018 Bare Costs Labor	Equipment	Total	Total Incl O&P
0010	**CHEMICAL TERMITE CONTROL**									
0020	Slab and walls, residential	1 Skwk	1200	.007	SF Flr.	.32	.35		.67	.88
0100	Commercial, minimum		2496	.003		.33	.17		.50	.62
0200	Maximum		1645	.005		.50	.25		.75	.93
0400	Insecticides for termite control, minimum		14.20	.563	Gal.	69	29.50		98.50	121
0500	Maximum		11	.727	"	118	38		156	189

For customer support on your Building Construction Costs with RSMeans data, call 800.448.8182.

639

31 32 Soil Stabilization

31 32 13 – Soil Mixing Stabilization

31 32 13.30 Calcium Chloride	Crew	Daily Output	Labor-Hours	Unit	Material	2018 Bare Costs Labor	Equipment	Total	Total Incl O&P
0010 **CALCIUM CHLORIDE**									
0020 Calcium chloride, delivered, 100 lb. bags, truckload lots				Ton	585			585	645
0030 Solution, 4 lb. flake per gallon, tank truck delivery				Gal.	1.51			1.51	1.66

31 33 Rock Stabilization

31 33 13 – Rock Bolting and Grouting

31 33 13.10 Rock Bolting

		Crew	Daily Output	Labor-Hours	Unit	Material	2018 Bare Costs Labor	Equipment	Total	Total Incl O&P
0010	**ROCK BOLTING**									
2020	Hollow core, prestressable anchor, 1" diameter, 5' long	2 Skwk	32	.500	Ea.	180	26		206	238
2025	10' long		24	.667		305	35		340	390
2060	2" diameter, 5' long		32	.500		685	26		711	790
2065	10' long		24	.667		1,225	35		1,260	1,400
2100	Super high-tensile, 3/4" diameter, 5' long		32	.500		48.50	26		74.50	93.50
2105	10' long		24	.667		130	35		165	196
2160	2" diameter, 5' long		32	.500		405	26		431	485
2165	10' long	↓	24	.667		690	35		725	815
4400	Drill hole for rock bolt, 1-3/4" diam., 5' long (for 3/4" bolt)	B-56	17	.941			43	85.50	128.50	159
4405	10' long		9	1.778			81	162	243	300
4420	2" diameter, 5' long (for 1" bolt)		13	1.231			56	112	168	208
4425	10' long		7	2.286			104	208	312	385
4460	3-1/2" diameter, 5' long (for 2" bolt)		10	1.600			73	145	218	270
4465	10' long	↓	5	3.200	↓		146	291	437	540

31 36 Gabions

31 36 13 – Gabion Boxes

31 36 13.10 Gabion Box Systems

		Crew	Daily Output	Labor-Hours	Unit	Material	2018 Bare Costs Labor	Equipment	Total	Total Incl O&P
0010	**GABION BOX SYSTEMS**									
0400	Gabions, galvanized steel mesh mats or boxes, stone filled, 6" deep	B-13	200	.280	S.Y.	19	12.25	2.95	34.20	43
0500	9" deep		163	.344		23.50	15	3.62	42.12	53
0600	12" deep		153	.366		31.50	16	3.86	51.36	64
0700	18" deep		102	.549		44.50	24	5.80	74.30	92
0800	36" deep	↓	60	.933	↓	75.50	41	9.85	126.35	156

31 37 Riprap

31 37 13 – Machined Riprap

31 37 13.10 Riprap and Rock Lining

		Crew	Daily Output	Labor-Hours	Unit	Material	2018 Bare Costs Labor	Equipment	Total	Total Incl O&P
0010	**RIPRAP AND ROCK LINING**									
0011	Random, broken stone									
0100	Machine placed for slope protection	B-12G	62	.258	L.C.Y.	30.50	12.40	14.05	56.95	67.50
0110	3/8 to 1/4 C.Y. pieces, grouted	B-13	80	.700	S.Y.	63	30.50	7.40	100.90	124
0200	18" minimum thickness, not grouted	"	53	1.057	"	19.15	46	11.15	76.30	103
0300	Dumped, 50 lb. average	B-11A	800	.020	Ton	25.50	.94	1.59	28.03	31
0350	100 lb. average		700	.023		25.50	1.07	1.82	28.39	31.50
0370	300 lb. average	↓	600	.027	↓	25.50	1.25	2.12	28.87	32

For customer support on your Building Construction Costs with RSMeans data, call 800.448.8182.

31 41 Shoring

31 41 13 – Timber Shoring

31 41 13.10 Building Shoring

		Crew	Daily Output	Labor-Hours	Unit	Material	2018 Bare Costs Labor	Equipment	Total	Total Incl O&P
0010	**BUILDING SHORING**									
0020	Shoring, existing building, with timber, no salvage allowance	B-51	2.20	21.818	M.B.F.	890	895	86	1,871	2,425
1000	On cribbing with 35 ton screw jacks, per box and jack	"	3.60	13.333	Jack	65	545	52.50	662.50	960
1100	Masonry openings in walls, see Section 02 41 19.16									

31 41 16 – Sheet Piling

31 41 16.10 Sheet Piling Systems

			Crew	Daily Output	Labor-Hours	Unit	Material	2018 Bare Costs Labor	Equipment	Total	Total Incl O&P
0010	**SHEET PILING SYSTEMS**										
0020	Sheet piling, 50,000 psi steel, not incl. wales, 22 psf, left in place		B-40	10.81	5.920	Ton	1,575	310	330	2,215	2,600
0100	Drive, extract & salvage	R314116-40		6	10.667		515	560	595	1,670	2,100
0300	20' deep excavation, 27 psf, left in place			12.95	4.942		1,575	259	275	2,109	2,450
0400	Drive, extract & salvage	R314116-45		6.55	9.771		515	510	545	1,570	1,975
0600	25' deep excavation, 38 psf, left in place			19	3.368		1,575	177	187	1,939	2,225
0700	Drive, extract & salvage			10.50	6.095		515	320	340	1,175	1,450
0900	40' deep excavation, 38 psf, left in place			21.20	3.019		1,575	158	168	1,901	2,175
1000	Drive, extract & salvage			12.25	5.224		515	274	291	1,080	1,325
1200	15' deep excavation, 22 psf, left in place			983	.065	S.F.	18.40	3.41	3.62	25.43	29.50
1300	Drive, extract & salvage			545	.117		5.80	6.15	6.55	18.50	23
1500	20' deep excavation, 27 psf, left in place			960	.067		23	3.49	3.71	30.20	35
1600	Drive, extract & salvage			485	.132		7.55	6.90	7.35	21.80	27
1800	25' deep excavation, 38 psf, left in place			1000	.064		34	3.35	3.56	40.91	46.50
1900	Drive, extract & salvage			553	.116		10.30	6.05	6.45	22.80	28
2100	Rent steel sheet piling and wales, first month					Ton	310			310	340
2200	Per added month						31			31	34
2300	Rental piling left in place, add to rental						1,150			1,150	1,275
2500	Wales, connections & struts, 2/3 salvage						480			480	525
2700	High strength piling, 60,000 psi, add						158			158	174
2800	65,000 psi, add						237			237	261
3000	Tie rod, not upset, 1-1/2" to 4" diameter with turnbuckle						2,075			2,075	2,275
3100	No turnbuckle						1,650			1,650	1,800
3300	Upset, 1-3/4" to 4" diameter with turnbuckle						2,375			2,375	2,625
3400	No turnbuckle						2,100			2,100	2,300
3600	Lightweight, 18" to 28" wide, 7 ga., 9.22 psf, and										
3610	9 ga., 8.6 psf, minimum					Lb.	.82			.82	.90
3700	Average						.88			.88	.97
3750	Maximum						1.05			1.05	1.16
3900	Wood, solid sheeting, incl. wales, braces and spacers,	R314116-40									
3910	drive, extract & salvage, 8' deep excavation		B-31	330	.121	S.F.	1.84	5.15	.67	7.66	10.60
4000	10' deep, 50 S.F./hr. in & 150 S.F./hr. out			300	.133		1.89	5.65	.73	8.27	11.50
4100	12' deep, 45 S.F./hr. in & 135 S.F./hr. out			270	.148		1.94	6.30	.81	9.05	12.60
4200	14' deep, 42 S.F./hr. in & 126 S.F./hr. out			250	.160		2	6.80	.88	9.68	13.50
4300	16' deep, 40 S.F./hr. in & 120 S.F./hr. out			240	.167		2.07	7.05	.92	10.04	14.05
4400	18' deep, 38 S.F./hr. in & 114 S.F./hr. out			230	.174		2.13	7.40	.96	10.49	14.65
4500	20' deep, 35 S.F./hr. in & 105 S.F./hr. out			210	.190		2.20	8.10	1.05	11.35	15.85
4520	Left in place, 8' deep, 55 S.F./hr.			440	.091		3.31	3.86	.50	7.67	10.05
4540	10' deep, 50 S.F./hr.			400	.100		3.48	4.24	.55	8.27	10.90
4560	12' deep, 45 S.F./hr.			360	.111		3.67	4.71	.61	8.99	11.90
4565	14' deep, 42 S.F./hr.			335	.119		3.89	5.05	.66	9.60	12.70
4570	16' deep, 40 S.F./hr.			320	.125		4.13	5.30	.69	10.12	13.40
4580	18' deep, 38 S.F./hr.			305	.131		4.41	5.55	.72	10.68	14.10
4590	20' deep, 35 S.F./hr.			280	.143		4.72	6.05	.79	11.56	15.30
4700	Alternate pricing, left in place, 8' deep			1.76	22.727	M.B.F.	745	965	125	1,835	2,425
4800	Drive, extract and salvage, 8' deep			1.32	30.303	"	660	1,275	167	2,102	2,850

31 41 Shoring

31 41 16 – Sheet Piling

31 41 16.10 Sheet Piling Systems	Crew	Daily Output	Labor-Hours	Unit	Material	2018 Bare Costs Labor	Equipment	Total	Total Incl O&P
5000 For treated lumber add cost of treatment to lumber									

31 43 Concrete Raising

31 43 13 – Pressure Grouting

31 43 13.13 Concrete Pressure Grouting

		Crew	Daily Output	Labor-Hours	Unit	Material	2018 Bare Costs Labor	Equipment	Total	Total Incl O&P
0010	**CONCRETE PRESSURE GROUTING**									
0020	Grouting, pressure, cement & sand, 1:1 mix, minimum	B-61	124	.323	Bag	19.20	13.70	2.59	35.49	45
0100	Maximum		51	.784	"	19.20	33.50	6.30	59	78.50
0200	Cement and sand, 1:1 mix, minimum		250	.160	C.F.	38.50	6.80	1.29	46.59	54.50
0300	Maximum		100	.400		57.50	17	3.22	77.72	93
0400	Epoxy cement grout, minimum		137	.292		755	12.40	2.35	769.75	850
0500	Maximum		57	.702		755	30	5.65	790.65	880
0700	Alternate pricing method: (Add for materials)									
0710	5 person crew and equipment	B-61	1	40	Day	1,700		320	2,020	2,925

31 45 Vibroflotation and Densification

31 45 13 – Vibroflotation

31 45 13.10 Vibroflotation Densification

		Crew	Daily Output	Labor-Hours	Unit	Material	2018 Bare Costs Labor	Equipment	Total	Total Incl O&P
0010	**VIBROFLOTATION DENSIFICATION**	R314513-90								
0900	Vibroflotation compacted sand cylinder, minimum	B-60	750	.075	V.L.F.		3.51	3.18	6.69	8.80
0950	Maximum		325	.172			8.10	7.35	15.45	20.50
1100	Vibro replacement compacted stone cylinder, minimum		500	.112			5.25	4.77	10.02	13.20
1150	Maximum		250	.224			10.50	9.55	20.05	26.50
1300	Mobilization and demobilization, minimum		.47	119	Total		5,600	5,075	10,675	14,100
1400	Maximum		.14	400	"		18,800	17,000	35,800	47,100

31 46 Needle Beams

31 46 13 – Cantilever Needle Beams

31 46 13.10 Needle Beams

		Crew	Daily Output	Labor-Hours	Unit	Material	2018 Bare Costs Labor	Equipment	Total	Total Incl O&P
0010	**NEEDLE BEAMS**									
0011	Incl. wood shoring 10' x 10' opening									
0400	Block, concrete, 8" thick	B-9	7.10	5.634	Ea.	53	227	33	313	440
0420	12" thick		6.70	5.970		64.50	240	35	339.50	475
0800	Brick, 4" thick with 8" backup block		5.70	7.018		64.50	282	41	387.50	545
1000	Brick, solid, 8" thick		6.20	6.452		53	260	37.50	350.50	495
1040	12" thick		4.90	8.163		64.50	330	48	442.50	625
1080	16" thick		4.50	8.889		87	360	52	499	700
2000	Add for additional floors of shoring	B-1	6	4		53	162		215	305

For customer support on your Building Construction Costs with RSMeans data, call 800.448.8182.

31 48 Underpinning

31 48 13 – Underpinning Piers

31 48 13.10 Underpinning Foundations	Crew	Daily Output	Labor-Hours	Unit	Material	2018 Bare Costs Labor	Equipment	Total	Total Incl O&P
0010 **UNDERPINNING FOUNDATIONS**									
0011 Including excavation,									
0020 forming, reinforcing, concrete and equipment									
0100 5' to 16' below grade, 100 to 500 C.Y.	B-52	2.30	24.348	C.Y.	335	1,125	262	1,722	2,350
0200 Over 500 C.Y.		2.50	22.400		300	1,050	241	1,591	2,175
0400 16' to 25' below grade, 100 to 500 C.Y.		2	28		370	1,300	300	1,970	2,700
0500 Over 500 C.Y.		2.10	26.667		350	1,225	287	1,862	2,575
0700 26' to 40' below grade, 100 to 500 C.Y.		1.60	35		405	1,625	375	2,405	3,300
0800 Over 500 C.Y.	▼	1.80	31.111		370	1,450	335	2,155	2,975
0900 For under 50 C.Y., add					10%	40%			
1000 For 50 C.Y. to 100 C.Y., add				▼	5%	20%			

31 52 Cofferdams

31 52 16 – Timber Cofferdams

31 52 16.10 Cofferdams

	Crew	Daily Output	Labor-Hours	Unit	Material	2018 Bare Costs Labor	Equipment	Total	Total Incl O&P
0010 **COFFERDAMS**									
0011 Incl. mobilization and temporary sheeting									
0080 Soldier beams & lagging H-piles with 3" wood sheeting									
0090 horizontal between piles, including removal of wales & braces									
0100 No hydrostatic head, 15' deep, 1 line of braces, minimum	B-50	545	.206	S.F.	8.90	10.20	4.81	23.91	31
0200 Maximum		495	.226		9.85	11.25	5.30	26.40	34
0400 15' to 22' deep with 2 lines of braces, 10" H, minimum		360	.311		10.45	15.45	7.30	33.20	43.50
0500 Maximum		330	.339		11.85	16.85	7.95	36.65	48
0700 23' to 35' deep with 3 lines of braces, 12" H, minimum		325	.345		13.65	17.10	8.05	38.80	50.50
0800 Maximum		295	.380		14.80	18.85	8.90	42.55	55
1000 36' to 45' deep with 4 lines of braces, 14" H, minimum		290	.386		15.30	19.15	9.05	43.50	56.50
1100 Maximum		265	.423		16.15	21	9.90	47.05	61
1300 No hydrostatic head, left in place, 15' deep, 1 line of braces, min.		635	.176		11.85	8.75	4.13	24.73	31
1400 Maximum		575	.195		12.70	9.65	4.56	26.91	34
1600 15' to 22' deep with 2 lines of braces, minimum		455	.246		17.75	12.20	5.75	35.70	45
1700 Maximum		415	.270		19.75	13.40	6.30	39.45	49.50
1900 23' to 35' deep with 3 lines of braces, minimum		420	.267		21	13.25	6.25	40.50	51
2000 Maximum		380	.295		23.50	14.65	6.90	45.05	55.50
2200 36' to 45' deep with 4 lines of braces, minimum		385	.291		25.50	14.45	6.80	46.75	58
2300 Maximum	▼	350	.320		29.50	15.90	7.50	52.90	65.50
2350 Lagging only, 3" thick wood between piles 8' OC, minimum	B-46	400	.120		1.97	5.50	.12	7.59	10.85
2370 Maximum		250	.192		2.96	8.80	.19	11.95	17.15
2400 Open sheeting no bracing, for trenches to 10' deep, min.		1736	.028		.89	1.27	.03	2.19	2.98
2450 Maximum	▼	1510	.032		.99	1.46	.03	2.48	3.39
2500 Tie-back method, add to open sheeting, add, minimum				▼				20%	20%
2550 Maximum								60%	60%
2700 Tie-backs only, based on tie-backs total length, minimum	B-46	86.80	.553	L.F.	15.50	25.50	.54	41.54	57
2750 Maximum		38.50	1.247	"	27.50	57	1.21	85.71	120
3500 Tie-backs only, typical average, 25' long		2	24	Ea.	680	1,100	23.50	1,803.50	2,475
3600 35' long	▼	1.58	30.380	"	910	1,400	29.50	2,339.50	3,200
6000 See also Section 31 41 16.10									

For customer support on your Building Construction Costs with RSMeans data, call 800.448.8182.

643

31 56 Slurry Walls

31 56 23 – Lean Concrete Slurry Walls

31 56 23.20 Slurry Trench

		Crew	Daily Output	Labor-Hours	Unit	Material	2018 Bare Costs Labor	2018 Bare Costs Equipment	Total	Total Incl O&P
0010	**SLURRY TRENCH**									
0011	Excavated slurry trench in wet soils									
0020	backfilled with 3,000 psi concrete, no reinforcing steel									
0050	Minimum	C-7	333	.216	C.F.	8.95	9.40	3.21	21.56	27.50
0100	Maximum		200	.360	"	15	15.65	5.35	36	46
0200	Alternate pricing method, minimum		150	.480	S.F.	17.90	21	7.15	46.05	59
0300	Maximum	↓	120	.600		27	26	8.90	61.90	79
0500	Reinforced slurry trench, minimum	B-48	177	.316		13.40	14.35	15.60	43.35	53.50
0600	Maximum	"	69	.812	↓	44.50	37	40	121.50	149
0800	Haul for disposal, 2 mile haul, excavated material, add	B-34B	99	.081	C.Y.		3.72	5.50	9.22	11.65
0900	Haul bentonite castings for disposal, add	"	40	.200	"		9.20	13.55	22.75	29

31 62 Driven Piles

31 62 13 – Concrete Piles

31 62 13.23 Prestressed Concrete Piles

		Crew	Daily Output	Labor-Hours	Unit	Material	2018 Bare Costs Labor	2018 Bare Costs Equipment	Total	Total Incl O&P
0010	**PRESTRESSED CONCRETE PILES**, 200 piles									
0020	Unless specified otherwise, not incl. pile caps or mobilization									
2200	Precast, prestressed, 50' long, cylinder, 12" diam., 2-3/8" wall	B-19	720	.089	V.L.F.	27	4.66	2.68	34.34	39.50
2300	14" diameter, 2-1/2" wall		680	.094		32	4.93	2.84	39.77	46.50
2500	16" diameter, 3" wall	↓	640	.100		41.50	5.25	3.01	49.76	57
2600	18" diameter, 3-1/2" wall	B-19A	600	.107		52	5.60	4.02	61.62	70.50
2800	20" diameter, 4" wall		560	.114		54.50	6	4.31	64.81	74
2900	24" diameter, 5" wall	↓	520	.123		73.50	6.45	4.64	84.59	96
3100	Precast, prestressed, 40' long, 10" thick, square	B-19	700	.091		14.30	4.79	2.75	21.84	26
3200	12" thick, square		680	.094		21.50	4.93	2.84	29.27	34.50
3400	14" thick, square		600	.107		24	5.60	3.21	32.81	38.50
3500	Octagonal		640	.100		28	5.25	3.01	36.26	42
3700	16" thick, square		560	.114		30.50	6	3.44	39.94	46.50
3800	Octagonal	↓	600	.107		33.50	5.60	3.21	42.31	49
4000	18" thick, square	B-19A	520	.123		41.50	6.45	4.64	52.59	60.50
4100	Octagonal	B-19	560	.114		44.50	6	3.44	53.94	62
4300	20" thick, square	B-19A	480	.133		47.50	7	5	59.50	69
4400	Octagonal	B-19	520	.123		50	6.45	3.71	60.16	69
4600	24" thick, square	B-19A	440	.145		54	7.60	5.50	67.10	77
4700	Octagonal	B-19	480	.133		59	7	4.02	70.02	80.50
4730	Precast, prestressed, 60' long, 10" thick, square		700	.091		15.05	4.79	2.75	22.59	27
4740	12" thick, square (60' long)		680	.094		22	4.93	2.84	29.77	35.50
4750	Mobilization for 10,000 L.F. pile job, add		3300	.019			1.02	.58	1.60	2.22
4800	25,000 L.F. pile job, add	↓	8500	.008	↓		.39	.23	.62	.86

31 62 16 – Steel Piles

31 62 16.13 Steel Piles

		Crew	Daily Output	Labor-Hours	Unit	Material	2018 Bare Costs Labor	2018 Bare Costs Equipment	Total	Total Incl O&P
0010	**STEEL PILES**									
0100	Step tapered, round, concrete filled									
0110	8" tip, 12" butt, 60 ton capacity, 30' depth	B-19	760	.084	V.L.F.	18.45	4.41	2.54	25.40	30
0120	60' depth with extension		740	.086		35	4.53	2.61	42.14	48.50
0130	80' depth with extensions		700	.091		54	4.79	2.75	61.54	70
0250	"H" Sections, 50' long, HP8 x 36		640	.100		17.05	5.25	3.01	25.31	30
0400	HP10 x 42		610	.105		20	5.50	3.16	28.66	34
0500	HP10 x 57		610	.105		27	5.50	3.16	35.66	41.50
0700	HP12 x 53	↓	590	.108	↓	25.50	5.70	3.27	34.47	40.50

For customer support on your Building Construction Costs with RSMeans data, call 800.448.8182.

31 62 Driven Piles

31 62 16 – Steel Piles

31 62 16.13 Steel Piles

		Crew	Daily Output	Labor-Hours	Unit	Material	2018 Bare Costs Labor	Equipment	Total	Total Incl O&P
0800	HP12 x 74	B-19A	590	.108	V.L.F.	35	5.70	4.09	44.79	52
1000	HP14 x 73		540	.119		35.50	6.20	4.47	46.17	53.50
1100	HP14 x 89		540	.119		42	6.20	4.47	52.67	61
1300	HP14 x 102		510	.125		48.50	6.60	4.73	59.83	68.50
1400	HP14 x 117		510	.125		55.50	6.60	4.73	66.83	76.50
1600	Splice on standard points, not in leads, 8" or 10"	1 Sswl	5	1.600	Ea.	103	87.50		190.50	257
1700	12" or 14"		4	2		143	109		252	335
1900	Heavy duty points, not in leads, 10" wide		4	2		189	109		298	385
2100	14" wide		3.50	2.286		230	125		355	455

31 62 19 – Timber Piles

31 62 19.10 Wood Piles

		Crew	Daily Output	Labor-Hours	Unit	Material	2018 Bare Costs Labor	Equipment	Total	Total Incl O&P
0010	**WOOD PILES**									
0011	Friction or end bearing, not including									
0050	mobilization or demobilization									
0100	Untreated piles, up to 30' long, 12" butts, 8" points	B-19	625	.102	V.L.F.	11.45	5.35	3.09	19.89	24.50
0200	30' to 39' long, 12" butts, 8" points		700	.091		11.45	4.79	2.75	18.99	23
0300	40' to 49' long, 12" butts, 7" points		720	.089		11.45	4.66	2.68	18.79	23
0400	50' to 59' long, 13" butts, 7" points		800	.080		13.60	4.19	2.41	20.20	24
0500	60' to 69' long, 13" butts, 7" points		840	.076		15.30	3.99	2.30	21.59	25.50
0600	70' to 80' long, 13" butts, 6" points		840	.076		17	3.99	2.30	23.29	27.50
0800	Treated piles, 12 lb./C.F.,									
0810	friction or end bearing, ASTM class B									
1000	Up to 30' long, 12" butts, 8" points	B-19	625	.102	V.L.F.	13.35	5.35	3.09	21.79	26.50
1100	30' to 39' long, 12" butts, 8" points		700	.091		14.50	4.79	2.75	22.04	26.50
1200	40' to 49' long, 12" butts, 7" points		720	.089		15.35	4.66	2.68	22.69	27
1300	50' to 59' long, 13" butts, 7" points		800	.080		17.90	4.19	2.41	24.50	29
1400	60' to 69' long, 13" butts, 6" points	B-19A	840	.076		22	3.99	2.87	28.86	33.50
1500	70' to 80' long, 13" butts, 6" points	"	840	.076		18.05	3.99	2.87	24.91	29
1600	Treated piles, C.C.A., 2.5 lb./C.F.									
1610	8" butts, 10' long	B-19	400	.160	V.L.F.	6.65	8.40	4.82	19.87	25.50
1620	11' to 16' long		500	.128		6.65	6.70	3.86	17.21	22
1630	17' to 20' long		575	.111		6.65	5.85	3.35	15.85	20
1640	10" butts, 10' to 16' long		500	.128		11.10	6.70	3.86	21.66	27
1650	17' to 20' long		575	.111		11.10	5.85	3.35	20.30	25
1660	21' to 40' long		700	.091		11.10	4.79	2.75	18.64	22.50
1670	12" butts, 10' to 20' long		575	.111		11.45	5.85	3.35	20.65	25.50
1680	21' to 35' long		650	.098		11.45	5.15	2.97	19.57	24
1690	36' to 40' long		700	.091		11.45	4.79	2.75	18.99	23
1695	14" butts, to 40' long		700	.091		14.50	4.79	2.75	22.04	26.50
1700	Boot for pile tip, minimum	1 Pile	27	.296	Ea.	39	15.20		54.20	67
1800	Maximum		21	.381		117	19.55		136.55	160
2000	Point for pile tip, minimum		20	.400		39	20.50		59.50	75.50
2100	Maximum		15	.533		140	27.50		167.50	198
2300	Splice for piles over 50' long, minimum	B-46	35	1.371		57.50	63	1.33	121.83	162
2400	Maximum		20	2.400		71.50	110	2.33	183.83	253
2600	Concrete encasement with wire mesh and tube		331	.145	V.L.F.	71.50	6.65	.14	78.29	89.50
2700	Mobilization for 10,000 L.F. pile job, add	B-19	3300	.019			1.02	.58	1.60	2.22
2800	25,000 L.F. pile job, add	"	8500	.008			.39	.23	.62	.86

31 62 Driven Piles

31 62 23 – Composite Piles

31 62 23.13 Concrete-Filled Steel Piles

	31 62 23.13 Concrete-Filled Steel Piles	Crew	Daily Output	Labor-Hours	Unit	Material	2018 Bare Costs Labor	Equipment	Total	Total Incl O&P
0010	**CONCRETE-FILLED STEEL PILES** no mobilization or demobilization									
2600	Pipe piles, 50' L, 8" diam., 29 lb./L.F., no concrete	B-19	500	.128	V.L.F.	21.50	6.70	3.86	32.06	38
2700	Concrete filled		460	.139		24	7.30	4.19	35.49	42.50
2900	10" diameter, 34 lb./L.F., no concrete		500	.128		24	6.70	3.86	34.56	41
3000	Concrete filled		450	.142		28.50	7.45	4.29	40.24	48
3200	12" diameter, 44 lb./L.F., no concrete		475	.135		29.50	7.05	4.06	40.61	48
3300	Concrete filled		415	.154		35	8.10	4.65	47.75	56
3500	14" diameter, 46 lb./L.F., no concrete		430	.149		31.50	7.80	4.48	43.78	52
3600	Concrete filled		355	.180		40	9.45	5.45	54.90	64.50
3800	16" diameter, 52 lb./L.F., no concrete		385	.166		38	8.70	5	51.70	60.50
3900	Concrete filled		335	.191		48.50	10	5.75	64.25	75.50
4100	18" diameter, 59 lb./L.F., no concrete		355	.180		43	9.45	5.45	57.90	67.50
4200	Concrete filled	▼	310	.206	▼	58.50	10.80	6.20	75.50	88
4400	Splices for pipe piles, stl., not in leads, 8" diameter	1 Sswl	5	1.600	Ea.	72	87.50		159.50	223
4410	10" diameter		4.75	1.684		81.50	92		173.50	240
4430	12" diameter		4.50	1.778		99.50	97		196.50	268
4500	14" diameter		4.25	1.882		136	103		239	320
4600	16" diameter		4	2		155	109		264	350
4650	18" diameter		3.75	2.133		226	117		343	440
4710	Steel pipe pile backing rings, w/spacer, 8" diameter		12	.667		8.60	36.50		45.10	69
4720	10" diameter		12	.667		11.30	36.50		47.80	72
4730	12" diameter		10	.800		13.60	43.50		57.10	86.50
4740	14" diameter		9	.889		15.15	48.50		63.65	96
4750	16" diameter		8	1		19.05	54.50		73.55	111
4760	18" diameter		6	1.333		20.50	73		93.50	142
4800	Points, standard, 8" diameter		4.61	1.735		85	95		180	249
4840	10" diameter		4.45	1.798		113	98.50		211.50	285
4880	12" diameter		4.25	1.882		158	103		261	340
4900	14" diameter		4.05	1.975		195	108		303	390
5000	16" diameter		3.37	2.374		279	130		409	515
5050	18" diameter		3.50	2.286		340	125		465	580
5200	Points, heavy duty, 10" diameter		2.90	2.759		232	151		383	500
5240	12" diameter		2.95	2.712		265	148		413	535
5260	14" diameter		2.95	2.712		277	148		425	545
5280	16" diameter		2.95	2.712		325	148		473	595
5290	18" diameter		2.80	2.857		410	156		566	705
5500	For reinforcing steel, add	▼	1150	.007	Lb.	.69	.38		1.07	1.38
5700	For thick wall sections, add				"	.60			.60	.66

31 63 Bored Piles

31 63 26 – Drilled Caissons

31 63 26.13 Fixed End Caisson Piles

	31 63 26.13 Fixed End Caisson Piles	Crew	Daily Output	Labor-Hours	Unit	Material	2018 Bare Costs Labor	Equipment	Total	Total Incl O&P
0010	**FIXED END CAISSON PILES** R316326-60									
0015	Including excavation, concrete, 50 lb. reinforcing									
0020	per C.Y., not incl. mobilization, boulder removal, disposal									
0100	Open style, machine drilled, to 50' deep, in stable ground, no									
0110	casings or ground water, 18" diam., 0.065 C.Y./L.F.	B-43	200	.240	V.L.F.	9.40	10.65	12.20	32.25	40
0200	24" diameter, 0.116 C.Y./L.F.		190	.253		16.80	11.20	12.85	40.85	49.50
0300	30" diameter, 0.182 C.Y./L.F.		150	.320		26.50	14.20	16.30	57	68.50
0400	36" diameter, 0.262 C.Y./L.F.	▼	125	.384	▼	38	17.05	19.55	74.60	89

For customer support on your Building Construction Costs with RSMeans data, call 800.448.8182.

31 63 26.13 Fixed End Caisson Piles	Crew	Daily Output	Labor-Hours	Unit	Material	2018 Bare Costs Labor	Equipment	Total	Total Incl O&P
0500	48" diameter, 0.465 C.Y./L.F.	B-43	100	.480	V.L.F.	67.50	21.50	24.50	113.50	134
0600	60" diameter, 0.727 C.Y./L.F.		90	.533		105	23.50	27	155.50	182
0700	72" diameter, 1.05 C.Y./L.F.		80	.600		152	26.50	30.50	209	241
0800	84" diameter, 1.43 C.Y./L.F.		75	.640		207	28.50	32.50	268	305
1000	For bell excavation and concrete, add									
1020	4' bell diameter, 24" shaft, 0.444 C.Y.	B-43	20	2.400	Ea.	53.50	106	122	281.50	355
1040	6' bell diameter, 30" shaft, 1.57 C.Y.		5.70	8.421		190	375	430	995	1,250
1060	8' bell diameter, 36" shaft, 3.72 C.Y.		2.40	20		450	885	1,025	2,360	2,975
1080	9' bell diameter, 48" shaft, 4.48 C.Y.		2	24		540	1,075	1,225	2,840	3,575
1100	10' bell diameter, 60" shaft, 5.24 C.Y.		1.70	28.235		635	1,250	1,450	3,335	4,175
1120	12' bell diameter, 72" shaft, 8.74 C.Y.		1	48		1,050	2,125	2,450	5,625	7,075
1140	14' bell diameter, 84" shaft, 13.6 C.Y.		.70	68.571		1,650	3,050	3,500	8,200	10,300
1200	Open style, machine drilled, to 50' deep, in wet ground, pulled									
1300	casing and pumping, 18" diameter, 0.065 C.Y./L.F.	B-48	160	.350	V.L.F.	9.40	15.85	17.25	42.50	53.50
1400	24" diameter, 0.116 C.Y./L.F.		125	.448		16.80	20.50	22	59.30	74
1500	30" diameter, 0.182 C.Y./L.F.		85	.659		26.50	30	32.50	89	110
1600	36" diameter, 0.262 C.Y./L.F.		60	.933		38	42.50	46	126.50	156
1700	48" diameter, 0.465 C.Y./L.F.	B-49	55	1.600		67.50	76.50	61	205	257
1800	60" diameter, 0.727 C.Y./L.F.		35	2.514		105	120	95.50	320.50	405
1900	72" diameter, 1.05 C.Y./L.F.		30	2.933		152	140	112	404	505
2000	84" diameter, 1.43 C.Y./L.F.		25	3.520		207	168	134	509	630
2100	For bell excavation and concrete, add									
2120	4' bell diameter, 24" shaft, 0.444 C.Y.	B-48	19.80	2.828	Ea.	53.50	128	139	320.50	405
2140	6' bell diameter, 30" shaft, 1.57 C.Y.		5.70	9.825		190	445	485	1,120	1,425
2160	8' bell diameter, 36" shaft, 3.72 C.Y.		2.40	23.333		450	1,050	1,150	2,650	3,375
2180	9' bell diameter, 48" shaft, 4.48 C.Y.	B-49	3.30	26.667		540	1,275	1,025	2,840	3,675
2200	10' bell diameter, 60" shaft, 5.24 C.Y.		2.80	31.429		635	1,500	1,200	3,335	4,325
2220	12' bell diameter, 72" shaft, 8.74 C.Y.		1.60	55		1,050	2,625	2,100	5,775	7,450
2240	14' bell diameter, 84" shaft, 13.6 C.Y.		1	88		1,650	4,200	3,350	9,200	11,900
2300	Open style, machine drilled, to 50' deep, in soft rocks and									
2400	medium hard shales, 18" diameter, 0.065 C.Y./L.F.	B-49	50	1.760	V.L.F.	9.40	84	67	160.40	212
2500	24" diameter, 0.116 C.Y./L.F.		30	2.933		16.80	140	112	268.80	355
2600	30" diameter, 0.182 C.Y./L.F.		20	4.400		26.50	210	167	403.50	535
2700	36" diameter, 0.262 C.Y./L.F.		15	5.867		38	280	223	541	715
2800	48" diameter, 0.465 C.Y./L.F.		10	8.800		67.50	420	335	822.50	1,075
2900	60" diameter, 0.727 C.Y./L.F.		7	12.571		105	600	480	1,185	1,550
3000	72" diameter, 1.05 C.Y./L.F.		6	14.667		152	700	560	1,412	1,850
3100	84" diameter, 1.43 C.Y./L.F.		5	17.600		207	840	670	1,717	2,250
3200	For bell excavation and concrete, add									
3220	4' bell diameter, 24" shaft, 0.444 C.Y.	B-49	10.90	8.073	Ea.	53.50	385	305	743.50	990
3240	6' bell diameter, 30" shaft, 1.57 C.Y.		3.10	28.387		190	1,350	1,075	2,615	3,475
3260	8' bell diameter, 36" shaft, 3.72 C.Y.		1.30	67.692		450	3,225	2,575	6,250	8,250
3280	9' bell diameter, 48" shaft, 4.48 C.Y.		1.10	80		540	3,825	3,050	7,415	9,775
3300	10' bell diameter, 60" shaft, 5.24 C.Y.		.90	97.778		635	4,675	3,725	9,035	11,900
3320	12' bell diameter, 72" shaft, 8.74 C.Y.		.60	147		1,050	7,000	5,575	13,625	18,000
3340	14' bell diameter, 84" shaft, 13.6 C.Y.		.40	220		1,650	10,500	8,375	20,525	27,000
3600	For rock excavation, sockets, add, minimum		120	.733	C.F.		35	28	63	84
3650	Average		95	.926			44	35.50	79.50	107
3700	Maximum		48	1.833			87.50	70	157.50	210
3900	For 50' to 100' deep, add				V.L.F.				7%	7%
4000	For 100' to 150' deep, add								25%	25%
4100	For 150' to 200' deep, add								30%	30%
4200	For casings left in place, add				Lb.	1.36			1.36	1.50

31 63 Bored Piles

31 63 26 – Drilled Caissons

31 63 26.13 Fixed End Caisson Piles

		Crew	Daily Output	Labor-Hours	Unit	Material	2018 Bare Costs Labor	Equipment	Total	Total Incl O&P
4300	For other than 50 lb. reinf. per C.Y., add or deduct				Lb.	1.26			1.26	1.39
4400	For steel I-beam cores, add	B-49	8.30	10.602	Ton	2,150	505	405	3,060	3,600
4500	Load and haul excess excavation, 2 miles	B-34B	178	.045	L.C.Y.		2.07	3.05	5.12	6.45
4600	For mobilization, 50 mile radius, rig to 36"	B-43	2	24	Ea.		1,075	1,225	2,300	2,975
4650	Rig to 84"	B-48	1.75	32			1,450	1,575	3,025	3,925
4700	For low headroom, add						50%			50%
4750	For difficult access, add						25%			25%
5000	Bottom inspection	1 Skwk	1.20	6.667	↓		350		350	535

31 63 26.16 Concrete Caissons for Marine Construction

		Crew	Daily Output	Labor-Hours	Unit	Material	2018 Bare Costs Labor	Equipment	Total	Total Incl O&P
0010	**CONCRETE CAISSONS FOR MARINE CONSTRUCTION**									
0100	Caissons, incl. mobilization and demobilization, up to 50 miles									
0200	Uncased shafts, 30 to 80 tons cap., 17" diam., 10' depth	B-44	88	.727	V.L.F.	24	37.50	23.50	85	111
0300	25' depth		165	.388		17.25	19.90	12.55	49.70	64
0400	80 to 150 ton capacity, 22" diameter, 10' depth		80	.800		30	41	26	97	126
0500	20' depth		130	.492		24	25.50	15.95	65.45	83
0700	Cased shafts, 10 to 30 ton capacity, 10-5/8" diam., 20' depth		175	.366		17.25	18.75	11.85	47.85	61
0800	30' depth		240	.267		16.10	13.70	8.65	38.45	48.50
0850	30 to 60 ton capacity, 12" diameter, 20' depth		160	.400		24	20.50	12.95	57.45	73
0900	40' depth		230	.278		18.60	14.30	9	41.90	52.50
1000	80 to 100 ton capacity, 16" diameter, 20' depth		160	.400		34.50	20.50	12.95	67.95	84.50
1100	40' depth		230	.278		32	14.30	9	55.30	67.50
1200	110 to 140 ton capacity, 17-5/8" diameter, 20' depth		160	.400		37	20.50	12.95	70.45	87.50
1300	40' depth		230	.278		34.50	14.30	9	57.80	70
1400	140 to 175 ton capacity, 19" diameter, 20' depth		130	.492		40.50	25.50	15.95	81.95	101
1500	40' depth	↓	210	.305	↓	37	15.65	9.85	62.50	76.50
1700	Over 30' long, L.F. cost tends to be lower									
1900	Maximum depth is about 90'									

31 63 29 – Drilled Concrete Piers and Shafts

31 63 29.13 Uncased Drilled Concrete Piers

		Crew	Daily Output	Labor-Hours	Unit	Material	2018 Bare Costs Labor	Equipment	Total	Total Incl O&P
0010	**UNCASED DRILLED CONCRETE PIERS**									
0020	Unless specified otherwise, not incl. pile caps or mobilization									
0050	Cast in place augered piles, no casing or reinforcing									
0060	8" diameter	B-43	540	.089	V.L.F.	4.20	3.94	4.53	12.67	15.60
0065	10" diameter		480	.100		6.70	4.44	5.10	16.24	19.70
0070	12" diameter		420	.114		9.40	5.05	5.80	20.25	24.50
0075	14" diameter		360	.133		12.70	5.90	6.80	25.40	30.50
0080	16" diameter		300	.160		17.10	7.10	8.15	32.35	38.50
0085	18" diameter	↓	240	.200	↓	21	8.85	10.20	40.05	48
0100	Cast in place, thin wall shell pile, straight sided,									
0110	not incl. reinforcing, 8" diam., 16 ga., 5.8 lb./L.F.	B-19	700	.091	V.L.F.	9.50	4.79	2.75	17.04	21
0200	10" diameter, 16 ga. corrugated, 7.3 lb./L.F.		650	.098		12.45	5.15	2.97	20.57	25
0300	12" diameter, 16 ga. corrugated, 8.7 lb./L.F.		600	.107		16.15	5.60	3.21	24.96	30
0400	14" diameter, 16 ga. corrugated, 10.0 lb./L.F.		550	.116		19	6.10	3.51	28.61	34.50
0500	16" diameter, 16 ga. corrugated, 11.6 lb./L.F.	↓	500	.128	↓	23.50	6.70	3.86	34.06	40
0800	Cast in place friction pile, 50' long, fluted,									
0810	tapered steel, 4,000 psi concrete, no reinforcing									
0900	12" diameter, 7 ga.	B-19	600	.107	V.L.F.	29.50	5.60	3.21	38.31	44.50
1000	14" diameter, 7 ga.		560	.114		32	6	3.44	41.44	48.50
1100	16" diameter, 7 ga.		520	.123		38	6.45	3.71	48.16	55.50
1200	18" diameter, 7 ga.	↓	480	.133	↓	44.50	7	4.02	55.52	64
1300	End bearing, fluted, constant diameter,									
1320	4,000 psi concrete, no reinforcing									

31 63 Bored Piles

31 63 29 – Drilled Concrete Piers and Shafts

31 63 29.13 Uncased Drilled Concrete Piers	Crew	Daily Output	Labor-Hours	Unit	Material	2018 Bare Costs Labor	Equipment	Total	Total Incl O&P
1340	12" diameter, 7 ga.	B-19	600	.107	V.L.F.	31	5.60	3.21	39.81	46
1360	14" diameter, 7 ga.		560	.114		39	6	3.44	48.44	55.50
1380	16" diameter, 7 ga.		520	.123		45	6.45	3.71	55.16	63.50
1400	18" diameter, 7 ga.		480	.133		49.50	7	4.02	60.52	70

31 63 29.20 Cast In Place Piles, Adds

		Crew	Daily Output	Labor-Hours	Unit	Material	2018 Bare Costs Labor	Equipment	Total	Total Incl O&P
0010	**CAST IN PLACE PILES, ADDS**									
1500	For reinforcing steel, add				Lb.	.96			.96	1.05
1700	For ball or pedestal end, add	B-19	11	5.818	C.Y.	150	305	175	630	835
1900	For lengths above 60', concrete, add	"	11	5.818	"	157	305	175	637	840
2000	For steel thin shell, pipe only				Lb.	1.50			1.50	1.65

For customer support on your Building Construction Costs with RSMeans data, call 800.448.8182.

Division Notes

	CREW	DAILY OUTPUT	LABOR-HOURS	UNIT	BARE COSTS				TOTAL INCL O&P
					MAT.	LABOR	EQUIP.	TOTAL	

Estimating Tips

32 01 00 Operations and Maintenance of Exterior Improvements

- Recycling of asphalt pavement is becoming very popular and is an alternative to removal and replacement. It can be a good value engineering proposal if removed pavement can be recycled, either at the project site or at another site that is reasonably close to the project site. Sections on repair of flexible and rigid pavement are included.

32 10 00 Bases, Ballasts, and Paving

- When estimating paving, keep in mind the project schedule. Also note that prices for asphalt and concrete are generally higher in the cold seasons. Lines for pavement markings, including tactile warning systems and fence lines, are included.

32 90 00 Planting

- The timing of planting and guarantee specifications often dictate the costs for establishing tree and shrub growth and a stand of grass or ground cover. Establish the work performance schedule to coincide with the local planting season. Maintenance and growth guarantees can add from 20%–100% to the total landscaping cost and can be contractually cumbersome. The cost to replace trees and shrubs can be as high as 5% of the total cost, depending on the planting zone, soil conditions, and time of year.

Reference Numbers

Reference numbers are shown at the beginning of some major classifications. These numbers refer to related items in the Reference Section. The reference information may be an estimating procedure, an alternate pricing method, or technical information.

Note: Not all subdivisions listed here necessarily appear. ■

Did you know?

RSMeans data is available through our online application with 24/7 access:

- Search for unit prices by keyword
- Leverage the most up-to-date data
- Build and export estimates

Try it free for 30 days!
www.rsmeans.com/2018freetrial

No part of this cost data may be reproduced, stored in a retrieval system, or transmitted in any form or by any means without prior written permission of Gordian.

32 01 13.61 Slurry Seal (Latex Modified)	Crew	Daily Output	Labor-Hours	Unit	Material	2018 Bare Costs Labor	Equipment	Total	Total Incl O&P
0010 **SLURRY SEAL (LATEX MODIFIED)**									
3780 Rubberized asphalt (latex) seal	B-45	5000	.003	S.Y.	1.24	.16	.16	1.56	1.78

32 01 13.64 Sand Seal

	Crew	Daily Output	Labor-Hours	Unit	Material	Labor	Equipment	Total	Total Incl O&P
0010 **SAND SEAL**									
2080 Sand sealing, sharp sand, asphalt emulsion, small area	B-91	10000	.006	S.Y.	1.49	.31	.22	2.02	2.34
2120 Roadway or large area	"	18000	.004	"	1.28	.17	.12	1.57	1.81

32 01 13.66 Fog Seal

	Crew	Daily Output	Labor-Hours	Unit	Material	Labor	Equipment	Total	Total Incl O&P
0010 **FOG SEAL**									
0012 Sealcoating, 2 coat coal tar pitch emulsion over 10,000 S.Y.	B-45	5000	.003	S.Y.	.85	.16	.16	1.17	1.36
0030 1,000 to 10,000 S.Y.	"	3000	.005		.85	.27	.27	1.39	1.63
0100 Under 1,000 S.Y.	B-1	1050	.023		.85	.93		1.78	2.35
0300 Petroleum resistant, over 10,000 S.Y.	B-45	5000	.003		1.32	.16	.16	1.64	1.87
0320 1,000 to 10,000 S.Y.	"	3000	.005		1.32	.27	.27	1.86	2.14
0400 Under 1,000 S.Y.	B-1	1050	.023		1.32	.93		2.25	2.86
0600 Non-skid pavement renewal, over 10,000 S.Y.	B-45	5000	.003		1.38	.16	.16	1.70	1.94
0620 1,000 to 10,000 S.Y.	"	3000	.005		1.38	.27	.27	1.92	2.21
0700 Under 1,000 S.Y.	B-1	1050	.023		1.38	.93		2.31	2.93
0800 Prepare and clean surface for above	A-2	8545	.003			.12	.02	.14	.20
1000 Hand seal asphalt curbing	B-1	4420	.005	L.F.	.63	.22		.85	1.03
1900 Asphalt surface treatment, single course, small area									
1901 0.30 gal./S.Y. asphalt material, 20 lb./S.Y. aggregate	B-91	5000	.013	S.Y.	1.32	.61	.44	2.37	2.87
1910 Roadway or large area		10000	.006		1.21	.31	.22	1.74	2.04
1950 Asphalt surface treatment, dbl. course for small area		3000	.021		2.98	1.02	.74	4.74	5.65
1960 Roadway or large area		6000	.011		2.68	.51	.37	3.56	4.13
1980 Asphalt surface treatment, single course, for shoulders		7500	.009		1.47	.41	.29	2.17	2.56

32 06 Schedules for Exterior Improvements
32 06 10 – Schedules for Bases, Ballasts, and Paving

32 06 10.10 Sidewalks, Driveways and Patios

	Crew	Daily Output	Labor-Hours	Unit	Material	Labor	Equipment	Total	Total Incl O&P
0010 **SIDEWALKS, DRIVEWAYS AND PATIOS** No base									
0020 Asphaltic concrete, 2" thick	B-37	720	.067	S.Y.	6.85	2.81	.21	9.87	12.05
0100 2-1/2" thick	"	660	.073	"	8.70	3.06	.23	11.99	14.50
0300 Concrete, 3,000 psi, CIP, 6 x 6 - W1.4 x W1.4 mesh,									
0310 broomed finish, no base, 4" thick	B-24	600	.040	S.F.	2.01	1.84		3.85	4.99
0350 5" thick		545	.044		2.68	2.03		4.71	6
0400 6" thick		510	.047		3.13	2.17		5.30	6.70
0450 For bank run gravel base, 4" thick, add	B-18	2500	.010		.43	.39	.02	.84	1.08
0520 8" thick, add	"	1600	.015		.87	.61	.03	1.51	1.92
0550 Exposed aggregate finish, add to above, minimum	B-24	1875	.013		.13	.59		.72	1.03
0600 Maximum	"	455	.053		.43	2.43		2.86	4.13
1000 Crushed stone, 1" thick, white marble	2 Clab	1700	.009		.48	.38		.86	1.10
1050 Bluestone	"	1700	.009		.20	.38		.58	.79
1700 Redwood, prefabricated, 4' x 4' sections	2 Carp	316	.051		4.77	2.57		7.34	9.15
1750 Redwood planks, 1" thick, on sleepers	"	240	.067		4.77	3.38		8.15	10.40
2250 Stone dust, 4" thick	B-62	900	.027	S.Y.	3.97	1.16	.19	5.32	6.35

32 06 10.20 Steps

	Crew	Daily Output	Labor-Hours	Unit	Material	Labor	Equipment	Total	Total Incl O&P
0010 **STEPS**									
0011 Incl. excav., borrow & concrete base as required									
0100 Brick steps	B-24	35	.686	LF Riser	17.25	31.50		48.75	66.50
0200 Railroad ties	2 Clab	25	.640		3.55	25.50		29.05	43

652

For customer support on your Building Construction Costs with RSMeans data, call 800.448.8182.

32 06 Schedules for Exterior Improvements

32 06 10 – Schedules for Bases, Ballasts, and Paving

32 06 10.20 Steps	Crew	Daily Output	Labor-Hours	Unit	Material	2018 Bare Costs Labor	Equipment	Total	Total Incl O&P
0300	Bluestone treads, 12" x 2" or 12" x 1-1/2"	B-24	30	.800	LF Riser	41.50	37		78.50	101
0500	Concrete, cast in place, see Section 03 30 53.40									
0600	Precast concrete, see Section 03 41 23.50									
4025	Steel edge strips, incl. stakes, 1/4" x 5"	B-1	390	.062	L.F.	4.54	2.49		7.03	8.80
4050	Edging, landscape timber or railroad ties, 6" x 8"	2 Carp	170	.094	"	2.36	4.77		7.13	9.85

32 11 Base Courses

32 11 23 – Aggregate Base Courses

32 11 23.23 Base Course Drainage Layers

		Crew	Daily Output	Labor-Hours	Unit	Material	2018 Bare Costs Labor	Equipment	Total	Total Incl O&P
0010	**BASE COURSE DRAINAGE LAYERS**									
0011	For roadways and large areas									
0050	Crushed 3/4" stone base, compacted, 3" deep	B-36C	5200	.008	S.Y.	2.71	.38	.74	3.83	4.38
0100	6" deep		5000	.008		5.40	.40	.77	6.57	7.40
0200	9" deep		4600	.009		8.15	.43	.84	9.42	10.50
0300	12" deep		4200	.010		10.85	.47	.92	12.24	13.65
0301	Crushed 1-1/2" stone base, compacted to 4" deep	B-36B	6000	.011		4.52	.51	.75	5.78	6.55
0302	6" deep		5400	.012		6.80	.57	.83	8.20	9.25
0303	8" deep		4500	.014		9.05	.68	1	10.73	12.10
0304	12" deep		3800	.017		13.55	.81	1.19	15.55	17.45
0350	Bank run gravel, spread and compacted									
0370	6" deep	B-32	6000	.005	S.Y.	3.60	.27	.36	4.23	4.75
0390	9" deep		4900	.007		5.40	.33	.44	6.17	6.95
0400	12" deep		4200	.008		7.20	.38	.51	8.09	9.10
6000	Stabilization fabric, polypropylene, 6 oz./S.Y.	B-6	10000	.002		.73	.10	.03	.86	.99
6900	For small and irregular areas, add						50%	50%		
7000	Prepare and roll sub-base, small areas to 2,500 S.Y.	B-32A	1500	.016	S.Y.		.79	.87	1.66	2.15
8000	Large areas over 2,500 S.Y.	"	3500	.007			.34	.37	.71	.92
8050	For roadways	B-32	4000	.008			.40	.54	.94	1.20

32 11 26 – Asphaltic Base Courses

32 11 26.19 Bituminous-Stabilized Base Courses

		Crew	Daily Output	Labor-Hours	Unit	Material	2018 Bare Costs Labor	Equipment	Total	Total Incl O&P
0010	**BITUMINOUS-STABILIZED BASE COURSES**									
0020	And large paved areas									
0700	Liquid application to gravel base, asphalt emulsion	B-45	6000	.003	Gal.	4.50	.13	.13	4.76	5.30
0800	Prime and seal, cut back asphalt		6000	.003	"	5.30	.13	.13	5.56	6.20
1000	Macadam penetration crushed stone, 2 gal./S.Y., 4" thick		6000	.003	S.Y.	9	.13	.13	9.26	10.25
1100	6" thick, 3 gal./S.Y.		4000	.004		13.50	.20	.20	13.90	15.35
1200	8" thick, 4 gal./S.Y.		3000	.005		18	.27	.27	18.54	20.50
8900	For small and irregular areas, add						50%	50%		

For customer support on your Building Construction Costs with RSMeans data, call 800.448.8182.

653

32 12 Flexible Paving

32 12 16 — Asphalt Paving

32 12 16.13 Plant-Mix Asphalt Paving

		Crew	Daily Output	Labor-Hours	Unit	Material	2018 Bare Costs Labor	Equipment	Total	Total Incl O&P
0010	**PLANT-MIX ASPHALT PAVING**									
0020	And large paved areas with no hauling included									
0025	See Section 31 23 23.20 for hauling costs									
0080	Binder course, 1-1/2" thick	B-25	7725	.011	S.Y.	5.30	.50	.35	6.15	7
0120	2" thick		6345	.014		7.05	.61	.42	8.08	9.15
0130	2-1/2" thick		5620	.016		8.80	.69	.48	9.97	11.25
0160	3" thick		4905	.018		10.60	.79	.55	11.94	13.45
0170	3-1/2" thick		4520	.019		12.35	.85	.60	13.80	15.55
0200	4" thick		4140	.021		14.10	.93	.65	15.68	17.65
0300	Wearing course, 1" thick	B-25B	10575	.009		3.50	.41	.28	4.19	4.77
0340	1-1/2" thick		7725	.012		5.90	.56	.38	6.84	7.70
0380	2" thick		6345	.015		7.90	.68	.46	9.04	10.25
0420	2-1/2" thick		5480	.018		9.75	.78	.53	11.06	12.50
0460	3" thick		4900	.020		11.60	.87	.60	13.07	14.80
0470	3-1/2" thick		4520	.021		13.65	.95	.65	15.25	17.15
0480	4" thick		4140	.023		15.60	1.04	.71	17.35	19.50
0500	Open graded friction course	B-25C	5000	.010		2.42	.43	.47	3.32	3.83
0800	Alternate method of figuring paving costs									
0810	Binder course, 1-1/2" thick	B-25	630	.140	Ton	65	6.10	4.27	75.37	85.50
0811	2" thick		690	.128		65	5.60	3.90	74.50	84.50
0812	3" thick		800	.110		65	4.82	3.37	73.19	82.50
0813	4" thick		900	.098		65	4.28	2.99	72.27	81.50
0850	Wearing course, 1" thick	B-25B	575	.167		71	7.45	5.10	83.55	95
0851	1-1/2" thick		630	.152		71	6.80	4.64	82.44	93.50
0852	2" thick		690	.139		71	6.20	4.24	81.44	92
0853	2-1/2" thick		765	.125		71	5.60	3.82	80.42	90.50
0854	3" thick		800	.120		71	5.35	3.66	80.01	90
1000	Pavement replacement over trench, 2" thick	B-37	90	.533	S.Y.	7.30	22.50	1.68	31.48	44
1050	4" thick		70	.686		14.40	29	2.16	45.56	62
1080	6" thick		55	.873		23	36.50	2.75	62.25	84

32 12 16.14 Asphaltic Concrete Paving

		Crew	Daily Output	Labor-Hours	Unit	Material	2018 Bare Costs Labor	Equipment	Total	Total Incl O&P
0011	**ASPHALTIC CONCRETE PAVING**, parking lots & driveways									
0015	No asphalt hauling included									
0018	Use 6.05 C.Y. per inch per M.S.F. for hauling									
0020	6" stone base, 2" binder course, 1" topping	B-25C	9000	.005	S.F.	1.82	.24	.26	2.32	2.65
0025	2" binder course, 2" topping		9000	.005		2.25	.24	.26	2.75	3.13
0030	3" binder course, 2" topping		9000	.005		2.65	.24	.26	3.15	3.56
0035	4" binder course, 2" topping		9000	.005		3.04	.24	.26	3.54	3.99
0040	1-1/2" binder course, 1" topping		9000	.005		1.63	.24	.26	2.13	2.44
0042	3" binder course, 1" topping		9000	.005		2.22	.24	.26	2.72	3.09
0045	3" binder course, 3" topping		9000	.005		3.08	.24	.26	3.58	4.04
0050	4" binder course, 3" topping		9000	.005		3.47	.24	.26	3.97	4.47
0055	4" binder course, 4" topping		9000	.005		3.90	.24	.26	4.40	4.93
0300	Binder course, 1-1/2" thick		35000	.001		.59	.06	.07	.72	.81
0400	2" thick		25000	.002		.76	.09	.09	.94	1.07
0500	3" thick		15000	.003		1.18	.14	.16	1.48	1.69
0600	4" thick		10800	.004		1.54	.20	.22	1.96	2.24
0800	Sand finish course, 3/4" thick		41000	.001		.31	.05	.06	.42	.48
0900	1" thick		34000	.001		.38	.06	.07	.51	.60
1000	Fill pot holes, hot mix, 2" thick	B-16	4200	.008		.81	.32	.13	1.26	1.51
1100	4" thick		3500	.009		1.18	.38	.16	1.72	2.05
1120	6" thick		3100	.010		1.59	.43	.18	2.20	2.59

For customer support on your Building Construction Costs with RSMeans data, call 800.448.8182.

32 12 Flexible Paving

32 12 16 – Asphalt Paving

32 12 16.14 Asphaltic Concrete Paving	Crew	Daily Output	Labor-Hours	Unit	Material	2018 Bare Costs Labor	Equipment	Total	Total Incl O&P
1140	Cold patch, 2" thick	B-51	3000	.016	S.F.	.92	.66	.06	1.64	2.08
1160	4" thick	↓	2700	.018	↓	1.75	.73	.07	2.55	3.11
1180	6" thick	▼	1900	.025	▼	2.72	1.03	.10	3.85	4.67

32 13 Rigid Paving

32 13 13 – Concrete Paving

32 13 13.25 Concrete Pavement, Highways

		Crew	Daily Output	Labor-Hours	Unit	Material	2018 Bare Costs Labor	Equipment	Total	Total Incl O&P
0010	**CONCRETE PAVEMENT, HIGHWAYS**									
0015	Including joints, finishing and curing									
0020	Fixed form, 12' pass, unreinforced, 6" thick	B-26	3000	.029	S.Y.	25.50	1.31	1	27.81	31
0100	8" thick		2750	.032		35	1.43	1.09	37.52	42
0110	8" thick, small area		1375	.064		35	2.85	2.19	40.04	45
0200	9" thick		2500	.035		40	1.57	1.20	42.77	47.50
0300	10" thick		2100	.042		44	1.87	1.43	47.30	52.50
0310	10" thick, small area		1050	.084		44	3.74	2.86	50.60	57
0400	12" thick		1800	.049		50.50	2.18	1.67	54.35	60.50
0410	Conc. pavement, w/jt., fnsh.& curing, fix form, 24' pass, unreinforced, 6"T		6000	.015		24.50	.65	.50	25.65	28.50
0430	8" thick		5500	.016		33.50	.71	.55	34.76	38
0440	9" thick		5000	.018		38	.79	.60	39.39	44
0450	10" thick		4200	.021		42	.93	.72	43.65	48.50
0460	12" thick		3600	.024		48.50	1.09	.83	50.42	56
0470	15" thick		3000	.029		63.50	1.31	1	65.81	72.50
0500	Fixed form 12' pass, 15" thick	▼	1500	.059	▼	64	2.62	2	68.62	76.50
0510	For small irregular areas, add				%	10%	100%	100%		
0520	Welded wire fabric, sheets for rigid paving 2.33 lb./S.Y.	2 Rodm	389	.041	S.Y.	1.36	2.25		3.61	4.92
0530	Reinforcing steel for rigid paving 12 lb./S.Y.		666	.024		6.05	1.31		7.36	8.65
0540	Reinforcing steel for rigid paving 18 lb./S.Y.	↓	444	.036		9.05	1.97		11.02	12.95
0620	Slip form, 12' pass, unreinforced, 6" thick	B-26A	5600	.016		25	.70	.55	26.25	29
0624	8" thick		5300	.017		34	.74	.59	35.33	39.50
0626	9" thick		4820	.018		39	.81	.64	40.45	44.50
0628	10" thick		4050	.022		43	.97	.77	44.74	49.50
0630	12" thick		3470	.025		49.50	1.13	.90	51.53	57
0632	15" thick		2890	.030		62.50	1.36	1.08	64.94	71.50
0640	Slip form, 24' pass, unreinforced, 6" thick		11200	.008		24.50	.35	.28	25.13	28
0644	8" thick		10600	.008		33	.37	.29	33.66	37
0646	9" thick		9640	.009		37.50	.41	.32	38.23	42.50
0648	10" thick		8100	.011		41.50	.48	.38	42.36	46.50
0650	12" thick		6940	.013		48	.57	.45	49.02	54.50
0652	15" thick	▼	5780	.015		60.50	.68	.54	61.72	68
0700	Finishing, broom finish small areas	2 Cefi	120	.133			6.35		6.35	9.40
1000	Curing, with sprayed membrane by hand	2 Clab	1500	.011	▼	1.09	.43		1.52	1.85
1650	For integral coloring, see Section 03 05 13.20									

For customer support on your Building Construction Costs with RSMeans data, call 800.448.8182.

655

32 14 Unit Paving

32 14 13 – Precast Concrete Unit Paving

32 14 13.13 Interlocking Precast Concrete Unit Paving

	Crew	Daily Output	Labor-Hours	Unit	Material	2018 Bare Costs Labor	Equipment	Total	Total Incl O&P
0010 **INTERLOCKING PRECAST CONCRETE UNIT PAVING**									
0020 "V" blocks for retaining soil	D-1	205	.078	S.F.	10.95	3.50		14.45	17.40

32 14 13.16 Precast Concrete Unit Paving Slabs

	Crew	Daily Output	Labor-Hours	Unit	Material	2018 Bare Costs Labor	Equipment	Total	Total Incl O&P
0010 **PRECAST CONCRETE UNIT PAVING SLABS**									
0710 Precast concrete patio blocks, 2-3/8" thick, colors, 8" x 16"	D-1	265	.060	S.F.	9.85	2.71		12.56	15
0750 Exposed local aggregate, natural	2 Bric	250	.064		9.20	3.22		12.42	15.05
0800 Colors		250	.064		9.15	3.22		12.37	15.05
0850 Exposed granite or limestone aggregate		250	.064		9.10	3.22		12.32	14.95
0900 Exposed white tumblestone aggregate		250	.064		9.85	3.22		13.07	15.80

32 14 13.18 Precast Concrete Plantable Pavers

	Crew	Daily Output	Labor-Hours	Unit	Material	2018 Bare Costs Labor	Equipment	Total	Total Incl O&P
0010 **PRECAST CONCRETE PLANTABLE PAVERS** (50% grass)									
0015 Subgrade preparation and grass planting not included									
0100 Precast concrete plantable pavers with topsoil, 24" x 16"	B-63	800	.050	S.F.	4.38	2.11	.22	6.71	8.25
0200 Less than 600 S.F. or irregular area	"	500	.080	"	4.38	3.37	.35	8.10	10.30
0300 3/4" crushed stone base for plantable pavers, 6" depth	B-62	1000	.024	S.Y.	4.22	1.05	.17	5.44	6.40
0400 8" depth		900	.027		5.60	1.16	.19	6.95	8.20
0500 10" depth		800	.030		7.05	1.31	.22	8.58	10
0600 12" depth		700	.034		8.40	1.50	.25	10.15	11.80
0700 Hydro seeding plantable pavers	B-81A	20	.800	M.S.F.	11.40	33.50	17.95	62.85	83.50
0800 Apply fertilizer and seed to plantable pavers	1 Clab	8	1	"	41.50	40		81.50	106

32 14 16 – Brick Unit Paving

32 14 16.10 Brick Paving

	Crew	Daily Output	Labor-Hours	Unit	Material	2018 Bare Costs Labor	Equipment	Total	Total Incl O&P
0010 **BRICK PAVING**									
0012 4" x 8" x 1-1/2", without joints (4.5 bricks/S.F.)	D-1	110	.145	S.F.	2.57	6.50		9.07	12.80
0100 Grouted, 3/8" joint (3.9 bricks/S.F.)		90	.178		2.08	7.95		10.03	14.50
0200 4" x 8" x 2-1/4", without joints (4.5 bricks/S.F.)		110	.145		2.40	6.50		8.90	12.65
0300 Grouted, 3/8" joint (3.9 bricks/S.F.)		90	.178		2.08	7.95		10.03	14.50
0455 Pervious brick paving, 4" x 8" x 3-1/4", without joints (4.5 bricks/S.F.)		110	.145		3.60	6.50		10.10	13.95
0500 Bedding, asphalt, 3/4" thick	B-25	5130	.017		.65	.75	.52	1.92	2.43
0540 Course washed sand bed, 1" thick	B-18	5000	.005		.35	.19	.01	.55	.70
0580 Mortar, 1" thick	D-1	300	.053		.65	2.39		3.04	4.37
0620 2" thick		200	.080		1.29	3.58		4.87	6.90
1500 Brick on 1" thick sand bed laid flat, 4.5/S.F.		100	.160		2.85	7.15		10	14.10
2000 Brick pavers, laid on edge, 7.2/S.F.		70	.229		4.46	10.25		14.71	20.50
2500 For 4" thick concrete bed and joints, add		595	.027		1.42	1.20		2.62	3.40
2800 For steam cleaning, add	A-1H	950	.008		.11	.34	.08	.53	.72

32 14 23 – Asphalt Unit Paving

32 14 23.10 Asphalt Blocks

	Crew	Daily Output	Labor-Hours	Unit	Material	2018 Bare Costs Labor	Equipment	Total	Total Incl O&P
0010 **ASPHALT BLOCKS**									
0020 Rectangular, 6" x 12" x 1-1/4", w/bed & neopr. adhesive	D-1	135	.119	S.F.	8.55	5.30		13.85	17.55
0100 3" thick		130	.123		12	5.50		17.50	21.50
0300 Hexagonal tile, 8" wide, 1-1/4" thick		135	.119		8.55	5.30		13.85	17.55
0400 2" thick		130	.123		12	5.50		17.50	21.50
0500 Square, 8" x 8", 1-1/4" thick		135	.119		8.55	5.30		13.85	17.55
0600 2" thick		130	.123		12	5.50		17.50	21.50
0900 For exposed aggregate (ground finish), add					.52			.52	.57
0910 For colors, add					.52			.52	.57

656

32 14 Unit Paving

32 14 40 – Stone Paving

32 14 40.10 Stone Pavers

		Crew	Daily Output	Labor-Hours	Unit	Material	2018 Bare Costs Labor	Equipment	Total	Total Incl O&P
0010	**STONE PAVERS**									
1100	Flagging, bluestone, irregular, 1" thick,	D-1	81	.198	S.F.	10.45	8.85		19.30	25
1150	Snapped random rectangular, 1" thick		92	.174		15.85	7.80		23.65	29.50
1200	1-1/2" thick		85	.188		19.05	8.45		27.50	34
1250	2" thick		83	.193		22	8.65		30.65	37.50
1300	Slate, natural cleft, irregular, 3/4" thick		92	.174		9.45	7.80		17.25	22.50
1350	Random rectangular, gauged, 1/2" thick		105	.152		20.50	6.85		27.35	33
1400	Random rectangular, butt joint, gauged, 1/4" thick	↓	150	.107	↓	22	4.78		26.78	32
1500	For interior setting, add								25%	25%
1550	Granite blocks, 3-1/2" x 3-1/2" x 3-1/2"	D-1	92	.174	S.F.	19.30	7.80		27.10	33
1600	4" to 12" long, 3" to 5" wide, 3" to 5" thick		98	.163		16.10	7.30		23.40	29
1650	6" to 15" long, 3" to 6" wide, 3" to 5" thick	↓	105	.152	↓	8.60	6.85		15.45	19.90

32 16 Curbs, Gutters, Sidewalks, and Driveways

32 16 13 – Curbs and Gutters

32 16 13.13 Cast-in-Place Concrete Curbs and Gutters

		Crew	Daily Output	Labor-Hours	Unit	Material	2018 Bare Costs Labor	Equipment	Total	Total Incl O&P
0010	**CAST-IN-PLACE CONCRETE CURBS AND GUTTERS**									
0290	Forms only, no concrete									
0300	Concrete, wood forms, 6" x 18", straight	C-2	500	.096	L.F.	2.93	4.73		7.66	10.40
0400	6" x 18", radius	"	200	.240	"	3.05	11.80		14.85	21.50
0402	Forms and concrete complete									
0404	Concrete, wood forms, 6" x 18", straight & concrete	C-2A	500	.096	L.F.	6.30	4.68		10.98	14.05
0406	6" x 18", radius		200	.240		6.45	11.70		18.15	25
0410	Steel forms, 6" x 18", straight		700	.069		5.25	3.34		8.59	10.80
0411	6" x 18", radius	↓	400	.120		5.40	5.85		11.25	14.80
0415	Machine formed, 6" x 18", straight	B-69A	2000	.024		3.68	1.05	.59	5.32	6.30
0416	6" x 18", radius	"	900	.053	↓	3.69	2.33	1.30	7.32	9
0421	Curb and gutter, straight									
0422	with 6" high curb and 6" thick gutter, wood forms									
0430	24" wide, 0.055 C.Y./L.F.	C-2A	375	.128	L.F.	16.05	6.25		22.30	27
0435	30" wide, 0.066 C.Y./L.F.		340	.141		17.70	6.90		24.60	30
0440	Steel forms, 24" wide, straight		700	.069		7.25	3.34		10.59	13
0441	Radius		500	.096		7.20	4.68		11.88	15.05
0442	30" wide, straight		700	.069		8.45	3.34		11.79	14.30
0443	Radius	↓	500	.096		8.30	4.68		12.98	16.25
0445	Machine formed, 24" wide, straight	B-69A	2000	.024		6.75	1.05	.59	8.39	9.70
0446	Radius		900	.053		6.75	2.33	1.30	10.38	12.40
0447	30" wide, straight		2000	.024		7.85	1.05	.59	9.49	10.90
0448	Radius	↓	900	.053	↓	7.85	2.33	1.30	11.48	13.60

32 16 13.23 Precast Concrete Curbs and Gutters

		Crew	Daily Output	Labor-Hours	Unit	Material	2018 Bare Costs Labor	Equipment	Total	Total Incl O&P
0010	**PRECAST CONCRETE CURBS AND GUTTERS**									
0550	Precast, 6" x 18", straight	B-29	700	.080	L.F.	9.35	3.50	1.24	14.09	16.90
0600	6" x 18", radius	"	325	.172	"	10.55	7.55	2.68	20.78	26

32 16 13.33 Asphalt Curbs

		Crew	Daily Output	Labor-Hours	Unit	Material	2018 Bare Costs Labor	Equipment	Total	Total Incl O&P
0010	**ASPHALT CURBS**									
0012	Curbs, asphaltic, machine formed, 8" wide, 6" high, 40 L.F./ton	B-27	1000	.032	L.F.	1.63	1.29	.32	3.24	4.12
0100	8" wide, 8" high, 30 L.F./ton		900	.036		2.18	1.43	.35	3.96	4.98
0150	Asphaltic berm, 12" W, 3" to 6" H, 35 L.F./ton, before pavement		700	.046		.04	1.84	.45	2.33	3.36
0200	12" W, 1-1/2" to 4" H, 60 L.F./ton, laid with pavement	B-2	1050	.038	↓	.02	1.53		1.55	2.37

32 16 13 – Curbs and Gutters

32 16 13.43 Stone Curbs	Crew	Daily Output	Labor-Hours	Unit	Material	2018 Bare Costs Labor	Equipment	Total	Total Incl O&P
0010 **STONE CURBS**									
1000 Granite, split face, straight, 5" x 16"	D-13	275	.175	L.F.	15.20	8.40	1.32	24.92	31
1100 6" x 18"	"	250	.192		19.95	9.20	1.46	30.61	37.50
1300 Radius curbing, 6" x 18", over 10' radius	B-29	260	.215	↓	24.50	9.40	3.34	37.24	45
1400 Corners, 2' radius	"	80	.700	Ea.	82	30.50	10.85	123.35	149
1600 Edging, 4-1/2" x 12", straight	D-13	300	.160	L.F.	7.60	7.70	1.21	16.51	21.50
1800 Curb inlets (guttermouth) straight	B-29	41	1.366	Ea.	182	59.50	21	262.50	315
2000 Indian granite (Belgian block)									
2100 Jumbo, 10-1/2" x 7-1/2" x 4", grey	D-1	150	.107	L.F.	8.35	4.78		13.13	16.50
2150 Pink		150	.107		8.40	4.78		13.18	16.55
2200 Regular, 9" x 4-1/2" x 4-1/2", grey		160	.100		4.45	4.48		8.93	11.75
2250 Pink		160	.100		5.85	4.48		10.33	13.30
2300 Cubes, 4" x 4" x 4", grey		175	.091		3.62	4.10		7.72	10.25
2350 Pink		175	.091		3.72	4.10		7.82	10.35
2400 6" x 6" x 6", pink	↓	155	.103	↓	12.70	4.62		17.32	21
2500 Alternate pricing method for Indian granite									
2550 Jumbo, 10-1/2" x 7-1/2" x 4" (30 lb.), grey				Ton	475			475	525
2600 Pink					490			490	540
2650 Regular, 9" x 4-1/2" x 4-1/2" (20 lb.), grey					315			315	345
2700 Pink					410			410	450
2750 Cubes, 4" x 4" x 4" (5 lb.), grey					440			440	485
2800 Pink					480			480	525
2850 6" x 6" x 6" (25 lb.), pink					495			495	545
2900 For pallets, add				↓	22			22	24

32 17 Paving Specialties

32 17 13 – Parking Bumpers

32 17 13.13 Metal Parking Bumpers

32 17 13.13 Metal Parking Bumpers	Crew	Daily Output	Labor-Hours	Unit	Material	2018 Bare Costs Labor	Equipment	Total	Total Incl O&P
0010 **METAL PARKING BUMPERS**									
0015 Bumper rails for garages, 12 ga. rail, 6" wide, with steel									
0020 posts 12'-6" OC, minimum	E-4	190	.168	L.F.	20.50	9.30	.52	30.32	38.50
0030 Average		165	.194		26	10.70	.60	37.30	46.50
0100 Maximum		140	.229		31	12.60	.70	44.30	55.50
0300 12" channel rail, minimum		160	.200		26	11.05	.62	37.67	47
0400 Maximum	↓	120	.267	↓	39	14.70	.82	54.52	67.50
1300 Pipe bollards, conc. filled/paint, 8' L x 4' D hole, 6" diam.	B-6	20	1.200	Ea.	610	52.50	15.60	678.10	765
1400 8" diam.		15	1.600		690	70	21	781	890
1500 12" diam.	↓	12	2		975	87.50	26	1,088.50	1,225
2030 Folding with individual padlocks	B-2	50	.800		199	32		231	268
8000 Parking lot control, see Section 11 12 13.10									
8900 Security bollards, SS, lighted, hyd., incl. controls, group of 3	L-7	.06	509	Ea.	39,200	24,800		64,000	80,500
8910 Group of 5	"	.04	683	"	65,500	33,200		98,700	122,500

32 17 13.16 Plastic Parking Bumpers

32 17 13.16 Plastic Parking Bumpers	Crew	Daily Output	Labor-Hours	Unit	Material	Labor	Equipment	Total	Total Incl O&P
0010 **PLASTIC PARKING BUMPERS**									
1200 Thermoplastic, 6" x 10" x 6'-0"	B-2	120	.333	Ea.	81	13.40		94.40	110

32 17 13.19 Precast Concrete Parking Bumpers

32 17 13.19 Precast Concrete Parking Bumpers	Crew	Daily Output	Labor-Hours	Unit	Material	Labor	Equipment	Total	Total Incl O&P
0010 **PRECAST CONCRETE PARKING BUMPERS**									
1000 Wheel stops, precast concrete incl. dowels, 6" x 10" x 6'-0"	B-2	120	.333	Ea.	54	13.40		67.40	80
1100 8" x 13" x 6'-0"	"	120	.333	"	62.50	13.40		75.90	89.50

For customer support on your Building Construction Costs with RSMeans data, call 800.448.8182.

32 17 Paving Specialties

32 17 13 – Parking Bumpers

32 17 13.26 Wood Parking Bumpers

	Crew	Daily Output	Labor-Hours	Unit	Material	2018 Bare Costs Labor	2018 Bare Costs Equipment	Total	Total Incl O&P
0010 **WOOD PARKING BUMPERS**									
0020 Parking barriers, timber w/saddles, treated type									
0100 4" x 4" for cars	B-2	520	.077	L.F.	3.07	3.10		6.17	8.10
0200 6" x 6" for trucks		520	.077	"	6.45	3.10		9.55	11.80
0600 Flexible fixed stanchion, 2' high, 3" diameter	↓	100	.400	Ea.	42.50	16.10		58.60	71.50

32 17 23 – Pavement Markings

32 17 23.13 Painted Pavement Markings

	Crew	Daily Output	Labor-Hours	Unit	Material	2018 Bare Costs Labor	2018 Bare Costs Equipment	Total	Total Incl O&P
0010 **PAINTED PAVEMENT MARKINGS**									
0020 Acrylic waterborne, white or yellow, 4" wide, less than 3,000 L.F.	B-78	20000	.002	L.F.	.10	.10	.03	.23	.29
0200 6" wide, less than 3,000 L.F.		11000	.004		.15	.18	.05	.38	.49
0500 8" wide, less than 3,000 L.F.		10000	.005		.20	.20	.05	.45	.58
0600 12" wide, less than 3,000 L.F.		4000	.012	↓	.30	.49	.13	.92	1.22
0620 Arrows or gore lines		2300	.021	S.F.	.20	.85	.22	1.27	1.76
0640 Temporary paint, white or yellow, less than 3,000 L.F.	↓	15000	.003	L.F.	.12	.13	.03	.28	.37
0660 Removal	1 Clab	300	.027			1.06		1.06	1.62
0680 Temporary tape	2 Clab	1500	.011		.50	.43		.93	1.20
0710 Thermoplastic, white or yellow, 4" wide, less than 6,000 L.F.	B-79	15000	.003		.35	.11	.08	.54	.65
0730 6" wide, less than 6,000 L.F.		14000	.003		.53	.12	.09	.74	.86
0740 8" wide, less than 6,000 L.F.		12000	.003		.70	.14	.10	.94	1.09
0750 12" wide, less than 6,000 L.F.		6000	.007	↓	1.03	.27	.21	1.51	1.79
0760 Arrows		660	.061	S.F.	.57	2.50	1.87	4.94	6.50
0770 Gore lines		2500	.016		.57	.66	.49	1.72	2.17
0780 Letters	↓	660	.061	↓	.57	2.50	1.87	4.94	6.50
1000 Airport painted markings									
1050 Traffic safety flashing truck for airport painting	A-2B	1	8	Day		355	188	543	740
1100 Painting, white or yellow, taxiway markings	B-78	4000	.012	S.F.	.33	.49	.13	.95	1.25
1110 with 12 lb. beads per 100 S.F.		4000	.012		.65	.49	.13	1.27	1.60
1200 Runway markings		3500	.014		.33	.56	.14	1.03	1.37
1210 with 12 lb. beads per 100 S.F.		3500	.014		.65	.56	.14	1.35	1.72
1300 Pavement location or direction signs		2500	.019		.33	.79	.20	1.32	1.78
1310 with 12 lb. beads per 100 S.F.		2500	.019	↓	.65	.79	.20	1.64	2.13
1350 Mobilization airport pavement painting	↓	4	12	Ea.		490	126	616	885
1400 Paint markings or pavement signs removal daytime	B-78B	400	.045	S.F.		1.85	.87	2.72	3.77
1500 Removal nighttime		335	.054	"		2.21	1.04	3.25	4.51
1600 Mobilization pavement paint removal	↓	4	4.500	Ea.		185	87	272	375

32 17 23.14 Pavement Parking Markings

	Crew	Daily Output	Labor-Hours	Unit	Material	2018 Bare Costs Labor	2018 Bare Costs Equipment	Total	Total Incl O&P
0010 **PAVEMENT PARKING MARKINGS**									
0790 Layout of pavement marking	A-2	25000	.001	L.F.		.04	.01	.05	.07
0800 Lines on pvmt., parking stall, paint, white, 4" wide	B-78B	400	.045	Stall	4.82	1.85	.87	7.54	9.05
0825 Parking stall, small quantities	2 Pord	80	.200		9.65	8.50		18.15	23.50
0830 Lines on pvmt., parking stall, thermoplastic, white, 4" wide	B-79	300	.133	↓	16	5.50	4.12	25.62	30.50
1000 Street letters and numbers	B-78B	1600	.011	S.F.	.72	.46	.22	1.40	1.73

32 17 26 – Tactile Warning Surfacing

32 17 26.10 Tactile Warning Surfacing

	Crew	Daily Output	Labor-Hours	Unit	Material	2018 Bare Costs Labor	2018 Bare Costs Equipment	Total	Total Incl O&P
0010 **TACTILE WARNING SURFACING**									
0100 Tactile warning tiles S.F.	2 Clab	400	.040	S.F.	17.25	1.59		18.84	21.50

For customer support on your Building Construction Costs with RSMeans data, call 800.448.8182.

659

32 18 Athletic and Recreational Surfacing

32 18 13 – Synthetic Grass Surfacing

32 18 13.10 Artificial Grass Surfacing

		Crew	Daily Output	Labor-Hours	Unit	Material	2018 Bare Costs Labor	Equipment	Total	Total Incl O&P
0010	**ARTIFICIAL GRASS SURFACING**									
0015	Not including asphalt base or drainage,									
0020	but including cushion pad, over 50,000 S.F.									
0200	1/2" pile and 5/16" cushion pad, standard	C-17	3200	.025	S.F.	5.25	1.32		6.57	7.75
0300	Deluxe		2560	.031		5.15	1.65		6.80	8.15
0500	1/2" pile and 5/8" cushion pad, standard		2844	.028		5.15	1.48		6.63	7.90
0600	Deluxe		2327	.034		5.15	1.81		6.96	8.45
0800	For asphaltic concrete base, 2-1/2" thick,									
0900	with 6" crushed stone sub-base, add	B-25	12000	.007	S.F.	1.67	.32	.22	2.21	2.57

32 18 16 – Synthetic Resilient Surfacing

32 18 16.13 Playground Protective Surfacing

		Crew	Daily Output	Labor-Hours	Unit	Material	2018 Bare Costs Labor	Equipment	Total	Total Incl O&P
0010	**PLAYGROUND PROTECTIVE SURFACING**									
0100	Resilient rubber surface, 4" thick, black	2 Skwk	300	.053	S.F.	13.90	2.79		16.69	19.55
0150	2" thick topping, colors	"	2800	.006		6.85	.30		7.15	8
0200	Wood chip mulch, 6" deep	1 Clab	300	.027		.79	1.06		1.85	2.49

32 18 23 – Athletic Surfacing

32 18 23.33 Running Track Surfacing

		Crew	Daily Output	Labor-Hours	Unit	Material	2018 Bare Costs Labor	Equipment	Total	Total Incl O&P
0010	**RUNNING TRACK SURFACING**									
0020	Running track, asphalt concrete pavement, 2-1/2"	B-37	300	.160	S.Y.	14.90	6.75	.50	22.15	27
0102	Surface, latex rubber system, 1/2" thick, black	B-20	115	.209		47.50	9.30		56.80	66.50
0152	Colors		115	.209		58	9.30		67.30	78
0302	Urethane rubber system, 1/2" thick, black		110	.218		35.50	9.75		45.25	54
0402	Color coating		110	.218		43.50	9.75		53.25	63

32 18 23.53 Tennis Court Surfacing

		Crew	Daily Output	Labor-Hours	Unit	Material	2018 Bare Costs Labor	Equipment	Total	Total Incl O&P
0010	**TENNIS COURT SURFACING**									
0020	Tennis court, asphalt, incl. base, 2-1/2" thick, one court	B-37	450	.107	S.Y.	33	4.49	.34	37.83	43.50
0200	Two courts		675	.071		18.80	2.99	.22	22.01	25.50
0300	Clay courts		360	.133		42.50	5.60	.42	48.52	55.50
0400	Pulverized natural greenstone with 4" base, fast dry		250	.192		39.50	8.10	.60	48.20	56.50
0800	Rubber-acrylic base resilient pavement		600	.080		65	3.37	.25	68.62	77
1000	Colored sealer, acrylic emulsion, 3 coats	2 Clab	800	.020		7.15	.80		7.95	9.05
1100	3 coats, 2 colors	"	900	.018		9.95	.71		10.66	12.05
1200	For preparing old courts, add	1 Clab	825	.010			.39		.39	.59
1400	Posts for nets, 3-1/2" diameter with eye bolts	B-1	3.40	7.059	Pr.	286	286		572	750
1500	With pulley & reel		3.40	7.059	"	860	286		1,146	1,375
1700	Net, 42' long, nylon thread with binder		50	.480	Ea.	262	19.45		281.45	320
1800	All metal		6.50	3.692	"	555	150		705	845
2000	Paint markings on asphalt, 2 coats	1 Pord	1.78	4.494	Court	226	191		417	535
2200	Complete court with fence, etc., asphaltic conc., minimum	B-37	.20	240		32,900	10,100	755	43,755	52,500
2300	Maximum		.16	300		67,000	12,600	945	80,545	94,000
2800	Clay courts, minimum		.20	240		36,800	10,100	755	47,655	56,500
2900	Maximum		.16	300		70,000	12,600	945	83,545	97,500

For customer support on your Building Construction Costs with RSMeans data, call 800.448.8182.

32 31 Fences and Gates

32 31 13 – Chain Link Fences and Gates

32 31 13.20 Fence, Chain Link Industrial

32 31 13.20 Fence, Chain Link Industrial	Crew	Daily Output	Labor-Hours	Unit	Material	2018 Bare Costs Labor	Equipment	Total	Total Incl O&P
0010 **FENCE, CHAIN LINK INDUSTRIAL**									
0011 Schedule 40, including concrete									
0020 3 strands barb wire, 2" post @ 10' OC, set in concrete, 6' H									
0200 9 ga. wire, galv. steel, in concrete	B-80C	240	.100	L.F.	19.80	4.14	.82	24.76	29
0248 Fence, add for vinyl coated fabric				S.F.	.68			.68	.75
0300 Aluminized steel	B-80C	240	.100	L.F.	22	4.14	.82	26.96	31
0301 Fence, wrought iron		240	.100		30	4.14	.82	34.96	40
0500 6 ga. wire, galv. steel		240	.100		25	4.14	.82	29.96	34.50
0600 Aluminized steel		240	.100		30.50	4.14	.82	35.46	40.50
0800 6 ga. wire, 6' high but omit barbed wire, galv. steel		250	.096		20	3.97	.78	24.75	29
0900 Aluminized steel, in concrete		250	.096		24	3.97	.78	28.75	33.50
0920 8' H, 6 ga. wire, 2-1/2" line post, galv. steel, in concrete		180	.133		32	5.50	1.09	38.59	44.50
0940 Aluminized steel, in concrete		180	.133		39	5.50	1.09	45.59	52.50
1100 Add for corner posts, 3" diam., galv. steel, in concrete		40	.600	Ea.	88.50	25	4.90	118.40	140
1200 Aluminized steel, in concrete		40	.600		88.50	25	4.90	118.40	140
1300 Add for braces, galv. steel		80	.300		39	12.40	2.45	53.85	64
1350 Aluminized steel		80	.300		50	12.40	2.45	64.85	76.50
1400 Gate for 6' high fence, 1-5/8" frame, 3' wide, galv. steel		10	2.400		208	99.50	19.60	327.10	400
1500 Aluminized steel, in concrete		10	2.400		209	99.50	19.60	328.10	405
2000 5'-0" high fence, 9 ga., no barbed wire, 2" line post, in concrete									
2010 10' OC, 1-5/8" top rail, in concrete									
2100 Galvanized steel, in concrete	B-80C	300	.080	L.F.	21	3.31	.65	24.96	28.50
2200 Aluminized steel, in concrete		300	.080	"	19.05	3.31	.65	23.01	26.50
2400 Gate, 4' wide, 5' high, 2" frame, galv. steel, in concrete		10	2.400	Ea.	219	99.50	19.60	338.10	415
2500 Aluminized steel, in concrete		10	2.400	"	197	99.50	19.60	316.10	390
3100 Overhead slide gate, chain link, 6' high, to 18' wide, in concrete		38	.632	L.F.	97	26	5.15	128.15	152
3110 Cantilever type, in concrete	B-80	48	.667		142	29.50	12.75	184.25	215
3120 8' high, in concrete		24	1.333		168	59	25.50	252.50	300
3130 10' high, in concrete		18	1.778		206	79	34	319	385
5000 Double swing gates, incl. posts & hardware, in concrete									
5010 5' high, 12' opening, in concrete	B-80C	3.40	7.059	Opng.	540	292	57.50	889.50	1,100
5020 20' opening, in concrete		2.80	8.571		605	355	70	1,030	1,275
5060 6' high, 12' opening, in concrete		3.20	7.500		470	310	61	841	1,050
5070 20' opening, in concrete		2.60	9.231		655	380	75.50	1,110.50	1,375
5080 8' high, 12' opening, in concrete	B-80	2.13	15.002		485	665	287	1,437	1,850
5090 20' opening, in concrete		1.45	22.069		785	980	420	2,185	2,800
5100 10' high, 12' opening, in concrete		1.31	24.427		840	1,075	465	2,380	3,100
5110 20' opening, in concrete		1.03	31.068		885	1,375	595	2,855	3,725
5120 12' high, 12' opening, in concrete		1.05	30.476		1,175	1,350	585	3,110	4,000
5130 20' opening, in concrete		.85	37.647		1,250	1,675	720	3,645	4,700
5190 For aluminized steel, add					20%				

32 31 13.25 Fence, Chain Link Residential

32 31 13.25 Fence, Chain Link Residential	Crew	Daily Output	Labor-Hours	Unit	Material	2018 Bare Costs Labor	Equipment	Total	Total Incl O&P
0010 **FENCE, CHAIN LINK RESIDENTIAL**									
0011 Schedule 20, 11 ga. wire, 1-5/8" post									
0020 10' OC, 1-3/8" top rail, 2" corner post, galv. stl. 3' high	B-80C	500	.048	L.F.	2	1.99	.39	4.38	5.65
0050 4' high		400	.060		7.05	2.48	.49	10.02	12.05
0100 6' high		200	.120		8.95	4.97	.98	14.90	18.50
0150 Add for gate 3' wide, 1-3/8" frame, 3' high		12	2	Ea.	76.50	83	16.30	175.80	228
0170 4' high		10	2.400		82.50	99.50	19.60	201.60	263
0190 6' high		10	2.400		103	99.50	19.60	222.10	286
0200 Add for gate 4' wide, 1-3/8" frame, 3' high		9	2.667		86	110	22	218	286
0220 4' high		9	2.667		92	110	22	224	292

32 31 Fences and Gates

32 31 13 – Chain Link Fences and Gates

32 31 13.25 Fence, Chain Link Residential

		Crew	Daily Output	Labor-Hours	Unit	Material	2018 Bare Costs Labor	Equipment	Total	Total Incl O&P
0240	6' high	B-80C	8	3	Ea.	116	124	24.50	264.50	345
0350	Aluminized steel, 11 ga. wire, 3' high		500	.048	L.F.	8.30	1.99	.39	10.68	12.60
0380	4' high		400	.060		7.75	2.48	.49	10.72	12.85
0400	6' high		200	.120	↓	10.65	4.97	.98	16.60	20.50
0450	Add for gate 3' wide, 1-3/8" frame, 3' high		12	2	Ea.	95.50	83	16.30	194.80	249
0470	4' high		10	2.400		95.50	99.50	19.60	214.60	278
0490	6' high		10	2.400		118	99.50	19.60	237.10	305
0500	Add for gate 4' wide, 1-3/8" frame, 3' high		10	2.400		99.50	99.50	19.60	218.60	282
0520	4' high		9	2.667		114	110	22	246	315
0540	6' high		8	3	↓	123	124	24.50	271.50	350
0620	Vinyl covered, 9 ga. wire, 3' high		500	.048	L.F.	7.40	1.99	.39	9.78	11.55
0640	4' high		400	.060		7.70	2.48	.49	10.67	12.75
0660	6' high		200	.120	↓	9.55	4.97	.98	15.50	19.15
0720	Add for gate 3' wide, 1-3/8" frame, 3' high		12	2	Ea.	89	83	16.30	188.30	242
0740	4' high		10	2.400		95	99.50	19.60	214.10	278
0760	6' high		10	2.400		113	99.50	19.60	232.10	298
0780	Add for gate 4' wide, 1-3/8" frame, 3' high		10	2.400		93.50	99.50	19.60	212.60	276
0800	4' high		9	2.667		97	110	22	229	298
0820	6' high	↓	8	3	↓	121	124	24.50	269.50	350
7076	Fence, for small jobs 100 L.F. fence or less w/or w/o gate, add				L.F.	20%				

32 31 13.26 Tennis Court Fences and Gates

		Crew	Daily Output	Labor-Hours	Unit	Material	2018 Bare Costs Labor	Equipment	Total	Total Incl O&P
0010	**TENNIS COURT FENCES AND GATES**									
0860	Tennis courts, 11 ga. wire, 2-1/2" post set									
0870	in concrete, 10' OC, 1-5/8" top rail									
0900	10' high	B-80	190	.168	L.F.	21.50	7.45	3.22	32.17	38.50
0920	12' high		170	.188	"	21.50	8.35	3.60	33.45	40
1000	Add for gate 4' wide, 1-5/8" frame 7' high		10	3.200	Ea.	229	142	61	432	535
1040	Aluminized steel, 11 ga. wire 10' high		190	.168	L.F.	20	7.45	3.22	30.67	37
1100	12' high		170	.188	"	22.50	8.35	3.60	34.45	41
1140	Add for gate 4' wide, 1-5/8" frame, 7' high		10	3.200	Ea.	240	142	61	443	545
1250	Vinyl covered, 9 ga. wire, 10' high		190	.168	L.F.	20.50	7.45	3.22	31.17	38
1300	12' high	↓	170	.188	"	24	8.35	3.60	35.95	42.50
1310	Fence, CL, tennis court, transom gate, single, galv., 4' x 7'	B-80A	8.72	2.752	Ea.	279	110	27.50	416.50	500
1400	Add for gate 4' wide, 1-5/8" frame, 7' high	B-80	10	3.200	"	296	142	61	499	610

32 31 13.33 Chain Link Backstops

		Crew	Daily Output	Labor-Hours	Unit	Material	2018 Bare Costs Labor	Equipment	Total	Total Incl O&P
0010	**CHAIN LINK BACKSTOPS**									
0015	Backstops, baseball, prefabricated, 30' wide, 12' high & 1 overhang	B-1	1	24	Ea.	2,625	970		3,595	4,350
0100	40' wide, 12' high & 2 overhangs	"	.75	32		6,925	1,300		8,225	9,575
0300	Basketball, steel, single goal	B-13	3.04	18.421		1,725	805	194	2,724	3,350
0400	Double goal	"	1.92	29.167	↓	2,125	1,275	310	3,710	4,600
0600	Tennis, wire mesh with pair of ends	B-1	2.48	9.677	Set	2,700	390		3,090	3,550
0700	Enclosed court	"	1.30	18.462	Ea.	9,150	750		9,900	11,300

32 31 13.53 High-Security Chain Link Fences, Gates and Sys.

		Crew	Daily Output	Labor-Hours	Unit	Material	2018 Bare Costs Labor	Equipment	Total	Total Incl O&P
0010	**HIGH-SECURITY CHAIN LINK FENCES, GATES AND SYSTEMS**									
0100	Fence, chain link, security, 7' H, standard FE-7, incl. excavation & posts	B-80C	480	.050	L.F.	45.50	2.07	.41	47.98	53.50
0200	Fence, barbed wire, security, 7' H, with 3 wire barbed wire arm	"	400	.060	"	7.60	2.48	.49	10.57	12.65
0300	Complete systems, including material and installation									
0310	Taunt wire fence detection system				M.L.F.				25,100	27,600
0410	Microwave fence detection system								41,300	45,400
0510	Passive magnetic fence detection system								19,500	21,400
0610	Infrared fence detection system								12,900	14,400
0710	Strain relief fence detection system				↓				25,100	27,600

For customer support on your Building Construction Costs with RSMeans data, call 800.448.8182.

32 31 Fences and Gates

32 31 13 – Chain Link Fences and Gates

32 31 13.53 High-Security Chain Link Fences, Gates and Sys.	Crew	Daily Output	Labor-Hours	Unit	Material	2018 Bare Costs Labor	Equipment	Total	Total Incl O&P
0810	Electro-shock fence detection system				M.L.F.				35,900	39,500
0910	Photo-electric fence detection system								16,300	18,000

32 31 13.80 Residential Chain Link Gate

		Crew	Daily Output	Labor-Hours	Unit	Material	Labor	Equipment	Total	Total Incl O&P
0010	**RESIDENTIAL CHAIN LINK GATE**									
0110	Residential 4' gate, single incl. hardware and concrete	B-80C	10	2.400	Ea.	123	99.50	19.60	242.10	310
0120	5'		10	2.400		133	99.50	19.60	252.10	320
0130	6'		10	2.400		143	99.50	19.60	262.10	330
0510	Residential 4' gate, double incl. hardware and concrete		10	2.400		219	99.50	19.60	338.10	415
0520	5'		10	2.400		234	99.50	19.60	353.10	430
0530	6'		10	2.400		273	99.50	19.60	392.10	475

32 31 13.82 Internal Chain Link Gate

		Crew	Daily Output	Labor-Hours	Unit	Material	Labor	Equipment	Total	Total Incl O&P
0010	**INTERNAL CHAIN LINK GATE**									
0110	Internal 6' gate, single incl. post flange, hardware and concrete	B-80C	10	2.400	Ea.	269	99.50	19.60	388.10	470
0120	8'		10	2.400		305	99.50	19.60	424.10	510
0130	10'		10	2.400		420	99.50	19.60	539.10	640
0510	Internal 6' gate, double incl. post flange, hardware and concrete		10	2.400		490	99.50	19.60	609.10	715
0520	8'		10	2.400		565	99.50	19.60	684.10	795
0530	10'		10	2.400		710	99.50	19.60	829.10	960

32 31 13.84 Industrial Chain Link Gate

		Crew	Daily Output	Labor-Hours	Unit	Material	Labor	Equipment	Total	Total Incl O&P
0010	**INDUSTRIAL CHAIN LINK GATE**									
0110	Industrial 8' gate, single incl. hardware and concrete	B-80C	10	2.400	Ea.	460	99.50	19.60	579.10	680
0120	10'		10	2.400		520	99.50	19.60	639.10	750
0510	Industrial 8' gate, double incl. hardware and concrete		10	2.400		715	99.50	19.60	834.10	960
0520	10'		10	2.400		815	99.50	19.60	934.10	1,075

32 31 13.88 Chain Link Transom

		Crew	Daily Output	Labor-Hours	Unit	Material	Labor	Equipment	Total	Total Incl O&P
0010	**CHAIN LINK TRANSOM**									
0110	Add for, single transom, 3' wide, incl. components & hardware	B-80C	10	2.400	Ea.	114	99.50	19.60	233.10	298
0120	Add for, double transom, 6' wide, incl. components & hardware	"	10	2.400	"	122	99.50	19.60	241.10	305

32 31 19 – Decorative Metal Fences and Gates

32 31 19.10 Decorative Fence

		Crew	Daily Output	Labor-Hours	Unit	Material	Labor	Equipment	Total	Total Incl O&P
0010	**DECORATIVE FENCE**									
5300	Tubular picket, steel, 6' sections, 1-9/16" posts, 4' high	B-80C	300	.080	L.F.	38	3.31	.65	41.96	47.50
5400	2" posts, 5' high		240	.100		42	4.14	.82	46.96	53
5600	2" posts, 6' high		200	.120		50.50	4.97	.98	56.45	64
5700	Staggered picket, 1-9/16" posts, 4' high		300	.080		38	3.31	.65	41.96	47.50
5800	2" posts, 5' high		240	.100		43	4.14	.82	47.96	54.50
5900	2" posts, 6' high		200	.120		50.50	4.97	.98	56.45	64
6200	Gates, 4' high, 3' wide	B-1	10	2.400	Ea.	345	97.50		442.50	530
6300	5' high, 3' wide		10	2.400		231	97.50		328.50	405
6400	6' high, 3' wide		10	2.400		281	97.50		378.50	460
6500	4' wide		10	2.400		330	97.50		427.50	515

32 31 26 – Wire Fences and Gates

32 31 26.10 Fences, Misc. Metal

		Crew	Daily Output	Labor-Hours	Unit	Material	Labor	Equipment	Total	Total Incl O&P
0010	**FENCES, MISC. METAL**									
0012	Chicken wire, posts @ 4', 1" mesh, 4' high	B-80C	410	.059	L.F.	3.50	2.42	.48	6.40	8.05
0100	2" mesh, 6' high		350	.069		3.93	2.84	.56	7.33	9.25
0200	Galv. steel, 12 ga., 2" x 4" mesh, posts 5' OC, 3' high		300	.080		2.70	3.31	.65	6.66	8.70
0300	5' high		300	.080		3.28	3.31	.65	7.24	9.35
0400	14 ga., 1" x 2" mesh, 3' high		300	.080		3.32	3.31	.65	7.28	9.35
0500	5' high		300	.080		4.44	3.31	.65	8.40	10.60

32 31 26.10 Fences, Misc. Metal

		Crew	Daily Output	Labor-Hours	Unit	Material	2018 Bare Costs Labor	Equipment	Total	Total Incl O&P
1000	Kennel fencing, 1-1/2" mesh, 6' long, 3'-6" wide, 6'-2" high	2 Clab	4	4	Ea.	495	159		654	790
1050	12' long		4	4		695	159		854	1,000
1200	Top covers, 1-1/2" mesh, 6' long		15	1.067		132	42.50		174.50	210
1250	12' long		12	1.333		185	53		238	285
1300	For kennel doors, see Section 08 31 13.40									
4490	Security fence, prison grade, set in concrete, 10' high	B-80	25	1.280	L.F.	61.50	57	24.50	143	181
4492	Security fence, prison grade, barbed wire, set in concrete, 10' high		22	1.455		62	64.50	28	154.50	197
4494	Security fence, prison grade, razor wire, set in concrete, 10' high		18	1.778		63	79	34	176	226
4500	Security fence, prison grade, set in concrete, 12' high		25	1.280		61.50	57	24.50	143	181
4600	16' high		20	1.600		79	71	30.50	180.50	229
4990	Security fence, prison grade, set in concrete, 10' high		25	1.280		61.50	57	24.50	143	181

32 31 26.20 Wire Fencing, General

		Crew	Daily Output	Labor-Hours	Unit	Material	2018 Bare Costs Labor	Equipment	Total	Total Incl O&P
0010	**WIRE FENCING, GENERAL**									
0015	Barbed wire, galvanized, domestic steel, hi-tensile 15-1/2 ga.				M.L.F.	130			130	142
0020	Standard, 12-3/4 ga.					146			146	160
0210	Barbless wire, 2-strand galvanized, 12-1/2 ga.					146			146	160
0500	Helical razor ribbon, stainless steel, 18" diam. x 18" spacing				C.L.F.	174			174	191
0600	Hardware cloth galv., 1/4" mesh, 23 ga., 2' wide				C.S.F.	34.50			34.50	38
0700	3' wide					26.50			26.50	29
0900	1/2" mesh, 19 ga., 2' wide					36			36	39.50
1000	4' wide					45			45	49.50
1200	Chain link fabric, steel, 2" mesh, 6 ga., galvanized					61			61	67
1300	9 ga., galvanized					54			54	59.50
1350	Vinyl coated					30			30	33
1360	Aluminized					184			184	203
1400	2-1/4" mesh, 11-1/2 ga., galvanized					42.50			42.50	46.50
1600	1-3/4" mesh (tennis courts), 11-1/2 ga. (core), vinyl coated					48			48	53
1700	9 ga., galvanized					85			85	93.50
2100	Welded wire fabric, galvanized, 1" x 2", 14 ga.	2 Carp	1600	.010	S.F.	.65	.51		1.16	1.49
2200	2" x 4", 12-1/2 ga.				C.S.F.	33			33	36

32 31 29 – Wood Fences and Gates

32 31 29.20 Fence, Wood Rail

		Crew	Daily Output	Labor-Hours	Unit	Material	2018 Bare Costs Labor	Equipment	Total	Total Incl O&P
0010	**FENCE, WOOD RAIL**									
0012	Picket, No. 2 cedar, Gothic, 2 rail, 3' high	B-1	160	.150	L.F.	8.10	6.10		14.20	18.15
0050	Gate, 3'-6" wide	B-80C	9	2.667	Ea.	78	110	22	210	277
0400	3 rail, 4' high		150	.160	L.F.	9.10	6.60	1.31	17.01	21.50
0500	Gate, 3'-6" wide		9	2.667	Ea.	95	110	22	227	295
0600	Open rail, rustic, No. 1 cedar, 2 rail, 3' high		160	.150	L.F.	6.05	6.20	1.22	13.47	17.40
0650	Gate, 3' wide		9	2.667	Ea.	83	110	22	215	282
0700	3 rail, 4' high		150	.160	L.F.	7.20	6.60	1.31	15.11	19.45
0900	Gate, 3' wide		9	2.667	Ea.	103	110	22	235	305
1200	Stockade, No. 2 cedar, treated wood rails, 6' high		160	.150	L.F.	9.10	6.20	1.22	16.52	21
1250	Gate, 3' wide		9	2.667	Ea.	97	110	22	229	298
1300	No. 1 cedar, 3-1/4" cedar rails, 6' high		160	.150	L.F.	19.25	6.20	1.22	26.67	32
1500	Gate, 3' wide		9	2.667	Ea.	227	110	22	359	440
1520	Open rail, split, No. 1 cedar, 2 rail, 3' high		160	.150	L.F.	6.05	6.20	1.22	13.47	17.40
1540	3 rail, 4' high		150	.160		8.10	6.60	1.31	16.01	20.50
3300	Board, shadow box, 1" x 6", treated pine, 6' high		160	.150		12.65	6.20	1.22	20.07	24.50
3400	No. 1 cedar, 6' high		150	.160		25	6.60	1.31	32.91	39
3900	Basket weave, No. 1 cedar, 6' high		160	.150		34	6.20	1.22	41.42	48.50
3950	Gate, 3'-6" wide	B-1	8	3	Ea.	234	122		356	445
4000	Treated pine, 6' high		150	.160	L.F.	16	6.50		22.50	27.50

For customer support on your Building Construction Costs with RSMeans data, call 800.448.8182.

32 31 Fences and Gates

32 31 29 – Wood Fences and Gates

32 31 29.20 Fence, Wood Rail

32 31 29.20 Fence, Wood Rail		Crew	Daily Output	Labor-Hours	Unit	Material	2018 Bare Costs Labor	Equipment	Total	Total Incl O&P
4200	Gate, 3'-6" wide	B-1	9	2.667	Ea.	178	108		286	360
5000	Fence rail, redwood, 2" x 4", merch. grade, 8'		2400	.010	L.F.	2.53	.41		2.94	3.40
5050	Select grade, 8'		2400	.010	"	5.55	.41		5.96	6.70
6000	Fence post, select redwood, earth packed & treated, 4" x 4" x 6'		96	.250	Ea.	14	10.15		24.15	31
6010	4" x 4" x 8'		96	.250		19.25	10.15		29.40	36.50
6020	Set in concrete, 4" x 4" x 6'		50	.480		22	19.45		41.45	53.50
6030	4" x 4" x 8'		50	.480		23	19.45		42.45	55
6040	Wood post, 4' high, set in concrete, incl. concrete		50	.480		14.20	19.45		33.65	45
6050	Earth packed		96	.250		17.20	10.15		27.35	34.50
6060	6' high, set in concrete, incl. concrete		50	.480		17.70	19.45		37.15	49
6070	Earth packed		96	.250		12.10	10.15		22.25	29

32 32 Retaining Walls

32 32 13 – Cast-in-Place Concrete Retaining Walls

32 32 13.10 Retaining Walls, Cast Concrete

32 32 13.10 Retaining Walls, Cast Concrete		Crew	Daily Output	Labor-Hours	Unit	Material	2018 Bare Costs Labor	Equipment	Total	Total Incl O&P
0010	**RETAINING WALLS, CAST CONCRETE**									
1800	Concrete gravity wall with vertical face including excavation & backfill									
1850	No reinforcing									
1900	6' high, level embankment	C-17C	36	2.306	L.F.	89	122	15.80	226.80	300
2000	33° slope embankment		32	2.594		103	137	17.80	257.80	345
2200	8' high, no surcharge		27	3.074		110	163	21	294	395
2300	33° slope embankment		24	3.458		133	183	23.50	339.50	450
2500	10' high, level embankment		19	4.368		157	231	30	418	560
2600	33° slope embankment		18	4.611		217	244	31.50	492.50	650
2800	Reinforced concrete cantilever, incl. excavation, backfill & reinf.									
2900	6' high, 33° slope embankment	C-17C	35	2.371	L.F.	80.50	125	16.25	221.75	298
3000	8' high, 33° slope embankment		29	2.862		93	151	19.65	263.65	355
3100	10' high, 33° slope embankment		20	4.150		121	219	28.50	368.50	500
3200	20' high, 500 lb./L.F. surcharge		7.50	11.067		360	585	76	1,021	1,375
3500	Concrete cribbing, incl. excavation and backfill									
3700	12' high, open face	B-13	210	.267	S.F.	40.50	11.65	2.81	54.96	65.50
3900	Closed face	"	210	.267	"	38	11.65	2.81	52.46	62.50
4100	Concrete filled slurry trench, see Section 31 56 23.20									

32 32 23 – Segmental Retaining Walls

32 32 23.13 Segmental Conc. Unit Masonry Retaining Walls

32 32 23.13 Segmental Conc. Unit Masonry Retaining Walls		Crew	Daily Output	Labor-Hours	Unit	Material	2018 Bare Costs Labor	Equipment	Total	Total Incl O&P
0010	**SEGMENTAL CONC. UNIT MASONRY RETAINING WALLS**									
7100	Segmental retaining wall system, incl. pins and void fill									
7120	base and backfill not included									
7140	Large unit, 8" high x 18" wide x 20" deep, 3 plane split	B-62	300	.080	S.F.	12.75	3.49	.58	16.82	20
7150	Straight split		300	.080		12.85	3.49	.58	16.92	20
7160	Medium, lt. wt., 8" high x 18" wide x 12" deep, 3 plane split		400	.060		6.30	2.62	.44	9.36	11.35
7170	Straight split		400	.060		9.65	2.62	.44	12.71	15.10
7180	Small unit, 4" x 18" x 10" deep, 3 plane split		400	.060		14.60	2.62	.44	17.66	20.50
7190	Straight split		400	.060		10.65	2.62	.44	13.71	16.15
7200	Cap unit, 3 plane split		300	.080		13.40	3.49	.58	17.47	20.50
7210	Cap unit, straight split		300	.080		13.40	3.49	.58	17.47	20.50
7250	Geo-grid soil reinforcement 4' x 50'	2 Clab	22500	.001		.68	.03		.71	.79
7255	Geo-grid soil reinforcement 6' x 150'	"	22500	.001		.54	.03		.57	.63

32 32 Retaining Walls

32 32 26 – Metal Crib Retaining Walls

32 32 26.10 Metal Bin Retaining Walls

		Crew	Daily Output	Labor-Hours	Unit	Material	2018 Bare Costs Labor	Equipment	Total	Total Incl O&P
0010	**METAL BIN RETAINING WALLS**									
0011	Aluminized steel bin, excavation									
0020	and backfill not included, 10' wide									
0100	4' high, 5.5' deep	B-13	650	.086	S.F.	28	3.77	.91	32.68	37.50
0200	8' high, 5.5' deep		615	.091		32	3.98	.96	36.94	42.50
0300	10' high, 7.7' deep		580	.097		35.50	4.22	1.02	40.74	47
0400	12' high, 7.7' deep		530	.106		38.50	4.62	1.11	44.23	50.50
0500	16' high, 7.7' deep		515	.109		40.50	4.75	1.15	46.40	53.50
0600	16' high, 9.9' deep		500	.112		43	4.90	1.18	49.08	56.50
0700	20' high, 9.9' deep		470	.119		48.50	5.20	1.26	54.96	63
0800	20' high, 12.1' deep		460	.122		43.50	5.30	1.28	50.08	57
0900	24' high, 12.1' deep		455	.123		46	5.40	1.30	52.70	60
1000	24' high, 14.3' deep		450	.124		54.50	5.45	1.31	61.26	69.50
1100	28' high, 14.3' deep	↓	440	.127	↓	56.50	5.55	1.34	63.39	72
1300	For plain galvanized bin type walls, deduct					10%				

32 32 29 – Timber Retaining Walls

32 32 29.10 Landscape Timber Retaining Walls

		Crew	Daily Output	Labor-Hours	Unit	Material	2018 Bare Costs Labor	Equipment	Total	Total Incl O&P
0010	**LANDSCAPE TIMBER RETAINING WALLS**									
0100	Treated timbers, 6" x 6"	1 Clab	265	.030	L.F.	2.47	1.20		3.67	4.55
0110	6" x 8"	"	200	.040	"	5.40	1.59		6.99	8.35
0120	Drilling holes in timbers for fastening, 1/2"	1 Carp	450	.018	Inch		.90		.90	1.37
0130	5/8"	"	450	.018	"		.90		.90	1.37
0140	Reinforcing rods for fastening, 1/2"	1 Clab	312	.026	L.F.	.35	1.02		1.37	1.95
0150	5/8"	"	312	.026	"	.55	1.02		1.57	2.16
0160	Reinforcing fabric	2 Clab	2500	.006	S.Y.	2.13	.26		2.39	2.73
0170	Gravel backfill		28	.571	C.Y.	16.80	23		39.80	53
0180	Perforated pipe, 4" diameter with silt sock	↓	1200	.013	L.F.	1	.53		1.53	1.91
0190	Galvanized 60d common nails	1 Clab	625	.013	Ea.	.18	.51		.69	.98
0200	20d common nails	"	3800	.002	"	.04	.08		.12	.17

32 32 36 – Gabion Retaining Walls

32 32 36.10 Stone Gabion Retaining Walls

		Crew	Daily Output	Labor-Hours	Unit	Material	2018 Bare Costs Labor	Equipment	Total	Total Incl O&P
0010	**STONE GABION RETAINING WALLS**									
4300	Stone filled gabions, not incl. excavation,									
4310	Stone, delivered, 3' wide									
4350	Galvanized, 6' high, 33° slope embankment	B-13	49	1.143	L.F.	57	50	12.05	119.05	152
4500	Highway surcharge		27	2.074		75.50	90.50	22	188	245
4600	9' high, up to 33° slope embankment		24	2.333		128	102	24.50	254.50	325
4700	Highway surcharge		16	3.500		132	153	37	322	420
4900	12' high, up to 33° slope embankment		14	4		200	175	42	417	530
5000	Highway surcharge	↓	11	5.091		189	223	53.50	465.50	605
5950	For PVC coating, add				↓	12%				

32 32 53 – Stone Retaining Walls

32 32 53.10 Retaining Walls, Stone

		Crew	Daily Output	Labor-Hours	Unit	Material	2018 Bare Costs Labor	Equipment	Total	Total Incl O&P
0010	**RETAINING WALLS, STONE**									
0015	Including excavation, concrete footing and									
0020	stone 3' below grade. Price is exposed face area.									
0200	Decorative random stone, to 6' high, 1'-6" thick, dry set	D-1	35	.457	S.F.	69	20.50		89.50	107
0300	Mortar set		40	.400		71	17.90		88.90	106
0500	Cut stone, to 6' high, 1'-6" thick, dry set		35	.457		71.50	20.50		92	110
0600	Mortar set	↓	40	.400	↓	72	17.90		89.90	107

For customer support on your Building Construction Costs with RSMeans data, call 800.448.8182.

32 32 Retaining Walls

32 32 53 – Stone Retaining Walls

32 32 53.10 Retaining Walls, Stone

	32 32 53.10 Retaining Walls, Stone	Crew	Daily Output	Labor-Hours	Unit	Material	2018 Bare Costs Labor	Equipment	Total	Total Incl O&P
0800	Random stone, 6' to 10' high, 2' thick, dry set	D-1	45	.356	S.F.	77.50	15.95		93.45	110
0900	Mortar set		50	.320		81	14.35		95.35	111
1100	Cut stone, 6' to 10' high, 2' thick, dry set		45	.356		78.50	15.95		94.45	111
1200	Mortar set	▼	50	.320	▼	81.50	14.35		95.85	112

32 33 Site Furnishings

32 33 33 – Site Manufactured Planters

32 33 33.10 Planters

	32 33 33.10 Planters	Crew	Daily Output	Labor-Hours	Unit	Material	2018 Bare Costs Labor	Equipment	Total	Total Incl O&P
0010	**PLANTERS**									
0012	Concrete, sandblasted, precast, 48" diameter, 24" high	2 Clab	15	1.067	Ea.	660	42.50		702.50	790
0100	Fluted, precast, 7' diameter, 36" high		10	1.600		1,575	64		1,639	1,850
0300	Fiberglass, circular, 36" diameter, 24" high		15	1.067		690	42.50		732.50	825
0320	36" diameter, 27" high		12	1.333		720	53		773	870
0330	33" high		15	1.067		725	42.50		767.50	865
0335	24" diameter, 36" high		15	1.067		425	42.50		467.50	535
0340	60" diameter, 39" high		8	2		1,300	79.50		1,379.50	1,575
0400	60" diameter, 24" high		10	1.600		1,150	64		1,214	1,350
0600	Square, 24" side, 36" high		15	1.067		765	42.50		807.50	910
0610	24" side, 27" high		12	1.333		835	53		888	1,000
0620	24" side, 16" high		20	.800		400	32		432	490
0700	48" side, 36" high		15	1.067		1,250	42.50		1,292.50	1,450
0900	Planter/bench, 72" square, 36" high		5	3.200		2,500	128		2,628	2,950
1000	96" square, 27" high		5	3.200		3,300	128		3,428	3,825
1200	Wood, square, 48" side, 24" high		15	1.067		1,625	42.50		1,667.50	1,875
1300	Circular, 48" diameter, 30" high		10	1.600		1,100	64		1,164	1,325
1500	72" diameter, 30" high		10	1.600		2,125	64		2,189	2,425
1600	Planter/bench, 72"	▼	5	3.200	▼	3,675	128		3,803	4,250

32 33 43 – Site Seating and Tables

32 33 43.13 Site Seating

	32 33 43.13 Site Seating	Crew	Daily Output	Labor-Hours	Unit	Material	2018 Bare Costs Labor	Equipment	Total	Total Incl O&P
0010	**SITE SEATING**									
0012	Seating, benches, park, precast conc., w/backs, wood rails, 4' long	2 Clab	5	3.200	Ea.	635	128		763	895
0100	8' long		4	4		1,075	159		1,234	1,425
0300	Fiberglass, without back, one piece, 4' long		10	1.600		1,775	64		1,839	2,050
0400	8' long		7	2.286		2,150	91		2,241	2,500
0500	Steel barstock pedestals w/backs, 2" x 3" wood rails, 4' long		10	1.600		1,350	64		1,414	1,600
0510	8' long		7	2.286		1,725	91		1,816	2,050
0515	Powder coated steel, 4" x 4" plastic slats, 6' long		8	2		545	79.50		624.50	720
0520	3" x 8" wood plank, 4' long		10	1.600		1,500	64		1,564	1,750
0530	8' long		7	2.286		1,500	91		1,591	1,800
0540	Backless, 4" x 4" wood plank, 4' square		10	1.600		1,250	64		1,314	1,475
0550	8' long		7	2.286		1,100	91		1,191	1,375
0560	Powder coated steel, with back and 2 anti-vagrant dividers, 6' long		8	2		1,175	79.50		1,254.50	1,425
0600	Aluminum pedestals, with backs, aluminum slats, 8' long		8	2		470	79.50		549.50	635
0610	15' long		5	3.200		1,000	128		1,128	1,325
0620	Portable, aluminum slats, 8' long		8	2		455	79.50		534.50	620
0630	15' long		5	3.200		460	128		588	700
0800	Cast iron pedestals, back & arms, wood slats, 4' long		8	2		435	79.50		514.50	600
0820	8' long		5	3.200		1,100	128		1,228	1,400
0840	Backless, wood slats, 4' long		8	2		585	79.50		664.50	765
0860	8' long	▼	5	3.200		1,075	128		1,203	1,400

For customer support on your Building Construction Costs with RSMeans data, call 800.448.8182.

32 33 Site Furnishings

32 33 43 – Site Seating and Tables

32 33 43.13 Site Seating	Crew	Daily Output	Labor-Hours	Unit	Material	2018 Bare Costs Labor	Equipment	Total	Total Incl O&P
1700	Steel frame, fir seat, 10' long	2 Clab	10	1.600	Ea.	395	64		459	530
2000	Benches, park, with back, galv. stl. frame, 4" x 4" plastic slats, 6' long	↓	7	2.286	↓	545	91		636	740

32 34 Fabricated Bridges

32 34 20 – Fabricated Pedestrian Bridges

32 34 20.10 Bridges, Pedestrian

		Crew	Daily Output	Labor-Hours	Unit	Material	2018 Bare Costs Labor	Equipment	Total	Total Incl O&P
0010	**BRIDGES, PEDESTRIAN**									
0011	Spans over streams, roadways, etc.									
0020	including erection, not including foundations									
0050	Precast concrete, complete in place, 8' wide, 60' span	E-2	215	.260	S.F.	146	14.15	7.85	168	192
0100	100' span		185	.303		160	16.45	9.10	185.55	213
0150	120' span		160	.350		174	19	10.55	203.55	233
0200	150' span		145	.386		181	21	11.65	213.65	245
0300	Steel, trussed or arch spans, compl. in place, 8' wide, 40' span		320	.175		121	9.50	5.25	135.75	154
0400	50' span		395	.142		108	7.70	4.27	119.97	136
0500	60' span		465	.120		108	6.55	3.63	118.18	133
0600	80' span		570	.098		129	5.35	2.96	137.31	154
0700	100' span		465	.120		181	6.55	3.63	191.18	213
0800	120' span		365	.153		229	8.35	4.62	241.97	270
0900	150' span		310	.181		244	9.80	5.45	259.25	290
1000	160' span		255	.220		244	11.90	6.60	262.50	294
1100	10' wide, 80' span		640	.088		129	4.75	2.64	136.39	153
1200	120' span		415	.135		167	7.30	4.07	178.37	200
1300	150' span		445	.126		187	6.85	3.79	197.64	221
1400	200' span	↓	205	.273		200	14.85	8.25	223.10	253
1600	Wood, laminated type, complete in place, 80' span	C-12	203	.236		87	11.85	2.44	101.29	116
1700	130' span	"	153	.314	↓	90.50	15.75	3.24	109.49	127

32 35 Screening Devices

32 35 16 – Sound Barriers

32 35 16.10 Traffic Barriers, Highway Sound Barriers

		Crew	Daily Output	Labor-Hours	Unit	Material	2018 Bare Costs Labor	Equipment	Total	Total Incl O&P
0010	**TRAFFIC BARRIERS, HIGHWAY SOUND BARRIERS**									
0020	Highway sound barriers, not including footing									
0100	Precast concrete, concrete columns @ 30' OC, 8" T, 8' H	C-12	400	.120	L.F.	169	6	1.24	176.24	197
0110	12' H		265	.181		254	9.10	1.87	264.97	296
0120	16' H		200	.240		340	12.05	2.48	354.53	395
0130	20' H	↓	160	.300		425	15.05	3.10	443.15	490
0400	Lt. wt. composite panel, cementitious face, st. posts @ 12' OC, 8' H	B-80B	190	.168		176	7.20	1.24	184.44	205
0410	12' H		125	.256		264	10.95	1.89	276.84	310
0420	16' H		95	.337		350	14.40	2.48	366.88	410
0430	20' H	↓	75	.427	↓	440	18.20	3.14	461.34	515

For customer support on your Building Construction Costs with RSMeans data, call 800.448.8182.

32 84 Planting Irrigation

32 84 23 – Underground Sprinklers

32 84 23.10 Sprinkler Irrigation System	Crew	Daily Output	Labor-Hours	Unit	Material	2018 Bare Costs Labor	Equipment	Total	Total Incl O&P
0010 SPRINKLER IRRIGATION SYSTEM									
0011 For lawns									
0800 Residential system, custom, 1" supply	B-20	2000	.012	S.F.	.25	.54		.79	1.10
0900 1-1/2" supply	"	1800	.013	"	.48	.60		1.08	1.44
1020 Pop up spray head w/risers, hi-pop, full circle pattern, 4"	2 Skwk	76	.211	Ea.	5.25	11		16.25	22.50
1030 1/2 circle pattern, 4"		76	.211		6.95	11		17.95	24.50
1040 6", full circle pattern		76	.211		11.70	11		22.70	29.50
1050 1/2 circle pattern, 6"		76	.211		11.70	11		22.70	29.50
1060 12", full circle pattern		76	.211		13.50	11		24.50	31.50
1070 1/2 circle pattern, 12"		76	.211		16.35	11		27.35	35
1080 Pop up bubbler head w/risers, hi-pop bubbler head, 4"		76	.211		5	11		16	22.50
1110 Impact full/part circle sprinklers, 28'-54' @ 25-60 psi		37	.432		19.30	22.50		41.80	55.50
1120 Spaced 37'-49' @ 25-50 psi		37	.432		24	22.50		46.50	61
1130 Spaced 43'-61' @ 30-60 psi		37	.432		67	22.50		89.50	109
1140 Spaced 54'-78' @ 40-80 psi		37	.432		116	22.50		138.50	163
1145 Impact rotor pop-up full/part commercial circle sprinklers									
1150 Spaced 42'-65' @ 35-80 psi	2 Skwk	25	.640	Ea.	16.75	33.50		50.25	70
1160 Spaced 48'-76' @ 45-85 psi	"	25	.640	"	20.50	33.50		54	74
1165 Impact rotor pop-up part. circle comm., 53'-75', 55-100 psi, w/accessories									
1180 Sprinkler, premium, pop-up rotator, 50'-100'	2 Skwk	25	.640	Ea.	96.50	33.50		130	158
1250 Plastic case, 2 nozzle, metal cover		25	.640		100	33.50		133.50	162
1260 Rubber cover		25	.640		103	33.50		136.50	165
1270 Iron case, 2 nozzle, metal cover		22	.727		146	38		184	219
1280 Rubber cover		22	.727		146	38		184	219
1282 Impact rotor pop-up full circle commercial, 39'-99', 30-100 psi									
1284 Plastic case, metal cover	2 Skwk	25	.640	Ea.	78.50	33.50		112	138
1286 Rubber cover		25	.640		103	33.50		136.50	166
1288 Iron case, metal cover		22	.727		133	38		171	206
1290 Rubber cover		22	.727		175	38		213	252
1292 Plastic case, 2 nozzle, metal cover		22	.727		101	38		139	170
1294 Rubber cover		22	.727		101	38		139	170
1296 Iron case, 2 nozzle, metal cover		20	.800		134	42		176	211
1298 Rubber cover		20	.800		139	42		181	217
1305 Electric remote control valve, plastic, 3/4"		18	.889		18.35	46.50		64.85	91
1310 1"		18	.889		25.50	46.50		72	99.50
1320 1-1/2"		18	.889		99.50	46.50		146	180
1330 2"		18	.889		119	46.50		165.50	202
1335 Quick coupling valves, brass, locking cover									
1340 Inlet coupling valve, 3/4"	2 Skwk	18.75	.853	Ea.	25	44.50		69.50	96
1350 1"		18.75	.853		33	44.50		77.50	105
1360 Controller valve boxes, 6" round boxes		18.75	.853		7.80	44.50		52.30	77
1370 10" round boxes		14.25	1.123		18.45	59		77.45	111
1380 12" square box		9.75	1.641		18.10	86		104.10	152
1388 Electromech. control, 14 day 3-60 min., auto start to 23/day									
1390 4 station	2 Skwk	1.04	15.385	Ea.	82	805		887	1,325
1400 7 station		.64	25		153	1,300		1,453	2,175
1410 12 station		.40	40		186	2,100		2,286	3,400
1420 Dual programs, 18 station		.24	66.667		219	3,500		3,719	5,600
1430 23 station		.16	100		241	5,225		5,466	8,300
1435 Backflow preventer, bronze, 0-175 psi, w/valves, test cocks									
1440 3/4"	2 Skwk	6	2.667	Ea.	89.50	140		229.50	315
1450 1"		6	2.667		103	140		243	325
1460 1-1/2"		6	2.667		273	140		413	515

32 84 Planting Irrigation

32 84 23 – Underground Sprinklers

32 84 23.10 Sprinkler Irrigation System	Crew	Daily Output	Labor-Hours	Unit	Material	2018 Bare Costs Labor	2018 Bare Costs Equipment	Total	Total Incl O&P
1470	2"	2 Skwk	6	2.667	Ea.	385	140		525	635
1475	Pressure vacuum breaker, brass, 15-150 psi									
1480	3/4"	2 Skwk	6	2.667	Ea.	27.50	140		167.50	244
1490	1"		6	2.667		75.50	140		215.50	297
1500	1-1/2"		6	2.667		86	140		226	310
1510	2"		6	2.667		147	140		287	375

32 91 Planting Preparation

32 91 13 – Soil Preparation

32 91 13.16 Mulching

		Crew	Daily Output	Labor-Hours	Unit	Material	2018 Bare Costs Labor	2018 Bare Costs Equipment	Total	Total Incl O&P
0010	**MULCHING**									
0100	Aged barks, 3" deep, hand spread	1 Clab	100	.080	S.Y.	3.80	3.19		6.99	9.05
0150	Skid steer loader	B-63	13.50	2.963	M.S.F.	420	125	12.90	557.90	670
0200	Hay, 1" deep, hand spread	1 Clab	475	.017	S.Y.	.45	.67		1.12	1.52
0250	Power mulcher, small	B-64	180	.089	M.S.F.	50	3.75	1.82	55.57	62.50
0350	Large	B-65	530	.030	"	50	1.27	.86	52.13	58
0400	Humus peat, 1" deep, hand spread	1 Clab	700	.011	S.Y.	2.96	.46		3.42	3.95
0450	Push spreader	"	2500	.003	"	2.96	.13		3.09	3.45
0550	Tractor spreader	B-66	700	.011	M.S.F.	330	.59	.35	330.94	360
0600	Oat straw, 1" deep, hand spread	1 Clab	475	.017	S.Y.	.56	.67		1.23	1.64
0650	Power mulcher, small	B-64	180	.089	M.S.F.	62	3.75	1.82	67.57	76
0700	Large	B-65	530	.030	"	62	1.27	.86	64.13	71.50
0750	Add for asphaltic emulsion	B-45	1770	.009	Gal.	5.80	.45	.45	6.70	7.60
0800	Peat moss, 1" deep, hand spread	1 Clab	900	.009	S.Y.	4.90	.35		5.25	5.95
0850	Push spreader	"	2500	.003	"	4.90	.13		5.03	5.60
0950	Tractor spreader	B-66	700	.011	M.S.F.	545	.59	.35	545.94	600
1000	Polyethylene film, 6 mil	2 Clab	2000	.008	S.Y.	.53	.32		.85	1.07
1100	Redwood nuggets, 3" deep, hand spread	1 Clab	150	.053	"	3.12	2.13		5.25	6.65
1150	Skid steer loader	B-63	13.50	2.963	M.S.F.	345	125	12.90	482.90	585
1200	Stone mulch, hand spread, ceramic chips, economy	1 Clab	125	.064	S.Y.	7	2.55		9.55	11.60
1250	Deluxe	"	95	.084	"	10.60	3.36		13.96	16.75
1300	Granite chips	B-1	10	2.400	C.Y.	72.50	97.50		170	228
1400	Marble chips		10	2.400		215	97.50		312.50	385
1600	Pea gravel		28	.857		111	34.50		145.50	175
1700	Quartz		10	2.400		190	97.50		287.50	355
1800	Tar paper, 15 lb. felt	1 Clab	800	.010	S.Y.	.48	.40		.88	1.14
1900	Wood chips, 2" deep, hand spread	"	220	.036	"	1.58	1.45		3.03	3.95
1950	Skid steer loader	B-63	20.30	1.970	M.S.F.	176	83	8.55	267.55	330

32 91 13.26 Planting Beds

		Crew	Daily Output	Labor-Hours	Unit	Material	2018 Bare Costs Labor	2018 Bare Costs Equipment	Total	Total Incl O&P
0010	**PLANTING BEDS**									
0100	Backfill planting pit, by hand, on site topsoil	2 Clab	18	.889	C.Y.		35.50		35.50	54
0200	Prepared planting mix, by hand	"	24	.667			26.50		26.50	40.50
0300	Skid steer loader, on site topsoil	B-62	340	.071			3.08	.51	3.59	5.25
0400	Prepared planting mix	"	410	.059			2.56	.42	2.98	4.35
1000	Excavate planting pit, by hand, sandy soil	2 Clab	16	1			40		40	60.50
1100	Heavy soil or clay	"	8	2			79.50		79.50	121
1200	1/2 C.Y. backhoe, sandy soil	B-11C	150	.107			4.99	2.08	7.07	9.85
1300	Heavy soil or clay	"	115	.139			6.50	2.72	9.22	12.85
2000	Mix planting soil, incl. loam, manure, peat, by hand	2 Clab	60	.267		44	10.65		54.65	64.50
2100	Skid steer loader	B-62	150	.160		44	7	1.16	52.16	60.50
3000	Pile sod, skid steer loader	"	2800	.009	S.Y.		.37	.06	.43	.64

For customer support on your Building Construction Costs with RSMeans data, call 800.448.8182.

32 91 Planting Preparation

32 91 13 - Soil Preparation

32 91 13.26 Planting Beds

		Crew	Daily Output	Labor-Hours	Unit	Material	2018 Bare Costs Labor	Equipment	Total	Total Incl O&P
3100	By hand	2 Clab	400	.040	S.Y.		1.59		1.59	2.43
4000	Remove sod, F.E. loader	B-10S	2000	.006			.29	.17	.46	.64
4100	Sod cutter	B-12K	3200	.005			.24	.39	.63	.79
4200	By hand	2 Clab	240	.067	↓		2.66		2.66	4.05

32 91 19 - Landscape Grading

32 91 19.13 Topsoil Placement and Grading

		Crew	Daily Output	Labor-Hours	Unit	Material	2018 Bare Costs Labor	Equipment	Total	Total Incl O&P
0010	**TOPSOIL PLACEMENT AND GRADING**									
0400	Spread from pile to rough finish grade, F.E. loader, 1-1/2 C.Y.	B-10S	200	.060	C.Y.		2.95	1.71	4.66	6.35
0500	Up to 200' radius, by hand	1 Clab	14	.571			23		23	34.50
0600	Top dress by hand, 1 C.Y. for 600 S.F.	"	11.50	.696	↓	28	27.50		55.50	72.50
0700	Furnish and place, truck dumped, screened, 4" deep	B-10S	1300	.009	S.Y.	3.47	.45	.26	4.18	4.80
0800	6" deep	"	820	.015	"	4.44	.72	.42	5.58	6.45

32 92 Turf and Grasses

32 92 19 - Seeding

32 92 19.13 Mechanical Seeding

		Crew	Daily Output	Labor-Hours	Unit	Material	2018 Bare Costs Labor	Equipment	Total	Total Incl O&P
0010	**MECHANICAL SEEDING**									
0020	Mechanical seeding, 215 lb./acre	B-66	1.50	5.333	Acre	560	274	165	999	1,200
0100	44 lb./M.S.Y.	"	2500	.003	S.Y.	.20	.16	.10	.46	.58
0101	44 lb./M.S.Y.	1 Clab	13950	.001	S.F.	.02	.02		.04	.05
0300	Fine grading and seeding incl. lime, fertilizer & seed,									
0310	with equipment	B-14	1000	.048	S.Y.	.45	2.02	.31	2.78	3.91
0400	Fertilizer hand push spreader, 35 lb./M.S.F.	1 Clab	200	.040	M.S.F.	9.85	1.59		11.44	13.30
0600	Limestone hand push spreader, 50 lb./M.S.F.		180	.044		4.65	1.77		6.42	7.80
0800	Grass seed hand push spreader, 4.5 lb./M.S.F.	↓	180	.044	↓	22	1.77		23.77	27
1000	Hydro or air seeding for large areas, incl. seed and fertilizer	B-81	8900	.003	S.Y.	.50	.13	.06	.69	.81
1100	With wood fiber mulch added	"	8900	.003	"	2	.13	.06	2.19	2.46
1300	Seed only, over 100 lb., field seed, minimum				Lb.	1.85			1.85	2.04
1400	Maximum					1.75			1.75	1.93
1500	Lawn seed, minimum					1.45			1.45	1.60
1600	Maximum				↓	2.50			2.50	2.75
1800	Aerial operations, seeding only, field seed	B-58	50	.480	Acre	600	21	63.50	684.50	760
1900	Lawn seed		50	.480		470	21	63.50	554.50	620
2100	Seed and liquid fertilizer, field seed		50	.480		675	21	63.50	759.50	845
2200	Lawn seed	↓	50	.480	↓	545	21	63.50	629.50	700

32 92 23 - Sodding

32 92 23.10 Sodding Systems

		Crew	Daily Output	Labor-Hours	Unit	Material	2018 Bare Costs Labor	Equipment	Total	Total Incl O&P
0010	**SODDING SYSTEMS**									
0020	Sodding, 1" deep, bluegrass sod, on level ground, over 8 M.S.F.	B-63	22	1.818	M.S.F.	246	76.50	7.90	330.40	395
0200	4 M.S.F.		17	2.353		240	99	10.25	349.25	425
0300	1,000 S.F.		13.50	2.963		290	125	12.90	427.90	525
0500	Sloped ground, over 8 M.S.F.		6	6.667		246	281	29	556	730
0600	4 M.S.F.		5	8		240	335	35	610	815
0700	1,000 S.F.		4	10		290	420	43.50	753.50	1,000
1000	Bent grass sod, on level ground, over 6 M.S.F.		20	2		257	84.50	8.70	350.20	420
1100	3 M.S.F.		18	2.222		271	93.50	9.65	374.15	450
1200	Sodding 1,000 S.F. or less		14	2.857		297	120	12.45	429.45	520
1500	Sloped ground, over 6 M.S.F.		15	2.667		257	112	11.60	380.60	465
1600	3 M.S.F.		13.50	2.963		271	125	12.90	408.90	500
1700	1,000 S.F.	↓	12	3.333	↓	297	140	14.50	451.50	555

For customer support on your Building Construction Costs with RSMeans data, call 800.448.8182.

32 93 Plants

32 93 10 – General Planting Costs

32 93 10.12 Travel

	Crew	Daily Output	Labor-Hours	Unit	Material	2018 Bare Costs Labor	Equipment	Total	Total Incl O&P
0010 **TRAVEL** add to all nursery items									
0015 10 to 20 miles one way, add				All				5%	5%
0100 30 to 50 miles one way, add				"				10%	10%

32 93 13 – Ground Covers

32 93 13.10 Ground Cover Plants

	Crew	Daily Output	Labor-Hours	Unit	Material	2018 Bare Costs Labor	Equipment	Total	Total Incl O&P
0010 **GROUND COVER PLANTS**									
0012 Plants, pachysandra, in prepared beds	B-1	15	1.600	C	79.50	65		144.50	187
0200 Vinca minor, 1 yr., bare root, in prepared beds		12	2	"	70	81		151	200
0600 Stone chips, in 50 lb. bags, Georgia marble		520	.046	Bag	4.80	1.87		6.67	8.15
0700 Onyx gemstone		260	.092		16	3.74		19.74	23.50
0800 Quartz		260	.092		16	3.74		19.74	23.50
0900 Pea gravel, truckload lots		28	.857	Ton	30	34.50		64.50	86.50

32 93 33 – Shrubs

32 93 33.10 Shrubs and Trees

	Crew	Daily Output	Labor-Hours	Unit	Material	2018 Bare Costs Labor	Equipment	Total	Total Incl O&P
0010 **SHRUBS AND TREES**									
0011 Evergreen, in prepared beds, B&B									
0100 Arborvitae pyramidal, 4'-5'	B-17	30	1.067	Ea.	103	47	21.50	171.50	209
0150 Globe, 12"-15"	B-1	96	.250		26	10.15		36.15	44
0300 Cedar, blue, 8'-10'	B-17	18	1.778		239	78.50	36	353.50	420
0500 Hemlock, Canadian, 2-1/2'-3'	B-1	36	.667		32.50	27		59.50	76.50
0550 Holly, Savannah, 8'-10' H		9.68	2.479		300	100		400	485
0600 Juniper, andorra, 18"-24"		80	.300		50	12.15		62.15	73.50
0620 Wiltoni, 15"-18"		80	.300		26.50	12.15		38.65	48
0640 Skyrocket, 4-1/2'-5'	B-17	55	.582		110	26	11.85	147.85	173
0660 Blue pfitzer, 2'-2-1/2'	B-1	44	.545		39.50	22		61.50	77
0680 Ketleerie, 2-1/2'-3'		50	.480		56	19.45		75.45	91
0700 Pine, black, 2-1/2'-3'		50	.480		62	19.45		81.45	98
0720 Mugo, 18"-24"		60	.400		56	16.20		72.20	86.50
0740 White, 4'-5'	B-17	75	.427		53	18.90	8.70	80.60	96.50
0800 Spruce, blue, 18"-24"	B-1	60	.400		69	16.20		85.20	101
0840 Norway, 4'-5'	B-17	75	.427		86	18.90	8.70	113.60	133
0900 Yew, denisforma, 12"-15"	B-1	60	.400		36.50	16.20		52.70	65
1000 Capitata, 18"-24"		30	.800		34	32.50		66.50	87
1100 Hicksi, 2'-2-1/2'		30	.800		103	32.50		135.50	164

32 93 33.20 Shrubs

	Crew	Daily Output	Labor-Hours	Unit	Material	2018 Bare Costs Labor	Equipment	Total	Total Incl O&P
0010 **SHRUBS**									
0011 Broadleaf Evergreen, planted in prepared beds									
0100 Andromeda, 15"-18", cont	B-1	96	.250	Ea.	33.50	10.15		43.65	52.50
0200 Azalea, 15"-18", cont		96	.250		30.50	10.15		40.65	49.50
0300 Barberry, 9"-12", cont		130	.185		18.70	7.50		26.20	32
0400 Boxwood, 15"-18", B&B		96	.250		44.50	10.15		54.65	64.50
0500 Euonymus, emerald gaiety, 12"-15", cont		115	.209		24.50	8.45		32.95	40
0600 Holly, 15"-18", B&B		96	.250		40	10.15		50.15	59.50
0900 Mount laurel, 18"-24", B&B		80	.300		72.50	12.15		84.65	98.50
1000 Paxistema, 9"-12" H		130	.185		22	7.50		29.50	35.50
1100 Rhododendron, 18"-24", cont		48	.500		38.50	20.50		59	73.50
1200 Rosemary, 1 gal. cont		600	.040		18.55	1.62		20.17	23
2000 Deciduous, planted in prepared beds, amelanchier, 2'-3', B&B		57	.421		123	17.05		140.05	161
2100 Azalea, 15"-18", B&B		96	.250		30.50	10.15		40.65	49
2300 Bayberry, 2'-3', B&B		57	.421		30	17.05		47.05	59
2600 Cotoneaster, 15"-18", B&B		80	.300		27	12.15		39.15	48.50

For customer support on your Building Construction Costs with RSMeans data, call 800.448.8182.

32 93 Plants

32 93 33 – Shrubs

32 93 33.20 Shrubs

	32 93 33.20 Shrubs	Crew	Daily Output	Labor-Hours	Unit	Material	2018 Bare Costs Labor	Equipment	Total	Total Incl O&P
2800	Dogwood, 3'-4', B&B	B-17	40	.800	Ea.	33	35.50	16.25	84.75	108
2900	Euonymus, alatus compacta, 15"-18", cont	B-1	80	.300		27	12.15		39.15	48
3200	Forsythia, 2'-3', cont	"	60	.400		17.80	16.20		34	44
3300	Hibiscus, 3'-4', B&B	B-17	75	.427		42	18.90	8.70	69.60	84
3400	Honeysuckle, 3'-4', B&B	B-1	60	.400		27.50	16.20		43.70	55
3500	Hydrangea, 2'-3', B&B	"	57	.421		31	17.05		48.05	60
3600	Lilac, 3'-4', B&B	B-17	40	.800		27.50	35.50	16.25	79.25	101
3900	Privet, bare root, 18"-24"	B-1	80	.300		15.15	12.15		27.30	35
4100	Quince, 2'-3', B&B	"	57	.421		29	17.05		46.05	58
4200	Russian olive, 3'-4', B&B	B-17	75	.427		25	18.90	8.70	52.60	65.50
4400	Spirea, 3'-4', B&B	B-1	70	.343		20.50	13.90		34.40	43.50
4500	Viburnum, 3'-4', B&B	B-17	40	.800		26	35.50	16.25	77.75	100

32 93 43 – Trees

32 93 43.20 Trees

	32 93 43.20 Trees		Crew	Daily Output	Labor-Hours	Unit	Material	2018 Bare Costs Labor	Equipment	Total	Total Incl O&P
0010	**TREES**										
0011	Deciduous, in prep. beds, balled & burlapped (B&B)										
0100	Ash, 2" caliper	G	B-17	8	4	Ea.	200	177	81.50	458.50	580
0200	Beech, 5'-6'	G		50	.640		210	28.50	13	251.50	288
0300	Birch, 6'-8', 3 stems	G		20	1.600		167	71	32.50	270.50	325
0500	Crabapple, 6'-8'	G		20	1.600		139	71	32.50	242.50	296
0600	Dogwood, 4'-5'	G		40	.800		140	35.50	16.25	191.75	225
0700	Eastern redbud, 4'-5'	G		40	.800		148	35.50	16.25	199.75	234
0800	Elm, 8'-10'	G		20	1.600		325	71	32.50	428.50	505
0900	Ginkgo, 6'-7'	G		24	1.333		148	59	27	234	283
1000	Hawthorn, 8'-10', 1" caliper	G		20	1.600		161	71	32.50	264.50	320
1100	Honeylocust, 10'-12', 1-1/2" caliper	G		10	3.200		206	142	65	413	510
1300	Larch, 8'	G		32	1		128	44.50	20.50	193	230
1400	Linden, 8'-10', 1" caliper	G		20	1.600		144	71	32.50	247.50	300
1500	Magnolia, 4'-5'	G		20	1.600		115	71	32.50	218.50	270
1600	Maple, red, 8'-10', 1-1/2" caliper	G		10	3.200		199	142	65	406	505
1700	Mountain ash, 8'-10', 1" caliper	G		16	2		177	88.50	40.50	306	375
1800	Oak, 2-1/2"-3" caliper	G		6	5.333		320	236	108	664	830
2100	Planetree, 9'-11', 1-1/4" caliper	G		10	3.200		266	142	65	473	580
2200	Plum, 6'-8', 1" caliper	G		20	1.600		79	71	32.50	182.50	230
2300	Poplar, 9'-11', 1-1/4" caliper	G		10	3.200		100	142	65	307	395
2500	Sumac, 2'-3'	G		75	.427		44.50	18.90	8.70	72.10	87
2700	Tulip, 5'-6'	G		40	.800		46	35.50	16.25	97.75	122
2800	Willow, 6'-8', 1" caliper	G		20	1.600		97	71	32.50	200.50	250

32 94 Planting Accessories

32 94 13 – Landscape Edging

32 94 13.20 Edging

	32 94 13.20 Edging	Crew	Daily Output	Labor-Hours	Unit	Material	2018 Bare Costs Labor	Equipment	Total	Total Incl O&P
0010	**EDGING**									
0050	Aluminum alloy, including stakes, 1/8" x 4", mill finish	B-1	390	.062	L.F.	2.21	2.49		4.70	6.25
0051	Black paint		390	.062		2.56	2.49		5.05	6.60
0052	Black anodized		390	.062		2.96	2.49		5.45	7.05
0100	Brick, set horizontally, 1-1/2 bricks per L.F.	D-1	370	.043		1.39	1.94		3.33	4.50
0150	Set vertically, 3 bricks per L.F.	"	135	.119		3.54	5.30		8.84	12.05
0200	Corrugated aluminum, roll, 4" wide	1 Carp	650	.012		2.19	.62		2.81	3.36
0250	6" wide	"	550	.015		2.74	.74		3.48	4.13

For customer support on your Building Construction Costs with RSMeans data, call 800.448.8182.

673

32 94 Planting Accessories

32 94 13 – Landscape Edging

32 94 13.20 Edging		Crew	Daily Output	Labor-Hours	Unit	Material	2018 Bare Costs Labor	Equipment	Total	Total Incl O&P
0600	Railroad ties, 6" x 8"	2 Carp	170	.094	L.F.	2.36	4.77		7.13	9.85
0650	7" x 9"		136	.118		2.62	5.95		8.57	12
0750	Redwood 2" x 4"		330	.048		2.26	2.46		4.72	6.25
0800	Steel edge strips, incl. stakes, 1/4" x 5"	B-1	390	.062		4.54	2.49		7.03	8.80
0850	3/16" x 4"	"	390	.062		3.59	2.49		6.08	7.75

32 94 50 – Tree Guying

32 94 50.10 Tree Guying Systems

		Crew	Daily Output	Labor-Hours	Unit	Material	Labor	Equipment	Total	Total Incl O&P
0010	**TREE GUYING SYSTEMS**									
0015	Tree guying including stakes, guy wire and wrap									
0100	Less than 3" caliper, 2 stakes	2 Clab	35	.457	Ea.	13.10	18.20		31.30	42.50
0200	3" to 4" caliper, 3 stakes	"	21	.762	"	19.85	30.50		50.35	68.50
1000	Including arrowhead anchor, cable, turnbuckles and wrap									
1100	Less than 3" caliper, 3" anchors	2 Clab	20	.800	Ea.	21	32		53	72
1200	3" to 6" caliper, 4" anchors		15	1.067		31.50	42.50		74	100
1300	6" caliper, 6" anchors		12	1.333		23.50	53		76.50	107
1400	8" caliper, 8" anchors		9	1.778		116	71		187	236

32 96 Transplanting

32 96 23 – Plant and Bulb Transplanting

32 96 23.23 Planting

		Crew	Daily Output	Labor-Hours	Unit	Material	Labor	Equipment	Total	Total Incl O&P
0010	**PLANTING**									
0012	Moving shrubs on site, 12" ball	B-62	28	.857	Ea.		37.50	6.20	43.70	64
0100	24" ball	"	22	1.091	"		47.50	7.90	55.40	81

32 96 23.43 Moving Trees

		Crew	Daily Output	Labor-Hours	Unit	Material	Labor	Equipment	Total	Total Incl O&P
0010	**MOVING TREES**, On site									
0300	Moving trees on site, 36" ball	B-6	3.75	6.400	Ea.		279	83.50	362.50	515
0400	60" ball	"	1	24	"		1,050	310	1,360	1,950

For customer support on your Building Construction Costs with RSMeans data, call 800.448.8182.

Estimating Tips
33 10 00 Water Utilities
33 30 00 Sanitary Sewerage Utilities
33 40 00 Storm Drainage Utilities

- Never assume that the water, sewer, and drainage lines will go in at the early stages of the project. Consider the site access needs before dividing the site in half with open trenches, loose pipe, and machinery obstructions. Always inspect the site to establish that the site drawings are complete. Check off all existing utilities on your drawings as you locate them. Be especially careful with underground utilities because appurtenances are sometimes buried during regrading or repaving operations. If you find any discrepancies, mark up the site plan for further research. Differing site conditions can be very costly if discovered later in the project.

- See also Section 33 01 00 for restoration of pipe where removal/replacement may be undesirable. Use of new types of piping materials can reduce the overall project cost. Owners/design engineers should consider the installing contractor as a valuable source of current information on utility products and local conditions that could lead to significant cost savings.

Reference Numbers
Reference numbers are shown at the beginning of some major classifications. These numbers refer to related items in the Reference Section. The reference information may be an estimating procedure, an alternate pricing method, or technical information.

Note: Not all subdivisions listed here necessarily appear. ■

No part of this cost data may be reproduced, stored in a retrieval system, or transmitted in any form or by any means without prior written permission of Gordian.

Note: Trade Service, in part, has been used as a reference source for some of the material prices used in Division 33.

Did you know?

RSMeans data is available through our online application with 24/7 access:

- Search for unit prices by keyword
- Leverage the most up-to-date data
- Build and export estimates

Try it free for 30 days!
www.rsmeans.com/2018freetrial

33 01 Operation and Maintenance of Utilities

33 01 10 – Operation and Maintenance of Water Utilities

33 01 10.10 Corrosion Resistance

33 01 10.10 Corrosion Resistance	Crew	Daily Output	Labor-Hours	Unit	Material	2018 Bare Costs Labor	Equipment	Total	Total Incl O&P
0010 **CORROSION RESISTANCE**									
0012 Wrap & coat, add to pipe, 4" diameter				L.F.	2.47			2.47	2.72
0040 6" diameter					3.71			3.71	4.08
0060 8" diameter					4.94			4.94	5.45
0100 12" diameter					7.40			7.40	8.15
0200 24" diameter					14.80			14.80	16.30
0500 Coating, bituminous, per diameter inch, 1 coat, add					.16			.16	.18
0540 3 coat					.48			.48	.53
0560 Coal tar epoxy, per diameter inch, 1 coat, add					.18			.18	.20
0600 3 coat					.54			.54	.59

33 01 30 – Operation and Maintenance of Sewer Utilities

33 01 30.72 Cured-In-Place Pipe Lining

33 01 30.72 Cured-In-Place Pipe Lining	Crew	Daily Output	Labor-Hours	Unit	Material	2018 Bare Costs Labor	Equipment	Total	Total Incl O&P
0010 **CURED-IN-PLACE PIPE LINING**									
0011 With cement incl. bypass & cleaning									
0020 Less than 10,000 L.F., urban, 6" to 10"	C-17E	130	.615	L.F.	9.55	32.50	.79	42.84	61
0200 24" to 36"		90	.889		15.25	47	1.14	63.39	90
0300 48" to 72"		80	1		24.50	53	1.28	78.78	109

33 05 Common Work Results for Utilities

33 05 07 – Trenchless Installation of Utility Piping

33 05 07.23 Utility Boring and Jacking

33 05 07.23 Utility Boring and Jacking	Crew	Daily Output	Labor-Hours	Unit	Material	2018 Bare Costs Labor	Equipment	Total	Total Incl O&P
0010 **UTILITY BORING AND JACKING**									
0011 Casing only, 100' minimum,									
0020 not incl. jacking pits or dewatering									
0100 Roadwork, 1/2" thick wall, 24" diameter casing	B-42	20	3.200	L.F.	117	144	57	318	415
0200 36" diameter		16	4		215	180	71	466	590
0300 48" diameter		15	4.267		297	192	76	565	705
0500 Railroad work, 24" diameter		15	4.267		117	192	76	385	505
0600 36" diameter		14	4.571		215	206	81.50	502.50	640
0700 48" diameter		12	5.333		297	240	95	632	800
0900 For ledge, add								20%	20%

33 05 07.36 Microtunneling

33 05 07.36 Microtunneling	Crew	Daily Output	Labor-Hours	Unit	Material	2018 Bare Costs Labor	Equipment	Total	Total Incl O&P
0010 **MICROTUNNELING**									
0011 Not including excavation, backfill, shoring,									
0020 or dewatering, average 50'/day, slurry method									
0100 24" to 48" outside diameter, minimum				L.F.				965	965
0110 Adverse conditions, add				"				500	500
1000 Rent microtunneling machine, average monthly lease				Month				97,500	107,000
1010 Operating technician				Day				630	705
1100 Mobilization and demobilization, minimum				Job				41,200	45,900
1110 Maximum				"				445,500	490,500

33 05 61 – Concrete Manholes

33 05 61.10 Storm Drainage Manholes, Frames and Covers

33 05 61.10 Storm Drainage Manholes, Frames & Covers	Crew	Daily Output	Labor-Hours	Unit	Material	2018 Bare Costs Labor	Equipment	Total	Total Incl O&P
0010 **STORM DRAINAGE MANHOLES, FRAMES & COVERS**									
0020 Excludes footing, excavation, backfill (See line items for frame & cover)									
0050 Brick, 4' inside diameter, 4' deep	D-1	1	16	Ea.	590	715		1,305	1,750
0100 6' deep		.70	22.857		835	1,025		1,860	2,500
0150 8' deep		.50	32		1,075	1,425		2,500	3,375
0200 For depths over 8', add		4	4	V.L.F.	88	179		267	370
0400 Concrete blocks (radial), 4' ID, 4' deep		1.50	10.667	Ea.	410	480		890	1,175

For customer support on your Building Construction Costs with RSMeans data, call 800.448.8182.

33 05 Common Work Results for Utilities

33 05 61 – Concrete Manholes

33 05 61.10 Storm Drainage Manholes, Frames and Covers

33 05 61.10 Storm Drainage Manholes, Frames and Covers	Crew	Daily Output	Labor-Hours	Unit	Material	2018 Bare Costs Labor	Equipment	Total	Total Incl O&P
0500	6' deep	D-1	1	16	Ea.	540	715		1,255	1,700
0600	8' deep		.70	22.857		675	1,025		1,700	2,325
0700	For depths over 8', add		5.50	2.909	V.L.F.	69	130		199	276
0800	Concrete, cast in place, 4' x 4', 8" thick, 4' deep	C-14H	2	24	Ea.	540	1,175	12.70	1,727.70	2,400
0900	6' deep		1.50	32		775	1,575	16.95	2,366.95	3,275
1000	8' deep		1	48		1,125	2,375	25.50	3,525.50	4,850
1100	For depths over 8', add		8	6	V.L.F.	123	296	3.18	422.18	590
1110	Precast, 4' ID, 4' deep	B-22	4.10	7.317	Ea.	845	345	47.50	1,237.50	1,500
1120	6' deep		3	10		1,050	470	65	1,585	1,925
1130	8' deep		2	15		1,200	705	97	2,002	2,500
1140	For depths over 8', add		16	1.875	V.L.F.	128	88	12.15	228.15	288
1150	5' ID, 4' deep	B-6	3	8	Ea.	1,750	350	104	2,204	2,575
1160	6' deep		2	12		2,100	525	156	2,781	3,275
1170	8' deep		1.50	16		2,550	700	208	3,458	4,075
1180	For depths over 8', add		12	2	V.L.F.	283	87.50	26	396.50	470
1190	6' ID, 4' deep		2	12	Ea.	2,500	525	156	3,181	3,725
1200	6' deep		1.50	16		2,950	700	208	3,858	4,525
1210	8' deep		1	24		3,625	1,050	310	4,985	5,950
1220	For depths over 8', add		8	3	V.L.F.	385	131	39	555	660
1250	Slab tops, precast, 8" thick									
1300	4' diameter manhole	B-6	8	3	Ea.	275	131	39	445	540
1400	5' diameter manhole		7.50	3.200		450	140	41.50	631.50	755
1500	6' diameter manhole		7	3.429		700	150	44.50	894.50	1,050
3800	Steps, heavyweight cast iron, 7" x 9"	1 Bric	40	.200		18.05	10.05		28.10	35.50
3900	8" x 9"		40	.200		21.50	10.05		31.55	39.50
3928	12" x 10-1/2"		40	.200		27	10.05		37.05	45
4000	Standard sizes, galvanized steel		40	.200		25	10.05		35.05	43
4100	Aluminum		40	.200		29	10.05		39.05	47.50
4150	Polyethylene		40	.200		29	10.05		39.05	47

33 05 63 – Concrete Vaults and Chambers

33 05 63.13 Precast Concrete Utility Structures

33 05 63.13 Precast Concrete Utility Structures	Crew	Daily Output	Labor-Hours	Unit	Material	2018 Bare Costs Labor	Equipment	Total	Total Incl O&P
0010	**PRECAST CONCRETE UTILITY STRUCTURES**, 6" thick									
0050	5' x 10' x 6' high, ID	B-13	2	28	Ea.	1,800	1,225	295	3,320	4,150
0100	6' x 10' x 6' high, ID		2	28		1,875	1,225	295	3,395	4,250
0150	5' x 12' x 6' high, ID		2	28		1,975	1,225	295	3,495	4,350
0200	6' x 12' x 6' high, ID		1.80	31.111		2,225	1,350	330	3,905	4,875
0250	6' x 13' x 6' high, ID		1.50	37.333		2,925	1,625	395	4,945	6,100
0300	8' x 14' x 7' high, ID		1	56		3,150	2,450	590	6,190	7,850
0350	Hand hole, precast concrete, 1-1/2" thick									
0400	1'-0" x 2'-0" x 1'-9", ID, light duty	B-1	4	6	Ea.	430	243		673	840
0450	4'-6" x 3'-2" x 2'-0", OD, heavy duty	B-6	3	8	"	1,525	350	104	1,979	2,325

33 05 97 – Identification and Signage for Utilities

33 05 97.05 Utility Connection

33 05 97.05 Utility Connection	Crew	Daily Output	Labor-Hours	Unit	Material	2018 Bare Costs Labor	Equipment	Total	Total Incl O&P
0010	**UTILITY CONNECTION**									
0020	Water, sanitary, stormwater, gas, single connection	B-14	1	48	Ea.	2,925	2,025	310	5,260	6,650
0030	Telecommunication	"	3	16	"	385	675	104	1,164	1,550

33 05 97.10 Utility Accessories

33 05 97.10 Utility Accessories	Crew	Daily Output	Labor-Hours	Unit	Material	2018 Bare Costs Labor	Equipment	Total	Total Incl O&P
0010	**UTILITY ACCESSORIES**									
0400	Underground tape, detectable, reinforced, alum. foil core, 2"	1 Clab	150	.053	C.L.F.	9	2.13		11.13	13.15
0500	6"	"	140	.057	"	36	2.28		38.28	43

33 11 Groundwater Sources

33 11 13 – Potable Water Supply Wells

33 11 13.10 Wells and Accessories

	33 11 13.10 Wells and Accessories	Crew	Daily Output	Labor-Hours	Unit	Material	2018 Bare Costs Labor	Equipment	Total	Total Incl O&P
0010	**WELLS & ACCESSORIES**									
0011	Domestic									
0100	Drilled, 4" to 6" diameter	B-23	120	.333	L.F.		13.40	22.50	35.90	45
0200	8" diameter	"	95.20	.420	"		16.90	28	44.90	57
0400	Gravel pack well, 40' deep, incl. gravel & casing, complete									
0500	24" diameter casing x 18" diameter screen	B-23	.13	308	Total	38,400	12,400	20,600	71,400	84,000
0600	36" diameter casing x 18" diameter screen		.12	333	"	39,600	13,400	22,400	75,400	88,500
0800	Observation wells, 1-1/4" riser pipe		163	.245	V.L.F.	20.50	9.90	16.45	46.85	55.50
0900	For flush Buffalo roadway box, add	1 Skwk	16.60	.482	Ea.	50.50	25		75.50	94
1200	Test well, 2-1/2" diameter, up to 50' deep (15 to 50 GPM)	B-23	1.51	26.490	"	800	1,075	1,775	3,650	4,450
1300	Over 50' deep, add	"	121.80	.328	L.F.	21.50	13.20	22	56.70	67.50
1500	Pumps, installed in wells to 100' deep, 4" submersible									
1510	1/2 HP	Q-1	3.22	4.969	Ea.	700	278		978	1,200
1520	3/4 HP		2.66	6.015		845	335		1,180	1,425
1600	1 HP		2.29	6.987		910	390		1,300	1,600
1700	1-1/2 HP	Q-22	1.60	10		1,800	560	310	2,670	3,150
1800	2 HP		1.33	12.030		1,725	675	375	2,775	3,325
1900	3 HP		1.14	14.035		2,225	785	435	3,445	4,100
2000	5 HP		1.14	14.035		3,050	785	435	4,270	5,000
3000	Pump, 6" submersible, 25' to 150' deep, 25 HP, 249 to 297 GPM		.89	17.978		1,975	1,000	555	3,530	4,300
3100	25' to 500' deep, 30 HP, 100 to 300 GPM		.73	21.918		2,250	1,225	680	4,155	5,075
8000	Steel well casing	B-23A	3020	.008	Lb.	1.28	.36	.85	2.49	2.88
9950	See Section 31 23 19.40 for wellpoints									
9960	See Section 31 23 19.30 for drainage wells									

33 11 13.20 Water Supply Wells, Pumps

	33 11 13.20 Water Supply Wells, Pumps	Crew	Daily Output	Labor-Hours	Unit	Material	2018 Bare Costs Labor	Equipment	Total	Total Incl O&P
0010	**WATER SUPPLY WELLS, PUMPS**									
0011	With pressure control									
1000	Deep well, jet, 42 gal. galvanized tank									
1040	3/4 HP	1 Plum	.80	10	Ea.	1,100	620		1,720	2,150
3000	Shallow well, jet, 30 gal. galvanized tank									
3040	1/2 HP	1 Plum	2	4	Ea.	895	249		1,144	1,350

33 14 Water Utility Transmission and Distribution

33 14 13 – Public Water Utility Distribution Piping

33 14 13.15 Water Supply, Ductile Iron Pipe

	33 14 13.15 Water Supply, Ductile Iron Pipe	Crew	Daily Output	Labor-Hours	Unit	Material	2018 Bare Costs Labor	Equipment	Total	Total Incl O&P
0010	**WATER SUPPLY, DUCTILE IRON PIPE** R331113-80									
0020	Not including excavation or backfill									
2000	Pipe, class 50 water piping, 18' lengths									
2020	Mechanical joint, 4" diameter	B-21A	200	.200	L.F.	30.50	10	1.82	42.32	50.50
2040	6" diameter		160	.250		37.50	12.50	2.27	52.27	62.50
2060	8" diameter		133.33	.300		52.50	15	2.73	70.23	83.50
2080	10" diameter		114.29	.350		69	17.50	3.18	89.68	106
2100	12" diameter		105.26	.380		88.50	18.95	3.45	110.90	129
2120	14" diameter		100	.400		93.50	19.95	3.64	117.09	137
2140	16" diameter		72.73	.550		103	27.50	5	135.50	160
2160	18" diameter		68.97	.580		127	29	5.25	161.25	189
2170	20" diameter		57.14	.700		137	35	6.35	178.35	211
2180	24" diameter		47.06	.850		163	42.50	7.75	213.25	253
3000	Push-on joint, 4" diameter		400	.100		21	4.99	.91	26.90	31.50
3020	6" diameter		333.33	.120		21.50	6	1.09	28.59	34.50
3040	8" diameter		200	.200		30.50	10	1.82	42.32	51

For customer support on your Building Construction Costs with RSMeans data, call 800.448.8182.

33 14 13.15 Water Supply, Ductile Iron Pipe

		Crew	Daily Output	Labor-Hours	Unit	Material	2018 Bare Costs Labor	2018 Bare Costs Equipment	Total	Total Incl O&P
3060	10" diameter	B-21A	181.82	.220	L.F.	46.50	11	2	59.50	70
3080	12" diameter		160	.250		49	12.50	2.27	63.77	75.50
3100	14" diameter		133.33	.300		49	15	2.73	66.73	79.50
3120	16" diameter		114.29	.350		56.50	17.50	3.18	77.18	92
3140	18" diameter		100	.400		63	19.95	3.64	86.59	103
3160	20" diameter		88.89	.450		65.50	22.50	4.09	92.09	111
3180	24" diameter		76.92	.520		86	26	4.73	116.73	139
8000	Piping, fittings, mechanical joint, AWWA C110									
8006	90° bend, 4" diameter	B-20A	16	2	Ea.	170	97		267	335
8020	6" diameter		12.80	2.500		255	121		376	465
8040	8" diameter		10.67	2.999		460	145		605	725
8060	10" diameter	B-21A	11.43	3.500		670	175	32	877	1,025
8080	12" diameter		10.53	3.799		880	190	34.50	1,104.50	1,300
8100	14" diameter		10	4		1,300	200	36.50	1,536.50	1,800
8120	16" diameter		7.27	5.502		1,625	275	50	1,950	2,275
8140	18" diameter		6.90	5.797		2,325	289	52.50	2,666.50	3,050
8160	20" diameter		5.71	7.005		2,800	350	63.50	3,213.50	3,700
8180	24" diameter		4.70	8.511		4,225	425	77.50	4,727.50	5,375
8200	Wye or tee, 4" diameter	B-20A	10.67	2.999		380	145		525	635
8220	6" diameter		8.53	3.751		570	182		752	900
8240	8" diameter		7.11	4.501		900	218		1,118	1,325
8260	10" diameter	B-21A	7.62	5.249		1,175	262	47.50	1,484.50	1,750
8280	12" diameter		7.02	5.698		1,950	285	52	2,287	2,625
8300	14" diameter		6.67	5.997		2,675	299	54.50	3,028.50	3,450
8320	16" diameter		4.85	8.247		3,125	410	75	3,610	4,125
8340	18" diameter		4.60	8.696		4,225	435	79	4,739	5,375
8360	20" diameter		3.81	10.499		6,125	525	95.50	6,745.50	7,650
8380	24" diameter		3.14	12.739		8,925	635	116	9,676	10,900
8450	Decreaser, 6" x 4" diameter	B-20A	14.22	2.250		232	109		341	420
8460	8" x 6" diameter	"	11.64	2.749		350	133		483	585
8470	10" x 6" diameter	B-21A	13.33	3.001		440	150	27.50	617.50	735
8480	12" x 6" diameter		12.70	3.150		575	157	28.50	760.50	900
8490	16" x 6" diameter		10	4		980	200	36.50	1,216.50	1,425
8500	20" x 6" diameter		8.42	4.751		1,525	237	43	1,805	2,075
8550	Piping, butterfly valves, cast iron									
8560	4" diameter	B-20	6	4	Ea.	405	179		584	720
8570	6" diameter	"	5	4.800		555	214		769	940
8580	8" diameter	B-21	4	7		720	325	32.50	1,077.50	1,325
8590	10" diameter		3.50	8		1,000	370	37	1,407	1,700
8600	12" diameter		3	9.333		1,375	430	43	1,848	2,225
8610	14" diameter		2	14		2,525	650	65	3,240	3,850
8620	16" diameter		2	14		3,925	650	65	4,640	5,350

33 14 13.25 Water Supply, Polyvinyl Chloride Pipe

		Crew	Daily Output	Labor-Hours	Unit	Material	2018 Bare Costs Labor	2018 Bare Costs Equipment	Total	Total Incl O&P
0010	**WATER SUPPLY, POLYVINYL CHLORIDE PIPE**									
0020	Not including excavation or backfill, unless specified									
2100	PVC pipe, Class 150, 1-1/2" diameter	Q-1A	750	.013	L.F.	.52	.83		1.35	1.83
2120	2" diameter		686	.015		.91	.91		1.82	2.37
2140	2-1/2" diameter		500	.020		1.18	1.25		2.43	3.18
2160	3" diameter	B-20	430	.056		1.59	2.49		4.08	5.55
3010	AWWA C905, PR 100, DR 25									
3030	14" diameter	B-20A	213	.150	L.F.	13.30	7.25		20.55	25.50
3040	16" diameter		200	.160		18.75	7.75		26.50	32

For customer support on your Building Construction Costs with RSMeans data, call 800.448.8182.

679

33 14 13.25 Water Supply, Polyvinyl Chloride Pipe		Crew	Daily Output	Labor-Hours	Unit	Material	2018 Bare Costs Labor	Equipment	Total	Total Incl O&P
3050	18" diameter	B-20A	160	.200	L.F.	23	9.70		32.70	39.50
3060	20" diameter		133	.241		28.50	11.65		40.15	48.50
3070	24" diameter		107	.299		43.50	14.45		57.95	69.50
3080	30" diameter		80	.400		70.50	19.35		89.85	107
3090	36" diameter		80	.400		110	19.35		129.35	151
3100	42" diameter		60	.533		133	26		159	185
3200	48" diameter		60	.533		173	26		199	229
4520	Pressure pipe Class 150, SDR 18, AWWA C900, 4" diameter		380	.084		2.78	4.07		6.85	9.20
4530	6" diameter		316	.101		5.10	4.90		10	13
4540	8" diameter		264	.121		8.40	5.85		14.25	18.15
4550	10" diameter		220	.145		12.75	7.05		19.80	24.50
4560	12" diameter	▼	186	.172	▼	16.90	8.35		25.25	31
8000	Fittings with rubber gasket									
8003	Class 150, DR 18									
8006	90° bend , 4" diameter	B-20	100	.240	Ea.	42.50	10.70		53.20	63.50
8020	6" diameter		90	.267		75.50	11.90		87.40	102
8040	8" diameter		80	.300		146	13.40		159.40	182
8060	10" diameter		50	.480		335	21.50		356.50	400
8080	12" diameter		30	.800		430	35.50		465.50	525
8100	Tee, 4" diameter		90	.267		58.50	11.90		70.40	82.50
8120	6" diameter		80	.300		131	13.40		144.40	165
8140	8" diameter		70	.343		204	15.30		219.30	248
8160	10" diameter		40	.600		665	27		692	770
8180	12" diameter		20	1.200		895	53.50		948.50	1,075
8200	45° bend, 4" diameter		100	.240		42.50	10.70		53.20	63.50
8220	6" diameter		90	.267		73.50	11.90		85.40	99
8240	8" diameter		50	.480		140	21.50		161.50	187
8260	10" diameter		50	.480		281	21.50		302.50	345
8280	12" diameter		30	.800		365	35.50		400.50	455
8300	Reducing tee 6" x 4"		100	.240		119	10.70		129.70	147
8320	8" x 6"		90	.267		199	11.90		210.90	237
8330	10" x 6"		90	.267		335	11.90		346.90	390
8340	10" x 8"		90	.267		350	11.90		361.90	405
8350	12" x 6"		90	.267		400	11.90		411.90	460
8360	12" x 8"		90	.267		415	11.90		426.90	480
8400	Tapped service tee (threaded type) 6" x 6" x 3/4"		100	.240		86.50	10.70		97.20	112
8430	6" x 6" x 1"		90	.267		86.50	11.90		98.40	114
8440	6" x 6" x 1-1/2"		90	.267		86.50	11.90		98.40	114
8450	6" x 6" x 2"		90	.267		86.50	11.90		98.40	114
8460	8" x 8" x 3/4"		90	.267		128	11.90		139.90	159
8470	8" x 8" x 1"		90	.267		128	11.90		139.90	159
8480	8" x 8" x 1-1/2"		90	.267		128	11.90		139.90	159
8490	8" x 8" x 2"		90	.267		128	11.90		139.90	159
8500	Repair coupling 4"		100	.240		23	10.70		33.70	42
8520	6" diameter		90	.267		35.50	11.90		47.40	57
8540	8" diameter		50	.480		86	21.50		107.50	127
8560	10" diameter		50	.480		181	21.50		202.50	232
8580	12" diameter		50	.480		264	21.50		285.50	325
8600	Plug end 4"		100	.240		21	10.70		31.70	39.50
8620	6" diameter		90	.267		36	11.90		47.90	57.50
8640	8" diameter		50	.480		61.50	21.50		83	100
8660	10" diameter		50	.480		92	21.50		113.50	134
8680	12" diameter	▼	50	.480	▼	113	21.50		134.50	157

33 14 Water Utility Transmission and Distribution

33 14 19 – Valves and Hydrants for Water Utility Service

33 14 19.30 Fire Hydrants	Crew	Daily Output	Labor-Hours	Unit	Material	2018 Bare Costs Labor	Equipment	Total	Total Incl O&P
0010 **FIRE HYDRANTS**									
0020 Mechanical joints unless otherwise noted									
1000 Fire hydrants, two way; excavation and backfill not incl.									
1100 4-1/2" valve size, depth 2'-0"	B-21	10	2.800	Ea.	2,375	130	12.95	2,517.95	2,825
1120 2'-6"		10	2.800		2,425	130	12.95	2,567.95	2,875
1140 3'-0"		10	2.800		2,475	130	12.95	2,617.95	2,925
1300 7'-0"		6	4.667		2,875	216	21.50	3,112.50	3,500
2400 Lower barrel extensions with stems, 1'-0"	B-20	14	1.714		350	76.50		426.50	500
2480 3'-0"	"	12	2		840	89.50		929.50	1,050

33 16 Water Utility Storage Tanks

33 16 11 – Elevated Composite Water Storage Tanks

33 16 11.50 Elevated Water Storage Tanks

								Total	Total Incl O&P
0010 **ELEVATED WATER STORAGE TANKS**									
0011 Not incl. pipe, pumps or foundation									
3000 Elevated water tanks, 100' to bottom capacity line, incl. painting									
3010 50,000 gallons				Ea.				185,000	204,000
3300 100,000 gallons								280,000	307,500
3400 250,000 gallons								751,500	826,500
3600 500,000 gallons								1,336,000	1,470,000
3700 750,000 gallons								1,622,000	1,783,500
3900 1,000,000 gallons								2,322,000	2,556,000

33 16 23 – Ground-Level Steel Water Storage Tanks

33 16 23.13 Steel Water Storage Tanks

								Total	Total Incl O&P
0010 **STEEL WATER STORAGE TANKS**									
0910 Steel, ground level, ht./diam. less than 1, not incl. fdn., 100,000 gallons				Ea.				202,000	244,500
1000 250,000 gallons								295,500	324,000
1200 500,000 gallons								417,000	458,500
1250 750,000 gallons								538,000	591,500
1300 1,000,000 gallons								558,000	725,500
1500 2,000,000 gallons								1,043,000	1,148,000
1600 4,000,000 gallons								2,121,000	2,333,000
1800 6,000,000 gallons								3,095,000	3,405,000
1850 8,000,000 gallons								4,068,000	4,475,000
1910 10,000,000 gallons								5,050,000	5,554,500
2100 Steel standpipes, ht./diam. more than 1,100' to overflow, no fdn.									
2200 500,000 gallons				Ea.				546,500	600,500
2400 750,000 gallons								722,500	794,500
2500 1,000,000 gallons								1,060,500	1,167,000
2700 1,500,000 gallons								1,749,000	1,923,000
2800 2,000,000 gallons								2,327,000	2,559,000

33 16 36 – Ground-Level Reinforced Concrete Water Storage Tanks

33 16 36.16 Prestressed Conc. Water Storage Tanks

								Total	Total Incl O&P
0010 **PRESTRESSED CONC. WATER STORAGE TANKS**									
0020 Not including fdn., pipe or pumps, 250,000 gallons				Ea.				299,000	329,500
0100 500,000 gallons								487,000	536,000
0300 1,000,000 gallons								707,000	807,500
0400 2,000,000 gallons								1,072,000	1,179,000
0600 4,000,000 gallons								1,706,000	1,877,000

33 16 Water Utility Storage Tanks

33 16 36 – Ground-Level Reinforced Concrete Water Storage Tanks

33 16 36.16 Prestressed Conc. Water Storage Tanks	Crew	Daily Output	Labor-Hours	Unit	Material	2018 Bare Costs Labor	Equipment	Total	Total Incl O&P
0700	6,000,000 gallons				Ea.				2,266,000	2,493,000
0750	8,000,000 gallons								2,924,000	3,216,000
0800	10,000,000 gallons								3,533,000	3,886,000

33 16 56 – Ground-Level Plastic Water Storage Cisterns

33 16 56.23 Plastic-Coated Fabric Pillow Water Tanks

		Crew	Daily Output	Labor-Hours	Unit	Material	2018 Bare Costs Labor	Equipment	Total	Total Incl O&P
0010	**PLASTIC-COATED FABRIC PILLOW WATER TANKS**									
7000	Water tanks, vinyl coated fabric pillow tanks, freestanding, 5,000 gallons	4 Clab	4	8	Ea.	3,650	320		3,970	4,475
7100	Supporting embankment not included, 25,000 gallons	6 Clab	2	24		14,100	955		15,055	17,000
7200	50,000 gallons	8 Clab	1.50	42.667		20,500	1,700		22,200	25,200
7300	100,000 gallons	9 Clab	.90	80		39,700	3,200		42,900	48,600
7400	150,000 gallons		.50	144		46,200	5,750		51,950	60,000
7500	200,000 gallons		.40	180		83,000	7,175		90,175	102,500
7600	250,000 gallons		.30	240		93,500	9,575		103,075	117,500

33 31 Sanitary Sewerage Piping

33 31 11 – Public Sanitary Sewerage Gravity Piping

33 31 11.15 Sewage Collection, Concrete Pipe

0010	**SEWAGE COLLECTION, CONCRETE PIPE**
0020	See Section 33 41 13.60 for sewage/drainage collection, concrete pipe

33 31 11.25 Sewage Collection, Polyvinyl Chloride Pipe

		Crew	Daily Output	Labor-Hours	Unit	Material	2018 Bare Costs Labor	Equipment	Total	Total Incl O&P
0010	**SEWAGE COLLECTION, POLYVINYL CHLORIDE PIPE**									
0020	Not including excavation or backfill									
2000	20' lengths, SDR 35, B&S, 4" diameter	B-20	375	.064	L.F.	1.64	2.86		4.50	6.15
2040	6" diameter		350	.069		3.49	3.06		6.55	8.50
2080	13' lengths, SDR 35, B&S, 8" diameter		335	.072		6.50	3.20		9.70	12.05
2120	10" diameter	B-21	330	.085		11.35	3.93	.39	15.67	18.90
2160	12" diameter		320	.088		12.90	4.05	.41	17.36	21
2200	15" diameter		240	.117		14.45	5.40	.54	20.39	24.50
4000	Piping, DWV PVC, no exc./bkfill., 10' L, Sch 40, 4" diameter	B-20	375	.064		3.74	2.86		6.60	8.45
4010	6" diameter		350	.069		8.10	3.06		11.16	13.60
4020	8" diameter		335	.072		12.75	3.20		15.95	18.90

33 34 Onsite Wastewater Disposal

33 34 13 – Septic Tanks

33 34 13.13 Concrete Septic Tanks

		Crew	Daily Output	Labor-Hours	Unit	Material	2018 Bare Costs Labor	Equipment	Total	Total Incl O&P
0010	**CONCRETE SEPTIC TANKS**									
0011	Not including excavation or piping									
0015	Septic tanks, precast, 1,000 gallon	B-21	8	3.500	Ea.	1,025	162	16.20	1,203.20	1,425
0060	1,500 gallon		7	4		1,600	185	18.50	1,803.50	2,075
0100	2,000 gallon		5	5.600		2,275	259	26	2,560	2,925
0200	5,000 gallon	B-13	3.50	16		6,350	700	169	7,219	8,200
0300	15,000 gallon, 4 piece	B-13B	1.70	32.941		21,500	1,450	585	23,535	26,400
0400	25,000 gallon, 4 piece		1.10	50.909		42,400	2,225	905	45,530	51,000
0500	40,000 gallon, 4 piece		.80	70		52,500	3,050	1,250	56,800	63,500
0520	50,000 gallon, 5 piece	B-13C	.60	93.333		60,500	4,075	3,125	67,700	76,000
0640	75,000 gallon, cast in place	C-14C	.25	448		73,500	21,500	103	95,103	113,500
0660	100,000 gallon	"	.15	747		91,000	35,900	172	127,072	154,500
1150	Leaching field chambers, 13' x 3'-7" x 1'-4", standard	B-13	16	3.500		500	153	37	690	825

33 34 Onsite Wastewater Disposal

33 34 13 – Septic Tanks

33 34 13.13 Concrete Septic Tanks

		Crew	Daily Output	Labor-Hours	Unit	Material	2018 Bare Costs Labor	2018 Bare Costs Equipment	Total	Total Incl O&P
1200	Heavy duty, 8' x 4' x 1'-6"	B-13	14	4	Ea.	284	175	42	501	620
1300	13' x 3'-9" x 1'-6"		12	4.667		1,250	204	49	1,503	1,750
1350	20' x 4' x 1'-6"		5	11.200		1,200	490	118	1,808	2,175
1400	Leaching pit, precast concrete, 3' diameter, 3' deep	B-21	8	3.500		725	162	16.20	903.20	1,050
1500	6' diameter, 3' section		4.70	5.957		815	276	27.50	1,118.50	1,350
2000	Velocity reducing pit, precast conc., 6' diameter, 3' deep		4.70	5.957		1,625	276	27.50	1,928.50	2,250

33 34 13.33 Polyethylene Septic Tanks

		Crew	Daily Output	Labor-Hours	Unit	Material	2018 Bare Costs Labor	2018 Bare Costs Equipment	Total	Total Incl O&P
0010	**POLYETHYLENE SEPTIC TANKS**									
0015	High density polyethylene, 1,000 gallon	B-21	8	3.500	Ea.	1,325	162	16.20	1,503.20	1,725
0020	1,250 gallon		8	3.500		1,200	162	16.20	1,378.20	1,575
0025	1,500 gallon		7	4		1,350	185	18.50	1,553.50	1,775

33 34 16 – Septic Tank Effluent Filters

33 34 16.13 Septic Tank Gravity Effluent Filters

		Crew	Daily Output	Labor-Hours	Unit	Material	2018 Bare Costs Labor	2018 Bare Costs Equipment	Total	Total Incl O&P
0010	**SEPTIC TANK GRAVITY EFFLUENT FILTERS**									
3000	Effluent filter, 4" diameter	1 Skwk	8	1	Ea.	39.50	52.50		92	124
3020	6" diameter		7	1.143		55	60		115	152
3030	8" diameter		7	1.143		250	60		310	365
3040	8" diameter, very fine		7	1.143		495	60		555	635
3050	10" diameter, very fine		6	1.333		237	70		307	365
3060	10" diameter		6	1.333		275	70		345	410
3080	12" diameter		6	1.333		650	70		720	820
3090	15" diameter		5	1.600		1,075	84		1,159	1,300

33 34 51 – Drainage Field Systems

33 34 51.10 Drainage Field Excavation and Fill

		Crew	Daily Output	Labor-Hours	Unit	Material	2018 Bare Costs Labor	2018 Bare Costs Equipment	Total	Total Incl O&P
0010	**DRAINAGE FIELD EXCAVATION AND FILL**									
2200	Septic tank & drainage field excavation with 3/4 C.Y. backhoe	B-12F	145	.110	C.Y.		5.30	4.62	9.92	13.10
2400	4' trench for disposal field, 3/4 C.Y. backhoe	"	335	.048	L.F.		2.29	2	4.29	5.65
2600	Gravel fill, run of bank	B-6	150	.160	C.Y.	16.80	7	2.08	25.88	31.50
2800	Crushed stone, 3/4"	"	150	.160	"	36	7	2.08	45.08	52.50

33 34 51.13 Utility Septic Tank Tile Drainage Field

		Crew	Daily Output	Labor-Hours	Unit	Material	2018 Bare Costs Labor	2018 Bare Costs Equipment	Total	Total Incl O&P
0010	**UTILITY SEPTIC TANK TILE DRAINAGE FIELD**									
0015	Distribution box, concrete, 5 outlets	2 Clab	20	.800	Ea.	92.50	32		124.50	151
0020	7 outlets		16	1		92.50	40		132.50	163
0025	9 outlets		8	2		550	79.50		629.50	725
0115	Distribution boxes, HDPE, 5 outlets		20	.800		75	32		107	131
0117	6 outlets		15	1.067		75	42.50		117.50	148
0118	7 outlets		15	1.067		75	42.50		117.50	148
0120	8 outlets		10	1.600		79	64		143	184
0240	Distribution boxes, outlet flow leveler	1 Clab	50	.160		2.22	6.40		8.62	12.15
0300	Precast concrete, galley, 4' x 4' x 4'	B-21	16	1.750		236	81	8.10	325.10	390
0350	HDPE infiltration chamber 12" H x 15" W	2 Clab	300	.053	L.F.	6.95	2.13		9.08	10.90
0351	12" H x 15" W end cap	1 Clab	32	.250	Ea.	18.65	9.95		28.60	35.50
0355	chamber 12" H x 22" W	2 Clab	300	.053	L.F.	6.45	2.13		8.58	10.35
0356	12" H x 22" W end cap	1 Clab	32	.250	Ea.	16.80	9.95		26.75	33.50
0360	chamber 13" H x 34" W	2 Clab	300	.053	L.F.	14.15	2.13		16.28	18.85
0361	13" H x 34" W end cap	1 Clab	32	.250	Ea.	50.50	9.95		60.45	70.50
0365	chamber 16" H x 34" W	2 Clab	300	.053	L.F.	19.05	2.13		21.18	24
0366	16" H x 34" W end cap	1 Clab	32	.250	Ea.	15.70	9.95		25.65	32.50
0370	chamber 8" H x 16" W	2 Clab	300	.053	L.F.	9.85	2.13		11.98	14.05
0371	8" H x 16" W end cap	1 Clab	32	.250	Ea.	11	9.95		20.95	27.50

33 41 Subdrainage

33 41 16 – Subdrainage Piping

33 41 16.25 Piping, Subdrainage, Corrugated Metal

33 41 16.25 Piping, Subdrainage, Corrugated Metal	Crew	Daily Output	Labor-Hours	Unit	Material	2018 Bare Costs Labor	Equipment	Total	Total Incl O&P
0010 **PIPING, SUBDRAINAGE, CORRUGATED METAL**									
0021 Not including excavation and backfill									
2010 Aluminum, perforated									
2020 6" diameter, 18 ga.	B-20	380	.063	L.F.	6.65	2.82		9.47	11.60
2200 8" diameter, 16 ga.	"	370	.065		8.25	2.90		11.15	13.50
2220 10" diameter, 16 ga.	B-21	360	.078		10.30	3.60	.36	14.26	17.25
2240 12" diameter, 16 ga.		285	.098		11.55	4.55	.45	16.55	20
2260 18" diameter, 16 ga.		205	.137		17.35	6.35	.63	24.33	29.50
3000 Uncoated galvanized, perforated									
3020 6" diameter, 18 ga.	B-20	380	.063	L.F.	6.45	2.82		9.27	11.40
3200 8" diameter, 16 ga.	"	370	.065		7.85	2.90		10.75	13.05
3220 10" diameter, 16 ga.	B-21	360	.078		8.30	3.60	.36	12.26	15.05
3240 12" diameter, 16 ga.		285	.098		9.25	4.55	.45	14.25	17.65
3260 18" diameter, 16 ga.		205	.137		14.20	6.35	.63	21.18	26
4000 Steel, perforated, asphalt coated									
4020 6" diameter, 18 ga.	B-20	380	.063	L.F.	6.65	2.82		9.47	11.60
4030 8" diameter, 18 ga.	"	370	.065		8.25	2.90		11.15	13.50
4040 10" diameter, 16 ga.	B-21	360	.078		10.40	3.60	.36	14.36	17.35
4050 12" diameter, 16 ga.		285	.098		11.65	4.55	.45	16.65	20.50
4060 18" diameter, 16 ga.		205	.137		18.90	6.35	.63	25.88	31.50

33 41 16.40 Piping, Subdrainage, Polyvinyl Chloride

0010 **PIPING, SUBDRAINAGE, POLYVINYL CHLORIDE**									
0020 Perforated, price as solid pipe, Section 33 31 13.25									

33 42 Stormwater Conveyance

33 42 11 – Stormwater Gravity Piping

33 42 11.40 Piping, Storm Drainage, Corrugated Metal

33 42 11.40 Piping, Storm Drainage, Corrugated Metal	Crew	Daily Output	Labor-Hours	Unit	Material	2018 Bare Costs Labor	Equipment	Total	Total Incl O&P
0010 **PIPING, STORM DRAINAGE, CORRUGATED METAL**									
0020 Not including excavation or backfill									
2000 Corrugated metal pipe, galvanized									
2020 Bituminous coated with paved invert, 20' lengths									
2040 8" diameter, 16 ga.	B-14	330	.145	L.F.	8.15	6.10	.95	15.20	19.35
2060 10" diameter, 16 ga.		260	.185		8.50	7.75	1.20	17.45	22.50
2080 12" diameter, 16 ga.		210	.229		10.55	9.60	1.49	21.64	28
2100 15" diameter, 16 ga.		200	.240		14.30	10.10	1.56	25.96	33
2120 18" diameter, 16 ga.		190	.253		18.70	10.65	1.64	30.99	38.50
2140 24" diameter, 14 ga.		160	.300		22.50	12.65	1.95	37.10	46.50
2160 30" diameter, 14 ga.	B-13	120	.467		27.50	20.50	4.92	52.92	67
2180 36" diameter, 12 ga.		120	.467		31	20.50	4.92	56.42	71
2200 48" diameter, 12 ga.		100	.560		48.50	24.50	5.90	78.90	96.50
2220 60" diameter, 10 ga.	B-13B	75	.747		74	32.50	13.25	119.75	146
2240 72" diameter, 8 ga.	"	45	1.244		88.50	54.50	22	165	205
2500 Galvanized, uncoated, 20' lengths									
2520 8" diameter, 16 ga.	B-14	355	.135	L.F.	7.55	5.70	.88	14.13	17.90
2540 10" diameter, 16 ga.		280	.171		8.65	7.20	1.12	16.97	21.50
2560 12" diameter, 16 ga.		220	.218		9.30	9.20	1.42	19.92	25.50
2580 15" diameter, 16 ga.		220	.218		11.70	9.20	1.42	22.32	28.50
2600 18" diameter, 16 ga.		205	.234		14.10	9.85	1.52	25.47	32
2620 24" diameter, 14 ga.		175	.274		21.50	11.55	1.79	34.84	43.50
2640 30" diameter, 14 ga.	B-13	130	.431		27	18.85	4.54	50.39	63.50
2660 36" diameter, 12 ga.		130	.431		32	18.85	4.54	55.39	68.50

For customer support on your Building Construction Costs with RSMeans data, call 800.448.8182.

33 42 11.40 Piping, Storm Drainage, Corrugated Metal

		Crew	Daily Output	Labor-Hours	Unit	Material	2018 Bare Costs Labor	Equipment	Total	Total Incl O&P
2680	48" diameter, 12 ga.	B-13	110	.509	L.F.	51.50	22.50	5.35	79.35	96.50
2690	60" diameter, 10 ga.	B-13B	78	.718		79.50	31.50	12.75	123.75	149
2780	End sections, 8" diameter	B-14	35	1.371	Ea.	66.50	57.50	8.95	132.95	170
2785	10" diameter		35	1.371		70	57.50	8.95	136.45	174
2790	12" diameter		35	1.371		103	57.50	8.95	169.45	210
2800	18" diameter		30	1.600		111	67.50	10.40	188.90	235
2810	24" diameter	B-13	25	2.240		207	98	23.50	328.50	405
2820	30" diameter		25	2.240		315	98	23.50	436.50	525
2825	36" diameter		20	2.800		475	122	29.50	626.50	745
2830	48" diameter		10	5.600		930	245	59	1,234	1,450
2835	60" diameter	B-13B	5	11.200		1,600	490	199	2,289	2,750
2840	72" diameter	"	4	14		1,975	610	249	2,834	3,375

33 42 11.60 Sewage/Drainage Collection, Concrete Pipe

		Crew	Daily Output	Labor-Hours	Unit	Material	2018 Bare Costs Labor	Equipment	Total	Total Incl O&P
0010	**SEWAGE/DRAINAGE COLLECTION, CONCRETE PIPE**									
0020	Not including excavation or backfill									
1000	Non-reinforced pipe, extra strength, B&S or T&G joints									
1010	6" diameter	B-14	265.04	.181	L.F.	7.55	7.60	1.18	16.33	21
1020	8" diameter		224	.214		8.30	9	1.40	18.70	24.50
1030	10" diameter		216	.222		9.20	9.35	1.45	20	26
1040	12" diameter		200	.240		10.75	10.10	1.56	22.41	29
1050	15" diameter		180	.267		15.10	11.20	1.74	28.04	35.50
1060	18" diameter		144	.333		18.25	14.05	2.17	34.47	44
1070	21" diameter		112	.429		19.50	18.05	2.79	40.34	52
1080	24" diameter		100	.480		20.50	20	3.12	43.62	57
2000	Reinforced culvert, class 3, no gaskets									
2010	12" diameter	B-14	150	.320	L.F.	12.70	13.45	2.08	28.23	36.50
2020	15" diameter		150	.320		16.85	13.45	2.08	32.38	41.50
2030	18" diameter		132	.364		20.50	15.30	2.37	38.17	49
2035	21" diameter		120	.400		25	16.85	2.60	44.45	56
2040	24" diameter		100	.480		30	20	3.12	53.12	67
2045	27" diameter	B-13	92	.609		43.50	26.50	6.40	76.40	95.50
2050	30" diameter		88	.636		49.50	28	6.70	84.20	104
2060	36" diameter		72	.778		68	34	8.20	110.20	136
2070	42" diameter	B-13B	72	.778		91.50	34	13.80	139.30	168
2080	48" diameter		64	.875		92.50	38.50	15.55	146.55	177
2090	60" diameter		48	1.167		174	51	20.50	245.50	293
2100	72" diameter		40	1.400		283	61	25	369	430
2120	84" diameter		32	1.750		315	76.50	31	422.50	500
2140	96" diameter		24	2.333		380	102	41.50	523.50	620
2200	With gaskets, class 3, 12" diameter	B-21	168	.167		13.95	7.70	.77	22.42	28
2220	15" diameter		160	.175		18.50	8.10	.81	27.41	33.50
2230	18" diameter		152	.184		23	8.55	.85	32.40	39
2240	24" diameter		136	.206		37	9.55	.95	47.50	56.50
2260	30" diameter	B-13	88	.636		57.50	28	6.70	92.20	112
2270	36" diameter	"	72	.778		77.50	34	8.20	119.70	146
2290	48" diameter	B-13B	64	.875		106	38.50	15.55	160.05	191
2310	72" diameter	"	40	1.400		300	61	25	386	450
2330	Flared ends, 12" diameter	B-21	31	.903	Ea.	257	42	4.18	303.18	350
2340	15" diameter		25	1.120		290	52	5.20	347.20	405
2400	18" diameter		20	1.400		340	65	6.50	411.50	480
2420	24" diameter		14	2		390	92.50	9.25	491.75	580
2440	36" diameter	B-13	10	5.600		815	245	59	1,119	1,325

33 42 11 – Stormwater Gravity Piping

33 42 11.60 Sewage/Drainage Collection, Concrete Pipe

		Crew	Daily Output	Labor-Hours	Unit	Material	2018 Bare Costs Labor	Equipment	Total	Total Incl O&P
3080	Radius pipe, add to pipe prices, 12" to 60" diameter				L.F.	50%				
3090	Over 60" diameter, add				"	20%				
3500	Reinforced elliptical, 8' lengths, C507 class 3									
3520	14" x 23" inside, round equivalent 18" diameter	B-21	82	.341	L.F.	42	15.80	1.58	59.38	72
3530	24" x 38" inside, round equivalent 30" diameter	B-13	58	.966		64.50	42	10.20	116.70	146
3540	29" x 45" inside, round equivalent 36" diameter		52	1.077		86	47	11.35	144.35	179
3550	38" x 60" inside, round equivalent 48" diameter		38	1.474		145	64.50	15.55	225.05	275
3560	48" x 76" inside, round equivalent 60" diameter		26	2.154		216	94	22.50	332.50	405
3570	58" x 91" inside, round equivalent 72" diameter	↓	22	2.545	↓	298	111	27	436	530
3780	Concrete slotted pipe, class 4 mortar joint									
3800	12" diameter	B-21	168	.167	L.F.	31	7.70	.77	39.47	46.50
3840	18" diameter	"	152	.184	"	35	8.55	.85	44.40	52.50
3900	Concrete slotted pipe, Class 4 O-ring joint									
3940	12" diameter	B-21	168	.167	L.F.	28	7.70	.77	36.47	43
3960	18" diameter	"	152	.184	"	32	8.55	.85	41.40	49

33 42 13 – Stormwater Culverts

33 42 13.15 Oval Arch Culverts

		Crew	Daily Output	Labor-Hours	Unit	Material	2018 Bare Costs Labor	Equipment	Total	Total Incl O&P
0010	**OVAL ARCH CULVERTS**									
3000	Corrugated galvanized or aluminum, coated & paved									
3020	17" x 13", 16 ga., 15" equivalent	B-14	200	.240	L.F.	13.20	10.10	1.56	24.86	31.50
3040	21" x 15", 16 ga., 18" equivalent		150	.320		17.55	13.45	2.08	33.08	42
3060	28" x 20", 14 ga., 24" equivalent		125	.384		22.50	16.15	2.50	41.15	52
3080	35" x 24", 14 ga., 30" equivalent	↓	100	.480		27	20	3.12	50.12	64
3100	42" x 29", 12 ga., 36" equivalent	B-13	100	.560		32.50	24.50	5.90	62.90	79
3120	49" x 33", 12 ga., 42" equivalent		90	.622		39.50	27	6.55	73.05	92
3140	57" x 38", 12 ga., 48" equivalent	↓	75	.747	↓	51	32.50	7.90	91.40	114
3160	Steel, plain oval arch culverts, plain									
3180	17" x 13", 16 ga., 15" equivalent	B-14	225	.213	L.F.	13.05	9	1.39	23.44	29.50
3200	21" x 15", 16 ga., 18" equivalent		175	.274		15.75	11.55	1.79	29.09	37
3220	28" x 20", 14 ga., 24" equivalent	↓	150	.320		23	13.45	2.08	38.53	48
3240	35" x 24", 14 ga., 30" equivalent	B-13	108	.519		28	22.50	5.45	55.95	71.50
3260	42" x 29", 12 ga., 36" equivalent		108	.519		38	22.50	5.45	65.95	82.50
3280	49" x 33", 12 ga., 42" equivalent		92	.609		45	26.50	6.40	77.90	97
3300	57" x 38", 12 ga., 48" equivalent		75	.747	↓	57.50	32.50	7.90	97.90	122
3320	End sections, 17" x 13"		22	2.545	Ea.	139	111	27	277	350
3340	42" x 29"	↓	17	3.294	"	375	144	35	554	665
3360	Multi-plate arch, steel	B-20	1690	.014	Lb.	1.29	.63		1.92	2.39

33 42 33 – Stormwater Curbside Drains and Inlets

33 42 33.13 Catch Basins

		Crew	Daily Output	Labor-Hours	Unit	Material	2018 Bare Costs Labor	Equipment	Total	Total Incl O&P
0010	**CATCH BASINS**									
0011	Not including footing & excavation									
1600	Frames & grates, C.I., 24" square, 500 lb.	B-6	7.80	3.077	Ea.	355	134	40	529	640
1700	26" D shape, 600 lb.		7	3.429		620	150	44.50	814.50	955
1800	Light traffic, 18" diameter, 100 lb.		10	2.400		203	105	31.50	339.50	420
1900	24" diameter, 300 lb.		8.70	2.759		197	120	36	353	440
2000	36" diameter, 900 lb.		5.80	4.138		640	181	54	875	1,050
2100	Heavy traffic, 24" diameter, 400 lb.		7.80	3.077		263	134	40	437	540
2200	36" diameter, 1,150 lb.		3	8		875	350	104	1,329	1,600
2300	Mass. State standard, 26" diameter, 475 lb.		7	3.429		645	150	44.50	839.50	985
2400	30" diameter, 620 lb.		7	3.429		340	150	44.50	534.50	650
2500	Watertight, 24" diameter, 350 lb.		7.80	3.077		365	134	40	539	650
2600	26" diameter, 500 lb.	↓	7	3.429		420	150	44.50	614.50	740

33 42 Stormwater Conveyance

33 42 33 – Stormwater Curbside Drains and Inlets

33 42 33.13 Catch Basins

		Crew	Daily Output	Labor-Hours	Unit	Material	2018 Bare Costs Labor	Equipment	Total	Total Incl O&P
2700	32" diameter, 575 lb.	B-6	6	4	Ea.	920	175	52	1,147	1,325
2800	3 piece cover & frame, 10" deep,									
2900	1,200 lb., for heavy equipment	B-6	3	8	Ea.	1,050	350	104	1,504	1,800
3000	Raised for paving 1-1/4" to 2" high									
3100	4 piece expansion ring									
3200	20" to 26" diameter	1 Clab	3	2.667	Ea.	178	106		284	355
3300	30" to 36" diameter	"	3	2.667	"	246	106		352	435
3320	Frames and covers, existing, raised for paving, 2", including									
3340	row of brick, concrete collar, up to 12" wide frame	B-6	18	1.333	Ea.	51.50	58	17.35	126.85	164
3360	20" to 26" wide frame		11	2.182		76.50	95.50	28.50	200.50	261
3380	30" to 36" wide frame		9	2.667		95	116	34.50	245.50	320
3400	Inverts, single channel brick	D-1	3	5.333		104	239		343	480
3500	Concrete		5	3.200		122	143		265	355
3600	Triple channel, brick		2	8		166	360		526	735
3700	Concrete		3	5.333		157	239		396	535

33 42 33.50 Stormwater Management

		Crew	Daily Output	Labor-Hours	Unit	Material	2018 Bare Costs Labor	Equipment	Total	Total Incl O&P
0010	**STORMWATER MANAGEMENT**									
0030	Add per S.F. of impervious surface	B-37	6000	.008	S.F.	2.14	.34	.03	2.51	2.90

33 52 Hydrocarbon Transmission and Distribution

33 52 13 – Liquid Hydrocarbon Piping

33 52 13.16 Gasoline Piping

		Crew	Daily Output	Labor-Hours	Unit	Material	2018 Bare Costs Labor	Equipment	Total	Total Incl O&P
0010	**GASOLINE PIPING**									
0020	Primary containment pipe, fiberglass-reinforced									
0030	Plastic pipe 15' & 30' lengths									
0040	2" diameter	Q-6	425	.056	L.F.	6.45	3.32		9.77	12.10
0050	3" diameter		400	.060		10	3.53		13.53	16.30
0060	4" diameter		375	.064		12.95	3.76		16.71	19.90
0100	Fittings									
0110	Elbows, 90° & 45°, bell ends, 2"	Q-6	24	1	Ea.	41.50	59		100.50	134
0120	3" diameter		22	1.091		51.50	64		115.50	154
0130	4" diameter		20	1.200		66	70.50		136.50	179
0200	Tees, bell ends, 2"		21	1.143		55.50	67		122.50	162
0210	3" diameter		18	1.333		60.50	78.50		139	185
0220	4" diameter		15	1.600		79.50	94		173.50	230
0230	Flanges bell ends, 2"		24	1		31.50	59		90.50	124
0240	3" diameter		22	1.091		37	64		101	137
0250	4" diameter		20	1.200		42.50	70.50		113	153
0260	Sleeve couplings, 2"		21	1.143		11.80	67		78.80	114
0270	3" diameter		18	1.333		16.80	78.50		95.30	136
0280	4" diameter		15	1.600		21.50	94		115.50	166
0290	Threaded adapters, 2"		21	1.143		16.95	67		83.95	120
0300	3" diameter		18	1.333		32	78.50		110.50	154
0310	4" diameter		15	1.600		35.50	94		129.50	182
0320	Reducers, 2"		27	.889		25.50	52.50		78	107
0330	3" diameter		22	1.091		26	64		90	125
0340	4" diameter		20	1.200		34.50	70.50		105	144
1010	Gas station product line for secondary containment (double wall)									
1100	Fiberglass reinforced plastic pipe 25' lengths									
1120	Pipe, plain end, 3" diameter	Q-6	375	.064	L.F.	25.50	3.76		29.26	33.50
1130	4" diameter		350	.069		30.50	4.03		34.53	39.50

For customer support on your Building Construction Costs with RSMeans data, call 800.448.8182.

687

33 52 13 – Liquid Hydrocarbon Piping

33 52 13.16 Gasoline Piping

		Crew	Daily Output	Labor-Hours	Unit	Material	2018 Bare Costs Labor	Equipment	Total	Total Incl O&P
1140	5" diameter	Q-6	325	.074	L.F.	34	4.34		38.34	44
1150	6" diameter	↓	300	.080	↓	37	4.70		41.70	47.50
1200	Fittings									
1230	Elbows, 90° & 45°, 3" diameter	Q-6	18	1.333	Ea.	135	78.50		213.50	266
1240	4" diameter		16	1.500		158	88		246	305
1250	5" diameter		14	1.714		173	101		274	345
1260	6" diameter		12	2		192	118		310	390
1270	Tees, 3" diameter		15	1.600		160	94		254	320
1280	4" diameter		12	2		191	118		309	385
1290	5" diameter		9	2.667		299	157		456	565
1300	6" diameter		6	4		355	235		590	745
1310	Couplings, 3" diameter		18	1.333		51.50	78.50		130	175
1320	4" diameter		16	1.500		113	88		201	257
1330	5" diameter		14	1.714		202	101		303	375
1340	6" diameter		12	2		299	118		417	505
1350	Cross-over nipples, 3" diameter		18	1.333		9.90	78.50		88.40	129
1360	4" diameter		16	1.500		12.05	88		100.05	146
1370	5" diameter		14	1.714		15	101		116	169
1380	6" diameter		12	2		18.05	118		136.05	197
1400	Telescoping, reducers, concentric 4" x 3"		18	1.333		43.50	78.50		122	166
1410	5" x 4"		17	1.412		90	83		173	224
1420	6" x 5"	↓	16	1.500	↓	221	88		309	375

33 52 16 – Gas Hydrocarbon Piping

33 52 16.20 Piping, Gas Service and Distribution, P.E.

		Crew	Daily Output	Labor-Hours	Unit	Material	2018 Bare Costs Labor	Equipment	Total	Total Incl O&P
0010	**PIPING, GAS SERVICE AND DISTRIBUTION, POLYETHYLENE**									
0020	Not including excavation or backfill									
1000	60 psi coils, compression coupling @ 100', 1/2" diameter, SDR 11	B-20A	608	.053	L.F.	.50	2.55		3.05	4.40
1010	1" diameter, SDR 11		544	.059		1.16	2.85		4.01	5.60
1040	1-1/4" diameter, SDR 11		544	.059		1.64	2.85		4.49	6.10
1100	2" diameter, SDR 11		488	.066		2.55	3.17		5.72	7.60
1160	3" diameter, SDR 11	↓	408	.078		5.35	3.80		9.15	11.65
1500	60 psi 40' joints with coupling, 3" diameter, SDR 11	B-21A	408	.098		7.30	4.90	.89	13.09	16.40
1540	4" diameter, SDR 11		352	.114		13	5.65	1.03	19.68	24
1600	6" diameter, SDR 11		328	.122		31	6.10	1.11	38.21	44.50
1640	8" diameter, SDR 11	↓	272	.147		44.50	7.35	1.34	53.19	61.50

33 52 16.23 Medium Density Polyethylene Piping

		Crew	Daily Output	Labor-Hours	Unit	Material	2018 Bare Costs Labor	Equipment	Total	Total Incl O&P
2010	**MEDIUM DENSITY POLYETHYLENE PIPING**									
2020	ASTM D2513, not including excavation or backfill									
2200	Butt fused pipe									
2205	80psi coils, butt fusion joint @ 100', 1/2" CTS diameter, SDR 7	B-22C	2050	.008	L.F.	.18	.36	.06	.60	.81
2210	1" CTS diameter, SDR 11.5		1900	.008		.25	.39	.06	.70	.94
2215	80psi coils, butt fusion joint @ 100', IPS 1/2" diameter, SDR 9.3		1950	.008		.26	.38	.06	.70	.94
2220	3/4" diameter, SDR 11		1950	.008		.33	.38	.06	.77	1.01
2225	1" diameter, SDR 11		1850	.009		.60	.40	.07	1.07	1.34
2230	1-1/4" diameter, SDR 11		1850	.009		.79	.40	.07	1.26	1.55
2235	1-1/2" diameter, SDR 11		1750	.009		.98	.42	.07	1.47	1.80
2240	2" diameter, SDR 11	↓	1750	.009		1.21	.42	.07	1.70	2.05
2245	80psi 40' lengths, butt fusion joint, IPS 3" diameter, SDR 11.5	B-22A	660	.061		2.22	2.79	1.04	6.05	7.80
2250	4" diameter, SDR 11.5	"	500	.080	↓	3.24	3.68	1.37	8.29	10.65
2400	Socket fused pipe									
2405	80psi coils, socket fusion coupling @ 100', 1/2" CTS diameter, SDR 7	B-20	1050	.023	L.F.	.20	1.02		1.22	1.78
2410	1" CTS diameter, SDR 11.5	↓	1000	.024		.28	1.07		1.35	1.95

For customer support on your Building Construction Costs with RSMeans data, call 800.448.8182.

33 52 16 – Gas Hydrocarbon Piping

33 52 16.23 Medium Density Polyethylene Piping	Crew	Daily Output	Labor-Hours	Unit	Material	2018 Bare Costs Labor	Equipment	Total	Total Incl O&P
2415	80psi coils, socket fusion coupling @ 100', IPS 1/2" diameter, SDR 9.3	B-20	1000	.024	L.F.	.28	1.07		1.35	1.94
2420	3/4" diameter, SDR 11		1000	.024		.35	1.07		1.42	2.03
2425	1" diameter, SDR 11		950	.025		.62	1.13		1.75	2.40
2430	1-1/4" diameter, SDR 11		950	.025		.82	1.13		1.95	2.62
2435	1-1/2" diameter, SDR 11		900	.027		1.01	1.19		2.20	2.93
2440	2" diameter, SDR 11		850	.028		1.24	1.26		2.50	3.30
2445	80psi 40' lengths, socket fusion coupling, IPS 3" diameter, SDR 11.5	B-21A	340	.118		2.70	5.85	1.07	9.62	13.05
2450	4" diameter, SDR 11.5	"	260	.154		4.12	7.70	1.40	13.22	17.65
2600	Compression coupled pipe									
2605	80psi coils, compression coupling @ 100', 1/2" CTS diameter, SDR 7	B-20	2250	.011	L.F.	.31	.48		.79	1.07
2610	1" CTS diameter, SDR 11.5		2175	.011		.51	.49		1	1.31
2615	80psi coils, compression coupling @ 100', IPS 1/2" diameter, SDR 9.3		2175	.011		.49	.49		.98	1.29
2620	3/4" diameter, SDR 11		2100	.011		.57	.51		1.08	1.41
2625	1" diameter, SDR 11		2025	.012		.86	.53		1.39	1.76
3000	Fittings, butt fusion									
3010	SDR 11, IPS unless noted CTS									
3100	Caps, 3/4" diameter	B-22C	28.50	.561	Ea.	2.99	26	4.25	33.24	47.50
3105	1" diameter		27	.593		3.06	27.50	4.49	35.05	50
3110	1-1/4" diameter		27	.593		4.53	27.50	4.49	36.52	51.50
3115	1-1/2" diameter		25.50	.627		3.67	29	4.75	37.42	53
3120	2" diameter		24	.667		4.89	30.50	5.05	40.44	58
3125	3" diameter		24	.667		12.10	30.50	5.05	47.65	66
3130	4" diameter		18	.889		22	41	6.75	69.75	94.50
3200	Reducers, 1" x 3/4" diameters		13.50	1.185		6.85	54.50	8.95	70.30	101
3205	1-1/4" x 1" diameters		13.50	1.185		7	54.50	8.95	70.45	101
3210	1-1/2" x 3/4" diameters		12.75	1.255		7.15	58	9.50	74.65	107
3215	1-1/2" x 1" diameters		12.75	1.255		7.15	58	9.50	74.65	107
3220	1-1/2" x 1-1/4" diameters		12.75	1.255		7.15	58	9.50	74.65	107
3225	2" x 1" diameters		12	1.333		7.90	61.50	10.10	79.50	114
3230	2" x 1-1/4" diameters		12	1.333		9.70	61.50	10.10	81.30	116
3235	2" x 1-1/2" diameters		12	1.333		9.95	61.50	10.10	81.55	116
3240	3" x 2" diameters		12	1.333		15.35	61.50	10.10	86.95	122
3245	4" x 2" diameters		9	1.778		21	82	13.45	116.45	163
3250	4" x 3" diameters		9	1.778		28	82	13.45	123.45	171
3300	Elbows, 90°, 3/4" diameter		14.25	1.123		4.92	52	8.50	65.42	94
3302	1" diameter		13.50	1.185		5.10	54.50	8.95	68.55	99
3304	1-1/4" diameter		13.50	1.185		8.60	54.50	8.95	72.05	103
3306	1-1/2" diameter		12.75	1.255		7.15	58	9.50	74.65	107
3308	2" diameter		12	1.333		12.20	61.50	10.10	83.80	119
3310	3" diameter		12	1.333		26	61.50	10.10	97.60	134
3312	4" diameter		9	1.778		35	82	13.45	130.45	178
3350	45°, 3" diameter		12	1.333		27	61.50	10.10	98.60	135
3352	4" diameter		9	1.778		37	82	13.45	132.45	180
3400	Tees, 3/4" diameter		9.50	1.684		14.20	77.50	12.75	104.45	149
3405	1" diameter		9	1.778		5.10	82	13.45	100.55	145
3410	1-1/4" diameter		9	1.778		11.05	82	13.45	106.50	152
3415	1-1/2" diameter		8.50	1.882		8.70	87	14.25	109.95	158
3420	2" diameter		8	2		13.60	92	15.15	120.75	173
3425	3" diameter		8	2		30	92	15.15	137.15	191
3430	4" diameter		6	2.667		39.50	123	20	182.50	254
3500	Tapping tees, high volume, butt fusion outlets									
3505	1-1/2" punch, 2" x 2" outlet	B-22C	11	1.455	Ea.	110	67	11	188	235
3510	1-7/8" punch, 3" x 2" outlet		11	1.455		110	67	11	188	235

33 52 16.23 Medium Density Polyethylene Piping		Crew	Daily Output	Labor-Hours	Unit	Material	2018 Bare Costs Labor	Equipment	Total	Total Incl O&P
3515	4" x 2" outlet	B-22C	9	1.778	Ea.	110	82	13.45	205.45	261
3550	For protective sleeves, add					2.90			2.90	3.19
3600	Service saddles, saddle contour x outlet, butt fusion outlets									
3602	1-1/4" x 3/4" outlet	B-22C	17	.941	Ea.	6.65	43.50	7.10	57.25	81.50
3604	1-1/4" x 1" outlet		17	.941		6.65	43.50	7.10	57.25	81.50
3606	1-1/4" x 1-1/4" outlet		16	1		6.65	46	7.55	60.20	86
3608	1-1/2" x 3/4" outlet		16	1		6.65	46	7.55	60.20	86
3610	1-1/2" x 1-1/4" outlet		15	1.067		6.65	49	8.05	63.70	91
3612	2" x 3/4" outlet		12	1.333		6.65	61.50	10.10	78.25	112
3614	2" x 1" outlet		12	1.333		6.65	61.50	10.10	78.25	112
3616	2" x 1-1/4" outlet		11	1.455		6.65	67	11	84.65	121
3618	3" x 3/4" outlet		12	1.333		6.65	61.50	10.10	78.25	112
3620	3" x 1" outlet		12	1.333		6.65	61.50	10.10	78.25	112
3622	3" x 1-1/4" outlet		11	1.455		6.65	67	11	84.65	121
3624	4" x 3/4" outlet		10	1.600		6.65	74	12.10	92.75	134
3626	4" x 1" outlet		10	1.600		6.65	74	12.10	92.75	134
3628	4" x 1-1/4" outlet		9	1.778		6.65	82	13.45	102.10	147
3685	For protective sleeves, 3/4" diameter outlets, add					.86			.86	.95
3690	1" diameter outlets, add					1.44			1.44	1.58
3695	1-1/4" diameter outlets, add					2.48			2.48	2.73
3700	Branch saddles, contour x outlet, butt outlets, round base									
3702	2" x 2" outlet	B-22C	11	1.455	Ea.	17.30	67	11	95.30	133
3704	3" x 2" outlet		11	1.455		18.80	67	11	96.80	135
3706	3" x 3" outlet		11	1.455		27	67	11	105	144
3708	4" x 2" outlet		9	1.778		21.50	82	13.45	116.95	163
3710	4" x 3" outlet		9	1.778		27	82	13.45	122.45	169
3712	4" x 4" outlet		9	1.778		36	82	13.45	131.45	179
3750	Rectangular base, 2" x 2" outlet		11	1.455		16.45	67	11	94.45	132
3752	3" x 2" outlet		11	1.455		17.90	67	11	95.90	134
3754	3" x 3" outlet		11	1.455		25.50	67	11	103.50	142
3756	4" x 2" outlet		9	1.778		20.50	82	13.45	115.95	162
3758	4" x 3" outlet		9	1.778		25.50	82	13.45	120.95	168
3760	4" x 4" outlet		9	1.778		31	82	13.45	126.45	174
4000	Fittings, socket fusion									
4010	SDR 11, IPS unless noted CTS									
4100	Caps, 1/2" CTS diameter	B-20	30	.800	Ea.	2.64	35.50		38.14	57.50
4105	1/2" diameter		28.50	.842		2.41	37.50		39.91	60
4110	3/4" diameter		28.50	.842		2.81	37.50		40.31	60.50
4115	1" CTS diameter		28	.857		6.90	38.50		45.40	66
4120	1" diameter		27	.889		3.18	39.50		42.68	64
4125	1-1/4" diameter		27	.889		3.32	39.50		42.82	64
4130	1-1/2" diameter		25.50	.941		4.10	42		46.10	68.50
4135	2" diameter		24	1		4.33	44.50		48.83	73
4140	3" diameter		24	1		19.30	44.50		63.80	89
4145	4" diameter		18	1.333		31.50	59.50		91	126
4200	Reducers, 1/2" x 1/2" CTS diameters		14.25	1.684		6	75.50		81.50	122
4202	3/4" x 1/2" CTS diameters		14.25	1.684		4.88	75.50		80.38	120
4204	3/4" x 1/2" diameters		14.25	1.684		6.10	75.50		81.60	122
4206	1" CTS x 1/2" diameters		14	1.714		5.65	76.50		82.15	123
4208	1" CTS x 3/4" diameters		13.50	1.778		6	79.50		85.50	128
4210	1" x 1/2" CTS diameters		13.50	1.778		7.45	79.50		86.95	129
4212	1" x 1/2" diameters		13.50	1.778		6	79.50		85.50	128
4214	1" x 3/4" diameters		13.50	1.778		7.90	79.50		87.40	130

For customer support on your Building Construction Costs with RSMeans data, call 800.448.8182.

33 52 Hydrocarbon Transmission and Distribution

33 52 16 – Gas Hydrocarbon Piping

33 52 16.23 Medium Density Polyethylene Piping	Crew	Daily Output	Labor-Hours	Unit	Material	2018 Bare Costs Labor	Equipment	Total	Total Incl O&P
4216	1" x 1" CTS diameters	B-20	13.50	1.778	Ea.	6.90	79.50		86.40	129
4218	1-1/4" x 1/2" CTS diameters		13.50	1.778		14.10	79.50		93.60	137
4220	1-1/4" x 1/2" diameters		13.50	1.778		8.65	79.50		88.15	131
4222	1-1/4" x 3/4" diameters		13.50	1.778		11.30	79.50		90.80	133
4224	1-1/4" x 1" CTS diameters		13.50	1.778		10.45	79.50		89.95	133
4226	1-1/4" x 1" diameters		13.50	1.778		9.80	79.50		89.30	132
4228	1-1/2" x 3/4" diameters		12.75	1.882		9.35	84		93.35	138
4230	1-1/2" x 1" diameters		12.75	1.882		8.25	84		92.25	137
4232	1-1/2" x 1-1/4" diameters		12.75	1.882		9.35	84		93.35	138
4234	2" x 3/4" diameters		12	2		9.15	89.50		98.65	146
4236	2" x 1" CTS diameters		12	2		9.80	89.50		99.30	147
4238	2" x 1" diameters		12	2		11.55	89.50		101.05	149
4240	2" x 1-1/4" diameters		12	2		10.75	89.50		100.25	148
4242	2" x 1-1/2" diameters		12	2		9.75	89.50		99.25	147
4244	3" x 2" diameters		12	2		12.10	89.50		101.60	149
4246	4" x 2" diameters		9	2.667		24.50	119		143.50	209
4248	4" x 3" diameters		9	2.667		28	119		147	213
4300	Couplings, 1/2" CTS diameter		15	1.600		1.83	71.50		73.33	111
4305	1/2" diameter		14.25	1.684		1.65	75.50		77.15	117
4310	3/4" diameter		14.25	1.684		2.23	75.50		77.73	117
4315	1" CTS diameter		14	1.714		3.06	76.50		79.56	120
4320	1" diameter		13.50	1.778		2.26	79.50		81.76	123
4325	1-1/4" diameter		13.50	1.778		2.68	79.50		82.18	124
4330	1-1/2" diameter		12.75	1.882		3	84		87	131
4335	2" diameter		12	2		3.36	89.50		92.86	140
4340	3" diameter		12	2		19.10	89.50		108.60	157
4345	4" diameter		9	2.667		35.50	119		154.50	221
4400	Elbows, 90°, 1/2" CTS diameter		15	1.600		3.42	71.50		74.92	113
4405	1/2" diameter		14.25	1.684		3.42	75.50		78.92	119
4410	3/4" diameter		14.25	1.684		4.05	75.50		79.55	119
4415	1" CTS diameter		14	1.714		3.81	76.50		80.31	121
4420	1" diameter		13.50	1.778		4.41	79.50		83.91	126
4425	1-1/4" diameter		13.50	1.778		4.93	79.50		84.43	126
4430	1-1/2" diameter		12.75	1.882		7.40	84		91.40	136
4435	2" diameter		12	2		8.65	89.50		98.15	146
4440	3" diameter		12	2		22	89.50		111.50	160
4445	4" diameter		9	2.667		57	119		176	245
4460	45°, 1" diameter		13.50	1.778		6.10	79.50		85.60	128
4465	2" diameter		12	2		11	89.50		100.50	148
4500	Tees, 1/2" CTS diameter		10	2.400		3.54	107		110.54	168
4505	1/2" diameter		9.50	2.526		3.11	113		116.11	175
4510	3/4" diameter		9.50	2.526		3.86	113		116.86	176
4515	1" CTS diameter		9.33	2.571		3.81	115		118.81	179
4520	1" diameter		9	2.667		4.52	119		123.52	187
4525	1-1/4" diameter		9	2.667		5.60	119		124.60	188
4530	1-1/2" diameter		8.50	2.824		8.95	126		134.95	203
4535	2" diameter		8	3		11.80	134		145.80	218
4540	3" diameter		8	3		44	134		178	254
4545	4" diameter		6	4		69	179		248	350
4600	Tapping tees, type II, 3/4" punch, socket fusion outlets									
4601	Saddle contour x outlet diameter, IPS unless noted CTS									
4602	1-1/4" x 1/2" CTS outlet	B-20	17	1.412	Ea.	11.95	63		74.95	110
4604	1-1/4" x 1/2" outlet		17	1.412		11.95	63		74.95	110

For customer support on your Building Construction Costs with RSMeans data, call 800.448.8182.

691

33 52 16.23 Medium Density Polyethylene Piping

		Crew	Daily Output	Labor-Hours	Unit	Material	2018 Bare Costs Labor	2018 Bare Costs Equipment	Total	Total Incl O&P
4606	1-1/4" x 3/4" outlet	B-20	17	1.412	Ea.	11.95	63		74.95	110
4608	1-1/4" x 1" CTS outlet		17	1.412		13.05	63		76.05	111
4610	1-1/4" x 1" outlet		17	1.412		13.05	63		76.05	111
4612	1-1/4" x 1-1/4" outlet		16	1.500		22	67		89	126
4614	1-1/2" x 1/2" CTS outlet		16	1.500		11.95	67		78.95	115
4616	1-1/2" x 1/2" outlet		16	1.500		11.95	67		78.95	115
4618	1-1/2" x 3/4" outlet		16	1.500		11.95	67		78.95	115
4620	1-1/2" x 1" CTS outlet		16	1.500		13.05	67		80.05	116
4622	1-1/2" x 1" outlet		16	1.500		13.05	67		80.05	116
4624	1-1/2" x 1-1/4" outlet		15	1.600		22	71.50		93.50	133
4626	2" x 1/2" CTS outlet		12	2		11.95	89.50		101.45	149
4628	2" x 1/2" outlet		12	2		11.95	89.50		101.45	149
4630	2" x 3/4" outlet		12	2		11.95	89.50		101.45	149
4632	2" x 1" CTS outlet		12	2		13.05	89.50		102.55	150
4634	2" x 1" outlet		12	2		13.05	89.50		102.55	150
4636	2" x 1-1/4" outlet		11	2.182		22	97.50		119.50	173
4638	3" x 1/2" CTS outlet		12	2		11.95	89.50		101.45	149
4640	3" x 1/2" outlet		12	2		11.95	89.50		101.45	149
4642	3" x 3/4" outlet		12	2		11.95	89.50		101.45	149
4644	3" x 1" CTS outlet		12	2		13.05	89.50		102.55	150
4646	3" x 1" outlet		12	2		13.05	89.50		102.55	150
4648	3" x 1-1/4" outlet		11	2.182		22	97.50		119.50	173
4650	4" x 1/2" CTS outlet		10	2.400		11.95	107		118.95	177
4652	4" x 1/2" outlet		10	2.400		11.95	107		118.95	177
4654	4" x 3/4" outlet		10	2.400		11.95	107		118.95	177
4656	4" x 1" CTS outlet		10	2.400		13.05	107		120.05	178
4658	4" x 1" outlet		10	2.400		13.05	107		120.05	178
4660	4" x 1-1/4" outlet	▼	9	2.667		22	119		141	206
4685	For protective sleeves, 1/2" CTS to 3/4" diameter outlets, add					.86			.86	.95
4690	1" CTS & IPS diameter outlets, add					1.44			1.44	1.58
4695	1-1/4" diameter outlets, add				▼	2.48			2.48	2.73
4700	Tapping tees, high volume, socket fusion outlets									
4705	1-1/2" punch, 2" x 1-1/4" outlet	B-20	11	2.182	Ea.	110	97.50		207.50	270
4710	1-7/8" punch, 3" x 1-1/4" outlet		11	2.182		110	97.50		207.50	270
4715	4" x 1-1/4" outlet	▼	9	2.667		110	119		229	305
4750	For protective sleeves, add				▼	2.90			2.90	3.19
4800	Service saddles, saddle contour x outlet, socket fusion outlets									
4801	IPS unless noted CTS									
4802	1-1/4" x 1/2" CTS outlet	B-20	17	1.412	Ea.	3.70	63		66.70	101
4804	1-1/4" x 1/2" outlet		17	1.412		3.70	63		66.70	101
4806	1-1/4" x 3/4" outlet		17	1.412		3.70	63		66.70	101
4808	1-1/4" x 1" CTS outlet		17	1.412		• 6.10	63		69.10	103
4810	1-1/4" x 1" outlet		17	1.412		6.10	63		69.10	103
4812	1-1/4" x 1-1/4" outlet		16	1.500		9.80	67		76.80	113
4814	1-1/2" x 1/2" CTS outlet		16	1.500		3.70	67		70.70	106
4816	1-1/2" x 1/2" outlet		16	1.500		3.70	67		70.70	106
4818	1-1/2" x 3/4" outlet		16	1.500		3.70	67		70.70	106
4820	1-1/2" x 1-1/4" outlet		15	1.600		9.80	71.50		81.30	120
4822	2" x 1/2" CTS outlet		12	2		3.70	89.50		93.20	140
4824	2" x 1/2" outlet		12	2		3.70	89.50		93.20	140
4826	2" x 3/4" outlet		12	2		3.70	89.50		93.20	140
4828	2" x 1" CTS outlet		12	2		6.10	89.50		95.60	143
4830	2" x 1" outlet		12	2		6.10	89.50		95.60	143

33 52 16.23 Medium Density Polyethylene Piping	Crew	Daily Output	Labor-Hours	Unit	Material	2018 Bare Costs Labor	Equipment	Total	Total Incl O&P
4832	2" x 1-1/4" outlet	B-20	11	2.182	Ea.	9.80	97.50		107.30	160
4834	3" x 1/2" CTS outlet		12	2		3.70	89.50		93.20	140
4836	3" x 1/2" outlet		12	2		3.70	89.50		93.20	140
4838	3" x 3/4" outlet		12	2		3.70	89.50		93.20	140
4840	3" x 1" CTS outlet		12	2		6.10	89.50		95.60	143
4842	3" x 1" outlet		12	2		6.10	89.50		95.60	143
4844	3" x 1-1/4" outlet		11	2.182		9.80	97.50		107.30	160
4846	4" x 1/2" CTS outlet		10	2.400		3.70	107		110.70	168
4848	4" x 1/2" outlet		10	2.400		3.70	107		110.70	168
4850	4" x 3/4" outlet		10	2.400		3.70	107		110.70	168
4852	4" x 1" CTS outlet		10	2.400		6.10	107		113.10	171
4854	4" x 1" outlet		10	2.400		6.10	107		113.10	171
4856	4" x 1-1/4" outlet		9	2.667		9.80	119		128.80	193
4885	For protective sleeves, 1/2" CTS to 3/4" diameter outlets, add					.86			.86	.95
4890	1" CTS & IPS diameter outlets, add					1.44			1.44	1.58
4895	1-1/4" diameter outlets, add					2.48			2.48	2.73
4900	Spigot fittings, tees, SDR 7, 1/2" CTS diameter	B-20	10	2.400		25.50	107		132.50	192
4901	SDR 9.3, 1/2" diameter		9.50	2.526		11.30	113		124.30	184
4902	SDR 10, 1-1/4" diameter		9	2.667		11.30	119		130.30	194
4903	SDR 11, 3/4" diameter		9.50	2.526		23.50	113		136.50	198
4904	2" diameter		8	3		21.50	134		155.50	229
4906	SDR 11.5, 1" CTS diameter		9.33	2.571		60	115		175	241
4907	3" diameter		8	3		61	134		195	272
4908	4" diameter		6	4		157	179		336	445
4921	90° elbows, SDR 7, 1/2" CTS diameter		15	1.600		13.45	71.50		84.95	124
4922	SDR 9.3, 1/2" diameter		14.25	1.684		5.80	75.50		81.30	121
4923	SDR 10, 1-1/4" diameter		13.50	1.778		5.90	79.50		85.40	128
4924	SDR 11, 3/4" diameter		14.25	1.684		6.60	75.50		82.10	122
4925	2" diameter		12	2		25	89.50		114.50	164
4927	SDR 11.5, 1" CTS diameter		14	1.714		33	76.50		109.50	153
4928	3" diameter		12	2		56.50	89.50		146	198
4929	4" diameter		9	2.667		121	119		240	315
4935	Caps, SDR 7, 1/2" CTS diameter		30	.800		4.46	35.50		39.96	59.50
4936	SDR 9.3, 1/2" diameter		28.50	.842		4.46	37.50		41.96	62.50
4937	SDR 10, 1-1/4" diameter		27	.889		4.62	39.50		44.12	65.50
4938	SDR 11, 3/4" diameter		28.50	.842		4.77	37.50		42.27	63
4939	1" diameter		27	.889		5.30	39.50		44.80	66.50
4940	2" diameter		24	1		5.70	44.50		50.20	74.50
4942	SDR 11.5, 1" CTS diameter		28	.857		18.70	38.50		57.20	79
4943	3" diameter		24	1		30.50	44.50		75	102
4944	4" diameter		18	1.333		29	59.50		88.50	123
4946	SDR 13.5, 4" diameter		18	1.333		73.50	59.50		133	172
4950	Reducers, SDR 10 x SDR 11, 1-1/4" x 3/4" diameters		13.50	1.778		15.25	79.50		94.75	138
4951	1-1/4" x 1" diameters		13.50	1.778		15.25	79.50		94.75	138
4952	SDR 10 x SDR 11.5, 1-1/4" x 1" CTS diameters		13.50	1.778		15.25	79.50		94.75	138
4953	SDR 11 x SDR 7, 3/4" x 1/2" CTS diameters		14.25	1.684		6.80	75.50		82.30	123
4954	SDR 11 x SDR 9.3, 3/4" x 1/2" diameters		14.25	1.684		6.80	75.50		82.30	123
4955	SDR 11 x SDR 10, 2" x 1-1/4" diameters		12	2		15.80	89.50		105.30	153
4956	SDR 11 x SDR 11, 1" x 3/4" diameters		13.50	1.778		7.85	79.50		87.35	130
4957	2" x 3/4" diameters		12	2		15.60	89.50		105.10	153
4958	2" x 1" diameters		12	2		15.70	89.50		105.20	153
4959	SDR 11 x SDR 11.5, 1" x 1" CTS diameters		13.50	1.778		8	79.50		87.50	130
4960	2" x 1" CTS diameters		12	2		15.60	89.50		105.10	153

33 52 16.23 Medium Density Polyethylene Piping	Crew	Daily Output	Labor-Hours	Unit	Material	2018 Bare Costs Labor	Equipment	Total	Total Incl O&P
4962	SDR 11.5 x SDR 11, 3" x 2" diameters	B-20	12	2	Ea.	21	89.50		110.50	159
4963	4" x 2" diameters		9	2.667		32	119		151	217
4964	SDR 11.5 x SDR 11.5, 4" x 3" diameters		9	2.667		34.50	119		153.50	220
6100	Fittings, compression									
6101	MDPE gas pipe, ASTM D2513/ASTM F1924-98									
6102	Caps, SDR 7, 1/2" CTS diameter	B-20	60	.400	Ea.	12.10	17.85		29.95	41
6104	SDR 9.3, 1/2" IPS diameter		58	.414		21.50	18.50		40	51.50
6106	SDR 10, 1/2" CTS diameter		60	.400		11.35	17.85		29.20	40
6108	1-1/4" IPS diameter		54	.444		67	19.85		86.85	105
6110	SDR 11, 3/4" IPS diameter		58	.414		20	18.50		38.50	50
6112	1" IPS diameter		54	.444		32.50	19.85		52.35	66
6114	1-1/4" IPS diameter		54	.444		70	19.85		89.85	108
6116	1-1/2" IPS diameter		52	.462		94.50	20.50		115	136
6118	2" IPS diameter		50	.480		83	21.50		104.50	124
6120	SDR 11.5, 1" CTS diameter		56	.429		21.50	19.15		40.65	53
6122	SDR 12.5, 1" CTS diameter		56	.429		24	19.15		43.15	55.50
6202	Reducers, SDR 7 x SDR 10, 1/2" CTS x 1/2" CTS diameters		30	.800		17.35	35.50		52.85	73.50
6204	SDR 9.3 x SDR 7, 1/2" IPS x 1/2" CTS diameters		29	.828		50.50	37		87.50	112
6206	SDR 11 x SDR 7, 3/4" IPS x 1/2" CTS diameters		29	.828		44	37		81	105
6208	1" IPS x 1/2" CTS diameters		27	.889		54	39.50		93.50	120
6210	SDR 11 x SDR 9.3, 3/4" IPS x 1/2" IPS diameters		29	.828		49	37		86	111
6212	1" IPS x 1/2" IPS diameters		27	.889		56.50	39.50		96	123
6214	SDR 11 x SDR 10, 2" IPS x 1-1/4" IPS diameters		25	.960		95.50	43		138.50	171
6216	SDR 11 x SDR 11, 1" IPS x 3/4" IPS diameters		27	.889		54	39.50		93.50	120
6218	1-1/4" IPS x 1" IPS diameters		27	.889		58	39.50		97.50	125
6220	2" IPS x 1-1/4" IPS diameters		25	.960		95.50	43		138.50	171
6222	SDR 11 x SDR 11.5, 1" IPS x 1" CTS diameters		27	.889		46	39.50		85.50	112
6224	SDR 11 x SDR 12.5, 1" IPS x 1" CTS diameters		27	.889		56	39.50		95.50	123
6226	SDR 11.5 x SDR 7, 1" CTS x 1/2" CTS diameters		28	.857		34	38.50		72.50	96
6228	SDR 11.5 x SDR 9.3, 1" CTS x 1/2" IPS diameters		28	.857		39.50	38.50		78	102
6230	SDR 11.5 x SDR 11, 1" CTS x 3/4" IPS diameters		28	.857		48	38.50		86.50	112
6232	SDR 12.5 x SDR 7, 1" CTS x 1/2" CTS diameters		28	.857		36	38.50		74.50	98
6234	SDR 12.5 x SDR 9.3, 1" CTS x 1/2" IPS diameters		28	.857		40.50	38.50		79	103
6236	SDR 12.5 x SDR 11, 1" CTS x 3/4" IPS diameters		28	.857		34.50	38.50		73	96.50
6302	Couplings, SDR 7, 1/2" CTS diameter		30	.800		12.65	35.50		48.15	68.50
6304	SDR 9.3, 1/2" IPS diameter		29	.828		23.50	37		60.50	82.50
6306	SDR 10, 1/2" CTS diameter		30	.800		12.40	35.50		47.90	68
6308	SDR 11, 3/4" IPS diameter		29	.828		24.50	37		61.50	83.50
6310	1" IPS diameter		27	.889		26.50	39.50		66	89.50
6312	SDR 11.5, 1" CTS diameter		28	.857		25.50	38.50		64	87
6314	SDR 12.5, 1" CTS diameter		28	.857		27.50	38.50		66	89
6402	Repair couplings, SDR 10, 1-1/4" IPS diameter		27	.889		66.50	39.50		106	134
6404	SDR 11, 1-1/4" IPS diameter		27	.889		68.50	39.50		108	136
6406	1-1/2" IPS diameter		26	.923		74	41		115	144
6408	2" IPS diameter		25	.960		87.50	43		130.50	162
6502	Elbows, 90°, SDR 7, 1/2" CTS diameter		30	.800		24.50	35.50		60	81.50
6504	SDR 9.3, 1/2" IPS diameter		29	.828		38	37		75	98
6506	SDR 10, 1-1/4" IPS diameter		27	.889		98.50	39.50		138	169
6508	SDR 11, 3/4" IPS diameter		29	.828		37.50	37		74.50	97.50
6510	1" IPS diameter		27	.889		33	39.50		72.50	97
6512	1-1/4" IPS diameter		27	.889		75	39.50		114.50	143
6514	1-1/2" IPS diameter		26	.923		146	41		187	223
6516	2" IPS diameter		25	.960		130	43		173	209

33 52 16.23 Medium Density Polyethylene Piping		Crew	Daily Output	Labor-Hours	Unit	Material	2018 Bare Costs Labor	Equipment	Total	Total Incl O&P
6518	SDR 11.5, 1" CTS diameter	B-20	28	.857	Ea.	41.50	38.50		80	105
6520	SDR 12.5, 1" CTS diameter		28	.857		53.50	38.50		92	118
6602	Tees, SDR 7, 1/2" CTS diameter		20	1.200		37	53.50		90.50	123
6604	SDR 9.3, 1/2" IPS diameter		19.33	1.241		45	55.50		100.50	134
6606	SDR 10, 1-1/4" IPS diameter		18	1.333		112	59.50		171.50	214
6608	SDR 11, 3/4" IPS diameter		19.33	1.241		57	55.50		112.50	147
6610	1" IPS diameter		18	1.333		48.50	59.50		108	145
6612	1-1/4" IPS diameter		18	1.333		87	59.50		146.50	187
6614	1-1/2" IPS diameter		17.33	1.385		210	62		272	325
6616	2" IPS diameter		16.67	1.440		138	64.50		202.50	250
6618	SDR 11.5, 1" CTS diameter		18.67	1.286		58.50	57.50		116	152
6620	SDR 12.5, 1" CTS diameter	▼	18.67	1.286	▼	59	57.50		116.50	153
8100	Fittings, accessories									
8105	SDR 11, IPS unless noted CTS									
8110	Protective sleeves, for high volume tapping tees, butt fusion outlets	1 Skwk	16	.500	Ea.	5.45	26		31.45	46
8120	For tapping tees, socket outlets, CTS, 1/2" diameter		18	.444		1.42	23.50		24.92	37
8122	1" diameter		18	.444		2.51	23.50		26.01	38.50
8124	IPS, 1/2" diameter		18	.444		1.96	23.50		25.46	37.50
8126	3/4" diameter		18	.444		2.51	23.50		26.01	38.50
8128	1" diameter		18	.444		3.42	23.50		26.92	39.50
8130	1-1/4" diameter		18	.444		4.31	23.50		27.81	40
8205	Tapping tee test caps, type I, yellow PE cap, "aldyl style"		45	.178		44	9.30		53.30	63
8210	Type II, yellow polyethylene cap		45	.178		44	9.30		53.30	63
8215	High volume, yellow polyethylene cap		45	.178		56.50	9.30		65.80	77
8250	Quick connector, female x female inlets, 1/4" N.P.T.		50	.160		18.30	8.40		26.70	33
8255	Test hose, 24" length, 3/8" ID, male outlets, 1/4" N.P.T.		50	.160		25	8.40		33.40	40.50
8260	Quick connector and test hose assembly	▼	50	.160		43.50	8.40		51.90	60.50
8305	Purge point caps, butt fusion, SDR 10, 1-1/4" diameter	B-22C	27	.593		32	27.50	4.49	63.99	81.50
8310	SDR 11, 1-1/4" diameter		27	.593		32	27.50	4.49	63.99	81.50
8315	2" diameter		24	.667		32.50	30.50	5.05	68.05	88
8320	3" diameter		24	.667		45.50	30.50	5.05	81.05	103
8325	4" diameter		18	.889		45	41	6.75	92.75	119
8340	Socket fusion, SDR 11, 1-1/4" diameter		27	.593		8.15	27.50	4.49	40.14	55.50
8345	2" diameter		24	.667		9.80	30.50	5.05	45.35	63.50
8350	3" diameter		24	.667		60	30.50	5.05	95.55	119
8355	4" diameter	▼	18	.889		44	41	6.75	91.75	118
8360	Purge test quick connector, female x female inlets, 1/4" N.P.T.	1 Skwk	50	.160		18.30	8.40		26.70	33
8365	Purge test hose, 24" length, 3/8" ID, male outlets, 1/4" N.P.T.		50	.160		25	8.40		33.40	40.50
8370	Purge test quick connector and test hose assembly	▼	50	.160		43.50	8.40		51.90	60.50
8405	Transition fittings, MDPE x zinc plated steel, SDR 7, 1/2" CTS x 1/2" MPT	B-22C	30	.533		25	24.50	4.04	53.54	69.50
8410	SDR 9.3, 1/2" IPS x 3/4" MPT		28.50	.561		36	26	4.25	66.25	83.50
8415	SDR 10, 1-1/4" IPS x 1-1/4" MPT		27	.593		38.50	27.50	4.49	70.49	89
8420	SDR 11, 3/4" IPS x 3/4" MPT		28.50	.561		18.85	26	4.25	49.10	65
8425	1" IPS x 1" MPT		27	.593		22.50	27.50	4.49	54.49	71.50
8430	1-1/4" IPS x 1-1/4" MPT		17	.941		42	43.50	7.10	92.60	121
8435	1-1/2" IPS x 1-1/2" MPT		25.50	.627		43	29	4.75	76.75	96.50
8440	2" IPS x 2" MPT		24	.667		49.50	30.50	5.05	85.05	107
8445	SDR 11.5, 1" CTS x 1" MPT	▼	28	.571	▼	31	26.50	4.33	61.83	79.50

33 52 Hydrocarbon Transmission and Distribution

33 52 16 – Gas Hydrocarbon Piping

33 52 16.26 High Density Polyethylene Piping	Crew	Daily Output	Labor-Hours	Unit	Material	2018 Bare Costs Labor	Equipment	Total	Total Incl O&P
2010	**HIGH DENSITY POLYETHYLENE PIPING**									
2020	ASTM D2513, not including excavation or backfill									
2200	Butt fused pipe									
2205	125psi coils, butt fusion joint @ 100', 1/2" CTS diameter, SDR 7	B-22C	2050	.008	L.F.	.13	.36	.06	.55	.75
2210	160psi coils, butt fusion joint @ 100', IPS 1/2" diameter, SDR 9		1950	.008		.18	.38	.06	.62	.85
2215	3/4" diameter, SDR 11		1950	.008		.26	.38	.06	.70	.94
2220	1" diameter, SDR 11		1850	.009		.41	.40	.07	.88	1.13
2225	1-1/4" diameter, SDR 11		1850	.009		.61	.40	.07	1.08	1.35
2230	1-1/2" diameter, SDR 11		1750	.009		1.06	.42	.07	1.55	1.89
2235	2" diameter, SDR 11		1750	.009		.94	.42	.07	1.43	1.75
2240	3" diameter, SDR 11	↓	1650	.010		1.54	.45	.07	2.06	2.45
2245	160psi 40' lengths, butt fusion joint, IPS 3" diameter, SDR 11	B-22A	660	.061		1.68	2.79	1.04	5.51	7.25
2250	4" diameter, SDR 11	"	500	.080	↓	2.76	3.68	1.37	7.81	10.15
2400	Socket fused pipe									
2405	125psi coils, socket fusion coupling @ 100', 1/2" CTS diameter, SDR 7	B-20	1050	.023	L.F.	.15	1.02		1.17	1.72
2410	160psi coils, socket fusion coupling @ 100', IPS 1/2" diameter, SDR 9		1000	.024		.20	1.07		1.27	1.86
2415	3/4" diameter, SDR 11		1000	.024		.28	1.07		1.35	1.94
2420	1" diameter, SDR 11		950	.025		.43	1.13		1.56	2.19
2425	1-1/4" diameter, SDR 11		950	.025		.63	1.13		1.76	2.41
2430	1-1/2" diameter, SDR 11		900	.027		1.08	1.19		2.27	3.01
2435	2" diameter, SDR 11		850	.028		.97	1.26		2.23	2.99
2440	3" diameter, SDR 11	↓	850	.028		1.65	1.26		2.91	3.75
2445	160psi 40' lengths, socket fusion coupling, IPS 3" diameter, SDR 11	B-21A	340	.118		1.96	5.85	1.07	8.88	12.25
2450	4" diameter, SDR 11	"	260	.154	↓	3.31	7.70	1.40	12.41	16.80
2600	Compression coupled pipe									
2610	160psi coils, compression coupling @ 100', IPS 1/2" diameter, SDR 9	B-20	1000	.024	L.F.	.41	1.07		1.48	2.10
2615	3/4" diameter, SDR 11		1000	.024		.50	1.07		1.57	2.20
2620	1" diameter, SDR 11		950	.025		.67	1.13		1.80	2.46
2625	1-1/4" diameter, SDR 11		950	.025		1.30	1.13		2.43	3.15
2630	1-1/2" diameter, SDR 11		900	.027		1.80	1.19		2.99	3.80
2635	2" diameter, SDR 11	↓	850	.028	↓	1.81	1.26		3.07	3.93
3000	Fittings, butt fusion									
3010	SDR 11, IPS unless noted CTS									
3105	Caps, 1/2" diameter	B-22C	28.50	.561	Ea.	6.60	26	4.25	36.85	51.50
3110	3/4" diameter		28.50	.561		3.53	26	4.25	33.78	48
3115	1" diameter		27	.593		4.07	27.50	4.49	36.06	51
3120	1-1/4" diameter		27	.593		4.23	27.50	4.49	36.22	51
3125	1-1/2" diameter		25.50	.627		4.82	29	4.75	38.57	54.50
3130	2" diameter		24	.667		4.74	30.50	5.05	40.29	58
3135	3" diameter		24	.667		7.15	30.50	5.05	42.70	60.50
3140	4" diameter		18	.889		15.10	41	6.75	62.85	86.50
3205	Reducers, 1/2" x 1/2" CTS diameters		14.25	1.123		12.25	52	8.50	72.75	102
3210	3/4" x 1/2" CTS diameters		14.25	1.123		8.80	52	8.50	69.30	98
3215	1" x 1/2" CTS diameters		13.50	1.185		13.65	54.50	8.95	77.10	108
3220	1" x 1/2" diameters		13.50	1.185		12.25	54.50	8.95	75.70	107
3225	1" x 3/4" diameters		13.50	1.185		13.50	54.50	8.95	76.95	108
3230	1-1/4" x 1" diameters		13.50	1.185		13.15	54.50	8.95	76.60	108
3232	1-1/2" x 3/4" diameters		12.75	1.255		9.75	58	9.50	77.25	110
3233	1-1/2" x 1" diameters		12.75	1.255		9.75	58	9.50	77.25	110
3234	1-1/2" x 1-1/4" diameters		12.75	1.255		9.75	58	9.50	77.25	110
3235	2" x 1" diameters		12	1.333		9.30	61.50	10.10	80.90	115
3240	2" x 1-1/4" diameters	↓	12	1.333		8.95	61.50	10.10	80.55	115

33 52 16 – Gas Hydrocarbon Piping

33 52 16.26 High Density Polyethylene Piping	Crew	Daily Output	Labor-Hours	Unit	Material	2018 Bare Costs Labor	Equipment	Total	Total Incl O&P
3245	2" x 1-1/2" diameters	B-22C	12	1.333	Ea.	9.30	61.50	10.10	80.90	115
3250	3" x 2" diameters		12	1.333		11.15	61.50	10.10	82.75	117
3255	4" x 2" diameters		9	1.778		14.45	82	13.45	109.90	156
3260	4" x 3" diameters		9	1.778		16.30	82	13.45	111.75	158
3305	Elbows, 90°, 3/4" diameter		14.25	1.123		6.50	52	8.50	67	95.50
3310	1" diameter		13.50	1.185		6.70	54.50	8.95	70.15	101
3315	1-1/4" diameter		13.50	1.185		7.10	54.50	8.95	70.55	101
3320	1-1/2" diameter		12.75	1.255		7.35	58	9.50	74.85	107
3325	2" diameter		12	1.333		8.30	61.50	10.10	79.90	114
3330	3" diameter		12	1.333		17.60	61.50	10.10	89.20	124
3335	4" diameter		9	1.778		22.50	82	13.45	117.95	164
3380	45°, 3/4" diameter		14.25	1.123		6.15	52	8.50	66.65	95
3385	1" diameter		13.50	1.185		6.15	54.50	8.95	69.60	100
3390	1-1/4" diameter		13.50	1.185		6.35	54.50	8.95	69.80	100
3395	1-1/2" diameter		12.75	1.255		8.25	58	9.50	75.75	108
3400	2" diameter		12	1.333		9.95	61.50	10.10	81.55	116
3405	3" diameter		12	1.333		17.40	61.50	10.10	89	124
3410	4" diameter		9	1.778		22.50	82	13.45	117.95	164
3505	Tees, 1/2" CTS diameter		10	1.600		8	74	12.10	94.10	135
3510	1/2" diameter		9.50	1.684		8.80	77.50	12.75	99.05	143
3515	3/4" diameter		9.50	1.684		6.65	77.50	12.75	96.90	140
3520	1" diameter		9	1.778		7.20	82	13.45	102.65	148
3525	1-1/4" diameter		9	1.778		7.95	82	13.45	103.40	149
3530	1-1/2" diameter		8.50	1.882		12.60	87	14.25	113.85	163
3535	2" diameter		8	2		10.30	92	15.15	117.45	169
3540	3" diameter		8	2		19.10	92	15.15	126.25	179
3545	4" diameter	▼	6	2.667	▼	29	123	20	172	242
3600	Tapping tees, type II, 3/4" punch, butt fusion outlets									
3601	Saddle contour x outlet diameter, IPS unless noted CTS									
3602	1-1/4" x 1/2" CTS outlet	B-22C	17	.941	Ea.	11.20	43.50	7.10	61.80	86.50
3604	1-1/4" x 1/2" outlet		17	.941		11.20	43.50	7.10	61.80	86.50
3606	1-1/4" x 3/4" outlet		17	.941		11.20	43.50	7.10	61.80	86.50
3608	1-1/4" x 1" outlet		17	.941		13.05	43.50	7.10	63.65	88.50
3610	1-1/4" x 1-1/4" outlet		16	1		23.50	46	7.55	77.05	104
3612	1-1/2" x 1/2" CTS outlet		16	1		11.20	46	7.55	64.75	91
3614	1-1/2" x 1/2" outlet		16	1		11.20	46	7.55	64.75	91
3616	1-1/2" x 3/4" outlet		16	1		11.20	46	7.55	64.75	91
3618	1-1/2" x 1" outlet		16	1		13.05	46	7.55	66.60	93
3620	1-1/2" x 1-1/4" outlet		15	1.067		23.50	49	8.05	80.55	109
3622	2" x 1/2" CTS outlet		12	1.333		11.20	61.50	10.10	82.80	117
3624	2" x 1/2" outlet		12	1.333		11.20	61.50	10.10	82.80	117
3626	2" x 3/4" outlet		12	1.333		11.20	61.50	10.10	82.80	117
3628	2" x 1" outlet		12	1.333		13.05	61.50	10.10	84.65	119
3630	2" x 1-1/4" outlet		11	1.455		23.50	67	11	101.50	140
3632	3" x 1/2" CTS outlet		12	1.333		11.20	61.50	10.10	82.80	117
3634	3" x 1/2" outlet		12	1.333		11.20	61.50	10.10	82.80	117
3636	3" x 3/4" outlet		12	1.333		11.20	61.50	10.10	82.80	117
3638	3" x 1" outlet		12	1.333		13.05	61.50	10.10	84.65	119
3640	3" x 1-1/4" outlet		11	1.455		23.50	67	11	101.50	140
3642	4" x 1/2" CTS outlet		10	1.600		11.20	74	12.10	97.30	139
3644	4" x 1/2" outlet		10	1.600		11.20	74	12.10	97.30	139
3646	4" x 3/4" outlet		10	1.600		11.20	74	12.10	97.30	139
3648	4" x 1" outlet	▼	10	1.600	▼	13.05	74	12.10	99.15	141

33 52 16.26 High Density Polyethylene Piping	Crew	Daily Output	Labor-Hours	Unit	Material	Labor	2018 Bare Costs Equipment	Total	Total Incl O&P
3650	4" x 1-1/4" outlet	B-22C	9	1.778	Ea.	23.50	82	13.45	118.95	165
3675	For protective sleeves, 1/2" to 3/4" diameter outlets, add					.86			.86	.95
3680	1" diameter outlets, add					1.44			1.44	1.58
3685	1-1/4" diameter outlets, add			↓		2.48			2.48	2.73
3700	Tapping tees, high volume, butt fusion outlets									
3705	1-1/2" punch, 2" x 2" outlet	B-22C	11	1.455	Ea.	110	67	11	188	235
3710	1-7/8" punch, 3" x 2" outlet		11	1.455		110	67	11	188	235
3715	4" x 2" outlet	↓	9	1.778	↓	110	82	13.45	205.45	261
3750	For protective sleeves, add					2.90			2.90	3.19
3800	Service saddles, saddle contour x outlet, butt fusion outlets									
3801	IPS unless noted CTS									
3802	1-1/4" x 3/4" outlet	B-22C	17	.941	Ea.	6.65	43.50	7.10	57.25	81.50
3804	1-1/4" x 1" outlet		16	1		6.65	46	7.55	60.20	86
3806	1-1/4" x 1-1/4" outlet		16	1		6.95	46	7.55	60.50	86.50
3808	1-1/2" x 3/4" outlet		16	1		6.65	46	7.55	60.20	86
3810	1-1/2" x 1-1/4" outlet		15	1.067		6.95	49	8.05	64	91.50
3812	2" x 3/4" outlet		12	1.333		6.65	61.50	10.10	78.25	112
3814	2" x 1" outlet		11	1.455		6.65	67	11	84.65	121
3816	2" x 1-1/4" outlet		11	1.455		6.95	67	11	84.95	122
3818	3" x 3/4" outlet		12	1.333		6.65	61.50	10.10	78.25	112
3820	3" x 1" outlet		11	1.455		6.65	67	11	84.65	121
3822	3" x 1-1/4" outlet		11	1.455		6.95	67	11	84.95	122
3824	4" x 3/4" outlet		10	1.600		6.65	74	12.10	92.75	134
3826	4" x 1" outlet		9	1.778		6.65	82	13.45	102.10	147
3828	4" x 1-1/4" outlet	↓	9	1.778		6.95	82	13.45	102.40	147
3880	For protective sleeves, 3/4" diameter outlets, add					.86			.86	.95
3885	1" diameter outlets, add					1.44			1.44	1.58
3890	1-1/4" diameter outlets, add					2.48			2.48	2.73
3905	Branch saddles, contour x outlet, 2" x 2" outlet	B-22C	11	1.455		18.65	67	11	96.65	135
3910	3" x 2" outlet		11	1.455		18.65	67	11	96.65	135
3915	3" x 3" outlet		11	1.455		25	67	11	103	142
3920	4" x 2" outlet		9	1.778		18.65	82	13.45	114.10	160
3925	4" x 3" outlet		9	1.778		25	82	13.45	120.45	167
3930	4" x 4" outlet		9	1.778		34	82	13.45	129.45	177
3980	Ball valves, full port, 3/4" diameter		14.25	1.123		59	52	8.50	119.50	153
3982	1" diameter		13.50	1.185		59	54.50	8.95	122.45	158
3984	1-1/4" diameter		13.50	1.185		59	54.50	8.95	122.45	158
3986	1-1/2" diameter		12.75	1.255		86	58	9.50	153.50	193
3988	2" diameter		12	1.333		135	61.50	10.10	206.60	254
3990	3" diameter		12	1.333		320	61.50	10.10	391.60	455
3992	4" diameter	↓	9	1.778	↓	420	82	13.45	515.45	600
4000	Fittings, socket fusion									
4010	SDR 11, IPS unless noted CTS									
4105	Caps, 1/2" CTS diameter	B-20	30	.800	Ea.	2.37	35.50		37.87	57
4110	1/2" diameter		28.50	.842		2.42	37.50		39.92	60
4115	3/4" diameter		28.50	.842		2.60	37.50		40.10	60.50
4120	1" diameter		27	.889		3.42	39.50		42.92	64.50
4125	1-1/4" diameter		27	.889		3.65	39.50		43.15	64.50
4130	1-1/2" diameter		25.50	.941		3.92	42		45.92	68.50
4135	2" diameter		24	1		3.97	44.50		48.47	72.50
4140	3" diameter		24	1		24	44.50		68.50	94.50
4145	4" diameter		18	1.333		41.50	59.50		101	137
4205	Reducers, 1/2" x 1/2" CTS diameters	↓	14.25	1.684	↓	6	75.50		81.50	122

For customer support on your Building Construction Costs with RSMeans data, call 800.448.8182.

33 52 16.26 High Density Polyethylene Piping	Crew	Daily Output	Labor-Hours	Unit	Material	2018 Bare Costs Labor	Equipment	Total	Total Incl O&P
4210	3/4" x 1/2" CTS diameters	B-20	14.25	1.684	Ea.	6.05	75.50		81.55	122
4215	3/4" x 1/2" diameters		14.25	1.684		6.05	75.50		81.55	122
4220	1" x 1/2" CTS diameters		13.50	1.778		5.95	79.50		85.45	128
4225	1" x 1/2" diameters		13.50	1.778		6.30	79.50		85.80	128
4230	1" x 3/4" diameters		13.50	1.778		3.96	79.50		83.46	125
4235	1-1/4" x 1/2" CTS diameters		13.50	1.778		5.95	79.50		85.45	128
4240	1-1/4" x 3/4" diameters		13.50	1.778		7.50	79.50		87	129
4245	1-1/4" x 1" diameters		13.50	1.778		5.75	79.50		85.25	127
4250	1-1/2" x 3/4" diameters		12.75	1.882		9.60	84		93.60	139
4252	1-1/2" x 1" diameters		12.75	1.882		8.25	84		92.25	137
4255	1-1/2" x 1-1/4" diameters		12.75	1.882		9.35	84		93.35	138
4260	2" x 3/4" diameters		12	2		9.10	89.50		98.60	146
4265	2" x 1" diameters		12	2		9.10	89.50		98.60	146
4270	2" x 1-1/4" diameters		12	2		9.10	89.50		98.60	146
4275	3" x 2" diameters		12	2		12.20	89.50		101.70	149
4280	4" x 2" diameters		9	2.667		24.50	119		143.50	209
4285	4" x 3" diameters		9	2.667		25.50	119		144.50	210
4305	Couplings, 1/2" CTS diameter		15	1.600		1.70	71.50		73.20	111
4310	1/2" diameter		14.25	1.684		1.70	75.50		77.20	117
4315	3/4" diameter		14.25	1.684		1.59	75.50		77.09	117
4320	1" diameter		13.50	1.778		1.59	79.50		81.09	123
4325	1-1/4" diameter		13.50	1.778		2.02	79.50		81.52	123
4330	1-1/2" diameter		12.75	1.882		2.47	84		86.47	131
4335	2" diameter		12	2		2.55	89.50		92.05	139
4340	3" diameter		12	2		11.15	89.50		100.65	148
4345	4" diameter		9	2.667		22	119		141	206
4405	Elbows, 90°, 1/2" CTS diameter		15	1.600		3.35	71.50		74.85	113
4410	1/2" diameter		14.25	1.684		3.42	75.50		78.92	119
4415	3/4" diameter		14.25	1.684		2.97	75.50		78.47	118
4420	1" diameter		13.50	1.778		3.77	79.50		83.27	125
4425	1-1/4" diameter		13.50	1.778		4.10	79.50		83.60	126
4430	1-1/2" diameter		12.75	1.882		7.30	84		91.30	136
4435	2" diameter		12	2		7.35	89.50		96.85	144
4440	3" diameter		12	2		23.50	89.50		113	162
4445	4" diameter		9	2.667		106	119		225	298
4450	45°, 3/4" diameter		14.25	1.684		6.95	75.50		82.45	123
4455	1" diameter		13.50	1.778		7.85	79.50		87.35	130
4460	1-1/4" diameter		13.50	1.778		6.60	79.50		86.10	128
4465	1-1/2" diameter		12.75	1.882		10.35	84		94.35	139
4470	2" diameter		12	2		7.70	89.50		97.20	144
4475	3" diameter		12	2		42.50	89.50		132	183
4480	4" diameter		9	2.667		55	119		174	243
4505	Tees, 1/2" CTS diameter		10	2.400		2.89	107		109.89	167
4510	1/2" diameter		9.50	2.526		2.95	113		115.95	175
4515	3/4" diameter		9.50	2.526		2.92	113		115.92	175
4520	1" diameter		9	2.667		3.65	119		122.65	186
4525	1-1/4" diameter		9	2.667		5.10	119		124.10	188
4530	1-1/2" diameter		8.50	2.824		8.95	126		134.95	203
4535	2" diameter		8	3		9.70	134		143.70	216
4540	3" diameter		8	3		36.50	134		170.50	246
4545	4" diameter		6	4		70.50	179		249.50	350
4600	Tapping tees, type II, 3/4" punch, socket fusion outlets									
4601	Saddle countour x outlet diameter, IPS unless noted CTS									

33 52 16.26 High Density Polyethylene Piping	Crew	Daily Output	Labor-Hours	Unit	Material	2018 Bare Costs Labor	Equipment	Total	Total Incl O&P
4602	1-1/4" x 1/2" CTS outlet	B-20	17	1.412	Ea.	11.20	63		74.20	109
4604	1-1/4" x 1/2" outlet		17	1.412		11.20	63		74.20	109
4606	1-1/4" x 3/4" outlet		17	1.412		11.20	63		74.20	109
4608	1-1/4" x 1" outlet		17	1.412		13.05	63		76.05	111
4610	1-1/4" x 1-1/4" outlet		16	1.500		23.50	67		90.50	128
4612	1-1/2" x 1/2" CTS outlet		16	1.500		11.20	67		78.20	114
4614	1-1/2" x 1/2" outlet		16	1.500		11.20	67		78.20	114
4616	1-1/2" x 3/4" outlet		16	1.500		11.20	67		78.20	114
4618	1-1/2" x 1" outlet		16	1.500		13.05	67		80.05	116
4620	1-1/2" x 1-1/4" outlet		15	1.600		23.50	71.50		95	135
4622	2" x 1/2" CTS outlet		12	2		11.20	89.50		100.70	148
4624	2" x 1/2" outlet		12	2		11.20	89.50		100.70	148
4626	2" x 3/4" outlet		12	2		11.20	89.50		100.70	148
4628	2" x 1" outlet		12	2		13.05	89.50		102.55	150
4630	2" x 1-1/4" outlet		11	2.182		23.50	97.50		121	175
4632	3" x 1/2" CTS outlet		12	2		11.20	89.50		100.70	148
4634	3" x 1/2" outlet		12	2		11.20	89.50		100.70	148
4636	3" x 3/4" outlet		12	2		11.20	89.50		100.70	148
4638	3" x 1" outlet		12	2		13.05	89.50		102.55	150
4640	3" x 1-1/4" outlet		11	2.182		23.50	97.50		121	175
4642	4" x 1/2" CTS outlet		10	2.400		11.20	107		118.20	176
4644	4" x 1/2" outlet		10	2.400		11.20	107		118.20	176
4646	4" x 3/4" outlet		10	2.400		11.20	107		118.20	176
4648	4" x 1" outlet		10	2.400		13.05	107		120.05	178
4650	4" x 1-1/4" outlet		9	2.667		23.50	119		142.50	208
4675	For protective sleeves, 1/2" to 3/4" diameter outlets, add					.86			.86	.95
4680	1" diameter outlets, add					1.44			1.44	1.58
4685	1-1/4" diameter outlets, add					2.48			2.48	2.73
4700	Tapping tees, high volume, socket fusion outlets									
4705	1-1/2" punch, 2" x 1-1/4" outlet	B-20	11	2.182	Ea.	110	97.50		207.50	270
4710	1-7/8" punch, 3" x 1-1/4" outlet		11	2.182		110	97.50		207.50	270
4715	4" x 1-1/4" outlet		9	2.667		110	119		229	305
4750	For protective sleeves, add					2.90			2.90	3.19
4800	Service saddles, saddle contour x outlet, socket fusion outlets									
4801	IPS unless noted CTS									
4802	1-1/4" x 1/2" CTS outlet	B-20	17	1.412	Ea.	5.90	63		68.90	103
4804	1-1/4" x 1/2" outlet		17	1.412		5.90	63		68.90	103
4806	1-1/4" x 3/4" outlet		17	1.412		5.90	63		68.90	103
4808	1-1/4" x 1" outlet		16	1.500		6.95	67		73.95	110
4810	1-1/4" x 1-1/4" outlet		16	1.500		9.80	67		76.80	113
4812	1-1/2" x 1/2" CTS outlet		16	1.500		5.90	67		72.90	109
4814	1-1/2" x 1/2" outlet		16	1.500		5.90	67		72.90	109
4816	1-1/2" x 3/4" outlet		16	1.500		5.90	67		72.90	109
4818	1-1/2" x 1-1/4" outlet		15	1.600		9.80	71.50		81.30	120
4820	2" x 1/2" CTS outlet		12	2		5.90	89.50		95.40	143
4822	2" x 1/2" outlet		12	2		5.90	89.50		95.40	143
4824	2" x 3/4" outlet		12	2		5.90	89.50		95.40	143
4826	2" x 1" outlet		11	2.182		6.95	97.50		104.45	157
4828	2" x 1-1/4" outlet		11	2.182		9.80	97.50		107.30	160
4830	3" x 1/2" CTS outlet		12	2		5.90	89.50		95.40	143
4832	3" x 1/2" outlet		12	2		5.90	89.50		95.40	143
4834	3" x 3/4" outlet		12	2		5.90	89.50		95.40	143
4836	3" x 1" outlet		11	2.182		6.95	97.50		104.45	157

For customer support on your Building Construction Costs with RSMeans data, call 800.448.8182.

33 52 16 – Gas Hydrocarbon Piping

33 52 16.26 High Density Polyethylene Piping	Crew	Daily Output	Labor-Hours	Unit	Material	2018 Bare Costs Labor	Equipment	Total	Total Incl O&P
4838	3" x 1-1/4" outlet	B-20	11	2.182	Ea.	9.80	97.50		107.30	160
4840	4" x 1/2" CTS outlet		10	2.400		5.90	107		112.90	171
4842	4" x 1/2" outlet		10	2.400		5.90	107		112.90	171
4844	4" x 3/4" outlet		10	2.400		5.90	107		112.90	171
4846	4" x 1" outlet		9	2.667		6.95	119		125.95	190
4848	4" x 1-1/4" outlet		9	2.667		9.80	119		128.80	193
4880	For protective sleeves, 1/2" to 3/4" diameter outlets, add					.86			.86	.95
4885	1" diameter outlets, add					1.44			1.44	1.58
4890	1-1/4" diameter outlets, add					2.48			2.48	2.73
4900	Reducer tees, 1-1/4" x 3/4" x 3/4" diameters	B-20	9	2.667		7.30	119		126.30	190
4902	1-1/4" x 3/4" x 1" diameters		9	2.667		9.50	119		128.50	192
4904	1-1/4" x 3/4" x 1-1/4" diameters		9	2.667		9.50	119		128.50	192
4906	1-1/4" x 1" x 3/4" diameters		9	2.667		9.50	119		128.50	192
4908	1-1/4" x 1" x 1" diameters		9	2.667		7.55	119		126.55	190
4910	1-1/4" x 1" x 1-1/4" diameters		9	2.667		9.50	119		128.50	192
4912	1-1/4" x 1-1/4" x 3/4" diameters		9	2.667		7.15	119		126.15	190
4914	1-1/4" x 1-1/4" x 1" diameters		9	2.667		7.25	119		126.25	190
4916	1-1/2" x 3/4" x 3/4" diameters		8.50	2.824		12.05	126		138.05	206
4918	1-1/2" x 3/4" x 1" diameters		8.50	2.824		12.05	126		138.05	206
4920	1-1/2" x 3/4" x 1-1/4" diameters		8.50	2.824		12.05	126		138.05	206
4922	1-1/2" x 3/4" x 1-1/2" diameters		8.50	2.824		12.05	126		138.05	206
4924	1-1/2" x 1" x 3/4" diameters		8.50	2.824		12.05	126		138.05	206
4926	1-1/2" x 1" x 1" diameters		8.50	2.824		12.05	126		138.05	206
4928	1-1/2" x 1" x 1-1/4" diameters		8.50	2.824		12.05	126		138.05	206
4930	1-1/2" x 1" x 1-1/2" diameters		8.50	2.824		12.05	126		138.05	206
4932	1-1/2" x 1-1/4" x 3/4" diameters		8.50	2.824		10.75	126		136.75	205
4934	1-1/2" x 1-1/4" x 1" diameters		8.50	2.824		10.75	126		136.75	205
4936	1-1/2" x 1-1/4" x 1-1/4" diameters		8.50	2.824		10.75	126		136.75	205
4938	1-1/2" x 1-1/4" x 1-1/2" diameters		8.50	2.824		12.05	126		138.05	206
4940	1-1/2" x 1-1/2" x 3/4" diameters		8.50	2.824		10.60	126		136.60	205
4942	1-1/2" x 1-1/2" x 1" diameters		8.50	2.824		10.60	126		136.60	205
4944	1-1/2" x 1-1/2" x 1-1/4" diameters		8.50	2.824		10.60	126		136.60	205
4946	2" x 1-1/4" x 3/4" diameters		8	3		11.90	134		145.90	218
4948	2" x 1-1/4" x 1" diameters		8	3		11.90	134		145.90	218
4950	2" x 1-1/4" x 1-1/4" diameters		8	3		12.20	134		146.20	218
4952	2" x 1-1/2" x 3/4" diameters		8	3		11.90	134		145.90	218
4954	2" x 1-1/2" x 1" diameters		8	3		11.90	134		145.90	218
4956	2" x 1-1/2" x 1-1/4" diameters		8	3		11.90	134		145.90	218
4958	2" x 2" x 3/4" diameters		8	3		11.75	134		145.75	218
4960	2" x 2" x 1" diameters		8	3		11.85	134		145.85	218
4962	2" x 2" x 1-1/4" diameters		8	3		12	134		146	218
8100	Fittings, accessories									
8105	SDR 11, IPS unless noted CTS									
8110	Flange adapters, 2" x 6" long	1 Skwk	32	.250	Ea.	15.25	13.10		28.35	37
8115	3" x 6" long		32	.250		18.85	13.10		31.95	40.50
8120	4" x 6" long		24	.333		24	17.45		41.45	52.50
8135	Backup flanges, 2" diameter		32	.250		19.45	13.10		32.55	41.50
8140	3" diameter		32	.250		26	13.10		39.10	48.50
8145	4" diameter		24	.333		51.50	17.45		68.95	83
8200	Tapping tees, test caps									
8205	Type II, yellow polyethylene cap	1 Skwk	45	.178	Ea.	44	9.30		53.30	63
8210	High volume, black polyethylene cap		45	.178		56.50	9.30		65.80	77
8215	Quick connector, female x female inlets, 1/4" N.P.T.		50	.160		18.30	8.40		26.70	33

33 52 16.26 High Density Polyethylene Piping	Crew	Daily Output	Labor-Hours	Unit	Material	2018 Bare Costs Labor	Equipment	Total	Total Incl O&P
8220	Test hose, 24" length, 3/8" ID, male outlets, 1/4" N.P.T.	1 Skwk	50	.160	Ea.	25	8.40		33.40	40.50
8225	Quick connector and test hose assembly	↓	50	.160	↓	43.50	8.40		51.90	60.50
8300	Threaded transition fittings									
8302	HDPE x MPT zinc plated steel, SDR 7, 1/2" CTS x 1/2" MPT	B-22C	30	.533	Ea.	29	24.50	4.04	57.54	73.50
8304	SDR 9.3, 1/2" IPS x 3/4" MPT		28.50	.561		37.50	26	4.25	67.75	85.50
8306	SDR 11, 3/4" IPS x 3/4" MPT		28.50	.561		20	26	4.25	50.25	66
8308	1" IPS x 1" MPT		27	.593		26	27.50	4.49	57.99	75
8310	1-1/4" IPS x 1-1/4" MPT		27	.593		38	27.50	4.49	69.99	88.50
8312	1-1/2" IPS x 1-1/2" MPT		25.50	.627		45	29	4.75	78.75	98.50
8314	2" IPS x 2" MPT		24	.667		50	30.50	5.05	85.55	108
8322	HDPE x MPT 316 stainless steel, SDR 11, 3/4" IPS x 3/4" MPT		28.50	.561		26	26	4.25	56.25	72.50
8324	1" IPS x 1" MPT		27	.593		27.50	27.50	4.49	59.49	76.50
8326	1-1/4" IPS x 1-1/4" MPT		27	.593		31	27.50	4.49	62.99	80.50
8328	2" IPS x 2" MPT		24	.667		41	30.50	5.05	76.55	98
8330	3" IPS x 3" MPT		24	.667		81	30.50	5.05	116.55	142
8332	4" IPS x 4" MPT		18	.889		107	41	6.75	154.75	188
8342	HDPE x FPT 316 stainless steel, SDR 11, 3/4" IPS x 3/4" FPT		28.50	.561		34	26	4.25	64.25	81.50
8344	1" IPS x 1" FPT		27	.593		45.50	27.50	4.49	77.49	96.50
8346	1-1/4" IPS x 1-1/4" FPT		27	.593		73.50	27.50	4.49	105.49	127
8348	1-1/2" IPS x 1-1/2" FPT		25.50	.627		72.50	29	4.75	106.25	129
8350	2" IPS x 2" FPT		24	.667		93	30.50	5.05	128.55	155
8352	3" IPS x 3" FPT		24	.667		178	30.50	5.05	213.55	249
8354	4" IPS x 4" FPT		18	.889		256	41	6.75	303.75	350
8362	HDPE x MPT epoxy carbon steel, SDR 11, 3/4" IPS x 3/4" MPT		28.50	.561		19.50	26	4.25	49.75	65.50
8364	1" IPS x 1" MPT		27	.593		21.50	27.50	4.49	53.49	70.50
8366	1-1/4" IPS x 1-1/4" MPT		27	.593		21.50	27.50	4.49	53.49	70.50
8368	1-1/2" IPS x 1-1/2" MPT		25.50	.627		25	29	4.75	58.75	76.50
8370	2" IPS x 2" MPT		24	.667		27	30.50	5.05	62.55	82
8372	3" IPS x 3" MPT		24	.667		41.50	30.50	5.05	77.05	98
8374	4" IPS x 4" MPT		18	.889		57.50	41	6.75	105.25	133
8382	HDPE x FPT epoxy carbon steel, SDR 11, 3/4" IPS x 3/4" FPT		28.50	.561		24.50	26	4.25	54.75	71
8384	1" IPS x 1" FPT		27	.593		27.50	27.50	4.49	59.49	76.50
8386	1-1/4" IPS x 1-1/4" FPT		27	.593		54	27.50	4.49	85.99	105
8388	1-1/2" IPS x 1-1/2" FPT		25.50	.627		58.50	29	4.75	92.25	114
8390	2" IPS x 2" FPT		24	.667		60.50	30.50	5.05	96.05	119
8392	3" IPS x 3" FPT		24	.667		103	30.50	5.05	138.55	166
8394	4" IPS x 4" FPT		18	.889		127	41	6.75	174.75	210
8402	HDPE x MPT poly coated carbon steel, SDR 11, 1" IPS x 1" MPT		27	.593		21.50	27.50	4.49	53.49	70.50
8404	1-1/4" IPS x 1-1/4" MPT		27	.593		28.50	27.50	4.49	60.49	78
8406	1-1/2" IPS x 1-1/2" MPT		25.50	.627		34	29	4.75	67.75	86.50
8408	2" IPS x 2" MPT		24	.667		38.50	30.50	5.05	74.05	95
8410	3" IPS x 3" MPT		24	.667		73	30.50	5.05	108.55	133
8412	4" IPS x 4" MPT	↓	18	.889		116	41	6.75	163.75	197
8602	Socket fused HDPE x MPT brass, SDR 11, 3/4" IPS x 3/4" MPT	B-20	28.50	.842		12	37.50		49.50	70.50
8604	1" IPS x 3/4" MPT		27	.889		14.50	39.50		54	76.50
8606	1" IPS x 1" MPT		27	.889		14	39.50		53.50	76
8608	1-1/4" IPS x 3/4" MPT		27	.889		15	39.50		54.50	77
8610	1-1/4" IPS x 1" MPT		27	.889		15	39.50		54.50	77
8612	1-1/4" IPS x 1-1/4" MPT		27	.889		21.50	39.50		61	84
8614	1-1/2" IPS x 1-1/2" MPT		25.50	.941		29.50	42		71.50	96.50
8616	2" IPS x 2" MPT		24	1		30	44.50		74.50	101
8618	Socket fused HDPE x FPT brass, SDR 11, 3/4" IPS x 3/4" FPT		28.50	.842		12	37.50		49.50	70.50
8620	1" IPS x 1/2" FPT		27	.889		12	39.50		51.50	73.50

33 52 Hydrocarbon Transmission and Distribution

33 52 16 – Gas Hydrocarbon Piping

33 52 16.26 High Density Polyethylene Piping

		Crew	Daily Output	Labor-Hours	Unit	Material	2018 Bare Costs Labor	Equipment	Total	Total Incl O&P
8622	1" IPS x 3/4" FPT	B-20	27	.889	Ea.	14.50	39.50		54	76.50
8624	1" IPS x 1" FPT		27	.889		14	39.50		53.50	76
8626	1-1/4" IPS x 3/4" FPT		27	.889		15	39.50		54.50	77
8628	1-1/4" IPS x 1" FPT		27	.889		14.50	39.50		54	76.50
8630	1-1/4" IPS x 1-1/4" FPT		27	.889		21.50	39.50		61	84
8632	1-1/2" IPS x 1-1/2" FPT		25.50	.941		30	42		72	97
8634	2" IPS x 1-1/2" FPT		24	1		30.50	44.50		75	102
8636	2" IPS x 2" FPT		24	1		32	44.50		76.50	103

33 71 Electrical Utility Transmission and Distribution

33 71 16 – Electrical Utility Poles

33 71 16.33 Wood Electrical Utility Poles

		Crew	Daily Output	Labor-Hours	Unit	Material	2018 Bare Costs Labor	Equipment	Total	Total Incl O&P
0010	**WOOD ELECTRICAL UTILITY POLES**									
0011	Excludes excavation, backfill and cast-in-place concrete									
6200	Electric & tel sitework, 20' high, treated wd., see Section 26 56 13.10	R-3	3.10	6.452	Ea.	189	375	42	606	815
6400	25' high		2.90	6.897		238	400	44.50	682.50	910
6600	30' high		2.60	7.692		340	445	50	835	1,100
6800	35' high		2.40	8.333		480	485	54	1,019	1,325
7000	40' high		2.30	8.696		640	505	56.50	1,201.50	1,525
7200	45' high		1.70	11.765		870	680	76	1,626	2,075
7400	Cross arms with hardware & insulators									
7600	4' long	1 Elec	2.50	3.200	Ea.	149	186		335	440
7800	5' long		2.40	3.333		165	194		359	470
8000	6' long		2.20	3.636		171	212		383	505

33 71 19 – Electrical Underground Ducts and Manholes

33 71 19.17 Electric and Telephone Underground

		Crew	Daily Output	Labor-Hours	Unit	Material	2018 Bare Costs Labor	Equipment	Total	Total Incl O&P
0010	**ELECTRIC AND TELEPHONE UNDERGROUND**									
0011	Not including excavation									
0200	backfill and cast in place concrete									
0400	Hand holes, precast concrete, with concrete cover									
0600	2' x 2' x 3' deep	R-3	2.40	8.333	Ea.	430	485	54	969	1,250
0800	3' x 3' x 3' deep		1.90	10.526		560	610	68	1,238	1,600
1000	4' x 4' x 4' deep		1.40	14.286		1,425	830	92.50	2,347.50	2,900
1200	Manholes, precast with iron racks & pulling irons, C.I. frame									
1400	and cover, 4' x 6' x 7' deep	B-13	2	28	Ea.	2,925	1,225	295	4,445	5,375
1600	6' x 8' x 7' deep		1.90	29.474		3,275	1,300	310	4,885	5,900
1800	6' x 10' x 7' deep		1.80	31.111		3,675	1,350	330	5,355	6,475
4200	Underground duct, banks ready for concrete fill, min. of 7.5"									
4400	between conduits, center to center									
4580	PVC, type EB, 1 @ 2" diameter	2 Elec	480	.033	L.F.	1.09	1.94		3.03	4.10
4600	2 @ 2" diameter		240	.067		2.18	3.88		6.06	8.20
4800	4 @ 2" diameter		120	.133		4.36	7.75		12.11	16.40
5000	2 @ 3" diameter		200	.080		3.63	4.66		8.29	10.95
5200	4 @ 3" diameter		100	.160		7.25	9.30		16.55	22
5400	2 @ 4" diameter		160	.100		3.61	5.80		9.41	12.70
5600	4 @ 4" diameter		80	.200		7.25	11.65		18.90	25.50
5800	6 @ 4" diameter		54	.296		10.85	17.25		28.10	38
6200	Rigid galvanized steel, 2 @ 2" diameter		180	.089		17.50	5.15		22.65	27
6400	4 @ 2" diameter		90	.178		35	10.35		45.35	54
6800	2 @ 3" diameter		100	.160		25	9.30		34.30	41.50

33 71 Electrical Utility Transmission and Distribution

33 71 19 – Electrical Underground Ducts and Manholes

33 71 19.17 Electric and Telephone Underground	Crew	Daily Output	Labor-Hours	Unit	Material	2018 Bare Costs Labor	Equipment	Total	Total Incl O&P
7000	4 @ 3" diameter	2 Elec	50	.320	L.F.	50	18.60		68.60	83
7200	2 @ 4" diameter		70	.229		37	13.30		50.30	61
7400	4 @ 4" diameter		34	.471		74.50	27.50		102	123
7600	6 @ 4" diameter		22	.727		112	42.50		154.50	187

33 81 Communications Structures

33 81 13 – Communications Transmission Towers

33 81 13.10 Radio Towers

		Crew	Daily Output	Labor-Hours	Unit	Material	2018 Bare Costs Labor	Equipment	Total	Total Incl O&P
0010	**RADIO TOWERS**									
0020	Guyed, 50' H, 40 lb. section, 70 MPH basic wind speed	2 Sswk	1	16	Ea.	2,775	875		3,650	4,475
0100	Wind load 90 MPH basic wind speed	"	1	16		3,675	875		4,550	5,450
0300	190' high, 40 lb. section, wind load 70 MPH basic wind speed	K-2	.33	72.727		9,525	3,775	720	14,020	17,300
0400	200' high, 70 lb. section, wind load 90 MPH basic wind speed		.33	72.727		15,800	3,775	720	20,295	24,100
0600	300' high, 70 lb. section, wind load 70 MPH basic wind speed		.20	120		25,300	6,225	1,200	32,725	39,100
0700	270' high, 90 lb. section, wind load 90 MPH basic wind speed		.20	120		27,400	6,225	1,200	34,825	41,400
0800	400' high, 100 lb. section, wind load 70 MPH basic wind speed		.14	171		38,200	8,900	1,700	48,800	58,000
0900	Self-supporting, 60' high, wind load 70 MPH basic wind speed		.80	30		4,500	1,550	298	6,348	7,775
0910	60' high, wind load 90 MPH basic wind speed		.45	53.333		5,250	2,775	530	8,555	10,800
1000	120' high, wind load 70 MPH basic wind speed		.40	60		9,925	3,125	595	13,645	16,500
1200	190' high, wind load 90 MPH basic wind speed		.20	120		28,300	6,225	1,200	35,725	42,400
2000	For states west of Rocky Mountains, add for shipping					10%				

For customer support on your Building Construction Costs with RSMeans data, call 800.448.8182.

Estimating Tips
34 11 00 Rail Tracks
This subdivision includes items that may involve either repair of existing or construction of new railroad tracks. Additional preparation work, such as the roadbed earthwork, would be found in Division 31. Additional new construction siding and turnouts are found in Subdivision 34 72. Maintenance of railroads is found under 34 01 23 Operation and Maintenance of Railways.

34 40 00 Traffic Signals
This subdivision includes traffic signal systems. Other traffic control devices such as traffic signs are found in Subdivision 10 14 53 Traffic Signage.

34 70 00 Vehicle Barriers
This subdivision includes security vehicle barriers, guide and guard rails, crash barriers, and delineators. The actual maintenance and construction of concrete and asphalt pavement are found in Division 32.

Reference Numbers
Reference numbers are shown at the beginning of some major classifications. These numbers refer to related items in the Reference Section. The reference information may be an estimating procedure, an alternate pricing method, or technical information.

Note: Not all subdivisions listed here necessarily appear. ■

Did you know?

RSMeans data is available through our online application with 24/7 access:

■ Search for unit prices by keyword

■ Leverage the most up-to-date data

■ Build and export estimates

Try it free for 30 days!
www.rsmeans.com/2018freetrial

No part of this cost data may be reproduced, stored in a retrieval system, or transmitted in any form or by any means without prior written permission of Gordian.

34 01 23 – Operation and Maintenance of Railways

34 01 23.51 Maintenance of Railroads		Crew	Daily Output	Labor-Hours	Unit	Material	2018 Bare Costs Labor	2018 Bare Costs Equipment	Total	Total Incl O&P
0010	**MAINTENANCE OF RAILROADS**									
0400	Resurface and realign existing track	B-14	200	.240	L.F.		10.10	1.56	11.66	17.05
0600	For crushed stone ballast, add	"	500	.096	"	13.55	4.04	.63	18.22	22

34 11 Rail Tracks

34 11 13 – Track Rails

34 11 13.23 Heavy Rail Track

		Crew	Daily Output	Labor-Hours	Unit	Material	Labor	Equipment	Total	Total Incl O&P
0010	**HEAVY RAIL TRACK**	R347216-10								
1000	Rail, 100 lb. prime grade				L.F.	31			31	34
1500	Relay rail				"	15.40			15.40	16.95

34 11 33 – Track Cross Ties

34 11 33.13 Concrete Track Cross Ties

		Crew	Daily Output	Labor-Hours	Unit	Material	Labor	Equipment	Total	Total Incl O&P
0010	**CONCRETE TRACK CROSS TIES**									
1400	Ties, concrete, 8'-6" long, 30" OC	B-14	80	.600	Ea.	151	25.50	3.91	180.41	209

34 11 33.16 Timber Track Cross Ties

		Crew	Daily Output	Labor-Hours	Unit	Material	Labor	Equipment	Total	Total Incl O&P
0010	**TIMBER TRACK CROSS TIES**									
1600	Wood, pressure treated, 6" x 8" x 8'-6", C.L. lots	B-14	90	.533	Ea.	56	22.50	3.47	81.97	99.50
1700	L.C.L. lots		90	.533		59	22.50	3.47	84.97	103
1900	Heavy duty, 7" x 9" x 8'-6", C.L. lots		70	.686		61.50	29	4.46	94.96	117
2000	L.C.L. lots	↓	70	.686	↓	61.50	29	4.46	94.96	117

34 11 33.17 Timber Switch Ties

		Crew	Daily Output	Labor-Hours	Unit	Material	Labor	Equipment	Total	Total Incl O&P
0010	**TIMBER SWITCH TIES**									
1200	Switch timber, for a #8 switch, pressure treated	B-14	3.70	12.973	M.B.F.	3,375	545	84.50	4,004.50	4,625
1300	Complete set of timbers, 3.7 MBF for #8 switch	"	1	48	Total	13,000	2,025	310	15,335	17,700

34 11 93 – Track Appurtenances and Accessories

34 11 93.50 Track Accessories

		Crew	Daily Output	Labor-Hours	Unit	Material	Labor	Equipment	Total	Total Incl O&P
0010	**TRACK ACCESSORIES**									
0020	Car bumpers, test	B-14	2	24	Ea.	3,425	1,000	156	4,581	5,475
0100	Heavy duty		2	24		6,500	1,000	156	7,656	8,850
0200	Derails hand throw (sliding)		10	4.800		1,425	202	31.50	1,658.50	1,900
0300	Hand throw with standard timbers, open stand & target		8	6	↓	1,525	253	39	1,817	2,100
2400	Wheel stops, fixed		18	2.667	Pr.	960	112	17.35	1,089.35	1,250
2450	Hinged	↓	14	3.429	"	1,300	144	22.50	1,466.50	1,675

34 11 93.60 Track Material

		Crew	Daily Output	Labor-Hours	Unit	Material	Labor	Equipment	Total	Total Incl O&P
0010	**TRACK MATERIAL**									
0020	Track bolts				Ea.	3.72			3.72	4.09
0100	Joint bars				Pr.	83.50			83.50	92
0200	Spikes				Ea.	2.01			2.01	2.21
0300	Tie plates				"	13.95			13.95	15.35

34 41 Roadway Signaling and Control Equipment

34 41 13 – Traffic Signals

34 41 13.10 Traffic Signals Systems	Crew	Daily Output	Labor-Hours	Unit	Material	2018 Bare Costs Labor	2018 Bare Costs Equipment	Total	Total Incl O&P
0010 **TRAFFIC SIGNALS SYSTEMS**									
0020 Component costs									
0600 Crew employs crane/directional driller as required									
1000 Vertical mast with foundation									
1010 Mast sized for single arm to 40'; no lighting or power function	R-11	.50	112	Signal	10,400	6,200	1,325	17,925	22,300
1100 Horizontal arm									
1110 Per linear foot of arm	R-11	50	1.120	Signal	224	62	13.25	299.25	355
1200 Traffic signal									
1210 Includes signal, bracket, sensor, and wiring	R-11	2.50	22.400	Signal	1,375	1,250	265	2,890	3,650
1300 Pedestrian signals and callers									
1310 Includes four signals with brackets and two call buttons	R-11	2.50	22.400	Signal	3,200	1,250	265	4,715	5,675
1400 Controller, design, and underground conduit									
1410 Includes miscellaneous signage and adjacent surface work	R-11	.25	224	Signal	25,000	12,400	2,650	40,050	49,100

34 71 Roadway Construction

34 71 13 – Vehicle Barriers

34 71 13.17 Security Vehicle Barriers

	Crew	Daily Output	Labor-Hours	Unit	Material	2018 Bare Costs Labor	2018 Bare Costs Equipment	Total	Total Incl O&P
0010 **SECURITY VEHICLE BARRIERS**									
0020 Security planters excludes filling material									
0100 Concrete security planter, exposed aggregate 36" diam. x 30" high	B-11M	8	2	Ea.	540	93.50	48	681.50	785
0200 48" diam. x 36" high		8	2		855	93.50	48	996.50	1,125
0300 53" diam. x 18" high		8	2		700	93.50	48	841.50	965
0400 72" diam. x 18" high with seats		8	2		1,975	93.50	48	2,116.50	2,350
0450 84" diam. x 36" high		8	2		2,150	93.50	48	2,291.50	2,575
0500 36" x 36" x 24" high square		8	2		475	93.50	48	616.50	720
0600 36" x 36" x 30" high square		8	2		620	93.50	48	761.50	875
0700 48" L x 24" W x 30" H rectangle		8	2		480	93.50	48	621.50	720
0800 72" L x 24" W x 30" H rectangle		8	2		555	93.50	48	696.50	805
0900 96" L x 24" W x 30" H rectangle		8	2		720	93.50	48	861.50	990
0950 Decorative geometric concrete barrier, 96" L x 24" W x 36" H		8	2		750	93.50	48	891.50	1,025
1000 Concrete security planter, filling with washed sand or gravel <1 C.Y.		8	2		34.50	93.50	48	176	233
1050 2 C.Y. or less	↓	6	2.667		68.50	125	64	257.50	335
1200 Jersey concrete barrier, 10' L x 2' by 0.5' W x 30" H	B-21B	16	2.500		380	109	31	520	615
1300 10 or more same site		24	1.667		380	72.50	20.50	473	550
1400 10' L x 2' by 0.5' W x 32" H		16	2.500		670	109	31	810	940
1500 10 or more same site		24	1.667		670	72.50	20.50	763	875
1600 20' L x 2' by 0.5' W x 30" H		12	3.333		760	145	41.50	946.50	1,100
1700 10 or more same site		18	2.222		760	96.50	27.50	884	1,025
1800 20' L x 2' by 0.5' W x 32" H		12	3.333		685	145	41.50	871.50	1,025
1900 10 or more same site		18	2.222		685	96.50	27.50	809	935
2000 GFRC decorative security barrier per 10' section including concrete		4	10		3,475	435	124	4,034	4,625
2100 Per 12' section including concrete	↓	4	10	↓	4,175	435	124	4,734	5,375
2210 GFRC decorative security barrier will stop 30 MPH, 4,000 lb. vehicle									
2300 High security barrier base prep per 12' section on bare ground	B-11C	4	4	Ea.	25.50	187	78	290.50	400
2310 GFRC barrier base prep does not include haulaway of excavated matl.									
2400 GFRC decorative high security barrier per 12' section w/concrete	B-21B	3	13.333	Ea.	5,450	580	165	6,195	7,050
2410 GFRC decorative high security barrier will stop 50 MPH, 15,000 lb. vehicle									
2500 GFRC decorative impaler security barrier per 10' section w/prep	B-6	4	6	Ea.	2,125	262	78	2,465	2,800
2600 Per 12' section w/prep	"	4	6	"	2,550	262	78	2,890	3,275
2610 Impaler barrier should stop 50 MPH, 15,000 lb. vehicle w/some penetr.									
2700 Pipe bollards, steel, concrete filled/painted, 8' L x 4' D hole, 8" diam.	B-6	10	2.400	Ea.	765	105	31.50	901.50	1,050

34 71 Roadway Construction

34 71 13 – Vehicle Barriers

34 71 13.17 Security Vehicle Barriers

		Crew	Daily Output	Labor-Hours	Unit	Material	2018 Bare Costs Labor	Equipment	Total	Total Incl O&P
2710	Schedule 80 concrete bollards will stop 4,000 lb. vehicle @ 30 MPH									
2800	GFRC decorative jersey barrier cover per 10' section excludes soil	B-6	8	3	Ea.	1,000	131	39	1,170	1,350
2900	Per 12' section excludes soil		8	3		1,200	131	39	1,370	1,575
3000	GFRC decorative 8" diameter bollard cover		12	2		375	87.50	26	488.50	575
3100	GFRC decorative barrier face 10' section excludes earth backing	↓	10	2.400	↓	950	105	31.50	1,086.50	1,250
3200	Drop arm crash barrier, 15,000 lb. vehicle @ 50 MPH									
3205	Includes all material, labor for complete installation									
3210	12' width				Ea.				64,500	71,000
3310	24' width								86,000	95,000
3410	Wedge crash barrier, 10' width								98,000	108,000
3510	12.5' width								108,000	119,000
3520	15' width								118,500	130,000
3610	Sliding crash barrier, 12' width								178,500	196,000
3710	Sliding roller crash barrier, 20' width								198,000	217,500
3810	Sliding cantilever crash barrier, 20' width				▼				198,000	217,500
3890	Note: Raised bollard crash barriers should be used w/tire shredders									
3910	Raised bollard crash barrier, 10' width				Ea.				38,800	42,900
4010	12' width								44,900	49,000
4110	Raised bollard crash barrier, 10' width, solar powered								46,900	52,000
4210	12' width								53,000	58,000
4310	In ground tire shredder, 16' width				▼				43,900	47,900

34 71 13.26 Vehicle Guide Rails

		Crew	Daily Output	Labor-Hours	Unit	Material	2018 Bare Costs Labor	Equipment	Total	Total Incl O&P
0010	**VEHICLE GUIDE RAILS**									
0012	Corrugated stl., galv. stl. posts, 6'-3" OC	B-80	850	.038	L.F.	26	1.67	.72	28.39	32
0200	End sections, galvanized, flared		50	.640	Ea.	100	28.50	12.25	140.75	166
0300	Wrap around end		50	.640	"	143	28.50	12.25	183.75	213
0400	Timber guide rail, 4" x 8" with 6" x 8" wood posts, treated		960	.033	L.F.	12.90	1.48	.64	15.02	17.10
0600	Cable guide rail, 3 at 3/4" cables, steel posts, single face		900	.036		11.35	1.58	.68	13.61	15.65
0700	Wood posts		950	.034		13.15	1.49	.64	15.28	17.40
0900	Guide rail, steel box beam, 6" x 6"		120	.267		36	11.85	5.10	52.95	63
1100	Median barrier, steel box beam, 6" x 8"	↓	215	.149		49	6.60	2.85	58.45	67
1400	Resilient guide fence and light shield, 6' high	B-2	130	.308	▼	20.50	12.40		32.90	41.50
1500	Concrete posts, individual, 6'-5", triangular	B-80	110	.291	Ea.	71	12.90	5.55	89.45	104
1550	Square	"	110	.291	"	76	12.90	5.55	94.45	109
2000	Median, precast concrete, 3'-6" high, 2' wide, single face	B-29	380	.147	L.F.	57.50	6.45	2.29	66.24	75.50
2200	Double face	"	340	.165	"	66	7.20	2.56	75.76	86.50
2400	Speed bumps, thermoplastic, 10-1/2" x 2-1/4" x 48" long	B-2	120	.333	Ea.	103	13.40		116.40	134
3030	Impact barrier, MUTCD, barrel type	B-16	30	1.067	"	395	44.50	18.10	457.60	520

34 71 19 – Vehicle Delineators

34 71 19.13 Fixed Vehicle Delineators

		Crew	Daily Output	Labor-Hours	Unit	Material	2018 Bare Costs Labor	Equipment	Total	Total Incl O&P
0010	**FIXED VEHICLE DELINEATORS**									
0020	Crash barriers									
0100	Traffic channelizing pavement markers, layout only	A-7	2000	.012	Ea.		.66	.03	.69	1.03
0110	13" x 7-1/2" x 2-1/2" high, non-plowable, install	2 Clab	96	.167		28	6.65		34.65	41
0200	8" x 8" x 3-1/4" high, non-plowable, install		96	.167		27	6.65		33.65	40
0230	4" x 4" x 3/4" high, non-plowable, install	↓	120	.133		2.50	5.30		7.80	10.85
0240	9-1/4" x 5-7/8" x 1/4" high, plowable, concrete pavmt.	A-2A	70	.343		4.58	14.20	4.16	22.94	31
0250	9-1/4" x 5-7/8" x 1/4" high, plowable, asphalt pavmt.	"	120	.200		3.66	8.30	2.42	14.38	19.25
0300	Barrier and curb delineators, reflectorized, 4" x 6"	2 Clab	150	.107		3.53	4.25		7.78	10.35
0310	3" x 5"	"	150	.107	▼	5.25	4.25		9.50	12.25
0500	Rumble strip, polycarbonate									
0510	24" x 3-1/2" x 1/2" high	2 Clab	50	.320	Ea.	9.35	12.75		22.10	29.50

For customer support on your Building Construction Costs with RSMeans data, call 800.448.8182.

34 72 Railway Construction

34 72 16 – Railway Siding

34 72 16.50 Railroad Sidings	Crew	Daily Output	Labor-Hours	Unit	Material	2018 Bare Costs Labor	Equipment	Total	Total Incl O&P
0010 **RAILROAD SIDINGS** R347216-10									
0800 Siding, yard spur, level grade									
0820 100 lb. new rail	B-14	57	.842	L.F.	125	35.50	5.50	166	197
1002 Steel ties in concrete, incl. fasteners & plates									
1020 100 lb. new rail	B-14	22	2.182	L.F.	189	92	14.20	295.20	365

34 72 16.60 Railroad Turnouts

	Crew	Daily Output	Labor-Hours	Unit	Material	2018 Bare Costs Labor	Equipment	Total	Total Incl O&P
0010 **RAILROAD TURNOUTS**									
2200 Turnout, #8 complete, w/rails, plates, bars, frog, switch point,									
2250 timbers, and ballast to 6" below bottom of ties									
2280 90 lb. rails	B-13	.25	224	Ea.	37,000	9,800	2,375	49,175	58,000
2290 90 lb. relay rails		.25	224		24,900	9,800	2,375	37,075	44,900
2300 100 lb. rails		.25	224		41,200	9,800	2,375	53,375	63,000
2310 100 lb. relay rails		.25	224		27,700	9,800	2,375	39,875	48,000
2320 110 lb. rails		.25	224		45,400	9,800	2,375	57,575	67,500
2330 110 lb. relay rails		.25	224		30,600	9,800	2,375	42,775	51,000
2340 115 lb. rails		.25	224		49,800	9,800	2,375	61,975	72,500
2350 115 lb. relay rails		.25	224		33,200	9,800	2,375	45,375	54,000
2360 132 lb. rails		.25	224		57,000	9,800	2,375	69,175	80,000
2370 132 lb. relay rails	▼	.25	224	▼	37,700	9,800	2,375	49,875	59,000

Division Notes

	CREW	DAILY OUTPUT	LABOR-HOURS	UNIT	BARE COSTS				TOTAL INCL O&P
					MAT.	LABOR	EQUIP.	TOTAL	

Estimating Tips

35 01 50 Operation and Maintenance of Marine Construction

Includes unit price lines for pile cleaning and pile wrapping for protection.

35 20 16 Hydraulic Gates

This subdivision includes various types of gates that are commonly used in waterway and canal construction. Various earthwork items and structural support are found in Division 31, and concrete work is found in Division 3.

35 20 23 Dredging

This subdivision includes barge and shore dredging systems for rivers, canals, and channels.

35 31 00 Shoreline Protection

This subdivision includes breakwaters, bulkheads, and revetments for ocean and river inlets. Additional earthwork may be required from Division 31 and concrete work from Division 3.

35 41 00 Levees

Contains information on levee construction, including the estimated cost of clay cone material.

35 49 00 Waterway Structures

This subdivision includes breakwaters and bulkheads for canals.

35 51 00 Floating Construction

This section includes floating piers, docks, and dock accessories. Fixed Pier Timber Construction is found in Section 06 13 33. Driven piles are found in Division 31, as well as sheet piling, cofferdams, and riprap.

Reference Numbers

Reference numbers are shown at the beginning of some major classifications. These numbers refer to related items in the Reference Section. The reference information may be an estimating procedure, an alternate pricing method, or technical information.

Note: Not all subdivisions listed here necessarily appear. ■

Did you know?

RSMeans data is available through our online application with 24/7 access:

■ Search for unit prices by keyword

■ Leverage the most up-to-date data

■ Build and export estimates

Try it free for 30 days!
www.rsmeans.com/2018freetrial

No part of this cost data may be reproduced, stored in a retrieval system, or transmitted in any form or by any means without prior written permission of Gordian.

35 24 Dredging

35 24 13 – Suction Dredging

35 24 13.13 Cutter Suction Dredging

	35 24 13.13 Cutter Suction Dredging	Crew	Daily Output	Labor-Hours	Unit	Material	2018 Bare Costs Labor	Equipment	Total	Total Incl O&P
0010	**CUTTER SUCTION DREDGING**									
0015	Add, Marine Equipment Rental, See Section 01 54 33.80									
1000	Hydraulic method, pumped 1,000' to shore dump, minimum	B-57	460	.104	B.C.Y.		4.83	3.97	8.80	11.65
1100	Maximum		310	.155			7.15	5.90	13.05	17.35
1400	Into scows dumped 20 miles, minimum		425	.113			5.20	4.30	9.50	12.65
1500	Maximum		243	.198			9.15	7.50	16.65	22
1600	For inland rivers and canals in South, deduct								30%	30%

35 24 23 – Clamshell Dredging

35 24 23.13 Mechanical Dredging

	35 24 23.13 Mechanical Dredging	Crew	Daily Output	Labor-Hours	Unit	Material	2018 Bare Costs Labor	Equipment	Total	Total Incl O&P
0010	**MECHANICAL DREDGING**									
0015	Add, Marine Equipment Rental, See Section 01 54 33.80									
0020	Dredging mobilization and demobilization, add to below, minimum	B-8	.53	121	Total	5,575	5,425	11,000	14,400
0100	Maximum	"	.10	640	"	29,600	28,800	58,400	76,500
0300	Barge mounted clamshell excavation into scows									
0310	Dumped 20 miles at sea, minimum	B-57	310	.155	B.C.Y.		7.15	5.90	13.05	17.35
0400	Maximum	"	213	.225	"		10.40	8.60	19	25.50
0500	Barge mounted dragline or clamshell, hopper dumped,									
0510	pumped 1,000' to shore dump, minimum	B-57	340	.141	B.C.Y.		6.55	5.40	11.95	15.80
0525	All pumping uses 2,000 gallons of water per cubic yard									
0600	Maximum	B-57	243	.198	B.C.Y.		9.15	7.50	16.65	22

35 51 Floating Construction

35 51 13 – Floating Piers

35 51 13.23 Floating Wood Piers

	35 51 13.23 Floating Wood Piers	Crew	Daily Output	Labor-Hours	Unit	Material	2018 Bare Costs Labor	Equipment	Total	Total Incl O&P
0010	**FLOATING WOOD PIERS**									
0020	Polyethylene encased polystyrene, no pilings included	F-3	330	.121	S.F.	30.50	6.30	1.50	38.30	45
0200	Pile supported, shore constructed, bare, 3" decking		130	.308		27.50	15.95	3.82	47.27	58
0250	4" decking		120	.333		31	17.25	4.13	52.38	64.50
0400	Floating, small boat, prefab, no shore facilities, minimum		250	.160		27	8.30	1.98	37.28	45
0500	Maximum		150	.267		58	13.80	3.31	75.11	88
0700	Per slip, minimum (180 S.F. each)		1.59	25.157	Ea.	5,525	1,300	310	7,135	8,400
0800	Maximum		1.40	28.571	"	9,000	1,475	355	10,830	12,500

For customer support on your Building Construction Costs with RSMeans data, call 800.448.8182.

Estimating Tips

Products such as conveyors, material handling cranes and hoists, as well as other items specified in this division, require trained installers. The general contractor may not have any choice as to who will perform the installation or when it will be performed. Long lead times are often required for these products, making early decisions in purchasing and scheduling necessary. The installation of this type of equipment may require the embedment of mounting hardware during construction of floors, structural walls, or interior walls/partitions. Electrical connections will require coordination with the electrical contractor.

Reference Numbers

Reference numbers are shown at the beginning of some major classifications. These numbers refer to related items in the Reference Section. The reference information may be an estimating procedure, an alternate pricing method, or technical information.

Note: Not all subdivisions listed here necessarily appear. ■

Did you know?

RSMeans data is available through our online application with 24/7 access:

■ Search for unit prices by keyword

■ Leverage the most up-to-date data

■ Build and export estimates

Try it free for 30 days!
www.rsmeans.com/2018freetrial

No part of this cost data may be reproduced, stored in a retrieval system, or transmitted in any form or by any means without prior written permission of Gordian.

41 21 Conveyors

41 21 23 – Piece Material Conveyors

41 21 23.16 Container Piece Material Conveyors

	Crew	Daily Output	Labor-Hours	Unit	Material	2018 Bare Costs Labor	Equipment	Total	Total Incl O&P
0010 **CONTAINER PIECE MATERIAL CONVEYORS**									
0020 Gravity fed, 2" rollers, 3" OC									
0050 10' sections with 2 supports, 600 lb. capacity, 18" wide				Ea.	515			515	565
0100 24" wide					580			580	640
0150 1,400 lb. capacity, 18" wide					685			685	755
0200 24" wide					580			580	640
0350 Horizontal belt, center drive and takeup, 60 fpm									
0400 16" belt, 26.5' length	2 Mill	.50	32	Ea.	3,550	1,700		5,250	6,350
0450 24" belt, 41.5' length		.40	40		5,475	2,100		7,575	9,100
0500 61.5' length		.30	53.333		7,700	2,825		10,525	12,600
0600 Inclined belt, 10' rise with horizontal loader and									
0620 End idler assembly, 27.5' length, 18" belt	2 Mill	.30	53.333	Ea.	7,700	2,825		10,525	12,600
0700 24" belt	"	.15	107	"	9,575	5,625		15,200	18,700
3600 Monorail, overhead, manual, channel type									
3700 125 lb./L.F.	1 Mill	26	.308	L.F.	21	16.25		37.25	46.50
3900 500 lb./L.F.	"	21	.381	"	20.50	20		40.50	52
4000 Trolleys for above, 2 wheel, 125 lb. capacity				Ea.	87.50			87.50	96
4200 4 wheel, 250 lb. capacity					345			345	380
4300 8 wheel, 500 lb. capacity					800			800	880

41 22 Cranes and Hoists

41 22 13 – Cranes

41 22 13.10 Crane Rail

	Crew	Daily Output	Labor-Hours	Unit	Material	2018 Bare Costs Labor	Equipment	Total	Total Incl O&P
0010 **CRANE RAIL**									
0020 Box beam bridge, no equipment included	E-4	3400	.009	Lb.	1.35	.52	.03	1.90	2.36
0200 Running track only, 104 lb. per yard		5600	.006	"	.67	.31	.02	1	1.27
0210 Running track only, 104 lb. per yard, 20' piece		160	.200	L.F.	23.50	11.05	.62	35.17	44.50

41 22 13.13 Bridge Cranes

	Crew	Daily Output	Labor-Hours	Unit	Material	2018 Bare Costs Labor	Equipment	Total	Total Incl O&P
0010 **BRIDGE CRANES**									
0100 1 girder, 20' span, 3 ton	M-3	1	34	Ea.	26,700	2,100	143	28,943	32,600
0125 5 ton		1	34		29,300	2,100	143	31,543	35,500
0150 7.5 ton		1	34		34,700	2,100	143	36,943	41,400
0175 10 ton		.80	42.500		46,200	2,625	179	49,004	55,000
0200 15 ton		.80	42.500		59,500	2,625	179	62,304	69,500
0225 30' span, 3 ton		1	34		27,700	2,100	143	29,943	33,800
0250 5 ton		1	34		30,600	2,100	143	32,843	36,900
0275 7.5 ton		1	34		36,500	2,100	143	38,743	43,400
0300 10 ton		.80	42.500		47,900	2,625	179	50,704	56,500
0325 15 ton		.80	42.500		62,500	2,625	179	65,304	72,500
0350 2 girder, 40' span, 3 ton	M-4	.50	72		45,300	4,375	365	50,040	57,000
0375 5 ton		.50	72		47,600	4,375	365	52,340	59,500
0400 7.5 ton		.50	72		52,000	4,375	365	56,740	64,500
0425 10 ton		.40	90		61,500	5,475	460	67,435	76,000
0450 15 ton		.40	90		83,500	5,475	460	89,435	100,500
0475 25 ton		.30	120		98,500	7,300	610	106,410	119,500
0500 50' span, 3 ton		.50	72		52,000	4,375	365	56,740	64,000
0525 5 ton		.50	72		54,000	4,375	365	58,740	66,500
0550 7.5 ton		.50	72		58,000	4,375	365	62,740	70,500
0575 10 ton		.40	90		67,000	5,475	460	72,935	82,500
0600 15 ton		.40	90		87,500	5,475	460	93,435	105,000
0625 25 ton		.30	120		103,000	7,300	610	110,910	125,000

For customer support on your Building Construction Costs with RSMeans data, call 800.448.8182.

41 22 Cranes and Hoists

41 22 13 – Cranes

41 22 13.19 Jib Cranes	Crew	Daily Output	Labor-Hours	Unit	Material	2018 Bare Costs Labor	Equipment	Total	Total Incl O&P
0010 **JIB CRANES**									
0020 Jib crane, wall cantilever, 500 lb. capacity, 8' span	2 Mill	1	16	Ea.	1,425	845		2,270	2,800
0040 12' span		1	16		1,550	845		2,395	2,950
0060 16' span		1	16		1,725	845		2,570	3,125
0080 20' span		1	16		2,200	845		3,045	3,625
0100 1,000 lb. capacity, 8' span		1	16		1,525	845		2,370	2,900
0120 12' span		1	16		1,650	845		2,495	3,050
0130 16' span		1	16		2,200	845		3,045	3,650
0150 20' span		1	16		2,575	845		3,420	4,050

41 22 23 – Hoists

41 22 23.10 Material Handling

41 22 23.10 Material Handling	Crew	Daily Output	Labor-Hours	Unit	Material	2018 Bare Costs Labor	Equipment	Total	Total Incl O&P
0010 **MATERIAL HANDLING**, cranes, hoists and lifts									
1500 Cranes, portable hydraulic, floor type, 2,000 lb. capacity				Ea.	4,200			4,200	4,625
1600 4,000 lb. capacity					4,825			4,825	5,300
1800 Movable gantry type, 12' to 15' range, 2,000 lb. capacity					4,225			4,225	4,625
1900 6,000 lb. capacity					6,400			6,400	7,025
2100 Hoists, electric overhead, chain, hook hung, 15' lift, 1 ton cap.					2,775			2,775	3,050
2200 3 ton capacity					3,825			3,825	4,200
2500 5 ton capacity					7,650			7,650	8,425
2600 For hand-pushed trolley, add					15%				
2700 For geared trolley, add					30%				
2800 For motor trolley, add					75%				
3000 For lifts over 15', 1 ton, add				L.F.	24.50			24.50	27
3100 5 ton, add				"	72			72	79
3110 Air powered hoist, 500 lb. capacity	2 Mill	2.67	5.993	Ea.	4,925	315		5,240	5,850
3300 Lifts, scissor type, portable, electric, 36" high, 2,000 lb.					4,025			4,025	4,425
3400 48" high, 4,000 lb.					4,575			4,575	5,025

For customer support on your Building Construction Costs with RSMeans data, call 800.448.8182.

Division Notes

	CREW	DAILY OUTPUT	LABOR-HOURS	UNIT	BARE COSTS				TOTAL INCL O&P
					MAT.	LABOR	EQUIP.	TOTAL	

Estimating Tips

This section involves equipment and construction costs for air noise and odor pollution control systems. These systems may be interrelated and care must be taken that the complete systems are estimated. For example, air pollution equipment may include dust and air-entrained particles that have to be collected. The vacuum systems could be noisy, requiring silencers to reduce noise pollution, and the collected solids have to be disposed of to prevent solid pollution.

Reference Numbers

Reference numbers are shown at the beginning of some major classifications. These numbers refer to related items in the Reference Section. The reference information may be an estimating procedure, an alternate pricing method, or technical information.

Note: Not all subdivisions listed here necessarily appear. ∎

Did you know?

RSMeans data is available through our online application with 24/7 access:

- Search for unit prices by keyword
- Leverage the most up-to-date data
- Build and export estimates

Try it free for 30 days!
www.rsmeans.com/2018freetrial

No part of this cost data may be reproduced, stored in a retrieval system, or transmitted in any form or by any means without prior written permission of Gordian.

44 11 16.10 Dust Collection Systems	Crew	Daily Output	Labor-Hours	Unit	Material	2018 Bare Costs Labor	Equipment	Total	Total Incl O&P
0010 **DUST COLLECTION SYSTEMS** Commercial/Industrial									
0120 Central vacuum units									
0130 Includes stand, filters and motorized shaker									
0200 500 CFM, 10" inlet, 2 HP	Q-20	2.40	8.333	Ea.	4,950	455		5,405	6,125
0220 1,000 CFM, 10" inlet, 3 HP		2.20	9.091		5,200	495		5,695	6,475
0240 1,500 CFM, 10" inlet, 5 HP		2	10		5,475	545		6,020	6,850
0260 3,000 CFM, 13" inlet, 10 HP		1.50	13.333		15,400	730		16,130	18,000
0280 5,000 CFM, 16" inlet, 2 @ 10 HP		1	20		16,400	1,100		17,500	19,700
1000 Vacuum tubing, galvanized									
1100 2-1/8" OD, 16 ga.	Q-9	440	.036	L.F.	3.37	1.96		5.33	6.70
1110 2-1/2" OD, 16 ga.		420	.038		3.70	2.05		5.75	7.20
1120 3" OD, 16 ga.		400	.040		4.72	2.15		6.87	8.50
1130 3-1/2" OD, 16 ga.		380	.042		6.70	2.27		8.97	10.80
1140 4" OD, 16 ga.		360	.044		6.85	2.39		9.24	11.20
1150 5" OD, 14 ga.		320	.050		12.35	2.69		15.04	17.65
1160 6" OD, 14 ga.		280	.057		13.80	3.08		16.88	19.90
1170 8" OD, 14 ga.		200	.080		21	4.31		25.31	29.50
1180 10" OD, 12 ga.		160	.100		40.50	5.40		45.90	52.50
1190 12" OD, 12 ga.		120	.133		53.50	7.20		60.70	70
1200 14" OD, 12 ga.		80	.200		61	10.75		71.75	83.50
1940 Hose, flexible wire reinforced rubber									
1956 3" diam.	Q-9	400	.040	L.F.	6.55	2.15		8.70	10.50
1960 4" diam.		360	.044		7.60	2.39		9.99	12
1970 5" diam.		320	.050		9.85	2.69		12.54	14.90
1980 6" diam.		280	.057		10.30	3.08		13.38	16.05
2000 90° elbow, slip fit									
2110 2-1/8" diam.	Q-9	70	.229	Ea.	10.85	12.30		23.15	30.50
2120 2-1/2" diam.		65	.246		15.20	13.25		28.45	36.50
2130 3" diam.		60	.267		20.50	14.35		34.85	44.50
2140 3-1/2" diam.		55	.291		25.50	15.65		41.15	52
2150 4" diam.		50	.320		31.50	17.25		48.75	61
2160 5" diam.		45	.356		60	19.15		79.15	95
2170 6" diam.		40	.400		82.50	21.50		104	124
2180 8" diam.		30	.533		157	28.50		185.50	216
2400 45° elbow, slip fit									
2410 2-1/8" diam.	Q-9	70	.229	Ea.	9.50	12.30		21.80	29
2420 2-1/2" diam.		65	.246		14	13.25		27.25	35.50
2430 3" diam.		60	.267		16.90	14.35		31.25	40.50
2440 3-1/2" diam.		55	.291		21.50	15.65		37.15	47.50
2450 4" diam.		50	.320		27.50	17.25		44.75	57
2460 5" diam.		45	.356		46.50	19.15		65.65	80
2470 6" diam.		40	.400		63	21.50		84.50	102
2480 8" diam.		35	.457		123	24.50		147.50	174
2800 90° TY, slip fit thru 6" diam.									
2810 2-1/8" diam.	Q-9	42	.381	Ea.	17.75	20.50		38.25	51
2820 2-1/2" diam.		39	.410		26.50	22		48.50	62.50
2830 3" diam.		36	.444		37	24		61	77
2840 3-1/2" diam.		33	.485		39	26		65	82.50
2850 4" diam.		30	.533		41.50	28.50		70	90
2860 5" diam.		27	.593		70.50	32		102.50	126
2870 6" diam.		24	.667		101	36		137	166
2880 8" diam., butt end					155			155	170
2890 10" diam., butt end					475			475	525

44 11 16.10 Dust Collection Systems		Crew	Daily Output	Labor-Hours	Unit	Material	2018 Bare Costs Labor	Equipment	Total	Total Incl O&P
2900	12" diam., butt end				Ea.	475			475	525
2910	14" diam., butt end					855			855	940
2920	6" x 4" diam., butt end					106			106	116
2930	8" x 4" diam., butt end					200			200	220
2940	10" x 4" diam., butt end					263			263	289
2950	12" x 4" diam., butt end					300			300	330
3100	90° elbow, butt end, segmented									
3110	8" diam., butt end, segmented				Ea.	157			157	172
3120	10" diam., butt end, segmented					390			390	430
3130	12" diam., butt end, segmented					495			495	545
3140	14" diam., butt end, segmented					620			620	680
3200	45° elbow, butt end, segmented									
3210	8" diam., butt end, segmented				Ea.	123			123	136
3220	10" diam., butt end, segmented					282			282	310
3230	12" diam., butt end, segmented					288			288	315
3240	14" diam., butt end, segmented					585			585	645
3400	All butt end fittings require one coupling per joint.									
3410	Labor for fitting included with couplings.									
3460	Compression coupling, galvanized, neoprene gasket									
3470	2-1/8" diam.	Q-9	44	.364	Ea.	11.40	19.55		30.95	42.50
3480	2-1/2" diam.		44	.364		11.40	19.55		30.95	42.50
3490	3" diam.		38	.421		16.40	22.50		38.90	52.50
3500	3-1/2" diam.		35	.457		18.35	24.50		42.85	57.50
3510	4" diam.		33	.485		19.90	26		45.90	62
3520	5" diam.		29	.552		22.50	29.50		52	70.50
3530	6" diam.		26	.615		26.50	33		59.50	80
3540	8" diam.		22	.727		48.50	39		87.50	113
3550	10" diam.		20	.800		71	43		114	144
3560	12" diam.		18	.889		88.50	48		136.50	171
3570	14" diam.		16	1		125	54		179	220
3800	Air gate valves, galvanized									
3810	2-1/8" diam.	Q-9	30	.533	Ea.	117	28.50		145.50	173
3820	2-1/2" diam.		28	.571		127	31		158	187
3830	3" diam.		26	.615		135	33		168	200
3840	4" diam.		23	.696		156	37.50		193.50	229
3850	6" diam.		18	.889		217	48		265	310

Division Notes

	CREW	DAILY OUTPUT	LABOR-HOURS	UNIT	BARE COSTS				TOTAL INCL O&P
					MAT.	LABOR	EQUIP.	TOTAL	

Estimating Tips

This division contains information about water and wastewater equipment and systems, which was formerly located in Division 44. The main areas of focus are total wastewater treatment plants and components of wastewater treatment plants. Also included in this section are oil/water separators for wastewater treatment.

Reference Numbers

Reference numbers are shown at the beginning of some major classifications. These numbers refer to related items in the Reference Section. The reference information may be an estimating procedure, an alternate pricing method, or technical information.

Note: Not all subdivisions listed here necessarily appear. ■

Did you know?

RSMeans data is available through our online application with 24/7 access:

■ Search for unit prices by keyword

■ Leverage the most up-to-date data

■ Build and export estimates

Try it free for 30 days!
www.rsmeans.com/2018freetrial

No part of this cost data may be reproduced, stored in a retrieval system, or transmitted in any form or by any means without prior written permission of Gordian.

46 07 53 – Packaged Wastewater Treatment Equipment

46 07 53.10 Biological Pkg. Wastewater Treatment Plants	Crew	Daily Output	Labor-Hours	Unit	Material	2018 Bare Costs Labor	Equipment	Total	Total Incl O&P
0010 **BIOLOGICAL PACKAGED WASTEWATER TREATMENT PLANTS**									
0011 Not including fencing or external piping									
0020 Steel packaged, blown air aeration plants									
0100 1,000 GPD				Gal.				55	60.50
0200 5,000 GPD								22	24
0300 15,000 GPD								22	24
0400 30,000 GPD								15.40	16.95
0500 50,000 GPD								11	12.10
0600 100,000 GPD								9.90	10.90
0700 200,000 GPD								8.80	9.70
0800 500,000 GPD								7.70	8.45
1000 Concrete, extended aeration, primary and secondary treatment									
1010 10,000 GPD				Gal.				22	24
1100 30,000 GPD								15.40	16.95
1200 50,000 GPD								11	12.10
1400 100,000 GPD								9.90	10.90
1500 500,000 GPD								7.70	8.45
1700 Municipal wastewater treatment facility									
1720 1.0 MGD				Gal.				11	12.10
1740 1.5 MGD								10.60	11.65
1760 2.0 MGD								10	11
1780 3.0 MGD								7.80	8.60
1800 5.0 MGD								5.80	6.70
2000 Holding tank system, not incl. excavation or backfill									
2010 Recirculating chemical water closet	2 Plum	4	4	Ea.	545	249		794	975
2100 For voltage converter, add	"	16	1		310	62		372	435
2200 For high level alarm, add	1 Plum	7.80	1.026		120	63.50		183.50	228

46 07 53.20 Wastewater Treatment System

	Crew	Daily Output	Labor-Hours	Unit	Material	2018 Bare Costs Labor	Equipment	Total	Total Incl O&P
0010 **WASTEWATER TREATMENT SYSTEM**									
0020 Fiberglass, 1,000 gallon	B-21	1.29	21.705	Ea.	4,325	1,000	101	5,426	6,375
0100 1,500 gallon	"	1.03	27.184	"	8,650	1,250	126	10,026	11,600

For customer support on your Building Construction Costs with RSMeans data, call 800.448.8182.

Estimating Tips

- When estimating costs for the installation of electrical power generation equipment, factors to review include access to the job site, access and setting up at the installation site, required connections, uncrating pads, anchors, leveling, final assembly of the components, and temporary protection from physical damage, such as environmental exposure.

- Be aware of the costs of equipment supports, concrete pads, and vibration isolators. Cross-reference them against other trades' specifications. Also, review site and structural drawings for items that must be included in the estimates.

- It is important to include items that are not documented in the plans and specifications but must be priced. These items include, but are not limited to, testing, dust protection, roof penetration, core drilling concrete floors and walls, patching, cleanup, and final adjustments. Add a contingency or allowance for utility company fees for power hookups, if needed.

- The project size and scope of electrical power generation equipment will have a significant impact on cost. The intent of RSMeans cost data is to provide a benchmark cost so that owners, engineers, and electrical contractors will have a comfortable number with which to start a project. Additionally, there are many websites available to use for research and to obtain a vendor's quote to finalize costs.

Reference Numbers

Reference numbers are shown at the beginning of some major classifications. These numbers refer to related items in the Reference Section. The reference information may be an estimating procedure, an alternate pricing method, or technical information.

Note: Not all subdivisions listed here necessarily appear. ■

Did you know?

RSMeans data is available through our online application with 24/7 access:

- Search for unit prices by keyword
- Leverage the most up-to-date data
- Build and export estimates

Try it free for 30 days!
www.rsmeans.com/2018freetrial

No part of this cost data may be reproduced, stored in a retrieval system, or transmitted in any form or by any means without prior written permission of Gordian.

Note: Trade Service, in part, has been used as a reference source for some of the material prices used in Division 48.

48 15 13.50 Wind Turbines and Components	Crew	Daily Output	Labor-Hours	Unit	Material	2018 Bare Costs Labor	Equipment	Total	Total Incl O&P
0010 **WIND TURBINES & COMPONENTS**									
0500 Complete system, grid connected									
1000 20 kW, 31' diam., incl. labor & material	G			System				49,900	49,900
1010 Enhanced	G							92,000	92,000
1500 10 kW, 23' diam., incl. labor & material	G							74,000	74,000
2000 2.4 kW, 12' diam., incl. labor & material	G			↓				18,000	18,000
2900 Component system									
3200 1,000 W, 9' diam.	G	2.05	3.902	Ea.	1,500	227		1,727	2,000
3400 Mounting hardware									
3500 30' guyed tower kit	G	2 Clab 5.12	3.125	Ea.	395	125		520	620
3505 3' galvanized helical earth screw	G	1 Clab 8	1		48	40		88	113
3510 Attic mount kit	G	1 Rofc 2.56	3.125		202	137		339	455
3520 Roof mount kit	G	1 Clab 3.41	2.346	↓	292	93.50		385.50	460
8900 Equipment									
9100 DC to AC inverter for, 48 V, 4,000 W	G	1 Elec 2	4	Ea.	2,275	233		2,508	2,850

For customer support on your Building Construction Costs with RSMeans data, call 800.448.8182.

Reference Section

All the reference information is in one section, making it easy to find what you need to know and easy to use the data set on a daily basis. This section is visually identified by a vertical black bar on the page edges.

In this Reference Section, we've included Equipment Rental Costs, a listing of rental and operating costs; Crew Listings, a full listing of all crews and equipment and their costs; Historical Cost Indexes for cost comparisons over time; City Cost Indexes and Location Factors for adjusting costs to the region you are in; Reference Tables, where you will find explanations, estimating information and procedures, and technical data; Change Orders, information on pricing changes to contract documents; and an explanation of all the Abbreviations in the data set.

Table of Contents

Construction Equipment Rental Costs   727

Crew Listings   739

Historical Cost Indexes   776

City Cost Indexes   777

Location Factors   820

Reference Tables   826

R01 General Requirements 826
R02 Existing Conditions 839
R03 Concrete 842
R04 Masonry 857
R05 Metals 859
R06 Wood, Plastics & Composites 866
R07 Thermal & Moisture Protection 866
R08 Openings 868
R09 Finishes 871
R13 Special Construction 874
R14 Conveying Equipment 876

Reference Tables (cont.)

R22 Plumbing 877
R23 Heating, Ventilating & Air Conditioning 878
R26 Electrical 879
R31 Earthwork 882
R32 Exterior Improvements 886
R33 Utilities 887
R34 Transportation 887

Change Orders   888

Square Foot Costs   891

Abbreviations   896

Estimating Tips

- This section contains the average costs to rent and operate hundreds of pieces of construction equipment. This is useful information when estimating the time and material requirements of any particular operation in order to establish a unit or total cost. Bare equipment costs shown on a unit cost line include not only rental, but also operating costs for equipment under normal use.

Rental Costs

- Equipment rental rates are obtained from the following industry sources throughout North America: contractors, suppliers, dealers, manufacturers, and distributors.

- Rental rates vary throughout the country, with larger cities generally having lower rates. Lease plans for new equipment are available for periods in excess of six months, with a percentage of payments applying toward purchase.

- Monthly rental rates vary from 2% to 5% of the purchase price of the equipment depending on the anticipated life of the equipment and its wearing parts.

- Weekly rental rates are about 1/3 of the monthly rates, and daily rental rates are about 1/3 of the weekly rate.

- Rental rates can also be treated as reimbursement costs for contractor-owned equipment. Owned equipment costs include depreciation, loan payments, interest, taxes, insurance, storage, and major repairs.

Operating Costs

- The operating costs include parts and labor for routine servicing, such as the repair and replacement of pumps, filters, and worn lines. Normal operating expendables, such as fuel, lubricants, tires, and electricity (where applicable), are also included.

- Extraordinary operating expendables with highly variable wear patterns, such as diamond bits and blades, are excluded. These costs can be found as material costs in the Unit Price section.

- The hourly operating costs listed do not include the operator's wages.

Equipment Cost/Day

- Any power equipment required by a crew is shown in the Crew Listings with a daily cost.

- This daily cost of equipment needed by a crew includes both the rental cost and the operating cost and is based on dividing the weekly rental rate by 5 (number of working days in the week), and then adding the hourly operating cost times 8 (the number of hours in a day). This "Equipment Cost/Day" is shown in the far right column of the Equipment Rental section.

- If equipment is needed for only one or two days, it is best to develop your own cost by including components for daily rent and hourly operating costs. This is important when the listed Crew for a task does not contain the equipment needed, such as a crane for lifting mechanical heating/cooling equipment up onto a roof.

- If the quantity of work is less than the crew's Daily Output shown for a Unit Price line item that includes a bare unit equipment cost, the recommendation is to estimate one day's rental cost and operating cost for equipment shown in the Crew Listing for that line item.

Mobilization, Demobilization Costs

- The cost to move construction equipment from an equipment yard or rental company to the job site and back again is not included in equipment rental costs listed in the Reference Section, nor in the bare equipment cost of any unit price line item, nor in any equipment costs shown in the Crew Listings.

- Mobilization (to the site) and demobilization (from the site) costs can be found in the Unit Price section.

- If a piece of equipment is already at the job site, it is not appropriate to utilize mobilization, demobilization costs again in an estimate. ∎

Did you know?

RSMeans data is available through our online application with 24/7 access:

- Search for unit prices by keyword
- Leverage the most up-to-date data
- Build and export estimates

Try it free for 30 days!
www.rsmeans.com/2018freetrial

No part of this cost data may be reproduced, stored in a retrieval system, or transmitted in any form or by any means without prior written permission of Gordian.

		UNIT	HOURLY OPER. COST	RENT PER DAY	RENT PER WEEK	RENT PER MONTH	EQUIPMENT COST/DAY
10	**0010**	**CONCRETE EQUIPMENT RENTAL** without operators R015433 -10						**10**
	0200	Bucket, concrete lightweight, 1/2 C.Y.	Ea.	.85	24.50	73	219	21.40
	0300	1 C.Y.		.95	28.50	85	255	24.60
	0400	1-1/2 C.Y.		1.20	38.50	115	345	32.60
	0500	2 C.Y.		1.30	46.50	140	420	38.40
	0580	8 C.Y.		6.35	263	790	2,375	208.80
	0600	Cart, concrete, self-propelled, operator walking, 10 C.F.		2.85	58.50	175	525	57.80
	0700	Operator riding, 18 C.F.		4.80	98.50	295	885	97.40
	0800	Conveyer for concrete, portable, gas, 16" wide, 26' long		10.60	130	390	1,175	162.80
	0900	46' long		11.00	155	465	1,400	181
	1000	56' long		11.15	163	490	1,475	187.20
	1100	Core drill, electric, 2-1/2 H.P., 1" to 8" bit diameter		1.56	58.50	175	525	47.50
	1150	11 H.P., 8" to 18" cores		5.40	115	345	1,025	112.20
	1200	Finisher, concrete floor, gas, riding trowel, 96" wide		9.65	148	445	1,325	166.20
	1300	Gas, walk-behind, 3 blade, 36" trowel		2.15	23	69	207	31
	1400	4 blade, 48" trowel		3.05	28	84	252	41.20
	1500	Float, hand-operated (Bull float), 48" wide		.08	13.65	41	123	8.85
	1570	Curb builder, 14 H.P., gas, single screw		13.90	275	825	2,475	276.20
	1590	Double screw		14.80	330	995	2,975	317.40
	1600	Floor grinder, concrete and terrazzo, electric, 22" path		2.95	183	550	1,650	133.60
	1700	Edger, concrete, electric, 7" path		1.12	56.50	170	510	42.95
	1750	Vacuum pick-up system for floor grinders, wet/dry		1.62	90	270	810	66.95
	1800	Mixer, powered, mortar and concrete, gas, 6 C.F., 18 H.P.		7.40	123	370	1,100	133.20
	1900	10 C.F., 25 H.P.		9.00	148	445	1,325	161
	2000	16 C.F.		9.35	172	515	1,550	177.80
	2100	Concrete, stationary, tilt drum, 2 C.Y.		7.40	242	725	2,175	204.20
	2120	Pump, concrete, truck mounted, 4" line, 80' boom		29.20	1,075	3,240	9,725	881.60
	2140	5" line, 110' boom		36.60	1,375	4,115	12,300	1,116
	2160	Mud jack, 50 C.F. per hr.		6.45	128	385	1,150	128.60
	2180	225 C.F. per hr.		8.80	147	440	1,325	158.40
	2190	Shotcrete pump rig, 12 C.Y./hr.		13.95	225	675	2,025	246.60
	2200	35 C.Y./hr.		15.65	242	725	2,175	270.20
	2600	Saw, concrete, manual, gas, 18 H.P.		5.40	46.50	140	420	71.20
	2650	Self-propelled, gas, 30 H.P.		7.55	70	210	630	102.40
	2675	V-groove crack chaser, manual, gas, 6 H.P.		1.80	18.35	55	165	25.40
	2700	Vibrators, concrete, electric, 60 cycle, 2 H.P.		.47	9	27	81	9.15
	2800	3 H.P.		.60	11.65	35	105	11.80
	2900	Gas engine, 5 H.P.		1.50	16.35	49	147	21.80
	3000	8 H.P.		2.00	16	48	144	25.60
	3050	Vibrating screed, gas engine, 8 H.P.		2.82	88	264	790	75.35
	3120	Concrete transit mixer, 6 x 4, 250 H.P., 8 C.Y., rear discharge		48.75	600	1,805	5,425	751
	3200	Front discharge		56.60	735	2,205	6,625	893.80
	3300	6 x 6, 285 H.P., 12 C.Y., rear discharge		55.65	700	2,095	6,275	864.20
	3400	Front discharge		57.95	735	2,210	6,625	905.60
20	**0010**	**EARTHWORK EQUIPMENT RENTAL** without operators R015433 -10						**20**
	0040	Aggregate spreader, push type, 8' to 12' wide	Ea.	2.60	26.50	80	240	36.80
	0045	Tailgate type, 8' wide		2.55	33.50	100	300	40.40
	0055	Earth auger, truck mounted, for fence & sign posts, utility poles		12.60	455	1,365	4,100	373.80
	0060	For borings and monitoring wells		42.60	695	2,090	6,275	758.80
	0070	Portable, trailer mounted		2.30	33	99	297	38.20
	0075	Truck mounted, for caissons, water wells		85.75	2,925	8,790	26,400	2,444
	0080	Horizontal boring machine, 12" to 36" diameter, 45 H.P.		22.70	197	590	1,775	299.60
	0090	12" to 48" diameter, 65 H.P.		30.60	340	1,020	3,050	448.80
	0095	Auger, for fence posts, gas engine, hand held		.45	6.35	19	57	7.40
	0100	Excavator, diesel hydraulic, crawler mounted, 1/2 C.Y. cap.		21.30	440	1,325	3,975	435.40
	0120	5/8 C.Y. capacity		28.60	580	1,740	5,225	576.80
	0140	3/4 C.Y. capacity		31.95	690	2,070	6,200	669.60
	0150	1 C.Y. capacity		39.90	705	2,115	6,350	742.20

01 54 33	Equipment Rental	UNIT	HOURLY OPER. COST	RENT PER DAY	RENT PER WEEK	RENT PER MONTH	EQUIPMENT COST/DAY
0200	1-1/2 C.Y. capacity	Ea.	47.45	855	2,570	7,700	893.60
0300	2 C.Y. capacity		54.15	1,025	3,095	9,275	1,052
0320	2-1/2 C.Y. capacity		78.90	1,300	3,900	11,700	1,411
0325	3-1/2 C.Y. capacity		113.65	2,150	6,430	19,300	2,195
0330	4-1/2 C.Y. capacity		143.70	2,650	7,925	23,800	2,735
0335	6 C.Y. capacity		181.40	3,400	10,180	30,500	3,487
0340	7 C.Y. capacity		175.20	3,125	9,365	28,100	3,275
0342	Excavator attachments, bucket thumbs		3.35	252	755	2,275	177.80
0345	Grapples		3.05	220	660	1,975	156.40
0346	Hydraulic hammer for boom mounting, 4000 ft lb.		13.20	365	1,095	3,275	324.60
0347	5000 ft lb.		15.50	450	1,350	4,050	394
0348	8000 ft lb.		22.85	655	1,970	5,900	576.80
0349	12,000 ft lb.		24.90	785	2,350	7,050	669.20
0350	Gradall type, truck mounted, 3 ton @ 15' radius, 5/8 C.Y.		44.95	850	2,550	7,650	869.60
0370	1 C.Y. capacity		60.20	1,275	3,805	11,400	1,243
0400	Backhoe-loader, 40 to 45 H.P., 5/8 C.Y. capacity		12.25	248	745	2,225	247
0450	45 H.P. to 60 H.P., 3/4 C.Y. capacity		17.80	283	850	2,550	312.40
0460	80 H.P., 1-1/4 C.Y. capacity		19.50	380	1,145	3,425	385
0470	112 H.P., 1-1/2 C.Y. capacity		31.90	605	1,820	5,450	619.20
0482	Backhoe-loader attachment, compactor, 20,000 lb.		6.25	148	445	1,325	139
0485	Hydraulic hammer, 750 ft lb.		3.55	102	305	915	89.40
0486	Hydraulic hammer, 1200 ft lb.		6.55	208	625	1,875	177.40
0500	Brush chipper, gas engine, 6" cutter head, 35 H.P.		9.15	110	330	990	139.20
0550	Diesel engine, 12" cutter head, 130 H.P.		24.00	335	1,005	3,025	393
0600	15" cutter head, 165 H.P.		26.05	400	1,200	3,600	448.40
0750	Bucket, clamshell, general purpose, 3/8 C.Y.		1.40	40	120	360	35.20
0800	1/2 C.Y.		1.50	48.50	145	435	41
0850	3/4 C.Y.		1.65	56.50	170	510	47.20
0900	1 C.Y.		1.70	61.50	185	555	50.60
0950	1-1/2 C.Y.		2.75	85	255	765	73
1000	2 C.Y.		2.90	93.50	280	840	79.20
1010	Bucket, dragline, medium duty, 1/2 C.Y.		.80	24.50	73	219	21
1020	3/4 C.Y.		.80	25.50	77	231	21.80
1030	1 C.Y.		.85	27	81	243	23
1040	1-1/2 C.Y.		1.30	41.50	125	375	35.40
1050	2 C.Y.		1.35	45	135	405	37.80
1070	3 C.Y.		2.10	65	195	585	55.80
1200	Compactor, manually guided 2-drum vibratory smooth roller, 7.5 H.P.		7.20	203	610	1,825	179.60
1250	Rammer/tamper, gas, 8"		2.25	46.50	140	420	46
1260	15"		2.50	53.50	160	480	52
1300	Vibratory plate, gas, 18" plate, 3000 lb. blow		2.15	24.50	73	219	31.80
1350	21" plate, 5000 lb. blow		2.60	33	99	297	40.60
1370	Curb builder/extruder, 14 H.P., gas, single screw		13.90	275	825	2,475	276.20
1390	Double screw		14.80	330	995	2,975	317.40
1500	Disc harrow attachment, for tractor		.47	78.50	235	705	50.75
1810	Feller buncher, shearing & accumulating trees, 100 H.P.		40.55	810	2,430	7,300	810.40
1860	Grader, self-propelled, 25,000 lb.		32.75	745	2,235	6,700	709
1910	30,000 lb.		32.35	640	1,925	5,775	643.80
1920	40,000 lb.		54.75	1,250	3,765	11,300	1,191
1930	55,000 lb.		67.80	1,650	4,975	14,900	1,537
1950	Hammer, pavement breaker, self-propelled, diesel, 1000 to 1250 lb.		28.40	460	1,380	4,150	503.20
2000	1300 to 1500 lb.		42.60	920	2,760	8,275	892.80
2050	Pile driving hammer, steam or air, 4150 ft lb. @ 225 bpm		11.80	565	1,695	5,075	433.40
2100	8750 ft lb. @ 145 bpm		14.30	805	2,410	7,225	596.40
2150	15,000 ft lb. @ 60 bpm		14.65	845	2,530	7,600	623.20
2200	24,450 ft lb. @ 111 bpm		15.65	935	2,800	8,400	685.20
2250	Leads, 60' high for pile driving hammers up to 20,000 ft lb.		3.60	84.50	253	760	79.40
2300	90' high for hammers over 20,000 ft lb.		5.35	148	444	1,325	131.60

For customer support on your Building Construction Costs with RSMeans data, call 800.448.8182.

729

01 54 33 | Equipment Rental

		UNIT	HOURLY OPER. COST	RENT PER DAY	RENT PER WEEK	RENT PER MONTH	EQUIPMENT COST/DAY	
2350	Diesel type hammer, 22,400 ft lb.	Ea.	18.25	470	1,405	4,225	427	20
2400	41,300 ft lb.		26.50	600	1,800	5,400	572	
2450	141,000 ft lb.		41.80	945	2,840	8,525	902.40	
2500	Vib. elec. hammer/extractor, 200 kW diesel generator, 34 H.P.		40.35	695	2,080	6,250	738.80	
2550	80 H.P.		70.10	1,000	3,000	9,000	1,161	
2600	150 H.P.		129.40	1,925	5,780	17,300	2,191	
2800	Log chipper, up to 22" diameter, 600 H.P.		46.10	660	1,980	5,950	764.80	
2850	Logger, for skidding & stacking logs, 150 H.P.		44.05	830	2,485	7,450	849.40	
2860	Mulcher, diesel powered, trailer mounted		17.10	220	660	1,975	268.80	
2900	Rake, spring tooth, with tractor		14.54	360	1,075	3,225	331.30	
3000	Roller, vibratory, tandem, smooth drum, 20 H.P.		7.65	150	450	1,350	151.20	
3050	35 H.P.		10.05	252	755	2,275	231.40	
3100	Towed type vibratory compactor, smooth drum, 50 H.P.		25.25	370	1,105	3,325	423	
3150	Sheepsfoot, 50 H.P.		25.35	370	1,115	3,350	425.80	
3170	Landfill compactor, 220 H.P.		71.35	1,575	4,710	14,100	1,513	
3200	Pneumatic tire roller, 80 H.P.		12.90	390	1,175	3,525	338.20	
3250	120 H.P.		19.45	645	1,930	5,800	541.60	
3300	Sheepsfoot vibratory roller, 240 H.P.		62.05	1,375	4,100	12,300	1,316	
3320	340 H.P.		83.10	2,025	6,075	18,200	1,880	
3350	Smooth drum vibratory roller, 75 H.P.		23.30	650	1,950	5,850	576.40	
3400	125 H.P.		27.60	730	2,195	6,575	659.80	
3410	Rotary mower, brush, 60", with tractor		19.00	340	1,025	3,075	357	
3420	Rototiller, walk-behind, gas, 5 H.P.		2.21	76.50	229	685	63.50	
3422	8 H.P.		2.80	87	261	785	74.60	
3440	Scrapers, towed type, 7 C.Y. capacity		6.30	122	365	1,100	123.40	
3450	10 C.Y. capacity		7.10	165	495	1,475	155.80	
3500	15 C.Y. capacity		7.60	192	575	1,725	175.80	
3525	Self-propelled, single engine, 14 C.Y. capacity		131.25	2,400	7,200	21,600	2,490	
3550	Dual engine, 21 C.Y. capacity		137.25	2,275	6,850	20,600	2,468	
3600	31 C.Y. capacity		184.55	3,500	10,500	31,500	3,576	
3640	44 C.Y. capacity		225.40	4,425	13,270	39,800	4,457	
3650	Elevating type, single engine, 11 C.Y. capacity		61.05	1,100	3,335	10,000	1,155	
3700	22 C.Y. capacity		114.65	2,275	6,835	20,500	2,284	
3710	Screening plant, 110 H.P. w/5' x 10' screen		20.55	385	1,160	3,475	396.40	
3720	5' x 16' screen		25.96	495	1,480	4,450	503.70	
3850	Shovel, crawler-mounted, front-loading, 7 C.Y. capacity		204.00	3,750	11,240	33,700	3,880	
3855	12 C.Y. capacity		332.00	5,200	15,567	46,700	5,769	
3860	Shovel/backhoe bucket, 1/2 C.Y.		2.70	71.50	215	645	64.60	
3870	3/4 C.Y.		2.75	78.50	235	705	69	
3880	1 C.Y.		2.85	88.50	265	795	75.80	
3890	1-1/2 C.Y.		3.00	102	305	915	85	
3910	3 C.Y.		3.35	137	410	1,225	108.80	
3950	Stump chipper, 18" deep, 30 H.P.		6.93	214	643	1,925	184.05	
4110	Dozer, crawler, torque converter, diesel 80 H.P.		25.05	440	1,320	3,950	464.40	
4150	105 H.P.		34.25	555	1,670	5,000	608	
4200	140 H.P.		42.20	845	2,540	7,625	845.60	
4260	200 H.P.		61.65	1,300	3,900	11,700	1,273	
4310	300 H.P.		82.45	1,950	5,845	17,500	1,829	
4360	410 H.P.		109.35	2,350	7,080	21,200	2,291	
4370	500 H.P.		137.25	2,875	8,600	25,800	2,818	
4380	700 H.P.		229.75	5,275	15,805	47,400	4,999	
4400	Loader, crawler, torque conv., diesel, 1-1/2 C.Y., 80 H.P.		29.60	570	1,715	5,150	579.80	
4450	1-1/2 to 1-3/4 C.Y., 95 H.P.		30.55	670	2,005	6,025	645.40	
4510	1-3/4 to 2-1/4 C.Y., 130 H.P.		47.50	1,000	2,995	8,975	979	
4530	2-1/2 to 3-1/4 C.Y., 190 H.P.		57.55	1,250	3,720	11,200	1,204	
4560	3-1/2 to 5 C.Y., 275 H.P.		71.55	1,475	4,435	13,300	1,459	
4610	Front end loader, 4WD, articulated frame, diesel, 1 to 1-1/4 C.Y., 70 H.P.		16.10	270	810	2,425	290.80	
4620	1-1/2 to 1-3/4 C.Y., 95 H.P.		19.35	315	940	2,825	342.80	

01 54 33 | Equipment Rental

		UNIT	HOURLY OPER. COST	RENT PER DAY	RENT PER WEEK	RENT PER MONTH	EQUIPMENT COST/DAY	
20	4650	1-3/4 to 2 C.Y., 130 H.P.	Ea.	20.50	380	1,140	3,425	392	**20**
	4710	2-1/2 to 3-1/2 C.Y., 145 H.P.		29.15	485	1,450	4,350	523.20	
	4730	3 to 4-1/2 C.Y., 185 H.P.		32.20	530	1,585	4,750	574.60	
	4760	5-1/4 to 5-3/4 C.Y., 270 H.P.		53.15	930	2,785	8,350	982.20	
	4810	7 to 9 C.Y., 475 H.P.		90.60	1,750	5,275	15,800	1,780	
	4870	9 to 11 C.Y., 620 H.P.		131.70	2,625	7,880	23,600	2,630	
	4880	Skid-steer loader, wheeled, 10 C.F., 30 H.P. gas		9.50	163	490	1,475	174	
	4890	1 C.Y., 78 H.P., diesel		18.35	365	1,090	3,275	364.80	
	4892	Skid-steer attachment, auger		.82	136	408	1,225	88.15	
	4893	Backhoe		.74	123	369	1,100	79.70	
	4894	Broom		.71	118	355	1,075	76.70	
	4895	Forks		.16	27	81	243	17.50	
	4896	Grapple		.69	115	346	1,050	74.70	
	4897	Concrete hammer		1.05	175	526	1,575	113.60	
	4898	Tree spade		.60	100	300	900	64.80	
	4899	Trencher		.69	115	344	1,025	74.30	
	4900	Trencher, chain, boom type, gas, operator walking, 12 H.P.		4.10	48.50	145	435	61.80	
	4910	Operator riding, 40 H.P.		16.30	345	1,030	3,100	336.40	
	5000	Wheel type, diesel, 4' deep, 12" wide		69.35	900	2,700	8,100	1,095	
	5100	6' deep, 20" wide		87.50	2,100	6,285	18,900	1,957	
	5150	Chain type, diesel, 5' deep, 8" wide		16.30	345	1,030	3,100	336.40	
	5200	Diesel, 8' deep, 16" wide		95.25	2,150	6,430	19,300	2,048	
	5202	Rock trencher, wheel type, 6" wide x 18" deep		43.65	925	2,775	8,325	904.20	
	5206	Chain type, 18" wide x 7' deep		106.10	3,050	9,160	27,500	2,681	
	5210	Tree spade, self-propelled		14.24	390	1,167	3,500	347.30	
	5250	Truck, dump, 2-axle, 12 ton, 8 C.Y. payload, 220 H.P.		23.95	245	735	2,200	338.60	
	5300	Three axle dump, 16 ton, 12 C.Y. payload, 400 H.P.		41.60	350	1,050	3,150	542.80	
	5310	Four axle dump, 25 ton, 18 C.Y. payload, 450 H.P.		50.00	510	1,530	4,600	706	
	5350	Dump trailer only, rear dump, 16-1/2 C.Y.		5.75	145	435	1,300	133	
	5400	20 C.Y.		6.20	163	490	1,475	147.60	
	5450	Flatbed, single axle, 1-1/2 ton rating		18.30	70	210	630	188.40	
	5500	3 ton rating		22.25	100	300	900	238	
	5550	Off highway rear dump, 25 ton capacity		61.80	1,425	4,245	12,700	1,343	
	5600	35 ton capacity		66.10	1,525	4,595	13,800	1,448	
	5610	50 ton capacity		81.20	1,675	5,060	15,200	1,662	
	5620	65 ton capacity		84.95	1,950	5,820	17,500	1,844	
	5630	100 ton capacity		119.60	2,850	8,560	25,700	2,669	
	6000	Vibratory plow, 25 H.P., walking		6.80	61.50	185	555	91.40	
40	0010	**GENERAL EQUIPMENT RENTAL** without operators							**40**
	0020	Aerial lift, scissor type, to 20' high, 1200 lb. capacity, electric	Ea.	3.40	51.50	155	465	58.20	
	0030	To 30' high, 1200 lb. capacity		3.85	68.50	205	615	71.80	
	0040	Over 30' high, 1500 lb. capacity		5.15	122	365	1,100	114.20	
	0070	Articulating boom, to 45' high, 500 lb. capacity, diesel		9.70	273	820	2,450	241.60	
	0075	To 60' high, 500 lb. capacity		13.70	470	1,410	4,225	391.60	
	0080	To 80' high, 500 lb. capacity		16.10	560	1,685	5,050	465.80	
	0085	To 125' high, 500 lb. capacity		18.40	780	2,335	7,000	614.20	
	0100	Telescoping boom to 40' high, 500 lb. capacity, diesel		11.60	315	950	2,850	282.80	
	0105	To 45' high, 500 lb. capacity		12.40	320	958	2,875	290.80	
	0110	To 60' high, 500 lb. capacity		16.20	540	1,625	4,875	454.60	
	0115	To 80' high, 500 lb. capacity		21.70	625	1,875	5,625	548.60	
	0120	To 100' high, 500 lb. capacity		28.80	805	2,420	7,250	714.40	
	0125	To 120' high, 500 lb. capacity		29.25	845	2,530	7,600	740	
	0195	Air compressor, portable, 6.5 CFM, electric		.91	13	39	117	15.10	
	0196	Gasoline		.66	19.65	59	177	17.10	
	0200	Towed type, gas engine, 60 CFM		9.30	51.50	155	465	105.40	
	0300	160 CFM		10.65	53.50	160	480	117.20	
	0400	Diesel engine, rotary screw, 250 CFM		12.05	118	355	1,075	167.40	
	0500	365 CFM		15.80	142	425	1,275	211.40	

Reference codes in table: R015433-10 (row 0020), R015433-15 (row 0070)

01 54 33 | Equipment Rental

		UNIT	HOURLY OPER. COST	RENT PER DAY	RENT PER WEEK	RENT PER MONTH	EQUIPMENT COST/DAY
0550	450 CFM	Ea.	19.70	177	530	1,600	263.60
0600	600 CFM		33.90	243	730	2,200	417.20
0700	750 CFM		34.10	252	755	2,275	423.80
0930	Air tools, breaker, pavement, 60 lb.		.55	10.35	31	93	10.60
0940	80 lb.		.55	10.65	32	96	10.80
0950	Drills, hand (jackhammer), 65 lb.		.65	17.65	53	159	15.80
0960	Track or wagon, swing boom, 4" drifter		54.80	925	2,775	8,325	993.40
0970	5" drifter		63.45	1,100	3,325	9,975	1,173
0975	Track mounted quarry drill, 6" diameter drill		104.15	1,650	4,945	14,800	1,822
0980	Dust control per drill		1.04	24.50	74	222	23.10
0990	Hammer, chipping, 12 lb.		.60	27	81	243	21
1000	Hose, air with couplings, 50' long, 3/4" diameter		.07	11.35	34	102	7.35
1100	1" diameter		.08	13	39	117	8.45
1200	1-1/2" diameter		.21	35	105	315	22.70
1300	2" diameter		.24	40	120	360	25.90
1400	2-1/2" diameter		.35	58.50	175	525	37.80
1410	3" diameter		.40	66.50	200	600	43.20
1450	Drill, steel, 7/8" x 2'		.08	13.65	41	123	8.85
1460	7/8" x 6'		.11	17.65	53	159	11.50
1520	Moil points		.03	4.67	14	42	3.05
1525	Pneumatic nailer w/accessories		.48	32	96	288	23.05
1530	Sheeting driver for 60 lb. breaker		.04	7.35	22	66	4.70
1540	For 90 lb. breaker		.15	9.65	29	87	7
1550	Spade, 25 lb.		.50	7.35	22	66	8.40
1560	Tamper, single, 35 lb.		.59	39.50	118	355	28.30
1570	Triple, 140 lb.		.89	59	177	530	42.50
1580	Wrenches, impact, air powered, up to 3/4" bolt		.45	13	39	117	11.40
1590	Up to 1-1/4" bolt		.55	23.50	71	213	18.60
1600	Barricades, barrels, reflectorized, 1 to 99 barrels		.03	5.35	16	48	3.45
1610	100 to 200 barrels		.02	4.13	12.40	37	2.65
1620	Barrels with flashers, 1 to 99 barrels		.04	6	18	54	3.90
1630	100 to 200 barrels		.03	4.80	14.40	43	3.10
1640	Barrels with steady burn type C lights		.05	8	24	72	5.20
1650	Illuminated board, trailer mounted, with generator		3.30	133	400	1,200	106.40
1670	Portable barricade, stock, with flashers, 1 to 6 units		.04	6	18	54	3.90
1680	25 to 50 units		.03	5.60	16.80	50.50	3.60
1685	Butt fusion machine, wheeled, 1.5 H.P. electric, 2" - 8" diameter pipe		2.63	167	500	1,500	121.05
1690	Tracked, 20 H.P. diesel, 4" - 12" diameter pipe		11.21	560	1,680	5,050	425.70
1695	83 H.P. diesel, 8" - 24" diameter pipe		49.46	2,525	7,560	22,700	1,908
1700	Carts, brick, gas engine, 1000 lb. capacity		2.95	61.50	185	555	60.60
1800	1500 lb., 7-1/2' lift		3.00	65	195	585	63
1822	Dehumidifier, medium, 6 lb./hr., 150 CFM		1.16	72.50	218	655	52.90
1824	Large, 18 lb./hr., 600 CFM		2.20	138	413	1,250	100.20
1830	Distributor, asphalt, trailer mounted, 2000 gal., 38 H.P. diesel		10.75	350	1,050	3,150	296
1840	3000 gal., 38 H.P. diesel		12.35	380	1,140	3,425	326.80
1850	Drill, rotary hammer, electric		1.12	27	81	243	25.15
1860	Carbide bit, 1-1/2" diameter, add to electric rotary hammer		.03	5	15	45	3.25
1865	Rotary, crawler, 250 H.P.		136.15	2,225	6,690	20,100	2,427
1870	Emulsion sprayer, 65 gal., 5 H.P. gas engine		2.77	103	309	925	83.95
1880	200 gal., 5 H.P. engine		7.30	172	515	1,550	161.40
1900	Floor auto-scrubbing machine, walk-behind, 28" path		5.41	350	1,055	3,175	254.30
1930	Floodlight, mercury vapor, or quartz, on tripod, 1000 watt		.46	22	66	198	16.90
1940	2000 watt		.63	27.50	82	246	21.45
1950	Floodlights, trailer mounted with generator, 1 - 300 watt light		3.60	76.50	230	690	74.80
1960	2 - 1000 watt lights		4.50	102	305	915	97
2000	4 - 300 watt lights		4.25	96.50	290	870	92
2005	Foam spray rig, incl. box trailer, compressor, generator, proportioner		23.78	515	1,545	4,625	499.25
2015	Forklift, pneumatic tire, rough terr, straight mast, 5000 lb, 12' lift, gas		19.00	212	635	1,900	279

For customer support on your Building Construction Costs with RSMeans data, call 800.448.8182.

01 54 33 | Equipment Rental

		UNIT	HOURLY OPER. COST	RENT PER DAY	RENT PER WEEK	RENT PER MONTH	EQUIPMENT COST/DAY	
40	2025	8000 lb., 12' lift	Ea.	22.75	283	850	2,550	352	40
	2030	5000 lb., 12' lift, diesel		15.70	237	710	2,125	267.60	
	2035	8000 lb., 12' lift, diesel		16.75	268	805	2,425	295	
	2045	All terrain, telescoping boom, diesel, 5000 lb., 10' reach, 19' lift		17.25	325	980	2,950	334	
	2055	6600 lb., 29' reach, 42' lift		21.10	380	1,140	3,425	396.80	
	2065	10,000 lb., 31' reach, 45' lift		23.65	490	1,475	4,425	484.20	
	2070	Cushion tire, smooth floor, gas, 5000 lb. capacity		8.25	76.50	230	690	112	
	2075	8000 lb. capacity		11.40	95	285	855	148.20	
	2085	Diesel, 5000 lb. capacity		7.75	83.50	250	750	112	
	2090	12,000 lb. capacity		12.05	130	390	1,175	174.40	
	2095	20,000 lb. capacity		17.00	165	495	1,475	235	
	2100	Generator, electric, gas engine, 1.5 kW to 3 kW		2.70	11.35	34	102	28.40	
	2200	5 kW		3.35	14.35	43	129	35.40	
	2300	10 kW		6.25	35	105	315	71	
	2400	25 kW		7.60	86.50	260	780	112.80	
	2500	Diesel engine, 20 kW		9.20	76.50	230	690	119.60	
	2600	50 kW		15.85	100	300	900	186.80	
	2700	100 kW		28.20	137	410	1,225	307.60	
	2800	250 kW		56.05	260	780	2,350	604.40	
	2850	Hammer, hydraulic, for mounting on boom, to 500 ft lb.		2.85	86.50	260	780	74.80	
	2860	1000 ft lb.		4.70	133	400	1,200	117.60	
	2900	Heaters, space, oil or electric, 50 MBH		1.47	8	24	72	16.55	
	3000	100 MBH		2.73	11.35	34	102	28.65	
	3100	300 MBH		7.84	40	120	360	86.70	
	3150	500 MBH		12.75	45	135	405	129	
	3200	Hose, water, suction with coupling, 20' long, 2" diameter		.02	3	9	27	1.95	
	3210	3" diameter		.03	4.33	13	39	2.85	
	3220	4" diameter		.03	5	15	45	3.25	
	3230	6" diameter		.11	17.65	53	159	11.50	
	3240	8" diameter		.28	46.50	140	420	30.25	
	3250	Discharge hose with coupling, 50' long, 2" diameter		.01	1.33	4	12	.90	
	3260	3" diameter		.01	2.33	7	21	1.50	
	3270	4" diameter		.02	3.67	11	33	2.35	
	3280	6" diameter		.06	9.35	28	84	6.10	
	3290	8" diameter		.24	40	120	360	25.90	
	3295	Insulation blower		.83	6	18	54	10.25	
	3300	Ladders, extension type, 16' to 36' long		.18	30	90	270	19.45	
	3400	40' to 60' long		.67	112	335	1,000	72.35	
	3405	Lance for cutting concrete		2.23	58.50	176	530	53.05	
	3407	Lawn mower, rotary, 22", 5 H.P.		1.15	25	75	225	24.20	
	3408	48" self-propelled		2.86	90	270	810	76.90	
	3410	Level, electronic, automatic, with tripod and leveling rod		1.05	70	210	630	50.40	
	3430	Laser type, for pipe and sewer line and grade		2.13	142	425	1,275	102.05	
	3440	Rotating beam for interior control		.90	60	180	540	43.20	
	3460	Builder's optical transit, with tripod and rod		.10	16.35	49	147	10.60	
	3500	Light towers, towable, with diesel generator, 2000 watt		4.25	96.50	290	870	92	
	3600	4000 watt		4.50	102	305	915	97	
	3700	Mixer, powered, plaster and mortar, 6 C.F., 7 H.P.		2.05	20.50	62	186	28.80	
	3800	10 C.F., 9 H.P.		2.20	33.50	100	300	37.60	
	3850	Nailer, pneumatic		.48	32	96	288	23.05	
	3900	Paint sprayers complete, 8 CFM		.94	62.50	188	565	45.10	
	4000	17 CFM		1.69	112	337	1,000	80.90	
	4020	Pavers, bituminous, rubber tires, 8' wide, 50 H.P., diesel		31.55	550	1,645	4,925	581.40	
	4030	10' wide, 150 H.P.		96.50	1,875	5,655	17,000	1,903	
	4050	Crawler, 8' wide, 100 H.P., diesel		87.60	2,025	6,105	18,300	1,922	
	4060	10' wide, 150 H.P.		104.50	2,350	7,015	21,000	2,239	
	4070	Concrete paver, 12' to 24' wide, 250 H.P.		87.45	1,625	4,875	14,600	1,675	
	4080	Placer-spreader-trimmer, 24' wide, 300 H.P.		117.20	2,375	7,115	21,300	2,361	

01 54 33 | Equipment Rental

		UNIT	HOURLY OPER. COST	RENT PER DAY	RENT PER WEEK	RENT PER MONTH	EQUIPMENT COST/DAY
4100	Pump, centrifugal gas pump, 1-1/2" diameter, 65 GPM	Ea.	3.90	53.50	160	480	63.20
4200	2" diameter, 130 GPM		5.00	63.50	190	570	78
4300	3" diameter, 250 GPM		5.15	63.50	190	570	79.20
4400	6" diameter, 1500 GPM		22.30	197	590	1,775	296.40
4500	Submersible electric pump, 1-1/4" diameter, 55 GPM		.41	17.65	53	159	13.90
4600	1-1/2" diameter, 83 GPM		.45	20.50	61	183	15.80
4700	2" diameter, 120 GPM		1.65	25.50	76	228	28.40
4800	3" diameter, 300 GPM		2.94	45	135	405	50.50
4900	4" diameter, 560 GPM		14.70	167	500	1,500	217.60
5000	6" diameter, 1590 GPM		21.94	218	655	1,975	306.50
5100	Diaphragm pump, gas, single, 1-1/2" diameter		1.12	54.50	164	490	41.75
5200	2" diameter		4.00	68.50	205	615	73
5300	3" diameter		4.05	68.50	205	615	73.40
5400	Double, 4" diameter		5.85	113	340	1,025	114.80
5450	Pressure washer 5 GPM, 3000 psi		3.95	53.50	160	480	63.60
5460	7 GPM, 3000 psi		4.90	63.50	190	570	77.20
5500	Trash pump, self-priming, gas, 2" diameter		3.80	23.50	70	210	44.40
5600	Diesel, 4" diameter		6.95	95	285	855	112.60
5650	Diesel, 6" diameter		16.90	167	500	1,500	235.20
5655	Grout pump		19.50	275	825	2,475	321
5700	Salamanders, L.P. gas fired, 100,000 BTU		2.93	14	42	126	31.85
5705	50,000 BTU		1.67	11.35	34	102	20.15
5720	Sandblaster, portable, open top, 3 C.F. capacity		.60	27	81	243	21
5730	6 C.F. capacity		1.00	40	120	360	32
5740	Accessories for above		.14	22.50	68	204	14.70
5750	Sander, floor		.77	17.65	53	159	16.75
5760	Edger		.52	15	45	135	13.15
5800	Saw, chain, gas engine, 18" long		1.80	22.50	67	201	27.80
5900	Hydraulic powered, 36" long		.80	66.50	200	600	46.40
5950	60" long		.80	68.50	205	615	47.40
6000	Masonry, table mounted, 14" diameter, 5 H.P.		1.32	56.50	170	510	44.55
6050	Portable cut-off, 8 H.P.		1.85	33.50	100	300	34.80
6100	Circular, hand held, electric, 7-1/4" diameter		.23	5	15	45	4.85
6200	12" diameter		.23	8	24	72	6.65
6250	Wall saw, w/hydraulic power, 10 H.P.		3.30	33.50	100	300	46.40
6275	Shot blaster, walk-behind, 20" wide		4.85	293	880	2,650	214.80
6280	Sidewalk broom, walk-behind		2.39	85	255	765	70.10
6300	Steam cleaner, 100 gallons per hour		3.35	80	240	720	74.80
6310	200 gallons per hour		4.40	96.50	290	870	93.20
6340	Tar kettle/pot, 400 gallons		15.15	76.50	230	690	167.20
6350	Torch, cutting, acetylene-oxygen, 150' hose, excludes gases		.45	15	45	135	12.60
6360	Hourly operating cost includes tips and gas		21.00				168
6410	Toilet, portable chemical		.13	22	66	198	14.25
6420	Recycle flush type		.16	27	81	243	17.50
6430	Toilet, fresh water flush, garden hose,		.19	32.50	97	291	20.90
6440	Hoisted, non-flush, for high rise		.16	26.50	79	237	17.10
6465	Tractor, farm with attachment		17.80	340	1,025	3,075	347.40
6480	Trailers, platform, flush deck, 2 axle, 3 ton capacity		1.60	21	63	189	25.40
6500	25 ton capacity		6.25	138	415	1,250	133
6600	40 ton capacity		8.00	193	580	1,750	180
6700	3 axle, 50 ton capacity		8.65	215	645	1,925	198.20
6800	75 ton capacity		10.90	285	855	2,575	258.20
6810	Trailer mounted cable reel for high voltage line work		5.79	276	827	2,475	211.70
6820	Trailer mounted cable tensioning rig		11.48	545	1,640	4,925	419.85
6830	Cable pulling rig		72.98	3,075	9,210	27,600	2,426
6850	Portable cable/wire puller, 8000 lb. max pulling capacity		3.72	167	502	1,500	130.15
6900	Water tank trailer, engine driven discharge, 5000 gallons		7.20	150	450	1,350	147.60
6925	10,000 gallons		9.70	207	620	1,850	201.60

40

01 54 33 | Equipment Rental

			UNIT	HOURLY OPER. COST	RENT PER DAY	RENT PER WEEK	RENT PER MONTH	EQUIPMENT COST/DAY	
40	6950	Water truck, off highway, 6000 gallons	Ea.	70.16	805	2,420	7,250	1,045	40
	7010	Tram car for high voltage line work, powered, 2 conductor		6.85	150	449	1,350	144.60	
	7020	Transit (builder's level) with tripod		.10	16.35	49	147	10.60	
	7030	Trench box, 3000 lb., 6' x 8'		.56	93.50	280	840	60.50	
	7040	7200 lb., 6' x 20'		.75	125	375	1,125	81	
	7050	8000 lb., 8' x 16'		1.08	180	540	1,625	116.65	
	7060	9500 lb., 8' x 20'		1.21	201	603	1,800	130.30	
	7065	11,000 lb., 8' x 24'		1.27	211	633	1,900	136.75	
	7070	12,000 lb., 10' x 20'		1.50	251	752	2,250	162.40	
	7100	Truck, pickup, 3/4 ton, 2 wheel drive		9.90	60	180	540	115.20	
	7200	4 wheel drive		10.20	75	225	675	126.60	
	7250	Crew carrier, 9 passenger		14.00	90	270	810	166	
	7290	Flat bed truck, 20,000 lb. GVW		14.90	130	390	1,175	197.20	
	7300	Tractor, 4 x 2, 220 H.P.		21.00	203	610	1,825	290	
	7410	330 H.P.		30.80	280	840	2,525	414.40	
	7500	6 x 4, 380 H.P.		35.15	325	975	2,925	476.20	
	7600	450 H.P.		43.30	395	1,185	3,550	583.40	
	7610	Tractor, with A frame, boom and winch, 225 H.P.		24.10	282	845	2,525	361.80	
	7620	Vacuum truck, hazardous material, 2500 gallons		12.85	305	910	2,725	284.80	
	7625	5000 gallons		13.11	425	1,270	3,800	358.90	
	7650	Vacuum, HEPA, 16 gallon, wet/dry		.90	18	54	162	18	
	7655	55 gallon, wet/dry		.81	27	81	243	22.70	
	7660	Water tank, portable		.74	123	370	1,100	79.90	
	7690	Sewer/catch basin vacuum, 14 C.Y., 1500 gallons		17.59	635	1,910	5,725	522.70	
	7700	Welder, electric, 200 amp		3.99	16.35	49	147	41.70	
	7800	300 amp		5.90	20	60	180	59.20	
	7900	Gas engine, 200 amp		9.10	24.50	74	222	87.60	
	8000	300 amp		10.35	26	78	234	98.40	
	8100	Wheelbarrow, any size		.06	10.65	32	96	6.90	
	8200	Wrecking ball, 4000 lb.		2.45	71.50	215	645	62.60	
50	0010	**HIGHWAY EQUIPMENT RENTAL** without operators R015433 -10							50
	0050	Asphalt batch plant, portable drum mixer, 100 ton/hr.	Ea.	85.49	1,500	4,505	13,500	1,585	
	0060	200 ton/hr.		97.81	1,600	4,800	14,400	1,742	
	0070	300 ton/hr.		116.21	1,875	5,625	16,900	2,055	
	0100	Backhoe attachment, long stick, up to 185 H.P., 10.5' long		.37	24.50	73	219	17.55	
	0140	Up to 250 H.P., 12' long		.41	27	81	243	19.50	
	0180	Over 250 H.P., 15' long		.56	37	111	335	26.70	
	0200	Special dipper arm, up to 100 H.P., 32' long		1.14	75.50	227	680	54.50	
	0240	Over 100 H.P., 33' long		1.42	94.50	284	850	68.15	
	0280	Catch basin/sewer cleaning truck, 3 ton, 9 C.Y., 1000 gal.		35.10	405	1,210	3,625	522.80	
	0300	Concrete batch plant, portable, electric, 200 C.Y./hr.		24.34	545	1,630	4,900	520.70	
	0520	Grader/dozer attachment, ripper/scarifier, rear mounted, up to 135 H.P.		3.15	61.50	185	555	62.20	
	0540	Up to 180 H.P.		4.10	91.50	275	825	87.80	
	0580	Up to 250 H.P.		5.70	145	435	1,300	132.60	
	0700	Pvmt. removal bucket, for hyd. excavator, up to 90 H.P.		2.10	56.50	170	510	50.80	
	0740	Up to 200 H.P.		2.25	71.50	215	645	61	
	0780	Over 200 H.P.		2.45	88.50	265	795	72.60	
	0900	Aggregate spreader, self-propelled, 187 H.P.		50.00	730	2,185	6,550	837	
	1000	Chemical spreader, 3 C.Y.		3.15	45	135	405	52.20	
	1900	Hammermill, traveling, 250 H.P.		68.23	2,200	6,620	19,900	1,870	
	2000	Horizontal borer, 3" diameter, 13 H.P. gas driven		5.50	56.50	170	510	78	
	2150	Horizontal directional drill, 20,000 lb. thrust, 78 H.P. diesel		27.50	680	2,045	6,125	629	
	2160	30,000 lb. thrust, 115 H.P.		33.65	1,050	3,135	9,400	896.20	
	2170	50,000 lb. thrust, 170 H.P.		48.35	1,325	4,005	12,000	1,188	
	2190	Mud trailer for HDD, 1500 gallons, 175 H.P., gas		24.10	158	475	1,425	287.80	
	2200	Hydromulcher, diesel, 3000 gallon, for truck mounting		16.35	253	760	2,275	282.80	
	2300	Gas, 600 gallon		7.40	103	310	930	121.20	
	2400	Joint & crack cleaner, walk behind, 25 H.P.		3.10	51.50	155	465	55.80	

01 54 33 | Equipment Rental

			UNIT	HOURLY OPER. COST	RENT PER DAY	RENT PER WEEK	RENT PER MONTH	EQUIPMENT COST/DAY	
50	2500	Filler, trailer mounted, 400 gallons, 20 H.P.	Ea.	8.40	218	655	1,975	198.20	**50**
	3000	Paint striper, self-propelled, 40 gallon, 22 H.P.		6.75	162	485	1,450	151	
	3100	120 gallon, 120 H.P.		18.90	405	1,220	3,650	395.20	
	3200	Post drivers, 6" I-Beam frame, for truck mounting		12.45	390	1,175	3,525	334.60	
	3400	Road sweeper, self-propelled, 8' wide, 90 H.P.		35.95	670	2,005	6,025	688.60	
	3450	Road sweeper, vacuum assisted, 4 C.Y., 220 gallons		56.05	655	1,960	5,875	840.40	
	4000	Road mixer, self-propelled, 130 H.P.		45.95	800	2,405	7,225	848.60	
	4100	310 H.P.		75.55	2,150	6,425	19,300	1,889	
	4220	Cold mix paver, incl. pug mill and bitumen tank, 165 H.P.		94.60	2,300	6,915	20,700	2,140	
	4240	Pavement brush, towed		3.40	96.50	290	870	85.20	
	4250	Paver, asphalt, wheel or crawler, 130 H.P., diesel		94.25	2,275	6,845	20,500	2,123	
	4300	Paver, road widener, gas, 1' to 6', 67 H.P.		46.65	940	2,825	8,475	938.20	
	4400	Diesel, 2' to 14', 88 H.P.		56.75	1,125	3,355	10,100	1,125	
	4600	Slipform pavers, curb and gutter, 2 track, 75 H.P.		56.30	1,200	3,615	10,800	1,173	
	4700	4 track, 165 H.P.		36.95	825	2,470	7,400	789.60	
	4800	Median barrier, 215 H.P.		57.45	1,275	3,805	11,400	1,221	
	4901	Trailer, low bed, 75 ton capacity		11.05	268	805	2,425	249.40	
	5000	Road planer, walk behind, 10" cutting width, 10 H.P.		2.50	33.50	100	300	40	
	5100	Self-propelled, 12" cutting width, 64 H.P.		8.00	115	345	1,025	133	
	5120	Traffic line remover, metal ball blaster, truck mounted, 115 H.P.		46.70	800	2,395	7,175	852.60	
	5140	Grinder, truck mounted, 115 H.P.		51.05	850	2,555	7,675	919.40	
	5160	Walk-behind, 11 H.P.		3.55	55	165	495	61.40	
	5200	Pavement profiler, 4' to 6' wide, 450 H.P.		218.90	3,450	10,350	31,100	3,821	
	5300	8' to 10' wide, 750 H.P.		336.50	4,550	13,635	40,900	5,419	
	5400	Roadway plate, steel, 1" x 8' x 20'		.09	14.35	43	129	9.30	
	5600	Stabilizer, self-propelled, 150 H.P.		41.50	680	2,045	6,125	741	
	5700	310 H.P.		77.65	1,825	5,485	16,500	1,718	
	5800	Striper, truck mounted, 120 gallon paint, 460 H.P.		48.50	505	1,510	4,525	690	
	5900	Thermal paint heating kettle, 115 gallons		7.73	26.50	80	240	77.85	
	6000	Tar kettle, 330 gallon, trailer mounted		11.65	60	180	540	129.20	
	7000	Tunnel locomotive, diesel, 8 to 12 ton		29.90	600	1,800	5,400	599.20	
	7005	Electric, 10 ton		28.40	685	2,060	6,175	639.20	
	7010	Muck cars, 1/2 C.Y. capacity		2.25	26	78	234	33.60	
	7020	1 C.Y. capacity		2.45	33.50	100	300	39.60	
	7030	2 C.Y. capacity		2.60	38.50	115	345	43.80	
	7040	Side dump, 2 C.Y. capacity		2.80	46.50	140	420	50.40	
	7050	3 C.Y. capacity		3.80	51.50	155	465	61.40	
	7060	5 C.Y. capacity		5.50	66.50	200	600	84	
	7100	Ventilating blower for tunnel, 7-1/2 H.P.		2.16	51.50	155	465	48.30	
	7110	10 H.P.		2.39	53.50	160	480	51.10	
	7120	20 H.P.		3.56	69.50	208	625	70.10	
	7140	40 H.P.		6.02	80	240	720	96.15	
	7160	60 H.P.		8.75	98.50	295	885	129	
	7175	75 H.P.		10.31	153	460	1,375	174.50	
	7180	200 H.P.		20.73	305	920	2,750	349.85	
	7800	Windrow loader, elevating		54.10	1,350	4,045	12,100	1,242	
60	0010	**LIFTING AND HOISTING EQUIPMENT RENTAL** without operators							**60**
	0150	Crane, flatbed mounted, 3 ton capacity [R015433-10]	Ea.	14.10	205	615	1,850	235.80	
	0200	Crane, climbing, 106' jib, 6000 lb. capacity, 410 fpm [R312316-45]		41.13	1,750	5,260	15,800	1,381	
	0300	101' jib, 10,250 lb. capacity, 270 fpm		48.18	2,225	6,670	20,000	1,719	
	0500	Tower, static, 130' high, 106' jib, 6200 lb. capacity at 400 fpm		45.23	2,025	6,080	18,200	1,578	
	0520	Mini crawler spider crane, up to 24" wide, 1990 lb. lifting capacity		12.54	755	2,265	6,800	553.30	
	0525	Up to 30" wide, 6450 lb. lifting capacity		14.57	840	2,520	7,550	620.55	
	0530	Up to 52" wide, 6680 lb. lifting capacity		23.17	1,325	3,960	11,900	977.35	
	0535	Up to 55" wide, 8920 lb. lifting capacity		25.87	1,500	4,500	13,500	1,107	
	0540	Up to 66" wide, 13,350 lb. lifting capacity		35.03	2,050	6,120	18,400	1,504	
	0600	Crawler mounted, lattice boom, 1/2 C.Y., 15 tons at 12' radius		37.10	885	2,660	7,975	828.80	
	0700	3/4 C.Y., 20 tons at 12' radius		49.46	1,100	3,320	9,950	1,060	

For customer support on your Building Construction Costs with RSMeans data, call 800.448.8182.

01 54 33 | Equipment Rental

		UNIT	HOURLY OPER. COST	RENT PER DAY	RENT PER WEEK	RENT PER MONTH	EQUIPMENT COST/DAY	
0800	1 C.Y., 25 tons at 12' radius	Ea.	65.95	1,375	4,100	12,300	1,348	60
0900	1-1/2 C.Y., 40 tons at 12' radius		66.45	1,400	4,190	12,600	1,370	
1000	2 C.Y., 50 tons at 12' radius		88.90	2,050	6,145	18,400	1,940	
1100	3 C.Y., 75 tons at 12' radius		76.00	1,825	5,500	16,500	1,708	
1200	100 ton capacity, 60' boom		86.05	1,975	5,920	17,800	1,872	
1300	165 ton capacity, 60' boom		105.00	2,325	6,970	20,900	2,234	
1400	200 ton capacity, 70' boom		140.85	3,125	9,385	28,200	3,004	
1500	350 ton capacity, 80' boom		182.50	4,125	12,375	37,100	3,935	
1600	Truck mounted, lattice boom, 6 x 4, 20 tons at 10' radius		37.84	1,300	3,900	11,700	1,083	
1700	25 tons at 10' radius		40.92	1,425	4,240	12,700	1,175	
1800	8 x 4, 30 tons at 10' radius		44.29	1,500	4,520	13,600	1,258	
1900	40 tons at 12' radius		47.09	1,575	4,720	14,200	1,321	
2000	60 tons at 15' radius		52.56	1,675	5,000	15,000	1,420	
2050	82 tons at 15' radius		58.46	1,775	5,340	16,000	1,536	
2100	90 tons at 15' radius		65.33	1,950	5,820	17,500	1,687	
2200	115 tons at 15' radius		73.60	2,175	6,500	19,500	1,889	
2300	150 tons at 18' radius		80.95	2,275	6,845	20,500	2,017	
2350	165 tons at 18' radius		85.95	2,425	7,260	21,800	2,140	
2400	Truck mounted, hydraulic, 12 ton capacity		31.00	415	1,240	3,725	496	
2500	25 ton capacity		37.45	485	1,455	4,375	590.60	
2550	33 ton capacity		50.70	890	2,675	8,025	940.60	
2560	40 ton capacity		51.00	905	2,710	8,125	950	
2600	55 ton capacity		56.60	900	2,705	8,125	993.80	
2700	80 ton capacity		77.85	1,500	4,475	13,400	1,518	
2720	100 ton capacity		76.80	1,525	4,595	13,800	1,533	
2740	120 ton capacity		101.90	1,825	5,460	16,400	1,907	
2760	150 ton capacity		107.85	1,975	5,960	17,900	2,055	
2800	Self-propelled, 4 x 4, with telescoping boom, 5 ton		15.00	232	695	2,075	259	
2900	12-1/2 ton capacity		20.45	335	1,000	3,000	363.60	
3000	15 ton capacity		33.95	520	1,560	4,675	583.60	
3050	20 ton capacity		25.40	615	1,840	5,525	571.20	
3100	25 ton capacity		37.30	615	1,850	5,550	668.40	
3150	40 ton capacity		44.10	635	1,910	5,725	734.80	
3200	Derricks, guy, 20 ton capacity, 60' boom, 75' mast		23.07	430	1,288	3,875	442.15	
3300	100' boom, 115' mast		36.55	735	2,210	6,625	734.40	
3400	Stiffleg, 20 ton capacity, 70' boom, 37' mast		25.74	555	1,670	5,000	539.90	
3500	100' boom, 47' mast		39.84	895	2,680	8,050	854.70	
3550	Helicopter, small, lift to 1250 lb. maximum, w/pilot		97.10	3,475	10,400	31,200	2,857	
3600	Hoists, chain type, overhead, manual, 3/4 ton		.15	.33	1	3	1.40	
3900	10 ton		.80	6	18	54	10	
4000	Hoist and tower, 5000 lb. cap., portable electric, 40' high		5.19	247	742	2,225	189.90	
4100	For each added 10' section, add		.12	19.35	58	174	12.55	
4200	Hoist and single tubular tower, 5000 lb. electric, 100' high		7.03	345	1,036	3,100	263.45	
4300	For each added 6'-6" section, add		.20	33.50	101	305	21.80	
4400	Hoist and double tubular tower, 5000 lb., 100' high		7.56	380	1,141	3,425	288.70	
4500	For each added 6'-6" section, add		.22	37	111	335	23.95	
4550	Hoist and tower, mast type, 6000 lb., 100' high		8.14	395	1,183	3,550	301.70	
4570	For each added 10' section, add		.14	22.50	68	204	14.70	
4600	Hoist and tower, personnel, electric, 2000 lb., 100' @ 125 fpm		17.23	1,050	3,150	9,450	767.85	
4700	3000 lb., 100' @ 200 fpm		19.70	1,200	3,570	10,700	871.60	
4800	3000 lb., 150' @ 300 fpm		21.85	1,325	4,000	12,000	974.80	
4900	4000 lb., 100' @ 300 fpm		22.62	1,350	4,080	12,200	996.95	
5000	6000 lb., 100' @ 275 fpm		24.32	1,425	4,270	12,800	1,049	
5100	For added heights up to 500', add	L.F.	.01	1.67	5	15	1.10	
5200	Jacks, hydraulic, 20 ton	Ea.	.05	2	6	18	1.60	
5500	100 ton		.40	12	36	108	10.40	
6100	Jacks, hydraulic, climbing w/50' jackrods, control console, 30 ton cap.		2.13	142	426	1,275	102.25	
6150	For each added 10' jackrod section, add		.05	3.33	10	30	2.40	

01 54 33 | Equipment Rental

		UNIT	HOURLY OPER. COST	RENT PER DAY	RENT PER WEEK	RENT PER MONTH	EQUIPMENT COST/DAY	
60	6300	50 ton capacity	Ea.	3.43	228	685	2,050	164.45	60
	6350	For each added 10' jackrod section, add		.06	4	12	36	2.90	
	6500	125 ton capacity		8.95	595	1,790	5,375	429.60	
	6550	For each added 10' jackrod section, add		.61	40.50	121	365	29.10	
	6600	Cable jack, 10 ton capacity with 200' cable		1.79	119	357	1,075	85.70	
	6650	For each added 50' of cable, add		.22	14.35	43	129	10.35	
70	0010	**WELLPOINT EQUIPMENT RENTAL** without operators	R015433 -10						70
	0020	Based on 2 months rental							
	0100	Combination jetting & wellpoint pump, 60 H.P. diesel	Ea.	15.83	350	1,057	3,175	338.05	
	0200	High pressure gas jet pump, 200 H.P., 300 psi	"	34.42	300	903	2,700	455.95	
	0300	Discharge pipe, 8" diameter	L.F.	.01	.57	1.71	5.15	.40	
	0350	12" diameter		.01	.84	2.53	7.60	.60	
	0400	Header pipe, flows up to 150 GPM, 4" diameter		.01	.52	1.56	4.68	.40	
	0500	400 GPM, 6" diameter		.01	.61	1.83	5.50	.45	
	0600	800 GPM, 8" diameter		.01	.84	2.53	7.60	.60	
	0700	1500 GPM, 10" diameter		.01	.89	2.66	8	.60	
	0800	2500 GPM, 12" diameter		.03	1.68	5.03	15.10	1.25	
	0900	4500 GPM, 16" diameter		.03	2.15	6.44	19.30	1.55	
	0950	For quick coupling aluminum and plastic pipe, add		.03	2.22	6.67	20	1.55	
	1100	Wellpoint, 25' long, with fittings & riser pipe, 1-1/2" or 2" diameter	Ea.	.07	4.44	13.31	40	3.20	
	1200	Wellpoint pump, diesel powered, 4" suction, 20 H.P.		7.07	203	609	1,825	178.35	
	1300	6" suction, 30 H.P.		9.51	252	756	2,275	227.30	
	1400	8" suction, 40 H.P.		12.87	345	1,036	3,100	310.15	
	1500	10" suction, 75 H.P.		19.01	405	1,211	3,625	394.30	
	1600	12" suction, 100 H.P.		27.56	645	1,930	5,800	606.50	
	1700	12" suction, 175 H.P.		39.50	710	2,130	6,400	742	
80	0010	**MARINE EQUIPMENT RENTAL** without operators	R015433 -10						80
	0200	Barge, 400 ton, 30' wide x 90' long	Ea.	18.05	1,150	3,455	10,400	835.40	
	0240	800 ton, 45' wide x 90' long		21.95	1,425	4,240	12,700	1,024	
	2000	Tugboat, diesel, 100 H.P.		28.80	228	685	2,050	367.40	
	2040	250 H.P.		54.40	410	1,225	3,675	680.20	
	2080	380 H.P.		122.45	1,225	3,685	11,100	1,717	
	3000	Small work boat, gas, 16-foot, 50 H.P.		12.50	63.50	190	570	138	
	4000	Large, diesel, 48-foot, 200 H.P.		74.90	1,300	3,930	11,800	1,385	

For customer support on your Building Construction Costs with RSMeans data, call 800.448.8182.

Crew No.	Bare Costs		Incl. Subs O&P		Cost Per Labor-Hour	
Crew A-1	**Hr.**	**Daily**	**Hr.**	**Daily**	**Bare Costs**	**Incl. O&P**
1 Building Laborer	$39.85	$318.80	$60.70	$485.60	$39.85	$60.70
1 Concrete Saw, Gas Manual		71.20		78.32	8.90	9.79
8 L.H., Daily Totals		$390.00		$563.92	$48.75	$70.49
Crew A-1A	**Hr.**	**Daily**	**Hr.**	**Daily**	**Bare Costs**	**Incl. O&P**
1 Skilled Worker	$52.35	$418.80	$80.15	$641.20	$52.35	$80.15
1 Shot Blaster, 20"		214.80		236.28	26.85	29.54
8 L.H., Daily Totals		$633.60		$877.48	$79.20	$109.69
Crew A-1B	**Hr.**	**Daily**	**Hr.**	**Daily**	**Bare Costs**	**Incl. O&P**
1 Building Laborer	$39.85	$318.80	$60.70	$485.60	$39.85	$60.70
1 Concrete Saw		102.40		112.64	12.80	14.08
8 L.H., Daily Totals		$421.20		$598.24	$52.65	$74.78
Crew A-1C	**Hr.**	**Daily**	**Hr.**	**Daily**	**Bare Costs**	**Incl. O&P**
1 Building Laborer	$39.85	$318.80	$60.70	$485.60	$39.85	$60.70
1 Chain Saw, Gas, 18"		27.80		30.58	3.48	3.82
8 L.H., Daily Totals		$346.60		$516.18	$43.33	$64.52
Crew A-1D	**Hr.**	**Daily**	**Hr.**	**Daily**	**Bare Costs**	**Incl. O&P**
1 Building Laborer	$39.85	$318.80	$60.70	$485.60	$39.85	$60.70
1 Vibrating Plate, Gas, 18"		31.80		34.98	3.98	4.37
8 L.H., Daily Totals		$350.60		$520.58	$43.83	$65.07
Crew A-1E	**Hr.**	**Daily**	**Hr.**	**Daily**	**Bare Costs**	**Incl. O&P**
1 Building Laborer	$39.85	$318.80	$60.70	$485.60	$39.85	$60.70
1 Vibrating Plate, Gas, 21"		40.60		44.66	5.08	5.58
8 L.H., Daily Totals		$359.40		$530.26	$44.92	$66.28
Crew A-1F	**Hr.**	**Daily**	**Hr.**	**Daily**	**Bare Costs**	**Incl. O&P**
1 Building Laborer	$39.85	$318.80	$60.70	$485.60	$39.85	$60.70
1 Rammer/Tamper, Gas, 8"		46.00		50.60	5.75	6.33
8 L.H., Daily Totals		$364.80		$536.20	$45.60	$67.03
Crew A-1G	**Hr.**	**Daily**	**Hr.**	**Daily**	**Bare Costs**	**Incl. O&P**
1 Building Laborer	$39.85	$318.80	$60.70	$485.60	$39.85	$60.70
1 Rammer/Tamper, Gas, 15"		52.00		57.20	6.50	7.15
8 L.H., Daily Totals		$370.80		$542.80	$46.35	$67.85
Crew A-1H	**Hr.**	**Daily**	**Hr.**	**Daily**	**Bare Costs**	**Incl. O&P**
1 Building Laborer	$39.85	$318.80	$60.70	$485.60	$39.85	$60.70
1 Exterior Steam Cleaner		74.80		82.28	9.35	10.29
8 L.H., Daily Totals		$393.60		$567.88	$49.20	$70.98
Crew A-1J	**Hr.**	**Daily**	**Hr.**	**Daily**	**Bare Costs**	**Incl. O&P**
1 Building Laborer	$39.85	$318.80	$60.70	$485.60	$39.85	$60.70
1 Cultivator, Walk-Behind, 5 H.P.		63.50		69.85	7.94	8.73
8 L.H., Daily Totals		$382.30		$555.45	$47.79	$69.43
Crew A-1K	**Hr.**	**Daily**	**Hr.**	**Daily**	**Bare Costs**	**Incl. O&P**
1 Building Laborer	$39.85	$318.80	$60.70	$485.60	$39.85	$60.70
1 Cultivator, Walk-Behind, 8 H.P.		74.60		82.06	9.32	10.26
8 L.H., Daily Totals		$393.40		$567.66	$49.17	$70.96
Crew A-1M	**Hr.**	**Daily**	**Hr.**	**Daily**	**Bare Costs**	**Incl. O&P**
1 Building Laborer	$39.85	$318.80	$60.70	$485.60	$39.85	$60.70
1 Snow Blower, Walk-Behind		70.10		77.11	8.76	9.64
8 L.H., Daily Totals		$388.90		$562.71	$48.61	$70.34

Crew No.	Bare Costs		Incl. Subs O&P		Cost Per Labor-Hour	
Crew A-2	**Hr.**	**Daily**	**Hr.**	**Daily**	**Bare Costs**	**Incl. O&P**
2 Laborers	$39.85	$637.60	$60.70	$971.20	$41.40	$62.80
1 Truck Driver (light)	44.50	356.00	67.00	536.00		
1 Flatbed Truck, Gas, 1.5 Ton		188.40		207.24	7.85	8.63
24 L.H., Daily Totals		$1182.00		$1714.44	$49.25	$71.44
Crew A-2A	**Hr.**	**Daily**	**Hr.**	**Daily**	**Bare Costs**	**Incl. O&P**
2 Laborers	$39.85	$637.60	$60.70	$971.20	$41.40	$62.80
1 Truck Driver (light)	44.50	356.00	67.00	536.00		
1 Flatbed Truck, Gas, 1.5 Ton		188.40		207.24		
1 Concrete Saw		102.40		112.64	12.12	13.33
24 L.H., Daily Totals		$1284.40		$1827.08	$53.52	$76.13
Crew A-2B	**Hr.**	**Daily**	**Hr.**	**Daily**	**Bare Costs**	**Incl. O&P**
1 Truck Driver (light)	$44.50	$356.00	$67.00	$536.00	$44.50	$67.00
1 Flatbed Truck, Gas, 1.5 Ton		188.40		207.24	23.55	25.91
8 L.H., Daily Totals		$544.40		$743.24	$68.05	$92.91
Crew A-3A	**Hr.**	**Daily**	**Hr.**	**Daily**	**Bare Costs**	**Incl. O&P**
1 Equip. Oper. (light)	$51.30	$410.40	$77.35	$618.80	$51.30	$77.35
1 Pickup Truck, 4x4, 3/4 Ton		126.60		139.26	15.82	17.41
8 L.H., Daily Totals		$537.00		$758.06	$67.13	$94.76
Crew A-3B	**Hr.**	**Daily**	**Hr.**	**Daily**	**Bare Costs**	**Incl. O&P**
1 Equip. Oper. (medium)	$53.75	$430.00	$81.05	$648.40	$49.88	$75.17
1 Truck Driver (heavy)	46.00	368.00	69.30	554.40		
1 Dump Truck, 12 C.Y., 400 H.P.		542.80		597.08		
1 F.E. Loader, W.M., 2.5 C.Y.		523.20		575.52	66.63	73.29
16 L.H., Daily Totals		$1864.00		$2375.40	$116.50	$148.46
Crew A-3C	**Hr.**	**Daily**	**Hr.**	**Daily**	**Bare Costs**	**Incl. O&P**
1 Equip. Oper. (light)	$51.30	$410.40	$77.35	$618.80	$51.30	$77.35
1 Loader, Skid Steer, 78 H.P.		364.80		401.28	45.60	50.16
8 L.H., Daily Totals		$775.20		$1020.08	$96.90	$127.51
Crew A-3D	**Hr.**	**Daily**	**Hr.**	**Daily**	**Bare Costs**	**Incl. O&P**
1 Truck Driver (light)	$44.50	$356.00	$67.00	$536.00	$44.50	$67.00
1 Pickup Truck, 4x4, 3/4 Ton		126.60		139.26		
1 Flatbed Trailer, 25 Ton		133.00		146.30	32.45	35.70
8 L.H., Daily Totals		$615.60		$821.56	$76.95	$102.69
Crew A-3E	**Hr.**	**Daily**	**Hr.**	**Daily**	**Bare Costs**	**Incl. O&P**
1 Equip. Oper. (crane)	$56.10	$448.80	$84.60	$676.80	$51.05	$76.95
1 Truck Driver (heavy)	46.00	368.00	69.30	554.40		
1 Pickup Truck, 4x4, 3/4 Ton		126.60		139.26	7.91	8.70
16 L.H., Daily Totals		$943.40		$1370.46	$58.96	$85.65
Crew A-3F	**Hr.**	**Daily**	**Hr.**	**Daily**	**Bare Costs**	**Incl. O&P**
1 Equip. Oper. (crane)	$56.10	$448.80	$84.60	$676.80	$51.05	$76.95
1 Truck Driver (heavy)	46.00	368.00	69.30	554.40		
1 Pickup Truck, 4x4, 3/4 Ton		126.60		139.26		
1 Truck Tractor, 6x4, 380 H.P.		476.20		523.82		
1 Lowbed Trailer, 75 Ton		249.40		274.34	53.26	58.59
16 L.H., Daily Totals		$1669.00		$2168.62	$104.31	$135.54

For customer support on your Building Construction Costs with RSMeans data, call 800.448.8182.

Crews - Standard

Crew A-3G

Crew A-3G	Hr.	Daily	Hr.	Daily	Bare Costs	Incl. O&P
1 Equip. Oper. (crane)	$56.10	$448.80	$84.60	$676.80	$51.05	$76.95
1 Truck Driver (heavy)	46.00	368.00	69.30	554.40		
1 Pickup Truck, 4x4, 3/4 Ton		126.60		139.26		
1 Truck Tractor, 6x4, 450 H.P.		583.40		641.74		
1 Lowbed Trailer, 75 Ton		249.40		274.34	59.96	65.96
16 L.H., Daily Totals		$1776.20		$2286.54	$111.01	$142.91

Crew A-3H

Crew A-3H	Hr.	Daily	Hr.	Daily	Bare Costs	Incl. O&P
1 Equip. Oper. (crane)	$56.10	$448.80	$84.60	$676.80	$56.10	$84.60
1 Hyd. Crane, 12 Ton (Daily)		628.60		691.46	78.58	86.43
8 L.H., Daily Totals		$1077.40		$1368.26	$134.68	$171.03

Crew A-3I

Crew A-3I	Hr.	Daily	Hr.	Daily	Bare Costs	Incl. O&P
1 Equip. Oper. (crane)	$56.10	$448.80	$84.60	$676.80	$56.10	$84.60
1 Hyd. Crane, 25 Ton (Daily)		759.40		835.34	94.92	104.42
8 L.H., Daily Totals		$1208.20		$1512.14	$151.03	$189.02

Crew A-3J

Crew A-3J	Hr.	Daily	Hr.	Daily	Bare Costs	Incl. O&P
1 Equip. Oper. (crane)	$56.10	$448.80	$84.60	$676.80	$56.10	$84.60
1 Hyd. Crane, 40 Ton (Daily)		1313.00		1444.30	164.13	180.54
8 L.H., Daily Totals		$1761.80		$2121.10	$220.22	$265.14

Crew A-3K

Crew A-3K	Hr.	Daily	Hr.	Daily	Bare Costs	Incl. O&P
1 Equip. Oper. (crane)	$56.10	$448.80	$84.60	$676.80	$52.35	$78.95
1 Equip. Oper. (oiler)	48.60	388.80	73.30	586.40		
1 Hyd. Crane, 55 Ton (Daily)		1353.00		1488.30		
1 P/U Truck, 3/4 Ton (Daily)		139.20		153.12	93.26	102.59
16 L.H., Daily Totals		$2329.80		$2904.62	$145.61	$181.54

Crew A-3L

Crew A-3L	Hr.	Daily	Hr.	Daily	Bare Costs	Incl. O&P
1 Equip. Oper. (crane)	$56.10	$448.80	$84.60	$676.80	$52.35	$78.95
1 Equip. Oper. (oiler)	48.60	388.80	73.30	586.40		
1 Hyd. Crane, 80 Ton (Daily)		2113.00		2324.30		
1 P/U Truck, 3/4 Ton (Daily)		139.20		153.12	140.76	154.84
16 L.H., Daily Totals		$3089.80		$3740.62	$193.11	$233.79

Crew A-3M

Crew A-3M	Hr.	Daily	Hr.	Daily	Bare Costs	Incl. O&P
1 Equip. Oper. (crane)	$56.10	$448.80	$84.60	$676.80	$52.35	$78.95
1 Equip. Oper. (oiler)	48.60	388.80	73.30	586.40		
1 Hyd. Crane, 100 Ton (Daily)		2144.00		2358.40		
1 P/U Truck, 3/4 Ton (Daily)		139.20		153.12	142.70	156.97
16 L.H., Daily Totals		$3120.80		$3774.72	$195.05	$235.92

Crew A-3N

Crew A-3N	Hr.	Daily	Hr.	Daily	Bare Costs	Incl. O&P
1 Equip. Oper. (crane)	$56.10	$448.80	$84.60	$676.80	$56.10	$84.60
1 Tower Crane (monthly)		1180.00		1298.00	147.50	162.25
8 L.H., Daily Totals		$1628.80		$1974.80	$203.60	$246.85

Crew A-3P

Crew A-3P	Hr.	Daily	Hr.	Daily	Bare Costs	Incl. O&P
1 Equip. Oper. (light)	$51.30	$410.40	$77.35	$618.80	$51.30	$77.35
1 A.T. Forklift, 31' reach, 45' lift		484.20		532.62	60.52	66.58
8 L.H., Daily Totals		$894.60		$1151.42	$111.83	$143.93

Crew A-3Q

Crew A-3Q	Hr.	Daily	Hr.	Daily	Bare Costs	Incl. O&P
1 Equip. Oper. (light)	$51.30	$410.40	$77.35	$618.80	$51.30	$77.35
1 Pickup Truck, 4x4, 3/4 Ton		126.60		139.26		
1 Flatbed Trailer, 3 Ton		25.40		27.94	19.00	20.90
8 L.H., Daily Totals		$562.40		$786.00	$70.30	$98.25

Crew A-3R

Crew A-3R	Hr.	Daily	Hr.	Daily	Bare Costs	Incl. O&P
1 Equip. Oper. (light)	$51.30	$410.40	$77.35	$618.80	$51.30	$77.35
1 Forklift, Smooth Floor, 8,000 Lb.		148.20		163.02	18.52	20.38
8 L.H., Daily Totals		$558.60		$781.82	$69.83	$97.73

Crew A-4

Crew A-4	Hr.	Daily	Hr.	Daily	Bare Costs	Incl. O&P
2 Carpenters	$50.70	$811.20	$77.20	$1235.20	$47.98	$72.82
1 Painter, Ordinary	42.55	340.40	64.05	512.40		
24 L.H., Daily Totals		$1151.60		$1747.60	$47.98	$72.82

Crew A-5

Crew A-5	Hr.	Daily	Hr.	Daily	Bare Costs	Incl. O&P
2 Laborers	$39.85	$637.60	$60.70	$971.20	$40.37	$61.40
.25 Truck Driver (light)	44.50	89.00	67.00	134.00		
.25 Flatbed Truck, Gas, 1.5 Ton		47.10		51.81	2.62	2.88
18 L.H., Daily Totals		$773.70		$1157.01	$42.98	$64.28

Crew A-6

Crew A-6	Hr.	Daily	Hr.	Daily	Bare Costs	Incl. O&P
1 Instrument Man	$52.35	$418.80	$80.15	$641.20	$50.42	$76.80
1 Rodman/Chainman	48.50	388.00	73.45	587.60		
1 Level, Electronic		50.40		55.44	3.15	3.46
16 L.H., Daily Totals		$857.20		$1284.24	$53.58	$80.27

Crew A-7

Crew A-7	Hr.	Daily	Hr.	Daily	Bare Costs	Incl. O&P
1 Chief of Party	$63.65	$509.20	$96.45	$771.60	$54.83	$83.35
1 Instrument Man	52.35	418.80	80.15	641.20		
1 Rodman/Chainman	48.50	388.00	73.45	587.60		
1 Level, Electronic		50.40		55.44	2.10	2.31
24 L.H., Daily Totals		$1366.40		$2055.84	$56.93	$85.66

Crew A-8

Crew A-8	Hr.	Daily	Hr.	Daily	Bare Costs	Incl. O&P
1 Chief of Party	$63.65	$509.20	$96.45	$771.60	$53.25	$80.88
1 Instrument Man	52.35	418.80	80.15	641.20		
2 Rodmen/Chainmen	48.50	776.00	73.45	1175.20		
1 Level, Electronic		50.40		55.44	1.58	1.73
32 L.H., Daily Totals		$1754.40		$2643.44	$54.83	$82.61

Crew A-9

Crew A-9	Hr.	Daily	Hr.	Daily	Bare Costs	Incl. O&P
1 Asbestos Foreman	$56.80	$454.40	$88.15	$705.20	$56.36	$87.49
7 Asbestos Workers	56.30	3152.80	87.40	4894.40		
64 L.H., Daily Totals		$3607.20		$5599.60	$56.36	$87.49

Crew A-10A

Crew A-10A	Hr.	Daily	Hr.	Daily	Bare Costs	Incl. O&P
1 Asbestos Foreman	$56.80	$454.40	$88.15	$705.20	$56.47	$87.65
2 Asbestos Workers	56.30	900.80	87.40	1398.40		
24 L.H., Daily Totals		$1355.20		$2103.60	$56.47	$87.65

Crew A-10B

Crew A-10B	Hr.	Daily	Hr.	Daily	Bare Costs	Incl. O&P
1 Asbestos Foreman	$56.80	$454.40	$88.15	$705.20	$56.42	$87.59
3 Asbestos Workers	56.30	1351.20	87.40	2097.60		
32 L.H., Daily Totals		$1805.60		$2802.80	$56.42	$87.59

Crew A-10C

Crew A-10C	Hr.	Daily	Hr.	Daily	Bare Costs	Incl. O&P
3 Asbestos Workers	$56.30	$1351.20	$87.40	$2097.60	$56.30	$87.40
1 Flatbed Truck, Gas, 1.5 Ton		188.40		207.24	7.85	8.63
24 L.H., Daily Totals		$1539.60		$2304.84	$64.15	$96.03

For customer support on your Building Construction Costs with RSMeans data, call 800.448.8182.

Crews - Standard

Crew A-10D

Crew No.	Bare Costs Hr.	Daily	Incl. Subs O&P Hr.	Daily	Cost Per Labor-Hour Bare Costs	Incl. O&P
2 Asbestos Workers	$56.30	$900.80	$87.40	$1398.40	$54.33	$83.17
1 Equip. Oper. (crane)	56.10	448.80	84.60	676.80		
1 Equip. Oper. (oiler)	48.60	388.80	73.30	586.40		
1 Hydraulic Crane, 33 Ton		940.60		1034.66	29.39	32.33
32 L.H., Daily Totals		$2679.00		$3696.26	$83.72	$115.51

Crew A-11

Crew No.	Bare Costs Hr.	Daily	Incl. Subs O&P Hr.	Daily	Cost Per Labor-Hour Bare Costs	Incl. O&P
1 Asbestos Foreman	$56.80	$454.40	$88.15	$705.20	$56.36	$87.49
7 Asbestos Workers	56.30	3152.80	87.40	4894.40		
2 Chip. Hammers, 12 Lb., Elec.		42.00		46.20	.66	.72
64 L.H., Daily Totals		$3649.20		$5645.80	$57.02	$88.22

Crew A-12

Crew No.	Bare Costs Hr.	Daily	Incl. Subs O&P Hr.	Daily	Cost Per Labor-Hour Bare Costs	Incl. O&P
1 Asbestos Foreman	$56.80	$454.40	$88.15	$705.20	$56.36	$87.49
7 Asbestos Workers	56.30	3152.80	87.40	4894.40		
1 Trk-Mtd Vac, 14 CY, 1500 Gal.		522.70		574.97		
1 Flatbed Truck, 20,000 GVW		197.20		216.92	11.25	12.37
64 L.H., Daily Totals		$4327.10		$6391.49	$67.61	$99.87

Crew A-13

Crew No.	Bare Costs Hr.	Daily	Incl. Subs O&P Hr.	Daily	Cost Per Labor-Hour Bare Costs	Incl. O&P
1 Equip. Oper. (light)	$51.30	$410.40	$77.35	$618.80	$51.30	$77.35
1 Trk-Mtd Vac, 14 CY, 1500 Gal.		522.70		574.97		
1 Flatbed Truck, 20,000 GVW		197.20		216.92	89.99	98.99
8 L.H., Daily Totals		$1130.30		$1410.69	$141.29	$176.34

Crew B-1

Crew No.	Bare Costs Hr.	Daily	Incl. Subs O&P Hr.	Daily	Cost Per Labor-Hour Bare Costs	Incl. O&P
1 Labor Foreman (outside)	$41.85	$334.80	$63.75	$510.00	$40.52	$61.72
2 Laborers	39.85	637.60	60.70	971.20		
24 L.H., Daily Totals		$972.40		$1481.20	$40.52	$61.72

Crew B-1A

Crew No.	Bare Costs Hr.	Daily	Incl. Subs O&P Hr.	Daily	Cost Per Labor-Hour Bare Costs	Incl. O&P
1 Labor Foreman (outside)	$41.85	$334.80	$63.75	$510.00	$40.52	$61.72
2 Laborers	39.85	637.60	60.70	971.20		
2 Cutting Torches		25.20		27.72		
2 Sets of Gases		336.00		369.60	15.05	16.56
24 L.H., Daily Totals		$1333.60		$1878.52	$55.57	$78.27

Crew B-1B

Crew No.	Bare Costs Hr.	Daily	Incl. Subs O&P Hr.	Daily	Cost Per Labor-Hour Bare Costs	Incl. O&P
1 Labor Foreman (outside)	$41.85	$334.80	$63.75	$510.00	$44.41	$67.44
2 Laborers	39.85	637.60	60.70	971.20		
1 Equip. Oper. (crane)	56.10	448.80	84.60	676.80		
2 Cutting Torches		25.20		27.72		
2 Sets of Gases		336.00		369.60		
1 Hyd. Crane, 12 Ton		496.00		545.60	26.79	29.47
32 L.H., Daily Totals		$2278.40		$3100.92	$71.20	$96.90

Crew B-1C

Crew No.	Bare Costs Hr.	Daily	Incl. Subs O&P Hr.	Daily	Cost Per Labor-Hour Bare Costs	Incl. O&P
1 Labor Foreman (outside)	$41.85	$334.80	$63.75	$510.00	$40.52	$61.72
2 Laborers	39.85	637.60	60.70	971.20		
1 Telescoping Boom Lift, to 60'		454.60		500.06	18.94	20.84
24 L.H., Daily Totals		$1427.00		$1981.26	$59.46	$82.55

Crew B-1D

Crew No.	Bare Costs Hr.	Daily	Incl. Subs O&P Hr.	Daily	Cost Per Labor-Hour Bare Costs	Incl. O&P
2 Laborers	$39.85	$637.60	$60.70	$971.20	$39.85	$60.70
1 Small Work Boat, Gas, 50 H.P.		138.00		151.80		
1 Pressure Washer, 7 GPM		77.20		84.92	13.45	14.80
16 L.H., Daily Totals		$852.80		$1207.92	$53.30	$75.50

Crew B-1E

Crew No.	Bare Costs Hr.	Daily	Incl. Subs O&P Hr.	Daily	Cost Per Labor-Hour Bare Costs	Incl. O&P
1 Labor Foreman (outside)	$41.85	$334.80	$63.75	$510.00	$40.35	$61.46
3 Laborers	39.85	956.40	60.70	1456.80		
1 Work Boat, Diesel, 200 H.P.		1385.00		1523.50		
2 Pressure Washers, 7 GPM		154.40		169.84	48.11	52.92
32 L.H., Daily Totals		$2830.60		$3660.14	$88.46	$114.38

Crew B-1F

Crew No.	Bare Costs Hr.	Daily	Incl. Subs O&P Hr.	Daily	Cost Per Labor-Hour Bare Costs	Incl. O&P
2 Skilled Workers	$52.35	$837.60	$80.15	$1282.40	$48.18	$73.67
1 Laborer	39.85	318.80	60.70	485.60		
1 Small Work Boat, Gas, 50 H.P.		138.00		151.80		
1 Pressure Washer, 7 GPM		77.20		84.92	8.97	9.86
24 L.H., Daily Totals		$1371.60		$2004.72	$57.15	$83.53

Crew B-1G

Crew No.	Bare Costs Hr.	Daily	Incl. Subs O&P Hr.	Daily	Cost Per Labor-Hour Bare Costs	Incl. O&P
2 Laborers	$39.85	$637.60	$60.70	$971.20	$39.85	$60.70
1 Small Work Boat, Gas, 50 H.P.		138.00		151.80	8.63	9.49
16 L.H., Daily Totals		$775.60		$1123.00	$48.48	$70.19

Crew B-1H

Crew No.	Bare Costs Hr.	Daily	Incl. Subs O&P Hr.	Daily	Cost Per Labor-Hour Bare Costs	Incl. O&P
2 Skilled Workers	$52.35	$837.60	$80.15	$1282.40	$48.18	$73.67
1 Laborer	39.85	318.80	60.70	485.60		
1 Small Work Boat, Gas, 50 H.P.		138.00		151.80	5.75	6.33
24 L.H., Daily Totals		$1294.40		$1919.80	$53.93	$79.99

Crew B-1J

Crew No.	Bare Costs Hr.	Daily	Incl. Subs O&P Hr.	Daily	Cost Per Labor-Hour Bare Costs	Incl. O&P
1 Labor Foreman (inside)	$40.35	$322.80	$61.45	$491.60	$40.10	$61.08
1 Laborer	39.85	318.80	60.70	485.60		
16 L.H., Daily Totals		$641.60		$977.20	$40.10	$61.08

Crew B-1K

Crew No.	Bare Costs Hr.	Daily	Incl. Subs O&P Hr.	Daily	Cost Per Labor-Hour Bare Costs	Incl. O&P
1 Carpenter Foreman (inside)	$51.20	$409.60	$78.00	$624.00	$50.95	$77.60
1 Carpenter	50.70	405.60	77.20	617.60		
16 L.H., Daily Totals		$815.20		$1241.60	$50.95	$77.60

Crew B-2

Crew No.	Bare Costs Hr.	Daily	Incl. Subs O&P Hr.	Daily	Cost Per Labor-Hour Bare Costs	Incl. O&P
1 Labor Foreman (outside)	$41.85	$334.80	$63.75	$510.00	$40.25	$61.31
4 Laborers	39.85	1275.20	60.70	1942.40		
40 L.H., Daily Totals		$1610.00		$2452.40	$40.25	$61.31

Crew B-2A

Crew No.	Bare Costs Hr.	Daily	Incl. Subs O&P Hr.	Daily	Cost Per Labor-Hour Bare Costs	Incl. O&P
1 Labor Foreman (outside)	$41.85	$334.80	$63.75	$510.00	$40.52	$61.72
2 Laborers	39.85	637.60	60.70	971.20		
1 Telescoping Boom Lift, to 60'		454.60		500.06	18.94	20.84
24 L.H., Daily Totals		$1427.00		$1981.26	$59.46	$82.55

Crew B-3

Crew No.	Bare Costs Hr.	Daily	Incl. Subs O&P Hr.	Daily	Cost Per Labor-Hour Bare Costs	Incl. O&P
1 Labor Foreman (outside)	$41.85	$334.80	$63.75	$510.00	$44.55	$67.47
2 Laborers	39.85	637.60	60.70	971.20		
1 Equip. Oper. (medium)	53.75	430.00	81.05	648.40		
2 Truck Drivers (heavy)	46.00	736.00	69.30	1108.80		
1 Crawler Loader, 3 C.Y.		1204.00		1324.40		
2 Dump Trucks, 12 C.Y., 400 H.P.		1085.60		1194.16	47.70	52.47
48 L.H., Daily Totals		$4428.00		$5756.96	$92.25	$119.94

Crew B-3A

Crew No.	Bare Costs Hr.	Daily	Incl. Subs O&P Hr.	Daily	Cost Per Labor-Hour Bare Costs	Incl. O&P
4 Laborers	$39.85	$1275.20	$60.70	$1942.40	$42.63	$64.77
1 Equip. Oper. (medium)	53.75	430.00	81.05	648.40		
1 Hyd. Excavator, 1.5 C.Y.		893.60		982.96	22.34	24.57
40 L.H., Daily Totals		$2598.80		$3573.76	$64.97	$89.34

For customer support on your Building Construction Costs with RSMeans data, call 800.448.8182.

Crew No.	Bare Costs Hr.	Daily	Incl. Subs O&P Hr.	Daily	Cost Per Labor-Hour Bare Costs	Incl. O&P
Crew B-3B					Bare Costs	Incl. O&P
2 Laborers	$39.85	$637.60	$60.70	$971.20	$44.86	$67.94
1 Equip. Oper. (medium)	53.75	430.00	81.05	648.40		
1 Truck Driver (heavy)	46.00	368.00	69.30	554.40		
1 Backhoe Loader, 80 H.P.		385.00		423.50		
1 Dump Truck, 12 C.Y., 400 H.P.		542.80		597.08	28.99	31.89
32 L.H., Daily Totals		$2363.40		$3194.58	$73.86	$99.83
Crew B-3C					Bare Costs	Incl. O&P
3 Laborers	$39.85	$956.40	$60.70	$1456.80	$43.33	$65.79
1 Equip. Oper. (medium)	53.75	430.00	81.05	648.40		
1 Crawler Loader, 4 C.Y.		1459.00		1604.90	45.59	50.15
32 L.H., Daily Totals		$2845.40		$3710.10	$88.92	$115.94
Crew B-4					Bare Costs	Incl. O&P
1 Labor Foreman (outside)	$41.85	$334.80	$63.75	$510.00	$41.21	$62.64
4 Laborers	39.85	1275.20	60.70	1942.40		
1 Truck Driver (heavy)	46.00	368.00	69.30	554.40		
1 Truck Tractor, 220 H.P.		290.00		319.00		
1 Flatbed Trailer, 40 Ton		180.00		198.00	9.79	10.77
48 L.H., Daily Totals		$2448.00		$3523.80	$51.00	$73.41
Crew B-5					Bare Costs	Incl. O&P
1 Labor Foreman (outside)	$41.85	$334.80	$63.75	$510.00	$44.11	$66.95
4 Laborers	39.85	1275.20	60.70	1942.40		
2 Equip. Oper. (medium)	53.75	860.00	81.05	1296.80		
1 Air Compressor, 250 cfm		167.40		184.14		
2 Breakers, Pavement, 60 lb.		21.20		23.32		
2 -50' Air Hoses, 1.5"		45.40		49.94		
1 Crawler Loader, 3 C.Y.		1204.00		1324.40	25.68	28.25
56 L.H., Daily Totals		$3908.00		$5331.00	$69.79	$95.20
Crew B-5A					Bare Costs	Incl. O&P
1 Labor Foreman (outside)	$41.85	$334.80	$63.75	$510.00	$44.31	$67.17
6 Laborers	39.85	1912.80	60.70	2913.60		
2 Equip. Oper. (medium)	53.75	860.00	81.05	1296.80		
1 Equip. Oper. (light)	51.30	410.40	77.35	618.80		
2 Truck Drivers (heavy)	46.00	736.00	69.30	1108.80		
1 Air Compressor, 365 cfm		211.40		232.54		
2 Breakers, Pavement, 60 lb.		21.20		23.32		
8 -50' Air Hoses, 1"		67.60		74.36		
2 Dump Trucks, 8 C.Y., 220 H.P.		677.20		744.92	10.18	11.20
96 L.H., Daily Totals		$5231.40		$7523.14	$54.49	$78.37
Crew B-5B					Bare Costs	Incl. O&P
1 Powderman	$52.35	$418.80	$80.15	$641.20	$49.64	$75.03
2 Equip. Oper. (medium)	53.75	860.00	81.05	1296.80		
3 Truck Drivers (heavy)	46.00	1104.00	69.30	1663.20		
1 F.E. Loader, W.M., 2.5 C.Y.		523.20		575.52		
3 Dump Trucks, 12 C.Y., 400 H.P.		1628.40		1791.24		
1 Air Compressor, 365 cfm		211.40		232.54	49.23	54.15
48 L.H., Daily Totals		$4745.80		$6200.50	$98.87	$129.18
Crew B-5C					Bare Costs	Incl. O&P
3 Laborers	$39.85	$956.40	$60.70	$1456.80	$46.25	$69.96
1 Equip. Oper. (medium)	53.75	430.00	81.05	648.40		
2 Truck Drivers (heavy)	46.00	736.00	69.30	1108.80		
1 Equip. Oper. (crane)	56.10	448.80	84.60	676.80		
1 Equip. Oper. (oiler)	48.60	388.80	73.30	586.40		
2 Dump Trucks, 12 C.Y., 400 H.P.		1085.60		1194.16		
1 Crawler Loader, 4 C.Y.		1459.00		1604.90		
1 S.P. Crane, 4x4, 25 Ton		668.40		735.24	50.20	55.22
64 L.H., Daily Totals		$6173.00		$8011.50	$96.45	$125.18
Crew B-5D					Bare Costs	Incl. O&P
1 Labor Foreman (outside)	$41.85	$334.80	$63.75	$510.00	$44.34	$67.24
4 Laborers	39.85	1275.20	60.70	1942.40		
2 Equip. Oper. (medium)	53.75	860.00	81.05	1296.80		
1 Truck Driver (heavy)	46.00	368.00	69.30	554.40		
1 Air Compressor, 250 cfm		167.40		184.14		
2 Breakers, Pavement, 60 lb.		21.20		23.32		
2 -50' Air Hoses, 1.5"		45.40		49.94		
1 Crawler Loader, 3 C.Y.		1204.00		1324.40		
1 Dump Truck, 12 C.Y., 400 H.P.		542.80		597.08	30.95	34.05
64 L.H., Daily Totals		$4818.80		$6482.48	$75.29	$101.29
Crew B-6					Bare Costs	Incl. O&P
2 Laborers	$39.85	$637.60	$60.70	$971.20	$43.67	$66.25
1 Equip. Oper. (light)	51.30	410.40	77.35	618.80		
1 Backhoe Loader, 48 H.P.		312.40		343.64	13.02	14.32
24 L.H., Daily Totals		$1360.40		$1933.64	$56.68	$80.57
Crew B-6A					Bare Costs	Incl. O&P
.5 Labor Foreman (outside)	$41.85	$167.40	$63.75	$255.00	$45.81	$69.45
1 Laborer	39.85	318.80	60.70	485.60		
1 Equip. Oper. (medium)	53.75	430.00	81.05	648.40		
1 Vacuum Truck, 5000 Gal.		358.90		394.79	17.95	19.74
20 L.H., Daily Totals		$1275.10		$1783.79	$63.76	$89.19
Crew B-6B					Bare Costs	Incl. O&P
2 Labor Foremen (outside)	$41.85	$669.60	$63.75	$1020.00	$40.52	$61.72
4 Laborers	39.85	1275.20	60.70	1942.40		
1 S.P. Crane, 4x4, 5 Ton		259.00		284.90		
1 Flatbed Truck, Gas, 1.5 Ton		188.40		207.24		
1 Butt Fusion Mach., 4"-12" diam.		425.70		468.27	18.19	20.01
48 L.H., Daily Totals		$2817.90		$3922.81	$58.71	$81.73
Crew B-6C					Bare Costs	Incl. O&P
2 Labor Foremen (outside)	$41.85	$669.60	$63.75	$1020.00	$40.52	$61.72
4 Laborers	39.85	1275.20	60.70	1942.40		
1 S.P. Crane, 4x4, 12 Ton		363.60		399.96		
1 Flatbed Truck, Gas, 3 Ton		238.00		261.80		
1 Butt Fusion Mach., 8"-24" diam.		1908.00		2098.80	52.28	57.51
48 L.H., Daily Totals		$4454.40		$5722.96	$92.80	$119.23
Crew B-7					Bare Costs	Incl. O&P
1 Labor Foreman (outside)	$41.85	$334.80	$63.75	$510.00	$42.50	$64.60
4 Laborers	39.85	1275.20	60.70	1942.40		
1 Equip. Oper. (medium)	53.75	430.00	81.05	648.40		
1 Brush Chipper, 12", 130 H.P.		393.00		432.30		
1 Crawler Loader, 3 C.Y.		1204.00		1324.40		
2 Chain Saws, Gas, 36" Long		92.80		102.08	35.20	38.72
48 L.H., Daily Totals		$3729.80		$4959.58	$77.70	$103.32

For customer support on your Building Construction Costs with RSMeans data, call 800.448.8182.

Crews - Standard

Crew B-7A

	Bare Costs Hr.	Daily	Incl. Subs O&P Hr.	Daily	Cost Per Labor-Hour Bare Costs	Incl. O&P
2 Laborers	$39.85	$637.60	$60.70	$971.20	$43.67	$66.25
1 Equip. Oper. (light)	51.30	410.40	77.35	618.80		
1 Rake w/Tractor		331.30		364.43		
2 Chain Saws, Gas, 18"		55.60		61.16	16.12	17.73
24 L.H., Daily Totals		$1434.90		$2015.59	$59.79	$83.98

Crew B-7B

	Bare Costs Hr.	Daily	Incl. Subs O&P Hr.	Daily	Cost Per Labor-Hour Bare Costs	Incl. O&P
1 Labor Foreman (outside)	$41.85	$334.80	$63.75	$510.00	$43.00	$65.27
4 Laborers	39.85	1275.20	60.70	1942.40		
1 Equip. Oper. (medium)	53.75	430.00	81.05	648.40		
1 Truck Driver (heavy)	46.00	368.00	69.30	554.40		
1 Brush Chipper, 12", 130 H.P.		393.00		432.30		
1 Crawler Loader, 3 C.Y.		1204.00		1324.40		
2 Chain Saws, Gas, 36" Long		92.80		102.08		
1 Dump Truck, 8 C.Y., 220 H.P.		338.60		372.46	36.22	39.84
56 L.H., Daily Totals		$4436.40		$5886.44	$79.22	$105.11

Crew B-7C

	Bare Costs Hr.	Daily	Incl. Subs O&P Hr.	Daily	Cost Per Labor-Hour Bare Costs	Incl. O&P
1 Labor Foreman (outside)	$41.85	$334.80	$63.75	$510.00	$43.00	$65.27
4 Laborers	39.85	1275.20	60.70	1942.40		
1 Equip. Oper. (medium)	53.75	430.00	81.05	648.40		
1 Truck Driver (heavy)	46.00	368.00	69.30	554.40		
1 Brush Chipper, 12", 130 H.P.		393.00		432.30		
1 Crawler Loader, 3 C.Y.		1204.00		1324.40		
2 Chain Saws, Gas, 36" Long		92.80		102.08		
1 Dump Truck, 12 C.Y., 400 H.P.		542.80		597.08	39.87	43.85
56 L.H., Daily Totals		$4640.60		$6111.06	$82.87	$109.13

Crew B-8

	Bare Costs Hr.	Daily	Incl. Subs O&P Hr.	Daily	Cost Per Labor-Hour Bare Costs	Incl. O&P
1 Labor Foreman (outside)	$41.85	$334.80	$63.75	$510.00	$46.21	$69.89
2 Laborers	39.85	637.60	60.70	971.20		
2 Equip. Oper. (medium)	53.75	860.00	81.05	1296.80		
1 Equip. Oper. (oiler)	48.60	388.80	73.30	586.40		
2 Truck Drivers (heavy)	46.00	736.00	69.30	1108.80		
1 Hyd. Crane, 25 Ton		590.60		649.66		
1 Crawler Loader, 3 C.Y.		1204.00		1324.40		
2 Dump Trucks, 12 C.Y., 400 H.P.		1085.60		1194.16	45.00	49.50
64 L.H., Daily Totals		$5837.40		$7641.42	$91.21	$119.40

Crew B-9

	Bare Costs Hr.	Daily	Incl. Subs O&P Hr.	Daily	Cost Per Labor-Hour Bare Costs	Incl. O&P
1 Labor Foreman (outside)	$41.85	$334.80	$63.75	$510.00	$40.25	$61.31
4 Laborers	39.85	1275.20	60.70	1942.40		
1 Air Compressor, 250 cfm		167.40		184.14		
2 Breakers, Pavement, 60 lb.		21.20		23.32		
2 -50' Air Hoses, 1.5"		45.40		49.94	5.85	6.43
40 L.H., Daily Totals		$1844.00		$2709.80	$46.10	$67.75

Crew B-9A

	Bare Costs Hr.	Daily	Incl. Subs O&P Hr.	Daily	Cost Per Labor-Hour Bare Costs	Incl. O&P
2 Laborers	$39.85	$637.60	$60.70	$971.20	$41.90	$63.57
1 Truck Driver (heavy)	46.00	368.00	69.30	554.40		
1 Water Tank Trailer, 5000 Gal.		147.60		162.36		
1 Truck Tractor, 220 H.P.		290.00		319.00		
2 -50' Discharge Hoses, 3"		3.00		3.30	18.36	20.19
24 L.H., Daily Totals		$1446.20		$2010.26	$60.26	$83.76

Crew B-9B

	Bare Costs Hr.	Daily	Incl. Subs O&P Hr.	Daily	Cost Per Labor-Hour Bare Costs	Incl. O&P
2 Laborers	$39.85	$637.60	$60.70	$971.20	$41.90	$63.57
1 Truck Driver (heavy)	46.00	368.00	69.30	554.40		
2 -50' Discharge Hoses, 3"		3.00		3.30		
1 Water Tank Trailer, 5000 Gal.		147.60		162.36		
1 Truck Tractor, 220 H.P.		290.00		319.00		
1 Pressure Washer		63.60		69.96	21.01	23.11
24 L.H., Daily Totals		$1509.80		$2080.22	$62.91	$86.68

Crew B-9D

	Bare Costs Hr.	Daily	Incl. Subs O&P Hr.	Daily	Cost Per Labor-Hour Bare Costs	Incl. O&P
1 Labor Foreman (outside)	$41.85	$334.80	$63.75	$510.00	$40.25	$61.31
4 Common Laborers	39.85	1275.20	60.70	1942.40		
1 Air Compressor, 250 cfm		167.40		184.14		
2 -50' Air Hoses, 1.5"		45.40		49.94		
2 Air Powered Tampers		56.60		62.26	6.74	7.41
40 L.H., Daily Totals		$1879.40		$2748.74	$46.98	$68.72

Crew B-10

	Bare Costs Hr.	Daily	Incl. Subs O&P Hr.	Daily	Cost Per Labor-Hour Bare Costs	Incl. O&P
1 Equip. Oper. (medium)	$53.75	$430.00	$81.05	$648.40	$49.12	$74.27
.5 Laborer	39.85	159.40	60.70	242.80		
12 L.H., Daily Totals		$589.40		$891.20	$49.12	$74.27

Crew B-10A

	Bare Costs Hr.	Daily	Incl. Subs O&P Hr.	Daily	Cost Per Labor-Hour Bare Costs	Incl. O&P
1 Equip. Oper. (medium)	$53.75	$430.00	$81.05	$648.40	$49.12	$74.27
.5 Laborer	39.85	159.40	60.70	242.80		
1 Roller, 2-Drum, W.B., 7.5 H.P.		179.60		197.56	14.97	16.46
12 L.H., Daily Totals		$769.00		$1088.76	$64.08	$90.73

Crew B-10B

	Bare Costs Hr.	Daily	Incl. Subs O&P Hr.	Daily	Cost Per Labor-Hour Bare Costs	Incl. O&P
1 Equip. Oper. (medium)	$53.75	$430.00	$81.05	$648.40	$49.12	$74.27
.5 Laborer	39.85	159.40	60.70	242.80		
1 Dozer, 200 H.P.		1273.00		1400.30	106.08	116.69
12 L.H., Daily Totals		$1862.40		$2291.50	$155.20	$190.96

Crew B-10C

	Bare Costs Hr.	Daily	Incl. Subs O&P Hr.	Daily	Cost Per Labor-Hour Bare Costs	Incl. O&P
1 Equip. Oper. (medium)	$53.75	$430.00	$81.05	$648.40	$49.12	$74.27
.5 Laborer	39.85	159.40	60.70	242.80		
1 Dozer, 200 H.P.		1273.00		1400.30		
1 Vibratory Roller, Towed, 23 Ton		423.00		465.30	141.33	155.47
12 L.H., Daily Totals		$2285.40		$2756.80	$190.45	$229.73

Crew B-10D

	Bare Costs Hr.	Daily	Incl. Subs O&P Hr.	Daily	Cost Per Labor-Hour Bare Costs	Incl. O&P
1 Equip. Oper. (medium)	$53.75	$430.00	$81.05	$648.40	$49.12	$74.27
.5 Laborer	39.85	159.40	60.70	242.80		
1 Dozer, 200 H.P.		1273.00		1400.30		
1 Sheepsft. Roller, Towed		425.80		468.38	141.57	155.72
12 L.H., Daily Totals		$2288.20		$2759.88	$190.68	$229.99

Crew B-10E

	Bare Costs Hr.	Daily	Incl. Subs O&P Hr.	Daily	Cost Per Labor-Hour Bare Costs	Incl. O&P
1 Equip. Oper. (medium)	$53.75	$430.00	$81.05	$648.40	$49.12	$74.27
.5 Laborer	39.85	159.40	60.70	242.80		
1 Tandem Roller, 5 Ton		151.20		166.32	12.60	13.86
12 L.H., Daily Totals		$740.60		$1057.52	$61.72	$88.13

Crew B-10F

	Bare Costs Hr.	Daily	Incl. Subs O&P Hr.	Daily	Cost Per Labor-Hour Bare Costs	Incl. O&P
1 Equip. Oper. (medium)	$53.75	$430.00	$81.05	$648.40	$49.12	$74.27
.5 Laborer	39.85	159.40	60.70	242.80		
1 Tandem Roller, 10 Ton		231.40		254.54	19.28	21.21
12 L.H., Daily Totals		$820.80		$1145.74	$68.40	$95.48

Crew No.	Bare Costs		Incl. Subs O&P		Cost Per Labor-Hour	
	Hr.	Daily	Hr.	Daily	Bare Costs	Incl. O&P
Crew B-10G						
1 Equip. Oper. (medium)	$53.75	$430.00	$81.05	$648.40	$49.12	$74.27
.5 Laborer	39.85	159.40	60.70	242.80		
1 Sheepsfoot Roller, 240 H.P.		1316.00		1447.60	109.67	120.63
12 L.H., Daily Totals		$1905.40		$2338.80	$158.78	$194.90
Crew B-10H	Hr.	Daily	Hr.	Daily	Bare Costs	Incl. O&P
1 Equip. Oper. (medium)	$53.75	$430.00	$81.05	$648.40	$49.12	$74.27
.5 Laborer	39.85	159.40	60.70	242.80		
1 Diaphragm Water Pump, 2"		73.00		80.30		
1 -20' Suction Hose, 2"		1.95		2.15		
2 -50' Discharge Hoses, 2"		1.80		1.98	6.40	7.04
12 L.H., Daily Totals		$666.15		$975.63	$55.51	$81.30
Crew B-10I	Hr.	Daily	Hr.	Daily	Bare Costs	Incl. O&P
1 Equip. Oper. (medium)	$53.75	$430.00	$81.05	$648.40	$49.12	$74.27
.5 Laborer	39.85	159.40	60.70	242.80		
1 Diaphragm Water Pump, 4"		114.80		126.28		
1 -20' Suction Hose, 4"		3.25		3.58		
2 -50' Discharge Hoses, 4"		4.70		5.17	10.23	11.25
12 L.H., Daily Totals		$712.15		$1026.22	$59.35	$85.52
Crew B-10J	Hr.	Daily	Hr.	Daily	Bare Costs	Incl. O&P
1 Equip. Oper. (medium)	$53.75	$430.00	$81.05	$648.40	$49.12	$74.27
.5 Laborer	39.85	159.40	60.70	242.80		
1 Centrifugal Water Pump, 3"		79.20		87.12		
1 -20' Suction Hose, 3"		2.85		3.13		
2 -50' Discharge Hoses, 3"		3.00		3.30	7.09	7.80
12 L.H., Daily Totals		$674.45		$984.76	$56.20	$82.06
Crew B-10K	Hr.	Daily	Hr.	Daily	Bare Costs	Incl. O&P
1 Equip. Oper. (medium)	$53.75	$430.00	$81.05	$648.40	$49.12	$74.27
.5 Laborer	39.85	159.40	60.70	242.80		
1 Centr. Water Pump, 6"		296.40		326.04		
1 -20' Suction Hose, 6"		11.50		12.65		
2 -50' Discharge Hoses, 6"		12.20		13.42	26.68	29.34
12 L.H., Daily Totals		$909.50		$1243.31	$75.79	$103.61
Crew B-10L	Hr.	Daily	Hr.	Daily	Bare Costs	Incl. O&P
1 Equip. Oper. (medium)	$53.75	$430.00	$81.05	$648.40	$49.12	$74.27
.5 Laborer	39.85	159.40	60.70	242.80		
1 Dozer, 80 H.P.		464.40		510.84	38.70	42.57
12 L.H., Daily Totals		$1053.80		$1402.04	$87.82	$116.84
Crew B-10M	Hr.	Daily	Hr.	Daily	Bare Costs	Incl. O&P
1 Equip. Oper. (medium)	$53.75	$430.00	$81.05	$648.40	$49.12	$74.27
.5 Laborer	39.85	159.40	60.70	242.80		
1 Dozer, 300 H.P.		1829.00		2011.90	152.42	167.66
12 L.H., Daily Totals		$2418.40		$2903.10	$201.53	$241.93
Crew B-10N	Hr.	Daily	Hr.	Daily	Bare Costs	Incl. O&P
1 Equip. Oper. (medium)	$53.75	$430.00	$81.05	$648.40	$49.12	$74.27
.5 Laborer	39.85	159.40	60.70	242.80		
1 F.E. Loader, T.M., 1.5 C.Y.		579.80		637.78	48.32	53.15
12 L.H., Daily Totals		$1169.20		$1528.98	$97.43	$127.42
Crew B-10O	Hr.	Daily	Hr.	Daily	Bare Costs	Incl. O&P
1 Equip. Oper. (medium)	$53.75	$430.00	$81.05	$648.40	$49.12	$74.27
.5 Laborer	39.85	159.40	60.70	242.80		
1 F.E. Loader, T.M., 2.25 C.Y.		979.00		1076.90	81.58	89.74
12 L.H., Daily Totals		$1568.40		$1968.10	$130.70	$164.01
Crew B-10P	Hr.	Daily	Hr.	Daily	Bare Costs	Incl. O&P
1 Equip. Oper. (medium)	$53.75	$430.00	$81.05	$648.40	$49.12	$74.27
.5 Laborer	39.85	159.40	60.70	242.80		
1 Crawler Loader, 3 C.Y.		1204.00		1324.40	100.33	110.37
12 L.H., Daily Totals		$1793.40		$2215.60	$149.45	$184.63
Crew B-10Q	Hr.	Daily	Hr.	Daily	Bare Costs	Incl. O&P
1 Equip. Oper. (medium)	$53.75	$430.00	$81.05	$648.40	$49.12	$74.27
.5 Laborer	39.85	159.40	60.70	242.80		
1 Crawler Loader, 4 C.Y.		1459.00		1604.90	121.58	133.74
12 L.H., Daily Totals		$2048.40		$2496.10	$170.70	$208.01
Crew B-10R	Hr.	Daily	Hr.	Daily	Bare Costs	Incl. O&P
1 Equip. Oper. (medium)	$53.75	$430.00	$81.05	$648.40	$49.12	$74.27
.5 Laborer	39.85	159.40	60.70	242.80		
1 F.E. Loader, W.M., 1 C.Y.		290.80		319.88	24.23	26.66
12 L.H., Daily Totals		$880.20		$1211.08	$73.35	$100.92
Crew B-10S	Hr.	Daily	Hr.	Daily	Bare Costs	Incl. O&P
1 Equip. Oper. (medium)	$53.75	$430.00	$81.05	$648.40	$49.12	$74.27
.5 Laborer	39.85	159.40	60.70	242.80		
1 F.E. Loader, W.M., 1.5 C.Y.		342.80		377.08	28.57	31.42
12 L.H., Daily Totals		$932.20		$1268.28	$77.68	$105.69
Crew B-10T	Hr.	Daily	Hr.	Daily	Bare Costs	Incl. O&P
1 Equip. Oper. (medium)	$53.75	$430.00	$81.05	$648.40	$49.12	$74.27
.5 Laborer	39.85	159.40	60.70	242.80		
1 F.E. Loader, W.M., 2.5 C.Y.		523.20		575.52	43.60	47.96
12 L.H., Daily Totals		$1112.60		$1466.72	$92.72	$122.23
Crew B-10U	Hr.	Daily	Hr.	Daily	Bare Costs	Incl. O&P
1 Equip. Oper. (medium)	$53.75	$430.00	$81.05	$648.40	$49.12	$74.27
.5 Laborer	39.85	159.40	60.70	242.80		
1 F.E. Loader, W.M., 5.5 C.Y.		982.20		1080.42	81.85	90.03
12 L.H., Daily Totals		$1571.60		$1971.62	$130.97	$164.30
Crew B-10V	Hr.	Daily	Hr.	Daily	Bare Costs	Incl. O&P
1 Equip. Oper. (medium)	$53.75	$430.00	$81.05	$648.40	$49.12	$74.27
.5 Laborer	39.85	159.40	60.70	242.80		
1 Dozer, 700 H.P.		4999.00		5498.90	416.58	458.24
12 L.H., Daily Totals		$5588.40		$6390.10	$465.70	$532.51
Crew B-10W	Hr.	Daily	Hr.	Daily	Bare Costs	Incl. O&P
1 Equip. Oper. (medium)	$53.75	$430.00	$81.05	$648.40	$49.12	$74.27
.5 Laborer	39.85	159.40	60.70	242.80		
1 Dozer, 105 H.P.		608.00		668.80	50.67	55.73
12 L.H., Daily Totals		$1197.40		$1560.00	$99.78	$130.00
Crew B-10X	Hr.	Daily	Hr.	Daily	Bare Costs	Incl. O&P
1 Equip. Oper. (medium)	$53.75	$430.00	$81.05	$648.40	$49.12	$74.27
.5 Laborer	39.85	159.40	60.70	242.80		
1 Dozer, 410 H.P.		2291.00		2520.10	190.92	210.01
12 L.H., Daily Totals		$2880.40		$3411.30	$240.03	$284.27
Crew B-10Y	Hr.	Daily	Hr.	Daily	Bare Costs	Incl. O&P
1 Equip. Oper. (medium)	$53.75	$430.00	$81.05	$648.40	$49.12	$74.27
.5 Laborer	39.85	159.40	60.70	242.80		
1 Vibr. Roller, Towed, 12 Ton		576.40		634.04	48.03	52.84
12 L.H., Daily Totals		$1165.80		$1525.24	$97.15	$127.10

For customer support on your Building Construction Costs with RSMeans data, call 800.448.8182.

Crew No.	Bare Costs		Incl. Subs O&P		Cost Per Labor-Hour	
	Hr.	Daily	Hr.	Daily	Bare Costs	Incl. O&P
Crew B-11A	Hr.	Daily	Hr.	Daily	Bare Costs	Incl. O&P
1 Equipment Oper. (med.)	$53.75	$430.00	$81.05	$648.40	$46.80	$70.88
1 Laborer	39.85	318.80	60.70	485.60		
1 Dozer, 200 H.P.		1273.00		1400.30	79.56	87.52
16 L.H., Daily Totals		$2021.80		$2534.30	$126.36	$158.39
Crew B-11B	Hr.	Daily	Hr.	Daily	Bare Costs	Incl. O&P
1 Equipment Oper. (light)	$51.30	$410.40	$77.35	$618.80	$45.58	$69.03
1 Laborer	39.85	318.80	60.70	485.60		
1 Air Powered Tamper		28.30		31.13		
1 Air Compressor, 365 cfm		211.40		232.54		
2 -50' Air Hoses, 1.5"		45.40		49.94	17.82	19.60
16 L.H., Daily Totals		$1014.30		$1418.01	$63.39	$88.63
Crew B-11C	Hr.	Daily	Hr.	Daily	Bare Costs	Incl. O&P
1 Equipment Oper. (med.)	$53.75	$430.00	$81.05	$648.40	$46.80	$70.88
1 Laborer	39.85	318.80	60.70	485.60		
1 Backhoe Loader, 48 H.P.		312.40		343.64	19.52	21.48
16 L.H., Daily Totals		$1061.20		$1477.64	$66.33	$92.35
Crew B-11J	Hr.	Daily	Hr.	Daily	Bare Costs	Incl. O&P
1 Equipment Oper. (med.)	$53.75	$430.00	$81.05	$648.40	$46.80	$70.88
1 Laborer	39.85	318.80	60.70	485.60		
1 Grader, 30,000 Lbs.		643.80		708.18		
1 Ripper, Beam & 1 Shank		87.80		96.58	45.73	50.30
16 L.H., Daily Totals		$1480.40		$1938.76	$92.53	$121.17
Crew B-11K	Hr.	Daily	Hr.	Daily	Bare Costs	Incl. O&P
1 Equipment Oper. (med.)	$53.75	$430.00	$81.05	$648.40	$46.80	$70.88
1 Laborer	39.85	318.80	60.70	485.60		
1 Trencher, Chain Type, 8' D		2048.00		2252.80	128.00	140.80
16 L.H., Daily Totals		$2796.80		$3386.80	$174.80	$211.68
Crew B-11L	Hr.	Daily	Hr.	Daily	Bare Costs	Incl. O&P
1 Equipment Oper. (med.)	$53.75	$430.00	$81.05	$648.40	$46.80	$70.88
1 Laborer	39.85	318.80	60.70	485.60		
1 Grader, 30,000 Lbs.		643.80		708.18	40.24	44.26
16 L.H., Daily Totals		$1392.60		$1842.18	$87.04	$115.14
Crew B-11M	Hr.	Daily	Hr.	Daily	Bare Costs	Incl. O&P
1 Equipment Oper. (med.)	$53.75	$430.00	$81.05	$648.40	$46.80	$70.88
1 Laborer	39.85	318.80	60.70	485.60		
1 Backhoe Loader, 80 H.P.		385.00		423.50	24.06	26.47
16 L.H., Daily Totals		$1133.80		$1557.50	$70.86	$97.34
Crew B-11N	Hr.	Daily	Hr.	Daily	Bare Costs	Incl. O&P
1 Labor Foreman (outside)	$41.85	$334.80	$63.75	$510.00	$47.26	$71.29
2 Equipment Operators (med.)	53.75	860.00	81.05	1296.80		
6 Truck Drivers (heavy)	46.00	2208.00	69.30	3326.40		
1 F.E. Loader, W.M., 5.5 C.Y.		982.20		1080.42		
1 Dozer, 410 H.P.		2291.00		2520.10		
6 Dump Trucks, Off Hwy., 50 Ton		9972.00		10969.20	183.96	202.36
72 L.H., Daily Totals		$16648.00		$19702.92	$231.22	$273.65
Crew B-11Q	Hr.	Daily	Hr.	Daily	Bare Costs	Incl. O&P
1 Equipment Operator (med.)	$53.75	$430.00	$81.05	$648.40	$49.12	$74.27
.5 Laborer	39.85	159.40	60.70	242.80		
1 Dozer, 140 H.P.		845.60		930.16	70.47	77.51
12 L.H., Daily Totals		$1435.00		$1821.36	$119.58	$151.78

Crew No.	Bare Costs		Incl. Subs O&P		Cost Per Labor-Hour	
Crew B-11R	Hr.	Daily	Hr.	Daily	Bare Costs	Incl. O&P
1 Equipment Operator (med.)	$53.75	$430.00	$81.05	$648.40	$49.12	$74.27
.5 Laborer	39.85	159.40	60.70	242.80		
1 Dozer, 200 H.P.		1273.00		1400.30	106.08	116.69
12 L.H., Daily Totals		$1862.40		$2291.50	$155.20	$190.96
Crew B-11S	Hr.	Daily	Hr.	Daily	Bare Costs	Incl. O&P
1 Equipment Operator (med.)	$53.75	$430.00	$81.05	$648.40	$49.12	$74.27
.5 Laborer	39.85	159.40	60.70	242.80		
1 Dozer, 300 H.P.		1829.00		2011.90		
1 Ripper, Beam & 1 Shank		87.80		96.58	159.73	175.71
12 L.H., Daily Totals		$2506.20		$2999.68	$208.85	$249.97
Crew B-11T	Hr.	Daily	Hr.	Daily	Bare Costs	Incl. O&P
1 Equipment Operator (med.)	$53.75	$430.00	$81.05	$648.40	$49.12	$74.27
.5 Laborer	39.85	159.40	60.70	242.80		
1 Dozer, 410 H.P.		2291.00		2520.10		
1 Ripper, Beam & 2 Shanks		132.60		145.86	201.97	222.16
12 L.H., Daily Totals		$3013.00		$3557.16	$251.08	$296.43
Crew B-11U	Hr.	Daily	Hr.	Daily	Bare Costs	Incl. O&P
1 Equipment Operator (med.)	$53.75	$430.00	$81.05	$648.40	$49.12	$74.27
.5 Laborer	39.85	159.40	60.70	242.80		
1 Dozer, 520 H.P.		2818.00		3099.80	234.83	258.32
12 L.H., Daily Totals		$3407.40		$3991.00	$283.95	$332.58
Crew B-11V	Hr.	Daily	Hr.	Daily	Bare Costs	Incl. O&P
3 Laborers	$39.85	$956.40	$60.70	$1456.80	$39.85	$60.70
1 Roller, 2-Drum, W.B., 7.5 H.P.		179.60		197.56	7.48	8.23
24 L.H., Daily Totals		$1136.00		$1654.36	$47.33	$68.93
Crew B-11W	Hr.	Daily	Hr.	Daily	Bare Costs	Incl. O&P
1 Equipment Operator (med.)	$53.75	$430.00	$81.05	$648.40	$46.13	$69.56
1 Common Laborer	39.85	318.80	60.70	485.60		
10 Truck Drivers (heavy)	46.00	3680.00	69.30	5544.00		
1 Dozer, 200 H.P.		1273.00		1400.30		
1 Vibratory Roller, Towed, 23 Ton		423.00		465.30		
10 Dump Trucks, 8 C.Y., 220 H.P.		3386.00		3724.60	52.94	58.23
96 L.H., Daily Totals		$9510.80		$12268.20	$99.07	$127.79
Crew B-11Y	Hr.	Daily	Hr.	Daily	Bare Costs	Incl. O&P
1 Labor Foreman (outside)	$41.85	$334.80	$63.75	$510.00	$44.71	$67.82
5 Common Laborers	39.85	1594.00	60.70	2428.00		
3 Equipment Operators (med.)	53.75	1290.00	81.05	1945.20		
1 Dozer, 80 H.P.		464.40		510.84		
2 Rollers, 2-Drums, W.B., 7.5 H.P.		359.20		395.12		
4 Vibrating Plates, Gas, 21"		162.40		178.64	13.69	15.06
72 L.H., Daily Totals		$4204.80		$5967.80	$58.40	$82.89
Crew B-12A	Hr.	Daily	Hr.	Daily	Bare Costs	Incl. O&P
1 Equip. Oper. (crane)	$56.10	$448.80	$84.60	$676.80	$47.98	$72.65
1 Laborer	39.85	318.80	60.70	485.60		
1 Hyd. Excavator, 1 C.Y.		742.20		816.42	46.39	51.03
16 L.H., Daily Totals		$1509.80		$1978.82	$94.36	$123.68
Crew B-12B	Hr.	Daily	Hr.	Daily	Bare Costs	Incl. O&P
1 Equip. Oper. (crane)	$56.10	$448.80	$84.60	$676.80	$47.98	$72.65
1 Laborer	39.85	318.80	60.70	485.60		
1 Hyd. Excavator, 1.5 C.Y.		893.60		982.96	55.85	61.44
16 L.H., Daily Totals		$1661.20		$2145.36	$103.83	$134.09

For customer support on your Building Construction Costs with RSMeans data, call 800.448.8182.

Crew No.	Bare Costs		Incl. Subs O&P		Cost Per Labor-Hour	

Crew B-12C

Crew B-12C	Hr.	Daily	Hr.	Daily	Bare Costs	Incl. O&P
1 Equip. Oper. (crane)	$56.10	$448.80	$84.60	$676.80	$47.98	$72.65
1 Laborer	39.85	318.80	60.70	485.60		
1 Hyd. Excavator, 2 C.Y.		1052.00		1157.20	65.75	72.33
16 L.H., Daily Totals		$1819.60		$2319.60	$113.72	$144.97

Crew B-12D	Hr.	Daily	Hr.	Daily	Bare Costs	Incl. O&P
1 Equip. Oper. (crane)	$56.10	$448.80	$84.60	$676.80	$47.98	$72.65
1 Laborer	39.85	318.80	60.70	485.60		
1 Hyd. Excavator, 3.5 C.Y.		2195.00		2414.50	137.19	150.91
16 L.H., Daily Totals		$2962.60		$3576.90	$185.16	$223.56

Crew B-12E	Hr.	Daily	Hr.	Daily	Bare Costs	Incl. O&P
1 Equip. Oper. (crane)	$56.10	$448.80	$84.60	$676.80	$47.98	$72.65
1 Laborer	39.85	318.80	60.70	485.60		
1 Hyd. Excavator, .5 C.Y.		435.40		478.94	27.21	29.93
16 L.H., Daily Totals		$1203.00		$1641.34	$75.19	$102.58

Crew B-12F	Hr.	Daily	Hr.	Daily	Bare Costs	Incl. O&P
1 Equip. Oper. (crane)	$56.10	$448.80	$84.60	$676.80	$47.98	$72.65
1 Laborer	39.85	318.80	60.70	485.60		
1 Hyd. Excavator, .75 C.Y.		669.60		736.56	41.85	46.03
16 L.H., Daily Totals		$1437.20		$1898.96	$89.83	$118.69

Crew B-12G	Hr.	Daily	Hr.	Daily	Bare Costs	Incl. O&P
1 Equip. Oper. (crane)	$56.10	$448.80	$84.60	$676.80	$47.98	$72.65
1 Laborer	39.85	318.80	60.70	485.60		
1 Crawler Crane, 15 Ton		828.80		911.68		
1 Clamshell Bucket, .5 C.Y.		41.00		45.10	54.36	59.80
16 L.H., Daily Totals		$1637.40		$2119.18	$102.34	$132.45

Crew B-12H	Hr.	Daily	Hr.	Daily	Bare Costs	Incl. O&P
1 Equip. Oper. (crane)	$56.10	$448.80	$84.60	$676.80	$47.98	$72.65
1 Laborer	39.85	318.80	60.70	485.60		
1 Crawler Crane, 25 Ton		1348.00		1482.80		
1 Clamshell Bucket, 1 C.Y.		50.60		55.66	87.41	96.15
16 L.H., Daily Totals		$2166.20		$2700.86	$135.39	$168.80

Crew B-12I	Hr.	Daily	Hr.	Daily	Bare Costs	Incl. O&P
1 Equip. Oper. (crane)	$56.10	$448.80	$84.60	$676.80	$47.98	$72.65
1 Laborer	39.85	318.80	60.70	485.60		
1 Crawler Crane, 20 Ton		1060.00		1166.00		
1 Dragline Bucket, .75 C.Y.		21.80		23.98	67.61	74.37
16 L.H., Daily Totals		$1849.40		$2352.38	$115.59	$147.02

Crew B-12J	Hr.	Daily	Hr.	Daily	Bare Costs	Incl. O&P
1 Equip. Oper. (crane)	$56.10	$448.80	$84.60	$676.80	$47.98	$72.65
1 Laborer	39.85	318.80	60.70	485.60		
1 Gradall, 5/8 C.Y.		869.60		956.56	54.35	59.78
16 L.H., Daily Totals		$1637.20		$2118.96	$102.33	$132.44

Crew B-12K	Hr.	Daily	Hr.	Daily	Bare Costs	Incl. O&P
1 Equip. Oper. (crane)	$56.10	$448.80	$84.60	$676.80	$47.98	$72.65
1 Laborer	39.85	318.80	60.70	485.60		
1 Gradall, 3 Ton, 1 C.Y.		1243.00		1367.30	77.69	85.46
16 L.H., Daily Totals		$2010.60		$2529.70	$125.66	$158.11

Crew B-12L	Hr.	Daily	Hr.	Daily	Bare Costs	Incl. O&P
1 Equip. Oper. (crane)	$56.10	$448.80	$84.60	$676.80	$47.98	$72.65
1 Laborer	39.85	318.80	60.70	485.60		
1 Crawler Crane, 15 Ton		828.80		911.68		
1 F.E. Attachment, .5 C.Y.		64.60		71.06	55.84	61.42
16 L.H., Daily Totals		$1661.00		$2145.14	$103.81	$134.07

Crew B-12M	Hr.	Daily	Hr.	Daily	Bare Costs	Incl. O&P
1 Equip. Oper. (crane)	$56.10	$448.80	$84.60	$676.80	$47.98	$72.65
1 Laborer	39.85	318.80	60.70	485.60		
1 Crawler Crane, 20 Ton		1060.00		1166.00		
1 F.E. Attachment, .75 C.Y.		69.00		75.90	70.56	77.62
16 L.H., Daily Totals		$1896.60		$2404.30	$118.54	$150.27

Crew B-12N	Hr.	Daily	Hr.	Daily	Bare Costs	Incl. O&P
1 Equip. Oper. (crane)	$56.10	$448.80	$84.60	$676.80	$47.98	$72.65
1 Laborer	39.85	318.80	60.70	485.60		
1 Crawler Crane, 25 Ton		1348.00		1482.80		
1 F.E. Attachment, 1 C.Y.		75.80		83.38	88.99	97.89
16 L.H., Daily Totals		$2191.40		$2728.58	$136.96	$170.54

Crew B-12O	Hr.	Daily	Hr.	Daily	Bare Costs	Incl. O&P
1 Equip. Oper. (crane)	$56.10	$448.80	$84.60	$676.80	$47.98	$72.65
1 Laborer	39.85	318.80	60.70	485.60		
1 Crawler Crane, 40 Ton		1370.00		1507.00		
1 F.E. Attachment, 1.5 C.Y.		85.00		93.50	90.94	100.03
16 L.H., Daily Totals		$2222.60		$2762.90	$138.91	$172.68

Crew B-12P	Hr.	Daily	Hr.	Daily	Bare Costs	Incl. O&P
1 Equip. Oper. (crane)	$56.10	$448.80	$84.60	$676.80	$47.98	$72.65
1 Laborer	39.85	318.80	60.70	485.60		
1 Crawler Crane, 40 Ton		1370.00		1507.00		
1 Dragline Bucket, 1.5 C.Y.		35.40		38.94	87.84	96.62
16 L.H., Daily Totals		$2173.00		$2708.34	$135.81	$169.27

Crew B-12Q	Hr.	Daily	Hr.	Daily	Bare Costs	Incl. O&P
1 Equip. Oper. (crane)	$56.10	$448.80	$84.60	$676.80	$47.98	$72.65
1 Laborer	39.85	318.80	60.70	485.60		
1 Hyd. Excavator, 5/8 C.Y.		576.80		634.48	36.05	39.66
16 L.H., Daily Totals		$1344.40		$1796.88	$84.03	$112.31

Crew B-12S	Hr.	Daily	Hr.	Daily	Bare Costs	Incl. O&P
1 Equip. Oper. (crane)	$56.10	$448.80	$84.60	$676.80	$47.98	$72.65
1 Laborer	39.85	318.80	60.70	485.60		
1 Hyd. Excavator, 2.5 C.Y.		1411.00		1552.10	88.19	97.01
16 L.H., Daily Totals		$2178.60		$2714.50	$136.16	$169.66

Crew B-12T	Hr.	Daily	Hr.	Daily	Bare Costs	Incl. O&P
1 Equip. Oper. (crane)	$56.10	$448.80	$84.60	$676.80	$47.98	$72.65
1 Laborer	39.85	318.80	60.70	485.60		
1 Crawler Crane, 75 Ton		1708.00		1878.80		
1 F.E. Attachment, 3 C.Y.		108.80		119.68	113.55	124.91
16 L.H., Daily Totals		$2584.40		$3160.88	$161.53	$197.56

Crew B-12V	Hr.	Daily	Hr.	Daily	Bare Costs	Incl. O&P
1 Equip. Oper. (crane)	$56.10	$448.80	$84.60	$676.80	$47.98	$72.65
1 Laborer	39.85	318.80	60.70	485.60		
1 Crawler Crane, 75 Ton		1708.00		1878.80		
1 Dragline Bucket, 3 C.Y.		55.80		61.38	110.24	121.26
16 L.H., Daily Totals		$2531.40		$3102.58	$158.21	$193.91

For customer support on your Building Construction Costs with RSMeans data, call 800.448.8182.

Crew No.	Bare Costs		Incl. Subs O&P		Cost Per Labor-Hour	

Crew B-12Y

	Hr.	Daily	Hr.	Daily	Bare Costs	Incl. O&P
1 Equip. Oper. (crane)	$56.10	$448.80	$84.60	$676.80	$45.27	$68.67
2 Laborers	39.85	637.60	60.70	971.20		
1 Hyd. Excavator, 3.5 C.Y.		2195.00		2414.50	91.46	100.60
24 L.H., Daily Totals		$3281.40		$4062.50	$136.72	$169.27

Crew B-12Z

	Hr.	Daily	Hr.	Daily	Bare Costs	Incl. O&P
1 Equip. Oper. (crane)	$56.10	$448.80	$84.60	$676.80	$45.27	$68.67
2 Laborers	39.85	637.60	60.70	971.20		
1 Hyd. Excavator, 2.5 C.Y.		1411.00		1552.10	58.79	64.67
24 L.H., Daily Totals		$2497.40		$3200.10	$104.06	$133.34

Crew B-13

	Hr.	Daily	Hr.	Daily	Bare Costs	Incl. O&P
1 Labor Foreman (outside)	$41.85	$334.80	$63.75	$510.00	$43.71	$66.35
4 Laborers	39.85	1275.20	60.70	1942.40		
1 Equip. Oper. (crane)	56.10	448.80	84.60	676.80		
1 Equip. Oper. (oiler)	48.60	388.80	73.30	586.40		
1 Hyd. Crane, 25 Ton		590.60		649.66	10.55	11.60
56 L.H., Daily Totals		$3038.20		$4365.26	$54.25	$77.95

Crew B-13A

	Hr.	Daily	Hr.	Daily	Bare Costs	Incl. O&P
1 Labor Foreman (outside)	$41.85	$334.80	$63.75	$510.00	$45.86	$69.41
2 Laborers	39.85	637.60	60.70	971.20		
2 Equipment Operators (med.)	53.75	860.00	81.05	1296.80		
2 Truck Drivers (heavy)	46.00	736.00	69.30	1108.80		
1 Crawler Crane, 75 Ton		1708.00		1878.80		
1 Crawler Loader, 4 C.Y.		1459.00		1604.90		
2 Dump Trucks, 8 C.Y., 220 H.P.		677.20		744.92	68.65	75.51
56 L.H., Daily Totals		$6412.60		$8115.42	$114.51	$144.92

Crew B-13B

	Hr.	Daily	Hr.	Daily	Bare Costs	Incl. O&P
1 Labor Foreman (outside)	$41.85	$334.80	$63.75	$510.00	$43.71	$66.35
4 Laborers	39.85	1275.20	60.70	1942.40		
1 Equip. Oper. (crane)	56.10	448.80	84.60	676.80		
1 Equip. Oper. (oiler)	48.60	388.80	73.30	586.40		
1 Hyd. Crane, 55 Ton		993.80		1093.18	17.75	19.52
56 L.H., Daily Totals		$3441.40		$4808.78	$61.45	$85.87

Crew B-13C

	Hr.	Daily	Hr.	Daily	Bare Costs	Incl. O&P
1 Labor Foreman (outside)	$41.85	$334.80	$63.75	$510.00	$43.71	$66.35
4 Laborers	39.85	1275.20	60.70	1942.40		
1 Equip. Oper. (crane)	56.10	448.80	84.60	676.80		
1 Equip. Oper. (oiler)	48.60	388.80	73.30	586.40		
1 Crawler Crane, 100 Ton		1872.00		2059.20	33.43	36.77
56 L.H., Daily Totals		$4319.60		$5774.80	$77.14	$103.12

Crew B-13D

	Hr.	Daily	Hr.	Daily	Bare Costs	Incl. O&P
1 Laborer	$39.85	$318.80	$60.70	$485.60	$47.98	$72.65
1 Equip. Oper. (crane)	56.10	448.80	84.60	676.80		
1 Hyd. Excavator, 1 C.Y.		742.20		816.42		
1 Trench Box		81.00		89.10	51.45	56.59
16 L.H., Daily Totals		$1590.80		$2067.92	$99.42	$129.25

Crew B-13E

	Hr.	Daily	Hr.	Daily	Bare Costs	Incl. O&P
1 Laborer	$39.85	$318.80	$60.70	$485.60	$47.98	$72.65
1 Equip. Oper. (crane)	56.10	448.80	84.60	676.80		
1 Hyd. Excavator, 1.5 C.Y.		893.60		982.96		
1 Trench Box		81.00		89.10	60.91	67.00
16 L.H., Daily Totals		$1742.20		$2234.46	$108.89	$139.65

Crew B-13F

	Hr.	Daily	Hr.	Daily	Bare Costs	Incl. O&P
1 Laborer	$39.85	$318.80	$60.70	$485.60	$47.98	$72.65
1 Equip. Oper. (crane)	56.10	448.80	84.60	676.80		
1 Hyd. Excavator, 3.5 C.Y.		2195.00		2414.50		
1 Trench Box		81.00		89.10	142.25	156.47
16 L.H., Daily Totals		$3043.60		$3666.00	$190.22	$229.13

Crew B-13G

	Hr.	Daily	Hr.	Daily	Bare Costs	Incl. O&P
1 Laborer	$39.85	$318.80	$60.70	$485.60	$47.98	$72.65
1 Equip. Oper. (crane)	56.10	448.80	84.60	676.80		
1 Hyd. Excavator, .75 C.Y.		669.60		736.56		
1 Trench Box		81.00		89.10	46.91	51.60
16 L.H., Daily Totals		$1518.20		$1988.06	$94.89	$124.25

Crew B-13H

	Hr.	Daily	Hr.	Daily	Bare Costs	Incl. O&P
1 Laborer	$39.85	$318.80	$60.70	$485.60	$47.98	$72.65
1 Equip. Oper. (crane)	56.10	448.80	84.60	676.80		
1 Gradall, 5/8 C.Y.		869.60		956.56		
1 Trench Box		81.00		89.10	59.41	65.35
16 L.H., Daily Totals		$1718.20		$2208.06	$107.39	$138.00

Crew B-13I

	Hr.	Daily	Hr.	Daily	Bare Costs	Incl. O&P
1 Laborer	$39.85	$318.80	$60.70	$485.60	$47.98	$72.65
1 Equip. Oper. (crane)	56.10	448.80	84.60	676.80		
1 Gradall, 3 Ton, 1 C.Y.		1243.00		1367.30		
1 Trench Box		81.00		89.10	82.75	91.03
16 L.H., Daily Totals		$2091.60		$2618.80	$130.72	$163.68

Crew B-13J

	Hr.	Daily	Hr.	Daily	Bare Costs	Incl. O&P
1 Laborer	$39.85	$318.80	$60.70	$485.60	$47.98	$72.65
1 Equip. Oper. (crane)	56.10	448.80	84.60	676.80		
1 Hyd. Excavator, 2.5 C.Y.		1411.00		1552.10		
1 Trench Box		81.00		89.10	93.25	102.58
16 L.H., Daily Totals		$2259.60		$2803.60	$141.22	$175.22

Crew B-13K

	Hr.	Daily	Hr.	Daily	Bare Costs	Incl. O&P
2 Equip. Opers. (crane)	$56.10	$897.60	$84.60	$1353.60	$56.10	$84.60
1 Hyd. Excavator, .75 C.Y.		669.60		736.56		
1 Hyd. Hammer, 4000 ft-lb		324.60		357.06		
1 Hyd. Excavator, .75 C.Y.		669.60		736.56	103.99	114.39
16 L.H., Daily Totals		$2561.40		$3183.78	$160.09	$198.99

Crew B-13L

	Hr.	Daily	Hr.	Daily	Bare Costs	Incl. O&P
2 Equip. Opers. (crane)	$56.10	$897.60	$84.60	$1353.60	$56.10	$84.60
1 Hyd. Excavator, 1.5 C.Y.		893.60		982.96		
1 Hyd. Hammer, 5000 ft-lb		394.00		433.40		
1 Hyd. Excavator, .75 C.Y.		669.60		736.56	122.33	134.56
16 L.H., Daily Totals		$2854.80		$3506.52	$178.43	$219.16

Crew B-13M

	Hr.	Daily	Hr.	Daily	Bare Costs	Incl. O&P
2 Equip. Opers. (crane)	$56.10	$897.60	$84.60	$1353.60	$56.10	$84.60
1 Hyd. Excavator, 2.5 C.Y.		1411.00		1552.10		
1 Hyd. Hammer, 8000 ft-lb		576.80		634.48		
1 Hyd. Excavator, 1.5 C.Y.		893.60		982.96	180.09	198.10
16 L.H., Daily Totals		$3779.00		$4523.14	$236.19	$282.70

For customer support on your **Building Construction Costs** with RSMeans data, call 800.448.8182.

747

Crew No.	Bare Costs		Incl. Subs O&P		Cost Per Labor-Hour	
Crew B-13N	Hr.	Daily	Hr.	Daily	Bare Costs	Incl. O&P
2 Equip. Opers. (crane)	$56.10	$897.60	$84.60	$1353.60	$56.10	$84.60
1 Hyd. Excavator, 3.5 C.Y.		2195.00		2414.50		
1 Hyd. Hammer, 12,000 ft-lb		669.20		736.12		
1 Hyd. Excavator, 1.5 C.Y.		893.60		982.96	234.86	258.35
16 L.H., Daily Totals		$4655.40		$5487.18	$290.96	$342.95
Crew B-14	Hr.	Daily	Hr.	Daily	Bare Costs	Incl. O&P
1 Labor Foreman (outside)	$41.85	$334.80	$63.75	$510.00	$42.09	$63.98
4 Laborers	39.85	1275.20	60.70	1942.40		
1 Equip. Oper. (light)	51.30	410.40	77.35	618.80		
1 Backhoe Loader, 48 H.P.		312.40		343.64	6.51	7.16
48 L.H., Daily Totals		$2332.80		$3414.84	$48.60	$71.14
Crew B-14A	Hr.	Daily	Hr.	Daily	Bare Costs	Incl. O&P
1 Equip. Oper. (crane)	$56.10	$448.80	$84.60	$676.80	$50.68	$76.63
.5 Laborer	39.85	159.40	60.70	242.80		
1 Hyd. Excavator, 4.5 C.Y.		2735.00		3008.50	227.92	250.71
12 L.H., Daily Totals		$3343.20		$3928.10	$278.60	$327.34
Crew B-14B	Hr.	Daily	Hr.	Daily	Bare Costs	Incl. O&P
1 Equip. Oper. (crane)	$56.10	$448.80	$84.60	$676.80	$50.68	$76.63
.5 Laborer	39.85	159.40	60.70	242.80		
1 Hyd. Excavator, 6 C.Y.		3487.00		3835.70	290.58	319.64
12 L.H., Daily Totals		$4095.20		$4755.30	$341.27	$396.27
Crew B-14C	Hr.	Daily	Hr.	Daily	Bare Costs	Incl. O&P
1 Equip. Oper. (crane)	$56.10	$448.80	$84.60	$676.80	$50.68	$76.63
.5 Laborer	39.85	159.40	60.70	242.80		
1 Hyd. Excavator, 7 C.Y.		3275.00		3602.50	272.92	300.21
12 L.H., Daily Totals		$3883.20		$4522.10	$323.60	$376.84
Crew B-14F	Hr.	Daily	Hr.	Daily	Bare Costs	Incl. O&P
1 Equip. Oper. (crane)	$56.10	$448.80	$84.60	$676.80	$50.68	$76.63
.5 Laborer	39.85	159.40	60.70	242.80		
1 Hyd. Shovel, 7 C.Y.		3880.00		4268.00	323.33	355.67
12 L.H., Daily Totals		$4488.20		$5187.60	$374.02	$432.30
Crew B-14G	Hr.	Daily	Hr.	Daily	Bare Costs	Incl. O&P
1 Equip. Oper. (crane)	$56.10	$448.80	$84.60	$676.80	$50.68	$76.63
.5 Laborer	39.85	159.40	60.70	242.80		
1 Hyd. Shovel, 12 C.Y.		5769.00		6345.90	480.75	528.83
12 L.H., Daily Totals		$6377.20		$7265.50	$531.43	$605.46
Crew B-14J	Hr.	Daily	Hr.	Daily	Bare Costs	Incl. O&P
1 Equip. Oper. (medium)	$53.75	$430.00	$81.05	$648.40	$49.12	$74.27
.5 Laborer	39.85	159.40	60.70	242.80		
1 F.E. Loader, 8 C.Y.		1780.00		1958.00	148.33	163.17
12 L.H., Daily Totals		$2369.40		$2849.20	$197.45	$237.43
Crew B-14K	Hr.	Daily	Hr.	Daily	Bare Costs	Incl. O&P
1 Equip. Oper. (medium)	$53.75	$430.00	$81.05	$648.40	$49.12	$74.27
.5 Laborer	39.85	159.40	60.70	242.80		
1 F.E. Loader, 10 C.Y.		2630.00		2893.00	219.17	241.08
12 L.H., Daily Totals		$3219.40		$3784.20	$268.28	$315.35

Crew No.	Bare Costs		Incl. Subs O&P		Cost Per Labor-Hour	
Crew B-15	Hr.	Daily	Hr.	Daily	Bare Costs	Incl. O&P
1 Equipment Oper. (med.)	$53.75	$430.00	$81.05	$648.40	$47.34	$71.43
.5 Laborer	39.85	159.40	60.70	242.80		
2 Truck Drivers (heavy)	46.00	736.00	69.30	1108.80		
2 Dump Trucks, 12 C.Y., 400 H.P.		1085.60		1194.16		
1 Dozer, 200 H.P.		1273.00		1400.30	84.24	92.66
28 L.H., Daily Totals		$3684.00		$4594.46	$131.57	$164.09
Crew B-16	Hr.	Daily	Hr.	Daily	Bare Costs	Incl. O&P
1 Labor Foreman (outside)	$41.85	$334.80	$63.75	$510.00	$41.89	$63.61
2 Laborers	39.85	637.60	60.70	971.20		
1 Truck Driver (heavy)	46.00	368.00	69.30	554.40		
1 Dump Truck, 12 C.Y., 400 H.P.		542.80		597.08	16.96	18.66
32 L.H., Daily Totals		$1883.20		$2632.68	$58.85	$82.27
Crew B-17	Hr.	Daily	Hr.	Daily	Bare Costs	Incl. O&P
2 Laborers	$39.85	$637.60	$60.70	$971.20	$44.25	$67.01
1 Equip. Oper. (light)	51.30	410.40	77.35	618.80		
1 Truck Driver (heavy)	46.00	368.00	69.30	554.40		
1 Backhoe Loader, 48 H.P.		312.40		343.64		
1 Dump Truck, 8 C.Y., 220 H.P.		338.60		372.46	20.34	22.38
32 L.H., Daily Totals		$2067.00		$2860.50	$64.59	$89.39
Crew B-17A	Hr.	Daily	Hr.	Daily	Bare Costs	Incl. O&P
2 Labor Foremen (outside)	$41.85	$669.60	$63.75	$1020.00	$42.95	$65.50
6 Laborers	39.85	1912.80	60.70	2913.60		
1 Skilled Worker Foreman (out)	54.35	434.80	83.20	665.60		
1 Skilled Worker	52.35	418.80	80.15	641.20		
80 L.H., Daily Totals		$3436.00		$5240.40	$42.95	$65.50
Crew B-17B	Hr.	Daily	Hr.	Daily	Bare Costs	Incl. O&P
2 Laborers	$39.85	$637.60	$60.70	$971.20	$44.25	$67.01
1 Equip. Oper. (light)	51.30	410.40	77.35	618.80		
1 Truck Driver (heavy)	46.00	368.00	69.30	554.40		
1 Backhoe Loader, 48 H.P.		312.40		343.64		
1 Dump Truck, 12 C.Y., 400 H.P.		542.80		597.08	26.73	29.40
32 L.H., Daily Totals		$2271.20		$3085.12	$70.97	$96.41
Crew B-18	Hr.	Daily	Hr.	Daily	Bare Costs	Incl. O&P
1 Labor Foreman (outside)	$41.85	$334.80	$63.75	$510.00	$40.52	$61.72
2 Laborers	39.85	637.60	60.70	971.20		
1 Vibrating Plate, Gas, 21"		40.60		44.66	1.69	1.86
24 L.H., Daily Totals		$1013.00		$1525.86	$42.21	$63.58
Crew B-19	Hr.	Daily	Hr.	Daily	Bare Costs	Incl. O&P
1 Pile Driver Foreman (outside)	$53.30	$426.40	$84.05	$672.40	$52.41	$81.27
4 Pile Drivers	51.30	1641.60	80.90	2588.80		
2 Equip. Oper. (crane)	56.10	897.60	84.60	1353.60		
1 Equip. Oper. (oiler)	48.60	388.80	73.30	586.40		
1 Crawler Crane, 40 Ton		1370.00		1507.00		
1 Lead, 90' High		131.60		144.76		
1 Hammer, Diesel, 22k ft-lb		427.00		469.70	30.13	33.15
64 L.H., Daily Totals		$5283.00		$7322.66	$82.55	$114.42

For customer support on your Building Construction Costs with RSMeans data, call 800.448.8182.

Crew No.	Bare Costs		Incl. Subs O&P		Cost Per Labor-Hour	

Crew B-19A

	Hr.	Daily	Hr.	Daily	Bare Costs	Incl. O&P
1 Pile Driver Foreman (outside)	$53.30	$426.40	$84.05	$672.40	$52.41	$81.27
4 Pile Drivers	51.30	1641.60	80.90	2588.80		
2 Equip. Oper. (crane)	56.10	897.60	84.60	1353.60		
1 Equip. Oper. (oiler)	48.60	388.80	73.30	586.40		
1 Crawler Crane, 75 Ton		1708.00		1878.80		
1 Lead, 90' High		131.60		144.76		
1 Hammer, Diesel, 41k ft-lb		572.00		629.20	37.68	41.45
64 L.H., Daily Totals		$5766.00		$7853.96	$90.09	$122.72

Crew B-19B

	Hr.	Daily	Hr.	Daily	Bare Costs	Incl. O&P
1 Pile Driver Foreman (outside)	$53.30	$426.40	$84.05	$672.40	$52.41	$81.27
4 Pile Drivers	51.30	1641.60	80.90	2588.80		
2 Equip. Oper. (crane)	56.10	897.60	84.60	1353.60		
1 Equip. Oper. (oiler)	48.60	388.80	73.30	586.40		
1 Crawler Crane, 40 Ton		1370.00		1507.00		
1 Lead, 90' High		131.60		144.76		
1 Hammer, Diesel, 22k ft-lb		427.00		469.70		
1 Barge, 400 Ton		835.40		918.94	43.19	47.51
64 L.H., Daily Totals		$6118.40		$8241.60	$95.60	$128.78

Crew B-19C

	Hr.	Daily	Hr.	Daily	Bare Costs	Incl. O&P
1 Pile Driver Foreman (outside)	$53.30	$426.40	$84.05	$672.40	$52.41	$81.27
4 Pile Drivers	51.30	1641.60	80.90	2588.80		
2 Equip. Oper. (crane)	56.10	897.60	84.60	1353.60		
1 Equip. Oper. (oiler)	48.60	388.80	73.30	586.40		
1 Crawler Crane, 75 Ton		1708.00		1878.80		
1 Lead, 90' High		131.60		144.76		
1 Hammer, Diesel, 41k ft-lb		572.00		629.20		
1 Barge, 400 Ton		835.40		918.94	50.73	55.81
64 L.H., Daily Totals		$6601.40		$8772.90	$103.15	$137.08

Crew B-20

	Hr.	Daily	Hr.	Daily	Bare Costs	Incl. O&P
1 Labor Foreman (outside)	$41.85	$334.80	$63.75	$510.00	$44.68	$68.20
1 Skilled Worker	52.35	418.80	80.15	641.20		
1 Laborer	39.85	318.80	60.70	485.60		
24 L.H., Daily Totals		$1072.40		$1636.80	$44.68	$68.20

Crew B-20A

	Hr.	Daily	Hr.	Daily	Bare Costs	Incl. O&P
1 Labor Foreman (outside)	$41.85	$334.80	$63.75	$510.00	$48.39	$73.22
1 Laborer	39.85	318.80	60.70	485.60		
1 Plumber	62.15	497.20	93.60	748.80		
1 Plumber Apprentice	49.70	397.60	74.85	598.80		
32 L.H., Daily Totals		$1548.40		$2343.20	$48.39	$73.22

Crew B-21

	Hr.	Daily	Hr.	Daily	Bare Costs	Incl. O&P
1 Labor Foreman (outside)	$41.85	$334.80	$63.75	$510.00	$46.31	$70.54
1 Skilled Worker	52.35	418.80	80.15	641.20		
1 Laborer	39.85	318.80	60.70	485.60		
.5 Equip. Oper. (crane)	56.10	224.40	84.60	338.40		
.5 S.P. Crane, 4x4, 5 Ton		129.50		142.45	4.63	5.09
28 L.H., Daily Totals		$1426.30		$2117.65	$50.94	$75.63

Crew B-21A

	Hr.	Daily	Hr.	Daily	Bare Costs	Incl. O&P
1 Labor Foreman (outside)	$41.85	$334.80	$63.75	$510.00	$49.93	$75.50
1 Laborer	39.85	318.80	60.70	485.60		
1 Plumber	62.15	497.20	93.60	748.80		
1 Plumber Apprentice	49.70	397.60	74.85	598.80		
1 Equip. Oper. (crane)	56.10	448.80	84.60	676.80		
1 S.P. Crane, 4x4, 12 Ton		363.60		399.96	9.09	10.00
40 L.H., Daily Totals		$2360.80		$3419.96	$59.02	$85.50

Crew B-21B

	Hr.	Daily	Hr.	Daily	Bare Costs	Incl. O&P
1 Labor Foreman (outside)	$41.85	$334.80	$63.75	$510.00	$43.50	$66.09
3 Laborers	39.85	956.40	60.70	1456.80		
1 Equip. Oper. (crane)	56.10	448.80	84.60	676.80		
1 Hyd. Crane, 12 Ton		496.00		545.60	12.40	13.64
40 L.H., Daily Totals		$2236.00		$3189.20	$55.90	$79.73

Crew B-21C

	Hr.	Daily	Hr.	Daily	Bare Costs	Incl. O&P
1 Labor Foreman (outside)	$41.85	$334.80	$63.75	$510.00	$43.71	$66.35
4 Laborers	39.85	1275.20	60.70	1942.40		
1 Equip. Oper. (crane)	56.10	448.80	84.60	676.80		
1 Equip. Oper. (oiler)	48.60	388.80	73.30	586.40		
2 Cutting Torches		25.20		27.72		
2 Sets of Gases		336.00		369.60		
1 Lattice Boom Crane, 90 Ton		1687.00		1855.70	36.58	40.23
56 L.H., Daily Totals		$4495.80		$5968.62	$80.28	$106.58

Crew B-22

	Hr.	Daily	Hr.	Daily	Bare Costs	Incl. O&P
1 Labor Foreman (outside)	$41.85	$334.80	$63.75	$510.00	$46.97	$71.48
1 Skilled Worker	52.35	418.80	80.15	641.20		
1 Laborer	39.85	318.80	60.70	485.60		
.75 Equip. Oper. (crane)	56.10	336.60	84.60	507.60		
.75 S.P. Crane, 4x4, 5 Ton		194.25		213.68	6.47	7.12
30 L.H., Daily Totals		$1603.25		$2358.07	$53.44	$78.60

Crew B-22A

	Hr.	Daily	Hr.	Daily	Bare Costs	Incl. O&P
1 Labor Foreman (outside)	$41.85	$334.80	$63.75	$510.00	$46.00	$69.98
1 Skilled Worker	52.35	418.80	80.15	641.20		
2 Laborers	39.85	637.60	60.70	971.20		
1 Equipment Operator, Crane	56.10	448.80	84.60	676.80		
1 S.P. Crane, 4x4, 5 Ton		259.00		284.90		
1 Butt Fusion Mach., 4"-12" diam.		425.70		468.27	17.12	18.83
40 L.H., Daily Totals		$2524.70		$3552.37	$63.12	$88.81

Crew B-22B

	Hr.	Daily	Hr.	Daily	Bare Costs	Incl. O&P
1 Labor Foreman (outside)	$41.85	$334.80	$63.75	$510.00	$46.00	$69.98
1 Skilled Worker	52.35	418.80	80.15	641.20		
2 Laborers	39.85	637.60	60.70	971.20		
1 Equip. Oper. (crane)	56.10	448.80	84.60	676.80		
1 S.P. Crane, 4x4, 5 Ton		259.00		284.90		
1 Butt Fusion Mach., 8"-24" diam.		1908.00		2098.80	54.17	59.59
40 L.H., Daily Totals		$4007.00		$5182.90	$100.18	$129.57

Crew B-22C

	Hr.	Daily	Hr.	Daily	Bare Costs	Incl. O&P
1 Skilled Worker	$52.35	$418.80	$80.15	$641.20	$46.10	$70.42
1 Laborer	39.85	318.80	60.70	485.60		
1 Butt Fusion Mach., 2"-8" diam.		121.05		133.16	7.57	8.32
16 L.H., Daily Totals		$858.65		$1259.95	$53.67	$78.75

Crew B-23

	Hr.	Daily	Hr.	Daily	Bare Costs	Incl. O&P
1 Labor Foreman (outside)	$41.85	$334.80	$63.75	$510.00	$40.25	$61.31
4 Laborers	39.85	1275.20	60.70	1942.40		
1 Drill Rig, Truck-Mounted		2444.00		2688.40		
1 Flatbed Truck, Gas, 3 Ton		238.00		261.80	67.05	73.75
40 L.H., Daily Totals		$4292.00		$5402.60	$107.30	$135.07

For customer support on your Building Construction Costs with RSMeans data, call 800.448.8182.

Crew No.	Bare Costs		Incl. Subs O&P		Cost Per Labor-Hour	
	Hr.	Daily	Hr.	Daily	Bare Costs	Incl. O&P
Crew B-23A						
1 Labor Foreman (outside)	$41.85	$334.80	$63.75	$510.00	$45.15	$68.50
1 Laborer	39.85	318.80	60.70	485.60		
1 Equip. Oper. (medium)	53.75	430.00	81.05	648.40		
1 Drill Rig, Truck-Mounted		2444.00		2688.40		
1 Pickup Truck, 3/4 Ton		115.20		126.72	106.63	117.30
24 L.H., Daily Totals		$3642.80		$4459.12	$151.78	$185.80

Crew No.	Bare Costs		Incl. Subs O&P		Cost Per Labor-Hour	
	Hr.	Daily	Hr.	Daily	Bare Costs	Incl. O&P
Crew B-23B						
1 Labor Foreman (outside)	$41.85	$334.80	$63.75	$510.00	$45.15	$68.50
1 Laborer	39.85	318.80	60.70	485.60		
1 Equip. Oper. (medium)	53.75	430.00	81.05	648.40		
1 Drill Rig, Truck-Mounted		2444.00		2688.40		
1 Pickup Truck, 3/4 Ton		115.20		126.72		
1 Centr. Water Pump, 6"		296.40		326.04	118.98	130.88
24 L.H., Daily Totals		$3939.20		$4785.16	$164.13	$199.38

Crew No.	Bare Costs		Incl. Subs O&P		Cost Per Labor-Hour	
	Hr.	Daily	Hr.	Daily	Bare Costs	Incl. O&P
Crew B-24						
1 Cement Finisher	$47.55	$380.40	$70.45	$563.60	$46.03	$69.45
1 Laborer	39.85	318.80	60.70	485.60		
1 Carpenter	50.70	405.60	77.20	617.60		
24 L.H., Daily Totals		$1104.80		$1666.80	$46.03	$69.45

Crew No.	Bare Costs		Incl. Subs O&P		Cost Per Labor-Hour	
	Hr.	Daily	Hr.	Daily	Bare Costs	Incl. O&P
Crew B-25						
1 Labor Foreman (outside)	$41.85	$334.80	$63.75	$510.00	$43.82	$66.53
7 Laborers	39.85	2231.60	60.70	3399.20		
3 Equip. Oper. (medium)	53.75	1290.00	81.05	1945.20		
1 Asphalt Paver, 130 H.P.		2123.00		2335.30		
1 Tandem Roller, 10 Ton		231.40		254.54		
1 Roller, Pneum. Whl., 12 Ton		338.20		372.02	30.60	33.66
88 L.H., Daily Totals		$6549.00		$8816.26	$74.42	$100.18

Crew No.	Bare Costs		Incl. Subs O&P		Cost Per Labor-Hour	
	Hr.	Daily	Hr.	Daily	Bare Costs	Incl. O&P
Crew B-25B						
1 Labor Foreman (outside)	$41.85	$334.80	$63.75	$510.00	$44.65	$67.74
7 Laborers	39.85	2231.60	60.70	3399.20		
4 Equip. Oper. (medium)	53.75	1720.00	81.05	2593.60		
1 Asphalt Paver, 130 H.P.		2123.00		2335.30		
2 Tandem Rollers, 10 Ton		462.80		509.08		
1 Roller, Pneum. Whl., 12 Ton		338.20		372.02	30.46	33.50
96 L.H., Daily Totals		$7210.40		$9719.20	$75.11	$101.24

Crew No.	Bare Costs		Incl. Subs O&P		Cost Per Labor-Hour	
	Hr.	Daily	Hr.	Daily	Bare Costs	Incl. O&P
Crew B-25C						
1 Labor Foreman (outside)	$41.85	$334.80	$63.75	$510.00	$44.82	$67.99
3 Laborers	39.85	956.40	60.70	1456.80		
2 Equip. Oper. (medium)	53.75	860.00	81.05	1296.80		
1 Asphalt Paver, 130 H.P.		2123.00		2335.30		
1 Tandem Roller, 10 Ton		231.40		254.54	49.05	53.95
48 L.H., Daily Totals		$4505.60		$5853.44	$93.87	$121.95

Crew No.	Bare Costs		Incl. Subs O&P		Cost Per Labor-Hour	
	Hr.	Daily	Hr.	Daily	Bare Costs	Incl. O&P
Crew B-25D						
1 Labor Foreman (outside)	$41.85	$334.80	$63.75	$510.00	$45.02	$68.28
3 Laborers	39.85	956.40	60.70	1456.80		
2.125 Equip. Oper. (medium)	53.75	913.75	81.05	1377.85		
.125 Truck Driver (heavy)	46.00	46.00	69.30	69.30		
.125 Truck Tractor, 6x4, 380 H.P.		59.52		65.48		
.125 Dist. Tanker, 3000 Gallon		40.85		44.94		
1 Asphalt Paver, 130 H.P.		2123.00		2335.30		
1 Tandem Roller, 10 Ton		231.40		254.54	49.10	54.01
50 L.H., Daily Totals		$4705.73		$6114.20	$94.11	$122.28

Crew No.	Bare Costs		Incl. Subs O&P		Cost Per Labor-Hour	
	Hr.	Daily	Hr.	Daily	Bare Costs	Incl. O&P
Crew B-25E						
1 Labor Foreman (outside)	$41.85	$334.80	$63.75	$510.00	$45.21	$68.54
3 Laborers	39.85	956.40	60.70	1456.80		
2.250 Equip. Oper. (medium)	53.75	967.50	81.05	1458.90		
.25 Truck Driver (heavy)	46.00	92.00	69.30	138.60		
.25 Truck Tractor, 6x4, 380 H.P.		119.05		130.96		
.25 Dist. Tanker, 3000 Gallon		81.70		89.87		
1 Asphalt Paver, 130 H.P.		2123.00		2335.30		
1 Tandem Roller, 10 Ton		231.40		254.54	49.14	54.05
52 L.H., Daily Totals		$4905.85		$6374.97	$94.34	$122.60

Crew No.	Bare Costs		Incl. Subs O&P		Cost Per Labor-Hour	
	Hr.	Daily	Hr.	Daily	Bare Costs	Incl. O&P
Crew B-26						
1 Labor Foreman (outside)	$41.85	$334.80	$63.75	$510.00	$44.60	$67.63
6 Laborers	39.85	1912.80	60.70	2913.60		
2 Equip. Oper. (medium)	53.75	860.00	81.05	1296.80		
1 Rodman (reinf.)	54.65	437.20	83.45	667.60		
1 Cement Finisher	47.55	380.40	70.45	563.60		
1 Grader, 30,000 Lbs.		643.80		708.18		
1 Paving Mach. & Equip.		2361.00		2597.10	34.15	37.56
88 L.H., Daily Totals		$6930.00		$9256.88	$78.75	$105.19

Crew No.	Bare Costs		Incl. Subs O&P		Cost Per Labor-Hour	
	Hr.	Daily	Hr.	Daily	Bare Costs	Incl. O&P
Crew B-26A						
1 Labor Foreman (outside)	$41.85	$334.80	$63.75	$510.00	$44.60	$67.63
6 Laborers	39.85	1912.80	60.70	2913.60		
2 Equip. Oper. (medium)	53.75	860.00	81.05	1296.80		
1 Rodman (reinf.)	54.65	437.20	83.45	667.60		
1 Cement Finisher	47.55	380.40	70.45	563.60		
1 Grader, 30,000 Lbs.		643.80		708.18		
1 Paving Mach. & Equip.		2361.00		2597.10		
1 Concrete Saw		102.40		112.64	35.31	38.84
88 L.H., Daily Totals		$7032.40		$9369.52	$79.91	$106.47

Crew No.	Bare Costs		Incl. Subs O&P		Cost Per Labor-Hour	
	Hr.	Daily	Hr.	Daily	Bare Costs	Incl. O&P
Crew B-26B						
1 Labor Foreman (outside)	$41.85	$334.80	$63.75	$510.00	$45.37	$68.75
6 Laborers	39.85	1912.80	60.70	2913.60		
3 Equip. Oper. (medium)	53.75	1290.00	81.05	1945.20		
1 Rodman (reinf.)	54.65	437.20	83.45	667.60		
1 Cement Finisher	47.55	380.40	70.45	563.60		
1 Grader, 30,000 Lbs.		643.80		708.18		
1 Paving Mach. & Equip.		2361.00		2597.10		
1 Concrete Pump, 110' Boom		1116.00		1227.60	42.92	47.22
96 L.H., Daily Totals		$8476.00		$11132.88	$88.29	$115.97

Crew No.	Bare Costs		Incl. Subs O&P		Cost Per Labor-Hour	
	Hr.	Daily	Hr.	Daily	Bare Costs	Incl. O&P
Crew B-26C						
1 Labor Foreman (outside)	$41.85	$334.80	$63.75	$510.00	$43.69	$66.29
6 Laborers	39.85	1912.80	60.70	2913.60		
1 Equip. Oper. (medium)	53.75	430.00	81.05	648.40		
1 Rodman (reinf.)	54.65	437.20	83.45	667.60		
1 Cement Finisher	47.55	380.40	70.45	563.60		
1 Paving Mach. & Equip.		2361.00		2597.10		
1 Concrete Saw		102.40		112.64	30.79	33.87
80 L.H., Daily Totals		$5958.60		$8012.94	$74.48	$100.16

Crew No.	Bare Costs		Incl. Subs O&P		Cost Per Labor-Hour	
	Hr.	Daily	Hr.	Daily	Bare Costs	Incl. O&P
Crew B-27						
1 Labor Foreman (outside)	$41.85	$334.80	$63.75	$510.00	$40.35	$61.46
3 Laborers	39.85	956.40	60.70	1456.80		
1 Berm Machine		317.40		349.14	9.92	10.91
32 L.H., Daily Totals		$1608.60		$2315.94	$50.27	$72.37

For customer support on your Building Construction Costs with RSMeans data, call 800.448.8182.

Crews - Standard

Crew No.	Bare Costs Hr.	Daily	Incl. Subs O&P Hr.	Daily	Cost Per Labor-Hour Bare Costs	Incl. O&P
Crew B-28	Hr.	Daily	Hr.	Daily	Bare Costs	Incl. O&P
2 Carpenters	$50.70	$811.20	$77.20	$1235.20	$47.08	$71.70
1 Laborer	39.85	318.80	60.70	485.60		
24 L.H., Daily Totals		$1130.00		$1720.80	$47.08	$71.70
Crew B-29	Hr.	Daily	Hr.	Daily	Bare Costs	Incl. O&P
1 Labor Foreman (outside)	$41.85	$334.80	$63.75	$510.00	$43.71	$66.35
4 Laborers	39.85	1275.20	60.70	1942.40		
1 Equip. Oper. (crane)	56.10	448.80	84.60	676.80		
1 Equip. Oper. (oiler)	48.60	388.80	73.30	586.40		
1 Gradall, 5/8 C.Y.		869.60		956.56	15.53	17.08
56 L.H., Daily Totals		$3317.20		$4672.16	$59.24	$83.43
Crew B-30	Hr.	Daily	Hr.	Daily	Bare Costs	Incl. O&P
1 Equip. Oper. (medium)	$53.75	$430.00	$81.05	$648.40	$48.58	$73.22
2 Truck Drivers (heavy)	46.00	736.00	69.30	1108.80		
1 Hyd. Excavator, 1.5 C.Y.		893.60		982.96		
2 Dump Trucks, 12 C.Y., 400 H.P.		1085.60		1194.16	82.47	90.71
24 L.H., Daily Totals		$3145.20		$3934.32	$131.05	$163.93
Crew B-31	Hr.	Daily	Hr.	Daily	Bare Costs	Incl. O&P
1 Labor Foreman (outside)	$41.85	$334.80	$63.75	$510.00	$42.42	$64.61
3 Laborers	39.85	956.40	60.70	1456.80		
1 Carpenter	50.70	405.60	77.20	617.60		
1 Air Compressor, 250 cfm		167.40		184.14		
1 Sheeting Driver		7.00		7.70		
2 -50' Air Hoses, 1.5"		45.40		49.94	5.50	6.04
40 L.H., Daily Totals		$1916.60		$2826.18	$47.91	$70.65
Crew B-32	Hr.	Daily	Hr.	Daily	Bare Costs	Incl. O&P
1 Laborer	$39.85	$318.80	$60.70	$485.60	$50.27	$75.96
3 Equip. Oper. (medium)	53.75	1290.00	81.05	1945.20		
1 Grader, 30,000 Lbs.		643.80		708.18		
1 Tandem Roller, 10 Ton		231.40		254.54		
1 Dozer, 200 H.P.		1273.00		1400.30	67.13	73.84
32 L.H., Daily Totals		$3757.00		$4793.82	$117.41	$149.81
Crew B-32A	Hr.	Daily	Hr.	Daily	Bare Costs	Incl. O&P
1 Laborer	$39.85	$318.80	$60.70	$485.60	$49.12	$74.27
2 Equip. Oper. (medium)	53.75	860.00	81.05	1296.80		
1 Grader, 30,000 Lbs.		643.80		708.18		
1 Roller, Vibratory, 25 Ton		659.80		725.78	54.32	59.75
24 L.H., Daily Totals		$2482.40		$3216.36	$103.43	$134.01
Crew B-32B	Hr.	Daily	Hr.	Daily	Bare Costs	Incl. O&P
1 Laborer	$39.85	$318.80	$60.70	$485.60	$49.12	$74.27
2 Equip. Oper. (medium)	53.75	860.00	81.05	1296.80		
1 Dozer, 200 H.P.		1273.00		1400.30		
1 Roller, Vibratory, 25 Ton		659.80		725.78	80.53	88.59
24 L.H., Daily Totals		$3111.60		$3908.48	$129.65	$162.85
Crew B-32C	Hr.	Daily	Hr.	Daily	Bare Costs	Incl. O&P
1 Labor Foreman (outside)	$41.85	$334.80	$63.75	$510.00	$47.13	$71.38
2 Laborers	39.85	637.60	60.70	971.20		
3 Equip. Oper. (medium)	53.75	1290.00	81.05	1945.20		
1 Grader, 30,000 Lbs.		643.80		708.18		
1 Tandem Roller, 10 Ton		231.40		254.54		
1 Dozer, 200 H.P.		1273.00		1400.30	44.75	49.23
48 L.H., Daily Totals		$4410.60		$5789.42	$91.89	$120.61

Crew No.	Bare Costs Hr.	Daily	Incl. Subs O&P Hr.	Daily	Cost Per Labor-Hour Bare Costs	Incl. O&P
Crew B-33A	Hr.	Daily	Hr.	Daily	Bare Costs	Incl. O&P
1 Equip. Oper. (medium)	$53.75	$430.00	$81.05	$648.40	$49.78	$75.24
.5 Laborer	39.85	159.40	60.70	242.80		
.25 Equip. Oper. (medium)	53.75	107.50	81.05	162.10		
1 Scraper, Towed, 7 C.Y.		123.40		135.74		
1.25 Dozers, 300 H.P.		2286.25		2514.88	172.12	189.33
14 L.H., Daily Totals		$3106.55		$3703.92	$221.90	$264.57
Crew B-33B	Hr.	Daily	Hr.	Daily	Bare Costs	Incl. O&P
1 Equip. Oper. (medium)	$53.75	$430.00	$81.05	$648.40	$49.78	$75.24
.5 Laborer	39.85	159.40	60.70	242.80		
.25 Equip. Oper. (medium)	53.75	107.50	81.05	162.10		
1 Scraper, Towed, 10 C.Y.		155.80		171.38		
1.25 Dozers, 300 H.P.		2286.25		2514.88	174.43	191.88
14 L.H., Daily Totals		$3138.95		$3739.55	$224.21	$267.11
Crew B-33C	Hr.	Daily	Hr.	Daily	Bare Costs	Incl. O&P
1 Equip. Oper. (medium)	$53.75	$430.00	$81.05	$648.40	$49.78	$75.24
.5 Laborer	39.85	159.40	60.70	242.80		
.25 Equip. Oper. (medium)	53.75	107.50	81.05	162.10		
1 Scraper, Towed, 15 C.Y.		175.80		193.38		
1.25 Dozers, 300 H.P.		2286.25		2514.88	175.86	193.45
14 L.H., Daily Totals		$3158.95		$3761.55	$225.64	$268.68
Crew B-33D	Hr.	Daily	Hr.	Daily	Bare Costs	Incl. O&P
1 Equip. Oper. (medium)	$53.75	$430.00	$81.05	$648.40	$49.78	$75.24
.5 Laborer	39.85	159.40	60.70	242.80		
.25 Equip. Oper. (medium)	53.75	107.50	81.05	162.10		
1 S.P. Scraper, 14 C.Y.		2490.00		2739.00		
.25 Dozer, 300 H.P.		457.25		502.98	210.52	231.57
14 L.H., Daily Totals		$3644.15		$4295.27	$260.30	$306.81
Crew B-33E	Hr.	Daily	Hr.	Daily	Bare Costs	Incl. O&P
1 Equip. Oper. (medium)	$53.75	$430.00	$81.05	$648.40	$49.78	$75.24
.5 Laborer	39.85	159.40	60.70	242.80		
.25 Equip. Oper. (medium)	53.75	107.50	81.05	162.10		
1 S.P. Scraper, 21 C.Y.		2468.00		2714.80		
.25 Dozer, 300 H.P.		457.25		502.98	208.95	229.84
14 L.H., Daily Totals		$3622.15		$4271.07	$258.73	$305.08
Crew B-33F	Hr.	Daily	Hr.	Daily	Bare Costs	Incl. O&P
1 Equip. Oper. (medium)	$53.75	$430.00	$81.05	$648.40	$49.78	$75.24
.5 Laborer	39.85	159.40	60.70	242.80		
.25 Equip. Oper. (medium)	53.75	107.50	81.05	162.10		
1 Elev. Scraper, 11 C.Y.		1155.00		1270.50		
.25 Dozer, 300 H.P.		457.25		502.98	115.16	126.68
14 L.H., Daily Totals		$2309.15		$2826.78	$164.94	$201.91
Crew B-33G	Hr.	Daily	Hr.	Daily	Bare Costs	Incl. O&P
1 Equip. Oper. (medium)	$53.75	$430.00	$81.05	$648.40	$49.78	$75.24
.5 Laborer	39.85	159.40	60.70	242.80		
.25 Equip. Oper. (medium)	53.75	107.50	81.05	162.10		
1 Elev. Scraper, 22 C.Y.		2284.00		2512.40		
.25 Dozer, 300 H.P.		457.25		502.98	195.80	215.38
14 L.H., Daily Totals		$3438.15		$4068.68	$245.58	$290.62

751

Crew No.	Bare Costs		Incl. Subs O&P		Cost Per Labor-Hour	
Crew B-33H	**Hr.**	**Daily**	**Hr.**	**Daily**	**Bare Costs**	**Incl. O&P**
.5 Laborer	$39.85	$159.40	$60.70	$242.80	$49.78	$75.24
1 Equipment Operator (med.)	53.75	430.00	81.05	648.40		
.25 Equipment Operator (med.)	53.75	107.50	81.05	162.10		
1 S.P. Scraper, 44 C.Y.		4457.00		4902.70		
.25 Dozer, 410 H.P.		572.75		630.02	359.27	395.19
14 L.H., Daily Totals		$5726.65		$6586.02	$409.05	$470.43
Crew B-33J	**Hr.**	**Daily**	**Hr.**	**Daily**	**Bare Costs**	**Incl. O&P**
1 Equipment Operator (med.)	$53.75	$430.00	$81.05	$648.40	$53.75	$81.05
1 S.P. Scraper, 14 C.Y.		2490.00		2739.00	311.25	342.38
8 L.H., Daily Totals		$2920.00		$3387.40	$365.00	$423.43
Crew B-33K	**Hr.**	**Daily**	**Hr.**	**Daily**	**Bare Costs**	**Incl. O&P**
1 Equipment Operator (med.)	$53.75	$430.00	$81.05	$648.40	$49.78	$75.24
.25 Equipment Operator (med.)	53.75	107.50	81.05	162.10		
.5 Laborer	39.85	159.40	60.70	242.80		
1 S.P. Scraper, 31 C.Y.		3576.00		3933.60		
.25 Dozer, 410 H.P.		572.75		630.02	296.34	325.97
14 L.H., Daily Totals		$4845.65		$5616.93	$346.12	$401.21
Crew B-34A	**Hr.**	**Daily**	**Hr.**	**Daily**	**Bare Costs**	**Incl. O&P**
1 Truck Driver (heavy)	$46.00	$368.00	$69.30	$554.40	$46.00	$69.30
1 Dump Truck, 8 C.Y., 220 H.P.		338.60		372.46	42.33	46.56
8 L.H., Daily Totals		$706.60		$926.86	$88.33	$115.86
Crew B-34B	**Hr.**	**Daily**	**Hr.**	**Daily**	**Bare Costs**	**Incl. O&P**
1 Truck Driver (heavy)	$46.00	$368.00	$69.30	$554.40	$46.00	$69.30
1 Dump Truck, 12 C.Y., 400 H.P.		542.80		597.08	67.85	74.64
8 L.H., Daily Totals		$910.80		$1151.48	$113.85	$143.94
Crew B-34C	**Hr.**	**Daily**	**Hr.**	**Daily**	**Bare Costs**	**Incl. O&P**
1 Truck Driver (heavy)	$46.00	$368.00	$69.30	$554.40	$46.00	$69.30
1 Truck Tractor, 6x4, 380 H.P.		476.20		523.82		
1 Dump Trailer, 16.5 C.Y.		133.00		146.30	76.15	83.77
8 L.H., Daily Totals		$977.20		$1224.52	$122.15	$153.07
Crew B-34D	**Hr.**	**Daily**	**Hr.**	**Daily**	**Bare Costs**	**Incl. O&P**
1 Truck Driver (heavy)	$46.00	$368.00	$69.30	$554.40	$46.00	$69.30
1 Truck Tractor, 6x4, 380 H.P.		476.20		523.82		
1 Dump Trailer, 20 C.Y.		147.60		162.36	77.97	85.77
8 L.H., Daily Totals		$991.80		$1240.58	$123.97	$155.07
Crew B-34E	**Hr.**	**Daily**	**Hr.**	**Daily**	**Bare Costs**	**Incl. O&P**
1 Truck Driver (heavy)	$46.00	$368.00	$69.30	$554.40	$46.00	$69.30
1 Dump Truck, Off Hwy., 25 Ton		1343.00		1477.30	167.88	184.66
8 L.H., Daily Totals		$1711.00		$2031.70	$213.88	$253.96
Crew B-34F	**Hr.**	**Daily**	**Hr.**	**Daily**	**Bare Costs**	**Incl. O&P**
1 Truck Driver (heavy)	$46.00	$368.00	$69.30	$554.40	$46.00	$69.30
1 Dump Truck, Off Hwy., 35 Ton		1448.00		1592.80	181.00	199.10
8 L.H., Daily Totals		$1816.00		$2147.20	$227.00	$268.40
Crew B-34G	**Hr.**	**Daily**	**Hr.**	**Daily**	**Bare Costs**	**Incl. O&P**
1 Truck Driver (heavy)	$46.00	$368.00	$69.30	$554.40	$46.00	$69.30
1 Dump Truck, Off Hwy., 50 Ton		1662.00		1828.20	207.75	228.53
8 L.H., Daily Totals		$2030.00		$2382.60	$253.75	$297.82

Crew No.	Bare Costs		Incl. Subs O&P		Cost Per Labor-Hour	
Crew B-34H	**Hr.**	**Daily**	**Hr.**	**Daily**	**Bare Costs**	**Incl. O&P**
1 Truck Driver (heavy)	$46.00	$368.00	$69.30	$554.40	$46.00	$69.30
1 Dump Truck, Off Hwy., 65 Ton		1844.00		2028.40	230.50	253.55
8 L.H., Daily Totals		$2212.00		$2582.80	$276.50	$322.85
Crew B-34I	**Hr.**	**Daily**	**Hr.**	**Daily**	**Bare Costs**	**Incl. O&P**
1 Truck Driver (heavy)	$46.00	$368.00	$69.30	$554.40	$46.00	$69.30
1 Dump Truck, 18 C.Y., 450 H.P.		706.00		776.60	88.25	97.08
8 L.H., Daily Totals		$1074.00		$1331.00	$134.25	$166.38
Crew B-34J	**Hr.**	**Daily**	**Hr.**	**Daily**	**Bare Costs**	**Incl. O&P**
1 Truck Driver (heavy)	$46.00	$368.00	$69.30	$554.40	$46.00	$69.30
1 Dump Truck, Off Hwy., 100 Ton		2669.00		2935.90	333.63	366.99
8 L.H., Daily Totals		$3037.00		$3490.30	$379.63	$436.29
Crew B-34K	**Hr.**	**Daily**	**Hr.**	**Daily**	**Bare Costs**	**Incl. O&P**
1 Truck Driver (heavy)	$46.00	$368.00	$69.30	$554.40	$46.00	$69.30
1 Truck Tractor, 6x4, 450 H.P.		583.40		641.74		
1 Lowbed Trailer, 75 Ton		249.40		274.34	104.10	114.51
8 L.H., Daily Totals		$1200.80		$1470.48	$150.10	$183.81
Crew B-34L	**Hr.**	**Daily**	**Hr.**	**Daily**	**Bare Costs**	**Incl. O&P**
1 Equip. Oper. (light)	$51.30	$410.40	$77.35	$618.80	$51.30	$77.35
1 Flatbed Truck, Gas, 1.5 Ton		188.40		207.24	23.55	25.91
8 L.H., Daily Totals		$598.80		$826.04	$74.85	$103.26
Crew B-34M	**Hr.**	**Daily**	**Hr.**	**Daily**	**Bare Costs**	**Incl. O&P**
1 Equip. Oper. (light)	$51.30	$410.40	$77.35	$618.80	$51.30	$77.35
1 Flatbed Truck, Gas, 3 Ton		238.00		261.80	29.75	32.73
8 L.H., Daily Totals		$648.40		$880.60	$81.05	$110.08
Crew B-34N	**Hr.**	**Daily**	**Hr.**	**Daily**	**Bare Costs**	**Incl. O&P**
1 Truck Driver (heavy)	$46.00	$368.00	$69.30	$554.40	$49.88	$75.17
1 Equip. Oper. (medium)	53.75	430.00	81.05	648.40		
1 Truck Tractor, 6x4, 380 H.P.		476.20		523.82		
1 Flatbed Trailer, 40 Ton		180.00		198.00	41.01	45.11
16 L.H., Daily Totals		$1454.20		$1924.62	$90.89	$120.29
Crew B-34P	**Hr.**	**Daily**	**Hr.**	**Daily**	**Bare Costs**	**Incl. O&P**
1 Pipe Fitter	$63.00	$504.00	$94.90	$759.20	$53.75	$80.98
1 Truck Driver (light)	44.50	356.00	67.00	536.00		
1 Equip. Oper. (medium)	53.75	430.00	81.05	648.40		
1 Flatbed Truck, Gas, 3 Ton		238.00		261.80		
1 Backhoe Loader, 48 H.P.		312.40		343.64	22.93	25.23
24 L.H., Daily Totals		$1840.40		$2549.04	$76.68	$106.21
Crew B-34Q	**Hr.**	**Daily**	**Hr.**	**Daily**	**Bare Costs**	**Incl. O&P**
1 Pipe Fitter	$63.00	$504.00	$94.90	$759.20	$54.53	$82.17
1 Truck Driver (light)	44.50	356.00	67.00	536.00		
1 Equip. Oper. (crane)	56.10	448.80	84.60	676.80		
1 Flatbed Trailer, 25 Ton		133.00		146.30		
1 Dump Truck, 8 C.Y., 220 H.P.		338.60		372.46		
1 Hyd. Crane, 25 Ton		590.60		649.66	44.26	48.68
24 L.H., Daily Totals		$2371.00		$3140.42	$98.79	$130.85

For customer support on your Building Construction Costs with RSMeans data, call 800.448.8182.

Crew No.	Bare Costs		Incl. Subs O&P		Cost Per Labor-Hour	
Crew B-34R	Hr.	Daily	Hr.	Daily	Bare Costs	Incl. O&P
1 Pipe Fitter	$63.00	$504.00	$94.90	$759.20	$54.53	$82.17
1 Truck Driver (light)	44.50	356.00	67.00	536.00		
1 Equip. Oper. (crane)	56.10	448.80	84.60	676.80		
1 Flatbed Trailer, 25 Ton		133.00		146.30		
1 Dump Truck, 8 C.Y., 220 H.P.		338.60		372.46		
1 Hyd. Crane, 25 Ton		590.60		649.66		
1 Hyd. Excavator, 1 C.Y.		742.20		816.42	75.18	82.70
24 L.H., Daily Totals		$3113.20		$3956.84	$129.72	$164.87

Crew B-34S	Hr.	Daily	Hr.	Daily	Bare Costs	Incl. O&P
2 Pipe Fitters	$63.00	$1008.00	$94.90	$1518.40	$57.02	$85.92
1 Truck Driver (heavy)	46.00	368.00	69.30	554.40		
1 Equip. Oper. (crane)	56.10	448.80	84.60	676.80		
1 Flatbed Trailer, 40 Ton		180.00		198.00		
1 Truck Tractor, 6x4, 380 H.P.		476.20		523.82		
1 Hyd. Crane, 80 Ton		1518.00		1669.80		
1 Hyd. Excavator, 2 C.Y.		1052.00		1157.20	100.82	110.90
32 L.H., Daily Totals		$5051.00		$6298.42	$157.84	$196.83

Crew B-34T	Hr.	Daily	Hr.	Daily	Bare Costs	Incl. O&P
2 Pipe Fitters	$63.00	$1008.00	$94.90	$1518.40	$57.02	$85.92
1 Truck Driver (heavy)	46.00	368.00	69.30	554.40		
1 Equip. Oper. (crane)	56.10	448.80	84.60	676.80		
1 Flatbed Trailer, 40 Ton		180.00		198.00		
1 Truck Tractor, 6x4, 380 H.P.		476.20		523.82		
1 Hyd. Crane, 80 Ton		1518.00		1669.80	67.94	74.74
32 L.H., Daily Totals		$3999.00		$5141.22	$124.97	$160.66

Crew B-34U	Hr.	Daily	Hr.	Daily	Bare Costs	Incl. O&P
1 Truck Driver (heavy)	$46.00	$368.00	$69.30	$554.40	$48.65	$73.33
1 Equip. Oper. (light)	51.30	410.40	77.35	618.80		
1 Truck Tractor, 220 H.P.		290.00		319.00		
1 Flatbed Trailer, 25 Ton		133.00		146.30	26.44	29.08
16 L.H., Daily Totals		$1201.40		$1638.50	$75.09	$102.41

Crew B-34V	Hr.	Daily	Hr.	Daily	Bare Costs	Incl. O&P
1 Truck Driver (heavy)	$46.00	$368.00	$69.30	$554.40	$51.13	$77.08
1 Equip. Oper. (crane)	56.10	448.80	84.60	676.80		
1 Equip. Oper. (light)	51.30	410.40	77.35	618.80		
1 Truck Tractor, 6x4, 450 H.P.		583.40		641.74		
1 Equipment Trailer, 50 Ton		198.20		218.02		
1 Pickup Truck, 4x4, 3/4 Ton		126.60		139.26	37.84	41.63
24 L.H., Daily Totals		$2135.40		$2849.02	$88.97	$118.71

Crew B-34W	Hr.	Daily	Hr.	Daily	Bare Costs	Incl. O&P
5 Truck Drivers (heavy)	$46.00	$1840.00	$69.30	$2772.00	$48.71	$73.48
2 Equip. Opers. (crane)	56.10	897.60	84.60	1353.60		
1 Equip. Oper. (mechanic)	56.30	450.40	84.90	679.20		
1 Laborer	39.85	318.80	60.70	485.60		
4 Truck Tractors, 6x4, 380 H.P.		1904.80		2095.28		
2 Equipment Trailers, 50 Ton		396.40		436.04		
2 Flatbed Trailers, 40 Ton		360.00		396.00		
1 Pickup Truck, 4x4, 3/4 Ton		126.60		139.26		
1 S.P. Crane, 4x4, 20 Ton		571.20		628.32	46.65	51.32
72 L.H., Daily Totals		$6865.80		$8985.30	$95.36	$124.80

Crew No.	Bare Costs		Incl. Subs O&P		Cost Per Labor-Hour	
Crew B-35	Hr.	Daily	Hr.	Daily	Bare Costs	Incl. O&P
1 Labor Foreman (outside)	$41.85	$334.80	$63.75	$510.00	$50.15	$76.02
1 Skilled Worker	52.35	418.80	80.15	641.20		
1 Welder (plumber)	62.15	497.20	93.60	748.80		
1 Laborer	39.85	318.80	60.70	485.60		
1 Equip. Oper. (crane)	56.10	448.80	84.60	676.80		
1 Equip. Oper. (oiler)	48.60	388.80	73.30	586.40		
1 Welder, Electric, 300 amp		59.20		65.12		
1 Hyd. Excavator, .75 C.Y.		669.60		736.56	15.18	16.70
48 L.H., Daily Totals		$3136.00		$4450.48	$65.33	$92.72

Crew B-35A	Hr.	Daily	Hr.	Daily	Bare Costs	Incl. O&P
1 Labor Foreman (outside)	$41.85	$334.80	$63.75	$510.00	$48.68	$73.83
2 Laborers	39.85	637.60	60.70	971.20		
1 Skilled Worker	52.35	418.80	80.15	641.20		
1 Welder (plumber)	62.15	497.20	93.60	748.80		
1 Equip. Oper. (crane)	56.10	448.80	84.60	676.80		
1 Equip. Oper. (oiler)	48.60	388.80	73.30	586.40		
1 Welder, Gas Engine, 300 amp		98.40		108.24		
1 Crawler Crane, 75 Ton		1708.00		1878.80	32.26	35.48
56 L.H., Daily Totals		$4532.40		$6121.44	$80.94	$109.31

Crew B-36	Hr.	Daily	Hr.	Daily	Bare Costs	Incl. O&P
1 Labor Foreman (outside)	$41.85	$334.80	$63.75	$510.00	$45.81	$69.45
2 Laborers	39.85	637.60	60.70	971.20		
2 Equip. Oper. (medium)	53.75	860.00	81.05	1296.80		
1 Dozer, 200 H.P.		1273.00		1400.30		
1 Aggregate Spreader		36.80		40.48		
1 Tandem Roller, 10 Ton		231.40		254.54	38.53	42.38
40 L.H., Daily Totals		$3373.60		$4473.32	$84.34	$111.83

Crew B-36A	Hr.	Daily	Hr.	Daily	Bare Costs	Incl. O&P
1 Labor Foreman (outside)	$41.85	$334.80	$63.75	$510.00	$48.08	$72.76
2 Laborers	39.85	637.60	60.70	971.20		
4 Equip. Oper. (medium)	53.75	1720.00	81.05	2593.60		
1 Dozer, 200 H.P.		1273.00		1400.30		
1 Aggregate Spreader		36.80		40.48		
1 Tandem Roller, 10 Ton		231.40		254.54		
1 Roller, Pneum. Whl., 12 Ton		338.20		372.02	33.56	36.92
56 L.H., Daily Totals		$4571.80		$6142.14	$81.64	$109.68

Crew B-36B	Hr.	Daily	Hr.	Daily	Bare Costs	Incl. O&P
1 Labor Foreman (outside)	$41.85	$334.80	$63.75	$510.00	$47.82	$72.33
2 Laborers	39.85	637.60	60.70	971.20		
4 Equip. Oper. (medium)	53.75	1720.00	81.05	2593.60		
1 Truck Driver (heavy)	46.00	368.00	69.30	554.40		
1 Grader, 30,000 Lbs.		643.80		708.18		
1 F.E. Loader, Crl, 1.5 C.Y.		645.40		709.94		
1 Dozer, 300 H.P.		1829.00		2011.90		
1 Roller, Vibratory, 25 Ton		659.80		725.78		
1 Truck Tractor, 6x4, 450 H.P.		583.40		641.74		
1 Water Tank Trailer, 5000 Gal.		147.60		162.36	70.45	77.50
64 L.H., Daily Totals		$7569.40		$9589.10	$118.27	$149.83

For customer support on your Building Construction Costs with RSMeans data, call 800.448.8182.

753

Crew B-36C	Hr.	Daily	Hr.	Daily	Bare Costs	Incl. O&P
1 Labor Foreman (outside)	$41.85	$334.80	$63.75	$510.00	$49.82	$75.24
3 Equip. Oper. (medium)	53.75	1290.00	81.05	1945.20		
1 Truck Driver (heavy)	46.00	368.00	69.30	554.40		
1 Grader, 30,000 Lbs.		643.80		708.18		
1 Dozer, 300 H.P.		1829.00		2011.90		
1 Roller, Vibratory, 25 Ton		659.80		725.78		
1 Truck Tractor, 6x4, 450 H.P.		583.40		641.74		
1 Water Tank Trailer, 5000 Gal.		147.60		162.36	96.59	106.25
40 L.H., Daily Totals		$5856.40		$7259.56	$146.41	$181.49

Crew B-36D	Hr.	Daily	Hr.	Daily	Bare Costs	Incl. O&P
1 Labor Foreman (outside)	$41.85	$334.80	$63.75	$510.00	$50.77	$76.72
3 Equip. Oper. (medium)	53.75	1290.00	81.05	1945.20		
1 Grader, 30,000 Lbs.		643.80		708.18		
1 Dozer, 300 H.P.		1829.00		2011.90		
1 Roller, Vibratory, 25 Ton		659.80		725.78	97.89	107.68
32 L.H., Daily Totals		$4757.40		$5901.06	$148.67	$184.41

Crew B-37	Hr.	Daily	Hr.	Daily	Bare Costs	Incl. O&P
1 Labor Foreman (outside)	$41.85	$334.80	$63.75	$510.00	$42.09	$63.98
4 Laborers	39.85	1275.20	60.70	1942.40		
1 Equip. Oper. (light)	51.30	410.40	77.35	618.80		
1 Tandem Roller, 5 Ton		151.20		166.32	3.15	3.46
48 L.H., Daily Totals		$2171.60		$3237.52	$45.24	$67.45

Crew B-37A	Hr.	Daily	Hr.	Daily	Bare Costs	Incl. O&P
2 Laborers	$39.85	$637.60	$60.70	$971.20	$41.40	$62.80
1 Truck Driver (light)	44.50	356.00	67.00	536.00		
1 Flatbed Truck, Gas, 1.5 Ton		188.40		207.24		
1 Tar Kettle, T.M.		129.20		142.12	13.23	14.56
24 L.H., Daily Totals		$1311.20		$1856.56	$54.63	$77.36

Crew B-37B	Hr.	Daily	Hr.	Daily	Bare Costs	Incl. O&P
3 Laborers	$39.85	$956.40	$60.70	$1456.80	$41.01	$62.27
1 Truck Driver (light)	44.50	356.00	67.00	536.00		
1 Flatbed Truck, Gas, 1.5 Ton		188.40		207.24		
1 Tar Kettle, T.M.		129.20		142.12	9.93	10.92
32 L.H., Daily Totals		$1630.00		$2342.16	$50.94	$73.19

Crew B-37C	Hr.	Daily	Hr.	Daily	Bare Costs	Incl. O&P
2 Laborers	$39.85	$637.60	$60.70	$971.20	$42.17	$63.85
2 Truck Drivers (light)	44.50	712.00	67.00	1072.00		
2 Flatbed Trucks, Gas, 1.5 Ton		376.80		414.48		
1 Tar Kettle, T.M.		129.20		142.12	15.81	17.39
32 L.H., Daily Totals		$1855.60		$2599.80	$57.99	$81.24

Crew B-37D	Hr.	Daily	Hr.	Daily	Bare Costs	Incl. O&P
1 Laborer	$39.85	$318.80	$60.70	$485.60	$42.17	$63.85
1 Truck Driver (light)	44.50	356.00	67.00	536.00		
1 Pickup Truck, 3/4 Ton		115.20		126.72	7.20	7.92
16 L.H., Daily Totals		$790.00		$1148.32	$49.38	$71.77

Crew B-37E	Hr.	Daily	Hr.	Daily	Bare Costs	Incl. O&P
3 Laborers	$39.85	$956.40	$60.70	$1456.80	$44.80	$67.79
1 Equip. Oper. (light)	51.30	410.40	77.35	618.80		
1 Equip. Oper. (medium)	53.75	430.00	81.05	648.40		
2 Truck Drivers (light)	44.50	712.00	67.00	1072.00		
4 Barrels w/Flasher		15.60		17.16		
1 Concrete Saw		102.40		112.64		
1 Rotary Hammer Drill		25.15		27.66		
1 Hammer Drill Bit		3.25		3.58		
1 Loader, Skid Steer, 30 H.P.		174.00		191.40		
1 Conc. Hammer Attach.		113.60		124.96		
1 Vibrating Plate, Gas, 18"		31.80		34.98		
2 Flatbed Trucks, Gas, 1.5 Ton		376.80		414.48	15.05	16.55
56 L.H., Daily Totals		$3351.40		$4722.86	$59.85	$84.34

Crew B-37F	Hr.	Daily	Hr.	Daily	Bare Costs	Incl. O&P
3 Laborers	$39.85	$956.40	$60.70	$1456.80	$41.01	$62.27
1 Truck Driver (light)	44.50	356.00	67.00	536.00		
4 Barrels w/Flasher		15.60		17.16		
1 Concrete Mixer, 10 C.F.		161.00		177.10		
1 Air Compressor, 60 cfm		105.40		115.94		
1 -50' Air Hose, 3/4"		7.35		8.09		
1 Spade (Chipper)		8.40		9.24		
1 Flatbed Truck, Gas, 1.5 Ton		188.40		207.24	15.19	16.71
32 L.H., Daily Totals		$1798.55		$2527.57	$56.20	$78.99

Crew B-37G	Hr.	Daily	Hr.	Daily	Bare Costs	Incl. O&P
1 Labor Foreman (outside)	$41.85	$334.80	$63.75	$510.00	$42.09	$63.98
4 Laborers	39.85	1275.20	60.70	1942.40		
1 Equip. Oper. (light)	51.30	410.40	77.35	618.80		
1 Berm Machine		317.40		349.14		
1 Tandem Roller, 5 Ton		151.20		166.32	9.76	10.74
48 L.H., Daily Totals		$2489.00		$3586.66	$51.85	$74.72

Crew B-37H	Hr.	Daily	Hr.	Daily	Bare Costs	Incl. O&P
1 Labor Foreman (outside)	$41.85	$334.80	$63.75	$510.00	$42.09	$63.98
4 Laborers	39.85	1275.20	60.70	1942.40		
1 Equip. Oper. (light)	51.30	410.40	77.35	618.80		
1 Tandem Roller, 5 Ton		151.20		166.32		
1 Flatbed Truck, Gas, 1.5 Ton		188.40		207.24		
1 Tar Kettle, T.M.		129.20		142.12	9.77	10.74
48 L.H., Daily Totals		$2489.20		$3586.88	$51.86	$74.73

Crew B-37I	Hr.	Daily	Hr.	Daily	Bare Costs	Incl. O&P
3 Laborers	$39.85	$956.40	$60.70	$1456.80	$44.80	$67.79
1 Equip. Oper. (light)	51.30	410.40	77.35	618.80		
1 Equip. Oper. (medium)	53.75	430.00	81.05	648.40		
2 Truck Drivers (light)	44.50	712.00	67.00	1072.00		
4 Barrels w/Flasher		15.60		17.16		
1 Concrete Saw		102.40		112.64		
1 Rotary Hammer Drill		25.15		27.66		
1 Hammer Drill Bit		3.25		3.58		
1 Air Compressor, 60 cfm		105.40		115.94		
1 -50' Air Hose, 3/4"		7.35		8.09		
1 Spade (Chipper)		8.40		9.24		
1 Loader, Skid Steer, 30 H.P.		174.00		191.40		
1 Conc. Hammer Attach.		113.60		124.96		
1 Concrete Mixer, 10 C.F.		161.00		177.10		
1 Vibrating Plate, Gas, 18"		31.80		34.98		
2 Flatbed Trucks, Gas, 1.5 Ton		376.80		414.48	20.08	22.09
56 L.H., Daily Totals		$3633.55		$5033.23	$64.88	$89.88

For customer support on your Building Construction Costs with RSMeans data, call 800.448.8182.

| Crew No. | Bare Costs | | Incl. Subs O&P | | Cost Per Labor-Hour | |

Crew B-37J

Crew B-37J	Hr.	Daily	Hr.	Daily	Bare Costs	Incl. O&P
1 Labor Foreman (outside)	$41.85	$334.80	$63.75	$510.00	$42.09	$63.98
4 Laborers	39.85	1275.20	60.70	1942.40		
1 Equip. Oper. (light)	51.30	410.40	77.35	618.80		
1 Air Compressor, 60 cfm		105.40		115.94		
1 -50' Air Hose, 3/4"		7.35		8.09		
2 Concrete Mixers, 10 C.F.		322.00		354.20		
2 Flatbed Trucks, Gas, 1.5 Ton		376.80		414.48		
1 Shot Blaster, 20"		214.80		236.28	21.38	23.52
48 L.H., Daily Totals		$3046.75		$4200.19	$63.47	$87.50

Crew B-37K

Crew B-37K	Hr.	Daily	Hr.	Daily	Bare Costs	Incl. O&P
1 Labor Foreman (outside)	$41.85	$334.80	$63.75	$510.00	$42.09	$63.98
4 Laborers	39.85	1275.20	60.70	1942.40		
1 Equip. Oper. (light)	51.30	410.40	77.35	618.80		
1 Air Compressor, 60 cfm		105.40		115.94		
1 -50' Air Hose, 3/4"		7.35		8.09		
2 Flatbed Trucks, Gas, 1.5 Ton		376.80		414.48		
1 Shot Blaster, 20"		214.80		236.28	14.67	16.14
48 L.H., Daily Totals		$2724.75		$3845.99	$56.77	$80.12

Crew B-38

Crew B-38	Hr.	Daily	Hr.	Daily	Bare Costs	Incl. O&P
1 Labor Foreman (outside)	$41.85	$334.80	$63.75	$510.00	$45.32	$68.71
2 Laborers	39.85	637.60	60.70	971.20		
1 Equip. Oper. (light)	51.30	410.40	77.35	618.80		
1 Equip. Oper. (medium)	53.75	430.00	81.05	648.40		
1 Backhoe Loader, 48 H.P.		312.40		343.64		
1 Hyd. Hammer (1200 lb.)		177.40		195.14		
1 F.E. Loader, W.M., 4 C.Y.		574.60		632.06		
1 Pvmt. Rem. Bucket		61.00		67.10	28.14	30.95
40 L.H., Daily Totals		$2938.20		$3986.34	$73.45	$99.66

Crew B-39

Crew B-39	Hr.	Daily	Hr.	Daily	Bare Costs	Incl. O&P
1 Labor Foreman (outside)	$41.85	$334.80	$63.75	$510.00	$42.09	$63.98
4 Laborers	39.85	1275.20	60.70	1942.40		
1 Equip. Oper. (light)	51.30	410.40	77.35	618.80		
1 Air Compressor, 250 cfm		167.40		184.14		
2 Breakers, Pavement, 60 lb.		21.20		23.32		
2 -50' Air Hoses, 1.5"		45.40		49.94	4.88	5.36
48 L.H., Daily Totals		$2254.40		$3328.60	$46.97	$69.35

Crew B-40

Crew B-40	Hr.	Daily	Hr.	Daily	Bare Costs	Incl. O&P
1 Pile Driver Foreman (outside)	$53.30	$426.40	$84.05	$672.40	$52.41	$81.27
4 Pile Drivers	51.30	1641.60	80.90	2588.80		
2 Equip. Oper. (crane)	56.10	897.60	84.60	1353.60		
1 Equip. Oper. (oiler)	48.60	388.80	73.30	586.40		
1 Crawler Crane, 40 Ton		1370.00		1507.00		
1 Vibratory Hammer & Gen.		2191.00		2410.10	55.64	61.20
64 L.H., Daily Totals		$6915.40		$9118.30	$108.05	$142.47

Crew B-40B

Crew B-40B	Hr.	Daily	Hr.	Daily	Bare Costs	Incl. O&P
1 Labor Foreman (outside)	$41.85	$334.80	$63.75	$510.00	$44.35	$67.29
3 Laborers	39.85	956.40	60.70	1456.80		
1 Equip. Oper. (crane)	56.10	448.80	84.60	676.80		
1 Equip. Oper. (oiler)	48.60	388.80	73.30	586.40		
1 Lattice Boom Crane, 40 Ton		1321.00		1453.10	27.52	30.27
48 L.H., Daily Totals		$3449.80		$4683.10	$71.87	$97.56

Crew B-41

Crew B-41	Hr.	Daily	Hr.	Daily	Bare Costs	Incl. O&P
1 Labor Foreman (outside)	$41.85	$334.80	$63.75	$510.00	$41.35	$62.91
4 Laborers	39.85	1275.20	60.70	1942.40		
.25 Equip. Oper. (crane)	56.10	112.20	84.60	169.20		
.25 Equip. Oper. (oiler)	48.60	97.20	73.30	146.60		
.25 Crawler Crane, 40 Ton		342.50		376.75	7.78	8.56
44 L.H., Daily Totals		$2161.90		$3144.95	$49.13	$71.48

Crew B-42

Crew B-42	Hr.	Daily	Hr.	Daily	Bare Costs	Incl. O&P
1 Labor Foreman (outside)	$41.85	$334.80	$63.75	$510.00	$45.08	$69.22
4 Laborers	39.85	1275.20	60.70	1942.40		
1 Equip. Oper. (crane)	56.10	448.80	84.60	676.80		
1 Equip. Oper. (oiler)	48.60	388.80	73.30	586.40		
1 Welder	54.65	437.20	89.35	714.80		
1 Hyd. Crane, 25 Ton		590.60		649.66		
1 Welder, Gas Engine, 300 amp		98.40		108.24		
1 Horz. Boring Csg. Mch.		448.80		493.68	17.78	19.56
64 L.H., Daily Totals		$4022.60		$5681.98	$62.85	$88.78

Crew B-43

Crew B-43	Hr.	Daily	Hr.	Daily	Bare Costs	Incl. O&P
1 Labor Foreman (outside)	$41.85	$334.80	$63.75	$510.00	$44.35	$67.29
3 Laborers	39.85	956.40	60.70	1456.80		
1 Equip. Oper. (crane)	56.10	448.80	84.60	676.80		
1 Equip. Oper. (oiler)	48.60	388.80	73.30	586.40		
1 Drill Rig, Truck-Mounted		2444.00		2688.40	50.92	56.01
48 L.H., Daily Totals		$4572.80		$5918.40	$95.27	$123.30

Crew B-44

Crew B-44	Hr.	Daily	Hr.	Daily	Bare Costs	Incl. O&P
1 Pile Driver Foreman (outside)	$53.30	$426.40	$84.05	$672.40	$51.32	$79.69
4 Pile Drivers	51.30	1641.60	80.90	2588.80		
2 Equip. Oper. (crane)	56.10	897.60	84.60	1353.60		
1 Laborer	39.85	318.80	60.70	485.60		
1 Crawler Crane, 40 Ton		1370.00		1507.00		
1 Lead, 60' High		79.40		87.34		
1 Hammer, Diesel, 15K ft.-lbs.		623.20		685.52	32.38	35.62
64 L.H., Daily Totals		$5357.00		$7380.26	$83.70	$115.32

Crew B-45

Crew B-45	Hr.	Daily	Hr.	Daily	Bare Costs	Incl. O&P
1 Equip. Oper. (medium)	$53.75	$430.00	$81.05	$648.40	$49.88	$75.17
1 Truck Driver (heavy)	46.00	368.00	69.30	554.40		
1 Dist. Tanker, 3000 Gallon		326.80		359.48		
1 Truck Tractor, 6x4, 380 H.P.		476.20		523.82	50.19	55.21
16 L.H., Daily Totals		$1601.00		$2086.10	$100.06	$130.38

Crew B-46

Crew B-46	Hr.	Daily	Hr.	Daily	Bare Costs	Incl. O&P
1 Pile Driver Foreman (outside)	$53.30	$426.40	$84.05	$672.40	$45.91	$71.33
2 Pile Drivers	51.30	820.80	80.90	1294.40		
3 Laborers	39.85	956.40	60.70	1456.80		
1 Chain Saw, Gas, 36" Long		46.40		51.04	.97	1.06
48 L.H., Daily Totals		$2250.00		$3474.64	$46.88	$72.39

Crew B-47

Crew B-47	Hr.	Daily	Hr.	Daily	Bare Costs	Incl. O&P
1 Blast Foreman (outside)	$41.85	$334.80	$63.75	$510.00	$44.33	$67.27
1 Driller	39.85	318.80	60.70	485.60		
1 Equip. Oper. (light)	51.30	410.40	77.35	618.80		
1 Air Track Drill, 4"		993.40		1092.74		
1 Air Compressor, 600 cfm		417.20		458.92		
2 -50' Air Hoses, 3"		86.40		95.04	62.38	68.61
24 L.H., Daily Totals		$2561.00		$3261.10	$106.71	$135.88

For customer support on your Building Construction Costs with RSMeans data, call 800.448.8182.

755

Crew B-47A

Crew No.	Bare Costs Hr.	Daily	Incl. Subs O&P Hr.	Daily	Cost Per Labor-Hour Bare Costs	Incl. O&P
1 Drilling Foreman (outside)	$41.85	$334.80	$63.75	$510.00	$48.85	$73.88
1 Equip. Oper. (heavy)	56.10	448.80	84.60	676.80		
1 Equip. Oper. (oiler)	48.60	388.80	73.30	586.40		
1 Air Track Drill, 5"		1173.00		1290.30	48.88	53.76
24 L.H., Daily Totals		$2345.40		$3063.50	$97.72	$127.65

Crew B-47C

Crew No.	Bare Costs Hr.	Daily	Incl. Subs O&P Hr.	Daily	Cost Per Labor-Hour Bare Costs	Incl. O&P
1 Laborer	$39.85	$318.80	$60.70	$485.60	$45.58	$69.03
1 Equip. Oper. (light)	51.30	410.40	77.35	618.80		
1 Air Compressor, 750 cfm		423.80		466.18		
2 -50' Air Hoses, 3"		86.40		95.04		
1 Air Track Drill, 4"		993.40		1092.74	93.97	103.37
16 L.H., Daily Totals		$2232.80		$2758.36	$139.55	$172.40

Crew B-47E

Crew No.	Bare Costs Hr.	Daily	Incl. Subs O&P Hr.	Daily	Cost Per Labor-Hour Bare Costs	Incl. O&P
1 Labor Foreman (outside)	$41.85	$334.80	$63.75	$510.00	$40.35	$61.46
3 Laborers	39.85	956.40	60.70	1456.80		
1 Flatbed Truck, Gas, 3 Ton		238.00		261.80	7.44	8.18
32 L.H., Daily Totals		$1529.20		$2228.60	$47.79	$69.64

Crew B-47G

Crew No.	Bare Costs Hr.	Daily	Incl. Subs O&P Hr.	Daily	Cost Per Labor-Hour Bare Costs	Incl. O&P
1 Labor Foreman (outside)	$41.85	$334.80	$63.75	$510.00	$43.21	$65.63
2 Laborers	39.85	637.60	60.70	971.20		
1 Equip. Oper. (light)	51.30	410.40	77.35	618.80		
1 Air Track Drill, 4"		993.40		1092.74		
1 Air Compressor, 600 cfm		417.20		458.92		
2 -50' Air Hoses, 3"		86.40		95.04		
1 Gunite Pump Rig		321.00		353.10	56.81	62.49
32 L.H., Daily Totals		$3200.80		$4099.80	$100.03	$128.12

Crew B-47H

Crew No.	Bare Costs Hr.	Daily	Incl. Subs O&P Hr.	Daily	Cost Per Labor-Hour Bare Costs	Incl. O&P
1 Skilled Worker Foreman (out)	$54.35	$434.80	$83.20	$665.60	$52.85	$80.91
3 Skilled Workers	52.35	1256.40	80.15	1923.60		
1 Flatbed Truck, Gas, 3 Ton		238.00		261.80	7.44	8.18
32 L.H., Daily Totals		$1929.20		$2851.00	$60.29	$89.09

Crew B-48

Crew No.	Bare Costs Hr.	Daily	Incl. Subs O&P Hr.	Daily	Cost Per Labor-Hour Bare Costs	Incl. O&P
1 Labor Foreman (outside)	$41.85	$334.80	$63.75	$510.00	$45.34	$68.73
3 Laborers	39.85	956.40	60.70	1456.80		
1 Equip. Oper. (crane)	56.10	448.80	84.60	676.80		
1 Equip. Oper. (oiler)	48.60	388.80	73.30	586.40		
1 Equip. Oper. (light)	51.30	410.40	77.35	618.80		
1 Centr. Water Pump, 6"		296.40		326.04		
1 -20' Suction Hose, 6"		11.50		12.65		
1 -50' Discharge Hose, 6"		6.10		6.71		
1 Drill Rig, Truck-Mounted		2444.00		2688.40	49.25	54.17
56 L.H., Daily Totals		$5297.20		$6882.60	$94.59	$122.90

Crew B-49

Crew No.	Bare Costs Hr.	Daily	Incl. Subs O&P Hr.	Daily	Cost Per Labor-Hour Bare Costs	Incl. O&P
1 Labor Foreman (outside)	$41.85	$334.80	$63.75	$510.00	$47.70	$72.80
3 Laborers	39.85	956.40	60.70	1456.80		
2 Equip. Oper. (crane)	56.10	897.60	84.60	1353.60		
2 Equip. Oper. (oilers)	48.60	777.60	73.30	1172.80		
1 Equip. Oper. (light)	51.30	410.40	77.35	618.80		
2 Pile Drivers	51.30	820.80	80.90	1294.40		
1 Hyd. Crane, 25 Ton		590.60		649.66		
1 Centr. Water Pump, 6"		296.40		326.04		
1 -20' Suction Hose, 6"		11.50		12.65		
1 -50' Discharge Hose, 6"		6.10		6.71		
1 Drill Rig, Truck-Mounted		2444.00		2688.40	38.05	41.86
88 L.H., Daily Totals		$7546.20		$10089.86	$85.75	$114.66

Crew B-50

Crew No.	Bare Costs Hr.	Daily	Incl. Subs O&P Hr.	Daily	Cost Per Labor-Hour Bare Costs	Incl. O&P
2 Pile Driver Foremen (outside)	$53.30	$852.80	$84.05	$1344.80	$49.63	$77.01
6 Pile Drivers	51.30	2462.40	80.90	3883.20		
2 Equip. Oper. (crane)	56.10	897.60	84.60	1353.60		
1 Equip. Oper. (oiler)	48.60	388.80	73.30	586.40		
3 Laborers	39.85	956.40	60.70	1456.80		
1 Crawler Crane, 40 Ton		1370.00		1507.00		
1 Lead, 60' High		79.40		87.34		
1 Hammer, Diesel, 15K ft.-lbs.		623.20		685.52		
1 Air Compressor, 600 cfm		417.20		458.92		
2 -50' Air Hoses, 3"		86.40		95.04		
1 Chain Saw, Gas, 36" Long		46.40		51.04	23.42	25.76
112 L.H., Daily Totals		$8180.60		$11509.66	$73.04	$102.76

Crew B-51

Crew No.	Bare Costs Hr.	Daily	Incl. Subs O&P Hr.	Daily	Cost Per Labor-Hour Bare Costs	Incl. O&P
1 Labor Foreman (outside)	$41.85	$334.80	$63.75	$510.00	$40.96	$62.26
4 Laborers	39.85	1275.20	60.70	1942.40		
1 Truck Driver (light)	44.50	356.00	67.00	536.00		
1 Flatbed Truck, Gas, 1.5 Ton		188.40		207.24	3.92	4.32
48 L.H., Daily Totals		$2154.40		$3195.64	$44.88	$66.58

Crew B-52

Crew No.	Bare Costs Hr.	Daily	Incl. Subs O&P Hr.	Daily	Cost Per Labor-Hour Bare Costs	Incl. O&P
1 Carpenter Foreman (outside)	$52.70	$421.60	$80.25	$642.00	$46.39	$70.32
1 Carpenter	50.70	405.60	77.20	617.60		
3 Laborers	39.85	956.40	60.70	1456.80		
1 Cement Finisher	47.55	380.40	70.45	563.60		
.5 Rodman (reinf.)	54.65	218.60	83.45	333.80		
.5 Equip. Oper. (medium)	53.75	215.00	81.05	324.20		
.5 Crawler Loader, 3 C.Y.		602.00		662.20	10.75	11.82
56 L.H., Daily Totals		$3199.60		$4600.20	$57.14	$82.15

Crew B-53

Crew No.	Bare Costs Hr.	Daily	Incl. Subs O&P Hr.	Daily	Cost Per Labor-Hour Bare Costs	Incl. O&P
1 Equip. Oper. (light)	$51.30	$410.40	$77.35	$618.80	$51.30	$77.35
1 Trencher, Chain, 12 H.P.		61.80		67.98	7.72	8.50
8 L.H., Daily Totals		$472.20		$686.78	$59.02	$85.85

Crew B-54

Crew No.	Bare Costs Hr.	Daily	Incl. Subs O&P Hr.	Daily	Cost Per Labor-Hour Bare Costs	Incl. O&P
1 Equip. Oper. (light)	$51.30	$410.40	$77.35	$618.80	$51.30	$77.35
1 Trencher, Chain, 40 H.P.		336.40		370.04	42.05	46.26
8 L.H., Daily Totals		$746.80		$988.84	$93.35	$123.61

Crew B-54A

Crew No.	Bare Costs Hr.	Daily	Incl. Subs O&P Hr.	Daily	Cost Per Labor-Hour Bare Costs	Incl. O&P
.17 Labor Foreman (outside)	$41.85	$56.92	$63.75	$86.70	$52.02	$78.54
1 Equipment Operator (med.)	53.75	430.00	81.05	648.40		
1 Wheel Trencher, 67 H.P.		1095.00		1204.50	116.99	128.69
9.36 L.H., Daily Totals		$1581.92		$1939.60	$169.01	$207.22

Crew B-54B

Crew No.	Bare Costs Hr.	Daily	Incl. Subs O&P Hr.	Daily	Cost Per Labor-Hour Bare Costs	Incl. O&P
.25 Labor Foreman (outside)	$41.85	$83.70	$63.75	$127.50	$51.37	$77.59
1 Equipment Operator (med.)	53.75	430.00	81.05	648.40		
1 Wheel Trencher, 150 H.P.		1957.00		2152.70	195.70	215.27
10 L.H., Daily Totals		$2470.70		$2928.60	$247.07	$292.86

Crew B-54C

Crew No.	Bare Costs Hr.	Daily	Incl. Subs O&P Hr.	Daily	Cost Per Labor-Hour Bare Costs	Incl. O&P
1 Laborer	$39.85	$318.80	$60.70	$485.60	$46.80	$70.88
1 Equipment Operator (med.)	53.75	430.00	81.05	648.40		
1 Wheel Trencher, 67 H.P.		1095.00		1204.50	68.44	75.28
16 L.H., Daily Totals		$1843.80		$2338.50	$115.24	$146.16

For customer support on your Building Construction Costs with RSMeans data, call 800.448.8182.

Crew No.	Bare Costs		Incl. Subs O&P		Cost Per Labor-Hour	
	Hr.	Daily	Hr.	Daily	Bare Costs	Incl. O&P
Crew B-54D						
1 Laborer	$39.85	$318.80	$60.70	$485.60	$46.80	$70.88
1 Equipment Operator (med.)	53.75	430.00	81.05	648.40		
1 Rock Trencher, 6" Width		904.20		994.62	56.51	62.16
16 L.H., Daily Totals		$1653.00		$2128.62	$103.31	$133.04
Crew B-54E	Hr.	Daily	Hr.	Daily	Bare Costs	Incl. O&P
1 Laborer	$39.85	$318.80	$60.70	$485.60	$46.80	$70.88
1 Equipment Operator (med.)	53.75	430.00	81.05	648.40		
1 Rock Trencher, 18" Width		2681.00		2949.10	167.56	184.32
16 L.H., Daily Totals		$3429.80		$4083.10	$214.36	$255.19
Crew B-55	Hr.	Daily	Hr.	Daily	Bare Costs	Incl. O&P
2 Laborers	$39.85	$637.60	$60.70	$971.20	$41.40	$62.80
1 Truck Driver (light)	44.50	356.00	67.00	536.00		
1 Truck-Mounted Earth Auger		758.80		834.68		
1 Flatbed Truck, Gas, 3 Ton		238.00		261.80	41.53	45.69
24 L.H., Daily Totals		$1990.40		$2603.68	$82.93	$108.49
Crew B-56	Hr.	Daily	Hr.	Daily	Bare Costs	Incl. O&P
1 Laborer	$39.85	$318.80	$60.70	$485.60	$45.58	$69.03
1 Equip. Oper. (light)	51.30	410.40	77.35	618.80		
1 Air Track Drill, 4"		993.40		1092.74		
1 Air Compressor, 600 cfm		417.20		458.92		
1 -50' Air Hose, 3"		43.20		47.52	90.86	99.95
16 L.H., Daily Totals		$2183.00		$2703.58	$136.44	$168.97
Crew B-57	Hr.	Daily	Hr.	Daily	Bare Costs	Incl. O&P
1 Labor Foreman (outside)	$41.85	$334.80	$63.75	$510.00	$46.26	$70.07
2 Laborers	39.85	637.60	60.70	971.20		
1 Equip. Oper. (crane)	56.10	448.80	84.60	676.80		
1 Equip. Oper. (light)	51.30	410.40	77.35	618.80		
1 Equip. Oper. (oiler)	48.60	388.80	73.30	586.40		
1 Crawler Crane, 25 Ton		1348.00		1482.80		
1 Clamshell Bucket, 1 C.Y.		50.60		55.66		
1 Centr. Water Pump, 6"		296.40		326.04		
1 -20' Suction Hose, 6"		11.50		12.65		
20 -50' Discharge Hoses, 6"		122.00		134.20	38.09	41.90
48 L.H., Daily Totals		$4048.90		$5374.55	$84.35	$111.97
Crew B-58	Hr.	Daily	Hr.	Daily	Bare Costs	Incl. O&P
2 Laborers	$39.85	$637.60	$60.70	$971.20	$43.67	$66.25
1 Equip. Oper. (light)	51.30	410.40	77.35	618.80		
1 Backhoe Loader, 48 H.P.		312.40		343.64		
1 Small Helicopter, w/ Pilot		2857.00		3142.70	132.06	145.26
24 L.H., Daily Totals		$4217.40		$5076.34	$175.72	$211.51
Crew B-59	Hr.	Daily	Hr.	Daily	Bare Costs	Incl. O&P
1 Truck Driver (heavy)	$46.00	$368.00	$69.30	$554.40	$46.00	$69.30
1 Truck Tractor, 220 H.P.		290.00		319.00		
1 Water Tank Trailer, 5000 Gal.		147.60		162.36	54.70	60.17
8 L.H., Daily Totals		$805.60		$1035.76	$100.70	$129.47
Crew B-59A	Hr.	Daily	Hr.	Daily	Bare Costs	Incl. O&P
2 Laborers	$39.85	$637.60	$60.70	$971.20	$41.90	$63.57
1 Truck Driver (heavy)	46.00	368.00	69.30	554.40		
1 Water Tank Trailer, 5000 Gal.		147.60		162.36		
1 Truck Tractor, 220 H.P.		290.00		319.00	18.23	20.06
24 L.H., Daily Totals		$1443.20		$2006.96	$60.13	$83.62

Crew No.	Bare Costs		Incl. Subs O&P		Cost Per Labor-Hour	
	Hr.	Daily	Hr.	Daily	Bare Costs	Incl. O&P
Crew B-60						
1 Labor Foreman (outside)	$41.85	$334.80	$63.75	$510.00	$46.98	$71.11
2 Laborers	39.85	637.60	60.70	971.20		
1 Equip. Oper. (crane)	56.10	448.80	84.60	676.80		
2 Equip. Oper. (light)	51.30	820.80	77.35	1237.60		
1 Equip. Oper. (oiler)	48.60	388.80	73.30	586.40		
1 Crawler Crane, 40 Ton		1370.00		1507.00		
1 Lead, 60' High		79.40		87.34		
1 Hammer, Diesel, 15K ft.-lbs.		623.20		685.52		
1 Backhoe Loader, 48 H.P.		312.40		343.64	42.59	46.85
56 L.H., Daily Totals		$5015.80		$6605.50	$89.57	$117.96
Crew B-61	Hr.	Daily	Hr.	Daily	Bare Costs	Incl. O&P
1 Labor Foreman (outside)	$41.85	$334.80	$63.75	$510.00	$42.54	$64.64
3 Laborers	39.85	956.40	60.70	1456.80		
1 Equip. Oper. (light)	51.30	410.40	77.35	618.80		
1 Cement Mixer, 2 C.Y.		204.20		224.62		
1 Air Compressor, 160 cfm		117.20		128.92	8.04	8.84
40 L.H., Daily Totals		$2023.00		$2939.14	$50.58	$73.48
Crew B-62	Hr.	Daily	Hr.	Daily	Bare Costs	Incl. O&P
2 Laborers	$39.85	$637.60	$60.70	$971.20	$43.67	$66.25
1 Equip. Oper. (light)	51.30	410.40	77.35	618.80		
1 Loader, Skid Steer, 30 H.P.		174.00		191.40	7.25	7.97
24 L.H., Daily Totals		$1222.00		$1781.40	$50.92	$74.22
Crew B-62A	Hr.	Daily	Hr.	Daily	Bare Costs	Incl. O&P
2 Laborers	$39.85	$637.60	$60.70	$971.20	$43.67	$66.25
1 Equip. Oper. (light)	51.30	410.40	77.35	618.80		
1 Loader, Skid Steer, 30 H.P.		174.00		191.40		
1 Trencher Attachment		74.30		81.73	10.35	11.38
24 L.H., Daily Totals		$1296.30		$1863.13	$54.01	$77.63
Crew B-63	Hr.	Daily	Hr.	Daily	Bare Costs	Incl. O&P
4 Laborers	$39.85	$1275.20	$60.70	$1942.40	$42.14	$64.03
1 Equip. Oper. (light)	51.30	410.40	77.35	618.80		
1 Loader, Skid Steer, 30 H.P.		174.00		191.40	4.35	4.79
40 L.H., Daily Totals		$1859.60		$2752.60	$46.49	$68.81
Crew B-63B	Hr.	Daily	Hr.	Daily	Bare Costs	Incl. O&P
1 Labor Foreman (inside)	$40.35	$322.80	$61.45	$491.60	$42.84	$65.05
2 Laborers	39.85	637.60	60.70	971.20		
1 Equip. Oper. (light)	51.30	410.40	77.35	618.80		
1 Loader, Skid Steer, 78 H.P.		364.80		401.28	11.40	12.54
32 L.H., Daily Totals		$1735.60		$2482.88	$54.24	$77.59
Crew B-64	Hr.	Daily	Hr.	Daily	Bare Costs	Incl. O&P
1 Laborer	$39.85	$318.80	$60.70	$485.60	$42.17	$63.85
1 Truck Driver (light)	44.50	356.00	67.00	536.00		
1 Power Mulcher (small)		139.20		153.12		
1 Flatbed Truck, Gas, 1.5 Ton		188.40		207.24	20.48	22.52
16 L.H., Daily Totals		$1002.40		$1381.96	$62.65	$86.37
Crew B-65	Hr.	Daily	Hr.	Daily	Bare Costs	Incl. O&P
1 Laborer	$39.85	$318.80	$60.70	$485.60	$42.17	$63.85
1 Truck Driver (light)	44.50	356.00	67.00	536.00		
1 Power Mulcher (Large)		268.80		295.68		
1 Flatbed Truck, Gas, 1.5 Ton		188.40		207.24	28.57	31.43
16 L.H., Daily Totals		$1132.00		$1524.52	$70.75	$95.28

For customer support on your Building Construction Costs with RSMeans data, call 800.448.8182.

Crews - Standard

Crew B-66	Hr.	Daily	Hr.	Daily	Bare Costs	Incl. O&P
1 Equip. Oper. (light)	$51.30	$410.40	$77.35	$618.80	$51.30	$77.35
1 Loader-Backhoe, 40 H.P.		247.00		271.70	30.88	33.96
8 L.H., Daily Totals		$657.40		$890.50	$82.17	$111.31

Crew B-67	Hr.	Daily	Hr.	Daily	Bare Costs	Incl. O&P
1 Millwright	$52.75	$422.00	$76.95	$615.60	$52.02	$77.15
1 Equip. Oper. (light)	51.30	410.40	77.35	618.80		
1 R.T. Forklift, 5,000 Lb., diesel		267.60		294.36	16.73	18.40
16 L.H., Daily Totals		$1100.00		$1528.76	$68.75	$95.55

Crew B-67B	Hr.	Daily	Hr.	Daily	Bare Costs	Incl. O&P
1 Millwright Foreman (inside)	$53.25	$426.00	$77.70	$621.60	$53.00	$77.33
1 Millwright	52.75	422.00	76.95	615.60		
16 L.H., Daily Totals		$848.00		$1237.20	$53.00	$77.33

Crew B-68	Hr.	Daily	Hr.	Daily	Bare Costs	Incl. O&P
2 Millwrights	$52.75	$844.00	$76.95	$1231.20	$52.27	$77.08
1 Equip. Oper. (light)	51.30	410.40	77.35	618.80		
1 R.T. Forklift, 5,000 Lb., diesel		267.60		294.36	11.15	12.27
24 L.H., Daily Totals		$1522.00		$2144.36	$63.42	$89.35

Crew B-68A	Hr.	Daily	Hr.	Daily	Bare Costs	Incl. O&P
1 Millwright Foreman (inside)	$53.25	$426.00	$77.70	$621.60	$52.92	$77.20
2 Millwrights	52.75	844.00	76.95	1231.20		
1 Forklift, Smooth Floor, 8,000 Lb.		148.20		163.02	6.17	6.79
24 L.H., Daily Totals		$1418.20		$2015.82	$59.09	$83.99

Crew B-68B	Hr.	Daily	Hr.	Daily	Bare Costs	Incl. O&P
1 Millwright Foreman (inside)	$53.25	$426.00	$77.70	$621.60	$57.06	$84.69
2 Millwrights	52.75	844.00	76.95	1231.20		
2 Electricians	58.20	931.20	87.00	1392.00		
2 Plumbers	62.15	994.40	93.60	1497.60		
1 R.T. Forklift, 5,000 Lb., gas		279.00		306.90	4.98	5.48
56 L.H., Daily Totals		$3474.60		$5049.30	$62.05	$90.17

Crew B-68C	Hr.	Daily	Hr.	Daily	Bare Costs	Incl. O&P
1 Millwright Foreman (inside)	$53.25	$426.00	$77.70	$621.60	$56.59	$83.81
1 Millwright	52.75	422.00	76.95	615.60		
1 Electrician	58.20	465.60	87.00	696.00		
1 Plumber	62.15	497.20	93.60	748.80		
1 R.T. Forklift, 5,000 Lb., gas		279.00		306.90	8.72	9.59
32 L.H., Daily Totals		$2089.80		$2988.90	$65.31	$93.40

Crew B-68D	Hr.	Daily	Hr.	Daily	Bare Costs	Incl. O&P
1 Labor Foreman (inside)	$40.35	$322.80	$61.45	$491.60	$43.83	$66.50
1 Laborer	39.85	318.80	60.70	485.60		
1 Equip. Oper. (light)	51.30	410.40	77.35	618.80		
1 R.T. Forklift, 5,000 Lb., gas		279.00		306.90	11.63	12.79
24 L.H., Daily Totals		$1331.00		$1902.90	$55.46	$79.29

Crew B-68E	Hr.	Daily	Hr.	Daily	Bare Costs	Incl. O&P
1 Struc. Steel Foreman (inside)	$55.15	$441.20	$90.15	$721.20	$54.75	$89.51
3 Struc. Steel Workers	54.65	1311.60	89.35	2144.40		
1 Welder	54.65	437.20	89.35	714.80		
1 Forklift, Smooth Floor, 8,000 Lb.		148.20		163.02	3.71	4.08
40 L.H., Daily Totals		$2338.20		$3743.42	$58.45	$93.59

Crew B-68F	Hr.	Daily	Hr.	Daily	Bare Costs	Incl. O&P
1 Skilled Worker Foreman (out)	$54.35	$434.80	$83.20	$665.60	$53.02	$81.17
2 Skilled Workers	52.35	837.60	80.15	1282.40		
1 R.T. Forklift, 5,000 Lb., gas		279.00		306.90	11.63	12.79
24 L.H., Daily Totals		$1551.40		$2254.90	$64.64	$93.95

Crew B-68G	Hr.	Daily	Hr.	Daily	Bare Costs	Incl. O&P
2 Structural Steel Workers	$54.65	$874.40	$89.35	$1429.60	$54.65	$89.35
1 R.T. Forklift, 5,000 Lb., gas		279.00		306.90	17.44	19.18
16 L.H., Daily Totals		$1153.40		$1736.50	$72.09	$108.53

Crew B-69	Hr.	Daily	Hr.	Daily	Bare Costs	Incl. O&P
1 Labor Foreman (outside)	$41.85	$334.80	$63.75	$510.00	$44.35	$67.29
3 Laborers	39.85	956.40	60.70	1456.80		
1 Equip. Oper. (crane)	56.10	448.80	84.60	676.80		
1 Equip. Oper. (oiler)	48.60	388.80	73.30	586.40		
1 Hyd. Crane, 80 Ton		1518.00		1669.80	31.63	34.79
48 L.H., Daily Totals		$3646.80		$4899.80	$75.97	$102.08

Crew B-69A	Hr.	Daily	Hr.	Daily	Bare Costs	Incl. O&P
1 Labor Foreman (outside)	$41.85	$334.80	$63.75	$510.00	$43.78	$66.22
3 Laborers	39.85	956.40	60.70	1456.80		
1 Equip. Oper. (medium)	53.75	430.00	81.05	648.40		
1 Concrete Finisher	47.55	380.40	70.45	563.60		
1 Curb/Gutter Paver, 2-Track		1173.00		1290.30	24.44	26.88
48 L.H., Daily Totals		$3274.60		$4469.10	$68.22	$93.11

Crew B-69B	Hr.	Daily	Hr.	Daily	Bare Costs	Incl. O&P
1 Labor Foreman (outside)	$41.85	$334.80	$63.75	$510.00	$43.78	$66.22
3 Laborers	39.85	956.40	60.70	1456.80		
1 Equip. Oper. (medium)	53.75	430.00	81.05	648.40		
1 Cement Finisher	47.55	380.40	70.45	563.60		
1 Curb/Gutter Paver, 4-Track		789.60		868.56	16.45	18.09
48 L.H., Daily Totals		$2891.20		$4047.36	$60.23	$84.32

Crew B-70	Hr.	Daily	Hr.	Daily	Bare Costs	Incl. O&P
1 Labor Foreman (outside)	$41.85	$334.80	$63.75	$510.00	$46.09	$69.86
3 Laborers	39.85	956.40	60.70	1456.80		
3 Equip. Oper. (medium)	53.75	1290.00	81.05	1945.20		
1 Grader, 30,000 Lbs.		643.80		708.18		
1 Ripper, Beam & 1 Shank		87.80		96.58		
1 Road Sweeper, S.P., 8' wide		688.60		757.46		
1 F.E. Loader, W.M., 1.5 C.Y.		342.80		377.08	31.48	34.63
56 L.H., Daily Totals		$4344.20		$5851.30	$77.58	$104.49

Crew B-70A	Hr.	Daily	Hr.	Daily	Bare Costs	Incl. O&P
1 Laborer	$39.85	$318.80	$60.70	$485.60	$50.97	$76.98
4 Equip. Oper. (medium)	53.75	1720.00	81.05	2593.60		
1 Grader, 40,000 Lbs.		1191.00		1310.10		
1 F.E. Loader, W.M., 2.5 C.Y.		523.20		575.52		
1 Dozer, 80 H.P.		464.40		510.84		
1 Roller, Pneum. Whl., 12 Ton		338.20		372.02	62.92	69.21
40 L.H., Daily Totals		$4555.60		$5847.68	$113.89	$146.19

Crew B-71	Hr.	Daily	Hr.	Daily	Bare Costs	Incl. O&P
1 Labor Foreman (outside)	$41.85	$334.80	$63.75	$510.00	$46.09	$69.86
3 Laborers	39.85	956.40	60.70	1456.80		
3 Equip. Oper. (medium)	53.75	1290.00	81.05	1945.20		
1 Pvmt. Profiler, 750 H.P.		5419.00		5960.90		
1 Road Sweeper, S.P., 8' wide		688.60		757.46		
1 F.E. Loader, W.M., 1.5 C.Y.		342.80		377.08	115.19	126.70
56 L.H., Daily Totals		$9031.60		$11007.44	$161.28	$196.56

For customer support on your Building Construction Costs with RSMeans data, call 800.448.8182.

Crew B-72

Crew No.	Bare Costs Hr.	Bare Costs Daily	Incl. Subs O&P Hr.	Incl. Subs O&P Daily	Cost Per Labor-Hour Bare Costs	Cost Per Labor-Hour Incl. O&P
1 Labor Foreman (outside)	$41.85	$334.80	$63.75	$510.00	$47.05	$71.26
3 Laborers	39.85	956.40	60.70	1456.80		
4 Equip. Oper. (medium)	53.75	1720.00	81.05	2593.60		
1 Pvmt. Profiler, 750 H.P.		5419.00		5960.90		
1 Hammermill, 250 H.P.		1870.00		2057.00		
1 Windrow Loader		1242.00		1366.20		
1 Mix Paver, 165 H.P.		2140.00		2354.00		
1 Roller, Pneum. Whl., 12 Ton		338.20		372.02	172.02	189.22
64 L.H., Daily Totals		$14020.40		$16670.52	$219.07	$260.48

Crew B-73

Crew No.	Bare Costs Hr.	Bare Costs Daily	Incl. Subs O&P Hr.	Incl. Subs O&P Daily	Cost Per Labor-Hour Bare Costs	Cost Per Labor-Hour Incl. O&P
1 Labor Foreman (outside)	$41.85	$334.80	$63.75	$510.00	$48.79	$73.80
2 Laborers	39.85	637.60	60.70	971.20		
5 Equip. Oper. (medium)	53.75	2150.00	81.05	3242.00		
1 Road Mixer, 310 H.P.		1889.00		2077.90		
1 Tandem Roller, 10 Ton		231.40		254.54		
1 Hammermill, 250 H.P.		1870.00		2057.00		
1 Grader, 30,000 Lbs.		643.80		708.18		
.5 F.E. Loader, W.M., 1.5 C.Y.		171.40		188.54		
.5 Truck Tractor, 220 H.P.		145.00		159.50		
.5 Water Tank Trailer, 5000 Gal.		73.80		81.18	78.51	86.36
64 L.H., Daily Totals		$8146.80		$10250.04	$127.29	$160.16

Crew B-74

Crew No.	Bare Costs Hr.	Bare Costs Daily	Incl. Subs O&P Hr.	Incl. Subs O&P Daily	Cost Per Labor-Hour Bare Costs	Cost Per Labor-Hour Incl. O&P
1 Labor Foreman (outside)	$41.85	$334.80	$63.75	$510.00	$48.59	$73.41
1 Laborer	39.85	318.80	60.70	485.60		
4 Equip. Oper. (medium)	53.75	1720.00	81.05	2593.60		
2 Truck Drivers (heavy)	46.00	736.00	69.30	1108.80		
1 Grader, 30,000 Lbs.		643.80		708.18		
1 Ripper, Beam & 1 Shank		87.80		96.58		
2 Stabilizers, 310 H.P.		3436.00		3779.60		
1 Flatbed Truck, Gas, 3 Ton		238.00		261.80		
1 Chem. Spreader, Towed		52.20		57.42		
1 Roller, Vibratory, 25 Ton		659.80		725.78		
1 Water Tank Trailer, 5000 Gal.		147.60		162.36		
1 Truck Tractor, 220 H.P.		290.00		319.00	86.80	95.48
64 L.H., Daily Totals		$8664.80		$10808.72	$135.39	$168.89

Crew B-75

Crew No.	Bare Costs Hr.	Bare Costs Daily	Incl. Subs O&P Hr.	Incl. Subs O&P Daily	Cost Per Labor-Hour Bare Costs	Cost Per Labor-Hour Incl. O&P
1 Labor Foreman (outside)	$41.85	$334.80	$63.75	$510.00	$48.96	$73.99
1 Laborer	39.85	318.80	60.70	485.60		
4 Equip. Oper. (medium)	53.75	1720.00	81.05	2593.60		
1 Truck Driver (heavy)	46.00	368.00	69.30	554.40		
1 Grader, 30,000 Lbs.		643.80		708.18		
1 Ripper, Beam & 1 Shank		87.80		96.58		
2 Stabilizers, 310 H.P.		3436.00		3779.60		
1 Dist. Tanker, 3000 Gallon		326.80		359.48		
1 Truck Tractor, 6x4, 380 H.P.		476.20		523.82		
1 Roller, Vibratory, 25 Ton		659.80		725.78	100.54	110.60
56 L.H., Daily Totals		$8372.00		$10337.04	$149.50	$184.59

Crew B-76

Crew No.	Bare Costs Hr.	Bare Costs Daily	Incl. Subs O&P Hr.	Incl. Subs O&P Daily	Cost Per Labor-Hour Bare Costs	Cost Per Labor-Hour Incl. O&P
1 Dock Builder Foreman (outside)	$53.30	$426.40	$84.05	$672.40	$52.29	$81.23
5 Dock Builders	51.30	2052.00	80.90	3236.00		
2 Equip. Oper. (crane)	56.10	897.60	84.60	1353.60		
1 Equip. Oper. (oiler)	48.60	388.80	73.30	586.40		
1 Crawler Crane, 50 Ton		1940.00		2134.00		
1 Barge, 400 Ton		835.40		918.94		
1 Hammer, Diesel, 15K ft.-lbs.		623.20		685.52		
1 Lead, 60' High		79.40		87.34		
1 Air Compressor, 600 cfm		417.20		458.92		
2 -50' Air Hoses, 3"		86.40		95.04	55.30	60.83
72 L.H., Daily Totals		$7746.40		$10228.16	$107.59	$142.06

Crew B-76A

Crew No.	Bare Costs Hr.	Bare Costs Daily	Incl. Subs O&P Hr.	Incl. Subs O&P Daily	Cost Per Labor-Hour Bare Costs	Cost Per Labor-Hour Incl. O&P
1 Labor Foreman (outside)	$41.85	$334.80	$63.75	$510.00	$43.23	$65.64
5 Laborers	39.85	1594.00	60.70	2428.00		
1 Equip. Oper. (crane)	56.10	448.80	84.60	676.80		
1 Equip. Oper. (oiler)	48.60	388.80	73.30	586.40		
1 Crawler Crane, 50 Ton		1940.00		2134.00		
1 Barge, 400 Ton		835.40		918.94	43.37	47.70
64 L.H., Daily Totals		$5541.80		$7254.14	$86.59	$113.35

Crew B-77

Crew No.	Bare Costs Hr.	Bare Costs Daily	Incl. Subs O&P Hr.	Incl. Subs O&P Daily	Cost Per Labor-Hour Bare Costs	Cost Per Labor-Hour Incl. O&P
1 Labor Foreman (outside)	$41.85	$334.80	$63.75	$510.00	$41.18	$62.57
3 Laborers	39.85	956.40	60.70	1456.80		
1 Truck Driver (light)	44.50	356.00	67.00	536.00		
1 Crack Cleaner, 25 H.P.		55.80		61.38		
1 Crack Filler, Trailer Mtd.		198.20		218.02		
1 Flatbed Truck, Gas, 3 Ton		238.00		261.80	12.30	13.53
40 L.H., Daily Totals		$2139.20		$3044.00	$53.48	$76.10

Crew B-78

Crew No.	Bare Costs Hr.	Bare Costs Daily	Incl. Subs O&P Hr.	Incl. Subs O&P Daily	Cost Per Labor-Hour Bare Costs	Cost Per Labor-Hour Incl. O&P
1 Labor Foreman (outside)	$41.85	$334.80	$63.75	$510.00	$40.96	$62.26
4 Laborers	39.85	1275.20	60.70	1942.40		
1 Truck Driver (light)	44.50	356.00	67.00	536.00		
1 Paint Striper, S.P., 40 Gallon		151.00		166.10		
1 Flatbed Truck, Gas, 3 Ton		238.00		261.80		
1 Pickup Truck, 3/4 Ton		115.20		126.72	10.50	11.55
48 L.H., Daily Totals		$2470.20		$3543.02	$51.46	$73.81

Crew B-78A

Crew No.	Bare Costs Hr.	Bare Costs Daily	Incl. Subs O&P Hr.	Incl. Subs O&P Daily	Cost Per Labor-Hour Bare Costs	Cost Per Labor-Hour Incl. O&P
1 Equip. Oper. (light)	$51.30	$410.40	$77.35	$618.80	$51.30	$77.35
1 Line Rem. (Metal Balls) 115 H.P.		852.60		937.86	106.58	117.23
8 L.H., Daily Totals		$1263.00		$1556.66	$157.88	$194.58

Crew B-78B

Crew No.	Bare Costs Hr.	Bare Costs Daily	Incl. Subs O&P Hr.	Incl. Subs O&P Daily	Cost Per Labor-Hour Bare Costs	Cost Per Labor-Hour Incl. O&P
2 Laborers	$39.85	$637.60	$60.70	$971.20	$41.12	$62.55
.25 Equip. Oper. (light)	51.30	102.60	77.35	154.70		
1 Pickup Truck, 3/4 Ton		115.20		126.72		
1 Line Rem.,11 H.P.,Walk Behind		61.40		67.54		
.25 Road Sweeper, S.P., 8' wide		172.15		189.37	19.38	21.31
18 L.H., Daily Totals		$1088.95		$1509.53	$60.50	$83.86

Crew B-78C

Crew No.	Bare Costs Hr.	Bare Costs Daily	Incl. Subs O&P Hr.	Incl. Subs O&P Daily	Cost Per Labor-Hour Bare Costs	Cost Per Labor-Hour Incl. O&P
1 Labor Foreman (outside)	$41.85	$334.80	$63.75	$510.00	$40.96	$62.26
4 Laborers	39.85	1275.20	60.70	1942.40		
1 Truck Driver (light)	44.50	356.00	67.00	536.00		
1 Paint Striper, T.M., 120 Gal.		690.00		759.00		
1 Flatbed Truck, Gas, 3 Ton		238.00		261.80		
1 Pickup Truck, 3/4 Ton		115.20		126.72	21.73	23.91
48 L.H., Daily Totals		$3009.20		$4135.92	$62.69	$86.17

For customer support on your Building Construction Costs with RSMeans data, call 800.448.8182.

Crews - Standard

Crew No.	Bare Costs Hr.	Daily	Incl. Subs O&P Hr.	Daily	Cost Per Labor-Hour Bare Costs	Incl. O&P
Crew B-78D	Hr.	Daily	Hr.	Daily	Bare Costs	Incl. O&P
2 Labor Foremen (outside)	$41.85	$669.60	$63.75	$1020.00	$40.72	$61.94
7 Laborers	39.85	2231.60	60.70	3399.20		
1 Truck Driver (light)	44.50	356.00	67.00	536.00		
1 Paint Striper, T.M., 120 Gal.		690.00		759.00		
1 Flatbed Truck, Gas, 3 Ton		238.00		261.80		
3 Pickup Trucks, 3/4 Ton		345.60		380.16		
1 Air Compressor, 60 cfm		105.40		115.94		
1 -50' Air Hose, 3/4"		7.35		8.09		
1 Breaker, Pavement, 60 lb.		10.60		11.66	17.46	19.21
80 L.H., Daily Totals		$4654.15		$6491.85	$58.18	$81.15
Crew B-78E	Hr.	Daily	Hr.	Daily	Bare Costs	Incl. O&P
2 Labor Foremen (outside)	$41.85	$669.60	$63.75	$1020.00	$40.57	$61.73
9 Laborers	39.85	2869.20	60.70	4370.40		
1 Truck Driver (light)	44.50	356.00	67.00	536.00		
1 Paint Striper, T.M., 120 Gal.		690.00		759.00		
1 Flatbed Truck, Gas, 3 Ton		238.00		261.80		
4 Pickup Trucks, 3/4 Ton		460.80		506.88		
2 Air Compressors, 60 cfm		210.80		231.88		
2 -50' Air Hoses, 3/4"		14.70		16.17		
2 Breakers, Pavement, 60 lb.		21.20		23.32	17.04	18.74
96 L.H., Daily Totals		$5530.30		$7725.45	$57.61	$80.47
Crew B-78F	Hr.	Daily	Hr.	Daily	Bare Costs	Incl. O&P
2 Labor Foremen (outside)	$41.85	$669.60	$63.75	$1020.00	$40.47	$61.59
11 Laborers	39.85	3506.80	60.70	5341.60		
1 Truck Driver (light)	44.50	356.00	67.00	536.00		
1 Paint Striper, T.M., 120 Gal.		690.00		759.00		
1 Flatbed Truck, Gas, 3 Ton		238.00		261.80		
7 Pickup Trucks, 3/4 Ton		806.40		887.04		
3 Air Compressors, 60 cfm		316.20		347.82		
3 -50' Air Hoses, 3/4"		22.05		24.25		
3 Breakers, Pavement, 60 lb.		31.80		34.98	18.79	20.67
112 L.H., Daily Totals		$6636.85		$9212.50	$59.26	$82.25
Crew B-79	Hr.	Daily	Hr.	Daily	Bare Costs	Incl. O&P
1 Labor Foreman (outside)	$41.85	$334.80	$63.75	$510.00	$41.18	$62.57
3 Laborers	39.85	956.40	60.70	1456.80		
1 Truck Driver (light)	44.50	356.00	67.00	536.00		
1 Paint Striper, T.M., 120 Gal.		690.00		759.00		
1 Heating Kettle, 115 Gallon		77.85		85.64		
1 Flatbed Truck, Gas, 3 Ton		238.00		261.80		
2 Pickup Trucks, 3/4 Ton		230.40		253.44	30.91	34.00
40 L.H., Daily Totals		$2883.45		$3862.68	$72.09	$96.57
Crew B-79A	Hr.	Daily	Hr.	Daily	Bare Costs	Incl. O&P
1.5 Equip. Oper. (light)	$51.30	$615.60	$77.35	$928.20	$51.30	$77.35
.5 Line Remov. (Grinder) 115 H.P.		459.70		505.67		
1 Line Rem. (Metal Balls) 115 H.P.		852.60		937.86	109.36	120.29
12 L.H., Daily Totals		$1927.90		$2371.73	$160.66	$197.64
Crew B-79B	Hr.	Daily	Hr.	Daily	Bare Costs	Incl. O&P
1 Laborer	$39.85	$318.80	$60.70	$485.60	$39.85	$60.70
1 Set of Gases		168.00		184.80	21.00	23.10
8 L.H., Daily Totals		$486.80		$670.40	$60.85	$83.80

Crew No.	Bare Costs Hr.	Daily	Incl. Subs O&P Hr.	Daily	Cost Per Labor-Hour Bare Costs	Incl. O&P
Crew B-79C	Hr.	Daily	Hr.	Daily	Bare Costs	Incl. O&P
1 Labor Foreman (outside)	$41.85	$334.80	$63.75	$510.00	$40.80	$62.04
5 Laborers	39.85	1594.00	60.70	2428.00		
1 Truck Driver (light)	44.50	356.00	67.00	536.00		
1 Paint Striper, T.M., 120 Gal.		690.00		759.00		
1 Heating Kettle, 115 Gallon		77.85		85.64		
1 Flatbed Truck, Gas, 3 Ton		238.00		261.80		
3 Pickup Trucks, 3/4 Ton		345.60		380.16		
1 Air Compressor, 60 cfm		105.40		115.94		
1 -50' Air Hose, 3/4"		7.35		8.09		
1 Breaker, Pavement, 60 lb.		10.60		11.66	26.34	28.97
56 L.H., Daily Totals		$3759.60		$5096.28	$67.14	$91.00
Crew B-79D	Hr.	Daily	Hr.	Daily	Bare Costs	Incl. O&P
2 Labor Foremen (outside)	$41.85	$669.60	$63.75	$1020.00	$40.93	$62.25
5 Laborers	39.85	1594.00	60.70	2428.00		
1 Truck Driver (light)	44.50	356.00	67.00	536.00		
1 Paint Striper, T.M., 120 Gal.		690.00		759.00		
1 Heating Kettle, 115 Gallon		77.85		85.64		
1 Flatbed Truck, Gas, 3 Ton		238.00		261.80		
4 Pickup Trucks, 3/4 Ton		460.80		506.88		
1 Air Compressor, 60 cfm		105.40		115.94		
1 -50' Air Hose, 3/4"		7.35		8.09		
1 Breaker, Pavement, 60 lb.		10.60		11.66	24.84	27.33
64 L.H., Daily Totals		$4209.60		$5733.00	$65.78	$89.58
Crew B-79E	Hr.	Daily	Hr.	Daily	Bare Costs	Incl. O&P
2 Labor Foremen (outside)	$41.85	$669.60	$63.75	$1020.00	$40.72	$61.94
7 Laborers	39.85	2231.60	60.70	3399.20		
1 Truck Driver (light)	44.50	356.00	67.00	536.00		
1 Paint Striper, T.M., 120 Gal.		690.00		759.00		
1 Heating Kettle, 115 Gallon		77.85		85.64		
1 Flatbed Truck, Gas, 3 Ton		238.00		261.80		
5 Pickup Trucks, 3/4 Ton		576.00		633.60		
2 Air Compressors, 60 cfm		210.80		231.88		
2 -50' Air Hoses, 3/4"		14.70		16.17		
2 Breakers, Pavement, 60 lb.		21.20		23.32	22.86	25.14
80 L.H., Daily Totals		$5085.75		$6966.60	$63.57	$87.08
Crew B-80	Hr.	Daily	Hr.	Daily	Bare Costs	Incl. O&P
1 Labor Foreman (outside)	$41.85	$334.80	$63.75	$510.00	$44.38	$67.20
1 Laborer	39.85	318.80	60.70	485.60		
1 Truck Driver (light)	44.50	356.00	67.00	536.00		
1 Equip. Oper. (light)	51.30	410.40	77.35	618.80		
1 Flatbed Truck, Gas, 3 Ton		238.00		261.80		
1 Earth Auger, Truck-Mtd.		373.80		411.18	19.12	21.03
32 L.H., Daily Totals		$2031.80		$2823.38	$63.49	$88.23
Crew B-80A	Hr.	Daily	Hr.	Daily	Bare Costs	Incl. O&P
3 Laborers	$39.85	$956.40	$60.70	$1456.80	$39.85	$60.70
1 Flatbed Truck, Gas, 3 Ton		238.00		261.80	9.92	10.91
24 L.H., Daily Totals		$1194.40		$1718.60	$49.77	$71.61
Crew B-80B	Hr.	Daily	Hr.	Daily	Bare Costs	Incl. O&P
3 Laborers	$39.85	$956.40	$60.70	$1456.80	$42.71	$64.86
1 Equip. Oper. (light)	51.30	410.40	77.35	618.80		
1 Crane, Flatbed Mounted, 3 Ton		235.80		259.38	7.37	8.11
32 L.H., Daily Totals		$1602.60		$2334.98	$50.08	$72.97

For customer support on your Building Construction Costs with RSMeans data, call 800.448.8182.

Crews - Standard

Crew B-80C

Crew No.	Hr.	Daily	Hr.	Daily	Bare Costs	Incl. O&P
2 Laborers	$39.85	$637.60	$60.70	$971.20	$41.40	$62.80
1 Truck Driver (light)	44.50	356.00	67.00	536.00		
1 Flatbed Truck, Gas, 1.5 Ton		188.40		207.24		
1 Manual Fence Post Auger, Gas		7.40		8.14	8.16	8.97
24 L.H., Daily Totals		$1189.40		$1722.58	$49.56	$71.77

Crew B-81

Crew No.	Hr.	Daily	Hr.	Daily	Bare Costs	Incl. O&P
1 Laborer	$39.85	$318.80	$60.70	$485.60	$46.53	$70.35
1 Equip. Oper. (medium)	53.75	430.00	81.05	648.40		
1 Truck Driver (heavy)	46.00	368.00	69.30	554.40		
1 Hydromulcher, T.M., 3000 Gal.		282.80		311.08		
1 Truck Tractor, 220 H.P.		290.00		319.00	23.87	26.25
24 L.H., Daily Totals		$1689.60		$2318.48	$70.40	$96.60

Crew B-81A

Crew No.	Hr.	Daily	Hr.	Daily	Bare Costs	Incl. O&P
1 Laborer	$39.85	$318.80	$60.70	$485.60	$42.17	$63.85
1 Truck Driver (light)	44.50	356.00	67.00	536.00		
1 Hydromulcher, T.M., 600 Gal.		121.20		133.32		
1 Flatbed Truck, Gas, 3 Ton		238.00		261.80	22.45	24.70
16 L.H., Daily Totals		$1034.00		$1416.72	$64.63	$88.55

Crew B-82

Crew No.	Hr.	Daily	Hr.	Daily	Bare Costs	Incl. O&P
1 Laborer	$39.85	$318.80	$60.70	$485.60	$45.58	$69.03
1 Equip. Oper. (light)	51.30	410.40	77.35	618.80		
1 Horiz. Borer, 6 H.P.		78.00		85.80	4.88	5.36
16 L.H., Daily Totals		$807.20		$1190.20	$50.45	$74.39

Crew B-82A

Crew No.	Hr.	Daily	Hr.	Daily	Bare Costs	Incl. O&P
2 Laborers	$39.85	$637.60	$60.70	$971.20	$45.58	$69.03
2 Equip. Opers. (light)	51.30	820.80	77.35	1237.60		
2 Dump Trucks, 8 C.Y., 220 H.P.		677.20		744.92		
1 Flatbed Trailer, 25 Ton		133.00		146.30		
1 Horiz. Dir. Drill, 20k lb. Thrust		629.00		691.90		
1 Mud Trailer for HDD, 1500 Gal.		287.80		316.58		
1 Pickup Truck, 4x4, 3/4 Ton		126.60		139.26		
1 Flatbed Trailer, 3 Ton		25.40		27.94		
1 Loader, Skid Steer, 78 H.P.		364.80		401.28	70.12	77.13
32 L.H., Daily Totals		$3702.20		$4676.98	$115.69	$146.16

Crew B-82B

Crew No.	Hr.	Daily	Hr.	Daily	Bare Costs	Incl. O&P
2 Laborers	$39.85	$637.60	$60.70	$971.20	$45.58	$69.03
2 Equip. Opers. (light)	51.30	820.80	77.35	1237.60		
2 Dump Trucks, 8 C.Y., 220 H.P.		677.20		744.92		
1 Flatbed Trailer, 25 Ton		133.00		146.30		
1 Horiz. Dir. Drill, 30k lb. Thrust		896.20		985.82		
1 Mud Trailer for HDD, 1500 Gal.		287.80		316.58		
1 Pickup Truck, 4x4, 3/4 Ton		126.60		139.26		
1 Flatbed Trailer, 3 Ton		25.40		27.94		
1 Loader, Skid Steer, 78 H.P.		364.80		401.28	78.47	86.32
32 L.H., Daily Totals		$3969.40		$4970.90	$124.04	$155.34

Crew B-82C

Crew No.	Hr.	Daily	Hr.	Daily	Bare Costs	Incl. O&P
2 Laborers	$39.85	$637.60	$60.70	$971.20	$45.58	$69.03
2 Equip. Opers. (light)	51.30	820.80	77.35	1237.60		
2 Dump Trucks, 8 C.Y., 220 H.P.		677.20		744.92		
1 Flatbed Trailer, 25 Ton		133.00		146.30		
1 Horiz. Dir. Drill, 50k lb. Thrust		1188.00		1306.80		
1 Mud Trailer for HDD, 1500 Gal.		287.80		316.58		
1 Pickup Truck, 4x4, 3/4 Ton		126.60		139.26		
1 Flatbed Trailer, 3 Ton		25.40		27.94		
1 Loader, Skid Steer, 78 H.P.		364.80		401.28	87.59	96.35
32 L.H., Daily Totals		$4261.20		$5291.88	$133.16	$165.37

Crew B-82D

Crew No.	Hr.	Daily	Hr.	Daily	Bare Costs	Incl. O&P
1 Equip. Oper. (light)	$51.30	$410.40	$77.35	$618.80	$51.30	$77.35
1 Mud Trailer for HDD, 1500 Gal.		287.80		316.58	35.98	39.57
8 L.H., Daily Totals		$698.20		$935.38	$87.28	$116.92

Crew B-83

Crew No.	Hr.	Daily	Hr.	Daily	Bare Costs	Incl. O&P
1 Tugboat Captain	$53.75	$430.00	$81.05	$648.40	$46.80	$70.88
1 Tugboat Hand	39.85	318.80	60.70	485.60		
1 Tugboat, 250 H.P.		680.20		748.22	42.51	46.76
16 L.H., Daily Totals		$1429.00		$1882.22	$89.31	$117.64

Crew B-84

Crew No.	Hr.	Daily	Hr.	Daily	Bare Costs	Incl. O&P
1 Equip. Oper. (medium)	$53.75	$430.00	$81.05	$648.40	$53.75	$81.05
1 Rotary Mower/Tractor		357.00		392.70	44.63	49.09
8 L.H., Daily Totals		$787.00		$1041.10	$98.38	$130.14

Crew B-85

Crew No.	Hr.	Daily	Hr.	Daily	Bare Costs	Incl. O&P
3 Laborers	$39.85	$956.40	$60.70	$1456.80	$43.86	$66.49
1 Equip. Oper. (medium)	53.75	430.00	81.05	648.40		
1 Truck Driver (heavy)	46.00	368.00	69.30	554.40		
1 Telescoping Boom Lift, to 80'		548.60		603.46		
1 Brush Chipper, 12", 130 H.P.		393.00		432.30		
1 Pruning Saw, Rotary		6.65		7.32	23.71	26.08
40 L.H., Daily Totals		$2702.65		$3702.68	$67.57	$92.57

Crew B-86

Crew No.	Hr.	Daily	Hr.	Daily	Bare Costs	Incl. O&P
1 Equip. Oper. (medium)	$53.75	$430.00	$81.05	$648.40	$53.75	$81.05
1 Stump Chipper, S.P.		184.05		202.46	23.01	25.31
8 L.H., Daily Totals		$614.05		$850.86	$76.76	$106.36

Crew B-86A

Crew No.	Hr.	Daily	Hr.	Daily	Bare Costs	Incl. O&P
1 Equip. Oper. (medium)	$53.75	$430.00	$81.05	$648.40	$53.75	$81.05
1 Grader, 30,000 Lbs.		643.80		708.18	80.47	88.52
8 L.H., Daily Totals		$1073.80		$1356.58	$134.22	$169.57

Crew B-86B

Crew No.	Hr.	Daily	Hr.	Daily	Bare Costs	Incl. O&P
1 Equip. Oper. (medium)	$53.75	$430.00	$81.05	$648.40	$53.75	$81.05
1 Dozer, 200 H.P.		1273.00		1400.30	159.13	175.04
8 L.H., Daily Totals		$1703.00		$2048.70	$212.88	$256.09

Crew B-87

Crew No.	Hr.	Daily	Hr.	Daily	Bare Costs	Incl. O&P
1 Laborer	$39.85	$318.80	$60.70	$485.60	$50.97	$76.98
4 Equip. Oper. (medium)	53.75	1720.00	81.05	2593.60		
2 Feller Bunchers, 100 H.P.		1620.80		1782.88		
1 Log Chipper, 22" Tree		764.80		841.28		
1 Dozer, 105 H.P.		608.00		668.80		
1 Chain Saw, Gas, 36" Long		46.40		51.04	76.00	83.60
40 L.H., Daily Totals		$5078.80		$6423.20	$126.97	$160.58

For customer support on your Building Construction Costs with RSMeans data, call 800.448.8182.

Crew No.	Bare Costs		Incl. Subs O&P		Cost Per Labor-Hour	

Crew B-88

Crew B-88	Hr.	Daily	Hr.	Daily	Bare Costs	Incl. O&P
1 Laborer	$39.85	$318.80	$60.70	$485.60	$51.76	$78.14
6 Equip. Oper. (medium)	53.75	2580.00	81.05	3890.40		
2 Feller Bunchers, 100 H.P.		1620.80		1782.88		
1 Log Chipper, 22" Tree		764.80		841.28		
2 Log Skidders, 50 H.P.		1698.80		1868.68		
1 Dozer, 105 H.P.		608.00		668.80		
1 Chain Saw, Gas, 36" Long		46.40		51.04	84.62	93.08
56 L.H., Daily Totals		$7637.60		$9588.68	$136.39	$171.23

Crew B-89	Hr.	Daily	Hr.	Daily	Bare Costs	Incl. O&P
1 Equip. Oper. (light)	$51.30	$410.40	$77.35	$618.80	$47.90	$72.17
1 Truck Driver (light)	44.50	356.00	67.00	536.00		
1 Flatbed Truck, Gas, 3 Ton		238.00		261.80		
1 Concrete Saw		102.40		112.64		
1 Water Tank, 65 Gal.		79.90		87.89	26.27	28.90
16 L.H., Daily Totals		$1186.70		$1617.13	$74.17	$101.07

Crew B-89A	Hr.	Daily	Hr.	Daily	Bare Costs	Incl. O&P
1 Skilled Worker	$52.35	$418.80	$80.15	$641.20	$46.10	$70.42
1 Laborer	39.85	318.80	60.70	485.60		
1 Core Drill (Large)		112.20		123.42	7.01	7.71
16 L.H., Daily Totals		$849.80		$1250.22	$53.11	$78.14

Crew B-89B	Hr.	Daily	Hr.	Daily	Bare Costs	Incl. O&P
1 Equip. Oper. (light)	$51.30	$410.40	$77.35	$618.80	$47.90	$72.17
1 Truck Driver (light)	44.50	356.00	67.00	536.00		
1 Wall Saw, Hydraulic, 10 H.P.		46.40		51.04		
1 Generator, Diesel, 100 kW		307.60		338.36		
1 Water Tank, 65 Gal.		79.90		87.89		
1 Flatbed Truck, Gas, 3 Ton		238.00		261.80	41.99	46.19
16 L.H., Daily Totals		$1438.30		$1893.89	$89.89	$118.37

Crew B-90	Hr.	Daily	Hr.	Daily	Bare Costs	Incl. O&P
1 Labor Foreman (outside)	$41.85	$334.80	$63.75	$510.00	$44.50	$67.39
3 Laborers	39.85	956.40	60.70	1456.80		
2 Equip. Oper. (light)	51.30	820.80	77.35	1237.60		
2 Truck Drivers (heavy)	46.00	736.00	69.30	1108.80		
1 Road Mixer, 310 H.P.		1889.00		2077.90		
1 Dist. Truck, 2000 Gal.		296.00		325.60	34.14	37.55
64 L.H., Daily Totals		$5033.00		$6716.70	$78.64	$104.95

Crew B-90A	Hr.	Daily	Hr.	Daily	Bare Costs	Incl. O&P
1 Labor Foreman (outside)	$41.85	$334.80	$63.75	$510.00	$48.08	$72.76
2 Laborers	39.85	637.60	60.70	971.20		
4 Equip. Oper. (medium)	53.75	1720.00	81.05	2593.60		
2 Graders, 30,000 Lbs.		1287.60		1416.36		
1 Tandem Roller, 10 Ton		231.40		254.54		
1 Roller, Pneum. Whl., 12 Ton		338.20		372.02	33.16	36.48
56 L.H., Daily Totals		$4549.60		$6117.72	$81.24	$109.25

Crew B-90B	Hr.	Daily	Hr.	Daily	Bare Costs	Incl. O&P
1 Labor Foreman (outside)	$41.85	$334.80	$63.75	$510.00	$47.13	$71.38
2 Laborers	39.85	637.60	60.70	971.20		
3 Equip. Oper. (medium)	53.75	1290.00	81.05	1945.20		
1 Roller, Pneum. Whl., 12 Ton		338.20		372.02		
1 Road Mixer, 310 H.P.		1889.00		2077.90	46.40	51.04
48 L.H., Daily Totals		$4489.60		$5876.32	$93.53	$122.42

Crew B-90C	Hr.	Daily	Hr.	Daily	Bare Costs	Incl. O&P
1 Labor Foreman (outside)	$41.85	$334.80	$63.75	$510.00	$45.50	$68.87
4 Laborers	39.85	1275.20	60.70	1942.40		
3 Equip. Oper. (medium)	53.75	1290.00	81.05	1945.20		
3 Truck Drivers (heavy)	46.00	1104.00	69.30	1663.20		
3 Road Mixers, 310 H.P.		5667.00		6233.70	64.40	70.84
88 L.H., Daily Totals		$9671.00		$12294.50	$109.90	$139.71

Crew B-90D	Hr.	Daily	Hr.	Daily	Bare Costs	Incl. O&P
1 Labor Foreman (outside)	$41.85	$334.80	$63.75	$510.00	$44.63	$67.62
6 Laborers	39.85	1912.80	60.70	2913.60		
3 Equip. Oper. (medium)	53.75	1290.00	81.05	1945.20		
3 Truck Drivers (heavy)	46.00	1104.00	69.30	1663.20		
3 Road Mixers, 310 H.P.		5667.00		6233.70	54.49	59.94
104 L.H., Daily Totals		$10308.60		$13265.70	$99.12	$127.55

Crew B-90E	Hr.	Daily	Hr.	Daily	Bare Costs	Incl. O&P
1 Labor Foreman (outside)	$41.85	$334.80	$63.75	$510.00	$45.39	$68.78
4 Laborers	39.85	1275.20	60.70	1942.40		
3 Equip. Oper. (medium)	53.75	1290.00	81.05	1945.20		
1 Truck Driver (heavy)	46.00	368.00	69.30	554.40		
1 Road Mixer, 310 H.P.		1889.00		2077.90	26.24	28.86
72 L.H., Daily Totals		$5157.00		$7029.90	$71.63	$97.64

Crew B-91	Hr.	Daily	Hr.	Daily	Bare Costs	Incl. O&P
1 Labor Foreman (outside)	$41.85	$334.80	$63.75	$510.00	$47.82	$72.33
2 Laborers	39.85	637.60	60.70	971.20		
4 Equip. Oper. (medium)	53.75	1720.00	81.05	2593.60		
1 Truck Driver (heavy)	46.00	368.00	69.30	554.40		
1 Dist. Tanker, 3000 Gallon		326.80		359.48		
1 Truck Tractor, 6x4, 380 H.P.		476.20		523.82		
1 Aggreg. Spreader, S.P.		837.00		920.70		
1 Roller, Pneum. Whl., 12 Ton		338.20		372.02		
1 Tandem Roller, 10 Ton		231.40		254.54	34.52	37.98
64 L.H., Daily Totals		$5270.00		$7059.76	$82.34	$110.31

Crew B-91B	Hr.	Daily	Hr.	Daily	Bare Costs	Incl. O&P
1 Laborer	$39.85	$318.80	$60.70	$485.60	$46.80	$70.88
1 Equipment Oper. (med.)	53.75	430.00	81.05	648.40		
1 Road Sweeper, Vac. Assist.		840.40		924.44	52.52	57.78
16 L.H., Daily Totals		$1589.20		$2058.44	$99.33	$128.65

Crew B-91C	Hr.	Daily	Hr.	Daily	Bare Costs	Incl. O&P
1 Laborer	$39.85	$318.80	$60.70	$485.60	$42.17	$63.85
1 Truck Driver (light)	44.50	356.00	67.00	536.00		
1 Catch Basin Cleaning Truck		522.80		575.08	32.67	35.94
16 L.H., Daily Totals		$1197.60		$1596.68	$74.85	$99.79

Crew B-91D	Hr.	Daily	Hr.	Daily	Bare Costs	Incl. O&P
1 Labor Foreman (outside)	$41.85	$334.80	$63.75	$510.00	$46.30	$70.08
5 Laborers	39.85	1594.00	60.70	2428.00		
5 Equip. Oper. (medium)	53.75	2150.00	81.05	3242.00		
2 Truck Drivers (heavy)	46.00	736.00	69.30	1108.80		
1 Aggreg. Spreader, S.P.		837.00		920.70		
2 Truck Tractors, 6x4, 380 H.P.		952.40		1047.64		
2 Dist. Tankers, 3000 Gallon		653.60		718.96		
2 Pavement Brushes, Towed		170.40		187.44		
2 Rollers Pneum. Whl., 12 Ton		676.40		744.04	31.63	34.80
104 L.H., Daily Totals		$8104.60		$10907.58	$77.93	$104.88

For customer support on your Building Construction Costs with RSMeans data, call 800.448.8182.

Crews - Standard

Crew No.		Bare Costs		Incl. Subs O&P		Cost Per Labor-Hour	
						Bare Costs	Incl. O&P
Crew B-92	Hr.	Daily	Hr.	Daily		Bare Costs	Incl. O&P
1 Labor Foreman (outside)	$41.85	$334.80	$63.75	$510.00		$40.35	$61.46
3 Laborers	39.85	956.40	60.70	1456.80			
1 Crack Cleaner, 25 H.P.		55.80		61.38			
1 Air Compressor, 60 cfm		105.40		115.94			
1 Tar Kettle, T.M.		129.20		142.12			
1 Flatbed Truck, Gas, 3 Ton		238.00		261.80		16.51	18.16
32 L.H., Daily Totals		$1819.60		$2548.04		$56.86	$79.63
Crew B-93	Hr.	Daily	Hr.	Daily		Bare Costs	Incl. O&P
1 Equip. Oper. (medium)	$53.75	$430.00	$81.05	$648.40		$53.75	$81.05
1 Feller Buncher, 100 H.P.		810.40		891.44		101.30	111.43
8 L.H., Daily Totals		$1240.40		$1539.84		$155.05	$192.48
Crew B-94A	Hr.	Daily	Hr.	Daily		Bare Costs	Incl. O&P
1 Laborer	$39.85	$318.80	$60.70	$485.60		$39.85	$60.70
1 Diaphragm Water Pump, 2"		73.00		80.30			
1 -20' Suction Hose, 2"		1.95		2.15			
2 -50' Discharge Hoses, 2"		1.80		1.98		9.59	10.55
8 L.H., Daily Totals		$395.55		$570.02		$49.44	$71.25
Crew B-94B	Hr.	Daily	Hr.	Daily		Bare Costs	Incl. O&P
1 Laborer	$39.85	$318.80	$60.70	$485.60		$39.85	$60.70
1 Diaphragm Water Pump, 4"		114.80		126.28			
1 -20' Suction Hose, 4"		3.25		3.58			
2 -50' Discharge Hoses, 4"		4.70		5.17		15.34	16.88
8 L.H., Daily Totals		$441.55		$620.63		$55.19	$77.58
Crew B-94C	Hr.	Daily	Hr.	Daily		Bare Costs	Incl. O&P
1 Laborer	$39.85	$318.80	$60.70	$485.60		$39.85	$60.70
1 Centrifugal Water Pump, 3"		79.20		87.12			
1 -20' Suction Hose, 3"		2.85		3.13			
2 -50' Discharge Hoses, 3"		3.00		3.30		10.63	11.69
8 L.H., Daily Totals		$403.85		$579.15		$50.48	$72.39
Crew B-94D	Hr.	Daily	Hr.	Daily		Bare Costs	Incl. O&P
1 Laborer	$39.85	$318.80	$60.70	$485.60		$39.85	$60.70
1 Centr. Water Pump, 6"		296.40		326.04			
1 -20' Suction Hose, 6"		11.50		12.65			
2 -50' Discharge Hoses, 6"		12.20		13.42		40.01	44.01
8 L.H., Daily Totals		$638.90		$837.71		$79.86	$104.71
Crew C-1	Hr.	Daily	Hr.	Daily		Bare Costs	Incl. O&P
3 Carpenters	$50.70	$1216.80	$77.20	$1852.80		$47.99	$73.08
1 Laborer	39.85	318.80	60.70	485.60			
32 L.H., Daily Totals		$1535.60		$2338.40		$47.99	$73.08
Crew C-2	Hr.	Daily	Hr.	Daily		Bare Costs	Incl. O&P
1 Carpenter Foreman (outside)	$52.70	$421.60	$80.25	$642.00		$49.23	$74.96
4 Carpenters	50.70	1622.40	77.20	2470.40			
1 Laborer	39.85	318.80	60.70	485.60			
48 L.H., Daily Totals		$2362.80		$3598.00		$49.23	$74.96
Crew C-2A	Hr.	Daily	Hr.	Daily		Bare Costs	Incl. O&P
1 Carpenter Foreman (outside)	$52.70	$421.60	$80.25	$642.00		$48.70	$73.83
3 Carpenters	50.70	1216.80	77.20	1852.80			
1 Cement Finisher	47.55	380.40	70.45	563.60			
1 Laborer	39.85	318.80	60.70	485.60			
48 L.H., Daily Totals		$2337.60		$3544.00		$48.70	$73.83

Crew No.		Bare Costs		Incl. Subs O&P		Cost Per Labor-Hour	
Crew C-3	Hr.	Daily	Hr.	Daily		Bare Costs	Incl. O&P
1 Rodman Foreman (outside)	$56.65	$453.20	$86.50	$692.00		$50.78	$77.38
4 Rodmen (reinf.)	54.65	1748.80	83.45	2670.40			
1 Equip. Oper. (light)	51.30	410.40	77.35	618.80			
2 Laborers	39.85	637.60	60.70	971.20			
3 Stressing Equipment		31.20		34.32			
.5 Grouting Equipment		79.20		87.12		1.73	1.90
64 L.H., Daily Totals		$3360.40		$5073.84		$52.51	$79.28
Crew C-4	Hr.	Daily	Hr.	Daily		Bare Costs	Incl. O&P
1 Rodman Foreman (outside)	$56.65	$453.20	$86.50	$692.00		$55.15	$84.21
3 Rodmen (reinf.)	54.65	1311.60	83.45	2002.80			
3 Stressing Equipment		31.20		34.32		.97	1.07
32 L.H., Daily Totals		$1796.00		$2729.12		$56.13	$85.28
Crew C-4A	Hr.	Daily	Hr.	Daily		Bare Costs	Incl. O&P
2 Rodmen (reinf.)	$54.65	$874.40	$83.45	$1335.20		$54.65	$83.45
4 Stressing Equipment		41.60		45.76		2.60	2.86
16 L.H., Daily Totals		$916.00		$1380.96		$57.25	$86.31
Crew C-5	Hr.	Daily	Hr.	Daily		Bare Costs	Incl. O&P
1 Rodman Foreman (outside)	$56.65	$453.20	$86.50	$692.00		$54.28	$82.60
4 Rodmen (reinf.)	54.65	1748.80	83.45	2670.40			
1 Equip. Oper. (crane)	56.10	448.80	84.60	676.80			
1 Equip. Oper. (oiler)	48.60	388.80	73.30	586.40			
1 Hyd. Crane, 25 Ton		590.60		649.66		10.55	11.60
56 L.H., Daily Totals		$3630.20		$5275.26		$64.83	$94.20
Crew C-6	Hr.	Daily	Hr.	Daily		Bare Costs	Incl. O&P
1 Labor Foreman (outside)	$41.85	$334.80	$63.75	$510.00		$41.47	$62.83
4 Laborers	39.85	1275.20	60.70	1942.40			
1 Cement Finisher	47.55	380.40	70.45	563.60			
2 Gas Engine Vibrators		51.20		56.32		1.07	1.17
48 L.H., Daily Totals		$2041.60		$3072.32		$42.53	$64.01
Crew C-7	Hr.	Daily	Hr.	Daily		Bare Costs	Incl. O&P
1 Labor Foreman (outside)	$41.85	$334.80	$63.75	$510.00		$43.44	$65.78
5 Laborers	39.85	1594.00	60.70	2428.00			
1 Cement Finisher	47.55	380.40	70.45	563.60			
1 Equip. Oper. (medium)	53.75	430.00	81.05	648.40			
1 Equip. Oper. (oiler)	48.60	388.80	73.30	586.40			
2 Gas Engine Vibrators		51.20		56.32			
1 Concrete Bucket, 1 C.Y.		24.60		27.06			
1 Hyd. Crane, 55 Ton		993.80		1093.18		14.86	16.34
72 L.H., Daily Totals		$4197.60		$5912.96		$58.30	$82.12
Crew C-7A	Hr.	Daily	Hr.	Daily		Bare Costs	Incl. O&P
1 Labor Foreman (outside)	$41.85	$334.80	$63.75	$510.00		$41.64	$63.23
5 Laborers	39.85	1594.00	60.70	2428.00			
2 Truck Drivers (heavy)	46.00	736.00	69.30	1108.80			
2 Conc. Transit Mixers		1811.20		1992.32		28.30	31.13
64 L.H., Daily Totals		$4476.00		$6039.12		$69.94	$94.36
Crew C-7B	Hr.	Daily	Hr.	Daily		Bare Costs	Incl. O&P
1 Labor Foreman (outside)	$41.85	$334.80	$63.75	$510.00		$43.23	$65.64
5 Laborers	39.85	1594.00	60.70	2428.00			
1 Equipment Operator, Crane	56.10	448.80	84.60	676.80			
1 Equipment Oiler	48.60	388.80	73.30	586.40			
1 Conc. Bucket, 2 C.Y.		38.40		42.24			
1 Lattice Boom Crane, 165 Ton		2140.00		2354.00		34.04	37.44
64 L.H., Daily Totals		$4944.80		$6597.44		$77.26	$103.09

For customer support on your Building Construction Costs with RSMeans data, call 800.448.8182.

Crew No.	Bare Costs		Incl. Subs O&P		Cost Per Labor-Hour	
	Hr.	Daily	Hr.	Daily	Bare Costs	Incl. O&P
Crew C-7C					$43.58	$66.17
1 Labor Foreman (outside)	$41.85	$334.80	$63.75	$510.00		
5 Laborers	39.85	1594.00	60.70	2428.00		
2 Equipment Operators (med.)	53.75	860.00	81.05	1296.80		
2 F.E. Loaders, W.M., 4 C.Y.		1149.20		1264.12	17.96	19.75
64 L.H., Daily Totals		$3938.00		$5498.92	$61.53	$85.92
Crew C-7D					$42.12	$64.04
1 Labor Foreman (outside)	$41.85	$334.80	$63.75	$510.00		
5 Laborers	39.85	1594.00	60.70	2428.00		
1 Equip. Oper. (medium)	53.75	430.00	81.05	648.40		
1 Concrete Conveyer		187.20		205.92	3.34	3.68
56 L.H., Daily Totals		$2546.00		$3792.32	$45.46	$67.72
Crew C-8					$44.32	$66.83
1 Labor Foreman (outside)	$41.85	$334.80	$63.75	$510.00		
3 Laborers	39.85	956.40	60.70	1456.80		
2 Cement Finishers	47.55	760.80	70.45	1127.20		
1 Equip. Oper. (medium)	53.75	430.00	81.05	648.40		
1 Concrete Pump (Small)		881.60		969.76	15.74	17.32
56 L.H., Daily Totals		$3363.60		$4712.16	$60.06	$84.15
Crew C-8A					$42.75	$64.46
1 Labor Foreman (outside)	$41.85	$334.80	$63.75	$510.00		
3 Laborers	39.85	956.40	60.70	1456.80		
2 Cement Finishers	47.55	760.80	70.45	1127.20		
48 L.H., Daily Totals		$2052.00		$3094.00	$42.75	$64.46
Crew C-8B					$43.03	$65.38
1 Labor Foreman (outside)	$41.85	$334.80	$63.75	$510.00		
3 Laborers	39.85	956.40	60.70	1456.80		
1 Equip. Oper. (medium)	53.75	430.00	81.05	648.40		
1 Vibrating Power Screed		75.35		82.89		
1 Roller, Vibratory, 25 Ton		659.80		725.78		
1 Dozer, 200 H.P.		1273.00		1400.30	50.20	55.22
40 L.H., Daily Totals		$3729.35		$4824.17	$93.23	$120.60
Crew C-8C					$43.78	$66.22
1 Labor Foreman (outside)	$41.85	$334.80	$63.75	$510.00		
3 Laborers	39.85	956.40	60.70	1456.80		
1 Cement Finisher	47.55	380.40	70.45	563.60		
1 Equip. Oper. (medium)	53.75	430.00	81.05	648.40		
1 Shotcrete Rig, 12 C.Y./hr		246.60		271.26		
1 Air Compressor, 160 cfm		117.20		128.92		
4 -50' Air Hoses, 1"		33.80		37.18		
4 -50' Air Hoses, 2"		103.60		113.96	10.44	11.49
48 L.H., Daily Totals		$2602.80		$3730.12	$54.23	$77.71
Crew C-8D					$45.14	$68.06
1 Labor Foreman (outside)	$41.85	$334.80	$63.75	$510.00		
1 Laborer	39.85	318.80	60.70	485.60		
1 Cement Finisher	47.55	380.40	70.45	563.60		
1 Equipment Oper. (light)	51.30	410.40	77.35	618.80		
1 Air Compressor, 250 cfm		167.40		184.14		
2 -50' Air Hoses, 1"		16.90		18.59	5.76	6.34
32 L.H., Daily Totals		$1628.70		$2380.73	$50.90	$74.40

Crew No.	Bare Costs		Incl. Subs O&P		Cost Per Labor-Hour	
	Hr.	Daily	Hr.	Daily	Bare Costs	Incl. O&P
Crew C-8E					$43.38	$65.61
1 Labor Foreman (outside)	$41.85	$334.80	$63.75	$510.00		
3 Laborers	39.85	956.40	60.70	1456.80		
1 Cement Finisher	47.55	380.40	70.45	563.60		
1 Equipment Oper. (light)	51.30	410.40	77.35	618.80		
1 Shotcrete Rig, 35 C.Y./hr		270.00		297.22		
1 Air Compressor, 250 cfm		167.40		184.14		
4 -50' Air Hoses, 1"		33.80		37.18		
4 -50' Air Hoses, 2"		103.60		113.96	11.98	13.18
48 L.H., Daily Totals		$2657.00		$3781.70	$55.35	$78.79
Crew C-10					$44.98	$67.20
1 Laborer	$39.85	$318.80	$60.70	$485.60		
2 Cement Finishers	47.55	760.80	70.45	1127.20		
24 L.H., Daily Totals		$1079.60		$1612.80	$44.98	$67.20
Crew C-10B					$42.93	$64.60
3 Laborers	$39.85	$956.40	$60.70	$1456.80		
2 Cement Finishers	47.55	760.80	70.45	1127.20		
1 Concrete Mixer, 10 C.F.		161.00		177.10		
2 Trowels, 48" Walk-Behind		82.40		90.64	6.09	6.69
40 L.H., Daily Totals		$1960.60		$2851.74	$49.02	$71.29
Crew C-10C					$44.98	$67.20
1 Laborer	$39.85	$318.80	$60.70	$485.60		
2 Cement Finishers	47.55	760.80	70.45	1127.20		
1 Trowel, 48" Walk-Behind		41.20		45.32	1.72	1.89
24 L.H., Daily Totals		$1120.80		$1658.12	$46.70	$69.09
Crew C-10D					$44.98	$67.20
1 Laborer	$39.85	$318.80	$60.70	$485.60		
2 Cement Finishers	47.55	760.80	70.45	1127.20		
1 Vibrating Power Screed		75.35		82.89		
1 Trowel, 48" Walk-Behind		41.20		45.32	4.86	5.34
24 L.H., Daily Totals		$1196.15		$1741.01	$49.84	$72.54
Crew C-10E					$44.98	$67.20
1 Laborer	$39.85	$318.80	$60.70	$485.60		
2 Cement Finishers	47.55	760.80	70.45	1127.20		
1 Vibrating Power Screed		75.35		82.89		
1 Cement Trowel, 96" Ride-On		166.20		182.82	10.06	11.07
24 L.H., Daily Totals		$1321.15		$1878.51	$55.05	$78.27
Crew C-10F					$44.98	$67.20
1 Laborer	$39.85	$318.80	$60.70	$485.60		
2 Cement Finishers	47.55	760.80	70.45	1127.20		
1 Telescoping Boom Lift, to 60'		454.60		500.06	18.94	20.84
24 L.H., Daily Totals		$1534.20		$2112.86	$63.92	$88.04
Crew C-11					$54.36	$87.40
1 Struc. Steel Foreman (outside)	$56.65	$453.20	$92.60	$740.80		
6 Struc. Steel Workers	54.65	2623.20	89.35	4288.80		
1 Equip. Oper. (crane)	56.10	448.80	84.60	676.80		
1 Equip. Oper. (oiler)	48.60	388.80	73.30	586.40		
1 Lattice Boom Crane, 150 Ton		2017.00		2218.70	28.01	30.82
72 L.H., Daily Totals		$5931.00		$8511.50	$82.38	$118.22

For customer support on your Building Construction Costs with RSMeans data, call 800.448.8182.

Crew No.	Bare Costs		Incl. Subs O&P		Cost Per Labor-Hour	

Crew C-12	Hr.	Daily	Hr.	Daily	Bare Costs	Incl. O&P
1 Carpenter Foreman (outside)	$52.70	$421.60	$80.25	$642.00	$50.13	$76.19
3 Carpenters	50.70	1216.80	77.20	1852.80		
1 Laborer	39.85	318.80	60.70	485.60		
1 Equip. Oper. (crane)	56.10	448.80	84.60	676.80		
1 Hyd. Crane, 12 Ton		496.00		545.60	10.33	11.37
48 L.H., Daily Totals		$2902.00		$4202.80	$60.46	$87.56

Crew C-13	Hr.	Daily	Hr.	Daily	Bare Costs	Incl. O&P
1 Struc. Steel Worker	$54.65	$437.20	$89.35	$714.80	$53.33	$85.30
1 Welder	54.65	437.20	89.35	714.80		
1 Carpenter	50.70	405.60	77.20	617.60		
1 Welder, Gas Engine, 300 amp		98.40		108.24	4.10	4.51
24 L.H., Daily Totals		$1378.40		$2155.44	$57.43	$89.81

Crew C-14	Hr.	Daily	Hr.	Daily	Bare Costs	Incl. O&P
1 Carpenter Foreman (outside)	$52.70	$421.60	$80.25	$642.00	$49.11	$74.54
5 Carpenters	50.70	2028.00	77.20	3088.00		
4 Laborers	39.85	1275.20	60.70	1942.40		
4 Rodmen (reinf.)	54.65	1748.80	83.45	2670.40		
2 Cement Finishers	47.55	760.80	70.45	1127.20		
1 Equip. Oper. (crane)	56.10	448.80	84.60	676.80		
1 Equip. Oper. (oiler)	48.60	388.80	73.30	586.40		
1 Hyd. Crane, 80 Ton		1518.00		1669.80	10.54	11.60
144 L.H., Daily Totals		$8590.00		$12403.00	$59.65	$86.13

Crew C-14A	Hr.	Daily	Hr.	Daily	Bare Costs	Incl. O&P
1 Carpenter Foreman (outside)	$52.70	$421.60	$80.25	$642.00	$50.54	$76.89
16 Carpenters	50.70	6489.60	77.20	9881.60		
4 Rodmen (reinf.)	54.65	1748.80	83.45	2670.40		
2 Laborers	39.85	637.60	60.70	971.20		
1 Cement Finisher	47.55	380.40	70.45	563.60		
1 Equip. Oper. (medium)	53.75	430.00	81.05	648.40		
1 Gas Engine Vibrator		25.60		28.16		
1 Concrete Pump (Small)		881.60		969.76	4.54	4.99
200 L.H., Daily Totals		$11015.20		$16375.12	$55.08	$81.88

Crew C-14B	Hr.	Daily	Hr.	Daily	Bare Costs	Incl. O&P
1 Carpenter Foreman (outside)	$52.70	$421.60	$80.25	$642.00	$50.42	$76.64
16 Carpenters	50.70	6489.60	77.20	9881.60		
4 Rodmen (reinf.)	54.65	1748.80	83.45	2670.40		
2 Laborers	39.85	637.60	60.70	971.20		
2 Cement Finishers	47.55	760.80	70.45	1127.20		
1 Equip. Oper. (medium)	53.75	430.00	81.05	648.40		
1 Gas Engine Vibrator		25.60		28.16		
1 Concrete Pump (Small)		881.60		969.76	4.36	4.80
208 L.H., Daily Totals		$11395.60		$16938.72	$54.79	$81.44

Crew C-14C	Hr.	Daily	Hr.	Daily	Bare Costs	Incl. O&P
1 Carpenter Foreman (outside)	$52.70	$421.60	$80.25	$642.00	$48.08	$73.11
6 Carpenters	50.70	2433.60	77.20	3705.60		
2 Rodmen (reinf.)	54.65	874.40	83.45	1335.20		
4 Laborers	39.85	1275.20	60.70	1942.40		
1 Cement Finisher	47.55	380.40	70.45	563.60		
1 Gas Engine Vibrator		25.60		28.16	.23	.25
112 L.H., Daily Totals		$5410.80		$8216.96	$48.31	$73.37

Crew C-14D	Hr.	Daily	Hr.	Daily	Bare Costs	Incl. O&P
1 Carpenter Foreman (outside)	$52.70	$421.60	$80.25	$642.00	$50.22	$76.39
18 Carpenters	50.70	7300.80	77.20	11116.80		
2 Rodmen (reinf.)	54.65	874.40	83.45	1335.20		
2 Laborers	39.85	637.60	60.70	971.20		
1 Cement Finisher	47.55	380.40	70.45	563.60		
1 Equip. Oper. (medium)	53.75	430.00	81.05	648.40		
1 Gas Engine Vibrator		25.60		28.16		
1 Concrete Pump (Small)		881.60		969.76	4.54	4.99
200 L.H., Daily Totals		$10952.00		$16275.12	$54.76	$81.38

Crew C-14E	Hr.	Daily	Hr.	Daily	Bare Costs	Incl. O&P
1 Carpenter Foreman (outside)	$52.70	$421.60	$80.25	$642.00	$49.07	$74.64
2 Carpenters	50.70	811.20	77.20	1235.20		
4 Rodmen (reinf.)	54.65	1748.80	83.45	2670.40		
3 Laborers	39.85	956.40	60.70	1456.80		
1 Cement Finisher	47.55	380.40	70.45	563.60		
1 Gas Engine Vibrator		25.60		28.16	.29	.32
88 L.H., Daily Totals		$4344.00		$6596.16	$49.36	$74.96

Crew C-14F	Hr.	Daily	Hr.	Daily	Bare Costs	Incl. O&P
1 Labor Foreman (outside)	$41.85	$334.80	$63.75	$510.00	$45.21	$67.54
2 Laborers	39.85	637.60	60.70	971.20		
6 Cement Finishers	47.55	2282.40	70.45	3381.60		
1 Gas Engine Vibrator		25.60		28.16	.36	.39
72 L.H., Daily Totals		$3280.40		$4890.96	$45.56	$67.93

Crew C-14G	Hr.	Daily	Hr.	Daily	Bare Costs	Incl. O&P
1 Labor Foreman (outside)	$41.85	$334.80	$63.75	$510.00	$44.54	$66.71
2 Laborers	39.85	637.60	60.70	971.20		
4 Cement Finishers	47.55	1521.60	70.45	2254.40		
1 Gas Engine Vibrator		25.60		28.16	.46	.50
56 L.H., Daily Totals		$2519.60		$3763.76	$44.99	$67.21

Crew C-14H	Hr.	Daily	Hr.	Daily	Bare Costs	Incl. O&P
1 Carpenter Foreman (outside)	$52.70	$421.60	$80.25	$642.00	$49.36	$74.88
2 Carpenters	50.70	811.20	77.20	1235.20		
1 Rodman (reinf.)	54.65	437.20	83.45	667.60		
1 Laborer	39.85	318.80	60.70	485.60		
1 Cement Finisher	47.55	380.40	70.45	563.60		
1 Gas Engine Vibrator		25.60		28.16	.53	.59
48 L.H., Daily Totals		$2394.80		$3622.16	$49.89	$75.46

Crew C-14L	Hr.	Daily	Hr.	Daily	Bare Costs	Incl. O&P
1 Carpenter Foreman (outside)	$52.70	$421.60	$80.25	$642.00	$46.99	$71.39
6 Carpenters	50.70	2433.60	77.20	3705.60		
4 Laborers	39.85	1275.20	60.70	1942.40		
1 Cement Finisher	47.55	380.40	70.45	563.60		
1 Gas Engine Vibrator		25.60		28.16	.27	.29
96 L.H., Daily Totals		$4536.40		$6881.76	$47.25	$71.69

Crew C-14M	Hr.	Daily	Hr.	Daily	Bare Costs	Incl. O&P
1 Carpenter Foreman (outside)	$52.70	$421.60	$80.25	$642.00	$48.72	$73.88
2 Carpenters	50.70	811.20	77.20	1235.20		
1 Rodman (reinf.)	54.65	437.20	83.45	667.60		
2 Laborers	39.85	637.60	60.70	971.20		
1 Cement Finisher	47.55	380.40	70.45	563.60		
1 Equip. Oper. (medium)	53.75	430.00	81.05	648.40		
1 Gas Engine Vibrator		25.60		28.16		
1 Concrete Pump (Small)		881.60		969.76	14.18	15.59
64 L.H., Daily Totals		$4025.20		$5725.92	$62.89	$89.47

Crew C-15	Hr.	Daily	Hr.	Daily	Bare Costs	Incl. O&P
1 Carpenter Foreman (outside)	$52.70	$421.60	$80.25	$642.00	$47.04	$71.23
2 Carpenters	50.70	811.20	77.20	1235.20		
3 Laborers	39.85	956.40	60.70	1456.80		
2 Cement Finishers	47.55	760.80	70.45	1127.20		
1 Rodman (reinf.)	54.65	437.20	83.45	667.60		
72 L.H., Daily Totals		$3387.20		$5128.80	$47.04	$71.23

Crew C-16	Hr.	Daily	Hr.	Daily	Bare Costs	Incl. O&P
1 Labor Foreman (outside)	$41.85	$334.80	$63.75	$510.00	$44.32	$66.83
3 Laborers	39.85	956.40	60.70	1456.80		
2 Cement Finishers	47.55	760.80	70.45	1127.20		
1 Equip. Oper. (medium)	53.75	430.00	81.05	648.40		
1 Gunite Pump Rig		321.00		353.10		
2 -50' Air Hoses, 3/4"		14.70		16.17		
2 -50' Air Hoses, 2"		51.80		56.98	6.92	7.61
56 L.H., Daily Totals		$2869.50		$4168.65	$51.24	$74.44

Crew C-16A	Hr.	Daily	Hr.	Daily	Bare Costs	Incl. O&P
1 Laborer	$39.85	$318.80	$60.70	$485.60	$47.17	$70.66
2 Cement Finishers	47.55	760.80	70.45	1127.20		
1 Equip. Oper. (medium)	53.75	430.00	81.05	648.40		
1 Gunite Pump Rig		321.00		353.10		
2 -50' Air Hoses, 3/4"		14.70		16.17		
2 -50' Air Hoses, 2"		51.80		56.98		
1 Telescoping Boom Lift, to 60'		454.60		500.06	26.32	28.95
32 L.H., Daily Totals		$2351.70		$3187.51	$73.49	$99.61

Crew C-17	Hr.	Daily	Hr.	Daily	Bare Costs	Incl. O&P
2 Skilled Worker Foremen (out)	$54.35	$869.60	$83.20	$1331.20	$52.75	$80.76
8 Skilled Workers	52.35	3350.40	80.15	5129.60		
80 L.H., Daily Totals		$4220.00		$6460.80	$52.75	$80.76

Crew C-17A	Hr.	Daily	Hr.	Daily	Bare Costs	Incl. O&P
2 Skilled Worker Foremen (out)	$54.35	$869.60	$83.20	$1331.20	$52.79	$80.81
8 Skilled Workers	52.35	3350.40	80.15	5129.60		
.125 Equip. Oper. (crane)	56.10	56.10	84.60	84.60		
.125 Hyd. Crane, 80 Ton		189.75		208.72	2.34	2.58
81 L.H., Daily Totals		$4465.85		$6754.13	$55.13	$83.38

Crew C-17B	Hr.	Daily	Hr.	Daily	Bare Costs	Incl. O&P
2 Skilled Worker Foremen (out)	$54.35	$869.60	$83.20	$1331.20	$52.83	$80.85
8 Skilled Workers	52.35	3350.40	80.15	5129.60		
.25 Equip. Oper. (crane)	56.10	112.20	84.60	169.20		
.25 Hyd. Crane, 80 Ton		379.50		417.45		
.25 Trowel, 48" Walk-Behind		10.30		11.33	4.75	5.23
82 L.H., Daily Totals		$4722.00		$7058.78	$57.59	$86.08

Crew C-17C	Hr.	Daily	Hr.	Daily	Bare Costs	Incl. O&P
2 Skilled Worker Foremen (out)	$54.35	$869.60	$83.20	$1331.20	$52.87	$80.90
8 Skilled Workers	52.35	3350.40	80.15	5129.60		
.375 Equip. Oper. (crane)	56.10	168.30	84.60	253.80		
.375 Hyd. Crane, 80 Ton		569.25		626.17	6.86	7.54
83 L.H., Daily Totals		$4957.55		$7340.77	$59.73	$88.44

Crew C-17D	Hr.	Daily	Hr.	Daily	Bare Costs	Incl. O&P
2 Skilled Worker Foremen (out)	$54.35	$869.60	$83.20	$1331.20	$52.91	$80.94
8 Skilled Workers	52.35	3350.40	80.15	5129.60		
.5 Equip. Oper. (crane)	56.10	224.40	84.60	338.40		
.5 Hyd. Crane, 80 Ton		759.00		834.90	9.04	9.94
84 L.H., Daily Totals		$5203.40		$7634.10	$61.95	$90.88

Crew C-17E	Hr.	Daily	Hr.	Daily	Bare Costs	Incl. O&P
2 Skilled Worker Foremen (out)	$54.35	$869.60	$83.20	$1331.20	$52.75	$80.76
8 Skilled Workers	52.35	3350.40	80.15	5129.60		
1 Hyd. Jack with Rods		102.25		112.47	1.28	1.41
80 L.H., Daily Totals		$4322.25		$6573.27	$54.03	$82.17

Crew C-18	Hr.	Daily	Hr.	Daily	Bare Costs	Incl. O&P
.125 Labor Foreman (outside)	$41.85	$41.85	$63.75	$63.75	$40.07	$61.04
1 Laborer	39.85	318.80	60.70	485.60		
1 Concrete Cart, 10 C.F.		57.80		63.58	6.42	7.06
9 L.H., Daily Totals		$418.45		$612.93	$46.49	$68.10

Crew C-19	Hr.	Daily	Hr.	Daily	Bare Costs	Incl. O&P
.125 Labor Foreman (outside)	$41.85	$41.85	$63.75	$63.75	$40.07	$61.04
1 Laborer	39.85	318.80	60.70	485.60		
1 Concrete Cart, 18 C.F.		97.40		107.14	10.82	11.90
9 L.H., Daily Totals		$458.05		$656.49	$50.89	$72.94

Crew C-20	Hr.	Daily	Hr.	Daily	Bare Costs	Incl. O&P
1 Labor Foreman (outside)	$41.85	$334.80	$63.75	$510.00	$42.80	$64.84
5 Laborers	39.85	1594.00	60.70	2428.00		
1 Cement Finisher	47.55	380.40	70.45	563.60		
1 Equip. Oper. (medium)	53.75	430.00	81.05	648.40		
2 Gas Engine Vibrators		51.20		56.32		
1 Concrete Pump (Small)		881.60		969.76	14.57	16.03
64 L.H., Daily Totals		$3672.00		$5176.08	$57.38	$80.88

Crew C-21	Hr.	Daily	Hr.	Daily	Bare Costs	Incl. O&P
1 Labor Foreman (outside)	$41.85	$334.80	$63.75	$510.00	$42.80	$64.84
5 Laborers	39.85	1594.00	60.70	2428.00		
1 Cement Finisher	47.55	380.40	70.45	563.60		
1 Equip. Oper. (medium)	53.75	430.00	81.05	648.40		
2 Gas Engine Vibrators		51.20		56.32		
1 Concrete Conveyer		187.20		205.92	3.73	4.10
64 L.H., Daily Totals		$2977.60		$4412.24	$46.52	$68.94

Crew C-22	Hr.	Daily	Hr.	Daily	Bare Costs	Incl. O&P
1 Rodman Foreman (outside)	$56.65	$453.20	$86.50	$692.00	$54.92	$83.82
4 Rodmen (reinf.)	54.65	1748.80	83.45	2670.40		
.125 Equip. Oper. (crane)	56.10	56.10	84.60	84.60		
.125 Equip. Oper. (oiler)	48.60	48.60	73.30	73.30		
.125 Hyd. Crane, 25 Ton		73.83		81.21	1.76	1.93
42 L.H., Daily Totals		$2380.53		$3601.51	$56.68	$85.75

Crew C-23	Hr.	Daily	Hr.	Daily	Bare Costs	Incl. O&P
2 Skilled Worker Foremen (out)	$54.35	$869.60	$83.20	$1331.20	$52.75	$80.52
6 Skilled Workers	52.35	2512.80	80.15	3847.20		
1 Equip. Oper. (crane)	56.10	448.80	84.60	676.80		
1 Equip. Oper. (oiler)	48.60	388.80	73.30	586.40		
1 Lattice Boom Crane, 90 Ton		1687.00		1855.70	21.09	23.20
80 L.H., Daily Totals		$5907.00		$8297.30	$73.84	$103.72

Crew C-23A	Hr.	Daily	Hr.	Daily	Bare Costs	Incl. O&P
1 Labor Foreman (outside)	$41.85	$334.80	$63.75	$510.00	$45.25	$68.61
2 Laborers	39.85	637.60	60.70	971.20		
1 Equip. Oper. (crane)	56.10	448.80	84.60	676.80		
1 Equip. Oper. (oiler)	48.60	388.80	73.30	586.40		
1 Crawler Crane, 100 Ton		1872.00		2059.20		
3 Conc. Buckets, 8 C.Y.		626.40		689.04	62.46	68.71
40 L.H., Daily Totals		$4308.40		$5492.64	$107.71	$137.32

For customer support on your Building Construction Costs with RSMeans data, call 800.448.8182.

Crew No.	Bare Costs		Incl. Subs O&P		Cost Per Labor-Hour	
	Hr.	Daily	Hr.	Daily	Bare Costs	Incl. O&P
Crew C-24					Bare Costs	Incl. O&P
2 Skilled Worker Foremen (out)	$54.35	$869.60	$83.20	$1331.20	$52.75	$80.52
6 Skilled Workers	52.35	2512.80	80.15	3847.20		
1 Equip. Oper. (crane)	56.10	448.80	84.60	676.80		
1 Equip. Oper. (oiler)	48.60	388.80	73.30	586.40		
1 Lattice Boom Crane, 150 Ton		2017.00		2218.70	25.21	27.73
80 L.H., Daily Totals		$6237.00		$8660.30	$77.96	$108.25
Crew C-25	Hr.	Daily	Hr.	Daily	Bare Costs	Incl. O&P
2 Rodmen (reinf.)	$54.65	$874.40	$83.45	$1335.20	$43.75	$69.38
2 Rodmen Helpers	32.85	525.60	55.30	884.80		
32 L.H., Daily Totals		$1400.00		$2220.00	$43.75	$69.38
Crew C-27	Hr.	Daily	Hr.	Daily	Bare Costs	Incl. O&P
2 Cement Finishers	$47.55	$760.80	$70.45	$1127.20	$47.55	$70.45
1 Concrete Saw		102.40		112.64	6.40	7.04
16 L.H., Daily Totals		$863.20		$1239.84	$53.95	$77.49
Crew C-28	Hr.	Daily	Hr.	Daily	Bare Costs	Incl. O&P
1 Cement Finisher	$47.55	$380.40	$70.45	$563.60	$47.55	$70.45
1 Portable Air Compressor, Gas		17.10		18.81	2.14	2.35
8 L.H., Daily Totals		$397.50		$582.41	$49.69	$72.80
Crew C-29	Hr.	Daily	Hr.	Daily	Bare Costs	Incl. O&P
1 Laborer	$39.85	$318.80	$60.70	$485.60	$39.85	$60.70
1 Pressure Washer		63.60		69.96	7.95	8.74
8 L.H., Daily Totals		$382.40		$555.56	$47.80	$69.44
Crew C-30	Hr.	Daily	Hr.	Daily	Bare Costs	Incl. O&P
1 Laborer	$39.85	$318.80	$60.70	$485.60	$39.85	$60.70
1 Concrete Mixer, 10 C.F.		161.00		177.10	20.13	22.14
8 L.H., Daily Totals		$479.80		$662.70	$59.98	$82.84
Crew C-31	Hr.	Daily	Hr.	Daily	Bare Costs	Incl. O&P
1 Cement Finisher	$47.55	$380.40	$70.45	$563.60	$47.55	$70.45
1 Grout Pump		321.00		353.10	40.13	44.14
8 L.H., Daily Totals		$701.40		$916.70	$87.67	$114.59
Crew C-32	Hr.	Daily	Hr.	Daily	Bare Costs	Incl. O&P
1 Cement Finisher	$47.55	$380.40	$70.45	$563.60	$43.70	$65.58
1 Laborer	39.85	318.80	60.70	485.60		
1 Crack Chaser Saw, Gas, 6 H.P.		25.40		27.94		
1 Vacuum Pick-Up System		66.95		73.64	5.77	6.35
16 L.H., Daily Totals		$791.55		$1150.79	$49.47	$71.92
Crew D-1	Hr.	Daily	Hr.	Daily	Bare Costs	Incl. O&P
1 Bricklayer	$50.30	$402.40	$77.00	$616.00	$44.80	$68.58
1 Bricklayer Helper	39.30	314.40	60.15	481.20		
16 L.H., Daily Totals		$716.80		$1097.20	$44.80	$68.58
Crew D-2	Hr.	Daily	Hr.	Daily	Bare Costs	Incl. O&P
3 Bricklayers	$50.30	$1207.20	$77.00	$1848.00	$46.34	$70.89
2 Bricklayer Helpers	39.30	628.80	60.15	962.40		
.5 Carpenter	50.70	202.80	77.20	308.80		
44 L.H., Daily Totals		$2038.80		$3119.20	$46.34	$70.89

Crew No.	Bare Costs		Incl. Subs O&P		Cost Per Labor-Hour	
	Hr.	Daily	Hr.	Daily	Bare Costs	Incl. O&P
Crew D-3					Bare Costs	Incl. O&P
3 Bricklayers	$50.30	$1207.20	$77.00	$1848.00	$46.13	$70.59
2 Bricklayer Helpers	39.30	628.80	60.15	962.40		
.25 Carpenter	50.70	101.40	77.20	154.40		
42 L.H., Daily Totals		$1937.40		$2964.80	$46.13	$70.59
Crew D-4	Hr.	Daily	Hr.	Daily	Bare Costs	Incl. O&P
1 Bricklayer	$50.30	$402.40	$77.00	$616.00	$45.05	$68.66
2 Bricklayer Helpers	39.30	628.80	60.15	962.40		
1 Equip. Oper. (light)	51.30	410.40	77.35	618.80		
1 Grout Pump, 50 C.F./hr.		128.60		141.46	4.02	4.42
32 L.H., Daily Totals		$1570.20		$2338.66	$49.07	$73.08
Crew D-5	Hr.	Daily	Hr.	Daily	Bare Costs	Incl. O&P
1 Bricklayer	50.30	402.40	77.00	616.00	50.30	77.00
8 L.H., Daily Totals		$402.40		$616.00	$50.30	$77.00
Crew D-6	Hr.	Daily	Hr.	Daily	Bare Costs	Incl. O&P
3 Bricklayers	$50.30	$1207.20	$77.00	$1848.00	$45.04	$68.92
3 Bricklayer Helpers	39.30	943.20	60.15	1443.60		
.25 Carpenter	50.70	101.40	77.20	154.40		
50 L.H., Daily Totals		$2251.80		$3446.00	$45.04	$68.92
Crew D-7	Hr.	Daily	Hr.	Daily	Bare Costs	Incl. O&P
1 Tile Layer	$46.85	$374.80	$69.35	$554.80	$42.02	$62.20
1 Tile Layer Helper	37.20	297.60	55.05	440.40		
16 L.H., Daily Totals		$672.40		$995.20	$42.02	$62.20
Crew D-8	Hr.	Daily	Hr.	Daily	Bare Costs	Incl. O&P
3 Bricklayers	$50.30	$1207.20	$77.00	$1848.00	$45.90	$70.26
2 Bricklayer Helpers	39.30	628.80	60.15	962.40		
40 L.H., Daily Totals		$1836.00		$2810.40	$45.90	$70.26
Crew D-9	Hr.	Daily	Hr.	Daily	Bare Costs	Incl. O&P
3 Bricklayers	$50.30	$1207.20	$77.00	$1848.00	$44.80	$68.58
3 Bricklayer Helpers	39.30	943.20	60.15	1443.60		
48 L.H., Daily Totals		$2150.40		$3291.60	$44.80	$68.58
Crew D-10	Hr.	Daily	Hr.	Daily	Bare Costs	Incl. O&P
1 Bricklayer Foreman (outside)	$52.30	$418.40	$80.05	$640.40	$49.50	$75.45
1 Bricklayer	50.30	402.40	77.00	616.00		
1 Bricklayer Helper	39.30	314.40	60.15	481.20		
1 Equip. Oper. (crane)	56.10	448.80	84.60	676.80		
1 S.P. Crane, 4x4, 12 Ton		363.60		399.96	11.36	12.50
32 L.H., Daily Totals		$1947.60		$2814.36	$60.86	$87.95
Crew D-11	Hr.	Daily	Hr.	Daily	Bare Costs	Incl. O&P
1 Bricklayer Foreman (outside)	$52.30	$418.40	$80.05	$640.40	$47.30	$72.40
1 Bricklayer	50.30	402.40	77.00	616.00		
1 Bricklayer Helper	39.30	314.40	60.15	481.20		
24 L.H., Daily Totals		$1135.20		$1737.60	$47.30	$72.40
Crew D-12	Hr.	Daily	Hr.	Daily	Bare Costs	Incl. O&P
1 Bricklayer Foreman (outside)	$52.30	$418.40	$80.05	$640.40	$45.30	$69.34
1 Bricklayer	50.30	402.40	77.00	616.00		
2 Bricklayer Helpers	39.30	628.80	60.15	962.40		
32 L.H., Daily Totals		$1449.60		$2218.80	$45.30	$69.34

Crew D-13

Crew No.	Bare Costs Hr.	Daily	Incl. Subs O&P Hr.	Daily	Cost Per Labor-Hour Bare Costs	Incl. O&P
1 Bricklayer Foreman (outside)	$52.30	$418.40	$80.05	$640.40	$48.00	$73.19
1 Bricklayer	50.30	402.40	77.00	616.00		
2 Bricklayer Helpers	39.30	628.80	60.15	962.40		
1 Carpenter	50.70	405.60	77.20	617.60		
1 Equip. Oper. (crane)	56.10	448.80	84.60	676.80		
1 S.P. Crane, 4x4, 12 Ton		363.60		399.96	7.58	8.33
48 L.H., Daily Totals		$2667.60		$3913.16	$55.58	$81.52

Crew D-14

Crew No.	Bare Costs Hr.	Daily	Incl. Subs O&P Hr.	Daily	Cost Per Labor-Hour Bare Costs	Incl. O&P
3 Bricklayers	$50.30	$1207.20	$77.00	$1848.00	$47.55	$72.79
1 Bricklayer Helper	39.30	314.40	60.15	481.20		
32 L.H., Daily Totals		$1521.60		$2329.20	$47.55	$72.79

Crew E-1

Crew No.	Bare Costs Hr.	Daily	Incl. Subs O&P Hr.	Daily	Cost Per Labor-Hour Bare Costs	Incl. O&P
1 Welder Foreman (outside)	$56.65	$453.20	$92.60	$740.80	$54.20	$86.43
1 Welder	54.65	437.20	89.35	714.80		
1 Equip. Oper. (light)	51.30	410.40	77.35	618.80		
1 Welder, Gas Engine, 300 amp		98.40		108.24	4.10	4.51
24 L.H., Daily Totals		$1399.20		$2182.64	$58.30	$90.94

Crew E-2

Crew No.	Bare Costs Hr.	Daily	Incl. Subs O&P Hr.	Daily	Cost Per Labor-Hour Bare Costs	Incl. O&P
1 Struc. Steel Foreman (outside)	$56.65	$453.20	$92.60	$740.80	$54.28	$86.84
4 Struc. Steel Workers	54.65	1748.80	89.35	2859.20		
1 Equip. Oper. (crane)	56.10	448.80	84.60	676.80		
1 Equip. Oper. (oiler)	48.60	388.80	73.30	586.40		
1 Lattice Boom Crane, 90 Ton		1687.00		1855.70	30.13	33.14
56 L.H., Daily Totals		$4726.60		$6718.90	$84.40	$119.98

Crew E-3

Crew No.	Bare Costs Hr.	Daily	Incl. Subs O&P Hr.	Daily	Cost Per Labor-Hour Bare Costs	Incl. O&P
1 Struc. Steel Foreman (outside)	$56.65	$453.20	$92.60	$740.80	$55.32	$90.43
1 Struc. Steel Worker	54.65	437.20	89.35	714.80		
1 Welder	54.65	437.20	89.35	714.80		
1 Welder, Gas Engine, 300 amp		98.40		108.24	4.10	4.51
24 L.H., Daily Totals		$1426.00		$2278.64	$59.42	$94.94

Crew E-3A

Crew No.	Bare Costs Hr.	Daily	Incl. Subs O&P Hr.	Daily	Cost Per Labor-Hour Bare Costs	Incl. O&P
1 Struc. Steel Foreman (outside)	$56.65	$453.20	$92.60	$740.80	$55.32	$90.43
1 Struc. Steel Worker	54.65	437.20	89.35	714.80		
1 Welder	54.65	437.20	89.35	714.80		
1 Welder, Gas Engine, 300 amp		98.40		108.24		
1 Telescoping Boom Lift, to 40'		282.80		311.08	15.88	17.47
24 L.H., Daily Totals		$1708.80		$2589.72	$71.20	$107.91

Crew E-4

Crew No.	Bare Costs Hr.	Daily	Incl. Subs O&P Hr.	Daily	Cost Per Labor-Hour Bare Costs	Incl. O&P
1 Struc. Steel Foreman (outside)	$56.65	$453.20	$92.60	$740.80	$55.15	$90.16
3 Struc. Steel Workers	54.65	1311.60	89.35	2144.40		
1 Welder, Gas Engine, 300 amp		98.40		108.24	3.08	3.38
32 L.H., Daily Totals		$1863.20		$2993.44	$58.23	$93.55

Crew E-5

Crew No.	Bare Costs Hr.	Daily	Incl. Subs O&P Hr.	Daily	Cost Per Labor-Hour Bare Costs	Incl. O&P
2 Struc. Steel Foremen (outside)	$56.65	$906.40	$92.60	$1481.60	$54.59	$87.92
5 Struc. Steel Workers	54.65	2186.00	89.35	3574.00		
1 Equip. Oper. (crane)	56.10	448.80	84.60	676.80		
1 Welder	54.65	437.20	89.35	714.80		
1 Equip. Oper. (oiler)	48.60	388.80	73.30	586.40		
1 Lattice Boom Crane, 90 Ton		1687.00		1855.70		
1 Welder, Gas Engine, 300 amp		98.40		108.24	22.32	24.55
80 L.H., Daily Totals		$6152.60		$8997.54	$76.91	$112.47

Crew E-6

Crew No.	Bare Costs Hr.	Daily	Incl. Subs O&P Hr.	Daily	Cost Per Labor-Hour Bare Costs	Incl. O&P
3 Struc. Steel Foremen (outside)	$56.65	$1359.60	$92.60	$2222.40	$54.53	$87.91
9 Struc. Steel Workers	54.65	3934.80	89.35	6433.20		
1 Equip. Oper. (crane)	56.10	448.80	84.60	676.80		
1 Welder	54.65	437.20	89.35	714.80		
1 Equip. Oper. (oiler)	48.60	388.80	73.30	586.40		
1 Equip. Oper. (light)	51.30	410.40	77.35	618.80		
1 Lattice Boom Crane, 90 Ton		1687.00		1855.70		
1 Welder, Gas Engine, 300 amp		98.40		108.24		
1 Air Compressor, 160 cfm		117.20		128.92		
2 Impact Wrenches		37.20		40.92	15.15	16.67
128 L.H., Daily Totals		$8919.40		$13386.18	$69.68	$104.58

Crew E-7

Crew No.	Bare Costs Hr.	Daily	Incl. Subs O&P Hr.	Daily	Cost Per Labor-Hour Bare Costs	Incl. O&P
1 Struc. Steel Foreman (outside)	$56.65	$453.20	$92.60	$740.80	$54.59	$87.92
4 Struc. Steel Workers	54.65	1748.80	89.35	2859.20		
1 Equip. Oper. (crane)	56.10	448.80	84.60	676.80		
1 Equip. Oper. (oiler)	48.60	388.80	73.30	586.40		
1 Welder Foreman (outside)	56.65	453.20	92.60	740.80		
2 Welders	54.65	874.40	89.35	1429.60		
1 Lattice Boom Crane, 90 Ton		1687.00		1855.70		
2 Welders, Gas Engine, 300 amp		196.80		216.48	23.55	25.90
80 L.H., Daily Totals		$6251.00		$9105.78	$78.14	$113.82

Crew E-8

Crew No.	Bare Costs Hr.	Daily	Incl. Subs O&P Hr.	Daily	Cost Per Labor-Hour Bare Costs	Incl. O&P
1 Struc. Steel Foreman (outside)	$56.65	$453.20	$92.60	$740.80	$54.35	$87.33
4 Struc. Steel Workers	54.65	1748.80	89.35	2859.20		
1 Welder Foreman (outside)	56.65	453.20	92.60	740.80		
4 Welders	54.65	1748.80	89.35	2859.20		
1 Equip. Oper. (crane)	56.10	448.80	84.60	676.80		
1 Equip. Oper. (oiler)	48.60	388.80	73.30	586.40		
1 Equip. Oper. (light)	51.30	410.40	77.35	618.80		
1 Lattice Boom Crane, 90 Ton		1687.00		1855.70		
4 Welders, Gas Engine, 300 amp		393.60		432.96	20.01	22.01
104 L.H., Daily Totals		$7732.60		$11370.66	$74.35	$109.33

Crew E-9

Crew No.	Bare Costs Hr.	Daily	Incl. Subs O&P Hr.	Daily	Cost Per Labor-Hour Bare Costs	Incl. O&P
2 Struc. Steel Foremen (outside)	$56.65	$906.40	$92.60	$1481.60	$54.53	$87.91
5 Struc. Steel Workers	54.65	2186.00	89.35	3574.00		
1 Welder Foreman (outside)	56.65	453.20	92.60	740.80		
5 Welders	54.65	2186.00	89.35	3574.00		
1 Equip. Oper. (crane)	56.10	448.80	84.60	676.80		
1 Equip. Oper. (oiler)	48.60	388.80	73.30	586.40		
1 Equip. Oper. (light)	51.30	410.40	77.35	618.80		
1 Lattice Boom Crane, 90 Ton		1687.00		1855.70		
5 Welders, Gas Engine, 300 amp		492.00		541.20	17.02	18.73
128 L.H., Daily Totals		$9158.60		$13649.30	$71.55	$106.64

Crew E-10

Crew No.	Bare Costs Hr.	Daily	Incl. Subs O&P Hr.	Daily	Cost Per Labor-Hour Bare Costs	Incl. O&P
1 Welder Foreman (outside)	$56.65	$453.20	$92.60	$740.80	$55.65	$90.97
1 Welder	54.65	437.20	89.35	714.80		
1 Welder, Gas Engine, 300 amp		98.40		108.24		
1 Flatbed Truck, Gas, 3 Ton		238.00		261.80	21.02	23.13
16 L.H., Daily Totals		$1226.80		$1825.64	$76.67	$114.10

Crew E-11

Crew No.	Bare Costs Hr.	Daily	Incl. Subs O&P Hr.	Daily	Cost Per Labor-Hour Bare Costs	Incl. O&P
2 Painters, Struc. Steel	$43.50	$696.00	$72.45	$1159.20	$44.54	$70.74
1 Building Laborer	39.85	318.80	60.70	485.60		
1 Equip. Oper. (light)	51.30	410.40	77.35	618.80		
1 Air Compressor, 250 cfm		167.40		184.14		
1 Sandblaster, Portable, 3 C.F.		21.00		23.10		
1 Set Sand Blasting Accessories		14.70		16.17	6.35	6.98
32 L.H., Daily Totals		$1628.30		$2487.01	$50.88	$77.72

For customer support on your Building Construction Costs with RSMeans data, call 800.448.8182.

Crew No.	Bare Costs		Incl. Subs O&P		Cost Per Labor-Hour	
	Hr.	Daily	Hr.	Daily	Bare Costs	Incl. O&P
Crew E-11A						
2 Painters, Struc. Steel	$43.50	$696.00	$72.45	$1159.20	$44.54	$70.74
1 Building Laborer	39.85	318.80	60.70	485.60		
1 Equip. Oper. (light)	51.30	410.40	77.35	618.80		
1 Air Compressor, 250 cfm		167.40		184.14		
1 Sandblaster, Portable, 3 C.F.		21.00		23.10		
1 Set Sand Blasting Accessories		14.70		16.17		
1 Telescoping Boom Lift, to 60'		454.60		500.06	20.55	22.61
32 L.H., Daily Totals		$2082.90		$2987.07	$65.09	$93.35
Crew E-11B	Hr.	Daily	Hr.	Daily	Bare Costs	Incl. O&P
2 Painters, Struc. Steel	$43.50	$696.00	$72.45	$1159.20	$42.28	$68.53
1 Building Laborer	39.85	318.80	60.70	485.60		
2 Paint Sprayer, 8 C.F.M.		90.20		99.22		
1 Telescoping Boom Lift, to 60'		454.60		500.06	22.70	24.97
24 L.H., Daily Totals		$1559.60		$2244.08	$64.98	$93.50
Crew E-12	Hr.	Daily	Hr.	Daily	Bare Costs	Incl. O&P
1 Welder Foreman (outside)	$56.65	$453.20	$92.60	$740.80	$53.98	$84.97
1 Equip. Oper. (light)	51.30	410.40	77.35	618.80		
1 Welder, Gas Engine, 300 amp		98.40		108.24	6.15	6.76
16 L.H., Daily Totals		$962.00		$1467.84	$60.13	$91.74
Crew E-13	Hr.	Daily	Hr.	Daily	Bare Costs	Incl. O&P
1 Welder Foreman (outside)	$56.65	$453.20	$92.60	$740.80	$54.87	$87.52
.5 Equip. Oper. (light)	51.30	205.20	77.35	309.40		
1 Welder, Gas Engine, 300 amp		98.40		108.24	8.20	9.02
12 L.H., Daily Totals		$756.80		$1158.44	$63.07	$96.54
Crew E-14	Hr.	Daily	Hr.	Daily	Bare Costs	Incl. O&P
1 Welder Foreman (outside)	$56.65	$453.20	$92.60	$740.80	$56.65	$92.60
1 Welder, Gas Engine, 300 amp		98.40		108.24	12.30	13.53
8 L.H., Daily Totals		$551.60		$849.04	$68.95	$106.13
Crew E-16	Hr.	Daily	Hr.	Daily	Bare Costs	Incl. O&P
1 Welder Foreman (outside)	$56.65	$453.20	$92.60	$740.80	$55.65	$90.97
1 Welder	54.65	437.20	89.35	714.80		
1 Welder, Gas Engine, 300 amp		98.40		108.24	6.15	6.76
16 L.H., Daily Totals		$988.80		$1563.84	$61.80	$97.74
Crew E-17	Hr.	Daily	Hr.	Daily	Bare Costs	Incl. O&P
1 Struc. Steel Foreman (outside)	$56.65	$453.20	$92.60	$740.80	$55.65	$90.97
1 Structural Steel Worker	54.65	437.20	89.35	714.80		
16 L.H., Daily Totals		$890.40		$1455.60	$55.65	$90.97
Crew E-18	Hr.	Daily	Hr.	Daily	Bare Costs	Incl. O&P
1 Struc. Steel Foreman (outside)	$56.65	$453.20	$92.60	$740.80	$54.87	$88.34
3 Structural Steel Workers	54.65	1311.60	89.35	2144.40		
1 Equipment Operator (med.)	53.75	430.00	81.05	648.40		
1 Lattice Boom Crane, 20 Ton		1083.00		1191.30	27.07	29.78
40 L.H., Daily Totals		$3277.80		$4724.90	$81.94	$118.12
Crew E-19	Hr.	Daily	Hr.	Daily	Bare Costs	Incl. O&P
1 Struc. Steel Foreman (outside)	$56.65	$453.20	$92.60	$740.80	$54.20	$86.43
1 Structural Steel Worker	54.65	437.20	89.35	714.80		
1 Equip. Oper. (light)	51.30	410.40	77.35	618.80		
1 Lattice Boom Crane, 20 Ton		1083.00		1191.30	45.13	49.64
24 L.H., Daily Totals		$2383.80		$3265.70	$99.33	$136.07

Crew No.	Bare Costs		Incl. Subs O&P		Cost Per Labor-Hour	
	Hr.	Daily	Hr.	Daily	Bare Costs	Incl. O&P
Crew E-20						
1 Struc. Steel Foreman (outside)	$56.65	$453.20	$92.60	$740.80	$54.33	$87.16
5 Structural Steel Workers	54.65	2186.00	89.35	3574.00		
1 Equip. Oper. (crane)	56.10	448.80	84.60	676.80		
1 Equip. Oper. (oiler)	48.60	388.80	73.30	586.40		
1 Lattice Boom Crane, 40 Ton		1321.00		1453.10	20.64	22.70
64 L.H., Daily Totals		$4797.80		$7031.10	$74.97	$109.86
Crew E-22	Hr.	Daily	Hr.	Daily	Bare Costs	Incl. O&P
1 Skilled Worker Foreman (out)	$54.35	$434.80	$83.20	$665.60	$53.02	$81.17
2 Skilled Workers	52.35	837.60	80.15	1282.40		
24 L.H., Daily Totals		$1272.40		$1948.00	$53.02	$81.17
Crew E-24	Hr.	Daily	Hr.	Daily	Bare Costs	Incl. O&P
3 Structural Steel Workers	$54.65	$1311.60	$89.35	$2144.40	$54.42	$87.28
1 Equipment Operator (med.)	53.75	430.00	81.05	648.40		
1 Hyd. Crane, 25 Ton		590.60		649.66	18.46	20.30
32 L.H., Daily Totals		$2332.20		$3442.46	$72.88	$107.58
Crew E-25	Hr.	Daily	Hr.	Daily	Bare Costs	Incl. O&P
1 Welder Foreman (outside)	$56.65	$453.20	$92.60	$740.80	$56.65	$92.60
1 Cutting Torch		12.60		13.86	1.58	1.73
8 L.H., Daily Totals		$465.80		$754.66	$58.23	$94.33
Crew E-26	Hr.	Daily	Hr.	Daily	Bare Costs	Incl. O&P
1 Struc. Steel Foreman (outside)	$56.65	$453.20	$92.60	$740.80	$56.01	$90.41
1 Struc. Steel Worker	54.65	437.20	89.35	714.80		
1 Welder	54.65	437.20	89.35	714.80		
.25 Electrician	58.20	116.40	87.00	174.00		
.25 Plumber	62.15	124.30	93.60	187.20		
1 Welder, Gas Engine, 300 amp		98.40		108.24	3.51	3.87
28 L.H., Daily Totals		$1666.70		$2639.84	$59.52	$94.28
Crew E-27	Hr.	Daily	Hr.	Daily	Bare Costs	Incl. O&P
1 Struc. Steel Foreman (outside)	$56.65	$453.20	$92.60	$740.80	$54.33	$87.16
5 Struc. Steel Workers	54.65	2186.00	89.35	3574.00		
1 Equip. Oper. (crane)	56.10	448.80	84.60	676.80		
1 Equip. Oper. (oiler)	48.60	388.80	73.30	586.40		
1 Hyd. Crane, 12 Ton		496.00		545.60		
1 Hyd. Crane, 80 Ton		1518.00		1669.80	31.47	34.62
64 L.H., Daily Totals		$5490.80		$7793.40	$85.79	$121.77
Crew F-3	Hr.	Daily	Hr.	Daily	Bare Costs	Incl. O&P
4 Carpenters	$50.70	$1622.40	$77.20	$2470.40	$51.78	$78.68
1 Equip. Oper. (crane)	56.10	448.80	84.60	676.80		
1 Hyd. Crane, 12 Ton		496.00		545.60	12.40	13.64
40 L.H., Daily Totals		$2567.20		$3692.80	$64.18	$92.32
Crew F-4	Hr.	Daily	Hr.	Daily	Bare Costs	Incl. O&P
4 Carpenters	$50.70	$1622.40	$77.20	$2470.40	$51.25	$77.78
1 Equip. Oper. (crane)	56.10	448.80	84.60	676.80		
1 Equip. Oper. (oiler)	48.60	388.80	73.30	586.40		
1 Hyd. Crane, 55 Ton		993.80		1093.18	20.70	22.77
48 L.H., Daily Totals		$3453.80		$4826.78	$71.95	$100.56
Crew F-5	Hr.	Daily	Hr.	Daily	Bare Costs	Incl. O&P
1 Carpenter Foreman (outside)	$52.70	$421.60	$80.25	$642.00	$51.20	$77.96
3 Carpenters	50.70	1216.80	77.20	1852.80		
32 L.H., Daily Totals		$1638.40		$2494.80	$51.20	$77.96

For customer support on your Building Construction Costs with RSMeans data, call 800.448.8182.

Crew No.	Bare Costs Hr.	Daily	Incl. Subs O&P Hr.	Daily	Cost Per Labor-Hour Bare Costs	Incl. O&P
Crew F-6						
2 Carpenters	$50.70	$811.20	$77.20	$1235.20	$47.44	$72.08
2 Building Laborers	39.85	637.60	60.70	971.20		
1 Equip. Oper. (crane)	56.10	448.80	84.60	676.80		
1 Hyd. Crane, 12 Ton		496.00		545.60	12.40	13.64
40 L.H., Daily Totals		$2393.60		$3428.80	$59.84	$85.72
Crew F-7						
2 Carpenters	$50.70	$811.20	$77.20	$1235.20	$45.27	$68.95
2 Building Laborers	39.85	637.60	60.70	971.20		
32 L.H., Daily Totals		$1448.80		$2206.40	$45.27	$68.95
Crew G-1						
1 Roofer Foreman (outside)	$45.95	$367.60	$77.35	$618.80	$41.06	$69.11
4 Roofers Composition	43.95	1406.40	73.95	2366.40		
2 Roofer Helpers	32.85	525.60	55.30	884.80		
1 Application Equipment		181.00		199.10		
1 Tar Kettle/Pot		167.20		183.92		
1 Crew Truck		166.00		182.60	9.18	10.10
56 L.H., Daily Totals		$2813.80		$4435.62	$50.25	$79.21
Crew G-2						
1 Plasterer	$46.45	$371.60	$69.70	$557.60	$42.02	$63.35
1 Plasterer Helper	39.75	318.00	59.65	477.20		
1 Building Laborer	39.85	318.80	60.70	485.60		
1 Grout Pump, 50 C.F./hr.		128.60		141.46	5.36	5.89
24 L.H., Daily Totals		$1137.00		$1661.86	$47.38	$69.24
Crew G-2A						
1 Roofer Composition	$43.95	$351.60	$73.95	$591.60	$38.88	$63.32
1 Roofer Helper	32.85	262.80	55.30	442.40		
1 Building Laborer	39.85	318.80	60.70	485.60		
1 Foam Spray Rig, Trailer-Mtd.		499.25		549.17		
1 Pickup Truck, 3/4 Ton		115.20		126.72	25.60	28.16
24 L.H., Daily Totals		$1547.65		$2195.49	$64.49	$91.48
Crew G-3						
2 Sheet Metal Workers	$59.80	$956.80	$91.25	$1460.00	$49.83	$75.97
2 Building Laborers	39.85	637.60	60.70	971.20		
32 L.H., Daily Totals		$1594.40		$2431.20	$49.83	$75.97
Crew G-4						
1 Labor Foreman (outside)	$41.85	$334.80	$63.75	$510.00	$40.52	$61.72
2 Building Laborers	39.85	637.60	60.70	971.20		
1 Flatbed Truck, Gas, 1.5 Ton		188.40		207.24		
1 Air Compressor, 160 cfm		117.20		128.92	12.73	14.01
24 L.H., Daily Totals		$1278.00		$1817.36	$53.25	$75.72
Crew G-5						
1 Roofer Foreman (outside)	$45.95	$367.60	$77.35	$618.80	$39.91	$67.17
2 Roofers Composition	43.95	703.20	73.95	1183.20		
2 Roofer Helpers	32.85	525.60	55.30	884.80		
1 Application Equipment		181.00		199.10	4.53	4.98
40 L.H., Daily Totals		$1777.40		$2885.90	$44.44	$72.15
Crew G-6A						
2 Roofers Composition	$43.95	$703.20	$73.95	$1183.20	$43.95	$73.95
1 Small Compressor, Electric		15.10		16.61		
2 Pneumatic Nailers		46.10		50.71	3.83	4.21
16 L.H., Daily Totals		$764.40		$1250.52	$47.77	$78.16

Crew No.	Bare Costs Hr.	Daily	Incl. Subs O&P Hr.	Daily	Cost Per Labor-Hour Bare Costs	Incl. O&P
Crew G-7						
1 Carpenter	$50.70	$405.60	$77.20	$617.60	$50.70	$77.20
1 Small Compressor, Electric		15.10		16.61		
1 Pneumatic Nailer		23.05		25.36	4.77	5.25
8 L.H., Daily Totals		$443.75		$659.57	$55.47	$82.45
Crew H-1						
2 Glaziers	$48.50	$776.00	$73.45	$1175.20	$51.58	$81.40
2 Struc. Steel Workers	54.65	874.40	89.35	1429.60		
32 L.H., Daily Totals		$1650.40		$2604.80	$51.58	$81.40
Crew H-2						
2 Glaziers	$48.50	$776.00	$73.45	$1175.20	$45.62	$69.20
1 Building Laborer	39.85	318.80	60.70	485.60		
24 L.H., Daily Totals		$1094.80		$1660.80	$45.62	$69.20
Crew H-3						
1 Glazier	$48.50	$388.00	$73.45	$587.60	$43.15	$65.90
1 Helper	37.80	302.40	58.35	466.80		
16 L.H., Daily Totals		$690.40		$1054.40	$43.15	$65.90
Crew H-4						
1 Carpenter	$50.70	$405.60	$77.20	$617.60	$47.04	$71.62
1 Carpenter Helper	37.80	302.40	58.35	466.80		
.5 Electrician	58.20	232.80	87.00	348.00		
20 L.H., Daily Totals		$940.80		$1432.40	$47.04	$71.62
Crew J-1						
3 Plasterers	$46.45	$1114.80	$69.70	$1672.80	$43.77	$65.68
2 Plasterer Helpers	39.75	636.00	59.65	954.40		
1 Mixing Machine, 6 C.F.		133.20		146.52	3.33	3.66
40 L.H., Daily Totals		$1884.00		$2773.72	$47.10	$69.34
Crew J-2						
3 Plasterers	$46.45	$1114.80	$69.70	$1672.80	$44.67	$66.84
2 Plasterer Helpers	39.75	636.00	59.65	954.40		
1 Lather	49.15	393.20	72.65	581.20		
1 Mixing Machine, 6 C.F.		133.20		146.52	2.77	3.05
48 L.H., Daily Totals		$2277.20		$3354.92	$47.44	$69.89
Crew J-3						
1 Terrazzo Worker	$47.00	$376.00	$69.55	$556.40	$43.05	$63.70
1 Terrazzo Helper	39.10	312.80	57.85	462.80		
1 Floor Grinder, 22" Path		133.60		146.96		
1 Terrazzo Mixer		177.80		195.58	19.46	21.41
16 L.H., Daily Totals		$1000.20		$1361.74	$62.51	$85.11
Crew J-4						
2 Cement Finishers	$47.55	$760.80	$70.45	$1127.20	$44.98	$67.20
1 Laborer	39.85	318.80	60.70	485.60		
1 Floor Grinder, 22" Path		133.60		146.96		
1 Floor Edger, 7" Path		42.95		47.24		
1 Vacuum Pick-Up System		66.95		73.64	10.15	11.16
24 L.H., Daily Totals		$1323.10		$1880.65	$55.13	$78.36

For customer support on your Building Construction Costs with RSMeans data, call 800.448.8182.

Crews - Standard

Crew J-4A

	Bare Costs Hr.	Daily	Incl. Subs O&P Hr.	Daily	Cost Per Labor-Hour Bare Costs	Incl. O&P
2 Cement Finishers	$47.55	$760.80	$70.45	$1127.20	$43.70	$65.58
2 Laborers	39.85	637.60	60.70	971.20		
1 Floor Grinder, 22" Path		133.60		146.96		
1 Floor Edger, 7" Path		42.95		47.24		
1 Vacuum Pick-Up System		66.95		73.64		
1 Floor Auto Scrubber		254.30		279.73	15.56	17.11
32 L.H., Daily Totals		$1896.20		$2645.98	$59.26	$82.69

Crew J-4B

	Bare Costs Hr.	Daily	Incl. Subs O&P Hr.	Daily	Bare Costs	Incl. O&P
1 Laborer	$39.85	$318.80	$60.70	$485.60	$39.85	$60.70
1 Floor Auto Scrubber		254.30		279.73	31.79	34.97
8 L.H., Daily Totals		$573.10		$765.33	$71.64	$95.67

Crew J-6

	Bare Costs Hr.	Daily	Incl. Subs O&P Hr.	Daily	Bare Costs	Incl. O&P
2 Painters	$42.55	$680.80	$64.05	$1024.80	$44.06	$66.54
1 Building Laborer	39.85	318.80	60.70	485.60		
1 Equip. Oper. (light)	51.30	410.40	77.35	618.80		
1 Air Compressor, 250 cfm		167.40		184.14		
1 Sandblaster, Portable, 3 C.F.		21.00		23.10		
1 Set Sand Blasting Accessories		14.70		16.17	6.35	6.98
32 L.H., Daily Totals		$1613.10		$2352.61	$50.41	$73.52

Crew J-7

	Bare Costs Hr.	Daily	Incl. Subs O&P Hr.	Daily	Bare Costs	Incl. O&P
2 Painters	$42.55	$680.80	$64.05	$1024.80	$42.55	$64.05
1 Floor Belt Sander		16.75		18.43		
1 Floor Sanding Edger		13.15		14.47	1.87	2.06
16 L.H., Daily Totals		$710.70		$1057.69	$44.42	$66.11

Crew K-1

	Bare Costs Hr.	Daily	Incl. Subs O&P Hr.	Daily	Bare Costs	Incl. O&P
1 Carpenter	$50.70	$405.60	$77.20	$617.60	$47.60	$72.10
1 Truck Driver (light)	44.50	356.00	67.00	536.00		
1 Flatbed Truck, Gas, 3 Ton		238.00		261.80	14.88	16.36
16 L.H., Daily Totals		$999.60		$1415.40	$62.48	$88.46

Crew K-2

	Bare Costs Hr.	Daily	Incl. Subs O&P Hr.	Daily	Bare Costs	Incl. O&P
1 Struc. Steel Foreman (outside)	$56.65	$453.20	$92.60	$740.80	$51.93	$82.98
1 Struc. Steel Worker	54.65	437.20	89.35	714.80		
1 Truck Driver (light)	44.50	356.00	67.00	536.00		
1 Flatbed Truck, Gas, 3 Ton		238.00		261.80	9.92	10.91
24 L.H., Daily Totals		$1484.40		$2253.40	$61.85	$93.89

Crew L-1

	Bare Costs Hr.	Daily	Incl. Subs O&P Hr.	Daily	Bare Costs	Incl. O&P
1 Electrician	$58.20	$465.60	$87.00	$696.00	$60.17	$90.30
1 Plumber	62.15	497.20	93.60	748.80		
16 L.H., Daily Totals		$962.80		$1444.80	$60.17	$90.30

Crew L-2

	Bare Costs Hr.	Daily	Incl. Subs O&P Hr.	Daily	Bare Costs	Incl. O&P
1 Carpenter	$50.70	$405.60	$77.20	$617.60	$44.25	$67.78
1 Carpenter Helper	37.80	302.40	58.35	466.80		
16 L.H., Daily Totals		$708.00		$1084.40	$44.25	$67.78

Crew L-3

	Bare Costs Hr.	Daily	Incl. Subs O&P Hr.	Daily	Bare Costs	Incl. O&P
1 Carpenter	$50.70	$405.60	$77.20	$617.60	$54.85	$83.16
.5 Electrician	58.20	232.80	87.00	348.00		
.5 Sheet Metal Worker	59.80	239.20	91.25	365.00		
16 L.H., Daily Totals		$877.60		$1330.60	$54.85	$83.16

Crew L-3A

	Bare Costs Hr.	Daily	Incl. Subs O&P Hr.	Daily	Bare Costs	Incl. O&P
1 Carpenter Foreman (outside)	$52.70	$421.60	$80.25	$642.00	$55.07	$83.92
.5 Sheet Metal Worker	59.80	239.20	91.25	365.00		
12 L.H., Daily Totals		$660.80		$1007.00	$55.07	$83.92

Crew L-4

	Bare Costs Hr.	Daily	Incl. Subs O&P Hr.	Daily	Bare Costs	Incl. O&P
2 Skilled Workers	$52.35	$837.60	$80.15	$1282.40	$47.50	$72.88
1 Helper	37.80	302.40	58.35	466.80		
24 L.H., Daily Totals		$1140.00		$1749.20	$47.50	$72.88

Crew L-5

	Bare Costs Hr.	Daily	Incl. Subs O&P Hr.	Daily	Bare Costs	Incl. O&P
1 Struc. Steel Foreman (outside)	$56.65	$453.20	$92.60	$740.80	$55.14	$89.14
5 Struc. Steel Workers	54.65	2186.00	89.35	3574.00		
1 Equip. Oper. (crane)	56.10	448.80	84.60	676.80		
1 Hyd. Crane, 25 Ton		590.60		649.66	10.55	11.60
56 L.H., Daily Totals		$3678.60		$5641.26	$65.69	$100.74

Crew L-5A

	Bare Costs Hr.	Daily	Incl. Subs O&P Hr.	Daily	Bare Costs	Incl. O&P
1 Struc. Steel Foreman (outside)	$56.65	$453.20	$92.60	$740.80	$55.51	$88.97
2 Structural Steel Workers	54.65	874.40	89.35	1429.60		
1 Equip. Oper. (crane)	56.10	448.80	84.60	676.80		
1 S.P. Crane, 4x4, 25 Ton		668.40		735.24	20.89	22.98
32 L.H., Daily Totals		$2444.80		$3582.44	$76.40	$111.95

Crew L-5B

	Bare Costs Hr.	Daily	Incl. Subs O&P Hr.	Daily	Bare Costs	Incl. O&P
1 Struc. Steel Foreman (outside)	$56.65	$453.20	$92.60	$740.80	$57.01	$88.11
2 Structural Steel Workers	54.65	874.40	89.35	1429.60		
2 Electricians	58.20	931.20	87.00	1392.00		
2 Steamfitters/Pipefitters	63.00	1008.00	94.90	1518.40		
1 Equip. Oper. (crane)	56.10	448.80	84.60	676.80		
1 Equip. Oper. (oiler)	48.60	388.80	73.30	586.40		
1 Hyd. Crane, 80 Ton		1518.00		1669.80	21.08	23.19
72 L.H., Daily Totals		$5622.40		$8013.80	$78.09	$111.30

Crew L-6

	Bare Costs Hr.	Daily	Incl. Subs O&P Hr.	Daily	Bare Costs	Incl. O&P
1 Plumber	$62.15	$497.20	$93.60	$748.80	$60.83	$91.40
.5 Electrician	58.20	232.80	87.00	348.00		
12 L.H., Daily Totals		$730.00		$1096.80	$60.83	$91.40

Crew L-7

	Bare Costs Hr.	Daily	Incl. Subs O&P Hr.	Daily	Bare Costs	Incl. O&P
2 Carpenters	$50.70	$811.20	$77.20	$1235.20	$48.67	$73.89
1 Building Laborer	39.85	318.80	60.70	485.60		
.5 Electrician	58.20	232.80	87.00	348.00		
28 L.H., Daily Totals		$1362.80		$2068.80	$48.67	$73.89

Crew L-8

	Bare Costs Hr.	Daily	Incl. Subs O&P Hr.	Daily	Bare Costs	Incl. O&P
2 Carpenters	$50.70	$811.20	$77.20	$1235.20	$52.99	$80.48
.5 Plumber	62.15	248.60	93.60	374.40		
20 L.H., Daily Totals		$1059.80		$1609.60	$52.99	$80.48

Crew L-9

	Bare Costs Hr.	Daily	Incl. Subs O&P Hr.	Daily	Bare Costs	Incl. O&P
1 Labor Foreman (inside)	$40.35	$322.80	$61.45	$491.60	$45.29	$70.16
2 Building Laborers	39.85	637.60	60.70	971.20		
1 Struc. Steel Worker	54.65	437.20	89.35	714.80		
.5 Electrician	58.20	232.80	87.00	348.00		
36 L.H., Daily Totals		$1630.40		$2525.60	$45.29	$70.16

Crew No.	Bare Costs		Incl. Subs O&P		Cost Per Labor-Hour	
	Hr.	**Daily**	**Hr.**	**Daily**	**Bare Costs**	**Incl. O&P**
Crew L-10						
1 Struc. Steel Foreman (outside)	$56.65	$453.20	$92.60	$740.80	$55.80	$88.85
1 Structural Steel Worker	54.65	437.20	89.35	714.80		
1 Equip. Oper. (crane)	56.10	448.80	84.60	676.80		
1 Hyd. Crane, 12 Ton		496.00		545.60	20.67	22.73
24 L.H., Daily Totals		$1835.20		$2678.00	$76.47	$111.58
Crew L-11	**Hr.**	**Daily**	**Hr.**	**Daily**	**Bare Costs**	**Incl. O&P**
2 Wreckers	$39.85	$637.60	$62.70	$1003.20	$46.77	$71.84
1 Equip. Oper. (crane)	56.10	448.80	84.60	676.80		
1 Equip. Oper. (light)	51.30	410.40	77.35	618.80		
1 Hyd. Excavator, 2.5 C.Y.		1411.00		1552.10		
1 Loader, Skid Steer, 78 H.P.		364.80		401.28	55.49	61.04
32 L.H., Daily Totals		$3272.60		$4252.18	$102.27	$132.88
Crew M-1	**Hr.**	**Daily**	**Hr.**	**Daily**	**Bare Costs**	**Incl. O&P**
3 Elevator Constructors	$82.20	$1972.80	$122.50	$2940.00	$78.09	$116.36
1 Elevator Apprentice	65.75	526.00	97.95	783.60		
5 Hand Tools		50.00		55.00	1.56	1.72
32 L.H., Daily Totals		$2548.80		$3778.60	$79.65	$118.08
Crew M-3	**Hr.**	**Daily**	**Hr.**	**Daily**	**Bare Costs**	**Incl. O&P**
1 Electrician Foreman (outside)	$60.20	$481.60	$90.00	$720.00	$61.51	$92.10
1 Common Laborer	39.85	318.80	60.70	485.60		
.25 Equipment Operator (med.)	53.75	107.50	81.05	162.10		
1 Elevator Constructor	82.20	657.60	122.50	980.00		
1 Elevator Apprentice	65.75	526.00	97.95	783.60		
.25 S.P. Crane, 4x4, 20 Ton		142.80		157.08	4.20	4.62
34 L.H., Daily Totals		$2234.30		$3288.38	$65.71	$96.72
Crew M-4	**Hr.**	**Daily**	**Hr.**	**Daily**	**Bare Costs**	**Incl. O&P**
1 Electrician Foreman (outside)	$60.20	$481.60	$90.00	$720.00	$60.93	$91.25
1 Common Laborer	39.85	318.80	60.70	485.60		
.25 Equipment Operator, Crane	56.10	112.20	84.60	169.20		
.25 Equip. Oper. (oiler)	48.60	97.20	73.30	146.60		
1 Elevator Constructor	82.20	657.60	122.50	980.00		
1 Elevator Apprentice	65.75	526.00	97.95	783.60		
.25 S.P. Crane, 4x4, 40 Ton		183.70		202.07	5.10	5.61
36 L.H., Daily Totals		$2377.10		$3487.07	$66.03	$96.86
Crew Q-1	**Hr.**	**Daily**	**Hr.**	**Daily**	**Bare Costs**	**Incl. O&P**
1 Plumber	$62.15	$497.20	$93.60	$748.80	$55.92	$84.22
1 Plumber Apprentice	49.70	397.60	74.85	598.80		
16 L.H., Daily Totals		$894.80		$1347.60	$55.92	$84.22
Crew Q-1A	**Hr.**	**Daily**	**Hr.**	**Daily**	**Bare Costs**	**Incl. O&P**
.25 Plumber Foreman (outside)	$64.15	$128.30	$96.60	$193.20	$62.55	$94.20
1 Plumber	62.15	497.20	93.60	748.80		
10 L.H., Daily Totals		$625.50		$942.00	$62.55	$94.20
Crew Q-1C	**Hr.**	**Daily**	**Hr.**	**Daily**	**Bare Costs**	**Incl. O&P**
1 Plumber	$62.15	$497.20	$93.60	$748.80	$55.20	$83.17
1 Plumber Apprentice	49.70	397.60	74.85	598.80		
1 Equip. Oper. (medium)	53.75	430.00	81.05	648.40		
1 Trencher, Chain Type, 8' D		2048.00		2252.80	85.33	93.87
24 L.H., Daily Totals		$3372.80		$4248.80	$140.53	$177.03
Crew Q-2	**Hr.**	**Daily**	**Hr.**	**Daily**	**Bare Costs**	**Incl. O&P**
2 Plumbers	$62.15	$994.40	$93.60	$1497.60	$58.00	$87.35
1 Plumber Apprentice	49.70	397.60	74.85	598.80		
24 L.H., Daily Totals		$1392.00		$2096.40	$58.00	$87.35

Crew No.	Bare Costs		Incl. Subs O&P		Cost Per Labor-Hour	
	Hr.	**Daily**	**Hr.**	**Daily**	**Bare Costs**	**Incl. O&P**
Crew Q-3						
1 Plumber Foreman (inside)	$62.65	$501.20	$94.35	$754.80	$59.16	$89.10
2 Plumbers	62.15	994.40	93.60	1497.60		
1 Plumber Apprentice	49.70	397.60	74.85	598.80		
32 L.H., Daily Totals		$1893.20		$2851.20	$59.16	$89.10
Crew Q-4	**Hr.**	**Daily**	**Hr.**	**Daily**	**Bare Costs**	**Incl. O&P**
1 Plumber Foreman (inside)	$62.65	$501.20	$94.35	$754.80	$59.16	$89.10
1 Plumber	62.15	497.20	93.60	748.80		
1 Welder (plumber)	62.15	497.20	93.60	748.80		
1 Plumber Apprentice	49.70	397.60	74.85	598.80		
1 Welder, Electric, 300 amp		59.20		65.12	1.85	2.04
32 L.H., Daily Totals		$1952.40		$2916.32	$61.01	$91.14
Crew Q-5	**Hr.**	**Daily**	**Hr.**	**Daily**	**Bare Costs**	**Incl. O&P**
1 Steamfitter	$63.00	$504.00	$94.90	$759.20	$56.70	$85.40
1 Steamfitter Apprentice	50.40	403.20	75.90	607.20		
16 L.H., Daily Totals		$907.20		$1366.40	$56.70	$85.40
Crew Q-6	**Hr.**	**Daily**	**Hr.**	**Daily**	**Bare Costs**	**Incl. O&P**
2 Steamfitters	$63.00	$1008.00	$94.90	$1518.40	$58.80	$88.57
1 Steamfitter Apprentice	50.40	403.20	75.90	607.20		
24 L.H., Daily Totals		$1411.20		$2125.60	$58.80	$88.57
Crew Q-7	**Hr.**	**Daily**	**Hr.**	**Daily**	**Bare Costs**	**Incl. O&P**
1 Steamfitter Foreman (inside)	$63.50	$508.00	$95.65	$765.20	$59.98	$90.34
2 Steamfitters	63.00	1008.00	94.90	1518.40		
1 Steamfitter Apprentice	50.40	403.20	75.90	607.20		
32 L.H., Daily Totals		$1919.20		$2890.80	$59.98	$90.34
Crew Q-8	**Hr.**	**Daily**	**Hr.**	**Daily**	**Bare Costs**	**Incl. O&P**
1 Steamfitter Foreman (inside)	$63.50	$508.00	$95.65	$765.20	$59.98	$90.34
1 Steamfitter	63.00	504.00	94.90	759.20		
1 Welder (steamfitter)	63.00	504.00	94.90	759.20		
1 Steamfitter Apprentice	50.40	403.20	75.90	607.20		
1 Welder, Electric, 300 amp		59.20		65.12	1.85	2.04
32 L.H., Daily Totals		$1978.40		$2955.92	$61.83	$92.37
Crew Q-9	**Hr.**	**Daily**	**Hr.**	**Daily**	**Bare Costs**	**Incl. O&P**
1 Sheet Metal Worker	$59.80	$478.40	$91.25	$730.00	$53.83	$82.13
1 Sheet Metal Apprentice	47.85	382.80	73.00	584.00		
16 L.H., Daily Totals		$861.20		$1314.00	$53.83	$82.13
Crew Q-10	**Hr.**	**Daily**	**Hr.**	**Daily**	**Bare Costs**	**Incl. O&P**
2 Sheet Metal Workers	$59.80	$956.80	$91.25	$1460.00	$55.82	$85.17
1 Sheet Metal Apprentice	47.85	382.80	73.00	584.00		
24 L.H., Daily Totals		$1339.60		$2044.00	$55.82	$85.17
Crew Q-11	**Hr.**	**Daily**	**Hr.**	**Daily**	**Bare Costs**	**Incl. O&P**
1 Sheet Metal Foreman (inside)	$60.30	$482.40	$92.00	$736.00	$56.94	$86.88
2 Sheet Metal Workers	59.80	956.80	91.25	1460.00		
1 Sheet Metal Apprentice	47.85	382.80	73.00	584.00		
32 L.H., Daily Totals		$1822.00		$2780.00	$56.94	$86.88
Crew Q-12	**Hr.**	**Daily**	**Hr.**	**Daily**	**Bare Costs**	**Incl. O&P**
1 Sprinkler Installer	$59.90	$479.20	$90.45	$723.60	$53.90	$81.40
1 Sprinkler Apprentice	47.90	383.20	72.35	578.80		
16 L.H., Daily Totals		$862.40		$1302.40	$53.90	$81.40

For customer support on your Building Construction Costs with RSMeans data, call 800.448.8182.

Crews - Standard

Crew No.	Bare Costs		Incl. Subs O&P		Cost Per Labor-Hour	
Crew Q-13	**Hr.**	**Daily**	**Hr.**	**Daily**	**Bare Costs**	**Incl. O&P**
1 Sprinkler Foreman (inside)	$60.40	$483.20	$91.20	$729.60	$57.02	$86.11
2 Sprinkler Installers	59.90	958.40	90.45	1447.20		
1 Sprinkler Apprentice	47.90	383.20	72.35	578.80		
32 L.H., Daily Totals		$1824.80		$2755.60	$57.02	$86.11
Crew Q-14	**Hr.**	**Daily**	**Hr.**	**Daily**	**Bare Costs**	**Incl. O&P**
1 Asbestos Worker	$56.30	$450.40	$87.40	$699.20	$50.67	$78.65
1 Asbestos Apprentice	45.05	360.40	69.90	559.20		
16 L.H., Daily Totals		$810.80		$1258.40	$50.67	$78.65
Crew Q-15	**Hr.**	**Daily**	**Hr.**	**Daily**	**Bare Costs**	**Incl. O&P**
1 Plumber	$62.15	$497.20	$93.60	$748.80	$55.92	$84.22
1 Plumber Apprentice	49.70	397.60	74.85	598.80		
1 Welder, Electric, 300 amp		59.20		65.12	3.70	4.07
16 L.H., Daily Totals		$954.00		$1412.72	$59.63	$88.30
Crew Q-16	**Hr.**	**Daily**	**Hr.**	**Daily**	**Bare Costs**	**Incl. O&P**
2 Plumbers	$62.15	$994.40	$93.60	$1497.60	$58.00	$87.35
1 Plumber Apprentice	49.70	397.60	74.85	598.80		
1 Welder, Electric, 300 amp		59.20		65.12	2.47	2.71
24 L.H., Daily Totals		$1451.20		$2161.52	$60.47	$90.06
Crew Q-17	**Hr.**	**Daily**	**Hr.**	**Daily**	**Bare Costs**	**Incl. O&P**
1 Steamfitter	$63.00	$504.00	$94.90	$759.20	$56.70	$85.40
1 Steamfitter Apprentice	50.40	403.20	75.90	607.20		
1 Welder, Electric, 300 amp		59.20		65.12	3.70	4.07
16 L.H., Daily Totals		$966.40		$1431.52	$60.40	$89.47
Crew Q-17A	**Hr.**	**Daily**	**Hr.**	**Daily**	**Bare Costs**	**Incl. O&P**
1 Steamfitter	$63.00	$504.00	$94.90	$759.20	$56.50	$85.13
1 Steamfitter Apprentice	50.40	403.20	75.90	607.20		
1 Equip. Oper. (crane)	56.10	448.80	84.60	676.80		
1 Hyd. Crane, 12 Ton		496.00		545.60		
1 Welder, Electric, 300 amp		59.20		65.12	23.13	25.45
24 L.H., Daily Totals		$1911.20		$2653.92	$79.63	$110.58
Crew Q-18	**Hr.**	**Daily**	**Hr.**	**Daily**	**Bare Costs**	**Incl. O&P**
2 Steamfitters	$63.00	$1008.00	$94.90	$1518.40	$58.80	$88.57
1 Steamfitter Apprentice	50.40	403.20	75.90	607.20		
1 Welder, Electric, 300 amp		59.20		65.12	2.47	2.71
24 L.H., Daily Totals		$1470.40		$2190.72	$61.27	$91.28
Crew Q-19	**Hr.**	**Daily**	**Hr.**	**Daily**	**Bare Costs**	**Incl. O&P**
1 Steamfitter	$63.00	$504.00	$94.90	$759.20	$57.20	$85.93
1 Steamfitter Apprentice	50.40	403.20	75.90	607.20		
1 Electrician	58.20	465.60	87.00	696.00		
24 L.H., Daily Totals		$1372.80		$2062.40	$57.20	$85.93
Crew Q-20	**Hr.**	**Daily**	**Hr.**	**Daily**	**Bare Costs**	**Incl. O&P**
1 Sheet Metal Worker	$59.80	$478.40	$91.25	$730.00	$54.70	$83.10
1 Sheet Metal Apprentice	47.85	382.80	73.00	584.00		
.5 Electrician	58.20	232.80	87.00	348.00		
20 L.H., Daily Totals		$1094.00		$1662.00	$54.70	$83.10
Crew Q-21	**Hr.**	**Daily**	**Hr.**	**Daily**	**Bare Costs**	**Incl. O&P**
2 Steamfitters	$63.00	$1008.00	$94.90	$1518.40	$58.65	$88.17
1 Steamfitter Apprentice	50.40	403.20	75.90	607.20		
1 Electrician	58.20	465.60	87.00	696.00		
32 L.H., Daily Totals		$1876.80		$2821.60	$58.65	$88.17

Crew No.	Bare Costs		Incl. Subs O&P		Cost Per Labor-Hour	
Crew Q-22	**Hr.**	**Daily**	**Hr.**	**Daily**	**Bare Costs**	**Incl. O&P**
1 Plumber	$62.15	$497.20	$93.60	$748.80	$55.92	$84.22
1 Plumber Apprentice	49.70	397.60	74.85	598.80		
1 Hyd. Crane, 12 Ton		496.00		545.60	31.00	34.10
16 L.H., Daily Totals		$1390.80		$1893.20	$86.92	$118.33
Crew Q-22A	**Hr.**	**Daily**	**Hr.**	**Daily**	**Bare Costs**	**Incl. O&P**
1 Plumber	$62.15	$497.20	$93.60	$748.80	$51.95	$78.44
1 Plumber Apprentice	49.70	397.60	74.85	598.80		
1 Laborer	39.85	318.80	60.70	485.60		
1 Equip. Oper. (crane)	56.10	448.80	84.60	676.80		
1 Hyd. Crane, 12 Ton		496.00		545.60	15.50	17.05
32 L.H., Daily Totals		$2158.40		$3055.60	$67.45	$95.49
Crew Q-23	**Hr.**	**Daily**	**Hr.**	**Daily**	**Bare Costs**	**Incl. O&P**
1 Plumber Foreman (outside)	$64.15	$513.20	$96.60	$772.80	$60.02	$90.42
1 Plumber	62.15	497.20	93.60	748.80		
1 Equip. Oper. (medium)	53.75	430.00	81.05	648.40		
1 Lattice Boom Crane, 20 Ton		1083.00		1191.30	45.13	49.64
24 L.H., Daily Totals		$2523.40		$3361.30	$105.14	$140.05
Crew R-1	**Hr.**	**Daily**	**Hr.**	**Daily**	**Bare Costs**	**Incl. O&P**
1 Electrician Foreman	$58.70	$469.60	$87.75	$702.00	$54.40	$81.33
3 Electricians	58.20	1396.80	87.00	2088.00		
2 Electrician Apprentices	46.55	744.80	69.60	1113.60		
48 L.H., Daily Totals		$2611.20		$3903.60	$54.40	$81.33
Crew R-1A	**Hr.**	**Daily**	**Hr.**	**Daily**	**Bare Costs**	**Incl. O&P**
1 Electrician	$58.20	$465.60	$87.00	$696.00	$52.38	$78.30
1 Electrician Apprentice	46.55	372.40	69.60	556.80		
16 L.H., Daily Totals		$838.00		$1252.80	$52.38	$78.30
Crew R-1B	**Hr.**	**Daily**	**Hr.**	**Daily**	**Bare Costs**	**Incl. O&P**
1 Electrician	$58.20	$465.60	$87.00	$696.00	$50.43	$75.40
2 Electrician Apprentices	46.55	744.80	69.60	1113.60		
24 L.H., Daily Totals		$1210.40		$1809.60	$50.43	$75.40
Crew R-1C	**Hr.**	**Daily**	**Hr.**	**Daily**	**Bare Costs**	**Incl. O&P**
2 Electricians	$58.20	$931.20	$87.00	$1392.00	$52.38	$78.30
2 Electrician Apprentices	46.55	744.80	69.60	1113.60		
1 Portable Cable Puller, 8000 lb.		130.15		143.16	4.07	4.47
32 L.H., Daily Totals		$1806.15		$2648.76	$56.44	$82.77
Crew R-2	**Hr.**	**Daily**	**Hr.**	**Daily**	**Bare Costs**	**Incl. O&P**
1 Electrician Foreman	$58.70	$469.60	$87.75	$702.00	$54.64	$81.79
3 Electricians	58.20	1396.80	87.00	2088.00		
2 Electrician Apprentices	46.55	744.80	69.60	1113.60		
1 Equip. Oper. (crane)	56.10	448.80	84.60	676.80		
1 S.P. Crane, 4x4, 5 Ton		259.00		284.90	4.63	5.09
56 L.H., Daily Totals		$3319.00		$4865.30	$59.27	$86.88
Crew R-3	**Hr.**	**Daily**	**Hr.**	**Daily**	**Bare Costs**	**Incl. O&P**
1 Electrician Foreman	$58.70	$469.60	$87.75	$702.00	$57.98	$86.82
1 Electrician	58.20	465.60	87.00	696.00		
.5 Equip. Oper. (crane)	56.10	224.40	84.60	338.40		
.5 S.P. Crane, 4x4, 5 Ton		129.50		142.45	6.47	7.12
20 L.H., Daily Totals		$1289.10		$1878.85	$64.45	$93.94

For customer support on your Building Construction Costs with RSMeans data, call 800.448.8182.

Crews - Standard

Crew No.	Bare Costs		Incl. Subs O&P		Cost Per Labor-Hour	

Crew R-4

	Hr.	Daily	Hr.	Daily	Bare Costs	Incl. O&P
1 Struc. Steel Foreman (outside)	$56.65	$453.20	$92.60	$740.80	$55.76	$89.53
3 Struc. Steel Workers	54.65	1311.60	89.35	2144.40		
1 Electrician	58.20	465.60	87.00	696.00		
1 Welder, Gas Engine, 300 amp		98.40		108.24	2.46	2.71
40 L.H., Daily Totals		$2328.80		$3689.44	$58.22	$92.24

Crew R-5

	Hr.	Daily	Hr.	Daily	Bare Costs	Incl. O&P
1 Electrician Foreman	$58.70	$469.60	$87.75	$702.00	$50.83	$76.65
4 Electrician Linemen	58.20	1862.40	87.00	2784.00		
2 Electrician Operators	58.20	931.20	87.00	1392.00		
4 Electrician Groundmen	37.80	1209.60	58.35	1867.20		
1 Crew Truck		166.00		182.60		
1 Flatbed Truck, 20,000 GVW		197.20		216.92		
1 Pickup Truck, 3/4 Ton		115.20		126.72		
.2 Hyd. Crane, 55 Ton		198.76		218.64		
.2 Hyd. Crane, 12 Ton		99.20		109.12		
.2 Earth Auger, Truck-Mtd.		74.76		82.24		
1 Tractor w/Winch		361.80		397.98	13.78	15.16
88 L.H., Daily Totals		$5685.72		$8079.41	$64.61	$91.81

Crew R-6

	Hr.	Daily	Hr.	Daily	Bare Costs	Incl. O&P
1 Electrician Foreman	$58.70	$469.60	$87.75	$702.00	$50.83	$76.65
4 Electrician Linemen	58.20	1862.40	87.00	2784.00		
2 Electrician Operators	58.20	931.20	87.00	1392.00		
4 Electrician Groundmen	37.80	1209.60	58.35	1867.20		
1 Crew Truck		166.00		182.60		
1 Flatbed Truck, 20,000 GVW		197.20		216.92		
1 Pickup Truck, 3/4 Ton		115.20		126.72		
.2 Hyd. Crane, 55 Ton		198.76		218.64		
.2 Hyd. Crane, 12 Ton		99.20		109.12		
.2 Earth Auger, Truck-Mtd.		74.76		82.24		
1 Tractor w/Winch		361.80		397.98		
3 Cable Trailers		635.10		698.61		
.5 Tensioning Rig		209.93		230.92		
.5 Cable Pulling Rig		1213.00		1334.30	37.17	40.89
88 L.H., Daily Totals		$7743.74		$10343.24	$88.00	$117.54

Crew R-7

	Hr.	Daily	Hr.	Daily	Bare Costs	Incl. O&P
1 Electrician Foreman	$58.70	$469.60	$87.75	$702.00	$41.28	$63.25
5 Electrician Groundmen	37.80	1512.00	58.35	2334.00		
1 Crew Truck		166.00		182.60	3.46	3.80
48 L.H., Daily Totals		$2147.60		$3218.60	$44.74	$67.05

Crew R-8

	Hr.	Daily	Hr.	Daily	Bare Costs	Incl. O&P
1 Electrician Foreman	$58.70	$469.60	$87.75	$702.00	$51.48	$77.58
3 Electrician Linemen	58.20	1396.80	87.00	2088.00		
2 Electrician Groundmen	37.80	604.80	58.35	933.60		
1 Pickup Truck, 3/4 Ton		115.20		126.72		
1 Crew Truck		166.00		182.60	5.86	6.44
48 L.H., Daily Totals		$2752.40		$4032.92	$57.34	$84.02

Crew R-9

	Hr.	Daily	Hr.	Daily	Bare Costs	Incl. O&P
1 Electrician Foreman	$58.70	$469.60	$87.75	$702.00	$48.06	$72.77
1 Electrician Lineman	58.20	465.60	87.00	696.00		
2 Electrician Operators	58.20	931.20	87.00	1392.00		
4 Electrician Groundmen	37.80	1209.60	58.35	1867.20		
1 Pickup Truck, 3/4 Ton		115.20		126.72		
1 Crew Truck		166.00		182.60	4.39	4.83
64 L.H., Daily Totals		$3357.20		$4966.52	$52.46	$77.60

Crew R-10

	Hr.	Daily	Hr.	Daily	Bare Costs	Incl. O&P
1 Electrician Foreman	$58.70	$469.60	$87.75	$702.00	$54.88	$82.35
4 Electrician Linemen	58.20	1862.40	87.00	2784.00		
1 Electrician Groundman	37.80	302.40	58.35	466.80		
1 Crew Truck		166.00		182.60		
3 Tram Cars		433.80		477.18	12.50	13.75
48 L.H., Daily Totals		$3234.20		$4612.58	$67.38	$96.10

Crew R-11

	Hr.	Daily	Hr.	Daily	Bare Costs	Incl. O&P
1 Electrician Foreman	$58.70	$469.60	$87.75	$702.00	$55.35	$83.01
4 Electricians	58.20	1862.40	87.00	2784.00		
1 Equip. Oper. (crane)	56.10	448.80	84.60	676.80		
1 Common Laborer	39.85	318.80	60.70	485.60		
1 Crew Truck		166.00		182.60		
1 Hyd. Crane, 12 Ton		496.00		545.60	11.82	13.00
56 L.H., Daily Totals		$3761.60		$5376.60	$67.17	$96.01

Crew R-12

	Hr.	Daily	Hr.	Daily	Bare Costs	Incl. O&P
1 Carpenter Foreman (inside)	$51.20	$409.60	$78.00	$624.00	$47.44	$72.73
4 Carpenters	50.70	1622.40	77.20	2470.40		
4 Common Laborers	39.85	1275.20	60.70	1942.40		
1 Equip. Oper. (medium)	53.75	430.00	81.05	648.40		
1 Steel Worker	54.65	437.20	89.35	714.80		
1 Dozer, 200 H.P.		1273.00		1400.30		
1 Pickup Truck, 3/4 Ton		115.20		126.72	15.78	17.35
88 L.H., Daily Totals		$5562.60		$7927.02	$63.21	$90.08

Crew R-13

	Hr.	Daily	Hr.	Daily	Bare Costs	Incl. O&P
1 Electrician Foreman	$58.70	$469.60	$87.75	$702.00	$56.37	$84.42
3 Electricians	58.20	1396.80	87.00	2088.00		
.25 Equip. Oper. (crane)	56.10	112.20	84.60	169.20		
1 Equipment Oiler	48.60	388.80	73.30	586.40		
.25 Hydraulic Crane, 33 Ton		235.15		258.67	5.60	6.16
42 L.H., Daily Totals		$2602.55		$3804.26	$61.97	$90.58

Crew R-15

	Hr.	Daily	Hr.	Daily	Bare Costs	Incl. O&P
1 Electrician Foreman	$58.70	$469.60	$87.75	$702.00	$57.13	$85.52
4 Electricians	58.20	1862.40	87.00	2784.00		
1 Equipment Oper. (light)	51.30	410.40	77.35	618.80		
1 Telescoping Boom Lift, to 40'		282.80		311.08	5.89	6.48
48 L.H., Daily Totals		$3025.20		$4415.88	$63.02	$92.00

Crew R-15A

	Hr.	Daily	Hr.	Daily	Bare Costs	Incl. O&P
1 Electrician Foreman	$58.70	$469.60	$87.75	$702.00	$51.02	$76.75
2 Electricians	58.20	931.20	87.00	1392.00		
2 Common Laborers	39.85	637.60	60.70	971.20		
1 Equip. Oper. (light)	51.30	410.40	77.35	618.80		
1 Telescoping Boom Lift, to 40'		282.80		311.08	5.89	6.48
48 L.H., Daily Totals		$2731.60		$3995.08	$56.91	$83.23

Crew R-18

	Hr.	Daily	Hr.	Daily	Bare Costs	Incl. O&P
.25 Electrician Foreman	$58.70	$117.40	$87.75	$175.50	$51.07	$76.35
1 Electrician	58.20	465.60	87.00	696.00		
2 Electrician Apprentices	46.55	744.80	69.60	1113.60		
26 L.H., Daily Totals		$1327.80		$1985.10	$51.07	$76.35

Crew R-19

	Hr.	Daily	Hr.	Daily	Bare Costs	Incl. O&P
.5 Electrician Foreman	$58.70	$234.80	$87.75	$351.00	$58.30	$87.15
2 Electricians	58.20	931.20	87.00	1392.00		
20 L.H., Daily Totals		$1166.00		$1743.00	$58.30	$87.15

For customer support on your Building Construction Costs with RSMeans data, call 800.448.8182.

Crews - Standard

Crew No.	Bare Costs		Incl. Subs O & P		Cost Per Labor-Hour	

Crew R-21	Hr.	Daily	Hr.	Daily	Bare Costs	Incl. O&P
1 Electrician Foreman	$58.70	$469.60	$87.75	$702.00	$58.21	$87.04
3 Electricians	58.20	1396.80	87.00	2088.00		
.1 Equip. Oper. (medium)	53.75	43.00	81.05	64.84		
.1 S.P. Crane, 4x4, 25 Ton		66.84		73.52	2.04	2.24
32.8 L.H., Daily Totals		$1976.24		$2928.36	$60.25	$89.28

Crew R-22	Hr.	Daily	Hr.	Daily	Bare Costs	Incl. O&P
.66 Electrician Foreman	$58.70	$309.94	$87.75	$463.32	$53.27	$79.64
2 Electricians	58.20	931.20	87.00	1392.00		
2 Electrician Apprentices	46.55	744.80	69.60	1113.60		
37.28 L.H., Daily Totals		$1985.94		$2968.92	$53.27	$79.64

Crew R-30	Hr.	Daily	Hr.	Daily	Bare Costs	Incl. O&P
.25 Electrician Foreman (outside)	$60.20	$120.40	$90.00	$180.00	$47.06	$71.05
1 Electrician	58.20	465.60	87.00	696.00		
2 Laborers (Semi-Skilled)	39.85	637.60	60.70	971.20		
26 L.H., Daily Totals		$1223.60		$1847.20	$47.06	$71.05

Crew R-31	Hr.	Daily	Hr.	Daily	Bare Costs	Incl. O&P
1 Electrician	$58.20	$465.60	$87.00	$696.00	$58.20	$87.00
1 Core Drill, Electric, 2.5 H.P.		47.50		52.25	5.94	6.53
8 L.H., Daily Totals		$513.10		$748.25	$64.14	$93.53

Crew W-41E	Hr.	Daily	Hr.	Daily	Bare Costs	Incl. O&P
.5 Plumber Foreman (outside)	$64.15	$256.60	$96.60	$386.40	$53.63	$81.04
1 Plumber	62.15	497.20	93.60	748.80		
1 Laborer	39.85	318.80	60.70	485.60		
20 L.H., Daily Totals		$1072.60		$1620.80	$53.63	$81.04

For customer support on your Building Construction Costs with RSMeans data, call 800.448.8182.

Historical Cost Indexes

The table below lists both the RSMeans® historical cost index based on Jan. 1, 1993 = 100 as well as the computed value of an index based on Jan. 1, 2018 costs. Since the Jan. 1, 2018 figure is estimated, space is left to write in the actual index figures as they become available through the quarterly *RSMeans Construction Cost Indexes*.

To compute the actual index based on Jan. 1, 2018 = 100, divide the historical cost index for a particular year by the actual Jan. 1, 2018 construction cost index. Space has been left to advance the index figures as the year progresses.

Year	Historical Cost Index Jan. 1, 1993 = 100		Current Index Based on Jan. 1, 2018 = 100		Year	Historical Cost Index Jan. 1, 1993 = 100	Current Index Based on Jan. 1, 2018 = 100		Year	Historical Cost Index Jan. 1, 1993 = 100	Current Index Based on Jan. 1, 2018 = 100	
	Est.	Actual	Est.	Actual		Actual	Est.	Actual		Actual	Est.	Actual
Oct 2018*					July 2003	132.0	61.2		July 1985	82.6	38.3	
July 2018*					2002	128.7	59.6		1984	82.0	38.0	
April 2018*					2001	125.1	58.0		1983	80.2	37.1	
Jan 2018*	215.8		100.0	100.0	2000	120.9	56.0		1982	76.1	35.3	
July 2017		213.6	99.0		1999	117.6	54.5		1981	70.0	32.4	
2016		207.3	96.1		1998	115.1	53.3		1980	62.9	29.1	
2015		206.2	95.6		1997	112.8	52.3		1979	57.8	26.8	
2014		204.9	94.9		1996	110.2	51.1		1978	53.5	24.8	
2013		201.2	93.2		1995	107.6	49.9		1977	49.5	22.9	
2012		194.6	90.2		1994	104.4	48.4		1976	46.9	21.7	
2011		191.2	88.6		1993	101.7	47.1		1975	44.8	20.8	
2010		183.5	85.0		1992	99.4	46.1		1974	41.4	19.2	
2009		180.1	83.5		1991	96.8	44.9		1973	37.7	17.5	
2008		180.4	83.6		1990	94.3	43.7		1972	34.8	16.1	
2007		169.4	78.5		1989	92.1	42.7		1971	32.1	14.9	
2006		162.0	75.1		1988	89.9	41.6		1970	28.7	13.3	
2005		151.6	70.3		1987	87.7	40.6		1969	26.9	12.5	
2004		143.7	66.6		1986	84.2	39.0		1968	24.9	11.5	

Adjustments to Costs

The "Historical Cost Index" can be used to convert national average building costs at a particular time to the approximate building costs for some other time.

Example:

Estimate and compare construction costs for different years in the same city.

To estimate the national average construction cost of a building in 1970, knowing that it cost $900,000 in 2018:

INDEX in 1970 = 28.7

INDEX in 2018 = 215.8

Note: The city cost indexes for Canada can be used to convert U.S. national averages to local costs in Canadian dollars.

Example:

To estimate and compare the cost of a building in Toronto, ON in 2018 with the known cost of $600,000 (US$) in New York, NY in 2018:

INDEX Toronto = 110.8

INDEX New York = 134.6

$$\frac{\text{INDEX Toronto}}{\text{INDEX New York}} \times \text{Cost New York} = \text{Cost Toronto}$$

$$\frac{110.8}{134.6} \times \$600,000 = .823 \times \$600,000 = \$493,908$$

The construction cost of the building in Toronto is $493,908 (CN$).

Time Adjustment Using the Historical Cost Indexes:

$$\frac{\text{Index for Year A}}{\text{Index for Year B}} \times \text{Cost in Year B} = \text{Cost in Year A}$$

$$\frac{\text{INDEX 1970}}{\text{INDEX 2018}} \times \text{Cost 2018} = \text{Cost 1970}$$

$$\frac{28.7}{215.8} \times \$900,000 = .133 \times \$900,000 = \$119,694$$

The construction cost of the building in 1970 was $119,694.

*Historical Cost Index updates and other resources are provided on the following website: **http://info.thegordiangroup.com/RSMeans.html**

For customer support on your Building Construction Costs with RSMeans data, call 800.448.8182.

How to Use the City Cost Indexes

What you should know before you begin

RSMeans City Cost Indexes (CCI) are an extremely useful tool for when you want to compare costs from city to city and region to region.

This publication contains average construction cost indexes for 731 U.S. and Canadian cities covering over 930 three-digit zip code locations, as listed directly under each city.

Keep in mind that a City Cost Index number is a percentage ratio of a specific city's cost to the national average cost of the same item at a stated time period.

In other words, these index figures represent relative construction factors (or, if you prefer, multipliers) for material and installation costs, as well as the weighted average for Total In Place costs for each CSI MasterFormat division. Installation costs include both labor and equipment rental costs. When estimating equipment rental rates only for a specific location, use 01 54 33 EQUIPMENT RENTAL COSTS in the Reference Section.

The 30 City Average Index is the average of 30 major U.S. cities and serves as a national average.

Index figures for both material and installation are based on the 30 major city average of 100 and represent the cost relationship as of July 1, 2017. The index for each division is computed from representative material and labor quantities for that division. The weighted average for each city is a weighted total of the components listed above it. It does not include relative productivity between trades or cities.

As changes occur in local material prices, labor rates, and equipment rental rates (including fuel costs), the impact of these changes should be accurately measured by the change in the City Cost Index for each particular city (as compared to the 30 city average).

Therefore, if you know (or have estimated) building costs in one city today, you can easily convert those costs to expected building costs in another city.

In addition, by using the Historical Cost Index, you can easily convert national average building costs at a particular time to the approximate building costs for some other time. The City Cost Indexes can then be applied to calculate the costs for a particular city.

Quick calculations

Location Adjustment Using the City Cost Indexes:

$$\frac{\text{Index for City A}}{\text{Index for City B}} \times \text{Cost in City B} = \text{Cost in City A}$$

Time Adjustment for the National Average Using the Historical Cost Index:

$$\frac{\text{Index for Year A}}{\text{Index for Year B}} \times \text{Cost in Year B} = \text{Cost in Year A}$$

Adjustment from the National Average:

$$\frac{\text{Index for City A}}{100} \times \text{National Average Cost} = \text{Cost in City A}$$

Since each of the other RSMeans data sets contains many different items, any *one* item multiplied by the particular city index may give incorrect results. However, the larger the number of items compiled, the closer the results should be to actual costs for that particular city.

The City Cost Indexes for Canadian cities are calculated using Canadian material and equipment prices and labor rates in Canadian dollars. Therefore, indexes for Canadian cities can be used to convert U.S. national average prices to local costs in Canadian dollars.

How to use this section

1. Compare costs from city to city.

In using the RSMeans Indexes, remember that an index number is not a fixed number but a ratio: It's a percentage ratio of a building component's cost at any stated time to the national average cost of that same component at the same time period. Put in the form of an equation:

$$\frac{\text{Specific City Cost}}{\text{National Average Cost}} \times 100 = \text{City Index Number}$$

Therefore, when making cost comparisons between cities, do not subtract one city's index number from the index number of another city and read the result as a percentage difference. Instead, divide one city's index number by that of the other city. The resulting number may then be used as a multiplier to calculate cost differences from city to city.

The formula used to find cost differences between cities for the purpose of comparison is as follows:

$$\frac{\text{City A Index}}{\text{City B Index}} \times \text{City B Cost (Known)} = \text{City A Cost (Unknown)}$$

In addition, you can use RSMeans CCI to calculate and compare costs division by division between cities using the same basic formula. (Just be sure that you're comparing similar divisions.)

2. Compare a specific city's construction costs with the national average.

When you're studying construction location feasibility, it's advisable to compare a prospective project's cost index with an index of the national average cost.

For example, divide the weighted average index of construction costs of a specific city by that of the 30 City Average, which = 100.

$$\frac{\text{City Index}}{100} = \% \text{ of National Average}$$

As a result, you get a ratio that indicates the relative cost of construction in that city in comparison with the national average.

3. Convert U.S. national average to actual costs in Canadian City.

$$\frac{\text{Index for Canadian City}}{100} \times \text{National Average Cost} = \text{Cost in Canadian City in \$ CAN}$$

4. Adjust construction cost data based on a national average.

When you use a source of construction cost data which is based on a national average (such as RSMeans cost data), it is necessary to adjust those costs to a specific location.

$$\frac{\text{City Index}}{100} \quad \times \quad \frac{\text{Cost Based on}}{\text{National Average Costs}} \quad = \quad \frac{\text{City Cost}}{\text{(Unknown)}}$$

5. When applying the City Cost Indexes to demolition projects, use the appropriate division installation index. For example, for removal of existing doors and windows, use the Division 8 (Openings) index.

What you might like to know about how we developed the Indexes

The information presented in the CCI is organized according to the Construction Specifications Institute (CSI) MasterFormat 2014 classification system.

To create a reliable index, RSMeans researched the building type most often constructed in the United States and Canada. Because it was concluded that no one type of building completely represented the building construction industry, nine different types of buildings were combined to create a composite model.

The exact material, labor, and equipment quantities are based on detailed analyses of these nine building types, and then each quantity is weighted in proportion to expected usage. These various material items, labor hours, and equipment rental rates are thus combined to form a composite building representing as closely as possible the actual usage of materials, labor, and equipment in the North American building construction industry.

The following structures were chosen to make up that composite model:

1. Factory, 1 story
2. Office, 2–4 stories
3. Store, Retail
4. Town Hall, 2–3 stories
5. High School, 2–3 stories
6. Hospital, 4–8 stories
7. Garage, Parking
8. Apartment, 1–3 stories
9. Hotel/Motel, 2–3 stories

For the purposes of ensuring the timeliness of the data, the components of the index for the composite model have been streamlined. They currently consist of:

- specific quantities of 66 commonly used construction materials;
- specific labor-hours for 21 building construction trades; and
- specific days of equipment rental for 6 types of construction equipment (normally used to install the 66 material items by the 21 trades.) Fuel costs and routine maintenance costs are included in the equipment cost.

Material and equipment price quotations are gathered quarterly from cities in the United States and Canada. These prices and the latest negotiated labor wage rates for 21 different building trades are used to compile the quarterly update of the City Cost Index.

The 30 major U.S. cities used to calculate the national average are:

Atlanta, GA	Memphis, TN
Baltimore, MD	Milwaukee, WI
Boston, MA	Minneapolis, MN
Buffalo, NY	Nashville, TN
Chicago, IL	New Orleans, LA
Cincinnati, OH	New York, NY
Cleveland, OH	Philadelphia, PA
Columbus, OH	Phoenix, AZ
Dallas, TX	Pittsburgh, PA
Denver, CO	St. Louis, MO
Detroit, MI	San Antonio, TX
Houston, TX	San Diego, CA
Indianapolis, IN	San Francisco, CA
Kansas City, MO	Seattle, WA
Los Angeles, CA	Washington, DC

What the CCI does not indicate

The weighted average for each city is a total of the divisional components weighted to reflect typical usage. It does not include the productivity variations between trades or cities.

In addition, the CCI does not take into consideration factors such as the following:

- managerial efficiency
- competitive conditions
- automation
- restrictive union practices
- unique local requirements
- regional variations due to specific building codes

City Cost Indexes

Table 1

DIVISION		UNITED STATES 30 CITY AVERAGE			ALABAMA ANNISTON 362			BIRMINGHAM 350 - 352			BUTLER 369			DECATUR 356			DOTHAN 363		
		MAT.	INST.	TOTAL	MAT.	INST.	TOTAL	MAT.	INST.	TOTAL	MAT.	INST.	TOTAL	MAT.	INST.	TOTAL	MAT.	INST.	TOTAL
015433	CONTRACTOR EQUIPMENT		100.0	100.0		104.4	104.4		104.5	104.5		101.9	101.9		104.4	104.4		101.9	101.9
0241, 31 - 34	SITE & INFRASTRUCTURE, DEMOLITION	100.0	100.0	100.0	86.8	92.4	90.7	93.0	92.8	92.8	99.8	87.9	91.5	86.4	91.9	90.3	97.5	88.2	91.0
0310	Concrete Forming & Accessories	100.0	100.0	100.0	88.7	66.5	69.6	92.5	67.5	70.9	85.3	67.1	69.6	95.0	62.6	67.1	93.9	68.2	71.7
0320	Concrete Reinforcing	100.0	100.0	100.0	87.2	72.0	79.5	95.5	71.9	83.6	92.3	72.2	82.2	89.4	68.9	79.1	92.3	71.6	81.9
0330	Cast-in-Place Concrete	100.0	100.0	100.0	91.6	67.5	82.5	102.4	68.4	89.5	89.3	67.9	81.2	100.5	67.4	87.9	89.3	67.6	81.1
03	CONCRETE	100.0	100.0	100.0	92.3	69.6	81.9	92.6	70.3	82.4	93.2	70.0	82.6	92.4	67.2	80.9	92.5	70.3	82.3
04	MASONRY	100.0	100.0	100.0	92.9	70.6	79.1	90.4	68.0	76.5	97.6	70.6	80.9	88.3	70.1	77.0	99.0	60.0	74.8
05	METALS	100.0	100.0	100.0	92.1	95.2	93.1	92.2	95.4	93.2	91.1	96.3	92.7	94.2	94.2	94.2	91.2	95.7	92.6
06	WOOD, PLASTICS & COMPOSITES	100.0	100.0	100.0	88.9	64.7	76.8	92.2	67.1	78.2	83.7	67.1	74.5	102.1	61.5	79.5	95.3	68.6	80.5
07	THERMAL & MOISTURE PROTECTION	100.0	100.0	100.0	95.4	65.2	82.6	98.3	67.3	85.2	95.4	67.8	83.8	97.3	67.0	84.5	95.4	64.9	82.5
08	OPENINGS	100.0	100.0	100.0	95.8	68.8	89.6	101.0	68.8	93.6	95.8	68.8	89.6	107.7	64.8	97.9	95.8	69.6	89.8
0920	Plaster & Gypsum Board	100.0	100.0	100.0	88.0	66.7	73.7	87.9	66.7	73.7	85.0	66.7	72.7	93.9	60.9	71.8	95.2	68.3	77.1
0950, 0980	Ceilings & Acoustic Treatment	100.0	100.0	100.0	77.6	66.7	70.2	82.6	66.7	71.9	77.6	66.7	70.2	82.9	60.9	68.1	77.6	68.3	71.3
0960	Flooring	100.0	100.0	100.0	80.5	82.5	81.0	92.3	75.9	87.8	84.9	82.5	84.2	87.0	82.5	85.7	90.0	84.1	88.3
0970, 0990	Wall Finishes & Painting/Coating	100.0	100.0	100.0	86.2	60.0	70.9	87.8	60.0	71.6	86.2	48.9	64.5	83.5	62.6	71.3	86.2	80.5	82.9
09	FINISHES	100.0	100.0	100.0	79.2	69.2	73.7	87.3	68.2	76.9	81.9	68.0	74.4	83.4	66.0	74.0	84.6	72.5	78.1
COVERS	DIVS. 10 - 14, 25, 28, 41, 43, 44, 46	100.0	100.0	100.0	100.0	71.6	93.7	100.0	84.9	96.6	100.0	78.5	95.2	100.0	70.8	93.5	100.0	73.0	94.0
21, 22, 23	FIRE SUPPRESSION, PLUMBING & HVAC	100.0	100.0	100.0	100.9	51.3	79.7	100.0	63.1	84.2	98.2	66.9	84.8	100.0	65.9	85.5	98.2	65.3	84.1
26, 27, 3370	ELECTRICAL, COMMUNICATIONS & UTIL.	100.0	100.0	100.0	96.4	60.2	77.6	98.1	61.5	79.1	98.5	67.4	82.3	93.9	66.0	79.4	97.2	76.1	86.3
MF2016	WEIGHTED AVERAGE	100.0	100.0	100.0	94.5	68.5	83.1	95.9	71.4	85.2	94.8	72.8	85.2	96.0	71.3	85.2	94.9	72.9	85.3

Table 2 — ALABAMA

DIVISION		EVERGREEN 364			GADSDEN 359			HUNTSVILLE 357 - 358			JASPER 355			MOBILE 365 - 366			MONTGOMERY 360 - 361		
		MAT.	INST.	TOTAL	MAT.	INST.	TOTAL	MAT.	INST.	TOTAL	MAT.	INST.	TOTAL	MAT.	INST.	TOTAL	MAT.	INST.	TOTAL
015433	CONTRACTOR EQUIPMENT		101.9	101.9		104.4	104.4		104.4	104.4		104.4	104.4		101.9	101.9		101.9	101.9
0241, 31 - 34	SITE & INFRASTRUCTURE, DEMOLITION	100.2	87.9	91.6	91.9	92.4	92.3	86.1	92.3	90.4	91.8	92.5	92.3	92.9	89.0	90.1	91.3	88.2	89.1
0310	Concrete Forming & Accessories	82.1	68.0	69.9	87.2	66.9	69.7	95.0	64.3	68.5	92.3	67.2	70.7	93.0	68.0	71.4	94.6	67.0	70.8
0320	Concrete Reinforcing	92.3	72.2	82.2	95.0	71.9	83.3	89.4	71.1	80.2	89.4	72.1	80.7	89.8	71.6	80.6	97.6	71.5	84.5
0330	Cast-in-Place Concrete	89.3	67.6	81.1	100.6	67.5	84.4	97.9	66.6	86.0	111.5	68.3	95.1	93.7	67.6	83.7	90.6	67.7	81.9
03	CONCRETE	93.5	70.3	82.9	96.8	69.7	84.4	91.2	68.1	80.6	100.5	70.2	86.6	89.1	70.2	80.5	87.4	69.8	79.4
04	MASONRY	97.6	70.1	80.6	86.8	66.4	74.2	89.6	65.0	74.4	84.5	71.7	76.6	95.8	60.0	73.6	93.1	60.5	72.9
05	METALS	91.2	96.2	92.7	92.2	95.5	93.2	94.2	94.9	94.4	92.1	95.5	93.2	93.1	95.5	93.8	92.2	95.0	93.1
06	WOOD, PLASTICS & COMPOSITES	80.3	68.6	73.8	92.9	67.1	78.5	102.1	64.3	81.1	99.2	67.1	81.4	93.9	68.6	79.9	92.3	67.1	78.3
07	THERMAL & MOISTURE PROTECTION	95.4	67.8	83.7	97.5	66.6	84.4	97.2	65.7	83.9	97.5	63.7	83.2	95.0	64.9	82.3	93.6	65.0	81.5
08	OPENINGS	95.8	69.6	89.8	104.2	68.8	96.0	107.4	67.0	98.2	104.1	68.8	96.0	98.7	69.6	92.0	97.5	68.8	90.9
0920	Plaster & Gypsum Board	84.4	68.3	73.6	86.7	66.7	73.3	93.9	63.8	73.7	90.7	66.7	74.6	92.1	68.3	76.1	88.8	66.7	74.0
0950, 0980	Ceilings & Acoustic Treatment	77.6	68.3	71.3	80.3	66.7	71.1	84.8	63.8	70.7	80.3	66.7	71.1	82.9	68.3	73.0	84.2	66.7	72.5
0960	Flooring	82.9	82.5	82.8	83.3	77.5	81.7	87.0	77.5	84.3	85.3	82.5	84.5	89.5	65.1	82.8	87.8	65.1	81.5
0970, 0990	Wall Finishes & Painting/Coating	86.2	48.9	64.4	83.5	60.0	69.8	83.5	62.7	71.4	83.5	60.0	69.8	89.4	48.9	65.8	86.5	60.0	71.1
09	FINISHES	81.3	68.8	74.5	81.3	68.1	74.1	83.8	66.4	74.3	82.3	69.5	75.3	84.7	65.2	74.1	85.7	65.7	74.8
COVERS	DIVS. 10 - 14, 25, 28, 41, 43, 44, 46	100.0	71.4	93.6	100.0	84.4	96.5	100.0	83.7	96.4	100.0	72.8	93.9	100.0	85.4	96.8	100.0	84.6	96.6
21, 22, 23	FIRE SUPPRESSION, PLUMBING & HVAC	98.2	66.7	84.7	102.1	62.9	85.3	100.0	62.2	83.8	102.1	63.0	85.4	99.8	66.7	85.6	99.9	65.5	85.2
26, 27, 3370	ELECTRICAL, COMMUNICATIONS & UTIL.	95.8	67.4	81.0	93.8	61.5	77.0	94.8	66.0	79.8	93.6	60.2	76.2	99.1	58.0	77.7	99.6	76.1	87.4
MF2016	WEIGHTED AVERAGE	94.5	72.6	84.9	96.1	71.0	85.1	96.0	70.7	84.9	96.5	71.2	85.5	95.4	70.2	84.4	94.8	72.3	85.0

Table 3

DIVISION		ALABAMA PHENIX CITY 368			SELMA 367			TUSCALOOSA 354			ALASKA ANCHORAGE 995 - 996			FAIRBANKS 997			JUNEAU 998		
		MAT.	INST.	TOTAL	MAT.	INST.	TOTAL	MAT.	INST.	TOTAL	MAT.	INST.	TOTAL	MAT.	INST.	TOTAL	MAT.	INST.	TOTAL
015433	CONTRACTOR EQUIPMENT		101.9	101.9		101.9	101.9		104.4	104.4		115.3	115.3		115.3	115.3		115.3	115.3
0241, 31 - 34	SITE & INFRASTRUCTURE, DEMOLITION	103.9	88.2	92.9	97.4	88.2	90.9	86.6	92.5	90.7	127.0	130.4	129.4	126.5	130.4	129.2	147.3	130.4	135.5
0310	Concrete Forming & Accessories	88.7	67.2	70.1	86.4	67.1	69.8	94.9	67.2	71.0	125.4	119.5	120.3	130.1	118.0	119.7	131.0	119.5	121.1
0320	Concrete Reinforcing	92.3	65.9	78.9	92.3	72.1	82.1	89.4	71.9	80.6	158.3	118.8	138.4	152.5	118.8	135.5	137.9	118.8	128.3
0330	Cast-in-Place Concrete	89.3	67.8	81.2	89.3	67.8	81.1	102.0	67.8	89.0	108.1	119.1	112.3	117.0	119.4	117.9	127.8	119.1	124.5
03	CONCRETE	96.2	68.9	83.7	92.0	70.0	81.9	93.1	69.9	82.5	117.6	118.4	118.0	110.2	117.9	113.7	123.5	118.4	121.2
04	MASONRY	97.6	70.6	80.8	100.9	70.6	82.1	88.5	66.8	75.1	175.5	121.0	141.7	185.3	121.0	145.4	165.8	121.0	138.0
05	METALS	91.1	93.6	91.9	91.1	95.6	92.6	93.5	95.6	94.1	120.0	104.9	115.4	121.2	105.0	116.2	117.0	104.9	113.3
06	WOOD, PLASTICS & COMPOSITES	88.6	67.0	76.6	85.7	67.1	75.4	102.1	67.1	82.6	117.8	117.9	117.9	129.1	115.8	121.7	124.7	117.9	120.9
07	THERMAL & MOISTURE PROTECTION	95.8	67.8	84.0	95.3	67.8	83.7	97.3	66.8	84.4	169.2	117.9	147.5	180.8	116.9	153.8	183.4	117.9	155.7
08	OPENINGS	95.8	67.1	89.2	95.8	68.8	89.6	107.4	68.8	98.6	128.9	117.3	126.3	134.8	116.4	130.6	129.7	117.3	126.8
0920	Plaster & Gypsum Board	89.3	66.6	74.1	87.3	66.7	73.5	93.9	66.7	75.6	143.2	118.3	126.5	166.9	116.1	132.8	147.9	118.3	128.0
0950, 0980	Ceilings & Acoustic Treatment	77.6	66.6	70.2	77.6	66.7	70.2	84.8	66.7	72.6	131.3	118.3	122.5	130.9	116.1	120.9	143.6	118.3	126.5
0960	Flooring	86.8	82.5	85.6	85.4	82.5	84.6	87.0	77.5	84.3	119.0	125.1	120.7	117.5	125.1	119.6	125.7	125.1	125.5
0970, 0990	Wall Finishes & Painting/Coating	86.2	80.5	82.9	86.2	60.0	70.9	83.5	60.0	69.8	103.4	115.3	110.3	106.4	119.9	114.3	108.9	115.3	112.6
09	FINISHES	83.5	71.4	76.9	82.1	69.2	75.1	88.6	68.2	75.3	125.6	120.4	122.8	126.1	119.8	122.6	128.4	120.4	124.1
COVERS	DIVS. 10 - 14, 25, 28, 41, 43, 44, 46	100.0	84.6	96.6	100.0	72.1	93.8	100.0	84.6	96.6	100.0	111.1	102.5	100.0	110.9	102.4	100.0	111.1	102.5
21, 22, 23	FIRE SUPPRESSION, PLUMBING & HVAC	98.2	65.4	84.2	98.2	65.5	84.2	100.0	68.3	86.5	100.8	106.4	103.2	100.6	110.1	104.7	100.6	106.4	103.1
26, 27, 3370	ELECTRICAL, COMMUNICATIONS & UTIL.	97.9	67.7	82.2	97.0	76.1	86.1	94.3	61.5	77.3	114.4	111.5	112.9	118.2	111.5	114.7	103.1	111.5	107.5
MF2016	WEIGHTED AVERAGE	95.4	72.7	85.4	94.6	73.6	85.4	96.0	72.3	85.6	118.9	114.8	117.1	120.1	115.3	118.0	118.8	114.8	117.0

For customer support on your Building Construction Costs with RSMeans data, call 800.448.8182.

Section 1

		ALASKA			ARIZONA														
		KETCHIKAN			CHAMBERS			FLAGSTAFF			GLOBE			KINGMAN			MESA/TEMPE		
DIVISION		999			865			860			855			864			852		
		MAT.	INST.	TOTAL	MAT.	INST.	TOTAL	MAT.	INST.	TOTAL	MAT.	INST.	TOTAL	MAT.	INST.	TOTAL	MAT.	INST.	TOTAL
015433	CONTRACTOR EQUIPMENT		115.3	115.3		90.7	90.7		90.7	90.7		92.0	92.0		90.7	90.7		92.0	92.0
0241, 31 - 34	SITE & INFRASTRUCTURE, DEMOLITION	183.9	130.4	146.4	72.9	94.9	88.3	92.6	94.9	94.2	105.2	95.9	98.7	72.9	94.8	88.3	94.5	95.9	95.5
0310	Concrete Forming & Accessories	121.3	119.5	119.7	97.2	68.7	72.6	102.4	68.7	73.3	93.4	68.7	72.1	95.5	68.6	72.3	96.5	68.8	72.6
0320	Concrete Reinforcing	113.8	118.8	116.3	100.0	81.8	90.9	99.8	81.8	90.8	110.6	81.9	96.1	100.2	81.8	90.9	111.3	81.9	96.5
0330	Cast-in-Place Concrete	236.9	119.1	192.1	89.7	68.4	81.6	89.8	68.4	81.6	82.7	68.1	77.2	89.4	68.4	81.4	83.4	68.1	77.6
03	CONCRETE	187.2	118.4	155.7	92.0	70.8	82.3	111.6	70.8	93.0	103.7	70.8	88.6	91.7	70.8	82.1	94.6	70.8	83.8
04	MASONRY	192.8	121.0	148.2	95.7	59.3	73.1	95.8	58.7	72.8	103.6	59.2	76.0	95.7	59.3	73.1	103.7	59.2	76.1
05	METALS	121.3	104.9	116.3	101.6	73.5	92.9	102.1	73.5	93.3	98.7	74.4	91.3	102.2	73.4	93.4	99.0	74.5	91.5
06	WOOD, PLASTICS & COMPOSITES	119.9	117.9	118.8	100.1	70.5	83.6	105.8	70.5	86.2	97.0	70.7	82.4	95.3	70.5	81.5	100.3	70.7	83.8
07	THERMAL & MOISTURE PROTECTION	185.6	117.9	157.0	93.6	70.1	83.6	95.4	69.8	84.6	99.9	68.8	86.7	93.5	69.9	83.5	99.7	68.5	86.5
08	OPENINGS	130.4	117.3	127.4	105.9	68.4	97.3	106.0	73.0	98.4	96.4	68.8	90.1	106.1	70.8	98.0	96.5	70.9	90.6
0920	Plaster & Gypsum Board	155.6	118.3	130.6	88.6	69.9	76.0	91.8	69.9	77.1	89.0	69.9	76.2	81.2	69.9	73.6	91.3	69.9	76.9
0950, 0980	Ceilings & Acoustic Treatment	124.2	118.3	120.2	105.7	69.9	81.6	106.6	69.9	81.9	95.0	69.9	78.1	106.6	69.9	81.9	95.0	69.9	78.1
0960	Flooring	117.5	125.1	119.6	91.1	60.1	82.5	93.5	60.1	84.3	102.7	60.1	90.9	89.8	60.1	81.6	104.6	60.1	92.3
0970, 0990	Wall Finishes & Painting/Coating	106.4	115.3	111.6	88.8	64.0	74.3	88.8	64.0	74.3	94.1	64.0	76.5	88.8	64.0	74.3	94.1	64.0	76.5
09	FINISHES	127.1	120.4	123.5	90.7	66.4	77.5	93.9	66.4	78.9	97.7	66.6	80.8	89.5	66.4	76.9	97.5	66.6	80.7
COVERS	DIVS. 10 - 14, 25, 28, 41, 43, 44, 46	100.0	111.1	102.5	100.0	84.1	96.5	100.0	84.0	96.4	100.0	84.4	96.5	100.0	84.0	96.4	100.0	84.4	96.5
21, 22, 23	FIRE SUPPRESSION, PLUMBING & HVAC	98.7	106.4	102.0	98.0	77.2	89.1	100.2	78.2	90.8	97.0	77.3	88.6	98.0	78.0	89.4	100.1	77.4	90.4
26, 27, 3370	ELECTRICAL, COMMUNICATIONS & UTIL.	118.2	111.5	114.7	103.2	86.2	94.4	102.2	63.1	81.9	99.0	66.4	82.0	103.2	66.4	84.1	95.9	66.4	80.5
MF2016	WEIGHTED AVERAGE	130.8	114.8	123.8	97.9	75.0	87.9	101.7	72.1	88.8	99.2	72.4	87.4	97.9	72.5	86.8	98.2	72.5	87.0

Section 2

		ARIZONA												ARKANSAS					
		PHOENIX			PRESCOTT			SHOW LOW			TUCSON			BATESVILLE			CAMDEN		
DIVISION		850, 853			863			859			856 - 857			725			717		
		MAT.	INST.	TOTAL	MAT.	INST.	TOTAL	MAT.	INST.	TOTAL	MAT.	INST.	TOTAL	MAT.	INST.	TOTAL	MAT.	INST.	TOTAL
015433	CONTRACTOR EQUIPMENT		93.0	93.0		90.7	90.7		92.0	92.0		92.0	92.0		92.2	92.2		92.2	92.2
0241, 31 - 34	SITE & INFRASTRUCTURE, DEMOLITION	95.1	94.5	94.7	79.8	94.8	90.3	107.5	95.9	99.4	90.1	95.9	94.2	73.6	90.1	85.2	82.5	90.1	87.8
0310	Concrete Forming & Accessories	101.4	72.9	76.9	98.7	68.7	72.8	100.5	68.8	73.1	96.9	68.8	72.7	83.8	57.4	61.1	81.0	57.4	60.7
0320	Concrete Reinforcing	109.2	79.3	94.2	99.8	81.8	90.8	111.3	81.9	96.5	92.4	81.8	87.1	90.9	65.5	78.1	94.9	65.2	79.9
0330	Cast-in-Place Concrete	83.0	70.7	78.3	89.7	68.3	81.6	82.8	68.1	77.2	85.9	68.1	79.1	68.9	75.8	71.5	80.4	75.8	78.7
03	CONCRETE	97.2	73.3	86.2	97.4	70.8	85.2	106.1	70.8	89.9	93.2	70.8	83.0	73.3	66.1	70.0	79.5	66.0	73.4
04	MASONRY	96.1	62.9	75.5	95.8	59.3	73.1	103.6	59.2	76.1	90.2	58.6	70.6	92.6	43.2	62.0	109.8	37.7	65.1
05	METALS	100.5	75.7	92.9	102.1	73.4	93.3	98.5	74.4	91.1	99.8	74.5	92.0	99.1	73.7	91.3	100.9	73.3	92.4
06	WOOD, PLASTICS & COMPOSITES	103.4	74.2	87.1	101.4	70.7	84.3	104.6	70.7	85.7	100.6	70.7	83.9	93.1	58.2	73.7	93.0	58.2	73.6
07	THERMAL & MOISTURE PROTECTION	100.9	70.4	88.0	94.1	70.2	84.0	100.1	68.6	86.8	100.8	68.3	87.1	103.1	59.0	84.4	96.4	57.5	79.9
08	OPENINGS	97.7	74.4	92.4	106.0	68.8	97.5	95.7	70.9	90.0	92.9	73.1	88.4	99.7	54.2	89.2	102.6	56.1	91.9
0920	Plaster & Gypsum Board	99.2	73.3	81.8	88.7	70.1	76.2	93.3	69.9	77.6	96.2	69.9	78.5	82.0	57.5	65.5	85.1	57.5	66.5
0950, 0980	Ceilings & Acoustic Treatment	105.6	73.3	83.8	104.9	70.1	81.4	95.0	69.9	78.1	95.8	69.9	78.4	86.7	57.5	67.0	82.1	57.5	65.5
0960	Flooring	106.7	60.1	93.8	92.0	60.1	83.2	106.6	60.1	93.7	95.7	60.1	85.8	84.5	54.2	76.1	92.0	36.3	76.6
0970, 0990	Wall Finishes & Painting/Coating	98.9	66.2	79.8	88.8	64.0	74.3	94.1	64.0	76.5	94.8	64.0	76.8	90.5	36.0	58.7	93.1	50.5	68.3
09	FINISHES	102.2	69.8	84.6	91.4	66.5	77.9	99.7	66.6	81.7	95.5	66.6	79.8	78.4	53.9	65.1	80.9	52.3	65.3
COVERS	DIVS. 10 - 14, 25, 28, 41, 43, 44, 46	100.0	86.5	97.0	100.0	84.1	96.5	100.0	84.4	96.5	100.0	84.4	96.5	100.0	67.2	92.7	100.0	67.2	92.7
21, 22, 23	FIRE SUPPRESSION, PLUMBING & HVAC	99.9	80.2	91.5	100.2	77.2	90.4	97.0	77.3	88.6	100.0	77.5	90.4	97.1	53.2	78.3	97.0	57.9	80.3
26, 27, 3370	ELECTRICAL, COMMUNICATIONS & UTIL.	101.4	63.1	81.5	101.9	66.4	83.4	96.2	66.4	80.7	98.1	61.0	78.8	95.5	62.6	78.4	94.4	59.0	76.0
MF2016	WEIGHTED AVERAGE	99.5	74.1	88.4	99.3	72.2	87.5	99.4	72.5	87.6	97.1	71.8	86.0	92.6	60.8	78.7	95.0	60.5	79.9

Section 3

		ARKANSAS																	
		FAYETTEVILLE			FORT SMITH			HARRISON			HOT SPRINGS			JONESBORO			LITTLE ROCK		
DIVISION		727			729			726			719			724			720 - 722		
		MAT.	INST.	TOTAL	MAT.	INST.	TOTAL	MAT.	INST.	TOTAL	MAT.	INST.	TOTAL	MAT.	INST.	TOTAL	MAT.	INST.	TOTAL
015433	CONTRACTOR EQUIPMENT		92.2	92.2		92.2	92.2		92.2	92.2		92.2	92.2		116.6	116.6		92.2	92.2
0241, 31 - 34	SITE & INFRASTRUCTURE, DEMOLITION	73.0	90.1	85.0	78.1	90.1	86.5	78.5	90.1	86.6	85.8	90.1	88.8	97.6	108.2	105.0	86.4	90.3	89.1
0310	Concrete Forming & Accessories	79.4	57.6	60.6	97.9	57.9	63.5	88.1	57.3	61.5	78.6	57.6	60.5	87.2	57.6	61.7	92.2	59.5	64.0
0320	Concrete Reinforcing	91.0	62.8	76.7	92.1	65.2	78.6	90.6	65.6	78.0	93.1	65.5	79.2	87.8	65.7	76.6	97.4	68.4	82.7
0330	Cast-in-Place Concrete	68.9	75.9	71.6	78.8	76.1	77.7	76.4	75.8	76.2	82.3	75.9	79.9	75.1	77.2	75.9	76.3	76.2	76.3
03	CONCRETE	73.0	65.7	69.7	80.1	66.3	73.8	79.7	66.0	73.4	82.8	66.2	75.2	77.3	67.7	72.9	81.0	67.6	74.9
04	MASONRY	84.0	42.7	58.4	90.9	63.4	73.8	93.0	44.3	62.8	82.3	42.9	57.9	85.2	40.8	57.6	86.7	67.2	74.6
05	METALS	99.1	72.9	91.0	101.4	75.2	93.3	100.2	73.6	92.1	100.9	73.8	92.6	95.6	89.1	93.6	101.6	76.8	94.0
06	WOOD, PLASTICS & COMPOSITES	89.7	58.2	72.2	109.4	58.2	80.9	98.9	58.2	76.3	90.4	58.2	72.5	97.3	58.6	75.7	93.9	59.7	74.9
07	THERMAL & MOISTURE PROTECTION	103.9	58.9	84.8	104.2	64.7	87.5	103.4	59.3	84.7	96.7	58.9	80.7	108.7	58.3	87.4	98.2	65.9	84.5
08	OPENINGS	99.7	56.1	89.7	101.7	56.6	91.4	100.5	53.1	89.6	102.6	56.0	91.9	105.4	57.2	94.3	95.9	58.2	87.2
0920	Plaster & Gypsum Board	81.4	57.5	65.3	88.3	57.5	67.6	87.3	57.5	67.3	83.8	57.5	66.1	96.1	57.5	70.1	95.4	59.0	71.0
0950, 0980	Ceilings & Acoustic Treatment	86.7	57.5	67.0	88.4	57.5	67.5	88.4	57.5	67.5	82.1	57.5	65.5	90.7	57.5	68.3	90.5	59.0	69.3
0960	Flooring	81.7	53.1	73.8	91.0	75.0	86.6	86.7	54.2	77.7	90.9	53.1	80.4	61.2	48.0	57.5	90.9	83.9	88.9
0970, 0990	Wall Finishes & Painting/Coating	90.5	48.2	65.8	90.5	50.8	67.3	90.5	31.1	55.9	93.1	50.5	68.3	80.1	44.1	59.1	88.8	50.5	66.5
09	FINISHES	77.5	54.9	65.2	81.8	60.1	70.0	80.5	53.3	65.7	80.7	55.3	66.9	76.2	53.7	64.0	87.8	62.8	74.2
COVERS	DIVS. 10 - 14, 25, 28, 41, 43, 44, 46	100.0	67.2	92.7	100.0	79.3	95.4	100.0	68.3	92.9	100.0	67.2	92.7	100.0	66.3	92.5	100.0	79.5	95.4
21, 22, 23	FIRE SUPPRESSION, PLUMBING & HVAC	97.2	62.3	82.3	100.3	49.2	78.4	97.1	49.4	76.7	97.0	53.5	78.4	100.6	53.2	80.4	99.9	53.6	80.1
26, 27, 3370	ELECTRICAL, COMMUNICATIONS & UTIL.	90.0	54.7	71.6	93.0	59.0	75.4	94.2	54.7	73.6	96.3	68.8	82.0	98.8	62.8	80.1	99.7	68.8	83.7
MF2016	WEIGHTED AVERAGE	91.5	61.6	78.5	95.0	63.1	81.0	93.9	58.9	78.6	94.3	62.0	80.1	94.5	63.7	81.0	95.3	66.6	82.7

For customer support on your Building Construction Costs with RSMeans data, call 800.448.8182.

City Cost Indexes

ARKANSAS / CALIFORNIA

DIVISION		PINE BLUFF 716 MAT.	INST.	TOTAL	RUSSELLVILLE 728 MAT.	INST.	TOTAL	TEXARKANA 718 MAT.	INST.	TOTAL	WEST MEMPHIS 723 MAT.	INST.	TOTAL	ALHAMBRA 917-918 MAT.	INST.	TOTAL	ANAHEIM 928 MAT.	INST.	TOTAL
015433	CONTRACTOR EQUIPMENT		92.2	92.2		92.2	92.2		93.5	93.5		116.6	116.6		95.8	95.8		101.0	101.0
0241, 31 - 34	SITE & INFRASTRUCTURE, DEMOLITION	87.8	90.2	89.4	74.9	90.1	85.5	98.9	92.3	94.3	104.6	108.5	107.3	101.0	109.1	106.7	102.0	109.9	107.5
0310	Concrete Forming & Accessories	78.4	58.2	61.0	84.4	57.3	61.0	84.1	57.5	61.1	92.6	57.9	62.7	116.4	134.6	132.1	104.9	138.5	133.9
0320	Concrete Reinforcing	94.8	68.1	81.3	91.5	65.4	78.4	94.4	65.5	79.8	87.8	63.9	75.7	104.8	128.4	116.7	94.8	128.1	111.6
0330	Cast-in-Place Concrete	82.3	76.2	80.0	72.2	75.8	73.6	89.7	75.9	84.5	78.7	77.3	78.2	82.8	129.0	100.4	87.9	132.7	104.9
03	CONCRETE	83.5	67.0	75.9	76.1	65.9	71.5	81.9	66.1	74.7	83.8	67.5	76.4	94.6	130.3	110.9	95.6	133.4	112.9
04	MASONRY	116.6	63.4	83.6	90.2	41.8	60.2	96.5	42.2	62.8	73.9	40.6	53.3	112.0	141.0	130.0	78.0	136.0	114.0
05	METALS	101.7	76.4	93.9	99.1	73.4	91.2	93.8	73.6	87.6	94.6	88.7	92.8	86.5	109.6	93.6	104.9	111.5	106.9
06	WOOD, PLASTICS & COMPOSITES	89.9	58.2	72.3	94.4	58.2	74.2	97.4	58.2	75.6	102.9	58.6	78.2	99.6	130.3	116.7	99.7	135.3	119.5
07	THERMAL & MOISTURE PROTECTION	96.7	64.6	83.1	104.1	58.6	84.8	97.4	58.7	81.0	109.2	58.2	87.6	98.9	130.1	112.1	106.2	134.4	118.1
08	OPENINGS	103.8	57.4	93.2	99.7	55.3	89.5	108.9	55.3	96.6	102.8	56.0	92.1	91.2	131.2	100.4	107.1	134.0	113.3
0920	Plaster & Gypsum Board	83.5	57.5	66.0	82.0	57.5	65.5	86.5	57.5	67.0	98.4	57.5	70.9	96.2	131.4	119.9	105.7	136.3	126.2
0950, 0980	Ceilings & Acoustic Treatment	82.1	57.5	65.5	86.7	57.5	67.0	85.5	57.5	66.6	88.8	57.5	67.7	110.3	131.4	124.5	111.1	136.3	128.1
0960	Flooring	90.6	75.0	86.3	84.1	54.2	75.8	92.8	45.8	79.8	63.4	48.0	59.2	102.9	116.0	106.5	97.4	118.6	103.3
0970, 0990	Wall Finishes & Painting/Coating	93.1	50.8	68.4	90.5	33.2	57.1	93.1	41.7	63.1	80.1	50.5	62.9	108.0	120.9	115.5	89.3	118.5	106.4
09	FINISHES	80.6	60.1	69.5	78.5	53.6	65.0	82.9	53.1	66.7	77.5	54.6	65.1	102.3	129.3	117.0	97.4	132.8	116.7
COVERS	DIVS. 10 - 14, 25, 28, 41, 43, 44, 46	100.0	79.3	95.4	100.0	67.2	92.7	100.0	67.2	92.7	100.0	68.1	92.9	100.0	119.7	104.4	100.0	120.9	104.6
21, 22, 23	FIRE SUPPRESSION, PLUMBING & HVAC	100.2	53.5	80.2	97.2	52.9	78.2	100.2	56.7	81.6	97.5	65.4	83.8	96.9	125.5	109.1	100.0	133.2	114.2
26, 27, 3370	ELECTRICAL, COMMUNICATIONS & UTIL.	94.6	64.4	78.9	93.0	54.7	73.1	96.2	56.4	75.5	100.3	64.6	81.7	117.8	126.8	122.5	91.1	108.6	100.2
MF2016	WEIGHTED AVERAGE	96.9	65.0	82.9	92.7	59.4	78.1	95.7	60.7	80.3	94.2	66.6	82.1	98.0	125.9	110.2	99.1	125.9	110.8

CALIFORNIA

DIVISION		BAKERSFIELD 932-933 MAT.	INST.	TOTAL	BERKELEY 947 MAT.	INST.	TOTAL	EUREKA 955 MAT.	INST.	TOTAL	FRESNO 936-938 MAT.	INST.	TOTAL	INGLEWOOD 903-905 MAT.	INST.	TOTAL	LONG BEACH 906-908 MAT.	INST.	TOTAL
015433	CONTRACTOR EQUIPMENT		98.9	98.9		102.7	102.7		98.7	98.7		98.9	98.9		97.8	97.8		97.8	97.8
0241, 31 - 34	SITE & INFRASTRUCTURE, DEMOLITION	99.6	107.6	105.2	107.4	109.4	108.8	113.2	106.2	108.3	104.1	107.0	106.1	91.7	105.8	101.6	99.1	105.9	103.8
0310	Concrete Forming & Accessories	104.7	134.4	130.3	115.2	165.0	158.2	113.6	150.3	145.2	102.1	149.9	143.3	108.3	134.9	131.2	103.3	135.0	130.6
0320	Concrete Reinforcing	90.0	128.0	109.2	90.6	130.1	110.5	102.9	129.5	116.3	75.5	129.1	102.5	94.7	128.4	111.7	93.9	128.4	111.3
0330	Cast-in-Place Concrete	89.3	131.6	105.4	117.4	133.0	123.3	95.4	129.0	108.2	96.9	128.8	109.0	80.8	131.6	100.1	92.0	131.7	107.1
03	CONCRETE	88.1	131.1	107.8	102.9	145.7	122.5	105.9	137.6	120.4	92.1	137.3	112.8	88.9	131.4	108.3	98.3	131.5	113.4
04	MASONRY	92.0	135.4	118.9	118.6	155.8	141.7	103.0	150.3	132.4	96.7	145.1	126.7	74.2	141.1	115.7	83.2	141.1	119.1
05	METALS	102.6	110.5	105.1	106.6	114.8	109.1	104.7	112.8	107.2	103.0	111.9	105.8	96.3	111.4	100.9	96.2	111.5	100.9
06	WOOD, PLASTICS & COMPOSITES	96.1	130.7	115.4	110.2	170.1	143.5	113.7	153.2	135.7	103.1	153.2	130.9	100.0	130.7	117.1	94.2	130.7	114.5
07	THERMAL & MOISTURE PROTECTION	102.6	124.5	111.9	102.7	153.7	124.3	110.0	145.8	125.1	94.0	133.0	110.5	103.4	130.2	114.7	103.7	131.8	115.6
08	OPENINGS	95.1	130.0	103.1	96.2	158.5	110.5	106.3	140.2	114.1	96.7	142.4	107.2	91.0	131.4	100.3	91.0	131.4	100.3
0920	Plaster & Gypsum Board	95.3	131.4	119.5	107.3	171.5	150.4	114.5	154.5	141.4	92.8	154.5	134.2	103.7	131.4	122.3	100.1	131.4	121.1
0950, 0980	Ceilings & Acoustic Treatment	97.9	131.4	120.5	107.5	171.5	148.4	114.5	154.5	141.5	95.3	154.5	135.2	109.2	131.4	124.2	109.2	131.4	124.2
0960	Flooring	100.9	116.0	105.1	116.9	136.3	122.3	101.2	130.6	109.4	103.5	114.8	106.6	109.0	116.0	110.9	105.7	116.0	108.6
0970, 0990	Wall Finishes & Painting/Coating	91.8	108.5	101.5	108.5	165.7	141.9	90.7	127.2	112.0	103.6	120.9	113.7	107.3	120.9	115.3	107.3	120.9	115.3
09	FINISHES	94.9	128.3	113.1	106.8	161.6	136.6	102.3	146.0	126.0	94.3	142.7	120.6	106.1	129.5	118.8	105.1	129.5	118.4
COVERS	DIVS. 10 - 14, 25, 28, 41, 43, 44, 46	100.0	112.1	102.7	100.0	131.3	107.0	100.0	128.1	106.2	100.0	128.0	106.2	100.0	120.6	104.6	100.0	120.6	104.6
21, 22, 23	FIRE SUPPRESSION, PLUMBING & HVAC	100.0	123.0	109.9	97.1	161.4	124.6	96.9	122.8	108.0	100.2	128.4	112.3	96.6	122.9	107.8	96.6	125.6	109.0
26, 27, 3370	ELECTRICAL, COMMUNICATIONS & UTIL.	106.3	105.1	105.6	102.5	159.0	131.9	97.6	120.7	109.6	96.5	108.5	102.7	100.5	126.8	114.2	100.2	126.8	114.1
MF2016	WEIGHTED AVERAGE	98.3	121.3	108.3	102.5	148.9	122.8	102.4	130.0	114.4	98.2	128.2	111.3	95.5	125.5	108.6	97.1	126.1	109.8

CALIFORNIA

DIVISION		LOS ANGELES 900-902 MAT.	INST.	TOTAL	MARYSVILLE 959 MAT.	INST.	TOTAL	MODESTO 953 MAT.	INST.	TOTAL	MOJAVE 935 MAT.	INST.	TOTAL	OAKLAND 946 MAT.	INST.	TOTAL	OXNARD 930 MAT.	INST.	TOTAL
015433	CONTRACTOR EQUIPMENT		103.6	103.6		98.7	98.7		98.7	98.7		98.9	98.9		102.7	102.7		97.8	97.8
0241, 31 - 34	SITE & INFRASTRUCTURE, DEMOLITION	97.7	111.1	107.1	109.7	106.3	107.3	104.3	106.5	105.8	97.3	107.6	104.5	112.3	109.4	110.3	104.3	105.7	105.3
0310	Concrete Forming & Accessories	106.3	138.9	134.4	103.8	150.0	143.7	100.3	149.9	143.1	113.8	134.3	131.5	104.4	165.0	156.7	105.3	138.6	134.0
0320	Concrete Reinforcing	95.4	128.6	112.1	102.9	128.9	116.0	106.6	128.8	117.8	90.4	128.0	109.4	92.7	130.1	111.6	88.8	128.1	108.6
0330	Cast-in-Place Concrete	83.4	132.6	102.1	106.6	128.9	115.1	95.5	128.8	108.1	83.8	131.6	102.0	111.3	133.0	119.6	97.6	132.0	110.7
03	CONCRETE	95.8	133.7	113.1	106.7	137.4	120.7	97.9	137.3	115.9	84.1	131.1	105.6	102.9	145.7	122.5	92.2	133.2	110.9
04	MASONRY	89.0	142.4	122.1	104.0	140.4	126.5	101.3	140.4	125.5	94.8	135.4	120.0	126.1	155.8	144.5	97.8	134.6	120.6
05	METALS	102.8	113.6	106.2	104.2	112.1	106.6	101.0	111.9	104.3	100.6	110.4	103.6	101.9	114.8	105.9	98.3	111.2	102.3
06	WOOD, PLASTICS & COMPOSITES	103.9	135.7	121.6	101.2	153.2	130.1	96.9	153.2	128.2	102.2	130.7	118.1	98.8	170.1	138.5	97.4	135.4	118.5
07	THERMAL & MOISTURE PROTECTION	99.5	134.1	114.1	109.4	135.1	120.3	108.9	132.4	118.9	99.9	122.2	109.3	101.4	153.7	123.6	103.2	132.5	115.6
08	OPENINGS	100.8	134.9	108.6	105.5	143.1	114.1	104.2	143.6	113.3	91.4	130.0	100.3	96.2	158.5	110.5	94.1	134.0	103.3
0920	Plaster & Gypsum Board	99.3	136.3	124.1	103.3	154.5	137.7	105.7	154.5	138.5	102.6	131.4	121.9	101.9	171.5	148.7	96.6	136.3	123.2
0950, 0980	Ceilings & Acoustic Treatment	116.2	136.3	129.7	113.6	154.5	141.2	111.1	154.5	140.4	96.7	131.4	120.1	103.3	171.5	149.3	99.2	136.3	124.2
0960	Flooring	106.9	118.6	110.1	96.9	121.2	103.6	97.3	121.0	103.9	104.5	116.0	107.7	109.7	136.3	117.1	96.7	118.6	102.7
0970, 0990	Wall Finishes & Painting/Coating	103.9	120.9	113.8	90.7	134.7	116.4	90.7	131.9	114.7	88.6	107.0	99.3	108.5	165.7	141.9	88.6	115.4	104.2
09	FINISHES	107.1	133.1	121.3	99.3	145.2	124.3	98.8	144.9	123.9	94.6	128.0	112.8	104.8	161.6	135.7	92.1	132.5	114.1
COVERS	DIVS. 10 - 14, 25, 28, 41, 43, 44, 46	100.0	118.7	104.2	100.0	128.1	106.2	100.0	128.0	106.2	100.0	112.1	102.7	100.0	131.3	107.0	100.0	121.1	104.7
21, 22, 23	FIRE SUPPRESSION, PLUMBING & HVAC	100.0	130.1	112.9	96.9	125.4	109.1	100.0	124.1	110.3	96.9	122.4	107.8	100.2	161.4	126.4	100.0	133.2	114.2
26, 27, 3370	ELECTRICAL, COMMUNICATIONS & UTIL.	98.9	130.6	115.4	94.5	118.3	106.9	96.8	107.3	102.3	95.0	107.9	101.7	101.7	159.0	131.5	100.4	117.2	109.1
MF2016	WEIGHTED AVERAGE	99.9	129.3	112.8	101.6	128.8	113.5	100.4	126.9	112.0	95.2	121.4	106.7	102.5	148.9	122.8	97.6	126.5	110.2

For customer support on your Building Construction Costs with RSMeans data, call 800.448.8182.

City Cost Indexes

CALIFORNIA

DIVISION		PALM SPRINGS 922			PALO ALTO 943			PASADENA 910 - 912			REDDING 960			RICHMOND 948			RIVERSIDE 925		
		MAT.	INST.	TOTAL	MAT.	INST.	TOTAL	MAT.	INST.	TOTAL	MAT.	INST.	TOTAL	MAT.	INST.	TOTAL	MAT.	INST.	TOTAL
015433	CONTRACTOR EQUIPMENT		99.8	99.8		102.7	102.7		95.8	95.8		98.7	98.7		102.7	102.7		99.8	99.8
0241, 31 - 34	SITE & INFRASTRUCTURE, DEMOLITION	93.4	107.9	103.5	103.7	109.4	107.7	97.7	109.1	105.7	133.9	106.3	114.5	111.6	109.4	110.1	100.5	107.9	105.7
0310	Concrete Forming & Accessories	101.7	134.9	130.3	102.6	164.7	156.1	105.4	134.5	130.5	104.3	149.7	143.5	117.8	164.6	158.2	105.4	138.5	133.9
0320	Concrete Reinforcing	108.3	128.2	118.3	90.6	129.6	110.3	105.8	128.4	117.2	127.7	128.9	128.3	90.6	129.6	110.3	105.2	128.1	116.7
0330	Cast-in-Place Concrete	83.8	132.6	102.4	99.3	132.9	112.1	78.6	128.9	97.7	115.6	128.9	120.6	114.2	132.8	121.3	91.2	132.7	106.9
03	CONCRETE	90.1	131.7	109.1	92.6	145.5	116.8	90.4	130.2	108.6	117.1	137.2	126.3	104.8	145.4	123.4	96.0	133.4	113.1
04	MASONRY	76.0	138.5	114.8	100.7	148.3	130.2	97.5	141.0	124.5	132.5	140.4	137.4	118.4	148.3	136.9	77.0	135.7	113.4
05	METALS	105.5	111.5	107.3	99.4	114.0	103.9	86.6	109.5	93.6	106.4	112.1	108.2	99.5	113.8	103.9	104.9	111.5	106.9
06	WOOD, PLASTICS & COMPOSITES	94.7	130.6	114.7	96.4	169.9	137.3	86.5	130.3	110.9	111.3	152.7	134.3	113.4	170.1	145.0	99.7	135.3	119.5
07	THERMAL & MOISTURE PROTECTION	105.5	133.0	117.4	96.0	152.2	122.4	98.5	130.1	111.9	130.6	135.1	132.5	104.9	149.3	121.6	106.4	133.4	117.8
08	OPENINGS	102.8	131.4	110.3	96.2	157.4	110.3	91.2	131.2	100.4	119.0	142.9	124.5	96.2	157.5	110.3	105.7	134.0	112.2
0920	Plaster & Gypsum Board	100.9	131.4	121.4	100.3	171.3	148.0	90.3	131.4	117.9	104.9	154.1	137.9	108.5	171.5	150.8	104.9	136.3	126.0
0950, 0980	Ceilings & Acoustic Treatment	107.7	131.4	123.7	101.4	171.3	148.5	110.3	131.4	124.5	141.6	154.1	150.0	101.4	171.5	148.7	116.3	136.3	129.7
0960	Flooring	99.4	122.3	105.7	108.2	136.3	116.0	96.4	116.0	101.8	91.7	124.9	100.9	119.5	136.3	124.2	101.0	118.6	105.9
0970, 0990	Wall Finishes & Painting/Coating	87.8	118.5	105.7	108.5	165.7	141.9	108.0	120.9	115.5	100.4	134.7	120.4	108.5	165.7	141.9	87.8	118.5	105.7
09	FINISHES	96.0	130.3	114.7	103.2	161.4	134.9	99.3	129.3	115.6	104.6	145.6	126.9	108.3	161.6	137.3	99.1	132.8	117.4
COVERS	DIVS. 10 - 14, 25, 28, 41, 43, 44, 46	100.0	120.4	104.5	100.0	131.3	107.0	100.0	119.7	104.4	100.0	128.0	106.2	100.0	131.2	106.9	100.0	120.9	104.6
21, 22, 23	FIRE SUPPRESSION, PLUMBING & HVAC	96.9	125.3	109.1	97.1	155.6	122.1	96.9	122.8	108.0	100.5	125.4	111.1	97.1	154.1	121.5	100.0	133.2	114.2
26, 27, 3370	ELECTRICAL, COMMUNICATIONS & UTIL.	94.1	110.8	102.8	101.6	169.0	136.6	114.7	126.8	121.0	99.0	118.3	109.1	102.2	136.0	119.8	90.9	108.0	99.8
MF2016	WEIGHTED AVERAGE	97.2	123.9	108.9	98.5	148.0	120.2	96.0	125.3	108.8	109.1	128.8	117.7	101.7	143.0	119.8	99.0	125.6	110.7

CALIFORNIA

DIVISION		SACRAMENTO 942, 956 - 958			SALINAS 939			SAN BERNARDINO 923 - 924			SAN DIEGO 919 - 921			SAN FRANCISCO 940 - 941			SAN JOSE 951		
		MAT.	INST.	TOTAL	MAT.	INST.	TOTAL	MAT.	INST.	TOTAL	MAT.	INST.	TOTAL	MAT.	INST.	TOTAL	MAT.	INST.	TOTAL
015433	CONTRACTOR EQUIPMENT		101.4	101.4		98.9	98.9		99.8	99.8		100.8	100.8		109.7	109.7		100.5	100.5
0241, 31 - 34	SITE & INFRASTRUCTURE, DEMOLITION	93.6	115.5	108.9	118.0	107.1	110.4	79.5	107.9	99.4	106.2	106.5	106.4	113.9	113.0	113.3	135.4	101.4	111.6
0310	Concrete Forming & Accessories	102.5	152.8	145.9	108.4	153.1	147.0	109.2	138.4	134.4	105.3	126.0	123.1	108.4	165.8	157.9	106.2	164.9	156.8
0320	Concrete Reinforcing	86.2	129.3	107.9	89.2	129.4	109.5	105.2	128.3	116.8	102.4	128.3	115.4	105.6	130.3	118.1	94.2	129.7	112.1
0330	Cast-in-Place Concrete	94.2	129.7	106.4	96.4	129.3	108.9	63.0	132.6	89.5	90.0	119.2	101.1	124.3	133.3	127.7	110.9	132.2	119.0
03	CONCRETE	93.9	138.7	114.4	101.1	139.0	118.4	71.0	133.4	99.5	100.6	123.2	110.9	114.2	146.7	129.1	103.9	145.7	123.0
04	MASONRY	101.6	146.1	129.2	94.9	146.5	126.9	83.3	138.5	117.5	89.8	132.5	116.3	135.9	156.5	148.7	131.2	149.8	142.7
05	METALS	97.6	108.5	101.0	103.2	113.2	106.3	104.9	111.6	107.0	102.8	113.0	106.0	107.4	122.2	112.0	99.3	119.1	105.4
06	WOOD, PLASTICS & COMPOSITES	92.2	156.5	128.0	101.8	156.3	132.1	103.2	135.3	121.1	97.9	122.6	111.6	101.1	170.1	139.5	109.7	169.9	143.2
07	THERMAL & MOISTURE PROTECTION	110.6	139.1	122.7	100.5	143.3	118.6	104.5	135.1	117.5	103.6	118.2	109.8	106.5	156.0	127.4	105.4	151.6	125.0
08	OPENINGS	109.6	145.4	117.8	95.3	150.9	108.1	102.8	134.6	110.1	100.1	125.8	106.0	102.2	158.5	115.2	95.7	158.4	110.1
0920	Plaster & Gypsum Board	97.6	157.7	137.9	97.6	157.7	138.0	106.3	136.3	126.4	96.2	123.0	114.2	99.9	171.5	148.0	101.8	171.5	148.6
0950, 0980	Ceilings & Acoustic Treatment	101.4	157.7	139.3	96.7	157.7	137.8	111.1	136.3	128.1	120.8	123.0	122.3	110.6	171.5	151.7	111.2	171.5	151.9
0960	Flooring	108.4	125.7	113.2	98.7	137.9	109.6	102.9	118.6	107.3	106.2	118.6	109.6	110.2	137.9	117.9	90.7	136.3	103.4
0970, 0990	Wall Finishes & Painting/Coating	105.2	134.7	122.4	89.4	165.7	133.9	87.8	120.9	107.1	105.6	120.9	114.5	106.4	175.4	146.7	91.0	165.7	134.6
09	FINISHES	102.1	148.0	127.1	94.0	153.3	126.3	97.4	132.9	116.7	107.0	123.7	116.1	106.4	162.9	137.2	97.8	161.4	132.4
COVERS	DIVS. 10 - 14, 25, 28, 41, 43, 44, 46	100.0	128.9	106.4	100.0	128.4	106.3	100.0	118.1	104.0	100.0	114.5	103.2	100.0	131.4	107.0	100.0	130.8	106.8
21, 22, 23	FIRE SUPPRESSION, PLUMBING & HVAC	100.1	128.1	112.1	96.9	134.5	113.0	96.9	129.9	111.1	99.9	126.5	111.3	100.0	189.7	138.4	100.0	166.0	128.3
26, 27, 3370	ELECTRICAL, COMMUNICATIONS & UTIL.	97.4	122.0	110.2	96.0	128.7	113.0	94.1	111.6	103.2	101.3	103.1	102.2	101.8	174.7	139.7	99.7	170.0	136.3
MF2016	WEIGHTED AVERAGE	100.0	131.6	113.8	98.8	134.8	114.6	94.8	125.7	108.4	101.0	119.5	109.1	106.3	158.5	129.1	102.4	150.4	123.4

CALIFORNIA

DIVISION		SAN LUIS OBISPO 934			SAN MATEO 944			SAN RAFAEL 949			SANTA ANA 926 - 927			SANTA BARBARA 931			SANTA CRUZ 950		
		MAT.	INST.	TOTAL	MAT.	INST.	TOTAL	MAT.	INST.	TOTAL	MAT.	INST.	TOTAL	MAT.	INST.	TOTAL	MAT.	INST.	TOTAL
015433	CONTRACTOR EQUIPMENT		98.9	98.9		102.7	102.7		101.8	101.8		99.8	99.8		98.9	98.9		100.5	100.5
0241, 31 - 34	SITE & INFRASTRUCTURE, DEMOLITION	110.0	107.7	108.4	109.5	109.4	109.4	106.3	115.2	112.5	91.8	107.9	103.0	104.3	107.7	106.7	135.0	101.3	111.4
0310	Concrete Forming & Accessories	115.5	136.8	133.9	108.4	164.9	157.1	112.9	164.9	157.7	109.4	134.7	131.2	106.1	138.6	134.1	106.2	153.5	147.0
0320	Concrete Reinforcing	90.4	128.1	109.4	90.6	129.8	110.4	91.4	129.9	110.8	108.9	128.2	118.6	88.8	128.1	108.6	116.2	129.4	122.9
0330	Cast-in-Place Concrete	103.6	131.8	114.3	110.6	132.9	119.1	128.6	132.0	129.9	80.5	132.5	100.2	97.2	131.9	110.4	110.2	131.1	118.1
03	CONCRETE	99.5	132.3	114.5	101.6	145.6	121.7	122.8	145.1	133.0	87.7	131.6	107.8	92.0	133.1	110.8	106.4	140.1	121.8
04	MASONRY	96.5	135.6	120.7	118.1	151.6	138.8	96.9	152.5	131.4	73.1	138.8	113.8	95.1	135.6	120.2	134.9	146.7	142.2
05	METALS	101.3	111.4	104.3	99.3	114.3	103.7	100.6	110.8	103.7	105.0	111.4	106.9	98.9	111.2	102.7	106.5	117.8	110.0
06	WOOD, PLASTICS & COMPOSITES	104.3	133.2	120.4	103.6	170.1	140.6	99.4	169.9	138.6	105.1	130.6	119.3	97.4	135.4	118.5	105.4	154.8	132.7
07	THERMAL & MOISTURE PROTECTION	100.7	131.7	113.8	101.0	152.4	122.7	105.3	151.9	125.0	105.8	133.7	117.6	100.1	132.6	113.9	105.0	145.8	122.2
08	OPENINGS	93.4	131.4	102.1	96.2	157.5	110.3	107.1	158.0	118.7	102.1	131.4	108.8	95.1	134.4	104.1	97.0	150.9	109.3
0920	Plaster & Gypsum Board	103.5	134.0	124.0	105.2	171.5	149.7	106.9	171.5	150.3	108.0	131.4	123.7	96.6	136.3	123.2	110.4	157.7	142.1
0950, 0980	Ceilings & Acoustic Treatment	96.7	134.0	121.9	101.4	171.5	148.7	108.2	171.5	150.9	111.1	131.4	124.8	99.2	136.3	124.2	111.9	157.7	142.8
0960	Flooring	105.4	116.0	108.4	112.3	136.3	118.9	124.7	130.6	126.3	103.5	116.0	107.0	97.8	118.6	103.6	94.7	137.9	106.7
0970, 0990	Wall Finishes & Painting/Coating	88.6	110.9	101.6	108.5	165.7	141.9	104.6	164.8	139.7	87.8	118.5	105.7	88.6	115.4	104.2	91.2	165.7	134.7
09	FINISHES	96.0	129.9	114.4	105.5	161.6	136.0	108.9	160.3	136.9	98.9	129.2	115.4	92.6	132.3	114.2	100.4	153.4	129.2
COVERS	DIVS. 10 - 14, 25, 28, 41, 43, 44, 46	100.0	127.6	106.1	100.0	131.3	107.0	100.0	130.6	106.8	100.0	120.4	104.5	100.0	118.3	104.1	100.0	128.7	106.4
21, 22, 23	FIRE SUPPRESSION, PLUMBING & HVAC	96.9	127.2	109.9	97.1	157.3	122.9	97.1	185.0	134.7	96.9	122.0	107.6	100.0	133.2	114.2	100.0	134.9	115.0
26, 27, 3370	ELECTRICAL, COMMUNICATIONS & UTIL.	95.0	112.5	104.1	101.6	161.1	132.6	98.8	123.8	111.8	94.1	110.0	102.4	93.9	114.8	104.8	98.9	128.7	114.4
MF2016	WEIGHTED AVERAGE	98.1	124.4	109.6	100.8	147.7	121.3	103.8	148.4	123.3	96.8	122.9	108.3	96.9	126.3	109.8	104.3	135.1	117.8

For customer support on your Building Construction Costs with RSMeans data, call 800.448.8182.

DIVISION		CALIFORNIA																COLORADO		
		SANTA ROSA			STOCKTON			SUSANVILLE			VALLEJO			VAN NUYS			ALAMOSA			
		954			952			961			945			913 - 916			811			
		MAT.	INST.	TOTAL	MAT.	INST.	TOTAL	MAT.	INST.	TOTAL	MAT.	INST.	TOTAL	MAT.	INST.	TOTAL	MAT.	INST.	TOTAL	
015433	CONTRACTOR EQUIPMENT		99.3	99.3		98.7	98.7		98.7	98.7		101.8	101.8		95.8	95.8		92.8	92.8	
0241, 31 - 34	SITE & INFRASTRUCTURE, DEMOLITION	106.1	106.5	106.4	104.0	106.5	105.8	142.3	106.2	117.0	94.8	115.3	109.1	116.1	109.1	111.2	143.5	88.4	104.9	
0310	Concrete Forming & Accessories	102.6	164.2	155.7	104.2	152.2	145.6	105.4	141.9	136.9	103.6	163.8	155.5	112.1	134.5	131.4	102.5	63.8	69.1	
0320	Concrete Reinforcing	103.8	130.0	117.0	106.6	128.8	117.8	127.7	128.8	128.3	92.5	129.8	111.3	105.8	128.4	117.2	107.9	67.8	87.7	
0330	Cast-in-Place Concrete	104.7	130.5	114.5	93.0	128.8	106.6	105.1	128.8	114.1	102.5	131.1	113.4	82.9	128.9	100.4	100.0	73.6	89.9	
03	CONCRETE	106.6	144.6	124.0	96.9	138.3	115.9	119.5	133.7	126.0	98.7	144.2	119.5	104.7	130.2	116.4	110.6	68.6	91.4	
04	MASONRY	101.8	153.3	133.7	101.2	140.4	125.5	131.1	140.4	136.8	73.7	151.4	121.9	112.0	141.0	130.0	130.4	60.9	87.3	
05	METALS	105.3	115.2	108.3	101.2	111.9	104.5	105.2	111.9	107.3	100.5	110.1	103.5	85.7	109.5	93.0	103.7	78.2	95.9	
06	WOOD, PLASTICS & COMPOSITES	96.1	169.6	137.0	102.3	156.3	132.3	113.0	142.3	129.3	89.3	169.9	134.1	94.6	130.3	114.5	97.8	64.1	79.0	
07	THERMAL & MOISTURE PROTECTION	106.3	152.0	125.6	109.4	134.0	119.8	132.3	131.9	132.1	103.3	151.8	123.8	99.7	130.1	112.6	104.3	69.0	89.4	
08	OPENINGS	103.7	158.3	116.3	104.2	143.2	113.2	119.9	137.1	123.9	108.9	156.3	119.8	91.0	131.2	100.3	96.4	67.0	89.6	
0920	Plaster & Gypsum Board	102.7	171.5	148.9	105.7	157.7	140.6	105.7	143.4	131.0	101.6	171.5	148.6	94.3	131.4	119.2	78.0	63.0	67.9	
0950, 0980	Ceilings & Acoustic Treatment	111.1	171.5	151.8	119.7	157.7	145.3	134.1	143.4	140.3	110.1	171.5	151.5	107.7	131.4	123.7	102.3	63.0	75.8	
0960	Flooring	99.9	128.7	107.9	97.3	121.0	103.9	92.1	124.9	101.2	118.0	136.3	123.1	99.8	116.0	104.3	109.3	73.0	99.2	
0970, 0990	Wall Finishes & Painting/Coating	87.8	155.6	127.4	90.7	128.3	112.6	100.4	134.7	120.4	105.6	164.8	140.1	108.0	120.9	115.5	103.1	61.9	79.1	
09	FINISHES	98.0	158.5	130.9	100.5	146.3	125.4	104.4	139.5	123.5	105.4	161.0	135.6	101.7	129.2	116.7	99.6	64.8	80.7	
COVERS	DIVS. 10 - 14, 25, 28, 41, 43, 44, 46	100.0	129.7	106.6	100.0	128.4	106.3	100.0	126.9	106.0	100.0	130.2	106.7	100.0	119.7	104.4	100.0	84.4	96.5	
21, 22, 23	FIRE SUPPRESSION, PLUMBING & HVAC	96.9	182.2	133.4	100.0	124.1	110.3	97.4	125.4	109.4	100.2	143.2	118.5	96.9	122.8	108.0	96.9	65.5	83.5	
26, 27, 3370	ELECTRICAL, COMMUNICATIONS & UTIL.	94.4	124.4	110.0	96.8	110.8	104.1	99.4	118.3	109.2	95.0	126.8	111.5	114.7	126.8	121.0	97.7	65.1	80.8	
MF2016	WEIGHTED AVERAGE	101.1	147.3	121.3	100.5	127.7	112.4	108.8	127.0	116.8	99.5	139.6	117.0	99.1	125.3	110.6	103.3	69.0	88.3	

DIVISION		COLORADO																	
		BOULDER			COLORADO SPRINGS			DENVER			DURANGO			FORT COLLINS			FORT MORGAN		
		803			808 - 809			800 - 802			813			805			807		
		MAT.	INST.	TOTAL	MAT.	INST.	TOTAL	MAT.	INST.	TOTAL	MAT.	INST.	TOTAL	MAT.	INST.	TOTAL	MAT.	INST.	TOTAL
015433	CONTRACTOR EQUIPMENT		94.7	94.7		92.6	92.6		99.1	99.1		92.8	92.8		94.7	94.7		94.7	94.7
0241, 31 - 34	SITE & INFRASTRUCTURE, DEMOLITION	97.3	95.7	96.2	99.1	90.7	93.2	104.2	102.4	102.9	136.6	88.4	102.8	109.6	95.6	99.8	99.5	95.4	96.6
0310	Concrete Forming & Accessories	103.0	74.9	78.8	93.8	60.3	64.9	102.0	63.8	69.1	108.6	63.6	69.8	100.6	69.9	74.1	103.5	53.3	62.2
0320	Concrete Reinforcing	97.7	67.9	82.7	96.9	67.9	82.3	96.9	68.0	82.3	107.9	67.8	87.7	97.8	67.9	82.7	97.9	67.9	82.8
0330	Cast-in-Place Concrete	106.0	74.2	93.9	108.9	73.3	95.4	113.6	72.7	98.0	115.0	73.5	99.2	119.7	72.8	101.9	104.0	72.7	92.1
03	CONCRETE	103.2	73.8	89.7	106.8	66.9	88.6	109.4	68.3	90.6	112.8	68.4	92.5	114.2	71.0	94.5	101.6	63.5	84.2
04	MASONRY	97.3	64.7	77.1	98.9	62.7	76.4	100.1	65.1	78.4	117.1	60.9	82.2	115.2	62.2	82.3	112.5	62.2	81.3
05	METALS	98.1	77.1	91.7	101.3	77.4	93.9	103.8	78.4	96.0	103.7	78.0	95.8	99.4	77.1	92.5	97.8	76.9	91.4
06	WOOD, PLASTICS & COMPOSITES	105.7	77.7	90.1	96.1	59.1	75.5	103.0	63.6	81.1	106.7	64.1	83.0	103.3	72.2	86.0	105.7	49.9	74.6
07	THERMAL & MOISTURE PROTECTION	98.9	73.5	88.1	99.6	70.7	87.4	97.4	71.6	86.5	104.3	69.0	89.4	99.2	71.8	87.6	98.8	69.3	86.3
08	OPENINGS	98.3	74.5	92.8	102.6	64.2	93.8	105.3	66.7	96.4	103.4	67.0	95.0	98.2	71.5	92.1	98.2	59.1	89.2
0920	Plaster & Gypsum Board	116.2	77.5	90.2	100.0	58.2	71.9	111.9	63.0	79.1	90.8	63.0	72.2	110.6	71.8	84.6	116.2	48.9	71.0
0950, 0980	Ceilings & Acoustic Treatment	90.2	77.5	81.6	97.9	58.2	71.1	102.5	63.0	75.9	102.3	63.0	75.8	90.2	71.8	77.8	90.2	48.9	62.4
0960	Flooring	111.8	73.0	101.0	102.8	73.0	94.5	109.0	73.0	99.0	114.7	73.0	103.1	107.7	73.0	98.1	112.3	73.0	101.4
0970, 0990	Wall Finishes & Painting/Coating	101.5	61.9	78.4	101.2	61.9	78.3	108.2	61.9	81.2	103.1	61.9	79.1	101.5	61.9	78.4	101.5	61.9	78.4
09	FINISHES	102.8	73.4	86.8	99.4	61.8	79.0	104.8	64.6	82.9	102.1	64.8	81.8	101.4	69.5	84.0	102.9	56.3	77.6
COVERS	DIVS. 10 - 14, 25, 28, 41, 43, 44, 46	100.0	85.8	96.8	100.0	83.2	96.3	100.0	83.5	96.3	100.0	84.4	96.5	100.0	84.4	96.5	100.0	81.9	96.0
21, 22, 23	FIRE SUPPRESSION, PLUMBING & HVAC	96.9	73.3	86.8	100.1	80.6	91.8	99.9	75.5	89.5	96.9	57.3	80.0	100.0	72.0	88.0	96.9	74.8	87.4
26, 27, 3370	ELECTRICAL, COMMUNICATIONS & UTIL.	95.2	79.5	87.1	98.2	75.8	86.5	99.8	79.6	89.3	97.2	53.1	74.3	95.2	79.5	87.1	95.5	79.4	87.2
MF2016	WEIGHTED AVERAGE	98.8	75.9	88.8	101.0	73.2	88.8	102.8	74.6	90.5	103.7	65.5	87.0	102.1	74.3	89.9	99.4	71.2	87.0

DIVISION		COLORADO																	
		GLENWOOD SPRINGS			GOLDEN			GRAND JUNCTION			GREELEY			MONTROSE			PUEBLO		
		816			804			815			806			814			810		
		MAT.	INST.	TOTAL	MAT.	INST.	TOTAL	MAT.	INST.	TOTAL	MAT.	INST.	TOTAL	MAT.	INST.	TOTAL	MAT.	INST.	TOTAL
015433	CONTRACTOR EQUIPMENT		96.0	96.0		94.7	94.7		96.0	96.0		94.7	94.7		94.4	94.4		92.8	92.8
0241, 31 - 34	SITE & INFRASTRUCTURE, DEMOLITION	153.2	96.4	113.4	110.6	95.7	100.2	135.9	96.6	108.4	95.9	95.6	95.7	146.3	92.4	108.5	127.3	88.4	100.0
0310	Concrete Forming & Accessories	99.5	53.2	59.6	96.1	59.6	64.6	107.5	74.3	78.9	98.5	73.9	77.3	99.0	53.5	59.8	104.1	60.3	66.4
0320	Concrete Reinforcing	106.7	67.8	87.1	97.9	67.9	82.8	107.0	67.9	87.3	97.7	67.9	82.7	106.5	67.8	87.0	103.1	67.9	85.3
0330	Cast-in-Place Concrete	99.9	72.4	89.5	104.1	74.1	92.7	110.7	73.9	96.7	100.0	72.8	89.7	100.0	73.0	89.7	99.2	73.6	89.5
03	CONCRETE	115.7	63.4	91.8	112.7	66.8	91.7	109.3	73.5	92.9	98.3	72.8	86.6	106.8	63.7	87.1	99.5	67.0	84.7
04	MASONRY	101.7	62.1	77.2	115.6	64.7	84.0	137.4	65.4	92.7	109.0	62.2	80.0	110.2	60.9	79.6	97.9	62.0	75.7
05	METALS	103.4	77.7	95.5	98.0	77.0	91.6	105.1	78.1	96.8	99.3	77.1	92.5	102.5	78.0	95.0	106.8	78.3	98.0
06	WOOD, PLASTICS & COMPOSITES	93.1	50.0	69.2	98.3	57.1	75.4	104.4	77.0	89.2	100.6	77.6	87.8	94.4	50.2	69.8	100.3	59.4	77.5
07	THERMAL & MOISTURE PROTECTION	104.2	67.9	88.8	99.9	71.2	87.8	103.3	72.2	90.1	98.5	72.4	87.4	104.4	67.5	88.8	102.8	69.1	88.5
08	OPENINGS	102.4	59.2	92.5	98.2	63.1	90.2	103.1	74.1	96.5	98.2	74.5	92.8	103.5	59.3	93.4	98.2	64.4	90.5
0920	Plaster & Gypsum Board	120.4	48.9	72.4	108.3	56.4	73.4	132.9	76.7	95.1	109.0	77.4	87.8	77.3	48.9	58.2	82.0	58.2	66.0
0950, 0980	Ceilings & Acoustic Treatment	101.5	48.9	66.0	90.2	56.4	67.4	101.5	76.7	84.7	90.2	77.4	81.6	102.3	48.9	66.3	110.0	58.2	75.1
0960	Flooring	108.4	73.0	98.6	102.5	73.0	94.3	114.0	73.0	102.6	106.5	73.0	97.2	112.0	73.0	101.1	110.6	73.0	100.1
0970, 0990	Wall Finishes & Painting/Coating	103.1	61.9	79.1	101.5	61.9	78.4	103.1	61.9	79.1	101.5	61.9	78.4	103.1	61.9	79.1	103.1	61.9	79.1
09	FINISHES	105.3	56.5	78.8	101.0	61.2	79.4	106.9	73.0	88.5	100.0	72.7	85.2	100.3	56.6	76.5	100.2	62.0	79.4
COVERS	DIVS. 10 - 14, 25, 28, 41, 43, 44, 46	100.0	82.2	96.0	100.0	83.5	96.3	100.0	86.0	96.9	100.0	85.0	96.7	100.0	82.5	96.1	100.0	83.9	96.4
21, 22, 23	FIRE SUPPRESSION, PLUMBING & HVAC	96.9	66.6	83.9	96.9	73.3	86.8	99.9	76.4	89.9	100.0	72.0	88.0	96.9	77.1	88.4	99.9	66.6	85.7
26, 27, 3370	ELECTRICAL, COMMUNICATIONS & UTIL.	94.7	53.1	73.1	95.5	79.5	87.2	96.9	53.1	74.1	95.2	79.5	87.1	96.9	53.1	74.1	97.7	66.2	81.3
MF2016	WEIGHTED AVERAGE	103.5	65.9	87.0	101.0	72.6	88.6	105.5	73.0	91.3	99.3	75.1	88.7	102.4	67.8	87.3	101.3	68.8	87.1

City Cost Indexes

		COLORADO			CONNECTICUT												
	DIVISION	SALIDA			BRIDGEPORT			BRISTOL			HARTFORD			MERIDEN			NEW BRITAIN		
		812			066			060			061			064			060		
		MAT.	INST.	TOTAL	MAT.	INST.	TOTAL	MAT.	INST.	TOTAL	MAT.	INST.	TOTAL	MAT.	INST.	TOTAL	MAT.	INST.	TOTAL
015433	CONTRACTOR EQUIPMENT		94.4	94.4		98.8	98.8		98.8	98.8		98.8	98.8		99.3	99.3		98.8	98.8
0241, 31 - 34	SITE & INFRASTRUCTURE, DEMOLITION	136.3	92.4	105.5	105.9	103.2	104.0	105.0	103.2	103.7	100.3	103.2	102.3	102.9	103.9	103.6	105.2	103.2	103.8
0310	Concrete Forming & Accessories	107.4	53.3	60.8	102.2	118.7	116.5	102.2	118.6	116.3	102.1	118.6	116.3	101.9	118.5	116.3	102.6	118.6	116.4
0320	Concrete Reinforcing	106.2	67.8	86.8	117.4	142.1	129.8	117.4	142.1	129.8	112.6	142.1	127.5	117.4	142.1	129.8	117.4	142.1	129.8
0330	Cast-in-Place Concrete	114.5	72.9	98.7	98.7	129.3	110.3	92.5	129.2	106.4	96.9	129.2	109.2	89.0	129.2	104.3	94.0	129.2	107.4
03	CONCRETE	107.8	63.6	87.6	96.2	125.6	109.7	93.3	125.5	108.0	93.5	125.6	108.2	91.6	125.5	107.1	94.0	125.5	108.4
04	MASONRY	138.9	60.9	90.5	113.9	130.1	124.0	105.1	130.1	120.6	104.4	130.1	120.3	104.8	130.1	120.5	107.5	130.1	121.5
05	METALS	102.2	77.8	94.7	96.2	116.4	102.4	96.2	116.2	102.4	101.2	116.3	105.8	93.6	116.2	100.6	92.7	116.3	100.0
06	WOOD, PLASTICS & COMPOSITES	101.3	50.2	72.9	110.3	116.7	113.9	110.3	116.7	113.9	96.7	116.7	107.8	110.3	116.7	113.9	110.3	116.7	113.9
07	THERMAL & MOISTURE PROTECTION	103.2	67.5	88.1	100.1	124.8	110.6	100.3	121.4	109.2	105.4	121.4	112.1	100.3	121.3	109.2	100.3	121.6	109.3
08	OPENINGS	96.5	59.3	87.9	95.5	122.4	101.7	95.5	122.4	101.7	94.0	122.4	100.5	97.8	122.4	103.4	95.5	122.4	101.7
0920	Plaster & Gypsum Board	77.6	48.9	58.3	115.6	116.8	116.4	115.6	116.8	116.4	103.9	116.8	112.5	116.8	116.8	116.8	115.6	116.8	116.4
0950, 0980	Ceilings & Acoustic Treatment	102.3	48.9	66.3	95.3	116.8	109.8	95.3	116.8	109.8	93.2	116.8	109.1	98.5	116.8	110.8	95.3	116.8	109.8
0960	Flooring	117.6	73.0	105.2	87.3	131.6	99.6	87.3	124.4	97.6	90.2	131.6	101.7	87.3	124.4	97.6	87.3	124.4	97.6
0970, 0990	Wall Finishes & Painting/Coating	103.1	61.9	79.1	92.3	128.8	113.6	92.3	128.8	113.6	94.4	128.8	114.5	92.3	128.8	113.6	92.3	128.8	113.6
09	FINISHES	101.0	56.6	76.8	91.4	121.4	107.7	91.4	120.2	107.0	93.6	121.4	108.7	92.2	120.2	107.4	91.4	120.2	107.0
COVERS	DIVS. 10 - 14, 25, 28, 41, 43, 44, 46	100.0	82.6	96.1	100.0	113.2	102.9	100.0	113.2	102.9	100.0	113.2	102.9	100.0	113.2	102.9	100.0	113.2	102.9
21, 22, 23	FIRE SUPPRESSION, PLUMBING & HVAC	96.9	62.9	82.3	100.1	118.8	108.1	100.1	118.7	108.1	100.0	118.8	108.0	97.0	118.7	106.3	100.1	118.7	108.1
26, 27, 3370	ELECTRICAL, COMMUNICATIONS & UTIL.	97.1	60.8	78.2	99.0	108.3	103.8	99.0	106.3	102.8	98.5	110.5	104.8	98.9	105.7	102.5	99.0	106.3	102.8
MF2016	WEIGHTED AVERAGE	103.0	65.8	86.7	98.5	118.4	107.2	97.7	117.9	106.5	98.4	118.6	107.3	96.6	117.8	105.9	97.4	117.9	106.3

		CONNECTICUT																
	DIVISION	NEW HAVEN			NEW LONDON			NORWALK			STAMFORD			WATERBURY			WILLIMANTIC		
		065			063			068			069			067			062		
		MAT.	INST.	TOTAL	MAT.	INST.	TOTAL	MAT.	INST.	TOTAL	MAT.	INST.	TOTAL	MAT.	INST.	TOTAL	MAT.	INST.	TOTAL
015433	CONTRACTOR EQUIPMENT		99.3	99.3		99.3	99.3		98.8	98.8		98.8	98.8		98.8	98.8		98.8	98.8
0241, 31 - 34	SITE & INFRASTRUCTURE, DEMOLITION	105.1	103.7	104.1	97.4	103.7	101.8	105.6	103.1	103.9	106.3	103.2	104.1	105.7	103.2	103.9	105.6	102.9	103.7
0310	Concrete Forming & Accessories	101.9	118.6	116.3	101.9	118.6	116.3	102.2	118.7	116.4	102.2	119.1	116.7	102.2	118.6	116.4	102.2	118.5	116.2
0320	Concrete Reinforcing	117.4	142.1	129.8	92.0	142.1	117.3	117.4	142.1	129.8	117.4	142.2	129.9	117.4	142.1	129.8	117.4	142.1	129.8
0330	Cast-in-Place Concrete	95.6	129.2	108.4	81.5	129.2	99.6	97.1	129.3	109.3	98.7	129.4	110.3	98.7	129.2	110.3	92.2	129.2	106.2
03	CONCRETE	106.6	125.5	115.3	82.3	125.5	102.1	95.4	125.6	109.2	96.2	125.8	109.8	96.2	125.6	109.6	93.1	125.5	107.9
04	MASONRY	105.4	130.1	120.7	103.7	130.1	120.1	104.8	130.1	120.5	105.6	130.1	120.8	105.6	130.1	120.8	105.0	130.1	120.6
05	METALS	92.9	116.3	100.1	92.7	116.3	99.9	96.2	116.4	102.4	96.2	116.8	102.5	96.2	116.3	102.4	96.0	116.2	102.2
06	WOOD, PLASTICS & COMPOSITES	110.3	116.7	113.9	110.3	116.7	113.9	110.3	116.7	113.9	110.3	116.7	113.9	110.3	116.7	113.9	110.3	116.7	113.9
07	THERMAL & MOISTURE PROTECTION	100.4	121.7	109.4	100.2	121.4	109.1	100.3	124.8	110.7	100.3	124.8	110.7	100.3	121.8	109.4	100.5	121.0	109.2
08	OPENINGS	95.5	122.4	101.7	97.9	122.4	103.5	95.5	122.4	101.7	95.5	122.4	101.7	95.5	122.4	101.7	97.9	122.4	103.5
0920	Plaster & Gypsum Board	115.6	116.8	116.4	115.6	116.8	116.4	115.6	116.8	116.4	115.6	116.8	116.4	115.6	116.8	116.4	115.6	116.8	116.4
0950, 0980	Ceilings & Acoustic Treatment	95.3	116.8	109.8	93.4	116.8	109.2	95.3	116.8	109.8	95.3	116.8	109.8	95.3	116.8	109.8	93.4	116.8	109.2
0960	Flooring	87.3	131.6	99.6	87.3	131.6	99.6	87.3	124.4	97.6	87.3	131.6	99.6	87.3	129.1	98.9	87.3	128.0	98.6
0970, 0990	Wall Finishes & Painting/Coating	92.3	128.8	113.6	92.3	128.8	113.6	92.3	128.8	113.6	92.3	128.8	113.6	92.3	128.8	113.6	92.3	128.8	113.6
09	FINISHES	91.4	121.4	107.7	90.6	121.4	107.3	91.4	120.2	107.1	91.5	121.4	107.7	91.3	121.0	107.4	91.1	120.8	107.3
COVERS	DIVS. 10 - 14, 25, 28, 41, 43, 44, 46	100.0	113.2	102.9	100.0	113.2	102.9	100.0	113.2	102.9	100.0	113.4	103.0	100.0	113.2	102.9	100.0	113.2	102.9
21, 22, 23	FIRE SUPPRESSION, PLUMBING & HVAC	100.1	118.8	108.1	97.0	118.7	106.3	100.1	118.8	108.1	100.1	118.8	108.1	100.1	118.8	108.1	100.1	118.6	108.0
26, 27, 3370	ELECTRICAL, COMMUNICATIONS & UTIL.	98.9	112.0	105.7	96.0	109.6	103.1	99.0	110.7	105.1	99.0	158.5	130.0	98.5	111.4	105.2	99.0	110.5	105.0
MF2016	WEIGHTED AVERAGE	98.9	118.9	107.6	94.7	118.5	105.1	98.0	118.6	107.0	98.1	125.5	110.1	98.1	118.7	107.1	97.9	118.5	106.9

		D.C.			DELAWARE									FLORIDA					
	DIVISION	WASHINGTON			DOVER			NEWARK			WILMINGTON			DAYTONA BEACH			FORT LAUDERDALE		
		200 - 205			199			197			198			321			333		
		MAT.	INST.	TOTAL	MAT.	INST.	TOTAL	MAT.	INST.	TOTAL	MAT.	INST.	TOTAL	MAT.	INST.	TOTAL	MAT.	INST.	TOTAL
015433	CONTRACTOR EQUIPMENT		106.4	106.4		122.9	122.9		122.9	122.9		123.1	123.1		101.9	101.9		94.8	94.8
0241, 31 - 34	SITE & INFRASTRUCTURE, DEMOLITION	104.0	96.5	98.7	104.1	114.8	111.6	103.8	114.8	111.5	105.0	115.2	112.1	102.4	88.7	92.8	94.2	76.1	81.5
0310	Concrete Forming & Accessories	99.2	72.3	76.0	95.9	102.3	101.4	98.3	102.3	101.8	96.7	102.3	101.5	97.3	63.6	68.3	95.8	56.5	61.9
0320	Concrete Reinforcing	100.6	87.1	93.8	105.3	114.3	109.9	100.8	114.3	107.6	105.3	114.3	109.9	92.6	65.9	79.2	92.3	57.5	74.7
0330	Cast-in-Place Concrete	109.9	81.0	98.9	101.0	107.7	103.5	88.3	107.7	95.7	98.2	107.7	101.8	88.8	65.8	80.0	88.6	62.7	78.7
03	CONCRETE	107.2	79.0	94.3	96.2	107.4	101.3	91.2	107.4	98.6	95.0	107.4	100.7	86.4	66.5	77.3	87.4	60.7	75.2
04	MASONRY	101.8	82.1	89.6	101.0	100.5	100.7	103.1	100.5	101.5	101.3	100.5	100.8	89.7	59.6	71.0	94.7	63.5	75.4
05	METALS	101.9	96.6	100.3	102.0	124.3	108.9	103.5	124.3	109.9	102.0	124.3	108.9	96.0	90.9	94.5	94.8	87.8	92.7
06	WOOD, PLASTICS & COMPOSITES	99.7	70.2	83.3	91.3	100.3	96.3	96.8	100.3	98.8	87.3	100.3	94.5	100.5	62.9	79.6	90.3	56.4	71.4
07	THERMAL & MOISTURE PROTECTION	99.8	85.6	93.8	100.7	113.2	106.0	105.0	113.2	108.5	100.1	113.2	105.6	97.2	65.8	83.9	98.4	62.2	83.1
08	OPENINGS	99.8	74.6	94.0	90.6	110.5	95.2	91.5	110.5	95.9	88.7	110.5	93.7	95.3	62.0	87.7	96.2	56.3	87.0
0920	Plaster & Gypsum Board	106.8	69.3	81.6	101.9	100.3	100.8	98.9	100.3	99.8	101.3	100.3	100.6	88.7	62.4	71.0	97.7	55.7	69.5
0950, 0980	Ceilings & Acoustic Treatment	99.2	69.3	79.0	94.2	100.3	98.3	96.1	100.3	98.9	91.6	100.3	97.5	83.2	62.4	69.2	87.2	55.7	65.9
0960	Flooring	97.4	79.0	92.3	87.5	111.5	94.1	86.5	111.5	93.4	88.0	111.5	94.5	96.8	58.0	86.0	94.1	78.5	89.8
0970, 0990	Wall Finishes & Painting/Coating	103.1	74.9	86.7	88.0	110.6	101.2	86.7	110.6	100.6	83.9	110.6	99.5	94.7	62.4	75.8	87.8	63.1	73.3
09	FINISHES	96.3	72.8	83.5	92.2	104.0	98.6	88.9	104.0	97.1	91.7	104.0	98.4	88.2	61.9	73.9	88.7	60.9	73.6
COVERS	DIVS. 10 - 14, 25, 28, 41, 43, 44, 46	100.0	95.9	99.1	100.0	106.2	101.4	100.0	106.2	101.4	100.0	106.2	101.4	100.0	82.6	96.1	100.0	81.6	95.9
21, 22, 23	FIRE SUPPRESSION, PLUMBING & HVAC	100.0	89.9	95.7	100.0	122.0	109.4	100.2	122.0	109.5	100.0	122.0	109.4	99.9	76.4	89.8	99.9	58.9	82.4
26, 27, 3370	ELECTRICAL, COMMUNICATIONS & UTIL.	100.6	102.8	101.7	97.3	112.7	105.3	98.9	112.7	106.1	97.3	112.7	105.3	92.7	62.2	76.8	94.6	69.4	81.4
MF2016	WEIGHTED AVERAGE	101.1	87.6	95.2	98.0	112.4	104.3	97.9	112.4	104.3	97.6	112.5	104.1	94.9	71.0	84.4	95.1	66.0	82.4

784

FLORIDA

DIVISION		FORT MYERS 339, 341			GAINESVILLE 326, 344			JACKSONVILLE 320, 322			LAKELAND 338			MELBOURNE 329			MIAMI 330 - 332, 340		
		MAT.	INST.	TOTAL	MAT.	INST.	TOTAL	MAT.	INST.	TOTAL	MAT.	INST.	TOTAL	MAT.	INST.	TOTAL	MAT.	INST.	TOTAL
015433	CONTRACTOR EQUIPMENT		101.9	101.9		101.9	101.9		101.9	101.9		101.9	101.9		101.9	101.9		94.8	94.8
0241, 31 - 34	SITE & INFRASTRUCTURE, DEMOLITION	105.9	88.2	93.5	110.0	88.5	95.0	102.4	88.5	92.7	107.8	88.7	94.4	109.1	88.9	94.9	95.5	76.1	81.9
0310	Concrete Forming & Accessories	91.7	56.5	61.3	92.5	56.6	61.5	97.1	56.6	62.1	88.0	62.5	66.0	93.7	63.8	67.9	100.0	56.2	62.3
0320	Concrete Reinforcing	93.3	56.0	74.5	98.2	58.4	78.1	92.6	58.2	75.3	95.6	76.5	86.0	93.6	65.9	79.7	99.1	56.9	77.8
0330	Cast-in-Place Concrete	92.4	62.6	81.1	101.9	64.0	87.5	89.6	63.9	79.9	94.5	65.3	83.4	107.0	65.9	91.4	85.6	62.5	76.8
03	CONCRETE	88.0	60.4	75.4	97.0	61.4	80.7	86.8	61.3	75.2	89.6	67.6	79.6	97.5	66.6	83.4	85.8	60.5	74.2
04	MASONRY	88.4	63.5	73.0	103.0	56.6	74.2	89.4	54.1	67.6	104.5	58.8	76.2	87.5	59.6	70.2	95.0	54.4	69.8
05	METALS	96.8	87.2	93.8	94.9	87.3	92.8	94.6	87.7	92.5	96.7	94.3	95.9	104.1	91.0	100.1	95.2	86.8	92.6
06	WOOD, PLASTICS & COMPOSITES	87.2	56.4	70.0	94.4	55.5	72.8	100.5	55.5	75.5	82.4	62.1	71.1	96.1	62.9	77.6	96.2	56.4	74.0
07	THERMAL & MOISTURE PROTECTION	98.2	62.5	83.1	97.6	60.9	82.1	97.5	60.2	81.7	98.1	62.6	83.0	97.7	64.6	83.7	98.2	59.7	81.9
08	OPENINGS	97.4	56.0	87.9	94.9	56.2	86.0	95.3	56.2	86.3	97.3	64.0	89.7	94.5	64.6	87.6	98.8	56.3	89.0
0920	Plaster & Gypsum Board	93.8	55.7	68.2	85.5	54.8	64.9	88.7	54.8	65.9	90.5	61.6	71.1	85.5	62.4	70.0	87.7	55.7	66.2
0950, 0980	Ceilings & Acoustic Treatment	82.7	55.7	64.5	77.6	54.8	62.2	83.2	54.8	64.0	82.7	61.6	68.5	82.3	62.4	68.9	86.8	55.7	65.8
0960	Flooring	91.1	76.7	87.1	94.2	58.0	84.2	96.8	58.0	86.0	88.9	58.0	80.3	94.4	58.0	84.3	93.0	58.0	83.3
0970, 0990	Wall Finishes & Painting/Coating	91.7	63.0	75.0	94.7	62.4	75.8	94.7	63.0	76.2	91.7	63.0	75.0	94.7	83.5	88.2	85.8	63.1	72.5
09	FINISHES	88.4	60.5	73.2	86.6	56.7	70.4	88.2	56.8	71.1	87.4	61.2	73.2	87.4	64.2	74.8	87.5	56.7	70.7
COVERS	DIVS. 10 - 14, 25, 28, 41, 43, 44, 46	100.0	78.4	95.2	100.0	80.8	95.7	100.0	79.0	95.3	100.0	80.5	95.7	100.0	82.6	96.1	100.0	81.6	95.9
21, 22, 23	FIRE SUPPRESSION, PLUMBING & HVAC	98.2	57.0	80.6	98.7	61.9	82.9	99.9	61.9	83.6	98.2	59.5	81.7	99.9	76.4	89.8	99.9	57.3	81.7
26, 27, 3370	ELECTRICAL, COMMUNICATIONS & UTIL.	96.4	62.9	79.0	92.9	59.0	75.3	92.4	64.1	77.7	94.8	63.4	78.5	93.7	61.0	76.7	98.2	73.3	85.2
MF2016	WEIGHTED AVERAGE	95.4	65.4	82.3	96.4	65.0	82.7	94.7	65.3	81.8	96.2	67.8	83.7	97.6	71.2	86.0	95.6	64.6	82.0

FLORIDA

DIVISION		ORLANDO 327 - 328, 347			PANAMA CITY 324			PENSACOLA 325			SARASOTA 342			ST. PETERSBURG 337			TALLAHASSEE 323		
		MAT.	INST.	TOTAL	MAT.	INST.	TOTAL	MAT.	INST.	TOTAL	MAT.	INST.	TOTAL	MAT.	INST.	TOTAL	MAT.	INST.	TOTAL
015433	CONTRACTOR EQUIPMENT		101.9	101.9		101.9	101.9		101.9	101.9		101.9	101.9		101.9	101.9		101.9	101.9
0241, 31 - 34	SITE & INFRASTRUCTURE, DEMOLITION	99.7	88.7	92.0	113.1	88.0	95.5	113.8	88.5	96.0	113.3	88.7	96.1	109.4	88.3	94.6	101.7	88.5	92.5
0310	Concrete Forming & Accessories	100.1	63.3	68.4	96.4	61.4	66.2	94.5	64.7	68.8	95.4	62.5	67.0	95.2	60.3	65.2	97.1	56.6	62.2
0320	Concrete Reinforcing	96.2	65.9	80.9	96.7	69.1	82.8	99.2	68.6	83.7	93.3	76.5	84.8	95.6	76.5	86.0	99.8	58.4	78.9
0330	Cast-in-Place Concrete	105.9	65.7	90.7	94.2	59.9	81.2	116.4	65.1	96.9	103.2	65.3	88.8	95.5	62.1	82.8	91.4	64.0	81.0
03	CONCRETE	94.1	66.3	81.4	95.1	64.0	80.9	104.9	67.2	87.7	95.2	67.6	82.6	91.1	65.5	79.4	88.3	61.4	76.0
04	MASONRY	95.3	59.6	73.2	94.1	58.6	72.1	114.3	58.7	79.8	92.0	58.8	71.4	144.1	53.2	87.7	91.7	56.6	69.9
05	METALS	95.2	90.6	93.8	95.8	92.1	94.6	96.8	91.4	95.1	98.7	94.3	97.4	97.6	93.2	96.3	92.3	87.9	91.0
06	WOOD, PLASTICS & COMPOSITES	89.8	62.9	74.9	99.3	65.4	80.4	97.3	65.4	79.5	98.1	62.1	78.1	91.7	62.1	75.2	96.0	55.5	73.5
07	THERMAL & MOISTURE PROTECTION	100.0	65.8	85.5	97.8	59.9	81.8	97.7	63.0	83.0	98.2	62.6	83.1	98.3	60.0	82.1	96.9	60.8	81.7
08	OPENINGS	98.1	62.0	89.8	93.3	64.0	86.6	93.3	64.0	86.6	99.5	64.0	91.3	96.1	63.6	88.6	100.1	56.2	90.0
0920	Plaster & Gypsum Board	94.7	62.4	73.0	87.8	64.9	72.4	95.5	64.9	75.0	94.7	61.6	72.5	96.1	61.6	72.9	93.3	54.8	67.4
0950, 0980	Ceilings & Acoustic Treatment	89.4	62.4	71.2	82.3	64.9	70.6	82.3	64.9	70.6	87.3	61.6	69.9	84.6	61.6	69.1	87.9	54.8	65.6
0960	Flooring	91.3	58.0	82.1	96.3	76.7	90.9	92.2	58.0	82.7	101.4	58.0	89.4	93.0	58.0	83.3	92.9	58.0	83.2
0970, 0990	Wall Finishes & Painting/Coating	93.4	63.0	75.7	94.7	60.6	74.8	94.7	62.4	75.8	96.3	63.0	76.9	91.7	63.6	75.3	91.9	63.0	75.1
09	FINISHES	90.4	61.9	74.9	88.8	64.3	75.5	88.4	63.1	74.6	94.3	61.2	76.3	89.9	59.9	73.6	89.8	56.8	71.8
COVERS	DIVS. 10 - 14, 25, 28, 41, 43, 44, 46	100.0	82.6	96.1	100.0	78.0	95.1	100.0	81.1	95.8	100.0	80.5	95.7	100.0	78.6	95.2	100.0	78.6	95.2
21, 22, 23	FIRE SUPPRESSION, PLUMBING & HVAC	100.0	57.8	81.9	99.9	51.9	79.4	99.9	63.0	84.1	99.9	59.5	82.6	99.9	56.6	81.4	99.9	67.8	86.2
26, 27, 3370	ELECTRICAL, COMMUNICATIONS & UTIL.	96.6	62.7	78.9	91.6	59.0	74.6	94.8	53.5	73.3	95.7	63.9	79.2	94.8	64.0	78.8	96.7	59.0	77.1
MF2016	WEIGHTED AVERAGE	96.8	67.0	83.8	96.2	65.1	82.5	98.8	67.1	84.9	98.1	67.9	84.9	99.0	65.9	84.5	95.6	66.2	82.7

		FLORIDA						GEORGIA											
DIVISION		TAMPA 335 - 336, 346			WEST PALM BEACH 334, 349			ALBANY 317, 398			ATHENS 306			ATLANTA 300 - 303, 399			AUGUSTA 308 - 309		
		MAT.	INST.	TOTAL	MAT.	INST.	TOTAL	MAT.	INST.	TOTAL	MAT.	INST.	TOTAL	MAT.	INST.	TOTAL	MAT.	INST.	TOTAL
015433	CONTRACTOR EQUIPMENT		101.9	101.9		94.8	94.8		95.7	95.7		94.5	94.5		96.4	96.4		94.5	94.5
0241, 31 - 34	SITE & INFRASTRUCTURE, DEMOLITION	109.9	88.7	95.1	91.1	76.1	80.6	104.3	79.0	86.6	102.6	95.1	97.3	99.4	95.3	96.5	95.6	95.4	95.5
0310	Concrete Forming & Accessories	98.3	62.4	67.4	99.4	56.3	62.2	94.0	68.9	72.4	91.4	45.1	51.5	96.0	71.7	75.1	92.5	72.9	75.6
0320	Concrete Reinforcing	92.3	76.5	84.3	94.9	56.9	75.7	88.1	71.9	79.9	102.0	65.6	83.6	101.2	72.0	86.5	102.4	68.1	85.1
0330	Cast-in-Place Concrete	93.4	65.3	82.7	84.2	62.5	76.0	88.3	69.0	81.0	109.3	69.9	94.3	112.7	71.4	97.0	103.3	70.2	90.7
03	CONCRETE	89.8	67.6	79.7	84.4	60.5	73.5	86.4	71.2	79.5	104.9	58.6	83.7	107.1	72.4	91.2	97.3	71.8	85.6
04	MASONRY	95.7	58.8	72.8	94.3	54.4	69.5	92.2	69.2	77.9	78.5	79.4	79.1	91.9	69.4	77.9	92.2	69.3	78.0
05	METALS	96.7	94.3	96.0	93.9	86.8	91.7	97.8	98.0	97.9	93.1	80.0	89.1	93.9	85.1	91.2	92.8	81.3	89.3
06	WOOD, PLASTICS & COMPOSITES	95.7	62.1	77.0	95.3	56.4	73.6	85.9	69.6	76.8	97.0	37.4	63.9	99.2	73.1	84.7	98.5	75.3	85.6
07	THERMAL & MOISTURE PROTECTION	98.6	62.6	83.3	98.1	61.3	82.5	97.3	68.9	85.3	97.9	70.2	86.2	99.4	72.4	88.0	97.7	71.7	86.7
08	OPENINGS	97.4	64.0	89.7	95.7	56.3	86.7	87.7	71.6	84.0	93.1	52.0	83.6	97.6	73.5	92.1	93.1	73.8	88.7
0920	Plaster & Gypsum Board	98.4	61.6	73.7	102.1	55.7	70.9	103.6	69.3	80.5	93.6	36.1	55.0	96.0	72.5	80.2	94.9	75.0	81.6
0950, 0980	Ceilings & Acoustic Treatment	87.2	61.6	69.9	82.7	55.7	64.5	83.8	69.3	74.0	98.8	36.1	56.6	91.6	72.5	78.7	99.8	75.0	83.1
0960	Flooring	94.1	58.0	84.1	95.9	58.0	85.4	96.9	67.4	88.7	95.8	86.6	93.3	98.8	67.4	90.1	96.0	67.4	88.1
0970, 0990	Wall Finishes & Painting/Coating	91.7	63.0	75.0	87.8	62.4	72.9	89.1	95.9	93.1	97.5	95.9	96.6	101.3	95.9	98.2	97.5	80.4	87.5
09	FINISHES	91.1	61.2	74.9	88.7	56.7	71.3	92.2	71.1	80.7	96.2	56.0	74.4	96.3	73.2	83.7	96.0	72.8	83.4
COVERS	DIVS. 10 - 14, 25, 28, 41, 43, 44, 46	100.0	80.5	95.7	100.0	81.6	95.9	100.0	84.9	96.6	100.0	81.2	95.8	100.0	86.0	96.9	100.0	85.3	96.7
21, 22, 23	FIRE SUPPRESSION, PLUMBING & HVAC	99.9	59.5	82.6	98.2	56.8	80.5	99.9	71.1	87.6	97.0	68.5	84.8	100.0	70.1	87.2	100.1	69.2	86.9
26, 27, 3370	ELECTRICAL, COMMUNICATIONS & UTIL.	94.5	62.2	77.7	95.6	69.4	81.9	95.0	63.6	78.7	99.1	67.5	82.7	98.5	71.4	84.4	99.7	70.5	84.5
MF2016	WEIGHTED AVERAGE	96.6	67.6	83.9	94.2	64.0	81.0	95.0	73.4	85.5	96.6	69.1	84.5	98.7	75.0	88.3	96.8	74.2	86.9

For customer support on your Building Construction Costs with RSMeans data, call 800.448.8182.

785

GEORGIA

DIVISION		COLUMBUS 318 - 319			DALTON 307			GAINESVILLE 305			MACON 310 - 312			SAVANNAH 313 - 314			STATESBORO 304		
		MAT.	INST.	TOTAL	MAT.	INST.	TOTAL	MAT.	INST.	TOTAL	MAT.	INST.	TOTAL	MAT.	INST.	TOTAL	MAT.	INST.	TOTAL
015433	CONTRACTOR EQUIPMENT		95.7	95.7		110.4	110.4		94.5	94.5		105.7	105.7		96.6	96.6		97.6	97.6
0241, 31 - 34	SITE & INFRASTRUCTURE, DEMOLITION	104.2	79.1	86.6	102.0	100.6	101.0	102.5	95.0	97.2	106.0	94.5	97.9	104.0	80.6	87.6	103.2	81.3	87.8
0310	Concrete Forming & Accessories	93.9	69.0	72.4	84.4	66.7	69.1	94.9	42.5	49.7	93.6	69.1	72.5	96.1	71.8	75.2	79.2	53.1	56.7
0320	Concrete Reinforcing	87.9	71.9	79.9	101.4	62.7	81.9	101.8	65.5	83.4	89.1	72.0	80.4	94.7	68.2	81.3	101.0	36.7	68.5
0330	Cast-in-Place Concrete	88.0	69.1	80.8	106.2	68.5	91.9	114.9	69.7	97.7	86.8	70.4	80.6	91.7	68.9	83.1	109.1	68.6	93.7
03	CONCRETE	86.2	71.3	79.4	104.1	68.5	87.8	106.7	57.3	84.1	85.8	71.8	79.4	87.7	71.9	80.4	104.5	57.8	83.2
04	MASONRY	92.2	69.2	77.9	79.6	78.7	79.1	87.0	79.4	82.3	104.5	69.3	82.7	87.2	69.2	76.0	81.6	79.4	80.2
05	METALS	97.4	98.3	97.7	94.1	94.7	94.3	92.4	79.4	88.4	93.1	98.4	94.7	94.2	96.4	94.9	97.6	84.9	93.7
06	WOOD, PLASTICS & COMPOSITES	85.9	69.6	76.8	79.7	68.3	73.3	100.8	34.7	64.0	92.6	69.6	79.8	90.3	73.6	81.0	73.6	49.3	60.1
07	THERMAL & MOISTURE PROTECTION	97.2	70.5	85.9	100.0	72.4	88.3	97.9	69.9	86.0	95.7	72.7	85.9	95.3	70.1	84.6	98.6	69.5	86.3
08	OPENINGS	87.7	71.6	84.0	94.1	68.3	88.2	93.1	50.5	83.3	87.5	71.6	83.8	94.4	72.9	89.4	95.0	50.5	84.8
0920	Plaster & Gypsum Board	103.6	69.3	80.5	82.4	67.8	72.6	95.9	33.3	53.9	109.5	69.3	82.5	98.9	73.4	81.8	82.3	48.4	59.6
0950, 0980	Ceilings & Acoustic Treatment	83.8	69.3	74.0	113.1	67.8	82.6	98.8	33.3	54.7	79.0	69.3	72.4	90.3	73.4	78.9	109.4	48.4	68.3
0960	Flooring	96.9	67.4	88.7	96.6	86.6	93.8	97.4	86.6	94.4	76.4	67.4	73.9	92.5	67.4	85.6	113.6	86.6	106.1
0970, 0990	Wall Finishes & Painting/Coating	89.1	95.9	93.1	87.6	73.2	79.2	97.5	95.9	96.6	91.1	95.9	93.9	88.3	87.9	88.1	95.3	73.2	82.4
09	FINISHES	92.1	71.1	80.7	105.3	71.6	87.0	96.9	54.4	73.8	81.9	71.1	76.0	92.1	72.6	81.5	108.8	60.6	82.6
COVERS	DIVS. 10 - 14, 25, 28, 41, 43, 44, 46	100.0	85.0	96.7	100.0	28.8	84.2	100.0	36.8	85.9	100.0	85.0	96.7	100.0	85.0	96.7	100.0	42.9	87.3
21, 22, 23	FIRE SUPPRESSION, PLUMBING & HVAC	100.0	67.4	86.1	97.1	63.2	82.6	97.0	68.3	84.7	100.0	70.5	87.4	100.1	68.3	86.6	97.6	69.1	85.4
26, 27, 3370	ELECTRICAL, COMMUNICATIONS & UTIL.	95.2	71.5	82.9	107.6	65.7	85.8	99.1	71.4	84.7	93.9	66.2	79.5	98.7	72.4	85.0	99.3	55.7	76.6
MF2016	WEIGHTED AVERAGE	94.9	73.8	85.7	98.3	72.2	86.9	97.2	67.6	84.2	93.8	75.1	85.6	95.4	74.4	86.2	98.7	66.3	84.5

DIVISION		GEORGIA VALDOSTA 316			WAYCROSS 315			HAWAII HILO 967			HONOLULU 968			STATES & POSS., GUAM 969			IDAHO BOISE 836 - 837		
		MAT.	INST.	TOTAL	MAT.	INST.	TOTAL	MAT.	INST.	TOTAL	MAT.	INST.	TOTAL	MAT.	INST.	TOTAL	MAT.	INST.	TOTAL
015433	CONTRACTOR EQUIPMENT		95.7	95.7		95.7	95.7		99.3	99.3		99.3	99.3		165.6	165.6		97.4	97.4
0241, 31 - 34	SITE & INFRASTRUCTURE, DEMOLITION	113.8	79.0	89.4	110.6	80.3	89.4	156.3	106.9	121.7	164.8	106.9	124.2	201.3	102.9	132.3	88.9	96.8	94.5
0310	Concrete Forming & Accessories	84.4	42.8	48.5	86.4	65.9	68.7	107.6	122.5	120.4	118.4	122.5	121.9	110.0	53.9	61.7	98.7	81.3	83.7
0320	Concrete Reinforcing	90.0	61.4	75.5	90.0	61.6	75.7	133.0	113.9	123.4	154.7	113.9	134.1	243.0	28.2	134.7	101.6	77.6	89.5
0330	Cast-in-Place Concrete	86.5	68.9	79.8	97.7	68.7	86.7	194.4	122.0	166.9	156.3	122.0	143.3	168.5	100.5	142.7	95.0	84.8	88.3
03	CONCRETE	91.1	57.5	75.7	94.1	67.7	82.0	151.7	119.8	137.1	142.6	119.8	132.2	153.7	66.6	113.9	96.1	82.0	89.7
04	MASONRY	97.6	79.4	86.3	98.5	79.4	86.7	153.6	121.4	133.6	140.0	121.4	128.5	217.7	36.2	105.2	125.6	89.0	102.9
05	METALS	97.0	94.1	96.1	96.0	89.7	94.1	112.9	102.3	109.6	125.9	102.3	118.7	145.3	77.1	124.4	108.1	82.1	100.1
06	WOOD, PLASTICS & COMPOSITES	74.1	34.5	52.0	75.7	66.4	70.5	114.9	122.4	119.1	129.4	122.4	125.5	126.7	56.3	87.5	92.4	79.9	85.5
07	THERMAL & MOISTURE PROTECTION	97.5	68.6	85.3	97.2	71.1	86.2	130.8	119.0	125.8	148.7	119.0	136.1	153.9	60.9	114.5	94.3	86.3	90.9
08	OPENINGS	84.3	49.2	76.2	84.6	64.2	79.9	117.0	118.9	117.5	131.5	118.9	128.6	121.6	45.9	104.3	97.1	74.3	91.9
0920	Plaster & Gypsum Board	95.7	33.2	53.7	95.7	66.0	75.8	108.1	122.9	118.0	151.3	122.9	132.2	216.4	44.6	101.0	92.4	79.4	83.7
0950, 0980	Ceilings & Acoustic Treatment	81.3	33.2	48.8	79.4	66.0	70.3	131.5	122.9	125.7	140.9	122.9	128.8	246.8	44.6	110.5	102.4	79.4	86.9
0960	Flooring	90.9	88.2	90.2	92.0	86.6	90.5	107.3	136.7	115.5	125.7	136.7	128.8	126.6	41.8	103.1	93.8	81.4	90.4
0970, 0990	Wall Finishes & Painting/Coating	89.1	95.9	93.1	89.1	73.2	79.8	95.6	137.2	119.9	103.4	137.2	123.1	100.4	33.0	61.1	93.6	41.3	63.1
09	FINISHES	89.8	54.6	70.6	89.3	70.6	79.2	108.7	127.1	118.7	124.8	127.1	126.1	178.9	50.4	109.0	92.7	77.2	84.3
COVERS	DIVS. 10 - 14, 25, 28, 41, 43, 44, 46	100.0	80.6	95.7	100.0	50.7	89.0	100.0	112.1	102.7	100.0	112.1	102.7	100.0	68.5	93.0	100.0	90.4	97.9
21, 22, 23	FIRE SUPPRESSION, PLUMBING & HVAC	100.0	70.6	87.4	98.1	65.4	84.1	100.5	109.4	104.3	100.5	109.4	104.3	103.5	34.9	74.2	100.1	74.9	89.3
26, 27, 3370	ELECTRICAL, COMMUNICATIONS & UTIL.	93.5	54.6	73.2	97.3	55.7	75.6	105.3	121.7	113.9	106.9	121.7	114.6	151.8	37.9	92.6	97.9	67.8	82.2
MF2016	WEIGHTED AVERAGE	95.1	67.3	82.9	95.3	69.5	84.0	116.9	116.0	116.5	121.1	116.0	118.9	139.1	53.4	101.6	100.5	79.8	91.4

DIVISION		IDAHO COEUR D'ALENE 838			IDAHO FALLS 834			LEWISTON 835			POCATELLO 832			TWIN FALLS 833			ILLINOIS BLOOMINGTON 617		
		MAT.	INST.	TOTAL	MAT.	INST.	TOTAL	MAT.	INST.	TOTAL	MAT.	INST.	TOTAL	MAT.	INST.	TOTAL	MAT.	INST.	TOTAL
015433	CONTRACTOR EQUIPMENT		92.2	92.2		97.4	97.4		92.2	92.2		97.4	97.4		97.4	97.4		105.2	105.2
0241, 31 - 34	SITE & INFRASTRUCTURE, DEMOLITION	87.9	90.9	90.0	86.5	96.8	93.7	94.7	91.8	92.7	89.9	96.8	94.7	97.0	96.6	96.7	95.7	100.5	99.1
0310	Concrete Forming & Accessories	108.9	79.6	83.6	92.8	80.1	81.8	113.9	81.6	86.0	98.8	80.9	83.4	100.0	79.9	82.7	83.4	119.9	114.9
0320	Concrete Reinforcing	110.0	99.2	104.6	103.8	77.6	90.6	110.0	99.3	104.6	102.1	77.6	89.7	104.1	77.6	90.7	96.1	105.6	100.9
0330	Cast-in-Place Concrete	97.8	82.6	92.0	86.2	83.2	85.0	101.7	86.0	95.7	93.0	84.7	89.9	95.5	83.1	90.8	97.7	117.5	105.2
03	CONCRETE	102.6	84.2	94.2	88.3	80.9	84.9	106.0	86.2	97.0	95.3	81.8	89.1	102.9	80.8	92.8	93.2	117.2	104.2
04	MASONRY	127.3	85.4	101.3	120.6	86.3	99.4	127.7	87.7	102.9	123.1	89.0	102.0	125.9	86.4	101.4	115.2	122.8	119.9
05	METALS	101.6	89.5	97.9	116.4	81.9	105.8	101.0	89.8	97.6	116.5	81.8	105.8	116.5	81.7	105.8	99.9	125.4	107.7
06	WOOD, PLASTICS & COMPOSITES	97.5	78.4	86.8	86.7	79.9	82.9	102.8	79.4	89.8	92.4	79.9	85.5	93.5	79.9	85.9	86.7	118.5	104.4
07	THERMAL & MOISTURE PROTECTION	147.3	83.8	120.4	93.5	76.4	86.3	147.6	85.5	121.4	94.0	77.2	86.9	94.9	83.3	90.0	96.9	115.8	104.9
08	OPENINGS	114.3	76.9	105.7	100.1	74.3	94.1	108.0	80.9	101.8	97.8	69.2	91.2	100.9	64.9	92.6	94.5	124.8	101.5
0920	Plaster & Gypsum Board	166.3	77.9	107.0	77.5	79.4	78.8	168.0	79.0	108.2	79.4	79.4	79.4	81.0	79.4	79.9	92.8	119.1	110.4
0950, 0980	Ceilings & Acoustic Treatment	136.6	77.9	97.1	103.2	79.4	87.1	136.6	79.0	97.8	110.0	79.4	89.4	105.7	79.4	88.0	89.2	119.1	109.3
0960	Flooring	132.6	81.4	118.4	93.4	81.4	90.1	136.1	81.4	120.9	97.1	81.4	92.7	98.3	81.4	93.7	84.6	126.3	96.2
0970, 0990	Wall Finishes & Painting/Coating	112.8	75.9	91.3	93.6	41.3	63.1	112.8	75.9	91.3	93.5	41.3	63.0	93.6	41.3	63.1	85.6	140.9	117.8
09	FINISHES	155.8	79.2	114.1	90.3	76.5	82.8	157.2	80.4	115.4	93.4	77.2	84.6	93.9	76.5	84.5	86.9	123.7	106.9
COVERS	DIVS. 10 - 14, 25, 28, 41, 43, 44, 46	100.0	95.0	98.9	100.0	89.5	97.7	100.0	95.9	99.1	100.0	90.4	97.9	100.0	89.5	97.7	100.0	107.6	101.7
21, 22, 23	FIRE SUPPRESSION, PLUMBING & HVAC	99.6	83.2	92.6	100.9	72.3	88.7	100.9	88.1	95.4	99.9	74.8	89.2	99.9	70.1	87.2	96.9	107.5	101.4
26, 27, 3370	ELECTRICAL, COMMUNICATIONS & UTIL.	90.2	86.1	88.1	89.9	68.2	78.6	88.4	82.6	85.4	95.4	69.7	82.0	91.3	68.2	79.3	95.6	95.4	95.5
MF2016	WEIGHTED AVERAGE	108.1	84.7	97.8	100.0	78.5	90.6	108.2	86.3	98.6	101.5	79.5	91.9	102.8	77.8	91.8	96.8	113.0	103.9

For customer support on your Building Construction Costs with RSMeans data, call 800.448.8182.

City Cost Indexes

ILLINOIS

DIVISION		CARBONDALE 629 MAT.	INST.	TOTAL	CENTRALIA 628 MAT.	INST.	TOTAL	CHAMPAIGN 618-619 MAT.	INST.	TOTAL	CHICAGO 606-608 MAT.	INST.	TOTAL	DECATUR 625 MAT.	INST.	TOTAL	EAST ST. LOUIS 620-622 MAT.	INST.	TOTAL
015433	CONTRACTOR EQUIPMENT		114.0	114.0		114.0	114.0		106.1	106.1		100.2	100.2		106.1	106.1		114.0	114.0
0241, 31 - 34	SITE & INFRASTRUCTURE, DEMOLITION	100.6	101.6	101.3	100.9	102.3	101.9	104.7	101.6	102.5	101.9	104.4	103.6	97.6	101.8	100.5	102.9	102.4	102.5
0310	Concrete Forming & Accessories	90.5	106.1	104.0	92.1	110.8	108.3	89.6	117.8	113.9	99.8	160.4	152.0	92.6	118.2	114.7	88.0	111.9	108.6
0320	Concrete Reinforcing	98.3	102.6	100.5	98.3	102.8	100.6	96.1	100.9	98.6	104.7	150.8	128.0	96.3	101.2	98.8	98.2	103.1	100.7
0330	Cast-in-Place Concrete	90.8	103.6	95.7	91.3	117.2	101.1	113.3	112.9	113.2	123.6	153.6	135.0	99.2	116.5	105.8	92.8	118.0	102.4
03	CONCRETE	83.1	106.3	93.7	83.5	113.2	97.1	105.5	113.8	109.3	109.9	155.7	130.8	93.5	115.2	103.4	84.6	114.0	98.0
04	MASONRY	78.1	109.9	97.8	78.1	112.4	99.4	139.8	121.5	128.5	102.6	164.9	141.2	73.8	122.8	104.2	78.3	117.2	102.4
05	METALS	97.8	130.5	107.9	97.9	132.9	108.6	99.9	121.7	106.6	97.9	146.9	113.0	101.6	121.2	107.6	99.0	133.3	109.5
06	WOOD, PLASTICS & COMPOSITES	90.9	102.7	97.5	93.3	108.3	101.6	93.2	117.1	106.5	104.3	160.0	135.3	91.2	117.1	105.6	88.4	108.9	99.8
07	THERMAL & MOISTURE PROTECTION	93.9	101.4	97.1	93.9	108.0	99.9	97.7	114.6	104.9	96.7	150.2	119.3	99.6	113.2	105.4	94.0	107.6	99.7
08	OPENINGS	89.4	115.7	95.4	89.4	118.8	96.1	95.1	119.4	100.7	101.9	170.3	117.6	100.5	119.5	104.9	89.4	119.1	96.2
0920	Plaster & Gypsum Board	97.3	102.9	101.1	98.3	108.6	105.2	94.7	117.6	110.1	103.3	161.7	142.5	99.7	117.6	111.7	96.3	109.2	105.0
0950, 0980	Ceilings & Acoustic Treatment	87.5	102.9	97.9	87.5	108.6	101.7	89.2	117.6	108.3	102.2	161.7	142.3	93.4	117.6	109.7	87.5	109.2	102.1
0960	Flooring	107.2	116.4	109.8	108.0	112.1	109.2	87.8	117.3	96.0	94.4	163.4	113.5	96.1	117.8	102.2	106.2	112.1	107.9
0970, 0990	Wall Finishes & Painting/Coating	94.5	104.8	100.5	94.5	115.2	106.6	85.6	117.6	104.2	101.4	162.5	137.0	86.5	111.6	101.1	94.5	115.2	106.6
09	FINISHES	91.6	106.6	99.7	92.0	111.1	102.4	88.8	118.3	104.8	99.3	162.1	133.4	91.4	118.1	105.9	91.2	112.4	102.7
COVERS	DIVS. 10 - 14, 25, 28, 41, 43, 44, 46	100.0	103.4	100.8	100.0	104.2	100.9	100.0	106.6	101.5	100.0	127.1	106.0	100.0	107.1	101.6	100.0	104.6	101.0
21, 22, 23	FIRE SUPPRESSION, PLUMBING & HVAC	96.9	105.8	100.7	96.9	94.1	95.7	96.9	105.1	100.4	100.0	138.4	116.4	100.1	98.2	99.3	100.0	97.7	99.0
26, 27, 3370	ELECTRICAL, COMMUNICATIONS & UTIL.	94.9	107.4	101.4	96.2	107.4	102.0	98.7	95.4	97.0	96.4	135.5	116.7	98.6	92.6	95.5	95.8	100.5	98.2
MF2016	WEIGHTED AVERAGE	93.2	108.8	100.0	93.4	108.8	100.1	100.3	110.6	104.8	100.8	145.8	120.5	97.3	109.0	102.4	94.4	109.4	100.9

ILLINOIS

DIVISION		EFFINGHAM 624 MAT.	INST.	TOTAL	GALESBURG 614 MAT.	INST.	TOTAL	JOLIET 604 MAT.	INST.	TOTAL	KANKAKEE 609 MAT.	INST.	TOTAL	LA SALLE 613 MAT.	INST.	TOTAL	NORTH SUBURBAN 600-603 MAT.	INST.	TOTAL
015433	CONTRACTOR EQUIPMENT		106.1	106.1		105.2	105.2		96.6	96.6		96.6	96.6		105.2	105.2		96.6	96.6
0241, 31 - 34	SITE & INFRASTRUCTURE, DEMOLITION	101.3	100.6	100.8	98.1	100.1	99.5	97.6	102.8	101.3	91.6	102.0	98.9	97.5	101.1	100.0	96.9	102.6	100.9
0310	Concrete Forming & Accessories	97.1	116.3	113.7	89.5	118.9	114.9	99.1	158.3	150.1	92.7	141.6	134.9	103.1	122.5	119.8	98.4	153.7	146.1
0320	Concrete Reinforcing	99.5	92.1	95.6	95.6	102.7	99.2	104.7	141.0	123.0	105.6	134.9	120.4	95.8	133.8	115.0	104.7	146.1	125.6
0330	Cast-in-Place Concrete	98.8	111.0	103.4	100.6	106.8	103.0	112.4	152.0	127.4	104.8	139.2	117.8	100.5	126.2	110.3	112.5	150.1	126.7
03	CONCRETE	94.3	110.8	101.8	96.0	112.6	103.6	102.8	152.2	125.4	96.7	139.1	116.1	96.8	126.5	110.4	102.8	150.4	124.6
04	MASONRY	82.0	115.2	102.6	115.4	120.6	118.6	101.8	163.4	140.0	98.1	147.8	128.9	115.4	145.7	134.2	98.6	159.1	136.2
05	METALS	98.7	114.1	103.4	99.9	122.9	107.0	95.9	138.4	108.9	95.9	133.9	107.5	100.0	143.1	113.2	97.0	140.1	110.2
06	WOOD, PLASTICS & COMPOSITES	93.5	117.1	106.6	93.0	118.6	107.3	102.9	157.3	133.2	96.1	139.8	120.4	107.7	119.5	114.3	101.3	152.6	129.8
07	THERMAL & MOISTURE PROTECTION	99.1	110.2	103.8	97.1	110.0	102.6	101.1	149.2	121.5	100.2	138.3	116.3	97.3	124.0	108.6	101.4	144.3	119.6
08	OPENINGS	95.3	116.0	100.0	94.5	117.3	99.7	99.5	165.7	114.7	92.7	153.2	106.6	94.5	133.5	103.5	99.6	163.8	114.3
0920	Plaster & Gypsum Board	99.6	117.6	111.7	94.7	119.3	111.2	96.1	159.0	138.3	94.1	141.0	125.6	101.7	120.1	114.1	99.1	154.1	136.0
0950, 0980	Ceilings & Acoustic Treatment	87.5	117.6	107.8	89.2	119.3	109.4	81.8	159.0	140.3	101.8	141.0	128.2	89.2	120.1	110.1	101.8	154.1	137.1
0960	Flooring	97.2	110.9	101.0	87.6	126.3	98.3	92.4	158.3	110.7	89.3	159.3	108.7	94.2	125.7	102.9	92.9	158.3	111.0
0970, 0990	Wall Finishes & Painting/Coating	86.5	109.3	99.8	85.6	99.9	93.9	93.5	170.4	138.4	93.5	140.9	121.2	85.6	139.1	116.8	95.5	159.5	132.9
09	FINISHES	90.6	115.7	104.3	88.2	119.2	105.1	94.7	160.6	130.6	93.2	145.2	121.4	91.1	123.5	108.7	95.4	156.0	128.3
COVERS	DIVS. 10 - 14, 25, 28, 41, 43, 44, 46	100.0	78.1	95.1	100.0	102.4	100.5	100.0	126.9	106.0	100.0	116.9	103.8	100.0	103.1	100.7	100.0	120.6	104.6
21, 22, 23	FIRE SUPPRESSION, PLUMBING & HVAC	97.0	104.2	100.1	96.9	105.2	100.5	100.0	135.4	115.2	96.9	129.7	111.0	96.9	123.8	108.4	99.9	133.2	114.2
26, 27, 3370	ELECTRICAL, COMMUNICATIONS & UTIL.	96.5	107.3	102.2	96.5	86.4	91.2	95.6	127.2	112.0	91.0	126.4	109.5	93.7	126.5	110.8	95.5	132.0	114.5
MF2016	WEIGHTED AVERAGE	95.8	108.7	101.5	97.4	108.9	102.5	98.8	142.0	117.7	95.6	133.5	112.2	97.6	126.5	110.3	98.9	140.6	117.1

ILLINOIS

DIVISION		PEORIA 615-616 MAT.	INST.	TOTAL	QUINCY 623 MAT.	INST.	TOTAL	ROCK ISLAND 612 MAT.	INST.	TOTAL	ROCKFORD 610-611 MAT.	INST.	TOTAL	SOUTH SUBURBAN 605 MAT.	INST.	TOTAL	SPRINGFIELD 626-627 MAT.	INST.	TOTAL
015433	CONTRACTOR EQUIPMENT		105.2	105.2		106.1	106.1		105.2	105.2		105.2	105.2		96.6	96.6		106.1	106.1
0241, 31 - 34	SITE & INFRASTRUCTURE, DEMOLITION	98.4	100.6	99.9	100.2	101.1	100.8	96.2	99.6	98.6	97.9	102.0	100.8	96.9	102.3	100.7	101.2	101.6	101.5
0310	Concrete Forming & Accessories	92.3	120.3	116.4	95.0	114.5	111.8	91.0	102.7	101.1	96.5	132.2	127.3	98.4	153.7	146.1	93.3	117.8	114.4
0320	Concrete Reinforcing	93.2	103.0	98.1	99.1	84.5	91.7	95.6	96.2	95.9	88.2	128.6	108.6	104.7	146.1	125.6	99.6	101.2	100.4
0330	Cast-in-Place Concrete	97.6	120.0	106.1	99.0	101.6	100.0	98.4	102.1	99.8	99.9	131.2	111.8	112.5	150.1	126.7	94.1	110.7	100.4
03	CONCRETE	93.3	117.8	104.5	93.9	105.5	99.2	94.0	102.5	97.8	94.0	131.7	111.2	102.8	150.4	124.6	91.7	113.1	101.4
04	MASONRY	114.5	123.6	120.2	105.4	114.0	110.8	115.2	101.9	106.9	89.1	139.1	120.1	98.6	159.1	136.2	84.6	121.6	107.5
05	METALS	102.6	124.2	109.3	98.7	112.4	102.9	99.9	118.7	105.7	102.6	140.4	114.2	97.0	140.1	110.2	99.1	121.3	105.9
06	WOOD, PLASTICS & COMPOSITES	100.5	118.6	110.6	91.1	117.1	105.5	94.6	102.2	98.8	100.5	129.6	116.7	101.3	152.6	129.8	92.2	117.1	106.0
07	THERMAL & MOISTURE PROTECTION	97.8	116.6	105.8	99.1	105.4	101.8	97.1	100.7	98.6	100.2	132.0	113.7	101.4	144.3	119.6	101.6	114.6	107.1
08	OPENINGS	100.6	124.1	106.0	96.0	113.6	100.0	94.5	106.7	97.3	100.6	140.6	109.7	99.6	163.8	114.3	98.8	118.8	103.4
0920	Plaster & Gypsum Board	98.5	119.3	112.4	98.3	117.6	111.3	94.7	102.4	99.9	98.5	130.5	120.0	99.1	154.1	136.0	100.8	117.6	112.1
0950, 0980	Ceilings & Acoustic Treatment	94.3	119.3	111.1	87.5	117.6	107.8	89.2	102.4	98.1	94.3	130.5	118.7	101.8	154.1	137.1	94.7	117.6	110.2
0960	Flooring	91.1	126.3	100.9	96.1	114.1	101.1	88.7	98.2	91.4	91.1	125.7	100.7	92.9	158.3	111.0	99.3	117.8	104.5
0970, 0990	Wall Finishes & Painting/Coating	85.6	140.9	117.8	86.5	115.2	103.2	88.7	99.9	93.9	85.6	149.2	122.7	95.5	159.5	132.9	84.7	115.2	102.5
09	FINISHES	90.8	123.9	108.8	90.1	115.7	104.0	88.4	101.4	95.5	90.8	132.6	113.6	95.4	156.0	128.3	96.9	118.1	108.4
COVERS	DIVS. 10 - 14, 25, 28, 41, 43, 44, 46	100.0	107.8	101.7	100.0	79.3	95.4	100.0	100.7	100.1	100.0	116.0	103.6	100.0	120.6	104.6	100.0	106.6	101.5
21, 22, 23	FIRE SUPPRESSION, PLUMBING & HVAC	100.0	102.7	101.1	97.0	101.5	98.9	96.9	100.1	98.3	100.1	118.6	108.0	99.9	133.2	114.2	100.0	103.0	101.3
26, 27, 3370	ELECTRICAL, COMMUNICATIONS & UTIL.	97.5	96.6	97.0	94.1	86.8	90.3	89.0	93.8	91.5	97.7	132.6	115.9	95.5	132.0	114.5	101.1	86.8	93.7
MF2016	WEIGHTED AVERAGE	99.3	112.2	104.9	96.7	103.9	99.8	96.4	101.9	98.8	98.2	128.3	111.4	98.9	140.6	117.1	97.8	108.8	102.6

For customer support on your Building Construction Costs with RSMeans data, call 800.448.8182.

City Cost Indexes

INDIANA

DIVISION		ANDERSON 460 MAT.	INST.	TOTAL	BLOOMINGTON 474 MAT.	INST.	TOTAL	COLUMBUS 472 MAT.	INST.	TOTAL	EVANSVILLE 476 - 477 MAT.	INST.	TOTAL	FORT WAYNE 467 - 468 MAT.	INST.	TOTAL	GARY 463 - 464 MAT.	INST.	TOTAL
015433	CONTRACTOR EQUIPMENT		96.7	96.7		83.4	83.4		83.4	83.4		113.9	113.9		96.7	96.7		96.7	96.7
0241, 31 - 34	SITE & INFRASTRUCTURE, DEMOLITION	97.9	93.4	94.7	87.8	92.1	90.8	84.4	91.9	89.7	93.4	121.8	113.3	98.9	93.2	94.9	98.5	97.1	97.5
0310	Concrete Forming & Accessories	94.6	81.2	83.0	99.0	81.6	84.0	93.2	79.1	81.0	92.6	82.3	83.7	92.7	74.8	77.3	94.7	114.8	112.1
0320	Concrete Reinforcing	90.1	85.7	87.8	87.1	85.3	86.2	87.5	85.4	86.4	95.4	78.6	86.9	90.1	77.1	83.5	90.1	113.1	101.7
0330	Cast-in-Place Concrete	100.6	78.7	92.3	99.1	77.8	91.0	98.7	75.7	89.9	94.7	87.6	92.0	106.9	75.0	94.8	105.1	116.3	109.4
03	CONCRETE	89.8	81.6	86.0	97.8	80.5	89.9	97.1	78.6	88.6	97.9	83.6	91.4	92.6	76.0	85.0	91.9	114.7	102.3
04	MASONRY	88.8	77.2	81.6	91.1	73.5	80.2	91.0	74.6	80.8	86.5	80.6	82.8	91.8	74.5	81.0	90.2	111.7	103.5
05	METALS	96.3	89.9	94.3	98.2	74.8	91.0	98.2	74.2	90.9	91.5	83.2	88.9	96.3	86.6	93.3	96.3	106.3	99.4
06	WOOD, PLASTICS & COMPOSITES	96.5	81.3	88.0	111.0	82.0	94.9	106.1	78.8	90.9	91.8	81.4	86.0	96.3	74.4	84.1	93.8	113.1	104.5
07	THERMAL & MOISTURE PROTECTION	110.1	77.9	96.5	96.6	78.9	89.1	96.0	78.5	88.6	100.9	84.8	94.1	109.8	74.0	94.7	108.5	108.2	108.4
08	OPENINGS	93.5	79.7	90.3	100.3	80.1	95.6	96.4	78.3	92.3	94.2	78.3	90.5	93.5	72.0	88.5	93.5	116.6	98.8
0920	Plaster & Gypsum Board	105.8	81.1	89.2	99.0	82.3	87.8	96.4	78.9	84.7	94.8	80.5	85.2	105.1	74.0	84.2	98.9	113.7	108.8
0950, 0980	Ceilings & Acoustic Treatment	89.7	81.1	83.9	79.8	82.3	81.5	79.8	78.9	79.2	83.6	80.5	81.5	89.7	74.0	79.1	89.7	113.7	105.9
0960	Flooring	94.0	79.3	89.9	98.5	83.7	94.4	93.3	83.7	90.6	93.0	75.5	88.1	94.0	73.2	88.2	94.0	116.8	100.3
0970, 0990	Wall Finishes & Painting/Coating	93.1	68.8	78.9	86.0	79.8	82.4	86.0	79.8	82.4	91.1	88.2	89.4	93.1	73.4	81.6	93.1	123.7	110.9
09	FINISHES	91.2	79.7	85.0	90.6	81.9	85.9	88.6	80.0	83.9	89.2	81.5	85.0	91.0	74.4	82.0	90.2	116.2	104.3
COVERS	DIVS. 10 - 14, 25, 28, 41, 43, 44, 46	100.0	91.0	98.0	100.0	87.5	97.2	100.0	87.2	97.1	100.0	95.1	98.9	100.0	90.5	97.9	100.0	107.2	101.6
21, 22, 23	FIRE SUPPRESSION, PLUMBING & HVAC	100.0	80.2	91.5	99.7	79.3	91.0	96.6	80.1	89.6	100.0	79.3	91.2	100.0	73.3	88.6	100.0	107.1	103.0
26, 27, 3370	ELECTRICAL, COMMUNICATIONS & UTIL.	87.9	87.8	87.9	99.9	88.1	93.8	99.2	87.7	93.2	96.0	86.0	90.8	88.6	76.5	82.3	99.4	114.1	107.0
MF2016	WEIGHTED AVERAGE	95.1	83.3	89.9	97.9	81.3	90.6	96.3	80.8	89.5	95.5	85.5	91.1	95.7	77.7	87.8	96.5	110.4	102.6

INDIANA

DIVISION		INDIANAPOLIS 461 - 462 MAT.	INST.	TOTAL	KOKOMO 469 MAT.	INST.	TOTAL	LAFAYETTE 479 MAT.	INST.	TOTAL	LAWRENCEBURG 470 MAT.	INST.	TOTAL	MUNCIE 473 MAT.	INST.	TOTAL	NEW ALBANY 471 MAT.	INST.	TOTAL
015433	CONTRACTOR EQUIPMENT		84.3	84.3		96.7	96.7		83.4	83.4		103.6	103.6		95.3	95.3		93.1	93.1
0241, 31 - 34	SITE & INFRASTRUCTURE, DEMOLITION	99.0	90.4	93.0	94.3	93.3	93.6	85.3	92.1	90.0	83.1	107.7	100.3	87.7	92.6	91.1	80.1	94.4	90.1
0310	Concrete Forming & Accessories	99.3	84.6	86.6	97.7	77.5	80.2	90.9	82.5	83.6	89.8	78.1	79.7	90.7	80.6	82.0	88.3	78.8	80.1
0320	Concrete Reinforcing	91.1	85.7	88.4	91.4	85.5	88.5	87.1	85.6	86.3	86.4	77.5	81.9	96.4	85.6	90.9	87.7	81.4	84.5
0330	Cast-in-Place Concrete	98.8	86.0	93.9	99.6	82.5	93.1	99.2	81.7	92.6	92.8	75.0	86.0	104.2	77.8	94.2	95.8	74.7	87.8
03	CONCRETE	95.7	84.7	90.7	86.9	81.2	84.3	97.3	82.3	90.4	90.5	77.5	84.6	96.2	81.1	89.3	95.8	78.1	87.7
04	MASONRY	90.5	79.2	83.5	88.5	75.2	80.3	96.5	77.1	84.5	75.7	73.3	74.2	92.9	77.3	83.2	82.3	69.5	74.4
05	METALS	97.1	75.3	90.4	92.7	89.6	91.8	96.6	75.1	90.0	93.3	85.3	90.8	100.0	89.9	96.9	95.2	81.8	91.1
06	WOOD, PLASTICS & COMPOSITES	98.8	85.1	91.2	99.5	76.2	86.5	103.3	83.0	92.0	90.1	77.9	83.3	105.1	80.8	91.6	92.0	80.1	85.4
07	THERMAL & MOISTURE PROTECTION	101.3	81.2	92.8	109.0	77.7	95.8	95.9	81.0	89.6	101.7	77.7	91.5	99.1	79.3	90.7	88.0	72.9	81.6
08	OPENINGS	99.7	81.8	95.6	89.0	76.9	86.2	94.9	80.6	91.6	96.1	74.7	91.2	93.7	79.4	90.4	93.7	78.8	90.3
0920	Plaster & Gypsum Board	94.7	85.0	88.2	110.4	75.8	87.2	94.0	83.3	86.8	73.6	77.9	76.5	94.8	81.1	85.6	92.5	80.0	84.1
0950, 0980	Ceilings & Acoustic Treatment	90.9	85.0	86.9	89.7	75.8	80.4	76.4	83.3	81.0	87.0	77.9	80.9	79.8	81.1	80.7	83.6	80.0	81.2
0960	Flooring	96.2	83.7	92.7	98.2	87.4	95.2	92.2	83.7	89.8	68.4	83.7	72.6	92.3	79.3	88.7	90.4	57.1	81.2
0970, 0990	Wall Finishes & Painting/Coating	98.6	79.8	87.6	93.1	71.0	80.2	86.0	85.1	85.4	86.8	74.5	79.6	86.0	68.8	76.0	91.1	68.2	77.8
09	FINISHES	94.6	84.1	88.9	93.0	78.4	85.0	87.3	83.1	85.0	79.2	79.0	79.1	87.8	79.1	83.1	88.3	73.7	80.4
COVERS	DIVS. 10 - 14, 25, 28, 41, 43, 44, 46	100.0	91.9	98.2	100.0	87.8	97.3	100.0	90.2	97.8	100.0	85.7	96.8	100.0	89.9	97.8	100.0	85.2	96.7
21, 22, 23	FIRE SUPPRESSION, PLUMBING & HVAC	99.9	80.8	91.7	96.9	80.2	89.8	96.6	79.4	89.3	97.6	75.9	88.3	99.7	80.1	91.3	96.9	77.0	88.4
26, 27, 3370	ELECTRICAL, COMMUNICATIONS & UTIL.	102.2	87.8	94.7	92.2	81.5	86.7	98.6	83.0	90.5	93.9	73.7	83.4	91.8	78.2	84.7	94.5	77.1	85.5
MF2016	WEIGHTED AVERAGE	98.3	83.2	91.7	93.5	81.7	88.3	96.0	81.6	89.7	92.8	79.6	87.0	96.3	81.7	89.9	94.0	78.0	87.0

INDIANA / IOWA

DIVISION		SOUTH BEND 465 - 466 MAT.	INST.	TOTAL	TERRE HAUTE 478 MAT.	INST.	TOTAL	WASHINGTON 475 MAT.	INST.	TOTAL	BURLINGTON 526 MAT.	INST.	TOTAL	CARROLL 514 MAT.	INST.	TOTAL	CEDAR RAPIDS 522 - 524 MAT.	INST.	TOTAL
015433	CONTRACTOR EQUIPMENT		108.4	108.4		113.9	113.9		113.9	113.9		101.5	101.5		101.5	101.5		98.2	98.2
0241, 31 - 34	SITE & INFRASTRUCTURE, DEMOLITION	97.4	93.9	94.9	95.2	122.2	114.1	95.0	122.4	114.2	99.5	96.6	97.5	88.3	97.4	94.7	100.7	96.1	97.5
0310	Concrete Forming & Accessories	94.5	79.6	81.6	93.4	80.8	82.6	94.3	82.5	84.1	93.0	82.5	83.9	81.3	80.6	80.7	98.7	87.9	89.4
0320	Concrete Reinforcing	89.4	86.2	87.8	95.4	85.7	90.5	88.2	85.1	86.7	91.7	86.1	88.9	92.4	82.9	87.6	92.4	82.7	87.5
0330	Cast-in-Place Concrete	99.1	81.5	92.4	91.7	82.4	88.2	99.8	87.8	95.3	107.6	57.4	88.5	107.6	58.9	89.1	107.9	85.3	99.3
03	CONCRETE	88.3	83.0	85.9	100.8	82.4	92.4	106.6	84.9	96.7	94.6	75.2	85.7	93.5	74.3	84.7	94.7	86.7	91.0
04	MASONRY	93.4	77.2	83.4	94.2	76.4	83.2	86.7	80.3	82.8	100.4	75.0	84.6	102.1	74.2	84.8	106.2	83.4	92.0
05	METALS	96.3	105.9	99.2	92.2	86.2	90.3	86.7	86.1	86.5	90.2	95.0	91.7	90.2	94.0	91.4	92.7	95.0	93.4
06	WOOD, PLASTICS & COMPOSITES	94.3	78.6	85.6	93.8	80.4	86.3	94.2	81.3	87.0	89.0	81.4	84.8	76.4	84.4	80.9	95.3	87.6	91.1
07	THERMAL & MOISTURE PROTECTION	101.2	82.0	93.1	101.0	83.0	93.4	100.9	84.5	93.9	104.2	78.5	93.3	104.5	73.4	91.4	105.2	83.5	96.0
08	OPENINGS	95.2	78.5	91.4	94.7	79.2	91.1	91.5	80.1	88.9	93.5	84.9	91.5	97.9	82.6	94.4	98.4	85.4	95.4
0920	Plaster & Gypsum Board	95.1	78.3	83.9	94.8	79.4	84.5	95.1	80.4	85.2	109.4	81.0	90.3	104.8	84.1	90.9	113.9	87.7	96.3
0950, 0980	Ceilings & Acoustic Treatment	91.6	78.3	82.7	83.6	79.4	80.8	79.4	80.4	80.1	98.0	81.0	86.5	98.0	84.1	88.7	100.6	87.7	91.9
0960	Flooring	91.6	91.1	91.5	93.0	80.0	89.4	93.9	81.1	90.3	94.1	73.0	88.3	88.2	84.1	87.1	107.9	89.1	102.7
0970, 0990	Wall Finishes & Painting/Coating	86.9	86.3	86.5	91.1	83.4	86.6	91.1	88.2	89.4	94.8	89.1	91.5	94.8	76.1	83.9	96.4	74.0	83.3
09	FINISHES	91.1	82.2	86.2	89.2	80.8	84.6	88.9	82.6	85.5	94.1	81.6	87.3	90.2	81.6	85.5	99.4	86.8	92.6
COVERS	DIVS. 10 - 14, 25, 28, 41, 43, 44, 46	100.0	92.1	98.2	100.0	92.4	98.3	100.0	95.1	98.9	100.0	91.7	98.1	100.0	64.9	92.2	100.0	93.7	98.6
21, 22, 23	FIRE SUPPRESSION, PLUMBING & HVAC	99.9	78.1	90.6	100.0	79.7	91.3	96.9	80.9	90.0	97.1	82.6	90.9	97.1	73.8	87.2	100.2	84.5	93.5
26, 27, 3370	ELECTRICAL, COMMUNICATIONS & UTIL.	100.1	88.6	94.1	94.3	88.1	91.1	94.8	86.0	90.2	100.8	76.3	88.0	101.4	81.2	90.9	98.4	79.6	88.6
MF2016	WEIGHTED AVERAGE	96.3	85.0	91.4	96.3	85.4	91.5	94.8	86.5	91.1	96.0	82.3	90.0	95.8	79.9	88.8	98.2	86.5	93.1

For customer support on your Building Construction Costs with RSMeans data, call 800.448.8182.

IOWA

DIVISION		COUNCIL BLUFFS 515			CRESTON 508			DAVENPORT 527 - 528			DECORAH 521			DES MOINES 500 - 503, 509			DUBUQUE 520		
		MAT.	INST.	TOTAL	MAT.	INST.	TOTAL	MAT.	INST.	TOTAL	MAT.	INST.	TOTAL	MAT.	INST.	TOTAL	MAT.	INST.	TOTAL
015433	CONTRACTOR EQUIPMENT		97.5	97.5		101.5	101.5		101.5	101.5		101.5	101.5		103.4	103.4		97.0	97.0
0241, 31 - 34	SITE & INFRASTRUCTURE, DEMOLITION	103.5	92.5	95.8	98.0	98.3	98.2	99.8	99.6	99.7	98.1	96.3	96.8	104.4	101.2	102.2	98.5	93.2	94.8
0310	Concrete Forming & Accessories	80.8	76.2	76.8	78.5	83.7	83.0	98.2	100.5	100.2	90.7	74.7	76.9	96.4	86.0	87.4	82.0	85.4	85.0
0320	Concrete Reinforcing	94.3	80.0	87.1	91.5	82.4	86.9	92.4	108.9	100.7	91.7	71.0	81.3	96.0	83.4	89.6	91.1	82.5	86.8
0330	Cast-in-Place Concrete	112.1	81.9	100.7	108.8	63.4	91.6	103.9	100.3	102.5	104.7	79.7	95.2	92.6	93.3	92.9	105.6	87.2	98.6
03	CONCRETE	96.9	79.8	89.1	93.9	77.2	86.3	92.7	102.2	97.1	92.6	76.7	85.3	87.5	89.0	88.2	91.6	86.2	89.2
04	MASONRY	107.2	77.5	88.8	101.9	72.4	83.6	102.8	96.1	98.7	122.3	69.7	89.7	89.0	86.0	87.1	107.1	71.8	85.2
05	METALS	97.7	93.1	96.3	94.5	93.5	94.2	92.7	106.3	96.9	90.3	88.6	89.8	100.5	100.7	100.6	91.2	94.0	92.1
06	WOOD, PLASTICS & COMPOSITES	75.3	75.6	75.5	72.1	84.4	78.9	95.3	99.9	97.9	86.1	74.3	79.5	91.4	84.4	87.5	76.9	86.2	82.0
07	THERMAL & MOISTURE PROTECTION	104.6	76.9	92.9	107.5	80.1	95.9	104.6	95.7	101.0	104.4	72.7	91.0	99.1	86.7	93.9	104.9	80.7	94.6
08	OPENINGS	97.4	78.7	93.1	105.0	84.4	100.3	98.4	100.5	98.9	96.5	77.4	92.1	96.9	87.5	94.7	97.5	86.6	95.0
0920	Plaster & Gypsum Board	104.8	75.3	85.0	102.5	84.1	90.2	113.9	100.0	104.6	108.1	73.7	85.0	94.5	84.1	87.5	104.8	86.2	92.3
0950, 0980	Ceilings & Acoustic Treatment	98.0	75.3	82.7	89.8	84.1	86.0	100.6	100.0	100.2	98.0	73.7	81.6	91.5	84.1	86.5	98.0	86.2	90.0
0960	Flooring	86.8	89.1	87.4	78.6	73.0	77.1	96.6	94.5	96.0	93.5	73.0	87.8	87.5	94.1	89.3	98.8	73.0	91.7
0970, 0990	Wall Finishes & Painting/Coating	90.9	74.0	81.0	84.3	77.7	80.4	94.8	94.6	94.7	94.8	85.0	89.1	85.8	85.0	85.3	95.6	82.0	87.7
09	FINISHES	90.8	78.2	84.0	83.4	81.5	82.4	95.9	98.8	97.5	93.7	75.6	83.9	89.5	87.0	88.1	94.7	82.9	88.3
COVERS	DIVS. 10 - 14, 25, 28, 41, 43, 44, 46	100.0	90.6	97.9	100.0	67.5	92.9	100.0	97.4	99.4	100.0	81.1	95.8	100.0	93.6	98.6	100.0	92.5	98.3
21, 22, 23	FIRE SUPPRESSION, PLUMBING & HVAC	100.2	74.8	89.3	96.9	81.7	90.4	100.2	97.8	99.2	97.1	73.1	86.9	99.9	83.9	93.0	100.2	78.7	91.0
26, 27, 3370	ELECTRICAL, COMMUNICATIONS & UTIL.	103.6	82.8	92.8	93.3	81.2	87.0	96.6	92.5	94.4	98.4	49.9	73.2	104.3	83.8	93.6	102.2	77.1	89.1
MF2016	WEIGHTED AVERAGE	99.0	81.1	91.2	96.2	82.1	90.0	97.3	98.6	97.9	96.8	74.0	86.8	97.2	88.6	93.5	97.3	82.8	91.0

IOWA

DIVISION		FORT DODGE 505			MASON CITY 504			OTTUMWA 525			SHENANDOAH 516			SIBLEY 512			SIOUX CITY 510 - 511		
		MAT.	INST.	TOTAL	MAT.	INST.	TOTAL	MAT.	INST.	TOTAL	MAT.	INST.	TOTAL	MAT.	INST.	TOTAL	MAT.	INST.	TOTAL
015433	CONTRACTOR EQUIPMENT		101.5	101.5		101.5	101.5		97.0	97.0		97.5	97.5		101.5	101.5		101.5	101.5
0241, 31 - 34	SITE & INFRASTRUCTURE, DEMOLITION	107.1	95.1	98.7	107.1	96.2	99.5	99.0	90.9	93.3	102.1	92.9	95.7	109.7	96.3	100.3	111.4	97.5	101.6
0310	Concrete Forming & Accessories	79.1	75.2	75.8	83.1	74.6	75.7	88.7	74.9	76.8	82.3	57.9	61.3	82.7	38.9	44.9	98.7	75.7	78.9
0320	Concrete Reinforcing	91.5	69.0	80.2	91.4	82.4	86.9	91.7	86.1	88.9	94.3	69.4	81.7	94.3	68.9	81.5	92.4	79.6	85.9
0330	Cast-in-Place Concrete	102.2	43.5	79.9	102.2	74.7	91.7	108.3	66.7	92.5	108.5	79.3	97.4	106.3	58.0	87.9	106.9	73.3	94.2
03	CONCRETE	89.5	64.2	77.9	89.8	77.0	83.9	94.2	75.0	85.4	94.4	68.8	82.7	93.4	52.7	74.8	94.0	76.5	86.0
04	MASONRY	100.8	51.7	70.3	114.2	68.4	85.8	103.6	55.1	73.5	106.8	73.2	86.0	126.0	52.0	80.1	100.0	69.6	81.1
05	METALS	94.6	87.6	92.4	94.6	93.4	94.2	90.2	94.8	91.6	96.7	88.1	94.1	90.4	87.1	89.4	92.7	92.5	92.6
06	WOOD, PLASTICS & COMPOSITES	72.5	84.4	79.1	76.3	74.3	75.2	83.5	81.2	82.2	76.9	52.8	63.5	77.6	35.7	54.3	95.3	75.7	84.4
07	THERMAL & MOISTURE PROTECTION	106.8	67.0	90.0	106.3	73.3	92.3	105.0	69.5	90.0	103.8	67.4	88.4	104.1	54.6	83.2	104.6	73.0	91.3
08	OPENINGS	99.0	70.3	92.4	91.4	80.5	88.9	97.9	81.7	94.2	89.1	59.0	82.2	94.6	43.3	82.8	98.4	77.1	93.5
0920	Plaster & Gypsum Board	102.5	84.1	90.2	102.5	73.7	83.2	105.7	81.0	89.1	104.8	51.9	69.2	104.8	34.1	57.3	113.9	75.2	87.9
0950, 0980	Ceilings & Acoustic Treatment	89.8	84.1	86.0	89.8	73.7	79.0	98.0	81.0	86.5	98.0	51.9	66.9	98.0	34.1	54.9	100.6	75.2	83.5
0960	Flooring	79.8	73.0	77.9	81.7	73.0	79.3	102.0	73.0	93.9	87.5	78.1	84.9	89.1	73.0	84.7	96.6	75.7	90.8
0970, 0990	Wall Finishes & Painting/Coating	84.3	85.0	84.7	84.3	85.0	84.7	95.6	85.0	89.4	90.9	63.7	75.0	94.8	59.8	74.4	94.8	69.5	80.0
09	FINISHES	85.3	77.1	80.8	85.9	75.3	80.2	95.9	76.0	85.1	90.9	60.9	74.6	93.4	45.6	67.4	97.4	74.9	85.2
COVERS	DIVS. 10 - 14, 25, 28, 41, 43, 44, 46	100.0	84.1	96.5	100.0	88.5	97.4	100.0	84.5	96.5	100.0	61.7	91.5	100.0	75.9	94.6	100.0	90.6	97.9
21, 22, 23	FIRE SUPPRESSION, PLUMBING & HVAC	96.9	70.6	85.7	96.9	79.5	89.5	97.1	72.7	86.7	97.1	87.1	92.8	97.1	66.3	84.0	100.2	79.5	91.4
26, 27, 3370	ELECTRICAL, COMMUNICATIONS & UTIL.	99.3	73.2	85.7	98.4	49.9	73.2	100.6	74.9	87.2	98.4	79.8	88.7	98.4	49.9	73.2	98.4	72.6	85.0
MF2016	WEIGHTED AVERAGE	95.9	73.0	85.9	95.8	76.0	87.1	96.7	76.2	87.7	96.4	76.5	87.7	97.1	61.0	81.3	97.9	79.2	89.7

IOWA / KANSAS

DIVISION		IOWA SPENCER 513			IOWA WATERLOO 506 - 507			KANSAS BELLEVILLE 669			KANSAS COLBY 677			KANSAS DODGE CITY 678			KANSAS EMPORIA 668		
		MAT.	INST.	TOTAL	MAT.	INST.	TOTAL	MAT.	INST.	TOTAL	MAT.	INST.	TOTAL	MAT.	INST.	TOTAL	MAT.	INST.	TOTAL
015433	CONTRACTOR EQUIPMENT		101.5	101.5		101.5	101.5		107.1	107.1		107.1	107.1		107.1	107.1		105.2	105.2
0241, 31 - 34	SITE & INFRASTRUCTURE, DEMOLITION	109.8	95.0	99.5	112.5	97.4	101.9	110.7	97.2	101.3	109.2	97.9	101.3	111.7	96.8	101.3	102.5	94.7	97.0
0310	Concrete Forming & Accessories	88.7	38.7	45.6	94.3	71.6	74.7	93.9	56.2	61.4	99.6	63.6	68.6	92.9	63.3	67.3	85.1	68.2	70.5
0320	Concrete Reinforcing	94.3	68.9	81.5	92.1	82.9	87.5	96.8	63.4	80.0	104.6	63.5	83.9	102.0	63.3	82.5	95.5	63.9	79.6
0330	Cast-in-Place Concrete	106.3	68.7	92.0	109.4	83.4	99.5	120.9	87.7	108.3	115.6	91.6	106.5	117.7	91.3	107.6	116.9	91.7	107.3
03	CONCRETE	93.8	56.3	76.7	95.3	78.7	87.7	111.5	70.0	92.2	108.9	74.7	93.3	110.3	74.4	93.9	104.1	76.9	91.6
04	MASONRY	126.0	52.0	80.1	101.5	77.7	86.7	92.0	55.2	69.2	102.2	61.5	77.0	112.5	59.9	79.9	97.9	68.9	79.9
05	METALS	90.4	87.0	89.3	97.0	94.7	96.3	98.7	86.4	95.5	99.1	87.3	95.5	100.6	86.0	96.1	98.5	87.9	95.2
06	WOOD, PLASTICS & COMPOSITES	83.4	35.7	56.9	89.3	67.4	77.1	98.2	53.6	73.4	107.0	60.1	80.9	99.0	60.1	77.4	89.6	65.9	76.4
07	THERMAL & MOISTURE PROTECTION	105.1	55.2	84.0	106.5	79.9	95.3	93.7	63.5	80.9	95.4	66.9	83.3	95.3	66.1	83.0	91.8	78.4	86.1
08	OPENINGS	105.9	43.3	91.5	91.7	76.7	88.3	98.2	55.9	88.5	103.4	63.3	94.2	103.4	59.5	93.3	96.1	65.2	89.0
0920	Plaster & Gypsum Board	105.7	34.1	57.6	111.0	66.7	81.2	95.4	52.4	66.5	100.6	59.1	72.7	95.0	59.1	70.9	92.7	65.1	74.2
0950, 0980	Ceilings & Acoustic Treatment	98.0	34.1	54.9	92.4	66.7	75.0	84.0	52.4	62.7	81.5	59.1	66.4	81.5	59.1	66.4	84.0	65.1	71.3
0960	Flooring	91.9	73.0	86.7	86.7	83.4	85.8	88.5	70.8	83.6	87.4	70.8	82.8	83.6	70.8	80.1	83.8	70.8	80.2
0970, 0990	Wall Finishes & Painting/Coating	94.8	59.8	74.4	84.3	85.0	84.7	84.5	57.6	68.8	91.7	57.6	71.8	91.7	57.6	71.8	84.5	57.6	68.8
09	FINISHES	94.3	44.4	67.2	89.4	74.3	81.2	86.6	57.8	71.0	86.3	63.3	73.8	84.5	63.3	72.9	83.8	66.7	74.5
COVERS	DIVS. 10 - 14, 25, 28, 41, 43, 44, 46	100.0	74.6	94.4	100.0	91.0	98.0	100.0	85.3	96.7	100.0	88.2	97.4	100.0	88.2	97.4	100.0	88.8	97.5
21, 22, 23	FIRE SUPPRESSION, PLUMBING & HVAC	97.1	70.7	85.9	100.0	83.1	92.8	96.9	71.7	86.1	96.9	72.5	86.5	100.0	72.5	88.2	96.9	75.9	87.9
26, 27, 3370	ELECTRICAL, COMMUNICATIONS & UTIL.	100.1	49.9	74.0	95.1	65.0	79.5	102.2	66.2	83.5	99.6	71.7	85.1	96.8	71.7	83.7	99.5	70.0	84.2
MF2016	WEIGHTED AVERAGE	98.7	62.1	82.7	97.2	80.2	89.7	99.1	69.9	86.3	99.8	73.5	88.3	100.9	72.9	88.7	97.4	75.8	87.9

KANSAS

DIVISION		FORT SCOTT 667			HAYS 676			HUTCHINSON 675			INDEPENDENCE 673			KANSAS CITY 660 - 662			LIBERAL 679		
		MAT.	INST.	TOTAL	MAT.	INST.	TOTAL	MAT.	INST.	TOTAL	MAT.	INST.	TOTAL	MAT.	INST.	TOTAL	MAT.	INST.	TOTAL
015433	CONTRACTOR EQUIPMENT		106.1	106.1		107.1	107.1		107.1	107.1		107.1	107.1		103.6	103.6		107.1	107.1
0241, 31 - 34	SITE & INFRASTRUCTURE, DEMOLITION	99.3	94.7	96.1	114.1	97.4	102.4	93.1	97.9	96.5	112.8	97.9	102.4	93.3	94.9	94.4	113.8	97.5	102.4
0310	Concrete Forming & Accessories	101.8	83.7	86.2	97.1	61.1	66.1	87.6	58.6	62.6	108.3	70.4	75.6	98.5	98.7	98.7	93.4	60.7	65.2
0320	Concrete Reinforcing	94.9	101.7	98.3	102.0	63.4	82.5	102.0	63.3	82.5	101.4	64.9	83.0	92.0	104.4	98.3	103.5	63.3	83.2
0330	Cast-in-Place Concrete	108.4	86.8	100.2	91.2	88.0	90.0	84.5	91.3	87.1	118.2	91.5	108.0	92.8	99.9	95.5	91.2	87.6	89.8
03	CONCRETE	99.3	88.6	94.4	100.4	72.3	87.6	83.7	72.3	78.5	111.4	77.9	96.1	91.4	100.6	95.6	102.3	71.9	88.4
04	MASONRY	99.3	55.9	72.4	111.6	55.3	76.6	102.1	59.9	75.9	99.3	64.6	77.8	99.8	98.2	98.8	110.3	53.6	75.1
05	METALS	98.4	100.3	99.0	98.7	87.0	95.1	98.5	86.0	94.7	98.4	86.5	94.8	106.0	106.3	106.1	99.0	85.8	94.9
06	WOOD, PLASTICS & COMPOSITES	108.2	90.6	98.4	103.9	60.1	79.5	94.1	53.8	71.7	117.3	69.0	90.4	104.1	98.6	101.0	99.6	60.1	77.6
07	THERMAL & MOISTURE PROTECTION	92.8	75.2	85.4	95.7	64.2	82.4	94.2	65.4	82.0	95.4	77.8	88.0	92.5	100.6	95.9	95.9	63.2	82.0
08	OPENINGS	96.1	87.5	94.1	103.3	59.5	93.3	103.3	56.0	92.4	101.1	64.0	92.6	97.4	98.0	97.6	103.4	59.5	93.3
0920	Plaster & Gypsum Board	97.7	90.4	92.8	97.9	59.1	71.8	94.0	52.6	66.2	107.5	68.2	81.1	91.4	98.6	96.2	95.6	59.1	71.1
0950, 0980	Ceilings & Acoustic Treatment	84.0	90.4	88.3	81.5	59.1	66.4	81.5	52.6	62.0	81.5	68.2	72.6	84.0	98.6	93.9	81.5	59.1	66.4
0960	Flooring	97.9	70.7	90.3	86.2	70.8	81.9	80.8	70.8	78.0	91.6	70.7	85.8	78.8	99.7	84.6	83.9	69.7	79.9
0970, 0990	Wall Finishes & Painting/Coating	86.1	81.0	83.1	91.7	57.6	71.8	91.7	57.6	71.8	91.7	57.6	71.8	91.5	102.4	97.8	91.7	57.6	71.8
09	FINISHES	89.0	82.2	85.3	86.0	61.7	72.8	82.0	59.6	69.8	88.6	69.1	78.0	84.1	99.3	92.4	85.3	61.4	72.3
COVERS	DIVS. 10 - 14, 25, 28, 41, 43, 44, 46	100.0	88.5	97.5	100.0	86.1	96.9	100.0	87.5	97.2	100.0	89.2	97.6	100.0	95.9	99.1	100.0	86.1	96.9
21, 22, 23	FIRE SUPPRESSION, PLUMBING & HVAC	96.9	69.0	85.0	96.9	69.3	85.1	96.9	70.4	85.6	96.9	72.9	86.6	99.9	98.8	99.4	96.9	67.9	84.5
26, 27, 3370	ELECTRICAL, COMMUNICATIONS & UTIL.	98.8	71.7	84.7	98.6	71.7	84.6	94.1	64.4	78.6	96.1	72.3	83.7	104.4	98.4	101.3	96.8	71.7	83.7
MF2016	WEIGHTED AVERAGE	97.3	79.1	89.3	99.1	71.3	86.9	95.0	70.5	84.3	99.6	75.6	89.1	98.4	99.3	98.8	99.0	70.6	86.6

DIVISION		KANSAS SALINA 674			TOPEKA 664 - 666			WICHITA 670 - 672			KENTUCKY ASHLAND 411 - 412			BOWLING GREEN 421 - 422			CAMPTON 413 - 414		
		MAT.	INST.	TOTAL	MAT.	INST.	TOTAL	MAT.	INST.	TOTAL	MAT.	INST.	TOTAL	MAT.	INST.	TOTAL	MAT.	INST.	TOTAL
015433	CONTRACTOR EQUIPMENT		107.1	107.1		105.2	105.2		107.1	107.1		100.3	100.3		93.1	93.1		99.5	99.5
0241, 31 - 34	SITE & INFRASTRUCTURE, DEMOLITION	102.0	97.4	98.8	96.9	94.5	95.2	98.0	98.6	98.4	115.9	82.6	92.6	80.2	94.2	90.0	89.3	95.5	93.6
0310	Concrete Forming & Accessories	89.4	56.3	60.9	97.2	69.0	72.8	95.7	56.1	61.6	85.8	95.9	94.5	84.7	79.2	79.9	87.7	83.2	83.8
0320	Concrete Reinforcing	101.4	63.4	82.2	91.6	90.6	91.1	99.5	101.3	100.4	89.2	94.7	92.0	86.5	79.1	82.8	87.3	95.2	91.3
0330	Cast-in-Place Concrete	102.1	88.0	96.8	97.5	89.9	94.6	95.5	76.3	88.2	87.2	97.1	91.0	86.5	72.5	81.2	96.5	70.6	86.7
03	CONCRETE	97.4	70.2	85.0	93.2	81.2	87.7	93.2	72.5	83.7	91.7	97.2	94.2	90.1	77.2	84.2	93.7	81.2	88.0
04	MASONRY	127.8	55.3	82.8	93.5	67.0	77.1	99.3	51.5	69.7	93.3	94.3	93.9	95.4	69.8	79.5	92.1	57.1	70.4
05	METALS	100.5	87.0	96.3	102.6	97.7	101.1	102.6	99.3	101.6	94.3	109.5	99.0	95.9	84.4	92.4	95.2	90.8	93.8
06	WOOD, PLASTICS & COMPOSITES	95.6	53.6	72.2	101.0	68.1	82.7	104.0	53.8	76.1	74.4	95.5	86.2	86.9	79.8	83.0	85.5	90.6	88.3
07	THERMAL & MOISTURE PROTECTION	94.8	63.5	81.6	96.5	77.7	88.6	94.1	60.6	79.9	92.0	91.7	91.9	88.0	79.2	84.2	101.0	70.9	88.3
08	OPENINGS	103.3	55.9	92.4	103.3	74.0	96.6	105.7	64.7	96.2	93.0	94.4	93.3	93.7	81.2	90.8	95.0	88.7	93.6
0920	Plaster & Gypsum Board	94.0	52.4	66.1	100.8	67.3	78.3	93.7	52.6	66.1	61.6	95.6	84.4	88.2	79.7	82.5	88.2	89.9	89.4
0950, 0980	Ceilings & Acoustic Treatment	81.5	52.4	61.9	88.7	67.3	74.3	85.3	52.6	63.3	78.9	95.6	90.2	83.6	79.7	81.0	83.6	89.9	87.9
0960	Flooring	82.2	70.8	79.0	89.6	70.8	84.4	92.6	70.8	86.6	73.6	85.9	77.0	88.3	64.4	81.7	90.4	65.7	83.5
0970, 0990	Wall Finishes & Painting/Coating	91.7	57.6	71.8	88.8	70.8	78.3	92.0	57.6	71.9	93.2	91.7	92.3	91.1	89.2	90.0	91.1	56.3	70.8
09	FINISHES	83.2	57.8	69.4	92.1	68.5	79.3	90.3	57.5	72.4	76.5	93.9	85.9	87.0	77.3	81.7	87.8	78.1	82.5
COVERS	DIVS. 10 - 14, 25, 28, 41, 43, 44, 46	100.0	86.1	96.9	100.0	83.3	96.3	100.0	85.4	96.8	100.0	91.7	98.1	100.0	87.4	97.2	100.0	49.5	88.8
21, 22, 23	FIRE SUPPRESSION, PLUMBING & HVAC	100.0	70.3	87.3	100.0	76.0	89.8	99.8	69.8	87.0	96.7	87.6	92.8	100.0	80.2	91.5	96.9	76.8	88.3
26, 27, 3370	ELECTRICAL, COMMUNICATIONS & UTIL.	96.5	74.1	84.8	103.2	73.7	87.8	100.2	74.1	86.6	92.3	93.2	92.8	94.8	78.9	86.5	92.4	93.1	92.8
MF2016	WEIGHTED AVERAGE	99.6	70.8	87.0	99.1	78.1	89.9	99.2	72.1	87.3	93.5	93.4	93.5	94.6	79.9	88.2	94.7	80.2	88.3

KENTUCKY

DIVISION		CORBIN 407 - 409			COVINGTON 410			ELIZABETHTOWN 427			FRANKFORT 406			HAZARD 417 - 418			HENDERSON 424		
		MAT.	INST.	TOTAL	MAT.	INST.	TOTAL	MAT.	INST.	TOTAL	MAT.	INST.	TOTAL	MAT.	INST.	TOTAL	MAT.	INST.	TOTAL
015433	CONTRACTOR EQUIPMENT		99.5	99.5		103.6	103.6		93.1	93.1		99.5	99.5		99.5	99.5		113.9	113.9
0241, 31 - 34	SITE & INFRASTRUCTURE, DEMOLITION	95.1	96.0	95.7	84.6	107.6	100.7	74.6	93.9	88.1	93.0	96.4	95.4	87.0	96.6	93.7	83.1	121.3	109.9
0310	Concrete Forming & Accessories	83.2	74.4	75.6	83.3	73.1	74.5	79.6	75.4	75.9	94.0	76.0	78.4	84.3	83.8	83.9	90.7	80.5	81.9
0320	Concrete Reinforcing	90.9	62.1	76.4	86.0	81.1	83.5	86.9	82.7	84.8	95.8	80.5	88.1	87.7	62.3	74.9	86.6	77.8	82.2
0330	Cast-in-Place Concrete	91.8	75.2	85.5	92.3	76.3	86.2	78.2	69.9	75.0	90.9	77.4	85.8	92.8	72.9	85.2	76.5	88.8	81.2
03	CONCRETE	87.1	73.1	80.7	92.1	76.5	84.9	82.3	75.2	79.1	89.3	77.7	84.0	90.6	76.6	84.2	87.3	83.2	85.4
04	MASONRY	90.0	62.6	73.0	106.9	75.5	87.1	79.2	64.4	70.0	85.5	73.3	77.9	90.8	59.2	71.2	99.0	82.1	88.5
05	METALS	93.9	78.4	89.1	93.2	89.9	92.2	95.1	85.1	92.0	96.6	86.2	93.4	95.2	78.8	90.1	86.4	84.6	85.9
06	WOOD, PLASTICS & COMPOSITES	72.1	75.0	73.7	83.7	70.2	76.1	82.4	77.5	79.7	89.6	75.0	81.5	82.7	90.6	87.1	89.5	79.4	83.9
07	THERMAL & MOISTURE PROTECTION	105.1	72.2	91.2	101.9	75.1	90.5	87.4	71.1	80.5	102.6	75.7	91.2	100.9	72.7	89.0	100.2	80.7	91.9
08	OPENINGS	89.4	62.5	83.2	97.0	74.1	91.7	93.7	75.6	89.6	95.3	77.2	91.2	95.4	80.7	92.0	91.9	79.8	89.1
0920	Plaster & Gypsum Board	93.1	73.9	80.2	70.8	70.0	70.2	87.5	77.3	80.6	92.5	73.9	80.0	87.5	89.9	89.1	91.5	78.4	82.7
0950, 0980	Ceilings & Acoustic Treatment	80.1	73.9	75.9	86.2	70.0	75.2	83.6	77.3	79.4	88.7	73.9	78.7	83.6	89.9	87.9	79.4	78.4	78.7
0960	Flooring	85.0	65.7	79.6	65.9	85.9	71.4	85.5	78.0	83.5	90.5	65.7	83.6	88.5	65.7	82.2	92.0	80.4	88.8
0970, 0990	Wall Finishes & Painting/Coating	90.9	64.8	75.7	86.8	74.4	79.5	91.1	70.1	78.9	93.2	71.2	80.4	91.1	56.3	70.8	91.1	90.7	90.9
09	FINISHES	83.2	71.9	77.0	78.1	75.0	76.4	85.7	75.0	79.9	89.9	73.6	81.0	87.0	78.6	82.4	87.1	81.8	84.2
COVERS	DIVS. 10 - 14, 25, 28, 41, 43, 44, 46	100.0	90.7	97.9	100.0	89.3	97.6	100.0	72.8	94.0	100.0	59.0	90.9	100.0	50.2	88.9	100.0	58.0	90.6
21, 22, 23	FIRE SUPPRESSION, PLUMBING & HVAC	97.0	76.4	88.2	97.7	79.0	89.7	97.0	77.1	88.5	100.0	80.2	91.5	96.9	77.6	88.7	97.0	79.0	89.3
26, 27, 3370	ELECTRICAL, COMMUNICATIONS & UTIL.	92.6	93.1	92.9	95.8	76.5	85.8	92.2	78.9	85.3	101.1	75.5	87.8	92.4	93.1	92.8	94.3	75.8	84.7
MF2016	WEIGHTED AVERAGE	92.7	77.7	86.2	94.7	80.5	88.5	91.5	77.2	85.3	96.0	78.5	88.3	94.1	78.7	87.4	92.4	83.0	88.3

For customer support on your Building Construction Costs with RSMeans data, call 800.448.8182.

KENTUCKY

DIVISION		LEXINGTON 403 - 405			LOUISVILLE 400 - 402			OWENSBORO 423			PADUCAH 420			PIKEVILLE 415 - 416			SOMERSET 425 - 426		
		MAT.	INST.	TOTAL	MAT.	INST.	TOTAL	MAT.	INST.	TOTAL	MAT.	INST.	TOTAL	MAT.	INST.	TOTAL	MAT.	INST.	TOTAL
015433	CONTRACTOR EQUIPMENT		99.5	99.5		93.1	93.1		113.9	113.9		113.9	113.9		100.3	100.3		99.5	99.5
0241, 31 - 34	SITE & INFRASTRUCTURE, DEMOLITION	97.2	97.8	97.6	91.6	94.2	93.4	93.3	122.0	113.4	85.9	121.2	110.6	127.4	81.8	95.5	79.6	96.0	91.1
0310	Concrete Forming & Accessories	94.2	76.7	79.1	94.7	78.4	80.7	89.2	80.5	81.7	87.2	82.3	82.9	94.5	78.1	80.3	85.3	74.8	76.3
0320	Concrete Reinforcing	99.8	83.0	91.3	95.8	82.9	89.3	86.6	79.8	83.1	87.1	81.0	84.0	89.7	94.5	92.1	86.9	62.3	74.5
0330	Cast-in-Place Concrete	93.9	80.8	88.9	90.9	71.7	83.6	89.2	88.3	88.9	81.5	84.5	82.6	95.9	92.3	94.5	76.5	92.1	82.4
03	CONCRETE	89.9	79.6	85.2	89.3	77.3	83.8	99.1	83.4	91.9	91.7	83.1	87.8	105.0	87.4	97.0	77.5	79.1	78.2
04	MASONRY	88.2	73.3	79.0	86.5	69.4	75.9	91.3	82.0	85.5	94.1	78.8	84.6	90.8	86.8	88.3	85.7	65.4	73.1
05	METALS	96.4	86.9	93.5	97.5	85.5	93.8	87.9	86.5	87.5	84.9	86.7	85.4	94.2	108.0	98.5	95.1	79.0	90.2
06	WOOD, PLASTICS & COMPOSITES	86.0	75.0	79.8	89.7	79.8	84.2	87.5	79.4	83.0	85.3	82.2	83.6	83.3	77.0	79.8	83.2	75.0	78.6
07	THERMAL & MOISTURE PROTECTION	105.3	76.5	93.1	102.4	74.2	90.5	100.9	80.1	92.1	100.3	84.7	93.7	92.8	82.6	88.5	100.2	73.2	88.8
08	OPENINGS	89.6	77.1	86.8	85.9	76.9	83.9	91.9	80.2	89.2	91.2	82.6	89.2	93.6	76.6	89.7	94.4	69.3	88.7
0920	Plaster & Gypsum Board	102.4	73.9	83.3	92.8	79.7	84.0	90.2	78.4	82.3	89.2	81.3	83.9	64.5	76.6	72.6	87.5	73.9	78.4
0950, 0980	Ceilings & Acoustic Treatment	83.5	73.9	77.0	88.7	79.7	82.6	79.4	78.4	78.7	79.4	81.3	80.7	78.9	76.6	77.3	83.6	73.9	77.1
0960	Flooring	89.9	65.7	83.2	89.8	64.4	82.7	91.3	64.4	83.9	90.3	80.4	87.5	77.9	65.7	74.5	88.9	65.7	82.4
0970, 0990	Wall Finishes & Painting/Coating	90.9	80.9	85.0	92.5	70.1	79.5	91.1	90.7	90.9	91.1	76.5	82.6	93.2	69.9	79.6	91.1	70.1	78.9
09	FINISHES	86.7	74.6	80.1	89.6	74.9	81.6	87.3	78.2	82.3	86.5	81.3	83.7	79.0	74.6	76.6	86.4	72.5	78.8
COVERS	DIVS. 10 - 14, 25, 28, 41, 43, 44, 46	100.0	91.8	98.2	100.0	89.3	97.6	100.0	96.9	99.3	100.0	88.1	97.3	100.0	48.3	88.5	100.0	90.7	97.9
21, 22, 23	FIRE SUPPRESSION, PLUMBING & HVAC	100.1	79.0	91.1	100.0	78.7	90.9	100.0	80.3	91.6	97.0	79.8	89.7	96.7	82.8	90.8	97.0	75.9	88.0
26, 27, 3370	ELECTRICAL, COMMUNICATIONS & UTIL.	95.2	78.9	86.7	101.1	78.9	89.5	94.3	75.8	84.7	96.5	78.1	87.0	95.1	66.7	80.3	92.8	93.1	93.0
MF2016	WEIGHTED AVERAGE	94.8	80.3	88.5	95.1	79.1	88.1	94.7	84.3	90.1	92.6	84.5	89.1	95.9	81.4	89.6	91.9	79.2	86.4

LOUISIANA

DIVISION		ALEXANDRIA 713 - 714			BATON ROUGE 707 - 708			HAMMOND 704			LAFAYETTE 705			LAKE CHARLES 706			MONROE 712		
		MAT.	INST.	TOTAL	MAT.	INST.	TOTAL	MAT.	INST.	TOTAL	MAT.	INST.	TOTAL	MAT.	INST.	TOTAL	MAT.	INST.	TOTAL
015433	CONTRACTOR EQUIPMENT		93.5	93.5		91.4	91.4		91.9	91.9		91.9	91.9		91.4	91.4		93.5	93.5
0241, 31 - 34	SITE & INFRASTRUCTURE, DEMOLITION	105.3	92.4	96.3	102.3	91.1	94.5	98.9	90.4	93.0	99.9	92.7	94.9	100.7	91.8	94.5	105.3	92.3	96.2
0310	Concrete Forming & Accessories	80.2	62.9	65.3	95.5	74.7	77.5	77.9	57.3	60.1	95.0	70.9	74.2	95.8	71.0	74.4	79.8	62.2	64.7
0320	Concrete Reinforcing	96.3	65.1	80.5	91.7	55.6	73.5	93.0	53.7	73.2	94.3	54.0	74.0	94.3	54.3	74.1	95.1	65.0	80.0
0330	Cast-in-Place Concrete	93.8	68.2	84.0	93.5	77.2	87.3	90.1	65.7	80.8	89.7	69.5	82.0	90.1	68.0	80.0	86.9	65.2	77.0
03	CONCRETE	87.1	65.8	77.3	88.8	72.5	81.4	86.9	60.4	74.8	87.8	67.8	78.7	90.1	68.0	80.0	86.9	65.2	77.0
04	MASONRY	114.2	62.9	82.4	87.1	64.9	73.4	91.7	54.7	68.8	91.7	65.3	75.4	91.1	69.7	77.8	108.7	61.6	79.5
05	METALS	92.5	75.1	87.2	96.6	70.4	88.5	87.6	69.1	81.9	86.9	69.8	81.7	86.9	70.2	81.8	92.5	75.0	87.1
06	WOOD, PLASTICS & COMPOSITES	92.9	61.6	75.5	102.1	76.5	87.9	85.6	56.8	69.6	105.4	71.3	86.4	103.6	71.3	85.6	92.3	61.6	75.2
07	THERMAL & MOISTURE PROTECTION	97.9	67.4	85.0	95.3	69.9	84.5	97.1	63.7	83.0	97.6	69.5	85.7	97.3	70.5	86.0	97.9	66.6	84.6
08	OPENINGS	110.5	63.8	99.8	99.7	70.4	93.0	95.5	57.0	86.6	99.1	64.9	91.3	99.1	64.9	91.3	110.5	62.2	99.4
0920	Plaster & Gypsum Board	84.0	60.9	68.5	99.3	76.3	83.9	101.2	56.0	70.9	109.0	70.9	83.4	109.0	70.9	83.4	83.7	60.9	68.4
0950, 0980	Ceilings & Acoustic Treatment	83.0	60.9	68.1	94.8	76.3	82.3	101.6	56.0	70.9	99.9	70.9	80.3	100.9	70.9	80.7	83.0	60.9	68.1
0960	Flooring	90.9	64.9	83.7	86.9	60.4	79.6	88.1	58.4	79.8	96.8	64.3	87.8	96.8	67.4	88.6	90.5	58.2	81.5
0970, 0990	Wall Finishes & Painting/Coating	93.1	62.5	75.2	89.4	62.5	73.7	94.9	62.5	76.0	94.9	62.5	76.0	94.9	62.5	76.0	93.1	62.5	75.2
09	FINISHES	82.0	62.8	71.6	89.7	71.2	79.6	91.4	57.6	73.0	94.8	69.0	80.8	95.0	69.7	81.2	81.8	61.1	70.6
COVERS	DIVS. 10 - 14, 25, 28, 41, 43, 44, 46	100.0	80.6	95.7	100.0	87.2	97.2	100.0	80.4	95.6	100.0	86.8	97.1	100.0	86.9	97.1	100.0	80.1	95.6
21, 22, 23	FIRE SUPPRESSION, PLUMBING & HVAC	100.2	65.4	85.3	100.0	66.4	85.6	97.1	63.0	82.5	100.2	66.6	85.8	100.2	66.6	85.8	100.2	64.6	85.0
26, 27, 3370	ELECTRICAL, COMMUNICATIONS & UTIL.	93.2	64.9	78.5	97.9	59.8	78.1	93.7	70.5	81.7	94.8	67.4	80.5	94.4	69.6	81.5	94.8	60.5	77.0
MF2016	WEIGHTED AVERAGE	96.9	68.2	84.3	96.3	70.1	84.8	93.2	65.1	80.9	94.9	70.1	84.0	95.1	70.9	84.5	96.7	66.9	83.7

LOUISIANA / MAINE

DIVISION		LOUISIANA NEW ORLEANS 700 - 701			SHREVEPORT 710 - 711			THIBODAUX 703			MAINE AUGUSTA 043			BANGOR 044			BATH 045		
		MAT.	INST.	TOTAL	MAT.	INST.	TOTAL	MAT.	INST.	TOTAL	MAT.	INST.	TOTAL	MAT.	INST.	TOTAL	MAT.	INST.	TOTAL
015433	CONTRACTOR EQUIPMENT		87.5	87.5		93.5	93.5		91.9	91.9		98.8	98.8		98.8	98.8		98.8	98.8
0241, 31 - 34	SITE & INFRASTRUCTURE, DEMOLITION	99.9	93.5	95.4	107.6	92.3	96.9	101.2	92.4	95.1	90.4	99.3	96.6	93.6	100.7	98.6	91.6	99.3	97.0
0310	Concrete Forming & Accessories	97.3	69.3	73.1	94.8	62.8	67.2	89.4	67.2	70.3	101.1	77.7	80.9	93.8	79.5	81.5	89.7	77.9	79.5
0320	Concrete Reinforcing	96.2	55.0	75.4	96.3	53.8	74.9	93.0	67.5	80.1	102.3	79.0	90.5	92.1	80.8	86.4	91.1	80.2	85.6
0330	Cast-in-Place Concrete	91.1	69.3	82.8	96.7	67.5	85.6	96.8	67.4	85.6	86.7	112.4	96.5	70.0	113.9	86.7	70.0	112.5	86.1
03	CONCRETE	93.1	66.7	81.0	90.5	63.6	78.2	91.7	67.6	80.7	88.1	90.5	89.2	81.7	92.1	86.5	81.7	90.8	85.9
04	MASONRY	96.9	60.6	74.4	98.1	58.9	73.8	114.7	53.9	77.0	102.7	66.9	80.5	118.3	95.5	104.1	125.5	86.9	101.5
05	METALS	96.8	62.7	86.3	96.6	70.8	88.7	87.6	68.9	81.9	101.1	89.5	97.6	91.0	92.7	91.5	89.4	91.2	89.9
06	WOOD, PLASTICS & COMPOSITES	103.7	71.4	85.7	99.1	62.4	78.6	87.0	65.1	79.4	98.6	76.5	86.3	97.9	76.5	86.0	92.5	76.5	83.6
07	THERMAL & MOISTURE PROTECTION	97.9	68.4	85.4	95.6	66.0	83.1	97.0	65.1	83.5	106.2	70.1	90.9	102.8	84.7	95.1	102.8	76.3	91.5
08	OPENINGS	103.1	65.3	94.5	105.4	59.0	94.8	100.0	59.4	90.7	100.6	80.3	96.0	98.0	80.3	94.0	98.0	80.3	93.9
0920	Plaster & Gypsum Board	98.6	70.9	80.0	93.5	61.7	72.2	102.5	68.2	79.5	108.2	75.4	86.2	112.7	75.4	87.7	108.5	75.4	86.3
0950, 0980	Ceilings & Acoustic Treatment	105.8	70.9	82.3	89.7	61.7	70.8	101.6	68.2	79.1	104.0	75.4	84.7	86.6	75.4	79.1	85.6	75.4	78.8
0960	Flooring	103.8	60.4	91.8	94.9	58.2	84.7	94.1	40.3	79.2	88.5	48.4	77.4	85.0	104.4	90.4	83.3	48.4	73.6
0970, 0990	Wall Finishes & Painting/Coating	99.0	62.5	77.7	85.6	62.5	72.1	96.1	62.5	76.5	90.5	93.0	91.9	90.7	93.0	92.0	90.7	93.0	92.0
09	FINISHES	101.3	67.1	82.7	88.7	61.7	74.0	93.7	61.6	76.3	93.1	72.7	82.0	87.4	84.8	86.0	86.1	72.7	78.8
COVERS	DIVS. 10 - 14, 25, 28, 41, 43, 44, 46	100.0	85.6	96.8	100.0	85.7	96.8	100.0	80.4	95.7	100.0	100.2	100.0	100.0	104.0	100.9	100.0	100.9	100.2
21, 22, 23	FIRE SUPPRESSION, PLUMBING & HVAC	100.1	64.0	84.7	99.9	64.8	84.9	97.1	64.5	83.2	100.0	71.9	88.0	100.2	73.0	88.5	97.1	72.0	86.3
26, 27, 3370	ELECTRICAL, COMMUNICATIONS & UTIL.	98.4	70.5	83.9	100.2	64.5	81.6	92.5	70.5	81.1	101.5	77.6	89.1	99.7	73.2	86.0	97.8	77.6	87.3
MF2016	WEIGHTED AVERAGE	98.7	68.4	85.5	97.9	66.7	84.2	95.6	67.4	83.3	98.4	79.8	90.3	95.8	85.0	91.1	94.8	82.2	89.3

For customer support on your Building Construction Costs with RSMeans data, call 800.448.8182.

City Cost Indexes

DIVISION		HOULTON 047 MAT.	INST.	TOTAL	KITTERY 039 MAT.	INST.	TOTAL	LEWISTON 042 MAT.	INST.	TOTAL	MACHIAS 046 MAT.	INST.	TOTAL	PORTLAND 040-041 MAT.	INST.	TOTAL	ROCKLAND 048 MAT.	INST.	TOTAL
015433	CONTRACTOR EQUIPMENT		98.8	98.8		98.8	98.8		98.8	98.8		98.8	98.8		98.8	98.8		98.8	98.8
0241, 31 - 34	SITE & INFRASTRUCTURE, DEMOLITION	93.4	99.3	97.5	82.9	99.3	94.4	91.3	100.7	97.9	92.8	99.3	97.3	91.2	100.7	97.8	89.3	99.3	96.3
0310	Concrete Forming & Accessories	97.6	77.7	80.4	87.8	78.2	79.6	99.4	79.6	82.3	94.7	77.7	80.0	101.1	79.6	82.5	95.7	77.7	80.2
0320	Concrete Reinforcing	92.1	78.7	85.3	85.3	80.3	82.8	113.1	80.8	96.8	92.1	78.7	85.3	102.3	80.8	91.4	92.1	78.7	85.3
0330	Cast-in-Place Concrete	70.0	112.4	86.1	73.3	112.6	88.2	71.4	113.9	87.6	70.0	112.4	86.1	86.7	113.9	97.0	71.5	112.4	87.0
03	CONCRETE	82.6	90.4	86.2	78.8	91.0	84.4	82.1	92.1	86.7	82.1	90.4	85.9	87.4	92.1	89.6	79.8	90.4	84.7
04	MASONRY	100.2	62.4	76.8	115.6	86.9	97.8	100.8	95.5	97.5	100.2	62.4	76.8	106.1	95.5	99.5	93.7	62.4	74.3
05	METALS	89.6	89.1	89.5	85.8	91.5	87.6	94.3	92.7	93.8	89.6	89.1	89.5	101.2	92.7	98.6	89.5	89.1	89.4
06	WOOD, PLASTICS & COMPOSITES	101.8	76.5	87.7	95.8	76.5	85.0	103.6	76.5	88.5	98.8	76.5	86.4	97.3	76.5	85.7	99.6	76.5	86.7
07	THERMAL & MOISTURE PROTECTION	102.9	69.9	88.9	102.4	75.4	91.0	101.2	84.7	95.0	102.8	69.1	88.6	106.5	84.7	97.2	102.5	69.1	88.4
08	OPENINGS	98.1	80.3	94.0	97.4	84.0	94.3	101.3	80.3	96.5	98.1	80.3	94.0	99.9	80.3	95.4	98.0	80.3	94.0
0920	Plaster & Gypsum Board	115.0	75.4	88.4	104.8	75.4	85.1	118.0	75.4	89.4	113.4	75.4	87.9	108.6	75.4	86.3	113.4	75.4	87.9
0950, 0980	Ceilings & Acoustic Treatment	85.6	75.4	78.8	99.5	75.4	83.3	95.1	75.4	81.8	85.6	75.4	78.8	102.2	75.4	84.1	85.6	75.4	78.8
0960	Flooring	86.2	45.2	74.9	85.4	50.4	75.7	87.8	104.4	92.4	85.4	45.2	74.2	89.2	104.4	93.4	85.8	45.2	74.5
0970, 0990	Wall Finishes & Painting/Coating	90.7	93.0	92.0	84.8	106.8	97.6	90.7	93.0	92.0	90.7	93.0	92.0	93.2	93.0	93.1	90.7	93.0	92.0
09	FINISHES	88.0	72.1	79.3	87.8	74.5	80.6	90.3	84.8	87.3	87.4	72.1	79.1	93.7	84.8	88.8	87.2	72.1	79.0
COVERS	DIVS. 10 - 14, 25, 28, 41, 43, 44, 46	100.0	100.9	100.2	100.0	100.9	100.2	100.0	104.1	100.9	100.0	100.9	100.2	100.0	104.1	100.9	100.0	100.9	100.2
21, 22, 23	FIRE SUPPRESSION, PLUMBING & HVAC	97.1	71.9	86.3	97.0	82.6	90.9	100.2	73.0	88.5	97.1	71.9	86.3	100.0	73.0	88.5	97.1	71.9	86.3
26, 27, 3370	ELECTRICAL, COMMUNICATIONS & UTIL.	101.7	77.6	89.2	96.0	77.6	86.4	101.7	77.6	89.2	101.7	77.6	89.2	103.8	77.6	90.2	101.6	77.6	89.1
MF2016	WEIGHTED AVERAGE	94.4	79.2	87.8	93.0	84.9	89.5	96.3	85.6	91.6	94.2	79.2	87.7	98.7	85.6	93.0	93.5	79.2	87.2

DIVISION		MAINE WATERVILLE 049 MAT.	INST.	TOTAL	ANNAPOLIS 214 MAT.	INST.	TOTAL	BALTIMORE 210-212 MAT.	INST.	TOTAL	COLLEGE PARK 207-208 MAT.	INST.	TOTAL	CUMBERLAND 215 MAT.	INST.	TOTAL	EASTON 216 MAT.	INST.	TOTAL
015433	CONTRACTOR EQUIPMENT		98.8	98.8		105.7	105.7		105.6	105.6		111.0	111.0		105.7	105.7		105.7	105.7
0241, 31 - 34	SITE & INFRASTRUCTURE, DEMOLITION	93.3	99.3	97.5	105.3	94.1	97.5	102.5	98.2	99.4	101.3	98.6	99.4	95.9	95.3	95.5	103.1	92.4	95.6
0310	Concrete Forming & Accessories	89.2	77.7	79.3	97.4	75.7	78.7	98.8	78.3	81.2	85.6	73.8	75.4	90.7	85.8	86.5	89.1	74.1	76.2
0320	Concrete Reinforcing	92.1	79.0	85.3	104.8	90.0	97.3	100.6	90.1	95.3	100.6	89.1	94.8	91.6	85.8	88.7	90.9	87.7	89.3
0330	Cast-in-Place Concrete	70.0	112.4	86.1	110.5	74.5	96.8	109.9	76.3	97.1	116.0	78.2	101.7	94.8	87.3	91.9	105.3	68.1	91.1
03	CONCRETE	83.1	90.5	86.5	98.2	79.3	89.5	107.2	80.7	95.1	106.9	79.7	94.4	89.6	87.6	88.7	97.6	75.8	87.6
04	MASONRY	110.8	66.9	83.6	99.5	70.6	81.6	101.8	70.6	82.5	115.6	76.5	91.4	101.0	87.9	92.9	115.2	61.1	81.6
05	METALS	89.6	89.5	89.6	102.1	105.0	103.0	101.7	99.9	101.1	88.5	107.3	94.3	98.1	105.0	100.2	98.4	101.0	99.2
06	WOOD, PLASTICS & COMPOSITES	92.0	76.5	83.4	94.2	77.4	84.8	101.2	80.7	89.8	83.4	71.7	76.9	90.2	84.9	87.2	88.3	80.9	84.1
07	THERMAL & MOISTURE PROTECTION	102.9	70.1	89.0	103.1	79.8	93.2	101.1	80.8	92.5	100.6	82.9	93.1	99.9	85.6	93.8	100.1	75.8	89.8
08	OPENINGS	98.1	80.3	94.0	97.3	83.0	94.0	99.8	84.8	96.4	95.2	79.6	91.6	100.1	87.8	97.3	98.3	84.3	95.1
0920	Plaster & Gypsum Board	108.5	75.4	86.3	100.1	77.0	84.6	104.3	80.1	88.1	98.7	71.0	80.1	105.7	84.7	91.6	105.7	80.6	88.9
0950, 0980	Ceilings & Acoustic Treatment	85.6	75.4	78.8	91.7	77.0	81.8	100.0	80.1	86.6	101.6	71.0	81.0	103.0	84.7	90.7	103.0	80.6	87.9
0960	Flooring	83.0	48.4	73.4	86.2	80.0	84.5	99.3	80.0	93.9	89.3	80.0	86.7	87.9	96.3	90.2	87.0	80.0	85.1
0970, 0990	Wall Finishes & Painting/Coating	90.7	93.0	92.0	87.3	76.5	81.0	102.9	76.5	87.5	104.1	74.3	86.7	97.3	88.5	92.2	97.3	74.3	83.9
09	FINISHES	86.2	72.7	78.8	86.1	76.2	80.7	99.8	78.2	88.1	92.8	74.1	82.6	94.1	88.0	90.8	94.3	75.7	84.2
COVERS	DIVS. 10 - 14, 25, 28, 41, 43, 44, 46	100.0	100.7	100.2	100.0	87.9	97.3	100.0	89.1	97.6	100.0	84.0	96.4	100.0	92.5	98.3	100.0	70.9	93.5
21, 22, 23	FIRE SUPPRESSION, PLUMBING & HVAC	97.1	71.9	86.3	100.1	82.1	92.4	100.1	82.2	92.4	97.0	86.1	92.3	96.8	74.9	87.5	96.8	75.8	87.8
26, 27, 3370	ELECTRICAL, COMMUNICATIONS & UTIL.	101.7	77.6	89.1	100.1	90.9	95.3	100.7	90.9	95.6	100.2	98.6	99.4	98.1	80.6	89.0	97.6	64.1	80.2
MF2016	WEIGHTED AVERAGE	94.7	79.8	88.2	98.9	84.2	92.4	101.4	84.7	94.1	97.9	86.8	93.0	96.9	86.3	92.3	98.6	76.6	88.9

DIVISION		ELKTON 219 MAT.	INST.	TOTAL	HAGERSTOWN 217 MAT.	INST.	TOTAL	SALISBURY 218 MAT.	INST.	TOTAL	SILVER SPRING 209 MAT.	INST.	TOTAL	WALDORF 206 MAT.	INST.	TOTAL	MASSACHUSETTS BOSTON 020-022,024 MAT.	INST.	TOTAL
015433	CONTRACTOR EQUIPMENT		105.7	105.7		105.7	105.7		105.7	105.7		102.4	102.4		102.4	102.4		103.0	103.0
0241, 31 - 34	SITE & INFRASTRUCTURE, DEMOLITION	89.8	93.9	92.6	94.0	95.2	94.8	103.1	92.4	95.6	89.7	90.0	89.9	96.1	90.1	91.9	95.5	102.1	100.1
0310	Concrete Forming & Accessories	94.8	90.7	91.3	90.0	82.5	83.5	102.8	52.6	59.5	94.1	73.1	75.9	101.3	70.5	74.8	101.6	139.0	133.8
0320	Concrete Reinforcing	90.9	113.2	102.2	91.6	85.8	88.7	90.9	67.3	79.0	99.4	89.0	94.1	100.0	89.0	94.4	101.1	147.6	124.6
0330	Cast-in-Place Concrete	85.3	74.0	81.0	90.3	87.3	89.2	105.3	66.1	90.4	118.8	78.6	103.5	133.1	78.6	112.4	96.0	139.2	112.4
03	CONCRETE	82.3	89.9	85.8	86.0	86.0	86.0	98.4	61.8	81.7	104.8	79.3	93.1	115.7	78.1	98.5	100.9	139.9	118.7
04	MASONRY	99.6	69.9	81.2	106.6	87.9	95.0	114.9	57.5	79.3	115.1	76.9	91.4	98.5	76.9	85.1	112.0	148.2	134.5
05	METALS	98.4	113.6	103.1	98.3	104.9	100.3	98.4	93.0	96.8	92.9	103.1	96.1	92.9	103.1	96.1	101.1	133.3	111.0
06	WOOD, PLASTICS & COMPOSITES	95.3	98.1	96.9	89.5	80.4	84.4	105.3	53.8	76.6	89.9	70.9	79.3	96.8	67.5	80.5	101.9	139.5	122.8
07	THERMAL & MOISTURE PROTECTION	99.6	81.6	92.0	100.3	86.7	94.5	100.3	71.1	87.9	104.3	88.0	97.4	104.8	87.6	97.5	106.0	137.6	119.3
08	OPENINGS	98.3	100.6	98.9	98.3	82.5	94.7	98.6	63.7	90.6	87.4	79.1	85.5	88.1	77.3	85.6	98.7	145.2	109.3
0920	Plaster & Gypsum Board	109.3	98.4	102.0	105.7	80.1	88.5	116.2	52.8	73.6	104.3	71.0	81.9	107.3	67.5	80.6	108.8	140.4	130.0
0950, 0980	Ceilings & Acoustic Treatment	103.0	98.4	99.9	103.9	80.1	87.9	103.0	52.8	69.2	110.1	71.0	83.7	110.1	67.5	81.4	109.4	140.4	130.3
0960	Flooring	89.5	80.0	86.9	87.5	96.3	89.9	93.4	80.0	89.7	95.4	80.0	91.2	99.2	80.0	93.8	95.3	163.7	114.3
0970, 0990	Wall Finishes & Painting/Coating	97.3	74.3	83.9	97.3	74.3	83.9	97.3	74.3	83.9	111.7	74.3	89.9	111.7	74.3	89.9	97.1	158.1	132.7
09	FINISHES	94.7	88.1	91.1	94.0	83.8	88.5	97.7	58.8	76.5	93.8	73.6	82.8	95.6	71.6	82.5	102.7	146.3	126.4
COVERS	DIVS. 10 - 14, 25, 28, 41, 43, 44, 46	100.0	58.1	90.7	100.0	92.0	98.2	100.0	79.7	95.5	100.0	84.4	96.5	100.0	84.0	96.4	100.0	116.7	103.7
21, 22, 23	FIRE SUPPRESSION, PLUMBING & HVAC	96.8	80.5	89.8	99.9	86.5	94.2	96.8	73.7	86.9	97.0	85.9	92.3	97.0	85.8	92.2	100.1	126.1	111.2
26, 27, 3370	ELECTRICAL, COMMUNICATIONS & UTIL.	99.3	89.1	94.0	97.9	80.6	88.9	96.4	63.5	79.3	97.5	98.6	98.1	95.1	98.6	96.9	101.4	129.8	116.2
MF2016	WEIGHTED AVERAGE	95.8	87.4	92.1	97.2	87.7	93.0	99.0	69.8	86.2	97.2	85.8	92.2	97.9	85.2	92.3	101.2	133.1	115.2

For customer support on your Building Construction Costs with RSMeans data, call 800.448.8182.

MASSACHUSETTS

DIVISION		BROCKTON 023 MAT.	INST.	TOTAL	BUZZARDS BAY 025 MAT.	INST.	TOTAL	FALL RIVER 027 MAT.	INST.	TOTAL	FITCHBURG 014 MAT.	INST.	TOTAL	FRAMINGHAM 017 MAT.	INST.	TOTAL	GREENFIELD 013 MAT.	INST.	TOTAL
015433	CONTRACTOR EQUIPMENT		101.4	101.4		101.4	101.4		102.3	102.3		98.8	98.8		100.7	100.7		98.8	98.8
0241, 31 - 34	SITE & INFRASTRUCTURE, DEMOLITION	94.0	103.8	100.9	84.2	103.8	97.9	93.0	103.9	100.6	85.3	103.7	98.2	82.0	103.6	97.1	88.9	102.5	98.4
0310	Concrete Forming & Accessories	98.6	125.5	121.8	96.3	124.9	120.9	98.6	125.1	121.4	92.9	125.1	120.7	100.3	125.5	122.0	91.3	109.4	106.9
0320	Concrete Reinforcing	102.0	147.3	124.9	81.8	124.4	103.3	102.0	124.5	113.4	84.6	147.0	116.1	84.6	147.1	116.1	88.3	120.1	104.3
0330	Cast-in-Place Concrete	89.4	138.6	108.1	74.3	138.4	98.6	86.5	138.9	106.4	78.7	138.5	101.4	78.7	138.7	101.5	80.9	122.3	96.6
03	CONCRETE	92.2	133.3	111.0	78.1	129.0	101.4	90.8	129.2	108.4	76.7	132.9	102.3	79.2	133.3	103.9	80.0	115.4	96.2
04	MASONRY	103.0	144.1	128.5	96.0	142.3	124.7	103.7	142.3	127.6	102.0	140.7	126.0	108.7	140.8	128.6	106.9	122.8	116.8
05	METALS	97.0	128.7	106.8	91.9	118.8	100.1	97.0	119.2	103.8	94.0	125.0	103.5	94.1	128.3	106.6	96.5	111.5	101.1
06	WOOD, PLASTICS & COMPOSITES	101.4	122.2	113.1	98.3	122.2	111.6	101.4	122.4	113.1	99.7	122.2	112.2	105.9	122.0	114.8	97.8	107.1	103.0
07	THERMAL & MOISTURE PROTECTION	100.1	132.9	114.0	99.0	129.5	111.9	100.0	127.8	111.8	100.8	126.7	111.7	100.8	131.5	113.8	100.8	110.2	104.8
08	OPENINGS	98.8	133.2	106.7	94.6	122.4	101.0	98.8	122.2	104.2	100.1	133.2	107.7	90.9	133.1	100.6	100.5	113.1	103.4
0920	Plaster & Gypsum Board	92.1	122.4	112.5	87.9	122.4	111.1	92.1	122.4	112.5	108.3	122.4	117.8	111.0	122.4	118.6	109.3	106.9	107.7
0950, 0980	Ceilings & Acoustic Treatment	108.7	122.4	117.9	93.3	122.4	112.9	108.7	122.4	117.9	93.1	122.4	112.9	93.1	122.4	112.9	101.6	106.9	105.2
0960	Flooring	90.8	163.7	111.0	88.7	163.7	109.5	89.9	163.7	110.4	86.6	163.7	108.0	88.3	163.7	109.2	85.8	140.4	101.0
0970, 0990	Wall Finishes & Painting/Coating	90.5	137.4	117.9	90.5	137.4	117.9	90.5	137.4	117.9	87.8	137.4	116.7	88.7	137.4	117.1	87.8	114.3	103.3
09	FINISHES	92.8	133.9	115.2	88.0	133.9	112.9	92.6	134.1	115.1	87.5	133.9	112.7	88.2	133.7	113.0	89.4	115.7	103.7
COVERS	DIVS. 10 - 14, 25, 28, 41, 43, 44, 46	100.0	115.1	103.4	100.0	115.1	103.4	100.0	115.7	103.5	100.0	109.7	102.2	100.0	114.7	103.3	100.0	105.3	101.2
21, 22, 23	FIRE SUPPRESSION, PLUMBING & HVAC	100.3	109.7	104.3	97.2	109.2	104.3	100.3	109.3	104.1	97.4	110.4	102.9	97.4	123.4	108.5	97.4	102.4	99.5
26, 27, 3370	ELECTRICAL, COMMUNICATIONS & UTIL.	100.1	103.2	101.7	97.5	103.2	100.4	100.0	103.2	101.6	101.4	102.5	101.9	98.1	129.3	114.3	101.4	98.2	99.7
MF2016	WEIGHTED AVERAGE	97.9	121.8	108.4	92.8	119.5	104.5	97.7	119.6	107.3	94.4	120.8	105.9	93.7	127.9	108.7	95.7	109.0	101.5

MASSACHUSETTS

DIVISION		HYANNIS 026 MAT.	INST.	TOTAL	LAWRENCE 019 MAT.	INST.	TOTAL	LOWELL 018 MAT.	INST.	TOTAL	NEW BEDFORD 027 MAT.	INST.	TOTAL	PITTSFIELD 012 MAT.	INST.	TOTAL	SPRINGFIELD 010 - 011 MAT.	INST.	TOTAL
015433	CONTRACTOR EQUIPMENT		101.4	101.4		101.4	101.4		98.8	98.8		102.3	102.3		98.8	98.8		98.8	98.8
0241, 31 - 34	SITE & INFRASTRUCTURE, DEMOLITION	90.3	103.8	99.8	94.1	103.8	100.9	93.0	103.8	100.5	91.4	103.9	100.2	94.0	102.2	99.7	93.5	102.5	99.8
0310	Concrete Forming & Accessories	90.9	124.9	120.2	101.5	125.8	122.4	98.0	125.7	121.9	98.6	125.2	121.5	98.0	107.8	106.4	98.2	109.4	107.9
0320	Concrete Reinforcing	81.8	124.4	103.3	104.7	139.0	122.3	105.6	138.7	122.3	102.0	124.5	113.4	87.7	120.8	104.4	105.6	120.1	111.4
0330	Cast-in-Place Concrete	81.6	138.4	103.2	91.0	138.8	109.2	82.8	138.7	104.0	76.2	138.9	100.0	90.3	119.5	101.4	86.1	122.3	99.9
03	CONCRETE	84.0	129.0	104.6	92.4	132.0	110.5	84.7	131.7	106.2	86.0	129.3	105.8	86.0	113.8	98.7	86.3	115.4	99.6
04	MASONRY	102.0	142.3	127.0	114.6	144.9	133.4	100.9	140.7	125.6	101.9	142.3	126.9	101.5	113.8	109.2	101.2	122.8	114.6
05	METALS	93.4	118.8	101.2	96.9	125.7	105.7	96.9	122.2	104.6	97.0	119.3	103.9	96.7	111.5	101.2	99.5	111.5	103.2
06	WOOD, PLASTICS & COMPOSITES	92.1	122.2	108.8	106.1	122.2	115.0	105.6	122.2	114.8	101.4	122.4	113.1	105.6	107.5	106.7	105.6	107.1	106.4
07	THERMAL & MOISTURE PROTECTION	99.5	129.5	112.2	101.5	133.1	114.8	101.2	131.9	114.7	99.9	127.8	111.7	101.3	106.9	103.7	101.2	110.2	105.0
08	OPENINGS	95.2	122.4	101.4	94.8	130.9	103.1	101.6	130.9	108.3	98.8	127.1	105.3	101.6	113.6	104.3	101.6	113.1	104.2
0920	Plaster & Gypsum Board	83.7	122.4	109.7	113.6	122.4	119.5	113.6	122.4	119.5	92.1	122.4	112.5	113.6	107.3	109.4	113.6	106.9	109.1
0950, 0980	Ceilings & Acoustic Treatment	101.9	122.4	115.7	103.5	122.4	116.3	103.5	122.4	116.3	108.7	122.4	117.9	103.5	107.3	106.1	103.5	106.9	105.8
0960	Flooring	86.0	163.7	107.6	88.8	163.7	109.6	88.8	163.7	109.6	89.9	163.7	110.4	89.2	132.9	101.4	88.2	140.4	102.7
0970, 0990	Wall Finishes & Painting/Coating	90.5	137.4	117.9	87.9	137.4	116.8	87.8	137.4	116.7	90.5	137.4	117.9	87.8	114.3	103.3	89.0	114.3	103.7
09	FINISHES	88.6	133.9	113.2	91.4	133.9	114.5	91.4	133.9	114.5	92.4	134.1	115.1	91.5	113.2	103.3	91.3	115.7	104.6
COVERS	DIVS. 10 - 14, 25, 28, 41, 43, 44, 46	100.0	115.1	103.4	100.0	115.2	103.4	100.0	115.2	103.4	100.0	115.7	103.5	100.0	103.7	100.8	100.0	105.3	101.2
21, 22, 23	FIRE SUPPRESSION, PLUMBING & HVAC	100.3	109.2	104.1	100.1	125.9	111.1	100.1	125.8	111.1	100.3	109.3	104.1	100.1	98.1	99.2	100.1	102.4	101.1
26, 27, 3370	ELECTRICAL, COMMUNICATIONS & UTIL.	97.9	103.2	100.6	100.5	129.8	115.7	100.9	126.0	114.0	100.0	103.2	102.1	100.9	98.2	99.5	101.0	98.2	99.5
MF2016	WEIGHTED AVERAGE	95.1	119.5	105.8	98.0	128.5	111.4	97.1	127.1	110.3	97.0	119.8	107.0	97.3	106.5	101.3	97.8	109.0	102.7

MASSACHUSETTS / MICHIGAN

DIVISION		WORCESTER 015 - 016 MAT.	INST.	TOTAL	ANN ARBOR 481 MAT.	INST.	TOTAL	BATTLE CREEK 490 MAT.	INST.	TOTAL	BAY CITY 487 MAT.	INST.	TOTAL	DEARBORN 481 MAT.	INST.	TOTAL	DETROIT 482 MAT.	INST.	TOTAL
015433	CONTRACTOR EQUIPMENT		98.8	98.8		112.1	112.1		98.9	98.9		112.1	112.1		112.1	112.1		96.3	96.3
0241, 31 - 34	SITE & INFRASTRUCTURE, DEMOLITION	93.4	103.7	100.6	82.2	96.7	92.3	92.1	85.8	87.7	73.6	96.3	89.5	81.9	96.8	92.3	97.3	100.7	99.7
0310	Concrete Forming & Accessories	98.6	125.1	121.5	96.4	106.3	105.0	95.2	80.9	82.9	96.5	82.6	84.5	96.3	107.0	105.5	99.3	107.0	105.9
0320	Concrete Reinforcing	105.6	147.0	126.5	98.4	104.5	101.5	88.6	80.9	84.7	98.4	103.7	101.1	98.4	104.6	101.5	99.3	106.5	102.9
0330	Cast-in-Place Concrete	85.6	138.5	105.7	88.4	101.5	93.3	85.3	95.8	89.3	84.6	87.7	85.7	86.4	102.3	92.5	103.2	101.9	102.7
03	CONCRETE	86.1	132.9	107.5	89.2	105.2	96.5	84.8	85.8	85.3	87.4	89.6	88.4	88.3	105.8	96.3	101.3	104.5	102.8
04	MASONRY	100.7	140.7	125.5	97.8	100.5	99.5	97.3	81.7	87.7	97.4	81.0	87.2	97.7	102.4	100.6	96.1	102.5	100.0
05	METALS	99.5	125.0	107.3	97.2	117.5	103.4	99.2	84.1	94.6	97.8	115.4	103.2	97.2	117.7	103.5	98.5	96.3	97.8
06	WOOD, PLASTICS & COMPOSITES	106.0	122.2	115.0	90.9	107.9	100.4	89.2	79.2	83.6	90.9	82.1	86.0	90.9	107.9	100.4	95.7	107.8	102.4
07	THERMAL & MOISTURE PROTECTION	101.2	126.7	112.0	106.1	102.7	104.6	97.5	81.9	90.9	103.3	84.5	95.4	104.3	106.2	105.1	103.5	106.7	104.9
08	OPENINGS	101.6	133.2	108.8	96.6	103.3	98.1	91.7	77.3	88.4	96.6	85.1	93.9	96.6	102.7	98.0	98.3	103.7	99.5
0920	Plaster & Gypsum Board	113.6	122.4	119.5	105.4	107.8	107.0	98.0	75.2	82.7	105.4	81.3	89.2	105.4	107.8	107.0	100.6	107.8	105.4
0950, 0980	Ceilings & Acoustic Treatment	103.5	122.4	116.3	88.9	107.8	101.6	82.4	75.2	77.5	89.8	81.3	84.0	88.9	107.8	101.6	96.9	107.8	104.3
0960	Flooring	88.8	163.7	109.6	90.1	113.4	96.6	88.3	74.5	84.4	90.1	81.3	87.7	89.6	107.4	94.6	97.6	107.4	100.3
0970, 0990	Wall Finishes & Painting/Coating	87.8	137.4	116.7	82.0	99.9	92.4	86.7	79.7	82.6	82.0	82.7	82.4	82.0	98.1	91.4	92.3	99.8	96.7
09	FINISHES	91.4	133.9	114.5	90.3	107.9	102.5	85.7	79.0	82.0	90.1	81.7	85.5	90.2	106.5	99.0	97.4	106.6	102.4
COVERS	DIVS. 10 - 14, 25, 28, 41, 43, 44, 46	100.0	109.7	102.2	100.0	97.1	99.4	100.0	97.6	99.5	100.0	91.2	98.0	100.0	97.6	99.5	100.0	102.2	100.6
21, 22, 23	FIRE SUPPRESSION, PLUMBING & HVAC	100.1	110.4	104.5	100.2	95.2	98.0	100.2	86.0	94.1	100.2	81.9	92.4	100.2	104.2	101.9	100.0	104.2	101.8
26, 27, 3370	ELECTRICAL, COMMUNICATIONS & UTIL.	101.0	102.5	101.7	96.3	100.9	98.7	93.5	82.1	87.6	95.5	87.1	91.1	96.3	101.5	99.0	99.0	101.4	100.3
MF2016	WEIGHTED AVERAGE	97.7	120.8	107.8	96.2	102.4	98.9	94.8	83.7	89.9	95.7	88.3	92.5	96.1	104.7	99.8	99.2	103.0	100.9

793

MICHIGAN

DIVISION		FLINT 484 - 485			GAYLORD 497			GRAND RAPIDS 493, 495			IRON MOUNTAIN 498 - 499			JACKSON 492			KALAMAZOO 491		
		MAT.	INST.	TOTAL	MAT.	INST.	TOTAL	MAT.	INST.	TOTAL	MAT.	INST.	TOTAL	MAT.	INST.	TOTAL	MAT.	INST.	TOTAL
015433	CONTRACTOR EQUIPMENT		112.1	112.1		106.8	106.8		98.9	98.9		93.6	93.6		106.8	106.8		98.9	98.9
0241, 31 - 34	SITE & INFRASTRUCTURE, DEMOLITION	71.5	96.2	88.8	86.6	82.6	83.8	91.0	85.7	87.3	95.1	91.6	92.7	109.1	85.1	92.3	92.4	85.8	87.7
0310	Concrete Forming & Accessories	99.2	85.8	87.6	93.9	73.2	76.0	93.8	79.4	81.4	85.7	79.8	80.6	90.5	84.1	85.0	95.2	80.8	82.8
0320	Concrete Reinforcing	98.4	104.1	101.3	82.6	89.0	85.8	93.0	80.7	86.8	82.5	85.7	84.1	80.2	104.0	92.2	88.6	78.7	83.6
0330	Cast-in-Place Concrete	89.0	90.2	89.4	85.1	82.2	84.0	89.5	94.0	91.2	100.7	70.8	89.3	84.9	92.1	87.6	87.0	95.8	90.3
03	CONCRETE	89.7	91.9	90.7	81.7	81.0	81.4	88.6	84.5	86.7	90.1	78.4	84.8	77.0	91.6	83.7	87.9	85.4	86.7
04	MASONRY	97.9	89.6	92.7	108.0	75.7	88.0	90.2	77.2	82.2	93.2	80.8	85.5	87.5	87.5	87.5	95.8	81.7	87.1
05	METALS	97.2	116.1	103.0	100.6	113.0	104.4	96.3	83.6	92.4	100.0	91.8	97.5	100.8	114.1	104.9	99.2	83.2	94.3
06	WOOD, PLASTICS & COMPOSITES	94.3	84.5	88.9	82.6	71.9	76.6	93.2	77.6	84.5	78.2	79.8	79.1	81.4	82.1	81.8	89.2	79.2	83.6
07	THERMAL & MOISTURE PROTECTION	103.7	88.7	97.3	96.2	70.8	85.4	98.8	74.1	88.3	100.0	76.4	90.0	95.4	89.0	92.7	97.5	81.9	90.9
08	OPENINGS	96.6	86.3	94.2	92.1	80.0	89.3	101.3	77.0	95.8	99.0	71.4	92.7	91.2	88.5	90.6	91.7	77.2	88.4
0920	Plaster & Gypsum Board	107.4	83.8	91.5	97.8	70.1	79.2	101.4	73.6	82.7	52.7	79.8	70.9	96.2	80.6	85.7	98.0	75.2	82.7
0950, 0980	Ceilings & Acoustic Treatment	88.9	83.8	85.4	81.5	70.1	73.8	95.1	73.6	80.6	80.5	79.8	80.0	81.5	80.6	80.9	82.4	75.2	77.5
0960	Flooring	90.1	90.9	90.3	81.8	90.5	84.2	90.3	80.5	87.6	100.0	92.4	97.9	80.5	83.1	81.2	88.3	74.5	84.4
0970, 0990	Wall Finishes & Painting/Coating	82.0	81.9	81.9	83.1	80.9	81.9	87.9	80.4	83.5	100.4	67.9	81.5	83.1	98.1	91.9	86.7	79.7	82.6
09	FINISHES	89.8	85.7	87.5	85.2	76.5	80.5	91.3	79.1	84.7	85.6	80.9	83.1	86.1	84.6	85.3	85.7	79.0	82.0
COVERS	DIVS. 10 - 14, 25, 28, 41, 43, 44, 46	100.0	92.5	98.3	100.0	80.5	95.7	100.0	97.3	99.4	100.0	89.0	97.6	100.0	93.7	98.6	100.0	97.6	99.5
21, 22, 23	FIRE SUPPRESSION, PLUMBING & HVAC	100.2	86.8	94.5	97.3	75.3	87.9	100.0	82.9	92.7	97.2	82.1	90.7	97.3	86.9	92.8	100.2	81.5	92.2
26, 27, 3370	ELECTRICAL, COMMUNICATIONS & UTIL.	96.3	94.1	95.2	91.7	75.9	83.5	98.5	87.6	92.8	97.4	81.3	89.0	95.2	99.8	97.6	93.4	78.3	85.5
MF2016	WEIGHTED AVERAGE	95.9	92.3	94.4	94.1	80.7	88.2	96.4	82.9	90.5	96.0	82.4	90.0	93.4	91.8	92.7	95.1	82.1	89.4

MICHIGAN / MINNESOTA

DIVISION		LANSING 488 - 489			MUSKEGON 494			ROYAL OAK 480, 483			SAGINAW 486			TRAVERSE CITY 496			BEMIDJI 566		
		MAT.	INST.	TOTAL	MAT.	INST.	TOTAL	MAT.	INST.	TOTAL	MAT.	INST.	TOTAL	MAT.	INST.	TOTAL	MAT.	INST.	TOTAL
015433	CONTRACTOR EQUIPMENT		112.1	112.1		98.9	98.9		91.7	91.7		112.1	112.1		93.6	93.6		100.3	100.3
0241, 31 - 34	SITE & INFRASTRUCTURE, DEMOLITION	94.1	96.5	95.8	90.1	85.7	87.0	86.1	95.8	92.9	74.6	96.2	89.8	80.8	91.1	88.0	96.1	99.6	98.6
0310	Concrete Forming & Accessories	93.3	78.4	80.5	95.5	79.2	81.5	91.7	106.6	104.6	96.4	83.4	85.2	85.7	72.9	74.7	85.0	85.2	85.1
0320	Concrete Reinforcing	101.2	103.8	102.5	89.2	80.6	84.9	89.2	90.0	89.6	98.4	103.6	101.1	83.7	77.9	80.8	94.7	98.7	96.6
0330	Cast-in-Place Concrete	99.3	89.0	95.4	85.0	93.9	88.4	77.3	102.0	86.7	87.3	87.6	87.4	78.5	78.5	78.5	102.1	107.1	104.0
03	CONCRETE	92.1	88.2	90.3	83.2	84.3	83.7	76.7	101.5	88.0	88.7	89.9	89.3	73.9	76.6	75.1	89.9	96.4	92.9
04	MASONRY	96.9	88.1	91.5	94.4	77.2	83.8	91.5	100.8	97.3	99.3	81.0	87.9	91.4	76.9	82.4	100.1	100.9	100.6
05	METALS	96.3	115.5	102.2	96.9	83.4	92.8	100.4	91.8	97.8	97.2	115.3	102.8	99.9	88.6	96.4	93.5	116.4	100.6
06	WOOD, PLASTICS & COMPOSITES	89.9	74.6	81.4	86.0	77.6	81.3	86.7	107.9	98.5	87.3	83.5	85.2	78.2	72.6	75.1	66.8	80.7	74.5
07	THERMAL & MOISTURE PROTECTION	101.7	86.1	95.1	96.4	74.4	87.1	101.9	102.6	102.2	104.6	84.8	96.2	99.0	68.4	86.0	106.1	91.9	100.1
08	OPENINGS	100.4	80.6	95.9	91.0	77.6	87.9	96.4	103.2	98.0	94.6	85.9	92.6	99.0	66.1	91.5	98.8	101.8	99.5
0920	Plaster & Gypsum Board	99.3	73.6	82.0	77.1	73.6	74.7	103.1	107.8	106.2	105.4	82.6	90.1	52.7	72.5	66.0	108.9	80.5	89.9
0950, 0980	Ceilings & Acoustic Treatment	88.7	73.6	78.5	83.2	73.6	76.7	88.1	107.8	101.4	88.9	82.6	84.7	80.5	72.5	75.1	125.3	80.5	95.1
0960	Flooring	91.9	83.1	89.5	87.0	76.4	84.1	87.2	107.4	92.8	90.1	81.3	87.7	100.0	90.5	97.4	90.6	90.4	90.6
0970, 0990	Wall Finishes & Painting/Coating	89.6	80.4	84.2	84.9	79.1	81.5	83.6	98.1	92.0	82.0	82.7	82.4	100.4	41.2	65.9	90.9	108.5	101.1
09	FINISHES	89.4	78.3	83.4	82.0	77.5	79.6	89.2	106.2	98.4	90.0	82.5	85.9	84.6	72.7	78.1	97.8	88.1	92.5
COVERS	DIVS. 10 - 14, 25, 28, 41, 43, 44, 46	100.0	91.7	98.2	100.0	97.3	99.4	100.0	102.4	100.5	100.0	91.4	98.1	100.0	87.2	97.2	100.0	95.3	99.0
21, 22, 23	FIRE SUPPRESSION, PLUMBING & HVAC	100.0	87.1	94.5	99.9	82.9	92.7	97.3	99.9	98.4	100.2	81.4	92.2	97.2	79.0	89.4	97.3	82.8	91.1
26, 27, 3370	ELECTRICAL, COMMUNICATIONS & UTIL.	97.9	92.5	95.1	93.8	77.0	85.1	98.3	101.4	99.9	94.7	89.2	91.8	93.2	75.9	84.2	104.4	104.2	104.3
MF2016	WEIGHTED AVERAGE	97.0	90.1	94.0	93.6	81.2	88.2	94.3	100.5	97.0	95.6	88.7	92.6	92.9	78.4	86.6	97.0	96.0	96.6

MINNESOTA

DIVISION		BRAINERD 564			DETROIT LAKES 565			DULUTH 556 - 558			MANKATO 560			MINNEAPOLIS 553 - 555			ROCHESTER 559		
		MAT.	INST.	TOTAL	MAT.	INST.	TOTAL	MAT.	INST.	TOTAL	MAT.	INST.	TOTAL	MAT.	INST.	TOTAL	MAT.	INST.	TOTAL
015433	CONTRACTOR EQUIPMENT		103.0	103.0		100.3	100.3		103.4	103.4		103.0	103.0		107.9	107.9		103.4	103.4
0241, 31 - 34	SITE & INFRASTRUCTURE, DEMOLITION	98.6	104.6	102.8	94.3	99.8	98.2	99.8	102.9	102.0	95.2	104.3	101.6	97.9	108.3	105.2	98.8	102.6	101.4
0310	Concrete Forming & Accessories	85.9	85.7	85.7	82.1	84.9	84.5	99.5	96.7	97.1	93.6	95.3	95.1	98.1	114.7	112.4	98.1	98.9	98.8
0320	Concrete Reinforcing	93.2	98.6	95.9	94.4	98.5	96.5	95.2	99.0	97.2	93.0	106.0	99.6	94.9	106.5	100.8	92.9	106.2	99.6
0330	Cast-in-Place Concrete	111.0	111.6	111.2	99.2	110.0	103.3	102.2	101.6	102.0	102.3	99.2	101.1	101.5	115.9	106.9	99.7	99.9	99.8
03	CONCRETE	93.8	98.2	95.8	87.6	97.3	92.0	95.1	99.9	97.3	89.4	99.6	94.1	98.0	114.3	105.5	91.5	101.6	96.2
04	MASONRY	125.1	106.4	113.5	124.5	103.7	111.6	99.2	110.1	106.0	112.9	104.3	107.6	105.4	118.5	113.5	98.8	107.8	104.4
05	METALS	94.7	116.0	101.2	93.5	115.7	100.3	101.2	118.8	106.6	94.5	119.7	102.2	101.2	124.0	108.2	100.9	122.0	107.4
06	WOOD, PLASTICS & COMPOSITES	83.6	77.8	80.4	64.1	78.0	71.8	93.1	93.6	93.3	92.5	93.3	93.0	95.7	112.1	104.8	95.7	97.4	96.6
07	THERMAL & MOISTURE PROTECTION	104.4	101.8	103.3	105.9	100.6	103.7	105.1	104.0	104.6	104.8	92.8	99.7	103.4	116.7	109.0	110.0	93.5	103.0
08	OPENINGS	85.8	100.2	89.1	98.8	100.3	99.1	105.8	106.0	105.8	90.3	112.2	95.3	101.3	122.6	106.2	100.7	113.5	103.6
0920	Plaster & Gypsum Board	94.8	77.8	83.3	107.9	77.8	87.7	92.7	93.9	93.5	99.4	93.7	95.6	100.8	112.7	108.8	103.6	97.8	99.7
0950, 0980	Ceilings & Acoustic Treatment	57.8	77.8	71.3	125.3	77.8	93.3	94.3	93.9	94.0	57.8	93.7	82.0	99.6	112.7	108.4	96.3	97.8	97.3
0960	Flooring	89.5	90.4	89.7	89.4	90.4	89.7	92.7	122.4	100.9	91.4	86.5	90.0	100.8	116.9	105.3	93.6	86.5	91.6
0970, 0990	Wall Finishes & Painting/Coating	85.3	108.5	98.8	90.9	87.2	88.7	85.5	109.2	99.3	96.2	102.6	100.0	100.9	129.5	117.6	85.7	102.6	95.6
09	FINISHES	81.8	88.0	85.2	97.1	85.5	90.8	90.5	102.2	96.8	83.6	94.7	89.6	99.8	116.7	109.0	91.4	97.5	94.7
COVERS	DIVS. 10 - 14, 25, 28, 41, 43, 44, 46	100.0	96.9	99.3	100.0	96.7	99.3	100.0	97.9	99.5	100.0	97.1	99.4	100.0	105.6	101.2	100.0	98.1	99.6
21, 22, 23	FIRE SUPPRESSION, PLUMBING & HVAC	96.5	86.3	92.1	97.3	85.3	92.2	100.0	96.3	98.4	96.5	86.4	92.2	99.9	108.4	103.6	100.0	93.9	97.4
26, 27, 3370	ELECTRICAL, COMMUNICATIONS & UTIL.	102.1	99.5	100.8	104.2	70.3	86.6	103.1	99.5	101.2	108.5	92.0	99.9	101.2	104.4	102.90	101.6	92.0	96.6
MF2016	WEIGHTED AVERAGE	95.9	97.5	96.6	97.8	92.0	95.3	99.8	102.7	101.1	96.0	98.0	96.9	100.4	113.0	105.9	98.8	100.9	99.7

For customer support on your Building Construction Costs with RSMeans data, call 800.448.8182.

DIVISION		MINNESOTA														MISSISSIPPI		
		SAINT PAUL 550 - 551			ST. CLOUD 563			THIEF RIVER FALLS 567			WILLMAR 562			WINDOM 561			BILOXI 395		
		MAT.	INST.	TOTAL	MAT.	INST.	TOTAL	MAT.	INST.	TOTAL	MAT.	INST.	TOTAL	MAT.	INST.	TOTAL	MAT.	INST.	TOTAL
015433	CONTRACTOR EQUIPMENT		103.4	103.4		103.0	103.0		100.3	100.3		103.0	103.0		103.0	103.0		102.5	102.5
0241, 31 - 34	SITE & INFRASTRUCTURE, DEMOLITION	96.0	103.6	101.3	93.7	105.6	102.1	95.1	99.4	98.1	93.2	104.6	101.2	87.0	103.9	98.8	102.9	89.7	93.6
0310	Concrete Forming & Accessories	97.1	114.4	112.0	83.3	113.7	109.5	85.6	84.2	84.4	83.1	89.8	88.9	87.3	85.8	86.0	93.4	66.9	70.5
0320	Concrete Reinforcing	95.5	106.5	101.1	93.2	105.8	99.6	94.8	98.4	96.6	92.8	105.6	99.3	92.8	105.0	99.0	83.1	50.1	66.4
0330	Cast-in-Place Concrete	101.2	114.4	106.2	98.0	114.1	104.1	101.2	83.5	94.5	99.5	82.2	92.9	86.1	87.3	86.5	114.5	67.1	96.5
03	CONCRETE	94.6	113.7	103.3	85.5	113.1	98.1	88.7	87.8	88.3	85.5	91.2	88.1	76.7	91.1	83.3	98.1	65.7	83.3
04	MASONRY	97.7	118.6	110.7	109.0	108.7	108.8	100.1	98.4	99.0	113.0	102.9	106.7	124.2	85.2	100.0	97.5	61.5	75.2
05	METALS	101.2	123.5	108.1	95.3	120.3	103.0	93.7	114.9	100.2	94.4	119.2	102.1	94.4	117.9	101.6	89.4	84.8	88.0
06	WOOD, PLASTICS & COMPOSITES	97.1	111.8	105.3	81.2	111.8	98.2	67.7	80.7	75.0	80.9	83.7	82.5	84.9	83.7	84.2	102.6	67.6	83.1
07	THERMAL & MOISTURE PROTECTION	106.0	116.3	110.4	104.5	106.0	105.1	106.9	87.4	98.6	104.3	97.9	101.6	104.3	83.6	95.5	94.8	63.9	81.7
08	OPENINGS	100.3	121.5	105.2	90.7	121.5	97.8	98.8	101.8	99.5	87.8	106.0	91.9	91.3	106.0	94.7	100.8	57.8	90.9
0920	Plaster & Gypsum Board	102.0	112.7	109.1	94.8	112.7	106.8	108.6	80.5	89.7	94.8	83.8	87.4	94.8	83.8	87.4	97.0	67.2	77.0
0950, 0980	Ceilings & Acoustic Treatment	95.8	112.7	107.2	57.8	112.7	94.8	125.3	80.5	95.1	57.8	83.8	75.4	57.8	83.8	75.4	90.2	67.2	74.7
0960	Flooring	87.7	116.9	95.8	86.3	90.4	87.4	90.3	90.4	90.3	87.6	90.4	88.4	90.2	90.4	90.2	88.6	57.5	80.0
0970, 0990	Wall Finishes & Painting/Coating	88.1	123.2	108.6	96.2	123.2	112.0	90.9	87.2	88.7	90.9	87.2	88.7	90.9	102.6	97.7	81.9	49.0	62.7
09	FINISHES	91.5	115.8	104.7	81.2	111.2	97.5	97.6	85.7	91.1	81.2	89.1	85.5	81.5	88.4	85.2	85.7	63.6	73.7
COVERS	DIVS. 10 - 14, 25, 28, 41, 43, 44, 46	100.0	102.8	100.6	100.0	102.6	100.6	100.0	95.2	98.9	100.0	97.5	99.4	100.0	94.3	98.7	100.0	72.7	93.9
21, 22, 23	FIRE SUPPRESSION, PLUMBING & HVAC	99.9	111.3	104.8	99.6	108.2	103.2	97.3	82.4	91.0	96.5	101.9	98.8	96.5	82.3	90.4	99.9	57.6	81.8
26, 27, 3370	ELECTRICAL, COMMUNICATIONS & UTIL.	101.6	113.2	107.6	102.1	113.2	107.9	101.6	70.2	85.3	102.1	79.7	90.5	108.5	92.0	99.9	100.0	56.7	77.5
MF2016	WEIGHTED AVERAGE	98.9	114.1	105.6	95.2	111.3	102.3	96.6	89.2	93.4	94.3	97.4	95.6	94.6	92.3	93.6	96.7	65.5	83.1

DIVISION		MISSISSIPPI																
		CLARKSDALE 386			COLUMBUS 397			GREENVILLE 387			GREENWOOD 389			JACKSON 390 - 392			LAUREL 394		
		MAT.	INST.	TOTAL	MAT.	INST.	TOTAL	MAT.	INST.	TOTAL	MAT.	INST.	TOTAL	MAT.	INST.	TOTAL	MAT.	INST.	TOTAL
015433	CONTRACTOR EQUIPMENT		102.5	102.5		102.5	102.5		102.5	102.5		102.5	102.5		102.5	102.5		102.5	102.5
0241, 31 - 34	SITE & INFRASTRUCTURE, DEMOLITION	97.3	88.3	91.0	101.7	89.2	92.9	102.8	89.5	93.5	100.1	88.1	91.7	98.0	89.5	92.1	106.6	88.4	93.9
0310	Concrete Forming & Accessories	84.9	45.7	51.1	82.3	47.6	52.4	81.7	64.5	66.9	94.0	45.5	52.2	90.7	65.6	69.1	82.4	61.8	64.6
0320	Concrete Reinforcing	100.7	64.8	82.6	89.0	35.9	62.2	101.2	51.5	76.1	100.7	44.0	72.1	92.2	51.1	71.5	89.6	32.9	61.0
0330	Cast-in-Place Concrete	102.2	50.8	82.7	116.8	56.1	93.7	105.3	57.2	87.0	109.9	50.4	87.3	99.8	65.7	86.8	114.2	52.1	90.6
03	CONCRETE	93.7	52.9	75.1	99.5	50.8	77.2	99.1	61.5	81.9	99.9	49.1	76.7	92.4	64.8	79.8	102.0	55.2	80.6
04	MASONRY	92.5	41.0	60.6	124.2	47.2	76.4	138.0	69.1	95.3	93.2	40.9	60.8	103.3	59.2	75.9	120.1	43.8	72.8
05	METALS	89.3	87.2	88.6	86.5	79.3	84.3	90.4	85.7	88.9	89.3	77.4	85.7	95.7	85.2	92.5	86.6	76.0	83.3
06	WOOD, PLASTICS & COMPOSITES	85.2	46.4	63.6	88.6	47.3	65.6	82.0	65.8	73.0	97.6	46.4	69.1	97.2	67.0	80.4	89.7	67.6	77.4
07	THERMAL & MOISTURE PROTECTION	95.8	49.5	76.2	94.8	54.5	77.7	96.3	63.8	82.5	96.2	51.8	77.4	93.1	62.4	80.1	94.9	55.5	78.2
08	OPENINGS	96.7	49.8	85.9	100.4	43.3	87.3	96.4	57.1	87.4	96.7	43.6	84.5	101.2	57.8	91.3	97.3	54.4	87.4
0920	Plaster & Gypsum Board	87.5	45.5	59.3	88.0	46.4	60.0	87.2	65.3	72.5	98.1	45.5	62.7	85.8	66.6	72.9	88.0	67.2	74.0
0950, 0980	Ceilings & Acoustic Treatment	84.7	45.5	58.2	84.9	46.4	58.9	88.5	65.3	72.9	84.7	45.5	58.2	90.8	66.6	74.5	84.9	67.2	72.9
0960	Flooring	95.7	47.6	82.3	82.9	53.4	74.7	94.0	47.6	81.1	101.4	47.6	86.5	85.9	57.5	78.1	81.6	47.6	72.1
0970, 0990	Wall Finishes & Painting/Coating	92.1	49.0	67.0	81.9	49.0	62.7	92.1	49.0	67.0	92.1	49.0	67.0	82.6	49.0	63.0	81.9	49.0	62.7
09	FINISHES	88.7	46.0	65.5	81.8	48.4	63.6	89.4	60.0	73.4	92.2	46.0	67.1	85.7	62.7	73.2	81.9	59.2	69.5
COVERS	DIVS. 10 - 14, 25, 28, 41, 43, 44, 46	100.0	49.8	88.8	100.0	50.8	89.1	100.0	71.8	93.7	100.0	49.8	88.8	100.0	71.9	93.7	100.0	35.4	85.6
21, 22, 23	FIRE SUPPRESSION, PLUMBING & HVAC	98.6	53.3	79.2	98.2	54.9	79.7	100.0	58.9	82.4	98.6	53.5	79.3	99.9	58.9	82.4	98.2	43.4	74.8
26, 27, 3370	ELECTRICAL, COMMUNICATIONS & UTIL.	95.6	43.9	68.7	97.5	46.6	71.0	95.6	56.7	75.4	95.6	41.2	67.3	101.5	56.7	78.2	99.0	56.7	77.0
MF2016	WEIGHTED AVERAGE	94.7	55.2	77.4	96.6	55.8	78.7	98.3	65.6	83.9	96.0	53.2	77.3	97.3	65.3	83.3	96.7	56.3	79.0

DIVISION		MISSISSIPPI									MISSOURI								
		MCCOMB 396			MERIDIAN 393			TUPELO 388			BOWLING GREEN 633			CAPE GIRARDEAU 637			CHILLICOTHE 646		
		MAT.	INST.	TOTAL	MAT.	INST.	TOTAL	MAT.	INST.	TOTAL	MAT.	INST.	TOTAL	MAT.	INST.	TOTAL	MAT.	INST.	TOTAL
015433	CONTRACTOR EQUIPMENT		102.5	102.5		102.5	102.5		102.5	102.5		110.5	110.5		110.5	110.5		104.8	104.8
0241, 31 - 34	SITE & INFRASTRUCTURE, DEMOLITION	94.8	88.2	90.2	98.8	89.7	92.4	94.8	88.2	90.2	89.1	94.9	93.1	90.7	94.8	93.6	105.8	92.9	96.8
0310	Concrete Forming & Accessories	82.3	47.0	51.9	79.6	65.3	67.3	82.2	47.4	52.2	93.6	91.5	91.8	86.4	78.3	79.4	82.9	96.1	94.2
0320	Concrete Reinforcing	90.1	34.5	62.1	89.0	51.1	69.9	98.5	44.1	71.0	106.8	96.7	101.7	108.1	78.4	93.1	98.0	101.3	99.7
0330	Cast-in-Place Concrete	101.3	50.3	81.9	108.4	66.7	92.6	102.2	71.0	90.3	91.4	98.5	94.1	90.4	88.1	89.6	94.8	89.0	92.6
03	CONCRETE	89.0	48.1	70.3	93.5	65.0	80.5	93.2	57.0	76.7	92.8	96.4	94.5	92.0	83.7	88.2	98.6	95.3	97.1
04	MASONRY	125.4	40.4	72.7	97.1	60.8	74.6	126.7	43.6	75.2	113.0	93.7	101.0	109.9	79.8	91.2	103.0	92.1	96.3
05	METALS	86.7	75.3	83.2	87.6	85.2	86.8	89.2	77.6	85.7	97.5	119.0	104.1	98.6	110.2	102.2	92.4	110.9	98.1
06	WOOD, PLASTICS & COMPOSITES	88.6	48.9	66.6	86.1	65.8	74.8	82.5	47.3	62.9	99.6	91.5	95.1	92.6	75.3	83.0	93.0	97.5	95.5
07	THERMAL & MOISTURE PROTECTION	94.3	51.9	76.3	94.5	63.0	81.1	95.8	53.1	77.7	100.1	99.3	99.8	99.6	85.4	93.6	92.5	94.5	93.3
08	OPENINGS	100.4	43.3	87.3	100.1	57.1	90.2	96.6	47.0	85.3	102.9	98.4	101.9	102.9	74.4	96.4	92.1	97.6	93.3
0920	Plaster & Gypsum Board	88.0	48.0	61.2	88.0	65.3	72.8	87.2	46.4	59.8	103.9	91.6	95.6	103.2	75.0	84.3	100.6	97.3	98.4
0950, 0980	Ceilings & Acoustic Treatment	84.9	48.0	60.1	86.8	65.3	72.3	84.7	46.4	58.8	91.4	91.6	91.5	91.4	75.0	80.3	91.1	97.3	95.3
0960	Flooring	82.9	47.6	73.1	81.5	57.5	74.9	94.2	47.6	81.3	95.6	98.5	96.4	92.2	87.2	90.8	94.7	104.7	97.5
0970, 0990	Wall Finishes & Painting/Coating	81.9	49.0	62.7	81.9	49.0	62.7	92.1	47.3	66.0	96.0	105.8	101.8	96.0	70.1	80.9	95.5	94.2	94.8
09	FINISHES	81.2	47.4	62.8	81.4	62.4	71.1	88.2	47.0	65.8	97.7	94.0	95.7	96.5	77.7	86.3	97.8	97.8	97.8
COVERS	DIVS. 10 - 14, 25, 28, 41, 43, 44, 46	100.0	52.8	89.5	100.0	72.3	93.8	100.0	50.8	89.1	100.0	82.7	96.2	100.0	92.8	98.4	100.0	83.4	96.3
21, 22, 23	FIRE SUPPRESSION, PLUMBING & HVAC	98.2	52.8	78.8	99.9	59.8	82.7	98.7	54.5	79.8	96.9	95.9	96.4	99.9	99.4	99.7	97.0	100.9	98.7
26, 27, 3370	ELECTRICAL, COMMUNICATIONS & UTIL.	96.1	46.8	70.5	99.0	58.7	78.1	95.3	46.7	70.0	101.8	79.6	90.3	101.7	101.0	101.3	95.9	77.9	86.5
MF2016	WEIGHTED AVERAGE	95.0	53.7	76.9	95.1	65.9	82.3	96.2	55.9	78.6	98.7	95.0	97.1	99.1	91.5	95.8	96.5	95.1	95.9

MISSOURI

DIVISION		COLUMBIA 652 MAT.	INST.	TOTAL	FLAT RIVER 636 MAT.	INST.	TOTAL	HANNIBAL 634 MAT.	INST.	TOTAL	HARRISONVILLE 647 MAT.	INST.	TOTAL	JEFFERSON CITY 650-651 MAT.	INST.	TOTAL	JOPLIN 648 MAT.	INST.	TOTAL
015433	CONTRACTOR EQUIPMENT		114.0	114.0		110.5	110.5		110.5	110.5		104.8	104.8		114.0	114.0		108.3	108.3
0241, 31 - 34	SITE & INFRASTRUCTURE, DEMOLITION	95.1	99.0	97.8	91.8	94.7	93.9	87.0	94.9	92.5	97.1	94.2	95.1	94.7	99.0	97.7	106.3	97.8	100.4
0310	Concrete Forming & Accessories	83.2	81.6	81.8	99.9	86.0	87.9	91.9	83.2	84.4	80.4	102.0	99.0	94.4	81.6	83.3	93.4	74.5	77.1
0320	Concrete Reinforcing	87.7	96.2	92.0	108.1	103.5	105.8	106.2	96.7	101.4	97.6	106.8	102.3	94.7	96.2	95.5	101.2	86.9	94.0
0330	Cast-in-Place Concrete	83.5	86.5	84.6	94.4	94.3	94.3	86.5	98.4	91.0	97.1	103.1	99.4	88.4	86.5	87.7	102.7	78.1	93.3
03	CONCRETE	80.7	87.8	83.9	95.7	93.7	94.8	89.2	92.7	90.8	94.6	103.9	98.9	87.4	87.8	87.6	97.4	79.2	89.1
04	MASONRY	144.4	87.5	109.1	110.2	78.5	90.5	104.7	93.7	97.9	97.4	102.3	100.4	102.0	87.5	93.0	96.3	81.3	87.0
05	METALS	102.6	118.1	107.4	97.4	120.9	104.6	97.5	118.6	104.0	92.9	114.6	99.5	102.2	118.1	107.1	95.6	101.0	97.2
06	WOOD, PLASTICS & COMPOSITES	85.6	79.1	82.0	107.7	85.2	95.2	97.9	80.8	88.4	89.9	101.6	96.4	97.8	79.1	87.4	103.5	73.0	86.5
07	THERMAL & MOISTURE PROTECTION	93.6	87.7	91.1	100.3	93.1	97.3	100.0	96.2	98.4	91.7	104.2	97.0	100.8	87.7	95.2	91.6	83.1	88.0
08	OPENINGS	99.6	84.1	96.0	102.9	97.0	101.6	102.9	85.0	98.8	92.1	104.5	95.0	97.6	84.1	94.5	93.2	78.0	89.7
0920	Plaster & Gypsum Board	93.1	78.6	83.3	109.8	85.1	93.2	103.6	80.6	88.1	96.2	101.5	99.8	96.8	78.6	84.6	107.0	72.1	83.6
0950, 0980	Ceilings & Acoustic Treatment	93.0	78.6	83.3	91.4	85.1	87.2	91.4	80.6	84.1	91.1	101.5	98.1	92.5	78.6	83.1	91.9	72.1	78.6
0960	Flooring	88.5	75.5	84.9	98.9	87.2	95.6	94.9	100.4	96.5	90.3	106.5	94.8	93.9	75.5	88.8	119.6	75.5	107.4
0970, 0990	Wall Finishes & Painting/Coating	92.4	81.7	86.2	96.0	75.0	83.8	96.0	98.4	97.4	99.8	107.6	104.3	86.0	81.7	83.5	95.1	79.3	85.9
09	FINISHES	85.9	79.9	82.7	99.7	83.9	91.1	97.3	86.7	91.6	95.5	103.4	99.8	91.2	79.9	85.1	104.7	75.4	88.8
COVERS	DIVS. 10 - 14, 25, 28, 41, 43, 44, 46	100.0	96.6	99.3	100.0	93.5	98.5	100.0	81.6	95.9	100.0	85.8	96.8	100.0	96.6	99.3	100.0	82.2	96.0
21, 22, 23	FIRE SUPPRESSION, PLUMBING & HVAC	99.9	99.9	99.9	96.9	99.0	97.8	96.9	99.9	98.2	96.9	102.7	99.4	99.9	100.0	100.0	100.1	71.8	88.0
26, 27, 3370	ELECTRICAL, COMMUNICATIONS & UTIL.	95.4	83.8	89.4	106.3	101.0	103.6	100.5	79.6	89.6	102.1	101.5	101.8	100.6	83.8	91.8	93.9	69.7	81.3
MF2016	WEIGHTED AVERAGE	98.1	92.5	95.7	99.6	95.7	97.9	97.6	93.6	95.9	95.9	102.8	98.9	97.8	92.5	95.5	97.8	79.6	89.8

MISSOURI

DIVISION		KANSAS CITY 640-641 MAT.	INST.	TOTAL	KIRKSVILLE 635 MAT.	INST.	TOTAL	POPLAR BLUFF 639 MAT.	INST.	TOTAL	ROLLA 654-655 MAT.	INST.	TOTAL	SEDALIA 653 MAT.	INST.	TOTAL	SIKESTON 638 MAT.	INST.	TOTAL
015433	CONTRACTOR EQUIPMENT		105.7	105.7		100.2	100.2		102.6	102.6		114.0	114.0		103.6	103.6		102.6	102.6
0241, 31 - 34	SITE & INFRASTRUCTURE, DEMOLITION	98.5	98.4	98.4	90.9	89.6	90.0	78.0	93.5	88.8	94.0	99.4	97.8	93.7	94.1	94.0	81.4	94.1	90.3
0310	Concrete Forming & Accessories	93.3	102.0	100.8	84.6	80.8	81.3	84.8	79.9	80.6	90.2	96.5	95.7	88.0	80.7	81.7	85.9	76.7	78.0
0320	Concrete Reinforcing	96.1	96.9	96.5	107.0	85.9	96.3	110.2	78.3	94.1	88.1	96.4	92.3	86.9	106.2	96.6	109.5	78.3	93.8
0330	Cast-in-Place Concrete	100.6	104.0	101.9	94.3	86.4	91.3	72.6	86.4	77.8	85.4	98.0	90.2	89.2	85.3	87.7	77.5	86.5	80.9
03	CONCRETE	97.4	102.4	99.7	107.7	84.9	97.3	82.9	83.1	83.0	82.3	98.5	89.7	96.0	88.0	92.3	86.6	81.7	84.4
04	MASONRY	103.4	102.3	102.7	116.8	87.6	98.7	108.2	76.1	88.3	117.4	88.2	99.3	123.8	84.1	99.2	107.6	76.1	88.2
05	METALS	102.6	109.1	104.6	97.2	102.8	98.9	97.8	99.5	98.3	102.1	118.7	107.2	100.9	111.9	104.3	98.2	99.7	98.7
06	WOOD, PLASTICS & COMPOSITES	98.5	101.5	100.2	85.3	78.8	81.7	84.4	79.9	81.9	92.9	98.1	95.8	86.0	79.1	82.1	85.9	75.4	80.1
07	THERMAL & MOISTURE PROTECTION	91.8	104.7	97.3	106.8	89.2	99.3	104.9	83.4	95.8	93.8	95.8	94.7	99.7	90.6	95.9	105.1	84.5	96.4
08	OPENINGS	100.1	101.9	100.5	108.2	82.2	102.2	109.3	76.9	101.9	99.6	94.6	98.4	104.5	88.7	100.9	109.3	74.4	101.3
0920	Plaster & Gypsum Board	102.5	101.5	101.8	98.9	78.6	85.3	99.3	79.7	86.1	95.1	98.2	97.2	88.8	78.6	81.9	100.9	75.0	83.5
0950, 0980	Ceilings & Acoustic Treatment	95.1	101.5	99.4	89.7	78.6	82.2	91.4	79.7	83.5	93.0	98.2	96.5	93.0	78.6	83.3	91.4	75.0	80.3
0960	Flooring	96.5	106.5	99.4	74.1	75.5	74.5	87.2	87.2	87.2	92.1	75.5	87.5	71.8	75.5	72.8	87.7	87.2	87.6
0970, 0990	Wall Finishes & Painting/Coating	98.4	107.6	103.7	92.0	81.9	86.1	91.2	70.1	78.9	92.4	94.3	93.5	92.4	81.9	86.3	91.2	70.1	78.9
09	FINISHES	99.1	103.4	101.4	96.4	79.8	87.4	96.2	79.6	87.2	87.4	92.7	90.3	84.7	79.1	81.7	96.8	77.3	86.2
COVERS	DIVS. 10 - 14, 25, 28, 41, 43, 44, 46	100.0	100.3	100.1	100.0	81.4	95.9	100.0	91.9	98.2	100.0	100.3	100.1	100.0	91.4	98.1	100.0	91.6	98.1
21, 22, 23	FIRE SUPPRESSION, PLUMBING & HVAC	99.9	102.6	101.1	96.9	99.3	97.9	96.9	96.8	96.8	96.9	100.4	98.4	96.8	97.4	97.0	96.9	96.8	96.9
26, 27, 3370	ELECTRICAL, COMMUNICATIONS & UTIL.	103.8	101.5	102.6	100.7	77.9	88.8	100.9	100.0	100.9	94.0	83.8	88.7	95.0	101.5	98.4	100.0	100.9	100.5
MF2016	WEIGHTED AVERAGE	100.3	102.7	101.4	101.1	88.5	95.6	97.5	89.7	94.1	96.2	96.9	96.5	98.6	93.2	96.2	98.0	89.1	94.1

MISSOURI / MONTANA

DIVISION		SPRINGFIELD 656-658 MAT.	INST.	TOTAL	ST. JOSEPH 644-645 MAT.	INST.	TOTAL	ST. LOUIS 630-631 MAT.	INST.	TOTAL	BILLINGS 590-591 MAT.	INST.	TOTAL	BUTTE 597 MAT.	INST.	TOTAL	GREAT FALLS 594 MAT.	INST.	TOTAL
015433	CONTRACTOR EQUIPMENT		106.1	106.1		104.8	104.8		111.1	111.1		100.6	100.6		100.3	100.3		100.3	100.3
0241, 31 - 34	SITE & INFRASTRUCTURE, DEMOLITION	96.4	96.6	96.5	100.8	92.0	94.6	96.0	99.4	98.4	90.7	97.6	95.6	95.9	96.4	96.3	99.5	97.4	98.0
0310	Concrete Forming & Accessories	96.1	76.6	79.3	92.2	90.2	90.5	97.1	104.1	103.2	94.7	67.7	71.4	82.8	67.5	69.6	94.6	66.9	70.7
0320	Concrete Reinforcing	84.6	95.8	90.2	94.9	106.4	100.7	98.8	104.9	101.9	95.4	80.9	88.1	103.5	80.8	92.1	95.4	80.8	88.1
0330	Cast-in-Place Concrete	90.7	77.5	85.7	95.4	100.1	97.1	102.1	102.8	102.3	108.9	71.3	94.7	119.9	71.1	101.4	126.7	71.2	105.6
03	CONCRETE	92.2	81.5	87.3	93.6	97.4	95.4	98.6	105.0	101.5	92.4	72.2	83.1	95.9	72.0	85.0	100.6	71.8	87.4
04	MASONRY	94.1	81.7	86.4	99.3	91.8	94.7	94.1	111.1	104.7	127.7	75.4	95.3	123.6	75.4	93.7	127.9	75.4	95.3
05	METALS	107.5	104.0	106.4	98.9	113.4	103.3	103.4	121.5	109.0	105.2	88.1	99.9	99.4	87.9	95.9	102.7	88.0	98.2
06	WOOD, PLASTICS & COMPOSITES	93.4	75.3	83.4	103.5	89.1	95.5	99.9	102.6	101.4	86.5	64.5	74.3	74.6	64.5	69.0	87.6	63.6	74.3
07	THERMAL & MOISTURE PROTECTION	97.9	78.4	89.6	92.1	91.6	91.9	97.9	107.1	101.8	107.6	71.1	92.2	107.4	71.0	92.0	108.0	70.6	92.2
08	OPENINGS	107.0	86.8	102.4	97.3	97.6	97.4	101.5	107.4	102.8	96.8	66.4	89.8	95.1	66.4	88.5	98.0	65.9	90.6
0920	Plaster & Gypsum Board	96.1	74.7	81.7	108.8	88.6	95.3	108.2	103.0	104.7	113.2	63.9	80.1	112.3	63.9	79.8	121.9	62.9	82.3
0950, 0980	Ceilings & Acoustic Treatment	93.0	74.7	80.7	99.6	88.6	92.2	93.1	103.0	99.8	91.9	63.9	73.0	97.8	63.9	74.9	99.5	62.9	74.9
0960	Flooring	91.1	75.5	86.8	99.9	104.7	101.2	98.0	100.4	98.7	93.7	79.7	89.8	91.1	79.7	87.9	97.9	79.7	92.9
0970, 0990	Wall Finishes & Painting/Coating	86.9	103.9	96.8	95.5	107.6	102.5	97.0	108.5	103.7	95.2	70.8	81.0	94.1	70.8	80.5	94.1	70.8	80.5
09	FINISHES	89.6	79.6	84.1	101.5	94.5	97.7	100.1	103.7	102.1	91.6	69.6	79.7	91.6	69.6	79.7	95.6	69.1	81.2
COVERS	DIVS. 10 - 14, 25, 28, 41, 43, 44, 46	100.0	94.0	98.7	100.0	97.4	99.4	100.0	100.9	100.5	100.0	92.2	98.3	100.0	92.2	98.3	100.0	92.1	98.2
21, 22, 23	FIRE SUPPRESSION, PLUMBING & HVAC	100.0	72.3	88.1	100.1	89.0	95.4	99.9	105.8	102.4	100.1	74.4	89.1	100.2	70.8	87.6	100.2	70.8	87.6
26, 27, 3370	ELECTRICAL, COMMUNICATIONS & UTIL.	99.1	71.7	84.8	102.1	77.9	89.5	104.3	101.0	102.6	98.6	72.0	84.8	105.4	71.2	87.6	98.0	71.2	84.1
MF2016	WEIGHTED AVERAGE	99.6	81.7	91.7	98.9	92.8	96.2	100.5	106.2	103.0	100.0	76.4	89.6	99.8	75.3	89.1	101.3	75.3	89.9

For customer support on your Building Construction Costs with RSMeans data, call 800.448.8182.

City Cost Indexes

MONTANA

DIVISION		HAVRE 595 MAT.	INST.	TOTAL	HELENA 596 MAT.	INST.	TOTAL	KALISPELL 599 MAT.	INST.	TOTAL	MILES CITY 593 MAT.	INST.	TOTAL	MISSOULA 598 MAT.	INST.	TOTAL	WOLF POINT 592 MAT.	INST.	TOTAL
015433	CONTRACTOR EQUIPMENT		100.3	100.3		100.3	100.3		100.3	100.3		100.3	100.3		100.3	100.3		100.3	100.3
0241, 31 - 34	SITE & INFRASTRUCTURE, DEMOLITION	102.8	96.5	98.4	90.0	96.4	94.5	86.6	96.3	93.4	92.5	96.4	95.2	79.6	96.2	91.2	108.8	96.5	100.2
0310	Concrete Forming & Accessories	76.4	67.4	68.6	97.5	67.4	71.6	85.7	67.5	70.0	93.2	67.5	71.0	85.7	67.4	69.9	86.4	66.8	69.5
0320	Concrete Reinforcing	104.3	80.8	92.5	110.2	79.6	94.8	106.2	85.6	95.8	104.0	80.8	92.3	105.2	85.6	95.3	105.4	80.1	92.7
0330	Cast-in-Place Concrete	129.1	71.1	107.1	93.7	71.1	85.1	104.1	71.1	91.6	114.0	71.1	97.7	88.4	71.1	81.8	127.7	71.1	106.2
03	CONCRETE	103.3	72.0	89.0	89.2	71.8	81.2	86.4	72.9	80.2	93.0	72.0	83.4	76.0	72.8	74.5	106.4	71.6	90.5
04	MASONRY	124.6	75.4	94.1	115.4	75.4	90.6	122.6	75.8	93.6	129.9	75.4	96.1	149.3	75.8	103.7	131.0	75.4	96.5
05	METALS	95.7	87.8	93.3	101.4	87.5	97.1	95.5	89.6	93.7	94.8	87.9	92.7	96.1	89.4	94.0	94.9	87.6	92.7
06	WOOD, PLASTICS & COMPOSITES	67.1	64.5	65.7	90.0	64.5	75.8	77.5	64.5	70.3	85.0	64.5	73.6	77.5	64.5	70.3	77.3	63.6	69.7
07	THERMAL & MOISTURE PROTECTION	107.8	66.2	90.2	102.4	71.0	89.1	107.0	73.2	92.7	107.3	66.3	90.0	106.6	70.6	91.3	108.3	66.2	90.5
08	OPENINGS	95.6	66.4	88.9	99.5	66.2	91.9	95.6	67.5	89.1	95.1	66.4	88.5	95.1	67.5	88.8	95.1	65.8	88.4
0920	Plaster & Gypsum Board	108.0	63.9	78.4	106.4	63.9	77.8	112.3	63.9	79.8	121.1	63.9	82.7	112.3	63.9	79.8	115.9	62.9	80.3
0950, 0980	Ceilings & Acoustic Treatment	97.8	63.9	74.9	96.8	63.9	74.6	97.8	63.9	74.9	96.1	63.9	74.4	97.8	63.9	74.9	96.1	62.9	73.7
0960	Flooring	88.6	79.7	86.1	98.9	79.7	93.6	92.8	79.7	89.2	97.7	79.7	92.8	92.8	79.7	89.2	94.3	79.7	90.2
0970, 0990	Wall Finishes & Painting/Coating	94.1	70.8	80.5	94.4	70.8	80.6	94.1	70.8	80.5	94.1	70.8	80.5	94.1	56.0	71.9	94.1	70.8	80.5
09	FINISHES	90.9	69.6	79.3	97.2	69.6	82.2	91.6	69.6	79.7	94.4	69.6	80.9	91.1	68.1	78.6	93.8	69.1	80.4
COVERS	DIVS. 10 - 14, 25, 28, 41, 43, 44, 46	100.0	92.2	98.3	100.0	92.2	98.3	100.0	92.2	98.3	100.0	92.2	98.3	100.0	92.2	98.3	100.0	92.1	98.2
21, 22, 23	FIRE SUPPRESSION, PLUMBING & HVAC	97.1	70.8	85.8	100.2	70.8	87.6	97.1	68.8	85.0	97.1	74.4	87.4	100.2	68.8	86.8	97.1	74.4	87.4
26, 27, 3370	ELECTRICAL, COMMUNICATIONS & UTIL.	98.0	71.2	84.1	104.6	71.2	87.2	102.2	68.7	84.8	98.0	76.6	86.8	103.2	67.7	84.8	98.0	76.6	86.8
MF2016	WEIGHTED AVERAGE	98.9	75.2	88.5	99.6	75.3	88.9	96.8	75.0	87.2	96.7	77.0	88.6	97.4	74.5	87.4	99.9	76.5	89.7

NEBRASKA

DIVISION		ALLIANCE 693 MAT.	INST.	TOTAL	COLUMBUS 686 MAT.	INST.	TOTAL	GRAND ISLAND 688 MAT.	INST.	TOTAL	HASTINGS 689 MAT.	INST.	TOTAL	LINCOLN 683 - 685 MAT.	INST.	TOTAL	MCCOOK 690 MAT.	INST.	TOTAL
015433	CONTRACTOR EQUIPMENT		98.2	98.2		105.2	105.2		105.2	105.2		105.2	105.2		105.2	105.2		105.2	105.2
0241, 31 - 34	SITE & INFRASTRUCTURE, DEMOLITION	99.4	101.2	100.6	100.8	95.2	96.9	105.6	95.2	98.3	104.4	95.2	97.9	91.1	95.2	94.0	102.7	95.2	97.4
0310	Concrete Forming & Accessories	86.3	55.8	60.0	95.2	74.8	77.6	94.8	71.2	74.4	98.0	74.3	77.6	93.7	76.4	78.8	91.6	55.9	60.9
0320	Concrete Reinforcing	111.2	88.0	99.5	97.2	86.9	92.0	96.6	76.3	86.4	96.6	73.4	84.9	95.9	76.5	86.2	103.4	69.7	86.4
0330	Cast-in-Place Concrete	107.8	77.3	96.2	110.5	80.9	99.2	117.0	80.3	103.0	117.0	62.2	96.2	91.8	80.4	87.5	116.6	58.6	94.6
03	CONCRETE	115.6	69.7	94.6	100.3	80.2	91.1	105.0	76.5	92.0	105.2	71.2	89.7	88.5	78.9	84.1	105.1	61.0	85.3
04	MASONRY	110.3	75.1	88.5	116.5	78.6	93.0	109.1	75.1	88.0	118.6	75.1	91.6	96.9	78.6	85.6	105.5	75.1	86.6
05	METALS	104.0	84.1	97.9	98.3	98.4	98.3	100.1	94.2	98.3	101.0	92.8	98.5	102.1	94.5	99.8	98.5	90.8	96.2
06	WOOD, PLASTICS & COMPOSITES	86.6	50.1	66.3	99.2	74.1	85.2	98.5	69.4	82.3	102.0	74.1	86.5	98.3	76.3	86.1	95.2	50.1	70.1
07	THERMAL & MOISTURE PROTECTION	102.5	65.9	87.0	101.7	80.5	92.7	101.8	79.1	92.2	101.9	78.7	92.1	98.2	80.9	90.9	97.5	76.1	88.4
08	OPENINGS	94.1	59.7	86.2	93.8	74.2	89.3	93.8	69.7	88.3	93.8	70.9	88.6	101.3	70.3	94.1	94.6	55.1	85.5
0920	Plaster & Gypsum Board	82.7	48.9	60.0	94.6	73.5	80.4	93.9	68.7	77.0	95.6	73.5	80.7	102.7	75.8	84.6	94.7	48.9	63.9
0950, 0980	Ceilings & Acoustic Treatment	93.2	48.9	63.3	85.9	73.5	77.5	85.9	68.7	74.3	85.9	73.5	77.5	95.1	75.8	82.1	89.2	48.9	62.0
0960	Flooring	89.7	83.4	88.0	82.4	89.7	84.5	82.2	83.4	82.5	83.5	83.4	83.5	93.7	89.7	92.6	88.5	83.4	87.1
0970, 0990	Wall Finishes & Painting/Coating	150.3	52.9	93.5	73.3	61.6	66.5	73.3	65.3	68.7	73.3	61.6	66.5	91.4	81.3	85.5	85.6	46.1	62.6
09	FINISHES	91.4	59.5	74.1	83.6	76.0	79.4	83.7	72.3	77.5	84.3	74.7	79.1	94.2	79.4	86.2	89.1	58.8	72.6
COVERS	DIVS. 10 - 14, 25, 28, 41, 43, 44, 46	100.0	60.9	91.3	100.0	84.1	96.5	100.0	89.0	97.5	100.0	84.1	96.5	100.0	89.7	97.7	100.0	60.3	91.2
21, 22, 23	FIRE SUPPRESSION, PLUMBING & HVAC	97.1	73.4	86.9	97.0	76.3	88.2	100.1	80.6	91.8	97.0	76.0	88.0	100.0	80.6	91.7	96.9	76.1	88.0
26, 27, 3370	ELECTRICAL, COMMUNICATIONS & UTIL.	93.5	65.6	79.0	93.7	80.2	86.7	92.4	65.7	78.5	91.8	81.4	86.4	101.4	65.7	82.8	95.5	65.6	80.0
MF2016	WEIGHTED AVERAGE	100.4	71.9	87.9	97.3	81.4	90.3	98.5	78.4	89.7	98.5	79.0	89.9	98.3	80.2	90.4	98.0	71.5	86.4

NEBRASKA / NEVADA

DIVISION		NORFOLK 687 MAT.	INST.	TOTAL	NORTH PLATTE 691 MAT.	INST.	TOTAL	OMAHA 680 - 681 MAT.	INST.	TOTAL	VALENTINE 692 MAT.	INST.	TOTAL	CARSON CITY 897 MAT.	INST.	TOTAL	ELKO 898 MAT.	INST.	TOTAL
015433	CONTRACTOR EQUIPMENT		94.4	94.4		105.2	105.2		94.4	94.4		97.9	97.9		97.4	97.4		97.4	97.4
0241, 31 - 34	SITE & INFRASTRUCTURE, DEMOLITION	83.6	94.3	91.1	103.9	95.2	97.8	90.9	94.3	93.3	87.3	100.3	96.4	89.2	97.8	95.2	71.8	96.1	88.8
0310	Concrete Forming & Accessories	81.9	75.5	76.4	94.1	70.3	73.6	93.3	76.1	78.5	82.7	55.5	59.1	105.0	79.7	83.2	110.2	77.4	82.0
0320	Concrete Reinforcing	97.3	64.6	80.8	102.8	73.3	87.9	95.9	76.5	86.1	103.4	64.2	83.6	99.0	113.3	106.2	106.2	91.7	98.9
0330	Cast-in-Place Concrete	111.1	61.5	92.2	116.6	63.3	96.3	92.1	78.2	86.8	102.8	56.3	85.1	102.5	82.7	94.9	99.2	74.1	89.7
03	CONCRETE	98.9	69.3	85.4	105.2	69.7	89.0	88.6	77.4	83.5	102.5	58.3	82.3	96.7	86.7	92.1	94.4	79.0	87.4
04	MASONRY	123.3	78.6	95.5	93.6	75.1	82.1	97.9	78.6	85.9	105.4	75.0	86.6	115.7	72.4	88.8	122.3	67.3	88.2
05	METALS	101.9	79.4	95.0	97.7	92.2	96.0	102.1	84.3	96.7	110.3	78.2	100.4	105.7	94.4	102.2	109.1	86.6	102.2
06	WOOD, PLASTICS & COMPOSITES	82.5	75.8	78.8	97.1	69.4	81.7	94.5	75.8	84.1	81.0	49.7	63.6	92.3	77.8	84.2	101.7	77.8	88.4
07	THERMAL & MOISTURE PROTECTION	101.6	79.2	92.1	97.4	78.2	89.3	98.1	80.3	90.6	98.3	75.0	88.4	105.7	80.4	95.0	102.3	71.8	89.4
08	OPENINGS	95.4	69.7	89.5	93.9	68.4	88.0	99.3	76.4	94.0	96.3	55.2	86.9	98.7	80.0	94.4	101.5	75.3	95.5
0920	Plaster & Gypsum Board	94.7	75.8	82.0	94.7	68.7	77.2	101.0	75.8	84.0	96.1	48.9	64.4	99.3	77.3	84.5	101.3	77.3	85.2
0950, 0980	Ceilings & Acoustic Treatment	99.8	75.8	83.6	89.2	68.7	75.4	97.2	75.8	82.8	105.6	48.9	67.4	101.4	77.3	85.1	100.1	77.3	84.7
0960	Flooring	102.0	89.7	98.6	89.6	83.4	87.9	94.3	89.7	93.0	114.6	83.4	105.9	96.0	71.0	89.0	101.0	71.0	92.6
0970, 0990	Wall Finishes & Painting/Coating	127.6	61.6	89.1	85.6	61.6	71.6	101.4	75.4	86.2	150.8	63.8	100.0	93.9	80.4	86.0	93.4	80.4	85.8
09	FINISHES	99.9	76.9	87.4	89.4	71.9	79.9	96.1	78.4	86.5	109.6	60.4	82.8	96.1	77.5	86.0	92.8	76.5	84.0
COVERS	DIVS. 10 - 14, 25, 28, 41, 43, 44, 46	100.0	86.1	96.9	100.0	62.4	91.6	100.0	88.7	97.5	100.0	58.3	90.7	100.0	95.2	98.9	100.0	90.5	97.9
21, 22, 23	FIRE SUPPRESSION, PLUMBING & HVAC	96.7	76.7	88.1	100.0	75.6	89.6	100.1	76.9	90.1	96.5	76.2	87.8	100.1	80.1	91.6	98.6	77.7	89.7
26, 27, 3370	ELECTRICAL, COMMUNICATIONS & UTIL.	92.7	83.9	88.1	93.8	65.6	79.2	102.7	83.9	92.9	91.2	66.1	78.2	103.2	93.2	98.0	100.4	93.2	96.7
MF2016	WEIGHTED AVERAGE	98.8	78.6	90.0	97.8	75.2	87.9	98.4	80.8	90.7	100.5	70.5	87.4	101.0	84.9	93.9	100.6	81.2	92.1

797

City Cost Indexes

Table 1

DIVISION		NEVADA — ELY 893 MAT.	INST.	TOTAL	LAS VEGAS 889-891 MAT.	INST.	TOTAL	RENO 894-895 MAT.	INST.	TOTAL	NEW HAMPSHIRE — CHARLESTON 036 MAT.	INST.	TOTAL	CLAREMONT 037 MAT.	INST.	TOTAL	CONCORD 032-033 MAT.	INST.	TOTAL
015433	CONTRACTOR EQUIPMENT		97.4	97.4		97.4	97.4		97.4	97.4		98.8	98.8		98.8	98.8		98.8	98.8
0241, 31 - 34	SITE & INFRASTRUCTURE, DEMOLITION	77.9	97.6	91.7	81.2	100.1	94.4	77.6	97.8	91.8	84.9	99.8	95.3	78.8	99.8	93.5	90.2	101.2	97.9
0310	Concrete Forming & Accessories	103.4	99.8	100.3	104.3	108.8	108.1	99.7	79.8	82.5	85.5	78.7	79.6	91.2	78.7	80.4	95.9	93.5	93.9
0320	Concrete Reinforcing	105.0	91.9	98.4	96.5	120.6	108.7	99.2	119.4	109.4	85.3	83.3	84.3	85.3	83.3	84.3	99.6	83.7	91.6
0330	Cast-in-Place Concrete	106.5	100.1	104.1	103.1	106.7	104.5	112.7	82.7	101.3	87.7	104.5	94.1	80.6	104.5	89.7	103.4	115.5	108.0
03	CONCRETE	102.6	98.3	100.6	97.9	109.5	103.2	102.5	87.8	95.8	87.3	88.7	88.0	80.2	88.7	84.1	94.6	99.3	96.7
04	MASONRY	127.9	78.1	97.0	114.8	99.2	105.1	121.5	72.4	91.0	95.6	77.4	84.3	95.9	77.4	84.4	104.0	100.1	101.6
05	METALS	109.1	89.7	103.1	117.6	101.6	112.7	110.8	96.4	106.4	93.0	89.3	91.9	93.0	89.3	91.9	99.6	90.6	96.9
06	WOOD, PLASTICS & COMPOSITES	92.6	101.0	97.3	90.7	106.9	99.7	87.0	77.8	81.9	94.2	83.2	88.1	100.0	83.2	90.7	94.9	92.7	93.7
07	THERMAL & MOISTURE PROTECTION	102.9	95.8	99.9	115.7	100.3	109.2	102.4	80.4	93.1	101.9	97.6	100.1	101.7	97.6	100.0	106.1	108.3	107.0
08	OPENINGS	101.4	88.2	98.4	100.4	111.9	103.0	99.2	81.4	95.1	97.3	82.9	94.0	98.4	82.9	94.8	95.7	88.1	94.0
0920	Plaster & Gypsum Board	96.7	101.1	99.7	91.9	107.2	102.1	87.0	77.3	80.5	104.8	82.4	89.7	106.1	82.4	90.2	107.3	92.1	97.1
0950, 0980	Ceilings & Acoustic Treatment	100.1	101.1	100.8	107.8	107.2	107.4	105.2	77.3	86.4	99.5	82.4	87.9	99.5	82.4	87.9	98.6	92.1	94.2
0960	Flooring	98.4	71.0	90.8	90.1	102.0	93.5	95.0	71.0	88.4	83.9	106.7	90.2	86.4	106.7	92.0	92.5	106.7	96.4
0970, 0990	Wall Finishes & Painting/Coating	93.4	117.1	107.2	95.9	114.9	107.0	93.4	80.4	85.8	84.8	94.1	90.2	84.8	94.1	90.2	88.8	94.1	91.9
09	FINISHES	92.0	96.9	94.7	90.5	108.3	100.1	90.5	77.5	83.4	86.6	85.4	85.9	87.1	85.4	86.2	90.6	95.7	93.3
COVERS	DIVS. 10 - 14, 25, 28, 41, 43, 44, 46	100.0	72.7	93.9	100.0	103.9	100.9	100.0	95.2	98.9	100.0	85.5	96.8	100.0	85.5	96.8	100.0	104.8	101.1
21, 22, 23	FIRE SUPPRESSION, PLUMBING & HVAC	98.6	99.5	99.0	100.3	100.7	100.5	100.2	80.1	91.6	97.0	60.1	81.2	97.0	60.1	81.2	100.0	85.2	93.7
26, 27, 3370	ELECTRICAL, COMMUNICATIONS & UTIL.	100.7	98.1	99.4	104.9	113.8	109.5	101.1	93.2	97.0	97.2	49.1	72.2	97.2	49.1	72.2	98.0	79.5	88.4
MF2016	WEIGHTED AVERAGE	101.9	94.1	98.5	103.1	105.2	104.0	101.8	85.3	94.6	94.3	76.3	86.4	93.5	76.3	86.0	97.9	94.2	95.5

Table 2

DIVISION		NEW HAMPSHIRE — KEENE 034 MAT.	INST.	TOTAL	LITTLETON 035 MAT.	INST.	TOTAL	MANCHESTER 031 MAT.	INST.	TOTAL	NASHUA 030 MAT.	INST.	TOTAL	PORTSMOUTH 038 MAT.	INST.	TOTAL	NEW JERSEY — ATLANTIC CITY 082, 084 MAT.	INST.	TOTAL
015433	CONTRACTOR EQUIPMENT		98.8	98.8		98.8	98.8		98.8	98.8		98.8	98.8		98.8	98.8		96.3	96.3
0241, 31 - 34	SITE & INFRASTRUCTURE, DEMOLITION	91.9	100.1	97.6	79.0	100.1	93.7	92.1	101.2	98.5	93.4	101.0	98.7	86.9	101.6	97.2	89.7	104.6	100.2
0310	Concrete Forming & Accessories	89.8	80.3	81.6	102.0	80.3	83.3	95.9	93.7	94.0	97.8	93.6	94.2	86.9	93.5	92.6	110.1	143.9	139.3
0320	Concrete Reinforcing	85.3	83.4	84.3	86.0	83.4	84.7	99.6	83.7	91.6	106.5	83.7	95.0	85.3	83.7	84.5	75.8	142.1	109.2
0330	Cast-in-Place Concrete	88.1	106.5	95.1	79.0	106.6	89.5	103.4	115.6	108.0	83.4	115.4	95.5	79.1	116.0	93.1	77.9	142.7	102.5
03	CONCRETE	87.0	90.2	88.4	79.7	90.2	84.5	94.5	99.4	96.8	87.0	99.3	92.6	79.8	99.5	88.8	84.1	141.5	109.2
04	MASONRY	98.4	80.7	87.4	108.5	80.7	91.2	100.7	100.1	100.4	100.7	100.0	100.3	96.0	100.3	98.7	111.3	140.6	129.5
05	METALS	93.7	89.7	92.5	93.8	89.7	92.5	101.2	90.8	98.0	99.3	90.3	96.6	95.3	91.5	94.2	96.5	115.7	102.4
06	WOOD, PLASTICS & COMPOSITES	98.4	83.2	89.9	109.2	83.2	94.7	95.0	92.7	93.7	107.6	92.7	99.3	95.4	92.7	93.9	118.0	144.6	132.8
07	THERMAL & MOISTURE PROTECTION	102.4	99.1	101.0	101.8	99.1	100.7	106.3	108.3	107.2	102.8	108.3	105.1	102.4	108.3	104.9	101.0	137.4	116.4
08	OPENINGS	96.0	86.5	93.8	99.3	82.9	95.5	96.7	91.8	95.5	100.6	91.1	98.4	100.9	81.2	96.4	96.8	140.3	106.8
0920	Plaster & Gypsum Board	105.1	82.4	89.8	118.9	82.4	94.4	107.3	92.1	97.1	114.5	92.1	99.5	104.8	92.1	96.3	116.0	145.4	135.7
0950, 0980	Ceilings & Acoustic Treatment	99.5	82.4	87.9	99.5	82.4	87.9	99.6	92.1	94.5	111.5	92.1	98.4	100.4	92.1	94.8	91.4	145.4	127.8
0960	Flooring	86.0	106.7	92.0	96.7	105.6	99.2	89.8	106.7	94.5	89.7	106.7	94.6	84.1	106.7	90.4	92.5	168.1	113.4
0970, 0990	Wall Finishes & Painting/Coating	84.8	108.1	98.4	84.8	94.1	90.2	90.8	108.1	100.9	84.8	106.9	97.7	84.8	106.9	97.7	83.0	145.9	119.6
09	FINISHES	88.4	87.8	88.0	91.9	86.0	88.7	91.9	97.2	94.8	93.1	97.1	95.3	87.4	97.1	92.7	89.7	149.9	122.4
COVERS	DIVS. 10 - 14, 25, 28, 41, 43, 44, 46	100.0	93.2	98.5	100.0	98.5	99.7	100.0	104.8	101.1	100.0	104.8	101.1	100.0	104.8	101.1	100.0	119.2	104.3
21, 22, 23	FIRE SUPPRESSION, PLUMBING & HVAC	97.0	63.3	82.6	97.0	69.8	85.4	100.1	85.2	93.7	100.1	85.2	93.7	100.1	85.2	93.7	99.8	131.7	113.4
26, 27, 3370	ELECTRICAL, COMMUNICATIONS & UTIL.	97.2	58.6	77.2	98.2	51.9	74.1	98.1	79.5	88.4	99.4	79.4	89.0	97.7	79.5	88.2	93.2	144.6	119.9
MF2016	WEIGHTED AVERAGE	94.8	79.6	88.1	94.9	79.8	88.3	98.3	92.7	95.9	97.7	92.6	95.5	95.1	92.4	93.9	95.7	134.7	112.8

Table 3

DIVISION		NEW JERSEY — CAMDEN 081 MAT.	INST.	TOTAL	DOVER 078 MAT.	INST.	TOTAL	ELIZABETH 072 MAT.	INST.	TOTAL	HACKENSACK 076 MAT.	INST.	TOTAL	JERSEY CITY 073 MAT.	INST.	TOTAL	LONG BRANCH 077 MAT.	INST.	TOTAL
015433	CONTRACTOR EQUIPMENT		96.3	96.3		98.8	98.8		98.8	98.8		98.8	98.8		96.3	96.3		95.9	95.9
0241, 31 - 34	SITE & INFRASTRUCTURE, DEMOLITION	90.9	104.6	100.5	96.5	106.3	103.4	100.3	106.3	104.5	97.5	106.3	103.6	88.7	106.2	100.9	92.5	105.9	101.9
0310	Concrete Forming & Accessories	101.3	143.8	137.9	99.4	146.5	140.0	111.6	146.5	141.7	99.4	146.4	139.9	103.7	146.4	140.6	104.2	145.6	139.9
0320	Concrete Reinforcing	99.5	132.6	116.2	78.1	150.5	114.6	78.1	150.5	114.6	78.1	150.5	114.6	101.3	150.5	126.1	78.1	150.5	114.6
0330	Cast-in-Place Concrete	75.6	140.8	100.3	85.2	136.1	104.5	73.2	136.1	97.1	83.2	136.1	103.3	66.4	136.1	92.9	73.8	140.8	99.3
03	CONCRETE	82.6	139.1	108.4	84.2	142.1	110.7	80.8	142.1	108.8	82.7	142.1	109.9	78.6	141.9	107.5	82.3	143.1	110.1
04	MASONRY	101.4	140.6	125.7	93.5	143.0	124.2	111.1	143.0	130.9	97.6	143.0	125.8	89.3	143.0	122.6	103.2	141.1	126.7
05	METALS	102.3	112.3	105.3	94.3	123.7	103.4	99.5	123.7	104.4	94.4	123.7	103.4	100.2	120.6	106.4	94.4	120.4	102.4
06	WOOD, PLASTICS & COMPOSITES	106.7	144.6	127.8	102.4	146.2	126.7	117.3	146.2	133.3	102.4	146.2	126.7	103.2	146.2	127.1	102.4	146.1	127.5
07	THERMAL & MOISTURE PROTECTION	100.8	138.4	116.8	108.7	136.4	120.4	108.9	136.4	120.5	108.4	135.6	119.9	108.1	136.4	120.1	108.3	135.2	119.7
08	OPENINGS	99.1	138.1	108.0	104.2	144.2	113.4	102.6	144.2	112.1	102.0	144.2	111.7	100.7	144.2	110.7	96.6	144.2	107.6
0920	Plaster & Gypsum Board	111.5	145.4	134.3	107.2	147.0	133.9	114.4	147.0	136.3	107.2	147.0	133.9	110.2	147.0	134.9	109.1	147.0	134.6
0950, 0980	Ceilings & Acoustic Treatment	101.0	145.4	130.9	87.8	147.0	127.7	89.7	147.0	128.3	87.8	147.0	127.7	97.3	147.0	130.8	87.8	147.0	127.7
0960	Flooring	88.6	168.1	110.6	81.8	191.9	112.4	86.9	191.9	116.0	81.8	191.9	112.4	82.9	191.9	113.1	83.1	187.0	111.9
0970, 0990	Wall Finishes & Painting/Coating	83.0	145.9	119.6	84.0	145.9	120.1	84.0	145.9	120.1	84.0	145.9	120.1	84.1	145.9	120.1	84.1	145.9	120.1
09	FINISHES	89.7	149.9	122.4	85.4	155.4	123.5	88.8	155.4	125.0	85.3	155.4	123.4	87.8	155.4	124.6	86.3	154.2	123.2
COVERS	DIVS. 10 - 14, 25, 28, 41, 43, 44, 46	100.0	119.2	104.3	100.0	131.8	107.1	100.0	131.8	107.1	100.0	131.8	107.1	100.0	131.8	107.1	100.0	119.4	104.3
21, 22, 23	FIRE SUPPRESSION, PLUMBING & HVAC	100.0	131.6	113.5	99.9	137.5	116.0	100.1	137.5	116.1	99.9	137.5	116.0	100.1	137.5	116.1	99.9	136.5	115.6
26, 27, 3370	ELECTRICAL, COMMUNICATIONS & UTIL.	97.7	139.5	119.4	94.8	142.9	119.8	95.4	142.9	120.1	94.8	140.9	118.8	99.5	140.9	121.0	94.4	137.2	116.7
MF2016	WEIGHTED AVERAGE	97.0	133.3	112.9	95.7	138.2	114.3	96.8	138.2	114.9	95.5	137.9	114.0	95.9	137.6	114.1	95.0	136.2	113.1

For customer support on your Building Construction Costs with RSMeans data, call 800.448.8182.

NEW JERSEY

DIVISION		NEW BRUNSWICK 088 - 089			NEWARK 070 - 071			PATERSON 074 - 075			POINT PLEASANT 087			SUMMIT 079			TRENTON 085 - 086		
		MAT.	INST.	TOTAL	MAT.	INST.	TOTAL	MAT.	INST.	TOTAL	MAT.	INST.	TOTAL	MAT.	INST.	TOTAL	MAT.	INST.	TOTAL
015433	CONTRACTOR EQUIPMENT		95.9	95.9		98.8	98.8		98.8	98.8		95.9	95.9		98.8	98.8		95.9	95.9
0241, 31 - 34	SITE & INFRASTRUCTURE, DEMOLITION	102.2	106.0	104.9	100.9	106.3	104.7	99.4	106.3	104.2	103.7	105.9	105.3	98.1	106.3	103.9	89.0	105.9	100.9
0310	Concrete Forming & Accessories	104.3	146.4	140.6	104.4	146.5	140.7	101.4	146.4	140.2	98.8	145.5	139.1	102.2	146.5	140.4	100.5	145.3	139.1
0320	Concrete Reinforcing	76.7	150.5	113.9	99.7	150.5	125.3	101.3	150.5	126.1	76.7	150.5	113.9	78.1	150.5	114.6	100.7	119.9	110.4
0330	Cast-in-Place Concrete	96.3	141.9	113.6	93.9	136.1	109.9	84.7	136.1	104.2	96.3	142.9	114.0	70.7	136.1	95.6	94.8	140.6	112.2
03	CONCRETE	97.5	143.8	118.7	91.6	142.1	114.7	87.1	142.1	112.2	97.2	143.8	118.5	78.2	142.1	107.4	91.9	137.6	112.8
04	MASONRY	109.7	143.0	130.4	96.6	143.0	125.4	94.0	143.0	124.4	98.0	141.1	124.7	96.6	143.0	125.4	101.9	141.1	126.2
05	METALS	96.6	120.5	103.9	102.0	123.7	108.7	95.5	123.7	104.2	96.6	120.3	103.9	94.3	123.7	103.4	102.0	109.1	104.2
06	WOOD, PLASTICS & COMPOSITES	111.5	146.1	130.8	100.0	146.2	125.7	105.0	146.2	127.9	104.1	146.1	127.5	106.1	146.2	128.4	93.5	146.1	122.8
07	THERMAL & MOISTURE PROTECTION	101.3	134.8	115.5	109.8	136.4	121.1	108.7	135.6	120.1	101.3	137.6	116.7	109.1	136.4	120.6	101.7	138.6	117.3
08	OPENINGS	92.0	144.2	104.0	100.6	144.2	110.6	107.3	144.2	115.8	93.8	144.2	105.3	108.3	144.2	116.5	98.5	136.7	107.3
0920	Plaster & Gypsum Board	113.3	147.0	136.0	102.5	147.0	132.4	110.2	147.0	134.9	108.4	147.0	134.3	109.1	147.0	134.6	103.4	147.0	132.7
0950, 0980	Ceilings & Acoustic Treatment	91.4	147.0	128.9	98.8	147.0	131.3	97.3	147.0	130.8	91.4	147.0	128.9	87.8	147.0	127.7	97.9	147.0	131.0
0960	Flooring	90.0	191.9	118.3	87.4	191.9	116.3	82.9	191.9	113.1	87.5	168.1	109.8	83.1	191.9	113.3	91.9	187.0	118.2
0970, 0990	Wall Finishes & Painting/Coating	83.0	145.9	119.6	85.4	145.9	120.6	84.0	145.9	120.1	83.0	145.9	119.6	84.0	145.9	120.1	86.6	145.9	121.2
09	FINISHES	89.6	155.3	125.4	90.7	155.4	125.9	88.0	155.4	124.6	88.2	150.9	122.3	86.4	155.4	123.9	92.4	154.2	126.0
COVERS	DIVS. 10 - 14, 25, 28, 41, 43, 44, 46	100.0	131.7	107.0	100.0	131.8	107.1	100.0	131.8	107.1	100.0	116.4	103.7	100.0	131.8	107.1	100.0	119.4	104.3
21, 22, 23	FIRE SUPPRESSION, PLUMBING & HVAC	99.8	137.5	115.9	100.0	137.5	116.1	100.1	137.5	116.1	99.8	136.5	115.5	99.9	137.5	116.0	100.2	136.2	115.6
26, 27, 3370	ELECTRICAL, COMMUNICATIONS & UTIL.	93.9	141.2	118.5	103.1	140.9	122.8	99.5	142.9	122.1	93.2	137.2	116.1	95.4	142.9	120.1	101.2	135.9	119.2
MF2016	WEIGHTED AVERAGE	97.4	137.8	115.1	99.1	137.9	116.1	97.4	138.2	115.3	96.8	135.8	113.9	98.1	138.2	114.3	98.5	133.9	114.0

DIVISION		NEW JERSEY VINELAND 080, 083			ALBUQUERQUE 870 - 872			CARRIZOZO 883			CLOVIS 881			FARMINGTON 874			GALLUP 873		
		MAT.	INST.	TOTAL	MAT.	INST.	TOTAL	MAT.	INST.	TOTAL	MAT.	INST.	TOTAL	MAT.	INST.	TOTAL	MAT.	INST.	TOTAL
015433	CONTRACTOR EQUIPMENT		96.3	96.3		111.0	111.0		111.0	111.0		111.0	111.0		111.0	111.0		111.0	111.0
0241, 31 - 34	SITE & INFRASTRUCTURE, DEMOLITION	93.9	104.7	101.4	92.3	102.4	99.4	112.0	102.4	105.3	98.8	102.4	101.3	98.8	102.4	101.3	108.5	102.4	104.2
0310	Concrete Forming & Accessories	96.0	144.0	137.4	99.0	64.1	68.9	97.1	64.1	68.6	97.1	64.0	68.5	99.1	64.1	68.9	99.1	64.1	68.9
0320	Concrete Reinforcing	75.8	135.4	105.9	97.0	71.0	83.9	109.7	71.0	90.2	110.9	71.0	90.8	106.1	71.0	88.4	101.5	71.0	86.1
0330	Cast-in-Place Concrete	84.0	140.9	105.6	93.6	70.1	84.7	94.7	70.1	85.3	94.6	70.0	85.3	94.5	70.1	85.2	88.9	70.1	81.8
03	CONCRETE	86.3	139.8	110.8	93.3	68.8	82.1	114.3	68.8	93.5	102.6	68.7	87.1	96.7	68.8	83.9	102.9	68.8	87.3
04	MASONRY	98.6	141.1	125.0	107.3	60.0	78.0	106.9	60.0	77.8	106.9	60.0	77.8	116.9	60.0	81.6	101.6	60.0	75.8
05	METALS	96.5	113.9	101.8	109.2	91.0	103.6	105.2	91.0	100.8	104.8	90.8	100.5	106.8	91.0	101.9	105.9	91.0	101.3
06	WOOD, PLASTICS & COMPOSITES	101.0	144.6	125.2	94.7	64.6	77.9	92.4	64.6	77.0	92.4	64.6	77.0	94.8	64.6	78.0	94.8	64.6	78.0
07	THERMAL & MOISTURE PROTECTION	100.8	137.6	116.4	96.9	70.8	85.8	99.5	70.8	87.3	98.2	70.8	86.6	97.1	70.8	86.0	98.2	70.8	86.6
08	OPENINGS	93.3	139.1	103.8	98.5	65.8	91.0	96.6	65.8	89.5	96.7	65.8	89.7	100.8	65.8	92.8	100.9	65.8	92.8
0920	Plaster & Gypsum Board	106.8	145.4	132.7	100.6	63.4	75.6	77.6	63.4	68.1	77.6	63.4	68.1	92.7	63.4	73.0	92.7	63.4	73.0
0950, 0980	Ceilings & Acoustic Treatment	91.4	145.4	127.8	99.3	63.4	75.1	102.3	63.4	76.1	102.3	63.4	76.1	97.1	63.4	74.4	97.1	63.4	74.4
0960	Flooring	86.6	168.1	109.2	87.7	69.5	82.7	98.5	69.5	90.4	98.5	69.5	90.4	89.3	69.5	83.8	89.3	69.5	83.8
0970, 0990	Wall Finishes & Painting/Coating	83.0	145.9	119.6	95.3	52.8	70.5	93.6	52.8	69.8	93.6	52.8	69.8	89.9	52.8	68.3	89.9	52.8	68.3
09	FINISHES	87.0	150.0	121.3	89.4	63.6	75.4	93.9	63.6	77.4	92.6	63.6	76.8	88.3	63.6	74.9	89.6	63.6	75.4
COVERS	DIVS. 10 - 14, 25, 28, 41, 43, 44, 46	100.0	119.4	104.3	100.0	84.0	96.5	100.0	84.0	96.5	100.0	84.0	96.5	100.0	84.0	96.5	100.0	84.0	96.5
21, 22, 23	FIRE SUPPRESSION, PLUMBING & HVAC	99.8	131.9	113.5	100.2	69.2	86.9	98.1	69.2	85.7	98.1	68.8	85.6	100.0	69.2	86.8	98.1	69.2	85.7
26, 27, 3370	ELECTRICAL, COMMUNICATIONS & UTIL.	93.2	144.6	119.9	88.1	88.5	88.3	91.7	88.5	90.0	89.4	88.5	88.9	86.2	88.5	87.4	85.6	88.5	87.1
MF2016	WEIGHTED AVERAGE	95.0	134.4	112.2	98.5	75.1	88.3	101.1	75.1	89.7	98.9	75.0	88.4	99.1	75.1	88.6	98.9	75.1	88.5

NEW MEXICO

DIVISION		LAS CRUCES 880			LAS VEGAS 877			ROSWELL 882			SANTA FE 875			SOCORRO 878			TRUTH/CONSEQUENCES 879		
		MAT.	INST.	TOTAL	MAT.	INST.	TOTAL	MAT.	INST.	TOTAL	MAT.	INST.	TOTAL	MAT.	INST.	TOTAL	MAT.	INST.	TOTAL
015433	CONTRACTOR EQUIPMENT		85.7	85.7		111.0	111.0		111.0	111.0		111.0	111.0		111.0	111.0		85.7	85.7
0241, 31 - 34	SITE & INFRASTRUCTURE, DEMOLITION	99.0	80.8	86.3	97.9	102.4	101.1	101.0	102.4	102.0	102.2	102.4	102.3	94.4	102.4	100.0	114.9	80.9	91.0
0310	Concrete Forming & Accessories	93.8	63.0	67.2	99.1	64.1	68.9	97.1	64.1	68.6	97.8	64.1	68.7	99.1	64.1	68.9	96.8	63.0	67.6
0320	Concrete Reinforcing	106.4	70.9	88.5	103.3	71.0	87.0	110.9	71.0	90.8	96.5	71.0	83.7	105.3	71.0	88.0	98.4	70.9	84.5
0330	Cast-in-Place Concrete	89.4	62.6	79.2	91.9	70.1	83.6	94.6	70.1	85.3	97.5	70.1	87.1	90.0	70.1	82.5	98.6	62.6	85.0
03	CONCRETE	81.3	65.3	74.0	94.3	68.8	82.6	103.4	68.8	87.6	93.5	68.8	82.2	93.1	68.8	82.0	87.1	65.3	77.1
04	MASONRY	102.2	59.6	75.8	101.9	60.0	75.9	118.3	60.0	82.1	94.9	60.0	73.2	101.8	60.0	75.9	99.0	59.6	74.6
05	METALS	103.6	83.0	97.2	105.6	91.0	101.1	106.1	91.0	101.4	102.7	91.0	99.1	105.9	91.0	101.3	105.5	83.0	98.6
06	WOOD, PLASTICS & COMPOSITES	81.4	63.5	71.4	94.8	64.6	78.0	92.4	64.6	77.0	93.9	64.6	77.6	94.8	64.6	78.0	85.9	63.5	73.5
07	THERMAL & MOISTURE PROTECTION	85.7	65.9	77.3	96.7	70.8	85.7	98.4	70.8	86.7	99.2	70.8	87.2	96.6	70.8	85.7	85.6	65.9	77.2
08	OPENINGS	92.3	65.2	86.1	97.3	65.8	90.1	96.6	65.8	89.5	100.0	65.8	92.2	97.1	65.8	89.9	90.7	65.2	84.9
0920	Plaster & Gypsum Board	75.9	63.4	67.5	92.7	63.4	73.0	77.6	63.4	68.1	107.8	63.4	78.0	92.7	63.4	73.0	94.1	63.4	73.5
0950, 0980	Ceilings & Acoustic Treatment	88.0	63.4	71.4	97.1	63.4	74.4	102.3	63.4	76.1	97.7	63.4	74.6	97.1	63.4	74.4	85.9	63.4	70.7
0960	Flooring	129.6	69.5	112.9	89.3	69.5	83.8	98.5	69.5	90.4	97.0	69.5	89.4	89.3	69.5	83.8	118.0	69.5	104.5
0970, 0990	Wall Finishes & Painting/Coating	82.8	52.8	65.3	89.9	52.8	68.3	93.6	52.8	69.8	95.9	52.8	70.8	89.9	52.8	68.3	82.3	52.8	65.1
09	FINISHES	102.6	62.7	80.9	88.2	63.6	74.8	92.7	63.6	76.9	96.6	63.6	78.6	88.1	63.6	74.8	100.0	62.8	79.8
COVERS	DIVS. 10 - 14, 25, 28, 41, 43, 44, 46	100.0	81.5	95.9	100.0	84.0	96.5	100.0	84.0	96.5	100.0	84.0	96.5	100.0	84.0	96.5	100.0	81.5	95.9
21, 22, 23	FIRE SUPPRESSION, PLUMBING & HVAC	100.4	68.9	86.9	98.1	69.2	85.7	99.9	69.2	86.8	100.1	69.2	86.9	98.1	69.2	85.7	98.1	68.9	85.6
26, 27, 3370	ELECTRICAL, COMMUNICATIONS & UTIL.	91.4	88.5	89.9	87.7	88.5	88.1	90.8	88.5	89.6	99.8	88.5	93.9	86.0	88.5	87.3	89.6	88.5	89.0
MF2016	WEIGHTED AVERAGE	96.4	71.7	85.6	97.2	75.1	87.5	100.3	75.1	89.3	99.1	75.1	88.6	96.8	75.1	87.3	96.6	71.8	85.7

For customer support on your Building Construction Costs with RSMeans data, call 800.448.8182.

DIVISION		NEW MEXICO TUCUMCARI 884			NEW YORK ALBANY 120 - 122			NEW YORK BINGHAMTON 137 - 139			NEW YORK BRONX 104			NEW YORK BROOKLYN 112			NEW YORK BUFFALO 140 - 142		
		MAT.	INST.	TOTAL	MAT.	INST.	TOTAL	MAT.	INST.	TOTAL	MAT.	INST.	TOTAL	MAT.	INST.	TOTAL	MAT.	INST.	TOTAL
015433	CONTRACTOR EQUIPMENT		111.0	111.0		115.8	115.8		121.0	121.0		106.9	106.9		112.3	112.3		98.9	98.9
0241, 31 - 34	SITE & INFRASTRUCTURE, DEMOLITION	98.4	102.4	101.2	83.9	104.7	98.5	96.2	93.2	94.1	92.6	113.7	107.4	120.9	125.5	124.1	99.6	101.6	101.0
0310	Concrete Forming & Accessories	97.1	64.0	68.5	99.3	105.9	105.0	100.6	90.7	92.0	99.1	193.1	180.2	106.7	182.5	172.1	102.3	120.3	117.8
0320	Concrete Reinforcing	108.7	71.0	89.7	103.7	119.5	111.7	100.4	104.2	102.3	96.4	172.4	134.8	101.8	222.6	162.8	108.6	115.3	111.9
0330	Cast-in-Place Concrete	94.6	70.0	85.3	79.7	116.1	93.5	105.1	107.6	106.1	86.4	173.4	119.5	107.2	171.7	131.7	108.3	125.3	114.8
03	CONCRETE	101.9	68.7	86.7	84.1	112.9	97.3	93.4	101.3	97.0	89.4	181.7	131.6	104.6	184.0	140.9	105.1	120.6	112.2
04	MASONRY	118.3	60.0	82.1	90.5	116.0	106.3	106.8	104.1	105.1	90.1	182.8	147.6	117.2	182.8	157.9	113.5	123.6	119.7
05	METALS	104.8	90.8	100.5	101.2	128.8	109.7	94.2	134.4	106.6	90.8	170.3	115.2	102.1	167.5	122.2	99.4	108.2	102.1
06	WOOD, PLASTICS & COMPOSITES	92.4	64.6	77.0	94.4	102.3	98.8	105.6	86.6	95.0	98.0	197.8	153.6	102.7	183.4	149.6	100.9	119.7	111.4
07	THERMAL & MOISTURE PROTECTION	98.2	70.8	86.6	106.6	110.3	108.2	107.1	95.0	102.0	104.4	168.9	131.7	107.3	166.6	132.4	104.8	113.3	108.4
08	OPENINGS	96.5	65.8	89.5	95.4	104.5	97.5	91.9	92.8	92.1	92.8	200.7	117.6	89.4	190.6	112.7	99.8	114.0	103.1
0920	Plaster & Gypsum Board	77.6	63.4	68.1	102.5	102.3	102.3	107.2	85.9	92.9	96.7	200.3	166.3	103.4	185.8	158.7	105.4	120.1	115.3
0950, 0980	Ceilings & Acoustic Treatment	102.3	63.4	76.1	99.6	102.3	101.4	94.1	85.9	88.6	85.2	200.3	162.8	89.9	185.8	154.5	97.3	120.1	112.7
0960	Flooring	98.5	69.5	90.4	90.6	111.1	96.3	99.4	99.2	99.4	92.6	189.9	119.6	106.3	189.9	129.5	95.2	118.5	101.7
0970, 0990	Wall Finishes & Painting/Coating	93.6	52.8	69.8	93.8	102.3	98.8	87.4	99.6	94.5	103.8	159.9	136.5	114.2	159.9	140.8	97.8	115.8	108.3
09	FINISHES	92.5	63.6	76.8	90.3	105.9	98.8	91.4	92.2	91.8	92.6	190.5	145.8	104.3	181.9	146.5	100.1	120.1	111.0
COVERS	DIVS. 10 - 14, 25, 28, 41, 43, 44, 46	100.0	84.0	96.5	100.0	102.0	100.4	100.0	98.8	99.7	100.0	137.9	108.4	100.0	135.5	108.0	100.0	108.1	101.8
21, 22, 23	FIRE SUPPRESSION, PLUMBING & HVAC	98.1	68.8	85.6	100.2	109.0	103.9	100.6	99.6	100.2	100.2	170.4	130.2	99.7	170.2	129.9	100.0	103.4	101.5
26, 27, 3370	ELECTRICAL, COMMUNICATIONS & UTIL.	91.7	88.5	90.0	95.5	110.9	103.5	99.4	100.8	100.1	94.2	181.5	139.6	99.1	181.5	141.9	98.9	106.0	102.6
MF2016	WEIGHTED AVERAGE	99.5	75.0	88.8	95.8	111.2	102.5	97.2	101.7	99.2	94.7	173.4	129.2	101.6	172.5	132.7	101.2	111.6	105.8

DIVISION		NEW YORK ELMIRA 148 - 149			NEW YORK FAR ROCKAWAY 116			NEW YORK FLUSHING 113			NEW YORK GLENS FALLS 128			NEW YORK HICKSVILLE 115, 117, 118			NEW YORK JAMAICA 114		
		MAT.	INST.	TOTAL	MAT.	INST.	TOTAL	MAT.	INST.	TOTAL	MAT.	INST.	TOTAL	MAT.	INST.	TOTAL	MAT.	INST.	TOTAL
015433	CONTRACTOR EQUIPMENT		123.0	123.0		112.3	112.3		112.3	112.3		115.8	115.8		112.3	112.3		112.3	112.3
0241, 31 - 34	SITE & INFRASTRUCTURE, DEMOLITION	102.2	92.8	95.6	124.2	125.5	125.1	124.2	125.5	125.1	74.3	104.2	95.3	113.8	124.6	121.3	118.4	125.5	123.4
0310	Concrete Forming & Accessories	83.5	96.0	94.3	93.1	193.0	179.2	97.0	193.0	179.8	84.8	96.3	94.8	89.6	162.1	152.1	97.0	193.0	179.8
0320	Concrete Reinforcing	108.1	108.3	108.2	101.8	222.7	162.8	103.5	222.7	163.6	99.3	111.3	105.4	101.8	222.5	162.7	101.8	222.7	162.8
0330	Cast-in-Place Concrete	98.3	106.7	101.5	115.9	171.8	137.1	115.9	171.8	137.1	77.0	110.8	89.8	98.5	168.2	125.0	102.7	171.8	131.7
03	CONCRETE	89.5	104.3	96.2	110.7	188.8	146.4	111.2	188.8	146.7	78.3	105.3	90.6	96.6	173.4	131.7	104.0	188.8	142.7
04	MASONRY	105.1	105.1	105.1	122.0	182.8	159.7	115.9	182.8	157.4	96.6	108.1	103.7	111.7	178.0	152.8	119.9	182.8	158.9
05	METALS	95.2	137.2	108.1	102.1	167.5	122.2	102.1	167.5	122.2	94.6	125.1	104.0	103.6	164.4	122.3	102.1	167.5	122.2
06	WOOD, PLASTICS & COMPOSITES	84.7	94.3	90.0	91.1	197.5	150.3	95.6	197.5	152.3	87.2	92.9	90.4	87.8	159.5	127.7	95.6	197.5	152.3
07	THERMAL & MOISTURE PROTECTION	103.2	95.6	100.0	107.2	168.2	133.0	107.2	168.2	133.0	99.7	105.3	102.1	106.9	161.4	130.0	107.0	168.2	132.9
08	OPENINGS	98.3	97.3	98.0	88.1	198.4	113.5	88.1	198.4	113.5	90.4	98.6	92.3	88.5	177.4	108.9	88.1	198.4	113.5
0920	Plaster & Gypsum Board	101.8	94.1	96.6	92.5	200.3	164.9	94.8	200.3	165.6	95.0	92.6	93.4	92.1	161.2	138.5	94.8	200.3	165.6
0950, 0980	Ceilings & Acoustic Treatment	98.5	94.1	95.5	79.7	200.3	161.0	79.7	200.3	161.0	85.6	92.6	90.3	78.7	161.2	134.4	79.7	200.3	161.0
0960	Flooring	86.5	102.4	90.9	101.1	189.9	125.7	102.6	189.9	126.9	82.3	111.1	90.3	100.0	176.4	121.2	102.6	189.9	126.9
0970, 0990	Wall Finishes & Painting/Coating	94.8	91.5	92.9	114.2	159.9	140.8	114.2	159.9	140.8	89.8	97.3	94.2	114.2	159.9	140.8	114.2	159.9	140.8
09	FINISHES	92.0	96.5	94.5	99.6	190.2	148.9	100.4	190.2	149.3	82.4	98.2	91.0	98.2	164.1	134.1	100.0	190.2	149.1
COVERS	DIVS. 10 - 14, 25, 28, 41, 43, 44, 46	100.0	97.9	99.5	100.0	137.1	108.3	100.0	137.1	108.3	100.0	98.1	99.6	100.0	124.9	105.5	100.0	137.1	108.3
21, 22, 23	FIRE SUPPRESSION, PLUMBING & HVAC	97.1	96.5	96.9	96.6	170.2	128.1	96.6	170.2	128.1	97.1	104.4	100.3	99.7	159.5	125.3	96.6	170.2	128.1
26, 27, 3370	ELECTRICAL, COMMUNICATIONS & UTIL.	97.7	103.8	100.9	105.9	181.5	145.2	105.9	181.5	145.2	90.7	110.9	101.2	98.6	142.4	121.4	97.6	181.5	141.2
MF2016	WEIGHTED AVERAGE	96.5	103.0	99.3	102.0	174.8	133.9	101.9	174.8	133.8	91.4	106.4	98.0	99.6	158.9	125.6	100.2	174.8	132.9

DIVISION		NEW YORK JAMESTOWN 147			NEW YORK KINGSTON 124			NEW YORK LONG ISLAND CITY 111			NEW YORK MONTICELLO 127			NEW YORK MOUNT VERNON 105			NEW YORK NEW ROCHELLE 108		
		MAT.	INST.	TOTAL	MAT.	INST.	TOTAL	MAT.	INST.	TOTAL	MAT.	INST.	TOTAL	MAT.	INST.	TOTAL	MAT.	INST.	TOTAL
015433	CONTRACTOR EQUIPMENT		91.9	91.9		112.3	112.3		112.3	112.3		112.3	112.3		106.9	106.9		106.9	106.9
0241, 31 - 34	SITE & INFRASTRUCTURE, DEMOLITION	103.7	93.5	96.6	145.6	120.7	128.1	122.0	125.5	124.5	140.6	121.2	127.0	97.5	109.9	106.2	97.3	109.9	106.1
0310	Concrete Forming & Accessories	83.6	89.1	88.3	86.2	128.1	122.4	101.3	193.0	180.4	93.3	128.1	123.3	89.4	140.1	133.1	104.5	140.1	135.2
0320	Concrete Reinforcing	108.3	114.5	111.4	99.8	156.7	128.5	101.8	222.7	162.8	99.0	156.7	128.1	95.2	170.9	133.4	95.3	170.9	133.4
0330	Cast-in-Place Concrete	102.0	85.3	95.7	103.9	145.2	119.6	110.6	171.8	133.8	97.3	145.2	115.5	96.5	150.7	117.1	96.5	150.7	117.1
03	CONCRETE	92.4	92.3	92.4	98.3	138.6	116.7	107.0	188.8	144.4	93.6	138.6	114.2	98.7	149.4	121.9	98.1	149.4	121.5
04	MASONRY	114.6	101.4	106.4	111.3	150.7	135.7	114.0	182.8	156.7	104.2	150.7	133.0	95.9	154.7	132.4	95.9	154.7	132.4
05	METALS	92.7	102.6	95.7	102.9	134.2	112.5	102.1	167.5	122.2	102.9	134.2	112.5	90.6	162.5	112.7	90.8	162.4	112.8
06	WOOD, PLASTICS & COMPOSITES	83.3	86.1	84.8	88.4	122.0	107.1	101.4	197.5	154.9	95.2	122.0	110.1	88.6	134.9	114.3	105.4	134.9	121.8
07	THERMAL & MOISTURE PROTECTION	102.7	94.0	99.0	123.4	143.5	131.9	107.2	168.2	133.0	123.1	143.5	131.7	105.4	146.6	122.8	105.4	146.6	122.9
08	OPENINGS	98.1	94.8	97.3	92.6	140.8	103.7	88.1	198.4	113.4	88.3	140.8	100.3	92.8	165.9	109.6	92.9	165.9	109.6
0920	Plaster & Gypsum Board	88.9	85.6	86.7	97.2	122.8	114.4	99.4	200.3	167.2	97.9	122.8	114.6	91.7	135.6	121.2	104.9	135.6	125.5
0950, 0980	Ceilings & Acoustic Treatment	95.1	85.6	88.7	76.3	122.8	107.6	79.7	200.3	161.0	76.3	122.8	107.6	83.5	135.6	118.6	83.5	135.6	118.6
0960	Flooring	89.0	102.4	92.7	98.7	162.9	116.5	104.3	189.9	128.0	101.1	162.9	118.2	84.6	173.4	109.3	92.0	173.4	114.6
0970, 0990	Wall Finishes & Painting/Coating	96.4	91.5	93.5	118.7	129.0	124.7	114.2	159.9	140.8	118.7	129.0	124.7	102.2	159.9	135.9	102.2	159.9	135.9
09	FINISHES	90.6	91.4	91.0	96.0	134.0	116.7	101.3	190.2	149.7	96.4	134.0	116.9	89.6	147.5	121.1	93.5	147.5	121.8
COVERS	DIVS. 10 - 14, 25, 28, 41, 43, 44, 46	100.0	96.6	99.3	100.0	115.9	103.5	100.0	137.1	108.3	100.0	115.9	103.5	100.0	119.9	104.4	100.0	116.8	103.7
21, 22, 23	FIRE SUPPRESSION, PLUMBING & HVAC	97.0	89.0	93.6	97.1	125.7	109.3	99.7	170.2	129.9	97.1	121.8	107.6	97.2	140.6	115.8	97.2	140.6	115.8
26, 27, 3370	ELECTRICAL, COMMUNICATIONS & UTIL.	96.7	93.8	95.2	92.0	121.0	107.1	98.1	181.5	141.4	92.0	121.0	107.1	92.4	149.6	122.2	92.4	149.6	122.2
MF2016	WEIGHTED AVERAGE	96.6	93.9	95.4	100.0	131.6	113.8	101.2	174.8	133.5	98.5	130.8	112.7	95.1	145.6	117.2	95.5	145.5	117.4

For customer support on your Building Construction Costs with RSMeans data, call 800.448.8182.

NEW YORK

DIVISION		NEW YORK 100 - 102			NIAGARA FALLS 143			PLATTSBURGH 129			POUGHKEEPSIE 125 - 126			QUEENS 110			RIVERHEAD 119		
		MAT.	INST.	TOTAL	MAT.	INST.	TOTAL	MAT.	INST.	TOTAL	MAT.	INST.	TOTAL	MAT.	INST.	TOTAL	MAT.	INST.	TOTAL
015433	CONTRACTOR EQUIPMENT		104.9	104.9		91.9	91.9		96.6	96.6		112.3	112.3		112.3	112.3		112.3	112.3
0241, 31 - 34	SITE & INFRASTRUCTURE, DEMOLITION	101.4	111.7	108.6	106.0	95.0	98.3	113.2	101.7	105.2	141.7	121.1	127.3	117.1	125.5	123.0	115.0	124.7	121.8
0310	Concrete Forming & Accessories	108.5	196.3	184.2	83.5	118.6	113.8	90.5	95.8	95.1	86.2	175.7	163.3	89.9	193.0	178.8	94.2	162.2	152.8
0320	Concrete Reinforcing	102.0	236.2	169.7	106.8	115.6	111.2	103.8	115.6	109.8	99.8	156.8	128.5	103.5	222.7	163.6	103.7	222.5	163.6
0330	Cast-in-Place Concrete	99.9	181.8	131.0	105.5	130.7	115.1	94.2	106.0	98.7	100.7	142.7	116.6	101.8	171.8	128.4	100.1	168.6	126.1
03	CONCRETE	103.3	195.8	145.6	94.6	121.5	106.9	90.9	102.5	96.2	95.8	159.3	124.8	99.8	188.8	140.5	97.2	173.6	132.1
04	MASONRY	103.5	189.8	157.0	121.8	130.5	127.2	90.9	101.5	97.5	103.8	145.5	129.7	108.1	182.8	154.4	117.5	178.4	155.3
05	METALS	102.0	169.5	122.8	95.3	104.3	98.0	98.8	101.6	99.7	102.9	135.3	112.9	102.1	167.5	122.2	104.1	164.5	122.7
06	WOOD, PLASTICS & COMPOSITES	102.4	197.9	155.5	83.2	113.4	100.0	93.6	92.4	93.0	88.4	188.0	143.9	87.9	197.5	148.9	92.7	159.5	129.9
07	THERMAL & MOISTURE PROTECTION	107.4	173.4	135.3	102.8	115.4	108.1	117.8	102.0	111.1	123.4	148.4	133.9	106.8	168.2	132.8	107.8	161.6	130.6
08	OPENINGS	95.8	201.1	119.9	98.1	110.0	100.9	98.1	99.5	98.5	92.7	177.3	112.1	88.1	198.4	113.4	88.5	177.4	108.9
0920	Plaster & Gypsum Board	102.2	200.3	168.1	88.9	113.7	105.6	114.8	91.7	99.3	97.2	190.6	159.9	92.1	200.3	164.8	93.3	161.2	138.9
0950, 0980	Ceilings & Acoustic Treatment	100.5	200.3	167.8	95.1	113.7	107.7	104.7	91.7	95.9	76.3	190.6	153.3	79.7	200.3	161.0	79.6	161.2	134.6
0960	Flooring	93.9	189.9	120.5	89.0	118.5	97.2	104.6	111.1	106.4	98.7	167.7	117.8	100.0	189.9	125.0	101.1	176.4	122.0
0970, 0990	Wall Finishes & Painting/Coating	100.5	164.8	138.0	96.4	116.5	108.1	114.4	99.5	105.7	118.7	129.0	124.7	114.2	159.9	140.8	114.2	159.9	140.8
09	FINISHES	97.5	192.8	149.3	90.7	118.6	105.9	94.7	98.2	96.6	95.8	172.5	137.5	98.6	190.2	148.4	98.8	163.9	134.2
COVERS	DIVS. 10 - 14, 25, 28, 41, 43, 44, 46	100.0	143.3	109.6	100.0	107.0	101.6	100.0	96.6	99.2	100.0	121.4	104.8	100.0	137.1	108.3	100.0	125.0	105.6
21, 22, 23	FIRE SUPPRESSION, PLUMBING & HVAC	100.1	174.1	131.8	97.0	107.7	101.6	97.1	104.5	100.3	97.1	125.3	109.1	99.7	170.2	129.9	99.9	159.6	125.5
26, 27, 3370	ELECTRICAL, COMMUNICATIONS & UTIL.	100.9	188.1	146.3	95.4	104.4	100.1	88.8	96.9	93.0	92.0	128.0	110.8	98.6	181.5	141.7	100.1	142.4	122.1
MF2016	WEIGHTED AVERAGE	100.6	178.2	134.6	97.6	111.9	103.9	96.6	100.9	98.5	99.2	142.4	118.1	99.6	174.8	132.6	100.4	159.0	126.1

NEW YORK

DIVISION		ROCHESTER 144 - 146			SCHENECTADY 123			STATEN ISLAND 103			SUFFERN 109			SYRACUSE 130 - 132			UTICA 133 - 135		
		MAT.	INST.	TOTAL	MAT.	INST.	TOTAL	MAT.	INST.	TOTAL	MAT.	INST.	TOTAL	MAT.	INST.	TOTAL	MAT.	INST.	TOTAL
015433	CONTRACTOR EQUIPMENT		119.8	119.8		115.8	115.8		106.9	106.9		106.9	106.9		115.8	115.8		115.8	115.8
0241, 31 - 34	SITE & INFRASTRUCTURE, DEMOLITION	94.1	108.8	104.4	84.6	104.7	98.7	101.5	113.7	110.1	94.5	109.2	104.8	94.8	103.5	100.9	73.2	103.1	94.1
0310	Concrete Forming & Accessories	99.8	101.5	101.3	101.1	105.9	105.3	88.9	182.9	169.9	97.8	147.6	140.7	99.7	93.4	94.2	100.8	90.5	91.9
0320	Concrete Reinforcing	110.7	108.4	109.5	98.3	119.5	109.0	96.4	222.7	160.1	95.3	156.9	126.4	101.4	104.2	102.8	101.4	103.5	102.5
0330	Cast-in-Place Concrete	97.7	107.7	101.5	92.9	116.1	101.7	96.5	173.5	125.8	93.6	146.5	113.7	97.6	107.7	101.4	89.3	106.3	95.8
03	CONCRETE	94.3	106.1	99.7	91.2	112.9	101.2	100.5	184.9	139.1	95.3	147.9	119.3	96.0	101.6	98.6	94.0	99.7	96.6
04	MASONRY	99.6	107.8	104.7	93.3	116.0	107.4	102.0	182.8	152.1	95.4	148.3	128.2	99.3	105.3	103.0	90.4	103.7	98.6
05	METALS	101.4	123.2	108.1	98.7	128.8	108.0	88.9	170.5	113.9	88.9	135.0	103.1	97.7	120.4	104.7	95.8	120.0	103.2
06	WOOD, PLASTICS & COMPOSITES	98.0	100.0	99.1	105.3	102.3	103.6	87.1	183.7	140.9	97.8	149.4	126.5	101.9	90.1	95.4	101.9	86.7	93.4
07	THERMAL & MOISTURE PROTECTION	102.5	103.6	102.9	101.2	110.3	105.1	104.8	167.3	131.3	105.3	145.0	122.1	101.7	98.8	100.5	90.1	98.7	93.7
08	OPENINGS	102.3	101.6	102.1	96.1	104.5	98.0	92.8	192.9	115.8	92.9	155.9	107.3	94.2	92.8	93.9	96.9	90.8	95.5
0920	Plaster & Gypsum Board	105.9	100.1	102.0	104.3	102.3	103.0	91.8	185.8	154.9	96.0	150.5	132.6	98.5	89.8	92.7	98.5	86.2	90.3
0950, 0980	Ceilings & Acoustic Treatment	97.0	100.1	99.1	92.4	102.3	99.1	85.2	185.8	153.0	83.5	150.5	128.7	94.1	89.8	91.2	94.1	86.2	88.8
0960	Flooring	90.1	111.0	95.9	89.5	111.1	95.5	88.1	189.9	116.3	88.1	181.4	114.0	88.9	97.6	91.3	86.9	97.7	89.9
0970, 0990	Wall Finishes & Painting/Coating	94.7	102.9	99.5	89.8	102.3	97.1	103.8	159.9	136.5	102.2	132.7	120.0	91.2	103.2	98.2	84.8	103.2	95.5
09	FINISHES	95.2	103.5	99.7	87.8	105.9	97.6	91.3	182.1	140.7	90.9	153.2	124.8	90.5	94.5	92.7	88.9	92.3	90.7
COVERS	DIVS. 10 - 14, 25, 28, 41, 43, 44, 46	100.0	102.0	100.5	100.0	102.0	100.4	100.0	136.3	108.1	100.0	119.5	104.3	100.0	99.0	99.8	100.0	93.5	98.6
21, 22, 23	FIRE SUPPRESSION, PLUMBING & HVAC	100.1	91.8	96.5	100.2	109.0	104.0	100.2	170.4	130.2	97.2	128.8	110.7	100.2	96.7	98.7	100.2	96.4	98.6
26, 27, 3370	ELECTRICAL, COMMUNICATIONS & UTIL.	102.7	94.9	98.6	94.7	110.9	103.1	94.2	181.5	139.6	99.0	121.0	110.5	99.4	106.7	103.2	97.5	106.7	102.3
MF2016	WEIGHTED AVERAGE	99.5	102.7	100.9	96.1	111.2	102.7	96.4	172.2	129.6	95.1	136.0	113.0	97.6	102.0	99.5	95.7	100.8	97.9

NEW YORK / NORTH CAROLINA

DIVISION		WATERTOWN 136			WHITE PLAINS 106			YONKERS 107			ASHEVILLE 287 - 288			CHARLOTTE 281 - 282			DURHAM 277		
		MAT.	INST.	TOTAL	MAT.	INST.	TOTAL	MAT.	INST.	TOTAL	MAT.	INST.	TOTAL	MAT.	INST.	TOTAL	MAT.	INST.	TOTAL
015433	CONTRACTOR EQUIPMENT		115.8	115.8		106.9	106.9		106.9	106.9		101.9	101.9		101.9	101.9		107.3	107.3
0241, 31 - 34	SITE & INFRASTRUCTURE, DEMOLITION	80.9	103.6	96.8	92.4	109.9	104.7	99.2	110.0	106.7	97.4	81.6	86.3	99.1	81.8	87.0	101.5	90.7	93.9
0310	Concrete Forming & Accessories	85.8	97.5	95.9	102.8	140.1	135.0	103.0	151.3	144.7	91.1	63.4	67.2	96.1	62.9	67.5	95.2	63.4	67.7
0320	Concrete Reinforcing	102.1	104.2	103.2	95.3	170.9	133.4	99.1	171.0	135.4	94.9	68.0	81.3	99.2	67.0	82.9	98.2	67.0	82.5
0330	Cast-in-Place Concrete	104.0	110.2	106.3	85.7	150.7	110.4	95.9	150.0	116.8	109.3	71.3	94.9	111.5	70.6	95.9	98.2	71.3	88.0
03	CONCRETE	106.4	104.4	105.5	89.1	149.4	116.7	98.1	154.6	124.0	99.5	68.6	85.4	99.2	68.0	84.9	94.2	68.4	82.4
04	MASONRY	91.6	109.3	102.6	94.8	154.7	131.9	98.6	154.7	133.4	90.1	63.7	73.7	93.9	62.3	74.3	86.6	63.7	72.4
05	METALS	95.8	120.4	103.4	90.5	162.5	112.6	98.6	163.0	118.4	97.8	91.7	95.9	98.6	91.4	96.4	113.7	91.3	106.8
06	WOOD, PLASTICS & COMPOSITES	84.4	94.2	89.9	103.1	134.9	120.8	103.0	149.4	128.8	90.7	61.8	74.7	91.2	61.8	74.8	94.9	61.8	76.5
07	THERMAL & MOISTURE PROTECTION	90.4	102.0	95.3	105.1	146.6	122.7	105.4	149.5	124.1	101.0	65.3	85.9	95.4	64.6	82.4	107.4	65.3	89.6
08	OPENINGS	96.9	97.3	97.0	92.9	165.9	109.6	96.5	173.7	114.2	93.7	62.9	86.6	98.6	62.7	90.3	100.8	62.7	92.0
0920	Plaster & Gypsum Board	89.7	94.0	92.6	98.9	135.6	123.6	102.3	150.5	134.7	106.6	60.6	75.7	101.6	60.6	74.1	98.7	60.6	73.1
0950, 0980	Ceilings & Acoustic Treatment	94.1	94.0	94.0	83.5	135.6	118.6	99.7	150.5	133.9	83.8	60.6	68.2	87.8	60.6	69.5	86.3	60.6	69.0
0960	Flooring	80.0	97.7	84.9	90.5	173.4	113.5	90.0	189.9	117.7	92.8	63.5	84.6	92.1	63.5	84.1	96.8	63.5	87.6
0970, 0990	Wall Finishes & Painting/Coating	84.8	97.2	92.0	102.2	159.9	135.9	102.2	159.9	135.9	104.5	59.7	78.4	97.0	59.7	75.3	98.8	59.7	76.0
09	FINISHES	86.2	97.0	92.1	91.7	147.5	122.0	95.7	159.3	130.3	90.1	62.7	75.2	90.0	62.4	75.0	89.7	62.7	75.0
COVERS	DIVS. 10 - 14, 25, 28, 41, 43, 44, 46	100.0	100.5	100.1	100.0	119.9	104.4	100.0	129.1	106.5	100.0	85.2	96.7	100.0	84.8	96.6	100.0	85.2	96.7
21, 22, 23	FIRE SUPPRESSION, PLUMBING & HVAC	100.2	91.4	96.5	100.4	140.6	117.6	100.4	140.8	117.7	100.5	62.6	84.3	99.9	63.1	84.1	100.5	62.6	84.3
26, 27, 3370	ELECTRICAL, COMMUNICATIONS & UTIL.	99.3	95.2	97.2	92.4	149.6	122.2	99.1	164.7	133.2	100.5	61.1	80.0	99.7	62.3	80.3	98.0	60.6	78.5
MF2016	WEIGHTED AVERAGE	97.3	100.7	98.8	94.6	145.6	116.9	98.9	150.8	121.6	97.6	68.3	84.8	98.1	68.3	85.0	100.2	68.9	86.5

City Cost Indexes

		NORTH CAROLINA																
	DIVISION	ELIZABETH CITY 279			FAYETTEVILLE 283			GASTONIA 280			GREENSBORO 270, 272 - 274			HICKORY 286			KINSTON 285		
		MAT.	INST.	TOTAL	MAT.	INST.	TOTAL	MAT.	INST.	TOTAL	MAT.	INST.	TOTAL	MAT.	INST.	TOTAL	MAT.	INST.	TOTAL
015433	CONTRACTOR EQUIPMENT		111.5	111.5		107.3	107.3		101.9	101.9		107.3	107.3		107.3	107.3		107.3	107.3
0241, 31 - 34	SITE & INFRASTRUCTURE, DEMOLITION	105.8	92.5	96.5	96.8	90.6	92.5	97.3	81.9	86.5	101.3	90.8	94.0	96.4	89.4	91.5	95.2	89.3	91.1
0310	Concrete Forming & Accessories	81.6	63.6	66.0	90.8	62.8	66.7	97.3	63.6	68.2	95.0	63.3	67.7	87.7	63.4	66.7	84.1	62.6	65.6
0320	Concrete Reinforcing	96.2	71.9	84.0	98.6	67.0	82.6	95.3	67.0	81.0	97.1	67.0	81.9	94.9	67.0	80.8	94.4	66.9	80.5
0330	Cast-in-Place Concrete	98.4	73.0	88.8	114.7	70.5	97.9	106.9	71.4	93.4	97.5	71.3	87.6	109.3	71.3	94.9	105.5	70.4	92.2
03	CONCRETE	94.4	70.0	83.2	101.4	67.9	86.1	98.0	68.6	84.5	93.6	68.4	82.1	99.3	68.4	85.2	96.1	67.8	83.2
04	MASONRY	98.8	62.3	76.2	93.2	62.3	74.1	94.5	63.7	75.4	82.5	63.7	70.9	78.9	63.7	69.5	85.8	62.3	71.2
05	METALS	100.0	94.0	98.2	117.8	91.3	109.7	98.3	91.6	96.3	106.2	91.3	101.7	97.8	91.3	95.8	96.6	91.1	94.9
06	WOOD, PLASTICS & COMPOSITES	80.1	62.5	70.3	90.0	61.8	74.3	98.5	61.8	78.1	94.6	61.8	76.4	85.9	61.8	72.5	83.0	61.8	71.2
07	THERMAL & MOISTURE PROTECTION	106.6	64.5	88.8	100.4	64.6	85.3	101.2	65.3	86.0	107.1	65.3	89.4	101.3	65.3	86.1	101.2	64.6	85.7
08	OPENINGS	97.6	64.1	89.9	93.8	62.7	86.7	97.5	62.7	89.5	100.8	62.7	92.0	93.8	62.7	86.6	93.9	62.7	86.7
0920	Plaster & Gypsum Board	92.3	60.6	71.0	110.8	60.6	77.1	112.7	60.6	77.7	100.2	60.6	73.6	106.6	60.6	75.7	106.4	60.6	75.7
0950, 0980	Ceilings & Acoustic Treatment	86.3	60.6	69.0	85.5	60.6	68.8	87.3	60.6	69.3	86.3	60.6	69.0	83.8	60.6	68.2	87.3	60.6	69.3
0960	Flooring	89.2	81.6	87.1	93.0	63.5	84.8	96.1	81.6	92.1	96.8	63.5	87.6	92.7	81.6	89.6	90.0	81.6	87.7
0970, 0990	Wall Finishes & Painting/Coating	98.8	59.7	76.0	104.5	59.7	78.4	104.5	59.7	78.4	98.8	59.7	76.0	104.5	59.7	78.4	104.5	59.7	78.4
09	FINISHES	86.9	66.5	75.8	91.1	62.4	75.5	92.6	66.4	78.3	90.0	62.7	75.1	90.3	66.4	77.3	89.9	66.0	76.9
COVERS	DIVS. 10 - 14, 25, 28, 41, 43, 44, 46	100.0	82.7	96.1	100.0	84.7	96.6	100.0	85.2	96.7	100.0	82.4	96.1	100.0	85.2	96.7	100.0	84.7	96.6
21, 22, 23	FIRE SUPPRESSION, PLUMBING & HVAC	97.3	60.5	81.6	100.2	61.9	83.9	100.5	61.5	83.8	100.4	62.6	84.2	97.4	61.5	82.0	97.4	60.3	81.5
26, 27, 3370	ELECTRICAL, COMMUNICATIONS & UTIL.	97.7	68.2	82.4	100.3	60.6	79.6	100.0	62.3	80.4	97.1	61.1	78.4	98.4	62.3	79.6	98.2	58.9	77.8
MF2016	WEIGHTED AVERAGE	97.2	70.4	85.5	101.3	68.4	86.9	98.4	68.7	85.4	98.6	68.9	85.6	96.1	69.3	84.4	95.8	68.2	83.7

		NORTH CAROLINA															NORTH DAKOTA		
	DIVISION	MURPHY 289			RALEIGH 275 - 276			ROCKY MOUNT 278			WILMINGTON 284			WINSTON-SALEM 271			BISMARCK 585		
		MAT.	INST.	TOTAL	MAT.	INST.	TOTAL	MAT.	INST.	TOTAL	MAT.	INST.	TOTAL	MAT.	INST.	TOTAL	MAT.	INST.	TOTAL
015433	CONTRACTOR EQUIPMENT		101.9	101.9		107.3	107.3		107.3	107.3		101.9	101.9		107.3	107.3		100.3	100.3
0241, 31 - 34	SITE & INFRASTRUCTURE, DEMOLITION	98.7	80.1	85.6	101.2	90.7	93.8	103.9	90.7	94.7	98.5	81.5	86.6	101.7	90.8	94.1	102.1	98.6	99.6
0310	Concrete Forming & Accessories	97.9	62.7	67.5	95.5	62.8	67.3	87.5	62.9	66.3	92.5	62.8	66.9	96.8	63.4	68.0	102.8	74.9	78.7
0320	Concrete Reinforcing	94.4	66.3	80.2	99.2	67.0	82.9	96.2	67.0	81.5	95.6	67.0	81.2	97.1	67.0	81.9	93.8	94.8	94.3
0330	Cast-in-Place Concrete	113.1	70.5	96.9	101.1	70.5	89.5	96.3	70.6	86.5	108.9	70.5	94.3	99.9	71.3	89.1	102.9	85.8	96.4
03	CONCRETE	102.4	67.7	86.6	92.5	67.9	81.3	95.1	68.0	82.7	99.4	67.9	85.0	94.9	68.4	82.8	92.5	82.8	88.1
04	MASONRY	82.0	62.3	69.8	80.9	62.3	69.4	76.1	62.3	67.6	79.4	62.3	68.8	82.8	63.7	71.0	103.6	83.3	91.0
05	METALS	95.6	91.1	94.2	98.7	91.3	96.5	99.2	91.4	96.8	97.3	91.3	95.5	103.4	91.3	99.7	103.2	93.7	100.3
06	WOOD, PLASTICS & COMPOSITES	99.1	61.7	78.3	92.7	61.8	75.5	86.6	61.8	72.8	92.5	61.8	75.5	94.6	61.8	76.4	97.2	70.7	82.5
07	THERMAL & MOISTURE PROTECTION	101.2	64.6	85.7	101.8	64.6	86.1	107.1	64.6	89.1	101.0	64.6	85.6	107.1	65.3	89.4	105.0	84.7	96.4
08	OPENINGS	93.7	62.4	86.5	99.1	62.7	90.7	96.8	62.7	89.0	93.9	62.7	86.7	100.8	62.7	92.0	106.5	81.0	100.7
0920	Plaster & Gypsum Board	111.9	60.5	77.4	92.6	60.6	71.1	94.2	60.6	71.7	108.7	60.6	76.4	100.2	60.6	73.6	102.8	70.3	81.0
0950, 0980	Ceilings & Acoustic Treatment	83.8	60.5	68.1	83.3	60.6	68.0	84.6	60.6	68.4	85.5	60.6	68.8	86.3	60.6	69.0	110.2	70.3	83.3
0960	Flooring	96.4	81.6	92.3	90.4	63.5	82.9	92.6	81.6	89.6	93.5	63.5	85.2	96.8	63.5	87.6	86.0	52.6	76.7
0970, 0990	Wall Finishes & Painting/Coating	104.5	59.7	78.4	93.8	59.7	73.9	98.8	59.7	76.0	104.5	59.7	78.4	98.8	59.7	76.0	91.2	60.1	73.1
09	FINISHES	92.1	65.9	77.9	88.4	62.4	74.2	88.0	66.0	76.0	91.0	62.4	75.4	90.0	62.7	75.1	96.0	69.6	81.7
COVERS	DIVS. 10 - 14, 25, 28, 41, 43, 44, 46	100.0	84.7	96.6	100.0	84.7	96.6	100.0	84.7	96.6	100.0	84.7	96.6	100.0	85.2	96.7	100.0	90.9	98.0
21, 22, 23	FIRE SUPPRESSION, PLUMBING & HVAC	97.4	60.7	81.7	99.9	61.9	83.7	97.3	60.6	81.6	100.5	61.9	84.0	100.4	62.6	84.2	100.1	76.9	90.2
26, 27, 3370	ELECTRICAL, COMMUNICATIONS & UTIL.	101.4	58.9	79.3	100.5	59.5	79.1	99.5	60.6	79.3	100.8	58.9	79.0	97.1	58.9	77.3	99.5	76.7	87.6
MF2016	WEIGHTED AVERAGE	96.9	67.5	84.0	96.9	68.3	84.4	96.3	68.6	84.2	97.2	67.5	84.2	98.4	68.7	85.4	100.3	81.4	92.0

		NORTH DAKOTA																	
	DIVISION	DEVILS LAKE 583			DICKINSON 586			FARGO 580 - 581			GRAND FORKS 582			JAMESTOWN 584			MINOT 587		
		MAT.	INST.	TOTAL	MAT.	INST.	TOTAL	MAT.	INST.	TOTAL	MAT.	INST.	TOTAL	MAT.	INST.	TOTAL	MAT.	INST.	TOTAL
015433	CONTRACTOR EQUIPMENT		100.3	100.3		100.3	100.3		100.3	100.3		100.3	100.3		100.3	100.3		100.3	100.3
0241, 31 - 34	SITE & INFRASTRUCTURE, DEMOLITION	106.9	98.6	101.1	114.4	98.6	103.3	97.6	98.6	98.3	110.4	98.6	102.1	105.9	98.6	100.8	108.0	98.6	101.4
0310	Concrete Forming & Accessories	99.3	74.8	78.1	89.1	74.8	76.8	96.9	75.0	78.0	92.8	74.8	77.3	90.7	74.7	76.9	88.8	74.8	76.8
0320	Concrete Reinforcing	95.3	94.7	95.0	96.2	94.8	95.5	95.4	95.1	95.3	93.9	94.7	94.3	95.9	95.1	95.5	97.2	94.8	96.0
0330	Cast-in-Place Concrete	123.1	85.8	108.9	111.4	85.8	101.6	100.8	85.8	95.1	111.3	85.8	101.6	121.5	85.7	107.9	111.3	85.8	101.6
03	CONCRETE	101.6	82.7	93.0	100.3	82.8	92.3	93.4	82.9	88.6	97.9	82.8	91.0	100.2	82.8	92.2	96.6	82.8	90.3
04	MASONRY	120.5	82.6	97.0	122.1	83.3	98.0	107.1	89.0	95.9	114.1	82.6	94.6	133.8	89.0	106.0	113.2	83.3	94.7
05	METALS	98.9	93.5	97.3	98.8	93.6	97.2	101.2	94.2	99.1	98.9	93.5	97.2	98.9	93.9	97.4	99.2	93.7	97.5
06	WOOD, PLASTICS & COMPOSITES	92.5	70.7	80.4	80.7	70.7	75.1	92.5	70.7	80.4	84.7	70.7	76.9	82.4	70.7	75.9	80.4	70.7	75.0
07	THERMAL & MOISTURE PROTECTION	105.7	84.5	96.7	106.3	84.7	97.1	105.9	86.7	97.8	105.9	84.5	96.9	105.5	86.7	97.6	105.7	84.7	96.8
08	OPENINGS	100.2	81.0	95.8	100.1	81.0	95.7	101.4	81.0	96.7	98.7	81.0	94.7	100.1	81.0	95.7	98.9	81.0	94.8
0920	Plaster & Gypsum Board	121.8	70.3	87.2	112.6	70.3	84.2	101.0	70.3	80.4	113.9	70.3	84.6	113.6	70.3	84.5	112.6	70.3	84.2
0950, 0980	Ceilings & Acoustic Treatment	108.4	70.3	82.7	108.4	70.3	82.7	99.9	70.3	79.9	108.4	70.3	82.7	108.4	70.3	82.7	108.4	70.3	82.7
0960	Flooring	92.6	52.6	81.5	86.1	52.6	76.8	96.1	52.6	84.1	88.0	52.6	78.2	86.8	52.6	77.3	85.8	52.6	76.6
0970, 0990	Wall Finishes & Painting/Coating	89.6	58.9	71.7	89.6	58.9	71.7	90.8	70.4	78.9	89.6	69.0	77.6	89.6	58.9	71.7	89.6	60.1	72.4
09	FINISHES	96.2	69.4	81.6	93.8	69.4	80.5	96.9	70.6	82.6	94.0	70.5	81.2	93.2	69.4	80.2	93.0	69.5	80.2
COVERS	DIVS. 10 - 14, 25, 28, 41, 43, 44, 46	100.0	90.8	98.0	100.0	90.9	98.0	100.0	90.9	98.0	100.0	90.9	98.0	100.0	90.8	98.0	100.0	90.9	98.0
21, 22, 23	FIRE SUPPRESSION, PLUMBING & HVAC	97.2	81.5	90.5	97.2	75.2	87.8	100.1	77.0	90.2	100.3	74.4	89.2	97.2	72.8	86.8	100.3	75.0	89.5
26, 27, 3370	ELECTRICAL, COMMUNICATIONS & UTIL.	97.4	48.0	71.7	105.8	71.7	88.0	103.5	73.3	87.8	100.8	63.0	81.1	97.4	48.0	71.7	103.8	76.7	89.7
MF2016	WEIGHTED AVERAGE	100.1	78.2	90.5	100.8	80.3	91.8	100.1	81.8	92.1	100.1	79.0	90.8	100.2	77.1	90.1	100.0	81.0	91.7

For customer support on your Building Construction Costs with RSMeans data, call 800.448.8182.

		NORTH DAKOTA			OHIO														
		WILLISTON			AKRON			ATHENS			CANTON			CHILLICOTHE			CINCINNATI		
	DIVISION	588			442 - 443			457			446 - 447			456			451 - 452		
		MAT.	INST.	TOTAL	MAT.	INST.	TOTAL	MAT.	INST.	TOTAL	MAT.	INST.	TOTAL	MAT.	INST.	TOTAL	MAT.	INST.	TOTAL
015433	CONTRACTOR EQUIPMENT		100.3	100.3		91.6	91.6		87.4	87.4		91.6	91.6		98.5	98.5		96.0	96.0
0241, 31 - 34	SITE & INFRASTRUCTURE, DEMOLITION	108.4	96.3	99.9	97.4	99.2	98.7	115.8	88.9	97.0	97.6	99.0	98.6	101.6	99.8	100.4	96.9	99.1	98.4
0310	Concrete Forming & Accessories	94.5	74.5	77.3	100.5	86.0	88.0	92.6	87.6	88.3	100.5	76.9	80.2	95.0	80.2	82.2	98.6	82.5	84.7
0320	Concrete Reinforcing	98.1	94.7	96.4	103.4	90.2	96.8	89.1	83.5	86.3	103.4	78.6	90.9	86.0	78.3	82.1	91.2	79.4	85.2
0330	Cast-in-Place Concrete	111.3	85.7	101.6	98.9	90.3	95.6	109.4	86.2	100.6	99.8	88.5	95.5	99.3	83.4	93.3	94.8	76.6	87.9
03	CONCRETE	97.8	82.6	90.9	98.3	87.6	93.4	101.1	85.9	94.1	98.7	80.9	90.6	95.5	81.4	89.1	94.6	80.1	88.0
04	MASONRY	107.4	83.3	92.5	93.5	91.6	92.3	80.1	81.2	80.8	94.1	79.1	84.8	87.3	91.2	89.7	90.5	78.0	82.7
05	METALS	99.0	93.3	97.3	96.8	80.1	91.7	103.2	77.5	95.3	96.8	74.8	90.0	95.2	85.7	92.3	97.5	83.8	93.3
06	WOOD, PLASTICS & COMPOSITES	86.2	70.7	77.6	107.2	85.0	94.8	86.4	90.6	88.8	107.5	75.2	89.6	99.7	77.0	87.1	99.4	83.9	90.7
07	THERMAL & MOISTURE PROTECTION	105.9	84.7	96.9	103.3	92.7	98.8	98.5	87.5	93.9	104.4	87.6	97.3	100.4	85.8	94.3	98.6	81.5	91.4
08	OPENINGS	100.2	81.0	95.8	109.6	85.3	104.0	97.1	84.9	94.3	103.3	73.2	96.4	88.7	74.4	85.4	96.6	80.1	92.8
0920	Plaster & Gypsum Board	113.9	70.3	84.6	103.3	84.5	90.7	95.6	90.2	92.0	104.3	74.4	84.2	98.3	76.7	83.8	97.8	83.8	88.4
0950, 0980	Ceilings & Acoustic Treatment	108.4	70.3	82.7	93.4	84.5	87.4	108.9	90.2	96.3	93.4	74.4	80.6	102.5	76.7	85.1	96.2	83.8	87.8
0960	Flooring	88.9	52.6	78.8	95.1	86.2	92.6	120.1	78.2	108.5	95.2	73.1	89.1	98.1	78.2	92.6	99.4	80.8	94.3
0970, 0990	Wall Finishes & Painting/Coating	89.6	58.9	71.7	98.2	95.0	96.3	100.7	88.8	93.7	98.2	72.2	83.0	97.9	80.1	87.5	97.5	74.6	84.1
09	FINISHES	94.2	69.4	80.7	97.4	86.6	91.5	101.5	86.5	93.3	97.6	75.0	85.3	98.4	79.3	88.0	97.7	81.6	88.9
COVERS	DIVS. 10 - 14, 25, 28, 41, 43, 44, 46	100.0	90.9	98.0	100.0	93.6	98.6	100.0	88.0	97.3	100.0	91.6	98.1	100.0	87.9	97.3	100.0	88.9	97.5
21, 22, 23	FIRE SUPPRESSION, PLUMBING & HVAC	97.2	75.2	87.8	100.0	89.0	95.3	96.9	76.3	88.1	100.0	78.0	90.6	97.4	94.6	96.2	99.9	77.1	90.2
26, 27, 3370	ELECTRICAL, COMMUNICATIONS & UTIL.	101.1	72.6	86.3	98.3	85.0	91.4	100.8	90.6	95.5	97.6	85.6	91.4	100.1	76.5	87.8	99.1	74.8	86.5
MF2016	WEIGHTED AVERAGE	99.3	80.2	90.9	99.7	88.2	94.7	99.1	83.7	92.4	99.1	81.0	91.1	96.2	86.1	91.7	97.7	80.9	90.3

		OHIO																	
		CLEVELAND			COLUMBUS			DAYTON			HAMILTON			LIMA			LORAIN		
	DIVISION	441			430 - 432			453 - 454			450			458			440		
		MAT.	INST.	TOTAL	MAT.	INST.	TOTAL	MAT.	INST.	TOTAL	MAT.	INST.	TOTAL	MAT.	INST.	TOTAL	MAT.	INST.	TOTAL
015433	CONTRACTOR EQUIPMENT		92.7	92.7		91.4	91.4		91.5	91.5		98.5	98.5		90.8	90.8		91.6	91.6
0241, 31 - 34	SITE & INFRASTRUCTURE, DEMOLITION	96.0	97.5	97.1	101.7	91.9	94.8	97.1	99.2	98.5	97.0	99.5	98.8	109.1	89.0	95.0	96.8	99.5	98.7
0310	Concrete Forming & Accessories	100.1	90.7	92.0	97.7	77.4	80.1	96.8	73.2	76.5	96.8	71.9	75.3	92.6	79.2	81.1	100.6	77.7	80.8
0320	Concrete Reinforcing	103.9	92.0	97.9	96.2	80.3	88.2	91.2	81.5	86.3	91.2	79.4	85.2	89.1	81.6	85.3	103.4	90.5	96.9
0330	Cast-in-Place Concrete	95.8	99.3	97.1	96.9	82.2	91.3	85.1	78.8	82.7	91.2	77.2	85.9	100.5	88.0	95.8	94.1	93.7	94.0
03	CONCRETE	98.5	93.4	96.2	96.2	79.6	88.6	86.9	76.6	82.2	89.8	75.7	83.4	94.0	82.7	88.8	96.0	85.2	91.1
04	MASONRY	100.0	99.2	99.5	91.9	86.5	88.5	86.3	77.2	80.6	86.7	79.4	82.2	108.7	78.6	90.0	90.2	94.6	92.9
05	METALS	98.4	84.9	94.3	97.5	80.8	92.4	96.7	77.8	90.9	96.8	86.9	93.7	103.2	80.7	96.3	97.4	81.5	92.5
06	WOOD, PLASTICS & COMPOSITES	101.0	88.3	93.9	99.4	76.1	86.4	103.5	71.7	85.8	102.1	69.0	83.7	86.4	78.7	82.1	107.2	73.4	88.4
07	THERMAL & MOISTURE PROTECTION	100.2	98.4	99.5	100.4	85.8	94.2	104.8	79.1	93.9	100.5	80.5	92.1	98.0	85.0	92.5	104.3	94.2	100.0
08	OPENINGS	97.9	87.5	95.5	99.7	74.4	93.9	95.8	72.8	90.5	93.5	71.5	88.4	97.1	75.9	92.2	103.3	78.9	97.7
0920	Plaster & Gypsum Board	103.2	87.7	92.8	97.2	75.5	82.6	100.1	71.3	80.8	100.1	68.5	78.9	95.6	78.0	83.8	103.3	72.6	82.7
0950, 0980	Ceilings & Acoustic Treatment	89.0	87.7	88.1	90.9	75.5	80.5	104.3	71.3	82.0	103.3	68.5	79.8	108.0	78.0	87.7	93.4	72.6	79.4
0960	Flooring	96.5	92.6	95.4	92.8	78.2	88.8	101.8	71.9	93.5	99.2	80.8	94.1	119.0	77.9	107.6	95.2	90.8	94.0
0970, 0990	Wall Finishes & Painting/Coating	101.1	95.0	97.5	100.4	80.1	88.5	97.9	71.6	82.6	97.9	71.0	82.2	100.7	71.9	83.9	98.2	95.0	96.3
09	FINISHES	97.8	91.2	94.2	93.9	77.3	84.8	99.9	72.3	84.8	98.9	72.8	84.7	100.5	78.0	88.3	97.4	81.0	88.5
COVERS	DIVS. 10 - 14, 25, 28, 41, 43, 44, 46	100.0	98.0	99.6	100.0	89.0	97.6	100.0	87.2	97.1	100.0	87.4	97.2	100.0	87.1	97.1	100.0	94.5	98.8
21, 22, 23	FIRE SUPPRESSION, PLUMBING & HVAC	100.0	93.0	97.0	99.9	85.7	93.9	100.8	75.7	90.1	100.5	77.5	90.7	96.9	83.1	91.0	100.0	90.0	95.7
26, 27, 3370	ELECTRICAL, COMMUNICATIONS & UTIL.	98.0	95.7	96.8	99.0	84.7	91.6	97.8	76.4	86.6	98.0	76.7	86.9	101.1	76.4	88.3	97.7	80.8	89.0
MF2016	WEIGHTED AVERAGE	98.9	93.4	96.5	98.1	83.3	91.6	96.7	77.9	88.5	96.6	79.4	89.1	99.4	81.1	91.4	98.6	86.9	93.5

		OHIO																	
		MANSFIELD			MARION			SPRINGFIELD			STEUBENVILLE			TOLEDO			YOUNGSTOWN		
	DIVISION	448 - 449			433			455			439			434 - 436			444 - 445		
		MAT.	INST.	TOTAL	MAT.	INST.	TOTAL	MAT.	INST.	TOTAL	MAT.	INST.	TOTAL	MAT.	INST.	TOTAL	MAT.	INST.	TOTAL
015433	CONTRACTOR EQUIPMENT		91.6	91.6		91.2	91.2		91.5	91.5		95.1	95.1		95.1	95.1		91.6	91.6
0241, 31 - 34	SITE & INFRASTRUCTURE, DEMOLITION	93.4	99.1	97.4	96.3	95.0	95.4	97.4	99.1	98.6	141.4	103.5	114.8	99.9	95.5	96.8	97.3	99.0	98.5
0310	Concrete Forming & Accessories	89.8	76.3	78.2	95.6	77.0	79.5	96.8	73.3	76.5	96.8	82.3	84.3	99.2	88.1	89.6	100.5	79.3	82.2
0320	Concrete Reinforcing	94.1	79.3	86.7	88.6	80.5	84.5	91.2	81.5	86.3	86.2	96.8	91.5	96.2	84.5	90.3	103.4	86.4	94.8
0330	Cast-in-Place Concrete	91.6	83.0	88.3	84.0	81.1	82.9	87.4	78.7	84.1	91.2	88.6	90.2	92.0	91.7	91.9	97.9	88.6	94.4
03	CONCRETE	90.6	78.9	85.2	83.4	78.9	81.4	88.0	76.6	82.8	87.2	86.5	86.9	91.0	88.7	89.9	97.8	83.4	91.2
04	MASONRY	92.6	91.7	92.0	93.5	89.2	90.8	86.7	76.5	80.5	80.8	91.3	87.3	99.9	90.7	94.2	93.8	87.3	89.7
05	METALS	97.6	76.0	91.0	96.5	78.9	91.1	96.7	77.9	90.9	93.1	82.2	89.7	97.3	85.9	93.8	96.8	78.2	91.1
06	WOOD, PLASTICS & COMPOSITES	94.9	73.4	83.0	96.6	76.1	85.2	105.0	71.7	86.5	92.1	80.0	85.4	100.5	88.0	93.5	107.2	78.2	91.1
07	THERMAL & MOISTURE PROTECTION	102.5	90.4	97.4	96.9	89.1	93.6	104.6	78.9	93.8	108.7	89.1	100.4	97.0	91.9	94.9	104.5	85.8	96.6
08	OPENINGS	103.7	73.2	96.7	93.4	71.5	88.4	94.0	72.8	89.1	93.7	82.0	91.0	96.2	85.1	93.6	103.3	78.6	97.6
0920	Plaster & Gypsum Board	96.9	72.6	80.6	98.5	75.5	83.1	100.1	71.3	80.8	97.6	79.1	85.1	100.5	87.7	91.9	103.3	77.5	86.0
0950, 0980	Ceilings & Acoustic Treatment	94.3	72.6	79.7	97.1	75.5	82.5	104.3	71.3	82.0	94.2	79.1	84.0	97.1	87.7	90.8	93.4	77.5	82.7
0960	Flooring	89.8	92.2	90.4	92.2	92.2	92.2	101.8	71.9	93.5	119.4	92.8	112.0	92.5	88.8	91.5	95.2	83.4	91.9
0970, 0990	Wall Finishes & Painting/Coating	98.2	76.5	85.5	104.9	76.5	88.3	97.9	71.6	82.6	118.5	88.8	101.2	104.9	87.3	94.6	98.2	82.0	88.7
09	FINISHES	94.8	78.6	86.0	95.3	79.5	86.7	99.9	72.2	84.8	112.9	84.3	97.4	96.0	88.0	91.6	97.5	79.9	87.9
COVERS	DIVS. 10 - 14, 25, 28, 41, 43, 44, 46	100.0	91.7	98.2	100.0	86.1	96.9	100.0	87.1	97.1	100.0	89.0	97.5	100.0	94.8	98.8	100.0	91.8	98.2
21, 22, 23	FIRE SUPPRESSION, PLUMBING & HVAC	96.9	89.0	93.5	96.9	85.1	91.9	100.8	81.1	92.4	97.3	93.3	95.6	100.0	93.5	97.2	100.0	83.8	93.1
26, 27, 3370	ELECTRICAL, COMMUNICATIONS & UTIL.	95.6	70.6	82.6	93.5	70.6	81.6	97.8	84.7	91.0	88.3	107.1	98.1	99.1	109.2	104.4	97.7	77.1	87.0
MF2016	WEIGHTED AVERAGE	96.8	82.8	90.7	94.4	81.6	88.8	96.7	80.2	89.5	96.2	91.8	94.3	97.5	93.1	95.6	98.9	83.3	92.1

DIVISION		OHIO ZANESVILLE 437 - 438 MAT.	INST.	TOTAL	OKLAHOMA ARDMORE 734 MAT.	INST.	TOTAL	CLINTON 736 MAT.	INST.	TOTAL	DURANT 747 MAT.	INST.	TOTAL	ENID 737 MAT.	INST.	TOTAL	GUYMON 739 MAT.	INST.	TOTAL
015433	CONTRACTOR EQUIPMENT		91.2	91.2		82.8	82.8		81.8	81.8		81.8	81.8		81.8	81.8		81.8	81.8
0241, 31 - 34	SITE & INFRASTRUCTURE, DEMOLITION	99.4	95.0	96.3	101.6	97.4	98.6	103.1	95.8	98.0	95.5	93.4	94.0	104.7	95.8	98.5	107.6	95.5	99.1
0310	Concrete Forming & Accessories	92.7	76.9	79.1	91.6	57.8	62.4	90.2	57.8	62.2	84.2	57.4	61.1	93.7	57.9	62.8	96.9	57.6	63.0
0320	Concrete Reinforcing	88.1	93.0	90.6	81.1	66.5	73.7	81.6	66.5	74.0	92.1	65.6	78.7	81.0	66.5	73.7	81.6	62.7	72.0
0330	Cast-in-Place Concrete	88.5	80.9	85.6	103.5	73.9	92.2	100.0	73.9	90.1	92.0	73.8	85.1	100.0	74.0	90.1	100.0	73.6	90.0
03	CONCRETE	87.0	81.1	84.3	89.8	64.7	78.3	89.1	64.7	78.0	85.7	64.3	75.9	89.6	64.8	78.3	92.1	63.9	79.2
04	MASONRY	91.2	77.7	82.8	95.8	57.9	72.3	119.7	57.9	81.4	88.7	64.8	73.8	101.5	57.9	74.5	98.0	56.6	72.4
05	METALS	98.0	83.6	93.6	102.0	61.3	89.5	102.1	61.3	89.6	91.3	60.9	82.0	103.6	61.4	90.7	102.6	59.1	89.3
06	WOOD, PLASTICS & COMPOSITES	92.4	76.1	83.3	104.1	57.0	77.9	103.1	57.0	77.5	91.5	57.0	72.3	105.5	57.0	79.0	110.0	57.0	80.5
07	THERMAL & MOISTURE PROTECTION	97.0	82.8	91.0	105.3	65.2	88.3	105.4	65.2	88.4	99.3	66.2	85.2	105.5	65.2	88.5	105.9	64.8	88.5
08	OPENINGS	93.4	78.1	89.9	103.2	57.0	92.6	103.2	57.0	92.6	96.9	56.7	87.7	104.4	57.3	93.6	103.3	56.1	92.5
0920	Plaster & Gypsum Board	95.2	75.5	82.0	88.9	56.4	67.0	88.5	56.4	66.9	77.9	56.4	63.4	89.2	56.4	67.1	89.2	56.4	67.1
0950, 0980	Ceilings & Acoustic Treatment	97.1	75.5	82.5	86.0	56.4	66.0	86.0	56.4	66.0	82.3	56.4	64.8	86.0	56.4	66.0	86.0	56.4	66.0
0960	Flooring	90.3	78.2	87.0	88.5	60.6	80.7	87.4	60.6	79.9	90.8	54.6	80.8	89.1	74.8	85.1	90.6	60.6	82.3
0970, 0990	Wall Finishes & Painting/Coating	104.9	80.1	90.4	84.0	45.1	61.3	84.0	45.1	61.3	90.9	45.1	64.2	84.0	45.1	61.3	84.0	44.2	60.8
09	FINISHES	94.5	77.0	85.0	81.8	55.9	67.7	81.6	55.9	67.6	82.5	54.8	67.4	82.2	58.8	69.5	83.1	57.2	69.0
COVERS	DIVS. 10 - 14, 25, 28, 41, 43, 44, 46	100.0	82.1	96.0	100.0	79.1	95.3	100.0	79.1	95.3	100.0	79.0	95.3	100.0	79.1	95.3	100.0	79.0	95.3
21, 22, 23	FIRE SUPPRESSION, PLUMBING & HVAC	96.9	90.9	94.5	97.2	67.8	84.7	97.2	67.8	84.7	97.2	67.8	84.6	100.3	67.9	86.4	97.2	65.6	83.7
26, 27, 3370	ELECTRICAL, COMMUNICATIONS & UTIL.	93.6	74.0	83.4	93.9	71.2	82.1	94.8	71.2	82.5	96.2	62.1	78.4	94.8	71.2	82.5	96.4	57.5	76.2
MF2016	WEIGHTED AVERAGE	95.0	82.5	89.5	96.7	66.7	83.6	97.9	66.6	84.2	93.2	65.6	81.1	98.3	67.0	84.6	97.8	63.8	82.9

DIVISION		OKLAHOMA LAWTON 735 MAT.	INST.	TOTAL	MCALESTER 745 MAT.	INST.	TOTAL	MIAMI 743 MAT.	INST.	TOTAL	MUSKOGEE 744 MAT.	INST.	TOTAL	OKLAHOMA CITY 730 - 731 MAT.	INST.	TOTAL	PONCA CITY 746 MAT.	INST.	TOTAL
015433	CONTRACTOR EQUIPMENT		82.8	82.8		81.8	81.8		93.5	93.5		93.5	93.5		83.1	83.1		81.8	81.8
0241, 31 - 34	SITE & INFRASTRUCTURE, DEMOLITION	100.6	97.4	98.3	88.7	95.5	93.5	90.0	92.8	91.9	90.3	92.7	92.0	99.0	97.9	98.2	96.0	95.8	95.9
0310	Concrete Forming & Accessories	96.9	57.9	63.3	82.3	43.2	48.6	95.1	57.6	62.7	99.5	57.4	63.2	94.3	57.9	62.9	90.6	57.7	62.3
0320	Concrete Reinforcing	81.2	66.5	73.8	91.8	65.2	78.4	90.3	66.5	78.3	91.2	64.9	77.9	89.7	66.5	78.0	91.2	66.5	78.7
0330	Cast-in-Place Concrete	96.7	74.0	88.1	80.7	73.5	78.0	84.6	74.9	80.9	85.6	74.8	81.5	96.7	74.0	88.1	94.5	73.9	86.7
03	CONCRETE	86.0	64.8	76.3	76.8	57.7	68.1	81.1	66.0	74.2	82.7	65.6	74.9	88.8	64.8	77.9	87.8	64.7	77.2
04	MASONRY	97.8	57.9	73.1	106.4	57.8	76.3	91.6	58.0	70.7	108.3	51.3	73.0	100.3	57.9	74.0	84.4	57.9	68.0
05	METALS	107.5	61.4	93.3	91.2	60.0	81.6	91.2	76.7	86.7	92.7	75.3	87.4	100.4	61.4	88.4	91.2	61.2	82.0
06	WOOD, PLASTICS & COMPOSITES	109.0	57.0	80.1	89.1	37.9	60.6	103.7	57.1	77.8	107.9	57.1	79.7	94.9	57.0	73.8	99.4	57.0	75.8
07	THERMAL & MOISTURE PROTECTION	105.3	65.2	88.3	98.9	62.1	83.3	99.3	65.2	84.8	99.6	62.9	84.1	96.2	65.5	83.2	99.4	65.5	85.1
08	OPENINGS	106.2	57.3	95.0	96.9	46.1	85.2	96.9	57.1	87.7	98.1	56.8	88.6	101.5	57.3	91.4	96.9	57.0	87.7
0920	Plaster & Gypsum Board	90.7	56.4	67.6	76.9	36.7	49.9	83.8	56.4	65.4	85.8	56.4	66.0	96.0	56.4	69.4	82.8	56.4	65.1
0950, 0980	Ceilings & Acoustic Treatment	92.8	56.4	68.2	82.3	36.7	51.6	82.3	56.4	64.8	90.9	56.4	67.6	92.6	56.4	68.1	82.3	56.4	64.8
0960	Flooring	90.9	74.8	86.5	89.7	60.6	81.6	97.0	54.6	85.2	99.4	35.1	81.6	88.2	74.8	84.5	93.9	60.6	84.7
0970, 0990	Wall Finishes & Painting/Coating	84.0	45.1	61.3	90.9	44.2	63.7	90.9	44.2	63.7	90.9	44.2	63.7	87.1	45.1	62.6	90.9	45.1	64.2
09	FINISHES	83.9	58.8	70.3	81.5	44.5	61.4	84.6	54.8	68.3	87.3	51.3	67.7	86.5	58.8	71.4	84.2	56.9	69.4
COVERS	DIVS. 10 - 14, 25, 28, 41, 43, 44, 46	100.0	79.1	95.3	100.0	77.0	94.9	100.0	79.5	95.4	100.0	79.5	95.4	100.0	79.1	95.3	100.0	79.1	95.3
21, 22, 23	FIRE SUPPRESSION, PLUMBING & HVAC	100.3	67.9	86.4	97.2	62.4	82.3	97.2	62.5	82.4	100.3	62.4	84.1	100.1	67.9	86.3	97.2	62.5	82.4
26, 27, 3370	ELECTRICAL, COMMUNICATIONS & UTIL.	96.4	69.4	82.4	94.7	64.5	79.0	96.0	64.6	79.7	94.3	70.9	82.1	101.7	71.2	85.8	94.3	67.6	80.4
MF2016	WEIGHTED AVERAGE	98.7	66.9	84.8	92.6	61.1	78.8	92.9	65.8	81.0	95.1	65.3	82.1	97.7	67.2	84.4	93.3	65.1	81.0

DIVISION		OKLAHOMA POTEAU 749 MAT.	INST.	TOTAL	SHAWNEE 748 MAT.	INST.	TOTAL	TULSA 740 - 741 MAT.	INST.	TOTAL	WOODWARD 738 MAT.	INST.	TOTAL	OREGON BEND 977 MAT.	INST.	TOTAL	EUGENE 974 MAT.	INST.	TOTAL
015433	CONTRACTOR EQUIPMENT		92.2	92.2		81.8	81.8		93.5	93.5		81.8	81.8		98.9	98.9		98.9	98.9
0241, 31 - 34	SITE & INFRASTRUCTURE, DEMOLITION	77.2	88.1	84.9	99.1	95.8	96.8	96.7	91.8	93.3	103.5	95.8	98.1	109.4	102.5	104.6	99.4	102.6	101.6
0310	Concrete Forming & Accessories	88.6	57.3	61.6	84.1	57.7	61.3	99.6	57.6	63.4	90.3	45.5	51.6	108.8	96.2	97.9	105.3	96.3	97.5
0320	Concrete Reinforcing	92.2	66.5	79.2	91.2	62.8	76.8	91.4	66.5	78.9	81.0	66.5	73.7	100.0	109.8	104.9	104.4	109.8	107.1
0330	Cast-in-Place Concrete	84.6	74.8	80.9	97.6	73.9	88.6	93.2	75.0	86.3	100.0	73.9	90.1	105.4	101.3	103.9	102.1	101.4	101.8
03	CONCRETE	83.2	65.8	75.3	89.4	64.0	77.8	87.9	66.1	77.9	89.4	59.2	75.6	103.8	99.9	102.0	95.6	99.9	97.6
04	MASONRY	91.9	58.0	70.9	107.8	57.9	76.8	92.5	58.0	71.1	91.0	57.9	70.5	104.0	100.8	102.0	101.0	100.8	100.9
05	METALS	91.2	76.4	86.7	91.1	59.9	81.5	95.8	76.8	89.9	102.2	61.2	89.6	96.3	94.5	95.7	97.0	94.6	96.3
06	WOOD, PLASTICS & COMPOSITES	96.1	57.1	74.4	91.3	57.0	72.2	107.1	57.1	79.3	103.3	40.5	68.3	101.8	95.1	98.1	97.8	95.1	96.3
07	THERMAL & MOISTURE PROTECTION	99.4	65.2	84.9	99.4	64.9	84.8	99.6	66.1	85.4	105.5	63.6	87.8	115.4	99.4	108.6	114.5	102.5	109.4
08	OPENINGS	96.9	57.1	87.7	96.9	56.1	87.5	99.8	57.1	90.0	103.2	47.9	90.5	101.2	98.9	100.6	101.4	98.9	100.9
0920	Plaster & Gypsum Board	80.9	56.4	64.4	77.9	56.4	63.4	85.8	56.4	66.0	88.5	39.4	55.5	115.2	94.8	101.5	113.2	94.8	100.8
0950, 0980	Ceilings & Acoustic Treatment	82.3	56.4	64.8	82.3	56.4	64.8	90.9	56.4	67.6	86.0	39.4	54.6	90.8	94.8	93.5	91.7	94.8	93.8
0960	Flooring	93.3	54.6	82.6	90.8	60.6	82.4	98.3	62.5	88.4	87.4	57.7	79.1	98.4	105.4	100.3	96.6	105.4	99.1
0970, 0990	Wall Finishes & Painting/Coating	90.9	45.1	64.2	90.9	44.2	63.7	90.9	44.2	63.7	84.0	45.1	61.3	90.9	72.8	80.3	90.9	72.8	80.3
09	FINISHES	82.3	54.9	67.4	82.7	55.8	68.1	87.1	56.3	70.4	81.6	45.6	62.0	96.1	95.1	95.6	94.5	95.1	94.8
COVERS	DIVS. 10 - 14, 25, 28, 41, 43, 44, 46	100.0	79.3	95.4	100.0	79.1	95.3	100.0	79.5	95.4	100.0	77.3	94.9	100.0	99.7	99.9	100.0	99.7	99.9
21, 22, 23	FIRE SUPPRESSION, PLUMBING & HVAC	97.2	62.4	82.4	97.2	67.8	84.6	100.3	64.3	84.9	97.2	67.8	84.7	97.0	102.2	99.2	100.1	102.2	101.0
26, 27, 3370	ELECTRICAL, COMMUNICATIONS & UTIL.	94.4	64.6	78.9	96.2	71.2	83.2	96.2	64.6	79.8	96.3	71.2	83.2	101.0	94.5	97.6	99.6	94.5	97.0
MF2016	WEIGHTED AVERAGE	92.5	65.4	80.6	94.7	66.3	82.3	96.0	66.4	83.0	96.7	63.8	82.3	100.0	98.7	99.4	99.1	98.8	99.0

For customer support on your Building Construction Costs with RSMeans data, call 800.448.8182.

OREGON

DIVISION		KLAMATH FALLS 976			MEDFORD 975			PENDLETON 978			PORTLAND 970-972			SALEM 973			VALE 979		
		MAT.	INST.	TOTAL	MAT.	INST.	TOTAL	MAT.	INST.	TOTAL	MAT.	INST.	TOTAL	MAT.	INST.	TOTAL	MAT.	INST.	TOTAL
015433	CONTRACTOR EQUIPMENT		98.9	98.9		98.9	98.9		96.4	96.4		98.9	98.9		98.9	98.9		96.4	96.4
0241, 31 - 34	SITE & INFRASTRUCTURE, DEMOLITION	113.4	102.5	105.8	107.1	102.5	103.9	106.8	95.5	98.9	102.1	102.6	102.4	93.9	102.5	99.9	94.0	95.4	95.0
0310	Concrete Forming & Accessories	101.7	95.9	96.7	100.8	95.9	96.5	102.2	96.2	97.1	106.4	96.3	97.7	105.9	96.2	97.5	108.7	95.1	97.0
0320	Concrete Reinforcing	100.0	109.7	104.9	101.7	109.7	105.8	99.2	109.8	104.5	105.2	109.8	107.5	113.1	109.8	111.4	96.7	109.6	103.2
0330	Cast-in-Place Concrete	105.4	101.2	103.8	105.4	101.2	103.8	106.2	100.2	103.9	104.9	101.4	103.6	96.4	101.3	98.3	83.6	102.2	90.6
03	CONCRETE	106.4	99.7	103.4	101.4	99.7	100.6	88.6	99.5	93.6	97.1	99.9	98.4	93.3	99.9	96.3	74.1	99.6	85.8
04	MASONRY	118.1	100.8	107.4	98.0	100.8	99.7	108.5	100.9	103.8	102.7	100.8	101.5	107.2	100.8	103.2	106.6	100.9	103.1
05	METALS	96.3	94.3	95.7	96.6	94.2	95.8	103.0	95.0	100.5	98.1	94.6	97.0	103.7	94.5	100.9	102.9	93.9	100.1
06	WOOD, PLASTICS & COMPOSITES	92.9	95.1	94.1	91.8	95.1	93.6	94.6	95.2	94.9	98.7	95.1	96.7	92.5	95.1	93.9	102.9	95.2	98.6
07	THERMAL & MOISTURE PROTECTION	115.6	95.2	107.0	115.2	95.2	106.8	108.0	95.8	102.9	114.4	99.4	108.1	110.5	99.4	105.8	107.4	93.4	101.5
08	OPENINGS	101.2	98.9	100.7	104.3	98.9	103.0	97.7	99.0	98.0	99.1	98.9	99.1	102.7	98.9	101.8	97.7	89.1	95.7
0920	Plaster & Gypsum Board	110.0	94.8	99.8	109.4	94.8	99.6	96.7	94.8	95.4	113.1	94.8	100.8	110.9	94.8	100.1	102.9	94.8	97.5
0950, 0980	Ceilings & Acoustic Treatment	98.4	94.8	96.0	105.1	94.8	98.2	65.0	94.8	85.1	93.6	94.8	94.4	99.6	94.8	96.4	65.0	94.8	85.1
0960	Flooring	95.2	105.4	98.1	94.6	105.4	97.6	66.1	105.4	77.0	94.5	105.4	97.5	98.0	105.4	100.0	68.2	105.4	78.5
0970, 0990	Wall Finishes & Painting/Coating	90.9	68.2	77.7	90.9	68.2	77.7	83.5	75.0	78.6	90.7	75.0	81.6	90.6	75.0	81.5	83.5	72.8	77.2
09	FINISHES	96.4	94.6	95.4	96.7	94.6	95.5	68.1	95.4	83.0	94.2	95.3	94.8	96.1	95.3	95.6	68.7	95.1	83.1
COVERS	DIVS. 10 - 14, 25, 28, 41, 43, 44, 46	100.0	99.6	99.9	100.0	99.6	99.9	100.0	99.5	99.9	100.0	99.7	99.9	100.0	99.7	99.9	100.0	99.9	100.0
21, 22, 23	FIRE SUPPRESSION, PLUMBING & HVAC	97.0	106.1	100.9	100.1	102.2	101.0	98.9	108.7	103.1	100.1	106.2	102.7	100.1	102.2	101.0	98.9	73.5	88.1
26, 27, 3370	ELECTRICAL, COMMUNICATIONS & UTIL.	99.7	77.5	88.2	103.1	77.5	89.8	92.2	96.3	94.3	99.8	103.5	101.7	107.2	94.5	100.6	92.2	67.7	79.4
MF2016	WEIGHTED AVERAGE	101.0	96.9	99.2	100.6	96.1	98.6	96.0	99.7	97.6	99.4	100.8	100.0	100.9	98.7	100.0	93.8	87.7	91.1

PENNSYLVANIA

DIVISION		ALLENTOWN 181			ALTOONA 166			BEDFORD 155			BRADFORD 167			BUTLER 160			CHAMBERSBURG 172		
		MAT.	INST.	TOTAL	MAT.	INST.	TOTAL	MAT.	INST.	TOTAL	MAT.	INST.	TOTAL	MAT.	INST.	TOTAL	MAT.	INST.	TOTAL
015433	CONTRACTOR EQUIPMENT		115.8	115.8		115.8	115.8		114.1	114.1		115.8	115.8		115.8	115.8		115.0	115.0
0241, 31 - 34	SITE & INFRASTRUCTURE, DEMOLITION	93.5	102.2	99.6	96.3	102.1	100.3	105.9	99.9	101.7	92.5	101.0	98.4	87.6	102.8	98.3	91.3	99.4	97.0
0310	Concrete Forming & Accessories	99.1	109.4	108.0	84.2	83.2	83.3	82.6	79.9	80.3	86.3	101.0	99.0	85.6	88.9	88.4	86.9	80.0	81.0
0320	Concrete Reinforcing	101.4	117.0	109.3	98.3	104.2	101.2	96.4	102.0	99.2	100.4	104.0	102.2	99.0	120.4	109.8	99.0	111.9	105.5
0330	Cast-in-Place Concrete	88.5	105.2	94.8	98.6	91.3	95.8	111.9	86.3	102.2	94.1	95.3	94.6	87.1	99.6	91.8	91.3	97.7	93.7
03	CONCRETE	90.8	110.1	99.6	86.8	91.2	88.8	103.3	87.6	96.1	92.7	100.6	96.3	78.9	99.5	88.3	94.5	93.3	93.9
04	MASONRY	94.3	97.1	96.0	97.8	87.0	91.1	105.7	83.7	92.0	95.2	82.1	87.1	100.0	96.7	97.9	100.0	82.3	89.0
05	METALS	98.0	123.2	105.7	92.0	115.6	99.3	97.2	114.0	102.3	95.8	115.4	101.8	91.7	123.4	101.5	95.5	117.8	102.4
06	WOOD, PLASTICS & COMPOSITES	101.3	111.6	107.0	80.5	81.3	80.9	84.0	79.1	81.3	86.9	106.1	97.6	81.9	85.6	84.0	88.2	79.2	83.2
07	THERMAL & MOISTURE PROTECTION	101.7	112.1	106.1	100.7	92.8	97.4	99.6	89.0	95.1	101.5	91.5	97.3	100.2	96.3	98.5	95.7	86.1	91.6
08	OPENINGS	94.2	109.9	97.8	87.9	85.6	87.4	96.3	84.5	93.6	94.1	99.4	95.3	87.9	96.1	89.8	91.4	84.7	89.9
0920	Plaster & Gypsum Board	96.8	111.8	106.9	88.5	80.7	83.2	100.5	78.5	85.8	89.6	106.2	100.7	88.5	85.2	86.3	104.3	78.5	87.0
0950, 0980	Ceilings & Acoustic Treatment	86.4	111.8	103.5	89.0	80.7	83.4	89.8	78.5	85.2	89.1	106.2	100.6	89.9	85.2	86.7	88.5	78.5	81.8
0960	Flooring	88.9	100.5	92.1	82.8	104.6	88.9	94.7	82.4	91.3	83.2	104.6	89.1	83.7	111.3	91.4	87.3	82.4	85.9
0970, 0990	Wall Finishes & Painting/Coating	91.2	106.5	100.1	86.8	111.0	100.9	96.7	111.0	105.1	91.2	105.3	99.4	86.8	111.0	100.9	88.6	105.3	98.3
09	FINISHES	88.8	108.0	99.2	86.5	89.2	88.0	99.4	83.3	90.6	86.6	102.9	95.5	86.5	94.4	90.8	86.5	82.7	84.5
COVERS	DIVS. 10 - 14, 25, 28, 41, 43, 44, 46	100.0	103.2	100.7	100.0	97.9	99.5	100.0	96.4	99.2	100.0	100.1	100.0	100.0	100.0	100.0	100.0	95.6	99.0
21, 22, 23	FIRE SUPPRESSION, PLUMBING & HVAC	100.2	112.0	105.3	99.7	85.6	93.7	96.9	82.3	90.6	97.2	94.4	96.0	96.7	97.1	96.8	97.2	88.1	93.3
26, 27, 3370	ELECTRICAL, COMMUNICATIONS & UTIL.	98.7	98.0	98.3	89.5	110.6	100.5	95.6	110.6	103.4	92.8	110.6	102.1	90.1	109.1	100.0	89.2	86.3	87.7
MF2016	WEIGHTED AVERAGE	96.5	107.7	101.4	93.2	95.1	94.0	98.7	92.2	95.8	94.8	100.2	97.2	91.4	101.5	95.8	94.4	90.8	92.8

PENNSYLVANIA

DIVISION		DOYLESTOWN 189			DUBOIS 158			ERIE 164-165			GREENSBURG 156			HARRISBURG 170-171			HAZLETON 182		
		MAT.	INST.	TOTAL	MAT.	INST.	TOTAL	MAT.	INST.	TOTAL	MAT.	INST.	TOTAL	MAT.	INST.	TOTAL	MAT.	INST.	TOTAL
015433	CONTRACTOR EQUIPMENT		93.5	93.5		114.1	114.1		115.8	115.8		114.1	114.1		115.0	115.0		115.8	115.8
0241, 31 - 34	SITE & INFRASTRUCTURE, DEMOLITION	107.5	89.2	94.7	110.8	100.0	103.2	93.2	102.4	99.6	101.8	101.9	101.9	91.4	100.3	97.6	86.9	101.5	97.1
0310	Concrete Forming & Accessories	83.3	126.4	120.5	82.1	80.7	80.9	98.4	87.0	88.5	88.8	92.8	92.2	98.6	87.5	89.0	80.9	86.7	85.9
0320	Concrete Reinforcing	98.1	124.0	111.2	95.7	104.1	99.9	100.4	105.2	102.8	95.7	120.1	108.0	108.4	114.5	111.5	98.4	113.0	105.8
0330	Cast-in-Place Concrete	83.6	133.8	102.6	107.9	94.7	102.9	96.9	93.3	95.5	103.8	99.2	102.0	89.8	100.9	94.0	83.6	99.1	89.5
03	CONCRETE	86.5	127.8	105.4	105.1	91.1	98.7	85.8	93.8	89.4	98.2	101.0	99.5	89.9	98.3	93.7	83.7	97.0	89.8
04	MASONRY	97.7	126.7	115.7	106.2	84.1	92.5	86.9	90.6	89.2	116.1	92.9	101.7	94.2	87.4	90.0	107.4	91.5	97.6
05	METALS	95.5	114.2	101.3	97.2	114.2	102.4	92.2	116.2	99.6	97.1	121.7	104.6	102.0	121.2	107.9	97.7	119.9	104.5
06	WOOD, PLASTICS & COMPOSITES	82.6	126.6	107.1	83.1	79.1	80.9	97.4	84.5	90.2	90.1	91.3	90.7	99.8	86.4	92.3	81.3	83.8	82.7
07	THERMAL & MOISTURE PROTECTION	99.2	127.8	111.3	99.9	91.2	96.2	101.2	92.1	97.4	99.5	95.8	97.9	98.6	103.9	100.8	101.0	100.8	100.9
08	OPENINGS	96.5	129.8	104.2	96.3	84.5	93.6	88.1	89.0	88.3	96.3	99.2	97.0	97.8	89.3	95.8	94.7	93.1	94.4
0920	Plaster & Gypsum Board	87.5	127.3	114.2	99.5	78.5	85.4	96.8	83.9	88.2	101.6	91.0	94.5	108.2	86.0	93.3	87.9	83.3	84.8
0950, 0980	Ceilings & Acoustic Treatment	85.6	127.3	113.7	98.8	78.5	85.2	86.4	83.9	84.7	98.0	91.0	93.3	97.0	86.0	89.6	87.4	83.3	84.6
0960	Flooring	73.8	147.3	94.2	94.5	104.6	97.3	89.6	104.6	93.7	98.8	82.4	94.2	89.8	95.6	91.4	80.4	94.4	84.3
0970, 0990	Wall Finishes & Painting/Coating	90.8	145.9	122.9	96.7	111.0	105.1	96.3	91.8	93.7	96.7	111.0	105.1	90.6	88.6	89.4	91.2	108.8	101.5
09	FINISHES	80.4	132.4	108.7	99.6	87.5	93.0	89.8	90.0	89.9	100.3	92.5	96.0	92.0	88.3	90.0	84.8	90.0	87.6
COVERS	DIVS. 10 - 14, 25, 28, 41, 43, 44, 46	100.0	113.0	102.9	100.0	96.9	99.3	100.0	99.5	99.9	100.0	100.4	100.1	100.0	97.7	99.5	100.0	96.8	99.3
21, 22, 23	FIRE SUPPRESSION, PLUMBING & HVAC	96.7	127.8	110.0	96.9	83.0	91.0	99.7	100.5	100.1	96.9	90.9	94.3	100.2	92.0	96.7	97.2	95.7	96.5
26, 27, 3370	ELECTRICAL, COMMUNICATIONS & UTIL.	92.3	127.0	110.3	96.2	110.6	103.7	91.1	92.4	91.8	96.2	110.6	103.7	95.9	88.7	92.2	93.6	91.7	92.6
MF2016	WEIGHTED AVERAGE	94.0	123.5	106.9	99.1	93.5	96.7	93.1	96.8	94.7	98.6	99.9	99.2	97.3	95.2	96.3	94.4	96.8	95.4

For customer support on your Building Construction Costs with RSMeans data, call 800.448.8182.

805

PENNSYLVANIA

DIVISION		INDIANA 157			JOHNSTOWN 159			KITTANNING 162			LANCASTER 175 - 176			LEHIGH VALLEY 180			MONTROSE 188		
		MAT.	INST.	TOTAL	MAT.	INST.	TOTAL	MAT.	INST.	TOTAL	MAT.	INST.	TOTAL	MAT.	INST.	TOTAL	MAT.	INST.	TOTAL
015433	CONTRACTOR EQUIPMENT		114.1	114.1		114.1	114.1		115.8	115.8		115.0	115.0		115.8	115.8		115.8	115.8
0241, 31 - 34	SITE & INFRASTRUCTURE, DEMOLITION	99.9	100.9	100.6	106.4	101.1	102.7	90.2	102.4	98.8	83.1	100.4	95.2	90.8	103.0	99.4	89.5	102.2	98.4
0310	Concrete Forming & Accessories	83.2	89.9	89.0	82.1	83.1	83.0	85.6	93.9	92.7	89.0	87.7	87.9	92.5	127.5	122.7	81.8	90.6	89.4
0320	Concrete Reinforcing	95.0	120.1	107.7	96.4	119.9	108.2	99.0	120.2	109.7	98.6	114.6	106.7	98.4	111.3	104.9	103.1	119.9	111.6
0330	Cast-in-Place Concrete	101.8	97.7	100.2	112.9	91.2	104.6	90.5	98.0	93.3	77.6	100.9	86.5	90.5	109.4	97.7	88.7	97.8	92.2
03	CONCRETE	95.7	99.1	97.3	104.3	93.7	99.5	81.4	101.1	90.4	83.1	98.4	90.1	89.9	118.8	103.1	88.6	99.6	93.6
04	MASONRY	102.2	92.4	96.1	103.1	87.3	93.3	102.6	92.3	96.2	106.1	88.9	95.4	94.3	109.9	104.0	94.3	95.5	95.0
05	METALS	97.2	121.2	104.6	97.2	119.8	104.1	91.8	122.4	101.2	95.5	121.4	103.4	97.6	120.9	104.8	95.9	123.0	104.2
06	WOOD, PLASTICS & COMPOSITES	84.7	88.8	87.0	83.1	81.2	82.0	81.9	93.6	88.4	90.9	86.4	88.4	93.2	130.3	113.8	82.0	87.3	85.0
07	THERMAL & MOISTURE PROTECTION	99.4	95.4	97.7	99.6	91.4	96.2	100.3	96.0	98.5	95.2	104.4	99.1	101.4	113.8	106.7	101.1	90.6	96.6
08	OPENINGS	96.3	93.5	95.7	96.3	89.3	94.7	87.9	100.5	90.8	91.4	89.3	90.9	94.7	118.9	100.3	91.4	94.4	92.1
0920	Plaster & Gypsum Board	100.8	88.4	92.5	99.3	80.7	86.8	88.5	93.4	91.8	106.2	86.0	92.6	90.5	131.0	117.7	88.3	86.9	87.4
0950, 0980	Ceilings & Acoustic Treatment	98.8	88.4	91.8	98.0	80.7	86.3	89.9	93.4	92.2	88.5	86.0	86.8	87.4	131.0	116.8	89.1	86.9	87.6
0960	Flooring	95.5	104.6	98.0	94.5	82.4	91.1	83.7	104.6	89.5	88.3	101.1	91.8	85.8	99.5	89.6	81.1	104.6	87.6
0970, 0990	Wall Finishes & Painting/Coating	96.7	111.0	105.1	96.7	111.0	105.1	86.8	111.0	100.9	88.6	88.6	88.6	91.2	62.5	74.4	91.2	108.8	101.5
09	FINISHES	99.2	94.3	96.5	99.0	85.5	91.6	86.6	97.2	92.4	86.5	89.4	88.1	87.0	116.8	103.2	85.6	94.2	90.3
COVERS	DIVS. 10 - 14, 25, 28, 41, 43, 44, 46	100.0	101.5	99.9	100.0	97.9	99.5	100.0	100.2	100.0	100.0	97.9	99.5	100.0	102.9	100.7	100.0	98.4	99.7
21, 22, 23	FIRE SUPPRESSION, PLUMBING & HVAC	96.9	83.3	91.1	96.9	82.1	90.6	96.7	93.3	95.2	97.2	92.3	95.1	97.2	113.0	103.9	97.2	98.6	97.8
26, 27, 3370	ELECTRICAL, COMMUNICATIONS & UTIL.	96.2	110.6	103.7	96.2	110.6	103.7	89.5	110.6	100.5	90.4	94.7	92.6	93.6	136.6	116.0	92.8	96.2	94.6
MF2016	WEIGHTED AVERAGE	97.4	97.8	97.6	98.7	94.7	96.9	91.9	101.2	95.9	93.2	96.4	94.6	94.9	117.4	104.7	93.8	99.5	96.3

PENNSYLVANIA

DIVISION		NEW CASTLE 161			NORRISTOWN 194			OIL CITY 163			PHILADELPHIA 190 - 191			PITTSBURGH 150 - 152			POTTSVILLE 179		
		MAT.	INST.	TOTAL	MAT.	INST.	TOTAL	MAT.	INST.	TOTAL	MAT.	INST.	TOTAL	MAT.	INST.	TOTAL	MAT.	INST.	TOTAL
015433	CONTRACTOR EQUIPMENT		115.8	115.8		100.1	100.1		115.8	115.8		100.3	100.3		102.6	102.6		115.0	115.0
0241, 31 - 34	SITE & INFRASTRUCTURE, DEMOLITION	88.0	102.7	98.3	97.0	100.2	99.2	86.6	101.2	96.8	98.8	100.6	100.1	106.0	100.4	102.1	86.1	100.6	96.3
0310	Concrete Forming & Accessories	85.6	94.6	93.4	84.1	126.7	120.8	85.6	93.4	92.3	101.5	141.9	136.4	99.0	98.9	99.0	80.6	88.9	87.8
0320	Concrete Reinforcing	97.8	101.9	99.8	99.5	126.6	113.2	99.0	97.7	98.3	103.3	126.7	115.1	96.9	120.6	108.8	97.8	114.5	106.2
0330	Cast-in-Place Concrete	87.9	99.2	92.2	83.4	134.1	102.6	85.3	97.6	90.0	91.2	137.7	108.9	111.3	99.9	107.0	82.7	101.3	89.8
03	CONCRETE	79.3	98.7	88.2	85.9	128.5	105.4	77.8	96.9	86.5	98.9	136.6	116.2	104.8	103.0	104.0	86.5	99.1	92.2
04	MASONRY	99.6	96.6	97.7	110.2	127.3	120.8	98.9	88.4	92.4	95.7	130.8	117.5	99.0	101.0	100.2	99.4	91.3	94.4
05	METALS	91.8	116.3	99.3	99.2	114.8	104.0	91.8	114.7	98.8	101.9	115.1	106.0	98.6	107.9	101.5	95.7	121.4	103.6
06	WOOD, PLASTICS & COMPOSITES	81.9	94.0	88.6	82.0	126.6	106.8	81.9	93.6	88.4	96.3	144.5	123.1	102.6	99.0	100.6	81.0	85.6	83.6
07	THERMAL & MOISTURE PROTECTION	100.2	95.7	98.3	104.6	128.2	114.6	100.1	93.2	97.2	101.5	136.9	116.5	99.8	99.1	99.5	95.3	103.1	98.6
08	OPENINGS	87.9	93.0	89.1	86.9	129.6	96.7	87.9	95.8	89.7	97.1	140.3	107.0	98.7	103.5	99.8	91.5	95.0	92.3
0920	Plaster & Gypsum Board	88.5	93.7	92.0	86.5	127.3	113.9	88.5	93.4	91.8	98.0	145.5	129.9	98.0	98.8	98.5	101.0	85.2	90.4
0950, 0980	Ceilings & Acoustic Treatment	89.9	93.7	92.5	91.0	127.3	115.4	89.9	93.4	92.2	101.3	145.5	131.1	92.6	98.8	96.8	88.5	85.2	86.3
0960	Flooring	83.7	111.3	91.4	81.8	147.3	99.9	83.7	106.1	89.9	96.7	147.3	110.8	104.0	107.7	105.0	84.3	101.1	89.0
0970, 0990	Wall Finishes & Painting/Coating	86.8	111.0	100.9	87.9	145.9	121.7	86.8	111.0	100.9	94.6	150.2	127.0	99.8	111.0	106.3	88.6	108.8	100.4
09	FINISHES	86.5	99.1	93.3	84.4	132.5	110.6	86.4	97.4	92.4	100.4	144.5	124.4	101.1	101.5	101.3	84.8	92.3	88.9
COVERS	DIVS. 10 - 14, 25, 28, 41, 43, 44, 46	100.0	100.8	100.2	100.0	113.2	102.9	100.0	100.1	100.0	100.0	120.0	104.4	100.0	101.8	100.4	100.0	95.2	98.9
21, 22, 23	FIRE SUPPRESSION, PLUMBING & HVAC	96.7	100.9	98.5	96.9	126.8	109.7	96.7	96.4	96.6	100.1	136.7	115.7	99.9	98.9	99.5	97.2	97.6	97.4
26, 27, 3370	ELECTRICAL, COMMUNICATIONS & UTIL.	90.1	100.7	95.6	93.7	150.0	123.0	91.7	109.1	100.8	100.1	155.2	128.7	98.5	111.6	105.3	88.8	91.0	89.9
MF2016	WEIGHTED AVERAGE	91.5	100.9	95.6	94.5	127.6	109.0	91.3	99.6	95.0	99.7	134.6	115.0	100.3	103.0	101.5	93.0	97.9	95.1

PENNSYLVANIA

DIVISION		READING 195 - 196			SCRANTON 184 - 185			STATE COLLEGE 168			STROUDSBURG 183			SUNBURY 178			UNIONTOWN 154		
		MAT.	INST.	TOTAL	MAT.	INST.	TOTAL	MAT.	INST.	TOTAL	MAT.	INST.	TOTAL	MAT.	INST.	TOTAL	MAT.	INST.	TOTAL
015433	CONTRACTOR EQUIPMENT		122.9	122.9		115.8	115.8		115.0	115.0		115.8	115.8		115.8	115.8		114.1	114.1
0241, 31 - 34	SITE & INFRASTRUCTURE, DEMOLITION	101.2	113.7	110.0	94.0	102.2	99.8	84.4	100.6	95.8	88.5	103.0	98.7	98.7	101.6	100.7	100.4	101.7	101.3
0310	Concrete Forming & Accessories	100.1	90.2	91.6	99.2	90.8	92.0	84.8	83.4	83.6	87.2	95.5	94.3	92.2	86.1	86.9	76.5	92.6	90.4
0320	Concrete Reinforcing	100.8	143.2	122.2	101.4	120.1	110.8	99.6	104.2	102.0	101.7	120.2	111.0	100.6	114.4	107.6	95.7	120.1	108.0
0330	Cast-in-Place Concrete	74.7	102.2	85.1	92.4	98.1	94.5	89.1	91.4	90.0	87.0	104.8	93.8	90.4	99.7	94.0	101.8	98.9	100.7
03	CONCRETE	84.8	105.0	94.0	92.6	99.8	95.9	92.8	91.3	92.1	87.4	104.2	95.1	89.9	97.2	93.3	95.3	100.7	97.8
04	MASONRY	99.2	96.3	96.8	94.6	97.9	96.6	100.3	87.5	92.3	92.4	109.7	103.1	99.9	84.4	91.3	117.8	92.8	102.3
05	METALS	99.5	132.9	109.7	100.0	123.3	107.1	95.6	115.8	101.8	97.7	123.6	105.7	95.4	121.1	103.3	96.9	121.2	104.4
06	WOOD, PLASTICS & COMPOSITES	99.4	86.4	92.2	101.3	87.3	93.5	88.7	81.3	84.5	87.7	87.3	87.5	88.9	85.6	87.1	77.6	91.2	85.2
07	THERMAL & MOISTURE PROTECTION	104.9	107.9	106.2	101.6	95.1	98.8	100.8	101.5	101.1	101.2	99.0	100.3	96.2	96.3	96.2	99.3	95.8	97.8
08	OPENINGS	91.2	102.2	93.8	94.2	94.4	94.2	91.2	85.6	89.9	94.7	97.3	95.3	91.6	92.2	91.7	96.3	98.7	96.8
0920	Plaster & Gypsum Board	96.3	86.0	89.4	98.5	86.9	90.7	90.4	80.7	83.9	89.2	86.9	87.6	100.2	85.2	90.1	97.7	91.0	93.2
0950, 0980	Ceilings & Acoustic Treatment	84.0	86.0	85.3	94.1	86.9	89.2	86.5	80.7	82.6	85.6	86.9	86.5	85.1	85.2	85.1	98.0	91.0	93.3
0960	Flooring	86.5	103.5	91.2	88.9	98.2	91.5	86.3	103.5	91.1	83.8	99.5	88.1	85.1	104.6	90.5	91.3	82.4	88.8
0970, 0990	Wall Finishes & Painting/Coating	86.7	106.5	98.2	87.2	113.2	104.0	91.2	111.0	102.8	91.2	108.8	101.5	88.6	108.8	100.4	96.7	108.9	103.8
09	FINISHES	86.1	92.9	89.8	90.5	93.8	92.3	86.2	89.1	87.8	85.7	96.5	91.6	85.7	91.2	88.7	97.3	92.4	94.6
COVERS	DIVS. 10 - 14, 25, 28, 41, 43, 44, 46	100.0	99.8	99.9	100.0	98.6	99.7	100.0	97.1	99.3	100.0	102.6	100.6	100.0	95.2	98.9	100.0	100.4	100.1
21, 22, 23	FIRE SUPPRESSION, PLUMBING & HVAC	100.2	111.1	104.8	100.2	98.8	99.6	97.2	91.6	94.8	97.2	104.8	100.4	97.2	89.6	93.9	96.9	85.6	92.1
26, 27, 3370	ELECTRICAL, COMMUNICATIONS & UTIL.	99.8	94.7	97.1	98.7	96.2	97.4	92.0	110.6	101.7	93.6	142.9	119.2	89.1	93.2	91.2	93.3	110.6	102.3
MF2016	WEIGHTED AVERAGE	96.1	105.2	100.1	97.2	100.0	98.4	94.4	96.5	95.3	94.3	110.4	101.3	93.9	95.1	94.4	97.6	98.6	98.1

For customer support on your Building Construction Costs with RSMeans data, call 800.448.8182.

City Cost Indexes

PENNSYLVANIA

DIVISION		WASHINGTON 153 MAT.	INST.	TOTAL	WELLSBORO 169 MAT.	INST.	TOTAL	WESTCHESTER 193 MAT.	INST.	TOTAL	WILKES-BARRE 186-187 MAT.	INST.	TOTAL	WILLIAMSPORT 177 MAT.	INST.	TOTAL	YORK 173-174 MAT.	INST.	TOTAL
015433	CONTRACTOR EQUIPMENT		114.1	114.1		115.8	115.8		100.1	100.1		115.8	115.8		115.8	115.8		115.0	115.0
0241, 31 - 34	SITE & INFRASTRUCTURE, DEMOLITION	100.5	102.0	101.6	96.0	101.6	99.9	102.9	101.1	101.6	86.5	102.2	97.5	89.3	101.7	98.0	86.9	100.4	96.3
0310	Concrete Forming & Accessories	83.3	94.6	93.1	85.8	85.9	85.9	90.8	127.3	122.3	89.8	89.4	89.4	88.8	86.8	87.1	84.0	87.7	87.1
0320	Concrete Reinforcing	95.7	120.4	108.2	99.6	119.9	109.8	98.5	143.9	121.4	100.4	120.1	110.3	99.8	114.5	107.2	100.6	114.6	107.6
0330	Cast-in-Place Concrete	101.8	99.3	100.8	93.3	92.7	93.1	92.3	135.2	108.6	83.6	97.9	89.0	76.5	93.6	83.0	83.1	100.9	89.9
03	CONCRETE	95.8	101.9	98.6	95.0	95.7	95.3	93.4	132.2	111.1	84.5	99.0	91.1	78.5	95.5	86.2	87.6	98.4	92.5
04	MASONRY	102.2	97.4	99.2	101.4	84.9	91.2	104.9	129.4	120.1	107.7	97.5	101.4	91.4	88.0	89.3	101.0	88.9	93.5
05	METALS	96.9	122.2	104.7	95.7	122.2	103.8	99.2	121.5	106.1	95.8	123.2	104.2	95.5	120.8	103.2	97.1	121.3	104.6
06	WOOD, PLASTICS & COMPOSITES	84.8	93.6	89.7	86.3	85.6	85.9	88.6	126.6	109.7	90.0	85.6	87.6	85.5	86.4	86.0	84.3	86.4	85.5
07	THERMAL & MOISTURE PROTECTION	99.4	97.7	98.6	101.8	88.8	96.3	105.0	124.7	113.3	101.0	94.7	98.4	95.6	90.8	93.6	95.3	104.4	99.2
08	OPENINGS	96.2	100.5	97.2	94.1	93.1	93.9	86.9	134.5	97.8	91.4	93.5	91.9	91.6	92.7	91.8	91.4	89.3	90.9
0920	Plaster & Gypsum Board	100.6	93.4	95.7	89.0	85.2	86.4	87.1	127.3	114.1	89.9	85.2	86.7	101.0	86.0	90.9	101.3	86.0	91.0
0950, 0980	Ceilings & Acoustic Treatment	98.0	93.4	94.9	86.5	85.2	85.6	91.0	127.3	115.4	89.1	85.2	86.4	88.5	86.0	86.8	87.6	86.0	86.5
0960	Flooring	95.6	82.4	91.9	82.8	101.1	87.9	84.6	147.3	102.0	84.6	98.2	88.4	84.0	91.7	86.2	85.6	101.1	89.9
0970, 0990	Wall Finishes & Painting/Coating	96.7	108.9	103.8	91.2	108.8	101.5	87.9	142.1	119.5	91.2	106.5	100.1	88.6	106.5	99.0	88.6	88.6	88.6
09	FINISHES	99.1	93.6	96.1	86.2	91.0	88.8	85.9	132.7	111.3	86.7	92.0	89.5	85.4	88.9	87.3	85.1	89.4	87.5
COVERS	DIVS. 10 - 14, 25, 28, 41, 43, 44, 46	100.0	100.7	100.2	100.0	92.9	98.4	100.0	111.2	102.5	100.0	98.2	99.6	100.0	97.5	99.5	100.0	97.8	99.5
21, 22, 23	FIRE SUPPRESSION, PLUMBING & HVAC	96.9	97.0	97.0	97.2	89.0	93.7	96.9	128.1	110.2	97.2	98.6	97.8	97.2	91.5	94.7	100.2	92.3	96.8
26, 27, 3370	ELECTRICAL, COMMUNICATIONS & UTIL.	95.6	110.6	103.4	92.8	93.6	93.2	93.6	109.4	101.9	93.6	91.7	92.6	89.6	75.9	82.4	90.4	81.2	85.6
MF2016	WEIGHTED AVERAGE	97.3	102.2	99.4	95.5	94.7	95.1	95.5	123.7	107.8	94.1	98.9	96.2	91.9	92.9	92.3	94.4	94.5	94.4

PUERTO RICO / RHODE ISLAND / SOUTH CAROLINA

DIVISION		SAN JUAN 009 MAT.	INST.	TOTAL	NEWPORT 028 MAT.	INST.	TOTAL	PROVIDENCE 029 MAT.	INST.	TOTAL	AIKEN 298 MAT.	INST.	TOTAL	BEAUFORT 299 MAT.	INST.	TOTAL	CHARLESTON 294 MAT.	INST.	TOTAL
015433	CONTRACTOR EQUIPMENT		87.1	87.1		101.1	101.1		101.1	101.1		106.8	106.8		106.8	106.8		106.8	106.8
0241, 31 - 34	SITE & INFRASTRUCTURE, DEMOLITION	131.9	89.9	102.5	88.7	102.8	98.6	90.6	102.8	99.1	128.2	89.6	101.2	123.3	89.0	99.3	108.2	89.3	95.0
0310	Concrete Forming & Accessories	91.0	21.7	31.3	98.6	119.9	117.0	99.9	119.9	117.1	95.7	65.0	69.3	94.7	39.8	47.4	93.9	65.0	69.0
0320	Concrete Reinforcing	189.9	18.4	103.4	102.0	114.6	108.4	104.8	114.6	109.7	95.8	62.6	79.0	94.8	26.6	60.4	94.6	66.1	80.2
0330	Cast-in-Place Concrete	95.0	32.8	71.4	72.7	118.7	90.2	93.2	118.7	102.9	90.9	67.6	82.1	90.9	67.2	81.9	106.7	67.6	91.8
03	CONCRETE	96.0	26.0	64.0	84.3	118.1	99.8	90.6	118.1	103.2	104.7	67.2	87.6	102.0	49.6	78.0	97.6	67.8	84.0
04	MASONRY	88.8	20.8	46.6	97.3	123.6	113.6	99.3	123.6	114.4	78.0	66.7	71.0	92.2	66.7	76.4	93.6	66.7	76.9
05	METALS	116.6	38.7	92.7	97.0	111.8	101.5	101.2	111.8	104.4	97.0	91.0	95.1	97.0	81.0	92.1	98.9	92.1	96.8
06	WOOD, PLASTICS & COMPOSITES	102.4	21.0	57.1	101.3	119.0	111.2	101.1	119.0	111.1	95.0	66.8	79.3	93.7	33.9	60.4	92.6	66.8	78.2
07	THERMAL & MOISTURE PROTECTION	130.6	25.9	86.3	99.6	117.7	107.2	103.1	117.7	109.3	99.1	66.3	85.2	98.8	58.1	81.5	97.9	65.9	84.3
08	OPENINGS	154.0	19.4	123.1	98.8	118.8	103.4	98.7	118.8	103.3	95.3	63.9	88.1	95.4	39.6	82.6	99.2	64.8	91.3
0920	Plaster & Gypsum Board	168.8	18.9	68.2	91.4	119.2	110.0	103.3	119.2	113.9	99.3	65.8	76.8	102.6	32.0	55.2	104.1	65.8	78.4
0950, 0980	Ceilings & Acoustic Treatment	250.5	18.9	94.4	96.7	119.2	111.8	101.6	119.2	113.4	86.4	65.8	72.5	89.0	32.0	50.5	89.0	65.8	73.3
0960	Flooring	203.1	21.3	152.6	89.9	127.6	100.3	87.3	127.6	98.5	95.9	94.9	95.6	97.3	80.1	92.6	97.0	81.6	92.7
0970, 0990	Wall Finishes & Painting/Coating	188.0	22.6	91.5	90.5	116.3	105.6	87.0	116.3	104.1	94.9	69.0	79.8	94.9	59.5	74.2	94.9	69.0	79.8
09	FINISHES	196.9	21.8	101.7	90.1	121.5	107.2	91.0	121.5	107.6	91.7	71.1	80.5	92.6	48.1	68.4	90.8	68.9	78.8
COVERS	DIVS. 10 - 14, 25, 28, 41, 43, 44, 46	100.0	19.8	82.1	100.0	107.5	101.7	100.0	107.5	101.7	100.0	72.1	93.8	100.0	76.9	94.9	100.0	72.0	93.8
21, 22, 23	FIRE SUPPRESSION, PLUMBING & HVAC	103.9	18.2	67.2	100.3	108.7	103.9	100.1	108.7	103.8	97.4	57.1	80.2	97.4	57.0	80.1	100.5	59.4	82.9
26, 27, 3370	ELECTRICAL, COMMUNICATIONS & UTIL.	124.8	17.5	69.0	100.9	96.6	98.6	101.1	96.6	98.7	100.9	65.3	82.4	104.6	32.6	67.2	103.0	60.1	80.7
MF2016	WEIGHTED AVERAGE	120.7	27.7	79.9	96.3	112.0	103.2	98.1	112.0	104.2	98.0	69.2	85.4	98.7	56.8	80.3	98.9	68.8	85.7

SOUTH CAROLINA / SOUTH DAKOTA

DIVISION		COLUMBIA 290-292 MAT.	INST.	TOTAL	FLORENCE 295 MAT.	INST.	TOTAL	GREENVILLE 296 MAT.	INST.	TOTAL	ROCK HILL 297 MAT.	INST.	TOTAL	SPARTANBURG 293 MAT.	INST.	TOTAL	ABERDEEN 574 MAT.	INST.	TOTAL
015433	CONTRACTOR EQUIPMENT		106.8	106.8		106.8	106.8		106.8	106.8		106.8	106.8		106.8	106.8		100.3	100.3
0241, 31 - 34	SITE & INFRASTRUCTURE, DEMOLITION	106.9	89.3	94.6	117.4	89.1	97.6	112.4	89.4	96.3	110.2	88.7	95.1	112.2	89.4	96.3	102.0	98.2	99.3
0310	Concrete Forming & Accessories	95.3	65.0	69.2	82.1	64.9	67.3	93.4	65.0	68.9	91.5	64.7	68.4	96.2	65.0	69.3	94.1	46.1	52.7
0320	Concrete Reinforcing	99.9	66.1	82.8	94.2	66.1	80.0	94.1	66.1	80.0	95.0	66.1	80.4	94.1	66.1	80.0	96.7	72.5	84.5
0330	Cast-in-Place Concrete	107.9	67.6	92.6	90.9	67.4	82.0	90.8	67.6	82.0	90.8	67.5	82.0	90.8	67.6	82.0	112.5	78.9	99.8
03	CONCRETE	95.8	67.8	83.0	96.5	67.6	83.3	95.1	67.8	82.6	93.0	67.6	81.4	95.3	67.8	82.7	99.3	63.6	83.0
04	MASONRY	87.3	66.7	74.5	78.0	66.7	71.0	75.6	66.7	70.1	99.6	66.7	79.2	78.0	66.7	71.0	113.6	73.4	88.7
05	METALS	96.1	92.1	94.9	97.8	91.5	95.9	97.8	92.1	96.0	97.0	91.8	95.4	97.8	92.1	96.0	100.9	83.0	95.4
06	WOOD, PLASTICS & COMPOSITES	93.1	66.8	78.5	79.9	64.9	72.7	94.5	66.8	78.1	90.9	64.6	77.5	96.1	66.8	79.8	92.4	34.7	60.3
07	THERMAL & MOISTURE PROTECTION	94.2	65.8	82.2	98.2	65.9	84.5	98.1	65.9	84.5	98.0	61.4	82.5	98.1	65.9	84.5	99.8	74.9	89.3
08	OPENINGS	95.4	64.8	88.4	95.4	64.8	88.4	95.3	64.8	88.3	95.4	64.8	88.3	95.4	64.8	88.3	95.7	41.4	83.2
0920	Plaster & Gypsum Board	94.8	65.8	75.3	93.6	65.8	74.9	98.4	65.8	76.5	98.0	65.8	76.4	101.0	65.8	77.4	112.0	33.3	59.1
0950, 0980	Ceilings & Acoustic Treatment	89.7	65.8	73.6	87.3	65.8	72.8	86.4	65.8	72.5	86.4	65.8	72.5	86.4	65.8	72.5	94.2	33.3	53.1
0960	Flooring	87.1	80.2	85.2	88.4	80.1	86.1	94.7	80.1	90.6	93.6	80.1	89.9	96.1	80.1	91.6	92.3	46.3	79.6
0970, 0990	Wall Finishes & Painting/Coating	89.0	69.0	77.3	94.9	69.0	79.8	94.9	69.0	79.8	94.9	69.0	79.8	94.9	69.0	79.8	90.4	34.9	58.0
09	FINISHES	86.8	68.6	76.9	87.6	68.6	77.3	89.6	68.6	78.2	88.9	68.6	77.9	90.3	68.6	78.5	91.4	43.0	65.1
COVERS	DIVS. 10 - 14, 25, 28, 41, 43, 44, 46	100.0	72.1	93.8	100.0	72.1	93.8	100.0	72.1	93.8	100.0	72.0	93.8	100.0	72.1	93.8	100.0	79.6	95.5
21, 22, 23	FIRE SUPPRESSION, PLUMBING & HVAC	100.1	58.7	82.4	100.5	58.7	82.7	100.5	58.7	82.7	97.4	57.1	80.2	100.5	58.7	82.7	100.2	54.8	80.8
26, 27, 3370	ELECTRICAL, COMMUNICATIONS & UTIL.	103.1	63.0	82.3	100.8	63.0	81.1	103.0	61.8	81.6	103.0	61.8	81.6	103.0	61.8	81.6	101.4	65.9	82.9
MF2016	WEIGHTED AVERAGE	96.9	69.1	84.7	97.0	69.0	84.8	97.1	68.9	84.8	97.1	68.3	84.5	97.3	68.9	84.9	99.7	64.5	84.3

For customer support on your Building Construction Costs with RSMeans data, call 800.448.8182.

SOUTH DAKOTA

DIVISION		MITCHELL 573			MOBRIDGE 576			PIERRE 575			RAPID CITY 577			SIOUX FALLS 570-571			WATERTOWN 572		
		MAT.	INST.	TOTAL	MAT.	INST.	TOTAL	MAT.	INST.	TOTAL	MAT.	INST.	TOTAL	MAT.	INST.	TOTAL	MAT.	INST.	TOTAL
015433	CONTRACTOR EQUIPMENT		100.3	100.3		100.3	100.3		100.3	100.3		100.3	100.3		101.3	101.3		100.3	100.3
0241, 31 - 34	SITE & INFRASTRUCTURE, DEMOLITION	98.8	98.1	98.3	98.8	98.2	98.3	100.4	97.6	98.4	100.3	97.6	98.4	93.4	100.0	98.1	98.7	98.1	98.3
0310	Concrete Forming & Accessories	93.3	46.2	52.7	84.4	46.6	51.8	96.5	47.5	54.2	101.5	59.1	64.9	100.7	64.5	69.5	81.2	44.9	49.9
0320	Concrete Reinforcing	96.2	72.7	84.3	98.6	72.7	85.6	98.6	97.0	97.8	90.4	97.2	93.8	96.8	97.3	97.0	93.5	41.1	67.1
0330	Cast-in-Place Concrete	109.4	54.8	88.6	109.4	79.0	97.8	104.2	79.8	94.9	108.5	80.1	97.7	92.5	80.8	88.1	109.4	57.9	89.8
03	CONCRETE	97.0	55.3	77.9	96.7	63.9	81.7	93.4	68.6	82.0	96.3	74.0	86.1	87.8	76.8	82.8	95.9	50.4	75.1
04	MASONRY	101.4	77.9	86.8	110.1	75.9	88.9	112.1	75.7	89.6	110.2	77.2	89.7	92.5	77.9	83.4	137.5	76.9	99.9
05	METALS	99.9	83.0	94.7	100.0	83.5	94.9	103.2	90.0	99.1	102.7	90.8	99.1	102.9	92.9	99.8	99.9	72.0	91.4
06	WOOD, PLASTICS & COMPOSITES	91.3	35.2	60.1	81.0	34.9	55.4	99.4	35.5	63.8	96.0	50.2	70.5	98.2	57.2	75.4	77.5	34.7	53.7
07	THERMAL & MOISTURE PROTECTION	99.6	76.2	89.7	99.7	78.8	90.8	101.9	75.7	90.8	100.3	80.4	91.9	98.4	81.9	91.4	99.4	75.7	89.4
08	OPENINGS	94.8	41.6	82.6	97.3	41.0	84.4	100.5	59.3	91.0	99.4	67.5	92.1	102.0	71.3	95.0	94.8	34.0	80.8
0920	Plaster & Gypsum Board	110.4	33.8	58.9	105.1	33.5	57.0	102.2	34.0	56.4	111.4	49.2	69.6	104.1	56.4	72.1	102.8	33.3	56.1
0950, 0980	Ceilings & Acoustic Treatment	89.9	33.8	52.1	94.2	33.5	53.3	94.2	34.0	53.6	95.9	49.2	64.4	89.7	56.4	67.2	89.9	33.3	51.7
0960	Flooring	91.9	46.3	79.3	87.5	49.1	76.9	93.3	33.0	76.6	91.7	76.8	87.6	95.8	77.8	90.8	86.1	46.3	75.1
0970, 0990	Wall Finishes & Painting/Coating	90.4	38.2	60.0	90.4	39.2	60.5	95.1	98.0	96.8	90.4	98.0	94.8	97.4	98.0	97.7	90.4	34.9	58.0
09	FINISHES	90.0	43.1	64.5	88.8	43.6	64.2	94.6	47.1	68.8	91.4	64.7	76.9	95.0	69.0	80.9	87.2	42.5	62.9
COVERS	DIVS. 10 - 14, 25, 28, 41, 43, 44, 46	100.0	79.6	95.5	100.0	79.5	95.4	100.0	81.4	95.9	100.0	83.0	96.2	100.0	83.9	96.4	100.0	44.8	87.7
21, 22, 23	FIRE SUPPRESSION, PLUMBING & HVAC	97.1	51.0	77.4	97.1	70.8	85.8	100.1	79.9	91.5	100.2	80.0	91.5	100.1	71.1	87.7	97.1	53.8	78.6
26, 27, 3370	ELECTRICAL, COMMUNICATIONS & UTIL.	99.7	65.9	82.1	101.4	42.0	70.5	104.1	50.4	76.1	97.9	50.4	73.2	102.5	65.9	83.5	98.9	46.1	71.4
MF2016	WEIGHTED AVERAGE	97.5	63.1	82.4	98.2	65.0	83.7	100.4	70.5	87.3	99.5	74.4	88.5	98.5	76.4	88.8	98.7	57.6	80.7

TENNESSEE

DIVISION		CHATTANOOGA 373 - 374			COLUMBIA 384			COOKEVILLE 385			JACKSON 383			JOHNSON CITY 376			KNOXVILLE 377 - 379		
		MAT.	INST.	TOTAL	MAT.	INST.	TOTAL	MAT.	INST.	TOTAL	MAT.	INST.	TOTAL	MAT.	INST.	TOTAL	MAT.	INST.	TOTAL
015433	CONTRACTOR EQUIPMENT		107.3	107.3		101.9	101.9		101.9	101.9		108.2	108.2		101.0	101.0		101.0	101.0
0241, 31 - 34	SITE & INFRASTRUCTURE, DEMOLITION	102.4	98.4	99.6	88.0	88.5	88.4	93.5	85.3	87.7	96.6	98.7	98.1	108.8	84.7	91.9	88.9	87.6	88.0
0310	Concrete Forming & Accessories	96.0	59.9	64.9	81.9	64.7	67.0	82.0	32.9	39.7	88.8	42.6	49.0	82.5	60.0	63.1	94.2	64.5	68.6
0320	Concrete Reinforcing	96.9	67.1	81.9	88.5	64.7	76.5	88.5	64.2	76.2	88.4	66.0	77.1	97.5	63.6	80.4	96.9	63.7	80.2
0330	Cast-in-Place Concrete	94.9	64.4	83.4	91.1	62.3	80.2	103.3	40.9	79.6	100.8	69.2	88.8	76.4	59.6	70.0	89.2	65.9	80.4
03	CONCRETE	93.2	64.5	80.1	92.1	65.6	80.0	102.2	43.7	75.4	93.0	58.2	77.1	103.2	62.2	84.5	90.8	66.5	79.7
04	MASONRY	99.5	58.7	74.2	116.4	56.4	79.2	111.3	39.8	67.0	116.7	42.7	70.8	113.5	42.1	69.2	77.9	52.8	62.4
05	METALS	96.9	88.9	94.5	90.2	89.6	90.0	90.3	87.6	89.5	92.6	89.6	91.7	94.1	86.7	91.9	97.4	87.2	94.3
06	WOOD, PLASTICS & COMPOSITES	104.1	58.9	78.9	72.3	66.4	69.0	72.5	30.8	49.3	88.5	41.5	62.3	77.4	65.6	70.8	90.4	65.6	76.6
07	THERMAL & MOISTURE PROTECTION	99.4	64.1	84.4	93.8	62.5	80.5	94.3	50.8	75.9	96.2	56.2	79.3	95.1	54.8	78.1	92.8	63.2	80.2
08	OPENINGS	101.2	60.4	91.8	92.4	55.5	83.9	92.4	36.7	79.6	99.6	45.8	87.2	97.5	62.6	89.5	94.7	58.0	86.3
0920	Plaster & Gypsum Board	78.9	58.3	65.1	85.4	66.0	72.3	85.4	29.4	47.8	87.3	40.4	55.8	99.1	65.2	76.3	105.9	65.2	78.5
0950, 0980	Ceilings & Acoustic Treatment	108.2	58.3	74.6	79.6	66.0	70.4	79.6	29.4	45.7	89.0	40.4	56.2	104.5	65.2	78.0	105.4	65.2	78.3
0960	Flooring	100.0	58.4	88.5	79.9	57.0	73.6	80.0	52.9	72.5	79.5	57.3	73.3	94.0	46.4	80.8	99.5	52.5	86.5
0970, 0990	Wall Finishes & Painting/Coating	96.9	62.2	76.7	83.0	59.0	69.0	83.0	59.0	69.0	84.6	59.0	69.7	93.7	61.4	74.9	93.7	60.2	74.2
09	FINISHES	96.6	59.4	76.3	84.6	62.7	72.7	85.0	37.9	59.4	84.9	45.5	63.5	100.0	58.0	77.2	93.6	61.8	76.3
COVERS	DIVS. 10 - 14, 25, 28, 41, 43, 44, 46	100.0	70.5	93.4	100.0	69.5	93.2	100.0	37.2	86.0	100.0	63.7	91.9	100.0	78.8	95.3	100.0	82.6	96.1
21, 22, 23	FIRE SUPPRESSION, PLUMBING & HVAC	100.2	59.7	82.9	98.0	75.9	88.6	98.0	69.8	86.0	100.1	61.1	83.4	99.9	57.3	81.7	99.9	63.0	84.1
26, 27, 3370	ELECTRICAL, COMMUNICATIONS & UTIL.	100.1	87.5	93.6	92.9	53.4	72.3	94.4	63.2	78.2	99.0	54.4	75.8	91.1	43.1	66.1	96.3	58.0	76.4
MF2016	WEIGHTED AVERAGE	98.6	70.3	86.2	94.2	68.4	82.9	95.6	57.7	79.0	97.1	60.6	81.1	98.8	60.3	82.0	95.3	66.2	82.5

TENNESSEE / TEXAS

DIVISION		MCKENZIE 382			MEMPHIS 375, 380 - 381			NASHVILLE 370 - 372			ABILENE 795 - 796			AMARILLO 790 - 791			AUSTIN 786 - 787		
		MAT.	INST.	TOTAL	MAT.	INST.	TOTAL	MAT.	INST.	TOTAL	MAT.	INST.	TOTAL	MAT.	INST.	TOTAL	MAT.	INST.	TOTAL
015433	CONTRACTOR EQUIPMENT		101.9	101.9		100.8	100.8		105.2	105.2		93.5	93.5		93.5	93.5		92.5	92.5
0241, 31 - 34	SITE & INFRASTRUCTURE, DEMOLITION	93.2	85.6	87.8	89.6	94.3	92.9	96.7	97.6	97.3	98.2	92.2	94.0	98.4	91.4	93.5	96.7	92.8	93.9
0310	Concrete Forming & Accessories	89.7	35.4	42.9	96.0	64.8	69.1	94.5	65.9	69.8	98.9	61.9	67.0	97.6	54.0	60.0	94.4	53.6	59.2
0320	Concrete Reinforcing	88.6	66.1	77.2	104.1	68.1	85.9	103.4	64.9	84.0	90.4	51.2	70.6	96.8	51.1	73.8	101.1	48.7	74.6
0330	Cast-in-Place Concrete	101.1	54.1	83.2	92.7	77.6	86.9	87.3	65.7	79.1	87.4	68.2	80.1	80.8	68.1	76.0	93.6	68.2	84.0
03	CONCRETE	100.7	49.7	77.4	95.6	70.9	84.3	94.0	67.0	81.6	82.8	63.0	73.7	82.3	59.4	71.8	88.6	58.7	74.9
04	MASONRY	115.4	47.0	73.0	98.0	57.8	73.1	84.6	57.8	68.0	95.1	63.2	75.3	94.5	62.7	74.8	97.3	63.2	76.1
05	METALS	90.3	88.9	89.9	99.8	84.4	95.1	96.9	85.9	93.5	107.8	70.7	96.4	102.7	70.5	92.8	97.2	69.0	88.5
06	WOOD, PLASTICS & COMPOSITES	80.9	33.1	54.3	98.0	65.1	79.7	91.2	66.7	77.6	104.6	63.7	81.8	103.2	53.3	75.4	93.1	52.4	70.5
07	THERMAL & MOISTURE PROTECTION	94.2	53.6	77.0	94.3	67.2	82.8	95.6	64.3	82.4	99.6	66.2	85.5	97.8	64.4	83.7	96.6	66.3	83.7
08	OPENINGS	92.4	38.7	80.1	97.7	61.8	89.4	96.0	63.4	88.5	102.2	59.1	92.3	103.4	53.3	91.9	102.8	50.6	90.8
0920	Plaster & Gypsum Board	88.3	31.8	50.4	90.4	64.4	72.9	91.6	66.0	74.4	84.9	63.1	70.2	94.1	52.4	66.1	88.6	51.6	63.7
0950, 0980	Ceilings & Acoustic Treatment	79.6	31.8	47.4	98.1	64.4	75.4	102.6	66.0	77.9	91.7	63.1	72.4	95.1	52.4	66.3	94.3	51.6	65.5
0960	Flooring	82.8	57.0	75.7	102.7	58.4	90.4	95.1	58.4	84.9	92.9	71.5	87.0	93.9	67.4	86.6	87.7	67.4	82.1
0970, 0990	Wall Finishes & Painting/Coating	83.0	46.2	61.5	91.3	59.6	72.8	100.7	70.3	83.0	91.0	54.0	69.4	86.6	54.0	67.6	92.4	46.2	65.5
09	FINISHES	86.2	39.3	60.7	98.1	62.5	78.8	98.0	64.6	79.8	83.7	63.1	72.5	90.6	56.1	71.8	89.6	54.7	70.6
COVERS	DIVS. 10 - 14, 25, 28, 41, 43, 44, 46	100.0	24.7	83.3	100.0	82.8	96.2	100.0	83.2	96.3	100.0	80.6	95.7	100.0	75.4	94.5	100.0	79.1	95.3
21, 22, 23	FIRE SUPPRESSION, PLUMBING & HVAC	98.0	61.0	82.2	100.0	71.3	87.7	100.0	76.6	90.0	100.3	49.3	78.5	99.9	53.2	79.9	100.0	65.2	85.2
26, 27, 3370	ELECTRICAL, COMMUNICATIONS & UTIL.	94.2	58.5	75.6	99.1	65.7	81.7	96.1	63.2	79.0	98.5	56.1	76.5	101.5	62.2	81.1	95.2	59.4	76.6
MF2016	WEIGHTED AVERAGE	95.8	56.8	78.7	98.4	70.7	86.3	96.7	71.7	85.8	97.7	62.6	82.3	97.6	62.2	82.1	96.7	62.7	81.8

For customer support on your Building Construction Costs with RSMeans data, call 800.448.8182.

TEXAS

DIVISION		BEAUMONT 776 - 777			BROWNWOOD 768			BRYAN 778			CHILDRESS 792			CORPUS CHRISTI 783 - 784			DALLAS 752 - 753		
		MAT.	INST.	TOTAL	MAT.	INST.	TOTAL	MAT.	INST.	TOTAL	MAT.	INST.	TOTAL	MAT.	INST.	TOTAL	MAT.	INST.	TOTAL
015433	CONTRACTOR EQUIPMENT		97.5	97.5		93.5	93.5		97.5	97.5		93.5	93.5		103.8	103.8		107.7	107.7
0241, 31 - 34	SITE & INFRASTRUCTURE, DEMOLITION	87.6	96.3	93.7	106.2	92.0	96.3	79.0	96.5	91.2	109.4	90.4	96.1	142.0	87.1	103.6	108.7	97.1	100.6
0310	Concrete Forming & Accessories	103.2	57.2	63.5	98.2	61.5	66.5	82.1	60.9	63.8	97.3	61.4	66.3	98.6	54.5	60.6	94.7	62.6	67.0
0320	Concrete Reinforcing	96.9	64.0	80.3	81.3	51.1	66.1	99.3	52.8	75.9	90.6	51.1	70.7	89.2	48.6	68.7	84.7	51.3	67.8
0330	Cast-in-Place Concrete	88.0	68.3	80.5	107.7	61.5	90.1	70.5	62.3	67.4	89.7	61.5	79.0	114.2	70.4	97.6	108.7	70.4	94.1
03	CONCRETE	90.9	63.2	78.3	94.5	60.5	78.9	76.0	60.9	69.1	90.5	60.4	76.7	96.5	61.0	80.3	99.7	64.9	83.8
04	MASONRY	102.5	64.3	78.8	128.3	63.2	87.9	142.8	63.2	93.4	99.2	62.7	76.6	85.5	63.3	71.7	100.0	61.8	76.3
05	METALS	93.5	77.2	88.5	95.5	70.3	87.8	93.3	72.9	87.0	105.2	70.2	94.5	93.5	84.7	90.8	97.6	83.5	93.3
06	WOOD, PLASTICS & COMPOSITES	114.0	56.6	82.0	104.7	63.7	81.8	79.1	62.1	69.7	104.0	63.7	81.6	122.5	53.9	84.3	95.3	64.0	77.9
07	THERMAL & MOISTURE PROTECTION	102.0	66.8	87.1	94.7	64.7	82.0	93.9	66.3	82.2	100.3	63.1	84.6	102.4	65.2	86.6	92.6	67.6	82.0
08	OPENINGS	94.9	57.8	86.4	98.0	59.1	89.1	95.3	58.2	86.8	97.4	59.1	88.6	103.9	51.6	91.9	101.2	59.3	91.6
0920	Plaster & Gypsum Board	97.1	55.9	69.4	85.3	63.1	70.4	85.0	61.5	69.3	84.5	63.1	70.1	96.1	52.9	67.1	93.8	63.1	73.2
0950, 0980	Ceilings & Acoustic Treatment	98.5	55.9	69.8	78.6	63.1	68.1	90.5	61.5	71.0	90.0	63.1	71.8	95.6	52.9	66.8	97.3	63.1	74.2
0960	Flooring	121.1	78.9	109.4	77.9	57.7	72.3	90.5	70.8	85.0	91.4	57.7	82.0	104.4	71.6	95.3	97.3	69.1	89.5
0970, 0990	Wall Finishes & Painting/Coating	88.7	56.3	69.8	89.7	54.0	68.9	86.8	60.0	71.2	91.0	54.0	69.4	104.9	46.2	70.7	102.2	54.0	74.1
09	FINISHES	94.7	60.6	76.2	76.5	60.3	67.7	81.8	62.3	71.2	84.1	60.3	71.1	98.2	56.5	75.5	97.4	62.8	78.6
COVERS	DIVS. 10 - 14, 25, 28, 41, 43, 44, 46	100.0	80.9	95.8	100.0	76.4	94.8	100.0	80.1	95.6	100.0	76.5	94.8	100.0	81.0	95.8	100.0	81.5	95.9
21, 22, 23	FIRE SUPPRESSION, PLUMBING & HVAC	100.2	64.3	84.8	97.1	47.3	75.8	97.1	63.8	82.8	97.2	53.2	78.4	100.2	50.7	79.0	99.9	60.6	83.1
26, 27, 3370	ELECTRICAL, COMMUNICATIONS & UTIL.	97.0	66.9	81.4	96.5	48.7	71.6	95.2	66.9	80.5	98.5	58.9	77.9	92.8	57.0	74.2	97.8	61.9	79.2
MF2016	WEIGHTED AVERAGE	96.5	68.0	84.0	96.9	60.2	80.8	94.0	67.3	82.3	97.6	62.7	82.3	98.7	62.4	82.8	99.2	67.6	85.4

TEXAS

DIVISION		DEL RIO 788			DENTON 762			EASTLAND 764			EL PASO 798 - 799, 885			FORT WORTH 760 - 761			GALVESTON 775		
		MAT.	INST.	TOTAL	MAT.	INST.	TOTAL	MAT.	INST.	TOTAL	MAT.	INST.	TOTAL	MAT.	INST.	TOTAL	MAT.	INST.	TOTAL
015433	CONTRACTOR EQUIPMENT		92.5	92.5		103.4	103.4		93.5	93.5		93.6	93.6		93.5	93.5		110.9	110.9
0241, 31 - 34	SITE & INFRASTRUCTURE, DEMOLITION	122.7	90.1	99.9	105.2	83.9	90.3	109.2	90.1	95.8	95.7	92.6	93.5	101.6	92.9	95.5	103.0	95.1	97.4
0310	Concrete Forming & Accessories	95.3	53.5	59.2	105.4	61.6	67.6	98.9	61.3	66.4	97.9	58.2	63.7	95.2	62.1	66.7	91.7	58.4	63.0
0320	Concrete Reinforcing	89.8	48.5	69.0	82.6	51.1	66.7	81.5	51.1	66.2	96.7	51.3	73.8	87.5	51.2	69.2	98.8	61.6	80.0
0330	Cast-in-Place Concrete	123.0	61.3	99.6	83.3	62.6	75.4	113.8	61.4	93.9	81.6	68.0	76.4	99.1	68.3	87.4	93.6	63.4	82.1
03	CONCRETE	116.3	56.3	88.8	74.2	62.0	68.6	99.1	60.3	81.4	82.7	61.2	72.9	89.6	63.1	77.5	92.1	62.8	78.7
04	MASONRY	100.5	63.1	77.3	137.5	61.8	90.6	96.9	63.2	76.0	85.9	64.4	72.6	95.9	61.8	74.7	98.6	63.3	76.7
05	METALS	93.3	68.2	85.6	95.2	86.0	92.3	95.3	70.1	87.6	102.7	70.1	92.7	97.1	70.9	89.1	94.9	92.9	94.3
06	WOOD, PLASTICS & COMPOSITES	100.9	52.8	74.1	116.1	63.8	87.0	111.8	63.7	85.0	88.7	58.5	71.9	95.1	63.7	77.6	96.5	58.5	75.3
07	THERMAL & MOISTURE PROTECTION	98.9	65.1	84.6	92.7	65.2	81.1	95.1	64.7	82.2	95.6	65.6	82.9	91.2	66.6	80.8	93.1	66.9	82.0
08	OPENINGS	99.2	50.1	87.9	115.6	58.7	102.5	70.6	59.1	67.9	98.6	54.0	88.3	99.7	59.1	90.4	99.4	58.3	90.0
0920	Plaster & Gypsum Board	92.4	51.9	65.2	89.9	63.1	71.9	85.3	63.1	70.4	92.3	57.7	69.1	95.5	63.1	73.7	91.6	57.7	68.8
0950, 0980	Ceilings & Acoustic Treatment	92.0	51.9	65.0	84.6	63.1	70.1	78.6	63.1	68.1	93.4	57.7	69.4	87.8	63.1	71.1	94.8	57.7	69.8
0960	Flooring	89.9	57.7	81.0	73.7	57.7	69.3	98.6	57.7	87.3	91.1	73.7	86.3	94.8	68.1	87.4	106.4	70.8	96.5
0970, 0990	Wall Finishes & Painting/Coating	92.7	44.5	64.6	99.7	51.1	71.4	91.1	54.0	69.5	95.5	49.9	68.9	92.0	54.1	69.9	97.5	56.1	73.4
09	FINISHES	90.8	52.8	70.1	75.7	60.1	67.2	83.3	60.3	70.8	89.3	60.0	73.4	89.0	62.4	74.6	91.6	59.7	74.3
COVERS	DIVS. 10 - 14, 25, 28, 41, 43, 44, 46	100.0	76.7	94.8	100.0	75.9	94.6	100.0	74.7	94.4	100.0	75.9	94.6	100.0	80.6	95.7	100.0	81.4	95.9
21, 22, 23	FIRE SUPPRESSION, PLUMBING & HVAC	97.0	61.2	81.7	97.1	58.1	80.4	97.1	47.2	75.8	99.9	66.3	85.5	100.0	57.7	81.9	97.0	63.9	82.8
26, 27, 3370	ELECTRICAL, COMMUNICATIONS & UTIL.	94.8	63.2	78.3	98.8	61.9	79.6	96.4	61.9	78.4	97.5	52.3	74.0	97.8	61.9	79.1	96.7	67.0	81.3
MF2016	WEIGHTED AVERAGE	99.5	62.9	83.4	96.8	65.3	83.0	93.6	61.8	79.7	96.0	64.7	82.3	96.6	65.1	82.8	96.2	69.0	84.3

TEXAS

DIVISION		GIDDINGS 789			GREENVILLE 754			HOUSTON 770 - 772			HUNTSVILLE 773			LAREDO 780			LONGVIEW 756		
		MAT.	INST.	TOTAL	MAT.	INST.	TOTAL	MAT.	INST.	TOTAL	MAT.	INST.	TOTAL	MAT.	INST.	TOTAL	MAT.	INST.	TOTAL
015433	CONTRACTOR EQUIPMENT		92.5	92.5		104.3	104.3		102.5	102.5		97.5	97.5		92.5	92.5		95.1	95.1
0241, 31 - 34	SITE & INFRASTRUCTURE, DEMOLITION	107.8	90.3	95.6	99.6	87.1	90.9	101.9	94.7	96.8	94.2	96.2	95.6	101.3	90.8	93.9	99.4	93.8	95.5
0310	Concrete Forming & Accessories	92.7	53.5	58.9	86.6	59.8	63.5	97.1	58.6	63.9	89.3	56.6	61.1	95.3	54.6	60.2	83.1	61.5	64.5
0320	Concrete Reinforcing	90.3	48.6	69.3	85.2	51.1	68.0	98.6	52.9	75.5	99.6	52.7	75.9	89.8	48.6	69.0	84.1	51.1	67.4
0330	Cast-in-Place Concrete	104.2	61.5	88.0	108.1	62.7	90.9	86.9	64.5	78.4	96.9	62.2	83.7	87.9	68.2	80.4	125.8	61.5	101.4
03	CONCRETE	93.6	56.4	76.6	93.4	61.1	78.6	91.3	61.0	77.4	98.4	58.9	80.3	88.0	59.2	74.8	110.5	60.4	87.6
04	MASONRY	107.9	63.2	80.2	155.5	61.8	97.4	97.5	63.3	76.4	141.0	63.2	92.8	93.8	63.2	74.8	151.7	61.7	95.9
05	METALS	92.7	68.7	85.4	95.1	84.6	91.9	97.6	78.4	91.7	93.2	72.6	86.9	95.8	69.1	87.6	88.3	69.3	82.5
06	WOOD, PLASTICS & COMPOSITES	100.0	52.8	73.7	85.5	61.3	72.0	101.9	58.7	77.9	87.9	56.6	70.5	100.9	53.8	74.7	80.3	63.8	71.1
07	THERMAL & MOISTURE PROTECTION	99.4	65.2	84.9	92.6	64.3	80.6	95.1	68.1	83.7	95.1	65.7	82.6	98.2	66.1	84.6	94.2	63.7	81.3
08	OPENINGS	98.3	50.8	87.4	97.8	57.8	88.6	102.5	56.3	91.9	95.3	55.1	86.1	99.2	51.4	88.2	88.4	58.7	81.6
0920	Plaster & Gypsum Board	91.4	51.9	64.9	85.2	60.3	68.5	90.8	57.7	68.6	89.0	55.9	66.7	93.2	52.9	66.2	84.2	63.1	70.0
0950, 0980	Ceilings & Acoustic Treatment	92.0	51.9	65.0	92.8	60.3	70.9	91.7	57.7	68.8	90.5	55.9	67.2	95.4	52.9	66.8	88.5	63.1	71.4
0960	Flooring	89.2	57.7	80.4	91.1	57.7	81.8	107.5	71.6	97.5	95.0	57.7	84.7	89.8	67.4	83.6	96.4	57.7	85.7
0970, 0990	Wall Finishes & Painting/Coating	92.7	46.2	65.6	96.3	54.0	71.6	95.4	58.3	73.8	86.8	58.3	70.1	92.7	46.2	65.6	87.0	51.1	66.0
09	FINISHES	89.4	53.0	69.6	91.9	58.9	74.0	97.3	60.6	77.3	84.6	56.6	69.4	89.9	55.5	71.2	97.3	60.0	77.0
COVERS	DIVS. 10 - 14, 25, 28, 41, 43, 44, 46	100.0	73.2	94.0	100.0	76.7	94.8	100.0	81.9	96.0	100.0	72.6	93.9	100.0	78.9	95.3	100.0	76.7	94.8
21, 22, 23	FIRE SUPPRESSION, PLUMBING & HVAC	97.1	62.6	82.3	96.9	58.9	80.7	100.1	65.2	85.2	97.1	63.7	82.8	100.1	61.3	83.5	96.9	57.5	80.0
26, 27, 3370	ELECTRICAL, COMMUNICATIONS & UTIL.	92.0	58.7	74.7	94.7	61.9	77.7	98.7	68.5	83.0	95.2	62.4	78.2	94.9	60.3	76.9	95.1	52.8	73.1
MF2016	WEIGHTED AVERAGE	96.1	62.6	81.4	98.6	65.2	84.0	98.3	68.0	85.0	97.4	65.1	83.3	96.1	63.5	81.8	98.9	62.8	83.1

For customer support on your Building Construction Costs with RSMeans data, call 800.448.8182.

809

TEXAS

DIVISION		LUBBOCK 793 - 794			LUFKIN 759			MCALLEN 785			MCKINNEY 750			MIDLAND 797			ODESSA 797		
		MAT.	INST.	TOTAL	MAT.	INST.	TOTAL	MAT.	INST.	TOTAL	MAT.	INST.	TOTAL	MAT.	INST.	TOTAL	MAT.	INST.	TOTAL
015433	CONTRACTOR EQUIPMENT		106.2	106.2		95.1	95.1		103.9	103.9		104.3	104.3		106.2	106.2		93.5	93.5
0241, 31 - 34	SITE & INFRASTRUCTURE, DEMOLITION	123.4	91.3	100.9	94.0	95.4	95.0	146.7	87.2	105.0	95.9	87.1	89.8	126.6	90.4	101.2	98.5	92.9	94.6
0310	Concrete Forming & Accessories	97.7	54.4	60.3	86.0	56.7	60.7	99.5	53.6	59.9	85.8	59.8	63.3	101.5	61.9	67.4	98.9	62.0	67.1
0320	Concrete Reinforcing	91.7	51.2	71.3	85.6	61.5	73.4	89.3	48.6	68.8	85.2	51.1	68.0	92.8	51.2	71.8	90.4	51.2	70.6
0330	Cast-in-Place Concrete	87.6	69.3	80.6	112.5	61.4	93.1	124.1	62.7	100.8	101.4	62.7	86.7	93.1	69.3	84.0	87.4	68.2	80.1
03	CONCRETE	81.7	61.1	72.2	101.8	59.9	82.7	104.2	57.9	83.0	88.4	61.1	75.9	85.7	64.4	76.0	82.8	63.1	73.8
04	MASONRY	94.5	62.8	74.8	116.7	63.1	83.4	100.1	63.3	77.3	166.9	61.8	101.7	111.1	62.8	81.1	95.1	62.7	75.0
05	METALS	111.7	86.5	104.0	95.2	72.4	88.2	93.2	84.5	90.5	95.0	84.6	91.8	109.9	86.4	102.7	107.2	70.8	96.0
06	WOOD, PLASTICS & COMPOSITES	103.8	53.4	75.8	87.6	56.8	70.4	121.2	52.9	83.2	84.6	61.3	71.6	108.5	63.8	83.6	104.6	63.7	81.8
07	THERMAL & MOISTURE PROTECTION	88.9	65.3	78.9	94.0	62.8	80.8	102.1	66.0	86.8	92.4	64.3	80.5	89.2	66.4	79.6	99.6	65.5	85.2
08	OPENINGS	109.8	53.4	96.9	67.2	57.7	65.0	103.1	50.1	90.9	97.8	57.7	88.6	111.3	59.2	99.3	102.2	59.1	92.3
0920	Plaster & Gypsum Board	85.1	52.4	63.1	83.2	55.9	64.8	96.8	51.9	66.7	85.2	60.3	68.5	87.0	63.1	70.9	84.9	63.1	70.2
0950, 0980	Ceilings & Acoustic Treatment	92.6	52.4	65.5	84.2	55.9	65.1	95.4	51.9	66.1	92.8	60.3	70.9	90.9	63.1	72.1	91.7	63.1	72.4
0960	Flooring	86.9	70.7	82.4	129.3	57.7	109.4	103.9	76.9	96.4	90.7	57.7	81.5	88.1	67.4	82.4	92.9	67.4	85.9
0970, 0990	Wall Finishes & Painting/Coating	101.4	54.0	73.8	87.0	54.0	67.7	104.9	44.5	69.7	96.3	54.0	71.6	101.4	54.0	73.8	91.0	54.0	69.4
09	FINISHES	86.1	56.9	70.2	106.1	56.6	79.2	98.5	56.8	75.8	91.6	58.9	73.8	86.8	62.3	73.5	83.7	62.2	72.0
COVERS	DIVS. 10 - 14, 25, 28, 41, 43, 44, 46	100.0	79.8	95.5	100.0	73.0	94.0	100.0	79.0	95.3	100.0	76.7	94.8	100.0	76.8	94.8	100.0	76.5	94.8
21, 22, 23	FIRE SUPPRESSION, PLUMBING & HVAC	99.7	49.3	78.2	96.9	60.4	81.3	97.1	50.7	77.2	96.9	58.9	80.7	96.6	49.4	76.4	100.3	55.3	81.0
26, 27, 3370	ELECTRICAL, COMMUNICATIONS & UTIL.	97.2	62.2	79.0	96.3	64.8	79.9	92.6	33.4	61.8	94.8	61.9	77.7	97.2	62.2	79.0	98.6	62.2	79.6
MF2016	WEIGHTED AVERAGE	99.3	63.4	83.6	95.7	64.9	82.2	99.6	58.6	81.7	98.4	65.2	83.9	100.0	64.8	84.6	97.6	64.5	83.1

TEXAS

DIVISION		PALESTINE 758			SAN ANGELO 769			SAN ANTONIO 781 - 782			TEMPLE 765			TEXARKANA 755			TYLER 757		
		MAT.	INST.	TOTAL	MAT.	INST.	TOTAL	MAT.	INST.	TOTAL	MAT.	INST.	TOTAL	MAT.	INST.	TOTAL	MAT.	INST.	TOTAL
015433	CONTRACTOR EQUIPMENT		95.1	95.1		93.5	93.5		93.3	93.3		93.5	93.5		95.1	95.1		95.1	95.1
0241, 31 - 34	SITE & INFRASTRUCTURE, DEMOLITION	99.5	94.1	95.7	102.4	92.5	95.4	101.1	93.5	95.8	90.3	92.3	91.7	88.3	96.6	94.1	98.6	93.9	95.3
0310	Concrete Forming & Accessories	77.6	61.6	63.8	98.5	53.8	59.9	93.6	54.2	59.6	101.9	53.5	60.2	92.7	62.2	66.4	87.6	61.7	65.3
0320	Concrete Reinforcing	83.5	51.1	67.1	81.2	51.2	66.1	96.0	48.6	72.1	81.4	48.6	64.8	83.3	51.2	67.1	84.1	51.1	67.5
0330	Cast-in-Place Concrete	102.8	61.5	87.1	101.6	68.2	88.9	86.2	69.6	79.9	83.2	61.4	74.9	103.6	68.3	90.2	123.5	61.6	100.0
03	CONCRETE	102.6	60.5	83.4	90.1	59.3	76.0	90.4	59.3	76.2	76.6	56.4	67.4	94.1	63.1	79.9	109.8	60.5	87.3
04	MASONRY	110.8	61.7	80.4	124.7	63.2	86.6	93.5	63.3	74.8	136.5	63.2	91.0	171.1	61.7	103.3	161.3	61.7	99.6
05	METALS	95.0	69.4	87.1	95.7	70.6	88.0	97.6	66.7	88.1	95.4	68.9	87.2	88.2	70.0	82.6	94.9	69.5	87.1
06	WOOD, PLASTICS & COMPOSITES	79.3	63.8	70.7	104.9	52.8	75.9	97.6	53.0	72.8	114.5	52.8	80.2	90.6	63.8	75.7	89.0	63.8	75.0
07	THERMAL & MOISTURE PROTECTION	94.6	64.8	82.0	94.5	65.7	82.3	93.0	67.5	82.2	94.1	64.7	81.7	93.7	66.1	82.0	94.4	64.8	81.9
08	OPENINGS	67.2	59.1	65.3	98.0	53.1	87.7	101.6	50.9	90.0	67.2	52.5	63.8	88.3	59.1	81.6	67.1	59.1	65.3
0920	Plaster & Gypsum Board	80.9	63.1	68.9	85.3	51.9	62.9	91.3	51.9	64.9	85.3	51.9	62.9	87.8	63.1	71.2	83.2	63.1	69.7
0950, 0980	Ceilings & Acoustic Treatment	84.2	63.1	70.0	78.6	51.9	60.6	94.5	51.9	65.8	78.6	51.9	60.6	88.5	63.1	71.4	84.2	63.1	70.0
0960	Flooring	120.2	57.7	102.8	78.0	57.7	72.3	101.6	71.6	93.3	100.2	57.7	88.4	104.8	67.4	94.4	131.5	57.7	111.0
0970, 0990	Wall Finishes & Painting/Coating	87.0	54.0	67.7	89.7	54.0	68.9	91.5	46.2	65.1	91.1	46.2	64.9	87.0	54.0	67.7	87.0	54.0	67.7
09	FINISHES	103.6	60.3	80.1	76.3	53.9	64.1	99.3	55.9	75.7	82.5	53.0	66.5	99.7	62.3	79.4	107.1	60.3	81.7
COVERS	DIVS. 10 - 14, 25, 28, 41, 43, 44, 46	100.0	76.7	94.8	100.0	79.2	95.4	100.0	79.2	95.4	100.0	75.3	94.5	100.0	80.9	95.7	100.0	80.8	95.7
21, 22, 23	FIRE SUPPRESSION, PLUMBING & HVAC	96.9	57.6	80.1	97.1	49.3	76.7	100.1	64.0	84.6	97.1	53.6	78.5	96.9	56.7	79.7	96.9	60.5	81.3
26, 27, 3370	ELECTRICAL, COMMUNICATIONS & UTIL.	92.6	51.1	71.0	100.4	57.6	78.2	97.3	62.0	79.0	97.6	54.9	75.4	96.2	57.6	76.1	95.1	56.7	75.1
MF2016	WEIGHTED AVERAGE	95.0	62.7	80.9	96.4	60.7	80.8	97.8	64.4	83.2	91.9	60.4	78.1	97.9	64.4	83.2	98.9	64.2	83.7

TEXAS / UTAH

DIVISION		VICTORIA 779			WACO 766 - 767			WAXAHACHIE 751			WHARTON 774			WICHITA FALLS 763			LOGAN 843		
		MAT.	INST.	TOTAL	MAT.	INST.	TOTAL	MAT.	INST.	TOTAL	MAT.	INST.	TOTAL	MAT.	INST.	TOTAL	MAT.	INST.	TOTAL
015433	CONTRACTOR EQUIPMENT		109.5	109.5		93.5	93.5		104.3	104.3		110.9	110.9		93.5	93.5		96.6	96.6
0241, 31 - 34	SITE & INFRASTRUCTURE, DEMOLITION	107.8	92.2	96.9	99.1	92.7	94.6	97.7	87.4	90.5	112.8	94.5	100.0	99.9	92.9	95.0	101.4	93.9	96.1
0310	Concrete Forming & Accessories	91.0	53.6	58.8	100.4	62.0	67.3	85.8	61.6	65.0	86.3	55.5	59.7	100.4	62.0	67.3	102.2	67.1	71.9
0320	Concrete Reinforcing	94.9	48.6	71.5	81.1	48.7	64.7	85.2	51.1	68.0	98.7	52.2	75.2	81.1	51.2	66.0	102.1	86.6	94.3
0330	Cast-in-Place Concrete	105.1	63.3	89.2	90.1	68.2	81.8	107.0	62.9	90.2	107.9	62.3	90.6	96.2	68.3	85.6	86.3	75.2	82.1
03	CONCRETE	99.4	58.3	80.6	83.0	62.6	73.7	92.4	62.0	78.5	103.5	59.4	83.3	85.9	63.1	75.5	103.3	73.8	89.8
04	MASONRY	116.5	63.3	83.5	94.9	63.2	75.2	156.0	61.9	97.6	99.8	63.2	77.1	95.4	62.7	75.1	108.4	67.9	83.3
05	METALS	93.4	87.8	91.7	97.7	69.4	89.0	95.1	84.9	92.0	94.9	88.8	93.0	97.6	70.8	89.4	107.9	84.0	100.6
06	WOOD, PLASTICS & COMPOSITES	99.9	52.9	73.7	112.8	63.7	85.4	84.6	63.9	73.1	90.0	55.2	70.6	112.8	63.7	85.4	84.2	67.1	74.6
07	THERMAL & MOISTURE PROTECTION	96.9	64.1	83.0	95.0	66.9	83.1	92.5	64.5	80.6	93.5	66.2	81.9	95.0	65.5	82.5	96.5	71.6	85.9
08	OPENINGS	99.3	52.1	88.4	78.7	58.5	74.1	97.8	59.2	89.0	99.4	54.2	89.0	78.7	59.1	74.2	92.4	66.9	86.5
0920	Plaster & Gypsum Board	88.5	51.9	63.9	85.9	63.1	70.5	85.5	63.1	70.4	87.3	54.3	65.1	85.9	63.1	70.5	77.8	66.2	70.0
0950, 0980	Ceilings & Acoustic Treatment	95.6	51.9	66.2	81.2	63.1	69.0	94.5	63.1	73.3	94.8	54.3	67.5	81.2	63.1	69.0	103.2	66.2	78.2
0960	Flooring	104.9	57.7	91.8	99.5	67.4	90.6	90.7	57.7	81.5	103.4	70.4	94.2	100.3	76.0	93.6	98.5	59.3	87.6
0970, 0990	Wall Finishes & Painting/Coating	97.7	58.3	74.7	91.1	54.0	69.4	96.3	54.0	71.6	97.5	58.3	74.6	93.5	54.0	70.4	93.6	56.7	72.1
09	FINISHES	89.3	54.4	70.3	83.3	62.2	71.9	92.2	60.5	74.9	90.8	57.9	72.9	83.8	64.0	73.0	93.4	64.0	77.4
COVERS	DIVS. 10 - 14, 25, 28, 41, 43, 44, 46	100.0	72.5	93.9	100.0	80.6	95.7	100.0	77.0	94.9	100.0	72.8	93.9	100.0	76.5	94.8	100.0	85.0	96.7
21, 22, 23	FIRE SUPPRESSION, PLUMBING & HVAC	97.1	63.6	82.8	100.2	59.1	82.6	96.9	52.5	77.9	97.0	61.0	81.6	100.2	52.7	79.9	99.9	70.0	87.1
26, 27, 3370	ELECTRICAL, COMMUNICATIONS & UTIL.	101.4	57.0	78.3	100.5	55.6	77.2	94.8	61.9	77.7	100.4	62.5	80.7	102.3	56.1	78.3	96.4	69.2	82.3
MF2016	WEIGHTED AVERAGE	98.3	64.9	83.7	93.6	64.4	80.8	98.5	64.3	83.5	98.2	66.2	84.2	94.2	63.3	80.7	100.2	72.9	88.3

For customer support on your Building Construction Costs with RSMeans data, call 800.448.8182.

		UTAH											VERMONT					
		OGDEN			PRICE			PROVO			SALT LAKE CITY			BELLOWS FALLS			BENNINGTON		
DIVISION		842, 844			845			846 - 847			840 - 841			051			052		
		MAT.	INST.	TOTAL	MAT.	INST.	TOTAL	MAT.	INST.	TOTAL	MAT.	INST.	TOTAL	MAT.	INST.	TOTAL	MAT.	INST.	TOTAL
015433	CONTRACTOR EQUIPMENT		96.6	96.6		95.6	95.6		95.6	95.6		96.6	96.6		98.8	98.8		98.8	98.8
0241, 31 - 34	SITE & INFRASTRUCTURE, DEMOLITION	88.8	93.9	92.4	98.8	92.1	94.1	97.5	92.2	93.8	88.5	93.8	92.2	92.2	101.5	98.7	91.5	101.5	98.5
0310	Concrete Forming & Accessories	102.2	67.1	71.9	104.6	66.9	72.1	103.9	67.1	72.1	104.5	67.1	72.2	93.8	106.3	104.6	91.6	106.2	104.1
0320	Concrete Reinforcing	101.7	86.6	94.1	109.4	86.6	97.9	110.4	86.6	98.4	103.9	86.6	95.2	80.9	81.4	81.1	80.9	81.4	81.1
0330	Cast-in-Place Concrete	87.6	75.2	82.9	86.4	75.2	82.1	86.4	75.2	82.2	95.7	75.2	88.0	90.7	115.5	100.1	90.7	115.5	100.1
03	CONCRETE	92.9	73.8	84.2	104.4	73.7	90.4	102.9	73.8	89.6	112.0	73.8	94.5	85.2	104.7	94.2	85.1	104.6	94.0
04	MASONRY	102.9	67.9	81.2	114.4	67.9	85.6	114.5	67.9	85.6	116.9	67.9	86.5	104.2	89.8	95.3	115.7	89.8	99.7
05	METALS	108.4	84.0	100.9	105.0	83.8	98.5	106.0	84.0	99.3	112.1	84.0	103.4	93.4	89.7	92.3	93.4	89.6	92.2
06	WOOD, PLASTICS & COMPOSITES	84.2	67.1	74.6	87.2	67.1	76.0	85.5	67.1	75.2	85.9	67.1	75.4	108.3	111.9	110.3	105.6	111.9	109.1
07	THERMAL & MOISTURE PROTECTION	95.3	71.6	85.3	98.3	71.6	87.0	98.4	71.6	87.0	102.8	71.6	89.6	98.0	92.6	95.7	98.0	92.6	95.7
08	OPENINGS	92.4	66.9	86.5	96.3	63.1	88.7	96.4	66.9	89.6	94.3	66.9	88.0	100.4	100.8	100.5	100.4	100.8	100.5
0920	Plaster & Gypsum Board	77.8	66.2	70.0	80.5	66.2	70.9	78.1	66.2	70.1	90.8	66.2	74.3	105.6	111.9	109.8	103.9	111.9	109.3
0950, 0980	Ceilings & Acoustic Treatment	103.2	66.2	78.2	103.2	66.2	78.2	103.2	66.2	78.2	95.5	66.2	75.8	93.8	111.9	106.0	93.8	111.9	106.0
0960	Flooring	96.4	59.3	86.1	99.7	59.3	88.5	99.4	59.3	88.2	100.4	59.3	89.0	91.1	104.5	94.8	90.4	104.5	94.3
0970, 0990	Wall Finishes & Painting/Coating	93.6	56.7	72.1	93.6	56.7	72.1	93.6	56.7	72.1	96.7	56.7	73.3	86.2	106.1	97.8	86.2	106.1	97.8
09	FINISHES	91.5	64.0	76.6	94.6	64.0	78.0	94.0	64.0	77.7	93.7	64.0	77.6	88.1	107.0	98.4	87.6	107.0	98.2
COVERS	DIVS. 10 - 14, 25, 28, 41, 43, 44, 46	100.0	85.0	96.7	100.0	85.0	96.7	100.0	85.0	96.7	100.0	85.0	96.7	100.0	95.7	99.0	100.0	95.6	99.0
21, 22, 23	FIRE SUPPRESSION, PLUMBING & HVAC	99.9	70.0	87.1	98.3	65.7	84.4	99.9	70.0	87.1	100.1	70.0	87.2	97.2	90.8	94.5	97.2	90.8	94.5
26, 27, 3370	ELECTRICAL, COMMUNICATIONS & UTIL.	96.7	69.2	82.4	101.6	69.2	84.7	97.0	69.2	82.5	99.3	69.2	83.6	105.2	84.0	94.1	105.1	57.5	80.4
MF2016	WEIGHTED AVERAGE	98.2	72.9	87.2	100.8	71.7	88.1	100.6	72.8	88.5	102.8	72.9	89.7	96.0	95.3	95.7	96.5	91.6	94.3

		VERMONT																
		BRATTLEBORO			BURLINGTON			GUILDHALL			MONTPELIER			RUTLAND			ST. JOHNSBURY		
DIVISION		053			054			059			056			057			058		
		MAT.	INST.	TOTAL	MAT.	INST.	TOTAL	MAT.	INST.	TOTAL	MAT.	INST.	TOTAL	MAT.	INST.	TOTAL	MAT.	INST.	TOTAL
015433	CONTRACTOR EQUIPMENT		98.8	98.8		98.8	98.8		98.8	98.8		98.8	98.8		98.8	98.8		98.8	98.8
0241, 31 - 34	SITE & INFRASTRUCTURE, DEMOLITION	93.1	101.5	99.0	96.9	101.4	100.0	91.1	96.9	95.2	95.4	101.4	99.6	95.6	101.4	99.7	91.2	100.5	97.7
0310	Concrete Forming & Accessories	94.0	106.3	104.6	97.3	82.6	84.6	92.0	99.4	98.4	97.3	105.3	104.2	94.3	82.6	84.2	90.7	99.8	98.6
0320	Concrete Reinforcing	80.0	81.4	80.7	99.4	81.2	90.2	81.6	81.2	81.4	99.4	81.3	90.2	101.1	81.2	91.1	80.0	81.3	80.7
0330	Cast-in-Place Concrete	93.5	115.5	101.9	108.2	114.7	110.7	87.9	106.3	94.9	108.1	114.7	110.6	88.8	114.7	98.6	87.9	106.5	94.9
03	CONCRETE	87.2	104.7	95.2	96.9	93.7	95.4	82.9	98.4	90.0	96.8	104.0	100.1	87.7	93.7	90.4	82.6	98.7	90.0
04	MASONRY	115.2	89.8	99.5	113.8	89.0	98.4	115.6	74.5	90.1	110.1	89.0	97.0	94.8	89.0	91.2	144.6	74.5	101.2
05	METALS	93.4	89.8	92.3	101.1	88.8	97.4	93.4	88.6	91.9	99.1	89.0	96.0	99.2	88.8	96.0	93.4	89.0	92.1
06	WOOD, PLASTICS & COMPOSITES	108.7	111.9	110.5	100.6	81.5	89.9	105.1	111.9	108.9	97.3	111.9	105.5	108.8	81.5	93.6	100.5	111.9	106.9
07	THERMAL & MOISTURE PROTECTION	98.1	92.6	95.8	105.1	88.9	98.2	97.9	85.6	92.7	105.0	92.2	99.6	98.2	88.9	94.3	97.8	85.6	92.6
08	OPENINGS	100.4	101.0	100.5	102.9	80.2	97.7	100.4	97.1	99.6	103.7	97.1	102.2	103.6	80.2	98.2	100.4	97.1	99.6
0920	Plaster & Gypsum Board	105.6	111.9	109.8	105.9	80.6	88.9	111.5	111.9	111.8	105.5	111.9	109.8	106.0	80.6	88.9	112.8	111.9	112.2
0950, 0980	Ceilings & Acoustic Treatment	93.8	111.9	106.0	100.8	80.6	87.1	93.8	111.9	106.0	101.0	111.9	108.3	98.4	80.6	86.4	93.8	111.9	106.0
0960	Flooring	91.2	104.5	94.9	97.3	104.5	99.3	94.0	104.5	96.9	98.4	104.5	100.1	91.1	104.5	94.8	97.1	104.5	99.2
0970, 0990	Wall Finishes & Painting/Coating	86.2	106.1	97.8	95.8	92.4	93.8	86.2	92.4	89.8	92.2	92.4	92.3	86.2	92.4	89.8	86.2	92.4	89.8
09	FINISHES	88.2	107.0	98.4	95.8	87.2	91.1	89.7	101.6	96.2	95.7	105.3	100.9	89.2	87.2	88.1	90.8	101.6	96.7
COVERS	DIVS. 10 - 14, 25, 28, 41, 43, 44, 46	100.0	95.7	99.0	100.0	92.0	98.2	100.0	90.4	97.9	100.0	95.3	99.0	100.0	92.0	98.2	100.0	90.4	97.9
21, 22, 23	FIRE SUPPRESSION, PLUMBING & HVAC	97.2	90.8	94.5	100.0	68.2	86.4	97.2	60.7	81.6	96.9	68.2	84.7	100.3	68.2	86.6	97.2	60.7	81.6
26, 27, 3370	ELECTRICAL, COMMUNICATIONS & UTIL.	105.1	84.0	94.1	105.0	55.6	79.3	105.2	57.5	80.4	104.3	55.6	79.0	105.2	55.6	79.4	105.2	57.5	80.4
MF2016	WEIGHTED AVERAGE	96.9	95.3	96.2	101.0	80.9	92.2	96.4	81.1	89.7	99.8	85.9	93.7	98.1	80.9	90.6	97.8	81.5	90.7

		VERMONT			VIRGINIA													
		WHITE RIVER JCT.			ALEXANDRIA			ARLINGTON			BRISTOL			CHARLOTTESVILLE			CULPEPER		
DIVISION		050			223			222			242			229			227		
		MAT.	INST.	TOTAL	MAT.	INST.	TOTAL	MAT.	INST.	TOTAL	MAT.	INST.	TOTAL	MAT.	INST.	TOTAL	MAT.	INST.	TOTAL
015433	CONTRACTOR EQUIPMENT		98.8	98.8		108.6	108.6		107.3	107.3		107.3	107.3		111.5	111.5		107.3	107.3
0241, 31 - 34	SITE & INFRASTRUCTURE, DEMOLITION	95.7	100.6	99.1	115.4	92.9	99.6	125.6	90.6	101.1	109.8	89.9	95.9	114.2	92.4	99.0	113.1	90.5	97.3
0310	Concrete Forming & Accessories	88.9	100.3	98.8	90.8	71.5	74.1	89.8	71.4	73.9	86.0	66.6	69.5	84.5	48.5	53.5	81.7	71.0	72.5
0320	Concrete Reinforcing	80.9	81.3	81.1	85.2	82.8	84.0	96.2	80.6	88.3	96.2	70.0	83.0	95.5	70.3	82.8	96.2	80.5	88.3
0330	Cast-in-Place Concrete	93.5	107.3	98.8	107.7	78.3	96.5	104.8	78.3	94.7	104.3	46.4	82.3	108.5	78.8	97.2	107.3	78.1	96.2
03	CONCRETE	88.7	99.2	93.5	100.0	77.3	89.6	104.5	76.9	91.9	100.8	62.1	83.1	101.5	65.0	84.8	98.5	76.6	88.5
04	MASONRY	129.7	76.0	96.4	91.2	73.7	80.3	106.3	73.7	86.1	95.3	47.3	65.5	120.2	56.9	80.9	107.2	73.7	86.4
05	METALS	93.4	89.0	92.1	103.4	100.5	102.5	102.0	99.6	101.3	100.8	93.8	98.7	101.1	95.8	99.5	101.2	98.8	100.5
06	WOOD, PLASTICS & COMPOSITES	102.6	111.9	107.8	94.6	69.7	80.7	91.3	69.7	79.2	84.5	73.2	78.2	83.1	41.6	60.0	81.9	69.7	75.1
07	THERMAL & MOISTURE PROTECTION	98.2	86.2	93.1	102.6	81.4	93.6	104.7	80.7	94.5	104.0	60.7	85.7	103.6	69.4	89.1	103.8	80.7	94.0
08	OPENINGS	100.4	97.1	99.6	98.5	74.1	92.9	96.6	73.6	91.3	99.6	67.1	92.2	97.8	55.8	88.2	98.1	73.6	92.5
0920	Plaster & Gypsum Board	102.6	111.9	108.8	103.9	68.7	80.3	100.6	68.7	79.2	97.1	72.3	80.4	97.1	39.1	58.2	97.3	68.7	78.1
0950, 0980	Ceilings & Acoustic Treatment	93.8	111.9	106.0	94.5	68.7	77.1	92.8	68.7	76.5	91.9	72.3	78.7	91.9	39.1	56.3	92.8	68.7	76.5
0960	Flooring	89.3	104.5	93.5	96.3	78.7	91.4	94.7	78.7	90.3	91.2	58.9	82.3	89.7	58.9	81.2	89.7	78.7	86.7
0970, 0990	Wall Finishes & Painting/Coating	86.2	92.4	89.8	114.7	75.0	91.5	114.7	75.0	91.5	101.5	56.6	75.3	101.5	60.0	77.3	114.7	60.0	82.8
09	FINISHES	87.5	102.0	95.4	94.4	72.4	82.4	94.3	72.4	82.4	91.0	65.3	77.1	90.6	49.5	68.3	91.3	70.7	80.1
COVERS	DIVS. 10 - 14, 25, 28, 41, 43, 44, 46	100.0	90.9	98.0	100.0	89.2	97.6	100.0	86.7	97.0	100.0	77.6	95.0	100.0	82.1	96.0	100.0	86.7	97.0
21, 22, 23	FIRE SUPPRESSION, PLUMBING & HVAC	97.2	61.4	81.9	100.4	85.3	93.9	100.4	85.2	93.9	97.3	47.7	76.1	97.3	70.4	85.8	97.3	70.6	85.9
26, 27, 3370	ELECTRICAL, COMMUNICATIONS & UTIL.	105.2	55.6	79.4	98.0	99.6	98.8	95.7	101.2	98.6	97.6	38.4	66.8	97.6	72.2	84.4	99.9	96.9	98.4
MF2016	WEIGHTED AVERAGE	97.7	81.6	90.7	99.8	84.7	93.2	100.7	84.5	93.7	98.6	60.5	82.0	99.8	69.3	86.5	99.1	80.5	91.0

City Cost Indexes

VIRGINIA

DIVISION		FAIRFAX 220-221 MAT.	INST.	TOTAL	FARMVILLE 239 MAT.	INST.	TOTAL	FREDERICKSBURG 224-225 MAT.	INST.	TOTAL	GRUNDY 246 MAT.	INST.	TOTAL	HARRISONBURG 228 MAT.	INST.	TOTAL	LYNCHBURG 245 MAT.	INST.	TOTAL
015433	CONTRACTOR EQUIPMENT		107.3	107.3		111.5	111.5		107.3	107.3		107.3	107.3		107.3	107.3		107.3	107.3
0241, 31 - 34	SITE & INFRASTRUCTURE, DEMOLITION	124.3	90.6	100.7	108.4	91.4	96.5	112.7	90.4	97.1	107.6	89.2	94.7	121.3	89.0	98.6	108.5	90.6	96.0
0310	Concrete Forming & Accessories	84.5	71.3	73.1	95.5	45.6	52.5	84.5	68.5	70.7	89.1	38.7	45.6	80.7	39.4	45.1	86.1	73.0	74.8
0320	Concrete Reinforcing	96.2	80.6	88.3	94.9	69.9	82.3	96.9	82.7	89.8	94.9	45.4	69.9	96.2	59.1	77.5	95.5	70.8	83.1
0330	Cast-in-Place Concrete	104.8	78.2	94.7	106.5	88.0	99.5	106.4	78.1	95.6	104.3	52.0	84.5	104.7	55.0	85.8	104.3	77.6	94.2
03	CONCRETE	104.2	76.8	91.7	97.2	66.6	83.2	98.3	75.9	88.0	99.4	46.6	75.3	102.0	50.6	78.5	99.3	75.7	88.5
04	MASONRY	106.2	73.7	86.0	104.3	52.1	71.9	106.3	69.6	83.6	96.1	55.7	71.1	104.2	60.3	77.0	111.7	65.8	83.2
05	METALS	101.3	99.5	100.7	98.0	91.8	96.1	101.2	100.3	100.9	100.8	76.1	93.2	101.1	88.8	97.3	101.0	95.9	99.5
06	WOOD, PLASTICS & COMPOSITES	84.5	69.7	76.2	93.8	40.2	64.0	84.5	67.7	75.1	87.1	31.9	56.4	81.0	35.5	55.7	84.5	74.6	79.0
07	THERMAL & MOISTURE PROTECTION	104.4	76.4	92.6	104.2	56.0	83.8	103.8	78.7	93.1	103.9	48.3	80.4	104.2	64.2	87.3	103.8	72.6	90.6
08	OPENINGS	96.6	73.6	91.3	97.6	45.3	85.6	97.8	72.0	91.9	99.6	31.6	84.0	98.1	45.3	86.0	98.1	67.9	91.2
0920	Plaster & Gypsum Board	97.3	68.7	78.1	106.4	37.7	60.3	97.3	66.7	76.7	97.1	29.9	52.0	97.1	33.6	54.5	97.1	73.7	81.4
0950, 0980	Ceilings & Acoustic Treatment	92.8	68.7	76.5	87.6	37.7	53.9	92.8	66.7	75.2	91.9	29.9	50.1	91.9	33.6	52.6	91.9	73.7	79.7
0960	Flooring	91.5	78.7	88.0	92.0	53.4	81.3	91.5	76.5	87.4	92.5	29.8	75.1	89.5	78.7	86.5	91.2	69.1	85.1
0970, 0990	Wall Finishes & Painting/Coating	114.7	75.0	91.5	99.7	59.8	76.4	114.7	59.6	82.6	101.5	32.7	61.3	114.7	50.5	77.3	101.5	56.6	75.3
09	FINISHES	92.9	72.4	81.7	90.5	47.3	67.0	91.8	68.3	79.0	91.2	35.0	60.7	91.7	46.2	67.0	90.8	70.6	79.9
COVERS	DIVS. 10 - 14, 25, 28, 41, 43, 44, 46	100.0	86.7	97.0	100.0	77.7	95.0	100.0	81.5	95.9	100.0	75.8	94.6	100.0	62.5	91.6	100.0	81.2	95.8
21, 22, 23	FIRE SUPPRESSION, PLUMBING & HVAC	97.3	85.2	92.1	97.4	46.9	75.8	97.3	83.2	91.3	97.3	67.4	84.5	97.3	68.1	84.8	97.3	70.7	85.9
26, 27, 3370	ELECTRICAL, COMMUNICATIONS & UTIL.	98.7	96.9	97.8	92.2	45.1	67.7	95.9	96.9	96.4	97.6	41.7	68.5	97.8	91.2	94.4	98.6	71.6	84.6
MF2016	WEIGHTED AVERAGE	100.0	83.8	92.9	97.4	58.6	80.4	98.7	82.2	91.5	98.5	56.0	79.9	99.5	67.3	85.4	99.2	75.2	88.7

VIRGINIA

DIVISION		NEWPORT NEWS 236 MAT.	INST.	TOTAL	NORFOLK 233-235 MAT.	INST.	TOTAL	PETERSBURG 238 MAT.	INST.	TOTAL	PORTSMOUTH 237 MAT.	INST.	TOTAL	PULASKI 243 MAT.	INST.	TOTAL	RICHMOND 230-232 MAT.	INST.	TOTAL
015433	CONTRACTOR EQUIPMENT		111.6	111.6		112.2	112.2		111.5	111.5		111.5	111.5		107.3	107.3		111.5	111.5
0241, 31 - 34	SITE & INFRASTRUCTURE, DEMOLITION	107.3	92.4	96.9	105.7	93.5	97.1	111.0	92.4	97.9	105.9	91.7	95.9	106.9	89.7	94.8	102.2	92.4	95.3
0310	Concrete Forming & Accessories	94.8	62.5	67.0	100.4	62.6	67.8	88.8	63.7	67.2	85.0	49.0	53.9	89.1	41.4	48.0	95.2	82.3	84.1
0320	Concrete Reinforcing	94.7	67.5	81.0	102.5	67.5	84.8	94.3	70.9	82.5	94.3	66.8	80.5	94.9	58.2	76.4	105.2	70.9	87.9
0330	Cast-in-Place Concrete	103.5	78.0	93.8	107.5	79.1	96.7	109.9	79.3	98.3	102.5	62.6	87.3	104.3	86.4	97.5	98.4	79.3	91.1
03	CONCRETE	94.5	70.5	83.6	96.5	71.0	84.8	99.7	72.1	87.1	93.4	59.0	77.7	99.4	62.2	82.4	92.2	80.6	86.9
04	MASONRY	98.6	64.5	77.4	96.6	64.5	76.7	112.9	64.5	82.9	104.6	49.4	70.4	90.8	56.6	69.6	97.3	64.5	77.0
05	METALS	100.3	95.1	98.7	102.0	95.2	99.9	98.0	96.8	97.7	99.3	93.6	97.5	100.9	87.4	96.7	102.0	97.0	100.5
06	WOOD, PLASTICS & COMPOSITES	92.8	61.4	75.3	96.1	61.4	76.8	85.2	62.4	72.5	82.0	46.9	62.5	87.1	33.5	57.3	95.9	87.4	91.2
07	THERMAL & MOISTURE PROTECTION	104.1	71.6	90.4	101.4	71.8	88.9	104.1	73.6	91.2	104.2	62.1	86.4	103.9	59.0	84.9	102.5	76.4	91.4
08	OPENINGS	98.0	60.3	89.3	96.2	61.6	88.2	97.3	67.3	90.4	98.0	49.9	87.0	99.7	38.3	85.6	100.2	81.1	95.8
0920	Plaster & Gypsum Board	107.4	59.5	75.2	100.3	59.5	72.9	100.7	60.5	73.7	101.2	44.6	63.2	97.1	31.5	53.1	101.4	86.2	91.2
0950, 0980	Ceilings & Acoustic Treatment	91.8	59.5	70.0	95.4	59.5	71.2	88.4	60.5	69.6	91.8	44.6	60.0	91.9	31.5	51.2	91.4	86.2	87.9
0960	Flooring	92.0	69.1	85.7	87.7	69.1	82.5	88.0	74.3	84.2	85.0	58.9	77.8	92.5	58.9	83.2	87.3	74.3	83.7
0970, 0990	Wall Finishes & Painting/Coating	99.7	60.0	76.5	97.4	60.0	75.6	99.7	60.0	76.5	99.7	60.0	76.5	101.5	50.5	71.8	92.7	60.0	73.6
09	FINISHES	91.3	63.0	75.9	90.7	63.0	75.6	88.9	64.6	75.7	88.4	50.5	67.8	91.2	42.9	64.9	91.0	79.4	84.7
COVERS	DIVS. 10 - 14, 25, 28, 41, 43, 44, 46	100.0	80.5	95.7	100.0	80.5	95.7	100.0	83.9	96.4	100.0	66.5	92.6	100.0	76.0	94.7	100.0	86.7	97.0
21, 22, 23	FIRE SUPPRESSION, PLUMBING & HVAC	100.5	64.0	84.9	100.1	67.3	86.1	97.4	69.8	85.6	100.5	64.3	85.1	97.3	66.2	84.0	100.1	69.8	87.1
26, 27, 3370	ELECTRICAL, COMMUNICATIONS & UTIL.	94.7	63.7	78.6	97.9	63.8	80.2	94.8	72.2	83.1	93.1	63.8	77.9	97.6	56.7	76.3	98.3	72.2	84.7
MF2016	WEIGHTED AVERAGE	98.2	70.4	86.0	98.5	71.3	86.6	98.3	73.9	87.6	97.7	64.3	83.1	98.2	62.7	82.7	98.5	77.9	89.5

VIRGINIA / WASHINGTON

DIVISION		ROANOKE 240-241 MAT.	INST.	TOTAL	STAUNTON 244 MAT.	INST.	TOTAL	WINCHESTER 226 MAT.	INST.	TOTAL	CLARKSTON 994 MAT.	INST.	TOTAL	EVERETT 982 MAT.	INST.	TOTAL	OLYMPIA 985 MAT.	INST.	TOTAL
015433	CONTRACTOR EQUIPMENT		107.2	107.2		111.5	111.5		107.3	107.3		91.5	91.5		102.2	102.2		102.2	102.2
0241, 31 - 34	SITE & INFRASTRUCTURE, DEMOLITION	107.5	90.6	95.7	110.7	91.3	97.1	119.9	90.5	99.3	107.5	89.6	94.9	92.9	110.9	105.5	93.0	110.9	105.5
0310	Concrete Forming & Accessories	95.2	73.1	76.1	88.8	50.2	55.5	82.9	69.5	71.4	111.4	67.0	73.1	116.2	102.6	104.5	105.3	102.4	102.8
0320	Concrete Reinforcing	95.9	70.8	83.2	95.5	70.0	82.7	95.6	82.8	89.1	112.0	98.3	105.1	108.9	111.8	110.4	113.8	111.8	112.8
0330	Cast-in-Place Concrete	118.5	87.8	106.8	108.5	88.6	100.9	104.7	67.1	90.5	87.8	84.1	86.4	98.6	109.1	102.6	90.8	109.0	97.7
03	CONCRETE	103.9	79.3	92.6	100.8	68.9	86.2	101.5	72.6	88.3	95.4	78.7	87.8	90.1	105.9	97.3	86.8	105.8	95.5
04	MASONRY	97.4	68.5	77.8	107.0	55.6	75.1	101.6	72.7	83.7	100.8	84.8	90.9	112.7	103.6	107.1	107.1	103.6	104.9
05	METALS	103.2	96.0	101.0	101.1	91.6	98.2	101.2	100.1	100.9	91.3	87.6	90.1	103.3	96.7	101.3	103.0	96.3	101.0
06	WOOD, PLASTICS & COMPOSITES	95.4	74.6	83.8	87.1	45.5	64.0	83.1	67.9	74.6	107.4	61.2	81.7	112.0	101.6	106.2	97.0	101.6	99.6
07	THERMAL & MOISTURE PROTECTION	103.7	76.1	92.0	103.5	57.5	84.1	104.3	78.3	93.3	158.3	81.7	125.8	111.9	105.8	109.3	111.6	101.9	107.5
08	OPENINGS	98.5	67.9	91.5	98.1	48.1	86.6	99.7	71.8	93.3	118.1	68.3	106.7	106.1	104.0	105.6	110.5	104.3	109.0
0920	Plaster & Gypsum Board	103.9	73.7	83.6	97.1	43.2	60.9	97.3	66.9	76.9	149.2	60.1	89.3	112.7	101.8	105.4	107.0	101.8	103.5
0950, 0980	Ceilings & Acoustic Treatment	94.5	73.7	80.5	91.9	43.2	59.1	92.8	66.9	75.3	108.8	60.1	75.9	105.7	101.8	103.1	105.7	101.8	103.1
0960	Flooring	96.3	69.1	88.8	92.1	35.0	76.2	91.0	78.7	87.6	86.7	81.9	85.3	105.2	101.5	104.2	93.1	101.5	95.4
0970, 0990	Wall Finishes & Painting/Coating	101.5	56.6	75.3	101.5	32.1	61.0	114.7	81.1	95.1	78.6	75.4	76.7	91.5	95.7	93.9	87.1	95.7	92.1
09	FINISHES	93.4	72.9	82.1	91.1	44.5	65.8	92.3	71.8	81.1	105.8	69.0	85.8	103.3	101.3	102.2	94.0	101.3	98.0
COVERS	DIVS. 10 - 14, 25, 28, 41, 43, 44, 46	100.0	81.1	95.8	100.0	78.8	95.3	100.0	86.1	96.9	100.0	92.9	98.4	100.0	100.0	100.0	100.0	102.5	100.6
21, 22, 23	FIRE SUPPRESSION, PLUMBING & HVAC	100.4	66.4	85.9	97.3	60.1	81.4	97.3	84.7	91.9	99.7	83.9	91.8	100.2	101.8	100.9	100.1	101.8	100.8
26, 27, 3370	ELECTRICAL, COMMUNICATIONS & UTIL.	97.6	57.8	76.9	96.6	72.2	83.9	96.3	96.9	96.6	86.6	98.0	92.6	108.2	98.4	103.1	106.5	98.3	102.2
MF2016	WEIGHTED AVERAGE	100.3	73.0	88.3	99.0	65.8	84.5	99.3	82.9	92.2	100.6	83.4	93.1	102.0	102.4	102.2	100.7	102.3	101.4

For customer support on your Building Construction Costs with RSMeans data, call 800.448.8182.

WASHINGTON

DIVISION		RICHLAND 993 MAT.	INST.	TOTAL	SEATTLE 980 - 981, 987 MAT.	INST.	TOTAL	SPOKANE 990 - 992 MAT.	INST.	TOTAL	TACOMA 983 - 984 MAT.	INST.	TOTAL	VANCOUVER 986 MAT.	INST.	TOTAL	WENATCHEE 988 MAT.	INST.	TOTAL
015433	CONTRACTOR EQUIPMENT		91.5	91.5		103.7	103.7		91.5	91.5		102.2	102.2		98.4	98.4		102.2	102.2
0241, 31 - 34	SITE & INFRASTRUCTURE, DEMOLITION	109.4	89.7	95.6	99.3	110.1	106.9	108.8	89.7	95.4	96.1	110.9	106.5	106.0	97.8	100.3	105.3	108.3	107.4
0310	Concrete Forming & Accessories	111.5	80.0	84.4	112.3	102.9	104.2	116.8	79.6	84.7	106.5	102.5	103.0	107.1	94.7	96.4	108.4	78.9	82.9
0320	Concrete Reinforcing	107.4	98.4	102.9	114.0	111.9	112.9	108.1	98.3	103.2	107.4	111.8	109.6	108.5	110.9	109.7	108.5	98.4	103.4
0330	Cast-in-Place Concrete	88.0	85.5	87.0	102.5	108.8	104.9	91.4	85.3	89.1	101.3	109.0	104.3	113.2	100.7	108.4	103.5	93.8	99.8
03	CONCRETE	95.0	85.1	90.5	98.2	106.2	101.9	97.1	84.8	91.5	91.7	105.8	98.2	101.1	99.4	100.3	98.8	87.4	93.6
04	MASONRY	101.5	86.8	92.4	119.1	103.6	109.5	102.0	86.8	92.6	112.2	103.6	106.9	111.7	87.8	96.9	115.6	84.4	96.3
05	METALS	91.7	88.2	90.6	103.7	99.9	102.5	94.0	87.8	92.1	105.1	96.3	102.4	102.7	96.4	100.8	102.5	87.9	98.0
06	WOOD, PLASTICS & COMPOSITES	107.6	77.0	90.6	107.3	101.6	104.1	116.2	77.0	94.4	101.5	101.6	101.6	93.0	93.9	93.5	103.2	76.8	88.5
07	THERMAL & MOISTURE PROTECTION	159.4	85.7	128.2	107.1	105.2	106.3	155.6	87.1	126.6	111.6	102.0	107.6	112.0	97.3	105.8	111.1	87.9	101.3
08	OPENINGS	116.1	75.9	106.9	106.5	104.9	106.1	116.7	79.5	108.2	106.9	104.3	106.3	103.2	98.7	102.2	106.4	76.7	99.5
0920	Plaster & Gypsum Board	149.2	76.3	100.2	108.6	101.8	104.0	139.6	76.3	97.1	108.8	101.8	104.1	106.9	94.1	98.3	111.7	76.3	87.9
0950, 0980	Ceilings & Acoustic Treatment	115.5	76.3	89.1	112.0	101.8	105.1	110.7	76.3	87.5	109.1	101.8	104.2	104.2	94.1	97.4	101.3	76.3	84.4
0960	Flooring	87.0	81.9	85.6	103.6	101.5	103.0	86.0	81.9	84.9	97.9	101.5	98.9	103.6	83.6	98.1	100.9	81.9	95.6
0970, 0990	Wall Finishes & Painting/Coating	78.6	79.9	79.3	105.4	95.7	99.7	78.7	79.9	79.4	91.5	95.7	93.9	94.0	75.2	83.0	91.5	74.1	81.4
09	FINISHES	107.4	79.3	92.1	106.9	101.3	103.9	105.0	79.3	91.0	101.4	101.3	101.3	99.2	90.1	94.2	102.1	77.9	88.9
COVERS	DIVS. 10 - 14, 25, 28, 41, 43, 44, 46	100.0	95.3	99.0	100.0	102.6	100.6	100.0	95.2	98.9	100.0	102.5	100.6	100.0	98.0	99.6	100.0	94.0	98.7
21, 22, 23	FIRE SUPPRESSION, PLUMBING & HVAC	100.8	112.6	105.9	100.0	113.5	105.8	100.8	86.6	94.7	100.2	101.8	100.9	100.3	101.5	100.8	97.1	94.2	95.8
26, 27, 3370	ELECTRICAL, COMMUNICATIONS & UTIL.	84.6	98.0	91.6	107.4	113.1	110.4	83.0	81.4	82.1	108.0	98.3	103.0	113.8	101.6	107.5	108.8	93.3	100.7
MF2016	WEIGHTED AVERAGE	101.1	92.7	97.4	103.6	107.3	105.2	101.4	85.0	94.2	102.4	102.3	102.3	103.3	97.2	100.6	102.6	89.5	96.9

DIVISION		WASHINGTON YAKIMA 989 MAT.	INST.	TOTAL	WEST VIRGINIA BECKLEY 258 - 259 MAT.	INST.	TOTAL	BLUEFIELD 247 - 248 MAT.	INST.	TOTAL	BUCKHANNON 262 MAT.	INST.	TOTAL	CHARLESTON 250 - 253 MAT.	INST.	TOTAL	CLARKSBURG 263 - 264 MAT.	INST.	TOTAL
015433	CONTRACTOR EQUIPMENT		102.2	102.2		107.3	107.3		107.3	107.3		107.3	107.3		107.3	107.3		107.3	107.3
0241, 31 - 34	SITE & INFRASTRUCTURE, DEMOLITION	98.6	109.4	106.2	101.5	91.8	94.7	101.5	91.8	94.7	107.7	91.7	96.5	100.7	92.8	95.2	108.5	91.7	96.7
0310	Concrete Forming & Accessories	107.0	98.0	99.2	86.9	90.2	89.7	86.0	89.9	89.3	85.5	86.4	86.3	97.7	91.3	92.2	83.0	86.4	85.9
0320	Concrete Reinforcing	108.0	98.5	103.2	91.8	89.4	90.6	94.5	70.8	82.6	95.1	94.3	94.7	99.3	89.7	94.4	95.1	97.3	96.2
0330	Cast-in-Place Concrete	108.2	84.6	99.3	104.4	93.1	100.1	102.0	92.9	98.6	101.7	93.4	98.5	101.0	94.5	98.5	111.5	91.6	103.9
03	CONCRETE	96.2	92.9	94.7	95.9	92.0	94.1	95.4	88.7	92.3	98.3	91.3	95.1	94.5	93.1	93.8	102.4	91.2	97.3
04	MASONRY	105.1	84.5	92.3	93.0	88.8	90.4	92.5	88.8	90.2	104.5	86.5	93.3	88.7	94.2	92.1	108.4	86.5	94.8
05	METALS	103.2	88.7	98.7	102.8	104.5	103.3	101.1	97.9	100.1	101.3	106.1	102.8	101.4	104.9	102.5	101.3	107.1	103.1
06	WOOD, PLASTICS & COMPOSITES	101.9	101.6	101.8	85.1	90.5	88.1	86.3	90.5	88.6	85.6	85.9	85.8	92.3	90.5	91.3	82.3	85.9	84.3
07	THERMAL & MOISTURE PROTECTION	111.7	88.9	102.1	106.5	88.6	98.9	103.7	88.6	97.3	104.0	87.6	97.1	103.1	90.6	97.8	103.9	87.3	96.9
08	OPENINGS	106.3	101.3	105.2	96.3	85.6	93.8	100.1	81.3	95.7	100.1	84.1	96.4	94.4	85.6	92.4	100.1	84.8	96.6
0920	Plaster & Gypsum Board	108.4	101.8	103.9	98.9	90.1	93.0	96.4	90.1	92.1	96.8	85.4	89.1	99.9	90.1	93.3	94.1	85.4	88.3
0950, 0980	Ceilings & Acoustic Treatment	103.2	101.8	102.2	84.1	90.1	88.1	90.2	90.1	90.1	91.9	85.4	87.5	95.2	90.1	91.7	91.9	85.4	87.5
0960	Flooring	98.8	81.9	94.1	91.6	100.6	94.1	88.9	100.6	92.2	88.7	96.5	90.8	94.0	100.6	95.8	87.6	96.5	90.1
0970, 0990	Wall Finishes & Painting/Coating	91.5	79.9	84.7	92.6	91.7	92.1	101.5	91.7	95.8	101.5	87.9	93.6	88.2	91.7	90.3	101.5	87.9	93.6
09	FINISHES	100.6	93.2	96.6	88.2	92.6	90.6	89.3	92.6	91.1	90.1	88.3	89.1	93.4	93.2	93.3	89.4	88.3	88.8
COVERS	DIVS. 10 - 14, 25, 28, 41, 43, 44, 46	100.0	99.3	99.8	100.0	93.8	98.6	100.0	93.8	98.6	100.0	92.4	98.3	100.0	94.1	98.7	100.0	92.4	98.3
21, 22, 23	FIRE SUPPRESSION, PLUMBING & HVAC	100.2	111.8	105.1	97.5	91.4	94.9	97.3	87.1	92.9	97.3	91.2	94.7	100.1	92.8	97.0	97.3	91.1	94.7
26, 27, 3370	ELECTRICAL, COMMUNICATIONS & UTIL.	111.0	98.1	104.3	94.6	87.0	90.7	96.6	87.0	91.6	98.0	91.9	94.8	99.7	88.8	94.0	98.0	91.9	94.8
MF2016	WEIGHTED AVERAGE	102.5	98.2	100.6	97.3	91.8	94.8	97.5	89.6	94.0	98.8	91.5	95.6	97.9	93.3	95.9	99.4	91.6	96.0

WEST VIRGINIA

DIVISION		GASSAWAY 266 MAT.	INST.	TOTAL	HUNTINGTON 255 - 257 MAT.	INST.	TOTAL	LEWISBURG 249 MAT.	INST.	TOTAL	MARTINSBURG 254 MAT.	INST.	TOTAL	MORGANTOWN 265 MAT.	INST.	TOTAL	PARKERSBURG 261 MAT.	INST.	TOTAL
015433	CONTRACTOR EQUIPMENT		107.3	107.3		107.3	107.3		107.3	107.3		107.3	107.3		107.3	107.3		107.3	107.3
0241, 31 - 34	SITE & INFRASTRUCTURE, DEMOLITION	105.2	91.7	95.7	106.2	92.8	96.8	117.2	91.8	99.4	105.5	92.5	96.4	102.5	92.6	95.5	110.7	92.7	98.1
0310	Concrete Forming & Accessories	84.9	86.3	86.1	98.6	91.0	92.1	83.3	89.5	88.6	87.0	77.7	79.0	83.3	86.5	86.1	87.6	86.8	86.9
0320	Concrete Reinforcing	95.1	94.1	94.6	93.1	93.6	93.3	95.1	70.6	82.7	91.8	94.2	93.0	95.1	97.3	96.2	94.5	94.3	94.4
0330	Cast-in-Place Concrete	106.5	92.4	101.1	113.9	98.4	108.0	102.1	92.8	98.6	109.4	86.9	100.8	101.7	93.4	98.5	104.0	93.5	100.0
03	CONCRETE	98.7	90.9	95.1	101.2	95.0	98.3	105.0	88.4	97.4	99.6	84.8	92.8	95.1	91.9	93.6	100.4	91.5	96.3
04	MASONRY	109.4	84.8	94.1	91.9	96.7	94.9	95.5	85.9	89.6	94.6	87.6	90.2	126.6	86.5	101.7	82.0	88.8	86.2
05	METALS	101.2	86.7	96.7	105.3	106.6	105.7	101.2	97.3	100.0	103.2	99.5	102.1	101.3	107.2	103.1	102.0	106.1	103.2
06	WOOD, PLASTICS & COMPOSITES	84.9	85.9	85.4	95.8	90.0	92.6	82.6	90.5	87.0	85.1	75.3	79.6	82.6	85.9	84.4	85.7	85.4	85.6
07	THERMAL & MOISTURE PROTECTION	103.7	87.0	96.6	106.6	91.7	100.3	104.6	87.8	97.5	106.7	83.0	96.7	103.8	87.6	96.9	103.8	88.4	97.3
08	OPENINGS	98.2	84.1	95.0	95.6	86.2	93.4	100.1	81.3	95.8	98.2	68.9	91.4	101.4	84.8	97.6	99.0	83.9	95.5
0920	Plaster & Gypsum Board	96.1	85.4	88.9	106.2	89.6	95.1	94.1	90.1	91.4	99.3	74.4	82.6	94.1	85.4	88.3	97.1	84.9	88.9
0950, 0980	Ceilings & Acoustic Treatment	91.9	85.4	87.5	85.9	89.6	88.4	91.9	90.1	90.7	85.9	74.4	78.2	91.9	85.4	87.5	91.9	84.9	87.2
0960	Flooring	88.4	100.6	91.8	99.0	109.2	101.8	87.7	100.6	91.3	91.6	96.6	93.0	87.7	96.5	90.2	91.6	97.2	93.2
0970, 0990	Wall Finishes & Painting/Coating	101.5	91.7	95.8	92.6	91.7	92.1	101.5	67.4	81.6	92.6	87.5	89.7	101.5	87.9	93.6	101.5	87.5	93.3
09	FINISHES	89.7	89.6	89.6	91.8	94.6	93.3	90.4	89.9	90.2	88.8	81.6	84.9	89.0	88.3	88.6	91.1	88.6	89.7
COVERS	DIVS. 10 - 14, 25, 28, 41, 43, 44, 46	100.0	92.4	98.3	100.0	93.9	98.6	100.0	67.7	92.8	100.0	84.0	96.4	100.0	86.7	97.0	100.0	93.0	98.4
21, 22, 23	FIRE SUPPRESSION, PLUMBING & HVAC	97.3	86.5	92.7	100.6	89.8	96.0	97.3	91.2	94.7	97.5	84.0	91.7	97.3	91.2	94.7	100.3	89.2	95.6
26, 27, 3370	ELECTRICAL, COMMUNICATIONS & UTIL.	98.0	88.8	93.2	98.2	91.0	94.4	94.2	87.0	90.5	100.2	76.0	87.6	98.1	91.9	94.9	98.1	78.3	87.8
MF2016	WEIGHTED AVERAGE	98.8	89.9	94.9	99.7	93.8	97.1	99.1	88.9	94.6	98.8	84.4	92.5	99.4	91.6	96.0	98.8	89.5	94.8

City Cost Indexes

		WEST VIRGINIA									WISCONSIN								
		PETERSBURG 268			ROMNEY 267			WHEELING 260			BELOIT 535			EAU CLAIRE 547			GREEN BAY 541 - 543		
DIVISION		MAT.	INST.	TOTAL	MAT.	INST.	TOTAL	MAT.	INST.	TOTAL	MAT.	INST.	TOTAL	MAT.	INST.	TOTAL	MAT.	INST.	TOTAL
015433	CONTRACTOR EQUIPMENT		107.3	107.3		107.3	107.3		107.3	107.3		101.9	101.9		102.7	102.7		100.3	100.3
0241, 31 - 34	SITE & INFRASTRUCTURE, DEMOLITION	101.7	92.5	95.3	104.6	92.6	96.2	111.4	92.6	98.2	95.8	106.6	103.4	97.0	104.3	102.1	100.3	99.8	99.9
0310	Concrete Forming & Accessories	86.4	77.8	79.0	82.5	78.2	78.8	89.1	86.8	87.1	96.4	92.3	92.9	95.0	105.8	104.4	103.6	105.6	105.3
0320	Concrete Reinforcing	94.5	94.1	94.3	95.1	94.3	94.7	93.9	97.3	95.6	95.5	131.6	113.7	91.4	110.6	101.1	89.4	106.3	97.9
0330	Cast-in-Place Concrete	101.7	87.0	96.1	106.5	87.1	99.1	104.0	96.6	101.2	106.4	99.1	103.6	100.2	99.8	100.0	103.6	103.8	103.7
03	CONCRETE	95.2	85.2	90.6	98.6	85.4	92.6	100.4	93.1	97.1	96.1	101.7	98.7	91.8	104.5	97.6	94.7	105.1	99.5
04	MASONRY	99.5	86.5	91.4	95.8	87.6	90.7	107.6	87.0	94.8	100.3	102.3	101.6	92.0	98.7	96.1	123.3	99.7	108.6
05	METALS	101.4	105.4	102.6	101.4	105.9	102.8	102.1	107.3	103.7	96.5	110.6	100.8	96.3	104.5	98.8	98.9	103.9	100.4
06	WOOD, PLASTICS & COMPOSITES	86.5	75.3	80.3	81.8	75.3	78.2	87.1	85.9	86.4	93.5	88.8	90.9	98.3	107.8	103.6	103.0	107.8	105.7
07	THERMAL & MOISTURE PROTECTION	103.8	82.9	94.9	103.9	83.9	95.4	104.2	88.8	97.7	100.7	94.0	97.3	103.5	98.9	101.6	105.8	100.8	103.7
08	OPENINGS	101.4	78.2	96.1	101.3	78.2	96.0	99.8	84.8	96.4	96.3	106.1	98.5	101.8	107.2	103.1	98.2	107.0	100.2
0920	Plaster & Gypsum Board	96.8	74.4	81.8	93.8	74.4	80.8	97.1	85.4	89.2	97.3	88.9	91.7	110.1	108.4	109.0	106.9	108.4	107.9
0950, 0980	Ceilings & Acoustic Treatment	91.9	74.4	80.1	91.9	74.4	80.1	91.9	85.4	87.5	86.3	88.9	88.1	90.3	108.4	102.5	84.6	108.4	100.7
0960	Flooring	89.5	96.5	91.4	87.5	98.5	90.6	92.5	96.5	93.6	91.3	125.2	100.7	80.8	112.6	89.6	96.5	121.4	103.4
0970, 0990	Wall Finishes & Painting/Coating	101.5	87.9	93.6	101.5	87.9	93.6	101.5	87.9	93.6	96.8	104.5	101.3	84.9	79.9	82.0	94.0	83.6	88.0
09	FINISHES	89.8	82.0	85.6	89.1	82.4	85.5	91.4	88.5	89.8	92.0	99.0	95.8	87.5	105.0	97.0	91.9	107.0	100.1
COVERS	DIVS. 10 - 14, 25, 28, 41, 43, 44, 46	100.0	55.1	90.0	100.0	91.2	98.0	100.0	87.8	97.3	100.0	94.1	98.7	100.0	96.4	99.2	100.0	96.3	99.2
21, 22, 23	FIRE SUPPRESSION, PLUMBING & HVAC	97.3	86.7	92.8	97.3	86.6	92.7	100.4	91.4	96.6	100.1	95.8	98.3	100.1	87.3	94.7	100.4	85.6	94.1
26, 27, 3370	ELECTRICAL, COMMUNICATIONS & UTIL.	101.1	76.0	88.0	100.4	76.0	87.7	95.4	91.9	93.6	99.3	85.2	91.9	104.0	84.8	94.0	98.6	81.3	89.6
MF2016	WEIGHTED AVERAGE	98.5	85.0	92.6	98.6	86.4	93.3	100.0	91.9	96.5	97.7	98.0	98.1	97.6	97.3	97.5	99.6	96.5	98.2

		WISCONSIN																	
		KENOSHA 531			LA CROSSE 546			LANCASTER 538			MADISON 537			MILWAUKEE 530, 532			NEW RICHMOND 540		
DIVISION		MAT.	INST.	TOTAL	MAT.	INST.	TOTAL	MAT.	INST.	TOTAL	MAT.	INST.	TOTAL	MAT.	INST.	TOTAL	MAT.	INST.	TOTAL
015433	CONTRACTOR EQUIPMENT		99.8	99.8		102.7	102.7		101.9	101.9		101.9	101.9		89.4	89.4		103.0	103.0
0241, 31 - 34	SITE & INFRASTRUCTURE, DEMOLITION	101.3	104.1	103.3	90.7	104.3	100.2	94.9	106.4	103.0	93.9	107.3	103.3	93.2	95.3	94.6	95.9	104.4	101.9
0310	Concrete Forming & Accessories	104.3	111.2	110.3	82.6	105.7	102.5	95.8	96.3	96.2	100.5	95.7	96.3	100.8	112.4	110.8	90.5	93.1	92.8
0320	Concrete Reinforcing	95.3	113.9	104.7	91.1	103.0	97.1	97.0	102.8	99.9	95.7	103.2	99.5	102.9	114.2	108.6	88.4	110.2	99.4
0330	Cast-in-Place Concrete	115.7	108.3	112.9	90.1	101.6	94.5	105.8	98.7	103.1	102.4	104.8	103.3	93.9	110.9	100.4	104.3	79.8	95.0
03	CONCRETE	101.0	110.4	105.4	83.4	103.8	92.7	95.9	98.4	97.0	93.7	100.3	96.7	95.7	111.2	102.8	89.8	91.9	90.7
04	MASONRY	97.8	110.7	105.8	91.1	98.7	95.8	100.3	95.9	97.6	95.6	99.6	98.1	105.8	114.7	111.3	118.1	95.7	104.2
05	METALS	97.4	105.5	99.9	96.2	102.5	98.1	93.9	99.3	95.5	98.3	100.6	99.0	97.9	97.4	97.8	96.6	103.5	98.7
06	WOOD, PLASTICS & COMPOSITES	97.3	110.8	104.8	84.3	107.8	97.4	92.8	95.7	94.4	93.9	93.4	93.6	98.8	111.3	105.7	87.8	91.7	90.0
07	THERMAL & MOISTURE PROTECTION	100.9	107.2	103.6	103.0	98.7	101.1	100.5	93.2	97.4	98.3	101.2	99.5	102.8	109.9	105.8	104.6	88.3	97.7
08	OPENINGS	90.9	110.8	95.5	101.8	100.2	101.4	92.3	93.5	92.5	102.3	98.3	101.3	101.0	111.1	103.3	87.9	91.8	88.8
0920	Plaster & Gypsum Board	87.3	111.6	103.6	105.0	108.4	107.3	96.5	96.1	96.2	104.3	93.6	97.1	100.3	111.6	107.9	95.6	92.0	93.2
0950, 0980	Ceilings & Acoustic Treatment	86.3	111.6	103.4	89.5	108.4	102.2	82.9	96.1	91.8	89.6	93.6	92.3	96.5	111.6	106.7	57.0	92.0	80.6
0960	Flooring	108.3	121.4	112.0	74.8	118.0	86.8	90.8	105.3	94.8	89.5	113.7	96.2	96.6	117.4	102.3	86.9	109.8	95.2
0970, 0990	Wall Finishes & Painting/Coating	108.0	118.8	114.3	84.9	81.2	82.8	96.8	102.1	99.9	90.1	104.5	98.5	106.8	123.4	116.5	96.2	85.4	89.9
09	FINISHES	96.8	114.3	106.3	84.3	106.3	96.3	91.1	99.0	95.4	91.7	99.6	96.0	100.5	114.7	108.2	82.4	95.7	89.7
COVERS	DIVS. 10 - 14, 25, 28, 41, 43, 44, 46	100.0	101.5	100.3	100.0	96.4	99.2	100.0	86.7	97.0	100.0	99.7	99.9	100.0	102.9	100.6	100.0	88.9	97.5
21, 22, 23	FIRE SUPPRESSION, PLUMBING & HVAC	100.3	97.0	98.9	100.1	87.3	94.6	97.0	87.1	92.8	100.0	96.7	98.6	99.9	105.5	102.3	96.5	87.4	92.6
26, 27, 3370	ELECTRICAL, COMMUNICATIONS & UTIL.	99.6	97.3	98.4	104.3	84.8	94.2	99.0	84.8	91.6	100.0	94.1	96.9	99.1	101.5	100.4	102.1	84.8	93.1
MF2016	WEIGHTED AVERAGE	98.4	105.0	101.3	95.9	96.9	96.4	96.0	93.9	95.1	98.0	98.9	98.4	99.3	106.6	102.5	95.6	92.7	94.3

		WISCONSIN																	
		OSHKOSH 549			PORTAGE 539			RACINE 534			RHINELANDER 545			SUPERIOR 548			WAUSAU 544		
DIVISION		MAT.	INST.	TOTAL	MAT.	INST.	TOTAL	MAT.	INST.	TOTAL	MAT.	INST.	TOTAL	MAT.	INST.	TOTAL	MAT.	INST.	TOTAL
015433	CONTRACTOR EQUIPMENT		100.3	100.3		101.9	101.9		101.9	101.9		100.3	100.3		103.0	103.0		100.3	100.3
0241, 31 - 34	SITE & INFRASTRUCTURE, DEMOLITION	92.1	99.4	97.2	85.8	106.2	100.1	95.6	108.1	104.3	104.1	99.5	100.9	92.6	104.1	100.7	87.9	100.7	96.9
0310	Concrete Forming & Accessories	87.1	89.6	89.3	88.5	91.5	91.1	96.8	111.3	109.3	84.9	90.3	89.6	88.6	89.6	89.5	86.6	105.0	102.4
0320	Concrete Reinforcing	89.6	98.5	94.1	97.1	102.9	100.0	95.5	114.0	104.8	89.7	96.8	93.3	88.4	106.9	97.8	89.7	102.8	96.3
0330	Cast-in-Place Concrete	96.1	94.2	95.4	90.8	99.3	94.0	104.3	108.1	105.8	108.9	96.9	104.4	98.3	101.9	99.6	89.6	95.8	91.9
03	CONCRETE	85.1	93.2	88.8	83.9	96.4	89.6	95.2	110.4	102.1	96.6	94.2	95.5	84.6	97.3	90.4	80.3	101.4	90.0
04	MASONRY	105.5	97.3	100.4	99.1	97.0	97.8	100.2	110.7	106.7	123.9	94.7	105.8	117.4	100.8	107.1	105.0	94.8	98.7
05	METALS	96.8	99.1	97.4	94.6	99.7	96.2	98.2	105.1	100.4	96.6	99.1	97.4	97.6	103.8	99.5	96.5	101.8	98.1
06	WOOD, PLASTICS & COMPOSITES	83.9	88.9	86.7	83.7	88.8	86.5	93.7	110.8	103.2	81.6	88.9	85.7	86.3	87.5	87.0	83.4	107.8	97.0
07	THERMAL & MOISTURE PROTECTION	104.8	82.2	95.2	99.9	88.9	95.3	100.8	106.8	103.3	105.7	82.7	96.0	104.2	89.4	98.0	104.6	84.2	96.0
08	OPENINGS	94.6	89.4	93.4	92.4	95.4	93.1	96.3	110.8	99.6	94.6	88.2	93.2	87.4	93.4	88.8	94.8	100.6	96.2
0920	Plaster & Gypsum Board	94.4	88.9	90.7	89.6	88.9	89.1	97.3	111.6	106.9	94.4	88.9	90.7	95.4	87.7	90.2	94.4	108.4	103.8
0950, 0980	Ceilings & Acoustic Treatment	84.6	88.9	87.5	85.4	88.9	87.8	86.3	111.6	103.4	84.6	88.9	87.5	57.8	87.7	78.0	84.6	108.4	100.7
0960	Flooring	87.6	121.4	97.0	87.2	114.8	94.9	91.3	121.4	99.6	87.0	111.7	93.9	90.7	127.2	100.8	87.5	111.7	94.2
0970, 0990	Wall Finishes & Painting/Coating	90.9	83.6	86.6	96.8	102.1	99.9	96.8	121.2	111.0	90.9	83.6	86.6	85.3	110.6	100.1	90.9	83.6	86.6
09	FINISHES	86.7	95.2	91.3	89.0	96.6	93.1	92.0	114.6	104.3	87.5	92.7	90.3	81.8	98.6	90.9	86.4	105.1	96.5
COVERS	DIVS. 10 - 14, 25, 28, 41, 43, 44, 46	100.0	86.4	97.2	100.0	87.4	97.2	100.0	100.5	100.3	100.0	87.6	97.2	100.0	87.2	97.2	100.0	96.5	99.2
21, 22, 23	FIRE SUPPRESSION, PLUMBING & HVAC	97.3	80.9	90.3	97.0	96.5	96.8	100.1	97.1	98.8	97.3	86.4	92.6	96.5	91.5	94.4	97.3	86.8	92.8
26, 27, 3370	ELECTRICAL, COMMUNICATIONS & UTIL.	102.6	76.3	88.9	102.7	94.1	98.2	99.1	103.5	101.4	102.0	77.6	89.3	106.9	98.7	102.6	103.8	80.2	91.5
MF2016	WEIGHTED AVERAGE	95.7	89.2	92.9	94.4	96.6	95.4	97.8	106.2	101.5	98.3	90.1	94.7	95.4	97.1	96.1	95.0	94.5	94.8

For customer support on your Building Construction Costs with RSMeans data, call 800.448.8182.

WYOMING

DIVISION		CASPER 826 MAT.	INST.	TOTAL	CHEYENNE 820 MAT.	INST.	TOTAL	NEWCASTLE 827 MAT.	INST.	TOTAL	RAWLINS 823 MAT.	INST.	TOTAL	RIVERTON 825 MAT.	INST.	TOTAL	ROCK SPRINGS 829-831 MAT.	INST.	TOTAL
015433	CONTRACTOR EQUIPMENT		97.4	97.4		97.4	97.4		97.4	97.4		97.4	97.4		97.4	97.4		97.4	97.4
0241, 31 - 34	SITE & INFRASTRUCTURE, DEMOLITION	100.6	94.7	96.5	96.1	94.7	95.1	88.0	94.2	92.4	102.6	94.2	96.7	95.8	94.1	94.6	92.2	94.2	93.6
0310	Concrete Forming & Accessories	100.5	55.2	61.4	101.0	68.4	72.9	92.2	73.8	76.4	96.3	74.1	77.1	91.3	62.9	66.8	98.3	63.9	68.7
0320	Concrete Reinforcing	111.9	85.2	98.4	107.4	85.3	96.2	115.3	85.0	100.0	115.0	85.0	99.9	116.1	85.0	100.4	116.1	83.2	99.5
0330	Cast-in-Place Concrete	104.4	77.5	94.2	98.6	77.6	90.6	99.6	76.5	90.8	99.6	76.7	90.9	99.6	76.5	90.8	99.6	76.5	90.8
03	CONCRETE	100.5	68.9	86.1	98.0	74.9	87.5	98.3	77.0	88.6	111.3	77.1	95.7	106.2	72.0	90.6	98.8	72.2	86.6
04	MASONRY	103.9	66.1	80.5	107.3	67.3	82.5	103.9	61.8	77.8	103.9	61.8	77.8	103.9	61.8	77.8	168.0	57.5	99.5
05	METALS	103.6	81.5	96.8	106.1	81.8	98.6	103.2	81.2	95.8	102.3	81.4	95.9	102.4	81.1	95.9	103.2	80.3	96.1
06	WOOD, PLASTICS & COMPOSITES	94.7	49.8	69.7	91.0	67.7	78.0	82.2	76.2	78.8	85.8	76.2	80.4	81.2	61.5	70.3	90.5	62.8	75.1
07	THERMAL & MOISTURE PROTECTION	102.5	67.1	87.5	97.1	69.4	85.4	98.7	66.3	85.0	100.3	77.9	90.8	99.7	64.8	84.9	98.8	67.9	85.7
08	OPENINGS	103.1	61.3	93.5	104.0	71.2	96.5	108.1	75.9	100.7	107.7	75.9	100.4	107.9	67.8	98.7	108.4	68.1	99.2
0920	Plaster & Gypsum Board	103.4	48.5	66.5	91.7	66.8	75.0	88.5	75.5	79.8	88.9	75.5	79.9	88.5	60.5	69.7	100.0	61.8	74.4
0950, 0980	Ceilings & Acoustic Treatment	103.1	48.5	66.3	98.7	66.8	77.2	100.6	75.5	83.7	100.6	75.5	83.7	100.6	60.5	73.6	100.6	61.8	74.5
0960	Flooring	105.8	73.6	96.9	105.0	73.6	96.3	98.7	47.1	84.4	101.6	47.1	86.5	98.2	63.7	88.7	103.6	47.1	88.1
0970, 0990	Wall Finishes & Painting/Coating	91.5	61.0	73.7	97.4	61.0	76.2	94.1	61.0	74.8	94.1	61.0	74.8	94.1	61.0	74.8	94.1	61.0	74.8
09	FINISHES	99.0	57.8	76.6	95.8	68.3	80.8	91.0	67.6	78.3	93.3	67.6	79.3	91.7	62.3	75.7	94.2	59.7	75.4
COVERS	DIVS. 10 - 14, 25, 28, 41, 43, 44, 46	100.0	84.5	96.6	100.0	86.5	97.0	100.0	94.5	98.8	100.0	94.5	98.8	100.0	80.5	95.7	100.0	84.7	96.6
21, 22, 23	FIRE SUPPRESSION, PLUMBING & HVAC	100.1	72.6	88.3	100.0	72.6	88.3	98.3	71.8	87.0	98.3	71.8	87.0	98.3	71.7	87.0	99.9	71.7	87.9
26, 27, 3370	ELECTRICAL, COMMUNICATIONS & UTIL.	101.9	63.7	82.0	104.0	64.6	83.5	102.7	63.3	82.2	102.7	63.3	82.2	102.7	62.4	81.8	100.8	80.9	90.5
MF2016	WEIGHTED AVERAGE	101.4	70.3	87.8	101.3	73.5	89.2	99.9	73.1	88.2	102.1	73.5	89.6	101.2	70.7	87.8	103.9	72.7	90.2

WYOMING / CANADA

DIVISION		SHERIDAN 828 MAT.	INST.	TOTAL	WHEATLAND 822 MAT.	INST.	TOTAL	WORLAND 824 MAT.	INST.	TOTAL	YELLOWSTONE NAT'L PA 821 MAT.	INST.	TOTAL	BARRIE, ONTARIO MAT.	INST.	TOTAL	BATHURST, NEW BRUNSWICK MAT.	INST.	TOTAL
015433	CONTRACTOR EQUIPMENT		97.4	97.4		97.4	97.4		97.4	97.4		97.4	97.4		103.0	103.0		102.8	102.8
0241, 31 - 34	SITE & INFRASTRUCTURE, DEMOLITION	96.2	94.7	95.1	93.0	94.2	93.8	90.2	94.2	93.0	90.3	94.2	93.0	119.2	99.9	105.7	107.0	95.7	99.1
0310	Concrete Forming & Accessories	99.0	63.7	68.6	94.1	50.6	56.6	94.1	63.9	68.0	94.2	63.9	68.1	124.2	86.3	91.5	103.6	60.6	66.5
0320	Concrete Reinforcing	116.1	85.0	100.4	115.3	84.4	99.8	116.1	85.0	100.4	118.1	85.0	101.4	175.3	89.7	132.1	137.7	59.4	98.2
0330	Cast-in-Place Concrete	103.0	77.5	93.3	103.9	76.4	93.5	99.6	76.5	90.8	99.6	76.5	90.8	158.9	86.5	131.4	117.7	59.1	95.4
03	CONCRETE	106.3	72.7	91.0	103.2	66.3	86.3	98.5	72.4	86.6	98.8	72.5	86.8	139.8	87.3	115.8	113.6	60.8	89.5
04	MASONRY	104.2	63.3	78.9	104.3	53.4	72.7	103.9	61.8	77.8	104.0	61.8	77.8	167.9	93.1	121.5	162.5	59.9	98.9
05	METALS	105.9	81.3	98.3	102.2	80.3	95.5	102.4	81.1	95.9	103.0	80.6	96.1	110.3	93.5	105.2	113.7	75.0	101.8
06	WOOD, PLASTICS & COMPOSITES	92.2	61.5	75.1	83.9	45.0	62.2	83.9	62.8	72.2	83.9	62.8	72.2	118.0	84.7	99.5	98.4	60.6	77.4
07	THERMAL & MOISTURE PROTECTION	100.0	67.6	86.3	99.0	60.3	82.6	98.8	66.3	85.0	98.2	66.3	84.7	113.8	90.3	103.9	109.4	61.3	89.0
08	OPENINGS	108.6	67.8	99.2	106.6	58.6	95.6	108.3	68.5	99.1	101.3	68.1	93.7	92.4	85.0	90.7	86.4	53.8	78.9
0920	Plaster & Gypsum Board	112.5	60.5	77.6	88.5	43.5	58.3	88.5	61.8	70.6	88.7	61.8	70.7	151.5	84.2	106.3	126.6	59.5	81.5
0950, 0980	Ceilings & Acoustic Treatment	103.5	60.5	74.5	100.6	43.5	62.1	100.6	61.8	74.5	101.5	61.8	74.7	95.3	84.2	87.9	112.5	59.5	76.7
0960	Flooring	102.5	63.7	91.7	100.3	46.3	85.3	100.3	47.1	85.5	100.3	47.1	85.5	118.1	92.3	110.9	98.6	43.8	83.4
0970, 0990	Wall Finishes & Painting/Coating	96.3	61.0	75.7	94.1	61.0	74.8	94.1	61.0	74.8	94.1	61.0	74.8	104.4	87.6	94.6	109.9	50.1	75.0
09	FINISHES	98.6	63.0	79.2	91.8	49.0	68.5	91.5	59.7	74.2	91.7	59.8	74.4	109.8	87.7	97.8	104.7	56.5	78.5
COVERS	DIVS. 10 - 14, 25, 28, 41, 43, 44, 46	100.0	85.8	96.8	100.0	83.5	96.3	100.0	79.3	95.4	100.0	79.4	95.4	139.2	67.6	123.3	131.1	60.2	115.4
21, 22, 23	FIRE SUPPRESSION, PLUMBING & HVAC	98.3	72.6	87.3	98.3	71.7	86.9	98.3	71.7	87.0	98.3	71.7	87.0	104.5	97.5	101.5	104.6	67.5	88.7
26, 27, 3370	ELECTRICAL, COMMUNICATIONS & UTIL.	105.3	62.4	83.0	102.7	80.9	91.4	102.7	80.9	91.4	101.6	90.9	96.1	116.1	87.6	101.3	112.0	59.0	84.4
MF2016	WEIGHTED AVERAGE	102.8	71.5	89.1	100.6	69.2	86.9	100.1	73.0	88.2	99.4	74.4	88.4	117.0	91.0	105.6	111.3	65.0	91.0

CANADA

DIVISION		BRANDON, MANITOBA MAT.	INST.	TOTAL	BRANTFORD, ONTARIO MAT.	INST.	TOTAL	BRIDGEWATER, NOVA SCOTIA MAT.	INST.	TOTAL	CALGARY, ALBERTA MAT.	INST.	TOTAL	CAP-DE-LA-MADELEINE, QUEBEC MAT.	INST.	TOTAL	CHARLESBOURG, QUEBEC MAT.	INST.	TOTAL
015433	CONTRACTOR EQUIPMENT		105.8	105.8		103.1	103.1		102.6	102.6		128.2	128.2		103.3	103.3		103.3	103.3
0241, 31 - 34	SITE & INFRASTRUCTURE, DEMOLITION	134.2	98.8	109.4	119.6	100.3	106.1	103.7	97.5	99.3	123.1	120.8	121.5	99.2	98.9	99.0	99.2	98.9	99.0
0310	Concrete Forming & Accessories	143.5	69.4	79.6	123.5	93.5	97.6	96.8	71.2	74.7	123.2	99.2	102.5	129.2	83.0	89.4	129.2	83.0	89.4
0320	Concrete Reinforcing	192.0	56.3	123.5	169.0	88.3	128.3	144.0	49.2	96.2	135.2	84.5	109.6	144.0	74.8	109.1	144.0	74.8	109.1
0330	Cast-in-Place Concrete	118.3	74.1	101.5	135.8	107.5	125.0	139.5	70.4	113.3	143.2	108.3	129.9	109.6	92.2	103.0	109.6	92.2	103.0
03	CONCRETE	127.0	69.7	100.8	127.0	97.6	113.6	125.3	68.0	99.1	129.2	100.5	116.1	111.7	85.2	99.6	111.7	85.2	99.6
04	MASONRY	221.4	63.3	123.3	168.8	97.4	124.5	164.2	68.4	104.8	200.4	90.6	132.3	164.9	79.6	112.0	164.9	79.6	112.0
05	METALS	133.5	80.4	117.2	112.4	94.3	106.8	111.7	77.6	101.2	129.5	104.8	121.9	110.8	86.9	103.5	110.8	86.9	103.5
06	WOOD, PLASTICS & COMPOSITES	150.9	70.0	105.9	119.8	92.0	104.3	90.7	70.6	79.5	96.4	98.8	97.8	130.5	82.8	103.9	130.5	82.8	103.9
07	THERMAL & MOISTURE PROTECTION	131.5	72.1	106.4	118.6	96.0	109.0	113.2	70.9	95.3	129.9	97.8	116.3	111.9	88.3	101.9	111.9	88.3	101.9
08	OPENINGS	102.2	63.0	93.2	90.3	91.1	90.5	84.5	64.4	79.9	83.5	87.8	84.4	91.6	76.0	88.0	91.6	76.0	88.0
0920	Plaster & Gypsum Board	113.1	68.8	83.3	116.6	91.7	99.9	122.4	69.8	87.1	121.5	97.9	105.6	144.8	82.2	102.8	144.8	82.2	102.8
0950, 0980	Ceilings & Acoustic Treatment	119.2	68.8	85.2	103.2	91.7	95.5	103.2	69.8	80.6	142.1	97.9	112.3	103.2	82.2	89.0	103.2	82.2	89.0
0960	Flooring	130.3	65.4	112.3	112.1	92.3	106.6	94.7	62.3	85.7	117.9	88.6	109.8	112.1	90.9	106.2	112.1	90.9	106.2
0970, 0990	Wall Finishes & Painting/Coating	117.0	56.5	81.8	109.0	96.2	101.5	109.0	62.0	81.6	117.1	111.2	113.7	109.0	87.4	96.4	109.0	87.4	96.4
09	FINISHES	118.4	67.6	90.8	106.3	93.9	99.6	101.1	68.8	83.6	122.2	99.1	109.6	109.4	85.2	96.3	109.4	85.2	96.3
COVERS	DIVS. 10 - 14, 25, 28, 41, 43, 44, 46	131.1	62.7	115.9	131.1	69.7	117.5	131.1	63.3	116.0	131.1	95.5	123.2	131.1	78.8	119.5	131.1	78.8	119.5
21, 22, 23	FIRE SUPPRESSION, PLUMBING & HVAC	104.7	82.2	95.1	104.6	100.4	102.8	104.6	82.8	95.3	105.5	93.3	100.3	105.1	87.7	97.6	105.1	87.7	97.6
26, 27, 3370	ELECTRICAL, COMMUNICATIONS & UTIL.	111.1	66.6	88.0	111.4	87.1	98.7	116.0	62.0	87.9	107.3	99.1	103.1	110.4	68.9	88.8	110.4	68.9	88.8
MF2016	WEIGHTED AVERAGE	123.6	73.8	101.8	114.3	94.9	105.8	112.3	73.4	95.3	119.5	98.8	110.4	111.7	83.6	99.4	111.7	83.6	99.4

City Cost Indexes

DIVISION		CHARLOTTETOWN, PRINCE EDWARD ISLAND			CHICOUTIMI, QUEBEC			CORNER BROOK, NEWFOUNDLAND			CORNWALL, ONTARIO			DALHOUSIE, NEW BRUNSWICK			DARTMOUTH, NOVA SCOTIA		
		MAT.	INST.	TOTAL	MAT.	INST.	TOTAL	MAT.	INST.	TOTAL	MAT.	INST.	TOTAL	MAT.	INST.	TOTAL	MAT.	INST.	TOTAL
015433	CONTRACTOR EQUIPMENT		120.7	120.7		103.4	103.4		104.0	104.0		103.1	103.1		102.6	102.6		102.9	102.9
0241, 31 - 34	SITE & INFRASTRUCTURE, DEMOLITION	133.6	108.5	116.0	103.9	98.6	100.2	138.7	96.1	108.8	117.7	99.7	105.1	102.8	95.5	97.7	124.9	97.7	105.9
0310	Concrete Forming & Accessories	118.4	55.4	64.1	131.9	91.0	96.7	120.7	79.3	85.0	121.3	86.5	91.3	102.4	60.8	66.5	110.5	71.2	76.6
0320	Concrete Reinforcing	152.2	48.0	99.6	103.9	94.2	99.0	176.3	50.4	112.8	169.0	88.0	128.2	147.3	59.4	102.9	184.1	49.2	116.0
0330	Cast-in-Place Concrete	149.0	59.1	114.9	110.8	97.0	105.6	137.7	67.0	110.8	122.1	97.9	112.9	115.9	59.1	94.3	132.9	70.4	109.2
03	CONCRETE	134.7	57.3	99.3	104.6	93.8	99.7	159.1	70.7	118.7	120.4	91.1	107.0	118.1	60.9	91.9	141.6	68.1	108.0
04	MASONRY	177.7	57.2	103.0	164.8	90.3	118.6	217.1	77.5	130.5	167.6	89.3	119.0	164.5	59.9	99.6	232.1	68.4	130.6
05	METALS	137.2	81.7	120.2	113.4	92.4	107.0	133.9	77.1	116.5	112.2	93.0	106.3	106.9	75.1	97.2	134.6	77.8	117.1
06	WOOD, PLASTICS & COMPOSITES	99.9	54.8	74.8	131.2	91.4	109.0	129.3	85.4	104.9	118.1	85.5	100.0	97.5	60.6	77.0	117.5	70.6	91.4
07	THERMAL & MOISTURE PROTECTION	128.5	59.1	99.1	109.8	98.0	104.8	136.2	69.2	107.9	118.4	90.5	106.6	116.2	61.3	92.9	133.2	71.0	106.9
08	OPENINGS	85.1	48.0	76.6	90.5	77.8	87.5	108.7	70.0	99.8	91.6	84.7	90.0	87.3	53.8	79.6	92.6	64.4	86.1
0920	Plaster & Gypsum Board	119.5	53.0	74.9	144.8	91.0	108.7	147.0	84.8	105.3	173.2	85.1	114.1	130.5	59.5	82.8	142.6	69.8	93.7
0950, 0980	Ceilings & Acoustic Treatment	128.0	53.0	77.4	111.7	91.0	97.8	120.0	84.8	96.3	105.7	85.1	91.8	106.5	59.5	74.8	126.9	69.8	88.4
0960	Flooring	106.3	58.4	93.0	114.3	90.9	107.8	112.4	52.4	95.7	112.1	90.9	106.2	100.6	67.1	91.3	107.3	62.3	94.8
0970, 0990	Wall Finishes & Painting/Coating	117.5	41.7	73.3	109.9	105.2	107.1	116.9	59.0	83.1	109.0	89.6	97.7	112.5	50.1	76.1	116.9	62.0	84.9
09	FINISHES	116.7	54.8	83.0	111.5	93.4	101.7	118.1	73.7	94.0	114.4	87.8	99.9	105.9	61.1	81.6	116.0	68.8	90.3
COVERS	DIVS. 10 - 14, 25, 28, 41, 43, 44, 46	131.1	60.4	115.4	131.1	82.5	120.3	131.1	63.1	116.0	131.1	67.3	116.9	131.1	60.1	115.3	131.1	63.3	116.1
21, 22, 23	FIRE SUPPRESSION, PLUMBING & HVAC	104.9	61.4	86.3	104.6	81.5	94.7	104.7	68.7	89.3	105.1	98.3	102.2	104.7	67.5	88.8	104.7	82.8	95.4
26, 27, 3370	ELECTRICAL, COMMUNICATIONS & UTIL.	114.3	50.1	80.9	108.7	85.6	96.7	108.8	54.4	80.5	111.7	88.0	99.4	113.9	55.8	83.6	113.0	62.0	86.5
MF2016	WEIGHTED AVERAGE	120.9	62.8	95.5	111.1	88.9	101.4	128.0	71.5	103.3	114.2	91.3	104.2	111.3	65.2	91.1	124.6	73.5	102.2

DIVISION		EDMONTON, ALBERTA			FORT MCMURRAY, ALBERTA			FREDERICTON, NEW BRUNSWICK			GATINEAU, QUEBEC			GRANBY, QUEBEC			HALIFAX, NOVA SCOTIA		
		MAT.	INST.	TOTAL	MAT.	INST.	TOTAL	MAT.	INST.	TOTAL	MAT.	INST.	TOTAL	MAT.	INST.	TOTAL	MAT.	INST.	TOTAL
015433	CONTRACTOR EQUIPMENT		128.5	128.5		105.5	105.5		120.7	120.7		103.3	103.3		103.3	103.3		118.3	118.3
0241, 31 - 34	SITE & INFRASTRUCTURE, DEMOLITION	129.3	121.0	123.5	124.8	101.0	108.1	113.6	110.3	111.3	99.0	98.9	99.0	99.5	98.9	99.1	107.8	109.8	109.2
0310	Concrete Forming & Accessories	119.7	99.2	102.0	121.4	91.7	95.7	120.2	61.4	69.5	129.2	82.9	89.3	129.2	82.9	89.3	117.4	85.0	89.4
0320	Concrete Reinforcing	135.7	84.5	109.9	156.5	84.3	120.1	151.1	59.6	105.0	152.2	74.8	113.1	152.2	74.7	113.1	152.2	72.3	111.9
0330	Cast-in-Place Concrete	147.7	108.4	132.7	180.3	103.5	151.1	116.8	59.5	95.0	108.1	92.2	102.1	111.7	92.2	104.3	106.9	86.0	98.9
03	CONCRETE	131.2	100.6	117.2	146.3	94.7	122.7	119.4	62.1	93.2	112.1	85.2	99.8	113.8	85.1	100.7	114.0	84.3	100.4
04	MASONRY	196.0	90.6	130.7	209.5	87.3	133.7	182.5	61.2	107.3	164.8	79.6	112.0	165.1	79.6	112.1	180.1	87.1	122.4
05	METALS	136.2	105.0	126.6	139.9	92.8	125.4	134.2	87.1	119.7	110.8	86.8	103.4	111.0	86.7	103.6	138.4	98.1	126.0
06	WOOD, PLASTICS & COMPOSITES	96.8	98.8	97.9	114.2	90.9	101.2	109.4	60.9	82.4	130.5	82.8	103.9	130.5	82.8	103.9	100.5	84.0	91.3
07	THERMAL & MOISTURE PROTECTION	129.7	97.8	116.2	127.1	93.9	113.0	123.7	61.3	97.3	111.9	88.3	101.9	111.9	86.7	101.2	129.0	86.1	110.8
08	OPENINGS	81.8	87.8	83.2	91.6	83.4	89.7	86.1	52.8	78.5	91.6	71.4	87.0	91.6	71.4	87.0	88.6	76.8	85.9
0920	Plaster & Gypsum Board	116.7	97.9	104.1	117.4	90.3	99.2	123.5	59.5	80.5	114.5	82.2	92.8	114.5	82.2	92.8	118.1	83.3	94.7
0950, 0980	Ceilings & Acoustic Treatment	137.2	97.9	110.7	110.8	90.3	97.0	133.9	59.5	83.7	103.2	82.2	89.0	103.2	82.2	89.0	128.5	83.3	98.0
0960	Flooring	114.0	88.6	107.0	112.1	88.6	105.6	109.5	70.0	98.6	112.1	90.9	106.2	112.1	90.9	106.2	99.6	83.8	95.2
0970, 0990	Wall Finishes & Painting/Coating	108.3	111.2	110.0	109.1	88.6	96.2	114.8	63.8	85.1	109.0	87.4	96.4	109.0	87.4	96.4	112.8	91.7	100.5
09	FINISHES	115.6	99.1	106.6	109.2	91.4	99.5	118.5	63.4	88.5	105.3	85.2	94.4	105.3	85.2	94.4	110.4	85.9	97.1
COVERS	DIVS. 10 - 14, 25, 28, 41, 43, 44, 46	131.1	95.5	123.2	131.1	92.3	122.5	131.1	60.7	115.5	131.1	78.8	119.5	131.1	78.8	119.5	131.1	68.7	117.2
21, 22, 23	FIRE SUPPRESSION, PLUMBING & HVAC	105.5	93.3	100.3	105.1	97.1	101.7	105.0	76.7	92.9	105.1	87.7	97.6	104.6	87.7	97.4	105.3	82.4	95.5
26, 27, 3370	ELECTRICAL, COMMUNICATIONS & UTIL.	110.3	99.1	104.5	105.6	81.8	93.3	111.9	73.6	92.0	110.4	68.9	88.8	111.0	68.9	89.1	109.4	87.0	97.7
MF2016	WEIGHTED AVERAGE	120.4	98.8	110.9	123.4	91.9	109.7	118.2	72.5	98.2	111.4	83.4	99.1	111.6	83.3	99.2	117.4	87.2	104.2

DIVISION		HAMILTON, ONTARIO			HULL, QUEBEC			JOLIETTE, QUEBEC			KAMLOOPS, BRITISH COLUMBIA			KINGSTON, ONTARIO			KITCHENER, ONTARIO		
		MAT.	INST.	TOTAL	MAT.	INST.	TOTAL	MAT.	INST.	TOTAL	MAT.	INST.	TOTAL	MAT.	INST.	TOTAL	MAT.	INST.	TOTAL
015433	CONTRACTOR EQUIPMENT		118.4	118.4		103.3	103.3		103.3	103.3		106.7	106.7		105.4	105.4		105.3	105.3
0241, 31 - 34	SITE & INFRASTRUCTURE, DEMOLITION	112.1	113.0	112.7	99.0	98.9	99.0	99.7	98.9	99.2	122.5	102.7	108.6	117.7	103.7	107.9	97.2	105.0	102.7
0310	Concrete Forming & Accessories	122.8	95.7	99.5	129.2	82.9	89.3	129.2	83.0	89.4	121.4	86.3	91.1	121.4	86.5	91.3	113.1	88.6	92.0
0320	Concrete Reinforcing	150.1	98.8	124.2	152.2	74.8	113.1	144.0	74.8	109.1	112.8	78.3	95.4	169.0	88.0	128.2	104.2	98.7	101.4
0330	Cast-in-Place Concrete	118.5	103.5	112.8	108.1	92.2	102.1	112.6	92.2	104.9	97.5	96.8	97.2	122.1	97.9	112.9	113.4	99.5	108.1
03	CONCRETE	116.3	98.6	108.7	112.1	85.2	99.8	113.2	85.2	100.4	121.2	88.9	106.4	122.3	91.1	108.0	100.5	94.4	97.7
04	MASONRY	170.9	101.7	128.0	164.8	79.6	112.0	165.2	79.6	112.1	172.2	87.6	119.7	174.7	89.3	121.8	145.6	99.0	116.8
05	METALS	131.7	106.4	123.9	111.0	86.8	103.6	111.0	86.9	103.6	112.8	89.2	105.6	113.5	92.9	107.2	121.0	97.5	113.8
06	WOOD, PLASTICS & COMPOSITES	99.9	94.7	97.0	130.5	82.8	103.9	130.5	82.8	103.9	102.5	84.5	92.5	118.1	85.6	100.0	107.6	86.4	95.8
07	THERMAL & MOISTURE PROTECTION	122.0	101.6	113.4	111.9	88.3	101.9	111.9	88.3	101.9	128.2	85.8	110.3	118.4	91.6	107.1	115.0	98.8	108.2
08	OPENINGS	85.6	93.8	87.5	91.6	71.4	87.0	91.6	76.0	88.0	88.4	82.5	87.1	91.6	84.4	90.0	83.1	87.4	84.1
0920	Plaster & Gypsum Board	121.1	94.1	103.0	114.5	82.2	92.8	144.8	82.2	102.8	102.2	83.7	89.8	176.2	85.2	115.1	105.6	86.0	92.4
0950, 0980	Ceilings & Acoustic Treatment	130.0	94.1	105.8	103.2	82.2	89.0	103.2	82.2	89.0	103.2	83.7	90.0	118.5	85.2	96.1	105.1	86.0	92.2
0960	Flooring	109.7	94.8	108.7	112.1	90.9	106.2	112.1	90.9	106.2	111.0	52.0	94.6	112.1	90.9	106.2	100.1	97.8	99.4
0970, 0990	Wall Finishes & Painting/Coating	102.9	105.4	104.3	109.0	87.4	96.4	109.0	87.4	96.4	109.0	79.9	92.0	109.0	82.8	93.7	103.1	95.1	98.4
09	FINISHES	113.5	97.1	104.6	105.3	85.2	94.4	109.4	85.2	96.3	105.8	79.6	91.5	117.3	87.1	100.9	99.1	90.5	94.4
COVERS	DIVS. 10 - 14, 25, 28, 41, 43, 44, 46	131.1	93.3	122.7	131.1	78.8	119.5	131.1	78.8	119.5	131.1	87.9	121.5	131.1	67.3	116.9	131.1	90.7	122.2
21, 22, 23	FIRE SUPPRESSION, PLUMBING & HVAC	105.5	92.2	99.8	104.6	87.7	97.4	104.6	87.7	97.4	104.6	91.2	98.9	105.1	98.4	102.2	104.8	90.9	98.8
26, 27, 3370	ELECTRICAL, COMMUNICATIONS & UTIL.	108.4	102.7	105.4	112.3	68.9	89.7	111.0	68.9	89.1	114.3	78.3	95.6	111.7	86.8	98.7	110.8	100.1	105.2
MF2016	WEIGHTED AVERAGE	116.0	99.6	108.8	111.5	83.4	99.2	111.9	83.6	99.5	114.1	87.2	102.3	115.3	91.4	104.8	109.1	95.1	103.0

For customer support on your Building Construction Costs with RSMeans data, call 800.448.8182.

City Cost Indexes

CANADA

DIVISION		LAVAL, QUEBEC			LETHBRIDGE, ALBERTA			LLOYDMINSTER, ALBERTA			LONDON, ONTARIO			MEDICINE HAT, ALBERTA			MONCTON, NEW BRUNSWICK		
		MAT.	INST.	TOTAL	MAT.	INST.	TOTAL	MAT.	INST.	TOTAL	MAT.	INST.	TOTAL	MAT.	INST.	TOTAL	MAT.	INST.	TOTAL
015433	CONTRACTOR EQUIPMENT		103.3	103.3		105.5	105.5		105.5	105.5		119.8	119.8		105.5	105.5		102.8	102.8
0241, 31 - 34	SITE & INFRASTRUCTURE, DEMOLITION	99.5	98.9	99.1	117.2	101.6	106.3	117.1	101.0	105.8	104.6	112.2	109.9	115.8	101.0	105.5	106.4	96.0	99.1
0310	Concrete Forming & Accessories	129.5	82.9	89.3	122.8	91.7	96.0	120.9	82.0	87.3	121.9	90.0	94.4	122.7	81.9	87.5	103.6	63.2	68.8
0320	Concrete Reinforcing	152.2	74.8	113.1	156.5	84.3	120.1	156.5	84.3	120.1	150.1	97.6	123.6	156.5	84.3	120.1	137.7	71.0	104.1
0330	Cast-in-Place Concrete	111.7	92.2	104.3	135.2	103.5	123.2	125.5	99.8	115.7	117.1	101.6	111.2	125.5	99.8	115.7	113.3	69.7	96.7
03	CONCRETE	113.8	85.2	100.7	125.1	94.8	111.2	120.3	89.1	106.0	115.6	96.2	106.7	120.4	89.0	106.1	111.5	68.0	91.6
04	MASONRY	165.1	79.6	112.1	183.2	87.3	123.7	164.1	80.9	112.5	179.6	99.9	130.2	164.1	80.9	112.5	162.1	61.4	99.7
05	METALS	110.9	86.8	103.5	134.6	92.8	121.8	112.8	92.7	106.6	131.7	107.7	124.3	113.0	92.6	106.7	113.7	85.9	105.2
06	WOOD, PLASTICS & COMPOSITES	130.6	82.8	104.0	117.6	90.9	102.7	114.2	81.2	95.8	106.2	87.7	95.9	117.6	81.2	97.3	98.4	62.6	78.4
07	THERMAL & MOISTURE PROTECTION	112.4	88.3	102.2	124.2	93.9	111.4	120.7	89.3	107.4	124.0	99.5	113.6	127.4	89.3	111.3	113.9	65.5	93.4
08	OPENINGS	91.6	71.4	87.0	91.6	83.4	89.7	91.6	78.0	88.5	80.4	88.8	82.4	91.6	78.0	88.5	86.4	61.5	80.7
0920	Plaster & Gypsum Board	114.8	82.2	92.9	106.7	90.3	95.7	102.7	80.3	87.6	123.0	86.9	98.8	105.0	80.3	88.4	126.6	61.5	82.8
0950, 0980	Ceilings & Acoustic Treatment	103.2	82.2	89.0	110.8	90.3	97.0	103.2	80.3	87.8	138.5	86.9	103.7	103.2	80.3	87.8	112.5	61.5	78.1
0960	Flooring	112.1	90.9	106.2	112.1	88.6	105.6	112.1	88.6	105.6	106.2	97.8	103.9	112.1	88.6	105.6	98.6	68.7	90.3
0970, 0990	Wall Finishes & Painting/Coating	109.0	87.4	96.4	108.9	100.8	104.2	109.1	78.5	91.3	109.3	102.0	105.0	108.9	78.5	91.2	109.9	51.4	75.8
09	FINISHES	105.3	85.2	94.4	106.8	92.4	99.0	104.9	82.6	92.7	116.2	92.3	103.2	105.0	82.6	92.8	104.7	62.9	82.0
COVERS	DIVS. 10 - 14, 25, 28, 41, 43, 44, 46	131.1	78.8	119.5	131.1	92.3	122.5	131.1	89.0	121.8	131.1	92.1	122.5	131.1	89.0	121.8	131.1	61.8	115.7
21, 22, 23	FIRE SUPPRESSION, PLUMBING & HVAC	104.8	87.7	97.5	104.9	93.8	100.2	105.1	93.7	100.2	105.5	89.4	98.6	104.6	90.5	98.6	104.6	69.9	89.8
26, 27, 3370	ELECTRICAL, COMMUNICATIONS & UTIL.	109.3	68.9	88.3	107.1	81.8	94.0	104.8	81.8	92.9	105.3	100.1	102.6	104.8	81.8	92.9	115.8	60.8	87.2
MF2016	WEIGHTED AVERAGE	111.5	83.4	99.2	118.3	91.4	106.5	112.7	88.1	101.9	115.6	97.0	107.5	112.9	87.4	101.7	111.5	69.3	93.0

CANADA

DIVISION		MONTREAL, QUEBEC			MOOSE JAW, SASKATCHEWAN			NEW GLASGOW, NOVA SCOTIA			NEWCASTLE, NEW BRUNSWICK			NORTH BAY, ONTARIO			OSHAWA, ONTARIO		
		MAT.	INST.	TOTAL	MAT.	INST.	TOTAL	MAT.	INST.	TOTAL	MAT.	INST.	TOTAL	MAT.	INST.	TOTAL	MAT.	INST.	TOTAL
015433	CONTRACTOR EQUIPMENT		123.5	123.5		101.7	101.7		102.9	102.9		102.8	102.8		103.5	103.5		105.3	105.3
0241, 31 - 34	SITE & INFRASTRUCTURE, DEMOLITION	114.0	113.5	113.6	117.1	95.1	101.7	119.1	97.7	104.1	107.0	95.7	99.1	134.5	99.6	110.0	108.3	103.9	105.2
0310	Concrete Forming & Accessories	124.4	91.6	96.1	106.3	57.2	64.0	110.5	71.2	76.6	103.6	60.8	66.7	144.3	84.0	92.3	118.5	88.7	92.8
0320	Concrete Reinforcing	126.8	94.3	110.4	110.4	63.4	86.7	176.3	49.2	112.2	137.7	59.4	98.2	207.8	87.6	147.1	165.2	93.6	129.1
0330	Cast-in-Place Concrete	130.9	99.1	118.8	105.8	62.7	86.1	132.9	70.4	109.2	117.7	59.1	95.4	128.0	84.7	111.5	131.1	87.2	114.4
03	CONCRETE	119.3	95.5	108.4	105.8	62.7	86.1	140.6	68.1	107.4	113.6	61.0	89.5	142.4	85.3	116.3	122.7	89.4	107.5
04	MASONRY	176.1	90.3	122.9	162.8	58.8	98.3	216.7	68.4	124.7	162.5	59.9	98.9	223.0	85.4	137.7	148.6	91.9	113.4
05	METALS	139.5	104.4	128.7	109.8	76.7	99.6	131.8	77.8	115.2	113.7	75.2	101.9	132.7	92.8	120.4	111.5	96.2	106.8
06	WOOD, PLASTICS & COMPOSITES	105.6	91.9	98.0	100.1	55.7	75.4	117.5	70.6	91.4	98.4	60.6	77.4	153.9	84.2	115.1	114.1	87.4	99.3
07	THERMAL & MOISTURE PROTECTION	119.2	98.2	110.3	111.3	63.0	90.9	133.2	71.0	106.9	113.9	61.3	91.6	139.8	86.8	117.4	116.0	91.1	105.4
08	OPENINGS	84.8	79.9	83.7	87.6	54.0	79.9	92.6	64.4	86.1	86.4	53.8	78.9	100.6	82.3	96.4	88.5	88.9	88.6
0920	Plaster & Gypsum Board	125.4	91.0	102.3	99.6	54.5	69.3	140.9	69.8	93.1	126.6	59.5	81.5	134.7	83.7	100.4	108.8	87.1	94.2
0950, 0980	Ceilings & Acoustic Treatment	140.4	91.0	107.1	103.2	54.5	70.3	119.2	69.8	85.9	112.5	59.5	76.7	119.2	83.7	95.3	101.7	87.1	91.8
0960	Flooring	110.0	93.1	105.3	102.4	56.7	89.7	107.3	62.3	94.8	98.6	67.1	89.9	130.3	90.9	119.4	102.9	100.3	102.2
0970, 0990	Wall Finishes & Painting/Coating	104.5	105.2	104.9	109.0	63.8	82.6	116.9	62.0	84.9	109.9	50.1	75.0	116.9	88.9	100.6	103.1	109.5	106.9
09	FINISHES	116.1	94.2	104.2	101.3	57.5	77.5	114.2	68.8	89.5	104.7	61.1	81.0	121.1	86.1	102.1	100.3	92.9	96.3
COVERS	DIVS. 10 - 14, 25, 28, 41, 43, 44, 46	131.1	83.7	120.6	131.1	59.8	115.3	131.1	63.3	116.1	131.1	60.2	115.4	131.1	66.1	116.7	131.1	90.4	122.1
21, 22, 23	FIRE SUPPRESSION, PLUMBING & HVAC	105.6	81.6	95.4	105.1	73.3	91.5	104.7	82.8	95.4	104.6	67.5	88.7	104.7	96.3	101.1	104.8	101.9	103.5
26, 27, 3370	ELECTRICAL, COMMUNICATIONS & UTIL.	108.4	85.7	96.6	113.3	59.5	85.3	109.1	62.0	84.6	111.4	59.0	84.1	109.4	87.9	98.2	111.9	90.3	100.7
MF2016	WEIGHTED AVERAGE	118.0	91.7	106.5	110.1	66.7	91.1	122.6	73.5	101.1	111.3	65.7	91.3	125.6	89.2	109.7	111.6	94.7	104.2

CANADA

DIVISION		OTTAWA, ONTARIO			OWEN SOUND, ONTARIO			PETERBOROUGH, ONTARIO			PORTAGE LA PRAIRIE, MANITOBA			PRINCE ALBERT, SASKATCHEWAN			PRINCE GEORGE, BRITISH COLUMBIA		
		MAT.	INST.	TOTAL	MAT.	INST.	TOTAL	MAT.	INST.	TOTAL	MAT.	INST.	TOTAL	MAT.	INST.	TOTAL	MAT.	INST.	TOTAL
015433	CONTRACTOR EQUIPMENT		121.6	121.6		103.0	103.0		103.1	103.1		105.4	105.4		101.7	101.7		106.7	106.7
0241, 31 - 34	SITE & INFRASTRUCTURE, DEMOLITION	108.4	114.4	112.6	119.2	99.7	105.6	119.6	99.7	105.7	118.3	98.5	104.5	112.6	95.3	100.5	126.0	102.7	109.7
0310	Concrete Forming & Accessories	122.2	90.0	94.4	124.2	82.5	88.2	123.5	85.0	90.3	122.9	68.9	76.3	106.3	57.0	63.8	111.6	81.3	85.4
0320	Concrete Reinforcing	130.3	97.6	113.8	175.3	89.7	132.1	169.0	88.1	128.2	156.5	56.3	105.9	115.3	63.3	89.1	112.8	78.3	95.4
0330	Cast-in-Place Concrete	126.9	104.4	118.4	158.9	80.6	129.2	135.8	86.2	116.9	125.5	73.4	105.7	110.3	66.9	93.8	122.2	96.8	112.5
03	CONCRETE	117.8	97.3	108.4	139.8	83.6	114.1	127.0	86.4	108.4	114.2	69.3	93.7	101.0	62.6	83.4	132.3	86.7	111.4
04	MASONRY	172.3	102.5	129.0	167.9	90.8	120.1	168.8	91.8	121.0	167.3	62.3	102.2	161.9	58.8	98.0	174.2	87.6	120.5
05	METALS	132.4	109.0	125.2	110.3	93.3	105.1	112.4	93.1	106.4	113.0	80.2	102.9	109.8	76.5	99.6	112.8	89.2	105.6
06	WOOD, PLASTICS & COMPOSITES	106.4	86.5	95.4	118.0	80.9	97.4	119.8	82.8	99.2	117.5	70.0	91.1	100.1	55.7	75.4	102.5	77.7	88.7
07	THERMAL & MOISTURE PROTECTION	130.7	100.6	118.0	113.8	87.6	102.7	118.6	92.4	107.5	111.8	71.6	94.8	111.2	61.9	90.3	122.0	85.0	106.4
08	OPENINGS	89.1	88.1	88.9	92.4	81.6	89.9	90.3	84.0	88.9	91.6	63.0	85.1	86.5	54.0	79.1	88.4	78.7	86.2
0920	Plaster & Gypsum Board	129.7	85.7	100.2	151.5	80.4	103.8	116.6	82.4	93.6	104.6	68.8	80.6	99.6	54.5	69.3	102.2	76.6	85.0
0950, 0980	Ceilings & Acoustic Treatment	134.2	85.7	101.5	95.3	80.4	85.3	103.2	82.4	89.1	103.2	68.8	80.0	103.2	54.5	70.3	103.2	76.6	85.3
0960	Flooring	102.3	93.2	99.8	118.1	92.3	110.9	112.1	90.9	106.2	112.1	65.4	99.2	102.4	56.7	89.7	107.7	71.2	97.6
0970, 0990	Wall Finishes & Painting/Coating	110.0	96.7	102.2	104.4	87.6	94.6	109.0	91.1	98.6	109.1	56.5	78.4	109.0	54.4	77.1	109.0	79.9	92.0
09	FINISHES	116.1	90.8	102.4	109.8	84.9	96.3	106.3	86.7	95.6	105.0	67.3	84.5	101.3	56.5	76.9	104.8	78.8	90.7
COVERS	DIVS. 10 - 14, 25, 28, 41, 43, 44, 46	131.1	90.7	122.1	139.2	66.4	123.0	131.1	67.5	117.0	131.1	62.3	115.8	131.1	59.8	115.3	131.1	87.2	121.4
21, 22, 23	FIRE SUPPRESSION, PLUMBING & HVAC	105.6	91.1	99.4	104.5	96.2	101.0	104.6	99.9	102.6	104.6	81.7	94.8	105.1	66.0	88.4	104.6	91.2	98.9
26, 27, 3370	ELECTRICAL, COMMUNICATIONS & UTIL.	108.5	100.9	104.5	117.6	86.7	101.5	111.4	87.5	99.0	113.0	57.5	84.1	113.3	59.5	85.3	111.4	78.3	94.2
MF2016	WEIGHTED AVERAGE	117.2	98.0	108.8	117.1	89.2	104.9	114.3	91.1	104.1	112.7	72.2	94.9	109.2	65.0	89.8	115.1	86.6	102.6

For customer support on your Building Construction Costs with RSMeans data, call 800.448.8182.

CANADA

DIVISION		QUEBEC CITY, QUEBEC			RED DEER, ALBERTA			REGINA, SASKATCHEWAN			RIMOUSKI, QUEBEC			ROUYN-NORANDA, QUEBEC			SAINT HYACINTHE, QUEBEC		
		MAT.	INST.	TOTAL	MAT.	INST.	TOTAL	MAT.	INST.	TOTAL	MAT.	INST.	TOTAL	MAT.	INST.	TOTAL	MAT.	INST.	TOTAL
015433	CONTRACTOR EQUIPMENT		124.2	124.2		105.5	105.5		124.2	124.2		103.3	103.3		103.3	103.3		103.3	103.3
0241, 31 - 34	SITE & INFRASTRUCTURE, DEMOLITION	115.4	113.3	113.9	115.8	101.0	105.5	127.4	115.0	118.7	99.3	98.5	98.8	99.0	98.9	99.0	99.5	98.9	99.1
0310	Concrete Forming & Accessories	122.5	91.7	95.9	136.9	81.9	89.5	124.8	95.6	99.6	129.2	91.0	96.3	129.2	82.9	89.3	129.2	82.9	89.3
0320	Concrete Reinforcing	129.3	94.3	111.7	156.5	84.3	120.1	147.1	91.7	119.1	108.6	94.2	101.3	152.2	74.8	113.1	152.2	74.8	113.1
0330	Cast-in-Place Concrete	132.7	98.9	119.9	125.5	99.8	115.7	144.3	100.7	127.7	113.7	97.0	107.4	108.1	92.2	102.1	111.7	92.2	104.3
03	CONCRETE	120.4	95.6	109.0	121.3	89.0	106.6	129.7	97.4	114.9	109.2	93.7	102.1	112.1	85.2	99.8	113.8	85.2	100.7
04	MASONRY	170.9	90.3	120.9	164.1	80.9	112.5	194.2	91.1	130.3	164.5	90.3	118.5	164.8	79.6	112.0	165.0	79.6	112.1
05	METALS	139.5	106.1	129.3	113.0	92.6	106.7	143.2	103.9	131.1	110.5	92.3	105.0	111.0	86.8	103.6	111.0	86.8	103.6
06	WOOD, PLASTICS & COMPOSITES	111.7	91.8	100.6	117.6	81.2	97.3	109.2	96.2	102.0	130.5	91.4	108.7	130.5	82.8	103.9	130.5	82.8	103.9
07	THERMAL & MOISTURE PROTECTION	115.8	98.4	108.4	137.7	89.3	117.2	132.7	90.0	114.7	111.9	98.0	106.0	111.9	88.3	101.9	112.2	88.3	102.1
08	OPENINGS	88.5	87.4	88.2	91.6	78.0	88.5	86.9	83.4	86.1	91.2	77.8	88.1	91.6	71.4	87.0	91.6	71.4	87.0
0920	Plaster & Gypsum Board	128.6	91.0	103.4	105.0	80.3	88.4	134.8	95.6	108.5	144.6	91.0	108.6	114.3	82.2	92.8	114.3	82.2	92.8
0950, 0980	Ceilings & Acoustic Treatment	137.2	91.0	106.1	103.2	80.3	87.8	141.5	95.6	110.6	102.3	91.0	94.7	102.3	82.2	88.8	102.3	82.2	88.8
0960	Flooring	110.8	93.1	105.7	114.5	88.6	107.3	113.4	97.8	109.1	113.4	90.9	107.2	112.1	90.9	106.2	112.1	90.9	106.2
0970, 0990	Wall Finishes & Painting/Coating	122.1	105.2	112.2	108.9	78.5	91.2	115.1	93.1	102.2	112.1	105.2	108.1	109.0	87.4	96.4	109.0	87.4	96.4
09	FINISHES	117.5	94.1	104.8	105.8	82.6	93.2	125.2	96.7	109.7	109.9	93.4	100.9	105.1	85.2	94.3	105.1	85.2	94.3
COVERS	DIVS. 10 - 14, 25, 28, 41, 43, 44, 46	131.1	83.4	120.5	131.1	89.0	121.8	131.1	71.8	117.9	131.1	82.5	120.3	131.1	78.8	119.5	131.1	78.8	119.5
21, 22, 23	FIRE SUPPRESSION, PLUMBING & HVAC	105.5	81.6	95.3	104.6	90.5	98.6	105.2	90.5	98.9	104.6	81.5	94.7	104.6	87.7	97.4	100.8	87.7	95.2
26, 27, 3370	ELECTRICAL, COMMUNICATIONS & UTIL.	109.6	85.7	97.2	104.8	81.8	92.9	111.4	96.6	103.7	111.0	85.6	97.8	111.0	68.9	89.1	111.6	68.9	89.4
MF2016	WEIGHTED AVERAGE	118.5	92.2	107.0	113.3	87.4	102.0	122.8	95.4	110.8	111.3	88.9	101.5	111.4	83.4	99.1	110.8	83.4	98.8

CANADA

DIVISION		SAINT JOHN, NEW BRUNSWICK			SARNIA, ONTARIO			SASKATOON, SASKATCHEWAN			SAULT STE MARIE, ONTARIO			SHERBROOKE, QUEBEC			SOREL, QUEBEC		
		MAT.	INST.	TOTAL	MAT.	INST.	TOTAL	MAT.	INST.	TOTAL	MAT.	INST.	TOTAL	MAT.	INST.	TOTAL	MAT.	INST.	TOTAL
015433	CONTRACTOR EQUIPMENT		102.8	102.8		103.1	103.1		101.5	101.5		103.1	103.1		103.3	103.3		103.3	103.3
0241, 31 - 34	SITE & INFRASTRUCTURE, DEMOLITION	107.1	97.4	100.3	118.2	99.8	105.3	114.1	97.9	102.7	107.9	99.4	101.9	99.5	98.9	99.1	99.7	98.9	99.2
0310	Concrete Forming & Accessories	123.5	66.3	74.1	122.2	91.9	96.1	106.0	95.1	96.6	111.8	87.9	91.2	129.2	82.9	89.3	129.2	83.0	89.4
0320	Concrete Reinforcing	137.7	71.6	104.4	120.7	89.5	104.9	117.9	91.6	104.6	108.4	88.2	98.2	152.2	74.8	113.1	144.0	74.8	109.1
0330	Cast-in-Place Concrete	115.9	71.4	99.0	125.3	94.9	115.5	119.1	98.1	111.1	112.6	84.8	102.0	111.7	92.2	104.3	112.6	92.2	104.9
03	CONCRETE	114.1	70.1	94.0	115.8	94.3	105.9	106.2	95.5	101.3	101.5	87.3	95.0	113.8	85.2	100.7	113.2	85.2	100.4
04	MASONRY	183.9	71.5	114.2	180.5	94.4	127.1	173.5	91.1	122.4	165.5	93.9	121.1	165.1	79.6	112.1	165.2	79.6	112.1
05	METALS	113.6	87.3	105.5	112.3	93.5	106.5	107.8	90.6	102.5	111.5	95.2	106.5	110.8	86.8	103.4	111.0	86.9	103.6
06	WOOD, PLASTICS & COMPOSITES	120.4	64.3	89.2	118.7	90.9	103.3	97.6	95.7	96.6	106.6	88.7	96.7	130.5	82.8	103.9	130.5	82.8	103.9
07	THERMAL & MOISTURE PROTECTION	114.2	72.4	96.5	118.7	96.0	109.1	112.3	89.2	102.5	117.4	91.2	106.3	111.9	88.3	101.9	111.9	88.3	101.9
08	OPENINGS	86.3	61.3	80.6	92.9	87.8	91.7	87.1	83.2	86.2	84.7	87.1	85.2	91.6	71.4	87.0	91.6	76.0	88.0
0920	Plaster & Gypsum Board	139.3	63.3	88.2	139.9	90.7	106.8	115.6	95.5	102.2	107.8	88.4	94.8	114.3	82.2	92.8	144.6	82.2	102.7
0950, 0980	Ceilings & Acoustic Treatment	117.6	63.3	81.0	107.4	90.7	96.1	123.5	95.6	104.7	103.2	88.4	93.2	102.3	82.2	88.8	102.3	82.2	88.8
0960	Flooring	110.1	68.7	98.6	112.1	99.5	108.6	105.0	97.8	103.0	105.2	96.4	102.7	112.1	90.9	106.2	112.1	90.9	106.2
0970, 0990	Wall Finishes & Painting/Coating	109.9	84.2	94.9	109.0	103.0	105.5	112.5	93.1	101.2	109.0	95.4	101.1	109.0	87.4	96.4	109.0	87.4	96.4
09	FINISHES	111.0	68.1	87.7	110.3	94.8	101.9	109.3	96.3	102.3	102.3	90.5	95.9	105.1	85.2	94.3	109.2	85.2	96.2
COVERS	DIVS. 10 - 14, 25, 28, 41, 43, 44, 46	131.1	63.0	116.0	131.1	68.8	117.3	131.1	71.0	117.8	131.1	89.5	121.9	131.1	78.8	119.5	131.1	78.8	119.5
21, 22, 23	FIRE SUPPRESSION, PLUMBING & HVAC	104.6	77.9	93.2	104.6	105.8	105.1	105.0	90.3	98.7	104.6	94.0	100.1	105.1	87.7	97.6	104.6	87.7	97.4
26, 27, 3370	ELECTRICAL, COMMUNICATIONS & UTIL.	118.5	86.3	101.7	114.3	89.7	101.5	112.6	96.6	104.3	113.0	87.9	100.0	111.0	68.9	89.1	111.0	68.9	89.1
MF2016	WEIGHTED AVERAGE	113.8	77.0	97.7	114.3	95.5	106.0	110.8	92.5	102.7	109.6	91.7	101.8	111.7	83.4	99.3	111.9	83.6	99.5

CANADA

DIVISION		ST CATHARINES, ONTARIO			ST JEROME, QUEBEC			ST JOHNS, NEWFOUNDLAND			SUDBURY, ONTARIO			SUMMERSIDE, PRINCE EDWARD ISLAND			SYDNEY, NOVA SCOTIA		
		MAT.	INST.	TOTAL	MAT.	INST.	TOTAL	MAT.	INST.	TOTAL	MAT.	INST.	TOTAL	MAT.	INST.	TOTAL	MAT.	INST.	TOTAL
015433	CONTRACTOR EQUIPMENT		103.1	103.1		103.3	103.3		121.7	121.7		103.1	103.1		102.9	102.9		102.9	102.9
0241, 31 - 34	SITE & INFRASTRUCTURE, DEMOLITION	97.4	101.4	100.2	99.0	98.9	99.0	121.6	111.7	114.7	97.5	101.0	99.9	129.3	95.0	105.3	114.8	97.7	102.9
0310	Concrete Forming & Accessories	111.1	94.9	97.1	129.2	82.9	89.3	122.8	88.0	92.8	107.1	89.9	92.3	110.8	55.1	62.7	110.5	71.2	76.6
0320	Concrete Reinforcing	105.1	98.7	101.9	152.2	74.8	113.1	161.2	85.0	122.8	106.0	102.5	104.3	173.8	48.0	110.3	176.3	49.2	112.2
0330	Cast-in-Place Concrete	108.3	101.4	105.6	108.1	92.2	102.1	140.6	101.1	125.6	109.1	98.0	104.9	126.2	57.7	100.1	102.5	70.4	90.4
03	CONCRETE	98.1	97.9	98.0	112.1	85.2	99.8	132.7	93.0	114.5	98.4	95.0	96.8	148.5	55.9	106.2	126.2	68.1	99.6
04	MASONRY	145.2	101.3	118.0	164.8	79.6	112.0	184.6	93.7	128.2	145.3	97.6	115.7	215.9	57.2	117.5	214.2	68.4	123.8
05	METALS	111.4	97.4	107.1	111.0	86.8	103.6	135.7	101.0	125.1	110.8	97.0	106.6	131.8	71.0	113.1	131.8	77.8	115.2
06	WOOD, PLASTICS & COMPOSITES	105.3	94.2	99.1	130.5	82.8	103.9	102.1	85.8	93.0	101.5	88.7	94.4	117.9	54.3	82.5	117.5	70.6	91.4
07	THERMAL & MOISTURE PROTECTION	115.0	103.0	109.9	111.9	88.3	101.9	139.2	98.0	121.8	114.5	97.5	107.3	132.6	60.1	101.9	133.2	71.0	106.9
08	OPENINGS	82.6	92.1	84.8	91.6	71.4	87.0	85.7	77.0	83.7	83.3	87.6	84.3	104.6	47.8	91.5	92.6	64.4	86.1
0920	Plaster & Gypsum Board	97.0	94.0	95.0	114.3	82.2	92.8	139.6	84.8	102.8	98.8	88.4	91.8	141.5	53.0	82.1	140.9	69.8	93.1
0950, 0980	Ceilings & Acoustic Treatment	101.7	94.0	96.5	102.3	82.2	88.8	129.6	84.8	99.4	96.6	88.4	91.1	119.2	53.0	74.6	119.2	69.8	85.9
0960	Flooring	98.9	94.5	97.7	112.1	90.9	106.2	111.1	54.2	95.3	97.1	96.4	96.9	107.4	58.4	93.8	107.3	62.3	94.8
0970, 0990	Wall Finishes & Painting/Coating	103.1	105.4	104.4	109.0	87.4	96.4	112.5	103.1	107.0	103.1	96.1	99.0	116.9	41.7	73.0	116.9	62.0	84.9
09	FINISHES	96.9	96.0	96.4	105.1	85.2	94.3	119.6	83.3	99.8	95.6	91.5	93.4	115.3	54.5	82.2	114.2	68.8	89.5
COVERS	DIVS. 10 - 14, 25, 28, 41, 43, 44, 46	131.1	70.5	117.7	131.1	78.8	119.5	131.1	69.7	117.5	131.1	90.8	122.2	131.1	59.4	115.2	131.1	63.3	116.1
21, 22, 23	FIRE SUPPRESSION, PLUMBING & HVAC	104.8	90.8	98.8	104.6	87.7	97.4	105.3	86.3	97.2	104.5	90.2	98.4	104.7	61.3	86.2	104.7	82.8	95.4
26, 27, 3370	ELECTRICAL, COMMUNICATIONS & UTIL.	112.4	100.9	106.4	111.7	68.9	89.4	109.3	85.9	97.1	112.5	101.3	106.7	108.2	50.1	78.0	109.1	62.0	84.6
MF2016	WEIGHTED AVERAGE	107.1	96.1	102.3	111.4	83.4	99.1	120.7	90.3	107.4	106.9	94.9	101.6	125.1	60.5	96.8	120.6	73.5	100.0

For customer support on your Building Construction Costs with RSMeans data, call 800.448.8182.

		CANADA																
	DIVISION	THUNDER BAY, ONTARIO			TIMMINS, ONTARIO			TORONTO, ONTARIO			TROIS RIVIERES, QUEBEC			TRURO, NOVA SCOTIA			VANCOUVER, BRITISH COLUMBIA		
		MAT.	INST.	TOTAL	MAT.	INST.	TOTAL	MAT.	INST.	TOTAL	MAT.	INST.	TOTAL	MAT.	INST.	TOTAL	MAT.	INST.	TOTAL
015433	CONTRACTOR EQUIPMENT		103.1	103.1		103.1	103.1		121.8	121.8		103.7	103.7		102.6	102.6		137.0	137.0
0241, 31 - 34	SITE & INFRASTRUCTURE, DEMOLITION	102.2	101.3	101.6	119.6	99.3	105.4	119.5	115.3	116.6	115.2	99.2	104.0	103.9	97.5	99.4	114.7	124.5	121.6
0310	Concrete Forming & Accessories	118.4	93.2	96.6	123.5	84.0	89.4	122.8	101.0	104.0	151.9	83.1	92.5	96.8	71.2	74.7	122.6	88.8	93.5
0320	Concrete Reinforcing	94.0	98.2	96.1	169.0	87.6	127.9	150.1	100.6	125.1	176.3	74.8	125.1	144.0	49.2	96.2	123.7	93.7	108.6
0330	Cast-in-Place Concrete	119.2	100.6	112.1	135.8	84.6	116.3	116.9	113.6	115.6	106.2	92.3	100.9	141.0	70.4	114.2	125.3	93.5	113.2
03	CONCRETE	105.4	96.7	101.4	127.0	85.3	107.9	115.6	105.9	111.1	128.5	85.3	108.7	126.0	68.0	99.5	118.7	92.7	106.8
04	MASONRY	145.8	101.5	118.3	168.8	85.4	117.1	170.9	108.4	132.1	218.4	79.6	132.3	164.3	68.4	104.8	170.9	88.0	119.5
05	METALS	111.4	96.4	106.8	112.4	92.7	106.3	132.3	109.7	125.3	131.0	87.1	117.5	111.7	77.6	101.2	139.8	114.4	131.1
06	WOOD, PLASTICS & COMPOSITES	114.1	91.5	101.6	119.8	84.1	99.9	105.5	98.5	101.6	168.1	82.8	120.7	90.7	70.6	79.5	100.9	86.8	93.1
07	THERMAL & MOISTURE PROTECTION	115.3	100.2	108.9	118.6	86.8	105.1	133.0	108.0	122.4	131.7	88.3	113.3	113.2	70.9	95.3	133.0	88.3	114.1
08	OPENINGS	81.9	90.0	83.8	90.3	82.3	88.5	84.1	98.0	87.3	102.2	76.0	96.2	84.5	64.4	79.9	83.0	87.2	84.0
0920	Plaster & Gypsum Board	124.9	91.3	102.3	116.6	83.7	94.5	119.4	98.1	105.1	169.3	82.2	110.8	122.4	69.8	87.1	114.2	85.4	94.8
0950, 0980	Ceilings & Acoustic Treatment	96.6	91.3	93.0	103.2	83.7	90.0	128.1	98.1	107.9	118.3	82.2	94.0	103.2	69.8	80.6	135.5	85.4	101.7
0960	Flooring	102.9	101.2	102.4	112.1	90.9	106.2	104.5	103.7	104.3	130.3	90.9	119.4	94.7	62.3	85.7	119.7	91.9	112.0
0970, 0990	Wall Finishes & Painting/Coating	103.1	97.1	99.6	109.0	88.9	97.3	100.8	109.5	105.9	116.9	87.4	99.7	109.0	62.0	81.6	110.8	89.2	98.2
09	FINISHES	101.2	95.0	97.9	106.3	86.1	95.3	110.6	102.3	106.1	124.8	85.2	103.3	101.1	68.8	83.6	119.1	89.3	102.9
COVERS	DIVS. 10 - 14, 25, 28, 41, 43, 44, 46	131.1	70.7	117.7	131.1	66.0	116.7	131.1	95.5	123.2	131.1	78.8	119.5	131.1	63.3	116.0	131.1	92.0	122.4
21, 22, 23	FIRE SUPPRESSION, PLUMBING & HVAC	104.8	91.0	98.9	104.6	96.3	101.1	105.4	99.0	102.7	104.7	87.7	97.5	104.6	82.8	95.3	105.5	78.6	94.0
26, 27, 3370	ELECTRICAL, COMMUNICATIONS & UTIL.	110.8	99.5	104.9	113.0	87.9	100.0	106.2	103.1	104.6	109.4	68.9	88.3	110.7	62.0	85.4	107.1	83.0	94.6
MF2016	WEIGHTED AVERAGE	108.3	95.4	102.7	114.4	89.2	103.4	115.9	104.2	110.8	123.3	83.6	105.9	111.9	73.4	95.1	118.0	91.2	106.3

		CANADA																	
	DIVISION	VICTORIA, BRITISH COLUMBIA			WHITEHORSE, YUKON			WINDSOR, ONTARIO			WINNIPEG, MANITOBA			YARMOUTH, NOVA SCOTIA			YELLOWKNIFE, NWT		
		MAT.	INST.	TOTAL	MAT.	INST.	TOTAL	MAT.	INST.	TOTAL	MAT.	INST.	TOTAL	MAT.	INST.	TOTAL	MAT.	INST.	TOTAL
015433	CONTRACTOR EQUIPMENT		109.7	109.7		130.0	130.0		103.1	103.1		123.9	123.9		102.9	102.9		129.9	129.9
0241, 31 - 34	SITE & INFRASTRUCTURE, DEMOLITION	125.5	106.8	112.4	146.2	116.0	125.0	93.8	101.3	99.0	117.4	111.2	113.1	118.9	97.7	104.1	154.5	122.1	131.8
0310	Concrete Forming & Accessories	110.8	87.6	90.8	133.6	59.2	69.5	118.4	91.4	95.1	125.4	66.7	74.8	110.5	71.2	76.6	135.9	79.3	87.1
0320	Concrete Reinforcing	115.3	93.5	104.3	159.8	65.3	112.1	103.0	97.5	100.2	142.1	61.3	101.3	176.3	49.2	112.2	156.6	67.9	111.9
0330	Cast-in-Place Concrete	123.4	90.8	111.0	179.5	73.1	139.1	110.9	102.1	107.6	144.8	73.4	117.6	131.5	70.4	108.3	177.1	90.8	144.3
03	CONCRETE	134.7	90.0	114.2	155.6	67.0	115.1	99.5	96.4	98.1	129.3	69.6	102.0	139.9	68.1	107.1	154.2	82.6	121.5
04	MASONRY	176.5	88.0	121.6	250.8	60.4	132.8	145.4	100.5	117.5	185.3	66.2	111.4	216.6	68.4	124.7	244.4	71.8	137.4
05	METALS	109.0	92.3	103.9	148.5	92.5	131.3	111.4	96.9	106.9	143.2	90.0	126.8	131.8	77.8	115.2	147.4	96.0	131.6
06	WOOD, PLASTICS & COMPOSITES	101.3	86.2	92.9	123.4	57.7	86.8	114.1	89.8	100.6	107.8	67.3	85.2	117.5	70.6	91.4	129.7	80.6	102.4
07	THERMAL & MOISTURE PROTECTION	122.2	89.3	108.3	142.8	64.9	109.9	115.1	99.6	108.5	125.3	71.0	102.3	133.2	71.0	106.9	144.1	81.4	117.5
08	OPENINGS	88.6	82.6	87.2	104.4	55.6	93.2	81.7	89.5	83.5	83.9	61.0	78.7	92.6	64.4	86.1	96.5	68.8	90.1
0920	Plaster & Gypsum Board	108.8	85.4	93.1	167.4	55.6	92.3	109.8	89.5	96.2	121.9	65.4	84.0	140.9	69.8	93.1	172.4	79.2	109.8
0950, 0980	Ceilings & Acoustic Treatment	105.6	85.4	92.0	161.4	55.6	90.1	96.6	89.5	91.8	143.8	65.4	91.0	119.2	69.8	85.9	155.1	79.2	103.9
0960	Flooring	110.4	71.2	99.5	125.4	58.3	106.8	102.9	98.5	101.7	109.0	70.4	98.3	107.3	62.3	94.8	123.4	87.6	113.5
0970, 0990	Wall Finishes & Painting/Coating	112.5	89.2	98.9	119.7	55.6	82.3	103.1	98.1	100.2	107.7	54.4	76.6	116.9	62.0	84.9	118.9	79.3	95.8
09	FINISHES	108.0	85.3	95.6	142.3	58.4	96.6	98.8	93.4	95.9	119.8	66.4	90.8	114.2	68.8	89.5	138.6	80.3	106.9
COVERS	DIVS. 10 - 14, 25, 28, 41, 43, 44, 46	131.1	67.4	116.9	131.1	62.6	115.9	131.1	70.1	117.6	131.1	64.7	116.4	131.1	63.3	116.1	131.1	66.1	116.7
21, 22, 23	FIRE SUPPRESSION, PLUMBING & HVAC	104.7	76.6	92.7	105.3	74.6	92.1	104.8	90.9	98.8	105.5	65.5	88.4	104.7	82.8	95.4	105.4	92.1	99.7
26, 27, 3370	ELECTRICAL, COMMUNICATIONS & UTIL.	112.8	82.4	97.0	132.0	60.3	94.7	115.6	101.0	108.0	112.8	65.1	88.0	109.1	62.0	84.6	121.5	81.8	100.8
MF2016	WEIGHTED AVERAGE	115.3	85.7	102.3	135.9	71.3	107.6	107.6	95.2	102.2	121.3	72.0	99.7	122.5	73.5	101.1	133.4	86.2	112.7

For customer support on your Building Construction Costs with RSMeans data, call 800.448.8182.

Location Factors - Commercial

Costs shown in RSMeans cost data publications are based on national averages for materials and installation. To adjust these costs to a specific location, simply multiply the base cost by the factor and divide by 100 for that city. The data is arranged alphabetically by state and postal zip code numbers. For a city not listed, use the factor for a nearby city with similar economic characteristics.

STATE/ZIP	CITY	MAT.	INST.	TOTAL
ALABAMA				
350-352	Birmingham	95.9	71.4	85.2
354	Tuscaloosa	96.0	72.3	85.6
355	Jasper	96.5	71.2	85.5
356	Decatur	96.0	71.3	85.2
357-358	Huntsville	96.0	70.7	84.9
359	Gadsden	96.1	71.0	85.1
360-361	Montgomery	94.8	72.3	85.0
362	Anniston	94.5	68.5	83.1
363	Dothan	94.9	72.9	85.3
364	Evergreen	94.5	72.6	84.9
365-366	Mobile	95.4	70.2	84.4
367	Selma	94.6	73.6	85.4
368	Phenix City	95.4	72.7	85.4
369	Butler	94.8	72.8	85.2
ALASKA				
995-996	Anchorage	118.9	114.8	117.1
997	Fairbanks	120.1	115.3	118.0
998	Juneau	118.8	114.8	117.0
999	Ketchikan	130.8	114.8	123.8
ARIZONA				
850,853	Phoenix	99.5	74.1	88.4
851,852	Mesa/Tempe	98.2	72.5	87.0
855	Globe	99.2	72.4	87.4
856-857	Tucson	97.1	71.8	86.0
859	Show Low	99.4	72.5	87.6
860	Flagstaff	101.7	72.1	88.8
863	Prescott	99.3	72.2	87.5
864	Kingman	97.9	72.5	86.8
865	Chambers	97.9	75.0	87.9
ARKANSAS				
716	Pine Bluff	96.9	65.0	82.9
717	Camden	95.0	60.5	79.9
718	Texarkana	95.7	60.7	80.3
719	Hot Springs	94.3	62.0	80.1
720-722	Little Rock	95.3	66.6	82.7
723	West Memphis	94.2	66.6	82.1
724	Jonesboro	94.5	63.7	81.0
725	Batesville	92.6	60.8	78.7
726	Harrison	93.9	58.9	78.6
727	Fayetteville	91.5	61.6	78.5
728	Russellville	92.7	59.4	78.1
729	Fort Smith	95.0	63.1	81.0
CALIFORNIA				
900-902	Los Angeles	99.9	129.3	112.8
903-905	Inglewood	95.5	125.5	108.6
906-908	Long Beach	97.1	126.1	109.8
910-912	Pasadena	96.0	125.3	108.8
913-916	Van Nuys	99.1	125.3	110.6
917-918	Alhambra	98.0	125.9	110.2
919-921	San Diego	101.0	119.5	109.1
922	Palm Springs	97.2	123.9	108.9
923-924	San Bernardino	94.8	125.7	108.4
925	Riverside	99.0	125.6	110.7
926-927	Santa Ana	96.8	122.9	108.3
928	Anaheim	99.1	125.9	110.8
930	Oxnard	97.6	126.5	110.2
931	Santa Barbara	96.9	126.3	109.8
932-933	Bakersfield	98.3	121.3	108.3
934	San Luis Obispo	98.1	124.4	109.6
935	Mojave	95.2	121.4	106.7
936-938	Fresno	98.2	128.2	111.3
939	Salinas	98.8	134.8	114.6
940-941	San Francisco	106.3	158.5	129.1
942,956-958	Sacramento	100.0	131.6	113.8
943	Palo Alto	98.5	148.0	120.2
944	San Mateo	100.8	147.7	121.3
945	Vallejo	99.5	139.6	117.0
946	Oakland	102.5	148.9	122.8
947	Berkeley	102.5	148.9	122.8
948	Richmond	101.7	143.0	119.8
949	San Rafael	103.8	148.4	123.3
950	Santa Cruz	104.3	135.1	117.8

STATE/ZIP	CITY	MAT.	INST.	TOTAL
CALIFORNIA (CONT'D)				
951	San Jose	102.4	150.4	123.4
952	Stockton	100.5	127.7	112.4
953	Modesto	100.4	126.9	112.0
954	Santa Rosa	101.1	147.3	121.3
955	Eureka	102.4	130.0	114.4
959	Marysville	101.6	128.8	113.5
960	Redding	109.1	128.8	117.7
961	Susanville	108.8	127.0	116.8
COLORADO				
800-802	Denver	102.8	74.6	90.5
803	Boulder	98.8	75.9	88.8
804	Golden	101.0	72.6	88.6
805	Fort Collins	102.1	74.3	89.9
806	Greeley	99.3	75.1	88.7
807	Fort Morgan	99.4	71.2	87.0
808-809	Colorado Springs	101.0	73.2	88.8
810	Pueblo	101.3	68.8	87.1
811	Alamosa	103.3	69.0	88.3
812	Salida	103.0	65.8	86.7
813	Durango	103.7	65.5	87.0
814	Montrose	102.4	67.8	87.3
815	Grand Junction	105.5	73.0	91.3
816	Glenwood Springs	103.5	65.9	87.0
CONNECTICUT				
060	New Britain	97.4	117.9	106.3
061	Hartford	98.4	118.6	107.3
062	Willimantic	97.9	118.5	106.9
063	New London	94.7	118.5	105.1
064	Meriden	96.6	117.8	105.9
065	New Haven	98.9	118.9	107.6
066	Bridgeport	98.5	118.4	107.2
067	Waterbury	98.1	118.7	107.1
068	Norwalk	98.0	118.6	107.0
069	Stamford	98.1	125.5	110.1
D.C.				
200-205	Washington	101.1	87.6	95.2
DELAWARE				
197	Newark	97.9	112.4	104.3
198	Wilmington	97.6	112.5	104.1
199	Dover	98.0	112.4	104.3
FLORIDA				
320,322	Jacksonville	94.7	65.3	81.8
321	Daytona Beach	94.9	71.0	84.4
323	Tallahassee	95.6	66.2	82.7
324	Panama City	96.2	65.1	82.5
325	Pensacola	98.8	67.1	84.9
326,344	Gainesville	96.4	65.0	82.7
327-328,347	Orlando	96.8	67.0	83.8
329	Melbourne	97.6	71.2	86.0
330-332,340	Miami	95.6	64.6	82.0
333	Fort Lauderdale	95.1	66.0	82.4
334,349	West Palm Beach	94.2	64.0	81.0
335-336,346	Tampa	96.6	67.6	83.9
337	St. Petersburg	99.0	65.9	84.5
338	Lakeland	96.2	67.8	83.7
339,341	Fort Myers	95.4	65.4	82.3
342	Sarasota	98.1	67.9	84.9
GEORGIA				
300-303,399	Atlanta	98.7	75.0	88.3
304	Statesboro	98.7	66.3	84.5
305	Gainesville	97.2	67.6	84.2
306	Athens	96.6	69.1	84.5
307	Dalton	98.3	72.2	86.9
308-309	Augusta	96.8	74.2	86.9
310-312	Macon	93.8	75.1	85.6
313-314	Savannah	95.4	74.4	86.2
315	Waycross	95.3	69.5	84.0
316	Valdosta	95.1	67.3	82.9
317,398	Albany	95.0	73.4	85.5
318-319	Columbus	94.9	73.8	85.7

For customer support on your Building Construction Costs with RSMeans data, call 800.448.8182.

Location Factors - Commercial

STATE/ZIP	CITY	MAT.	INST.	TOTAL	STATE/ZIP	CITY	MAT.	INST.	TOTAL
HAWAII					**KANSAS (CONT'D)**				
967	Hilo	116.9	116.0	116.5	678	Dodge City	100.9	72.9	88.7
968	Honolulu	121.1	116.0	118.9	679	Liberal	99.0	70.6	86.6
STATES & POSS.					**KENTUCKY**				
969	Guam	139.1	53.4	101.6	400-402	Louisville	95.1	79.1	88.1
					403-405	Lexington	94.8	80.3	88.5
IDAHO					406	Frankfort	96.0	78.5	88.3
832	Pocatello	101.5	79.5	91.9	407-409	Corbin	92.7	77.7	86.2
833	Twin Falls	102.8	77.8	91.8	410	Covington	94.7	80.5	88.5
834	Idaho Falls	100.0	78.5	90.6	411-412	Ashland	93.5	93.4	93.5
835	Lewiston	108.2	86.3	98.6	413-414	Campton	94.7	80.2	88.3
836-837	Boise	100.5	79.8	91.4	415-416	Pikeville	95.9	81.4	89.6
838	Coeur d'Alene	108.1	84.7	97.8	417-418	Hazard	94.1	78.7	87.4
					420	Paducah	92.6	84.5	89.1
ILLINOIS					421-422	Bowling Green	94.6	79.9	88.2
600-603	North Suburban	98.9	140.6	117.1	423	Owensboro	94.7	84.3	90.2
604	Joliet	98.8	142.0	117.7	424	Henderson	92.4	83.0	88.3
605	South Suburban	98.9	140.6	117.1	425-426	Somerset	91.9	79.2	86.4
606-608	Chicago	100.8	145.8	120.5	427	Elizabethtown	91.5	77.2	85.3
609	Kankakee	95.6	133.5	112.2					
610-611	Rockford	98.2	128.3	111.4	**LOUISIANA**				
612	Rock Island	96.4	101.9	98.8	700-701	New Orleans	98.7	68.4	85.5
613	La Salle	97.6	126.5	110.3	703	Thibodaux	95.6	67.4	83.3
614	Galesburg	97.4	108.9	102.5	704	Hammond	93.2	65.1	80.9
615-616	Peoria	99.3	112.2	104.9	705	Lafayette	94.9	70.1	84.0
617	Bloomington	96.8	113.0	103.9	706	Lake Charles	95.1	70.9	84.5
618-619	Champaign	100.3	110.6	104.8	707-708	Baton Rouge	96.3	70.1	84.8
620-622	East St. Louis	94.4	109.4	100.9	710-711	Shreveport	97.9	66.7	84.2
623	Quincy	96.7	103.9	99.8	712	Monroe	96.7	66.9	83.7
624	Effingham	95.8	108.7	101.5	713-714	Alexandria	96.9	68.2	84.3
625	Decatur	97.3	109.0	102.4					
626-627	Springfield	97.8	108.8	102.6	**MAINE**				
628	Centralia	93.4	108.8	100.1	039	Kittery	93.0	84.9	89.5
629	Carbondale	93.2	108.8	100.0	040-041	Portland	98.7	85.6	93.0
					042	Lewiston	96.3	85.6	91.6
INDIANA					043	Augusta	98.4	79.8	90.3
460	Anderson	95.1	83.3	89.9	044	Bangor	95.8	85.0	91.1
461-462	Indianapolis	98.3	83.2	91.7	045	Bath	94.8	82.2	89.3
463-464	Gary	96.5	110.4	102.6	046	Machias	94.2	79.2	87.7
465-466	South Bend	96.3	85.0	91.4	047	Houlton	94.4	79.2	87.8
467-468	Fort Wayne	95.7	77.7	87.8	048	Rockland	93.5	79.2	87.2
469	Kokomo	93.5	81.7	88.3	049	Waterville	94.7	79.8	88.2
470	Lawrenceburg	92.8	79.6	87.0					
471	New Albany	94.0	78.0	87.0	**MARYLAND**				
472	Columbus	96.3	80.8	89.5	206	Waldorf	97.9	85.2	92.3
473	Muncie	96.3	81.7	89.9	207-208	College Park	97.9	86.8	93.0
474	Bloomington	97.9	81.3	90.6	209	Silver Spring	97.2	85.8	92.2
475	Washington	94.8	86.5	91.1	210-212	Baltimore	101.4	84.7	94.1
476-477	Evansville	95.5	85.5	91.1	214	Annapolis	98.9	84.2	92.4
478	Terre Haute	96.3	85.4	91.5	215	Cumberland	96.9	86.3	92.3
479	Lafayette	96.0	81.6	89.7	216	Easton	98.6	76.6	88.9
					217	Hagerstown	97.2	87.7	93.0
IOWA					218	Salisbury	99.0	69.8	86.2
500-503,509	Des Moines	97.2	88.6	93.5	219	Elkton	95.8	87.4	92.1
504	Mason City	95.8	76.0	87.1					
505	Fort Dodge	95.9	73.0	85.9	**MASSACHUSETTS**				
506-507	Waterloo	97.2	80.2	89.7	010-011	Springfield	97.8	109.0	102.7
508	Creston	96.2	82.1	90.0	012	Pittsfield	97.3	106.5	101.3
510-511	Sioux City	97.9	79.2	89.7	013	Greenfield	95.7	109.0	101.5
512	Sibley	97.1	61.0	81.3	014	Fitchburg	94.4	120.8	105.9
513	Spencer	98.7	62.1	82.7	015-016	Worcester	97.7	120.8	107.8
514	Carroll	95.4	79.9	88.8	017	Framingham	93.7	127.9	108.7
515	Council Bluffs	99.0	81.1	91.2	018	Lowell	97.1	127.1	110.3
516	Shenandoah	96.4	76.5	87.7	019	Lawrence	98.0	128.5	111.4
520	Dubuque	97.3	82.8	91.0	020-022, 024	Boston	101.2	133.1	115.2
521	Decorah	96.8	74.0	86.8	023	Brockton	97.9	121.8	108.4
522-524	Cedar Rapids	98.2	86.5	93.1	025	Buzzards Bay	92.8	119.5	104.5
525	Ottumwa	96.7	76.2	87.7	026	Hyannis	95.1	119.5	105.8
526	Burlington	96.0	82.3	90.0	027	New Bedford	97.0	119.8	107.0
527-528	Davenport	97.3	98.6	97.9					
					MICHIGAN				
KANSAS					480,483	Royal Oak	94.3	100.5	97.0
660-662	Kansas City	98.4	99.3	98.8	481	Ann Arbor	96.2	102.4	98.9
664-666	Topeka	99.1	78.1	89.9	482	Detroit	99.2	103.0	100.9
667	Fort Scott	97.3	79.1	89.3	484-485	Flint	95.9	92.3	94.4
668	Emporia	97.4	75.8	87.9	486	Saginaw	95.6	88.7	92.6
669	Belleville	99.1	69.9	86.3	487	Bay City	95.7	88.3	92.5
670-672	Wichita	99.2	72.1	87.3	488-489	Lansing	97.0	90.1	94.0
673	Independence	99.6	75.6	89.1	490	Battle Creek	94.8	83.7	89.9
674	Salina	99.6	70.8	87.0	491	Kalamazoo	95.1	82.1	89.4
675	Hutchinson	95.0	70.5	84.3	492	Jackson	93.4	91.8	92.7
676	Hays	99.1	71.3	86.9	493,495	Grand Rapids	96.4	82.9	90.5
677	Colby	99.8	73.5	88.3	494	Muskegon	93.6	81.2	88.2

For customer support on your Building Construction Costs with RSMeans data, call 800.448.8182.

STATE/ZIP	CITY	MAT.	INST.	TOTAL
MICHIGAN (CONT'D)				
496	Traverse City	92.9	78.4	86.6
497	Gaylord	94.1	80.7	88.2
498-499	Iron Mountain	96.0	82.4	90.0
MINNESOTA				
550-551	Saint Paul	98.9	114.1	105.6
553-555	Minneapolis	100.4	113.0	105.9
556-558	Duluth	99.8	102.7	101.1
559	Rochester	98.8	100.9	99.7
560	Mankato	96.0	98.0	96.9
561	Windom	94.6	92.3	93.6
562	Willmar	94.3	97.4	95.6
563	St. Cloud	95.2	111.3	102.3
564	Brainerd	95.9	97.5	96.6
565	Detroit Lakes	97.8	92.0	95.3
566	Bemidji	97.0	96.0	96.6
567	Thief River Falls	96.6	89.2	93.4
MISSISSIPPI				
386	Clarksdale	94.7	55.2	77.4
387	Greenville	98.3	65.6	83.9
388	Tupelo	96.2	55.9	78.6
389	Greenwood	96.0	53.2	77.3
390-392	Jackson	97.3	65.3	83.3
393	Meridian	95.1	65.9	82.3
394	Laurel	96.7	56.3	79.0
395	Biloxi	96.7	65.5	83.1
396	McComb	95.0	53.7	76.9
397	Columbus	96.6	55.8	78.7
MISSOURI				
630-631	St. Louis	100.5	106.2	103.0
633	Bowling Green	98.7	95.0	97.1
634	Hannibal	97.6	93.6	95.9
635	Kirksville	101.1	88.5	95.6
636	Flat River	99.6	95.7	97.9
637	Cape Girardeau	99.1	91.5	95.8
638	Sikeston	98.0	89.1	94.1
639	Poplar Bluff	97.5	89.7	94.1
640-641	Kansas City	100.3	102.7	101.4
644-645	St. Joseph	98.9	92.8	96.2
646	Chillicothe	96.5	95.1	95.9
647	Harrisonville	95.9	102.8	98.9
648	Joplin	97.8	79.6	89.8
650-651	Jefferson City	97.8	92.5	95.5
652	Columbia	98.1	92.5	95.7
653	Sedalia	98.6	93.2	96.2
654-655	Rolla	96.2	96.9	96.5
656-658	Springfield	99.6	81.7	91.7
MONTANA				
590-591	Billings	100.0	76.4	89.6
592	Wolf Point	99.9	76.5	89.7
593	Miles City	97.8	76.7	88.6
594	Great Falls	101.3	75.3	89.9
595	Havre	98.9	75.2	88.5
596	Helena	99.6	75.3	88.9
597	Butte	99.8	75.3	89.1
598	Missoula	97.4	74.5	87.4
599	Kalispell	96.8	75.0	87.2
NEBRASKA				
680-681	Omaha	98.4	80.8	90.7
683-685	Lincoln	98.3	80.2	90.4
686	Columbus	97.3	81.4	90.3
687	Norfolk	98.8	78.6	90.0
688	Grand Island	98.5	78.4	89.7
689	Hastings	98.5	79.0	89.9
690	McCook	98.0	71.5	86.4
691	North Platte	97.8	75.2	87.9
692	Valentine	100.5	70.5	87.4
693	Alliance	100.4	71.9	87.9
NEVADA				
889-891	Las Vegas	103.1	105.2	104.0
893	Ely	101.9	94.1	98.5
894-895	Reno	101.8	85.3	94.6
897	Carson City	101.0	84.9	93.9
898	Elko	100.6	81.2	92.1
NEW HAMPSHIRE				
030	Nashua	97.7	92.6	95.5
031	Manchester	98.3	92.7	95.9

STATE/ZIP	CITY	MAT.	INST.	TOTAL
NEW HAMPSHIRE (CONT'D)				
032-033	Concord	97.9	92.4	95.5
034	Keene	94.8	79.6	88.1
035	Littleton	94.9	79.8	88.3
036	Charleston	94.3	76.3	86.4
037	Claremont	93.5	76.3	86.0
038	Portsmouth	95.1	92.4	93.9
NEW JERSEY				
070-071	Newark	99.1	137.9	116.1
072	Elizabeth	96.8	138.2	114.9
073	Jersey City	95.9	137.6	114.1
074-075	Paterson	97.4	138.2	115.3
076	Hackensack	95.5	137.9	114.0
077	Long Branch	95.0	136.2	113.1
078	Dover	95.7	138.2	114.3
079	Summit	95.8	138.2	114.3
080,083	Vineland	95.0	134.4	112.2
081	Camden	97.0	133.3	112.9
082,084	Atlantic City	95.7	134.7	112.8
085-086	Trenton	98.5	133.9	114.0
087	Point Pleasant	96.8	135.8	113.9
088-089	New Brunswick	97.4	137.8	115.1
NEW MEXICO				
870-872	Albuquerque	98.5	75.1	88.3
873	Gallup	98.9	75.1	88.5
874	Farmington	99.1	75.1	88.6
875	Santa Fe	99.1	75.1	88.6
877	Las Vegas	97.2	75.1	87.5
878	Socorro	96.8	75.1	87.3
879	Truth/Consequences	96.6	71.8	85.7
880	Las Cruces	96.4	71.7	85.6
881	Clovis	98.9	75.0	88.4
882	Roswell	100.3	75.1	89.3
883	Carrizozo	101.1	75.1	89.7
884	Tucumcari	99.5	75.0	88.8
NEW YORK				
100-102	New York	100.6	178.2	134.6
103	Staten Island	96.4	172.2	129.6
104	Bronx	94.7	173.4	129.2
105	Mount Vernon	95.1	145.6	117.2
106	White Plains	94.6	145.6	116.9
107	Yonkers	98.9	150.8	121.6
108	New Rochelle	95.5	145.5	117.4
109	Suffern	95.1	136.0	113.0
110	Queens	99.6	174.8	132.6
111	Long Island City	101.2	174.8	133.5
112	Brooklyn	101.6	172.5	132.7
113	Flushing	101.9	174.8	133.8
114	Jamaica	100.2	174.8	132.9
115,117,118	Hicksville	99.6	158.9	125.6
116	Far Rockaway	102.0	174.8	133.9
119	Riverhead	100.4	159.0	126.1
120-122	Albany	95.8	111.2	102.5
123	Schenectady	96.1	111.2	102.7
124	Kingston	100.0	131.6	113.8
125-126	Poughkeepsie	99.2	142.4	118.1
127	Monticello	98.5	130.8	112.7
128	Glens Falls	91.4	106.4	98.0
129	Plattsburgh	96.6	100.9	98.5
130-132	Syracuse	97.6	102.0	99.5
133-135	Utica	95.7	100.8	97.9
136	Watertown	97.3	100.7	98.8
137-139	Binghamton	97.2	101.7	99.2
140-142	Buffalo	101.2	111.6	105.8
143	Niagara Falls	97.6	111.9	103.9
144-146	Rochester	99.5	102.7	100.9
147	Jamestown	96.6	93.9	95.4
148-149	Elmira	96.5	103.0	99.3
NORTH CAROLINA				
270,272-274	Greensboro	98.6	68.9	85.6
271	Winston-Salem	98.4	68.7	85.4
275-276	Raleigh	96.9	68.3	84.4
277	Durham	100.2	68.9	86.5
278	Rocky Mount	96.3	68.6	84.2
279	Elizabeth City	97.2	70.4	85.5
280	Gastonia	98.4	68.7	85.4
281-282	Charlotte	98.1	68.3	85.0
283	Fayetteville	101.3	68.4	86.9
284	Wilmington	97.2	67.5	84.2
285	Kinston	95.8	68.2	83.7

For customer support on your Building Construction Costs with RSMeans data, call 800.448.8182.

Location Factors - Commercial

STATE/ZIP	CITY	MAT.	INST.	TOTAL
NORTH CAROLINA (CONT'D)				
286	Hickory	96.1	69.3	84.4
287-288	Asheville	97.6	68.3	84.8
289	Murphy	96.9	67.5	84.0
NORTH DAKOTA				
580-581	Fargo	100.1	81.8	92.1
582	Grand Forks	100.1	79.0	90.8
583	Devils Lake	100.1	78.2	90.5
584	Jamestown	100.2	77.1	90.1
585	Bismarck	100.3	81.4	92.0
586	Dickinson	100.8	80.3	91.8
587	Minot	100.0	81.0	91.7
588	Williston	99.3	80.2	90.9
OHIO				
430-432	Columbus	98.1	83.3	91.6
433	Marion	94.4	81.6	88.8
434-436	Toledo	97.5	93.1	95.6
437-438	Zanesville	95.0	82.5	89.5
439	Steubenville	96.2	91.8	94.3
440	Lorain	98.6	86.9	93.5
441	Cleveland	98.9	93.4	96.5
442-443	Akron	99.7	88.2	94.7
444-445	Youngstown	98.9	83.3	92.1
446-447	Canton	99.1	81.0	91.1
448-449	Mansfield	96.8	82.8	90.7
450	Hamilton	96.6	79.4	89.1
451-452	Cincinnati	97.7	80.9	90.3
453-454	Dayton	96.7	77.9	88.5
455	Springfield	96.7	80.2	89.5
456	Chillicothe	96.2	86.1	91.7
457	Athens	99.1	83.7	92.4
458	Lima	99.4	81.1	91.4
OKLAHOMA				
730-731	Oklahoma City	97.7	67.2	84.4
734	Ardmore	96.7	66.7	83.6
735	Lawton	98.7	66.9	84.8
736	Clinton	97.9	66.6	84.2
737	Enid	98.3	67.0	84.6
738	Woodward	96.7	63.8	82.3
739	Guymon	97.8	63.8	82.9
740-741	Tulsa	96.0	66.4	83.0
743	Miami	92.9	65.8	81.0
744	Muskogee	95.1	65.3	82.1
745	McAlester	92.6	61.1	78.8
746	Ponca City	93.3	65.1	81.0
747	Durant	93.2	65.6	81.1
748	Shawnee	94.7	66.3	82.3
749	Poteau	92.5	65.4	80.6
OREGON				
970-972	Portland	99.4	100.8	100.0
973	Salem	100.9	98.7	100.0
974	Eugene	99.1	98.8	99.0
975	Medford	100.6	96.1	98.6
976	Klamath Falls	101.0	96.9	99.2
977	Bend	100.0	98.7	99.4
978	Pendleton	96.0	99.7	97.6
979	Vale	93.8	87.7	91.1
PENNSYLVANIA				
150-152	Pittsburgh	100.3	103.0	101.5
153	Washington	97.3	102.2	99.4
154	Uniontown	97.6	98.6	98.1
155	Bedford	98.7	92.2	95.8
156	Greensburg	98.6	99.9	99.2
157	Indiana	97.4	97.8	97.6
158	Dubois	99.1	93.5	96.7
159	Johnstown	98.7	94.7	96.9
160	Butler	91.4	101.5	95.8
161	New Castle	91.5	100.9	95.6
162	Kittanning	91.9	101.2	95.9
163	Oil City	91.3	99.6	95.0
164-165	Erie	93.1	96.8	94.7
166	Altoona	93.2	95.1	94.0
167	Bradford	94.8	100.2	97.2
168	State College	94.4	96.5	95.3
169	Wellsboro	95.5	94.7	95.1
170-171	Harrisburg	97.3	95.2	96.3
172	Chambersburg	94.4	90.8	92.8
173-174	York	94.4	94.5	94.4
175-176	Lancaster	93.2	96.4	94.6

STATE/ZIP	CITY	MAT.	INST.	TOTAL
PENNSYLVANIA (CONT'D)				
177	Williamsport	91.9	92.8	92.3
178	Sunbury	93.9	95.1	94.4
179	Pottsville	93.0	97.9	95.1
180	Lehigh Valley	94.9	117.4	104.7
181	Allentown	96.5	107.7	101.4
182	Hazleton	94.4	96.8	95.4
183	Stroudsburg	94.3	110.4	101.3
184-185	Scranton	97.2	100.0	98.4
186-187	Wilkes-Barre	94.1	98.9	96.2
188	Montrose	93.8	99.5	96.3
189	Doylestown	94.0	123.5	106.9
190-191	Philadelphia	99.7	134.6	115.0
193	Westchester	95.5	123.7	107.8
194	Norristown	94.5	127.6	109.0
195-196	Reading	96.1	105.2	100.1
PUERTO RICO				
009	San Juan	120.7	27.7	79.9
RHODE ISLAND				
028	Newport	96.3	112.0	103.2
029	Providence	98.1	112.0	104.2
SOUTH CAROLINA				
290-292	Columbia	96.9	69.1	84.7
293	Spartanburg	97.3	68.9	84.9
294	Charleston	98.9	68.8	85.7
295	Florence	97.0	69.0	84.8
296	Greenville	97.1	68.9	84.8
297	Rock Hill	97.1	68.3	84.5
298	Aiken	98.0	69.2	85.4
299	Beaufort	98.7	56.8	80.3
SOUTH DAKOTA				
570-571	Sioux Falls	98.5	76.4	88.8
572	Watertown	98.7	57.6	80.7
573	Mitchell	97.5	63.1	82.4
574	Aberdeen	99.7	64.5	84.3
575	Pierre	100.4	70.5	87.3
576	Mobridge	98.2	65.0	83.7
577	Rapid City	99.5	74.4	88.5
TENNESSEE				
370-372	Nashville	96.7	71.7	85.8
373-374	Chattanooga	98.6	70.3	86.2
375,380-381	Memphis	98.4	70.7	86.3
376	Johnson City	98.8	60.3	82.0
377-379	Knoxville	95.3	66.2	82.5
382	McKenzie	95.8	56.8	78.7
383	Jackson	97.1	60.6	81.1
384	Columbia	94.2	68.4	82.9
385	Cookeville	95.6	57.7	79.0
TEXAS				
750	McKinney	98.4	65.2	83.9
751	Waxahachie	98.5	64.3	83.5
752-753	Dallas	99.2	67.6	85.4
754	Greenville	98.6	65.2	84.0
755	Texarkana	97.9	64.4	83.2
756	Longview	98.9	62.8	83.1
757	Tyler	98.9	64.2	83.7
758	Palestine	95.0	62.7	80.9
759	Lufkin	95.7	64.9	82.2
760-761	Fort Worth	96.6	65.1	82.8
762	Denton	96.8	65.3	83.0
763	Wichita Falls	94.2	63.3	80.7
764	Eastland	93.6	61.8	79.7
765	Temple	91.9	60.4	78.1
766-767	Waco	93.6	64.4	80.8
768	Brownwood	96.9	60.2	80.8
769	San Angelo	96.4	60.7	80.8
770-772	Houston	98.3	68.0	85.0
773	Huntsville	97.4	65.1	83.3
774	Wharton	98.2	66.2	84.2
775	Galveston	96.2	69.0	84.3
776-777	Beaumont	96.5	68.0	84.0
778	Bryan	94.0	67.3	82.3
779	Victoria	98.3	64.9	83.7
780	Laredo	96.1	63.5	81.8
781-782	San Antonio	97.8	64.4	83.2
783-784	Corpus Christi	98.7	62.4	82.8
785	McAllen	99.6	58.6	81.7
786-787	Austin	96.7	62.7	81.8

823

Location Factors - Commercial

STATE/ZIP	CITY	MAT.	INST.	TOTAL
TEXAS (CONT'D)				
788	Del Rio	99.5	62.9	83.4
789	Giddings	96.1	62.6	81.4
790-791	Amarillo	97.6	62.2	82.1
792	Childress	97.6	62.7	82.3
793-794	Lubbock	99.3	63.4	83.6
795-796	Abilene	97.7	62.6	82.3
797	Midland	100.0	64.8	84.6
798-799,885	El Paso	96.0	64.7	82.3
UTAH				
840-841	Salt Lake City	102.8	72.9	89.7
842,844	Ogden	98.2	72.9	87.2
843	Logan	100.2	72.9	88.3
845	Price	100.8	71.7	88.1
846-847	Provo	100.6	72.8	88.5
VERMONT				
050	White River Jct.	97.7	81.6	90.7
051	Bellows Falls	96.0	95.3	95.7
052	Bennington	96.5	91.6	94.3
053	Brattleboro	96.9	95.3	96.2
054	Burlington	101.0	80.9	92.2
056	Montpelier	99.8	85.9	93.7
057	Rutland	98.1	80.9	90.6
058	St. Johnsbury	97.8	81.5	90.7
059	Guildhall	96.4	81.1	89.7
VIRGINIA				
220-221	Fairfax	100.0	83.8	92.9
222	Arlington	100.7	84.5	93.7
223	Alexandria	99.8	84.7	93.2
224-225	Fredericksburg	98.7	82.2	91.5
226	Winchester	99.3	82.9	92.2
227	Culpeper	99.1	80.5	91.0
228	Harrisonburg	99.5	67.3	85.4
229	Charlottesville	99.8	69.3	86.5
230-232	Richmond	98.5	77.9	89.5
233-235	Norfolk	98.5	71.3	86.6
236	Newport News	98.2	70.4	86.0
237	Portsmouth	97.7	64.3	83.1
238	Petersburg	98.3	73.9	87.6
239	Farmville	97.4	58.6	80.4
240-241	Roanoke	100.3	73.0	88.3
242	Bristol	98.6	60.5	82.0
243	Pulaski	98.2	62.7	82.7
244	Staunton	99.0	65.8	84.5
245	Lynchburg	99.2	75.2	88.7
246	Grundy	98.5	56.0	79.9
WASHINGTON				
980-981,987	Seattle	103.6	107.3	105.2
982	Everett	102.0	102.4	102.2
983-984	Tacoma	102.4	102.3	102.3
985	Olympia	100.7	102.3	101.4
986	Vancouver	103.3	97.2	100.6
988	Wenatchee	102.6	89.5	96.9
989	Yakima	102.5	98.2	100.6
990-992	Spokane	101.4	85.0	94.2
993	Richland	101.1	92.7	97.4
994	Clarkston	100.6	83.4	93.1
WEST VIRGINIA				
247-248	Bluefield	97.5	89.6	94.0
249	Lewisburg	99.1	88.9	94.6
250-253	Charleston	97.9	93.3	95.9
254	Martinsburg	98.8	84.4	92.5
255-257	Huntington	99.7	93.8	97.1
258-259	Beckley	97.3	91.8	94.8
260	Wheeling	100.0	91.9	96.5
261	Parkersburg	98.8	89.5	94.8
262	Buckhannon	98.8	91.5	95.6
263-264	Clarksburg	99.4	91.6	96.0
265	Morgantown	99.4	91.6	96.0
266	Gassaway	98.8	89.9	94.9
267	Romney	98.6	86.4	93.3
268	Petersburg	98.5	85.0	92.6
WISCONSIN				
530,532	Milwaukee	99.3	106.6	102.5
531	Kenosha	98.4	105.0	101.3
534	Racine	97.8	106.2	101.5
535	Beloit	97.7	98.6	98.1
537	Madison	98.0	98.9	98.4

STATE/ZIP	CITY	MAT.	INST.	TOTAL
WISCONSIN (CONT'D)				
538	Lancaster	96.0	93.9	95.1
539	Portage	94.4	96.6	95.4
540	New Richmond	95.6	92.7	94.3
541-543	Green Bay	99.6	96.5	98.2
544	Wausau	95.0	94.5	94.8
545	Rhinelander	98.3	90.1	94.7
546	La Crosse	95.9	96.9	96.4
547	Eau Claire	97.6	97.3	97.5
548	Superior	95.4	97.1	96.1
549	Oshkosh	95.7	89.2	92.9
WYOMING				
820	Cheyenne	101.3	73.5	89.2
821	Yellowstone Nat'l Park	99.4	74.4	88.4
822	Wheatland	100.6	69.2	86.9
823	Rawlins	102.1	73.5	89.6
824	Worland	100.1	73.0	88.2
825	Riverton	101.2	70.7	87.8
826	Casper	101.4	70.3	87.8
827	Newcastle	99.9	73.1	88.2
828	Sheridan	102.8	71.5	89.1
829-831	Rock Springs	103.9	72.7	90.2
CANADIAN FACTORS (reflect Canadian currency)				
ALBERTA				
	Calgary	119.5	98.8	110.4
	Edmonton	120.4	98.8	110.9
	Fort McMurray	123.4	91.9	109.7
	Lethbridge	118.3	91.4	106.5
	Lloydminster	112.7	88.1	101.9
	Medicine Hat	112.9	87.4	101.7
	Red Deer	113.3	87.4	102.0
BRITISH COLUMBIA				
	Kamloops	114.1	87.2	102.3
	Prince George	115.1	86.6	102.6
	Vancouver	118.0	91.2	106.3
	Victoria	115.3	85.7	102.3
MANITOBA				
	Brandon	123.6	73.8	101.8
	Portage la Prairie	112.7	72.2	94.9
	Winnipeg	121.3	72.0	99.7
NEW BRUNSWICK				
	Bathurst	111.3	65.0	91.0
	Dalhousie	111.3	65.2	91.1
	Fredericton	118.2	72.5	98.2
	Moncton	111.5	69.3	93.0
	Newcastle	111.3	65.7	91.3
	St. John	113.8	77.0	97.7
NEWFOUNDLAND				
	Corner Brook	128.0	71.5	103.3
	St. Johns	120.7	90.3	107.4
NORTHWEST TERRITORIES				
	Yellowknife	133.4	86.2	112.7
NOVA SCOTIA				
	Bridgewater	112.3	73.4	95.3
	Dartmouth	124.6	73.5	102.2
	Halifax	117.4	87.2	104.2
	New Glasgow	122.6	73.5	101.1
	Sydney	120.6	73.5	100.0
	Truro	111.9	73.4	95.1
	Yarmouth	122.5	73.5	101.1
ONTARIO				
	Barrie	117.0	91.0	105.6
	Brantford	114.3	94.9	105.8
	Cornwall	114.2	91.3	104.2
	Hamilton	116.0	99.6	108.8
	Kingston	115.3	91.4	104.8
	Kitchener	109.1	95.1	103.0
	London	115.6	97.0	107.5
	North Bay	125.6	89.2	109.7
	Oshawa	111.6	94.7	104.2
	Ottawa	117.2	98.0	108.8
	Owen Sound	117.1	89.2	104.9
	Peterborough	114.3	91.1	104.1
	Sarnia	114.3	95.5	106.0
ONTARIO (CONT'D)				

For customer support on your Building Construction Costs with RSMeans data, call 800.448.8182.

STATE/ZIP	CITY	MAT.	INST.	TOTAL
	Sault Ste. Marie	109.6	91.7	101.8
	St. Catharines	107.1	96.1	102.3
	Sudbury	106.9	94.9	101.6
	Thunder Bay	108.3	95.4	102.7
	Timmins	114.4	89.2	103.4
	Toronto	115.9	104.2	110.8
	Windsor	107.6	95.2	102.2
PRINCE EDWARD ISLAND				
	Charlottetown	120.9	62.8	95.5
	Summerside	125.1	60.5	96.8
QUEBEC				
	Cap-de-la-Madeleine	111.7	83.6	99.4
	Charlesbourg	111.7	83.6	99.4
	Chicoutimi	111.1	88.9	101.4
	Gatineau	111.4	83.4	99.1
	Granby	111.6	83.3	99.2
	Hull	111.5	83.4	99.2
	Joliette	111.9	83.6	99.5
	Laval	111.5	83.4	99.2
	Montreal	118.0	91.7	106.5
	Quebec City	118.5	92.2	107.0
	Rimouski	111.3	88.9	101.5
	Rouyn-Noranda	111.4	83.4	99.1
	Saint-Hyacinthe	110.8	83.4	98.8
	Sherbrooke	111.7	83.4	99.3
	Sorel	111.9	83.6	99.5
	Saint-Jerome	111.4	83.4	99.1
	Trois-Rivieres	123.3	83.6	105.9
SASKATCHEWAN				
	Moose Jaw	110.1	66.7	91.1
	Prince Albert	109.2	65.0	89.8
	Regina	122.8	95.4	110.8
	Saskatoon	110.8	92.5	102.7
YUKON				
	Whitehorse	135.9	71.3	107.6

825

R011105-05   Tips for Accurate Estimating

1. Use pre-printed or columnar forms for orderly sequence of dimensions and locations and for recording telephone quotations.

2. Use only the front side of each paper or form except for certain pre-printed summary forms.

3. Be consistent in listing dimensions: For example, length x width x height. This helps in rechecking to ensure that, the total length of partitions is appropriate for the building area.

4. Use printed (rather than measured) dimensions where given.

5. Add up multiple printed dimensions for a single entry where possible.

6. Measure all other dimensions carefully.

7. Use each set of dimensions to calculate multiple related quantities.

8. Convert foot and inch measurements to decimal feet when listing. Memorize decimal equivalents to .01 parts of a foot (1/8″ equals approximately .01′).

9. Do not "round off" quantities until the final summary.

10. Mark drawings with different colors as items are taken off.

11. Keep similar items together, different items separate.

12. Identify location and drawing numbers to aid in future checking for completeness.

13. Measure or list everything on the drawings or mentioned in the specifications.

14. It may be necessary to list items not called for to make the job complete.

15. Be alert for: Notes on plans such as N.T.S. (not to scale); changes in scale throughout the drawings; reduced size drawings; discrepancies between the specifications and the drawings.

16. Develop a consistent pattern of performing an estimate. For example:
    a. Start the quantity takeoff at the lower floor and move to the next higher floor.
    b. Proceed from the main section of the building to the wings.
    c. Proceed from south to north or vice versa, clockwise or counterclockwise.
    d. Take off floor plan quantities first, elevations next, then detail drawings.

17. List all gross dimensions that can be either used again for different quantities, or used as a rough check of other quantities for verification (exterior perimeter, gross floor area, individual floor areas, etc.).

18. Utilize design symmetry or repetition (repetitive floors, repetitive wings, symmetrical design around a center line, similar room layouts, etc.). Note: Extreme caution is needed here so as not to omit or duplicate an area.

19. Do not convert units until the final total is obtained. For instance, when estimating concrete work, keep all units to the nearest cubic foot, then summarize and convert to cubic yards.

20. When figuring alternatives, it is best to total all items involved in the basic system, then total all items involved in the alternates. Therefore you work with positive numbers in all cases. When adds and deducts are used, it is often confusing whether to add or subtract a portion of an item; especially on a complicated or involved alternate.

For customer support on your Building Construction Costs with RSMeans data, call 800.448.8182.

R011105-50 Metric Conversion Factors

Description: This table is primarily for converting customary U.S. units in the left hand column to SI metric units in the right hand column. In addition, conversion factors for some commonly encountered Canadian and non-SI metric units are included.

	If You Know		Multiply By		To Find
Length	Inches	x	25.4[a]	=	Millimeters
	Feet	x	0.3048[a]	=	Meters
	Yards	x	0.9144[a]	=	Meters
	Miles (statute)	x	1.609	=	Kilometers
Area	Square inches	x	645.2	=	Square millimeters
	Square feet	x	0.0929	=	Square meters
	Square yards	x	0.8361	=	Square meters
Volume (Capacity)	Cubic inches	x	16,387	=	Cubic millimeters
	Cubic feet	x	0.02832	=	Cubic meters
	Cubic yards	x	0.7646	=	Cubic meters
	Gallons (U.S. liquids)[b]	x	0.003785	=	Cubic meters[c]
	Gallons (Canadian liquid)[b]	x	0.004546	=	Cubic meters[c]
	Ounces (U.S. liquid)[b]	x	29.57	=	Milliliters[c, d]
	Quarts (U.S. liquid)[b]	x	0.9464	=	Liters[c, d]
	Gallons (U.S. liquid)[b]	x	3.785	=	Liters[c, d]
Force	Kilograms force[d]	x	9.807	=	Newtons
	Pounds force	x	4.448	=	Newtons
	Pounds force	x	0.4536	=	Kilograms force[d]
	Kips	x	4448	=	Newtons
	Kips	x	453.6	=	Kilograms force[d]
Pressure, Stress, Strength (Force per unit area)	Kilograms force per square centimeter[d]	x	0.09807	=	Megapascals
	Pounds force per square inch (psi)	x	0.006895	=	Megapascals
	Kips per square inch	x	6.895	=	Megapascals
	Pounds force per square inch (psi)	x	0.07031	=	Kilograms force per square centimeter[d]
	Pounds force per square foot	x	47.88	=	Pascals
	Pounds force per square foot	x	4.882	=	Kilograms force per square meter[d]
Flow	Cubic feet per minute	x	0.4719	=	Liters per second
	Gallons per minute	x	0.0631	=	Liters per second
	Gallons per hour	x	1.05	=	Milliliters per second
Bending Moment Or Torque	Inch-pounds force	x	0.01152	=	Meter-kilograms force[d]
	Inch-pounds force	x	0.1130	=	Newton-meters
	Foot-pounds force	x	0.1383	=	Meter-kilograms force[d]
	Foot-pounds force	x	1.356	=	Newton-meters
	Meter-kilograms force[d]	x	9.807	=	Newton-meters
Mass	Ounces (avoirdupois)	x	28.35	=	Grams
	Pounds (avoirdupois)	x	0.4536	=	Kilograms
	Tons (metric)	x	1000	=	Kilograms
	Tons, short (2000 pounds)	x	907.2	=	Kilograms
	Tons, short (2000 pounds)	x	0.9072	=	Megagrams[e]
Mass per Unit Volume	Pounds mass per cubic foot	x	16.02	=	Kilograms per cubic meter
	Pounds mass per cubic yard	x	0.5933	=	Kilograms per cubic meter
	Pounds mass per gallon (U.S. liquid)[b]	x	119.8	=	Kilograms per cubic meter
	Pounds mass per gallon (Canadian liquid)[b]	x	99.78	=	Kilograms per cubic meter
Temperature	Degrees Fahrenheit	(F-32)/1.8		=	Degrees Celsius
	Degrees Fahrenheit	(F+459.67)/1.8		=	Degrees Kelvin
	Degrees Celsius	C+273.15		=	Degrees Kelvin

[a]The factor given is exact
[b]One U.S. gallon = 0.8327 Canadian gallon
[c]1 liter = 1000 milliliters = 1000 cubic centimeters
 1 cubic decimeter = 0.001 cubic meter

[d]Metric but not SI unit
[e]Called "tonne" in England and "metric ton" in other metric countries

For customer support on your Building Construction Costs with RSMeans data, call 800.448.8182.

R011105-60   Weights and Measures

Measures of Length
1 Mile = 1760 Yards = 5280 Feet
1 Yard = 3 Feet = 36 inches
1 Foot = 12 Inches
1 Mil = 0.001 Inch
1 Fathom = 2 Yards = 6 Feet
1 Rod = 5.5 Yards = 16.5 Feet
1 Hand = 4 Inches
1 Span = 9 Inches
1 Micro-inch = One Millionth Inch or 0.000001 Inch
1 Micron = One Millionth Meter + 0.00003937 Inch

Surveyor's Measure
1 Mile = 8 Furlongs = 80 Chains
1 Furlong = 10 Chains = 220 Yards
1 Chain = 4 Rods = 22 Yards = 66 Feet = 100 Links
1 Link = 7.92 Inches

Square Measure
1 Square Mile = 640 Acres = 6400 Square Chains
1 Acre = 10 Square Chains = 4840 Square Yards =
  43,560 Sq. Ft.
1 Square Chain = 16 Square Rods = 484 Square Yards =
  4356 Sq. Ft.
1 Square Rod = 30.25 Square Yards = 272.25 Square Feet = 625 Square
  Lines
1 Square Yard = 9 Square Feet
1 Square Foot = 144 Square Inches
An Acre equals a Square 208.7 Feet per Side

Cubic Measure
1 Cubic Yard = 27 Cubic Feet
1 Cubic Foot = 1728 Cubic Inches
1 Cord of Wood = 4 x 4 x 8 Feet = 128 Cubic Feet
1 Perch of Masonry = 16½ x 1½ x 1 Foot = 24.75 Cubic Feet

Avoirdupois or Commercial Weight
1 Gross or Long Ton = 2240 Pounds
1 Net or Short Ton = 2000 Pounds
1 Pound = 16 Ounces = 7000 Grains
1 Ounce = 16 Drachms = 437.5 Grains
1 Stone = 14 Pounds

Power
1 British Thermal Unit per Hour = 0.2931 Watts
1 Ton (Refrigeration) = 3.517 Kilowatts
1 Horsepower (Boiler) = 9.81 Kilowatts
1 Horsepower (550 ft-lb/s) = 0.746 Kilowatts

Shipping Measure
For Measuring Internal Capacity of a Vessel:
  1 Register Ton = 100 Cubic Feet

For Measurement of Cargo:
  Approximately 40 Cubic Feet of Merchandise is considered a Shipping
  Ton, unless that bulk would weigh more than 2000 Pounds, in which case
  Freight Charge may be based upon weight.

40 Cubic Feet = 32.143 U.S. Bushels = 31.16 Imp. Bushels

Liquid Measure
1 Imperial Gallon = 1.2009 U.S. Gallon = 277.42 Cu. In.
1 Cubic Foot = 7.48 U.S. Gallons

R011110-10   Architectural Fees

Tabulated below are typical percentage fees by project size, for good professional architectural service. Fees may vary from those listed depending upon degree of design difficulty and economic conditions in any particular area.

Rates can be interpolated horizontally and vertically. Various portions of the same project requiring different rates should be adjusted proportionately. For alterations, add 50% to the fee for the first $500,000 of project cost and add 25% to the fee for project cost over $500,000.

Architectural fees tabulated below include Structural, Mechanical and Electrical Engineering Fees. They do not include the fees for special consultants such as kitchen planning, security, acoustical, interior design, etc.

Civil Engineering fees are included in the Architectural fee for project sites requiring minimal design such as city sites. However, separate Civil Engineering fees must be added when utility connections require design, drainage calculations are needed, stepped foundations are required, or provisions are required to protect adjacent wetlands.

Building Types	Total Project Size in Thousands of Dollars						
	100	250	500	1,000	5,000	10,000	50,000
Factories, garages, warehouses, repetitive housing	9.0%	8.0%	7.0%	6.2%	5.3%	4.9%	4.5%
Apartments, banks, schools, libraries, offices, municipal buildings	12.2	12.3	9.2	8.0	7.0	6.6	6.2
Churches, hospitals, homes, laboratories, museums, research	15.0	13.6	12.7	11.9	9.5	8.8	8.0
Memorials, monumental work, decorative furnishings	—	16.0	14.5	13.1	10.0	9.0	8.3

For customer support on your Building Construction Costs with RSMeans data, call 800.448.8182.

R011110-30   Engineering Fees

Typical **Structural Engineering Fees** based on type of construction and total project size. These fees are included in Architectural Fees.

Type of Construction	Total Project Size (in thousands of dollars)			
	$500	$500-$1,000	$1,000-$5,000	Over $5000
Industrial buildings, factories & warehouses	Technical payroll times 2.0 to 2.5	1.60%	1.25%	1.00%
Hotels, apartments, offices, dormitories, hospitals, public buildings, food stores		2.00%	1.70%	1.20%
Museums, banks, churches and cathedrals		2.00%	1.75%	1.25%
Thin shells, prestressed concrete, earthquake resistive		2.00%	1.75%	1.50%
Parking ramps, auditoriums, stadiums, convention halls, hangars & boiler houses		2.50%	2.00%	1.75%
Special buildings, major alterations, underpinning & future expansion	▼	Add to above 0.5%	Add to above 0.5%	Add to above 0.5%

For complex reinforced concrete or unusually complicated structures, add 20% to 50%.

Typical **Mechanical and Electrical Engineering Fees** are based on the size of the subcontract. The fee structure for both is shown below. These fees are included in Architectural Fees.

Type of Construction	Subcontract Size							
	$25,000	$50,000	$100,000	$225,000	$350,000	$500,000	$750,000	$1,000,000
Simple structures	6.4%	5.7%	4.8%	4.5%	4.4%	4.3%	4.2%	4.1%
Intermediate structures	8.0	7.3	6.5	5.6	5.1	5.0	4.9	4.8
Complex structures	10.1	9.0	9.0	8.0	7.5	7.5	7.0	7.0

For renovations, add 15% to 25% to applicable fee.

R012153-60   Security Factors

Contractors entering, working in, and exiting secure facilities often lose productive time during a normal workday. The recommended allowances in this section are intended to provide for the loss of productivity by increasing labor costs. Note that different costs are associated with searches upon entry only and searches upon entry and exit. Time spent in a queue is unpredictable and not part of these allowances. Contractors should plan ahead for this situation.

Security checkpoints are designed to reflect the level of security required to gain access or egress. An extreme example is when contractors, along with any materials, tools, equipment, and vehicles, must be physically searched and have all materials, tools, equipment, and vehicles inventoried and documented prior to both entry and exit.

Physical searches without going through the documentation process represent the next level and take up less time.

Electronic searches—passing through a detector or x-ray machine with no documentation of materials, tools, equipment, and vehicles—take less time than physical searches.

Visual searches of materials, tools, equipment, and vehicles represent the next level of security.

Finally, access by means of an ID card or displayed sticker takes the least amount of time.

Another consideration is if the searches described above are performed each and every day, or if they are performed only on the first day with access granted by ID card or displayed sticker for the remainder of the project. The figures for this situation have been calculated to represent the initial check-in as described and subsequent entry by ID card or displayed sticker for up to 20 days on site. For the situation described above, where the time period is beyond 20 days, the impact on labor cost is negligible.

There are situations where tradespeople must be accompanied by an escort and observed during the work day. The loss of freedom of movement will slow down productivity for the tradesperson. Costs for the observer have not been included. Those costs are normally born by the owner.

829

General Requirements — R0121 Allowances

R012157-20 Construction Time Requirements

The table below lists the construction durations for various building types along with their respective project sizes and project values. Design time runs 25% to 40% of construction time.

Building Type	Size S.F.	Project Value	Construction Duration
Industrial/Warehouse	100,000	$8,000,000	14 months
	500,000	$32,000,000	19 months
	1,000,000	$75,000,000	21 months
Offices/Retail	50,000	$7,000,000	15 months
	250,000	$28,000,000	23 months
	500,000	$58,000,000	34 months
Institutional/Hospitals/Laboratory	200,000	$45,000,000	31 months
	500,000	$110,000,000	52 months
	750,000	$160,000,000	55 months
	1,000,000	$210,000,000	60 months

General Requirements — R0129 Payment Procedures

R012909-80 Sales Tax by State

State sales tax on materials is tabulated below (5 states have no sales tax). Many states allow local jurisdictions, such as a county or city, to levy additional sales tax.

Some projects may be sales tax exempt, particularly those constructed with public funds.

State	Tax (%)	State	Tax (%)	State	Tax (%)	State	Tax (%)
Alabama	4	Illinois	6.25	Montana	0	Rhode Island	7
Alaska	0	Indiana	7	Nebraska	5.5	South Carolina	6
Arizona	5.6	Iowa	6	Nevada	6.85	South Dakota	5
Arkansas	6.5	Kansas	6.5	New Hampshire	0	Tennessee	7
California	7.25	Kentucky	6	New Jersey	7	Texas	6.25
Colorado	2.9	Louisiana	4	New Mexico	5.125	Utah	5.95
Connecticut	6.35	Maine	5.5	New York	4	Vermont	6
Delaware	0	Maryland	6	North Carolina	4.75	Virginia	5.3
District of Columbia	5.75	Massachusetts	6.25	North Dakota	5	Washington	6.5
Florida	6	Michigan	6	Ohio	5.75	West Virginia	6
Georgia	4	Minnesota	6.875	Oklahoma	4.5	Wisconsin	5
Hawaii	4	Mississippi	7	Oregon	0	Wyoming	4
Idaho	6	Missouri	4.225	Pennsylvania	6	Average	5.11%

Sales Tax by Province (Canada)

GST - a value-added tax, which the government imposes on most goods and services provided in or imported into Canada. PST - a retail sales tax, which five of the provinces impose on the prices of most goods and some services. QST - a value-added tax, similar to the federal GST, which Quebec imposes. HST - Three provinces have combined their retail sales taxes with the federal GST into one harmonized tax.

Province	PST (%)	QST (%)	GST(%)	HST(%)
Alberta	0	0	5	0
British Columbia	7	0	5	0
Manitoba	8	0	5	0
New Brunswick	0	0	0	15
Newfoundland	0	0	0	15
Northwest Territories	0	0	5	0
Nova Scotia	0	0	0	15
Ontario	0	0	0	13
Prince Edward Island	0	0	0	15
Quebec	0	9.975	5	0
Saskatchewan	6	0	5	0
Yukon	0	0	5	0

R012909-85 Unemployment Taxes and Social Security Taxes

State unemployment tax rates vary not only from state to state, but also with the experience rating of the contractor. The federal unemployment tax rate is 6.0% of the first $7,000 of wages. This is reduced by a credit of up to 5.4% for timely payment to the state. The minimum federal unemployment tax is 0.6% after all credits.

Social security (FICA) for 2018 is estimated at time of publication to be 7.65% of wages up to $127,200.

R012909-86 Unemployment Tax by State

Information is from the U.S. Department of Labor, state unemployment tax rates.

State	Tax (%)	State	Tax (%)	State	Tax (%)	State	Tax (%)
Alabama	6.74	Illinois	7.75	Montana	6.12	Rhode Island	9.79
Alaska	5.4	Indiana	7.474	Nebraska	5.4	South Carolina	5.46
Arizona	8.91	Iowa	8	Nevada	5.4	South Dakota	9.5
Arkansas	6.0	Kansas	7.6	New Hampshire	7.5	Tennessee	10.0
California	6.2	Kentucky	10.0	New Jersey	5.8	Texas	7.5
Colorado	8.9	Louisiana	6.2	New Mexico	5.4	Utah	7.2
Connecticut	6.8	Maine	5.4	New York	8.5	Vermont	8.4
Delaware	8.0	Maryland	7.50	North Carolina	5.76	Virginia	6.27
District of Columbia	7	Massachusetts	11.13	North Dakota	10.72	Washington	5.7
Florida	5.4	Michigan	10.3	Ohio	8.7	West Virginia	7.5
Georgia	5.4	Minnesota	9.0	Oklahoma	5.5	Wisconsin	12.0
Hawaii	5.6	Mississippi	5.4	Oregon	5.4	Wyoming	8.8
Idaho	5.4	Missouri	9.75	Pennsylvania	10.89	Median	7.47%

831

For customer support on your Building Construction Costs with RSMeans data, call 800.448.8182.

R012909-90   Overtime

One way to improve the completion date of a project or eliminate negative float from a schedule is to compress activity duration times. This can be achieved by increasing the crew size or working overtime with the proposed crew.

To determine the costs of working overtime to compress activity duration times, consider the following examples. Below is an overtime efficiency and cost chart based on a five, six, or seven day week with an eight through twelve hour day. Payroll percentage increases for time and one half and double times are shown for the various working days.

Days per Week	Hours per Day	Production Efficiency					Payroll Cost Factors	
		1st Week	2nd Week	3rd Week	4th Week	Average 4 Weeks	@ 1-1/2 Times	@ 2 Times
5	8	100%	100%	100%	100%	100%	1.000	1.000
	9	100	100	95	90	96	1.056	1.111
	10	100	95	90	85	93	1.100	1.200
	11	95	90	75	65	81	1.136	1.273
	12	90	85	70	60	76	1.167	1.333
6	8	100	100	95	90	96	1.083	1.167
	9	100	95	90	85	93	1.130	1.259
	10	95	90	85	80	88	1.167	1.333
	11	95	85	70	65	79	1.197	1.394
	12	90	80	65	60	74	1.222	1.444
7	8	100	95	85	75	89	1.143	1.286
	9	95	90	80	70	84	1.183	1.365
	10	90	85	75	65	79	1.214	1.429
	11	85	80	65	60	73	1.240	1.481
	12	85	75	60	55	69	1.262	1.524

R013113-40   Builder's Risk Insurance

Builder's risk insurance is insurance on a building during construction. Premiums are paid by the owner or the contractor. Blasting, collapse and underground insurance would raise total insurance costs.

For customer support on your Building Construction Costs with RSMeans data, call 800.448.8182.

R013113-50   General Contractor's Overhead

There are two distinct types of overhead on a construction project: Project overhead and main office overhead. Project overhead includes those costs at a construction site not directly associated with the installation of construction materials. Examples of project overhead costs include the following:

1. Superintendent
2. Construction office and storage trailers
3. Temporary sanitary facilities
4. Temporary utilities
5. Security fencing
6. Photographs
7. Cleanup
8. Performance and payment bonds

The above project overhead items are also referred to as general requirements and therefore are estimated in Division 1. Division 1 is the first division listed in the CSI MasterFormat but it is usually the last division estimated. The sum of the costs in Divisions 1 through 49 is referred to as the sum of the direct costs.

All construction projects also include indirect costs. The primary components of indirect costs are the contractor's main office overhead and profit. The amount of the main office overhead expense varies depending on the following:

1. Owner's compensation
2. Project managers' and estimators' wages
3. Clerical support wages
4. Office rent and utilities
5. Corporate legal and accounting costs
6. Advertising
7. Automobile expenses
8. Association dues
9. Travel and entertainment expenses

These costs are usually calculated as a percentage of annual sales volume. This percentage can range from 35% for a small contractor doing less than $500,000 to 5% for a large contractor with sales in excess of $100 million.

R013113-55   Installing Contractor's Overhead

Installing contractors (subcontractors) also incur costs for general requirements and main office overhead.

Included within the total incl. overhead and profit costs is a percent mark-up for overhead that includes:

1. Compensation and benefits for office staff and project managers
2. Office rent, utilities, business equipment, and maintenance
3. Corporate legal and accounting costs

4. Advertising
5. Vehicle expenses (for office staff and project managers)
6. Association dues
7. Travel, entertainment
8. Insurance
9. Small tools and equipment

For customer support on your Building Construction Costs with RSMeans data, call 800.448.8182.

R013113-60   Workers' Compensation Insurance Rates by Trade

The table below tabulates the national averages for workers' compensation insurance rates by trade and type of building. The average "Insurance Rate" is multiplied by the "% of Building Cost" for each trade. This produces the "Workers' Compensation" cost by % of total labor cost, to be added for each trade by building type to determine the weighted average workers' compensation rate for the building types analyzed.

Trade	Insurance Rate (% Labor Cost) Range	Average	% of Building Cost Office Bldgs.	Schools & Apts.	Mfg.	Workers' Compensation Office Bldgs.	Schools & Apts.	Mfg.
Excavation, Grading, etc.	2.7 % to 20.1%	8.5%	4.8%	4.9%	4.5%	0.41%	0.42%	0.38%
Piles & Foundations	5.3 to 29.8	13.4	7.1	5.2	8.7	0.95	0.70	1.17
Concrete	4.1 to 28.0	11.8	5.0	14.8	3.7	0.59	1.75	0.44
Masonry	3.9 to 49.3	13.8	6.9	7.5	1.9	0.95	1.04	0.26
Structural Steel	5.3 to 59.1	21.2	10.7	3.9	17.6	2.27	0.83	3.73
Miscellaneous & Ornamental Metals	3.3 to 24.4	10.6	2.8	4.0	3.6	0.30	0.42	0.38
Carpentry & Millwork	4.4 to 32.4	13.0	3.7	4.0	0.5	0.48	0.52	0.07
Metal or Composition Siding	5.5 to 107.2	19.0	2.3	0.3	4.3	0.44	0.06	0.82
Roofing	5.5 to 120.3	29.0	2.3	2.6	3.1	0.67	0.75	0.90
Doors & Hardware	3.2 to 32.4	11.0	0.9	1.4	0.4	0.10	0.15	0.04
Sash & Glazing	4.7 to 25.5	12.1	3.5	4.0	1.0	0.42	0.48	0.12
Lath & Plaster	3.0 to 31.6	10.7	3.3	6.9	0.8	0.35	0.74	0.09
Tile, Marble & Floors	2.7 to 18.3	8.7	2.6	3.0	0.5	0.23	0.26	0.04
Acoustical Ceilings	2.4 to 46.3	8.5	2.4	0.2	0.3	0.20	0.02	0.03
Painting	3.3 to 38.8	11.2	1.5	1.6	1.6	0.17	0.18	0.18
Interior Partitions	4.4 to 32.4	13.0	3.9	4.3	4.4	0.51	0.56	0.57
Miscellaneous Items	2.3 to 97.7	11.2	5.2	3.7	9.7	0.58	0.42	1.09
Elevators	1.3 to 13.7	4.7	2.1	1.1	2.2	0.10	0.05	0.10
Sprinklers	2.0 to 15.5	6.7	0.5	—	2.0	0.03	—	0.13
Plumbing	1.7 to 14.0	6.3	4.9	7.2	5.2	0.31	0.45	0.33
Heat., Vent., Air Conditioning	3.3 to 17.8	8.3	13.5	11.0	12.9	1.12	0.91	1.07
Electrical	1.9 to 11.6	5.2	10.1	8.4	11.1	0.53	0.44	0.58
Total	1.3 % to 120.3%	—	100.0%	100.0%	100.0%	11.71%	11.15%	12.52%
			Overall Weighted Average	11.79%				

Workers' Compensation Insurance Rates by States

The table below lists the weighted average Workers' Compensation base rate for each state with a factor comparing this with the national average of 11.8%.

State	Weighted Average	Factor	State	Weighted Average	Factor	State	Weighted Average	Factor
Alabama	15.0%	127	Kentucky	10.4%	88	North Dakota	6.2%	53
Alaska	10.4	88	Louisiana	18.7	158	Ohio	7.2	61
Arizona	9.6	81	Maine	10.4	88	Oklahoma	8.9	75
Arkansas	7.0	59	Maryland	11.3	96	Oregon	9.3	79
California	22.2	188	Massachusetts	11.2	95	Pennsylvania	21.1	179
Colorado	7.5	64	Michigan	8.2	69	Rhode Island	13.7	116
Connecticut	17.5	148	Minnesota	16.9	143	South Carolina	16.5	140
Delaware	13.9	118	Mississippi	11.8	100	South Dakota	11.8	100
District of Columbia	9.1	77	Missouri	12.4	105	Tennessee	8.6	73
Florida	11.1	94	Montana	8.8	75	Texas	6.6	56
Georgia	31.9	270	Nebraska	13.5	114	Utah	7.4	63
Hawaii	8.5	72	Nevada	7.5	64	Vermont	10.9	92
Idaho	9.4	80	New Hampshire	12.0	102	Virginia	6.9	58
Illinois	21.1	179	New Jersey	14.8	125	Washington	9.1	77
Indiana	4.1	35	New Mexico	13.3	113	West Virginia	4.5	38
Iowa	13.7	116	New York	19.2	163	Wisconsin	12.2	103
Kansas	6.5	55	North Carolina	15.8	134	Wyoming	5.6	47
			Weighted Average for U.S. is	11.8% of payroll = 100%				

The weighted average skilled worker rate for 35 trades is 11.8%. For bidding purposes, apply the full value of Workers' Compensation directly to total labor costs, or if labor is 38%, materials 42% and overhead and profit 20% of total cost, carry 38/80 x 11.8% = 6.0% of cost (before overhead and profit)

into overhead. Rates vary not only from state to state but also with the experience rating of the contractor.

Rates are the most current available at the time of publication.

For customer support on your Building Construction Costs with RSMeans data, call 800.448.8182.

General Requirements   R0131 Project Management & Coordination

R013113-80   Performance Bond

This table shows the cost of a performance bond for a construction job scheduled to be completed in 12 months. Add 1% of the premium cost per month for jobs requiring more than 12 months to complete. The rates are "standard" rates offered to contractors that the bonding company considers financially sound and capable of doing the work. Preferred rates are offered by some bonding companies based upon financial strength of the contractor. Actual rates vary from contractor to contractor and from bonding company to bonding company. Contractors should prequalify through a bonding agency before submitting a bid on a contract that requires a bond.

Contract Amount	Building Construction Class B Projects			Highways & Bridges					
				Class A New Construction			Class A-1 Highway Resurfacing		
First $ 100,000 bid	$25.00 per M			$15.00 per M			$9.40 per M		
Next 400,000 bid	$ 2,500	plus	$15.00	per M	$ 1,500	plus	$10.00	per M	$ 940	plus	$7.20	per M
Next 2,000,000 bid	8,500	plus	10.00	per M	5,500	plus	7.00	per M	3,820	plus	5.00	per M
Next 2,500,000 bid	28,500	plus	7.50	per M	19,500	plus	5.50	per M	15,820	plus	4.50	per M
Next 2,500,000 bid	47,250	plus	7.00	per M	33,250	plus	5.00	per M	28,320	plus	4.50	per M
Over 7,500,000 bid	64,750	plus	6.00	per M	45,750	plus	4.50	per M	39,570	plus	4.00	per M

General Requirements   R0151 Temporary Utilities

R015113-65   Temporary Power Equipment

Cost data for the temporary equipment was developed utilizing the following information.

1) Re-usable material-services, transformers, equipment and cords are based on new purchase and prorated to three projects.
2) PVC feeder includes trench and backfill.
3) Connections include disconnects and fuses.

4) Labor units include an allowance for removal.
5) No utility company charges or fees are included.
6) Concrete pads or vaults are not included.
7) Utility company conduits not included.

For customer support on your Building Construction Costs with RSMeans data, call 800.448.8182.

R015423-10   Steel Tubular Scaffolding

On new construction, tubular scaffolding is efficient up to 60' high or five stories. Above this it is usually better to use a hung scaffolding if construction permits. Swing scaffolding operations may interfere with tenants. In this case, the tubular is more practical at all heights.

In repairing or cleaning the front of an existing building the cost of tubular scaffolding per S.F. of building front increases as the height increases above the first tier. The first tier cost is relatively high due to leveling and alignment.

The minimum efficient crew for erecting and dismantling is three workers. They can set up and remove 18 frame sections per day up to 5 stories high. For 6 to 12 stories high, a crew of four is most efficient. Use two or more on top and two on the bottom for handing up or hoisting. They can

also set up and remove 18 frame sections per day. At 7' horizontal spacing, this will run about 800 S.F. per day of erecting and dismantling. Time for placing and removing planks must be added to the above. A crew of three can place and remove 72 planks per day up to 5 stories. For over 5 stories, a crew of four can place and remove 80 planks per day.

The table below shows the number of pieces required to erect tubular steel scaffolding for 1000 S.F. of building frontage. This area is made up of a scaffolding system that is 12 frames (11 bays) long by 2 frames high.

For jobs under twenty-five frames, add 50% to rental cost. Rental rates will be lower for jobs over three months duration. Large quantities for long periods can reduce rental rates by 20%.

Description of Component	Number of Pieces for 1000 S.F. of Building Front	Unit
5' Wide Standard Frame, 6'-4" High	24	Ea.
Leveling Jack & Plate	24	
Cross Brace	44	
Side Arm Bracket, 21"	12	
Guardrail Post	12	
Guardrail, 7' section	22	
Stairway Section	2	
Stairway Starter Bar	1	
Stairway Inside Handrail	2	
Stairway Outside Handrail	2	
Walk-Thru Frame Guardrail	2	

Scaffolding is often used as falsework over 15' high during construction of cast-in-place concrete beams and slabs. Two foot wide scaffolding is generally used for heavy beam construction. The span between frames depends upon the load to be carried with a maximum span of 5'.

Heavy duty shoring frames with a capacity of 10,000#/leg can be spaced up to 10' O.C. depending upon form support design and loading.

Scaffolding used as horizontal shoring requires less than half the material required with conventional shoring.

On new construction, erection is done by carpenters.

Rolling towers supporting horizontal shores can reduce labor and speed the job. For maintenance work, catwalks with spans up to 70' can be supported by the rolling towers.

R015423-20   Pump Staging

Pump staging is generally not available for rent. The table below shows the number of pieces required to erect pump staging for 2400 S.F. of building

frontage. This area is made up of a pump jack system that is 3 poles (2 bays) wide by 2 poles high.

Item	Number of Pieces for 2400 S.F. of Building Front	Unit
Aluminum pole section, 24' long	6	Ea.
Aluminum splice joint, 6' long	3	
Aluminum foldable brace	3	
Aluminum pump jack	3	
Aluminum support for workbench/back safety rail	3	
Aluminum scaffold plank/workbench, 14" wide x 24' long	4	
Safety net, 22' long	2	
Aluminum plank end safety rail	2	

The cost in place for this 2400 S.F. will depend on how many uses are realized during the life of the equipment.

R015433-10   Contractor Equipment

Rental Rates shown elsewhere in the book pertain to late model high quality machines in excellent working condition, rented from equipment dealers. Rental rates from contractors may be substantially lower than the rental rates from equipment dealers depending upon economic conditions; for older, less productive machines, reduce rates by a maximum of 15%. Any overtime must be added to the base rates. For shift work, rates are lower. Usual rule of thumb is 150% of one shift rate for two shifts; 200% for three shifts.

For periods of less than one week, operated equipment is usually more economical to rent than renting bare equipment and hiring an operator.

Costs to move equipment to a job site (mobilization) or from a job site (demobilization) are not included in rental rates, nor in any Equipment costs on any Unit Price line items or crew listings. These costs can be found elsewhere. If a piece of equipment is already at a job site, it is not appropriate to utilize mob/demob costs in an estimate again.

Rental rates vary throughout the country with larger cities generally having lower rates. Lease plans for new equipment are available for periods in excess of six months with a percentage of payments applying toward purchase.

Rental rates can also be treated as reimbursement costs for contractor-owned equipment. Owned equipment costs include depreciation, loan payments, interest, taxes, insurance, storage, and major repairs.

Monthly rental rates vary from 2% to 5% of the cost of the equipment depending on the anticipated life of the equipment and its wearing parts. Weekly rates are about 1/3 the monthly rates and daily rental rates are about 1/3 the weekly rates.

The hourly operating costs for each piece of equipment include costs to the user such as fuel, oil, lubrication, normal expendables for the equipment, and a percentage of the mechanic's wages chargeable to maintenance. The hourly operating costs listed do not include the operator's wages.

The daily cost for equipment used in the standard crews is figured by dividing the weekly rate by five, then adding eight times the hourly operating cost to give the total daily equipment cost, not including the operator. This figure is in the right hand column of the Equipment listings under Equipment Cost/Day.

Pile Driving rates shown for the pile hammer and extractor do not include leads, cranes, boilers or compressors. Vibratory pile driving requires an added field specialist during set-up and pile driving operation for the electric model. The hydraulic model requires a field specialist for set-up only. Up to 125 reuses of sheet piling are possible using vibratory drivers. For normal conditions, crane capacity for hammer type and size is as follows.

Crane Capacity	Hammer Type and Size		
	Air or Steam	Diesel	Vibratory
25 ton	to 8,750 ft.-lb.		70 H.P.
40 ton	15,000 ft.-lb.	to 32,000 ft.-lb.	170 H.P.
60 ton	25,000 ft.-lb.		300 H.P.
100 ton		112,000 ft.-lb.	

Cranes should be specified for the job by size, building and site characteristics, availability, performance characteristics, and duration of time required.

Backhoes & Shovels rent for about the same as equivalent size cranes but maintenance and operating expenses are higher. The crane operator's rate must be adjusted for high boom heights. Average adjustments: for 150' boom add 2% per hour; over 185', add 4% per hour; over 210', add 6% per hour; over 250', add 8% per hour and over 295', add 12% per hour.

Tower Cranes of the climbing or static type have jibs from 50' to 200' and capacities at maximum reach range from 4,000 to 14,000 pounds. Lifting capacities increase up to maximum load as the hook radius decreases.

Typical rental rates, based on purchase price, are about 2% to 3% per month.

Erection and dismantling run between 500 and 2000 labor hours. Climbing operation takes 10 labor hours per 20' climb. Crane dead time is about 5 hours per 40' climb. If crane is bolted to side of the building add cost of ties and extra mast sections. Climbing cranes have from 80' to 180' of mast while static cranes have 80' to 800' of mast.

Truck Cranes can be converted to tower cranes by using tower attachments. Mast heights over 400' have been used.

A single 100' high material **Hoist and Tower** can be erected and dismantled in about 400 labor hours; a double 100' high hoist and tower in about 600 labor hours. Erection times for additional heights are 3 and 4 labor hours per vertical foot respectively up to 150', and 4 to 5 labor hours per vertical foot over 150' high. A 40' high portable Buck hoist takes about 160 labor hours to erect and dismantle. Additional heights take 2 labor hours per vertical foot to 80' and 3 labor hours per vertical foot for the next 100'. Most material hoists do not meet local code requirements for carrying personnel.

A 150' high **Personnel Hoist** requires about 500 to 800 labor hours to erect and dismantle. Budget erection time at 5 labor hours per vertical foot for all trades. Local code requirements or labor scarcity requiring overtime can add up to 50% to any of the above erection costs.

Earthmoving Equipment: The selection of earthmoving equipment depends upon the type and quantity of material, moisture content, haul distance, haul road, time available, and equipment available. Short haul cut and fill operations may require dozers only, while another operation may require excavators, a fleet of trucks, and spreading and compaction equipment. Stockpiled material and granular material are easily excavated with front end loaders. Scrapers are most economically used with hauls between 300' and 1-1/2 miles if adequate haul roads can be maintained. Shovels are often used for blasted rock and any material where a vertical face of 8' or more can be excavated. Special conditions may dictate the use of draglines, clamshells, or backhoes. Spreading and compaction equipment must be matched to the soil characteristics, the compaction required and the rate the fill is being supplied.

R015433-15   Heavy Lifting

Hydraulic Climbing Jacks

The use of hydraulic heavy lift systems is an alternative to conventional type crane equipment. The lifting, lowering, pushing, or pulling mechanism is a hydraulic climbing jack moving on a square steel jackrod from 1-5/8" to 4" square, or a steel cable. The jackrod or cable can be vertical or horizontal, stationary or movable, depending on the individual application. When the jackrod is stationary, the climbing jack will climb the rod and push or pull the load along with itself. When the climbing jack is stationary, the jackrod is movable with the load attached to the end and the climbing jack will lift or lower the jackrod with the attached load. The heavy lift system is normally operated by a single control lever located at the hydraulic pump.

The system is flexible in that one or more climbing jacks can be applied wherever a load support point is required, and the rate of lift synchronized.

Economic benefits have been demonstrated on projects such as: erection of ground assembled roofs and floors, complete bridge spans, girders and trusses, towers, chimney liners and steel vessels, storage tanks, and heavy machinery. Other uses are raising and lowering offshore work platforms, caissons, tunnel sections and pipelines.

837

R015436-50 Mobilization

Costs to move rented construction equipment to a job site from an equipment dealer's or contractor's yard (mobilization) or off the job site (demobilization), are not included in the rental or operating rates, nor in the equipment cost on a unit price line or in a crew listing. These costs can be found consolidated in the Mobilization section of the data and elsewhere in particular site work sections. If a piece of

equipment is already on the job site, it is not appropriate to include mob/demob costs in a new estimate that requires use of that equipment. The following table identifies approximate sizes of rented construction equipment that would be hauled on a towed trailer. Because this listing is not all-encompassing, the user can infer as to what size trailer might be required for a piece of equipment not listed.

3-ton Trailer	20-ton Trailer	40-ton Trailer	50-ton Trailer
20 H.P. Excavator	110 H.P. Excavator	200 H.P. Excavator	270 H.P. Excavator
50 H.P. Skid Steer	165 H.P. Dozer	300 H.P. Dozer	Small Crawler Crane
35 H.P. Roller	150 H.P. Roller	400 H.P. Scraper	500 H.P. Scraper
40 H.P. Trencher	Backhoe	450 H.P. Art. Dump Truck	500 H.P. Art. Dump Truck

For customer support on your Building Construction Costs with RSMeans data, call 800.448.8182.

R024119-10   Demolition Defined

Whole Building Demolition - Demolition of the whole building with no concern for any particular building element, component, or material type being demolished. This type of demolition is accomplished with large pieces of construction equipment that break up the structure, load it into trucks and haul it to a disposal site, but disposal or dump fees are not included. Demolition of below-grade foundation elements, such as footings, foundation walls, grade beams, slabs on grade, etc., is not included. Certain mechanical equipment containing flammable liquids or ozone-depleting refrigerants, electric lighting elements, communication equipment components, and other building elements may contain hazardous waste, and must be removed, either selectively or carefully, as hazardous waste before the building can be demolished.

Foundation Demolition - Demolition of below-grade foundation footings, foundation walls, grade beams, and slabs on grade. This type of demolition is accomplished by hand or pneumatic hand tools, and does not include saw cutting, or handling, loading, hauling, or disposal of the debris.

Gutting - Removal of building interior finishes and electrical/mechanical systems down to the load-bearing and sub-floor elements of the rough building frame, with no concern for any particular building element, component, or material type being demolished. This type of demolition is accomplished by hand or pneumatic hand tools, and includes loading into trucks, but not hauling, disposal or dump fees, scaffolding, or shoring. Certain mechanical equipment containing flammable liquids or ozone-depleting refrigerants, electric lighting elements, communication equipment components, and other building elements may contain hazardous waste, and must be removed, either selectively or carefully, as hazardous waste, before the building is gutted.

Selective Demolition - Demolition of a selected building element, component, or finish, with some concern for surrounding or adjacent elements, components, or finishes (see the first Subdivision (s) at the beginning of appropriate Divisions). This type of demolition is accomplished by hand or pneumatic hand tools, and does not include handling, loading,

storing, hauling, or disposal of the debris, scaffolding, or shoring. "Gutting" methods may be used in order to save time, but damage that is caused to surrounding or adjacent elements, components, or finishes may have to be repaired at a later time.

Careful Removal - Removal of a piece of service equipment, building element or component, or material type, with great concern for both the removed item and surrounding or adjacent elements, components or finishes. The purpose of careful removal may be to protect the removed item for later re-use, preserve a higher salvage value of the removed item, or replace an item while taking care to protect surrounding or adjacent elements, components, connections, or finishes from cosmetic and/or structural damage. An approximation of the time required to perform this type of removal is 1/3 to 1/2 the time it would take to install a new item of like kind (see Reference Number R220105-10). This type of removal is accomplished by hand or pneumatic hand tools, and does not include loading, hauling, or storing the removed item, scaffolding, shoring, or lifting equipment.

Cutout Demolition - Demolition of a small quantity of floor, wall, roof, or other assembly, with concern for the appearance and structural integrity of the surrounding materials. This type of demolition is accomplished by hand or pneumatic hand tools, and does not include saw cutting, handling, loading, hauling, or disposal of debris, scaffolding, or shoring.

Rubbish Handling - Work activities that involve handling, loading or hauling of debris. Generally, the cost of rubbish handling must be added to the cost of all types of demolition, with the exception of whole building demolition.

Minor Site Demolition - Demolition of site elements outside the footprint of a building. This type of demolition is accomplished by hand or pneumatic hand tools, or with larger pieces of construction equipment, and may include loading a removed item onto a truck (check the Crew for equipment used). It does not include saw cutting, hauling or disposal of debris, and, sometimes, handling or loading.

R024119-20   Dumpsters

Dumpster rental costs on construction sites are presented in two ways:

The cost per week rental includes the delivery of the dumpster; its pulling or emptying once per week, and its final removal. The assumption is made that the dumpster contractor could choose to empty a dumpster by simply bringing in an empty unit and removing the full one. These costs also include the disposal of the materials in the dumpster.

The alternate pricing can be used when actual planned conditions are not approximated by the weekly numbers. For example, these lines can be used when a dumpster is needed for 4 weeks and will need to be emptied 2 or 3 times per week. Conversely the alternate pricing lines can be used when a dumpster will be rented for several weeks or months but needs to be emptied only a few times over this period.

R024119-30   Rubbish Handling Chutes

To correctly estimate the cost of rubbish handling chute systems, the individual components must be priced separately. First choose the size of the system; a 30-inch diameter chute is quite common, but the sizes range from 18 to 36 inches in diameter. The 30-inch chute comes in a standard weight and two thinner weights. The thinner weight chutes are sometimes chosen for cost savings, but they are more easily damaged.

There are several types of major chute pieces that make up the chute system. The first component to consider is the top chute section (top intake hopper) where the material is dropped into the chute at the highest point. After determining the top chute, the intermediate chute pieces called the regular chute sections are priced. Next, the number of chute control door sections (intermediate intake hoppers) must be determined. In the more complex systems, a chute control door section is provided at each floor level. The last major component to consider is bolt down frames; these are usually provided at every other floor level.

There are a number of accessories to consider for safe operation and control. There are covers for the top chute and the chute door sections. The top

chute can have a trough that allows for better loading of the chute. For the safest operation, a chute warning light system can be added that will warn the other chute intake locations not to load while another is being used. There are dust control devices that spray a water mist to keep down the dust as the debris is loaded into a dumpster. There are special breakaway cords that are used to prevent damage to the chute if the dumpster is removed without disconnecting from the chute. There are chute liners that can be installed to protect the chute structure from physical damage from rough abrasive materials. Warning signs can be posted at each floor level that is provided with a chute control door section.

In summary, a complete rubbish handling chute system will include one top section, several intermediate regular sections, several intermediate control door (intake hopper) sections and bolt down frames at every other floor level starting with the top floor. If so desired, the system can also include covers and a light warning system for a safer operation. The bottom of the chute should always be above the Dumpster and should be tied off with a breakaway cord to the Dumpster.

839

R026510-20   Underground Storage Tank Removal

Underground storage tank removal can be divided into two categories: non-leaking and leaking. Prior to removing an underground storage tank, tests should be made, with the proper authorities present, to determine whether a tank has been leaking or the surrounding soil has been contaminated.

To safely remove liquid underground storage tanks:

1. Excavate to the top of the tank.
2. Disconnect all piping.
3. Open all tank vents and access ports.
4. Remove all liquids and/or sludge.
5. Purge the tank with an inert gas.
6. Provide access to the inside of the tank and clean out the interior using proper personal protective equipment (PPE).
7. Excavate soil surrounding the tank using proper PPE for on-site personnel.
8. Pull and properly dispose of the tank.
9. Clean up the site of all contaminated material.
10. Install new tanks or close the excavation.

R028213-20   Asbestos Removal Process

Asbestos removal is accomplished by a specialty contractor who understands the federal and state regulations regarding the handling and disposal of the material. The process of asbestos removal is divided into many individual steps. An accurate estimate can be calculated only after all the steps have been priced.

The steps are generally as follows:

1. Obtain an asbestos abatement plan from an industrial hygienist.
2. Monitor the air quality in and around the removal area and along the path of travel between the removal area and transport area. This establishes the background contamination.
3. Construct a two part decontamination chamber at entrance to removal area.
4. Install a HEPA filter to create a negative pressure in the removal area.
5. Install wall, floor and ceiling protection as required by the plan, usually 2 layers of fireproof 6 mil polyethylene.
6. Industrial hygienist visually inspects work area to verify compliance with plan.
7. Provide temporary supports for conduit and piping affected by the removal process.
8. Proceed with asbestos removal and bagging process. Monitor air quality as described in Step #2. Discontinue operations when contaminate levels exceed applicable standards.
9. Document the legal disposal of materials in accordance with EPA standards.
10. Thoroughly clean removal area including all ledges, crevices and surfaces.
11. Post abatement inspection by industrial hygienist to verify plan compliance.
12. Provide a certificate from a licensed industrial hygienist attesting that contaminate levels are within acceptable standards before returning area to regular use.

R028319-60   Lead Paint Remediation Methods

Lead paint remediation can be accomplished by the following methods.

1. Abrasive blast
2. Chemical stripping
3. Power tool cleaning with vacuum collection system
4. Encapsulation
5. Remove and replace
6. Enclosure

Each of these methods has strengths and weaknesses depending on the specific circumstances of the project. The following is an overview of each method.

1. **Abrasive blasting** is usually accomplished with sand or recyclable metallic blast. Before work can begin, the area must be contained to ensure the blast material with lead does not escape to the atmosphere. The use of vacuum blast greatly reduces the containment requirements. Lead abatement equipment that may be associated with this work includes a negative air machine. In addition, it is necessary to have an industrial hygienist monitor the project on a continual basis. When the work is complete, the spent blast sand with lead must be disposed of as a hazardous material. If metallic shot was used, the lead is separated from the shot and disposed of as hazardous material. Worker protection includes disposable clothing and respiratory protection.

2. **Chemical stripping** requires strong chemicals to be applied to the surface to remove the lead paint. Before the work can begin, the area under/adjacent to the work area must be covered to catch the chemical and removed lead. After the chemical is applied to the painted surface it is usually covered with paper. The chemical is left in place for the specified period, then the paper with lead paint is pulled or scraped off. The process may require several chemical applications. The paper with chemicals and lead paint adhered to it, plus the containment and loose scrapings collected by a HEPA (high efficiency particulate air filter) vac, must be disposed of as hazardous material. The chemical stripping process usually requires a neutralizing agent and several wash downs after the paint is removed. Worker protection includes a neoprene or other compatible protective clothing and respiratory

protection with face shield. An industrial hygienist is required intermittently during the process.

3. **Power tool cleaning** is accomplished using shrouded needle blasting guns. The shrouding with different end configurations is held up against the surface to be cleaned. The area is blasted with hardened needles and the shroud captures the lead with a HEPA vac and deposits it in a holding tank. An industrial hygienist monitors the project. Protective clothing and a respirator are required until air samples prove otherwise. When the work is complete the lead must be disposed of as hazardous material.

4. **Encapsulation** is a method that leaves the well bonded lead paint in place after the peeling paint has been removed. Before the work can begin, the area under/adjacent to the work must be covered to catch the scrapings. The scraped surface is then washed with a detergent and rinsed. The prepared surface is covered with approximately 10 mils of paint. A reinforcing fabric can also be embedded in the paint covering. The scraped paint and containment must be disposed of as hazardous material. Workers must wear protective clothing and respirators.

5. **Removing and replacing** are effective ways to remove lead paint from windows, gypsum walls and concrete masonry surfaces. The painted materials are removed and new materials are installed. Workers should wear a respirator and tyvek suit. The demolished materials must be disposed of as hazardous waste if it fails the TCLP (toxicity characteristic leachate process) test.

6. **Enclosure** is the process that permanently seals lead painted materials in place. This process has many applications such as covering lead painted drywall with new drywall, covering exterior construction with tyvek paper then residing, or covering lead painted structural members with aluminum or plastic. The seams on all enclosing materials must be securely sealed. An industrial hygienist monitors the project, and protective clothing and a respirator are required until air samples prove otherwise.

All the processes require clearance monitoring and wipe testing as required by the hygienist.

For customer support on your Building Construction Costs with RSMeans data, call 800.448.8182.

R031113-10   Wall Form Materials

Aluminum Forms

Approximate weight is 3 lbs. per S.F.C.A. Standard widths are available from 4" to 36" with 36" most common. Standard lengths of 2', 4', 6' to 8' are available. Forms are lightweight and fewer ties are needed with the wider widths. The form face is either smooth or textured.

Metal Framed Plywood Forms

Manufacturers claim over 75 reuses of plywood and over 300 reuses of steel frames. Many specials such as corners, fillers, pilasters, etc. are available. Monthly rental is generally about 15% of purchase price for first month and 9% per month thereafter with 90% of rental applied to purchase for the first month and decreasing percentages thereafter. Aluminum framed forms cost 25% to 30% more than steel framed.

After the first month, extra days may be prorated from the monthly charge. Rental rates do not include ties, accessories, cleaning, loss of hardware or freight in and out. Approximate weight is 5 lbs. per S.F. for steel; 3 lbs. per S.F. for aluminum.

Forms can be rented with option to buy.

Plywood Forms, Job Fabricated

There are two types of plywood used for concrete forms.

1. Exterior plyform is completely waterproof. This is face oiled to facilitate stripping. Ten reuses can be expected with this type with 25 reuses possible.
2. An overlaid type consists of a resin fiber fused to exterior plyform. No oiling is required except to facilitate cleaning. This is available in both high density (HDO) and medium density overlaid (MDO). Using HDO, 50 reuses can be expected with 200 possible.

Plyform is available in 5/8" and 3/4" thickness. High density overlaid is available in 3/8", 1/2", 5/8" and 3/4" thickness.

5/8" thick is sufficient for most building forms, while 3/4" is best on heavy construction.

Plywood Forms, Modular, Prefabricated

There are many plywood forming systems without frames. Most of these are manufactured from 1-1/8" (HDO) plywood and have some hardware attached. These are used principally for foundation walls 8' or less high. With care and maintenance, 100 reuses can be attained with decreasing quality of surface finish.

Steel Forms

Approximate weight is 6-1/2 lbs. per S.F.C.A. including accessories. Standard widths are available from 2" to 24", with 24" the most common. Standard lengths are from 2' to 8', with 4' the most common. Forms are easily ganged into modular units.

Forms are usually leased for 15% of the purchase price per month prorated daily over 30 days.

Rental may be applied to sale price, and usually rental forms are bought. With careful handling and cleaning 200 to 400 reuses are possible.

Straight wall gang forms up to 12' x 20' or 8' x 30' can be fabricated. These crane handled forms usually lease for approx. 9% per month.

Individual job analysis is available from the manufacturer at no charge.

R031113-40   Forms for Reinforced Concrete

Design Economy

Avoid many sizes in proportioning beams and columns.

From story to story avoid changing column dimensions. Gain strength by adding steel or using a richer mix. If a change in size of column is necessary, vary one dimension only to minimize form alterations. Keep beams and columns the same width.

From floor to floor in a multi-story building vary beam depth, not width, as that will leave the slab panel form unchanged. It is cheaper to vary the strength of a beam from floor to floor by means of a steel area than by 2" changes in either width or depth.

Cost Factors

Material includes the cost of lumber, cost of rent for metal pans or forms if used, nails, form ties, form oil, bolts and accessories.

Labor includes the cost of carpenters to make up, erect, remove and repair, plus common labor to clean and move. Having carpenters remove forms minimizes repairs.

Improper alignment and condition of forms will increase finishing cost. When forms are heavily oiled, concrete surfaces must be neutralized before finishing. Special curing compounds will cause spillages to spall off in first frost. Gang forming methods will reduce costs on large projects.

Materials Used

Boards are seldom used unless their architectural finish is required. Generally, steel, fiberglass and plywood are used for contact surfaces. Labor on plywood is 10% less than with boards. The plywood is backed up with

2 x 4's at 12" to 32" O.C. Walers are generally 2 - 2 x 4's. Column forms are held together with steel yokes or bands. Shoring is with adjustable shoring or scaffolding for high ceilings.

Reuse

Floor and column forms can be reused four or possibly five times without excessive repair. Remember to allow for 10% waste on each reuse.

When modular sized wall forms are made, up to twenty uses can be expected with exterior plyform.

When forms are reused, the cost to erect, strip, clean and move will not be affected. 10% replacement of lumber should be included and about one hour of carpenter time for repairs on each reuse per 100 S.F.

The reuse cost for certain accessory items normally rented on a monthly basis will be lower than the cost for the first use.

After the fifth use, new material required plus time needed for repair prevent the form cost from dropping further; it may go up. Much depends on care in stripping, the number of special bays, changes in beam or column sizes and other factors.

Costs for multiple use of formwork may be developed as follows:

2 Uses	3 Uses	4 Uses
$\dfrac{\text{(1st Use + Reuse)}}{2} = \text{avg. cost/2 uses}$	$\dfrac{\text{(1st Use + 2 Reuses)}}{3} = \text{avg. cost/3 uses}$	$\dfrac{\text{(1st use + 3 Reuses)}}{4} = \text{avg. cost/4 uses}$

For customer support on your Building Construction Costs with RSMeans data, call 800.448.8182.

R031113-60   Formwork Labor-Hours

Item	Unit	Hours Required			Total Hours	Multiple Use		
		Fabricate	Erect & Strip	Clean & Move	1 Use	2 Use	3 Use	4 Use
Beam and Girder, interior beams, 12" wide	100 S.F.	6.4	8.3	1.3	16.0	13.3	12.4	12.0
Hung from steel beams		5.8	7.7	1.3	14.8	12.4	11.6	11.2
Beam sides only, 36" high		5.8	7.2	1.3	14.3	11.9	11.1	10.7
Beam bottoms only, 24" wide		6.6	13.0	1.3	20.9	18.1	17.2	16.7
Box out for openings		9.9	10.0	1.1	21.0	16.6	15.1	14.3
Buttress forms, to 8' high		6.0	6.5	1.2	13.7	11.2	10.4	10.0
Centering, steel, 3/4" rib lath			1.0		1.0			
3/8" rib lath or slab form			0.9		0.9			
Chamfer strip or keyway	100 L.F.		1.5		1.5	1.5	1.5	1.5
Columns, fiber tube 8" diameter			20.6		20.6			
12"			21.3		21.3			
16"			22.9		22.9			
20"			23.7		23.7			
24"			24.6		24.6			
30"			25.6		25.6			
Columns, round steel, 12" diameter			22.0		22.0	22.0	22.0	22.0
16"			25.6		25.6	25.6	25.6	25.6
20"			30.5		30.5	30.5	30.5	30.5
24"			37.7		37.7	37.7	37.7	37.7
Columns, plywood 8" x 8"	100 S.F.	7.0	11.0	1.2	19.2	16.2	15.2	14.7
12" x 12"		6.0	10.5	1.2	17.7	15.2	14.4	14.0
16" x 16"		5.9	10.0	1.2	17.1	14.7	13.8	13.4
24" x 24"		5.8	9.8	1.2	16.8	14.4	13.6	13.2
Columns, steel framed plywood 8" x 8"			10.0	1.0	11.0	11.0	11.0	11.0
12" x 12"			9.3	1.0	10.3	10.3	10.3	10.3
16" x 16"			8.5	1.0	9.5	9.5	9.5	9.5
24" x 24"			7.8	1.0	8.8	8.8	8.8	8.8
Drop head forms, plywood		9.0	12.5	1.5	23.0	19.0	17.7	17.0
Coping forms		8.5	15.0	1.5	25.0	21.3	20.0	19.4
Culvert, box			14.5	4.3	18.8	18.8	18.8	18.8
Curb forms, 6" to 12" high, on grade		5.0	8.5	1.2	14.7	12.7	12.1	11.7
On elevated slabs		6.0	10.8	1.2	18.0	15.5	14.7	14.3
Edge forms to 6" high, on grade	100 L.F.	2.0	3.5	0.6	6.1	5.6	5.4	5.3
7" to 12" high	100 S.F.	2.5	5.0	1.0	8.5	7.8	7.5	7.4
Equipment foundations		10.0	18.0	2.0	30.0	25.5	24.0	23.3
Flat slabs, including drops		3.5	6.0	1.2	10.7	9.5	9.0	8.8
Hung from steel		3.0	5.5	1.2	9.7	8.7	8.4	8.2
Closed deck for domes		3.0	5.8	1.2	10.0	9.0	8.7	8.5
Open deck for pans		2.2	5.3	1.0	8.5	7.9	7.7	7.6
Footings, continuous, 12" high		3.5	3.5	1.5	8.5	7.3	6.8	6.6
Spread, 12" high		4.7	4.2	1.6	10.5	8.7	8.0	7.7
Pile caps, square or rectangular		4.5	5.0	1.5	11.0	9.3	8.7	8.4
Grade beams, 24" deep		2.5	5.3	1.2	9.0	8.3	8.0	7.9
Lintel or Sill forms		8.0	17.0	2.0	27.0	23.5	22.3	21.8
Spandrel beams, 12" wide		9.0	11.2	1.3	21.5	17.5	16.2	15.5
Stairs			25.0	4.0	29.0	29.0	29.0	29.0
Trench forms in floor		4.5	14.0	1.5	20.0	18.3	17.7	17.4
Walls, Plywood, at grade, to 8' high		5.0	6.5	1.5	13.0	11.0	9.7	9.5
8' to 16'		7.5	8.0	1.5	17.0	13.8	12.7	12.1
16' to 20'		9.0	10.0	1.5	20.5	16.5	15.2	14.5
Foundation walls, to 8' high		4.5	6.5	1.0	12.0	10.3	9.7	9.4
8' to 16' high		5.5	7.5	1.0	14.0	11.8	11.0	10.6
Retaining wall to 12' high, battered		6.0	8.5	1.5	16.0	13.5	12.7	12.3
Radial walls to 12' high, smooth		8.0	9.5	2.0	19.5	16.0	14.8	14.3
2' chords		7.0	8.0	1.5	16.5	13.5	12.5	12.0
Prefabricated modular, to 8' high		—	4.3	1.0	5.3	5.3	5.3	5.3
Steel, to 8' high		—	6.8	1.2	8.0	8.0	8.0	8.0
8' to 16' high		—	9.1	1.5	10.6	10.3	10.2	10.2
Steel framed plywood to 8' high		—	6.8	1.2	8.0	7.5	7.3	7.2
8' to 16' high		—	9.3	1.2	10.5	9.5	9.2	9.0

843

For customer support on your **Building Construction Costs** with RSMeans data, call 800.448.8182.

R032110-10 Reinforcing Steel Weights and Measures

Bar Designation No.**	Nominal Weight Lb./Ft.	U.S. Customary Units Nominal Dimensions*			SI Units Nominal Dimensions*			
		Diameter in.	Cross Sectional Area, in.^{2}	Perimeter in.	Nominal Weight kg/m	Diameter mm	Cross Sectional Area, cm^2	Perimeter mm
3	.376	.375	.11	1.178	.560	9.52	.71	29.9
4	.668	.500	.20	1.571	.994	12.70	1.29	39.9
5	1.043	.625	.31	1.963	1.552	15.88	2.00	49.9
6	1.502	.750	.44	2.356	2.235	19.05	2.84	59.8
7	2.044	.875	.60	2.749	3.042	22.22	3.87	69.8
8	2.670	1.000	.79	3.142	3.973	25.40	5.10	79.8
9	3.400	1.128	1.00	3.544	5.059	28.65	6.45	90.0
10	4.303	1.270	1.27	3.990	6.403	32.26	8.19	101.4
11	5.313	1.410	1.56	4.430	7.906	35.81	10.06	112.5
14	7.650	1.693	2.25	5.320	11.384	43.00	14.52	135.1
18	13.600	2.257	4.00	7.090	20.238	57.33	25.81	180.1

* The nominal dimensions of a deformed bar are equivalent to those of a plain round bar having the same weight per foot as the deformed bar.
** Bar numbers are based on the number of eighths of an inch included in the nominal diameter of the bars.

R032110-20 Metric Rebar Specification - ASTM A615-81

Grade 300 (300 MPa* = 43,560 psi; +8.7% vs. Grade 40)				
Grade 400 (400 MPa* = 58,000 psi; –3.4% vs. Grade 60)				
Bar No.	Diameter mm	Area mm^2	Equivalent in.^{2}	Comparison with U.S. Customary Bars
10M	11.3	100	.16	Between #3 & #4
15M	16.0	200	.31	#5 (.31 in.^{2})
20M	19.5	300	.47	#6 (.44 in.^{2})
25M	25.2	500	.78	#8 (.79 in.^{2})
30M	29.9	700	1.09	#9 (1.00 in.^{2})
35M	35.7	1000	1.55	#11 (1.56 in.^{2})
45M	43.7	1500	2.33	#14 (2.25 in.^{2})
55M	56.4	2500	3.88	#18 (4.00 in.^{2})

* MPa = megapascals

For customer support on your Building Construction Costs with RSMeans data, call 800.448.8182.

R032110-40 Weight of Steel Reinforcing Per Square Foot of Wall (PSF)

Reinforced weights: The table below suggests the weights per square foot for reinforcing steel in walls. Weights are approximate and will be the same for all grades of steel bars. For bars in two directions, add weights for each size and spacing.

C/C Spacing in Inches	#3 Wt. (PSF)	#4 Wt. (PSF)	#5 Wt. (PSF)	#6 Wt. (PSF)	#7 Wt. (PSF)	#8 Wt. (PSF)	#9 Wt. (PSF)	#10 Wt. (PSF)	#11 Wt. (PSF)
						Bar Size			
2"	2.26	4.01	6.26	9.01	12.27				
3"	1.50	2.67	4.17	6.01	8.18	10.68	13.60	17.21	21.25
4"	1.13	2.01	3.13	4.51	6.13	8.10	10.20	12.91	15.94
5"	.90	1.60	2.50	3.60	4.91	6.41	8.16	10.33	12.75
6"	.752	1.34	2.09	3.00	4.09	5.34	6.80	8.61	10.63
8"	.564	1.00	1.57	2.25	3.07	4.01	5.10	6.46	7.97
10"	.451	.802	1.25	1.80	2.45	3.20	4.08	5.16	6.38
12"	.376	.668	1.04	1.50	2.04	2.67	3.40	4.30	5.31
18"	.251	.445	.695	1.00	1.32	1.78	2.27	2.86	3.54
24"	.188	.334	.522	.751	1.02	1.34	1.70	2.15	2.66
30"	.150	.267	.417	.600	.817	1.07	1.36	1.72	2.13
36"	.125	.223	.348	.501	.681	.890	1.13	1.43	1.77
42"	.107	.191	.298	.429	.584	.753	.97	1.17	1.52
48"	.094	.167	.261	.376	.511	.668	.85	1.08	1.33

R032110-50 Minimum Wall Reinforcement Weight (PSF)

This table lists the approximate minimum wall reinforcement weights per S.F. according to the specification of .12% of gross area for vertical bars and .20% of gross area for horizontal bars.

Location	Wall Thickness	Bar Size	Horizontal Steel Spacing C/C	Sq. In. Req'd per S.F.	Total Wt. per S.F.	Bar Size	Vertical Steel Spacing C/C	Sq. In. Req'd per S.F.	Total Wt. per S.F.	Horizontal & Vertical Steel Total Weight per S.F.
Both Faces	10"	#4	18"	.24	.89#	#3	18"	.14	.50#	1.39#
	12"	#4	16"	.29	1.00	#3	16"	.17	.60	1.60
	14"	#4	14"	.34	1.14	#3	13"	.20	.69	1.84
	16"	#4	12"	.38	1.34	#3	11"	.23	.82	2.16
	18"	#5	17"	.43	1.47	#4	18"	.26	.89	2.36
One Face	6"	#3	9"	.15	.50	#3	18"	.09	.25	.75
	8"	#4	12"	.19	.67	#3	11"	.12	.41	1.08
	10"	#5	15"	.24	.83	#4	16"	.14	.50	1.34

R032110-70 Bend, Place, and Tie Reinforcing

Placing and tying by rodmen for footings and slabs run from nine hrs. per ton for heavy bars to fifteen hrs. per ton for light bars. For beams, columns, and walls, production runs from eight hrs. per ton for heavy bars to twenty hrs. per ton for light bars. The overall average for typical reinforced concrete buildings is about fourteen hrs. per ton. These production figures include the time for placing accessories and usual inserts, but not their material cost (allow 15% of the cost of delivered bent rods). Equipment handling is necessary for the larger-sized bars so that installation costs for the very heavy bars will not decrease proportionately.

Installation costs for splicing reinforcing bars include allowance for equipment to hold the bars in place while splicing as well as necessary scaffolding for iron workers.

R032110-80 Shop-Fabricated Reinforcing Steel

The material prices for reinforcing, shown in the unit cost sections of the data set, are for 50 tons or more of shop-fabricated reinforcing steel and include:
1. Mill base price of reinforcing steel
2. Mill grade/size/length extras
3. Mill delivery to the fabrication shop
4. Shop storage and handling
5. Shop drafting/detailing
6. Shop shearing and bending
7. Shop listing
8. Shop delivery to the job site

Both material and installation costs can be considerably higher for small jobs consisting primarily of smaller bars, while material costs may be slightly lower for larger jobs.

R032205-30   Common Stock Styles of Welded Wire Fabric

This table provides some of the basic specifications, sizes, and weights of welded wire fabric used for reinforcing concrete.

	New Designation	Old Designation		Steel Area per Foot				Approximate Weight per 100 S.F.	
	Spacing — Cross Sectional Area (in.) — (Sq. in. 100)	Spacing — Wire Gauge (in.) — (AS & W)		Longitudinal		Transverse			
				in.	cm	in.	cm	lbs	kg
Rolls	6 x 6 — W1.4 x W1.4	6 x 6 — 10 x 10		.028	.071	.028	.071	21	9.53
	6 x 6 — W2.0 x W2.0	6 x 6 — 8 x 8	1	.040	.102	.040	.102	29	13.15
	6 x 6 — W2.9 x W2.9	6 x 6 — 6 x 6		.058	.147	.058	.147	42	19.05
	6 x 6 — W4.0 x W4.0	6 x 6 — 4 x 4		.080	.203	.080	.203	58	26.91
	4 x 4 — W1.4 x W1.4	4 x 4 — 10 x 10		.042	.107	.042	.107	31	14.06
	4 x 4 — W2.0 x W2:0	4 x 4 — 8 x 8	1	.060	.152	.060	.152	43	19.50
	4 x 4 — W2.9 x W2.9	4 x 4 — 6 x 6		.087	.227	.087	.227	62	28.12
	4 x 4 — W4.0 x W4.0	4 x 4 — 4 x 4		.120	.305	.120	.305	85	38.56
Sheets	6 x 6 — W2.9 x W2.9	6 x 6 — 6 x 6		.058	.147	.058	.147	42	19.05
	6 x 6 — W4.0 x W4.0	6 x 6 — 4 x 4		.080	.203	.080	.203	58	26.31
	6 x 6 — W5.5 x W5.5	6 x 6 — 2 x 2	2	.110	.279	.110	.279	80	36.29
	4 x 4 — W1.4 x W1.4	4 x 4 — 4 x 4		.120	.305	.120	.305	85	38.56

NOTES: 1. Exact W—number size for 8 gauge is W2.1
2. Exact W—number size for 2 gauge is W5.4

The above table was compiled with the following excerpts from the WRI Manual of Standard Practices, 7th Edition, Copyright 2006. Reproduced with permission of the Wire Reinforcement Institute, Inc.:

1. Chapter 3, page 7, Table 1 Common Styles of Metric Wire Reinforcement (WWR) With Equivalent US Customary Units
2. Chapter 6, page 19, Table 5 Customary Units
3. Chapter 6, Page 23, Table 7 Customary Units (in.) Welded Plain Wire Reinforcement
4. Chapter 6, Page 25 Table 8 Wire Size Comparison
5. Chapter 9, Page 30, Table 9 Weight of Longitudinal Wires Weight (Mass) Estimating Tables
6. Chapter 9, Page 31, Table 9M Weight of Longitudinal Wires Weight (Mass) Estimating Tables
7. Chapter 9, Page 32, Table 10 Weight of Transverse Wires Based on 62″ lengths of transverse wire (60″ width plus 1″ overhand each side)
8. Chapter 9, Page 33, Table 10M Weight of Transverse Wires

Concrete

R0330 Cast-In-Place Concrete

R033053-50   Industrial Chimneys

Foundation requirements in C.Y. of concrete for various sized chimneys.

Size Chimney	2 Ton Soil	3 Ton Soil	Size Chimney	2 Ton Soil	3 Ton Soil	Size Chimney	2 Ton Soil	3 Ton Soil
75' x 3'-0″	13 C.Y.	11 C.Y.	160' x 6'-6″	86 C.Y.	76 C.Y.	300' x 10'-0″	325 C.Y.	245 C.Y.
85' x 5'-6″	19	16	175' x 7'-0″	108	95	350' x 12'-0″	422	320
100' x 5'-0″	24	20	200' x 6'-0″	125	105	400' x 14'-0″	520	400
125' x 5'-6″	43	36	250' x 8'-0″	230	175	500' x 18'-0″	725	575

For customer support on your Building Construction Costs with RSMeans data, call 800.448.8182.

R033105-10 Proportionate Quantities

The tables below show both quantities per S.F. of floor areas as well as form and reinforcing quantities per C.Y. Unusual structural requirements would increase the ratios below. High strength reinforcing would reduce the steel weights. Figures are for 3000 psi concrete and 60,000 psi reinforcing unless specified otherwise.

Type of Construction	Live Load	Span	Per S.F. of Floor Area				Per C.Y. of Concrete		
			Concrete	Forms	Reinf.	Pans	Forms	Reinf.	Pans
Flat Plate	50 psf	15 Ft.	.46 C.F.	1.06 S.F.	1.71lb.		62 S.F.	101 lb.	
		20	.63	1.02	2.40		44	104	
		25	.79	1.02	3.03		35	104	
	100	15	.46	1.04	2.14		61	126	
		20	.71	1.02	2.72		39	104	
		25	.83	1.01	3.47		33	113	
Flat Plate (waffle construction) 20" domes	50	20	.43	1.00	2.10	.84 S.F.	63	135	53 S.F.
		25	.52	1.00	2.90	.89	52	150	46
		30	.64	1.00	3.70	.87	42	155	37
	100	20	.51	1.00	2.30	.84	53	125	45
		25	.64	1.00	3.20	.83	42	135	35
		30	.76	1.00	4.40	.81	36	160	29
Waffle Construction 30" domes	50	25	.69	1.06	1.83	.68	42	72	40
		30	.74	1.06	2.39	.69	39	87	39
		35	.86	1.05	2.71	.69	33	85	39
		40	.78	1.00	4.80	.68	35	165	40
Flat Slab (two way with drop panels)	50	20	.62	1.03	2.34		45	102	
		25	.77	1.03	2.99		36	105	
		30	.95	1.03	4.09		29	116	
	100	20	.64	1.03	2.83		43	119	
		25	.79	1.03	3.88		35	133	
		30	.96	1.03	4.66		29	131	
	200	20	.73	1.03	3.03		38	112	
		25	.86	1.03	4.23		32	133	
		30	1.06	1.03	5.30		26	135	
One Way Joists 20" Pans	50	15	.36	1.04	1.40	.93	78	105	70
		20	.42	1.05	1.80	.94	67	120	60
		25	.47	1.05	2.60	.94	60	150	54
	100	15	.38	1.07	1.90	.93	77	140	66
		20	.44	1.08	2.40	.94	67	150	58
		25	.52	1.07	3.50	.94	55	185	49
One Way Joists 8" x 16" filler blocks	50	15	.34	1.06	1.80	.81 Ea.	84	145	64 Ea.
		20	.40	1.08	2.20	.82	73	145	55
		25	.46	1.07	3.20	.83	63	190	49
	100	15	.39	1.07	1.90	.81	74	130	56
		20	.46	1.09	2.80	.82	64	160	48
		25	.53	1.10	3.60	.83	56	190	42
One Way Beam & Slab	50	15	.42	1.30	1.73		84	111	
		20	.51	1.28	2.61		68	138	
		25	.64	1.25	2.78		53	117	
	100	15	.42	1.30	1.90		84	122	
		20	.54	1.35	2.69		68	154	
		25	.69	1.37	3.93		54	145	
	200	15	.44	1.31	2.24		80	137	
		20	.58	1.40	3.30		65	163	
		25	.69	1.42	4.89		53	183	
Two Way Beam & Slab	100	15	.47	1.20	2.26		69	130	
		20	.63	1.29	3.06		55	131	
		25	.83	1.33	3.79		43	123	
	200	15	.49	1.25	2.70		41	149	
		20	.66	1.32	4.04		54	165	
		25	.88	1.32	6.08		41	187	

847

R033105-10   Proportionate Quantities (cont.)

4000 psi Concrete and 60,000 psi Reinforcing—Form and Reinforcing Quantities per C.Y.					
Item	Size	Forms	Reinforcing	Minimum	Maximum
Columns (square tied)	10″ x 10″	130 S.F.C.A.	#5 to #11	220 lbs.	875 lbs.
	12″ x 12″	108	#6 to #14	200	955
	14″ x 14″	92	#7 to #14	190	900
	16″ x 16″	81	#6 to #14	187	1082
	18″ x 18″	72	#6 to #14	170	906
	20″ x 20″	65	#7 to #18	150	1080
	22″ x 22″	59	#8 to #18	153	902
	24″ x 24″	54	#8 to #18	164	884
	26″ x 26″	50	#9 to #18	169	994
	28″ x 28″	46	#9 to #18	147	864
	30″ x 30″	43	#10 to #18	146	983
	32″ x 32″	40	#10 to #18	175	866
	34″ x 34″	38	#10 to #18	157	772
	36″ x 36″	36	#10 to #18	175	852
	38″ x 38″	34	#10 to #18	158	765
	40″ x 40″	32	#10 to #18	143	692

Item	Size	Form	Spiral	Reinforcing	Minimum	Maximum
Columns (spirally reinforced)	12″ diameter	34.5 L.F.	190 lbs.	#4 to #11	165 lbs.	1505 lb.
		34.5	190	#14 & #18	—	1100
	14″	25	170	#4 to #11	150	970
		25	170	#14 & #18	800	1000
	16″	19	160	#4 to #11	160	950
		19	160	#14 & #18	605	1080
	18″	15	150	#4 to #11	160	915
		15	150	#14 & #18	480	1075
	20″	12	130	#4 to #11	155	865
		12	130	#14 & #18	385	1020
	22″	10	125	#4 to #11	165	775
		10	125	#14 & #18	320	995
	24″	9	120	#4 to #11	195	800
		9	120	#14 & #18	290	1150
	26″	7.3	100	#4 to #11	200	729
		7.3	100	#14 & #18	235	1035
	28″	6.3	95	#4 to #11	175	700
		6.3	95	#14 & #18	200	1075
	30″	5.5	90	#4 to #11	180	670
		5.5	90	#14 & #18	175	1015
	32″	4.8	85	#4 to #11	185	615
		4.8	85	#14 & #18	155	955
	34″	4.3	80	#4 to #11	180	600
		4.3	80	#14 & #18	170	855
	36″	3.8	75	#4 to #11	165	570
		3.8	75	#14 & #18	155	865
	40″	3.0	70	#4 to #11	165	500
		3.0	70	#14 & #18	145	765

For customer support on your Building Construction Costs with RSMeans data, call 800.448.8182.

R033105-10   Proportionate Quantities (cont.)

		3000 psi Concrete and 60,000 psi Reinforcing—Form and Reinforcing Quantities per C.Y.					
Item	Type	Loading	Height	C.Y./L.F.	Forms/C.Y.	Reinf./C.Y.	
Retaining Walls	Cantilever	Level Backfill	4 Ft.	0.2 C.Y.	49 S.F.	35 lbs.	
			8	0.5	42	45	
			12	0.8	35	70	
			16	1.1	32	85	
			20	1.6	28	105	
		Highway Surcharge	4	0.3	41	35	
			8	0.5	36	55	
			12	0.8	33	90	
			16	1.2	30	120	
			20	1.7	27	155	
		Railroad Surcharge	4	0.4	28	45	
			8	0.8	25	65	
			12	1.3	22	90	
			16	1.9	20	100	
			20	2.6	18	120	
	Gravity, with Vertical Face	Level Backfill	4	0.4	37	None	
			7	0.6	27	↓	
			10	1.2	20		
		Sloping Backfill	4	0.3	31		
			7	0.8	21		
			10	1.6	15	↓	

		Live Load in Kips per Linear Foot						
	Span	Under 1 Kip		2 to 3 Kips		4 to 5 Kips		6 to 7 Kips	
		Forms	Reinf.	Forms	Reinf.	Forms	Reinf.	Forms	Reinf.
Beams	10 Ft.	—	—	90 S.F.	170 #	85 S.F.	175 #	75 S.F.	185 #
	16	130 S.F.	165 #	85	180	75	180	65	225
	20	110	170	75	185	62	200	51	200
	26	90	170	65	215	62	215	—	—
	30	85	175	60	200	—	—	—	—

Item	Size	Type	Forms per C.Y.	Reinforcing per C.Y.
Spread Footings	Under 1 C.Y.	1,000 psf soil	24 S.F.	44 lbs.
		5,000	24	42
		10,000	24	52
	1 C.Y. to 5 C.Y.	1,000	14	49
		5,000	14	50
		10,000	14	50
	Over 5 C.Y.	1,000	9	54
		5,000	9	52
		10,000	9	56
Pile Caps (30 Ton Concrete Piles)	Under 5 C.Y.	shallow caps	20	65
		medium	20	50
		deep	20	40
	5 C.Y. to 10 C.Y.	shallow	14	55
		medium	15	45
		deep	15	40
	10 C.Y. to 20 C.Y.	shallow	11	60
		medium	11	45
		deep	12	35
	Over 20 C.Y.	shallow	9	60
		medium	9	45
		deep	10	40

849

R033105-10   Proportionate Quantities (cont.)

			3000 psi Concrete and 60,000 psi Reinforcing — Form and Reinforcing Quantities per C.Y.			
Item	Size	Pile Spacing	50 T Pile	100 T Pile	50 T Pile	100 T Pile
Pile Caps (Steel H Piles)	Under 5 C.Y.	24" O.C.	24 S.F.	24 S.F.	75 lbs.	90 lbs.
		30"	25	25	80	100
		36"	24	24	80	110
	5 C.Y. to 10 C.Y.	24"	15	15	80	110
		30"	15	15	85	110
		36"	15	15	75	90
	Over 10 C.Y.	24"	13	13	85	90
		30"	11	11	85	95
		36"	10	10	85	90

	8" Thick		10" Thick		12" Thick		15" Thick	
Height	Forms	Reinf.	Forms	Reinf.	Forms	Reinf.	Forms	Reinf.
7 Ft.	81 S.F.	44 lbs.	65 S.F.	45 lbs.	54 S.F.	44 lbs.	41 S.F.	43 lbs.
8		44		45		44		43
9		46		45		44		43
10		57		45		44		43
12		83		50		52		43
14		116		65		64		51
16				86		90		65
18						106		70

(Basement Walls)

R033105-20   Materials for One C.Y. of Concrete

This is an approximate method of figuring quantities of cement, sand and coarse aggregate for a field mix with waste allowance included.

With crushed gravel as coarse aggregate, to determine barrels of cement required, divide 10 by total mix; that is, for 1:2:4 mix, 10 divided by 7 = 1-3/7 barrels. If the coarse aggregate is crushed stone, use 10-1/2 instead of 10 as given for gravel.

To determine tons of sand required, multiply barrels of cement by parts of sand and then by 0.2; that is, for the 1:2:4 mix, as above, 1-3/7 x 2 x .2 = .57 tons.

Tons of crushed gravel are in the same ratio to tons of sand as parts in the mix, or 4/2 x .57 = 1.14 tons.

1 bag cement = 94#	1 C.Y. sand or crushed gravel = 2700#	1 C.Y. crushed stone = 2575#
4 bags = 1 barrel	1 ton sand or crushed gravel = 20 C.F.	1 ton crushed stone = 21 C.F.

Average carload of cement is 692 bags; of sand or gravel is 56 tons.

Do not stack stored cement over 10 bags high.

R033105-30   Metric Equivalents of Cement Content for Concrete Mixes

94 Pound Bags per Cubic Yard	Kilograms per Cubic Meter	94 Pound Bags per Cubic Yard	Kilograms per Cubic Meter
1.0	55.77	7.0	390.4
1.5	83.65	7.5	418.3
2.0	111.5	8.0	446.2
2.5	139.4	8.5	474.0
3.0	167.3	9.0	501.9
3.5	195.2	9.5	529.8
4.0	223.1	10.0	557.7
4.5	251.0	10.5	585.6
5.0	278.8	11.0	613.5
5.5	306.7	11.5	641.3
6.0	334.6	12.0	669.2
6.5	362.5	12.5	697.1

a. If you know the cement content in pounds per cubic yard, multiply by .5933 to obtain kilograms per cubic meter.

b. If you know the cement content in 94 pound bags per cubic yard, multiply by 55.77 to obtain kilograms per cubic meter.

For customer support on your Building Construction Costs with RSMeans data, call 800.448.8182.

R033105-40 Metric Equivalents of Common Concrete Strengths
(to convert other psi values to megapascals, multiply by 0.006895)

U.S. Values psi	SI Values Megapascals	Non-SI Metric Values kgf/cm²*
2000	14	140
2500	17	175
3000	21	210
3500	24	245
4000	28	280
4500	31	315
5000	34	350
6000	41	420
7000	48	490
8000	55	560
9000	62	630
10,000	69	705

* kilograms force per square centimeter

R033105-50 Quantities of Cement, Sand and Stone for One C.Y. of Concrete per Various Mixes

This table can be used to determine the quantities of the ingredients for smaller quantities of site mixed concrete.

Concrete (C.Y.)	Mix = 1:1:1-3/4			Mix = 1:2:2.25			Mix = 1:2.25:3			Mix = 1:3:4		
	Cement (sacks)	Sand (C.Y.)	Stone (C.Y.)	Cement (sacks)	Sand (C.Y.)	Stone (C.Y.)	Cement (sacks)	Sand (C.Y.)	Stone (C.Y.)	Cement (sacks)	Sand (C.Y.)	Stone (C.Y.)
1	10	.37	.63	7.75	.56	.65	6.25	.52	.70	5	.56	.74
2	20	.74	1.26	15.50	1.12	1.30	12.50	1.04	1.40	10	1.12	1.48
3	30	1.11	1.89	23.25	1.68	1.95	18.75	1.56	2.10	15	1.68	2.22
4	40	1.48	2.52	31.00	2.24	2.60	25.00	2.08	2.80	20	2.24	2.96
5	50	1.85	3.15	38.75	2.80	3.25	31.25	2.60	3.50	25	2.80	3.70
6	60	2.22	3.78	46.50	3.36	3.90	37.50	3.12	4.20	30	3.36	4.44
7	70	2.59	4.41	54.25	3.92	4.55	43.75	3.64	4.90	35	3.92	5.18
8	80	2.96	5.04	62.00	4.48	5.20	50.00	4.16	5.60	40	4.48	5.92
9	90	3.33	5.67	69.75	5.04	5.85	56.25	4.68	6.30	45	5.04	6.66
10	100	3.70	6.30	77.50	5.60	6.50	62.50	5.20	7.00	50	5.60	7.40
11	110	4.07	6.93	85.25	6.16	7.15	68.75	5.72	7.70	55	6.16	8.14
12	120	4.44	7.56	93.00	6.72	7.80	75.00	6.24	8.40	60	6.72	8.88
13	130	4.82	8.20	100.76	7.28	8.46	81.26	6.76	9.10	65	7.28	9.62
14	140	5.18	8.82	108.50	7.84	9.10	87.50	7.28	9.80	70	7.84	10.36
15	150	5.56	9.46	116.26	8.40	9.76	93.76	7.80	10.50	75	8.40	11.10
16	160	5.92	10.08	124.00	8.96	10.40	100.00	8.32	11.20	80	8.96	11.84
17	170	6.30	10.72	131.76	9.52	11.06	106.26	8.84	11.90	85	9.52	12.58
18	180	6.66	11.34	139.50	10.08	11.70	112.50	9.36	12.60	90	10.08	13.32
19	190	7.04	11.98	147.26	10.64	12.36	118.76	9.84	13.30	95	10.64	14.06
20	200	7.40	12.60	155.00	11.20	13.00	125.00	10.40	14.00	100	11.20	14.80
21	210	7.77	13.23	162.75	11.76	13.65	131.25	10.92	14.70	105	11.76	15.54
22	220	8.14	13.86	170.05	12.32	14.30	137.50	11.44	15.40	110	12.32	16.28
23	230	8.51	14.49	178.25	12.88	14.95	143.75	11.96	16.10	115	12.88	17.02
24	240	8.88	15.12	186.00	13.44	15.60	150.00	12.48	16.80	120	13.44	17.76
25	250	9.25	15.75	193.75	14.00	16.25	156.25	13.00	17.50	125	14.00	18.50
26	260	9.64	16.40	201.52	14.56	16.92	162.52	13.52	18.20	130	14.56	19.24
27	270	10.00	17.00	209.26	15.12	17.56	168.76	14.04	18.90	135	15.02	20.00
28	280	10.36	17.64	217.00	15.68	18.20	175.00	14.56	19.60	140	15.68	20.72
29	290	10.74	18.28	224.76	16.24	18.86	181.26	15.08	20.30	145	16.24	21.46

851

R033105-65   Field-Mix Concrete

Presently most building jobs are built with ready-mixed concrete except at isolated locations and some larger jobs requiring over 10,000 C.Y. where land is readily available for setting up a temporary batch plant.

The most economical mix is a controlled mix using local aggregate proportioned by trial to give the required strength with the least cost of material.

R033105-70   Placing Ready-Mixed Concrete

For ground pours allow for 5% waste when figuring quantities.

Prices in the front of the data set assume normal deliveries. If deliveries are made before 8 A.M. or after 5 P.M. or on Saturday afternoons add 30%. Negotiated discounts for large volumes are not included in prices in front of the data set.

For the lower floors without truck access, concrete may be wheeled in rubber-tired buggies, conveyer handled, crane handled or pumped. Pumping is economical if there is top steel. Conveyers are more efficient for thick slabs.

At higher floors the rubber-tired buggies may be hoisted by a hoisting tower and wheeled to the location. Placement by a conveyer is limited to three floors and is best for high-volume pours. Pumped concrete is best when the building has no crane access. Concrete may be pumped directly as high as thirty-six stories using special pumping techniques. Normal maximum height is about fifteen stories.

The best pumping aggregate is screened and graded bank gravel rather than crushed stone.

Pumping downward is more difficult than pumping upward. The horizontal distance from pump to pour may increase preparation time prior to pour. Placing by cranes, either mobile, climbing or tower types, continues as the most efficient method for high-rise concrete buildings.

R033105-85   Lift Slabs

The cost advantage of the lift slab method is due to placing all concrete, reinforcing steel, inserts and electrical conduit at ground level and in reduction of formwork. Minimum economical project size is about 30,000 S.F. Slabs may be tilted for parking garage ramps.

It is now used in all types of buildings and has gone up to 22 stories high in apartment buildings. The current trend is to use post-tensioned flat plate slabs with spans from 22' to 35'. Cylindrical void forms are used when deep slabs are required. One pound of prestressing steel is about equal to seven pounds of conventional reinforcing.

To be considered cured for stressing and lifting, a slab must have attained 75% of design strength. Seven days are usually sufficient with four to five days possible if high early strength cement is used. Slabs can be stacked using two coats of a non-bonding agent to insure that slabs do not stick to each other. Lifting is done by companies specializing in this work. Lift rate is 5' to 15' per hour with an average of 10' per hour. Total areas up to 33,000 S.F. have been lifted at one time. 24 to 36 jacking columns are common. Most economical bay sizes are 24' to 28' with four to fourteen stories most efficient. Continuous design reduces reinforcing steel cost. Use of post-tensioned slabs allows larger bay sizes.

For customer support on your Building Construction Costs with RSMeans data, call 800.448.8182.

R033543-10   Polished Concrete Floors

A polished concrete floor has a glossy mirror-like appearance and is created by grinding the concrete floor with finer and finer diamond grits, similar to sanding wood, until the desired level of reflective clarity and sheen are achieved. The technical term for this type of polished concrete is bonded abrasive polished concrete. The basic piece of equipment used in the polishing process is a walk-behind planetary grinder for working large floor areas. This grinder drives diamond-impregnated abrasive discs, which progress from coarse- to fine-grit discs.

The process begins with the use of very coarse diamond segments or discs bonded in a metallic matrix. These segments are coarse enough to allow the removal of pits, blemishes, stains, and light coatings from the floor surface in preparation for final smoothing. The condition of the original concrete surface will dictate the grit coarseness of the initial grinding step which will generally end up being a three- to four-step process using ever finer grits. The purpose of this initial grinding step is to remove surface coatings and blemishes and to cut down into the cream for very fine aggregate exposure, or deeper into the fine aggregate layer just below the cream layer, or even deeper into the coarse aggregate layer. These initial grinding steps will progress up to the 100/120 grit. If wet grinding is done, a waste slurry is produced that must be removed between grit changes and disposed of properly. If dry grinding is done, a high performance vacuum will pick up the dust during grinding and collect it in bags which must be disposed of properly.

The process continues with honing the floor in a series of steps that progresses from 100-grit to 400-grit diamond abrasive discs embedded in a plastic or resin matrix. At some point during, or just prior to, the honing step, one or two coats of stain or dye can be sprayed onto the surface to give color to the concrete, and two coats of densifier/hardener must be applied to the floor surface and allowed to dry. This sprayed-on densifier/hardener will penetrate about 1/8″ into the concrete to make the surface harder, denser and more abrasion-resistant.

The process ends with polishing the floor surface in a series of steps that progresses from resin-impregnated 800-grit (medium polish) to 1500-grit (high polish) to 3000-grit (very high polish), depending on the desired level of reflective clarity and sheen.

The Concrete Polishing Association of America (CPAA) has defined the flooring options available when processing concrete to a desired finish. The first category is aggregate exposure, the grinding of a concrete surface with bonded abrasives, in as many abrasive grits necessary, to achieve one of the following classes:

©Concrete Polishing Association of America "Glossary."

A. Cream – very little surface cut depth; little aggregate exposure

B. Fine aggregate (salt and pepper) – surface cut depth of 1/16″; fine aggregate exposure with little or no medium aggregate exposure at random locations

C. Medium aggregate – surface cut depth of 1/8″; medium aggregate exposure with little or no large aggregate exposure at random locations

D. Large aggregate – surface cut depth of 1/4″; large aggregate exposure with little or no fine aggregate exposure at random locations

The second CPAA defined category is reflective clarity and sheen, the polishing of a concrete surface with the minimum number of bonded abrasives as indicated to achieve one of the following levels:

1. Ground – flat appearance with none to very slight diffused reflection; none to very low reflective sheen; using a minimum total of 4 grit levels up 100-grit

2. Honed – matte appearance with or without slight diffused reflection; low to medium reflective sheen; using a minimum total of 5 grit levels up to 400-grit

3. Semi-polished – objects being reflected are not quite sharp and crisp but can be easily identified; medium to high reflective sheen; using a minimum total of 6 grit levels up to 800-grit

4. Highly-polished – objects being reflected are sharp and crisp as would be seen in a mirror-like reflection; high to highest reflective sheen; using a minimum total of up to 8 grit levels up to 1500-grit or 3000-grit

The CPAA defines reflective clarity as the degree of sharpness and crispness of the reflection of overhead objects when viewed 5′ above and perpendicular to the floor surface. Reflective sheen is the degree of gloss reflected from a surface when viewed at least 20′ from and at an angle to the floor surface. These terms are relatively subjective. The final outcome depends on the internal makeup and surface condition of the original concrete floor, the experience of the floor polishing crew, and the expectations of the owner. Before the grinding, honing, and polishing work commences on the main floor area, it might be beneficial to do a mock-up panel in the same floor but in an out of the way place to demonstrate the sequence of steps with increasingly fine abrasive grits and to demonstrate the final reflective clarity and reflective sheen. This mock-up panel will be within the area of, and part of, the final work.

853

R034105-30   Prestressed Precast Concrete Structural Units

Type	Location	Depth	Span in Ft.		Live Load Lb. per S.F.
Double Tee	Floor	28" to 34"	60 to 80		50 to 80
	Roof	12" to 24"	30 to 50		40
	Wall	Width 8'	Up to 55' high		Wind
Multiple Tee	Roof	8" to 12"	15 to 40		40
	Floor	8" to 12"	15 to 30		100
Plank	Roof		Roof	Floor	
		4"	13	12	40 for Roof
		6"	22	18	
or	or	8"	26	25	
		10"	33	29	100 for Floor
	Floor	12"	42	32	
Single Tee	Roof	28"	40		
		32"	80		
		36"	100		40
		48"	120		
AASHTO Girder	Bridges	Type 4	100		
		5	110		Highway
		6	125		
Box Beam	Bridges	15"	40		
		27"	to		Highway
		33"	100		

The majority of precast projects today utilize double tees rather than single tees because of speed and ease of installation. As a result casting beds at manufacturing plants are normally formed for double tees. Single tee projects will therefore require an initial set up charge to be spread over the individual single tee costs.

For floors, a 2" to 3" topping is field cast over the shapes. For roofs, insulating concrete or rigid insulation is placed over the shapes.

Member lengths up to 40' are standard haul, 40' to 60' require special permits and lengths over 60' must be escorted. Excessive width and/or length can add up to 100% on hauling costs.

Large heavy members may require two cranes for lifting which would increase erection costs by about 45%. An eight man crew can install 12 to 20 double tees, or 45 to 70 quad tees or planks per day.

Grouting of connections must also be included.

Several system buildings utilizing precast members are available. Heights can go up to 22 stories for apartment buildings. The optimum design ratio is 3 S.F. of surface to 1 S.F. of floor area.

For customer support on your Building Construction Costs with RSMeans data, call 800.448.8182.

R034136-90 Prestressed Concrete, Post-Tensioned

In post-tensioned concrete the steel tendons are tensioned after the concrete has reached about 3/4 of its ultimate strength. The cableways are grouted after tensioning to provide bond between the steel and concrete. If bond is to be prevented, the tendons are coated with a corrosion-preventative grease and wrapped with waterproofed paper or plastic. Bonded tendons are usually used when ultimate strength (beams & girders) are controlling factors.

High strength concrete is used to fully utilize the steel, thereby reducing the size and weight of the member. A plasticizing agent may be added to reduce water content. Maximum size aggregate ranges from 1/2" to 1-1/2" depending on the spacing of the tendons.

The types of steel commonly used are bars and strands. Job conditions determine which is best suited. Bars are best for vertical prestresses since

they are easy to support. The trend is for steel manufacturers to supply a finished package, cut to length, which reduces field preparation to a minimum.

Bars vary from 3/4" to 1-3/8" diameter. The table below gives time in labor-hours per tendon for placing, tensioning and grouting (if required) a 75' beam. Tendons used in buildings are not usually grouted; tendons for bridges usually are grouted. For strands the table indicates the labor-hours per pound for typical prestressed units 100' long. Simple span beams usually require one- end stressing regardless of lengths. Continuous beams are usually stressed from two ends. Long slabs are poured from the center outward and stressed in 75' increments after the initial 150' center pour.

Labor Hours per Tendon and per Pound of Prestressed Steel						
Length	100' Beam		75' Beam		100' Slab	
Type Steel	Strand		Bars		Strand	
Diameter	0.5"		3/4"	1-3/8"	0.5"	0.6"
Number	4	12	1	1	1	1
Force in Kips	100	300	42	143	25	35
Preparation & Placing Cables	3.6	7.4	0.9	2.9	0.9	1.1
Stressing Cables	2.0	2.4	0.8	1.6	0.5	0.5
Grouting, if required	2.5	3.0	0.6	1.3		
Total Labor Hours	8.1	12.8	2.3	5.8	1.4	1.6
Prestressing Steel Weights (Lbs.)	215	640	115	380	53	74
Labor-hours per Lb. Bonded	0.038	0.020	0.020	0.015		
Non-bonded					0.026	0.022

Flat slab construction — 4000 psi concrete with span-to-depth ratio between 36 and 44. Two way post-tensioned steel averages 1.0 lb. per S.F. for 24' to 28' bays (usually strand) and additional reinforcing steel averages .5 lb. per S.F.

Pan and joist construction — 4000 psi concrete with span-to-depth ratio between 28 to 30. Post-tensioned steel averages .8 lb. per S.F. and reinforcing steel about 1.0 lb. per S.F. Placing and stressing average 40 hours per ton of total material.

Beam construction — 4000 to 5000 psi concrete. Steel weights vary greatly.

Labor cost per pound goes down as the size and length of the tendon increase. The primary economic consideration is the cost per kip for the member.

Post-tensioning becomes feasible for beams and girders over 30' long; for continuous two-way slabs over 20' clear; and for transferring upper building loads over longer spans at lower levels. Post-tension suppliers will provide engineering services at no cost to the user. Substantial economies are possible by using post-tensioned lift slabs.

For customer support on your Building Construction Costs with RSMeans data, call 800.448.8182.

Concrete — R0345 Precast Architectural Concrete

R034513-10 Precast Concrete Wall Panels

Panels are either solid or insulated with plain, colored or textured finishes. Transportation is an important cost factor. Prices shown in the unit cost section of the data set are based on delivery within 50 miles of a plant including fabricators' overhead and profit. Engineering data is available from fabricators to assist with construction details. Usual minimum job size for economical use of panels is about 5000 S.F. Small jobs can double the prices shown. For large, highly repetitive jobs, deduct up to 15% from the prices shown.

2″ thick panels cost about the same as 3″ thick panels, and maximum panel size is less. For building panels faced with granite, marble or stone, add the material prices from those unit cost sections to the plain panel price shown. There is a growing trend toward aggregate facings and broken rib finishes rather than plain gray concrete panels.

No allowance has been made in the unit cost section for supporting steel framework. On one story buildings, panels may rest on grade beams and require only wind bracing and fasteners. On multi-story buildings panels can span from column to column and floor to floor. Plastic-designed steel-framed structures may have large deflections which slow down erection and raise costs.

Large panels are more economical than small panels on a S.F. basis. When figuring areas include all protrusions, returns, etc. Overhangs can triple erection costs. Panels over 45′ have been produced. Larger flat units should be prestressed. Vacuum lifting of smooth finish panels eliminates inserts and can speed erection.

Concrete — R0347 Site-Cast Concrete

R034713-20 Tilt Up Concrete Panels

The advantage of tilt up construction is in the low cost of forms and the placing of concrete and reinforcing. Panels up to 75′ high and 5-1/2″ thick have been tilted using strongbacks. Tilt up has been used for one to five story buildings and is well-suited for warehouses, stores, offices, schools and residences.

The panels are cast in forms on the floor slab. Most jobs use 5-1/2″ thick solid reinforced concrete panels. Sandwich panels with a layer of insulating materials are also used. Where dampness is a factor, lightweight aggregate is used. Optimum panel size is 300 to 500 S.F.

Slabs are usually poured with 3000 psi concrete which permits tilting seven days after pouring. Slabs may be stacked on top of each other and are separated from each other by either two coats of bond breaker or a film of polyethylene. Use of high early-strength cement allows tilting two days after a pour. Tilting up is done with a roller outrigger crane with a capacity of at least 1-1/2 times the weight of the panel at the required reach. Exterior precast columns can be set at the same time as the panels; interior precast columns can be set first and the panels clipped directly to them. The use of cast-in-place concrete columns is diminishing due to shrinkage problems. Structural steel columns are sometimes used if crane rails are planned. Panels can be clipped to the columns or lowered between the flanges. Steel channels with anchors may be used as edge forms for the slab. When the panels are lifted the channels form an integral steel column to take structural loads. Roof loads can be carried directly by the panels for wall heights to 14′.

Requirements of local building codes may be a limiting factor and should be checked. Building floor slabs should be poured first and should be a minimum of 5″ thick with 100% compaction of soil or 6″ thick with less than 100% compaction.

Setting times as fast as nine minutes per panel have been observed, but a safer expectation would be four panels per hour with a crane and a four-man setting crew. If a crane erects from inside a building, some provision must be made to get the crane out after walls are erected. Good yarding procedure is important to minimize delays. Equalizing three-point lifting beams and self-releasing pick-up hooks speed erection. If panels must be carried to their final location, setting time per panel will be increased and erection costs may approach the erection cost range of architectural precast wall panels. Placing panels into slots formed in continuous footers will speed erection.

Reinforcing should be with #5 bars with vertical bars on the bottom. If surface is to be sandblasted, stainless steel chairs should be used to prevent rust staining.

Use of a broom finish is popular since the unavoidable surface blemishes are concealed.

Precast columns run from three to five times the C.Y. price of the panels only.

Concrete — R0352 Lightweight Concrete Roof Insulation

R035216-10 Lightweight Concrete

Lightweight aggregate concrete is usually purchased ready mixed, but it can also be field mixed.

Vermiculite or Perlite comes in bags of 4 C.F. under various trade names. The weight is about 8 lbs. per C.F. For insulating roof fill use 1:6 mix. For a structural deck use 1:4 mix over gypsum boards, steeltex, steel centering, etc., supported by closely spaced joists or bulb trees. For structural slabs use 1:3:2 vermiculite sand concrete over steeltex, metal lath, steel centering, etc., on joists spaced 2′-0″ O.C. for maximum L.L. of 80 P.S.F. Use same mix

for slab base fill over steel flooring or regular reinforced concrete slab when tile, terrazzo or other finish is to be laid over.

For slabs on grade use 1:3:2 mix when tile, etc., finish is to be laid over. If radiant heating units are installed use a 1:6 mix for a base. After coils are in place, cover with a regular granolithic finish (mix 1:3:2) to a minimum depth of 1-1/2″ over top of units.

Reinforce all slabs with 6 × 6 or 10 × 10 welded wire mesh.

For customer support on your Building Construction Costs with RSMeans data, call 800.448.8182.

Masonry   R0401 Maintenance of Masonry

R040130-10   Cleaning Face Brick

On smooth brick a person can clean 70 S.F. an hour; on rough brick 50 S.F. per hour. Use one gallon muriatic acid to 20 gallons of water for 1000

S.F. Do not use acid solution until wall is at least seven days old, but a mild soap solution may be used after two days.

Time has been allowed for cleanup in brick prices.

Masonry   R0405 Common Work Results for Masonry

R040513-10   Cement Mortar (material only)

Type N - 1:1:6 mix by volume. Use everywhere above grade except as noted below. - 1:3 mix using conventional masonry cement which saves handling two separate bagged materials.

Type M - 1:1/4:3 mix by volume, or 1 part cement, 1/4 (10% by wt.) lime, 3 parts sand. Use for heavy loads and where earthquakes or hurricanes may occur. Also for reinforced brick, sewers, manholes and everywhere below grade.

Mix Proportions by Volume and Compressive Strength of Mortar

Where Used	Mortar Type	Allowable Proportions by Volume				Compressive Strength @ 28 days
		Portland Cement	Masonry Cement	Hydrated Lime	Masonry Sand	
Plain Masonry	M	1	1	—	6	
		1	—	1/4	3	2500 psi
	S	1/2	1	—	4	
		1	—	1/4 to 1/2	4	1800 psi
	N	—	1	—	3	
		1	—	1/2 to 1-1/4	6	750 psi
	O	—	1	—	3	
		1	—	1-1/4 to 2-1/2	9	350 psi
	K	1	—	2-1/2 to 4	12	75 psi
Reinforced Masonry	PM	1	1	—	6	2500 psi
	PL	1	—	1/4 to 1/2	4	2500 psi

Note: The total aggregate should be between 2.25 to 3 times the sum of the cement and lime used.

The labor cost to mix the mortar is included in the productivity and labor cost of unit price lines in unit cost sections for brickwork, blockwork and stonework.

The material cost of mixed mortar is included in the material cost of those same unit price lines and includes the cost of renting and operating a 10 C.F. mixer at the rate of 200 C.F. per day.

There are two types of mortar color used. One type is the inert additive type with about 100 lbs. per M brick as the typical quantity required. These colors are also available in smaller-batch-sized bags (1 lb. to 15 lb.) which can be placed directly into the mixer without measuring. The other type is premixed and replaces the masonry cement. Dark green color has the highest cost.

R040519-50   Masonry Reinforcing

Horizontal joint reinforcing helps prevent wall cracks where wall movement may occur and in many locations is required by code. Horizontal joint reinforcing is generally not considered to be structural reinforcing and an unreinforced wall may still contain joint reinforcing.

Reinforcing strips come in 10′ and 12′ lengths and in truss and ladder shapes, with and without drips. Field labor runs between 2.7 to 5.3 hours per 1000 L.F. for wall thicknesses up to 12″.

The wire meets ASTM A82 for cold drawn steel wire and the typical size is 9 ga. sides and ties with 3/16″ diameter also available. Typical finish is mill galvanized with zinc coating at .10 oz. per S.F. Class I (.40 oz. per S.F.) and Class III (.80 oz. per S.F.) are also available, as is hot dipped galvanizing at 1.50 oz. per S.F.

857

R042110-10   Economy in Bricklaying

Have adequate supervision. Be sure bricklayers are always supplied with materials so there is no waiting. Place experienced bricklayers at corners and openings.

Use only screened sand for mortar. Otherwise, labor time will be wasted picking out pebbles. Use seamless metal tubs for mortar as they do not leak or catch the trowel. Locate stack and mortar for easy wheeling.

Have brick delivered for stacking. This makes for faster handling, reduces chipping and breakage, and requires less storage space. Many dealers will deliver select common in 2′ × 3′ × 4′ pallets or face brick packaged. This affords quick handling with a crane or forklift and easy tonging in units of ten, which reduces waste.

Use wider bricks for one wythe wall construction. Keep scaffolding away from the wall to allow mortar to fall clear and not stain the wall.

On large jobs develop specialized crews for each type of masonry unit.

Consider designing for prefabricated panel construction on high rise projects.

Avoid excessive corners or openings. Each opening adds about 50% to the labor cost for area of opening.

Bolting stone panels and using window frames as stops reduce labor costs and speed up erection.

R042110-20   Common and Face Brick

Common building brick manufactured according to ASTM C62 and facing brick manufactured according to ASTM C216 are the two standard bricks available for general building use.

Building brick is made in three grades: SW, where high resistance to damage caused by cyclic freezing is required; MW, where moderate resistance to cyclic freezing is needed; and NW, where little resistance to cyclic freezing is needed. Facing brick is made in only the two grades SW and MW. Additionally, facing brick is available in three types: FBS, for general use; FBX, for general use where a higher degree of precision and lower permissible variation in size than FBS are needed; and FBA, for general use to produce characteristic architectural effects resulting from non-uniformity in size and texture of the units.

In figuring the material cost of brickwork, an allowance of 25% mortar waste and 3% brick breakage was included. If bricks are delivered palletized

with 280 to 300 per pallet, or packaged, allow only 1-1/2% for breakage. Packaged or palletized delivery is practical when a job is big enough to have a crane or other equipment available to handle a package of brick. This is so on all industrial work but not always true on small commercial buildings.

The use of buff and gray face is increasing, and there is a continuing trend to the Norman, Roman, Jumbo and SCR brick.

Common red clay brick for backup is not used that often. Concrete block is the most usual backup material with occasional use of sand lime or cement brick. Building brick is commonly used in solid walls for strength and as a fire stop.

Brick panels built on the ground and then crane erected to the upper floors have proven to be economical. This allows the work to be done under cover and without scaffolding.

R042110-50   Brick, Block & Mortar Quantities

Running Bond							For Other Bonds Standard Size Add to S.F. Quantities in Table to Left			
Number of Brick per S.F. of Wall - Single Wythe with 3/8″ Joints					C.F. of Mortar per M Bricks, Waste Included					
Type Brick	Nominal Size (incl. mortar)			Modular Coursing	Number of Brick per S.F.	3/8″ Joint	1/2″ Joint	Bond Type	Description	Factor
	L	H	W							
Standard	8 x 2-2/3 x 4			3C=8″	6.75	8.1	10.3	Common	full header every fifth course	+20%
Economy	8 x 4 x 4			1C=4″	4.50	9.1	11.6		full header every sixth course	+16.7%
Engineer	8 x 3-1/5 x 4			5C=16″	5.63	8.5	10.8	English	full header every second course	+50%
Fire	9 x 2-1/2 x 4-1/2			2C=5″	6.40	550 # Fireclay	—	Flemish	alternate headers every course	+33.3%
Jumbo	12 x 4 x 6 or 8			1C=4″	3.00	22.5	29.2		every sixth course	+5.6%
Norman	12 x 2-2/3 x 4			3C=8″	4.50	11.2	14.3	Header = W x H exposed		+100%
Norwegian	12 x 3-1/5 x 4			5C=16″	3.75	11.7	14.9	Rowlock = H x W exposed		+100%
Roman	12 x 2 x 4			2C=4″	6.00	10.7	13.7	Rowlock stretcher = L x W exposed		+33.3%
SCR	12 x 2-2/3 x 6			3C=8″	4.50	21.8	28.0	Soldier = H x L exposed		—
Utility	12 x 4 x 4			1C=4″	3.00	12.3	15.7	Sailor = W x L exposed		-33.3%

Concrete Blocks Nominal Size		Approximate Weight per S.F.		Blocks per 100 S.F.	Mortar per M block, waste included	
		Standard	Lightweight		Partitions	Back up
2″	x 8″ x 16″	20 PSF	15 PSF	113	27 C.F.	36 C.F.
4″		30	20		41	51
6″		42	30		56	66
8″		55	38		72	82
10″		70	47		87	97
12″		85	55		102	112

Brick & Mortar Quantities
©Brick Industry Association. 2009 Feb. Technical Notes on
Brick Construction 10:
    Dimensioning and Estimating Brick Masonry. Reston (VA): BIA. Table 1
    Modular Brick Sizes and Table 4 Quantity Estimates for Brick Masonry.

Masonry

R0422 Concrete Unit Masonry

R042210-20 Concrete Block

The material cost of special block such as corner, jamb and head block can be figured at the same price as ordinary block of equal size. Labor on specials is about the same as equal-sized regular block.

Bond beams and 16″ high lintel blocks are more expensive than regular units of equal size. Lintel blocks are 8″ long and either 8″ or 16″ high.

Use of a motorized mortar spreader box will speed construction of continuous walls.

Hollow non-load-bearing units are made according to ASTM C129 and hollow load-bearing units according to ASTM C90.

Metals

R0505 Common Work Results for Metals

R050516-30 Coating Structural Steel

On field-welded jobs, the shop-applied primer coat is necessarily omitted. All painting must be done in the field and usually consists of red oxide rust inhibitive paint or an aluminum paint. The table below shows paint coverage and daily production for field painting.

See Division 05 05 13.50 for hot-dipped galvanizing and Division 09 97 13.23 for field-applied cold galvanizing and other paints and protective coatings.

See Division 05 01 10.51 for steel surface preparation treatments such as wire brushing, pressure washing and sand blasting.

Type Construction	Surface Area per Ton	Coat	One Gallon Covers		In 8 Hrs. Person Covers		Average per Ton Spray	
			Brush	Spray	Brush	Spray	Gallons	Labor-hours
Light Structural	300 S.F. to 500 S.F.	1st	500 S.F.	455 S.F.	640 S.F.	2000 S.F.	0.9 gals.	1.6 L.H.
		2nd	450	410	800	2400	1.0	1.3
		3rd	450	410	960	3200	1.0	1.0
Medium	150 S.F. to 300 S.F.	All	400	365	1600	3200	0.6	0.6
Heavy Structural	50 S.F. to 150 S.F.	1st	400	365	1920	4000	0.2	0.2
		2nd	400	365	2000	4000	0.2	0.2
		3rd	400	365	2000	4000	0.2	0.2
Weighted Average	225 S.F.	All	400	365	1350	3000	0.6	0.6

R050521-20 Welded Structural Steel

Usual weight reductions with welded design run 10% to 20% compared with bolted or riveted connections. This amounts to about the same total cost compared with bolted structures since field welding is more expensive than bolts. For normal spans of 18′ to 24′ figure 6 to 7 connections per ton.

Trusses — For welded trusses add 4% to weight of main members for connections. Up to 15% less steel can be expected in a welded truss compared to one that is shop bolted. Cost of erection is the same whether shop bolted or welded.

General — Typical electrodes for structural steel welding are E6010, E6011, E60T and E70T. Typical buildings vary between 2# to 8# of weld rod per

ton of steel. Buildings utilizing continuous design require about three times as much welding as conventional welded structures. In estimating field erection by welding, it is best to use the average linear feet of weld per ton to arrive at the welding cost per ton. The type, size and position of the weld will have a direct bearing on the cost per linear foot. A typical field welder will deposit 1.8# to 2# of weld rod per hour manually. Using semiautomatic methods can increase production by as much as 50% to 75%.

R050523-10 High Strength Bolts

Common bolts (A307) are usually used in secondary connections (see Division 05 05 23.10).

High strength bolts (A325 and A490) are usually specified for primary connections such as column splices, beam and girder connections to columns, column bracing, connections for supports of operating equipment or of other live loads which produce impact or reversal of stress, and in structures carrying cranes of over 5-ton capacity.

Allow 20 field bolts per ton of steel for a 6 story office building, apartment house or light industrial building. For 6 to 12 stories allow 25 bolts per ton, and above 12 stories, 30 bolts per ton. On power stations, 20 to 25 bolts per ton are needed.

R051223-10   Structural Steel

The bare material prices for structural steel, shown in the unit cost sections of the data set, are for 100 tons of shop-fabricated structural steel and include:

1. Mill base price of structural steel
2. Mill scrap/grade/size/length extras
3. Mill delivery to a metals service center (warehouse)
4. Service center storage and handling
5. Service center delivery to a fabrication shop
6. Shop storage and handling
7. Shop drafting/detailing
8. Shop fabrication
9. Shop coat of primer paint
10. Shop listing
11. Shop delivery to the job site

In unit cost sections of the data set that contain items for field fabrication of steel components, the bare material cost of steel includes:

1. Mill base price of structural steel
2. Mill scrap/grade/size/length extras
3. Mill delivery to a metals service center (warehouse)
4. Service center storage and handling
5. Service center delivery to the job site

R051223-20   Steel Estimating Quantities

One estimate on erection is that a crane can handle 35 to 60 pieces per day. Say the average is 45. With usual sizes of beams, girders, and columns, this would amount to about 20 tons per day. The type of connection greatly affects the speed of erection. Moment connections for continuous design slow down production and increase erection costs.

Short open web bar joists can be set at the rate of 75 to 80 per day, with 50 per day being the average for setting long span joists.

After main members are calculated, add the following for usual allowances: base plates 2% to 3%; column splices 4% to 5%; and miscellaneous details 4% to 5%, for a total of 10% to 13% in addition to main members.

The ratio of column to beam tonnage varies depending on type of steels used, typical spans, story heights and live loads.

It is more economical to keep the column size constant and to vary the strength of the column by using high strength steels. This also saves floor space. Buildings have recently gone as high as ten stories with 8" high strength columns. For light columns under W8X31 lb. sections, concrete filled steel columns are economical.

High strength steels may be used in columns and beams to save floor space and to meet head room requirements. High strength steels in some sizes sometimes require long lead times.

Round, square and rectangular columns, both plain and concrete filled, are readily available and save floor area, but are higher in cost per pound than rolled columns. For high unbraced columns, tube columns may be less expensive.

Below are average minimum figures for the weights of the structural steel frame for different types of buildings using A36 steel, rolled shapes and simple joints. For economy in domes, rise to span ratio = .13. Open web joist framing systems will reduce weights by 10% to 40%. Composite design can reduce steel weight by up to 25% but additional concrete floor slab thickness may be required. Continuous design can reduce the weights up to 20%. There are many building codes with different live load requirements and different structural requirements, such as hurricane and earthquake loadings, which can alter the figures.

Structural Steel Weights per S.F. of Floor Area

Type of Building	No. of Stories	Avg. Spans	L.L. #/S.F.	Lbs. Per S.F.	Type of Building	No. of Stories	Avg. Spans	L.L. #/S.F.	Lbs. Per S.F.
Steel Frame Mfg.	1	20'x20'	40	8	Apartments	2-8	20'x20'	40	8
		30'x30'		13		9-25			14
		40'x40'		18	Office	to 10	Various	80	10
Parking garage	4	Various	80	8.5		20			18
Domes (Schwedler)*	1	200'	30	10		30			26
		300'		15		over 50			35

For customer support on your Building Construction Costs with RSMeans data, call 800.448.8182.

R051223-25   Common Structural Steel Specifications

ASTM A992 (formerly A36, then A572 Grade 50) is the all-purpose carbon grade steel widely used in building and bridge construction.

The other high-strength steels listed below may each have certain advantages over ASTM A992 structural carbon steel, depending on the application. They have proven to be economical choices where, due to lighter members, the reduction of dead load and the associated savings in shipping cost can be significant.

ASTM A588 atmospheric weathering, high-strength, low-alloy steels can be used in the bare (uncoated) condition, where exposure to normal atmosphere causes a tightly adherant oxide to form on the surface, protecting the steel from further oxidation. ASTM A242 corrosion-resistant, high-strength, low-alloy steels have enhanced atmospheric corrosion resistance of at least two times that of carbon structural steels with copper, or four times that of carbon structural steels without copper. The reduction or elimination of maintenance resulting from the use of these steels often offsets their higher initial cost.

Steel Type	ASTM Designation	Minimum Yield Stress in KSI	Shapes Available
Carbon	A36	36	All structural shape groups, and plates & bars up through 8″ thick
	A529	50	Structural shape group 1, and plates & bars up through 2″ thick
High-Strength Low-Alloy Quenched & Self-Tempered	A913	50	All structural shape groups
		60	
		65	
		70	
High-Strength Low-Alloy Columbium-Vanadium	A572	42	All structural shape groups, and plates & bars up through 6″ thick
		50	All structural shape groups, and plates & bars up through 4″ thick
		55	Structural shape groups 1 & 2, and plates & bars up through 2″ thick
		60	Structural shape groups 1 & 2, and plates & bars up through 1-1/4″ thick
		65	Structural shape group 1, and plates & bars up through 1-1/4″ thick
High-Strength Low-Alloy Columbium-Vanadium	A992	50	All structural shape groups
Weathering High-Strength Low-Alloy	A242	42	Structural shape groups 4 & 5, and plates & bars over 1-1/2″ up through 4″ thick
		46	Structural shape group 3, and plates & bars over 3/4″ up through 1-1/2″ thick
		50	Structural shape groups 1 & 2, and plates & bars up through 3/4″ thick
Weathering High-Strength Low-Alloy	A588	42	Plates & bars over 5″ up through 8″ thick
		46	Plates & bars over 4″ up through 5″ thick
		50	All structural shape groups, and plates & bars up through 4″ thick
Quenched and Tempered Low-Alloy	A852	70	Plates & bars up through 4″ thick
Quenched and Tempered Alloy	A514	90	Plates & bars over 2-1/2″ up through 6″ thick
		100	Plates & bars up through 2-1/2″ thick

R051223-30   High Strength Steels

The mill price of high strength steels may be higher than A992 carbon steel but their proper use can achieve overall savings through total reduced weights. For columns with L/r over 100, A992 steel is best; under 100, high strength steels are economical. For heavy columns, high strength steels are economical when cover plates are eliminated. There is no economy using high strength steels for clip angles or supports or for beams where deflection governs. Thinner members are more economical than thick.

The per ton erection and fabricating costs of the high strength steels will be higher than for A992 since the same number of pieces, but less weight, will be installed.

861

R051223-35   Common Steel Sections

The upper portion of this table shows the name, shape, common designation and basic characteristics of commonly used steel sections. The lower portion explains how to read the designations used for the above illustrated common sections.

Shape & Designation	Name & Characteristics	Shape & Designation	Name & Characteristics
W	W Shape / Parallel flange surfaces	MC	Miscellaneous Channel / Infrequently rolled by some producers
S	American Standard Beam (I Beam) / Sloped inner flange	L	Angle / Equal or unequal legs, constant thickness
M	Miscellaneous Beams / Cannot be classified as W, HP or S; infrequently rolled by some producers	T	Structural Tee / Cut from W, M or S on center of web
C	American Standard Channel / Sloped inner flange	HP	Bearing Pile / Parallel flanges and equal flange and web thickness

Common drawing designations follow:

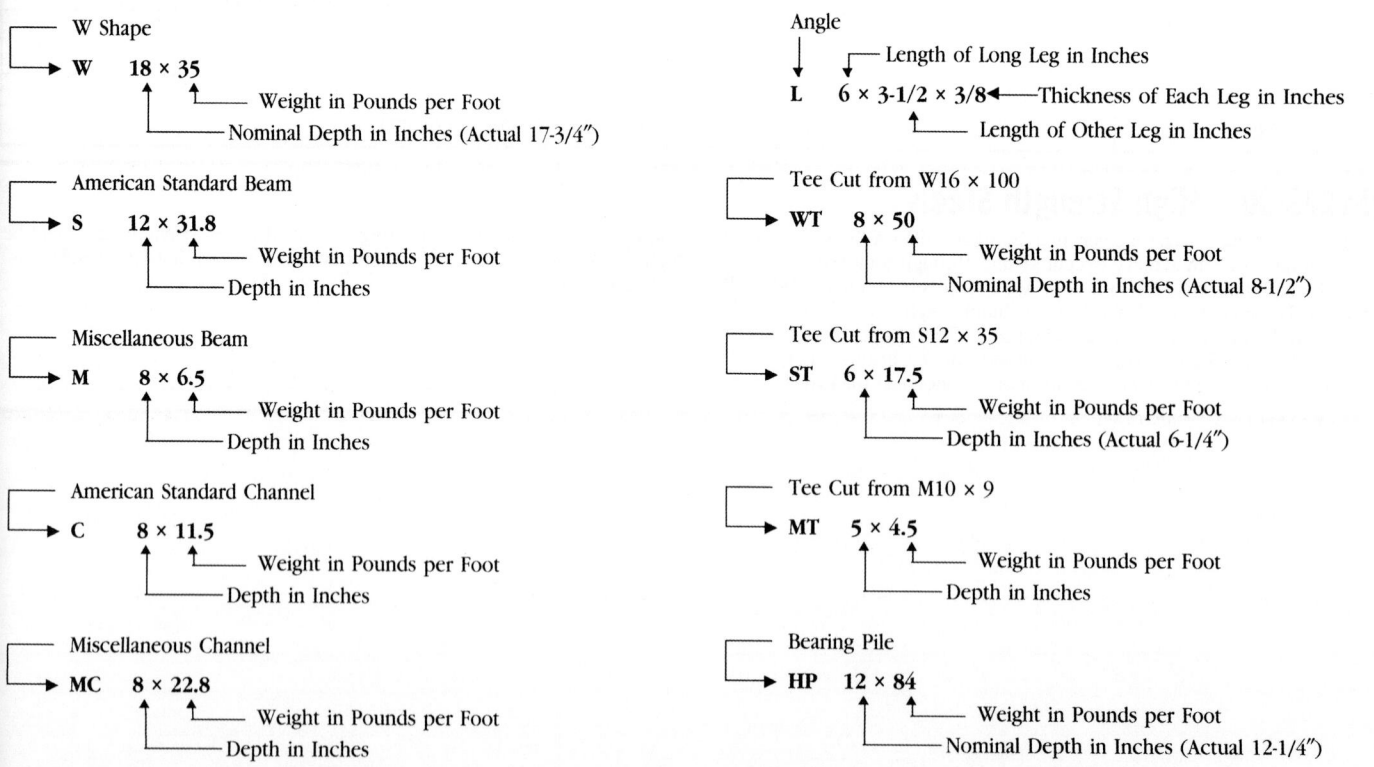

W Shape

W 18 × 35
— Weight in Pounds per Foot
— Nominal Depth in Inches (Actual 17-3/4″)

American Standard Beam

S 12 × 31.8
— Weight in Pounds per Foot
— Depth in Inches

Miscellaneous Beam

M 8 × 6.5
— Weight in Pounds per Foot
— Depth in Inches

American Standard Channel

C 8 × 11.5
— Weight in Pounds per Foot
— Depth in Inches

Miscellaneous Channel

MC 8 × 22.8
— Weight in Pounds per Foot
— Depth in Inches

Angle

L 6 × 3-1/2 × 3/8 — Length of Long Leg in Inches / Thickness of Each Leg in Inches / Length of Other Leg in Inches

Tee Cut from W16 × 100

WT 8 × 50
— Weight in Pounds per Foot
— Nominal Depth in Inches (Actual 8-1/2″)

Tee Cut from S12 × 35

ST 6 × 17.5
— Weight in Pounds per Foot
— Depth in Inches (Actual 6-1/4″)

Tee Cut from M10 × 9

MT 5 × 4.5
— Weight in Pounds per Foot
— Depth in Inches

Bearing Pile

HP 12 × 84
— Weight in Pounds per Foot
— Nominal Depth in Inches (Actual 12-1/4″)

For customer support on your Building Construction Costs with RSMeans data, call 800.448.8182.

R051223-45 Installation Time for Structural Steel Building Components

The following tables show the expected average installation times for various structural steel shapes. Table A presents installation times for columns, Table B for beams, Table C for light framing and bolts, and Table D for structural steel for various project types.

Table A

Description	Labor-Hours	Unit
Columns		
Steel, Concrete Filled		
3-1/2" Diameter	.933	Ea.
6-5/8" Diameter	1.120	Ea.
Steel Pipe		
3" Diameter	.933	Ea.
8" Diameter	1.120	Ea.
12" Diameter	1.244	Ea.
Structural Tubing		
4" x 4"	.966	Ea.
8" x 8"	1.120	Ea.
12" x 8"	1.167	Ea.
W Shape 2 Tier		
W8 x 31	.052	L.F.
W8 x 67	.057	L.F.
W10 x 45	.054	L.F.
W10 x 112	.058	L.F.
W12 x 50	.054	L.F.
W12 x 190	.061	L.F.
W14 x 74	.057	L.F.
W14 x 176	.061	L.F.

Table B

Description	Labor-Hours	Unit	Labor-Hours	Unit
Beams, W Shape				
W6 x 9	.949	Ea.	.093	L.F.
W10 x 22	1.037	Ea.	.085	L.F.
W12 x 26	1.037	Ea.	.064	L.F.
W14 x 34	1.333	Ea.	.069	L.F.
W16 x 31	1.333	Ea.	.062	L.F.
W18 x 50	2.162	Ea.	.088	L.F.
W21 x 62	2.222	Ea.	.077	L.F.
W24 x 76	2.353	Ea.	.072	L.F.
W27 x 94	2.581	Ea.	.067	L.F.
W30 x 108	2.857	Ea.	.067	L.F.
W33 x 130	3.200	Ea.	.071	L.F.
W36 x 300	3.810	Ea.	.077	L.F.

Table C

Description	Labor-Hours	Unit
Light Framing		
Angles 4" and Larger	.055	lbs.
Less than 4"	.091	lbs.
Channels 8" and Larger	.048	lbs.
Less than 8"	.072	lbs.
Cross Bracing Angles	.055	lbs.
Rods	.034	lbs.
Hanging Lintels	.069	lbs.
High Strength Bolts in Place		
3/4" Bolts	.070	Ea.
7/8" Bolts	.076	Ea.

Table D

Description	Labor-Hours	Unit	Labor-Hours	Unit
Apartments, Nursing Homes, etc.				
1-2 Stories	4.211	Piece	7.767	Ton
3-6 Stories	4.444	Piece	7.921	Ton
7-15 Stories	4.923	Piece	9.014	Ton
Over 15 Stories	5.333	Piece	9.209	Ton
Offices, Hospitals, etc.				
1-2 Stories	4.211	Piece	7.767	Ton
3-6 Stories	4.741	Piece	8.889	Ton
7-15 Stories	4.923	Piece	9.014	Ton
Over 15 Stories	5.120	Piece	9.209	Ton
Industrial Buildings				
1 Story	3.478	Piece	6.202	Ton

R051223-50 Subpurlins

Bulb tee subpurlins are structural members designed to support and reinforce a variety of roof deck systems such as precast cement fiber roof deck tiles, monolithic roof deck systems, and gypsum or lightweight concrete over formboard. Other uses include interstitial service ceiling systems, wall panel systems, and joist anchoring in bond beams. See the Unit Price section for pricing on a square foot basis at 32-5/8" O.C. Maximum span is based on a 3-span condition with a total allowable vertical load of 40 psf.

R051223-80   Dimensions and Weights of Sheet Steel

| Gauge No. | Approximate Thickness | | | | Weight | | |
| | Inches (in fractions) | Inches (in decimal parts) | | Millimeters | | | per Square |
	Wrought Iron	Wrought Iron	Steel	Steel	per S.F. in Ounces	per S.F. in Lbs.	Meter in Kg.
0000000	1/2"	.5	.4782	12.146	320	20.000	97.650
000000	15/32"	.46875	.4484	11.389	300	18.750	91.550
00000	7/16"	.4375	.4185	10.630	280	17.500	85.440
0000	13/32"	.40625	.3886	9.870	260	16.250	79.330
000	3/8"	.375	.3587	9.111	240	15.000	73.240
00	11/32"	.34375	.3288	8.352	220	13.750	67.130
0	5/16"	.3125	.2989	7.592	200	12.500	61.030
1	9/32"	.28125	.2690	6.833	180	11.250	54.930
2	17/64"	.265625	.2541	6.454	170	10.625	51.880
3	1/4"	.25	.2391	6.073	160	10.000	48.820
4	15/64"	.234375	.2242	5.695	150	9.375	45.770
5	7/32"	.21875	.2092	5.314	140	8.750	42.720
6	13/64"	.203125	.1943	4.935	130	8.125	39.670
7	3/16"	.1875	.1793	4.554	120	7.500	36.320
8	11/64"	.171875	.1644	4.176	110	6.875	33.570
9	5/32"	.15625	.1495	3.797	100	6.250	30.520
10	9/64"	.140625	.1345	3.416	90	5.625	27.460
11	1/8"	.125	.1196	3.038	80	5.000	24.410
12	7/64"	.109375	.1046	2.657	70	4.375	21.360
13	3/32"	.09375	.0897	2.278	60	3.750	18.310
14	5/64"	.078125	.0747	1.897	50	3.125	15.260
15	9/128"	.0713125	.0673	1.709	45	2.813	13.730
16	1/16"	.0625	.0598	1.519	40	2.500	12.210
17	9/160"	.05625	.0538	1.367	36	2.250	10.990
18	1/20"	.05	.0478	1.214	32	2.000	9.765
19	7/160"	.04375	.0418	1.062	28	1.750	8.544
20	3/80"	.0375	.0359	.912	24	1.500	7.324
21	11/320"	.034375	.0329	.836	22	1.375	6.713
22	1/32"	.03125	.0299	.759	20	1.250	6.103
23	9/320"	.028125	.0269	.683	18	1.125	5.490
24	1/40"	.025	.0239	.607	16	1.000	4.882
25	7/320"	.021875	.0209	.531	14	.875	4.272
26	3/160"	.01875	.0179	.455	12	.750	3.662
27	11/640"	.0171875	.0164	.417	11	.688	3.357
28	1/64"	.015625	.0149	.378	10	.625	3.052

Reference Tables

R053100-10   Decking Descriptions

General - All Deck Products

A steel deck is made by cold forming structural grade sheet steel into a repeating pattern of parallel ribs. The strength and stiffness of the panels are the result of the ribs and the material properties of the steel. Deck lengths can be varied to suit job conditions, but because of shipping considerations, are usually less than 40 feet. Standard deck width varies with the product used but full sheets are usually 12", 18", 24", 30", or 36". The deck is typically furnished in a standard width with the ends cut square. Any cutting for width, such as at openings or for angular fit, is done at the job site.

The deck is typically attached to the building frame with arc puddle welds, self-drilling screws, or powder or pneumatically driven pins. Sheet to sheet fastening is done with screws, button punching (crimping), or welds.

Composite Floor Deck

After installation and adequate fastening, a floor deck serves several purposes. It (a) acts as a working platform, (b) stabilizes the frame, (c) serves as a concrete form for the slab, and (d) reinforces the slab to carry the design loads applied during the life of the building. Composite decks are distinguished by the presence of shear connector devices as part of the deck. These devices are designed to mechanically lock the concrete and deck together so that the concrete and the deck work together to carry subsequent floor loads. These shear connector devices can be rolled-in embossments, lugs, holes, or wires welded to the panels. The deck profile can also be used to interlock concrete and steel.

Composite deck finishes are either galvanized (zinc coated) or phosphatized/painted. Galvanized deck has a zinc coating on both the top and bottom surfaces. The phosphatized/painted deck has a bare (phosphatized) top surface that will come into contact with the concrete. This bare top surface can be expected to develop rust before the concrete is placed. The bottom side of the deck has a primer coat of paint.

A composite floor deck is normally installed so the panel ends do not overlap on the supporting beams. Shear lugs or panel profile shapes often prevent a tight metal to metal fit if the panel ends overlap; the air gap caused by overlapping will prevent proper fusion with the structural steel supports when the panel end laps are shear stud welded.

Adequate end bearing of the deck must be obtained as shown on the drawings. If bearing is actually less in the field than shown on the drawings, further investigation is required.

Roof Deck

A roof deck is not designed to act compositely with other materials. A roof deck acts alone in transferring horizontal and vertical loads into the building frame. Roof deck rib openings are usually narrower than floor deck rib openings. This provides adequate support of the rigid thermal insulation board.

A roof deck is typically installed to endlap approximately 2" over supports. However, it can be butted (or lapped more than 2") to solve field fit problems. Since designers frequently use the installed deck system as part of the horizontal bracing system (the deck as a diaphragm), any fastening substitution or change should be approved by the designer. Continuous perimeter support of the deck is necessary to limit edge deflection in the finished roof and may be required for diaphragm shear transfer.

Standard roof deck finishes are galvanized or primer painted. The standard factory applied paint for roof decks is a primer paint and is not intended to weather for extended periods of time. Field painting or touching up of abrasions and deterioration of the primer coat or other protective finishes is the responsibility of the contractor.

Cellular Deck

A cellular deck is made by attaching a bottom steel sheet to a roof deck or composite floor deck panel. A cellular deck can be used in the same manner as a floor deck. Electrical, telephone, and data wires are easily run through the chase created between the deck panel and the bottom sheet.

When used as part of the electrical distribution system, the cellular deck must be installed so that the ribs line up and create a smooth cell transition at abutting ends. The joint that occurs at butting cell ends must be taped or otherwise sealed to prevent wet concrete from seeping into the cell. Cell interiors must be free of welding burrs, or other sharp intrusions, to prevent damage to wires.

When used as a roof deck, the bottom flat plate is usually left exposed to view. Care must be maintained during erection to keep good alignment and prevent damage.

A cellular deck is sometimes used with the flat plate on the top side to provide a flat working surface. Installation of the deck for this purpose requires special methods for attachment to the frame because the flat plate, now on the top, can prevent direct access to the deck material that is bearing on the structural steel. It may be advisable to treat the flat top surface to prevent slipping.

A cellular deck is always furnished galvanized or painted over galvanized.

Form Deck

A form deck can be any floor or roof deck product used as a concrete form. Connections to the frame are by the same methods used to anchor floor and roof decks. Welding washers are recommended when welding a deck that is less than 20 gauge thickness.

A form deck is furnished galvanized, prime painted, or uncoated. A galvanized deck must be used for those roof deck systems where a form deck is used to carry a lightweight insulating concrete fill.

For customer support on your Building Construction Costs with RSMeans data, call 800.448.8182.

R061110-30   Lumber Product Material Prices

The price of forest products fluctuates widely from location to location and from season to season depending upon economic conditions. The bare material prices in the unit cost sections of the data set show the National Average material prices in effect Jan. 1 of this data year. It must be noted that lumber prices in general may change significantly during the year.

Availability of certain items depends upon geographic location and must be checked prior to firm-price bidding.

Wood, Plastics & Comp. | R0616 Sheathing

R061636-20   Plywood

There are two types of plywood used in construction: interior, which is moisture-resistant but not waterproofed, and exterior, which is waterproofed.

The grade of the exterior surface of the plywood sheets is designated by the first letter: A, for smooth surface with patches allowed; B, for solid surface with patches and plugs allowed; C, which may be surface plugged or may have knot holes up to 1″ wide; and D, which is used only for interior type plywood and may have knot holes up to 2-1/2″ wide. "Structural Grade" is specifically designed for engineered applications such as box beams. All CC & DD grades have roof and floor spans marked on them.

Underlayment-grade plywood runs from 1/4″ to 1-1/4″ thick. Thicknesses 5/8″ and over have optional tongue and groove joints which eliminate the need for blocking the edges. Underlayment 19/32″ and over may be referred to as Sturd-i-Floor.

The price of plywood can fluctuate widely due to geographic and economic conditions.

Typical uses for various plywood grades are as follows:

AA-AD Interior — cupboards, shelving, paneling, furniture

BB Plyform — concrete form plywood

CDX — wall and roof sheathing

Structural — box beams, girders, stressed skin panels

AA-AC Exterior — fences, signs, siding, soffits, etc.

Underlayment — base for resilient floor coverings

Overlaid HDO — high density for concrete forms & highway signs

Overlaid MDO — medium density for painting, siding, soffits & signs

303 Siding — exterior siding, textured, striated, embossed, etc.

Thermal & Moist. Protec. | R0731 Shingles & Shakes

R073126-20   Roof Slate

16″, 18″ and 20″ are standard lengths, and slate usually comes in random widths. For standard 3/16″ thickness use 1-1/2″ copper nails. Allow for 3% breakage.

Thermal & Moist. Protec. | R0751 Built-Up Bituminous Roofing

R075113-20   Built-Up Roofing

Asphalt is available in kegs of 100 lbs. each; coal tar pitch in 560 lb. kegs. Prepared roofing felts are available in a wide range of sizes, weights and characteristics. However, the most commonly used are #15 (432 S.F. per roll, 13 lbs. per square) and #30 (216 S.F. per roll, 27 lbs. per square).

Inter-ply bitumen varies from 24 lbs. per sq. (asphalt) to 30 lbs. per sq. (coal tar) per ply, MF4@ 25%. Flood coat bitumen also varies from 60 lbs. per sq. (asphalt) to 75 lbs. per sq. (coal tar), MF4@ 25%. Expendable equipment (mops, brooms, screeds, etc.) runs about 16% of the bitumen cost. For new, inexperienced crews this factor may be much higher.

A rigid insulation board is typically applied in two layers. The first is mechanically attached to nailable decks or spot or solid mopped to non-nailable decks; the second layer is then spot or solid mopped to the first layer. Membrane application follows the insulation, except in protected membrane roofs, where the membrane goes down first and the insulation on top, followed with ballast (stone or concrete pavers). Insulation and related labor costs are NOT included in prices for built-up roofing.

For customer support on your Building Construction Costs with RSMeans data, call 800.448.8182.

Thermal & Moist. Protec.　　**R0752 Modified Bituminous Membrane Roofing**

Reference Tables

R075213-30　Modified Bitumen Roofing

The cost of modified bitumen roofing is highly dependent on the type of installation that is planned. Installation is based on the type of modifier used in the bitumen. The two most popular modifiers are atactic polypropylene (APP) and styrene butadiene styrene (SBS). The modifiers are added to heated bitumen during the manufacturing process to change its characteristics. A polyethylene, polyester or fiberglass reinforcing sheet is then sandwiched between layers of this bitumen. When completed, the result is a pre-assembled, built-up roof that has increased elasticity and weatherability. Some manufacturers include a surfacing material such as ceramic or mineral granules, metal particles or sand.

The preferred method of adhering SBS-modified bitumen roofing to the substrate is with hot-mopped asphalt (much the same as built-up roofing). This installation method requires a tar kettle/pot to heat the asphalt, as well as the labor, tools and equipment necessary to distribute and spread the hot asphalt.

The alternative method for applying APP and SBS modified bitumen is as follows. A skilled installer uses a torch to melt a small pool of bitumen off the membrane. This pool must form across the entire roll for proper adhesion. The installer must unroll the roofing at a pace slow enough to melt the bitumen, but fast enough to prevent damage to the rest of the membrane.

Modified bitumen roofing provides the advantages of both built-up and single-ply roofing. Labor costs are reduced over those of built-up roofing because only a single ply is necessary. The elasticity of single-ply roofing is attained with the reinforcing sheet and polymer modifiers. Modifieds have some self-healing characteristics and because of their multi-layer construction, they offer the reliability and safety of built-up roofing.

R078413-30　Firestopping

Firestopping is the sealing of structural, mechanical, electrical and other penetrations through fire-rated assemblies. The basic components of firestop systems are safing insulation and firestop sealant on both sides of wall penetrations and the top side of floor penetrations.

Pipe penetrations are assumed to be through concrete, grout, or joint compound and can be sleeved or unsleeved. Costs for the penetrations and sleeves are not included. An annular space of 1″ is assumed. Escutcheons are not included.

A metallic pipe is assumed to be copper, aluminum, cast iron or similar metallic material. An insulated metallic pipe is assumed to be covered with a thermal insulating jacket of varying thickness and materials.

A non-metallic pipe is assumed to be PVC, CPVC, FR Polypropylene or similar plastic piping material. Intumescent firestop sealants or wrap strips are included. Collars on both sides of wall penetrations and a sheet metal plate on the underside of floor penetrations are included.

Ductwork is assumed to be sheet metal, stainless steel or similar metallic material. Duct penetrations are assumed to be through concrete, grout or joint compound. Costs for penetrations and sleeves are not included. An annular space of 1/2″ is assumed.

Multi-trade openings include costs for sheet metal forms, firestop mortar, wrap strips, collars and sealants as necessary.

Structural penetrations joints are assumed to be 1/2″ or less. CMU walls are assumed to be within 1-1/2″ of the metal deck. Drywall walls are assumed to be tight to the underside of metal decking.

Metal panel, glass or curtain wall systems include a spandrel area of 5′ filled with mineral wool foil-faced insulation. Fasteners and stiffeners are included.

867

Openings | R0813 Metal Doors

R081313-20 Steel Door Selection Guide

Standard steel doors are classified into four levels, as recommended by the Steel Door Institute in the chart below. Each of the four levels offers a range of construction models and designs to meet architectural requirements for preference and appearance, including full flush, seamless, and stile & rail. Recommended minimum gauge requirements are also included.

For complete standard steel door construction specifications and available sizes, refer to the Steel Door Institute Technical Data Series, ANSI A250.8-98 (SDI-100), and ANSI A250.4-94 Test Procedure and Acceptance Criteria for Physical Endurance of Steel Door and Hardware Reinforcements.

Level		Model	Construction	For Full Flush or Seamless		
				Min. Gauge	Thickness (in)	Thickness (mm)
I	Standard Duty	1	Full Flush			
		2	Seamless	20	0.032	0.8
II	Heavy Duty	1	Full Flush			
		2	Seamless	18	0.042	1.0
III	Extra Heavy Duty	1	Full Flush			
		2	Seamless			
		3	*Stile & Rail	16	0.053	1.3
IV	Maximum Duty	1	Full Flush			
		2	Seamless	14	0.067	1.6

*Stiles & rails are 16 gauge; flush panels, when specified, are 18 gauge

Openings | R0851 Metal Windows

R085123-10 Steel Sash

An ironworker crew will erect 25 S.F. or 1.3 sash unit per hour, whichever is less.

A mechanic will point 30 L.F. per hour.

A painter will paint 90 S.F. per coat per hour.

A glazier production depends on light size.

Allow 1 lb. special steel sash putty per 16" x 20" light.

Openings | R0852 Wood Windows

R085216-10 Window Estimates

To ensure a complete window estimate, be sure to include the material and labor costs for each window, as well as the material and labor costs for an interior wood trim set.

For customer support on your Building Construction Costs with RSMeans data, call 800.448.8182.

R087110-10   Hardware Finishes

This table describes hardware finishes used throughout the industry. It also shows the base metal and the respective symbols in the three predominate systems of identification. Many of these are used in pricing descriptions in Division Eight.

US"	BMHA*	CDN^	Base	Description
US P	600	CP	Steel	Primed for Painting
US 1B	601	C1B	Steel	Bright Black Japanned
US 2C	602	C2C	Steel	Zinc Plated
US 2G	603	C2G	Steel	Zinc Plated
US 3	605	C3	Brass	Bright Brass, Clear Coated
US 4	606	C4	Brass	Satin Brass, Clear Coated
US 5	609	C5	Brass	Satin Brass, Blackened, Satin Relieved, Clear Coated
US 7	610	C7	Brass	Satin Brass, Blackened, Bright Relieved, Clear Coated
US 9	611	C9	Bronze	Bright Bronze, Clear Coated
US 10	612	C10	Bronze	Satin Bronze, Clear Coated
US 10A	641	C10A	Steel	Antiqued Bronze, Oiled and Lacquered
US 10B	613	C10B	Bronze	Antiqued Bronze, Oiled
US 11	616	C11	Bronze	Satin Bronze, Blackened, Satin Relieved, Clear Coated
US 14	618	C14	Brass/Bronze	Bright Nickel Plated, Clear Coated
US 15	619	C15	Brass/Bronze	Satin Nickel, Clear Coated
US 15A	620	C15A	Brass/Bronze	Satin Nickel Plated, Blackened, Satin Relieved, Clear Coated
US 17A	621	C17A	Brass/Bronze	Nickel Plated, Blackened, Relieved, Clear Coated
US 19	622	C19	Brass/Bronze	Flat Black Coated
US 20	623	C20	Brass/Bronze	Statuary Bronze, Light
US 20A	624	C20A	Brass/Bronze	Statuary Bronze, Dark
US 26	625	C26	Brass/Bronze	Bright Chromium
US 26D	626	C26D	Brass/Bronze	Satin Chromium
US 20	627	C27	Aluminum	Satin Aluminum Clear
US 28	628	C28	Aluminum	Anodized Dull Aluminum
US 32	629	C32	Stainless Steel	Bright Stainless Steel
US 32D	630	C32D	Stainless Steel	Stainless Steel
US 3	632	C3	Steel	Bright Brass Plated, Clear Coated
US 4	633	C4	Steel	Satin Brass, Clear Coated
US 7	636	C7	Steel	Satin Brass Plated, Blackened, Bright Relieved, Clear Coated
US 9	637	C9	Steel	Bright Bronze Plated, Clear Coated
US 5	638	C5	Steel	Satin Brass Plated, Blackened, Bright Relieved, Clear Coated
US 10	639	C10	Steel	Satin Bronze Plated, Clear Coated
US 10B	640	C10B	Steel	Antique Bronze, Oiled
US 10A	641	C10A	Steel	Antiqued Bronze, Oiled and Lacquered
US 11	643	C11	Steel	Satin Bronze Plated, Blackened, Bright Relieved, Clear Coated
US 14	645	C14	Steel	Bright Nickel Plated, Clear Coated
US 15	646	C15	Steel	Satin Nickel Plated, Clear Coated
US 15A	647	C15A	Steel	Nickel Plated, Blackened, Bright Relieved, Clear Coated
US 17A	648	C17A	Steel	Nickel Plated, Blackened, Relieved, Clear Coated
US 20	649	C20	Steel	Statuary Bronze, Light
US 20A	650	C20A	Steel	Statuary Bronze, Dark
US 26	651	C26	Steel	Bright Chromium Plated
US 26D	652	C26D	Steel	Satin Chromium Plated

* - BMHA Builders Hardware Manufacturing Association
" - US Equivalent
^ - Canadian Equivalent
Japanning is imitating Asian lacquer work

For customer support on your Building Construction Costs with RSMeans data, call 800.448.8182.

R088110-10   Glazing Productivity

Some glass sizes are estimated by the "united inch" (height + width). The table below shows the number of lights glazed in an eight-hour period by the crew size indicated, for glass up to 1/4″ thick. Square or nearly square lights are more economical on a S.F. basis. Long slender lights will have a high S.F. installation cost. For insulated glass reduce production by 33%. For 1/2″ float glass reduce production by 50%. Production time for glazing with two glaziers per day averages: 1/4″ float glass 120 S.F.; 1/2″ float glass 55 S.F.; 1/2″ insulated glass 95 S.F.; 3/4″ insulated glass 75 S.F.

Glazing Method	United Inches per Light							
	40″	60″	80″	100″	135″	165″	200″	240″
Number of Men in Crew	1	1	1	1	2	3	3	4
Industrial sash, putty	60	45	24	15	18	—	—	—
With stops, putty bed	50	36	21	12	16	8	4	3
Wood stops, rubber	40	27	15	9	11	6	3	2
Metal stops, rubber	30	24	14	9	9	6	3	2
Structural glass	10	7	4	3	—	—	—	—
Corrugated glass	12	9	7	4	4	4	3	—
Storefronts	16	15	13	11	7	6	4	4
Skylights, putty glass	60	36	21	12	16	—	—	—
Thiokol set	15	15	11	9	9	6	3	2
Vinyl set, snap on	18	18	13	12	12	7	5	4
Maximum area per light	2.8 S.F.	6.3 S.F.	11.1 S.F.	17.4 S.F.	31.6 S.F.	47 S.F.	69 S.F.	100 S.F.

For customer support on your Building Construction Costs with RSMeans data, call 800.448.8182.

R092000-50   Lath, Plaster and Gypsum Board

Gypsum board lath is available in 3/8″ thick × 16″ wide × 4′ long sheets as a base material for multi-layer plaster applications. It is also available as a base for either multi-layer or veneer plaster applications in 1/2″ and 5/8″ thick–4′ wide × 8′, 10′ or 12′ long sheets. Fasteners are screws or blued ring shank nails for wood framing and screws for metal framing.

Metal lath is available in diamond mesh patterns with flat or self-furring profiles. Paper backing is available for applications where excessive plaster waste needs to be avoided. A slotted mesh ribbed lath should be used in areas where the span between structural supports is greater than normal. Most metal lath comes in 27″ × 96″ sheets. Diamond mesh weighs 1.75, 2.5 or 3.4 pounds per square yard, slotted mesh lath weighs 2.75 or 3.4 pounds per square yard. Metal lath can be nailed, screwed or tied in place.

Many **accessories** are available. Corner beads, flat reinforcing strips, casing beads, control and expansion joints, furring brackets and channels are some examples. Note that accessories are not included in plaster or stucco line items.

Plaster is defined as a material or combination of materials that when mixed with a suitable amount of water, forms a plastic mass or paste. When applied to a surface, the paste adheres to it and subsequently hardens, preserving in a rigid state the form or texture imposed during the period of elasticity.

Gypsum plaster is made from ground calcined gypsum. It is mixed with aggregates and water for use as a base coat plaster.

Vermiculite plaster is a fire-retardant plaster covering used on steel beams, concrete slabs and other heavy construction materials. Vermiculite is a group name for certain clay minerals, hydrous silicates or aluminum, magnesium and iron that have been expanded by heat.

Perlite plaster is a plaster using perlite as an aggregate instead of sand. Perlite is a volcanic glass that has been expanded by heat.

Gauging plaster is a mix of gypsum plaster and lime putty that when applied produces a quick drying finish coat.

Veneer plaster is a one or two component gypsum plaster used as a thin finish coat over special gypsum board.

Keenes cement is a white cementitious material manufactured from gypsum that has been burned at a high temperature and ground to a fine powder. Alum is added to accelerate the set. The resulting plaster is hard and strong and accepts and maintains a high polish, hence it is used as a finishing plaster.

Stucco is a Portland cement based plaster used primarily as an exterior finish.

Plaster is used on both interior and exterior surfaces. Generally it is applied in multiple-coat systems. A three-coat system uses the terms scratch, brown and finish to identify each coat. A two-coat system uses base and finish to describe each coat. Each type of plaster and application system has attributes that are chosen by the designer to best fit the intended use.

Gypsum Plaster Quantities for 100 S.Y.	2 Coat, 5/8″ Thick		3 Coat, 3/4″ Thick		
	Base	Finish	Scratch	Brown	Finish
	1:3 Mix	2:1 Mix	1:2 Mix	1:3 Mix	2:1 Mix
Gypsum plaster	1,300 lb.		1,350 lb.	650 lb.	
Sand	1.75 C.Y.		1.85 C.Y.	1.35 C.Y.	
Finish hydrated lime		340 lb.			340 lb.
Gauging plaster		170 lb.			170 lb.

Vermiculite or Perlite Plaster Quantities for 100 S.Y.	2 Coat, 5/8″ Thick		3 Coat, 3/4″ Thick		
	Base	Finish	Scratch	Brown	Finish
Gypsum plaster	1,250 lb.		1,450 lb.	800 lb.	
Vermiculite or perlite	7.8 bags		8.0 bags	3.3 bags	
Finish hydrated lime		340 lb.			340 lb.
Gauging plaster		170 lb.			170 lb.

Stucco–Three-Coat System Quantities for 100 S.Y.	On Wood Frame	On Masonry
Portland cement	29 bags	21 bags
Sand	2.6 C.Y.	2.0 C.Y.
Hydrated lime	180 lb.	120 lb.

For customer support on your Building Construction Costs with RSMeans data, call 800.448.8182.

R092910-10   Levels of Gypsum Drywall Finish

In the past, contract documents often used phrases such as "industry standard" and "workmanlike finish" to specify the expected quality of gypsum board wall and ceiling installations. The vagueness of these descriptions led to unacceptable work and disputes.

In order to resolve this problem, four major trade associations concerned with the manufacture, erection, finish, and decoration of gypsum board wall and ceiling systems developed an industry-wide *Recommended Levels of Gypsum Board Finish.*

The finish of gypsum board walls and ceilings for specific final decoration is dependent on a number of factors. A primary consideration is the location of the surface and the degree of decorative treatment desired. Painted and unpainted surfaces in warehouses and other areas where appearance is normally not critical may simply require the taping of wallboard joints and 'spotting' of fastener heads. Blemish-free, smooth, monolithic surfaces often intended for painted and decorated walls and ceilings in habitated structures, ranging from single-family dwellings through monumental buildings, require additional finishing prior to the application of the final decoration.

Other factors to be considered in determining the level of finish of the gypsum board surface are (1) the type of angle of surface illumination (both natural and artificial lighting), and (2) the paint and method of application or the type and finish of wallcovering specified as the final decoration. Critical lighting conditions, gloss paints, and thin wall coverings require a higher level of gypsum board finish than heavily textured surfaces which are subsequently painted or surfaces which are to be decorated with heavy grade wall coverings.

The following descriptions were developed by the Association of the Wall and Ceiling Industries-International (AWCI), Ceiling & Interior Systems Construction Association (CISCA), Gypsum Association (GA), and Painting and Decorating Contractors of America (PDCA) as a guide.

Level 0: Used in temporary construction or wherever the final decoration has not been determined. Unfinished. No taping, finishing or corner beads are required. Also could be used where non-predecorated panels will be used in demountable-type partitions that are to be painted as a final finish.

Level 1: Frequently used in plenum areas above ceilings, in attics, in areas where the assembly would generally be concealed, or in building service corridors and other areas not normally open to public view. Some degree of sound and smoke control is provided; in some geographic areas, this level is referred to as "fire-taping," although this level of finish does not typically meet fire-resistant assembly requirements. Where a fire resistance rating is required for the gypsum board assembly, details of construction should be in accordance with reports of fire tests of assemblies that have met the requirements of the fire rating acceptable.

All joints and interior angles shall have tape embedded in joint compound. Accessories are optional at specifier discretion in corridors and other areas with pedestrian traffic. Tape and fastener heads need not be covered with joint compound. Surface shall be free of excess joint compound. Tool marks and ridges are acceptable.

Level 2: It may be specified for standard gypsum board surfaces in garages, warehouse storage, or other similar areas where surface appearance is not of primary importance.

All joints and interior angles shall have tape embedded in joint compound and shall be immediately wiped with a joint knife or trowel, leaving a thin coating of joint compound over all joints and interior angles. Fastener heads and accessories shall be covered with a coat of joint compound. Surface shall be free of excess joint compound. Tool marks and ridges are acceptable.

Level 3: Typically used in areas receiving heavy texture (spray or hand applied) finishes before final painting, or where commercial-grade (heavy duty) wall coverings are to be applied as the final decoration. This level of finish should not be used where smooth painted surfaces or where lighter weight wall coverings are specified. The prepared surface shall be coated with a drywall primer prior to the application of final finishes.

All joints and interior angles shall have tape embedded in joint compound and shall be immediately wiped with a joint knife or trowel, leaving a thin coating of joint compound over all joints and interior angles. One additional coat of joint compound shall be applied over all joints and interior angles. Fastener heads and accessories shall be covered with two separate coats of joint compound. All joint compounds shall be smooth and free of tool marks and ridges. The prepared surface shall be covered with a drywall primer prior to the application of the final decoration.

Level 4: This level should be used where residential grade (light duty) wall coverings, flat paints, or light textures are to be applied. The prepared surface shall be coated with a drywall primer prior to the application of final finishes. Release agents for wall coverings are specifically formulated to minimize damage if coverings are subsequently removed.

The weight, texture, and sheen level of the wall covering material selected should be taken into consideration when specifying wall coverings over this level of drywall treatment. Joints and fasteners must be sufficiently concealed if the wall covering material is lightweight, contains limited pattern, has a glossy finish, or has any combination of these features. In critical lighting areas, flat paints applied over light textures tend to reduce joint photographing. Gloss, semi-gloss, and enamel paints are not recommended over this level of finish.

All joints and interior angles shall have tape embedded in joint compound and shall be immediately wiped with a joint knife or trowel, leaving a thin coating of joint compound over all joints and interior angles. In addition, two separate coats of joint compound shall be applied over all flat joints and one separate coat of joint compound applied over interior angles. Fastener heads and accessories shall be covered with three separate coats of joint compound. All joint compounds shall be smooth and free of tool marks and ridges. The prepared surface shall be covered with a drywall primer like Sheetrock® First Coat prior to the application of the final decoration.

Level 5: The highest quality finish is the most effective method to provide a uniform surface and minimize the possibility of joint photographing and of fasteners showing through the final decoration. This level of finish is required where gloss, semi-gloss, or enamel is specified; when flat joints are specified over an untextured surface; or where critical lighting conditions occur. The prepared surface shall be coated with a drywall primer prior to the application of the final decoration.

All joints and interior angles shall have tape embedded in joint compound and be immediately wiped with a joint knife or trowel, leaving a thin coating of joint compound over all joints and interior angles. Two separate coats of joint compound shall be applied over all flat joints and one separate coat of joint compound applied over interior angles. Fastener heads and accessories shall be covered with three separate coats of joint compound.

A thin skim coat of joint compound shall be trowel applied to the entire surface. Excess compound is immediately troweled off, leaving a film or skim coating of compound completely covering the paper. As an alternative to a skim coat, a material manufactured especially for this purpose may be applied such as Sheetrock® Tuff-Hide primer surfacer. The surface must be smooth and free of tool marks and ridges. The prepared surface shall be covered with a drywall primer prior to the application of the final decoration.

For customer support on your Building Construction Costs with RSMeans data, call 800.448.8182.

Finishes	R0966 Terrazzo Flooring

R096613-10 Terrazzo Floor

The table below lists quantities required for 100 S.F. of 5/8" terrazzo topping, either bonded or not bonded.

Description	Bonded to Concrete 1-1/8" Bed, 1:4 Mix	Not Bonded 2-1/8" Bed and 1/4" Sand
Portland cement, 94 lb. Bag	6 bags	8 bags
Sand	10 C.F.	20 C.F.
Divider strips, 4' squares	50 L.F.	50 L.F.
Terrazzo fill, 50 lb. Bag	12 bags	12 bags
15 Lb. tarred felt		1 C.S.F.
Mesh 2 x 2 #14 galvanized		1 C.S.F.
Crew J-3	0.77 days	0.87 days

2' × 2' panels require 1.00 L.F. divider strip per S.F.

3' × 3' panels require 0.67 L.F. divider strip per S.F.

4' × 4' panels require 0.50 L.F. divider strip per S.F.

5' × 5' panels require 0.40 L.F. divider strip per S.F.

6' × 6' panels require 0.33 L.F. divider strip per S.F.

Finishes	R0972 Wall Coverings

R097223-10 Wall Covering

The table below lists the quantities required for 100 S.F. of wall covering.

Description	Medium-Priced Paper	Expensive Paper
Paper	1.6 dbl. rolls	1.6 dbl. rolls
Wall sizing	0.25 gallon	0.25 gallon
Vinyl wall paste	0.6 gallon	0.6 gallon
Apply sizing	0.3 hour	0.3 hour
Apply paper	1.2 hours	1.5 hours

Most wallpapers now come in double rolls only.

To remove old paper, allow 1.3 hours per 100 S.F.

873

Finishes · R0991 Painting

R099100-10 Painting Estimating Techniques

Proper estimating methodology is needed to obtain an accurate painting estimate. There is no known reliable shortcut or square foot method. The following steps should be followed:

- List all surfaces to be painted, with an accurate quantity (area) of each. Items having similar surface condition, finish, application method and accessibility may be grouped together.
- List all the tasks required for each surface to be painted, including surface preparation, masking, and protection of adjacent surfaces. Surface preparation may include minor repairs, washing, sanding and puttying.
- Select the proper Means line for each task. Review and consider all adjustments to labor and materials for type of paint and location of work. Apply the height adjustment carefully. For instance, when applying the adjustment for work over 8' high to a wall that is 12' high, apply the adjustment only to the area between 8' and 12' high, and not to the entire wall.

When applying more than one percent (%) adjustment, apply each to the base cost of the data, rather than applying one percentage adjustment on top of the other.

When estimating the cost of painting walls and ceilings remember to add the brushwork for all cut-ins at inside corners and around windows and doors as a LF measure. One linear foot of cut-in with a brush equals one square foot of painting.

All items for spray painting include the labor for roll-back.

Deduct for openings greater than 100 SF or openings that extend from floor to ceiling and are greater than 5' wide. Do not deduct small openings.

The cost of brushes, rollers, ladders and spray equipment are considered part of a painting contractor's overhead, and should not be added to the estimate. The cost of rented equipment such as scaffolding and swing staging should be added to the estimate.

R099100-20 Painting

Item	Coat	One Gallon Covers			In 8 Hours a Laborer Covers			Labor-Hours per 100 S.F.		
		Brush	Roller	Spray	Brush	Roller	Spray	Brush	Roller	Spray
Paint wood siding	prime	250 S.F.	225 S.F.	290 S.F.	1150 S.F.	1300 S.F.	2275 S.F.	.695	.615	.351
	others	270	250	290	1300	1625	2600	.615	.492	.307
Paint exterior trim	prime	400	—	—	650	—	—	1.230	—	—
	1st	475	—	—	800	—	—	1.000	—	—
	2nd	520	—	—	975	—	—	.820	—	—
Paint shingle siding	prime	270	255	300	650	975	1950	1.230	.820	.410
	others	360	340	380	800	1150	2275	1.000	.695	.351
Stain shingle siding	1st	180	170	200	750	1125	2250	1.068	.711	.355
	2nd	270	250	290	900	1325	2600	.888	.603	.307
Paint brick masonry	prime	180	135	160	750	800	1800	1.066	1.000	.444
	1st	270	225	290	815	975	2275	.981	.820	.351
	2nd	340	305	360	815	1150	2925	.981	.695	.273
Paint interior plaster or drywall	prime	400	380	495	1150	2000	3250	.695	.400	.246
	others	450	425	495	1300	2300	4000	.615	.347	.200
Paint interior doors and windows	prime	400	—	—	650	—	—	1.230	—	—
	1st	425	—	—	800	—	—	1.000	—	—
	2nd	450	—	—	975	—	—	.820	—	—

Special Construction · R1311 Swimming Pools

R131113-20 Swimming Pools

Pool prices given per square foot of surface area include pool structure, filter and chlorination equipment, pumps, related piping, ladders/steps, maintenance kit, skimmer and vacuum system. Decks and electrical service to equipment are not included.

Residential in-ground pool construction can be divided into two categories: vinyl lined and gunite. Vinyl lined pool walls are constructed of different materials including wood, concrete, plastic or metal. The bottom is often graded with sand over which the vinyl liner is installed. Vermiculite or soil cement bottoms may be substituted for an added cost.

Gunite pool construction is used both in residential and municipal installations. These structures are steel reinforced for strength and finished with a white cement limestone plaster.

Municipal pools will have a higher cost because plumbing codes require more expensive materials, chlorination equipment and higher filtration rates.

Municipal pools greater than 1,800 S.F. require gutter systems to control waves. This gutter may be formed into the concrete wall. Often a vinyl/stainless steel gutter or gutter/wall system is specified, which will raise the pool cost.

Competition pools usually require tile bottoms and sides with contrasting lane striping, which will also raise the pool cost.

Special Construction | R1331 Fabric Structures

R133113-10 Air Supported Structures

Air supported structures are made from fabrics that can be classified into two groups: temporary and permanent. Temporary fabrics include nylon, woven polyethylene, vinyl film, and vinyl coated dacron. These have lifespans that range from five to fifteen plus years. The cost per square foot includes a fabric shell, tension cables, primary and back-up inflation systems and doors. The lower cost structures are used for construction shelters, bulk storage and pond covers. The more expensive are used for recreational structures and warehouses.

Permanent fabrics are teflon coated fiberglass. The life of this structure is twenty plus years. The high cost limits its application to architectural designed structures which call for a clear span covered area, such as stadiums and convention centers. Both temporary and permanent structures are available in translucent fabrics which eliminate the need for daytime lighting.

Areas to be covered vary from 10,000 S.F. to any area up to 1000 feet wide by any length. Height restrictions range from a maximum of 1/2 of the width to a minimum of 1/6 of the width. Erection of even the largest of the temporary structures requires no more than a week.

Centrifugal fans provide the inflation necessary to support the structure during application of live loads. Airlocks are usually used at large entrances to prevent loss of static pressure. Some manufacturers employ propeller fans which generate sufficient airflow (30,000 CFM) to eliminate the need for airlocks. These fans may also be automatically controlled to resist high wind conditions, regulate humidity (air changes), and provide cooling and heat.

Insulation can be provided with the addition of a second or even third interior liner, creating a dead air space with an "R" value of four to nine. Some structures allow for the liner to be collapsed into the outer shell to enable the internal heat to melt accumulated snow. For cooling or air conditioning, the exterior face of the liner can be aluminized to reflect the sun's heat.

R133113-90 Seismic Bracing

Sometimes referred to as anti-sway bracing, this support system is required in earthquake areas. The individual components must be assembled to make a required system.

Example				Additionally, height factors must be taken into account. Add the following percentages to labor for elevated installations:	
	C-Clamp 3/8" rod	2 ea.		15' to 20' high	10%
	Rod, continuous thread 3/8"	10 L.F.		21' to 25' high	20%
	Field, weld 1"	2 ea.		26' to 30' high	30%
				31' to 35' high	40%
				36' to 40' high	50%
				41' to 50' high	60%

Special Construction | R1334 Fabricated Engineered Structures

R133419-10 Pre-Engineered Steel Buildings

These buildings are manufactured by many companies and normally erected by franchised dealers throughout the U.S. The four basic types are: Rigid Frames, Truss type, Post and Beam and the Sloped Beam type. The most popular roof slope is a low pitch of 1" in 12". The minimum economical area of these buildings is about 3000 S.F. of floor area. Bay sizes are usually 20' to 24' but can go as high as 30' with heavier girts and purlins. Eave heights are usually 12' to 24' with 18' to 20' most typical.

Material prices shown in the Unit Price section are bare costs for the building shell only and do not include floors, foundations, anchor bolts, interior finishes or utilities. Costs assume at least three bays of 24' each, a 1" in 12" roof slope, and they are based on a 30 psf roof load and a 20 psf wind load (wind load is a function of wind speed, building height, and terrain characteristics; this should be determined by a Registered Structural Engineer) and no unusual requirements. Costs include the structural frame, 26 ga. non-insulated colored corrugated or ribbed roofing and siding panels, fasteners, closures, trim and flashing but no allowance for insulation, doors, windows, skylights, gutters or downspouts. Very large projects would generally cost less for materials than the prices shown. For roof panel substitutions and wall panel substitutions, see appropriate Unit Price sections.

Conditions at the site, weather, shape and size of the building, and labor availability will affect the erection cost of the building.

R133423-30 Dome Structures

Steel — The four types are Lamella, Schwedler, Arch and Geodesic. For maximum economy, the rise should be about 15 to 20% of the diameter. Most common diameters are in the 200' to 300' range. Lamella domes weigh about 5 P.S.F. of floor area less than Schwedler domes. The Schwedler dome weight in lbs. per S.F. approaches .046 times the diameter. Domes below 125' diameter weigh .07 times the diameter and the cost per ton of steel is higher. See R051223-20 for estimating weight.

Wood — Small domes are of sawn lumber. Larger ones are laminated. In larger sizes, triaxial and triangular cost about the same; radial domes cost more. Radial domes are economical in the 60' to 70' diameter range. The most economical range of all types is 80' to 200' diameters. Diameters can run over 400'. All costs are quoted above the foundation. Prices include 2" decking and a tension tie ring in place.

Plywood — Stock prefab geodesic domes are available with diameters from 24' to 60'.

Fiberglass — Aluminum framed translucent sandwich panels with spans from 5' to 45' are commercially available.

Aluminum — Stressed skin aluminum panels form geodesic domes with spans ranging from 82' to 232'. An aluminum space truss, triangulated or nontriangulated, with aluminum or clear acrylic closure panels can be used for clear spans of 40' to 415'.

875

R142000-10   Freight Elevators

Capacities run from 2,000 lbs. to over 100,000 lbs. with 3,000 lbs. to 10,000 lbs. most common. Travel speeds are generally lower and the control is less intricate than on passenger elevators. Costs in the Unit Price sections are for hydraulic and geared elevators.

R142000-20   Elevator Selective Costs   See R142000-40 for cost development.

	Passenger		Freight		Hospital	
A. Base Unit	Hydraulic	Electric	Hydraulic	Electric	Hydraulic	Electric
Capacity	1,500 lb.	2,000 lb.	2,000 lb.	4,000 lb.	4,000 lb.	4,000 lb.
Speed	100 F.P.M.	200 F.P.M.	100 F.P.M.	200 F.P.M.	100 F.P.M.	200 F.P.M.
#Stops/Travel Ft.	2/12	4/40	2/20	4/40	2/20	4/40
Push Button Oper.	Yes	Yes	Yes	Yes	Yes	Yes
Telephone Box & Wire	"	"	"	"	"	"
Emergency Lighting	"	"	No	No	"	"
Cab	Plastic Lam. Walls	Plastic Lam. Walls	Painted Steel	Painted Steel	Plastic Lam. Walls	Plastic Lam. Walls
Cove Lighting	Yes	Yes	No	No	Yes	Yes
Floor	V.C.T.	V.C.T.	Wood w/Safety Treads	Wood w/Safety Treads	V.C.T.	V.C.T.
Doors, & Speedside Slide	Yes	Yes	Yes	Yes	Yes	Yes
Gates, Manual	No	No	No	No	No	No
Signals, Lighted Buttons	Car and Hall	Car and Hall	Car and Hall	Car and Hall	Car and Hall	Car and Hall
O.H. Geared Machine	N.A.	Yes	N.A.	Yes	N.A.	Yes
Variable Voltage Contr.	"	"	N.A.	"	"	"
Emergency Alarm	Yes	"	Yes	"	Yes	"
Class "A" Loading	N.A.	N.A.	"	"	N.A.	N.A.

R142000-30   Passenger Elevators

Electric elevators are used generally but hydraulic elevators can be used for lifts up to 70' and where large capacities are required. Hydraulic speeds are limited to 200 F.P.M. but cars are self leveling at the stops. On low rises, hydraulic installation runs about 15% less than standard electric types but on higher rises this installation cost advantage is reduced. Maintenance of hydraulic elevators is about the same as the electric type but the underground portion is not included in the maintenance contract.

In electric elevators there are several control systems available, the choice of which will be based upon elevator use, size, speed and cost criteria. The two types of drives are geared for low speeds and gearless for 450 F.P.M. and over.

The tables on the preceding pages illustrate typical installed costs of the various types of elevators available.

R142000-40   Elevator Cost Development

To price a new car or truck from the factory, you must start with the manufacturer's basic model, then add or exchange optional equipment and features. The same is true for pricing elevators.

Requirement: One-passenger elevator, five-story hydraulic, 2,500 lb. capacity, 12' floor to floor, speed 150 F.P.M., emergency power switching and maintenance contract.

Example:

Description	Adjustment
A. Base Elevator: Hydraulic Passenger, 1500 lb. Capacity, 100 fpm, 2 Stops, Standard Finish	1 Ea.
B. Capacity Adjustment (2,500 lb.)	1 Ea.
C. Excess Travel Adjustment: 48' Total Travel (4 x 12') minus 12' Base Unit Travel =	36 V.L.F.
D. Stops Adjustment: 5 Total Stops minus 2 Stops (Base Unit) =	3 Stops
E. Speed Adjustment (150 F.P.M.)	1 Ea.
F. Options:	
1. Intercom Service	1 Ea.
2. Emergency Power Switching, Automatic	1 Ea.
3. Stainless Steel Entrance Doors	5 Ea.
4. Maintenance Contract (12 Months)	1 Ea.
5. Position Indicator for main floor level (none indicated in Base Unit)	1 Ea.

For customer support on your Building Construction Costs with RSMeans data, call 800.448.8182.

Conveying Equipment   R1432 Moving Walks

R143210-20   Moving Ramps and Walks

These are a specialized form of conveyor 3' to 6' wide with capacities of 3,600 to 18,000 persons per hour. Maximum speed is 140 F.P.M. and normal incline is 0° to 15°.

Local codes will determine the maximum angle. Outdoor units would require additional weather protection.

Plumbing   R2201 Operation & Maintenance of Plumbing

R220102-20   Labor Adjustment Factors

Labor Adjustment Factors are provided for Divisions 21, 22, and 23 to assist the mechanical estimator account for the various complexities and special conditions of any particular project. While a single percentage has been entered on each line of Division 22 01 02.20, it should be understood that these are just suggested midpoints of ranges of values commonly used by mechanical estimators. They may be increased or decreased depending on the severity of the special conditions.

The group for "existing occupied buildings" has been the subject of requests for explanation. Actually there are two stages to this group: buildings that are existing and "finished" but unoccupied, and those that also are occupied. Buildings that are "finished" may result in higher labor costs due to the

workers having to be more careful not to damage finished walls, ceilings, floors, etc. and may necessitate special protective coverings and barriers. Also corridor bends and doorways may not accommodate long pieces of pipe or larger pieces of equipment. Work above an already hung ceiling can be very time consuming. The addition of occupants may force the work to be done on premium time (nights and/or weekends), eliminate the possible use of some preferred tools such as pneumatic drivers, powder charged drivers, etc. The estimator should evaluate the access to the work area and just how the work is going to be accomplished to arrive at an increase in labor costs over "normal" new construction productivity.

R220105-10   Demolition (Selective vs. Removal for Replacement)

Demolition can be divided into two basic categories.

One type of demolition involves the removal of material with no concern for its replacement. The labor-hours to estimate this work are found under "Selective Demolition" in the Fire Protection, Plumbing and HVAC Divisions. It is selective in that individual items or all the material installed as a system or trade grouping such as plumbing or heating systems are removed. This may be accomplished by the easiest way possible, such as sawing, torch cutting, or sledge hammering as well as simple unbolting.

The second type of demolition is the removal of some items for repair or replacement. This removal may involve careful draining, opening of unions,

disconnecting and tagging electrical connections, capping pipes/ducts to prevent entry of debris or leakage of the material contained as well as transporting the item away from its in-place location to a truck/dumpster. An approximation of the time required to accomplish this type of demolition is to use half of the time indicated as necessary to install a new unit. For example: installation of a new pump might be listed as requiring 6 labor-hours so if we had to estimate the removal of the old pump we would allow an additional 3 hours for a total of 9 hours. That is, the complete replacement of a defective pump with a new pump would be estimated to take 9 labor-hours.

Plumbing   R2211 Facility Water Distribution

R221113-50   Pipe Material Considerations

1. Malleable fittings should be used for gas service.
2. Malleable fittings are used where there are stresses/strains due to expansion and vibration.
3. Cast fittings may be broken as an aid to disassembling heating lines frozen by long use, temperature and minerals.
4. A cast iron pipe is extensively used for underground and submerged service.
5. Type M (light wall) copper tubing is available in hard temper only and is used for nonpressure and less severe applications than K and L.

Domestic/Imported Pipe and Fittings Costs

The prices shown in this publication for steel/cast iron pipe and steel, cast iron, and malleable iron fittings are based on domestic production sold at the normal trade discounts. The above listed items of foreign manufacture may be available at prices 1/3 to 1/2 of those shown. Some imported items after minor machining or finishing operations are being sold as domestic to further complicate the system.

6. Type L (medium wall) copper tubing, available hard or soft for interior service.
7. Type K (heavy wall) copper tubing, available in hard or soft temper for use where conditions are severe. For underground and interior service.
8. Hard drawn tubing requires fewer hangers or supports but should not be bent. Silver brazed fittings are recommended, but soft solder is normally used.
9. Type DMV (very light wall) copper tubing designed for drainage, waste and vent plus other non-critical pressure services.

Caution: Most pipe prices in this data set also include a coupling and pipe hangers which for the larger sizes can add significantly to the per foot cost and should be taken into account when comparing "book cost" with the quoted supplier's cost.

877

R235616-60   Solar Heating (Space and Hot Water)

Collectors should face as close to due South as possible, but variations of up to 20 degrees on either side of true South are acceptable. Local climate and collector type may influence the choice between east or west deviations. Obviously they should be located so they are not shaded from the sun's rays. Incline collectors at a slope of latitude minus 5 degrees for domestic hot water and latitude plus 15 degrees for space heating.

Flat plate collectors consist of a number of components as follows: Insulation to reduce heat loss through the bottom and sides of the collector. The enclosure which contains all the components in this assembly is usually weatherproof and prevents dust, wind and water from coming in contact with the absorber plate. The cover plate usually consists of one or more layers of a variety of glass or plastic and reduces the reradiation by creating an air space which traps the heat between the cover and the absorber plates.

The absorber plate must have a good thermal bond with the fluid passages. The absorber plate is usually metallic and treated with a surface coating which improves absorptivity. Black or dark paints or selective coatings are used for this purpose, and the design of this passage and plate combination helps determine a solar system's effectiveness.

Heat transfer fluid passage tubes are attached above and below or integral with an absorber plate for the purpose of transferring thermal energy from the absorber plate to a heat transfer medium. The heat exchanger is a device for transferring thermal energy from one fluid to another.

Piping and storage tanks should be well insulated to minimize heat losses.

Size domestic water heating storage tanks to hold 20 gallons of water per user, minimum, plus 10 gallons per dishwasher or washing machine. For domestic water heating an optimum collector size is approximately 3/4 square foot of area per gallon of water storage. For space heating of residences and small commercial applications the collector is commonly sized between 30% and 50% of the internal floor area. For space heating of large commercial applications, collector areas less than 30% of the internal floor area can still provide significant heat reductions.

A supplementary heat source is recommended for Northern states for December through February.

The solar energy transmission per square foot of collector surface varies greatly with the material used. Initial cost, heat transmittance and useful life are obviously interrelated.

Heating, Ventilating & A.C. R2360 Central Cooling Equipment

R236000-20   Air Conditioning Requirements

BTUs per hour per S.F. of floor area and S.F. per ton of air conditioning.

Type of Building	BTU/Hr per S.F.	S.F. per Ton	Type of Building	BTU/Hr per S.F.	S.F. per Ton	Type of Building	BTU/Hr per S.F.	S.F. per Ton
Apartments, Individual	26	450	Dormitory, Rooms	40	300	Libraries	50	240
Corridors	22	550	Corridors	30	400	Low Rise Office, Exterior	38	320
Auditoriums & Theaters	40	300/18*	Dress Shops	43	280	Interior	33	360
Banks	50	240	Drug Stores	80	150	Medical Centers	28	425
Barber Shops	48	250	Factories	40	300	Motels	28	425
Bars & Taverns	133	90	High Rise Office—Ext. Rms.	46	263	Office (small suite)	43	280
Beauty Parlors	66	180	Interior Rooms	37	325	Post Office, Individual Office	42	285
Bowling Alleys	68	175	Hospitals, Core	43	280	Central Area	46	260
Churches	36	330/20*	Perimeter	46	260	Residences	20	600
Cocktail Lounges	68	175	Hotel, Guest Rooms	44	275	Restaurants	60	200
Computer Rooms	141	85	Corridors	30	400	Schools & Colleges	46	260
Dental Offices	52	230	Public Spaces	55	220	Shoe Stores	55	220
Dept. Stores, Basement	34	350	Industrial Plants, Offices	38	320	Shop'g. Ctrs., Supermarkets	34	350
Main Floor	40	300	General Offices	34	350	Retail Stores	48	250
Upper Floor	30	400	Plant Areas	40	300	Specialty	60	200

*Persons per ton

12,000 BTU = 1 ton of air conditioning

For customer support on your Building Construction Costs with RSMeans data, call 800.448.8182.

R260519-90 Wire

Wire quantities are taken off by either measuring each cable run or by extending the conduit and raceway quantities times the number of conductors in the raceway. Ten percent should be added for waste and tie-ins. Keep in mind that the unit of measure of wire is C.L.F. not L.F. as in raceways so the formula would read:

$$\frac{\text{(L.F. Raceway x No. of Conductors) x 1.10}}{100} = \text{C.L.F.}$$

Price per C.L.F. of wire includes:
1. Setting up wire coils or spools on racks
2. Attaching wire to pull in means
3. Measuring and cutting wire
4. Pulling wire into a raceway
5. Identifying and tagging

Price does not include:
1. Connections to breakers, panelboards, or equipment
2. Splices

Job Conditions: Productivity is based on new construction to a height of 15' using rolling staging in an unobstructed area. Material staging is assumed to be within 100' of work being performed.

Economy of Scale: If more than three wires at a time are being pulled, deduct the following percentages from the labor of that grouping:

4-5 wires	25%
6-10 wires	30%
11-15 wires	35%
over 15	40%

If a wire pull is less than 100' in length and is interrupted several times by boxes, lighting outlets, etc., it may be necessary to add the following lengths to each wire being pulled:

Junction box to junction box	2 L.F.
Lighting panel to junction box	6 L.F.
Distribution panel to sub panel	8 L.F.
Switchboard to distribution panel	12 L.F.
Switchboard to motor control center	20 L.F.
Switchboard to cable tray	40 L.F.

Measure of Drops and Riser: It is important when taking off wire quantities to include the wire for drops to electrical equipment. If heights of electrical equipment are not clearly stated, use the following guide:

	Bottom A.F.F.	Top A.F.F.	Inside Cabinet
Safety switch to 100A	5'	6'	2'
Safety switch 400 to 600A	4'	6'	3'
100A panel 12 to 30 circuit	4'	6'	3'
42 circuit panel	3'	6'	4'
Switch box	3'	3'6"	1'
Switchgear	0'	8'	8'
Motor control centers	0'	8'	8'
Transformers - wall mount	4'	8'	2'
Transformers - floor mount	0'	12'	4'

879

R260519-92   Minimum Copper and Aluminum Wire Size Allowed for Various Types of Insulation

	Minimum Wire Sizes								
	Copper		Aluminum			Copper		Aluminum	
Amperes	THW THWN or XHHW	THHN XHHW *	THW XHHW	THHN XHHW *	Amperes	THW THWN or XHHW	THHN XHHW *	THW XHHW	THHN XHHW *
15A	#14	#14	#12	#12	195	3/0	2/0	250kcmil	4/0
20	#12	#12	#10	#10	200	3/0	3/0	250kcmil	4/0
25	#10	#10	#10	#10	205	4/0	3/0	250kcmil	4/0
30	#10	#10	# 8	# 8	225	4/0	3/0	300kcmil	250kcmil
40	# 8	# 8	# 8	# 8	230	4/0	4/0	300kcmil	250kcmil
45	# 8	# 8	# 6	# 8	250	250kcmil	4/0	350kcmil	300kcmil
50	# 8	# 8	# 6	# 6	255	250kcmil	4/0	400kcmil	300kcmil
55	# 6	# 8	# 4	# 6	260	300kcmil	4/0	400kcmil	350kcmil
60	# 6	# 6	# 4	# 6	270	300kcmil	250kcmil	400kcmil	350kcmil
65	# 6	# 6	# 4	# 4	280	300kcmil	250kcmil	500kcmil	350kcmil
75	# 4	# 6	# 3	# 4	285	300kcmil	250kcmil	500kcmil	400kcmil
85	# 4	# 4	# 2	# 3	290	350kcmil	250kcmil	500kcmil	400kcmil
90	# 3	# 4	# 2	# 2	305	350kcmil	300kcmil	500kcmil	400kcmil
95	# 3	# 4	# 1	# 2	310	350kcmil	300kcmil	500kcmil	500kcmil
100	# 3	# 3	# 1	# 2	320	400kcmil	300kcmil	600kcmil	500kcmil
110	# 2	# 3	1/0	# 1	335	400kcmil	350kcmil	600kcmil	500kcmil
115	# 2	# 2	1/0	# 1	340	500kcmil	350kcmil	600kcmil	500kcmil
120	# 1	# 2	1/0	1/0	350	500kcmil	350kcmil	700kcmil	500kcmil
130	# 1	# 2	2/0	1/0	375	500kcmil	400kcmil	700kcmil	600kcmil
135	1/0	# 1	2/0	1/0	380	500kcmil	400kcmil	750kcmil	600kcmil
150	1/0	# 1	3/0	2/0	385	600kcmil	500kcmil	750kcmil	600kcmil
155	2/0	1/0	3/0	3/0	420	600kcmil	500kcmil		700kcmil
170	2/0	1/0	4/0	3/0	430		500kcmil		750kcmil
175	2/0	2/0	4/0	3/0	435		600kcmil		750kcmil
180	3/0	2/0	4/0	4/0	475		600kcmil		

*Dry Locations Only

Notes:

1. Size #14 to 4/0 is in AWG units (American Wire Gauge).
2. Size 250 to 750 is in kcmil units (Thousand Circular Mils).
3. Use next higher ampere value if exact value is not listed in table.
4. For loads that operate continuously increase ampere value by 25% to obtain proper wire size.
5. Table R260519-92 has been written for estimating only. It is based on an ambient temperature of 30°C (86° F). For ambient temperature other than 30°C (86° F), ampacity correction factors will be applied.

Reprinted with permission from NFPA 70-2014, *National Electrical Code®*, Copyright © 2013, National Fire Protection Association, Quincy, MA. This reprinted material is not the complete and official position of the NFPA on the referenced subject, which is represented solely by the standard in its entirety. NFPA 70®, *National Electrical Code* and *NEC®* are registered trademarks of the National Fire Protection Association, Quincy, MA.

R260533-22   Conductors in Conduit

The table below lists the maximum number of conductors for various sized conduit using THW, TW or THWN insulations.

Copper Wire Size	1/2"			3/4"			1"			1-1/4"			1-1/2"			2"			2-1/2"			3"		3-1/2"		4"	
	TW	THW	THWN	TW	THW	THWN	TW	THW	THWN	TW	THW	THWN	TW	THW	THWN	TW	THW	THWN	TW	THW	THWN	THW	THWN	THW	THWN	THW	THWN
#14	9	6	13	15	10	24	25	16	39	44	29	69	60	40	94	99	65	154	142	93		143		192			
#12	7	4	10	12	8	18	19	13	29	35	24	51	47	32	70	78	53	114	111	76	164	117		157			
#10	5	4	6	9	6	11	15	11	18	26	19	32	36	26	44	60	43	73	85	61	104	95	160	127		163	
#8	2	1	3	4	3	5	7	5	9	12	10	16	17	13	22	28	22	36	40	32	51	49	79	66	106	85	136
#6		1	1		2	4		4	6		7	11		10	15		16	26		23	37	36	57	48	76	62	98
#4		1	1		1	2		3	4		5	7		7	9		12	16		17	22	27	35	36	47	47	60
#3		1	1		1	1		2	3		4	6		6	8		10	13		15	19	23	29	31	39	40	51
#2		1	1		1	1		2	3		4	5		5	7		9	11		13	16	20	25	27	33	34	43
#1					1	1		1	1		3	3		4	5		6	8		9	12	14	18	19	25	25	32
1/0					1	1		1	1		2	3		3	4		5	7		8	10	12	15	16	21	21	27
2/0					1	1		1	1		1	2		3	3		5	6		7	8	10	13	14	17	18	22
3/0					1	1		1	1		1	1		2	3		4	5		6	7	9	11	12	14	15	18
4/0						1		1	1		1	1		1	2		3	4		5	6	7	9	10	12	13	15
250 kcmil								1	1		1	1		1	1		2	3		4	4	6	7	8	10	10	12
300								1	1		1	1		1	1		2	3		3	4	5	6	7	8	9	11
350									1		1	1		1	1		1	2		3	3	4	5	6	7	8	9
400											1	1		1	1		1	1		2	3	4	5	5	6	7	8
500											1	1		1	1		1	1		1	2	3	4	4	5	6	7
600												1		1	1		1	1		1	1	3	3	4	4	5	5
700														1	1		1	1		1	1	2	3	3	4	4	5
750														1	1		1	1		1	1	2	2	3	3	4	4

Reprinted with permission from NFPA 70-2014, *National Electrical Code®*, Copyright © 2013, National Fire Protection Association, Quincy, MA. This reprinted material is not the complete and official position of the NFPA on the referenced subject, which is represented solely by the standard in its entirety.

For customer support on your Building Construction Costs with RSMeans data, call 800.448.8182.

R312316-40   Excavating

The selection of equipment used for structural excavation and bulk excavation or for grading is determined by the following factors.
1. Quantity of material
2. Type of material
3. Depth or height of cut
4. Length of haul
5. Condition of haul road
6. Accessibility of site
7. Moisture content and dewatering requirements
8. Availability of excavating and hauling equipment

Some additional costs must be allowed for hand trimming the sides and bottom of concrete pours and other excavation below the general excavation.

Number of B.C.Y. per truck = 1.5 C.Y. bucket × 8 passes = 12 loose C.Y.

$$= 12 \times \frac{100}{118} = 10.2 \text{ B.C.Y. per truck}$$

Truck Haul Cycle:

Load truck, 8 passes	=	4 minutes
Haul distance, 1 mile	=	9 minutes
Dump time	=	2 minutes
Return, 1 mile	=	7 minutes
Spot under machine	=	1 minute
		23 minute cycle

Add the mobilization and demobilization costs to the total excavation costs. When equipment is rented for more than three days, there is often no mobilization charge by the equipment dealer. On larger jobs outside of urban areas, scrapers can move earth economically provided a dump site or fill area and adequate haul roads are available. Excavation within sheeting bracing or cofferdam bracing is usually done with a clamshell and production

When planning excavation and fill, the following should also be considered.
1. Swell factor
2. Compaction factor
3. Moisture content
4. Density requirements

A typical example for scheduling and estimating the cost of excavation of a 15′ deep basement on a dry site when the material must be hauled off the site is outlined below.

Assumptions:
1. Swell factor, 18%
2. No mobilization or demobilization
3. Allowance included for idle time and moving on job
4. No dewatering, sheeting, or bracing
5. No truck spotter or hand trimming

Fleet Haul Production per day in B.C.Y.

$$4 \text{ trucks} \times \frac{50 \text{ min. hour}}{23 \text{ min. haul cycle}} \times 8 \text{ hrs.} \times 10.2 \text{ B.C.Y.}$$

$$= 4 \times 2.2 \times 8 \times 10.2 = 718 \text{ B.C.Y./day}$$

is low, since the clamshell may have to be guided by hand between the bracing. When excavating or filling an area enclosed with a wellpoint system, add 10% to 15% to the cost to allow for restricted access. When estimating earth excavation quantities for structures, allow work space outside the building footprint for construction of the foundation and a slope of 1:1 unless sheeting is used.

For customer support on your Building Construction Costs with RSMeans data, call 800.448.8182.

R312316-45 Excavating Equipment

The table below lists theoretical hourly production in C.Y./hr. bank measure for some typical excavation equipment. Figures assume 50 minute hours, 83% job efficiency, 100% operator efficiency, 90° swing and properly sized hauling units, which must be modified for adverse digging and loading conditions. Actual production costs in the front of the data set average about 50% of the theoretical values listed here.

Equipment	Soil Type	B.C.Y. Weight	% Swell	1 C.Y.	1-1/2 C.Y.	2 C.Y.	2-1/2 C.Y.	3 C.Y.	3-1/2 C.Y.	4 C.Y.
Hydraulic Excavator	Moist loam, sandy clay	3400 lb.	40%	165	195	200	275	330	385	440
"Backhoe"	Sand and gravel	3100	18	140	170	225	240	285	330	380
15' Deep Cut	Common earth	2800	30	150	180	230	250	300	350	400
	Clay, hard, dense	3000	33	120	140	190	200	240	260	320
	Moist loam, sandy clay	3400	40	170	245	295	335	385	435	475
				(6.0)	(7.0)	(7.8)	(8.4)	(8.8)	(9.1)	(9.4)
	Sand and gravel	3100	18	165	225	275	325	375	420	460
Power Shovel				(6.0)	(7.0)	(7.8)	(8.4)	(8.8)	(9.1)	(9.4)
Optimum Cut (Ft.)	Common earth	2800	30	145	200	250	295	335	375	425
				(7.8)	(9.2)	(10.2)	(11.2)	(12.1)	(13.0)	(13.8)
	Clay, hard, dense	3000	33	120	175	220	255	300	335	375
				(9.0)	(10.7)	(12.2)	(13.3)	(14.2)	(15.1)	(16.0)
	Moist loam, sandy clay	3400	40	130	180	220	250	290	325	385
				(6.6)	(7.4)	(8.0)	(8.5)	(9.0)	(9.5)	(10.0)
	Sand and gravel	3100	18	130	175	210	245	280	315	375
Drag Line				(6.6)	(7.4)	(8.0)	(8.5)	(9.0)	(9.5)	(10.0)
Optimum Cut (Ft.)	Common earth	2800	30	110	160	190	220	250	280	310
				(8.0)	(9.0)	(9.9)	(10.5)	(11.0)	(11.5)	(12.0)
	Clay, hard, dense	3000	33	90	130	160	190	225	250	280
				(9.3)	(10.7)	(11.8)	(12.3)	(12.8)	(13.3)	(12.0)

Equipment	Soil Type	B.C.Y. Weight	% Swell	Wheel Loaders				Track Loaders		
				3 C.Y.	4 C.Y.	6 C.Y.	8 C.Y.	2-1/4 C.Y.	3 C.Y.	4 C.Y.
	Moist loam, sandy clay	3400	40	260	340	510	690	135	180	250
Loading Tractors	Sand and gravel	3100	18	245	320	480	650	130	170	235
	Common earth	2800	30	230	300	460	620	120	155	220
	Clay, hard, dense	3000	33	200	270	415	560	110	145	200
	Rock, well-blasted	4000	50	180	245	380	520	100	130	180

For customer support on your Building Construction Costs with RSMeans data, call 800.448.8182.

R312319-90   Wellpoints

A single stage wellpoint system is usually limited to dewatering an average 15' depth below normal ground water level. Multi-stage systems are employed for greater depth with the pumping equipment installed only at the lowest header level. Ejectors with unlimited lift capacity can be economical when two or more stages of wellpoints can be replaced or when horizontal clearance is restricted, such as in deep trenches or tunneling projects, and where low water flows are expected. Wellpoints are usually spaced on 2-1/2' to 10' centers along a header pipe. Wellpoint spacing, header size, and pump size are all determined by the expected flow as dictated by soil conditions.

In almost all soils encountered in wellpoint dewatering, the wellpoints may be jetted into place. Cemented soils and stiff clays may require sand wicks about 12" in diameter around each wellpoint to increase efficiency and eliminate weeping into the excavation. These sand wicks require 1/2 to 3 C.Y. of washed filter sand and are installed by using a 12" diameter steel casing and hole puncher jetted into the ground 2' deeper than the wellpoint. Rock may require predrilled holes.

Labor required for the complete installation and removal of a single stage wellpoint system is in the range of 3/4 to 2 labor-hours per linear foot of header, depending upon jetting conditions, wellpoint spacing, etc.

Continuous pumping is necessary except in some free draining soil where temporary flooding is permissible (as in trenches which are backfilled after each day's work). Good practice requires provision of a stand-by pump during the continuous pumping operation.

Systems for continuous trenching below the water table should be installed three to four times the length of expected daily progress to ensure uninterrupted digging, and header pipe size should not be changed during the job.

For pervious free draining soils, deep wells in place of wellpoints may be economical because of lower installation and maintenance costs. Daily production ranges between two to three wells per day, for 25' to 40' depths, to one well per day for depths over 50'.

Detailed analysis and estimating for any dewatering problem is available at no cost from wellpoint manufacturers. Major firms will quote "sufficient equipment" quotes or their affiliates will offer lump sum proposals to cover complete dewatering responsibility.

Description for 200' System with 8" Header		Quantities
Equipment & Material	Wellpoints 25' long, 2" diameter @ 5' O.C.	40 Each
	Header pipe, 8" diameter	200 L.F.
	Discharge pipe, 8" diameter	100 L.F.
	8" valves	3 Each
	Combination jetting & wellpoint pump (standby)	1 Each
	Wellpoint pump, 8" diameter	1 Each
	Transportation to and from site	1 Day
	Fuel for 30 days x 60 gal./day	1800 Gallons
	Lubricants for 30 days x 16 lbs./day	480 Lbs.
	Sand for points	40 C.Y.
Labor	Technician to supervise installation	1 Week
	Labor for installation and removal of system	300 Labor-hours
	4 Operators straight time 40 hrs./wk. for 4.33 wks.	693 Hrs.
	4 Operators overtime 2 hrs./wk. for 4.33 wks.	35 Hrs.

R312323-30   Compacting Backfill

Compaction of fill in embankments, around structures, in trenches, and under slabs is important to control settlement. Factors affecting compaction are:

1. Soil gradation
2. Moisture content
3. Equipment used
4. Depth of fill per lift
5. Density required

Production Rate:

$$\frac{1.75' \text{ plate width} \times 50 \text{ F.P.M.} \times 50 \text{ min./hr.} \times .67' \text{ lift}}{27 \text{ C.F. per C.Y.}} = 108.5 \text{ C.Y./hr.}$$

Production Rate for 4 Passes:

$$\frac{108.5 \text{ C.Y.}}{4 \text{ passes}} = 27.125 \text{ C.Y./hr.} \times 8 \text{ hrs.} = 217 \text{ C.Y./day}$$

Example:

Compact granular fill around a building foundation using a 21" wide x 24" vibratory plate in 8" lifts. Operator moves at 50 F.P.M. working a 50 minute hour to develop 95% Modified Proctor Density with 4 passes.

For customer support on your Building Construction Costs with RSMeans data, call 800.448.8182.

R314116-40   Wood Sheet Piling

Wood sheet piling may be used for depths to 20′ where there is no ground water. If moderate ground water is encountered Tongue & Groove sheeting will help to keep it out. When considerable ground water is present, steel sheeting must be used.

For estimating purposes on trench excavation, sizes are as follows:

Depth	Sheeting	Wales	Braces	B.F. per S.F.
To 8′	3 x 12′s	6 x 8′s, 2 line	6 x 8′s, @ 10′	4.0 @ 8′
8′ x 12′	3 x 12′s	10 x 10′s, 2 line	10 x 10′s, @ 9′	5.0 average
12′ to 20′	3 x 12′s	12 x 12′s, 3 line	12 x 12′s, @ 8′	7.0 average

Sheeting to be toed in at least 2′ depending upon soil conditions. A five person crew with an air compressor and sheeting driver can drive and brace 440 SF/day at 8′ deep, 360 SF/day at 12′ deep, and 320 SF/day at 16′ deep.

For normal soils, piling can be pulled in 1/3 the time to install. Pulling difficulty increases with the time in the ground. Production can be increased by high pressure jetting.

R314116-45   Steel Sheet Piling

Limiting weights are 22 to 38#/S.F. of wall surface with 27#/S.F. average for usual types and sizes. (Weights of piles themselves are from 30.7#/L.F. to 57#/L.F. but they are 15″ to 21″ wide.) Lightweight sections 12″ to 28″ wide from 3 ga. to 12 ga. thick are also available for shallow excavations. Piles may be driven two at a time with an impact or vibratory hammer (use vibratory to pull) hung from a crane without leads. A reasonable estimate of the life of steel sheet piling is 10 uses with up to 125 uses possible if a vibratory hammer is used. Used piling costs from 50% to 80% of new piling depending on location and market conditions. Sheet piling and H piles

can be rented for about 30% of the delivered mill price for the first month and 5% per month thereafter. Allow 1 labor-hour per pile for cleaning and trimming after driving. These costs increase with depth and hydrostatic head. Vibratory drivers are faster in wet granular soils and are excellent for pile extraction. Pulling difficulty increases with the time in the ground and may cost more than driving. It is often economical to abandon the sheet piling, especially if it can be used as the outer wall form. Allow about 1/3 additional length or more for toeing into ground. Add bracing, waler and strut costs. Waler costs can equal the cost per ton of sheeting.

R314513-90   Vibroflotation and Vibro Replacement Soil Compaction

Vibroflotation is a proprietary system of compacting sandy soils in place to increase relative density to about 70%. Typical bearing capacities attained will be 6000 psf for saturated sand and 12,000 psf for dry sand. Usual range is 4000 to 8000 psf capacity. Costs in the front of the data set are for a vertical foot of compacted cylinder 6′ to 10′ in diameter.

Vibro replacement is a proprietary system of improving cohesive soils in place to increase bearing capacity. Most silts and clays above or below the water table can be strengthened by installation of stone columns.

The process consists of radial displacement of the soil by vibration. The created hole is then backfilled in stages with coarse granular fill which is thoroughly compacted and displaced into the surrounding soil in the form of a column.

The total project cost would depend on the number and depth of the compacted cylinders. The installing company guarantees relative soil density of the sand cylinders after compaction and the bearing capacity of the soil after the replacement process. Detailed estimating information is available from the installer at no cost.

For customer support on your Building Construction Costs with RSMeans data, call 800.448.8182.

Earthwork — R3163 Drilled Caissons

R316326-60   Caissons

The three principal types of caissons are:

(1) Belled Caissons, which except for shallow depths and poor soil conditions, are generally recommended. They provide more bearing than shaft area. Because of its conical shape, no horizontal reinforcement of the bell is required.

(2) Straight Shaft Caissons are used where relatively light loads are to be supported by caissons that rest on high value bearing strata. While the shaft is larger in diameter than for belled types this is more than offset by the savings in time and labor.

(3) Keyed Caissons are used when extremely heavy loads are to be carried. A keyed or socketed caisson transfers its load into rock by a combination of end-bearing and shear reinforcing of the shaft. The most economical shaft often consists of a steel casing, a steel wide flange core and concrete. Allowable compressive stresses of .225 f'c for concrete, 16,000 psi for the wide flange core, and 9,000 psi for the steel casing are commonly used. The usual range of shaft diameter is 18″ to 84″. The number of sizes specified for any one project should be limited due to the problems of casing and auger storage. When hand work is to be performed, shaft diameters should not be less than 32″. When inspection of borings is required a minimum shaft diameter of 30″ is recommended. Concrete caissons are intended to be poured against earth excavation so permanent forms, which add to cost, should not be used if the excavation is clean and the earth is sufficiently impervious to prevent excessive loss of concrete.

Soil Conditions for Belling		
Good	**Requires Handwork**	**Not Recommended**
Clay	Hard Shale	Silt
Sandy Clay	Limestone	Sand
Silty Clay	Sandstone	Gravel
Clayey Silt	Weathered Mica	Igneous Rock
Hard-pan		
Soft Shale		
Decomposed Rock		

Exterior Improvements — R3292 Turf & Grasses

R329219-50   Seeding

The type of grass is determined by light, shade and moisture content of soil plus intended use. Fertilizer should be disked 4″ before seeding. For steep slopes disk five tons of mulch and lay two tons of hay or straw on surface per acre after seeding. Surface mulch can be staked, lightly disked or tar emulsion sprayed. Material for mulch can be wood chips, peat moss, partially rotted hay or straw, wood fibers and sprayed emulsions. Hemp seed blankets with fertilizer are also available. For spring seeding, watering is necessary. Late fall seeding may have to be reseeded in the spring. Hydraulic seeding, power mulching, and aerial seeding can be used on large areas.

For customer support on your Building Construction Costs with RSMeans data, call 800.448.8182.

R331113-80   Piping Designations

There are several systems currently in use to describe pipe and fittings. The following paragraphs will help to identify and clarify classifications of piping systems used for water distribution.

Piping may be classified by schedule. Piping schedules include 5S, 10S, 10, 20, 30, Standard, 40, 60, Extra Strong, 80, 100, 120, 140, 160 and Double Extra Strong. These schedules are dependent upon the pipe wall thickness. The wall thickness of a particular schedule may vary with pipe size.

Ductile iron pipe for water distribution is classified by Pressure Classes such as Class 150, 200, 250, 300 and 350. These classes are actually the rated water working pressure of the pipe in pounds per square inch (psi). The pipe in these pressure classes is designed to withstand the rated water working pressure plus a surge allowance of 100 psi.

The American Water Works Association (AWWA) provides standards for various types of **plastic pipe**. C-900 is the specification for polyvinyl chloride (PVC) piping used for water distribution in sizes ranging from 4″ through 12″. C-901 is the specification for polyethylene (PE) pressure pipe, tubing and fittings used for water distribution in sizes ranging from 1/2″ through 3″. C-905 is the specification for PVC piping sizes 14″ and greater.

PVC pressure-rated pipe is identified using the standard dimensional ratio (SDR) method. This method is defined by the American Society for Testing and Materials (ASTM) Standard D 2241. This pipe is available in SDR numbers 64, 41, 32.5, 26, 21, 17, and 13.5. A pipe with an SDR of 64 will have the thinnest wall while a pipe with an SDR of 13.5 will have the thickest wall. When the pressure rating (PR) of a pipe is given in psi, it is based on a line supplying water at 73 degrees F.

The National Sanitation Foundation (NSF) seal of approval is applied to products that can be used with potable water. These products have been tested to ANSI/NSF Standard 14.

Valves and strainers are classified by American National Standards Institute (ANSI) Classes. These Classes are 125, 150, 200, 250, 300, 400, 600, 900, 1500 and 2500. Within each class there is an operating pressure range dependent upon temperature. Design parameters should be compared to the appropriate material dependent, pressure-temperature rating chart for accurate valve selection.

Transportation   R3472 Railway Construction

R347216-10   Single Track R.R. Siding

The costs for a single track RR siding in the Unit Price Section include the components shown in the table below.

Description of Component	Qty. per L.F. of Track	Unit
Ballast, 1-1/2″ crushed stone	.667	C.Y.
6″ x 8″ x 8′-6″ Treated timber ties, 22″ O.C.	.545	Ea.
Tie plates, 2 per tie	1.091	Ea.
Track rail	2.000	L.F.
Spikes, 6″, 4 per tie	2.182	Ea.
Splice bars w/ bolts, lock washers & nuts, @ 33′ O.C.	.061	Pair
Crew B-14 @ 57 L.F./Day	.018	Day

R347216-20   Single Track, Steel Ties, Concrete Bed

The costs for a R.R. siding with steel ties and a concrete bed in the Unit Price section include the components shown in the table below.

Description of Component	Qty. per L.F. of Track	Unit
Concrete bed, 9′ wide, 10″ thick	.278	C.Y.
Ties, W6x16 x 6′-6″ long, @ 30″ O.C.	.400	Ea.
Tie plates, 4 per tie	1.600	Ea.
Track rail	2.000	L.F.
Tie plate bolts, 1″, 8 per tie	3.200	Ea.
Splice bars w/bolts, lock washers & nuts, @ 33′ O.C.	.061	Pair
Crew B-14 @ 22 L.F./Day	.045	Day

For customer support on your Building Construction Costs with RSMeans data, call 800.448.8182.

Change Orders

Change Order Considerations

A change order is a written document usually prepared by the design professional and signed by the owner, the architect/engineer, and the contractor. A change order states the agreement of the parties to: an addition, deletion, or revision in the work; an adjustment in the contract sum, if any; or an adjustment in the contract time, if any. Change orders, or "extras", in the construction process occur after execution of the construction contract and impact architects/engineers, contractors, and owners.

Change orders that are properly recognized and managed can ensure orderly, professional, and profitable progress for everyone involved in the project. There are many causes for change orders and change order requests. In all cases, change orders or change order requests should be addressed promptly and in a precise and prescribed manner. The following paragraphs include information regarding change order pricing and procedures.

The Causes of Change Orders

Reasons for issuing change orders include:

- Unforeseen field conditions that require a change in the work
- Correction of design discrepancies, errors, or omissions in the contract documents
- Owner-requested changes, either by design criteria, scope of work, or project objectives
- Completion date changes for reasons unrelated to the construction process
- Changes in building code interpretations, or other public authority requirements that require a change in the work
- Changes in availability of existing or new materials and products

Procedures

Properly written contract documents must include the correct change order procedures for all parties—owners, design professionals, and contractors—to follow in order to avoid costly delays and litigation.

Being "in the right" is not always a sufficient or acceptable defense. The contract provisions requiring notification and documentation must be adhered to within a defined or reasonable time frame.

The appropriate method of handling change orders is by a written proposal and acceptance by all parties involved. Prior to starting work on a project, all parties should identify their

authorized agents who may sign and accept change orders, as well as any limits placed on their authority.

Time may be a critical factor when the need for a change arises. For such cases, the contractor might be directed to proceed on a "time and materials" basis, rather than wait for all paperwork to be processed—a delay that could impede progress. In this situation, the contractor must still follow the prescribed change order procedures including, but not limited to, notification and documentation.

Lack of documentation can be very costly, especially if legal judgments are to be made, and if certain field personnel are no longer available. For time and material change orders, the contractor should keep accurate daily records of all labor and material allocated to the change.

Owners or awarding authorities who do considerable and continual building construction (such as the federal government) realize the inevitability of change orders for numerous reasons, both predictable and unpredictable. As a result, the federal government, the American Institute of Architects (AIA), the Engineers Joint Contract Documents Committee (EJCDC), and other contractor, legal, and technical organizations have developed standards and procedures to be followed by all parties to achieve contract continuance and timely completion, while being financially fair to all concerned.

Pricing Change Orders

When pricing change orders, regardless of their cause, the most significant factor is when the change occurs. The need for a change may be perceived in the field or requested by the architect/engineer *before* any of the actual installation has begun, or may evolve or appear *during* construction when the item of work in question is partially installed. In the latter cases, the original sequence of construction is disrupted, along with all contiguous and supporting systems. Change orders cause the greatest impact when they occur *after* the installation has been completed and must be uncovered, or even replaced. Post-completion changes may be caused by necessary design changes, product failure, or changes in the owner's requirements that are not discovered until the building or the systems begin to function.

Specified procedures of notification and record keeping must be adhered to and enforced regardless of the stage of construction: *before, during,* or *after* installation. Some bidding documents anticipate change orders by requiring that unit prices including overhead and profit percentages—for additional as well as deductible changes—be listed. Generally these unit prices do not fully take into account the ripple effect, or impact on other trades, and should be used for general guidance only.

When pricing change orders, it is important to classify the time frame in which the change occurs. There are two basic time frames for change orders: *pre-installation change orders,* which occur before the start of construction, and *post-installation change orders,* which involve reworking after the original installation. Change orders that occur between these stages may be priced according to the extent of work completed using a combination of techniques developed for pricing *pre-* and *post-installation* changes.

Factors To Consider When Pricing Change Orders

As an estimator begins to prepare a change order, the following questions should be reviewed to determine their impact on the final price.

General

- Is the change order work pre-installation or post-installation?

  Change order work costs vary according to how much of the installation has been completed. Once workers have the project scoped in their minds, even though they have not started, it can be difficult to refocus. Consequently they may spend more than the normal amount of time understanding the change. Also, modifications to work in place, such as trimming or refitting, usually take more time than was initially estimated. The greater the amount of work in place, the more reluctant workers are to change it. Psychologically they may resent the change and as a result the rework takes longer than normal. Post-installation change order estimates must include demolition of existing work as required to accomplish the change. If the work is performed at a later time, additional obstacles, such as building finishes, may be present which must be protected. Regardless of whether the change occurs

For customer support on your Building Construction Costs with RSMeans data, call 800.448.8182.

pre-installation or post-installation, attempt to isolate the identifiable factors and price them separately. For example, add shipping costs that may be required pre-installation or any demolition required post-installation. Then analyze the potential impact on productivity of psychological and/or learning curve factors and adjust the output rates accordingly. One approach is to break down the typical workday into segments and quantify the impact on each segment.

Change Order Installation Efficiency

The labor-hours expressed (for new construction) are based on average installation time, using an efficiency level. For change order situations, adjustments to this efficiency level should reflect the daily labor-hour allocation for that particular occurrence.

- *Will the change substantially delay the original completion date?*

A significant change in the project may cause the original completion date to be extended. The extended schedule may subject the contractor to new wage rates dictated by relevant labor contracts. Project supervision and other project overhead must also be extended beyond the original completion date. The schedule extension may also put installation into a new weather season. For example, underground piping scheduled for October installation was delayed until January. As a result, frost penetrated the trench area, thereby changing the degree of difficulty of the task. Changes and delays may have a ripple effect throughout the project. This effect must be analyzed and negotiated with the owner.

- *What is the net effect of a deduct change order?*

In most cases, change orders resulting in a deduction or credit reflect only bare costs. The contractor may retain the overhead and profit based on the original bid.

Materials

- *Will you have to pay more or less for the new material, required by the change order, than you paid for the original purchase?*

The same material prices or discounts will usually apply to materials purchased for change orders as new construction. In some

instances, however, the contractor may forfeit the advantages of competitive pricing for change orders. Consider the following example:

A contractor purchased over $20,000 worth of fan coil units for an installation and obtained the maximum discount. Some time later it was determined the project required an additional matching unit. The contractor has to purchase this unit from the original supplier to ensure a match. The supplier at this time may not discount the unit because of the small quantity, and he is no longer in a competitive situation. The impact of quantity on purchase can add between 0% and 25% to material prices and/or subcontractor quotes.

- *If materials have been ordered or delivered to the job site, will they be subject to a cancellation charge or restocking fee?*

Check with the supplier to determine if ordered materials are subject to a cancellation charge. Delivered materials not used as a result of a change order may be subject to a restocking fee if returned to the supplier. Common restocking charges run between 20% and 40%. Also, delivery charges to return the goods to the supplier must be added.

Labor

- *How efficient is the existing crew at the actual installation?*

Is the same crew that performed the initial work going to do the change order? Possibly the change consists of the installation of a unit identical to one already installed; therefore, the change should take less time. Be sure to consider this potential productivity increase and modify the productivity rates accordingly.

- *If the crew size is increased, what impact will that have on supervision requirements?*

Under most bargaining agreements or management practices, there is a point at which a working foreman is replaced by a nonworking foreman. This replacement increases project overhead by adding a nonproductive worker. If additional workers are added to accelerate the project or to perform changes while maintaining the schedule, be sure to add additional supervision time if warranted. Calculate the

hours involved and the additional cost directly if possible.

- *What are the other impacts of increased crew size?*

The larger the crew, the greater the potential for productivity to decrease. Some of the factors that cause this productivity loss are: overcrowding (producing restrictive conditions in the working space) and possibly a shortage of any special tools and equipment required. Such factors affect not only the crew working on the elements directly involved in the change order, but other crews whose movements may also be hampered. As the crew increases, check its basic composition for changes by the addition or deletion of apprentices or nonworking foreman, and quantify the potential effects of equipment shortages or other logistical factors.

- *As new crews, unfamiliar with the project, are brought onto the site, how long will it take them to become oriented to the project requirements?*

The orientation time for a new crew to become 100% effective varies with the site and type of project. Orientation is easiest at a new construction site and most difficult at existing, very restrictive renovation sites. The type of work also affects orientation time. When all elements of the work are exposed, such as concrete or masonry work, orientation is decreased. When the work is concealed or less visible, such as existing electrical systems, orientation takes longer. Usually orientation can be accomplished in one day or less. Costs for added orientation should be itemized and added to the total estimated cost.

- *How much actual production can be gained by working overtime?*

Short term overtime can be used effectively to accomplish more work in a day. However, as overtime is scheduled to run beyond several weeks, studies have shown marked decreases in output. The following chart shows the effect of long term overtime on worker efficiency. If the anticipated change requires extended overtime to keep the job on schedule, these factors can be used as a guide to predict the impact on time and cost. Add project overhead, particularly supervision, that may also be incurred.

Days per Week	Hours per Day	Production Efficiency					Payroll Cost Factors	
		1st Week	2nd Week	3rd Week	4th Week	Average 4 Weeks	@ 1-1/2 Times	@ 2 Times
5	8	100%	100%	100%	100%	100%	100%	100%
	9	100	100	95	90	96.25	105.6	111.1
	10	100	95	90	85	92.50	110.0	120.0
	11	95	90	75	65	81.25	113.6	127.3
	12	90	85	70	60	76.25	116.7	133.3
6	8	100	100	95	90	96.25	108.3	116.7
	9	100	95	90	85	92.50	113.0	125.9
	10	95	90	85	80	87.50	116.7	133.3
	11	95	85	70	65	78.75	119.7	139.4
	12	90	80	65	60	73.75	122.2	144.4
7	8	100	95	85	75	88.75	114.3	128.6
	9	95	90	80	70	83.75	118.3	136.5
	10	90	85	75	65	78.75	121.4	142.9
	11	85	80	65	60	72.50	124.0	148.1
	12	85	75	60	55	68.75	126.2	152.4

Effects of Overtime

Caution: Under many labor agreements, Sundays and holidays are paid at a higher premium than the normal overtime rate.

The use of long-term overtime is counterproductive on almost any construction job; that is, the longer the period of overtime, the lower the actual production rate. Numerous studies have been conducted, and while they have resulted in slightly different numbers, all reach the same conclusion. The figure above tabulates the effects of overtime work on efficiency.

As illustrated, there can be a difference between the *actual* payroll cost per hour and the *effective* cost per hour for overtime work. This is due to the reduced production efficiency with the increase in weekly hours beyond 40. This difference between actual and effective cost results from overtime work over a prolonged period. Short-term overtime work does not result in as great a reduction in efficiency and, in such cases, effective cost may not vary significantly from the actual payroll cost. As the total hours per week are increased on a regular basis, more time is lost due to fatigue, lowered morale, and an increased accident rate.

As an example, assume a project where workers are working 6 days a week, 10 hours per day. From the figure above (based on productivity studies), the average effective productive hours over a 4-week period are:

$$0.875 \times 60 = 52.5$$

Depending upon the locale and day of week, overtime hours may be paid at time and a half or double time. For time and a half, the overall (average) *actual* payroll cost (including regular and overtime hours) is determined as follows:

$$\frac{40 \text{ reg. hrs.} + (20 \text{ overtime hrs.} \times 1.5)}{60 \text{ hrs.}} = 1.167$$

Based on 60 hours, the payroll cost per hour will be 116.7% of the normal rate at 40 hours per week. However, because the effective production (efficiency) for 60 hours is reduced to the equivalent of 52.5 hours, the effective cost of overtime is calculated as follows:

For time and a half:

$$\frac{40 \text{ reg. hrs.} + (20 \text{ overtime hrs.} \times 1.5)}{52.5 \text{ hrs.}} = 1.33$$

The installed cost will be 133% of the normal rate (for labor).

Thus, when figuring overtime, the actual cost per unit of work will be higher than the apparent overtime payroll dollar increase, due to the reduced productivity of the longer work week. These efficiency calculations are true only for those cost factors determined by hours worked. Costs that are applied weekly or monthly, such as equipment rentals, will not be similarly affected.

Equipment

- *What equipment is required to complete the change order?*

Change orders may require extending the rental period of equipment already on the job site, or the addition of special equipment brought in to accomplish the change work. In either case, the additional rental charges and operator labor charges must be added.

Summary

The preceding considerations and others you deem appropriate should be analyzed and applied to a change order estimate. The impact of each should be quantified and listed on the estimate to form an audit trail.

Change orders that are properly identified, documented, and managed help to ensure the orderly, professional, and profitable progress of the work. They also minimize potential claims or disputes at the end of the project.

Back by customer demand!

You asked and we listened. For customer convenience and estimating ease, we have made the 2018 Project Costs available for download at **www.RSMeans.com/2018books**. You will also find sample estimates, an RSMeans data overview video and book registration form to receive quarterly data updates throughout 2018.

Estimating Tips

- The cost figures available in the download were derived from hundreds of projects contained in the RSMeans database of completed construction projects. They include the contractor's overhead and profit. The figures have been adjusted to January of the current year.

- These projects were located throughout the U.S. and reflect a tremendous variation in square foot (S.F.) costs. This is due to differences, not only in labor and material costs, but also in individual owners' requirements. For instance, a bank in a large city would have different features than one in a rural area. This is true of all the different types of buildings analyzed. Therefore, caution should be exercised when using these Project Costs. For example, for courthouses, costs in the database are local courthouse costs and will not apply to the larger, more elaborate federal courthouses.

- None of the figures "go with" any others. All individual cost items were computed and tabulated separately. Thus, the sum of the median figures for plumbing, HVAC, and electrical will not normally total up to the total mechanical and electrical costs arrived at by separate analysis and tabulation of the projects.

- Each building was analyzed as to total and component costs and percentages. The figures were arranged in ascending order with the results tabulated as shown. The 1/4 column shows that 25% of the projects had lower costs and 75% had higher. The 3/4 column shows that 75% of the projects had lower costs and 25% had higher. The median column shows that 50% of the projects had lower costs and 50% had higher.

- Project Costs are useful in the conceptual stage when no details are available. As soon as details become available in the project design, the square foot approach should be discontinued and the project priced as to its particular components. When more precision is required, or for estimating the replacement cost of specific buildings, the current edition of *RSMeans Square Foot Costs* should be used.

- In using the figures in this section, it is recommended that the median column be used for preliminary figures if no additional information is available. The median figures, when multiplied by the total city construction cost index figures (see City Cost Indexes) and then multiplied by the project size modifier at the end of this section, should present a fairly accurate base figure, which would then have to be adjusted in view of the estimator's experience, local economic conditions, code requirements, and the owner's particular requirements. There is no need to factor the percentage figures, as these should remain constant from city to city.

- The editors of this data would greatly appreciate receiving cost figures on one or more of your recent projects, which would then be included in the averages for next year. All cost figures received will be kept confidential, except that they will be averaged with other similar projects to arrive at square foot cost figures for next year.

See the website above for details and the discount available for submitting one or more of your projects.

Did you know?

RSMeans data is available through our online application with 24/7 access:

- Search for unit prices by keyword
- Leverage the most up-to-date data
- Build and export estimates

Try it free for 30 days!
www.rsmeans.com/2018freetrial

No part of this cost data may be reproduced, stored in a retrieval system, or transmitted in any form or by any means without prior written permission of Gordian.

50 17 00 | Project Costs		UNIT	UNIT COSTS			% OF TOTAL		
			1/4	MEDIAN	3/4	1/4	MEDIAN	3/4
01	**0000**	**Auto Sales with Repair**	S.F.							**01**
	0100	Architectural		98.50	110	119	58%	64%	67%
	0200	Plumbing		8.25	8.65	11.55	4.84%	5.20%	6.80%
	0300	Mechanical		11.05	14.80	16.35	6.40%	8.70%	10.15%
	0400	Electrical		17	21	26.50	9.05%	11.70%	15.90%
	0500	Total Project Costs		165	173	177			
02	**0000**	**Banking Institutions**	S.F.							**02**
	0100	Architectural		149	183	222	59%	65%	69%
	0200	Plumbing		6	8.35	11.60	2.12%	3.39%	4.19%
	0300	Mechanical		11.90	16.45	19.40	4.41%	5.10%	10.75%
	0400	Electrical		29	35	54	10.45%	13.05%	15.90%
	0500	Total Project Costs		247	278	340			
03	**0000**	**Court House**	S.F.							**03**
	0100	Architectural		78.50	78.50	78.50	54.50%	54.50%	54.50%
	0200	Plumbing		2.96	2.96	2.96	2.07%	2.07%	2.07%
	0300	Mechanical		18.55	18.55	18.55	12.95%	12.95%	12.95%
	0400	Electrical		24	24	24	16.60%	16.60%	16.60%
	0500	Total Project Costs		143	143	143			
04	**0000**	**Data Centers**	S.F.							**04**
	0100	Architectural		177	177	177	68%	68%	68%
	0200	Plumbing		9.70	9.70	9.70	3.71%	3.71%	3.71%
	0300	Mechanical		24.50	24.50	24.50	9.45%	9.45%	9.45%
	0400	Electrical		23.50	23.50	23.50	9%	9%	9%
	0500	Total Project Costs		261	261	261			
05	**0000**	**Detention Centers**	S.F.							**05**
	0100	Architectural		164	174	184	52%	53%	60.50%
	0200	Plumbing		17.35	21	25.50	5.15%	7.10%	7.25%
	0300	Mechanical		22	31.50	37.50	7.55%	9.50%	13.80%
	0400	Electrical		36	42.50	55.50	10.90%	14.85%	17.95%
	0500	Total Project Costs		278	293	345			
06	**0000**	**Fire Stations**	S.F.							**06**
	0100	Architectural		97	120	174	49%	55.50%	63%
	0200	Plumbing		9.45	13.30	15.75	4.86%	5.80%	6.30%
	0300	Mechanical		12.80	18.65	24	5.45%	8.25%	9.70%
	0400	Electrical		21.50	27.50	32.50	8.35%	12.80%	14.70%
	0500	Total Project Costs		195	222	294			
07	**0000**	**Gymnasium**	S.F.							**07**
	0100	Architectural		82	108	108	57%	64.50%	64.50%
	0200	Plumbing		2.01	6.60	6.60	1.58%	3.48%	3.48%
	0300	Mechanical		3.09	28	28	2.42%	14.65%	14.65%
	0400	Electrical		10.10	19.60	19.60	7.95%	10.35%	10.35%
	0500	Total Project Costs		128	189	189			
08	**0000**	**Hospitals**	S.F.							**08**
	0100	Architectural		99.50	164	178	43%	47.50%	48%
	0200	Plumbing		7.30	13.95	30	6%	7.45%	7.65%
	0300	Mechanical		48.50	54.50	71	14.20%	17.95%	23.50%
	0400	Electrical		22	44	57.50	10.95%	13.75%	16.85%
	0500	Total Project Costs		233	345	375			
09	**0000**	**Industrial Buildings**	S.F.							**09**
	0100	Architectural		42	66.50	216	46%	54%	56.50%
	0200	Plumbing		1.62	6.10	12.30	2%	3.06%	6.30%
	0300	Mechanical		4.47	8.50	40.50	4.77%	5.55%	14.80%
	0400	Electrical		6.80	7.80	65	7.85%	13.55%	16.20%
	0500	Total Project Costs		74.50	97	400			
10	**0000**	**Medical Clinics & Offices**	S.F.							**10**
	0100	Architectural		79.50	118	152	48.50%	55.50%	62.50%
	0200	Plumbing		8.10	12.20	19.65	4.44%	6.40%	8.05%
	0300	Mechanical		13.70	25	42.50	8.30%	11.05%	17.25%
	0400	Electrical		18.60	25	36	8.55%	12.10%	14.65%
	0500	Total Project Costs		156	208	272			

For customer support on your Building Construction Costs with RSMeans data, call 800.448.8182.

		50 17 00	Project Costs	UNIT	UNIT COSTS			% OF TOTAL			
					1/4	MEDIAN	3/4	1/4	MEDIAN	3/4	
11	0000	**Mixed Use**		S.F.							11
	0100	Architectural			85	120	182	45.50%	52.50%	70%	
	0200	Plumbing			5.70	8.70	11	2.84%	3.44%	4.18%	
	0300	Mechanical			14.05	36.50	65	8.05%	13.75%	20%	
	0400	Electrical			14.85	36.50	49	8.30%	12.95%	15.80%	
	0500	Total Project Costs			179	201	320				
12	0000	**Multi-Family Housing**		S.F.							12
	0100	Architectural			70.50	106	158	56.50%	61.50%	66.50%	
	0200	Plumbing			5.90	10.20	13.90	4.81%	6.30%	7.40%	
	0300	Mechanical			6.60	8.80	35.50	4.92%	6.90%	11.20%	
	0400	Electrical			9.20	14.35	17.40	6.20%	8.45%	10.25%	
	0500	Total Project Costs			104	194	233				
13	0000	**Nursing Home & Assisted Living**		S.F.				.			13
	0100	Architectural			66.50	87	116	51.50%	56.50%	69%	
	0200	Plumbing			6.75	10.45	11.70	6.05%	7.40%	8.80%	
	0300	Mechanical			5.90	7.55	17.05	4.04%	6.70%	9.55%	
	0400	Electrical			9.75	12.80	21.50	7%	10.05%	12.20%	
	0500	Total Project Costs			114	135	179				
14	0000	**Office Buildings**		S.F.							14
	0100	Architectural			90.50	120	168	56.50%	62%	71.50%	
	0200	Plumbing			4.73	7.05	11.45	2.62%	3.48%	5.85%	
	0300	Mechanical			10.25	15.80	23.50	5.50%	8.20%	11.10%	
	0400	Electrical			11.85	20	31.50	7.75%	10%	12.70%	
	0500	Total Project Costs			150	186	262				
15	0000	**Parking Garage**		S.F.							15
	0100	Architectural			29.50	36	37.50	70%	79%	88%	
	0200	Plumbing			.97	1.01	1.90	2.05%	2.70%	2.83%	
	0300	Mechanical			.75	1.16	4.39	2.11%	3.62%	3.81%	
	0400	Electrical			2.58	2.83	5.90	5.30%	6.35%	7.95%	
	0500	Total Project Costs			36	43.50	47				
16	0000	**Parking Garage/Mixed Use**		S.F.							16
	0100	Architectural			95.50	104	106	61%	62%	65.50%	
	0200	Plumbing			3.06	4.01	6.15	2.47%	2.72%	3.66%	
	0300	Mechanical			13.10	14.75	21	7.80%	13.10%	13.60%	
	0400	Electrical			13.70	19.70	20.50	8.20%	12.65%	18.15%	
	0500	Total Project Costs			156	162	168				
17	0000	**Police Stations**		S.F.							17
	0100	Architectural			108	121	152	49%	56.50%	61%	
	0200	Plumbing			14.25	17.10	17.20	5.05%	5.55%	9.05%	
	0300	Mechanical			32.50	45	46.50	13%	14.55%	16.55%	
	0400	Electrical			24.50	26.50	28	9.15%	12.10%	14%	
	0500	Total Project Costs			202	248	282				
18	0000	**Police/Fire**		S.F.							18
	0100	Architectural			105	105	320	55.50%	66%	68%	
	0200	Plumbing			8.40	8.70	32	5.45%	5.50%	5.55%	
	0300	Mechanical			12.85	20.50	73.50	8.35%	12.70%	12.80%	
	0400	Electrical			14.65	18.70	84	9.50%	11.75%	14.55%	
	0500	Total Project Costs			154	159	580				
19	0000	**Public Assembly Buildings**		S.F.							19
	0100	Architectural			119	148	200	58%	61.50%	64%	
	0200	Plumbing			5.65	7.40	11.45	2.60%	3.32%	4.01%	
	0300	Mechanical			14.95	23	33	6.90%	9.15%	12.75%	
	0400	Electrical			19.70	24.50	38.50	8.65%	10.75%	13%	
	0500	Total Project Costs			188	242	335				
20	0000	**Recreational**		S.F.							20
	0100	Architectural			103	162	219	53.50%	59%	66%	
	0200	Plumbing			8.40	15.30	20	3.12%	4.76%	7.40%	
	0300	Mechanical			12.65	19.25	30.50	5.35%	7.90%	12.75%	
	0400	Electrical			14.15	24.50	33	7.35%	8.50%	10.65%	
	0500	Total Project Costs			183	272	375				

		50 17 00 | Project Costs	UNIT	UNIT COSTS			% OF TOTAL			
				1/4	MEDIAN	3/4	1/4	MEDIAN	3/4	
21	0000	**Restaurants**	S.F.							21
	0100	Architectural		50.50	179	231	57.50%	60.50%	78%	
	0200	Plumbing		9.45	29.50	37	7.35%	7.75%	8.95%	
	0300	Mechanical		7.25	18.30	44.50	6.50%	10.80%	18.30%	
	0400	Electrical		11.65	21.50	48.50	6.75%	8.45%	12.45%	
	0500	Total Project Costs	↓	65	283	400				
22	0000	**Retail**	S.F.							22
	0100	Architectural		52	57.50	104	60%	60%	64.50%	
	0200	Plumbing		5.35	8.05	10.15	5.05%	6.70%	9%	
	0300	Mechanical		4.88	7	8.60	5.05%	6.15%	6.60%	
	0400	Electrical		6.85	10.95	17.60	7.90%	10.15%	12.25%	
	0500	Total Project Costs	↓	87	106	173				
23	0000	**Schools**	S.F.							23
	0100	Architectural		88.50	114	152	52.50%	57%	61.50%	
	0200	Plumbing		7.10	9.85	14.40	3.75%	4.82%	7%	
	0300	Mechanical		16.85	23	35.50	9.50%	12.25%	14.55%	
	0400	Electrical		16.65	23	29	9.45%	11.40%	13.05%	
	0500	Total Project Costs	↓	151	206	272				
24	0000	**University, College & Private School Classroom & Admin Buildings**	S.F.							24
	0100	Architectural		115	142	179	50.50%	55%	59.50%	
	0200	Plumbing		6.55	10.15	14.35	2.74%	4.30%	6.35%	
	0300	Mechanical		24.50	35.50	43	10.10%	12.15%	14.70%	
	0400	Electrical		18.55	26	31.50	7.65%	9.50%	11.55%	
	0500	Total Project Costs	↓	191	264	350				
25	0000	**University, College & Private School Dormitories**	S.F.							25
	0100	Architectural		75	132	140	54.50%	65%	68.50%	
	0200	Plumbing		9.95	14.05	21	6.45%	7.30%	9.15%	
	0300	Mechanical		4.46	18.95	30	4.13%	9%	12.05%	
	0400	Electrical		5.30	18.40	28	4.75%	7.35%	12.30%	
	0500	Total Project Costs	↓	111	211	249				
26	0000	**University, College & Private School Science, Eng. & Lab Buildings**	S.F.							26
	0100	Architectural		129	137	178	50.50%	56.50%	58%	
	0200	Plumbing		8.90	9.30	24.50	3.29%	3.95%	8.40%	
	0300	Mechanical		40.50	64	64	13.20%	21.50%	23.50%	
	0400	Electrical		26	30	35.50	8.95%	9.70%	12.85%	
	0500	Total Project Costs	↓	265	271	295				
27	0000	**University, College & Private School Student Union Buildings**	S.F.							27
	0100	Architectural		69.50	102	102	51%	59.50%	59.50%	
	0200	Plumbing		5	23	23	4.27%	11.45%	11.45%	
	0300	Mechanical		11.25	29.50	29.50	9.60%	14.55%	14.55%	
	0400	Electrical		15.40	25.50	25.50	12.80%	13.15%	13.15%	
	0500	Total Project Costs	↓	117	201	201				
28	0000	**Warehouses**	S.F.							28
	0100	Architectural		44	68.50	162	61.50%	67.50%	72%	
	0200	Plumbing		2.28	4.88	9.40	2.82%	3.72%	5%	
	0300	Mechanical		2.70	15.40	24	4.56%	8.20%	10.70%	
	0400	Electrical		4.91	18.45	30.50	7.50%	10.10%	18.30%	
	0500	Total Project Costs	↓	65.50	116	226				

For customer support on your Building Construction Costs with RSMeans data, call 800.448.8182.

Square Foot Project Size Modifier

One factor that affects the S.F. cost of a particular building is the size. In general, for buildings built to the same specifications in the same locality, the larger building will have the lower S.F. cost. This is due mainly to the decreasing contribution of the exterior walls plus the economy of scale usually achievable in larger buildings. The Area Conversion Scale shown below will give a factor to convert costs for the typical size building to an adjusted cost for the particular project.

Example: Determine the cost per S.F. for a 152,600 S.F. Multi-family housing.

$$\frac{\text{Proposed building area} = 152{,}600 \text{ S.F.}}{\text{Typical size from below} = 76{,}300 \text{ S.F.}} = 2.00$$

Enter Area Conversion scale at 2.0, intersect curve, read horizontally the appropriate cost multiplier of .94. Size adjusted cost becomes .94 x $194.00 = $182.36 based on national average costs.

Note: For Size Factors less than .50, the Cost Multiplier is 1.1
For Size Factors greater than 3.5, the Cost Multiplier is .90

The Square Foot Base Size lists the median costs, most typical project size in our accumulated data, and the range in size of the projects.

The Size Factor for your project is determined by dividing your project area in S.F. by the typical project size for the particular Building Type. With this factor, enter the Area Conversion Scale at the appropriate Size Factor and determine the appropriate cost multiplier for your building size.

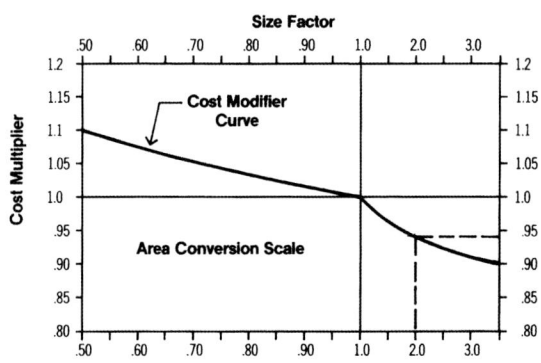

System	Median Cost (Total Project Costs)	Typical Size Gross S.F. (Median of Projects)	Typical Range (Low – High) (Projects)
Auto Sales with Repair	$173.00	25,300	8,200 – 28,700
Banking Institutions	278.00	26,300	3,300 – 28,700
Detention Centers	293.00	42,000	12,300 – 183,300
Fire Stations	222.00	14,600	6,300 – 29,600
Hospitals	345.00	137,500	54,700 – 410,300
Industrial Buildings	$97.00	16,900	5,100 – 200,600
Medical Clinics & Offices	208.00	5,500	2,600 – 327,000
Mixed Use	201.00	49,900	14,400 – 49,900
Multi-Family Housing	194.00	76,300	12,500 – 1,161,500
Nursing Home & Assisted Living	135.00	16,200	1,500 – 242,600
Office Buildings	186.00	10,000	1,100 – 930,000
Parking Garage	43.50	174,600	99,900 – 287,000
Parking Garage/Mixed Use	162.00	5,300	5,300 – 318,000
Police Stations	248.00	15,400	15,400 – 31,600
Public Assembly Buildings	242.00	30,500	2,200 – 235,300
Recreational	272.00	2,300	1,500 – 223,800
Restaurants	283.00	6,100	5,500 – 42,000
Retail	106.00	28,700	5,800 – 61,000
Schools	206.00	30,000	5,500 – 410,800
University, College & Private School Classroom & Admin Buildings	264.00	89,200	9,400 – 196,200
University, College & Private School Dormitories	211.00	50,800	1,500 – 126,900
University, College & Private School Science, Eng. & Lab Buildings	271.00	39,800	36,000 – 117,600
Warehouses	116.00	2,100	600 – 303,800

895

A	Area Square Feet; Ampere	Brk., brk	Brick	Csc	Cosecant
AAFES	Army and Air Force Exchange Service	brkt	Bracket	C.S.F.	Hundred Square Feet
ABS	Acrylonitrile Butadiene Stryrene; Asbestos Bonded Steel	Brs.	Brass	CSI	Construction Specifications Institute
		Brz.	Bronze		
A.C., AC	Alternating Current;	Bsn.	Basin	CT	Current Transformer
	Air-Conditioning;	Btr.	Better	CTS	Copper Tube Size
	Asbestos Cement;	BTU	British Thermal Unit	Cu	Copper, Cubic
	Plywood Grade A & C	BTUH	BTU per Hour	Cu. Ft.	Cubic Foot
ACI	American Concrete Institute	Bu.	Bushels	cw	Continuous Wave
ACR	Air Conditioning Refrigeration	BUR	Built-up Roofing	C.W.	Cool White; Cold Water
ADA	Americans with Disabilities Act	BX	Interlocked Armored Cable	Cwt.	100 Pounds
AD	Plywood, Grade A & D	°C	Degree Centigrade	C.W.X.	Cool White Deluxe
Addit.	Additional	c	Conductivity, Copper Sweat	C.Y.	Cubic Yard (27 cubic feet)
Adh.	Adhesive	C	Hundred; Centigrade	C.Y./Hr.	Cubic Yard per Hour
Adj.	Adjustable	C/C	Center to Center, Cedar on Cedar	Cyl.	Cylinder
af	Audio-frequency	C-C	Center to Center	d	Penny (nail size)
AFFF	Aqueous Film Forming Foam	Cab	Cabinet	D	Deep; Depth; Discharge
AFUE	Annual Fuel Utilization Efficiency	Cair.	Air Tool Laborer	Dis., Disch.	Discharge
AGA	American Gas Association	Cal.	Caliper	Db	Decibel
Agg.	Aggregate	Calc	Calculated	Dbl.	Double
A.H., Ah	Ampere Hours	Cap.	Capacity	DC	Direct Current
A hr.	Ampere-hour	Carp.	Carpenter	DDC	Direct Digital Control
A.H.U., AHU	Air Handling Unit	C.B.	Circuit Breaker	Demob.	Demobilization
A.I.A.	American Institute of Architects	C.C.A.	Chromate Copper Arsenate	d.f.t.	Dry Film Thickness
AIC	Ampere Interrupting Capacity	C.C.F.	Hundred Cubic Feet	d.f.u.	Drainage Fixture Units
Allow.	Allowance	cd	Candela	D.H.	Double Hung
alt., alt	Alternate	cd/sf	Candela per Square Foot	DHW	Domestic Hot Water
Alum.	Aluminum	CD	Grade of Plywood Face & Back	DI	Ductile Iron
a.m.	Ante Meridiem	CDX	Plywood, Grade C & D, exterior glue	Diag.	Diagonal
Amp.	Ampere			Diam., Dia	Diameter
Anod.	Anodized	Cefi.	Cement Finisher	Distrib.	Distribution
ANSI	American National Standards Institute	Cem.	Cement	Div.	Division
		CF	Hundred Feet	Dk.	Deck
APA	American Plywood Association	C.F.	Cubic Feet	D.L.	Dead Load; Diesel
Approx.	Approximate	CFM	Cubic Feet per Minute	DLH	Deep Long Span Bar Joist
Apt.	Apartment	CFRP	Carbon Fiber Reinforced Plastic	dlx	Deluxe
Asb.	Asbestos	c.g.	Center of Gravity	Do.	Ditto
A.S.B.C.	American Standard Building Code	CHW	Chilled Water; Commercial Hot Water	DOP	Dioctyl Phthalate Penetration Test (Air Filters)
Asbe.	Asbestos Worker				
ASCE	American Society of Civil Engineers	C.I., CI	Cast Iron	Dp., dp	Depth
A.S.H.R.A.E.	American Society of Heating, Refrig. & AC Engineers	C.I.P., CIP	Cast in Place	D.P.S.T.	Double Pole, Single Throw
		Circ.	Circuit	Dr.	Drive
ASME	American Society of Mechanical Engineers	C.L.	Carload Lot	DR	Dimension Ratio
		CL	Chain Link	Drink.	Drinking
ASTM	American Society for Testing and Materials	Clab.	Common Laborer	D.S.	Double Strength
		Clam	Common Maintenance Laborer	D.S.A.	Double Strength A Grade
Attchmt.	Attachment	C.L.F.	Hundred Linear Feet	D.S.B.	Double Strength B Grade
Avg., Ave.	Average	CLF	Current Limiting Fuse	Dty.	Duty
AWG	American Wire Gauge	CLP	Cross Linked Polyethylene	DWV	Drain Waste Vent
AWWA	American Water Works Assoc.	cm	Centimeter	DX	Deluxe White, Direct Expansion
Bbl.	Barrel	CMP	Corr. Metal Pipe	dyn	Dyne
B&B, BB	Grade B and Better; Balled & Burlapped	CMU	Concrete Masonry Unit	e	Eccentricity
		CN	Change Notice	E	Equipment Only; East; Emissivity
B&S	Bell and Spigot	Col.	Column	Ea.	Each
B.&W.	Black and White	CO₂	Carbon Dioxide	EB	Encased Burial
b.c.c.	Body-centered Cubic	Comb.	Combination	Econ.	Economy
B.C.Y.	Bank Cubic Yards	comm.	Commercial, Communication	E.C.Y	Embankment Cubic Yards
BE	Bevel End	Compr.	Compressor	EDP	Electronic Data Processing
B.F.	Board Feet	Conc.	Concrete	EIFS	Exterior Insulation Finish System
Bg. cem.	Bag of Cement	Cont., cont	Continuous; Continued, Container	E.D.R.	Equiv. Direct Radiation
BHP	Boiler Horsepower; Brake Horsepower	Corkbd.	Cork Board	Eq.	Equation
		Corr.	Corrugated	EL	Elevation
B.I.	Black Iron	Cos	Cosine	Elec.	Electrician; Electrical
bidir.	bidirectional	Cot	Cotangent	Elev.	Elevator; Elevating
Bit., Bitum.	Bituminous	Cov.	Cover	EMT	Electrical Metallic Conduit; Thin Wall Conduit
Bit., Conc.	Bituminous Concrete	C/P	Cedar on Paneling		
Bk.	Backed	CPA	Control Point Adjustment	Eng.	Engine, Engineered
Bkrs.	Breakers	Cplg.	Coupling	EPDM	Ethylene Propylene Diene Monomer
Bldg., bldg	Building	CPM	Critical Path Method		
Blk.	Block	CPVC	Chlorinated Polyvinyl Chloride	EPS	Expanded Polystyrene
Bm.	Beam	C.Pr.	Hundred Pair	Eqhv.	Equip. Oper., Heavy
Boil.	Boilermaker	CRC	Cold Rolled Channel	Eqlt.	Equip. Oper., Light
bpm	Blows per Minute	Creos.	Creosote	Eqmd.	Equip. Oper., Medium
BR	Bedroom	Crpt.	Carpet & Linoleum Layer	Eqmm.	Equip. Oper., Master Mechanic
Brg., brng.	Bearing	CRT	Cathode-ray Tube	Eqol.	Equip. Oper., Oilers
Brhe.	Bricklayer Helper	CS	Carbon Steel, Constant Shear Bar Joist	Equip.	Equipment
Bric.	Bricklayer			ERW	Electric Resistance Welded

For customer support on your Building Construction Costs with RSMeans data, call 800.448.8182.

Abbreviations

E.S.	Energy Saver	H	High Henry	Lath.	Lather
Est.	Estimated	HC	High Capacity	Lav.	Lavatory
esu	Electrostatic Units	H.D., HD	Heavy Duty; High Density	lb.; #	Pound
E.W.	Each Way	H.D.O.	High Density Overlaid	L.B., LB	Load Bearing; L Conduit Body
EWT	Entering Water Temperature	HDPE	High Density Polyethylene Plastic	L. & E.	Labor & Equipment
Excav.	Excavation	Hdr.	Header	lb./hr.	Pounds per Hour
excl	Excluding	Hdwe.	Hardware	lb./L.F.	Pounds per Linear Foot
Exp., exp	Expansion, Exposure	H.I.D., HID	High Intensity Discharge	lbf/sq.in.	Pound-force per Square Inch
Ext., ext	Exterior; Extension	Help.	Helper Average	L.C.L.	Less than Carload Lot
Extru.	Extrusion	HEPA	High Efficiency Particulate Air	L.C.Y.	Loose Cubic Yard
f.	Fiber Stress		Filter	Ld.	Load
F	Fahrenheit; Female; Fill	Hg	Mercury	LE	Lead Equivalent
Fab., fab	Fabricated; Fabric	HIC	High Interrupting Capacity	LED	Light Emitting Diode
FBGS	Fiberglass	HM	Hollow Metal	L.F.	Linear Foot
F.C.	Footcandles	HMWPE	High Molecular Weight	L.F. Hdr	Linear Feet of Header
f.c.c.	Face-centered Cubic		Polyethylene	L.F. Nose	Linear Foot of Stair Nosing
f'c.	Compressive Stress in Concrete;	HO	High Output	L.F. Rsr	Linear Foot of Stair Riser
	Extreme Compressive Stress	Horiz.	Horizontal	Lg.	Long; Length; Large
F.E.	Front End	H.P., HP	Horsepower; High Pressure	L & H	Light and Heat
FEP	Fluorinated Ethylene Propylene	H.P.F.	High Power Factor	LH	Long Span Bar Joist
	(Teflon)	Hr.	Hour	L.H.	Labor Hours
F.G.	Flat Grain	Hrs./Day	Hours per Day	L.L., LL	Live Load
F.H.A.	Federal Housing Administration	HSC	High Short Circuit	L.L.D.	Lamp Lumen Depreciation
Fig.	Figure	Ht.	Height	lm	Lumen
Fin.	Finished	Htg.	Heating	lm/sf	Lumen per Square Foot
FIPS	Female Iron Pipe Size	Htrs.	Heaters	lm/W	Lumen per Watt
Fixt.	Fixture	HVAC	Heating, Ventilation & Air-	LOA	Length Over All
FJP	Finger jointed and primed		Conditioning	log	Logarithm
Fl. Oz.	Fluid Ounces	Hvy.	Heavy	L-O-L	Lateralolet
Flr.	Floor	HW	Hot Water	long.	Longitude
Flrs.	Floors	Hyd.; Hydr.	Hydraulic	L.P., LP	Liquefied Petroleum; Low Pressure
FM	Frequency Modulation;	Hz	Hertz (cycles)	L.P.F.	Low Power Factor
	Factory Mutual	I.	Moment of Inertia	LR	Long Radius
Fmg.	Framing	IBC	International Building Code	L.S.	Lump Sum
FM/UL	Factory Mutual/Underwriters Labs	I.C.	Interrupting Capacity	Lt.	Light
Fdn.	Foundation	ID	Inside Diameter	Lt. Ga.	Light Gauge
FNPT	Female National Pipe Thread	I.D.	Inside Dimension; Identification	L.T.L.	Less than Truckload Lot
Fori.	Foreman, Inside	I.F.	Inside Frosted	Lt. Wt.	Lightweight
Foro.	Foreman, Outside	I.M.C.	Intermediate Metal Conduit	L.V.	Low Voltage
Fount.	Fountain	In.	Inch	M	Thousand; Material; Male;
fpm	Feet per Minute	Incan.	Incandescent		Light Wall Copper Tubing
FPT	Female Pipe Thread	Incl.	Included; Including	M²CA	Meters Squared Contact Area
Fr	Frame	Int.	Interior	m/hr.; M.H.	Man-hour
F.R.	Fire Rating	Inst.	Installation	mA	Milliampere
FRK	Foil Reinforced Kraft	Insul., insul	Insulation/Insulated	Mach.	Machine
FSK	Foil/Scrim/Kraft	I.P.	Iron Pipe	Mag. Str.	Magnetic Starter
FRP	Fiberglass Reinforced Plastic	I.P.S., IPS	Iron Pipe Size	Maint.	Maintenance
FS	Forged Steel	IPT	Iron Pipe Threaded	Marb.	Marble Setter
FSC	Cast Body; Cast Switch Box	I.W.	Indirect Waste	Mat; Mat'l.	Material
Ft., ft	Foot; Feet	J	Joule	Max.	Maximum
Ftng.	Fitting	J.I.C.	Joint Industrial Council	MBF	Thousand Board Feet
Ftg.	Footing	K	Thousand; Thousand Pounds;	MBH	Thousand BTU's per hr.
Ft lb.	Foot Pound		Heavy Wall Copper Tubing, Kelvin	MC	Metal Clad Cable
Furn.	Furniture	K.A.H.	Thousand Amp. Hours	MCC	Motor Control Center
FVNR	Full Voltage Non-Reversing	kcmil	Thousand Circular Mils	M.C.F.	Thousand Cubic Feet
FVR	Full Voltage Reversing	KD	Knock Down	MCFM	Thousand Cubic Feet per Minute
FXM	Female by Male	K.D.A.T.	Kiln Dried After Treatment	M.C.M.	Thousand Circular Mils
Fy.	Minimum Yield Stress of Steel	kg	Kilogram	MCP	Motor Circuit Protector
g	Gram	kG	Kilogauss	MD	Medium Duty
G	Gauss	kgf	Kilogram Force	MDF	Medium-density fibreboard
Ga.	Gauge	kHz	Kilohertz	M.D.O.	Medium Density Overlaid
Gal., gal.	Gallon	Kip	1000 Pounds	Med.	Medium
Galv., galv	Galvanized	KJ	Kilojoule	MF	Thousand Feet
GC/MS	Gas Chromatograph/Mass	K.L.	Effective Length Factor	M.F.B.M.	Thousand Feet Board Measure
	Spectrometer	K.L.F.	Kips per Linear Foot	Mfg.	Manufacturing
Gen.	General	Km	Kilometer	Mfrs.	Manufacturers
GFI	Ground Fault Interrupter	KO	Knock Out	mg	Milligram
GFRC	Glass Fiber Reinforced Concrete	K.S.F.	Kips per Square Foot	MGD	Million Gallons per Day
Glaz.	Glazier	K.S.I.	Kips per Square Inch	MGPH	Million Gallons per Hour
GPD	Gallons per Day	kV	Kilovolt	MH, M.H.	Manhole; Metal Halide; Man-Hour
gpf	Gallon per Flush	kVA	Kilovolt Ampere	MHz	Megahertz
GPH	Gallons per Hour	kVAR	Kilovar (Reactance)	Mi.	Mile
gpm, GPM	Gallons per Minute	KW	Kilowatt	MI	Malleable Iron; Mineral Insulated
GR	Grade	KWh	Kilowatt-hour	MIPS	Male Iron Pipe Size
Gran.	Granular	L	Labor Only; Length; Long;	mj	Mechanical Joint
Grnd.	Ground		Medium Wall Copper Tubing	m	Meter
GVW	Gross Vehicle Weight	Lab.	Labor	mm	Millimeter
GWB	Gypsum Wall Board	lat	Latitude	Mill.	Millwright
				Min., min.	Minimum, Minute

897

Misc.	Miscellaneous	PCM	Phase Contrast Microscopy	SBS	Styrene Butadiere Styrene
ml	Milliliter, Mainline	PDCA	Painting and Decorating	SC	Screw Cover
M.L.F.	Thousand Linear Feet		Contractors of America	SCFM	Standard Cubic Feet per Minute
Mo.	Month	P.E., PE	Professional Engineer;	Scaf.	Scaffold
Mobil.	Mobilization		Porcelain Enamel;	Sch., Sched.	Schedule
Mog.	Mogul Base		Polyethylene; Plain End	S.C.R.	Modular Brick
MPH	Miles per Hour	P.E.C.I.	Porcelain Enamel on Cast Iron	S.D.	Sound Deadening
MPT	Male Pipe Thread	Perf.	Perforated	SDR	Standard Dimension Ratio
MRGWB	Moisture Resistant Gypsum	PEX	Cross Linked Polyethylene	S.E.	Surfaced Edge
	Wallboard	Ph.	Phase	Sel.	Select
MRT	Mile Round Trip	P.I.	Pressure Injected	SER, SEU	Service Entrance Cable
ms	Millisecond	Pile.	Pile Driver	S.F.	Square Foot
M.S.F.	Thousand Square Feet	Pkg.	Package	S.F.C.A.	Square Foot Contact Area
Mstz.	Mosaic & Terrazzo Worker	Pl.	Plate	S.F. Flr.	Square Foot of Floor
M.S.Y.	Thousand Square Yards	Plah.	Plasterer Helper	S.F.G.	Square Foot of Ground
Mtd., mtd., mtd	Mounted	Plas.	Plasterer	S.F. Hor.	Square Foot Horizontal
Mthe.	Mosaic & Terrazzo Helper	plf	Pounds Per Linear Foot	SFR	Square Feet of Radiation
Mtng.	Mounting	Pluh.	Plumber Helper	S.F. Shlf.	Square Foot of Shelf
Mult.	Multi; Multiply	Plum.	Plumber	S4S	Surface 4 Sides
MUTCD	Manual on Uniform Traffic Control	Ply.	Plywood	Shee.	Sheet Metal Worker
	Devices	p.m.	Post Meridiem	Sin.	Sine
M.V.A.	Million Volt Amperes	Pntd.	Painted	Skwk.	Skilled Worker
M.V.A.R.	Million Volt Amperes Reactance	Pord.	Painter, Ordinary	SL	Saran Lined
MV	Megavolt	pp	Pages	S.L.	Slimline
MW	Megawatt	PP, PPL	Polypropylene	Sldr.	Solder
MXM	Male by Male	P.P.M.	Parts per Million	SLH	Super Long Span Bar Joist
MYD	Thousand Yards	Pr.	Pair	S.N.	Solid Neutral
N	Natural; North	P.E.S.B.	Pre-engineered Steel Building	SO	Stranded with oil resistant inside
nA	Nanoampere	Prefab.	Prefabricated		insulation
NA	Not Available; Not Applicable	Prefin.	Prefinished	S-O-L	Socketolet
N.B.C.	National Building Code	Prop.	Propelled	sp	Standpipe
NC	Normally Closed	PSF, psf	Pounds per Square Foot	S.P.	Static Pressure; Single Pole; Self-Propelled
NEMA	National Electrical Manufacturers	PSI, psi	Pounds per Square Inch		
	Assoc.	PSIG	Pounds per Square Inch Gauge	Spri.	Sprinkler Installer
NEHB	Bolted Circuit Breaker to 600V.	PSP	Plastic Sewer Pipe	spwg	Static Pressure Water Gauge
NFPA	National Fire Protection Association	Pspr.	Painter, Spray	S.P.D.T.	Single Pole, Double Throw
NLB	Non-Load-Bearing	Psst.	Painter, Structural Steel	SPF	Spruce Pine Fir; Sprayed
NM	Non-Metallic Cable	P.T.	Potential Transformer		Polyurethane Foam
nm	Nanometer	P. & T.	Pressure & Temperature	S.P.S.T.	Single Pole, Single Throw
No.	Number	Ptd.	Painted	SPT	Standard Pipe Thread
NO	Normally Open	Ptns.	Partitions	Sq.	Square; 100 Square Feet
N.O.C.	Not Otherwise Classified	Pu	Ultimate Load	Sq. Hd.	Square Head
Nose.	Nosing	PVC	Polyvinyl Chloride	Sq. In.	Square Inch
NPT	National Pipe Thread	Pvmt.	Pavement	S.S.	Single Strength; Stainless Steel
NQOD	Combination Plug-on/Bolt on	PRV	Pressure Relief Valve	S.S.B.	Single Strength B Grade
	Circuit Breaker to 240V.	Pwr.	Power	sst, ss	Stainless Steel
N.R.C., NRC	Noise Reduction Coefficient/	Q	Quantity Heat Flow	Sswk.	Structural Steel Worker
	Nuclear Regulator Commission	Qt.	Quart	Sswl.	Structural Steel Welder
N.R.S.	Non Rising Stem	Quan., Qty.	Quantity	St.; Stl.	Steel
ns	Nanosecond	Q.C.	Quick Coupling	STC	Sound Transmission Coefficient
NTP	Notice to Proceed	r	Radius of Gyration	Std.	Standard
nW	Nanowatt	R	Resistance	Stg.	Staging
OB	Opposing Blade	R.C.P.	Reinforced Concrete Pipe	STK	Select Tight Knot
OC	On Center	Rect.	Rectangle	STP	Standard Temperature & Pressure
OD	Outside Diameter	recpt.	Receptacle	Stpi.	Steamfitter, Pipefitter
O.D.	Outside Dimension	Reg.	Regular	Str.	Strength; Starter; Straight
ODS	Overhead Distribution System	Reinf.	Reinforced	Strd.	Stranded
O.G.	Ogee	Req'd.	Required	Struct.	Structural
O.H.	Overhead	Res.	Resistant	Sty.	Story
O&P	Overhead and Profit	Resi.	Residential	Subj.	Subject
Oper.	Operator	RF	Radio Frequency	Subs.	Subcontractors
Opng.	Opening	RFID	Radio-frequency Identification	Surf.	Surface
Orna.	Ornamental	Rgh.	Rough	Sw.	Switch
OSB	Oriented Strand Board	RGS	Rigid Galvanized Steel	Swbd.	Switchboard
OS&Y	Outside Screw and Yoke	RHW	Rubber, Heat & Water Resistant;	S.Y.	Square Yard
OSHA	Occupational Safety and Health		Residential Hot Water	Syn.	Synthetic
	Act	rms	Root Mean Square	S.Y.P.	Southern Yellow Pine
Ovhd.	Overhead	Rnd.	Round	Sys.	System
OWG	Oil, Water or Gas	Rodm.	Rodman	t.	Thickness
Oz.	Ounce	Rofc.	Roofer, Composition	T	Temperature; Ton
P.	Pole; Applied Load; Projection	Rofp.	Roofer, Precast	Tan	Tangent
p.	Page	Rohe.	Roofer Helpers (Composition)	T.C.	Terra Cotta
Pape.	Paperhanger	Rots.	Roofer, Tile & Slate	T & C	Threaded and Coupled
P.A.P.R.	Powered Air Purifying Respirator	R.O.W.	Right of Way	T.D.	Temperature Difference
PAR	Parabolic Reflector	RPM	Revolutions per Minute	TDD	Telecommunications Device for
P.B., PB	Push Button	R.S.	Rapid Start		the Deaf
Pc., Pcs.	Piece, Pieces	Rsr	Riser	T.E.M.	Transmission Electron Microscopy
P.C.	Portland Cement; Power Connector	RT	Round Trip	temp	Temperature, Tempered, Temporary
P.C.F.	Pounds per Cubic Foot	S.	Suction; Single Entrance; South	TFFN	Nylon Jacketed Wire

For customer support on your Building Construction Costs with RSMeans data, call 800.448.8182.

Abbreviations

TFE	Tetrafluoroethylene (Teflon)	U.L., UL	Underwriters Laboratory	w/	With
T. & G.	Tongue & Groove;	Uld.	Unloading	W.C., WC	Water Column; Water Closet
	Tar & Gravel	Unfin.	Unfinished	W.F.	Wide Flange
Th., Thk.	Thick	UPS	Uninterruptible Power Supply	W.G.	Water Gauge
Thn.	Thin	URD	Underground Residential	Wldg.	Welding
Thrded	Threaded		Distribution	W. Mile	Wire Mile
Tilf.	Tile Layer, Floor	US	United States	W-O-L	Weldolet
Tilh.	Tile Layer, Helper	USGBC	U.S. Green Building Council	W.R.	Water Resistant
THHN	Nylon Jacketed Wire	USP	United States Primed	Wrck.	Wrecker
THW.	Insulated Strand Wire	UTMCD	Uniform Traffic Manual For Control	WSFU	Water Supply Fixture Unit
THWN	Nylon Jacketed Wire		Devices	W.S.P.	Water, Steam, Petroleum
T.L., TL	Truckload	UTP	Unshielded Twisted Pair	WT., Wt.	Weight
T.M.	Track Mounted	V	Volt	WWF	Welded Wire Fabric
Tot.	Total	VA	Volt Amperes	XFER	Transfer
T-O-L	Threadolet	VAT	Vinyl Asbestos Tile	XFMR	Transformer
tmpd	Tempered	V.C.T.	Vinyl Composition Tile	XHD	Extra Heavy Duty
TPO	Thermoplastic Polyolefin	VAV	Variable Air Volume	XHHW	Cross-Linked Polyethylene Wire
T.S.	Trigger Start	VC	Veneer Core	XLPE	Insulation
Tr.	Trade	VDC	Volts Direct Current	XLP	Cross-linked Polyethylene
Transf.	Transformer	Vent.	Ventilation	Xport	Transport
Trhv.	Truck Driver, Heavy	Vert.	Vertical	Y	Wye
Trlr	Trailer	V.F.	Vinyl Faced	yd	Yard
Trlt.	Truck Driver, Light	V.G.	Vertical Grain	yr	Year
TTY	Teletypewriter	VHF	Very High Frequency	Δ	Delta
TV	Television	VHO	Very High Output	%	Percent
T.W.	Thermoplastic Water Resistant	Vib.	Vibrating	~	Approximately
	Wire	VLF	Vertical Linear Foot	∅	Phase; diameter
UCI	Uniform Construction Index	VOC	Volatile Organic Compound	@	At
UF	Underground Feeder	Vol.	Volume	#	Pound; Number
UGND	Underground Feeder	VRP	Vinyl Reinforced Polyester	<	Less Than
UHF	Ultra High Frequency	W	Wire; Watt; Wide; West	>	Greater Than
U.I.	United Inch			Z	Zone

899

A

Abandon catch basin . . . 29
Abatement asbestos . . . 41
  equipment asbestos . . . 40
ABC extinguisher portable . . . 394
  extinguisher wheeled . . . 394
Aboveground storage tank concrete . . . 534
Abrasive aluminum oxide . . . 50
  floor . . . 80
  floor tile . . . 336
  silicon carbide . . . 50, 80
  terrazzo . . . 347
  tile . . . 334
  tread . . . 336
ABS DWV pipe . . . 503
Absorber shock . . . 507
Absorption testing . . . 15
  water chiller . . . 555
A/C packaged terminal . . . 559
Accelerator sprinkler system . . . 488
Access card key . . . 608
  control card type . . . 608
  control fingerprint scanner . . . 608
  control video camera . . . 609
  door and panel . . . 278
  door basement . . . 84
  door duct . . . 540
  door fire rated . . . 278
  door floor . . . 278
  door metal . . . 278
  door roof . . . 256
  door stainless steel . . . 278
  floor . . . 352
  flooring . . . 352
  road and parking area . . . 22
Accessory anchor bolt . . . 64
  bathroom . . . 334, 388, 390
  door . . . 310
  drainage . . . 253
  drywall . . . 333
  duct . . . 539
  fireplace . . . 391
  gypsum board . . . 333
  masonry . . . 99
  plaster . . . 326
  rebar . . . 68
  reinforcing . . . 68
  steel wire rope . . . 136
  toilet . . . 388
Accordion door . . . 275
Acid etch finish concrete . . . 81
  proof floor . . . 342, 349
  resistant pipe . . . 525
Acoustic ceiling board . . . 338
Acoustical barrier . . . 356
  batt . . . 355
  block . . . 108
  booth . . . 470
  ceiling . . . 338, 339
  door . . . 283
  enclosure . . . 470, 472
  folding door . . . 386
  folding partition . . . 386
  glass . . . 315
  metal deck . . . 143
  panel . . . 386
  panel ceiling . . . 338
  partition . . . 385, 387
  room . . . 461
  sealant . . . 261, 331
  treatment . . . 355
  wall tile . . . 339
Acrylic floor . . . 349
  plexiglass . . . 313
  rubber roofing . . . 247
  sign . . . 378
  wall coating . . . 370
  wallcovering . . . 353
  wood block . . . 343
ADA compliant drinking fount. . . . 522
  compliant fixture support . . . 520
  compliant water cooler . . . 523
Adhesive cement . . . 346
  EPDM . . . 225
  neoprene . . . 225
  PVC . . . 225
  removal floor . . . 82
  roof . . . 243
  wallpaper . . . 353
Adjustable astragal . . . 309
  jack post . . . 128
Adjustment factor labor . . . 492, 566
Admixture cement . . . 80
  concrete . . . 50
Adobe brick . . . 111
Aeration sewage . . . 722
Aerator . . . 524
Aerial photography . . . 14
  seeding . . . 671
  survey . . . 28
Agent form release . . . 51
Aggregate exposed . . . 81, 652
  lightweight . . . 51
  masonry . . . 97
  spreader . . . 728, 735
  stone . . . 672
  testing . . . 15
Agricultural equipment . . . 433
Aid staging . . . 21
Air balancing . . . 529
  barrier curtain . . . 543
  cleaner electronic . . . 548
  compressor . . . 731
  compressor control system . . . 532
  compressor dental . . . 428
  compressor portable . . . 731
  conditioner cooling & heating . . . 559
  conditioner direct expansion . . . 561
  conditioner fan coil . . . 561
  conditioner gas heat . . . 558
  conditioner general . . . 528
  conditioner packaged term. . . . 559
  conditioner receptacle . . . 581
  conditioner removal . . . 528
  conditioner rooftop . . . 558
  conditioner self-contained . . . 559
  conditioner thru-wall . . . 559
  conditioner ventilating . . . 538
  conditioner wiring . . . 583
  conditioning . . . 878
  conditioning computer . . . 559
  conditioning fan . . . 541
  conditioning ventilating . . 531, 538, 539, 541, 542, 547, 555, 557, 559
  control . . . 534
  cooled belt drive condenser . . . 554
  cooled condensing unit . . . 554
  cooled direct drive condenser . . . 555
  curtain . . . 543
  entraining agent . . . 50
  filter . . . 547, 548
  filter mechanical media . . . 548
  filter permanent washable . . . 547
  filter roll type . . . 547
  filter washable . . . 547
  filtration . . . 40
  filtration charcoal type . . . 548
  handling troffer . . . 594
  hose . . . 732
  lock . . . 464
  make-up unit . . . 558
  powered hoist . . . 715
  quality . . . 413
  register . . . 545
  return grille . . . 545
  sampling . . . 40
  separator . . . 534
  spade . . . 732
  supply pneumatic control . . . 532
  supply register . . . 545
  supported building . . . 463
  supported storage tank cover . . . 463
  supported structure demo . . . 454
  supported structures . . . 875
  tube system . . . 482
  unit make-up . . . 558
  vent automatic . . . 534
  vent convector or baseboard . . . 561
  vent roof . . . 257
  wall . . . 387
Air-compressor mobilization . . . 615
Aircraft cable . . . 138
Airfoil fan centrifugal . . . 542
Air-handling unit . . . 557, 561
Airless sprayer . . . 40
Airplane hangar . . . 470
Airport painted marking . . . 659
Air-source heat pump . . . 560
Alarm device . . . 611
  exit control . . . 609
  panel and device . . . 611
  residential . . . 582
  valve sprinkler . . . 486
All fuel chimney . . . 549
Alley bowling . . . 425
Alloy steel chain . . . 162
Altar . . . 432
Alteration fee . . . 10
Alternating tread ladder . . . 155
Alternative fastening . . . 231
Aluminum astragal . . . 308
  bench . . . 667
  blind . . . 436
  cable tray . . . 576
  ceiling tile . . . 339
  column . . . 164, 212
  column base . . . 161, 164
  column cap . . . 164
  column cover . . . 161
  commercial door . . . 270
  conduit . . . 573
  coping . . . 117
  cross . . . 432
  curtain wall glazed . . . 288
  diffuser perforated . . . 544
  directory board . . . 377
  dome . . . 464
  door . . . 265, 284
  door double acting swing . . . 285
  downspout . . . 251
  drip edge . . . 255
  ductwork . . . 538
  edging . . . 673
  entrance . . . 287
  expansion joint . . . 255
  extrusion . . . 135
  fence . . . 661, 662
  flagpole . . . 400
  flashing . . . 239, 249, 509
  foil . . . 229, 232, 356
  framing . . . 135
  grating . . . 157
  gravel stop . . . 251
  grille . . . 545
  gutter . . . 253
  handrail . . . 352
  joist shoring . . . 61
  louver . . . 315, 546
  louver door . . . 316
  Mansard . . . 238
  mesh grating . . . 159
  nail . . . 171
  operating louver . . . 546
  oxide abrasive . . . 50
  pipe . . . 684
  plank grating . . . 157
  plank scaffold . . . 21
  register . . . 545
  reglet . . . 254
  rivet . . . 125
  roof . . . 237
  roof panel . . . 237
  salvage . . . 493
  sash . . . 289
  screen/storm door . . . 265
  screw . . . 239
  security driveway gate . . . 282
  service entrance cable . . . 571
  shape structural . . . 135
  sheet metal . . . 251
  shelter . . . 470
  shingle . . . 233
  siding . . . 238
  siding paint . . . 360
  siding panel . . . 238
  sign . . . 378
  sliding door . . . 279
  steeple . . . 399
  storefront . . . 288
  storm door . . . 265
  structural . . . 135
  threshold . . . 305
  tile . . . 236, 333
  transom . . . 287
  trench cover . . . 159
  tube frame . . . 285
  weatherstrip . . . 308
  weld rod . . . 127
  window . . . 289
  window demolition . . . 264
  window impact resistant . . . 289
Aluminum-framed door . . . 287
Ammeter . . . 584
Analysis petrographic . . . 15
  sieve . . . 15
Anchor bolt . . . 64, 98
  bolt accessory . . . 64
  bolt screw . . . 69
  bolt sleeve . . . 64
  brick . . . 98
  buck . . . 98
  bumper . . . 407
  chemical . . . 88, 121
  dovetail . . . 67
  epoxy . . . 88, 121
  expansion . . . 121
  framing . . . 172
  hollow wall . . . 121
  lead screw . . . 122
  machinery . . . 67
  masonry . . . 98
  metal nailing . . . 121
  nailing . . . 121
  nylon nailing . . . 121
  partition . . . 98
  plastic screw . . . 122
  rafter . . . 172
  rigid . . . 98
  roof safety . . . 21
  screw . . . 69, 121
  sill . . . 172
  steel . . . 98

For customer support on your Building Construction Costs with RSMeans data, call 800.448.8182.

Index

stone 98
super high-tensile 640
toggle-bolt 121
wedge 122
Anechoic chamber 461
Angle corner 129
  curb edging 129
  fiberglass 216
  valve bronze 493
  valve plastic 496
Anti-ligature locksets 304
Anti-siphon device 506
Anti-slip floor treatment 366
Apartment call system 605
  hardware 299
Appliance 412
  plumbing 512, 513
  residential 412, 582
Approach ramp 352
Arch culvert oval 686
  laminated 193
  radial 193
Architectural fee 10, 828
Area clean-up 43
  window well 400
Ark 432
Armored cable 569
Arrow 659
Articulating boom lift 731
Artificial grass surfacing 660
  turf 660
Asbestos abatement 41
  abatement equipment 40
  area decontamination 43
  demolition 42
  disposal 43
  encapsulation 44
  remediation plan/method 40
  removal 42
  removal process 840
Ash conveyor 430
  receiver 445
  urn 445
Ashlar stone 114
  veneer 112
Asphalt base sheet 243
  block 656
  block floor 656
  coating 224
  concrete 614
  curb 657
  cutting 34
  distributor 732
  felt 243, 244
  flashing 249
  flood coat 243
  mix design 15
  pavement 660
  paver 733
  paving 654
  plant portable 735
  primer 222, 346
  roof cold-applied built-up 243
  roof shingle 233
  sealcoat rubberized 652
  sheathing 192
  shingle 233
  sidewalk 652
  testing 15
Asphaltic binder 654
  concrete 654
  concrete paving 654
  cover board 230
  emulsion 670
  pavement 654
  wearing course 654

Aspirator dental 428
Astragal adjustable 309
  aluminum 308
  exterior moulding 308
  magnetic 308
  molding 206
  one piece 308
  overlapping 309
  rubber 308
  split 309
  steel 308
Astronomy observation dome ... 471
Athletic backstop 662
  carpet indoor 351
  equipment 424
  paving 660
  pole 425
  post 427
Atomizer water 41
Attachment ripper 735
Attachments skid-steer 731
Attic stair 414
Audio masking 472
Audiometric room 461
Auditorium chair 450
Auger boring 29
  earth 728
  hole 28
Augering 614
Auto park drain 509
Autoclave 427
  aerated concrete block 104
Automatic fire-suppression 489
  flush 519
  opener 299
  opener commercial 299
  opener industrial 299
  operator 299
  teller 411
  wall switch 584
  washing machine 413
Automotive equipment 406
  lift 481
  spray painting booth 406
Autopsy equipment 430
Auto-scrub machine 732
Awning canvas 398
  fabric 398
  window 290, 398
  window aluminum 289
  window metal-clad 292
  window vinyl clad 293

B

Backer rod 260
  rod polyethylene 64
Backerboard 224
  cementitious 329
Backfill 628, 629
  compaction 884
  dozer 628
  planting pit 670
  structural 628
  trench 619, 621
Backflow preventer 506
Backhoe 623, 729
  bucket 730
  extension 735
Backsplash 441
  countertop 442
Backstop athletic 662
  baseball 662
  basketball 424, 662
  chain link 662

electric 424
  handball court 426
  squash court 426
  tennis court 662
Backup block 104
Backwater valve 511
Baffle roof 257
Bag disposable 41
  glove 42
Baked enamel door 268
  enamel frame 267
Balanced entrance door 287
Balancing air 529
  water 530
Balcony fire escape 154
Bale hay 639
Baler 431
Ball valve bronze 494
  valve plastic 496
  valve polypropylene 496
  wrecking 735
Ballast fixture 595
  railroad 706
Ballistic door wood 282
Baluster wood stair 209
Balustrade painting 367
Bamboo flooring 341
Band joist framing 148
  riser 422
Bank counter 411
  equipment 411
  run gravel 614
  window 411
Banquet booth 447
Baptistry 432
Bar bell 423
  chair reinforcing 68
  front 471
  grab 389
  grating tread 160
  grouted 73
  joist painting 141
  masonry reinforcing 98
  restaurant 447
  tie reinforcing 68
  towel 389
  tub 389
  ungrouted 73
  Zee 333
Barbed wire fence 661
Barber chair 409
  equipment 409
Barge construction 738
Bariatric equipment 428
  lift 428
Bark mulch 670
  mulch redwood 670
Barn 433
Barrel bolt 299
Barrels flasher 732
  reflectorized 732
Barricade 23, 708
  flasher 732
  tape 24
Barrier acoustical 356
  and enclosure 23
  crash 708
  delineator 708
  dust 23
  highway sound traffic 668
  impact 708
  median 708
  moisture 225
  parking 659
  plenum 356
  security 408

separation 41, 46
  slipform paver 736
  weather 232
  X-ray 473
Base cabinet 439
  carpet 351
  ceramic tile 334
  column 172
  course 653
  course drainage layer 653
  gravel 653
  masonry 110
  molding 196
  plate column 134
  quarry tile 336
  resilient 345
  road 653
  screed 69
  sheet 243, 244
  sheet asphalt 243
  sink 439
  stabilizer 736
  stone 114, 653
  terrazzo 347, 348
  vanity 440
  wall 344
  wood 196
Baseball backstop 662
  scoreboard 424
Baseboard demolition 170
  heat 561
  heat electric 564
  heater electric 564
Basement bulkhead stair 84
Basement/utility window 290
Baseplate scaffold 19
  shoring 19
Basic finish material 318
Basketball backstop 424, 662
  equipment 424
  scoreboard 424
Batch trial 15
Bath paraffin 428
  steam 462
  whirlpool 428, 459
Bathroom 516
  accessory 334, 388, 390
  exhaust fan 543
  faucet 517
  fixture 515, 516
  heater & fan 543
  soaking 516
Bathtub bar 389
  enclosure 391
  removal 493
  residential 516
Batt acoustical 355
  insulation 227
  thermal 355
Bay wood window double hung .. 293
Beacon warning 597
Bead blast demo 319
  board insulation 460
  casing 333
  corner 326, 333
  parting 207
Beam & girder framing 181
  and girder formwork 52
  bond 97, 105
  bondcrete 328
  bottom formwork 52
  castellated 134
  ceiling 212
  concrete 75
  concrete placement 78
  concrete placement grade .. 79

decorative	212	Blanket curing	83	Bluegrass sod	671	auger	29
drywall	329	insulation	530	Bluestone	112	cased	28
fiberglass wide flange	216	insulation wall	227	sidewalk	652	horizontal	676
fireproofing	257	sound attenuation	355	sill	116	machine horizontal	728
formwork side	53	Blast curtain	437	step	653	Borosilicate pipe	524
hanger	172	demo bead	319	Board & batten siding	240	Borrow	614, 629
laminated	193	floor shot	319	batten fence	664	Bosun's chair	22
load-bearing stud wall box	144	Blaster shot	734	bulletin	376	Bottle storage	415
mantel	212	Blasting	622	ceiling	338	vial	47
plaster	327	cap	623	control	377, 423	Boulder excavation	623
precast	84	mat	623	directory	377, 378	Boundary and survey marker	28
precast concrete	84	water	95	dock	407	Bow & bay metal-clad window	293
precast concrete tee	84	Bleacher	466	drain	225	window	293
reinforcing	71	outdoor	466	gypsum	329	Bowl toilet	515
reinforcing bolster	68	stadium	466	high abuse gypsum	332	Bowling alley	425
side formwork	52	telescoping	450	insulation	226	Bowstring truss	194
soldier	643	Blind exterior	214	insulation foam	226	Box & wiring device	575
test	15	structural bolt	126	paneling	209	beam load-bearing stud wall	144
wood	181, 189	vertical	437	partition gypsum	322	buffalo	678
Bearing pad	126	vinyl	214	partition NLB gypsum	321	culvert formwork	54
Bed molding	199	window	436	sheathing	191	distribution	683
Bedding brick	656	wood	438	siding wood	240	electrical	575
pipe	629	Block acoustical	108	valance	440	electrical pull	575
Belgian block	658	asphalt	656	verge	205	flow leveler distribution	683
Bell	423	autoclave concrete	104	Boat work	738	locker	394
& spigot pipe	508	backup	104	Body harness	21	mail	396
bar	423	Belgian	658	Boiler	549	pull	575
caisson	647	bituminous	656	control system	531	safety deposit	411
church	423	Block, brick and mortar	858	demolition	528	stair	209
signal	611	Block cap	109	electric	549	steel mesh	640
system	591	chimney	104	electric steam	549	storage	17
Bellow expansion joint	534	column	105	gas fired	549	termination	605
Belt material handling	714	concrete	104-108, 859	gas/oil combination	550	trench	735
Bench	667	concrete bond beam	105	general	492	utility	677
aluminum	667	concrete exterior	107	hot water	549, 550	vent	99
fiberglass	667	corner	109	insulation	531	Boxed header/beam	148
folding	450	decorative concrete	105	mobilization	615	header/beam framing	148
greenhouse	465	filler	368	oil fired	550	Boxing ring	424
park	667	floor	344	removal	528	Brace cross	130
planter	667	floor wood	343	steam	549, 550	Bracing	180
plastic	396	glass	110	Bollard	128	framing	151
player	668	glazed	109	pipe	658	let-in	180
wood	667	glazed concrete	109	security	707	load-bearing stud wall	143
work	422	grooved	105	Bolster beam reinforcing	68	metal	180
Bentonite	225, 644	ground face	106	slab reinforcing	68	metal joist	147
panel waterproofing	225	hexagonal face	106	Bolt & Hex nut steel	123	roof rafter	151
Berm pavement	657	high strength	107	anchor	64, 98	shoring	19
road	657	insulation	105	barrel	299	support seismic	496
Bevel glass	311	insulation insert concrete	105	blind structural	126	wood	180
siding	240	lightweight	108	flush	299	Bracket roof	21
Bicycle rack	425, 451	lintel	108	high strength	124	scaffold	19
trainer	423	manhole	677	remove	120	sidewall	21
Bi-fold door	269, 274	partition	108	steel	172	Braided bronze hose	505
Bimodal MDPE, ASTM D2513	688	patio	114, 244	Bolt-on circuit-breaker	587	Branch circuit fusible switch	586
Bin part	397	profile	106	Bond beam	97, 105	Brass hinge	306
retaining wall metal	666	reflective	110	performance	14, 835	pipe	500
Binder asphaltic	654	scored	106	Bondcrete	328	salvage	493
Biological safety cabinet	422	scored split face	106	Bonding agent	50	screw	171
Biometric identity access	608	sill	109	surface	97	valve	493
Bi-passing closet door	274	slotted	108	Bookcase	397	Brazed connection	573
Birch exterior door	277	slump	106	Bookshelf	449	Break glass station	611
hollow core door	275	solar screen	109	Boom lift articulating	731	tie concrete	81
molding	206	split rib	106	lift telescoping	731	Breaker circuit	591
paneling	208	wall removal	31	Booster fan	542	pavement	732
solid core door	274	Blocking carpentry	180	Boot pile	645	vacuum	506
wood frame	213	steel	180	Booth acoustical	470	Breeching draft inducer	548
Bird control film	314	wood	180, 184	banquet	447	insulation	531
control netting	401	Blood pressure unit	427	fixed	447	Brick	102
Bit core drill	90	Bloodbank refrigeration	421	mounted	448	adobe	111
Bituminous block	656	Blower insulation	733	painting	406	anchor	98
coating	224, 676	pneumatic tube	482	portable	470	bedding	656
concrete curb	657	Blown-in cellulose	229	ticket	470	Brick, block and mortar	858
paver	733	fiberglass	229	Border light	597	Brick cart	732
Bituminous-stabilized base		insulation	229	Bored pile	646	catch basin	676
course	653	Blueboard	328	Borer horizontal	735	chimney simulated	391
Blackboard	386	partition	322	Boring and exploratory drilling	28	cleaning	95

For customer support on your Building Construction Costs with RSMeans data, call 800.448.8182.

common . . . 102
common building . . . 102
concrete . . . 109
demolition . . . 96, 318
economy . . . 102
edging . . . 673
engineer . . . 102
face . . . 102, 858
floor . . . 342, 656
flooring miscellaneous . . . 342
manhole . . . 676
molding . . . 203, 206
oversized . . . 101
paving . . . 656
removal . . . 30
sand . . . 51
saw . . . 94
sidewalk . . . 656
sill . . . 116
simulated . . . 117
stair . . . 112
step . . . 652
structural . . . 102
testing . . . 15
veneer . . . 99
veneer demolition . . . 96
veneer thin . . . 101
vent . . . 546
vent box . . . 99
wall . . . 111
wall panel . . . 111
wash . . . 95
Bricklaying . . . 858
Bridge . . . 668
concrete . . . 668
crane . . . 714
pedestrian . . . 668
sidewalk . . . 20
Bridging . . . 181
framing . . . 151
joist . . . 140
load-bearing stud wall . . . 144
metal joist . . . 147
roof rafter . . . 151
Broiler . . . 417
Bronze angle valve . . . 493
ball valve . . . 494
body strainer . . . 535
cross . . . 432
globe valve . . . 495
letter . . . 378
plaque . . . 378
push-pull plate . . . 305
swing check valve . . . 494
valve . . . 493
Broom cabinet . . . 439
finish concrete . . . 80
sidewalk . . . 734
Brown coat . . . 347
Brownstone . . . 114
Brush chipper . . . 729
clearing . . . 616
cutter . . . 729, 730
mowing . . . 616
Bubbler . . . 522
drinking . . . 523
Buck anchor . . . 98
rough . . . 183
Bucket backhoe . . . 730
clamshell . . . 729
concrete . . . 728
dragline . . . 729
excavator . . . 735
pavement . . . 735
Buffalo box . . . 678
Buggy concrete . . . 79, 728
Builder's risk insurance . . . 832
Building air supported . . . 463
component deconstruction . . . 36
deconstruction . . . 35
demo footing & foundation . . . 31
demolition . . . 30
demolition selective . . . 31
directory . . . 377, 605
greenhouse . . . 465
hangar . . . 470
insulation . . . 229
model . . . 10
moving . . . 38
paper . . . 232
permit . . . 14
portable . . . 17
prefabricated . . . 465
relocation . . . 38
shoring . . . 641
sprinkler . . . 488
temporary . . . 17
tension . . . 464
Built-in range . . . 412
shower . . . 517
Built-up roof . . . 243
roofing . . . 866
roofing component . . . 243
roofing system . . . 243
Bulb tee sub-purlin . . . 87
Bulb tee subpurlin . . . 135
Bulk bank measure excavating . . . 623
dozer excavating . . . 624
drilling and blasting . . . 622
scraper excavation . . . 625
storage dome . . . 469
Bulkhead door . . . 278
moveable . . . 458
Bulkhead/cellar door . . . 278
Bulldozer . . . 628, 730
Bullet resistant window . . . 411
Bulletin board . . . 376
board cork . . . 376
Bulletproof glass . . . 315
Bullfloat concrete . . . 728
Bullnose block . . . 107
Bumper car . . . 658, 659
dock . . . 407
door . . . 300, 304
parking . . . 659
railroad . . . 706
wall . . . 304, 388
Buncher feller . . . 729
Bunk house trailer . . . 17
Burglar alarm indicating panel . . . 609
Burial cell . . . 38
Burlap curing . . . 83
rubbing concrete . . . 81
Bus duct cable tap box . . . 589
copper . . . 589
fitting . . . 589
plug in . . . 589
switch . . . 590
tap box . . . 589
Bush hammer . . . 81
hammer finish concrete . . . 81
Butt fusion machine . . . 732
Butterfly valve . . . 528, 679
Butyl caulking . . . 261
expansion joint . . . 255
flashing . . . 250
rubber . . . 260
waterproofing . . . 225

C

Cab finish elevator . . . 478
Cabinet & casework
paint/coating . . . 363
base . . . 439
broom . . . 439
corner base . . . 439
corner wall . . . 439
current transformer . . . 590
defibrillator . . . 393
demolition . . . 170
electric . . . 590
electrical hinged . . . 575
filler . . . 440
fire equipment . . . 393
hardboard . . . 439
hardware . . . 440
hinge . . . 440
hose rack . . . 393
hospital . . . 442
hotel . . . 390
key . . . 310
kitchen . . . 439
kitchen metal . . . 442
laboratory . . . 441
medicine . . . 390
metal . . . 442
nurse station . . . 442
oven . . . 439
pharmacy . . . 422
portable . . . 415
school . . . 441
shower . . . 390
stain . . . 363
storage . . . 441
strip . . . 605
transformer . . . 590
valve . . . 393
varnish . . . 363
wall . . . 439, 441, 442
wardrobe . . . 397
Cable aircraft . . . 138
armored . . . 569
control . . . 573
copper . . . 569, 573
copper shielded . . . 569
crimp termination . . . 570
electric . . . 569, 570
electrical . . . 572, 573
guide rail . . . 708
jack . . . 738
MC . . . 570
medium-voltage . . . 569
mineral insulated . . . 571
pulling . . . 734
PVC jacket . . . 573
railing . . . 164
safety railing . . . 138
sheathed nonmetallic . . . 571
sheathed Romex . . . 571
tap box bus duct . . . 589
tensioning . . . 734
termination . . . 570
trailer . . . 734
tray . . . 576
tray aluminum . . . 576
tray ladder type . . . 576
wire THWN . . . 573
Cable/wire puller . . . 734
Cafe door . . . 276
Caisson bell . . . 647
concrete . . . 646
displacement . . . 648
foundation . . . 646
Caissons . . . 886
Calcium chloride . . . 50, 640
Call system apartment . . . 605
system emergency . . . 605
system nurse . . . 606
Camera TV closed circuit . . . 609
Canopy . . . 398
entrance . . . 399, 467
entry . . . 398
framing . . . 127
framing wood . . . 186
metal . . . 398
patio/deck . . . 398
Cant roof . . . 186, 244
Cantilever needle beam . . . 642
retaining wall . . . 665
Canvas awning . . . 398
Cap blasting . . . 623
block . . . 109
concrete placement pile . . . 79
dowel . . . 72
pile . . . 77
post . . . 172
Capacitor indoor . . . 593
Car bumper . . . 658, 659
tram . . . 735
Carbon dioxide extinguisher . . . 489
monoxide detector . . . 610
Carborundum rub finish . . . 81
Card catalog file . . . 449
key access . . . 608
Carillon . . . 423
Carousel compactor . . . 431
Carpentry finish . . . 439
Carpet base . . . 351
computer room . . . 352
felt pad . . . 350
floor . . . 351
maintenance . . . 318
nylon . . . 351
pad . . . 350
pad commercial grade . . . 350
padding . . . 350
removal . . . 318
sheet . . . 351
stair . . . 351
tile . . . 350
tile removal . . . 318
urethane pad . . . 350
wool . . . 351
Carport . . . 399
Carrel . . . 449
Carrier ceiling . . . 325
channel . . . 339
fixture support . . . 520
Carrier/support fixture . . . 520
Cart brick . . . 732
concrete . . . 79, 728
disposal . . . 431
food delivery . . . 417
mounted extinguisher . . . 394
Carving stone . . . 114
Case display . . . 441
exhibit . . . 441
good metal . . . 446
good wood . . . 446
refrigerated . . . 414
Cased boring . . . 28
Casement window . . . 289-291, 293
window metal-clad . . . 292
window vinyl . . . 296
Casework custom . . . 439
demo . . . 436
demolition . . . 170
educational . . . 441
ground . . . 188
hospital . . . 441

For customer support on your Building Construction Costs with RSMeans data, call 800.448.8182.

metal 310, 441
painting 363
varnish 363
Cash register 409
Casing bead 333
molding 197
Cast in place concrete 75, 78
in place concrete curb 657
in place pile 648
in place pile add 649
in place terrazzo 347
iron bench 667
iron damper 392
iron grate cover 159
iron manhole cover 686
iron pipe 508
iron pull box 575
iron radiator 561
iron trap 508
trim lock 303
Castellated beam 134
Caster scaffold 19
Casting construction 161
fiberglass 216
Cast iron column base 161
weld rod 127
Catch basin 686
basin brick 676
basin cleaner 735
basin precast 677
basin removal 29
basin vacuum 735
door 440
Cathodic protection 471
Catwalk scaffold 20
Caulking 261
lead 508
oakum 508
polyurethane 261
sealant 261
Cavity wall 111
wall grout 98
wall insulation 228, 229
wall reinforcing 99
Cedar closet 208
fence 664
paneling 209
post 213
roof deck 190
roof plank 190
shingle 234
siding 240
Ceiling 338
acoustical 338, 339
beam 212
board 338
board acoustic 338
board fiberglass 338
bondcrete 328
bracing seismic 339
carrier 325
demolition 318
diffuser 544
drywall 330
fan 541
framing 182
furring 188, 322, 325
heater 413
insulation 226, 229
lath 325, 326
linear metal 340
luminous 338, 599
molding 199
painting 367, 368
panel 338
panel metal 339, 340

plaster 327
polystyrene 460
register 545
stair 414
support 127
suspended 325, 338
suspension system 339
tile 338, 339
translucent 340
wood 340
woven wire 384
Cell equipment 433
prison 471
Cellar door 278
wine 415
Cellular concrete 75
decking 143
Cellulose blown-in 229
insulation 228
Cement adhesive 346
admixture 80
color 97
content 850
flashing 222
grout 98, 642
Keenes 327
liner 676
masonry 96
masonry unit 104, 106-108
mortar 335, 336, 857
parging 224
plaster 460
Portland 51
terrazzo Portland 347
testing 15
underlayment self-leveling 88
vermiculite 258
Cementitious backerboard 329
waterproofing 225
wood fiber plank 87
Center bulb waterstop 63
Central vacuum 430
vacuum unit 718
Centrifugal airfoil fan 542
liquid chiller 555
pump 734
type HVAC fan 542
Ceramic coating 371
mulch 670
tile 334, 335
tile accessory 333
tile countertop 443
tile demolition 319
tile floor 335
tile repair 333
tile waterproofing membrane . . 337
tiling 333-337
Certification welding 16
Chain core rope 408
fall hoist 737
hoist 715
hoist door 283
link backstop 662
link fence 24, 661, 664
link fence high-security 662
link fence paint 358
link gate 662
link gate industrial 663
link gate internal 663
link gate residential 663
link industrial fence 661
link transom 663
saw 734
steel 162
trencher 621, 731
Chair 448

auditorium 450
barber 409
bosun's 22
dental 428
hydraulic 428
life guard 458
lift 481
locker bench 396
molding 206
movie 420
office 447
portable 422
rail demolition 170
seating 450
Chalkboard 374
electric 375
fixed 374
freestanding 376
liquid chalk 374
metal 375, 386
portable 376
sliding 375
swing leaf 375
wall hung 374
Chamber anechoic 461
decontamination 41, 46
echo 472
infiltration 683
Chamfer strip 60
Channel carrier 339
curb edging 129
frame 267
furring 322, 333
grating tread 160
metal frame 267
siding 240
steel 326
Channelizing traffic 708
Charcoal type air filter 548
Charge disposal 43
powder 125
Charging desk 449
Check floor 305
rack clothes 409
swing valve 494
valve 486
valve steel 529
Checkered plate 159
plate cover 159
plate platform 159
Checking bag garment 409
Checkout counter 409
scanner 409
supermarket 409
Checkroom equipment 409
Chemical anchor 88, 121
cleaning masonry 95
dry extinguisher 393
removal graffiti 370
spreader 735
termite control 639
toilet 734
water closet 722
Chemical-resistant quarry tiling . . 337
Chest 448
Chilled water coil 560
Chiller centrifugal type water 555
water 555
Chimney 101
accessory 391
all fuel 549
all fuel vent 549
block 104
brick 101
demolition 95
fireplace 391

flue 116
foundation 75
metal 391
positive pressure 549
screen 391
simulated brick 391
vent 548
Chip quartz 51
silica 51
white marble 51
wood 670
Chipper brush 729
log 730
stump 730
Chipping hammer 732
stump 615
Chloride accelerator concrete . . . 78
Chlorination system 524
Chlorosulfunated polyethylene . . 38
Church bell 423
equipment 432
equipment demolition 404
pew 450
Chute 482
linen 482
package 482
refuse 482
system 33
Cinema equipment 420
Circuit breaker 591
bolt-on 587
feeder section 586
Circular rotating entrance door . . 287
saw 734
Circulating pump 536
Clamp water pipe ground 573
Clamshell 623, 712
bucket 729
Clapboard painting 359
Classroom heating & cooling 561
seating 450
Clay fire 116
tennis court 660
tile 235
tile coping 117
Clean control joint 63
room 460
tank 39
Cleaner catch basin 735
crack 735
steam 734
Cleaning & disposal equipment . 419
brick 95
face brick 857
masonry 94
metal 120
up 25
Cleanout door 392
floor 498
pipe 498
PVC 498
tee 498
Clean-up area 43
Clear and grub site 615
grub 615
Clearing brush 616
selective 616
site 615
Clevis hook 162
Climber mast 21
Climbing crane 736
hydraulic jack 737
Clip plywood 172
wire rope 136
Clock equipment 606
timer 582

For customer support on your Building Construction Costs with RSMeans data, call 800.448.8182.

wall . . . 445
Closed circuit television system . 609
circuit TV . . . 411
deck grandstand . . . 465
Closer concealed . . . 300
door . . . 300
door electric automatic . . . 301
door electro magnetic . . . 300
electronic . . . 301
floor . . . 300
holder . . . 300
Closet cedar . . . 208
door . . . 269, 274
pole . . . 206
rod . . . 397
water . . . 515, 518
Closing coordinator door . . . 305
Closure trash . . . 451
Cloth hardware . . . 664
Clothes check rack . . . 409
dryer commercial . . . 410
CMU . . . 104
pointing . . . 94
Coal tar pitch . . . 222, 243
Coat brown . . . 347
glaze . . . 367
hook . . . 389
rack . . . 397
rack & wardrobe . . . 397
scratch . . . 347
Coated roofing foamed . . . 247
Coating bituminous . . . 224, 676
ceramic . . . 371
concrete wall . . . 370
dry erase . . . 371
elastomeric . . . 370
epoxy . . . 81, 371, 676
exterior steel . . . 371
flood . . . 244
intumescent . . . 370
polyethylene . . . 676
reinforcing . . . 72
roof . . . 222
rubber . . . 226
silicone . . . 226
spray . . . 224
structural steel . . . 859
stucco wall . . . 370
trowel . . . 224
wall . . . 370
water repellent . . . 226
waterproofing . . . 224
Cock gas . . . 493
Coffee urn . . . 416
Cofferdam . . . 643
excavation . . . 624
Cohesion testing . . . 15
Coil cooling . . . 560
flanged . . . 560
Coiling door and grille . . . 280
grill . . . 280
Coiling-type counter door . . . 280
Coin operated gate . . . 408
CO_2 extinguisher . . . 394
fire extinguishing system . . . 489
Coin-operated washer & dryer . . . 410
Cold mix paver . . . 736
roofing . . . 243
storage door . . . 281
storage room . . . 460
Collection box lobby . . . 396
concrete pipe sewage . . . 682
sewage/drainage . . . 685
system dust . . . 718
Collector solar energy system . . . 553
Colonial door . . . 274, 277
wood frame . . . 213
Color floor . . . 81
wheel . . . 597
Coloring concrete . . . 51
Column . . . 212
accessory formwork . . . 60
aluminum . . . 164, 212
base . . . 172
base aluminum . . . 161, 164
base cast iron . . . 161
base plate . . . 134
block . . . 105
bondcrete . . . 328
brick . . . 101
cap aluminum . . . 164
concrete . . . 75
concrete block . . . 105
concrete placement . . . 78
cover . . . 161
cover aluminum . . . 161
cover stainless . . . 161
demolition . . . 96
drywall . . . 329
fireproof . . . 257
formwork . . . 53
framing . . . 182
lally . . . 128
laminated wood . . . 194
lath . . . 325
lightweight . . . 128
ornamental . . . 164
plaster . . . 327
plywood formwork . . . 53
precast . . . 84
precast concrete . . . 84
reinforcing . . . 71
removal . . . 96
round fiber tube formwork . . . 53
round fiberglass formwork . . . 53
round steel formwork . . . 53
steel framed formwork . . . 54
stone . . . 114
structural . . . 128
structural shape . . . 129
wood . . . 182, 189, 212
Combination device . . . 579
storm door . . . 274, 277
Comfort station . . . 469
Command dog . . . 24
Commercial automatic opener . . 299
dishwasher . . . 419
floor door . . . 278
folding partition . . . 386
gas water heater . . . 514
greenhouse . . . 465
half glass door . . . 268
lavatory . . . 518
mail box . . . 396
metal door . . . 269
oil-fired water heater . . . 514
refrigeration . . . 414, 415, 460
safe . . . 410
toilet accessory . . . 388
water heater . . . 513, 514
water heater electric . . . 513
Commissioning . . . 26
Common brick . . . 102
building brick . . . 102
nail . . . 171
Communicating lockset . . . 302, 303
Communication optical fiber . . . 604
system . . . 605
Compact fill . . . 629
Compaction . . . 639
backfill . . . 884
soil . . . 628
structural . . . 639
test Proctor . . . 16
vibroflotation . . . 885
Compactor . . . 431
earth . . . 729
landfill . . . 730
residential . . . 412
waste . . . 431
Compartment toilet . . . 380
Compensation workers' . . . 12
Compliant drinking fountain
ADA . . . 522
fixture support ADA . . . 520
water cooler ADA . . . 523
Composite decking . . . 185, 218
decking woodgrained . . . 218
door . . . 268, 276
fabrication . . . 218
insulation . . . 231
metal deck . . . 141
rafter . . . 185
railing . . . 219
railing encased . . . 219
Composition floor . . . 350
flooring . . . 343, 349
flooring removal . . . 319
Compound dustproofing . . . 50
Compressed air equipment
automotive . . . 406
Compression seal . . . 260
Compressive strength . . . 15
Compressor air . . . 731
tank mounted . . . 532
Computer air conditioning . . . 559
floor . . . 352
room carpet . . . 352
Concealed closer . . . 300
Concrete . . . 847-851
acid etch finish . . . 81
admixture . . . 50
asphalt . . . 614
asphaltic . . . 654
beam . . . 75
block . . . 104-108, 859
block autoclave . . . 104
block back-up . . . 104
block bond beam . . . 105
block column . . . 105
block decorative . . . 105
block demolition . . . 32
block exterior . . . 107
block foundation . . . 107
block glazed . . . 109
block grout . . . 98
block insulation . . . 229
block insulation insert . . . 105
block interlocking . . . 107
block lintel . . . 108
block painting . . . 368
block partition . . . 108
block planter . . . 656
break tie . . . 81
brick . . . 109
bridge . . . 668
broom finish . . . 80
bucket . . . 728
buggy . . . 79, 728
bullfloat . . . 728
burlap rubbing . . . 81
bush hammer finish . . . 81
caisson . . . 646
cart . . . 79, 728
cast-in-place . . . 75, 78
catch basin . . . 676
cellular . . . 75
chloride accelerator . . . 78
coloring . . . 51
column . . . 75
conversion . . . 851
conveyer . . . 728
coping . . . 117
core drilling . . . 89
crane and bucket . . . 78
cribbing . . . 665
curb . . . 657
curing . . . 83, 655
curing compound . . . 83
cutout . . . 32
cylinder . . . 15
dampproofing . . . 52
demo . . . 29
demolition . . . 32, 50
direct chute . . . 79
distribution box . . . 683
drilling . . . 90
equipment rental . . . 728
fiber reinforcing . . . 78
field mix . . . 852
finish . . . 655
float finish . . . 80, 81
floor edger . . . 728
floor grinder . . . 728
floor patching . . . 50
floor staining . . . 341
floors . . . 853
form liner . . . 58
forms . . . 842
furring . . . 187
granolithic finish . . . 87
grinding . . . 82
grout . . . 98
hand hole . . . 703
hand mix . . . 77
hand trowel finish . . . 80
hardener . . . 80
hydrodemolition . . . 29
impact drilling . . . 90
integral color . . . 50
integral topping . . . 87
integral waterproofing . . . 51
joist . . . 75
lance . . . 733
lift slab . . . 852
lightweight . . . 75, 78, 856
lightweight insulating . . . 87
machine trowel finish . . . 80
manhole . . . 677, 703
materials . . . 850
membrane curing . . . 83
mix design . . . 15
mixer . . . 728
mixes . . . 850, 851
non-chloride accelerator . . . 78
pan tread . . . 80
paper curing . . . 83
patio block . . . 656
pavement highway . . . 655
paver . . . 733
pier and shaft drilled . . . 648
pile prestressed . . . 644
pipe . . . 685
pipe removal . . . 29
placement beam . . . 78
placement column . . . 78
placement footing . . . 79
placement slab . . . 79
placement wall . . . 79
placing . . . 78, 852
plantable paver precast . . . 656
planter . . . 667
polishing . . . 82
post . . . 708

For customer support on your Building Construction Costs with RSMeans data, call 800.448.8182.

pressure grouting	642	in slab PVC	576
prestressed	855	in trench electrical	576
prestressed precast	854	in trench steel	576
processing	82	intermediate steel	574
pump	728	raceway	573
pumped	78	rigid in slab	576
ready mix	78	rigid steel	574
reinforcing	71	Cone traffic	23
removal	29	Confessional	432
retaining wall	665	Connection brazed	573
retaining wall segmental	665	fire-department	486
retarder	78	motor	578
sand	51	standpipe	486
sandblast finish	81	utility	677
saw	728	Connector flexible	505
saw blade	89	gas	505
scarify	319	joist	172
septic tank	682	timber	172
shingle	235	water copper tubing	505
short load	78	wire	570
sidewalk	652	Consolidation test	16
slab	75	Construction aid	18
slab conduit	576	barge	738
slab cutting	89	casting	161
slab X-ray	16	management fee	10
small quantity	77	photography	14
spreader	733	sump hole	627
stair	77	temporary	23
stair finish	81	time	10, 829, 830
stengths	851	Containment hazardous waste	38
tank	681	Contaminated soil	39
tank aboveground storage	534	Contingency	11
testing	15	Continuous hinge	307
tie railroad	706	Contractor overhead	25
tile	235	scale	402
tilt-up	856	Control & motor starter	588
topping	87	air	534
track cross tie	706	board	377, 423
trowel	728	cable	573
truck	728	damper volume	539
truck holding	78	hand/off/automatic	592
unit paving slab precast	656	joint	63, 99
utility vault	677	joint clean	63
vertical shoring	61	joint PVC	99
vibrator	728	joint rubber	99
volumetric mixed	77	joint sawn	63
wall	76	mirror traffic	401
wall coating	370	radiator supply	535
wall demolition	31	station	592
wall panel tilt-up	86	station stop/start	592
wall patching	50	system air compressor	532
water reducer	78	system boiler	531
water storage tank	681	system cooling	531
wheeling	79	system electronic	531
winter	78	system pneumatic	531
Concrete-filled steel pile	646	system split system	531
Condensate pump	537	system ventilator	531
removal pump	537	tower	471
return pump	537	valve heating	535
steam meter	537	Controller time system	606
Condenser	554	Convection oven	417
air cooled belt drive	554	Convector hydronic heating	561
air cooled direct drive	555	Conveyor	714
Condensing unit air cooled	554	ash	430
Conditioner general air	528	door	543
Conductive floor	347, 349	material handling	714
rubber & vinyl flooring	347	Cooking equipment	412, 417
terrazzo	349	range	412
Conductor	569	Cooler	461
& grounding	569-572	beverage	415
Conductors	881	water	523
Conduit & fitting flexible	578	Cooling coil	560
aluminum	573	control system	531
concrete slab	576	tower	557
electrical	573	tower fiberglass	557
high installation	574	tower forced-draft	557
tower louver	546	sink	516
tower stainless	557	solid surface	443
Coping	113, 117	stainless steel	436, 442
aluminum	117	Course drainage layer base	653
clay tile	117	wearing	654
concrete	117	Court air supported	463
removal	96	handball	462
terra cotta	103, 117	paddle tennis	426
Copper bus duct	589	racquetball	462
cable	569, 573	squash	462
downspout	252	Cove base ceramic tile	335
drum trap	509	base terrazzo	348
DWV tubing	500	molding	199
flashing	249, 250, 509	Cover aluminum trench	159
gravel stop	251	board asphaltic	230
gutter	253	board gypsum	230
pipe	500	cast iron grate	159
reglet	254	checkered plate	159
rivet	125	column	161
roof	248	manhole	686
salvage	493	pool	458
shielded cable	569	stadium	464
wall covering	353	stair tread	24
wire	572	tank	463
Core drill	15, 728	trench	159
drill bit	90	walkway	399
drilling concrete	89	Covering wall	873
testing	15	CPVC pipe	504
Coreboard	331	valve	496
Cork floor	342	Crack cleaner	735
insulation	460	filler	736
tile	342	Crane	715
tile flooring	342	and bucket concrete	78
wall covering	352	bridge	714
wall tile	352	climbing	736
wallpaper	352	crawler	736
Corner base cabinet	439	crew daily	18
bead	326, 333	hydraulic	715, 737
block	109	jib	715
guard	129, 387	material handling	715
protection	387	mobilization	615
wall cabinet	439	overhead bridge	714
Cornerboard PVC	217	rail	714
Cornice drain	511	scale	402
molding	199, 202	tower	18, 736
painting	367	truck mounted	737
stone	114	Crash barrier	708
Corridor sign	378	Crawler crane	736
Corrosion resistance	676	drill rotary	732
resistant pipe	524	shovel	730
Corrugated metal pipe	684	Crematory	430
pipe	684	Crew daily crane	18
roof tile	235	forklift	18
siding	237-239	survey	25
steel pipe	684	Cribbing concrete	665
Cost control	14	Crimp termination cable	570
mark-up	13	Critical path schedule	14
Cot prison	448	Cross brace	130
Counter bank	411	wall	432
checkout	409	Crown molding	199, 206
door	280	Crushed stone	614
door coiling-type	280	stone sidewalk	652
flashing	254	CSPE roof	245
hospital	441	Cubicle	384
top	442	curtain	384
top demolition	170	detention	433
Countertop backsplash	442	shower	516
ceramic tile	443	toilet	382
engineered stone	444	track & hardware	384
granite	443	Cultured stone	117
laboratory	443	Culvert end	685
laminated	442	pipe	686
maple	442	reinforced	685
marble	443	Cupola	399
plastic-laminate-clad	442	Cupric oxychloride floor	349
postformed	443	Curb	657

906

and gutter 657
asphalt 657
builder 728
concrete 657
edging 129, 658
edging angle 129
edging channel 129
extruder 729
granite 113, 658
inlet 658
precast 657
removal 29
roof 186
seal 652
slipform paver 736
terrazzo 347, 348
Curing blanket 83
compound concrete 83
concrete 83, 655
concrete membrane 83
concrete paper 83
concrete water 83
paper 232
Current transformer 584
transformer cabinet 590
Curtain air 543
air barrier 543
blast 437
cubicle 384
damper fire 539
divider 425
gymnasium divider 425
rod 388
sound 472
stage 423
track 423
vinyl 388
wall 288
wall glass 288
wall glazed aluminum 288
Custom architectural door 273
casework 439
Cut stone curb 658
stone trim 114
Cutoff pile 614
Cutout counter 443
demolition 32
slab 32
Cutter brush 729, 730
Cutting asphalt 34
block 442
concrete floor saw 89
concrete slab 89
concrete wall saw 89
masonry 34
steel 122
torch 122, 734
wood 34
Cylinder concrete 15
lockset 302
recore 302, 303

D

Daily crane crew 18
Dairy case 414
Damper 539
assembly splitter 539
fire curtain 539
fireplace 392
foundation vent 99
multi-blade 539
volume-control 539
Dampproofing 225
Darkroom 461
door 282
equipment 410
lead lined 461
revolving 461
Dasher hockey 459
ice rink 459
Day gate 283
Daylighting sensor 584
Dead lock 303
Deadbolt 303, 305
Deadlocking latch 303
Deaerator 534
Deciduous shrub 672
Deck drain 509
edge form 143
framing 184, 215
metal floor 143
roof 190, 192
slab form 142
wood 190
Decking 865
cellular 143
composite 185, 218
floor 141-143
form 142
roof 142
woodgrained composite 218
Deconstruction building 35
building component 36
material handling 37
Decontamination asbestos area 43
chamber 41, 46
enclosure 41, 43, 46
equipment 41
Decorative beam 212
block 105
fence 663
Decorator device 579
switch 579
Deep freeze 412
seal trap 509
therapy room 472
Deep-longspan joist 138
Defibrillator cabinet 393
Dehumidifier 732
Delicatessen case 414
Delineator barrier 708
Delivery charge 22
Deluge sprinkler monitoring panel 488
valve assembly sprinkler 489
Demo air supported structure 454
casework 436
concrete 29
footing & foundation building 31
garden house 454
geodesic dome 454
greenhouse 454
hangar 454
lightning protection 454
pre-engineered steel building 455
roof window 265
silo 455
skylight 265
sound control 455
special purpose room 455
specialty 374
storage tank 456
swimming pool 456
tension structure 457
thermal & moisture protection 222
X-ray/radio freq protection 457
Demolish decking 167
remove pavement and curb 29
Demolition 839
air compressor 404
asbestos 42
bank equipment 404
barber equipment 404
baseboard 170
boiler 528
bowling alley 405
brick 96, 318
brick veneer 96
cabinet 170
casework 170
ceiling 318
central vacuum 404
ceramic tile 319
checkout counter 404
chimney 95
church equipment 404
column 96
commercial dishwasher 405
concrete 32, 50
concrete block 32
concrete wall 31
cooking equipment 405
crematory 404
darkroom equipment 405
dental equipment 405
detection equipment 404
disappearing stairway 405
dishwasher hood 405
disposal 32
door 264
drywall 318, 319
ductwork 528
electrical 567
equipment 404
explosive/implosive 31
fencing 30
fire protection system 405
fireplace 96
flashing 42
flooring 318
food case 404
food delivery cart 405
food dispensing equip 405
food preparation equip 405
food storage equipment 405
food storage shelving 405
framing 167
fume hood 405
garbage disposal 405
glass 264
granite 96
gutter 222
gutting 34
hammer 729
health club equipment 405
hood & ventilation equip 405
HVAC 528
hydraulic gate 404
ice machine 405
implosive 31
joist 167
lath 318
laundry equipment 404
library equipment 404
loading dock equipment 404
lubrication equip auto 404
masonry 32, 95, 319
medical equipment 405
metal 120
metal stud 319
millwork 170
millwork & trim 170
mold contaminated area 46
movable wall 319
movie equipment 404
paneling 170
parking equipment 404
partition 319
pavement 29
physical therapy 405
plaster 318, 319
plenum 319
plumbing 493
plywood 318, 319
post 168
rafter 168
railing 170
residential water heater 405
roofing 42, 223
saw cutting 34
selective building 31
sewage pumping system 404
sewage treatment 404
shooting range 405
site 29
sound system 404
spray paint booth 404
stage equipment 404
steel 120
steel window 264
stucco 319
subfloor 319
surgical light 405
terra cotta 320
terrazzo 319
tile 318
torch cutting 35
trim 170
truss 170
vault day gate 404
vault door and frame 404
vocational shop equip 405
wall 319
wall and partition 319
waste handling equipment 404
wastewater treatment 405
water softener 405
whirlpool bath 405
window 264
window & door 264
wood 318
wood framing 166
Demountable partition 385
Dental chair 428
equipment 428
metal interceptor 510
office equipment 428
Dentil crown molding 200
Depository night 411
Derail railroad 706
Derrick crane guyed 737
crane stiffleg 737
Desk 448
Detection intrusion 609
leak 610
probe leak 610
system 609
tape 677
Detector carbon monoxide 610
explosive 608
infrared 609
metal 608
motion 609
smoke 610
temperature rise 611
ultrasonic motion 609
Detention cubicle 433
equipment 433
Detour sign 24
Developing tank 410
Device & box wiring 575
alarm 611

For customer support on your Building Construction Costs with RSMeans data, call 800.448.8182.

anti-siphon 506
  combination 579
  decorator 579
  GFI 580
  receptacle 580
  residential 578
  wiring 575, 591
Dewater 626, 627
  pumping 627
Dewatering 627, 884
  equipment 738
Diamond lath 325
  plate tread 160
Diaper station 388
Diaphragm pump 734
Diesel generator 592
  generator set 592
  hammer 730
  tugboat 738
Diesel-engine-driven generator . . 592
Diffuser ceiling 544
  opposed blade damper 544
  perforated aluminum 544
  rectangular 545
  steel 545
  T-bar mount 545
Digital movie equipment 420
Dimensions and weights of
sheet steel 864
Dimmer switch 579, 591
Direct chute concrete 79
  expansion A/C 561
Directional drill horizontal 735
  sign 380
Directory 377
  board 377, 378
  building 377, 605
Disappearing stair 414
Disc harrow 729
Discharge hose 733
Dishwasher commercial 419
  residential 412
Diskette safe 410
Dispenser hot water 513
  napkin 389
  soap 389
  ticket 406
  toilet tissue 389
  towel 388
Dispensing equipment food 418
Dispersion nozzle 489
Displacement caisson 648
Display case 441
Disposable bag 41
Disposal 31
  asbestos 43
  cart 431
  charge 43
  field 682
  garbage 412, 419
  or salvage value 35
  waste 45, 431
Distiller medical 427
  water 427
Distribution box 683
  box concrete 683
  box flow leveler 683
  box HDPE 683
  pipe water 678, 679
  switchboard 586
Distributor asphalt 732
Ditching 627, 644
Divider curtain 425
  strip terrazzo 347
Diving board 458
  stand 458

Dock board 407
  board magnesium 407
  bumper 407
  equipment 407
  floating 712
  leveler 407
  light 408
  loading 407
  seal 407
  shelter 407
  truck 407
Dog command 24
Dome 401, 469
  drain 511
  geodesic 464
  security 401
  structures 875
Domed metal-framed skylight . . . 298
Door & window interior
paint 364, 365
  accessory 310
  accordion 275
  acoustical 283
  acoustical folding 386
  air lock 464
  aluminum 265, 284
  aluminum commercial 270
  aluminum double acting
  swing 285
  aluminum louver 316
  aluminum-framed 287
  aluminum-framed entrance . . . 287
  and grille coiling 280
  and panel access 278
  and window material 264
  balanced entrance 287
  bell residential 581
  bi-fold 269, 274
  bi-passing closet 274
  birch exterior 277
  birch hollow core 275
  birch solid core 274
  blind 214
  bulkhead 278
  bulkhead/cellar 278
  bullet resistant 411
  bumper 300, 304
  cafe 276
  casing PVC 217
  catch 440
  chain hoist 283
  cleanout 392
  closer 300
  closet 269, 274
  closing coordinator 305
  cold storage 281
  colonial 274, 277
  combination storm 274, 277
  commercial floor 278
  commercial half glass 268
  composite 268, 276
  conveyor 543
  counter 280
  custom architectural 273
  darkroom 282
  darkroom revolving 282
  demolition 264
  double 269
  double acting swing 285
  duct access 540
  dutch 274
  dutch oven 392
  electric automatic closer 301
  electro magnetic closer 300
  entrance 265, 274, 277, 286
  entrance strip 285

exterior 276
exterior pre-hung 277
fiberglass 276, 284
fire 270, 273, 281, 611
fire rated access 278
fireplace 392
flexible 285
floor 279
flush 271, 274
flush wood 271
folding 386
folding accordion 275
frame 266, 267
frame exterior wood 213
frame grout 98
frame guard 388
frame interior 213
frame lead lined 473
french exterior 276
garage 283, 284
glass 265, 279, 286, 287
hand carved 270
handle 440
hangar 470
hardware 299
hardware accessory 305
hollow metal 267
hollow metal exterior 269
industrial 280
interior pre-hung 277
jamb molding 207
kennel 279
kick plate 307
knob 303
labeled 268, 272
louver 315
louvered 275, 277
mahogany 270
metal 265, 269, 284, 467
metal access 278
metal fire 268
metal toilet component 382
mirror 313
molding 207
moulded 273
opener 264, 281, 284, 299
operator 264, 299, 300
overhead 283, 284
overhead commercial 283
panel 273, 274
paneled 269
partition 386
passage 275, 277
patio 280
plastic laminate 271
plastic-laminate toilet 383
prefinished 271
pre-hung 276, 277
protector 300
refrigerator 281
release 605
removal 264
residential 213, 269, 274
residential garage 284
residential steel 269
residential storm 265
revolving 282, 287, 464
revolving entrance 287
rolling 280, 288
roof 256
sauna 462
seal 407
sectional 283
sectional overhead 284
security vault 283
shower 390

sidelight 269, 276
sill 207, 213, 305
sliding aluminum 279
sliding glass 279, 280
sliding glass vinyl clad 279
sliding panel 288
sliding vinyl clad 279
special 278, 283
stain 363
stainless steel 277
stainless steel (access) 278
steel . . . 267, 268, 283, 284, 467, 868
steel louver 315
steel louver fire-rated 316
stop 286, 304
storm 265
swinging glass 287
switch burglar alarm 609
threshold 214
torrified 275
traffic 285
varnish 363
vault 283
vertical lift 284
weatherstrip 309
weatherstrip garage 309
wire partition 384
wood 270, 271, 273
wood ballistic 282
wood double acting swing 285
wood fire 272
wood panel 273
wood storm 277
Doorbell system 591
Dormer gable 187
Dormitory furniture 448
Double acting swing door 285
  hung aluminum sash window . 289
  hung bay wood window 293
  hung steel sash window 290
  hung window 294
  hung wood window 291
  precast concrete tee beam 85
  wall pipe 687
  weight hinge 306
Dovetail anchor 67
Dowel cap 72
  reinforcing 71
  sleeve 72
Downspout 251, 253, 399
  aluminum 251
  copper 252
  demolition 222
  elbow 252
  lead coated copper 252
  steel 252
  strainer 252
Dozer 730
  backfill 628
  excavation 624
Draft inducer breeching 548
Draft-induction fan 548
Dragline 623, 712
  bucket 729
Drain 508, 511
  board 225
  deck 509
  dome 511
  facility trench 511
  floor 509
  main 511
  pipe 627
  roof 511
  sanitary 509
  scupper 511
  sediment bucket 509

For customer support on your Building Construction Costs with RSMeans data, call 800.448.8182.

shower 508
trench 511
Drainage accessory 253
field 683
pipe 524, 682, 684, 685
trap 508, 509
Drapery hardware 437
Drawer kitchen 439
pass-thru 433
security 433
track 440
type cabinet 441
Dredge 712
hydraulic 712
Dressing unit 448
Drill auger 29
console dental 428
core 15, 728
earth 28, 646
hole 640
quarry 732
rig 28
rig mobilization 615
rock 28, 622
shop 422
steel 732
track 732
wood 172, 666
Drilled concrete pier and shaft 648
concrete pier uncased 648
pier 647
Drilling and blasting bulk 622
and blasting rock 622
concrete 90
concrete impact 90
horizontal 676
plaster 322
quarry 622
rig 728
rock bolt 622
steel 122
Drinking bubbler 523
fountain 522, 523
fountain deck rough-in 523
fountain floor rough-in 523
fountain support 520
fountain wall rough-in 522
Drip edge 255
edge aluminum 255
Dripproof motor 602
Driven pile 644
Driver post 736
sheeting 732
Drive-up window 411
Driveway 652
gate security 282
removal 29
security gate 282
security gate opener 283
security gate solar panel 283
Drum trap copper 509
Dry erase coating 371
fall painting 369
pipe sprinkler head 488
wall leaded 473
Drycleaner 409
Dryer commercial clothes 410
hand 389
industrial 409
receptacle 581
residential 413
vent 413
Dry-pipe sprinkler system 488
Dry-type transformer 585
Drywall 329
accessory 333
column 329
cutout 32
demolition 318, 319
finish 872
frame 266
gypsum 329, 332
gypsum high abuse 332
high abuse 332
nail 171
painting 367, 368
partition 322
partition NLB 321
prefinished 330
removal 319
screw 333
Duck tarpaulin 23
Duct accessory 539
connection fabric flexible 538
electric 577, 589
electrical underfloor 577
fire rated blanket 530
fitting trench 577
fitting underfloor 577
flexible 540
flexible insulated 540
flexible noninsulated 540
furnace 551
grille 352
heater electric 560
humidifier 564
HVAC 538
insulation 530
liner 541
liner non-fibrous 541
mechanical 538
silencer 540
steel trench 577
thermal insulation 530
trench 577
underfloor 577
underground 703
utility 703
Ductile iron fitting 679
iron fitting mech joint 679
iron grooved joint 502, 503
iron pipe 678
Ductwork 538
aluminum 538
demolition 528
fabric coated flexible 540
fabricated 538
fiberglass 539
galvanized 538
metal 538
rectangular 538
rigid 538
Dumbbell 423
waterstop 62
Dumbwaiter electric 476
manual 476
Dump charge 34
truck 731
Dumpster 33
Dumpsters 839
Dump truck off highway 731
Duplex receptacle 591
Dust barrier 23
collection system 718
collector shop 422
Dustproofing 81
compound 50
Dutch door 274
oven door 392
DWV pipe ABS 503
PVC pipe 504, 682
tubing copper 500

E

Earth auger 728
compactor 729
drill 28, 646
rolling 639
scraper 730
vibrator 619, 628, 639
Earthwork 617
equipment 628
equipment rental 728
haul 629
Eave overhang 467
Ecclesiastical equipment 432
Echo chamber 472
Economy brick 102
Edge drip 255
Edger concrete floor 728
Edging 673
aluminum 673
curb 129, 658
Educational casework 441
T.V. studio 609
Efficiency flush valve high 519
Efflorescence testing 15
Effluent-filter septic system 683
EIFS 231
Ejector pump 510
Elastomeric coating 370
liquid flooring 349
membrane 334
roof 247
sheet waterproofing 225
waterproofing 225
Elbow downspout 252
pipe 502
Electric appliance 412
backstop 424
ballast 595
baseboard heater 564
boiler 549
cabinet 590
cable 569, 570
capacitor 593
chalkboard 375
commercial water heater 513
duct 577, 589
duct heater 560
dumbwaiter 476
feeder 589
fixture 594
fixture exterior 598
furnace 551
generator 733
heater 413
heating 549, 564
hinge 306
hoist 715
lamp 599
log 392
metallic tubing 574
meter 584
motor 602
panelboard 587
pool heater 550
service 591
stair 414
switch 591, 592
traction freight elevator 476
utility 704
vehicle charging 593
water cooler 523
water heater residential 513
Electrical & telephone undrgrnd 703
box 575
cable 572, 573
conduit 573
demolition 567
fee 10
laboratory 421
metering smart 584
pole 703
tubing 573
wire 572
Electrically tinted window film 314
Electricity temporary 17
Electronic air cleaner 548
closer 301
control system 531
directory board 377
markerboard 376
Electrostatic painting 367
Elevated conduit 574
floor 75
installation add 492
pipe add 492
slab 75
slab formwork 54
slab reinforcing 71
water storage tank 681
water tank 681
Elevating scraper 625
Elevator 476, 876
cab finish 478
control 479
electric traction freight 476
fee 10
freight 476
hydraulic freight 477
option 479
passenger 476
residential 477
shaft wall 321
Embossed print door 273
Emergency call system 605
equipment laboratory 421
eyewash 522
lighting 596
lighting unit 596
shower 522
Employer liability 12
EMT 574
Emulsion adhesive 346
asphaltic 670
pavement 652
penetration 653
sprayer 732
Encapsulation asbestos 44
pipe 44
Encased railing 219
Encasement pile 645
Encasing steel beam formwork 52
Enclosed bus assembly 589
Enclosure acoustical 470, 472
bathtub 391
commercial telephone 380
decontamination 41, 43, 46
shower 390
swimming pool 465
telephone 380
End culvert 685
grain block floor 343
Endwall overhang 467
Energy circulator air solar 553
saving lighting device 584
system controller solar 553
system heat exchanger solar 553
Engineer brick 102
Engineered stone countertop 444
Engineering fee 10, 829
Engraved panel signage 378
Entrance aluminum 287

and storefront 287
canopy 399, 467
door 265, 274, 277, 286
door aluminum-framed 287
door fiberous glass 276
floor mat 445
frame 213
lock 305
screen 381
strip 285
Entry canopy 398
EPDM adhesive 225
flashing 250
roofing 245
Epoxy anchor 88, 121
coated reinforcing 72
coating 81, 371, 676
fiberglass wound pipe 524
floor 349, 350
grating 217
grout 334, 336, 642
lined silo 470
terrazzo 348, 349
wall coating 370
welded wire 73
Equipment 409, 712, 728
agricultural 433
athletic 424
automotive 406
bank 411
barber 409
basketball 424
cell 433
checkroom 409
cinema 420
cleaning & disposal 419
clock 606
darkroom 410
demolition 404
dental 428
dental office 428
detention 433
earthwork 628
ecclesiastical 432
exercise 424
fire hose 487
fire-extinguishing 489
food preparation 416
foundation formwork 55
gymnasium 423, 424
health club 423
hospital 427
ice skating 459
installation 405
insulation 531
insurance 12
laboratory 421, 441
laboratory emergency 421
laundry 409, 410
loading dock 407
lube 406
lubrication 406
medical 427-430
medical sterilizing 427
mobilization 648
pad 76
parking 406
parking collection 406
parking control 406
parking gate 406
parking ticket 406
playfield 425
playground 425, 426, 662
refrigerated storage 415
refrigeration 459
rental 728, 837, 838
rental concrete 728
rental earthwork 728
rental general 731
rental highway 735
rental lifting 736
rental marine 738
rental wellpoint 738
safety eye/face 522
security 433
security and vault 410
shop 422
stage 422
swimming pool 458
theater and stage 422
waste handling 430, 431
Erosion control synthetic 639
Escalator 480
Escape fire 154
Estimate electrical heating 564
Estimates window 868
Estimating 826
Evergreen shrub 672
tree 672
Exam equipment medical 427
light 427
Excavating bulk bank measure 623
bulk dozer 624
equipment 883
trench 618
utility trench 621
Excavation 618, 623, 882
boulder 623
bulk scraper 625
cofferdam 624
dozer 624
hand 619, 621, 622, 628
machine 628
planting pit 670
scraper 625
septic tank 683
structural 622
tractor 619
trench 618, 621
Excavator bucket 735
hydraulic 728
Exchanger heat 554
Exercise equipment 424
ladder 424
rope 424
weight 423
Exhaust hood 413, 421
vent 546
Exhauster roof 542
Exhibit case 441
Exit control alarm 609
device panic 301
light 596
lighting 596
Expansion anchor 121
joint 63, 255, 327, 534
joint assembly 262
joint bellow 534
joint butyl 255
joint floor 262
joint neoprene 255
joint roof 255
shield 121
tank 535
tank steel 535
Expense office 13
Explosion proof fixture 597
lighting 597
motor 602
Explosive 623
detection equipment 608
detector 608
Explosive/implosive demolition 31
Exposed aggregate 81, 652
aggregate coating 370
Extension backhoe 735
ladder 733
Exterior blind 214
concrete block 107
door 276
door frame 213
door hollow metal 269
fixture lighting 598
floodlamp 599
insulation 231
insulation finish system 231
LED fixture 598
lighting fixture 598
molding 200
moulding astragal 308
paint door & window 361
plaster 328
pre-hung door 277
PVC molding 217
residential door 274
roll-up shutter 398
shutter 214
siding painting 359
signage 378
steel coating 371
surface preparation 356
tile 335
trim 200
trim paint 362
wood door frame 213
wood frame 213
Extinguisher cart mounted 394
chemical dry 393
CO_2 394
fire 393, 489
FM200 fire 489
installation 394
portable ABC 394
pressurized fire 394
standard 393
wheeled ABC 394
Extra work 13
Extruder curb 729
Extrusion aluminum 135
Eye bolt screw 69
wash fountain 522
wash portable 522
Eye/face wash safety equipment 522
Eyewash emergency 522
safety equipment 522

F

Fabric awning 398
flashing 249
pillow water tank 682
stabilization 653
stile 308
structure 463
waterproofing 224
welded wire 72
wire 664
Fabricated ductwork 538
Fabrication composite 218
Fabric-backed flashing 250
Face brick 102, 858
brick cleaning 857
wash fountain 522
Facial scanner unit 608
Facility trench drain 511
Facing panel 115
stone 114
tile structural 103
Factor security 11
Fall painting dry 369
Fan air conditioning 541
bathroom exhaust 543
booster 542
ceiling 541
centrifugal type HVAC 542
coil air conditioning 561
draft-induction 548
HVAC axial flow 541
induced draft 548
in-line 542
kitchen exhaust 543
paddle 582
propeller exhaust 543
residential 582
roof 542
utility set 541, 542
vaneaxial 541
ventilation 582
wall 543
wall exhaust 542
wiring 582
Farm type siding 238
Fascia board demolition 167
metal 251
PVC 217
wood 186, 203
Fast food equipment 416
Fastener timber 171
wood 171
Fastening alternative 231
Faucet & fitting 517, 519
bathroom 517
laundry 517
lavatory 517
Fee architectural 10
engineering 10
Feeder electric 589
section branch circuit 586
section circuit breaker 586
section frame 586
section switchboard 586
stock 433
Feller buncher 729
Felt 244
asphalt 243, 244
carpet pad 350
tarred 244
waterproofing 224
Fence 358
aluminum 661, 662
and gate 661
board batten 664
cedar 664
chain link 24, 664
chain link industrial 661
chain link residential 661
decorative 663
helical topping 664
mesh 661
metal 661
misc metal 663
picket paint 358
plywood 24
security 664
snow 39
steel 661, 663
temporary 24
tennis court 662
treated pine 664
tubular 663
wire 24, 663
wood rail 664
wrought iron 661

For customer support on your Building Construction Costs with RSMeans data, call 800.448.8182.

Fencing demolition 30
wire 664
Fertilizer 671
Fiber cement siding 242
optic cable 604
reinforcing concrete 78
steel 74
synthetic 74
Fiberboard insulation 230
Fiberglass angle 216
area wall cap 400
bench 667
blown-in 229
casting 216
ceiling board 338
cooling tower 557
cross 432
door 276, 284
ductwork 539
flagpole 400
flat sheet 216
floor grating 217
formboard 59
grating 216
handrail 216
insulation 226, 227, 229, 468
panel 23, 238, 355
panel skylight 469
planter 667
reinforced ceiling 460
reinforced plastic panel 354
round bar 216
round tube 216
shade 438
single hung window 298
square bar 216
square tube 216
stair tread 216
steeple 399
tank 532
threaded rod 216
trash receptacle 451
trench drain 511
wall lamination 371
waterproofing 224
wide flange beam 216
window 298
wool 228
Field disposal 682
drainage 683
mix concrete 852
office 17
office expense 17
personnel 12
sample 47
seeding 671
Fieldstone 112
Fill 628, 629
by borrow 629
by borrow & utility bedding . . 629
floor 75
gravel 614, 629
Filler block 368
cabinet 440
crack 736
joint 260
strip 309
Fillet welding 123
Film bird control 314
equipment 410, 420
polyethylene 670
safety 314
security 314
Filter air 548
grille 545
mechanical media 547
stone 640
swimming pool 524
Filtration air 40
equipment 458
Fin tube radiation 562
Fine grade 671
Finish carpentry 439
concrete 655
floor 344, 369
grading 618
Keenes cement 327
lime 97
nail 171
refrigeration 460
wall 353
Finishing floor concrete 80
wall concrete 81
Fir column 212
floor 343
molding 206
roof deck 190
roof plank 190
Fire brick 116
call pullbox 611
clay 116
damper curtain type 539
door 270, 273, 281, 611
door frame 266
door metal 268
door wood 272
equipment cabinet 393
escape 154
escape balcony 154
escape stair 155
extinguisher 393, 489
extinguisher portable 393
extinguishing system 489
horn 611
hose 487
hose adapter 487
hose equipment 487
hose nozzle 487
hose rack 487
hose storage cabinet 393
hose valve 487
hydrant 681
hydrant building 486
hydrant remove 29
protection 393
pump 490
rated blanket duct 530
rated tile 103
resistant drywall 330, 331
resistant glass 311
resistant wall 321
retardant pipe 525
Firebrick 116
Fire-department connection 486
Fire-extinguishing equipment . . . 489
Fireplace accessory 391
box 117
built-in 392
chimney 391
damper 392
demolition 96
door 392
form 391
free standing 392
mantel 212
mantel beam 212
masonry 117
prefabricated 391, 392
Fireproofing 257
plaster 257
plastic 258
sprayed cementitious 257
Fire-rated door steel louver 316
Firestop wood 183
Firestopping 258, 867
Fire-suppression automatic 489
Fitting bus duct 589
ductile iron 679
grooved joint 501
grooved joint pipe 502
poke-thru 574
PVC 680
underfloor duct 578
waterstop 63
Fixed blade louver 546
booth 447
chalkboard 374
end caisson pile 646
roof tank 533
tackboard 376
vehicle delineator 708
Fixture ballast 595
bathroom 515, 516
carrier/support 520
electric 594
explosion proof 597
fluorescent 594
incandescent 595
incandescent vaportight 595
interior light 594
plumbing . . 510, 515, 522, 536, 678
removal 493
residential 581
sodium high pressure 599
vandalproof 595
Flagging 342, 657
slate 657
Flagpole 400
aluminum 400
fiberglass 400
foundation 401
steel internal halyard 400
structure mounted 401
wall 401
wood 400
Flange beam fiberglass wide 216
Flanged coil 560
Flasher barrels 732
barricade 732
Flashing 249-251
aluminum 239, 249, 509
asphalt 249
butyl 250
cement 222
copper 249, 250, 509
counter 254
demolition 42
EPDM 250
fabric 249
fabric-backed 250
laminated sheet 250
masonry 249
mastic-backed 250
membrane 244
metal 468
neoprene 250
paperbacked 250
plastic sheet 250
PVC 250
self-adhering 251
sheet metal 249
stainless 249
valley 233
vent 509
Flat seam sheet metal roofing . . . 248
sheet fiberglass 216
Flatbed truck 731, 735
truck crane 736
Flexible conduit & fitting 578
connector 505
door 285
duct 540
duct connection fabric 538
ductwork fabric coated 540
insulated duct 540
metal hose 505
noninsulated duct 540
plastic netting 401
sign 378
Float finish concrete 80, 81
glass 310
Floater equipment 12
Floating dock 712
floor 341
pin 306
roof tank 533
wood pier 712
Flood coating 244
Floodlamp exterior 599
Floodlight LED 599
pole mounted 599
trailer 732
tripod 732
Floor 342
abrasive 80
access 352
acid proof 342
acrylic 349
adhesive removal 82
asphalt block 656
athletic resilient 347
brick 342, 656
carpet 351
ceramic tile 335
check 305
cleaning 25
cleanout 498
closer 300
color 81
composition 350
concrete finishing 80
conductive 347, 349
cork 342
decking 141-143
door 279
door commercial 278
drain 509
elevated 75
end grain block 343
epoxy 349, 350
expansion joint 262
fill 75
fill lightweight 76
finish 369
flagging 657
framing removal 32
grating 217
grating aluminum 157
grating fiberglass 217
grating steel 157
grinding 82
hardener 50
hatch 279
heating radiant 562
honing 82
insulation 226
marble 114
mastic 349
mat 445
mat entrance 445
nail 171
neoprene 349
oak 343
paint 366

paint & coating interior 365
paint removal 82
parquet 343
pedestal 352
plank 189
plate stair 154
plywood 190
polyacrylate 348
polyester 350
polyethylene 351
polyurethane 349, 350
portable 422
quarry tile 336
refrigeration 460
register 545
removal 31
rubber 346, 660
rubber and vinyl sheet 346
sander 734
scupper 511
sealer 51
slate 115
sleeper 186
stain 366
subfloor 190
terrazzo 347, 873
tile 348
tile or terrazzo base 347
tile terrazzo 348
topping 81, 87
transition strip 350
treatment anti-slip 366
underlayment 191
varnish 366, 369
vinyl 346
vinyl sheet 346
wood 343
wood athletic 344
wood composition 343
Flooring bamboo 341
composition 343, 349
conductive rubber & vinyl 347
demolition 318
elastomeric liquid 349
masonry 342
miscellaneous brick 342
quartz 350
treatment 341
wood strip 343
Flow meter 536
Flue chimney 116
chimney metal 548
liner 101, 116
prefab metal 548
screen 391
tile 116
Fluid applied membrane
air barrier 233
heat transfer 553
Fluorescent fixture 594
Fluoroscopy room 472
Flush automatic 519
bolt 299
door 271, 274
high efficiency toilet (het) 519
metal door 269
tube framing 285
valve 519
wood door 271
Flying truss shoring 61
FM200 fire extinguisher 489
Foam board insulation 226
core DWV ABS pipe 504
insulation 229
pipe covering 499
roofing 247

spray rig 732
Foamed coated roofing 247
in place insulation 228
Foam-water system component .. 489
Fog seal 652
Foil aluminum 229, 232, 356
metallic 229
Folder laundry 409
Folding accordion door 275
accordion partition 386
bench 450
door 386
door shower 390
gate 385
table 447
Food delivery cart 417
dispensing equipment 418
mixer 416
preparation equipment 416
service equipment 417
storage metal shelving 416
warmer 418
Foot valve 496
Football goalpost 427
scoreboard 424
Footing concrete placement 79
formwork continuous 55
formwork spread 56
keyway 55
keyway formwork 55
reinforcing 71
removal 31
spread 76
Forced-draft cooling tower 557
Forest stewardship council 180
Forged valve 529
Forklift 732
crew 18
Form deck edge 143
decking 142
fireplace 391
liner concrete 58
material 842
release agent 51
Formblock 105
Formboard fiberglass 59
roof deck 59
wood fiber 59
Forms 842
concrete 842
Formwork beam and girder 52
beam bottom 52
beam side 52
box culvert 54
column 53
column accessory 60
column plywood 53
column round fiber tube 53
column round fiberglass 53
column round steel 53
column steel framed 54
continuous footing 55
elevated slab 54
encasing steel beam 52
equipment foundation 55
footing keyway 55
gang wall 58
gas station 56
gas station island 56
girder 52
grade beam 56
hanger accessory 60
insert concrete 67
insulating concrete 59
interior beam 52
labor hours 843

light base 56
mat foundation 56
oil 62
patch 62
pile cap 55
plywood 52, 54
radial wall 57
retaining wall 58
shoring concrete 61
side beam 53
sign base 56
slab blockout 56
slab box out opening 55
slab bulkhead 55, 56
slab curb 55, 56
slab edge 55, 56
slab flat plate 54
slab haunch 76
slab joist dome 54
slab joist pan 54
slab on grade 56
slab screed 57
slab thickened edge 76
slab trench 57
slab turndown 76
slab void 55, 57
slab with drop panel 54
sleeve 61
snap-tie 61
spandrel beam 52
spread footing 56
stair 59
stake 62
steel framed plywood 58
upstanding beam 53
wall 57
wall accessory 62
wall boxout 57
wall brick shelf 57
wall bulkhead 57
wall buttress 57
wall corbel 57
wall lintel 58
wall pilaster 58
wall plywood 57
wall prefab plywood 58
wall sill 58
wall steel framed 58
Foundation caisson 646
chimney 75
concrete block 107
flagpole 401
mat 76
mat concrete placement 79
pile 648
scale pit 402
underpin 643
vent 99
wall 107
Fountain 459
drinking 522, 523
eye wash 522
face wash 522
indoor 459
lighting 524
outdoor 459
pump 524
wash 519
water pump 524
yard 459
Frame baked enamel 267
door 266, 267
drywall 266
entrance 213
exterior wood door 213
fire door 266

grating 159
labeled 266
metal 266
metal butt 267
scaffold 19
shoring 61
steel 266
trench grating 159
welded 266
window 292
Framing aluminum 135
anchor 172
band joist 148
beam & girder 181
boxed header/beam 148
bracing 151
bridging 151
canopy 127
ceiling 182
column 182
deck 184, 215
demolition 167
heavy 189
joist 182
laminated 193
ledger 186
lightweight 129
lightweight angle 129
lightweight channel 130
lightweight junior beam 130
lightweight tee 130
lightweight zee 130
load-bearing stud partition 145
load-bearing stud wall 145
metal joist 149
metal roof parapet 151
miscellaneous wood 183
NLB partition 323
open web joist wood 182
opening 467
pipe support 131
porch 184
removal 166
roof 186
roof metal rafter 151
roof metal truss 153
roof rafter 185
roof soffit 152
roof truss 193
sill & ledger 186
sleeper 186
slotted channel 130
soffit & canopy 186
stair stringer 183
steel 133
steel tee 130
steel zee 130
suspended ceiling 182
timber 189
treated lumber 187
tube 285
wall 187
web stiffener 149
window 467
window wall 285
wood 180
wood beam & girder 181
wood joist 182
wood plate 187
wood sleeper 186
wood soffit & canopy 187
wood stair stringer 183
wood stub wall 187
Freestanding chalkboard 376
Freezer deep 412
Freezer 414, 415, 461

For customer support on your Building Construction Costs with RSMeans data, call 800.448.8182.

Freight elevator 476
   elevator hydraulic 477
   elevators 876
French exterior door 276
Friction pile 645, 648
Frieze PVC 218
Front end loader 623
   vault 283
FRP panel 354
Fryer 417
Fuel storage tank 533
   tank 532
Full vision glass 311
Fume hood 421
Furnace duct 551
   electric 551
   gas 551
   gas fired 551
   hot air 551
   oil fired 551
   wall 552
Furnishing library 449
   site 451
Furniture 446
   dormitory 448
   hospital 449
   hotel 447
   library 449
   office 446, 447
   restaurant 447
   school 448
Furring and lathing 322
   ceiling 188, 322, 325
   channel 322, 333
   concrete 187
   masonry 187
   metal 322
   steel 322
   wall 187, 323
   wood 187
Fusible link closer 300
   switch branch circuit 586
Fusion machine butt 732

G

Gabion box 640
   retaining wall stone 666
Gable dormer 187
Galley septic 683
Galvanized ductwork 538
   plank grating 158
   reinforcing 72
   roof 237
   steel reglet 254
   welded wire 73
Galvanizing 371
   lintel 131
   metal in field 371
   metal in shop 120
Gang wall formwork 58
Gantry crane 715
Garage 469
   door 283, 284
   door residential 284
   door weatherstrip 309
   public parking 469
   residential 469
Garbage disposal 412, 419
Garden house 470
   house demo 454
Garment checking bag 409
Gas cock 493
   connector 505
   fired boiler 549

fired furnace 551
fired infrared heater 551
fired space heater 552
furnace 551
generator set 592
heat air conditioner 558
incinerator 430
log 392
pipe 688
station formwork 56
station island formwork 56
station tank 532
stop valve 493
vent 548
water heater commercial 514
water heater instantaneous ... 513
water heater residential 513
water heater tankless 513
Gas-fired duct heater 551
   heating 551, 552
Gasket joint 260
   neoprene 260
Gas/oil combination boiler 550
Gasoline generator 593
   piping 687
   tank 533
Gate chain link 662
   day 283
   fence 661
   folding 385
   opener driveway security 283
   parking 406
   security 385
   security driveway 282
   slide 661
   swing 661
   valve 528, 529
   valve soldered 494
Gauging plaster 327
General air conditioner 528
   boiler 492
   contractor's overhead 833
   equipment rental 731
   fill 629
Generator diesel 592
   diesel-engine-driven 592
   electric 733
   emergency 592
   gasoline 593
   natural gas 592
   set 592
   steam 427
Geodesic dome 464
   dome demo 454
   hemisphere greenhouse 465
Geo-grid 665
GFI receptacle 580
Girder formwork 52
   reinforcing 71
   wood 181, 189
Girt steel 134
Glass 310, 312, 313
   acoustical 315
   and glazing 315
   bead 285
   bevel 311
   block 110
   bulletin board 376
   bulletproof 315
   curtain wall 288
   demolition 264
   door 265, 279, 286, 287
   door astragal 309
   door shower 390
   door sliding 280
   door swinging 287

fiber rod reinforcing 72
fire resistant 311
float 310
full vision 311
heat reflective 312
insulating 311
laminated 315
lead 473
lined water heater 413
low emissivity 311
mirror 313, 389
mosaic 336
mosaic sheet 336
obscure 312
patterned 312
pipe 524
reduce heat transfer 311
reflective 313
safety 311
sandblast 311
sheet 312
shower stall 390
solar film 314
spandrel 312
tempered 310
tile 313
tinted 310
window 312
window wall 286
wire 312
Glassware sterilizer 421
   washer 421
Glaze coat 367
Glazed aluminum curtain wall ... 288
   block 109
   brick 102
   ceramic tile 334
   concrete block 109
   wall coating 370
Glazing 310, 311
   application 311
   plastic 313
   polycarbonate 313
   productivity 870
Globe valve 529
   valve bronze 495
Glove bag 42
   box 421
Glued laminated 193
Goalpost 427
   football 427
   soccer 427
Golf shelter 426
   tee surface 351
Gore line 659
Grab bar 389
Gradall 623, 729
Grade beam concrete placement . 79
   beam formwork 56
   fine 671
Grader motorized 729
Grading 617, 635
   rough 617
   site 617
Graffiti chemical removal 370
   resistant treatment 369
Grandstand 465, 466
Granite 112
   building 112
   chip 670
   conductive floor 349
   countertop 443
   curb 113, 658
   demolition 96
   Indian 658
   paver 113

paving block 657
reclaimed or antique 113
sidewalk 657
Granolithic finish concrete 87
Grass cloth wallpaper 353
   lawn 671
   seed 671
   surfacing artificial 660
Grating aluminum 157
   aluminum floor 157
   aluminum mesh 159
   aluminum plank 157
   area wall 400
   area way 400
   fiberglass 216
   fiberglass floor 217
   floor 217
   frame 159
   galvanized plank 158
   plank 158
   stainless bar 158
   stainless plank 158
   stair 154
   steel 157
   steel floor 157
   steel mesh 159
Gravel base 653
   fill 614, 629
   pack well 678
   pea 670
   roof 51
   stop 251
Gravity retaining wall 665
Grease interceptor 510
   trap 510
Green roof membrane 236
   roof soil mixture 236
   roof system 236
Greenhouse 465
   air supported 463
   commercial 465
   cooling 465
   demo 454
   geodesic hemisphere 465
   residential 465
Grid spike 172
Griddle 417
Grill coiling 280
Grille air return 545
   aluminum 545
   decorative wood 212
   duct 352
   filter 545
   painting 366
   plastic 545
   side coiling 280
   top coiling 280
   window 294
Grinder concrete floor 728
   shop 422
   system pump 510
Grinding concrete 82
   floor 82
Grooved block 105
   joint ductile iron 502, 503
   joint fitting 501
   joint pipe 501
Ground 188
   box hydrant 507
   clamp water pipe 573
   cover plant 672
   face block 105
   fault protection 584
   rod 573
   socket 425
   water monitoring 16

For customer support on your Building Construction Costs with RSMeans data, call 800.448.8182.

wire 573
Grounding 573
& conductor 569-572
wire brazed 573
Group shower 519
wash fountain 519
Grout 97
cavity wall 98
cement 98, 642
concrete 98
concrete block 98
door frame 98
epoxy 334, 336, 642
metallic non-shrink 88
non-metallic non-shrink 88
pump 734
tile 334
topping 88
wall 98
Grouted bar 73
strand 73
Guard corner 129, 387
gutter 254
house 469
lamp 600
service 24
snow 256
wall 129, 388
wall & corner 387
window 162
Guardrail scaffold 19
temporary 24
Guide rail 708
rail cable 708
rail removal 30
rail timber 708
rail vehicle 708
Guide/guard rail 708
Gunite 82
drymix 82
mesh 82
Gutter 253, 468
aluminum 253
copper 253
demolition 222
guard 254
lead coated copper 253
monolithic 657
stainless 253
steel 253
strainer 254
valley 468
vinyl 253
wood 253
Gutting 34
demolition 34
Guyed derrick crane 737
tower 704
Gym floor underlayment 344
mat 424
Gymnasium divider curtain 425
equipment 423, 424
floor 344, 351
Gypsum block demolition 32
board 329
board accessory 333
board high abuse 332
board leaded 473
board partition 322
board partition NLB 321
board removal 319
board system 320
cover board 230
drywall 329, 332
fabric wallcovering 353
high abuse drywall 332
lath 325
lath nail 171
partition 385
partition NLB 322
plaster 327
restoration 318
roof deck 87
shaft wall 321
sheathing 193
sound dampening panel 331
underlayment poured 88
wallboard repair 318
weatherproof 193

H

H pile 644
Hair interceptor 510
Half round molding 206
Halotron 394
Hammer bush 81
chipping 732
demolition 729
diesel 730
drill rotary 732
hydraulic 729, 733
pile 729
pile mobilization 615
vibratory 730
Hammermill 735
Hand carved door 270
clearing 616
dryer 389
excavation 619, 621, 622, 628
hole 677
hole concrete 703
scanner unit 608
split shake 235
trowel finish concrete 80
Handball court 462
court backstop 426
Handball/squash court 426
Handicap lever 305
opener 299
ramp 76
Handle door 440
Handling material 25, 715
waste 430, 431
Hand/off/automatic control 592
Handrail 388
aluminum 352
fiberglass 216
wood 206, 211
Hangar 470
demo 454
Hanger accessory formwork 60
beam 172
joist 172
Hanging lintel 130
wire 339
Hardboard cabinet 439
molding 208
overhead door 284
paneling 208
tempered 208
underlayment 191
Hardener concrete 80
floor 50
Hardware 299
apartment 299
cabinet 440
cloth 664
door 299
drapery 437
finishes 869
motel/hotel 299
window 299
Hardwood carrel 449
floor stage 422
grille 212
Harness body 21
Harrow disc 729
Hasp 306
Hat and coat strip 389
rack 397
Hatch floor 279
roof 256
smoke 256
Haul earthwork 629
Hauling 629
cycle 629
Haunch slab 76
Hay 670
bale 639
Hazardous waste cleanup 40
waste containment 38
waste disposal 40
waste handling 431
HDPE 38
distribution box 683
infiltration chamber 683
Head sprinkler 488
Header load-bearing stud wall 144
pipe wellpoint 738
wood 187
Header/beam boxed 148
Headrail metal toilet component 383
plastic-laminate toilet 383
Health club equipment 423
Hearth 117
Heat baseboard 561
electric baseboard 564
exchanger 554
exchanger plate-type 554
exchanger shell-type 554
greenhouse 465
pump 560
pump air-source 560
pump residential 583
recovery 430
reflective glass 312
temporary 52
therapy 428
transfer fluid 553
transfer package 514
Heated pad 52
Heater & fan bathroom 543
electric 413
floor mounted space 552
gas fired infrared 551, 552
gas residential water 513
gas water 514
gas-fired duct 551
infrared 551
oil fired space 552
sauna 462
space 733
swimming pool 550
terminal 551
tubular infrared 552
unit 552, 561
warm air 551
water 413, 583
water residential electric 513
Heating & cooling classroom 561
control valve 535
electric 549, 564
estimate electrical 564
gas-fired 551, 552
hot air 551, 561
hydronic 550, 561
insulation 499, 531
kettle 736
solar 553
steam-to-water 554
subsoil 459
Heavy construction 668
duty shoring 19
framing 189
lifting 837
rail track 706
timber 189
Helicopter 737
Hemlock column 212
Hex bolt steel 123
Hexagonal face block 106
High abuse drywall 332
build coating 371
chair reinforcing 68
efficiency flush valve 519
efficiency toilet (het) flush 519
efficiency urinal 519
installation conduit 574
pressure HDPE, ASTM D2513 696
rib lath 325
strength block 107
strength bolt 124, 859
strength decorative block 105
strength steel 134, 861
High-security chain link fence 662
Highway equipment rental 735
paver 657
sign 380
Hinge 306
brass 306
cabinet 440
continuous 307
electric 306
hospital 307
paumelle 306
prison 306
residential 306
school 307
security 306
special 306
stainless steel 306
steel 306
wide throw 307
Hip rafter 186
Hockey dasher 459
scoreboard 424
Hoist 714
air powered 715
chain 715
chain fall 737
electric 715
overhead 715
personnel 737
tower 737
Holder closer 300
screed 69
Holding tank 722
Holdown 172
Hole drill 640
Hollow core door 271, 275
metal door 267
metal frame 266
metal stud partition 326
precast concrete plank 83
wall anchor 121
Honed block 107
Honing floor 82
Hood and ventilation equip. 547
exhaust 413, 421
fume 421
range 413
Hook clevis 162

For customer support on your Building Construction Costs with RSMeans data, call 800.448.8182.

coat 389
robe 389
Hopper refuse 482
Horizontal borer 735
boring 676
boring machine 728
directional drill 735
drilling 676
Horn fire 611
Hose adapter fire 487
air 732
bibb sillcock 517
braided bronze 505
discharge 733
equipment 487
fire 487
metal flexible 505
nozzle 487
rack 487
rack cabinet 393
suction 733
valve cabinet 393
water 733
Hospital cabinet 441, 442
casework 441
door hardware 299
equipment 427
furniture 449
hinge 307
kitchen equipment 418
partition 384
tip pin 306
Hot air furnace 551
air heating 551, 561
water boiler 549, 550
water dispenser 513
water heating 549, 561
water-steam exchange 514, 554
water-water exchange 554
Hotel cabinet 390
furniture 447
lockset 302, 303
House garden 470
guard 469
safety flood shut-off 495
telephone 605
Housewrap 232
Hubbard tank 428
Humidification equipment 465
Humidifier 414, 564
duct 564
room 564
Humus peat 670
HVAC axial flow fan 541
demolition 528
duct 538
equipment insulation 531
louver 546
piping specialty 534, 536
power circulator & ventilator 543
Hydrant building fire 486
fire 681
ground box 507
removal 29
remove fire 29
wall 507
water 507
Hydrated lime 97
Hydraulic chair 428
crane 715, 737
dredge 712
excavator 728
hammer 729, 733
jack 737
jack climbing 737
jacking 837
lift 481
seeding 671
Hydrodemolition 29
concrete 29
Hydromulcher 735
Hydronic heating 550, 561
heating control valve 535
heating convector 561
Hypalon neoprene roofing 247

I

Ice machine 419
rink dasher 459
skating equipment 459
Icemaker 412, 419
I-joist wood & composite 193
Impact barrier 708
wrench 732
Impact-resistant aluminum window 289
Implosive demolition 31
Incandescent fixture 595
Incinerator gas 430
municipal 430
waste 430
Inclined ramp 481
Incubator 421
Indian granite 658
Indicating panel burglar alarm 609
Indirect-fired water chiller 555
Indoor athletic carpet 351
fountain 459
Induced draft fan 548
Industrial address system 605
chain link gate 663
chimneys 846
door 278, 280
dryer 409
equipment installation 405
folding partition 386
lighting 594
railing 156
safety fixture 522
window 290
Inert gas 39
Infiltration chamber 683
chamber HDPE 683
Infrared broiler 417
detector 609
heater 551
heater gas fired 551, 552
heater tubular 552
Inlet curb 658
In-line fan 542
Insecticide 639
Insert concrete formwork 67
slab lifting 69
Inspection technician 16
Installation add elevated 492
extinguisher 394
Instantaneous gas water heater 513
Instrument switchboard 584
Insulated glass spandrel 312
panel 231
precast wall panel 86
Insulating concrete formwork 59
glass 311
sheathing 190
Insulation 226, 229
batt 227
blanket 530
blower 733
blown-in 229
board 226
boiler 531
breeching 531
building 229
cavity wall 228
ceiling 226, 229
cellulose 228
composite 231
duct 530
duct thermal 530
equipment 531
exterior 231
fiberglass 226, 227, 229, 468
finish system exterior 231
floor 226
foam 229
foamed in place 228
heating 499, 531
HVAC equipment 531
insert concrete block 105
isocyanurate 226
loose fill 228
masonry 229
mineral fiber 228
pipe 499
polystyrene 226, 228, 460
reflective 229
refrigeration 460
removal 42, 223
rigid 226
roof 230, 468
roof deck 230
shingle 233
spray 229
sprayed-on 229
subsoil 459
vapor barrier 232
vermiculite 228
wall 226, 468
wall blanket 227
Insurance 12, 832
builder risk 12
equipment 12
public liability 12
Intake-exhaust louver 546
vent 547
Integral color concrete 50
topping concrete 87
waterproofing 52
waterproofing concrete 51
Interceptor 510
grease 510
hair 510
metal recovery 510
oil 510
Intercom 605
Interior beam formwork 52
door frame 213
LED fixture 595
light fixture 594
lighting 594, 595
paint door & window 363
paint wall & ceiling 368
planter 450
pre-hung door 277
residential door 274
shutter 438
wood door frame 213
Interlocking concrete block 107
precast concrete unit 656
Intermediate metallic conduit 574
Internal chain link gate 663
Interval timer 579
Intrusion detection 609
system 609
Intumescent coating 370
Invert manhole 687
Inverted bucket steam trap 535
Iron alloy mechanical joint pipe 524
body valve 528
Ironer laundry 410
Ironing center 391
Ironspot brick 342
Irrigation system sprinkler 669
Isocyanurate insulation 226
IV track system 384

J

Jack cable 738
hydraulic 737
ladder 21
post adjustable 128
pump 19
roof 186
screw 641
Jackhammer 623, 732
Jacking 676
Jet water system 678
Jeweler safe 411
Jib crane 715
Job condition 11
Jockey pump fire 490
Joint assembly expansion 262
control 63, 99
expansion 63, 255, 327, 534
filler 260
gasket 260
push-on 678
reinforcing 98
roof 255
sealant replacement 222
sealer 260, 261
Jointer shop 422
Joist bridging 140
concrete 75
connector 172
deep-longspan 138
demolition 167
framing 182
hanger 172
longspan 139
metal framing 149
open web bar 140
precast concrete 85
removal 167
truss 141
web stiffener 149
wood 182, 193
Jumbo brick 102
Junction box 574
Jute mesh 639

K

Keenes cement 327
Kennel door 279
fence 664
Ketone ethylene ester roofing 246
Kettle 417
heating 736
tar 734, 736
Key cabinet 310
keeper 396
Keyless lock 302
Keyway footing 55
Kick plate 307
plate door 307
Kiln vocational 422
King brick 102
Kiosk 470

For customer support on your Building Construction Costs with RSMeans data, call 800.448.8182.

Kitchen appliance 412
cabinet 439
equipment 417
equipment fee 10
exhaust fan 543
metal cabinet 442
sink 516
sink faucet 517
sink residential 516
unit commercial 414
K-lath 326
Knob door 303
Kraft paper 232

L

Labeled door 268, 272
frame 266
Labor adjustment factors 492, 566
formwork 843
modifier 492
Laboratory analytical service 47
cabinet 441
casework metal 441
countertop 443
equipment 421, 441
safety equipment 421
sink 421
table 421
test equipment 421
Ladder alternating tread 155
exercise 424
extension 733
inclined metal 155
jack 21
monkey 425
rolling 20
ship 155
swimming pool 458
towel 389
type cable tray 576
vertical metal 155
Lag screw 125
screw shield 121
Lagging 643
Lally column 128
Laminated beam 193
countertop 442
epoxy & fiberglass 371
framing 193
glass 315
glued 193
lead 473
roof deck 190
sheet flashing 250
veneer member 194
wood 193
Lamp electric 599
guard 600
LED 600
metal halide 600
post 163, 597
Lampholder 591
Lamphouse 420
Lance concrete 733
Landfill compactor 730
Landing metal pan 154
stair 348
Landscape fee 10
surface 351
Laser level 733
Latch deadlocking 303
set 302, 303
Lateral arm awning retractable 398
Latex caulking 261
underlayment 344
Lath demolition 318
gypsum 325
metal 320, 325
Lath, plaster and gypsum board 871
Lath rib 325
Lathe shop 422
Lattice molding 206
Lauan door 275
Laundry equipment 409, 410
faucet 517
folder 409
ironer 410
presser 410
sink 517
spreader 409
Lava stone 113
Lavatory commercial 518
faucet 517
pedestal type 515
removal 493
residential 515
sink 515
support 520
vanity top 515
wall hung 515
Lawn grass 671
mower 733
seed 671
Lazy Susan 439
Leaching field chamber 683
pit 683
Lead barrier 356
caulking 508
coated copper downspout 252
coated copper gutter 253
coated downspout 252
flashing 249
glass 473
gypsum board 473
lined darkroom 461
lined door frame 473
paint encapsulation 44
paint remediation 44
paint remediation methods 841
paint removal 45
plastic 473
roof 248
salvage 493
screw anchor 122
sheet 472
shielding 473
testing 44
Leads pile 729
Leak detection 610
detection probe 610
detection tank 610
Lean-to type greenhouse 465
Lectern 432
Lecture hall seating 450
LED fixture exterior 598
fixture interior 595
floodlight 599
lamp 600
lighting parking 598
luminaire roadway 598
Ledger framing 186
wood 186
Lens movie 420
Let-in bracing 180
Letter sign 378
slot 396
Level laser 733
Leveler dock 407
truck 407
Leveling jack shoring 19
Lever handicap 305
Lexan 313
Liability employer 12
insurance 833
Library equipment demolition 404
furnishing 449
furniture 449
shelf 449
Life guard chair 458
Lift 481
automotive 481
bariatric 428
hydraulic 481
scissor 715, 731
slab 77, 852
wheelchair 481
Lifter platform 407
Lifting equipment rental 736
Light base formwork 56
border 597
dental 428
dock 408
exam 427
exit 596
fixture interior 594, 597
fixture troffer 595
loading dock 408
nurse call 606
pole 703
pole aluminum 597
pole steel 598
shield 708
stand 40
strobe 597
support 128
temporary 17
tower 733
underwater 458
Lighting darkroom 410
emergency 596
exit 596
explosion proof 597
exterior fixture 598
fixture exterior 598
fountain 524
incandescent 595
industrial 594
interior 594, 595
outlet 581
pole 597
residential 581
stage 597
strip 594
surgical 427
temporary 17
theatrical 597
unit emergency 596
Lightning protection demo 454
suppressor 578
Lightweight aggregate 51
angle framing 129
block 108
channel framing 130
column 128
concrete 75, 78, 856
floor fill 76
framing 129
insulating concrete 87
junior beam framing 130
natural stone 113
tee framing 130
Zee framing 130
Lime 97
finish 97
hydrated 97
Limestone 113, 671
coping 117
Line gore 659
remover traffic 736
Linen chute 482
collector 482
wallcovering 353
Liner cement 676
duct 541
flue 101, 116
non-fibrous duct 541
pipe 676
Link transom chain 663
Lint collector 409
Lintel 113
block 108
concrete block 108
galvanizing 131
hanging 130
precast concrete 86
steel 130
Liquid chiller centrifugal 555
chiller screw 556
Load test pile 614
Load-bearing stud partition framing 145
stud wall bracing 143
stud wall bridging 144
stud wall framing 145
stud wall header 144
Loader front end 623
skidsteer 731
tractor 730
vacuum 41
wheeled 730
windrow 736
Loading dock 407
dock equipment 407
dock light 408
Loam 614, 617
Lobby collection box 396
Lock electric release 302
entrance 305
keyless 302
time 283
tubular 303
Locker metal 394
plastic 395
steel 394
wall mount 395
wire mesh 394
Locking receptacle 581
Lockset communicating 302, 303
cylinder 302
hotel 302, 303
mortise 303
Locksets anti-ligature 304
Locomotive tunnel 736
Log chipper 730
electric 392
gas 392
skidder 730
Longspan joist 139
Loose fill insulation 228
Louver 315
aluminum 315, 546
aluminum operating 546
coating 546
cooling tower 546
door 315
fixed blade 546
HVAC 546
intake-exhaust 546
midget 315
mullion type 546
redwood 316
ventilation 316

For customer support on your Building Construction Costs with RSMeans data, call 800.448.8182.

wall ... 316
wood ... 212
Louvered door ... 275, 277
Lowbed trailer ... 736
Low-voltage silicon rectifier ... 591
switching ... 591
switchplate ... 591
transformer ... 591
Lube equipment ... 406
Lubrication equipment ... 406
Lug terminal ... 570
Lumber ... 181
core paneling ... 208
plastic ... 215
product prices ... 866
recycled plastic ... 214
structural plastic ... 215
treated ... 186
Luminaire roadway ... 598
Luminous ceiling ... 338, 599

M

Macadam ... 653
penetration ... 653
Machine auto-scrub ... 732
excavation ... 628
screw ... 125
trowel finish concrete ... 80
welding ... 735
X-ray ... 608
Machinery anchor ... 67
Magazine shelving ... 449
Magnesium dock board ... 407
oxychloride ... 258
Magnetic astragal ... 308
motor starter ... 588
particle test ... 16
Mahogany door ... 270
Mail box ... 396
box call system ... 605
box commercial ... 396
slot ... 396
Main drain ... 511
office expense ... 13
Maintenance carpet ... 318
railroad ... 706
railroad track ... 706
Make-up air unit ... 558
Mall front ... 288
Management fee construction ... 10
stormwater ... 687
Manhole ... 676
brick ... 676
concrete ... 677, 703
cover ... 686
electric service ... 703
frame and cover ... 686
invert ... 687
raise ... 687
removal ... 29
step ... 677
Man-made soil mix ... 236
Mansard aluminum ... 238
Mantel beam ... 212
fireplace ... 212
Manual dumbwaiter ... 476
Map rail ... 376
Maple countertop ... 442
floor ... 344
Marble ... 114
chip ... 670
chip white ... 51
coping ... 117
countertop ... 443
floor ... 114
screen ... 382
shower stall ... 390
sill ... 116
soffit ... 114
stair ... 114
synthetic ... 342
tile ... 342
Marina small boat ... 712
Marine equipment rental ... 738
Marker boundary and survey ... 28
Markerboard electronic ... 376
Mark-up cost ... 13
Masking ... 95
Mason scaffold ... 19
Masonry accessory ... 99
aggregate ... 97
anchor ... 98
base ... 110
brick ... 101
cement ... 96
cleaning ... 94
color ... 97
cutting ... 34
demolition ... 32, 95, 319
fireplace ... 117
flashing ... 249
flooring ... 342
furring ... 187
insulation ... 229
manhole ... 677
nail ... 171
painting ... 368
panel ... 111
panel pre-fabricated ... 111
pointing ... 94
reinforcing ... 98, 857
removal ... 30, 96
restoration ... 97
saw ... 734
sawing ... 94
selective demolition ... 95
sill ... 116
stabilization ... 94
step ... 652
testing ... 15
toothing ... 32
ventilator ... 99
wall ... 104, 666
wall tie ... 98
waterproofing ... 97
Mass notification system ... 611
Massage table ... 424
Mast climber ... 21
Mastic floor ... 349
Mastic-backed flashing ... 250
Mat blasting ... 623
concrete placement foundation . 79
floor ... 445
foundation ... 76
foundation formwork ... 56
gym ... 424
wall ... 424
Material door and window ... 264
handling ... 25, 715
handling belt ... 714
handling conveyor ... 714
handling deconstruction ... 37
handling system ... 714
removal/salvage ... 35
Materials concrete ... 850
MC cable ... 570
Meat case ... 414
Mechanical ... 877
dredging ... 712
duct ... 538
equipment demolition ... 528
fee ... 10
media filter ... 547
seeding ... 671
Median barrier ... 708
precast ... 708
Medical distiller ... 427
equipment ... 427-430
exam equipment ... 427
sterilizer ... 427
sterilizing equipment ... 427
waste cart ... 431
waste disposal ... 431
waste sanitizer ... 431
X-ray ... 429
Medicine cabinet ... 390
Medium-voltage cable ... 569
Membrane elastomeric ... 334
flashing ... 244
protected ... 247
roofing ... 243, 247
waterproofing ... 224
Mercury vapor lamp ... 600
Mesh fence ... 661
gunite ... 82
partition ... 384
security ... 326
stucco ... 326
Metal bin retaining wall ... 666
bookshelf ... 449
bracing ... 180
butt frame ... 267
cabinet ... 442
cabinet school ... 441
canopy ... 398
case good ... 446
casework ... 310, 441
ceiling linear ... 340
ceiling panel ... 339, 340
chalkboard ... 375, 386
chimney ... 391
cleaning ... 120
deck ... 142
deck acoustical ... 143
deck composite ... 141
deck ventilated ... 143
demolition ... 120
detector ... 608
door ... 265, 269, 284, 467
door frame ... 467
door residential ... 269
ductwork ... 538
facing panel ... 239
fascia ... 251
fence ... 661
fire door ... 270
flashing ... 468
flexible hose ... 505
floor deck ... 143
flue chimney ... 548
frame ... 266
frame channel ... 267
framing parapet ... 151
furring ... 322
halide lamp ... 600
in field galvanizing ... 371
in field painting ... 371
in shop galvanizing ... 120
interceptor dental ... 510
joist bracing ... 147
joist bridging ... 147
joist framing ... 149
laboratory casework ... 441
ladder inclined ... 155
lath ... 320, 325
locker ... 394
molding ... 326
nailing anchor ... 121
overhead door ... 284
paint & protective coating ... 120
pan ceiling ... 338
pan landing ... 154
pan stair ... 154
parking bumper ... 658
pipe ... 684
pipe removal ... 493
plate stair ... 154
pressure washing ... 120
rafter framing ... 151
recovery interceptor ... 510
roof ... 237
roof parapet framing ... 151
roof truss ... 134
sandblasting ... 120
sash ... 290
screen ... 290
sheet ... 248
shelf ... 396
shelving food storage ... 416
shingle ... 234
siding ... 237, 238
sign ... 378
soffit ... 242
steam cleaning ... 120
stud ... 320
stud demolition ... 319
stud NLB ... 323
support assembly ... 322
threshold ... 305
tile ... 333
toilet component door ... 382
toilet component headrail ... 383
toilet component panel ... 382
toilet component pilaster ... 382
toilet partition ... 380
trash receptacle ... 445
truss framing ... 153
water blasting ... 120
window ... 289, 290, 469
wire brushing ... 120
Metal-clad dbl. hung window ... 292
window bow & bay ... 293
window picture & sliding ... 292
Metal-framed skylight ... 298
Metallic conduit intermediate ... 574
foil ... 229
hardener ... 80
non-shrink grout ... 88
Meter electric ... 584
flow ... 536
steam condensate ... 537
venturi flow ... 536
water supply ... 505
water supply domestic ... 505
Metric conversion ... 851
conversion factors ... 827
rebar specs ... 844
Microphone ... 605
Microtunneling ... 676
Microwave oven ... 412
Mill construction ... 189
extra steel ... 134
Millwork ... 196
& trim demolition ... 170
demolition ... 170
Mineral fiber ceiling ... 338, 339
fiber insulation ... 228
fiberboard panel ... 355
insulated cable ... 571
roof ... 248
Minor site demolition ... 29
Mirror ... 313, 389

For customer support on your Building Construction Costs with RSMeans data, call 800.448.8182.

door . . . 313
glass . . . 313, 389
plexiglass . . . 313
wall . . . 313
Miscellaneous painting . . . 358, 363
Mix design asphalt . . . 15
planting pit . . . 670
Mixed bituminous conc. plant . . . 614
Mixer concrete . . . 728
food . . . 416
mortar . . . 728, 733
plaster . . . 733
road . . . 736
Mixes concrete . . . 850, 851
Mixing valve . . . 517
Mobile shelving . . . 449
X-ray . . . 429
Mobilization . . . 615, 648
air-compressor . . . 615
equipment . . . 642, 648
or demobilization . . . 22
Model building . . . 10
Modification to cost . . . 11
Modified bitumen roof . . . 244
bitumen roofing . . . 867
bituminous barrier sheet . . . 233
bituminous membrane SBS . . . 245
Modifier labor . . . 492
Modular office system . . . 385
playground . . . 426
Modulus of elasticity . . . 15
Moil point . . . 732
Moisture barrier . . . 225
content test . . . 16
Mold abatement . . . 46
abatement work area . . . 46
contaminated area demolition . . . 46
Molding base . . . 196
bed . . . 199
birch . . . 206
brick . . . 206
casing . . . 197
ceiling . . . 199
chair . . . 206
cornice . . . 199, 202
cove . . . 199
crown . . . 199
dentil crown . . . 200
exterior . . . 200
hardboard . . . 208
metal . . . 326
pine . . . 202
soffit . . . 207
trim . . . 206
window and door . . . 207
wood transition . . . 344
Money safe . . . 411
Monitor support . . . 128
Monitoring sampling . . . 47
Monkey ladder . . . 425
Monolithic gutter . . . 657
terrazzo . . . 347
Monorail . . . 714
Monument survey . . . 28
Mop holder strip . . . 389
roof . . . 244
sink . . . 518
Mortar . . . 97
admixture . . . 97
Mortar, brick and block . . . 858
Mortar cement . . . 335, 336, 857
masonry cement . . . 97
mixer . . . 728, 733
pigment . . . 97
Portland cement . . . 97
restoration . . . 97
sand . . . 97
testing . . . 15
thinset . . . 336
Mortise lockset . . . 303
Mortuary equipment . . . 430
refrigeration . . . 430
Mosaic glass . . . 336
Moss peat . . . 670
Motel/hotel hardware . . . 299
Motion detector . . . 609
Motor . . . 602
connection . . . 578
dripproof . . . 602
electric . . . 602
explosion proof . . . 602
starter . . . 588
starter & control . . . 588
starter enclosed & heated . . . 588
starter magnetic . . . 588
starter w/circuit protector . . . 588
starter w/fused switch . . . 588
support . . . 128
Motorized grader . . . 729
roof . . . 465
Moulded door . . . 273
Mounted booth . . . 448
Mounting board plywood . . . 192
Movable louver blind . . . 438
office partition . . . 385
wall demolition . . . 319
Moveable bulkhead . . . 458
Movie equipment . . . 420
equipment digital . . . 420
lens . . . 420
projector . . . 420
screen . . . 419
Moving building . . . 38
ramp . . . 481
ramps and walks . . . 877
shrub . . . 674
stair & walk . . . 480
structure . . . 38
tree . . . 674
walk . . . 481
Mower lawn . . . 733
Mowing brush . . . 616
Muck car tunnel . . . 736
Mud pump . . . 728
sill . . . 186
trailer . . . 735
Mulch bark . . . 670
ceramic . . . 670
stone . . . 670
Mulcher power . . . 730
Mulching . . . 670
Mullion type louver . . . 546
vertical . . . 286
Multi-blade damper . . . 539
Multi-channel rack enclosure . . . 604
Multizone air conditioner
rooftop . . . 558
Municipal incinerator . . . 430
Muntin window . . . 294
Mushroom ventilator stationary . . . 547
Music room . . . 461
Mylar tarpaulin . . . 23

N

Nail . . . 171
common . . . 171
lead head . . . 472
stake . . . 62
Nailer pneumatic . . . 732, 733
wood . . . 183
Nailing anchor . . . 121
Napkin dispenser . . . 389
Natural fiber wall-covering . . . 353
gas generator . . . 592
Needle beam cantilever . . . 642
Neoprene adhesive . . . 225
expansion joint . . . 255
flashing . . . 250
floor . . . 349
gasket . . . 260
roof . . . 247
waterproofing . . . 225
Net safety . . . 18
tennis court . . . 660
Netting bird control . . . 401
flexible plastic . . . 401
Newel wood stair . . . 210
Newspaper rack . . . 449
Night depository . . . 411
No hub pipe . . . 508
Non-chloride accelerator conc. . . . 78
Non-destructive testing . . . 16
Non-metallic non-shrink grout . . . 88
Non-removable pin . . . 306
Norwegian brick . . . 102
Nosing rubber . . . 345
safety . . . 345
stair . . . 345, 347
Nozzle dispersion . . . 489
fire hose . . . 487
fog . . . 487
playpipe . . . 487
Nurse call light . . . 606
call system . . . 606
speaker station . . . 606
station cabinet . . . 441, 442
Nursery item travel . . . 672
Nursing home bed . . . 449
Nut remove . . . 120
Nylon carpet . . . 351
nailing anchor . . . 121

O

Oak door frame . . . 213
floor . . . 343
molding . . . 206
paneling . . . 208
threshold . . . 214
Oakum caulking . . . 508
Obscure glass . . . 312
Observation well . . . 678
Occupancy sensor . . . 584
Offhighway dumptruck . . . 731
Office & storage space . . . 17
chair . . . 447
expense . . . 13
field . . . 17
floor . . . 352
furniture . . . 446, 447
partition . . . 385
partition movable . . . 385
safe . . . 410
system modular . . . 386
trailer . . . 17
Oil fired boiler . . . 550
fired furnace . . . 551
fired space heater . . . 552
formwork . . . 62
interceptor . . . 510
storage tank . . . 533
Oil-filled transformer . . . 585
Oil-fired water heater . . . 514
water heater commercial . . . 514
water heater residential . . . 514
Olive knuckle hinge . . . 306
Omitted work . . . 13
One piece astragal . . . 308
One-way vent . . . 257
Onyx . . . 347
Open web bar joist . . . 140
Opener automatic . . . 299
door . . . 264, 281, 284, 299
driveway security gate . . . 283
handicap . . . 299
industrial automatic . . . 299
Opening framing . . . 467
roof frame . . . 130
Operable partition . . . 387
Operating room equipment . . . 429
Operator automatic . . . 299
Option elevator . . . 479
Ornamental aluminum rail . . . 164
column . . . 164
glass rail . . . 164
railing . . . 164
steel rail . . . 164
wrought iron rail . . . 164
OSB faced panel . . . 188
OSHA testing . . . 43
Outdoor bleacher . . . 466
fountain . . . 459
Outlet box steel . . . 574
lighting . . . 581
Outrigger wall pole . . . 401
Oval arch culvert . . . 686
Oven . . . 412, 417
cabinet . . . 439
convection . . . 417
microwave . . . 412
Overbed table . . . 449
Overhang eave . . . 467
endwall . . . 467
Overhaul . . . 34
Overhead & profit . . . 13
bridge crane . . . 714
commercial door . . . 283
contractor . . . 25, 833
door . . . 283, 284
hoist . . . 715
Overlapping astragal . . . 309
Overlay face door . . . 271, 272
Overpass . . . 668
Oversized brick . . . 101
Overtime . . . 12, 13
Oxygen lance cutting . . . 35

P

P trap . . . 509
trap running . . . 508
Package chute . . . 482
receiver . . . 411
Packaged terminal A/C . . . 559
Packaging waste . . . 43
Pad bearing . . . 126
carpet . . . 350
commercial grade carpet . . . 350
equipment . . . 76
heated . . . 52
vibration . . . 126
Padding carpet . . . 350
Paddle blade air circulator . . . 543
fan . . . 582
tennis court . . . 426
Paint & coating . . . 356
& coating interior floor . . . 365
& protective coating metal . . . 120
aluminum siding . . . 360
and protective coating . . . 120

chain link fence ... 358
door & window exterior ... 361
door & window interior ... 363-365
encapsulation lead ... 44
exterior miscellaneous ... 359
fence picket ... 358
floor ... 366
floor concrete ... 365, 366
floor wood ... 366
interior miscellaneous ... 366
remediation lead ... 44
removal ... 45
removal floor ... 82
removal lead ... 45
siding ... 360
sprayer ... 733
striper ... 736
trim exterior ... 362
wall & ceiling interior ... 367, 368
wall masonry exterior ... 362
Paint/coating cabinet & casework ... 363
Painted marking airport ... 659
pavement marking ... 659
Painting ... 874
balustrade ... 367
bar joist ... 141
booth ... 406
casework ... 363
ceiling ... 367, 368
clapboard ... 359
concrete block ... 368
cornice ... 367
decking ... 359
drywall ... 367, 368
electrostatic ... 367
exterior siding ... 359
grille ... 366
masonry ... 368
metal in field ... 371
miscellaneous ... 358, 363
parking stall ... 659
pavement ... 659
pipe ... 366
plaster ... 367, 368
railing ... 359
reflective ... 659
shutter ... 359
siding ... 359
stair stringer ... 359
steel ... 371
steel siding ... 359
stucco ... 359
swimming pool ... 458
temporary road ... 659
tennis court ... 660
thermoplastic ... 659
trellis/lattice ... 359
trim ... 366
truss ... 367
wall ... 360, 368, 369
window ... 364
Palladian window ... 294
Pallet rack ... 396
Pan shower ... 249
slab ... 75
stair metal ... 154
tread concrete ... 80
Panel acoustical ... 386
and device alarm ... 611
board residential ... 578
brick wall ... 111
ceiling ... 338
ceiling acoustical ... 338
door ... 273, 274
facing ... 115
fiberglass ... 23, 238, 355
fiberglass refrigeration ... 460
FRP ... 354
insulated ... 231
masonry ... 111
metal facing ... 239
metal toilet component ... 382
mineral fiberboard ... 355
OSB faced ... 188
plastic-laminate toilet ... 383
portable ... 386
precast concrete double wall ... 85
prefabricated ... 231
sandwich ... 238
shearwall ... 188
sound ... 472
sound absorbing ... 355
sound dampening gypsum ... 331
spandrel ... 312
steel roofing ... 237
structural ... 188
structural insulated ... 188
system ... 209
vision ... 433
wall ... 117
wood folding ... 438
woven wire ... 384
Panelboard ... 587
electric ... 587
w/circuit breaker ... 587
Paneled door ... 269
pine door ... 275
Paneling ... 208
birch ... 208
board ... 209
cedar ... 209
cutout ... 32
demolition ... 170
hardboard ... 208
plywood ... 208
redwood ... 209
wood ... 208
Panelized shingle ... 234
Panic exit device ... 301
Paper building ... 232
sheathing ... 232
Paperbacked flashing ... 250
Paperhanging ... 352
Paperholder ... 389
Paraffin bath ... 428
Parallel bar ... 424, 428
Parapet metal framing ... 151
Parging cement ... 224
Park bench ... 667
Parking barrier ... 659
barrier precast ... 658
bumper ... 659
bumper metal ... 658
bumper plastic ... 658
bumper wood ... 659
collection equipment ... 406
control equipment ... 406
equipment ... 406
garage public ... 469
gate ... 406
gate equipment ... 406
LED lighting ... 598
lot paving ... 654
marking pavement ... 659
stall painting ... 659
ticket equipment ... 406
Parquet floor ... 343
wood ... 343
Part bin ... 397
Particle board underlayment ... 191
core door ... 271
Parting bead ... 207
Partition ... 384, 386
acoustical ... 385, 387
anchor ... 98
block ... 108
blueboard ... 322
bulletproof ... 411
concrete block ... 108
demolition ... 319
demountable ... 385
door ... 386
drywall ... 322
folding accordion ... 386
folding leaf ... 387
framing load-bearing stud ... 145
framing NLB ... 323
gypsum ... 385
hospital ... 384
mesh ... 384
movable office ... 385
NLB drywall ... 321
NLB gypsum ... 322
office ... 385
operable ... 387
plaster ... 320
portable ... 387
refrigeration ... 460
shower ... 114, 390
sliding ... 387
steel ... 385, 387
support ... 127
thin plaster ... 322
tile ... 103
toilet ... 114, 380, 381
toilet stone ... 382
wall ... 321
wall NLB ... 321
wire ... 384
wire mesh ... 384
wood frame ... 183
woven wire ... 384
Passage door ... 275, 277
Passenger elevator ... 476
elevators ... 876
Pass-thru drawer ... 433
Patch core hole ... 15
formwork ... 62
roof ... 222
Patching asphalt ... 614
concrete floor ... 50
concrete wall ... 50
Patient care equipment ... 428
nurse call ... 606
Patio ... 653
block ... 114, 244
block concrete ... 656
door ... 280
Patio/deck canopy ... 398
Patterned glass ... 312
Paumelle hinge ... 306
Pavement ... 656
asphalt ... 660
asphaltic ... 654
berm ... 657
breaker ... 732
bucket ... 735
demolition ... 29
emulsion ... 652
highway concrete ... 655
marking ... 658
marking painted ... 659
painting ... 659
parking marking ... 659
planer ... 736
profiler ... 736
replacement ... 654
sealer ... 660
slate ... 115
widener ... 736
Paver asphalt ... 733
bituminous ... 733
cold mix ... 736
concrete ... 733
floor ... 342
highway ... 657
roof ... 257
shoulder ... 736
tile exposed aggregate ... 656
Paving asphalt ... 654
asphaltic concrete ... 654
athletic ... 660
block granite ... 657
brick ... 656
parking lot ... 654
Pea gravel ... 670
stone ... 51
Peastone ... 672
Peat humus ... 670
moss ... 670
Pedestal floor ... 352
type lavatory ... 515
type seating ... 450
Pedestrian bridge ... 668
Peephole ... 304
Pegboard ... 208
Penetration macadam ... 653
test ... 15
Penthouse roof louver ... 546
Perforated aluminum pipe ... 684
pipe ... 684
PVC pipe ... 684
Performance bond ... 14, 835
Perlite insulation ... 226
plaster ... 327
sprayed ... 371
Permit building ... 14
Personal respirator ... 40
Personnel field ... 12
hoist ... 737
protection ... 18
Petrographic analysis ... 15
Pew church ... 450
sanctuary ... 450
PEX pipe ... 504
tubing ... 504, 562
tubing fitting ... 563
Pharmacy cabinet ... 422
Phone booth ... 380
Photoelectric control ... 584
Photography ... 14
aerial ... 14
construction ... 14
time lapse ... 14
Photoluminescent signage ... 379
Physician's scale ... 427
PIB roof ... 246
Picket railing ... 156
Pickup truck ... 735
Pick-up vacuum ... 728
Picture window ... 291
window aluminum sash ... 289
window steel sash ... 290
window vinyl ... 297
Pier brick ... 101
Pigment mortar ... 97
Pilaster metal toilet component ... 382
plastic-laminate toilet ... 383
toilet partition ... 382
wood column ... 212
Pile ... 648
boot ... 645
bored ... 646

For customer support on your Building Construction Costs with RSMeans data, call 800.448.8182.

cap 77
cap concrete placement 79
cap formwork 55
cutoff 614
driven 644
driving 837, 838
driving mobilization 615
encasement 645
foundation 648
friction 645, 648
H 643, 644
hammer 729
high strength 641
leads 729
lightweight 641
load test 614
mobilization hammer 615
pipe 648
point 645, 646
point heavy duty 646
precast 644
prestressed 644, 648
sod 670
splice 645, 646
steel 644
steel sheet 641
step tapered 644
testing 614
timber 645
treated 645
wood 645
wood sheet 641
Piling sheet 641, 885
special cost 614
Pillow tank 682
Pin powder 125
Pine door 274
door frame 213
fireplace mantel 212
floor 344
molding 202
roof deck 190
shelving 397
siding 240
stair tread 209
Pipe & fitting 520
& fitting backflow preventer . . 506
& fitting backwater valve 511
& fitting bronze 495
& fitting copper 500
& fitting grooved joint 502
& fitting hydrant 507
& fitting iron body 529
& fitting polypropylene . . 496, 525
& fitting PVC 503
& fitting steel 501
& fitting trap 509
acid resistant 525
add elevated 492
aluminum 684
and fittings 877
bedding 629
bedding trench 621
bollard 658
brass 500
cast iron 508
cleanout 498
concrete 685
copper 500
corrosion resistant 524
corrugated 684
corrugated metal 684
covering 499
covering fiberglass 499
CPVC 504
culvert 686

double wall 687
drain 627
drainage 524, 682, 684, 685
ductile iron 678
DWV PVC 504, 682
elbow 502
encapsulation 44
epoxy fiberglass wound 524
fire retardant 525
fitting grooved joint 502
foam core DWV ABS 504
gas 688
glass 524
grooved joint 501
insulation 499
insulation removal 42
iron alloy mechanical joint . . 524
liner 676
metal 684
no hub 508
painting 366
perforated aluminum 684
PEX 504
pile 648
plastic 503, 504
polyethylene 688
polypropylene 525
proxylene 525
PVC 503, 679, 682
rail aluminum 156
rail galvanized 156
rail stainless 156
rail steel 156
rail wall 156
railing 156
reinforced concrete 685
relay 627
removal 29
removal metal 493
removal plastic 493
sewage 682, 684, 685
sewage collection PVC 682
shock absorber 507
single hub 508
sleeve plastic 61
soil 508
stainless steel 503
steel 501, 684
subdrainage 684
support framing 131
tee 502
water 678
weld joint 501
wrapping 676
Piping designations 887
Piping gas service polyethylene . 688
gasoline 687
specialty HVAC 534, 536
storm drainage 684
Pit excavation 622
leaching 683
scale 402
sump 627
test 28
Pitch coal tar 243
emulsion tar 652
pocket 251
Pivoted window 290
Placing concrete 78, 852
reinforcment 845
Plain tube framing 285
Plan remediation 40, 44
Planer pavement 736
shop 422
Plank floor 189
grating 158

hollow precast concrete 83
precast concrete nailable 84
precast concrete roof 83
precast concrete slab 83
roof 190
scaffolding 20
Plant and bulb transplanting . . . 674
and planter 450
bed preparation 670
ground cover 672
mixed bituminous concrete . . 614
screening 730
Plantation shutter 438
Planter 450, 667
bench 667
concrete 667
fiberglass 667
interior 450
Planting 674
Plant-mix asphalt paving 654
Plaque bronze 378
Plaster 320
accessory 326
beam 327
ceiling 327
cement 460
column 327
cutout 32
demolition 318, 319
drilling 322
gauging 327
ground 188
gypsum 327
mixer 733
painting 367, 368
partition 320
partition thin 322
perlite 327
soffit 327
thincoat 329
Venetian 328
vermiculite 327
wall 327
Plasterboard 329
Plastic angle valve 496
ball valve 496
bench 396
faced hardboard 208
fireproofing 258
glazing 313
grille 545
laminate door 271
lead 473
locker 395
lumber 215
lumber structural 215
matrix terrazzo 348
parking bumper 658
pipe 503, 504
pipe removal 493
railing 217
screw anchor 122
sheet flashing 250
sign 378
skylight 298
toilet compartment 383
toilet partition 382
trench drain 511
valve 496
window 295
Plastic-laminate toilet
compartment 381
toilet component 383
toilet door 383
toilet headrail 383
toilet panel 383

toilet pilaster 383
Plastic-laminate-clad countertop . 442
Plate checkered 159
roadway 736
shear 172
steel 131
stiffener 131
wall switch 591
wood 187
Plate-type heat exchanger 554
Platform checkered plate 159
lifter 407
telescoping 422
tennis 426
trailer 734
Plating zinc 171
Player bench 668
Playfield equipment 425
Playground equipment . 425, 426, 662
modular 426
protective surfacing 660
slide 426
surface 351
whirler 426
Plenum barrier 356
demolition 319
Plexiglass 313
acrylic 313
mirror 313
Plow vibrator 731
Plug in bus duct 589
in circuit breaker 590
in switch 590
wall 99
Plugmold raceway 591
Plumbing 506
appliance 512, 513
demolition 493
fixture 510, 515, 522, 536, 678
fixture removal 493
laboratory 421
Plywood 866
clip 172
demolition 318, 319
fence 24
floor 190
formwork 52, 54
formwork steel framed 58
joist 193
mounting board 192
paneling 208
sheathing roof & wall 192
shelving 397
sidewalk 24
siding 240
sign 378
soffit 207
subfloor 190
underlayment 191
Pneumatic control system 531
nailer 732, 733
tube 411
tube system 482
Pocket door 299
door frame 214
pitch 251
Point heavy duty pile 646
moil 732
pile 645, 646
Pointing CMU 94
masonry 94
Poisoning soil 639
Poke-thru fitting 574
Pole aluminum light 597
athletic 425
closet 206

For customer support on your Building Construction Costs with RSMeans data, call 800.448.8182.

cross arm	703
electrical	703
light	703
lighting	597
portable decorative	408
steel light	598
telephone	703
utility	597, 703
wood	703
wood electrical utility	703
Polished concrete	853
Polishing concrete	82
Polyacrylate floor	348
terrazzo	348
Polycarbonate glazing	313
Polyester floor	350
room darkening shade	439
Polyethylene backer rod	64
coating	676
film	670
floor	351
pipe	688
pool cover	458
septic tank	683
tarpaulin	23
waterproofing	224, 225
Polymer trench drain	511
Polyolefin roofing thermoplastic	246
Polypropylene pipe	525
shower	516
siding	242
valve	496
Polystyrene blind	214
ceiling	460
insulation	226, 228, 460
Polysulfide caulking	261
Polyurethane caulking	261
floor	349, 350
varnish	371
Polyvinyl chloride (PVC)	38
soffit	207, 242
tarpaulin	23
Polyvinyl-chloride roof	246
Pool accessory	458
cover	458
cover polyethylene	458
filtration swimming	524
heater electric	550
swimming	457, 458
Porcelain tile	334
Porch framing	184
Portable air compressor	731
asphalt plant	735
booth	470
building	17
cabinet	415
chalkboard	376
eye wash	522
fire extinguisher	393
floor	344
panel	386
partition	387
scale	402
stage	422
Portland cement	51
cement terrazzo	347
Positive pressure chimney	549
Post and panel signage	379
athletic	427
cap	172
cedar	213
concrete	708
demolition	168
driver	736
fence	661
lamp	163, 597

pedestrian traffic control	408
recreational	425
shore	61
sign	380
tennis court	660
Postal specialty	396
Postformed countertop	443
Post-tensioned concrete	855
slab on grade	73
Potable water softener	512
water treatment	512
Potters wheel	422
Poured gypsum underlayment	88
Powder actuated tool	125
charge	125
pin	125
Power equipment	835
mulcher	730
temporary	17
trowel	728
wiring	591
Preaction valve cabinet	488
Preblast survey	623
Precast beam	84
bridge	668
catch basin	677
column	84
concrete beam	84
concrete channel slab	83
concrete column	84
concrete joist	85
concrete lintel	86
concrete nailable plank	84
concrete plantable paver	656
concrete roof plank	83
concrete slab plank	83
concrete stair	84
concrete tee beam	84
concrete tee beam double	85
concrete tee beam quad	85
concrete tee beam single	85
concrete unit paving slab	656
concrete wall	86
concrete wall panel tilt-up	86
concrete window sill	86
conrete	856
coping	117
curb	657
median	708
members	854
parking barrier	658
pile	644
receptor	390
septic tank	682
tee	85
terrazzo	348
wall	856
wall panel	86
wall panel insulated	86
Pre-engineered steel building	466
steel building demo	455
steel buildings	875
Prefab metal flue	548
Prefabricated building	465
comfort station	469
fireplace	391
Pre-fabricated masonry panel	111
Prefabricated panel	231
wood stair	209
Prefinished door	271
drywall	330
floor	343
hardboard paneling	208
shelving	397
Preformed roof panel	237
roofing & siding	237

Pre-hung door	276, 277
Preparation exterior surface	356
interior surface	357
plant bed	670
Presser laundry	410
Pressure grouting cement	642
reducing valve water	495
regulator	505
regulator steam	505
relief valve	495
switch switchboard	586
valve relief	495
wash	357
washer	734
washing metal	120
Pressurized fire extinguisher	394
Prestressed concrete	855
concrete pile	644
pile	644, 648
precast concrete	854
Prestressing steel	73
Preventer backflow	506
Prices lumber products	866
Prime coat	653
Primer asphalt	222, 346
steel	371
Prison cell	471
cot	448
hinge	306
toilet	433
Processing concrete	82
Proctor compaction test	16
Produce case	414
Product piping	687
Productivity glazing	870
Profile block	106
Profiler pavement	736
Progress schedule	14
Project overhead	13
sign	25
Projected window aluminum	289
window steel	290
window steel sash	290
Projection screen	375, 419, 420
Projector movie	420
Propeller exhaust fan	543
unit heater	562
Property line survey	28
Protected membrane	247
Protection corner	387
fire	393
slope	639
stile	308
temporary	25
termite	639
winter	23, 52
worker	41
Protector door	300
Proxylene pipe	525
P&T relief valve	495
PTAC unit	559
Public address system	605
Pull box	575
box electrical	575
door	440
plate	305
Pulpit church	432
Pump	537, 678
centrifugal	734
circulating	536
concrete	728
condensate	537
condensate removal	537
condensate return	537
contractor	627
diaphragm	734

fire	490
fire jockey	490
fountain	524
general utility	508
grinder system	510
grout	734
heat	560
in-line centrifugal	536
jack	19
mud	728
operator	627
sewage ejector	510
shallow well	678
shotcrete	728
staging	18, 836
submersible	512, 678, 734
sump	413, 511
trash	734
water	536, 678, 734
water supply well	678
wellpoint	738
Pumped concrete	78
Pumping	627
dewater	627
Purlin roof	189
steel	134
Push plate	304
Push-on joint	678
Push-pull plate	304
Putlog scaffold	20
Putting surface	351
Puttying	356
PVC adhesive	225
blind	437
cleanout	498
conduit in slab	576
control joint	99
cornerboard	217
door casing	217
fascia	217
fitting	680
flashing	250
frieze	218
gravel stop	251
molding exterior	217
pipe	503, 679, 682
pipe perforated	684
rake	218
roof	246
sheet	225, 351
siding	241
soffit	218
trim	217
underground duct	703
valve	496
waterstop	62

Q

Quad precast concrete tee beam	85
Quarry drill	732
drilling	622
tile	336
tiling	337
tiling chemical-resistant	337
Quarter round molding	207
Quartz	670
chip	51
flooring	350
Quoin	113

R

Raceway	574, 576

For customer support on your Building Construction Costs with RSMeans data, call 800.448.8182.

conduit 573
plugmold 591
surface 590
trench duct 577
wiremold 590
Rack bicycle 425, 451
coat 397
hat 397
hose 487
pallet 396
Racquetball court 462
Radial arch 193
wall formwork 57
Radiant floor heating 562
Radiation fin tube 562
Radiator cast iron 561
supply control 535
thermostat control system 532
Radio frequency security shielding 341
frequency shielding 474
tower 704
Radiography test 16
Radiology equipment 429
Rafter anchor 172
composite 185
demolition 168
framing metal 151
metal bracing 151
metal bridging 151
tie 186
wood 185, 186
Rail aluminum pipe 156
crane 714
crash 388
dock shelter 407
galvanized pipe 156
guide 708
guide/guard 708
map 376
ornamental aluminum 164
ornamental glass 164
ornamental steel 164
ornamental wrought iron 164
stainless pipe 156
steel pipe 156
trolley 388
wall pipe 156
Railing cable 164
church 432
composite 219
demolition 170
encased 219
encased composite 219
industrial 156
ornamental 164
picket 156
pipe 156
plastic 217
wood 206, 209, 211
wood stair 210
Railroad ballast 706
bumper 706
concrete tie 706
derail 706
maintenance 706
siding 709, 887
tie 653, 674
tie step 652
timber switch tie 706
timber tie 706
track accessory 706
track heavy rail 706
track maintenance 706
track material 706
track removal 30
turnout 709
wheel stop 706
Raise manhole 687
manhole frame 687
Raised floor 352
Rake PVC 218
tractor 730
Rammer/tamper 729
Ramp approach 352
handicap 76
moving 481
temporary 22
Ranch plank floor 344
Range cooking 412
hood 413
receptacle 581, 591
restaurant 417
shooting 425
Ratio water cement 15
Razor wire 664
Reach-in refrigeration 415
Reading table 449
Ready mix concrete 78, 852
Rebar accessory 68
Receiver ash 445
trash 445
Receptacle air conditioner 581
device 580
dryer 581
duplex 591
GFI 580
locking 581
range 581, 591
telephone 581
television 581
trash 451
waste 390
weatherproof 581
Receptor precast 390
shower 390, 516, 519
terrazzo 390
Recessed mat 445
Reciprocating water chiller 555
Recirculating chemical toilet 722
Reclaimed or antique granite 113
Recorder videotape 609
Recore cylinder 302, 303
Recovery heat 430
Recreational post 425
Rectangular diffuser 545
ductwork 538
Rectifier low-voltage silicon 591
Recycled plastic lumber 214
rubber tire tile 445
Red bag 431
Reduce heat transfer glass 311
Redwood bark mulch 670
cupola 399
louver 316
paneling 209
siding 240
wine cellar 415
Refinish floor 344
textured ceiling 318
Reflective block 110
glass 313
insulation 229
painting 659
sign 380
Reflectorized barrels 732
Refrigerant removal 528
Refrigerated case 414
storage equipment 415
wine cellar 415
Refrigeration 460
bloodbank 421
commercial 414, 415
equipment 459
floor 460
insulation 460
mortuary 430
panel fiberglass 460
partition 460
reach-in 415
residential 412
walk-in 460
Refrigerator door 281
Refuse chute 482
hopper 482
Register air supply 545
cash 409
return 546
steel 545
wall 546
Reglet aluminum 254
galvanized steel 254
Regulator pressure 505
steam pressure 505
Reinforced concrete pipe 685
culvert 685
plastic panel fiberglass 354
PVC roof 246
Reinforcement 845
welded wire 846
Reinforcing 845
accessory 68
bar chair 68
bar splicing 70
bar tie 68
beam 71
chair subgrade 69
coating 72
column 71
concrete 71
dowel 71
elevated slab 71
epoxy coated 72
footing 71
galvanized 72
girder 71
glass fiber rod 72
high chair 68
joint 98
masonry 98
metric 844
slab 71
sorting 71
spiral 71
steel 844, 845
steel fiber 74
synthetic fiber 74
testing 15
tie wire 69
wall 71
Relay pipe 627
Release door 605
Relief valve P&T 495
valve self-closing 495
valve temperature 495
vent ventilator 547
Relining sewer 676
Remediation plan 40, 44
plan/method asbestos 40
Remote power pack 584
Removal air conditioner 528
asbestos 42
bathtub 493
block wall 31
boiler 528
catch basin 29
concrete 29
concrete pipe 29
curb 29
driveway 29
fixture 493
floor 31
guide rail 30
hydrant 29
insulation 42, 223
lavatory 493
masonry 30, 96
paint 45
pipe 29
pipe insulation 42
plumbing fixture 493
railroad track 30
refrigerant 528
shingle 223
sidewalk 30
sink 493
sod 671
steel pipe 29
stone 30
stump 616
tank 39
tree 615, 616
urinal 493
utility line 29
vat 42
water closet 493
water fountain 493
water heater 493
water softener 493
window 264
Removal/salvage material 35
Remove bolt 120
nut 120
Rendering 10
Renovation tread cover 161
Rental equipment 728
Repellent water 368
Replacement joint sealant 222
pavement 654
Resaturant roof 222
Residential alarm 582
appliance 412, 582
application 578
bathtub 516
chain link gate 663
closet door 269
device 578
dishwasher 412
door 213, 269, 274
door bell 581
dryer 413
elevator 477
fan 582
fixture 581
folding partition 386
garage 469
gas water heater 513
greenhouse 465
gutting 34
heat pump 583
hinge 306
kitchen sink 516
lavatory 515
lighting 581
oil-fired water heater 514
overhead door 284
panel board 578
refrigeration 412
roof jack 543
service 578
sink 516
smoke detector 582
stair 209
storm door 265

For customer support on your Building Construction Costs with RSMeans data, call 800.448.8182.

switch 578
transition 543
wall cap 543
wash bowl 515
washer 413
water heater 413, 514, 583
water heater electric 513
wiring 578, 582
Resilient base 344, 345
floor athletic 347
pavement 660
Resistance corrosion 676
Respirator 41
personal 40
Resquared shingle 234
Restaurant furniture 447
range 417
Restoration gypsum 318
masonry 97
mortar 97
window 45
Retaining wall 77, 667
wall cast concrete 665
wall concrete segmental 665
wall formwork 58
wall segmental 665
wall stone 666
wall stone gabion 666
wall timber 666
Retarder concrete 78
vapor 232
Retractable lateral arm awning 398
Return register 546
Revolving darkroom 461
dome 471
door 282, 287, 464
door darkroom 282
entrance door 287
Rewind table 420
Rib lath 325
Ribbed siding 238
waterstop 62
Ridge board 186
cap 233, 237
flashing 468
roll 238
shingle slate 234
vent 256
Rig drill 28
Rigid anchor 98
conduit in trench 576
in slab conduit 576
insulation 226
joint sealant 261
metal-framed skylight 298
Ring boxing 424
split 172
toothed 173
Ripper attachment 735
Riprap and rock lining 640
Riser pipe wellpoint 738
rubber 345
stair 348
terrazzo 348
wood stair 210
River stone 51
Rivet 125
aluminum 125
copper 125
stainless 125
steel 125
tool 125
Road base 653
berm 657
mixer 736
sign 380
sweeper 736
temporary 22
Roadway LED luminaire 598
luminaire 598
plate 736
Robe hook 389
Rock bolt drilling 622
bolting 640
drill 28, 622
removal 622
trencher 731
Rod backer 260
closet 397
curtain 388
ground 573
shower 388
tie 130, 641
weld 127
Roll ridge 238
roof 243
roofing 248
type air filter 547
Roller compaction 628
sheepsfoot 628, 639, 730
tandem 730
vibratory 730
Rolling door 280, 288
earth 639
ladder 20
service door 281
tower scaffold 21
Roll-up exterior shutter 398
shutter 398
Romex copper 571
Roof adhesive 243
aluminum 237
baffle 257
beam 193
bracket 21
built-up 243
cant 186, 244
clay tile 235
coating 222
cold-applied built-up asphalt 243
copper 248
CSPE 245
curb 186
deck 190
deck formboard 59
deck gypsum 87
deck insulation 230
deck laminated 190
deck wood 190
decking 142
drain 511
elastomeric 247
exhauster 542
expansion joint 255
fan 542
fiberglass 238
fill 856
frame opening 130
framing 186
framing removal 32
gravel 51
hatch 256
hatch removal 223
insulation 230, 468
jack 186
jack residential 543
joint 255
lead 248
membrane green 236
metal 237
metal rafter framing 151
metal tile 236
metal truss framing 153
mineral 248
modified bitumen 244
mop 244
nail 171
panel aluminum 237
panel preformed 237
patch 222
paver 257
paver and support 257
PIB 246
polyvinyl-chloride 246
purlin 189
PVC 246
rafter bracing 151
rafter bridging 151
rafter framing 185
reinforced PVC 246
resaturant 222
roll 243
safety anchor 21
sheathing 192
sheet metal 248
shingle 233
shingle asphalt 233
slate 234, 866
soffit framing 152
soil mixture green 236
specialty prefab 251
stainless steel 248
steel 237
stressed skin 131
system green 236
thermoplastic polyolefin 246
tile 235
TPO 246
truss 193, 194
vent 256, 469
ventilator 257
walkway 244
window demo 265
zinc 248
Roofing & siding preformed 237
built-up 866
cold 243
demolition 42, 223
EPDM 245
flat seam sheet metal 248
ketone ethylene ester 246
membrane 243, 247
modified bitumen 244
roll 248
single-ply 245
SPF 247
system built-up 243
Rooftop air conditioner 558
multizone air conditioner 558
Room acoustical 461
audiometric 461
clean 460
humidifier 564
Rope decorative 408
exercise 424
safety line 21
steel wire 137
Rotary crawler drill 732
hammer drill 732
Rototiller 730
Rough buck 183
grading 617
grading site 617
stone wall 112
Rough-in drinking fountain deck 523
drinking fountain floor 523
drinking fountain wall 522
sink countertop 516
sink raised deck 516
sink service floor 518
sink service wall 518
tub 516
Round bar fiberglass 216
diffuser 545
rail fence 664
table 447
tube fiberglass 216
Rowing machine 423
Rub finish carborundum 81
Rubber and vinyl sheet floor 346
astragal 308
base 345
coating 226
control joint 99
dock bumper 407
floor 346, 660
floor tile 346
nosing 345
pavement 660
pipe insulation 499
riser 345
sheet 346
stair 345
threshold 305
tile 347
waterproofing 225
waterstop 63
Rubberized asphalt sealcoat 652
Rubbing wall 81
Rubbish handling 33
handling chutes 839
Rumble strip 708
Run gravel bank 614
Running track 344
track surfacing 660
trap 508
Rupture testing 15

S

Safe commercial 410
diskette 410
office 410
Safety cabinet biological 422
deposit box 411
equipment laboratory 421
eye/face equipment 522
film 314
fixture industrial 522
flood shut off house 495
glass 311
line rope 21
net 18
nosing 345
railing cable 138
shower 522
shut off flood whole house 495
switch 592
water and gas shut off 495
Salamander 734
Sales tax 12, 830, 831
Salvage or disposal value 35
Sample field 47
Sampling air 40
Sanctuary pew 450
Sand 97
brick 51
concrete 51
fill 614
screened 97
seal 652
Sandblast finish concrete 81
glass 311

For customer support on your Building Construction Costs with RSMeans data, call 800.448.8182.

Index

masonry 95
Sandblasting equipment 734
metal 120
Sander floor 734
Sanding 356
floor 344
Sandstone 114
flagging 657
Sandwich panel 238
wall panel 239
Sanitary base cove 335
drain 509
Sash 468
aluminum 289
metal 290
prefinished storm 469
security 290
steel 290
wood 291
Sauna 462
door 462
Saw blade concrete 89
brick 94
chain 734
circular 734
concrete 728
cutting concrete floor 89
cutting concrete wall 89
cutting demolition 34
cutting slab 89
masonry 734
shop 422
table 422
Sawing masonry 94
Sawn control joint 63
SBS modified bituminous membrane 245
Scaffold aluminum plank 21
baseplate 19
bracket 19
caster 19
catwalk 20
frame 19
guardrail 19
mason 19
putlog 20
rolling tower 21
specialty 20
stairway 19
wood plank 19, 21
Scaffolding 836
plank 20
tubular 19
Scale 402
commercial 402
contractor 402
crane 402
physician's 427
pit 402
portable 402
truck 402
warehouse 402
Scanner checkout 409
unit facial 608
Scarify concrete 319
Schedule 14
board 377
critical path 14
progress 14
School cabinet 441
door hardware 300
equipment 424
furniture 448
hinge 307
panic device door hardware 300
T.V. 609
Scissor gate 385
lift 715, 731
Scoreboard baseball 424
basketball 424
football 424
hockey 424
Scored block 106
split face block 106
Scraper earth 730
elevating 625
excavation 625
self propelled 626
Scratch coat 347
Screed base 69
Screed, gas engine, 8HP vibrating 728
Screed holder 69
Screen chimney 391
entrance 381
fence 664
metal 290
molding 206
projection 375, 419, 420
security 163, 290
sight 386
squirrel and bird 391
urinal 382
window 290, 295
wood 295
Screened sand 97
Screening plant 730
Screen/storm door aluminum 265
Screw aluminum 239
anchor 69, 121
anchor bolt 69
brass 171
drywall 333
eye bolt 69
jack 641
lag 125
liquid chiller 556
machine 125
sheet metal 171
steel 171
water chiller 556
wood 171
Scroll water chiller 555
Scrub station 429
Scupper drain 511
floor 511
Seal compression 260
curb 652
dock 407
door 407
fog 652
grout 333
pavement 652
security 308
Sealant 224
acoustical 261, 331
caulking 261
replacement joint 222
rigid joint 261
tape 261
Sealcoat 652
Sealer floor 51
joint 260, 261
pavement 660
Seamless floor 349
Seating church 432
movie 420
Secondary treatment plant 722
Sectional door 283
overhead door 284
Security and vault equipment 410
barrier 408
bollard 707
dome 401
door and frame 282
drawer 433
driveway gate 282
driveway gate aluminum 282
driveway gate steel 282
driveway gate wood 282
equipment 433
factor 11
fence 664
film 314
gate 385
gate driveway 282
gate opener driveway 283
hinge 306
mesh 326
planter 707
sash 290
screen 163, 290
seal 308
turnstile 408
vault door 283
vehicle barrier 707
X-ray equipment 608
Sediment bucket drain 509
strainer Y valve 496
Seeding 671, 886
See-saw 426
Segmental concrete retaining wall 665
retaining wall 665
Seismic bracing 875
bracing support 496
ceiling bracing 339
Selective clearing 616
demolition equipment 404
demolition fencing 30
demolition masonry 95
tree removal 616
Self-adhering flashing 251
Self-closing relief valve 495
Self-contained air conditioner 559
Self-propelled scraper 626
Self-supporting tower 704
Sentry dog 24
Separation barrier 41, 46
Separator air 534
Septic galley 683
system 682
system chamber 683
system effluent-filter 683
tank 682
tank concrete 682
tank polyethylene 683
tank precast 682
Service door 280
electric 591
entrance cable aluminum 571
laboratory analytical 47
residential 578
sink 518
sink faucet 520
station equipment 406
Sewage aeration 722
collection concrete pipe 682
collection PVC pipe 682
ejector pump 510
municipal waste water 722
pipe 682, 684, 685
treatment plant 722
Sewage/drainage collection 685
Sewer relining 676
Shade 438
polyester room darkening 439
Shaft wall 321
Shake wood 235
Shape structural aluminum 135
Shear connector welded 126
plate 172
test 16
wall 192
Shearwall panel 188
Sheathed nonmetallic cable 571
Romex cable 571
Sheathing 191, 192
asphalt 192
board 191
gypsum 193
insulating 190
paper 232
roof 192
roof & wall plywood 192
wall 192
Sheepsfoot roller 628, 639, 730
Sheet carpet 351
glass 312
glass mosaic 336
lead 472
metal 248
metal aluminum 251
metal flashing 249
metal roof 248
metal screw 171, 239
modified bituminous barrier 233
piling 641, 885
steel pile 644
Sheeting 641
driver 732
open 643
tie back 643
wale 641
wood 627, 641, 885
Shelf bathroom 389
bin 397
library 449
metal 396
Shellac door 363
Shell-type heat exchanger 554
Shelter aluminum 470
dock 407
golf 426
rail dock 407
temporary 52
Shelving 397
food storage metal 416
mobile 449
pine 397
plywood 397
prefinished 397
refrigeration 461
steel 396
storage 396
wood 397
Shield expansion 121
lag screw 121
light 708
Shielded copper cable 569
Shielding lead 473
radio frequency 474
radio frequency security 341
Shift work 12
Shingle 233
aluminum 233
asphalt 233
cedar 234
concrete 235
metal 234
panelized 234
removal 223
roof 233
slate 234

For customer support on your Building Construction Costs with RSMeans data, call 800.448.8182.

steel 234
strip 233
wood 234
Ship ladder 155
Shock absorber 507
  absorber pipe 507
  absorbing door 264, 285
Shooting range 425
Shop drill 422
  equipment 422
Shore post 61
Shoring 641, 642
  aluminum joist 61
  baseplate 19
  bracing 19
  concrete formwork 61
  concrete vertical 61
  flying truss 61
  frame 61
  heavy duty 19
  leveling jack 19
  slab 20
  steel beam 61
  temporary 641
Short load concrete 78
Shot blast floor 319
  blaster 734
Shotcrete 83
  pump 728
Shoulder paver 736
Shovel 623
  crawler 730
Shower arm 517
  built-in 517
  by-pass valve 517
  cabinet 390
  cubicle 516
  door 390
  drain 508
  emergency 522
  enclosure 390
  glass door 390
  group 519
  pan 249
  partition 114, 390
  polypropylene 516
  receptor 390, 516, 519
  rod 388
  safety 522
  stall 516
  surround 391
Shower/tub control set 517
Shower-tub valve spout set 517
Shredder 431
  compactor 431
Shrinkage test 15
Shrub and tree 672
  broadleaf evergreen 672
  deciduous 672
  evergreen 672
  moving 674
Shut off flood whole house
safety 495
  safety water and gas 495
  water heater safety 495
Shutter 438
  exterior 214
  exterior roll-up 398
  interior 438
  plantation 438
  wood 214
  wood interior 438
Side coiling grille 280
Sidelight 213
  door 269, 276
Sidewalk 342, 656

asphalt 652
brick 656
bridge 20
broom 734
concrete 652
driveway and patio 652
removal 30
temporary 24
Sidewall bracket 21
Siding aluminum 238
  bevel 240
  cedar 240
  fiber cement 242
  fiberglass 238
  metal 237, 238
  nail 171
  paint 360
  painting 359
  panel aluminum 238
  plywood 240
  polypropylene 242
  railroad 709, 887
  redwood 240
  removal 223
  ribbed 238
  stain 240
  steel 239
  vinyl 241
  wood 240
  wood board 240
Sieve analysis 15
Sign 25, 378, 380
  acrylic 378
  aluminum 378
  base formwork 56
  corridor 378
  detour 24
  directional 380
  flexible 378
  highway 380
  letter 378
  metal 378
  plastic 378
  plywood 378
  post 380
  project 25
  reflective 380
  road 380
  stainless 378
  street 378
  traffic 380
Signage custom wayfinding 379
  engraved panel 378
  exterior 378
  photoluminescent 379
  post and panel 379
Signal bell 611
  system traffic 707
Silencer duct 540
Silica chip 51
Silicon carbide abrasive 50, 80
Silicone 260
  caulking 261
  coating 226
  water repellent 226
Sill 113
  & ledger framing 186
  anchor 172
  block 109
  door 207, 213, 305
  masonry 116
  mud 186
  precast concrete window 86
  quarry tile 336
  stone 112, 115
  window 292

wood 186
Sillcock hose bibb 517
Silo 470
  demo 455
Silt fence 639
Simulated brick 117
  stone 117, 118
Single hub pipe 508
  hung aluminum sash window 289
  hung aluminum window 289
  precast concrete tee beam 85
  zone rooftop unit 558
Single-ply roofing 245
Sink barber 409
  base 439
  countertop 516
  countertop rough-in 516
  faucet kitchen 517
  kitchen 516
  laboratory 421
  laundry 517
  lavatory 515
  mop 518
  raised deck rough-in 516
  removal 493
  residential 516
  service 518
  service floor rough-in 518
  service wall rough-in 518
  support 520
Site clear and grub 615
  clearing 615
  demolition 29
  furnishing 451
  grading 617
  improvement 667, 669
  preparation 28
  rough grading 617
Skidder log 730
Skidsteer attachments 731
  loader 731
Skylight demo 265
  domed metal-framed 298
  fiberglass panel 469
  metal-framed 298
  plastic 298
  removal 223
  rigid metal-framed 298
Skyroof 288
Slab blockout formwork 56
  box out opening formwork 55
  bulkhead formwork 55, 56
  concrete 75
  concrete placement 79
  curb formwork 55, 56
  cutout 32
  edge form 143
  edge formwork 55, 56
  elevated 75
  flat plate formwork 54
  haunch 76
  haunch formwork 76
  joist dome formwork 54
  joist pan formwork 54
  lift 77, 852
  lifting insert 69
  on grade 76
  on grade formwork 56
  on grade post-tensioned 73
  on grade removal 30
  pan 75
  precast concrete channel 83
  reinforcing 71
  reinforcing bolster 68
  saw cutting 89
  screed formwork 57

shoring 20
stamping 82
textured 77
thickened edge 76
thickened edge formwork 76
trench formwork 57
turndown 76
turndown formwork 76
void formwork 55, 57
waffle 75
with drop panel formwork 54
Slate 115
  flagging 657
  pavement 115
  roof 234, 866
  shingle 234
  sidewalk 657
  sill 116
  stair 115
  tile 343
Slatwall 209, 354
Sleeper floor 186
  framing 186
  wood 186
Sleeve anchor bolt 64
  dowel 72
  formwork 61
  plastic pipe 61
Slide gate 661
  playground 426
  swimming pool 458
Sliding aluminum door 279
  chalkboard 375
  door shower 390
  glass door 279
  glass vinyl clad door 279
  mirror 390
  panel door 288
  partition 387
  vinyl clad door 279
  window 292
  window aluminum 289
Slipform paver barrier 736
  paver curb 736
Slope protection 639
  protection rip-rap 640
Slot letter 396
Slotted block 108
  channel framing 130
  pipe 686
Slump block 106
Slurry seal (latex modified) 652
  trench 644
Smart electrical metering 584
Smoke detector 610
  hatch 256
  vent 256
  vent chimney 548
Snap-tie formwork 61
Snow fence 39
  guard 256
Soaking bathroom 516
  tub 516
Soap dispenser 389
  holder 390
Soccer goalpost 427
Socket ground 425
  wire rope 136
Sod 671
  tennis court 660
Sodding 671
Sodium high pressure fixture 599
  low pressure fixture 598
Soffit & canopy framing 186
  drywall 329
  framing wood 186

For customer support on your Building Construction Costs with RSMeans data, call 800.448.8182.

marble	114
metal	242
molding	207
plaster	327
plywood	207
polyvinyl	207, 242
PVC	218
steel	467
stucco	328
wood	207
Softener water	512
Soil compaction	628
decontamination	39
mix man-made	236
pipe	508
poisoning	639
sample	29
stabilization	640
tamping	628
test	16
treatment	639
Solar energy	553
energy circulator air	553
energy system collector	553
energy system controller	553
energy sys. heat exchanger	553
film glass	314
heating	553, 878
heating system	553
panel driveway security gate	283
screen block	109
Soldier beam	643
Solid surface countertop	443
wood door	273
Sorting reinforcing	71
Sound absorbing panel	355
attenuation blanket	355
control demo	455
curtain	472
dampening panel gypsum	331
movie	420
panel	472
proof enclosure	461
system	605
system speaker	605
Source heat pump water	560
Space heater	733
heater floor mounted	552
office & storage	17
Spade air	732
tree	731
Spandrel beam formwork	52
glass	312
Spanish roof tile	235
Speaker movie	420
sound system	605
station nurse	606
Special construction	472
door	278, 283
hinge	306
nurse call system	606
purpose room demo	455
Specialty	389
demo	374
piping HVAC	534, 536
scaffold	20
telephone	380
Specific gravity	16
gravity testing	15
Speed bump	708
SPF roofing	247
Spike grid	172
unit traffic	407
Spinner ventilator	546
Spiral reinforcing	71
stair	163, 209

Spire church	399
Splice pile	645, 646
Splicer movie	420
Splicing reinforcing bar	70
Split astragal	309
rail fence	664
rib block	106
ring	172
system control system	531
Splitter damper assembly	539
Spotlight	597
Spotter	635
Spray coating	224
insulation	229
painting booth automotive	406
rig foam	732
substrate	43
Sprayed cementitious frprfng	257
Sprayed-on insulation	229
Sprayer airless	40
emulsion	732
paint	733
Spread footing	76
footing formwork	56
Spreader aggregate	728, 735
chemical	735
concrete	733
laundry	409
Spring bolt astragal	308
bronze weatherstrip	309
hinge	307
Sprinkler head	488
irrigation system	669
monitoring panel deluge	488
system	488
system accelerator	488
system dry-pipe	488
underground	669
Square bar fiberglass	216
tube fiberglass	216
Squash court	462
court backstop	426
Stabilization fabric	653
masonry	94
soil	640
Stabilizer base	736
Stacked bond block	106
Stadium	465
bleacher	466
cover	464
Stage curtain	423
equipment	422
equipment demolition	404
lighting	597
portable	422
Staging aid	21
pump	18
swing	22
Stain cabinet	363
door	363
floor	366
lumber	194
siding	240
truss	367
Staining concrete floor	341
Stainless bar grating	158
column cover	161
cooling tower	557
duct	538
flashing	249
gutter	253
plank grating	158
reglet	254
rivet	125
screen	381
sign	378

steel corner guard	387
steel cot	448
steel countertop	436, 442
steel cross	432
steel door	277
steel door and frame	277
steel downspout	252
steel gravel stop	251
steel hinge	306
steel pipe	503
steel roof	248
steel shelf	389
steel storefront	288
weld rod	127
Stair & walk moving	480
basement	209
basement bulkhead	84
brick	112
carpet	351
ceiling	414
climber	481
concrete	77
disappearing	414
electric	414
finish concrete	81
fire escape	155
floor plate	154
formwork	59
grating	154
landing	348
marble	114
metal pan	154
metal plate	154
nosing	345, 347
pan tread	80
part wood	209
precast concrete	84
prefabricated wood	209
railroad tie	653
removal	169
residential	209
riser	348
riser vinyl	345
rubber	345
slate	115
spiral	163, 209
stringer	348
stringer framing	183
stringer wood	183
temporary protection	24
terrazzo	347
tread	112, 160, 345
tread and riser	345
tread fiberglass	216
tread insert	62
tread terrazzo	348
tread tile	336
wood	209
Stairlift wheelchair	481
Stairway door hardware	300
scaffold	19
Stairwork handrail	211
Stake formwork	62
nail	62
subgrade	69
Stall shower	516
toilet	380-382
type urinal	518
urinal	518
Stamping slab	82
texture	81
Standard extinguisher	393
Standing seam	248
Standpipe connection	486
steel	681
Starter board & switch	586, 587

motor	588
Station control	592
diaper	388
hospital	428
transfer	431
weather	474
Stationary ventilator	547
ventilator mushroom	547
Steam bath	462
bath residential	462
boiler	549, 550
boiler electric	549
clean masonry	95
cleaner	734
cleaning metal	120
condensate meter	537
humidifier	564
jacketed kettle	417
pressure valve	535
regulator pressure	505
trap	535
Steamer	417
Steam-to-water heating	554
Steel anchor	98
astragal	308
ballistic door	282
beam shoring	61
beam W-shape	131
bin wall	666
blocking	180
bolt	172
bolt & HEX nut	123
bridge	668
bridging	181
building components	863
building pre-engineered	466
chain	162
channel	326
conduit in slab	576
conduit in trench	576
conduit intermediate	574
conduit rigid	574
corner guard	129
cutting	122
demolition	120
diffuser	545
dome	469
door	267, 268, 283, 284, 467, 868
door and frame stainless	277
door residential	269
downspout	252
drill	732
drilling	122
edging	653, 674
estimating	860
expansion tank	535
fence	661, 663
fiber	74
fiber reinforcing	74
flashing	249
frame	266
framing	133
furring	322
girt	134
grating	157
gravel stop	251
gutter	253
HEX bolt	123
high strength	134
hinge	306
lath	325
lintel	130
locker	394
louver door	315
member structural	131
mesh box	640

For customer support on your Building Construction Costs with RSMeans data, call 800.448.8182.

Index

mesh grating ... 159
mill extra ... 134
outlet box ... 574
painting ... 371
partition ... 385, 387
pile ... 644
pile concrete-filled ... 646
pile sheet ... 644
pipe ... 501, 684
pipe corrugated ... 684
pipe removal ... 29
plate ... 131
prestressing ... 73
primer ... 371
project structural ... 133
purlin ... 134
register ... 545
reinforcement ... 845
reinforcing ... 844, 845
rivet ... 125
roof ... 237
roofing panel ... 237
salvage ... 493
sash ... 290, 868
screw ... 171
sections ... 862
security driveway gate ... 282
sheet pile ... 641
shelving ... 396
shingle ... 234
siding ... 239
siding demolition ... 224
silo ... 470
standpipe ... 681
structural ... 860
stud NLB ... 323
tank ... 535
tank above ground ... 533
testing ... 15
tower ... 704
underground duct ... 703
underground storage tank ... 532
valve ... 529
water storage tank ... 681
water tank ground level ... 681
weld rod ... 127
well casing ... 678
window ... 289, 290, 468
window demolition ... 264
wire rope ... 137
wire rope accessory ... 136
Steeple ... 399
aluminum ... 399
tip pin ... 306
Step ... 380
bluestone ... 653
brick ... 652
manhole ... 677
masonry ... 652
railroad tie ... 652
stone ... 113
tapered pile ... 644
Sterilizer barber ... 409
dental ... 428
glassware ... 421
medical ... 427
Stiffener joist web ... 149
plate ... 131
Stiffleg derrick crane ... 737
Stile fabric ... 308
protection ... 308
Stock feeder ... 433
Stockpiling of soil ... 617
Stone aggregate ... 672
anchor ... 98
ashlar ... 114
base ... 114, 653
cast ... 117
column ... 114
crushed ... 614
cultured ... 117
curb cut ... 658
fill ... 614
filter ... 640
floor ... 114
gabion retaining wall ... 666
ground cover ... 672
mulch ... 670
paver ... 113, 657
pea ... 51
removal ... 30
retaining wall ... 666
river ... 51
sill ... 112, 115
simulated ... 117, 118
step ... 113
stool ... 115
toilet compartment ... 382
tread ... 112
trim cut ... 114
wall ... 112, 666
Stool cap ... 207
doctor ... 428
stone ... 115
window ... 114, 116
Stop door ... 207
gravel ... 251
Stop/start control station ... 592
Storage bottle ... 415
box ... 17
cabinet ... 441
dome ... 469
dome bulk ... 469
metal shelving food ... 416
room cold ... 460
shelving ... 396
tank ... 532, 681
tank concrete aboveground ... 534
tank cover ... 463
tank demo ... 456
tank steel underground ... 532
tank underground ... 532
Storefront aluminum ... 288
Storm door ... 265
door residential ... 265
drainage manhole frame ... 676
drainage piping ... 684
sash prefinished ... 469
window ... 295
window aluminum residential ... 295
Stormwater management ... 687
Stove ... 393, 417
woodburning ... 393
Strainer bronze body ... 535
downspout ... 252
gutter ... 254
roof ... 252
wire ... 252
Y type bronze body ... 535
Y type iron body ... 536
Strand grouted ... 73
ungrouted ... 73
Stranded wire ... 572
Strap tie ... 173
Straw ... 670
bale construction ... 471
Street sign ... 378
Streetlight ... 598
Strength compressive ... 15
Stressed skin roof ... 131
Stringer stair ... 348
stair terrazzo ... 348
Strip cabinet ... 605
chamfer ... 60
entrance ... 285
filler ... 309
floor ... 343
footing ... 76
lighting ... 594
rumble ... 708
shingle ... 233
soil ... 617
Striper paint ... 736
Stripping topsoil ... 617
Strobe light ... 597
Structural aluminum ... 135
backfill ... 628
brick ... 102
column ... 128
compaction ... 639
excavation ... 622
fabrication cost ... 134
face tile ... 103
facing tile ... 103
fee ... 10
framing ... 861
insulated panel ... 188
panel ... 188
shape column ... 129
steel ... 860
steel member ... 131
steel project ... 133
tile ... 103
welding ... 123, 134
Structure fabric ... 463
moving ... 38
tension ... 464
Stub switchboard ... 589
Stucco ... 328
demolition ... 319
mesh ... 326
painting ... 359
wall coating ... 370
Stud demolition ... 169
metal ... 320
NLB metal ... 323
NLB steel ... 323
NLB wall ... 323
partition ... 183, 326
partition framing load-bearing ... 145
wall ... 183, 187, 322
wall blocking load-bearing ... 144
wall box beam load-bearing ... 144
wall bracing load-bearing ... 143
wall framing load-bearing ... 145
wall header load-bearing ... 144
wall wood ... 321
welded ... 126
wood ... 187, 320
Stump chipper ... 730
chipping ... 615
removal ... 616
Subcontractor O&P ... 13
Subdrainage pipe ... 684
system ... 684
Subfloor ... 190
adhesive ... 191
demolition ... 319
plywood ... 190
wood ... 191
Subgrade reinforcing chair ... 69
stake ... 69
Submersible pump ... 512, 678, 734
sump pump ... 512
Sub purlin bulb tee ... 87
bulb tee ... 135
Subpurlins ... 863
Subsoil heating ... 459
Subsurface exploration ... 28
Suction hose ... 733
Sump hole construction ... 627
pit ... 627
pump ... 413, 511
pump submersible ... 512
Super high-tensile anchor ... 640
Supermarket checkout ... 409
scanner ... 409
Supply ductile iron pipe water ... 678
Support carrier fixture ... 520
ceiling ... 127
drinking fountain ... 520
framing pipe ... 131
lavatory ... 520
light ... 128
monitor ... 128
motor ... 128
partition ... 127
sink ... 520
urinal ... 520
water closet ... 520
water cooler ... 521
X-ray ... 128
Suppressor lightning ... 578
Surface bonding ... 97
landscape ... 351
playground ... 351
preparation exterior ... 356
preparation interior ... 357
raceway ... 590
Surfacing ... 652
Surfactant ... 43
Surgery equipment ... 429
table ... 427
Surgical lighting ... 427
Surround shower ... 391
tub ... 391
Surveillance system TV ... 609
Survey aerial ... 28
crew ... 25
monument ... 28
preblast ... 623
property line ... 28
stake ... 25
topographic ... 28
Suspended ceiling ... 325, 338
ceiling framing ... 182
tile ... 339
Suspension mounted heater ... 552
system ceiling ... 339
Sweeper road ... 736
Swell testing ... 15
Swimming pool ... 457, 458
pool blanket ... 458
pool demo ... 456
pool enclosure ... 465
pool equipment ... 458
pool filter ... 524
pool filtration ... 524
pool hydraulic lift ... 459
pool ladder ... 458
pool painting ... 458
pool ramp ... 459
pools ... 874
Swing ... 426
check steel valve ... 529
check valve ... 494, 528
check valve bronze ... 494
clear hinge ... 307
gate ... 661
staging ... 22
Switch box ... 574
bus duct ... 590
decorator ... 579
dimmer ... 579, 591

927

For customer support on your Building Construction Costs with RSMeans data, call 800.448.8182.

electric 591, 592
general duty 592
residential 578
safety 592
toggle 591
Switchboard distribution 586
electric 592
feeder section 586
incoming 586
instrument 584
pressure switch 586
stub 589
w/CT compartment 586
Switching low-voltage 591
Switchplate low-voltage 591
Synthetic erosion control 639
fiber 74
fiber reinforcing 74
marble 342
turf 351
System chamber septic 683
component foam-water 489
control 531
fire extinguishing 489
grinder pump 510
septic 682
sprinkler 488, 489
subdrainage 684
tube 482
T.V. 604
T.V. surveillance 609
UHF 604
VHF 604

T

Table folding 447
laboratory 421
massage 424
overbed 449
physical therapy 428
reading 449
rewind 420
round 447
saw 422
surgery 427
Tackboard fixed 376
Tactile warning surfacing 659
Tamper 619, 732
Tamping soil 628
Tandem roller 730
Tank above ground steel 533
aboveground storage conc. . . . 534
clean 39
cover 463
darkroom 410
developing 410
disposal 39
expansion 535
fiberglass 532
fixed roof 533
gasoline 533
holding 722
horizontal above ground 533
Hubbard 428
leak detection 610
pillow 682
removal 39
septic 682
steel 535
steel underground storage 532
storage 532, 681
testing 16
water 681, 735
water storage solar 553

Tankless gas water heater 513
Tap box bus duct 589
Tape barricade 24
detection 677
sealant 261
temporary 659
underground 677
Tar kettle 734, 736
paper 670
pitch emulsion 652
roof 222
Target range 425
Tarpaulin 23
duck 23
Mylar 23
polyethylene 23
polyvinyl 23
Tarred felt 244
Tax 12
sales 12
social security 12
unemployment 12
Taxes 831, 832
T-bar mount diffuser 545
Teak floor 343
molding 207
paneling 208
Technician inspection 16
Tee cleanout 498
framing steel 130
pipe 502
precast 85
precast concrete beam 84
Telephone enclosure 380
enclosure commercial 380
house 605
manhole 703
pole 703
receptacle 581
specialty 380
Telescoping bleacher 450
boom lift 731
platform 422
Television equipment 605
receptacle 581
system 604
Teller automatic 411
window 411
Temperature relief valve 495
rise detector 611
Tempered glass 310
glass greenhouse 465
hardboard 208
Tempering valve 495
valve water 495
Temporary barricade 23
building 17
construction 17, 23
electricity 17
facility 23
fence 24
guardrail 24
heat 52
light 17
lighting 17
power 17
protection 25
ramp 22
road 22
road painting 659
shelter 52
shoring 641
tape 659
toilet 734
utility 17
Tennis court air supported 463

court backstop 662
court clay 660
court fence 662, 664
court fence and gate 662
court net 660
court painting 660
court post 660
court sod 660
court surface 350
court surfacing 660
Tensile test 15
Tension structure 464
structure demo 457
Terminal A/C packaged 559
air conditioner packaged 559
heater 551
lug 570
Termination box 605
cable 570
Termite control chemical 639
protection 639
Terne coated flashing 249
Terra cotta 103, 104
cotta coping 103, 117
cotta demolition 32, 320
cotta tile 104
Terrazzo 347
abrasive 347
base 347, 348
conductive 349
cove base 348
curb 347, 348
demolition 319
epoxy 348, 349
floor 347, 873
floor tile 348
monolithic 347
plastic matrix 348
polyacrylate 348
precast 348
receptor 390
riser 348
stair 347
Venetian 347, 348
wainscot 347, 348
Test beam 15
equipment laboratory 421
load pile 614
moisture content 16
pile load 614
pit 28
soil 16
ultrasonic 16
well 678
Testing 15
lead 44
OSHA 43
pile 614
sulfate soundness 15
tank 16
Textile wall covering 353
Texture stamping 81
Textured ceiling refinish 318
slab 77
Theater and stage equipment 422
Theatrical lighting 597
Thermal & moisture
protection demo 222
batt 355
Thermometer 421
Thermoplastic 313
painting 659
polyolefin roof 246
polyolefin roofing 246
Thermostat control sys. radiator . 532
integral 564

wire 583
Thickened edge slab 76
Thimble wire rope 136
Thin brick veneer 101
plaster partition 322
Thincoat plaster 329
Thinning tree 615
Thin-set ceramic tile 334
mortar 336
Thin-set tile 334, 335
Threaded rod fiberglass 216
Threshold 305
aluminum 305
door 214
stone 114
wood 207
Thru-wall air conditioner 559
Ticket booth 470
dispenser 406
Tie back sheeting 643
rafter 186
railroad 653, 674
rod 130, 641
strap 173
wall 98
wire reinforcing 69
Tier locker 394
Tile 333, 337, 346
abrasive 334
aluminum 236, 333
carpet 350
ceiling 338, 339
ceramic 334, 335
clay 235
concrete 235
cork 342
cork wall 352
demolition 318
exposed aggregate paver 656
exterior 335
fire rated 103
floor 348
flooring cork 342
flue 116
glass 313, 336
grout 333, 334
marble 342
metal 333
or terrazzo base floor 347
partition 103
porcelain 334
quarry 336
recycled rubber tire 445
roof 235
roof metal 236
rubber 347
slate 343
stainless steel 333
stair tread 336
structural 103
terra cotta 104
vinyl composition 346
wall 334
waterproofing membrane
ceramic 337
window sill 336
Tiling ceramic 333-337
quarry 337
Tilt-up concrete 856
concrete wall panel 86
precast concrete wall panel 86
Timber connector 172
fastener 171
framing 189
guide rail 708
heavy 189

For customer support on your Building Construction Costs with RSMeans data, call 800.448.8182.

laminated . . . 193
pile . . . 645
retaining wall . . . 666
switch tie . . . 706
switch tie railroad . . . 706
tie railroad . . . 706
Time lapse photography . . . 14
lock . . . 283
system controller . . . 606
Timer clock . . . 582
interval . . . 579
Tin ceiling . . . 340
Tinted glass . . . 310
window film electrically . . . 314
Toaster . . . 417
Toggle switch . . . 591
Toggle bolt anchor . . . 121
Toilet . . . 722
accessory . . . 388
accessory commercial . . . 388
bowl . . . 515
chemical . . . 734
compartment . . . 380
compartment plastic . . . 383
compartment plastic-laminate . . . 381
compartment stone . . . 382
partition . . . 114, 380, 381
partition removal . . . 320
prison . . . 433
stall . . . 380-382
stone partition . . . 382
temporary . . . 734
tissue dispenser . . . 389
Tool powder actuated . . . 125
rivet . . . 125
Toothed ring . . . 173
Toothing masonry . . . 32
Top coiling grille . . . 280
demolition counter . . . 170
dressing . . . 671
vanity . . . 444
Topographical survey . . . 28
Topping concrete . . . 87
epoxy . . . 349
floor . . . 81, 87
grout . . . 88
Topsoil . . . 614, 617
placement and grading . . . 671
stripping . . . 617
Torch cutting . . . 122, 734
cutting demolition . . . 35
Torrified door . . . 275
Towel bar . . . 389
dispenser . . . 388
Tower control . . . 471
cooling . . . 557
crane . . . 18, 736, 837, 838
hoist . . . 737
light . . . 733
radio . . . 704
TPO roof . . . 246
Track accessory . . . 706
accessory railroad . . . 706
curtain . . . 423
drawer . . . 440
drill . . . 732
heavy rail railroad . . . 706
material railroad . . . 706
running . . . 344
traverse . . . 437
Tractor . . . 628
loader . . . 730
rake . . . 730
truck . . . 735
Traffic barrier highway sound . . . 668
channelizing . . . 708
cone . . . 23
control mirror . . . 401
door . . . 285
line . . . 659
line remover . . . 736
sign . . . 380
signal system . . . 707
spike unit . . . 407
Trailer bunk house . . . 17
floodlight . . . 732
lowbed . . . 736
mud . . . 735
office . . . 17
platform . . . 734
truck . . . 734
water . . . 734
Trainer bicycle . . . 423
Tram car . . . 735
Transceiver . . . 604
Transfer station . . . 431
Transformer . . . 585
& bus duct . . . 585, 590
cabinet . . . 590
current . . . 584
dry-type . . . 585
fused potential . . . 584
low-voltage . . . 591
oil-filled . . . 585
Transition molding wood . . . 344
residential . . . 543
strip floor . . . 350
Translucent ceiling . . . 340
Transom aluminum . . . 287
lite frame . . . 267
window . . . 294
Transplanting plant and bulb . . . 674
Trap cast iron . . . 508
deep seal . . . 509
drainage . . . 508, 509
grease . . . 510
inverted bucket steam . . . 535
rock surface . . . 81
steam . . . 535
Trash closure . . . 451
compactor . . . 419
pump . . . 734
receiver . . . 445
receptacle . . . 451
receptacle fiberglass . . . 451
receptacle metal . . . 445
Travel nursery item . . . 672
Traverse . . . 437
track . . . 437
Travertine . . . 114
Tray cable . . . 576
Tread abrasive . . . 336
and riser stair . . . 345
bar grating . . . 160
channel grating . . . 160
cover renovation . . . 161
diamond plate . . . 160
insert stair . . . 62
rubber . . . 345
stair . . . 112, 160, 345
stair pan . . . 80
stone . . . 114, 115
vinyl . . . 345
wood stair . . . 211
Treadmill . . . 424
Treated lumber . . . 186
lumber framing . . . 187
pile . . . 645
pine fence . . . 664
Treatment acoustical . . . 355
equipment . . . 428
flooring . . . 341
plant secondary . . . 722
plant sewage . . . 722
potable water . . . 512
Tree deciduous . . . 673
evergreen . . . 672
guying system . . . 674
moving . . . 674
removal . . . 615, 616
removal selective . . . 616
spade . . . 731
thinning . . . 615
Trench backfill . . . 619, 621
box . . . 735
cover . . . 159
disposal field . . . 683
drain . . . 511
drain fiberglass . . . 511
drain plastic . . . 511
drain polymer . . . 511
duct . . . 577
duct fitting . . . 577
duct raceway . . . 577
duct steel . . . 577
excavating . . . 618
excavation . . . 618, 621
grating frame . . . 159
slurry . . . 644
utility . . . 621
Trencher chain . . . 621, 731
rock . . . 731
wheel . . . 731
Trenching . . . 622, 627
Trial batch . . . 15
Trim demolition . . . 170
exterior . . . 200, 201
exterior paint . . . 362
molding . . . 206
painting . . . 366
PVC . . . 217
tile . . . 335
wood . . . 203
Triple weight hinge . . . 306
Tripod floodlight . . . 732
Troffer air handling . . . 594
light fixture . . . 595
Trolley rail . . . 388
Trowel coating . . . 224
concrete . . . 728
power . . . 728
Truck concrete . . . 728
crane flatbed . . . 736
dock . . . 407
dump . . . 731
flatbed . . . 731, 735
holding concrete . . . 78
leveler . . . 407
loading . . . 622
mounted crane . . . 737
pickup . . . 735
scale . . . 402
tractor . . . 735
trailer . . . 734
vacuum . . . 735
winch . . . 735
Truss bowstring . . . 194
demolition . . . 170
framing metal . . . 153
joist . . . 141
metal roof . . . 134
painting . . . 367
plate . . . 173
roof . . . 193, 194
stain . . . 367
varnish . . . 367
Tub bar . . . 389
rough-in . . . 516
soaking . . . 516
surround . . . 391
Tube framing . . . 285
pneumatic . . . 411
system . . . 482
system air . . . 482
system pneumatic . . . 482
Tubing copper . . . 500
electric metallic . . . 574
electrical . . . 573
fitting PEX . . . 563
PEX . . . 504, 562
Tubular fence . . . 663
lock . . . 303
scaffolding . . . 19
Tugboat diesel . . . 738
Tumbler holder . . . 389
Tunnel locomotive . . . 736
muck car . . . 736
ventilator . . . 736
Turbine wind . . . 724
Turf artificial . . . 660
synthetic . . . 351
Turnbuckle wire rope . . . 137
Turndown slab . . . 76
Turned column . . . 212
Turnout railroad . . . 709
Turnstile . . . 408
security . . . 408
TV closed circuit camera . . . 609
system . . . 604
Two piece astragal . . . 309
Tying wire . . . 61

U

UHF system . . . 604
Ultrasonic cleaner . . . 429
motion detector . . . 609
test . . . 16
Uncased drilled concrete pier . . . 648
Underdrain . . . 677
Undereave vent . . . 316
Underfloor duct . . . 577
duct electrical . . . 577
duct fitting . . . 577
Underground duct . . . 703
electrical & telephone . . . 703
sprinkler . . . 669
storage tank . . . 532
storage tank removal . . . 840
storage tank steel . . . 532
tape . . . 677
Underlayment gym floor . . . 344
hardboard . . . 191
latex . . . 344
self-leveling cement . . . 88
Underpin foundation . . . 643
Underwater light . . . 458
Undisturbed soil . . . 16
Unemployment tax . . . 12
Ungrouted bar . . . 73
strand . . . 73
Unit air-handling . . . 557, 561
commercial kitchen . . . 414
heater . . . 552, 561
heater propeller . . . 562
Upstanding beam formwork . . . 53
Urethane foam . . . 355
insulation . . . 460
wall coating . . . 371
Urinal . . . 518
high efficiency . . . 519
removal . . . 493
screen . . . 382

For customer support on your Building Construction Costs with RSMeans data, call 800.448.8182.

stall 518
stall type 518
support 520
wall hung 518
waterless 518
watersaving 518
Urn ash 445
Utensil washer medical 427
washer-sanitizer 421
Utility accessory 677
box 677
connection 677
duct 703
electric 704
line removal 29
pole 597, 703
pole wood electrical 703
set fan 541, 542
sitework 703
temporary 17
trench 621
trench excavating 621

V

Vacuum breaker 506
catch basin 735
central 430
cleaning 430
loader 41
pick-up 728
truck 735
unit central 718
wet/dry 735
Valance board 440
Valley flashing 233
gutter 468
rafter 186
Valve 528
assembly sprinkler deluge 489
backwater 511
brass 493
bronze 493
bronze angle 493
butterfly 528, 679
cabinet 393
check 486
CPVC 496
fire hose 487
flush 519
foot 496
forged 529
gas stop 493
gate 528, 529
globe 529
heating control 535
hot water radiator 535
iron body 528
mixing 517
plastic 496
plastic angle 496
polypropylene 496
polypropylene ball 496
PVC 496
relief pressure 495
shower by-pass 517
soldered gate 494
spout set shower-tub 517
sprinkler alarm 486
steam pressure 535
steel 529
swing check 494, 528
swing check steel 529
tempering 495
water pressure 495

Y sediment strainer 496
Vandalproof fixture 595
Vaneaxial fan 541
Vanity base 440
top 444
top lavatory 515
Vapor barrier 232, 461
barrier sheathing 192
retarder 232
Vaportight fixture incandescent . 595
Varnish cabinet 363
casework 363
door 363
floor 366, 369
polyurethane 371
truss 367
VAT removal 42
Vault day gate demolition 404
door 283
front 283
Vaulting side horse 424
VCT removal 319
Vehicle barrier security 707
charging electric 593
guide rail 708
Veneer ashlar 112
brick 99
core paneling 208
granite 112
member laminated 194
removal 96
Venetian plaster 328
terrazzo 347, 348
Vent automatic air 534
box 99
brick 546
chimney 548
chimney all fuel 549
convector or baseboard air ... 561
dryer 413
exhaust 546
flashing 509
foundation 99
gas 548
intake/exhaust 547
metal chimney 548
one-way 257
ridge 256
ridge strip 316
roof 256, 469
smoke 256
Ventilated metal deck 143
Ventilating air conditioning . 531, 538,
    539, 541, 542, 547, 555, 557, 559
Ventilation equip. and hood .. 547
fan 582
louver 316
Ventilator control system 531
masonry 99
mushroom stationary 547
relief vent 547
roof 257
spinner 546
stationary mushroom 547
tunnel 736
Venturi flow meter 536
Verge board 205
Vermiculite cement 258
insulation 228
plaster 327
Vertical blind 437
lift door 284
metal ladder 155
Very low density polyethylene . 38
VHF system 604
Vial bottle 47

Vibrating screed, gas engine,
8HP 728
Vibration pad 126
Vibrator concrete 728
earth 619, 628, 639
plate 628
plow 731
Vibratory hammer 730
roller 730
Vibroflotation 642
compaction 885
Video surveillance 609
Videotape recorder 609
Vinyl blind 214
casement window 296
clad premium window 294
coated fence 662
composition floor 346
composition tile 346
curtain 388
downspout 252
faced wallboard 329
floor 346
gutter 253
lead barrier 356
sheet floor 346
siding 241
stair riser 345
stair tread 345
tread 345
wall coating 371
wallpaper 353
window 296
window double hung 296
window single hung 295
Vision panel 433
Vocational kiln 422
shop equipment 422
Voltmeter 584
Volume control damper 539
Volume-control damper 539

W

Waffle slab 75
Wainscot ceramic tile 334
molding 207
quarry tile 336
terrazzo 347, 348
Wale 641
sheeting 641
Walk 652
moving 481
Walk-in refrigeration 460
Walkway cover 399
roof 244
Wall & corner guard 387
accessory formwork 62
aluminum 238
and partition demolition 319
base 344
blocking load-bearing stud ... 144
box beam load-bearing stud ... 144
boxout formwork 57
bracing load-bearing stud 143
brick 111
brick shelf formwork 57
bulkhead formwork 57
bumper 304, 388
buttress formwork 57
cabinet 439, 441, 442
canopy 398
cap residential 543
cast concrete retaining 665
cavity 111

ceramic tile 334
clock 445
coating 370
concrete 76
concrete finishing 81
concrete placement 79
concrete segmental retaining .. 665
corbel formwork 57
covering 873
covering cork 352
covering textile 353
cross 432
curtain 288
cutout 32
demolition 319
drywall 329
exhaust fan 542
fan 543
finish 353
flagpole 401
forms 842
formwork 57
foundation 107
framing 187
framing load-bearing stud 145
framing removal 32
furnace 552
furring 187, 323
grout 98
guard 129, 388
header load-bearing stud 144
heater 413
hung lavatory 515
hung urinal 518
hydrant 507
insulation 226, 468
lath 325, 326
lintel formwork 58
louver 316
masonry 104, 666
mat 424
mirror 313
NLB partition 321
painting 360, 368, 369
panel 117
panel brick 111
panel precast 86
panel precast concrete double . 85
panel woven wire 384
paneling 208
partition 321
pilaster formwork 58
plaster 327
plug 99
plywood formwork 57
precast concrete 86, 856
prefab plywood formwork 58
register 546
reinforcing 71, 845
retaining 77, 667
rubbing 81
screen 386
shaft 321
shear 192
sheathing 192
sill formwork 58
steel bin 666
steel framed formwork 58
stone 112, 666
stone gabion retaining 666
stucco 328
stud 183, 187, 320, 322
stud NLB 323
switch plate 591
tie 98
tie masonry 98

930

tile 334
tile cork 352
window 286, 288
wood stud 321
Wallboard repair gypsum 318
Wallcovering 353
acrylic 353
gypsum fabric 353
Wall-covering natural fiber 353
Wallguard 387
Wallpaper 873
cork 352
grass cloth 353
vinyl 353
Walnut floor 343
Wardrobe 397
cabinet 397
Warehouse 463
scale 402
Warm air heater 551
Warning beacon 597
surfacing tactile 659
Wash bowl residential 515
brick 95
fountain 519
fountain group 519
pressure 357
safety equipment eye/face 522
Washable air filter 547
air filter permanent 547
Washer 173
& dryer coin-operated 410
and extractor 410
commercial 410
pressure 734
residential 413
Washer-sanitizer utensil 421
Washing machine automatic 413
Waste cart medical 431
cleanup hazardous 40
compactor 431
disposal 45, 431
disposal hazardous 40
handling 430, 431
handling equipment 430, 431
incinerator 430
packaging 43
receptacle 390
Wastewater treatment system 722
Watchdog 24
Watchman service 24
Water atomizer 41
balancing 530
blasting 95
blasting metal 120
cement ratio 15
chiller 555
chiller absorption 555
chiller centrifugal type 555
chiller indirect-fired 555
chiller reciprocating 555
chiller screw 556
chiller scroll 555
closet 515, 518
closet chemical 722
closet removal 493
closet support 520, 521
coil chilled 560
cooler 523
cooler electric 523
cooler support 521
copper tubing connector 505
curing concrete 83
dispenser hot 513
distiller 427
distribution pipe 678, 679
effect testing 15
fountain removal 493
hammer arrester 507
heater 413, 583
heater commercial 513, 514
heater gas 514
heater gas residential 513
heater oil-fired 514
heater removal 493
heater residential 413, 514, 583
heater residential electric 513
heater safety shut off 495
heating hot 549, 561
hose 733
hydrant 507
pipe 678
pipe ground clamp 573
pressure reducing valve 495
pressure relief valve 495
pressure valve 495
pump 536, 678, 734
pump fire 490
pump fountain 524
pumping 627
reducer concrete 78
repellent 368
repellent coating 226
repellent silicone 226
softener 512
softener potable 512
softener removal 493
source heat pump 560
storage solar tank 553
supply domestic meter 505
supply ductile iron pipe 678
supply meter 505
supply PVC pipe 679
supply well pump 678
tank 681, 735
tank elevated 681
tank ground level steel 681
tempering valve 495
trailer 734
treatment 722
well 678
Waterless urinal 518
Waterproofing 225
butyl 225
cementitious 225
coating 224
demolition 224
elastomeric 225
elastomeric sheet 225
integral 52
masonry 97
membrane 224
neoprene 225
rubber 225
Watersaving urinal 518
Waterstop center bulb 63
dumbbell 62
fitting 63
PVC 62
ribbed 62
rubber 63
Water-water exchange hot 554
Watt meter 584
Wayfinding signage custom 379
Wearing course 654
Weather barrier 232
barrier or wrap 232
station 474
Weatherproof receptacle 581
Weatherstrip 281
aluminum 308
door 309
spring bronze 309
window 310
zinc 309
Weatherstripping 309
Web stiffener framing 149
Wedge anchor 122
Weight exercise 423
lifting multi station 424
Weights and measures 828
Weld joint pipe 501
rod 127
rod aluminum 127
rod cast iron 127
rod stainless 127
rod steel 127
Welded frame 266
shear connector 126
structural steel 859
stud 126
wire epoxy 73
wire fabric 72, 846
wire galvanized 73
Welder arc 422
Welding certification 16
fillet 123
machine 735
structural 123, 134
Well 627
& accessory 678
area window 400
casing steel 678
gravel pack 678
pump 678
pump shallow 678
water 678
Wellpoint 627
discharge pipe 738
equipment rental 738
header pipe 738
pump 738
riser pipe 738
Wellpoints 884
Wet/dry vacuum 735
Wheel color 597
guard 388
potter's 422
stop railroad 706
trencher 731
Wheelbarrow 735
Wheelchair lift 481
stairlift 481
Wheeled loader 730
Whirler playground 426
Whirlpool bath 428, 459
Wide flange beam fiberglass 216
throw hinge 307
Widener pavement 736
Winch truck 735
Wind turbine 724
Window 289
& door demolition 264
aluminum 289
aluminum awning 289
aluminum projected 289
aluminum residential storm 295
aluminum sliding 289
and door molding 207
awning 290, 398
bank 411
basement/utility 290
blind 436
bow & bay metal-clad 293
bullet resistant 411
casement 289-291, 293
casement wood 291
demolition 264
double hung 294
double hung aluminum sash 289
double hung bay wood 293
double hung steel sash 290
double hung vinyl 296
double hung wood 291
drive-up 411
estimates 868
fiberglass 298
fiberglass single hung 298
frame 292
framing 467
glass 312
grille 294
guard 162
hardware 299
impact resistant aluminum 289
industrial 290
metal 289, 290, 469
metal-clad awning 292
metal-clad casement 292
metal-clad double hung 292
muntin 294
painting 364
Palladian 294
picture 291
picture & sliding metal-clad 292
pivoted 290
plastic 295
precast concrete sill 86
removal 264
restoration 45
sash wood 292
screen 290, 295
sill 116
sill marble 114
sill tile 336
single hung aluminum 289
single hung aluminum sash 289
single hung vinyl 295
sliding 292
sliding wood 292
steel 289, 290, 468
steel projected 290
steel sash picture 290
steel sash projected 290
stool 114, 116
storm 295
teller 411
transom 294
trim set 207
vinyl 296
vinyl casement 296
vinyl clad awning 293
vinyl clad premium 294
wall 286
wall aluminum 286
wall framing 285
weatherstrip 310
well area 400
wood 290
wood awning 290
wood bow 293
wood double hung 291
Windrow loader 736
Wine cellar 415
Winter concrete 78
protection 23, 52
Wire 879, 880
brushing metal 120
connector 570
copper 572
electrical 572
fabric welded 72
fence 24, 663
fencing 664

For customer support on your Building Construction Costs with RSMeans data, call 800.448.8182.

Index

glass 312
ground 573
hanging 339
mesh 328, 664
mesh locker 394
mesh partition 384
partition 384
razor 664
reinforcement 846
rope clip 136
rope socket 136
rope thimble 136
rope turnbuckle 137
strainer 252
stranded 572
thermostat 583
THW 572
THWN-THHN 572
tying 61
Wiremold raceway 590
Wireway 576
Wiring air conditioner 583
device 575, 591
device & box 575
fan 582
method 585
power 591
residential 578, 582
Wood & composite I-joist 193
awning window 290
ballistic door 282
base 196
beam 181, 189
beam & girder framing 181
bench 667
blind 438
block floor 343
block floor demolition 319
blocking 180, 184
bow window 293
bracing 180
bridge 668
cabinet school 441
canopy framing 186
case good 446
ceiling 340
chip 670
column 182, 189, 212
composition floor 343
cupola 399
cutting 34

deck 190
demolition 318
dome 469
door 270, 271, 273
door frame exterior 213
door frame interior 213
door residential 274
double acting swing door 285
double hung window 291
electrical utility pole 703
fascia 186, 203
fastener 171
fiber formboard 59
fiber plank cementitious 87
fiber sheathing 192
fiber underlayment 191
firestop 183
flagpole 400
floor 343
floor demolition 319
folding panel 438
folding partition 386
framing 180
framing demolition 166
framing miscellaneous 183
framing open web joist 182
framing plate 187
furring 187
girder 181
gutter 253
handrail 206, 211
header 187
interior shutter 438
joist 182, 193
joist framing 182
laminated 193
ledger 186
louver 212
nailer 183
overhead door 284
panel door 273
paneling 208
parking bumper 659
parquet 343
partition 183
pier floating 712
pile 645
plank scaffold 19, 21
planter 667
plate 187
pole 703

rafter 185, 186
rail fence 664
railing 206, 209, 211
roof deck 190
roof truss 193
sash 291
screen 295
screw 171
security driveway gate 282
shade 438
shake 235
sheet piling 885
sheeting 627, 641, 885
shelving 397
shingle 234
shutter 214
sidewalk 652
siding 240
siding demolition 224
sill 186
sleeper 186
sleeper framing 186
soffit 207
soffit & canopy framing 187
soffit framing 186
stair 209
stair baluster 209
stair newel 210
stair part 209
stair railing 210
stair riser 210
stair stringer 183
stair stringer framing 183
stair tread 211
storm door 277
strip flooring 343
stub wall framing 187
stud 187, 320
subfloor 191
threshold 207
trim 203
veneer wall covering 354
window 290
window casement 291
window demolition 264
window double hung bay 293
window sash 292
window sliding 292
Woodburning stove 393
Wool carpet 351
fiberglass 228

Work boat 738
extra 13
Worker protection 41
Workers' compensation 12, 834
Woven wire partition 384
Wrapping pipe 676
Wrecking ball 735
Wrench impact 732
Wrestling mat 424
Wrought iron fence 661

X

X-ray barrier 473
concrete slab 16
dental 429
equipment 429, 608
equipment security 608
machine 608
medical 429
mobile 429
protection 472
support 128
X-ray/radio freq prot. demo 457

Y

Y sediment strainer valve 496
type bronze body strainer 535
type iron body strainer 536
Yard fountain 459
Yellow pine floor 344

Z

Z bar suspension 339
Zee bar 333
framing steel 130
Zinc divider strip 347
plating 171
roof 248
terrazzo strip 347
weatherstrip 309

For customer support on your Building Construction Costs with RSMeans data, call 800.448.8182.

Division Notes

	CREW	DAILY OUTPUT	LABOR-HOURS	UNIT	BARE COSTS				TOTAL INCL O&P
					MAT.	LABOR	EQUIP.	TOTAL	

Division Notes

		CREW	DAILY OUTPUT	LABOR-HOURS	UNIT	BARE COSTS				TOTAL INCL O&P
						MAT.	LABOR	EQUIP.	TOTAL	

Division Notes

	CREW	DAILY OUTPUT	LABOR-HOURS	UNIT	BARE COSTS				TOTAL INCL O&P
					MAT.	LABOR	EQUIP.	TOTAL	

Division Notes

		CREW	DAILY OUTPUT	LABOR-HOURS	UNIT	BARE COSTS				TOTAL INCL O&P
						MAT.	LABOR	EQUIP.	TOTAL	

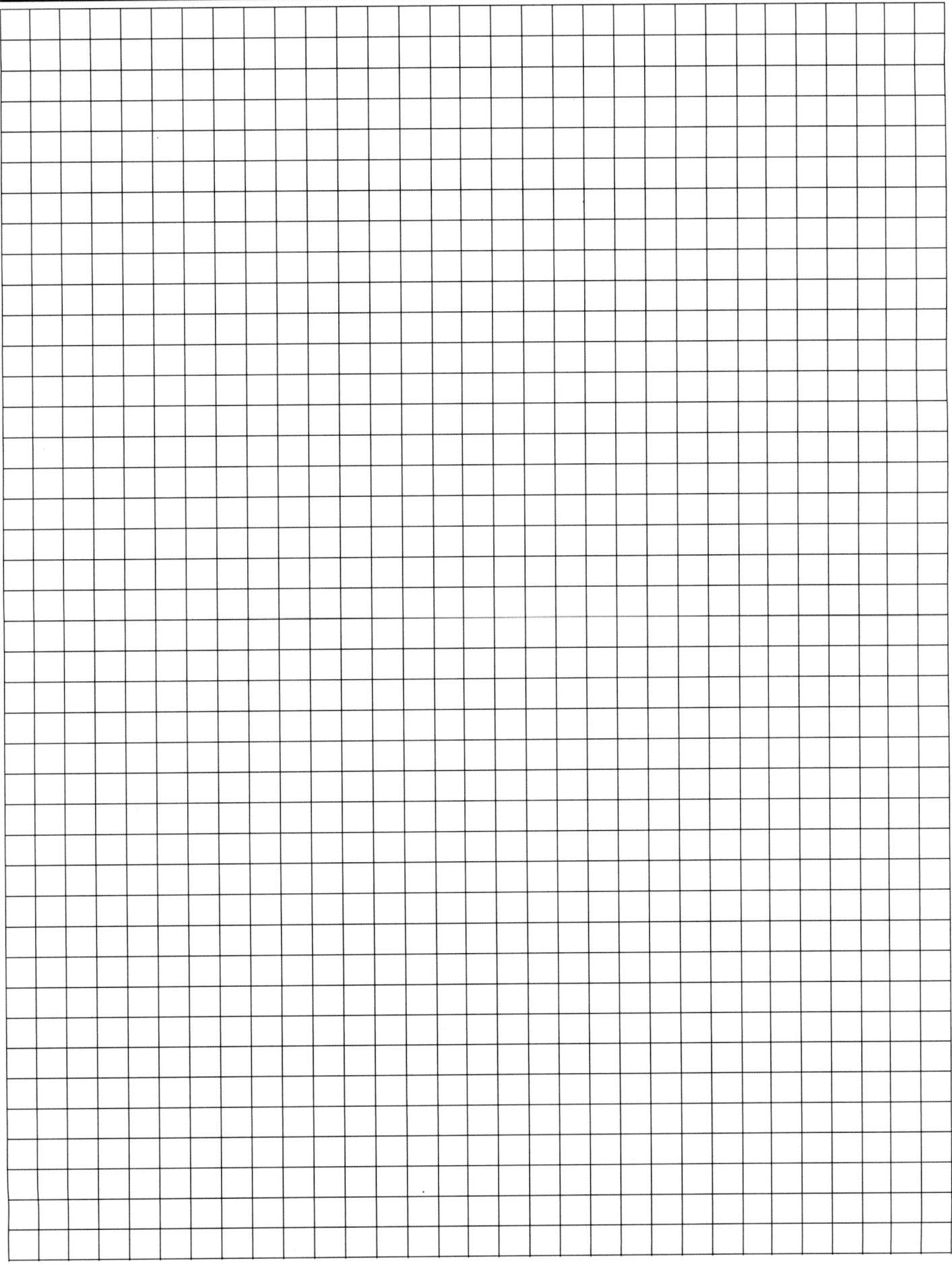

Other Data & Services

A tradition of excellence in construction cost information
and services since 1942

Table of Contents
Annual Cost Guides
Online Estimating Solution
Seminars and Professional Development

For more information visit our website at www.RSMeans.com

Unit prices according to the latest MasterFormat

Cost Data Selection Guide

The following table provides definitive information on the content of each cost data publication. The number of lines of data provided in each unit price or assemblies division, as well as the number of crews, is listed for each data set. The presence of other elements such as reference tables, square foot models, equipment rental costs, historical cost indexes, and city cost indexes, is also indicated. You can use the table to help select the RSMeans data set that has the quantity and type of information you most need in your work.

Unit Cost Divisions	Building Construction	Mechanical	Electrical	Commercial Renovation	Square Foot	Site Work Landsc.	Green Building	Interior	Concrete Masonry	Open Shop	Heavy Construction	Light Commercial	Facilities Construction	Plumbing	Residential
1	590	411	428	530	0	524	200	331	473	589	527	273	1065	421	178
2	777	280	86	733	0	991	207	399	214	776	733	481	1220	287	274
3	1688	340	230	1081	0	1480	986	354	2034	1688	1690	482	1788	316	389
4	960	21	0	920	0	725	180	615	1158	928	615	533	1175	0	447
5	1901	158	155	1093	0	852	1799	1106	729	1901	1037	979	1918	204	746
6	2453	18	18	2111	0	110	589	1528	281	2449	123	2141	2125	22	2661
7	1596	215	128	1634	0	580	763	532	523	1593	26	1329	1697	227	1049
8	2140	80	3	2733	0	255	1140	1813	105	2142	0	2328	2966	0	1552
9	2107	86	45	1931	0	309	455	2193	412	2050	15	1756	2356	54	1521
10	1090	17	10	685	0	234	32	899	136	1090	34	589	1181	237	224
11	1097	201	166	541	0	135	56	925	29	1064	0	231	1117	164	110
12	548	0	2	299	0	219	147	1551	14	515	0	273	1574	23	217
13	744	149	158	253	0	366	125	254	78	720	267	109	760	115	104
14	273	36	0	223	0	0	0	257	0	273	0	12	293	16	6
21	130	0	41	37	0	0	0	296	0	130	0	121	668	688	259
22	1165	7557	160	1226	0	1572	1063	849	20	1154	1681	875	7505	9414	719
23	1194	6995	581	938	0	157	898	787	38	1177	110	886	5235	1917	480
25	0	0	14	14	0	0	0	0	0	0	0	0	0	0	0
26	1512	491	10456	1293	0	811	644	1159	55	1438	600	1360	10237	399	636
27	94	0	447	101	0	0	0	71	0	94	39	67	388	0	56
28	143	79	223	124	0	0	28	97	0	127	0	70	209	57	41
31	1510	733	610	806	0	3265	288	7	1217	1455	3276	604	1569	660	613
32	836	49	8	905	0	4472	353	405	314	808	1889	440	1751	142	487
33	1246	1076	534	252	0	3021	38	0	237	523	3058	128	1698	2085	154
34	107	0	47	4	0	190	0	0	31	62	212	0	136	0	0
35	18	0	0	0	0	327	0	0	0	18	442	0	84	0	0
41	62	0	0	33	0	8	0	22	0	61	31	0	68	14	0
44	75	79	0	0	0	0	0	0	0	0	0	0	75	75	0
46	23	16	0	0	0	274	261	0	0	23	264	0	33	33	0
48	10	0	38	2	0	0	23	0	0	10	17	10	23	0	10
Totals	26089	19087	14588	20502	0	20877	10275	16450	8098	24858	16686	16077	50914	17570	12933

Assem Div	Building Construction	Mechanical	Electrical	Commercial Renovation	Square Foot	Site Work Landscape	Assemblies	Green Building	Interior	Concrete Masonry	Heavy Construction	Light Commercial	Facilities Construction	Plumbing	Asm Div	Residential
A		15	0	188	164	577	598	0	0	536	571	154	24	0	1	378
B		0	0	848	2554	0	5661	56	329	1976	368	2094	174	0	2	211
C		0	0	647	954	0	1334	0	1642	146	0	844	255	0	3	588
D		1057	941	712	1859	72	2538	330	825	0	0	1345	1105	1088	4	851
E		0	0	86	261	0	301	0	5	0	0	258	5	0	5	391
F		0	0	0	114	0	143	0	0	0	0	114	0	0	6	357
G		527	447	318	312	3377	792	0	0	534	1349	205	293	677	7	307
															8	760
															9	80
															10	0
															11	0
															12	0
Totals		1599	1388	2799	6218	4026	11367	386	2801	3192	2288	5014	1856	1765		3923

Reference Section	Building Construction Costs	Mechanical	Electrical	Commercial Renovation	Square Foot	Site Work Landscape	Assem.	Green Building	Interior	Concrete Masonry	Open Shop	Heavy Construction	Light Commercial	Facilities Construction	Plumbing	Resi.
Reference Tables	yes	yes	yes	yes	no	yes	yes	yes	yes	yes	yes	yes	yes	yes	yes	yes
Models					111			25					50			28
Crews	578	578	578	556		578		578	578	578	555	578	555	556	578	555
Equipment Rental Costs	yes	yes	yes	yes		yes		yes	yes	yes	yes	yes	yes	yes	yes	yes
Historical Cost Indexes	yes	yes	yes	yes	yes	yes	yes	yes	yes	yes	yes	yes	yes	yes	yes	no
City Cost Indexes	yes	yes	yes	yes	yes	yes	yes	yes	yes	yes	yes	yes	yes	yes	yes	yes

Online Estimating Solution *(side margin)*

Unit prices according to the latest MasterFormat *(side margin)*

Our Online Estimating Solution

Competitive Cost Estimates Made Easy

Our online estimating solution is a web-based service that provides accurate and up-to-date cost information to help you build competitive estimates or budgets in less time.

Quick, intuitive, easy to use and automatically updated, you'll gain instant access to hundreds of thousands of material, labor, and equipment costs from RSMeans' comprehensive database, delivering the information you need to build budgets and competitive estimates every time.

With our online estimating solutions, you can perform quick searches to locate specific costs and adjust costs to reflect prices in your geographic area. Tag and store your favorites for fast access to your frequently used line items and assemblies and clone estimates to save time. System notifications will alert you as updated data becomes available. This data is automatically updated throughout the year.

Our visual, interactive estimating features help you create, manage, save and share estimates with ease! You'll enjoy increased flexibility with customizable advanced reports. Easily edit custom report templates and import your company logo onto your estimates.

	Core	Advanced	Complete
Unit Prices	✓	✓	✓
Assemblies	⊘	✓	✓
Sq Foot Models	⊘	⊘	✓
Editable Sq Foot Models	⊘	⊘	✓
Editable Assembly Components	⊘	✓	✓
Custom Cost Data	⊘	✓	✓
User Defined Components	⊘	✓	✓
Advanced Reporting & Customization	⊘	✓	✓
Union Labor Type	✓	✓	✓

Continue to check our website at www.RSMeans.com for more product offerings.

Estimate with Precision

Find everything you need to develop complete, accurate estimates.
- Verified costs for construction materials
- Equipment rental costs
- Crew sizing, labor hours and labor rates
- Localized costs for U.S. and Canada

Save Time & Increase Efficiency

Make cost estimating and calculating faster and easier than ever with secure, online estimating tools.
- Quickly locate costs in the searchable database
- Create estimates in minutes with RSMeans cost lines
- Tag and store favorites for fast access to frequently used items

Improve Planning & Decision-Making

Back your estimates with complete, accurate and up-to-date cost data for informed business decisions.
- Verify construction costs from third parties
- Check validity of subcontractor proposals
- Evaluate material and assembly alternatives

Increase Profits

Use our online estimating solution to estimate projects quickly and accurately, so you can gain an edge over your competition.
- Create accurate and competitive bids
- Minimize the risk of cost overruns
- Reduce variability

For more information visit our website at www.RSMeans.com

Access the data online

Search for unit prices by keyword | Leverage the most up-to-date data | Build and export estimates

Try it free for 30 days   www.rsmeans.com/2018freetrial

2018 Seminar Schedule   📞 877-620-6245

Note: call for exact dates, locations, and details as some cities are subject to change.

Location	Dates	Location	Dates
Seattle, WA	January and August	San Francisco, CA	June
Dallas/Ft. Worth, TX	January	Bethesda, MD	June
Austin, TX	February	El Segundo, CA	August
Anchorage, AK	March and September	Dallas, TX	September
Las Vegas, NV	March	Raleigh, NC	October
New Orleans, LA	March	Salt Lake City, UT	October
Washington, DC	April and September	Baltimore, MD	November
Phoenix, AZ	April	Orlando, FL	November
Toronto	May	San Diego, CA	December
Denver, CO	May	San Antonio, TX	December

Gordian also offers a suite of online RSMeans data self-paced offerings. Check our website www.RSMeans.com for more information.

Self-Paced Professional Development Courses

Training on how to use RSMeans data and estimating tools, as well as Professional Development courses on industry topics are now offered in a convenient self-paced format. These courses are on-demand and allow more flexibility to learn around your busy schedule, while saving the cost of travel and time.

Current course offerings include:

Facilities Construction Estimating—our best-selling live class now available as an on-demand training course! Let the subject matter experts of construction estimating—the RSMeans Engineering Staff—walk you through the basics and much more of estimating for renovation and facilities construction.

RSMeansOnline.com Training—learn the ins and outs of the flagship delivery method of RSMeans data!

The Construction Process—how much do you and your team really know about the ins and outs of the "contract-side" of a construction project? This self-paced course will clarify best practices for items such as schedules, change orders, and project closeout.

These self-paced training courses can be completed over the course of 45 days and are comprised of multiple lessons with documentation, video presentation, software simulation, assessment quizzes and certificate of completion.

Site Work Estimating with RSMeans data

This new one-day program focuses directly on site work costs, a unique portion of most construction projects that often is the wild card in determining whether you have developed a good estimate or not. Accurately scoping, quantifying, and pricing site preparation, underground utility work, and improvements to exterior site elements are often the most difficult estimating tasks on any project. The program takes the participant from preparing a never-developed site through underground utility installation, pad preparation, paving and sidewalks, and landscaping. Attendees will use the full array of site work cost data through the RSMeans online program and participate in exercises to strengthen their estimating skills.

Some of what you'll learn:

- Evaluation of site work and understanding site scope of work.
- Site work estimating topics including: site clearing, grading, excavation, disposal and trucking of materials, erosion control devices, backfill and compaction, underground utilities, paving, sidewalks, fences & gates, and seeding & planting.
- Unit price site work estimates—Correct use of RSMeans site work cost data to develop a cost estimate.
- Using and modifying assemblies—Save valuable time when estimating site work activities using custom assemblies.

Who should attend: Engineers, contractors, estimators, project managers, owner's representatives, and others who are concerned with the proper preparation and/or evaluation of site work estimates.

Please bring a laptop with ability to access the internet.

Visit RSMeans.com/Online for more details on data titles in online format.

For more information visit our website at www.RSMeans.com

Professional Development

Training for our Online Estimating Solution

Construction estimating is vital to the decision-making process at each state of every project. Our online solution works the way you do. It's systematic, flexible and intuitive. In this one-day class you will see how you can estimate any phase of any project faster and better.

Some of what you'll learn:
- Customizing our online estimating solution
- Making the most of RSMeans "Circle Reference" numbers
- How to integrate your cost data
- Generating reports, exporting estimates to MS Excel, sharing, collaborating and more

Also offered as a self-paced or on-site training program!

Maintenance & Repair Estimating for Facilities

This two-day course teaches attendees how to plan, budget, and estimate the cost of ongoing and preventive maintenance and repair for existing buildings and grounds.

Some of what you'll learn:
- The most financially favorable maintenance, repair, and replacement scheduling and estimating
- Auditing and value engineering facilities
- Preventive planning and facilities upgrading
- Determining both in-house and contract-out service cost
- Annual, asset-protecting M&R plan

Who should attend: facility managers, maintenance supervisors, buildings and grounds superintendents, plant managers, planners, estimators, and others involved in facilities planning and budgeting.

Facilities Construction Estimating

In this two-day course, professionals working in facilities management can get help with their daily challenges to establish budgets for all phases of a project.

Some of what you'll learn:
- Determining the full scope of a project
- Identifying the scope of risks and opportunities
- Creative solutions to estimating issues
- Organizing estimates for presentation and discussion
- Special techniques for repair/remodel and maintenance projects
- Negotiating project change orders

Who should attend: facility managers, engineers, contractors, facility tradespeople, planners, and project managers.

Practical Project Management for Construction Professionals

In this two-day course, acquire the essential knowledge and develop the skills to effectively and efficiently execute the day-to-day responsibilities of the construction project manager.

Some of what you'll learn:
- General conditions of the construction contract
- Contract modifications: change orders and construction change directives
- Negotiations with subcontractors and vendors
- Effective writing: notification and communications
- Dispute resolution: claims and liens

Who should attend: architects, engineers, owners' representatives, and project managers.

Construction Cost Estimating: Concepts and Practice

This one-day introductory course to improve estimating skills and effectiveness starts with the details of interpreting bid documents and ends with the summary of the estimate and bid submission.

Some of what you'll learn:
- Using the plans and specifications to create estimates
- The takeoff process—deriving all tasks with correct quantities
- Developing pricing using various sources; how subcontractor pricing fits in
- Summarizing the estimate to arrive at the final number
- Formulas for area and cubic measure, adding waste and adjusting productivity to specific projects
- Evaluating subcontractors' proposals and prices
- Adding insurance and bonds
- Understanding how labor costs are calculated
- Submitting bids and proposals

Who should attend: project managers, architects, engineers, owners' representatives, contractors, and anyone who's responsible for budgeting or estimating construction projects.

Mechanical & Electrical Estimating

This two-day course teaches attendees how to prepare more accurate and complete mechanical/electrical estimates, avoid the pitfalls of omission and double-counting, and understand the composition and rationale within the RSMeans mechanical/electrical database.

Some of what you'll learn:
- The unique way mechanical and electrical systems are interrelated
- M&E estimates—conceptual, planning, budgeting, and bidding stages
- Order of magnitude, square foot, assemblies, and unit price estimating
- Comparative cost analysis of equipment and design alternatives

Who should attend: architects, engineers, facilities managers, mechanical and electrical contractors, and others who need a highly reliable method for developing, understanding, and evaluating mechanical and electrical contracts.

Visit RSMeans.com/Online for more details on data titles in online format.

For more information visit our website at www.RSMeans.com

Unit Price Estimating

This interactive two-day seminar teaches attendees how to interpret project information and process it into final, detailed estimates with the greatest accuracy level.

The most important credential an estimator can take to the job is the ability to visualize construction and estimate accurately.

Some of what you'll learn:
- Interpreting the design in terms of cost
- The most detailed, time-tested methodology for accurate pricing
- Key cost drivers—material, labor, equipment, staging, and subcontracts
- Understanding direct and indirect costs for accurate job cost accounting and change order management

Who should attend: corporate and government estimators and purchasers, architects, engineers, and others who need to produce accurate project estimates.

Life Cycle Cost Estimating for Facility Asset Managers

Life Cycle Cost Estimating will take the attendee through choosing the correct RSMeans database to use and then correctly applying RSMeans data to their specific life cycle application. Conceptual estimating through RSMeans new building models, conceptual estimating of major existing building projects through RSMeans renovation models, pricing specific renovation elements, estimating repair, replacement and preventive maintenance costs today and forward up to 30 years will be covered.

Some of what you'll learn:
- Cost implications of managing assets
- Planning projects and initial & life cycle costs
- How to use RSMeans data online

Who should attend: facilities owners and managers and anyone involved in the financial side of the decision making process in the planning, design, procurement, and operation of facility real assets.

Please bring a laptop with ability to access the internet.

Training for our CD Estimating Solution

This one-day course helps users become more familiar with the functionality of the CD. Each menu, icon, screen, and function found in the program is explained in depth. Time is devoted to hands-on estimating exercises.

Some of what you'll learn:
- Searching the database using all navigation methods
- Exporting RSMeans data to your preferred spreadsheet format
- Viewing crews, assembly components, and much more
- Automatically regionalizing the database

This training session requires you to bring a laptop computer to class.

When you register for this course you will receive an outline for your laptop requirements.

Also offered as a self-paced or on-site training program!

Assessing Scope of Work for Facilities Construction Estimating

This two-day practical training program addresses the vital importance of understanding the scope of projects in order to produce accurate cost estimates for facility repair and remodeling.

Some of what you'll learn:
- Discussions of site visits, plans/specs, record drawings of facilities, and site-specific lists
- Review of CSI divisions, including means, methods, materials, and the challenges of scoping each topic
- Exercises in scope identification and scope writing for accurate estimating of projects
- Hands-on exercises that require scope, take-off, and pricing

Who should attend: corporate and government estimators, planners, facility managers, and others who need to produce accurate project estimates.

Facilities Estimating Using the CD

This two-day class combines hands-on skill-building with best estimating practices and real-life problems. You will learn key concepts, tips, pointers, and guidelines to save time and avoid cost oversights and errors.

Some of what you'll learn:
- Estimating process concepts
- Customizing and adapting RSMeans cost data
- Establishing scope of work to account for all known variables
- Budget estimating: when, why, and how
- Site visits: what to look for and what you can't afford to overlook
- How to estimate repair and remodeling variables

This training session requires you to bring a laptop computer to class.

Who should attend: facility managers, architects, engineers, contractors, facility tradespeople, planners, project managers, and anyone involved with JOC, SABRE, or IDIQ.

Building Systems and the Construction Process

This one-day course was written to assist novices and those outside the industry in obtaining a solid understanding of the construction process - from both a building systems and construction administration approach.

Some of what you'll learn:
- Various systems used and how components come together to create a building
- Start with foundation and end with the physical systems of the structure such as HVAC and Electrical
- Focus on the process from start of design through project closeout

This training session requires you to bring a laptop computer to class.

Who should attend: building professionals or novices to help make the crossover to the construction industry; suited for anyone responsible for providing high level oversight on construction projects.

Professional Development

Visit RSMeans.com/Online for more details on data titles in online format.

For more information visit our website at www.RSMeans.com

Registration Information

Register early and save up to $100!

Register 30 days before the start date of a seminar and save $100 off your total fee. Note: This discount can be applied only once per order. It cannot be applied to team discount registrations or any other special offer.

How to register

By Phone
Register by phone at 877-620-6245

Online
Register online at
www.RSMeans.com/products/seminars.aspx

Note: Purchase Orders or Credits Cards are required to register.

Two-day seminar registration fee - $1,045.

One-Day Construction Cost Estimating or Building Systems and the Construction Process - $630.

Government pricing

All federal government employees save off the regular seminar price. Other promotional discounts cannot be combined with the government discount.

Team discount program

For over five attendee registrations. Call for pricing: 781-422-5115

Refund policy

Cancellations will be accepted up to ten business days prior to the seminar start. There are no refunds for cancellations received later than ten working days prior to the first day of the seminar. A $150 processing fee will be applied for all cancellations. Written notice of the cancellation is required. Substitutions can be made at any time before the session starts. No-shows are subject to the full seminar fee.

Note: Pricing subject to change.

AACE approved courses

Many seminars described and offered here have been approved for 14 hours (1.4 recertification credits) of credit by the AACE International Certification Board toward meeting the continuing education requirements for recertification as a Certified Cost Engineer/Certified Cost Consultant.

AIA Continuing Education

We are registered with the AIA Continuing Education System (AIA/CES) and are committed to developing quality learning activities in accordance with the CES criteria. Many seminars meet the AIA/CES criteria for Quality Level 2. AIA members may receive 14 learning units (LUs) for each two-day RSMeans course.

Daily course schedule

The first day of each seminar session begins at 8:30 a.m. and ends at 4:30 p.m. The second day begins at 8:00 a.m. and ends at 4:00 p.m. Participants are urged to bring a hand-held calculator since many actual problems will be worked out in each session.

Continental breakfast

Your registration includes the cost of a continental breakfast and a morning and afternoon refreshment break. These informal segments allow you to discuss topics of mutual interest with other seminar attendees. (You are free to make your own lunch and dinner arrangements.)

Hotel/transportation arrangements

We arrange to hold a block of rooms at most host hotels. To take advantage of special group rates when making your reservation, be sure to mention that you are attending the RSMeans Institute data seminar. You are, of course, free to stay at the lodging place of your choice. (Hotel reservations and transportation arrangements should be made directly by seminar attendees.)

Important

Class sizes are limited, so please register as soon as possible.

Professional Development